The Welfare of Animals

Animal Welfare

VOLUME 8

Series Editor

Clive Phillips, *Professor of Animal Welfare, Centre for Animal Welfare and Ethics, School of Veterinary Science, University of Queensland, Australia*

Titles published in this series:

Volume 1: The Welfare of Horses
Natalie Waran
ISBN 1-4020-0766-3

Volume 2: The Welfare of Laboratory Animals
Eila Kaliste
ISBN 1-4020-2270-0

Volume 3: The Welfare of Cats
Irene Rochlitz
ISBN 978-1-4020-3226-4

Volume 4: The Welfare of Dogs
Kevin Stafford
ISBN 978-1-4020-4361-1

Volume 5: The Welfare of Cattle
Jeffrey Rushen, Anne Marie de Passillé,
Marina A.G. von Keyserlingk and
Daniel M. Weary
ISBN 978-1-4020-6557-6

Volume 6: The Welfare of Sheep
Cathy M. Dwyer
ISBN 978-1-4020-8552-9

Volume 7: The Welfare of Pigs
Jeremy N. Marchant-Forde
ISBN 978-1-4020-8908-4

The Welfare of Animals

The Silent Majority

Clive Phillips

University of Queensland, Centre for Animal Welfare and Ethics,
Australia

Springer

Prof. Clive Phillips
University of Queensland
School of Veterinary Science
Centre for Animal Welfare & Ethics
Australia

ISBN: 978-1-4020-9218-3 e-ISBN: 978-1-4020-9219-0

DOI: 10.1007/978-1-4020-9219-0

Library of Congress Control Number: 2008935890

Printed on acid-free paper

9 8 7 6 5 4 3 2 1

springer.com

To my late father, Michael Phillips

Foreword

The Welfare of Animals is an exciting book that will stimulate and provoke its readers. It describes many problems faced by animals – those we use for food, for pleasure or in research, and those simply but harshly affected by shrinking habitats in the face of the ever-growing human population. And yet it is not a depressing read. It focuses not only on the difficulties that animals face, but on their capacity for free-choice, for joy and excitement, and on the possible ways in which the planet can be shared between species if only we take the time and trouble to think more carefully about the impact of our actions.

Clive Phillips moved from the United Kingdom to take up a Foundation Chair in Animal Welfare at the University of Queensland, becoming Australia's first Professor of Animal Welfare in 2003. This cultural leap, combined with his travels in countries like Malaysia and Borneo, permits him the unique and broad perspective that forms the backbone of this book. Eschewing the normal territory patrolled by the animal scientist (explaining the physiological basis of the stress response or causation of abnormal behaviour), Clive ventures into jungles and deserts, city centres and tribal homelands, and presents a book that remarkably and successfully combines travel-diary, nature notes, social and cultural history. By some sleight of hand, Clive throws all of these disparate elements into the air and emerges with an insightful analysis of how political, economic, religious and psychological truths impact on the animals whose planet we share. The portrayal of animals in art and religion, for example, is integrated with a consideration of how such depictions are associated with shifting perceptions about the place of animals within a moral framework. Parts of the book remind me of meditative passages in the Snow Leopard, written by Peter Matthiessen, as he accompanied the zoologist George Schaller through Nepal to study the behaviour of the blue sheep found in the Himalayas. Other sections remind me of the lectures given by the semi-fictional character Elizabeth Costello (created by the Nobel-prize winning author J.M. Coetzee) as she agonises over our catastrophic and industrial-scale disregard for animals, or of the art book *Animals and Men* by Kenneth Clark. The difference is that Clive's forays into history, art, religion or culture are always informed by his deep knowledge of the biology and psychology of the animals he is writing about.

This book provides readers with a vast array of original material. As such it will form a vitally important resource and text book for students and members of the public with an interest in animals from almost any perspective. The scale and intensity of animal use across the world is documented, and the rise of animal welfare and animal rights organisations is charted and analysed. In addition, the growth of animal welfare as a scientific discipline is given a place as a piece of modern cultural history in its own right. The political context in which animal welfare is considered in different countries is compared. Funding from industry or charitable sources is hugely important in advancing the field, but sometimes conflicts of interest can arise. It is therefore crucial that at least some animal welfare funding comes from government research councils whose primary interest is the promotion of good science. I do hope that Clive's exposition of the politics of animal welfare funding acts as a spur to governments worldwide to invest more in animal welfare science, even if this threatens the position of my own University as the top publisher in the field (Table 8.3)!

Clive Phillips has gathered some unpalatable facts and presented them at a crucial time. The human population has risen to 6.8 billion (www.optimumpopulation.org). The development of livestock agriculture is seen as a route out of poverty and towards increased resilience for some of the world's poorest people (UN World Food Programme) at the same time as demand for animal produce shoots up amongst increasingly wealthy urban populations. Clive estimates that, on average, each one of us is responsible for the life (or death) of some 18 animals per year. Generally, as humans, we acknowledge responsibility for the dog that shares our hearth and, gradually, this concept of responsibility is being extended to the animals we farm, or use in scientific procedures. But billions more sentient creatures remain truly forgotten: the wild-caught fish, the animals killed by our cars (and the offspring they will never return to), those forced out of their habitats, or killed by our own pet cats. Each and every one of these creatures is impacted by the way we choose to live our lives. The importance of this book is that it presents some new ways of thinking about how we can manage these problems without totally disregarding the feelings of our fellow species. This book matters. Thank you, Clive, for writing it.

Bristol, UK Christine Nicol

Animal Welfare Series Preface

Animal welfare is attracting increasing interest worldwide, especially in developed countries where the knowledge and resources are available to provide better management systems for farm animals, as well as companion, zoo and laboratory animals. The key requirements for adequate food, water, a suitable environment, companionship and health are important for animals kept for all of these purposes.

There has been increased attention given to farm animal welfare in the West in recent years. This derives largely from the fact that the relentless pursuit of financial reward and efficiency, to satisfy market demands, has led to the development of intensive animal production systems that challenge the conscience of many consumers in those countries.

In developing countries, human survival is still a daily uncertainty, so that provision for animal welfare has to be balanced against human welfare. Animal welfare is usually a priority only if it supports the output of the animal, be it food, work, clothing, sport or companionship. In principle the welfare needs of both humans and animals can be provided for, in both developing and developed countries, if resources are properly husbanded. In reality, however, the inequitable division of the world's riches creates physical and psychological poverty for humans and animals alike in many parts of the world. Livestock are the world's biggest land users (FAO, 2002) and the farmed animal population is increasing rapidly to meet the needs of an expanding human population. This results in a tendency to allocate fewer resources to each animal and to value individual animals less, particularly in the case of farmed pigs where herds of several thousand are not uncommon. In these circumstances, the importance of each individual's welfare is diminished.

Increased attention to welfare issues is just as evident for companion, laboratory, wild and zoo animals. Of increasing importance is the ethical management of breeding programmes, since genetic manipulation is more feasible, but there is less public tolerance of the deliberate breeding of animals with genetic abnormalities. However, the quest for producing novel genotypes has fascinated breeders for centuries. Dog and cat breeders have produced a variety of extreme forms with adverse effects on their welfare, but nowadays the quest is pursued in the laboratory, where the mouse is genetically manipulated with equally profound effects.

The intimate connection between animals and humans that was once so essential in husbandry is rare nowadays, having been superseded by technologically efficient production systems where animals on farms and in laboratories are tended by increasingly few humans in the drive to enhance labour efficiency. With today's busy lifestyle, companion animals too may suffer from reduced contact with humans, although their value in providing companionship, particularly for groups such as the elderly, is increasingly recognised. Consumers also rarely have any contact with the animals that produce their food.

In this estranged, efficient world, people struggle to find the moral imperatives to determine the level of welfare that they should afford to animals within their charge. Some, in particular many companion animal owners, aim for what they believe to be the highest levels of welfare provision, while others, deliberately or through ignorance, keep animals in impoverished conditions where their health and wellbeing is severely compromised. Today's multiplicity of moral codes for animal care and use are derived from a broad range of cultural influences, including media reports of animal abuse, guidelines on ethical consumption and campaigning and lobbying groups.

This series has been designed to help contribute towards a culture of respect for animals and their welfare by producing academic texts discussing the provision for the welfare of the major animal species that are managed and cared for by humans. They are not detailed blue-prints for the management of each species, rather they describe and consider the major welfare concerns, often in relation to the wild progenitors of the managed animals. Welfare is considered in relation to the animal's needs, concentrating on nutrition, behaviour, reproduction and the physical and social environment. Economic effects of animal welfare provision are also considered where relevant, as are key areas where further research is required.

With the growing pace of knowledge in this new area of research, it is hoped that this series will provide a timely and much-needed set of texts for researchers, lecturers, practitioners, and students. My thanks are particularly due to the publishers for their support, and to the authors and editors for their hard work in producing the texts on time and in good order.

Clive Phillips
Series Editor
Professor of Animal Welfare and Director,
Centre for Animal Welfare and Ethics,
School of Veterinary Science,
University of Queensland,
Australia

Reference

Food and Agriculture Organisation (2002). http://www.fao.org/ag/aga/index_en.htm

Preface

Like most people, I learnt about caring for animals when I was young. My experiences shaped my views on animal welfare, and this book is an attempt to share these views. My earliest animal experiences were mostly of pets, and I laughed with them, played with them and cried when they died like most of you, and it was in my early years that I was taught a sense of respect for wild animals. Once a week I visited my grandparents' house and garden, where a chaffinch regularly flew onto their lounge window sill and entered the room to take some nuts which had been put out for him. Whenever it appeared at the windowsill there was an immediate call for the children to be quiet in order that it could enter and get his reward without fear. For those precious few seconds we sat motionless and enjoyed the presence of that bird. In that one small action of crossing the windowsill, the bird had bridged that great divide, from being a totally wild bird, to entering into a 'contract' with humans, that could have been the first step on the way to domestication. One year the bird no longer came to the window, which made us sad, but I had received an early lesson on respect for wild animals and the joy that they can give us.

Now, after some years as a student of animals, and the ways that they behave, feed, reproduce and interact with humans, this book offers some thoughts on their welfare. Many people are working tirelessly to improve animal welfare with little thanks or praise and it is hoped that this will provide them with inspiration. Some work in animal industries, others in charity organizations, shelters, wildlife parks and educational establishments. Working *for* animals, as opposed to *with* animals, requires courage and those involved often face criticisms from those who seek to get the maximum from their animals for the smallest input. Many whose livelihoods depend on using animals are averse to any change that will affect the short-term profitability of their enterprise, although by improving animal welfare they are guaranteeing their future markets and the ethical acceptability of their business in the long-term.

Animal welfare scientists are few in number, and their work may not be accepted as well as those working in traditional animal disciplines, such as zoology, animal production science and agriculture. Some in the animal production sciences may believe that showing kindness to animals is a sign of weakness. Animal welfare may not be regarded as a rigorous scientific discipline, and this is in part because simple measurements are often most effective in any relatively new science, including animal welfare. It would not be good use of research funds to engage in detailed reductionist analysis of just a few problems when simple measurements are lacking in most branches of the science. For example, we could go to great lengths to find criteria that prove when an animal is malnourished, but is this really useful when animals are dying in drought conditions? Another issue relates to the type of measurements taken, and how an animal behaves is a key indicator of welfare, yet behaviour measurements are regarded by many scientists as lacking rigour and not as scientific as physiological or anatomical measurements. However, few could doubt that animal welfare science often requires more careful thought than other disciplines in framing the paradigm that is to be tested.

Animal welfare science is different because the beneficiaries are animals not humans. Most related fields – animal, crop, food and soil sciences, and even sometimes veterinary medicine – aim to improve the efficiency, and hence profitability, of the industry for the benefit of humans. Animal welfare science, however, aims to improve an animal's quality of life, which is an ambitious goal since we can often only guess at the impact on the animal's welfare. Such an intangible goal is too obscure for many traditional scientists, but those involved in animal welfare must have faith that their work is valuable to the animals. They must believe that addressing the animal welfare problems in the world today is a necessity for the future moral health of society, and they will require the courage of their convictions to pursue this goal. Furthermore, the significant advances in animal welfare science made in the last quarter of a century should encourage new scientists to take up this challenge – to provide the information necessary to meet the public demand for improved animal welfare. If scientists fail to provide this information, the politicians will have only opinions on which to base the improved standards that the public require. Most codes of practice for the welfare of animals are currently based on unproven beliefs, and they often lack credibility with those directly involved in the animal industries as a result. For every expert with one opinion there is another with a counter argument. Basing standards on whim or fancy will ensure that they will be challenged forcefully and this will hinder animal welfare reform.

Politicians must recognize that science cannot solve all of the animal welfare problems quickly. Patience will be required in building the research base with which to tackle welfare issues, since in many countries there are few, if any, scientists devoted to this topic. In addition we must recognize that for many closely involved in the animal industries, providing for animal welfare is as much an art as a science – an art that has been handed down over the centuries and which is closely connected with the different religious and philosophical approaches to animals that are possessed by the various cultures. As Gandhi said, "there should be no science without humanity" (Gandhi, 1948). Many different types of people have responsibility for animal welfare and we must understand and recognize the importance of all of their views. For those directly involved, stockmanship – the art of looking after animals – is commonly recognized as a more important influence on welfare than where they are kept, what they are fed etc. (AWC, 2002; Cross et al., 2008a). The art of providing for animal welfare is the cement that binds the scientific building blocks together, that enables scientific advances to be implemented with the support of society.

What does the future hold for animal welfare movements worldwide? We live in an increasingly egalitarian age, when the rights of animals and women, as well as the mentally retarded, racial minorities, gay and disabled people are recognized as having been too often ignored in the past. Sometimes there are concerns, especially amongst the elders in the community, that the pace of change and development is too rapid, faster than humankind can cope with in social terms. Hence there is concern that traditions are being eroded, that there are undetected dangers facing those embracing modern technologies. However, it is this rapid pace of technological change that has allowed us to progress from being concerned primarily about survival of the human race to our being able to control the survival and welfare of all the species on earth. Adequate welfare provision for all is within our sights and this rapid change in emphasis is now essential, because focusing only on human welfare, when we have developed such powerful tools to manipulate and potentially benefit animal and plant life, would be a certain means of rapid destruction of both humans and animals. Hence improving animal welfare is not just a desirable option, it is essential for the wellbeing of future societies.

I hope this book will encourage us all to reflect on animal welfare – how it was in the past, how it is changing and how we want it to change in the future. Remembering that animal welfare is not just a scientific discipline, but has a strong humanitarian component as well, we must allow our approach to animal welfare to recognize cultural, gender and social differences. Reflection is what separates us out as a species, the power of rational thought that has served us so well in the past. We have overcome dictators, with ambitions to conquer the world, diseases that threaten to wipe out our species, and we are now tackling both environmental change, that theatens to erode our standard of living everywhere, and poverty that threatens the existence of the poor and needy in the world today. We will overcome these challenges with ingenuity, correct actions and careful planning. But we often forget our biggest responsibility: to recognize that our actions have a major impact on the welfare of animals. Increasingly humans are cognizant of this fact and are taking action to improve animal welfare throughout the globe. After social movements to prevent the abuse of children, racial minorities, disabled people and homosexuals, to be in the midst of a social revolution that recognizes the need to improve animal welfare is truly a privilege.

These are ethical issues which need debating, but most welfare problems are easy for all to see. 'So far as the animals are concerned it matters not what we think or feel but what we do' (Webster, 1995). If we do not act, it is not just the animals that will suffer; our lives will be impoverished because the human-animal bond is crucial to our existence. It is incumbent on all of us to do everything in our power to improve the welfare of animals, so that future generations cannot say that we stood idly by whilst animals were exploited at the expense of their welfare.

Brisbane, QLD Clive Phillips

Acknowledgements

A preliminary version of part of Chapter 7 was published as follows: Phillips, C.J.C. 2005. How animal welfare science assists in defining cruelty to animals. *Proceedings of the 2005 RSPCA Australia Scientific Seminar*, Canberra, 22 February 2005, pp 29–37. http://www.rspca.org.au/events/seminars05_proc.PDF. It was also published in Veterinaria.org (URL: http://veterinaria.org/revistas/redvet and http://www.veterinaria.org). A preliminary version of the section on pain and cruelty in Chapter 6 was published in the Proceedings of the Animal Welfare Science Summit – Pain and Pain Management (http://www.daff.gov.au/animal-plant-health/welfare/aaws/summit) and will be published in a forthcoming volume in the OIE Technical Series.[1] The section on animal welfare in veterinary education in this chapter was initially presented to a conference on Humane Education at Griffiths University, Brisbane, the proceedings of which will be published electronically.

I am grateful to the Australian Animal Welfare Strategy for cover photography, to the late David Piggins for helpful discussions on the chapter on Empathy, Martin Phillips for a tour of London art galleries, Donald Broom for the opportunity to spend time in his research group during the preparation of the book and for information on the early history of the International Society for Applied Ethology, Jia Meng for data on companion animal populations, Andrew Tribe for information on zoo animal populations, Zainal Zafari for assistance with my visit to Krau Wildlife Reserve, Malaysia, Humrawali Khan and Shokri Jusoh for assisting with my visit to Borneo. I acknowledge the Department of Agriculture, Fisheries and Forestry, Adrianne Yzerman and Kerrie Rubie for photographs used on the front cover. The views expressed in the book are entirely my own.

[1] OIE Technical Series, Vol. 10, 2009, "Scientific assessment and management of animal pain"

Contents

Chapter 1
Definitions and Concepts of Animal Welfare

Defining animal welfare, coping with the environment, feelings and experiences – a variable scale for experiences – accounting good and bad experiences – naturalness of an animal's environment – animal needs and desires.

Introduction

Managing animal welfare is one of the most contentious societal issues of today, so it is important to be clear what is understood by the term. However, in choosing a definition that best encapsulates our interest in the topic, we need to be aware of who we are defining it for. Lawyers may need a more precise definition than the general public, who need a definition in terms that can easily be understood. Scientists require a definition in terms that can be measured, and animal managers or keepers require one that is practically relevant and can be easily applied to their animals.

The word 'welfare' comes from the Old Norse word *velferth*, derived from the words meaning *good* (val in Old Norse) and *travel* (fara in Old Norse). A similar word, *wohlfahrt*, is used today in the German language. A commutation of this concept is the popular English word used when people part – farewell. In the Romantic languages the concept is rather different, being based on 'being good' (bienestar in Spanish, bien-être in French and bemestar in Portuguese). This variation is now supported in the United States, where the word 'well-being' is predominantly used, rather than 'welfare', because of potential confusion with the welfare state for people. Thus the Northern European concept is one of the 'state whilst travelling', indicating an assessment over a period of time, whereas the Southern European, and now American, concept is one of an assessment at one point in time. Some people in Europe and Australia now advocate using 'well-being' rather than 'welfare' because of the supposedly negative connotations associated with 'welfare'. This is a similar problem to that of finding a suitable word to represent negroid people in predominantly white societies, and the solution is not to regularly change to new words when the conventional ones begin to be used in a derogatory fashion

C. Phillips, *The Welfare of Animals*, Animal Welfare 8,

DOI 10.1007/978-1-4020-9219-0_1, © Springer Science+Business Media B.V. 2009

by some members of society, but to ensure that the word is clearly and openly defined for all.

In defining the welfare of any animal, we are essentially trying to answer the question, 'How good is an animal's state in its environment?' This has several components, first, the animal's environment; second, the animal's awareness of its environment; third, the animal's internal state, based on its genetics and past and present experiences; and fourth, the animal's awareness of its internal state.

Coping with the Environment

The most widely quoted definition of animal welfare is also one of the oldest – 'the state of an animal with regard to its ability to cope with its environment' (Broom, 1986). This adequately addresses at least the first of these four components, and probably the second (depending on what constitutes a 'state'). However, the response to internal influences, such as the presence of a tumour, which will reduce the animal's welfare, is not addressed, and neither is the animal's awareness of its internal state. Another problem in relation to Donald Broom's definition is that 'coping' is an emotive word which suggests a negative overall attitude in the animal, whereas it should be positive or at least neutral. 'Coping' accords well with the view that life is difficult, miserable or laden with original sin, which is a view favoured by religious leaders and was probably proposed in an attempt to persuade people to behave more altruistically, so that they can have increased happiness or a better afterlife. The ability to thrive, rather than cope, is a better description of an animal's response to their environment.

Broom's definition focuses on dealing with unpleasant experiences, and an outside observer could be forgiven for believing that 'all of creation groans with pain', a sentiment expressed by the apostle Paul (Paul, 1994). The concept is perpetuated elsewhere – four of the Five Freedoms advocated by many as a good description of welfare needs (Webster, 1995) are freedoms from unpleasant experiences (1. hunger/thirst, 2. discomfort, 3. pain/injury/disease, 4. fear/distress), and only one addresses positive welfare (freedom to perform normal behaviour). In reality animal welfare should be a construct of both positive and negative affect.[1] Pain and other negative affects evolved to guard us from danger, and equally important are the positive affects that attract us to things that will improve our lives. The primary affects are probably common to all higher animals and humans. Negative affects include pain, fear, terror, hatred, distress, dissmell,[2] anguish and disgust (Tomkins, 1963; 1991) and positive affects includes interest, excitement, strength, enthusiasm, pride, alertness, inspiration, determination, attentiveness and activity.

---

[1] An affect is a brief biological, innate, instinctive response to a stimulus

[2] The negative affect of experiencing noxious odours

Animal's Feelings

Many consider that welfare relates principally to an animal's feelings (e.g. Duncan and Fraser, 1997). Under this hypothesis, an animal that is not sentient cannot experience poor welfare. Therefore an animal which is not consciously aware and responds purely automatically, i.e. without recourse to reason, learned responses or experiences, cannot be said to have any degree of well-being (Piggins and Phillips, 1998). This definition favours animals with advanced reasoning, such as ourselves, which have developed these abilities because of the way in which we function in the environment. However, it is important to realise that most of our behaviour, and that of other 'higher' animals, is automatic.

In a 'feelings' approach to welfare, it is reasonable to propose that the gradations of sentience that exist in animals should be taken into account by providing gradations of welfare standards, according to the sentience level of the species. This is very different from Broom's definition, proposing that welfare is the animal's state in respect to its coping with the environment, in which case it matters not whether the animal is sentient or not. The animal may not perceive that it is in a poor state if it is not sentient, but in terms of man's provision of reasonable standards for all animals that he manages, it should not matter whether the animal perceives the standards or not. The perception will not affect the major goal of animal life, to thrive in an environment and to be able to continue existing there through reproduction (hence the suggestion that 'thriving' is a better description than 'coping' above). Focusing on just an animal's feelings implies that just one mechanism for the pursuit of this goal[3] – stimuli perception and processing – is dominant over any others, such as reproduction, autonomic responses to stimuli or the destruction of competitive species in that environment. All of these will contribute to the animal's welfare, not just the feelings.

Although defining welfare by an animal's feelings is more related to psychological affect, we do not know, and may never know, much about what it actually feels like to be an animal. Longuet-Higgins (1994) has commented that "in practice Westerners attribute consciousness to other species only when they can imagine 'what it feels like' to be" that species. Therefore, in assessing animal welfare we tend to rely on psychological affects that we know are supported by several of the usual welfare indicators – behaviour, physiology, animal choices, longevity, production and reproduction.

Another problem with defining welfare from feelings or sentience is that it is almost impossible to measure an animal's feelings with any degree of accuracy, and this is likely to be the case for at least the foreseeable future. I am usually, but not always, aware of my own feelings, I recognise that others claim to have similar feelings and therefore probably theirs are similar to mine. However,

[3] A goal that can simply be described as of the preservation of life in a specific environment

I can only occasionally impute the existence of feelings in other animals and have little confidence in describing them. People may agree on the feelings that they believe animals are displaying (Wemelsfelder et al., 2001), but it does not mean that they are correct, because the species barrier between humans and animals may contribute a systematic error in their assessment. Indeed Griffin (1976) has argued that ..... 'It is not necessary to assume that such mental experiences (of animals) are at all similar to those which a person might have under analogous circumstances'. We have little reason to suspect that an animal's feelings may be accurately inferred from our own experiences, their sensory apparatus is usually very different from our own and their processing capacities equally so. For example, as humans, we use almost one fifth of our energy intake supporting the activities of our brain (Kurpad et al., 1994), but it would be maladaptive for other animals to devote such a large proportion of the nutrients purely for processing if this was not required for their occupation of their particular ecological niche. Therefore to base assessment of welfare purely on an animal's feelings is speciesist, since it favours animals such as us with well-developed cognitive capacity. Longuet-Higgins (1994) believed that this argument even invalidated any "scientific explanation of consciousness", which "is a logical absurdity because it would entail explaining the existence of observers in terms of their own observations".

Feelings are transient and relate primarily to short-term considerations of animal well-being and not long-term welfare assessment. For example, an animal may select food items that are desirable in the short-term, especially if they are carbohydrate-rich, because they provide a pleasurable feeling and adaptive advantage in the short-term (Matsuno and Thibault, 1995), but in the long-term exclusive selection of these items may reduce its welfare by causing disease, such as dental decay or obesity. Some events, which may damage an animal's welfare, such as removal of a limb under anaesthetics, will not result in any adverse feelings at the time but may damage an animal's welfare in the long-term. However, it is evident that a long-term summation of feelings could give, in an ideal world, a useful measure of just one component of welfare.

With our current knowledge, we can only rely on expert opinion or folk knowledge for an estimation of the relative feelings of different animals. For example, a survey of nearly 500 students has shown that animals are perceived as having the following hierarchy of sentience: monkey > dog > newborn baby > fox > pig > chicken > rat > fish (Phillips and McCulloch, 2005). However even this does not provide an objective assessment of sentience levels, since there were marked differences between students of different nationalities in their attribution of sentience to different species, demonstrating that human perception is inaccurate. When the questions were classified as providing information on the students' attitudes towards suffering during life, which is more concerned with animal welfare, or reverence for life, which is more concerned with an animal's rights, it was clear that Asian students had less concern for animal suffering during life than the European students, although both had similar reverence for animal life. Our views on animal's feelings are therefore not constant, but determined by experience and culture.

When assessing the welfare of an animal, we develop a hierarchy of similarities. For example, I understand my behaviour better than your behaviour, which I understand better than another primate's behaviour, which I understand better than that of a fish. To produce this hierarchy we have to make simple comparisons in our mind, in this case (1) Your behaviour is more similar to mine than another primate's, (2) a primate's behaviour is more similar to mine than a fish, (3) Your behaviour is more similar to mine than a fish. The hierarchy therefore in terms of understanding feelings is my behaviour> another person's behaviour>another primate's behaviour>a fish's behaviour.

Experiences

A third possible definition of an animal's welfare is its 'state with respect to the quality and quantity of its experiences'. This has been considered recently (Bracke et al., 1999), although experiences were dismissed as too difficult to measure. However, an experience can be considered to be 'an event or a series of events participated in or lived through' and does not include attributed mental states. The difference between experiences and feelings is shown by a simple situation where a man hits a dog. The dog experiences being hit, it feels pain. The experiences can be accurately described, for example the number of times that it happened and the severity. The feelings cannot be accurately described, they may be inferred from behavioural responses, or even physiological, but we have little certainty that we know what the dog feels.

The sequence of human interpretation of stimuli are as follows:

Environmental and Internal stimuli presented
⇓
Events experienced
⇓
Sensation
⇓ ↘
Perception → Processing
⇓ ↗
Cognition

Defining welfare as the animal's state with regard to its responses to the environment includes some processing, as does the approach of defining welfare by 'feelings'. However, if we define welfare by experiences, we omit processing and can be sure that at least some are definable and measurable with a reasonable degree of accuracy. Whether such experiences contribute positively to animal welfare can be determined experimentally, by offering animals choices or by monitoring their responses to specific imposed situations, or by soliciting expert or even public opinion. The approach will vary with the situation.

The introduction of stimulus processing into experience assessment is difficult, but a scale of increasing complexity for human stimulus reception and processing has been proposed by Young (1994):

1) *Phenomenal awareness*: the experience of seeing, hearing, touching and so on.
2) *Access awareness*: where stored information is brought to mind
3) *Monitoring:* including awareness of our own actions and their effects, and monitoring perceptual information for discrepancies with current plans and hypotheses.
4) *Executive awareness:* awareness of our goals and intentions

However, although we can be certain that animals possess phenomenal and access awareness, we cannot be sure of the extent of their monitoring awareness, and even whether they have any executive awareness at all. Griffin (1976) also attempted to produce an awareness processing scale, this time for animals, and again progressing from the simple to the complex forms of processing (Table 1.1).

It is possible to combine Young and Griffin's awareness scales in relation to visual stimuli (Piggins and Phillips, 1998, Table 1.2). This demonstrates that the boundaries between sensation, perception and cognition are not as fixed as traditionally assumed. Some of the research necessary to determine the visual

Table 1.1 An awareness processing scale (Griffin, 1976)

Processing by animal	Complexity
Pattern recognition	Simple
Neural template formation	
Search image	
Develop affect	↓
Expectancy	
Covert verbal behaviour	
Develop internal image	
Conceptualisation	↓
Understanding	
Intention	
Feeling	
Mental experience	↓
Thought	
Choice	
Free will	
Consciousness	Complex

Table 1.2 Scheme for visual awareness assessment based on Young (1994) and Griffin (1976) with examples (Piggins and Phillips, 1998)

Basis of assessment			Purpose	Examples
Sensation/ Perception/ Cognition	Young (1994)	Griffin (1976)		
Sensation	Phenomenal awareness	–	Detection	Reflexive Absolute threshold Simple Reaction Time Motion Sensory memory Colour vision Brightness Contrast Sensitivity
Sensation/ perception	Access awareness	–	Discrimination	Acuity Complex reaction time Optic flow Difference threshold Imagery
			Depth	Binocular Monocular Size constancy (perceptual) Perspective (perceptual) Motion parrallax (sensory)
Perception/ Cognition 'Intelligent perception'	Monitoring awareness	Pattern recognition	Recognition organisation	Gestalts Pattern Form Object Short term, working memory Events Location (Ego centrality) Hallucination Mental rotation
Cognition	Monitoring Executive	–	Evaluation	Imagery Long term memory and stored knowledge Experience

Table 1.2 (continued)

Basis of assessment			Purpose	Examples
Sensation/ Perception/ Cognition	Young (1994)	Griffin (1976)		
				Problem solving (e.g. tool use)
				Decision making
				Stroop effort
				Mental representation
			Meaning	Volition, Intent
				Self awareness and that of others
				Mental states
				Creativity

experiences of some animals has been attempted, most notably for their sensory capacities, but much remains to be undertaken. Visual awareness is one of the better understood forms of awareness (Lomas et al., 1998), other forms of awareness such as auditory (Heffner, 1998), dietary (Forbes, 1998) and olfactory (Sommerville and Broom, 1998) awareness are much less well understood.

Quantifying Experiences – An Accounting Perspective

Animal welfare should be considered over as long a time frame as possible. If the immediate effects of, for example, invasive procedures are considered, the wrong conclusions about their impact on welfare can be gained if they are performed to safeguard the animal's longterm welfare. So for example, the removal of skin from the hindquarters of sheep in the mulesing operation, although appearing inhumane, may actually improve welfare by preventing the sheep from being attached by flies. Veterinary operations are therefore often painful but may improve an animal's welfare in the long term. It is important to consider an animal's life as comprised of many experiences, which can simplistically be considered good or bad. If only the quality of experiences is considered, then an animal with just one good experience and no bad experiences would have the same welfare as an animal with many good experiences (at the same level) and no bad experiences. Thus the *number* of each type of experience is important as well as its quality. This can be expressed mathematically by a simple formula for animal welfare, as the sum of different good and bad experiences. To arrive at this formula, we must first imagine that every animal undergoes different types of good and bad experiences, which can be labeled G_1, G_2 and up to the almost innumerable type of experiences, which can be called G_n, and also B_1, B_2, B_n etc. Each type of experience has to be qualified by the number of times that it happens, which mathematically can be described by the coefficient of G_n, we can call this g_1 for experience G_1, g_2 for G_2 and g_n for G_n etc.

So, the equation for animal welfare becomes:

$$W = g_1G_1 + g_2G_2.....g_nG_n - b_1B_1 - b_2B_2.......b_nB_n$$

Where

W = welfare of an animal
G_n = type of good experience
g_n = the number of G_n good experiences
B_n = type of bad experience
b_n = the number of B_n bad experiences
n = number 1 to ∞

This equation can be summarized as:

$$W = \sum_{n=1}^{\infty} g_nG_n - \sum_{n=1}^{\infty} b_nB_n$$

Information on the impact of experiences on welfare is limited, especially in the long term, and if necessary the animal's lifetime. However, if we accept that human responsibility to animals includes a positive welfare provision, the above approach can be used to enumerate the relative impact of good and bad experiences in the animal's lifetime. Human responsibilities in this respect are considered in Chapters 4 and 6. Using this method, we can develop the tools to determine lifetime welfare, for example does an animal that has good free range conditions for most of its life and then travels a long way to slaughter have better welfare than an animal that has intensive housing conditions for most of its life but only a short journey to slaughter? Can high welfare during rearing be offset by poor welfare at slaughter? A potential flaw with this approach is that exposure to some bad experiences may be beneficial for young animals as it prepares them to deal with potentially worse experiences in later life, or similar experiences at times when their behaviour is less malleable and they are less able to cope than in infancy. Another potential issue is that members of the public usually have certain standards below which a practice is unacceptable.

The different welfare experiences cannot be added unless the animal's perception of the experience can be quantified. Welfare is therefore not absolute but affected by the animal's response to the experiences. Welfare perception by the animal is not probably the same as welfare perception by the human. Like humans, animals make decisions to optimise their welfare over time (Abeyesinghe et al., 2005), even though their ability to do this is probably not as accurate as in humans. For example, chickens can show restraint and delay taking a food reward if it is larger than an immediate reward (e.g. Abeyesinghe et al., 2005). However, the increase in size of the reward has to be substantial and the delay short. If offered the choice of a slightly increased reward size after a significant delay, animals become impulsive and take the smaller reward. Although direct comparisons have not been made, humans, and especially adults would be expected to show more restraint.

The total welfare perception 'P' by the animal at a particular time is a function of the sum (Σ) of these good and bad experiences, taking into account the influence of past experiences. This can be incorporated by adding another coefficient, *x* for good experiences and *y* for bad experiences:

$$P = \sum_{n=1}^{\infty} x_n g_n G_n - \sum_{n=1}^{\infty} y_n b_n B_n$$

Where x_n = influence of past good experiences on perceived welfare
And y_n = influence of past bad experiences on perceived welfare

In many cases x_n and $y_n < 1$, so the perception of individual good and bad experiences diminishes with repetition, i.e. the animal habituates. However, there may be situations where x and y > 1, i.e. individual past experiences potentiate the perceived welfare impact of any individual experience to a greater-than-normal response. For example, the impact of sticking spears into cattle during a bullfight is probably incremental until the animal is worn down and the matedor can get close enough to kill the animal with a sword. The adverse effect of each successive spear will reduce welfare in ever increasing amounts. Thus the marginal response is likely to depend on the previous level of experience. The physiology of pain potentiation of this nature is beginning to be understood (McKenna and Melzack, 2001).

Under this model, our contract with animals could be to provide a certain ratio of good to bad experiences, presumably greater than 1:1, for a predetermined period of time. Thus premature slaughter would be not only an ethical issue, but a welfare issue too, because the total number of good experiences would be reduced. This model of animal welfare accords with public sentiment, who believe that killing an animal reduces its welfare.

The direction of change in the quality of the experiences over a lifetime may be influential in determining welfare perception by the animal. An animal that starts off life with bad experiences that gradually progress to good experiences may perceive its welfare to be better overall than one which starts off life with good experiences but these gradually worsen over time. Many animals naturally experience a worsening of their experiences over time, as their bodily functions and capabilities for sentience, repair and activity decline with age. Others experience a reduction in welfare because of the way we manage them, for example, beef cattle are often raised on rangeland, where there is plentiful space, nutrition from their mother and freedom to perform most natural behaviours. Then for the final few months of their short lives they are transferred to feedlots, where there is limited space, often hot conditions and a novel social structure quite different to the natural matriarchal grouping that they experienced on the rangeland. There are further complications, for although most lay people perceive that the provision of grazing for cattle is good for their welfare (Hemsworth et al., 1995), expert opinion suggests that the risk of parasitism on rangeland in some regions is a significant problem (Cross et al., 2008b). An alternative system would remove calves from their mothers and rear them intensively, a process called early weaning. This may reduce welfare because the calves lose the assistance of their mother, which for many young mammals cushions them from the realities and responsibilities of self-determination during suckling period. The intensification of agriculture in the late 20th C led to a trend for early weaning in calves, at least in the dairy sector, for economic expediency (Heleski et al., 2006a). Recently, in Nordic countries in particular, where animal welfare is a high priority, there have been attempts to develop systems where nurse cows take over the milking cow's responsibilities for rearing the calf (Vaarst et al., 2001). In most developing countries, the advantages of keeping the cow and calf together are well understood, and it even improves the health of both, providing the calf gets adequate milk (Margerison et al., 2002; Phillips and Sorensen, 1993).

Naturalness

A fourth way of defining animal welfare, which is closest to the position held by many members of the general public (Lassen et al., 2006), is that the key element is whether animals are living in a natural environment and in particular whether they have the ability to perform natural behaviour. Many people have a strong respect for, and affinity with, nature (Kellert and Wilson, 1984), and this may have conferred an adaptive advantage in the past and even today. Over the period of our evolution, humans that understood nature better would have had an improved ability to find the best food and the location of safe refuge away from predators and would have been generally better adapted to their environment. This relationship with nature is explored in the next chapter.

Animal Needs and Desires

Animals have both 'needs', that are essential for life to be sustained in the long term, and 'desires', the satisfaction of which will improve the quality of life, but they are not essential. 'Needs' relate to both life preservation and reproduction (Bracke et al., 1999). For life preservation, an adequate supply of food and water and a good health status are most likely to be limiting. For reproduction, the total environment must be suitable for it to be worthwhile for the animal to expend resources. Thus factors such as temperature stress, social stress etc may limit reproduction but are not necessarily life threatening in the short term, but may threaten the maintenance of the species in that environment. Animal welfare can be estimated using an Animal Needs Index that has been created by Italian researchers (Napolitano et al., 2007).

Animal's desires include such resources as companionship, space, and probably variety in the diet. There is a tendency for scientists to anthropomorphically concentrate on the 'desires', perhaps because this is a major pre-occupation for humans, whose needs are nearly always adequately met without thinking about them. For example, humans are rarely sufficiently hungry or thirsty that their survival or ability to reproduce is threatened. Animals' needs are not often adequately considered, for example, the availability and quality of the food and water supply receives little attention from a welfare standpoint (see Chapter 7), whereas adequate space is usually considered one of the most important influences on welfare. Another reason that 'needs' are not considered so much for animals is because they seriously impact on the profitability of keeping animals; 'desires' are unlikely to have such a major effect. The economic impacts of different nutritional strategies and some health factors are major and well known. Desires, such as a suitable stocking density, can still have some impact on profitability, for example if dairy cow cleanliness is affected, as somatic cell counts may increase at high stocking densities in buildings (Arave et al., 1974). However, overall the economic impact of desires is likely to be less than needs.

Chapter 2
Mankind's Relationship to Animals in Nature

Animals in art and prose – changing attitudes to animals – benefits of a close connection with the natural world – comparing the behaviour and welfare of animals kept in captivity and the wild – observing the behaviour of animals in the Malaysian jungle

Introduction

Man's relationship to nature is of fundamental importance to the well-being of both individuals and society. Many people relate to nature more closely than to fellow humans or the artificial products of human civilisations. Who has not marvelled at natural wonders, admired the awesome power displayed by nature's forces, such as a waterfall in full flood, and felt inspired by the beauty of the animals and plants that inhabit our natural environment? As technological advances have diminished human reliance on the natural world, we have adapted our relationship with animals to concentrate more on the mental benefits of close contact with animals and plants, for example from ownership of companion animals, with less reliance on the physical benefits. Even the group of physicists, mathematicians and chemists that designed the first atom bomb took strong support from nature and were inspired by it (Fiege, 2007). Man's relationship with animals is still one of the most important components of moral behaviour, and to understand man's current position it is important to see how this has developed over the centuries. Nowhere is this better illustrated than in art, prose and religious beliefs, which provide us with a permanent record of the way in which man has changed his attitudes to animals over time. The following section chronologically charts man's changing attitude to animals through these media.

Changes in Attitude to Animals Over Time

Palaeolithic Period

The earliest evidence that we have of man's relationship with animals in nature comes from the prehistoric art of the Palaeolithic period. These show that man

C. Phillips, *The Welfare of Animals*, Animal Welfare 8,
DOI 10.1007/978-1-4020-9219-0_2, © Springer Science+Business Media B.V. 2009

was reliant on animals and probably in awe of them. Cave paintings were discovered recently in caves at Chauvet Pont d'Arc in the Ardeche region of France that date from about 29,500 B.C., when modern *Homo sapiens* was just emerging as a successful world coloniser, and Neanderthal man was in decline (Valladas et al., 2001). Like many later cave paintings (Garfinkel, 2006), these are deep inside the cave labyrinth, out of reach of natural light, in a part of the cave complex that was not inhabited, suggesting that they were not used for decoration of living quarters but for magical or religious purposes (Lewis-Williams, 1997). Because the paintings are layered on top of each other, it appears that it was the act of drawing them, rather than the end product that was the main purpose. The animal profiles depicted at Chauvet include bison, rhinoceros, panther, bears, horse, deer, lions and an owl. Many abstract symbols were also drawn but the significance of these is unclear (Anon, 2007c). Paintings of similar antiquity have been found in aboriginal Australia, again layered on top of each other and commonly depicting food animals so that hunting techniques could be improved.

The palaeolithic cave paintings have a sense of nobility (Lorblanchet, 2007), and although killing is often depicted, no sense of hatred is evident. In the Chauvet paintings, about 60% of the animals are those that would have been dangerous to man, if mammoths are included. Because men hunted as a pack and were reliant on others to help them kill animals that were individually much stronger than each human, the outcome must have been far from certain. The actions and behaviour of the hunted animals were unpredictable and the hunt would only be successful if the men acted in unison, with clear thinking, self assurance and calmness. This would have encouraged a respect, even reverence, for the animals, so that by supplications to the animal gods before the hunt the hunters could build up their confidence for an audacious attack. Group activities, such as chanting, may have assisted the bonding process to ensure cohesion in the hunt (Thackeray, 2005). We can picture the scene before the hunt. The men in the tribe had learned of the whereabouts of a herd of mammoths, they withdrew to an inner part of the caves, away from the distractions of the women and children. They carried a torch to light their way. Then deep in the cave system, they sat and watched as the artist of the group drew a huge mammoth, with symbols to represent aspects of the hunt that did not need to be drawn in detail, perhaps the position of people, or the approach route. Then they chanted songs about the hunt that they were planning for the next day, focusing on the dangerous activities ahead, and they danced in unison to bond together. Later groups may have used hallucinogenic drugs to enhance the spiritual nature of the process, but it is not clear whether the early hunters had the necessary botanical knowledge.

There is little evidence of any transition in the European cave paintings over time, except where there were changes in the flora and fauna. Australian cave paintings progressed from the Bradshaw human figures, which are similar to many early African paintings, to more complex drawings of animal outlines with hatched shading to allow a solid image to be depicted without using too much paint. If an accurate depiction of the painter's subject was the desired aim, one would expect a transition from crude paintings in early periods to more sophisticated ones later. This is not generally apparent, except perhaps in Australia, which supports the contention that it was the act of drawing them that was their purpose. These paintings are to be found in almost all parts of the world, and apart from differences in the animals depicted due to geographical variation, they are remarkably similar in form, shape and size. It appears that prehistoric man had a sense of reverence for at least the large mammals that he hunted and that this helped him to co-operate in slaughtering them. Some birds are depicted, and marine mammals are common in sites near the sea, but overall there is a focus on the most dangerous animals, which suggests a magical purpose to the drawings.

Later rock artists appeared to use their artform to express their wishes for human society, such as maintaining male dominance over women (Whitley, 1994). They also used sculpture for animal imagery, but the nomadic nature of the human population in the Palaeolithic period would have made this of limited value. The earliest known animal sculptures, a bird and a horse carved in ivory and found in Germany, suggest that such figurative art emerged about 30,000 years ago (Conard, 2003). Somewhat later (20–19,000 BC) is a small stone figure of a flying swan, probably an amulet. The reliance of the ancient nomadic people on animals is demonstrated by their use of an Animal Style of art, which employed animal images in weaponry, clothes, ornaments and jewellery (Carter, 1957). These images were not detailed but portrayed the most prominent features of the animals only. Use of the imagery is suspected to have been a means of protecting the people from evil, with the animals believed to have magical power (Mundkur, 1984).

Animals at this time were, therefore, more than just *things*, they were perceived to have spirits. The animal teleos, or individual identity, was held in high regard by the people that hunted them. At this time in our history, humans were therefore totally dependent on animals for their livelihood. Nowadays, most animals are partly dependent on people for their livelihood, and it behooves us to honour the opportunity that they gave our forebears to develop into the human society that dominates the world today.

Neolithic Period and Early Civilisations

As man started to develop closer associations with animals and to cultivate plants in the Neolithic period, there was more opportunity to construct artifacts to represent their relationship with animals. Many of the earliest sculptures, created 30–25,000 years ago, represented not animals but pregnant humans (Anon, 2007d), evidence of a desire for fertility that would have helped the population to replicate. As settled agriculture developed, the prospect of misuse of animal and plant resources became possible, and even likely. Many early civilisations failed when their use of natural resources became unsustainable or climate change forced new directions in agricultural techniques. There is evidence of this in the livestock farming in the Middle East. There sheep and goats were first domesticated about 6–8,000 BC (Hatziminaoglou and Boyazoglu, 2004), in a region called the 'Fertile Crescent', but now that land is unable to support even enough animals to feed the local population. To fill this gap a buoyant trade in live sheep between Australia and the Middle East has been established. So now on the wharf in Western Australia you can often see large numbers of sheep waiting to travel by ship to the Middle East, leaving one fragile environment that could easily lose its ability to produce sheep in a sustainable way to another fragile environment, the Middle East, where the water available for agricultural production has reduced very significantly since prehistoric times (Araus et al., 1999). Other problems in the region which probably were associated with the collapse of agriculture in the region are salt accumulation from repeated irrigation of the land and forest destruction (Nissenbaum, 1994). History has shown us that truly sustainable agriculture is hard to achieve (Brown, 2007; Gintzburger et al., 2005), particularly in relation to water use, and one of the lessons from this period of early agricultural development is that we must ensure that animal production in fragile environments does not permanently damage the environment for future generations.

The Garden of Eden and its meaning to humanity The Garden of Eden is believed to have existed about 4,000 BC (Hill, 2000), 2–4,000 years after mankind first started to develop domestic sheep and goat breeds in the Fertile Crescent. It is tempting to think that it was man's changing relationship with nature, as a result of lessons learnt from over-exploitation of plant and animal resources, which was the reason for this story and similar ones in other religions. Carlson referred to the allegorical significance of this story when she wrote "today we use animals as symbols of a long-lost Eden, a connection to a religious 'wild'" (Carlson, 2002). Man's relationship with animals was changing from one in which he merely co-existed with animals, exploiting them where he could, to one in which he was responsible for managing them. He was developing the knowledge required to use the animals and plants judiciously, through agriculture. Although the Fertile Crescent was well supplied with all the necessities for domestic life and it was here that most of our current plant and animal species were domesticated (Diamond, 1997), the problems of drought, salinity and deforestation challenged farmers in the region in ways that are familiar to farmers pursuing agricultural sustainability today. The Fertile Crescent was a good testing ground for man's skills in sustainable agricultural management, and after several thousand years of developing systems of animal agriculture, this story warns us to respect the forces of nature.

In the allegorical story, man's relationship with nature was potentially the source of all wrong-doings, since his relationships with animals (the serpent) and plants (the apple) were fundamental to his knowledge of right and wrong. He learned directly from nature, in the form of the serpent, about right and wrong. He learnt that he was different from the rest of the animal kingdom (because he was naked) and was initially ashamed of this fact and covered it up. He therefore shrank from his knowledge of good and evil, and not only did he cover his nakedness, he hid in the garden. The task required of him, to understand what is right and what is wrong and behave correctly according to this code, was initially too great. He was tasked (by god according to the story) with having dominion over the animals, managing them, making him distinctly different to all other animals. This is an enormous responsibility, and even if we don't believe that a god required us to take it on, it is clear that we have that responsibility now. Our impact on the plant and animal kingdoms' of the world is indisputable. Apart from all the animals that we manage directly, the farm, companion, research, sport, work and zoo animals, our impact on the natural environment is becoming increasingly evident. Sometimes we maintain or recreate ecosystems of a type we prefer and feel comfortable with. Or they may be from a particular period in history, as in the preservation of small farms and their integral connection to nature in British national parks. But our influence is far greater than that. We encroach on natural habitats to convert land to agricultural or housing use and we contaminate the atmosphere with gaseous pollutants, which eventually changes the environment worldwide. The influence of man on the atmosphere is not just greenhouse gases, heavy metals like lead have been emitted into the atmosphere and have been found in polar icecaps. Accepting that we have this influence on nature is a major responsibility, and one that, like Adam, we may find difficult to accept. As a society, we should respect those who take on that responsibility and manage animals well – farmers, staff in animal shelters, zoos, sanctuaries, animal laboratories etc.

We have learnt a lot about animal management since the Neolithic period, and important milestones, such as the simultaneous discovery of evolutionary principles by Darwin and Wallace 150 years ago, can now be viewed as increasing our knowledge to assist us to manage nature. The religiously inclined may believe that it was assisting God's purpose for man, rather than any challenge to his supremacy, which Darwin was so afraid of.

Early civilizations The development of early civilisations in the Middle East and Europe maintained a close contact with animals, firstly as objects to be hunted and secondly as gods, for example the cat in Egypt. However, they added new uses of animals, as livestock and pets, for guarding and as hunters. The ancient Mesopotamian civilisations were the first to develop urban societies, but the inhabitants still clearly relied much on animals. Art of this region has survived in the form of sculptures, ranging from the 26th C BC to the 6th C BC. Animals sculpted included bulls, sheep and horses. Sculpted dogs, believed to be for guarding purposes, have been found that are at least 15th C BC (Fitzwilliam Museum, Cambridge, UK). Assyrian cultures favoured depictions of bulls that had been stabbed, demonstrating a celebration of the animals' death that contrasts with many of the images of the prehistoric cave paintings. Later the Romans developed the practice of sacrificing bulls both before battle to cleanse the troops and prepare them for the task ahead and also after, to celebrate victory. Our best record of literature of attitudes to animals at this time is the Old Testament, in which it is evident that animals played a major role in human society. However, they were beginning to become items of property, as evidenced by their use in sacrifice to appease gods. They were still revered, such as the Golden Calf that was carried with the Israelites in their travels in the Middle East.

Egyptian civilisations had many gods, most of whom had an animal form in at least some part of their body, and which was fundamental to the life of the everyday people. Many were half human, half animal – chimeras – which were common in other ancient cultures. Early Egyptian civilisations (about 7,000 BCE) commonly used animals in their artwork: baboons, perhaps as a fertility symbol, frogs, which symbolised the inundation of the Nile, and boars. The greatest evidence of animal symbolism comes from the late Egyptian to Roman period of 4,380 to 4,200 BCE, when animals were extensively depicted in their art. There were many representations of dogs, for hunting and guarding property, hippopotami, often shown being hunted, scorpions, which represented kings because they could both attack enemies and defend their territory, and cats, which were reared to be mummified at an early age. Some animals were believed to be representations of gods when they had died, and the coffins of mummified humans were adorned with much animal art, especially cattle, ibises and dogs.

The Etruscans of Italian Tuscany developed a vibrant agronomy, growing flax, grain and olives. There was less reliance on animal art because of this, but there were still the mythical centaurs, swans, lions, oxen and horses used to embellish everyday items, especially pots. The artifacts from this period in the Dachian civilisation of the nearby land that is now Romania also regularly depicted centaurs, chimeras and animals. As early as 7–11 C BC when the Etruscan civilisation started, there was less animal use because of the thriving plant-based agriculture had been developed in the near perfect climate. Later, it is interesting that few Italian saints are associated with animals, in comparison with for example, the Celtic saints (Preece and Fraser, 2000).

Greek and Roman civilisations The ancient civilisations of Greece had a close connection with animals. Their confident approach to depictions of animals in art in the Classical period reflects the self-assured nature that they displayed to their interaction with the natural world. Even the constellations of the sky were represented in animal forms. The Greeks greatly respected beauty, and like the Palaeolithic hunters, they depicted the large mammals in much of their art, particularly horses and bulls. The horse, from which they derived substantial benefit, was idealized in their art as a creature of great beauty. The art historian Kenneth Clark described their relationship with these animals that even now we consider beautiful:

> Primitive man's admiration for the beauty and strength of animals never died out, and in more evolved societies became the inspiration of great works of art. From about the year 1,500 BC sculptors have found in certain animals a grace, a delicate balance and a smooth relationship of the part to the whole which we have come to describe as beautiful. An undefinable word; but since it expresses the difference between our

> feelings when we look at a gazelle or a horse and those aroused by a crocodile or a wart-hog, it is worth preserving . . .as the description of a precious human experience. . ...The Greeks . . .had one superb vehicle for their sense of beauty, the horse. . . ..No wonder artists have been inspired by horses. The splendid curves of energy – the neck and the rump, united by the passive curve of the belly, and capable of infinite variation, from calm to furious strength – are without question the most satisfying piece of formal relationship in nature: so much so that good photographs of horses have the same effect on me as works of art (and of course a well-bred horse is to some extent the result of art) (Clark, 1977).

The extent to which the Greeks valued the beauty of horses is shown by their regular use of horse images on their coinage. The depiction of the horse in art not only emphasized their grace and fine lines, so expertly painted by artists such as George Stubbs, Leonardo da Vinci, Géricault and Degas, it also drew attention to their energy and at times ferocity.

The Middle Ages

The mediaeval period of the 5 – 14/15 C AD was characterised by much use of animal imagery, particularly on jewelry. Most of the people had close contact with animals, and in particular depended on them for food and tillage of their land. The pervasive nature of animal imagery at this time also extended to domestic artefacts, where even the jugs were often in the form of animals (aquamaniles).

It was during this period that the confidence of the European population was shaken by the ravages of plague and famine. The people turned to their God for help, hoping for appeasement through the construction of grandiose cathedrals, and the Christian church grew rapidly in power and influence. The respectful relationship of Europeans with animals was evident in the gargoyles that adorned the cathedrals.

Religion therefore played an important part in the lives of everyday people, and animal imagery was much used in ecclesiastical architecture and literature. In the absence of printing presses, monks were tasked with transcribing books, and they often illustrated the texts with depictions of animals in majestic or humorous scenes, adding interest and amusement. It was in this era that the signs of the zodiac were devised to provide shorthand to the months of the year.

The mediaeval period left a significant store of animal art in adorned books. The monks who created the animal art had better living standards than the majority of the population and they devoted themselves to two things, God and nature. The best example of their close relationship with nature is the Bestiary, a much used book with depictions of animals, mostly exotic or grotesque in nature, in symbolic pictorial stories. It was used during their long hours of prayer and reading. The stories drew theological or spiritual conclusions from observation of the natural world. The text derived from a 2nd C AD Christian book, Physiologus (meaning 'naturalist'). The animals used to embellish the manuscripts often came from the Roman and Greek literature of the early Christian church. Many were of North African origin. Their strange nature, with combinations of body parts from several species, also relates partly to the lack of direct experience of the animal on the part of the artist.

The classical attitudes to animals of the ancient civilisations clearly influenced the Middle Age animal art. For example, the respect shown for the horse in the Romano-Greek era, which verged on idolatry, was further developed in the Middle Ages by Europeans in the form of the unicorn. This was a mythical horned horse, which had probably been imagined by the explorers of the newly discovered African and Asian lands when they saw horses with riders brandishing spears or lances. Despite its large horn, a defensive weapon, it was portrayed as submissive in the presence of even the most vulnerable of humans, the young virgin, and was frequently depicted resting its head in the lap of a virgin. This may have symbolized the submission of the human male to the feminine nature of woman, but probably the true thoughts of the original unicorn artists will never be known. It later came to symbolize the risen Christ and thus the sacred character of the natural world. It was believed that a unicorn could purify water by dipping its horn into it, which symbolised Christ's purification of the world after the fall of mankind in the Garden of Eden. So prevalent was the image of the unicorn in mediaeval Europe, when later European explorers saw rhinoceroses in Africa, they mistook them for unicorns.

The monks' rudimentary knowledge of animal breeding, and the possibility to create different animal forms by cross-breeding, may also have driven them to depict fantastical animals. Some well-known northern European animals were drawn in forms that the artists must have known did not exist. The artists, who were mainly mediaeval monks, probably had a considerable interest in, and perhaps knowledge about the breeding of animals. In mediaeval times an extension of their knowledge of the inheritance of characteristics of farm animals to wild animals is a possible explanation for the fanciful animals that adorn the Bestiary. Several centuries later, another monk, the now celebrated Austrian, Gregor Mendel, discovered the mathematical principles of genetic inheritance through his careful studies with peas.

However, there is another reason for the mysticism, which may have prevailed in the early days of bestiary drawing. In the early Christian church these drawings were allegorical and there was a need for symbolism in the days of Christian persecution. It also represents the uncertainty that surrounded the church at the time, with a rejection of traditional paganism by many and a search for mystical and fantastical theories. Such new religious theories were explored through the depiction of animals in strange forms and situations, thereby demonstrating the closeness of the artists at least to the natural world. This symbolic use of animals was also evident about a millennium earlier in the biblical Old Testament book of Job, in which it was written 'Ask the beasts and they shall teach thee, and the fowls of the air, and they shall teach thee' (Job, Chapter 12, Vs 7).

Not all the mediaeval depictions of animals were of fantastical animals. The monks often drew pictures in their manuscripts of animals with which they interacted every day, such as fish or wild boar, sometimes to decorate an insertion of text, sometimes for no other apparent purpose than to provide a pleasing appearance to a page or to satisfy the artist's need to communicate his feelings for the natural world. These animal 'doodles' may mean little to us now, but it is symptomatic of the close communion with nature of the monks of the mediaeval period, that animals were commonly depicted in their texts. In some mediaeval texts and ecclesiastical frescoes, animals are put in humorous scenes, clearly for the amusement of the reader or viewer (Figure 2.1). Later, after the arrival of the printing press, such embellishment of the text with animals disappeared. It may have been unnecessarily cumbersome to set up on the printing presses, or it may be that writers did not wish to reveal such intimate aspects of their character for mass readership. Later, some bestiaries were defaced to remove the humourous images, such as the horned devils and people's bottoms in the Macclesfield Psalter of 1330, probably by those that equated laughter with evil in the Puritanical spirit that became popular through much of Europe in later years.

In addition to gargoyles and bestiaries, in England and other prosperous regions of Europe at the time, church effigies and brasses provide some material to enlighten us on attitudes to animals at the time. Memorials often included dogs lying at the feet of the deceased, but this was not because of their being required to accompany their owner to the next life, not to indicate the submissiveness of the dogs, nor even because of their emotional significance in the life of the owner, but rather because in colder regions of Europe at time, people slept with dogs at their feet for warmth and protection. The importance of some animals of the mediaeval period for the livelihood of the general populace is demonstrated by sheep depicted at the feet of wool producers (Fig. 2.2). However, lions were also commonly depicted in this position, illustrating the

Fig. 2.1 Execution of the cat by a procession of rats and mice. Detail from the Cathedral of Tarragona, Spain (Evans, 1896)

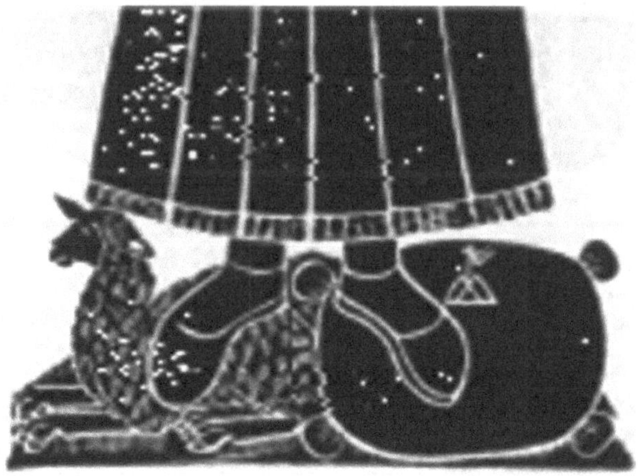

Fig. 2.2 Footrest from brass to unknown wool man, c. 1485, Northleach, Gloucestershire (http://www.mbs-brasses.co.uk/pic_lib/Picture_Library-Wool_and_Cloth_Trades.htm, accessed December, 2006)

reverence for this animal as a symbol of courage and strength, as in Richard the Lionheart. The lion has been the symbol of English kings since that time. In addition to the lion, many other animals had symbolic characteristics, as developed in heraldry. In Christianity, birds were the symbol of the risen Christ, and were also protectors against the plague. A mythical bird Charadrius could take any disease from humans. Doves came to represent the Holy Spirit, a symbolism that has been maintained to this day. Dragons represented the devil, and a red dragon symbolised the British people before the invasions by the Anglo-Saxons, which were represented by the white dragon. The red dragon is still used by the Welsh people as a symbol of their Celtic origins. Another fanciful animal, the basilisk, could kill by its smell, glance or hissing sound, and the manticore, with the face of a man, body of a lion and tail of a scorpion, had so seductive a voice of siren song that it could tempt Christians away from their life's journey in prayer and reflection.

The early Christian churches and manuscripts in England were therefore much adorned with animal art, mostly of the allegorical kind, some of which was copied from the Mediterranean animals, but some depicted northern European animals. In addition to that remaining today – the gargoyles, the illustrated texts with bestiaries etc. – the walls of the churches and cathedrals would have been painted with images that have long since been eroded away. The people of the Middle Ages, especially the monks, therefore had close contact with animals and they depicted them regularly in their art. The fact that they chose to illustrate books, adorn churches and probably their homes with them, suggests that they understood how the natural world functioned, and they believed that it had a moral perspective that humans needed to learn and understand.

Animals were also much used in sport. In one game, the village idiot was encouraged to bite off the head of a sparrow, which usually resisted fiercely. Many other barbaric rituals masqueraded as sports, such as cockfighting, dogfighting and bearbaiting. Clearly animals were used both in the positive and negative aspects of mediaeval life.

The Renaissance

In contrast to the allegorical use of animals in the Middle Ages, the Renaissance movement, which originated in Italy in the late 14th C, was a worldly movement, enjoying life's pleasures and celebrating them in natural art. Renaissance art had less reliance on animals than previous cultures, but it was sometimes characterized by life-like depictions of animals in dramatic poses, displaying the ferocity that reminded people that nature could not be easily tamed. Animal art of the period was characterised by depictions of one of the most graceful animals ever painted: the horse, most notably by Leonardo da Vinci, who painted horses for almost his entire artistic career. He had a flair for, and interest in, depicting the anatomy of horses with great accuracy. The confidence with which he drew his horses, that may have surpassed even Stubbs in detail, was symptomatic of the enlightenment that pervaded this era. The contrast between the direct approach to animals and nature in Classical and Renaissance periods and the mythical approach of the northern Europeans in the Middle Ages may simply reflect the later development of civilization in northern, compared with southern Europe. It was not until the Victorian period that the northern Europeans came to portray dominance of humans over nature in art, but when they did it was with great romanticism, reflecting the confident mood of the people at this time of rapid progress.

The employment of any frivolity, such as the inclusion of animals to enrich religious texts, was greatly frowned upon in English texts of this period, and especially in the King James' version of the Book of Common Prayer, produced in 1603. Such heavyweight texts were needed to confirm the King's authority and to bring a sense of order at a time when England was troubled by civil war and there was general unease in the population. Despite the demise of the Stuart kings, the Book of Common Prayer continues to be used in some English churches until this day, reflecting the popularity of the ancient prose.

The first appearance of animal activism occurred at this time, encapsulated in a painting in the Hermitage museum of St Petersburg, *Portrait of a Hunter* by Paulus Potter (Fig. 2.3). In this picture hunting scenes in the peripheral paintings are juxtaposed with two main scenes that depict, first, a bear and two wolves hauling a hunter before a tribunal consisting of an elephant, a ram, a leopard, and a lion. A fox holds down the bill of indictment with his paw. The hunter's cringing dogs are dragged behind him, by a bear and a boar. The second scene shows the hunter being roasted on a spit by his former victims, while his dog auxiliaries are hoisted skyward on a rope. This reversal of roles was more famously portrayed in prose in George Orwell's Animal Farm, first

Fig. 2.3 Portrait of a hunter, Paulus Potter, Hermitage Museum, St Petersburg. c. 1650

published in 1945. Paulus Potter was a prolific painter of animals in the Golden Age of Dutch painting in the early 17th C. He wandered the countryside observing how animals behaved and was unusual because he was one of the first accomplished artists who made animals the focus of his paintings, rather than an accompaniment. His meadows were enlivened with frogs, lizards, poppies and butterflies, and he was more attuned to nature's moods and the timeless harmony of animal, landscape, and weather than any of his contemporaries, or perhaps predecessors. He died of tuberculosis at the tragically young age of 28.

In the 17th and 18th C, one of the most common depictions of animals was in family portraits, which usually had at least one dog in accompaniment. Edward Landseer, George Stubbs and others portrayed animals in bold, dramatic postures and settings, but under the control of humans, adding to the belief that animals were there to be managed. Frequently, characteristics of animals that people admired and desired were portrayed, such as the ferocity of the lions that Stubbs often depicted. Alternatively the animals were depicted dutifully mirroring the family head's gaze and expressions. Stubbs also painted cattle, such as the Lincolnshire Ox (Walker Art Gallery, National Museums, Liverpool, 1790), which was probably approximately 50% bigger than life size, by comparison to the human standing beside it. Other painters of agricultural animals also exaggerated body size, so that they appear to have impossibly small legs that could not support them.

Victorian Times

The historical art period that represents the strongest affinity with nature is probably the post-industrial revolution era in northern Europe. This may reflect nervousness towards industrialization, with its reliance on man's discoveries, perhaps also a reaction to the way in which modern societies had distanced themselves from nature in towns and cities, perhaps a concern about the pollution that began to pervade these areas or perhaps increased leisure time and money to enjoy the beauty of nature. Nowhere was this better represented than in the art of the day, in which the viewer was encouraged to see the countryside as innocent, beautiful, traditional and untainted with modern development, and generally a happy place to be in.

The European Pre-Raphaelite Brotherhood, founded in 1848, was at the forefront of the Victorian Romantic movement. A similar movement of romantic naturalism re-established nature as an acceptable subject of moral philosophy (Kalenda, 2005). In Victorian art, in contrast to the dull images of classical 19th C nudes, Constable's countryside scenes or even some of Stubbs' animals, the seven brethren of the Pre-Raphaelite movement depicted nature as romantic, with sensuous women and bucolic reposes of farm workers in natural settings. They used bright colours, with many shades of greens. Often there are small groups of animals in the fields, with bountiful supplies of fodder and shade from trees. They are depicted lying and resting, like those tending them; hence the life of the grazing animal appears idyllic, like that of their carers. This suggests that the artists believed that animals had to be happy for their carers to be happy. They could have depicted the animals hard at work grazing, while their carers rested. This contrasts with the paintings of Constable and his followers 50 years earlier, where rural scenes were less warming to the viewer, often with dark skies, distant animals grazing and the focus was on the harshness of rural life at the time. The Industrial Revolution had started and people were alerted to its potentially powerful effects by artists such as Turner, who did not espouse the classical style of the early 19th C, and painted neo-industrial scenes, such as the first steamships, with dramatic sunsets that are evocative of the dawn of industrialization. In contrast to this, fifty years after Constable and Turner, in the mid-late Victorian period, the Pre-Raphaelite brethren attempted to draw the city dwellers' attention to the beauty of the natural world that they had left behind.

The brethren began their work in Oxford, but continued in London, where the rapid growth of the Victorian city must have incited reminiscences of the pleasures that nature could bring. They promoted painting directly from nature (i.e. outdoors), being true to it, with a great attention to, and reverence for, the detail of nature, and they encouraged people to look at nature with pleasure and to appreciate wild places. Typical scenes included a few cattle or sheep, being tended by a small group of children or a family group, who were resting in the shade of a tree in a well-wooded valley. There was never a suggestion of failing crops, starving cattle or a poverty stricken rural population.

The reality of rural life was quite different from the pre-Raphaelite's romantic perceptions. By the 1870s England was in the grip of a severe agricultural depression, characterized by low wages, intense competition between farmers for markets in the face of cheap imported food, and rural depopulation fuelled by an increasingly mechanized agriculture. Still the landscape painters continued to depict idyllic rural scenes. This was the biggest period of rural change of the millennium, with the proportion of the population living in cities increasing from 20% in 1801 to 80% in 1911 in Britain. The expansion of Victorian cities was accompanied by the development of strict moral codes, necessary because of the poor working and living conditions. Fear of disease, poverty and a people's revolution similar to that in France added to the misery of city dwellers. At this time, the rural idyll depicted by such painters as the brethren's founders, Dante Rosetti and Benjamin Leader, attracted great attention in city exhibitions. The idea of family groups with their accompanying animals, which could easily gain a living from a countryside resplendent with abundant resources, must have given hope to many. The reality was that landowners operated strict control of their animal and plant resources, with gaming laws preventing families from supplementing their meager diet with even rabbits or pigeons, and some even preventing the collection of berries and nuts on their properties.

By contrast, the Australian paintings of the late 19th and early 20th C agricultural pioneers were typically more honest, showing large mobs of cattle in a wide open, but still beautiful landscape, traveling on a dusty road to slaughter. Meanwhile in Europe visiting the wild places became popular, which was facilitated by the growth of the railways in the mid to late 19th C. For example, the mountains and lakes of England and Wales became increasingly fashionable for tourists as the Industrial Revolution progressed (Trevelyan, 1949), suggesting that people working in stressful conditions needed contact with nature to recuperate. A typical picture of the day exists in the Bristol Art gallery and depicts the true story of a climber lost off Striding Edge on Helvellyn in the English Lake District, with the faithful dog found beside the body some months later. Fidelity, courage, strength, beauty were all favourite topics for Victorian artists painting animals.

Literature too embraced the benefits of a harmonious union between humans and animals. At the end of the Victorian era a major series of animal stories, the Jungle Books, were written by Rudyard Kipling, which used animals to portray moral stories to children. The canniness of the fox, the independence of the cat, and the sociability of the dog were all used to explain to children how they should behave. In these days, regular contact with animals was normal for most children, and the tradition of using animals to instruct children has continued to this day. However, it is debatable how long this will continue, because children will increasingly fail to recognise the animal characteristics that their forebears knew so well. There may be a stronger focus on animals known to city dwellers – domesticated pets and animals, such as foxes, that have adapted to city life.

Recent Times

In modern art, animals are less frequently depicted compared with most earlier art, showing that we have become more remote from contact with animals. Artists prefer to use manufactured items, especially those in every day use, and put them into unusual contexts to surprise the viewer. War scenes were commonly depicted in the first half of the 20th C, as artists wanted to depict the horrors that they perceived (for example Guernica by Picasso), in contrast to the patriotic and heroic messages of governments of the day. A rare but evocative first World War animal painting depicts a faithful dog, looking forlornly out of the window, with an obituary on the wall, and widow's knitting and reading book on the windowsill ('Will he come home', Bristol Art Gallery). Elsewhere few animals were depicted in art, which was preoccupied with war and the last vestiges of the Industrial Revolution. The motorcar had begun to have an impact on society, man could fly in machines and the Victorian fascination with the natural world must have appeared old-fashioned and irrelevant in the machine age.

In the mid to late 20th C, the public became captivated by the domestication processes that are at the centre of our relationship with nature. Genetic manipulation by humans had been a concern since Darwin's time, and the scientists had begun to consider the mechanisms of domestication. Jack London famously explored the relationship between wolf and dog in his influential books 'Call of the Wild' and 'White Fang', but was criticized for humanizing canines (London, 1939). In the 1960s, after the ravages of the wars of the first half of the century, there was a time of exploration of nature's beauty, perhaps as an antidote. The public were captivated by the work of Joy and George Adamson in Africa, in the books Born and Living Free and a film of the same name, in which they rescued orphaned lion cubs and cheetahs, hand raised them and reintroduced them to the wild. There was resurgence in art of the painting of animals in wild African settings, led by David Shepherd. People were for the first time beginning to question the ethics of rearing wild animals in captivity, and whether the domestication process changed animal form and function. Further afield the public became able to view the natural world from the comfort of their living room, through the endeavours of companies such as the British Broadcasting Corporation and their Bristol wildlife unit. Presenters such as David Attenborough, Johnny Morris and David Bellamy became household names and icons for their portrayal of the natural world in the media. Underwater, Jacques Cousteau began to reveal a world that people had only had very limited understanding and knowledge of before television. Children's television regularly used animal stories to reinforce the belief there is a hierarchy of care for animals, with mammals at the top and animals such as fish and invertebrates at the bottom (Paul, 1996). The use of farm animals in products such as meat and wool was rarely portrayed, thus avoiding the paradox of caring attitudes towards animals that are ultimately going to be killed for our benefit.

More recently, there has been a resurgence of interest in the beauty of animals and in their behaviour, which coincided with a growing realisation amongst scientists that the natural world was at risk from man's activities, and again a period of rapid urbanisation, as it had in Victorian times. The benefits of close communion with nature, for example taking 'green' exercise (walking or cycling in parks, conservation work etc) have now since been shown to be particularly beneficial for both self esteem and mental health (Pretty et al., 2007). Scientific studies have demonstrated that people living in buildings surrounded by green parks are less likely to suffer from stress, are less aggressive and more likely to engage in physical activities and have increased social cohesion (see Groenewegen et al., 2006). People increasingly understand that they need some close contact with nature and describe it as beautiful, soothing or edifying, not just green surroundings but also animals in natural surroundings. Despite this, where animals are used in art today, they are often included symbolically, or in an attempt to shock, a notion which pervades modern culture. Perhaps this is in response to a living style that is too protected, and as most people are free from involvement with wars, protected from the major infectious diseases, and enjoying unprecedented high standards of living, some have turned to nature to shock, terrify and amuse them. The depiction of animals in graphic or horrific images, or pickled as laboratory specimens, reminds the viewer that the natural world can still arouse our passions that have been dulled by the sterility of modern living. This was a technique used earlier by Stubbs, in his many paintings of lions attacking horses (an event he is believed to have witnessed), which had the power to alert our primitive instincts of fear and alarm. In the carefully constructed artificial world that many live in today, particularly those in cities, people appear to welcome this reminder, perhaps reflecting a primitive need for exposure to mock danger, so that we are prepared when real danger occurs.

More recently, in Australia at least there has been evidence of resurgence in interest in communion with nature through art, and renewed interest in the ancient aboriginal paintings of natural subjects is confirmation of this. The relationship which the traditional aboriginal peoples of Australia had, and to some extent still have, with nature has emerged to become respected by many in the western world. This recent reverence for that intimate relationship between man and animals was first evident when the cave paintings of horses, primitive cattle and others were discovered in the middle of the twentieth century in France. It may reflect nervousness that we are losing touch with nature, a concern that we are damaging the environment, beginning with issues such as the damaging effects of DDT, progressing to widespread habitat destruction and more recently pollution of the atmosphere with greenhouse gases. A return to nature is evident in today's Green movement, and the popularity of aboriginal art and its communion with nature is evidence of this movement.

Conclusions on Changes in Attitudes to Animals Over Time

It is often hard to discern the artist's message concerning animals, but it ranges from predominantly veneration in prehistoric times and early civilisations through to symbolism, which was particularly evident in the Middle Ages, respect and adoration, most famously in the Industrial Revolution era, and finally to attempts to shock and disturb, which has been used throughout history, but are probably the most common reason for including animals in modern art.

From the Palaeolithic period, when animals were the dominant subject in art, and in many cases the only subject, there has been a transition to animals becoming one of the less common subjects in the most fashionable art of today. This reflects the place of animals in society, from humans relying heavily on animals for their food supply to their having little or no physical need to have direct contact with animals. In between there were periods in early civilisations when most art had some connection with animals, but equally there were civilisations, both ancient and modern, whose art had little association with animals. Not all observers agree with this trend, Fraser (2005) believes that animals have received *increased* attention in literature, the visual arts and philosophy over time, at least since the beginning of the 18th C, and attributes this to increased scientific knowledge about animals and the transition from contact mainly with farm animals to contact mainly with companion animals. The best way to test the hypothesis is to visit national art galleries, which usually chart the development of the nation's art chronologically. For example, a visit to the national art museum in Bucharest, Romania, readily demonstrates that there was a transition from regular use of animals as centaurs, chimeras and agricultural animals in their early civilizations to much less frequent use by the second millennium AD. Indeed only a handful of the 1000 or so paintings from the 14th to 20th C on display include animals. Most of the bucolic images, that were almost as popular in this Latin country as they were for the Pre-Raphaelites, illustrate cropped fields and rustic buildings, with people enjoying the fresh air, good food and open space. The archetypal family portraits that were more common in northern Europe, with a father, typically standing and the tallest, a mother, often sitting, some children and one or two dogs are largely absent. A notable exception was a single painting of people and animals enjoying the forest, of which there is much in Romania, with the animals playing musical instruments, dancing etc, in a way that is reminiscent of the mediaeval bestiary of England, described earlier. Where animals are depicted in the countryside, and horses and cattle are included sometimes, they are not a central part of the picture, but are there to demonstrate their usefulness to agriculture. Animals of traction have pained expressions, in contrast to the humans, and other animals accompanying them. This simple examination of a country's national art treasures can therefore reveal the changes in attitude towards animals, at times showing close reliance on animals but also often demonstrating a utilitarian approach to animals. Nowadays the continued popularity of household pets in these countries perhaps demonstrates their strong need for animal companionship.

The trend in British art over the last two hundred years to depict agricultural animals and their carers in harmony with nature is usually far from reality. The pressure on farmers to produce cheap food resulted in some unethical animal and plant management practices emerging that necessitated legislation to control the worst examples of animal cruelty, as well as requiring long working hours by the farmer and produced a feeling of isolation in many. The amalgamation of farms to make economically viable units has led to farmers being estranged through rural depopulation. The level of suicides amongst farmers is one of the highest of any profession. Technological developments have changed agriculture from an art to a science, leading Rollin (2006) to lament that *husbandry* is not taught in agricultural colleges any more, having been replaced by animal *science* courses. Nevertheless, the bucolic images continue, to tempt us into thinking that there is a better world away from the cities. The emerging focus on high quality food production both draws on our desire to support this rural idyll and may be instrumental in reversing some of the damage that has been done to rural life by commercial pressures.

Over time, the proportion of the population that has a close connection with animals has declined. In hunter-gather societies, most men would have had involvement with the hunt, and women and children would be involved in looking after companion animals and preparing food animals for eating. Nowadays in developed countries, only a small proportion of the population have a direct connection with animals through their work, perhaps no more than 10% in total. A large proportion of the population has companion animals, but this is voluntary. It conceivably represents an attempt to maintain a connection with the natural world, to bring it into our homes, where we can wonder at it in comfort and free from danger. There is increasing interest in keeping exotic, dangerous animals as pets, which demonstrates the sense of satisfaction that we have in mastering nature.

The Benefits of Naturalness

There has been a focus in art over the centuries to depict mainly the animals that are of benefit to us as beautiful, and these are animals that we empathise with most. Often in art, animals were depicted in a wild setting and both animals and nature were romanticized. This was particularly seen in animals that are beneficial to us. Rarely has the rat or the snake been depicted as a beautiful animal, on the contrary the snake is usually associated with the guile and cunning that was first displayed in the biblical Garden of Eden. The curving lines of the snake are more likely to evoke feelings of fear than beauty, despite the fact that they possess features that might be considered attractive in human empathy – curvaceous, slim and slow, gracious movement. Such features are admired in other humans and some animals, such as horses, that are routinely depicted as beautiful in art. Fears of snakes and spiders are inherited characteristics (Kendler et al., 1993; Davey et al., 1993), demonstrating that our response to nature is not simplified to certain forms and features, but is complex and to some extent dependent on our genetic makeup.

Not only do the public view animals that are beneficial to themselves as more beautiful but also more sentient. In the ranking of sentience by university students described earlier (monkey > dog > newborn baby > fox > pig > chicken > rat > fish, Phillips and McCullough, 2005), the physiological validity of this ranking is dubious and there is no evidence that the dog is actually any more sentient than the rat. However, they were ranked at the opposite ends of the spectrum because the dog provides many benefits and the rat is generally viewed as a pest. Furthermore, these students believed that more sentient animals deserved better welfare, so perception of sentience may have a major influence on welfare standards. This shows the danger of relying on people's opinion to set welfare standards.

The feeling of beauty and wonder when we view the nature at its best is probably generated biologically in a similar way to the feeling of wonder when we see new technology that can help us in our daily lives, or the image of an attractive person. However, this perception of the sanctity of nature is something longer lasting, something that transcends the wonder of modern technology. Who has not stood in the nave of a big cathedral and wondered at the magnificence of the structure above? However, how more awesome is it to stand at the foot of great forest trees and wonder at nature's ability to create such an impressive structure. One of the characteristics that evokes that sense of wonder is probably the uniqueness of the feat, and this might lead us sometimes to be more inspired by man's 'cathedrals'. However, one can pass a beautiful natural scene every day and still feel respect for it and a calmness that is generated by the sight. The same cannot be said to be true for the sense of wonder on appreciating the possibilities offered for our lives by modern technology, whose attractiveness rarely survives the first few encounters. The delight of learning that a small machine like an Ipod, which you can keep in your pocket, can store all your favourite music is short-lived. No sooner has one wonder of modern technology delighted us than another captures our attention, as we search for that quick solution to correct the deficiencies in our lives.

There is likely to be adaptive benefit to such a sense of beauty, which is a feeling that has evolved to keep us knowledgeable about the natural world, developed after millions of years of dependence on nature. However, the same sense of wonder can be gained from watching the moon light up the waters of the sea, or the grandeur of a great mountain. Such attraction to scenes like these could have had positional benefit in the past, allowing man to accurately pinpoint his location in relation to his home territory. If this is true, we would expect such feelings to have developed in other animals.

Can we expect the same of animals – would they benefit from a close relationship with nature? Clearly they do not appear to use art to portray any positive feelings about nature, although sometimes their constructions, such as the nest of a bower bird, portray a sense of order and invoke a feeling of wonder that may incline us to believe that it is art. In contrast to this, the activities of the elephants that are persuaded to daub paint on canvases which are sold to tourists in Asia are not art, but merely the performance of a reward-driven behaviour. Even though they don't usually portray their feelings through the sort of art that we know and understand, we have reason to expect that animals derive similar benefit from a close connection to nature. As well as us, animals are likely to benefit from being able to identify the natural features of the environment and return to safe places at times of danger, find food sources more readily and obtain shelter when needed. If a latent need to be close to nature is present in humans *and* animals, it is likely that both derive mental satisfaction from a more natural environment – a need that may be partially satisfied if the enrichment in a cage is natural rather than artificial.

There have been a few experiments designed to test this hypothesis. One such was an attempt to discover whether rabbits, a natural grazing animal, prefer to eat grass or whether they would be satisfied with an artificial food mix that is commonly offered to caged rabbits (Leslie et al., 2004). The rabbits showed no clear preference for grass, although this could be explained because the mix could be eaten faster, and most prey animals like to consume their food as quickly as possible so that they can retreat to safety. Other students in my group were unable to demonstrate any benefit of, or serious interest in, natural enrichment (foliage and tree branches) for gliders (Greer, 2006) or the scents of favourite plants for squirrel monkeys (Carling, 2005) in a zoo environment. In another experiment, my research group did find that cattle function more efficiently (as in circumnavigating an obstacle faster) when their environment is bathed in green light, compared with red or blue light (Phillips and Lomas, 2001), suggesting that the light under trees might be more attractive. However, this could just be because this is in the middle of their visual spectrum, where acuity would be expected to be increased. This is a limited set of experiments, and there is much more to be done to test the hypothesis, but we can only conclude that so far there is very little evidence on whether animals benefit from being kept in a more natural environment.

If the hypothesis is supported, that animals do have a sense of beauty or respect for natural things, compared with man-made, then it may well be of benefit to the animals in zoos to provide tree branches for animals to climb, rather than alkathene pipes, or foliage to eat rather than pelleted food. Cages are often made to look natural for the benefit of the viewing public, who equate it with better welfare. This could be tested and adopted if found beneficial for animals in other intensive management situations – farmed and laboratory animals in particular. What features of the natural environment would be beneficial to add to the environment of animals housed in intensive environments? Should it be green, but this might not be appropriate for a desert animal, and to what extent are these enrichment forms species specific? There are opportunities for research to determine the responses to natural or unnatural enrichment, but the difficulties centre on how to measure the responses. Some experts consider that it is not important that enrichment mimics the situation in the wild (naturalism), rather that the animal can perform similar tasks to those that it would perform in the wild (functionalism) (Swaisgood et al., 2003).

There are alternative hypotheses to the benefit of a sense of beauty that man has when he is in touch with nature. In the Christian religion, followers are encouraged to view nature as the work of God, for example in the Psalms, which encourage reverence for God and a belief in his powers in nature. Furthermore, if a god has instilled in man a sense of respect for nature, and in the Christian religion at least, has ordained man to manage and look after animals, then a sense of respect for nature, and a feeling of pleasure when we are in contact with it and it is correctly managed, would be a significant step towards achieving this goal. We may feel a sense of awe when we see a magnificent mountain, but when we see nature destroyed, such as when we come across animals killed on the road, or the mountain is transformed into piles of waste stones or slate by open-caste mining, we feel a sense of loss or shame. In the past many works of art were created by people supposedly inspired by their religion, but as Dawkins has argued, this is not necessarily evidence that a god exists, rather that the artists were following the dominant convictions of the time (Dawkins, 2006).

If our sense of wonder at nature was simply a feeling of nostalgia, a yearning to return to the times when man *was* in close contact with nature, it is unclear what benefit would derive from such a feeling. There can be no doubt that man's inventions, his construction of an artificial world around him, have benefited his survival. They have enabled him to colonise the planet in even the most hostile of regions, to live in relative comfort, with increased longevity and improved quality of life. And yet man still benefits mentally and to some extent physically from close contact with nature. The close and positive relationship with animals benefits people as much as it does the animals. The advantages of a close relationship between animals and their owners are emphasized in books on animal management, e.g. English et al. (1992), and they provide an altruistic reason for improving animal welfare, which is often referred to in prose, since people looking after animals well are themselves enriched by the experience. Conversely people that are cruel to animals are considered outcasts by society. For example, the poet William Blake emphasized the antisocial nature of ill-treatment of animals:

> He who shall hurt the little wren
> Shall never be beloved by men
> He who the Ox to wrath has movd
> Shall never be by Woman lov'd
>
> Blake, 1803

Having exhorted those who might be considering animal ill-treatment not to, Blake then encouraged people not to harm animals for fear of the wrath that might be upon them if they did:

Kill not the moth or butterfly
For the Last Judgement draweth nigh

Furthermore those who perpetrate cruelty to animals may not be at peace with themselves, let alone other men. Yet cruelty continues for several reasons. One is that it can become a form of redirected aggression. About one half of prisoners convicted of animal abuse are motivated by anger, the rest being motivated by the need for sex or to impress or imitate others (Hensley and Tallichet, 2005). When confronted with aggression from other humans, an individual may turn to animals, and particularly companion animals, to release their own aggressive impulses, because the chances of retaliation are less. A second reason is because animal managers are confronted with diverse ethical dilemmas, such as whether to place personal ethical standards, which require that he provides for his family, above the welfare of the animals in his charge. Similarly, provision of high standards of animal care may conflict with minimizing the environmental impact of a farm, an important consideration in relation to free range systems of production. A third reason is the desensitization of animal managers to the plight of animals in their charge. It seems likely that this is most common when animals are only in the care of the manager for a brief time (such as abattoir workers). Those with long term care responsibilities, such as companion animal or guide dog owners or managers of farm animals that are used to produce milk, are less likely to become desensitized to pain and cruelty.

Life in a Natural Setting

Some people might imagine that for animals in the wild there are stresses untold, which reduces their welfare compared with the husbanded animals. However, it is wrong to imagine that grazing impala on the plains of the Serengeti, with a crouched lion just a few metres away from them, suffer prolonged stress. They know their escape capabilities and can judge their flight distance very effectively. There is a similarly relationship between the wolves and caribou of the northern Canadian territories (Mowat, 1963), where wolves will periodically test the fitness of fawns and old does by making them run, since in these animals there is a greater proportion of injured, malformed or inferior animals. According to Mowat (1963), the caribou herd respond by sheltering these animals in the centre of their herds. The young wolf pups are taught to test weak animals in this way and ignore the majority of the herd, who are free to graze within a few metres of the wolves. Population density is mainly controlled by birth rate, which increases in times of plenty. In times of food shortage the biggest direct killer is not starvation but disease, with malnourished animals quickly succumbing to rabies, distemper or mange. The major welfare influence has probably been from humans, even in this remote area, since wolves have been poisoned with strychnine because of their supposed remorseless killing of caribou.

Hence there is little evidence that animals in nature are constantly at risk of predation or that this causes prolonged stress, which would not in any case be adaptive for survival. The stress reaction evolved to cope with short term danger and the increased metabolic rate and other physiological adaptations caused by stress would not be adaptive in the long term. Being stressed is less efficient metabolically but it places the animal in a position of readiness to cope with danger. Darwin epitomized this view when he wrote 'We may console ourselves with the full belief, that the war of nature is not incessant, that no fear is felt, that death is generally prompt, and that the vigorous, the healthy, and the happy survive and multiply' (Darwin, 1859). His words suggest that he believed that it was maladaptive for animals to be constantly stressed by the presence of predators, and hence the flight or fight response is not usually continually activated.

Although they may not be stressed by the presence of predators, the optimisation of population size in relation to food resources means that wild animals are often less well nourished than their captive counterparts, which can reduce their welfare. Their typical longevity is usually less than their captive counterparts, perhaps only one half, with not only reduced plane of nutrition, but also limited possibilities for medication, in the event of sickness, and exposure to climatic extremes. However, some animals are particularly difficult to keep in captivity, such as elephants. Mean longevity of zoo elephants is only about 20 years, compared with 70 years in the wild (Wiese and Willis, 2004). Foot problems, caused by inadequate exercise, moist substrate on the floor, and exacerbated by obesity, together with circulatory disorders, account for the majority of premature deaths in zoos. In addition elephants will rarely breed in zoos (Clubb and Mason, 2003). Thus the achievement of potential lifespan in captivity depends on man's willingness to provide suitable living conditions.

Exposure to climatic extremes can cause major loss of life in both wildlife and free range livestock. The following passage by the wife of one of the pioneering Kenyan game wardens describes the devastating impact of drought on wildlife.

> the rains had been disappointing and insufficient to promote much regeneration of the shrubs favoured by rhino. Every bush was browsed almost to the ground, leaving only the hardwood. Lack of water elsewhere in the Park forced the elephant herds on to the river, and the vegetation suffered still further. Patrols brought in distressing reports of rhino dying daily.......... the plight of the rhino in the area was indeed pitiful and the reports had in no way been exaggerated. We saw several rhino, all pathetically weak and covered in black patches, and came upon one, which, unable to stand, was lying in the blazing sun and had only just enough strength to snort faintly and toss its head as we approached. We . . . tried to encourage it to eat, but it had lost the desire to live and died shortly afterwards......Further along, we came across a baby rhino standing sadly beside its mother's carcass, nudging it at intervals and obviously puzzled at getting no response. The rangers quickly surrounded this little calf, who, although small and

> helpless, courageously refused to leave his mother's body and stood to face what he believed to be his enemies. The gallant behaviour of this baby rhino moved me to such an extent that I wished with all my heart that he might be spared and that we would be able to rear him. But another look at his wasted body, made me realise that there was little hope and this proved to be the case for, although we took him back to camp and did our best to save him, he died that night.
>
> On another occasion, we found a rhino lying on the banks of the river in the last stages of exhaustion, while a host of vultures were tearing the living flesh off its hind quarters. It was too feeble to keep them off and could only lie there and endure the torture in silence. A merciful bullet brought its suffering to end.
>
> Every day brought fresh examples of the appalling suffering which these unfortunate creatures were enduring. One particular incident which upset us greatly was the death of an old female rhino, who was well known to us for she possessed a pair of unusually long horns. She had been trying to reach some green leaves growing from a branch of a tree overhanging the river bank and had lost her footing and fallen into a pile of driftwood below. We found her fairly wedged between two logs with her head only a few feet from the edge of the river. It was obvious that she had been in this position from many hours, if not days, and it is not difficult to imagine the torment she had undergone dying an agonising death of thirst with cool water running by just out of reach. When we found her she was still alive and while.... the Rangers tried to free her with the help of an axe, I dipped my sweater into the river and squeezed the water though her parched lips. She gave a couple of weak gulps but again we were too late , and, with a heartrending sigh, she died a few minutes later. (Sheldrick, 1966).

In contrast to the suffering of wild animals under such conditions, farm animals are usually offered supplementary feed or moved to better conditions in such circumstances.. The delicate balance in the natural world is well understood by those managing game and national parks. Although they can do little to influence the forces of nature, or the balance of wildlife, they understand the importance of maintaining these reserves as a sanctuary for wildlife, when the forces of modern population pressure and the ensuing agricultural development bring even more hardship to wild animals attempting to gain succour from the land. This is how Sheldrick describes the objectives of one of the African game parks:

> the foundations of the Park have been laid with infinite care, patience and endurance, by a handful of dedicated men, not for material gain, but simply out of a deep rooted and sincere love of animals. It was as though the Creator, conscious of the threat to so many of his creatures in a fast changing world, called upon them to establish an island, where His animals can enjoy the freedom craved for so desperately by man himself, but often denied by him to his four-legged neighbours; a sanctuary where these have the right to live their lives in peace, and in doing so, can bring enjoyment to hundreds of people..... I have come to look upon them, not as four-legged machines put here for the benefit of mankind, but as creatures with as much right to enjoy the world God gave them as we have. (Sheldrick, 1966).

Although there are examples like the one above of occasional widespread slaughter and suffering of native animals, for the most part they are well fed and healthy. This can be illustrated by comparing the welfare of wild cats, which are native to most of the world, feral cats, which live in most peri-urban districts, and domesticated cats. The wild cats have evolved over millennia to their environment, and they are usually well fed. Evolving as an animal that thrives particularly well in desert environment, they are well prepared for hot, dry conditions. Nearly all are free from diseases for the majority of their life, because of their large home range and low stocking density, which reduces the chance of high parasite populations to challenge the cats. The fluctuations in the numbers of their prey would be less than for feral cats, and natural selection would quickly remove any animals that became sick and vulnerable. The genetic variation is much greater than for feral and domestic cats, enabling an effective immune response to disease challenges in at least some animals (O'Brien et al., 2006). Wild cats are capable of being carriers of exotic diseases, such as the Feline Immunodeficiency Virus or feline homologue of the human HIV, without major suffering. Such diseases cause serious clinical symptoms in domestic cats (O'Brien et al., 2006). There is no clear evidence of greater susceptibility to disease of feral cats or domestic cats in shelters (Case et al., 2006).

As well as having limited genetic diversity to cope with disease, feral cats in peri-urban districts are subjected to a fluctuating food supply, often based around fast food outlets. They scavenge and may have to resort to consuming unnatural 'food' items, such as plastic bags. They are often in bad condition, emaciated and with skin disorders and parasites. The disease status can be influenced by the health status of the prey animals, particularly in areas where wild cats are not native. So, for example in New Zealand it has been found that 38% of feral cats are infected with the rabbit haemorrhagic disease virus (Henning et al., 2006). However, they have freedom to roam, which is often restricted in domesticated cats. The latter are fed a highly nutritious diet, but high nutrient intake coupled with inadequate opportunities to exercise may lead to health complications such as diabetes. Usually they have to change from being nocturnal to diurnal to match their owners (although they tend to revert back to nocturnality in old age). They are often kept permanently indoors, especially if the owners live in an apartment, as is increasingly popular. Although artificial breeding has produced domesticated cats that appear to have less need for access to outside areas than wild cats, they are not yet well adapted to the extremes of intensive human existence, in small apartments several floors from the ground. Their desire to perform natural behaviours is thwarted by their environment. Most people would agree that, of these three different types of cat, the welfare is best for the native animal.

Charles Darwin believed that nature prepares animals better for environmental challenges than artificial breeding: 'Man selects only for his own good, Nature only for that of the being which she tends' (Darwin, 1857). Thus, it is likely that valuable information can be attained from studying the wild relatives of domesticated animals, especially in relation to their behavioural needs. For example the Gaur cattle of Asia (*Bos gaurus gaurus*) could inform us about the behavioural needs of domestic cattle. Gaur cattle are one of the last remnants of wild cattle with a similar genotype to our modern domesticated cattle and despite the obvious merit in studying their behaviour, few attempts have been made to do so. The extent to which animals can perform natural behaviour could be a better indicator of welfare than an anthropomorphic assessment of the conditions of the animals.

An example of the difficulties of an anthropomorphic assessment of welfare is the early weaning of dairy calves from their mother. Most people would believe that removing the offspring at one day of age would greatly reduced the welfare of both cow and calf, even if it does continue to be fed milk, although this time reconstituted from powdered sources. Research shows that stress levels experienced by the cow following separation do not support any contention of a major welfare impact (Hopster et al., 1995). However, it is necessary to consider what opportunities for close bonding between cow and calf have been forgone by the separation, emphasizing not the negative impact of the separation, but the absence of positive events.

Animals in the Wild, What Can They Tell Us About the Needs of Domestic Animals?

How important is the domestication process in determining an animal's behavioural needs? To answer that question, we must study the behaviour of domestic animals in wild and semi-extensive conditions and compare it to the behaviour of wild progenitors of domestic animals in the same environments. For example, there are opportunities to observe the behaviour of domestic cattle under natural conditions, such as at the Chillingham estate in northern England. Parkland cattle such as these were typically introduced to British stately homes to enhance the aesthetic qualities of the landscape in the 18th and 19th C, and have often been resident as a herd for several hundred years. A point of concern is that the opening of the estates to visitors in recent times has meant that the cattle can no longer be considered entirely free from human contact (Ritvo, 1992). One point of interest of such herds is their natural configuration of mixed sex groups. Unlike most domestic herds, where male and female cattle are segregated, these mixed sex groups typically adopt a matriarchal herd structure with groups of 10 to 20 animals being led by a dominant cow, and bulls that are evicted from the herd at puberty (Hall and Hall, 1988). This structure mirrors that of wild cattle herds, but is the behaviour of wild and domestic cattle similar? The behaviour of domestic cattle is well understood (Phillips, 2002), but opportunities to observe the behaviour of wild cattle have been rare.

I have been fortunate to study this in the central highlands of Malaysia, where there exists one of the last remaining groups of wild cattle that are close relatives of the domesticated *Bos taurus* and *Bos indicus* cattle used in farming systems today. Deep in the highland jungle, there are several hundred *Bos Gaurus* (Gaur) cattle, which have lived in this ancient habitat for many millennia. Most of Malaysia is covered with date palm or rubber plantation, but the Highland regions are difficult to cultivate and there are a remnants of rainforest that preserve an important habitat for tropical fauna, including the Gaur cattle, or *seladang* in Malaysian (Solti et al., 2000). Cattle exist there in small groups of about half a dozen individuals, led by a dominant female with a dominant male close by, but outside the matriarchal group (Fig. 2.4). The only predator is the tiger, which will occasionally take small calves, but when there is the threat of attack the members of the herd will form a stockade facing outwards and the male will join the group as the primary defence unit. Bull threatening behaviour is similar to that seen in domesticated bulls: shaking of the head, pawing the ground, snorting and adopting a threatening 'intention to charge' posture. This demonstrates that the behaviour of wild cattle under threat of attack is similar to that of domestic cattle. In this and other behaviours, the differences between wild cattle behaviour and that of domestic cattle appear due mainly to the environments in which they are kept, and not their genetic constitution. There are differences in morphology between wild cattle and today's domestic cattle, with the former being deeper bodied, with pronounced strengthening of the anterior vertebrae to enhance the ability of bulls to clash heads and withstand the impact of charging. This deeper body structure causes the vocalisations to lower in pitch compared with domestic cattle. Apart from these differences in morphology, it is clearly possible to use observation of wild cattle behaviour as an indicator of the behavioural needs of domestic cattle.

To find out more about the behaviour of Gaur cattle, I trekked in the jungle with local rangers and camped at the logging stations, using a four-wheel-drive vehicle to scour the logging roads and those at the edge of the jungle near plantations, to try to find the tracks of the wild Gaur cattle. We found old cattle tracks, as well as tigers, wild boar, deer and tapir, but local villagers told us that the logging activities were disturbing the cattle too much and they had retreated to the higher lands. At the junction between the jungle and the oil palm plantation, where the cattle had previously been observed entering the plantation to forage at night, a double stranded high-voltage electric fence had been

Fig. 2.4 Family group of Bos gaurus cattle, Krau Wildlife Reserve, Malaysia

erected to keep both cattle and elephants from damaging the palm trees. Elephants are helpful to the Gaur cattle in clearing areas of the forest to allow shoots, grasses and other diverse young, nutritious vegetation to grow, but the size of this wildlife reserve was too small to sustain a large number of elephants.

We found recent cattle tracks at a salt lick, but even there the cattle were too elusive, being now very wary of human presence, because of the intrusion of loggers and local people. The local Orang Asli people were causing more disturbance in the forest now that they have motorbikes to go to the villages for the foodstuffs that they cannot get from the forest, such as sugar and rice. However, their natural coexistence with other forest life was evident, and they were still using blowpipes to secure monkeys for food, trapping birds by putting sticky substances on tree branches and collecting rattan, which they sell for furniture making. They had little to do with the cattle, and it was clear that these cattle were very shy of humans. We returned to the park headquarters, with no direct sightings, but evidence of cattle activity in the park. Fortunately, there was a captive Gaur cattle breeding programme at the park headquarters, with ten large paddocks of 1–2 hectares each, and three to five Gaur cattle in each. Some animals had been in this programme for as long as 25 years, demonstrating that in this case the lifespan of wild cattle was well in excess of that of domesticated cattle, which are usually only kept for four to five years if they are dairy cows and one to three years if they are being reared for beef. Observing wild cattle in these large paddocks was ideal as it resembled their natural forest habitat, but the animals were close enough for observation.

The cattle being primarily nocturnal, I visited the animals at all times of the day and night to obtain an accurate picture of their natural behaviour. By night they foraged continuously, taking mainly small twigs and leaves from tree material cut locally (the paddocks were not large enough to provide sufficient browse material for the cattle on a permanent basis). Their diet was markedly different from domesticated cattle, which are primarily offered pasture grass. It is still possible to see domestic cattle browsing the lower branches of orchards and trees in mixed tree/grass systems, demonstrating that grazing is not their only method of food procurement. The wild Gaur cattle only grazed if there was a shortage of tree fodder. The paddocks being relatively small meant that the grass was quite short, but of good nutritive value.

Another evident difference between the behaviour of wild and domestic cattle was the nocturnal habit of wild cattle, feeding by night and resting and ruminating during the day. This may have been a strategy that evolved to limit activity during the hot periods of the day in the tropical environment, but it may also limit the predation risk to young calves, since they can lie out of sight during the day. Cattle have large eyes with a reflective layer, tapetum, on the retina, which gives them good night vision (Lomas et al., 1998). Domestic cattle are also active at night, rarely sleeping, but their carers generally only see them during the day.

Heat stress resistance was enhanced in the Gaur cattle by the production of sebum, which reflects the heat, something which we no longer see in domestic cattle in colder climates, which developed long coats in the domestication process in the northern latitudes. The sebum also deters flies from attacking the cattle's skin (Warnes, 1995). In hot climates, such as in Israel, cattle have reverted to a shiny, sleek coat which reflects the heat, compared with the long, hairy coats of cold climate cattle. The sebum production of the Gaur cattle was most evident in the mature animals, which are more susceptible to heat stress, and not in the calves. Suckling naturally lasted for nine months, compared to just a day or two in most dairy cows, after which the calves are offered powdered milk for just six weeks.

Reproductive behaviour in the Gaur cattle was limited to contact between the males and females during copulation, with little evidence of a prolonged courtship. Copulation started soon after the cow had given birth and usually led to a second pregnancy, which progressed whilst she suckled her first calf. There was no apparent seasonality in their reproduction, as expected in this equatorial latitude. By contrast, domestic cows often aggregate into sexually active groups during their oestrus and engage in homosexual courtship behaviour: mounting, sniffing and licking the anogenital region and rubbing their chins on each other's rumps. This is more pronounced in intensively-managed housed cows, rather than in cows outdoors, which suggests that it may be partly a response to the stress of the intensive environment, a phenomenon that we have observed in other mammals (Feige et al., 2007). It is also possible that humans selected for this behaviour during the domestication process, when bulls were probably shared between several families. A distinctive behaviour such as mounting would enable the cowmen to identify when their animals were ready to be served by the village bull.

The final difference that I observed between domestic and wild cattle was in their lying behaviour. When domestic cattle lie down, they usually tuck their head back towards their thorax, which may be to protect it from being trampling in a crowded environment (Phillips, 2002). The wild cattle always lay asleep with their head fully outstretched.

In all other respects, the behaviour of the wild cattle matched that of domestic cattle very closely. Tails swishing to remove flies, herding behaviour, cleaning the nostrils with the tongue and many other behaviours were all identical. These captive wild cattle adjusted to friendly human presence quite readily and would allow themselves to be touched, and stroked by people, and they recognised familiar individuals. Therefore, some behaviours have changed, because of the different circumstances of domestic cattle, but the innate motivations are mostly the same. It is most often the stressful conditions of the housed most after environment that requires cattle to change their behaviour. Their lying stalls, or cubicles, are often cramped and they can have difficulty changing position and standing up and lying down. In a cubicle house, they may be confronted by more dominant cows that can be aggressive, so subordinate cows stand half inside their lying stalls, to get some protection from other cows. Much more could be done to examine the behaviour of wild cattle to assist us in understanding the behaviour of domestic cattle, especially as the wild Gaur cattle of the Malaysian highlands are threatened with extinction. There are also Gaur cattle in India, but they are mostly semi-domesticated.

Observing animals in the wild therefore reinforces concerns about housing them in small, confined spaces. Another animal that lives in the Malaysian jungle that has controversially been brought into confined spaces, this time for public viewing, is the Asian elephant. The circumstances of elephants in zoos present a number of welfare concerns, of which the lack of space and the absence of natural foraging behaviour are the most serious. These problems lead to low reproductive rates and the display of abnormal behaviours such as rocking and swaying (Wilson et al., 2004).

In relation to reproduction, studies of the elephants in North American and European zoos (where there are several hundred altogether and it is possible to evaluate their reproductive success) have shown that the reproductive rate will not sustain even the zoo population, let alone providing elephants to return to the wild (Clubb and Mason, 2003). One reason for the low reproductive rate is the small number of animals that exist in each zoo, which limits breeding opportunities, and another is likely to be the inadequate conditions in which they are often kept in comparison to their natural environment. The size and cost of maintaining an elephant in a zoo means that there are rarely more than two or three animals together. Introduction of new animals causes stress (Schmid et al., 2001). This contrasts with the situation in the wild, where the animals live in matriarchal groups of 10–30 animals, with several generations together and long-lasting social relationships (Schulte, 2000). In rangeland conditions some much larger herds of 50 to 100 animals form. Bulls are evicted from the matriarchal group at puberty and naturally live an isolated existence, although they sometimes roam the forest in a bachelor group of two or three animals.

In the wild, the distances covered by both the matriarchal group and the bulls are considerable. These animals, by virtue of their size and feeding habits, damage the trees in their habitat, and cannot afford to stay in one place too long. Home ranges vary from 10 to 800 km^2 (Dolmia et al., 2007; Shannon et al., 2006), but in zoos they are in enclosures typically of only a few hundred square metres, so that they can be readily seen by the public. Minimum recommended size requirements for enclosures, which are about 100–200 m^2 per elephant, relate more to what is possible than any considerations of the animal's behavioural needs.

It is difficult to recreate an elephant's environmental needs in a small zoo enclosure. Their natural environment is highly complex, and in the case of the Asian elephants, they may obtain food from almost all strata of the jungle: grasses and herbs from the floor, fruit, roots, leaves from bushes and shrubs, as well as small trees, some of which may be knocked over so that they can feed from the floor. Sometimes they will even stand on their hind legs to reach for tallest browse material. They spend about 16–17 hours per day looking for and eating food, whereas in zoos a nutrient rich diet is provided that they usually consume in about 10 hours (Clubb and Mason, 2003). As a result of their environment being so restricted, they develop repeated, stereotyped behaviours: rocking or swaying repetitively, raising and lowering their legs or stimulating their mouths with their trunk, particularly before they are fed, handled or trained. These stereotyped behaviours have recently been linked to increased levels of the stress hormone cortisol and seem to function as a mechanism to cope with the stress of their environment (Wilson et al., 2004).

The training methods present another serious welfare issue, with the animals initially needing to be 'broken' by their keeper, so that they can be chained by the leg both during the night and when procedures are carried out on the elephants. The methods used in training include electric goads or long metal rods to control the animals' movement and isolation in between training sessions, so that the animal comes to value its moments of freedom and becomes responsive to the demands of the trainer. Close attachments can form, but the movement of elephants in captive breeding programmes makes these long-term attachments difficult. Regular movement of bull elephants around zoos for breeding purposes, although mimicking the roving behaviour of the bull in the wild, leads to serious welfare problems, because zoos will usually only provide small, restricted enclosures for the bulls to live in for the short time that they are required to serve the females. Not surprisingly, reproductive behaviour in such circumstances is very limited and there is a likelihood that zoos will now use artificial insemination to overcome this problem (Andrabi and Maxwell, 2007; Hermes et al., 2007). However, the bulls will still have to travel, because bulls' semen is not usually viable after freezing. The insemination procedure is much more difficult than domestic animals like cattle or sheep, and calf mortality is high. Hence breeding success rate, even with artificial insemination, has been disappointingly low.

In my visit to the central highlands of Malaysia, I was able to see first hand the problems facing the Asian elephant, such as enclosing the oil palm plantations with electric fences so that they cannot forage there at night, logging of at least the biggest trees in the forest and greater human activity. Our riverside camp was visited by a herd of elephants at 5 a.m., and I wondered then how long they would survive there with the extent of the logging and other human activities. Controlling the expansion of the human population in south-east Asia and the human activity in the forest, including logging, is the only way to ensure the survival of this species. Every child learns about elephants in story books and wants to see one, but confinement of these most intelligent and demanding of animals in small spaces in zoos is less desirable than in situ conservation, in particular addressing the problems in the South East Asian region. If in the face of continued human population and agricultural expansion in the region the in situ conservation ultimately proves impossible, then facilities with adequate conditions for elephant welfare should be provided in zoos and wildlife parks.

My next opportunity to compare wild animals' behaviour directly with those in captivity was again in Malaysia, this time in Sarawak on the island of Borneo. I visited the Bako National Park, where there are about 200 proboscis monkeys – an endangered species only found on the island of Borneo. Named for their exceptionally long nose, especially in the male, these large ginger-coloured monkeys were of particular interest because much less is known about their behaviour than more common primate species. The best time to see them was at dawn, so I went to the mangrove trees that they often visited at about 6 am at the end of their daily migration from the hilly hinterland. At 7.10 am the first of the monkeys came crashing through the trees overhead. Clearly this daily migration to the beach could not be mimicked in captive proboscis monkeys. Later I observed another troupe more closely as it was settled eating near the beach. Dripping with exertion and tense with excitement, I was able to film these creatures for several minutes before they made their way back into the impenetrable jungle. I reflected on why these animals should inspire such a feeling of awe in the wild, when an encounter in the zoo would be only mildly stimulating. Was it respect, even jealousy, for the natural setting in which they were living, for the freedom that they exhibited, for their apparent love of life? Was it their rarity or their apparent control of their surroundings? Was it their unusual form, the vivid ginger-red colour of their coat? It mattered not, I came away enriched by the experience.

Later on this trip, I saw orangutans in captive and semi-wild conditions. Just outside the capital of Sarawak, Kuching, there is an orang-utan sanctuary called Semangok. Here orangutans that had been kept locally as pets were rescued and prepared for rehabilitation, by giving them restricted access to the jungle. They could return for food if they wished, but most chose to move away and colonise the surrounding areas. The offering of food provided a natural transition between captivity and the wild. The behaviour of the rehabilitated orangutans did not resemble the complexity of their wild relatives in respect of nest building, feeding and other complex behaviour patterns. Wild orangutans carefully bend branches to make their nest for the night, orang utans from captivity were comparatively naïve about nest building, frequently breaking branches and often having to reuse their nest because of the construction difficulties. Despite the fact that they would have been taught these behaviours as infants from their parents, the trauma that they experienced at the hands of their captors appeared to reduce their ability to perform complex behaviours. Animals that have been severely stressed can only make simple choices.

I was reminded of the stress imposed by captivity when we moved on to see the sanctuary at Matang (just inland from Semangok), which first receives the orangutans, performs any veterinary treatment and offers a home to those that are unsuitable for rehabilitation. Although relatively large and enriched compared with many European enclosures for orangutans, their enclosure obviously did not provide enough stimulation for these highly intelligent animals. The single orang-utan in one enclosure had only the visitors to amuse her, and it was clear that she wanted to make eye contact with each new arrival to the viewing platform. What differences could she recognise – gender, colour, clothing? Recognition of gender was very likely, as this one was reported to have been violent to female visitors after she had had a baby, perhaps suspecting that they might challenge her for the baby. The contrast in behaviour from the orangutans that I had seen earlier was striking, here was an animal obviously bored and probably stressed. She vomited into a bottle and then drank the vomit, a common abnormal behaviour in captive great apes. She played with the many plastic bottles that the viewers had thrown into her enclosure, probably in an attempt to enrich her life.

Experiences such as this, seeing orang utans in wild and captive conditions, emphasize the difficulties in keeping cognitively advanced species in captivity. Often the space and enrichment requirements for adequate welfare are not compatible with allowing the viewing public an adequate experience of seeing the animals close up. These brief experiences of natural and captive behaviour of animals in the tropics served to reinforce my belief that much more has to be done to improve the life of captive animals. Not just in zoos and sanctuaries, the impact of captivity can be equally great on domestic animals. I once had to compare the behaviour of a herd of supposedly wild cattle at Chillingham in Northumberland with that of a 'tame' dairy herd for a documentary film. It transpired that the 'wild' Chillingham cattle were much safer to approach and appeared more contented than the 'tame' ones, probably because there were regular visitors to their park. The dairy herd's Friesian bull was particularly fierce, alone in a small pen, and was passed by many cows daily on their way to milking, which probably contributed to his dangerous demeanour. All animals, both wild and domesticated need the provision of a suitable environment, and learning about the behaviour of animals in a natural setting can prove invaluable in learning about their behavioural needs in captivity.

Chapter 3
Empathy Towards Animals

Moral behaviour – empathy, sympathy and feelings – empathy and domestication – gender issues – empathy and animal welfare – empathy and learning

Introduction

Moral behaviour involves actions to improve the welfare of others (Kurtines and Gewirtz, 1984). It is one of many values that motivate people (Schwartz, 2007) and contains elements of universalism (social justice, equality), benevolence, tradition, conformity and security. Other values directly promote self interest: power, achievement, hedonism, stimulation (leading an exciting and varied life), self-direction (ability to determine one's future). The balance between self interest and moral related behaviour is likely to determine an individual's propensity towards caring for others, including animals.

A perspective of our need to improve animal welfare can be gained by trying to understand how humans acquired the strong interspecies empathy that is at the centre of our concern for animals (Hoffman, 1987). Empathic theory originated with nineteenth century German aesthetes, who believed that humans can derive feelings from objects, particularly architectural or natural, for which they used the word *Einfühlung* or 'feeling into' (Wispe, 1987; Taylor, 1994). For example, considering the shape of columns in a Greek temple, short, fat columns would make them feel uncomfortable, as if they reflected their own feelings for people of that shape, whereas tall, slender columns of the same proportion as elegant humans would engender a comfortable feeling.

Although the term 'empathy' was first used by Titchener (1924) to mean 'humanising objects, ... reading or feeling ourselves into them', its common use now relates more to 'an emotional response that stems from another's emotional state or condition and that is congruent with the other's emotional state or situation' (Eisenberg and Strayer, 1987) or 'the capacity to experience the feelings of another person or an animal, cognitively and emotionally' (O'Connell, 1995). Empathy assumes knowledge of the feelings of others and the ability to relate behaviour to feelings; it also involves the ability to attribute

C. Phillips, *The Welfare of Animals*, Animal Welfare 8,
DOI 10.1007/978-1-4020-9219-0_3, © Springer Science+Business Media B.V. 2009

mental states and perceive the feelings of others (Anon, 1987). I have already argued (Chapter 1) that we cannot literally experience the feelings of others, and that even imagining the feelings of animals will be immeasurably more difficult than humans. Thus empathy can only involve an assumed emotional state.

Empathy must be distinguished from sympathy, which is the heightened awareness of the suffering of another as something to be alleviated (Wispe, 1986). Sympathy involves an urge to take mitigating action, whereas empathy does not. Together empathy and sympathy are responsible for the caring attitude that we have for animals that we come into contact with. Their evolution could be due to utilitarian benefits to humans, but there are clear and frequent instances where caring for animals does not produce a net welfare advantage to the individuals involved. Self sacrifice for a drowning pet is one such example. The possibility of reciprocal altruism cannot be dismissed. Caring for pets is linked to caring for humans in some studies (see review by Paul, 2000) and may be considered a form of generalisation, but in one study the likelihood of owning a pet has been shown to be *inversely* related to the size of a woman's friend networks (Ory and Goldberg, 1984). Clearly generalisation does occur, as for instance in the generalisation of the empathy shown by young children towards their pets to siblings (Poresky, 1996), but it is not a universal phenomenon.

Empathy and Domestication

Caring for other species is not unique to humans, many animals in symbiotic or commensal relationships with other species have to protect their partners in the relationship (Wiese, 1996), and social species such as chimpanzees can show empathy across a wide range of circumstances (O'Connell, 1995). Plutchnik (1987) describes the survival value of empathy in animals as group behaviour, imitation and bonding, but he uses a broader definition of empathy ('the capacity for participating in, or a vicarious experiencing of another's feelings, volitions, or ideas and sometimes another's movements to the point of executing bodily movements resembling his'). In the narrower definition of empathy given earlier, these benefits of empathy have yet to be demonstrated in relation to human:animal interactions. Only bonding would seem to clearly offer survival benefit in the 'domestic alliance' that Coppinger and Smith (1983) propose is responsible for the evolution of human:animal symbiotic relationships. Initially, they argue, opportunistic plants which benefited from increased soil fertility colonised the regions around human settlements. Productive animals that required the nutrient-rich plants and could tolerate human presence became associated with the settlements, but animals and humans could only co-exist if the animals' behaviour, needs and disease status could be managed by humans. Farmsteads where the humans could recognise the animals' needs and respond more rapidly were more successful, and thus the ability to respond to animals' needs evolved.

It is easier to demonstrate interspecial empathy in humans than other primates (Chalmeau et al., 1997), and the development of empathy appears to lie on an evolutionary continuum (Brothers, 1990). There is circumstantial evidence that human sympathy towards animals may have evolved into a 'psychological need', which if thwarted could lead to dissatisfaction, stress and even violent and irrational behaviour (Groves, 1995). Human empathic feelings are not generalised to all species – they do not normally extend, for example, to potentially dangerous species, such as snakes or spiders. However, this is unlikely be true of professionals in the field, herpetologists and arachnologists, suggesting that there is substantial variation between individuals in the extent of their empathic feelings. The ability of humans to interact with and manage a large number of animal species for our own benefit (e.g. for food, protection and clothing) may indicate why we empathise with so many animal species. An ability to recognise, and ameliorate where possible, the suffering of animals in their charge would have benefited humans throughout the long history of dependency on animals, thereby giving reward to the evolution of empathy through reciprocal altruism (Leak and Christopher, 1982). For example, an ability to recognize and treat livestock diseases in the early domestication period would have enhanced the survival rate of both the animals and humans which depended on them. Evidence for the adaptive significance of empathy is chiefly that those animals that give us the most benefit evoke considerable empathy, particularly farm animals and companion animals (Serpell, 1986). However, animals such as primates that are phylogenetically close to us also evoke empathy, particularly if they are juveniles, suggesting that there is some generalisation of the human:offspring bond. Size of the animal also appears to be influential, with very large mammals such as whales evoking particularly high levels of empathy (Howard and Parsons, 2006). These stereotypes appear to be perpetuated in societial teaching, most recently through the medium of television (Paul, 1996).

Empathy in humans probably first appeared before domestication, since it is apparent in some other primates such as chimpanzees (O'Connell, 1995). A primitive form of empathy, termed an emotional contagion, has even been detected in mice (Miller, 2006). Miller found that the response of mice to painful stimuli was greater if they were with a familiar mouse that was also given the stimuli than if they were with an unfamiliar mouse given the same stimuli. The inference is that the mice were able to share the experiences of other mice and that they were more influenced by the experiences of familiar mice. Other research with rodents has identified that the amygdala in the brain is activated during the transfer of this information (Knapska et al., 2006).

It is often reported that empathy is an innate characteristic in humans (Stein, 1964; Kohut, 1977; Wispe, 1987), but in the certain knowledge that it can be influenced experientially, it is unclear whether nature or nurture is the greater influence. A well-studied instance of human:animal empathy relates to the evolution of dogs. Initially, commensal relationships between hunting men and dogs in the pursuit of their prey are likely to have promoted caring attitudes by humans towards their dogs. The affiliative behaviour that companion dogs normally demonstrate towards their owners suggests that this attitude may well have been reciprocated. The process of domestication relies on the selection of behaviour patterns that are more prevalent in juveniles than adults – acceptance of novel situations and objects, receipt of food items from conspecifics, and incomplete behaviour patterns e.g. mock rather than injurious fighting, oestrous display in cattle (Woodgush, 1983). Hence, if the morphological characteristics of juveniles are related to their (juvenile) behaviour patterns (the neotenization hypothesis, Geist, 1971), the preservation of juvenile features by selective breeding of domesticated animals will have value in the maintenance of suitable behavioural patterns (Price, 1984). This is exemplified by the selection of two major types of working dogs used by livestock farmers – conducting and guarding dogs. The former are used to move sheep by guiding them as if they were stalking prey. The latter protect livestock by virtue of their size but do not stalk them. Both types of dog develop these behaviours instinctively, and Coppinger et al. (1987) suggests that the absence of predatory behaviour in the guarding dogs is evidence of selection for retarded motor pattern development (neoteny).

Gender Issues

The greater concern for animals by women than men in some societies (Herzog et al., 1991; Hills, 1995; Phillips and McCulloch, 2005; (Daly and Morton, 2006; Howard and Parsons, 2006) suggests that this trait may have been selected for post animal domestication, since in societies with domesticated animals women were usually involved with looking after animals in the farmstead and the men were responsible for other tasks such as hunting or fishing. Alternatively, the gender effect could be related to the greater responsibility that women have for the care of their young, and there is evidence that such interspecial generalisation does occur (Poresky, 1990). The greater intuitive care devoted by women to the young, which is logical since they have invested more than men in each individual offspring (Alcock, 1989), may cause them to seek to adopt the role of animal carers. Hence girls are more likely than boys to want to keep pets (Rost and Hartmann, 1994). Not only do women have a greater level of concern for animals, they also credit cats and dogs with a greater understanding of their owners and stronger feelings of love and compassion for them, compared with men (Vitulli, 2006). This suggests that women are able to enter into a deeper relationship with animals than men.

At the other end of the scale, animal phobias are more common in women than men (Fredrikson et al., 1996). They are at least partly genetically determined (Kendler et al., 1993; Davey et al., 1993), which provides indirect evidence that our attitudes to animals have a genetic component and, therefore, *ceteris paribus* an adaptive significance.

Nationality Issues

International comparisons suggest that it is in those areas where human populations traditionally depended most on animals for food that people show the greatest concern for animal welfare. There is in Europe at least and probably also in North America, an increase in concern for animals in more northerly latitudes, where humans relied more on animals for food (Curtis and Guither, 1983). Concern also tends to be greater in Europe than in Asia, but this could relate to the recent intensification of animal production in Europe (Phillips and McCulloch, 2005). This paradigm does not include recent migrations and changes in the geographical spread of animal husbandry. For example, the widespread use of the Americas for cattle raising does not necessarily equate with an increase in concern for animal welfare in these regions. However, the management of animals for meat production in this region is quite different to traditional methods of animal management for food production that have prevailed over the last 5,000 years or so. Contact with individuals is minimized and excessive attention to the welfare of individual animals would often be contrary to the efficient economic management of the farm unit. In addition, the relationship between the dependency on animals and concern for them does not easily extend to the animal keepers on modern farms, in fact in one study farmers, who spend more of their time with animals than the general public, demonstrated less empathy (Hills, 1995). This is possibly because they are subject to conflicts of interests, in which their livelihood and ability to care for other animals are reduced if they show too much empathy to particular animals in their care. Alternatively empathic feelings may become habituated over time in farmers, particularly as the supposed reward of reciprocity (Leak and Christopher, 1982) may be of little benefit to the farmer, or may not be evident in modern intensive farming situations. However, a recent study shows that many Australian livestock farmers believe that their sensitivity to animal welfare has increased, not decreased, over time (Phillips and Phillips, unpublished data). This may be because of increased responsibility in their position, a general time trend in attitudes to animal welfare, an increased realization of their own frailty over time or because of a reduced willingness to exploit animals to make a living as they age.

Empathy and Animal Welfare

The relation between human empathy to animals and their welfare is important in relation to its role in the development of sympathetic attitudes. We care most for animals that give, and more importantly gave, us the most benefit. For example, legal protection was introduced much earlier for animals such as horses, from which we derive benefit, than for vermin and other animals that may carry disease, with which we also have regular contact but few empathic feelings. In 1641, the Massachusetts Bay Colony in the USA drafted one of the first laws which forbade cruelty to farm animals including horses (Stull, 1996). Our attitudes to animals therefore depend on their value to us, as evidenced by the anthropogenic decline in predatory and other dangerous animals in the wild, and the simultaneous proliferation of animals such as cattle that are suitable for food production (Coppinger and Smith, 1983; Rifkin, 1994). There is no lack of logic in our attitudes to different types of animals, since this approach represents the best human survival strategy. There has been a global pattern of animal differentiation, into those that are, and those that are not, useful to us, with survival of the latter jeopardised to expand the range of the former (Coppinger and Smith, 1983). However, with rapid change in global living conditions, climate and opportunities for the animal management, we must remember that we will in future benefit if we have preserved as much of the genetic diversity that we have inherited as possible. The pressure from human population expansion, climate change, resource exhaustion encourages a focus on systems of animal production that are of benefit to us now, with disregard for future opportunities.

Recently the ethical and environmental logic of our extensive use of traditionally beneficial animals has been questioned (e.g. Rifkin, 1994). The proliferation of beneficial animal species, however desirable in terms of providing food that is highly nutritious and therefore attractive, is not essential, since we now have the technology to produce enough food for the human population by agronomic means alone. The use of prime agricultural land for producing food from ruminants in particular may be less efficient than producing the equivalent amount of nutrients from crops, but the greater processing costs of crops compared with animal products is often not considered in this comparison (Blaxter, 1995). The environmental concerns focus on the facts that animals are often concentrated into small areas, making it difficult to dispose of their waste, that they utilise land in developing countries that has until recently been afforested and that they contribute to atmospheric pollution (Phillips and Sorensen, 1993; Phillips, 1994).

Empathy and Learning

Some aspects of empathy, such as imitative behaviour, are undoubtedly influenced by experience (Wilson, 1975), and some argue that parental and cultural influences i.e. learning, are primarily responsible for the observed differences in empathy to the various animal species, rather than genetic influences. There are three reasons why this is unlikely. First, other personality traits are known to be strongly influenced by genetic inheritance (e.g. Goldsmith, 1983; Carey and DiLalla, 1994), so it is to be expected by analogy that a trait with such obvious survival value as caring for animals would also have a genetic component. Second, empathy for animals is a trait observed at a very early age in children, and this cannot be explained satisfactorily by such environmental influences as pet presence (Poresky and Hendrix, 1990). Third, within a population the strongest concern for the welfare of farm animals is to be seen in urban dwellers, which have little direct contact with farm animals and therefore less opportunity to learn about attitudes to them (Hills, 1995).

Further evidence for a relationship between the benefit derived from animals and the degree of empathy shown by humans is provided by a detailed examination of the different benefits. Benefits from animals are derived principally from the provision of food and other commodities, such as clothing materials, and this is mainly from farmed animals. Additional benefits are derived from symbiotic relationships, as in animals kept for companionship, draught and transport purposes. Finally some benefits are obtained from a semi-parasitic relationship between man and animals, such as when they are kept for sport. The greatest material benefit derives from the provision of food. Even though most people in industrialised countries are not in regular contact with animals farmed for food, the concern for their welfare, in particular pigs and poultry, is greater than for other anthropogenic risks to animal welfare, such as destroying natural habitats, hunting or the endangering of wild animals through the use of chemicals in agriculture (Fölsch, 1984). Hence we show the greatest concern for the animals from which we derive most benefit.

Conclusions on Empathy

Empathy is the ability to infer and experience what we understand to be the feelings of others, which is not unique to humans but is most evident in higher order animals. Evidence for an adaptive significance of animal empathy comes from the fact that we often focus our empathy on animals which give us benefit, either physical as in farm animals, or social as in companion animals. Such empathic reactions will benefit us through enhanced bonding and recognition of animal needs, as well as symbiotic activities such as hunting. Other influences on the level of empathy include phylogenetic similarity to humans and size of the target animal. The greater level of empathic emotions for animals in women than men suggests that there is a generalisation from empathic childcare emotions, which may have developed post domestication, when women looked after animals in the homestead. Several other lines of evidence support the view that empathic responses to animal emotions were influenced by the need to look after domesticated animals.

The evidence for empathy being an inherited trait in humans is strong, but definite proof is likely to remain elusive. If empathy evolved in response to the symbiotic relationship between humans and domesticated animals, we may conclude the following about the implications for our attitude towards the welfare of animals: we direct our attention most to those animals that are currently useful to us, but must also consider the relationship that we want to have with animals in the future. It was proposed 50 years ago that empathy provides a reliable basis for consensus in moral judgement (Hume, 1957), but in view of the influences of experience on empathic values this would seem unwise. Some argue that we cannot logically take the view that all animals should equally be given the benefit of our care (Allott, 1991). Currently our concern to improve the welfare of animals is primarily directed at those animals that benefit us (Leak and Christopher, 1982), which has adaptive advantage for these species. As well as more tangible methods of improving welfare, this could take the form of educating children to enhance empathic attitudes (Hills, 1995) and instructing those using, or directly benefiting from animals on how they should be treated sympathetically (e.g. Wiebers et al., 1994). The evolution of animal empathy was probably also influenced by the need to manage animals successfully, which included understanding their feelings and taking appropriate action when necessary.

Chapter 4
Animal Welfare and Animal Rights

The evolution of moral standards – major influences on our attitudes to animals – animal welfare and animal rights – animal morality – cruelty towards animals

The Evolution of Standards Supporting Moral Behaviour Towards Animals

Northern European countries have been pre-eminent in establishing a legal framework for ethical standards in the latter part of the last millennium. This was in part because this region was a focus of industrial development at this time, together with the opportunities that the increased wealth generated. The towns and cities that had developed in the industrial nations were overcrowded, disease-ridden and with little space for penal settlements. In these urban centres, legislation had to be enacted to protect the underprivileged minorities, who were in danger of exploitation in these less moralistic societies than those in rural areas, which had evolved their own methods for maintaining moral standards. The call for an improvement in social standards in European countries was led by public demand, politicians, protestors and activists, who became aware that city dwellers needed protection through legislation in a way that village dwellers in the past did not. Closeknit village communities could protect the less privileged members of their society through mutual consent, admonition of miscreants by elders in the society and the threat of being ostracized from the society for misdemeanors.

Starting with the rights of the common over (Bill of Rights in 1789 in the USA and the abolition of slavery in the UK in 1833), there have been major initiatives to improve standards for children[1] (1889), women (1870–1920),[2]

[1] The Society for the Prevention of Cruelty to Children was founded in the UK in 1889

[2] Like animals, up until this time women were viewed largely as property. The 20th C attitudes towards women contrasts sharply with earlier attitudes, for example a 12th C BC Mesopotamian law determined that "when she deserves it, a man may pull out the hair of his wife, mutilate or twist her ears, with no liability attaching to him" (Starr, 1973)

C. Phillips, *The Welfare of Animals*, Animal Welfare 8,
DOI 10.1007/978-1-4020-9219-0_4, © Springer Science+Business Media B.V. 2009

ethnic minorities (1960s–1980s),[3] homosexuals (1960s–1990s)[4] and disabled people (1970s–2000s)[5]. Having championed the rights of disadvantaged people, society's attention increasingly focused on animals in the 20th C. Although the Society for the Protection of Cruelty to Animals was founded in the United Kingdom in 1824, major activities to advance the welfare of animals did not begin in earnest until the late 20th C, in approximately 1980. The 20th C focus of attention on animal welfare can be traced to 1965 when the Brambell Committee pronounced on farm animal welfare issues in the United Kingdom, but it took about a further 15 years before animal welfare became a major subject of concern in this country, and by the late 1980s there was an exponential increase in animal welfare research (see Fig. 8.1).

With all of these social movements, many relevant activities have occurred outside of these time periods, but the movements are periods when major advances were demanded and made, particularly in Western Society. This social development was unprecedented in world history, indeed in the urbanization of early civilizations laws were introduced that legitimized a much more hierarchical society than today, where the punishment for crimes depended on the status of the victims[6] (Starr, 1973). Typically there were four classes of human beings: the king, who was considered of divine origin, the upper classes, commoners and slaves. Like the slaves, women were considered to be property.

It was not just fortuitous that the modern era ushered in a period of major social development. The period of increasing prosperity since the mid 20th C has introduced opportunities for widespread availability of welfare support (universalism,[7] Schwartz, 2007) that would have been hitherto impossible. For example, if Russell and Burch had been born in the 19th instead of the 20th C, their attempts to reform laboratory animal welfare would have been futile, since the use of animals for laboratory research was in its infancy. The major social developments have been led by people who developed the right ideas at the right time. Even Darwin encapsulated the thoughts and ideas that were beginning to form in the minds of the British public at the time. He

---

[3] The American Black Power movement began in the 1960s and progressed to black independent electoral activism during the 1970s. Countryman MJ (2006) "From protest to politics"–Community control and black independent politics in Philadelphia, 1965–1984. *Journal of Urban History* **32**, 813–861.

[4] Often considered to commence with the Stonewall Riots in the USA in 1969.

[5] The UN General Assembly adopted a Declaration of the Rights of Disabled Persons in 1975. Since that time there have been many major initiatives in different regions to advance the cause of disabled people.

[6] Punishments for violence were decreed by the King of Babylon in approximately the year 1800 BC as follows: "If a noble has broken another noble's bone, they shall break his bone. If he has destroyed the eye of a commoner, or has broken the bone of a commoner, he shall pay one mina of silver. If he has destroyed the eye of the noble's slave or broken the bone of a noble's slave, he shall pay one of less value." (Starr, 1973).

[7] Defined in social science as the motivation to promote the welfare of others, but taken here to mean the motivation to make welfare support available to all sectors of society

developed and led the reform of ideas on evolution, but there were other scientists proposing the same ideas at the same time.

In recent times the northern European nations have been the most influential reformers of societal standards. Even to this day, new standards in animal welfare devised in Europe are often extended to former Anglo-Saxon colonies, such as Australia, New Zealand, the United States and Canada several years later. The concerns are spreading worldwide, with societies to protect animal welfare becoming established or strengthened in most areas of the world.

Given that most social movements of the last two centuries have usually lasted between 25 and 50 years, we can expect that there will continue to be a major emphasis on animal rights and welfare improvement for several more decades, and probably it will last until standards have considerably improved. Given the contentious nature of our moral stance on animal welfare issues, it was perhaps inevitable that all animal rights activity would become synonymous with extremist views, in just the same way as those directly involved in the female emancipation movement were branded extremists. In reality, most members of the public acknowledge that animals should have some rights and recognize that this will lead to an improvement in welfare. However, they would not necessarily hold the extremist view that animals cannot be used by humans. A mutually symbiotic relationship between animals and man is, and will continue to be, accepted by most people, even after the 40–50 years or so of welfare reforms that can be anticipated. Such a relationship acknowledges that man dictates, and to some extent restricts, the basic freedoms of animals, but also assures a life that is reasonably well provided for, at least in terms of nutrition, safety and health care.

The animal rights advocacy framework has been not only growing but also becoming more sophisticated, and is part of the general evolution in social cause support groups. In the US these are doubling in size every twenty years, partly due to disillusion with political forces, and in particular the large size of the electoral unit. Communication with members is greatly facilitated by the internet (Lewis, 2005). The major activist organizations have membership lists of millions of supporters and very significant budgets. They employ many well-trained scientists to research campaigns, so that the organisation is well prepared when the campaign starts. Campaigns are focused on achievable targets, often involving groups in society that are susceptible to pressure. Typical weaknesses that can be exploited include the belief by young school children that animals used for fast food production are unhappy, the guilt of house spouses if they purchase fast food in preference to spending time preparing quality meals for their family, and the teenagers' fear that meat would make them smell unattractive or that milk would cause acne. Campaigns are often run in militaristic style, with victories heralded on the website. Bequests are still the major source of funding, but increasingly industry is targeted for support, and a seal of approval by the activist group may assist sales as well as helping advertising. By contrast the target animal industries have smaller budgets and employ fewer researchers to defend their practices.

In addition to the legitimate non-government organisations, there are also animal activist groups, that support illegal acts, although they usually require that these should not harm people or animals. These might include arson, harassment, vandalism, animal release and even bombing. Because the members of such societies engage in illegal activity, they do not have leaders but active spokespeople. Similarly, for legal protection they are not a club or an organization that people can join, but a concept that is realized only when an action takes place using the society's name. They aim to liberate animals from enclosed situations, such as laboratories, intensive meat animal farms, fur farms, etc, and place them in homes where they may live out their natural lives. They also seek to inflict economic damage on those who profit from using animals, and to make the public aware of the circumstances in which the animals are kept. The societies increasingly focus on electronic civil disobedience, such as frequent e-mails or telephone calls to those involved in the animal industries. They may identify a network of companies associated with a target organization, and try to persuade them to withdraw their support for the company. Whilst few people would condone the illegal nature of the activities of some members of these societies, it must be remembered that in the past activists of this nature have often illegally protested against activities that seemed acceptable at the time, but eventually come to be viewed as unacceptable to society at large.

Slow responses, bureaucracy and congestion in the legislative channels encourage members of the public to support groups engaged in direct action. Although the activities of some of the larger societies are across all the major animal use industries, the food sector is an increasingly popular target. This is partly because of our strong sense of empathy with farm animals that provide us with food and many other commodities (see Chapter 3), and partly because the food industry is now dominated by a small number of integrated, multinational companies (making them easy targets and creating the possibility of a domino effect within the industry).

The mode of action of the social activist groups is changing. Traditionally they simply lobbied parliament, which would then regulate industry. However, nowadays activist groups manufacture an issue (which is given a catchy slogan), create a public debate around the issue and make someone within the sector responsible. A viable alternative to the practice in question must be available and the transition must be achievable. The company is then forced to pursue this in order not to lose public support.

The most popular targets are practices that are unnatural, cruel, the result of human greed and displaying a lack of human care. These will attract far more concern than natural events that challenge the animals' welfare, such as drought. Consider the livestock export industry, sending about 6 million sheep from Australia to the Middle East and about one million cattle each year in large vessels. Such long distance transport is easy portray to the public as unnatural, as the animals are kept on large vessels for up to two weeks. Even before entering the ship, stock are transported to the port in vehicles, in which there may be bruising to the limbs, or animals may lie down and be unable to get up (Jarvis et al., 1996). At the wharf, where loading is often in full view of the public, electronic goads may be used to handle the animals during loading, which stresses them (Warner et al., 2007). Once on board, the biggest problem for cattle leaving Australia in winter is that they still have their winter coat, which renders them more susceptible to heat stress when they enter the summer of the northern hemisphere. To counter this problem, the Australian government has stopped cattle being shipped from the southern ports of Australia in winter (Anon, 2006). The industry has also developed a computer model that estimates the impact of the type of cattle, their degree of acclimatization and the anticipated weather conditions during the voyage on heat stress risk, and adjusts the stocking density of cattle or sheep on the ship accordingly (Stacey, 2003). A high risk of heat stress results in stocking density being relaxed, allowing the animals to lose heat more easily. The most serious problem facing sheep on the ship is inappetence, with animals that are too fat being particularly prone to anorexia, in which apparently fit and healthy animals give up eating (Higgs et al., 1991). Fat animals are able to do without food for a few days more easily than thin ones, but then they permanently lose the desire to eat and may die as a result. Although the numbers of Australian sheep rejected are not known, in England over the last eight years approximately 0.02% or 2755 animals were rejected at the place of origin (out of total shipment of almost 200,000 sheep, 500 cattle and 150 pigs, traveling from England to the European continent in 50 shipments each year, DEFRA, 2006).

As I have argued elsewhere (Phillips, 2005b), it is possible to consider such long distance transport necessary because the livestock are reared in extensive rangeland conditions, far from the centres of human population. However, with such large numbers of animals and over the considerable period of time that they are transported, even on the best shipments there will be some mortalities *en route*. Defendants often point to the reduction in mortality in recent years (Norris and Gorman, 2007), now down to about 0.9% for sheep, but the animal welfare activists counter this by saying that 0.9% of 50,000 sheep is still an average of 450 deaths on each voyage. This argument appeals to the public perception that all animal's lives are to be respected and valued. The trade is also easily portrayed as a result of human greed, since the city-dwelling public is inclined to believe that livestock farmers are profiting excessively from the trade. Finally the lack of human care has been prominently exploited in the video footage of the handling of the animals in the recipient countries, some of which showed cattle having their leg tendons cut to stop them running away in an Egyptian abattoir (Animals Australia, 2008). The battle for the hearts and minds of the public continues.

Increasingly, the activist groups work through networks of supporters that are created on the internet. Direct assistance for lobbying is provided, which is a powerful influence on politicians in their decision making. Support for school activities is common, because this will influence opinion makers of the future. Information packs to assist teachers are made available, although in many countries industry has retaliated by attempting to reach into schools with counter views.

The high profile of the activists' campaigns ensures public support, which generates funds for the next campaign, and so on. In these campaigns, the focus is usually on a small part of the industry or one practice within it, such as the recent effective action against the mulesing[8] of sheep, rather than targeting the whole industry, even though many of the campaigners will be against the use of animals for food. The response of industry should be measured and considered, but it is often more of a knee-jerk reaction, in part because they see a dramatic shift in public support and are unsure what effect this will have on their business. They recognize that the financial impact of the adverse publicity is likely to exceed the direct losses caused by any change in practice. Often the media attention is very one-sided, and it is clear that the media have devised their story to appeal to public sentiment, without considering that industry may have a valid counter-story. The major risk if considered action is not taken is that management of the animal industries by referendum evolves, rather than by considered government.

Influences on Concern for Animal Welfare

To properly understand how the animal welfare movement is strengthening and expanding, it is necessary to consider the main influences, or drivers, for this movement. Concern for animal welfare is growing in most regions of the world, that much is evident from the attention paid to it by the media, the growing volume of scientific research in animal welfare (Fig. 8.1) and the increasing attention paid to animal welfare issues by governments around the world. It is important for the animal industries, and in particular the livestock industries, to be able to predict future changes in concern for animal welfare in order to manage the scale and direction of the industry, and most importantly, to provide the type of production system with which the public feel comfortable. Rapid changes may have detrimental effects on animal welfare, for example if live export of cattle from the northern parts of Australia were banned, they would be taken to southern states for sale and the lower price achieved would probably reduce the farmers' ability to feed their cattle adequately, at least in the short term. Different regions of the world will have their own unique consideration for animal welfare issues, and the many drivers of concern for these issues will have levels of importance that are peculiar to the region. Therefore, it is essential to understand not only how animal welfare concern is changing globally, but also the influences that are likely to be the most important in different regions.

One of the key influences is affluence, with people having more disposable income being more likely to purchase products from a high animal welfare system (Rahmann, 2000). Most regions of the world have experienced

[8] Surgical removal of loose skin from the hind quarters of sheep to prevent flies laying their eggs in the moist folds.

increasing affluence in the last 50 years – a result of economic growth and development and exploitation of many of the world's resources. The major exception to this is Africa, and in addition to this, there have been some serious but mostly temporary setbacks to economic growth in the former communist countries. As well as people having more money to spend on welfare-friendly products, countries that have experienced high economic growth have sought to increase the profitability of their animal industries by intensifying the production systems. Thus it may be difficult to discern whether the increased purchase of welfare-friendly animal products in the affluent countries derives from the increased concern about the animal production system or increased disposable income to spend on these products.

As well as there being increased disposable income, with the widespread adoption of capitalist financial policies, many countries have witnessed a greater divergence of income in their population. This is particularly relevant to animal welfare purchases if there is a threshold above which people start to spend more on welfare-friendly products, after they have satisfied their own material needs. There is no direct evidence of this to date, but if it is determined to be an economic phenomenon, we might expect some exponential growth in spending on welfare-friendly products as incomes increase, dependent on the difference in cost of welfare-friendly products relative to conventionally produced goods. This divergence in disposable income varies between regions. It has been most pronounced recently in the former communist countries, emerging from the conversion to capitalism, and least marked in countries with a high level of social responsibility, such as Australia, New Zealand and the Scandinavian countries.

As well as affluence, different regions of the world display varying levels of concern for animal welfare depending on their cultural heritage. As outlined previously, a preliminary survey has suggested that students from some Asian countries had less concern for animal welfare than those from Europe or America, but they all had similar levels of concern for animal rights (Phillips and McCulloch, 2005). This is likely to be due to their different cultural heritage and the levels of education of people within these countries. Levels of concern for animal welfare issues tend to increase with the level of education (Poss and Bader, 2007). The increase in educational standards in South and East Asia, following on from a period of significant economic growth, may be one reason for the growth in animal welfare interests in this area. The growth in awareness of animal welfare issues was originally a phenomenon largely confined to the Anglo-Saxon regions of the world, northern Europe, the northern sector of North America, and to a lesser extent, Australia and New Zealand. However, the movement appears to be spreading, partly driven by new national and international standards from the European Union.

Some cultural traditions towards animals derive from religious teachings, others from the different practices in relation to animal management that have evolved over the last few centuries. There is much debate about the influence of religion on concerns surrounding animal welfare (Lindeman and Vaananen,

2000), but the low number of adherents to religious creeds and practices in much of the Anglo-Saxon world suggests that religion is not a direct major driver of concern, rather an indirect driver through former influences on culture and beliefs (Li, 2000). An absence of direct influence of religion is suggested by reduced levels of concern for animal welfare in Asian people compared with northern Europeans (Phillips and McCulloch, 2005), even though the Buddhist and Moslem religions, which predominate in Asia, contain more teaching and instruction on the good management of animals than the teachings of Christ, which are officially followed in northern Europe (see Chapter 6). Furthermore, the small proportion of the population that is confirmed adherents to religious faith of their country in northern European countries may have increased the need for the development of codes of practice and legislation in these countries.

Coupled with increasing affluence in nearly all first world countries, there has been an increasing trend towards urbanization in all regions of the world. For many in developed regions of the world this brings greater affluence, and so the two influences on animal welfare are intertwined. But urbanization also brings removal from day-to-day contact with farming practices, and this loss of regular contact with the farming industry results in people becoming more sensitive to farm animal cruelty. Standards for keeping animals may be derived more from their companion animals, than food animals. Probably regular exposure to food animals in the rural population brings about a desire to support the rural industries and community, even if people are not directly involved in farming.

Urbanisation brings opportunities of choice for consumers of animal products. City dwellers use their affluence to eat out more in restaurants and fast food outlets, and this brings the opportunity to purchase high quality meals from welfare-friendly items without requiring more time and knowledge on the part of the consumer in the preparation process. Hence expenditure on food is increasing, and because food consumption is a major source of pleasure, we may expect the provision of welfare-friendly products through commercial outlets to increase substantially in future years (Denton et al., 1999).

Gender is also a major influence on animal welfare concerns. Females have greater levels of concern for animal welfare issues than males (Phillips and McCullough, 2005), although this may not extend to food purchasing habits (Lindeman and Vaananen, 2000). This greater concern of women may derive from the close relationship between the women and animals at the homestead during human evolution (see Chapter 3). There may also be some generalisation, from the more caring attitude that women have towards children than men, as a result of their greater investment in the reproductive process. Up until recently, women have always been the major food purchasers in shops and markets, however, with female emancipation having been a major force in the last century, and many women leaving the home to work, we should not assume that the majority of purchases of food items will be by women in the future. Animal welfare concern is very much related to gender. The personality traits usually associated with masculinity (adventure-seeking, aggression and dominance) are in contradiction to the caring, compassionate nature of the human feminine nature that is often associated with concern for animal welfare. The association between femininity and animal welfare attitude is explored by Laurie Carlson in her book *Cattle – an Informal Social History,* in which she contrasts the femininity of cow keeping with the masculinity of men hunting wild animals in prehistory (Carlson, 2002). The link between animal welfare and gender was also advocated by Mahatma Gandhi, who indirectly corrected the masculine human personality characteristics to a lack of care for animals, which he considered to be an indicator of moral backwardness[9](Gandhi, 1927).

Animal Welfare vs Animal Rights

There is a fundamental distinction between animal welfare – the quality and quantity of an animal's experiences – and animal rights – man's duty to exercise morally correct behaviour in relation to animals. An animal's welfare is a scientific absolute, evaluated on a continuous scale from low to high. By contrast, animals' rights are determined by beliefs and their existence is even denied by many individuals, religious groups and most legislation, for which animals are just property. Extreme advocates of animal rights usually believe that the life and integrity of individuals is of paramount importance and cannot be sacrificed for the benefit of humans or other animals. However, some have argued more generally that 'rights' are unsuitable to be the building blocks of society, because they are firstly illogical in some instances (why do wild animals have fewer rights than farm animals, for example?) and they are secondly, focused on ourselves and those injustices that are uncomfortable for us to live with (Bagaric, 2006). Bargaric argues that focusing on the consequences of our actions would be more logical. However, public sentiment is usually supportive of increasing both the standards of animal welfare and the level of rights afforded to animals, although they are not always willing to pay the extra cost. Prescribing rights creates a sphere where we can live comfortably without being confronted by hardship and cruelty. However, with increased communication around the globe, we must remember that images and actions from places other than our immediate neighbourhood may intrude on our daily lives.

In some cases, the entitlement to life proposed by many animal rightists may conflict with the animal welfare advocates, who frequently espouse a utilitarian view. For example, animal welfare advocates may believe that animals may legitimately be sacrificed in experimentation to find a technique for improving animal health, if it improves the welfare of other animals sufficiently, whereas animal rightists would oppose such sacrifice. Mahatma Gandhi was one of

---

[9] 'The greatness of a nation and its moral progress can be judged by the way its animals are treated' M. Gandhi.

many animal rights advocates who did not believe in utilitarianism.[10] Animal rights do not have to be extreme, they could simply relate to a right not to be maltreated to a certain degree, rather than a right to a life or the maintenance of integrity. This is enshrined in much new legislation concerning people's duty of care to animals. In this less extreme rights philosophy, the type of beneficiary may determine the level of an animal's rights. If the perceived beneficiary of, for example, animal experimentation is a member of the family of the aggrieved animal, its right to avoidance of the experimentation would in most people's view be less than if it was merely conducted to be a benefit for another animal species. Many people still take an anthropocentric view and consider that a benefit to humans is the most worthy and justifiable reason for animal experimentation. Preservation of similar genetics is the primary force at work and is likely to be more engrained in our attitudes for the purposes of adaptive evolution. Some even argue that this attitude prevails within our own species: countries where the inhabitants are ethnically pure, such as in Scandinavia, are more likely to have an extensive human welfare support systems (Ragin, 1994; Rojas, 1999; Kildal and Kuhnle, 2002; Bay et al., 2007). People in the Scandinavian countries have the highest expectations for the welfare of their animals of any in Europe, and they also recognize that provision for welfare is at a high level in their country (Eurobarometer, 2007), suggesting that high expectations can produce improved welfare outcome.

The animal rights movement is likely to grow in rapidly developing countries such as Australia, due to increased urbanization and the increasing spread of American ideologies, of which animal rights philosophy is one. The modern animal welfare movement originated in Europe, and hence the two most influential regions of the world on Australian ideology, the USA and Europe, both have strong considerations for animals, although from different perspectives. These influences are likely to continue and even strengthen due to the dominance of the USA and Europe in world affairs.

Western Attitudes Towards Animal Welfare

Australia is a high income, urbanised society, and both the affluence and the degree of urbanisation tend to increase the strength of concern for animal welfare and rights. The recent history of colonising land that is marginal and climatically challenged has led to speculation that the competition created between livestock and native fauna in such environments is too great, and that these areas should be left for native fauna (Higgins et al., 2002). This movement has been growing in parallel with the animal welfare movement, and both movements are increasing as people have more money to spend on

[10] 'I do not believe in the doctrine of the greatest good of the greatest number. The only real, dignified, human doctrine is the greatest good of all' (Gandhi, undated).

food that is produced to high standards of welfare, as well as safety and environmental sustainability. Increasingly, there will be less acceptance of farming practices that do not conform to the high ideals of a largely urbanised society.

The emergence of animal rights considerations has been a gradual development over the last 150 years, beginning with the Darwinian acceptance that there are no fundamental biological differences between animals and man, and that man evolved from animal progenitors. In fact, this fundamental challenge to anthropocentrism was much more revolutionary than the gradual acceptance that animals have rights, which is emerging today. Extreme animal rights advocates today form a body of highly dedicated and determined individuals, who are resolute in their pursuance of rights for animals and who usually occupy a more radical position in their beliefs about animal rights than the general public. History suggests that these people will be seen in the future as the social reformers of their day, in the same way that female emancipators and slave trade campaigners are now viewed as a necessary part of the social evolution of our democracies. As they represent an extreme sector of the population's views on animals' rights, we do not have to expect that all of their beliefs will become incorporated as a societal norm, for instance opposing the killing of animals for meat. Although they often support the most extreme positions, e.g. veganism, they usually expect to persuade government and individuals to support some of their more moderate demands, for instance the banning of cages or stalls for sows to be held in during pregnancy. Their message is appealing to the media, containing the classically attractive elements of first, cruelty or at least antisocial behaviour towards a defenceless being; second, domesticated animals; and third, social reform: an attack on the landowners, who are perceived to be rich. There is also the possibility of redistribution of their wealth to the land poor members of the public, if the landowners can be persuaded or forced to adopt more costly, welfare-friendly practices at no extra charge to the consumer. Over the last 50 years, industrialisation of animal agriculture has dramatically reduced the price of animal products, but sometimes at the expense of the quality of life of the animals. However, with man's inherent love of aesthetics and the natural world, we readily empathise with animals that are kept in unsuitable facilities, and act to eliminate unacceptable practices.

Addressing the Animal Rights Issues

In the long term, the best way to address the issues posed by animal rights extremists is through scientific investigation to find suitable alternatives to the systems that are the subject of the criticism. In addition, knowing how the public perceive animal practices and educating farmers about how new practices can improve animal welfare is essential. Most farmers would be very happy to use more or better resources than they currently do to improve the welfare of their animals, however, their system of production has to be economic. The welfare state of farm animals is therefore at least partly the product of consumer buying habits. Developing accurate information for consumers on welfare status is essential.

Anticipating the activists' next focus of attention will enable an effective public education response to be mounted at the right time, but in reality research should be in place and information programmes available for all the major animal practices that are suspected of presenting a challenge to animal welfare. Australia has some different welfare issues to Europe, such as poor feed availability for rangeland stock, which is generally not so much of a problem in Europe. This relates to the more marginal land that is used for livestock rearing in Australia, compared with Europe. The extensive nature of Australian livestock farming brings threats of food shortages, but also opportunities to inform the public that these practices are conducive to fulfilling the animal's natural requirements for space and a natural social order, in comparison with European production systems that more commonly include intensive animal housing, which is often crowded and does not respect the mother-offspring relationship. Such advantages are actively promoted by New Zealand, but Australia has to first address the worldwide view that its livestock farming is inherently cruel because of the invasive practices that it employs. The unacceptability of invasive practices to animal rights activists is founded in their belief in maintaining the integrity of the animal. Therefore practices such as dehorning, tail docking, branding and mulesing are all seen as undesirable, even if the consequences of not doing the practice brings greater harm, or a risk of greater harm. Some would even take this view if not doing the practice produces a negative welfare situation overall, such as when a sheep is struck by flies because it has not had the folds of skin removed from the hind quarters in the mulesing operation. There is an urgent need for critical evaluation of the impact of invasive practices on whole-of-life welfare, including disease evaluation that incorporates assessment of the duration of the disease and the severity.

Free choice of the animal to select the optimum environment and diet are also perceived by many as desirable, since *our* ability to exercise free choice is one of our most valued resources. There is currently little evidence that animals do or do not value this free choice. Research on diet selection by sheep suggests that they make limited use of pre-consumption information that might direct choices, rather regular information processing during feeding informs choices directly (Illius et al., 1992). Furthermore, when eating an ideal diet, they will regularly return to sample an inadequate diet. This could be because they want to confirm that the diet is inadequate, but it could also be to maintain a microflora in the gut that is capable of digesting both an ideal and inadequate diet, in case they only have access to the latter in future. Thus it is beneficial for animals to experience good and bad circumstances, particularly as they develop, so that they can learn to cope with the bad circumstances.

Morality Towards Animals

Although most people agree that conditions and facilities for animals are improving, although perhaps not as rapidly as for their human counterparts (see Chapter 1), largely as a result of technological advances, there is limited evidence of improvements in our moral attitudes to standards for animal welfare. An animal's welfare is a function of both the level of welfare that we are able to afford and the level that we aspire to provide for them. The higher the aspirations, the more we are likely to sacrifice luxury goods for our own welfare, to make provision for the welfare of animals in our care. The improvements that we see in animal welfare in recent decades appear to derive largely from improved wealth and opportunity to provide better conditions for animals. Thus we are more able to keep them closer to the standard which we would aspire to, as a result of technological advances and increased affluence.

Some evidence that we may be aspiring to better welfare standards for animals comes from increased activity by nongovernmental organizations, increased concern by the public, in the media and more legislation and codes However, we also see increased activity in many other areas of social awareness and this may merely be a reflection of more responsive governments that listen to such pressure. Contrary evidence, that moral attitudes towards animals are not improving, derives largely from the reports of animal abuse and the minor offences against animals that we all witness in our everyday life. Whilst there has not been any formal documentation of the frequency of such abuses, it is likely that in the past they occurred regularly but were not reported or of interest. In Australia at least, most abuse (80%) does not involve deliberate cruelty on the part of the perpetrator but is the result of ignorance, poverty or adherence to tradition (Green and Gullone, 2005). Deliberate animal abuse may be as low as 0.6 % of cases seen by veterinarians (Sharpe and Wittum, 1999) and may be motivated by aggression, or an outlet for sex or attention seeking[11] (Munro and Thrusfield, 2001b). Only about half of one sample of small animal veterinarians have had to deal with abuse, mostly in dogs (Munro and Thrusfield, 2001a), yet they are ill-prepared by their training (Landau, 1999). Of increasing concern is the fact that abuse of animals links directly to abusive attitudes and bullying of children and women, but veterinarians rarely address this associated human abuse (Henry and Sanders, 2007). We shall return to animal abuse when considering cruelty.

Looking at societal changes generally, it might be suspected that the pressures of modern living would cause some people to vent their frustrations on animals. Man's humanity to his fellow man appears at first sight to have

[11] A Munchausen syndrome by proxy syndrome has been observed in about 2% of animal abuse cases, involving attention-seeking behaviour by the owner, apparently fictitious clinical signs and injury (Munro and Thrusfield, 2001b)

deteriorated considerably in the first half of the 20th century. Although accurate statistics are difficult to find, historians are agreed that there was a considerable escalation in the proportion of the world population killed in wars in the 20th C (about 5%), indeed there has been a progressive increase since the advent of modern warfare in the 16th C (Roland, 2007). Undoubtedly much of the increase was due to the increased effectiveness of weapons and the conflict between global empires rather than individual countries, rather than a decline in moral standards. Another reason for the increase in 20th C lies in the emergence of eugenics movement, which was at least partly responsible for the most expensive war yet, the Second World War, claiming over 50 million lives (White, 2005). This movement can be traced to America's attempt to suppress the coloured people at the turn of the century (Black, 2003), and before this to the views of Sir Francis Galton, who took the works of his cousin, Darwin, one stage further to advocate selective breeding of humans for the purposes of genetic improvement (Allen, 2002).

Fortunately, although the eugenics movement continues in a variety of forms, the rate of death from war declined substantially in the second half of the century, which some attribute to the development of nuclear weapons and the reluctance to use them because of the scale of the consequences[12] (Roland, 2007). More likely it was due to a growing responsibility for, and ability to control, world events. In addition, following the atrocities committed by totalitarian governments with misguided objectives, there was a marked change in the latter part of the 20th century towards more democratic government. This has brought the opportunity for more social responsibility in governments and also for the people to demonstrate their concern for the less fortunate members of society, including the animals in our care. As a result of these democratic changes, we now see active animal welfare organisations around the globe, enabling people to express their intentions to treat animals better.

Thus it has taken time for man to start to develop a responsible attitude to managing change after he realised that he could control and manipulate his own destiny, not just his own genetic destiny and that of the animals within his care, but it is increasingly realised that management of the global environment is man's responsibility too. There will be further brutal acts by dictators and probably more genocide, but the seed has been sown for major and prolonged social improvements that will enable us to care better for the animals within our charge. We must never forget that, as Lord Acton said 'power tends to corrupt, and absolute power corrupts absolutely'. However, the worldwide move towards representative government which emerged in the late 20th century, and most notably in the former communist countries, must give us optimism that multicultural societies will thrive and flourish in the future, embracing a multitude of attitudes towards animals.

---

[12] The same argument was made for dynamite when it was invented (Roland, 2007)

Cruelty to Animals

Cruelty to an animal usually has the elements of an intentional act towards an animal, or willful neglect, that causes unnecessary suffering, in that it affects their life, health or comfort (American Humane, 2003; LLL, 2008). It can include mental harm and reasonable apprehension or fear of physical or mental harm. It does not include acts which merely offend or can be assumed to hurt the feelings or brief displays of passion.

Widespread and regrettably sometimes glamorized cruelty to animals has been a scar on man's attempts to create civilized societies for several thousand years. However, when considering primeval man's hunting practices, it may be difficult at first sight to differentiate cruelty from natural predator behaviour. In prehistoric times, men engaged regularly in hunting forays, where an animal would be injured or killed by a range of projectiles, or it could be surrounded and driven into a pit, and then gradually beaten or speared to death, perhaps with a degree of torture, in the same way that many predators will play with their prey before the kill. Is this any different to the slaughter of large animals by the ancient Romans, surrounded by an audience of appreciative citizens? The major difference is the necessity of the act, which affects the motivation. This fundamentally alters the acceptability of the practices, since the Roman games were a spectacle designed to appeal to the crowd's sense of satisfaction in animal suffering, in much the same way that an animal sacrificed in the bullring delights the audience nowadays. The motivation for primeval man was survival, for which he had to procure food. The feelings of satisfaction in suffering have no place in modern society, and many great civilizations have thrived in the past without such activity. Hence the British government recently made the traditional practice of hunting foxes with dogs illegal (Hunting Act, 2004), because the event is a spectacle for the enjoyment of both those engaged in the sport and other viewers.

Early hunting activities did not always celebrate the animal suffering in the way that the Roman games did. The ancient Egyptians rarely depicted hunting scenes in their art, and did not engage in the mass slaughter of animals for pleasure. When hunting scenes were painted, in the earliest Egyptian cliff engravings of the 8th millenium BC, they conveyed more of a feeling of grace rather than ferocity (Andrews, 2005). The Assyrian kings were keen on hunting lions (British Museum, 2007) in a ritualized chase, with caged lions released so that the king and his followers could shoot at them with arrows. This has the element of intention that is necessary in cruelty and also some of the danger associated with the Spanish bullfight. Early reliefs show lions being driven towards the king's men in chariots, sometimes attacking the chariots (Anon, 2008b). Victorian pictures of hunting, where the squire (who often commissioned the painting) sits aloft an elegant horse surrounded by well-groomed and obedient dogs, convey an appearance of grace and elegance in the same way as the Egyptian engravings. The emphasis is on mastery of nature, rather than the suffering involved in animal sacrifice. Similarly the ancient Greeks did not glorify the torture and slaughter of animals in the way the Romans did, although they did sacrifice animals to appease the gods, in particular bulls (Thomas, 2003). Earlier civilizations had glorified hunting, which in essence was not dissimilar from bull-fighting.

Animal sacrifice, the ritual killing of an animal as part of a religion, was a unique form of cruelty which featured strongly in many early civilizations. The ritual element lent acceptability to a cruel practice. It is inconceivable to imagine that such ceremonial sacrifice of animals would be tolerated by modern society, yet it is perpetuated in pagan ceremonies in many countries, where chickens or other farm animals are regularly slaughtered. The Romans sacrificed vast numbers of animals to appease the gods, even accepting that there may be some inaccuracies in the historical records. Animals' heads were first sprinkled with bread and wine and then disemboweled for inspection of the quality of their internal organs, before having their throats cut (Anon, 2008a). Any imperfections led to other animals being used. Bulls were particularly favoured, which probably stems from their being an object of reverence both for their ferocious nature and displays of sexual activity. The scale of their sacrifice may be a reflection of the Romans' strong belief in the need for the support of the gods. Bullfighting in the Iberian peninsular, France and South America persists, but perhaps surprisingly it did not evolve from the Greek and Roman bull sacrifice, but was introduced more recently into Spain by the Moors, who established bull-fights in the ancient Roman amphitheatres of Andalusia. The bull-fight, unlike the ritualized pagan slaughter of animals, cock-fighting or fox hunting, assumes some of its appeal because of the danger that men face when fighting the animals. This supplication to the hot-blooded nature of the Latin temperament has ensured its recent survival in the face of mounting criticism from animal welfare supporters. However, despite the claims of bull-fighting supporters, the practice has not continued for centuries but in many cases achieved a resurgence in popularity in the 1970s and 80s (FAACE, 2001).

As the bull was revered in both the ancient Egyptian and Roman civilizations, so too were other animals revered by ancient people: lions by the Assyrians, and more recently bears by the Siberian Samoyede nomads. The Samoyede peoples of Siberia were one of the last hunter-gatherer societies to survive and illustrate the close relationship between hunters and their animals. They relied on hunting deer, wolves, bears and squirrels for meat and hides and led a largely nomadic existence as recently as the Middle Ages, living in the tundra region where no agriculture was possible (Newell, undated). They kept domesticated reindeer, some of whom they venerated as gods, and these were decorated and given special treatment. Eating the flesh of the reindeer was rare, and forbidden in the case of the sacred animals, but the does were used as decoys to lure wild stags to be killed for meat and hides. Even more recently than this, a similar close and respectful relationship between men and cattle was found in the Nuer tribesmen of the

Southern Sudan by a famous anthropologist working in the middle of the last century (Evans-Pritchard, 1940).

Thus the traditional relationship between primitive man and animals was one of respect and veneration on the part of man, with some animals benefiting from the care that humans gave to them in a symbiotic relationship that came to represent the domestic contract. This contract, which primitive people employed for keeping animals when they started domesticating them about 12 C ago, provided for a good standard of nutrition, health and safety for the animals, but also dictated their longevity, their relationship with their offspring and other animals that they kept and the utilization of products from the animals, such as milk. As humans have developed, in terms of the conditions under which they live, it is to be expected that they should seek to keep their animals in better conditions. Because of this, cruelty to animals is increasingly recognised as unacceptable by most members of society. Yet as society becomes more sophisticated and complex, attempts to impose new standards of morality towards animals are often ignored and flagrantly flouted by a minority of individuals.

Animal welfare activists may have a different perspective on animal cruelty to those directly involved in the animal industries, in that they focus on invasive procedures, such as mulesing in sheep or dehorning or castration in cattle, rather than longer term problems of poor stockmanship, long distance transport, inadequate food supplies or lack of shelter. This may be due to a lack of knowledge, as in the public, who may change their views on a practice when they have more knowledge. For example, with limited knowledge the public generally disapprove of mechanical harvesting of poultry, but when informed about the welfare impact in comparison with harvesting by people, they frequently change their mind (Delezie et al., 2006). The least informed are young female, urban members of society, who also display the greatest concern about animal welfare issues (Delezie et al., 2007).

Those more directly involved in farming recognize that invasive practices prevent some far more damaging welfare problems, such as mulesing sheep to prevent them from being attacked by flies, or dehorning cattle to prevent them from injuring each other. There is little evidence that farmers become immune to the short-term pain that animals suffer during and after these operations, because of their regular exposure to them. However, although such people are usually doing their best for the animals, with societal standards for animal welfare increasing at a relatively rapid rate, some systems of production may be accepted by the producers but unacceptable to the public, and even to the consumers of the products. Hence it may be necessary for systems to change, which requires producers to be flexible. The long time that it takes to learn how to manage an animal production system mitigates against sudden change, but it is the mark of an advanced, responsible society that it helps producers to make the necessary changes if they are drastic. Financial incentives (golden handshakes) for those leaving the industry and support for those adopting new, more suitable methods of production are two methods of facilitating the adoption of systems of production that are in accord with public demand.

Action on animal cruelty depends on being able to measure the risk of it happening. Theoretically the risk to welfare of cruelty events can be evaluated mathematically by determining two factors – the severity of the event and the likelihood of it happening. Systems of risk management that employ Hazard Analysis and Critical Control Point are now being advocated to deal with animal welfare issues (Hegelund and Sorensen, 2007). There is often some understanding of the severity of different events, but little understanding of the risk factors for the occurrence. One such risk factor is abuse to humans, since criminals that have committed such offences often admit to having abused animals as well (Arluke et al., 1999). It is not clear whether this is a causal relationship, but the fact that the two co-exist should encourage protection societies for humans and animals to collaborate in identification of offenders.

Whilst we have no direct evidence that the levels of cruelty inflicted on animals are actually increasing, we might anticipate that the changing demographics in modern society are likely to have an impact. The majority of the perpetrators of cruelty are male (Herzog, 2007) and likely to be without employment or a good education. Both young boys, eager to demonstrate their bravado to their peers by abusing animals, especially cats, and teenage or adult males, who may find cruelty to animals, especially dogs, to be an easy form of release for their pent-up aggression, are commonly involved (Arluke and Luke, 1997). Cruelty may offer a form of amusement, when there is little else to occupy them. The pressures of modern society, and the increased number of people living alone without the stability of the traditional family unit, are likely to increase the number of people wishing to harm animals. Although this cannot be condoned, understanding the aggressive impulses of the young male, and how best to disburse these through peaceful means, would help to deal with the problem of cruelty to animals.

Incidences of extreme cruelty are entering the media headlines with increasing regularity, demonstrating that the phenomenon creates a sense of outrage in the law-abiding members of society. The type of human abuse with which animal abuse is often linked is usually wife battering or child beating (Ascione et al., 2007). Although this is increasingly speculated upon, the evidence supporting a causal link is still equivocal. About 25% of aggressive criminals, when questioned in prisons, self report that they have performed violent acts to five or more animals previously. Non-aggressive criminals self-report that about 5% of them conducted violence to animals, and this compares with an incidence in the general population approaching 0% (Frank Ascione, personal communication). One of the most brutal killers in recent times, Saddam Hussein, was also reputed to have been an animal abuser as a teenager. According to his Press Secretary, "he used to put an iron bar on the fire and make it red hot, and when he saw an animal passing he would run out and stab it in the stomach" (Simpson, 2004). He had had a turbulent childhood, and showed a penchant for violence throughout his life. These reports give some support for the contention that childhood violence towards animals will eventually lead, at least in a proportion of the perpetrators, to violence towards fellow human beings. However, care in interpretation is warranted because aggressive criminals may be likely to inflate their reputation for aggression by claiming to have abused animals in the past, since it can be done with much less risk than claiming aggression towards humans. Alternatively aggressive criminals may even under report, due to the stigma attached to being a violent criminal in prison. The determination of any link between animal and human violence is therefore difficult, because only a very minute proportion of the population are ever convicted of either offence. As only about one quarter of aggressive criminals report having abused animals, it seems likely that animal abuse does not lead to human abuse in many cases. However, it should be treated, not just for the offence against animals that it represents, but more generally, as evidence of psychological disturbance in the mind of the perpetrator. The involvement of social care workers should be sought as soon as a case is recognised.

The role of the veterinarians in reporting animal cruelty is crucial. As only about one half of veterinarians will have to deal with animal abuse cases, they have little opportunity to develop the correct responses. Most likely, the first response will be to counsel the client, and then the dilemma will emerge as to whether to preserve client confidentiality and perhaps maintain harmony in a household if the veterinarian was called in by someone other than the animal abuser, or whether to report the incident to the police or the animal cruelty inspectorate. Ethicists usually come down on the side of the animal (Rollin, 2006). In the United Kingdom, newly qualified veterinarians, on entry to their controlling body or college at the end of their study swear an oath[13] that requires them to put the welfare of animals in their charge as the first priority for the duration of their career.

In other countries, veterinarians may be legally required to report animal cruelty to the police. This dilemma will face an average veterinarian only rarely, but their decision will be very important for the welfare of animals that come into contact with the potential animal abuser. Inevitably, there is much abuse that is unreported, and even within a family situation, aggression towards a pet may be seen as a more acceptable form of aggression than between two people. For the veterinarian, the issue is a classical dilemma, since they have responsibilities to their clients, to their profession, to animals, to society and to themselves, which can easily present conflicts of interest, particularly between clients and animals.

The most commonly abused animals are cats and dogs, which is no doubt in part due to the fact that they are the most common family pets. We may speculate that cats in particular often appear quite fearless in their approach to other animals and humans, an attribute which has probably been selected for

---

[13] "I promise, above all, that I will pursue the work of my profession with uprightness of conduct and that my constant endeavour will be to ensure the welfare of animals committed to my care."

over the course of their evolution as a solitary hunter. This may bring them into conflict with the dominant member of a household, who may feel that his position at the head of a household is threatened. Tensions in the modern household today are more likely to arise from the pressures of both parents having to work longer hours than previously. The sorts of injuries sustained by both cats and dogs include burns, lacerations, gunshot wounds, poisoning, injury to genitalia, bruising and fractures (McGuinness et al., 2005). In Australia there is believed to have been a recent increase in antipathy towards cats as an introduced species (PIAS, undated). Cats readily become feral and threaten native fauna, in particular birds and small mammals. Such concerns may appear to legitimise the disposal and sometimes brutal killing of members of the introduced species. A similar stigmatisation has condemned many cane toads in Australia, with some environmental agencies actively encouraging people to put these animals to death (Beatty, undated). The impact of such apparently benevolent action by members of the public on population dynamics is negligible. In this, as in other areas, animal welfare may be in direct conflict with environmentalism. During wet weather, when the toads emerge onto the roads, many car drivers will attempt to drive over them, which in their minds is legitimized because of the environmental cause.

Another form of animal abuse that appears to be increasing is that of bestiality. As modern society becomes more complex and pressured and we put up social barriers to protect our integrity, more people live alone rather than in the traditional family. There appears to have been an increase in the tendency for people to seek sexual gratification from animals rather than fellow human beings, perhaps due to the aforementioned social changes, perhaps due to the availability of images on the internet (Mehta, 2001). At least there is greater awareness of this problem, and of concern is the possible link to violence towards people (Hensley et al., 2006). In an age when the barriers towards sexual activities of homosexuals, transsexuals and transvestites are being broken down in the name of freedom of expression, the stigma attached to bestiality is inevitably being challenged (Landry, 2001). However, although the debate will continue about whether the moral depravity involved in sexual acts with animals will facilitate the perpetrator engaging in similar acts with humans, perhaps even infants, the adverse effects on animal welfare are obvious, particularly if the animal is small and helpless.

There is an urgent need for more research to investigate deliberate animal cruelty, because of the outrage caused in the general public when serious incidences arise, and because the small proportion of cases that come to the notice of the general public are undoubtedly an indication that there is a much greater problem of deliberate cruelty towards animals that needs addressing. Increased vigilance on the part of veterinarians, animal cruelty inspectors, the police and the general public will assist in the raising of levels of awareness in the general public, which is likely to further increase the stigma attached to animal cruelty so that people are aware of its social unacceptability.

Cruelty Legislation

Traditional animal welfare legislation relies on preventing cruelty to animals, which as well as the definition in scientific terms (see introduction to Cruelty, above) has a specific legal context. Definitions of cruelty have evolved over the last two centuries, being originally taken to mean 'the unnecessary abuse of any animal'. In the early twentieth century, cruelty legislation was defined in the United Kingdom largely for the purposes of preventing horses being mistreated in London, and therefore contained a catalogue of potential abuses of such animals (a person being guilty of 'cruelty' if they [cruelly] beat, kick, ill-treat, override, over-drive, overload, torture, unfuriate, or terrify any animal). The most important offence under which most people are prosecuted in the United Kingdom today is when they 'wantonly or unreasonably do or omit any act causing unnecessary suffering to any animal'. Two key elements are the necessity of intentionality, which was not present in the first definition, and the possibility that unnecessary suffering can be caused by either commission *or* omission. Other pioneering twentieth century British legislation relates to specific acts of suffering that may not necessarily require commission: enabling fighting between animals, administering injurious drugs, operating without due care and humanity, and tethering horses (Radford, 2001). It should be noted that cruelty generally refers just to the negative treatment of animals, whereas 'animal welfare' also has positive elements within it (Radford, 2001). This deficiency is increasingly being taken into account in new welfare legislation, which identifies that people in charge of an animal have a duty of care towards it (e.g. Queensland Animal Care and Protection Act, 2001).

The legislative difficulties that animal welfare science can address are principally in the determination of when unnecessary suffering has been inflicted. Assuming that the suffering is both unnecessary and inflicted voluntarily, the difficulty often confronting the courts is whether significant pain[14] and distress[15] has been caused. The Queensland Animal Care and Protection Act defines cruelty both by the causation of unjustifiable, unnecessary and unreasonable pain, and by specific instances of cruelty that include beating an animal to cause pain, abusing, terrifying, tormenting or overworking it. Cruelty during transport, killing and the inflicting of injury are qualified by the Act, by being required to be inappropriate or unreasonable.

The number of cruelty complaints to the RSPCA, Australia, is approximately 45000/year, which results in about 330 prosecutions and 200 convictions annually (RSPCA, 2002a). Most complaints are received for dogs (15,000), livestock (7,000) and cats (5500). Wildlife and horses each receive about 3,100

---

[14] Defined as an aversive sensory experience caused by actual or potential injury that elicits protective motor and vegetative reactions, results in learned avoidance, and may modify species specific behaviour, including social behaviour (Broom, 1992)

[15] Defined as a severe stress response accompanied by behavioural signs suggesting that the animal finds it unpleasant (Ewbank, 1992)

complaints and 1,800 for birds. The different reporting systems and data collection methods in the different Australian States and Territories make comparison over time difficult.

Both physiological, disease and behavioural evidence can be used to determine that pain and distress has been inflicted. The courts tend to rely more on physical evidence (disease) where deliberate cruelty is inflicted by commission. However, there are many cases of omission which inflict cruelty, for example about 300 prosecutions for abandonment are brought each year by the RSPCA in the United Kingdom. In such cases physiological evidence is preferred to behavioural, despite the fact that behavioural responses are more directly involved in the definitions of pain and distress. The reasons for this are (1) samples for chemical analysis of physiological parameters are relatively easily taken, and (2) normal (reference) ranges for physiological parameters are well known and are generally within a narrow range. Animal behaviour is considerably more variable than physiology, because homeostasis confines the optimum function of the body's metabolites to a narrow range for its own protection. Physiological parameters that indicate abnormal function are difficult to identify, in part because behavioural modification protects the animal's physiology. For example, animals that are deficient in a particular nutrient will often develop an appetite for abnormal food items that will satisfy their need, as in bone chewing in phosphorus-deficient cattle. Demonstrating that animals are severely undernourished is difficult because the physiological indicators that are utilized in starvation cases mainly refer to short-term undernutrition (Table 4.1). The reference range of these indicators is easy to establish, but when my research team investigated the values obtained for two herds of cattle suffering from prolonged undernutrition, none of them were consistently outside the reference range, despite the fact that several of the animals had died (Agenäs et al., 2006).

In such instances, behavioural changes, such as the development of a depraved appetite and lethargy are more likely to be of value in indicating pain and distress, but they are difficult to monitor and present to a court of law. In other cases the incidence of abnormal behaviour can be taken to indicate the existence of pain and distress. Stereotyped behaviour that serves no obvious

Table 4.1 Reference ranges for serum nutrition status indicators in adequately-nourished beef cattle and values for two undernourished herds (Agenäs et al. 2006)

Indicator	Reference range	Undernourished herds	
		1	2
Albumin, g/l	25.0–44.4	36.1	27.1
β-hydroxybutyrate, mmol/l	0.12–0.61	0.31	0.29
Creatinine, μmol/l	110–225	157	81
Fructosamine, μmol/l	183–365	248	367
Globulin, g/l	27.2–49.2	36.6	34.9
NEFA, μmol/l	176–1317	467	343
Urea, mmol/l	1.88–7.00	4.3	6.5

purpose other than to provide a distraction for the animal concerned is one possible indicator of mental suffering. Broom and Johnson (1993) suggest that stereotypies performed for 40% of the time indicate very poor welfare, those performed for just 5% of the time indicating some adverse effects on welfare. However, the incidence of stereotypy performance varies significantly between animal species and between individuals within a species, with most people acknowledging that the animal probably derives some psychological benefit from performing the stereotypy. Furthermore, it is not clear whether the absence of stereotypy performance in species such as cattle demonstrates a greater ability to cope with a difficult environment. Ruminant animals in particular show few oral stereotypies, even when placed in severely confined conditions that would elicit prolonged stereotypy performance in pigs, for example. This may be because they already spend more than one half of their day in chewing activity, either when consuming food or ruminating. Does this legitimize such conditions for ruminant animals? Using abnormal behaviour as indicators of pain and distress may therefore be feasible for 'extrovert' species such as pigs and chickens that readily perform such behaviours, but of less value for 'introvert' species, such as ruminants. In stressful situations it is not adaptive for wild cattle or sheep to draw attention to themselves, as the best defence for a vulnerable animal from attack by predators is hiding. Wild boar, however, scatter to their den in times of danger and jungle fowl fly into the trees, so warning others would be of adaptive advantage. Hence abnormal behaviour, including vocalization, is a common part of the adaptation of some, but not all animals to stress and danger.

An alternative way in which animal welfare science reduces cruelty is through improved codes of practice, assurance schemes and directly into commercial practice. Industry funds a significant amount of animal welfare science, but is most likely to adopt changes to normal practice when there is little adverse commercial impact or there is a positive benefit.

Concluding Remarks

Animal welfare and animal rights movements are distinct, but linked in many people's minds. The growing public concern for these issues will pose a significant financial threat to the animal industries unless consumers can be persuaded to pay for animal output produced to a higher standard of welfare than at present. The cost of welfare improvements can eventually be passed on to the consumer, but in the short term intense worldwide competition keeps the price low and discourages some farmers from adopting novel, welfare-friendly practices. Greater understanding of how to measure welfare and develop improved systems is essential before audit systems for animal welfare can be effectively used in industry. A failure to improve welfare in this way and preserve animal rights will lead to a greater demand for legislation, which will confine the industry to prescribed legitimate practices.

Chapter 5
Welfare Assessment

Welfare perception – positive and negative welfare components – developing useful measures – legislation and audits

Introduction

The welfare of an animal, or a group of animals, is a complex concept and therefore assessment is not simple, and there are no perfect, instantaneous measures. Expert opinion may be all that is available, with little or no scientific evaluation, but even though experts may agree, their assessment can be wrong or exaggerated. To take an example, how do we assess whether it is wrong for cows to be kept in dirty conditions, and in extreme cases this might be buildings where so much faeces and urine have collected on the floor that their walking becomes difficult (see Phillips and Morris, 2000, for evidence of this)? First, the effects on the animal's welfare have to be assessed, but by whom and over what period of time? We can measure the choices that cows make when offered the opportunity to be in clean or dirty conditions, but this will not necessarily tell us *how much* they prefer the clean conditions, if they do. We can make cows work to obtain the preferred environment, for example by having to walk further to get to it. We can ask people their perception of the situation, and if they understand it in reasonable detail they may automatically assume that it is damaging to a cow's welfare to be kept in dirty conditions, as it is likely to soil her coat and makes the floor slippery to walk on. This may be unnecessarily anthropomorphic, as we ourselves would not like to live under such conditions and we would be particularly concerned about the spread of diseases. However, the limited amount of research with dairy cows kept under these conditions suggests that they quickly learn to tolerate the presence of faeces in the passageways of their buildings and when given a choice of clean or dirty passageways, they will not actively avoid them after just a short period of exposure (Phillips and Morris, 2000; Phillips et al., 2000). When given the opportunity, cattle *will* avoid faecal deposits (Whistance et al., 2007), but there is only limited evidence so far that their welfare is adversely affected by being in dirty passageways.

C. Phillips, *The Welfare of Animals*, Animal Welfare 8,
DOI 10.1007/978-1-4020-9219-0_5,

An opinion based solely on the available scientific evidence might therefore conclude that such conditions do not adversely affect the cows' welfare. However, cattle may not be able to foresee the long-term consequences of their choices, and the avoidance of faeces will help to protect animals from acquiring parasites, in particular stomach worms, and potentially dangerous bacteria, such *Escherichia coli*, and those causing tuberculosis and paratuberculosis. Such diseases are rare, but this scenario emphasises that there may be risks that the cow cannot foresee, which her genetically-controlled behaviour does not recognise or respond to. The cow will only learn to associate the presence of faeces with consequentially poor welfare if the two are closely related in time, and there may need to be repeated events for the relationship to be learnt.

Therefore, there are some benefits to expert assessment, which may make it more useful than direct choices posed to the animal. However, the extent to which assessment should depend on perceived values or observed behaviour is partly dependent on whether the expert opinion recognises real dangers to the animal. Expert opinion may be essentially folk knowledge, which may be based in reality, but could also be based on hearsay or speculation. People are particularly concerned about issues that would bother them, rather than the animal, and they may exaggerate the perceived welfare impact, so in the case of animals living in or near their own faeces it may pose only a remote risk of contamination if they are given the recommended anti-parasite medication. People are most concerned if the animals themselves have no control over their environment, for example in hot conditions cattle may choose to shelter under a tree if they have the choice. In requiring animals to have some freedom of choice, people acknowledge that animals often know better than humans what is good for them. People are even more concerned if others that they don't know and trust have control over their animals, for example, when animals are sent overseas for slaughter. These are some of the public's simple rules for assessing animal welfare: trust the animals' ability to choose their best environment and trust people that are looking after them to do the same if, and only if, they are people like themselves.

We often rely on politicians to implement our concerns about animal welfare, but in wishing to please their electorate they are likely to have divided loyalties. Relatively large numbers of the general public make some form of direct representation about animal welfare issues to politicians, compared to other social issues, so they feel obliged to act.[1] However, too great a level of government control over animal welfare in the primary industries, especially in countries in which these are economically important, will reduce their profitability and potentially cause problems of rural

---

[1] Quote from Euro-MP Neil Parish 'Animal welfare is the issue raised most frequently in my mailbag'. "MP to support animals". South-West Farmer, Thursday, February 1, 2007. http://www.southwestfarmer.co.uk/mostpopular.var.1163285.mostviewed.mp_to_support_animals.php

depopulation, as well as impacting on the general economy of the country. Countries with high welfare standards will always be under threat of importation of animal products from countries with lower standards, unless there is restriction of trade on the grounds of differing animal welfare standards. Similarly, although too great a level of control over pet owners will not affect national prosperity, it may alienate some members of this sector of the population.

The public are particularly suspicious about the extent to which farmers and laboratory scientists provide adequate care for their animals because of their vested financial interest, especially following the recent intensification of the animal industries (see Chapter 9). The lack of contact with individual animals and the pressures on managers to produce financial returns to their investors may have led some to become desensitised to animal suffering. In the case of farmers an additional constraint is the severe economic competition which has prevailed in times of food surpluses. This may encourage some to place personal ethical responsibilities to their family above their professional responsibilities to animals. Farmers are often not collectively organized to obtain an adequate return from the retailers of their products, some of which may take advantage of this, so that they can reduce prices to the public and improve the competitiveness of their products.

The general public appears to be increasingly less well informed about the management of animals in rural districts, as fewer people live and work there than previously and there is limited information provided to consumers on production methods (Duffy et al., 2005). For example, in Australia, 92% of people live in cities (compared with 50% worldwide) (Newman, 2006). Whereas only a few decades ago most Australians had a strong connection with activities in the countryside through their parents or grandparents, this is increasingly no longer the case, and people rely more on the media for information. 'Abuse of animals' stories are popular with the public and appeal to their concern for animals.

Another concern relating to assessment of welfare by the public is that their views differ depending on their culture, country of origin etc. In the multinational survey of students of many different nationalities previously referred to (Phillips and McCulloch, 2005), in which students from Asian countries appeared to have less concern for suffering of animals during life than reverence for the life of animals, whereas those from European countries generally had more concern for suffering during life, this may be because welfare provision is worse in European than Asian countries. In addition across the nations, women have a more caring attitude towards animals than men, and hence are more often employed as animal carers (see Chapter 3). In the former Communist states of Eastern Europe, where division of labour on the large collective and state farms was at a more advanced level than in the West, women would usually be given the tasks relating to animal caring, and men were primarily involved in mechanical aspects of the farm's work, driving tractors, working as engineers etc.

Welfare Perception

Welfare perception by humans is therefore influenced by many factors, including cultural traditions, gender, intelligence level, probably human genetics and possibly age. There may be a distinct difference between the perceived and actual animal welfare. Both will be relevant for welfare assessment, but the former will be most useful to understand the public position on welfare requirements and the latter for objective improvement. As society progresses, the perception of the desirable animal welfare state will change, and it is likely that there will be greater emphasis on equity in provision for animal welfare. Currently very different standards are aimed for, depending on the type of animal. Greater equity would be a mark of a more caring society, representing societal progress, whereas focusing on traditional attitudes to animals that derive from the benefits that they produce ignores the responsibility that we have to manage all animals. For example, rats used to be a major cause of disease, infesting crops and offering no benefit to human society. They were universally reviled and where possible exterminated. Now that their antihuman activities have, in most developed societies at least, been controlled, their benefits to society as companions or laboratory animals are beginning to be recognized.

Positive and Negative Welfare Components

Animal welfare can be measured in terms of good and bad experiences, as outlined in Chapter 1. In terms of good experiences, happiness is a major goal for all living beings, as numerous spiritual leaders over the centuries have taught, perhaps most notably the current Dallai Llama (Mehrotra, 2005). Human and animal happiness are both dependent on the balance between perceived negative and positive experiences, but for humans with their complex cognitive abilities there is the opportunity to alter the perception of any event from negative to positive just by training the mind. It is likely that the opportunity for animals to train themselves, or be trained, to increase their level of happiness by freeing their mind from worry, hatred or other negative emotions is more limited than for humans. Nevertheless, companion animals will often be comforted by their owners, providing reassurance that they should not be frightened, for example in a thunder storm. The benefits of complementary therapy for animals, including relaxation techniques, such as through touch, are evident for humans and may also be applicable to animals but are rarely explored scientifically. Cats and dogs are often patted and stroked to enhance the bond with humans and calm them, and sometimes cattle stockmen will also use contact positively in this way. Animal physiotherapy is now adopting a more universal application, rather than just for veterinary medicine. Animals that suffer from anxiety, such as dogs separated from their owners, probably would benefit just as much as us from relaxation therapy.

The impacts of diet on animal welfare are also starting to be explored. A high protein diet, long recognised to stimulate boxers to be more aggressive, has some of the same mood enhancing effects in the common dairy cow (Phillips and Kitwood, 2003). Conversely diets that are deficient in essential nutrients may stimulate animals to fight over food, or develop exploratory feeding habits in an attempt to rectify the deficiency. Odours may influence the mood of animals, as it does in humans, and beneficial effects of lavender straw have been observed in reducing travel sickness in pigs (Bradshaw et al., 1998). Some odours, such as citronella oil, are noxious to animals and are now used to control barking behaviour in dogs, with a collar emitting a short burst of the oil every time a dog barks (Steiss et al., 2007).

Some scientists are beginning to question whether there should be more emphasis on the creation of positive welfare states, instead of focusing on avoiding negative welfare. For example, Yeates and Main (2008) recently suggested that more attempts should be made to extend welfare assessment to indicators of positive affect, or emotion, recognising that they largely concentrate on negative emotion at present. The reason that they concentrate on the negative elements may be partly because the public are better able to empathise with animal's negative experiences. Many would agree that we owe animals a life with avoidance of the most serious negative emotions, but that there is less moral imperative to encourage us to create experiences likely to result in positive emotions. However, a major common theme underpinning most religions, and hence moral imperatives, in the world today is the golden rule which says that we should treat others in a way that we would like them to treat us. This does not distinguish between positive and negative consequences of our actions. It does not suggest that treating others badly is any more important than not treating them well.

Nevertheless, most research has been conducted on negative aspects of welfare and the several different methods of measuring welfare allows us to be confident that some practices do indeed cause negative emotion. So animals are likely to respond to a practice which induces negative emotions with negative behaviour responses (such as abnormal behaviours, stereotypies and avoidance behaviour), increased disease incidence, reduced production and reproductive rate, reduced longevity and adverse effects on physiology. For example, a lame dairy cow will have behavioural indicators that she is experiencing negative emotions – she will limp, in order to withhold pressure on her diseased claw and will lie down for a long time (O'Callaghan et al., 2003). She also will eat less and produce less milk (Bach et al., 2007), have a reduced life expectancy and is less likely to become pregnant (Bicalho et al., 2007; Melendez et al., 2003). Her nutrient status, as evidenced by her body condition, is likely to be low (Garbarino et al., 2004), and physiological measures could detect the metabolic consequences of the lameness (high cortisol concentrations, adverse effects on reproductive and nutritional hormones, for example) (El-Ghoul and Hofmann,

2002). The tools for welfare assessment all suggest that the cow is being negatively affected by the lameness. More specifically it is now possible to distinguish which forms of negative emotion are associated with specific behavioural, physiological and immunological changes. In cats, stimulating different areas of the hypothalamus can induce different forms of negative emotion, which appear to represent restlessness, defensive attack, retreat and biting attack (Mori et al., 2001). The first three all have similar behavioural components, but at different levels, and are associated with elevated cortisol, but they are different from biting attacks, which have different behavioural components and during which cortisol is not elevated. Defensive attack and restlessness are associated with increased immunocompetence, but not the other negative traits. It is this sort of information that is needed to assess the welfare impact of negative emotions, and it may ultimately make the assessment of welfare from experiences that are classified as good or bad, or positive and negative, appear too simplistic.

We can have less confidence that supposedly positive emotions are beneficial for the animal, rather than just neutral. For example, animal play is often used to infer positive affect, yet it is now believed that social play can switch rapidly from positive to negative affect even within a bout (Burgdorf et al., 2006). It is difficult to ascribe a common purpose to play, with often disparate characteristics and different affective properties. For some aspects of welfare, there is an obvious continuum, such as in nutrition, which includes both positive and negative emotions. We feel good when we eat to satiate hunger, which is related to the stress responses abating, and we feel bad when we need to eat, mainly because of physiologically-induced stress associated with this state (Adam and Epel, 2007). However, for other welfare measures, such as the thermal environment, it is not necessarily the case that increasing provision of the resource will increase the positive emotion resulting from it. Moving from low temperatures to a satisfactory temperature improves welfare, but increasing temperature still further will return welfare to a low level. There is good reason for addressing positive and negative affect separately – they are not just the opposite ends of a cognitive continuum, even though negative welfare is often inversely correlated with positive welfare measures. Further evidence that positive and negative affect are not diametrically opposed comes from depressed humans, who respond physiologically in a different way to normal humans on presentation of pictures suggesting negative emotion, but both groups respond similarly to pictures suggesting neutral or positive emotion (Abler et al., 2007). Physiologically the negative emotion is clearly dominated by amygdala activity, whereas the brain centres responsible for most positive affects have yet to be identified (Garolera et al., 2007).

Until we understand positive emotions better, we remain compelled to focus on welfare indicators that suggest negative emotions, because there is general agreement that these impact on welfare. However, because of the inverse correlation between many negative emotions and the productivity of animal units, systems of animal management have been developed that largely prevent animals experiencing major negative emotions. Controlled environments, used especially for pig and poultry production and laboratory animals, attempt to prevent extremes of temperature, to control infectious diseases and avoid major social challenges. While effectively minimising negative emotions, they do little to foster positive emotions, and if the trend towards welfare improvement continues it will be increasingly important that we include positive emotions in welfare assessment schemes, examining the opportunities for play, environmental exploration, satiation following eating, free choice etc.

One positive emotion, happiness, has been quite extensively studied in humans because of its obvious relevance to life satisfaction. It has been scientifically researched by Richard Layard of London University (Layard, 2005), and his findings potentially have some important implications for animal welfare assessment. Layard provides evidence for two compelling arguments:

1: Most people in developed countries of the world have not experienced an increase in happiness over the last the 40 years, despite increased personal wealth
2: At any one point in time, rich people are happier than very poor people

These can only be reconciled by accepting that above a certain base income level, which Layard estimates is probably about US $20 k, people only strive to gain more resources in order to elevate their status. [2] However, it is impossible for everyone to gain increased status, so if becoming happier is our goal we would be better off changing our lifestyle to adopt other established techniques of achieving this – altruistic deeds, religious pursuits, calming exercises that reduce negative emotions etc. By doing this everyone could be happier, not just the privileged few of high status.

It is likely that the same principles apply in animal societies that humans manage, in which a higher status does not necessarily confer successful reproduction. Like humans, it is likely that above a certain level of resources, animals only compete to elevate themselves in the dominance hierarchy and increase their chance of their genes surviving through increased reproduction. Therefore, above a minimum level of resources, being dominant through having access to more resources is more important to animals than the resources themselves.

In an attempt to measure human happiness, quality of life surveys have broadened the types of resources that are normally included in any measures of welfare. A popular Quality of Life measurement index is one developed by Mercer Human Resource Consulting (MHRC, 2007), which takes into account the following key indicators in determining the best place for humans to live in:

---

[2] They are probably genetically programmed to do this, since it would have had adaptive advantage in the processes of evolutionary selection, with higher status people successfully rearing more offspring. Nowadays it no longer has adaptive benefit – wealthier people do not necessarily rear more offspring successfully, and in modern society benefits to that society are no longer gained by proliferating one's genes to the greatest extent. So fecundity is greatest in poor countries today (Aarssen, 2005).

- Political and social environment (political stability, crime, law enforcement, etc)
- Economic environment (currency exchange regulations, banking services, etc)
- Socio-cultural environment (censorship, limitations on personal freedom, etc)
- Medical and health considerations (medical supplies and services, infectious diseases, sewage, waste disposal, air pollution, etc)
- Schools and education (standard and availability of schools, etc)
- Public services and transportation (electricity, water, public transport, traffic congestion, etc)
- Recreation (restaurants, theatres, cinemas, sports and leisure, etc)
- Consumer goods (availability of food/daily consumption items, cars, etc)
- Housing (housing, household appliances, furniture, maintenance services, etc)
- Natural environment (climate, record of natural disasters)

This assessment focuses on the quality of the resources offered to individuals in different locations. Quality of life surveys can also focus on the individual's ability to utilize such resources, which can then be used to prioritise health care provision by public services. The Mercer scale can be adapted to provide a scale to determine an animal's quality of life, which would be wider ranging than conventional animal welfare assessments.

- Political environment – consistency and quality of management, availability of personal choice
- Economic environment – economic provision for animal care, including provision for emergencies
- Socio-cultural environment – companionship with suitable conspecifics, or failing that similar species
- Medical and health considerations – veterinary care
- Education – provisions for training and development, availability of parental care
- Transportation – transport facilities and availability of personnel to maintain facilities
- Recreation – environmental enrichment
- Consumer goods – availability of food, water etc
- Housing – quality of accommodation offered to animals
- Natural environment – climate, natural disaster frequency

Welfare Assessment

Welfare assessment can be based on scientific research, public opinion or the opinion of experts. Scientific research is slow to provide the answers to welfare questions, usually taking several decades, whereas public opinion can change quite quickly, often in response to media releases, but also in the long-term in response to changing societal standards. However, scientific research is invaluable in setting standards because it is objective and untainted by anthropomorphic attitudes. Often science is needed to provide the welfare assessment, but it is vital to understand public opinion as well because this will dictate the level of provision for the animal to an acceptable standard. Hence we might use science to evaluate an animal's responses to a particular practice, say vehicular transport, but then we need public opinion to say what is acceptable once we know how the animals respond. Public opinion is not usually particularly valuable for detailed welfare assessment, because the public do not have sufficient knowledge to make such an assessment, and they are open to persuasion by welfare activist groups. Nevertheless, scientists must recognise that it is usually public pressure that most often brings about changes in animal management systems and that their role is a supportive one, not a decision-making one.

The opinion of experts can be rapidly gathered; it represents an informed opinion and is often based on scientific principles (see examples of indices devised from expert opinion by Whay et al., 2003; Rousing et al., 2007). However, it may be biased if it comes from those integrally involved in industry or from academics dependent on industry funding or goodwill for their work. Most codes of practice for welfare assessment are based primarily on expert opinion, and that is why they are not usually enshrined in law, because scientific evidence is not available to provide definitive proof of welfare status. Over time, more codes of practice will become based on scientific evidence and more will then be able to be legally enforced.

The Centre for Animal Welfare and Ethics at the University of Queensland has initiated a series of welfare assessments based on expert opinion, as a first stage to developing robust standards. These all assume that different aspects of welfare provision are at least partly exchangeable. So if food availability was sub-standard in a particular animal keeping practice, this can be at least partially compensated by improving another attribute, such as space availability. The currency adopted for exchange of welfare attributes is the Importance attached to each, as determined by the experts. Indices of performance that can be used to compare the welfare level of different animal systems are being produced. These are being constructed for the welfare of farm livestock on ships, great apes, especially orang utans, chimpanzees and gorillas, and for elephants in captivity. The first step for the researcher constructing an index is to identify who the experts are. This might include veterinarians, keepers of the animals, managers of the animal facility, scientists studying the species, knowledgeable animal welfare organisation representatives, those who transport the animals, and any other stakeholders or interest groups with a detailed knowledge of managing the species. Secondly a small group is interviewed, that is usually one to two individuals nominated by relevant societies representing an interest group. The aim in this part of the process is to identify the principle welfare resources that the species needs. If these were set by the researcher, the questionnaire to finally determine the importance of each resource would be biased by their choice and description of resources. The output is a list of the key welfare components that can be elaborated upon in the questionnaire to the different interest groups. Often these are based loosely around the Five Freedoms, that are now commonly used as a basis for welfare assessment (Webster et al., 2004):

- Freedom from hunger and thirst
- Freedom from discomfort
- Freedom from pain, injury and disease
- Freedom to express most normal behaviour
- Freedom from fear and distress

Having derived this framework, a group of about 10–20 welfare indicators are chosen on the basis of their being most popular with the stakeholder representatives, their practicality to be measured and their perceived relationship to animal welfare. They usually include resources like space availability, dietary adequacy, frequency of feeding etc. Care has to be taken that welfare impact is not counted twice, with four of the freedoms primarily indicating feelings and one (freedom to express most normal behaviour) indicating an expression of the feelings externally. Suitable levels are chosen, usually two to four per welfare indicator, in conjunction with those directly involved in managing animals in the systems that are the focus of the study. So, for space availability for chimpanzees, the levels could be providing enough space for individuals to escape from dominant animals all of the time, most of the time, or not at all. Another welfare component could be access to an outdoor enclosure, with the levels being all of the time, some of the time or never. For stocking density of animals in transport, we could choose enough space for the animal to perform most normal behaviours, enough to turn around and enough to stand up and lie down. This approach recognises the difficulties in putting figures to many components because of differences in size and breed of the animals and quality of space. These are then entered into a questionnaire, which is available on the worldwide web, as this potentially allows large numbers of experts to contribute to the construction of the welfare index. Typically this will be several hundred and could run into thousands, but responses can be weighted according to an individual's level of experience. Running the questionnaire on a computer allows questions to be tailored to a respondent's interests, producing an adaptive questionnaire, so if two components are rated similarly and of high importance, the computer will cease asking about components that it already has been told were rated unimportant by the respondent and begin trying to differentiate between the two similar components. The respondent is asked questions in the following form: if all else was equal, which of the following two welfare components is more important, or which of the following scenarios is more acceptable from a welfare perspective: component x at level 1 or component y at level 2? Respondents are also asked which is the preferable of two scenarios, each with the same two components but at different levels, such as scenario 1 with animals having enough space to avoid dominant animals most of the time but no access to an outdoor enclosure, compared with scenario 2, where animals have insufficient space to avoid dominant animals but do have complete access to an outdoor enclosure. The questions are manufactured by the computer to focus on welfare components that the respondent is rating of similar value. Conjoint questions of this nature, whilst appearing difficult to answer and sometimes rather contrived, are a powerful tool to elicit detailed information on the respondent's preferences. Armed with ratings for the perceived importance of the different levels of each welfare resource and the perceived relative importance of the different resources, these can be simply compiled into a mathematical index for use in the field.

After it is formulated, it is important to test the accuracy of the welfare index. So for an index for zoos, for example, each enclosure can be rated for the different welfare indicators, either by a visiting assessor or in a questionnaire sent to the zoo director, and the total added to provide an overall score for the zoo for the particular animal species. Comparing different zoos' performance will allow assessors to determine which welfare indicators are presenting the most difficulty in achieving a reasonable score. It is important to modify the index if it is considered that there is scientific evidence that refutes the experts' opinions. If there was no clear consensus on whether a particular component is important or not, or which level is best for the animals, it might be dropped until clear evidence becomes available. Finally, surveys of consumers can be used to determine how much people would pay for the animals to be provided with higher welfare. For example, how much would people pay to enter a zoo where animals are being kept at a higher point on the welfare index? How much more would people pay for meat products from animals kept at a higher welfare? Such information could be compared to the cost of providing the extra facilities, or even to determine the most cost effective way to improve the welfare of the animals. In this way, zoo directors, farm managers and even animal transporters can objectively determine the best way to improve the welfare of animals in their custody. It may then be possible to make an economic argument for the improvement of animal welfare, if the public survey indicates that people would prefer to pay more to access the product (zoo visit/foodstuff etc) if animal welfare is at a higher level.

These indices can be used in practice to assess either individual animal welfare, or more normally, the welfare of a group of animals, for example in a farm or a zoo. League tables will encourage competition to improve welfare standards, just as tables for individual farm productivity used to be constructed to encourage high production in dairy cow herds.

Legislation and Audits

Despite the goodwill of many animal managers towards the animals in their care, the conflicting ethical responsibilities that they are faced with often means that legislation and audits are needed to achieve minimum standards required by the public. Given the strength of public opinion today, we may expect that the animal welfare and rights movements will not diminish until there is a fundamental change in provision for improved conditions for animals, which may take several decades of legislation. Slavery did not disappear entirely following the 19th C campaigns, but the proportion of the world population that were slaves diminished due to new legislation, leaving the only remaining incidences of slavery as covert operations, and this state remains today (Walvin, 2007). The same is likely to happen in the animal welfare movement, new legislation will substantially improve the welfare of animals, but some problems will continue, particularly in fields that are unsuitable for legislation. Animal welfare legislation can be based on expert or public opinion, but it is likely to be more credible and long-lasting if it is based on scientific data. This can be provided by physiological or behavioural information collected from animals, and also the preferences of animals that are given choices. The preferences that they display will indicate the extent of their feelings about a particular resource, particularly if the strength of their preferences is measured by requiring them to work to gain access to the resource. When setting standards, legislators prefer to use evidence of physiological impact on an animal, rather than preferences, which may indicate a difference in mentality rather than health.

An alternative, which is likely to be preferable to legislation for industry, is the development of a system of industry-led audits or accreditation/certification programmes. This has had some success in Europe, for example of Swedish pig producers (Bruckmeier and Prutzer, 2007), with legislation being reserved for the most severe welfare problems. Given the strength of feeling by the general public, accreditation schemes that simply provide a rubber stamp for the status quo in the industry will be only temporarily credible. Thus audits must be sufficiently robust to ensure that standards are improved, preferably to levels acceptable to the public, although the possibility remains to convince them that other scientific alternatives are more desirable. This may mean pressure on some farmers to leave the industry, allowing the best farmers to remain. With a better understanding of animal welfare, it should be possible to devise audits that allow a variety of routes to a common endpoint – a healthy, happy animal. This requires knowledge of the relative merits and demerits of specific practices – for example, how do hot and cold branding affect welfare, and how severe is the problem, in the animal's perception, compared with tail docking? Until we know the answers to questions such as these, based on scientific data, we can do no more than rely on experts' opinion.

The best audits will allow farmers to trade welfare impacts, allowing a long journey to slaughter, for example, to occur only if the animal has been reared in benign, free range conditions with adequate food and social resources. Such exchange is only possible if a fully numerical audit is devised. So it is not only necessary to know that transport is to a high standard, but to allocate numerical values to each component of the practice. These may be based on scores by auditors or direct measurements. Often the measures chosen will not be ideal in terms of relation to end products. For example, the potential for cattle on ships to develop heat stress is known to be high when they are in hot ambient temperatures. The most appropriate biological measure is probably a panting score, which relates directly to the animal's apparent suffering due to heat stress (Mader and Davis, 2002). However, assessment is subjective, and repeatability both within and between individuals is likely to be low. Using it to assess welfare would present problems of both measurement and interpretation. A more precise animal measurement, which does not relate so directly to the animal's suffering, is respiratory rate, but this suffers from the problem that it does not relate linearly to ambient temperature (Brown-Brandl et al., 2006). Even this would be difficult to apply on ships, because it is unclear who could measure it and on which animals. An audit would typically in this situation fall back on wet bulb temperature measurements, which are repeatable, fast and cannot be manipulated. These do not relate so well to animal discomfort, because wind speed cannot be taken into consideration easily, but they would still allow standards to be improved so that severe heat stress events are prevented. The greatest risks are when the ship docks in port, as the ventilating effect of open sea breezes is lost. Minimising the time in port will reduce the likelihood that heat stress could occur. This illustrates the difficulties in deciding which measures to include in audits or welfare assessment schemes. Under conditions where animals can be more easily monitored, such as laying hens, the strong and well understood relationship between animal measures and their welfare suggests that these can be used more frequently (Mollenhorst et al., 2005).

Chapter 6
Managing Animal Welfare and Rights

Religious and historical perspectives – recent developments of attitudes – modern management of animal welfare – animal's right to life and welfare – animal sacrifice – animal slaughter – pain – improving animal welfare in developed and developing countries – treatment of animals by indigenous people

Introduction

Animal welfare and rights advocacy are two of the most pervasive influences of our time, but they are viewed as a threat by many in the animal industries because the changes sought by proponents of this movement are likely to reduce the profitability of animal enterprises. Effective animal business management, including the welfare of the animals, requires an understanding of, and ability to predict the standards expected by advocates, consumers and the public. In many cases these standards are derived from religious and historical perspectives. In this chapter welfare and rights management is considered from a variety of different perspectives, including that of the general public, those involved in teaching animal managers, and traditional societies.

Religious and Historical Perspectives

Of the four major religions in the world, Christianity, Mohamedanism, Hinduism and Buddhism, all have different perspectives on the management of animals. In terms of the number of adherents to the major faiths, one could add Chinese Traditional or Folk Religion as a fifth major religion, and because of its historical significance Judaism warrants inclusion as a sixth.

Christianity

The dominant Christian view of the management of animals is that God ordained that man should have dominion over, or rule them. At the start of

C. Phillips, *The Welfare of Animals*, Animal Welfare 8,
DOI 10.1007/978-1-4020-9219-0_6, © Springer Science+Business Media B.V. 2009

the Bible, we are told that "God said, 'Let us make humankind in our own image . . . and let them have dominion over the fish of the sea, the birds of the air, and over the cattle, and over all the wild animals of the earth, and over every creeping thing that creeps upon the earth." (Genesis Chapter 1, verse 26). Although there may be some doubt about the extent to which the Hebrew word for 'dominion', rādâ, means either lordship/mastery or careful husbandry, the intention of the unknown author of Genesis is clear (Preece and Fraser, 2000). Humans are believed to be made in god's image and have the responsibility to manage all living things. Nevertheless, some critics of Christian attitudes to animals, including celebrated ethicists such as Peter Singer (2005), have used this text to suggest that the dominant Christian approach to animals is one of enforced servitude. However, clarification of the Christian ideology is presented later in the Old Testament of the Bible, particularly in relation to the treatment of livestock, which as Preece and Fraser (2000) describe, reads in part like a husbandry manual for livestock. This was because it was particularly relevant to the pastoral society for which it was written. This is just one instance of how the messages of the ancient scriptures, in this case the bible, should be considered in the context in which they were written, and to gain a correct understanding of attitudes of the time it is often necessary to consider texts from several books, which may in the first instance appear conflicting (Regan, 1990; Preece and Fraser, 2000). The writer of the Genesis account, which was probably written in approximately 1,400 BC, obviously had no knowledge of the scientific principles of human evolution, and the text usefully exhorts us to consider our responsibility to look after animals, a responsibility that when fulfilled brings a sense of satisfaction that the long-term health of the animal kingdom is assured.

Given that humans have, according to the bible, been empowered to manage the animal kingdom, probably the most contentious message of the bible is that there is a hierarchy in the animal kingdom, described by a Psalmist's supplication to god as follows:

> 'You made us a little lower than yourself, and you have crowned us with glory and honour. You let us rule everything your hands have made. And you put all of it under our power – the sheep and the cattle, and every wild animal, the birds in the sky, the fish in the sea, and all ocean creatures' (Psalm 8, verses 7–9).

This approach is supported by many in the Western world, and even Tom Regan, the animal rights advocate and staunch opponent of speciesism, believes that god has given man the authority to manage animals (Regan, 1990). Proof of our responsibility will never be found, and indeed in Aristotle's view should not be necessary for educated people (Regan, 1990).

Regardless of whether the power is divinely given or not, and there is remarkably little evidence on which to base any decision, it has become abundantly clear in recent times that we *are* responsible for virtually all the animals on earth. Our activities touch the lives of almost every animal on land and in the sea. Whereas in the past, we might have thought that our impact on some ecosystems, in particular the marine ones, was limited, it has recently become clear that we have had an even greater impact than on many terrestrial ecosystems. Clearly for some animals our influence is remote, and not a direct action, occurring only through our influence on the ecosystem in which they live. This could be through the harvesting of some animals or plants or the pollution of the environment. As well as the influence of harvesting of wild animals, it is now evident we are now having a serious impact on the global atmospheric environment through our use of fossil fuels in particular, which may be faster than the ability of wild animals to adapt to the changing conditions, with some species facing extinction as a result. In relation to environmental pollution, human influences continue to be discovered that were hitherto unexpected. Deer grazing on lichen and natural grasses in the Austrian Alps have been found to have a high prevalence of cadmium toxicity in their kidneys, because this element can travel from the industrial areas of central Europe to the alpine grazing pastures (Beiglbock et al., 2002). Similar instances of nephropathology have now been discovered in a number of European birds and mammals (see review by Phillips and Prankel, 2008).

The concept of human differentiation from animals is further developed in the story of the Garden of Eden, which suggests that humans are different from animals because they have a knowledge of good and evil,[1] or right or wrong. This knowledge was imparted by god, the story tells us, since man was made in god's image. In the story, man's role of managing nature is a highly demanding one, and for this reason knowledge of right and wrong is essential. People need to regularly consider these responsibilities and discuss together how best to foster good animal welfare. Prayer is one method advanced by the Christian church to provide guidance and may be useful for people with responsibility for managing animals, especially any sick animals within their charge. In a survey of American owners of cats and dogs diagnosed with cancer, prayer was used by almost one half of those surveyed, mainly in an attempt to improve their animal's welfare (Lana et al., 2006). After supplements, it was the second most commonly used form of 'alternative or complementary treatment'. There is little evidence of its efficacy in animals, but there have been several meta-analyses[2] of reports of the effects of intercessory prayer for the health of the human subjects (who do not know that they are being prayed for)

[1] Many people reject the concept of good and evil because of the way in which the message was used to guide behaviour in the past. The threat of going to hell because of wrongdoing was used to frighten people into correct or right behaviour. The concept of the devil, as the antidote to god, was used by many, including religious leaders, to externalise wrongdoing, as a means of putting the blame onto some other being, so that we do not become weighed down with the consequences of our actions. However, rejecting this concept fails to acknowledge that religions are useful in offering opportunities to learn from one's mistakes, receiving forgiveness following acknowledgement and confession of wrongdoing, and encouragement to improve on subsequent occasions.

[2] A review, usually with statistical analysis, of all the published or available literature on a topic

(Abbot, 2000; Astin et al., 2000; Hodge, 2007; Masters and Spielmans, 2007). One (Masters and Spielmans, 2007) suggests that there is no benefit to the subjects, and the rest suggest small benefits or they are inconclusive (Abbot, 2000; Astin et al., 2000; Hodge, 2007). In the only study of the effects of intercessory prayer on the healing of animals, in this case self-inflicted wounds following excessive self-grooming in primates, the group of primates that were prayed for healed much more quickly than the control without prayers (Lesniak, 2006). The biological mechanism was explored and the primates that were prayed for had increased red blood cell, haemaglobin and haematocrit concentrations and they groomed themselves less. While it is not good science to compare the results of meta-analyses of several tens of trials involving many humans, with a preliminary study involving just 22 primates, it should be noted that in all of the meta-analyses some of the included studies generated significant benefits of prayer to humans. The difficulties of setting up double-blind trials of this nature may explain some of the ambiguity in results and also explains why praying for animals was the subject of the one study cited. One meta-analysis generating inconclusive results of the effects of prayer on human healing noted that the best designed studies yielded positive results (Abbot, 2000; Hodge, 2007).

Although the direct effects of prayer are unclear, increased hope and decreased anxiety are two main ways in which prayer could benefit those with close bonds to their animals. However, these effects have not yet been demonstrated in people's attitudes towards animals, even though there is evidence that prayer can reduce generally reduce anxiety (Arias et al., 2006; Coruh et al., 2005), especially if the prayer involves actively seeking god's assistance (Harris et al., 2005). Conversely, concentration on an animal's health problem can exacerbate anxiety (Hill and Pargament, 2003).

Although the Old Testament initially emphasizes the responsibility of humans to have dominion over animals, later we see a different view of man's relationship with animals emerging. This is evident after the fall of Adam and Eve from their special relationship with god, and states that man should have a covenant with animals, created by god. In the biblical book of Hosea, written as if god was instructing the people, it is said that '(I) will make for you a covenant..... with the wild animals, the birds of the air, and the creeping things of the ground, and I will abolish ... war from the land, and you shall lie down in safety' (Hosea, Chapter 2, Verse 18). The emphasis was on creating harmony with nature and the environment, with man's safety secured. This covenant could have been to protect humans from wild animals, but the fact that birds were included, which were not a threat to man, suggests that the covenant was more universal, with humans entering into an agreement with animals that was ordained by god. This book of the Bible was written about 700 BC, and demonstrates the gradual transition towards stewardship in the Christian faith.

Although people were exhorted to enter into this agreement in Old Testament times, still later in the Bible, in the New Testament, we get renewed confirmation of the Christian belief in man's superiority to animals. For example, the Gospel writer Matthew wrote that if a man could rescue a sheep from a pit, how much better it would be to rescue a man from a pit (Matthew Chapter 12, verses 11–12). Elsewhere in the New Testament there are few references to the way in which we should manage animals, indeed Jesus had remarkably little to say on the matter. He did contentiously cast out demons from a wild man, presumably a mentally disturbed man, and apparently sent them into a herd of about 2000 pigs, which then killed themselves by rushing down a steep hill into the sea (Matthew Chapter 8, verses 30–32, Mark Chapter 5, verses 11–13 and Luke Chapter 8, verses 32–33). Whilst we can perhaps explain the pigs' behaviour as being prompted by their fear of the 'wild man', which may have caused a stampede down the hill, there is no doubt that the gospel writers thought this an acceptable way to treat a madman – to pass on his problems to 'lesser' beings.

Although there was much teaching on the management of animals in the Old Testament, there is very little in the New Testament. In the latter, Jesus sometimes referred to animals in his parables, particularly sheep, as well as occasional references to wild birds, poultry and goats (Matthews, 1937). The different emphasis in the two Testaments is probably because in Old Testament times there was a need for instruction about animal management in a predominantly pastoral society (Preece and Fraser, 2000), but in New Testament times there was a major need for instruction about man's relationship to his fellow man. The Romans had occupied Israel and brought with them challenging and aggressive attitudes towards people of other nations in the occupied territories. Jesus' principle message was one of non-violence and love to all people, similar to Gandhi's message during the British occupation of India. Both men gave their life to their cause and subsequently became revered, but it is symptomatic of the fundamental differences between the Hindu and Christian faiths that Gandhi's teaching included a message of respect for animals, but Jesus' did not. Indeed Jesus is reputed to have said "Look at the birds in the sky; .... Are not you more important than they?" (Matthew Chapter 6 vs 26). The Christian faith has therefore consistently emphasized man's superiority to animals, but also the responsibility to manage them well.

Furthermore, Christianity tolerates meat eating in a way which most of the Eastern religions do not. However, there are prophecies in the Bible that indicate that ultimately it is expected that the consumption of meat will cease, even in predatory animals. 'The wolf and the lamb will graze together, and the lion will eat straw like the ox; and dust will be the serpent's food. They will do no evil or harm in all My holy mountain' (Isaiah, Chapter 65, vs 25), and 'Wolves will live with lambs. Leopards will lie down with goats. Calves, young lions, and year-old lambs will be together, and little children will lead them' (Isaiah, Chapter 11 vs 6). In this last prophecy the responsibility of humankind, even young children, for the animals of the world is recognized. Although biologically implausible, these prophecies represent the Christian objective that ultimately there will be complete harmony between living creatures, even between predator and prey. Prophecies of harmony in the world are not confined to religious texts but are common in the writings and sayings of society's cultural icons. For example, in the most popular song of recent times, Imagine, John Lennon wrote: 'You may say I'm a dreamer, but I'm not the only one. I hope some day you will join us, when the world will be as one'.

The Islamic Faith

The religious texts of the Islamic faith contained many references to good management of animals, in particular cattle, but it is assumed, as in the Old Testament writings, that animals were made available for human use and benefit. For example, it is written in the Qur'an, that Allah says "And cattle He has created for you. From them you drive wont and numerous benefits and of their meat, you eat" (16: 5–8). Like Jesus, Mohammed ate meat and when one of his companions wanted to give up meat, he was told that this was wrong. Ritual slaughter of livestock was normal on several feast occasions in the year, in particular the feast of Eid-ul-Adha, which still entails dividing the carcase of a sheep or cattle into one third portions to be distributed to the poor, neighbours and the members of the household managing the slaughter, in memory of Abraham's generous offer of his son when God asked for a sacrifice. Nowadays although the principle of dividing the animal is preserved, the slaughter process is often delegated to abattoir workers.

It is often assumed that Muslims have little concern for animals because they do not usually allow animals to be stunned before slaughter and because there are many widely cited examples of animal cruelty in Muslim countries, such as the recent publicity given to the cutting of the tendons of cattle in Egyptian abattoirs to immobolise them before slaughter (Sidhom, 2003). However, it must be remembered that many Christian countries engaged in similar abuses of animals when they were at the same stage of development as many of the Muslim countries are today, so it is not necessarily the religion that affects animal welfare, but the stage of economic development.

One of the least troublesome examples of animal slaughter that I have witnessed was by the Bedouin Muslims deep in the Negev Desert in Israel. Accompanied by an Israeli colleague from Beersheva University on the edge of the desert, I visited Bedouin encampments and settlements to see how they kept their animals, mainly sheep, goats, horses and camels, which are all well adapted to cope with the hot, dry conditions of the region. Camels and horses roamed close to the tents, while their owners went about their daily business, such as working on Israel's intensive fruit farms. We were often welcomed into the Bedouin's tents and invited to drink tea and eat the unleavened bread and meats that they cooked over an open fire at the front of the tent. On one occasion we arrived at a small village on the edge of the desert just as the preparations were being made for a feast. A key part of this was the killing of a sheep to feed one of the extended families. We were invited to watch the head of the household preparing to slaughter the sheep. Knives were sharpened, a clean area prepared and the man then calmly entered the flock, that was kept in a modest shelter adjacent to his house. These animals were used to the presence of people and did not associate human presence with anything painful or distressing. Children played in the streets by the animals and helped to look after them. They were taught to respect the animals and not to taunt or abuse them, a central tenet of the teaching of the Qur'an. The animals for their part learnt that humans were in control, and that to resist would lead to punishment, but that humans provided them with food, water and protection.

The head of the household entered the pen and extracted a large sheep, which offered little or no resistance as it was led outside. Then at the corner of the shelter, facing Mecca and holding the sheep on the ground by one horn, with the other hand he reached for the knife and then drew it quickly across the animal's neck. Any attempt by the animal to struggle was stifled by a second man kneeling gently on the animal's back. As the slaughterer slit the sheep's throat, he said the usual prayer to God, "Bismillah Allahu akbar" (in the name of God, God is great), and then both men held the animal until it had expired. In the event, the animal did not offer any perceptible struggle, or any noise, and in less than a minute it was respectfully transferred from the land of the living to that of the dead. It lay on the floor in a pool of blood. Almost immediately, it was hung from the eaves of the shelter and stripped of its skin and internal organs by the slaughterer putting his foot in the space between the skin and the carcase and gradually easing the two apart by pushing downwards with his foot. Three small boys watched the spectacle with interest before returning to their play. Gradually the slaughterer dissected the muscles from the carcase and put them into a large pan of boiling water, where the women were busy adding vegetables, herbs and spices. There was no shame in the process; I was invited to photograph the entire procedure and even to partake in the feast, which unfortunately we did not have time for. Nigel Brown, the British vet in the Middle East who was employed by the Australian meat and livestock corporation to assist in promoting good welfare for the Australian livestock that are shipped to that region, tells of similar experiences. He testifies that he has seen some of the best and worst examples of animal welfare in the Middle East.

Good animal slaughter practice is one of the central beliefs of Muslims in relation to the management of their their animals, and it is therefore ironic that Westerners often single out the failure of Muslims to stun animals before slaughter as evidence that their treatment of animals is inhumane. Muslims believe that animals must die in the quickest and most painless way possible, using methods prescribed in the Qur'an and Hadith (the sayings of the prophet Mohammed). Contrasting the welfare implications of the death of a sheep that I had just witnessed being slaughtered in the Muslim way with the lengthy gathering of livestock from the field, waiting to be transported, often without food or water, and then lingering in abattoir lairage before finally being killed for sheep in Western countries, it seemed evident that Muslim slaughter can be just as humane as that in Christian countries.

When visiting Muslim countries, it is easy to see examples of poor animal welfare (as it is in Western countries). Cattle in the Nile delta confined to walking all day in a small circle around a shadoof, blindfolded so that they do not get distracted, clearly indicates poor welfare. Such practices have remained unchanged for centuries, and the Western industrial revolution of the last two hundred years has yet to have a major impact on livestock management practices. It is not unusual to come across a few buffalo housed in the basement of a house in the middle of Cairo being kept for milk production. These are likely to be permanently tethered in a hot, windowless clay building, fed on a small amount of rice straw and cottonseed waste, leading to an apparently poor welfare status. But can we be sure that these buffalos' welfare is worse than intensively farmed dairy cows? They are fed regularly and they are likely to have a close relationship with their keeper. Unlike the dairy cows, they don't have to fight for their daily ration, they are not struggling to stand upright on a slippery floor, their manager probably talks nicely to them when they come to milk them, in contrast to the dairy manager, who ushers them into a courtyard with an electrified fence that moves gradually forward behind them, forcing them into the milking parlour. The Egyptian system has evolved over centuries, with the manager dominant to his cows, but respecting them. The intensive dairy farm is a recent invention, devised to minimise labour use, but also meaning that each animal has a respectful, but probably also necessarily fearful relationship with its manager to enable him to control large numbers of animals.

Perhaps the worst examples of poor welfare is where developing countries are persuaded, often by aid agencies, that intensive animal production is the solution to the country's burgeoning hunger for meat and milk products. A typical example was witnessed in Malaysia, when I saw their attempt to build a Western-style milking parlour, but because of space constraints the entry and exit ramps were so steep that many animals slipped on them. When I visited, a cow was giving birth on a concrete pad, watched by all the other cows. The calves were taken to narrow pens for rearing, in which they could neither turn around nor reach down to groom themselves. Sustainability and good welfare are just as necessary in developing country animal production as they are in the industrialised countries.

Judaism

Judaism has similar origins to Christianity and the Muslim faith, all three originating with Abraham, of Old Testament fame. Jewish principles are enshrined in Talmudic literature and include 'tsa'ar ba'alei chayim' or the mandate to avoid causing unnecessary harm to animals. Traditionally, Judaism has several rituals requiring animal slaughter, such as the Feast of the Passover. Hence meat consumption is accepted by most adherents of the faith, although some argue that it was appropriate at the time of the Old Testament teachings, but is not necessary today.

Hinduism

India has followers of several related faiths; Hindus, Hare Krishna, Jains and Sikhs, all with variations on common central beliefs. The Hindu religion has the most followers and has numerous gods, many of which are animals or animal-related. Followers are encouraged to have reverence for the animal gods, and this translates to care for the animals themselves. There are gods for rats, tigers and almost all common Indian animals, and the divinity is closely allied with the natural world. Encouraging good treatment of animals by their association with gods has benefited animal welfare, and in recent years has aided the conservation of endangered animals, such as elephants.Because animal welfare attitudes in India have a firm basis in their religious teaching, the animal welfare laws are well established, however, implementation cannot always be assured.

The elephant god is one of the most popular, and for this reason many elephants are paraded in religious festivals. Elephants for this purpose are kept in sanctuaries, which are supported and visited by the public. Elephants may be donated or loaned to the sanctuaries, which then hire the animals out for the many festivals taking place in Spring time. They are not usually slaughtered and when they die naturally they are buried.

Cattle are also sacred and their five major products – milk, curd, butter, urine and faeces – are believed to have cleansing properties. They are still allowed to roam the streets in much of northern India, scavenging for any biodegradable material. Unfortunately this often leads to plastic consumption and a pot-bellied, gaunt appearance. In some of the more affluent parts of India, cattle have been largely removed from the streets and refuse collection instigated. Severe fines are levied on motorists that hit cattle on the streets, and most cattle are allowed to live out their natural life. Sometimes members of other faiths, especially Muslims, are employed to slaughter cattle, thereby at least partially externalizing the ethical impact of meat consumption for Hindus.

Hinduism encourages vegetarianism but does not insist on it. Indeed meat consumption in the Indian subcontinent is increasing in line with the growing affluence. The Indian sacred book, *The Laws of Manu V, 45–52* states that 'Meat cannot be obtained without injury to animals, and the slaughter of animals obstructs the way to Heaven; let him therefore shun the use of meat.' Hindus also believe that what they eat has a strong bearing on their well-being.

The Vedic texts are the set of instructions on which the religion was founded over 3000 years ago. Although these espouse the vegetarian concept as desirable, they also acknowledge that animal sacrifice has a role in producing meat for consumption. It was long after the Vedic texts and the development of Hinduism that vegetarianism became common. Ahimsa, or non-violence, is a guiding principle afforded to humans and animals alike.

Hindus believe both in hierarchies of humans in society, the caste system, and that humans are superior to animals. They believe that their spirit came from a series of migrations from animal to animal, finally ending up in humans, which have the ability to use their reason to attain the Ultimate Truth. Thus it is possible to end the cycle of rebirth and attain the Kingdom of God. However, those less fortunate or badly behaved are punished by being reborn as animals.

Buddhism

Like Hinduism, Buddhism evolved in India, but its followers separated because they were unhappy with the caste system. Thus, although Buddhism shares many beliefs with Hinduism, it differs in that it openly espouses equality of humans and animals. The Tibetan Buddhist spiritual teacher, the Dalai Llama, advocates that all sentient beings should be considered equal, the inference being that they should be treated equally (Mehrotra, 2005). Accordingly, affording such equal treatment to humans and animals is necessitated by the interdependence of the human and animal worlds, and the Dalai Llama counsels that 'the interests and well-being of human beings is dependent upon the well-being of animals living on the same planet' (Mehrotra, 2005).

In the Buddhist faith, all animals and humans alike have the potential to become Buddhas, or enlightened beings. Buddhists pursue ahimsa, or an absence of harm or death to any other being, and an absence of suffering, even to the smallest of animals. Like Hindus, Buddhists believe that they are reborn as either other humans or as animals, and hence all animals and humans are interrelated. As in Hinduism, it is recognized that animals suffer at the hands of humans, so it is considered a penalty for leading a bad life to be reborn as an animal. Humans are therefore exhorted to be kind to animals because of the possibility that they might be reborn as one. Reincarnation as an animal is undesirable, or in the words of the Dalai Llama, 'is extremely miserable' (Mehrota, 2005). This belief evolved when many animals managed by man did indeed have a worse existence than their masters. Both were struggling to survive and humans were able to take advantage of their position of control by exploiting them. Even though many of the worst instances of animal exploitation have been brought under control, the threat of rebirth as an animal is still used to exhort Buddhists today to abstain from specified non-virtuous actions. These are taking life, taking what is not given (to them), abusive sexual behaviour, lying, slander, harsh words, idle gossip, covetousness, malice and wrong views. In Buddhism, humans may even be reborn as many small animals at one time, hence even small insects are to be respected. For this reason, and because life is to be revered, killing another animal is a major sin. This contrasts with the Christian belief of original sin and the belief that all beings are sinful, unless redeemed by god. Even unborn foetuses are the subject of Buddhist compassion, and Buddhists must prevent others from killing animals, thereby liberating them. Buddhists also have a strong recognition of the sanctity of nature, and many Buddhist temples in east Asia exploit this belief by selling animals, mainly small birds, to people to be released from captivity into the wild. In one survey, about 30% of the population of Taiwan purchased animals for release into the wild (Severinghaus and Chi, 1999).

Religious Unification on Attitudes Towards Animals

As the world becomes increasingly united through common media, travel and trade, there is a need to develop a common philosophy in relation to the welfare of animals. Globalisation trends will increasingly bring people of different cultural and religious backgrounds into contact, and this will lead to conflict unless an agreed common approach is developed. Although the positions of the different world religions on the treatment of animals may at first sight seem to be diametrically opposed, they are not necessarily mutually exclusive, provided that the scriptures are viewed as instructional and allegorical, not literally the word of god that must be obeyed at all times. Religions have to be allowed to develop and adapt to changing circumstances.

The position of the Abrahamic faiths, Christianity, the Islamic faith and Judaism, that humans are given the responsibility to manage animals, through their dominion over animals, is not necessarily incompatible with the position of the Eastern religions that animals and humans are equal. An analogous situation is the treatment of children by adults. The responsibility of adults to manage children does not confer superiority, which has increasingly come to be recognised in the new social ethic that has evolved in the last century, with legal protection of children and affirmation of their rights. The biblical position that man was superior to animals should be seen as outdated and a new social ethic established that if man is to gain benefit from animals he must engage in a mutually-beneficial contract where animals are not penalised. Such unwritten contracts are not unique in human society; many professions have similar constraints on a person's living and the employee often has no more opportunity to relinquish the 'contract' than animals have.

No-one can deny that we do have the ability to manage animals, since our actions affect nearly all animals, wild or domesticated. Whether that ability is given by an all-powerful being, or god, or whether we assumed the right by virtue of our greater intellect and managerial ability, is controversial. Statistically, we appear to be outliers in the animal kingdom. Our use of tools and ability to harness technology, our cognitive power, the complexity of social relationships and our language ability are all considerably in excess of those of even the most closely related primates, so that for a long time it was even disputed whether these skills were possessed by any other animals. We also have a unique ability to represent things and events pictorially (Valladas et al., 1992), and our advanced capacity to place events in a contextual timeframe may be the key to our high level of consciousness (Humphrey, 2006).

This cognitive divide between ourselves and other animals is often attributed to brain size, rather than any unique abilities of the human brain (Kirkcaldie and Kitchener, 2007). Even the size of the human neocortex fits into a comparative (exponential) relationship between species evolution and brain size (Kirkcaldie and Kitchener, 2007). Our brain is about 3 times bigger in relation to our size than other primates, which in turn are about twice as big as other mammals (Schoenemann, 2006).

There are two pieces of evidence that suggest that our animal management skills were acquired by natural selection, rather than by divine intervention, which could still be part of intelligent design. First, as previously mentioned, the human brain, although unique in its cognitive ability, appears to be the pinnacle of cognitive evolution. As you progress through the phylogenetic tree, from vertebrates to mammals to primates to man, the brain becomes increasingly large in relation to body size. Second, there is fossil evidence that the human brain increased in size, if not gradually, in a punctuated manner, with no evidence of a single, large increase (Schoenemann, 2006). Although the emergence of a dominant, intelligent animal is without precedent in prehistory on this planet, it seems likely that the social skills that we have evolved to exist as a complex society also benefit us by enabling us to manage animals. Thus, although the debate on whether our peculiar faculties derive from a god or evolution has continued for over 125 years with no clear winner (Bleckmann, 2006), the scientific evidence favours the evolutionary perspective. However, Intelligent Design supporters argue that humans possess souls created in the image and likeness of god, giving them a unique sense of self-awareness (Moreland, 2001). On this science must remain agnostic at present, although future neuropsychological research may eventually enlighten us.

The position of the Eastern religions on reincarnation of humans as animals at first sight appears directly opposed to the Christian position. However, we now know that higher mammals share most of their DNA, and whereas the literal interpretation of the scriptures implies that individuals will be reborn as specific animals, a common position would be that our DNA is shared with other animals, as we are all inter-related. When we die, our DNA is perpetuated most directly in our children, but because much of it is shared with other animals, the evolution of animal life on our planet involves a constant selection of animals with the most suitable DNA for the occupied niche. Seeing ourselves as part of that common evolution of genetic material allows us to believe in our continued contribution to life on earth. We saw earlier how attitudes towards animal management in the Christian religion have evolved over time, and a common approach to reincarnation may involve this more liberal interpretation. In other respects the scriptures, most of which are derived from the teachings of inspired prophets several thousand years ago, contain many messages that are equally relevant today as when they were written.

Recent Development of Attitudes to Animals

As the human race has advanced, there has been an increasing trend to view animals as part of man's custodianship of the earth and less to view them as needing to be suppressed or in a combative role. This may be in part because we are gaining greater controlling ability in relation to our management of animals. The suppression began to end in Victorian times, when there was a focus was on understanding animals, by recording and classifying them in great scientific detail – the first stages in management of animals and nature. Expeditions to explore the world's fauna were funded by geographical societies, such as Darwin's famous voyage to South America on the Beagle. Sometimes this new-found scientific purpose was used as a front for colonising expeditions, which sought to seize territory. Sometimes it even hampered the expeditions, as was the case with Scott's bid to be the first to the South Pole in 1910. On this occasion, the time devoted to scientific discovery and experimentation was at least partly responsible for Amundsen claiming the prize. The acquisition of knowledge of the natural world was well established in British explorations, mainly to satisfy an ever curious and knowledge-hungry British public. The British government had supported expeditions of discovery for over 300 years and now, following the Industrial Revolution, the British public was eager for links to the natural world. The prospect of 'better' worlds, with exotic fauna and flora, were appealing to a population seeing their country being buried in smoke, pollution and hardship. Advances in news reporting meant that the British explorations in foreign lands were rapidly conveyed back to a public eager for news of the great adventures.

The emphasis on gaining control of new conquests in the Victorian era by establishing an ordered system of mapping and recording was applied just as avidly to the natural world as it was to their territorial conquests. A systematic method for the classification of animals and plants had been first proposed in the mid 18th C by Carl Linnaeus, and by the late 19th C approximately 1,000 new species were being described by British explorers and scientists every year (Ritvo, 2001). The mapping of animal species in Victorian times was comparable to the mapping of the animal genomes that is currently underway, marking the establishment of a knowledge base before major advances in plant and animal management and manipulation could be made. We can look forward to major advances in animal management over the next century as a result of our recently enhanced understanding of genetic effects on animal form and function.

The rudiments of breeding management for improved cattle and sheep genotypes had been established as early as the late 18th C, but in Victorian times the same techniques were applied to the increasingly popular companion animals, especially dogs. Before the Victorian era contact with nature was confined to situations in which man could be expected to benefit financially. A geographical text from the early 19th C describes how nature was necessarily 'subservient to the distribution and industry of mankind' (Pinkerton, 1807). In the late 20th and early 21st C, the ability to genetically manipulate animals led to an accelerated potential to develop new genotypes, which was utilised most in medical research, primarily with rodents, but is just now beginning to have a major impact on farm and companion animals. In future the emphasis will be on developing animals that fit better into their environment, thereby avoiding some of the damage done by, for example, developing dogs with extreme features, such as short noses, leading to respiratory problems, overlong backs, unnaturally short legs etc.

Pet ownership thrived in the 19th and 20th C, although signs of decline in the cat population were beginning to be seen in Australia in the late 20th C (Baldock et al., 2001), as a result of lifestyle changes and concern for native fauna. Standards for pedigree companion animal breeds were established in the same way as they had been for livestock. The Kennel Club of England was established in 1873, and was followed by the National Cat Club in 1874. Champion dogs of highly desired breeds, such as St Bernards, were worth considerable sums, often in excess of 1,000 pounds, equivalent to many tens of thousands of pounds nowadays. The welfare of companion animals became an issue at the same time as rising ownership. In 1868 Queen Victoria exhorted her subjects to be more concerned for animal welfare, complaining that 'the English are inclined to be more cruel to animals than some other civilised nations are' (Hibbert, 1984). It is not clear what her evidence was.

The Victorian era also saw the widespread establishment of menageries in Europe to allow the public to view exotic animals. In London, Regent's Park Zoo was established in 1828, amidst fear that England was falling 'behind' other European countries in exhibiting animals to the public. Some menagarists indulged the public blood lust by offering live prey. Concern about this practice was expressed by the newly established Royal Society for the Protection of Animals (founded in 1824), as much for the moral wellbeing of spectators as for the welfare of the animals. Contact with animals by viewers was encouraged, with viewers taking umbrellas and sticks into the zoo to provoke the animals through the bars of the cages. This accentuated the belief that the animals were aggressive, and that the viewer could share a similar, if less extreme, experience to the explorers that collected the animals. Food was often provided for the visitors to offer to the animals. However, attitudes began to change, and the historian Harriet Ritvo, described the position thus 'Gradually administrators began to understand wild fauna as a valuable resource in need of husbanding, rather than a source of diversion (and sometimes a nuisance) whose disappearance was an inevitable by-product of the progress of civilisation. . . . . . The perceived balance between humans and the natural world had definitely shifted. The role of the people was now to protect.' (Ritvo, 2001). However, Anglocentism was embraced by colonisers in many parts of the world, and the introduction of British fauna, such as rabbits and foxes, is rued to this day in colonies such as Australia.

Modern Management of Animal Welfare

Following the Victorian era of discovery, and the tumultuous war years of the first half of the 20th C, which heralded an era of peace and development, there has been a quest to improve the welfare of animals in at least the developed regions of the world. Affluent societies worldwide are requiring better conditions for managed animals, and most have introduced codes of welfare for the major animal species, supported by appropriate legislation (Fraser, 2006). The codes attempt to protect animals from suffering and cruelty, and often extend to a duty of care that animal owners have towards their charges. They are more effective than extensive legislation, which can only protect against the worst instances of cruelty. Many animal industries are characterized by their diversity and in modern Western societies the marketplace usually requires products from animals kept under a variety of different welfare standards. A minority of people will choose to purchase products from animals kept at a very high level of welfare, most people will purchase products from animals kept under normal conditions, and it is conceivable that only a few people would, if they were allowed, purchase products from animals kept under very poor welfare, assuming that there is a direct relationship between welfare level and cost of production (Fig. 6.1). The shape of this curve will differ between animal products and populations. If the majority of the population are of the opinion that animals should not be kept in systems where the welfare is very poor, then this practice is usually prohibited by law (Fig. 6.2). People in developed countries are increasingly demanding that food items that they export and import are produced to at least the same standards as foods that are produced and consumed in their home country. Hence the welfare of livestock exported from Australia to developing countries is scrutinized closely by the welfare lobby group and must be to a very high standard.

Welfare decisions may be based on individual experiences or on the sum of experiences. In the former case, individual events may be deemed to be too severe, particularly if they offend the majority of the population. The degree of

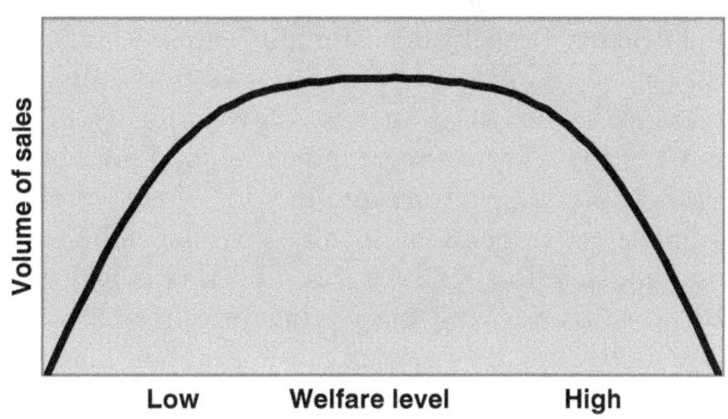

Fig. 6.1 Hypothetical changes in volume of sales of animal products with welfare level in a market economy

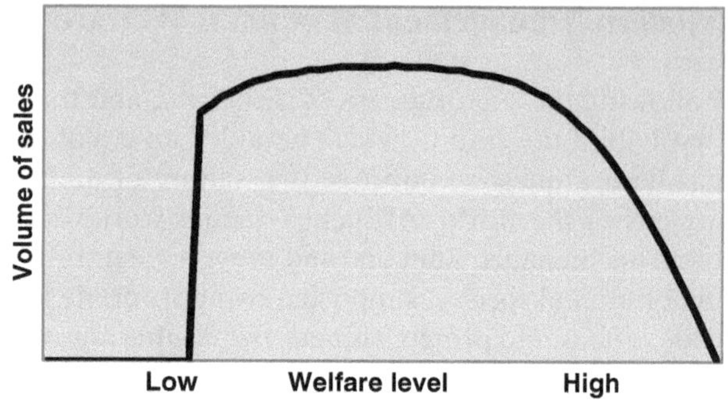

Fig. 6.2 Hypothetical changes in volume of sales of animal products with welfare level, in a market economy with a restriction on sales of animal products with low animal welfare

offence caused will depend on the essentiality of the experience to the quality of life of the animal, for example experiences such as the dehorning of cattle, which are largely for the animal's benefit because it prevents injury during fighting, may cause less offence than procedures which are arguably less severe, such as transport or hormone treatment, but are solely for the benefit of consumers. 'Welfare' experiences are traded by humans all the time in their own lives, and therefore it is logical to allow the same for animals. A typical human scenario might be 'I am buying a meal for dinner that I know is unhealthy, but it tastes good, and I need to improve my wellbeing after a bad morning'. The implication is that the person wishes to forego resources (long term health status, capital) in the short-term to achieve a rapid resumption of their welfare state.

Welfare can be measured from events as they happen or the resultant outcome on the animal, such as the final body weight or condition score of cattle after a period in a feedlot. Outcome-based measures could potentially provide a flexibility of approaches, which would be useful for on-farm assessments, and they are more likely to be directly related to welfare, compared to resource-based measures (Botreau et al., 2007; Edwards, 2007). However, it is hard to identify suitable indicators for welfare, especially if access to the animals can only be gained some time after the experiences have occurred. For example, the welfare of sheep transported by ship can be prescribed in the form of direct influences of resources provided for the animals during the journey (e.g. stocking density, temperature, humidity, noise levels), or the state of the animals at the end of the journey (e.g. live weight, coat soiling, behaviour, skin elasticity – a measure of dehydration status). We already know the relationship between most of these key resource indicators and possible animal measures, for example between temperature and the risk of heat stress (Marai et al., 2007). Not all animals will respond in the same way, depending on their physiology, morphology and previous experiences, and this is a disadvantage of using resource-based measures, but employing resource-based measures is often the only possibility on the basis of cost, reliability, repeatability and acceptability. Metabolic 'markers' of welfare state have proved equally elusive, as attempts to identify metabolic indicators of undernutrition have shown, because the animal's homeostasis maintains most metabolites at 'normal' levels (Agenäs et al, 2006). A combination of measures is likely to be needed in most circumstances, and it is likely that a prescriptive resource-based set of measures will be used in most instances for the forseeable future.

The minimum level of care afforded to animals managed by humans could be that which they would get in the wild, assuming that there is a niche for them to occupy. There may be wild relatives of our domesticated animals, whose welfare we can assess in pristine habitats to provide benchmarks for their domesticated relatives. In this way our contract with animals would have at least a neutral and preferably positive effect on the animals. Alternatively, for animals that are managed and used (including consumed) in their country of origin, the level could be as close as possible to that afforded to the minimum level that humans in that environment experience. However, this suggests that most humans will be better cared for than animals, alluding to human supremacy, and would invoke a charge of speciesism, that is so fiercely opposed by philosophers such as Tom Regan (1990).

An alternative approach is to use the democratic processes in a country to determine minimum welfare standards. Armed with useful measures of animal welfare, minimum standards for animal managers could be established in a democratic society according to the majority view. Some practices will be deemed by a majority to be unacceptable, for example the deliberate mutilation of animals for pleasure, as in bullfighting. Others might be seen as acceptable if the purpose is to secure animal welfare in the future. For example, removal of horns in young calves with the application of anaesthetic should involve little pain (Sylvester et al., 2004), and will prevent injury later during fighting. Other operations are more contentious: the mulesing operation in sheep (removal of flaps of skin in the region of the vulva of sheep) is conducted in order to protect the animal from blowfly infestation (James, 2006). From a utilitarian approach there are major welfare benefits for the few animals that have avoided a blowfly infestation by having the operation, whilst for the majority of animals the impact is negative, since only a small proportion would suffer a blow fly infestation. For some animals the impact would be profoundly negative, since there are risks associated with the operation, in particular a mulesing-induced flystrike. Perhaps the public would decide that if most animals had no benefit, the operation is too painful to allow, and that if the risks of flystrike are very high on some farms then sheep should not be kept there.

In the long term, human society will not be at peace with itself until sentient animals, wherever possible, are offered as good conditions for their welfare as humans. In support of this, the previous social movements have strived for nothing less than to improve the opportunities of less fortunate members of society, including children, women, disabled people, ethnic minorities etc. so that they equal that of the most privileged, in particular the healthy adult males. As with less fortunate members of society, conditions for animals do not need to be the same or even similar, but appropriate. The facilities provided to all sentient species, including humans, therefore need to be tailored to their needs

(Bartussek, 1999). Suitable rather than equivalent levels of care should be the aim, and consideration needs to be given to other ethical issues than welfare which are involved in our management of animals: genetic modification, premature slaughter, speciesism, altering the animal's integrity, reproductive manipulation, habitat destruction etc.

Many animals are treated worse than humans, and this is tolerated if not approved by society. As evidence of this, comparison with animal conditions is used as a means to express o concern for human standards, saying that they are no better than those offered to animals. 'Brutal' treatment, the 'cattle class' in aeroplanes, living rooms being a 'pig sty' all demonstrate that we recognise that animals are treated worse than humans. Whilst conditions are still improving to an acceptable level for many humans it is inevitable that animals will be less well provided for. The greater the human deprivation, the more likely it is that conditions for animals will be poor. However, as conditions for humans rise above the threshold for happiness that Layard (2005) enumerated, animals will be increasingly better cared for. There are even instances of animals being offered better conditions than humans, some cats and dogs for example, whose owners afford them all the luxuries that many humans aspire to. Other highly valued animals, such as those with rare genetics that are used for breeding animals for sale, may be kept in superior conditions to many humans, who are not only constrained by the captivity of their immediate environment, they also have to spend up to 50% of their time (or perhaps 80% of their time awake) working to be able to support themselves and dependent offspring. Elephants kept in Indian sanctuaries for religious purposes are usually given plenty of food, companionship, space, good veterinary care, and there are several mahouts to look after each one. They have to parade in festivals for several months and then have at least six months without work.

The public concern about animals kept in small enclosures with limited stimulation is not always extended to humans that have to endure similar conditions, but the impact of living in small confined conditions is little understood in either. Familiarity with the environment is of obvious survival value and therefore can be comforting, and living in one small room enables the occupier to come to know their surroundings in intricate detail. However, humans have considerable cognitive ability, the suppression of which could lead to boredom. This is also recognised in many animals that are kept in solitary conditions, especially those in zoos and non-domesticated animals. At the opposite end of the spectrum, are animals that find their environment too difficult to cope with, leading to anxiety, which may be associated with self-directed behaviour, sometimes harmful. Other symptoms known in humans are rapid switching between tasks, tachycardia, tightness of the chest because of shallow, rapid breathing, over-oxygenation of the blood, leading sometimes to dizziness and panic attacks. Many animals also suffer from social anxiety, but particularly the gregarious ones. Sheep, for example will demonstrate a severe stress response when they are isolated (Degabriele and Fell, 2001), and dogs with a strong bond to their owners suffer anxiety when separated from them (Houpt et al., 2007).

Despite these problems, there can be little doubt that conditions for animals have improved in many situations in recent years. For example, livestock mortality in the late 18th C shipments from the Cape, Calcutta or the west coast of America to Australia averaged about 50% (Peel, 1986), whereas today for the shipments from Australia to the Middle East it is just 1% for sheep (Higgs et al., 1999) and 0.1% for cattle. Carlson has provided graphic descriptions of the cruelty inflicted on cattle shipped to England only just over one hundred years ago, demonstrating how transport conditions have improved over the course of the last century (Carlson, 2002). The cattle were given scant food and water and were continually prodded and made to move to keep them alive. On arrival, since only live cattle were paid for, hot paraffin was sometimes poured into their ear canal, which was stuffed with hay and then set alight in order to incite the near-dead animals to move. In this case, standards for animals have undoubtedly improved, but the trade is still regarded as cruel by many (RSPCA, 2006). This demonstrates the rapid improvement in welfare standards expected by the general public.

Animal feeding too has seen many improvements in recent years. In the early 20th C, the ability of farmers to keep their cattle alive over winter in cold climates such as in northern Britain was often limited by their stocks of conserved fodder, in particular hay, and most animals would lose weight. Some would even die in a hard winter. In milder climates standing fodder or foggage could be used, and it still is in many countries, but in Britain snow cover limited this option. Nowadays, with fodder production vastly improved due to mechanisation of the process and fertilization of the soil, such malnutrition is rare (Phillips, 2001). Even in Australian drought conditions the ability of farmers to keep their animals alive by either bringing conserved feed onto the farm or sending cattle away to areas where feed is available is much improved over the last 50 years. Farmers' ability to manage their feedstocks has improved, with consequential benefits to animal welfare (Hogan, 1996). In other animal industries, nutrition has improved in parallel with improvements in human nutrition. Diets are available for companion animals that will not only optimise their growth, they can correct for diseases and enhance the animal's welfare (Diez and Istasse, 1995).

The driver for improvements in living conditions around the globe is partly the new social ethic, described in Chapter 4, but it is also new technology, which has been developing at an ever increasing pace. Having been at the mercy of nature for so many millennia, we are at last learning how to manage the planet and its animals and plants. The nirvana, the attainment of good living standards for all sentient animals and people alike, will take many hundreds of years to reach. Current improvements in animal welfare should not lead to complacency, but neither is it correct to say that deteriorating animal welfare standards through intensification are the main driver for increasing animal welfare concern, as proposed by Rollin (2006). In many fields of society, post war generations have been both spectacularly ambitious and achieving, and in animal welfare there have been many improvements, and we can still anticipate future benefits to animals, particularly as we develop better tools to manage their genetics.

Animal's Right to Life and Welfare

The animal rights philosopher Tom Regan (1983) believes that some animals are sufficiently similar to us, in that they show evidence of sentience, that they should be afforded special status, termed by Regan the 'subject-of-a-life'. This makes them eligible for certain rights, such as life, freedom from hunger, fear etc and other important aspects of welfare provision. In support of this concept, the great apes have been the subject of an attempt by a group of philosophers to afford them the legal status of humans, principally because of their rational thought powers (Singer and Cavalieri, 1993). This has been partially achieved in Spain, although it seems likely that they will still legally be kept in captivity (GAP, 2008). Other animals who do not demonstrate sentience are ineligible for such considerations (Regan, 1983). The division between sentient and non-sentient animals is frequently used in setting standards for animal management (e.g. Australian Code of Practice for the Care and Use of Animals for Scientific Purposes, 2004), but it is difficult to imagine how a distinct division between those with and without this capacity can be scientifically justified. A graded scale of sentience is more defendable. Others argue that animals can only have these rights if they claim and accept the rights and the responsibilities that accompany them, and because the animals are in most cases managed by man they cannot demonstrate the free will necessary to assume responsibilities (Seamer, 1998).

The validity of using sentience as a criterion for assigning welfare benefits depends on how animal welfare is defined. If it is defined as the animal's feelings then it must be essential for an animal to have the power of sentience in order to have the opportunity to have good feelings rather than bad. However, another key criterion for attributing our welfare concern is the animals' role in their ecosystem. Some are essential members and hold a key role in the ecosystems managed by humans. Others are not, and as stated previously, the right to life is not absolute. All animals are interdependent in the living ecosystem, and they are not all equal. For example, it must be considered whether an animal is native to the habitat or introduced, and if the latter how long ago was it introduced? Maintaining a high proportion of native species preserves stability and diversity and helps to limit the rate of change in ecosystems. However, it is not just the status of the animal itself but the interdependencies with other animals that are important. Large predators have been largely eliminated from the Australian landscape, so the Australian dingo appears to have a role to play despite its relatively recent introduction about 6000 years ago (Savolainen et al., 2004).

The available habitat and the species' impact on ecosystems have also to be considered, with African elephants being controlled even though they are native because of their destructive effects on local fauna. Finally the use of the species for human purposes must be considered.

Animals' relationship to humans and the human ethical responsibility to end animal suffering must also be taken into account (Albright, 2002). The annual Japanese whale cull of 860 animals evokes much greater public outrage than for the annual Australian kangaroo cull of 3 million animals (RSPCA, 2002b). All animals have their part to play in the ecosystem, but for some species that part will include preparing the way for others. That is the nature of evolution. Who determines whether an animal species has a major part to play, whether its welfare should be preserved at the expense of others? Generally human society takes this responsibility, but society's attitudes are changing to become more inclusive, with more concern for the animals that have previously attracted little attention. Society will sometimes get it wrong, but we must accept that all animal life is part of a dynamic ecosystem and not a mass of individual entities. Humans have been called upon in the past to sacrifice their life, or part of it, for the benefit of others, most notably in conflicts. Sometimes this is in error, but humans learn from the error of past mistakes, that is part of our contribution. Evidence for the integrated nature of human society is to be found in the many examples where humans willingly sacrifice themselves for the benefit of others. Animals do exactly the same, most famously the lemmings of Scandinavia, who sacrifice themselves approximately every four years for the benefit of the next generation, thereby depriving the animals that prey on them, snowy owls, long-tailed skuas, arctic foxes and stoats, of their sustenance and limiting their population for the future benefit of the species (Wang and Kuang, 2007). Altruism is not unique to humans, nor does it have to be reciprocated to be of genetic benefit.

A key moral issue is whether the rights or welfare of individual animals can be sacrificed for the benefit of other conspecifics or even humans. Tom Regan's philosophy places an emphasis on the rights of individuals, which cannot be forfeited for the benefit of others (Regan, 1983). The opposing (utilitarian) view is that the rights of an individual can be sacrificed if it brings overall benefit (or increased happiness) to the population. Although the latter is a form of trade off that happens all the time in human society, there has been a movement in the last century to diminish the responsibility of individuals to society. The sacrifice of millions of young men in the First World War for the benefit of civilized European society went almost unquestioned at the time, but it is doubtful whether it would be morally acceptable nowadays. Society's boundaries are expanding with globalization, and with this the traditional allegiance of the individual to their country is diminishing. However, with this changing perception of human responsibilities has come the recognition that an individual animal's rights are also important. The question of degree is important, and few would argue that a mild injustice to an individual should not be tolerated if it brings considerable benefit to many others. Such is the essence of altruism, which may actually benefit the individual, since he or she will gains a sense of satisfaction in helping the community and rewards if such assistance is reciprocated. This good feeling probably evolved in communities that benefit from individuals acting for the common good. However, even those that espouse a utilitarian approach to animal rights, such as Peter Singer, do not accept that a major loss of rights, such as the right to life in farm or laboratory animals, is acceptable for the benefit of others (Singer, 1975). Singer also argues that the use of farm or laboratory animals brings about a major cost to the animals, but the benefit to humans is only minor. A key moral issue is therefore how much should individuals be prepared to sacrifice for the benefit of others? Is it just sufficient for them to gain benefit of belonging to a close community, or should it be sufficient for others to gain benefit at their expense?

The right to life is one of the most fundamental rights, yet it is dependent on the use of animals, for example being controlled by humans for many farm and laboratory animals. Farm animals kept for the production of meat usually lose their right to life after they have reached about half of their mature size. Dairy cows kept for milk live as long as they are economically producing milk, which is usually until they reach the age of about five, considerably less than their potential longevity, which is about 25 years. In southern Mediterranean countries there is a tradition of killing food animals at a very young age, when their muscles are tender. Many animals, such as lambs, calves and piglets are killed when they are still consuming milk, directly from the mother in the case of lambs and piglets. In northern Europe farm animals are slaughtered at an older age, because they then have more fat in their bodies, and there was traditionally a need for the inhabitants of these colder climes to consume meat with a higher fat content in the past.

Laboratory animals rarely reach senescence, indeed they are hardly ever used for more than one experiment. Furthermore, the repetition of experimental procedures on an animal is not advised by some authorities due to the potential cumulative effects of the experiments on the animal's welfare (Australian Code of Practice for the Care and Use of Animals for Scientific Purposes, 2004).

Companion animals will often live to senescence, and indeed because of the strong bond established with the owner, the life of pets is often maintained even if the quality of life is severely reduced. However, this desire sometimes contrasts with that of the attendant veterinarians, who will often counsel that it is kinder to destroy an animal whose quality of life is reduced than keep it alive. Perhaps because of their unwillingness to allow animals to suffer, euthanasia is advocated, although we do not fully understand the animal's ability to preserve its mental wellbeing in the face of physical disabilities. The advantage to an individual's genes of preserving life, even in the face of severe physical problems is considerable, as long as it could potentially breed. Hence it is likely that animals share the human desire for longevity.

Nevertheless, there is a distinctly different attitude to the right to life in humans and animals. Some humans would accept voluntary euthanasia as octogenarians but as teenagers it would be considered wrong by nearly everyone, with a severe stigma attached. A long lifespan in humans is heralded with a sense of achievement in obituaries. In animals the attitude to preservation of life depends on species, situation and the owner or manager's cultural background and religious beliefs (Phillips and McCulloch, 2005).

Animal Sacrifice

Although the deliberate taking of life before natural senescence occurs may seem by some an infringement of an animal's rights, it is deeply embedded in human society. Indeed in the ancient biblical period before Moses' time the taking of an animals' life was believed to be necessary to maintain a good relationship with god. The scriptures do not state whether this was believed at the time to be fundamental to our primeval nature or ordained by god. The latter belief prevailed in some scholars as recently as the late Victorian period (Smith, 1880). Sacrifice by the Israelites was probably originally borrowed from neighbouring countries, especially Egypt, where it was part of their religious practices. However, the biblical explanation of sacrifice by the Israelites as a ritual practice to cement the covenant between god and humans differs from the beliefs of neighbouring peoples, who viewed their gods as being angry and jealous, requiring sacrifice to appease them.

In the biblical writings after the time of Moses, the story gradually unfolds that sacrifice, especially of animals, was for the atonement of the sins of humans, and it is clear in the New Testament that this was its principle purpose. Biblical texts also speak of the sacrifice of praise, thanksgiving, charity and devotion, which were offered to god and 'with which he is well pleased' (Hebrews Chapter 13, vs 15–16). There is considerably less emphasis on religious sacrifice of animals in the New than the Old Testament, because in the former it is proposed that Jesus Christ became the 'sacrificial lamb' in a single act of atonement for man's sins through his death. Nevertheless, the idea prevailed that humans were sinful and that the sins could be offloaded onto sacrificial animals or other humans, rendering the people pure and holy. Such ideas persisted into the Middle Ages, when animals were killed to atone for particular crimes. The practice of sacrifice at religious festivals still persists in the Middle East (Alboga, 2003), but there is less emphasis on appeasement of god. Even in Western society the consumption of lamb remains traditional at Easter, when Christians remember that Jesus became the sacrificial lamb. Also in the Christian religion the idea developed that believers could atone for their sins by confessing them to god and their priest and repenting for them. Now that animal welfare has become a major societal concern, this is a more acceptable way of dealing with immoral behaviour and the priest replaces the sacrificial 'scapegoat'.

In Old Testament times sacrifice was both public and private. Public sacrifice might involve, for example, the slaughter of two goats for the people and one bullock for the priest on a day of atonement (Leviticus Chapter 16). This would encourage the people to be reverent to both god and the priests. Although it was considered to be essentially a peace offering in the Old Testament era after Moses, it was also used in supplication for benefits, such as clement weather. Private sacrifice was also accepted by the law, which guided and limited the practice. Hence, in ancient Judaea, although sacrifice of lambs aged about one year was common, no lamb under eight days of age was allowed to be killed (Leviticus, Ch 23, 27), (Smith, 1880). Prevalent as it was in many ancient religions, it is likely that requiring humans through private sacrifice to be willing to forsake the things that were most precious to them, i.e. their animals and in Abraham's case even his own child, enabled the priests to maintain a degree of control over the people. In the case of animals, the priests' power was only over the life and death of the animal, not its use for human consumption. Controlling the latter would have severely constrained the food supply for any society, so it is entirely logical that animal sacrifice became associated with religious festivals.

In the Muslim religion, animals are sacrificed at festival times to share between relatives, neighbouring families and the poor (Alboga, 2003). This is a logical development because it would bond people together, preserve the population and also because there is too much in one animal for one family. The Koran advocates such sharing of larger animals:

> 'We have made the camels a part of God's rites. They are of much use to you. Pronounce over them the name of God as you draw them up in a line and slaughter them; and when they have fallen to the ground eat of their flesh and feed the uncomplaining beggar and the demanding suppliant. Thus have We subjected them to your service, so that you may give thanks.
>
> The Koran, Pilgrimage, 22, 35–37 (1990)

Cattle were commonly used for sacrifice in Muslim society, and the Koran requires that the name of God should be invoked whenever cattle are offered for sacrifice. This should be done by priests only after the cattle have been used by humans for other useful purposes, such as producing fuel in the form of dung and working to till the fields. The scriptures remind the people that the cattle are a gift from God (Koran, Pilgrimage, 22, 32–35).

> 'In the cattle, you have but an example of Our power. You eat their flesh, and gain other benefits from them besides. By them, as the ships that sail the sea, you are carried.'
>
> Koran, The Believers, 23, 21–23

Like Muslims, Hindus still regularly practise animal sacrifice in India, although those in Western countries have difficulty in obtaining permission (Smith, 2000). In India chiefly goats and chickens are sacrificed, and the practice is often managed by the temples. In Buddhist cultures it is much rarer, but still exists in the form of externalizing punishment for people's crimes in Sri Lanka (Feddema, 1995).

Animal Slaughter

Animals are killed mainly for meat production, other 'products' such as wool, dung for fuel, companionship and sport being derived from live animals. The choice of animals for meat production is principally driven by the ease and safety of keeping them. Animals that are polygynous, precocious and prolific are favoured. Herbivores are both less likely to transmit zoonotic diseases than carnivores and more efficient users of land, and hence cattle, sheep, deer and other herbivores are popular. Omnivores, such as pigs and chickens are also popular food animals but are fed mainly on plant rather than animal products, especially after the emergence of zoonotic BSE in cattle after they were fed animal products.

Animals used for meat production mostly possess high levels of sentience and there is therefore considerable public concern to make the killing process as quick and painless as possible. People demonstrate considerable empathy in their high level of interest in this process. Preventing awareness of their fate is a major concern, but there is only limited evidence of awareness of the death process in animals, even 'higher' animals such as elephants (Bradshaw, 2004), dogs and primates. Indeed fear of death is almost certainly greater in humans than in animals because of a greater ability to anticipate and imagine the event, and the concern for an afterlife. Animals do not apparently plan for the future to the same extent as humans and therefore probably do not demonstrate a long-term fear of death (Hui et al., 2006).

Awareness of death encourages the development of religious beliefs, which are obviously most evident so far in man. The anxiety created by the desire to maximize productive life, may be controlled by a belief in another life, be it the immortal soul residing in heaven in the Christian religion or the rebirth in the form of other animals or humans in the eastern religions. Whilst it could be argued that the major cognitive powers of man have allowed for considerable inventiveness in religious beliefs, there are some who argue that the rudiments of religious belief exist in higher animals, in that they perform some moral behaviours, by avoiding incest for example (Broom, 2003).

Methods used for animal slaughter range from the purely physical, such as when piglets are killed by swinging them hard against a wall or the floor, to the chemical, such as when animals are injected with a lethal dose of sedative or deprived of oxygen. The need for rapid slaughter of large numbers of animals in a painless manner has led to the search for chemical methods that can be applied to a roomful of animals, without the need to treat each animal separately. Increasing the carbon dioxide concentration is one method favoured by some for the slaughter of laboratory and farm animals, even though animals will suffer pain for about 10 seconds before becoming unconscious if they are rapidly exposed to high concentrations (Hawkins et al., 2006). A gradual fill may reduce the pain sensation but increase breathlessness. Humans, who metabolise carbon dioxide in just the same way as farm animals, have reported severe respiratory trauma when accidentally exposed to the gas (Hawkins et al., 2006). Considerable concern rightly exists amongst laboratory and farm animal scientists that this constitutes unnecessary suffering but other methods, such as lethal injection are time consuming and not practical for rapid euthanasia of large numbers of animals.

The death process, even if relatively quick, commands more attention than long-term suffering. The few seconds before death, when the pain is probably severe and exacerbated by anxiety, represent perhaps 0.003% if the animal's life, yet they receive more attention than the keeping of laboratory animals in enclosures that prevent them performing natural behaviour throughout their short lives. By focusing concern on the animals' death, people may be displaying remorse for the killing of these animals.

There are other anomalies in animal death that appear speciesist. Little is known about the duration and extent of suffering in wild animals, for example the kangaroos and camels that are slaughtered in Australia to manage their population (Ford, 1986), or seals that are slaughtered for their pelts (Ambrose, 1992). By contrast, even though the suffering of farm and laboratory animals during death appears shorter and less severe, it is much researched and also closely monitored for quality control.

Expectations for the normal longevity of an animal will depend on the environment in which it is kept, its genotype and management. Longevity in sexually reproducing animals is determined principally by the need to exchange genes on a regular basis, in order to accommodate both environmental change and the dynamic population of other species that compete with them. In asexually-reproducing animals reproduction provides the opportunity to rapidly increase the population, with large numbers of offspring produced to potentially colonise new habitats or confirm occupancy of existing ones. Across species, natural longevity tends to be greater for larger animals, because of the more protracted investment in growth of the animal.

Sexual reproduction therefore evolved because the ability to change conferred advantage in a competitive world. The environment constantly changes and it is advantageous for animals to be able to adapt to meet those changes. A fundamental central question is why the world needs to change. If a supreme being established the physical laws and designed a world which humans evolved to manage, with increasing effectiveness, why did the being not design a perfect world in the first place, with the physical and biological systems in harmony. The answer may lie in our construct of good and evil (see above, Religious and Historical Perspectives in this chapter), which we have developed to allow us to manage biological systems. Pain and cruelty are only associated with evil because they have been useful constructs in the past to steer people towards sustainable management systems. We may already exist in a perfect management system and have all the tools for sustainable management of our animals and plants. Our biological systems and the evolutionary principles that allow them to adapt to novel circumstances are designed to optimize sustainable use of the planetary resources. Many philosophers, including Goethe, Liebniz and

Spinoza, have argued that each animal species is in itself an instance of perfected design, even if others have erroneously argued that perfection increases as we move up the species hierarchy (Preece and Fraser, 2000). Any pain or suffering that we experience, or which animals experience, appears wrong to us because of our desire for self-preservation, whereas we should see ourselves as just part of the much larger spectrum of life on the planet, interconnected and interdependent (Dawkins, 2006). Therefore we should not *expect* to live for any particular period of time, nor could any animal within our charge, but we should maximise the contribution of each individual to the spectrum of life, in order to sustain life most effectively. This may mean not slaughtering animals for food at a very early age, not sacrificing animals or humans unnecessarily – in other words behaving altruistically.

Pain

Pain has been described as "an aversive sensory experience caused by actual or potential injury that elicits protective motor and vegetative reactions, results in learned avoidance, and may modify species specific behaviour, including social behaviour" (Zimmerman, 1986). It therefore fulfills a vital function of alerting an animal to potential or actual injury, so that they can take evasive action. Without the feeling of pain animals would die at an early age. If we acknowledge the welfare of an animal to be a function of its experiences, positive and negative, then pain obviously has a direct input into welfare, because it exists both during and after a negative experience. However, it may help to avert further negative experiences through the animal's evasive action, which will have a positive impact on lifetime welfare. The exhibition of pain, therefore, is not necessarily indicative of an overall negative impact on welfare.

Pain is one of a suite of primary negative affects,[3] all of which are probably common to higher animals and humans. They include fear, terror, hatred, distress, dissmell,[4] anguish and disgust (Tomkins, 1963; 1991). Secondary negative emotions include shame and guilt, which in the view of humans at least may be more common in humans than other animals.[5] These affects often co-exist and one may predispose to another, so in the case of pain, the additional negative affects are most commonly fear and distress.

The pain response to physical injury is not a 'straight through' biological mechanism, where the affective response is directly proportional to the tissue damage, but it is influenced by the subject's level of attention, anxiety, suggestion and prior experience (Melzack, 1987). Up until the 1950s it was believed that pain responses were only a function of the somatic neuronal pathways,

---

[3] A brief biological, innate, instinctive response to a stimulus

[4] The negative affect of experiencing noxious odours

[5] Legally animals cannot experience guilt since they are recognized only as property in most parts of the world

which could be best controlled by neurosurgery, at least experimentally (Melzack, 1993). After this date, psychological influences, such as those referred to above, that had previously been dismissed as *reactions* to pain began to be recognized as pain modulators, and it became clear that a 'top-down' approach was necessary in examining pain responses. Furthermore, Melzack and Wall (1965) suggested that the transmission of nerve impulses from afferent fibres to spinal cord transmission cells is modulated by a spinal gating mechanism in the dorsal horn, which can be influenced by signals from the brain. They proposed that if the number of nerve impulses is more than a critical value, pain is experienced. They found that rubbing the affected area closed the gate, whereas pinching opened it. This began two decades of research to elucidate the mechanisms involved in brain regulation of pain responses.

Surveys of Attitudes to Pain as a Component of Animal Welfare

We have little direct evidence of the importance of pain in animal welfare, other than the extent of scientific research, compared to other areas of welfare, and attitudes of stakeholders in the animal industries. By the year 2005, the number of scientific journal articles that had been published on pain and animal welfare (4793) was considerably greater than those on the other Freedoms commonly recognized as the major components of welfare (normal behaviour, 2,599; fear/stress 1,286; hunger and thirst 1,247 and discomfort 486) (Phillips, 2005a).

Most surveys of attitudes to pain have investigated it in relation to animal welfare. Several surveys have investigated attitudes in American university staff and students concerned with animals in veterinary or animal science programs, who will be, or are leaders in the field of animal management and health treatment. When animal science faculty members in a variety of American universities were surveyed on their attitudes towards animal welfare, nearly all (97%) believed that animals should have freedom from unnecessary pain (Heleski et al., 2004). They also believed that animals should be free from other negative affect, such as thirst, hunger, fear or distress. However, when asked about specific practices that are commonly used in the animal production industries and are known to cause pain, such as castration without anaesthetic use (Taylor and Weary, 2000) or beak trimming in poultry, approximately 70% of the staff believed that these practices did not warrant concern. This agreed with the majority view that current farm practices did not require any change, or that changes should be minor. Later, in a direct comparison of veterinary and animal science staff, Heleski et al. (2006) concluded that the former had greater levels of concern for farm animal welfare, especially for meat birds, beef and sheep. Both groups of staff strongly believed that farm animals should be free from pain and discomfort, with no difference between disciplines.

Amongst tertiary level students of veterinary medicine and other disciplines, there is a belief that some domesticated species are more capable of experiencing emotions than others (Levine et al., 2005; Phillips and McCulloch, 2005). Specifically, students of veterinary medicine in an American college believed that dogs and cats were more capable in this respect than farm animals, with a particularly low rating for poultry (Levine et al., 2005). In contrast to the animal science faculty members, most of the students considered normal practices with farm animals that are known to cause pain, such as branding with a hot iron, castration by banding or at a very young age and surgical procedures, to be inhumane. Amongst students of all disciplines and of a mix of nationalities, there was a belief that species have differing ability to experience sentience, and the dog receives a particularly high rating, despite an absence of anatomical or physiological evidence to support this belief (sentience ratings: monkey > dog > newborn baby > fox > pig > chicken > rat > fish, Phillips and McCulloch, 2005). In a further American survey, animal science students also believed that there are species differences in the capacity to feel pain, with chickens apparently having less capacity than other farm animal species (Heleski and Zanella, 2006). Approximately 50% of students believed that farm animals experienced pain in a similar way to humans. What is most concerning is that these differences in attributed sentience levels influence the students' attitudes to practices that cause pain or cruelty, with a more tolerant attitude if the animals were believed to be less sentient (Phillips and McCulloch, 2005). Some differences between students of different nationalities towards sentience were also detected, for example pigs and poultry were attributed high levels of sentience by students from southeast Asia, and Chinese students attributed particularly high levels of sentience to rats and fish. In an international comparison of 15 different nations (Pifer et al., 1994), members of the public were asked to agree or disagree with the statement *'Scientists should be allowed to do research that causes pain and injury to animals like dogs and chimpanzees if it produces new information about human health problems.'* The intensity of opposition ranged from low levels in Japan (42%) and the USA (42%) to high levels in France (68%) and West Germany (60%), with Great Britain, Italy and Spain being intermediate.

In Pifer et al.'s study, gender also affected the attitude to animal research, with women being more opposed than men in all 15 nations. Gender also influences attitudes to practices that affect animal welfare, with women having greater concern than men (Pifer et al., 1994; Heleski et al., 2005; Phillips and McCulloch, 2005), but men and women have similar attitudes to the sentience capabilities of different species (Phillips and McCulloch, 2005).

Evaluating Pain Responses as a Component of Animal Welfare

The science of determining that an animal is in pain has been successfully developed through the self-administration of analgesics (e.g. Sneddon et al., 2003). Self-selection of analgesics by chickens has demonstrated that lame birds are experiencing pain (Danbury et al., 2000). More detailed quantification of pain is needed, in order to compare welfare impact of different husbandry procedures. It is possible to determine the severity of short-term pain responses from the levels of stress invoked and the longevity of the response, and by multiplying these two an estimate of welfare impact is possible (as has been proposed by Broom, 1999, for pest animals). Hence, responses to mastitis could theoretically be evaluated by measuring the duration and the degree of pyrexia. However, mean disease durations and the stress responses over the time have received little attention, so caution is required in any interpretation of such models. There are exceptions, for example the mean duration of lameness in dairy cows is approximately three months (Phillips, 1990). Evaluating the welfare implications of pain responses in this way does not take into account the long term increase in avoidance of humans and possible anxiety/fear that may develop in the animals concerned.

Pain also has the potential to cause anxiety and fear in humans, and pain therapy aims to dissociate pain from such conditioning factors (Pruimboom and van Dam, 2006). Asmundson and Hadjistavropoulos (2007) have demonstrated physiologically, by looking at the time taken to react to words, that human patients with high fear of pain also have a generalized anxiety towards their health and even non-threatening issues. However, Wilson et al. (2007) have suggested that individual differences in baseline anxiety levels do not modulate pain responses in rodents, and it is therefore unclear whether animals respond cognitively to pain in the same way as humans.

Conclusions on Pain

Pain is perceived as one of the most important topics in animal welfare, both in terms of the research focus and effect on animal welfare. There is major concern amongst stakeholders in the farming industries that it should be controlled when invasive practices are performed (Cross et al., 2008a, b). However, those closely associated with the animal industries may become habituated to the pain that animals experience when they are treated and accept it as a normal part of the system. Scientific methods of demonstrating pain in animals have improved considerably, but quantifying pain is difficult, especially given the paucity of information on responses in animals. Significant pain is likely to induce secondary negative affects, such as fear and anxiety, but this has rarely been investigated in animals.

We need to understand both people's opinions on the importance of pain as a welfare issue and the extent to which animals can experience pain better. Whilst we can theoretically compare different experiences in terms of the pain produced, it will be much harder to compare the importance of different negative emotions for animal welfare.

Improving Animal Welfare in Developed and Developing Countries

There have been some substantial lapses in standards along the general path of improvement of animal welfare in recent times. The emergence of intensive 'factory' farming practices in industrialized countries in the 1960s to increase food production, after the shortages of the 1940s and 1950s, represented an unacceptable shift to prioritising human welfare at the expense of animals in many people's view. This was mainly a reaction to the threat of starvation and food shortages endured during and after the Second World War, particularly in the United Kingdom, but it was also a continuation of the process of intensification begun in the mid 18th C. At that time British breeders began to develop specialised livestock for different purposes, meat and wool production in particular, because of the need to intensify production for an expanding population. The population expansion at the start of the Industrial Revolution heralded an increase in demand for animal products. The intensification of livestock production in Britain in the mid-late 20th C met with fierce opposition at times (e.g. Harrison, 1964), but because of the economic consequences its reversal has been slow and somewhat limited. The widespread adoption of intensive production in the developed and now increasingly in the developing world had been assured by the greater efficiency of labour use and often superior profits that it generated, compared to traditional techniques. The argument has been commonly made that animal welfare in intensive production systems is good because the animals are growing, lactating or reproducing effectively, without considering the impact on the animals themselves. Even before the major era of intensification, a prominent agricultural lecturer reported that 'cattle are equally well suited to the extensive ranching methods of the remote and indifferent grazing regions, where the animals fend for themselves, and the intensive methods of densely populated countries, where they seldom, and sometimes never, leave the stall' (Shanahan, 1925). It is true that they can be kept in either situation, but their greater suitability for extensive grazing systems than indoor management is apparent from the health and behaviour consequences of the latter (Phillips, 2001, 2002; Phillips and Sorensen, 1993). It is increasingly accepted that some forms of farming allow for economic production but do not adequately provide for animal welfare (Appleby et al., 2003).

Animal production has not always been intensifying and we have a tendency to regard traditional systems as extensive and modern ones as intensive. This is not always the case. Sheep farming is nowadays generally regarded as one of the most extensive forms of animal farming, but accounts from biblical times show that the production systems employed then were labour intensive, with each animal being known by name and led out to pasture each day by a shepherd calling them individually (Smith, 1880).

The response of the public to intensive animal production has been varied. In the United States, intensification was rapid in the pig and poultry industries, with multi-tier buildings and many of the tasks automated, from egg collection to 'spent hen' harvesting in the poultry industry. Only recently has there been opposition from the American public to unacceptable animal management practices. Such opposition is only just beginning to emerge in developing countries, and many do not have the democratic procedures to allow the people's wishes to be considered. The last century was dominated by a greater struggle than ever before to achieve democracy, the struggle against totalitarianism being one of the most protracted and difficult. In the words of Margaret Thatcher, 'Civilization has its ebbs and flows, but if we look at the history of the last five hundred years, whether in the field of art, science, technology, religious tolerance or in the practice of politics, the conscious inspiration of it all has been the belief and practice of freedom under law; freedom disciplined by morality, under the law perceived to be just' (Thatcher, 1986). This moral freedom is now being exercised in many developed countries, and includes improvements in animal welfare.

Improvement of animal welfare on farms has required the industrialized farm model that evolved in post-war Europe to be challenged and amended to take into account the people's demand that animals be treated with respect and greater concern for their welfare. The intensive route to food production has significant risks: antibiotics will often be routinely administered because the high stocking density of the animals means that they have to be protected artificially from infectious diseases (Ravindran et al., 2006); food supplies may be exhausted if insufficient reserves are held; selection for rapid growth and high milk production can increase susceptibility to disease and reductions in genetic diversity through intensive breeding practices may limit future production capacity. The recent trend to use animal feed for biofuel production is likely to have a more significant and rapid impact on intensive animal production than animal welfare concerns. In the medium term future, the keeping of farm animals is likely to be relegated in future to marginal land, instead of using prime land that might be needed for fuel (biodiesel or ethanol) or human food production, and animal food products will be used more as a luxury food item by some people, rather than the main item of most meals for most people. Much of this will be necessitated by the need to use very large areas of land to produce human food or fuel, extracted from potatoes, sugar cane and other suitable crops, in order to reduce reliance on fossil fuels, particularly those from the Middle East (Mattison and Norris, 2007). Relegated to marginal land, animal products will be more likely to come from cattle or sheep than pigs or poultry. Producing in a less intensive way involves the farmer in less risk and is more sustainable (Phillips and Sorensen, 1993). At present farmers in many countries, including Australia, England and other highly developed nations, receive government aid if there are adverse climatic or other unforeseen events, in order to preserve food production capacity. In future additional government assistance is likely to be offered for the establishment of sustainable farming systems, incorporating agroforestry, recycling of resources and integrated systems mixing crops and animal farming.

It is evident that the majority of the population in Europe wants improvements in animal welfare (Eurobarometer, 2007), and the activities of the non-government organisations (NGOs) against the corporate animal industries has been increasingly intense. Recently, the growth of these NGOs, as a result of public donations, and the support of many prominent people in the media, as well as some industry, has resulted in significant changes in industrial practices. A recent example is the outlawing of the mulesing operation in Australian sheep. This operation to remove loose skin on the hindquarters of sheep in order to prevent flies laying their eggs there has been necessitated by the breeding of wool sheep with folds of skin in this region of their body, to increase wool yield. The outlawing of the operation was achieved by public pressure, channelled through one of the NGOs (Flugge, 2004). Such a freedom for NGOs to pressurize industry would not be possible under a totalitarian regime, and our hope for the future of relationships between animals and man should surely be encouraged by the democratic government systems that were established in much of the world in the last century (UNESCO, 2001). Using these effectively requires those in the NGO to have a good understanding of the major welfare issues on farms (Cross et al., 2008a, b) and it may initially bring conflict with industry. However, in the long term after product prices have stabilized to take into account increased costs of production, farmers will be more satisfied from keeping animals that are in a high state of welfare. Demonstrating that animal welfare is at a high level is also good for public moral and the country's international image.

In developing countries, although the livestock industries are often integral to the nation's wellbeing, with animals produced largely for local consumption, it is not fair to the human inhabitants to require that the animals receive the same level of welfare as those in developed countries. Therefore while a marked difference exists in human living standards between world regions, truly international animal welfare standards that are equivalent in different countries and anything more than very basic minimum standards are not possible or just whilst there are major inequities between the living standards for humans. The subjection of animals to inadequate standards is still morally wrong, but a wrong from which the perpetrators can be excused (Regan, 1990). If, however, animals are exported from a developing country to a developed country, public sentiment in the latter may require that living standards for the animals in the developing country are increased to that of the animals in the developed country, the cost of which has to be met by the consumer. Therefore, if the animals are destined for export to another country, the level of welfare afforded to them should be at least that in their country of destination. Consumers in developed countries can then be assured that their animal products are from systems with similar welfare standards, regardless of place of origin. In the case of developed countries exporting animal products to developing countries, it will be necessary to take into account the standards demanded by the population in the developed country, and they may not wish animals to be exported to countries where they are not managed well and where the journey is long and arduous.

A high level of animal welfare can lead to improved people-animal and people-people relationships and better social cohesion. Improved people-animal relationships are evident at farm level, since stockpeople will work better if they have the resources at their disposal to feed and care for the animals in their charge to a high standard (Hemsworth and Coleman, 1998). They are likely to spend more time visiting the stock out of normal working hours to make sure that the animals are well cared for, to take a more active interest and pride in their job and to spend time updating themselves on the latest techniques of animal keeping, from which the animals may benefit.

To improve animal welfare standards, better documentation will be required, preferably in the form of a Welfare Record for all animals. This could include a health record, psychosocial record, information on the animal's response to humans/handling and any experimental procedures that have been conducted. Much of this information is already collected, but there is a need to standardise across species and management systems, to ensure a more equitable treatment of all animals in different management systems. In France for example, farmers are required to keep a register of veterinary treatment, which will provide information on the health of animals on the farm, but also enable potentially hazardous or misused drugs, such as antibiotics, to be known for each animal. The register contains information which includes a farm description, a livestock fact sheet, a record of all animal movements on and off the farm, a comprehensive health record and a record of all veterinary visits. Such a register should remain the responsibility of the animal's owner throughout its life and must accompany any animal if it is moved from the farm.

The Treatment of Animals by Indigenous People

When considering animal welfare and rights, it is important to reflect on the traditional relationship between indigenous people and their animals. Although many have emphasized the closeness of this relationship (e.g. Serpell, 1986), nowadays poorly developed indigenous communities are often chided for their ill-treatment of animals. In part this is because of the Western concept of animal welfare is different to their own and in part it may be because they do not use Western methods to manage health and reproduction in their animals. For example, dogs and cats may be communally owned and be evident in the streets, rather than having an individual owner who locks the animal in his house or garden when he goes to work each day. Modern-day indigenous societies are sometimes characterised by a breakdown of the traditional tribal management system, which uses Elders in the form of a local council presiding over the village people. Sometimes the poor living conditions for the people and lack of opportunities for work means that companion animals are badly cared for. In the past it was probably worse in many primitive societies. In the middle of the last century the American writer Joseph Furnas described how Polynesian and

Melanesian people treated their animals when he visited them: 'Hawaiians .... when collecting dogs for a feast, ...let them lie moaning for days with muzzles tied up and legs broken and tied over their backs ......Hawaiian women flocked to see cattle slaughtered as a good show. Melanesia was no better. The Trobrianders singed hogs alive as lingeringly as possible in order to enjoy their screams. South Sea missionaries have accomplished little, and I can find no indication that they have tried hard to encourage humanity towards animals among their converts.' (Furnas, 1948). It is ironic that the terms humanity and humane treatment are often used to describe gentle treatment, when in reality it is only the human population that perpetuate and derive pleasure from torture of animals in this way. Many predators will play with their prey before killing it, thus affording practice in chasing and managing the prey, but no such benefit appears to derive from deliberate cruelty to animals by humans, other than as an outlet for stress.

A recent case of recurring animal ill-treatment in Australia was that of horses on Palm Island, just off the coast of Queensland. There are about 200 horses on the island that were originally brought there to help work the cattle. The island is inhabited by about 1600 indigenous people that were deported there almost 100 years ago, because of the problems that they posed to the society that the new settlers were trying to create. Cattle and pig farms were established on the island to give them work, but fell into disrepair and were discontinued as the locals became increasingly reliant on social support in the second part of the 20th century. Nowadays, the island population is about 70% unemployed and although subject to state legislation, the main governance is by a local council of Elders, which struggles to maintain law and order in the poor socioeconomic conditions on the island. Although many westerners would consider life on a tropical island in northern Queensland to be idyllic, there is actually little to occupy the villagers, who are hampered by their remoteness from the mainland. There are strong tribal factions, as originally about 40 clans contributed to the island population. A large part of the island was cleared for cattle grazing, and after they were removed the pasture was available for the horse population. As the youths of the island have little to occupy them and no possibilities of a career or societal structure to give moral guidance, horses are a major recreational facility for 8-14 year old boys and girls. Lacking the funds for saddles or horse tackle, they ride bareback, and improvise for the bridle and bit, using wire or rope, which cuts into the horses' mouths, and can lead to paralysis of the lower lip. Excessive bareback riding causes saddle sores and of the 200 horses, only a proportion will allow themselves to be ridden, perhaps 30 or 40, and most of these have saddle sores. The sores may be initiated by the horses biting each other, particularly stallions, since there is no reproductive control or gelding of the stallions, and they comprise about 40% of the island's horse population. There are also reports of cruelty to the horses, such as using barbed wire to inflict injuries on the horses (Wilson, 2003).

The youths muster the horses in an attempt to ride them, and many horses seek refuge on the high ground. Some seek refuge in the sea but their movement is restricted, and the youths can readily jump on their backs. Their sandy, nylon shorts provide an abrasive surface to create the saddle sores, like sandpaper. Once the riders are on the horses, they may ride them to exhaustion, over a whole day or more. A proportion of the horses become lame, which makes them even easier to mount.

The major lesson is that welfare is often reduced through ignorance and neglect, and is related to human social problems. Children need to be taught how to treat animals, they need to be gainfully occupied during free time. They need structure in their society to know where their position and their goals lie. They need training and instruction, in this case in horse management, but also in all aspects of life, and often the level of education is pitifully inadequate. In Australia horses have held a valued place in aboriginal society, so, for example. there has for a long time been a taboo against eating them. Many aboriginal workers are highly valued for both their horsemanship skills and their skill in managing cattle. This respect for the animals needs to be conveyed to the young members of society in a structured teaching programme. In this instance a serious animal welfare issue is interwoven into societal problems, with racial issues, democracy and governance being at the heart of the problem. Attempts to solve the problem have resulted in horses being removed, impounded when their saddle sores become too bad, which usually means that the youths work even harder to gain new recruits from the horse population, so an even greater number of animals suffer. These problems have to be addressed from all angles, by social workers, educators and animal welfare staff, but with the active involvement of the tribal Elders.

Chapter 7
Teaching Animal Welfare

Animal welfare knowledge – veterinary training in animal welfare – animal ethical considerations in veterinary medicine – decision making for the treatment of animals by veterinarians

Animal welfare knowledge, correctly disseminated to those directly managing animals, can often achieve improvements in animal welfare much faster in the short term than research. This knowledge is potentially disseminated in schools, universities and in adult education. At present the emphasis is on university education, with animal welfare being taught in veterinary, animal science and agriculture courses. Agricultural colleges often run animal care and stockmanship courses, which aim to improve animal management, even though they often don't specifically consider animal welfare. Much of the responsibility for animal welfare lies directly with animal carers. The declining attractiveness in many countries of agricultural animal management jobs is a serious cause for concern in relation to animal welfare improvement, and is partly due to the greater attractiveness of jobs in cities. In addition to this concern, the increased size of industrialised livestock farms means that the contact between stockperson and animal is limited. No longer can the conscience of the stockperson be relied upon to prevent the ill-treatment of animals. This is causing a greater reliance on regulatory control, but because of the diverse nature of farming enterprises, it is difficult to apply. The cost is ultimately borne by the consumer in many cases, and the industry progression to industrialised systems should be considered in the light of the cost of regulating them from a welfare and environmental perspective. The decline in attractiveness of farm stockperson positions and farm veterinary practices in many parts of the world is a source for concern, which needs to be considered in the light of the cost of regulating standards, as well as the reduced attractiveness to stockpeople and vets of working in industrialized systems of animal production.

Veterinarians are often at the vanguard of animal welfare concerns, and they are held in high regard by members of society (Seabrook and Wilkinson, 2000). They have a particular role by virtue of their clinical skills, but their diagnosis and treatment of diseases and correct use of prophylaxis are all key elements

C. Phillips, *The Welfare of Animals*, Animal Welfare 8,
DOI 10.1007/978-1-4020-9219-0_7, © Springer Science+Business Media B.V. 2009

of their profession. The role of veterinarians is being strengthened by new legislation and the codes, particularly concerning the duty of care, as opposed to cruelty being the main basis for prosecutions. In teaching, as well, there's also a key role for veterinarians, with about 5% of graduates entering this profession. Some will be involved in teaching other veterinarians, and veterinary nursing students, but many will be involved in agriculture or animal science programmes. Veterinary graduates also enter research programmes, but because of the long training that they have already undergone, regrettably few enter doctoral research training. Veterinary graduates play a key role in government, and animal welfare issues may be included in statutory work, such as amending, redefining and administering legislation and codes of practice. In the recently emerging activities of the World Animal Health Organisation in establishing global animal welfare standards, veterinarians have played a prominent role in guiding and governing welfare practices.

Veterinarians will be also involved in more minor roles – the ethical approval of research, for example. Australian Animal Ethics Committees all have to have a (Category A) veterinary member. Many veterinarians become involved in legal cases, providing expert witness and opinion.

The activities of veterinarians are not solely defined by the universities that provide their training, because the practice of veterinarians is in many countries governed by legislation through Act of Parliament. Thus a statutory body exists in most countries – in Australia, the Veterinary Surgeons Board. In the United Kingdom, students are admitted to the Royal College of Veterinary Surgeons (RCVS) at or soon after graduation, in a ceremony that recognises that welfare is of paramount importance in their qualification. Graduating students swear an oath, which acknowledges the prime importance of their welfare responsibilities:

> "I promise, above all, that I will pursue the work of my profession with uprightness of conduct and that my constant endeavour will be to ensure the welfare of animals committed to my care."

The United Kingdom's Royal College of Veterinary Surgeons has also defined essential competencies, including the recognition of clinical signs of specified diseases and effective treatment. Several essential competencies relate directly to welfare and ethics (Table 7.1). Students, on graduating, have to be aware of their ethical responsibilities and the emotional climate in which they function, which will govern the treatment that they give. They have to be aware of the ethical codes, of their own personal limitations, and must seek treatment from elsewhere if they feel that they can't conduct the necessary surgery. They have to be aware of legislation relating to welfare and should promote welfare. They should euthanase with sensitivity to the feelings of the owner of the animal, address and implement welfare records and advise on accepted welfare standards.

Table 7.1 Core competencies established by the Royal College of Veterinary Surgeons (RCVS) that relate to animal welfare and ethics

• Being aware of ethical responsibilities
• Awareness of 'emotional climate'
• Ethical codes
• Personal limitations re treatments
• Legislation relating to welfare
• Promotion of welfare
• Euthanasing animals with sensitivity to feelings of owner
• Assessing and implementing welfare records
• Advising on accepted welfare standards

There are many skills required of veterinarians and welfare concern is central to their profession. The key skills are disease detection and reporting, including surgical treatment, drug therapy, epidemiology, population medicine and preventive medicine. In addition, they must understand animal behaviour, nutrition and reproduction sufficiently well to be able to advise how to prevent or cure disease. They have to make treatment decisions in accord with ethical norms, and they have to know and apply relevant animal law and codes of practice. They should handle their patient and the owner with care and diplomacy, particularly in relation to small animal practice, and must often manage their own business.

Veterinarians often have to give guidance on ethical issues, and in doing so they must balance the requirements and demands of a number of different interest groups. They have responsibilities to their clients, to their patients, to the public, in managing the welfare of animals in the community, and to other veterinarians, in terms of maintaining professional standards and upholding the reputation of the profession.

Teaching Veterinarians About Animal Welfare

Almost the entire veterinary course could be said to be aimed at improving animal welfare, and preserving welfare should be a central tenet in all the core courses in veterinary training (anatomy, physiology, biochemistry, pharmacology, pathology, biology of diseases, epidemiology, oncology, reproduction, animal husbandry, animal behaviour, nutrition, species medicine and clinical practice). The only parts of the course that are not directly concerned with aspects of welfare are some zoonotic diseases, with the aim being to uphold human, rather than animal health, and some toxic substances such as cadmium, which rarely presents problems to domestic animals but can present problems for the terminal consumer, humans, who are therefore subject to the greatest problems of accumulation.

In addition to the core courses, veterinarians may study professional practice, which is usually focused on client and business management communication, skills, etc. This may include aspects of animal welfare management. Other optional subjects include wildlife and exotic animal medicine, poultry medicine and advanced nutrition[1]. Animal welfare is taught as a separate subject in many veterinary colleges, and this is likely to cover some of the theoretical background – ethical approaches to managing animals, which will influence animal welfare concerns, the means of measuring animal welfare, contrasting physiological and behavioural methods, and relations between the major animal diseases and welfare. It is useful, but less essential to highlight the major animal welfare issues in each animal management system. Much of this can be taught in other subjects, parasitology for example, but it is helpful also to have a broad-ranging perspective which combines the considerations of animal welfare, economics and environmental sustainability for each system. Animal handling and transport can also be included. The objective should be to enhance the students' welfare knowledge so that their clinical and paraclinical skills can be used to best effect.

The veterinarian qualifies with clinical, para-clinical and associated clinical skills. Many, but not all of the clinical skills are unique to veterinarians, and they are derived from a sound medical knowledge. As in human medicine, the amount of knowledge is increasing very rapidly, and the increased expectations of owners, particularly of companion animals, encourages the introduction of advanced medical science into the program. Para-clinical skills, such as in animal behaviour, epidemiology, production medicine, nutrition are also growing in importance. Associated skills include health and welfare management, economics, animal ethics, role of animals in society, client management, etc. In this field, the level of skills will probably depend on the level of interest of the student and the particular interests of the university faculty.

There is a transition from farm to small animal practice and usually between two thirds and three quarters of graduates will small animal practice. The remainder mostly enter large animal practice. Small animal practitioners have particular need of para-clinical and associated skills, especially animal behaviour, nutrition of small animals and client management skills, because the problems that they have to deal with may relate to the client's management of the animals.

The recent 'gender switch' in veterinary education is likely to have an impact on welfare management. The majority of all veterinarians practising in developed countries will soon be female. At the beginning of the 1980s, there were approximately 92% male and 8% female students on veterinary courses, and now it's almost reversed, approximately 20%–30% male and 70%–80% female, in most of the developed world. The reasons are numerous, including

[1] RCVS specifies that their veterinarians do not have to be advanced nutritionists, in particular in relation to dairy cow nutrition, since there are others who will take this role.

females performing better academically in late teens, an across-discipline tendency for more females to enter university, the job characteristics of low salary and long hours being more likely to be accepted by women, the transition from farm to small animal work and females being often more patient with animals than males. Women are keener to work with small animals, especially if this includes regular working hours, and they particularly value good workplace relationships. There is little truth to the traditional image of women struggling to cope with an extended calving, and with their smaller hands than men, they may be better at maneuvering calves and lambs in a malpresentation than men. Women tend to be more concerned about animal welfare than men, in a survey of attitudes of international male and female students towards welfare issues, females rated the issues on average 4.0 out of 7 and males only 3.6, with 1 representing little concern and 7 a major concern (Phillips and McCulloch, 2005). The increasing science content of veterinary medicine courses may be more attractive to males. Subjects such as oncology were barely considered 20 years ago, whereas today it is an important part of small animal medicine.

Teaching correct attitudes to animals is a vital part of veterinary education. A sound understanding of animal ethics is particularly needed to ensure that the most appropriate decisions are taken in treating animals (Rollin, 2006). Terminal surgery practicals are considered by many to be essential, because they expose students to surgical practices which would be difficult to achieve using alternative teaching methods. Some universities utilise dogs from local pounds that would otherwise be euthanized, but students may have concerns about dogs being used for their benefit in this way, or about becoming desensitised to unnecessary death. There is a risk of stress and trauma to some of the students, particularly if they start with an apparently healthy dog, which is then surgically operated upon and euthanased by the students. The advantages and disadvantages of this particular use of dogs should be assessed by animal ethics committees that are usually responsible for approving the practicals. If students are suffering stress and trauma, veterinary faculties should be offering counselling facilities before, during and after a particular practice. Because of these controversies, many universities use pigs and sheep for this purpose, which could be considered speciesist, or arrange for cadavers to be donated from local veterinary practices or pounds. Guided teaching of veterinarians in commercial practices, shelters or pounds already occurs, but could be used to a greater extent. Learning on surplus animals or animals in developing countries presents another alternative. Other alternatives to using live animals for teaching, such as videos, models and computer simulations are being developed, and the American Veterinary Medical Association (AVMA) and the RCVS have already accredited some veterinary schools with ‘no harm or kill’ policies.

Another ethical issue relating to veterinary practice is that some students, particularly those from non-Christian countries, may have concerns about studying and treating food production animals. Muslims and Hindus are likely to have concerns about pig and beef cattle medicine, respectively. Students should have the opportunity to have their ethical objections considered and the different persuasions of students acknowledged, as they are sincerely held and it takes courage for students to stand up against a majority view. There may be concerns amongst some students about visiting abattoirs, a topic which has been increasing in importance in the veterinary curriculum. There are many different ways in which students can be taught about the public health implications of veterinary medicine. Vegetarian students of veterinary medicine have to learn to trim chickens beaks, castrate lambs and calves and conduct other practices that allow the animals to be kept in intensive management systems. Veterinary students have a range of ethical persuasions, from those that are unwilling to treat farm animals, but accept the value of animals as companions for example, to those that treat animals as commodities and are focused mainly on developing a good clinical skills base. Although an industry body ensuring sound practice and learning is a major advantage for any profession with such responsibilities, the imposition of a compulsory framework of learning, which includes practices that some of the veterinary students do not agree with, is a major disadvantage of accreditation. Some of the more contentious practices should be optional and universities should be encouraged to offer the students choice and accredit them only in practices that they wish to study. They could then be licensed to practice only in areas that they are accredited in, and this accreditation could be acquired during their training degree, or indeed afterwards in dedicated courses. Such diversification would lead to more dedicated and skilled veterinarians, which would be a benefit for animal welfare. Specialisation is already possible through advanced courses, such as in exotic animal disease (EAD) recognition, in which case it is acknowledged that only veterinarians with the relevant postgraduate training should be involved in EAD diagnosis (Kerwick et al., 2008). The veterinary accreditation bodies should give careful consideration to allowing students to specialise in areas of animal medicine that are most appropriate to them, rather than requiring them to be able and by inference, willing, to treat all types of animals.

The veterinary profession is therefore changing rapidly. There are many more women, which may be beneficial because they tend to be more caring to animals than men, and there is a greater focus on companion animals. And the long-term trend is for an increasing demand for science in the course, in response to public demand. Today's students are likely to be concerned about the ethics of animal use.

Decision Making for the Treatment of Animals by Veterinarians

After graduation veterinary students will be faced with many ethical dilemmas, not least the decision of whether to treat an animal or not. This is likely to be made on many grounds, including the probability of the treatment being effective, the expected lifespan of the animal, the cost of treatment, the value of the animal, the ability of the client to pay, and other risks, such as environmental or human health that are involved in the treatment. The major rule in decision making is to maximise utility. Sometimes the decision may be too hard, and the client may choose to avoid treatment in order to minimise their regret if things go wrong. If an animal is afflicted with a serious ailment, it is usually not seen as the owner's fault. However, if the owner opts to allow the animal to undergo major surgery, for example, then there would substantial and sometimes unbearable regret at having committed the animal to the surgery if it results in a worse welfare than if no treatment had been made. The risk of this may be particularly unbearable if the net result of surgery may shorten the animal's life if it goes wrong. The owner 'plays safe' by allowing the animal to live out its natural life, that way he or she absolves himself or herself of guilt, even if the outcome for the animal is likely to be worse than if some treatment is attempted. The owner may also feel that the decision is too big to make, and may need professional guidance, in part to absolve himself or herself of guilt if it goes wrong. Such circumstances may result in irrational decisions being taken by owners if no guidance is given, and it is necessary to consider whether the owner is the best person to make the judgement. In the same way that parents may make irrational judgements over whether their children should be medically treated, owners may need to be counselled by the veterinarian over whether their pets should be treated.

Some owners, particularly of companion animals, may go to the other extreme and feel that their animal's illness is their fault, and then they will go to any lengths to have the animal treated in the hope that a cure will be found. They cannot bring themselves to face the demise of the animal and again irrational decisions can be made, that veterinarians should counsel against. If there is little hope of drugs producing a cure in a specific case, there may still be a case for veterinarians advising treatment on the grounds that the experience gained may assist in future when animals are afflicted with the same disease.

Owners may feel that giving their animals at least a chance of continued life is a risk worth taking, and this makes them feel better. In this circumstance they must balance the risk of the utility to the animal with the utility of the risk that they might acquire through this course of action. In all events the veterinarians should counsel their client to choose in the best interests of the patient.

Chapter 8
Animal Welfare Science

Historical development – growth in animal welfare science – impact of animal welfare science on animal welfare – independence of research effort – industry funding of animal welfare research – Relationships between farm profitability and animal welfare: the lessons for animal welfare research

Historical Development

Compared with most other scientific disciplines, animal welfare is a relatively new science. Whilst philosophers and ethicists have debated the way in which animals should be treated for thousands of years, it is only in the last thirty years that a scientific approach has become properly established. There had been limited research in animal welfare over the course of the 20th C in particular humane killing methods, experimental techniques, pain control, treatment and prevention of painful diseases, and the University Federation for Animal Welfare had, amongst others, promoted a scientific approach to animal welfare since the 1920s (UFAW, undated). However, in the late 20th C the establishment of animal welfare positions in universities, journals focusing on animal welfare and courses teaching animal welfare as a science gave credibility to the contention that animal welfare could be considered a science. Furthermore, although agricultural research had attempted to improve conditions for farm animals for over one hundred years, the objective was to increase profit from animals, rather than improving their welfare *per se*.

Many of the principles for animal welfare reform were set in the mid 1960s, with the publication of Ruth Harrison's Animal Machines in 1965, and the Brambell report, which established the Five Freedom's in the same year (Brambell, 1965). However, it was not until the early 1980s that animal welfare science really began in earnest, fostered by a small group of British scientists, the foremost of which were David Woodgush, Donald Broom and Marian Dawkins. The forum for such scientists to meet and discuss their new interest was the Society for Veterinary Ethology,[1] which had been started by the Edinburgh veterinarians Andrew Fraser,

[1] The science of animal behaviour

C. Phillips, *The Welfare of Animals*, Animal Welfare 8,
DOI 10.1007/978-1-4020-9219-0_8, © Springer Science+Business Media B.V. 2009

Alex Brownlea and Bill Jackson in the late 1960s. Although originally confined to veterinarians, the SVE was soon opened up to other interested scientists in the early 1970s, principally David Woodgush and later, Donald Broom and Ian Duncan. An early task was to convince people that animal welfare was a legitimate scientific discipline, particularly in some of the prestigious British universities where many of the key scientists held academic positions. Pain, anger and other terms that might imply emotion were rarely used at this stage (Broom, personal communication). Later the SVE became the International Society for Applied Ethology, the major international society dedicated to animal welfare issues, despite its obvious focus on animal behaviour.

The foundation of animal welfare science in ethology has had a profound impact on the way in which the science has developed. Although many would see components of veterinary medicine, animal physiology and animal psychology as being of at least equal importance, the fact that animal welfare science began with a strong connection to ethology has unfortunately sometimes distanced it from other relevant disciplines, in particular veterinary medicine. Ownership of the new discipline should not be claimed by any branch of science, and the dedication of the early founders of this should be respected for the breadth of scientific disciplines which they encouraged, not just their former interests. Thus, Donald Broom embraced agricultural science, animal physiology, philosophy and veterinary medicine in his research, Marian Dawkins has vigorously pursued animal psychology and cognition, and undoubtedly David Woodgush would have done the same, were it not for his early death. Since the pioneers developed and formulated the new discipline, the science has grown with the support of governments in 2005, non-government organizations and philanthropists. However, the number of scientists worldwide teaching and carrying out research in animal welfare on a permanent or semi-permanent basis was still less than one hundred in 2005, comprising 19 professors of animal welfare, mostly in veterinary schools, 12 other professors who mainly work in animal welfare and 45 other active scientists (Broom, 2005). There are also philosophers teaching ethics of animal use and veterinarians and animal scientists teaching animal care/husbandry. Most of these 76 scientists working directly in animal welfare were appointed after 1995, and the majority of the appointments, for example 13 out of the 19 professors of animal welfare, are in the English-speaking countries: the United Kingdom, Canada, the United States of America, New Zealand and Australia. The concentration of animal welfare professors in veterinary schools results from a desire by many, including politicians, to closely connect animal welfare science to animal health. However, although most research by veterinary faculty members is aimed at diseases that adversely affect animal welfare, there is a separate, more holistic focus for animal welfare research in the 21st C, which embraces sociology (understanding attitudes to animal welfare), philosophy (examining the moral basis for welfare provision), nutrition and reproduction (as major animal needs), fundamental biology (for strategic advances) and psychology (for the mental state of animals). Veterinary schools are traditionally narrowly focused on clinical disorders and pathology, especially infectious diseases, which is understandable because even this is a large remit given the number of species covered. Ultimately animal welfare study will probably stand alone as a university faculty, and attempts to centralize research in one or two locations in each country are a step in this direction.

Growth in Animal Welfare Science

As public concern for the welfare of animals has grown, there has been an increase in attention given to the science, which is seen by many as the best solution to the increasingly entrenched positions occupied by the animal rights lobby and those who make their living from animal management. Care must be taken that the animal industries do not hide behind the need for scientific evidence before making changes, because this takes many years to gather, and in the absence of robust science, there still may be a justification for change on the basis of public opinion, or the opinion of those knowledgeable of the industry.

The scale of the recent increase in science is evident from the increase in the number of publications on animal welfare or wellbeing in scientific journals, reviewed for the RSPCA in 2005 (Phillips, 2005a) (Fig. 8.1). Of course, there have been many articles written that are of relevance to animal welfare, but

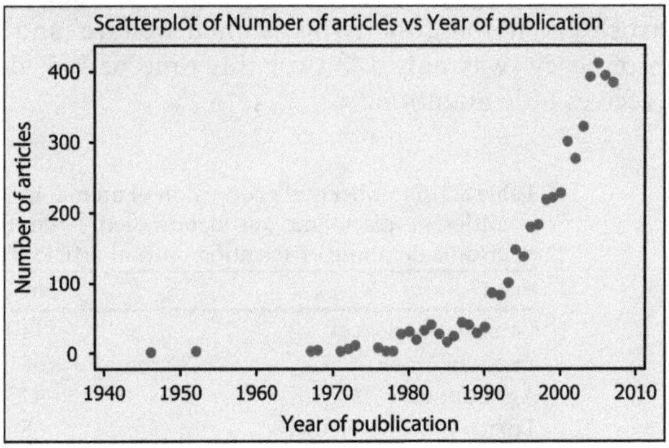

Fig. 8.1 Number of animal welfare articles[2] in the electronic database of scientific journal articles, Web of Science, 1946–2005

[2] All articles containing 'animal welfare' in the abstract or key words in the electronic database of publications in scientific journals, Web of Science. Note that in America the term 'wellbeing' is sometimes used in preference to 'welfare', but in this search only nine articles contained 'animal wellbeing' and not 'animal welfare'. In addition some articles address animal welfare without specifically mentioning the term, but this survey demonstrates the increase in popularity in animal welfare science as an entity.

which do not explicitly mention the term 'animal welfare' or 'animal wellbeing', but there is still an underlying trend of increasing popularity of the discipline, and the graph shows an exponential increase starting about 1987. The reasons for the increase in public concern for animal welfare are much debated (e.g. by Rollin, 2003), and these have been considered in chapter 6.

The different research disciplines represented by articles on animal welfare, as classified by the Web of Science, are listed in Table 8.1. Despite the origin of the animal welfare science being in animal ethology, veterinary sciences are responsible for over half of the publications, with zoology and domestic animal science (including ethology) accounting for most of the rest. A small minority are devoted to ethical and economic issues.

The number of publications in the areas defined by the Five Freedoms, that are commonly used to describe animal welfare needs, is presented in Table 8.2. Since veterinary sciences are well represented in animal welfare publications, it is not surprising that the majority of publications are devoted to *health aspects*, and to a lesser extent *behaviour*. Investigations of *fear* and *discomfort* are few, perhaps because they are less easily identified as serious welfare problems in animals, compared to disease and behaviour. *Nutrition* is also not well researched in relation to animal welfare, despite the long standing record of farm animal nutritional investigation in the countries leading animal welfare research. This may be partly because its impact on animal welfare is less obvious than disease, but also because farm animals are generally well fed or they would be unproductive, albeit often with foods that differ markedly from those that they evolved to consume. The other major animal 'need' is reproduction, although it is partially covered under 'normal behaviour'. The number of articles containing the terms 'animal welfare' and 'reproduction, breeding or prolificacy' was only 188 over this time period, demonstrating that the area receives little attention.

Table 8.1 Number and proportion of animal welfare scientific papers in different disciplines, as determined from the Web of Science electronic database of scientific journal articles from 1945–2005

Field	No.	%
Veterinary sciences	1110	55.5
Zoology	481	24.1
Agriculture	473	23.7
Toxicology and medicine	80	4.0
Environment	79	3.9
Food science and technology	69	3.5
Behavioural sciences	57	2.9
Ethics	52	2.6
History and philosophy of science	45	2.2
Psychology, biology	33	1.7
Ecology	26	1.3
Other	157	7.8

Table 8.2 Number and proportion of animal welfare scientific papers in the Five Freedoms (and respective key words), as determined from the Web of Science electronic database of scientific journal articles from 1945–2005

Freedom from pain/injury (pain, injury, disease, health)	4793
Freedom to perform normal behaviour (behaviour, behavior)	2599
Freedom from fear and distress (fear, distress, stress)	1286[3]
Freedom from hunger and thirst (nutrition, nutrient, food, eat, hunger, thirst, drink)	1247
Freedom from discomfort (discomfort, comfort, space)	486[4]

The language of most (85%) articles on 'animal welfare' was, as expected, English, however, the significant proportion of articles in German (12%) is worthy of note, given that the proportion of articles in other disciplines that are in German is less: nutrition (4.6%), animal reproduction (2.5%), toxicology (2.0%). This shows a focus of attention to animal welfare in the German-speaking countries.

There are many journals in which animal welfare articles were published (Table 8.3), with the majority being in two journals specializing in animal welfare and ethology: Animal Welfare and Applied Animal Behaviour Science. Veterinary journals published a significant number of animal welfare articles, but the number in traditional animal science journals (Journal of Animal

Table 8.3 Number and proportion of animal welfare scientific papers in different journals, as determined from the Web of Science electronic database of scientific journal articles from 1945–2005

Journal	No. papers	%
Animal Welfare	354	17.7
Applied Animal Behaviour Science	120	6.0
Veterinary Record	91	4.5
Deutsche Tierarztliche Wochenschrift	84	4.2
Alternatives to Laboratory Animals	82	4.1
Journal of the American Veterinary Medical Association	42	2.1
Acta Agriculturae Scandinavica	41	2.1
Australian Veterinary Journal	40	2.0
Journal of Agricultural and Environmental Ethics	39	1.9
Alternativen zu Tierexperimenten	36	1.8
Laboratory Animals	35	1.8
Journal of Animal Science	34	1.7
Zuchtungskunde	29	1.5
Animal Science	27	1.4
Livestock Production Science	27	1.4
Other	170	7.7

[3] Only 424 if stress omitted
[4] Only 114 if space omitted

Science, Animal Science, Livestock Production Science etc.) was small. This may reflect a reluctance of some editors in these traditionally production-orientated journals to accept scientific papers on the new science of animal welfare or a reluctance of authors to relate their work to animal welfare in the titles or abstracts of the article.

Many organizations were responsible for conducting the research in animal welfare (Table 8.4). The preponderance of British and other northern European institutions is evident, as well as the lead taken by universities in this field, as compared with research institutions. The only major institution in the southern hemisphere publishing in this field was the University of Massey in New Zealand, although this has now been joined by a substantial commitment to animal welfare research in Australia. There has therefore been a more progressive attitude to animal welfare science in northern Europe than elsewhere (Broom, 1992). Conditions, attitudes and types of production systems are very different in the two hemispheres, and this will determine the optimum welfare status for the animals.

The types of animals that were the subject of the animal welfare research are listed in Table 8.5. The major focus was on farm animals, with cattle, pigs and poultry featuring most strongly. The quantity of research on cattle and sheep is surprisingly large, given that most public concern is focused on the pig and poultry industries because of the intensive housing systems and because of their size compared with, for example, the cattle and sheep industries (see Chapter 8). The focus on farm animals, compared to other types of animals, may reflect the fact that research facilities and personnel were already established for the purposed of increasing productivity in Northern European countries, particularly after the Second World War when these countries had been subjected to food shortages. The changed directive of many of these research units is away from research focused on increasing farm productivity towards sustainable

Table 8.4 Number and proportion of animal welfare scientific papers from different organisations, as determined from the Web of Science electronic database of scientific journal articles from 1945–2005

Organisation	No. papers	%
Univ. Bristol	68	3.4
Univ. Edinburgh	49	2.5
Univ. Utrecht	47	2.4
Royal Vet & Agr Univ.	42	2.1
Univ. Oxford	38	1.9
Swedish Univ. Agr. Sci.	34	1.7
Univ. Cambridge	33	1.7
Danish Inst. Agr. Sci.	30	1.5
Scottish Agr. Coll.	29	1.5
INRA, France	25	1.2
Massey Univ.	24	1.2
Other	222	10.3

Table 8.5 Number and proportion of animal welfare scientific papers using the different animal types,[5] as determined from the Web of Science electronic database of scientific journal articles from 1945–2005

Animal type	Number of articles
Food animals	
Cattle	557
Pigs	506
Poultry	413
Sheep	270
Goats	21
Total	1767
Entertainment animals	
Racing	457
Zoo	78
Circus	8
Rodeo	1
Total	542
Experimental animals	
Mouse	114
Rat	27
Total	442
Companion animals	
Horse	109
Dog	78
Rabbit	45
Cat	39
Hamster/gerbil/guinea pig	27
Total	298
Fibre animals	
Fur	69
Mink	61
Wool	16
Total	137
Wild animals	116

[5] Words used to define animal types, in conjunction with the generic term
Cattle: calf, steer, bull, cow
Pig: piglet, swine, sow, boar
Sheep: lamb, ewe, ram, wether
Poultry: chicken, hen, cockerel, chick
Goat: kid, nanny, billy, caprine
Cat: kitten
Dog: puppy, hound
Horse: equine, stallion, mare, colt (not racing/race)
Rabbit: doe, buck, bunny
Hamster, gerbil, guinea pig
Mouse: mice, murine

farming systems that do not damage the environment and take account of consumer demands for improved welfare of the animals. Such established research capabilities have not been available for other types of animal use, such as zoo or companion animals, even though the public have considerable concerns about the welfare of these animals. The research capability in farm animals has matched the concerns by industry that consumers may be influenced in their buying habits by welfare issues, and therefore more research funding is provided by farming than other animal industries. Hence the attention paid to farm animals by the scientific research community is greater than that paid to entertainment, experimental, companion, fibre and wild animals together (Table 8.5). Of the other areas, only racing animals commanded significant attention, which was predominantly research on horses.

Identifying the most important animal welfare problems that industries have, does not necessarily indicate where changes should be made, since the most serious problems may have greatest commercial impact. For example, the perceived biggest improvement in the welfare of cattle might be reducing the duration of transport, but this would render production in remote areas difficult, therefore more achievable goals, such as improving nutrition, might be more appropriate targets. The research sponsored by the farming industry is therefore largely directed at making the greatest improvement in welfare with minimal impact on production efficiency. Some topics that deserve research attention because of their significance to welfare may be avoided because of the potential impact on production efficiency. However, if the public find a particular practice offensive, for example mulesing of sheep in Australia, there may be sudden changes demanded following campaigns by the animal activist groups. Industry must be proactive in researching alternatives to any practice that is likely to have a major impact on welfare, as well as being aware of public concerns and how they change over time. Regrettably, only a small minority of scientists are proactive in animal welfare research, attempting to work on areas that *they* believe could bring dividends, because they are reliant on research funding from industry. This reflects the change in Western universities from curiosity-led research to that driven by ambitious goals for the scientists, in terms of research grant income, and full cost recovery by the university. This progress will not necessarily lead to the most rapid improvement of animal welfare, because strategic research is likely to be scarce, with short term welfare problems dominating the research conducted. Research on some issues may be so far from identifying a solution that industry is unwilling to commit funding. Such is likely to be the case in the under-researched Freedoms identified in Table 8.2 or species identified in Table 8.5.

Animal welfare research has the potential to generate further concerns by the public, if evaluation of welfare issues demonstrates greater impact on welfare than hitherto believed. Therefore industries may avoid important areas because any changes recommended following such research could reduce the profitability of the animal enterprise. Unscrupulous industries may even attempt to use research to justify existing practices. An element of animal welfare research, at least, should be under the control of bodies other than those industries that are directly involved, for example, government, the universities and animal welfare organizations. Governments should be aware of significant welfare problems that industries avoid because of their commercial significance or the considerable funds required to provide solutions. They should maintain an overview and some control over funds provided for animal welfare scientists from the industrial sectors, particularly if they are matching the industry funding with public money.

Impact of Animal Welfare Science on Animal Welfare

Despite this significant and continued increase in scientific activity in the field of animal welfare, it would be rash to assume that this is having a major impact on animal welfare. Our attitude towards animal welfare is as much governed by our gender, culture and education as it is by the science that has been conducted to evaluate and improve standards of animal care (Thompson et al., 2007). In the short term, giving advice to those looking after animals is likely to bring about the greatest improvement (Anderson, 2007), however, in the long term a systematic and prolonged research effort will produce significant dividends (Blokhuis et al., 2000). The areas in which research is most likely to have a significant impact are amending or redefining legislation, improving the codes of practice or assurance schemes, and improving management practices involved in the husbanding of animals. It will also sometimes have an indirect impact by informing the public and research funders of latest developments in issues, which may then translate into additional funds or concern expressed by the public.

Independence of Research Effort

The animal industries are constantly being challenged by welfare activists. However, although their concern is well meant, they have different views on what constitutes the most serious welfare issues, focusing on invasive practices, compared to those directly involved in the industry, who focus on longer term issues (Wojciechowska et al., 2006; Cross et al., 2008a, b). It is likely that stakeholders in the industry are better informed on the most pertinent issues to challenge animals within their care, although it is possible that some of them become inured to the pain and stress involved in the most invasive procedures over time. The ultimate test of whether practices are harmful to animals must be by well-planned, co-ordinated research, and in many countries welfare research in specific industry sectors is now managed by a research and development corporation funded by industry, often by levies raised from industry members. These research and development corporations may receive at least matched funding from government as part of the effort to develop a state-of-the-art industry. However, many welfare activists' perceived improved welfare systems for animal management would be expensive both to effect the change and to maintain the system into the future, and so the animal industries have a vested interest in procuring research that demonstrates that their practices do not actually cause damage or harm to the animals. Researchers may be under pressure from their university or institute to generate research income from external sources and should have strict ethical standards to preserve their integrity as animal welfare scientists. The following principles are a minimum set of ethical guidelines for animal welfare scientific research work:

- The principle research activities should have animal welfare and/or ethics as the primary focus
- The results of research should be in the public domain, particularly if supported by public money, and if necessary after a time for industry to prepare for any necessary changes
- Research should not be conducted if it causes animals to suffer unnecessarily
- An independent, evidence-based approach to welfare issues is essential
- Publicity for the research should be approved by the scientific leader, and if necessary by the sponsor of the research

Industry Funding of Animal Welfare Research

There is concern nowadays that industry is corrupting university research, which may be relevant to animal welfare considerations:

> We must ensure that some areas of science remain free of commercialization. It must not be a requirement that all research 'pays'. The independence of science has values that are measurable in ways that are not financial. Scientists themselves have to speak out when commerce corrodes disinterest. Then we can renew the presumption that scientists are honest, disinterested and incorruptible.
>
> Kennedy et al. (2007).

Scientists used to be highly respected for their integrity. In the days of Victorian scientific endeavour, the financial rewards to scientists were small. Most had other means of support, and it was considered an honourable pursuit for clergymen or explorers. Darwin waited until the last possible moment to publish his most famous treatise on evolution (Darwin, 1857), knowing that he was not alone in his discovery. The delay was due to his sense of responsibility to the population of the United Kingdom. With admirable foresight, he knew that his results had profound implications for religious believers and possibly beyond, and he wanted to delay the onset of the attention that he knew would detract from his work. His motivation was clearly to advance science, not his own scientific standing. However, nowadays the system of rewarding scientists only if they produce major publications has fostered some unethical practices, such as repeat publishing of the same results.

In many developed countries, industry funds a significant proportion of animal welfare research. Some of this may be to increase profits, for example by control of diseases or an economically viable increase in productivity due to alleviation of stress. However, welfare research could also decrease profitability of farming units, for example a discovery that farm animals need more space, even though they do not produce more product, would potentially improve welfare and decrease profits if implemented. Industry is most likely to fund research in which both animal welfare and industry profitability are likely to be increased. They are less likely to fund work in which profit is reduced if welfare is increased, for example research on stocking density of animals during transport. This will be increasingly unlikely to occur as scientists develop all the methods of improving welfare that do not involve reductions in profit. Of course even if profit is reduced in the short term, in the long term a better market may be accessible if welfare is improved, such as to consumers paying more to purchase products from animals kept in high welfare systems.

Second, industry may fund research because changes to the production systems are being demanded by animal welfare activists, arguing that such changes cannot be made without scientific evaluations of welfare impacts. A major research programme can take ten years to complete, but is necessary to provide information for an informed decision. Thus in some countries, particularly those in which the industries have major economic significance, governments and industry may be reluctant to use expert opinion to establish standards, when no scientific information is available. European research on animal welfare is the most comprehensive, but is often seen as irrelevant outside of this region, because it was not conducted in the climate/situation of the country in question.

Relationships Between Farm Profitability and Animal Welfare: the Lessons for Animal Welfare Research

The level of animal welfare on production units is significantly affected by the quality of animal management, which also will have an important influence on the profitability of the unit. The most important animal management influences on welfare are the quality of stockpersonship, the infrastructure for animal keeping (buildings, handling facilities etc.), and the resources available for providing for the daily requirements of the animals, in particular food, water and veterinary attention. The industrialisation of animal food production industries over the last 50 years has made it imperative that there is a legislative framework for the maintenance of adequate animal welfare, as well as regular inspection of animal units, but this is a last resort, to deal with extreme cases of animal mismanagement. Regulatory control is particularly difficult because of the diverse nature of the systems used to produce animals, and hence the various interest groups are partly dependent on animal welfare being improved by good husbandry and the improved profitability that will normally ensue. On a broad scale, husbandry extremes do lead to direct welfare correlates, for example a diseased animal will be less profitable than a healthy one, taking into account loss of productivity and treatment costs. However, there are numerous examples of systems or components of systems that are both unsupportive of good welfare and yet return good profits to the producer. Battery cages for chickens and veal calf production are two of the best known examples, but there are also systems that regularly experience breakdowns in animal welfare when adequate resources and environmental conditions are not provided for the animals, such as barn and free range systems of egg production or feedlots for beef cattle. These still may be profitable systems at all other times. Furthermore, there are numerous practices that are the most economic option, but they do not necessarily lead to good welfare, such as early weaning of pigs and calves. Finally there are many practices that lead to a temporary reduction in welfare but in the long term will increase welfare as well as preserving the economic output of the system, such as some invasive procedures and preventative medicine, which entails catching and treating the animal, with all the associated stress. This emphasizes the need to research welfare measures over a long duration, and preferably the life time of the animal. The adverse effects of long-term stress on immunocompetence, particularly leucocytes, can clearly be demonstrated (Lin et al., 2003; Stoyanchev et al., 2007; Tuchscherer and Manteuffel, 2000). However, although they could adversely affect animal health, the hygienic conditions operated in many forms of intensive animal production, with limited exposure to external pathogens, mean that often animal health is not be affected.

The relationship between management practices and welfare is therefore not simple, but it is the belief of many that substantial improvements in animal welfare could be obtained on most animal farms by better research directed at management. Attention to detail, good observational skills, a thorough knowledge of animal keeping technologies and a dedication to the job are all essential attributes of good stockpeople that will improve conditions for animals in farms (Hemsworth and Coleman, 1998).

Chapter 9
The Scale and Intensity of the World's Animal Industries

Intensity of animal industries and corporate interests – Differences between intensive and extensive animal industries – Scale of the animal industries – Intensification of animal production for food – Changes in companion animal management

Intensity of Animal Industries and Corporate Interests

As a result of industrialization of agriculture following the Second World War, the profitability of farming animals for food was generally greater in the second half of the 20th C than the first (Clements, 2006). Although profit per farm increased over this period, it has been argued that decreased profit per animal may have resulted in reduced animal welfare (Fraser, 2005). Fraser argues that pressure from retailers reduces the product price, which reduces profit per animal and forces the farmer to reduce inputs. He cites two examples of welfare apparently being worse in low input outdoor pig production systems than intensive indoor production: first, a survey that found more lameness and shorter longevity of outdoor sows in Croatia, compared with indoor sows (Cox and Bilkei, 2004), and second, the trend towards higher levels of neonatal mortality in oudoor piglets, compared with those reared indoors (Kerr et al., 1988), which was also observed in Cox and Bilkei's survey. However, these are only components of the system and if, as well as sow lameness and piglet mortality, freedom of movement, freedom from pain and injury and opportunities for social interaction were considered, it is likely that the outdoor system would be judged to lead to better welfare. In addition there is probably considerable variation between farms in their welfare provision, particularly in outdoor systems with variable environmental conditions and a need for advanced management skills.

The increasing net worth of farms since the Second World War offers farmers the opportunity to invest to improve welfare. This is unlikely to involve reversion to traditional systems, which were usually based on low levels of mechanical input. Technical advances have enabled animal management to be automated and labour input to be reduced.

C. Phillips, *The Welfare of Animals*, Animal Welfare 8,
DOI 10.1007/978-1-4020-9219-0_9, © Springer Science+Business Media B.V. 2009

Intensification has been most actively embraced by corporate enterprise, which for example used their corporate investment potential to adopt intensive housing effectively in the latter part of the last century. This led to increased profitability and the emergence of a new industrial farming sector, based on corporate enterprise. This sector is now expanding rapidly, at the expense of family businesses, and it brings a new economic imperative to the animal industries. Even if automation can effectively improve animal management, the profit-driven enterprises can pose a challenge to animal welfare in the size of the operations, the potential lack of close contact between stockperson and animals, and the reliance on housing systems with relatively little space for its occupants. There is no universal truism that intensive systems are associated with low welfare and extensive systems with high, but if these challenges are not addressed an intensive system will inevitably lead to poor welfare. However, there are also occasions when large-scale corporate-owned livestock farms can utilize capital resources to sustain an enterprise when confronted with difficulties, such as drought, disease challenges etc, in a way that smaller family owned units could not. Multiple sites in a corporate beef enterprise, for example, facilitates the movement of cattle from one site to another when drought persists, when such a move from a privately-owned property would be more difficult, requiring negotiations to arrange agistment for the stock and the capital for the move.

Differences Between Intensive and Extensive Animal Industries

The principal differences between extensive and intensive animal industries are difficult to define. The most significant is that capital utilization is greater in the intensive system. The need for large amounts of capital to be injected into intensive systems tends to concentrate them into the hands of industry, which often has other assets and available finance. Small family operations, which were traditionally the normal systems of animal management in most countries, are increasingly rare. Occasionally this trend has been temporarily reversed, such as when the collective and state farms of the former communist countries of Eastern Europe and Asia were dismantled in the 1990s, and the land returned to the multiple previous owners (Rizov, 2003) However, following the rapid political change in these regions, small farms have exchanged hands regularly to make larger, more viable units. Extensive livestock production, in particular sheep, was a common path chosen by families with limited capital to start farming, but the low returns have resulted in farms rapidly being sold and combined into larger units.

Intensive units are more likely to be at the forefront of technology and engage in entrepreneurial activity in order to maximize profit, but they are also more likely to be indebted than small-holder farms. The latter are characterized by the multiple functions livestock serve, the integrated nature of livestock activities, the multiple objectives of the producers, as well as poor infrastructure, markets, and access to information at the community level (McDermott et al., 1999). The reduced reliance on labour inputs and increased mechanization in intensive units could be viewed as more efficient, but has the disadvantage of a reduced ability to employ people in sometimes deprived rural districts or forcing them to look for non-farm income to alleviate poverty (McDermott et al., 1999). The increased labour efficiency is likely to result in weaker bonds between the animals and the stockmen, although it is sometimes argued that employing labour-saving technology, for example automatic milking for dairy cows, allows labour to be more effectively used in assessing the stock and detecting problems earlier.

Intensive units aim to maximize animal output per unit of food, and often offer highly processed energy and protein rich food to achieve this, which does not support normal feeding behaviour. The concentrated energy and protein feeds could alternatively be used for human consumption, whereas in extensive livestock units the feed is more likely to be pasture based and produced in areas that could not grow arable crops needed for human consumption. Intensive units often use animals that have been bred for high productivity, which further necessitates them being fed an energy and protein rich diet.

Scale of the Animal Industries

If intensification is reducing the welfare of animals, compared with the improved extensive systems that are available nowadays, attention for welfare reform should be focused on the large scale intensive industries, as well as those increasing in scale and those increasing in intensity. Although the global position is not always well understood, Table 9.1 represents an attempt to quantify the scale of different animal enterprises where there is a direct impact on animals.[1] On a global scale, the most numerous land-based animal industry is animals used for farming, with many more pigs and poultry than cattle and sheep, and the number of pig and poultry units expanding in response to the demand for more meat in developing countries (Table 9.2). Pigs and poultry are more easily kept in intensive units and they utilize high energy and protein feeds more efficiently than the ruminants like cattle and sheep, which are better suited to utilizing rough grazing pastures.

Ruth Harrison, in her influential book *Animal Machines* (Harrison, 1964), as well as drawing attention to the industrialization of poultry farms, also commented that farm animals were much more numerous than companion or laboratory animals and therefore deserved more attention (Broom, 2005). As

[1] There are indirect impacts, such as disease introduction or environmental change, which are almost impossible to quantify. Those presented are impacts that are the direct and immediate result of human activities.

Table 9.1 Estimated world population of animals in the major sectors, whose welfare is directly and immediately impacted upon by humans (see text for sources)

Animal group	Estimated population
Farm animals	
Ruminants	3 billion
Pigs/poultry	17 billion
Companion animals[2]	1–2 billion
Laboratory animals	50–70 million
Zoo animals[3]	1 million
Work/entertainment animals	100 million
Wild animals (annual mortality)	
Wild caught fish	70 billion
Bycatch	10 billion
Cat kills	12 billion
Road kills	1 billion
Death by habitat loss	Unknown
Total	Approximately 120 billion

well as numbers of animals involved, it is important to consider the extent of human involvement. For those whose focus is animal rights, interest will be focused on animals that are managed by humans, such as farm animals, and less on wild-caught animals, such as fish, that are affected by humans only

Table 9.2 World livestock populations in 1992 and 2001. From CAB International (2002)

	Population 1992	Population 2001	Percentage change
Donkey	43.7×10^6	42.0×10^6	– 2.29
Water buffalo	153.1×10^6	167.6×10^6	+ 9.50
Cattle	1302.8×10^6	1360.1×10^6	+ 4.41
Goat	597.6×10^6	693.5×10^6	+ 16.0
Horse	60.3×10^6	58.3×10^6	– 3.40
Pig	869.6×10^6	927.7×10^6	+ 6.68
Sheep	1161.3×10^6	1059.1×10^6	– 9.80
Camels	18.2×10^6	19.1×10^6	+ 5.17
Other camelids	5.41×10^6	6.20×10^6	+ 14.5
Mules	15.1×10^6	13.5×10^6	– 10.4
Rabbits	519.7×10^6	481.6×10^6	– 7.33
Rodents	$22.9 x 10^6$	14.9×10^6	– 34.9
Chickens	11500×10^6	14700×10^6	+ 27.3
Ducks	652.1×10^6	915.4×10^6	+ 40.4
Turkeys	223.4×10^6	241.4×10^6	+ 8.04
Geese	163.5×10^6	240.6×10^6	+ 47.2
Beehives	57.9×10^6	58.9×10^6	+ 1.68

[2] A restricted definition of companion is used here, including animals that provide friendship, but excluding those, such as ornamental fish, that are kept primarily for display purposes.
[3] The number of wild animals in 'rescue and rehabilitation' centres probably greatly exceeds the numbers in zoos, but would be difficult to quantify.

transiently at the end of their life. However, the impact of humans is not always negative; some animals are well cared for and derive much benefit from this symbiotic relationship.

Agricultural Animals

Cattle and sheep farms are usually less intensively managed than pig and poultry farms, with the latter usually housed in controlled environments. However, there has been a trend towards intensification of cattle production through the expansion of feedlot finishing in many of the major producing countries, with only a temporary retardation as a result of increased cost of high energy foods, because of their potential for biofuel production. Poultry units intensified in the latter part of the 20th C in response to demand for cheap meat and eggs and appear to have reached a point where further intensification, especially genetic increases in growth rate, will reduce the welfare of the bird, for example by increasing skeletal disorders (Tatara, 2006). The number of birds kept worldwide has expanded and also the size of individual farms. Some farms now house more than four million hens, and the majority of poultry production is in the hands of a small number of companies in countries such as the US, which have intensified the most. Here just 60 companies own nearly 80% of the country's birds, and many of these have fully integrated production systems, controlling the process from production of the food, growing the bird to slaughter, processing and marketing of the finished product. Increasingly, these companies operate internationally, taking advantage of reduced labour costs abroad to outsource some of their activities (USDA, undated). The intensification is driven by economies of scale that allow both greater mechanization and reduced purchase costs, of for example feed supplies, due the high volume being supplied. Modern egg processing machines can handle over a hundred thousand eggs per hour, which can only be provided by a large layer flock. To justify expenditure on such automation, US companies therefore have to manage several million birds (United Egg, 2007).

Some developing countries are only now establishing intensive farm animal production systems, which are replacing extensive, small-scale production in villages, even though they are not necessarily more biologically efficient or sustainable (Verhulst, 1993). However, in response to animal welfare concerns in developed countries, and strong economic competition between new entrants to the industry, extensification of individual farms, particularly in less profitable upland livestock farms seems likely (Matthews et al., 2006). There is much criticism of intensive animal food production industries, and a general belief that we should return to more extensive systems. This would probably benefit animal welfare in many cases, but we should not imagine that the extensive systems of producing food today are the same as the extensive systems used in the past. Indeed it is probably wrong to say that the movement from extensive to intensive management in the latter half of the 20th C necessarily caused a reduction in animal welfare, a theory advanced by Rollin (2006). In the past, animal farming was often devastated by problems such as high mortality (as a result of failure to control virulent diseases, like rinderpest and tuberculosis in cattle), chronic undernutrition during winter or drought (because foods could not be adequately stored), and cramped housing conditions, that caused the rapid spread of infectious disease, such as tuberculosis. So, for example, the mortality of the livestock on the First Fleet sent from England to Australia at the end of the 18th C was about 50% (see page 111). This may be an extreme example, but in most other areas of production agriculture the major welfare indicators have improved. Hence the new extensive systems, that for example meet the demands of the public for more space for their animals, should only be introduced if welfare is improved. In some circumstances, such as the transition from a battery cage system for laying hens to a free range system in a harsh climate, the welfare of some birds may actually be reduced by extensification Feather pecking and even cannabilism are common in free range systems involving up to 18% of birds in some genotypes (Kjaer and Sorensen, 2002; Sedlackova et al., 2004), and the birds may make little use of the outside area if it is too hot or cold or there is insufficient cover (Dawkins et al., 2003).

Intensification of agricultural production brings many threats, not just to animal welfare, due to the concentration of wastes onto small sites, which has led to the bizarre spectacle of manure being transported from northern Holland, where pig farms are sited so that they can benefit from cheap grain near the ports, to southern Holland, where the soil can tolerate the manure without toxic substances, such as copper, accumulating to dangerous levels (Westhoek et al., 2004). Another threat from intensive production is the reduction in biodiversity, as companies use only animals with genetics suitable for the economic climate of today. Short-term profit is maintained but the gene pool is reduced and changes in the future, for example a demand for a return to more extensive production, would leave the industry woefully lacking in suitable animals from which to choose. A most extreme example of this is poultry production, where fewer than two dozen companies worldwide control the production of the breeding (grandparent) stock (Hoffmann, 2005). These birds are developed for high egg production or muscle growth from energy-rich, cereal-based diets. If the cereals were required exclusively for human consumption, these birds would be unsuitable for less concentrated diets.

Most people do not support intensive production, for example a survey of Australians living in Canberra that showed that nearly three quarters of the public would support banning cages for poultry (Greens, 2007), but this often does not translate into purchasing preferences, when it comes to choosing from a variety of products (Heleski et al., 2006). There are several possible reasons. First the public do not yet understand that banning cages will inevitably increase cost. Second, the food purchasers in the family are prepared to sacrifice the moral standards provided for the birds for the financial benefit of their family. In the early days of intensification, the public could be excused for not knowing how chickens were kept in the new 'battery' farms. Nowadays, with greater publicity, particularly by high profile personalities, more use of video technology and major campaigns by the activist groups, few can be unaware that intensive production is believed by at least activists to be damaging the animals' welfare. A third possibility that food purchasing habits do not always agree with people's sentiment about production systems is that many people's food consumption habits tend to be conservative, emulating their parents' diet (Montonen et al., 2005). Changes in dietary preferences take time to evolve, but the recent increase in sales of food from animals with improved welfare standards is evidence that purchasing habits may indeed be changing (Hoogland et al., 2005).

Companion Animals

It is difficult to define the exact group of animals commonly referred to as companion animals. In essence the term refers to animals with which something is shared. This could just be time, in which case non-domesticated animals such as birds and fish would be included (CAWC, 2003). However, if companion animals are providing friendship they would include mainly domesticated animals, especially cats, dogs and horses, although there undoubtedly many individual examples of captive wild animals providing friendship in special circumstances. For the most part, fish, reptiles and possibly birds are kept for other reasons, the beauty and extraordinary nature of their bodies and songs, their behaviour and in the case of some birds, their ability to mimic humans. Determining populations of non-domesticated companion animals is difficult and has necessitated that a restricted estimate of numbers, assuming that the global dog and cat populations to each be approximately 40%, with 20% other animals, principally horses. Others estimate the proportions as 36% dogs, 32% cats and 32% other for the US (AVMA, undated; Belotto and Silva, 2006). About 80% of the cats and dogs worldwide are strays (Butcher, 2006; Davies, 2006).

Total worldwide companion animal populations are much less than for farm animals, estimated to be 150 million in the United States, 500 million in Europe and 100 million in the China (WSPA, undated, data from 2002, the European Union, 2005, and Li, 2004, respectively). In China, the population of companion animals is predicted to expand rapidly because of strict control of the number of human offspring in families and the people's growing wealth. The official press agency of the government predicted a five-fold increase between 2004 and 2009 (Li, 2004). In many countries there is a change in the management of companion animals, which resembles the intensification of the farm animal production. It takes the form of less contact with the owners and less space for living, and arises because first, within a family it is more likely that both parents will be working than in the past, and second, people are moving into high density housing with no garden or land for the animals to use, which means that they are often confined indoors all day by themselves. In addition there is increasingly less opportunity to exercise dogs off leash, with small areas being devoted to such activity in the major towns and cities and restrictions being placed on use of open land and beaches.

The large numbers of animals euthanased in shelters each year continues to be a major cause for concern. Estimates in the United States vary from 3–4 million (HSUS, 2005) to 10 million (AH, 1997) cats and dogs euthanased annually.

Laboratory Animals

The number of laboratory animals worldwide is hard to estimate because countries record the different species used in different ways and some record none at all. A few record all animals used. Clearly the total number used annually is much less than the number of companion or farm animals. Recent estimates suggest that Australia,[4] UK, Canada and Italy respectively use annually approximately 5.8, 2.6, 1.7 and 0.9 million vertebrate animals annually for research, with a total for 15 European Union member states of 11.6 million animals in 1996 (Bayvel, 2004; Gauthier, 2004; Passantino et al., 2004; Anon, 1996). In Japan the data collected from universities, institutes and laboratory testing companies suggested that just over 10 million animals were used in 1995, but the number recorded had supposedly dropped to 5.6 million by 1998, mainly because concerns about adverse publicity caused many institutes involved in vaccine development to withhold information on the numbers of animals that they were using. In addition the Japanese government prevented stray dogs and cats in pounds from being used in research. In the USA only the number of non-human primates, cats, dogs, rabbits, hamsters and guinea-pigs used for research is recorded, which is perhaps only 4% of the total, that must include many mice and rats. Thus the total use of these six species in the USA, currently about 0.8 million per year, suggests a total vertebrate use of approximately 20 million per year.

Developing countries, particularly in south-east Asia, are increasing their animal research, partly because they are becoming more involved in medical research and partly because some scientists are moving there because stringent regulations are making it difficult to work freely and rapidly in developed countries where the public are more vocal about the animals' welfare. In some countries, such as the United Kingdom, governments are attempting to limit numbers to assuage public concerns, and there are reports of declining numbers of animals being used for research being since the 1980s in the UK, since about 1985 in the US and since the early 1990s in Canada (Gauthier, 2004). One

---

[4] In Australia the definition of animals used is likely to be more broad ranging than other countries, with a greater chance of double counting of animals by different institutions.

analyst believes that animal use for experimentation increased exponentially between 1910 until 1970 due to the development of biomedical disciplines, then declined due to greater public awareness, increased legislation and better quality of animals used in laboratories until the mid 1990s (Baumans, 2004). After this time, he believes that it has increased due to the increased use of large numbers of animals for genetic modification experiments. Although his sources of information are not clear, his total use of animals for research, at 75–100 million per year is similar, but slightly higher, to the estimate presented above. The United Kingdom publishes annual statistics on animal use, and these have recently shown small (1–2%) increases annually (Hudson, 2007). This is believed to be due to large-scale genetic experiments.

The most common laboratory animal is still the mouse, accounting for about two thirds of all procedures (Hudson, 2007). The number used increased in the 1980s due to their extensive use in molecular biology experiments, then in the 1990s decreased as in vitro models became more popular. More recently, in many institutions the use of mice for the production of transgenic mutants has been expanding due to technical developments (Gauthier, 2004). In this research thousands of animals are used in individual experiments in the hope that a few valuable mutants will result from genetic modifications. Previously only a few tens or hundreds of animals would be used for a single experiment.

Zoo Animals

Zoo animals are less numerous than most other forms of animal use, there being 1200 core zoos worldwide, with an estimated 1 million captive animals in total, and about 3000 vertebrate species exhibited in total (IUDZG/CBSG, 1993). Many of these contribute to the International Species Information System (ISIS), which involves 613 institutions from 70 countries on six continents. Members keep and share standardized information on more than 1.8 million zoological specimens of 10,000 taxa, but this includes invertebrates (World Association of Zoos and Aquariums, 2005). The number of animals worldwide may be increasing as zoos attempt to keep self-sustainable numbers for captive breeding and release programmes. However, there is also public pressure on zoos to treat animals well, which may result in some zoos keeping fewer species with additional space and enrichment for each. There are also many animals in sanctuaries and shelters, apart from companion animals, which it would be difficult to quantify because most are small and unregulated.

Utility Animals

Utility animals, for work or entertainment, number about 100 million, which are mainly horses used for agricultural work. The mechanization of the agricultural industries in recent years is reducing the number of animals used for work, principally horses, but also some donkeys and cattle (Table 9.2). There are smaller numbers of animals used for entertainment, but the welfare and ethical impact can be extreme, for example the approximately 10,000 bulls that are slaughtered annually in bull fights worldwide (Catan, 2007). In addition to the obvious cruelty of the fight itself, the animals are selected at an early age by challenging them to determine their levels of aggression. They are kept on extensive pasturelands, so that they have little contact with humans, and their reaction to the matedor in the fight is then all the more valiant if they have not been subjugated by humans before. In some countries, such as Portugal and France, the bulls are fought but not killed in the ring, although they are slaughtered immediately afterwards. Extending the period of severe pain and cruel treatment in this way probably has an overall negative effect on the animal's welfare.

Wild Animals

Wild animals obviously number many billions, but only a proportion has their welfare impacted by man. This includes animals killed or maimed by vehicles, hunted animals, wild caught fish and those whose habitat has been affected or even destroyed by man.

Road kills are mainly mammals and birds, many of which are killed while feeding off other kills. Reptiles and amphibians are underrepresented. In Britain alone, an estimated 50,000 badgers, 100,000 foxes and 10 million birds are killed or maimed on the roads each year (Born Free Foundation, 2007). There are also many cats and dogs maimed or killed in road traffic accidents. In one Australian survey kangaroos were killed at a rate of 0.03 deaths/km/day on a major road, attracted to it by the proliferation of food supply on the verges of the road (Klocker et al., 2006). Most of these are killed at night, because the kangaroos freeze when spotlighted by a car's headlights.

To get a global figure for road kills, it is possible to relate the numbers killed in Britain to the distance vehicles travel on the roads. The car population in Britain is approximately 33 million, and vehicles travel about 500 billion km annually (Optimum Population Trust, 2007). Scaling this up worldwide, the number of vehicles is between 600 (Anon, 2007b) and 750 billion (Optimum Population Trust, 2007), and they are driven about 6500 billion kilometers annually. By this crude method of estimating the worldwide number killed or maimed on roads annually, it is probably about 130 million birds and 2 million mammals. However, in the US alone it has been estimated that about 365 million vertebrates are run over each year (Anon, 2007a), suggesting that the British figures are an underestimate. Also, with population growth and the number of cars increasing at about 2% annually, it is predicted that the number of car kilometers traveled annually worldwide could reach 70,000 billion by

2050, which would increase annual road kills/maims to at least 1.4 billion birds and 22 million mammals. In the face of this uncertainty on road kill statistics, it would seem likely that at least 1 billion vertebrates are currently killed annually on roads. In addition to the animal welfare issues that surround every road accident to an animal, the trauma and financial consequences to owners when domestic animals are killed or maimed on the roads is very significant. Animals that survive become nervous and frightened of cars; their owners let them outside less; most owners consider the emotional consequences to be severe, in comparison with the less serious financial consequences (Rochlitz, 2004).

Many countries sanction widespread slaughter of feral animals, in the belief that it helps to control the population. They may be slaughtered by shooting, hunting, mustering or the administration of poisons. Although statistics are hard to obtain, it is likely that several million rabbits, at least a million pigs, and smaller numbers of goats, horses, buffalo, donkeys, camels, foxes and wild cats are killed annually in Australia. The systematic killing of wild animals for food is perhaps most advanced in the case of kangaroo culling in Australia. Approximately 3 million are harvested annually, from a quota of almost 6 million (1.5% of the population)(RSPCA, 2002b). Commercial shooters are licensed and in some states the kangaroos must be shot in the head if they are to be sold commercially. The greatest concern is for the welfare of the young at foot and pouch young that will die when their mother is shot. Nevertheless, this probably represents a less significant impact on welfare than kangaroos that are maimed by vehicles which do not stop (RSPCA, 2002b).

Although these numbers of animals may seem considerable, they are small compared with the numbers of wild fish caught for human consumption. The commercial fishing quota is declining, but is still very substantial. United Kingdom commercial fishermen alone caught over one million tonnes of sea fish in 1997 (Parnell et al., 2000). Worldwide the total tonnage of wild caught fish, including shell fish, in 2001 was 92 million tonnes, five times greater than in 1950 (Vannuccini, 2003). At an average weight of 1.3 kg each (Karpov and Albin, 1995), there are approximately 70 billion fish caught annually. Most of these die by asphyxiation on the deck of the ship. The most numerous species are anchovy, pollock and mackerel. This does not include the bycatch, fish that are too small or of the wrong species for consumption, which may be discarded, made into fish paste or fishmeal for animal food. The weight of fish and other animals returned as bycatch is about 8% of the fish actually harvested, i.e. about 8 million tones globally each year (Kellcher, 2005), but they are smaller than the main catch, so the number of animals affected is greater. Bycatch species that are returned to the sea have a high mortality rate, as much as 50% for some species. Some progress in reducing the bycatch was made when driftnets were banned internationally in 1992. The impact of the bycatch on the ecology of the area and the biology of the species concerned is considerable (Dayton et al., 1996). Some countries are beginning to implement sustainable fishing policies that include a guiding principle that there will be no discards. It is not just the land animals affected by habitat destruction, but trawler fishing has enabled considerable numbers of fish to be harvested and has been very damaging to the marine environment, especially where the nets are dragged along the bottom of the ocean. In addition, the welfare impact of commercial fishing is not just to the fish, since porpoises and other mammals are caught in the nets. Sea birds, especially diving birds, are killed by gill nets, and trawling along the bottom of the oceans has produced widespread destruction that must be affecting the entire ecosystem. Other human activities impact on the welfare of marine life. The underwater noise generated by shipping, and in particular the naval forces, is suspected of interfering with the navigation and communication systems of the higher mammals, such as whales and dolphins.

Although commercial fishing is declining, recreational fishing and fish farming are increasing. Recreational fishing has been increasing as people in developed countries have more leisure time, and in developed countries such as Australia and Canada it is estimated that approximately 17–25% of the population participate in this sport (Hardy-Smith, P. personal communication; Anon, 2008c). In Florida alone, the number of angling trips has increased from 2 to 5 million per annum over the last 20 years (Florida Fish and Wildlife Conservation Commission, 2007). The major welfare impacts are the pain induced by the hook and the pain endured between capture and death. The latter depends on species, with eels for example it is particularly slow. As well as utilization of fish for food and recreation, there are fish used for ornamental purposes and fish held in aquaria in restaurants before being killed and served fresh to the customers, a growing trend in many regions of the world.

In addition to animals that are deliberately killed by man, there is secondary killing by animals owned and managed by humans. The world domestic cat population, at approximately 0.5 billion, is responsible for the killing of several billion wild animals, mainly birds, rodents and amphibians each year. In the United States, it is estimated that there are approximately 90 million domestic cats and a similar number of feral ones. These together kill hundreds of millions of birds, and more than a billion small mammals, such as rabbits, squirrels, and chipmunks, each year (ABC, 2007). The killing of wildlife by cats has received much adverse publicity in Australia, particularly because they kill some endangered native animals such as tree frogs. It is estimated that the average Australian household cat kills 25 creatures a year; a total of 100 million creatures every year in the entire country (WIRES, 2007). Feral cats eat the equivalent of 7 bush rats each week, over 400 million creatures a year. The adverse publicity given to hunting by cats in Australia is believed to be partly responsible for the declining cat population. In Australia there are approximately 3 million pet cats and 12 million feral cats, giving an annual total slaughter of perhaps 0.4 billion native animals (Queensland Parks and Wildlife Services, personal communication). Scaling this slaughter up to worldwide populations, there could be about 12 billion vertebrates killed each year by cats. Roughly 60% to 70% of the wildlife that cats kill is small mammals; 20% to 30% are birds; and up to 10% are amphibians, reptiles, and insects (ABC, 2007).

The number of wild animals whose welfare is affected by humans or their companions is very considerable and probably much greater than that of the next largest sector, the farm animals.

The Human Footprint on the 'Silent Majority' of Animals

The annual population of sentient animals whose welfare is directly affected by man, at about 120 billion, is clearly a majority compared with the global human population of c. 6.8 billion. Each person on the planet potentially affects the welfare of about 18 animals each year. Given the significant overall impact of our actions on animal welfare, it is not surprising that animal welfare activists worldwide are increasingly concerned about this 'silent majority' of animals whose welfare needs better protection. We have the potential, the knowledge and the resources to manage animals in a better way, and the constant plea from the activists is that our 'footprint' on the animal kingdom is considerable and we need to make sure that it is a more positive one. In terms of the subject of attention, the focus on farm animals is logical if both the number of animals affected and the considerable welfare impact of many husbandry practices are taken into consideration. However, more attention should probably be given to wild animals, particularly in the marine environment, even though the welfare impact is not yet well understood.

Intensification of Animal Production for Food

In order to focus on the most important issues in animal welfare, it is important to consider not just of the size of the animal sectors and the welfare impact of human management practices, but also how the different groups are changing over time.

Agriculture has undergone the most rapid period of intensification of any of the animal industries, because of the opportunities provided by mechanization and the economic benefit to consumers of increased efficiency. This has been a progressive change since the start of settled agriculture, but it accelerated in the latter half of the 20th C to match the escalating growth in population and personal wealth, which created a strong demand for high quality food products from animals. England was one of the first countries to intensify its agriculture and it is worthwhile considering the pressures that prompted this small country to seek to change its systems of production in response to public pressure for cheap meat and milk.

The industrial revolution, which started in the 18th C, was a major stimulus to the early mechanization of agriculture, which in turn became the key to increased output (Crafts, 1985). One of the precursors to this in England was a shortage of land. The British government passed several Enclosure Acts in the late 18th and early 19th C, which restricted the rights of the people to graze their livestock on common land. Enclosing this land enabled it to be used to grow more cereals to feed the expanding population. At the same time, land use was intensified through mechanization, and Britain became a major exporter of agricultural machinery in Victorian times. The recipient countries were mainly countries that were in the process of colonial development, such as India, which were encouraged to pay by supplying food back to the home country. Even this did not keep pace with increasing demand in Britain. By the start of the Second World War, Britain was only about 40% self sufficient for food production, with a population approximately half that of today. British agriculture was in a moribund state, with a shortage of labour after the losses of manpower in the First World Way, derelict land and a market undermined by cheap imports from overseas. The Depression of the late 1920s did not have an impact on agriculture until later, and grain prices were at their lowest level in 1934. During the Second World War, the poor state of British agriculture was recognized by the Germans as the Achilles heel of the island people. Indeed, the German U-boats nearly succeeded in starving Britain into submission by preventing shipments of food from crossing the Atlantic. Food imports from overseas, that had hitherto been taken for granted, dried up because of the blockade. The shortage of food supplies led to rations on staple foods being imposed in 1940, initially just bacon, butter and sugar, but then all meat and flour.

The farmers of the day rose to the challenge, with assistance from the 'Land Girls' and prisoners in the later years of the war, and the mentality of intensifying food production started to be instilled into the island race. The principle strategy for increasing food production involved ploughing up grazing lands for the production of cereal crops, thereby utilizing fertility that had accumulated in the pre-war years, when much land was fallow or underutilized. Farmers were given quotas of crops to grow, including such staple foods as potatoes, but also cereals that were grown for livestock to increase their milk or meat output. The nature of British farming was changing rapidly. In the pre-war years, cows were usually kept at pasture during winter. Hence farms had to be kept in permanent pasture to withstand the pressure of the cows' hooves during the wet months of winter. More productive temporary leys were badly damaged during wet weather. With increased land required for crop production, farmers began to keep their cows inside during winter and feed them hay and cereals. The cows were tethered throughout this period, even though the restriction on movement could make them lame.

After the war, the shortage of food persisted for several years in Britain, and food rations were not lifted until 1953–4. Trading industrial goods for food from the colonies was rare in post-war austerity. At the final lifting of meat rations in 1954, prices escalated because of limited supply. Most of the prisoners of war and land girls had left the land, and many of the demobbed men from the armed forces went to farm, because of the shortage of other jobs. These men were not trained in agriculture and were often reluctant to accept advice from labourers. In times of such rapid change new farming methods often evolve.

Intensification continued through the 1960s, with the introduction of group housing for pigs and cages for laying hens. In the late 1970s, government moved to ensure that Britain would never again be vulnerable to food shortages by investing heavily in agricultural research and encouraging farmers, through grants, to increase the intensity of their production systems. A comprehensive agricultural development and advisory service was established, which had evolved from the War Agricultural Committee that controlled production during the war years and immediately afterwards. Agricultural education thrived. Government grants were made available to improve farms, for example by removing hedges between fields, providing housing for animals and access roads for hill stock. The emphasis was on increasing production with little consideration for either the environmental or animal welfare consequences of farming methods.

In the 1960s intensive housing units were first developed so that animals could be more productive, but it was not until the 1970s and 80s that most farms began to adopt them. This included cages arranged in 'batteries' or rows of similar units for laying hens, in which the hens were grouped 5–6 in each cage, with the eggs rolling out of the cage for easier collection, and the faeces falling through the wire floor. Intensive breeding and fattening units were constructed for pigs, with farrowing crates to restrain the sows, creep feeding for the piglets and verandah units (with indoor and outdoor accommodation) to fatten them into porkers or baconers. More sheep and cattle were kept indoors during the winter, which gave better control of the feeding and management, but also gave rise to lameness and behaviour problems because of lack of space. Male cattle that had previously been raised in the fields, after castration to control their aggression, began to be kept indoors without being castrated. This increased their growth rate, even though it meant that potentially aggressive and dangerous animals were being raised in small, confined spaces. They were prone to riding one another and developed other sexual behaviour abnormalities, but there was not sufficient interest in animal welfare issues at the time for any control to be considered.

In addition to intensive housing, high energy and protein feeds were manufactured from cereals and other quality feed sources for feeding to the farm animals, which responded by growing faster, producing more milk and laying more eggs. Piglets were weaned from their mother after just three weeks, so that more than two litters could be obtained from each sow every year. Cows were fed concentrated energy and protein supplements that could bypass their rumen and increase production still further. For winter feeding hay came to be replaced by grass conserved by a natural process of acidification – silage – which had higher feeding value because it could be cut at a young and green and transported directly into storage systems. Previously hay had had to be cut at a mature stage and then dried in the field to ensure that it would not go mouldy when formed into stacks. Ensiling grass and other crops was made possible by mechanized harvesting, necessary to transport the wet grass, storage in pits and towers and distribution by machine to the animals.

Dairy cows were an exception to the intenstification drive, as they were not as intensively managed in the post war period as some had been in early Victorian times. Before railways were developed to transport the milk to cities from rural areas, there were cow keepers in the all the major centres of population. These usually had 8–10 cows each, housed in cellars, tied day and night and fed hay and concentrates (Lea, 2005). They were often the older cows whose milk was no longer good for cheese production. They were not mated, so after about a year their lactation had declined to an uneconomic level and they were walked out of the cellar, for the first time for a year, to be slaughtered. Such city production continues in some developing countries today.

During the industrial revolution, as well as milk produced in the cities, cheese production was developed in the western parts of Britain, where the grass grew well and the product could be stored before being taken to the cities for sale. Stocking rates at pasture in the mid 19th C were typically about 2–3 acres per cow, each animal producing approximately 250 lbs of cheese per year. By contrast in the 1980s, with inexpensive fertilizer to increase grass growth and supplementary feeding of concentrates, cows were stocked at about one per acre and each animal was expected to produce over 1000 lbs of cheese per year, a tenfold increase in output per acre compared with 130 years ago. Multinational fertilizer companies were producing large quantities of artificial nitrogen fertilizer to allow farmers to sustain the high stocking rates, and farmers were being encouraged to spread up to 3–400 kg of nitrogen on each hectare of land. However, the high stocking density caused damage to the land in winter, particularly in wet conditions, and winter housing became preferred for high productivity. Intensive stocking in summer resulted in significant health problems from parasites on the pasture, which could be easily transmitted from one animal to another. The winter housing also produced many health problems: lameness from standing on concrete all day, mastitis from lying in bedding contaminated with faeces and swollen joints from lying in uncomfortable stalls, as well preventing the normal foraging behaviour of the cows. Hence this tenfold increase in productivity per acre in little over 100 years was achieved at the expense of the welfare of the cows, but it was a profitable system of production, as long as nitrogen fertilizer was cheaply available, and it reduced the milk price to the consumer and increased profit for farmers. Government advisers were actively involved in the intensification of British dairy farms, attempting to assist farmers to get the most from their land, whilst ignoring the cost to the cows.

The development of milking parlours in the mid 20th C allowed cows to be free in their shed during winter, instead of being tied in individual stalls. Dairy cows that had been milked in their stalls came to be milked in these parlours, usually twice a day, but sometimes three times if milk prices were good. Milking machines of the sort that are used in parlours were first developed at the start of the 20th C, but the economic climate was not right for their widespread adoption until the 1960s. Before this the farmer had had to take the milking unit from one cow to the next in a byre, or mobile milking station, which could be transported to the field where the cows were grazing. The development of parlours – static milking stations to which cows were taken – probably represented a welfare benefit of mechanization for housed cows, since it enabled the cows to have more freedom of movement to display natural behaviour. More machinery was needed in the parlour than in a portable unit, but these became increasingly sophisticated until about 100 cows could be milked per hour, thus reducing the labour requirement for this task.

In early 20th C Britain, the government controlled the markets for most of the major animal products, for example the Milk Marketing Board, which was established in 1933 and offered a guaranteed price for all the milk that farmers could produce. This incentive for expansion of output continued until Britain became self-sufficient for milk (with an agreed importation of New Zealand dairy products included) in the 1980s. At this time, dairy cows were often fed so much cereal-derived concentrate, in an attempt to maximize milk production, that they developed digestive upsets and the so-called 'production diseases' of acetonaemia, fatty liver syndrome and low milk fat syndrome. Such was the drive to increase milk yield per cow that the Milk Marketing Board produced league tables of farms in each region that produced the most milk. Some control of digestive upsets was achieved by feeding the cereals in small amounts regularly over the day, by mixing it into a complete diet with forages, or by rationing cows by providing them with an electronic key to control their access to the feeder. If the cereals were over-processed or were fed in large amounts at milking, cows got acidosis – acidic conditions in their rumen – which badly affected the micro-organisms in their digestive system and could result in the cow's death. The new feeding methods for high milk yields were pioneered by Professor Boutflour at the Royal Agricultural College in the late 1940s, but did not come to be in widespread use for dairy cows until the early 1980s. The intensification of the production of beef cattle came earlier, driven by inexpensive availability of calves and cereals and growing demand for beef in the early 1970s. Some farmers began to fatten cattle intensively indoors on a diet of just cereals, despite the metabolic disturbances and ill-health that often caused welfare problems (Preston and Willis, 1974). The calves grew very fast, and were allowed to eat as much concentrated food as they wanted, so that by nine months of age they were ready for slaughter, instead of the usual 18–24 months if they were fed a grass-based diet.

The driving forces behind the intensification of animal production in the late 20th C were clearly not aimed at improving animal welfare. Most of the impact of the intensification had adverse effects on welfare, since animals were pushed to their limits metabolically. Furthermore, ill health and boredom often resulted when the animals were kept in systems that only aimed to maximize the rate of output of animal product, be it muscle growth, milk or egg yield or reproduction. The biological system was often stretched to the limit by focusing on one aspect of production, in just the same way that an athlete pushes elements of his or her body. Dairy cows developed massive udders, which could only last a few years before the suspensor ligaments gave way and the cows had to be slaughtered (Brade, 2005). The bodies of beef cattle (Freudenberg et al., 2007), pigs (Barnett et al., 2001) and poultry (Oviedo-Rondon et al., 2006) grew so fast that their legs could not support them, and developed joint problems. Sheep were developed that could produce so many lambs that the mothers could not adequately rear them (Davis et al., 1993). Farming came to be managed more by big businesses, often with integrated chains that controlled all aspects from the growing of the animal food, to the slaughter in the abattoir, and the family farms began to decline.

The reason that these changes were not stopped was principally because animal welfare was, and to some extent still is, just a small part of the economic and ethical package in which farmers function. At the time of the recent major intensification of the British animal production industries, there were more pressing issues to deal with, national food security and human welfare in particular, so that animal welfare was barely a consideration when the movement was in full force. Later, when the threat to national food security had abated and human welfare was not at risk, animal welfare became a key consideration and developed into the focus of attention that it now holds.

Farming Skills

Farming is a diverse operation, taking into account the varied conditions, climatic, topographical and economic, of each unit. Families have benefited in the past from handing down local knowledge about farming practices from one generation to the next. Good stockpersonship is recognized as one of the most important influences on animal welfare. Any period of rapid change, such as the entry into farming of the demobbed soldiers after the Second World War or the intensification of the late 20th C, is likely to lead to some people managing animals badly, until the necessary skills have been learnt. For many centuries it was expected that the sons of farmers would take over the family farm when the father became too old to do the job, or they would take over a similar farm in the locality. In rangelands animals usually stayed on the farm when ownership was transferred from one farmer to the next, to maintain continuity and because the animals were adapted to the territory. The same was essentially true of the farming families. The opportunities for children with a farming background to enter other types of work were limited, and the standard of husbandry was maintained at a high level by the accumulated knowledge. Such knowledge is important for the management of the stock. It can be learned, but farmers' offspring learn most of what they need to know about managing the animals at home before they even leave conventional school.

In former times, the importance of this knowledge was recognized in leasing agreements for farms. In mid Victorian England, during the (first) agricultural revolution, a typical mixed farm would be about 50 acres, with approximately six cows, six steers, a few calves, four horses, three pigs, eight sheep, hens, ducks and geese and a total net value of just a few hundred pounds. Most farms were tenanted, a legacy of the feudal system that had predominated in mediaeval times, and three people could be named in the lease, usually a farmer and two of his sons, ensuring that farms could be transferred to the next generation easily. At this time, an initial entry charge of £ 110 was payable by the first tenant, which was the equivalent of about 50 cows, and then a small rent of about £ 8–10 per year had to be paid. To transfer the tenancy to the second person cost the family only their single, best beast, or increasingly as the century progressed a small monetary sum was required, about £ 3–5. Thus the system ensured that farms and the knowledge that had accrued were maintained between the generations. The landlord had much greater control over his tenants than today: a set number of days had to be worked free of charge for the landlord or a fine was levied, and in times of war every tenant had to provide a man and/or a horse, depending on the size of the tenancy. The stock that the tenant kept and the crops that he grew were dictated by the landlord and sometimes a proportion of the products had to be surrendered annually.

Life was hard for the 18th C farmer and most of his possessions, such as furniture and linen, were handmade on the farm using locally grown materials. His life was integrally connected with his animals; recreation included cock fighting, hare coursing etc, which today are considered cruel, but his interest in these pursuits reflected the constant struggle with nature that characterized his life and that of his family. At the same time, the bond between stockperson and the animals in his care was strong and engendered the empathetic attitudes to the stock that are often absent in large, intensive units today. This was vividly illustrated in Thomas Hardy's celebrated novel 'Far from the Madding Crowd', when the shepherd, Oak, saw all his sheep lying dead at the foot of a cliff after being chased by dogs: "Oak was an intensely humane man. . .A shadow in his life had always been that his sheep ended up as mutton – that a day came and found every shepherd an arrant traitor to his defenceless sheep. His first feeling now was one of pity for the untimely fate of these gentle ewes and their unborn lambs." (Hardy, 1902). Such views typify the strong bond between stock and stockman that developed when they spend long hours with their animals.

In the last quarter of the 20th C, the traditional system of farms passing between the generations and being managed by the whole family was breaking down. There was an increased standard of education amongst farmers, providing increased job flexibility. Families became smaller and there were fewer people on the farm to take care of the animals. On dairy farms, looking after the calves, or on sheep farms, the orphan lambs, had often been assigned to the farmer's wife or children. Then economic pressures for cheap food led to the farmer's wife often working away from the farm, and even the farmer having to take part-time work.

Whereas previously it was expected that a farming family would endure for several generations, by the late 20th C it had become common for farms to change hands several times within a generation. People now enter and leave jobs frequently, and since fewer people are needed in agricultural work and the cities offer the prospect of higher incomes and more leisure opportunities, rural depopulation has been a major problem worldwide (Westhoek et al., 2006). Agricultural training programmes have suffered from a lack of interested students (Pongratz and Schmitt, 1990), and more people are entering the industry in later years as part of a lifestyle change after working in cities. Such 'hobby farmers' are often lacking in knowledge about animal husbandry, but they usually have money to spend on the farm, including maintaining the health of their animals in conjunction with veterinarians, and their ethical standards may be higher than those of young, ambitious farmers who might be prepare to sacrifice animal welfare to build up the economic viability of their farm (Holloway, 2001).

The period of rapid agricultural intensification in Britain after the Second World Way was already coming to an end by the mid 1980s because the country was becoming self sufficient for most commodities. As a member of the European Community since 1972, Britain under the Thatcher government of the 1980s took the stance that agricultural subsidies, which continental European countries benefited from much more than Britain, must be reduced if the Community was to grow economically. Mrs Thatcher also fought hard to control the agricultural surpluses that were being produced as a result of the subsidies – which in her words demonstrated 'the Mad Hatter economics of the Common Agricultural Policy' (Thatcher, 1995). Subsidies were strongly favoured by the French and the Germans because of their large rural population that benefited from them.

For a period, the investment in intensive animal production units that had occurred during the 1970s and 80s continued to pay economic dividends, but as animal houses wore out and public demand for less intensively produced animal products increased, there was a gradual trend to replace these units with less intensive ones. So from the early 1990s until today, there has been a gradual movement towards less extensive animal production. This is not a reversal to production as it was in the 1930s, because the buoyant economy in Britain has created a population that is prepared and able to pay for high quality food. Governmental support is increasingly directed towards environmental management, rather than food production. Previously support for farmers depended on their stock numbers, so they were encouraged to overstock their farms, which then might not be able to produce enough food for all the animals.

Locally-produced food is now favored, and local marketing and farmers' markets, where the food is sold direct to the public, have gained in popularity. Although originally a small niche market, the local production of food to high environmental and animal welfare standards has only become possible after the liberation of the market place from centralized control, achieved by the Thatcher government of the 1980s. In less than 50 years, British agriculture had been transformed from almost total state control, with enforced cultivation of land, the grading of farms by the War Agricultural Committee according to production potential and centralized machinery stores operated by the

Committee, to a free market system that allowed farmers to produce according to the demands of the population. Such changes were not confined to the heavily industrialized countries. Similar changes in agricultural policy occurred in New Zealand in the 1980s, which although initially causing hardship for farmers, eventually produced an agriculture that was able to respond to market demand, including the demand for high welfare products. New Zealand dairy products are now marketed abroad as coming from grazing cows that are kept in a high state of welfare, which are recognized as a marketing asset.

As some cultures move rapidly towards high animal welfare provision, others struggle to evolve from their ancient past. Pressures on cultures to develop must be considered in the light of other changes taking place, principally pressure from a population that is three times larger than when our grandparents were born, and pressure to preserve and improve the environment. Aboriginal cultures still exist in Australia that survive mainly on 'bush tucker', comprising the numerous fruits and nuts of the forest, but also wallaby, goanna, and other small mammals that can be easily harvested. These communities mainly exist on the islands off northern Australia, and the harder they are to reach, the slower the adoption of western culture. Traditional animal management practices are under threat, such as the harvesting of sea turtles. Australian law requires that such animals should be stunned before slaughter, whereas the Aboriginal method of killing is wringing the neck or smashing the head with a rock. Exceptions to the laws are made for Aboriginal communities, but there is considerable pressure for common standards for all Australians. The apparent neglect of cats, dogs and horses in Aboriginal communities attracts the accusatorial eye of westerners, but derives mainly from a lack of ownership of the animals by individuals, in favour of communal ownership, and a reluctance to use modern veterinary medicine to treat skin diseases and other non-lethal ailments. Aboriginals reliant on bush foods have less footprint on animal welfare. The average Western person eats over two tons of fat, nearly two tons of protein and eight tons of cholesterol in his lifetime. Bush animals do not have their welfare interfered with by man, only the time of death. The impact of the Aborigine is much less, but the balance of nature is fragile – commercialization of such bush tucker harvesting practices would potentially destroy their symbiotic relationship with animals.

Industrial Impacts on Animal Health and Welfare

There are many short-term activities on farms that challenge animal welfare, and these often receive the most attention from animal welfare activists, transport of animals for example. These events are finite, but what is often not realized is that the industrial revolution has left a legacy of polluted land that will have an impact on animal health and welfare for centuries and which is very difficult to redress. Some of the issues have been recognized and addressed, for example the influence of PCBs in the environment on the viability of birds' eggs (Fernie et al., 2000), but many are only just becoming evident. The accumulation of heavy metals on pasture land has been a recent cause for concern (Wilkinson et al., 2003), with much criticism of the former Communist countries of Eastern Europe for heavy emissions. Cadmium accumulation from smelter emissions and phosphate fertilizers has recently become a cause for concern, and the longevity of both cadmium and lead in the soil is considerable, particularly lead which will not be leached for hundreds of years. Both of these metals have potentially serious adverse effects on grazing animals. Rumen fermentation of pasture grass is disrupted, kidney function is likely to be impaired in animals surviving to maturity (such as horses), because of the long half life of the metals in the body (30 years for cadmium). Many domestic grazing animals do not live long enough for their health to be affected, but wild animals such as deer are increasingly found to have kidney function damaged by cadmium (see review by Phillips and Prankel, 2008).

Changes in Companion Animal Management

Because of the strong emotional bond with companion animals, people often display more concern for their welfare than farm, laboratory, wild or other animals. Indeed demands for increased welfare standards for farm animals may be influenced by the members of the public's increasing concern for the welfare of companion animals. Although such generalization of attention to companion animals to concern for farm animals may be true in many cases, there are noted exceptions. Ernest Hemingway's affection for cats was well known, but he delighted in the spectacle of the bull fight and his favourite pursuits were big game hunting and deep sea fishing.

The keeping of companion animals has been subject to different patterns of change in the various parts of the world, which means that there is no overall pattern of increasing intensity as there is with farm animals in most parts of the world. However, there is one almost universal truism, that the trend towards high density living is forcing dog and cat owners to leave their animals indoors for longer, and in some cases permanently. In heavily industrialized or regulated countries such as Japan and Australia, there are limited off-leash areas to allow the dog to run free.

Dog keeping has become specialized as the industry has grown over time. It is now recognized that there are many different reasons that people keep dogs and these need to be taken into account by matching the dog's characteristics to the owner's needs when shelters rehome dogs or puppies are purchased. These include companionship alternatives, in particular child, partner or friend substitute; exercise for the owner; protection of the owner's property, or their person (people also may own dogs to help them avoid contact with other people); breeding, either to sell puppies or to hire out a stud dog; showing, to be good examples of a breed, to bring monetary reward or recognition to the owner; an outlet for nurturance instincts (humans, especially females, are believed to have an innate need to nurture others); sport/recreation (dog sports are becoming more popular and include agility and obedience trials, lure coursing, endurance competitions, fly ball and scavenger hunts); utility (dogs often assist their owners, such as guide dogs for the blind, deaf or disabled, herding of farm animals, or act as security guards, and the keen olfactory sense of dogs is used in immigration controls); social connectivity facilitators (some people like to keep a dog to help them to integrate into society, they may facilitate conversation when walking) and finally animals may be kept as a status symbol (some persons seek to enhance their reputation through animal ownership, perhaps because of its beauty, value, danger or ferocity).

As society becomes more complex and people adapt to high density living, there is a need for everyone to conform to a sustainable, recognised method of keeping companion animals. Hence individuals should not be allowed to keep large numbers of pets, treat them badly or ignore their needs. A uniform set of standards has rarely been available, let alone enforced in the past, but nowadays it is more likely that the behaviour of an individual dog or cat will impact on people around them. Regulatory control of the number of pets is now common in Australia, but the specified constraints, for example animal numbers per household, will depend on the locality. In Brisbane the maximum number of cats or dogs allowed is three per household (BCC, 2003), on the Gold Coast of Australia the number varies with the size of the property, if 600 m^2 or less only one dog is allowed, otherwise two, and only two cats are allowed on each property (GCCC, 2007). Birds are similarly regulated, a property of less than 300 m^2 can only accommodate four small birds, e.g. budgerigars. A larger property may be used to keep twenty small birds and four big birds, such as Galahs, whereas only one is allowed on properties of up to 4000 m^2. Thus there are increasingly common regulations on companion animal ownership that prevent them from being kept in very intensive conditions.

Conclusions

Intensification has affected both the farm and companion animal industries, but intensification of interaction with wild animals, although not considered here also has had a major impact on welfare. The number of animals whose welfare is affected by humans is much greater than the number of people on the planet, with each person potentially directly impacting on the welfare of about 20 animals each year of their life, emphasizing the considerable responsibility that we all hold for management of animals on the planet.

Intensification has had negative effects on many aspects of animal welfare, even though it is not the only reason that people are increasingly concerned about welfare issues. Animal welfare concern is greatest and advances are most easily made when the economic conditions are favourable and the technical knowledge is available. The potential to improve animal welfare in intensive units exists, but to date the main focus in intensive units has been to increase productivity of farm animals.

Chapter 10
Animals in Research

Historical background – ethical assessment of animal use in research – ethical dilemmas – genetic modification of organisms – xenotransplantation

Historical Background

Concern for the use of animals in research has been evident since they first began to be used for this purpose in modern Europe, beginning in the late 17th C (Pocard, 1999). An emerging view of some philosophers, most notably Descartes, was that animals were just machines. This was not a common belief, but was used by a minority of scientists to justify their use of animals for research (Preece and Fraser, 2000). Many scientists at this time, including Isaac Newton and Robert Boyle, opposed the treatment of animals for such research, as did many philosophers and writers. The concern was strongest in the United Kingdom, where the experimentation was most advanced and the populace had a burgeoning concern for animal welfare. It was here that the first legislation in the world to protect animals in research was enacted in 1876, which was utilized through most of the 20th C (Pocard, 1999).

Ethical Assessment of Animal Use in Research

As a result of expansion of medical science, animals are now used extensively for research on a world-wide basis. The welfare standards and ethical issues surrounding the use of these animals are a focus of attention for activist groups, particularly where the use of the animals is for research that is not directly aimed at improving the health of humans or animals, such as the testing of cosmetics. The response of most governments in developed countries, and an increasing number in developing countries, is to have the animal experimentation monitored by either government agencies or institutional bodies or a combination of the two. This formal ethical assessment of the use of animals for research is a response to the concern and has evolved in parallel with

C. Phillips, *The Welfare of Animals*, Animal Welfare 8,

DOI 10.1007/978-1-4020-9219-0_10,

assessment of the use of humans for medical research. Regulation of animal use in research is generally more advanced than that for humans in research, probably because of the activities of animal advocates against animal experimentation (Schuppli and Mcdonald, 2005). In the United Kingdom and Germany there has been a major reliance on government monitoring, through the Home Office in Britain, which is now augmented by institutional bodies which provide a preliminary assessment of the ethical viability of the research (the Ethical Review Process) (Bradshaw, 2002). The Ethical Review Panels of institutions must include a named veterinarian and representatives from the Animal Care and Welfare Officers. By contrast, in Australia, Canada and the US there has been primary reliance on institutional monitoring, although in Australia this is now augmented by government auditing at a State level (Schuppli and Fraser, 2007). Many developing countries are now establishing institutional monitoring of this type. Systems of institutional monitoring, which can include members of the public, and is accompanied by government auditing of the process in some instances, has emerged as the most popular system, probably because the major cost and responsibility is then passed back to the organization conducting the research. Comparing the United Kingdom and Australia, the systems of assessment were until recently quite different, with the UK relying almost totally on government inspection and Australia relying on institutional monitoring (Bradshaw, 2002). Both have broadened the scope of the monitoring and now use institutional and government bodies for approval. In developing countries, animal research standards are usually voluntary and individual institutions may introduce their own standards, administered by an ethics committee, for example in some South Korean universities. Some countries, such as Singapore and Iran, are consulting the Australian Code of Practice for the Care and Use of Animals for Scientific Purposes (2004) and adopting the institution-led model for ethical assessment. The development of adequate assessment methods is particularly important in developing countries, as increasingly more research is conducted there to benefit from the low costs of labour and laboratories, as well as there being fewer, if any, legal restrictions. In the past, developing countries have trained many of their best scientists in the Western countries, some of whom are now continuing their work at home. Unless these countries adopt similar requirements for research work, so that it lies within acceptable ethical standards, there will be an added attraction to working in developing countries, of reduced regulatory control. However, it is necessary to understand the cultural differences in attitudes to the use of animals in research in these countries, as the concerns of the European public may not be shared in Asia, for example (for evidence of this, see Phillips and McCulloch, 2005).

Although government monitoring is cumbersome, time-consuming and expensive, it should guarantee high standards of assessment that are repeatable. Government inspectors are trained in the administration of government regulations and therefore can be held accountable for the quality of the work that they do in monitoring projects. They understand the ethical principles upon which decisions are made. A key principle is that there should be substantial gain to either animals or humans if any animals are to suffer as a result of the experimentation: a utilitarian perspective. Research aimed at cosmetic or frivolous changes in animals is unlikely to be allowed, for, example breeding animals to have extraordinary features purely for the amusement of the public. The inspectors also recognize that sometimes the impact on the animals will be too severe, regardless of the benefits to humans or animals: an animal rights perspective.

In contrast to regulation by government bodies, institutional monitoring depends largely on untrained individuals, whose beliefs are likely to be governed by their upbringing and cultural identities. A study of the views of animal ethics committee members in Canada has revealed significant differences in attitudes and approaches of individual members (Schuppli and Fraser, 2005), for example on whether it is the role of committee members to address reduction in the number of animals requested to be used in experiments. Some members felt that it was not their role, that numbers were irrelevant providing that the optimum procedures were conducted. Some members believed that they had insufficient expertise to suggest changes in numbers; others believed that scientific review by grant-awarding bodies adequately fulfilled the process, and others believed that the cost of using animals in experiments would naturally keep numbers to a minimum. Few believed that it was should ever be necessary to recommend increased as well as decreased numbers to avoid animal wastage, which is the reality of inadequate experimental design. Refinement of techniques, one of the three R's,[1] was very poorly understood, with a variety of views as to what it meant. Some members employed a scale of sentience in their own minds, which for example made it preferable to use a hamster than a rabbit, which in turn was preferable to using a cat for research. Such review practices are evidence of a lack of understanding of the complexities of ethical review by the panel, and in this case directly contradict biological evidence, for example that a cat is no more sentient than a rabbit. Avoidance of pain is clearly not uniformly addressed, or even advocated. Environmental enrichment is not supported by many scientists, because of the effects on experimental results (Hubrecht, 2000), even though there is considerable evidence that abnormal results will result from situations where animals lack a suitable environment (Poole, 1997).

Such discrepancies between members are understandable, since they come to the committees with varied backgrounds, but it is apparent that members may have widespread misconceptions which can only be addressed by a rigorous training. Regrettably, the review by Schuppli and Fraser (2005) does not engender confidence in the process. Different outcomes may emerge from different committees and there is currently little attempt to harmonise the

[1] Replacement, Reduction and Refinement, advocated as a framework for the improvement of laboratory animal research and hence of central importance to the ethical review process (Russell and Burch, 1959)

processes or results. This could be done by training or standardisation of recruitment procedures for committee members, and also by a reviewing body. The lack of training of committee members is evident; some institutions provide training to new researchers, but this is often superficial and covers only the basics of ethical practice in animal research. A few institutions sponsor committee members to attend relevant conferences, but in Schuppli and Fraser's study, a lack of familiarity with guidelines on research ethics was particularly evident, with some committees sanctioning LD 50 tests,[2] for example, in contradiction to international guidelines.

The method of recruitment is crucial to ensuring that committee members represent the public's point of view (Schuppli and Fraser, 2007). However, all too often it is hard to find volunteers for these positions, and people with extreme views and vested interests may find their way onto the committees. Payment may bring people to the committees for the wrong reasons. Volunteers can be chosen as representatives of the major stakeholders in the process: animal laboratory staff, animal scientists, veterinarians and animal welfare organizations, but these may not reflect the public viewpoint. They may be inherently more caring, having been attracted to work with animals, but they may also have become inured to some of the more invasive practices. Members of the general public may also serve on institutional committees, and are likely to have a wide variety of opinions, depending on their upbringing and circumstances. Adding one or two members of the public to the committee may make the decision-making process more difficult. They will probably have more disparate views than the scientists or veterinarians on the committee, who have all been through similar training programmes (Schuppli and Fraser, 2007). Most committees place considerable importance on the views of those supposed to represent informed public members when it comes to ethical decisions. However, if discussion focuses on the scientific conduct of the experiment, as often it does in institutional committees, the lay members can feel ostracized because of their lack of experience in experimentation (Schuppli and Fraser, 2007).

Australian animal ethics committees contain one representative from each of the following groups – animal scientists, veterinarians, animal welfare organizations and the public. The administering body for the committees, the National Health and Medical Research Council, stipulates that members of the last two categories should not be numerically outnumbered by the animal scientists and veterinarians. There is no official representation from animal laboratory staff, although these are often co-opted as non-voting members or in attendance. Because of the lay membership on the committees, applications by scientists to conduct work must be written in plain English. Inevitably, much of the discussion in these committees focuses on the science to be undertaken,

---

[2] An index of toxicity (*lethal dose 50%*), the amount of the substance that kills 50% of the test population of experimental animals when administered as a single dose

for example the impact of specific procedures on the welfare of the animals. The chairman is often also an institutional animal scientist, although it is generally advocated that they should ideally be from another discipline (Australian Code of Practice for the Care and Use of Animals for Scientific Purposes, 2004; Schuppli and Fraser, 2007). The scientific discussion helps to refine the experimental protocol, so that the work is scientifically sound and imposes the minimum welfare cost to the animals. However, there is much less, if any consideration of whether the benefit of conducting the work, to humans, animals or theoretical knowledge, is justifiable on ethical grounds, i.e. the benefit outweighs the cost to the animals.

Some guidance is given in the Australian Code of Practice for the Care and Use of Animals for Scientific Purposes (2004) on making decisions on the ethical acceptability of research work, suggesting that the benefits to humans or animals must outweigh the cost to the animals involved in the experimentation. This therefore advocates that the decision should be made on utilitarian grounds, i.e. favouring the outcome which produces the greatest good for the greatest number of individuals. However, there are concerns that utilitarianism may support some unnatural decisions, that members of the public would not make, especially major impacts on the welfare of individuals for a relatively minor benefit for a large number of animals. Most people would prefer to preserve the welfare of the experimental animals, even if it meant that an opportunity to improve welfare of a large number of animals had to be foregone. Many people with no vested interest in the process would take the view that all sentient animals are the subject of a valuable life, and that welfare cannot be compromised significantly for the benefit of other animals or humans. The contrast between this view and the views of some scientists, who might wish to make sure that the path to experimentation is easy, could prevent non-scientists from expressing their views openly. Alternatively assessors who are scientists may be influenced by professional competitiveness, which is targeted at specific individuals, or alternatively could be keen to defend the rights of scientists and animal experimentation. Some veterinarians, because of their training, are less likely to acknowledge mental than physical suffering in experimental animals (physical symptoms of disease tend to be considered more important than the mental aspects of animal suffering in veterinary courses, Rollin, 2006).

Volunteer assessors from outside the institution, who represent animal welfare organisations or the general public, may be concerned if the institution profits from the work undertaken, and this may influence their judgement on applications. Sometimes, institutions take on the role of assessing applications from other nearby small companies or research establishments. These companies may have only a profit motive, unlike academic establishments which usually have a learning objective. Some assessors within the university system may dislike research which is conducted purely for profit and be less likely to approve it. Institutional rivalries may surface as well. If a university is assessing applications for other organisations, it is essential to ensure that the latter are inspected regularly by the university's or government authorities and that they are subjected to the same rigorous process as those within the university, especially if the university benefits financially for providing ethical approval for the animal research. The training of research personnel in small establishments may be more difficult to achieve than in a large university. If a company is profiting significantly from the research, institutional members may feel that the process is a misuse of the panel's time, which is supposed to be supporting the university. Scientist members may consider that the company is providing competition to their own research, or those of colleagues within the university. From the company's point of view, especially if they are entrepreneurial companies set up with venture capital, they may be using more novel technology, which they believe the universities are not adequately able to assess, xenotransplantation for example. A different attitude towards scientific research often pervades the two types of establishment. In some countries, institutions are establishing their own committees, and representation must be fair and open to scrutiny, as well as the activities of the committee.

Ethical Dilemmas

Some ethical concerns on the use of animals for research have their parallel in the ethics of experimentation involving humans. Increasingly, human medical research is being taken to developing countries, where patients will be willing to take unproven medication if there is a chance that they will be cured (Abbas, 2007; Kemp, 1996). By conventional ethical standards, the researchers should offer the control patients the best possible treatment that has been medically proven. However, this adds considerably to the cost and researchers accept that they may improve the lives of future users of the drug if they identify a useful treatment, even if the control patients have no benefit. The moral dilemma is whether an ethical panel assessing the morality of such experimentation should support the rights of the patients to receive best treatment or the rights of a future population to effective medicine? If they insist that the control treatment should be best available practice, the work may never be done, because it is too expensive. Similarly control animals in case:control studies should theoretically be given best practice treatment, as long as it is based on scientific evidence. This becomes an issue when panels have to assess applications in which the objective is to improve the conditions of animals housed in less than optimum conditions, such as some zoo animals, animals in live export etc. Although we may know little about the actual husbandry standards offered to such animals, requests to learn about the conditions and seek to improve them may not be favourably judged by those that are fundamentally opposed to such practices.

In this era of globalisation, there is a likelihood that the more contentious and risk-taking animal research will migrate from developed countries, with their increasingly stringent standards, to developing countries. The best solution to this dilemma is extensive collaboration between developing and developed countries in producing and administering the standards for animal research. Those few people championing the pursuance of ethical standards for animals in research in developing countries should be given every support possible from the developed countries, because they are at the vanguard of animal protection, whereas in developed countries animal advocates already receive considerable public and other support.

A suitable compromise might accept that utilitarian values are useful to judge most applications (Schuppli et al., 2004; Schuppli and Fraser, 2005), so that allocating some animals to be transported in a ship would be acceptable even if no improvement in their conditions was being tested, as long as animals would benefit in the future from the knowledge gained. However, there are extremes examples of offences to animals which cannot be allowed under any circumstances, whatever the benefit to humans or animals (Schuppli et al., 2004). Little is known about precisely which these offences are, but they probably include severe mutilation and other affronts to an animal's integrity, vivisection, trials that result in the painful death of animals and trials that involve the creation of extreme anxiety or fear in animals.

Much of the laboratory animal research is far from public view, unlike the rearing of farm animals or the management of companion animals, which is exposed to many. The public rely on the appraisal systems for the maintenance of their ethical standards. Without adequate and regular appraisal of experimentation by trained government assessors, there is a possibility that some institutional appraisal will be superficial and without the knowledge of trained officials. People on institutional assessment panels usually act voluntarily, and many are dedicated in their service to the committee. However, the system will only work if the institutions support the appraisal panels, with training, secretariat etc, so that each application can be rigorously and promptly considered.

Another ethical dilemma is whether the committees should be primarily dealing improving the design and analysis of experiments, assuming that their objectives are ethically valid. Usually it is rare for members of the panel to have received training in ethical decision making in relation to animal experimentation, but they will be faced with ethical dilemmas, such as applications using modern genetic technologies with large numbers of animals, only a small minority of which will produce successful mutants. Such research is justifiable from an economic and scientific perspective, because the mutants produced may be extremely valuable for medicine, but with the loss rate being much greater than in traditional experiments, the ethical validity of the work should be carefully considered. Once an ethical case has been accepted, there should be careful scrutinisation of the number of animals actually used and destroyed after the experiment, but unfortunately statistical justification is often rare or vague (Schuppli and Fraser, 2005). The increased use of animals for genetic modification, largely rodents, demonstrates a reversal of the trend towards reduced numbers of animals used for laboratory research at the beginning of the 21st C (Schuppli et al., 2004).

Many experiments do not use the correct number of animals to test their hypotheses (McCance, 1995). Two thirds of articles published in the Australian Veterinary Journal, for example, show statistical flaws, and nearly 10% used too few animals to prove or disprove their hypothesis (McCance, 1995). The work may still be useful to be published if repeated experiments can be linked through a statistical combination of several experiments, or meta-analysis (Phillips, 2005b). However, inadequate attention to statistical design leads to wasteful use of the animals in research.

The process of ethical approval by institutional panels is time-consuming and sometimes underfunded, relying largely on volunteer labour and a skeleton staff for administrative and clerical matters. However, it has the advantage over governmental processes in that some recognition is given to the views of the public and those members of activist organizations concerned about the welfare of research animals. The failure of some institutional ethical review processes to include members of the public or activist groups, such as the new ethical review system started in Iran (S. Aldavood, personal communication), will need to be addressed if it is to be credible internationally.

The processes can be accelerated by using documented standard procedures, which can be referred to in applications. In theory, this should give more time for discussion of the ethical merits of the work, but in practice it is likely that this will still be assumed to be acceptable. There's also a risk that the assessors will be desensitised to the procedures by seeing them referred to just as a code or reference number, and they may not adequately consider the relevance of the procedures to the specific experiments being evaluated. If they are used, these documented procedures must be regularly reviewed, so that when an improved technique becomes available its use is rapidly made known.

After these general considerations concerning the welfare of laboratory animals, it is pertinent to consider two contentious and relatively new areas of research that are likely to have a major impact on animal welfare.

Genetic Modification of Organisms

Genetic modification of animals has been pursued by man ever since they were first domesticated (Uzogara, 2000). Initially, the objective was to select animals that were best suited to the environment. In the last 50 years, however, with the industrialization of livestock production, the objective has moved rapidly towards economic goals, with the focus on increased productivity. Although genetic modification is not new, the speed with which changes can be introduced has been accelerating and the knowledge base has increased. Animal modifications are now conducted with some understanding of the changes at gene level, whereas in the past selection was based on phenotype alone. As the genetic constitution was unknown, progress was slow, but the phenotype could be expected to lie somewhere between the most extreme expression of the selected trait and the normal phenotype of the population. The traits selected for were usually multilocus and therefore extreme results were rare. However, now that the genes themselves are deliberately targeted, and the expression is often improperly understood, extreme results are more common (Sillence, 2004). Hence the research can be conducted with a danger of producing phenotypes that could potentially release unwanted genes into the environment. As the precise functioning of the genes is often uncertain, and the modifications are targeted at an array of possible genes, the animals produced could be at risk of congenital welfare problems. Some will have high morbidity, and be susceptible to a variety of physiological complications. In addition, the very low success rate of many genetic modification programs, for mice at least, gives cause for concern about the ethics of the procedure. Sometimes, in large experiments with several hundreds of mice, the offspring will all be euthanased or they may not reach maturity, because of malfunctions and morphological complications, or because they failed to produce any suitable modification and are redundant for the experimental purposes. The standard production of GM mice in the laboratory therefore poses a major ethical dilemma as to whether large numbers of animals should be used in a production process with high mortality rates.

If the production of GM animals for laboratory research is contentious, so too is their utilisation in agriculture. Genetic modification of crops that are produced to be resistance to specific diseases or to be able to withstand pesticides and herbicides, to avoid the crop being contaminated with pests and weeds, respectively, is less morally questionable (Knight, 2007). An ability to tolerate pesticides and herbicides may actually reduce the volume of these chemicals required (Uzogara, 2000). These objectives may be laudable, but the long-term impact on the native flora and fauna is unclear. The impact in particular, on soil micro-organisms, which are at the start of the food chain, has received inadequate attention (Toro et al., 1998). Although most investigations have found little evidence of danger to humans, animals or micro-organisms of the production of genetically-modified crops (Toro et al., 1998), experimentation at Cornell University with the ecologically-valuable Monarch butterflies demonstrated the potential for their larvae to be killed by genetically-modified corn (Dively et al., 2004).

Genetical modification of sentient animals is more contentious, and early experimentation demonstrated the potential for welfare problems, because of the uncertainty of the phenotype. Some animals were genetically modified for increased growth and had problems with their leg joints, because farm animals have already been selected for rapid growth and other productive traits. Selection for cattle with a double muscling gene, which has a high prevalence in the Belgian Blue breed, directs growth preferentially to muscle and away from fat deposition and basic organs (Clinquart et al., 1998). The size of these animals and their high level of muscularity make them difficult to join with conventional cattle breeds without producing large foetuses, which require parturition by Caesarean section (Webster, 2002). Nevertheless, the search for genes connected with increased growth and production has accelerated in the last

30 years. The resulting animals are showing susceptibility to welfare problems: broiler chickens have high levels of congenital leg disorders (Bessei, 2006), many pigs are lame (Barnett et al., 2001) and the reproductive rate in extreme dairy type cows has declined (Roman et al., 1999; Shook, 2006).

There is now an increasing emphasis amongst farmers to breed for better disease resistance in their livestock, for example to enable cows to produce 10,000 litres per year or more without succumbing to lameness (Distl, 1999). Despite considerable improvements in treatment (Bolgov et al., 2002), the prevalence of mastitis is at the same level as 50 years ago. This is because many dairy cows are kept in intensive housing systems, where they come into contact with their excreta, that harbours the bacteria causing some of the most serious types of mastitis. In addition, walking on concrete all day can lead to damage to the laminar structure of the hooves, making walking painful (da Silva et al., 2004). The potential exists to genetically improve the hoof laminar structure in cows or to change the management system, but concrete is the normal material for floors because of its durability and ready availability.

There are therefore many ethical issues associated with genetical modification of animals: the welfare of all the laboratory animals used to develop them; the impact on humans or farm, companion or wild animals of consuming food from genetically modified animals, including the potential transmission of bacterial resistance to the animals themselves or their gut microflora, and finally the risk to the welfare of farm animals when using genetic modification to enhance production. However, there are potential welfare benefits from judicious use of this technology, by breeding animals that are more resistant to disease for example, or breeding animals better able to cope with heat, nutritional or other stresses.

Xenotransplantation

One of the new potential uses of laboratory animals, which illustrates the emerging ethical dilemmas now facing researchers, governments and the public, is xenotransplantation, the transplantation of animal tissues, cells or organs to humans. At a time when there is major shortage of organs for transplantation, it offers a hope of meeting the human demand, as well as potentially contributing to cell therapy, for example in renal dialysis. However, as well as presenting serious ethical issues, the transplantation procedures are technically difficult, mainly because of the rejection of the transplanted tissue or cells by the recipient (Yang and Sykes, 2007). This rejection is often very rapid, causing a hyperacute immune response, and in the case of major organ transplantation it is usual for the recipient to survive only a few days. Some improvements in controlling the immune response are emerging, particularly eliminating rejection by genetic modification of the source animals (Groth, 2007). However, the technique also poses dangers in relation to the emergence of new diseases and transmission of existing diseases between species. This risk to the animal and human environment is heightened by the fact that the immune system of the recipient is heavily suppressed in an attempt to control the rejection problems.

Many western governments have prohibited animal to human transplantation, at least until the science is better understood. This is because, although most patients in need of a transplant would accept one from an animal, they are unlikely to understand the risks associated with this action (Ellison, 2006). Public opinion is less supportive (Deschamps et al., 2000). Most countries, including Australia, allow animal to animal transplantation, so that scientists can explore the techniques before they are used on humans. Animal to human cadaver transplantation is another possible method of improving the technique (Siepe et al., 2007). There is a common concern for both animal welfare and the environment as a result of the potential emergence of novel pathogens in this process. The most common source animal is the pig. The pig is omnivorous, like humans, and many of its organs are of approximately the same size as those of humans, although the heart poses some challenges because of the quadrupedal gait of pigs (Siepe et al., 2007). It is routinely kept for food production, and so may be seen as more ethically acceptable than animals that have to be captured from the wild before surrendering their organs, or animals, such as the dog, from which we derive emotional comfort. The pig as a domesticated animal has become partly adapted to intensive management systems and a considerable amount of research has been conducted on the improvement of the welfare of pigs in intensive management systems. Methodology has been prepared to assess porcine welfare accurately (e.g. API, 2004). However, when they are kept as source animals, pigs need to be confined in very clean, bioexclusion environments to control most pathogens (Tucker et al., 2002). No access to soil for rooting and nesting behaviour is possible, unless it is sterilised, and supplying toys to enrich their environment also has to be strictly controlled. The need to keep their environment scrupulously clean means that the housing systems are sterile, featureless enclosures, which leads to welfare problems such as fighting and tail biting. Even with all these precautions, pigs routinely carry some microorganisms that could be conveyed to humans during transplantation. The porcine endogenous retrovirus (PERV) is carried by most pigs, and at least theoretically has the potential to infect humans (Boneva and Folks, 2004). Recipients that become infected with PERV could develop novel diseases, particularly because the organism is changing from one species to another, and these may be capable of causing a worldwide epidemic. Thus recipients of transplants from pigs would have to agree to regular monitoring of their health status and this could lead to potential complications if they disappear or emigrate. They could not themselves donate their own organs for fear of transmission of any novel diseases. The risk has not yet been fully evaluated and appears considerable, although in animal-to-animal transplants the recipients can be closely monitored and contact with other animals limited. Such control would be very difficult or impossible to achieve with human recipients.

Some people find that xenotransplantation is contrary to the values that require the maintenance of the integrity of the animals (Thompson, 1997). There may be ethical problems for the human recipients relating to their religious beliefs (Hagelin et al., 2001), since Muslims are not allowed to eat the flesh of the pig and using their tissue could be equally abhorrent. There could also be a psychological stigma surrounding the human identity of recipients, who are essentially hybrid animals or chimeras (Modell, 2007). These were commonly depicted in the art and prose of many cultures, occasionally as graceful animals, such as the centaur, but usually as demonic animals or those meant to amuse, particularly in mediaeval manuscripts (Powell, 2004). Such chimeras are now beginning to be possible with genetic engineering and are raising people's concerns about the resulting 'animals'. However, the use of animal tissues in human medicine has been tolerated for many years without complaint, and it is likely that some of the fears surrounding the ethics of chimeras are unjustified.

Whilst there is no doubting that it is very tragic to see thousands of people dying whilst waiting for organ transplants, xenotransplantation is unlikely to offer an easy solution. The alternatives of artificial organs, stem cell transplants or medical campaigns to prove improve awareness of health issues probably offer more cost-effective and immediate control measures and a greater likelihood of success. The challenge of transplanting organs to humans will probably attract scientists to pursue this endeavour for years to come, but whether it should be a major recipient of government funds is highly debatable because of the ethical and welfare concerns. Possibilities remain that animal cells may be useful for transfusion procedures in renal or hepatic dialysis in the case of kidney or liver failure, respectively. But even in this situation animal cells are likely to be recognised as foreign material by the recipient and an immune reaction initiated. Attempts to encapsulate the source cells in an inert material have not been very successful. Another problem concerning rejection relates to the alpha galactose molecules on the surface of most mammalian cells, which cause immune rejection in the initial stages (Kiessling, 2002). It may be possible to genetically modify pigs to remove these molecules but the impacts on other aspects of porcine physiology are unknown.

The widespread use of pigs as source animals for transplanting organs would undoubtedly cause problems to many in the world's population that avoid pig meat and other products for religious reasons, in particular, the Muslim and Jewish people. Buddhists and Hindus would probably avoid this technology, in all probability, because of their belief that animal should not be used for human benefit, so the technology seems mainly targeted at the world's Christian population (Hagelin et al., 2001). The Buddhist belief in the transmutation of humans into animals between one life and the next might make the practice more acceptable for some. However, many people within Christian societies have developed serious concerns for the welfare of pigs. This together with concern for the potential threat to the environment of novel pathogens and suspicion with modern scientific developments suggests that there may be limited acceptance of the techniques (Glass et al., 2005). Many argue that if there is a substantial need for organ transplants (and of this there is no doubt), then this would be more ethically met from the human rather than animal population. The limited number of organs available from humans is largely because people have to opt into schemes for organ donation. Some progressive countries such as Belgium have adopted a policy that assumes people will allow use of their organs after their death, unless they opt out (Roels, 1999). Thus people have to positively state that they will not donate their organs, and availability has dramatically increased.

Chapter 11
Future Developments in Animal Welfare

Regulatory control – farm animals – companion animals – wild animals – the scope of animal welfare concerns

Having considered the development of animal welfare in the past and the present, it is pertinent to reflect on how it is going to develop in the future. The significant changes in attitudes to animals in recent centuries suggest that over the course of the current century, attitudes will continue to change, probably stimulating an accelerated improvement in animal welfare. There will be many technical developments that facilitate improvements in animal welfare, some of which are generic to all animals. For example, there is likely to be more use of systems of electronic surveillance of animals, coupled with automatic provision of resources in response to demand or need.

Over the course of the century, it is probable that regulatory control of animal welfare will increase. Codes of practice will become more prescriptive, specifying facilities required for animals in detail. There will be increased emphasis on the establishment and maintenance of global standards for animal welfare. These are starting to be formulated by consensus between co-operating countries, such as in the European Union. However there will be a major role for international organizations with responsibility for animal welfare, in particular the World Organisation for Animal Health (OIE). This organization started in 1924 and now has representation from 172 countries. It adopted animal welfare as part of its portfolio in 2001. It has a particular focus on harmonizing animal welfare standards internationally, whilst recognizing cultural differences in attitudes to animals. International campaign activity is mainly the domain of the World Society for Protection of Animals, which started in 1981 and currently has about 700 member societies. Animal welfare campaigns are also becoming global activities, for example the recent campaign, orchestrated in the United States, to persuade worldwide clothing manufacturers not to buy wool from Australian sheep farmers still practising mulesing of their animals. It is also possible that more countries will give sentient animals legal status, following a declaration by the United Nations that animals are sentient beings, which means that their current status as property is inappropriate (WSPA, 2007).

C. Phillips, *The Welfare of Animals*, Animal Welfare 8,
DOI 10.1007/978-1-4020-9219-0_11, © Springer Science+Business Media B.V. 2009

Religious guidance on our treatment of animals has been influential in the past, and will probably still be important for the coming century. However, it will have to evolve to take account of changing human attitudes and aspirations. Guidance on the management of animals that is relevant today is generally not available in the ancient scriptures, because these were written for people living at different times in very different circumstances. The Islamic restrictions on stunning animals before slaughter, for example, appear outdated. However, being deeply embedded in their beliefs, development of alternative stunning procedures that are reversible appear urgent.[1] The religious leaders must address people's concerns about the welfare of animals, and those in the Abrahamic faiths that ignore this issue avoid the fundamental doctrine that humans have responsibility to look after animals. Indeed the failure of these faiths to adequately address such issues may even be partly responsible for the rise in popularity of the major Eastern religions, which place animals and humans on an equal basis. As discussed earlier, this is not necessarily incompatible with humans assuming responsibility for animals, in just the same way as children are our equals but we have responsibility for them. A just society does not allow the abuse of children or animals, and the popular demand for improved standards will probably grow over the course of the century.

Farm Animals

Meat consumption may decline in developed countries because of concerns about the cost, consumer health, animal welfare and the environment, and the meat that is consumed will be in smaller portions and with little fat. However, meat consumption is likely to increase in many rapidly developing countries, including India and China, because it was previously unaffordable to many people. An increasing number of people in developed countries are questioning methods used for food production from animals, but these doubts do not emanate from the traditional moral leaders in society, but from changing attitudes in a wealthier, better fed and materially richer society. In a recent Australian survey nearly one half of the population were either eating or contemplating eating a predominately plant-based diet (Lea et al., 2006), a trend which some believe represents a more moral approach to eating behaviour (Berndsen and van der Pligt, 2004). If this trend continues this pattern of behaviour will have significant consequences for our requirements for farmed livestock. Consumers are now better educated and require more information on the items available for purchase. More comprehensive and informative food labeling is starting to fulfil that need, together with an ability to trace products from the farm to the supermarket, using electronic coding. To avoid

[1] Conventional stunning is irreversible and animals that have been stunned are not therefore killed by a knife cut to the throat, which the Islamic Qur'an requires.

unnecessary labeling complexity, the branding of food products by certifying bodies is gaining in popularity, although many consumers do not yet recognize the certification schemes (Hoogland et al., 2007). Flexible schemes will allow the consumer to select goods produced to the welfare standard that they wish to support, but pressure from the major retailers could improve farm animal welfare standards relatively rapidly. Consumers are likely to increasingly support organic farming methods, which represent a certified standard of production in all the major areas of concern – animal welfare, food safety and care for the environment. Eventually organic products may be universally accepted in retail outlets, with little demand for conventional products, making the retailers the arbiters of animal welfare standards, in conjunction with the certifying bodies.

Competition for the consumer market will increasingly be between global companies, rather than between countries (Toulmin, 1999). The globalization of the world's food markets and harmonization of diet across different cultures will see more animal products exported from countries that can produce them efficiently (Van Dooren, 2006). However, this centralization can have large effects locally, for example waste concentration. Transport of the product over long distances is likely to reduce efficiency. The concern that people have for the welfare impact of long distance transport of live animals is likely to mean that the products, rather than the animals, will be transported wherever possible. The harmonization of the world's cultures, as a result of migration and internationalization of the media, may diminish the demand for live animals to be transported overseas for religious slaughter. More production will be based in developing countries because of the cheap labour available there, but there is already some harmonization of labour costs and standards between developed and developing countries (Frenkel, 2001). India, which has the largest cattle population of any country, allows most cattle to scavenge off refuse in the streets. The refuse may have alternative uses, such as for biofuel production, and it is likely that India will remove cattle from the streets into farms, where their production can be controlled.

Companion Animals

Humans are a very social species and the likely continued rise in popularity of urban living to limit transport time and to increase the opportunities for entertainment and social intercourse, will be at the cost of further estrangement from nature. When people are estranged from nature they express their desire for a natural environment in their art, poetry and music. Urban living and female emancipation has brought opportunities for both members of partnerships to work full-time, and many are choosing to do so to acquire the high standard of living that modern technology offers. This style of living does not easily allow the inclusion of the traditional dog or cat in the household. Dogs are social animals in need of care, attention and exercise, and they require a considerable time commitment. The emergence of daytime animal care facilities, in its infancy today, is likely to accelerate over the course of the century to meet the demands of modern lifestyles. Cats are also social and they like to hunt. Many welfare agencies now recommend that cats should be kept indoors all day, to protect wildlife (e.g. HSUS, 2008), but if they only interact with their owners for a few hours each day this will be seen as providing insufficient stimulation and activity.

Stray dogs and cats are a cause for concern to many because of their perceived welfare, the spread of disease and the risks to humans and wildlife, therefore compulsory control over their breeding activity is likely to be more common. In the more progressive countries only desexed cats and dogs will be available to the general public, with breeding licences available on application for those that can offer suitable facilities. In the longer term, other less demanding types of companion animal are likely to increase in popularity. Aquaria fish are already growing rapidly in popularity (e.g. NSW DPI, undated), offering attractive animals in a natural setting, that only need a small expenditure of time to be maintained. They do not provide the physical contact of cats and dogs, but it will be increasingly recognized that this can only be obtained alongside strict enforcement of minimum standards for exercise, health provisions and containment of these two species. Health care for companion animals can expect to adopt many of the advances in human medicine, which coupled with breeding for healthier animals, will allow them to live longer and healthier lives. Alongside the medical advances that will improve humans' quality of life, there will be more emphasis on people taking responsibility for a good diet, appropriate exercise and the morality of their behaviour. Animals will always be important contributors to the maintenance of our physical and mental health. However, to achieve this there will be an increased need for skilled animal experts to provide guidance on the management of animals and their interaction with the environment that they share with us.

Wild Animals

Wild animals will be increasingly valued as a vital component in the natural areas in which we relax. Ecotourism will help to preserve this future. However, in regions where the combination of natural areas and urban living is under strain, such as in developing countries with rapidly populations, the survival and welfare of wild animals is likely to become an increasing focus of concern in developed countries. Habitat destruction and global climate change (Marai et al., 2007) will the two major challenges, although the impact of tourism will also be an increasing concern in many regions. The accelerated rate of climate change as a result of human industrial activity may exceed the ability of animals to change their species characteristics. In addition, the possibilities of mass migration to more suitable regions have been diminished for most of the larger animals by the transition of natural habitat into farming land. The possibility of extinction of wild animal species, partly as a result of global climate change, will increasingly become a reality, which will need to be addressed by politicians, biologists, animal welfare scientists, geographers and a range of scientists from other disciplines. There will be a major emphasis on conservation of biodiversity and maintenance of sustainable ecosystems, with special consideration given to maintaining native animal species that are adapted to the environment.

There is increased awareness of the impact of fishing practices, and the ability to inform the public of the issues through the modern media is likely to lead to more attention being paid to the welfare of fish, not just commercial but also recreational fishing (Davie and Kopf, 2006). Now that it is realized that fish feel pain in similar ways to other animals (Nordgreen et al., 2007), there are many changes to recreational fishing practice proposed that could reduce the welfare impact (Davie and Kopf, 2006).

As wild animals and their habitat become increasingly threatened, their preservation and reproduction in wildlife parks and zoos will be more valued. These will be responsible for holding stocks of endangered animals and for rehabilitating animals into suitable environments when they are available. These sanctuaries can also receive the public to be educated about animals in nature, to learn to respect the animal's form and function and to assist in the process through their donations, however, this should never be their main function. Adopting animals is a good way of ensuring active involvement of the public, which fulfills the need that many have to look after nature. In children it teaches them to care for animals and that they can make a difference. However, welfare standards for animals in many sanctuaries need addressing, and it is likely that global standards will emerge.

Living standards for animals in zoos should ideally be as good as those living standards of the people that visit. Clearly many zoos have a long way to go, but this transition is essential if they are to have a good future.

The Scope of Animal Welfare Concerns

Over the course of the century, it is likely that the emphasis of the public's concern for animals will move away from the current focus on sentient animals that are useful to humans, towards a more holistic view of our responsibilities towards animals, embracing concern for all animals, and extending the focus to genetic integrity, longevity, relationship to the environment, teleos and animal uses. The combination of wellbeing and longevity indicators to create overall indicators of animal welfare would bring the animal sector more in line with human indicators of wellbeing, which usually unite the two components (Perenboom et al., 2004).

Animal welfare science will continue to develop and will increasingly attract top scientists, as well as leading academics from other areas: philosophy, theology and bioethics in particular. Other countries are likely to follow the United Kingdom's lead establishing tertiary level courses in animal welfare, producing well trained individuals that will contribute to the development of this new science. Progress will continue until there is appropriate attention to the welfare of all animals. Legislative control of animal welfare will almost certainly increase, supported by detailed Codes of Practice. Licensing of animal industry operators, and in particular farm workers and managers, is likely in some developed countries within 50 years.

There may be increased conflict between supporters of improved welfare standards for animals, welfare standards for humans and the welfare of the environment. However, true progress will only be made when the integrated nature of these three objectives is realised and pursued by all. According to the Confucian proverb, we can gain wisdom in several ways, through experience and imitation, but the noblest way is through reflection. Our reflections on the welfare of that silent majority, the animals with whom we share the planet, will guide our thoughts, words and, most importantly, our deeds.

References

Aarssen, L.W. 2005. Why is fertility lower in wealthier countries? The role of relaxed fertility-selection. Population and Development Review 31: 113.

Abbas, E.E. 2007. Industry-sponsored research in developing countries. Contemporary Clinical Trials 28: 677–683.

Abbot, N.C. 2000. Healing as a therapy for human disease: A systematic review. Journal of Alternative and Complementary Medicine 6: 159–169.

Abeyesinghe, S.M., Nicol, C.J. et al. 2005. Can domestic fowl, Gallus gallus domesticus, show self-control? Animal Behaviour 70: 1–11.

Abler, B., Erk, S. et al. 2007. Anticipation of aversive stimuli activates extended amygdala in unipolar depression. Journal of Psychiatric Research 41: 511–522.

Adam, T.C. and Epel, E.S. 2007. Stress, eating and the reward system. Physiology & Behavior 91: 449–458.

Agenäs, S., Heath, M.F., Nixon, R.M., Wilkinson J.M. and Phillips, C.J.C. 2003. A meta-regression and field study aiming to evaluate frequently analysed metabolites as indicators of undernutrition in cattle. Report to the RSPCA, UK, July 2003.

Agenäs, S., Heath, M.F., Nixon, R.M., Wilkinson, J.M. and Phillips, C.J.C. 2006. Indicators of undernutrition in cattle. Animal Welfare 15: 149–160.

Alboga, M.A.B. 2003. Ritual slaughter: Animal friendly slaughter and protection of animals in Islam. Deutsche Tierarztliche Wochenschrift 110: 189–192.

Albright, K. M. 2002. The extension of legal rights to animals under a caring ethic: An ecofeminist exploration of Steven Wise's Rattling the cage. Natural Resources Journal 42: 915–937.

Alcock, J. 1989. Animal Behaviour and Evolutionary Approach. Sinauer Associates, Inc., Sunderland, U.S.A.

Allen, G. 2002. The measure of a Victorian polymath: Pulling together the strands of Francis Galton's legacy to modern biology. Nature 145: 19–20.

Allott, R. 1991. Objective morality. Journal of Social and Biological Structures 4: 455–471.

Ambrose, P. 1992. Conflict over Canadian Seal Cull. Marine Pollution Bulletin 24: 226–226.

American Bird Conservancy. 2007. Domestic cat predation on birds and other wildlife. http://www.abcbirds.org/cats/factsheets/predation.pdf. Accessed February 2007.

American Humane, 1997. Animal Shelter Euthanasia. http://www.americanhumane.org/site/PageServer?pagename = nr_fact_sheets_animal_euthanasia. Accessed 21 January, 2008.

American Humane, 2003. Animal cruelty statutes by State. americanhumane.org/site/DocServer/ta_cruelty_stats_2003.pdf?docID = 442

American Veterinary Medicine Association AVMA undated. Companion Animal/Exotic. http://www.avma.org/membshp/marketstats/comp exotic.asp. Accessed 17 July, 2006.

Anderson, L. C. 2007. Institutional and IACUC responsibilities for animal care and use education and training programs. ILAR Journal 48: 90–95.

Andrabi, S.M.H. and Maxwell, W.M.C. 2007. A review on reproductive biotechnologies for conservation of endangered mammalian species. Animal Reproduction Science 99: 223–243.

Andrews, M. 2005. An Introduction to Egyptian Art. http://touregypt.net/featurestories/artoverview.htm. Accessed 7 January, 2008.

Animal Welfare Centre (AWC). 2002. Final lists of animal welfare issues for each animal industry. In 'Proceedings of a Workshop to Identify Animal Welfare Issues with Animal Industries', pp 5–29. Animal Welfare Centre, Werribee, Australia.

Animals Australia, 2007. Change is Needed. Video. http://www.animalsaustralia.org/media/videos.php?vid = changeisneeded&native = 1. Accessed 2 January, 2008.

Anon, 1987. Empathy, pp 220–221. In: R.L. Gregory Ed. Oxford Companion to the Mind. Oxford University Press, UK.

Anon, 1996. Second report on the statistics on the number of animals used for experimental and other scientific purposes in the member states of the European Union. Commission Communication to the Council and the Parliament. Com 99 191. http://ec.europa.eu/environment/docum/99191_en.htm, accessed January, 2007.

Anon, 2006. Australian Standards for the Export of Livestock. http://www.daff.gov.au/content/output.cfm?ObjectID = E23635D3-2B6F-48D9-BFF4C8F52C618BAF. Accessed 1 March, 2006.

Anon, 2007a. www.hen.org/servlets/hcp. Accessed January, 2007.

Anon, 2007b. http://rqriley.com/images. Accessed January, 2007.

Anon, 2007c. The Cave Paintings and Rock Art of Chauvet. http://www.bradshawfoundation.com/chauvet/page10.php. Accessed October, 2008.

Anon, 2007d. The Venus of Willendorf, Fertility Symbol, Pre-Historic Sculpture, 30000-25000 BC. http://www.art.com/asp/sp-asp/_/pd–11725851/The_Venus_of_Willendorf_Fertility_Symbol_PreHistoric_Sculpture_3000025000_BC_Front_View.htm. Accessed October, 2007.

Anon, 2008a. Prayer and Sacrifice. www.roman-empire.net/religion/sacrifice.html. Accessed, 8 January, 2008.

Anon, 2008b. Ashurbanipal II Killing Lions, the Palace of Ashurbanipal II,Nimrud, Iraq, Neo-Assyrian, 668–627 BC. http://www.allposters.com/-sp/Ashurbanipal-II-Killing-Lions-the-Palace-of-Ashurbanipal-II-Nimrud-Iraq-Neo-Assyrian-668-627-BC-Posters_i1951619_.htm. Accessed 8 January, 2008.

Anon, 2008c. Recreational Fishing. http://www.ec.gc.ca/nature/highlights/bc.htm, accessed December, 2008.

Appleby, M.C., Cutler, N. et al. 2003. What price cheap food? Journal of Agricultural & Environmental Ethics 16: 395–408.

Araus, J.L., Febrero, A. et al. 1999. Crop water availability in early agriculture: evidence from carbon isotope discrimination of seeds from a tenth millennium BP site on the Euphrates. Global Change Biology 5: 201–212.

Arave, C.W., Albright, J.L. and Sinclair, C.L. 1974. Behaviour, milk yield and leucocytes of dairy cows in reduced space and isolation. Journal of Dairy Science 59: 974–985.

Arias, A. J., Steinberg, K. et al. 2006. Systematic review of the efficacy of meditation techniques as treatments for medical illness. Journal of Alternative and Complementary Medicine 12: 817–832.

Arluke, A. and Luke, C. 1997. Physical cruelty toward animals in Massachusetts, 1975–1996. Society & Animals 5: 195–204.

Arluke, A., Levin, J. et al. 1999. The relationship of animal abuse to violence and other forms of antisocial behavior. Journal of Interpersonal Violence 14: 963–975.

Ascione, F.R., Weber, C.V. et al. 2007. Battered pets and domestic violence – Animal abuse reported by women experiencing intimate violence and by nonabused women. Violence against Women 13: 354–373.

Asmundson, G.J.G. and Hadjistavropoulos, H.D. 2007. Is high fear of pain associated with attentional biases for pain-related or general threat? A categorical reanalysis. Journal of Pain 8: 11–18.

Astin, J.A., Harkness, E. et al. 2000. The efficacy of distant healing: A systematic review of randomized trials. Annals of Internal Medicine 132: 903–910.

Australian Code of Practice for the Care and Use of Animals for Scientific Purposes, 2004. Seventh edition. National Health and Medical Research Council, Canberra. http://www.health.gov.au/nhmrc/issues/pdf/animcode.pdf, accessed August, 2007.

Australian Pork Industry. Pigs: Welfare Audit for the Pork Industry. http://www.apl.au.com/media/Pork%20Industry%20Welfare%20Audit.pdf. 2004.

Bach, A., Dinares, M. et al. 2007. Associations between lameness and production, feeding and milking attendance of Holstein cows milked with an automatic milking system. Journal of Dairy Research 74: 40–46.

Bagaric, M. 2006. How to Live: Being Happy and Dealing with Moral Dilemmas, University Press of America, Lanham, MD, USA.

Baldock, F.C., Alexander, L. and More, S.J. 2001. Estimated and predicted changes in the cat population of Australian households from 1979 to 2005. Australian Veterinary Journal 81: 289–292.

Barnett, J.L., Hemsworth, P.H. et al. 2001. A review of the welfare issues for sows and piglets in relation to housing. Australian Journal of Agricultural Research 52: 1–28.

Bartussek, H. 1999. A review of the animal needs index ANI for the assessment of animals' well-being in the housing systems for Austrian proprietary products nad legislation. Livestock Production Science 61: 179–192.

Baumans, V. 2004. Use of animals in experimental research: an ethical dilemma? Gene Therapy 11: S64–S66.

Bay, A.H., Hellevik, O. et al. 2007. Does immigration undermine welfare state support? Tidsskrift for Samfunnsforskning 48: 377–408.

Bayvel, D. 2004. Animal use statistics: the Australian and New Zealand experience and perspective. Alternatives to Laboratory Animals, 32, Supplement 1, 287–292.

Beatty, M. Undated. RSPCA recognises cane toad menace. Media release. http://www.rspcaqld.org.au/news/mediareleases/cane_toad_menace.pdf

Beiglbock, C., Steinbeck, T., Tataruch, F. and Rug, T. 2002. Environmental cadmium induces histopathological changes in kidneys of roe deer. Environmental Toxicology and Chemistry 21: 1811–1816.

Belotto, A.J. and Silva, D.F.D. 2006. Major issues affecting companion animals worldwide. Proceedings of the Symposium of the World Society for the Protection of Animals, London, July, 2006.

Berndsen, M. and van der Pligt, J. 2004. Ambivalence towards meat. Appetite 42: 71–78.

Bessei, W. 2006. Welfare of broilers: a review. Worlds Poultry Science Journal 62: 455–466.

Bicalho, R. C., Vokey, F. et al. 2007. Visual locomotion scoring in the first seventy days in milk: impact on pregnancy and survival. Journal of Dairy Science 90: 4586–4591.

Black, E. 2003. War Against the Weak: Eugenics and America's Campaign to Create a Master Race. Four Walls Eight Windows, New York, USA.

Blake, W. 1803. Auguries of Innocence. In: Nicholson & Lee Eds. The Oxford Book of English Mystical Verse. 1917. Oxford, England.

Blaxter, K.L. 1995. From Dearth to Plenty: The Modern Revolution in Food Production. 296 pp. Cambridge University Press, Cambridge.

Bleckmann, C.A. 2006. Evolution and creationism in Science: 1880–2000. Bioscience 56: 151–158.

Blokhuis, H.J., Ekkel, E.D. et al. 2000. Farm animal welfare research in interaction with society. Veterinary Quarterly 22: 217–222.

Bolgov, A.E., Karmanova, E.P. et al. 2002. The influence of different factors on resistance of dairy cows to mastitis. Journal of Animal and Feed Sciences 11: 237–254.

Boneva, R.S. and Folks, T.M. 2004. Xenotransplantation and risks of zoonotic infections. Annals of Medicine 36: 504–517.

Born Free Foundation, 2007. Roadkill Survey. www.bornfree.org.uk/ukroad.htm, accessed January, 2007.

Botreau, R., Bonde, M. et al. 2007. Aggregation of measures to produce an overall assessment of animal welfare. Part 1: a review of existing methods. Animal 1: 1179–1187.

Bracke, M.B.M., Spruijt, B.M. et al. 1999. Overall animal welfare reviewed. Part 3: Welfare assessment based on needs and supported by expert opinion. Netherlands Journal of Agricultural Science 47: 307–322.

Brade, W. 2005. Longevity and reasons for culling of milk yield controlled Holstein cows: Consequences for breeding? Praktische Tierarzt 86: 658.

Bradshaw, I.G.A. 2004. Not by bread alone: symbolic loss, trauma, and recovery in elephant communities. Society & Animals 12: 143–158.

Bradshaw, R.H. 2002. The ethical review process in the UK and Australia: The Australian experience of improved dialogue and communication. Animal Welfare 11: 141–156.

Bradshaw, R.H., Marchant, J.N. et al. 1998. Effects of lavender straw on stress and travel sickness in pigs. Journal of Alternative and Complementary Medicine 4: 271–275.

Brambell, F.W.R. 1965. Report of the Technical Committee to Enquire into the Welfare of Animals Kept under Intensive Livestock Husbandry Systems. HMSO Cmnd. 2836, London.

Brisbane City Council BCC , 2003. Animal Laws and Permits. http://www.brisbane.qld.gov.au/BCC:BASE:544886598:pc = PC_1826#dogs. Accessed 21 January, 2008.

British Museum, 2007. Assyria, Lion Hunts, Room 10a. http://www.britishmuseum.org/explore/galleries/middle_east/room_10a_assyria_lion_hunts.aspx

Broom, D.M. 1986. Indicators of poor welfare. British Veterinary Journal 142: 524–526.

Broom, D.M. 1992. Animal welfare: its scientific measurement and current relevance to animal husbandry in Europe. In: C.J.C. Phillips and D. Piggins Eds. Farm Animals and the Environment, pp 245–253. CAB International, Wallingford.

Broom, D.M. 1999. The welfare of vertebrate pests in relation to their management. In: D.P.F. Cowand Ed. Advances in Vetebrate Pest Management, pp. 309–329. Fũrth, Finlander Verlad.

Broom, D.M. 2003. The Evolution of Morality and Religion. Cambridge University Press, Cambridge, UK.

Broom, D.M. 2005. Animal welfare education: development and prospects. Journal of Veterinary Medical Education 32: 438–441.

Broom D.M. and Johnson, K.G. 1993. Stress and Animal Welfare. Chapman and Hall, London.

Brothers, L. 1990. The neural basis of primate social communication. Special issue: Empathy. Motivation and Emotion 14: 81–91.

Brown, D.F. 2007. The spatial and temporal dimensions of a rural landscape: the Yucatec Maya k'ax. Canadian Geographer-Geographe Canadien 51: 91–108.

Brown-Brandl, T.M., Eigenberg, R.A. et al. 2006. Heat stress risk factors of feedlot heifers. Livestock Science 105: 57–68.

Bruckmeier, K. and Prutzer, M. 2007. Swedish pig producers and their perspectives on animal welfare: a case study. British Food Journal 109: 906–918.

Burgdorf, J., Panksepp, J. et al. 2006. Regional brain cholecystokinin changes as a function of rough-and-tumble play behavior in adolescent rats. Peptides 27: 172–177.

Butcher, R. 2006. Companion animal issues: solutions. Proceedings of the Symposium of the World Society for the Protection of Animals, London, July, 2006.

CAB International. 2002. Animal Health and Production Compendium. CAB International, Wallingford, UK.

Carey, G. and DiLalla, D.L. 1994. Personality and psychopathology: genetic perspectives. Journal of Abnormal Psychology 103: 32–43.

Carling, L. 2005. Squirrel Monkey Behaviour Enrichment. Honours' thesis, University of Queensland, Australia.

Carlson, L.W. 2002. Cattle: An Informal Social History. National Book Networks, Lanham, USA.

Carter, D. 1957. The Symbol of the Beast – the Animal Style Art of Eurasia. Ronald Press Co., New York.

Case, J.B., Chomel, B. et al. 2006. Serological survey of vector-borne zoonotic pathogens in pet cats and cats from animal shelters and feral colonies. Journal of Feline Medicine and Surgery 8: 111–117.

Catan, T. 2007. Bullfighting in dock after 14-year old is badly gored. The Times, 18 April, 2007, 33 pp.

Chalmeau, R., Visalberghi, E. and Gallo, A. 1997. Capuchin monkeys, Cebus apella, fail to understand a cooperative task. Animal Behaviour 54: 1215–1225.

Clark, A.C. 1977. Animals and Men, their Relationship as Reflected in Western Art from Prehistory to the Present Day. William Morrow and Company, New York.

Clements, R.J. 2006. Farrer oration – shaking windows; rattling walls. Tropical Grasslands 40: 129–136.

Clinquart, A., Hornick, J.L. et al. 1998. The influence of double muscling on production and quality of meat in Belgian Blue cattle. Productions Animales 11: 285–297.

Clubb, R. and Mason, G. 2003: A review of the welfare of zoo elephants in Europe. Report commissioned by the RSPCA. University of Oxford, Animal Behaviour Research Group, Oxford.

Companion Animal Welfare Council, CAWC, 2003. Report on the Welfare of Non-domesticated Animals kept for Companionship. CAWC, Sidmouth, Devon.

Conard, N. J. 2003. Palaeolithic ivory sculptures from southwestern Germany and the origins of figurative art. Nature 426: 830–832.

Coppinger, R., Glendinning, E, Torop, E, Mattay, C, Sutherland, M and Smith, C. 1987. Degree of behavioural neoteny differentiates canid polymorphs. Ethology 75: 89–108.

Coppinger, R.F. and Smith, C.K. 1983. The domestication of evolution. Environmental Conservation 10: 283–292.

Coruh, B. A., Ayele, H. et al. 2005. Does religious activity improve health outcomes? A critical review of the recent literature. Explore – the Journal of Science and Healing 1: 186–191.

Countryman, M.J. 2006. From protest to politics – Community control and black independent politics in Philadelphia, 1965–1984. Journal of Urban History 32: 813–861.

Cox, B. and Bilkei, G. 2004. Lifetime reproductive performance of sows kept indoors and outdoors in Croatia. Veterinary Record 154: 569–570.

Crafts, N. 1985. British Economic Growth during the Industrial Revolution. Oxford: Clarendon Press.

Cross, N., Wojciechowska, J., Meng, J. and Phillips, C.J.C. 2008a. Perceptions of the importance of different welfare issues in Australian beef cattle production. Submitted for publication.

Cross, N., Wojciechowska, J., Meng, J. and Phillips, C.J.C. 2008b. Perceptions of the importance of different welfare issues in Australian sheep and goat production. Submitted for publication.

Curtis, S.E. and Guither, H.D. 1983. Animal welfare, an international perspective. In Baker, F.D. ed. Beef Cattle Science Handbook, pp 1187–1191. Westview Press, Boulder, Colorado, USA.

da Silva, L.A.F., Fioravanti, M.C.S. et al. 2004. Foot diseases in dairy cows: association with clinical mastitis, metrites and predisposed factors. Pesquisa Veterinaria Brasileira 24: 217–222.

Daly, B. and Morton, L.L. 2006. An investigation of human-animal interactions and empathy as related to pet preference, ownership, attachment, and attitudes in children. Anthrozoos 19: 113–127.

Danbury, T.C., Weeks, C.A., Chambers, J.P., Waterman-Pearson, A.E., Kestin, S.C. 2000. Self-selection of the analgesic drug carprofen by lame broiler chickens. Veterinary Record 146: 307.

Darwin, C. 1859. On the Orgin of Species by Means of Natural Selection. First edition. John Murray Publishers, London.
Davey, G.C., Forster, L., and Mayhew, G. 1993. Familial resemblances in disgust sensitivity and animal phobias. Behavioural Research Therapy 31: 41–50.
Davie, P.S. and Kopf, R.K. 2006. Physiology, behaviour and welfare of fish during recreational fishing and after release. New Zealand Veterinary Journal 54: 161–172.
Davies, P. 2006. Speech of the 10 million signatures petition – Animals matter to me. Proceedings of the Symposium of the World Society for the Protection of Animals, London, July, 2006.
Davis, G.H., Dodds, K.G. et al. 1993. Liveweight, fleece weight and prolificacy of Romney ewes carrying the Inverdale prolificacy gene Fecxi located on the X-chromosome. Livestock Production Science 34 1–2: 83–91.
Dawkins, M.S., Cook, P.A.et al. 2003. What makes free-range broiler chickens range? In situ measurement of habitat preference. Animal Behaviour 66: 151–160.
Dawkins, R. 2006. The God Delusion. Bantam Press, London.
Dayton L. Alverson, D.L., Freeberg, M.H., Murawski, S.A. and Pope, J.G. 1996. A global assessment of fisheries bycatch and discards. FAO Fisheries Technical Paper 339, Food and Agriculture Organisation, Rome.
Degabriele, R. and Fell, L.R. 2001. Changes in behaviour, cortisol and lymphocyte types during isolation and group confinement of sheep. Immunology and Cell Biology 79: 583–589.
Delezie, E., Verbeke, W. et al. 2006. Consumers' preferences toward techniques for improving manual catching of poultry. Poultry Science 85: 2019–2027.
Delezie, E., Verbeke, W. et al. 2007. Consumer perception versus scientific evidence about alternatives for manual catching of broilers in Belgium. Poultry Science 86: 413–419.
Denton, F.T., Mountain, D.C. et al. 1999. Age, trend, and cohort effects in a macro model of Canadian expenditure patterns. Journal of Business & Economic Statistics 17: 430–443.
Deschamps, J.Y., Chaillous, L. et al. 2000. Acceptability of xenograft by patients with type 1 diabetes mellitus and the general French population. Pathologie Biologie 48: 419–421.
Diamond, J. 1997. Guns, Germs and Steel: The Fates of Human Societies. W.W. Norton, USA.
Diez, M. and Istasse, L. 1995. Commercial dog food – characteristics and use in different physiological states. Annales De Medecine Veterinaire 139: 309–316.
Distl, O. 1999. Breeding for soundness of feet and legs in dairy cattle. Zuchtungskunde 71: 446–458.
Dively, G.P., Rose, R. et al. 2004. Effects on monarch butterfly larvae Lepidoptera: Danaidae after continuous exposure to Cry1Ab-expressing corn during anthesis. Environmental Entomology 33: 1116–1125.
Dolmia, N.M., Calenge, C. et al. 2007. Preliminary observations of elephant Loxodonta africana, Blumenbach movements and home range in Zakouma National Park, Chad. African Journal of Ecology 45: 594–598.
Duffy, R., Fearne, A. et al. 2005. Reconnection in the UK food chain - bridging the communication gap between food chain producers and consumers. British Food Journal 107: 17–33.
Duncan, I.J.H. and Fraser, D. 1997. Understanding animal welfare. In: M.C. Appleby and B.O. Hughes Ed. Animal Welfare, pp 19–31. CABI, Oxford, UK.
Dwyer, C.M. 2004. How has the risk of predation shaped the behavioural responses of sheep to fear and distress? Animal Welfare 13: 269–281.
Edwards, S.A. 2007. Experimental welfare assessment and on-farm application. Animal Welfare 16: 111–115.
Eisenberg, N. and Strayer, J. 1987. Critical issues in the study of empathy. In N. Eisenberg and J. Strayer ed. Empathy and Its Development, pp 3–16. Cambridge University Press, Cambridge.
El-Ghoul, W. and Hofmann, W. 2002. Influence of claw diseases of varying degrees on measurable stress reaction with special consideration of cortisol and lactate in bovine blood serum. Praktische Tierarzt 83: 354.

Ellison, T. 2006. Xenotransplantation – ethics and regulation. Xenotransplantation 13: 505–509.

English, P., Gurgess, G., Segundo, R. and Dunne, J. 1992. Stockmanship. Improving the Care of the Pig and other Livestock. Farming Press, Ipswich.

Eurobarometer, 2007. Attitudes of EU citizens towards Animal Welfare. russels. http://ec.europa.eu/public_opinion/archives/ebs/ebs_270_en.pdf. Retrieved 13 November 2004.

Evans, E.P. 1896. Animal Symbolism in Ecclesiastical Architecture. 105 pp. Heinemann, London.

Evans-Pritchard, E.E. 1940. The Nuer: A Description of the Modes of Livelihood and Political Institutions of a Nilotic People. Oxford University Press, Oxford.

Ewbank, R. 1992. Stress: a general overview. In: C.J.C. Phillips and D. Piggins Eds. Farm Animals and the Environment, pp 255–262. CAB International, Wallingford.

Feddema, J.P.H. 1995. The lesser violence of animal sacrifice – a somewhat hidden and overlooked ignored reality in Sinhala Buddhism. Anthropos 90: 133–148.

Feige, S., Nilsson, K., Phillips, C.J.C. and Johnston, S. 2007. Heterosexual and homosexual behaviour and vocalisations in captive female koalas (Phascolarctos cinereus). Applied Animal Behaviour Science 103: 131–145.

Fernie, K.J., Bortolotti, G.R., Smits, J.E., Wilson, J., Drouillard, K.G. and Bird, D.M. 2000. Changes in egg composition of American kestrels exposed to dietary polychlorinated biphenyls. Journal of Toxicology and Environmental Health – Part A 60: 291–303.

Fiege, M. 2007. The atomic scientists, the sense of wonder, and the bomb. Environmental History 12: 578–613.

Florida Fish and Wildlife Conservation Commission, 2007. http://research.myfwc.com/deatures/view_articlw.asp?id = 19870, accessed January, 2007.

Flugge, M. 2004. Australian wool and sheep industry – leaders in sheep care pdf. Australian Wool Innovation Ltd. http://www.woolisbest.com/documents/wool_industry%20update_Nov_2004.pdfAccessed 15 January, 2008.

Fölsch, D.W. 1984. Determinants on the man-animal relationship. In Unshelm, J., van Putten, G. and Zeeb, K. ed. Proceedings of the International Congress on Applied Ethology in Farm Animals, pp 257–263. International Congress on Applied Ethology in Farm Animals, Kiel, Germany.

Food and Agriculture Organisation (2002). http://www.fao.org/ag/aga/index_en.htm.

Forbes, M.J. 1998. Dietary awareness. In special issue on 'Animal Awareness, Concepts and Implications for Domesticated Animals' ed D. Piggins and C.J.C. Phillips. Applied Animal Behaviour Science 57: 287–298.

Ford, J. 1986. The great kangaroo cull begins. New Scientist 109 1494: 27–27.

Fraser, D. 2006. Animal welfare assurance programs in food production: a framework for assessing the options. Animal Welfare 15: 93–104.

Fraser, D. 2005. Animal welfare and the intensification of animal production. Food and Agricultural Organisation Readings in Ethics, 2. FAO, Rome.

Fredrikson, M., Annas, P., Fischer, H. and Wik, G. 1996. Gender and age differences in the prevalence of specific fears and phobias. Behavioural Research Therapy 34: 33–39.

Freedom Against Animal Cruelty in Europe FAACE, 2001. Frequently asked questions. http://www.faace.co.uk/faqs2.htm. Accessed 8 January, 2008.

Frenkel, S.J. 2001. Globalization, athletic footwear commodity chains and employment relations in China. Organization Studies 22: 531–562.

Freudenberg, K., Plefka, C. et al. 2007. Osteochondrosis dissecans in a growing bull. Tieraerztliche Praxis Ausgabe Grosstiere Nutztiere 35: 11.

Furnas, J.C. 1948. Anatomy of Paradise: Hawaii and the Islands of the South Pacific. 542 pp. Sloan Association, New York.

Gandhi, M.K. 1927. An Autobiography. Penguin Books, Hamondsworth, England.

Gandhi, M.K. 1948. The Seven Blunders of the World. List given by Mahatma Gandhi to his grandson Arun Gandhi, written on a piece of paper, on their final day together, shortly before his assassination. http://en.wikipedia.org/wiki/Seven_Blunders_of_the_World, accessed 31/01/07

Gandhi, M.K. undated. http://en.wikiquote.org/wiki/Mahatma_Gandhi, accessed 31/01/07

Garbarino, E.J., Hernandez, J.A. et al. 2004. Effect of lameness on ovarian activity in postpartum Holstein cows. Journal of Dairy Science 87: 4123–4131.

Garfinkel, A.P. 2006. Paradigm shifts, rock art studies, and the Coso sheep cult of eastern California. North American Archaeologist 27: 203–244.

Garolera, M., Coppola, R. et al. 2007. Amygdala activation in affective priming: a magnetoencephalogram study. Neuroreport 18: 1449–1453.

Gauthier, C. 2004. Overview and analysis of animal use in North America. Alternatives to Laboratory Animals, 32: Supplement 1, 275–285.

Geist, V. 1971. Mountain Sheep: A Study in Behaviour and Evolution. University of Chicago Press, Chicago.

Gintzburger, G., Le Houerou, H.N. et al. 2005. The steppes of Middle Asia: Post-1991 agricultural and rangeland adjustment. Arid Land Research and Management 19: 215–239.

Glass, C.A., Hutchinson, W.G. et al. 2005. Measuring the value to the public of pig welfare improvements: a contingent valuation approach. Animal Welfare 14: 61–69.

Gold Coast City Council. 2007. Dog Laws: How many Dogs Can you Keep? http://www.goldcoast.qld.gov.au/t_standard.aspx?pid = 22. Accessed 21 January, 2008

Goldsmith, H.H. 1983. Genetic influences on personality from infancy to adulthood. Child Development 54: 331–355

Grandin, T. 1993. Management and economic factors of handling and transport. In: T. Grandin ed Livestock Handling and Transport. CAB International, Wallingford, UK.

Great Ape Project 2008. GAP, 2008. Spanish Congress Announces Support for the Great Ape Project and Great Apes. http://www.greatapeproject.org/news.php. Accessed July, 2008.

Green, P.C. and Gullone, E. 2005. Knowledge and attitudes of Australian veterinarians to animal abuse and human interpersonal violence. Australian Veterinary Journal 83: 619–625.

Greens, 2007. Greens put the chicken before the egg. http://www.act.greens.org.au/988. Accessed 21 January, 2008.

Greer, T. 2006. Observations of the behaviour of the mahogany glider (Petaurus gracilis). Honours' thesis, University of Queensland, Australia.

Griffin, D.R., 1976. The Question of Animal Awareness. Rockefeller University Press, New York.

Groenewegen, P.P., den Berg, A.E. et al. 2006. Vitamin G: effects of green space on health, well-being, and social safety. British Medical Council Public Health Report No. 6.

Groth, C.G. 2007. Prospects in xenotransplantation: a personal view. Transplantation Proceedings 39: 685–687.

Groves, J.M. 1995. Learning to feel – the neglected sociology of social movements. Sociological Review 43: 435–461.

Hagelin, J., Hau, J. et al. 2001. Religious beliefs and opinions on clinical xenotransplantation – a survey of university students from Kenya, Sweden and Texas. Clinical Transplantation 15: 421–425.

Hall, S.J.G. and Hall, J.G. 1988. Inbreeding and population-dynamics of the Chillingham cattle Bos taurus. Journal of Zoology 216: 479–493.

Hardy, J. 1874. Far from the Madding Crowd, Cornhill Magazine, London.

Harris, J.I., Schoneman, S.W. et al. 2005. Preferred prayer styles and anxiety control. Journal of Religion & Health 44: 403–412.

Harrison, R. 1964. Animal Machines. Robert Cunningham and Sons Ltd, London.

Hatziminaoglou, Y. and Boyazoglu, J. 2004. The goat in ancient civilisations: from the Fertile Crescent to the Aegean Sea. Small Ruminant Research 51: 123–129.

Hawkins, P., Playle, L, Golledge, H., Leach, M., Banzett, R. et al. 2006. Newcastle Consensus Meeting on Carbon Dioxide Euthanasia of Laboratory Animals. 27, 28th February, 2006.

University of Newcastle upon Tyne. http://www.nc3rs.org.uk/downloaddoc.asp?id=416&page=292&skin=0. Accessed 21 December, 2007.

Heffner, H. 1998. Auditory awareness. In special issue on 'Animal Awareness, Concepts and Implications for Domesticated Animals' ed D. Piggins and C.J.C. Phillips. Applied Animal Behaviour Science 57, 259–268.

Hegelund, L. and Sorensen, J.T. 2007. Developing a HACCP-like system for improving animal health and welfare in organic egg production – based on an expert panel analysis. Animal 1: 1018–1025.

Heleski, C.R., and Zanella, A.J. 2006. Animal science student attitudes to farm animal welfare. Anthrozoos 19: 3–16.

Heleski, C.R., Mertig, A.G. and Zanella, A.J. 2004. Assessing attitudes toward farm animal welfare: A national survey of animal science faculty members. Journal of Animal Science 82: 2806–2814

Heleski, C.R., Mertig, A.G., and Zanella, A.J. 2005. Results of a national survey of US veterinary college faculty regarding attitudes toward farm animal welfare. Journal of the American Veterinary Medicine Association 226: 1538–1546.

Heleski, C.R., Mertig, A.G., and Zanella, A.J. 2006. Stakeholder attitudes toward farm animal welfare. Anthrozoos 19: 290–307.

Hemsworth, P. H., Barnett, J. L. et al. 1995. The welfare of extensively managed dairy-cattle – a review. Applied Animal Behaviour Science 42: 161–182.

Hemsworth, P.H. and Coleman, G.J. 1998. Human Livestock Interactions: The Stockperson and the Productivity and Welfare of Intensely Farmed Animals. CAB International, Wallingford, Oxon, UK.

Henning, J., Davies, P.R. et al. 2006. Seropositivity to rabbit haemorrhagic disease virus in non-target mammals during periods of viral activity in a population of wild rabbits in New Zealand. Wildlife Research 33: 305–311.

Henry, B.C. and Sanders, C.E. 2007. Bullying and animal abuse: Is there a connection? Society & Animals 15: 107–126.

Hensley, C. and Tallichet, S.E. 2005. Animal cruelty motivations – assessing demographic and situational influences. Journal of Interpersonal Violence 20: 1429–1443.

Hensley, C., Tallichet, S.E. et al. 2006. Exploring the possible link between childhood and adolescent bestiality and interpersonal violence. Journal of Interpersonal Violence 21: 910–923.

Hermes, R., Goritz, F. et al. 2007. Assisted reproduction in female rhinoceros and elephants – current status and future perspective. Reproduction in Domestic Animals 42: 33–44.

Herzog, H.A. 2007. Gender differences in human-animal interactions: a review. Anthrozoos 20: 7–21.

Herzog, H.A., Betchart, N.S. and Pittman, R.B. 1991. Gender sex role orientation and attitudes towards animals. Anthrozoös 4: 184–191.

Hibbert C. (editor) 1984. Queen Victoria in Her Lletters and Journals. Hammondsworth, UK.

Higgins, K.F., Naugle, D.E. et al. 2002. A case study of changing land use practices in the Northern Great Plains, USA: an uncertain future for waterbird conservation. Waterbirds 25: 42–50.

Higgs, A., Norris, R.T. et al. 1999. Mortality of sheep exported by sea: evidence of similarity by farm group and of regional differences. Australian Veterinary Journal 77: 729–733.

Higgs, A.R.B., Norris, R.T. and Richards, R.B. 1991. Season, age and adiposity influence death rates in sheep exported by sea. Australian Journal of Agricultural Research 42: 205–214.

Hill, C.A. 2000. The Garden of Eden: a modern landscape. Perspectives on Science and Christian Faith 52: 31–46.

Hill, P.C. and Pargament, K.I. 2003. Advances in the conceptualization and measurement of religion and spirituality – implications for physical and mental health research. American Psychologist 58: 64–74.

Hills, A.M. 1995. Empathy and belief in the mental experience of animals. Anthrozoös 8: 132–142.

Hodge, D.R. 2007. A systematic review of the empirical literature on intercessory prayer. Research on Social Work Practice 17: 174–187.

Hoffman, M.L. 1987. Empathy, its limitations, and its role in a comprehensive moral theory. In W.M. Kurtines and J.L. Gewitz Eds. Morality, Moral Behaviour, and Moral Development, pp. 283–302. New York: John Wiley.

Hoffmann, I. 2005. Research and investment in poultry genetic resources – challenges and options for sustainable use. Worlds Poultry Science Journal 61: 57–70.

Hogan, J.P. 1996. Options for manipulating nutrition if feed supply is immutable. Australian Journal of Agricultural Research 47: 289–305.

Hogan, J.P., Petherick, J.C. and Phillips, C.J.C. 2007. The nutritional impact on sheep and cattle of feed and water deprivation prior to and during transport. Nutrition Research Reviews 20: 1–13.

Holloway, L. 2001. Pets and protein: placing domestic livestock on hobby-farms in England and Wales. Journal of Rural Studies 17: 293–307.

Hoogland, C.T., de Boer, J. et al. 2005. Transparency of the meat chain in the light of food culture and history. Appetite 45: 15–23.

Hoogland, C.T., de Boer, J. et al. 2007. Food and sustainability: Do consumers package information and do customers recognize, understand and value production standards? Appetite 49: 47–57.

Hopster, H., Oconnell, J.M. et al. 1995. Acute effects of cow-calf separation on heart-rate, plasma cortisol and behavior in multiparous dairy cows. Applied Animal Behaviour Science 44: 1–8.

Houpt, K.A., Goodwin, D. et al. 2007. Proceedings of a workshop to identify dog welfare issues in the US, Japan, Czech Republic, Spain and the UK. Applied Animal Behaviour Science 106: 221–233.

Howard, C. and Parsons, E.C.M. 2006. Attitudes of Scottish city inhabitants to cetacean conservation. Biodiversity and Conservation 15: 4335–4356.

Hubrecht, R. 2000. Advancing rabbit and rodent housing and husbandry research. In Progress in the Reduction, Refinement and Replacement of Animal Experimentation. Proc. 3rd World Congress on Alternatives and Animal Use in the Life Sciences, Bologna, Italy ed. M. Balls, A.-M. van Zeller, and M.E. Halder, pp 1257–1267. Amsterdam, Elsevier Science B.V.

Hudson, M. 2007. Why do the numbers of laboratory animal procedures conducted continue to rise? An analysis of the Home Office Statistics of Scientific Procedures on Living Animals: Great Britain 2005. Alternatives to Laboratory Animals 35: 177–187.

Hui, V.K.Y., Bond, M.H. et al. 2006. General beliefs about the world as defensive mechanisms against death anxiety. Omega-Journal of Death and Dying 54: 199–214.

Humane Society of the United States (HSUS). 2005. [http://www.hsus.org/pets/issues_affecting_our_pets/pet_overpopulation_and_ownership_statistics/hsus_pet_overpopulation_estimates.html] cited by White, 2005

Humane Society of the United States (HSUS). 2008. Keeping your cat happy indoors. hsus.org/pets/pet_care/cat_care/keeping_your_cat_happy_indoors.html, accessed 30 January, 2008.

Hume, D. 1957. An Inquiry Concerning the Principle of Morals Vol. 4. Liberal Arts Press, New York original work 1751.

Humphrey, N. 2006. Seeing Red: A Study in Consciousness. The Belknap Press of Harvard University Press, Cambridge, Massachusetts, USA.

Hunting Act, 2004. http://www.opsi.gov.uk/acts/acts2004/ukpga_20040037_en_1. Accessed 8 January, 2008.

Illius, A.W., Clark, D.A. et al. 1992. Discrimination and patch choice by sheep grazing grass-clover swards. Journal of Animal Ecology 61: 183–194.

IUDZG/CBSG IUCN/SSC 1993. The World Zoo Conservation Strategy: The Role of Zoos and Aquaria of the World in Global Conservation. The Chicago Zoological Society, Chicago.

James, P.J. 2006. Genetic alternatives to mulesing and tail docking in sheep: a review. Australian Journal of Experimental Agriculture 46: 1–18.

Jarvis, A.M., Harrington, D.W.J. and Cockram, M.S. 1996. Effects of source and lairage on some behavioural and biochemical measurements of feed restriction and dehydration in cattle at a slaughterhouse. Applied Animal Behaviour Science 50: 83–94.

Kalenda, C. 2005. Ecological ethics: origins, principles, and applications. Problemos 68: 135–147.

Karpov, K. and Albin, D.P. 1995. The marine recreational fishery in northern and central California. Fish Bulletin, 176. State of California, the Resources Agency, Department of Fish and Game. http://www.psmfc.org/~wvanbusk/pub/bull176/bull176.htm, accessed January 2007.

Kellcher, K. 2005. Discards in the world's marine fisheries – an update. FAO Fisheries Technical Paper no. 470. Food and Agriculture Organisation of the United Nations, Rome.

Kellert, S.R. and Wilson, E.O. 1984. The Biophilia Hypothesis. Shearwater Book, Island Press, USA.

Kemp, J. 1996. Collecting medical specimens in South America: a dilemma in medical ethics. Anthropological Quarterly 69: 142–148.

Kendler, K.S., Neale, M.C., Kessler, R.C., Heath, A.C. and Eaves, L.J. 1993. Major depression and phobias: the genetic and environmental sources of comorbidity. Psychological Medicine 23: 361–371.

Kennedy, H., Little, M., Pell, G., Sommerville, H.D. and Jacobsen, L. 2007. Ethically Challenged – Big Questions for Science. Miegunyah Press, Melbourne.

Kerr, S.G.C., Woodgush, D.G.M., Moser, H. and Whittemore, C.T. 1988. Enrichment of the production environment and the enhancement of welfare through the Edinburgh Family Pen System of Pig Production. Research and Development in Agriculture 5: 171–186.

Kerwick, C., Meers, J. and Phillips, C.J.C. 2008. The training of veterinary personnel for effective identification and diagnosis of an exotic animal disease. Journal of Veterinary Medical Education, 35, 255–262.

Kiessling, L.L. 2002. Investigating and manipulating responses to the major antigen in xenotransplantation: The galactosyl-alpha 1-3 galactose epitope. Abstracts of Papers of the American Chemical Society 224: U259–U259.

Kildal, N. and Kuhnle, S. 2002 The Principle of Universalism: Tracing a Key Concept in the Scandinavian Welfare Model. Paper presented at the 9th Basic Income Economic Network International Congress, Geneva, 12–14 September. http://www.etes.ucl.ac.be/BIEN/Resources/Congress2002.htm, accessed 30 January, 2008.

Kirkcaldie, M.T.K. and Kitchener, P.D. 2007. When brains expand: mind and the evolution of cortex. Acta Neuropsychiatrica 19: 139–148.

Kjaer, J.B. and Sorensen, P. 2002. Feather pecking and cannibalism in free-range laying hens as affected by genotype, dietary level of methionine plus cystine, light intensity during rearing and age at first access to the range area. Applied Animal Behaviour Science 76: 21–39.

Klocker, U., Croft, D.B. et al. 2006. Frequency and causes of kangaroo-vehicle collisions on an Australian outback highway. Wildlife Research 33: 5–15.

Knapska, E., Nikolaev, E. et al. 2006. Between-subject transfer of emotional information evokes specific pattern of amygdala activation. Proceedings of the National Academy of Sciences of the United States of America 103: 3858–3862.

Knight, A. 2007. Intervening effects of knowledge, morality, trust, and benefits on support for animal and plant biotechnology applications. Risk Analysis 27: 1553–1563.

Kohut, H. 1977. The Restoration of the Self. International Universities Press: New York.

Kurpad, A.V., Khan, K., Calder, A.G. et al. 1994. Muscle and whole body metabolism after norepinephrine. American Journal of Physiology 266: E877–E884.

Kurtines, W.M., and Gewirtz, J.L. eds. 1984. Morality, Moral Behavior, and Moral Development. New York: John Wiley.

Lana, S.E., Kogan, L.R. et al. 2006. The use of complementary and alternative therapies in dogs and cats with cancer. Journal of the American Animal Hospital Association 42: 361–365.

Landau, R.E. 1999. A survey of teaching and implementation: the veterinarian's role in recognizing and reporting abuse. Journal of the American Veterinary Medical Association 215: 328–331.

Landry, D. 2001. Horsy and persistently queer: Imperialism, feminism and bestiality. Textual Practice 15: 467–485.

Lassen, J., Sandoe, P. et al. 2006. Happy pigs are dirty! Conflicting perspectives on animal welfare. Livestock Science 103: 221–230.

Layard, R. 2005. Happiness: Lessons from a New Science. Penguin Press, London.

Lea, E. J., Crawford, D. et al. 2006. Consumers' readiness to eat a plant-based diet. European Journal of Clinical Nutrition 60: 342–351.

Lea, J. 2005. Down the Cobbled Street, Memories of a Cheshire Farmer. Isis Publishing Ltd., Oxford.

Leak, G.K. and Christopher, S.B. 1982. Empathy from an evolutionary perspective. Journal for the Theory of Social Behaviour 12: 79–82.

Lectric Law Library's Lexicon LLL 2008. Cruelty. http://www.lectlaw.com/def/c332.htm, accessed January, 2008.

Leslie, T.K., Dalton, L. and Phillips, C.J.C. 2004. The preference of domestic rabbits for pasture and concentrate feeds. Animal Welfare 13: 57–62.

Lesniak, K.T. 2006. The effect of intercessory prayer on wound healing in nonhuman primates. Alternative Therapies in Health and Medicine 12: 42–48.

Levine, F.D., Mills, D.S. and Houpt, K.A. Attitudes of veterinary students at one US college towards factors relating to farm animal welfare. Journal of Veterinary Medical Education 32: 481–490.

Lewis, J.E. 2005. Statement of John E. Lewis, Deputy Assistant Director, Counterterrorism Division, Federal Bureau of Investigation, before the Senate Committee on Environment and Public Works, May 18, 2005. www.fbi.gov/congress/congress05/lewis051805.htm. Accessed 2 January, 2008.

Lewis-Williams, J.D. 1997. Agency, art and altered consciousness: a motif in French Quercy upper Palaeolithic parietal art. Antiquity 71: 810–830.

Li, C.H. 2000. A union of Christianity, humanity, and philanthropy: The Christian tradition and the prevention of cruelty to animals in nineteenth-century England. Society & Animals 8: 265–285.

Li, Z.-Y. 2004. Pet Related Trade has Exceed Ten Billion Yuan in China XinHua online. 03 November, 2004. http://news.xinhuanet.com/fortune/2004-11/03/content_2173277.htm. Retrieved 12 September 2006

Lin, W.J., Wang, W.W. et al. 2003. New animal model of emotional stress: Behavioral, neuroendocrine and immunological consequences. Chinese Science Bulletin 48: 1565–1568.

Lindeman, M. and Vaananen, M. 2000. Measurement of ethical food choice motives. Appetite 34: 55–59.

Lomas, C.A., Piggins, D. and Phillips, C.J.C. 1998. Visual awareness. In special issue on 'Animal Awareness, Concepts and Implications for Domesticated Animals' ed D. Piggins and C.J.C. Phillips. Applied Animal Behaviour Science 57, 247–257.

London, J. 1939. Jack London and His Times. New York, Doubleday.

Longuet-Higgins, C. 1994. Guest editorial: 150 words on consciousness Perception 23: 1265–1266.

Lorblanchet, M. 2007. The origin of art. Diogenes 54: 98.

Mader, T.L. and Davis, M.S. 2002. Wind speed and solar radiation corrections for the temperature–humidity index. American Meteorological Society. 15th Conference Biometereology and Aerobiology, Joint with 16th International Congress on Biometereology 2002.
Marai, I.F.M., El-Darawany, A.A. et al. 2007. Physiological traits as affected by heat stress in sheep – A review. Small Ruminant Research 71 1–3: 1–12.
Margerison, J.K., Phillips, C.J.C. and Preston, T.R. 2002. Restricted suckling of tropical dairy cows by their own calf or other cows' calves. Journal of Animal Science 80: 1663–1670.
Masters, K.S. and Spielmans, G.I. 2007. Prayer and health: review, meta-analysis, and research agenda. Journal of Behavioral Medicine 30: 329–338.
Matsuno, A.Y. and Thibault, L. 1995. Effect of sucrose and fructose macronutrient diets on feeding-behavior of rats. Physiology & Behavior 58: 1277–1280.
Matthews, B. 1937. The World in which Jesus Lived. pp 49–53. Oxford University Press, London.
Matthews, K.B., Wright, I.A. et al. 2006. Assessing the options for upland livestock systems under CAP reform: Developing and applying a livestock systems model within whole-farm systems analysis. Agricultural Systems 90 1–3: 32–61.
Mattison, E.H.A. and Norris, K. 2007. Intentions of UK farmers toward biofuel crop production: Implications for policy targets and land use change. Environmental Science & Technology 41: 5589–5594.
McCance, I. 1995. Assessment of statistical procedures used in papers in the Australian Veterinary Journal. Australian Veterinary Journal 72: 322–328.
McDermott, J.J., Randolph, T.F. et al. 1999. The economics of optimal health and productivity in smallholder livestock systems in developing countries. Revue Scientifique et Technique de l'Office International des Epizooties 18: 399–424.
McGuinness, K., Allen, M. et al. 2005. Non-accidental injury in companion animals in the Republic of Ireland. Irish Veterinary Journal 58: 392–396.
McKenna, J.E. and Melzack, R. 2001. Blocking NMDA receptors in the hippocampal dentate gyrus with AP5 produces analgesia in the formalin pain test. Experimental Neurology 172: 92–99.
Mehrotra, R. editor. 2005. The Essential Dalai Lama, His Important Teachings. Penguin Books, New York, USA.
Mehta, M.D. 2001. Pornography in Usenet: A study of 9,800 randomly selected images. Cyberpsychology & Behavior 4: 695–703.
Melendez, P., Bartolome, J. et al. 2003. The association between lameness, ovarian cysts and fertility in lactating dairy cows. Theriogenology 59 3–4: 927–937.
Melzack R. and Wall, P.D. 1965. Pain mechanisms: a new theory. Science 150: 971–979.
Melzack, R. 1987, Pain. In The Oxford Companion to the Mind, by O. L. Zangwill Author, Richard L. Gregory Editor. Oxford University Press, Oxford, United Kingdom.
Melzack, R. 1993. Pain: past, present and future. Canadian Journal of Experimental Psychology 47: 615–629.
Mercer Human Resources Consulting MHRC , 2007. World-wide quality of living survey http://www.mercerhr.com/pressrelease/details.jhtml/dynamic/idContent/117310. Accessed 1 August, 2007.
Miller, G. 2006. Animal behavior – signs of empathy seen in mice. Science 312: 1860–1861.
Modell, S.M. 2007. Approaching religious guidelines for chimera policymaking. Zygon 42: 629–641.
Mollenhorst, H., Rodenburg, T.B. et al. 2005. On-farm assessment of laying hen welfare: a comparison of one environment-based and two animal-based methods. Applied Animal Behaviour Science 90 3–4: 277–291.
Montonen, J., Knekt, P. et al. 2005. Dietary patterns and the incidence of type 2 diabetes. American Journal of Epidemiology 161: 219–227.
Moreland, J.P. 2001. Intelligent design psychology and evolutionary psychology: a comparison of rival paradigms. Journal of Psychology and Theology 29: 361–377.
Mori, Y., Ma, J.Y. et al. 2001. Hypothalamically induced emotional behavior and immunological changes in the cat. Psychiatry and Clinical Neurosciences 55: 325–332.

Mowatt, F. 1963. Never Cry Wolf. McClelland and Stewart Ltd., Toronto.
Mundkur, B. 1984. The bicephalic animal style in northern Eurasian religious art and its Western Hemispheric analogs. Current Anthropology 25: 451–482.
Munro, H.M.C. and Thrusfield, M.V. 2001a. 'Battered pets': features that raise suspicion of non-accidental injury. Journal of Small Animal Practice 42: 218–226.
Munro, H.M.C. and Thrusfield, M.V. 2001b. 'Battered pets': Munchausen syndrome by proxy and factitious illness by proxy. Journal of Small Animal Practice 42: 385–389.
Napolitano, F., De Rosa, G. et al. 2007. Bridging consumer perception and on-farm assessment of animal welfare. Animal Welfare 16: 249–253.
New South Wales, Department of Primary Industries NSW DPI, undated. Aquarium fish. http://www.fisheries.nsw.gov.au/aquatic_habitats/aquatic_habitats/aquarium_fish. Accessed 30 January, 2008.
Newell, G.A. undated. Samoyede; the People and their Dog. http://www.samoyed.org/samoyede_people.html. Accessed 8 January, 2008.
Newman, M. 2006. City Living. SASI Group University of Sheffield and Mark Newman University of Michigan http://www.sasi.group.shef.ac.uk/worldmapper/posters/worldmapper_map189_ver5.pdf. Accessed 9 January, 2008.
Nissenbaum, A. 1994. Sodom, Gomorrah and the other lost cities of the plain – a climatic perspective. Climatic Change 26: 435–446.
Nordgreen, J., Horsberg, T.E. et al. 2007. Somatosensory evoked potentials in the telencephalon of Atlantic salmon (Salmo salar) following galvanic stimulation of the tail. Journal of Comparative Physiology A – Neuroethology, Sensory, Neural and Behavioral Physiology 193: 1235–1242.
Norris, R.T. and Gorman, G.J. 2007. National Livestock Exports Mortality Summary 2006. 36 pp. Report B.LIV.0241 to Livecorp/Meat and Livestock Australia MLA. MLA, Sydney, Australia.
O'Connell, S.M. 1995. Empathy in chimpanzees – evidence for theory of mind. Primates 36: 397–410.
O'Brien, S.J., Troyer, J.L. et al. 2006. Plagues and adaptation: Lessons from the Felidae models for SARS and AIDS. Biological Conservation 131: 255–267.
O'Callaghan, K.A., Cripps, P.J. et al. 2003. Subjective and objective assessment of pain and discomfort due to lameness in dairy cattle. Animal Welfare 12: 605–610.
Office of the Queensland Parliamentary Counsel (OQPC). 2001. Animal Care and Protection Act 2001. Reprint 2A. OQPC, Brisbane.
Optimum Population Trust, 2007. Transport. www.optimumpopulation.org/opt.more.traffic.html. Accessed January, 2007
Ory, M.G. and Goldberg, E.L. 1984. An epidemiological study of pet ownership in the community. In: R.K. Anderson, B.L. Hart and L.A. Hart Eds. The Pet Connection: its Influence on Our Health and Quality of Life, pp. 320–330. Minneapolis: CENSHARE, University of Minneapolis.
Oviedo-Rondon, E.O., Ferket, P.R. et al. 2006. Understanding long bone development in broilers and turkeys. Avian and Poultry Biology Reviews 17: 77–88.
Parnell, S., Kettle, C. and Williamson, J. 2000. Sea fisheries – a renewable resource. http://helios.bto.ed.ac.uk/resman/nrm/Projects%201999/sea_fisheries.htm, accessed January 2007.
Passantino, A., Panichi, M., Russo, M.R. and Passantino, M. 2004. Ethics for research using animals in Italy. Alternatives to Laboratory Animals, 32, Supplement 1, 303–306.
Paul, 1994. Letter to the Romans, 8: 22. Good News Bible, pp 197. Harper Collins, London.
Paul, E. S. 1996. The representation of animals on children's television. Anthrozoos 9: 169–181.
Paul, E.S. 2000. Love of Pets and Love of People. In: A.L. Podberscek, E.S. Paul and J.A. Serpell Eds. Companion Animals and Us, pp 168–186. Cambridge University Press, Cambridge.
Peel, L.J. 1986. History of the Australian pastoral industries. Chapter 2 In: G. Alexander and O.B. Williams Eds. The Pastoral Industries of Australia, pp 41. Sydney University Press, Sydney.

Perenboom, R. J. M., L. M. Van Herten, et al. 2004. Trends in life expectancy and wellbeing. Social Indicators Research 65: 227–244.

Petcare Information & Advisory Service (PIAS). undated. The metropolitan cat. A survey of the population characteristics and hunting behaviour of the domestic cat in Australia. http://www.petnet.com.au/reark/reark.html. accessed 8 January, 2008.

Peters, J., Helmer, D. et al. 1999. Early animal husbandry in the Northern Levant. Paleorient 25, 27–47.

Phillips, C.J.C. 1990. Adverse effects on reproductive performance and lameness of feeding grazing dairy cows partially on silage indoors. Journal of Agricultural Science, Cambridge, 115, 253–258.

Phillips, C.J.C. 1994. El futuro rol de los animales de crianza en el medio ambiente rural. (The future role of farm animals in the rural environment). In Proceedings of the First Buiatrics' Conference of Latin America, Osorno, Chile, 25–27 November, 1993, pp.11–21.

Phillips, C.J.C. 2001. Principles of Cattle Production, 286 pp. CAB International, Wallingford.

Phillips, C.J.C. 2002. Cattle Behaviour and Welfare, 264 pp. Blackwell's Scientific, Oxford.

Phillips, C.J.C. 2004. Animal welfare matters: editorial. Australian Veterinary Journal 82: 63–64.

Phillips, C.J.C. 2005a. How animal welfare science assists in defining cruelty to animals. Proceedings of the 2005 RSPCA Australia Scientific Seminar, Canberra, 22 February 2005, pp 29–37. http://www.rspca.org.au/events/seminars05_proc.PDF

Phillips, C.J.C. 2005b. Meta-analysis – a systematic and quantitative review of animal experiments to maximise the information derived. Animal Welfare 14: 333–338.

Phillips, C.J.C. 2008b. The welfare of livestock during sea transport. In: Appleby, M., Cussen, V., Garces, L., Lambert, L.A. and Turner, J. Ed., Long Distance Transport and Welfare of Farm Animals. pp 137–156. CAB International, Wallingford.

Phillips, C.J.C. and Kitwood, S.E. 2003. The effects of feed supplements on the production, metabolism and behaviour of dairy cows fed low levels of concentrates during lactation. Journal of Applied Research in Veterinary Medicine, 1, 187–195.

Phillips, C.J.C. and Lomas, C.A. 2001. The perception of color by cattle and its influence on behavior. Journal of Dairy Science 84: 801–813.

Phillips, C. J. C. and McCulloch, S. 2005. Attitudes of students of different nationalities towards animal sentience and the use of animals in society, with implications for animal use in education. Journal of Biological Education 40: 17–24.

Phillips, C.J.C. and Morris, I.D. 2000. The locomotion of dairy cows on concrete floors that are dry, wet or covered with a slurry of excreta. Journal of Dairy Science 83: 1767–1772.

Phillips, C.J.C., Morris, I.D. and Lomas, C.A. 2000. The locomotion of cattle in passageways with different light intensities. Animal Welfare 9: 421–431.

Phillips and Prankel, C.J.C. 2008. A review of the effects of cadmium on the welfare of animals. In: Encyclopedia of Environmental Health, in press. Elsevier Press.

Phillips, C.J.C. and Sorensen, J.T. 1993. Sustainability in cattle production systems. Journal of Agricultural and Environmental Ethics 6: 61–73.

Phillips, C.J.C. and Youssef , M.Y.I. 2003. The effects of previous grazing experience and ewe presence on the response to novel grass species by weaned lambs. Animal Science 77: 335–341.

Pifer L., Shimizu K., and Pifer R. 1994. Public attitudes towards animal research: some international comparisons. Society and Animals 2: 95–113.

Piggins, D. and Phillips, C.J.C. 1998. Awareness in domesticated animals – concepts and definitions. In special issue on 'Animal Awareness, Concepts and Implications for Domesticated Animals' ed D. Piggins and C.J.C. Phillips. Applied Animal Behaviour Science 57: 181–200.

Pinkerton, J.M. 1807. Modern Geography. A Description of the Empires, Kingdoms, States and Colonies with the Oceans, Seas and Isles in all the most Recent Discoveries and Political Alteration. Volume 2, Cadell and Davies: Longman, Hust, Rees and Orme, London.

Plutchnik, R. 1987. Evolutionary basis of empathy. In N. Eisenberg and J. Strayer Ed. Empathy and Its Development pp 38–46. Cambridge University Press, Cambridge.

Pocard, M. 1999. Origin of animal experimentation legislation in the 19th century. Annales De Chirurgie 53: 627–631.

Pongratz, H. and Schmitt, M. 1990. Prospects for Professional-Training in Farming. Berichte Uber Landwirtschaft 68: 567–581.

Poole, T. 1997. Happy animals make good science. Laboratory Animals 31: 116–124.

Popkin, B.M. and S.F. Du 2003. Dynamics of the nutrition transition toward the animal foods sector in China and its implications: A worried perspective. Journal of Nutrition 133: 3898S–3906S.

Poresky, R.H. 1990. The young children's empathy measure: reliability, validity and effects of companion animal bonding. Psychological Reports 66: 931–936.

Poresky, R.H. 1996. Companion animals and other factors affecting young children's development. Anthrozoös 4: 159–168.

Poresky, R.H. and Hendrix, C. 1990. Differential effects of pet presence and pet bonding on young children. Psychological Reports 67: 51–54.

Poss, J.E. and Bader, J.O. 2007. Attitudes toward companion animals among Hispanic residents of a Texas border community. Journal of Applied Animal Welfare Science 10: 243–253.

Powell, D. 2004. Chimera contemporary: The enduring art of the composite beast. Leonardo 37: 332–340.

Preece, R. and Fraser, D. 2000. The status of animals in biblical and Christian thought: a study in colliding values. Society and Animals 8: 245–263.

Preston, T.R. and Willis, M.S. 1974. Intensive Beef Production. Second Edition. Pergamon Press, Oxford.

Pretty, J., Peacock, J. et al. 2007. Green exercise in the UK countryside: Effects on health and psychological well-being, and implications for policy and planning. Journal of Environmental Planning and Management 50: 211–231.

Price, E.O. 1984. Behavioural aspects of animal domestication. Quarterly Review in Biology 59: 1–32.

Pruimboom, L. and van Dam, A.C. 2006. Chronic pain: A non-use disease. Medical Hypotheses 68: 506–511.

Radford, M. 2001. Animal Welfare Law in Britain. Oxford University Press, Oxford.

Ragin, C. 1994. A qualitative comparative analysis of pension systems, In: T. Janoski and A.M. Hicks Ed. The Comparative Political Economy of the Welfare State. Cambridge University Press, Cambridge.

Rahmann, G. 2000. Marketing analysis for meat from nature conservation in the Biosphere Reserve Rhon. Berichte Uber Landwirtschaft 78: 248–261.

Ravindran, V., Thomas, D.V. et al. 2006. Performance and welfare of broilers as affected by stocking density and zinc bacitracin supplementation. Animal Science Journal 77: 110–116.

Regan, T. 1983. The Case for Animal Rights. Paperback Edition. University of California Press, Berkeley, California, USA.

Regan, T. 1990. Christianity and animal rights: the challenge and promise. In: C. Birch, W. Eaken and J.B. McDaniel Eds. Liberating Life: Contemporary Approaches in Ecological Theology. Chapter 6, pp 73–87. Orbis Books, Maryknoll, New York.

Rifkin, J. 1994. Beyond Beef: The Rise and Fall of Cattle Culture, 353 pp. Thorsons, London.

Ritvo, H. 1992. Race, breed, and myths of origin – Chillingham cattle as ancient Britons. Representations 39: 1–22.

Ritvo, H. 2001. Ch 12. The natural world. In: J.M MacKenzie Ed. The Victorian Vision, pp 281–295. Victoria & Albert Publications, London.

Rizov, M. 2003. An essay on the agricultural production organization in former communist countries. Social Science Journal 40: 665–672.

Rochlitz, I. 2004. The effects of road traffic accidents on domestic cats and their owners. Animal Welfare 13: 51–55

Roels, L. 1999. Opt out registers for organ donation have existed in Belgium since 1981. British Medical Journal 318: 399–399.
Rojas, M. 1999 The Rise and Fall of the Swedish Model. Social Market Foundation, London
Roland, A. 2007. Technology and War. American Diplomacy 2: http://www.unc.edu/depts/diplomat/AD_Issues/amdipl_4/roland.html. Accessed December, 2007.
Rollin, B.E. 2003. Foreward. In: S.J. Armstrong and R.G. Botzler Eds. The Animal Ethics Reader, Routledge, New York.
Rollin, B.E. 2006. An Introductory to Veterinary Medical Ethics. Theory and Cases. Second Edition. Blackwell Publishing, Oxford.
Roman, R.M., Wilcox, C.J. et al. 1999. Genetic trends for milk yield of Jerseys and correlated changes in productive and reproductive performance. Journal of Dairy Science 82: 196–204.
Rost, D.H. and Hartmann, A. 1994. Children and their pets. Anthrozoös 7: 242–254.
Rousing, T., Jakobsen, I.A., Hindhede, J., Klaas, I.C., Bonde, M., and Rensen, J.T. 2007. Evaluation of a welfare indicator protocol for assessing animal welfare in AMS herds: researcher, production advisor and veterinary practitioner opinion. Animal Welfare 16: 213–216.
Royal Society for the Protection of Cruelty to Animals RSPCA 2002a. RSPCA Australia National Statistics 2001–2002 http://www.rspca.org.au/resource/stats2002.pdf. Accessed January, 2008.
Royal Society for the Protection of Cruelty to Animals RSPCA 2002b. Kangaroo shooting compliance. A survey of the extent of compliance with the requirements of the Code of Practice for the Humane Shooting of Kangaroos. Department of Environment and Water Resources, Australian Government, Canberra.
Royal Society for the Protection of Cruelty to Animals, RSPCA 2006. RSPCA research confirms public believe live export is cruel. www.rspca.org.au/media releases/MRShow.asp, accessed January 2007.
Russell, W.M.S. and Burch, R.L. 1959. Principles of Humane Experimental Technique. Methuen, London.
Savolainen, P., Leitner, T. et al. 2004. A detailed picture of the origin of the Australian dingo, obtained from the study of mitochondrial DNA. Proceedings of the National Academy of Sciences of the United States of America 101: 12387–12390.
Schmid, J., Heistermann, M. et al. 2001. Introduction of foreign female Asian elephants (Elephas maximus) into an existing group: Behavioural reactions and changes in cortisol levels. Animal Welfare 10: 357–372.
Schoenemann, P.T. 2006. Evolution of the size and functional areas of the human brain. Annual Review of Anthropology 35: 379–406.
Schulte, B. A. 2000. Social structure and helping behavior in captive elephants. Zoo Biology 19: 447–459.
Schuppli, C.A. and Fraser, D. 2005. The interpretation and application of the Three R's by Animal Ethics Committee members. Alternatives to Laboratory Animals 33: 487–500.
Schuppli, C.A. and Fraser, D. 2007. Factors influencing the effectiveness of research ethics committees. Journal of Medical Ethics: 33: 294–301.
Schuppli, C.A. and McDonald, M. 2005. Contrasting modes of governance for the protection of humans and animals in Canada: lessons for reform. Health Law Review 13: 97–106.
Schuppli, C.A., Fraser, D. and McDonald, M. 2004. Expanding the 3Rs to meet new challenges in humane animal education. Alternatives to Laboratory Animals 32: 525–532.
Schwartz, S. H. 2007. Universalism values and the inclusiveness of our moral universe. Journal of Cross-Cultural Psychology 38: 711–728.
Seabrook, M.F. and Wilkinson, J.M. 2000. Stockpersons' attitudes to the husbandry of dairy cows. Veterinary Record 147: 157–160.
Seamer, J.H. 1998. Human stewardship and animal welfare. Applied Animal Behaviour Science 59: 201–205.

Sedlackova, M., Bilcik, B. et al. 2004. Feather pecking in laying hens: Environmental and endogenous factors. Acta Veterinaria Brno 73: 521–531.
Serpell, J. 1986. In the Company of Animals, pp 66–68. Blackwell, Oxford.
Severinghaus, L.L. and Chi, L. 1999. Prayer animal release in Taiwan. Biological Conservation 89: 301–304.
Shanahan, E.W. 1925. The world's cattle situation and prospects. In: G.F. Finlay Ed Cattle Breeding, pp 482–488. Oliver and Boyd, Edinburgh.
Shannon, G., Page, B. et al. 2006. African elephant home range and habitat selection in Pongola Game Reserve, South Africa. African Zoology 41: 37–44.
Sharpe, M.S. and Wittum, T.E. 1999. Veterinarian involvement in the prevention and intervention of human violence and animal abuse: A survey of small animal practitioners. Anthrozoos 12: 97–104.
Sheldrick, D. 1966. The Orphans of Tsavo, Pp 90–92 and 159. Collins and Harvill Press, London.
Shook, G.E. 2006. Major advances in determining appropriate selection goals. Journal of Dairy Science 89: 1349–1361.
Sidhom, P. 2003. Welfare of cattle transported from Australia to Egypt. Letter to the Editor. Australian Veterinary Journal 81: 364–365.
Siepe, M., Martin, J. et al. 2007. Anatomical study on the surgical technique used for xenotransplantation: Porcine hearts into humans. Journal of Surgical Research 143: 211–215.
Sillence, M.N. 2004. Technologies for the control of fat and lean deposition in livestock. Veterinary Journal 167: 242–257.
Simpson, J. 2004. The Wars against Saddam, pp 55. Pan Books, London.
Singer, P. 1975. Animal Liberation. Avon Books, New York, USA.
Singer, P. (ed.) 2005. In Defense of Animals: the Second Wave. Blackwell Publishing, Oxford.
Singer, P. and Cavalieri, P. Editors 1993. The Great Ape Project: Equality beyond Humanity. Fourth Estate publishing, London, England.
Smith W., 1880. A Concise Dictionary of the Bible, Its Antiquities, Biography, Geography and Natural History. Murray, London.
Smith, F.M. 2000. Indra goes west: Report on a Vedic soma sacrifice in London in July 1996 (regarding Western European hierarchies and modes of sectarian Hindu worship). History of Religions 39: 247–267.
Sneddon, L.U., Braithwaite, V.A. and Gentle, M.J. 2003. Novel object test: examining nociception and fear in the rainbow trout. Journal of Pain 4: 431–440.
Solti, L., Crichton, E.G. et al. 2000. Economical and ecological importance of indigenous livestock and the application of assisted reproduction to their preservation. Theriogenology 53: 149–162.
Sommerville, B.A. and Broom, D.M. 1998. Olfactory awareness. In special issue on 'Animal Awareness, Concepts and Implications for Domesticated Animals' ed D. Piggins and C.J.C. Phillips. Applied Animal Behaviour Science 57: 269–286.
Stacey, C. 2003. Development of a heat stress risk management model. Livecorp Report no. 116 to Meat & Livestock Australia, North Sydney NSW.
Starr, C.G. 1973. Early Man, Prehistory and the Civilizations of the Ancient Near East. New York, Oxford University Press.
Stein, E. 1964. On the Problem of Empathy. Martinus Nijhoff, The Hague.
Steiss, J.E., Schaffer, C. et al. 2007. Evaluation of plasma cortisol levels and behavior in dogs wearing bark control collars. Applied Animal Behaviour Science 106: 96–106.
Stoyanchev, K., Sotirov, L. et al. 2007. Study on the level of natural humoral immunity in turkey-broilers with muscular dystrophy, reared under conditions of either animal welfare or stress. Revue de Medecine Veterinaire 158: 314–319.
Stull, C.L. 1996. History of US equine welfare and legislation. Pferdeheilkunde 4: 391–392.
Swaisgood R.R., Ellis S., Forthman D.L., Shepherdson, D.J. 2003. Commentary: Improving well-being for captive giant pandas: Theoretical and practical issues. Zoo Biology 22: 347–354.

Sylvester, S.P., Stafford, K.J. et al. 2004. Behavioural responses of calves to amputation dehorning with and without local anaesthesia. Australian Veterinary Journal 82: 697–700.
Tatara, M.R. 2006. Current methods for in vivo assessment of the skeletal system in poultry. Medycyna Weterynaryjna 62: 266–269.
Taylor, A.A., and Weary, D.M. 2000. Vocal responses of piglets to castration: identifying procedural sources of pain. Applied Animal Behaviour Science 70: 17–26.
Taylor, K.F. 1994. Architectural empathy and the renaissance of French classicism. Journal of the Society of Architectural Historians 53: 239–242.
Thackeray, J.F. 2005. The wounded roan: a contribution to the relation of hunting and trance in southern African rock art. Antiquity 79: 5–18.
Thatcher, M. 1986. In Defence of Freedom, Speeches on Britain's Relations with the World, 1976–1986, pp 115. Aurum Press, London.
Thatcher, M. 1995. Margaret Thatcher, the Downing Street Years, pp 733. Harper Collins Publishers, London.
The Koran, 1990. Translated with notes by N.J. Dawood. Fifth Revised Edition. Penguin Books, London.
Thomas, J. 2003. The symbolic interpretation of rock engravings from Mont Bego. Anthropologie 107: 271–290.
Thompson, P.B. 1997. Ethics and the genetic engineering of food animals. Journal of Agricultural & Environmental Ethics 10: 1–23.
Thompson, P., Harris, C. et al. 2007. Livestock welfare product claims: The emerging social context. Journal of Animal Science 85: 2354–2360.
Titchener, E. 1924. A Textbook of Psychology. Macmillan, New York.
Tomkins, S. 1963. Affect, Imagery, Consciousness Volume 2: The Negative Affects.Springer, New York.
Tomkins, S. 1991. Affect, Imagery, Consciousness Volume 3: The Negative Affects: Anger and Fear. Springer, New York.
Toro, M., Azcon, R. et al. 1998. The use of isotopic dilution techniques to evaluate the interactive effects of Rhizobium genotype, mycorrhizal fungi, phosphate-solubilizing rhizobacteria and rock phosphate on nitrogen and phosphorus acquisition by Medicago sativa. New Phytologist 138: 265–273.
Toulmin, S. 1999. The ambiguities of globalization. Futures 31: 905–912.
Trevelyan, G.M. 1949. An Autobiography and Other Eessays. Longman Green, London.
Tuchscherer, M. and Manteuffel, G. 2000. The effect of psycho stress on the immune system. Another reason for pursuing animal welfare Review. Archiv Fur Tierzucht-Archives of Animal Breeding 43: 547–560.
Tucker, A., Belcher, C. et al. 2002. The production of transgenic pigs for potential use in clinical xenotransplantation: microbiological evaluation. Xenotransplantation 9: 191–202.
UNESCO, 2001. Unesco seminar: Democracy and World Governance in the 21st Century, Porto Alegre Brazil, 29–30 January 2001 http://www.unesco.org/most/wsfunesco.htm, accessed 15 January, 2008.
United Egg, 2007. US Egg Industry. http://www.unitedegg.org/useggindustry_generalstats.aspx. Accessed 20 January, 2007
United States Department of Agriculture USDA , undated. Poultry Production in the United States Economic Research Service/USDA; Broiler Farms. Organization, Management, and Performance/ AIB-748. http://www.ers.usda.gov/publications/aib748/aib748b.pdf. Accessed 20 January 2008.
University Federation for Animal Welfare (UFAW). Undated. A History of Improving Animal Welfare. http://www.ufaw.org.uk/history.php. Accessed 15 January, 2008.
Uzogara, S.G. 2000. The impact of genetic modification of human foods in the 21st century: A review. Biotechnology Advances 18: 179–206.
Vaarst, M., Jensen, M.B. et al. 2001. Behaviour of calves at introduction to nurse cows after the colostrum period. Applied Animal Behaviour Science 73: 27–33.

Valladas, H., Cachier, H. et al. 1992. Direct Radiocarbon-Dates for Prehistoric Paintings at the Altamira, El-Castillo and Niaux Caves. Nature 357: 68–70.

Valladas, H., Tisnerat-Laborde, N. et al. 2001. Radiocarbon AMS dates for paleolithic cave paintings. Radiocarbon 43 (2B): 977–986.

Van Dooren, R. 2006. La Laguna: of exporting jeans and changing labour relations. Tijdschrift Voor Economische En Sociale Geografie 97: 480–490.

Vannuccini, S. 2003. Overview of fish production, utilization, consumption and trade. Food and Agricultural Organization of the United Nations, Fishery Information, Data and Statistics Unit. FAO, Rome.

Verhulst, A. 1993. Lessons from field experiences in the development of monogastric animal production, 271 pp. FAO Animal Production and Health Papers – 107. FAO, Rome.

Vitulli, W.F. 2006. Attitudes toward empathy in domestic dogs and cats. Psychological Reports 99: 981–991.

Walvin, J. 2007. Abolishing the slave trade. www.history.ac.uk/ihr/Focus/Slavery/articles/walvin.html. Accessed, 9 January, 2008.

Wang, H. and Kuang, Y. 2007. Alternative models for cyclic lemming dynamics. Mathematical Biosciences and Engineering 4: 85–99.

Warnes, M.L. 1995. Field studies on the effect of cattle skin secretion on the behavior of tsetse. Medical and Veterinary Entomology 9: 284–288.

Warner, R.F., Ferguson, D.M., Cottrell, J.J., and Knee, B.W. 2007. Acute stress induced by the preslaughter use of electric prodders causes tougher beef meat. Australian Journal of Experimental Agriculture, 47:782–788.

Webster A.J.F. 1995. Animal Welfare: A Cool Eye towards Eden. Oxford, U.K., Blackwell

Webster, A.J.F. 2002. Rendering unto Caesar: welfare problems in Belgian Blue cattle. Veterinary Journal 163: 228–229.

Webster, A.J.F., Main, D.C.J. et al. 2004. Welfare assessment: indices from clinical observation. Animal Welfare 13: S93–S98.

Wemelsfelder F., Hunter T.E.A., Mendl M.T., and Lawrence A.B. 2001. Assessing the 'whole animal': a free choice profiling approach. Animal Behaviour 62: 209–220.

Westhoek, H.J., van den Berg, M. et al. 2006. Scenario development to explore the future of Europe's rural areas. Agriculture Ecosystems & Environment 114: 7–20.

Westhoek, H., van den Berg, R. et al. 2004. Economic and environmental effects of the manure policy in The Netherlands: synthesis of integrated ex-post and ex-ante evaluation. Water Science and Technology 49: 109–116.

Whay, H.R., Main, D.C.J., Green, L.E. and Webster, A.J.F. 2003. Animal-based measures for the assessment of welfare state of dairy cattle, pigs and laying hens: consensus of expert opinion. Animal Welfare 12: 205–217.

Whistance, L., Arney, D.R., Sinclair, L.A. and Phillips, C.J.C. 2007. Defaecation behaviour of dairy cows housed in straw yards or cubicle systems. Applied Animal Behaviour Science 105: 14–25.

Whitley, D.S. 1994. By the hunter, for the gatherer – art, social relations and subsistence change in the prehistoric Great-Basin. World Archaeology 25: 356–373.

Wiebers, D.O., Barron, R.A., Leaning, J. and Ascione, F.R. 1994. Ethics and animal issues in United States medical education. Medical Education 28: 517–523.

Wiese, M. 1996. Searching for the biological roots of self-care and care behaviour. 1. The evolutional importance of self-care and care. Pflege 9: 105–112.

Wiese, R. J. and Willis, K. 2004: Calculation of longevity and life expectancy in captive elephants. Zoo Biology 23: 365–373.

Wildlife Information and Rescue Service. 2007. Protect your cat, protect your wildlife. http://members.tripod.com/wires_msc/page8.html. Accessed February, 2007.

Wilkinson, J.M., Hill, J. and Phillips, C.J.C. 2003. The accumulation of potentially toxic metals by grazing ruminants. Proceedings of the Nutrition Society 62, 267–277.

Wilson, A. 2003. RSPCA baulks on horse cruelty , The Australian, 29 July 2003, pp 3.

Wilson, E.O. 1975. Sociobiology: The New Synthesis. Belknap, Cambridge, MA.

Wilson, H.D., Boyette-Davis, J. and Fuchs, P.N. 2007. The relationship between basal level of anxiety and the affective response to inflammation. Physiology and Behaviour 90: 506–511.

Wilson, M.L., Bloomsmith, M.A. et al. 2004. Stereotypic swaying and serum cortisol concentrations in three captive African elephants (Loxodonta Africana). Animal Welfare 13: 39–43.

Wispe, L. 1986. The distinction between sympathy and empathy: to call forth a concept, a word is needed. Journal of Personality and Social Psychology 50: 314–321.

Wispe, L. 1987. History of the concept of empathy. In N. Eisenberg and J. Strayer ed., Empathy and its Development, pp 17–37. Cambridge University Press, Cambridge.

Wojciechowska, J., Meng, J., Cross, N. and Phillips, C.J.C.2006. Industry perceptions of animal welfare in the red meat industry. Report to Meat and Livestock Australia, Sydney, Australia

Woodgush, D.G.M. 1983. Elements of Ethology. Chapman and Hall, London.

World Association of Zoos and Aquariums 2005. Building a Future for Wildlife, World Zoo and Aquarium Conservation Strategy. http://www.waza.org/conservation/wzacs.php. Accessed February 2007.

World Society for the Protection of Animals (WSPA). 2007. Animals matter to me. http://www.animalsmatter.org/downloads/UDAW_Text_2005.pdf. Accessed 21 January, 2007.

Yang, Y.G. and Sykes, M. 2007. Xenotransplantation: current status and a perspective on the future. Nature Reviews Immunology 7: 519–531.

Yeates, J. and Main, D.C. 2008. Assessment of positive welfare: a review. The Veterinary Journal, 175, 293–300.

Young, A.W., 1994. Neuropschology of awareness. Chapter 8 In: A. Revonsuo and M. Kamppinnen, Consciousness in Philosophy and Cognitive Neuroscience. Erlbaum, Hillsdale.

Zimmerman, M. 1986. Behavioural investigations of pain in animals. In: Duncan I.J.H. and Molony, V. Eds Assessing Pain in Farm Animals, pp 30–35. Commission of the European Communities, Luxembourg.

Index

Kolibri 360 GmbH
Ferdinand-Jühlke-Straße 7
99095 Erfurt, Deutschland
produktsicherheit@kolibri360.de

Max-Planck-Institut für ausländisches
öffentliches Recht und Völkerrecht

Beiträge zum ausländischen öffentlichen Recht und Völkerrecht

Begründet von Viktor Bruns

Herausgegeben von
Rudolf Bernhardt · Karl Doehring
Jochen Abr. Frowein

Band 81

Völkerrecht als Rechtsordnung
Internationale Gerichtsbarkeit
Menschenrechte

Festschrift für Hermann Mosler

Herausgegeben von
Rudolf Bernhardt · Wilhelm Karl Geck
Günther Jaenicke · Helmut Steinberger

Springer-Verlag Berlin Heidelberg New York 1983

ISBN-13: 978-3-642-68694-8 e-ISBN-13: 978-3-642-68692-4
DOI: 10.1007/978-3-642-68692-4

CIP-Kurztitelaufnahme der Deutschen Bibliothek
Völkerrecht als Rechtsordnung – internationale Gerichtsbarkeit – Menschenrechte :
Festschr. für Hermann Mosler / hrsg. von Rudolf Bernhardt ... –
Berlin ; Heidelberg ; New York : Springer, 1983.
(Beiträge zum ausländischen öffentlichen Recht und Völkerrecht ; Bd. 81)
ISBN-13: 978-3-642-68694-8

NE: Bernhardt, Rudolf [Hrsg.]; Mosler, Hermann: Festschrift; GT

Softcover reprint of the hardcover 1st edition 1983

2140/3130-543210

Festschrift für Hermann Mosler

Hermann Mosler
zum 26. Dezember 1982

Diese Festschrift zum 70. Geburtstag von Hermann Mosler trägt die Überschrift: Völkerrecht als Rechtsordnung – Internationale Gerichtsbarkeit – Menschenrechte. Es sind drei Themenbereiche, die das wissenschaftliche Werk und das praktische Wirken Hermann Moslers seit Jahrzehnten kennzeichnen.

Seit der 1912 in Hennef im Rheinland Geborene nach Abschluß des Studiums 1937 in das Kaiser-Wilhelm-Institut für ausländisches öffentliches Recht und Völkerrecht in Berlin eintrat, blieb er mit diesem Institut und dem späteren Max-Planck-Institut sowie mit den dort gepflegten Wissenschaftszweigen aufs engste verbunden. In den schweren Jahren bis zum Ende des Zweiten Weltkriegs legte Hermann Mosler in dem Berliner Institut, von dessen Mitarbeitern die meisten dem Ungeist der Zeit trotzten, die Grundlagen für sein weiteres wissenschaftliches Werk. Nach dem Krieg folgten in raschem Wechsel neue Aufgaben: 1946 die Habilitation in Bonn bei Richard Thoma, 1949 die erste Professur in Frankfurt, bald darauf zugleich die Leitung der Rechtsabteilung des neuen Auswärtigen Amtes der Bundesrepublik Deutschland. 1954 folgte die Übernahme der Leitung des Max-Planck-Instituts (als Nachfolger von Carl Bilfinger) und einer Professur an der Ruprecht-Karl-Universität in Heidelberg. 1959 wurde Hermann Mosler zum Richter am Europäischen Gerichtshof für Menschenrechte gewählt, 1976 wurde er Richter am Internationalen Gerichtshof.

Vita contemplativa und *vita activa* sind in Hermann Mosler eine ungewöhnlich glückliche Verbindung eingegangen. Die Wissenschaft stand im Vordergrund, aber man brauchte und braucht den Professor des Völkerrechts, des deutschen öffentlichen Rechts und der öffentlichrechtlichen Rechtsvergleichung immer wieder auch in der Praxis.

Das Kaiser-Wilhelm-Institut in Berlin hat den jungen Hermann Mosler geprägt, er selbst hat das Nachfolgeinstitut in Heidelberg geprägt. Die Herausgeber dieser Festschrift sind ihm teils schon in Berlin, teils in Frankfurt begegnet, um dann in den Heidelberger Jahren erneut mit ihm zusammenzuarbeiten. In Heidelberg hat Hermann Mosler ihre – und zahlreiche weitere – Habilitationen gefördert und betreut.

Wir sind mit vielen früheren und jetzigen Mitarbeitern des Instituts der Überzeugung, daß das Institut zu den in fachlicher und menschlicher Hinsicht herausragenden Forschungsstätten gehört, und das ist für die Nachkriegszeit im wesentlichen Hermann Mosler zu verdanken. Viele Bände der Zeitschrift und der Beiträge zum ausländischen öffentlichen Recht und Völkerrecht sowie der Fontes Iuris Gentium sind während der Leitung des Instituts durch Hermann Mosler erschienen. Große rechtsvergleichende und völkerrechtliche Kolloquien haben Gelehrte aus allen Ländern der Welt nach Heidelberg geführt. Viele Hochschullehrer in der Bundesrepublik und im Ausland, viele Praktiker, die in hohen Positionen, etwa als Rechtsberater und Botschafter, die internationalen Beziehungen der Bundesrepublik pflegen, sind durch die Schule des Instituts und damit die Schule Hermann Moslers gegangen. Er selbst fand bei allen Aufgaben immer wieder Zeit und Kraft für eigene wissenschaftliche Arbeiten. Sie sind im Anhang zu dieser Festschrift zusammengestellt. Von ihnen seien hier nur die beiden Vorlesungsreihen in der Haager Académie de Droit international und die entsprechenden Publikationen erwähnt. Die erste befaßte sich 1957 mit der innerstaatlichen Anwendung des Völkerrechts, die zweite durchmißt unter dem Titel »The International Society as a Legal Community« in einem großen Wurf die wichtigsten Bereiche des Völkerrechts.

Ungewöhnlich sind auch, als Folge dieser Leistungen, die Ehrungen durch wissenschaftliche Gremien und Institutionen. Dem Institut de Droit international gehört Hermann Mosler seit 1957 an. 1972 erfolgte die Wahl zum Korrespondierenden Mitglied der Österreichischen Akademie der Wissenschaften. An seinem 70. Geburtstag ist Hermann Mosler seit wenigen Wochen Präsident der Heidelberger Akademie der Wissenschaften. Ehrendoktorate der Freien Universität Brüssel (1969) und der Universität des Saarlandes in Saarbrücken (1982) kamen hinzu. Die Deutsche Gesellschaft für Völkerrecht wählte Hermann Mosler für einige Jahre zu ihrem Vorsitzenden, das Auswärtige Amt der Bundesrepublik Deutschland berief ihn in seinen Völkerrechtswissenschaftlichen Beirat.

1959 begann das faszinierende und erfolgreiche Experiment des Europäischen Gerichtshofs für Menschenrechte. Hermann Mosler gehörte zu den ersten Richtern; später war er – zeitweise als Vizepräsident – einer der dienstältesten Richter, bis er 1980/81 freiwillig wegen der Belastungen durch das Richteramt in Den Haag zurücktrat. Der Straßburger Menschenrechtsgerichtshof hatte in den ersten Jahren seines Bestehens nur wenige Fälle zu entscheiden, aber das änderte sich. Die letzten Entscheidungen, an denen Hermann Mosler noch mitwirkte, ergingen 1981 und tragen die amtlichen Zahlen 42 und 43. Die zunehmende Belastung durch die Straßburger Richtertätigkeit hat Hermann Mosler neben seinen zahlreichen anderen Verpflichtungen mit gleichbleibender Pflichtauffassung getragen. Der Menschenrechtsgerichtshof hat ein Stück europäischer Völkerrechtsgeschichte geschrieben, und Hermann Mosler hat daran tatkräftig und überzeugend mitgewirkt. Daß das mit der Akribie des Wissenschaftlers und dem Verantwortungsbewußtsein dessen geschah, der sich dem Staat und Europa gleichermaßen verpflichtet weiß, ergibt sich aus den Entscheidungen des Gerichtshofes ebenso wie aus den zahlreichen wissenschaftlichen Abhandlungen Hermann Moslers zum Schutz der Menschenrechte.

Ende 1975, als die Arbeit Hermann Moslers ihren Höhepunkt erreicht zu haben schien, wählten ihn Generalversammlung und Sicherheitsrat der Vereinten Nationen zum Richter am Internationalen Gerichtshof. Vor ihm war nur einmal ein Deutscher, Walter Schücking, von 1931–1935 ständiges Mitglied des Weltgerichtshofes gewesen. Die Wahl Hermann Moslers wenige Jahre nach dem Beitritt der Bundesrepublik Deutschland zu den Vereinten Nationen war ein Zeichen für das internationale Ansehen, das der Gewählte besaß. Seit 1976 und weit über den 70. Geburtstag hinaus gehört Hermann Mosler dem Gerichtshof an, zweifellos auch hier als außerordentlich aktives Mitglied.

In einer Festschrift stehen meist das äußere Wirken und die wissenschaftlichen Leistungen im Vordergrund, und so ist es auch hier. Aber es darf nicht unerwähnt bleiben, daß mit dem öffentlichen Wirken des Jubilars stets menschliche Wärme, Verständnis für andere und Gastfreundschaft einhergingen. Hierin haben Hermann und Anne Mosler gemeinsam und einzeln hohe Maßstäbe gesetzt. Auch hierfür danken die Autoren und die Herausgeber dieser Festschrift, sie sprechen dabei für viele andere.

RUDOLF BERNHARDT WILHELM KARL GECK
GÜNTHER JAENICKE HELMUT STEINBERGER

Inhaltsverzeichnis

Wesen und Wirkung des »Konsens« im Völkerrecht

HANS BALLREICH

I. Zur Fragestellung

1. In seinem ideenreichen und gründlichen Werk zum Verständnis der internationalen Gesellschaft als Rechtsgemeinschaft hat *Hermann Mosler*, dem diese Betrachtung in herzlicher Verbundenheit gewidmet ist, nur wenig zum Problem des »Consensus« bzw. der sogenannten »Konsens-Resolution« ausgesagt[1]. Er bemerkt lediglich, eher beiläufig, daß trotz aller in den Satzungen internationaler Organisationen enthaltener Bestimmungen über Abstimmungen und die erforderlichen Mehrheiten sich daneben ein anderes, ein mehr informelles Verfahren ausgebildet habe, dessen man sich bediene, wenn dem förmlichen Verfahren widrige Umstände entgegenstehen. In kontroversen und delikaten Angelegenheiten bestehe heute die Tendenz, dann, wenn durch eine förmliche Abstimmung ein in der Sache bestehender Dissens offenkundig werden könne, deshalb das kompetente internationale Organ Schaden nehmen würde, an Stelle der förmlichen Abstimmung die Feststellung eines bestehenden Consensus durch den Vorsitzenden treten zu lassen. »Consensus cannot, however, have a binding effect on the members of the organization«, sagt dazu *Mosler* abschließend[2].

2. Als *Mosler* Jahre vor dem Erscheinen als Buch seine Arbeit niedergeschrieben hat[3], war es eine offene Frage, ob der Weg der Konsensfeststellung, auf den ersten Blick erkennbar mit mannigfachen Problemen behaftet, sich einführen und nennenswerte praktische Bedeutung erlangen würde. Inzwischen ist er aus dem Verfahren der Willensbildung insbesondere der internationalen Organisationen nicht mehr wegdenkbar. Der Weg der Konsens-

---

[1] *Hermann Mosler*, The International Society as a Legal Community (Alphen aan den Rijn 1980).

[2] *Mosler*, a.a.O., S. 210.

[3] Das Buch beruht auf einer vor der Akademie für internationales Recht, Den Haag, gehaltenen Vorlesung. Vgl. RdC 140 (1974 IV), S. 1 ff.

feststellung wird heute auch dann gewählt, wenn es sich keineswegs um delikate Fragen handelt oder um solche, die selbst bei Aufwendung einiger Geduld nicht in förmlicher Beschlußfassung abklärbar wären. Es ist vielmehr ein Mittel geworden, um das Verfahren der Meinungsbildung zu beschleunigen, dabei bestehende Diskrepanzen eher zu umgehen, als sie, wie es besser zu einem Rechtsbildungsprozeß passen würde, auszudiskutieren und so am Ende vielleicht doch eine für alle Beteiligte akzeptable, wirklich einvernehmliche Lösung zu finden.

3. Bei dieser Lage der Dinge lohnt es sich, auf das Problem der Konsensbildung, seine Bedeutung für das Völkerrecht näher einzugehen, zumal eine soeben erschienene verdienstvolle Arbeit neu deutlich gemacht hat, daß schon vom Begrifflichen her gesehen leicht Mißverständnisse entstehen können[4]. Das Problem Konsensbildung und -feststellung liegt heute noch immer in einer Grauzone, die aufzuhellen man sich bemühen muß, zumal, wie schon gesagt, Konsensfeststellung und »Konsensresolution« in der Staatenpraxis beträchtliche Bedeutung erlangt haben. Ein Verzicht auf diesen nach klassischen Maßstäben ungewöhnlichen Weg der Rechtsschöpfung kann wohl nicht mehr angenommen werden[5].

II. Vorfragen des völkerrechtlichen Konsensproblems

1. Der Begriff Konsens (lat. consensus = Übereinstimmung) hatte lange Zeit in der Rechtssprache und in anderen Sprachbereichen keine aktuelle Bedeutung. Zutreffend weist z. B. *Ulrich Scheuner* für das Recht darauf hin, daß der Begriff in keinem neueren Werk der Staatslehre vorkommt, daß er in keinem Register verzeichnet ist (etwa bei *Krüger, Zippelius, Herzog* u. a.). Er findet sich (so *Scheuner*) auch nicht im Ev. Staatslexikon (2. Aufl. 1955), das er als eher »zu gründlich« kennzeichnet[6]. Der Begriff fehlt auch, darf ergänzt werden, in der sonst so zuverlässigen Encyclopaedia Britannica. Daß »Konsens« sogar im Ev. Staatslexikon nicht aufgeführt ist, ist schon deshalb erstaunlich, weil dieser Begriff in der Entwicklung der evangelischen Kirche in Verbindung mit den Versuchen, unterschiedliche Lehrmeinungen auszugleichen, schon früh eine Rolle gespielt hat. Die Bezugnahme auf eine eher theologische Fragestellung zeigt aber bereits, daß es sich bei »Konsens« nicht um

---

[4] *Bruno Simma,* Zur völkerrechtlichen Bedeutung von Resolutionen der UN-Generalversammlung, in: R. Bernhardt u. a. (Hrsg.), Völkerrecht und Außenpolitik, Fünftes deutsch-polnisches Juristen-Kolloquium (Baden-Baden 1981).

[5] *Amadou-Mahtar McBow,* La pratique du consensus comme moyen de prise de décision est irremplaçable, in: ONU-Chronique, VII/1979, S. 94 f., der sich insbes. mit der Unverzichtbarkeit des *consensus* und den dafür maßgeblichen Gründen auseinandersetzt.

[6] *Ulrich Scheuner,* Konsens und Pluralismus als verfassungsrechtliches Problem, in: G. Jakobs (Hrsg.), Rechtsgestaltung und Konsens (Berlin 1976).

einen eigentlichen Rechtsbegriff handelt. Tatsächlich hat der Begriff in den verschiedenen fachlichen Bereichen ganz unterschiedliche Sinngehalte erlangt. Die »Brockhaus-Enzyklopädie« (1968) zählt z. B. unter dem Stichwort »Consensus« als maßgebliche Anwendungsfelder dieses Begriffs auf: Philosophie, katholische Dogmatik, katholisches Kirchenrecht, evangelische Konsens-Auffassung und schließlich auch das Zivilrecht [7]. Vom Verfassungsrecht und vom Völkerrecht ist keine Rede.

2. Das zeigt bereits, daß der Begriff aus seinem jeweiligen Umfeld, also z. B. aus dem Zivilrecht, neuerdings auch aus dem Staatsrecht, dem Völkerrecht usw., aus den diesen Rechtsmassen zugehörenden Besonderheiten verstanden werden muß. So bedeutet im *Zivilrecht* Konsens »das durch Willenserklärung der Kontrahenten bewirkte Einverständnis« (*Flume*). Das impliziert, daß volle Übereinstimmung der Beteiligten bestehen muß, was bei etwaigen Abstimmungen über den Gegenstand *Einstimmigkeit* bedeutet. Da sich im Zivilrecht die Partner als Gleichgestellte begegnen, entspricht das Erfordernis der völligen Übereinstimmung der Natur der Sache. Dissens schließt, als Gegenbegriff von Konsens, den vollen Konsens aus, macht die rechtliche Einbindung des Dissentierenden unmöglich.

3. In *Staatslehre* und *Staatsrecht* weicht das Konsenskonzept von dem des Zivilrechts ab. Es wurzelt auch nicht in diesem, sondern in der *Politikwissenschaft* [8]. Sie geht von der *Autonomie des Individuums* aus, das durch politische Willensakte die rechtsstaatliche Demokratie erzeugt. Diese politische Zwecksetzung trägt unter anderem das dem modernen Parlamentarismus zugehörige *Prinzip der Majorisierung,* der Unterwerfung der überstimmten Minderheit unter den politischen Willen der Mehrheit. Das ist ein allgemeiner Verfassungsgrundsatz, lehrt man, der im Staat von einem allgemeinen Konsens, von einem *consensus omnium* getragen wird. Bestünde die Möglichkeit der Majorisierung nicht, wie es im Zivilrecht der Fall ist, wäre der demokratische Staat funktionsunfähig. Ziel des im Konsens geeinten politischen Willens ist aber gerade der funktionsfähige Staat [9].

Interessant ist bei diesem Konsens-Konzept, daß es nicht wirklich von allen einem Staat Zugehörenden getragen werden muß, wie es im Zivilrecht der Fall ist, sondern im Grunde ein bloß fiktiver Konsens ist, bei dem durchaus abweichende Meinungen möglich sind, ohne daß dadurch der Konsens

---

[7] Hinsichtlich des Konsenses in weiteren Bereichen des Rechts vgl. *Jakobs,* a.a.O., mit Beiträgen von *Graue* (Vertragsschluß durch Konsens?), *Mayer-Maly* (Die Bedeutung des Konsens in privatrechtsgestaltender Sicht), *Podlech* (Die Bedeutung des Konsenses der Beteiligten beim Strafprozeß).

[8] Dazu u. a. *Rudolf Wildenmann,* Macht und Konsens als Problem der Innen- und Außenpolitik (Frankfurt, Bonn 1963).

[9] *Ulrich Scheuner,* Das Mehrheitsprinzip in der Demokratie, Rheinisch-Westfälische Akademie der Wissenschaften (1973), S. 176 ff.; die Problematik des Mehrheitsentscheides, der selbst eine mathematische Seite zugehört, wurde schon sehr früh gesehen. Vgl. *Marquis de Condorcet,* Essai sur l'application de l'analyse à probabilité des décisions rendues à la pluralité des voix (Paris 1785).

als solcher in dieser besonderen »politischen« Auffassung illusorisch werden würde. Solche möglichen Abweichungen sind unbeachtlich.

4. Die Idee der auf Konsens beruhenden, also damit rechtlich zulässigen Majorisierung wird immer wieder, wenngleich unzulässig, auf das Völkerrecht erstreckt. Verkannt wird dabei, daß dann das Bedürfnis nach einer Majorisierung aus dem Völkerrecht selber heraus, aus dem Staatenwillen, aus den aktuellen strukturellen Gegebenheiten der Völkerrechtsgemeinschaft herleitbar sein müßte. Aber die Völkerrechtsgemeinschaft ist, jedenfalls gegenwärtig noch, weit davon entfernt, eine dem Staat vergleichbare Verdichtung erreicht zu haben, so daß sich ein Bedürfnis nach einer Majorisierung, ein entsprechender Staatenwille auch nicht ausgebildet hat. Anders als im Staat kann von einem die Majorisierung tragenden Konsens in der Völkerrechtsgemeinschaft nicht gesprochen werden [10].

5. Auch andere Elemente des modernen Verfassungsstaates, die wie der Staat selber auf einem allgemeinen Konsens beruhen, sind dem Raum des Völkerrechts fremd, in diesen nicht einführbar. Das weist erneut aus, daß die jeweilige Erscheinungsform eines Konsenses aus seinem ihm eigenen Raum verstanden werden muß. Der »permanente Konsens« (»perpetuierliche Konsens«) etwa ist im Grunde die von *Rudolf Smend* beschriebene »*Integration*« im Staat, ein Prozeß zwischen den Individuen und ihren Gruppen, den es so im internationalen Raum nicht gibt, nicht geben kann, weil dort die Individuen noch immer durch den Staat mediatisiert sind, am Rechtsbildungsprozeß unmittelbar nicht teilnehmen.

6. Aus dieser Position des Individuums wurde gelegentlich abgeleitet, daß es im Völkerrecht Konsens nicht geben kann, weil dieser als Summe von Individualwillen nur im personalen Verband ausbildbar ist, dessen oberste Ausprägung der Staat ist. Dem steht aber schon die Tatsache entgegen, daß nicht nur die Verträge des Zivilrechts, sondern auch die des Völkerrechts in ihrem Kern auf willentlicher Übereinstimmung, auf Konsens beruhen. Daß die Vertragspartner von unterschiedlicher Qualität sind, nämlich einmal Individuen und einmal Staaten (und andere Völkerrechtssubjekte), ist unbeachtlich. Konsens gibt es also auch im Völkerrecht, allerdings hat erst in neuester Zeit das völkerrechtliche Konsensproblem aus dem Wesen der internationalen Gemeinschaft und dem Völkerrecht selber heraus neue, ihm gemäße besondere Aspekte erhalten. Um sie wird es in den nachfolgenden Abschnitten dieser Betrachtung gehen.

7. Mindestens eine Parallele im Konsensproblem des Staatsrechts und dem des Völkerrechts gibt es aber doch, nämlich die der da wie dort notwendigen Konkretisierung. Ohne sie würde der Konsens Fiktion bleiben. Sie erfolgt im Staat in vielfältiger Weise, durch Wahlen und Abstimmungen, durch die Presse und alle anderen Meinungsbildner, die ja auch immer Spie-

---

[10] Daher betont *Mosler* immer wieder, daß trotz aller eingetretener Wandlungen das Völkerrecht auch gegenwärtig noch *Koordinationsrecht* ist.

gel der Meinungen sind. Auch die Bürgerinitiativen, sofern sie nicht das Chaos zum Ziel haben und sich so außerhalb der Verfassung stellen, was selten ist, gehören hierher.

8. Wie immer der Konsens im Staat konkretisiert werden mag, niemals ist er selber Quelle von unmittelbar anwendbarem Recht. Selbst dann, wenn durch Verweisungen auf »Treu und Glauben«, auf die »Verkehrssitte« ein nur im allgemeinen Rechtsbewußtsein verankerter Grundsatz, eine entsprechende Verhaltensweise als rechtlich beachtlich erklärt wird, gelangt im konkreten Fall als Recht erst das zur Anwendung, was gestützt auf eine solche Ermächtigung der Richter als anwendbar erkennt. Hier zeigt es sich, daß dann, wenn aus einem tatsächlich oder angeblich bestehenden Konsens konkrete Folgerungen gezogen werden sollen, das Verfahren der Feststellung des Konsenses, seine Umsetzung in eine anwendbare Norm, von entscheidender Bedeutung ist. Es ist, kann man sagen, um so differenzierter, je allgemeiner der bestehende Konsens ist. Das komplizierte, in viele und vielerlei Verfahrensschritte gegliederte Gesetzgebungsverfahren, dem ja immer ein bestehender Konsens der einzelnen Bürger zugrunde liegen sollte, belegt das. Weitere Beispiele erübrigen sich hier.

9. Einige Feststellungen sind noch zur Art der Verwendung der üblich gewordenen Begriffe »Konsens«, *»consensus«* und »Konsens-Resolution« erforderlich.

a) *»Konsens«* meint primär echte Willensübereinstimmung, also Einstimmigkeit der Beteiligten, Freiheit von an sich möglichem Dissens. Aber im Staatsrecht hat sich das in der oben geschilderten Weise verändert, so daß es maßgeblich darauf ankommt, in welchem Zusammenhang der Begriff gebraucht wird. Auch im Völkerrecht gibt es gewisse Tendenzen, die darauf abzielen, den Begriff vom Erfordernis der Übereinstimmung aller zu lösen, die es für ausreichend halten, wenn die Großen, die in der internationalen Gemeinschaft maßgeblichen Mächte, in ihrem Willen übereinstimmen, sich dafür der Argumentation aus dem Staatsrecht bedienen. »Maßgeblich« in diesem Sinn soll sein, wer nicht ausgelassen, nicht übergangen werden kann, ohne daß dadurch dem internationalen System Schaden zugefügt wird. Ein solcher Schaden wäre es auch, wenn der den angeblichen Konsens konkretisierende Beschluß usw. wegen der Nichtberücksichtigung eines Großen ineffektiv bleiben würde.

Dieses auf eine Majorisierung der Kleinen hinauslaufende Konzept hat sich allerdings im Völkerrecht keineswegs schon durchgesetzt. Immer noch wird – vermutlich weit überwiegend – gemeint, daß Konsens volle Übereinstimmung bedeute, also die zwischen den Parteien übereinstimmende Willensgrundlage eines völkerrechtlichen Vertrages sei.

b) Die Begriffe *consensus* und *Konsens-Resolution* sind von der Sache her im Grunde deckungsgleich. Der Begriff *consensus* wird regelmäßig dann gebraucht, wenn im Schoße einer internationalen Organisation, bei einer internationalen Konferenz ein Ergebnis ohne die eigentlich gebotene förmliche

Abstimmung, durch bloße Feststellung von Übereinstimmung durch den Vorsitzenden, unter Umständen ganz in der Nähe einer bloßen Übereinstimmungsbehauptung gelegen, erfolgt. Die *Konsens-Resolution* ist dann die Einbindung, die Ausfertigung also des durch *consensus* gewonnenen Ergebnisses in der Gestalt, die für die förmlich gefaßten Beschlüsse vorgesehen ist, z.B. für die Beschlüsse der UN-Generalversammlung, wobei regelmäßig darauf hingewiesen wird, daß der Inhalt durch *consensus* zustande gekommen ist.

10. Vom Konsens ist die *Akklamation* zu unterscheiden. Bei ihr erfolgt Abstimmung (Wahl) durch Zuruf. Sie hat also ein aktives Verhalten als Voraussetzung, während ein Konsens durch bloß passives Hinnehmen ausgedrückt werden kann. Die Akklamation spielt auch in der Praxis der internationalen Organisationen eine Rolle, interessiert hier aber nicht weiter.

III. Die Konsens-Resolutionen internationaler Organisationen

1. Ohne alle vorausgehenden theoretischen Erwägungen zur Frage des Konsenses im Völkerrecht [11] hat das Problem im Schoße der internationalen Organisationen, vorab der Vereinten Nationen, eine rein pragmatische Regelung gefunden. Auf sie bezieht sich *Hermann Mosler* in seiner eingangs mitgeteilten Stellungnahme zum Konsens-Problem: Ein Organ einer internationalen Organisation steht aus politischen und anderen Gründen unter Erfolgszwang, kann das Scheitern einer positiven Willensbildung, wie es bei Anwendung des Regelverfahrens vorauszusehen ist, nicht hinnehmen, muß das Ziel also auf einem Umweg bzw. mit Hilfe eines in der maßgeblichen Satzung nicht vorgesehenen Auswegs erreichen. Dafür wurde der Weg der Konsensbildung und Konsensfeststellung ausgebildet, der heute als *»consensus«* eingeführt ist.

---

[11] Auch in der Literatur wurde zunächst die Konsensfrage selbst von namhaften Autoren nicht beachtet. Beispiele dafür sind u.a. *Max Sørensen,* Manuel of Public International Law (New York 1968); *D. W. Bowett,* The Law of International Institutions (2. Aufl., London 1970); *Ingrid Detter,* Law Making of International Organizations (Stockholm 1965); *Ignaz Seidl-Hohenveldern,* Das Recht der internationalen Organisationen einschließlich der supranationalen Gemeinschaften (Köln, Berlin 1967); *derselbe,* Völkerrecht (3. Aufl., Köln 1974). Vgl. dagegen (ohne Lehrbücher) die allerdings oft nur knappen einschlägigen Darlegungen bei (Beispiele) *Suzanne Bastid,* Observations sur la pratique du consensus, in: Multitudo legum ius unum, Festschrift für Wilhelm Wengler, Bd. 1 (Berlin 1973), S. 1 ff.; *Albert Bleckmann,* Die Aufgaben einer Methodenlehre des Völkerrechts, Probleme der Rechtsquellenlehre im Völkerrecht (Heidelberg, Karlsruhe 1978); *Ingrid Detter,* Essays on the Law of Treaties (Stockholm, London 1967); *Günther Jaenicke,* Die Dritte Seerechtskonferenz der Vereinten Nationen, ZaöRV 38 (1978), S. 438 ff., 445 ff.; *Wilfried C. Jenks,* Law in the World Community (London 1967); *de Lacharrière,* Consensus et Nations Unies, Annuaire français de droit international 1968, S. 9 ff.; *Christian Tomuschat,* Die Charta der wirtschaftlichen Rechte und Pflichten der Staaten, Zur Geltungskraft von Resolutionen der UN-Generalversammlung, ZaöRV 36 (1976), S. 444 ff.

2. Bei bestimmten regelungsbedürftigen Materien hätte man vielleicht die Ausbildung eines auf fortschreitender Verdichtung der internationalen Beziehungen beruhenden »historischen Konsenses« abwarten können, also die nach längerer Entwicklung sich ergebende einhellige Meinung, daß ein bestimmtes politisches Konzept auch rechtlich verbindlich sei. Ein Beispiel für eine solche Entwicklung ist die *Überwindung des Kolonialismus.* In der UN-Charta war dieses Problem ausgespart oder seine Regelung vergessen worden. Bald wurde aber klar, daß das überlieferte Kolonialsystem nicht fortgesetzt werden kann. Das wurde nicht nur Meinung der Staaten, sondern über alle Staatsgrenzen hinweg Meinung der Bevölkerungen, der einzelnen Menschen. Ergebnis war die nach 1960 forciert einsetzende, von einem allgemeinen Konsens getragene Dekolonisierung [12], die zur Bildung von zahlreichen souveränen Neustaaten führte, die sämtlich UN-Mitglieder wurden, so daß die Zahl der Mitgliedstaaten, die bei der Gründung bei rund 50 gelegen hatte, heute auf rund 160 angewachsen ist, darunter allerdings zahlreiche »Ministaaten«, über deren Staatsqualität man durchaus streiten könnte. Aber das ist in diesem Zusammenhang unbeachtlich.

3. Diese über eine längere Zeitspanne sich vollziehende Meinungsbildung, die dann durch Effektuierung ihre Bestätigung als Konsens erfährt, ist keineswegs unproblematisch. Wann ist aus einer bloßen Konsensbehauptung ein tatsächlicher, ein rechtlich relevanter Konsens geworden, ist die Frage, die kaum einmal so eindeutig wie im Fall der Dekolonisierung beantwortet werden kann. Dafür ist die rechtliche Qualifizierung der »Apartheid« kennzeichnend. Schon frühzeitig ist gesagt worden, Apartheid sei mit rechtsstaatlichen Grundsätzen unvereinbar. Bis 1960 war es herrschende, insbesondere von den westlichen Industriestaaten vertretene Meinung, daß daher Apartheid – lediglich – zu mißbilligen sei. Nach 1960, nach der Verschiebung der Mehrheitsverhältnisse in der UN-Generalversammlung aufgrund der Dekolonisierung bildete sich die Meinung aus und wurde schließlich – wie es schien – maßgeblich, daß Apartheid als schwere Menschenrechtsverletzung ein internationales Verbrechen sei. Es schien sich ein entsprechender Konsens gebildet zu haben. Als aber dann versucht wurde, Südafrika wegen der Apartheidpolitik aus den UN auszuschließen, scheiterte das Verfahren im Sicherheitsrat, dessen nach der UN-Charta erforderliche Zustimmung zum Ausschluß nicht zu erlangen war. Bestand dennoch ein rechtlich beachtlicher Konsens, war die Frage. Man wird das schon deshalb verneinen müssen, weil sonst die für die Rechtsbildung außerordentlich wichtigen Verfahrensvorschriften der UN-Charta ihren Sinn verlieren würden [13]. Daß Verfahrensvorschriften im Prozeß

---

[12] Zur Dekolonisierung z. B. *Alfred Verdross/Bruno Simma,* Universelles Völkerrecht, Theorie und Praxis (Berlin 1976), insbes. S. 254 ff., 485 ff.

[13] Dazu eingehend *Eckart Klein,* Zur Beschränkung von Mitgliedschaftsrechten in den Vereinten Nationen, Eine Untersuchung zum Südafrika-Beschluß der Vereinten Nationen vom 12. November 1974, Vereinte Nationen, Jg. 1975, S. 51 ff.

der Rechtsneuschöpfung einen besonderen Rang haben, warum das so ist, wird später noch näher darzutun sein.

4. Um eine ganz andere Art von Konsens handelt es sich bei dem dann seit der Mitte der sechziger Jahre zuerst im Schoß der UN-Generalversammlung ausgebildeten Konsens, international eingeführtem Sprachgebrauch folgend, hinfort *consensus* genannt. Bei ihm liegt, das ist zu zeigen, das Schwergewicht auf dem Verfahren.

Ziel ist es, in einem konkreten Rechtsschöpfungsverfahren eine allgemeine Willensübereinstimmung aller Beteiligten bezogen auf einen bestimmten Zweck zu erzielen, also insofern einen echten Konsens herzustellen. Dabei geht es dem Wesen des Völkerrechts gemäß nicht darum, einen solchen der Bevölkerungen zu erreichen, sondern zwischen den am internationalen Rechtsschöpfungsprozeß unmittelbar Beteiligten, also vor allem den Staaten. Dabei genügt nach dem Willen der UN-Charta ein Mehrheitsentscheid (je nach Bedeutung der Sache einfache oder qualifizierte Mehrheit) [14], was natürlich *in concreto* Einstimmigkeit nicht ausschließt. Aber gerade darin steckt das Problem. Einstimmigkeit ist bei Fragen von einiger Bedeutung nicht erreichbar, und ein Mehrheitsbeschluß (einfach oder qualifiziert) bleibt ineffektiv, wenn er nicht von den wirklich potenten Staaten mitgetragen wird. Der Ausweg *consensus* aus diesem Dilemma besteht nun darin, daß eine Sache bis zu einer gewissen Reife erörtert wird und schließlich der Konferenzvorsitzende im geeigneten Zeitpunkt (der dann eingetreten ist, wenn aus welchen Gründen auch immer nennenswerte Einwendungen und Vorbehalte nicht mehr zu erwarten sind) bestehenden *consensus* feststellt. Eine förmliche Abstimmung findet also nicht statt. Ein solcher *consensus* gilt als für die Beteiligten verbindlich, wobei ihm regelmäßig der Rang einer einstimmigen Beschlußfassung zuerkannt wird, der z. B. im Fall der UN-Generalversammlung in der Form der Resolution im Sinne des Art. 18 der UN-Charta ergeht [15].

5. Grundsätzlich kann ein solcher *consensus* keinen über eine einstimmig beschlossene förmliche Resolution hinausgehenden Rechtswert haben [16]. Man könnte umgekehrt sogar eher fragen, ob die Bildung von *consensus* überhaupt zulässig sein kann, wenn die Charta das Verfahren der Willensbildung in der UN-Generalversammlung genau festlegt, von der Zulässigkeit der

---

[14] UN-Charta Art. 10 ff., 18.

[15] Dabei handelt es sich naturgemäß nur um eine Rang-Behauptung ohne eigentliche rechtliche Fundierung. Aus der UN-Charta selbst ist für einstimmig gefaßte Beschlüsse kein höherer Rang herleitbar als für mehrheitliche gem. Art. 18 der Charta. Zu bedenken ist weiter, daß naturgemäß von Einstimmigkeit nur gesprochen werden könnte, wenn auch förmlich abgestimmt worden wäre. Eine Quasi-Einstimmigkeit gibt es nicht.

[16] *Simma* (Anm. 4), S. 51: »Die Resolutionen als solche können niemals über den Schatten ihres bloßen Empfehlungscharakters springen. ... Die Deutung der Verabschiedung von Generalversammlungsresolutionen als Abschluß völkerrechtlicher *Verträge* ist grundsätzlich verfehlt«.

Feststellung des *consensus* ohne förmliche Abstimmung in der Charta keine Rede ist. Aber das Verfahren, einmal praktiziert, ist ohne Widerspruch hingenommen worden, so daß solche Zweifel wohl kein Gewicht mehr haben können. Zudem ist schon lange anerkannt, daß die Charta wie auch jede andere Satzung einer internationalen Organisation durch die Art der Handhabung weiterbildbar ist. Insofern bestehen also kaum rechtlich beachtliche Bedenken.

6. Mehr Bedeutung hat jedoch die Frage der zulänglichen Bevollmächtigung der Staatenvertreter. Deckt die bloße Verhandlungsmacht, die ja die Regel ist, die Hinnahme der Konsensfeststellung? Ist das Teil der Verhandlungsführung oder rückt das als Zustimmung durch Schweigen[17] so nahe an den eigentlichen Vertragsschluß, daß Abschlußvollmacht vorliegen müßte? Aber eine Resolution, gleichgültig ob sie auf förmlicher Abstimmung oder *consensus* beruht, kann nicht einem Vertragsschluß gleichgestellt werden. Die Hinnahme einer *consensus*-Feststellung ist daher durch die erteilte Vertretungs- bzw. Verhandlungsvollmacht gedeckt. Sie ermöglicht auch Widerspruch oder Vorbehalt, um das Zustandekommen eines *consensus* zu verhindern.

7. Bei *consensus*-Feststellung ergibt sich für jeden beteiligten Staat, so wie bei jedem förmlichen Organbeschluß, z. B. bei jeder Resolution der UN-Generalversammlung, die Notwendigkeit der Bestimmung einer etwaigen Bindungswirkung[18]. Es stellt sich also die ganz allgemeine und immer wieder diskutierte Frage des Rechtswerts von Resolutionen schlechthin, mit der Besonderheit, daß *consensus* den *einstimmig* beschlossenen Resolutionen gleichgesetzt wird, von denen immer wieder behauptet wird, daß sie, trotz des Wortlautes des Art. 18 UN-Charta, für die Beteiligten verbindlich sind.

Wenn nun ein Staat (sein Bevollmächtigter also) bei der *consensus*-Feststellung die Bedeutung der Sache verkennt, die eventuell sich ergebende Beeinträchtigung seiner wirklichen Interessen falsch einschätzt und deshalb auf einen Widerspruch verzichtet, gilt er trotz eventueller späterer Einwendungen als durch den *consensus* gebunden. Man könnte meinen, daß dann der gleiche Effekt eintritt, wie im Fall des Mehrheitsbeschlusses, wo der überstimmte Staat sich auf den bloßen Empfehlungscharakter beruft und Erfüllung in

---

[17] Im Bereich des Völkerrechts als Folge des Souveränitätsgrundsatzes galt »Schweigen« stets als Ablehnung. Erst in unserer Zeit, als rechtliche Folge der eingetretenen Verdichtung der Staatenbeziehungen, auch durch die Ausbildung eines besonderen internationalen Organisationsrechts ist das Verständnis des Schweigens als Zustimmung möglich geworden.

[18] Dazu u. a. *Bleckmann* (Anm. 11), *Tomuschat* (Anm. 11), *Eckart Wehser,* Die Bindungswirkung der Empfehlungen der Vollversammlung der Vereinten Nationen, in: The Law of the United Nations, Thesaurus Acroasium, 1976, Bd. 2 (Thessaloniki, 1976), S. 371–394. Die bei einer solchen Prüfung unter Umständen sich ergebende Erkenntnis, daß die Resolution positiviertes Gewohnheitsrecht ist bzw. enthält, begründet nicht die Verbindlichkeit der Resolution, sondern macht diese zu einer Verweisung auf anwendbares Gewohnheitsrecht.

rechtlich zulässiger Weise verweigert, so daß auch ein im *consensus* ohne seinen eigentlichen Willen, ein eher durch ein Versehen gebundener Staat die Erfüllung ablehnt. Aber hier setzen dann eigentümliche psychologische Mechanismen ein, die regelmäßig doch ein anderes Ergebnis bewirken. Darin ist begründet, daß heute nicht nur in den Vereinten Nationen, sondern in allen internationalen Organisationen und Konferenzen Konsensbildung und Konsensfeststellung als nicht mehr verzichtbar gelten[19], sehr viel häufiger in Anspruch genommen werden, als das aus dem Wesen der Sache herzuleiten ist.

Tatsächlich ist es häufig so, daß ein beteiligter Staat einer *consensus*-Feststellung nicht widerspricht, obwohl er bei förmlicher Abstimmung gegenstimmen würde, und daß dann, wenn einmal *consensus* widerspruchslos festgestellt worden ist, der eigentlich in seinem wirklichen Wollen opponierende Staat die einmal getroffene Feststellung als doch auch für sich maßgeblich hinnimmt. Bei dieser Verhaltensweise, die bei diesem Verfahren einkalkuliert ist, spielen vielerlei Faktoren eine Rolle, so z. B. daß ein doch noch negierender Staat leicht als unsolidarisch, als Obstruktion betreibend propagandistisch in der internationalen Meinung diffamiert werden könnte. Beim heute bereits eingetretenen Ausmaß der internationalen Verdichtung nimmt das kein Staat gerne in Kauf, wenn es nicht um seine wirklich existentiellen Interessen geht.

8. Diese aus der internationalen Konferenzdiplomatie hergeleitete Erfahrung zeigt bereits, daß *consensus* nicht Konsens im Sinn einer willentlichen Übereinstimmung aller Beteiligter bedeuten muß, daß völkerrechtlicher *consensus* nichts mit Konsens im Sinn des Zivilrechts zu tun hat, wo Deckungsgleichheit (also Konsens ohne Dissens), der Willen aller Beteiligter das Wesen der Sache ausmacht. Es wäre aber gewiß falsch zu sagen, daß wie beim fiktiven Basiskonsens des Staatsrechts oder beim Partialkonsens (etwa hinsichtlich der föderalistischen Staatsstruktur) auch im Völkerrecht Unbestimmtheit hinsichtlich der tatsächlichen Willen der Partner des *consensus* akzeptabel sein müsse. Sie ist ja, so wird gemeint, beim Verfahren der *consensus*-Feststellung im Grunde unvermeidbar, ist positiv gewendet einer der Anlässe des Ausweichens in den *consensus,* also hinzunehmen, wenn man überhaupt zu einem Ergebnis kommen wolle.

Weil das so ist, begnügt man sich beim *consensus* sozusagen mit einem Verfahrenstrick. Übereinstimmung, tatsächliche oder doch behauptete, muß nur im Zeitpunkt der Konsens-Feststellung durch den Vorsitzenden bestehen und ist selbst dann problematisch. Sie bedeutet allenfalls einen Querschnitt in einem dynamischen Prozeß, der dadurch relativ stabilisiert wird, daß man unter Berufung auf den Vertrauensgrundsatz, auf Treu und Glauben usw. die Hinnahme der einmal – eigentlich in einem Eilverfahren – erfolgten Feststellung des bestehenden *consensus* als verbindlich ansieht. Diese Verbindlichkeit

---

[19] Dazu *McBow* (Anm. 5).

steht natürlich, wie schon gesagt, immer unter dem Vorbehalt, daß sie nicht mehr sein kann, keinen höheren Rechtswert haben kann, als der Beschluß, die Resolution, die im satzungsgemäßen, also im förmlichen Verfahren mit Mehrheitsfeststellung gefaßt worden ist.

9. Von Anfang an war klar, daß ein *consensus* nicht leichter zu erzielen ist, auch keine größere Erfüllungserwartung hat als ein mehrheitlich gefaßter Beschluß, wenn es um die Regelung konkreter Einzelheiten geht, womöglich gar um die Bildung von Recht, das als solches innerstaatlich unmittelbar anwendbar sein könnte. Es besteht daher die Tendenz, *consensus* vor allem zur Festlegung allgemeiner Grundsätze anzustreben, denen bei Licht besehen schon wegen ihrer inhaltlichen Unbestimmtheit keine eigentliche rechtliche Relevanz zukommen kann. Geht es dagegen um konkrete Regelungen, wird doch meist das herkömmliche Instrumentarium, mit förmlicher Beschlußfassung, in Anspruch genommen.

Kennzeichnend dafür ist, wie es zu den anfänglichen Regelungen mit dem Ziel der Ausbildung einer *Neuen Weltwirtschaftsordnung* gekommen ist[20].

Am Anfang steht die »*Declaration on the Establishment of a New International Economic Order*«, inhaltlich eine ganz allgemeine Absichtserklärung, aus der unmittelbar rechtliche Wirkungen nicht hergeleitet werden können und die ohne Abstimmung, als »Konsens-Resolution« also, festgestellt worden ist[21]. Gerade die Unbestimmtheit machte den Durchgang ohne jeden Widerstand möglich, obwohl ja die Errichtung einer Neuen Weltwirtschaftsordnung die unmittelbaren Interessen eines jeden in die Weltwirtschaft eingebundenen Staates berühren muß. Im gleichen Verfahren kam es zum zwar schon bestimmteren, aber doch auch bloß Absichten fixierenden »*Programme of Action on the Establishment of a New International Economic Order*«. Die dann schon als rechtliche Grundlage der Bemühungen um eine Neue Weltwirtschaftsordnung gedachte »*Charter of Economic Rights and Duties of States*«, deshalb inhaltlich schon konkret gehalten, wenn auch gewiß nicht in jeder Hinsicht ausreichend, ist schon nicht mehr als Konsens-Resolution, sondern im regulären Verfahren mit Abstimmung gemäß der UN-Charta beschlossen worden[22]. Es wurde, darf man sagen, erst der Bereich »Neuordnung der Weltwirtschaft« mit sehr allgemeinen Erklärungen, denen zu widersprechen man keinen Grund sah, aufgebrochen, wofür *consensus* ausreichte, und dann in mehrheitlicher Abstimmung eine rechtliche Präzisierung von sehr bemerkenswerten Konsequenzen vorgenommen, wobei man für den reibungslosen

---

[20] Allgemein dazu *Bergsten*, The Future of the International Economic Order: An Agenda for Research (1) 1973; *Edwin P. Reubens* (Hrsg.), The Callenge of the New International Economic Order (Boulder, Colorado 1981).

[21] *Edward McWhinney*, The World Court and the Contemporary International Law-Marking Process (Alphen aan den Rijn 1979), S. 130 ff.

[22] A/RES/3281 (XXIX) 12 Dec. 1974; Die Beschlußfassung erfolgte mit 120 Für und 6 Gegenstimmen, bei 10 Enthaltungen.

Durchgang dieser Resolution auch geltend machen konnte, daß sie in ihren Prinzipien ja schon durch *consensus* gebilligt sei. Natürlich wird so die prinzipielle Entscheidungsfreiheit eines Staates nicht tangiert, handelt es sich dabei eher um eine Frage des Taktierens, aber auch solches Taktieren ist eine Frage des Verfahrens insgesamt. Wie schon gesagt worden ist, im Zusammenhang mit dem *consensus* spielen Verfahrensfragen eine Rolle, deren Bedeutung man auf keinen Fall unterschätzen darf.

10. Wenn ein Verfahren mit dem Ziel eines *consensus* stattfinden soll, kommt es, wie ebenfalls die Praxis zeigt, ganz entscheidend auf seine Vorbereitung, auf die Verfahrenszwischenstufen an. Deshalb haben die der Verhandlung im Plenum vorausgehenden Ausschußsitzungen, die Beratungen und Meinungsbekundungen der als Ratgeber zugezogenen Experten an Bedeutung ganz außerordentlich gewonnen. In diesen Gremien hat sich der Grundsatz ausgebildet, daß einer in einer Gruppe ausgebildeten und festgestellten Meinung zu einer Sache durch diese Gruppe oder ein Gruppenmitglied nachträglich nicht mehr widersprochen wird, wie immer das rechtlich zu erklären sein mag[23]. Das bedeutet, daß in das Plenum festgefügte Positionen eingebracht werden, die leichter ausgleichbar sind, als wenn *ab ovo* kontrovers verhandelt werden müßte. Dadurch wird die Möglichkeit der Feststellung eines *consensus* ganz außerordentlich erleichtert. Hinzu kommt, daß zwar im Plenum im Prinzip jeder Partner seine Meinung noch ändern, seine Position wechseln kann, daß aber ein solcher Wechsel ganz ungewöhnlich, vielleicht sogar als Verletzung einer Solidaritätsverpflichtung unzulässig ist, jedenfalls dann, wenn Einbindungen in eine Gruppe bestehen. All das bestätigt die oben schon vertretene Auffassung, daß die Bildung und Feststellung eines *consensus* ganz maßgeblich eine Verfahrensfrage ist. Das Verfahren würde aber nicht funktionieren, wenn die Verfahrensvorstufen, die Beratungen in den Ausschüssen usw. nicht gut ausgebaut wären und in konsequenter Weise eingesetzt würden.

11. Wann *consensus* erstrebt wird, wie in der Konferenzpraxis verfahren wird, welche rechtlichen Fragen sich ergeben, ist deutlicher als das an Hand allgemeiner Erwägungen möglich sein kann, an einem konkreten Beispiel verstehbar zu machen. Ein solches kann gerade aktuellster Staaten- bzw. Konferenzpraxis entnommen werden, nämlich der Konferenz zur Reform der Pariser Verbandsübereinkunft, die nach langen Vorbereitungen im Februar 1980 in Genf eröffnet worden ist und deren Abschluß gegenwärtig noch nicht abzusehen ist.

---

[23] Das Verfahren der vorbereitenden Meinungsbildung und auch Abstimmung in Gruppen wurde in der UNCTAD ausgebildet und ist von anderen internationalen Einrichtungen entsprechender Interessengruppierung übernommen worden. Es gibt die A-, B-, C- und D-Gruppe, wobei die Entwicklungsländer (»Gruppe der 77«) die A-Gruppe bilden, die westlichen Industrieländer die B-Gruppe usw. Vgl. dazu *Autar Krishna Koul,* The Legal Framework of UNCTAD in World Trade (Leyden, Bombay 1977), S. 49 ff.

IV. Consensus im Anwendungsfall

1. Die Pariser Verbandsübereinkunft zum Schutz des gewerblichen Eigentums (PVÜ), in der Ursprungsfassung aus dem Jahr 1883 stammend, eine der großen Konventionen des Funktionalismus des letzten Drittels des vorigen Jahrhunderts und nach mehreren Revisionen noch heute geltend, ist gemäß ihrem Art. 18 der Stockholmer, also der letzten, 1967 erfolgten Neufassung, von Zeit zu Zeit zur Anpassung an neu entstandene Bedürfnisse einer Revision zu unterziehen. Ein solches Revisionsverfahren läuft gegenwärtig[24]. Mit ihm soll insbesondere den Bedürfnissen der Entwicklungsländer, wie sie erst in neuester Zeit klar erkannt worden sind, besser genügt werden. Eine erste Verhandlungsrunde fand 1980 in Genf statt, ihr folgte 1981 eine weitere in Nairobi, die nächste ist für 1982 wieder in Genf in Aussicht genommen. Schon dieser bisherige Ablauf zeigt, daß besondere Schwierigkeiten bestehen, denn alle vorausgehenden Revisionen, selbst die umfangreiche und komplizierte von 1967 in Stockholm, mit der verbunden die Gründung der Weltorganisation für geistiges Eigentum (WIPO) erfolgte, konnten in einem Konferenzdurchgang abgewickelt werden.

2. Schon bei den Konferenzvorbereitungen im Rahmen der dafür zuständigen Weltorganisation für geistiges Eigentum (mit Sitz in Genf) zeigte es sich, daß von vordringlicher Bedeutung die Frage sein wird, ob für eine Neufassung Einstimmigkeit der an der Konferenz teilnehmenden und abstimmenden Staaten erzielt werden muß oder ob ein Mehrheitsentscheid – zunächst einmal gleichgültig mit welchem Mehrheitsverhältnis – ausreichend sein kann[25]. Maßgebliches Motiv der Forderung nach dem Mehrheitsentscheid war der Umstand, daß bei gegenwärtig konventionsrechtlich gebotener Einstimmigkeit ein einziges Partnerland der PVÜ (Verbandsland des Pariser Verbandes) eine von allen anderen für erforderlich gehaltene Revision verhindern kann.

Das ist im Prinzip richtig, wäre aber mit dem Geist der PVÜ, der in ihr für den nicht zustimmenden Staat angelegten Ausweichsmöglichkeiten unvereinbar. Nach jeder Neufassung gilt nämlich die vorausgehende Fassung als neben der Neufassung anwendbar weiter. Ein Staat, der eine Neuregelung nicht will, kann den entsprechenden Beschluß, ohne zur Wahrung seiner Interessen widersprechen zu müssen, passieren lassen. Er wird an die Neufassung nicht gebunden, er bleibt Partnerstaat der vorausgehenden Fassung, bleibt weiter im Besitz der aus ihr herleitbaren rechtlichen Position, auch aufgrund entsprechender konventionsrechtlicher Verknüpfungen in Bezie-

---

[24] *Hans Peter Kunz-Hallstein,* Die Genfer Konferenz zur Revision der Pariser Verbandsübereinkunft, Gewerblicher Rechtsschutz und Urheberrecht, Internationaler Teil (GRUR Int.) 1981, S. 137 ff.

[25] *Hans Ballreich/Hans Peter Kunz-Hallstein,* Zur Frage der Reform des für Revisionen der Pariser Verbandsübereinkunft geltenden Einstimmigkeitsprinzips, GRUR Int. 1977, S. 251 ff.

hung auf die Partner der Neufassung, bleibt also Mitglied des Pariser Verbandes. Würde ein Verbandsland gegen den Willen aller anderen Partner widersprechen, wenngleich seine Interessen dabei nicht beeinträchtigt werden, wäre das wohl ein Fall von schierer Obstruktion. Das ist aber hier hinsichtlich der möglichen rechtlichen Folgen nicht auszudeuten.

Der Pariser Verband beruht von Anfang an auf dem Prinzip der ungefähren Interessenübereinstimmung der in ihm verbundenen Länder (Staaten), in dem Obstruktion sinnlos, daher nicht zu erwarten ist, in dem ein einmal als den neueren Bedürfnissen besser genügender Weg erkannt, auch durchsetzbar ist, zumal für den, der doch nicht zu folgen vermag, die geschilderte Ausweichmöglichkeit besteht. Für ihn gibt es aus seiner Interessenlage also keinen Anlaß, den neuen Weg zu blockieren.

3. Dennoch hat gleich in der ersten Verhandlungsrunde die Frage der Willensbildung, des Abstimmungsverfahrens, eine zentrale Rolle gespielt[26]. Das Einstimmigkeitsprinzip fand beträchtliche Gegnerschaft. Argumente, wie sie gerade für seine Beibehaltung geschildert worden sind, aus denen sich sogar seine Bedeutung für die Elastizität des Verbandes ergibt, wurden nicht mehr erörtert. Die stimmungsmäßige Abkehr der meisten Staatenvertreter vom Einstimmigkeitsprinzip fand vor allem auch darin eine Stütze, daß das einst für den Völkerbund maßgebliche Einstimmigkeitsprinzip wegen seiner ihm zugeschriebenen Ineffektivität für die Vereinten Nationen vom Mehrheitsprinzip abgelöst worden ist, daß die Wiener Vertragsrechtskonvention von 1969 (WVK) für die Willensbildung bei internationalen Konferenzen zur Vorbereitung multilateraler Verträge mehrheitliche Beschlußfassung vorsieht (Art. 9 II WVK)[27].

4. Aber all diese neueren Regelungen betreffen nicht ältere internationale Instrumente, üben keine generell-obligatorische Wirkung aus und haben keine rückwirkende Kraft. Das bedeutet, daß dann, wenn man den Übergang zum Mehrheitsprinzip bei der Willensbildung im Rahmen der PVÜ für erforderlich gehalten hätte, im Rahmen des normalen Revisionsverfahrens nach den Regeln der PVÜ selber, das ihr zugehörige Einstimmigkeitsprinzip hätte ändern, durch irgendein Mehrheitsprinzip hätte ersetzen müssen. In den Vorverhandlungen wurden Möglichkeiten von einfacher Mehrheit bis zur 9/10-Mehrheit als erstrebenswert bedacht, was immerhin ausweist, daß man jedenfalls nicht der Meinung war, die Abstimmungsmodalitäten wären kraft sich überlagernden Völkerrechts anderer Quellen als der PVÜ bereits im Sinn des notwendigen bzw. des zulässigen Mehrheitsentscheids überlagert.

5. Jedoch hielt man das konventionsgemäße Verfahren (einstimmige Beschlußfassung, Unterzeichnung, Ratifikation) überwiegend für zu langsam, den modernen Bedürfnissen nicht mehr adäquat, vor allem deshalb, weil da-

---

[26] *Kunz-Hallstein* (Anm. 24), S. 140 f.

[27] Konvention über das Recht der Verträge vom 23. 5. 1969, engl./frz. Fassung, ZaöRV 29 (1969), S. 711 ff.

mit ja die Mehrheitsregelung erst ab nächster Revisionskonferenz, also zu einem jetzt noch nicht absehbaren Zeitpunkt, gegriffen hätte. So einigte man sich dahingehend, daß die Abstimmungsmodalitäten als Verfahrensfragen in einer der Konvention selbst nicht zugehörenden Verfahrensordnung zu regeln seien, daß darüber »Konsensresolution« zulässig sei, zu der es dann auch gekommen ist. Dabei zeigte sich aber sofort die ganze Schwäche eines solchen »dynamischeren« Weges. Als es zur Konsensfeststellung durch den Vorsitzenden im Plenum kam, erkannten die US-Delegierten nicht bzw. verspätet die Tragweite der auf dem Wege des Konsens zu treffenden Feststellung (andere Staatenvertreter auch nicht, darf man annehmen), widersprachen also nicht bzw. verspätet, galten nach Meinung des Plenums daher als durch *consensus* gebunden, drangen also mit ihren nachträglichen Einwendungen nicht mehr durch.

6. Wäre konventionsrechtsgemäß, entsprechend der klassischen Lehre verfahren worden, hätte es zu diesem Dilemma nicht kommen können. Über den tatsächlichen Willen der Beteiligten, dann durch förmliche Abstimmung festgestellt, hätte kein Zweifel bestanden. Man macht dagegen zwar geltend, daß auch so kein Schaden entstehe, weil die USA am Ende, falls sie ihre Auffassung nicht wieder ändern würden, die Ratifikation verweigern könnten. Aber Sinn einer PVÜ-Revision ist es, in eine Neufassung möglichst alle, jedenfalls alle potenten Verbandsländer einzubinden und nicht durch Verfahrenszüge, die nach herkömmlicher Ansicht konventionswidrig sind, die Verbandseinheit als Rechtseinheit zu beeinträchtigen. Müßten solche Fragen nämlich ausverhandelt werden, weil sonst die Abstimmungsreife nicht gegeben wäre, würden diese Verhandlungen als Ausweichmöglichkeit nicht durch verfrühte Feststellung eines *consensus* mit allen psychologischen Implikationen schon in einem Vorstadium beendet, wären doch die Chancen groß, einen für alle Partner gangbaren Weg zu finden. Das würde dann mehr Zeit in Anspruch nehmen, dafür aber auch wirkliche Übereinstimmung bewirken. Verfahrensbeschleunigung um nahezu jeden Preis dürfte im Rechtsschöpfungsprozeß nicht der Weisheit letzter Schluß sein.

7. Das Verfahren, wie es in geschilderter Weise für eine Teilfrage auf der PVÜ-Revisionskonferenz in Genf 1980 durchgezogen worden ist, macht mehrere Probleme deutlich, so z. B. das, wer eigentlich auf einer solchen Konferenz bei der *consensus*-Bildung mitwirkungsberechtigt ist, nachdem auch Nichtverbandsländer der PVÜ teilnahmeberechtigt sind[28]. Solchen Fragen soll aber hier nicht nachgegangen werden. Zu fragen ist aber doch noch, ob es mit dem Grundsatz der Bindung an einen geltenden Vertrag, mit dem zu den tragenden Prinzipien des Völkerrechts gehörenden Satz *pacta sunt servanda* zu vereinbaren ist, daß eine aus einem geltenden Vertrag, für alle Part-

---

[28] Die üblich gewordene Einladung von Nichtmitgliedern bei internationalen Konferenzen beruht auf A/RES/366 (IV) »Rules for the Calling of International Conferences of States«.

ner verbindliche Regelung einfach auf die Seite geschoben, wegen tatsächlicher oder vermeintlicher neuer Bedürfnisse bedenkenlos durch eine ganz andere, aus dem konkreten Vertrag jedenfalls nicht herzuleitende Verfahrensweise ersetzt wird, ohne daß dabei die dafür einschlägigen Vorschriften des alten Vertrages beachtet werden. Dieses Problem spielt auch bei einem anderen, mit *consensus* nicht zu verwechselnden Konsens-Konzept eine zentrale Rolle.

V. Zum Versuch eines neuen Konsens-Konzepts

1. Anknüpfend an schon früher angestellte Betrachtungen zum Konsens-Problem hat soeben *Bruno Simma* ein Konsens-Konzept vorgelegt, das sich deutlich von dem des *consensus* der oben geschilderten Art unterscheidet, den *Simma* als eher »nebulös« kennzeichnet[29].

Nach *Simma* ist das Völkerrecht reines *Konsensrecht.* Eine heteronome Rechtsetzung gibt es nicht. Es existiert kein zentrales Gesetzgebungsorgan. Zwar haben sich, sagt er, gewisse formalisierte Verfahren der Rechtsschöpfung ausgebildet, die in Art. 38 IGH-Statut ihren Ausdruck gefunden haben (vor allem Vertrag, Gewohnheitsrecht, allgemeine Rechtsgrundsätze), aber diese Möglichkeiten gründen sämtlich im *übergeordneten Prinzip des Konsens,* der willentlichen Übereinstimmung der Staaten, sind immer nur dessen Ausdruck. Dieser Konsens der Partner, der Völkerrechtssubjekte, kann auf höchst unterschiedliche Weise zum Ausdruck gebracht werden. Die in Art. 38 IGH-Statut genannten Möglichkeiten sind nicht erschöpfend gemeint. Als »Herren des Völkerrechts« bestimmen nämlich die Staaten, denen die Rechtsbildung und Weiterbildung obliegt, einvernehmlich und autonom, auf welche Weise die bestehende Übereinstimmung zum Ausdruck gebracht werden soll. Eine Bindung an bestimmte Formen bzw. Verfahren, etwa die des Art. 38 IGH-Statut, gibt es im Völkerrecht seinem Wesen gemäß nicht. Wörtlich heißt es: »Die Erzeugung des Völkerrechts ist also nicht in bestimmten formalisierten Gestalten erstarrt, sondern befindet sich gewissermaßen in einem flüssigen Aggregatzustand«. Aus dieser Gesamtmasse gewinnen die Staaten Völkerrecht, unter Umständen in den Verfahren des Art. 38 IGH-Statut, also auf dem Wege einer formalisierten Rechtsetzung, aber im Prinzip doch gestützt auf die grundsätzliche Formfreiheit. Sie können daher beim Rechtsschöpfungsakt, wegen auf ihrer im Rahmen des gegenseitigen Einvernehmens bestehenden vollen Dispositionsfreiheit, sich auch einer ganz formlosen Rechtserzeugung bedienen, »indem sie in einem ständigen Ringen, durch Anerkennung, Duldung und Bestreitung von Ansprüchen und Situationen sowie durch formlose Abmachungen, also in einem ›process of contin-

[29] *Simma* (Anm. 4), S. 58 ff.

uous interaction, of continuous demand and response‹ (zitiert *Simma*[30]) das geltende Völkerrecht nicht nur feststellen, sondern auch weiterbilden«.

2. Dieser Vorgang, der allerdings an Formlosigkeit nichts zu wünschen übrig läßt, denn nichts ist fixiert, nichts geregelt, geordnet (was eine zwingende Konsequenz der im Völkerrecht maßgeblichen Formfreiheit sein soll), ist der »*formlose Konsens*«, der als solcher die eigentliche Quelle des Völkerrechts bildet.

Es gibt demnach, kann man sagen, wenngleich solche Bilder immer problematisch sind, verschiedene Rechtsschöpfungsstufen, deren unterste, im Sinn einer Basis, der völlig formlose Konsens darstellt. »Er führte« – es muß, schon um Mißverständnisse des *Simma*schen Konzepts auszuschließen, seine eigene Textfassung zitiert werden – »in der Geburtsstunde des Völkerrechts *uno actu* mit und durch die gegenseitige Anerkennung der Flächenstaaten als souveräne Herrschaftsverbände zur Herausbildung der ursprünglichen staatlichen Grundrechte sowie zur Anerkennung von Verträgen, Gewohnheitsrecht und allgemeinen Rechtsgrundsätzen als weiterer, stärker formgebundener Erzeugungsarten des Völkerrechts«. Das sind dann die jeweils in einem gewissen Maße formalisierten, den nur formlos ausgebildeten Konsens überlagernden Stufen, bzw., besser gesagt, Verfahrensweisen, die sich aber so noch nicht in vollem Maße mit dem decken müssen, was Art. 38 IGH-Statut sagt. Es könnten also zwischen dem, was hier unter »formlosem Konsens« verstanden werden soll, und den Möglichkeiten des Art. 38 IGH-Statut noch zahlreiche andere, kaum formalisierte vorausgehende Stufen möglich sein. Wie diese auszusehen hätten, bleibt offen; das kann hier auch dahingestellt bleiben.

3. Wesentlicher Gedanke zum Verhältnis dieser Quellen bzw. Stufen zueinander ist es, daß sie nicht als »historische Abfolge« zu verstehen sind, daß also dann, wenn der formlose Konsens sozusagen seine Aufgabe als Nährboden weiterer Rechtsschöpfung erfüllt hat, nun also mehr oder weniger formalisierte Verfahren eingreifen, damit der formlose Konsens nicht etwa entfunktionalisiert wird, sondern fortbestehend die förmlichen Verfahren sogar überlagert und damit weiter seinen Rang als *die* originäre Quelle des Völkerrechts behält[31]. Daraus wird dann eine bemerkenswerte Konsequenz gezogen: Formlos erzeugtes Völkerrecht kann zwar durch eine formalisierte Rechtsquelle geändert werden, da aber diese in ihrer Anwendung zu einer Auseinandersetzung zwischen den beteiligten Staaten führen kann, besteht wiederum die Möglichkeit einer Neuregelung durch formlosen Konsens, wobei, wird zugegeben, die Grenze zwischen der formalisierten Rechtserzeugung in den Bahnen des Art. 38 IGH-Statut und der formlosen Erzeugung von Völkerrecht fließend ist.

---

[30] Nach *McDougal,* The Hydrogen Bomb Tests and the International Law of the Sea, AJIL 49 (1955), S. 336.

[31] *Simma* (Anm. 4), S. 58/59.

4. Die Schwierigkeit dieser Grenzbestimmung beruht vor allem darauf, daß auch die »formalisierten Verfahrensweisen« im Sinn von Art. 38 IGH-Statut ihre besonderen Verfahrensprobleme in sich bergen, wie *Simma* anerkennt. Verträge können eben formfrei, eventuell nur durch konkludente Handlung abgeschlossen werden, auch »Schweigen« auf eine Willensbekundung kann unter gewissen Umständen als »Zustimmung« verstanden werden. Welche Verfahrensproblematik dann bei der Bildung von Gewohnheitsrecht besteht, nicht minder bei der Bestimmung des Bestandes allgemeiner Rechtsgrundsätze (alles ausgebildet aufgrund von »Verfahren« im Sinn des Art. 38 IGH-Statut), braucht hier keine weitere Darlegung. Ein »Vertrag« (so *Simma*) beruht ebenso wie der »formlose Konsens« auf Übereinstimmung, auf Konsens im Wortsinn, aber, meint er weiter, wenn auch jeder Vertrag auf einem Konsens beruht, nur durch einen solchen zustande kommen kann, ist der Konsens (als eigentliche Quelle des Völkerrechts zu verstehen, wie oben näher dargelegt) gegenüber dem Vertrag der weitere Begriff, so daß nicht jeder Konsens auch zu einem Vertrag führen muß. Ein Vertrag beruht in diesem Sinn immer auf einer (formellen oder formlosen) Willenseinigung, während der Konsens auch aus einem bloß passiven Verhalten, der Bereitschaft, sich einer bestimmten Ordnung zu fügen, erschlossen werden kann. Und weiter heißt es – nun in bezug auf die Bildung von Völkergewohnheitsrecht – »eine Norm des Völkergewohnheitsrechts etwa kann nicht nur durch eine als Recht anerkannte allgemeine Übung, sondern auch in der Weise begründet werden, daß die Staaten eine bestimmte allgemeine Regelung oder Praxis widerspruchslos hinnehmen«. Als Beleg dafür dient z. B. der *Fisheries* Case, bei dem der IGH gesagt hat, daß allgemeine Duldung (*tolérance générale*) einer einseitigen Festlegung der inneren Küstenmeergrenze deren Anerkennung durch die Staatengemeinschaft als im Einklang mit dem Völkerrecht stehend bedeute [32]. Diese dem eigentlichen Vertrag sozusagen vorgelagerte, aber ebenfalls auf Konsens beruhende Rechtsmasse ist, wenn *Simma* hier richtig verstanden wird, als Gewohnheitsrecht zu qualifizieren, obwohl es an einer *anhaltenden* Übung gemäß der klassischen Vorstellung der Gewohnheitsrechtsbildung fehlt, einer Übung, die als rechtlich geboten anerkannt worden ist.

5. Voraussetzung einer solchen Rechtsbildung soll allerdings sein, daß ein *echter* Konsens vorliegt, »Er darf . . . mit keinen Willensmängeln behaftet sein«. Seine Ermittlung muß mit großer Vorsicht erfolgen, »da das Verfahren der formlosen Rechtserzeugung die Gefahr in sich birgt, selbst leitende Grundsätze, auf denen das Völkerrecht der Vereinten Nationen beruht, aufzulockern« (*Simma*). Es geht also darum, einen bestehenden, in eine bestimmte Richtung gehenden, auf einen bestimmten Effekt ausgerichteten Staatenwillen exakt, frei von etwaigen Zweifeln, auszumachen. Nochmals: Konsens nach diesem Konzept ist Willensübereinstimmung zweier oder

---

[32] *South West Africa* Cases, ICJ Pleadings, Bd. 9 (1966), S. 345.

mehrerer Partner, wobei hinsichtlich des Bestandes des mängelfreien Willens volle Klarheit herstellbar sein muß, volle Einigkeit bestehen muß.

6. Nachdrücklich wird betont, daß ein Konsens nicht bloß durch eine *expressis verbis* erfolgende Willensbekundung gebildet werden kann, sondern daß er auch aus einem bloß passiven Verhalten, aus der erkennbaren Bereitschaft, sich einer bestimmten Ordnung zu fügen, erschlossen werden kann. Gerade dieser Weg soll deutlich von dem des Vertrages unterschieden werden. Glaubt man also, das Bestehen eines Vertrages verneinen zu müssen, ist immer noch die Möglichkeit offen, daß eine derartige stillschweigende, aber doch auf entsprechender – einseitiger – Willensbildung beruhende Einbindung mit allen rechtlichen Konsequenzen erfolgt ist. Ob das geschehen ist, muß die Prüfung ausweisen.

7. Diese Art von Konsens ist nach *Simma* streng vom »nebulösen Konzept« eines *consensus* zu scheiden, zu dessen Charakterisierung er auf den *South West Africa* Case verweist, wo von den Klägern als Anspruchsgrundlage ein bestehender *consensus* behauptet worden ist, wesensmäßig bestimmt als »an overwhelming majority, a convergence of international opinion, a predominance of view ... considerably more than a simple majority, but something less than unanimity«. Tatsächlich kann man meinen, daß bei dieser Auffassung Recht eher aus einer Stimmungslage als aus einer eigentlichen Rechtsüberzeugung mit Bindungswillen der Beteiligten gewonnen werden soll. Solche Gedanken sind heute in der Literatur nicht mehr selten und sind auch dann im Spiel, wenn die Staatenvertreter auf internationalen Konferenzen meinen, auf eine förmliche Abstimmung verzichten zu sollen, durch *consensus* leichter zu einem Ergebnis kommen zu können, wie oben näher ausgeführt worden ist.

8. Zur Verdeutlichung und um mögliche Mißverständnisse zu vermeiden, seien hier die für das geschilderte neue Konsens-Konzept maßgeblichen Thesen nochmals zusammengefaßt:

a) Völkerrecht ist Konsensrecht, beruht also nur auf übereinstimmendem Parteiwillen.

b) Konsensbildung erfolgt formlos. Der formlose Konsens ist die Basis allen Völkerrechts, aus dem im Prinzip alles weitere Völkerrecht herleitbar ist.

c) Art. 38 IGH-Statut enthält in formalisierten Erzeugungsarten gewonnene bzw. zu gewinnende Regeln des Völkerrechts. Die dort genannten Arten und Verfahrensweisen seiner Erzeugung (Vertrag, Gewohnheitsrecht, allgemeine Rechtsgrundsätze) sind keine erschöpfende Aufzählung. Andere Möglichkeiten sind ausbildbar, bzw. bedürfen als Konsequenz der grundsätzlichen Formlosigkeit erst gar keiner besonderen Ausbildung.

d) Der »formlose Konsens« (als Basis allen Völkerrechts) ist nicht bloße Ausgangsstufe in einem als *historischem Ablauf* zu verstehenden Prozeß der Ausbildung von Völkerrecht, weshalb er mit den Möglichkeiten des Art. 38 IGH-Statut als gegenstandslos geworden oder doch zur Ruhe gekommen anzusehen wäre, sondern er hat neben den formalisierten Quellen weiter Wirksamkeit. Eine durch förmlichen Vertrag geregelte Materie bleibt daher stets für eine Neuordnung durch formlosen Konsens offen.

e) Bestehender Konsens ist immer Voraussetzung eines förmlichen Vertragsschlusses. Ohne Konsens gibt es keinen Vertrag. Es kann aber einen rechtlich relevanten, einem Vertrag *nachfolgenden* Konsens geben, auch für den gleichen Regelungsgegenstand, der nicht zu einem Vertrag führt, aber doch völkerrechtlich verbindlich ist. Der formlose Konsens reicht also weiter als ein Vertrag.

f) Ein formloser Konsens vermag wegen des Grundsatzes der Formfreiheit der Völkerrechtsschöpfung einen vorausgehenden völkerrechtlichen Vertrag obsolet zu machen, zu ändern oder zu ersetzen.

g) Der formlose Konsens beruht auf *Willensübereinstimmung* der an seiner Ausbildung Beteiligten. Die Willenbildung muß fehlerfrei erfolgt sein. Willensmängel machen den Konsens unwirksam bzw. verhindern sein Zustandekommen.

h) Da auch völkerrechtliche Verträge formlos abgeschlossen werden können, bereitet die Abgrenzung von formlosem Konsens und formlos abgeschlossenem Vertrag unter Umständen Schwierigkeiten. Das zwingt zur außerordentlich sorgsamen Prüfung, ob wirklich Konsens – frei von Willensmängeln – vorliegt. Das ist schon deshalb unerläßlich, weil sonst unter Berufung auf angeblich bestehenden formlosen Konsens selbst tragende Grundsätze des Völkerrechts (»des Völkerrechts der Vereinten Nationen«) ausgehöhlt, ihrer Tragfähigkeit beraubt werden könnten.

i) Die Möglichkeit des formlosen Konsens macht u.a. spontane Rechtsbildung, die Ausbildung von Völkergewohnheitsrecht ohne vorausgehende langanhaltende Übung in der Anwendung als Recht verstehbar (*instant law* u.a.).

k) Der *Positivismus klassischer Prägung,* mit dem dieses Konzept wohl nicht zu vereinbaren wäre, ist überwunden. Heute bekennt sich niemand mehr zu ihm.

VI. Würdigung und Kritik

1. Die bis in die neuere Zeit gängige Meinung, Kennzeichen des Politischen sei seine *Dynamik,* Kennzeichen des Rechts dagegen seine *Statik,* wird heute nicht mehr akzeptiert. Man weiß und anerkennt, daß dem Recht eine erhebliche Dynamik innewohnt, ihm innewohnen muß, weil es die Aufgabe hat, einer sich schnell verändernden Wirklichkeit rechtlich zu genügen, sie zu erfassen, zu ordnen, verstehbar zu machen. So ist die Regelungssubstanz »Recht« heute, wo Veränderungen sich schnell vollziehen, in hohem Maße selber dynamisch geworden. Dabei ist allerdings dieser Dynamik dort eine Grenze gesetzt, wo das Recht durch sie seine Fähigkeit verlieren würde, in der Fülle der

sozialen Prozesse stabilisierend zu wirken. Rechtsregeln, die dem nicht mehr genügen können, sind kein Recht mehr, sondern allenfalls Widerspiegelung faktischer Geschehensabläufe. Die Substanz »Recht«, gleich welcher Ausprägung, voran aber das Völkerrecht, weil bei ihm die Koppelung an das Politische besonders intensiv ist, wäre in ihrem Wesen getroffen. Das sollte man bei der Einführung neuer Verfahrensweisen der Rechtsgewinnung, bei der Ausbildung neuer Theorien, nie aus dem Auge verlieren, es stets als unerläßliches Korrektiv zu Hilfe nehmen. Daran ist auch *consensus* und der *kontinuierlich mögliche formlose Konsens* der obigen Schilderung zu messen.

2. Wenn neue Wege gegangen werden sollen, stellt sich vorab stets die Frage, ob es der Neuerung tatsächlich bedarf. Für das Institut des *consensus* wird das, wie eingangs schon gesagt, deshalb oft bejaht, weil damit in internationalen Beschlußgremien selbst in »hoffnungslosen« Fällen unter Umständen noch ein Ergebnis erzielt werden kann, weil damit die Funktionsfähigkeit teurer internationaler Einrichtungen auszuweisen ist, vor allem aber weil damit relativ schnelle Ergebnisse zu erzielen sind.

Aber gerade der Gesichtspunkt der Schnelligkeit weckt Zweifel. Das mühsame Sichdahinschleppen der PVÜ-Revisionskonferenz, die früher in einer Session abzuwickeln war, wurde schon genannt. Statistische Auswertungen der zeitlichen Abläufe der Konferenzen, die mit förmlicher Abstimmung endeten, in Beziehung gesetzt zu denen, an deren Ende *consensus* festgestellt wurde, gibt es nicht oder sie sind nicht zugänglich. Alles spricht aber dafür, daß zwar die Beratungen im Plenum bis zur Feststellung eines etwaigen *consensus* relativ schnell über die Bühne gehen, zumal regelmäßig im Plenum auf rechtliche, ökonomische, technische Einzelheiten nicht mehr eingegangen wird, insofern auf die vorangegangenen Ausschuß-, Experten- usw. Beratungen verwiesen wird. Dort findet aber meist ein außerordentlich zeitraubendes Ringen statt, zumal dann, wenn Übereinstimmung in Blöcken bzw. Gruppen als Vorstufe einer Einigung für erforderlich gehalten wird[33]. Sieht man das Gesamtverfahren als Einheit, betrachtet man also Beratungen in Ausschüssen und im Plenum auch hinsichtlich des zeitlichen Ablaufes als zusammengehörig, dann dürfte kaum einmal von einer Zeitersparnis die Rede sein können. Es besteht also auch kein echter zeitlicher Druck auf die Partner. Dann ist aber ein Verzicht auf die klärende förmliche Abstimmung ein echter Verlust, Quelle mannigfacher Unklarheiten.

3. Der Gesichtspunkt »Zeit«, wenn auch mit anderer Akzentuierung, spielt auch beim Weg des formlosen Konsens eine Rolle. Während zunächst kaum rechtlich deutlich wird, wozu dieses Verfahren eigentlich taugen soll, wo sein wirklicher Nutzen liegt, wird schließlich gesagt, daß mit ihm die Ausbildung von *instant law*, also von spontan entstehendem Gewohnheitsrecht verstehbar wird. Dabei bleibt dann aber die Frage offen, ob es denn wirklich des Verständnisses gewisser neuer Erscheinungen als *instant law* be-

---

[33] Vgl. oben S. 12.

darf. Ist das wirklich alles »Recht«, was aus aktuellem Anlaß schnell hingezaubert und als Recht behauptet wird? Ist es nicht eine im Bereich des Faktischen liegende bloße Verhaltensweise, die naturgemäß schnell auch wieder änderbar sein muß und auch – formlos – geändert wird? Darauf sollen hier keine Antworten versucht werden.

4. Rechtlich gravierender ist es, daß im Schoß einer internationalen Organisation bei *consensus* die in den jeweiligen Satzungen enthaltenen Verfahrensregeln für die Beschlußfassung unbeachtet bleiben. Aber auch das wäre nur bei den ersten entsprechenden Vorgängen noch von praktischem Interesse gewesen. Damals wurde gegen das Verfahren kein Widerstand geleistet, nun – nach bald 20 Jahren ständiger Praxis – ist es eingeführt, als zulässig anerkannt. Man kann sich nur noch darüber wundern, was die Theorie der *implied powers,* aus dem US-Verfassungsrecht im Grunde in unzulässiger Weise übernommen, im Raum des internationalen Organisationsrechts alles zu leisten, zu decken hat.

Dabei wird im Grunde vom Recht her gesehen kein Gewinn erzielt. Oben wurde schon betont, daß die Konsens-Resolution rechtlich keinen höheren Rang haben kann als die im förmlichen Verfahren mehrheitlich beschlossene Resolution. Wenn eine solche Resolution aber mehr bringen soll als ein bloßes Bekenntnis zu einem politischen usw. Prinzip, dann sollte man auf sie verzichten, weil sie im Raum des Rechts kaum etwas bewirken wird, die Staaten sich dann doch auf die prinzipielle Unverbindlichkeit, auf den Empfehlungscharakter berufen werden. Das Bedenkliche liegt dann im Propagandistischen. Da in der Öffentlichkeit, auch in der internationalen Öffentlichkeit, kaum noch jemand verbindliches Recht und unverbindliche Empfehlung in der Fülle der Betätigungen internationaler Einrichtungen zu unterscheiden vermag, fällt es leicht, einen Staat, der seinen eigenen Weg gehen will, als unsolidarischen Rechtsverweigerer zu disqualifizieren.

5. Ein Staat kann sich gegen das Zustandekommen eines *consensus,* der ihm solchermaßen politisch unangenehm werden könnte, nicht dadurch schützen, daß er die Vollmacht seines Unterhändlers so begrenzt, daß von ihr die Zustimmung zu einem *consensus* nicht gedeckt ist. Denn bei der *consensus*-Feststellung bedarf es keiner ausdrücklichen Zustimmung. Widerspruchslose Hinnahme reicht für das Zustandekommen aus. Dem Unterhändler aufzugeben, grundsätzlich jedem *consensus* zu widersprechen, wäre extrem unbefriedigend, da das Verfahren ja auch einmal seinen guten Sinn haben kann. Regel sollte aber werden und der Auftrag sollte entsprechend lauten, daß stets dann zu widersprechen ist und so der *consensus* zu verhindern ist, wenn hinsichtlich seines Inhalts, seiner möglichen Folgen auch nur leichte Zweifel bestehen. Denn das gilt als gesichert und allgemein anerkannt: durch ausdrücklichen Widerspruch bei der *consensus*-Feststellung kommt er nicht zustande.

6. Beim formlosen, auf völliger Willensübereinstimmung beruhenden und jederzeit ausbildbaren Konsens ist es zentrale Frage: Kann im formalisierten Verfahren geschaffenes Völkerrecht, z. B. ein Vertrag, der auch Vor-

schriften darüber enthält, wie – in welchem Verfahren – er abänderbar ist, ohne Rücksicht auf diese Vorschriften jederzeit durch bloßen formlosen Konsens abgeändert werden? Die schlichte Bejahung dieser Frage, wie sie erfolgt ist, mit der Begründung, die Staaten seien als Schöpfer des Rechts auch seine Herren und könnten es willentlich jederzeit und ohne Rücksicht auf bestehende vertragliche Festlegungen ändern, ist extrem unbefriedigend. Sicher, so erreicht man eine erstaunliche Dynamik des Rechts, aber in Wirklichkeit hat man dann die Grenze dessen, was dem Recht angesichts seiner Stabilisierungsaufgabe zugemutet werden kann, überschritten. Die sich so ergebende Fluktuation des Rechts ist nicht tragbar. Das Konzept verstößt überdies gegen einen Basissatz des Völkerrechts, dessen Preisgabe das Ende des Rechts bewirken würde: die Parteien sind an das von ihnen gesetzte Recht (konkret also an den Vertrag mit seinen Änderungsklauseln) selber gebunden[34].

7. Diese Erwägungen sind keine theoretische Spielerei. Oben wurde für den Fall der PVÜ-Revisionskonferenz gezeigt, daß man die aus Art. 18 der Konvention herleitbare Verfahrensregelung für das Revisionsverfahren herausgebrochen und *per consensus* durch ein ganz anderes Verfahren ersetzt hat. Diesen Vorgang würde das Konzept des jederzeitigen formlosen Konsens voll verstehbar machen, denn ehe *per consensus* ein anderes Verfahren möglich war, mußte der in der PVÜ verankerte, das Revisionsverfahren betreffende Parteiwille obsolet sein. Wie das ohne förmliches Änderungsverfahren geschehen kann, erklärt das Konzept des immer ausbildbaren formlosen Konsenses als Ausdruck eines voll übereinstimmenden Parteiwillens. Wie schwierig es *realiter* wäre, eine solcher Übereinstimmung, frei von jeder Fiktion festzustellen, hat die PVÜ-Revisionskonferenz in ihrem bisherigen Verlauf bereits deutlich ausgewiesen. Auch mit diesem Konsens-Konzept ist dem Völkerrecht als stabilisierendem Faktor im internationalen Beziehungssystem nicht gedient.

8. Das von *Simma* entwickelte Konzept des formlosen Konsenses, der auch eine *expressis verbis* geltende vertragliche Regelung ohne Beachtung der Revisionsvorschriften dieses Vertrages zu ersetzen vermag, bringt keine Lösung. Gewiß kann sich ein stillschweigender, jedenfalls formloser Konsens dahingehend ausbilden, daß die positiven Vertragsbestimmungen den Bedürfnissen der Partner nicht mehr genügen, daß sie geändert oder aufgegeben werden sollten, aber diese Art von Konsens ist erst Voraussetzung der Rechtsbildung, nicht selber schon anwendbares Recht. Um diesen Effekt bewirken zu können, bedarf es, wie oben schon betont worden ist, der Konkretisierung, der näheren inhaltlichen Bestimmung, des dahingehend übereinstimmenden und ausgedrückten Willens, daß ein solcher Konsens als Recht

[34] Es handelt sich, wie *Mosler* sagen würde (Anm. 1, S. 134 f.), um eines der *Prinzipien* des Rechts, gegen die nicht verstoßen werden darf, weil sonst das Recht selber Schaden nehmen würde.

gelten soll[35]. Verzichtet man darauf ohne Not, wird das Gefüge »Recht« zerstört, selbst wenn dann andere Regelungssysteme eingreifen sollten[36]. Sie alle würden niemals das Recht in seinem Rang wirklich ersetzen können.

[35] Vgl. oben S. 18.
[36] Politische Verhaltensmuster, internationale Sitte, Moral u.a.m.

La personalidad jurídica internacional

JULIO A. BARBERIS

Introducción

El lenguaje es un instrumento del cual nos valemos para transmitir a nuestros semejantes ideas, deseos, opiniones, sentimientos, emociones, órdenes, sugerencias. Su estructura es muy sutil y compleja, y su riqueza es tal que pone a nuestra disposición numerosísimos términos para expresar los más variados matices de nuestros pensamientos.

A veces, en la lectura o en la conversación, hallamos una palabra que no conocemos y cuyo sentido no resulta aclarado por el contexto. Debemos entonces solicitar o buscar la *definición* de dicha palabra. Definir un término es explicar su significación.

Algunas palabras pueden tener más de un significado y puede ocurrir que en un contexto no aparezca claro cuál de ellos se pretende atribuir a un término. Se dice entonces que dicho término es ambiguo. La ambigüedad puede dar origen a falacias consistentes en utilizar en un razonamiento términos cuyos significados cambian. Las ambigüedades pueden ser eliminadas definiendo con precisión los términos que se utilizan.

La ambigüedad no conduce sólo a razonamientos falaces, sino que provoca a veces discusiones o controversias verbales. Se plantean a menudo diferencias de opinión entre dos personas que, en realidad, constituyen un desacuerdo aparente pues se debe a que utilizan un mismo término con diferentes significaciones. En el plano jurídico, algunas polémicas famosas en el derecho civil de ciertos países, como la planteada entre quienes afirman que la locación es un contrato y quienes replican que se trata de un derecho real, o la disputa acerca de si la posesión es un hecho o un derecho, se deben en gran medida, si no enteramente, a que no se ha precisado el sentido de los términos empleados.

Otra fuente de diferencias aparentes está dada por la circunstancia de que en el lenguaje se utiliza el mismo término para designar a un objeto determinado y a la palabra que designa dicho objeto. Pueden plantearse dudas en un contexto sobre si alguien utiliza un término para referirse a la palabra en sí (p. ej., la palabra »perro«) o al objeto designado por ella (un perro). En el derecho de gentes, algunas controversias doctrinales se deben a esta causa. Así, la discusión existente acerca de si una norma consuetudinaria internacional puede ser creada sólo mediante una repetición de hechos »positivos« o también a través de hechos »negativos« (abstenciones) es una controversia que se sitúa en el plano del lenguaje y que no hace al de la realidad de las cosas[1].

Pero la ambigüedad no es la única fuente de dificultades en el uso del lenguaje. En muchas oportunidades, se conoce el significado de un término, pero se presentan dudas en cuanto a los límites de su aplicabilidad. Se dice entonces que el término es vago. Por ejemplo el adjetivo »rico« se aplica indudablemente a una persona que posee una renta mensual de dos millones de francos suizos, tiene varias mansiones en distintos países de Europa y una extensión considerable de campo en Brasil. Es éste un caso típico o paradigmático. Pero podemos fácilmente imaginar otros casos de personas con un determinado patrimonio en que sea difícil decidir si les es aplicable o no el término »rico«. Y lo que ocurre con esta palabra, sucede con muchos otros adjetivos: alto, inteligente, amplio, etc. Los sustantivos presentan también esta característica. Un edificio de cien pisos es indudablemente un rascacielos, pero ¿cuántos pisos ha de tener como mínimo un edificio para ser denominado así?

La realidad de las cosas ofrece casos a los que puede aplicarse un término sin duda alguna, son los casos típicos. Pero, se presenta además un número indeterminado de casos que se hallan en una zona gris, en un halo de sombra, en que resulta difícil decidir si un cierto término les corresponde o no.

La aplicación de las normas jurídicas plantea también esta dificultad pues dichas normas están expresadas en palabras de nuestro lenguaje. Por ejemplo, cuando un código penal prevé cierta sanción para quien roba a otro valiéndose de un »arma«, ¿debe ser considerado tal un revólver de juguete o un trozo de caño de hierro? Igualmente, los códigos civiles establecen la distinción entre los contratos de compra y venta y de donación. Si alguien vende su casa por cien mil dólares o si la regala a una sociedad de beneficencia, no hay duda en cuanto al contrato que celebra. Pero, ¿realiza verdaderamente una venta si vende la casa por dos mil dólares, o por cien, o por veinte? ¿Dónde está el límite entre la venta a precio vil y la donación simulada?

En una cierta medida, la vaguedad de los términos puede ser eliminada tratando de precisar o detallar su significado, añadiendo para ello algunos otros términos a la definición dada. Pero, como los nuevos términos que se introducen en la definición contienen también cierta dosis de vaguedad, resulta imposible trazar siempre con toda precisión el límite de los casos a los

---

[1] Para el análisis de esta cuestión, ver *Barberis,* Fuentes del Derecho internacional (La Plata 1973), pp. 37 ss. y 40 ss.

que corresponde aplicar un cierto término. Este fenómeno es llamado *textura abierta del lenguaje.*

Las ciencias sociales se valen generalmente del lenguaje natural para expresar sus proposiciones. En el lenguaje jurídico se parte a menudo de las palabras de uso común. Se utilizan términos corrientes y así, se habla de buques, salario, lesiones, frutos, etc. Pero, dado que estos términos son muy imprecisos en cuanto a su significación, los juristas han ido determinando con detalle las características definitorias de cada palabra y los límites de su aplicación.

Cuando se explica el significado que a cada palabra le asignan las personas que hablan un lenguaje determinado, se dice que se da su definición informativa o lexicográfica. Este tipo de definiciones describen un uso determinado y, por consiguiente, pueden ser verdaderas o falsas.

Por el contrario, cuando alguien introduce en un lenguaje por primera vez un término nuevo, tiene entera libertad de estipular el significado que él tendrá. Las definiciones de estos términos nuevos, que se realizan mediante asignación de un significado, se llaman definiciones estipulativas. Por ejemplo, los términos que introduce Edmund Husserl en su Fenomenología para describir los distintos elementos de una vivencia son definidos estipulativamente. Cuando el matemático Kasner advirtió que necesitaba referirse frecuentemente en sus trabajos a la décima potencia de diez, preguntó a su nieto pequeño cómo la designaría y éste la llamó »googol«. Este término está incorporado hoy al lenguaje matemático. Las definiciones estipulativas no tienen un carácter informativo, sino directivo. Constituyen una directiva de cómo utilizar un término nuevo. Por lo tanto, ellas no pueden ser ni verdaderas ni falsas.

En la tarea de descripción y sistematización del derecho, el jurista trata de valerse de conceptos lo más precisos posibles. Toma como base las palabras del lenguaje común, pero las modifica al determinar sus características definitorias y fijar los límites de su aplicación. Es decir, el jurista utiliza términos del lenguaje común pero les asigna cierta significación específica, o sea, los define estipulativamente. Por ello, es posible encontrar en las obras de diversos juristas que términos tales como responsabilidad, sanción, derecho subjetivo, etc., no tienen siempre el mismo significado. No se trata aquí de que existe una definición verdadera para cada concepto, sino que todas ellas son igualmente válidas. El jurista tiene libertad para acuñar los conceptos con los que va a describir las normas jurídicas, como la tiene cualquier otro científico, un botánico o un geólogo, para definir los términos de que se valdrá para describir su objeto de conocimiento.

Cada autor determina las características definitorias y los límites de aplicación de los conceptos que emplea teniendo en cuenta la utilidad de la definición adoptada, su simplicidad y su exactitud. De esta manera, trata de eliminar la ambigüedad y, en la medida de lo posible, de reducir la vaguedad de los términos.

I. El concepto de sujeto en derecho internacional

Los autores ofrecen distintas definiciones de sujeto de derecho de gentes. En este estudio se exponen primeramente dos concepciones que se caracterizan por su rigurosidad y originalidad en la materia y que han hallado aceptación por parte de la doctrina. Son ellas la teoría pura del derecho y la que se puede denominar teoría de la responsabilidad. Luego, se propone un concepto de sujeto que permite, a nuestro modo de ver, superar las críticas dirigidas a las teorías anteriores.

1. La teoría pura del derecho

Es una noción comúnmente aceptada por la teoría general del derecho que existen determinados ámbitos dentro de los cuales las normas jurídicas son válidas, o sea, son tenidas por tales. El ámbito de validez espacial de una norma jurídica indica el espacio en el cual dicha norma rige. El período de tiempo en el que la norma es válida se lo llama ámbito de validez temporal. El llamado generalmente contenido de la norma, esto es, aquella conducta que está prescripta como permitida, prohibida u obligatoria, es el ámbito material de la norma. Por último, el ámbito personal de la norma jurídica está dado por las personas cuyas conductas son reguladas por dicha norma[2].

El concepto de sujeto de derecho se halla vinculado con la noción de ámbito de validez personal de la norma jurídica[3]. Cuando se expresa que un individuo es sujeto de una obligación jurídica, esto significa que una conducta de ese individuo aparece descripta por el orden jurídico como obligatoria. Igualmente, cuando se afirma que un individuo tiene cierta competencia jurídica para crear normas, se quiere indicar que determinadas conductas de ese individuo, según el orden jurídico, participan en la creación de normas jurídicas. En general, cuando se afirma que el orden jurídico otorga personalidad jurídica a un individuo, esto significa que dicho orden toma la conducta de ese individuo como contenido de derechos y obligaciones[4].

De esto resulta que, para que un individuo sea sujeto de derecho, es suficiente que una norma del orden jurídico prevea una conducta suya como contenido de un derecho o de una obligación jurídica.

---

[2] Conf. *Kelsen,* General Theory of Law and State (Cambridge [Mass.] 1946), pp. 42 ss.; *id.,* Théorie du droit international public, RdC 84 (1953 III), pp. 62 ss.; *id.,* Reine Rechtslehre (2 a. ed., Wien 1960), pp. 9 ss.

[3] Conf. *Kelsen,* Théorie du droit international public, RdC 84 (1953 III), p. 66.

[4] »Wenn man sagt, die Rechtsordnung verleihe einem Menschen Rechtspersönlichkeit, bedeutet das nur, daß die Rechtsordnung das Verhalten eines Menschen zum Inhalt von Pflichten und Rechten macht« (*Kelsen,* Reine Rechtslehre, pp. 193–194).

Es de señalar que el término »derechos« está usado en el sentido de »derechos subjetivos« (Conf. *Kelsen, op. cit.,* p. 177). En cuanto al concepto de »derecho subjetivo«, ver *Kelsen, op. cit.,* pp. 130 ss.

Una vez expuesta esta idea básica, *Kelsen* la aplica tanto a las personas físicas como a las personas jurídicas, según la denominación de la teoría tradicional.

Se expresó ya que el derecho considera a alguien como sujeto en tanto y en cuanto alguna conducta suya aparece como contenido de un derecho o de una obligación. Pues bien, los seres humanos tendrán personalidad jurídica en la medida en que ello ocurra. La extensión de esta personalidad puede ser más o menos amplia y de hecho así sucede en los distintos órdenes jurídicos. Por ejemplo, una persona por nacer tiene generalmente una personalidad muy reducida que se limita a la posibilidad de adquirir bienes por donación o herencia. Una persona mayor de edad goza, en principio, de una personalidad plena. En algunos países, los extranjeros poseen una personalidad jurídica algo limitada pues no son titulares de ciertos derechos.

Dado que lo que interesa a la teoría pura del derecho son las normas jurídicas en que se mencionan conductas de un individuo determinado como contenido de derechos u obligaciones, *Kelsen* define a la persona física como un conjunto de normas que tienen como contenido derechos y obligaciones de un hombre determinado[5]. Dicho en otros términos, la persona física es la expresión de la unidad de las normas que prescriben como jurídicamente permitidas, prohibidas u obligatorias las conductas de un mismo individuo. Obsérvese que en la teoría tradicional se afirma que la persona *posee* derechos u obligaciones, haciendo así una distinción entre el ser humano, por una parte, y sus derechos y obligaciones, por la otra. Por el contrario, la teoría pura afirma que la persona física *es* un conjunto de normas que tienen como contenido derechos y obligaciones de un mismo ser humano.

Respecto de las llamadas personas jurídicas, *Kelsen* efectúa un razonamiento análogo[6]. Una sociedad es sujeto de un orden jurídico en la medida en que sea titular de derechos u obligaciones.

La persona jurídica actúa a través de individuos que se desempeñan como órganos de aquélla. El estatuto de la entidad determina en cada caso qué actos u omisiones de sus órganos son atribuídos (*zugeschrieben*) a la persona jurídica como habiendo sido realizados por ella. Los individuos que representan la persona jurídica son órganos de ella en la medida en que sus actos son regulados por el estatuto de aquélla. El estatuto de la persona jurídica constituye un orden jurídico parcial.

Este orden jurídico parcial se encuentra en una determinada relación con el orden jurídico del Estado. Cuando, por ejemplo, el derecho del Estado prescribe que una persona jurídica está obligada a hacer algo, aquél deja librado al estatuto de la persona jurídica que indique qué individuo o miembro de ella,

---

[5] Conf. *Kelsen,* General Theory of Law and State, p. 95; *id.,* Reine Rechtslehre, p. 178.

[6] *Kelsen,* General Theory of Law and State, pp. 96 ss.; *id.,* Théorie du droit international public, RdC 84 (1953 III), pp. 67 ss.; *id.,* Reine Rechtslehre, pp. 178 ss.

actuando como su órgano, deberá cumplir la obligación. El estatuto de la persona jurídica precisará si ello es función del gerente, del secretario o de otro de sus miembros. En este caso, el derecho del Estado determina el ámbito de validez material de la norma (o sea, *qué* se debe hacer), en tanto que el ámbito de validez personal (o sea, *quién* lo debe hacer) es indicado por el estatuto de la persona jurídica. Igualmente, cuando el orden jurídico estatal atribuye un derecho a la persona jurídica, aquél deja librado al estatuto de la entidad la determinación de qué individuo, actuando como su órgano, ejercerá o hará valer el derecho atribuído. Respecto de los individuos que en estos casos actúan como órganos de la persona jurídica, se puede decir que su conducta se encuentra regulada directamente por el estatuto de la persona jurídica y de modo indirecto por el orden jurídico estatal.

Así como para la teoría pura del derecho la persona física es la expresión de la unidad de las normas que regulan la conducta de un mismo individuo, la persona jurídica es la expresión de la unidad de las normas que regulan la conducta de los miembros de dicha persona en cuanto tales. Expresado en otros términos, la persona jurídica está constituída por las normas que componen su estatuto. Obsérvese que, en la teoría tradicional, la persona jurídica *posee* su estatuto, estableciéndose así una distinción entre la entidad y las normas que la regulan. Por el contrario, la teoría pura afirma que la persona jurídica *es* su estatuto.

Kelsen analiza en detalle los derechos y los deberes de la persona jurídica y muestra que, en todo caso, ellos pueden ser explicados como derechos y deberes colectivos de los individuos que componen dicha persona jurídica. Para la teoría pura, la persona jurídica es una construcción auxiliar de la ciencia del derecho y no del derecho mismo[7]. El jurista puede prescindir de ella y describir los derechos y los deberes de la persona jurídica como derechos y deberes colectivos de sus miembros. Sin embargo, su empleo es útil pues simplifica en gran medida la descripción del derecho positivo. El concepto puede ser utilizado teniendo en cuenta que, en última instancia, todo derecho o deber es siempre el de un hombre ya que el orden jurídico sólo regula conductas humanas.

La teoría pura aplica los conceptos expuestos al ámbito del derecho internacional[8]. Es preciso distinguir aquí los casos en que una conducta humana se halla regulada por el derecho de gentes de manera directa o indirecta. *Kelsen* estima que existen situaciones en que la conducta de un individuo es regulada directamente por el orden internacional como permitida, prohibida u

---

[7] ». . . diese Personifikation und ihr Ergebnis, der Hilfsbegriff der juristischen Person, ist ein Produkt der das Recht beschreibenden Wissenschaft, nicht des Rechts« (Reine Rechtslehre, p. 194).

[8] *Kelsen,* Principles of International Law (New York 1952), pp. 96 ss.; *id.,* Théorie du droit international public, RdC 84 (1953 III), pp. 66 ss. y 85 ss.; *id.,* Reine Rechtslehre, pp. 325 y 326.

obligatoria. En la mayoría de los casos, sin embargo, no ocurre así pues el derecho internacional se refiere directamente a entidades tales como el Estado, la Iglesia, las organizaciones internacionales, etc. A su vez, los ordenamientos jurídicos de estas entidades se refieren de manera directa a los individuos. Estos últimos, por lo tanto, son regidos indirectamente por el derecho internacional. El derecho de gentes regula en estos casos la conducta de los individuos de la misma manera en que lo hace el orden jurídico de un Estado respecto de los miembros de una persona jurídica.

Las normas del derecho internacional, como cualquier norma jurídica, regulan siempre conductas humanas, pero lo hacen de un modo directo o indirecto. La teoría pura considera sujetos del orden jurídico internacional a toda entidad o individuo que sea destinatario directo de una norma de dicho orden. Cuando se dice que una entidad o persona jurídica es destinataria directa de una norma de derecho de gentes significa que los individuos que la componen tienen un derecho o un deber colectivos y que sus conductas están reguladas sólo indirectamente por el orden jurídico internacional.

2. La teoría de la responsabilidad

La teoría que aquí se denomina de la responsabilidad ha sido expuesta principalmente por *Eustathiades* y por *Wengler,* si bien hay algunas diferencias entre ambos autores.

Eustathiades ha desarrollado su concepción en el curso dictado en la Academia de Derecho Internacional de La Haya sobre los sujetos del derecho de gentes y la responsabilidad internacional[9]. Este autor considera como sujeto de derecho internacional a todo aquel que se halla al menos en una de las dos situaciones siguientes: *i)* ser titular de un derecho y poder hacerlo valer mediante reclamación internacional, o *ii)* ser titular de un deber jurídico y tener la capacidad de cometer un delito internacional[10]. Los dos casos tienen en común, según *Eustathiades,* el elemento de la responsabilidad internacional. En el primero, se trata de quien puede invocar y hacer valer la responsabilidad internacional; en el segundo, se trata de quien puede asumirla.

Conviene precisar ahora los términos empleados por el autor.

En cuanto a la primera situación, no existe diferencia con lo enseñado por la teoría pura del derecho porque ser titular de un derecho y poder hacerlo valer mediante reclamación es lo que *Kelsen* denomina ser titular de un derecho subjetivo[11].

Respecto del segundo caso, *Eustathiades* entiende por capacidad de cometer un delito el hecho de ser responsable, o sea, sufrir sus consecuen-

---

[9] *Eustathiades,* Les sujets du Droit international et la responsabilité internationale – Nouvelles tendances, RdC 84 (1953 III), pp. 397 ss.

[10] *Eustathiades, op. cit.,* RdC 84 (1953 III), pp. 414–415.

[11] *Kelsen,* Reine Rechtslehre, p. 140. En cuanto al derecho subjetivo en derecho internacional, ver *op. cit.,* p. 325.

cias [12]. Por lo tanto, esto se da cuando alguien no cumple con el deber jurídico y debe asumir las consecuencias de su incumplimiento.

Wengler ha desarrollado la teoría de la responsabilidad más ampliamente [13]. A diferencia de *Eustathiades*, el profesor de Berlín distingue claramente entre acto antijurídico o delito, por una parte, y responsabilidad, por la otra [14]. El acto antijurídico es el antecedente de una sanción, en tanto que la responsabilidad consiste en ser destinatario de ella, independientemente del hecho de haber sido o no el autor del acto antijurídico. Esta distinción es importante porque se presentan casos en que no existe identidad entre el autor del acto antijurídico y el destinatario de la sanción.

Después de efectuar una serie de consideraciones acerca de las razones por las cuales el derecho utiliza a veces la técnica de convertir en destinatario de una sanción a quien no es el autor del acto ilícito, *Wengler* examina quiénes son destinatarios de sanciones en el orden internacional, o sea, quiénes son responsables. En este examen comprueba, por ejemplo, que no sólo el Estado, sino también una fracción del Estado puede ser objeto de represalia [15] y afirma que dicha fracción posee entonces la calidad de sujeto de derecho internacional [16]. También analiza el derecho de guerra y estima que todo individuo o comunidad que sea objeto de una sanción en este ámbito, es sujeto de derecho internacional [17].

Wengler generaliza luego este análisis y llega a la conclusión que sujeto de derecho de gentes es quien es responsable internacionalmente, o sea, todo aquel que puede ser destinatario de una sanción internacional [18].

3. El sujeto como destinatario efectivo de un derecho o de una obligación en el plano internacional

Según se ha expuesto precedentemente, para la teoría pura, es sujeto de derecho todo aquel cuya conducta es regulada directamente por el orden jurídico como contenido de una obligación o de un derecho. *Wengler*, por el contrario, identifica el sujeto de derecho con el responsable, o sea, con el destinatario de una sanción.

Tal como se expresó al comienzo de este estudio, el jurista es libre de adoptar el concepto de sujeto que desee; sólo debe ser consecuente con él en su descripción del derecho. Sin embargo, en la práctica, el jurista elige los

---

[12] *Eustathiades, op. cit.*, RdC 84 (1953 III), pp. 416 y 424.

[13] *Wengler*, Der Begriff des Völkerrechtssubjektes im Lichte der politischen Gegenwart, Die Friedens-Warte (FW) 51 (1951–53), pp. 113 ss.

[14] ». . . das menschliche Objekt der sogenannten Unrechtsfolge und der Urheber des Unrechtstatbestandes können verschieden sein, ohne daß dem normativ Wesentlichen etwas gestört wird« (*Wengler*, FW 51, p. 114).

[15] *Wengler*, FW 51, p. 122.

[16] *Wengler*, FW 51, pp. 123 y 124.

[17] *Wengler*, FW 51, p. 125.

[18] *Wengler*, FW 51, pp. 128, 141 y 142.

conceptos de que se ha de servir guiándose por criterios de utilidad, simplicidad y exactitud. Teniendo en cuenta estos criterios, conviene efectuar un breve comentario de las dos teorías expuestas.

En cuanto a la concepción de *Wengler,* este autor admite que el derecho es siempre regulación de conducta humana[19]. Sin embargo, él adopta un concepto de sujeto que puede comprender a individuos cuyas conductas no se encuentran reguladas jurídicamente. En efecto, el derecho regula determinadas conductas como permitidas, prohibidas u obligatorias. Es decir, cuando un individuo está jurídicamente autorizado a hacer algo, o le está prohibido determinado acto o está obligado a cierta prestación, su conducta, como hombre, se encuentra regulada por el orden jurídico. Pero, cuando la norma jurídica lo hace destinatario de una sanción, no puede decirse que el derecho regula su conducta como hombre. La sanción es un acto coactivo que la autoridad impone al individuo *independientemente* de su voluntad. Por lo tanto, no se trata de un acto humano del individuo, no puede decirse en este caso que el derecho regula la conducta del sancionado. Este último es sólo un objeto.

Ahora bien, es posible imaginar, y en la práctica así sucede, que una norma jurídica prevea sanciones para ciertos individuos, o sea, que los haga responsables, sin atribuirles ningún derecho u obligación. En este caso, según la concepción de *Wengler,* los individuos serían sujetos de derecho sin que su conducta aparezca regulada por la norma jurídica. En esta hipótesis, la teoría de la responsabilidad podría responder que, cuando el derecho establece en una norma que, si un individuo *A* no realiza cierto acto, el individuo *F* será sancionado, la conducta de *F* se encuentra regulada jurídicamente. Así, dirá que, al existir esa norma, *F* tratará de influir de alguna manera para que *A* realice el acto prescripto por aquélla[20]. Pero, es preciso reconocer que, aún cuando esta posibilidad de influencia en la conducta de otro se dé en la realidad, no es una regulación de conducta humana establecida por la norma jurídica. Por ello, no parece útil adoptar un concepto de sujeto de derecho que pueda conducir a este resultado.

La concepción de la teoría pura, por el contrario, puede servir como punto de partida para una definición de sujeto. Sin embargo, el concepto acuñado por su fundador posee cierta dosis de vaguedad que hace que, al ser aplicado en la práctica, resulte difícil determinar en algunos casos si se está en presencia de un sujeto de derecho de gentes o no.

En primer término, es necesario precisar algo que en distintos pasajes de las obras de *Kelsen* no está bien especificado. Se trata de saber si sujeto de derecho es aquel cuya conducta está prevista en el orden jurídico como contenido de derechos *y* obligaciones, o basta sólo que lo sea de derechos *u* obligaciones. En el presente estudio se adopta la segunda alternativa y se con-

---

[19] *Wengler,* FW 51, p. 114.
[20] *Wengler,* FW 51, p. 117.

sidera sujeto de derecho internacional a todo aquel cuya conducta está prevista directamente por el derecho de gentes, al menos, como contenido de un derecho o de una obligación[21].

En segundo lugar, se debe analizar más detalladamente qué se entiende por ser destinatario directo de una norma internacional que otorga un derecho o impone una obligación. El titular del derecho u obligación puede no ser quien figura como tal en la letra de un tratado, sino que es quien efectivamente hace valer el derecho o asume la obligación. Si una organización aparece como titular de una obligación o de un derecho internacional en un tratado, pero esa obligación o ese derecho es efectivamente asumido por otro ente, aquélla carecerá de personalidad internacional y será tenida sólo como un órgano de este último, que será el verdadero sujeto internacional.

Un caso de esta naturaleza que se presenta en la práctica es el de los miembros de un Estado federal. Si el miembro *S* de un Estado federal *B* contrae obligaciones en el plano internacional y resulta que, en todos los casos, las obligaciones de *S* son asumidas en última instancia por *B*, no puede hablarse de obligaciones internacionales propias de *S*. Igualmente, si los derechos de *S* se traducen en derechos que, en todo caso, hace valer *B* en el plano internacional, tampoco podría hablarse de derechos internacionales de *S*. Por lo tanto, si se da que las obligaciones y los derechos de la provincia, Land o cantón *S* en el plano internacional se reducen, en todos los casos, a obligaciones y derechos del Estado federal *B*, se sigue entonces que *S* no tiene personalidad en derecho de gentes, sino que es un mero órgano de *B*.

Por esta razón, determinar si una provincia, un cantón o un Land es sujeto de derecho de gentes equivale a precisar si existen casos en que sus obligaciones o sus derechos en el plano internacional son distintos de los que, en última instancia, es titular el Estado federal. Si esta diferencia existe, el Estado miembro es sujeto de derecho internacional. Si, por el contrario, los derechos y las obligaciones del Estado miembro son en todo caso derechos y obligaciones del Estado federal, aquél es sólo un órgano de este último pues todas las consecuencias jurídicas de sus actos en el orden internacional son atribuídas al Estado federal.

Se debe puntualizar que aquí se habla de obligación jurídica y no de responsabilidad. Quien asume efectivamente una obligación es quien cumple la prestación y, en caso de incumplimiento, paga las indemnizaciones, intereses o perjuicios correspondientes.

En consecuencia de lo expuesto, se puede definir el sujeto de derecho internacional como aquel cuya conducta está prevista directa y efectivamente por el derecho de gentes, al menos, como contenido de un derecho o de una obligación.

---

[21] Comp. *Anzilotti,* Corso di Diritto Internazionale (3a. ed., Roma 1928) t. 1, p. 112; *Guggenheim,* Traité de Droit international public (Genève 1953) t. 1, p. 171; *Verdross/Simma,* Universelles Völkerrecht (Berlin 1976), p. 200.

4. La calidad y el contenido de la personalidad internacional

La calidad de sujeto no depende de la cantidad de derechos y obligaciones de que una entidad es titular. La designación es válida tanto para quien sólo goza de un derecho adjudicado por una norma consuetudinaria como para un Estado soberano. Por ello, del hecho de que una entidad es titular de modo directo y efectivo de un derecho o de una obligación en el orden internacional puede deducirse que dicha entidad es un sujeto de derecho de gentes. Pero, de la circunstancia de que alguien es sujeto de derecho de gentes no puede deducirse que es titular de un determinado derecho u obligación [22].

La categoría de sujeto de derecho es una sistematización efectuada por la ciencia del derecho. Esta realiza un análisis del derecho positivo y, como consecuencia de él, determina en cada caso qué entes poseen personalidad internacional. Una parte de la doctrina estima, por el contrario, que existen una o varias normas del derecho de gentes que atribuyen a ciertas entidades personalidad internacional [23]. Algunos autores que sostienen esta tesis afirman también que la atribución de subjetividad internacional trae aparejada la concesión de determinados derechos.

La existencia de normas que indican que ciertas entidades poseen personalidad internacional y de otras normas que confieren a todo sujeto determinados derechos, no ha sido probada en el derecho internacional positivo. La práctica internacional no permite deducir la existencia de normas con este contenido. el derecho sólo confiere personalidad a un ente atribuyéndole derechos o deberes, pero no diciendo que es un sujeto de derecho.

La doctrina que se examina parece incurrir en una confusión entre el derecho positivo y la ciencia del derecho. Una cosa son las normas en vigor que conforman un orden jurídico y otra son la descripción y la sistematización que el jurista hace de dichas normas. Esta distinción entre el objeto de conocimiento y la ciencia que describe dicho objeto aparece más clara en las ciencias naturales. Una cosa es el reino animal y otra la zoología, que describe y clasifica los animales. La naturaleza no determina, por ejemplo, que los quirópteros, los cetáceos y los ungulados pertenecen a la clase de los mamíferos. Esto lo elabora el hombre de ciencia tomando como punto de partida la realidad natural. Ocurre lo mismo en el ámbito jurídico. La categoría de sujeto es una elaboración de los juristas que describen el derecho, y no del derecho mismo.

---

[22] Un ejemplo de razonamiento erróneo lo ofrece la resolución de la Asamblea Parlamentaria Europea del 19. 11. 1960 (Journal officiel des Communautés européennes, 1960, p. 1496/60). Esta resolución parte de la premisa que las Comunidades europeas poseen personalidad jurídica internacional para llegar a la conclusión que gozan del derecho de legación activo y pasivo.

[23] Ver en este sentido: *Balladore Pallieri,* Diritto Internazionale Pubblico (8 a. ed., Milano 1962), pp. 105 ss.; *Diez de Velasco,* Instituciones de Derecho Internacional Público (5 a. ed., Madrid 1980) t. 1, pp. 179 y 180.

Sujeto de derecho es la denominación que los juristas dan a los destinatarios de derechos u obligaciones en las condiciones indicadas. El derecho no puede prescribir en una norma jurídica que »*X* es sujeto de derecho«, porque carecería de sentido y sería irrelevante. Aún cuando un orden jurídico contuviera una norma que dijera que »*X* es sujeto de derecho«, *X* no sería sujeto si ese ordenamiento no le atribuyera, al menos, un derecho o una obligación. Las normas jurídicas no pueden imponer al jurista que considere sujeto a quien no es titular de ningún derecho ni de ninguna obligación.

Los sujetos internacionales son muy distintos entre sí y la cantidad de derechos y obligaciones de que son titulares es también muy diversa. En un conocido pasaje de la opinión consultiva sobre la reparación de los daños sufridos al servicio de las Naciones Unidas, el Tribunal Internacional de Justicia expresó en este sentido:

> »Los sujetos de derecho, en un sistema jurídico, no son necesariamente idénticos en cuanto a su naturaleza o a la extensión de sus derechos«[24].

La cantidad o extensión de los derechos y obligaciones de cada sujeto internacional sólo puede precisarse realizando un análisis de las normas del derecho de gentes.

II. La comunidad jurídica internacional

La comunidad internacional evoluciona a través de los siglos y se modifica constantemente. La característica principal de su evolución en las últimas décadas es la notable ampliación del número de sus miembros[25].

La doctrina del derecho de gentes ha tenido en ciertas épocas dificultad en reconocer la realidad de la comunidad jurídica internacional y describirla tal como se presenta. A veces, ello se ha debido a la carencia de elementos de una teoría general del derecho, a concepciones *a priori,* a la tendencia de querer ajustarse a categorías jurídicas del derecho privado o simplemente a razones ideológicas o políticas. En los parágrafos siguientes se analizan algunas formas cómo se ha concebido la comunidad jurídica internacional.

1. La uniformidad de los sujetos del derecho internacional

Algunas teorías han elaborado una visión uniforme de la comunidad internacional y han pretendido que todos sus miembros pertenecen a una misma especie.

---

[24] »The subjects of law in any legal system are not necessarily identical in their nature or in the extent of their rights« (C.I.J., Recueil 1949, p. 178). El Tribunal invocó este precedente en su opinión consultiva sobre el Sahara occidental (C.I.J., Recueil 1975, p. 63).

[25] Ver en este sentido *Mosler,* Die Erweiterung des Kreises der Völkerrechtssubjekte, ZaöRV 22 (1962), pp. 1 ss.; *Truyol y Serra,* La sociedad internacional (2a. ed., Madrid 1977), pp. 81 ss. y 101 ss.

Una primera corriente en este sentido fue la tesis que, en otra época, sostuvo que los Estados eran los únicos sujetos del derecho de gentes[26]. Esta opinión ejerció cierta influencia en el Tribunal permanente de Justicia internacional quien, en el caso del *Lotus,* consideró al derecho internacional como el que rige las relaciones entre Estados independientes[27].

La comunidad internacional sería, según esta concepción, una comunidad de Estados. Cada uno de ellos tendría un ordenamiento jurídico propio y, a su vez, todos estarían vinculados por el derecho de gentes. El derecho interno de los Estados y el derecho interestatal configurarían el cuadro completo del orden jurídico universal. Así lo reconoce el Tribunal permanente en el caso de los empréstitos serbios emitidos en Francia cuando afirma que »todo contrato que no es un contrato entre Estados como sujetos del derecho internacional, tiene su fundamento en una ley nacional« [28].

Esta teoría uniforme de la comunidad internacional modificaba, en cierta medida, la realidad. Ante cualquier entidad no estatal que surgía, sus sostenedores adoptaban la alternativa siguiente: o le negaban personalidad internacional, o la consideraban un Estado. En este sentido, se puede recordar que ciertas comisiones internacionales administradoras de ríos fueron definidas como »Estados fluviales« [29].

Sin embargo, en su momento se advirtió que la teoría no constituía una descripción fiel de la realidad internacional y fue abandonada, habiendo sido los juristas soviéticos sus últimos sostenedores.

La segunda corriente tendiente a uniformar la comunidad internacional es aquella que sostiene que los únicos sujetos del derecho de gentes son los individuos. Esta tesis prescinde del Estado y de cualquier tipo de sociedad y sólo considera a los hombres. »Hay que desterrar definitivamente de la jurisprudencia los conceptos metafísicos de sustancia, de sujeto de derecho, de derecho subjetivo, fuente de controversias interminables, agotadoras y estériles«, escribe *León Duguit* en el prefacio de una de sus obras[30]. Y refiriéndose específicamente al derecho de gentes, expresa: »los sujetos de derecho internacional no son los Estados, sino los individuos miembros de los Estados« [31]. Entre los autores partidarios de esta doctrina se hallan también *Krabbe*[32] y *Scelle*[33].

---

[26] Ver, por ejemplo, *von Ullman,* Völkerrecht (2 a. ed., Tübingen 1908), p. 88; *von Liszt,* Das Völkerrecht (10 a. ed., Berlin 1915), p. 47.

[27] C.P.J.I., Série A, No. 10, p. 18.

[28] »Tout contrat qui n'est pas un contrat entre des Etats en tant que sujets du droit international a son fondement dans une loi nationale« (C.P.J.I., Série A, Nos. 20/21, p. 41).

[29] Ver, por ejemplo, *von Holtzendorff,* Rumäniens Uferrechte an der Donau (Leipzig 1883), p. 107.

[30] *Duguit,* Traité de Droit Constitutionnel (2 a. ed., Paris 1921) t. 1, p. VI. Ver también *op. cit.,* t. 1, pp. 319 ss. y 385 ss.

[31] *Duguit, op. cit.,* t. 1, p. 561.

[32] *Krabbe,* Die moderne Staats-Idee (2 a. ed., Den Haag 1919), pp. 272 ss.

[33] *Scelle,* Précis de Droit des gens (Paris 1932) t. 1, pp. 42, 43 y 49.

No existe actualmente ninguna concepción jurídica seria que niegue que el derecho regula siempre conductas humanas. Se trata además de una proposición ya aceptada en la época en que los autores mencionados publicaron sus obras. Pero, esta teoría no parece distinguir con claridad algo que se ha precisado anteriormente en este estudio y que consiste en la diferencia que hay entre aquellos casos en que una norma jurídica regula directamente una conducta humana y cuando lo hace indirectamente[34]. Se ha expresado ya que es posible describir en proposiciones jurídicas los derechos y las obligaciones de un Estado prescindiendo enteramente de este concepto y formulándolos como derechos y obligaciones colectivos de los individuos que lo integran. Estos derechos y obligaciones colectivos son diferentes de los derechos y obligaciones de que un individuo puede ser destinatario directo por parte de una norma jurídica. Esta diferencia es esencial y no se trata de ninguna ficción que pueda ser eliminada mediante un análisis jurídico realista. Si esta distinción no se tiene en cuenta, la descripción del derecho de gentes que se haga no será correcta. Si, por el contrario, esta distinción se realiza, será posible entonces efectuar una descripción fiel del derecho internacional recurriendo sólo a individuos como destinatarios de sus normas. Pero, en este caso, se podrá comprobar que la descripción que efectúa un jurista partidario de esta teoría y la que hace otro que se sirve de los conceptos de Estado, organización internacional, etc., serán idénticas y sólo diferirán en el lenguaje empleado. En efecto, cuando este último se refiera a un derecho del Estado, aquél hablará de un derecho colectivo de ciertos individuos y deberá recurrir a una larga y complicada formulación para explicar el modo y las condiciones de su ejercicio, pero las dos descripciones serán igualmente válidas.

2. La aparición de nuevas entidades y el recurso a un tercer orden jurídico

La doctrina no ha sabido siempre interpretar las modificaciones que continuamente sufre la comunidad internacional, en particular cuando se trata de comprobar si ciertas entidades no estatales poseen personalidad jurídica internacional. En este sentido, los autores han elaborado en reiteradas oportunidades teorías que postulan la existencia de otro orden jurídico distinto del derecho de gentes y de los ordenamientos nacionales. Así, ante la presencia de una entidad determinada, se afirma que se encuentra regulada por un derecho especial, que no es el internacional ni el derecho interno estatal. De esta manera, la entidad en cuestión pasa a ser sujeto de ese nuevo orden jurídico y queda excluída de la comunidad internacional y de los ordenamientos nacionales.

---

[34] Sobre esta cuestión ver particularmente *Kelsen,* Zur Grundlegung der Völkerrechtslehre, Österreichische Zeitschrift für öffentliches Recht (ÖZöR) 1 (1946–48), pp. 69 ss.

El recurso a un tercer orden jurídico es utilizado desde principios de siglo para explicar la situación de la Iglesia Católica. Según una teoría, las relaciones entre ella y los Estados estarían regidas por un ordenamiento especial, que sería distinto del derecho de gentes[35]. Por consiguiente, la Iglesia Católica no sería un sujeto de derecho internacional, sino sólo de ese tercer orden jurídico.

Como consecuencia de los crímenes de guerra cometidos durante la segunda conflagración mundial surgió la idea de un »derecho humano«, distinto del derecho internacional y del derecho interno de los Estados, para justificar el castigo individual de sus autores[36]. Este tercer orden jurídico sancionaría ciertos delitos llamados de »lesa humanidad«.

Otra tesis sostiene que ciertos acuerdos entre Estados y empresas extranjeras, que una parte considerable de la doctrina denomina convenios cuasi-internacionales (*quasi-völkerrechtliche Verträge*), no pertenecen ni al derecho de gentes ni al orden jurídico de los Estados[37].

El derecho interno de las organizaciones internacionales fue considerado también, en un principio, como un tercer ordenamiento jurídico[38].

Los acuerdos celebrados por las Iglesias evangélicas, particularmente en Alemania, que reciben el nombre de *Staatskirchenverträge,* han sido interpretados como que configuran un derecho distinto de los ordenamientos jurídicos internos y del internacional[39].

---

[35] *de Taube,* La situation internationale actuelle du Pape et l'idée d'un ›droit entre pouvoirs‹ (jus inter potestates), Archiv für Rechts- und Wirtschaftsphilosophie 1 (1907/1908), pp. 368 y 510 ss.; *Quadri,* Diritto Internazionale Pubblico (5 a. ed., Napoli 1968), p. 510; *Berber,* Lehrbuch des Völkerrechts (2 a. ed., München 1975) t. 1, pp. 163 y 164.

[36] *Nuvolone,* La punizione dei crimini di guerra e le nuove esigenze giuridiche (Roma 1945), pp. 43 ss., 59 ss. y 73 ss.

[37] Ver, en este sentido, las publicaciones siguientes de *Verdross,* Protection of Private Property under Quasi-International Agreements, Varia juris gentium – Liber amicorum aangeboden aan Jean Pierre Adrien François (Leiden 1959), pp. 355 ss.; The Status of Foreign Private Interests Stemming from Economic Development Agreements with Arbitration Clauses, ÖZöR 9 (1958–1959), pp. 449 ss.; Zwei Schweizer Schiedssprüche über quasi-völkerrechliche Verträge, Schweizerisches Jahrbuch für intern. Recht 21 (1964), pp. 15 ss.; Gibt es Verträge, die weder dem innerstaatlichen Recht noch dem Völkerrecht unterliegen?, Zeitschrift für Rechtsvergleichung 6 (1965), pp. 129 ss.

[38] Refiriéndose a las normas creadas por las organizaciones internacionales, escribe *von Liszt:* »Sie bilden eine selbständige Gruppe von Rechtssätzen zwischen dem staatlichen Recht und Völkerrecht« (*op. cit.,* nota 26, p. 48).

[39] *Huber,* Verträge zwischen Staat und Kirche im Deutschen Reich (Breslau 1930), pp. 83, 84 y 85.; *Hollerbach,* Verträge zwischen Staat und Kirche in der Bundesrepublik Deutschland (Frankfurt am Main 1965), pp. 100 y 101.; *Grundmann,* Vertragskirchenrecht, en Evangelisches Staatslexikon (2 a. ed., Stuttgart, Berlin 1975), columna 2760.

Es de señalar que *Hollerbach* modificó su opinión en estudios posteriores.

La existencia de un tercer orden jurídico se ha mencionado asimismo como posibilidad para solucionar las cuestiones planteadas acerca de las normas que rigen los acuerdos transfronterizos[40] y las luchas de los llamados »movimientos de liberación nacional«[41].

La cuestión del recurso a un tercer orden jurídico para explicar ciertas situaciones que se presentan puede ser analizada desde un punto de vista práctico y particular o desde otro teórico y general. El análisis práctico consiste en examinar cada caso particular y verificar si realmente existen normas que no pertenecen ni al derecho de gentes ni a los ordenamientos internos. El análisis teórico consiste en determinar si, según la estructura del derecho internacional y de los órdenes internos, es posible la existencia de ese tercer orden jurídico.

Para dar una respuesta desde el punto de vista teórico general, conviene examinar las hipótesis de un tercer orden jurídico que pueden plantearse.

Ocurre que en cierto momento aparece una entidad que se coloca, si se permite la expresión, »al lado« de los miembros de la comunidad internacional. En general, se trata de una entidad nueva, aunque puede tratarse también de un sujeto ya existente del derecho de gentes que, por razones particulares, pretende abandonar la comunidad internacional y situarse al margen de ella. Por hipótesis, se supone que la entidad en cuestión mantiene relaciones *jurídicas* con los miembros de la comunidad internacional, porque si así no fuera, no podría pensarse en la existencia de un tercer orden *jurídico.*

Las razones por las cuales una entidad puede ser considerada al margen de la comunidad internacional son diversas. Así, una corriente antirreligiosa inspira, sin duda, la idea de marginar a la Iglesia Católica del derecho de gentes. En cuanto a los convenios cuasi-internacionales, cierta ideología se resiste a ver que las empresas multinacionales puedan ocupar un lugar, por reducido que fuere, entre los sujetos internacionales. Ciertos movimientos de liberación nacional han pretendido estar regidos por un ordenamiento jurídico especial para no verse obligados a cumplir con las normas sobre derecho de guerra y derecho humanitario. Puede pensarse también que un Estado que disponga de recursos cuantiosos y de un poder extraordinario puede pretender marginarse de la sociedad internacional y querer imponer las reglas jurídicas que lo vincularán con los miembros de ella. En resumen, las razones por las cuales una entidad es marginada o pretende marginarse de la sociedad internacional son ideológicas. En unos casos, es la comunidad internacional que se resiste a admitir en su seno a la nueva entidad y, en otros, es la entidad que por propia conveniencia pretende mantenerse fuera de aquélla. Pero, tanto en uno como en otro caso, se acude al recurso de calificar las relaciones

---

[40] *Witmer,* Grenznachbarliche Zusammenarbeit (Zürich 1979), pp. 186 y 187.

[41] Conférence d'experts gouvernementaux sur la réaffirmation et le développement du droit international humanitaire applicable dans les conflits armés (Genève, 24 mai–12 juin 1971), Rapport sur les travaux de la Conférence, p. 61, parágrafo 319.

jurídicas de la entidad marginada con los sujetos de la sociedad internacional como pertenecientes a un tercer orden jurídico, distinto del derecho de gentes y del derecho interno de los miembros de dicha sociedad.

La relación jurídica de la entidad que se halla »al lado« de la comunidad internacional con los sujetos de ésta puede tener la característica de una relación de subordinación o de coordinación.

Una relación de subordinación puede darse, por ejemplo, si una entidad con un poder muy superior al de los miembros de la comunidad internacional se erige en legislador y establece las normas jurídicas que regirán sus relaciones para con cada uno de dichos miembros. Otro ejemplo de subordinación se presenta si los miembros de la comunidad internacional imponen a la entidad sus propias reglas jurídicas que regirán sus relaciones para con ella. En el primer ejemplo, el derecho de gentes habrá desaparecido para transformarse en el derecho interno de un Estado o Imperio mundial. En el segundo, las normas que regulan la relación con la entidad marginada pertenecen al derecho interno de los distintos miembros de la comunidad internacional.

Una relación de coordinación se presenta si los miembros de la comunidad internacional, mediante acuerdos con la entidad marginada (que podrán llamarse »acuerdos especiales« para distinguirlos de los tratados del derecho de gentes), establecen normas para regir sus vinculaciones recíprocas. En este caso, las normas así establecidas formarán parte, indudablemente, del derecho internacional debido a su modo de creación. Sólo un prejuicio ideológico podrá impedir que estas normas sean incluídas en el derecho de gentes. El jurista, que debe describir el derecho prescindiendo de intereses políticos e ideologías, deberá incluir en el ámbito del derecho internacional, que es el derecho coordinador de las distintas jurisdicciones, las relaciones con la entidad marginada. Si estas relaciones tienen un contenido distinto, y quizás contradictorio, respecto del derecho de gentes hasta ese momento vigente entre los demás miembros de la comunidad internacional, deberá explicar esto como si se tratara de un derecho regional o mediante otra forma, pero no podrá prescindir de esas normas en su descripción del derecho internacional.

Según resulta del análisis precedente, las relaciones jurídicas entre los sujetos de la comunidad internacional y una entidad »marginada« pueden configurar una relación de subordinación o de coordinación. Si se trata de una relación de subordinación, las normas pertenecerán al derecho interno de un Estado mundial que se crea o al derecho interno de los miembros de la comunidad internacional. Si se trata de una relación de coordinación, las normas pertenecerán al derecho internacional. Hay, pues, una alternativa: subordinación o coordinación, que se traduce en derecho interno o internacional.

La hipótesis de un tercer orden jurídico queda enteramente excluída. La relación entre dos entidades, o se da en un plano de igualdad (coordinación), o se da en un plano de desigualdad (subordinación), *tertius non datur*. La hipótesis de un tercer orden jurídico es imposible.

3. La comunidad jurídica internacional no constituye un sujeto de derecho de gentes

El orden jurídico de cada Estado precisa, mediante el otorgamiento de obligaciones y derechos, quiénes son los sujetos de dicho orden.

Pero, además existen normas que atribuyen conductas al orden jurídico estatal, o sea, hay actos que se imputan al Estado como habiendo sido realizados por él. Son los actos que cumplen los órganos del Estado. En el orden jurídico estatal, el legislador dicta leyes en nombre del Estado, los jueces pronuncian y ejecutan sentencias en nombre del Estado, la administración pública actúa también en su nombre. Por consiguiente, en el ordenamiento interno de cada Estado, además de las personas físicas y jurídicas que son sujetos de derecho, el Estado mismo es también un sujeto.

Lo que aquí se expresa del Estado puede decirse igualmente de los ordenamientos jurídicos de otros sujetos del derecho de gentes, como la Iglesia Católica, las organizaciones internacionales, la comunidad beligerante, etc.

Al igual que los órdenes jurídicos internos, el derecho internacional determina quiénes son sus sujetos, a través del otorgamiento de derechos y obligaciones. Pero, el derecho de gentes es un orden jurídico descentralizado y no existe en él ninguna norma que atribuya determinados actos al orden internacional. No hay ningún ente que actúe como órgano del ordenamiento internacional, no se atribuyen a él derechos ni obligaciones. O sea, el orden jurídico internacional no es un sujeto, la comunidad internacional no posee personalidad en el plano del derecho de gentes.

Dado que el orden internacional carece de subjetividad y que no hay órganos ejecutivos, ni legislativos, ni judiciales cuyos actos sean atribuídos al orden, los miembros de la comunidad internacional suelen crear, según sus necesidades y para determinados casos, entidades llamadas a cumplir ciertas funciones respecto sólo de quienes les han dado origen. Por ejemplo, un número de Estados puede crear un tribunal internacional para resolver los litigios que se susciten entre ellos, otros Estados pueden establecer una organización internacional para administrar la navegación y la explotación de una cuenca hídrica internacional, etc. Pero, ninguna de las entidades así creadas personifica el orden jurídico internacional. Ello no se da ni siquiera en el caso de la Organización de las Naciones Unidas.

Ante la ausencia de personalidad en la comunidad internacional, algunos autores han pretendido hallar que la »humanidad« es sujeto de derecho de gentes[42]. Su personalidad surgiría principalmente de los artículos 1 y 5 del tratado sobre exploración y utilización del espacio ultraterrestre, incluso la Luna y otros cuerpos celestes (27. 1. 1967), y de la resolución 2749 (XXV) de la Asamblea General de la O.N.U. sobre los fondos marinos[43].

---

[42] Ver, por ejemplo, *Marcoff*, Traité de Droit international public de l'espace (Fribourg, Genève, Paris, New York 1973), pp. 272 ss.

[43] Ver también los arts. 136 y 137 del proyecto de convención sobre el derecho del mar (texto oficioso), O.N.U., doc. A/CONF.62/WP.10/Rev. 3 (22. 9. 1980).

Sin embargo, el análisis de estos textos no permite llegar a la conclusión que la »humanidad« sea destinataria directa y efectiva de derechos u obligaciones internacionales. Es cierto que se hace mención de la »humanidad« en ellos, pero una mera declaración o enunciado no basta para erigirla en sujeto de derecho. En cuanto a los fondos marinos, la tercera Conferencia de las Naciones Unidas sobre el derecho del mar prevé en el proyecto de convención una cierta organización internacional y derechos y deberes para los Estados, pero no para la »humanidad«. Otro tanto hace el tratado de 1967 sobre el espacio ultraterrestre que se limita a considerar a los astronautas como »enviados de la humanidad«, sin adjudicar a esta última derechos ni obligaciones.

4. La comunidad jurídica internacional actual

Una vez precisado el concepto de sujeto de derecho de gentes, el procedimiento a seguir para identificar los miembros de la comunidad jurídica internacional actual consiste en analizar a la luz de dicho concepto cuáles son las entidades que se ajustan a él. El examen de cada sujeto en particular excede indudablemente el marco de este estudio. Sin embargo, resulta interesante efectuar aquí algunas consideraciones generales.

Algunos sujetos del derecho internacional pertenecen a una categoría o género. Así, los Estados soberanos o las organizaciones internacionales constituyen categorías que comprenden un número considerable de sujetos. Pero, otros sujetos son únicos, configuran casos particulares que no pueden ser clasificados en categorías. La Ciudad del Vaticano o el Comité internacional de la Cruz Roja son ejemplos de ello.

Todos los sujetos del derecho internacional, por el hecho de serlo, se encuentran regulados, en alguna medida, por ese orden jurídico. Pero hay algunos de ellos, como los Estados soberanos, que se hallan sujetos sólo al derecho de gentes. Se los puede denominar sujetos independientes. Otros, si bien se hallan regulados por el derecho de gentes, dependen también en cierta medida del orden jurídico de otro sujeto internacional. Se los puede denominar entonces sujetos dependientes. Hay sujetos internacionales que dependen de un Estado soberano, como el reino de Bután, el principado de Mónaco, el sultanato de Brunei, la Unión Postal Universal o la Conferencia de Derecho Internacional Privado de La Haya [44]. Hay otros que dependen de la Iglesia Católica, como es el caso de la Ciudad del Vaticano y de la Soberana Orden Militar de Malta; o de una organización internacional, como la Corporación Financiera Internacional [45]; o de varios sujetos simultáneamente, como los Valles de Andorra [46].

---

[44] Bután, Mónaco y Brunei reconocen cierta dependencia respecto de la India, Francia y Gran Bretaña respectivamente. Las organizaciones internacionales citadas reconocen cierta dependencia de Suiza y los Países Bajos respectivamente.

[45] Esta depende del Banco Internacional para la Reconstrucción y el Desarrollo.

La actual comunidad jurídica internacional está integrada por un conjunto muy heterogéneo de sujetos. Cada uno dispone de su ordenamiento jurídico interno. No sólo los Estados tienen su orden interno, sino también los demás sujetos: el derecho melitense es el orden jurídico de la Soberana Orden Militar de Malta, el derecho canónico es el de la Iglesia Católica, las organizaciones internacionales poseen el suyo, las comunidades beligerantes tienen también su ordenamiento jurídico, etc. Estos órdenes jurídicos internos se relacionan con el derecho de gentes, que cumple primordialmente una función de coordinación de aquéllos.

Resulta conveniente subrayar que el ordenamiento jurídico universal está compuesto por el derecho de gentes y por los órdenes jurídicos de los diversos sujetos internacionales. Fuera de estos órdenes no existen otras normas jurídicas. Ciertas teorías, sin embargo, afirman que las hay y dan como ejemplos los llamados contratos *in vacuo* y una supuesta *lex mercatoria* que regularía ciertos aspectos del comercio internacional[47].

Las personas reciben siempre la facultad de contratar de una ley interna. Cada orden jurídico determina el grado de autonomía de que dispondrán las personas para concertar sus contratos. En general, en los países de Occidente la autonomía de la voluntad es muy amplia. Pero, se dará siempre que cuando alguien concierta un contrato, lo hace en virtud de una ley que le da capacidad para ello y esa ley es la fuente de validez del contrato. No puede haber un contrato sin una norma anterior que otorgue a un individuo la capacidad de contratar, y esto es válido también para los llamados contratos *in vacuo.*

Se afirma también que una *lex mercatoria,* creada mediante costumbre, regiría algunos contratos comerciales internacionales. Las mismas razones ya indicadas acerca de los presuntos contratos *in vacuo* se aplican igualmente a la supuesta creación de normas consuetudinarias fuera del derecho de gentes y de los órdenes internos. La costumbre es un método de creación de normas jurídicas, o sea, es una fuente de derecho, como lo es también el contrato. Y así como no pueden existir normas contractuales *in vacuo,* tampoco puede haber normas consuetudinarias en el vacío. Las distintas normas que componen la llamada *lex mercatoria,* en la medida en que se trata verdaderamente de normas jurídicas, encuentran su fuente de validez en el derecho de gentes o en un orden jurídico interno.

Según lo expuesto en el parágrafo II.2. y en el presente, se puede, pues, expresar que toda norma jurídica pertenece, o al derecho de gentes, o al orden jurídico de un sujeto del derecho de gentes.

---

[46] Andorra depende en cierta medida de la Iglesia Católica y de Francia pues son los ordenamientos jurídicos de estos dos sujetos (el derecho canónico y el derecho francés) los que determinan la designación de los Copríncipes.

[47] Para un análisis de estas teorías, ver *Reimann,* Zur Lehre vom »rechtsordnungslosen« Vertrag (Bonn 1970).

La mise en oeuvre d'un recours concernant les droits de l'homme dans le domaine relevant de la compétence de l'UNESCO

SUZANNE BASTID

1. Dans la *Contribution de la France à la consultation des Etats membres en vue de la préparation du plan à moyen terme de l'UNESCO pour 1984–1989,* le chapitre intitulé »Les droits de l'homme et la paix«, marque que pour la France le respect et la promotion des droits de l'homme est le premier de tous les problèmes généraux que le monde d'aujourd'hui pose à l'UNESCO. Relevant des événements récents qui révèlent des tendances à faire fi de la vie humaine et de la dignité de l'homme, à revenir à des formes d'intolérance et de racisme, il est affirmé qu'à cette montée des périls »le second plan à moyen terme doit opposer un ensemble de mesures propres à prolonger et à renforcer l'action entreprise depuis près de trente-cinq ans«. Après avoir rappelé l'action normative de l'UNESCO dans ce domaine, ce document ajoute: »Consolider l'acquis, c'est aussi et sans doute surtout faire passer dans la *réalité vécue* les droits déjà consacrés par la Déclaration Universelle et par les Pactes, il y va de l'autorité des droits de l'homme. Si leur proclamation devient sans effet sur la condition humaine, l'opinion s'en détachera et n'y verra que des formules creuses. De là l'intérêt que la France n'a cessé de porter aux travaux du Conseil Exécutif pour l'examen des cas de violation des droits de l'homme qui lui sont soumis. Cette action, discrète mais déjà efficace, doit trouver son prolongement dans le second plan« [1].

La compétence du Conseil Exécutif concernant les droits de l'homme a déjà fait l'objet de plusieurs études [2], on se bornera ici à exposer les condi-

[1] Contribution de la France, juin 1981, par. 20–23.

[2] *Philip Alston,* UNESCO Procedure for Dealing with Human Rights Violations, Santa Clara Law Review, Vol. 20, pp. 665–696, article écrit dans le cadre du programme de recherches de la Société antiesclavagiste pour la protection des droits de l'homme de Londres; *Shellos F. Perrakis,* L'UNESCO et les droits de l'homme, La procédure d'examen des communications adressées à l'UNESCO pour violation des droits de l'homme relevant de sa compétence, Cahiers de droit international 4, Democritos Université de Thrace, Ed. Sakkoulas, (Komotini 1980); *Charles Rousseau,* Chronique, R.G.D.I.P. 83 (1979), p. 200.

tions dans lesquelles a été établi le régime actuel ainsi que ses caractères juridique propres.

2. Dans les mois qui ont précédé la XIXe Conférence Générale de l'UNESCO, qui devait avoir lieu à Nairobi, le Président de la Commission française, Jean Thomas, a montré l'intérêt que pourrait avoir une initiative française dans le domaine des droits de l'homme si celle-ci pouvait se traduire par des effets pratiques sur la condition humaine. L'attention a été alors attirée par Madame de Brissac, qui avait une expérience directe du fonctionnement de l'UNESCO, sur le fait que des plaintes concernant les droits de l'homme ne faisaient pas l'objet d'une procédure bien établie permettant une vérification des faits et aboutissant à porter remède à la situation en question[3]. On a relevé par ailleurs que la Conférence générale procédait fréquemment à des »condamnations« péremptoires prononcées sans garanties de procédure ni réel examen des situations litigieuses, et ceci alors qu'aucun des organes de l'UNESCO n'est habilité à condamner ni à s'ériger en juridiction. L'entrée en vigueur des Pactes internationaux devrait confirmer la nécessité d'un renforcement des mécanismes existants, dans la mesure où plusieurs de leurs dispositions concernent directement l'UNESCO.

Le groupe d'étude de la Commission française chargé du problème a examiné le mécanisme établi par le Protocole à la Convention concernant la lutte contre la discrimination dans le domaine de l'enseignement et le système de la Commission de conciliation et de bons offices chargée de rechercher la solution de différends qui naîtraient entre Etats parties à la convention. Ce système comporte un organisme de conciliation et non une juridiction appelée à prononcer une condamnation[4]. Bien qu'en fait la Commission n'ait pas fonctionné, l'intérêt d'une procédure de conciliation a été souligné, afin d'aider les Etats à améliorer les situations qui leur sont reprochées, comme aussi a été marquée la préoccupation d'éviter que les violations des droits de l'homme ne soient dénoncées en fonction de critères politiques au gré de majorités circonstancielles et par conséquent de manière sélective.

Le groupe d'étude a marqué nettement qu'il ne s'agissait pas seulement de considérer les plaintes relatives aux conventions établies dans le cadre de l'UNESCO, mais toutes celles qui concernent les droits relevant de la compétence de l'UNESCO.

La Commission française pour l'UNESCO est arrivée à la conclusion qu'il convenait que la France dépose un projet de résolution, modeste sans doute, mais engageant l'Organisation dans une étude préalable concernant la

---

[3] Sur cette procédure menée devant le Comité sur les conventions et recommandations dans le domaine de l'éducation et sur ses lacunes voir par. 109 & s. document 102 EX/19 et sur son historique 100 EX/CR/2. Elle a été adoptée par le Conseil exécutif à sa 77e session (décision 77.EX/8.3) et confirmée à la 98e session (décisions 98.EX/9.4,9.5,9.6).

[4] *S. Bastid,* Une nouvelle Commission de conciliation?, in: Mélanges offerts à Henri Rolin (1964), pp. 1–12.

possibilité de procédures comparables dans leur principe à celles du B.I.T. Suivant ces suggestions, le Gouvernement français a déposé un projet de résolution qui le rattachait aux décisions antérieurement prises par le Conseil exécutif (cf. note 3):

»La Conférence générale . . .
5. Prenant acte des décisions 98 EX/9.4, 9.5 et 9.6 et 99 EX/9.4, 9.5 respectivement adoptées par le Conseil exécutif lors de sa 98e et de sa 99e session,
6. Soucieuse de mettre l'Organisation en mesure d'exercer aussi pleinement que possible sa fonction en matière de droits de l'homme, telle qu'elle résulte des principes qui inspirent son Acte constitutif,
7. *Invite le Conseil exécutif et le Directeur général a étudier les mesures propres à améliorer l'examen des cas dont l'UNESCO peut être saisie en ce qui concerne l'exercice des droits de l'homme dans les domaines relevant de sa compétence*«.

3. A la Conférence de Nairobi en 1976, ce projet a donné lieu à une longue discussion. Le projet français visait les *cas*. La délégation de l'U.R.S.S. voulait remplacer ce terme par *questions*. Plusieurs positions se sont fait jour: volonté d'éviter une formule trop abstraite et sentiment que l'UNESCO était mieux préparée pour les cas individuels, opinion suivant laquelle le terme »questions« couvrait les cas individuels et les englobait, enfin suggestion de juxtaposer »cas et questions« en tant que compromis: l'examen des cas pouvant conduire à l'examen plus large des questions.

C'est finalement cette dernière idée, dans une formule proposée par le Brésil, qui a été acceptée par la France. La partie opérationnelle se lit donc comme suit:

»*Invite* le Conseil exécutif et le Directeur général à *étudier les procédures qu'il conviendrait de suivre dans l'examen des cas et des questions dont l'UNESCO pourrait être saisie en ce qui concerne l'exercice des droits de l'homme dans les domaines relevant de sa compétence, afin de rendre son action plus efficace*«.

Ce texte, adopté à l'unanimité moins une abstention, est devenu la résolution 19 C/6.113.

La partie opérationnelle se retrouve dans la résolution générale (12.1.9) adoptée sur la proposition du Groupe de rédaction et de négociation. Cette dernière résolution contient d'autres dispositions qui doivent être citées:

»Constatant . . . que les violations des droits de l'homme dans les domaines de la compétence de l'UNESCO sont de plus en plus fréquentes, et font l'objet de *nombreuses plaintes* adressées à l'organisation,
Prenant acte du fait que, dans son introduction au débat de politique générale (19 C/Inf. 12) le Directeur général a demandé à propos des problèmes relatifs aux droits de l'homme qu'on lui assure à l'avenir, dans le

cadre des compétences de l'UNESCO, les *moyens moraux* qui *lui permettent d'agir plus efficacement pour la sauvegarde de ces droits...*
Rappelle que l'UNESCO *n'est pas un organe judiciaire international* et que, conformément au paragraphe 3 de l'article premier de son Acte constitutif, elle doit *éviter toute intervention dans les affaires intérieures des Etats membres*«.

Enfin est rappelée la nécessité de coordination avec l'action des Nations Unies.

Cette »grande« résolution complétait sous plusieurs aspects la résolution 19 C/6.113:
- constatation du nombre des plaintes;
- constatation que les possibilités d'action sont actuellement limitées;
- demande du Directeur général tendant à être autorisé à agir;
- limitation des droits reconnues à l'Organisation par rapport aux Etats. De plus rapport devait être fait à la 20e session de la Conférence.

4. La mise en oeuvre de ces résolutions supposait une action concertée du Conseil Exécutif et du Directeur général. Un rôle important a été joué par le Directeur de la division des droits de l'homme, Karel Vasak[5], dont la connaissance approfondie des recours ouverts en matière de droit de l'homme a été largement mise à profit dès ce stade.

Le 7 avril 1977 une importante étude a été présentée pour favoriser l'examen de la question posée par la Conférence; elle reproduisait en annexe une étude antérieure en date du 26 août 1976 préparée à l'intention du Comité sur les conventions et recommandations dans le domaine de l'éducation qui avait été invité par le Conseil exécutif »à revoir les procédures qu'il applique à l'heure actuelle, y compris ses méthodes de travail et la présentation de rapports au Conseil exécutif en vue de recommander des améliorations dans le cas où cela est nécessaire«.

Ces documents font apparaître une évolution institutionnelle très curieuse en même temps que sont fournies des informations sur le nombre de plaintes qui ont été présentées concernant des droits de la compétence de l'UNESCO.

Lors de la 29e session du Conseil exécutif, la décision 29 EX/11.3 constate qu'en »l'état actuel des textes constitutionnels et réglementaires de l'Organisation, aucune suite ne peut être donnée à des communications émanant de particuliers ou d'associations qui invoquent la violation de certains droits de l'homme ...«[6]. Il est décidé cependant que le Directeur transmettra les plaintes au Président du Conseil Exécutif qui les examinera en consultation

---

[5] Devenu ultérieurement directeur des affaires juridiques, Karel Vasak reste chargé des questions concernant les recours devant le Conseil exécutif.

[6] Décision 29 EX/11.3; 77 EX/8.3 par. 4; 93 EX/8.2 par. 7 et 98 EX/9.4, 9.5, 9.6 par. 11.

avec les autres membres du Bureau. En fait, le Bureau a progressivement cessé de prendre connaissance des communications. A sa 34e session, le Conseil exécutif invitait le Directeur général à étudier »les mesures qui pourraient être proposées en vue d'assurer la possibilité d'un recours devant une instance internationale en cas de violation des droits éducatifs et culturels de l'homme«.

A la 38e session, le Conseil exécutif, tout en réaffirmant l'importance et l'intérêt de la question, décidait de renvoyer toute décision en attendant l'adoption par l'Assemblée générale des textes des pactes internationaux sur les droits de l'homme.

La question n'a été reprise qu'en 1967 à la 77e session du Conseil exécutif. On a pu relever que de 1960 à 1967, soixante-douze plaintes avaient été transmises au Président du Conseil exécutif fondées pour la plupart sur un article précis de la Déclaration universelle ou qui signalaient des faits paraissant contraires aux buts et aux principes de l'UNESCO. Sur la base d'une étude faite par le secrétariat une décision a été prise (77 EX/29) sur la *»Procédure concernant la suite à donner aux communications relatives à des cas particuliers et invoquant les droits de l'homme dans les domaines de l'éducation, de la science et de la culture«*.

Il est stipulé que la procédure sera analogue à celle qui est prévue par la résolution 728 du Conseil économique et social et que le Directeur général devra »conformément à cette procédure porter les communications en question à la connaissance du Comité spécial sur la discrimination dans le domaine de l'enseignement«.

Cette décision est importante à deux points de vue: d'une part se référant au systéme établi par le Conseil économique et social elle va servir de base à l'élaboration de lettres modèles destinées à l'auteur de la communication et à l'Etat membre mis en cause. Ces lettres indiquent certains éléments procéduraux et notamment la faculté pour l'Etat de présenter une réponse à la plainte dont la teneur lui est communiquée. La transmission d'une communication à l'Etat concerné implique non seulement que l'Etat sera informé de l'identité de l'auteur, mais des détails des »cas particuliers« dont il s'agit, y compris l'identité des personnes mentionnées dans la communication comme ayant fait l'objet des violations alléguées.

Par ailleurs la décision du Conseil exécutif désigne un organe subsidiaire, *le Comité spécial sur la discrimination dans le domaine de l'enseignement*, à la connaissance duquel le Directeur général devra porter les communications et la réponse de l'Etat. C'est certainement là une étape essentielle. Le Conseil exécutif décide »d'élargir à cette fin le mandat du Comité«.

Ce Comité a été créé en 1965 par le Conseil exécutif à la suite de la résolution 13.C.16.1 concernant la présentation par les Etats des rapports périodique sur l'application de la convention et de la recommandation concernant la lutte contre la discrimination dans le domaine de l'enseignement. La dénomination du Comité a été modifiée lors de la 75e session pour devenir

Comité spécial sur la discrimination dans le domaine de l'enseignement[7]. Il est alors composé de 12 membres. Sa compétence sera double: examen de rapports périodiques sur l'application des conventions et recommandations lorsque l'examen de ces rapports est confié au Conseil exécutif, examen des cas concernant l'exercice des droits de l'homme dans les domaines de la compétence de l'UNESCO.

Ainsi c'est un organe composé de représentants de gouvernements désignés parmi les membres du Conseil qui va être saisi par le secrétariat d'un document comprenant les textes des communications du requérant et du gouvernement concerné précédés d'un résumé succinct préparé par le secrétariat. Si le Comité décidait de maintenir le cas à son ordre du jour, le secrétariat en informait l'auteur de la communication ainsi que le gouvernement concerné en leur demandant éventuellement des renseignements complémentaires en vue de permettre au Comité de poursuivre l'examen.

Le nombre de communications placées devant le Comité a d'abord été très faible (8 jusqu'en 1974). Il a considérablement augmenté ultérieurement (20 en 1974, 16 en 1975, 29 en 1976). Bien qu'aucune règle de procédure concernant le Comité n'ait été prévue, celui-ci a examiné les communications en séance privée et s'est borné tout d'abord à mentionner dans un rapport au Conseil exécutif qu'il avait pris connaissance des communications et des réponses des Etats. A la suite d'une décision prise par le Conseil exécutif 93/EX.8.2, le Comité, à dater d'avril 1974, a présenté un rapport distinct plus détaillé faisant état de chacune des communications reçues avec la mention de leurs auteurs, ainsi que des informations fournies par le gouvernement concerné dont le nom était également mentionné. Le rapport contenait également *in fine* les conclusions auxquelles le Comité était parvenu à l'issue de ses travaux[8].

Le Comité a été amené dans son rapport à souligner que la procédure pouvait être prolongée pour demander des précisions supplémentaires, soit aux auteurs de la communications, soit au gouvernement en cause. Il a pris d'autres initiatives: transmission aux auteurs des communications de copie de ses rapports, audition des représentants des Etats non représentés au Conseil, etc.

La question de l'examen par le Conseil du rapport du Comité en séance publique a posé des questions difficiles et la note du secrétariat du 26 août 1976 faisait diverses suggestions pour limiter le débat aux conclusions et propositions de caractère général. Cette même note soulignait une évolution importante de la conception que le Comité se faisait de son rôle ne se bornant plus aux communications reçues par l'Organisation sans sollicitation de sa part, mais recherchant des informations supplémentaires suivant des règles à définir, éventuellement auditions de témoins, et même investigations *in situ*.

[7] 70 EX/déc. 5.2.1; 71 EX/déc. 3.2; 75/EX. déc. 6 II.
[8] 100 EX/CR/2 par. 30.

5. *L'étude des procédures qu'il conviendrait de suivre dans l'examen des cas et des questions dont l'UNESCO pourrait être saisie en ce qui concerne l'exercice des droits de l'homme dans les domaines relevant de sa compétence, afin de rendre son action plus efficace,* présentée le 7 avril 1977 par le Directeur général[9] constitue un exposé systématique des diverses procédures internationales existantes pour l'examen des communications relatives aux droits de l'homme dans le cadre de l'O.N.U., de l'O.I.T. et des organisations régionales. On peut y relever spécialement ce qui concerne la procédure utilisant la communication comme élément d'information permettant d'établir un rapport sur la situation *générale* des droits de l'homme dans le pays concerné telle qu'elle résulte des résolutions 1235 et 1503 de l'Ecosoc établies à la suite des demandes de la Commission des droits de l'homme tendant à l'examen des renseignements concernant les violations flagrantes des droits de l'homme contenues dans les communications relevées sur la liste dressée en application de la résolution 728 F (XXVIII), celle-là même que mentionne la décision 77 EX/29.

Les systèmes déjà adoptés par l'UNESCO sont rappelés, qu'il s'agisse des communications liées à une convention ou au jeu de la décision 77 EX/8.3 du Conseil exécutif. Sont présentées enfin des considérations pouvant servir de base à la discussion du Conseil exécutif[10]: Il s'agit de clarifier et de consolider les procédures existantes et par ailleurs de les renforcer et de les approfondir. On notera en particulier les points suivants: l'énoncé des bases juridiques directes de la compétence en matière de droits de l'homme: Acte constitutif (art. 1.1), conventions et recommandations, décisions de la Conférence générale et du Conseil exécutif complétant les bases constitutionnelles et conventionnelles, l'idée que la recherche de nouvelles procédures d'examen des communications doit tenir compte des limites à l'action de l'Organisation et de l'interdiction d'intervention en aucune matière relevant de la juridiction intérieure des Etats membres, l'idée que les méthodes d'action de l'Organisation sont essentiellement des méthodes de promotion. »L'action de promotion des droits de l'homme relevant de la compétence de l'UNESCO, si elle ne peut transformer l'Organisation en un tribunal international, pourrait néanmoins lui permettre d'*agir par voie d'étude, d'examen, d'enquête et de conciliation*«.

Sont également explorés la détermination des droits de l'homme relevant de l'UNESCO, les problèmes concernant la procédure à suivre, qu'il s'agisse des cas ou des questions, la solution à donner à l'affaire, le problème de la coordination avec les autres procédures d'examen des communications.

Ce document répondait à la mission confiée au Directeur général et au Conseil exécutif par la Conférence générale mais il correspondait également aux problèmes dont le Comité sur les conventions et recommandations dans le domaine de l'éducation avait été chargé par le Conseil exécutif.

---

[9] 102 EX/19, 7 avril 1977.
[10] *Ibid.*, par. 125 à 162.

6. Le Conseil exécutif devant établir une décision appliquant les résolutions de Nairobi, agissait dans le cadre des fonctions qui lui sont imparties suivant l'art. V.5 b de la Constitution de l'UNESCO, fonctions *d'exécution* du programme adopté par la Conférence.

A cette fin, ayant mis le point à l'ordre du jour de la 102e session, le Conseil a décidé de procéder à un examen préalable du document 102 EX/19. Un débat important est intervenu sur les diverses questions présentées dans ce document. On peut avoir une idée de la diversité des opinions énoncées en se référant au résumé analytique des discussions que le Conseil exécutif a prié le Directeur général d'établir et qui était destiné à tous les membres du Conseil[11]. Ce débat a porté sur les bases juridiques de la compétence de l'UNESCO, la détermination des droits de l'homme relevant de sa compétence, mais on ne peut manquer de relever la place donnée par plusieurs membres du Conseil au problème des violations massives et flagrantes des droits de l'homme comme aux aspirations à une transformation fondamentale du milieu international et à un nouvel ordre économique international.

Cependant à l'issue du débat le Conseil décidait de créer un Groupe de travail chargé de procéder à une étude approfondie de la question et de lui faire rapport à sa 103e session. Ce Groupe composé de treize membres, sous la présidence du Président du Conseil exécutif, avait un mandat qui apparaît comme charnière entre le débat général et l'élaboration d'une décision mettant en application les résolutions de la Conférence générale. Il s'agissait en somme pour lui de »décanter« la discussion antérieure et de présenter des solutions possibles. Il devait »soumettre à une étude approfondie le document 102 EX/19, le résumé analytique« et les commentaires écrits que les membres du Conseil exécutif avaient dû présenter et qu'en fait, sept d'entre eux avaient présentés. D'autre part il avait pour mission: »identifier les points d'accord et de désaccord et s'efforcer, en se fondant dans la mesure du possible sur un consensus, de réduire les divergences«.

Enfin il était indiqué que le rapport à présenter au Conseil devait contenir »des suggestions concernant les procédures à suivre à l'avenir (en proposant plusieurs variantes chaque fois que cela sera nécessaire)«.

Ainsi une méthode d'action était tracée et l'objectif du consensus à atteindre se profilait.

Le Groupe de travail s'est réuni en août 1977. Il ressort du rapport[12] que plusieurs membres se sont fait assister, voire remplacer, par des personnalités possédant une compétence spéciale dans ce domaine. C'est ainsi que le professeur T. Buergenthal a siégé à la place du membre américain du Conseil exécutif[13].

---

[11] 103 EX/18, 16 juin 1977.

[12] 103 EX/19, 19 août 1977.

[13] Le groupe composé suivant le principe de représentation géographique ne comportait pas de membre français. M. François Valéry n'a été élu que le 12 septembre 1977 pour remplacer Jean de Broglie décédé tragiquement.

Le rapport de ce Groupe est un document d'un grand intérêt: il témoigne du sérieux de la discussion et de l'importance des problèmes abordés mais force est de constater qu'il ne fait état d'un consensus que sur l'idée que la compétence de l'Organisation en matière de droits de l'homme ne l'autorise pas à prendre des mesures d'ordre contraignant (par. 14) et que l'esprit de dialogue et non de mise en accusation et de confrontation doit présider à l'accomplissement des tâches de l'organe compétent (par. 44). Par contre, le problème des »questions« a été l'occasion d'un débat difficile qu'il s'agisse de les définir ou même de déterminer le rôle de l'Organisation. On a pu avoir à la lecture de ce rapport, le sentiment que des difficultés considérables devaient être surmontées pour mettre à exécution les objectifs de Nairobi et que, faute de mieux, la décision 77 EX 8.3 continuerait à s'appliquer. Cependant, quelques mois plus tard, un projet de décision approuvé par consensus devait être présenté par le Groupe de travail [14].

7. Le Conseil exécutif ayant pris note du rapport 103 EX/19, a prié en effet le Groupe de travail de se réunir à nouveau en janvier 1978 afin d'établir un rapport final. Le Groupe de travail disposait des observations du Comité sur les conventions et recommandations dans le domaine de l'éducation [15]. Ce document contient une partie sur les questions de procédure et de méthodes de travail où sont examinées la recevabilité des communications, l'identité des auteurs, la présentation de renseignements complémentaires. Certaines observations concernaient l'avenir et une modification éventuelle du rôle du Comité: il était suggéré qu'il se borne désormais à connaître des cas où sont invoqués des »violations massives et flagrantes« des droits de l'homme dans le domaine de l'UNESCO. C'était reprendre la thèse soutenue par l'U.R.S.S. à Nairobi.

Le Groupe de travail a reçu également des projets de rapport émanant du Président du Conseil exécutif, des membres russe, polonais, américain, ainsi que les amendements présentés par d'autres membres.

Le Groupe de travail a adopté un projet de décision approuvé par consensus. Il a été entendu qu'aucun des douze membres du Groupe de travail ne présenterait d'amendements ou remettrait en cause les principes de base devant le Conseil exécutif. Les membres du Groupe se sont engagés à défendre durant le Conseil exécutif, l'ensemble des dispositions adoptés.

L'attitude du Groupe de travail impliquait que le texte élaboré était un compromis entre des positions très différentes au départ mais un compromis tel qu'il ne pouvait y être porté atteinte sans compromettre le résultat recherché finalement.

On doit relever ici le rôle décisif joué par les résolutions de Nairobi qui en partant de l'insuffisance du système pratiqué, créait pour le Conseil exécutif l'obligation de faire quelque chose de plus pour les communications

---

[14] 104 EX/3, 3 mars 1978.

[15] 103 EX/17, partie II.

concernant les cas comme les questions, d'où, au moment décisif, la nécessité d'en arriver à des concessions réciproques dont l'équilibre ne pouvait être remis en question. Mais l'acquis de l'UNESCO, quelles qu'en soient les faiblesses, allait jouer également un rôle important: en effet un schéma organique existait qui s'était déjà essayé à ce genre de problèmes, et qui, en ayant mesuré les insuffisances du cadre existant, était préparé à agir sur une base juridique plus solide avec des compétences mieux définies. La situation était en fait la même, qu'il s'agisse du Directeur général, du Comité sur les conventions et recommandations dans le domaine de l'éducation – qui allait être dénommé ultérieurement Comité sur les conventions et recommandations par la décision 104 EX 13.3 – et du Conseil exécutif.

Cet ensemble de circonstances a conduit à un système sans précédent dans le domaine des droits de l'homme, système fondé sur la compétence constitutionnelle de l'Organisation internationale et sur l'accord réalisé au sein des organes de celle-ci, accord obtenu à la suite de négociations mais sans donner la forme conventionnelle aux documents établis. Le rapport du Groupe de travail en date du 3 mars 1978 lorsqu'il a été connu des autres membres du Conseil exécutif, a soulevé des critiques concernant la forme mais aussi des questions de fond telle la disposition relative aux recours internes; quoiqu'il en soit, l'accord relatif au consensus a été respecté et la décision adoptée par le Conseil exécutif au cours de sa cent quatrième session reproduit intégralement le texte du Groupe de travail[16].

8. Pour l'essentiel cette décision s'inspire de la procédure de la résolution 1503 du Conseil économique et social et du système de la liberté syndicale. Elle part de la distinction faite à Nairobi entre les *cas* individuels et les *questions*. S'agissant de celles-ci une distinction est faite entre les violations massives, systématiques ou flagrantes qui sont la conséquence d'une politique pratiquée en droit ou en fait par un Etat comme l'apartheid, et celles qui sont »la conséquence d'une accumulation de cas individuels qui constituent un ensemble concordant«. Cette distinction est exposée explicitement au point 10. Elle réapparaît dans plusieurs textes. Toutefois ce n'est qu'au niveau du Conseil qu'une différence de procédure est faite expressément.

Après avoir rappelé les fondements juridiques de la décision: Acte constitutif, résolutions de l'Assemblée générale, le texte mentionne la disposition interdisant d'intervenir en aucune matière relevant essentiellement de la juridiction intérieure des Etats ainsi que les compétences spécifiques de l'UNESCO en matière de droits de l'homme qui commandent une action dans un esprit de coopération internationale, »l'UNESCO ne peut pas jouer le rôle d'un organisme judiciaire international«.

La décision s'attache ensuite au rôle important du Directeur général – rôle qui est évoqué en général et indépendamment de la procédure alors organisée. Elle se réfère d'autre part au rôle antérieurement imparti au Comité

[16] 104 EX/déc. 3.3.

sur les conventions et recommandations dans le domaine de l'éducation en matière de droits de l'homme et décide que le Comité, désormais désigné sous le vocable de »Comité sur les conventions et recommandations«, examinera les communications concernant les cas et les questions de violation des droits de l'homme conformément aux conditions et procédures énoncées dans la décision sur ce point. Ainsi sa compétence est affirmée de façon générale.

Les conditions de recevabilité sont énumérées sous dix points, parmi celles-ci, il faut relever la possibilité d'une communication d'une personne, d'un groupe de personnes ou organisation non gouvernementale «qui a une connaissance digne de foi« des violations.

La procédure est définie avec le souci de préciser le rôle du Directeur général pour assurer le respect des conditions et la transmission de la communication tout d'abord au Gouvernement concerné puis au Comité avec la réponse du Gouvernement.

Il est prévu que la communication doit indiquer si un effort a été fait afin d'épuiser les voies de recours interne disponibles concernant les faits qui constituent l'objet de la communication, ainsi que les résultats éventuels de tels efforts. Cette référence originale à l'épuisement des voies de recours interne a suscité d'assez vives réactions. On a pu faire valoir qu'elle est plus raisonnable que la mise en application des exceptions tirées de la nature et de l'efficacité des recours.

Le rôle du Comité – agissant en séance privée – est enfin défini: décision sur la recevabilité, débat possible avec les représentants des gouvernements concernés, possibilité d'entendre le requérant, possibilité de rechercher des informations supplémentaires.

Le Comité peut décider l'irrecevabilité, le rejet au fond ou poursuivre l'examen de la communication de façon »à contribuer à faire prévaloir une solution amiable destinée à favoriser la promotion des droits de l'homme«.

Le Comité présente au Conseil exécutif des rapports confidentiels contenant tous renseignements appropriés résultant de l'examen des communications et des recommandations, soit d'une manière générale, soit quant à la suite à donner aux communications soumises à son examen.

Le Conseil décide d'examiner ces rapports confidentiels en séance privée et de leur donner toute la suite qui paraît nécessaire.

Les communications qui attestent l'existence d'une question font l'objet d'une procédure spéciale. Les questions relatives à des violations massives, systématiques ou flagrantes des droits de l'homme et des libertés fondamentales »devraient être examinées par le Conseil exécutif et la Conférence générale en séance publique«. La décision cite, entre autres, les questions qui résultent de politiques d'agression, d'ingérences dans les affaires intérieures d'un Etat, etc.

Ce dernier texte qui s'inspire des termes de la résolution 1503, se rattache elle aussi à des résolutions de la Conférence générale (18 C/9.1 par. 4).

9. Depuis que cette décision capitale a été adoptée, le Conseil exécutif a pris régulièrement acte du rapport du Comité concernant les communications concernant »les cas«. En 1979 seulement, le Conseil exécutif a mentionné les »questions« à côté des »cas«.

Le Conseil exécutif a également demandé au Directeur général de donner suite aux recommandations contenues dans le rapport du Comité [17] et l'année suivante il a fait siens les *voeux* du Comité qui y sont exprimés [18] sans que soient rendus publics les termes du rapport du Comité.

Il faut noter que la Conférence générale a, lors de sa XXe session, *pris acte* de l'adoption de cette nouvelle procédure, invité le Directeur général à faire en sorte qu'elle fonctionne au mieux [19]. Lors de la XXIe session, la résolution *Contribution de l'UNESCO à la paix* s'est formellement référée aux »progrès accomplis« grâce à la nouvelle procédure d'examen, procédure qui permet de »donner suite aux pétitions individuelles« et qui permet et facilite un dialogue permanent avec les Etats membres concernés par les communications parvenues au secrétariat.

Les termes de ces résolutions méritent d'être relevés car ils marquent que la décision du Conseil exécutif adoptée par consensus est acceptée comme correspondant aux objectifs définis en 1976 par la Conférence générale. Ainsi s'affirme la portée juridique de ce texte. On est dans une situation où la réaction des Etats membres consolide un système qui oblige les organes de l'organisation dans le cadre de leur compétence, mais »affecte« également les Etats membres puis qu'il crée un titre à leur égard, à tout le moins le droit de leur demander de s'expliquer.

10. Quant à l'application pratique du système il faut relever que le Comité qui compte actuellement 25 membres [20] a été présidé, à une exception près, par un membre du Conseil exécutif d'Amérique latine. Désigné par le Conseil le choix de ce président est vraisemblablement d'une grande importance. De même le secret des délibérations et des conclusions est essentiel pour éviter une politisation des débats, politisation dont on peut imaginer le risque s'agissant de communications concernant certaines personnalités dont le sort a été abondamment discuté par les medias. Le nombre de communications s'est accru. Certaines affaires ont connu une heureuse issue: le pianiste Estraila a pu quitter l'Uruguay. Il semble que le sort du mathématicien Masera pourrait donner lieu à de nouvelles initiatives du Comité alors que dans le cadre du Comité des droits de l'homme de l'O.N.U. sur la base du Pacte des droits civils et politiques aucun résultat n'a été obtenu. A cet égard, il faut relever que le problème des recours parallèles n'a pas été posé dans des termes conflictuels et qu'il s'agit là d'un élément important du système.

---

[17] 108 EX/8.1.

[18] 110 EX/9.1.

[19] Résolution générale, Résolution sur le respect des droits de l'homme.

[20] Sa composition dépend du Conseil exécutif suivant l'art. V.17 du règlement intérieur.

Il faut noter également que le Conseil exécutif n'a pas estimé qu'il était saisi de »questions« dont il aurait à débattre en séance publique à la suite de communications. A vrai dire on peut penser que l'identification même des »questions« est susceptible de discussions et que les termes de la décision 104 EX/3.3 qui s'y réfère expressément à diverses reprises ne sont pas de nature à éliminer les controverses qui se sont exprimées lors des travaux préparatoires ou les discussions qui ont eu lieu dans d'autres enceintes. Divers critères sont évidemment concevables: durée des violations, nombre des personnes en cause, gravité des violations notamment en rapport avec le maintien de la paix et de la sécurité internationales, etc.

On constate que le texte actuel ne contient aucune disposition particulière de procédure s'agissant des questions quant au rôle du Directeur général ou du Comité. Il semble donc que ce n'est qu'à l'issue de son examen que le Comité devrait prendre position comme il semble l'avoir fait au moins une fois. Mais la rédaction du par. 18 de la décision 104 EX/3.3 laisse alors une large mesure d'appréciation au Conseil exécutif sur le sort à réserver au rapport du Comité.

Il y a tout lieu de penser, d'autre part, que dans l'application de la décision 104 EX/3.3, le Comité s'est trouvé en présence de problèmes non réglés par un texte et qu'une jurisprudence est en cours de formation. Si l'on pense à la pratique de la Commission des droits de l'homme du Conseil de l'Europe on peut imaginer qu'il y a eu lieu de préciser les cas et les conditions dans lesquels des allégations parvenues à l'UNESCO peuvent être écartées *a priori*. D'autre part, le problème de la coopération du gouvernement concerné par une communication a inévitablement posé des questions concrètes eu égard à la volonté de rechercher le dialogue avec lui, volonté qui doit sans doute être affectée par le comportement même de ce gouvernement.

Au stade actuel de cette expérience il est probable que le souci d'informer ou le désir de recevoir plus d'informations de la part de ceux qui se consacrent à la protection des droits de l'homme doivent tenir compte de la relative nouveauté du système et de l'importance essentielle de préserver une zone convenable de discrétion afin que peu à peu s'affirme au sein même du Comité et du Conseil exécutif une tradition suffisamment solide sur laquelle puissent s'appuyer ceux qui ont la redoutable responsabilité d'oeuvrer dans ce cadre.

Die unbeschränkte und die beschränkte Einkommensteuer – ein Vergleich

Zugleich ein Beitrag zu den Grundlagen des deutschen Internationalen Einkommensteuerrechts

HERMANN-WILFRIED BAYER

I. Einleitung

Wer es unternimmt, zwei Steuern miteinander zu vergleichen, stellt sich eine Aufgabe, deren Lösung ebenso schwierig wie unabweisbar ist. Die Schwierigkeiten sind spätestens seit der Zeit eines *Hensel*[1] und *Markull*[2] zur Genüge bekannt. Sie lassen sich, in einem Satze zusammengefaßt, damit umschreiben, daß es an einem mehr oder weniger verbindlichen Kanon von Merkmalen fehlt, nach denen sich bestimmt, ob zwei Steuern gleich oder verschieden sind[3]. Es wird also, etwas konkreter ausgedrückt, immer solche geben, die Beifall mit der Auffassung ernten, daß zwei Steuern stets schon dann als im Rechtssinne gleich zu gelten haben, wenn nur die »Quellen der wirtschaftlichen Leistungsfähigkeit«, an die angeknüpft wird, im einen oder im anderen Falle dieselben sind[4]. Mit dem gleichen oder einem ähnlichen Maß an Recht

---

[1] *Albert Hensel,* Steuerrecht (2. Aufl. 1927), S. 22; *derselbe,* Reichsfinanzhof und Staatsrecht, Steuer und Wirtschaft (StuW) 1928 I, Sp. 1127 ff., 1166.

[2] *Wilhelm Markull,* Gleichartige Steuern, Vierteljahresschrift für Steuer- und Finanzrecht (VjSchrStuFinR) 4 (1930), S. 535 ff.

[3] Vgl. schon *Wilhelm Markull,* Kommentar zum Gesetz über den Finanzausgleich (1923), S. 125, der in bezug auf das Gleichartigkeitsproblem von einer »völligen Zerfahrenheit der steuerlichen Terminologie und Systematik« spricht; ferner jetzt *Dieter Bökelmann,* Die örtlichen Steuern und das Gleichartigkeitsverbot in Art. 105 Abs. 2a Grundgesetz (1974), S. 61 ff.

[4] Vgl. dazu schon BVerfG vom 10. 5. 1962, Entscheidungen des Bundesverfassungsgerichts (BVerfGE) 14, 76 ff., 91 betr. die Württ.-Hohenzollerische Einwohnersteuer; ferner BVerfG vom 12. 10. 1978, BVerfGE 49, 343 ff., 355 betr. die Schleswig-Holsteinische Wohnbauabgabe; noch deutlicher allerdings *Klaus Tipke,* Über die Gleichartigkeit von Steuern, StuW 1975 S. 242 ff., 245; demgegenüber aber zu Recht kritisch *Hartmut Hahn,* Gleichheit und Gleichartigkeit im Steuerrecht, Deutsches Steuer-Recht (DStR) 1980, S. 215 ff., 218.

werden andere behaupten können, daß das Ergebnis des Vergleichs ausschließlich davon abhängen darf, ob die gesetzlichen Tatbestände, auf denen die Besteuerung im Einzelfall beruht, mehr gemeinsames oder mehr verschiedenes haben [5]. Den einen wie den anderen werden schließlich all die mit einer mehr oder weniger offenen Ablehnung begegnen, die für sich selbst, wiederum nicht ganz zu Unrecht, in Anspruch nehmen, daß sich die Gleichheit oder Verschiedenheit zweier Steuern nach der Art und Weise bestimmt, wie die für die Durchführung der Besteuerung verantwortlichen Finanzbehörden das Verfahren der Steuererhebung betreiben [6].

Der Umstand, daß der Vergleich zweier Steuern mit fast unüberwindlichen Schwierigkeiten verbunden ist, ändert nichts daran, daß er von Fall zu Fall vorgenommen werden muß. Denn davon, ob zwei Steuern gleich oder verschieden sind, hängt nicht nur das rein wissenschaftliche Urteil darüber ab, welchen Standort die einzelne Steuer im systematischen Gesamtaufbau der Steuern hat [7]. Die Konsequenzen, die mit einer Feststellung wie dieser verbunden sind, sind vielmehr auch von ganz praktischer Art. Es mag genügen, dazu mehr beispielhaft ins Gedächtnis zu rufen, daß es nichts anderes als die »Gleichartigkeit« zweier Steuern, in diesem Falle der Speiseeis- und der Umsatzsteuer, war, die das Bundesverfassungsgericht (BVerfG) in seinem bekannten »Hessischen Speiseeissteuer-Urteil« dazu zwang, von der Verfassungswidrigkeit der einen der beiden zu sprechen [8].

Wer zwei Steuern miteinander vergleichen will, muß die gesetzlichen *Grundtatbestände* miteinander vergleichen, bei deren Vorliegen im Einzelfall die Steuerschuld, sei es der einen, sei es der anderen Art, dem Grunde nach entsteht [9]. Er muß sich also in einem ersten Schritt die Frage vorlegen, ob die

---

[5] Statt aller: *Vogel/Walter*, Kommentar zum Bonner Grundgesetz (Zweitbearbeitung Art. 105 GG (1971)), Rz. 103 zu Art. 105: »Möglich ist deshalb nur ein Gesamtvergleich der gesetzlichen Steuertatbestände im weitesten Sinne; d. h. der Gesamtheit aller Voraussetzungen, von denen Entstehung und Höhe der Steuerpflicht abhängen«.

[6] Vgl. dazu wiederum BVerfG vom 12.10.1978, BVerfGE 49, 343 ff., 355. Ein Beispiel wäre etwa die Lohnsteuer, und zwar in ihrer Eigenschaft als im sog. Abzugsverfahren (vgl. § 38 EStG 1979) erhobene Einkommensteuer.

[7] Vgl. dazu etwa *Wilhelm Wengler*, Beiträge zum Problem der internationalen Doppelbesteuerung (1935), S. 142 ff.; ferner schon, wenn auch nicht sehr klar, *Friedrich Geyler*, Steuerliche Mehrfachbelastungen und ihre normative Abwehr, Bd. 1 (1931), S. 90 ff.

[8] BVerfG vom 23. 7. 1963, BVerfGE 16, 306 ff., 328; vgl. auch BVerwG vom 26. 7. 1979, Entscheidungen des Bundesverwaltungsgerichts (BVerwGE) 58, 230 ff., 241 ff. betr. die Überlinger Zweitwohnungsteuer, und dazu jetzt *Hermann- Wilfried Bayer* (Hrsg.), Die Zweitwohnungsteuer. Eine Dokumentation (1982), Rz. 17.

[9] Zum Begriff des Grundtatbestandes: *Hermann-Wilfried Bayer*, Grundbegriffe des Steuerrechts, Juristische Arbeitsblätter (JA) Sonderheft 15 (1977), Rz. 120; ferner *derselbe*, Die Liebhaberei im Steuerrecht (1981), S. 6 ff., 158 mit weiteren Nachweisen; mit Bezug auf das im Text Gesagte auch schon Reichsfinanzhof vom 20. 10. 1923, Entscheidungen des Reichsfinanzhofs (RFHE) 13, 28 ff., 31: »Der Große Senat hat schon in der Entscheidung vom 1. Mai 1922 . . . ausgesprochen, daß die äußerliche

persönlichen Verhältnisse, die im Rahmen der zu vergleichenden Tatbestände dem *Steuersubjekt* das Gepräge geben, die gleichen oder verschiedene sind. Zeigt sich, um nur die eine der beiden an sich denkbaren Möglichkeiten herauszugreifen, daß die Verhältnisse insoweit *verschieden* liegen, so darf dies, wenn auch noch nicht als Beweis, so doch jedenfalls als ein erstes sicheres Indiz dafür gelten, daß die Steuern, auf die sich der Vergleich als solcher bezieht, verschiedene sind.

Jeder einzelne steuerliche Grundtatbestand ist von Haus aus so aufgebaut, daß er die Entstehung einer Steuerschuld dem Grunde nach außer von dem Vorhandensein eines Steuersubjekts auch noch von dem eines *Steuergegenstandes* oder, wie man besser und deutlicher sagt, einer *steuerbaren Handlung* abhängig macht[10]. Der Vergleich, der soeben ins Auge gefaßt worden ist, trägt diesem Umstand nur für den Fall Rechnung, daß er sich in einem zweiten Schritt mit der Frage befaßt, ob sich diejenigen mehr *sachlich*-handlungsbezogenen Verhältnisse, denen innerhalb der zu vergleichenden Tatbestände das Merkmal der steuerbaren Handlung entspricht, als gleich oder verschieden darstellen. Tritt zu der Feststellung, daß die persönlichen Verhältnisse verschieden sind, die hinzu, daß für die sachlichen Verhältnisse dasselbe gilt, so ist endgültig klar, daß nicht die Gleichheit, sondern die Verschiedenheit das den verglichenen Steuern eigentümliche ist. Selbst für den Fall, daß die persönlichen Verhältnisse nicht verschiedene, sondern *gleiche* sind, ändert sich hieran bei Verschiedenheit der sachlichen Verhältnisse freilich nichts. Gesetzt, die Sache liegt so, macht sich nämlich der Umstand bemerkbar, daß die *sachlichen* Verhältnisse, aufs Ganze gesehen, diejenigen sind, die über das *höhere Maß* an Beweiskraft verfügen. Oder, wie es das Bundesverwaltungsgericht (BVerwG) schon in seinem »Jagdsteuer- Urteil« vom 8. November 1957 ausgedrückt hat: »Für die Gleichartigkeit von Steuern ist entscheidend der Steuergegenstand«[11]. Die Steuern sind also auch in einem solchen Fall zwei voneinander *verschiedene*.

Das bis jetzt Gesagte bedeutet nicht, daß es im einzelnen Falle »verboten« wäre, dem Vergleich zweier Steuern neben dem Vergleich der ihre

---

Bezeichnung des unmittelbaren Gegenstandes der Besteuerung nicht maßgebend ist und auch bei einer Verschiedenheit des als steuerpflichtig bezeichneten Tatbestandes eine Gleichartigkeit der Steuern bestehen kann. Es kommt, wie gesagt, nur auf die wirtschaftliche *Grundlage* der Steuer an« (Hervorhebung vom Verf.); ähnlich jetzt BVerfGE 16, 64 ff., 75: »Dabei kann es jedoch nicht allein auf eine formale, äußerliche Abweichung der steuerlichen Anknüpfungsmerkmale ankommen«; zum Ganzen auch *Markull*(Anm. 2), S. 593 f.; *Bökelmann* (Anm. 3), S. 91 ff.

[10] Vgl. dazu *Bayer,* Grundbegriffe (Anm. 9), Rz. 57 ff.

[11] BVerwG vom 8. 11. 1957, BVerwGE 5, 339 ff., 344; vgl. auch schon *Markull* (Anm. 2), S. 595: »Im Steuergegenstand ist sicherlich das Innerste und Eigenste der Steuer, jener Punkt erreicht, in dem sie sich von allen anderen Steuern artbestimmend unterscheidet«; zu Unrecht insoweit kritisch *Vogel/Walter* (Anm. 5), Rz. 102 zu Art. 105; im Ergebnis wie der Text dagegen schon *Bayer,* Die Liebhaberei (Anm. 9), S. 159.

Grundtatbestände begründenden Merkmale auch noch den Vergleich weiterer Umstände zugrunde zu legen. Wer es aus seiner Sicht für angezeigt hält, den Vorgang des Vergleichs auch noch auf solche Faktoren wie etwa den Steuermaßstab [12], den Steuersatz [13] oder das Steuerverfahren [14] auszudehnen, mag dies also gern tun. Er sollte sich allerdings nichts vormachen. Aus Gründen, die in der Natur der Sache liegen [15], wird es ihm nämlich nicht möglich sein, an dem noch einmal etwas zu ändern, was der Vergleich der den Grundtatbeständen eigentümlichen Merkmale ohnehin schon an Ergebnissen gezeitigt hat. Oder, um das »Jagdsteuer-Urteil« des BVerwG noch einmal, und zwar nunmehr vollständiger, zu zitieren: »Für die Gleichartigkeit von Steuern ist entscheidend der Steuergegenstand, *hilfsweise* der Besteuerungsmaßstab, die subjektive Steuerpflicht und die wirtschaftliche Steuerwirkung« [16].

Nachdem bis hierher einige allgemeine Grundlagen gelegt worden sind, soll im folgenden der Versuch unternommen werden, die *unbeschränkte* und die *beschränkte Einkommensteuer* miteinander zu vergleichen. Eine Festschrift, die *Hermann Mosler,* dem großen Lehrer des internationalen Rechts und der Rechtsvergleichung, gewidmet ist, scheint dem Verf. dafür der richtige Ort. Denn wenn es überhaupt einen Umstand gibt, der den Vergleich zwischen der unbeschränkten und der beschränkten Einkommensteuer von dem zwischen allen anderen Arten von Steuern unterscheidet, dann ist es der, daß mit ihm auf ein Verhältnis von »internationalem Charakter« Bezug genommen wird.

Eine Bemerkung mehr terminologischer Art sei dem weiteren noch vorausgeschickt. Dem deutschen Einkommensteuerrecht sind die Tatbestände der unbeschränkten und der beschränkten Einkommensteuer (ESt) seit inzwischen beinahe einhundert Jahren vertraut [17]. Es hat sich dabei mehr und mehr

---

[12] Vgl. dazu nochmals BVerfG vom 12. 10. 1978, BVerfGE 49, 343 ff., 355, wo aus dem »steuerbegründenden Tatbestand« neben dem »Steuergegenstand« ausdrücklich auch der »Steuermaßstab« als ein geeigneter Vergleichsmaßstab herausgehoben wird.

[13] Vgl. dazu etwa Bundesfinanzhof vom 24. 1. 1952, Entscheidungen des Bundesfinanzhofs (BFHE) 56, 184 ff., 189: ». . . die Hebesätze . . . sind für die Grundsteuer und die Gewerbesteuer typisch. Demnach ist die bayerische Baunotabgabe nicht als Grundsteuer anzusprechen«; skeptisch gegenüber dem Steuersatz als Vergleichsmaßstab dagegen schon *Markull* (Anm. 2), S. 594.

[14] Vgl. oben Anm. 6.

[15] Was den Steuermaßstab anbetrifft, so sei an dieser Stelle nur angedeutet, daß er schon deshalb keinen eigenen Vergleichsmaßstab bilden kann, weil sich in ihm, wenn auch auf numerische Weise, nur das noch einmal widerspiegelt, was auch schon dem Steuersubjekt und dem Steuergegenstand das Gepräge gibt; vgl. *Bayer,* Grundbegriffe (Anm. 9), Rz. 90 ff. sowie unten S. 70 f.

[16] A.a.O. (Anm. 11). Hervorhebung vom Verf.

[17] Der Sache nach sind die unbeschränkte und die beschränkte ESt bereits Gegenstand der §§ 1 und 2 Preuß. Einkommensteuergesetz (EStG) vom 24. 6. 1891 (Ges.-Sammlung 1891 S. 175); vgl. dazu *B. Fuisting,* Die Preußischen direkten Steuern, Bd. 1 (7. Aufl. 1907), Anm. 1 zu § 1. – Die Begriffe der »unbeschränkten« und der

eingebürgert, denjenigen, der den Tatbestand der unbeschränkten ESt-pflicht erfüllt, als Steuer*inländer*, und denjenigen, der den Tatbestand der beschränkten ESt-pflicht erfüllt, als Steuer*ausländer* zu bezeichnen[18]. Die nachfolgenden Ausführungen halten daran fest, und zwar unbeschadet der Vorbehalte, zu denen die Festlegung auf ganz bestimmte Begriffe gerade an einer Stelle wie dieser zwingt.

II. Das einkommensteuerrechtliche Personenrecht

1. Der Steuerinländer

Nach dem deutschen Einkommensteuerrecht (EStR) zeichnet sich der Steuerinländer in persönlicher Hinsicht zunächst einmal dadurch aus, daß er eine *natürliche* Person oder ein *Mensch* ist[19]. Das gilt zumindest dann, wenn man, wie es hier geschehen soll, das deutsche EStR mit denjenigen einkommensteuerrechtlichen Gesetzgebungsakten gleichsetzt, die seit dem Inkrafttreten des ersten reichseinheitlichen deutschen EStG im Jahre 1920 Geltung haben[20]. Mögen die Dinge früher auch in gewisser Weise etwas anders gelegen haben, so ist also heute jedenfalls ausgeschlossen, daß ein Steuerinländer »von Person« auch für den Fall ESt schulden kann, daß er die Rechtsnatur nicht eines Menschen, sondern einer Kapitalgesellschaft, einer Personengesellschaft des Bürgerlichen Rechts oder des Handelsrechts oder gar irgendeiner sonstigen Art von Personenvereinigung hat. Die persönlichen Verhältnisse des Steuerinländers sind, modern – deutsch – einkommensteuerrechtlich gesprochen, stets und ohne jede Ausnahme »menschliche« Verhältnisse[21].

Das deutsche EStR geht, was die persönlichen Verhältnisse des Steuerinländers anbetrifft, weiterhin davon aus, daß dieser ein *privates Leben führt*[22]. Der Begriff der privaten Lebensführung läßt sich dabei, bezogen auf den einkommensteuerrechtlichen Steuerinländer, ebenso wenig definieren wie in an-

---

»beschränkten ESt-pflicht« tauchen dagegen im Gesetz erstmals mit §§ 2 und 3 EStG vom 10. 8. 1925 (Reichsgesetzblatt [RGBl.] I, S. 189) auf. – Zur Geschichte der deutschen ESt, nicht zuletzt unter Berücksichtigung der Rechtsvergleichung, neuerdings: *Bernhard Großfeld,* Die Einkommensteuer (1981), S. 26 ff.

[18] Statt aller: *Helmut Debatin,* in: Korn/Dietz/Debatin, Doppelbesteuerung – Kommentar (Loseblatt-Sammlung, Stand: 1. Oktober 1981), Vorb. System. Teil, Rz. 23; vgl. auch schon *derselbe,* Konzeptionen zur Steuerpflicht, Finanz-Rundschau (FR) 1969, S. 277 ff. – Der Vorschlag von *Edgar Herzfeld,* Probleme des internationalen Steuerrechts, VjSchrStuFinR 1 (1927), S. 422 ff., 423, anstatt von »Steuerausländern« von »Außenwohnern« oder nur von »Fremden« zu sprechen, hat sich nicht durchgesetzt.

[19] Vgl. § 1 Abs. 1 Satz 1 EStG 1979 (Bundesgesetzblatt [BGBl.] I, S. 721).

[20] Vgl. § 2 EStG vom 29. 3. 1920 (RGBl. S. 359), anders noch etwa § 1 Ziff. 4 und 5 Preuß. EStG 1891 (Anm. 17).

[21] Statt aller: *Enno Becker,* Die Grundlagen der Einkommensteuer (1940), S. 89.

[22] Zum folgenden ausführlich *Bayer,* Die Liebhaberei (Anm. 9), S. 21 ff.

derem Zusammenhang. Das, was gemeint ist, wird indessen auch schon dadurch deutlich, daß man sich einerseits die typischen Formen einer Betätigung »in der Privatsphäre«, nämlich vor allem das *Wohnen*, ferner aber auch das Essen, Trinken und Schlafen sowie nicht zuletzt das Steuerzahlen[23] vergegenwärtigt, um dem dann andererseits die Betätigung »in der Erwerbssphäre«, nämlich das Leben »im Beruf« und ähnliches, gegenüberzustellen. Wer im einkommensteuerrechtlichen Sinne Steuerinländer sein will, muß also in jedem Fall als »Privater« in die Erscheinung treten.

Für den Steuerinländer ist nach dem deutschen EStR in persönlicher Hinsicht schließlich charakteristisch, daß für ihn die *Bundesrepublik Deutschland* den Mittelpunkt seiner privaten Lebensführung bildet[24] oder der hier, wie es gegenwärtig in § 1 Abs. 1 EStG 1979 heißt, »einen Wohnsitz« oder seinen »gewöhnlichen Aufenthalt« hat. Jemand braucht sich, um den diesbezüglichen Anforderungen zu genügen, im Einzelfall selbstverständlich nicht stets und ohne jede Unterbrechung innerhalb des Bundesgebietes zu bewegen. Es steht ihm frei, sich schon aus privaten und erst recht aus beruflichen Gründen in das Ausland zu begeben. Die Verhältnisse müssen nur so liegen, daß er in dem Sinne in der Bundesrepublik Deutschland »sitzt«, daß er hierhin immer wieder und in nicht allzu großen Abständen »zurückkehrt«[25]: Die Eigenart des Steuerinländers besteht an dieser Stelle, anders ausgedrückt, darin, daß die Bundesrepublik Deutschland für ihn das ist, was in der älteren Lehre und Rechtsprechung auch als der »dauernde Stützpunkt für die persönliche Existenz«[26] oder, einfacher, als die »Heimat«[27] bezeichnet worden ist.

Der Kreis der persönlichen Umstände, die im vorliegenden Zusammenhang Erwähnung verdienen, mag sich schließen, indem noch kurz unterstrichen wird, daß es zumindest seit dem EStG 1925 keinen deutsch-einkommensteuerrechtlichen Satz mehr gibt, der verlangen würde, daß der Steuerinländer die deutsche Staatsangehörigkeit besitzt[28]. Obwohl die, die als Steuer-

---

[23] Vgl. dazu schon *Karl Neumeyer,* Die Staatsangehörigkeit als Anknüpfung im Internationalen Recht, Niemeyers Zeitschrift für Internationales Recht 32 (1925), S. 1 ff., 10: »Steuerzahlung ist eine wirtschaftliche Angelegenheit, die als solche ihren Mittelpunkt am Ort der *bürgerlichen* Betätigung hat« (Hervorhebung vom Verf.); im Ergebnis ebenso auch BFH vom 22. 7. 1966, BFHE 86, 482 ff., 489.

[24] Statt aller: *Herrmann-Heuer,* Einkommensteuergesetz (18. Aufl. 1977), Rz. 2 zu § 1.

[25] In diesem Sinne bereits *Georg Strutz,* Kommentar zum Einkommensteuergesetz vom 10. August 1925, Bd. 1 (1927), Anm. 21 zu § 2; ausführlich zu dem im Text Gesagten auch *Arnim Spitaler,* Das Doppelbesteuerungsproblem bei den direkten Steuern (1936), S. 569 ff.; sowie *Becker* (Anm. 21), S. 176 ff.

[26] Vgl. *Strutz,* a.a.O., Anm. 21 zu § 2 mit weiteren Nachweisen.

[27] Vgl. etwa *Ernst Isay,* Internationales Finanzrecht (1934), S. 75 f.

[28] Ausnahme: § 1 Abs. 2 EStG 1979 (»Auslandsbeamtenklausel«), dazu *Hermann-Wilfried Bayer,* Das Völkerrecht in der Rechtsprechung des Bundesfinanzhofs, StuW 1981, S. 61 ff., 70 f., mit weiteren Nachweisen. – Auf dem Staatsangehörig-

inländer deutsche Einkommensteuer schulden, in der Mehrzahl der Fälle Deutsche sein werden, ist es demnach ebenso möglich, daß sie Inhaber einer fremden Staatsangehörigkeit oder staatenlos sind.

2. Der Steuerausländer

Die persönlichen Verhältnisse des Steuerausländers sind in den Augen des deutschen EStR einmal dadurch gekennzeichnet, daß er eine *natürliche* Person und damit ein *Mensch* ist. Die Dinge liegen beim Steuerausländer insoweit ebenso wie beim Steuerinländer. Das oben dazu Gesagte gilt sinngemäß.

Das deutsche EStR geht darüber hinaus von der Annahme aus, daß der Steuerausländer *kein privates Leben* führt. Der Steuerausländer mag sich also *tatsächlich* in dieser oder jener Form wie ein Privatmann gerieren. Das EStR sieht davon ab und behandelt ihn so, als ob er, wenn überhaupt, nur über eine *Erwerbssphäre* verfügte. In Anlehnung an eine Formulierung, die von *Debatin* stammt, kann man auch sagen, daß das deutsche EStR, was die persönlichen Verhältnisse des Steuerausländers anbetrifft, auf einer »gewissen Entpersönlichung« beruht[29].

Wer von einem Recht wie dem deutschen EStR so angesehen wird, als wenn er keine Privatsphäre hat, muß im gleichen Zuge so behandelt werden, als wenn er *nirgendwo* einen *Wohnsitz* hätte. Denn das eine ist Ausdruck des anderen[30]. Nach dem, was soeben gesagt worden ist, ist es deshalb auch nur folgerichtig, daß das deutsche EStR seit langem, gegenwärtig in der Gestalt des § 1 Abs. 3 EStG 1979, eine Vorschrift enthält, die es als ein persönliches Merkmal der Steuerausländer bezeichnet, daß sie »im Inland *weder einen Wohnsitz* noch *ihren gewöhnlichen Aufenthalt* haben«. Der Gesetzgeber hätte die Sache, um die es geht, freilich noch sehr viel schärfer getroffen, wenn er eine Formulierung gefunden hätte, die es *verboten* hätte, aus dem Erfordernis des *Fehlens* eines Wohnsitzes oder gewöhnlichen Aufenthalts im Inland auf die Notwendigkeit des *Vorhandenseins* eines Wohnsitzes oder gewöhnlichen

---

keitsprinzip beruhend zuletzt § 2 Abs. 1 Ziff. 1 EStG 1920 (Anm. 20), anders dann § 2 EStG 1925 (Anm. 17), dazu *Herbert Dorn,* Das Recht der internationalen Doppelbesteuerung, VjSchrStuFinR 1 (1927), S. 189 ff., 199 f., der den Wandel so erklärt: »Zunächst war die Besteuerung auf die politischen Beziehungen des Staates zur Person, die Staatsangehörigkeit, gegründet ... Diese subjektive politische Beziehung ist mit dem Anwachsen der die Staatsgrenzen sprengenden Privatwirtschaften zurückgetreten. In der weitaus überwiegenden Zahl knüpfen die Steuern jetzt an wirtschaftliche Beziehungen der Person zum Gebiete an ... Dies gilt für die Einkommensbesteuerung der meisten Staaten«; ferner *Neumeyer* (Anm. 23), S. 1 ff.

[29] In: Korn/Dietz/Debatin (Anm. 18), Rz. 24; ferner BFH vom 30. 11. 1966, BFHE 88, 378 ff., 380.

[30] Vgl. dazu *Ernst Blumenstein,* System des Steuerrechts (3. Aufl. 1971), S. 52 f., der von dem »Individuum«, das im obigen Text als Steuerausländer bezeichnet wird, als von jemandem spricht, der »nicht mit seiner Person als solcher, *wie beim Wohnsitz«,* der inländischen Gebietshoheit unterworfen ist (Hervorhebung vom Verf.).

Aufenthalts im *Ausland* zu schließen[31]. Es wird eines der Anliegen künftiger Einkommensteuerreformen sein müssen, hier klärend zu wirken[32].

Ebenso wenig wie beim Steuerinländer tritt beim Steuerausländer zu dem bis jetzt Gesagten noch hinzu, daß er im Besitz irgendeiner bestimmten Staatsangehörigkeit ist. Dem Umstand, daß jemand Steuerausländer ist, steht, um nur diesen einen Fall zu nehmen, also nicht entgegen, daß er, staatsangehörigkeitsrechtlich gesehen, *Deutscher* ist[33].

3. Ergebnis

Vergleicht man, wie es hier geschehen ist, die unbeschränkte und die beschränkte ESt an Hand der ihren Grundtatbeständen zugrunde liegenden *persönlichen* Verhältnisse, so läßt sich erkennen, daß beide schon insoweit offenbar eher *verschieden* als gleich sind. Der Umstand, daß sowohl im Falle der unbeschränkten wie in dem der beschränkten ESt eine *natürliche* Person vorhanden sein muß, ändert daran nichts. Als schwerwiegender muß nämlich die Tatsache gelten, daß allein von dem, der als Steuerinländer den Tatbestand der *unbeschränkten* ESt erfüllt, auch in dem Sinne ein Person-Sein gefordert wird, daß er als eine Person *mit Privatleben* in die Erscheinung tritt, und noch dazu einem solchen, das im Gebiet der *Bundesrepublik Deutschland* seinen Mittelpunkt hat. Es trifft im übrigen, um nur das noch hinzuzufügen, im wesentlichen das gleiche, wenn das Hessische FG das hier in Rede stehende Verhältnis in seinem Urteil vom 18. Mai 1971 so charakterisiert: »Beide Besteuerungsformen beruhen auf unterschiedlichen Voraussetzungen, die so erheblich sind, daß verschiedene rechtliche Folgerungen möglich sind: *Allein* der Umstand, ob jemand im Inland den Mittelpunkt seines *Lebens* (Wohnsitz, gewöhnlichen Aufenthalt) hat, rechtfertigt generell eine unterschiedliche einkommensteuerrechtliche Behandlung«[34].

---

[31] Unzutreffend also *Joachim Lang*, Systematisierung der Steuervergünstigungen (1974), S. 48, der davon ausgeht, daß »das Rechtsinstitut der beschränkten Steuerpflicht« »an den ausländischen Wohnsitz« anknüpft; ansatzweise wie im Text dagegen Niedersächs. Finanzgericht (FG) vom 18. 4. 1977, Entscheidungen der Finanzgerichte (EFG) 1978 Nr. 126, wo bezüglich der »beschränkten Steuerpflicht« darauf abgehoben wird, daß jemand »*keine* Beziehungen zum Inland hatte bzw. aufrechterhält« (Hervorhebung vom Verf.); vgl. auch BFH vom 23. 5. 1973, BFHE 111, 33 ff., 36.

[32] Zu denken wäre insoweit an eine Rückkehr zu § 2 Abs. 2 EStG 1920 (Anm. 20), wo die Steuerausländer noch als »natürliche Personen *ohne Rücksicht* auf Staatsangehörigkeit, Wohnsitz oder Aufenthalt« bezeichnet werden.

[33] Vgl. BFH vom 30. 10. 1973, BFHE 111, 235 ff., 241.

[34] EFG 1971 Nr. 514 (Hervorhebung vom Verf.); ferner schon FG Bremen vom 8. 9. 1960, EFG 1961 Nr. 174: »Nach dem System des EStG ist für die Entscheidung, ob unbeschränkte oder beschränkte ESt-pflicht besteht, nicht davon auszugehen, ob sich die Steuerpflicht auf sämtliche oder nur auf einen Teil der Einkünfte erstreckt, sondern davon, ob der Stpfl. einen Wohnsitz oder seinen gewöhnlichen Aufenthalt im Inland hat oder nicht«.

III. Das sachliche Einkommensteuerrecht

1. Der Steuerinländer

Der Steuerinländer zeichnet sich nach dem deutschen EStR in sachlicher Hinsicht zunächst einmal dadurch aus, daß er einer *Erwerbstätigkeit* nachgeht. Der Gesetzgeber läßt dabei allerdings seit jeher offen, was unter einer »Erwerbstätigkeit« oder, wie es im Preußischen EStG von 1891 hieß, unter einer »Gewinn bringenden Beschäftigung« zu verstehen ist[35]. Aus der Gesamtheit aller Umstände kann indessen geschlossen werden[36], daß der Begriff der Erwerbstätigkeit in erster Linie die vier herkömmlichen Formen einer *Berufstätigkeit*[37], nämlich den Betrieb einer Land- und Forstwirtschaft, eines Gewerbes, einer selbständigen und einer nichtselbständigen Arbeit umfaßt, weiter die beiden »*Neben*«-Erwerbstätigkeitsarten der Überlassung von Kapitalvermögen und der Überlassung von Grundvermögen zur Nutzung einschließt und endlich auch noch insoweit gilt, wie »sonstige«, den bis jetzt genannten »*ähnliche*« Tätigkeiten auf dem Spiele stehen. Der Steuerinländer ist also, um ein Beispiel zu geben, »sachlich« einkommensteuerpflichtig, gleichviel, ob er ein großes Unternehmen betreibt, ein einzelnes Grundstück an einen Dritten vermietet oder gar nur auf Dauer den Pfand für eine Sammlung leerer Coca-Cola-Flaschen entgegennimmt[38].

Soll jemand aus sachlichen Gründen Steuerinländer im Sinne des deutschen EStR sein, so ist darüber hinaus erforderlich, daß er seiner Erwerbstätigkeit *irgendwo* nachgeht. Es gilt das genaue Gegenteil dessen, was oben über die private Lebensführung des *Steuerausländers* gesagt worden ist. Die Verhältnis in seinem Urteil vom 18. Mai 1971 so charakterisiert: »Beide Besteu-Gebiet der Bundesrepublik Deutschland, so doch jedenfalls im *Ausland* der Ort für eine Betätigung innerhalb der »Erwerbssphäre« vorhanden ist. Wo

---

[35] Vgl. § 7 Ziff. 2 Preuß. EStG (Anm. 17). – Seit dem EStG 1925 (Anm. 17) ist die allgemeinste gesetzliche Umschreibung der sachlichen ESt-pflicht die der »Leistung« (§ 6 Abs. 1 Ziff. 8) oder auch die der »Tätigkeit« (§ 41 Abs. 1 Ziff. 1). Zum Ganzen im übrigen jetzt § 2 Abs. 1 EStG 1979.

[36] Grundlegend zum folgenden: *Hermann-Wilfried Bayer,* Die allgemeinen Grundsätze des Einkommensteuerrechts, JA 1979, S. 140 ff., 196 ff.; *derselbe,* Die Liebhaberei (Anm. 9), S. 11 ff., 39 ff.; ferner *Harald Schulz,* Grundlagen und System der Einkommensbesteuerung (3. Aufl. 1981), Tz. 59 f., 131 ff.; und dazu *Thomas Birtel,* Betriebs-Berater (BB) 1982, S. 1142 f. (Buchbesprechung). Vgl. auch Art. 3 Ziff. 1 des Schweizerischen Wehrsteuerbeschlusses vom 9. 12. 1940 i.d.F. vom 28. 6. 1971, wonach zur »Entrichtung der Wehrsteuer« die »natürlichen Personen« verpflichtet sind, »a) die in der Schweiz ihren Wohnsitz haben, b) die sich in der Schweiz aufhalten und hier eine Erwerbstätigkeit ausüben . . .« (hier zitiert nach *Heinz Masshardt,* Kommentar zur eidgenössischen Wehrsteuer 1971–1982, Ausgabe 1972 (1972)).

[37] Vgl. zu diesem Begriff etwa § 6 Abs. 1 Ziff. 3 EStG 1925 (Anm. 17).

[38] Beispiel nach BFH vom 6. 6. 1973, BFHE 110, 22 ff.

die »Einkunftsquelle« genau liegt, ist, um mit dem Bundesfinanzhof (BFH) zu sprechen, aus dem Blickwinkel des deutschen EStR »unerheblich« [39].

2. Der Steuerausländer

Für den Steuerausländer ist nach dem sachlichen deutschen EStR zunächst einmal charakteristisch, daß er einer *Erwerbstätigkeit* nachgeht. Der Begriff der Erwerbstätigkeit, der damit auch an dieser Stelle vorausgesetzt wird, ist im wesentlichen derselbe wie oben [40]. Es erübrigen sich also besondere Ausführungen darüber, daß der Steuerausländer keineswegs bloß dann eine deutsche ESt zu schulden braucht, wenn er sich etwa einem Gewerbebetrieb oder einer der drei anderen Formen einer »Haupterwerbstätigkeit« bzw. eines »Berufes« widmet.

Die sachlichen Verhältnisse des Steuerausländers werden im deutschen EStR außerdem durch den Umstand bestimmt, daß er die *Bundesrepublik Deutschland* zum Ort seiner Erwerbstätigkeit macht [41]. Die Anforderungen, die das im einzelnen mit sich bringt, können hier auf sich beruhen. Es ist nur wichtig, sich zu vergegenwärtigen, daß das, was oben über die Bundesrepublik Deutschland in ihrer Eigenschaft als Wohnsitz eines Steuerinländers gesagt worden ist, im vorliegenden Zusammenhang mit der Maßgabe gilt, daß an die Stelle des Wohnsitzes die *Betriebsstätte* tritt. Soll ein Steuerausländer dem Grunde nach zur Zahlung von ESt verpflichtet sein, muß sich von ihm also sagen lassen, daß die von ihm ausgeübte Erwerbstätigkeit mit dem Bundesgebiet durch eine Betriebsstätte oder ähnliches *verbunden* ist. Oder, um das FG Düsseldorf mit seinem Urteil vom 26. Mai 1977 zu Wort kommen zu lassen: »Als ... Merkmal dieser Grenzziehung ergibt sich, daß es wesentlich darauf ankommt, daß die Einkommensquelle mit dem Inland in einem besonderen *örtlichen und wirtschaftlichen Zusammenhang* steht« [42].

---

[39] Vgl. BFH vom 29. 10. 1974, BFHE 114, 76 ff., 78; ferner schon BFH vom 29. 10. 1970, BFHE 101, 91 ff., 93.

[40] Vgl. § 49 Abs. 1 EStG 1979; ferner schon BFH vom 7. 7. 1971, BFHE 103, 153 ff., 154; anders dagegen etwa Hess. FG vom 18. 5. 1971, EFG 1971 Nr. 514.

[41] Vgl. dazu schon § 2 Abs. 2 EStG 1920 (Anm. 20), wo ausdrücklich von »einer im Inland ausgeübten Erwerbstätigkeit« die Rede ist. Die Vorschrift des § 1 Abs. 3 EStG 1979 verwendet statt dessen den Begriff der »inländischen Einkünfte«, verfälscht damit aber insofern den wahren Sachverhalt, als »Einkünfte« eine Größe sind, die schon von Haus aus nicht an einen bestimmten Ort gebunden sein kann; vgl. dazu BFH vom 29. 10. 1970, BFHE 101, 91 ff., 93, wo das »Einkommen« zutreffend als »nicht gebietsgebunden« bezeichnet wird; ferner bereits *Spitaler* (Anm. 25), S. 424: »Eine Lagebestimmung seines (d. h.: des Ackers) wirtschaftlichen Ertrages versuchen zu wollen, ist eine unlösbare Aufgabe; denn etwas Unkörperliches, wie es der wirtschaftliche (nicht technische) Ertrag ist, hat keine körperliche Lage«.

[42] EFG 1977 Nr. 597 (Hervorhebung vom Verf.). – Die Kritik, die *Klaus Vogel*, Der räumliche Anwendungsbereich der Verwaltungsrechtsnormen (1965), S. 114 ff., an dem dem Text zugrunde liegenden Konzept einer »Territorialität« der beschränkten ESt übt, trifft die Sache nicht.

3. Ergebnis

Hatten sich oben unter II. bereits erste deutliche Anzeichen dafür ergeben, daß die unbeschränkte und die beschränkte ESt zwei voneinander verschiedene Steuern sind, so werden diese jetzt zur Gewißheit. Wer immer noch behaupten wollte, daß die unbeschränkte und die beschränkte ESt im Rechtssinne nicht verschieden, sondern gleich sind, könnte sich zwar, was den vorliegenden Zusammenhang anbetrifft, auf den Umstand berufen, daß der steuerliche Grundtatbestand im einen wie im anderen Fall die Ausübung einer *Erwerbstätigkeit* verlangt. Er würde aber, und das erscheint mindestens ebenso gewichtig, aus dem Auge lassen, daß im Falle der *beschränkten*, anders als dem der unbeschränkten ESt noch *hinzukommen* muß, daß die Ausübung der Erwerbstätigkeit mit dem Gebiet der Bundesrepublik Deutschland *örtlich verknüpft* ist.

Die je *eigene* Art, die die unbeschränkte und die beschränkte ESt hiernach besitzen[43], läßt sich noch auf zwei anderen Wegen zum Ausdruck bringen. Man kann sich einmal daran erinnern, daß herkömmlicherweise all die, aber auch nur die Steuern als Personal- oder *Personensteuern* bezeichnet werden, deren Grundtatbestand an irgendwelche besonderen persönlichen Verhältnisse des jeweiligen Steuersubjekts anknüpft[44]. Tut man dies im Zusammenhang mit dem hier behandelten Thema, so liegt auf der Hand, daß die Art der *unbeschränkten* ESt auf jeden Fall die einer *Personen-* und die Art der beschränkten ESt auf jeden Fall die einer *Nicht-Personensteuer* ist[45].

Es läßt sich aber zweitens auch »umgekehrt« verfahren und darauf abheben, daß Steuern immer dann und nur dann als *Objeksteuern* in die Erscheinung treten, wenn ihr Steuergegenstand, wie insbesondere im Falle der Grund- und der Gewerbesteuer, so ausgestaltet ist, daß eine *räumliche* Verknüpfung mit dem *Inland* bestehen muß[46]. Auch der Schluß, der sich, so gesehen, an dieser Stelle ergibt, ist klar. Er kann nämlich nur in der Feststellung

---

[43] Anders aber *Becker* (Anm. 21), S. 151: Keine »einschneidende rechtliche Verschiedenheit«; aus neuerer Zeit ferner *Horst Endriss,* Ist die Unterscheidung zwischen unbeschränkter und beschränkter Steuerpflicht noch zeitgemäß?, FR 1968, S. 338 ff.

[44] Statt aller *Hermann-Wilfried Bayer,* Art. Personensteuer, Handwörterbuch des Steuerrechts (2. Aufl. 1981), S. 1083 f.

[45] Vgl. etwa *Helmut Debatin,* Die beschränkte Steuerpflicht bei der Einkommen- und Körperschaftsteuer, BB 1969, S. 1015 ff., 1015: »Geht man davon aus, daß die unbeschränkte Steuerpflicht die Grundform der Einkommen- und Körperschaftsteuer als Personensteuer bildet, so . . .«; aus dem älteren Schrifttum *Geyler,* (Anm. 7), S. 533 ff.; bezüglich der beschränkten ESt abweichend etwa *Eberhard Littmann,* Das Einkommensteuerrecht (13. Aufl. 1981), Rz. 6 zu § 1.

[46] Vgl. bezüglich der Gewerbesteuer statt aller BFH vom 4. 9. 1962, BFHE 75, 679 ff., 681: »Das Gewerbesteuergesetz befaßt sich für jeden Steuerpflichtigen erkennbar nur mit der Besteuerung im Inland erzielter Erträge und inländischer Kapitalien (§ 2 Abs. 1 Sätze 1 und 2 GewStG)«; ähnlich neuerdings auch BFH vom 10. 7. 1974, BFHE 113, 242 ff., 245.

liegen, daß die *beschränkte* ESt, was auch immer sonst noch ihre Art sein mag, eine *Objekt-* und die *unbeschränkte* ESt eine *Nicht-Objektsteuer* ist[47].

IV. Sonstige Gesichtspunkte

Schon im Eingangsteil dieser Studie ist darauf hingewiesen worden, daß der gesetzliche Steuermaßstab und das Steuerverfahren zwei Merkmale sind, die in aller Regel die Fähigkeit haben, das Ergebnis, zu dem der Vergleich zwischen zwei Steuern an Hand ihrer gesetzlichen Grundtatbestände führt, noch zu bekräftigen. Das Ziel der nachfolgenden, mehr thesenhaften Bemerkungen soll es sein, unter Beweis zu stellen, daß das auch für das Ergebnis des Vergleichs zwischen den hier ins Auge gefaßten Steuern gilt.

1. Das einkommensteuerrechtliche Steuermaßstabrecht

Die Verschiedenheit zwischen der unbeschränkten und der beschränkten ESt ist oben einmal mit dem Hinweis darauf gekennzeichnet worden, daß der einkommensteuerrechtliche Grundtatbestand im Falle der unbeschränkten ESt auf der Erwerbstätigkeit *als solcher* und im Falle der beschränkten auf der im *Bundesgebiet* ausgeübten Erwerbstätigkeit beruht. Im Rahmen des einkommensteuerrechtlichen Steuermaßstabrechts findet sich schon diese Art der Verschiedenheit wieder. Man braucht sich, um das zu erkennen, nur diejenigen seiner Vorschriften vor Augen zu führen, die bestimmen, daß bei der Ermittlung der Höhe der Steuerschuld im Falle der unbeschränkten ESt von dem »*Gesamtbetrag* der Einkünfte« und im Falle der beschränkten von den »*inländischen* Einkünften« auszugehen sein soll[48]. Denn es bedarf wohl kaum irgendeiner weiteren Erörterung, daß in bezug auf den »Gesamtbetrag der Einkünfte« in dem gleichen Sinne eine Erwerbstätigkeit *als solche* vorausgesetzt sein muß, wie eine »inländische Einkunft« nicht ohne eine im *Bundesgebiet* ausgeübte Erwerbstätigkeit gedacht sein kann.

Als zweites Merkmal der Verschiedenheit im hier erörterten Sinne ist zuvor der Umstand bezeichnet worden, daß im Falle der unbeschränkten ESt

---

[47] Statt aller BFH vom 12. 4. 1978, BFHE 125, 157 ff., 160: »Bei beschränkter Steuerpflicht ist darauf abzustellen, daß sich diese Einkünfte vom Inland aus gesehen darstellen. Auf die Person oder die Organisationsform des Steuerpflichtigen kommt es wegen des objektsteuerähnlichen Charakters der beschränkten Steuerpflicht nicht an (sog. isolierende Betrachtungsweise)«; aus dem Schrifttum etwa *Debatin* (Anm. 45), S. 1015; *derselbe,* in: Korn/Dietz/Debatin (Anm. 18), Rz. 24; *Lang* (Anm. 31), S. 48; *Bayer* (Anm. 44), S. 1084; stark einschränkend dagegen noch *Becker* (Anm. 21), S. 150 f.

[48] Vgl. § 2 Abs. 3 EStG 1979 einerseits, § 1 Abs. 3 in Verbindung mit § 49 Abs. 1 EStG 1979 andererseits. – Die lange Zeit praktizierte Übung (vgl. erstmals § 2 Abs. 1 EStG 1920 [Anm. 20]), die Erstreckung der unbeschränkten ESt auf das »gesamte Einkommen« ausdrücklich im Gesetz zu betonen (sog. Welteinkommensprinzip), ist vernünftigerweise fallengelassen worden.

die natürliche Person in ihrer Eigenschaft als *Privater* und im Falle der beschränkten ESt die natürliche Person *als solche* die ESt dem Grunde nach schuldet. Das einkommensteuerrechtliche Steuermaßstabrecht ist dazu angetan, auch diese Form der Unterscheidung zu bestätigen, und zwar, um es kurz zu machen, in der Gestalt, daß es die Befugnis, den »Gesamtbetrag der Einkünfte« zum Zwecke der endgültigen Ermittlung der Höhe der Steuerschuld um ganz bestimmte »Aufwendungen für die *Lebensführung*«, nämlich die Sonderausgaben und die außergewöhnlichen Belastungen, zu kürzen, denjenigen natürlichen Personen vorbehält, die die Eigenschaft von unbeschränkt Einkommensteuerpflichtigen und damit die von *Privaten* haben [49].

2. Das einkommensteuerrechtliche Verfahrensrecht

Die beiden Merkmale, die der Unterscheidung zwischen der unbeschränkten und der beschränkten ESt bis hierher das Gepräge gegeben haben, sind schließlich auch ein Bestandteil des einkommensteuerrechtlichen Verfahrensrechts. Aus dem Blickwinkel des einkommensteuerrechtlichen Verfahrensrechts ist die Unterscheidung nach Maßgabe der natürlichen Person als *Privatem* und der natürlichen Person als *solcher* so etwa eine ganz selbstverständliche Voraussetzung all derjenigen Bestimmungen, die, wie heute etwa § 26 Abs. 1 EStG 1979, besagen, daß das Recht der Zusammenveranlagung solchen natürlichen Personen vorbehalten sein soll, die je für sich unbeschränkt einkommensteuerpflichtige »Ehegatten« und d. h. *Private* sind. Und was dieselbe Unterscheidung, jetzt nur unter Zugrundelegung der Kriterien der Erwerbstätigkeit als *solcher* einerseits und der Erwerbstätigkeit *im Bundesgebiet* andererseits anbetrifft, so kommt diese im Rahmen des einkommensteuerrechtlichen Verfahrensrechts zumindest an der Stelle zum Ausdruck, wo die Finanzbehörden dazu angehalten werden, die Steuer im Falle der *beschränkten* ESt auf dem Wege des Abzugs an der *im Bundesgebiet* belegenen Quelle zu erheben [50].

3. Die einkommensteuerrechtlichen Rechtfertigungsgründe

Dem bis jetzt Gesagten sei noch ein letzter Gesichtspunkt angefügt. Das einschlägige Schrifttum hat immer wieder die Ansicht vertreten, daß die unbeschränkte und die beschränkte ESt sich dadurch unterschieden, daß die unbeschränkte ESt dazu diene, dem Gedanken der Besteuerung nach der *Leistungsfähigkeit* zum Durchbruch zu verhelfen, während das Ziel der beschränkten ESt darin bestehe, den Gedanken der Besteuerung nach der *Äquivalenz* zu

---

[49] Vgl. § 2 Abs. 4 EStG 1979 einerseits, § 50 Abs. 1 EStG 1979 andererseits; dazu auch *Debatin,* Konzeptionen (Anm. 18), S. 279, sowie Niedersächs. FG vom 2. 5. 1978, EFG 1978 Nr. 463.

[50] Vgl. § 50 a Abs. 4 EStG 1979, dazu auch *Debatin* (Anm. 45), S. 1016.

verwirklichen[51]. Hebt man auf der Grundlage dieser Untersuchung darauf ab, daß die je eigene Art der unbeschränkten und der beschränkten ESt in erster Linie darin liegt, daß die ihnen eigenen Grundtatbestände auf der Gegenüberstellung der Erwerbstätigkeit *als solcher* einerseits und der Erwerbstätigkeit *im Bundesgebiet* andererseits beruhen, so läßt sich dem folgen. Denn für diesen Fall kann man in der Tat behaupten, daß die *unbeschränkte* ESt nichts anderes als ein Ausdruck für die mit einer Erwerbstätigkeit als solcher erlangte wirtschaftliche *Leistungsfähigkeit* des Einzelnen und die *beschränkte* ESt nichts anderes als ein Ausdruck für die mit einer Erwerbstätigkeit im Bundesgebiet erlangten Leistungen der *Bundesrepublik Deutschland* ist[52]. Die Dinge liegen dagegen eher umgekehrt, wenn man die je eigene Art der unbeschränkten und der beschränkten ESt mehr darin sieht, daß die ihnen entsprechenden Grundtatbestände im ersten Falle die natürliche Person *als Private* und im zweiten die natürliche Person *als solche* voraussetzen. Der Annahme, daß zwischen der unbeschränkten und der beschränkten ESt der oben genannte Unterschied besteht, steht, *so gesehen,* nämlich schon der Umstand entgegen, daß zumindest die natürliche Person *als solche* ein Ausdruck für *gar nichts* und nicht etwa ein solcher für irgendwelche seitens der Bundesrepublik Deutschland erbrachten *Leistungen* ist[53].

---

[51] Statt aller: *Volker Kluge,* Das Internationale Steuerrecht der Bundesrepublik (1976), S. 35. – Das Unterscheidungsmerkmal wird freilich einmal von denen geleugnet, die sowohl die unbeschränkte als auch die beschränkte ESt als Ausdruck einer, wenn auch »abgestuften« Äquivalenz betrachten; vgl. dazu grundlegend *Georg Schanz,* Zur Frage der Steuerpflicht, Finanzarchiv (FA) 9 (1893), S. 365 ff.; ferner, diesem folgend, etwa *Dorn* (Anm. 28), S. 208 ff.; *Spitaler* (Anm. 25), S. 425 ff.; *Blumenstein* (Anm. 30), S. 47; *Jörg M. Mössner,* Der Begriff des internationalen Steuerrechts in der neueren Literatur, Österreichische Zeitschrift für öffentliches Recht 25 (1974), S. 255 ff., 285; *Walter Rudolf,* Über Territorialgrenzen der Steuergesetze, in: Festschrift für Johannes Bärmann zum 70. Geburtstag (1975), S. 769 ff., 780 ff. – Umgekehrt ist dasselbe Unterscheidungsmerkmal für die nicht existent, die in der unbeschränkten wie in der beschränkten ESt den Ausdruck einer, wenn auch »abgestuften« Leistungsfähigkeit sehen; vgl. dazu schon *Becker* (Anm. 21), S. 169, und jetzt vor allem BVerfG vom 12. 10. 1976, BVerfGE 43, 1 ff., 9 betr. die Verfassungsmäßigkeit des § 50 Abs. 1 EStG 1961: »Der beschränkt Steuerpflichtige wird daher im Inland nicht [erg.: wie der unbeschränkt Steuerpflichtige] im Rahmen seiner *vollen* (vom Gesamteinkommen abhängenden) Leistungsfähigkeit zur Einkommensteuer herangezogen« (Hervorhebung vom Verf.); vorher auch BFH vom 16. 8. 1963, BFHE 77, 456 ff.; sowie BFH vom 10. 10. 1973, BFHE 110, 414 ff., 415.

[52] Ebenso schon *Bayer* (Anm. 28), S. 73 mit Anm. 94; ferner Niedersächs. FG vom 18. 4. 1977, EFG 1978 Nr. 126, wo bezüglich der »Steuergerechtigkeit« in ihrem Verhältnis zur beschränkten Steuerpflicht ausdrücklich auf die Inanspruchnahme der »staatlichen Einrichtungen der Bundesrepublik Deutschland« abgestellt wird; vgl. ähnlich auch Schleswig-Holsteinisches FG vom 24. 3. 1970, EFG 1970 Nr. 471.

[53] Folgt man dem im Text Gesagten, ist die Unterscheidung der unbeschränkten und der beschränkten ESt nach Maßgabe der Person also nur aus Zweckmäßigkeitserwägungen zu erklären; so freilich ausdrücklich BFH vom 5. 12. 1963, BFHE 78, 379 ff., 381: »Der Gesetzgeber bestimmt nach seinem Ermessen, welche Merkmale die unbeschränkte Einkommensteuerpflicht begründen. Er hat sich dabei an das zu halten, was im Rahmen des Zweckmäßigen, Vernünftigen und Gerechten vertretbar ist«.

V. Schlußbemerkung

Das steuerrechtliche Schrifttum hat es sich unter Federführung von *Klaus Vogel* mehr und mehr angewöhnt, die Probleme, zu deren Behandlung einerseits das Recht der unbeschränkten und andererseits das Recht der beschränkten ESt Anlaß gibt, unter dem Axiom einer rechtlichen *Einheit* zu betrachten[54]. Es gibt zwar, wie man hinzufügen muß, kaum einen Autor, der das ausdrücklich so auch aussprechen würde. Die Tatsache als solche bleibt davon jedoch unberührt. Wenn auch eher unter dem Etikett des Rechts der einkommensteuerrechtlichen »Sach-«[55] bzw. des Rechts der einkommensteuerrechtlichen »Rechtsanwendungs-«, »Kollisions-« oder »Grenznorm«[56] als unter dem des Rechts der unbeschränkten bzw. der beschränkten ESt wird also behauptet, daß es irgendwelche Möglichkeiten, zu Unterscheidungen zu gelangen, einfach nicht gäbe[57].

Die vorausgegangenen Überlegungen sind vielleicht dazu angetan, ein insoweit schiefes Bild gerade zu rücken. Denn wenn sich aus ihnen etwas mit hinreichender Deutlichkeit ergeben hat, dann ist es doch wohl der Umstand, daß die unbeschränkte und die beschränkte ESt zwei Steuern sind, die nicht auf den gleichen, sondern auf ganz und gar *verschiedenen* rechtlichen Voraussetzungen beruhen[58]. Oder, wie man auch sagen könnte: Das, was die unbeschränkte und die beschränkte ESt gemeinsam haben, geht kaum über die Äußerlichkeit des Namensteiles »Einkommensteuer« hinaus.

---

[54] *Vogel* (Anm. 42), S. 298 ff.; vorher schon *Ottmar Bühler,* Von der Struktur des internationalen Steuer- und Wirtschaftsrechts, Neue Juristische Wochenschrift 1956, S. 889 ff., 890; ferner etwa, wenn auch nicht ganz widerspruchsfrei, *Mössner* (Anm. 51), S. 273, 277; *Lang* (Anm. 31), S. 30 ff.

[55] Vgl. zum Begriff der steuerrechtlichen Sachnorm etwa *Isay* (Anm. 27), S. 4 f.; *Bühler* (Anm. 54), S. 890; *Wolfgang Arendt,* Zum Qualifikationsproblem im deutschen Internationalen Steuerrecht, StuW 1959, Sp. 381 ff., 386.

[56] Vgl. zu diesen Begriffen etwa *Bühler* (Anm. 54), S. 890; *Vogel* (Anm. 42), S. 153 ff.; aus früherer Zeit auch schon *Isay*(Anm. 27), S. 17 ff.; *Karl Neumeyer,* Internationales Verwaltungsrecht, Bd. 4 (1936), S. 115 ff.; *Wengler* (Anm. 7), S. 1 ff. – Die im Text unterstellte Gleichsetzung als solche klingt sowohl bei *Vogel* (Anm. 42), S. 299 Anm. 6, als auch bei *Mössner* (Anm. 51), S. 270, an.

[57] Statt aller: *Vogel* (Anm. 42), S. 312.

[58] In der Sache damit übereinstimmend früher schon *Isay* (Anm. 27), S. 9 f.; *Neumeyer* (Anm. 56), S. 115 ff.; *Wengler* (Anm. 7), S. 4, 87; ferner BFH vom 19. 2. 1960, BFHE 70, 527 ff., 530: »Die Besteuerung der beschränkt Steuerpflichtigen erfolgt nach zum Teil *wesentlich anderen Grundsätzen* als die der unbeschränkt Steuerpflichtigen« (Hervorhebung vom Verf.).

Internationaler Menschenrechtsschutz und nationaler Gestaltungsspielraum

Rudolf Bernhardt

Auf Grund seiner langjährigen Erfahrung als Richter des Europäischen Gerichtshofes für Menschenrechte hat *Hermann Mosler* ein Grundproblem, dem in dieser Skizze nachgegangen werden soll, wie folgt formuliert [1]:

> »Das schwierigste Problem, dem sich Kommission und Gerichtshof in gleicher Weise gegenübergestellt sehen, ist die Abgrenzung der von der Konvention geschützten Rechte von den im nationalen Zuständigkeitsbereich verbliebenen Zuständigkeiten. Sie haben dafür folgende Methoden entwickelt: die im Verhältnis zu den Definitionen und Kategorien der innerstaatlichen Rechtsordnungen autonomen Definitionen der Konvention und die Abgrenzung zwischen dem einheitlichen Standard, den die Konvention für alle Mitgliedstaaten fordert, und den Unterschieden, die innerhalb des nationalen Beurteilungsspielraums bestehen dürfen«.

Es ist mit Sicherheit zu erwarten, daß diese Problematik auch in Zukunft und in mannigfaltigen Konstellationen eine gewichtige Rolle spielen wird. Eine erneute Betrachtung zu Ehren von *Hermann Mosler* erscheint daher sinn- und reizvoll.

I

Vorangestellt werden soll eine allgemeine Betrachtung: Das viel diskutierte Verhältnis des Völkerrechts zum Landesrecht bedarf generell einer Überprüfung im Hinblick auf internationale Menschenrechtskonventionen, und zwar besonders dann, wenn diese Konventionen internationale Organe mit der Kontrolle ihrer Einhaltung betrauen.

---

[1] *H. Mosler,* Der Europäische Gerichtshof für Menschenrechte nach zwanzig Jahren, in: Recht als Prozeß und Gefüge, Festschrift für Hans Huber (1981), S. 595 ff., 600.

Das Besondere an den Menschenrechtspakten der Nachkriegszeit ist das Bemühen, die staatlichen Organe in ihrem Verhalten gegenüber den eigenen Staatsangehörigen in Pflicht zu nehmen und an internationale Standards zu binden. Insbesondere im Fremdenrecht und im Minderheitenrecht gab es schon zuvor explizite völkerrechtliche Verpflichtungen zur Respektierung von Individuen und Gruppen. Aber nunmehr soll die Staatsgewalt generell bei hoheitlichen Maßnahmen auf ihrem Staatsgebiet gebunden und beschränkt werden. Das kann kaum ohne Folgewirkungen für die innerstaatliche Bedeutung der Menschenrechtskonventionen und der Rechtsprechung etwaiger Konventionsorgane bleiben. Zwar folgt die Staatenpraxis weiterhin grundsätzlich dem traditionellen Konzept, daß Menschenrechtskonventionen nicht anders als andere völkerrechtliche Verträge zu bewerten sind und wie diese zur innerstaatlichen Geltung eines staatlichen Anwendungsbefehls – einer Transformation oder einer Vollzugsanordnung – bedürfen, aber es zeigen sich doch Besonderheiten.

In den Staaten, in denen völkerrechtliche Verträge allgemein innerstaatlich anwendbar sind, in denen also Verträge auf Grund eines wie auch immer ausgestalteten Anwendungsbefehls ohne besonderes Ausführungsgesetz innerstaatlich verbindlich und von den staatlichen Organen zu respektieren sind, ist die innerstaatliche Geltung und Anwendung der Menschenrechtspakte an sich nichts Neues. Aber die Intensität der Einwirkungen auf die nationalen Rechtsordnungen ist in der Regel erheblich größer als in anderen Bereichen. Es handelt sich ja regelmäßig um Grundrechtskataloge, die gewichtige Teile der staatlichen Rechtsordnung und der staatlichen Organisation betreffen, etwa die Polizeigewalt und das Gerichtswesen. Auch müssen oder sollten die staatlichen Organe beachten, daß die Garantien der internationalen Pakte in ihren Formulierungen wie in ihrem sachlichen Gehalt als völkerrechtliche Verpflichtungen von den gewohnten nationalen Standards abweichen können; darauf ist zurückzukommen. Schließlich ist anzunehmen, daß die Rechtsprechung der jeweiligen Konventionsorgane bei der innerstaatlichen Anwendung der betreffenden Konvention zu beachten ist, das heißt konkret, daß die Rechtsprechung der Europäischen Menschenrechtskommission und des Europäischen Menschenrechtsgerichtshofes von den nationalen Organen einschließlich der Gerichte bei der Auslegung und Anwendung der Konvention zu beachten ist. Wie das dogmatisch zu begründen ist[2], hängt nicht zuletzt von den Rechtsprinzipien des nationalen Rechts ab, am ehesten bietet sich wohl die Konstruktion an, daß der einmal erteilte Anwendungsbefehl für die Konvention auch die verbindlichen Aussprüche der Konventionsorgane erfaßt.

Nach ganz überwiegender Auffassung sind die Staaten nach wie vor völkerrechtlich nicht gehalten, die Menschenrechtspakte selbst in das nationale

---

[2] Allgemein zur Problematik: *C. Schreuer*, Die Behandlung internationaler Organakte durch staatliche Gerichte (1977), insbes. S. 173 ff.

Recht aufzunehmen oder zu überführen. Die Staaten sind verpflichtet, dafür zu sorgen, daß im Ergebnis das nationale Recht und seine Anwendung den Konventionen entsprechen, aber die Wege, die zu diesem Ergebnis eingeschlagen werden können, sind nicht verbindlich vorgeschrieben. Daher handeln die Mitgliedstaaten der Europäischen Menschenrechtskonvention, die dieser die innerstaatliche Anwendung versagen, deshalb nicht konventionswidrig. Aber auch für diese Staaten kann im Zuge der weiteren Entwicklung eine Frage größere Bedeutung erlangen, die in neuerer Zeit verschiedentlich im Vereinigten Königreich aufgeworfen wurde. Britische Gerichte und Autoren haben diskutiert, ob nicht die Europäische Menschenrechtskonvention und ihre Auslegung durch den Menschenrechtsgerichtshof auch ohne formelle innerstaatliche Verbindlichkeit als Argumente und Faktoren für die Auslegung des nationalen Rechts heranzuziehen sind[3]. Obwohl nicht unmittelbar verbindlich, könnte so die Konvention zur Konkretisierung des nationalen Rechts dienen. Das ist eine bemerkenswerte Überlegung, die zeigt, daß die traditionelle Trennung von Völkerrecht und Landesrecht – die bei näherem Zusehen nie exakt beachtet wurde, man denke etwa an die interne Anwendung von Völkergewohnheitsrecht oder an die Rechtsprechung nationaler Prisengerichte zum völkerrechtlichen Prisenrecht – möglicherweise relativiert werden muß.

Wie fragwürdig eine rigorose Trennung von Völkerrecht und Landesrecht ist, zeigt schließlich das Rechtssschutzsystem der Europäischen Menschenrechtskonvention. Die Anerkennung der Individualbeschwerde nach Art. 25 EMRK ist offensichtlich ganz unabhängig davon, ob die Konvention in einem Staat innerstaatlich direkt anwendbar ist. Wenn einmal die Erklärung gegenüber dem Generalsekretär des Europarates abgegeben wurde, kann sich der Einzelne mit seiner Beschwerde unmittelbar an die Menschenrechtskommission wenden, und die Frage der innerstaatlichen Geltung der Konvention in dem betreffenden Staat spielt in dem Verfahren vor den Konven-

---

[3] Die Rechtsprechung britischer Gerichte betont zwar nach wie vor, daß völkerrechtliche Verträge innerstaatlich nicht verbindlich sind, es finden sich aber Äußerungen, nach denen das nationale Recht wenn möglich so ausgelegt und angewandt werden sollte, daß die Europäische Menschenrechtskonvention und ihre Auslegung durch den Menschenrechtsgerichtshof respektiert werden; vgl. *Attorney General* v. *the B.B.C.*, [1980] 3 All England Law Reports, S. 161, 178; *Shearing Chemicals* v. *Hawkman*, [1981] 2 Weekly Law Reports, S. 848, 861; auf beide Entscheidungen wird in einer – wohl noch nicht publizierten – Entscheidung des High Court of Justice of the Isle of Man vom 6. 10. 1981 zustimmend Bezug genommen. Vgl. andererseits die Notes in The Law Quarterly Review 98 (1982), S. 183 ff., mit Hinweis auf die Entscheidung *Kaur* v. *Lord Advocate*, 1981 Scots Law Times 322. Siehe im übrigen an neueren Stellungnahmen etwa *W. Finnie*, The European Convention on Human Rights, Journal of the Law Society of Scotland 25 (1980), S. 434 ff.; *P. J. Duffy*, English Law and the European Convention on Human Rights, The International and Comparative Law Quarterly 29 (1980), S. 585 ff.; *E. H. Riedel*, Die Habeas Corpus-Akte, Europäische Grundrechte-Zeitschrift (EuGRZ) 1980, S. 192 ff., 197 f.

tionsorganen keine Rolle, sondern nur bei der materiellen Prüfung, ob das interne Recht im Einklang mit der Konvention steht.

II

Noch eine weitere grundsätzliche Betrachtung ist voranzustellen. Es ist davon auszugehen, daß alle Menschenrechtskonventionen völkerrechtlich verbindliche Mindestgarantien einführen wollen, zugleich aber keine umfassende Revision oder Reform der innerstaatlichen Rechtsordnungen verlangen. Beide Teile dieser Aussage ergeben sich aus der Natur und der Interpretation der Konventionen. Die Partner wollen, wie schon die gewählte Vertragsform zeigt, eine völkerrechtlich verbindliche Standardisierung des Menschenrechtsschutzes. Es kann aber nicht unterstellt werden, daß sie über partielle und punktuelle Änderungen ihrer innerstaatlichen Rechtsordnungen hinaus diese generell in Frage stellen und reformieren wollten. Eine derartige Absicht kann wohl keinem der Partner unterstellt werden, zudem gebietet es die souveräne Gleichheit aller Staaten, ihre innerstaatlichen Rechtsordnungen grundsätzlich als gleichwertig anzusehen. Dieses Spannungsverhältnis zwischen effektivem völkerrechtlichem Menschenrechtsschutz und nationaler Autonomie bei der Ausgestaltung der innerstaatlichen Rechtsordnungen ist eine Grundgegebenheit, die zu den Menschenrechtspakten geführt hat und diese ständig begleitet.

Je größer der Kreis der Vertragspartner und je inhomogener ihre Zusammensetzung ist, um so skeptischer ist die Aussicht zu beurteilen, zu einheitlichen Auslegungen und zu umfassend wirksamen Garantien zu kommen. Das zeigt sich bei den Menschenrechtspakten der Vereinten Nationen, deren Ratifikation durch manche Staaten angesichts der tatsächlichen innerstaatlichen Praxis entweder nur als Lippenbekenntnis oder als Zeichen dafür verstanden werden kann, daß dieselben Formulierungen ganz unterschiedlich verstanden und ausgelegt werden. Die weitere Entwicklung muß ziegen, ob die Vorschriften der UN-Pakte weitgehend papierene Formelkompromisse ohne sachliche Auswirkungen bleiben (oder in der Realität zu solchen werden) oder ob es doch gelingt, den hehr klingenden Garantien zur Wirksamkeit zu verhelfen. Dabei dürfte die Arbeit des Menschenrechtskomitees nach dem UN-Pakt über staatsbürgerliche und politische Rechte eine wichtige Rolle spielen.

Diese skeptischen Betrachtungen sind nicht angebracht bei der Europäischen Menschenrechtskonvention. (Für die Interamerikanische Konvention steht die Probe noch aus[4].) Das hat mehrere Gründe, von denen den gemeinsamen Grundauffassungen der beteiligten Staaten und der Existenz von Kon-

---

[4] Eine Sachentscheidung des Interamerikanischen Menschenrechtsgerichtshofs liegt noch nicht vor.

ventionsorganen mit wirksamen Kontrollbefugnissen eine besondere Bedeutung zukommt. Zur relativen Homogenität gehört das gemeinsame Bekenntnis der Partnerstaaten zu Demokratie und Rechtsstaat; das zeitweilige Ausscheiden Griechenlands ist ein Beleg dafür, daß allenfalls eine durch Notstand verursachte und vorübergehende, aber keine verfestigte Diktatur unter der Menschenrechtskonvention hingenommen werden kann. Nichts anderes kann für ein rechtsstaatswidriges Willkürregiment jeglicher Ausprägung gelten.

Die Konventionsorgane – Menschenrechtskommission und Menschenrechtsgerichtshof – haben schließlich durch ihre Rechtsprechung dafür gesorgt, daß die Europäische Konvention zu einem wichtigen und wirksamen völkerrechtlichen Instrument geworden ist. Diese Rechtsprechung ist nicht selten weitergegangen, als dies bei Schaffung der Konvention 1950 erwartet werden konnte. Die Tatsache, daß inzwischen 16 der 20 Vertragsstaaten die Individualbeschwerde und 18 Staaten die Zuständigkeit des Gerichtshofes anerkannt haben, ist ein Beleg dafür, daß die Tätigkeit der Konventionsorgane im großen und ganzen positiv bewertet wird und die Staaten bereit sind, den Konventionsorganen zu folgen. Dieses Ergebnis konnte wohl nur dadurch erreicht werden, daß sowohl die in der Konvention enthaltenen Garantien angewandt und ausgebaut als auch der notwendige Raum staatlicher Autonomie bei der Ausgestaltung der innerstaatlichen Rechtsordnung respektiert wurden.

III

Aus dem Ziel der Europäischen Menschenrechtskonvention, für die Bürger und Bewohner der Vertragsstaaten einen wirksamen Grundrechtsschutz zu schaffen und diesen internationaler Kontrolle zu unterstellen, folgt mit Notwendigkeit, daß die Vorschriften der Konvention »europäisch«, d.h. über die Staatsgrenzen hinweg einheitlich ausgelegt werden müssen und daß keine einzige Vorschrift den Staaten zur einseitigen und beliebigen Auslegung und Anwendung überlassen bleibt. Wenn es etwa um die Frage geht, ob »une accusation en matière pénale« (»criminal charge«) vorliegt und deshalb nach Art. 6 Abs. 1 »un tribunal indépendant et impartial« (»an independent and impartial tribunal«) entscheiden muß, so bleibt es nicht dem Belieben der Staaten überlassen, autonom über das Vorliegen einer strafrechtlichen Anklage[5] oder über die Qualifizierung einer Person oder eines Gremiums als Gericht[6] zu entscheiden, sondern es müssen europäische Standards gefunden

[5] Hierzu Cour européenne des Droits de l'Homme (CEDH) (*Engel*-Fall), Urteil vom 8. 6. 1976, série A n° 22, S. 34 (mit Hinweisen auf die vorherige Rechtsprechung).

[6] *Ibid.*, S. 28, mit Hinweis auf das Urteil *De Wilde* vom 18. 6. 1971, série A n° 12, S. 41.

und entwickelt werden. Dabei ist sicher insoweit eine Rückkoppelung zu den nationalen Rechtsordnungen angebracht, als diesen allgemeine Rechtsgrundsätze und gemeinsame Rechtsüberzeugungen entnommen werden können, aber das Wesentliche ist, daß nicht eine einzelne staatliche Rechtsordnung maßgeblich ist, sondern daß sich aus dem Ineinandergreifen der Menschenrechtskonvention und der gemeinsamen Rechtsanschauungen der Mitgliedstaaten eine europäische Menschenrechtsordnung bildet. Die Zusammensetzung der Konventionsorgane garantiert hierbei, daß die Rechtsanschauungen der Mitgliedstaaten wirklich zusammenfließen.

Europäische Standards lassen sich nicht von heute auf morgen und nicht für alle Bestimmungen der Konvention gleichermaßen und in kurzen Zeiträumen entwickeln. In allen Vorschriften der Konvention finden sich Begriffe und Regeln, die der Konkretisierung und der Entfaltung bedürfen. Der Begriff des Gesetzes, der in mehreren Artikeln auftaucht, verweist nicht einfach auf den einzelstaatlichen Gesetzesbegriff, sondern verlangt nach einer europäischen Definition oder Umschreibung des Gesetzes, dem nationale Vorschriften gerecht werden müssen. Die Begriffe »Tortur«, »inhumane und entwürdigende Behandlung« (Art. 3), »Sklaverei«, »Abhängigkeit« (Art. 4), »Freiheitsentzug«, »Verurteilung« (Art. 5), »Unschuldsvermutung« (Art. 6), »Privatleben«, »Familie« (Art. 8), »Gedanken-, Gewissens- und Religionsfreiheit« (Art. 9), »Meinungsfreiheit« (Art. 10), »Eigentum« (Art. 1 des 1. Zusatzprotokolls), »Erziehung« (Art. 2 des Protokolls) und viele andere Begriffe der Konvention bedürfen der europäischen Ausfüllung und Bestimmung. Dabei kann es nicht darum gehen, den optimalen oder den dem jeweiligen Betrachter optimal erscheinenden Schutz des Einzelnen zu erreichen, gleichsam einen Maximalstandard zu entwickeln, sondern es muß so viel an europäischen Gemeinsamkeiten festgelegt werden, wie dies einerseits zum Schutz der Menschen notwendig, andererseits mit grundlegenden kulturellen und traditionellen Vorstellungen der Mitgliedstaaten vereinbar ist. In manchen, vielleicht den meisten Fällen wird sich der europäische Standard auf Mindestanforderungen beschränken müssen. Die fortbestehenden Besonderheiten und Unterschiede der staatlichen Rechtsordnungen sind an diesen Mindestanforderungen zu messen, und sie können nur dann beanstandet und als konventionswidrig verworfen werden, wenn sie den Mindeststandard nicht erreichen.

IV

Besonders schwierige Probleme werfen Beschränkungen der Menschenrechte auf. Mit wenigen Ausnahmen, zu denen etwa das Torturverbot des Art. 3 gehört, sind die Grund- und Menschenrechte ja nicht unbeschränkt garantiert, sondern sie können von den Staaten vielfältig eingegrenzt werden. Die Europäische Menschenrechtskonvention umschreibt diese Einschränkungsmöglichkeiten in unterschiedlicher Weise. Ähnliche Formeln werden in den Abs. 2 der Art. 8–11 EMRK verwandt[7], danach sind Einschränkungen im wesentlichen unter drei Voraussetzungen möglich: Die Beschränkung muß im *Gesetz* vorgesehen sein, sie muß (zur Erreichung bestimmter legitimer Ziele) *notwendig* sein, und Maßstab für die Notwendigkeit sind die Erfordernisse einer *demokratischen Gesellschaft*. An anderen Stellen der Konvention werden andere Formeln gebraucht, so ist in Art. 2 Abs. 2 von »absolut notwendig« die Rede, andererseits spricht Art. 1 Abs. 2 des 1. Zusatzprotokolls von Gesetzen, die »die Staaten für notwendig halten«.

Hier soll nicht auf die Einzelheiten dieser Klauseln eingegangen, sondern nur das gemeinsame Grundproblem aufgezeigt werden: Wie weit reichen die Kontrollbefugnisse der Konventionsorgane einerseits und die Beurteilungs- und Gestaltungsmöglichkeiten der Staaten andererseits? Prüfen die Menschenrechtskommission und der Menschenrechtsgerichtshof in vollem Umfang, was im Einzelfall in einer demokratischen Gesellschaft notwendig ist, oder kontrollieren sie nur die Einhaltung bestimmter Grenzen?

Hierzu ist von den Konventionsorganen[8] die Doktrin von der *marge d'appréciation*, der *margin of appreciation* der Staaten entwickelt worden[9]. Den Staaten soll ein Beurteilungs- und Ermessensspielraum (die Unterscheidung dieser Begriffe im nationalen Recht läßt sich kaum *tel quel* auf die europäische Ebene übertragen) zustehen, nur die Beachtung gewisser Grenzen soll von den Konventionsorganen kontrolliert werden können. *Hermann Mosler* hat zutreffend darauf hingewiesen, daß die *marge d'appréciation* zwei Seiten oder Bestandteile hat: »... den Spielraum der nationalen Organe der Rechtsetzung, Verwaltung und Rechtsprechung zur Bewertung einer Situation

---

[7] Dazu *U. Hoffmann-Remy*, Die Möglichkeiten der Grundrechtseinschränkung nach den Art. 8–11 Abs. 2 der Europäischen Menschenrechtskonvention (1976).

[8] Belege aus der Rechtsprechung des Menschenrechtsgerichtshofes werden im folgenden angeführt; die Menschenrechtskommission hat in letzter Zeit etwa in der Entscheidung 8440/78, *Christians against Racism and Fascism* v. *U.K.*, Stellung genommen, vgl. European Commission of Human Rights, Decisions and Reports 21, S. 138 ff., 149 ff.

[9] Vgl. hierzu etwa *C. C. Morrison*, Margin of Appreciation in European Human Rights Law, Human Rights Journal 6 (1973), S. 263 ff.; *Sir Humphrey Waldock*, Die Wirksamkeit des Systems der Europäischen Menschenrechtskonvention, EuGRZ 1979, S. 599 ff.; *R. Sapienza*, La libertà d'espressione nella Convenzione Europea dei Diritti dell'Uomo: Il caso Sunday Times, Rivista di Diritto Internazionale 1981, S. 43 ff., 47, 54 ff. – *C. S. Feingold*, The Doctrine of Margin of Appreciation and the European Convention on Human Rights, Notre Dame Lawyer 53 (1977/78), S. 90–106, kommt auf Grund einer unzulänglichen Analyse der Rechtsprechung der Konventionsorgane zu dem unhaltbaren Ergebnis: »By applying the doctrine of margin of appreciation to Articles 8 through 11, the Commission and the Court have extended to Member States *unlimited* discretion to restrict the enumerated rights [Hervorhebung im Original]. In effect, Member States may impose restrictions without regard to whether such restraints are necessary in a democratic society. Thus, the doctrine has been used to circumvent the express requirements of the Convention«.

oder eines Sachverhalts und den Ermessensspielraum der nationalen Behörden zur Entscheidung über die anzuwendenden Maßnahmen«[10]. Die Staaten haben darüber zu befinden, ob die öffentliche Sicherheit und Ordnung oder der Schutz der Gesundheit und der Moral Einschränkungen der Individualrechte erfordern, und sie entscheiden weiter auch darüber, welche Maßnahmen »notwendig« sind. Der staatliche Beurteilungs- und Gestaltungsspielraum ist jedoch nicht unbegrenzt, sondern er findet seine Grenze an den Anforderungen der Konvention, und die Beachtung der Grenze wird von den Konventionsorganen kontrolliert.

Diese Doktrin erscheint zutreffend, denn sie wird der zuvor umschriebenen Forderung gerecht, den internationalen Menschenrechtsschutz und die nationalen Belange und Unterschiede gleichermaßen zu beachten. Die Richtigkeit der Doktrin garantiert natürlich nicht ihre richtige Anwendung im Einzelfall. Hier liegen außerordentlich schwierige Probleme, die kaum durch allgemeine Formeln, sondern nur im Wege fallweiser und sachgerechter Einzelentscheidungen zu lösen sind. Dabei können die Konventionsorgane die Entwicklung des öffentlichen Bewußtseins in den Konventionsstaaten nicht unberücksichtigt lassen. Auch die – ebenfalls von *Hermann Mosler* hervorgehobene[11] – evolutive Auslegung der Konvention ist wohl nur in dem Umfang zu rechtfertigen und erfolgversprechend, wie sie sich im Einklang mit den sich wandelnden Wertvorstellungen in den demokratischen Staaten Europas befindet.

Die Spruchtätigkeit der Straßburger Organe zur staatlichen *marge d'appréciation* ist mehrfach dargestellt worden[12] und kann hier nicht im einzelnen wiedergegeben werden. Lediglich einige Beispiele sollen das Problem und die gefundenen Ergebnisse illustrieren.

Das Urteil vom 1. Juli 1961 im *Lawless*-Fall soll erst später kurz erörtert werden. Es wird zwar zu Recht als Ausgangspunkt der Rechtsprechung zur nationalen *marge d'appréciation* angesehen, aber es betrifft doch mit Art. 15 der Konvention eine spezielle Problematik.

Mehr oder weniger deutlich hat der Europäische Menschenrechtsgerichtshof in einer Reihe von Entscheidungen den nationalen Spielraum bei der Ausgestaltung des Sozialbereichs, in dem die Menschenrechte der Kon-

---

[10] *Mosler* (Anm. 1), S. 601.

[11] A.a.O., S. 602; *derselbe*, Problems of Interpretation in the Case Law of the European Court of Human Rights, in: Essays on the Development of the International Legal Order in memory of Haro F. van Panhuys (1980), S. 149 ff., 158 ff.; vgl. weiter *J. Abr. Frowein*, Die Europäische Menschenrechtskonvention in der neueren Praxis der Europäischen Kommission und des Europäischen Gerichtshofs für Menschenrechte, EuGRZ 1980, S. 231 ff., 234 f.; *W. J. Ganshof van der Meersch*, La garantie des droits de l'homme et la Cour européenne de Strasbourg, Journal des Tribunaux, Jg. 101 (1982), S. 102 ff., 112 ff.; *derselbe*, Die Bezugnahme auf das innerstaatliche Recht der Vertragsstaaten in der Rechtsprechung des Europäischen Gerichtshofs für Menschenrechte, EuGRZ 1981, S. 481 ff., 486 ff.

[12] Vgl. die in Anm. 9 genannte Literatur.

vention tangiert werden können, hervorgehoben. Das gilt etwa für das Urteil im *Belgischen Sprachenstreit*[13], das Urteil in der *Affaire Syndicat National de la police belge*[14] und das Urteil im *Engel*-Fall[15]. Dabei ist anzumerken, daß das Gleichbehandlungsgebot des Art. 14 der Konvention in diesem Zusammenhang eine wichtige Rolle spielt und nach der Rechtsprechung nur dann verletzt sein soll, wenn die Staaten keine vernünftigen Gründe für Differenzierungen anführen können.

In der *Handyside*-Entscheidung[16] hat der Gerichtshof die These vom nationalen Gestaltungsspielraum nachdrücklich unterstrichen und dazu ausgeführt:

> »... on ne peut dégager du droit interne des divers Etats contractants une notion européenne uniforme de la ›morale‹. L'idée que leurs lois respectives se font des exigences de cette dernière varie dans le temps et l'espace, spécialement à notre époque caractérisée par une évolution rapide et profonde des opinions en la matière. Grâce à leurs contacts directs et constants avec les forces vives de leur pays, les autorités de l'Etat se trouvent en principe mieux placées que le juge international pour se prononcer sur le contenu précis de ces exigences comme sur la ›nécessité‹ d'une ›restriction‹ ou ›sanction‹ destinée à y répondre. La Cour note à cette occasion que si l'adjectif ›nécessaire‹, au sens de l'article 10 § 2, n'est pas synonyme d'›indispensable‹ (comp., aux articles 2 § 2 et 6 § 1, les mots ›absolument nécessaire‹ et ›strictement nécessaire‹ et, à l'article 15 § 1, le membre de phrase ›dans la stricte mesure où la situation l'exige‹), il n'a pas non plus la souplesse de termes tels qu'›admissible‹, ›normal‹ (comp. l'article 4 § 3), ›utile‹ (comp. le premier alinéa de l'article 1 du Protocole n° 1), ›raisonnable‹ (comp. les articles 5 § 3 et 6 § 1) ou ›opportun«. Il n'en appartient pas moins aux autorités nationales de juger, au premier chef, de la réalité du besoin social impérieux qu'implique en l'occurrence le concept de ›nécessité‹.
>
> Dès lors, l'article 10 § 2 réserve aux Etats contractants une marge d'appréciation. Il l'accorde à la fois au législateur national (›prévues par la loi‹) et aux organes, notamment judiciaires, appelés à interpréter et appliquer les lois en vigueur (arrêt Engel et autres du 8 juin 1976, série A n° 22, pp. 41–42, § 100; comp., pour l'article 8 § 2, l'arrêt De Wilde, Ooms et Versyp du 18 juin 1971, série A n° 12, pp. 45–46, § 93, et l'arrêt Golder du 21 février 1975, série A n° 18, pp. 21–22, § 45).
>
> L'article 10 § 2 n'attribue pas pour autant aux Etats contractants un pouvoir d'appréciation illimité. Chargée, avec la Commission, d'assurer le

---

[13] CEDH, Urteil vom 23. 7. 1968, série A n° 6, S. 32, 34 f., 56, auch S. 90 (Sondervotum).

[14] CEDH, Urteil vom 27. 10. 1975, série A n° 19, S. 20, 22.

[15] CEDH, Urteil vom 8. 6. 1976, série A n° 22, S. 25, 31, 41 f.

[16] CEDH, Urteil vom 7. 12. 1976, série A n° 24, S. 22 f., auch S. 26, 28, dazu das Sondervotum Mosler, S. 32 f.

respect de leurs engagements (article 19), la Cour a compétence pour statuer par un arrêt définitif sur le point de savoir si une ›restriction‹ ou ›sanction‹ se concilie avec la liberté d'expression telle que la protège l'article 10. La marge nationale d'appréciation va donc de pair avec un contrôle européen«.

Die jüngste Rechtsprechung des Menschenrechtsgerichtshofs folgt im großen und ganzen dem früher eingeschlagenen Weg. Im *Young*-Fall[17] ging es darum, ob eine bestimmte Ausgestaltung des *closed shop*-Systems, nämlich die Entlassung von Arbeitnehmern, die vor der Einführung des Systems eingestellt und wegen ihrer späteren Weigerung, einer Gewerkschaft beizutreten, entlassen worden waren, mit der Konvention vereinbar ist. Der Gerichtshof hat dies verneint, er hatte dabei zur Notwendigkeit der Regelung in einer demokratischen Gesellschaft Stellung zu nehmen. Bei der mit großer Mehrheit (18:3) gefällten Entscheidung hat der Gerichtshof darauf hingewiesen, daß die Notwendigkeit der Maßnahmen weder von den staatlichen Organen noch von den gesellschaftlichen Institutionen und Kräften nachdrücklich und überzeugend behauptet und begründet worden sei. Insgesamt enthielten die getroffenen Maßnahmen und die zugrundeliegenden Vorschriften unverhältnismäßige und nicht notwendige Beschränkungen der Individualrechte. Die Schlußfolgerung des Gerichtshofs lautete[18]:

»Even making due allowance for a State's ›margin of appreciation‹ . . ., the Court thus finds that the restriction complained of were not ›necessary in a democratic society‹, as required by paragraph 2 of Article 11«.

In der *Dudgeon*-Entscheidung[19] ging es um die Strafbarkeit bestimmter sexueller Verhaltensweisen in Nordirland. In der Argumentation der Beteiligten spielte auch die Frage eine Rolle, ob die Strafbarkeit ein und desselben Verhaltens nur in Nordirland, nicht aber in anderen Teilen desselben Staates als »notwendig« angesehen werden könne; das hat der Gerichtshof für möglich gehalten. Das zeigt, daß die Notwendigkeit einer Maßnahme *in a democratic society* nicht für alle Staaten und Staatsteile einheitlich zu beurteilen ist, daß es also insoweit keinen einheitlichen europäischen Standard gibt. Der Gerichtshof hat weiter geprüft, ob die Notwendigkeit der beanstandeten Strafbestimmungen bejaht werden könne, er hat in diesem Zusammenhang ausgeführt[20]:

»Notwithstanding the margin of appreciation left to the national authorities, it is for the Court to make the final evaluation as to whether the reasons it has found to be relevant were sufficient in the circum-

---

[17] CEDH, Urteil *Young, James und Webster* vom 13. 8. 1981, série A n° 44.
[18] *Ibid.*, S. 26.
[19] CEDH, Urteil *Dudgeon* vom 22. 10. 1981, série A n° 45.
[20] *Ibid.*, S. 23.

stances, in particular whether the interference complained of was proportionate to the social need claimed for it . . .«.

In diesen beiden zuletzt genannten Entscheidungen ist letztlich eine Überschreitung des Spielraums nationaler Gestaltungsmöglichkeiten angenommen worden, aber die grundsätzliche Existenz der staatlichen *margin of appreciation* wurde nicht in Zweifel gezogen.

Bei der Prüfung, welche Menschenrechtsbeschränkungen zulässig sind, müssen nicht nur der jeweilige Lebenssachverhalt und die gesellschaftliche und politische Situation im jeweiligen Staat oder Staatsteil, sondern auch die unterschiedlichen Formulierungen der Konvention beachtet werden. Das »absolut notwendig« in Art. 2 Abs. 2 erfordert eine intensivere Prüfung als die einfache »Notwendigkeit« in den Abs. 2 der Art. 8–11; es handelt sich nicht um mehr oder weniger zufällige Unterschiede in den Formulierungen, sondern die Tötung eines Menschen (Art. 2 Abs. 2) soll offensichtlich nur im Extremfall zulässig sein, wenn keine anderen Mittel zum Schutz hoher Gemeinschaftsgüter zur Verfügung stehen. Die Kontrollbefugnisse der Konventionsorgane dürften hier weiter reichen als in anderen Fällen, wobei jedoch auch hier eine *marge d'appréciation* der staatlichen Organe kaum ganz ausgeschlossen werden kann, etwa bei der Einschätzung, welches Ausmaß ein Aufruhr angenommen hat und welche Mittel zu seiner Unterdrückung unbedingt erforderlich erschienen.

Wenn andererseits Art. 1 des 1. Zusatzprotokolls im Rahmen des Eigentumsschutzes vom Recht der Staaten spricht »de mettre en vigueur les lois *qu'ils jugent nécessaires*«, so zeigt der Vertragstext an, daß in erster Linie das Urteil der staatlichen Organe entscheiden soll. Doch auch hier kann kaum angenommen werden, daß der staatliche Beurteilungs- und Ermessensspielraum unbegrenzt ist.

V

In Literatur und Rechtsprechung wird zu Recht darauf hingewiesen, daß die Grenzziehung zwischen dem nationalen Gestaltungsspielraum einerseits und der internationalen Kontrolle andererseits schon in dem ersten Fall vor dem Europäischen Menschenrechtsgerichtshof eine Rolle spielte und daß die weitere Rechtsprechung die der *Lawless*-Entscheidung[21] zugrundegelegte Doktrin nur weiter entwickelt und entfaltet hat. Wenn dieser Ausgangsfall erst am Ende unserer Skizze gewürdigt wird, so hat das sachliche Gründe. Soweit bei den einzelnen Menschenrechtsgarantien der Konvention Einschränkungen implizit oder kraft ausdrücklicher Vorschrift zulässig sind, handelt es sich um die Anwendung der Konvention »im Alltag«, im stetig sich entwickelnden und verändernden Gemeinschaftsleben und in der sich

---

[21] CEDH, *Lawless*-Fall, Urteil vom 1. 7. 1961, série A n° 3, insbes. S. 55, 56.

wandelnden staatlichen Rechtsordnung. Hier stellt sich ständig die Frage, wie weit der einzelstaatliche Gestaltungsspielraum reicht und wo die Kontrolle der Konventionsorgane wirksam werden kann.

Die Notstandsklausel des Art. 15 EMRK[22] hat eine andere Zielsetzung. An sich garantierte und unter normalen Bedingungen nicht weiter einschränkbare Menschenrechte sollen »en cas de guerre ou en cas d'autre danger public menaçant la vie de la nation« zur staatlichen Disposition stehen, aber nur »dans la stricte mesure où la situation l'exige« (und soweit keine anderen völkerrechtlichen Verpflichtungen entgegenstehen). Diese Formeln bieten besondere Schwierigkeiten. Wann liegt eine Gefahr für das Leben der Nation vor? Welche Maßnahmen gebietet die jeweilige Situation zwingend? Wenn die legalen und demokratisch bestellten Organe eines Staates das Vorliegen der Voraussetzungen und die Notwendigkeit der getroffenen Maßnahmen mit plausiblen Gründen bejahen, dürfte es für eine internationale Instanz nicht leicht sein, das Gegenteil festzustellen. Schon die Tatsachenfeststellung, die in vielen Fällen erst *post festum* möglich ist, bereitet besondere Schwierigkeiten, nicht weniger schwierig ist die Bewertung der Tatsachen. Hier spricht viel dafür, daß die Beurteilung der Situation in erster Linie den staatlichen Organen obliegt. Aber das heißt nicht, daß die internationale Kontrolle durch die Konventionsorgane ganz ausfällt. Vielmehr werden sie zunächst von den staatlichen Organen eine zuverlässige Darstellung der Fakten und einleuchtende Begründungen für die getroffenen Maßnahmen verlangen können; den Staaten obliegt es, ihr Vorgehen zu rechtfertigen, nicht im Sinne einer Beweisführungslast, sondern im Sinne einer Pflicht zur plausiblen Begründung ihres Vorgehens. Weiter aber sind die Konventionsorgane berechtigt und verpflichtet, die Einhaltung der Grenzen der staatlichen Notstandsbefugnisse zu kontrollieren. Damit haben wir es erneut, wenn auch in einer besonderen Ausprägung, mit der Grenzziehung zwischen nationaler Kompetenz und internationaler Kontrolle zu tun.

Der Europäische Gerichtshof für Menschenrechte hat in zwei Entscheidungen zu diesem Problem Stellung genommen.

Im *Lawless*-Urteil vom 1. Juli 1961 verwendet der Gerichtshof den Ausdruck *marge d'appréciation* noch nicht, in der Sache aber betont er einerseits seine Befugnis oder Aufgabe, das Vorliegen der Voraussetzungen der Notstandsklausel in Art. 15 der Konvention zu prüfen, er deutet aber zugleich an, daß die irische Regierung aus den gegebenen Umständen vernünftigerwei-

---

22 Allgemein hierzu und zu ähnlichen Klauseln etwa *J. F. Hartman*, Derogation from Human Rights Treaties in Public Emergencies, Harvard International Law Journal 1981, S. 1 ff.; *T. Buergenthal*, To Respect and to Ensure: State Obligations and Permissible Derogations, in: L. Henkin, The International Bill of Rights (1981), S. 72 ff. Schon früher *J. Velu*, Le contrôle des organes prévus par la Convention européenne des droits de l'homme sur le but, le motif et l'objet des mesures d'exception dérogeant à cette Convention, in: Mélanges offerts à Henri Rolin (1964), S. 463 ff.

se gewisse Folgerungen ziehen konnte. Es heißt in der Entscheidung unter anderem:

>». . . il appartient à la Cour de vérifier si les conditions énumérées à l'article 15 pour l'exercice du droit exceptionnel de dérogation étaient réunies dans le cas présent . . .«.
>». . . l'existence à cette époque d'un ›danger public menaçant la vie de la nation‹ a pu être raisonnablement déduite par le Gouvernement irlandais de la conjonction de plusieurs éléments constitutifs . . .«.

Noch klarer hat der Gerichtshof seine Auffassung im Urteil vom 18. Januar 1978 im Fall *Irland* gegen *Vereinigtes Königreich*[23] dargelegt:

>»Les limites du pouvoir de contrôle de la Cour (arrêt du 23 juillet 1968 sur le fond de l'affaire ›linguistique belge‹, série A n° 6, p. 35, § 10 in fine; arrêt Handyside zu 7 décembre 1976, série A n° 24, p. 22, § 48) se manifestent avec une clarté particulière dans le domaine de l'article 15.
>Il incombe d'abord à chaque Etat contractant, responsable de ›la vie de (sa) nation‹, de déterminer si un ›danger public‹ la menace et, dans l'affirmative, jusqu'où il faut aller pour essayer de le dissiper. En contact direct et constant avec les réalités pressantes du moment, les autorités nationales se trouvent en principe mieux placées que le juge international pour se prononcer sur la présence de pareil danger comme sur la nature et l'étendue de dérogations nécessaires pour le conjurer. L'article 15 § 1 leur laisse en la matière une large marge d'appréciation.
>Les Etats ne jouissent pas pour autant d'un pouvoir illimité en ce domaine. Chargée, avec la Commission, d'assurer le respect de leurs engagements (article 19), la Cour a compétence pour décider s'ils ont excédé la ›stricte mesure‹ des exigences de la crise (arrêt Lawless du 1er juillet 1961, série A n° 3, p. 55, § 22, et pp. 57–59, §§ 36–38). La marge nationale d'appréciation s'accompagne donc d'un contrôle européen«.

VI

Der Umfang des staatlichen Gestaltungsspielraums im Menschenrechtsbereich muß jeweils im konkreten Zusammenhang bestimmt werden, und die Konventionsorgane können nur den Einzelfall entscheiden. Aber wir haben es zugleich mit einer Grundfrage der internationalen Entwicklung zu tun. Die demokratischen Staaten Europas wollen im gegenwärtigen Stand der Entwicklung einerseits ihre staatliche Identität, ihre gesellschaftlichen Strukturen und ihre Traditionen bewahren, andererseits bekennen sie sich zu gemeinsamen Werten beim Schutz ihrer Bürger, und sie haben internationalen

---

[23] CEDH, série A n° 25, S. 78 f., auch S. 81, 82, 84.

Organen eine wirksame Kontrolle aufgegeben und anvertraut. Es kann daher weder darum gehen, über die Menschenrechte uniforme Lebensverhältnisse und Regeln in den beteiligten Staaten zu erzwingen, noch kann es darum gehen, gemeinsame Verpflichtungen und Worte, eben diejenigen der Menschenrechtskonvention, folgenlos zu lassen. Die Staaten, ihre Bürger und die Konventionsorgane stehen gleichermaßen vor der Aufgabe, mehr Gemeinsamkeiten im Bereich der Menschenrechte – und darüber hinaus – zu schaffen, ohne die Eigenheiten jeder Nation zu vergessen. Hinter dem scheinbar technischen oder positivrechtlichen Problem der Abgrenzung von internationaler Bindung und staatlichem Gestaltungsspielraum steht eine große politische und gesellschaftliche Aufgabe. Es ist zu hoffen, daß sie bewältigt werden wird.

Die Praxis des Völkergewohnheitsrechts als konsekutive Rechtsetzung

Albert Bleckmann

Am 27. Februar 1981 hat *Hermann Mosler* anläßlich der Einführung der neuen Direktoren des Max-Planck-Instituts für ausländisches öffentliches Recht und Völkerrecht eine Abschiedsrede »Rückblick und Ausblick« gehalten, in welcher er betonte, daß es zwar keine Methode oder Schule des Max-Planck-Instituts gebe und geben könne, wohl aber ein besonderes Interesse an allen methodischen Problemen. Dieses methodische und vor allem dogmatische Interesse kommt denn auch in den Schriften des Jubilars sehr deutlich zum Ausdruck. Ich meine deshalb, gerade mit einem Beitrag zum Methoden- und Rechtsquellenproblem im Völkerrecht dem Anliegen des Jubilars nahezukommen.

I. Übung und Rechtsüberzeugung als Voraussetzungen der Bildung von Völkergewohnheitsrecht

Liest man die Literatur zum Völkergewohnheitsrecht, hat man den Eindruck, die Übung und die Rechtsüberzeugung der Staaten flössen aus jeweils verschiedenen Akten. So soll nach der wohl herrschenden Lehre Völkergewohnheitsrecht meist dadurch entstehen, daß sich zunächst einmal eine rein faktische Übung bilde und später die Rechtsüberzeugung hinzuträte [1].

Nach dieser Lehre wäre zunächst das rein faktische Handeln als die für die Bildung von Völkergewohnheitsrecht notwendige Übung zu untersu-

[1] Zu den verschiedenen Formen der historischen Entwicklung von Völkergewohnheitsrecht vgl. *Alfred Verdross*, Entstehungsweisen und Geltungsgrund des universellen völkerrechtlichen Gewohnheitsrechts, ZaöRV 29 (1969), S. 635, sowie *Alfred Verdross/Bruno Simma*, Universelles Völkerrecht (1976), S. 276 ff. Generell gesehen erscheint es mir allerdings dogmatisch bedenklich, verschiedene Entstehungsweisen des Völkergewohnheitsrechts anzuerkennen; die Lehre und Rechtsprechung gehen vielmehr einmütig von einheitlichen Bedingungen der Entstehung von Gewohnheitsrecht aus.

chen; anschließend müßte man sich davon unterschiedlichen Akten wie etwa den diplomatischen Noten der Außenämter oder den Urteilen der nationalen Gerichte zuwenden, um die Rechtsüberzeugung zu begründen. Konkret gesprochen reicht es damit für die Bildung von Völkergewohnheitsrecht nach dieser Ansicht nicht aus, wenn die Regierungen in diplomatischen Noten eine bestimmte völkerrechtliche Auffassung zum Mindeststandard der Fremden vertreten; erforderlich wäre vielmehr auch, daß die Fremden in diesen Staaten dieser Rechtsauffassung entsprechend tatsächlich auch behandelt werden.

Wenn ich auch keineswegs bestreiten will, daß die meisten Rechtssätze des Völkergewohnheitsrechts auf diese Weise entstanden sind und heute noch auf diese Weise entstehen können, gibt diese Sicht doch den heutigen Stand der Entwicklung von Völkergewohnheitsrecht in den meisten Fällen nicht richtig wieder [2].

Das zeigt sich deutlich, wenn man die Präjudizien näher ins Auge faßt, die heute für die Bildung von Völkergewohnheitsrecht herangezogen werden: In der Tat stützt man sich dabei meist auf diplomatische Noten, auf Entscheidungen internationaler und nationaler Gerichte und auf die Materialien der innerstaatlichen Gesetzgebungsverfahren [3]. In all diesen Fällen stellt man fest, daß die Praxis gerade nicht in einem rein faktischen Handeln, sondern in *Rechtsakten* besteht, welche die Rechtslage im nationalen oder im internationalen Rechtsraum entweder feststellen oder gestalten.

In diesen Rechtsakten findet man in den meisten Fällen die Behauptung eines abstrakten Rechtssatzes des Völkergewohnheitsrechts [4]. Auch insoweit will ich nicht behaupten, daß die Rechtsüberzeugung sich nicht auch auf die Subsumtion eines Einzelfalls unter einen völkerrechtlichen Rechtssatz oder die Behauptung beziehen kann, ein bestimmter Einzelfall sei in einer bestimmten Weise völkerrechtlich zu beurteilen. Ich möchte nur hervorheben, daß die Rechtsüberzeugung in der überwiegenden Mehrheit aller Fälle in der Behauptung eines abstrakten Rechtssatzes des Völkergewohnheitsrechts besteht.

Daraus ergibt sich aber, daß die Völkerrechtspraxis sich in den weitaus meisten Fällen in der *Behauptung eines abstrakten Rechtssatzes des Völkergewohnheitsrechts* erschöpft.

Diese Behauptung wird dabei meist in völkerrechtlichen oder innerstaatlichen Rechtsakten oder Verfahren – etwa im Rahmen eines innerstaatlichen oder völkerrechtlichen Rechtsstreits – geäußert. Von irgendeiner *tatsächlichen*

[2] Eine Durchsicht des klassischen Völkerrechtslehrbuchs von *Verdross/Simma* (a.a.O.) zeigt, daß die Lehre nur auf die Rechtssatzbehauptungen der Praxis abstellt.

[3] Vgl. hierzu und zum folgenden die methodisch besonders wichtigen Immunitätsentscheidungen des Bundesverfassungsgerichts, Entscheidungen des Bundesverfassungsgerichts (BVerfGE) 46, 342.

[4] Hierzu und zum folgenden vgl. *Albert Bleckmann,* Zur Feststellung und Auslegung von Völkergewohnheitsrecht, ZaöRV 37 (1977), S. 504 ff.

Übung wird diese Erklärung in der Regel nicht begleitet. Die Erklärung bezieht sich meist auch nicht auf ein vorangehendes tatsächliches Handeln, und sie wird meist durch ein solches Handeln im innerstaatlichen oder völkerrechtlichen Rechtsraum auch nicht durchgeführt. Hält man an der Voraussetzung der Übung fest, könnte ein solches Handeln nur in der Tatsache gesehen werden, daß diese Erklärung in einem Rechtsakt oder in bezug auf einen Rechtsstreit geäußert worden ist. Damit wird der Begriff der Übung aber sehr weit ausgedehnt. Auf der anderen Seite wird ersichtlich, daß dann Übung und Äußerung der Rechtsüberzeugung nicht, wie die Lehre vorauszusetzen scheint, in zwei unterschiedlichen Akten zu suchen sind, sondern in einem einzigen Akt zusammenfallen.

Um auch hierfür ein Beispiel zu bieten, wird auf die zahlreichen Urteile der nationalen Gerichte zur Immunität für fremde Staaten verwiesen[5]. Hierbei handelt es sich um innerstaatliche Rechtsakte, die das Rechtsverhältnis zwischen einem fremden Staat und einem Individuum regeln. Sieht man solche Urteile als die nach der herrschenden Lehre erforderliche Übung an, ist nicht zu übersehen, daß die erforderliche Rechtsüberzeugung – nämlich die aktuelle Stellungnahme des Gerichts zur Immunität fremder Staaten – in demselben Urteil zum Ausdruck gelangt, das die Übung darstellt.

Noch schwieriger ist die angeblich notwendige Übung zu ermitteln, wenn es sich um die Stellungnahme eines Staates zu einem ihn berührenden völkerrechtlichen Streit, etwa zu einem Grenzstreit, handelt. Hier ist mit Ausnahme der Erklärungen der Außenminister in diplomatischen Noten überhaupt keine tatsächliche Praxis festzustellen. Die Noten stellen auch insoweit keine Übung dar, als es sich nicht um Rechtsakte handelt, die eine Rechtslage verbindlich feststellen oder regeln und insoweit als Handeln aufgefaßt werden können. Es handelt sich nur um Erklärungen zur Rechtslage und damit eigentlich nur um das Äußern einer Rechtsüberzeugung und nicht um eine Übung. Und dennoch werden solche Akte als Völkergewohnheitsrecht begründende, mit Rechtsüberzeugung gesetzte Staatenpraxis verstanden. Das kann doch nur bedeuten, daß grundsätzlich nur die Rechtsüberzeugung, die Behauptung abstrakter Rechtssätze des Völkergewohnheitsrechts und die Subsumtion des Einzelfalls unter diesen Rechtssatz, für die Bildung von Völkergewohnheitsrecht erforderlich ist: Auf die Übung als tatsächliches Handeln wird in all diesen Fällen verzichtet. *Die faktische Übung kann damit aber generell nicht als eine notwendige Voraussetzung der Bildung von Völkergewohnheitsrecht angesehen werden.*

Wie wir oben ausgeführt haben, gibt es neben den Praxisfällen, in denen nur die Existenz eines Rechtssatzes behauptet wird, Fälle der rein tatsächlichen Übung ohne Rechtssatzbehauptung[6]. Hier kann die Rechtsüberzeu-

---

[5] Als Beispiel vgl. die Immunitätsentscheidungen des Bundesverfassungsgerichts (oben Anm. 3).

[6] Vgl. *Bleckmann,* a.a.O. (Anm. 4).

gung häufig nur durch Rückgriff auf neben der Übung stehende Akte bewiesen werden. Manchmal kann man die Rechtsüberzeugung in diesen Fällen aber aus der tatsächlichen Praxis selbst ableiten. So ist davon auszugehen, daß die Staaten sich an das Völkerrecht halten wollten. Wenn sie also positiv handeln, muß von ihrer stillschweigenden Behauptung ausgegangen werden, dieses Handeln entspreche dem Völkerrecht. Wie die Rechte begründende ausdrückliche Rechtsbehauptung ist dieses tatsächliche *positive* Handeln aber nur dahin auszulegen, daß der betreffende Staat so handeln *darf,* nicht aber, daß er auch so handeln *muß.* Aus einer *Unterlassung* kann man dagegen auf die erforderliche Rechtspflicht zum Unterlassen nicht schließen.

Da wir nun gesehen haben, daß die reine Rechtsbehauptung auch ohne tatsächliche Übung Völkergewohnheitsrecht begründet und es für alle Fälle der Bildung von Völkergewohnheitsrecht nur einheitliche Regeln geben kann, muß davon ausgegangen werden, daß – auch wenn die Staaten rein tatsächlich ohne ausdrückliche Rechtsbehauptung handeln – nicht eigentlich die tatsächliche Übung, sondern die stillschweigende Rechtsbehauptung das entscheidende Element für die Bildung von Völkergewohnheitsrecht ist. Daß diese Rechtsbehauptung nur stillschweigend ist und sich nur auf den konkreten Einzelfall bezieht, daß der abstrakte Rechtssatz in solchen Fällen also durch Induktion, durch Generalisierung der Einzelfälle, entwickelt werden muß, ändert an diesem Ergebnis nichts.

Aus diesen Feststellungen ergeben sich eine Anzahl von Rechtsfolgen:

1. In der Literatur ist umstritten, ob bei den Voraussetzungen der Bildung von Völkergewohnheitsrecht der Übung oder der Rechtsüberzeugung das Übergewicht zufällt. Die klassische Theorie, welche die Übung als auf den Abschluß des völkerrechtlichen Vertrages gerichtete Willenserklärung verstand[7], sah das Hauptelement der Bildung von Völkergewohnheitsrecht naturgemäß in der Rechtsüberzeugung. Die neuere Vertrauensschutztheorie[8] muß dagegen die tatsächliche Übung in den Vordergrund schieben[9]. Dasselbe gilt von Theoretikern wie *Guggenheim,* der wegen der Schwierigkeit der Feststellung des subjektiven Elements des Völkergewohnheitsrechts annahm, auf das subjektive Element ganz verzichten zu müssen[10].

Die Entscheidung über diese Frage kann im Gegensatz zur Auffassung der Literatur nicht von der Theorie abhängig gemacht werden, der man zu folgen bereit ist. Eine Lösung kann vielmehr nur gefunden werden, wenn man auf der einen Seite den Inhalt der Praxis genau analysiert und auf der anderen Seite untersucht, auf welche Voraussetzungen der Bildung von Völ-

---

[7] Hierzu und zum folgenden vgl. *Albert Bleckmann,* Die Aufgaben einer Methodenlehre des Völkerrechts (1978), S. 20 ff.

[8] Vgl. *Jörg P. Müller,* Vertrauensschutz im Völkerrecht (Beiträge zum ausländischen öffentlichen Recht und Völkerrecht, Bd. 56) (1971), S. 77 ff.

[9] Vgl. *Bleckmann,* a.a.O. (Anm. 7).

[10] Zur Auffassung von *Guggenheim* vgl. *Karol Wolfke,* Custom in Present International Law (1964), S. 53.

kergewohnheitsrecht die Rechtsprechung, Praxis und Lehre bei der Feststellung konkreter Regeln des Völkergewohnheitsrechts und nicht nur bei der abstrakten Darstellung der Voraussetzungen der Bildung von Völkergewohnheitsrecht tatsächlich abstellen: Auch die zur Bindungswirkung von Völkergewohnheitsrecht aufgestellten Theorien müssen sich nach dem tatsächlichen Verhalten der Staaten, der Gerichte und der Lehre richten und dürfen nicht aus abstrakten, *a priori* festgelegten Prämissen abgeleitet werden. Solche abstrakten Theorien dürfen nur dann bemüht werden, wenn die Praxis keine Anhaltspunkte zur Lösung der konkreten Rechtsfragen bietet. Auch dann dürfen die Rechtsfolgen der Theorie zumindest in allen wesentlichen Punkten der Auffassung der Praxis nicht widersprechen. Vielleicht wird man auch voraussetzen müssen, daß umgekehrt diese Theorie auch durch die konkreten Ergebnisse der Praxis getragen wird, sich aus der Praxis ergibt, ohne daß die Theorie selbst in der Praxis behauptet zu werden braucht. Denn nur in einem solchen Fall stellt die Theorie sich als aus dem Völkergewohnheitsrecht abgeleitetes und damit als verbindliches Völkerrecht dar.

Methodisch gesehen begehen die Untersuchungen der Lehre über das Völkergewohnheitsrecht also den Fehler, daß sie für die Frage, wie Völkergewohnheitsrecht entsteht, entweder eine Auslegung des Art. 38 des IGH-Statuts versuchen oder sich auf ausdrückliche Aussagen des StIGH und des IGH zu den Voraussetzungen des Völkergewohnheitsrechts berufen[11]. Methodisch richtiger erscheint mir, den Inhalt der Praxis zu analysieren, welche Praxis, Lehre und Rechtsprechung in der tatsächlichen Begründung konkreter Rechtssätze des Völkergewohnheitsrechts als für die Bildung von Gewohnheitsrecht relevant ansehen. Es wird sich zeigen, daß erhebliche Unterschiede zwischen den ausdrücklich vertretenen Theorien und der tatsächlichen Praxis bestehen. Mangels theoretischer Behandlung der folgenden Probleme stützt sich dieser Aufsatz also auf die tatsächliche Behandlung der Praxis in Lehre, Rechtsprechung und Praxis. So wird etwa auf die Tatsache abgestellt, daß die Staatenpraxis absolut bindet, auch wenn die Staaten sich bei ihrer Formulierung irrten oder unter dem Vorwand der Rechtsfeststellung tatsächlich völlig neue Rechtssätze des Völkergewohnheitsrechts behaupten.

Auf dieser Analyse der Fakten muß anschließend eine neue Dogmatik der Rechtsquellenlehre aufgebaut werden. Nur so ist es nämlich möglich, die zahlreichen Lücken allmählich zu schließen, welche die heutige Lehre vom Völkergewohnheitsrecht offen läßt. Dabei kann ich natürlich noch keine völlig fertigen Theorien, sondern nur einen Anfang hierfür bieten, Ansätze, die sich erst in einer allgemeinen Diskussion entwickeln können. Hier wie in zahlreichen anderen Bereichen muß dem Mangel an Dogmatik der meist nur pragmatisch verfahrenden heutigen Völkerrechtswissenschaft entgegengewirkt werden.

---

[11] So verfahren etwa *Wolfke,* a.a.O., sowie *Julio A. Barberis,* Fuentes del derecho internacional (1973), S. 66, und *Max Sørensen,* Les sources du droit international (1946), S. 106.

Die obige Analyse hat nun gezeigt, daß man insoweit der Auffassung der Vertragstheorie folgen muß: Das eigentliche Element der von Rechtsüberzeugung getragenen Praxis ist die Behauptung der Existenz eines bestimmten Rechtssatzes als Völkergewohnheitsrecht. Nur mit einer solchen Auffassung lassen sich der Praxis wie erforderlich und üblich bestimmte abstrakte Rechtssätze des Völkergewohnheitsrechts entnehmen, während man bei der Abstellung auf die faktische Übung den abstrakten Völkerrechtssatz nur durch Generalisierung der Einzelfälle induzieren könnte. Damit gewänne man aber eine hinreichende Gewißheit der Existenz der allgemeinen Rechtssätze im Völkerrecht nicht. Man würde damit auch verkennen, daß die Praxis gerade die Existenz der allgemeinen Normen behauptet.

Die herrschende Lehre, welche die Bildung von allgemeinen Rechtssätzen als eine induzierende Generalisierung der auf die Einzelfälle gerichteten Rechtsüberzeugung erklärt [12], übersieht damit den Hauptfall der Praxis, nämlich die Tatsache, daß die Praxis selbst die Existenz abstrakter Rechtssätze des Völkergewohnheitsrechts behauptet. Damit wird auch die häufig vertretene Auffassung widerlegt, die Rechtsüberzeugung der Praxis sei immer nur schwer nachzuweisen [13]. Da die Praxis in fast allen Fällen die Existenz abstrakter Rechtssätze des Völkergewohnheitsrechts behauptet, ist diese Rechtsüberzeugung vielmehr sehr leicht festzustellen.

Auf der anderen Seite führt die obige Erklärung der Dinge zu einer völlig verfehlten Sicht der Problematik der Auslegung des Völkergewohnheitsrechts. Wenn, wie die herrschende Lehre meint, die Rechtsüberzeugung immer nur auf den konkreten Einzelfall gerichtet wäre, würde sich das Völkergewohnheitsrecht zunächst nur auf die Feststellung der Rechtswidrigkeit oder Rechtmäßigkeit des Einzelfalls beziehen. Damit gäbe es das Problem der Auslegung abstrakter Rechtssätze des Völkergewohnheitsrechts nicht: Es handelte sich bei der Frage nach der Anwendung des Völkergewohnheitsrechts auf den Einzelfall immer nur um ein Problem der Feststellung des Völkergewohnheitsrechts [14]. Wenn dagegen die Praxis die Existenz eines abstrakten Rechtssatzes des Völkergewohnheitsrechts behauptet, kann dieser Rechtssatz aus der Praxis unmittelbar abgeleitet werden. Dann stellt sich natürlich auch das Problem der Auslegung dieser abstrakten Rechtssätze [15]. Da diese Rechtssätze nur in wenigen Praxisfällen auf bestimmte konkrete Einzelfälle bezogen worden sind, entsteht nur wenig Völkergewohnheitsrecht, das sich neben den abstrakten Rechtssätzen auf die konkreten Einzelfälle bezieht.

---

[12] Dazu *Adolf Schüle,* Methoden der Völkerrechtswissenschaft, Berichte der deutschen Gesellschaft für Völkerrecht Heft 3 (1959), S. 1 ff.

[13] So *Hans Kelsen,* Théorie du droit coutumier, Revue internationale 1939, S. 263, sowie *Sørensen,* a.a.O. (Anm. 11).

[14] Diese weit verbreitete falsche Auffassung geht auf *Felix Somlo,* Juristische Grundlehre (2. Aufl. 1927), S. 273, zurück: Vgl. *Karl Larenz,* Methodenlehre der Rechtswissenschaft (3. Aufl. 1975), S. 345.

[15] Vgl. *Bleckmann,* a.a.O. (Anm. 4).

Nur insoweit handelt es sich um die Feststellung von konkretem Gewohnheitsrecht. Soweit dagegen neue Fälle zu prüfen sind, die in der Praxis bisher noch nicht gelöst worden sind, müssen die allgemeinen Rechtssätze des Völkergewohnheitsrechts auf den konkreten Einzelfall angewendet und folglich ausgelegt werden.

Die Rechtslage ist damit ähnlich wie im britischen Präjudizienrecht, wenn auch die Bindung an Völkergewohnheitsrecht in der Regel nicht durch ein einzelnes Präjudiz, sondern erst infolge einer Mehrheit von Präzedenzfällen entsteht. Denn auch im britischen Präjudizienrecht ist die nachfolgende Entscheidung nicht nur an die Entscheidung des Einzelfalls, sondern auch durch die *ratio decidendi* der Entscheidung, d. h. aber an seine Rechtssatzbehauptung gebunden[16].

Mit den vorangehenden Feststellungen will ich auch nicht behaupten, daß im Völkergewohnheitsrecht eine induzierende Generalisierung niemals notwendig wird. Solche Induktionsschlüsse sind vielmehr einmal dann notwendig, wenn die Praxis abweichend von der Regel die Existenz eines abstrakten Rechtssatzes des Völkergewohnheitsrechts nicht behauptet, sondern sich auf die Behauptung der Rechtswidrigkeit oder Rechtmäßigkeit des Einzelfalls beschränkt. Auf der anderen Seite wird eine solche Induzierung auch erforderlich, wenn die Präzedenzfälle die Existenz eines bestimmten allgemeinen Rechtssatzes behaupten. Dann stellt sich nämlich das Problem der Auslegung, der Anwendung der allgemeinen Regeln auf einen neuen Einzelfall, und dabei kann man einerseits die klassischen Auslegungsmittel auf die abstrakte Regel anwenden; zum anderen muß man aber auch versuchen, im Wege der Analogie die Einzelfallentscheidung auf neue Fälle anzuwenden. Und schließlich meine ich, daß, wenn auch die Behauptung eines allgemeinen Rechtssatzes selbst zu Völkergewohnheitsrecht erstarken kann, dieser allgemeine Rechtssatz in der notwendig vor allem auf den jeweiligen Einzelfall gerichteten Entscheidung der Staaten noch verfehlt werden kann; das eigentliche Schwergewicht der richterlichen Bindung liegt dann, bis die Praxis den allgemeinen Rechtssatz hinreichend absichert, wie beim Präjudizienrecht in der Entscheidung des konkreten Einzelfalls.

Der im Endergebnis bindende allgemeine Rechtssatz des Völkergewohnheitsrechts ist also einerseits durch einen Vergleich der verschiedenen Formulierungen des Rechtssatzes in den unterschiedlichen Rechtssatzbehauptungen der Praxis, andererseits durch eine induzierende Generalisierung der Einzelfälle zu suchen.

Und umgekehrt müssen bei der Konkretisierung dieses allgemeinen Rechtssatzes auf neue Streitfälle einerseits die abstrakten klassischen Auslegungsregeln auf den abstrakten Rechtssatz angewendet werden; zur Kontrolle und Bestätigung ist aber stets auch die induzierende Generalisierung der entschiedenen Einzelfälle zu versuchen.

---

[16] Vgl. zum britischen Präjudizienrecht *Holdsworth,* A History of English Law, Bd. 12 (1967), S. 146 ff.

Daß die Rechtsüberzeugung und nicht die Übung die eigentliche Bindungsquelle des Völkergewohnheitsrechts darstellt, ergibt sich nicht nur aus der tatsächlichen Übung der Praxis, Rechtsprechung und Lehre bei der Entwicklung konkreter Rechtssätze des Völkergewohnheitsrechts, sondern auch aus theoretischen Erwägungen. Dabei kann dahingestellt bleiben, ob Recht immer nur aus dem Willen eines Trägers der Rechtsetzungsgewalt entstehen kann: In der Tat könnte man davon ausgehen, daß das Völkerrecht die Bindungswirkung nur der tatsächlichen Staatenübung ausdrücklich anordnet; diese Anordnung könnte so etwa aus dem Vertrauensschutzprinzip des Völkerrechts abgeleitet werden. Im Endergebnis kann diese Frage dahingestellt bleiben, weil auch bei einer solchen Konstruktion praktisch gesehen sich die allgemeinen Rechtssätze des Völkerrechts nur aus den ausdrücklichen oder hinter der tatsächlichen Praxis entstehenden stillschweigenden Rechtssatzbehauptungen der Staaten ableiten lassen. Auch wenn die Staaten keine ausdrücklichen Rechtssätze des Völkergewohnheitsrechts behaupten, kann die tatsächliche Übung nämlich nur als Rohmasse des gesuchten Völkerrechtssatzes angesehen werden, aus der durch die Beschränkung der Sachverhalte auf die wesentlichen Elemente der Rechtssatz herausgefiltert werden muß, der hinter der tatsächlichen Übung steht: Wird das Hauptgewicht der Völkergewohnheitsrechtsbindung auf die tatsächliche Praxis gelegt, wird diese Induzierung des Rechtssatzes aus der Übung als eigentliche Rechtsquelle des Völkergewohnheitsrechts angesehen und die Tatsache vernachlässigt, daß Praxis, Rechtsprechung und Lehre den Rechtssatz meist unmittelbar aus den abstrakten Rechtssatzbehauptungen der Staaten ableiten.

Überdies halten die Praxis, Rechtsprechung und Lehre neben der tatsächlichen Übung mit Ausnahme von *Guggenheim* einmütig an dem zweiten Erfordernis der Rechtsüberzeugung für die Bildung von Völkergewohnheitsrecht fest. Diese zweite Voraussetzung kann nur die Funktion haben, die Bindungswirkung des Völkergewohnheitsrechts zu begründen. Dann kann diese Bindungswirkung nicht aus einem auf die Bindung der tatsächlichen Übung gerichteten Rechtssatz des Völkerrechts wie etwa das Vertrauensschutzprinzip, sondern nur aus dem subjektiven Element der Rechtsüberzeugung abgeleitet werden, was immer man darunter auch verstehen mag.

Wenn so die Bindungswirkung des Völkergewohnheitsrechts auf das subjektive Element zurückzuführen ist, kann die Funktion der tatsächlichen Übung nur darin bestehen, die notwendige *Ernstlichkeit* der Rechtsüberzeugung zu sichern. Selbst wenn nun aber diese Beweissicherungsfunktion eine wichtige Rolle spielen mag, tritt diese Aufgabe doch notwendig hinter die Rolle der Rechtsüberzeugung zurück, der die wesentlichste Funktion bei der Bildung von Völkergewohnheitsrecht zufallen muß. Denn auch die Ableitung allgemeiner Rechtssätze aus der Reduzierung des Sachverhalts der tatsächlichen Übung auf die wesentlichsten Elemente bindet nach Ansicht der Praxis, Rechtsprechung und Lehre nur, weil im Bewußtsein der Staaten hinter der tatsächlichen Übung ein bestimmter abstrakter Rechtssatz steht.

2. Überspitzt gesagt, setzen die Praxis, Rechtsprechung und Lehre deshalb für die Bildung von Völkergewohnheitsrecht nicht die beiden Elemente der Rechtsüberzeugung und der Übung, sondern als einziges Element die Rechtsüberzeugung voraus, die in der Behauptung eines abstrakten Rechtssatzes des Völkergewohnheitsrechts besteht. Es fragt sich damit, welches Gewicht der eventuellen Übung noch zufallen kann:

a) In den durchaus meisten Fällen wird die Übung neben der Rechtsüberzeugung nur verlangt, um klarzustellen, daß die Behauptung der Rechtsüberzeugung in bestimmten Einzeldokumenten zum Ausdruck gelangen muß und daß der Rechtssatz während einer längeren Zeitdauer, gleichförmig und von hinreichend zahlreichen Staaten behauptet worden sein muß. In dieser Form ist die Übung heute kein neben die Behauptung des Rechtssatzes tretendes selbständiges Element der Bildung von Völkergewohnheitsrecht mehr; die entsprechenden Voraussetzungen können vielmehr auch unmittelbar auf die Behauptung der Rechtssätze bezogen werden, die gleichförmig sind, eine gewisse Dauer aufweisen und von hinreichend zahlreichen Staaten erklärt sein muß.

b) Die tatsächliche Übung könnte aber einen hierüber hinaus reichenden Inhalt haben. Insbesondere könnte sie sicherstellen, daß die Staaten die betreffende Rechtsüberzeugung tatsächlich auch besitzen. Die *Ernstlichkeit* der Rechtsüberzeugung käme dann darin zum Ausdruck, daß die Staaten in ihrem konkreten Handeln den behaupteten Rechtssatz auch tatsächlich anwenden. Dabei kann das Handeln der Staaten sicherlich nicht nur in einem faktischen Handeln, etwa im Abschuß von das Gebiet unerlaubt überfliegenden fremden Militärmaschinen zum Ausdruck gelangen. Auch rein feststellende oder gestaltende Rechtsakte wie etwa die Urteile der nationalen Gerichte und der Parlamente stellen vielmehr eine solche faktische Übung dar.

Fragt man nach der Notwendigkeit einer solchen faktischen Übung neben der Behauptung eines abstrakten Rechtssatzes, wird man feststellen, daß in der Literatur jedenfalls eine von der Rechtsbehauptung verschiedene faktische Übung nicht verlangt wird: Auch ein Gesetz und ein Urteil sowie eine diplomatische Note, welche die Rechtsbehauptung enthalten, sind vielmehr als faktische Übung angesehen worden. Werden solche Präzedenzfälle angeführt, wird in der Tat niemals weiter nach einem dahinterstehenden faktischen Handeln der Staaten gefragt.

Es fragt sich damit nur noch, ob es sich bei den Akten, in denen die Rechtsbehauptung zum Ausdruck kommt, selbst um eine faktische Übung in diesem Sinne handeln muß. Handelt es sich mit anderen Worten etwa bei einer Rede eines Ministers, die sicherlich kein faktisches Handeln in diesem obigen Sinne darstellt, um eine mit Rechtsüberzeugung gesetzte Praxis?

Die Berichterstattung der Zeitschrift für ausländisches öffentliches Recht und Völkerrecht über die deutsche Völkerrechtspraxis bezieht auch derartige Reden mit ein. Wenn man die Funktion der Übung berücksichtigt, die Ernstlichkeit der abstrakten Rechtsüberzeugung nachzuweisen, wird man an einem solchen Ergebnis zweifeln können. In der Tat steht der Bundesminister, wenn er eine solche Rede hält, nicht unbedingt in derselben politischen und rechtlichen Verantwortlichkeit, wie wenn er an eine fremde Regierung eine diplomatische Note sendet oder wenn ein Gericht einen internationalen oder nationalen Rechtsstreit entscheidet. Trotzdem meine ich, daß man eine solche Rede auch als für die Bildung von Völkergewohnheitsrecht relevante Praxis ansehen muß, man aber bei der dann notwendigen Abwägung das Gewicht einer solchen Rede nicht überbewerten darf [17].

Ich meine schließlich auch, daß für die Feststellung der Ernstlichkeit die faktische Übung nicht ausreicht, insbesondere dann nicht, wenn in den Erklärungen und in der Übung die Existenz eines Rechtssatzes behauptet wird, welche den betreffenden Staat begünstigt. In der Tat meine ich in meinem Buch über die »Funktionen der Lehre im Völkerrecht« gezeigt zu haben, daß es sowohl für die Ernstlichkeit als auch für die Gerechtigkeit der Staatenpraxis auf das Vorliegen bestimmter institutioneller Garantien ankommt [18]. So kann etwa die Ernstlichkeit der Rechtsbehauptung ebensowenig wie der Versuch eines generellen Interessenausgleichs bestritten werden, wenn ein Staat ohne Druck durch einen anderen Staat die Existenz eines Rechtssatzes behauptet, der gegen seine Interessen spricht.

Allgemein stellt sich dabei aber die weiter unten näher behandelte Frage, ob es auf einen solchen Beweis der Ernstlichkeit einer Behauptung von Völkergewohnheitsrecht überhaupt ankommt, ob nicht aus Gründen des Vertrauensschutzes und der Verantwortlichkeit der Staaten jeder Staat an seine Rechtsbehauptungen auch dann gebunden ist, wenn es sich nur um Lippenbekenntnisse handelt. In der Tat könnte die Ernstlichkeit einer Rechtsbehauptung nur widerlegt werden, wenn man die Motive des betreffenden Staates untersucht; die Lehre ist aber einmütig der Auffassung, daß es für die Rechtsüberzeugung auf die Motive der Staaten nicht ankommt.

c) Ich komme als zu dem Ergebnis, daß in der ganz überwiegenden Mehrheit der Fälle die Behauptung eines Rechtssatzes nicht nur das entscheidende, sondern das einzige Element der Bildung von Völkergewohnheitsrecht ist.

d) Fraglich bleibt dann nur die Lösung der Fälle, in denen das faktische Handeln der Staaten von der erklärten Rechtsüberzeugung abweicht. In solchen Fällen könnten aus Vertrauensschutzgründen die Staaten an ihrem Lippenbekenntnis festhalten und dieses faktische Handeln unbeachtet lassen. Die Rechtslage ist dann genauso, wie wenn man aus der Tatsache, daß in der Verfassung zahlreicher Staaten bestimmte Grundrechte verankert sind, einen allgemeinen Rechtsgrundsatz des Europäischen Gemeinschaftsrechts ableitet, obwohl diese Grundrechte sich in der gerichtlichen Praxis der Staaten nicht durchsetzen.

---

[17] Zur Gewichtung der Praxis vgl. *Albert Bleckmann,* Völkergewohnheitsrecht trotz widersprüchlicher Praxis?, ZaöRV 36 (1976), S. 374 ff.

[18] *Albert Bleckmann,* Die Funktionen der Lehre im Völkerrecht (1982).

II. Die Behauptung eines Rechtssatzes als Rechtsetzung

1. Die herrschende Lehre ist der Auffassung, es komme für die Bildung von Völkergewohnheitsrecht auf die Rechtsüberzeugung der Staaten an.

In dieser Sicht wird den Staaten die Funktion eines internationalen Gerichts zugesprochen. Mangels obligatorischer internationaler Gerichtsbarkeit im Völkerrecht wird die richterliche Funktion dezentralisiert von den staatlichen Streitparteien wahrgenommen. *Die Staaten stellen im konkreten Streitfall die völkerrechtliche Rechtsgrundlage wie ein internationaler Richter fest und sprechen die daraus fließenden Rechtsfolgen aus* (Rechtsprechungsfunktion der Staatenpraxis).

Legt man dieses Bild zugrunde, wird mit der Abstellung auf die Rechtsüberzeugung selbst die Funktion der internationalen und der nationalen Gerichtsbarkeit, wie sie in der nationalen Methodendiskussion gesehen wird[19], völlig verkannt.

In dieser Sicht hätten die Staaten nur das vorher bestehende Recht und hieraus durch anerkannte Methoden abgeleitete Rechtssätze auf den Streitfall anzuwenden. Dabei handelt es sich um einen mehr oder weniger automatischen Vorgang der Rechtsanwendung. Irgendein Ermessens- oder Beurteilungsspielraum stünde dem Richter und folglich auch den Staaten nicht zu. Es handelt sich folglich um eine reine Erkenntnis vorherbestehenden Rechts und nicht um Rechtsschöpfung.

Die neuere Methodendiskussion hat aufgedeckt, daß es sich bei diesem Bild um eine reine Fiktion handelt. In Wirklichkeit ist der Richter in den meisten Fällen kein Subsumtionsautomat. Er entwickelt die Obersätze häufig selbständig. Die Subsumtion besteht häufig in einer Wertung des Sachverhalts im Lichte des so gewonnenen Obersatzes. Diese für das innerstaatliche Recht treffende Charakterisierung der Rolle des Richters, der sich mehr und mehr der Funktion eines Gesetzgebers annähert, gilt vielleicht für das Völkerrecht und insbesondere für das Völkergewohnheitsrecht noch verstärkt:

Häufig liegt vor der Entscheidung der Streitparteien über den konkreten Fall nur wenig gleichförmige Praxis vor. Die Entscheidung, daß diese Praxis für die Existenz von Völkergewohnheitsrecht ausreicht, kann nur im Wege einer Wertung gewonnen werden. Dasselbe gilt für die Aufgabe, aus den voneinander abweichenden in der Praxis behaupteten Rechtssätzen den »einzigen« Rechtssatz des Völkergewohnheitsrechts zu entwickeln. Meist ist dieser Rechtssatz nur auf bestimmte einzelne Fälle angewendet worden. Die Subsumtion eines neuen Falles unter diesen Rechtssatz setzt deshalb häufig wiederum eine wertende Stellungnahme voraus.

In einer solchen Situation übernimmt der Richter notwendigerweise *Rechtsetzungsfunktionen.* Er muß entscheiden, ob der behauptete Rechtssatz

[19] Vgl. dazu *Larenz,* a.a.O. (Anm. 14), sowie neuerdings *René A. Rhinow,* Rechtssetzung und Methodik (1979), und *Hans-Martin Pawlowski,* Methodenlehre für Juristen (1981).

generalisierungsfähig ist, in das System und die Strukturen der Völkerrechtsordnung paßt, dem Gerechtigkeitsideal und den Allgemeininteressen der Völkerrechtsgemeinschaft entspricht etc.

Dabei weicht die Funktion eines den konkreten Streit entscheidenden Staates von der eines internationalen Richters in einem entscheidenden Punkt ab. Der betreffende Staat wird, wenn das Völkerrecht sich noch nicht hinreichend verdichtet hat und mehrere vertretbare Lösungen offenläßt, sich nicht für die *»einzig richtige«*, sondern für die Lösung entscheiden, die seinen *nationalen Interessen* am besten entspricht. Das entspricht nicht nur einer möglichen verfassungsrechtlichen Verpflichtung dem Volk gegenüber, sondern auch der Funktion der internationalen Rechtsentwicklung, die versuchen muß, die nationalen Interessen aller Staaten miteinander auszugleichen: und hierfür ist erforderlich, daß die Staaten diese Interessen in die internationale Auseinandersetzung zunächst einmal einbringen.

Die Annäherung der Funktion der Staaten bei der Praxissetzung an die Funktion eines internationalen oder nationalen Richters widerspricht aus einem zweiten Grund der tatsächlichen Rechtslage. Es ist zwar richtig, daß die Parteien eines Rechtsstreits, wenn sie den Streit auf dem Boden des geltenden Rechts lösen wollen, nur das vorherbestehende Recht anwenden und in dem durch die anerkannten Methoden des Völkerrechts gezogenen Rahmen weiterentwickeln dürfen. Diese der richterlichen Funktion entsprechende Staatenfunktion wird aber durch die Tatsache überlagert, daß die Praxis eben nicht nur eine Lösung des vorhergehenden Rechtsstreites sucht, *sondern gleichzeitig für die Zukunft geltendes Recht setzen soll.* In der Praxis entscheiden die Staaten nämlich nicht nur über den zwischen ihnen schwebenden konkreten Rechtsstreit; die Praxis stellt gleichzeitig vielmehr Präzedenzfälle für die Entwicklung künftigen Völkergewohnheitsrechts. In dieser zweiten Funktion sind die Staaten vom vorherbestehenden Recht aber stärker befreit als in ihrer ersten Funktion. Sie können folglich auch ohne Rücksicht auf das vorherbestehende Recht neues Völkergewohnheitsrecht behaupten. Dieses neue Völkergewohnheitsrecht kann zwar von einem internationalen Gericht noch nicht zur Lösung des betreffenden konkreten Streitfalls herangezogen werden. Trotz der Abweichung von geltendem Recht handelt es sich aber für die Zukunft um relevante Staatenpraxis, die, wenn die anderen Staaten ihr folgen, neues Völkergewohnheitsrecht bildet. Die Tatsache, daß selbst eine völlige Abweichung vom geltenden Recht für die Zukunft zu beachten ist, zeigt, daß es sich bei der Praxis eben nicht nur um die Ausübung einer richterlichen Funktion, sondern gerade auch um Rechtsetzung handelt.

Die für das Völkergewohnheitsrecht relevante Praxis hat damit eine Doppelfunktion. Der konkrete Rechtsstreit kann nur durch die Anwendung des vorherbestehenden Rechts geklärt werden. Insoweit überwiegt bei der Feststellung der Rechtslage durch die Staaten die richterliche Funktion. Auf der anderen Seite ist die Staatenpraxis aber auch auf die zukünftige Rechtsentwicklung gerichtet. Insoweit können die den Streit entscheidenden Staaten die Existenz neuen, von dem bisherigen Gewohnheitsrecht abweichenden Völkergewohnheitsrechts behaupten. Auch diese Praxis ist für die Feststellung des Völkergewohnheitsrechts in zukünftigen Streitfällen relevant. Wenn auch schon der richterlichen Funktion der Staaten ein rechtsschöpferisches Element zugrunde liegt, tritt bei dieser zweiten Funktion das gesetzgeberische Element der Völkergewohnheitsrechtspraxis in den Vordergrund.

Im nationalen Recht wie im Völkerrecht besteht eine gewisse Tendenz, diesen rechtsschöpferischen Charakter der richterlichen Tätigkeit wie der staatlichen Aufgabe bei der Praxissetzung zu vernachlässigen. Im staatlichen Rechtsraum wird diese Aufgabe des Richters meist bewußt verkannt, um den Schwierigkeiten der verfassungsrechtlichen Legitimation der richterlichen Rechtsschöpfung auszuweichen. Dabei kann man im nationalen Rechtsraum zu akzeptablen Ergebnissen nur dann gelangen, wenn man die tatsächliche Rechtsschöfung durch die Richter anerkennt und anschließend versucht, diese Rechtsschöpfung durch die Verfassung zu legitimieren und damit gleichzeitig zu begrenzen. Im Völkerrechtsraum besteht eine solche Legitimationsaufgabe in weit geringerem Maße als im nationalen Recht. Eine Beschränkung der Rechtsschöpfung der Staaten bei der Praxissetzung ist nur erforderlich, um sicherzustellen, daß die Staaten bei der völkerrechtlichen Beurteilung des Sachverhalts, also in ihrer richterlichen Funktion, sich an das vorhergehende Völkerrecht gehalten haben. Auf der anderen Seite ist gerade die Anerkennung der rechtsschöpferischen Funktion der Staaten bei der Praxissetzung erforderlich, um die notwendige Entwicklungsfähigkeit des Völkerrechts, die Anpassung des Völkerrechts an neue Bedürfnisse, Werte und Lagen zu sichern.

2. In noch stärkerem Maße als die Rechtsprechung, enthält deshalb die Rechtsüberzeugung bei der Bildung von Völkergewohnheitsrecht sehr starke Elemente einer dezentralisierten Rechtsetzung: Die Staaten erkennen zwar auf der einen Seite das vorherbestehende Völkergewohnheitsrecht an, bilden es auf der anderen Seite aber auch stets selbst weiter, so daß neben die Überzeugung von vorherbestehendem Recht der Wille zur Setzung neuen Rechts tritt.

Diese Deutung der für die Bildung von Völkergewohnheitsrecht notwendigen Rechtsüberzeugung als Rechtsetzung drängt sich geradezu auf, wenn man mit den vorangehenden Darlegungen berücksichtigt, daß nach der Praxis, Rechtsprechung und Lehre nur die *Rechtsbehauptung,* nicht aber die nach unseren allgemeinen Voraussetzungen vom Gewohnheitsrecht geradezu lebenswichtige *Übung* für die Bildung von Völkergewohnheitsrecht erforderlich ist. Reduziert sich die für das Völkergewohnheitsrecht notwendige Praxis auf Rechtsbehauptungen, können diese Rechtsbehauptungen eben nur eine konsekutive Rechtsetzung darstellen: Reduziert sich eine Rechtsquelle auf ein subjektives Element, kommt es notwendig auf den Rechtsetzungswillen und nicht auf die Rechtserkenntnis an.

Im Gegensatz zur Rechtserkenntnis scheint mir bei der Praxis des Völkergewohnheitsrechts dieses Rechtsetzungselement das Rechtserkenntniselement sogar zu überwiegen.

Das zeigt sich insbesondere beim beginnenden Völkergewohnheitsrecht. In der Tat hat die Lehre bisher vergeblich versucht, die Rechtsüberzeugung im ersten Praxisfall zu erklären, bei dem es definitionsgemäß weder Völkergewohnheitsrecht und nicht einmal beginnendes Völkergewohnheitsrecht gab[20]. In diesem, aber auch in allen anderen Fällen, in denen die Praxis über das vorherbestehende Völkergewohnheitsrecht hinausgeht[21], befindet sich der Staat – geht man von der Interpretation der Rechtsüberzeugung als Rechtserkenntnis aus – in einem Irrtum. Aufgrund der allgemeinen Irrtumslehre sollte man deshalb meinen, eine solche Rechtserkenntnis müsse rechtlich unbeachtlich bleiben; das liegt insbesondere dann nahe, wenn man der Vertragstheorie des Völkergewohnheitsrechts folgt[22]. Sieht man dagegen in der Rechtsüberzeugung ein der richterlichen Rechtserkenntnis verwandtes Element, müßte man, wenn jede vorherige Praxis fehlt, unter Umständen auf einen *excès de pouvoir* des Richters schließen, der die Verbindlichkeit des betreffenden Urteils ebenfalls aufhebt.

Gerade in der Tatsache, daß in diesen wie in ähnlichen Fällen die Staatenpraxis trotz dieser »Irrtümer« absolut verbindlich bleibt, möchte ich den Beweis dafür sehen, daß die Rechtsüberzeugung – der Tatsache entsprechend, daß es sich beim Völkergewohnheitsrecht um eine echte Rechtsquelle und nicht nur um eine Quelle der Rechtserkenntnis handelt – im wesentlichen den Willen zur Rechtsetzung und weniger eine Rechtsfeststellung enthält.

Neben dem Völkerrechtsvertrag kennt das Völkerrecht mit dem Gewohnheitsrecht also eine zweite Quelle der Rechtsetzung, die dezentralisiert durch nacheinander folgende Rechtsbehauptungen zahlreicher Staaten zustande kommt, wenn die Zahl der Rechtsbehauptungen gleichförmig und hinreichend dicht ist.

Die Interpretation der Rechtsüberzeugung als Rechtsetzung erklärt auch die dynamische Entwicklung des Völkergewohnheitsrechts, das durchaus nicht so starr ist, wie die Lehre manchmal behauptet: Durch jeden Akt der Staatenpraxis wird der abstrakte Rechtssatz des Völkergewohnheitsrechts und seine Konkretisierung im Einzelfall weiterentwickelt.

Dabei ist es auch möglich, daß ein Staat die Existenz eines abstrakten Rechtssatzes des Völkergewohnheitsrechts behauptet, der von den bisherigen Rechtsbehauptungen völlig abweicht. Ginge man davon aus, daß die Rechtsüberzeugung nur eine Rechtserkenntnis beinhaltet, müßte eine solche von

---

[20] Dieses Problem wurde von *Kelsen*, a.a.O. (Anm. 12), entdeckt. Vgl. dazu auch *Barberis*, a.a.O. (Anm. 11), und *Sørensen*, a.a.O. (Anm. 11).

[21] Eigentlich stellt jede Anwendung des Gewohnheitsrechts auf neue, noch nicht entschiedene Fälle eine Rechtsschöpfung dar.

[22] Dazu *Bleckmann*, a.a.O. (Anm. 7).

der Rechtserkenntnis völlig abweichende Rechtsbehauptung offensichtlich völkerrechtlich irrelevant sein. Wie die Behandlung solcher Fälle durch die Lehre zeigt, handelt es sich aber auch hier um relevante Praxis der Staaten, die sogar nicht völkerrechtswidrig ist, weil nicht die Behauptung eines abstrakten Völkerrechtssatzes, sondern erst die konkrete gegen das Völkerrecht verstoßende Behandlung eines fremden Staates gegen das Völkerrecht verstößt. Die neue Rechtsbehauptung trägt allerdings erst dann einen neuen Rechtssatz des Völkergewohnheitsrechts, wenn die übrige Staatenpraxis diesem Präzedenzfall folgt.

Die Behauptung, daß ein einzelnes Präjudiz der Staatenpraxis bindet, muß in einem gewissen Sinne relativiert werden. Meines Erachtens hat jeder Fall der Staatenpraxis einen gewissen Bindungswert, selbst wenn er allein noch kein Völkergewohnheitsrecht begründet. Die Rechtsetzung erfolgt auf der Ebene des Völkergewohnheitsrechts durch konsekutive Einzelakte, die jeweils einen eigenen Bindungswert besitzen. Mit jedem weiteren entschiedenen Fall nähert sich die Praxis dem Schwellenwert, in dem sie hinreichend dicht ist, um die Aussage zu tragen, sie repräsentiere alle Staaten der Welt und stelle folglich universelles Völkergewohnheitsrecht dar. *Völkergewohnheitsrecht entsteht so durch die allmähliche Kumulierung des Bindungswertes der einzelnen Praxisfälle.*

Dieser Bindungswert der einzelnen Praxisfälle kann dazu führen, daß sie in Verbindung mit der Lehre, mit der Rechtsvergleichung, mit Vernunftgründen Rechtssätze des Völkerrechts begründeten, die eindeutig keiner der Einzelrechtsquellen des Art. 368 des IGH-Statuts zugeordnet werden können, aber Völkerrecht darstellen, wenn man das Kästchendenken in der Rechtsquellenlehre überwindet.

Allerdings besitzen auch gewisse Gerichte eine Rechtsetzungsfunktion, die deutlicher von der Rechtsfeststellungsfunktion geschieden ist als bei der Staatenpraxis. Im Interesse der Rechtssicherheit entscheiden so das Bundesverfassungsgericht[23] und der Gerichtshof der Europäischen Gemeinschaften[24] manchmal einen Rechtsstreit an Hand der vorherigen Norm A, sprechen dabei aber gleichzeitig aus, daß sie zukünftige Fälle an Hand der weitergehenden Norm B entscheiden werden. Ein solches Vorgehen erscheint mir im Interesse der Rechtssicherheit und der notwendigen Entwicklung und Anpassung des Völkerrechts an neue Umstände auch bei der Staatenpraxis möglich: Die Staaten können auf den konkreten Rechtsstreit dann in ihrer Rechtsfeststellungsfunktion den vorherbestehenden Rechtssatz A anwenden und gleichzeitig feststellen, daß sie in Zukunft den Rechtssatz B anwenden werden.

---

[23] Zur Anwendung des Gesetzesvorbehalts auf das »besondere Gewaltverhalten« des Strafgefangenen vgl. so BVerfGE 33,1.

[24] Zur zukünftigen Anwendung des Art. 117 EWG-Vertrag vgl. EuGH, *Defrenne*-Fall, Slg. 1978, S. 1365.

Wie bei den nationalen und internationalen Gerichten wird aber in der Staatenpraxis auch insoweit die Rechsetzungsfunktion durch die Rechtsfeststellungsfunktion überlagert: Relevante Staatenpraxis liegt nur vor, wenn die Staaten zumindest *behaupten,* der neue Rechtssatz B ergebe sich aus der Entwicklung des Völkergewohnheitsrechts, wenn die Staaten sich also auf eine Rechtsfeststellung berufen. Wenn sie dagegen ausdrücklich behaupten, der Rechtssatz B ergebe sich zwar *nicht* aus dem vorhergehenden Recht, sie wollten diesen Rechtssatz aber in Zukunft kraft ihrer Rechtsetzungsgewalt anwenden, handelt es sich um für Völkergewohnheitsrecht relevante Praxis wohl nicht. Das zeigt die enge Verbindung zwischen der Rechtsfeststellung und der Rechtsetzungsfunktion der Staaten auf. Wenn es auch nicht erforderlich ist, daß sich der behauptete Rechtssatz des Völkergewohnheitsrechts aus der vorherigen Praxis tatsächlich ergibt, darf der Staat doch wie ein Gericht den Rechtssatz nicht völlig frei von der vorherigen Praxis setzen wollen.

In diesem Punkt unterscheidet sich meine Theorie von der Vertragstheorie des Völkergewohnheitsrechts. Nach der Vertragstheorie braucht der Staat seine Rechtsbehauptung gerade nicht in die Form einer Rechtsfeststellung zu kleiden; der völkerrechtliche Vertrag kann nämlich subjektiv und objektiv gesehen völlig neues Recht setzen.

Angesichts der Behandlung der Praxis durch Praxis, Rechtsprechung und Lehre wird man kaum behaupten können, daß nur die Praxis für die Bildung von Völkergewohnheitsrecht relevant ist, welche ihre Lösung mit *vertretbaren* Methoden aus dem vorherigen Recht ableitet, die also selbst vertretbar ist. Mir ist in der Tat kein Fall bekannt, in welchem die Rechtsprechung, Lehre oder Praxis in dieser Weise die Lösungen der Praxis korrigiert haben. Da die Staatenpraxis die eigentliche Grundlage des Völkerrechts ist, es also auf die Bewertung der Rechtslagen durch die Staaten ankommt, kann diese staatliche Beurteilung der Rechtslage durch die Lehre oder Rechtsprechung nicht korrigiert werden. Eine solche Korrektur ist nämlich nur hinsichtlich der auf den konkreten Streitfall bezogenen Rechtsfeststellungs-, nicht hinsichtlich der auf die Zukunft bezogenen Rechtsetzungsfunktion möglich. Gerade hierin kommt der Rechtsquellen- und Rechtsetzungscharakter des Völkergewohnheitsrechts zum Ausdruck.

Auch diese Behauptung soll durch einen Beispielsfall verdeutlicht werden: Nehmen wir an, zu Beginn der Entwicklung der beschränkten Staatenimmunität habe in einem konkreten Rechtsstreit der Staat A die Theorie der unbeschränkten, der Staat B die Theorie der beschränkten Staatenimmunität vertreten. Für den konkreten Rechtsstreit, auf den die Rechtsfeststellungsfunktion der Rechtsüberzeugung ausgerichtet ist, binden dann diese Theorien den internationalen Richter nicht. Dieselben Rechtssatzbehauptungen der Staaten A und B bilden aber für spätere Rechtsstreitigkeiten zwischen denselben oder anderen Staaten für die Entwicklung von Völkergewohnheitsrecht relevante und damit für die Praxis und Rechtsprechung verbindliche, durch den Rückgriff auf die Methoden des Völkerrechts oder die vorherige

Praxis nicht korrigierbare Praxis. Darin kommt die Rechtsetzungsfunktion der Staatenpraxis zum Ausdruck.

Der Unterschied der Rechtsetzungs- von der Rechtsfeststellungsfunktion und das Übergewicht der Rechtsetzungsfunktion bei der Praxis der Staaten kommt auch in der Tatsache zum Ausdruck, daß mit Ausnahme des entschiedenen Rechtsstreits Urteile der nationalen und internationalen Gerichte in anderen als den entschiedenen Fällen nicht unbedingt binden, sondern als Präjudiz vernachlässigt werden können, wenn sie das richtige Recht verfehlen. Die Praxis der Staaten ist dagegen, wenn sie zu Völkergewohnheitsrecht auch erst mit der notwendigen Dichte der Staatenpraxis wird, als Präjudiz für das Völkergewohnheitsrecht absolut verbindlich; sie kann nicht durch Anwendung der gesicherten Methoden des Völkerrechts korrigiert werden. Sie bindet ebenso wie ein Gesetz oder ein Vertrag, Rechtsakte, die, selbst wenn sie auf die Kodifikation des vorhergehenden Rechts gerichtet sind, absolut binden, selbst wenn sie das vorherige Recht nicht richtig widergeben.

Bei der Setzung der Praxis ist den Staaten darüber hinaus bewußt, daß sie neben der auf die Lösung des gegenwärtigen Falles gerichteten Rechtsfeststellungsfunktion eine auf zukünftige Fälle bezogene Rechtsetzungsfunktion ausüben. Denn bei der Auswahl des den rechtlich vertretbaren Lösungen an Hand ihres nationalen Interesses berücksichtigen sie stets auch, ob der von ihnen so behauptete Rechtssatz ihrem Interesse auch für zukünftige Fälle entspricht. Diese Ausrichtung auf die Zukunft wird überaus deutlich bei den Großmächten, weil bei ihren zahlreichen Außenbeziehungen die Möglichkeit einer Entstehung ähnlicher Fallkonstellationen wächst.

Dabei ist aber zu berücksichtigen, daß beide so aufgezeigten Funktionen in engem Zusammenhang miteinander stehen: Die auf die Zukunft gerichtete Rechtsetzungsfunktion kann nicht in der Weise von der auf die Gegenwart gerichteten Feststellungsfunktion getrennt werden, daß im Rahmen jeder Funktion unterschiedliche Rechtssätze behauptet werden. Auch die Rechtsetzungsfunktion wirkt so auf die Feststellungsfunktion ein, und umgekehrt zieht die Feststellungsfunktion der Rechtsetzungsfunktion enge, wenn auch nicht rechtliche, so doch faktische Grenzen, weil eine Lösung von Rechtsstreitigkeiten nicht erreicht werden kann, wenn die behaupteten Rechtssätze des Völkerrechts zu weit vom vorher geltenden Recht abweichen.

Allerdings liegt in dieser Begrenzung der Rechtsetzungs- durch die Rechtsfeststellungsfunktion ein gewisses Problem. Einerseits darf man nämlich die Rechtsetzungsfunktion nicht so eng fassen, daß sie die notwendige Anpassung des Völkergewohnheitsrechts an neue Interessen, Werte und Tatsachen stört. Wenn auf der anderen Seite andere Staaten der frei, d.h. ohne Bezug auf eine Rechtsfeststellung geäußerten Praxis des ersten Staates folgen, kann man schwerlich von der Irrelevanz der späteren Praxis ausgehen, weil sie sich auf angeblich irrelevante Praxis des Völkerrechts, also nicht auf eine Feststellung des Völkergewohnheitsrechts stützt. Das zeigt die Notwendigkeit, die Begrenzung der Rechtsetzungs- durch die Rechtsfeststellungsfunktion stark zu lockern.

Die in der konsekutiven Praxis geäußerte Rechtsetzung wird von Lehre, Rechtsprechung und Praxis also nur beachtet, soweit sie die Behauptung einer Rechtsfeststellung enthält. Dabei handelt es sich um für die Bildung von Völkergewohnheitsrecht relevante Praxis nur dann, wenn diese Rechtsfeststellung sich auf bestimmte Quellen des Völkerrechts bezieht[25]. So können die Staaten zunächst einmal behaupten, es handle sich bei der behaupteten Norm um Völkergewohnheitsrecht, das aus einer bestimmten Staatenpraxis abgeleitet wird. Um für das Völkergewohnheitsrecht relevante Praxis handelt es sich ferner dann, wenn die Staaten sich auf die Existenz eines allgemeinen Rechtsgrundsatzes des Völkerrechts beziehen. Zweifelhafter ist die Lage, wenn die Rechtsbehauptung sich auf die Auslegung eines völkerrechtlichen Vertrages bezieht. Hier zieht das heutige Recht die Praxis nur als Mittel der Auslegung des Vertrages heran. Meines Erachtens kann aber nicht zweifelhaft sein, daß eine solche Praxis die Vertragsstaaten, wenn auch möglicherweise nicht Drittstaaten bindendes Völkergewohnheitsrecht begründen kann[26]. Für die Bildung von Völkergewohnheitsrecht relevante Praxis wird ferner dann gesetzt, wenn die Staaten den Rechtssatz aufgrund von anerkannten oder evidenten Methoden aus dem Völkergewohnheitsrecht, dem Völkervertragsrecht und aus allgemeinen Rechtsgrundsätzen ableiten, etwa eine Analogie zu einem Satz des Völkergewohnheitsrechts ziehen. Um für die Bildung von Völkergewohnheitsrecht relevante Praxis dürfte es sich schließlich dann handeln, wenn die Staaten versuchen, eine Norm der Moral oder Sitte in eine Norm des Völkergewohnheitsrechts zu verwandeln, oder wenn sie sich nur auf Vernunftgründe oder auf die Lehre stützten. Hier sind allerdings die Grenzen der Rechtsfeststellung erreicht. In der Tat kann man der Auffassung sein, daß es sich, wenn man sich nur auf Sitte oder Moral, auf die Vernunft oder auf die Lehre stützt, nicht mehr um eine Rechtsfeststellung handelt, weil diese Normen keine Rechtsquellenqualität im Völkerrecht besitzen. Auf der anderen Seite ist heute mehr oder weniger anerkannt, daß eine Norm der Moral durch Praxis in Völkergewohnheitsrecht verwandelt werden kann[27]. Theoretisch stehen hier zwei Möglichkeiten der Erklärung offen: Entweder ist man der Auffassung, es handle sich hier um Rechtsquellen des Völkerrechts in einem weiteren Sinn, dann kann man an der Behauptung festhalten, die Praxis sei nur als Rechtsfeststellung relevant. Oder aber, man ist der Auffassung, es handle sich hier nicht um Rechtsquellen; dann muß man anerken-

---

[25] Das kommt in der Rechtsprechung des IGH nicht deutlich genug zum Ausdruck, nach welcher sich die Rechtsüberzeugung nur auf die Existenz von *law* richten muß, vgl. *Verdross/Simma,* a.a.O. (Anm. 1), S. 283 ff.

[26] Vgl. *Albert Bleckmann,* Zur Funktion europäischen Gewohnheitsrechts im Europäischen Gemeinschaftsrecht, EuR 1981, S. 101 ff.

[27] Vgl. die Auffassung von *De Visscher* bei *Sørensen,* a.a.O. (Anm. 11), S. 116.

nen, daß im Rahmen der für Gewohnheitsrecht relevanten Praxis die Rechtsetzung über die Grenzen der Rechtsfeststellung hinausgehen kann.

III. Die Rechtsetzungskompetenz

1. Schon wenn wir davon ausgingen, es handle sich bei der Rechtsüberzeugung nur um eine Form der Rechtserkenntnis, müßten wir, weil diese Rechtserkenntnis zu Völkergewohnheitsrecht führen soll, nach der Kompetenz der die Praxis setzenden Organe fragen. Dies gilt um so mehr, wenn wir davon ausgehen, daß es sich bei der Rechtsüberzeugung in erster Linie um eine Rechtsetzungsfunktion handelt. Rechtsetzung ist immer nur dann verbindlich, wenn sie von einem hierfür zuständigen Organ ausgeht.

Die Vertragstheorie hat diese Notwendigkeit der Kompetenzfeststellung gesehen: Da es sich bei der Praxis des Völkergewohnheitsrechts um Angebote zum Abschluß eines völkerrechtlichen Vertrages handelt, darf nur die Praxis der nach Völkerrecht und nationalem Recht zum Vertragsschluß zuständigen Außenminister verbindlich sein[28]. Mit der Aufgabe dieser Theorie haben die Lehre und Rechtsprechung auch die notwendige Kompetenzfeststellung über Bord geworfen: Alle staatlichen und internationalen Organe dürfen nach dieser impliziten Auffassung Praxis setzen.

Bei der Frage nach der völkergewohnheitsrechtlichen Rechtsetzungskompetenz ist zwischen einer völkerrechtlichen und einer nationalen Zuständigkeit zu entscheiden. Völkerrechtlich gesehen liegt die Kompetenz zur Schaffung universellen Völkergewohnheitsrechts bei der gesamten Staatengemeinschaft, nicht bei den jeweils entscheidenden fünf bis zwanzig oder mehr Staaten, welche die Staatenpraxis setzten. Allerdings ist nach der Regel über die Bildung von Völkergewohnheitsrecht nicht erforderlich, daß alle Staaten der Welt eine entsprechende Praxis gesetzt haben: Bei hinreichend enger Staatenpraxis wird vielmehr von einem für die Gesamtstaatengemeinschaft repräsentativen Willen nur weniger Staaten ausgegangen. Neben die Frage nach der völkerrechtlichen Kompetenz tritt die nach der nationalen Kompetenz, d.h. das Problem, welche staatlichen Organe in welchem Verfahren zur Setzung relevanter Staatenpraxis zuständig sind.

2. Da das Völkerrecht die Kompetenz zur dezentralisierten Rechtsetzung nicht selbst bestimmt, muß es auf das nationale Recht verweisen. Auf den ersten Blick ist das Problem, daß das nationale Recht die Kompetenz zur Rechtsetzung im Wege des Völkergewohnheitsrechts offensichtlich nicht zu bestimmen scheint. Dabei fragt es sich in zahlreichen Bereichen, ob die Regeln über die Bildung des Völkergewohnheitsrechts nicht den Regeln der nationalen Hierarchie und der nationalen Rechtsquellenbildung geradezu ins Gesicht schlagen:

---

[28] Vgl. *Bleckmann,* a.a.O. (Anm. 7).

a) Zunächst kann auch die Praxis erstinstanzlicher nationaler Untergerichte Völkergewohnheitsrecht begründen. Das ist sogar häufig in den Bereichen der Fall, in denen wie bei der Zwangsvollstreckung (gegen fremde Staaten) nur die erstinstanzlichen Gerichte zuständig sind. In diesen Fällen bestehen vom nationalen Recht her gesehen gegen die Bildung von Völkergewohnheitsrecht Bedenken nicht. Etwas anderes gilt, wenn in der betreffenden Materie auch die obersten Gerichte zuständig sind. Hier kann die den nationalen Regeln widersprechende Situation entstehen, daß die Praxis der Untergerichte Völkergewohnheitsrecht setzt, an das auch die Obergerichte gebunden sind – eine für das nationale Recht nicht akzeptierbare Situation.

b) Die Regierungen – sagen wir einmal der westlichen Staatengruppe – könnten, um die parlamentarische Zustimmung zu einem entsprechenden Völkerrechtsvertrag zu umgehen, etwa auf internationalen Konferenzen eine übereinstimmende Rechtsbehauptung aufstellen, die in Völkergewohnheitsrecht erwächst, ohne daß die Parlamente zu diesem Völkergewohnheitsrecht Stellung beziehen können. Auch hier liegt eine klare Umgehung des nationalen Verfassungsrechts vor.

c) Vom nationalen Recht aus noch unhaltbarer wird die Rechtslage, wenn dieselben Rechtsbehauptungen von Abgeordneten der nationalen Parlamente einzelner Staaten aufgestellt werden, die im Gegensatz zur Exekutive nach nationalem Recht als Einzelne keinerlei Rechtsetzungsbefugnis besitzen: Auch diese Rechtsbehauptungen sind nach der herrschenden Lehre Praxis, die bei der Entwicklung von Völkergewohnheitsrecht zu beachten ist.

d) Das zeigt schon, daß die Regeln über die Bildung von Völkergewohnheitsrecht sich in einem bestimmten Umfang nach den nationalen gemeinsamen Verfassungsregeln richten müssen, um zu verhindern, daß auch innerstaatlich wirkende Regeln gebildet werden, die gegen das Verfassungsrecht der Staaten und vor allem auch gegen die nationale Hierarchie der Staatsorgane verstoßen.

e) Daß die innerstaatlichen Regeln bei der Bildung von Völkergewohnheitsrecht mit berücksichtigt werden müssen, zeigt schließlich auch der Fall, in welchem die Praxis – sagen wir des Parlaments, des Außenamts, des Bundesverfassungsgerichts und des Bundesgerichtshofs – voneinander abweicht. Wenn dies in verschiedenen Staaten der Fall ist, wird man nicht ohne weiteres davon ausgehen können, daß widersprüchliches Völkergewohnheitsrecht entsteht, so daß grundsätzlich von der Bildung von Völkergewohnheitsrecht nicht ausgegangen werden kann.

In der Tat muß man davon ausgehen, daß völkerrechtlich gesehen jedem Staat nur eine einzige Stellungnahme zum Völkergewohnheitsrecht zugeschrieben werden kann: Die Staaten sprechen nach außen mit einer einzigen Stimme. Nach außen verbindlich kann nur die Äußerung sein, die entweder im Verhältnis nach außen abgegeben wird – das wäre die Rechtsauffassung des Auswärtigen Amtes – oder die nach innerstaatlichem Recht die höhere

Verbindlichkeit beanspruchen kann – das wäre die im Rahmen des Art. 100 II GG abgegebene Stellungnahme des Bundesverfassungsgerichts.

3. Es muß also darauf ankommen, wie das innerstaatliche Recht die Kompetenz zur völkergewohnheitsrechtlichen Rechtsetzung bestimmt. Dabei kann es nicht auf die Kompetenz zur Setzung innerstaatlichen Rechts ankommen, weil die Außenkompetenzen sich nicht nach den innerstaatlichen Kompetenzen ausrichten. Es muß vielmehr versucht werden, die Kompetenz zur völkergewohnheitsrechtlichen Rechtsetzung selbständig zu bestimmen. Diese Kompetenz kann man auffinden, wenn man berücksichtigt, daß die Feststellung von Völkergewohnheitsrecht meist nur eine Vorfrage zu einer innerstaatlichen Rechtsfeststellung oder Rechtsetzung ist.

Da die Rechtsetzung des Völkergewohnheitsrechts im Rahmen einer Rechtsfeststellung erfolgt, kommt es auf die staatlichen Vorschriften an, welche den Gerichten, Behörden, Parlamenten usw. inzidentell die Feststellung des Völkergewohnheitsrechts gestatten.

a) So sind die innerstaatlichen Gerichte befugt, im Rahmen innerstaatlicher Rechtsstreitigkeiten notwendige völkerrechtliche Vorfragen zu klären. In diesem Rahmen müssen sie folglich auch befugt sein, Rechtssätze des Völkergewohnheitsrechts zu behaupten. Die Kompetenz zur Entscheidung über den innerstaatlichen Rechtsstreit beinhaltet die Kompetenz zur Rechtsetzung im Völkerrecht im Rahmen des Völkergewohnheitsrechts.

Dabei kann man offensichtlich nur auf die generelle Kompetenz der Gerichte zur Feststellung des Völkergewohnheitsrechts in einem schwebenden Verfahren, nicht auf die sachliche, örtliche und funktionelle Zuständigkeit zur Entscheidung des innerstaatlichen konkreten Streitfalls abstellen, weil diese innerstaatlichen Zuständigkeiten relativ schwer erkennbar und völkerrechtlich offensichtlich irrelevant sind. Auch kommt es möglicherweise nicht darauf an, ob die Feststellung des Völkergewohnheitsrechts im konkreten Rechtsstreit eine notwendige Vorfrage darstellt oder ob das Gericht nur ein *obiter dictum* zum Völkergewohnheitsrecht erlassen hat.

Eine völkerrechtlich irrelevante Praxis liegt dagegen vor, wenn ein Gericht eine völkerrechtliche Stellungnahme außerhalb seiner eigentlichen Rechtsprechungsfunktionen abgegeben hat.

b) Das Parlament ist innerstaatlich nach Völkerrecht zuständig, den innerstaatlichen Vollzug des Völkerrechts zu regeln. Die für diese in diesem Rahmen erlassenen Gesetze und parlamentarischen Stellungnahmen stellen verbindliche Praxis des betreffenden Staates dar. Auch wenn ein außerhalb dieses Rahmens ergangenes Gesetz möglicherweise völkerrechtliche Rechte von Drittstaaten verletzt, muß das Parlament diese Fragen prüfen und kann hierzu rechtlich verbindliche Stellungnahmen abgeben. Mangels innerstaatlicher Kompetenz müssen dagegen Stellungnahmen des Parlaments und einzelner Abgeordneter außerhalb dieses Rahmens unbeachtet bleiben.

c) Die Regierungen und Außenämter sind nach dem nationalen Verfassungsrecht für die auswärtige Gewalt zuständig. Sie dürfen deshalb auch innerhalb und außerhalb eines konkreten Rechtsstreits mit einem Drittstaat völkerrechtlich verbindliche Praxis setzen. Die verfassungsrechtlichen Grenzen sind dort erreicht, wo die Regierungen mehrerer Staaten Praxis setzen, um die Zustimmungsbedürftigkeit eines entsprechenden Völkerrechtsvertrages zu umgehen. Hier verstößt die Praxis offensichtlich gegen das Verfassungsrecht und ist deshalb auch völkerrechtlich unverbindlich.

d) Die Länder besitzen außenpolitische Befugnisse nur im Rahmen des Art. 32 III GG. Soweit sie in diesem Rahmen völkerrechtliche Rechtsbehauptungen aufstellen, ist diese Praxis auch völkerrechtlich verbindlich. Völkerrechtliche Rechtssatzbehauptungen der Länder außerhalb dieses Rahmens und der Gemeinden und Anstalten des öffentlichen Rechts im allgemeinen sind völkerrechtlich dagegen unverbindlich.

4. Bei einer Verweisung des Völkerrechts auf das nationale Verfassungs- und Gesetzesrecht muß der Vertrauensschutz des Völkerrechts berücksichtigt werden, weil es sich hier um einen allgemeinen Rechtsgrundsatz handelt oder weil die Analogie zu Art. 46 der Wiener Vertragsrechtskonvention greift: Wenn die nationalen Verfassungsregelungen zu kompliziert und Verstöße nicht evident sind, muß der völkerrechtliche Rechtsverkehr sich darauf verlassen können, daß das betreffende Organ rechtsverbindliche Praxis gesetzt hat. In der Tat muß diese Äußerung eines unzuständigen Organs dem betreffenden Heimatstaat zugerechnet werden, wenn er es unterlassen hat, diese Äußerung etwa einer Gemeinde im Völkerrechtsraum richtigzustellen.

Völkerrecht und Bundesstaat

Gedanken zu einem juristischen Gegenverkehr

MICHAEL BOTHE

I. Einführung

Die Beziehungen zwischen Bundesstaatsrecht und Völkerrecht sind eines der wesentlichen wissenschaftlichen Interessengebiete des Jubilars, auf dem er sowohl dem Völkerrecht als auch dem Staatsrecht wesentliche Impulse gegeben hat[1]. Darüber hinaus verdanken wir ihm Anstöße, über Fragen der gegenseitigen Beeinflussung von Staats- und Völkerrecht nachzudenken, die staatsrechtliche Rechtsvergleichung für das Völkerrecht[2] und das Völkerrecht für das Verständnis des Staatsrechts fruchtbar zu machen[3]. Diesem Ziel sind auch die nachfolgenden Erwägungen gewidmet.

Das Problem der Anwendung von Völkerrecht im Verhältnis zwischen Gliedstaaten eines Bundesstaates ist eine alte Fragestellung der Staats- und Völkerrechtswissenschaft[4]. Die bundesstaatliche Verfassung wird meist als

---

[1] Vgl. z. B. *Mosler*, Die völkerrechtliche Wirkung bundesstaatlicher Verfassungen, Eine Untersuchung zum Völkerrecht und zum vergleichenden Verfassungsrecht, in: Festschrift für Richard Thoma (1950), S. 129–172.

[2] Zusammenfassend vgl. etwa *Mosler*, Das Heidelberger Kolloquium über Gerichtsschutz gegen die Exekutive, in: Gerichtsschutz gegen die Exekutive – Judicial Protection against the Executive – La protection juridictionnelle contre l'exécutif (Beiträge zum ausländischen öffentlichen Recht und Völkerrecht, Bd. 52) (1969), S. XIII–XXII (insbes. S. XIV f., XXI f.).

[3] So insbesondere in den Arbeiten zur »Harmonie von Völkerrecht und nationalem Recht«, vgl. *Mosler,* Die Gewährleistung des Völkerrechts durch die nationale Verfassung, dargestellt am Grundgesetz für die Bundesrepublik Deutschland, in: Problèmes contemporains de droit comparé, Bd. 1 (Tokio 1962), S. 161–178; *derselbe,* Das Völkerrecht in der Praxis der deutschen Gerichte, Juristische Studiengesellschaft Karlsruhe, Schriftenreihe Heft 32/33 (1957).

[4] *Triepel*, Völkerrecht und Landesrecht (1899), S. 186 f. und 198 ff.; *Bolle,* Das interkantonale Recht, Die völkerrechtlichen Beziehungen der Kantone unter sich (1907), S. 2 ff.; *Müller,* Die Beziehungen der Gliedstaaten im Bundesstaatsrecht (1936), insbes. S. 28 ff., 55 ff., 91 ff.; *Verdross,* Die Völkerrechtssubjektivität der

eine erschöpfende, alle Beziehungen der Glieder untereinander regelnde Ordnung angesehen. »Wenn daher«, so formuliert *Mosler* die herrschende Lehre, »Völkerrechtsnormen zwischen Gliedstaaten angewendet werden, wie es in Bundesstaaten allgemein üblich ist, so füllen sie die Lücken des Bundesverfassungsrechts aus und sind als bundesverfassungsrechtlich rezipiertes Recht anzusehen. Ihr Geltungsgrund innerhalb des Bundesstaates ist nicht das Völkerrecht, sondern die Bundesverfassungsordnung« [5].

Kann sich die wissenschaftliche Fragestellung mit dieser Antwort – ihre Richtigkeit unterstellt – zufriedengeben? Ist nicht vielmehr weiter zu prüfen, warum und unter welchen Umständen Völkerrecht im Verhältnis zwischen Gliedstaaten als Entscheidungsmaßstab herangezogen wird?

Weiteren Aufschluß über dieses Problem ist vielleicht aus der Tatsache zu gewinnen, daß es sich hier sozusagen nicht um eine rechtliche Einbahnstraße, sondern um einen Gegenverkehr handelt. Denn in der Praxis der Rechtsanwendung kommt es nicht nur zur Heranziehung von Völkerrecht bei der Lösung von Rechtsfragen zwischen Gliedstaaten von Bundesstaaten, sondern umgekehrt zur Verwendung bundesstaatlicher Präzedenzfälle im Verhältnis zwischen souveränen Staaten [6]. Ein Paradebeispiel ist hier das völkerrechtliche Nachbarrecht, insbesondere das Recht der grenzüberschreitenden Umweltbelastungen. Die grundlegenden wissenschaftlichen Arbeiten hierzu stützen sich allesamt – zunächst ohne dies dogmatisch zu rechtfertigen – auf Entscheidungen des Supreme Court der USA und des schweizerischen Bundesgerichts, die das Verhältnis zwischen Gliedstaaten betreffen [7]. *Handl* macht die Problematik dieses Verfahrens deutlicher, wenn er es vorsichtig, wohl etwas übervorsichtig, wie folgt rechtfertigt [8]:

> »[T]his study at times draws heavily on quasi-international decisions, i.e. on judicial precedents in inter-state disputes, in support of its general

---

Gliedstaaten der Sowjetunion, ÖZöR 1 (1946), S. 212 ff.; *Cowles*, International Law as Applied between Subdivisions of Federations, RdC 74 (1949 I), S. 654–754; *Krüger*, Völkerrecht im Bundesstaat, in: Um Recht und Gerechtigkeit, Festgabe für Erich Kaufmann (1950), S. 239–248; *Bleckmann*, Völkerrecht im Bundesstaat?, Gedanken zum zweiten Coburg-Urteil des Bundesverfassungsgerichts vom 30. Januar 1973, Schweiz. Jahrbuch für internationales Recht 29 (1973), S. 9–48.

[5] A.a.O. (Anm. 1), S. 137, Anm. 3.

[6] *Cowles* (Anm. 4), S. 740 ff.; das wird auch ansatzweise gesehen von *Bleckmann* (Anm. 4), S. 20.

[7] *Wildhaber*, Die Öldestillerieanlage Sennwald und das Völkerrecht der grenzüberschreitenden Luftverschmutzung, Schweiz. Jahrbuch für internationales Recht 31 (1975), S. 97–120 (S. 104 f., 116 f.); vgl. auch *Bothe*, International Legal Problems of Industrial Siting in Border Areas and National Envrionmental Policies, in: Transfrontier Pollution and the Role of States, OECD 1981, S. 79–97 (S. 83 f; 89 ff.).

[8] *Handl*, The Principle of »Equitable Use« as Applied to Internationally Shared Natural Resources: Its Role in Resolving Potential International Disputes over Transfrontier Pollution, in: Transfrontier Pollution and the Role of States, OECD 1981, S. 98–121 (100).

conclusions. Given the abundance of U.S. law and the relative scarcity of other national decisions, the former has been principally relied upon for substantiating merely points of law which seemed already well developed on an international level. In other words, special care was taken not to rely on decisions in which constitutional considerations concerning, in particular, the question of relief, prevailed; and, in any event, to use them only in a supplementary function«.

Die Entwicklung völkerrechtlicher Aussagen aus bundesstaatlichen, insbesondere amerikanischen und schweizerischen Präzedenzfällen ist keine Erfindung der Wissenschaft. Sie kann sich vielmehr auf schiedsgerichtliche Präzedenzfälle berufen. So argumentiert Huber im *Las Palmas*-Schiedsspruch [9]:

> »[T]his principle [Gebietserwerb durch unbestrittene länger dauernde Hoheitsausübung] has further been recognized in more than one federal state, where a jurisdiction is established in order to apply, as need arises, rules of international law to the interstate relations of the states members«.

Hier wird auf den wichtigen Gesichtspunkt der wechselseitigen Beeinflussung von Völkerrecht und Bundesstaatsrecht angespielt. Ein Ausgangspunkt für die Übernahme bundesstaatlicher Präzedenzfälle ins Völkerrecht ist ferner einer der grundlegenden Schiedssprüche zu Fragen grenzüberschreitender Umweltbelastungen, der *Trail Smelter*-Fall [10]. Die Aussagekraft dieses Schiedsspruchs für das Völkerrecht wird gelegentlich bezweifelt, da das Schiedsgericht nach dem Kompromiß auch zur Anwendung amerikanischen Rechts befugt war. Letzteres ist zwar richtig, aber das Gericht wollte seine wesentlichen Aussagen zur Zulässigkeit grenzüberschreitender Luftverschmutzung erklärtermaßen als völkerrechtliche verstanden wissen. Das Gericht rechtfertigt sein Verfahren wie folgt [11]:

> »There are, however, as regards both air pollution and water pollution, certain decisions of the Supreme Court of the United States which may legitimately be taken as guide in this field of international law, for it is reasonable to follow by analogy, in international cases, precedents established by that court in dealing with controversies between States of the Union or with other controversies concerning the quasi-sovereign rights of such States, where no contrary rule prevails in international law and no reason for rejecting such precedents can be adduced from the limitations of sovereignty inherent in the Constitution of the United States«.

---

[9] AJIL 22 (1928), S. 867 ff. (876 f.).

[10] Reports of International Arbitral Awards (RIAA) Bd. 3, S. 1906; vgl. dazu *A. P. Rubin*, Pollution by Analogy, The Trail Smelter Arbitration, Oregon Law Review 50 (1971), S. 259–282 (266 ff.).

[11] A.a.O., S. 1964.

Das Schiedsgericht gewinnt also für das Völkerrecht Erkenntnisse aus dem innerstaatlichen Recht. Der Weg, auf dem dies normalerweise geschieht, ist Art. 38 Abs. 1 (c) IGH-Statut: über die allgemeinen von Staaten anerkannten Rechtsgrundsätze. Die Ermittlung solcher allgemeiner Rechtsgrundsätze bedeutet die Übertragung innerstaatlich geltender Prinzipien in das Völkerrecht[12]. Sieht man einmal von der möglichen Besonderheit bilateral oder regional geltender Rechtsgrundsätze ab, so erfüllen nur solche Rechtsgrundsätze die Erfordernisse, die Art. 38 Abs. 1 (c) an die Rechtsquelle stellt, die universal, d. h. jedenfalls von einer Vielzahl von Staaten anerkannt sind[13]. Bei der hier in Frage stehenden Argumentation geht es aber nur um die Übertragung von Rechtsgrundsätzen, die im Recht eines oder einiger weniger Staaten vorgefunden werden, nämlich in Staaten mit einer bundesstaatlichen Verfassungsordnung. Streng genommen ist also das Vorgehen des Schiedsgerichts im *Trail Smelter*-Fall nicht eine Anwendung des Art. 38 Abs. 1 (c) IGH-Statut. Daß in der Sache aber die Anwendung allgemeiner Rechtsgrundsätze durch internationale Gerichte kaum von dem Vorgehen des Schiedsgerichts zu unterscheiden ist, wird noch zu zeigen sein.

Das Schiedsgericht bezeichnet seine Heranziehung des amerikanischen Rechts für das Völkerrecht als einen Analogieschluß, eine Formel, die auch in der Debatte um die Anwendung von Völkerrecht im Bundesstaat auftaucht[14]. Die Voraussetzungen und Grenzen eines solchen Analogieschlusses müssen darum im folgenden näher untersucht werden. Eine genauere Erklärung der dargestellten Argumentationsmuster muß aber noch umfassender aussehen, muß die Erklärungsmöglichkeiten vollständiger ausleuchten. Es gibt hier im Grunde zwei Alternativen: Einmal könnte es sich bei den entschiedenen Fällen doch um solche aus einer einheitlichen Rechtsordnung handeln, nämlich aus dem Völkerrecht. Dann ist die Verwendung von Entscheidungen, die im Verhältnis zwischen Gliedstaaten eines Bundesstaates ergangen sind, für das Verhältnis zwischen voll-souveränen Staaten und umgekehrt ein normaler Fall der Anwendung von Präzedenzfällen, einer Hilfsquelle nach § 38 Abs. 1 (d) IGH-Statut. Die entscheidende Frage ist dabei, ob und inwieweit entgegen der oben dargestellten herrschenden Lehre doch eine Geltung von Völkerrecht im Verhältnis zwischen den Gliedstaaten eines Bundesstaates möglich ist. Zum anderen kann es sich um die Übertragung von Regelungsmustern einer Rechtsordnung in die andere handeln. Deren Möglichkeit wäre dann grundsätzlich zu klären.

---

[12] *Verdross/Simma*, Universelles Völkerrecht (2. Aufl. 1981), S. 309 ff.

[13] Vgl. *Bothe*, Die Bedeutung der Rechtsvergleichung in der Praxis internationaler Gerichte, ZaöRV 36 (1976), S. 280–299 (282).

[14] *Bleckmann* (Anm. 4), S. 43.

II. Geltung von Völkerrecht im Bundesstaat?

Die These, daß im Verhältnis zwischen Gliedstaaten eines Bundesstaates Völkerrecht nicht anwendbar sein könnte, beruht auf der Annahme, daß begriffsnotwendig eine bundesstaatliche Verfassung die Beziehungen zwischen den Gliedstaaten untereinander sowie zwischen den Gliedstaaten und dem Bund umfassend und unter Ausschluß des Völkerrechts regeln müsse. Dies setzt voraus, daß sich der staatsrechtlich organisierte vom völkerrechtlich organisierten, der Bundesstaat vom Staatenbund trennscharf abgrenzen läßt, daß fließende Übergänge, Gemengelagen zwischen staatsrechtlicher und völkerrechtlicher Regelung unmöglich sind. Nun können wir solche Gemengelagen zwischen Staats- und Völkerrecht eben doch in gewissen Fällen zweifelsfrei ausmachen, z. B. im Britischen Commonwealth oder im Deutschen Reich nach 1648[15]. Der Streit um die Zuordnung der Souveränität im Bundesstaat (Calhoun und Webster in USA, Seydel und Treitschke im Deutschen Reich)[16], wenn er auch heute überwunden ist, läßt doch an der Berechtigung allzu scharfer begrifflicher Ableitungen zweifeln. Die richtige Frage lautet darum wohl nicht, ob Völkerrecht in den Beziehungen zwischen Gliedstaaten eines Bundesstaates gelten kann, sondern ob es gilt[16a]. Das kann nur aus einer genaueren Analyse der jeweiligen bundesstaatlichen Ordnung festgestellt werden. Die Tatsache, daß Gerichte völkerrechtliche Normen zwischen den Gliedstaaten eines Bundesstaates angewandt haben, reicht allerdings zum Nachweis der Geltung von Völkerrecht nicht aus. Diese Anwendung läßt sich auch, wie oben erwähnt, mit Analogie oder ähnlich erklären. Was den deutschen Bundesstaat angeht, so wurde für das Bismarck-Reich überwiegend eine Gemengelage von Staats- und Völkerrecht angenommen[17]. Erst zur Weimarer Zeit wurde eine Tendenz zur Ausschaltung des Völkerrechts erkennbar. Die Äußerungen des Staatsgerichtshofs für das Deutsche Reich zu dieser Frage sind nicht ganz eindeutig[18], werden aber teil-

---

[15] Vgl. dazu *Randelzhofer*, Völkerrechtliche Aspekte des Heiligen Römischen Reiches nach 1648 (1967), insbes. S. 299 f.

[16] Vgl. dazu *Deuerlein*, Föderalismus, Die historischen und philosophischen Grundlagen des föderativen Prinzips (1972), S. 55, 155 ff.

[16a] Zutreffend *Wolff*, Die Rechtsbeziehungen zwischen den Ländern der Bundesrepublik Deutschland (Diss. Marburg 1968), S. 79.

[17] *Müller* (Anm. 4), S. 31 mit weiteren Nachweisen. Vgl. auch *Rudolf*, Zum System der staatlichen Rechtsordnungen in der Bundesrepublik Deutschland, Die Öffentliche Verwaltung 19 (1966), S. 73–77, 75.

[18] Vgl. *Müller* (Anm. 4), S. 32; *Bothe*, L'application des règles du droit international public aux relations entre les Etats Membres d'un Etat fédéral: le cas de la République Fédérale d'Allemagne, in: Bothe/Vinuesa (Hrsg.), International Law and Municipal Law (1982), S. 207–212 (209).

weise in dem Sinne verstanden, daß das Gericht von einer Ausschließlichkeit des staatsrechtlichen Bandes zwischen den Ländern ausging[19]. Das Bundesverfassungsgericht hat sich im *Coburg*-Urteil 1973 eindeutig im Sinne einer Nichtgeltung von Völkerrecht ausgesprochen, hat dabei aber durchaus völkerrechtliche Normen (die *clausula rebus sic stantibus*) als Entscheidungsmaßstab herangezogen[20]. Die klare Aussage des *Coburg*-Urteils wurde allerdings später im Urteil über den Grundlagenvertrag wieder relativiert, eine der vielen Ungereimtheiten des letztgenannten Urteils[21].

Auch in der Schweiz geht jedenfalls die ältere Lehre von einer Gemengelage von Staats- und Völkerrecht aus[22]. In neueren Abhandlungen über das interkantonale Recht findet sich jedoch die These in dieser Form nicht mehr[23]. Was den amerikanischen Bundesstaat angeht, so ist die Bewertung der einschlägigen Entscheidungen des Supreme Court aus dem letzten und der ersten Hälfte dieses Jahrhunderts unklar und umstritten[24]. Man wird der neuesten Analyse hierzu von *Bleckmann* wohl insoweit zustimmen müssen, daß gewisse *dicta*, die die Annahme einer Geltung von Völkerrecht nahelegen, nicht eindeutig sind und zum Nachweis dieser These letztlich nicht ausreichen. Die Formel für die Bezeichnung der zwischengliedstaatlichen Rechtsordnung, die sich schließlich durchgesetzt hat, ist die des *interstate common law*[25], ein Ausdruck, der sehr plastisch die Besonderheiten dieser Rechtsbeziehungen zwischen den Gliedstaaten widerspiegelt: als *interstate law* ist er dem Völkerrecht verwandt, ist jedoch formal nicht oder nicht notwendigerweise zum Völkerrecht, sondern eher zum staatlichen Recht zu zählen.

---

[19] In diesem Sinne interpretiert offenbar das Bundesverfassungsgericht den Staatsgerichtshof, Entscheidungen des Bundesverfassungsgerichts (BVerfGE) 34, 216 (231). So auch *Schneider*, Verträge zwischen Gliedstaaten im Bundesstaat, Veröffentlichungen der Vereinigung der Deutschen Staatsrechtslehrer (VVDStRL) Heft 19, S. 1–85 (14); *Rill*, Gliedstaatsverträge (1972), S. 150.

[20] BVerfGE 34, 216.

[21] Neue Juristische Wochenschrift (NJW) 1973, S. 1539. Kritisch *Bleckmann* (Anm. 4), S. 11 Anm. 2.

[22] *Bolle* (Anm. 4), S. 2 ff., unter Bezugnahme auf *Tripel*; *Huber*, The International Law of Switzerland (Swiss Interstate Law), AJIL 3 (1909), S. 62–98 (65 f.).

[23] Vgl. etwa die vorsichtigen Ausführungen von *Schaumann*, Verträge zwischen Gliedstaaten in Bundesstaaten, VVDStRL Heft 19, S. 86–125 (119 ff.), ferner *Kehrli*, Interkantonales Konkordatsrecht (1968), S. 121.

[24] *Bleckmann* (Anm. 4), S. 12 mit weiteren Nachweisen. Vgl. dazu *Müller* (Anm. 4), S. 42 ff.

[25] So die Formel des Supreme Court in *Kansas* v. *Colorado*, 27 S.Ct. 655 (1907), S. 667. Vgl. dazu *Caldwell*, The Settlement of Interstate Disputes, AJIL 14 (1920), S. 38–69 (66 ff.); *Scott*, Kansas v. Colorado Revisited, AJIL 52 (1958), S. 432–454; *Müller* (Anm. 4), S. 47.

III. Der »Analogieschluß« vom Völkerrecht zum Bundesstaatsrecht und umgekehrt

1. Die Übertragung von Regelungen einer Rechtsordnung in die andere

Es geht bei den hier erörterten Fragen um die Füllung von Lücken einer Rechtsordnung durch Übernahme von Problemlösungen aus anderen Rechtsordnungen.

Diese Übertragung von Problemlösungen einer Rechtsordnung in eine andere, die in den oben erörterten Fällen als Analogieschluß bezeichnet wird, ist sicherlich kein klassischer Fall des Analogieschlusses, wie er in der juristischen Methodenlehre behandelt wird, nämlich die Übertragung der Lösung eines ausdrücklich geregelten Falles auf einen nicht ausdrücklich geregelten innerhalb einer Rechtsordnung[26]. Solche Analogieschlüsse werden gelegentlich mit dem in den ausdrücklichen Regelungen zutage getretenen Gemeinwillen gerechtfertigt[27]. Eine Übertragung von Lösungen aus einer Rechtsordnung in die andere trägt *diese* Rechtfertigung selbstverständlich nicht. Lückenfüllung, insbesondere solche durch richterliche Rechtsfortbildung, ist aber nicht auf solche Ableitungen aus einer wie auch immer konstruierbaren *volonté générale* bestimmter Rechtsgemeinschaften beschränkt. Rechtsfortbildung und Lückenfüllung durch Herausarbeiten von typischen Interessenkonstellationen, von Falltypen, denen gewisse innere Sachgesetzlichkeiten zukommen, ist eine notwendige Aufgabe des Rechtsanwenders[28]. Ob man dieses Verfahren nun in jedem Fall Analogieschluß nennt, ist eine eher zweitrangige terminologische Frage[29]. Anschauungsmaterial für dieses Verfahren zu bieten, ist eine wesentliche Aufgabe der Rechtsvergleichung[30]. In der Praxis vieler Staaten ist es durchaus üblich, daß die Gerichte Lösungen, die sie bei der Analyse der Rechtsordnungen anderer Staaten vorfinden, für die Entscheidung entsprechender Fragen der eigenen Rechtsordnung verwenden[31], und zwar nicht nur dort, wo die fremde Rechtsordnung aus historischen Gründen

---

[26] *Larenz*, Methodenlehre der Rechtswissenschaft (3. Aufl. 1975), S. 366 ff.; *Enneccerus/Nipperdey*, Allgemeiner Teil des bürgerlichen Rechts (15. Aufl. 1959), 1. Halbbd., S. 339 ff.; *Engisch*, Einführung in das juristische Denken (5. Aufl. 1956), S. 142 ff.

[27] *Enneccerus/Nipperdey*, a.a.O., S. 341.

[28] *Enneccerus/Nipperdey*, a.a.O., S. 343.

[29] In der deutschen juristischen Methodenlehre erscheint der Analogieschluß eher unter der gesetzesimmanenten Rechtsfortbildung, stellt also nur einen sehr begrenzten Ausschnitt der Lösungsmöglichkeiten bei Feststellung von Regelungslücken dar. Vgl. *Enneccerus/Nipperdey*, a.a.O., S. 342; *Larenz* (Anm. 26), S. 366 ff.

[30] *Enneccerus/Nipperdey*, a.a.O., S. 343.

[31] Aus dem hier noch zu erörternden Bereich des Bundesstaatsrechts vgl. etwa Entscheidungen des Reichsgerichts in Zivilsachen 112, Anhang, S. 28 (Bezugnahme des Staatsgerichtshofs für das Deutsche Reich auf das schweizerische Bundesgericht).

in der eigenen ganz oder teilweise, manchmal auch nur vermeintlich, rezipiert worden ist, so beim Fortgelten und Fortwirken englischen Rechts in USA[32], bei der Übernahme (?) nordamerikanischen verfassungsrechtlichen Gedankenguts durch die lateinamerikanischen Staaten[33], beim »Export« der privatrechtlichen Kodifikationen Europas[34]. In aller Regel spricht man allerdings bei einer solchen Fortentwicklung des Rechts unter Berücksichtigung ausländischer Modelle nicht von Analogie, der gedankliche Vorgang ist aber jedenfalls ähnlich. Es wird je ein Lebenssachverhalt, je eine bestimmte gesellschaftliche Situation in der einen und in der anderen Rechtsordnung (bei der klassischen Gesetzesanalogie: im geregelten und im nicht geregelten Lebensbereich) darauf untersucht, ob sie genügend gemeinsame Merkmale aufweisen, um die Regelung des einen Bereichs auch für den anderen angemessen erscheinen zu lassen. Dieses Problem der Übertragbarkeit ist ein zentrales Problem der praktischen Rechtsvergleichung, es existiert aber ähnlich auch bei der Übertragung von Prinzipien des innerstaatlichen Rechts in das Völkerrecht als allgemeine Rechtsgrundsätze im Sinne von Art. 38 Abs. 1 (c) IGH-Statut[35].

Um eine solche Übertragung von Lösungen aus einer Rechtsordnung zum Zwecke der Lückenfüllung und Rechtsfortbildung in die andere geht es auch bei den hier untersuchten Fällen der Anwendung von Völkerrecht im Bundesstaat[36] und von Bundesstaatsrecht in den internationalen Beziehungen.

2. Die Übertragbarkeit von Lösungen aus einer Rechtsordnung in die andere

a) Die Staatsqualität von Gliedstaaten eines Bundesstaates

Eine wesentliche Voraussetzung der Übertragung rechtlicher Lösungen aus einer Rechtsordnung in die andere ist die Vergleichbarkeit des geregelten Sachverhalts.

---

[32] *Blumenwitz,* Einführung in das Anglo-Amerikanische Recht (2. Aufl. 1976), S. 18 ff.

[33] Vgl. *Sato Cárdenas,* Influencia de la Independencia de los Estados Unidos en la Constitución de las Naciones Latinoamericanas (1979); vgl. dazu aus der Rechtsprechung der argentinischen Corte Suprema etwa den Fall *S. A. Merck Química Argentina* c. *Nación Argentina,* Fallos 211, S. 162 ff., abgedruckt bei Bothe/Vinuesa (Hrsg.) (Anm. 18), S. 231 ff. (234 f.).

[34] Zur Rezeption des Code civil *Zweigert/Kötz,* Einführung in die Rechtsvergleichung auf dem Gebiete des Privatrechts, Bd. 1 (1971), S. 106 ff., zur internationalen Ausstrahlung der deutschen Zivilistik und das BGB sowie des schweizerischen ZGB, ebenda, S. 178 ff., 185 ff.

[35] *Hailbronner,* Ziele und Methoden völkerrechtlich relevanter Rechtsvergleichung, ZaöRV 36 (1976), S. 190–226 (205 ff.); *Bothe,* Die Bedeutung der Rechtsvergleichung in der Praxis internationaler Gerichte, ebenda, S. 280–299 (294ff.).

[36] *Rill,* Gliedstaatsverträge (1972), S. 155 ff.

Dies sei für die hier zur Debatte stehende Frage Bundesstaatsrecht – Völkerrecht an vier Problembereichen gezeigt: die Abgrenzung des Staatsgebiets, das Recht grenzüberschreitender Umweltbelastungen, die Einschränkung von Handelshindernissen und die Geltung bzw. Bestandskraft von Verträgen.

Ein *tertium comparationis*, das für die Übertragbarkeit rechtlicher Regelungen in die eine oder andere Richtung von ausschlaggebender Bedeutung ist, ist, so banal es klingen mag, die Staatsqualität der Gliedstaaten eines Bundesstaates[37]. Sie sind zwar keine Staaten im Sinne des Völkerrechts, d.h. keine geborenen Völkerrechtssubjekte[38], da sie, trotz der gegenteiligen verbalen Beteuerung der schweizerischen Bundesverfassung[39], nicht souverän, nicht unabhängig werden. Dennoch ist ihre Bezeichnung als Staaten ja durchaus sinnvoll.

Von den vier Elementen der völkerrechtlichen Staatsdefinition besitzen sie jedenfalls zwei (Staatsgebiet und Staatsgewalt), in unterschiedlichem Maße auch ein drittes, das Staatsvolk[40]. Sie haben eine – vorbehaltlich der Befugnisse des Gesamtstaates – umfassende und ausschließliche Regelungsbefugnis für ihr Gebiet, die in aller Regel auch so viel politisches Gewicht hat, daß sich der Gliedstaat qualitativ von einer lokalen Gebietskörperschaft unterscheidet. Deswegen können zwischen den Gliedstaaten rechtlich zu regelnde Interessenkonflikte auftreten, die denen zwischen souveränen Staaten gleichen.

Dies wird von den Gerichten auch deutlich so gesehen. Das schweizerische Bundesgericht stellt in seiner Entscheidung *Aargau* gegen *Solothurn*[41] darauf ab, daß sich die beiden Kantone »als staatliche Organismen«, nicht wie Privatpersonen gegenübertreten. Der Begriff der Souveränität (im Sinne der schweizerischen Bundesverfassung) und Gebietshoheit bedeute jedenfalls für das Verhältnis nach außen, gegenüber anderen Staaten, daß der Kanton »bestimmte absolute, von den anderen Staaten unbedingt anzuerkennende Rechte« besitze, »darunter namentlich das Recht der unbeschränkten Herrschaft über Land und Leute«. Dies erscheint in der Entscheidung zwar als Folge, ist aber auch Voraussetzung der Anwendung von Völkerrecht, hier der Anwendung von Grundsätzen des völkerrechtlichen Nachbarrechts, auf die sich das Bundesgericht zur Begründung seiner Entscheidung (Verbot einer grenzüberschreitenden Immission) ausdrücklich beruft.

---

[37] Vgl. *Schaumann*, Verträge zwischen Gliedstaaten im Bundesstaat, VVDStRL Heft 19, S. 86–125 (124 f.).

[38] Zur Unterscheidung zwischen ursprünglichen und abgeleiteten Völkerrechtssubjekten vgl. *Mosler*, Die Erweiterung des Kreises der Völkerrechtssubjekte, ZaöRV 22 (1962), S. 1–48 (22 ff.).

[39] Art. 3 Bundesverfassung.

[40] Die Problematik der Landesangehörigkeit kann hier nicht eingehender erörtert werden.

[41] Entscheidungen des Schweizerischen Bundesgerichts (BGE) 26 I, 444 (1900), S. 449 ff.

In den USA wird vom Supreme Court bei Entscheidungen von Streitigkeiten zwischen Gliedstaaten immer wieder betont, daß solche Streitigkeiten eher denen zwischen souveränen Staaten als solchen zwischen Privatpersonen gleichen:

> »It may be imagined that a nuisance might be created by a state upon a navigable river like the Danube, which would amount to a *casus belli* for a state lower down, unless removed« [42].
> »If the two states were absolutely independent nations (the dispute) would be settled by treaty or by force« [43].

Solche Streitigkeiten werden von denen zwischen Individuen unterschieden:

> »[I]t does not follow that every matter which would warrant a resort to equity to one citizen against another in the same jurisdiction would warrant an interference by this court with the action of a state« [44].
> »The states, by entering the Union, did not sink to the position of private owners, subject to one system of private law. This court has not quite the same freedom to balance the harm that will be done by an injunction against that of which the plaintiff complains, that it would have in deciding between two subjects of a single political power« [45].

Die Gliedstaaten werden als »quasi-sovereign« bezeichnet [46], sie besitzen »full powers of local sovereignty« [47], die Handlung eines Staates wird als »reasonable exercise of its sovereignty« bezeichnet [48].

Die Folgerungen, die der Supreme Court für das anwendbare Recht aus dieser Tatsache zieht, sind allerdings nicht ganz eindeutig [49]. Während sich einige Hinweise auf Völkerrecht finden, scheinen es letztlich doch mehr allgemeine Grundsätze der *equity* zu sein, die den Entscheidungen zugrunde liegen [50]. Es sind aber eben dieselben Fälle, die das Schiedsgericht im *Trail Smel-*

---

[42] *Missouri* v. *Illinois*, 26 S.Ct. 268 (1906), S. 269.

[43] *Kansas* v. *Colorado*, 27 S.Ct. 655 (1907), S. 668.

[44] *Missouri* v. *Illinois* (Anm. 42), S. 270.

[45] *Georgia* v. *Tennessee Copper*, 27 S.Ct. 618 (1907), 619.

[46] *Georgia* v. *Tennessee Copper Co.*, a.a.O., S. 619; *North Dakota* v. *Minnesota*, 44 S.Ct. 138 (1923), S. 139.

[47] *Kansas* v. *Colorado* (Anm. 43), S. 666.

[48] *Kansas* v. *Colorado*, a.a.O., S. 669.

[49] Viel zitiert ist vor allem der Satz aus *Kansas* v. *Colorado*, a.a.O., S. 656: »Sitting, as it were, as an international as well as a domestic tribunal, we apply Federal law, state law, and international law, as the exigencies of the particular case may demand . . .«.

[50] Insoweit zutreffend *Bleckmann* (Anm. 4), S. 32 f. Teilweise ist *Bleckmanns* Argumentation allerdings unhaltbar, so wenn er meint, der Supreme Court habe im Jahre 1907 schon deshalb kein völkerrechtliches Wasserrecht anwenden können, weil es das 1907 noch gar nicht gab!

ter-Fall dann als Basis seines »Analogieschlusses« für den Nachweis von Völkerrecht anführt [51].

Der Rückgriff auf Regeln des innerstaatlichen Rechts zum Nachweis allgemeiner Rechtsgrundsätze im Sinne der Art. 38 Abs. 1 (c) IGH-Statut ist ein dem Völkerrechtler geläufiger Vorgang. Dabei handelt es sich meist um Regeln des Privatrechts, sogenannte *private law analogies* [52]. Das entscheidende Problem ist dabei die Frage, ob die mittels solcher »Analogien« zu regelnden Fragen der internationalen Beziehungen privatrechtlichen Beziehungen vergleichbar sind, d. h. ob die privatrechtliche Regelung sinnvoll auf die internationale Ebene übertragbar ist [53]. Gerade dieses Problem stellt sich bei den hier untersuchten Fällen der innerstaatlichen Regelung des Verhältnisses zwischen Gliedstaaten eines Bundesstaates nicht in gleichem Maße. Hier ist die geregelte Frage der internationalen eher vergleichbar, das Ergebnis eher übertragbar [54]. Die Problematik für Art. 38 Abs. 1 (c) IGH-Statut liegt, wie bereits erwähnt, in der mangelnden »Allgemeinheit« des Rechtssatzes.

b) Einzelne Fallgruppen

Diese allgemeinen Erwägungen zur Übertragbarkeit rechtlicher Lösungen zwischen bundesstaatlichen Ordnungen und internationalen Beziehungen besagen aber noch nichts Endgültiges für die Vergleichbarkeit der zu regelnden Rechtsverhältnisse bei einzelnen Fallgruppen. Darauf ist nun noch näher einzugehen.

aa) Grenzfragen

Völkerrechtliche Grundsätze sind für die Grenzregelung zwischen Gliedstaaten eines Bundesstaates in zwei verschiedenen Zusammenhängen bedeutsam geworden. Einmal geht es um allgemeine Grundsätze der Gebietsabgrenzung, wobei sich insbesondere die Frage der Übertragbarkeit zivilrechtlicher Grundsätze über Eigentumserwerb an Grundstücken auf staats- bzw. völkerrechtlichen Gebietserwerb stellt. Hier hat das Bundesstaatsrecht die gleiche Frage zu lösen wie das Völkerrecht. So führt der amerikanische Supreme Court in einem Grenzstreit zwischen Michigan und Wisconsin aus [55]:

---

[51] RIAA Bd. 3, insbes. S. 1964 f. An der Übertragbarkeit der Rechtsprechung des Supreme Court auf die Ebene des Völkerrechts zweifelt jedoch *Scott* (Anm. 25), S. 453.

[52] Grundlegend dazu *H. Lauterpacht*, Private Law Sources and Analogies of International Law (1927; unveränderter Neudruck 1970).

[53] *Lauterpacht*, a.a.O., S. 71 ff.; *Bothe* (Anm. 13), S. 296, 298 mit weiteren Nachweisen; *Hailbronner*, Ziele und Methoden völkerrechtlich relevanter Rechtsvergleichung, ZaöRV 36 (1976), S. 190–226 (211).

[54] *Rill* (Anm. 36), S. 158 f.

[55] *Michigan* v. *Wisconsin*, 46 S.Ct. 290 (1926), S. 294.

»That rights of the character here claimed may be acquired, on the one hand and lost on the other, by open, long-continued, and uninterrupted possession of territory, is a doctrine not confined to individuals, but applicable to sovereign nations as well (folgen Nachweise aus der Literatur zum Völkerrecht) and, a fortiori to the quasi sovereign states of the Union«.

Die Grundsätze, um die es hier geht, sind etwa die der Ersitzung, *acquiescence, prescription*[56], Anwachsung und Abwachsung (*avulsion, accretion*[57]), aber auch Fragen der Auslegung von Grenzverträgen[58].

Ein ganz anderer Zusammenhang, in dem Völkerrecht für die Grenzziehung zwischen Gliedstaaten eines Bundesstaates von Bedeutung werden kann, ist das Prinzip der geschichtlich hergebrachten Grenzen. Zur Bestimmung der Grenzen zwischen Gliedstaaten muß unter Umständen auf die historische Entwicklung dieser Grenzen zu einer Zeit zurückgegriffen werden, zu der diese Staaten noch unabhängige, souveräne Staaten waren[59]. Dabei sind dann die zu der fraglichen Zeit geltenden Völkerrechtsnormen zu beachten. Dies ist z.B. der Grund, weshalb der Supreme Court das im Völkerrecht geltende Thalweg-Prinzip auf die Grenze zwischen New Jersey and Delaware anwendet[60].

Keine Frage der Grenzziehung, sondern eine solche der grenzüberschreitenden Ausübung bestimmter Hoheitsbefugnisse sind sogenannte Staatsservituten. Das schweizerische Bundesgericht hat die Möglichkeit solcher Staatsservituten im Verhältnis zwischen den Kantonen unter Berufung auf völkerrechtliche Literatur bejaht[61].

bb) Grenzüberschreitende Immissionen und Ressourcennutzung

Die zweite, bereits eingehend behandelte Fallgruppe ist diejenige der grenzüberschreitenden Immissionen und Ressourcennutzung (vor allem Was-

---

[56] Vgl. auch *Oklahoma* v. *Texas*, 47 S.Ct. 9 (1926); *Virginia* v. *Tennessee*, 13 S.Ct. 728 (1893), für einen Fall der Nichtübertragung zivilrechtlicher Grundsätze auf das Verhältnis zwischen Staaten vgl. jedoch *Rhode Island* v. *Massachusetts*, 15 Pet. 233 (1841), S. 272 f.

[57] *Nebraska* v. *Iowa*, 12 S.Ct. 396 (1892), insbes. S. 397: »These propositions, which are universally recognized as correct where the boundaries of private property touch on streams, are in like manner recognized where the boundaries between states or nations are ... found in running water.« Zum Nachweis der fraglichen Prinzipien bezieht sich der Supreme Court einmal auf völkerrechtliche Literatur, verfolgt die Prinzipien aber zurück auf ihre Ursprünge im römischen Recht und weist ihre Geltung im englischen und amerikanischen Recht nach.

[58] *Alabama* v. *Georgia*, 23 How. 505 (1860). Der Supreme Court bezieht sich hier zu Auslegungsfragen eingehend auf *Vattel*, aber auch auf *Grotius*.

[59] Vgl. aus der Rechtsprechung des schweizerischen Bundesgerichts: *Schaffhausen* gegen *Zürich*, BGE 23. II, 1405; *Zürich* gegen *Schaffhausen*, BGE 33. I, 537; *Graubünden* gegen *Tessin*, BGE 18, 673.

[60] *New Jersey* v. *Delaware*, 54 S.Ct. 407 (1933).

[61] *Luzern* gegen *Aargau*, BGE 8, 43 (56).

sernutzung) mit grenzüberschreitender Wirkung. Die Ähnlichkeit der Fallkonstellationen liegt hier auf der Hand. Die Behandlung unter Zugrundelegung privatrechtlicher Regeln des Immissionsschutzes (im angelsächsischen Bereich *trespass* und *nuisance*) wird hier von den Gerichten nicht als angemessen angesehen, jedenfalls nicht ohne Modifikationen. Die amerikanischen Gerichte haben immer wieder betont, daß hier Grundsätze des *common law* für das Verhältnis zwischen Privatpersonen nicht unverändert übernommen werden könnten [62]. Das schweizerische Bundesgericht und der Staatsgerichtshof für das Deutsche Reich haben bei solchen Fällen Völkerrecht angewandt [63]. Das ist der Bereich, in dem das Recht der Bundesstaaten die Völkerrechtsentwicklung stark beeinflußt hat und noch beeinflußt [64].

Die moderne Entwicklung dieses Fragenkreises in bundesstaatlichen Ordnungen ist sehr stark geprägt von Gedanken der Kooperation. Es gibt in Bundesstaaten eine umfangreiche Vertragspraxis der gemeinschaftlichen Verwaltung grenzüberschreitender Wasserressourcen und der Eindämmung grenzüberschreitender Umweltbelastungen durch grenzüberschreitende Planung [65]. Während in den USA diese Praxis offenbar eher als eine Frage der Opportunität angesehen wird [66], spielt in der Bundesrepublik Deutschland hier zunehmend der Gedanke einer aus dem Grundsatz der Bundestreue folgenden Kooperationspflicht eine Rolle [67]. Anderseits sind im internationalen Bereich Fortschritte in bezug auf eine allgemeine Anerkennung solcher Pflichten zu Kooperation und Mitwirkung an gemeinsamer Ressourcenverwaltung [68] (die über die wohl anerkannte Pflicht zu Information

---

[62] Siehe oben Text zu Anm. 42–45.

[63] *Württemberg und Preußen ./. Baden*, RGZ 116, Anhang S. 18, 27 ff. (Donauversinkung); *Bremen ./. Preußen, Thüringen und Braunschweig*, RGZ 121, Anhang, S. 1, 6 (Weserverschmutzung).

[64] Siehe oben Text zu Anm. 8–11. Zurückhaltend insofern *Scott* (Anm. 25), S. 453.

[65] Nachweise bei *Bothe* (Anm. 7), S. 90 f.

[66] Vgl. etwa die Ausführungen des Supreme Court in *New York* v. *New Jersey*, 41 S.Ct. 492 (1918), S. 498: »We cannot withhold the *suggestion*, inspired by the consideration of this case, that the grave problem of sewage disposal presented by the large and growing populations living on the shores of New York Bay is one *more likely* to be *wisely solved* by co-operative study and by conferences and mutual concession on the part of representatives of the states so vitally interested in it than by proceedings in any court however constituted« (Hervorhebung des Verf.). Vgl. ferner *Washington* v. *Oregon*, 29 S.Ct. 631 (1909), S. 633; *Minnesota* v. *Wisconsin*, 40 S.Ct. 313 (1920), S. 319; *Hinderlider* v. *La Plata*, 58 S.Ct. 803 (1938), S. 808 f.

[67] Vgl. etwa BVerfGE 1, 117 (131); 33, 165 (194).

[68] Das allgemein genannte Vorbild für solche Regelung ist das Regime der Great Lakes. Weitere Nachweise bei *Bothe* (Anm. 7), S. 91 f.; *derselbe*, The Trends in Both National and International Politics for Achieving a Unification of Standards in Pollution Matters, Zeitschrift für Umweltpolitik 1979, S. 293–312 (299 ff.); *derselbe*, Transfrontier Environmental Management, in: Bothe (Hrsg.), Trends in Environmental Policy and Law (1980), S. 391–401.

und Konsultation hinausgehen) nur sehr zögerlich[69]. Der UNEP-Bericht über »Shared Natural Resources«[70] ist mit seinen diesbezüglichen Formulierungen sehr zurückhaltend.

Dies führt zu der Frage, wieweit die geschilderte innerstaatliche Praxis zur Begründung internationaler Kooperationspflichten herangezogen werden kann. Es stellt sich damit die Frage, inwieweit die rechtliche Einbindung der Gliedstaaten in ein gesamtstaatliches Gemeininteresse nicht eine unterschiedliche Lösung dieser Fragen im Bundesstaatsrecht und im Völkerrecht erfordert[71]. Unter diesem Gesichtspunkt ist schon die zitierte Rechtsprechung des Staatsgerichtshofs kritisiert[72], sind Zweifel an der Übertragbarkeit der wasserrechtlichen Entscheidungen des Supreme Court zum *equitable apportionment* in das Völkerrecht geäußert worden[73]. Darauf wird noch zurückzukommen sein[74].

cc) Handelsfreiheit und staatlicher Rechtsgüterschutz

Ein anderer einschlägiger Fragenkreis, der allerdings in der Debatte um die hier untersuchten Fragen bislang kaum Beachtung gefunden hat, ist das Spannungsverhältnis von Regeln über Handelsfreiheit auf der einen und Schutz anderer Rechtsgüter durch (Glied-)Staaten auf der anderen Seite. Es geht um die Frage, inwieweit (glied)staatliche Regelungen zulässig sind, die negative (Neben?)Wirkungen auf die rechtlich gebotene Freiheit des Handels zwischen den (Glied-)Staaten haben. Diese Frage stellt sich (wie zu zeigen sein wird: in vergleichbarer Weise)[75] in einigen Bundesstaaten, in den Europäischen Gemeinschaften und weltweit im GATT.

Die Freiheit des Handels zwischen den amerikanischen Staaten zu sichern, ist eines der Grundanliegen der amerikanischen Bundesverfassung[76]. Die Tatsache, daß dieses Anliegen in einer Vorschrift über Bundeskompetenzen, der sogenannten *commerce clause*, formuliert ist, hat zu einigen juristischen Schwierigkeiten geführt. Es ist aber jedenfalls seit langem unbestritten, daß aus der *commerce clause* auch ein Verbot gliedstaatlicher Maßnahmen folgt, die den Handel zwischen den Gliedstaaten über Gebühr beeinträchtigen. Umstritten ist bis heute immer wieder die Grenze zwischen zulässiger

---

[69] Vgl. dazu auch *Riphagen*, The International Concern for the Environment as Expressed in the Concept of the »Common Heritage of Mankind« and of »Shared Natural Resources«, in: Bothe (Hrsg.) (Anm. 68), S. 343–362.

[70] International Legal Materials (ILM) 17 (1978), S. 1098.

[71] Vgl. *Krüger* (Anm. 4), S. 245.

[72] *Müller* (Anm. 4), S. 95 ff.

[73] *Scott* (Anm. 25), S. 453.

[74] Siehe unten IV. 2. d.

[75] Vgl. dazu neuerdings *T. Sandalow/E. Stein* (Hrsg.), Courts and Free Markets, 2 Bde. (Oxford University Press, 1982).

[76] *Lenaerts*, Distribution of Powers in American Federalism, Jahrbuch für öffentliches Recht 30 (1981), S. 569 ff. mit weiteren Nachweisen.

und unzulässiger Beeinträchtigung des Handels im Einzelfall[77]. Mit durchaus wechselnden Begründungen war es hier die Tendenz der Rechtsprechung, gliedstaatliche Maßnahmen, die die Gerichte, an ihrer Spitze der Supreme Court, als eine vernünftige (*reasonable*) Regelung zum Schutz legitimer gliedstaatlicher Interessen ansahen, als nicht verboten anzusehen; gliedstaatliche Maßnahmen, die in den Augen der Richter diesem Test nicht standhielten, wurden als verfassungswidrig angesehen. Dieser Grundgedanke wurde in der Rechtsprechung verfeinert. Von besonderem Interesse sind hier zwei Ansätze: Nichtdiskriminierung und Interessenabwägung. Ein wesentlicher Test der Zulässigkeit gliedstaatlicher Maßnahmen ist, daß sie nicht den zwischengliedstaatlichen Handel diskriminieren, daß sie *interstate* und *intrastate commerce* in gleicher Weise treffen[78]. Dabei wird auch *disguised discrimination* geprüft. Ist eine gliedstaatliche Regelung nicht diskriminierend, werden gliedstaatliche Regelungsinteressen gegen das Bundesinteresse an der Freiheit des Handels zwischen den Staaten abgewogen. Als Beispiel sei eine Entscheidung zitiert, die eine Regelung des Staates Arizona über die Höchstlänge von Eisenbahnzügen betraf (vorgeblich zum Zwecke der Einschränkung von Unfallgefahren, in Wahrheit wohl eine gewerkschaftlich veranlaßte »Arbeitsbeschaffungsmaßnahme«, da mehr und kürzere Züge mehr Personal erfordern)[79]:

> »Here we conclude that the state goes too far. Its regulation of train length, admittedly obstructive to interstate train operation, and having a seriously adverse effect on transportation efficiency and economy, *passes beyond what is plainly essential for safety* since it does not appear that it will lessen rather than increase the danger of accident. Its attempted regulation of interstate trains cannot establish nation-wide control such as is essential to the maintenance of an efficient transportation system, which Congress alone can prescribe. The state interest cannot be preserved at the expense of the national interest by an enactment which regulates interstate train lengths without securing such control, which is a *matter of national concern. To this the interest of the state here asserted is subordinate*« (Hervorhebung vom Verf.).

Der wesentliche Punkt der Interessenabwägung ist hier offenbar der Gedanke der Erforderlichkeit. Wenn der Staat auch nicht gehindert ist, Maßnahmen zum Schutze ihm schutzwürdig erscheinender Rechtsgüter zu ergreifen, so darf er dabei den freien Handel jedenfalls nicht mehr beeinträchtigen, als es zum Schutze dieser Rechtsgüter unbedingt erforderlich ist. Das Gericht

---

77 Die Entwicklung ist im einzelnen dargestellt bei *Lenaerts*, a.a.O.

78 Vgl. aus der neueren Rechtsprechung *Minnesota* v. *Clover Leaf Creamery Co.*, 101 S.Ct. 715 (1981), S. 728: »This statute is therefore unlike statutes discriminating against interstate commerce, which we have consistently struck down« (folgen Nachweise).

79 *Southern Pacific Co.* v. *Arizona*, 65 S.Ct. 1515 (1945).

prüft also auch, ob weniger belastende andere Maßnahmen (*less restrictive alternatives*) zum Schutze dieser Rechtsgüter möglich sind[80].

Zu den Gedanken der Erforderlichkeit kommt der der Verhältnismäßigkeit. Eine den zwischenstaatlichen Handel belastende Maßnahme ist auch dann verfassungswidrig, wenn ihre Schädlichkeit für den Handel außer Verhältnis zu dem Nutzen für das zu schützende Rechtsgut steht:

> ». . . the controlling question is whether the incidental burden imposed on interstate commerce . . . is clearly excessive in relation to the putative local benefits«[81].

Diese Argumentationsmuster des verfassungsrechtlichen Ausgleichs beim Schutz konfligierender Rechtsgüter, die der Supreme Court übrigens auch im Bereich des Grundrechtsschutzes verwendet[82], sind dem deutschen Verfassungsjuristen vor allem aus dem zuletzt genannten Bereich vertraut[83]. Sie finden sich aber auch in der Rechtsprechung des Gerichtshofs der Europäischen Gemeinschaften (EuGH), und zwar gerade bei Fragen des Verhältnisses von Handelsfreiheit und mitgliedstaatlichem Rechtsgüterschutz.

Der EWG-Vertrag verbietet nicht nur die klassischen und offenkundigen Handelsbarrieren, Zölle und mengenmäßigen Beschränkungen; nach Art. 30 EWG-Vertrag sind auch verboten »Maßnahmen mit gleicher Wirkung wie mengenmäßige Beschränkungen«. Dies sind nach der Definition des EuGH, nunmehr ständige Rechtsprechung, alle staatlichen Maßnahmen, die sich mittelbar oder unmittelbar auf den Handel zwischen den Mitgliedstaaten auswirken[84]. Damit ist Art. 30 eine entscheidende Garantie der Freiheit des Handels zwischen den Mitgliedstaaten. Allerdings macht diese Garantie den Schutz wichtiger Rechtsgüter durch die Mitgliedstaaten nicht unmöglich. Art. 36 sieht Ausnahmen zum Schutze bestimmter Rechtsgüter ausdrücklich vor. Offenbar ist der EuGH nicht der Auffassung, daß der Katalog der in Art. 36 genannten Rechtsgüter abschließend ist, daß staatliche Maßnahmen zum Schutze anderer als der in Art. 36 genannten Rechtsgüter stets unter das Verbot des Art. 30 fallen. Ohne Art. 36 zu erwähnen, hat der EuGH mehrfach Rechtsgüter als mögliche Schutzobjekte staatlicher Regelungen genannt, die nicht ausdrücklich in Art. 36 erwähnt werden[85], es sei denn, man subsumiert sie unter dem umfassenden Begriff der öffentlichen Ordnung und Sicherheit. Unzulässig sind solche Maßnahmen allerdings dann, wenn sie den Handel zwischen den Mitgliedstaaten (offen oder versteckt) diskriminieren[86].

---

[80] *Dean Milk Co.* v. *City of Madison*, 71 S.Ct. 295 (1951).

[81] *Minnesota* v. *Clover Leaf Creamery Co.*, 101 S.Ct. 715 (1981), S. 728.

[82] Vgl. die Nachweise bei *Lenaerts* (Anm. 76), S. 600 f.

[83] *Hesse*, Grundzüge des Verfassungsrechts der Bundesrepublik Deutschland (13. Aufl. 1981), S. 127 f.

[84] Grundlegend das Urteil *Dassonville*, Sammlung der Rechtsprechung des Gerichtshofes der Europäischen Gemeinschaften (EuGHE) 1974, 837.

[85] EuGHE 1979, 649 (*Rewe*), 1980, 2071, 2078 (*Gilli und Andres*).

Ferner wird verlangt, daß die staatliche Norm »notwendig ist, um zwingenden Erfordernissen in bezug auf (die zu schützenden Rechtsgüter) gerecht zu werden«[87]. Das Verhältnismäßigkeitsprinzip hat der EuGH in anderem Zusammenhang angewandt[88]. Es hätte sicherlich auch hier seinen Platz. Dies sind die gleichen Gedanken und Erwägungen, die auch für die dargestellte Rechtsprechung des Supreme Court bestimmend sind. Die gleichen Regelungsprinzipien finden sich nun auch in den einschlägigen Regeln des GATT wieder, Art. III, XX GATT, sowie des im Rahmen der Tokio-Runde ausgehandelten »Agreement on Technical Barriers to Trade«[89]. Art. III GATT schreibt Inländerbehandlung, mit anderen Worten Nichtdiskriminierung für gewisse staatliche handelsrelevante Regelungen vor, Art. XX läßt gewisse Ausnahmen für den Schutz bestimmter Rechtsgüter zu, wiederum allerdings mit dem Vorbehalt, daß diese kein Mittel zu einer »willkürlichen und ungerechtfertigten Diskriminierung« zwischen Ländern, bei denen gleiche Bedingungen vorliegen, oder zu einer verschleierten Beschränkung des internationalen Handels sind. Das Abkommen über nicht-tarifäre Handelshindernisse (Art. 2) schreibt unter anderem vor, daß in bezug auf technische Regeln und Standards Importwaren wie Waren inländischen Ursprungs zu behandeln sind (Nichtdiskriminierung im Sinne von Inländerbehandlung). Ferner müssen die Vertragsparteien sicherstellen, daß solche Regeln und Standards nicht die Wirkung haben, »unnötige« Behinderungen des internationalen Handels zu schaffen. Auch hier treffen wir also die bekannten Prinzipien: Verbot der Diskriminierung und Beschränkung staatlicher Regelungen auf das zum Rechtsgüterschutz Erforderliche.

Diese Parallelentwicklung von Bundesstaatsrecht, EG-Recht und Völkerrecht läßt den Schluß zu, daß es sich um eine Fallkonstellation handelt, in denen es auf den verschiedenen Ebenen eine vergleichbare Problemlage, vergleichbare Sachzwänge gibt. Dies legt es nahe, bei der Lösung von Einzelfragen in diesem Bereich auf einer Ebene die Erfahrungen und Präzedenzfälle der anderen Ebene zu berücksichtigen. Insbesondere könnte die noch am wenigsten durch Rechtsprechung ausdifferenzierte Rechtsordnung, die des GATT, hier *case law* des Supreme Court und des EuGH zur Lösung ihrer Fragen fruchtbar machen.

dd) Das Recht zwischenstaatlicher Verträge

Ein ganz anderer Bereich, wo die Frage des Schlusses von einer auf die andere Rechtsordnung zur Debatte steht, ist das Recht der Verträge. Das völker-

---

[86] Ein Beispiel für eine verschleierte Diskriminierung findet sich in EuGHE 1980, 2299 (*Kommission./. Französische Republik*, »Werbung für alkoholische Getränke«).

[87] EuGHE 1980, 2071 (2078).

[88] Vgl. das Urteil *Bela-Mühle*, Europäische Grundrechte-Zeitschrift (EuGRZ) 1977, S. 402.

[89] ILM 1979, S. 1079.

rechtliche Vertragsrecht mußte sich zunächst durch Rückgriff auf Konzepte des innerstaatlichen Vertragsrechts entwickeln. Das Grundkonzept des Vertrages als Schaffung rechtlicher Verpflichtungen durch übereinstimmende Willenserklärungen der Parteien ist dem Privatrecht und dem Völkerrecht gemeinsam [90]. Von daher waren die *private law analogies* für das völkerrechtliche Vertragsrecht durchaus angemessen. Das völkerrechtliche Vertragsrecht hat sich aber im Laufe der Zeit eigenständig entwickelt, da und insoweit diesbezüglich in den internationalen Beziehungen Probleme auftauchen, die sich bei privatrechtlichen Verträgen nicht oder nicht in gleicher Weise stellen: Problem des typischen zwischenstaatlichen Abschlußverfahrens, Vorbehalte bei multilateralen Verträgen, in gewissem Maße auch Willensmängel und Beendigungsgründe [91], schließlich die Fragen der innerstaatlichen Wirkung völkerrechtlicher Verträge. Für die hier untersuchte Frage des Schlusses von den Beziehungen zwischen Gliedstaaten eines Bundesstaates auf die zwischen souveränen Staaten und umgekehrt ist zunächst festzustellen, daß Verträge zwischen Gliedstaaten eines Bundesstaates, was z. B. Fragen des Abschlußverfahrens angeht, den völkerrechtlichen Verträgen ähnlicher sind als privatrechtlichen Kontrakten. Wir haben auch hier häufig mehrphasige Abschlußverfahren mit Beteiligung der staatlichen Parlamente. Auch bei den zwischengliedstaatlichen Verträgen stellt sich z. B. die Frage ihres Verhältnisses zum (späteren oder früheren) innerstaatlichen Gesetz. Dies ist der innere Grund dafür, daß Fragen des zwischengliedstaatlichen Vertragsrechts von den Gerichten häufig durch Rückgriff auf völkerrechtliche Normen gelöst wurden [92].

aaa) Vertragsauslegung

So sagt das schweizerische Bundesgericht noch in einer Entscheidung aus dem Jahre 1974 zur Auslegung sogenannter Konkordate:

> »Auf interkantonale Verträge sind, soweit nicht nach Bundesrecht, Gewohnheitsrecht oder Vereinbarung etwas anderes gilt, die Grundsätze des Völkerrechts anwendbar« [93].

Es stellt dann ab auf die *travaux préparatoires*, »soweit sie den Willen der vertragschließenden Kantone klar erkennen lassen« [94], sowie auf den Vertragszweck [95]. Verzicht auf Befugnisse sei einschränkend auszulegen [96].

---

[90] *Lauterpacht* (Anm. 52), S. 156 ff.

[91] Dazu *Lauterpacht* (Anm. 52), S. 161 ff.

[92] *Bothe* (Anm. 18), S. 210 ff.

[93] *Kuster u. a. ./. Regierungsrat des Kantons Schwyz*, BGE 100 Ia, 418 (423).

[94] Vgl. dazu *Bernhardt*, Die Auslegung völkerrechtlicher Verträge (Beiträge zum ausländischen öffentlichen Recht und Völkerrecht, Bd. 40) (1963), S. 109 ff.

[95] Dazu *Bernhardt*, a.a.O., S. 88 ff.

[96] Kritisch hierzu *Bernhardt*, a.a.O., S. 143 ff.

Auch vom Supreme Court werden bei der Auslegung innerbundesstaatlicher Verträge völkerrechtliche Grundsätze herangezogen [97].

bbb) Verträge, die gegen die gliedstaatliche Verfassung verstoßen

Bei der Frage der Mängel des Vertragsschlusses stellt sich insbesondere das Problem der völkerrechtlichen Wirkung verfassungswidriger Verträge bzw. der zwischengliedstaatlichen Wirkung landesverfassungswidriger Verträge, wobei zu unterscheiden ist zwischen der Verletzung von Kompetenzvorschriften beim Abschluß der Verträge und die Unvereinbarkeit des Vertragsinhalts mit einer Verfassungsnorm. Im Völkerrecht kann nunmehr wohl als herrschend bezeichnet werden die Evidenztheorie, eine modifizierte Irrelevanztheorie, wie sie in Art. 46 der Wiener Konvention über das Recht der Verträge von 1969 [98] ihren Ausdruck gefunden hat.

Das Bundesverwaltungsgericht hat in seiner Entscheidung zum Staatsvertrag über die Vergabe von Studienplätzen [99], den der Bayerische Verfassungsgerichtshof für landesverfassungswidrig erklärt hatte (ein Fall der inhaltlichen Unvereinbarkeit), sein Ergebnis, daß der Vertrag zwischen den Ländern trotzdem gelte, zwar allein auf verfassungsrechtliche Erwägungen gestützt, hat aber eine völkerrechtliche Argumentation unterstützend herangezogen:

> »Die aus dem Verfassungsgrundsatz der Bundestreue folgende Konsequenz der Weitergeltung der Bonus-malus-Regelung ohne Rücksicht auf die Entscheidung des Landesverfassungsgerichts stellt eine mögliche Konsequenz bei einer entsprechenden Konfliktsituation im Völkerrecht dar, worauf hier nur unterstützend hingewiesen werden soll. Da in der Bundesrepublik Deutschland die staatsrechtlichen Beziehungen zwischen den Gliedstaaten ausschließlich durch das geltende Verfassungsrecht bestimmt werden, bleibt insoweit für die – an sich mögliche – Anwendung von Völkerrecht kein Raum ... Dies schließt aber nicht aus, das aus dem Grundsatz der Bundestreue gefundene Ergebnis anhand völkerrechtlicher Regeln zu überprüfen«.

Es folgen Erwägungen zur völkerrechtlichen Lage. Das Gericht fährt fort:

> »Mehr als der Hinweis, daß auch im Verhältnis von Staaten, die nicht bundesstaatlich verbunden sind, in der entsprechenden Konfliktsituation die Weitergeltung des Vertrages trotz innerverfassungsrechtlicher Mängel grundsätzlich ... angenommen wird, läßt sich dem Völkerrecht für die Lösung des Konflikts in bundesstaatlicher Sicht nicht entnehmen«.

---

[97] Vgl. oben Anm. 58.

[98] ZaöRV 29 (1969), S. 711 ff.

[99] NJW 1977, S. 66 (69). Vgl. zum Folgenden auch *Wolff* (Anm. 16a), S. 131 ff.

In einer ähnlichen Konfliktsituation (allerdings ein Fall der Verletzung von Kompetenzvorschriften) hat der U.S. Supreme Court im Ergebnis genauso entschieden, jedoch fehlt ein ausdrücklicher Rückgriff aufs Völkerrecht. Der Oberste Gerichtshof des Staates West Virginia hatte einen zwischengliedstaatlichen Vertrag als unverbindlich angesehen, da die Exekutive dieses Staates aus landesverfassungsrechtlichen Gründen nicht hätte abschließen dürfen [100]. Der Supreme Court hat diese Erwägung zurückgewiesen, den Vertrag als verbindlich angesehen und das Urteil aus West Virginia aufgehoben [101]. Interessant sind die verschiedenen für dieses Ergebnis gegebenen Begründungen. Die von Richter Frankfurter verfaßte Mehrheitsmeinung stellt darauf ab, daß die Verfassung von West Virginia nicht den Inhalt habe, den das gliedstaatliche Gericht ihr gegeben habe, daß der Vertrag also in Wahrheit nicht landesverfassungswidrig sei. Dagegen wendet Richter Reed in seiner *concurring opinion* ein, daß der Supreme Court zu einer eigenständigen Auslegung der gliedstaatlichen Verfassung nicht befugt sei. Er stützt das Ergebnis vielmehr auf die *supremacy clause* der Bundesverfassung, den Vorrang des Bundesrechts [102]. Der Richter Jackson liefert dagegen in seiner *concurring opinion* eine Begründung, die sehr stark an die der völkerrechtlichen Evidenztheorie erinnert [103], die ja eine eingeschränkte Irrelevanztheorie ist. Der mögliche verfassungsrechtliche Mangel sei für die anderen Staaten und den Kongreß, der dem Vertrag zugestimmt habe, nicht erkennbar gewesen [104]:

> »(West Virginia) may not apply retroactively that interpretation [daß der Vertrag gegen seine Verfassung verstoße] to place an unforeseeable construction upon what the other States to this compact were entitled to believe was a fully authorized act«.

Einem solchen Verhalten stünde das Prinzip des *estoppel* entgegen. Dieser Gesichtspunkt hat auch ganz entscheidende Bedeutung für die völkerrechtliche Regel, daß jedenfalls nicht erkennbare verfassungsrechtliche Mängel der Abschlußkompetenz der Staatsorgane für die Gültigkeit des Vertrages unerheblich sind [105].

---

100 Vgl. zu diesem Verfahren *Abel*, Ohio Valley Panorama, West Virginia Law Review 32 (1946/47), S. 203–231; *Zimmermann/Wendell*, The Interstate Compact and Dyer v. Sims, Columbia Law Review 51 (1951), S. 931–950.

101 *Dyer* v. *Sims*, 71 S.Ct. 557 (1951).

102 A.a.O., S. 563.

103 Vgl. dazu kritisch *Geck*, Die völkerrechtlichen Wirkungen verfassungswidriger Verträge (Beiträge zum ausländischen öffentlichen Recht und Völkerrecht, Bd. 38) (1963), insbes. S. 416 ff.; *derselbe*, The Conclusion of Treaties in Violation of the Internal Law of a Party, ZaöRV 27 (1967), S. 429–450.

104 A.a.O., S. 564.

105 Eingehend dazu *J. P. Müller*, Vertrauensschutz im Völkerrecht (Beiträge zum ausländischen öffentlichen Recht und Völkerrecht, Bd. 56) (1971), S. 191 ff.

In der Schweiz wird nach der herrschenden Lehre die Frage, unter bewußter Abkehr von völkerrechtlichen Grundsätzen, umgekehrt entschieden. Ein Konkordat kommt gültig nur zustande, wenn beim Abschluß kantonale Kompetenzvorschriften nicht verletzt werden [106].

ccc) Der Rang zwischengliedstaatlicher Verträge im gliedstaatlichen Recht

Ein weiteres Problem, das mit dem soeben erörterten zusammenhängt, ist das der innergliedstaatlichen Geltung bzw. des innergliedstaatlichen Ranges der Verträge zwischen Gliedstaaten. Das schweizerische Bundesgericht löst diese Frage im Sinne einer Parallelität zur Rechtslage bei völkerrechtlichen Verträgen [107].

> »Nach allgemein geltenden Rechtsgrundsätzen geht das Konkordatsrecht, als gemeinsam vereinbartes Recht, dem Recht jedes einzelnen der am Konkordat teilnehmenden Kantone vor, ähnlich wie völkerrechtliche Verträge dem Landesrecht vorgehen (folgen Nachweise). Auf jeden Fall geht Konkordatsrecht dem kantonalen Recht vor, das nicht auf Verfassungsstufe steht«.

In USA ist die Rechtsprechung uneinheitlich. Eine Reihe von Entscheidungen sehen die *interstate compacts* als Bundesrecht an, das durch die *supremacy clause* der Bundesverfassung Vorrang vor Landesrecht genießt. Ansatzpunkt dieser Sichtweise ist die Tatsache, daß *compacts* der Zustimmung des Kongresses bedürfen, die in Form eines Bundesgesetzes ergeht [108]. Andere Entscheidungen stützen die Bindungswirkung der Verträge auf die sogenannte *contract clause* (Art. I sect. 10: No State shall . . . pass any . . . Law impairing the Obligation of Contracts . . .) [109]. In jedem Fall liegt eine bundesverfassungsrechtliche Garantie der Bestandskraft solcher Verträge vor, die ihnen Vorrang vor dem Landesrecht, wohl auch dem Landesverfassungsrecht verschafft. Auf völkerrechtliche Gedanken wird hier nicht bezug genommen.

In Deutschland hat der Staatsgerichtshof für das Deutsche Reich die Frage nur kurz gestreift [110]:

---

[106] *Schaumann* (Anm. 37), S. 110 f. Vgl. *Neuenburg* gegen *Bern*, BGE 40. I, 550; *Engel* gegen *Zürich*, BGE 40. I, 389.

[107] Dabei bestehen diesbezüglich allerdings auch in der Schweiz gewisse Unklarheiten. Der Vorrang des Völkerrechts vor dem Landesrecht, von dem die hier zitierte Textstelle ausgeht, wurde vom schweizerischen Bundesgericht in einigen Fällen bejaht, in anderen verneint. Vgl. dazu *Müller/Wildhaber*, Praxis des Völkerrechts (1977), S. 121 ff.

[108] Vgl. Richter Reed in *Dyer* v. *Sims* (Anm. 101); vgl. ferner *Pennsylvania* v. *Wheeling & B. Bridge Co.*, 13 How. 518 (1852).

[109] *Green* v. *Biddle*, 8 Wheat. 1 (1823); S. 92 *Virginia* v. *West Virginia*, 246 U.S. 565 (1918).

[110] RGZ 121, Anhang, S. 7.

»Was die Landesgesetze angeht, so können durch diese die Normen des zwischenstaatlichen Rechts nicht außer Kraft gesetzt werden, wie keiner näheren Darlegung bedarf«.

Dies scheint auf einen Vorrang der Verträge auch gegenüber späterem Landesrecht zu deuten. In dem Streit um den Staatsvertrag über die Vergabe von Studienplätzen argumentierte dann der Bayerische Verfassungsgerichtshof auf der Basis der Gleichbehandlung der zwischengliedstaatlichen Verträge mit völkerrechtlichen Verträgen [111]. Ansatzpunkt seiner verfassungsgerichtlichen Kontrolle ist der Zustimmungsakt des bayerischen Landtages. Ist dieser verfassungswidrig, ist der Vertrag innerstaatlich nicht anwendbar. Dies ist beim völkerrechtlichen Vertrag nach der deutschen verfassungsrechtlichen Rechtslage auch unbestritten [112]. Für den zwischengliedstaatlichen Vertrag sieht der Bayerische Verfassungsgerichtshof die Lage genauso. Das Bundesverwaltungsgericht folgt ihm hierin nicht. Es stellt nicht nur die Fortgeltung des Vertrages zwischen den Ländern fest, sondern ist darüber hinaus der Auffassung, daß aus Bundesrecht, nämlich aus dem Grundsatz des bundesfreundlichen Verhaltens, eine Verpflichtung Bayerns folge, den Vertrag weiter zu erfüllen [113]. Damit müßte eben auch bayerisches Verfassungsrecht im innerbayerischen Bereich weichen. Wenn sich das Bundesverwaltungsgericht zur Unterstützung seiner Auffassung auf die entsprechende Lage im Völkerrecht beruft [114], so trifft das allerdings nur für die Geltung des Vertrages zwischen den Partnern, nicht für die Geltung im innerstaatlichen Bereich zu. Einschränkend ist zu bemerken, daß das Bundesverwaltungsgericht die hier angeschnittenen Fragen nicht umfassend, sondern nur für einen engen Bereich entschieden hat, nämlich für den Fall, daß der geltend gemachte Verstoß gegen die Landesverfassung ein Grundrecht betrifft, das parallel durch Bundes- und Landesverfassung garantiert wird und über dessen Tragweite das Bundesverwaltungsgericht infolgedessen selbst abschließend entscheiden kann. Gleichfalls nicht entschieden ist die Frage des Verhältnisses von zwischengliedstaatlichem Vertrag und späterem Landesgesetz. In der Konsequenz der Entscheidung des bayerischen Verfassungsgerichtshofes und einer Gleichbehandlung mit dem völkerrechtlichen Vertrag läge es, die Gültigkeit des späteren Landesgesetzes anzunehmen [115]. Aus der Entscheidung des Bundesverwaltungsgerichts folgt andererseits, daß ein bewußter Bruch der vertraglichen Vereinbarung durch ein vertragswidriges Gesetz als Verstoß gegen die Bundestreue als nichtig zu behandeln wäre.

---

[111] Bayerischer Verfassungsgerichtshof, NJW 1975, 1733 (1934).

[112] *Boehmer,* Der völkerrechtliche Vertrag im deutschen Recht (Beiträge zum ausländischen öffentlichen Recht und Völkerrecht, Bd. 43) (1965), S. 72 f.

[113] NJW 1977, S. 66.

[114] Vgl. Text zu Anm. 99.

[115] *Schneider* (Anm. 19), S. 15 f., unter ausdrücklichem Hinweis auf die Parallele beim völkerrechtlichen Vertrag.

ddd) Vertragsbeendigung und clausula rebus sic stantibus

Eine Frage, für die völkerrechtliche Erwägungen in der Praxis eine große Rolle gespielt haben, ist die der Vertragsbeendigung [115a] und hier insbesondere die der *clausula rebus sic stantibus* [116]. Sowohl der Staatsgerichtshof für das Deutsche Reich [117] als auch das Bundesverfassungsgericht [118] haben die *clausula* als auf Verträge zwischen Gliedstaaten anwendbar angesehen. Während das Bundesverfassungsgericht aber die *clausula* kurz und knapp (ohne jeden Beleg, ohne jedes Zitat) zu einer logischen Konsequenz des Prinzips der Bundestreue und damit zu einer Norm des Verfassungsrechts erklärt, macht sich der Staatsgerichtshof mehr Mühe. Es zeigt zunächst die grundsätzliche Geltung der *clausula* im Völkerrecht, weist dabei aber auch auf die verschiedenen Auffassungen über ihre Tragweite hin. Mit den allgemeinen völkerrechtlichen Aussagen konnte der Staatsgerichtshof in der Tat für die Lösung des Falles, in dem es um die Grenzen der *clausula* ging, wenig anfangen. Deshalb tat er das, was eine reine völkerrechtliche Streitentscheidung auch hätte tun müssen: er ging zurück zu den Quellen des Vertragsrechts im Privatrecht und fand hier den Ansatz der Lösung [119]. So gesehen, ist diese Entscheidung des Staatsgerichtshofs durchaus ein Beleg für die Übertragbarkeit völkerrechtlicher Lösung auf das Bundesstaatsrecht [120], und die hier konkret gefundenen Ergebnisse sind ins Völkerrecht rückübertragbar.

Auch das schweizerische Bundesgericht hat sich mit dem Problem der Geltung der *clausula* für interkantonale Verträge befaßt. Es hat unter Berufung auf völkerrechtliche Literatur die Geltung dieser Regel angenommen, das Vorliegen der Voraussetzungen ihrer Anwendung aber im konkreten Fall recht knapp verneint [121].

115a In seiner Entscheidung über die Kündigung des Staatsvertrages über den Norddeutschen Rundfunk, in der es u.a. um die Notwendigkeit der Beteiligung des schleswig-holsteinischen Parlaments ging, berief sich das Bundesverwaltungsgericht (BVerwG) nicht ausdrücklich darauf, daß diese Frage genauso wie bei völkerrechtlichen Verträgen zu behandeln sei, stützt aber seine Auffassung, das Parlament sei nicht zu beteiligen, auf die einschlägige Literatur zum völkerrechtlichen Vertrag (Deutsches Verwaltungsblatt 1980, S. 845 ff.).

116 Vgl. dazu auch *Lauterpacht* (Anm. 52), S. 167 ff.

117 *Bremen ./. Preußen*, RGZ 112, Anhang, S. 21 (28 ff.).

118 BVerfGE 34, 216 (231 ff.).

119 A.a.O., S. 31 f.

120 Zweifelnd *Bleckmann* (Anm. 4), S. 38.

121 *Luzern* gegen *Aargau*, BGE 8, 43 (57 f.) (1882).

IV. Die Bedeutung der Unterschiede zwischen Bundesstaatsrecht und Völkerrecht

1. Das Überspielen der Unterschiede zwischen Völkerrecht und Bundesstaatsrecht durch Argumente a maiore ad minus

Die Grenzen der Übertragbarkeit der Lösungen einer Rechtsordnung auf die andere, die in den zitierten Ausführungen des Bundesverwaltungsgerichts[122] anklingen, werden von den Gerichten häufig mit einer »erst-recht«-Argumentation umgangen[123]. So betont der Staatsgerichtshof für das Deutsche Reich in seiner Entscheidung zur Weser-Verschmutzung[124]:

> »Hierbei ist nach den vom Staatsgerichtshof festgestellten Grundsätzen von jedem deutschen Lande zu verlangen, daß es die Interessen von Angehörigen des anderen Landes, wenn auch nicht in gleichem Maße wie die Interessen der eigenen Landesangehörigen, so doch jedenfalls stärker berücksichtigt als die anderer, nichtdeutscher Staaten«.

Das Völkerrecht wird hier also mit anderen Worten zum Nachweis eines Mindeststandards der Rücksichtnahme im Nachbarverhältnis zwischen Gliedstaaten herangezogen. Wieweit die offenbar als völkerrechtlich zulässig vorausgesetzte Schlechterbehandlung ausländischer Interessen gegenüber inländischen heute noch dem geltenden Völkerrecht entspricht, ist eine andere Frage[125]. Noch deutlicher wird diese Auffassung von Völkerrecht als bundesstaatlichem Mindeststandard in dem folgenden Zitat[126]:

> »Das deutsche Staatsrecht wird in Fragen wie der hier vorliegenden, maßgebend allein beherrscht von der Tatsache der Verbindung der deutschen Bundesstaaten (Länder) zum Deutschen Reich und der für diese Verbindung gesetzten Normierung, der Reichsverfassung. Wenn aber die letztere, wie im Streitfalle, keinerlei Anhaltspunkte gibt, wie die entstandenen Streitfragen zu lösen sind, so wird der Schluß nicht von der Hand zu weisen sein, daß diejenige Rücksichtnahme auf die Interessen des Gegenkontrahenten eines Staatsvertrages, die im Völkerrecht jedem Staat zugemutet wird, auch im Verbande des Deutschen Reichs nicht als unbillig und einer rechtlichen Grundlage entbehrend angesehen werden kann«.

---

[122] Text zu Anm. 99.

[123] Vgl. dazu *Schneider* (Anm. 19), S. 14.

[124] RGZ 121, Anhang, S. 8.

[125] Sie widerspricht sicherlich dem von der OECD postulierten Prinzip der Nichtdiskriminierung (Resolution des OECD Rates vom 14. 11. 1974: Principles Concerning Transfrontier Pollution, Text in: OECD and the Environment, 1976, S. 52 ff.). Vgl. dazu International Law Association, Committee on Legal Aspects of the Conservation of the Environment, in: ILA, Report of the Fifty-Eighths Conference, 1978, S. 383 ff. (408).

[126] RGZ 112, Anhang, S. 28.

Dieser Schluß von einer Rechtsordnung auf die andere läßt sich auch umkehren, ist in der Tat vom Schiedsgericht im *Trail Smelter*-Fall umgedreht worden: Wenn schon im Verhältnis zwischen Gliedstaaten nur eine beschränkte Pflicht zur Verhinderung grenzüberschreitender Immissionen besteht, dann erst recht im Völkerrecht [127]:

> »What is true between the States of the Union is, at least, equally true concerning the relations between the United States and the Dominion of Canada«.

In ähnlicher Weise, allerdings gedanklich präziser, argumentiert M. Huber im *Las Palmas*-Fall [128]. Nachdem er darauf verwiesen hat, daß das Prinzip des Gebietserwerbs durch Ersitzung auch in bundesstaatlichen Präzedenzfällen anerkannt sein, fährt er fort:

> »This is the more significant, in that it might well be conceived that in a federal state possessing a complete judicial system for interstate matters – far more than in the domain of international relations properly so-called – there should be applied to territorial questions the principle that, failing any specific provision of the law to the contrary, a *jus in re* once lawfully acquired shall prevail over *de facto* possession however well established«.

Wenn also schon die entwickeltere, richterliche Rechtsdurchsetzungsmöglichkeiten anbietende Rechtsordnung in gewissen Fällen wie bei der Ersitzung die normative Kraft des Faktischen anerkennt, dann muß erst recht das Völkerrecht dies tun. Hier sind also die Strukturunterschiede zwischen Völkerrecht und Bundesstaatsrecht kein Hindernis für die Übertragung von Konfliktlösungen von einer rechtlichen Ebene auf die andere, diese Unterschiede verstärken sogar die Berechtigung dieser Übertragung.

So einleuchtend diese Argumentation in den konkret zitierten Fällen auch sein mag, sie darf nicht unkritisch verallgemeinert werden. Zutreffend weist das Bundesverwaltungsgericht auf die Grenzen einer solchen »erst-recht«-Argumentation hin [129]:

> »Zwar ließe sich aus einer Weitergeltung des Vertrags zwischen Völkerrechtspartnern schließen, daß auch die enger verbundenen Länder eines Bundesstaates an einen einmal geschlossenen Vertrag gebunden bleiben müssen; dieser Überlegung steht aber der gegenläufige Gesichtspunkt gegenüber, daß die engere Verbundenheit auch eine größere Rücksichtnahme aufeinander und damit auch auf innerstaatliche Erfüllungsschwierigkeiten gebietet«.

Hier steht wieder das ewige Dilemma der Wahl zwischen Analogie und *argumentum a contrario* im Raum. Es läßt sich nur durch eine sorgfältige Ana-

---

[127] A.a.O. (Anm. 10), S. 1964.
[128] Oben Anm. 9.
[129] A.a.O. (Anm. 99), S. 69.

lyse der Eigenheiten zu vergleichender Sachverhalte und der involvierten Interessen lösen.

2. Einzelne Unterschiede zwischen Bundesstaatsrecht und Völkerrecht

Für die hier untersuchten Fallgruppen muß darum versucht werden, etwas genauer der Frage nachzugehen, welche Unterschiede der Rechtsordnung bzw. der politischen, wirtschaftlichen und sozialen Bedingungen der Übertragung von Ergebnissen aus einer Rechtsordnung in die andere entgegenstehen könnten.

a) Unterschiede des Entwicklungsstandes der Rechtsordnung

Die Übertragung von Ergebnissen aus einer stark entwickelten, ausdifferenzierten Rechtsordnung in eine weniger entwickelte ist sicherlich ein Problem. Aber letztlich ist gerade dies der Sinn der hier untersuchten Übertragungsvorgänge, aus einer Rechtsordnung, die mit der Lösung bestimmter Fragen schon weiter fortgeschritten ist als eine andere, Gewinn für Problemlösungen in der letzteren zu ziehen. Gäbe es keine Unterschiede in der Entwicklung der Rechtsordnungen, würden sich die hier untersuchten Fragen gar nicht erst stellen. Das Problem des unterschiedlichen Entwicklungsstandes der zu vergleichenden Rechtsordnungen ist seinerseits differenziert zu betrachten.

b) Die Bedeutung der stärkeren Entwicklung des Rechtsschutzes im innerstaatlichen Bereich

Ein wesentlicher Strukturunterschied zwischen der Situation im Bundesstaat und der in den internationalen Beziehungen besteht wie erwähnt in der stärkeren Entwicklung gerichtlichen Rechtsschutzes innerhalb des Staates. Deshalb ist die Ausübung des »diplomatischen« Schutzes im Verhältnis zwischen Gliedstaaten eines Bundesstaates grundsätzlich nicht möglich, können auch die Gerichte nicht von einem Gliedstaat mit der Begründung angerufen werden, ein anderer Gliedstaat verletze individuelle Rechte der Angehörigen des ersteren Staates. Der Gliedstaat kann solche individuellen Rechte nicht geltend machen, sondern nur die Verletzung eigener staatsrechtlicher Rechtspositionen, etwa seiner Territorialhoheit. So führt der Staatsgerichtshof für das Deutsche Reich aus[130]:

> »(Der württembergische Staat) muß es seinen Staatsangehörigen überlassen, die Rechte, die sie aus (der Reichsgewerbeordnung) glauben herleiten zu können, vor den verfassungsmäßig dazu berufenen Behörden ... geltend zu machen. Im Verhältnis von Staat zu Staat unmittelbar ihre Durchsetzung zu erzwingen, ist nicht angängig«.

---

[130] RGZ 116, Anhang, S. 27 (Donauversinkung).

Ähnlich unterscheidet der amerikanische Supreme Court[131]:

> »The right of a state as parens patriae to bring suit to protect the *general* comfort, health, or property rights of its inhabitants threatened by the proposed or continued action of another state by prayer for injunction is to be differentiated from its *lost power as a sovereign* to present and enforce *individual* claims of its citizens as their trustee against a sister state«.

Der entscheidende Gesichtspunkt für den Verlust der staatlichen Befugnis zur diplomatischen Protektion ist nach Auffassung des Supreme Court die Möglichkeit des individuellen Rechtsschutzes[132]:

> ». . . the giving of the direct remedy to the citizien himself was equivalent to taking away any indirect remedy he might otherwise have claimed, through the intervention of his state, upon any principle of the law of nations«.

Die Rechtsprechung des schweizerischen Bundesgerichts geht in die gleiche Richtung[133].

Man wird sich allerdings fragen müssen, wieweit hier nicht eine Tendenz zur Angleichung der Situation im Bundesstaatsrecht und im Völkerrecht in dem Sinne zu erkennen ist, daß der diplomatische Schutz auch im Völkerrecht zurücktritt. Traditionell tritt der diplomatische Schutz nur vorläufig zurück, soweit das Erfordernis der Erschöpfung des innerstaatlichen Rechtsweges eingreift. Es ist aber eine alte Forderung der lateinamerikanischen Staaten (Calvo-Doktrin), den diplomatischen Schutz stärker zurückzudrängen und das Individuum auf innerstaatlichen Rechtsschutz zu verweisen. Auch neuere Entwicklungen im Bereich des internationalen Umweltschutzrechtes gehen in die Richtung einer stärkeren Betonung individuellen (grenzüberschreitenden) gerichtlichen Rechtsschutzes gegenüber der Durchsetzung individueller Rechte durch die Regierung des Heimatstaates[134].

c) Bundesstaatliches »Fremdenrecht«

Mit der Frage des »diplomatischen Schutzes« hängt die Frage des »bundesstaatlichen Fremdenrechts« zusammen. Hier geht es um die Frage, ob und inwieweit Gliedstaaten eines Bundesstaates die »Angehörigen« eines anderen

---

[131] *North Dakota* v. *Minnesota*, 44 S.Ct. 138 (1923), S. 140 (Hervorhebung vom Verf.).

[132] *New Hampshire* v. *Louisiana*, 108 U.S. 76 (1883). Nach dieser Entscheidung ist die Befugnis der Staaten zur diplomatischen Protektion auch nicht dadurch wieder aufgelebt, daß das XI. Amendment diese individuelle Rechtsschutzmöglichkeit wieder weitgehend beseitigt hat.

[133] Vgl. die Nachweise bei *Bleckmann* (Anm. 4), S. 18.

[134] Vgl. dazu *Bothe*, Transfrontier Environmental Management (Anm. 68), S. 391–401 (396 ff.).

Gliedstaates gegenüber Landeskindern unterschiedlich behandeln können [135]. Völkerrechtliche Inländerbehandlungsgebote oder fremdenrechtlicher Mindeststandard bieten keine vernünftige Antwort auf diese Frage. Sie wird meist unter dem Gesichtspunkt des Gleichheitssatzes diskutiert. Es wird gefragt, inwieweit die Zugehörigkeit einer Person zu einem anderen Gliedstaat im Rahmen des Gleichheitssatzes ein zulässiges Differenzierungskriterium darstellt. Insoweit liegt die Frage noch parallel zum Problem der Ausländereigenschaft als Differenzierungskriterium. Der Angehörige eines anderen Gliedstaates ist aber sicherlich »gleicher« als der Ausländer. Darum sind auch für das völkerrechtliche Fremdenrecht sinnvolle Schlüsse aus dem bundesstaatlichen Fremdenrecht nicht möglich.

d) Unterschiede der Loyalitätspflichten zwischen den Staaten

Die andersartige Behandlung der Fragen des »Fremdenrechts« im Bundesstaat und im Völkerrecht ist sicherlich ein Fall, in dem die Tatsache ausschlaggebend ist, daß das rechtliche Loyalitätsband zwischen den Gliedstaaten eines Bundesstaates dichter, intensiver ist als das zwischen unabhängigen Staaten. In dieser stärkeren Intensität des staatsrechtlichen Bandes wird häufig ein wesentlicher Unterschied zwischen der bundesstaatlichen und der völkerrechtlichen Ebene gesehen [136]. Es muß aber davor gewarnt werden, vorschnell Unterschiede der rechtlichen Verbundenheit festzustellen und daraus auf die Unübertragbarkeit von Lösungen einer Rechtsordnung auf die andere zu schließen. Die zunehmende Interdependenz der Staaten, die Zersetzung ihrer Hülle der Souveränität durch eine Fülle nicht durch Staaten vermittelter grenzüberschreitender Interaktion läßt heute das Verhältnis zwischen Staaten dem zwischen Gliedstaaten vergleichbarer erscheinen, als man es vielleicht früher angenommen hätte [137]. Es fällt auf, daß auch solche Entscheidungen, die die stärkere Bindung im zwischengliedstaatlichen Verhältnis betonen [138], im Ergebnis gerade keine vom Völkerrecht abweichende Ergebnisse erzielen.

Die gesteigerte Loyalität im Bundesstaat (Bundestreue) steht einer Übertragbarkeit der Regelungsmuster nur begrenzt im Wege. Auch dem Völker-

---

[135] Nachweise für die Rechtslage in der Bundesrepublik Deutschland, der Schweiz und den USA bei *Bothe*, Die Kompetenzstruktur des modernen Bundesstaates in rechtsvergleichender Sicht (Beiträge zum ausländischen öffentlichen Recht und Völkerrecht, Bd. 69) (1977), S. 131 ff.

[136] *Müller* (Anm. 4), S. 80 ff.

[137] So beruht die ablehnende Haltung *Müllers* (Anm. 4), S. 80 ff., gegen die Anwendbarkeit völkerrechtlicher Prinzipien im Bundesstaat klar auf einer Völkerrechtskonzeption, die heute nicht mehr zu vertreten ist. Dennoch will übrigens *Müller* die Anwendung bundesstaatlicher Prinzipien im Völkerrecht nicht völlig ausschließen, a.a.O., S. 91 ff.

[138] So nicht nur das BVerfG (Anm. 20) und das BVerwG (Anm. 99), sondern auch der Staatsgerichtshof für das Deutsche Reich, 112, Anhang, S. 1 (28); 121, Anhang, S. 1 (8).

recht ist der heutige Gedanke eines dem Individualinteresse übergeordneten Gemeinschaftsinteresses nicht fremd. Das wird deutlich in der Entwicklung des Menschenrechtsschutzes, des humanitären Völkerrechts, im Prinzip der Solidarität, wie es im Recht internationaler Hilfsaktionen und in den Bestrebungen zu einer neuen Weltwirtschaftsordnung, was immer man im einzelnen davon halten mag, immer mehr zur Verwirklichung drängt, in Konzepten wie *common heritage of mankind* und *shared natural resources.* Verstärkt ist dies in Europa zu beobachten, im Konzept einer europäischen Menschenrechtsordnung, im Gedanken einer Gemeinschaftstreue in Analogie zur Bundestreue.

Das moderne Völkerrecht kennt auch Kooperations- und Rücksichtnahmepflichten[138a]. Die souveränen Staaten sind keine erratischen Blöcke, die scharf und säuberlich getrennt nebeneinander existieren. Diesem Befund steht auch die Tatsache nicht entgegen, daß die Betonung der staatlichen Souveränität in der Dritten Welt fröhliche Urständ feiert. Dies erklärt sich aus dem Bestreben dieser Staaten, überkommene Dependenzen abzubauen, sowie aus Unterentwicklung der Interdependenz in der Dritten Welt, die aus der historischen Ausrichtung dieser Staaten auf die »erste Welt« folgt. Aber auch hier zeigen sich ökologische und wirtschaftliche Interdependenzen, die zur Entwicklung von grenzüberschreitender Kooperation geführt haben, soziale Interdependenzen durch Wanderarbeit u. a. m. Der Befund einer gesteigerten Permeabilität des Staates gilt in zunehmendem Maße auch für die Dritte Welt. Dieses Bild der offenen Staatlichkeit läßt sich am wenigsten im Ost-West-Verhältnis diagnostizieren. Diese Unterschiede der Interdependenz und der politischen Rahmenbedingungen für Kooperation sind für die Frage der Übertragbarkeit bundesstaatlicher Lösungen in das Völkerrecht und umgekehrt genauso ein Problem wie es die Unterschiedlichkeit der bundesstaatlichen Ordnungen untereinander ist. Deshalb sind für die beschriebene vertikale Rechtsvergleichung auf der völkerrechtlichen Ebene besonders geeignet Regionen stärkerer politischer Homogenität (Westeuropa) oder Sektoren starker Interdependenz ohne Rücksicht auf politische Verhältnisse (Umweltschutz, internationaler Handel), auf seiten der Bundesstaaten solche mit vergleichsweise starker Dezentralisierung (USA).

e) Bundesrechtliche Sicherungen der Bestandskraft zwischengliedstaatlicher Verträge

Eine Frage, bei der die Dichte des rechtlichen Bandes zwischen Gliedstaaten wohl doch Unterschiede zum Völkerrecht begründet, ist die der Bestandskraft zwischengliedstaatlicher Verträge.

Bei der Frage der Gültigkeit landesverfassungswidriger Verträge ist in der Schweiz offenbar unbestritten, daß jedenfalls die Verletzung kantonal-ver-

---

[138a] *Neuhold,* Die Pflicht zur Zusammenarbeit zwischen den Staaten: moralisches Postulat oder völkerrechtliche Norm?, in: Ius humanitatis, Festschrift zum 90. Geburtstag von Alfred Verdross (1980), S. 575–606.

fassungsrechtlicher Kompetenzvorschriften den Vertragsbeitritt mit Wirkung nach außen unwirksam machen. Die Entscheidung des Bundesverwaltungsgerichts[139] würde dem für die Bundesrepublik nicht entgegenstehen, denn sie betrifft keine Verletzung von Kompetenzvorschriften, sondern die materielle Unvereinbarkeit des Vertrages mit Grundrechtsbestimmungen der Landesverfassung. In dem vom U.S. Supreme Court entschiedenen Fall *Dyer* v. *Sims*[140] ging es um Vorschriften über die Einhaltung von Verpflichtungen zu Lasten des Staatshaushaltes, also wohl Normen, die dem Bereich der Kompetenzvorschriften zuzurechnen wären. Hier wird man wohl aus der Homogenität der Rechtsordnungen im Bundesstaat grundsätzlich den Gliedstaaten zumuten können, von der Verfassungsordnung der anderen Staaten Kenntnis zu nehmen, worauf auch das Bundesverwaltungsgericht hinweist[141]. Auch in der deutschsprachigen und amerikanischen Literatur wird auf diesen Gesichtspunkt hingewiesen[142]. Insofern ist die Lage doch anders als bei völkerrechtlichen Verträgen, wo diese Homogenität und rechtliche Nähe zwischen den Vertragspartnern nicht besteht. Allerdings kommt hier eine Korrekturerwägung zum Tragen, die letztlich die Entscheidung des U.S. Supreme Court doch rechtfertigt und die letztlich auch hinter der Entscheidung des Bundesverwaltungsgerichts steht. In beiden Fällen wurde von dem Gliedstaat, der seine Verpflichtung negieren wollte, die Verletzung verfassungsrechtlicher Normen geltend gemacht, die den Vertragspartnern gemeinsam waren, die im Falle des Bundesverwaltungsgerichts gar eine Entsprechung in der Bundesverfassung hatten. Der Gliedstaat stützte sich hier jeweils unter Berufung auf ein Urteil seines höchsten Gerichts nachträglich auf eine von der der übrigen Partner abweichende Auslegung gemeinsamer verfassungsrechtlicher Normen. Um derartiges auszuschließen, nahmen in beiden Fällen die Bundesgerichte (im Falle *Dyer* v. *Sims* die Mehrheit des Gerichts, für die der angesehene Richter Frankfurter sprach) die Kompetenz in Anspruch, über die Auslegung solcher gemeinsamer Vorschriften zu entscheiden. Dies ist meines Erachtens letztlich ein sachlich gerechtfertigter Gesichtspunkt, wenn dieses Vorgehen auch von den Vorschriften über die gerichtliche Kompetenz her etwas schwierig zu rechtfertigen ist. Damit ist der Verfassungsverstoß nicht irrelevant, er wird vom Bundesgericht eben nur in eigener Kompetenz geprüft. Ganz klar in diese Richtung geht die Praxis des schweizerischen Bundesgerichts, das in Fällen, in denen die Ungültigkeit eines Konkordats wegen Verstoßes gegen eine Kantonsverfassung geltend gemacht wird, das Vorliegen eines solchen Verstoßes stets uneingeschränkt nachprüft[143]. Daß die Lage grundverschieden ist von derjenigen im Völker-

---

[139] Siehe oben Anm. 99.

[140] Oben Anm. 101.

[141] Siehe Text zu Anm. 129.

[142] *Rill* (Anm. 36), S. 369; *Abel* (Anm. 100), S. 193 f.

[143] *Neuenburg* gegen *Bern*, BGE 40. I, 580; *Engel* gegen *Zürich*, BGE 40. I, 389; zu den innerkantonalen Kompetenzproblemen vgl. *R. J. Schweizer*, Kantonale Kompe-

recht, wo eine solche eigenständige Prüfung der Lage nach staatlichem Verfassungsrecht grundsätzlich nicht möglich ist (mögliche Ausnahmen im Bereich der Menschenrechte können hier nicht vertieft werden), leuchtet ohne weiteres ein.

Ist ein Vertrag aber einmal verfassungsgemäß zustande gekommen, so steht er unter dem Schutz der übergeordneten Rechtsordnung, d.h. der Rechtsordnung des Bundes, wie auch immer man diesen Befund dogmatisch konstruiert[144]. Dies ist der innere Grund für die in USA und der Schweiz herrschende Praxis[145], daß jedenfalls der unmittelbar anwendungsfähige zwischengliedstaatliche Vertrag auch späterem Gliedstaatenrecht vorgeht[146]. Hier ist das bundesstaatliche Band eben doch dichter als das völkerrechtliche, es nimmt die Gültigkeit vertragsverletzender Normen nicht hin, braucht sie angesichts der Existenz einschlägiger Rechtsbehelfe auch praktisch nicht hinzunehmen.

V. Schlußfolgerungen

Die vorstehenden Erwägungen haben sich bewußt weitgehend darauf beschränkt, der Rechtsprechung »aufs Maul zu schauen«. Denn es ging nicht um die mit dogmatischen Erwägungen für die meisten heutigen Bundesstaaten recht überzeugend negativ zu beantwortende Frage nach der »Geltung« von Völkerrecht im Bundesstaat. Es ging vielmehr um den sachgerechten Einsatz bestimmter Argumentationsstrukturen zur Lösung typischer Konflikte in verschiedenen Rechtsordnungen. Hierfür bietet die Rechtsprechung immer noch das beste Anschauungsmaterial, haben Richter meistens ein besseres Gespür als ihre dogmatischen Kritiker. Es konnte dabei gezeigt werden, daß die Übertragung von Lösungsmustern aus der bundesstaatlichen auf die völkerrechtliche Ebene und umgekehrt mit Hilfe einer vertikalen Rechtsvergleichung durchaus hilfreich ist. Freilich stellen sich auch Probleme der Übertragbarkeit, die denen vergleichbar sind, die sich der horizontalen Rechtsvergleichung stellen.

Die Kriterien für eine Übertragung der Lösungsmuster lassen sich nur aus einer näheren Analyse der Interessenlage im konkreten Fall unter sozial- und rechtswissenschaftlichen Gesichtspunkten gewinnen. Eine solche pflegt, mehr oder weniger deutlich gemacht, mehr oder weniger reflektiert, auch den Entscheidungen in der Rechtsprechung zugrunde zu liegen. Allgemeine Aus-

---

tenzordnung und interkantonale Vereinbarungen, Festgabe zum Schweizerischen Juristentag (1973), S. 131–157.

[144] Für USA siehe oben Text zu Anm. 108–109. Zur Bundesrepublik Deutschland und Österreich *Rill* (Anm. 36), S. 491 ff.

[145] Oben Anm. 107–109.

[146] Daß *Rill* (Anm. 36) diese Konsequenz letztlich doch nicht zieht, vermag nicht zu überzeugen.

sagen dazu sind nur begrenzt möglich. Immerhin läßt sich soviel sagen, daß die Analogie der Situation darin besteht, daß sich gleichgeordnete staatliche Einheiten gegenüberstehen. Liegt eine ausdrückliche Regelung durch Bundesverfassung oder Bundesgesetz vor, so wird eine Übertragbarkeit nicht in Betracht kommen. Sinnvoll für die Übertragung sind nur solche Regelungsmuster, die allgemeine und verallgemeinerungsfähige Formen des Interessenausgleichs zwischen gleichgeordneten staatlichen Einheiten darstellen. Deshalb sind auch einzelne Bundesstaaten mehr oder weniger für die Vergleichung geeignet. Es kommt hier auf die Regelung im einzelnen Fall an.

Wie ist diese Übertragung von Problemlösungen aus einer Rechtsordnung in die andere in das jeweilige System der Rechtsquellen einzuordnen. Die Beispiele, die hier vorgestellt wurden, waren aus mehreren Gründen aus der Rechtsprechung gewählt. Die »Zulässigkeit« solcher Übertragung ist in einem nicht geringen Teil gleichbedeutend mit der Frage nach den zulässigen Erwägungen richterlicher Rechtsfortbildung. Solche Übertragungen, »Analogien«, sind Elemente der Rationalisierung richterlicher Konfliktlösung bei Versagen der sozusagen näherliegenden Rechtserkenntnisquellen, insbesondere des geschriebenen Rechts. Für das innerstaatliche Recht ergeben sich hier keine grundsätzlichen theoretischen Schwierigkeiten [147], für das Völkerrecht mag man einen Augenblick angesichts des Textes des Art. 38 IGH-Statut zögern. Es wurde dargelegt, daß ein Übertragungsvorgang der hier beschriebenen Art nicht durch Art. 38 Abs. 1 (c) IGH-Statut gedeckt wäre [148]. Kann diese Bestimmung den internationalen Richter bei der Suche nach sachgerechten Vorbildern in innerstaatlichen Rechtsordnungen auf solche Lösungen beschränken, die universell vorzufinden sind? Schon bei der praktischen Anwendung des Art. 38 Abs. 1 (e) fällt auf, daß eine nachgewiesene universelle Rechtsvergleichung als Grundlage richterlicher Entscheidungen die große Ausnahme ist, daß der internationale Richter auf der Suche nach Sachgerechtigkeit bei der Rechtsvergleichung eher eklektisch vorzugehen pflegt [149]. Das drängt natürlich die auch in anderen Zusammenhängen gestellte Frage auf, wieweit Art. 38 die Rechtserkenntnisquellen für den internationalen Richter wirklich erschöpfend regelt[150]. Diese Frage kann hier nicht abschließend behandelt werden. Die Übernahme sachgerechter Lösungen aus einigen wenigen innerstaatlichen Rechtsordnungen, oder auch nur aus einer, läßt sich wohl, ohne dem Text Gewalt anzutun, in Art. 38 Abs. 1 (d) als sog. Hilfsquelle verorten. Sind nicht solche innerstaatlichen Lösungsmuster jedenfalls den »teachings of the most qualified publicists« gleichzuachten?

Allerdings ist im Völkerrecht die Leitfunktion richterlicher Rechtsfortbildung heute weniger bedeutsam als in vielen innerstaatlichen Rechtsord-

---

[147] Vgl. oben III. 1.
[148] Oben Text zu Anm. 13.
[149] *Bothe* (Anm. 13), S. 283 ff.
[150] *Verdross/Simma* (Anm. 12), S. 258 f.

nungen. Muß der nicht-richterliche Rechtsanwender im Völkerrecht deswegen auf den Gebrauch solcher Kriterien der Sachgerechtigkeit, die ihm der Blick auf bundesstaatliche Problemlösungen liefert, verzichten? Wohl kaum. Auch bei Rechtsanwendung durch andere relevante Akteure der internationalen Ordnung kann der Blick auf die bundesstaatliche Ordnung Orientierungsdaten liefern, Orientierungsdaten für die Staatenpraxis. Der Blick auf den Bundesstaat kann helfen, sachgerechte Lösungen akzeptabel zu machen, ist damit ein Element der Bildung von Konsens, der letztlich die Grundlage der zwischenstaatlichen Rechtsordnung ist.

Zum »Recht auf Leben« aus nationaler und internationaler Sicht

Karl Doehring

I. Allgemeine Betrachtung

Mit der Darbietung der folgenden Überlegungen ist nicht beabsichtigt, in die Tiefen der Probleme einzudringen, die sich aus der Einordnung des Lebens als Rechtswert unter rechts- und staatsphilosophischen Gesichtspunkten ergeben. Es geht, wie zu zeigen ist, um ein bescheideneres Ziel, nämlich um die Frage, wie das »Recht auf Leben« in die Notwendigkeiten der zu praktizierenden Rechtsanwendung einzufügen ist.

Der Satz »Das Leben ist der Güter höchstes nicht« sei grundgesetzwidrig, schreibt *Walter Leisner*[1]. Diese bemerkenswerte Aussage führt zu Konsequenzen bei der Beurteilung von Rechtslagen, die im einzelnen noch betrachtet werden sollen. Das Bundesverfassungsgericht war etwas vorsichtiger, wenn es feststellte, daß das Leben innerhalb der grundgesetzlichen Ordnung »einen« Höchstwert darstelle und daß die entsprechende Schutzverpflichtung des Staates besonders ernst genommen werden müsse, weil eben das menschliche Leben »einen« Höchstwert bedeute[2]. Die Frage, ob man der einen oder der anderen Auffassung folgt – das Leben als »der« Höchstwert oder das Leben als »ein« Höchstwert der Rechtsordnung – kann und darf aber nicht unbeantwortet bleiben. Im einen Falle wäre eine Rechtsgüterabwägung – Leben gegen ein anderes hochwertiges Rechtsgut – verfassungswidrig, im anderen Falle nicht nur zulässig, sondern auch geboten. Wie verunsichert Lehre und Rechtsprechung auf diesem Gebiet sind, zeigt sich vielfältig; es sei nur erinnert an die Diskussionen über den sog. gezielten Todesschuß der Polizei etwa zur Geiselbefreiung, an die Grenzen des Notwehrrechts und vor allem des Nothilferechts bei Angriffen auf andere Rechtsgüter durch die Polizei oder

---

[1] *W. Leisner,* Das Lebensrecht, hrsg. von der Niedersächsischen Landeszentrale für politische Bildung (Hannover 1976), S. 25.

[2] So die Entscheidung des BVerfG im Falle *Schleyer* vom 15. 10. 1977, Entscheidungen des Bundesverfassungsgerichts (BVerfGE) 46, 160 ff.

auch durch Privatpersonen, an die Grenzen des Einsatzes von Polizei und Feuerwehr bei eigener Lebensgefährdung, an den Lebenseinsatz von Soldaten im Kriege, die Auslieferung bei drohender Todesstrafe und den Schwangerschaftsabbruch. Es ließen sich sicherlich noch weitere Problemsituationen andeuten. In allen diesen Fällen ist die Beantwortung der Frage nach dem Stellenwert des Rechts auf Leben in einer positivierten Rechtsordnung unausweichbar. Wenn, wie behauptet wird, nur Lebensschutz gegen Lebensschutz abgewogen werden darf[3] und also der Lebensschutz jedem anderen Rechtsgut vorgeordnet wird, müßten auch die noch zu zeigenden Konsequenzen akzeptiert werden. Es entsteht aber der Eindruck, daß die Folgerungen aus dieser zum Teil so idealistischen Ansicht doch wohl nicht völlig durchdacht wurden.

Im internationalen Recht steht es nicht anders. Die Europäische Menschenrechtskonvention[4] spricht von dem Recht jedes Menschen auf Leben und zählt dann auf, wann doch Einschränkungen zulässig seien (Art. 2 EMRK). Der Stellenwert dieses Grundrechts im Rahmen der Konvention und gemessen an anderen Rechtsgütern scheint so durch eine erschöpfende *lex specialis* geklärt, doch werden dadurch Unklarheiten nicht völlig vermieden, denn – um nur ein Beispiel zu nennen – die Anwendung der Notwehr soll nur dann zulässig und rechtmäßig sein, wenn es um die Verteidigung »eines Menschen gegenüber rechtswidriger Gewaltanwendung« geht; ist nun Gewaltanwendung »gegenüber einem Menschen« auch eine solche, die ihm gewaltsam eine Sachherrschaft entziehen will? Zwar ist *Karl Josef Partsch* zu folgen, wenn er ausführt, Art. 2 EMRK wende sich nur an die Staatsgewalt[5]. Doch bleibt dann auch von diesem Spezialisten die Frage ungelöst, ob die Staatsgewalt nur und ausschließlich im Rahmen ihrer Schutzausübung Leben vernichten darf, wenn Leben bedroht ist, oder ob andere Rechtsgüter auch durch Eingriffe in das Leben des Angreifers geschützt werden dürfen, etwa die körperliche Unversehrtheit oder hohe Sachwerte.

Der Pakt der Vereinten Nationen über bürgerliche und politische Rechte[6] zeigt ebenfalls kein klares Bild. Abgesehen von eingehenden Regelungen über Zulässigkeit und Vollzug der Todesstrafe enthält Art. 6 des Paktes nur zwei Hinweise, die für eine Güterabwägung bedeutsam sein können. So wird

---

[3] *G. Dürig* (in: Maunz/Dürig/Herzog/Scholz, Grundgesetz, Kommentar, zu Art. 2, Rdn. 13) führt aus, daß dann, wenn der Staat rechtmäßig einen Lebenseinsatz fordern oder eine Lebensvernichtung als rechtmäßig anerkennen könne, es sich nur um Fälle handeln dürfe, in denen »Leben gegen Leben« stehe. Dabei soll nicht verkannt werden, daß spätere Ausführungen wesentliche Differenzierungen aufweisen; es geht hier nur um die grundsätzliche Aussage.

[4] Konvention zum Schutz der Menschenrechte und Grundfreiheiten vom 4. 11. 1950, Bundesgesetzblatt (BGBl.) 1952 II, S. 686, 953 mit Änderungen.

[5] *K. J. Partsch,* Die Rechte und Pflichten der Europäischen Menschenrechtskonvention, in: Die Grundrechte, Bd. 1, 1. Hlbbd. (1966), S. 336.

[6] Internationaler Pakt über bürgerliche und politische Rechte vom 19. 12. 1966, BGBl. 1973 II, S. 1534.

das Recht auf Leben als ein *erent right* bezeichnet, was in der üblichen deutschen Übersetzung als »angeborenes Recht« charakterisiert wird. Man kann wohl annehmen, daß dieser Terminologie eine naturrechtliche Vorstellung zugrunde lag, wonach das Recht auf Leben jenseits jeder Positivierung Bestand habe. Doch auch das Naturrecht (welches?) sagt nichts Zwingendes aus, oder besser gesagt, gibt unterschiedliche Aussagen darüber, ob das Recht auf Leben nicht doch Grenzen aufweise[7]. Um diese Grenzen aber geht es, wenn es dann heißt, daß niemand »willkürlich« seines Lebens beraubt werden dürfe; hier nun wieder scheint das Tor zur Güterabwägung breit geöffnet.

Man kann wohl auch von einem weltweiten Pakt nichts anderes erwarten, da die Kultur- und Rechtsordnungen der Völkergemeinschaft auf diesem Gebiet eine außerordentliche Heterogenität aufweisen, die sich zwischen religiösem Lebensschutz auch der Tierwelt bis zur gnadenlosen Vernichtung des ideologisch andersdenkenden Menschen bewegt. In großer Selbstaufopferung und Selbstgefährdung werden Menschen aus den Meeren aufgefischt, während sowohl »Befreiungskämpfer« als auch um die Erhaltung ihrer Macht besorgte Regierungen vor der Ausrottung von Frauen und Kindern und vor Völkermord nicht zurückschrecken. Gleichzeitig beraten internationale Kommissionen über den Begriff der Willkür. Während in Frankreich die Todesstrafe abgeschafft wurde[8], ist ihre Vollstreckung in anderen Teilen der Welt zum Exzeß angewachsen; die Rechtfertigungen in beiden Fällen beruhen letztlich auf Irrationalität. In der Bartholomäusnacht läßt Conrad Ferdinand Meyer die Flußgöttin der Steinfrau die Frage stellen, warum die Menschen sich morden. Die Antwort lautet, weil sie sich nicht einigen können über den Weg zur Seligkeit. Die Übertragung dieses Bildes auf weite Teile der internationalen Welt fällt nicht schwer. Auch die Privilegierung der Kämpfer für das Selbstbestimmungsrecht, wie sie sich in internationalen Dokumenten findet[9], scheint von einer Allgemeinauffassung, daß das Leben nur zur Erhaltung anderen Lebens vernichtet werden dürfe, weit entfernt. Hierauf ist zurückzukommen, aber auch hier schon sei ein Hinweis gegeben. Es scheint auch die Art des Lebens zu sein, die zur Lebensvernichtung und zur Gefährdung des eigenen Lebens der abwehrbereiten Menschen berechtigen soll. Zwar wird der Nachweis versucht, daß nahezu jede der bestehenden Kultur- und Rechtsordnungen den Grundsatz, wonach Leben nur zur Lebenserhaltung geopfert werden darf, auf der Grundlage ihrer eigenen Konzeption bejahen müßte[10]; doch bleibt auch das zweifelhaft. Die auf christlicher Rechts-

[7] *K. Doehring,* Das Widerstandsrecht des Grundgesetzes und das überpositive Recht, Der Staat 8 (1969), S. 429 ff.

[8] Loi no 81-908 portant abolition de la peine de mort, J.O., 113e Année – No 238, 10. 10. 1981, p. 2759.

[9] Definition des Angriffskrieges, Generalversammlung der UN vom 14. 12. 1974, Res. 3314 (XXIX), Art. 7.

[10] *Leisner,* a.a.O. (Anm. 1), S. 17 ff.

kultur beruhenden Systeme, so wird ausgeführt, sähen sich diesem Grundsatz verpflichtet, weil das individuelle Leben und nicht der Bestand einer Gemeinschaft von Individuen einen »unvergleichlichen« Wert bedeute. Fraglich ist aber, ob nicht doch auch bei dieser Sicht die transzendentale Blickrichtung christlicher Auffassung zu kurz kommt. Ist christliches Martyrium wirklich beschränkt auf das Ziel, hiermit nur anderes irdisches Leben zu schützen? Vielleicht ist das doch dann nicht der Fall, wenn man der Auffassung folgt, daß die Vollendung göttlicher Wertordnung das irdische Leben nicht zum höchsten Gut erhebt. So bleiben Zweifel an dieser Schlußfolgerung. Das gleiche gilt für den Marxismus, wenn auch wohl eher im Hinblick auf ein Auseinanderfallen von Theorie und Praxis. Richtig wird gesagt, daß die Ablehnung transzendentaler Ziele notwendig das individuelle irdische Leben zum höchsten Gut erheben müsse [11]. Kommunistische Ideologien und ihre Praxis haben aber durchaus auch nicht gezögert, das Lebensopfer, insoweit irrational, für ihre Ziele zu fordern, nämlich für ein »besseres« Leben, nicht für die Lebenserhaltung schlechthin [12].

Die These, daß Lebensverlust und Lebensgefährdung ausschließlich dann gefordert werden können, wenn mit ihnen Lebenserhaltung bezweckt ist, scheint in dieser Art einer rigorosen Formel nicht durchhaltbar. Daß die Logik als eine der wissenschaftlichen Erkenntnisquellen diese These nicht stützt, bedarf keiner besonderen Erläuterung, denn das wertende Setzen von Prämissen geht jeder Logik voraus und diese können daher selbst nicht als logisch notwendig nachgewiesen werden. Doch auch die Empirie, als die andere wissenschaftliche Erkenntnisquelle, stützt diese These nicht. Bevor auf Einzelheiten der nationalen und der internationalen Rechtsordnungen einzugehen ist, seien noch einige allgemeine Betrachtungen angestellt, denn diese könnten zur Aufhellung problematischer Situationen beitragen.

Auf eine in gewisser Weise »logische« Schlußfolgerung bei Beurteilung des Lebens als eines der höchsten Rechtsgüter scheint das Bundesverfassungsgericht sich berufen zu wollen, wenn ausgeführt wird, es müsse nicht näher begründet werden, daß das Leben innerhalb der grundgesetzlichen Ordnung einen Höchstwert darstelle, denn »es ist die vitale Basis der Menschenwürde und die Voraussetzung aller anderen Grundrechte« [13]. Der Hinweis, daß das Leben die Basis der Menschenwürde darstelle, könnte darauf hindeuten, daß auch um letzterer willen das Leben zu schützen sei. Dann aber könnte wiederum auch die Menschenwürde selbst als höherrangig gedacht werden, wenn gerade um ihretwillen das Leben zu schützen ist. Die zweite Aussage, der Bestand des Lebens bilde die »Voraussetzung« aller anderen Grund-

---

[11] Dazu *Leisner*, a.a.O. (Anm. 1), S. 19.

[12] So heißt es im Kommunistischen Manifest: ». . . Die Kommunisten verschmähen es, ihre Ansichten und Absichten zu verheimlichen. Sie erklären es offen, daß ihre Zwecke nur erreicht werden können durch den gewaltsamen Umsturz aller bisherigen Gesellschaftsordnungen. . .«.

[13] BVerfG vom 25. 2. 1975, BVerfGE 39, 42.

rechte, führt ebenfalls zu keiner eindeutigen Bestimmung seines Stellenwertes. Das Leben ist sicherlich die »faktische« Voraussetzung der Grundrechtsanwendung, aber ebenso sicher ist, daß auch nur gegenüber einem Lebenden Grundrechte verletzt werden können, was gleichermaßen nur die Feststellung einer Tatsache bedeutet. Sicherlich kann ein Toter seine Meinung unter der Garantie des Art. 5 GG nicht mehr äußern, aber eine Leiche zu foltern, ist ebenfalls unmöglich. Auch in der Entscheidung anläßlich der Entführung von Hanns-Martin Schleyer hat das Bundesverfassungsgericht – sicherlich weise – keine endgültige Aussage über den Stellenwert des Lebens in der Rechtsordnung abgegeben. Dort ist ausgeführt, daß das Leben als ein höchstes Rechtsgut durch die Staatsgewalt zu schützen sei, daß sein Schutz aber auch ohne Rechtsmißbrauch unterlassen werden dürfe, wenn anders der »effektive Schutz« der Bürger unmöglich werde[14]; auch und gerade die Bürger zu schützen, sei Aufgabe der Staatsgewalt gemäß Art. 2 Abs. 2 GG. Da anzunehmen war, daß freigepreßte Terroristen wiederum das Leben der Bürger bedrohen würden, war eine weitere Abwägung nicht notwendig und hätte nur ein *obiter dictum* bedeutet. Die Frage, ob andere Rechtsgüter dem Lebensrecht in der Güterabwägung gleichgestellt werden dürfen (welche?), bleibt offen.

Die Einstufung des Lebens als den höchsten Wert der Rechtsordnung kann nur unter der Prämisse standhalten, wonach jeder andere Rechtswert letztlich und ausschließlich im Konfliktsfalle dem individuellen Leben unterzuordnen ist, nicht schon der Ermöglichung des Bestandes einer Gemeinschaft und der von ihr gewünschten Lebensform. Hierzu gelangt auch eine extreme Sicht, wie *Walter Leisner* sie in voller Konsequenz bietet[15]. Da es sich hierbei aber nur bei Anerkennung einer mit Mitteln der Logik nicht zu widerlegenden Prämisse um eine zwingende Schlußfolgerung handeln kann, die Anerkennung dieser Prämisse also vorausgesetzt wird, soll doch ein kurzer Blick auf die Empirie als Erkenntnisquelle gerichtet werden. Niemand wird leugnen, daß im Laufe der Rechtsgeschichte zur Opferung individuellen Lebens auch andere Rechtsgüter als *iusta causa* gewertet wurden. Mit Ehrfurcht berichtet man von Helden, die nicht nur für die Erhaltung individuellen Lebens ihr eigenes Leben opferten, sondern auch für ein »lebenswürdiges« Leben der Gemeinschaft oder auch für die Aufrechterhaltung einer Rechtsordnung, deren Zerbrechen durchaus nicht auch gleichzeitig Lebensvernichtung bedeutet hätte, sondern vielleicht »nur« Sklaverei, Ausrottung eines Glaubens oder Auferlegung von – nach Auffassung der Betroffenen – unwürdigen Lebensbedingungen. Ähnlich ist es wohl gemeint, wenn Friedrich von Schiller (grundgesetzwidrig?) das Leben nicht als »der Güter höchstes« bezeichnen läßt. Streik der unterdrückten Arbeitnehmer etwa – auch in Erwartung von lebensgefährdenden Konsequenzen – sollte Lebensbedingungen und nicht nur Lebenserhaltung erzeugen. Der Übergang von der These, daß Le-

---

[14] BVerfG vom 15. 10. 1977, BVerfGE 46, 160 ff.

[15] *Leisner,* a.a.O. (Anm. 1), insbes. S. 25 und S. 37.

bensopfer nur anderes Leben schützen darf, zur These, daß das Leben nicht *per se* und unabdingbar das höchste Rechtsgut darstellen muß, wird hergestellt durch den Übergang von der These unbedingter Betonung der Individualrechte zur These von der Schutzwürdigkeit überindividueller Werte. Die Extreme aber sind wohl niemals durchgehalten worden. Sie sind aber verbindbar durch eine Synthese, welche die Antithesen »Individualschutz« und »Gemeinschaftsschutz« überwölbt; sie lautet: Ohne Schutz der Rechtsordnung – deren Inhalt hier zunächst dahingestellt sein mag – ist kein Rechtsgut, weder der Individualschutz noch der Gemeinschaftsschutz, zu sichern. Welcher Wert auch immer als der »höchste« angesehen wird und wie auch immer die Abwägung innerhalb einer Rechtsgemeinschaft erfolgt, die Wertbestimmung wird sinnlos, wenn das »Recht dem Unrecht weicht«, d.h. wenn diejenige Ordnung nicht eingehalten wird, die zum Schutz der – postulierten – Werte eingerichtet wurde. Unter dieser Perspektive sollen nun Einzelprobleme aufgezeigt werden.

II. Einzelfragen

1. Notwehr und Nothilfe im nationalen und internationalen Recht

Die Abwehr von Angriffen auf Sachgüter durch Privatpersonen mit dem Mittel der Tötung gilt als gerechtfertigt, falls andere Abwehr wegen der Intensität des Angriffs und der Beachtlichkeit des Sachwerts nicht erfolgreich sein kann [16]. Nach jedenfalls überwiegender Auffassung hat auch die Europäische Menschenrechtskonvention das geltende deutsche Strafrecht in dieser Beziehung nicht geändert [17]. Das soll hier nicht nochmals ausgeführt werden. Es wäre theoretisch denkbar, daß der Angegriffene dem Angreifer, der nicht sein Leben bedroht, sondern nur stärker ist, alles hinzugeben hätte, weil das Leben auch des Angreifers nicht vernichtet werden darf [18]. *Walter Leisner* meint auch, das sei unter der Herrschaft des Grundgesetzes geltendes Recht, und die Europäische Menschenrechtskonvention, so wie er sie auslegt, habe deswegen die deutsche Rechtsordnung nicht ändern können, weil diese selbst schon diesem Grundsatz entspreche [19]. Würde wirklich dieses Ergebnis dem

---

[16] Über die Erforderlichkeit der Verteidigung siehe *K. Lackner,* Strafgesetzbuch (14. Aufl. 1981), S. 167.

[17] *Lackner,* a.a.O., S. 168.

[18] Zu dieser Schlußfolgerung müßten jedenfalls diejenigen gelangen, die, wie *Leisner,* a.a.O. (Anm. 1), S. 34, der EMRK entnehmen, daß lebensbedrohende Notwehr zum Sachgüterschutz nicht zulässig sei und daß entweder die EMRK das deutsche Strafrecht verändert habe oder, entgegen der herrschenden Lehre, dieser Grundsatz ohnehin im deutschen Recht schon enthalten ist. So *Dürig,* a.a.O. (Anm. 3), Rdn. 62 zu Art. 1 GG.

[19] Die wohl herrschende Auffassung ist richtig wiedergegeben von *H. Guradze,* Die Europäische Menschenrechtskonvention, Kommentar (1968), zu Art. 2 Ziff. 9.

geltenden Recht entsprechen, wäre die Sachnotwehr durch Tötung des Angreifers nicht gerechtfertigt, sondern bestenfalls entschuldigt. Warum die herrschende Lehre an diesem Rechtfertigungsgrund aber festhält, obwohl das Leben »eines« der höchsten Rechtsgüter ist, erklärt sich daraus, daß die Aufhebung dieses Abwehrrechts Gemeinschaftsleben unmöglich machen könnte. Hier zeigt sich die genannte Synthese. Das *zoon politikon* könnte nur in Vereinzelung leben: eine *contradictio in adjecto.* Würde den Einzelnen dieses Recht genommen, müßte der allein zugelassene Schutz durch eine allgegenwärtige Staatsmacht ausgeübt werden, was wiederum das Ende individuellen Privatlebens bedeuten würde. Es ist wohl nicht notwendig, das weiter auszuführen. Der brutale Räuber würde bei physischer Unterlegenheit des zu Beraubenden kein Risiko eingehen; sein Leben wäre grundsätzlich geschützt und seine Beute so unter Umständen auch. Ein ernstlicher Streit besteht in dieser Frage wohl auch nicht, obwohl die These von der objektiven Wertordnung der Grundrechte (Drittwirkung) verbunden mit der These von der Höchstwertigkeit des Lebens zu diesem Ergebnis kommen müßte.

Uneinigkeit besteht aber bei der Frage nach den Grenzen hoheitlichen, insbesondere polizeilichen Eingriffs. Das Problem, ob die Polizei bei gezieltem Todesschuß, etwa zur Geiselbefreiung, dienstlich oder privat handelt, ob also eine Unterscheidung gemacht werden darf zwischen dienstlicher Kompetenzausübung und strafrechtlichem Rechtfertigungsgrund, ist von *Peter Lerche* und *Paul Kirchhof* eingehend behandelt worden[20]; das soll hier nicht wiederholt werden, zumal die Ergebnisse dieser Untersuchungen überzeugen. Weniger behandelt wurde die Frage nach der Zulässigkeit polizeilichen Lebenseingriffes bei reinem Sachgüterschutz. Es wird behauptet, Art. 2 EMRK lasse den Sachgüterschutz durch hoheitliche Gewalt nicht mit dem Mittel der Lebensvernichtung zu[21]. Das ergibt sich durchaus nicht mit zwingender Schlußfolgerung. Der englische Text lautet, die Lebensvernichtung des Angreifers sei zulässig, »in defence of any person from unlawful violance«, im französischen Text heißt es, »pour assurer la défence de toutes personnes contre la violence illégale«. Die Wegnahme privaten Eigentums unter Gewaltandrohung und auch unter einer solchen, die das Leben des Eigentümers nicht bedroht, sondern nur seinen Widerstand brechen will, kann durchaus als »Gewalt gegen eine Person« bezeichnet werden. Diese Auslegung des Art. 2 EMRK liegt um so näher, als sogar Flucht und Widerstand gegen Festnahme mit Lebensvernichtung verhindert werden dürfen (Art. 2 Abs. 2 b EMRK).

Die deutsche Gesetzgebung gibt hier eine recht klare Auskunft; fraglich könnte nur sein, ob die dort getroffene Regelung mit dem Grundgesetz über-

[20] *P. Lerche,* Der gezielte tödliche Schuß nach künftigem einheitlichem Polizeirecht – Zum Verhältnis hoheitlicher Eingriffsbefugnisse zu den allgemeinen Notrechten, in: Um Recht und Freiheit, Festschrift für Friedrich August von der Heydte (1977), S. 1033 ff.; *P. Kirchhof,* Polizeiliche Eingriffsbefugnisse und private Nothilfe, NJW 1978, S. 969 ff.

[21] Hinweise bei *Lackner,* a.a.O. (Anm. 16), S. 168.

einstimmt. Das Gesetz über den unmittelbaren Zwang bei Ausübung öffentlicher Gewalt durch Vollzugsbeamte des Bundes[22] – hier beispielhaft angeführt und im folgenden als UZwG bezeichnet – schränkt das Recht auf Leben des Art. 2 Abs. 2 GG ausdrücklich in § 3 ein. Der Schußwaffengebrauch ist gemäß § 10 zulässig, wenn ein Verbrechen verhindert werden kann. Das Verbrechen des Raubes (§ 249 Strafgesetzbuch), dessen Tatbestandsmerkmal »mit Gewalt« sich nicht als Leibes- oder Lebensgefahr darstellen muß, dürfte also mit Schußwaffengebrauch verhindert werden; das gleiche gilt u.a. auch für die Brandstiftung. Die Abwehr hat, wie jede polizeiliche Maßnahme, den Grundsatz der Verhältnismäßigkeit zu beachten, aber im Extremfall ist die Tötung des Angreifers rechtmäßig. Wäre das Leben – auch des Angreifers – nicht »eines«, sondern »das« höchste Rechtsgut des Grundgesetzes, wäre das UZwG verfassungswidrig. Nicht jedes Verbrechen dürfte von der Hoheitsgewalt auch durch Tötung des Angreifers verhindert werden, sondern nur ein solches, dessen Durchführung selbst auf die Lebensvernichtung des Angegriffenen gerichtet ist. Schon die Drohung etwa, Geiseln nicht zu töten, sondern »nur« zu verstümmeln (Ohrenabschneiden im Falle Getty, Italien), würde den Todesschuß unzulässig machen, um so mehr die Verhinderung des »einfachen« Raubes, der Brandstiftung oder des Sprengstoffverbrechens, falls es im konkreten Fall nur Sachgüter bedroht. Wäre es richtig, daß nur der Lebensschutz die Tötung rechtfertigen könnte, bedürfte es des Gesetzesvorbehalts in Art. 2 Abs. 2 Satz 2 GG insoweit nicht, als diese Vorschrift die Einschränkungsmöglichkeit auch des Rechts auf Leben vorsieht; denn wenn von zwei Leben notwendig nur eines erhalten werden kann, wäre angesichts der von niemandem geleugneten Schutzpflicht der Staatsgewalt[23] die Unterlassung der Lebensrettung ihrerseits ein »Eingriff« in das Lebensrecht. Der Gesetzesvorbehalt erhält einen Sinn nur dann, wenn er nicht etwas Selbstverständliches legalisieren soll, wenn es also um einen »Eingriff« als Alternative zum »Nichteingriff« geht. Welches Leben in diesen Fällen notwendiger Kollision von zwei Lebensrechten schutzwürdiger ist, kann sich nur nach dem Grundsatz bestimmen, daß das Recht dem Unrecht nicht zu weichen hat. Der Gesetzesvorbehalt bezüglich des Lebens in Art. 2 Abs. 2 GG kann sich so sinnvoll nur auf den Schutz der körperlichen Unversehrtheit und den Sachgüterschutz beziehen und ist in UZwG verfassungsgemäß verwendet. Wenn es im Gesetz heißt, daß der Waffengebrauch nur dazu dienen darf, »angriffsunfähig« zu machen (§ 12), so bedeutet das, daß das in bezug auf seine Erforderlichkeit mildeste Mittel einzusetzen ist. Falls aber in einer besonderen Situation nur die Tötung »angriffsunfähig« machen kann, wäre auch sie legal. Der bis zu seiner eigenen letzten Lebenssekunde auf Erfolg sei-

[22] Gesetz über den unmittelbaren Zwang bei Ausübung öffentlicher Gewalt durch Vollzugsbeamte des Bundes vom 10. 3. 1961, BGBl. I, S. 165 ff.

[23] *I. von Münch*, Grundgesetz – Kommentar, Bd. 1 (2. Aufl. 1981), Rdn. 46 zu Art. 2, mit Nachweisen der Rechtsprechung des BVerfG.

nes Handelns bedachte Fanatiker ist eine auch in der modernen Welt durchaus bekannte Erscheinung.

Das UZwG verbietet den Schußwaffengebrauch, wenn erkennbar Unbeteiligte gefährdet würden (§ 12), es sei denn, bei zulässigem Einschreiten gegen eine Menschenmenge (§ 10) sei das unvermeidbar. Im Extremfall bedeutet das, daß dann, wenn die Vernichtung auch hoher Sachgüter droht (Sprengung der Münchner Pinakothek – Beispiel von *P. Lerche*[24]), auf den Täter nicht geschossen werden darf, wenn ein Unbeteiligter getroffen werden könnte. Der Grundsatz, das Recht müsse dem Unrecht nicht weichen, kann ein anderes Ergebnis auch nicht fordern, denn der Unbeteiligte ist nicht »im Unrecht«. Es stellt sich nur die Frage, ob das Grundgesetz dieses Ergebnis fordert oder ein anderes ebenfalls zuläßt. Der Gesetzgeber hat – weise – im UZwG den Waffengebrauch nur verboten, wenn »mit hoher Wahrscheinlichkeit« ein Unbeteiligter gefährdet würde, und also in Kauf genommen, daß nicht jede vorstellbare Gefährdung vermieden werden muß. Der Gesetzesvorbehalt des Art. 2 Abs. 2 GG scheint auch diese gesteigerte Einschränkung zu decken. Die These allerdings, daß Lebensgefährdung nur zur Lebensrettung in Kauf genommen werden darf, steht mit dem Erfordernis der nur »hohen Wahrscheinlichkeit« nicht in Übereinstimmung.

Man wird zu dem Schluß kommen dürfen, daß das Grundgesetz eine klare Aussage in bezug auf das Verhältnis von Sachgüterschutz und Lebensgefährdung nicht enthält, sondern dem Gesetzgeber einen Ermessensspielraum läßt. Zwar ist das Leben als Rechtsgut im Grundgesetz nun genannt, was unter der Herrschaft der Weimarer Verfassung nicht der Fall war und sich aus den Erfahrungen der nationalsozialistischen Zeit erklärt[25]. Aber ebenso wie andere sog. absolute Grundrechte dem Gemeinschaftsvorbehalt unterliegen[26], kann wohl auch das Lebensrecht nicht jeglichem Gemeinschaftsvorbehalt entzogen sein und sich nicht gegen jedes andere notwehrfähige Rechtsgut in voller Unbedingtheit behaupten. Soweit das Grundgesetz keine eindeutige Aussage enthält, ist es Sache des politischen Parlaments, die ihm als schutzwürdig erscheinenden Werte zu bestimmen und ihr Verhältnis zueinander zu regeln.

Notwehr und Nothilfe im internationalen Recht haben einen vielfältigen Aspekt, denn es kann sich sowohl um die Rechtsverteidigung durch das Individuum, durch Menschengruppen und auch um eine solche durch den Staat selbst handeln.

Das individuelle Recht zur Notwehr im Rahmen der Europäischen Menschenrechtskonvention wurde schon im Hinblick auf die Lebensgefährdung durch Abwehrhandlung charakterisiert. Hierauf soll nicht nochmals zurück-

---

[24] *Lerche*, a.a.O. (Anm. 20), S. 1048.

[25] Dazu *Dürig*, a.a.O. (Anm. 3), Rdn. 8 zu Art. 2 Abs. II GG.

[26] Vgl. statt vieler Entscheidungen BVerfG vom 1. 8. 1978, BVerfGE 49, 24 (53 ff.).

gekommen werden; die lateinamerikanische Menschenrechtskonvention mag auch außer Betracht bleiben, weil sie unseren Staat nicht bindet. Der Pakt der Vereinten Nationen über bürgerliche und politische Rechte[27] sagt in Art. 6 nicht mehr, als daß ein *inherent right* auf das Leben bestehe, daß dieses gesetzlich zu schützen sei und Willkür bei Lebensvernichtung nicht zulässig sei. Daß dieser Bestimmung für die Frage der Güterabwägung bei der Ausübung der Notwehr nichts Näheres zu entnehmen ist, wurde schon aufgezeigt, insbesondere daß die These, wonach Tötung nur zum Schutz anderen Lebens und nicht zum Schutz von körperlicher Unversehrtheit oder zum Sachgüterschutz zulässig ist, dort keine zwingende Rechtsgrundlage findet.

Aus der Charta der Vereinten Nationen (Art. 2 Ziff. 4) könnte entnommen werden, daß auf der Grundlage des strikten Gewaltverbots ein Staat unter keinen Umständen berechtigt sein soll, das Leben der Angehörigen eines anderen Staates zu bedrohen oder zu vernichten. Das gilt jedenfalls für den rechtswidrigen Aggressor. Unklar aber bleibt, ob und welche inhaltlichen Grenzen bei der staatlichen Notwehr gegenüber rechtswidrigem Angriff anderer Staaten bestehen. Art. 51 der UN-Charta spricht nur von dem *inherent right* zur Abwehr eines bewaffneten Angriffs durch andere Staaten. Auch die Aggressionsdefinition[28] der Generalversammlung der Vereinten Nationen verweist auf die Anwendung und die Androhung mit bewaffneter Gewalt. Die Vernichtung des Lebens fremder Staatsbürger scheint also nur dann als rechtmäßig angesehen zu werden, wenn sie zum Schutze des Lebens der eigenen Staatsbürger erfolgt. So einfach ist jedoch die Rechtslage nicht, denn die Aggressionsdefinition läßt Ausnahmen zu. Es heißt dort, daß es immer rechtmäßig sei, das Selbstbestimmungsrecht mit Gewaltmaßnahmen durchzusetzen, die sonst unzulässig wären und eine verbotene Aggression bedeuten würden (Art. 7)[29]. Die Resolution ist formell nicht bindend, aber das Selbstbestimmungsrecht hat in der Charta der Vereinten Nationen eine legale Grundlage und könnte als Escapeklausel auch für Art. 51 der Charta gelten, abgesehen von seiner vielfältig zum Ausdruck gebrachten Rechtsqualität in anderen internationalen Dokumenten.

Da es hier um das Leben als eines der höchsten zu schützenden Rechtsgüter geht, erhebt sich die Frage, ob auch das Selbstbestimmungsrecht gewaltsam und also mit der Möglichkeit der Lebensvernichtung nur dann durchgeführt werden darf, wenn der Mißachter des Selbstbestimmungsrechts seinerseits Leben der Gruppenmitglieder vernichtet. Eine solche Begrenzung scheint jedoch der überwiegenden Auffassung in der Völkergemeinschaft

---

[27] Oben Anm. 6.

[28] Oben Anm. 9.

[29] Die Bestimmung lautet: »Keine Bestimmung dieser Definition, insbesondere Artikel 3, kann in irgendeiner Weise das aus der Charta hergeleitete Recht auf Selbstbestimmung, Freiheit und Unabhängigkeit von Völkern beeinträchtigen . . .«; der in bezug genommene Art. 3 enthält die Aufzählung der als verboten geltenden Angriffshandlungen.

nicht zu entsprechen [30]. Die zur Ausübung des Selbstbestimmungsrechts berechtigende Situation setzt, bei allen Zweifeln hinsichtlich ihrer Abgrenzung, eine die Menschenrechte unerträglich verletzende Diskriminierung der Gruppe voraus [31]. Dieser Standard sollte auch hoch angesetzt werden. Jedoch muß diese unerträgliche Diskriminierung nicht unbedingt in der Tötung der Gruppenmitglieder bestehen. Gegen Völkermord [32] besteht selbstverständlich ein Abwehrrecht im Sinne des *inherent right* auf Notwehr. Faktische Sklaverei, Heiratsverbot, Sprachverbot, Kulturvernichtung, völliger Eigentumsentzug oder permanente Gruppeninternierung sind aber Beispiele dafür, daß das Selbstbestimmungsrecht auch ohne Lebensvernichtung der Gruppe zu gewaltsamer Aktion berechtigen könnte. Es zeigt sich so, daß auch hier die These, wonach Leben nur zur Rettung anderen Lebens vernichtet werden darf, zweifelhaft wird und wohl in Extremfällen nicht standhält. Allgemein kann wohl gesagt werden, daß das geltende Völkerrecht der Verteidigung von fundamentalen Menschenrechten bisher nicht eine Grenze gesetzt hat, wonach nur Völkermord lebensgefährdende Abwehr rechtfertigt.

2. Der Einsatz des Lebens von Soldaten, Polizisten und Vollzugsbeamten nationaler und internationaler Instanzen

Die Behauptung, daß Lebensopfer nur zur Lebensrettung verlangt werden dürfe, hat die Frage entstehen lassen, ob das mit der Forderung in Übereinstimmung steht, die vom Soldaten den Lebenseinsatz im Kriege verlangt [33]. Die Situationen des militärischen Einsatzes können so vielfältig sein, die Distanz zwischen Lebenseinsatz des Soldaten, seiner Lebensbedrohung, der Lebensvernichtung des militärischen Gegners und der Lebensgefahr nicht unmittelbar Beteiligter kann so groß sein, daß von der Unmittelbarkeit der Notwehr nicht mehr gesprochen werden kann. Man versucht – es wäre frivol, von einer Eselsbrücke zu sprechen –, diese Denkschwierigkeit aufzulösen, indem man den Staat selbst als in einer Notwehrlage befindlich ansieht, die zur Aufrechnung von Leben gegen Leben führt. Richtig wird auch erkannt, daß die Qualifikation des Krieges als »gerecht« hier nicht weiterhelfen kann. Das Völkerrecht trennt mit gutem Grund die rechtliche Beurteilung des *ius ad bellum* von derjenigen des *ius in bello*; der Teilnehmer an einem verbotenen Angriffskrieg gilt nicht als Mörder, sondern als regulärer Kombattant und kann persönlich nicht für die Mißachtung des *ius ad bellum* haftbar gemacht

---

[30] *I. Seidl-Hohenveldern,* Völkerrecht (4. Aufl. 1980), Rdn. 1158.

[31] Zu der gesamten Frage *K. Doehring,* Das Selbstbestimmungsrecht der Völker als Grundsatz des Völkerrechts, Berichte der Deutschen Gesellschaft für Völkerrecht, Heft 14 (1974).

[32] In der Konvention über die Verhütung und Bestrafung des Völkermordes vom 9. 12. 1948, BGBl. 1954 II, S. 730, heißt es in Art. 1, daß Völkermord als ein »Verbrechen gemäß internationalem Recht« anzusehen sei, woraus ein Notwehrrecht mit Selbstverständlichkeit sich ergibt.

[33] *Dürig,* a.a.O. (Anm. 3), Rdn. 19, 20.

werden[34]. Die Forderung an den Soldaten, sein Leben einzusetzen, und die Berechtigung des Soldaten, das Leben des Gegners zu vernichten, bedürfen nicht der Grundlegung durch den Satz, daß Leben nur gegen Leben eingesetzt werden darf. Verteidigungsfähig sind hier durchaus auch überindividuelle Rechtsgüter wie die Souveränität des Staates, die Freiheit zur Gestaltung der eigenen Rechtsordnung, der Schutz der Bevölkerung nicht nur gegen Lebensbedrohung, sondern auch gegen Fremdherrschaft unter nicht gewünschten Lebensbedingungen. Das entspricht dem geltenden Völkerrecht. Wäre es anders, müßte der Aufforderung des einen Staates an den anderen, man möge einer bestimmten politischen Auffassung folgen, Gewalt werde nur bei Weigerung eingesetzt, nachgegeben werden, denn ein unbedingtes Gebot, Leben nur zu gefährden, um Leben zu retten, würde kein anderes Ergebnis zulassen.

Der in Staaten westlich-liberaler Demokratie und vor allem in der Bundesrepublik Deutschland feststellbare Trend zur Ausweitung des Rechts auf Kriegsdienstverweigerung[35] ist aufschlußreich auch für die hier zu behandelnden Fragen. Der Kriegsdienstverweigerer beruft sich darauf, daß sein Gewissen ihm das Töten des Gegners auch im Kriege verbiete. Konsequenterweise müßte er in Kauf nehmen, daß, unter dem Gesichtspunkt der Lebenserhaltung, die Androhung der Gewalt das Nachgeben gebietet, will er sich nicht selbst doch und letztlich auf die »Gewissenlosigkeit« seiner Mitbürger verlassen. Moderne Slogans, »Frieden schaffen ohne – eigene – Waffen«, führen auch zu dieser Konsequenz. Wenn es richtig wäre, daß der Lebenseinsatz für den Soldaten im Kriegszustand und die Bedrohung des Lebens der Angreifer sich nur mit dem Notwehrrecht zum Lebensschutz rechtfertigen könnte, müßte die Abschaffung des Rechts zur Kriegsdienstverweigerung unproblematisch sein, denn bei bekannten Befragungen pflegt der Kriegsdienstverweigerer eigene Tötungshandlungen dennoch nicht auszuschließen, wenn sein eigenes Leben oder das seiner Angehörigen bedroht ist; die Befragungskommission mutet ihm eine gegenteilige Entscheidung auch nicht zu. Die Rechtfertigung der Tötung des Gegners im Kriege mit der Begründung, hier stehe Leben gegen Leben im Sinne einer »privaten« Notwehrsituation, erscheint höchst zweifelhaft[36].

Die Probleme des das eigene Leben gefährdenden Einsatzes von Polizisten oder auch Feuerwehrleuten zeigen kein anderes Bild. Soll von ihnen ebenfalls nur verlangt werden dürfen, das eigene Leben dann zu riskieren, wenn es gilt, anderes Leben zu retten? Mit Gewalttätigkeit verbundene Demonstrationen und Unruhen haben – so sei unterstellt – bisher nicht das Ziel

---

[34] Zahlreiche Nachweise bei *F. Berber,* Lehrbuch des Völkerrechts, Bd. 2 (Kriegsrecht) (1962), S. 57 Anm. 3.

[35] *D. Blumenwitz,* Wehrpflicht und Ersatzdienst (1978) Teil III, S. 343 ff., gibt tabellenmäßige Übersichten zur Entwicklung in der Bundesrepublik, deren Trend sich bis heute noch wesentlich verstärkt hat.

[36] Zur Rsprechung in diesen Fällen *U. Hemmrich,* Grundgesetz – Kommentar (hrsg. von I. von Münch), a.a.O. (Anm. 23), zu Art. 4, Rdn. 33b.

gehabt, Leben zu zerstören, aber doch Kraftwagen in Brand zu setzen, Banken zu beschädigen, Schaufenster einzuschlagen und Geschäfte zu plündern. Hiergegen müßte der Einsatz von Polizei und Feuerwehr dann verboten sein, wenn er das Leben der Vollzugsbeamten bedroht, denn es geht »nur« um Sachschaden. Die Feuerwehr müßte z. B. den Kölner Dom gegebenenfalls abbrennen lassen und dürfte nur die Umgebung sichern, falls Löscharbeiten lebensgefährdend sein könnten. Die Beispiele ließen sich beliebig vermehren; sie sollten zeigen, wie vorsichtig man auch mit Abgrenzungen umgehen muß, die zunächst eine gewisse Evidenz für sich beanspruchen könnten.

Der Satz »Das Leben ist der Güter höchstes nicht« sei grundgesetzwidrig, kann nicht unwidersprochen bleiben. Das Grundgesetz enthält ihn ebensowenig wie das Völkerrecht. Es ist auch nicht anzunehmen, daß das Grundgesetz eine gesetzgeberische Entscheidung zulassen würde, wonach das Lebensrecht sich gegen jedes andere Rechtsgut durchzusetzen habe und also unbedingten Schutz genießt. Soll wirklich trotz der in Art. 1 GG enthaltenen Versicherung, die Würde des Menschen sei »unantastbar« und sie zu schützen sei eine Verpflichtung aller Staatsgewalt, die Erniedrigung des Menschen, etwa durch physische oder psychische Folterung, nicht mit Lebensvernichtung des Rechtsbrechers verhindert werden dürfen, wenn jede andere Maßnahme versagen würde? Und erhält der Gesetzesvorbehalt des Art. 2 Abs. 2 GG, auf das Lebensrecht bezogen, nicht erst so seinen Sinn? Eine Überbetonung der Schutzwürdigkeit des individuellen Lebens als ausschließlich höchstem Wert der Rechtsordnung, dem jeder andere Wert im Konfliktsfalle zu weichen hätte, macht eine jede Gemeinschaft abwehrunfähig, sowohl im nationalen als auch im internationalen Recht. Die abwehrunfähige Gemeinschaft kann dann auch das Individuum und sein Leben nicht mehr schützen. Da das Individuum sich selbst nicht gegen stärkere Individuen, stärkere Gemeinschaften oder auch Naturgewalten zu schützen vermag, kann die Überbetonung seines Wertes und der Schutzwürdigkeit seines Lebens letztlich auch zu seiner Schutzlosigkeit führen. Das individuelle Leben mag dann als »der Güter höchstes« bezeichnet werden – aber es wird nicht »überleben«. Gemeinschaftsschutz und Individualschutz sind keine Alternativen, sondern bedingen sich gegenseitig im Rahmen einer rechtlich geregelten Ordnung. Diese Ordnung sollte weder lauten »Recht ist, was dem Volke nützt«, noch sollte sie den Wert des individuellen Lebens jeder Gemeinschaftsnotwendigkeit vorordnen. In einer so verstandenen Synthese der gegenseitigen Abhängigkeit sollte der Satz gelten, daß das Recht – beruhend auf der Erkenntnis der Gemeinschaftsbezogenheit des Menschen – dem Unrecht nicht zu weichen habe.

The Limits of the Right of Intervention in a Case before the International Court of Justice

TASLIM O. ELIAS

I. Introduction

The Statute of the Court, under the general heading of »Procedure« (Chapter 3), provides, among other things, for the right of intervention by States in proceedings before the Court, in certain circumstances, in two types of specific situation.

Article 62 provides as follows:

(1) Should a State consider that it has an interest of a legal nature which may be affected by the decision in the case, it may submit a request to the Court to be permitted to intervene.

(2) It shall be for the Court to decide upon this request.

Article 63 provides as follows:

(1) Whenever the construction of a Convention to which States other than those concerned in the case are parties is in question, the Registrar shall notify all such States forthwith.

(2) Every State so notified has the right to intervene in the proceedings; but if it uses this right, the construction given by the Judgment will be equally binding upon it.

Despite these statutory provisions, the literature on the subject is sparse[1], and State practice is relatively even more sparse. There have been only three occasions on which interventions have fallen to be considered by the Court: once by the Permanent Court of International Justice, and twice by the International Court of Justice.

In the first decided case in which the issue arose, the case of the S. S. *»Wimbledon«*[2], the Permanent Court of International Justice observed that

[1] Apart from incidental and scattered references in a few pages to be found in journal articles and monographs, there would understandably appear to be as yet no autonomous studies on the subject of intervention before the International Court of Justice.

[2] 1923, P.C.I.J., Series A, No. 1, pp. 9–13.

the Statute of the Court provides for »two sets of circumstances and two different forms in which intervention is possible on the part of States which are not from the outset concerned in a suit brought before it«. It pointed out that the first form is dealt with in Article 62 of the Statute and Articles 58 and 59 of the then Rules of Court, and that »it is based on an interest of a legal nature advanced by the intervening party, and the Court should only admit such intervention if, in its opinion, the existence of this interest is sufficiently demonstrated«. The Court pointed out, however, that »when the object of the suit before the Court is the interpretation of an international Convention, any State which is a party to this Convention has, under Article 63 of the Statute, the right to intervene in the proceedings instituted by others and, should it make use of the right thus accorded, the construction given by the Judgment of the Court will be equally binding upon it as upon the original applicant parties«.

In the case itself, Great Britain, France, Italy and Japan brought proceedings asking the Court to decide whether the German authorities were within their rights in refusing to the S. S. *Wimbledon*, on 21 March 1921, free access to the Kiel Canal and, if it considered it necessary, to determine the damages due for the prejudice caused to the vessel by reason of this refusal.

The Polish Government, in a note of 22 May 1953, informed the Court that it desired permission to intervene in the suit on the side of the four applicant States under Article 62 of the Statute and Articles 58 and 59 of the then Rules of Court; but, although Article 63 was not expressly referred to, the note cited in support of the application Poland's participation in the Treaty of Versailles, in particular Article 380 thereof, in which the rights and interests of Poland were guaranteed, pointing out that the refusal to allow the S. S. *Wimbledon* access to the Kiel Canal was a violation of that guarantee. In a further communication made to the Court at the public sitting on 25 June, it would seem that the Polish Government abandoned its exclusive claim in the earlier note, and seemed intent on availing itself of the right conferred upon it, as a party to the Treaty of Versailles, by Article 63 of the Statute, and did »not insist that the grounds submitted by it as justification for intervention under Article 62 should be taken into consideration«. In these circumstances, the Court observed as follows: »The attitude thus adopted renders it unnecessary for the Court to consider and satisfy itself whether Poland's intervention in the suit before it is justified by an interest of a legal nature, within the meaning of Article 62 of the Statute. It will suffice for the Court to note that in this case the interpretation of certain clauses of the Treaty of Versailles is involved in the suit and that the Polish Republic is one of the States which are parties to this treaty«. The Court added, rather significantly: »In view of the facts established above, which are conclusive, and of the statements made at the hearing by the representatives of the applicant Powers, who left the matter to the decision of the Court, *the Court records* that the Polish Government intends to avail itself of the right to intervene conferred upon it by Article 63 of the Statute«. The Court therefore accepted the intervention of Poland in the case of the S. S. *Wimbledon* under Article 63.

The second occasion on which the issue of intervention proceedings came before the Court, which was also the first occasion that the matter was dealt with by the International Court of Justice, as distinct from the Permanent Court of International Justice, but nevertheless under the same Articles of the Statute of the Court, was in the *Haya de la Torre*[3] case, which itself arose out of the asylum dispute[4] between Colombia and Peru. In the instant case, the Government of Cuba, availing itself of the right which Article 63 of the Statute of the Court confers on States parties to a Convention, filed a Declaration of Intervention with the Court, attaching thereto a Memorandum in which it stated its views with regard to the interpretation of the Havana Convention of 1928, which it had ratified, and also its general attitude towards asylum. This Memorandum was considered by the Court as constituting the Written Observations of the Government of Cuba, which Article 63, paragraph 4 of the then Rules of Court required that the Government of Cuba should submit. The Government of Peru, however, contended that the intervention of the Government of Cuba was inadmissible on two grounds: *(a)* that it was out of time, and *(b)* that the Declaration, as well as the Memorandum, did not constitute »an intervention in the true meaning of the term, but an attempt by a third State to appeal against the Judgment delivered by the Court earlier in the case between the two contesting Parties«. On this point, the Court observed that »every intervention is incidental to the proceedings in a case«, and that »it follows, therefore, that a declaration filed as an intervention only acquires that character in law if it actually relates to the subject-matter of the pending proceedings. The subject-matter of the present case differs from that of the case which was terminated by the Judgment of 20 November 1950«. This was because it concerned the surrender of Haya de la Torre to the Peruvian authorities, a point not dealt with in that earlier Judgment. The Court was of the view that the arguments set out in the Memorandum appeared to be an attempt to raise a matter which was already *res judicata* as already decided in the earlier case[5]; but Cuba argued, at the public hearing, that the intervention was based on the fact that the Court was required to interpret a new aspect of the Havana Convention, an aspect not dealt with in the earlier Judgment of the Court. The Court accordingly granted intervention to Cuba in these words: »Reduced in this way, and operating within these limits, the intervention of the Government of Cuba conforms to the conditions of Article 63 of the Statute, and the Court, having deliberated on the matter, decided on 16 May to admit the intervention

---

[3] Judgment [of 13 June 1951], I.C.J. Reports 1951, p. 71.
[4] Asylum, Judgment, I.C.J. Reports 1950, p. 266.
[5] Colombia took the line that Cuba was entitled to intervene: *ibid.*, p. 74.

in pursuance of paragraph 2 of Article 66 of the Rules of Court«. Thus, again, the Court found itself called upon to deal with the issue of intervention within the context of Article 63 of the Statute of the Court, not of Article 62.

We shall now turn to the third example of intervention before the Court, which was the second such case to arise. This occurred in the *Nuclear Tests*[6] case, where the Government of Fiji submitted an Application to the Court, seeking to intervene in the dispute between France, on the one hand, and Australia and New Zealand, on the other, concerning the holding of atmospheric tests of nuclear weapons by the French Government in the Pacific Ocean. In view of the fact that, by an order dated 20 June 1973, the Court had decided that the written proceedings in the case itself should first be addressed to the questions of the jurisdiction of the Court to entertain the dispute between Australia and France, and of the admissibility of Australia's Application, the Court decided to defer its consideration of the Application of the Government of Fiji for permission to intervene in the proceedings until it had pronounced upon the questions to which the pleadings were to be addressed. In other words, the Court postponed consideration of Fiji's Application for Intervention until the merits of the substantive dispute had been disposed of. After hearing that case, and after coming to the conclusion that the case itself had become moot, so that it was unnecessary to adjudicate upon it, the Court decided that Fiji's Application need no longer be considered. The question of intervention was not raised in the Court's Judgment, but a number of separate and dissenting opinions were submitted by some Judges, most of which suggested that, before an intending intervener could come to the Court for permission under Article 62 of the Statute of the Court, it should establish a jurisdictional link between itself and the parties to the main case. In one of the opinions, it was argued that it could be regarded as an oversight on the part of those reviewing the Statute of the Court in 1922 that they did not specify the relevance of Article 36 of the Statute in relation to their formulation of Article 62; and also that such a requirement was necessary, especially where the intervener (Fiji, in this case) wanted to intervene on the side of certain parties (Australia and New Zealand) against another (France) with regard to a mutual concern (shared by Fiji with Australia and New Zealand regarding the atomic tests in the Pacific Ocean). We shall consider this point later, in the analysis of Article 62 below.

A fourth occasion would almost have arisen had Albania decided, in the *Monetary Gold Removed from Rome in 1943*[7] case, to intervene in the case between Italy, on the one hand, and France, the United Kingdom and the Unit-

---

[6] (*Australia* v. *France*), Application to Intervene, Order of 12 July 1973, I.C.J. Reports 1973, p. 320.

[7] Judgment, I.C.J. Reports 1954, p. 19.

ed States of America, on the other. The Governments of France, the United Kingdom and the United States, to which the implementation of the reparations agreement had been entrusted, decided that the gold in Rome should be delivered to the United Kingdom, unless, within a certain time- limit, Italy or Albania applied to the Court, requesting it to adjudicate on their respective rights. Albania, however, took no action. Italy, on the other hand, made an Application to the Court, and later raised the preliminary question as to whether the Court had jurisdiction to adjudicate upon the validity of the Italian claim against Albania. Since Albania had not consented to come to Court, the Court found that it could not attempt to deal with a dispute between Albania and Italy, and that it was therefore unable to decide the question submitted to it. The gold issue touched and concerned the interests of Albania in the way contemplated by Article 62 (1) of the Statute, and it seemed that intervention of Albania would have been permitted had she sought it, since Albania would have been able to show that she had »an interest of a legal nature in the subject-matter of the dispute« between Italy and the other three States. It is significant to note the concluding remarks made by the Court on this point: »Albania has not submitted a request to the Court to be permitted to intervene. In the present case, Albania's legal interests would not only be affected by a decision, but would form the very subject-matter of the decision. In such a case, the Statute cannot be regarded, by implication, as authorizing proceedings to be continued in the absence of Albania« [8]. Thus, the Court fell just short of inviting Albania to intervene in that case.

II. Argument that Article 62 should not be Read in Isolation

This may be so, but reading it within the context of the Statute as a whole, including Article 36, must involve not only reading Article 62 (1) above as subject to Article 36, but also to reading it as intended to be regarded as apart from and independent of Article 36. If this were not so, and Article 62 (1) were to be read subject to Article 36, what would have been more natural than to make the issue of intervention subject to compulsory jurisdiction in, say, a subsection (7) of Article 36? Indeed, a cross-reference to Article 36 might have been included in Article 62, making it clear that compulsory jurisdiction as envisaged in Article 36 is really intended to apply in the case of contentious proceedings no less than in the case of intervention. For example, Article 53 of the Statute, dealing with the problem of the non-appearing defendant, contains a specific reference to Articles 36 and 37 in these words: »The Court must, before doing so, satisfy itself, not only that it has

[8] *Ibid.*, p. 32, last para.

jurisdiction in accordance with Articles 36 and 37, but also that the claim is well-founded in fact and law«[9]. It is clearly no answer to say that it is because the defendant State is absent in such a case that this specific reference to Articles 36 and 37 has been inserted. The Statute really requires the issue of jurisdiction and stipulates it expressly.

Now, it may be asked why some such clarification or emendation was not made until the unsatisfactory attempt, to be found in our present Article 81 (2) *(c)* of the 1978 Rules of Court. There had been at least three occasions for having done so, but nothing had in fact been changed. The first was during the most interesting debate on the whole subject, when the 1920 Committee of Jurists considered whether or not to include the requirement of jurisdictional link in intervention proceedings[10]. The second notable occasion was during the formulation of the General Act of Pacific Settlement of 1928 (the so-called Kellogg-Briand Pact) on the whole issue of pacific settlement of disputes in the contemplation of judicial or arbitral proceedings. The issue of compulsory jurisdiction was discussed without any specific mention of any link with intervention proceedings, which were no doubt considered as incidental to judicial or arbitral settlement of disputes. The third occasion for an emendation of the specific provision of Article 62 (1) of the Statute with regard to the insertion of the requirement of jurisdictional link was during the overall reconsideration of the Statute of the Court in 1945, including in particular the question of the compulsory jurisdiction of the Court. The Conference seriously reviewed the whole question and, in relation to intervention proceedings, all it finally decided to do was to delete the words »as a third party« from the text as it then stood, leaving the rest of Article 62 (1) as it now stands in the Statute[11]. No other change was made to the text of Article 62; and no attempt was made to align it with Article 36 with regard to jurisdiction of the Court. Article 62 was left severely alone to continue its inexorable course as it had been left in consequence of the 1922 debate in the Judges Rules Committee[12].

---

[9] Wherever the Statute requires to be specific, it often does so by cross-references; e.g., reference in Article 31 (6) to Articles 2, 17 (2), 20 and 24, emphasizing the conditions necessary to be fulfilled by *ad hoc* judges in order to entitle them to be put on terms of complete equality with their colleagues.

[10] See *Continental Shelf (Tunisia/Libyan Arab Jamahiriya),* Application to Intervene, Judgment, I.C.J. Reports 1981, p. 3, para. 21 for the origin of the idea.

[11] *Ibid.*, para. 22.

[12] The three following points were discussed: *(a)* Lord Finlay's suggestion (see Lord Finlay's Opinion as to whether and how far a would-be intervener should be allowed to inspect the pleadings of the parties to the principal case). *(b)* Have third parties interested in a case the right of intervention only when the original parties have accepted the compulsory jurisdiction of the Court? *(c)* Is there any difference with regard to this question between Articles 62 and 63 of the Statute? All three questions were finally answered by saying that each should be decided on its merits as it arose. See pp. 24–28 and p. 46 for the opposing views of Moore and others and

It has since then been left to the Court to apply it as it deems fit in each particular case, without making a jurisdictional link a *sine qua non* in every case of intervention. It may be pointed out in parenthesis that, in the argument of those who insist that the requirement of jurisdictional link is *implicit* in Article 62 (1) of the Statute there would seem to lie an assimilation of the intervention proceedings with the issue of compulsory jurisdiction which contentious proceedings necessarily imply. Intervention, it is true, involves an incidental proceeding to that in the main case[13], but it need not necessarily follow exactly the same route, since the objects and goals may, and indeed often do, differ.

It has again been said that not to insist on the jurisdictional link might scare States away from the Court. There is some force in this argument but, as it will be shown presently, it is a double-edged sword. It is true that when States accept the jurisdiction of the Court in a particular case they have mainly themselves in contemplation and do not normally anticipate what must seem to them an intrusion or interference in their *inter se* dispute. This is especially so in the case of two or more parties agreeing, by means of a *compromis*, to submit their particular dispute to the Court. The tendency to regard such a procedure as private or exclusive to themselves is natural and understandable. However, States must realize that, in the circumstances of present-day international life, third parties are more likely than not to show or assert a genuine interest in the subject-matter of their dispute, which was by inadvertence not contemplated by the parties themselves when the dispute was first formulated. States must also make due allowance for the possibility of third-party intervention whenever they are about to embark upon submitting their dispute to judicial settlement; States must be prepared to regard intervention as a predictable contingency of international life in the rapidly increasing inter-dependence of State relations. So intermingled have the economic, technological, technical and geographical interests of States become today that they must expect the possibility that their bilateral disputes might impinge upon the interests of a third State, even tangentially; it all depends on the circumstances of each particular situation or case. Therefore, it behoves States contemplating litigation to consider third-party intervention as a possibility and, in any case, not always to be regarded as an unfriendly act.

---

those of Anzilotti as to whether or not a jurisdictional link must be established before an intending intervener could be permitted to intervene. (Prelim. Session, 16th Meeting, held at the Peace Palace, The Hague, on 23 February 1922, at 10.30 a.m.; M. Loder, President, in the chair.)

The conclusion was: »The President stated that he could not take a vote upon a proposal the effect of which would be to limit the right of intervention (as prescribed in Article 62) to such States as had accepted compulsory jurisdiction. If a proposal in this sense were adopted, it would be contrary to the Statute.« (See Prelim. Session, 17th Meeting, 24 February 1922, p. 96).

[13] The I.C.J. observed in the *Haya de la Torre* case: »every intervention is incidental to the proceedings in a case« (I.C.J. Reports 1951, p. 16).

They must bear in mind that contemporary international law permits third States to intervene in the dispute between two or more States, in certain well-defined circumstances, as in Articles 62 and 63 of the Statute.

The other aspect of the requirement of a jurisdictional link which might hamper the development of international law is that strict insistence on it as a *sine qua non* in every case may mean that States are likely to abstain from coming to the Court even where they have interests worth submitting for consideration within the meaning of Article 62. It may, of course, be contended that the remedy is for all States to accept the compulsory jurisdiction of the Court; and that that would encourage wider participation in international judicial activities generally. But that stage has not yet been reached and, in the mean time, intervention proceedings should not be unduly hampered by an inflexible requirement of jurisdictional link in judicial proceedings.

III. Is there a Contradiction between Article 62 of the Statute and Rule 81 (2) (c)?

If carefully considered, Rule 81 (2) *(c)* would not seem to contradict or go beyond Article 62 of the Statute. Indeed, it cannot do so. It must be seen as an attempt to clarify it on the issue of the requirement of jurisdiction where that question is a relevant consideration in a particular case. It does not lay down any invariable rule that jurisdiction must be established between the intervening State and the two or more States parties to the original dispute. Two other factors, namely, whether a would-be intervener has an interest of a legal nature that may be affected by the Court's decision in the main case, or whether the object of the proposed intervention is relevant and appropriate to the subject-matter of the principal dispute, might enable the Court to determine whether or not intervention should be permitted in a particular situation. Rule 81 (2) *(c)* spells out these two requirements admirably.

The requirement in Article 62 that the intervening State should show *(a)* that it has an interest of a legal nature, and *(b)* that that interest may be affected by the decision of the Court in the dispute between the two original parties to the case is a crucial one. It is the one most distinguishing feature between Article 62 and Article 63. Intervention under Article 63 is open to all those States that can show that the construction of an international convention, to which they are all parties, is involved; no other requirement need be fulfilled before an intervening State can participate in such proceedings before the Court. Intervention under Article 63 is, on this basis, automatic for the State intending to intervene. Under Article 62, however, a legal interest, in the sense of legitimate, relevant and genuine interest, must be proved by a State desiring to intervene; furthermore, such a State must specify the way and manner that the decision in the substantive case *may* affect its own alleged interest, that is, »a right which is a legally protected interest«[14]. Mere concern, however greatly or genuinely expressed or shown, is not enough.

Views vary as to whether, apart from this basic distinction, intervention may be, or is indeed, open to States under both Articles 62 and 63, irrespective of whether they are or are not signatories to the Statute of the Court or member States of the United Nations. The point deserves further study before we can be categorical about it. It has also been questioned why indeed no jurisdictional link is demanded under Article 63, but such a link is insisted on in the case of Article 62. Surely, the mere fact of participation by a would-be intervener in a multilateral convention should not necessarily be regarded as a valid substitute, in Article 63, for the compulsory jurisdiction of the Court[15]? All this is not to deny that Article 62 altogether and in every case does not require a jurisdictional link. It means that the opposite case is not as obvious as some would argue.

IV. Is a Would-be Intervener a Party in the Case?

In the course of the Judges' deliberations on the Rules in 1922, the issue was raised as to the procedural capacity in which a would-be intervener could be regarded as standing in the case. Lord Philimore averred that, in the English common law, an intervener could, at least at that time, only associate itself with the defendant. On the other hand, Judge Loder pointed out that intervention, as a right, could, in Dutch law, be exercised both on the side of the plaintiff and of the defendant. Thereafter, the issue was not pursued, although the juridical nature of the question would seem to be recognized.

The relative position of the intervener, it was suggested, might be one of three kinds: an intervening State might be a co-plaintiff with one of the two or more original parties to the main dispute; it might claim certain exclusive rights; or it might intervene as a replacement of one of the parties so that the displaced party would withdraw[16]. The third possibility seems to be so rare

[14] See, for example, *Roscoe Pound*'s Interpretations of Legal History, Ch. 7; *id.*, Individual Interests of Substance, 59 Harvard Law Review (1945); and *id.*, Jurisprudence (1959), Vol. 3, Part 4, for stimulating analyses and classifications of legally protected interests.

[15] For a summary account of the history of that evolution, see paras. 21–23 of the *Continental Shelf (Tunisia/Libyan Arab Jamahiriya)*, Application to Intervene, Judgment, I.C.J. Reports 1981, p. 3, at pp. 14 and 15.

[16] »There are three possibilities: a party may wish to take sides with the plaintiff or the defendant; a party may claim certain exclusive rights; or a party may request that one of the two contesting States should withdraw on the ground that it is not the real *dominus* of the right which it claims. In this latter case intervention tends to become exclusion, but as a rule a State is content to take joint action with one of the parties: should this be allowed? The Committee replies in the affirmative, but on condition that an interest of a legal nature is involved. The Court is to decide whether

that it has not been recorded in the known history of intervention, whether in international arbitral adjudication or in judicial settlement. What is of interest is to enquire whether an intervener has more cause to establish a jurisdictional link when seeking to intervene in either of the two capacities of plaintiff or defendant than if both the original parties are in the same interest in the original dispute. The point has yet to be decided in a concrete case. But it has nevertheless been argued that in so-called quasi-advisory proceedings like those in the *North Sea Continental Shelf* cases [17], where the adversary character does not seem to be prominent between the two or more original parties, there is no necessity for the intervener to establish a jurisdictional link. Indeed, in the *Continental Shelf (Tunisia/Libyan Arab Jamahiriya)*, Application to Intervene, Judgment [18], to be discussed below, one of the counsel for Malta, Mr. Bathurst, put forward this argument, to which there was no clear reply during the public hearings.

Another related question is to consider whether there is such a thing as a non-party intervener; that is to say, whether a would-be intervener can, in a sense, be regarded as entitled to be accepted or treated as a party without incurring any of the obligations involved in participation in the proceedings. The idea was considered for a while during the 1972 revision of the Rules of Court, but it would seem to have been abandoned, as there is no mention of it in the adopted text of the Rules in the 1978 version. The truth is that it is ludicrous to accept the existence of such an enigma [19]. The suggestion most closely bordering on such acceptance was that Malta might be conceded the right to take part in the proceedings between Tunisia and Libya without being bound by the decision of the Court in the case. The Maltese representatives, however, repudiated such an attribution to their legal argument, which firmly denied any desire to play the part of onlooker in the main case should intervention be permitted. The question remains as to precisely how far Malta should be deemed to have agreed to be bound in the light of Article 59 of the Statute and of the limited scope of the stated object of intervention.

One last point is to consider who it is that really makes the decision contemplated in Article 62 (2) on the crucial question as to whether or not to

---

the interest is legitimate and consequently whether the intervention is admissible. To refuse all right of intervention might have unfortunate results. The essential point is to limit it to cases in which an interest of a legal nature can be shown, so that political intervention will be excluded, and to give the Court the right of decision.« [Twenty-eighth Meeting (Private), held at the Peace Palace, The Hague, on 20 July 1920. Committee of Judges, Minutes, pp. 11 and 12].

[17] I.C.J. Reports 1968, p. 9; I.C.J. Reports 1969, p. 3.

[18] I.C.J. Reports 1981, p. 3.

[19] It may be recalled that Sir Cecil Hurst, President of the Court, once observed »according to Article 62 of the Statute, the English text of which was particularly clear, it was only if the Court allowed the request of the State desiring to intervene that the State became a party.« (*P.C.I.J., Series D, No. 2,* 3rd Addendum, p. 306; *Travaux préparatoires* of the Rules).

permit intervention. It has been argued that the Court would not appear clearly as the arbiter under Article 62 (2) if one or other(s) of the original parties objected to the application to intervene, since in that case whatever decision was awarded by the Court might be regarded as having been influenced by the view that had prevailed. Unless both parties together refuse the intervener's application, while the Court accepts it, the matter is not free from controversy. On the whole, however, the Court should not concern itself unduly, even if, in the case of differences between the original parties, it reaches a final verdict which happens to agree with the view of one party; more important is the reasoning with which the verdict is supported by the Court.

V. The Last Phase

The case between Tunisia and Libya may now be briefly considered.

Under a Special Agreement (*compromis*), Libya and Tunisia submitted their dispute to the Court, asking it to define: »What are the principles and rules of international law which may be applied for the delimitation of the area of the continental shelf appertaining to the Republic of Tunisia and the area of the continental shelf appertaining to the Socialist People's Libyan Arab Jamahiriya and, in rendering its decision, to take account of equitable principles and the relevant circumstances which characterize the area, as well as the recent trends admitted at the Third Conference on the Law of the Sea.

Also, the Court is further requested to specify precisely the practical way in which the aforesaid principles and rules apply in this particular situation so as to enable the experts of the two countries to delimit those areas without any difficulties« [20].

The Special Agreement clearly stated that the Court was not to determine or demarcate the boundaries of the continental shelves appertaining to each adjacent State, but only to lay down general guidelines which their respective experts would apply in drawing the relevant boundary lines. If both States should be unable to carry out an acceptable delimitation, they would ask the Court for clarification. The Court should base its guidance upon equitable principles, not upon the principle of equidistance as enunciated in Article 6 of the 1958 Continental Shelf Convention. The Court was to take into account recent rules developed by the current Third Law of the Sea Conference.

After their written pleadings had been closed, Malta, which had pending a continental shelf boundary dispute with Libya, brought to the Court an

[20] Special Agreement between the Republic of Tunisia and the Socialist People's Libyan Arab Jamahiriya for the submission of the question of the Continental Shelf between the two countries to the International Court of Justice (notified to the Court on 1 December 1978), Art. 1, at p. 11.

Application for permission to intervene under Article 62 of the Statute. Malta made the following preliminary observation:

»In Malta's case there is a continental shelf boundary with both Libya and Tunisia, and the boundaries between all three States *converge at a single, as yet undetermined, point.* Given the proximity of Libya, Tunisia and Malta, the ›principles and rules of international law‹ applicable to the delimitation of the Libya/Tunisia boundary are bound to be relevant to the delimitation of the Malta/Libya and Malta/Tunisia boundaries. Furthermore, there is a substantial probability that many of the ›relevant circumstances‹ – geographic, geologic, geomorphic, economic, and other – which affect the determination of the boundary between Libya and Tunisia would also be relevant to the determination of Malta's boundaries with those two States. The Court's treatment of such factors in the *Libya/Tunisia* case is thus bound to affect the treatment of the same factors in a subsequent case involving Malta's boundaries« [21].

The principal object of Malta's intervention as confirmed in its Application was »to enable Malta to submit its views to the Court on the issues raised in the pending case, before the Court has given its decision in that case« [22]. Malta stressed, in addition, that it was not its object »by way, or in the course of intervention in the *Libya/Tunisia* case to obtain any form of ruling or decision from the Court concerning its continental shelf boundaries with either or both of those countries« [23].

Thus, Malta merely wanted to submit its views to the Court, and did not seek any determination of any legal right or interest if permitted to intervene in the main dispute between Tunisia and Libya. Instead, Malta insisted that the very purpose of the principal proceedings in the case was to secure a statement from the Court of what the appropriate law is, arguing that it is not justifiable to suggest that Malta's object in seeking to intervene should be »more exact, more precise or more operative in formal terms« [24] than the object of the Parties in the principal proceedings.

In sum, the stated object of Malta was not to assert its own particular legal interests as against those of Libya and Tunisia, but merely for it to be permitted to appear as a kind of participating but not *amicus curiae* in the principal proceedings, without incurring any of the obligations of a real litigant in the case. Clearly, this is not contemplated by Article 62 of the Statute. For the purpose of this particular provision, an intervener must be more

---

[21] Application for Permission to Intervene by the Government of Malta (filed in the Registry of the Court on 30 January 1981) in the case concerning the *Continental Shelf (Tunisia/Libyan Arab Jamahiriya),* Article 10, pp. 4–6.

[22] *Ibid.,* para. 20.

[23] *Ibid.,* para. 22.

[24] *Ibid.,* para. 26.

than a participating spectator in the arena of the proceedings between the two principal parties.

* * *

Malta also introduced a new dimension into the proceedings by nominating its own *ad hoc* judge for the purpose of taking part in the intervention proceedings, arguing that Malta was entitled, under Article 31, paragraph 5 of the Statute, to have an *ad hoc* judge in the same way as Libya and Tunisia, who had each appointed one.

Malta maintained that intervention proceedings are in themselves a distinct »case before the Court« for the purpose of Article 31, and that paragraph 5 applies. Both Libya and Tunisia are »in the same interest« by reference to the identity or near identity of their »conclusions«. The Court considered the submission and, without calling in the two *ad hoc* judges of Libya and Tunisia to participate, decided that Malta, not yet being a party to the case until permitted to intervene, was not entitled to have an *ad hoc* judge at that stage of the proceedings.

The Court then heard the respective representatives of Malta, Libya and Tunisia on the Application for permission to intervene, after having obtained the objections of both Libya and Tunisia. Both Libya and Tunisia argued that Malta must show a jurisdictional link between itself and each of them as a condition of its Application being granted. They also pointed out that the stated object of the intervention was vague and not sufficiently precise to satisfy Article 62 on the issue of the requirement that the would-be intervener must establish that it has an interest of a legal nature in the subject-matter of the dispute between the parties. Malta, for its part, strove to prove the required legal interest by showing that its continental shelf was likely to be affected by the decision of legally applicable principles to be laid down by the Court. It feared that the Court's decision might suggest the drawing of boundary lines that might impinge upon its own continental shelves. It was not enough, in Malta's submission, to point to Article 59 of the Statute as a safeguard of its legal interests; this was because the ambit of the Court's decision might embrace wider rules or principles than the *dispositif* in such a way as to prejudice Malta's legal interests. The Court also considered that the real object of Malta's proposed intervention was unclear and seemed to represent no more than the general interest which all States might have in the – undoubtedly interesting – disputations on the developing continental shelf and the attendant laws and concepts. This did not seem to the Court to be within the purview of Article 62 of the Statute. The Court felt that Malta's Application lacked the true purpose of intervention which the founding fathers of the Statute assumed to be at the heart of the matter, namely, a concrete, legally protected interest of the would-be intervener which may be affected by the Court's decision. On the whole, the Court was not convinced that Malta's Application satisfied the requirements of a true intervention under Article 62 of the Statute.

Having reached this conclusion on the nature and character of Malta's Application, the Court did not think that, in the present state of Article 62, it would be necessary to attempt to enquire whether, in every case of intervention under this particular provision, an intervener need establish a jurisdictional link. The learned arguments addressed to the Court by counsel from all sides concerning the 1920 Committee of Jurists and the Rules Committee of Judges of 1922 remained inconclusive, leaving the Court to make its main basis Article 62 of the Statute, as well as the special circumstances of each particular case, in considering any application by a State to intervene under Article 62.

On this first genuine occasion, in the *Tunisia/Libya* case, when an application by Malta was based directly on Article 62 alone, the Court found itself constrained to refuse the permission sought for the reasons already stated.

Sind die Mitgliedstaaten der Europäischen Gemeinschaft noch Herren der Verträge?

Zum Verhältnis von Europäischem Gemeinschaftsrecht und Völkerrecht

ULRICH EVERLING

Hermann Mosler, der an der Entstehung des Europäischen Gemeinschaftsrechts wesentlich beteiligt war[1], hat sich von Anfang an um die Deutung dieser neuen Rechtsordnung bemüht[2]. Dabei hat er das Gemeinschaftsrecht mit der Europäischen Menschenrechtskonvention und anderen Abkommen als ein besonderes »Europarecht« zusammengefaßt, und dieses hat er als Ausdruck eines der europäischen Rechtsgemeinschaft gemeinsamen *ordre public* im Sinne der für sie existenznotwendigen, als Normen geltenden Rechtswerte verstanden[3].

Dieser weite Begriff des Europarechts dürfte dem heutigen Stand kaum noch entsprechen; die qualitative und quantitative Entwicklung des Gemeinschaftsrechts erfordert eine besondere Behandlung in Forschung und Lehre. Wichtiger als diese letztlich zweitrangige Frage der begrifflichen Abgrenzung ist *Moslers* Versuch, das Gemeinschaftsrecht in den Gesamtzusammenhang der Rechtsordnung einzugliedern.

Wenn man das Gemeinschaftsrecht als Ausdruck eines besonderen, durch eine europäische Rechtsgemeinschaft geprägten Wertbewußtseins begreift, weist man ihm eine eigene Stellung sowohl gegenüber den nationalen Rechtsordnungen, also dem Recht ihrer Mitgliedstaaten, als auch gegenüber dem allgemeinen Völkerrecht als der von der internationalen Rechtsgemeinschaft getragenen umfassenden Rechtsordnung zu. Das Verhältnis zum na-

---

[1] Vgl. *H. Mosler*, Die Entstehung des Modells supranationaler und gewaltenteilender Staatenverbindungen in den Verhandlungen über den Schuman-Plan, in: Festschrift für Walter Hallstein (Frankfurt 1966), S. 355 ff.

[2] So bereits *H. Mosler*, Der Vertrag über die EGKS, Entstehung und Qualifizierung, ZaöRV 14 (1951/52), S. 1 ff.

[3] *H. Mosler*, Begriff und Gegenstand des Europarechts, ZaöRV 28 (1968), S. 481 ff.

tionalen Recht ist häufig behandelt worden und wohl weitgehend geklärt. Dem Verhältnis zum Völkerrecht soll hier nachgegangen werden, und zwar vor allem unter dem Gesichtspunkt, ob und wie weit neben dem Gemeinschaftsrecht die Regeln des Völkerrechts innerhalb der Europäischen Gemeinschaft[4] angewandt werden können.

Diese Problematik kann hier nicht erschöpfend behandelt werden, insbesondere kann auf die rechtliche Wirkung der allgemeinen Regeln des Völkerrechts und der von der Gemeinschaft oder ihren Mitgliedstaaten geschlossenen Verträge und der Mitgliedschaft in internationalen Organisationen im Gemeinschaftsrecht nicht näher eingegangen werden. Es soll vielmehr die engere Frage herausgegriffen werden, wie weit das Völkerrecht noch zwischen den Mitgliedstaaten der Gemeinschaft gilt, und zwar einerseits innerhalb ihres Kompetenzbereichs und andererseits außerhalb, aber im Zusammenhang oder mit Bezug auf die Gemeinschaft.

Dies führt zu der Frage, ob die Mitgliedstaaten noch zur Verfügung über die Gemeinschaft berechtigt, also nach einem geflügelten Wort »Herren der Verträge« sind. Sie soll, obwohl gerade unlängst wieder jeder Zweifel daran als »Illusion« und als »unrealistisch« bezeichnet wurde[5], erneut zur Diskussion gestellt werden.

1. Das Gemeinschaftsrecht schließt als Sonderregelung die Anwendung des Völkerrechts aus, soweit es Regeln für die Beziehungen der Mitgliedstaaten untereinander enthält. Darüber besteht Einigkeit, wenn auch die Begründung im einzelnen schwankt. So heißt es meist, daß das Gemeinschaftsrecht als besonderes Völkerrecht dem allgemeinen vorgehe und als Recht einer internationalen Organisation oder eines besonderen Staatenzusammenschlusses insoweit, als es Regelungen treffe, für das allgemeine Völkerrecht keinen Raum lasse[6].

Umgekehrt scheint vielfach angenommen zu werden, daß die Regeln des Völkerrechts und insbesondere die Wiener Vertragsrechtskonvention insoweit innerhalb der Gemeinschaft für die Beziehungen der Mitgliedstaaten untereinander gelten, als das Gemeinschaftsrecht keine besonderen Regelungen enthalte, also etwa für Fragen wie Repressalien, Schadensersatz bei Vertragsverletzungen oder Kündigung[7].

---

[4] Unter der Bezeichnung »Gemeinschaft« werden im folgenden, entsprechend dem politischen Sinn, die drei rechtlich weiterhin selbständigen Europäischen Gemeinschaften zusammengefaßt; Besonderheiten von EGKS- und EAG-Vertrag werden ausgeklammert.

[5] So *M. Zuleeg*, Neuere Literatur zum Europarecht, Archiv des öffentlichen Rechts 106 (1981), S. 136.

[6] Vgl. etwa *A. Bleckmann*, Die Rechtsnatur des Europäischen Gemeinschaftsrechts, Die Öffentliche Verwaltung (DÖV) 1978, S. 391 ff., ferner *derselbe*, Europarecht (2. Aufl., Köln 1978), S. 93 ff.

[7] Dies ist bisher wenig behandelt worden; vgl. vor allem *Bleckmann*, a.a.O., DÖV 1978, S. 391 ff.; *M. Zuleeg*, Der Bestand der Europäischen Gemeinschaft, in: Gedächtnisschrift für Christoph Sasse, Bd. 1 (Baden-Baden 1981), S. 55 ff.; ablehnend

Dies wird oft aus begrifflichen Klassifizierungen abgeleitet, die nichts weiter als eine *petitio principii* sind, wie etwa daß die Gemeinschaft eine »völkerrechtliche Staatengemeinschaft besonderer Art« sei, daß das Gemeinschaftsrecht »völkerrechtlichen Charakter« habe oder daß sich in ihm »staatliche mit völkerrechtlichen Elementen verbinden«[8]. Teilweise wird auch angeführt, daß sich die Gemeinschaft noch nicht von der Grundlage der völkerrechtlichen Gründungsverträge gelöst habe[9], was sich manche auch erst dann vorstellen können, wenn die Gemeinschaft zu einem Staat geworden ist[10].

Eine Betrachtung, die sich an vorgefaßten Begriffen[11] oder an der traditionellen Aufteilung von Völkerrecht und Landesrecht[12] orientiert, vermag das Problem jedoch nicht zu lösen. Grundlage der Gemeinschaftsrechtsordnung sind zwar die zwischen den Mitgliedstaaten geschlossenen völkerrechtlichen Verträge in ihrer gegenwärtigen, sich aus den Beitritts- und Änderungsverträgen ergebenden Fassung. Aber ein solcher Gesamtakt, der eine vom Gründungsvertrag unabhängige Rechtspersönlichkeit schafft, ist eine bekannte Erscheinung im Recht[13]; daraus ergibt sich noch nichts für den Charakter des Gemeinschaftsrechts und darüber, welche Regeln innerhalb der

---

L.-J. Constantinesco, Das Recht der Europäischen Gemeinschaften, Bd. 1 (Baden-Baden 1977), S. 180 ff.

[8] So vor allem *Bleckmann* in seinen verschiedenen Beiträgen, vgl. Anm. 6 sowie: Die Position des Völkerrechts im inneren Rechtsraum der Europäischen Gemeinschaften, Monismus oder Dualismus der Rechtsordnungen?, Jahrbuch für internationales Recht (JIR) 1975, S. 300 ff.; *derselbe*, Zur Funktion des Gewohnheitsrechts im Europäischen Gemeinschaftsrecht, Europarecht 1981, S. 101 ff., 108; *derselbe*, Nationales und europäisches Souveränitätsverständnis, Strukturalistisches Modelldenken im Europäischen Gemeinschaftsrecht und im Völkerrecht, in: G. Ress (Hrsg.), Souveränitätsverständnis in den Europäischen Gemeinschaften (Baden-Baden 1980), S. 11 ff.

[9] So etwa *R. Bernhardt*, Das Recht der Europäischen Gemeinschaften zwischen Völkerrecht und staatlichem Recht, in: Festschrift für Rudolf Bindschedler (Bern 1980), S. 229 ff.

[10] Vgl. etwa *C. Tomuschat*, Bonner Kommentar zum Grundgesetz (Zweitbearbeitung 1981), Art. 24 Rn. 48; daß die Gemeinschaft kein Staat ist, entspricht der einhelligen Meinung, wenn auch die Begründung differiert; überzeugend neuerdings *E. Grabitz*, in: Grabitz/Läufer, Das Europäische Parlament (Bonn 1980), S. 345 ff., 356; vgl. ferner (entgegen dem Eindruck von *Tomuschat*, a.a.O., Rn. 47) *U. Everling*, Vom Zweckverband zur Europäischen Union – Überlegungen zur Struktur der Europäischen Gemeinschaft, in: Festschrift für Hans Peter Ipsen (Tübingen 1977), S. 595 ff., 607 f.

[11] *Bleckmann* (Anm. 6) DÖV 1978, S. 396, und Souveränitätsverständnis, S. 42, 60, versucht sich – gewissermaßen prophylaktisch – gegen den Vorwurf der Begriffsjurisprudenz zu wehren, indem er auf die Ordnungsfunktion der Begriffe verweist. Diese erfordert jedoch, zunächst die Wirklichkeit zu analysieren und sie sodann begrifflich einzuordnen.

[12] So im Ergebnis *Bernhardt* (Anm. 9) in seiner abgewogenen Studie. Für eine differenzierte Betrachtung auch *Ress* (Anm. 8), S. 11 ff.

[13] Für die Gemeinschaftsverträge vgl. *H. P. Ipsen*, Über Supranationalität, in: Festschrift für Ulrich Scheuner (Berlin 1973), S. 211 ff., 220 f.; *Constantinesco* (Anm. 7), S. 317 f.

Gemeinschaft gelten. Vielmehr muß gefragt werden, ob sich die Gemeinschaftsrechtsordnung nach Struktur, Inhalt und Umfang bereits als ein derart eigenständiges und vervollständigtes Normensystem darstellt, daß sie die Anwendung ergänzender Regelungen des allgemeinen Völkerrechts ausschließt.

2. Den besonderen Charakter des Gemeinschaftsrechts als eigene Rechtsordnung hat der Gerichtshof bereits 1964 in einer heute als klassisch zu bezeichnenden Formulierung [14] umschrieben, der das Bundesverfassungsgericht weitgehend gefolgt ist [15]. Die Elemente, die er schon damals als konstitutiv genannt hat, bilden die Grundlage seiner ganzen weiteren Rechtsprechung: Gemeinschaft auf unbegrenzte Zeit, eigene Organe, Rechts- und Geschäftsfähigkeit, internationale Handlungsfähigkeit, echte Hoheitsrechte, Verbindlichkeit für die Mitgliedstaaten und Verbindlichkeit für die Bürger.

Auf der Basis dieser grundsätzlichen Einschätzung hat sich die Rechtsordnung der Gemeinschaft in den letzten Jahrzehnten trotz der permanenten Krisen, die ihr Bild in der Öffentlichkeit verdunkeln, in eindrucksvoller Weise entwickelt.

Die Rechtsordnung der Gemeinschaft gilt heute für weite Bereiche, die bisher der staatlichen Regelung unterlagen, und bewirkt dadurch eine enge Verflechtung der Mitgliedstaaten sowie ihrer Unternehmen und Bürger. Vor allem das Wirtschaftsrecht wird zunehmend vom Gemeinschaftsrecht erfaßt oder wenigstens beeinflußt [16]. Das gilt besonders für sämtliche Regelungen, die den Warenverkehr zwischen den Mitgliedstaaten regeln oder sich auf ihn auswirken. Alle Vorschriften des nationalen Rechts über die Produktion und Vermarktung von Waren werden mehr oder weniger stark betroffen. Die Beispiele reichen vom Lebensmittelrecht über das technische Recht bis zum gewerblichen Rechtsschutz. Diese Vorschriften sind Gegenstand einer zwar langsam, aber ständig fortschreitenden Rechtsangleichung. Sie erstreckt sich auch auf Bereiche des Zivilrechts wie des Gesellschaftsrechts und auf das Steuerrecht.

Das Gemeinschaftsrecht regelt ferner die Freizügigkeit der Arbeitnehmer und das damit zusammenhängende Recht der sozialen Sicherheit sowie das Niederlassungsrecht und die Koordinierung des Gewerbe- und Berufsrechts bis hin zum Bank- und Versicherungsrecht sowie zum Recht der öffentlichen Aufträge. Im Rahmen der Agrarpolitik wurde ein umfassender Regelungskomplex mit mehreren tausend Rechtsvorschriften jährlich geschaffen. Im Bereich der Industrie- und Energiepolitik bestehen weniger Regelungen, aber vor allem das Wettbewerbsrecht einschließlich der Beihilferegeln ist ein

---

[14] Urteil vom 15. 7. 1964, Rs. 6/64, *Costa/ENEL*, Slg. S. 1251 ff.; diese Rechtsprechung wurde wiederholt bekräftigt, etwa Urteil vom 9. 3. 1978, Rs. 106/77, *Simmenthal*, Slg. S. 629 ff.

[15] Beschluß vom 18. 10. 1967, BVerfGE 22, 293 ff.

[16] Vgl. dazu die zusammenfassenden Überblicke bei *B. Beutler/R. Bieber/J. Pipkorn/J. Streil*, Die Europäische Gemeinschaft, Rechtsordnung und Politik (Baden-Baden 1979).

wichtiger, vom Gemeinschaftsrecht erfaßter Bereich, und Regionalpolitik, Forschungspolitik sowie die allgemeine Wirtschafts- und Währungspolitik sind wenigstens Gegenstand von gemeinsamen Aktionen oder Koordinierungen.

Die Außenbeziehungen sind durch Zollrecht und Außenhandelsrecht umfassend geregelt. Ein kaum noch übersehbares Netz von Assoziationen und ähnlichen, auf Abbau von Handelshemmnissen und Zusammenarbeit ausgerichteten Verträgen erstreckt sich vom Nordkap über ganz Westeuropa und Afrika bis vor die Tore Südafrikas und in die Karibik. Sie werden ergänzt durch zahlreiche Abkommen mit anderen Staaten und insbesondere Entwicklungsländern und vor allem durch gemeinschaftliches oder wenigstens abgestimmtes Verhalten in internationalen Organisationen.

Auf diese Weise ist trotz aller Krisen, Widerstände und Rückschläge fast unbemerkt ein überaus umfangreiches Regelungssystem entstanden, das nicht mit den traditionellen internationalen Organisationen auf eine Stufe gestellt werden kann. Die Aufgaben der Gemeinschaft sind nicht mehr klar begrenzt, sie werden durch eine extensive Auslegung der Art. 100 und 235 EWG-Vertrag ständig erweitert und lassen kaum noch einen staatlichen Bereich unberührt [17]. Damit wird aber die Quantität zu einem Problem der Qualität.

3. Die Qualität dieser Rechtsordnung und insbesondere ihre Struktur wird entscheidend durch die Rechtsprechung des Gerichtshofs bestimmt. Die geringe Normdichte, die zunächst wegen des Rahmen- und Plancharakters der Verträge bestand, hat den Gerichtshof nicht etwa veranlaßt, die innergemeinschaftliche Regelung auch nur in beschränktem Umfang dem Völkerrecht zu überlassen, sondern er hat die Rechtsordnung aus sich heraus entwickelt und ausgebildet, und zwar weitgehend gegen den Widerstand der Mitgliedstaaten [18].

Das Vorgehen des Gerichtshofs und sein Beitrag zur Entstehung eines in sich kohärenten Systems sind bereits mehrfach eingehend analysiert worden [19]. Auch hierzu können nur einige Stichworte genannt werden.

Zunächst ist der Grundsatz des Vorrangs des Gemeinschaftsrechts vor jedem nationalen Recht gleich welcher Qualität zu erwähnen, der sich – mit den bekannten Zweifeln im Grundrechtsbereich – allgemein durchgesetzt

---

[17] Vgl. *Everling* (Anm. 10), ferner: Die allgemeine Ermächtigung der Europäischen Gemeinschaft zur Zielverwirklichung nach Art. 235 EWG-Vertrag, Europarecht 1976/Sonderheft, S. 2 ff., 17 f.

[18] Dazu *E. Stein*, Lawyers, Judges and the Making of a Constitution, in: Festschrift für Konrad Zweigert (Tübingen 1981), S. 771 ff.; auch der Verfasser, der in einigen der von *Stein* angeführten Verfahren als Vertreter der Bundesregierung aufgetreten ist, muß das heute begrüßen.

[19] Vgl. vor allem *J. Schwarze*, Die Befugnis zur Abstraktion im Europäischen Gemeinschaftsrecht (Baden-Baden 1976), insbes. S. 105 ff.; ferner *R. Lecourt*, L'Europe des juges (Bruxelles 1976); *H. Kutscher*, Über den Gerichtshof der Europäischen Gemeinschaft, Europarecht 1981, S. 392 ff., jeweils mit Nachweisen zum folgenden.

hat. Ferner hat der Gerichtshof weiten Teilen des Gemeinschaftsrechts direkte Wirkung zuerkannt und damit den Einzelnen die Möglichkeit eröffnet, sich gegenüber dem nationalen Recht auf Gemeinschaftsrecht zu berufen. Die Skala reicht von dem Verbot der Diskriminierungen und Einfuhrhindernisse über weitestgehende Gleichbehandlung und Aufenthaltsrechte der Arbeitnehmer und Selbständigen sowie den Grundsatz der Lohngleichheit von Mann und Frau bis zu begünstigenden, nicht ausführungsbedürftigen Bestimmungen von Richtlinien, die ein Mitgliedstaat nicht rechtzeitig in nationales Recht umgesetzt hat.

Der Gerichtshof hat mit seiner Rechtsprechung die Vertragsregeln erläutert und praktikabel gemacht, so etwa das Zollrecht, das Agrarrecht mit seinem System des Währungsausgleichs, das Recht der sozialen Sicherheit und nicht zuletzt das europäische Beamtenrecht. Er hat eine reiche Rechtsprechung zur Frage normativen Unrechts und zur Frage des Schadensersatzes und der Rückforderung ungerechtfertigter Zahlungen entwickelt. Schließlich hat er den Schutz von Grundrechten im Gemeinschaftsrecht übernommen und die Rechtsakte der Gemeinschaft an allgemeinen Rechtsgrundsätzen, vor allem am Prinzip der Proportionalität und der Gleichheit gemessen.

Dabei hat der Gerichtshof eine teleologische und dynamische Methode der Auslegung angewendet, die sich von der üblichen Auslegung völkerrechtlicher Verträge mindestens insofern unterscheidet, als die den Verträgen innewohnende Finalität ohne Rücksicht auf das spätere Verhalten der Vertragspartner, deren gemeinsames Handeln im Rat sogar der Kontrolle unterworfen ist, zugrunde gelegt wird [20]. Jedenfalls in dieser Hinsicht, also bei der Anwendung und Auslegung des Gemeinschaftsrechts durch den Gerichtshof, sind die Mitgliedstaaten deshalb nicht mehr »Herren der Verträge« [21]. Allerdings muß einschränkend hinzugefügt werden, daß der Gerichtshof etwaige Wandlungen des internationalen Systems oder des allgemeinen politischen Verständnisses von der Gemeinschaft und ihren Zielen – bei aller Distanz gegenüber tagespolitischen Strömungen – auf Dauer ebensowenig ignorieren kann wie der nationale Richter Wandlungen des allgemeinen Rechtsbewußtseins und seiner sozialen Grundlagen [22].

4. Bei einer so vielfältig entwickelten Rechtsordnung, die aufgrund eines nahezu perfekten Rechtsschutzsystems vom Gerichtshof ständig weiter ausgebaut wird, bleibt für die Anwendung des Völkerrechts einerseits und des

---

[20] Vgl. die Analyse der Rechtsprechung durch *H. Kutscher*, Thesen zu den Methoden der Auslegung des Gerichtshofs aus der Sicht eines Richters, Begegnung von Justiz und Hochschule 27./28. September 1976 (Luxemburg 1976), ferner *R. Bernhardt*, Zur Auslegung des europäischen Gemeinschaftsrechts, in: Festschrift für Hans Kutscher (Baden-Baden 1981), S. 17 ff.

[21] So *Bernhardt*, a.a.O., S. 21; in der Sache bereits *Kutscher*, a.a.O., S. 33.

[22] Vgl. *A. Bleckmann,* Teleologie und dynamische Auslegung des Europäischen Gemeinschaftsrechts, Europarecht 1979, S. 239 ff., 258.

nationalen Rechts andererseits Raum nur insoweit, als es sich aus den wechselseitigen Verflechtungen ergibt.

Das Gemeinschaftsrecht ist sicher nicht geschlossen in dem Sinne, daß seine Regeln isoliert stehen und keiner Ergänzung bedürfen. Es weist schon wegen der begrenzten Kompetenz der Gemeinschaft und ihrer schrittweisen Entwicklung erhebliche Lücken auf. Der Gerichtshof nimmt dies teilweise bewußt – etwa bei unvollkommener Rechtsangleichung – hin, bis der Gemeinschaftsgesetzgeber tätig geworden ist[23]. In der Tat wäre es eigenartig, wenn die Gemeinschaft sich unitarischer als die Bundesstaaten verhalten würde, die unterschiedliche Regelungen ihrer Glieder in erheblichem Umfang tolerieren. Gemeinschaftsrechtsordnung und nationale Rechtsordnung sind in vielfältiger Weise aufeinander bezogen[24].

Ebenso ist die Gemeinschaftsrechtsordnung auch zum Völkerrecht hin offen. Für die Staaten wird seit langem diskutiert, daß die strikte Trennung von Völkerrecht und Landesrecht, wie sie seit *Triepel* vorherrscht, der Wirklichkeit nicht mehr gerecht wird. Internationale Lagen wirken vielfältig auf das interne Recht und die verfassungsmäßige Willensbildung ein; man hat von einer »Internationalisierung der Verfassung« gesprochen[25].

Die Gemeinschaft kann sich dieser Entwicklung nicht entziehen. Durch die von ihr geschlossenen völkerrechtlichen Verträge und durch die Verträge ihrer Mitgliedstaaten, die sie teils ablösen, teils respektieren muß, sowie aufgrund ihrer eigenen Mitgliedschaft in internationalen Organisationen neben oder anstelle der Mitgliedstaaten und aufgrund ihres Auftretens in Organisationen, in denen die Mitgliedstaaten Mitglieder geblieben sind, beeinflußt das Völkerrecht das Gemeinschaftsrecht in vielfältiger Form[26].

Aber auch die allgemeinen Grundsätze des Völkerrechts müssen im Rahmen des Gemeinschaftsrechts etwa wie nach Art. 25 GG unmittelbar und mit Vorrang gegenüber dem Gemeinschaftsrecht angewandt werden[27]. Das

---

[23] *K. Mortelmans*, Les lacunes provisoires en droit communautaire, Cahiers de droit européen 1981, S. 410 ff.

[24] Vgl. *P. Pescatore*, Das Zusammenwirken der Gemeinschaftsrechtsordnung mit den nationalen Rechtsordnungen, Europarecht 1970, S. 307.

[25] Dazu die zusammenfassende Darstellung bei *W. Fiedler*, Auswärtige Gewalt und Verfassungsgewichtung, in: Festschrift für Hans-Jürgen Schlochauer (Berlin 1981), S. 57 ff. mit Nachweisen, ferner *C. Tomuschat*, Der Verfassungsstaat im Geflecht der internationalen Beziehungen, Veröffentlichungen der Vereinigung der deutschen Staatsrechtslehrer (VVDStL) H. 36 (1978), S. 7 ff.; *R. Bernhardt*, Verfassungsrecht und internationale Lagen, DÖV 1977, S. 457 ff.

[26] Vgl. etwa *H. Krück*, Völkerrechtliche Verträge im Recht der Europäischen Gemeinschaften (Beiträge zum ausländischen öffentlichen Recht und Völkerrecht, Bd. 70) (Berlin 1977); *E. U. Petersmann*, Auswärtige Gewalt, Völkerrechtspraxis und Völkerrechtsbindungen der Europäischen Wirtschaftsgemeinschaft, ZaöRV 35 (1975), S. 213 ff.

[27] Vgl. *K. M. Meessen*, The Application of Rules of Public International Law within Community Law, Common Market Law Review (CMLR) 1976, S. 485 ff., 497 ff. Abweichend *Bleckmann* (Anm. 8) JIR 1975, S. 300 ff.

folgt aus der allgemein völkerrechtsfreundlichen Haltung des EWG-Vertrages, wie sie sich etwa in Art. 110 und Art. 228 ff. zeigt, aber auch aus einem aus der gemeinsamen Praxis der Mitgliedstaaten herzuleitenden allgemeinen Rechtsgrundsatz des Gemeinschaftsrechts[28].

Daraus ergibt sich aber noch nichts für die Geltung der Völkerrechtsregeln innerhalb des Verfassungssystems der Gemeinschaft. Im Gegenteil, die Gemeinschaftsrechtsordnung ist dem Völkerrecht gegenüber nicht deshalb offen, weil sie auf völkerrechtsvertraglicher Grundlage beruht, sondern weil sie als Verfassungsordnung wie jedes Recht in den internationalen Zusammenhang gestellt ist. Gerade dadurch beweist sie ihre Eigenständigkeit.

5. Die Gemeinschaftsrechtsordnung muß nach ihrer umfassenden quantitativen und qualitativen Ausgestaltung also als eigenständige Rechtsordnung begriffen werden, die sich von ihrer völkerrechtsvertraglichen Grundlage in einem erheblichen Ausmaß gelöst hat und über ein eigenes, nach eigenen Grundsätzen auszulegendes Verfassungssystem verfügt.

Dieses durch die Rechtsprechung des Gerichtshofs näher ausgeformte und vervollständigte Verfassungssystem stellt die Regeln zur Verfügung, die innerhalb der Gemeinschaft anzuwenden sind. Sie bestimmen das Verhältnis der Gemeinschaft zu ihren Mitgliedstaaten, so etwa die Verpflichtungen der Mitgliedstaaten aus den Rechtsakten und Finanzregelungen und umgekehrt auch ihre Rechte aus diesen, ferner die Mitwirkung in den Gemeinschaftsorganen sowie die Aufsichtsbefugnisse der Kommission. Sie regeln ferner das Verhältnis der Mitgliedstaaten untereinander, vor allem ihr Zusammenwirken in den Organen und ihre gegenseitigen Rechte und Pflichten. Sie enthalten insgesamt gesehen ein ganzes Geflecht von Bindungen, die nicht mehr ohne weiteres abzustreifen sind, weil sie in der tatsächlichen Entwicklung bereits zu einer Vielzahl von wechselseitigen Beziehungen und Abhängigkeiten auf politischer, wirtschaftlicher und gesellschaftlicher Ebene geführt haben.

Diese Einschätzung wird durch die ständigen offenen und latenten Krisen, von denen die Gemeinschaft betroffen wird, und durch deren weitgehende Handlungsunfähigkeit gegenüber den gegenwärtigen politischen Konflikten und wirtschaftlichen Schwierigkeiten nicht in Frage gestellt. In den Mitgliedstaaten haben wir uns längst an langwierige Entscheidungsprozesse gewöhnt, an Pattsituationen zwischen verschiedenen Institutionen wie Bundestag und Bundesrat und an Widerstände gesellschaftlicher Gruppen, die Zweifel an der Regierbarkeit der modernen Industriegesellschaften aufkommen lassen. Auf der Gemeinschaftsebene vervielfachen sich diese Probleme noch, weil hier nur beschränkt von einem gemeinsamen Grundkonsens, wie er etwa

---

[28] Von der Geltung allgemeiner Grundsätze des Völkerrechts zu unterscheiden ist die Geltung allgemeiner Grundsätze des Gemeinschaftsrechts; dazu *K. M. Meessen*, Zur Theorie allgemeiner Rechtsgrundsätze des internationalen Rechts: Der Nachweis allgemeiner Rechtsgrundsätze des Europäischen Gemeinschaftsrechts, JIR 1974, S. 283 ff.

den Bundesstaat trägt, ausgegangen werden kann. Der Konsens muß in der Gemeinschaft in jeder Einzelsituation wieder neu gesucht werden.

Die dazu erforderlichen langwierigen Auseinandersetzungen sagen aber nichts über Realität und Bestand des Systems aus. Die sogenannten Krisen müssen vielmehr als Ausdruck des Spannungsverhältnisses angesehen werden, das Systemen mit mehrstufiger Gliederung von Entscheidungsebenen immanent ist[29]. Nicht zuletzt daran, daß dieses Spannungsverhältnis in der Praxis immer wieder aufgelöst werden konnte, zeigt sich die Bestandskraft des in vielfachen Verflechtungen auf allen Ebenen verwirklichten politischen, wirtschaftlichen, gesellschaftlichen Systems der Gemeinschaft, das sich bereits in rechtlicher Form weitgehend verfestigt hat.

Ein derartig ausgebildetes Verfassungssystem enthält aber in sich selbst die Regeln, die sein Funktionieren gewährleisten. Nicht das Völkerrecht, sondern allein die aus den Rechtssätzen und Zielen der Gemeinschaft entwickelten Regeln bestimmen das Verhältnis der Gemeinschaft zu ihren Mitgliedstaaten und das Verhältnis ihrer Mitgliedstaaten untereinander. Vor allem ist die Anwendung der Wiener Vertragsrechtskonvention innerhalb der Gemeinschaft ausgeschlossen. Das muß für zwei Fragenkreise, die in jüngster Zeit erörtert worden sind, besonders erläutert werden, nämlich für Reaktionen auf Vertragsverletzungen einzelner Mitgliedstaaten und für die Möglichkeit der Mitgliedstaaten, aus der Gemeinschaft auszuscheiden.

6. Für Vertragsverletzungen eines Mitgliedstaates sieht der Vertrag ausdrücklich Regelungen vor. Die Kommission und die anderen Mitgliedstaaten können nach Art. 169, 170 EWG-Vertrag Klage erheben. Ferner kann sich die Partei eines Rechtsstreits vor einem nationalen Gericht gegen vertragswidrige Rechtsvorschriften eines Mitgliedstaates wehren, was zu einer Überprüfung im Vorlageverfahren nach Art. 177 EWG-Vertrag führt.

Im ersten Fall endet der Rechtsstreit vor dem Gerichtshof mit der Feststellung der Vertragsverletzung, im zweiten mit einer Vertragsinterpretation, aus der sich die Vertragsverletzung indirekt ergibt. Für den ersten sagt Art. 171 EWG-Vertrag ausdrücklich, daß der Mitgliedstaat die Maßnahmen zu ergreifen hat, die sich aus dem Urteil ergeben; im zweiten folgt ein solcher Ausspruch erst, wenn ein Vertragsverletzungsverfahren nachgeschoben wird, aber auch ohne dieses muß die Verpflichtung des Mitgliedstaates, sein Recht zu ändern, angenommen werden. Wenn ein Mitgliedstaat die erforderlichen Maßnahmen nicht ergreift, um einem Urteil nachzukommen, so besitzt die Gemeinschaft keine Möglichkeiten, ihn zum Handeln zu zwingen[30]. Sanktionen gegenüber Mitgliedstaaten sind – abgesehen vom engen Bereich des

---

[29] *U. Everling*, Überlegungen zum Fortgang der Europäischen Integration, Europa-Archiv 1979, S. 737.

[30] Zur Problematik vgl. *U. Everling*, Das Europäische Gemeinschaftsrecht im Spannungsfeld von Politik und Wirtschaft, in: Festschrift für Hans Kutscher (Baden-Baden 1981), S. 155 ff., 183 ff.

Montanvertrages unter den in dessen Art. 88 vorgesehenen strengen Bedingungen – nicht vorgesehen. Dem Bundeszwang vergleichbare Regelungen bestehen nicht, und erst recht gibt es keine Gemeinschaftsgewalt[31].

Erst in einer solchen Lage, also wenn das im Vertrag vorgesehene Verfahren durchgeführt worden ist und die Befolgung eines Urteils verweigert wird, könnte sich die Frage stellen, ob ein Mitgliedstaat auf Regeln des Völkerrechts zurückgreifen kann, um dem vertragswidrig handelnden Staat gegenüber Repressalien anzuwenden oder Schadensersatz zu verlangen[32]. Das wäre jedoch nicht mit der Rechts- und Verfassungsordnung der Gemeinschaft vereinbar. Diese läßt nach ihrer umfassenden Ausgestaltung keine Ergänzung durch andere Ordnungsprinzipien zu[33]. Sie geht davon aus, daß die wechselseitige Verflechtung und das gemeinsame Interesse am Bestand der Gemeinschaft den Respekt des Gemeinschaftsrechts sichert, so wie es schließlich auch immer wieder geschehen ist[34].

Die Gemeinschaftsrechtsordnung beruht nicht auf dem Prinzip gegenseitiger Vertragsleistungen zwischen den Mitgliedstaaten, das Gegen- oder Ausgleichsmaßnahmen zwischen diesen zulassen könnte, sondern enthält ein ganzes Geflecht von Regelungskomplexen mit vielfältigen Bezügen aller Mitgliedstaaten untereinander. Vor allem aber begründen die Gemeinschaftsregelungen unmittelbare Rechte der Bürger, die nicht mehr von der Vermittlung durch die Mitgliedstaaten abhängig sind[35]. Wenn ein Mitgliedstaat einem anderen gegenüber, durch den er sich geschädigt glaubt, im Wege der Retorsion Gemeinschaftsbestimmungen nicht anwenden würde, so würde er mit dem schädigenden Mitgliedstaat zugleich dessen Bürger treffen, die über eigene, nicht mehr vom Staat abhängige und daher auch nicht mehr diesem gegenüber entziehbare Rechte im Rahmen der Gemeinschaft verfügen.

---

[31] Vgl. *E. Grabitz*, Der Verfassungsstaat in der Gemeinschaft, Deutsches Verwaltungsblatt (DVBl.) 1977, S. 786 ff., ferner *derselbe* (Anm. 10), S. 186.

[32] Dafür vor allem *A. Bleckmann*, Zwangsmittel im Gemeinsamen Markt, Recht der Internationalen Wirtschaft, Außenwirtschaftsdienst des Betriebs-Beraters (RIW/AWD) 1978, S. 91 ff., der sie jedoch nur als »allerletztes Mittel« und nur »im europäischen Gemeinschaftsinteresse« anwenden will – aber dessen Wahrnehmung dürfen sich einzelne Mitgliedstaaten nicht anmaßen. Anders *Zuleeg* (Anm. 7), S. 63, der einseitige Maßnahmen der Mitgliedstaaten ablehnt.

[33] So hat der Gerichtshof stets den Versuch von Mitgliedstaaten zurückgewiesen, ihre Vertragsverletzung als Antwort auf eine Vertragsverletzung anderer Mitgliedstaaten oder des Rates, insbesondere dessen Untätigkeit, zu rechtfertigen; vgl. Urteil vom 14. 12. 1971, Rs. 7/71, *Kommission/Frankreich* (Versorgungsagentur), Slg. S. 1003.

[34] Vgl. zur Problematik *C. D. Ehlermann*, Die Verfolgung von Vertragsverletzungen der Mitgliedstaaten durch die Kommission, in: Festschrift für Hans Kutscher (Baden-Baden 1981), S. 135 ff., mit Hinweisen auf den Schaffleischfall, Urteil vom 25. 9. 1979, Rs. 232/78, *Kommission/Frankreich*, Slg. S. 2729. Dazu *G. Meier*, Die Europäische Rechtsgemeinschaft am Scheideweg, RIW/AWD 1981, S. 246.

[35] Aus diesem Grunde kommt auch ein diplomatischer Schutz, auf den *Bleckmann* (Anm. 32), S. 94, Ansprüche stützen will, nicht in Betracht.

Gegenmaßnahmen können deshalb nicht mehr auf der Grundlage des Völkerrechts in der Hand der einzelnen Mitgliedstaaten liegen. Es fragt sich lediglich, ob der Gerichtshof etwa aus allgemeinen Rechtsgrundsätzen des Gemeinschaftsrechts herleiten könnte, daß Gemeinschaftssanktionen gegenüber einem widersätzlichen Mitgliedstaat, etwa der Entzug von Mitteln aus einem Gemeinschaftsfonds, die Leistung von Schadensersatz an die Gemeinschaft oder andere Mitgliedstaaten[36] oder als letzte Konsequenz der Ausschluß durch die anderen Mitgliedstaaten[37], zulässig sind. Hierzu sei – ohne rechtliche Stellungnahme – lediglich bemerkt, daß dies wiederum zu Gegenreaktionen und Zahlungsverweigerungen führen könnte, und ob bei einer solchen Eskalation die Gemeinschaft letztlich am längeren Hebel sitzen würde, erscheint mindestens fraglich.

7. Für die Beendigung der Mitgliedschaft können die Regeln des Völkerrechts ebenfalls nicht maßgebend sein. Die Mitgliedstaaten sind politisch, wirtschaftlich und rechtlich vielfältig in das Gesamtsystem der Gemeinschaft eingebunden; ihre Bürger haben eigene, einklagbare Rechte erworben. Eine derartige wechselseitige Verflechtung, wie sie dargestellt wurde, schließt ein einseitiges Ausscheiden wegen veränderter Umstände oder gar aufgrund einer einseitigen Kündigung aus[38].

Man hat versucht, das Gegenteil aus dem Wortlaut des Art. 240 EWG-Vertrag herzuleiten. Danach soll die Formulierung, daß der Vertrag auf »unbegrenzte Zeit« gilt, als »nicht festgelegt« in dem Sinne verstanden werden, daß die Geltungsdauer offen sei, also jederzeit nach den Regeln der Wiener Vertragsrechtskonvention beendet werden könne[39]. Aber abgesehen davon, daß eine solche Wortinterpretation das Gegenteil der vom Gerichtshof geforderten und angewandten teleologischen Auslegung ist, entspricht sie nicht der dargelegten Struktur der Gemeinschaft.

---

[36] Dafür *Ehlermann* (Anm. 34), S. 151 f., unter Hinweis auf das Urteil vom 7. 2. 1973, Rs. 39/72, *Kommission/Italien*, Slg. S. 101 ff.

[37] Dafür *Zuleeg* (Anm. 7), S. 63 f.; vgl. ferner *G. Meier*, Die Beendigung der Mitgliedschaft in der Europäischen Gemeinschaft, Neue Juristische Wochenschrift (NJW) 1974, S. 391.

[38] *H. P. Ipsen*, Europäisches Gemeinschaftsrecht (Tübingen 1972), S. 100 f.; *Constantinesco* (Anm. 7), S. 180 f.; *Beutler* (Anm. 16), S. 61. Mit Einschränkungen: *M. Hilf*, in: Groeben-Boeckh-Thiesing, Kommentar zum EWG-Vertrag (2. Aufl. 1974), Art. 240 Anm. 6.2, sowie *Meier* (Anm. 37). Für ein Recht zur Kündigung und Anrufung der *clausula rebus sic stantibus: P. Dagtoglou*, Recht auf Rücktritt von den Römischen Verträgen?, in: Festschrift für Ernst Forsthoff (München 1972), S. 72 ff.; *K. Doehring*, Die Wiedervereinigung Deutschlands und die europäische Integration als Inhalt der Präambel des Grundgesetzes, DVBl. 1979, S. 639; *L. M. Bentivoglio*, in: R. Quadri/R. Monaco/A. Trabucchi, Trattato Istitutivo della Comunità Economica Europea, Bd. 3 (Mailand 1965), Art. 240 Anm. 1; *A. Gauland*, Noch einmal: Die Beendigung der Mitgliedschaft in der Europäischen Gemeinschaft, NJW 1974, S. 1034. Mit Einschränkungen *Zuleeg* (Anm. 7), S. 60 ff.

[39] So vor allem *Dagtoglou*, a.a.O.; ferner *S. Malawer*, International Law, European Community and the Rule of Reason, Journal of World Trade Law 1974, S.17 ff.

Die Gemeinschaft ist nach ihrem Gesamtcharakter, wie er sich aus dem dargestellten Verfassungssystem und der bereits erreichten tatsächlichen Verflechtung auf der Ebene von Politik, Wirtschaft und Gesellschaft ergibt, und nach ihrer in der Präambel ausgedrückten Zielsetzung, einen immer engeren Zusammenschluß der Völker Europas herbeizuführen, auf Dauer angelegt. Der Gerichtshof hat bereits im Jahre 1964 die unbegrenzte Dauer als Argument für den besonderen Charakter des Gemeinschaftsrechts angeführt, also auch im Sinne einer endgültigen Bindung verstanden [40]. Die geschilderte weitere Ausgestaltung der Gemeinschaftsrechtsordnung und deren Anwendung in der Praxis bestätigen diese Auslegung. Kündigung und Geltendmachung der *clausula rebus sic stantibus* sind demnach ausgeschlossen [41].

Dagegen läßt sich auch nicht einwenden, daß ein Mitgliedstaat nicht gegen seinen Willen in der Gemeinschaft gehalten werden könne und daß niemand seinen Austritt verhindern könne [42]. Das ist sicher richtig, aber wer so argumentiert, verwechselt Recht und Macht. Gegen Vertragsbruch ist die Gemeinschaft ebensowenig geschützt wie die Staaten gegen Revolutionen, und sie besitzt keine Machtinstrumente, keine Gemeinschaftsgewalt, um sich dagegen zu wehren. Ein Staat, der einseitig aus der Gemeinschaft ausscheidet, negiert damit die Geltung des Rechts, das Grundlage des Zusammenlebens der Völker ist [43].

In diesem Zusammenhang wird vielfach auf die Erklärung zur Wiedervereinigung hingewiesen, die von der deutschen Delegation bei den Vertragsverhandlungen abgegeben wurde [44]. Ob es sich dabei wirklich um einen Vorbehalt handelt, der ein einseitiges Ausscheiden ermöglicht [45], oder nicht vielmehr um den Hinweis auf eine gegebenenfalls erforderliche Gesamtüberprü-

---

[40] Vgl. Urteil vom 15. 7. 1964, Rs. 6/64, *Costa/ENEL*, Slg. S. 1251; dazu oben bei Anm. 14.

[41] Ein einvernehmliches Ausscheiden mit Zustimmung der anderen Mirgliedstaaten, wie es etwa *Beutler, Meier* und *Zuleeg* (Anm. 38) für zulässig halten, ist in Wahrheit eine Vertragsänderung und als solche zu behandeln; vgl. unten Ziffer 9 und 10.

[42] So etwa *M. Zuleeg*, Der Beitritt Griechenlands zur Europäischen Gemeinschaft, politische und rechtliche Probleme, Hellenic Review of International Relations 1 (1980/1), S. 357 ff., 360, der ein einseitiges Ausscheiden zwar für einen Vertragsbruch, aber dennoch für einen wirksamen Rechtsakt hält.

[43] Auch die Vorgänge um Grönland im Anschluß an die den Verbleib in der Gemeinschaft ablehnende Volksabstimmung vom 23. 2. 1982 sprechen nicht gegen diese Auslegung. Ein etwaiges Ausscheiden Grönlands wäre den Vorgängen um die sog. überseeischen Länder und Hoheitsgebiete im Sinne der Art. 131 ff. EWG-Vertrag nach Erlangung der Selbständigkeit vergleichbar. Zum neuen Statut Grönlands vgl. *I. Foiglul*, Home Rule in Greenland: A Framework for Local Autonomy, CMLR 1980, S. 91 ff.

[44] Zur Problematik ausführlich *Doehring* (Anm. 38), S. 633; *E. Wohlfarth*, in: E. Wohlfarth/U. Everling/H.-J. Glaesner/R. Sprung, Die Europäische Wirtschaftsgemeinschaft, Kommentar zum Vertrag (Berlin 1960), Art. 227 Anm. 5, und Art. 240 Anm. 2; *Zuleeg* (Anm. 6), S. 61; *Dagtoglou* (Anm. 38), S. 92.

[45] So *Wohlfarth*, a.a.O.

fung durch die Vertragspartner, mag dahingestellt bleiben. Denn jedenfalls wäre eine solche Lage nur bei umstürzenden Vorgängen von nahezu revolutionärem Ausmaß denkbar, die einen Wandel aller bestehenden Bündnissysteme zur Folge hätten. Unter den auf absehbare Zeit bestehenden Verhältnissen ist eine Annäherung der beiden deutschen Staaten nur bei Fortbestand ihrer beiderseitigen Bindungen im Rahmen gesamteuropäischer Entwicklungen vorstellbar[46]. Auch unter diesem Gesichtspunkt ist also die Endgültigkeit der Bindungen in der Gemeinschaft nicht in Frage gestellt.

8. Es stellt sich jedoch die weitere Frage, wie weit die Mitgliedstaaten auch bei ihrem Handeln außerhalb der Gemeinschaft, soweit sich dieses auf die Gemeinschaft bezieht, durch die Gemeinschaftsverträge gebunden sind. Dabei ist davon auszugehen, daß die Regeln des Völkerrechts zwischen den Mitgliedstaaten sicher insoweit weiter anzuwenden sind, als es um Materien geht, die nicht durch die Gemeinschaftsverträge erfaßt sind. So gibt es weite Bereiche, etwa der Verteidigungspolitik und der inneren und äußeren Sicherheit, in denen das Völkerrecht und insbesondere die Wiener Vertragsrechtskonvention zwischen den EG-Staaten uneingeschränkt gelten.

Daraus folgt aber noch nicht, daß die Mitgliedstaaten außerhalb des eigentlichen Geltungsbereichs des Gemeinschaftsrechts von dessen Bindungen völlig frei sind. Das Gemeinschaftsrecht wirkt auch auf das außergemeinschaftliche Verhalten der Mitgliedstaaten ein.

Das gilt vor allem für die Grauzone, die sich in immer größerem Umfang um die Verträge herum bildet und in der die Mitgliedstaaten zwar nicht auf der Grundlage der Bestimmungen der Verträge, aber inhaltlich in engem Zusammenhang mit diesen und zur Verwirklichung ihrer Ziele handeln. In den sechziger Jahren handelte es sich vorwiegend um Beschlüsse der im Rat vereinigten Vertreter, deren Bindungswirkung nie ganz geklärt wurde. Seit der Gipfelkonferenz von Den Haag und den Entschließungen über die Wirtschafts- und Währungsunion weichen die Mitgliedstaaten in noch größerem Umfang in vom Vertrag nicht als Rechtsakte qualifizierte Willensäußerungen aus. Zu erwähnen sind vor allem die Entschließungen des Europäischen Rates über vielfältige anstehende Themen von der Konjunkturpolitik bis zum Finanzausgleich für Großbritannien, ferner die zahlreichen Entschließungen des Rates und die Beschlüsse der Europäischen Politischen Zusammenarbeit, die in einem unmittelbaren sachlichen Zusammenhang mit der Gemeinschaft und ihren Zielen stehen.

Derartigen Entschließungen fehlt in aller Regel der rechtliche Bindungswille; als »politische Verpflichtungen« begründen sie keine direkte rechtliche Wirkung. Man hat deshalb gefragt, ob diese »nicht-rechtlichen Normen«, die

---

[46] Dazu *P. Bender*, Das Ende des ideologischen Zeitalters. Die Europäisierung Europas (Berlin 1981), S. 234 ff.; *derselbe,* Europa und die Deutschen. Die Entschärfung eines Problems, Deutschland Archiv 1981, S. 251; ferner *Constantinesco* (Anm. 7), S. 182 f.

immerhin gewisse Erwartungen für das künftige Verhalten wecken, nach neueren Tendenzen im Völkerrecht als *soft-law* zu beurteilen sind, was mit Recht verneint wird[47].

Der Schlüssel zur Beurteilung dieser Akte muß trotz des Fehlens einer unmittelbaren Rechtsgrundlage im Gemeinschaftsrecht gesucht werden. Die Verpflichtung zur Gemeinschaftstreue nach Art. 5 EWG-Vertrag bindet die Mitgliedstaaten auch bei ihren Aktivitäten außerhalb des Kompetenzbereichs der Gemeinschaft, soweit sie mit dieser in Zusammenhang stehen und für die Verwirklichung ihrer Ziele von Bedeutung sind[48]. Wie weit diese Bindung reicht, insbesondere ob und in welchem Umfang deshalb Beschlüsse dieser Art doch gewisse rechtliche Wirkungen entfalten können, kann hier nicht weiter untersucht werden[49]. Es kommt nur darauf an zu zeigen, daß das Gemeinschaftsrecht Ausstrahlungen über seinen eigentlichen Bereich hinaus entfalten und die Mitgliedstaaten in ihrem Handeln auch über den Kompetenzbereich der Gemeinschaft hinaus erfassen kann.

9. Die Mitgliedstaaten sind demnach zwar Schöpfer und Träger der Gemeinschaft, aber sie sind doch zugleich auch vielfältig in sie eingebunden. Können sie in dieser Lage noch beliebig über ihr Geschöpf verfügen? Es stellt sich damit die im Thema aufgezeichnete Frage, ob die Mitgliedstaaten die Gemeinschaft durch einen völkerrechtlichen Vertrag, den *actus contrarius* zum Gründungsvertrag, von außen her beliebig und ohne Bindung an die Ziele der Verträge und die in ihnen vorgesehenen Verfahren ändern oder gar aufheben können.

Die Frage wird meist unter Hinweis auf die politischen Realitäten und den fehlenden Staatscharakter der Gemeinschaft bejaht[50]. Dabei wird vielfach auf die Auffassungen aus der Zeit der Ausarbeitung des EWG-Vertrages zurückgegriffen[51], in der die wechselseitige Verflechtung in der Tat noch ge-

---

[47] Vgl. *M. Bothe*, »Soft Law« in den Europäischen Gemeinschaften?, in: Festschrift für Hans-Jürgen Schlochauer (Berlin 1981), S. 761 ff., auf den wegen aller Nachweise verwiesen wird; ferner *O. Jacot-Guillarmod*, Droit communautaire et droit international public (Genf 1979), S. 5 ff., 36 ff.

[48] Abweichend *Bothe*, a.a.O.; vgl. aber etwa *Jacot-Guillarmod*, a.a.O., S. 11 ff.

[49] Der Verfasser beabsichtigt, dazu eine gesonderte Untersuchung in der Gedächtnisschrift für L.-J. Constantinesco vorzulegen.

[50] *Bernhardt* (Anm. 9), S. 234 f.; *Bleckmann* (Anm. 8), Europarecht 1981, S. 114; *Constantinesco* (Anm. 7), S. 180; *A. Matthias*, Integrationsprobleme innerhalb der Europäischen Gemeinschaft, Jahrbuch des öffentlichen Rechts 29 (1980), S. 85 ff.; *W. Meng*, Das Recht der Internationalen Organisationen – eine Entwicklungsstufe des Völkerrechts (1979), S. 120 ff.; *Zuleeg* (Anm. 7), S. 59; zurückhaltender *Hilf* (Anm. 38), Art. 240 Anm. 6. – Ablehnend *Ipsen* (Anm. 38), S. 211, 228, 232; *Beutler* (Anm. 16), S. 61; *M. Sørensen*, »Eigene Rechtsordnungen«, Skizze zu einigen systemanalytischen Betrachtungen über ein Problem der internationalen Organisationen, in: Festschrift für Hans Kutscher (Baden-Baden 1981), S. 415 ff., 431; *O. Jacot-Guillarmod* (Anm. 47), S. 11 ff.; *G. Nicolaysen*, Europäisches Gemeinschaftsrecht (Stuttgart 1979), S. 24; *P. Pescatore*, Les répartitions des compétences entre la Communauté et ses Etats membres, in: Actes du sixième Colloque à Liège (1973), S. 61 ff., 82.

ring und die Entwicklung nicht vorauszusehen war. Aber der Rückgriff auf die Gründerzeit besagt ebensowenig wie der Hinweis auf die Änderung des Montanvertrages im Zusammenhang mit dem Abschluß des Saarvertrages im Jahre 1956[52], zumal in der späteren Praxis derartige »Jugendsünden« sorgfältig vermieden wurden. Die These von der Verfügungsmacht der Mitgliedstaaten ist auch nicht durch die Einsetzung des Europäischen Rates bekräftigt worden[53]. Im Gegenteil zeigt die Formulierung der Entschließung der Regierungschefs von 1974[54], nach der der Europäische Rat »als Rat der Gemeinschaft und im Rahmen der Politischen Zusammenarbeit« zusammentritt, daß die Regierungschefs die Gemeinschaftsregelungen respektiert und sich nicht als »Herren« geriert haben[55].

Schon frühzeitig wurde darauf hingewiesen, daß es für die Beurteilung der Frage auf die weitere Entwicklung ankomme, also auf die sich aus langwieriger gemeinsamer Erfüllung der überstaatlichen Aufgaben ergebenden gesellschaftlichen und politischen »haltgebenden Elemente im Außerrechtlichen« und auf die »Unterwerfung des Staatslebens unter eine durch die Gemeinschaften fortschreitend zu verwirklichende faktische Zwangsläufigkeit, die weniger und weniger reversibel und weniger und weniger einer Alternativentscheidung seitens der Mitgliedstaaten ausgesetzt ist«[56]. Wie weit dieser Prozeß fortgeschritten ist, muß an Hand der Entwicklung, die skizziert wurde, beantwortet werden.

In der Literatur liegen dazu nur vorsichtige Äußerungen vor, wenn auch das Bewußtsein dafür, daß sich die Gemeinschaft der Schwelle der Irreversibilität ihrer Existenz nähert, wächst[57]. Aus dem geschilderten Stand der Ent-

---

[51] *K. Carstens*, Die kleine Revision des Vertrages über die Europäische Gemeinschaft für Kohle und Stahl, ZaöRV 21 (1961), S. 1 ff., 8, mit Hinweisen auf die ältere Literatur zum Montanvertrag; ferner *U. Scheuner*, Die Rechtsetzungsbefugnis internationaler Gemeinschaften, in: Festschrift für Alfred Verdross (Wien 1960), S. 229 ff., 237; *Wohlfarth* (Anm. 44), Art. 240 Anm. 1; anders damals wohl nur *W. Thieme*, Das Grundgesetz und die öffentliche Gewalt internationaler Gemeinschaften, VVDStL H. 18 (1960), S. 50 ff. 72.

[52] Vgl. *Zuleeg* (Anm. 7), S. 58.

[53] Vgl. *H. P. Ipsen*, Marginalien zum Europäischen Rat, Europarecht 1981, S. 309 ff., 329.

[54] Vgl. Europa-Archiv 1975, S. D 41 ff.

[55] Vgl. *U. Everling*, Die Europäische Gemeinschaft nach der zweiten Pariser Konferenz, Europa-Archiv 1975, S. 59 ff., 64; *W. Wessels*, Der Europäische Rat (Bonn 1980), S. 120 ff.

[56] *W. v. Simson*, Der politische Wille als Gegenstand der europäischen Gemeinschaft, in: Festschrift für Otto Riese (Karlsruhe 1964), S. 83 ff. Auf die Entwicklung stellt auch *Hilf* (Anm. 38) ab.

[57] Vgl. etwa *K. Doehring*, Das Staatsrecht der Bundesrepublik Deutschland (2. Aufl. 1980), S. 79 ff. *W. v. Simson* scheint insoweit noch zu zögern; vgl. seinen Beitrag: Wachstumsprobleme einer europäischen Verfassung, in: Festschrift für Hans Kutscher (Baden-Baden 1981), S. 481 ff. Die Überlegungen von *Bleckmann* (Anm. 8) Souveränitätsverständnis, S. 61, dürften wohl hypothetisch gemeint sein, nämlich für den Fall der Verwendung des Verfassungsbegriffs als »Integrationsreserve«.

wicklung wird man aber nunmehr Folgerungen ziehen müssen. Das Gemeinschaftsrecht hat quantitativ und qualitativ vielfältige Bindungen zwischen der Gemeinschaft und den Mitgliedstaaten geschaffen. Diese sind als Glieder der Gemeinschaft, die deren Rechtshoheit unterliegen, nicht mehr in der Lage, beliebig Herrschaft über sie auszuüben. Vor allem entspricht dieser rechtlichen Bindung auch eine Verflechtung auf politischer, wirtschaftlicher und gesellschaftlicher Ebene[58], die es ausschließt, den Mitgliedstaaten noch die volle Verfügungsgewalt zuzusprechen.

Das Ausmaß dieser Verflechtung zeigt sich täglich an den vielfältigen Formen der Zusammenarbeit und Abhängigkeit, die im Bewußtsein der Öffentlichkeit freilich teilweise von den täglichen Querelen verdeckt werden. Allerdings ist nicht zweifelsfrei, ob in allen Mitgliedstaaten die Entwicklung gleich weit fortgeschritten ist. Vor allem für Frankreich, in dem das traditionelle Verständnis der Nation fortwirkt, und für die neuen Mitgliedstaaten wie Großbritannien, in denen die Anpassungsprozesse erst begonnen haben, mögen Zweifel bestehen[59].

10. Fragt man nach den Grenzen der Verfügungsmacht, so muß man sich auf die Herkunft der Gemeinschaft besinnen. Schon frühzeitig wurde darauf hingewiesen, daß der Akt, durch den die nach Art. 24 GG ermächtigte Bundesregierung gemeinsam mit anderen Regierungen, die auf vergleichbare Weise ermächtigt sind, eine zwischenstaatliche Einrichtung gründet, Verfassungsgebung ist[60]. Die Staaten wurden also bei der Gründung der Gemeinschaft als *pouvoir constituant* tätig, nur als solche könnten sie deshalb auch künftig noch Verfügungsgewalt über die Gemeinschaft ausüben.

Daraus folgt zunächst, daß Vertragsänderungen, die sich im Rahmen der bisherigen Vertragsziele und -grundlagen halten, den dafür im Vertrag vorgesehenen Verfahren unterliegen, denn dafür konstituiert sich nicht die verfassunggebende Gewalt. Alle Vertragsänderungen üblicher Art müssen daher im Verfahren nach Art. 236 EWG-Vertrag beschlossen werden, also unter Mitwirkung der Kommission und des Europäischen Parlaments[61].

---

[58] Vgl. etwa *C. J. Friedrich*, Europa im Werden (Bonn 1972).

[59] Für Frankreich vgl. z. B. *J. Rideau/P. Gerbet/M. Torelli/R.-M. Chevallier*, La France et les Communautés Européennes (Paris 1975); *J.-P. Jacqué*, La Souveraineté française et l'élection du Parlement européen au suffrage universel direct, in: Ress (Hrsg.) (Anm. 8), S. 71 ff.; *E.-W. Fuss*, Souveränitätsdenken als Fessel des Integrationsfortschritts, Die Judikatur des französischen Verfassungsrats zur europäischen Einigung, DVBl. 1980, S. 98 ff.; für Großbritannien vgl. *R. Mayne*, Das »britische Problem« der Europäischen Gemeinschaft, Europa-Archiv 1980, S. 337 ff.

[60] So vor allem *J. H. Kaiser*, Bewahrung und Veränderung demokratischer und rechtsstaatlicher Verfassungsstruktur in den internationalen Gemeinschaften, VVDStL H. 23 (1966), S. 1 ff., 18, ferner *derselbe*, Zur gegenwärtigen Differenzierung von Recht und Staat – staatstheoretische Lehren der Integration, Österr. Zeitschrift für öffentliches Recht 10 (1959/60), S. 413 ff.

[61] *Constantinesco* (Anm. 7), S. 192 ff.; *C. v. Arnim*, in: Groeben/Boeckh/Thiesing (Anm. 38), Art. 236 Anm. I 1; *Jacot-Guillarmod* (Anm. 47), S. 14. Anders *Zuleeg*

Weiter folgt daraus, daß die Mitgliedstaaten als Träger und Gründer der Gemeinschaft in diese von außen nur noch unter den außergewöhnlichen Umständen eingreifen können, unter denen sich – bei aller gebotenen Vorsicht mit solchen Vergleichen – im Staat eine verfassunggebende Gewalt konstituieren würde, also bei Wandlungen mit revolutionärem Ausmaß[62]. Bei der Gemeinschaft würden solche Voraussetzungen vor allem dann vorliegen, wenn umstürzende Zusammenbrüche oder Neuorientierungen des internationalen Systems ihrem Fortbestand die Grundlage entziehen würden oder wenn umwälzende Vorgänge in einzelnen Mitgliedstaaten, etwa die grundlegende Änderung ihrer politischen Systeme, die Zielsetzung der Gemeinschaft, zur Erhaltung von Friede und Freiheit in Westeuropa und in der Welt beizutragen, ernsthaft gefährden würden. Im letzteren Fall käme auch ein Ausschluß eines Mitgliedstaates durch die anderen in Betracht.

Fraglich ist dabei aber, was heute als *pouvoir constituant* der Gemeinschaft anzusehen wäre. Dabei kann maßgebend nicht mehr der Gründungsakt sein[63], sondern es kommt auf den Gesamtcharakter der durch ihn geschaffenen und bis heute fortentwickelten Gemeinschaft an. Überwiegend wird angenommen, daß die Gemeinschaftsgewalt ihre Legitimität immer noch über die Mitgliedstaaten vermittelt erhält[64]. Ob das aber nach der bisherigen Entwicklung noch in vollem Umfang zutrifft, erscheint doch zweifelhaft. Jedenfalls kann man sich dafür nicht auf die Errichtung des Europäischen Rates berufen, denn die in ihm vereinigten Regierungschefs vermögen sich schon wegen ihrer ganz unterschiedlichen verfassungsmäßigen Stellung im Verhältnis zu ihren Staatsvölkern nicht als Träger der Gemeinschaft zu etablieren[65].

Im Gegenteil, es ist an den Staaten und ihren Regierungschefs vorbei eine unmittelbare Verbindung zwischen den Bürgern und der Gemeinschaft entstanden, die nicht mehr ignoriert werden kann. Die unmittelbare Wirkung der wichtigsten Rechtssätze des Gemeinschaftsrechts, die uneinge-

---

(Anm. 6), S. 59 f.; *M. Deliege-Sequaris*, Révision des traités européens en dehors des procédures prévues, Cah. dr. eur. 1980, S. 539 ff.; gegen letztere *J. V. Louis*, ebenda, S. 553 ff.

[62] Zur verfassunggebenden Gewalt vgl. *K. Stern*, Das Staatsrecht der Bundesrepublik Deutschland, Bd. 1 (München 1977), S. 118 ff., sowie *G. Burdeau*, Traité de science politique, Bd. 4 (2. Aufl. 1969), n° 296 ff., mit weiteren Nachweisen.

[63] So nahm *J. H. Kaiser* (Anm. 60) noch an, daß die Mitgliedstaaten, und zwar deren durch die Parlamente handelnden Völker (vgl. Art. 137 EWG-Vertrag) als Constituante tätig würden; ebenso *P. Pescatore*, Droit International et Droit communautaire, essai de réflexion comparative, Collection des Conférences européennes n° 5 (Nancy 1969), S. 21.

[64] Vgl. *Ipsen* (Anm. 38), S. 70 ff. und *passim*, ferner *derselbe* (Anm. 53), S. 333; *Constantinesco* (Anm. 6), S. 313 ff.; *R. Herzog*, Zur Legitimation der Willensbildung in den Gemeinschaften, in: Die institutionelle Entwicklung der Europäischen Gemeinschaften in den siebziger Jahren (Kölner Schriften zum Europarecht, Bd. 22) (Köln 1973), S. 35 ff.

[65] *Ipsen* (Anm. 53), S. 333 f., scheint dagegen hier Ansätze zur Legitimierung zu sehen. Vgl. dazu schon *Everling* (Anm. 10), S. 610.

schränkte Gleichstellung bei Freizügigkeit und wirtschaftlicher Betätigung, die Gewährleistung grundrechtlicher Positionen und der umfassende Rechtsschutz haben dazu geführt, daß die Mitgliedstaaten ihr Vermittlungsmonopol für den Volkswillen, der auch in der Gemeinschaft letzte Grundlage aller Hoheitsgewalt darstellt, nicht mehr uneingeschränkt besitzen. Diese Entwicklung hat einen weiteren Impuls durch die Direktwahl der Abgeordneten des Europäischen Parlaments erhalten. Sie hat der Gemeinschaft sicher noch keine volle eigene Legitimität verschafft, aber doch Ansätze dazu gebracht[66]. Wenn es auch bis zur Bildung eines echten »Gemeinschaftsbewußtseins« noch ein weiter Weg ist, so können doch diese Ansätze nicht ohne rechtliche Bedeutung sein.

Die Mitgliedstaaten werden deshalb ihre Völker nicht nur über ihre nationalen Parlamente, sondern auch auf Gemeinschaftsebene in geeigneter Form einbeziehen müssen, wenn sie die Gemeinschaft kraft ihrer Verfügungsgewalt als *pouvoir constituant* in Zielsetzung, grundsätzlicher Ausgestaltung und Zusammensetzung ändern oder gar aufheben wollen. Wie diese Beteiligung aussehen muß, hängt vom Stand der Integration ab. Mindestanforderung wäre wohl die Einhaltung der Formen des Art. 236 des EWG-Vertrages, aber darüber hinaus kämen bei weiteren Fortschritten je nach Bedeutung des Falls eine Mitentscheidung des Europäischen Parlaments oder einer besonderen Versammlung und in letzter Konsequenz eine Volksabstimmung in Betracht.

11. Mit dieser These wird der Gemeinschaft und ihrem Recht eine Selbständigkeit zugesprochen, die sie abhebt von der bisherigen mittelbaren Verankerung im staatlichen Recht einerseits und im Völkerrecht andererseits. Und dabei wird davon ausgegangen, daß die Gemeinschaft die Schwelle zum Staat auch weiterhin nicht überschreitet und die Staatlichkeit ihrer Mitgliedstaaten in der Substanz nicht in Frage stellt[67].

Wie eine solche Entwicklung theoretisch zu deuten ist, muß hier offen bleiben. Immerhin sei soviel bemerkt, daß sich das Gemeinschaftsrecht bei dieser Bewertung durchaus in das moderne Verständnis von der Völkerrechtsordnung einfügt, wie es *Hermann Mosler* unlängst dargestellt hat[68]. Es muß davon ausgegangen werden, daß über den Staaten als den vom Volkswillen legitimierten politischen Einheiten mit Gewaltmonopol, die nach wie vor primäre Entscheidungsträger sind, eine Reihe von teils übereinander, teils

---

[66] Zur Problematik *Grabitz* (Anm. 10), S. 360, 371 ff.; ferner *H.-J. Glaesner*, Die Legitimität der Gemeinschaft, in: Gedächtnisschrift für Christoph Sasse (Baden-Baden 1980), S. 73 ff.

[67] Dazu würde die zunehmende rechtliche und tatsächliche Verflechtung nicht ausreichen, sondern es wäre ein konstitutiver Begründungsakt einer europäischen verfassunggebenden Gewalt notwendig, der den Rahmen des Art. 24 GG überschreiten würde; vgl. *P. Badura*, Verfassung und Verfassungsgesetz, in: Festschrift für U. Scheuner (Berlin 1973), S. 19 ff., 38 ff.

[68] *H. Mosler*, Völkerrecht als Rechtsordnung, ZaöRV 36 (1976), S. 6 ff.

nebeneinander geordneten Rechtsordnungen bis hin zum allgemeinen Völkerrecht bestehen, deren Grad an Regelungsintensität jeweils unterschiedlich ist[69]. Ihre Rechtfertigung erhält diese Hierarchie von ihrer Verankerung im Bewußtsein der Menschen, die die verschiedenen Rechtsordnungen tragen. Bei den Europäischen Gemeinschaften erreicht diese einen solchen Grad von Dichte und Unmittelbarkeit, daß sich ihr Bestand bis zu einem gewissen Grade vom Willen der sie tragenden Staaten zu lösen beginnt. Aber die Schwelle zum Staat wird damit noch nicht überschritten, und wenig spricht dafür, daß dies in absehbarer Zeit geschieht oder auch nur angestrebt wird. Wahrscheinlich handelt es sich bei der Europäischen Gemeinschaft in ihrer jetzigen Gestalt um eine Organisationsform, die der gegenwärtigen geschichtlichen Situation Westeuropas angemessen ist und mit der die Mitgliedstaaten, die in erster Linie Diener und nicht Herren der Verträge sein sollen, bis auf weiteres leben müssen und können.

---

[69] Vgl. dazu *Sørensen* (Anm. 50), S. 415 ff.

L'interaction entre la dimension internationale et la dimension nationale de la Convention européenne des Droits de l'Homme: Notions autonomes et effet direct

DIMITRIOS EVRIGENIS

J'ai choisi de parler[1] de la Convention européenne des Droits de l'Homme pour la raison que cet instrument représente, par excellence, un terrain juridique de convergence européenne qui se manifeste sur le plan à la fois national, international et communautaire. En laissant de côté le traitement qui est réservé à la Convention dans le cadre de l'ordre juridique communautaire et, en particulier, dans celui de la jurisprudence communautaire, ainsi que les tentatives récentes de construction d'un rattachement plus substantiel et plus élaboré entre le système de la Convention et la Communauté, je me limiterai à traiter brièvement deux questions de caractère général qui se situent sur ce qui est devenu courant d'appeler la dimension internationale et la dimension nationale de la Convention européenne des Droits de l'Homme: Il s'agit, d'une part, des *notions autonomes*[2] et, d'autre part, de l'*effet direct* de la Convention. Bien que traditionnellement considérées comme attributs respectivement de la dimension internationale et de la dimension nationale de la Convention, ces questions n'en finissent pas moins par constituer une démonstration de la relativité de la distinction entre ces deux dimensions et, inversément, une preuve de l'interdépendance et de l'interaction qui existe entre elles.

* * *

[1] Le texte reproduit l'essentiel d'une conférence donnée à Bruxelles, le 18 novembre 1981, sous les auspices du Centre interuniversitaire de droit comparé.

[2] *Hermann Mosler* a largement consacré ses réflexions à la question des notions autonomes. Voir, entre autres, ses études: Rechtsvergleichung vor völkerrechtlichen Gerichten, in: Internationale Festschrift für Alfred Verdross (1971), p. 381 ss.; L'influence du droit national sur la Convention européenne des droits de l'Homme, in: Miscellanea W. J. Ganshof van der Meersch, t. 1 (1972), p. 521 ss.; Problems of Interpretation in the Case Law of the European Court of Human Rights, in: Essays on the Development of the International Legal Order in memory of Haro F. van Panhuys (1980), p. 149 ss.

Les *notions autonomes*, tout d'abord. Je commencerai par formuler certaines précisions sur ce sujet. J'aborderai ensuite la question de savoir comment déterminer le contenu d'une notion autonome. Et je finirai avec quelques réflexions sur la mission que ces notions sont appelées à remplir.

Dans un langage laïc on pourrait décrire le phénomène des notions autonomes par un adage construit: Homophonie n'est pas forcément synonymie. En d'autres termes, identité terminologique ne signifie pas nécessairement identité conceptuelle s'il s'agit de termes énoncés et fonctionnant dans des contextes systématiques différents. Exemple: »droit civil« dans l'art. 6 para. 1 de la Convention européenne des Droits de l'Homme ne s'identifie pas conceptuellement avec »droit civil« dans tel système juridique national ou tel autre instrument international. Ces affirmations ont l'air de véritables truismes. En effet, elles n'expriment qu'une règle élémentaire de l'interprétation juridique dès que celle-ci dépasse les limites d'une simple exégèse et tient compte – pour rappeler les termes de la Convention de Vienne sur le droit des traités – de l'objet, du but et du contexte de la notion juridique à interpréter. On pourrait affirmer, sans toujours quitter le terrain des tautologies, que toute notion juridique est autonome dans la mesure de l'autonomie dont jouit le système ou le contexte normatif dont elle fait partie.

En deuxième lieu, il faut observer que les notions autonomes ne sont pas un phénomène de la logique juridique limité, comme on est parfois enclin à le penser, au seul droit international conventionnel. Qu'on songe, par exemple, à l'autonomie dont jouissent, dans certains systèmes juridiques nationaux, les notions du droit commercial par rapport à celles homophones du droit civil ou les notions patrimoniales du droit fiscal par rapport à celles terminologiquement identiques ou similaires du droit privé. Le droit communautaire, de son côté, qui par son évolution transcende et dépasse la catégorie du droit international conventionnel, fournit un autre exemple de terrain juridique fertile en notions autonomes. Toujours est-il que c'est sur le plan du droit des traités proprement dit que la technique des notions autonomes a été développée et affinée. Et c'est sur ce dernier plan que se pose la question de l'autonomie des notions de la Convention européenne des Droits de l'Homme.

En troisième lieu, il est nécessaire de définir dans le phénomène de la notion autonome le concept d'autonomie. Autonomie d'une notion contenue dans un traité international par rapport à quoi? Une réponse facile et logiquement impeccable serait d'affirmer qu'il s'agit d'une autonomie globale et absolue. La notion autonome est autonome par rapport à toute autre notion terminologiquement identique ou similaire nonobstant le système juridique dont cette dernière fait partie. Il n'en est pas moins vrai qu'une telle réponse resterait largement évasive sur le plan pratique car l'autonomie d'une notion contractuelle a besoin d'être précisée et mise en oeuvre surtout par rapport à des notions terminologiquement identiques ou similaires qui existent dans un *milieu juridique proche*. Par milieu juridique proche, j'entends l'ensemble des contextes juridiques systématiques avec lesquels l'instrument auquel appartient la notion autonome à interpréter présente des liens juridiques. Tels sont les systèmes juridiques nationaux des parties contractantes, mais également le droit international général, ainsi que d'autres conventions internationales en vigueur entre les mêmes parties contractantes. On peut illustrer cette proposition avec un exemple: L'autonomie d'une notion contenue dans la Convention européenne des Droits de l'Homme se pose effectivement dans les rapports de la Convention avec le droit national des Etats contractants, avec le droit international général et avec d'autres instruments internationaux en vigueur entre ces Etats, cette dernière catégorie devant comprendre également le droit communautaire. En particulier, entre la Convention et le droit communautaire il existe une réclamation réciproque d'autonomie, même si la Convention devait être considérée, selon une certaine doctrine jurisprudentielle, comme en quelque sorte intégrée dans le droit communautaire. Les juridictions de Strasbourg et de Luxembourg ont, par exemple, interprété de façon sensiblement différente la notion de »morale« (art. 10 para. 2 de la Convention et 36 du Traité CEE) dans les arrêts *Handyside*[3] et *Henn*[4]. Les rapports conceptuels réciproques, dont il est question, méritent bien un approfondissement doctrinal.

Une quatrième précision: Comment distinguer dans un instrument international entre notions autonomes et notions non-autonomes? La Cour de Strasbourg éprouve fréquemment le besoin d'indiquer expressément que telle notion qu'elle interprète est autonome. J'y vois une précision donnée dans un but déclaratoire, plutôt qu'une distinction. Lorsque la Cour s'abstient de qualifier d'autonome une notion de la Convention, ceci ne signifie pas que cette notion n'est pas nécessairement autonome. Il existe d'ailleurs d'autres expressions jurisprudentielles courantes pour indiquer le caractère autonome d'une notion, telles que, par exemple, les formules »au sens de la Convention«, »dans le contexte de la Convention«. Car finalement la question se pose de savoir si la distinction entre notions autonomes et non-autonomes de la Convention n'est pas un faux dilemme. Toute notion contenue dans un instrument international ne devrait-elle pas avoir un caractère autonome? Je répondrais par l'affirmative à cette question, avec la précision suivante: pour toute notion contenue dans un instrument international il existe une présomption d'autonomie, présomption pourtant réfragable en ce sens que le traité peut aussi contenir de notions qui sont définies par renvoi direct au droit national. Les notions pour lesquelles le traité renvoie directement au droit national ne sont pas autonomes puisque leur contenu est puisé dans un autre contexte juridique. Mais là aussi, il faut nuancer l'affirmation. Le renvoi au droit national est opéré par le traité lui-même et, par conséquent, condi-

---

[3] Arrêt du 7 décembre 1976, Série A, vol. 24 (1976).

[4] Arrêt du 14 décembre 1979, *Regina* c. *Henn et Darby*, affaire 34/79, Recueil 1979. 3795.

tionné par celui-ci. La non-autonomie d'une notion définie par renvoi direct au droit national est donc relative. Prenons deux exemples: il est admis que lorsqu'un traité utilise la notion de nationalité d'une personne physique il renvoie au droit de l'Etat intéressé. Mais là aussi, dans une matière ressortissant, par excellence, de la souveraineté nationale, telle que la nationalité, la définition de ce qui est entendu par »nationalité« relève largement du traité lui-même, bien que la constatation de l'existence d'une certaine nationalité au chef d'un individu se fasse par application du droit national correspondant. Deuxième exemple: selon plusieurs dispositions de la Convention européenne des Droits de l'Homme, pour qu'une restriction nationale à un droit fondamental soit compatible avec la Convention, elle doit être »prévue par la loi«, à savoir la loi nationale de l'Etat introduisant la restriction (voir les para. 2 des art. 8, 9, 10 et 11 de la Convention). Le renvoi au droit national est ici clair et certain. Il n'en reste pas moins qu'il est fait en vertu et en fonction de la formule »prévue par la loi« qui fait partie de la Convention et qui conditionne le sens qu'il faut attribuer à ce renvoi au droit national[5].

Sans entrer dans un examen plus poussé de cette question, je voudrais conclure que le caractère soit autonome, soit non-autonome d'une notion contenue dans un traité international comporte un élément important de relativité. Comme les notions non-autonomes ne sont qu'en principe non-autonomes et par conséquent elles contiennent, elles aussi, un degré d'autonomie, ainsi les notions autonomes ne sont qu'en principe autonomes et par conséquent elles contiennent, elles aussi, un élément de non-autonomie. Dans ce climat de relativité on pourrait, peut-être, reprendre ici la célèbre formule de *G. Orwell* en disant qu'il existe des notions autonomes qui sont plus autonomes que d'autres.

Je voudrais illustrer le caractère relatif de l'autonomie des notions autonomes en essayant d'apporter une dernière précision à la question. Comment définir une notion autonome? Où puiser son contenu conceptuel? Si autonomie signifie qu'une notion n'est pas synonyme, donc conceptuellement identique à une autre notion terminologiquement identique appartenant à un autre contexte juridique, il n'en reste pas moins que son contenu s'alimente également à l'aide des éléments puisés dans ces autres contextes. Les sources disponibles pour la définition d'une notion autonome contenue dans un traité sont le droit international général, le droit national des Parties contractantes, les principes généraux du droit. La méthode comparative, c'est évident, est ici un instrument de travail dont on ne saurait trop souligner l'importance.

Les notions autonomes sont appelées à jouer un rôle essentiel dans la mise en oeuvre d'un instrument international. Ce rôle est double: Tout d'abord, les notions autonomes sont un facteur d'*uniformité d'interprétation*. C'est leur fonction principale. Détachées surtout du contexte juridique na-

---

[5] Voir l'arrêt *Sunday Times* du 26 avril 1979, Série A, vol. 30 (1979), §§ 46 ss., p. 30 ss.

tional, elles réalisent une uniformité internationale au-dessus de la diversité nationale. Le procédé logique développé en l'espèce mérite notre attention. Si les notions autonomes puisent dans une certaine mesure leur contenu dans le droit des Etats contractants, elles y reviennent non seulement comme définitions uniformes des engagements internationaux de ces Etats mais, dans les cas où la convention s'incorpore dans l'ordre juridique national, également comme définitions uniformes qui, émanant de la convention internationale, sont elles-mêmes incorporées dans l'ordre juridique national. Si le concept autonome se situe sur la dimension internationale du traité et fonctionne à partir de celle-ci, il n'en influence pas moins de façon substantielle l'application de la convention sur la dimension nationale. Cet échange conceptuel réciproque met en relief l'interdépendance et l'interaction entre les deux dimensions, internationale et nationale, de la convention.

Mais, à côté de leur fonction d'uniformisation, les notions autonomes sont appelées à jouer un autre rôle, également important: elles sont, par excellence, l'outil d'une *interprétation créatrice*. Affranchies de la servitude d'une dépendance absolue des conceptions juridiques nationales, parfois sclérosées et en tout cas particularistes, mais tout en pouvant mettre en valeur l'évolution et l'épanouissement de ces conceptions, elles peuvent devenir l'instrument d'une jurisprudence adaptée aux transformations constantes des données sociales. La jurisprudence de la Cour européenne des Droits de l'Homme sur la notion de »droits et obligations de caractère civil« de l'art. 6 para. 1 de la Convention européenne des Droits de l'Homme, exprimée dans les arrêts *Ringeisen*[6], *König*[7] et *Le Compte, Van Leuven et De Meyere*[8], nonobstant la critique que l'on a cru pouvoir lui adresser quant au fond, est un exemple représentatif d'une telle utilisation des notions autonomes.

* * *

Je disais tout à l'heure que la fonction uniformatrice des notions autonomes s'exerce de façon beaucoup plus efficace lorsque la Convention, et par conséquent ses notions autonomes, sont incorporées dans l'ordre juridique des pays contractants. On pourrait à cet égard parler d'un effet direct de l'*autonomie* de la notion autonome. Nous voici donc dans l'antichambre de la deuxième question que je voudrais traiter, celle de l'*effet direct* de la Convention. Mon objectif est modeste. Je vais rappeler les données fondamentales du problème, tel qu'il se pose eu égard à la Convention européenne des Droits de l'Homme, et formuler, par la suite, quelques observations.

L'effet direct est une des conséquences juridiques que les dispositions des traités internationaux sont susceptibles de déployer dans l'ordre juridique des

[6] Arrêt du 16 juillet 1971, Série A, 1971.

[7] Arrêt du 28 juin 1978, Série A, vol. 27 (1978).

[8] Arrêt du 23 juin 1981, Série A, vol. 43 (1981). Voir surtout les §§ 86 ss., de l'arrêt *König*.

Etats contractants. En ce sens, il relève de cet ensemble d'attributs qu'on appelle le statut des traités en droit national, statut principalement exprimé par l'incorporation – ou la non-incorporation – du traité dans l'ordre juridique national et le rang qui est reconnu aux dispositions du traité dans l'échelle hiérarchique des normes composant le système juridique national.

Incorporation, rang, effet direct sont ainsi les concepts techniques qui expriment pour l'essentiel le statut d'un instrument international dans l'ordre juridique interne, mais dont les rapports mutuels logiques et hiérarchiques ne sont pas clairement définis. D'ailleurs, le cadre dans lequel ces concepts se concrétisent étant celui de l'ordre juridique des Etats contractants, il s'ensuit que leurs modalités sont largement déterminées en fonction des conceptions particulières prévalant en la matière dans ces ordres. C'est ainsi qu'au-delà de certaines considérations de base, il n'existe pas une seule doctrine et pratique uniformes de l'effet direct valables pour l'ensemble des ordres juridiques nationaux concernés. Il est intéressant de rappeler à cet égard que le problème de l'effet direct a reçu dans le cadre du droit communautaire des solutions qui ne sont pas toujours comparables à celles adoptées sur le plan des traités internationaux traditionnels. Cette différenciation s'explique largement par des particularités de l'ordre juridique communautaire. De surcroît, les procédures coercitives communautaires, juridictionnelles ou autres, parviennent, en principe, à rappeler à l'ordre les systèmes juridiques nationaux non ou mal disposés à reconnaître d'effet direct aux règles de droit communautaire considérées comme directement applicables. La situation est différente en ce qui concerne la Convention européenne des Droits de l'Homme. Ici, la souveraineté nationale et à travers elle les particularités juridiques de chaque Etat contractant restent largement intactes. Le statut juridique de la Convention dans l'ordre juridique interne diffère d'un pays contractant à l'autre. L'éventail des solutions est trop large et trop nuancé pour que l'on puisse le présenter ici d'un bout à l'autre. J'essaierai néanmoins d'en donner une esquisse sommaire.

En laissant de côté pour le moment la question de savoir si l'effet direct d'un traité suppose nécessairement, comme l'affirme l'opinion dominante, son incorporation dans l'ordre interne, je crois qu'on peut distinguer, en matière de statut interne de la Convention et selon les Etats contractants, les variations suivantes: Incorporation ou non-incorporation de la Convention. Rang: niveau supraconstitutionnel, constitutionnel, interprétation de la Constitution conforme à la Convention, niveau situé entre la Constitution et la loi, niveau de loi avec interprétation de la loi conforme à la Convention, niveau simple de loi. Ces variations doivent être vues sur la toile de fond que constitue le système national de contrôle – ou de non-contrôle – de la constitutionnalité ou, dans un sens plus général, le système national de solutions des conflits hiérarchiques entre normes juridiques. Dans la détermination des rapports entre les dispositions matérielles de la Convention et le droit national, il faut également tenir compte de l'art. 60 de la Convention, à savoir de la clause dite »de la disposition la plus favorable« ou »clause de surenchère«, qui accorde la priorité au droit national lorsque celui-ci est plus généreux que la Convention dans la protection des Droits de l'Homme.

C'est dans ce cadre aussi bien général, qu'individuel pour chaque Etat contractant, que se pose le problème de l'*effet direct* des dispositions de la Convention. En traçant les grandes lignes de ce cadre d'accueil national, je n'ai pas voulu répondre implicitement à la question de savoir si l'effet direct est declenché par un impératif du droit international ou bien du droit interne. Avec ses origines aux dicta du Chief Justice Marshall dans l'arrêt *Foster* v. *Nelson* (1829) et à l'arrêt *Dantzig* de la Cour Permanente de Justice internationale, le débat a bondi de nos jours. Le fondement juridique de l'effet direct s'analyse, à mon sens, en un processus de coordination entre les deux dimensions, internationale et nationale, dans lequel la priorité logique revient à la dimension internationale et la priorité hiérarchique à la dimension nationale ou, au moins, est déterminée par celle-ci. Si, par conséquent, l'effet direct des dispositions d'un traité dépend largement des attitudes nationales, il est exact de parler, en matière de Convention européenne des Droits de l'Homme, d'un effet direct *territorialement limité*. Mais dans la mesure où l'effet direct est conditionné par la dimension internationale il faudrait tenir compte, eu égard à la Convention, de deux facteurs: Premièrement, de la volonté contractuelle exprimée dans les dispositions de l'instrument. On connaît à cet égard l'argument tiré du libellé de l'article premier de la Convention – tel qu'il a été finalement rédigé sur proposition d'Henri Rolin – selon lequel les Etats contractants ne »s'engageraient pas à reconnaître«, mais »ils reconnaîtraient« (*reconnaissent/shall secure*) les droits consacrés par la Convention. On connaît également l'argument tiré de l'art. 13 de la Convention qui prévoit le droit à l'octroi d'un recours effectif national contre les violations de la Convention. La Cour européenne des Droits de l'Homme a, dans certains arrêts et plus clairement dans celui rendu en l'affaire *Irlande* c. *Royaume Uni*[9], insisté sur la volonté contractuelle qui se dégage de l'article premier en affirmant que »l'intention des Etats contractants se reflète avec une fidélité particulière là où la Convention a été incorporée à l'ordre juridique interne«.

Deuxièmement, dans la mesure où la volonté des Etats contractants exprimée dans la Convention est d'importance à cet égard, l'interprétation par la Cour de Strasbourg des dispositions de la Convention est également un facteur à prendre en considération. Dans l'arrêt *Van Oosterwijck*, la Cour a tenu à préciser qu'à son avis, l'art. 8 de la Convention (droit au respect de la vie privée), est directement applicable, mais elle a ajouté que tel était aussi l'avis des juridictions de l'Etat défendeur, ainsi que d'autres Etats contractants[10]. Pareilles affirmations ont trait au délicat problème des qualités intrinsèques que

---

[9] Arrêt du 18 janvier 1978, Série A, vol. 25 (1978), § 279, p. 91.
[10] Arrêt du 6 novembre 1980, Série A, vol. 40 (1981), § 33, p. 16 et les références.

doit réunir la disposition d'un traité – clarté, précision et caractère direct du message juridique qu'elle contient – pour qu'elle puisse déployer un effet direct. Je me contente de rappeler ce problème.

Si les facteurs auxquels je viens de me référer préparent sur le terrain de la dimension internationale l'effet direct d'une disposition de la Convention, c'est ensuite à la dimension nationale qu'il appartient de mettre en oeuvre cet effet. C'est dans cette phase que les particularités nationales jaillissent. En effet, c'est selon les attitudes nationales particulières que les rapports entre la dimension internationale et la dimension nationale peuvent devenir des rapports de coordination ou de subordination d'une dimension à l'autre ou vice versa.

Je voudrais conclure avec quelques considérations générales, mais aussi avec quelques interrogations.

Première considération: Dans la mesure où l'effet direct dans l'ordre national d'une disposition de la Convention fait partie de la volonté contractuelle effective, pourrait-on dire qu'un Etat qui n'incorpore pas la Convention ou qui ne reconnaît pas d'effet direct à telle ou telle de ses dispositions ne se conforme pas entièrement aux obligations internationales qui lui incombent en vertu de cet instrument ? La Cour paraît avoir répondu par la négative à cette question dans l'arrêt *Irlande* c. *Royaume-Uni*, lorsqu'elle a observé que l'incorporation de la Convention dans l'ordre juridique national est la preuve d'une »fidélité particulière« à la volonté contractuelle [11]. Les Etats qui n'incorporent pas la Convention dans leur droit ne se voient pas décerner cette prime de »fidélité particulière«, mais ils ne se trouvent cependant pas en violation de leurs engagements internationaux.

Deuxième considération: L'incorporation est-elle une condition préalable nécessaire de l'effet direct? Une réponse affirmative à cette question paraît s'imposer par la logique juridique elle-même. Pourtant, il faut encore une fois nuancer. Je veux citer un cas jurisprudentiel très récent: Dans son arrêt *Tyrer* [12], la Cour européenne des Droits de l'Homme a jugé que la fustigation (*birching*) pratiquée en tant que sanction criminelle dans l'Ile de Man était une peine dégradante et par conséquent incompatible avec l'art. 3 de la Convention. L'arrêt fut accueilli avec une certaine désapprobation dans l'Ile, où, paraît-il, une partie, peut-être la majorité, de l'opinion publique est en faveur du maintien de ce châtiment corporel. En juin 1981, les juridictions de l'Ile ont imposé la même peine à un anglais de seize ans et demi accusé d'actes de violence illégale. Le condamné saisit la juridiction d'appel qui rendit son arrêt le 6 octobre 1981 [13]. Le texte de cet arrêt s'offre à une lecture

[11] Voir la référence à la note 9.

[12] Arrêt du 25 avril 1978, Série A, vol. 26 (1978).

[13] *Teare* v. *O'Callaghan*, High Court of Justice of the Isle of Man, Staff of Government Division, Criminal Jurisdiction, Judgement of October 6, 1981 (reasons delivered by B. A. Hytner, Q. C., Judge of Appeal).

extrêmement intéressante. Le juge Hytner qui l'a prononcé a cassé le jugement de première instance en tenant compte de l'arrêt *Tyrer* de la Cour de Strasbourg. Bien que conscient de la non-incorporation de la Convention dans son droit interne, il a cru devoir interpréter ce droit de façon conforme à la Convention, telle que celle-ci a été interprétée par la Cour européenne des Droits de l'Homme. Cette *treaty-friendly interpretation* n'est-elle pas la consécration d'un certain effet direct sans incorporation? Ici, la dimension nationale conserve tout son poids mais la dimension internationale n'en reste pas moins importante, bien que le maillon de l'incorporation faisant défaut, la chaîne juridique entre les deux dimensions reste formellement rompue. Dans son arrêt, le juge Hytner précise que les engagements internationaux l'obligeaient, à défaut d'une action du législateur national, à suivre une certaine »politique judiciaire«, une *sentencing policy*. C'est par cette technique, fondée sur le principe de droit interne d'une interprétation du droit national conforme à la Convention (*konventionskonforme Gesetzesauslegung, treaty-friendly interpretation*) que la Convention peut parfois déployer un effet direct qu'on pourrait qualifier d'effet direct *indirect*.

La politique judiciaire appliquée par le juge britannique nous amène à notre dernière considération: La technique de l'effet direct contient virtuellement un puissant élément politique. Malgré les définitions juridiques du principe de l'effet direct, des conditions et des conséquences de son application, il est évident qu'en matière de traités internationaux, plus ou moins traditionnels, l'ordre juridique national retient entre ses mains un large pouvoir d'appréciation, donc un pouvoir politique, appartenant au judiciaire ou à l'exécutif, de barrer aux traités la route vers l'effet direct. Comme l'incorporation et le rang sont en matière de traités internationaux des armes à double tranchant dans les mains de l'Etat, il en est de même de la technique de l'effet direct par laquelle on peut aussi bien réserver aux traités un accueil généreux dans l'ordre juridique national, que leur en interdire pratiquement l'accès. Des développements récents sur le plan communautaire font surgir les mêmes dilemmes et, peut-être, les mêmes embarras. Après avoir suivi une trajectoire plutôt sereine, la doctrine communautaire de l'effet direct semble maintenant traverser une zone de turbulence avec les affaires *Polydor*[14], *Kupferberg*[15] et *Pabst*[16] où il est question de l'effet direct des dispositions d'accords mixtes de la Communauté (accord avec le Portugal, accord d'association CEE/Grèce). Il est intéressant de noter que le Directeur général du Service juridique de la Commission M. Ehlermann, en introduisant devant la Cour de Justice des Communautés la thèse de la Commission en l'affaire *Polydor*, a insisté sur l'aspect politique du problème et fit observer que la reconnaissance ou non, dans la Communauté, d'effet direct à telle ou telle dis-

---

[14] Affaire 270/80, arrêt du 9 février 1982.

[15] Affaire 104/81.

[16] Affaire 17/81, arrêt du 29 avril 1982.

position d'accords mixtes se ramène à un »choix de politique juridique«. Si l'on est en droit d'affirmer que la question se pose de façon différente sur le plan de la Convention européenne des Droits de l'Homme par rapport à des accords de coopération économique, il n'en reste pas moins qu'un équilibre acceptable entre appréciation utilitaire des engagements internationaux et leur respect effectif à travers la reconnaissance, dans l'ordre juridique national, de leur effet direct, bref un juste équilibre entre réalisme politique et civisme international est partout difficile à atteindre. Pour le juriste, en tout cas, ces problèmes laissent du bon pain sur la planche.

Some Reflections on the European Convention on Human Rights – and on Human Rights

SIR GERALD FITZMAURICE

Note: I had originally intended, as my contribution to Judge Mosler's Festschrift, to offer some comments on the well-known decision of the European Court of Human Rights in the Sunday Times (Thalidomide) case, delivered in April 1979, in which we both participated though of different persuasions. It seemed to me that this might be acceptable, seeing that neither of us is any longer a member of the Court; that some time had now elapsed since the decision was given; and that it had only been reached by a narrow vote of eleven to nine, which could easily – had things fallen out differently – have been the other way round, or have resulted in a ten-ten tie, in which case the outcome would have been no different – but would have been narrower still, depending on the President's casting vote. Moreover, this was a case in which I had not followed my usual practice when in the minority, of delivering an individual dissenting or separate Opinion, and had renounced doing so (although I had prepared one) in order not to detract from the weight of what would otherwise be a single unanimous minority view, subscribed to by nine Judges of the Court.

However, when I came to write the present contribution I found it would be necessary to preface my intended remarks with some general observations about human rights as such, about the character of the European Convention on Human Rights, and about Strasbourg's approach to its interpretation. By the time I had done that, I found that »the tail had wagged the dog« and that I had neither space, time, nor energy for what I had originally intended which, accordingly, must await another occasion – if there ever is one.

1. Because of the way in which the European Convention on Human Rights is drafted, it lends itself to a method of interpretation that makes it possible to apply certain of its provisions in a manner that intermixes two conceptually quite distinct categories of acts, both treated as offences against Human Rights, and that deals with them as being all on the same plane – thereby degrading what are truly matters of human rights in the proper his-

torical sense of the term to the level of what are mere matters of socio-economics, while exalting the latter to a stratum of ideas that is not intrinsically theirs. For instance, the interdiction of torture is unquestionably a matter of human rights *stricto sensu*, whereas registration of the births of illegitimate or legitimate children is not, – it is a matter of civic rights. To treat such radically different orders of questions as all coming under the same umbrella is to introduce a distortion into the heart of the Convention – a distortion which a more restrained interpretation of its provisions than has in fact occurred could have done something to remedy.

2. Worse still, not only does this approach tend to an assimilation of what are basically separate categories of acts or measures, as one and all involving violations of the Convention, but, in addition, all violations of the Convention are treated as being in principle of equal gravity. Any municipal system of law distinguishes between serious crimes and lesser offences as regards the type and degree of sentences or punishments to be passed or inflicted. But, whether it is a case of torture – or whether it is a case of a mere infraction of civil rights which the Convention has been interpreted as protecting – all that the Court can do, if it finds a violation, is to declare it, – i.e. find that the Government or authority concerned is in breach of the Convention. True, by Article 50 – but only in those cases where the internal law of the offending State does not allow of adequate reparation being made – the Court is then »if necessary« to »afford just satisfaction to the injured party«. But what »just satisfaction« can there be to compensate for being killed, or maimed for life, or spending twenty years in a concentration camp?

3. In practice, the Court can sometimes decree a species of *restitutio in integrum*, or specific performance, where for instance an individual has been removed from his post, or excluded from some facility, in breach of a provision of the Convention, – but otherwise, all it can award is monetary damages as compensation, the amount of which *can* be varied accordingly to the gravity of the offences – but there still remain offences for which no monetary award can compensate. Such a situation is of course to a considerable extent inevitable where the offending parties are States that cannot be coerced. Even so, the total failure of the Convention to distinguish between what is serious and what – relatively speaking – is not, constitutes a grave flaw in the system it establishes – a flaw which the adopted method of interpretation has done little to lessen.

4. If one considered the development of the concept of human rights in the light of the great historic instruments and declarations that have testified to its character and course over the last 500 years, such as those listed in the footnote below[1], it is evident that although there are occasional references to

---

[1] Alfonso IX of Spain's Confirmation (1188) of the series of rights enacted by the Cortes of Leon; Magna Carta (1215); the English Petition of Rights (1628) and Bill of Right (1689); the Habeas Corpus Act of 1640 and the much older common-

lesser matters, their main preoccupation has, *grosso modo*, been with the larger issues of human rights and fundamental freedoms; the right to life, liberty and security of the person; prohibition of slavery or other forms of servitude or forced labour; freedom of conscience, religion and worship; dignity of the person; freedom from arbitrary arrest and imprisonment *incommunicado*: government by consent and limitations on absolute sovereignity; equality and the right to a fair trial before the law: immunity from persecution and cruel or inhuman, or degrading, treatment.

5. In the aftermath of the War of 1939–1945, and given the nature of the events of the previous decade or so, and of those taking place during the war itself over much of Europe and under a variety of political and social regimes – it was above all on the cruel or the inhuman and degrading character of the measures and practices concerned, together with the extermination, mass killings, deportations and quasi-enslavement in prison and concentration camps, that were chiefly in the public mind when questions of violations of human rights and fundamental freedoms came up, – and certainly in the minds of governments in framing such instruments as the European Convention. These were the things that must cease and not be allowed to recur, – and in this context it was not of political, civil[2], social, or economic rights that most people or governments were thinking. Naturally these other rights had to be protected too, but this should be as a separate exercise.

6. Although the distinction is considerably blurred, its existence in principle was recognised and reflected in the work of the United Nations in the human rights field. The Preamble to the Charter, after reaffirming »faith in fundamental human rights« and »the dignity and worth of the human person« spoke generally amongst its »Purposes and Principles«, of »promoting and encouraging respect for human rights and fundamental freedoms for all, without distinction as to race, sex, language or religion«, and Article 13 directed the Assembly to initiate studies and make recommendations for, *inter alia*, the same purposes – the definition of its further responsibilities, functions and powers in that regard being relegated to Chapters IX and X of the Charter. Of these two, Chapter X is not relevant (it sets up the Economic and Social Council of the U.N.), but Chapter IX is, and its Article 55

---

law Writ; the Virginia Declaration of Rights and the American Declaration of Independence (1776), together with the early amendments to the United States Constitution; the French Declaration of the Rights of Man (1789); the »Declaration of the International rights of Man« adopted by the *Institut de Droit International* in 1929; and then, at the end of the second World War, the Nuremberg Charter, and the Law of the Allied Control Council defining Crimes against Humanity – the last two arising directly out of the occurrences of the War and immediate pre-war periods.

[2] Political and civil rights can of course be, and often are, seen as covering fundamental human rights and freedoms. But having regard to the other and much more common uses of the terms »political« and »civil«, these are not apt terms for describing concepts that, as will be mentioned later, derive essentially from the natural law postulate of what is »inalienable« from the notion of man.

places in separate sub-paragraphs, on the one hand social, economic, cultural and educational matters (heads (a) and (b)) and, on the other (head (c)) »universal respect for, and observances of, human rights and fundamental freedoms for all . . . etc.«; and this dichotomy reappeared in the work done by, or under the *aegis* of, the Assembly in carrying out its mandate.

7. The Universal Declaration of Human Rights adopted by the Assembly in December 1948, did indeed embrace both categories: although some two thirds of its main provisions treated of what may be called fundamental human rights – or human rights proper – the rest concerned such socio-economic matters as employment, remuneration, trade unions, health care, social security services, education and cultural activities, copyright, etc. But the Declaration was not, as such, legally binding, and was in due course (1966) to be followed up by instruments that were to become so on receiving the required number of signatures and ratifications. Apart from an Optional Protocol containing some rudimentary machinery for supervision and control, these instruments consisted of two Covenants that (always with some blurring) dealt separately with the two categories of rights involved, the first on »Economic, Social and Cultural Rights« – the second on »Civil and Political Rights«. It was this second Covenant that dealt, very nearly exclusively, with fundamental human rights and freedoms proper, although certain provisions, such as those on freedom of movement, the right to enter and leave a country, the expulsion of aliens, non-imprisonment for breach of contract, etc. could be regarded as doubtful candidates for figuring under that rubric. Be that as it may, the conceptual distinction was in substance recognised and given effect to.

8. It was no doubt partly as an acknowledgment of this (if only to a limited extent) mixed content of the second Covenant, that it was entitled »Covenant on Civil and Political Rights«, although in fact this was not an apt description (note 2 *supra*) and did not – at least to the man in the street – convey the idea of an instrument devoted to fundamental human rights and freedoms, – and all this is borne out by the fact that in practice the Covenant is invariably known as the UN Covenant on Human Rights and not by its actual formal title.

9. But there were deeper reasons both for the separation into two Covenants and for the titles given them, inappropriate though one of them was – which leads in turn to reasons yet more profound. In regard to the basic human rights and freedoms, there is first of all the point that it is essentially governments and governmental authorities – in other words the executive arms of the State – which are primarily responsible both for respecting human rights and also for denying them, and for the malpractices and breaches of them that occur. It is essentially in his capacity as a civilian, as a citizen and member of the body politic, that the ordinary man is both the object of these rights and the victim of any violation of them. It is in this sense that the notion of civil and political rights has some relevance. Primary

responsibility with respect to the basic human rights and freedoms does not normally rest on parliaments, legislatures, or private bodies or entities – nor yet on individuals, unless executive officers, or acting under government orders. It may to some extent rest on the courts or law, but these usually administer the law more or less correctly[3], and it is but seldom that the law as such is in open contradiction with basic human rights[3]. At all events in European countries, the law in these regards is *on paper* orthodox and even benign, – what counts is the way in which it is administered by the courts, but – above all by the executive. Indeed, the law is often silent from a belief that to speak would be superfluous. Few systems specifically forbid acts of torture or inhuman treatment because these would automatically infringe more general laws and be punished by the courts. Yet they occur.

10. Different is the case of those rights, equally figuring as »human rights« in the Universal Declaration, but subsequently correctly assembled in a separate Covenant under the label of »Economic, Social and Cultural Rights«. This is quite a different order of thing, as can be seen from the brief summary given in the footnote below[4]. With one or two possible exceptions[5], there is virtually nothing in this Covenant that is fundamental to the concept of a human being in the sense that without them he would be, as it were, dehumanised, or lead an existence that would be, for him, futile – such things as the reasonable expectation of life not arbitrarily or artificially cut short, liberty without licence, freedom of thought and conscience, dignity of the person and equality before the law, humane treatment, etc. These are things that are in some sort inherent in the idea of Man, whether the law grants them or not – and even if it does not – and even if it purports to deny them. But the great majority of the matters with which the Economic, Social

[3] In spite of the language of Art. 1 of the European Convention, which might appear to warrant a finding that a country's laws, as such, and in principle, were contrary to the Convention, the better view seems to be that any breach of the latter arises from the particular act performed in the application of the law. Of course a law which, for instance, forbade all worship in church or – still more – legalized slavery, might seem *ipso jure* to be peccant. Yet strictly it would not be until people were physically prevented from going to church or were forcibly taken into slavery that a concrete breach would result. However, this view may be difficult to reconcile with Art. 1, although where the courts have administered the law honestly and correctly according to its own terms, it may also be difficult to say that there is a breach of the Convention unless that breach resides *en soi*, and in the very terms that the courts have applied.

[4] The right to work and related matters, wages, conditions of work, holidays, leisure, trade unions and related matters, social security and insurance, family law, standards of living, food production and distribution, health, hygiene, control of disease, medical care, compulsory education, parental freedom over choice of schools, cultural, artistic, creative and scientific activities.

[5] Such matters as the right of peoples to self determination, the free disposal of their wealth and natural resources, and related matters, with which the Covenant also deals, belong to yet another and quite different category, and no attempt is made to consider them here.

and Cultural Covenant is concerned have no immanence in the notion of Man, – they are man-*made*, and also the work, less of governments, than of parliaments, legislatures, commissions and committees, institutions, writers and publicists – translated into, and taking the form of social and economic legislation. But there is no need to legislate to confer the right to life and liberty, for example: legislation to prevent or punish the unlawful *taking away of it*, suffices in itself to show that it already exists. On the other hand there is no intrinsic right to e.g. social security benefits, – only the legislature can confer these[6]. Consequently there comes about a sort of debasing of the currency in connection with human rights if those that are man-made are equated with those that have their roots in nature.

11. This leads straight to the heart of the analysis that must precede consideration of the actual text of the European Convention. Nothing is more striking, when contemplating the terms in which the fundamental human rights and freedoms have been described historically, than the type of characterization given them, and the epithets or adjectival phrases used for that purpose. These are underlined in the following passages: the famous words of the archetypal American Declaration of Independence, in which it was held to be a *»self-evident«* truth »that all men . . . are endowed by their Creator with certain *inalienable rights*, that among these are Life, Liberty and the Pursuit of Happiness«: the words of the Virginia Declaration of Rights 1776, namely that all men are *by nature* equally free and independent, and have certain *inherent rights* of which . . . they cannot . . . *deprive or divest their posterity: Blackstone*, in his Commentaries on the Laws of England, speaks of the »absolute rights of every Englishman as they are *founded in nature* and reason«: the *immemorial* rights of Englishmen (Petition of Right, Bill of Rights): the 1929 Resolution of the *Institut de Droit International* spoke of the »declaration of rights written into many constitutions and especially into the (18th century) American and French constitutions, as being *»ordained* not only for the citizen *but for man*«, – such rights, the Resolution said, demanded to be »preserved from all infringement on the part of the State« – in other words they were »entrenched« rights deriving from a law above the State – natural law. The State might refuse to accord them *de facto*, it could not destroy man's entitlement to them *jure naturale*.

12. This kind of language, held over three centuries[7] – »self-evident«, »inalienable«, »by nature«, »inherent rights« of which no one can »deprive or

---

[6] The argument that such rights do inherently exist, legislation being required only to provide the mechanism for implementing them, may be noted; but it is weak, for it still leaves unanswered the question of why some »rights« are thus implemented and others not, – the decision is man-made, not inherent in the nature of things. More interesting is it to ask whether the quality of inherency cannot come to attach to what was in origin a deliberate creation.

[7] And much earlier, for the notion of rights derived from a *jus naturale* goes back to antiquity.

divest ... posterity«, »absolute rights«, »founded in nature and reason«, »immemorial rights«, »ordained ... for man«, »to be preserved from all (State) infringement«, – such words and phrases testify to a deep seated and persistent feeling that certain rights are so fundamental, so founded in nature and *la condition humaine*, as to constitute a different order of right – not placed there but self-existent »before the foundation of the world«, and on another plane from that occupied by the ordinary everyday run of rights deriving from man-made laws. It is an essentially instinctive, almost inarticulate feeling of the kind described in a recent philosophical work[8] as »that kind of life which human beings lack unless they feel more than they can say, and grasp more than they can explain«. Another philosopher, in the same vein but closer to the concrete aspects of our theme, says[9]:

> »The torturer whose activities are owed to the profound belief that he is on the side of history is ... not more engaging than the torturer who is only there to earn a living wage. Still more to the point ... it can hardly be that there is more to be said for the ... acts of the first torturer«[10].

This passage, even if in a somewhat grimly humerous way, brings out the, in principle, »absolute« character of the prohibition of torture and analogous treatment, duly given effect to (see paragraphs 16 and note 17 below) in the European Convention on Human Rights, to which it is now time to come.

13. The European Convention, as compared with previous instruments on human rights, introduced two important innovations, both well known but also having a special significance in the present context that is generally recognised. The first is the system of the »right of individual petition« to the European Commission at Strasbourg on the part of nationals of States accepting that system, as a considerable number of European States have done. This meant that, for the first time[11], a *quantity* (for so it has proved) of *individual*,

---

[8] Moral Luck by *Bernard Williams* (Cambridge University Press) – »moral luck« – for instance two friends out shooting shoot into a wood at what they think is a deer. It happens to be the keeper. One of the two hits the keeper and kills him, the other misses. The first is convicted of causing death by gross negligence, – the other is held to have committed no offence. Moral luck! – or is he, at least morally, guilty too? – (advantage of being a poor shot, penalties of being a good one!).

[9] *T. Honderich* in an article in the New Statesman for 15 January 1982, p. 17.

[10] However this is not the real dilemma. The real dilemma is exemplified by the man who tortures in order to obtain information that will prevent the destruction of a whole city: the old dilemma of the end and the means.

[11] The United Nations Covenant only provided for the reference of *State against State* complaints to a Human Rights Committee, and then only as between States having both specially accepted that system. No right of complaint on the part of *individuals* was provided for in the Covenant itself, except under an annexed Optional Protocol which has received few acceptances, – and even under the Protocol the Human Rights Committee only has powers of investigation and report.

non government sponsored, cases – most of them indeed arraigning the individual's *own* government – would come forward to Strasbourg for action in accordance with the Convention. The second innovation was the provision of machinery for the processing of complaints – if possible their settlement – but, if ultimately necessary, their adjudication by a Court empowered to give a judicial decision binding (even if adverse) on the government against which the original complaint had been made. This also was quite new [12], as was the possibility it created of what was in effect an appeal to an *international* tribunal against the decision of even the highest *national* ones [13].

14. The consequences of these innovations were potentially momentous, not only because there would now be something like a flood [14] of concrete cases reaching a human rights organization comprising institutions empowered to give authoritative interpretations and, in the last resort, legally binding decisions – but also for another important reason. This was that the substantive (i.e. non-procedural) provisions of the European Convention – the provisions specifying the rights to be protected – were, in their essentials, taken – often word for word and always in echoing language – from the corresponding provisions of the United Nations Covenant [15]. It results from this that if ever a World Court of Human Rights is set up, it will have at its disposal a large body of precedent deriving from the work of the European Court and Commission. The importance of the character of the approach of these bodies to the interpretation of the clauses of the European Convention is therefore manifest.

15. Turning now to the actual articles of the European Convention, it is noticeable that although, as just stated above, they echo and often virtually reproduce the language of corresponding provisions of the United Nations Covenant, they do so much less amply and elaborately; and also that many of

---

[12] The United Nations Universal Declaration on Human Rights was declatory only, not a legally binding instrument. The Covenant was binding, (on coming into force) for the States ratifying it, but (see note 11 above) contained no machinery for enforcement or even for furnishing legally authoritative interpretations of its provisions.

[13] The European Court – correctly of course – has always maintained that it had no power actually to reverse the *ipsissimum* decision of a national Court. But the effect, though not in form and not identical, is very similar in practice, since the Government concerned becomes bound to remedy the situation created by the decision of its Courts.

[14] It is probably not realised outside Human Rights circles that the decisions of the European Court constitute only the tip of the iceberg of the cases reaching Strasbourg. Many get no further than the Commission of Human Rights, and many not even as far as that, being rejected *in limine* on various grounds.

[15] The first of the four Protocols to the Convention – the only one that is in force – deals with property, and educational and voting rights. The last echoes a provision of the U.N. Covenant on Human Rights. The second reflects provisions of the Covenant on Economic, Social and Cultural Rights, while the first figures solely in the Universal Declaration.

the provisions of the Covenant are not reproduced at all. This suggests that those who framed the European Convention were resolved upon a more sober and restrained formulation of the rights involved than the markedly idealistic and even high-flown expression given to them – or many of them – in the Universal Declaration and the Covenant[16].

16. The articles of the European Convention that deal with the most fundamental of human rights and freedoms – those that would in the past have earned the name of »self-evident«, »inherent« »inalienable«, are Articles 2–6 and 9, – the right to life; prohibition of torture or degrading treatment or punishment; freedom from slavery, servitude or forced labour; freedom from arbitrary arrest or imprisonment (i.e. the right to liberty and security of the person); the right to a fair trial and public hearing at law, and the presumption of innocence; and freedom of thought, conscience, and religious worship. Less fundamental – or not on quite the same basic level – are the contents of Articles 8, 10 & 12 – respect for private and family life, home and correspondence; the right to freedom of expression; and the right to marry and found a family. This last right cannot exactly be denied the appelation of »human« but is more in the nature of a social right, and in any case is somewhat tautologous or truistical in the European context. Also, one can think of more fundamental aspects of the relationship of the sexes than actual marriage. Finally, there are Articles 7 & 11, respectively concerning non-retroactivity of laws, of freedom of association, and the right to form and join trade unions – neither of these, in our view, basic human rights, but important rights of a civil or civic character. All the Articles concerned, except No. 3 (torture and inhuman or degrading treatment), contain important qualifications in the interests of law and order, defence, public policy, public safety, etc.; and one provision, Article 15, permits derogations from the Convention (in effect its temporary suspension) to be made under certain conditions in case of war or national emergency[17]. However, for present purposes, it is not necessary for any of these qualifications or possibilities of derogation to be considered here. Of the remaining Articles, nos. 1 and 13–18, which are regulatory in character, Article 1 has been briefly noticed in note 3. Mention of Article 13 is made in note 21, and Article 14 is discussed in paragraphs 29–31 below.

17. Somewhat of a mixed bunch, as they say, both in substance and otherwise. But the substantive articles all have one common characteristic, that they are to be interpreted strictly, though for reasons of a different order

[16] Compare for instance Arts. 2, 3, 6, 9 & 12 of the Convention with, respectively, Arts. 6, 7, 14, 18 & 23 of the Covenant; and as examples of matters dealt with in the Covenant but not in the Convention, see Arts. 1–3, 11–13, 16, 20, 23–25 & 27 of the former.

[17] However, no derogations are permitted from Art. 2 (right to life) except as resulting from lawful acts of war; Art. 3 (torture, etc.); Art. 4, para. 1 (slavery or servitude), and 7 (non-retroactivity of laws).

and with a different aim. Those that are fundamental are to be interpreted strictly just because they are such, and no derogations from them can be permitted unless expressly provided for, – and those that are less fundamental, and particularly if not fundamental at all even if important, are to be strictly construed in order to avoid making them a vehicle for the introduction into the Convention – by way of interpretation – of rights and obligations not in the nature of human rights proper, and not originally intended to be there. This has certainly happened; but it is not proposed here to discuss particular cases, and it will suffice to indicate the type of case in which it seems that the Convention has been, or is liable to be, misapplied.

18. No particular difficulty of principle has arisen in regard to the really fundamental articles (right to life, liberty and humane treatment), except insofar as the interpretation of particular terms such as »torture« »inhuman treatment« and »degrading treatment« have involved problems. At least on the question of »torture« the Court has shown a commendable reserve, – for here the danger (often overlooked) that the very abhorrence felt for this odious practice, and the eagerness to condemn any manifestation of it, may lead to that serious debasing of the currency of a vital concept that has been remarked on earlier. If every case of the infliction of a relatively slight degree (and we stress the word »relatively«) of physical pain or discomfort, or of mental stress, is to be characterized as »torture« and as a breach of the Convention, *on that ground* – (it may still be a breach on other grounds) – then one must ask what is left in order to give what should be their true weight, and true condemnation, to such horrendous examples as were instanced by the present writer in his separate opinions in certain cases before the Court [18]?

19. The same applies, if to a lesser extent, to such concepts as those of inhuman or degrading treatment or punishment. There are many kinds, there are many degrees, of inhuman or degrading treatment, and they are not all on the same level of detestability and culpability. The trouble is that the Convention does not, any more than does the Covenant, afford any basis for distinguishing between acts which are so greatly removed from one another in degree that it becomes wrong to regard them as being all the same in kind. Yet the Convention leaves little room for escape [19], and this places the Court in much the same position as the 18th and early 19th century juries in England who, because the punishment for stealing a sheep was – at least on paper, and sometimes in reality – death, refused to convict at all – (and there is certainly one present day analogy to this) [20].

[18] In particular the *Northern Ireland* & *Tyrer* cases, both 1978.

[19] In the atmosphere of the period (see paragraph 5 above) this may well have been deliberate.

[20] Modern juries, often themselves consisting of car owners, will, not infrequently, refuse to condemn for motoring offences, on the basis of »There, but for the grace of God, go I«.

20. So far so good – if it is good. But we now come to the Articles which, from the standpoint of the present essay, are the real, potential and sometimes actual, offenders. Apart from Article 6 which is of a different character and commented on in a note[21], and Article 9 (freedom of conscience and worship, which need not be discussed for present purposes), these Articles are nos. 8 and 10 (rights to respect for private, family, and home life, and the right to freedom of expression). There is of course no fault whatever to be found in principle with these Articles as such. But consider the kind of practices which really motivated their inclusion, such as, amongst others, the 4.0 a.m. knock on the door, getting children to spy on their parents, obtaining information from one spouse by threatening harm to the other, telephone-tapping, holding up of correspondence, harassment in leaving or entering the house, cutting off electricity, water or other essential services, newspaper censorship, proscription for publishing views disagreeable to the authorities, or being in possession of what they consider to be subversive literature, etc., etc. But »etcetera, etcetera« was never intended to bring within the ambit of these provisions many of the actual matters which it has been sought to argue that they cover – on occasion successfully, on occasion not – (sometimes it almost seems as if despair at finding any genuine grievance against a government must inspire some of the petty complaints taken to Strasbourg, – for the governments habitually arraigned there are usually precisely those which in general are least guilty of serious offences against genuine human rights).

21. This is not to say that there are not many things of importance amongst those that go to Strasbourg. There are, but that is not the point in the present context. The point is, are they matters of »human rights« within the proper conception of that term – or are they simply matters of civic, social or economic concern? Consider some concrete cases that have arisen: in the case of a man turned into woman, or woman into man after medical treatment and operations, can the refusal of the authorities to register the artificially produced sex, as dating from birth, rank as a violation of human rights? – for the *genes* can never be altered, the »man« can never produce a child or the »woman« father one. But the real point is, should this type of

---

21 There is nothing wrong with the substance of this Article (fair trial). But it is liable to be misapplied through ignoring the governing words of its opening: *»In the determination of«* – a litigant's »civil rights and obligations or of any criminal charge«, there shall be a right to a fair trial, etc. This means that *where* a remedy exists under the national law concerned, it must be afforded promptly by means of a fair and public hearing. It does not mean that national law must *provide* a remedy for each and every complaint that anyone cares to make. No system of law could do that – see the writer's separate opinion in the *Golder* case (1975). In so far as the Convention obliges the States Parties to provide a remedy, it does so by Article 13, not Article 6, and Article 13 is confined to remedies for infractions of the »rights and freedoms set forth in this Convention«. It creates no obligation to provide a universal set of remedies for everything.

complaint have been received at all at Strasbourg as a matter of human rights? Or again, is a refusal to re-register a legitimated child as having been born legitimate a violation of any genuine human right? It may be that such refusals would be wrong and that, if correct under the law as it stands, the law ought to be changed, – but is it as a matter of respect for human rights that this should be done, or merely as a recognition of a change in the climate of opinion regarding matters of sex? The Strasbourg Court was never instituted in order to act as a sort of general law-reformer, but to protect genuine human rights, and these are not such.

22. Exactly the same queries arise in regard to a recent case of an alleged breach of the Convention stemming from laws (subsequently changed) penalizing homosexual acts, even if between adults and in private. Did such a law, and any prosecutions taking place under it, really constitute a violation of human rights? It would surely be going too far to postulate that anything that involves a human being must be a »human« matter, and therefore a matter of »Human Rights« (capitals and all), – for this would simply be a play on words, and on that basis there would be little limit to what could be brought within the concept, thereby completely emasculating it, and degrading what should be the very serious matter of violating human rights down to the level of an ordinary offence or breach of the law.

23. It should also not be forgotten that in the other language of the Convention – French – the term Human Rights is rendered »Droits de l'Homme«, the Rights of Man – in some ways a better term, because it shows immediately that certain refusals, denials, deprivations or penalizations, even if not legally or morally justifiable, cannot without absurdity be glorified into acts contrary to the Rights of Man (or Woman).

24. The error involved in what jurists call a category error, or error of classification. Objecting to such errors and to decisions arrived at on the basis of them, is not merely formalistic. Persisted in, they lead in the end to unacceptable distortions; and it may be asked how it is that in the field of human rights they are so liable to be made. Bearing in mind that these questions arise on the basis of concrete allegations of breaches of the Convention, the answer lies partly in the wording of certain of its provisions that facilitates and even suggests the making of such allegations, and partly in the general approach or attitude that has been adopted concerning the interpretation and application of that wording, – because if the approach takes the form of regarding the language as presenting a good opportunity for plausibly imparting to the Convention a scope which it is virtually certain its originators never even thought of, much less intended, then wording of this kind provides a highly fertile seed-bed for such an exercise. Take for instance the phrases of Article 8 to the effect that »Everyone has the right to respect for his private life«, »... to respect for his ... family life«, »... for his ... home«. What is there, not falling under the head of public activities, that cannot colourably be brought within these categories? And what is meant by

»respect« for private life, for family life, for the home. We shall soon see complaints at Strasbourg based on the failure of the State and/or its Courts to stop the neighbours from disturbing the night with television noises: – lack of respect for private life, family life and the home, all in one breath – what a congeries! It may well be that this would amount to a lack of respect for these things – personally one could think so – but would it be the kind of lack of respect that the Convention ought to be envisaged as encompassing or was intended to encompass? Above all would it be a violation of the »Rights of Man« (or Woman)? Surely not. Such matters belong to the order of community rights, or rights which a citizen has as a member of the community – (no neighbours, no nuisances). They are out of place considered as human rights. There can be no »absolute«, »self-evident«, »inherent« objections to failing to proscribe musical instruments[22] or manipulate the register of births, for instance, as there is to torture, inhuman or degrading treatment, or unlawful deprivations of life or liberty. Yet in one fairly recent case, the request presented at Strasbourg (and to a not inconsiderable extent responded to there) was one that virtually involved judicially enacting a whole new code of family law as being what Article 8 required by way of respect for family life, – and therefore (here of course is the logical *non-sequitur*) as amounting to a violation of human right if this code should not be adhered to.

25. In short, some of the phraseology of the Convention cannot be taken literally or it will let in too much and falsify and distort the whole concept of fundamental human rights. The contrast which the right to marry (Article 12 of the Convention) affords to the case of laws on homosexuality, even if certain of these laws are wrong and illiberal, is instructive, – for although there may be a number of grounds on which marriage between consenting adults can legitimately be prohibited, or be deemed a nullity[23], no one can doubt that such a prohibition imposed by means of differences in race or colour (and though the prohibition might be perfectly lawful under the relevant national law) would be a violation of human rights quite *apart* from its incompatibility with Article 12. In the case of fundamental human rights, the obligation to respect them is not created, but only registered, by covenant or convention, – and it is because it antecedently exists as a matter of natural law, that the violation of it is formally prohibited by man-made law.

26. In the same sort of way, the right to »freedom of expression« and, *inter alia*, to »impart information and ideas without interference by public authority« (Article 10) has been made the vehicle in more than one case for complaints about the suppression of pornographic material by a government. Even if such a suppression is possibly undesirable, can it be a violation of

---

[22] One can imagine some readers saying that there both can and should be!

[23] Marriage within the prohibited degrees of affinity; one or both parties already married, etc.

human rights? Is there a *human right* to publish pornographic material? Of course this is not quite the way in which the matter arises at Strasbourg, – but it is highly illuminating to consider how it does arise. The flaw in the process of reasoning lies in the literal answer given to the initial question »has there been an interference by a public authority with freedom of expression«. To this, the answer immediately given is »Yes«, on the basis that since the government concerned did, incontrovertibly in point of fact, suppress the material, there has, physically, so to speak, been an interference with »freedom of expression«. Hence, so it is said, *prima facie* and in principle there has been a breach of the Convention! It is never asked at that stage whether the suppression of pornographic material (irrespective of whether it is a breach of the Convention, and even if it is a breach – in the *literal* sense) can, without absurdity, be considered a violation of human rights. If that question were asked[24], and at that stage, more time might be given to asking what freedom of expression was intended to mean in a human right context, and it might be realized that since the suppression of pornographic material can by no stretch of imagination be regarded as a violation of human rights, since there can be no human right to publish pornographic material – (there may be a civil law right but if there is, that is a separate and different matter), the complaint should be rejected *in limine litis* without further enquiry. What actually happens is that – it having been decided that *prima facie* there has been a breach of the Convention – the equiry proceeds to the question of whether the breach is as it were excusable, i.e. whether the act occasioning it was justifiable by reason of the case coming under one of the instances listed in paragraph 2 of Article 10 as permitting interference with freedom of expression. This of course makes it possible in practice to reject unmeritorious complaints. But it should never come to that, partly because there is a danger that the particular case may just fail technically to come under one of the permitted exceptions, so that there is then no alternative to holding that there has been a breach of the Convention although it is not a breach of any genuine human right (and this has happened), – but also, and more important, because this entire method of interpretation is flawed by two serious vices.

27. The lesser vice lies in the failure to consider this type of article in the Convention as an integrated whole, consisting at one and the same time of the paragraph that specifies the right and the paragraph that specifies the exceptions to it, so that these are *from the start* considered together, instead of successively and in isolation[25]. It is wrong to hold, first, that in principle there has been a breach of the Convention, and then to hold, as a sort of after

[24] *Mea culpa*! – for in neither of the cases of this kind occurring in his time did the writer take this point. »Mais le temps porte conseil«.

[25] Here at least (see previous footnote) the writer did in one of his separate opinions draw attention to this anomaly.

thought, that, because of an exception, the act concerned does not, when all is said and done, amount to a breach. The real truth of course is, that in these circumstances *there never was a breach at all.* The situation is reminiscent of Queen Victoria's indignant refusal to accept a verdict of »not guilty« in the case of a lunatic who had shot at her. Her point was that he *had* shot at her. The verdict of course meant that if he was insane he could not be guilty in law. This is supposed to have been the origin, at the Queen's insistence, of the revised verdict that long persisted in English practice, of »guilty but insane«!

28. The other and much more serious vice consists in the initial approach of considering only (à la Queen Victoria) whether there has *factually* been an interference or failure to »respect«, instead of considering whether that interference or failure – supposing it to have occurred factually – was of the *kind*, was of the order of occurrence, or touched upon a matter, to which a violation of human rights should properly be regarded as belonging. It would then be seen, or should be seen, that however undesirable or reprehensible some interferences, etc. might be, they were just not on that level at all. If European Governments want them to be subject to international adjudication and condemnation, this should be done by means of a separate Convention on social, not human, rights, instead of sliding them in by the back door of an instrument never intended to give them house-room, – and thereby (for this will surely happen ultimately) bringing the whole Convention into disrepute.

29. There remains one more Article for examination in the present context, though different in character, – Article 14, which is considered at Strasbourg in virtually every case that comes up. Yet its sphere of application, though wide, is not unlimited. What it provides is that »the enjoyment of the rights and freedoms« specified (»set forth«) in the Convention is to be secured to the beneficiaries »without discrimination« on any ground such as sex, race, colour, language, religion, etc. (and there follows a comprehensive list of possible bases of discrimination, all of which come under the ban). Nevertheless it is evident that what has to be secured without discrimination is not just anything, but the »enjoyment of the rights and freedoms« specified (»set forth«) in the Convention. Thus, even if a government is duly affording these rights and freedoms to those within its jurisdiction, it will still be in breach of the Convention if it does so in a discriminatory manner, e.g. by affording them to some but not others, or to all but more amply or beneficially to some than to others. On the other hand, if what is involved – if what is said to have been afforded in a discriminatory manner – is *not* one of the rights and freedoms provided for by the Convention, but something outside it, then Article 14 has no field of play, and cannot apply, – indeed the Convention itself does not apply, and there can be no breach of it or of Article 14. In short, this provision is not one of the substantive articles of the Convention creating a separate and independent (human) right not to be dis-

criminated against. It is essentially an adjectival clause. Its real purpose was simply to prevent governments from by-passing their obligations under the Convention by ostensibly granting the rights and freedoms concerned, but doing so in such a discriminatory way that the object would be frustrated, or frustrated for substantial elements of the population.

30. It follows that there are really only two types of cases in which it is legitimate to have recourse to Article 14 in order to see whether there has been a breach of it. *The first* is when it has been held that one of the substantive provisions of the Convention is *applicable*, but that there has been no breach of it, i.e. that one of the Convention's rights or freedoms *is* involved but has been duly granted. It is then legitimate to go on and consider whether it has nevertheless been granted in a discriminatory way. Secondly, where, again, it has been found that one of the substantive provisions is involved, and that there *has* been a breach of it, it may then be legitimate – though strictly it is unnecessary – to enquire under Article 14 whether, additionally, there has been a second breach of the Convention by reason of some discriminatory element that is involved.

31. What can never be legitimate, if the language of Article 14 is to be respected, is to start by holding that there has been no breach of the Convention *because* the act complained of did not relate to one of the conventional rights or freedoms – i.e. that no substantive provision of the Convention was *applicable*, so none could be breached – but *then* to go on to hold that because the act complained of had discriminatory aspects, a breach of Article 14 has occurred, – for this is to set up that Article as an independent substantive provision prohibiting any form of discrimination in all circumstances[26], irrespective of whether it takes place in the granting of the rights and freedoms set forth in the Convention or not[27]. The distinction in question may be a subtle one - yet it is real, and such as should appeal to the juridically minded. In actual fact however, although the point has more than once been drawn attention to[28], these considerations tend to be shrugged off. What actually exists at Strasbourg is a strong propensity to have habitual, quasi-automatic, almost endemic recourse to Article 14 as if it were a provision embodying a right (separate from and independent of the others »set forth« in the Convention) not to be discriminated against.

32. *In conclusion*, it is worth registering that to a considerable extent, the anomalies to which this essay has sought to draw attention could in practice be reduced or neutralised by giving greater play to a doctrine which has been

---

[26] »Discrimination« in the context means of course *arbitrary* discrimination, not grounded in any recognised basis justifying or calling for it. Otherwise it is differentiation, not discrimination. This is always understood as underlying any prohibition of discrimination. For instance minors are (in their own interests) subjected to certain disabilities that adults are not, – but that is not »discrimination«.

[27] Compare in this respect Art. 26 of the U.N. Covenant.

[28] Certainly by the present writer.

applied in certain cases, that of the »margin of appreciation« or license that should be afforded to governments before they are found to be in breach of the Convention. It is not every niggling negligence or trifling transgression that the Convention was intended to repress, but acts, measures and practices of much more serious import. In the same way – in regard to the action of the Courts of Law – use should be made of the public international law principle that governments do not incur responsibility for the mere error of their Courts, even if it results in an injustice. All Courts are fallible. The responsibility of the State only arises where the decision or action in question is so manifestly erroneous or unjust that no competent court acting honestly, would have been likely to give that decision or take that action.

However, an enquiry into all this must be for another day, – perhaps!

Le Conseil constitutionnel et les engagements internationaux de la France

MICHEL FROMONT

La Constitution française du 4 obtobre 1958 est à certains égards fidèle à l'ouverture au droit international qui caractérisait la Constitution de la IVème République. Elle rappelle en premier lieu son attachement aux principes posés dans le Préambule de la Constitution du 27 octobre 1946 et plus particulièrement aux quatorzième et quinzième alinéas: »La République française, fidèle à ses traditions, se conforme aux règles du droit public international. Elle n'entreprendra aucune guerre dans des vues de conquête et n'emploiera jamais sa force contre la liberté d'aucun peuple. Sous réserve de réciprocité, la France consent aux limitations de souveraineté nécessaires à l'organisation et à la défense de la paix«. En second lieu, la Constitution de 1958 reprend l'essentiel des principes de l'applicabilité immédiate et de la supériorité des traités internationaux sur les lois nationales posés par la Constitution de 1946[1]. Mais précisément sur ce point, elle lui apporte des retouches assez importantes.

En effet, le Général de Gaulle était attaché passionnément à l'idée d'indépendance nationale. Ce terme apparaît d'ailleurs à deux reprises à propos des attributions du Président de la République, auxquelles le Général avait accordé une particulière attention[2]. Bien plus, la notion de souveraineté na-

---

[1] Art. 26 de la Constitution du 27 octobre 1946: »Les traités diplomatiques régulièrement ratifiés et publiés ont force de loi dans le cas même où ils seraient contraires à des lois françaises, sans qu'il soit besoin pour en assurer l'application d'autres dispositions législatives que celles qui auraient été nécessaires pour assurer leur ratification«. Art. 28, lère phase: »Les traités diplomatiques régulièrement ratifiés et publiés ayant une autorité supérieure à celle des lois internes, leurs dispositions ne peuvent être abrogés, modifiées ou suspendues qu' à la suite d'une dénonciation régulière, notifiée par voie diplomatique«.

[2] Art. 5, al. 2: »Il [le Président de la République] est le garant de l'indépendance nationale, de l'intégrité du territoire et du respect des traités«. Art. 16, al. 1: »Lorsque les institutions de la République, l'indépendance de la Nation, l'intégrité de son territoire ou l'exécution de ses engagements internationaux sont menacées d'une manière grave et immédiate. . ., le Président de la République prend les mesures exigées par les circonstances. . .«.

tionale, traditionnelle en droit constitutionnel français, a désormais un sens plus étendu: au lieu d'exprimer simplement la vocation de la nation à désigner ses dirigeants, elle signifie désormais le droit de la nation à gérer ses affaires elle-même tant sur le plan interne (principe démocratique) que sur le plan externe (principe de l'indépendance nationale). Chargée de ce sens nouveau, elle occupe une place éminente dans le texte constitutionnel puisqu'elle est mentionnée dans le préambule dans les termes suivants: «Le peuple français proclame solennellement son attachement aux Droits de l'Homme et aux principes de la souveraineté nationale tels qu'ils ont été définis par la Déclaration de 1789, confirmée et complétée par le préambule de la Constitution de 1946«.

Cette inspiration nationaliste a eu pour effet de restreindre quelque peu l'ouverture du droit constitutionnel français au droit international.

D'une part, la Constitution de 1958 a prévu que les engagements internationaux reconnus contraires aux principes qu'elle pose ne peuvent devenir définitifs qu'après révision de la Constitution. Le soin de constater la contrariété avec la Constitution incombe en premier lieu au Président de la République, »garant de l'indépendance nationale«; mais en dernier ressort, il incombe à un organe institué en 1958 et chargé d'assumer le respect de la constitution, le Conseil constitutionnel[3].

D'autre part, la Constitution de 1958 a certes maintenu le principe de l'incontestabilité des engagements internationaux de la France qui sont devenus définitifs; mais elle a subordonné la supériorité des traités sur les lois nationales postérieures à leur application réciproque par toutes les parties[4]. A l'égard de ces engagements internationaux définitifs, le Conseil constitutionnel joue également un rôle important quoique moindre: selon sa jurisprudence, il veille en effet au respect de leur incontestabilité bien qu'il n'assure pas lui-même l'application du principe de leur supériorité sur les lois. Il le fait à l'occasion du contrôle qu'il exerce sur les engagements internationaux subséquents ou les lois ultérieures.

Tant à l'égard des engagements internationaux non définitifs qu'à l'égard de ceux qui sont devenus définitifs, le Conseil constitutionnel a développé une jurisprudence qui s'efforce de tenir la balance égale entre le maintien de la souveraineté nationale et le développement de la coopération internationale[5].

---

[3] Art. 54 de la Constitution: »Si le Conseil constitutionnel, saisi par le Président de la République, par le Premier Ministre ou par le Président de l'une ou l'autre assemblée, a déclaré qu'un engagement international comporte une clause contraire à la Constitution, l'autorisation de le ratifier ou de l'approuver ne peut intervenir qu'après la révision de la Constitution«. Sur le Conseil constitutionnel, voir *Fromont,* Der Conseil constitutionnel, Zeitschrift für Politik 1979, p. 151.

[4] Art. 55 de la Constitution: »Les traités ou accords régulièrement ratifiés ou approuvés ont, dès leur publication, une autorité supérieure à celle des lois, sous réserve, pour chaque accord ou traité, de son application par l'autre partie«.

[5] Bibliographie: *Nguyen Quoc Dinh,* Le Conseil constitutionnel français et les règles du droit international public, R.G.D.I.P. 1976, p. 1001; *Blumann,* L'article 54 de

Nous aurons l'occasion de montrer que malgré des principes de base différents et des attributions peu comparables, le Conseil constitutionnel a élaboré une jurisprudence qui n'est pas très éloignée de celle du Bundesverfassungsgericht.

I. Le contrôle exercé sur les engagements internationaux non définitifs

Le compromis réalisé en 1958 entre l'ouverture vers la société internationale et le maintien de la souveraineté nationale a très largement inspiré la jurisprudence du Conseil constitutionnel. Si celui-ci a interprété de façon extensive les textes relatifs à ses attributions (A), il a en revanche interprété les principes constitutionnels que doivent respecter les engagements internationaux de façon assez équilibrée (B).

A. Les engagements internationaux soumis au Conseil constitutionnel

Le contrôle de la constitutionnalité des engagements internationaux n'est prévu expressément qu'à l'art. 54 de la Constitution. Mais le Conseil constitutionnel a accepté d'exercer également ce contrôle dans le cadre du contrôle de la constitutionnalité des lois institué par l'art. 61, al. 2, de la Constitution. Tandis que le premier se situe normalement avant l'intervention du Parlement, le second se situe entre celle-ci et l'acte gouvernemental de ratification ou d'approbation.

a) Le contrôle exercé sur l'engagement international lui-même

L'art. 54 de la Constitution dispose: »si le Conseil constitutionnel, saisi par le Président de la République par le Premier Ministre ou par le Président de l'une ou l'autre assemblée, a déclaré qu'un engagement international comporte une clause contraire à la Constitution, l'autorisation de le ratifier ou de l'approuver ne peut intervenir qu'après la révision de la Constitution«[6].

Cette disposition affirme très clairement le caractère préventif de ce contrôle puisque celui-ci doit intervenir entre la signature de l'engagement et l'autorisation parlementaire de le rendre définitif. Un tel système contraste avec la solution allemande selon laquelle la demande de contrôle abstrait des normes ne peut être dirigée contre un traité international qu'après le vote de

---

la Constitution et le contrôle de la constitutionnalité des traités en France, R.G.D.I.P. 1978, p. 537; *Favoreu,* Le Conseil constitutionnel et le droit international, A.F.D.I. 1977, p. 95; *Luchaire,* Le Conseil constitutionnel (Paris 1980), pp. 225–265; *De Villiers,* Norme constitutionnelle et norme internationale, Revue administrative 1981, p. 143.

[6] Sur l'art. 54 de la Constitution, voir *Blumann,* L'article 54 de la Constitution et le contrôle de la constitutionnalité des traités en France, R.G.D.I.P. 1978, pp. 537 s.

la loi d'autorisation par le Parlement[7]. Il est à cet égard pleinement compatible avec le principe de la supériorité du droit international sur le droit interne: le Conseil constitutionnel n'est appelé à exercer son contrôle que sur des projets d'engagements internationaux et, en cas de contrariété avec la Constitution, le Pouvoir exécutif a le choix entre la renonciation à rendre définitif l'engagement souscrit et le déclenchement d'une procédure de révision de la Constitution. En fait, dans le premier cas où la procédure de l'art. 54 a été mise en oeuvre, la saisine a eu lieu après le dépôt du projet de loi portant autorisation, mais avant son examen en séance publique (décision du 19 juin 1970 relative au budget des Communautés européennes), et dans le second cas, elle a eu lieu avant le dépôt du projet de loi portant autorisation (décision du 30 décembre 1976 relative à l'élection au suffrage universel direct de l'assemblée des Communautés européennes). Mais dans les deux cas, le Conseil constitutionnel a statué avant que le Parlement n'engage l'examen des textes en séance publique.

Le principe de la saisine du Conseil constitutionnel avant que n'intervienne l'autorisation parlementaire entraîne une conséquence importante: seuls les engagements internationaux soumis à autorisation parlementaire peuvent être déférés au Conseil constitutionnel. Or, selon l'art. 53 de la Constitution, seuls certains traités ou accords doivent recueillir l'assentiment du Parlement avant d'être ratifiés ou approuvés: ce sont »les traités de paix, les traités de commerce, les traités ou accords relatifs à l'organisation internationale, ceux qui engagent les finances de l'Etat, ceux qui modifient des dispositions de nature législative, ceux qui sont relatifs à l'état des personnes, ceux qui comportent cession, échange ou adjonction de territoire«. Ainsi, le 19 juin 1970, le Conseil constitutionnel s'est prononcé sur le traité signé à Luxembourg le 22 avril 1970 qui portait modification des dispositions budgétaires du traité instituant les Communautés européennes et qui était donc de façon incontestable »un traité relatif à l'organisation internationale«.

Mais faut-il interpréter l'art. 54 comme limitant les possibilités de saisine du Conseil constitutionnel aux seuls traités et accords internationaux pour lesquels une autorisation parlementaire est nécessaire? Le Conseil constitutionnel ne l'a pas pensé et bien qu'il ne se soit pas clairement exprimé sur ce point, il semble qu'il étende sa compétence aux engagements internationaux de toute sorte pour lesquels le Gouvernement sollicite l'autorisation du Parlement. C'est ainsi que le 19 juin 1970[8], le Conseil constitutionnel a accepté d'examiner s'il y avait contradiction entre la Constitution et la décision du Conseil des Communautés européennes relative au

---

[7] Entscheidungen des Bundesverfassungsgerichts (BVerfGE) 5, 28.

[8] C.C. 19 juin 1970, Rec. p. 15; J.C.P. 1970, I, 2354, note *D. Ruzié;* R.D.P. 1971, p. 172, note *Emeri* et *Gautron*; R.G.D.I. P. 1971, p. 23, note X; et *Favoreu/ Philip,* Les grandes décisions du Conseil constitutionnel (lère éd., Paris 1975), n° 23, p. 257 (noter que la décision n'est pas commentée dans la deuxième éd. de ce recueil).

remplacement des contributions financières des Etats membres par des ressources propres aux Communautés, après avoir relevé que »l'adoption des dispositions prévues par cette décision« est »subordonnée, conformément à l'art. 53, à l'intervention d'une loi«; selon cette décision, le Conseil serait donc compétent pour connaître des décisions des organisations internationales qui doivent être approuvées par le Parlement français. Le Conseil constitutionnel semble être allé encore plus loin dans sa décision du 30 décembre 1976 [9]: il a accepté de faire porter son examen sur la décision du Conseil des Communautés européennes relative à l'élection de l'Assemblée des Communautés au suffrage universel direct et à l'acte qui y est annexé sans donner la moindre justification sauf l'affirmation qu'il s'agit d'un engagement international: est-ce simplement parce que la décision était un engagement relatif à l'organisation internationale et que le Gouvernement était donc tenu de le soumettre au Parlement ou bien est-ce parce que la décision a été prise à l'unanimité par le Conseil des Communautés européennes et qu'elle présente de ce fait un certain caractère contractuel ou encore parce que le Gouvernement avait spontanément demandé au Parlement de lui donner son approbation? Le commentateur est condamné à faire des hypothèses. Ce qui est incontestable c'est que les seconde et troisième interprétations conduiraient à une nouvelle extension des attributions du Conseil constitutionnel.

Selon l'art. 54, seules quatre personnes peuvent saisir le Conseil constitutionnel d'un engagement international non définitif: le Président de la République, le Premier Ministre, le Président de l'Assemblée nationale et le Président du Sénat. Dans les deux cas d'application de l'art. 54 qui se sont présentés jusqu'à maintenant, les auteurs de la saisine ont été respectivement le Premier Ministre en 1970 et le Président de la République en 1976; tous deux désiraient faire cesser une querelle entre les parlementaires de la majorité au sujet de la compatibilité avec la Constitution française des décisions relatives respectivement au budget et à l'assemblée des Communautés européennes. On notera que les quatre titulaires du droit de saisine appartiennent en général à la même majorité, celle qui est au pouvoir; toutefois le Président du Sénat a appartenu à l'opposition de 1962 à 1974 et il y appartient de nouveau depuis 1981. Cette remarque explique cependant pourquoi le Conseil constitutionnel a également accepté d'être saisi d'un engagement international au titre de l'art. 61.

---

[9] C.C. 30 décembre 1976, Rec. p. 15; J.C.P. 1978, I, 2895, note *De Villiers* et II, 18704 note *C. Franck;* D. 1977, J. p. 201, note *L. Hamon;* R.D.P. 1977, p. 129, chronique de *Favoreu* et *Philip;* Cahiers de droit européen, 1977, p. 458, note *J. Boulouis,* Revue trimestrielle de droit européen 1977, p. 665, commentaire *Kovar* et *Simon* et p. 695, article *Darras* et *Pirotte;* J.D.I. 1977, p. 66, note *D. Ruzié;* A.F.D.I. 1976, p. 787, observations *Cocâtre-Ziglien* et p. 805, note *Coussirat-Coustère.* En langue allemande: EuR 1977, p. 51, note *Bieber.*

b) Le contrôle exercé sur la loi d'autorisation

Selon l'art. 61, al. 2, »les lois peuvent être déférées au Conseil constitutionnel avant leur promulgation par le Président de la République, le Premier Ministre, le Président de l'Assemblée nationale, le Président du Sénat, soixante députés ou soixante sénateurs« afin que soit examinée leur conformité à la Constitution.

Dans sa décision des 29–30 décembre 1976, le Conseil constitutionnel avait déjà indiqué dans un *obiter dictum* que »toute modification des traités« serait susceptible de donner lieu à l'application »tant des articles figurant au titre VI (et donc notamment de l'article 54) que de l'article 61 de la Constitution«. De fait, dans sa décision du 17 juillet 1980[10] le Conseil constitutionnel a déclaré recevable une demande émanant de soixante députés qui était dirigée contre la Convention franco-allemande additionnelle à la Convention européenne d'entraide judiciaire en matière pénale du 20 avril 1959; il s'est contenté de dire »qu'une telle demande doit s'entendre comme concernant la loi autorisant la ratification et entraîne, par voie de conséquence, l'examen de la Convention franco-allemande additionnelle signée le 24 octobre 1974«.

Certains auteurs avaient douté de cette possibilité de saisine, faisant valoir que la loi autorisant la ratification d'une convention internationale n'était pas une véritable loi, compte tenu de ce qu'elle n'est qu'un acte d'autorisation, que les parlementaires ne disposent d'ailleurs d'aucun pouvoir d'amendement et qu'en revanche le pouvoir exécutif n'est pas tenu de promulguer la loi d'autorisation dans les quinze jours qui suivent son adoption par les deux chambres. Mais d'autres ont soutenu que la loi autorisant la ratification d'un traité était peu différente d'une loi de finances, c'est-à-dire d'une loi autorisant les recettes et les dépenses de l'Etat, ou encore d'une loi d'habilitation, c'est-à-dire d'une loi autorisant le gouvernement à prendre par voie de règlement des mesures qui sont normalement du domaine de la loi (art. 38). Or personne n'a jamais contesté que de telles lois puissent faire l'objet d'une saisine au titre de l'art. 61 de la Constitution. Il semble que le Conseil constitutionnel se soit rallié à cette seconde opinion qui présente l'avantage de donner le droit de saisine aux parlementaires, il est vrai, à une date plus proche de l'engagement définitif. Il s'agit donc d'une nouvelle extension des compétences du Conseil constitutionnel.

Dans le cadre de l'art. 61 comme dans celui de l'art. 54, le contrôle ne peut être que préventif contrairement à la procédure allemande du contrôle abstrait des normes. Certes en matière de traités internationaux, la Cour constitutionnelle fédérale a admis que la loi d'autorisation puisse (et non pas doive) lui être déférée dès son adoption par le Bundestag[11]. Néanmoins, même

---

[10] C.C. 17 juillet 1980, Rec. p. 36; R.D.P. 1980, p. 1640, observations *Favoreu*.

[11] BVerfGE 5, 28.

lorsque la saisine intervient avant que l'engagement ne devienne définitif, le jugement de la Cour intervient le plus souvent après qu'il le soit devenu. Cela s'est produit notamment lors de l'affaire relative au traité fondamental entre les deux Etats allemands [12]; ce qui pose évidemment de graves problèmes tant au juge constitutionnel qu'au gouvernement responsable des relations extérieures de son pays.

A cet égard, le système français de vérification de la constitutionnalité des engagements internationaux réalise un meilleur compromis entre les exigences de la protection de la Constitution nationale et celles de la participation au développement de la société internationale. C'est ce même sens du compromis qui caractérise les positions adoptées par le Conseil constitutionnel sur les problèmes de fond.

B. Les principes constitutionnels appliqués par le Conseil constitutionnel

Le Conseil constitutionnel a posé en principe qu'il ne pouvait apprécier les engagements internationaux de la France qu'au regard de la Constitution. En particulier, il ne peut pas apprécier leur compatibilité avec les engagements internationaux antérieurs. C'est ainsi qu'à propos de la Convention franco-allemande additionnelle à la Convention européenne d'entraide judiciaire en matière pénale, il a déclaré: »il n'appartient pas au Conseil constitutionnel, lorsqu'il est saisi en application de l'article 61 de la Constitution, d'apprécier la conformité d'un engagement international aux stipulations d'un traité ou d'un accord international; dès lors, il n'y a pas lieu d'examiner la conformité de la Convention franco-allemande aux stipulations de la Convention européenne« [13].

Les principes constitutionnels que le Conseil constitutionnel a jusqu' à présent appliqués sont principalement ceux relatifs à la souveraineté nationale et à ses limites; toutefois la protection des prérogatives du Parlement et des droits de l'homme a parfois été prise en considération par le juge constitutionnel français.

a) Le principe de souveraineté nationale et ses limites

Dans l'ensemble, la jurisprudence du Conseil constitutionnel reflète assez fidèlement le compromis réalisé par la Constitution de 1958 entre les exigences de l'indépendance nationale et celles de la participation de la France à la vie internationale et plus particulièrement au fonctionnement des organisations internationales.

Cette jurisprudence s'est développée principalement à propos des modifications apportées aux institutions communautaires en 1970 et 1976. La

[12] BVerfGE 35, 193 et 257; BVerfGE 36, 1.
[13] C.C. 17 juillet 1980 (références à la note 10).

première affaire, jugée le 19 juin 1970, concernait en particulier le traité, signé à Luxembourg le 22 avril 1970, portant modification de certaines dispositions budgétaires des traités communautaires [14]. Le Conseil constitutionnel se borna à constater que le principe de souveraineté nationale n'était pas mis en cause par le traité soumis à son examen, car celui-ci »ne contenait que des dispositions relatives au fonctionnement interne des communautés, modifiant la répartition des compétences entre les divers organes de celles-ci et n'affectant pas l'équilibre des relations entre les Communautés européennes, d'une part, et les Etats membres, d'autre part«. A cet égard, la seconde décision du Conseil constitutionnel est beaucoup plus riche [15]. Délibérée le 30 décembre 1976, elle concernait la décision du Conseil des Communautés européennes du 20 septembre 1976, relative à l'élection de l'Assemblée au suffrage universel direct, ainsi que l'acte qui y était annexé. Avant d'apprécier ces textes au regard de la constitution, le Conseil constitutionnel expose en termes généraux ce qu'exige le respect du principe de souveraineté nationale: »Si le préambule de la Constitution de 1946, confirmé par celui de la Constitution de 1958, dispose que, sous réserve de réciprocité, la France consent aux limitations de souveraineté nécessaires à l'organisation et à la défense de la paix, aucune disposition de nature constitutionnelle n'autorise les transferts de tout ou partie de la souveraineté nationale à quelque organisation internationale que ce soit«. Cette formule n'est pas très claire et elle a fait l'objet de commentaires souvent divergents. Ce qui est incontestable, c'est que le Conseil constitutionnel considère implicitement que le respect de la souveraineté nationale est la règle et que sans révision de la constitution, il ne peut y être dérogé que dans les conditions définies par le préambule de 1946. Or précisément, ce préambule ne prévoit que de simples limitations en vue de la défense de la paix et sous réserve de réciprocité. Reste à savoir ce qu'il faut entendre par limitations de souveraineté. Le Conseil constitutionnel a répondu à cette question en les opposant aux transferts de souveraineté qui seraient formellement prohibés. Mais que faut-il entendre par là? C'est là que les interprétations des commentateurs divergent. Pour notre part, nous pencherions pour la thèse du professeur *Jean Boulouis*: selon lui, la distinction »ne vise qu'à faire obstacle au transfert compris comme le déplacement du

---

[14] Conseil constitutionnel 19 juin 1970, Rec. p. 15; J.C.P. 1970, I. 2354, observations *D. Ruzié;* R.D.P. 1971, p. 172, observations *C. Emeri* et *J. C. Gautron;* R.G.D.I.P. 1971, p. 239, observations X.

[15] Conseil constitutionnel 29–30 décembre 1976, Rec. p. 15; D. 1977, J., p. 201, note *L. Hamon;* J.C.P. 1977, II, 18704, note *C. Franck,* J.C.P. 1978, I, 2895, observations *M. De Villiers;* R.D.P., 1977, p. 129, observations *L. Favoreu* et *L. Philip;* Pouvoirs, 1977, n° 1, 312, observations *P. Avril* et *J. Gicquel;* Revue trimestrielle de droit européen 1977, p. 666, observations *R. Kovar* et *D. Simon;* Cahiers de droit européen 1977, p. 458, note *J. Boulouis;* Journal de droit international 1977, p. 66, note *D. Ruzié;* A.F.D.I. 1976, p. 805, observations *Cocâtre-Zilgien;* A.F.D.I. 1977, p. 965, note *J. L. Lachaume.*

plan national au plan international de la souveraineté; autrement dit, à s'opposer à ce que l'organisation internationale puisse se considérer comme directement détentrice d'un titre qui n'appartient qu'aux Etats«. De fait, le Conseil constitutionnel a estimé que »l'élection au suffrage universel direct des représentants des peuples des Etats membres à l'Assemblée des communautés européennes n'a pour effet de créer ni une souveraineté, ni des institutions dont la nature serait incompatible avec le respect de la souveraineté nationale«. Certes les hommes politiques hostiles à l'élection au suffrage universel de l'Assemblée européenne avaient fait valoir que précisément l'élection au suffrage universal avait pour effet de créer un organe ayant pour le moins vocation à exercer une partie de la souveraineté française. Mais le Conseil constitutionnel a rejeté très nettement cette argumentation en faisant valoir que »l'acte soumis à son examen est relatif à l'élection d'une assemblée qui n'appartient pas à l'ordre institutionnel de la République française et qui ne participe pas à l'exercice de la souveraineté nationale«, tout en précisant que cette Assemblée »demeure composée de représentants de chacun des peuples de ces Etats«.

Le Conseil constitutionnel a également apprécié les actes prévoyant l'élection au suffrage universel de l'Assemblée européenne au regard d'un attribut traditionnel de la souveraineté nationale en droit public français, à savoir l'indivisibilité de la République française qui est consacrée par l'art. 2 de la Constitution [16]. En effet, parmi les critiques adressées à l'élection de l'Assemblée européenne au suffrage universel direct, figurait celle-ci: l'élection des représentants du peuple français se fera dans le cadre de grandes circonscriptions électorales qui auront ainsi une existence internationale et apparaîtront donc comme des démembrements de la République française. Le Conseil constitutionnel a réfuté cette argumentation en déclarant que les termes de »procédure électorale uniforme dont il est fait mention à l'article 7 de l'acte... ne sauraient être interprétés comme pouvant permettre qu'il soit porté atteinte au principe d'indivisibilité de la République«. En interprétant l'acte de façon à ce qu'il ne soit pas en contrariété avec la Constitution, le Conseil imposait ainsi au législateur français l'obligation de ne prévoir qu'une seule circonscription pour l'ensemble de la République; cette obligation fut d'ailleurs respectée par la loi française du 7 juillet 1977 [17].

Le principe de la souveraineté nationale a également été invoqué à l'occasion de la Convention franco-allemande additionnelle à la Convention eu-

---

[16] Art. 2, al. 1: »La France est une République indivisible...«.

[17] Il convient de noter que le principe d'indivisibilité a joué également un rôle à l'occasion de l'interprétation de l'art. 53, al. 3, de la Constitution selon lequel »nulle cession, nul échange, nulle adjonction de territoire n'est valable sans le consentement des populations intéressées«. En effet, le 30 décembre 1975, le Conseil constitutionnel a jugé que l'exigence d'une consultation des populations s'appliquait également au cas d'accession à l'indépendance d'un territoire d'outre-mer faisant partie de la République française (en l'espèce, l'archipel des Comores, à l'exception de l'Ile de Mayotte).

ropéenne d'entraide judiciaire en matière pénale. Les auteurs de la saisine, soixante députés communistes, soutinrent que cette Convention portait atteinte à la souveraineté nationale en ce que les autorités judiciaires françaises seraient pratiquement tenues de déférer aux demandes d'entraide judiciaire présentées par la République fédérale d'Allemagne. Le Conseil a rejeté cette argumentation en relevant, tout d'abord que l'entraide judiciaire pouvait, conformément à la Convention européenne, être refusée si la demande se référait à des infractions politiques ou si son exécution était de nature à porter atteinte aux intérêts essentiels du pays. Puis, il a admis que le principe de souveraineté nationale implique que »les autorités judiciaires françaises, telles qu'elles sont définies par la loi française, sont seules compétentes pour accomplir en France, dans les formes prescrites par cette loi, les actes qui peuvent être demandés par une autorité étrangère au titre de l'entraide judiciaire en matière pénale«; mais il a considéré, sans d'ailleurs le démontrer, que la Convention franco-allemande ne porte aucune atteinte à cette règle.

Si le respect de la souveraineté nationale constitue la question majeure soulevée par les engagements internationaux devant le Conseil constitutionnel, d'autres problèmes ont néanmoins été abordés dans les affaires qui lui ont été soumises.

b) Le respect des droits de l'homme et des droits du Parlement

Un seul droit de l'homme a été jusqu'à présent invoqué devant le Conseil constitutionnel à l'encontre d'un engagement international: le droit d'asile, qui est reconnu par l'alinéa 4 du préambule de la Constitution de 1946[18]. Dans l'affaire relative à la Convention franco-allemande additionnelle à la Convention européenne d'entraide judiciaire en matière pénale, le Conseil constitutionnel a en effet constaté »qu'aucune disposition de la Convention additionnelle n'ouvre la possibilité de porter atteinte du droit d'asile, tel qu'il est proclamé par le Préambule de la Constitution du 27 octobre 1946, réaffirmé par celui de la Constitution du 4 octobre 1958«.

Par ailleurs, l'application de l'art. 53 de la Constitution qui subordonne la ratification ou l'approbation de certains traités ou accords à une autorisation du Parlement a soulevé des difficultés à propos des dispositions des lois budgétaires par lesquelles le Parlement autorise le Gouvernement à accorder des prêts à des Gouvernements étrangers. Dans l'affaire jugée le 30 décembre 1975[19], les députés socialistes faisaient valoir que ces autorisations étaient globales et ne comportaient donc pas une approbation explicite des différents accords financiers conclus avec les Gouvernements étrangers alors que

---

[18] »Tout homme persécuté en raison de son action en faveur de la liberté a droit d'asile sur les territoires de la République«.

[19] Conseil constitutionnel 30 décembre 1975, Rec. p. 28; D. 1976, J. p. 461, note *L. Hamon;* J.C.P. 1976, II, 18368, note *J. M. Bollé;* A.J.D.A. 1976, p. 308, note *Py;* R.D.P. 1976, p. 995; J.D.I. 1976, p. 405, note *D. Ruzié.*

l'art. 53 de la Constitution subordonne à l'autorisation du Parlement l'approbation des accords »qui engagent les finances de l'Etat«. La réponse du Conseil constitutionnel est assez peu claire. Il semble qu'il n'ait pas vu dans ces accords de véritables engagements internationaux au sens de l'art. 53 de la Constitution, mais de simples décisions financières régies par les dispositions constitutionnelles applicables en matière budgétaire. Ainsi, c'est en se fondant sur le respect des pouvoirs budgétaires du Parlement que le Conseil constitutionnel déclare conformes à la Constitution les disposition budgétaires qui étaient contestées.

Ainsi, jusqu' à présent, les principes constitutionnels appliqués par le Conseil constitutionnel se ramènent principalement au principe de la souveraineté nationale; de fait, le respect de l'indépendance nationale a été un thème majeur de la vie politique française depuits trente ans. A cet égard, il est remarquable que le Bundesverfassungsgericht ait de son côté apprécié la constitutionnalité des traités qui lui étaient soumis par référence aux principes de la réunification de l'Allemagne et de la libre détermination de son peuple[20] et qu'il ait dû comme le Conseil constitutionnel français concilier les exigences contradictoires de l'unité de la nation allemande et de la coopération internationale. L'écart entre les jurisprudences constitutionnelles française et allemande est donc moins considérable que pouvaient le faire penser les différences de technique procédurale.

Une remarque équivalente pourra êtra faite à propos des solutions françaises et allemandes relatives aux engagements internationaux définitifs.

II. Le Conseil constitutionnel et les engagements internationaux définitifs

Depuis 1975, le Conseil constitutionnel a développé progressivement une jurisprudence qui peut être aujourd'hui synthétisée de la façon suivante. Il refuse de contrôler si un engagement international ou une loi est contraire aux engagements internationaux définitifs malgré le principe de supériorité de ces derniers; il considère, en effet, que les actes contraires aux engagements internationaux sont néanmoins conformes à la constitution (A). Inversement, il refuse de contrôler la conformité à la constitution des mesures d'application des engagements internationaux, ce qui conduit à admettre leur supériorité de fait sur la constitution (B).

A. La conformité à la Constitution des actes contraires à des engagements internationaux définitifs

Le problème du respect des engagements internationaux ne se pose pas dans les mêmes termes selon qu'il s'agit d'engagements internationaux sub-

---

[20] BVerfGE 1, 396 (30. 7. 1952); BVerfGE 4, 157 (4. 5. 1955); BVerfGE 36, 1 (31. 7. 1973).

séquents qui ne sont pas encore entrés en vigueur, ou de lois votées ultérieurement par les assemblées parlementaires, mais non encore promulguées.

a) Le refus de contrôler le respect des engagements internationaux définitifs par un engagement ultérieur subséquent

En 1980, le Conseil constitutionnel a été saisi d'un recours dirigé par des parlementaires communistes contre la Convention franco-allemande additionnelle à la Convention européenne d'entraide judiciaire pénale[21].

Les auteurs de la saisine faisaient en particulier valoir que »la convention franco-allemande serait contraire à la Constitution en tant qu'elle dérogerait à une convention en vigueur, à savoir la Convention européenne d'entraide judiciaire en matière pénale, du 20 avril 1959«. A cet argument, le juge constitutionnel français répondit: »il n'appartient pas au Conseil constitutionnel, lorsqu'il est saisi en application de l'art. 61 de la Constitution, d'apprécier la conformité d'un engagement international aux stipulations d'un traité ou d'un accord international; dès lors, il n'y a pas lieu d'examiner la conformité de la Convention franco-allemande aux stipulations de la Convention européenne«.

Cette argumentation n'est pas fondée sur des considérations de droit international (liberté des Etats de conclure des traités internationaux dérogatoires à des traités antérieurs sous réserve des cas définis par la Convention de Vienne du 23 mai 1969), mais sur l'interprétation de la Constitution française. Plus précisément, le Conseil constitutionnel invoque expressément l'art. 61 qui lui donne pour mission de contrôler la conformité à la constitution de divers actes étatiques, notamment celle des lois autorisant la ratification ou l'approbation d'un traité ou d'un accord international. Implicitement, il considère donc que la conformité à un engagement international ne peut pas être identifiée à la conformité à la Constitution nonobstant le quatorzième alinéa du Préambule de la Constitution de 1946 selon lequel »la République française, fidèle à ses traditions, se conforme aux règles du droit public international«. Devant le laconisme du Conseil constitutionnel, le commentateur est réduit à faire des hypothèses: le Conseil constitutionnel a-t-il refusé cette identification parce que le droit international ne comprendrait pas la règle *pacta sunt servanda* compte tenu des restrictions qui lui sont généralement apportées et qui ont été codifiées par la Convention de Vienne ou plus simplement a-t-il considéré que la règle *pacta sunt servanda* ne rentrait pas dans le champ d'application du quatorzième alinéa du Préambule de la Constitution de 1946 compte tenu de ce que les textes constitutionnels français ont toujours consacré des dispositions particulières à la place des traités internationaux dans l'ordre juridique français?

---

[21] Conseil constitutionnel 17 juillet 1980, Rec. p. 36; R.D.P. 1980, p. 1640, observations *Favoreu;* R.G.D.I.P. 1981, p. 202, note *Vallée.*

En tout cas, nous devons rejeter l'interprétation donnée par les deux auteurs cités à la note 21: selon ceux-ci, le Conseil constitutionnel aurait simplement étendu au cas des engagements internationaux sa jurisprudence relative aux rapports entre les traités et les lois. Cette extension nous semble exclue dans la mesure où cette jurisprudence est fondée sur l'art. 55 de la Constitution qui ne nous semble pas pouvoir régir les rapports entre un engagement international définitif et un engagement international subséquent, mais seulement les rapports entre un engagement international définitif et une loi subséquente ainsi que nous allons le montrer.

b) Le refus de contrôler le respect des engagements internationaux définitifs par une loi subséquente

L'art. 55 de la Constitution dispose: »Les traités ou accords régulièrement ratifiés ou approuvés ont, dès leur publication, une autorité supérieure à celle des lois, sous réserve, pour chaque accord ou traité, de son application par l'autre partie«.

Le Conseil constitutionnel a interprété pour la première fois cette disposition dans sa célèbre décision du 15 janvier 1975 relative à la loi sur l'interruption volontaire de grossesse[22].

Les auteurs de la saisine, des députés gaullistes, faisaient valoir que la loi française était contraire à l'art. 2 de la Convention européenne de sauvegarde des droits de l'homme[23] et qu'elle était donc inconstitutionnelle en vertu de l'art. 55 de la Constitution. Mais le Conseil constitutionnel a refusé d'exercer ce contrôle et a affirmé: »il n'appartient pas au Conseil constitutionnel, lorsqu'il est saisi en application de l'art. 61 de la Constitution, d'examiner la conformité d'une loi aux stipulations d'un traité ou d'un accord international«.

L'argumentation développée par le Conseil pour justifier son refus est d'une logique irréfutable. Elle peut se résumer de la façon suivante: l'art. 55 n'instaure pas une supériorité générale et absolue du traité sur la loi, car l'existence de la réserve de réciprocité suppose précisément que des lois contraires à un traité puissent être appliquées par le juge lorsqu'un autre Etat partie au traité ne le respecte pas. La supériorité du traité sur la loi n'est donc

---

[22] Rec. p. 19; D. 1975, J., p. 529, note *L. Hamon;* J.C.P. 1975, II, 18030, note *E. M. Bey;* G.P. 1976, 1, p. 25, note *A. Pellet;* R.D.P. 1975, pp. 185 et 1335, observations *L. Favoreu* et *L. Philip;* A.J.D.A. 1975, p. 134, note *Rivero;* R.T.D.E. 1975, p. 49, observations *M. Lagrange;* Revue du marché commun 1975, p. 69; note *G. Druesne;* Cahiers de droit européen 1975, p. 875, note *J. Rideau;* R.G.D.I.P. 1975, p. 1070, note *C. Franck;* A.F.D.I. 1975, p. 859, observations *Nguyen Quoc Dinh;* Revue critique de droit international privé 1975, p. 606, note *P. Lagarde;* J.D.I. 1975, p. 249, note *D. Ruzié;* R.I.D.C. 1975, p. 875, note *J. Robert;* Annales de Clermont, 1976, p. 207, observations *Montgroux.*

[23] Art. 2 de la Convention européenne: »Le droit de toute personne à la vie est protégé par la loi. La mort ne peut être infligée à quiconque intentionnellement, sauf en exécution d'une sentence capitale«.

que »relative et contingente«. Or précisément les décisions prises par le Conseil constitutionnel à l'égard d'une loi »revètent un caractère absolu et définitif« puisqu'elles »font obstacle à la promulgation et à la mise en application de toute disposition déclarée inconstitutionnelle«. En d'autres termes, la contrariété d'une loi avec un traité ne peut pas affecter sa validité, mais seulement son applicabilité.

Bien des auteurs français ont cependant reproché au Conseil constitutionnel d'avoir poussé jusqu' à l'absurde la logique inhérente à l'art. 55 de la Constitution. En particulier, la clause de réciprocité est tout à fait injustifiée pour les traités relatifs aux droits de l'homme: en effet, en ratifiant de tels traités, les Etats ne contractent pas des obligations au profit des autres Etats ou de leurs ressortissants, mais en réalité au profit de leurs propres nationaux. Il aurait donc été préférable que le Conseil constitutionnel réserve le cas des traités protecteurs des droits de l'homme en faisant valoir que ceux-ci ont une valeur absolue qui ne doit pas dépendre du comportement des autres Etats. Certains auteurs vont même jusqu' à soutenir que la même attitude devrait être adoptée pour les traités multilatéraux qui instituent des organisations internationales ayant leurs propres procédures pour sanctionner le manquement des Etats membres. Mais en réalité, les inconvénients pratiques de cette jurisprudence sont très réduits dans ce dernier cas du fait même que les manquements éventuels de la France à de tels traités ne resteraient pas sans sanction.

En prenant cette position, le Conseil constitutionnel confiait implicitement au juge ordinaire le soin de trancher les conflits entre la loi et le traité. C'est d'ailleurs ainsi que l'a entendu la Cour de cassation qui, peu de temps après, n'hésitait pas à écarter l'application d'une loi contraire au traité instituant la Communauté économique européenne[24]. Malheureusement le Conseil d'Etat n'a pas suivi cet exemple, arguant que le refus d'appliquer une loi contraire à un traité aboutirait à contrôler la constitutionnalité de celle-ci, tâche qui serait réservée exclusivement au Conseil constitutionnel[25].

Quoiqu'il en soit, le refus de censurer une loi contraire à un traité a été renouvelé par le Conseil constitutionnel à plusieurs reprises. Le traité invoqué était selon les cas soit la Convention européenne des droits de l'homme[26] soit une convention de l'Organisation internationale du travail[27].

---

[24] Cour de cassation, Ch. mixte, *Société des cafés Jacques Vabre,* 24 mai 1975, Rec. 1975, p. 500, conclusion *Touffait;* D. 1975, p. 497, conclusions *Touffait;* G.P. 306, R.T.D.E. 1975, p. 337, conclusions *Touffait;* R.G.D.I.P. 1976, p. 960, note *Rousseau;* Cahier de droit européen 1975, p. 655, note *Kovar;* R.D.P. 1975, p. 1335, observations *Favoreu* et *Philip;* A.J.D.A. 1975, p. 567, note *J. Boulouis;* J.D.I. 1975, p. 802, note *D. Ruzié;* R.C.D.I. 1976, p. 347, note *J. Foyer* et *D. Holleaux.*

[25] Conseil d'Etat, 22 octobre 1979, *Union démocratique du travail,* Rec. p. 383; R.D.P. 1980, p. 531, conclusions *M. D. Hagelsteen;* A.J.D.A. 1980, n° 1, p. 19, Chronique *Robineau* et *Feffer* et p. 39, note *B. G.*

[26] Conseil constitutionnel 18 janvier 1978, Rec. p. 21; R.D.P. 1978, p. 834, observations *Favoreu.*

Cette jurisprudence du Conseil constitutionnel n'apparaît pas très éloignée des solutions du droit allemand, du moins quant aux résultats: dans les deux pays, la supériorité des traités sur les lois n'est pas pleinement assurée, même si, dans la pratique, elle est admise dans la quasi-totalité des cas. En France, les traités sont théoriquement supérieurs aux lois, mais le juge constitutionnel laisse au juge ordinaire le soin d'appliquer ce principe; en Allemagne, traités et lois sont placés en principe sur un pied d'égalité, mais l'idée d'ouverture au droit international *(Völkerrechtsfreundlichkeit der nationalen Rechtsordnung)* assure aux traités internationaux une supériorité de fait sur les lois.

La même remarque vaut pour les mesures d'application des traités internationaux: les deux juges constitutionnels admettent très largement la supériorité de celles-ci sur la constitution même s'il ne l'admettent pas complètement.

B. L'incontestabilité des mesures d'application des engagements internationaux définitifs

L'incontestabilité des mesures d'application des engagements internationaux sur la Constitution n'est pas explicitement formulée par le Conseil constitutionnel et ne l'est qu'assez rarement par la doctrine [28]. C'est pourquoi il est nécessaire de démontrer l'existence de ce principe avant d'étudier les problèmes posés par sa mise en oeuvre.

a) L'existence du principe

Le problème des pouvoirs du juge constitutionnel à l'égard des mesures d'application des traités internationaux s'est posé dès 1970 au sujet de la décision du Conseil des Communautés européennes du 21 avril 1970 relative au remplacement des contributions financières des Etats membres par des ressources propres aux Communautés. Ne contenait-elle pas des dispositions contraires à la Constitution française? Or, dans sa décision du 19 juin 1970 [29], le Conseil constitutionnel releva tout d'abord que les trois traités institutifs des Communautés ont prévu expressément le passage progressif d'un système de contribution des Etats membres à un régime de ressources propres et que ceux-ci »ont été régulièrement ratifiés et publiés et sont, dès lors, entrés dans le champ d'application de l'article 55 de la Constitution«, puis il déclara: »La décision du 21 avril 1970. . . a le caractère d'une mesure d'application des dispositions susrappelées des traités instituant les Communautés européennes. . .; en conséquence, elle n'est pas en contradiction avec la Constitution«.

---

[27] Conseil constitutionnel 20 juillet 1977, Rec. p. 39; R.D.P. 1978, p. 827, observations *Favoreu* (convention n° 95).

[28] Voir toutefois, dans une large mesure, *L. Philip*, R.D.P. 1977, p. 473, à propos de la décision du Conseil constitutionnel du 30 décembre 1977.

[29] Pour les références, voir *supra* note 14.

Cette argumentation du Conseil constitutionnel peut, nous semble-t-il, être comprise de la façon suivante: les traités institutifs sont entrés dans le champ d'application de l'art. 55 de la Constitution, c'est-à-dire qu'ils ne sont plus susceptibles d'être contestés devant le juge constitutionnel et qu'ils doivent donc produire tous leurs effets conformément à l'art. 55. Par voie de conséquence, les mesures d'application de ces traités doivent bénéficier du même régime, sinon l'autorité des traités ne serait pas pleinement respectée. Nous ne cachons pas qu'une telle interprétation est assez hardie dans la mesure où l'art. 55 affirme seulement la supériorité des traités sur les lois, et non celle des mesures d'application; néanmoins, il est incontestable que dans l'espèce, la décision communautaire d'avril 1970 n'a pas été déclarée en contradiction avec la Constitution bien qu'elle restreignît les pouvoirs fiscaux et budgétaires du Parlement français.

Sept ans plus tard, le Conseil constitutionnel adopta une argumentation beaucoup plus nette à propos, non plus d'une décision des Communautés européennes, mais d'une loi française prise en application d'une décision communautaire. Dans les deux affaires jugées le 30 décembre 1977 [30], les auteurs de la saisine, des parlementaires de l'opposition, contestaient la conformité à la Constitution de lois de finances qui précisaient les modalités de recouvrement ou autorisaient la perception de prélèvements institués par des règlements des Communautés européennes (prélèvement de coresponsabilité sur le lait et cotisation à la production d'isoglucose). Le Conseil constitutionnel admit très clairement que ces règlements communautaires avaient pour effet de restreindre les compétences du Parlement français: »dans le cas du règlement du 17 mai 1977, le Parlement n'avait pas à intervenir dans la détermination de l'assiette et du taux de la cotisation et il lui revenait seulement de régler les modalités de recouvrement non fixées par le règlement; les répercussions de la répartition des compétences ainsi opérées entre les institutions communautaires et les autorités nationales au regard tant des conditions d'exercice de la souveraineté nationale que du jeu des règles de l'article 34 de la Constitution relative au domaine de la loi ne sont que la conséquence d'engagements internationaux souscrits par la France qui sont entrés dans le champ d'application de l'article 55 de la Constitution; dans ces conditions, les dispositions de la loi de finances rectificative pour 1977 soumise à l'examen du Conseil constitutionnel ne sont contraires à aucune règle ni à aucun principe ayant valeur constitutionnelle« [31]. Cette seconde décision est par-

---

[30] Conseil constitutionnel 30 décembre 1977 (décisions n°s 77–89 D.C. et 77–90 D. C.), Rec. 44 et 46; R.D.P. 1979, p. 468, observations *L. Philip;* R.T.D.E. 1979, p. 142, note *G. Isaac* et *J. Molinier*.

[31] Dans l'affaire relative à la loi autorisant le Gouvernement français à augmenter la quote-part de la France au Fonds Monétaire International jugée le 22 avril 1978, la question posée était moins claire: les auteurs de la saisine faisaient valoir que cette loi était inconstitutionnelle parce qu'elle n'aurait été qu'une mesure d'application d'une résolution du Conseil des Gouverneurs et que cette dernière ne lierait pas la France

ticulièrement nette: des mesures d'application des engagements internationaux peuvent écarter l'application des règles posées par la Constitution française.

Toutefois, la référence à l'art. 55 de la Constitution pourrait faire penser que cette supériorité est subordonnée au respect de l'engagement par les autres parties. De fait, ce fut l'argumentation développée par des parlementaires de l'opposition dans l'affaire jugée le 30 décembre 1980 par le Conseil constitutionnel[32]: ceux-ci soutinrent en effet que la loi harmonisant les impôts français sur les boissons avec le droit communautaire était contraire à la Constitution parce que d'autres Etats membres ne respectaient pas non plus le traité de Rome sur ce point et qu'ainsi le Parlement aurait méconnu la règle de réciprocité posée par l'art. 55 de la Constitution. Mais le Conseil constitutionnel a rejeté cette argumentation, car, selon lui, »si elle affecte la supériorité des traités ou accords sur les lois, (la règle de réciprocité) n'est pas une condition de conformité des lois à la Constitution«; elle »ne s'oppose (donc) pas à ce que la loi édicte des mesures ayant pour objet d'harmoniser la législation nationale avec les dispositions découlant d'un traité, alors même que celles-ci ne seraient pas appliquées par l'ensemble des parties signataires«.

Mais, si le principe a ainsi une existence incontestable, il n'en comporte pas moins des limites qui ont pour effet de laisser au Conseil constitutionnel une certaine marge d'appréciation.

b) Les limites du principe

Selon la jurisprudence du Conseil constitutionnel, le principe de l'incontestabilité des mesures d'application des engagements internationaux souffre deux limites.

La première limite découle de la logique même du principe: seuls les actes qui mettent en application un engagement international antérieur bénéficient de l'incontestabilité, ce qui amène le Conseil constitutionnel à vérifier si un lien rattache l'acte contesté à un engagement international antérieur et si par conséquent cet acte est conforme aux dispositions contenues dans cet engagement. Par exemple, dans les deux décisions du 30 décembre 1977 relatives aux taxes européennes[33], le Conseil montre que celles-ci trouvent leur fondement dans le traité instituant la Communauté économique européenne: elles ont été créées par des règlements du Conseil qui ont été pris

---

faute d'avoir fait l'objet d'une approbation expresse de sa part. Le Conseil répondit que cette résolution modifiant les statuts du Fonds s'imposait à la France en vertu de l'art. XVII de ceux-ci et que la loi d'autorisation n'était donc pas contraire à la Constitution (Conseil constitutionnel 29 avril 1978, Rec. p. 23; R.D.P. 1979, p. 494, observations *Philip;* J.D.I. 1979, p. 577, note *D. Ruzié*).

[32] Conseil constitutionnel 30 décembre 1980, Rec. p. 53; R.G.D.I.P. 1981, p. 601, note *Decaux*.

[33] Voir les références à la note 30.

en vertu de l'art. 145 du traité et »dont la force obligatoire n'est pas subordonnée à une intervention des autorités des Etats membres« (art. 189 du traité). Ainsi le Conseil constitutionnel s'assure que l'acte soumis à son examen (qu'il s'agisse de l'acceptation d'une décision d'organisation internationale ou d'une loi française) est bien la conséquence d'engagements antérieurs[34]. Il ne s'est abstenu de le faire que dans une décision, il est vrai, importante, celle des 29 et 30 décembre 1976 relative à la décision du Conseil des Communautés européennes du 20 septembre 1976 prévoyant l'élection de l'assemblée au suffrage universel direct: il a admis implicitement que ce n'était pas une mesure d'application des traités institutifs, peut-être parce qu'elle y dérogeait en renonçant à imposer aux Etats une procédure électorale uniforme.

Si cette limite apportée au principe de l'incontestabilité des mesures d'application des engagements internationaux est parfaitement logique, il n'en est pas de même pour l'autre: en effet, l'étude de la jurisprudence montre que la nature de mesure d'application de l'acte soumis à son contrôle ne conduit pas toujours le Conseil constitutionnel à rejeter tous les griefs d'inconstitutionnalité: il accepte parfois d'examiner certains d'entre eux. Par exemple, dans l'affaire jugée le 19 juin 1970[35], le grief de violation du principe de la souveraineté nationale est examiné puisque le Conseil relève que la mesure d'application contestée "ne peut porter atteinte, ni par sa structure, ni par son importance, aux conditions essentielles d'exercice de la souveraineté«; de même, dans celle jugée le 30 décembre 1980[36], le Conseil examine si la loi française prise en application d'une décision communautaire »édicterait une règle rétroactive en méconnaissance du principe posé par l'article 8 de la Déclaration des droits de l'homme et du citoyen«. Cependant cette jurisprudence n'est pas constante comme le montre la décision du 30 décembre 1977[37] qui conclut à la constitutionnalité de la loi soumise à son contrôle dès lors qu'elle est »la conséquence d'engagements internationaux souscrits par la France«. Il est difficile de systématiser une jurisprudence qui se trouve encore en formation. Toutefois cette attitude du Conseil constitutionnel français est à rapprocher de celle du Bundesverfassungsgericht qui, le 29 mai 1974, a reconnu la supériorité des règlements communautaires sur la Constitution allemande à l'exception toutefois des dispositions relatives au droits fon-

---

[34] Cette vérification est d'ailleurs assez superficielle comme le montre la décision du 30 décembre 1977 par laquelle le Conseil constitutionnel a rejeté les critiques dirigées contre les mesures d'application du règlement instituant une taxe sur l'isoglucose, alors que ce règlement communautaire a été précisément jugé contraire au traité par la Cour de justice des Communautés une année plus tard (C.J.C.E. 25 octobre 1978, affaires jointes 103 et 145/77, R.T.D.E. 1979, p. 131). C'est d'ailleurs pourquoi le Conseil constitutionnel ne semble pas tenu de recourir dans de tels cas à la procédure de renvoi préjudiciel prévue par l'art. 177 du traité C. E. E.

[35] Voir les références à la note 8.

[36] Voir les références à la note 32.

[37] Voir les références à la note 30.

damentaux (du moins pour quelque temps)[38]. Il semble que dans les deux pays le juge constitutionnel manifeste quelque hésitation à admettre pleinement la supériorité du droit international en ce domaine.

Conclusion

L'institution d'une justice constitutionnelle en France a eu pour effet de donner au droit constitutionnel français une précision et une rigueur qu'il n'avait jamais eues. A l'imprécision des règles et à l'absence de sanction des violations de la Constitution ont succédé les contraintes de règles bien définies et de sanctions efficaces. De ce fait les relations entre le droit constitutionnel et le droit international sont techniquement plus difficiles à aménager que par le passé. De plus, un nationalisme, modéré il est vrai, a inspiré incontestablement le texte de la Constitution du 4 octobre 1958, ce qui a quelque peu atténué l'ouverture au droit international qu'avait réalisée la Constitution du 27 octobre 1946.

Malgré ces difficultés qui expliquent bien des tâtonnements, le juge constitutionnel français a su tirer profit de la caractéristique essentielle du système français de contrôle de constitutionnalité, à savoir son caractère préventif, pour élaborer une jurisprudence qui soit adaptée aux exigences de la participation d'un pays à la vie internationale. C'est elle en particulier qui permet de justifier les dérogations aux règles constitutionnelles françaises lorsque celles-ci ne sont que la conséquence d'engagements précédemment souscrits par la France. C'est elle également qui permet de limiter la portée de la règle de réciprocité qui a été introduite de façon si malencontreuse en 1958 dans le droit public français. Dans le domaine des relations internationales comme d'ailleurs dans les autres domaines où il intervient, le Conseil constitutionnel est ainsi parvenu à maintenir un équilibre assez satisfaisant entre les tendances contradictoires qui dominent le droit constitutionnel français.

---

[38] BVerfGE, 37, 271.

Die Verpflichtungen erga omnes im Völkerrecht und ihre Durchsetzung

JOCHEN ABR. FROWEIN

1. Einführung

Die Entwicklung des Völkerrechts seit 1945 trägt ambivalente Züge. Nach dem Zusammenbruch der europäischen Ordnung im zweiten Weltkrieg war es natürlich, daß eine Rückbesinnung auf die Grundlagen der zwischenstaatlichen Rechtsordnung, der ihr vorausliegenden Wertentscheidungen und ihrer Ausrichtung auf den Menschen in der Völkerrechtstheorie einsetzte [1]. Die Qualität der Staatengesellschaft als Rechtsgemeinschaft hat der Jubilar eindrucksvoll seinem Verständnis der Entwicklung des Völkerrechts zugrunde gelegt [2]. Die Anerkennung der Menschenrechte in der Satzung der Vereinten Nationen, die Verabschiedung der völkerrechtlich nicht verbindlichen, wohl aber erheblichen Allgemeinen Erklärung der Menschenrechte von 1948, vor allem aber der Abschluß einer Vielzahl von die Rechte des einzelnen Menschen schützenden Verträgen, an ihrer Spitze der Pakte von 1966, scheinen den Befund einer Rückkehr des Völkerrechts zu eindeutigen Wertgrundlagen zu stützen. Dazu gehört auch die Festlegung des Verbotes der Gewaltanwendung zwischen Staaten in Art. 2 Ziff. 4 der UN-Satzung und der Einrichtung eines Verfahrens zur Friedenssicherung in deren VII. Kapitel, das die Wiederholung von 1939 verhindern sollte. Es war nur konsequent, wenn sich aus dieser Entwicklung die Notwendigkeit einer Anerkennung der wichtigsten dieser Grundlagen der Völkerrechtsverfassung als zwingendes Recht in der Literatur durchsetzte [3] und in der Staatenpraxis Anerkennung fand [4].

---

[1] *H. Mosler,* Völkerrecht als Rechtsordnung, ZaöRV 36 (1976), S. 6, 30 f., 36 f. und *passim.*

[2] *H. Mosler,* The International Society as a Legal Community, RdC 140 (1974 IV); revidierte Ausgabe 1980.

[3] Dazu *H. Mosler,* Ius cogens im Völkerrecht, Schweizerisches Jahrbuch für internationales Recht 25 (1968), S. 9 ff.

[4] Vor allem in Art. 53 und 64 Wiener Vertragsrechtskonvention (WVRK).

Freilich ist dies nur die eine, bei beschränkter Betrachtung eher täuschende Seite. Das Völkerrecht befindet sich als Rechtsordnung in einer Krise, die tiefgreifender und qualitativ ungleich gefährlicher sein dürfte als alle vorhergehenden. Trotz der sehr erfolgreichen Kodifikationstätigkeit, die zur Klärung und Fortbildung des Völkerrechts in nicht mehr hinwegzudenkender Weise beiträgt, besteht über die Grundlagen der Völkerrechtsordnung ein tiefer Dissens. *Wilhelm Grewe* hat darauf hingewiesen, daß der Charakter der Völkerrechtsordnung sich wegen der bestehenden Dissense über Grundbegriffe nicht aus den Normen der UN-Charta oder der Fülle der Texte völkerrechtlichen oder quasi-völkerrechtlichen Inhalts ergründen lasse, die in dieser deklarations- und kodifikationsfreudigen Periode produziert worden sind. Mehr denn je komme es darauf an, die Realitäten der Staatenpraxis, das tatsächliche Verhalten der Mächte und seine Auswirkungen für die Völkerrechtsordnung ins Auge zu fassen und systematisch zu bewerten [5]. Während das europäische Völkerrecht bis zum ersten Weltkrieg unangefochten von der Vorherrschaft des europäisch-amerikanischen Rechtsdenkens lebte, die zwischen den Kriegen nur von der isolierten Sowjetunion in Frage gestellt wurde, besteht seit 1945 ein Antagonismus des Völkerrechtsverständnisses, der sich auf die neu entstandene dritte Welt ausgedehnt hat [6]. Weder die Bedeutung des Gewaltverbotes noch die Auslegung der Menschenrechtspakte entziehen sich der Auseinandersetzung, die die Grundlagen des Völkerrechts ergriffen hat.

Wenn in einer derartigen Situation die Begriffe des *ius cogens* oder der Verpflichtungen *erga omnes* im Völkerrecht Anerkennung finden, so wird es genauer und nüchterner Prüfung bedürfen, um festzustellen, inwieweit hiermit Rechtsfolgen verbunden sind, deren Beachtung als relativ gesichert bezeichnet werden kann. *R.-J. Dupuy* hat davor gewarnt, sich durch den Begriff der *communauté internationale* über den konfliktgeladenen Charakter der heutigen Völkerrechtsordnung täuschen zu lassen [7]. Das ist gerade in einem Zusammenhang zu beachten, in dem es auch um die Wirkungen von Rechtsnormen geht, die in der Definition des *ius cogens* in Art. 53 WVRK von der »internationalen Staatengemeinschaft in ihrer Gesamtheit« als so wichtig angesehen werden, daß von ihnen nicht abgewichen werden darf.

Die genaue Beziehung zwischen *ius cogens* und Verpflichtungen *erga omnes* kann hier zunächst offen bleiben. Mit *H. Mosler* sollte der Begriff des *ius*

---

[5] *W. Grewe,* Über den Gesamtcharakter der jüngsten Epoche der Völkerrechtsgeschichte, in: Staatsrecht-Völkerrecht-Europarecht, Festschrift für H.-J. Schlochauer (1981), S. 304.

[6] Vgl. *U. Scheuner,* Solidarität unter den Nationen als Grundsatz in der gegenwärtigen internationalen Gemeinschaft, in: Recht im Dienst des Friedens, Festschrift für E. Menzel (1975), S. 251 ff.; mit Beispielen *J. A. Frowein*, Der Beitrag der internationalen Organisationen zur Entwicklung des Völkerrechts, ZaöRV 36 (1976), S. 147, 156 ff.; *W. Grewe,* Spiel der Kräfte in der Weltpolitik (1970), S. 388 ff., 398.

[7] *R.-J. Dupuy,* Leçon Inaugurale, Collège de France 22. 2. 1980 (1980), S. 9, 24 ff., und *passim.*

cogens auf das durch Vertrag nicht abänderbare Recht beschränkt werden[8]. Der Begriff der Verpflichtung *erga omnes,* die aus denselben Rechtsnormen entstehen kann, legt das Gewicht auf die entstehende Rechts- und Pflichtenbeziehung und wirft damit die Frage der Durchsetzung auf.

2. Die Verpflichtungen erga omnes in der Rechtsprechung des IGH

Es wird wohl nie aufgeklärt werden, wie der Internationale Gerichtshof in seinem Urteil vom 5. Februar 1970 in der Sache *Barcelona Traction* zu der jedenfalls auch für Verfahrensbeteiligte überraschenden Erörterung des unterschiedlichen Charakters von völkerrechtlichen Verpflichtungen gekommen ist. Der Gerichtshof hat hier bekanntlich erstmals ausdrücklich anerkannt, daß es Verpflichtungen gegenüber der internationalen Gemeinschaft als ganzer und gegenüber einzelnen Staaten gebe[9]. Die gewählte Formulierung legt die Annahme nahe, daß die Verabschiedung der Wiener Vertragsrechtskonvention im Jahr davor mit ihren Bestimmungen über das *ius cogens* von erheblichem Einfluß gewesen ist[10]. Die Beachtung der Verpflichtungen *erga omnes* ist nach dem Urteil des Gerichtshofes »the concern of all states«, und alle Staaten haben ein rechtliches Interesse an ihrer Durchsetzung[11]. Was das prozessual bedeutet, bedarf späterer Erörterung[12]. Im Augenblick interessiert nur die konzeptionelle Basis der Argumentation.

Als Verpflichtungen, die zu dieser Kategorie zu rechnen sind, werden das Verbot der Aggression, des Genozids sowie die Verpflichtungen genannt, die aus den Prinzipien und Regeln über die grundlegenden Rechte der menschlichen Person (*basic rights of the human person*) einschließlich des Schutzes vor Sklaverei und rassischer Diskriminierung folgen[13]. Die Verbote der Aggression und des Genozids werden als einigermaßen klar umgrenzt angesehen werden können, was freilich nicht ausschließt, daß die Auseinandersetzungen im Einzelfall um die Anwendung erheblich sein können. Es dürfte aber wichtig sein, daß es sich hier um Grundnormen der Völkerrechtsgemeinschaft handelt, deren Charakter als zwingend jedenfalls nach dem zweiten Weltkrieg nicht mehr ernsthaft in Zweifel gezogen werden kann. Wesentlich weniger eindeutig erscheint der Verweis auf die *basic rights of the human person.* Die Verwendung des Wortes *basic* an Stelle von *fundamental* dürfte zeigen, daß

---

[8] *Mosler* (Anm. 3), S. 26 und *passim.*

[9] ICJ Reports 1970, S. 32, Par. 33, 34.

[10] In der *sep. op.* von Ammoun wird die Wiener Vertragsrechtskonvention in unmittelbarem Zusammenhang mit dem Konzept des *ius cogens* erwähnt, freilich bei einer etwas anderen Erörterung, a.a.O., S. 304.

[11] A.a.O., S. 32, Par. 33.

[12] Vgl. unten S. 256.

[13] A.a.O. (Anm. 9), S. 32, Par. 34.

man hier an den wichtigsten Kern der Menschenrechte, nicht dagegen an die in den internationalen Menschenrechtskonventionen auch geschützten liberalen Freiheitsrechte gedacht hat. Dafür spricht auch die ausdrückliche Erwähnung des Schutzes gegen Sklaverei und Rassendiskriminierung. Freilich dürfte es nicht leicht sein, die Grenze hier aufzufinden [14].

In dem Urteil vom 24. Mai 1980 in der *Teheraner Geiselaffäre* hat der IGH die Freiheitsentziehung als mit den Prinzipien der UN-Satzung und den *fundamental principles* in der Allgemeinen Erklärung der Menschenrechte unvereinbar bezeichnet [15]. Er hat dann den Verstoß gegen die Gesamtheit der Regeln des Gesandtschafts- und Konsularrechts, dessen *fundamental character* besonders betont werden müsse, unterstrichen [16]. Dieser Teil des Urteils wird mit zwei Sätzen abgeschlossen, die hier wörtlich wiedergegeben werden sollen [17]:

> »Therefore in recalling yet again the extreme importance of the principles of law which it is called upon to apply in the present case, the Court considers it to be its duty to draw the attention of the entire international community, of which Iran itself has been a member since time immemorial, to the irreparable harm that may be caused by events of the kind now before the Court. Such events cannot fail to undermine the edifice of law carefully constructed by mankind over a period of centuries, the maintenance of which is vital for the security and well-being of the complex international community of the present day, to which it is more essential than ever that the rules developed to ensure the ordered progress of relations between its members should be constantly and scrupulously respected«.

Hier stellt der Gerichtshof nicht nur fest, daß der Völkerrechtsverstoß in diesem Fall die ganze Völkerrechtsgemeinschaft angeht, sondern fordert die Staatengemeinschaft förmlich dazu auf, sich dieses Vorgangs anzunehmen und zu reagieren. Ohne daß der Internationale Gerichtshof es ausdrücklich sagte, erscheint es unzweifelhaft, daß er in den Verpflichtungen, die mit der Geiselnahme von diplomatischem und konsularischem Personal verletzt worden sind, solche sieht, die »towards the international community as a whole« im Sinne des *Barcelona Traction*-Urteils bestehen und damit »the concern of all States« sind [18].

Der Internationale Gerichtshof stand bisher noch nicht vor der Notwendigkeit, genauer zu prüfen, welche Maßnahmen nicht unmittelbar betroffene

---

[14] Dazu *F. A. Mann,* The Doctrine of Jus Cogens in International Law, in: Festschrift für U. Scheuner (1973), S. 399, 408.

[15] ICJ Reports 1980, S. 42, Par. 91.

[16] A.a.O., S. 42, Par. 91.

[17] A.a.O., S. 42 f., Par. 92.

[18] ICJ Reports 1970, S. 32, Par. 33.

Staaten in Fällen von Verstößen gegen Völkerrechtsnormen ergreifen können, die die Staatengemeinschaft insgesamt betreffen. Immerhin scheint der Gerichtshof bestimmte, auch rechtlich relevante Maßnahmen nicht nur für möglich, sondern im Fall der Geiseln in Teheran sogar für geboten zu halten. Im *Barcelona Traction*-Urteil findet sich eine interessante, wenn auch nicht leicht verständliche Andeutung über die Möglichkeiten von Staaten, auf Verletzungen von Menschenrechten zu reagieren. Im Zusammenhang mit der Möglichkeit des Staates, bei Verletzung seiner Rechte gegen einen anderen Staat vorzugehen, verweist der Gerichtshof zunächst auf seine früheren Ausführungen über die Menschenrechte und erklärt, daß zu ihnen auch der Schutz gegen *denial of justice* gehöre. Dann folgt der Satz: »However, on the universal level, the instruments which embody human rights do not confer on States the capacity to protect the victims of infringements of such rights irrespective of their nationality«. Das sei nur im regionalen Rechtskreis anders, was mit einem Verweis auf das Staatenbeschwerdeverfahren in Art. 24 der Europäischen Konvention für Menschenrechte begründet wird [19].

Diese Passage des Urteils ist sehr unterschiedlich interpretiert worden [20]. Es ist nicht deutlich, ob der Gerichtshof hier auch die UN-Menschenrechtspakte gemeint hat, die 1970 verabschiedet, aber nicht in Kraft waren [21]. Bei der sehr generellen Formulierung dürfte einiges für diese Annahme sprechen. Darauf ist noch zurückzukommen [22]. Von größerer Bedeutung ist aber die Frage, ob hiermit die Aussage zu der Wirkung der Verpflichtungen *erga omnes* eingeschränkt werden sollte. Dagegen dürfte sprechen, daß in dem genannten Par. 91 des Urteils einfach von *human rights* die Rede ist, zu denen auch der Schutz gegen *denial of justice* gerechnet wird, während in Par. 34 die *basic rights of the human person* als Verpflichtungen *erga omnes* begründend bezeichnet werden. Insofern könnte hier der engere Kreis der elementaren Menschenrechte als stärker geschützt angesehen werden, als die in die Menschenrechtspakte einbezogenen freiheitlich-rechtsstaatlichen Gewährleistungen, zu denen auch das Verbot der Rechtsverweigerung gehört. Freilich ist diese Auslegung wiederum zweifelhaft, weil eigenartigerweise Par. 34, der die Verpflichtungen *erga omnes* exemplifiziert, seinerseits den Satz enthält, die den Verpflichtungen entsprechenden Rechte des Schutzes, also der Durchsetzung, seien teilweise allgemeines Völkerrecht, teilweise seien sie »by international instruments of a universal or quasi-universal character« übertragen [23].

---

[19] A.a.O., S. 47, Par. 91.

[20] Vgl. etwa *B. Simma,* Fragen der zwischenstaatlichen Durchsetzung vertraglich vereinbarter Menschenrechte, in: Staatsrecht-Völkerrecht-Europarecht, Festschrift für H.-J. Schlochauer (1981), S. 635 ff.

[21] So *J. A. Frowein,* The Interrelationship between the Helsinki Final Act, the International Covenants on Human Rights, and the European Convention on Human Rights, in: T. Buergenthal (ed.), Human Rights, International Law and the Helsinki Accord (1977), S. 78 ff.

[22] Vgl. unten S. 256.

[23] A.a.O. (Anm. 9), S. 32, Par. 34.

Auch hier dürfte jedenfalls mit an die UN-Pakte gedacht sein[24]. Aber man wird diesen nicht klaren Verweis wohl kaum als Rücknahme der ausdrücklich getroffenen Feststellung werten dürfen, daß Verpflichtungen *erga omnes* »a legal interest in their protection« für alle Staaten begründen.

Es ist nicht möglich, in der Frage der Reaktionen auf Verletzungen von *erga omnes* geltenden Verpflichtungen eindeutigere Auskunft aus den Separatvoten zum Urteil in der Sache *Barcelona Traction* zu erhalten. Sir Gerald Fitzmaurice scheint es für einen Zirkelschluß zu halten, von der Frage, gegenüber wem eine bestimmte Verpflichtung besteht, das Recht, Ansprüche geltend zu machen, abhängen zu lassen[25]. Riphagen kritisiert in seiner abweichenden Meinung die vom Gerichtshof getroffene Unterscheidung generell, da das Völkerrecht die Rechte der Staaten schütze, indem es anderen Staaten Pflichten auferlege, weil es ein übergeordnetes Interesse der Staatengemeinschaft sei, daß diese Rechte beachtet würden[26]. Er legt außerdem dar, es sei eindeutig falsch, ein *ius standi* jedes Staates anzunehmen, wo die territoriale Integrität oder die politische Unabhängigkeit eines Staates anders als durch bewaffneten Angriff verletzt werde oder wo der Staatsangehörige eines anderen Staates Opfer einer Verletzung seiner individuellen Freiheitsrechte werde[27]. Offenbar will Riphagen hier nur bei einem bewaffneten Angriff eine Reaktion von Drittstaaten zulassen. In diesem Fall ist sie freilich durch das Recht zur kollektiven Selbstverteidigung in der UN-Satzung eindeutig gerechtfertigt.

3. Die Entwicklung in Völkerrechtslehre und International Law Commission

Die Vorstellung, bei bestimmten Völkerrechtsverletzungen könne auch ein nicht unmittelbar geschädigter Staat sich als verletzt ansehen und Repressalien ergreifen, hat eine durchaus beachtliche Tradition in der Rechtslehre. Als Voraussetzung für meist als »Intervention« bezeichnete Maßnahmen von Staaten, die selbst nicht unmittelbar betroffen sind, wurde häufig eine offene oder durch die übertretene Norm besonders qualifizierte Völkerrechtsverletzung angesehen[28]. Besonders charakteristisch ist die Position von *Heffter,* der als Voraussetzung für ein derartiges Eingreifen nennt, daß der andauernden Völkerrechtsverletzung ein Ende gesetzt oder Verstöße gegen Menschlichkeit und Gerechtigkeit unterdrückt würden. Er fügt hinzu: »En ce cas les Etats ne font que remplir une mission commune qui leur est tracée naturellement. Or-

[24] Dazu *Frowein* (Anm. 21).
[25] A.a.O. (Anm. 9), S. 66, Par. 4.
[26] A.a.O., S. 339, Par. 8.
[27] A.a.O., S. 340, Par. 8.
[28] Eingehend *M. Akehurst,* Reprisals by Third States, British Year Book of International Law 44 (1970), S. 1 ff.

ganes suprêmes et multiples de l'humanité, ils sont appelés à en faire respecter les lois partout où elles sont violées«[29]. Die Problematik der humanitären Intervention vermengt sich hier mit der Frage nach der Berechtigung von Gegenmaßnahmen dritter Staaten gegen Völkerrechtsverletzungen.

Es ist freilich auch leicht, eine Vielzahl von Gegenstimmen bis in die neueste Zeit aufzufinden. *G. F. von Martens* verneinte die Möglichkeit von Repressalien zugunsten dritter Staaten unter Berufung auf *Vattel* und *Grotius* generell[30]. Das wird in der deutschen Literatur bis in die Zeit nach dem zweiten Weltkrieg wohl zur herrschenden Auffassung, wobei häufig die Möglichkeit von Verpflichtungen gegenüber der Völkerrechtsgemeinschaft allgemein geleugnet wird[31]. Erst in neuester Zeit scheint sich hier eine Veränderung der Auffassung anzudeuten[32]. *Bindschedler* hat sich kürzlich dafür ausgesprochen, daß ein Staat intervenieren könne, um Rechte aus einer allgemein geltenden Norm des Gewohnheitsrechts durchzusetzen, da ein »Rechtsanspruch auf Einhaltung des allgemeinen Rechtssatzes durch alle Angehörigen der fraglichen Rechtsgemeinschaft« bestehe[33]. In der sowjetischen Völkerrechtslehre hat *Tunkin* seit 1963 vor allem für das Gewaltverbot, aber auch für andere Normen die Möglichkeit bejaht, daß andere als direkt betroffene Staaten das Recht zu Gegenmaßnahmen haben[34]. Besonders eingehend hat sich *Wengler* 1964 der Problematik angenommen. Ohne sich mit Literatur oder Praxis näher auseinanderzusetzen, sieht er es als rechtspolitisch unentbehrlich an, daß bei »gewissen Normen des allgemeinen Völkerrechts« eine universale Aktivlegitimation für die völkerrechtliche Durchsetzung besteht. Er nennt als Beispiel die Verletzung der Menschenrechte und fügt hinzu, die Entwick-

---

[29] *A.-G. Heffter,* Le droit international public de l'Europe (1866), S. 211; vgl. auch *H. Mosler,* Die Intervention im Völkerrecht (1937), besonders S. 63 ff.

[30] *G. F. v. Martens,* Einleitung in das positive Europäische Völkerrecht (1796), S. 295, Par. 256. Interessant ist die von *Martens,* a.a.O., anerkannte Ausnahme für die Schweizer Kantone, die auch für die Staaten der Europäischen Gemeinschaft bei einer stärker zum Staatenbund tendierenden Entwicklung bedeutsam werden könnte.

[31] So *A. Schüle,* Völkerrechtliches Delikt, in: Strupp-Schlochauer, Wörterbuch des Völkerrechts, Bd. 1 (1960), S. 330; *I. von Münch,* Das völkerrechtliche Delikt (1963), S. 137.

[32] Zweifelnd *K. Ipsen,* in: Menzel/Ipsen, Völkerrecht (2. Aufl. 1979), S. 348 f.; siehe auch *M. Bothe,* International Obligations, Means to Secure Performance, in: Bernhardt (ed.), Encyclopedia of Public International Law, Instalment 1 (1981), S. 104 ff.

[33] *R. L. Bindschedler,* Der Schutz der Menschenrechte und das Verbot der Einmischung, in: Staatsrecht-Völkerrecht-Europarecht, Festschrift für H.-J. Schlochauer (1981), S. 179, 182. Ebenso unter »Intervention« *Oppenheim/Lauterpacht,* International Law, Bd. 1 (8. Aufl. 1955), S. 308. Vgl. auch *B. Bollecker-Stern,* Le préjudice dans la théorie de la responsabilité internationale (1973), S. 50 ff., 57 ff..

[34] *G. I. Tunkin,* Das Völkerrecht der Gegenwart (1963), S. 231 ff., vor allem S. 234 f.; zur Diskussion der Problematik des *ius cogens* in der sowjetischen Doktrin *T. Schweisfurth,* Sozialistisches Völkerrecht? (Beiträge zum ausländischen öffentlichen Recht und Völkerrecht, Bd. 73) (1979), S. 257 und *passim.*

lung gehe ferner dort zur Aktivlegitimation aller Staaten zur Anwendung von Repressalien bei Verletzungen des allgemeinen Völkerrechts, wo ein universales Interesse an der Durchsetzung der Befolgung einer abstrakten Norm vermutet werde[35].

Von großer Bedeutung für die Frage, die hier zur Diskussion steht, sind die Arbeiten in der International Law Commission zur Staatenverantwortlichkeit. Dabei hat vor allem Ago als Special Rapporteur die Problematik erörtert. In seinem Bericht von 1976 findet sich eine genaue Analyse der Entwicklung in dem Abschnitt über den Inhalt der verletzten völkerrechtlichen Verpflichtung[36]. Er geht von der Frage aus, ob alle Verletzungen des Völkerrechts dasselbe »Régime« der Staatenverantwortlichkeit auslösen oder ob hier Unterscheidungen notwendig sind. Die Frage sei bis zum Ende des zweiten Weltkrieges kaum als relevant angesehen worden, habe danach aber zunehmend Bedeutung gewonnen[37]. Nach der Anerkennung des *ius cogens* sieht er es als Widerspruch an, wenn der Verantwortungszusammenhang nur zwischen dem Staat, der die Norm verletzt, und dem direkt verletzten Staat bestehe[38]. Nach einer Analyse der Äußerungen von Staatenvertretern und der Literatur kommt er zu dem Vorschlag, zwischen dem völkerrechtlichen Delikt und dem völkerrechtlichen Verbrechen zu unterscheiden[39]. Das Verbrechen wird in Anlehnung an die Diskussionen über das *ius cogens* umschrieben. Die International Law Commission hat die Unterscheidung von Ago aufgenommen und in Art. 19 ihres Entwurfs zwischen völkerrechtlichen Delikten und Verbrechen unterschieden[40]. Die Diskussion hatte eine weitgehende Übereinstimmung mit der Haltung von Ago gezeigt[41]. Nur vereinzelt wurde Kritik an der Einführung des Begriffes *crime* geübt[42]. Sehr kritisch äußerte sich allerdings der Amerikaner Kearney, der es als nicht annehmbar bezeichnete, nicht verletzten dritten Staaten ein Eingriffsrecht zu geben. Das sei ein Verstoß gegen die UN-Satzung und komme der Erlaubnis für große Staaten gleich, kleine einzuschüchtern und in ihre Angelegenheiten zu intervenieren[43].

Der von der International Law Commission vorläufig angenommene Art. 19 definiert die völkerrechtlichen Verbrechen sehr viel vorsichtiger, als es der Berichterstatter vorgesehen hatte[44]. Wichtig erscheint die Beschrän-

---

[35] *W. Wengler*, Völkerrecht, Bd. 1 (1964), S. 579 ff., insbes. 580 f.
[36] Yearbook of the ILC 1976 II, S. 24 ff.
[37] A.a.O., S. 29 ff.
[38] A.a.O., S. 32.
[39] A.a.O., S. 54.
[40] A.a.O., S. 95.
[41] A.a.O., S. 67 ff.
[42] Vor allem von Quentin-Baxter, a.a.O., S. 78 ff.
[43] A.a.O., S. 76 f.
[44] Der Text lautet: (Article 19) »International crimes and international delicts. 1. An act of a State which constitutes a breach of an international obligation is an in-

kung auf den *serious breach*. Freilich ist klar, daß es sich um eine rein beispielhafte Aufzählung handelt. Auch bei anderen Delikten kann auf Grund der Formulierung die Auffassung vertreten werden, daß sie in diese Kategorie gehören. Die Gefahren für die Rechtssicherheit, auf die Kearney als Mitglied der International Law Commission hingewiesen hatte, sind damit nicht gebannt[45]. Entscheidend wird es sein, welche Konsequenzen sich für die Möglichkeiten der Reaktionen dritter Staaten hieraus ergeben. Darüber besteht bisher offenbar weder in der International Law Commission noch unter den Staaten, die zu Art. 19 Stellung genommen haben, Einigkeit. Deutlich erscheint nur, daß die Frage, inwieweit Gegenmaßnahmen dritter Staaten bei der Verletzung grundlegender Normen des Völkerrechts gerechtfertigt sind und welche Grenzen sie zu beachten haben, zu einem zentralen Problem der weiteren Erörterung über die Staatenverantwortlichkeit werden wird.

Der jetzt zuständige Berichterstatter Riphagen hat am 1. April 1980 einen vorläufigen Bericht vorgelegt, in dem er drei Parameter der Rechtsbeziehungen aus einer Völkerrechtsverletzung unterscheidet. Das sind 1. die neuen Verpflichtungen des die Norm verletzenden Staates, 2. die neuen Rechte des verletzten Staates und 3. die Position dritter Staaten in Beziehung zu der Völkerrechtsverletzung[46]. Für den dritten Parameter geht er von der Grundregel aus, daß die rechtliche Beziehung nur zwischen Verletzer und Verletz-

---

ternationally wrongful act, regardless of the subject-matter of the obligation breached. 2. An internationally wrongful act which results from the breach by a State of an international obligation so essential for the protection of fundamental interests of the international community that its breach is recognized as a crime by that community as a whole, constitutes an international crime. 3. Subject to paragraph 2, and on the basis of the rules of international law in force, an international crime may result, inter alia, from: (a) a serious breach of an international obligation of essential importance for the maintenance of international peace and security, such as that prohibiting aggression; (b) a serious breach of an international obligation of essential importance for safeguarding the right of self-determination of peoples, such as that prohibiting the establishment or maintenance by force of colonial domination; (c) a serious breach on a widespread scale of an international obligation of essential importance for safeguarding the human being, such as those prohibiting slavery, genocide and apartheid; (d) a serious breach of an international obligation of essential importance for the safeguarding and preservation of the human environment, such as those prohibiting massive pollution of the atmosphere or of the seas. 4. Any internationally wrongful act which is not an international crime in accordance with paragraph 2 constitutes an international delict.«

Vgl. dazu *C. Dominicé*, Die internationalen Verbrechen und deren rechtliches Regime, in: Völkerrecht und Rechtsphilosophie, Festschrift für S. Verosta (1980), S. 227 ff.

[45] So sehr deutlich der schwedische Delegierte Danelius im 6. Komitee der Generalversammlung A/C.6/35/SR. 49, S. 2: »In reality, however, judgements on the question whether on obligation was essential for the protection of fundamental interests of the international community must necessarily be subjective and political. The distinction made in draft article 19 between different categories of obligations did not seem useful and could create considerable difficulties in practice«.

[46] A/CN.4/330, S. 13.

tem besteht. Er erkennt aber drei Ausnahmen an, wovon die einzige echte Verletzungen einer Verpflichtung betrifft, die nicht nur das Interesse eines Staates angeht[47]. Riphagen scheint freilich jedenfalls normalerweise eine kollektive Entscheidung, wohl der zuständigen Organe der Vereinten Nationen, als Voraussetzung für rechtliche Maßnahmen von Drittstaaten für notwendig zu halten, obwohl er einräumt, daß die Frage zweifelhaft ist[48]. In der Diskussion dieses Berichtes ist dem insbesondere von dem Amerikaner Schwebel widersprochen worden, der kurz zuvor das Teheraner Geiseldrama erwähnt und gewiß seine Position auch in diesem Zusammenhang formuliert hat[49].

Riphagen selbst hat das Problem, diesmal als Vertreter der Niederlande im 6. Ausschuß, dahin formuliert, daß noch vor kurzer Zeit Rechtsbeziehungen nur zwischen dem verletzten Staat und dem Verletzer angenommen worden seien, aber die Welt auch heute noch weit von einer Rechtsordnung sei, nach der jede Völkerrechtsverletzung ein Bruch der internationalen Rechtsordnung als solcher sei und Reaktionen der internationalen Gemeinschaft ermögliche. Das heutige Regime der Staatenverantwortlichkeit befinde sich irgendwo zwischen diesen Extremen und sei daher naturgemäß differenziert[50]. Der Vertreter der Bundesrepublik Deutschland erkannte die Möglichkeit an, daß ein Drittstaat eine »nicht-neutrale Position« gegenüber bestimmten Völkerrechtsverletzungen einnehme, und hielt es für möglich, daß darin Art. 19 des Entwurfs der International Law Commission seine Rechtfertigung finde. Er umschrieb die fraglichen Normen als solche, auf deren Einhaltung die Staatengemeinschaft ein *vested interest* habe[51].

Dieser Befund zeigt, daß nicht nur in der völkerrechtlichen Literatur, sondern vor allem im Rahmen der Kodifikation und Fortbildung des Völkerrechts die vom Internationalen Gerichtshof im *Barcelona Traction*-Urteil getroffene Unterscheidung zwischen völkerrechtlichen Verpflichtungen, die *erga omnes* wirken und deren Durchsetzung allen Staaten möglich ist, und einfachen Pflichten, die jeweils im bilateralen Verhältnis durchzusetzen sind, als grundsätzlich akzeptiert erscheint. Das heißt freilich nicht, daß über die Konsequenzen dieser Aussage wirklich Klarheit bestünde.

4. Beispiele aus der Staatenpraxis

Die neuere Staatenpraxis kennt einige eindeutige Beispiele, in denen Drittstaaten massiven Völkerrechtsverletzungen rechtlich mit Sanktionen begegnet sind. Dabei erscheinen die Fälle der Anwendung von Art. 51 der UN-Satzung, des Rechts der kollektiven Selbstverteidigung, nicht relevant, da hier

[47] A.a.O., S. 25.
[48] A.a.O., S. 29 f., vor allem Par. 71 und 72.
[49] Yearbook ILC 1980 I, S. 93, die Erwähnung der Geiselnahme S. 91 f.
[50] Sitzung 10. 11. 1981, A/C.6/36/SR. 45, S. 5.
[51] Sitzung 11. 11. 1981, A/C.6/35/SR. 45, S. 5.

eine Rechtfertigung für bewaffnete Maßnahmen zur Unterstützung des Angegriffenen in der UN-Satzung gegeben ist[52]. Ebensowenig ist die Anwendung von Sanktionsmaßnahmen auf Grund von Beschlüssen des Sicherheitsrates im Rahmen von Kapitel VII einzubeziehen. Hier enthält die freilich unvollkommene Verfassung der Staatengemeinschaft eindeutige Regeln[53].

Der wohl bisher klarste Fall, in dem dritte Staaten Sanktionen wegen einer Verletzung des Völkerrechts ergriffen haben, die sie nicht unmittelbar betraf, ist die Geiselnahme in der amerikanischen Botschaft in Teheran. Am 22. April 1980 beschlossen die Außenminister der Europäischen Gemeinschaft im Rahmen der Europäischen Politischen Zusammenarbeit, Maßnahmen zu ergreifen, um gegen den Iran Sanktionen zu verhängen[54]. Am 18. Mai wurde ein förmlicher Sanktionsbeschluß gefaßt, der vor allem die Stornierung nach dem 4. November 1979, dem Zeitpunkt der Geiselnahme, abgeschlossener Verträge betraf[55]. Die Maßnahmen sind von den Mitgliedstaaten, allerdings mit gewissen Unterschieden, durchgeführt worden[56]. Besonders bedeutsam ist die in den Beschlüssen enthaltene Begründung. Es wird dabei auf die Gefährdung des Friedens und der Sicherheit in der Welt hingewiesen und auf die im Sicherheitsrat wegen des Vetos der Sowjetunion nicht angenommene Resolution Bezug genommen, die Sanktionen verhängen sollte[57]. Außerdem wird ausdrücklich betont, daß diese Situation für die gesamte Völkergemeinschaft Anlaß zur Besorgnis sei[58]. Zu diesem Zeitpunkt lag bereits die Entscheidung des Internationalen Gerichtshofs über eine einstweilige Anordnung vor, in der die Bedeutung der Völkerrechtsnormen über den Schutz von Diplomaten und Konsuln für die Staatengemeinschaft vom Gerichtshof klar herausgestellt worden war[59]. Er hatte dem Iran die sofortige Freilassung aufgegeben. Allerdings wird auf die Entscheidung des Internationalen Gerichtshofs an keiner Stelle der Sanktionsbeschlüsse Bezug genommen. Nicht sicher erscheint, ob in Ausführung der Beschlüsse völkerrechtliche Verpflichtungen gegenüber dem Iran tatsächlich nicht beachtet worden sind, aber die Mitgliedstaaten der Europäischen Gemeinschaft haben das Recht dazu in Anspruch genommen.

---

[52] Dazu *J. Delbrück,* Collective Self-defense, in: Bernhardt (ed.), Encyclopedia of Public International Law, Instalment 3 (1982), S. 114 ff.

[53] Siehe *J. A. Frowein,* Friedenssicherung durch die Vereinten Nationen, in: Die Vereinten Nationen und die Mitarbeit der Bundesrepublik Deutschland, hrsg. von U. Scheuner und B. Lindemann (1973), S. 45–67; *M. Schaefer,* Die Funktionsfähigkeit des Sicherheitsmechanismus der Vereinten Nationen (Beiträge zum ausländischen öffentlichen Recht und Völkerrecht, Bd. 77) (1981).

[54] Bulletin der EG 1980 Nr. 4, S. 26 f.

[55] A.a.O., S. 28.

[56] A.a.O., 1980 Nr. 5, S. 28 ff.; vgl. auch *Lindemeyer,* Recht der Internationalen Wirtschaft 1981, S. 10; *Enderlein,* ebenda, 1980, S. 453.

[57] Bulletin der EG 1980 Nr. 4, S. 26, 28.

[58] A.a.O., S. 27.

[59] ICJ Reports 1979, S. 16 f.

Ein weiteres Beispiel, das hier zu nennen ist, betrifft die Reaktion von Drittstaaten auf Flugzeugentführungen, die am 17. Juli 1978 und erneut am 20. Juli 1981 zwischen den zehn Staats- und Regierungschefs der wichtigsten Wirtschafts- und Handelsstaaten vereinbart wurde. Nach dem Wortlaut der Erklärung vom 17. Juli 1978 werden die Staaten ihre gemeinsamen Anstrengungen verstärken, um den internationalen Terrorismus zu bekämpfen. Wenn ein Staat die Auslieferung oder die strafrechtliche Verfolgung von Flugzeugentführern verweigert oder das Flugzeug nicht zurückgibt, so werden danach Maßnahmen getroffen werden, um alle Flüge in dieses Land und aus diesem Land oder Flüge seiner Fluggesellschaft zu unterbinden [60]. Es ist deutlich, daß hier an eine sofortige Suspendierung der Rechte aus bestehenden Luftverkehrsabkommen gedacht ist. Ohne nähere Begründung wird offenbar davon ausgegangen, daß die Nichtverfolgung von Flugzeugentführungen oder gar die indirekte Unterstützung ein Verstoß gegen die *erga omnes* bestehende Pflicht von Staaten ist, die Sicherheit des internationalen Luftverkehrs nicht nur nicht selbst zu gefährden, sondern auch wirksam gegen Zuwiderhandlungen einzuschreiten. Dafür läßt sich auf die vielfältigen Resolutionen und Verträge verweisen, die sich gegen Geiselnahme, Terrorismus und Flugzeugentführung aussprechen und eine Zusammenarbeit der Staaten zu ihrer Bekämpfung für notwendig erklären [61]. Man wird in diesem Zusammenhang auch zu berücksichtigen haben, daß Art. 14 der Konvention über die Hohe See die Staaten zu einer engen Zusammenarbeit zur Unterbindung der Piraterie verpflichtet. Dazu gehört nach Art. 15 auch die Flugzeugentführung über der Hohen See. Es spricht vieles dafür, hierin eine von der Staatengemeinschaft insgesamt anerkannte und gewohnheitsrechtlich verfestigte Verpflichtung mit Wirkung *erga omnes* zum Schutz des See- und Luftverkehrs zu sehen. Ein Unterschied zwischen Flugzeugentführungen über Land oder See kann nicht sinnvoll gemacht werden. Man wird hier auch nicht ein Problem im Rahmen des Gewaltverbotes der UN-Charta zu sehen haben, da die Unterbindung des Luftverkehrs keine Gewaltanwendung im Sinne von Art. 2 Ziff. 4 der Satzung ist [62]. Im Grunde setzt sich die alte Vorstellung vom Piraten als *hostis humani generis* gegenüber den Terroristen durch und führt nicht nur zu einer Begründung der staatlichen Jurisdiktion zur Aburtei-

---

[60] Bulletin der Bundesregierung 19. 7. 1978, Nr. 80, S. 766, und 24. 7. 1981, Nr. 71, S. 617.

[61] Vgl. dazu *T. Oppermann,* Der Beitrag des Internationalen Rechts zur Bekämpfung des Internationalen Terrorismus, in: Staatsrecht-Völkerrecht-Europarecht, Festschrift für H.-J. Schlochauer (1981), S. 495 ff.; siehe auch *T. Stein,* Die Bekämpfung des Terrorismus im Rahmen der Mitgliedstaaten der Europäischen Gemeinschaften, ZaöRV 40 (1980), S. 319 ff., und *K. W. Platz,* Internationale Konvention gegen Geiselnahmen, ZaöRV 40 (1980), S. 276 ff.

[62] Anders wohl *Oppermann,* a.a.O., S. 512, der sich aber dennoch für die Zulässigkeit ausspricht.

lung, sondern auch dazu, daß gegen Staaten vorgegangen werden kann, die mit diesem *hostis* gemeinsame Sache machen [63].

Ein weiteres Beispiel, das hier einbezogen werden muß, ergibt sich aus den Diskussionen um Sanktionsmaßnahmen in Zusammenhang mit der Einführung des Kriegsrechts in Polen am 13. Dezember 1981. Insbesondere amerikanische Erklärungen zeigen, daß die Anwendung von Repressalien gegen die Sowjetunion wegen des Vorgehens in Polen rechtlich als zulässig angesehen wurde [64]. Dabei wird unterstellt, daß die Ereignisse in Polen nicht ohne sowjetischen Druck entstanden sind. Der drohende Truppenaufmarsch an der polnischen Grenze lange vor dem 13. Dezember 1981 mußte hier in der Tat als eindeutiger Verstoß gegen die grundlegende Norm des Art. 2 Ziff. 4 der UN-Satzung angesehen werden, der gerade auch die Drohung mit Gewalt, die gegen die politische Unabhängigkeit eines Staates gerichtet ist, erfaßt [65]. Daß es sich bei Art. 2 Ziff. 4 um eine der wichtigsten *erga omnes* wirkenden Normen des Rechts der Vereinten Nationen, aber auch des allgemeinen Völkerrechts handelt, kann nicht zweifelhaft sein. Allerdings ist die Verordnung des Rates der Europäischen Gemeinschaft vom 15. März 1982 zur Änderung der Einfuhrregelung für bestimmte Waren mit Ursprung in der UdSSR so gefaßt, daß er als bloße Retorsion nicht in entstandene Rechte eingreift [66].

Es ist nicht ohne Interesse festzustellen, daß die Beispiele aus der Staatenpraxis, die die Verhängung von Repressalien durch Drittstaaten betreffen, alle sehr jungen Datums sind. Obwohl das Problem, wie sich aus der Entwicklung der Meinungen in der Literatur zeigte, schon früher durchaus bekannt war, hat erst in neuester Zeit die Erkenntnis, daß das Sanktionssystem der Vereinten Nationen weithin funktionsunfähig ist, dazu geführt, daß Drittstaaten bei Verstößen gegen grundlegende Normen der Völkerrechtsordnung wie der Gefährdung des Friedens auch rechtliche Reaktionen prüfen und teilweise zur Anwendung bringen.

5. Die Bedeutung der UN-Satzung und der Regeln über Verletzungen multilateraler Verträge

Da die hier untersuchte Frage der Verpflichtungen *erga omnes* und ihrer Durchsetzung in Zusammenhang mit der Übernahme von Ersatzfunktionen

---

[63] Ähnlich *Oppermann,* a.a.O., S. 507.

[64] Dazu vor allem die Erklärung des Präsidenten vom 29. 12. 1981, nach der alle Flüge von Aeroflot in die USA suspendiert wurden, Europa-Archiv 1982, S. D 163.

[65] In der amerikanischen Erklärung wird der angewendete Druck ausdrücklich erwähnt, a.a.O., S. D 163.

[66] ABl. EG L 72 16. 3. 1982, S. 15. Vgl. jetzt auch Verordnung vom 16. 4. 1982, ABL. EG L 102 vom 16. 4. 1982, S. 1, 3 mit Sanktionen gegen Argentinien wegen der »Invasion der Falkland-Inseln«. Wiederum ist unklar, ob nur Retorsionen oder auch Repressalien ins Auge gefaßt sind.

für den weitgehenden Ausfall des UN-Systems durch die dezentralisierte Staatengemeinschaft gesehen werden muß, erscheint es angemessen, dem Verhältnis der Problematik zur UN-Satzung noch einige Aufmerksamkeit zu widmen. Die Satzung der Vereinten Nationen verkörpert das gemeinsame Interesse der Staaten an der Erhaltung des Friedens, aber auch an der Gewährleistung des Selbstbestimmungsrechts, des Schutzes der Menschenrechte und der internationalen Zusammenarbeit. Sie ist damit der wichtigste Ausdruck des heutigen internationalen *ordre public*[67]. Es ist charakteristisch für das UN-System, daß nicht nur die Prinzipien formuliert, sondern auch ihre Wahrung bestimmten Organen der Staatengemeinschaft übertragen worden ist. Das gilt besonders für die Friedenswahrung nach den Kapiteln VI und VII der Satzung.

Es wäre durchaus denkbar, daß die UN-Satzung die Wahrung des Staatengemeinschaftsinteresses abschließend regelte und damit jedenfalls einzelnen Staaten die Möglichkeit der Durchsetzung von Verpflichtungen *erga omnes* unmöglich machte. Art. 51 zeigt allerdings mit der Anerkennung des naturgegebenen Rechts der individuellen, aber vor allem auch der kollektiven Selbstverteidigung, daß für den wichtigsten Fall der Verletzung des Staatengemeinschaftsinteresses eine Reaktion von Drittstaaten, nämlich die militärische Hilfeleistung für den Angegriffenen, ausdrücklich in das System aufgenommen worden ist. Wenn die militärische Gewaltanwendung gegen den Friedensbrecher als Reaktion eines Drittstaates gestattet ist, so ist *a maiore ad minus* zu schließen, daß jede andere Reaktion, etwa die Vorenthaltung völkerrechtlich gebotener Leistungen, ebenfalls gerechtfertigt ist. Daß die Staatenpraxis diese Haltung bestätigt, hat sich gezeigt[68]. Keinerlei Regelung enthält die Satzung zur Verletzung anderer Grundprinzipien, etwa des Schutzes der Menschenrechte oder des Rechts der Selbstbestimmung. Daraus wird man aber nicht schließen dürfen, die Charta habe hier eine Aussage gegen jede Berücksichtigung des formulierten Interesses der Staatengemeinschaft getroffen. Eher dürfte das Fehlen von Vorschriften in der Satzung als Verweis auf die Lage nach allgemeinem Völkerrecht zu verstehen sein.

Beachtung verdient im vorliegenden Zusammenhang die Regelung der Durchsetzung des gemeinsamen Interesses bei multilateralen Verträgen. Durch ihren Abschluß haben die betroffenen Staaten sich zu einer Vertragsgemeinschaft zusammengeschlossen, die insgesamt an der Einhaltung des Vertrages interessiert sein muß. Art. 60 WVRK sieht bei multilateralen Verträgen Suspendierung und Beendigung des Vertrages als Reaktion auf eine erhebliche Verletzung durch eine Partei nur durch einstimmige Entscheidung aller anderen Parteien vor[69]. Daraus folgt, daß eine oder einzelne Parteien hier von einer derartigen Reaktion ausgeschlossen sind. Für den Fall, daß eine Partei durch die Verletzung besonders betroffen ist (*specially affected*), aber

[67] Zum *ordre public* im Völkerrecht siehe *Mosler* (Anm. 2), S. 17 ff.
[68] Vgl. oben S. 250 ff.
[69] Art. 60 Abs. 2, a WVRK.

nur dann, kann sie den Vertrag zwischen sich und dem Verletzer suspendieren. Im übrigen besteht diese Möglichkeit nur, wenn die Verletzung nach der Natur des Vertrages die Position jeder Partei für die weitere Erfüllung radikal verändert[70]. Riphagen hat auf diese Regelung mit der Notwendigkeit kollektiver Entscheidungen hingewiesen, wenn nicht unmittelbar betroffene Drittstaaten gegen eine auch ihnen gegenüber obliegende Verpflichtung vorgehen wollen[71]. Er vergleicht die Verträge, deren Nichterfüllung durch eine Partei die rechtliche Lage jeder anderen Partei im Sinne von Art. 60 Abs. 2, c verändert, mit den Verpflichtungen *erga omnes*[72].

Die Frage der Reaktionsmöglichkeit nicht unmittelbar betroffener Drittstaaten gegenüber Verletzungen eines multilateralen Vertrages hat zu einer Kontroverse im Hinblick auf die Verträge zum Schutz der Menschenrechte geführt[73]. Dabei geht es vor allem darum, ob die Einrichtung bestimmter, bei den UN-Pakten sehr unvollkommener Überwachungsverfahren die Durchsetzung der Vertragspflichten zwischen den Mitgliedstaaten ausschließt oder jedenfalls dem etwa mit einem förmlichen Protest wegen der Nichteinhaltung des Vertrages konfrontierten Staat die Möglichkeit gibt, auf die Priorität des Überwachungsverfahrens zu verweisen. Der Verfasser hat sich im letzteren Sinn geäußert[74], während etwa *Simma* dezidiert die Meinung vertritt, zur Sicherung der Erfüllung der Verpflichtungen aus Menschenrechtsverträgen stünden neben den dort vereinbarten Verfahren auch alle anderen Selbsthilfemittel zur Verfügung, die nach geltendem Völkerrecht zur Durchsetzung von Rechtsansprüchen zulässig sind[75]. Es geht hier nicht darum, diese Frage eingehend weiter zu verfolgen. Sie erscheint aber bedeutsam als Beispiel für die Grundproblematik der Durchsetzung von Verpflichtungen *erga omnes*.

Die in den UN-Pakten über bürgerliche und politische sowie über wirtschaftliche, soziale und kulturelle Rechte gewährleisteten Individualrechte sind in hohem Maße von der ideologischen Grundauseinandersetzung betroffen, die das heutige Völkerrecht durchzieht. Das hat sich etwa in den Berichten der sich als sozialistisch verstehenden Staaten an das Menschenrechtskomitee des Paktes über bürgerliche und politische Rechte sehr klar gezeigt[76].

---

[70] Art. 60 Abs. 2, b, c WVRK.

[71] Riphagen, A/CN. 4/330, S. 26 f.

[72] A.a.O., S. 27.

[73] Dazu jetzt eingehend *B. Simma,* Fragen der zwischenstaatlichen Durchsetzung vertraglich vereinbarter Menschenrechte, in: Staatsrecht-Völkerrecht-Europarecht, Festschrift für H. J. Schlochauer (1981), S. 635 ff.

[74] *Frowein* (Anm. 21), S. 78 ff.

[75] A.a.O. (Anm. 73), S. 647 f. und *passim.*

[76] Vgl. dazu *M. Nowak*, The Effectiveness of the International Covenant on Civil and Political Rights – Stocktaking after the first eleven sessions of the UN-Human Rights Committee, Human Rights Law Journal 1 (1980), S. 136 ff.; *derselbe,* Die Durchsetzung des Internationalen Paktes über bürgerliche und politische Rechte, EuGRZ 7 (1980), S. 532 ff.

Es war bei Abschluß dieser Verträge bekannt. Man wird sich, ehe man zu dem Schluß kommt, daß etwa Repressalien wegen des angeblichen Verstoßes gegen die Verträge möglich sind, genau die Konsequenzen einer solchen Auffassung klar machen müssen. Der Abschluß der Menschenrechtspakte hätte zu einem von Anfang an eindeutigen Repressaliengrund geführt, da etwa die in Art. 19 des Paktes über bürgerliche und politische Rechte gewährleistete Meinungs- und Pressefreiheit in vielen sich sozialistisch verstehenden Staaten nicht in einer Weise besteht, die auch nur im Kern dem westlichen Verständnis nahekommt[77]. Das gilt auch für viele andere Staaten, die den Pakt ratifiziert haben. Ähnliches kann für das Recht auf Arbeit in Art. 6 des Paktes über wirtschaftliche und soziale Rechte im umgekehrten Sinne gesagt werden, da hier das Verständnis in Staaten mit freier Wirtschaftsordnung völlig von dem in Planwirtschaften abweicht. Die Zulässigkeit der Repressalie als Mittel der Austragung dieser Streitfragen würde eine erhebliche Rechtsunsicherheit in das gegenwärtige Völkerrecht hereintragen. Von daher dürfte viel für die vom Internationalen Gerichtshof in dem *Barcelona Traction*-Urteil wohl auch in Bezug auf die Menschenrechtspakte formulierte Ansicht sprechen, daß ein Schutzrecht nicht unabhängig von der Staatsangehörigkeit der Betroffenen übertragen werde[78]. Das dürfte auch der folgende indirekte Hinweis auf Art. 24 der Europäischen Menschenrechtskonvention zeigen, der jedem Mitgliedstaat *ipso iure* ein Recht zuweist, gegen einen anderen Mitgliedstaat eine Beschwerde wegen Verletzung der Konvention anzustrengen, während Art. 41 des UN-Paktes über bürgerliche und politische Rechte nur nach einer ausdrücklichen Unterwerfung des betroffenen Staates ein solches Recht gewährt[79].

Damit wird nicht bezweifelt, daß ein Vertrag von der Art der Pakte gegenseitige Verpflichtungen erzeugt. Es geht allein darum, wie diese Verpflichtungen durchgesetzt werden können. Wenn der Vertrag dafür Verfahren festlegt, dürfte ihnen Priorität zukommen. Der Europäische Gerichtshof für Menschenrechte hat für die Europäische Menschenrechtskonvention festgestellt, daß sie über gegenseitige Verpflichtungen hinaus objektive Verpflichtungen festlegt, die dem System gemeinsamer Durchsetzung unterliegen. Art. 24 ermögliche jedem Staat die Durchsetzung, ohne daß er ein besonderes Interesse zu zeigen habe, etwa die Verletzung der Rechte eines eigenen Staatsangehörigen[80]. Diese Formulierung scheint dafür zu sprechen, daß die gegenseitigen Rechte nicht so verstanden werden, daß sie eine Berechtigung zur Durchsetzung ohne spezifisches Interesse abgesehen von der in Art. 24 vorgesehenen Beschwerde gewähren würden[81].

---

[77] Vgl. *J. A. Frowein*, Das Problem des grenzüberschreitenden Informationsflusses und des »domaine réservé«, Berichte der Deutschen Gesellschaft für Völkerrecht, Heft 19 (1979), S. 20 ff.

[78] Vgl. oben S. 245.

[79] Auch dann sind nur *communications* möglich.

[80] Case of *Ireland* v. *The United Kingdom*, Series A, Bd. 25, 1978, S. 90 f.

Gewiß sind die hier erörterten Fragen von der Durchsetzung allgemeiner, *erga omnes* wirkender Verpflichtungen des Völkerrechts zu unterscheiden. Aber es ist doch bedeutsam, wie die Menschenrechtsschutzverträge, die weltweit und von der Zielsetzung her *erga omnes* Menschenrechte schützen, die Durchsetzung regeln. Die offenbar relativ vorsichtige Gewährung von Möglichkeiten der Durchsetzung im bilateralen Verhältnis kann kaum ganz unberücksichtigt bleiben, wenn jetzt auf Grund der bisherigen Befunde der Versuch gemacht wird, die Durchsetzung der Verpflichtungen *erga omnes* insgesamt etwas weiter aufzuhellen.

6. Die heutige Völkerrechtsordnung und die Durchsetzung von Verpflichtungen erga omnes

Jede Rechtsordnung wird von der gesellschaftlichen Entwicklung der von ihr verfaßten Rechtsgemeinschaft ihrerseits geprägt [82]. Das Völkerrecht wird von der Geschichte der Staatenbeziehungen und dem weltpolitischen Zustand in besonderem Maße beeinflußt. Das Völkerrecht seit 1945 kann von der ideologischen Teilung der Welt und der Existenz von zwei Supermächten nicht absehen [83]. *Hermann Mosler*, dem diese Zeilen gewidmet sind, betont die Notwendigkeit, daß die Rechtsgemeinschaft sich den Veränderungen der Staatengesellschaft anpaßt [84]. Auch die politischen Doktrinen, die auf die Staatengemeinschaft einwirken, können nicht unberücksichtigt bleiben, wie *Mosler* darlegt, und er erwähnt dann die Teilung Europas in Einflußzonen und auch die sogenannte Breschnew-Doktrin [85]. Es stellt sich die grundsätzliche Frage, inwieweit diese für das Völkerrecht sicherlich sehr relevanten Tatsachen auch in dem hier vorliegenden Zusammenhang berücksichtigt werden müssen. Man mag dem entgegnen, daß hierin eine unzulässige Vermischung von Recht und Politik liege und das Recht gegenüber der Politik abdanke. Indessen kann die Problematik mit so einfachen Kategorien nicht völlig erfaßt werden. Wenn das labile System des Völkerrechts auf der Existenz der heutigen Staatengemeinschaft aufbaut, dann bedarf jedenfalls ein Ergebnis genauer Begründung, das von der Struktur dieser Staatengemeinschaft absieht. Es ist nicht abwegig, die Frage zu stellen, welches die Konsequenzen für die Völ-

---

[81] Freilich brauchte sich der Gerichtshof mit dieser Frage nicht zu befassen. Art. 63 der Europäischen Menschenrechtskonvention schließt andere Streitschlichtungsverfahren zwischen den Parteien ausdrücklich aus.

[82] »The law can never be oblivious to the changes in life, circumstance and community standards in which it functions«, P. Jessup, *diss. op.*, South West Africa cases (*Ethiopia* v. *South Africa; Liberia* v. *South Africa*) Second Phase, ICJ Reports 1966, S. 439.

[83] Vgl. *Mosler* (Anm. 2) (rev. Ausgabe 1980), S. 6 ff., 13.

[84] A.a.O., S. 20.

[85] A.a.O., S. 21 f.

kerrechtsordnung sind, wenn Drittstaaten diese oder jene Möglichkeit zur Durchsetzung der *erga omnes* wirkenden Völkerrechtsnormen haben.

Die Verletzung des Gewaltverbotes ist ein eindeutiger Fall. Bei einem bewaffneten Angriff rechtfertigt Art. 51 der UN-Satzung die kollektive Selbstverteidigung und damit auch weniger scharfe Abwehrmaßnahmen dritter Staaten [86]. Aber auch die Drohung mit Gewalt kann mit Gegenmaßnahmen dritter Staaten beantwortet werden, wenn der Sanktionsmechanismus der UN-Satzung versagt. Freilich muß hier wieder auf die Gefahren geachtet werden, die durch eine nicht ausreichend objektiv abgesicherte und stark ideologisierte Anwendung von einschneidenden Repressalien gegen Gewaltandrohung hervorgerufen werden können. Bei der Weite des Spektrums möglicher Drohmaßnahmen und der häufig unterschiedlichen Würdigung ist die Gefahr von Fehlentscheidungen hier nicht von der Hand zu weisen. Eine Eskalation kann durch derartige Reaktionen auf als Drohung eingeordnete Handlungen leicht eintreten. Von besonderer Bedeutung ist daher der für die Anwendung von Repressalien generell geltende scharfe Maßstab der Verhältnismäßigkeit [87].

Nur selten wird die Lage bei Verletzungen von Normen des allgemeinen Völkerrechts außerhalb des Gewaltverbotes so eindeutig sein wie bei der Geiselnahme amerikanischer Diplomaten in Teheran [88]. Hier waren fundamentale Normen des Völkerrechts mißachtet worden, es gab keinerlei auch nur erwägenswerte Rechtfertigung, der Internationale Gerichtshof hatte gesprochen [89]. Hier waren Repressalien dritter Staaten zur Durchsetzung der *erga omnes* wirkenden Verpflichtungen zulässig.

Besondere Schwierigkeiten ergeben sich, wo der Verstoß gegen *erga omnes* wirkende Völkerrechtsnormen die Menschenrechte betrifft. Selbst wenn die Aussage des Internationalen Gerichtshofs im *Barcelona Traction*-Urteil bezüglich der *basic rights of the human person* nur auf die elementaren Menschenrechte bezogen wird, so ist beim Zustand der Staatengesellschaft die Gefahr einer erheblichen Rechtsunsicherheit nicht zu verkennen, wenn man bei ihrer Verletzung Repressalien für zulässig hält. Das Argument von *Akehurst,* daß bei Auseinandersetzungen häufig beide Seiten sich gegenseitig Verstöße gegen das Völkerrecht vorwerfen und Drittstaaten im Zweifel ihre Verbündeten unterstützen würden, erscheint hier besonders zutreffend [90]. Andererseits besteht die Gefahr, daß der Schutz der Menschenrechte ohne Durchsetzungsmittel wertlos bleibt. Daß Drittstaaten bei Verletzungen der Menschenrechte ihre Position formulieren, den Rechtsbrecher anprangern, die Rechtfertigung verlangen, ist im Sinne des Völkerrechts und der internationalen Ordnung zu

---

86 Vgl. oben S. 254.

87 Dazu *Verdross/Simma,* Universelles Völkerrecht (1976), S. 652 ff.

88 Vgl. oben S. 251.

89 Vgl. oben S. 251.

90 A.a.O. (Anm. 28), S. 15.

begrüßen. Repressalien werden von ihnen nur angewendet werden können, wenn die Verletzung ein hohes Maß an Eindeutigkeit erreicht. Auch wird hier der kollektiven Entscheidung von Staatengemeinschaftsorganen oder Staatengruppen gegenüber isolierten Maßnahmen im Prinzip der Vorrang gebühren müssen.

Wo sich der Völkerrechtsbruch klar gegen ein internationales Rechtsgut wie den diplomatischen Verkehr, den Seeverkehr, den Luftverkehr richtet, hat die geschilderte Praxis gezeigt, daß die Staatengemeinschaft sich als solche betroffen sieht. Sie kann hier auch reagieren. Dabei sind ebenfalls kollektive Maßnahmen eher zu rechtfertigen, die freilich durchaus von einer Staatengruppe vorgenommen werden können. Insofern erscheinen verhältnismäßige Repressalien, wie sie in der genannten Erklärung der Staats- und Regierungschefs zu Flugzeugentführungen angekündigt worden sind, vom Völkerrecht voll gedeckt[91].

Bisher hat noch kein Staat den Versuch gemacht, *erga omnes* bestehende Verpflichtungen ohne ein spezielles eigenes Interesse gerichtlich durchzusetzen. *F. A. Mann* hat auf die erstaunlichen Konsequenzen, die sich bei einer Zulässigkeit der gerichtlichen Durchsetzung für die Unterwerfungserklärungen nach Art. 36 des IGH-Statuts ergäben, im Anschluß an das Urteil im Fall *Barcelona Traction* hingewiesen[92]. Das einzige hier immerhin zu nennende Beispiel ist der freilich vergebliche, aber nur durch die Stimme des Präsidenten abgewiesene Versuch von Liberia und Äthiopien, die Mandatsbestimmungen für das damalige Mandatsgebiet Südwestafrika gegenüber Südafrika gerichtlich zu erzwingen[93]. Hier lag immerhin in Gestalt der Völkerbundssatzung und der in ihrem Kontext getroffenen Mandatsregelung ein Vertragsband vor, dem die Kläger Äthiopien und Liberia angehört hatten. Einige dissentierende Voten erklären die Durchsetzung der Menschenrechte daher auch vor Gericht unabhängig von jedem weiteren rechtlichen Interesse bei vorhandener Zuständigkeit des IGH für möglich[94]. Die durch die Stimme des Präsidenten hergestellte Mehrheit verneinte dagegen ausdrücklich die Existenz der *actio popularis* im Völkerrecht[95]. Es erscheint zumindest sehr zweifelhaft, ob sich daran durch das Urteil in der Sache *Barcelona Traction* etwas geändert hat. Die gemeinsame abweichende Meinung von vier Richtern zu dem Urteil über die Atomversuche im Pazifik stellt fest, daß das Problem der *actio popu-*

---

[91] Vgl. oben S. 252.

[92] A.a.O. (Anm. 14), S. 406; vgl. auch *Dominicé* (Anm. 44), S. 247; *J. Juste Ruiz,* Los obligaciones »erga omnes« en derecho internacional público, in: Estudios de Derecho Internacional, Homenaje al Profesor Miaja de la Muela, Bd. 1 (1979), S. 219, 230–233; *I. Seidl-Hohenveldern,* Actio Popularis im Völkerrecht?, Comunicazioni e Studi 14 (1975), S. 803, 813.

[93] South West Africa cases (*Ethiopia* v. *South Africa*; *Liberia* v. *South Africa*), Second Phase, ICJ Reports 1966, p. 6.

[94] Tanaka, a.a.O., S. 252 ff.; Jessup, a.a.O., S. 373 ff., 423 ff.

[95] A.a.O., S. 47.

laris im Völkerrecht kontrovers sei, will es aber von der konkreten Norm abhängig machen, inwieweit ein Staat das Recht hat, die Einhaltung vor Gericht individuell durchzusetzen [96]. Es erscheint zutreffend, daß hier vor allem Unterschiede aus der Schutzrichtung der Rechtsnorm entnommen werden müssen. Je stärker eine Norm das gemeinsame Interesse der Staatengemeinschaft schützt, desto weniger wird ein nicht betroffener Staat sie allein durchsetzen können. Andererseits wird aber auch dort die gerichtliche Geltendmachung durch den nicht selbst betroffenen Drittstaat unzulässig sein, wo etwa der unmittelbar betroffene Staat auf Grund freier Entscheidung nichts gegen die ihm gegenüber begangene Verletzung unternimmt. Hier wäre die Geltendmachung der auf Grund der Norm entstandenen konkreten Rechtsbeziehung durch einen Drittstaat ein Eingriff in die Rechte der unmittelbar Beteiligten.

Die Frage nach möglichen Reaktionen von nicht unmittelbar betroffenen Staaten auf Verletzungen von Völkerrechtsnormen, die Verpflichtungen *erga omnes* erzeugen, hat noch eine weitere Dimension. Die Drittstaaten können die Verletzung zum Anlaß für Maßnahmen nehmen, die sich nicht unmittelbar gegen den verletzenden Staat richten, sondern die völkerrechtliche Wirkung der Verletzungshandlung betreffen. Die Nichtanerkennung gewaltsam herbeigeführter Gebietsveränderungen, durch die Declaration on Friendly Relations heute ausdrücklich als verpflichtend bezeichnet, ist eines der wichtigsten Beispiele [97]. Aber auch die Nichtanerkennung von Verletzungen des Völkerrechts in der eigenen Rechtsordnung, etwa bei der Anwendung fremden Rechts durch Gerichte, ist hier zu nennen [98].

Schließlich muß auf eine wichtige Konsequenz der Regelungen der Wiener Vertragsrechtskonvention über die Nichtigkeit von Verträgen in diesem Zusammenhang aufmerksam gemacht werden. Zwischen den in Art. 46–53 der Konvention geregelten Nichtigkeitsgründen besteht ein wichtiger Unterschied. Während Irrtum, arglistige Täuschung und Bestechung nur dem betroffenen Staat das Recht zur Anfechtung geben, wirken der Zwang gegenüber dem Staat oder seinem Vertreter sowie der Verstoß gegen eine Norm des zwingenden Völkerrechts absolut. In den letzten Fällen tritt die Nichtigkeit *ipso iure* und von Anfang an ein, während im übrigen der betroffene Staat frei entscheiden kann, ob er den Vertrag aufrecht erhalten oder anfechten will. Wie sich der Kommentierung zum Entwurf der International Law Commission entnehmen läßt, ist die Unterscheidung sehr bewußt getroffen worden. Der Zwang gegen den Staat durch Drohung oder Anwendung von

---

[96] ICJ Reports 1974, S. 370.

[97] In der Declaration 2625 (XXV) heißt es zum Gewaltverbot: »No territorial acquisition resulting from the threat or use of force shall be recognized as legal«. Dazu *J. A. Frowein*, Freundschaft und Zusammenarbeit über den Staaten, Europa-Archiv 28 (1973), S. 70.

[98] Die *adv. op.* des IGH zu Namibia sieht in diesem Sonderfall sogar eine Verpflichtung zur Nichtanerkennung, ICJ Reports 1971, S. 54 ff.

Gewalt muß nach Ansicht der Kommission zur absoluten Nichtigkeit führen, weil es sich hierbei um Verletzungen von Völkerrechtsnormen handelt, deren Einhaltung alle Staaten angeht [99].

Freilich wird die Unterscheidung zwischen den beiden Fällen durch die Verfahrensbestimmungen zur Geltendmachung der Nichtigkeitsgründe relativiert. Nach Art. 65 und 66 muß das Vorliegen eines Nichtigkeitsgrundes in der dort vorgesehenen Weise dem anderen Staat notifiziert werden. Das gilt auch für die Fälle, wo ein absoluter Nichtigkeitsgrund geltend gemacht wird. Damit ist der Unterschied zwischen den vernichtbaren Verträgen, die einer förmlichen Anfechtung bedürfen, und den anderen, die absolut nichtig sind, für die Vertragsparteien weitgehend aufgehoben. Sie müssen die Feststellung der Nichtigkeit in dem genannten Verfahren abwarten, bevor sie die Frage als geklärt ansehen können. Das zeigt auch die jetzige Fassung des Art. 69, der entgegen dem Entwurf der International Law Commission, nach der die Bestimmungen eines nichtigen Vertrages keine Gültigkeit haben sollten, nun festlegt: »Ein Vertrag, dessen Ungültigkeit auf Grund dieses Übereinkommens festgestellt wird, ist nichtig« [100]. Das scheint zu zeigen, daß die Feststellung nach der Konvention Voraussetzung für die Nichtigkeit ist.

Es ist aber zweifelhaft, ob dieses auch für dritte Staaten und internationale Organisationen gilt. Nur wenn man das annehmen könnte, hätte die Unterscheidung zwischen absoluten und nur durch Anfechtung geltend zu machenden Nichtigkeitsgründen keinen Sinn mehr. Die Möglichkeit der Geltendmachung einer Nichtigkeit des Vertrages wegen Verstoßes gegen einen absoluten Nichtigkeitsgrund, insbesondere gegen *ius cogens,* durch dritte Staaten oder internationale Organisationen würde dagegen die getroffene Unterscheidung rechtfertigen. Äußerungen von *Rosenne* und *Verosta*, die beide an der Konferenz teilgenommen haben, deuten in diese Richtung [101].

Davon geht auch die Staatenpraxis aus. Der Rechtsberater des State Department hat den Vertrag zwischen Afghanistan und der Sowjetunion von 1978 für den Fall als nichtig bezeichnet, daß er die Gewaltanwendung im Verstoß gegen Art. 2 der UN-Satzung erlauben sollte [102]. Der Internationale Gerichtshof hat keine Bedenken gehabt zu prüfen, ob Notenwechsel zwischen Island und dem Vereinigten Königreich bzw. der Bundesrepublik nichtig waren, weil sie unter Androhung oder Anwendung von Gewalt zustande

---

[99] Yearbook ILC 1966 II, S. 246 f.

[100] Die Vorschrift lautete im Entwurf: »Die Bestimmungen eines nichtigen Vertrags haben keine Gültigkeit«. Zu der auf französischen Antrag zurückgehenden Änderung vgl. *S. Verosta*, Die Vertragsrechts-Konferenz der Vereinten Nationen 1968/69 und die Wiener Konvention über das Recht der Verträge, ZaöRV 29 (1969), S. 634, 689.

[101] *Verosta,* a.a.O., S. 690; vor allem *S. Rosenne,* The Settlement of Treaty Disputes under the Vienna Convention of 1969, ZaöRV 31 (1971), S. 1, 35 mit Anm. 97, 54 mit Anm. 168.

[102] AJIL 74 (1980), S. 419.

gekommen seien. Der Gerichtshof bezeichnete es als unzweifelhaft, daß Verträge dieser Art, wie es Art. 52 WVRK anerkenne, nichtig sind[103].

Hier zeigt sich, daß die Wiener Vertragsrechtskonvention bei Verträgen, die gegen absolute Nichtigkeitsgründe verstoßen, ein fein austariertes System gewählt hat. Die einseitige Geltendmachung von Nichtigkeitsgründen durch eine Vertragspartei muß immer in dem Verfahren der Art. 65 und 66 überprüft werden. Der andere Vertragspartner kann auf dieses Verfahren verweisen. Wo aber die Nichtigkeit von Drittstaaten oder in einer internationalen Organisation behauptet wird, kann für die absoluten Nichtigkeitsgründe nicht auf das Fehlen einer Entscheidung im Verfahren nach Art. 65 und 66 verwiesen werden. Hier wird der Völkerrechtsgemeinschaft, aber auch einzelnen Drittstaaten eine Reaktionsmöglichkeit gegenüber Verträgen zugesprochen, die die Grundlagen der Völkerrechtsordnung mißachten.

Es hat sich gezeigt, daß Staaten, die nicht unmittelbar durch einen Völkerrechtsbruch betroffen sind, als Mitglieder der Staatengemeinschaft unter bestimmten Voraussetzungen durchaus rechtlich in der Lage sind, auf die Verletzung von Völkerrechtsnormen zu reagieren, die Verpflichtungen *erga omnes* erzeugen. Freilich wird eine vorsichtige und den Einzelfall sehr genau prüfende Beurteilung immer notwendig sein, um die Zulässigkeit der einzelnen Reaktion zu ermitteln.

Dabei muß in der heterogenen und auch in den Grundfragen zum Teil antagonistischen Völkerrechtsgemeinschaft des ausgehenden 20. Jahrhunderts die Verantwortung der für die Staatengemeinschaft handelnden Drittstaaten sowohl auf das jeweils betroffene Rechtsgut als auch auf die Struktur der Völkergemeinschaft ausgerichtet sein. Nur so wird diese als Rechtsgemeinschaft im Sinne des Jubilars gestärkt werden[104].

---

103 »There can be little doubt, as is implied in the Charter of the United Nations and recognized in Article 52 of the Vienna Convention on the Law of Treaties, that under contemporary international law an agreement concluded under the threat or use of force is void«, ICJ Reports 1973, S. 59.

104 The International Society as a Legal Community (1980), S. 312.

Réflexions sur les restrictions à l'exercice des droits de l'homme dans la jurisprudence de la Cour européenne de Strasbourg

WALTER J. GANSHOF VAN DER MEERSCH

Le juge *Hermann Mosler* a été, depuis sa fondation, l'un des artisans de la jurisprudence de la Cour européenne des droits de l'homme. Il en a été aussi l'un des commentateurs les plus autorisés[1]. On ne saurait aujourd'hui lui rendre d'hommage plus justifié que de consacrer cette brève étude, qui lui est dédiée, à l'un des aspects les plus délicats du système de protection des droits et libertés, institué par la Convention de Sauvegarde, signée à Rome le 4 novembre 1950.

Nous avons en vue de confronter les principes qui gouvernent l'interprétation par la Cour des dispositions garantissant les droits et libertés proclamés par la Convention aux règles qui, dans le second paragraphe des art. 8, 9, 10 et 11 de celle-ci, établissent des *restrictions* à ces droits et libertés.

On sait que les auteurs de la Convention n'ont pas adopté dans l'élaboration de celle-ci la technique des déclarations de droits et des chartes et constitutions de droit interne, qui se bornent à proclamer le principe des droits et des libertés, laissant au juge le soin et la charge de déterminer le champ de leurs applications[2].

La Convention européenne des droits de l'homme compte des dispositions de caractère absolu. C'est le cas de l'art. 3, qui dispose que: »Nul ne peut être soumis à la torture ni à des peines ou traitements inhumains ou dégradants«. C'est le cas aussi du paragraphe 1er de l'art. 4 qui déclare que:

---

[1] *H. Mosler,* Problems of Interpretation in the Case Law of the European Court of Human Rights, in: Essays on the Development of the Legal Order, in memory of H. F. van Panhuys (1980), pp. 149–167. Du même auteur, Der Europäische Gerichtshof für Menschenrechte nach zwanzig Jahren, in: Recht als Prozeß und Gefüge, Festschrift für Hans Huber (Bern 1981), pp. 595–607.

[2] Le droit constitutionnel interne connaît aussi la garantie conditionnée du droit qu'elle proclame. L'expropriation pour cause d'utilité publique en est, sans doute, l'exemple le plus général.

»Nul ne peut être tenu en esclavage ni en servitude«. Le droit protégé lui-même n'est pas exprimé. Seule apparaît l'interdiction d'y porter atteinte dans les conditions décrites.

Sans doute la Commission et la Cour se trouvent-elles, en pareil cas, dans l'obligation d'interpréter la notion-même de ce qui est interdit: la »torture«, les »peines ou traitements inhumains ou dégradants«, l'»esclavage«, la »servitude«, mais l'interdiction est absolue. Il n'y a ni conditions ni exceptions. Si les manières d'agir ou les actes interdits sont accomplis ou les situations prescrites néanmoins réalisées, il y a violation de la Convention avec la conséquence que l'intégrité de la personne physique, la dignité de l'homme, sa liberté d'aller et de venir et le droit de ne pas être tenu en totale dépendance, tous droits que les actions ou les situations expressément mentionnées dans les art. 3 et 4 sont de nature à violer, bénéficient, dans cette mesure, de la protection instituée par la Convention.

Mais il est, dans ce système aux multiples facettes et aux techniques diverses, d'autres droits et libertés, énoncés comme tels, mais au respect intégral desquels il est des limitations ou des restrictions explicitement prévues. Dans certaines situations mentionnées mais non décrites, ils perdent leur nature de droits protégés de manière absolue. Les auteurs de la Convention ont entrepris la tâche difficile de privilégier dans l'intérêt général, un certain nombre de données et de situations en spécifiant les conditions dans lesquelles l'*exercice* de ces droits pouvait être limité ou restreint[3]. La justification de ces exceptions ou dérogations ressortit aux obligations inhérentes à l'organisation de la vie sociale et à la confrontation du droit ou de la liberté fondamentale et de certains aspects de l'intérêt général.

Les Hautes Parties contractantes ont, pour certains droits et libertés[4], tracé, dans un second paragraphe, des limites à la garantie qu'ils proclamaient et entendaient assurer. Ces droits sont »le respect de (la) vie privée et familiale«, du »domicile« et de la »correspondance« (art. 8); la »liberté de pensée, de conscience et de religion« (art. 9); la »liberté d'expression« (art. 10); la »liberté de réunion pacifique« et la »liberté d'association« (art. 11). Mais ces limitations ou restrictions laissent subsister le droit; elles n'en restreignent que l'exercice[5].

---

[3] La distinction entre »limitation« et »restriction« nous paraît revêtir un caractère artificiel.

[4] Nous ne nous référons, dans cette brève étude, qu'aux art. 8, 9, 10 et 11 de la Convention pour ne pas l'alourdir. Mais on observera que la situation est la même pour l'art. 2, par. 3, du 4ème Protocole additionnel, qui concerne le droit pour quiconque se trouve régulièrement sur le territoire d'un Etat d'y circuler librement et d'y choisir librement sa résidence (par. 1) et le droit pour toute personne de quitter n'importe quel pays, y compris le sien. Voir aussi l'art. 2 de la Convention, en ses deux paragraphes; l'art. 4, en son par. 3; l'art. 5 en son ler par.; l'art. 1, 2ème alinéa du 1er Protocole additionnel.

[5] C. E. D. H. *Affaire linguistique belge*, 23 juillet 1968, Série A, p. 32 § 5.

Les exceptions à l'exercice garanti du droit ou de la liberté revêtent diverses formes, adaptées au domaine auquel elles apportent des dérogations. Il s'agit d'*ingérences* de l'autorité admises dans le domaine garanti au par. 1er, dans le cas des art. 8 et 10 de la Convention. Des *restrictions* peuvent, dans des conditions déterminées, être apportées à l'exercice des libertés garanties aux articles 9, 10 et 11. Enfin, des *formalités,* des *conditions* et des *sanctions* peuvent être imposées à la liberté d'expression, qui fait l'objet de l'article 10.

Les limitations au plein exercice des droits et libertés garantis au 1er par. de ces articles sont explicitement qualifiées et justifiées par référence à des intérêts de la vie étatique, de la vie sociale ou des droits d'autrui au sein de la société. Ces intérêts couvrent des matières étendues. Certains peuvent affecter plusieurs droits et libertés et sont mentionnés dans plus d'une disposition. C'est le cas de la »sécurité nationale« (art. 8, 9, 10 et 11); de la »sûreté« ou de la »sécurité publiques«[6] (art. 8, 9 et 11); de la »défense« ou de la »protection de l'ordre« (art. 8, 9, 10 et 11); de la »protection de la santé« ou de la »morale publiques« (art. 8, 9, 10 et 11); de la »protection des droits et libertés d'autrui« (art. 8, 9, 10 et 11); de la »prévention du crime« ou »des infractions pénales« (art. 8 et 10). D'autres objectifs, qui peuvent eux-mêmes être des droits collectifs ou individuels, ou des intérêts seulement, ne peuvent légitimement affecter qu'une seule liberté; c'est le cas du »bien-être économique du pays« (art. 8); de la volonté d' »empêcher la divulgation d'informations confidentielles« (art. 10); de l' »autorité et l'impartialité du pouvoir judiciaire« (art. 10).

L'examen auquel la Cour est appelée à procéder, en vue de déterminer si l'acte de l'autorité que le requérant prétend violer le droit en cause est justifié par la défense ou le respect de l'un de ces objectifs, comporte nécessairement l'interprétation de la notion même de celui-ci et souvent l'appréciation de la mesure dans laquelle il s'impose.

La frontière entre le droit ou la liberté dont l'exercice est placé sous la souvegarde de la Convention et donc soustrait à la souveraineté de l'Etat et la part qui demeure du ressort de l'autorité nationale, en vue d'apprécier la nature et les exigences des objectifs eu égard auxquels l'exercice du droit est limité, est l'un des problèmes les plus difficiles, sans cesse renouvelé, qui se pose à la Commission et à la Cour.

Sans doute la Cour demeure-t-elle, pour l'application des dispositions qui énoncent les droits et libertés proclamées par la Convention et organisent son système de garantie, dans le cadre des règles d'interprétation proposées à l'art. 31 de la Convention de Vienne sur le droit des Traités. La Cour s'y est référée longuement dans l'arrêt *Golder* c. *le Royaume-Uni*[7]. Elle a égard à la »*bonne foi*« et aux »*termes dans leur contexte*«. Mais, elle attache au »*but*«

[6] On observera qu'il ne règne pas une grande rigueur dans le choix des termes. Voir aussi le § 3 de l'art. 2 du 4e Protocole additionnel.

[7] C. E. D. H. 21 février 1975, Série A, n° 18, p. 14, § 29 et 30.

qu'elle lie à l' »*objet*« une importance primordiale et les fait peser fermement dans son raisonnement, guidé de manière générale par une philosophie progressiste. Celle-ci est justifiée par la matière-même qui fait l'objet de son examen et de sa décision: la *sauvegarde* des droits de l'homme et des libertés fondamentales.

Le »but du traité est le moyen de réaliser son ›objet‹«. Les deux éléments sont confondus dans la Convention et forment une pièce maitresse des principes d'interprétation de la Cour. Ils apparaissent dans le préambule de la Convention [8]. Aux termes de ce préambule le but de la Convention est de réaliser non seulement la »sauvegarde des droits de l'homme«, mais leur »*développement*«. Il y est question aussi de »prendre les *premières* mesures propres à assurer la garantie collective de certains des droits énoncés dans la Déclaration universelle«. Une volonté progressiste analogue apparaît dans le préambule du statut du Conseil de l'Europe, au sein duquel la Convention fut conçue et conclue; il se réfère, en effet, à la volonté de »sauvegarder et de faire triompher *progressivement*« un idéal, qui comporte notamment les »principes de liberté individuelle, de liberté politique et de prééminence du droit«. De part et d'autre, le préambule guide, par les aspirations qu'il révèle clairement, le juge dans la voie d'une interprétation téléologique des termes de la Convention [9].

Cette interprétation finaliste s'observe de manière très générale dans la jurisprudence de la Cour, dont on ne saurait donner ici que quelques exemples. Ainsi, dans l'affaire *Golder* c. *le Royaume-Uni* [10], la Cour a déduit le refus du droit d'accès à un tribunal, qui n'est pas mentionné explicitement dans la Convention, de l'interdiction faite au détenu de consulter un avocat. »En matière civile la prééminence du droit ne se conçoit guère sans la possibilité d'accéder aux tribunaux«, affirme l'arrêt (§ 34). »Le principe selon lequel une contestation civile doit pouvoir être portée devant un juge compte au nombre des principes fondamentaux de droit universellement reconnus« (§ 35). La doctrine a progressé encore dans l'affaire *Airey* c. *la République d'Irlande* [11] où la Cour a invoqué l'art. 6 § 1er de la Convention aux fins de garantir le droit à l'assistance judiciaire en vue d'intenter une action en séparation de corps, alors que le droit à l'assistance judiciaire n'existe pas pour les litiges civils en droit irlandais.

Les normes dans lesquelles la garantie des droits de l'homme s'exprime ne sauraient être statiques. L'interprétation de la Cour, là où il s'agit d'assurer la garantie de l'un des droits ou de l'une des libertés proclamés dans la

---

[8] C. E. D. H. »Le préambule d'un traité forme partie intégrante du contexte (. . .), il offre d'ordinaire une grande utilité pour la détermination de l'objet et du but de l'instrument à interpréter«. *Golder* c. *le Royaume-Uni,* Série A, p. 16, § 14, déja cité.

[9] Voir notre étude: La garantie des droits de l'homme et la Cour européenne de Strasbourg, Journ. des Tribunaux 1982, pp. 110 ss.

[10] C. E. D. H. 21 février 1975, Série A, n° 18.

[11] C. E. D. H. 9 octobre 1979, Série A, n° 32.

Convention, n'est pas seulement téléologique. Elle est aussi évolutive. La jurisprudence de la Cour – comme celle de la Commission, du reste – présente très nettement ce caractère. Le juge international européen prend en considération les transformations acquises sur le terrain des droits de l'homme dans la société. Il n'est, d'autre part, pas non plus indifférent à l'évolution du droit dans les Etats membres du Conseil de l'Europe. La Cour, qui a la charge d'appliquer une Convention largement issue de l'inspiration, sinon de la conjugaison des systèmes constitutionnels des Etats contractants, ne saurait prendre de décision sans approches comparatives sur le terrain de l'évolution du droit public interne des Etats membres, qui vont devoir s'aligner sur sa décision et participer à la création d'un système unifié.

Dans l'affaire *Marckx* c. *l'Etat belge* (Série A, n° 31), les requérantes, une mère et sa fille née en dehors du mariage, se plaignaient des dispositions du Code civil relatives au mode légal d'établissement de la filiation maternelle »naturelle« et des effets de ce système sur la »famille« et les droits patrimoniaux tant de l'enfant que de la mère. Elles soutenaient que ces dispositions constituaient une »ingérence« non justifiée (art. 8, § 2) dans leur »vie privée et familiale« . . . dont l'art. 8 de la Convention garantit le respect.

La Cour décida que la législation critiquée était contraire à la Convention, tant en ce qui concerne la filiation que les droits patrimoniaux: la Convention »doit s'interpréter à la lumière des conditions d'aujourd'hui«. L'arrêt relève aussi que: »Le droit interne de la grande majorité des Etats membres du Conseil de l'Europe a évolué et continue d'évoluer, corrélativement avec les instruments internationaux pertinents, vers la consécration juridique intégrale de l'adage *mater semper certa est*« (§ 41).

A l'appui de cette interprétation évolutive, la Cour s'est référée à l'évolution du droit au sein des Etats membres du Conseil de l'Europe[12].

L'affaire *Guzzardi* c. *la République italienne* (affaire dite des *mafiosi*) concernait le droit à la liberté et la régularité des conditions de détention (Convention, art. 5). L'intéressé avait été condamné pour complicité dans l'enlèvement d'un otage libéré après paiement d'une rançon, mais seule était en cause devant la Cour la privation de liberté au cours de l'instruction préparatoire, qui fut fort longue.

En vertu de l'art. 272, al. 1er du Code de procédure pénale italienne, la détention préventive ne peut dépasser 2 ans. A l'expiration de ce terme, l'instruction n'étant pas terminée et devant être longue encore, Guzzardi, qui

---

[12] Dans l'exposé des motifs d'un projet de loi déposé par le gouvernement belge lui-même assez paradoxalement au Sénat le 15 février 1978 (Doc. parl. Sénat, 305 n° 1, sess. 1977–1978, pp. 1 à 28) il était fait état de ce que »ces dernières années, plusieurs pays de l'Europe occidentale, parmi lesquels la République fédérale d'Allemagne, la Grande-Bretagne, les Pays-Bas, la France, l'Italie et la Suisse, ont adopté une législation nouvelle, bouleversant l'économie traditionnelle du droit de la filiation et instaurant une égalité quasi complète entre les enfants légitimes et les enfants naturels«.

avait de lourds antécédents judiciaires, fut placé sous »surveillance spéciale« pour 3 ans à l'île de l'Asinara, ce qui implique un état de liberté, mais de liberté surveillée. Or, le séjour dans cette île répondait à un total isolement et à des conditions qui étaient celles d'un »camp de concentration«. Elles étaient, dans la réalité des faits, une prolongation déguisée de la détention préventive, arrivée à son terme.

La Cour constata qu'il y avait violation de l'art. 5 de la Convention, le séjour de Guzzardi à l'île de l'Asinara ne correspondant à aucune des exceptions »légales« au droit à la liberté [13].

Dans l'affaire *Dudgeon* c. *le Royaume-Uni,* le requérant se plaignait de risquer, aux termes de la législation en vigueur en Irlande du Nord, des poursuites pénales en raison de son comportement homosexuel, au mépris de l'art. 8 de la Convention, qui garantit le droit au respect de la vie privée. Une législation datant de 1861 et de 1885, ainsi que des »précédents« jurisprudentiels de *common law* condamnaient ces pratiques.

Par son arrêt du 22 octobre 1981 (Série A, n° 45), la Cour décida que l'intéressé avait subi et continuait à subir une atteinte, non justifiée aux termes du § 2 de l'art. 8 de la Convention, à son droit au respect de sa vie privée. Elle s'est exprimée comme suit: »On comprend mieux aujourd'hui le comportement homosexuel qu'à l'époque de l'adoption de cette législation et on témoigne de plus de tolérance: dans la grande majorité des Etats membres du Conseil de l'Europe, on a cessé de croire que les pratiques de ce genre appellent par elles-mêmes une répression pénale; la législation interne y a subi sur ce point une nette évolution que la Cour ne saurait négliger«. (§§ 58 et 60).

* * *

Le système des restrictions au plein exercice des droits et libertés énoncés aux art. 8, 9, 10 et 11 de la Convention ne se borne pas à privilégier certains objectifs d'intérêt général à ce point justifiés en principe qu'ils font reculer, en certains de ses éléments, la garantie du droit proclamé. La Cour n'est pas libre d'apprécier sans réserve si ces causes d'ingérence, de restriction ou de sanction, inscrites au second paragraphe de l'article, s'imposent »légitimement«.

Il existe des limites à ces limites, en ce sens que ces objectifs d'intérêt public ne valent que s'ils répondent à trois conditions impératives dont le juge doit apprécier l'existence. La restriction au plein exercice du droit ou de la liberté en cause doit être *»prévue par la loi«;* elle doit être *»nécessaire«*; cette nécessité doit s'apprécier eu égard au critère d'une *»société démocratique«* [14]. Si la première de ces conditions répond à des critères objectifs, et si la troisième,

---

[13] C. E. D. H., 6 novembre 1980, Série A, n° 39.

[14] Cons. *G. Cohen Jonathan,* Cour eur. des droits de l'homme. Chr. jurispr., Cah. dr. eur. 1980, pp. 463 ss. Voir *infra,* p. 273.

tout en offrant un large horizon à l'interprétation, comporte aussi des éléments pouvant être objectivement déterminés, le caractère »nécessaire« de l'exception ouvre au juge un vaste champ de réflexion d'où doit être écarté tout élément d'appréciation subjective.

L'application du second paragraphe de chacun des art. 8, 9, 10 et 11 de la Convention[15] comporte donc pour la Cour une double obligation d'interprétation: une première dans le contrôle de la situation en cause, au regard de l'objectif dans l'intérêt duquel l'exercice du droit ou de la liberté sont limités; une seconde dans le contrôle de l'existence des trois conditions auxquelles la légitimité de cet objectif est soumise.

En tous cas, l'interprétation de la Convention et donc celle des restrictions ou limitations qu'elle prévoit trouve dans l'art. 17 un terme infranchissable: l'interprétation ne saurait impliquer le droit »d'accomplir un acte visant à la destruction des droits et libertés reconnus dans la (...) Convention ou à des limitations plus amples de ces droits ou libertés que celles prévues à (la) Convention«[16].

Les exigences des objectifs de nature à justifier une restriction à l'exercice des droits et libertés dont la garantie est proclamée ne sauraient être déterminées de manière rigoureuse à la lumière de la seule logique et de la syntaxe. Les limites en sont fréquemment incertaines. Elles varient avec les situations de fait au sein de la vie sociale.

En ce qui concerne ces objectifs et les conditions auxquelles il peut en être tenu compte, la jurisprudence de la Cour – comme celle de la Commission – reconnaît une *marge d'appréciation* aux autorités nationales de l'Etat. Cette marge d'appréciation, qui est inhérente à la nature même des objectifs explicitement mentionnés dans la Convention, est reconnue et appréciée par la Cour en tenant compte de la situation concrète que crée, dans chaque cas, la mesure contestée et de la part de discrétion que comporte l'appréciation de l'autorité nationale[17]; il s'agit là d'éléments intimement associés qu'il serait artificiel de vouloir isoler. La Cour reconnaît la marge d'appréciation à la fois au législateur national et aux organes, notamment judiciaires, appelés à interpréter et appliquer les lois en vigueur[18].

A plusieurs reprises, la Cour a rappelé quelle est la nature et quelles sont les limites de cette marge d'appréciation, eu égard sans doute aux droits ou

---

[15] La situation est identique en ce qui concerne le § 3 de l'art. 2 du 4e Protocole additionnel.

[16] Voir *F. G. Jacobs*, »Restrictions à l'exercice des droits et libertés garantis par la Convention européenne des droits de l'homme«. Coll. international, Rome, novembre 1975, éd. Conseil de l'Europe (Strasbourg 1975), p. 3.

[17] Cons. *Mosler*, Der Europäische Gerichtshof (note 1), pp. 600 ss.

[18] C. E. D. H. *Engel*, 8 juin 1976, Série A, n° 22, pp. 41–42, § 100; *Handyside*, 7 décembre 1976, Série A, n° 24, p. 22, § 48; *Sunday Times*, 26 avril 1979, Série A, p. 36, § 59.

aux libertés en cause, mais aussi à l'objectif dont ces droits et libertés vont devoir subir les effets restrictifs [19].

Déjà dans l'*affaire linguistique belge* la Cour a dit qu'elle »ne saurait ignorer les données de droit et de fait caractérisant la vie de la société dans l'Etat qui, en qualité de Partie Contractante, répond de la mesure contestée. Ce faisant, elle ne saurait se substituer aux autorités nationales compétentes, faute de quoi elle perdrait de vue le caractère subsidiaire du mécanisme international de garantie collective instauré par la Convention« [20].

Dans l'affaire *Klass,* il étati reproché à la République fédérale d'Allemagne d'avoir violé notamment l'art. 8 de la Convention, qui garantit le respect de la vie privée, en instituant législativement et en appliquant un système de surveillance de la correspondance et des télécommunications. La Cour, ici encore, a fait application de la doctrine de la marge d'appréciation dont l'Etat dispose dans certains états dérogatoires. Elle s'est exprimée dans les termes suivants: »Les sociétés démocratiques se trouvent menacées de nos jours par des formes très complexes d'espionnage et par le terrorisme, de sorte que l'Etat doit être capable, pour combattre efficacement ces menaces, de surveiller en secret les éléments subversifs opérant sur son territoire. La Cour doit donc admettre que l'existence de dispositions législatives accordant des pouvoirs de surveillance secrète de la correspondance, des envois postaux et des télécommunications est, devant une situation exceptionnelle, nécessaire dans une société démocratique à la sécurité nationale et/ou à la défense de l'ordre et à la prévention des infractions pénales« [21].

La Cour va s'étendre davantage sur la nature et la portée de la marge d'appréciation et sur ses limites dans l'arrêt *Handyside* [22], où il s'agissait du § 2 de l'art. 10 de la Convention relatif à la liberté d'expression dans le cas d'une diffusion, notamment aux abords des écoles, du livre obscène »*The little red Schoolbook*«. Elle va, en s'exprimant de la manière suivante, énoncer dans cet arrêt, au regard du cas d'espèce, les lignes principales de sa doctrine de la marge d'appréciation des autorités nationales:

> »La Cour relève que le mécanisme de sauvegarde instauré par la Convention revêt un caractère subsidiaire par rapport aux systèmes nationaux de garantie des droits de l'homme. La Convention confie en premier lieu à chacun des Etats contractants le soin d'assurer la joissance des droits et

---

[19] Voir sur la »marge d'appréciation de l'autorité nationale«: W. Ganshof van der Meersch. La référence au droit interne des Etats contractants dans la jurisprudence de la Cour européenne des droits de l'homme. Rev. intern. de droit comparé (Paris) 1980, pp. 317 ss., spéc. pp. 329 ss.

[20] C. E. D. H. 23 juillet 1968. A. pp. 34–35 § 10. Voir aussi arrêt *De Wilde, Ooms et Versyp,* 18 juin 1971, Série A, p. 45 § 93. Cons. C. E. D. H. *Delcourt,* 17 janvier 1970, Série A, § 36.

[21] C. E. D. H. 6 septembre 1978, Série A, n° 28, p. 23, § 49.

[22] C. E. D. H. 7 décembre 1976. Série A, n° 24. Voir aussi *Sunday Times*. Série A, n° 30, p. 36 § 59.

libertés qu'elle consacre. Les institutions créées par elle y contribuent de leur côté, mais elles n'entrent en jeu que par la voie contentieuse et après épuisement des voies de recours internes. . .« (§ 48).

»L'article 10 § 2 n'attribue pas pour autant aux Etats contractants un pouvoir d'appréciation illimité. Chargée, avec la Commission, d'assurer le respect de leurs engagements (article 19), la Cour a compétence pour statuer par un arrêt définitif sur le point de savoir si une »restriction« ou »sanction« se concilie avec la liberté d'expression telle que la protège l'article 10. La marge nationale d'appréciation va donc de pair avec un contrôle européen. Celui-ci concerne à la fois la finalité de la mesure litigieuse et sa ›nécessité‹. Il porte tant sur la loi de base que sur la décision l'appliquant, même quand elle émane d'une juridiction indépendante«. (§ 49).

La Cour a nuancé sa pensée, en relevant que la marge d'appréciation de l'autorité nationale n'a pas toujours ni la même étendue ni la même autorité, dans l'arrêt *Sunday Times*[23], où il était question des restrictions et sanctions dont la liberté d'expression est susceptible »pour garantir l'autorité et l'impartialité du pouvoir judiciaire«. Après avoir dit que »le *pouvoir national d'appréciation*[24] n'a pas une ampleur identique pour chacun des buts énumérés à l'article 10 § 2«, la Cour compare avec une autre affaire qu'elle a eu à juger[25] et qui concernait la »protection de la morale«. Se référant aux constations de l'arrêt dans cette affaire, elle rappelle que les exigences de la morale varient, selon les Etats contractants, dans le temps et l'espace, et ajoute que: »Il n'en va pas exactement de même de la notion, beaucoup plus objective, d'»autorité« du pouvoir judiciaire (. . .). Elle se reflète dans une série de clauses de la Convention, dont l'article 6, qui n'ont pas d'équivalent pour la »morale«. Et la Cour conclut, élevant cette conclusion au niveau d'une règle d'interprétation: »A une liberté d'appréciation moins discrétionnaire correspond donc ici un contrôle européen plus étendu« (§ 59).

* * *

a) Le contrôle de la Cour portera donc d'abord sur la question de savoir si la mesure contestée prise par l'Etat est de celles qui sont »*prévues par la loi*«. En présence du sens apparemment restrictif du mot »loi«, elle s'est longuement étendue sur la question. Elle l'a fait sans doute eu égard au cas d'espèce qui lui était soumis, mais elle a été au-delà de cette technique à laquelle généralement elle se tient, dans l'arrêt *Sunday Times* c. *le Royaume Uni*[26], qui

[23] Série A, n° 30, déjà cité.
[24] »*The domestic power of appreciation*«. Le texte anglais de l'arrêt fait foi. La traduction française est fidèle; mais on peut douter de ce que la marge d'appréciation de l'autorité nationale soit un »pouvoir«.
[25] C. E. D. H. *Handyside* (note 18), p. 22 § 48.
[26] C. E. D. H. 26 avril 1979, Série A, n° 30.

mettait fin à une affaire dans laquelle étaient en cause le délit et al procédure du »*contempt of court*«, institution dont le fondement se trouve dans la *common law*. Il s'agissait en l'espèce de l'interdiction à peine de sanction pénale du projet de publication d'un article dans lequel l'auteur se prononçait sur l'existence de la faute d'une société, alors que la question était soumise au juge. Elle mettait en cause la liberté d'expression que garantit l'art. 10 de la Convention. Le choix de cet arrêt dans la jurisprudence de la Cour nous a paru s'imposer en raison de ce que, dans cette affaire où la question se présentait de manière fort complexe, la Cour a longuement exposé ce qu'était, selon elle, la portée des termes »prévue par la loi«.

La Cour commence par écarter l'interprétation étroite selon laquelle le mot »loi« exigerait une règle de droit écrit; tout le droit de la *common law* est ici en cause. Elle s'exprime sur ce point en ces termes: »La Cour constate que dans ›prévue par la loi‹ le mot ›loi‹ englobe à la fois le droit écrit et le droit non écrit. Elle n'attache donc pas ici d'importance au fait que le *contempt of court* est une création de la *common law* et non de la législation. On irait manifestement à l'encontre de l'intention des auteurs de la Convention si l'on disait qu'une restriction imposée en vertu de la *common law* n'est pas ›prévue par la loi‹ au seul motif qu'elle ne ressort d'aucun texte législatif: on priverait un Etat de *common law* partie à la Convention, de la protection de l'article 10 § 2 et l'on frapperait à la base son système juridique«. . . (§ 47 al. 1er).

Après avoir relevé qu'en l'espèce il était soutenu qu'en raison de ce que ici les règles de la *common law* étaient incertaines et que, par conséquent, le principe général de la sécurité juridique s'opposait à une interprétation qui aurait permis de considérer que le *contempt of court* était prévu par la loi, la Cour poursuit, en se référant au texte anglais dans cette cause où celui-ci faisait foi: »L'expression *›prescribed by law‹* apparaît au paragraphe 2 des articles 9, 10 et 11 de la Convention, son équivalent en français étant chaque fois ›prévues par la loi‹. Cependant, alors que la même expression française se retrouve aux articles 8 § 2 de la Convention, 1 du Protocole n° 1 et 2 du Protocole n° 4, la version anglaise la rend respectivement par *›in accordance with the law‹*, *›provided for by law‹* et *›in accordance with law‹*). Placée ainsi devant des textes d'un même traité normatif faisant également foi mais ne concordant pas entièrement, la Cour doit les interpréter d'une manière qui les concilie dans la mesure du possible et soit la plus propre à atteindre le but et réaliser l'objet de ce traité (arrêt *Wemhoff* du 27 juin 1968, Série A, n° 7, p. 23, § 8, et article 33 § 4 de la Convention de Vienne du 23 mai 1969 sur le droit des traités)« (§ 48). On le voit, la Cour rappelle ici sa doctrine fondamentale d'interprétation téléologique.

Elle va ensuite analyser les éléments qui interviennent dans la locution, prise dans son ensemble, lorsque l'on se trouve, comme en l'espèce, en présence d'une règle de *common law* et que, donc, la précision des »précédents« doit entrer en ligne de compte pour interpréter leur autorité au regard de la règle de droit en cause: »Aux yeux de la Cour, les deux conditions suivantes comptent parmi celles qui se dégagent des mots ›prévues par la loi‹. Il faut d'abord que la ›loi‹ soit suffisamment accessible: le citoyen doit pouvoir disposer de renseignements suffisants, dans les circonstances de la cause, sur les normes juridiques applicables à un cas donné. En second lieu, on ne peut considérer comme une ›loi‹ qu'une norme énoncée avec assez de précision pour permettre au citoyen de régler sa conduite; en s'entourant au besoin de conseils éclairés, il doit être à même de prévoir, à un degré raisonnable dans les circonstances de la cause, les conséquences de nature à dériver d'un acte déterminé. Elles n'ont pas besoin d'être prévisibles avec une certitude absolue: l'expérience la révèle hors d'atteinte. En outre la certitude, bien que hautement souhaitable, s'accompagne parfois d'une rigidité excessive; or le droit doit savoir s'adapter aux changements de situation. Aussi beaucoup de lois se servent-elles, par la force des choses, de formules plus ou moins vagues dont l'interprétation et l'application dépendent de la pratique« (§ 49).

Donc »accessibilité« et »précision«; dès lors »prévisibilité« de la règle sous sa forme de *précédent*. La première de ces conditions était d'autant plus délicate à établir que si la *House of Lords* avait, par un jugement unanime, décidé qu'en l'espèce il y avait *contempt of court*, la doctrine des cinq *Law Lords* variait quant aux motifs.

Après avoir dit que le »principe des prévisions«, c'est-à-dire de la volonté d'influer sur la décision du juge, avait été formulé avec assez de précision pour permettre aux requérants de prévoir au degré voulu, les conséquences de nature à découler de la publication du projet d'article et avoir, d'autre part, renvoyé à des précédents concluants dans la *common law*, en ce qui concerne le »principe du jugement prématuré«, pour conclure que les intéressés avaient pu prévoir »à un degré raisonnable« que le projet de l'article en cause risquait de se heurter à ce principe, la Cour a conclu que l'ingérence dans l'exercice de la liberté d'expression était, en l'espèce, »prévue par la loi«, au sens de l'art. 10 de la Convention.

b) Les restrictions au plein exercice du droit ou de la liberté en cause, en vue de donner satisfaction aux objectifs d'intérêt général aux droits d'autrui, sont soumises, dans le par. 2 des art. 8, 9, 10 et 11, en termes identiques, au caractère *nécessaire* de ces mesures. Cette nécessité est liée, on l'a vu, au critère de la »société démocratique«[27].

L'appréciation de ce caractère de nécessité offre de grandes difficultés. Elle suppose, d'une part, une vue sur le caractère impérieux de l'objectif d'intérêt général envisagé, ainsi que sur le régime de droit interne et la vie sociale au sein de l'Etat en cause et, d'autre part, une prise de position sur l'étendue et la mesure respectives de la marge d'appréciation de l'autorité nationale au regard du pouvoir de décision du juge international. C'est sans doute celle des

---

[27] Voir aussi C. E. D. H. *De Wilde, Ooms et Versyp*, 18 juin 1971, Série A, p. 45 § 93; *Handyside*, n° 24, 7 décembre 1976 (note 18), p. 21, § 44. Voir ci-dessus note 14.

trois conditions où la marge d'appréciation de l'Etat est la plus sensible en raison de la meilleure connaissance qu'a l'autorité nationale du milieu social.

Dans l'arrêt *Handyside,* la Cour s'est exprimée de la manière la plus complète, au regard d'un cas d'espèce, qui mettait en cause la liberté d'expression, l'application du par. 2 de l'art. 10 de la Convention, sur la manière selon laquelle se pose pour elle l'appréciation de la *nécessité* des restrictions à l'exercice du droit ou de la liberté en cause[28]: »Grâce à leurs contacts directs et constants avec les forces vives de leur pays« a dit la Cour »les autorités de l'Etat se trouvent en principe mieux placées que le juge international pour se prononcer sur (. . .) la ›nécessité‹ d'une ›restriction‹ ou de ›sanction‹ destinées à y répondre. La Cour note à cette occasion que si l'adjectif ›nécessaire‹, au sens de l'article 10 § 2, n'est pas synonyme d'›indispensable‹ (comp., aux articles 2 § 2 et 6 § 1, les mots ›absolument nécessaire‹ et ›strictement nécessaire‹ et, à l'article 15 § 1, le membre de phrase ›dans la stricte mesure où la situation l'exige‹), il n'a pas non plus la souplesse de termes tels qu'›admissible‹, ›normal‹ (comp. l'article 4 § 3), ›utile‹ (comp. le premier alinéa de l'article 1 du Protocole n° 1), ›raisonnable‹ (comp. les articles 5 § 3 et 6 § 1) ou ›opportun‹. Il n'en appartient pas moins aux autorités nationales de juger, au premier chef, de la réalité du besoin social impérieux qu'implique en l'occurrence le concept de ›nécessité‹. . .«.

Dans l'arrêt *Sunday Times*[29], la Cour précise que le caractère »nécessaire« doit ressortir des conditions dans lesquelles se présente l'ingérence dans le cas d'espèce dont elle est saisie et non point de la législation seulement. Elle s'est exprimée dans les termes suivants: »Pour apprécier si l'ingérence incriminée se fondait sur des motifs ›suffisants‹ qui la rendaient ›nécessaire dans une société démocratique‹, il faut donc tenir compte de tout aspect de l'affaire relevant de l'intérêt public« (§ 65). Et elle poursuit: »Il ne suffit pas que l'ingérence dont il s'agit se classe parmi celle des exceptions énumérées à l'article 10 § 2 (. . .); il ne suffit pas davantage qu'elle ait été imposée parce que son objet se rangeait dans telle catégorie ou tombait sous le coup d'une règle juridique formulée en termes généraux ou absolus: la Cour doit s'assurer qu'il était nécessaire d'y recourir eu égard aux faits et circonstances de la cause précise pendante devant elle« (§ 65).

c) Reste la qualification de *»société démocratique«*, critère qui trouve son origine dans l'art. 9, par. 2 de la Déclaration universelle. La Cour ne s'est pas contentée ici de se reporter, comme référence au système socio-politique de la »société démocratique«, au texte de la constitution des Etats concernés. Elle a donné de l'un des éléments fondamentaux d'une société de cette nature une notion à la fois vivante et progressiste, qui mérite sans doute d'être reproduite[30]. Après avoir dit que: »Son rôle de surveillance commande à la

---

[28] C. E. D. H. *Handyside,* Série A, n° 24 (note 18), p. 22 § 48. Voir aussi: *Sunday Times,* Série A (note 18), pp. 35–36 § 59.

[29] C. E. D. H., Série A, n° 30 (note 18), p. 41, § 65.

[30] C. E. D. H., *Handyside* (note 18), § 49.

Cour de prêter une extrême attention aux principes propres à la Société démocratique«; elle a poursuivi en disant que: »La liberté d'expression constitue l'un des fondements essentiels de pareille société, l'une des conditions primordiales de son progrès et de l'épanouissement de chacun (. . .); elle vaut non seulement pour les informations ou idées accueillies avec faveur ou considérées comme inoffensives ou indifférentes, mais aussi pour celles qui heurtent, choquent ou inquiètent l'Etat ou une fraction quelconque de la population. Ainsi le veulent le pluralisme, la tolérance et l'esprit d'ouverture sans lesquels il n'y a pas de »société démocratique«.

* * *

Peut-être n'est-il pas sans intérêt de compléter ces quelques réflexions consacrées au problème des restrictions admises au plein exercice des droits et libertés proclamés aux art. 8, 9, 10 et 11 de la Convention, en donnant quelques aperçus de la jurisprudence de la Cour concernant l'autre volet du système »à deux temps« que consacre la Convention, c'est-à-dire les objectifs privilégiés susceptibles de justifier une restriction *lato sensu* à l'exercice des droits et libertés en cause. On choisira, à titre d'exemples, deux des objectifs dans l'intérêt desquels des restrictions à la liberté d'expression sont prévues au second paragraphe de l'art. 10: *»la protection de la morale«* et la garantie de *l'autorité et (de) l'impartialité du pouvoir judiciaire«*.

a) C'est encore dans l'affaire *Handyside,* où des *»sanctions«* pénales avaient été prises par l'autorité judiciaire nationale pour la diffusion d'un livre obscène destiné aux élèves des écoles, que la Cour a mis en lumière certains caractères propres à l'expression *»la protection de la morale«* dans le cadre de la Convention, en disant que: »on ne peut dégager du droit interne des divers Etats contractants« une notion européenne uniforme de la »morale«. L'idée que leurs lois respectives se font des exigences de cette dernière varie dans le temps et l'espace, spécialement à notre époque caractérisée par une évolution rapide et profonde des opinions en la matière. Grâce à leurs contacts directs et constants avec les forces vives de leur pays, les autorités de l'Etat se trouvent en principe mieux placées que le juge international pour se prononcer sur le contenu précis de ces exigences« [31].

Après avoir analysé les faits, et avoir dit qu'elle »n'a point pour tâche de se substituer aux juridictions internes compétentes, mais d'apprécier, sous l'angle de l'article 10, les décisions qu'elles ont rendues dans l'exercice de leur pouvoir d'appréciation« (§ 50), la Cour a conclu, en ce qui concerne les mesures prises par l'autorité judiciaire »pour la protection de la morale«, que: »malgré la diversité et l'évolution constante des conceptions éthiques et éducatives au Royaume-Uni, les magistrats anglais étaient en droit de croire à l'époque, dans l'exercice de leur pouvoir d'appréciation [32], que le *Schoolbook*

---

[31] C. E. D. H. (note 18), § 48.

[32] On voit que la Cour reconnaît ici une *»marge d'appréciation«* particulièrement large à l'autorité nationale.

aurait des répercussions néfastes sur la moralité de beaucoup des enfants et adolescents qui le liraient« (§ 52). Il est certain que cette dernière considération a joué un rôle déterminant dans l'acceptation par la Cour de l'exception à la liberté d'expression fondée sur la protection de la morale.

L'interprétation de l'objectif que constitue la *»protection de la morale«* a aussi fait l'objet de l'arrêt de la Cour dans l'affaire *Dudgeon* c. *le Royaume-Uni*. Le requérant se plaignait dans cette affaire d'être toujours sous la menace, en vertu de la législation en vigueur en Irlande du Nord, de poursuites pénales à cause de son comportement homosexuel, et cela au mépris de l'article 8 de la Convention, qui garantit le droit au respect de la vie privée. Une législation datant de 1861 et de 1885, ainsi que des »précédents« jurisprudentiels de *common law* condamnaient pareilles pratiques.

Par son arrêt du 22 octobre 1981 [33], la Cour décida que le requérant subissait une atteinte, non justifiée par la protection de la morale, à son droit au respect de sa vie privée. Elle s'est exprimée comme suit: »On comprend mieux aujourd'hui le comportement homosexuel qu'à l'époque de l'adoption de cette législation et on témoigne de plus de tolérance: dans la grande majorité des Etats membres du Conseil de l'Europe, on a cessé de croire que les pratiques de ce genre appellent par elles-mêmes une répression pénale, la législation interne y a subi sur ce point une nette évolution que la Cour ne saurait négliger (. . .)«. Elle a conclu en ces termes, après avoir relevé qu'aucune poursuite n'a été engagée en Irlande du Nord pour actes d'homosexualité commis de leur plein gré et en privé par des hommes de plus de 21 ans, capables d'y consentir: »On ne saurait, dès lors, parler d'un besoin social impérieux d'ériger de tels actes en infractions, faute d'une justification suffisante, déterminée par le risque de nuire à des individus vulnérables à protéger ou par des répercussions sur la collectivité« (§ 60). L'arrêt porte, on le voit, par le rejet de la restriction prévue au § 2 dans l'intérêt de la »protection de la morale« le caractère évolutif très affirmé de l'interprétation de la Convention.

b) La Cour a eu à connaître, dans l'affaire *Sunday Times* c. *le Royaume-Uni* [34] de l'objectif d'intérêt national qu'est la *»garantie (de) l'autorité (. . .) du pouvoir judiciaire«* [35]. Il s'agissait, en l'espèce, de poursuites pénales du chef du délit de *contempt of court,* que le requérant soutenait violer la liberté d'expression. Au sujet de l'autorité du pouvoir judiciaire, la Cour s'exprima de la manière suivante:

»Les termes ›pouvoir judiciaire‹ (›*judiciary*‹) recouvrent l'appareil de la justice ou le secteur judiciaire du pouvoir autant que les juges en leur qualité officielle. Quant à l'expression ›autorité du pouvoir judiciaire‹, elle reflète notamment l'idée que les tribunaux constituent les organes appropriés pour ap-

---

[33] C. E. D. H., Série A, n° 45.
[34] C. E. D. H. 26 avril 1979, Série A, n° 30.
[35] Art. 10 § 2.

préciser les droits et obligations juridiques et statuer sur les différends y relatifs, que le public les considère comme tels et que leur aptitude à s'acquitter de cette tâche lui inspire du respect et de la confiance« (§ 55, al. 2).

La Cour poursuit son raisonnement en se référant au »rapport Philimore«[36] sur l'institution du *contempt of court,* où il est exposé que ce délit vise pour l'essentiel des types de conduite ayant trait, soit à la position des juges, soit au fonctionnement des tribunaux et de l'appareil de la justice«. Elle relève que »la garantie de l'autorité et l'impartialité du pouvoir judiciaire figure donc parmi ses fins« (§ 55, al. 3), et que: »l'ingérence dans l'exercice de la liberté d'expression du requérant répondait donc (en principe) à un but légitime au regard de l'article 10 § 24« (§ 57, al. 5).

Elle considérera néanmoins qu'il y a violation de la liberté d'expression. Après avoir dit qu'elle »doit s'assurer qu'il était nécessaire de recourir (à cette ingérence) eu égard aux faits et circonstances de la cause (...) pendante devant elle«, elle conclut que: »Eu égard à l'ensemble des circonstances de la cause et sur la base de la démarche définie (...) ci-dessus, (...) l'ingérence incriminée ne correspondait pas à un besoin social assez impérieux pour primer l'intérêt public s'attachant à la liberté d'expression au sens où l'entend la Convention« (§ 67)[37].

* * *

La Cour a le devoir de donner plein effet à la Convention. Sa jurisprudence présente, on l'a vu, un caractère à la fois téléologique et évolutif en s'inspirant de l'objet et du but de la Convention. Elle puise dans la nature même des droits de l'homme et des libertés fondamentales dont elle doit assurer le respect la justification d'une interprétation naturellement sensible à leur développement et aux progrès de l'évolution de la »société démocratique« à laquelle la Convention se réfère.

Si ce caractère téléologique et évolutif de la jurisprudence de la Cour apparaît dans l'interprétation progressiste qu'elle donne aux dispositions qui ont pour objet la *garantie même des droits* proclamés par la Convention dans un système de normes minimales, la situation ne saurait être caractérisée de manière aussi absolue en ce qui concerne les dispositions qui énoncent des *restrictions* à l'exercice de ces droits, en général prévues en vue du respect d'objectifs de caractère et d'intérêt public national.

Le caractère d'applicabilité directe de la plupart des dispositions de droit national de la Convention et leur incorporation dans le droit interne d'un grand nombre des Etats contractants et adhérents fait du juge national l'un

---

[36] Commission, instituée le 8 juin 1971 par le *Lord Chancellor* et le *Lord Advocate* et chargée sous la présidence de Lord Justice Phillimore, de dire si, selon elle, le droit du *contempt,* dans son ensemble appelait des modifications.

[37] Voir *supra* les considérations relatives au caractère *»nécessaire«* des restrictions ou sanctions pour qu'elles soient légitimes au sens de la Convention.

des juges de l'application de la Convention. D'autre part, le système même de la Convention implique une certaine collaboration avec le droit interne des Etats et les juridictions de ceux-ci. Cette collaboration s'exprime par la marge d'appréciation qu'ils conservent dans le champ même des droits et libertés, inscrits dans la Convention et dans le 1er et le 4ème de ses Protocoles.

On observe au surplus que la Cour tient compte d'un certain droit commun, qui se dégage, dans une large homogénéité, de l'ensemble des droits de l'homme des Etats contractants, qui ont inspiré la Convention et forment dans ce domaine aujourd'hui un véritable ordre public européen [38]. C'est ainsi que ce n'est pas un paradoxe de voir la Cour, bien qu'exerçant son contrôle sur les actes du législateur et du juge national dont elle apprécie la conformité à la Convention, se référer parfois, à l'appui de son raisonnement, au droit interne des Etats membres du Conseil de l'Europe, au sein duquel la Convention fut élaborée [38]. Il y a là, dans le cadre du respect de la Convention, un certain jeu de réciprocité.

Si les limitations, restrictions et sanctions que la Convention apporte, spécialement dans le second paragraphe des art. 8, 9, 10 et 11, aux droits et libertés qu'elle proclame dans le premier des paragraphes de ces articles sont, en principe, l'objet d'une interprétation restrictive, encore faut-il, dans chaque cas, tenir compte de la forme affirmative ou négative de la rédaction de la disposition limitative.

Mais, quelle que soit la forme sous laquelle apparaît cette limitation, elle n'est, dans l'interprétation que lui donne la Cour, qu'un autre langage pour assurer la garantie de ces droits et libertés dans le même esprit progressiste. L'interprétation restrictive d'une exception ou d'une dérogation rejoint, dans son principe, l'interprétation large de la règle à laquelle il est dérogé.

A plusieurs reprises, la Cour a exprimé l'idée que les exceptions aux limitations à l'exercice des droits et libertés que la Convention a pour objet de sauvegarder doivent en principe recevoir une interprétation restrictive. C'est là une constante logique de sa doctrine d'interprétation à laquelle les autorités nationales des Etats contractants et adhérents se doivent d'être attentifs.

Elle y a insisté fermement dans l'arrêt *Klass* [39], où les requérants invoquaient la violation de l'art. 8 de la Convention, qui garantit le respect de la vie privée et familiale et celle de l'art. 6, dans ses dispositions relatives aux garanties judiciaires, en raison de l'usage en République Fédérale d'Allemagne d'écoutes téléphoniques clandestines [40], dans l'instruction d'actes de ter-

[38] Voir notre étude, La référence au droit interne des Etats contractants dans la jurisprudence de la Cour européenne des droits de l'homme, Revue internationale de droit comparé (Paris) 1980, pp. 317–335, spec. pp. 322 ss.

[39] C. E. D. H., 6 septembre 1978, Série A, n° 28, p. 21, § 42.

[40] La Cour, après avoir »déclaré« que les faits en cause étaient, dans le cas de l'espèce, »nécessaires dans une société démocratique et dans la situation actuelle, à la sécurité nationale comme à la défense de l'ordre et à la prévention des infractions pénales« (§ 68), a dit à l'unanimité qu'il n'y avait eu ni violation de l'art. 6, ni violation de l'art. 8 de la Convention (p. 33).

rorisme. Après avoir situé la portée de l'art. 8 dans ses données essentielles au regard des faits de la cause, la Cour, rappelant que: »La principale question qui se posait en l'occurrence sur le terrain de l'article 8 consistait à savoir si les termes du paragraphe 2 justifiaient l'ingérence ainsi constatée« a poursuivi en disant que »Ménageant une exception à un droit garanti par la Convention, ce paragraphe appelle une interprétation étroite«, mais que, »Caractéristique de l'Etat policier, le pouvoir de surveiller en secret les citoyens n'est tolérable d'après la Convention que dans la mesure strictement nécessaire à la sauvegarde des institutions démocratiques.« (§ 42).

Dans l'arrêt *Golder*[41] déjà, où était en cause le refus du droit d'accès à un tribunal, déduit de l'interdiction faite au requérant de consulter un avocat, la Cour avait mis en lumière le caractère restrictif du par. 2 de l'art. 8, qui ressortit a sa rédaction même: »Il ne peut y avoir ingérence (. . .) *que* pour autant que (. . .)«[42]. Dans le même arrêt, la Cour, à propos de l'interprétation du même art. 8 de la Convention, avait déclaré que celui-ci »ne laisse pas place à l'idée de *limitations implicites*«[42], en réponse au gouvernement qui soutenait que les restrictions du par. 2 n'étaient pas elles-mêmes limitatives et pouvaient être étendues. La Cour avait écarté ce raisonnement tendant à voir introduire la théorie des compétences implicites dans la jurisprudence de la Cour, là où celle-ci a la charge d'interpréter les dispositions limitatives à l'exercice des droits et libertés proclamés par la Convention.

Dans l'arrêt *Sunday Times*[43], la Cour, comparant la nature et la portée de la liberté elle-même, proclamée dans le premier paragraphe de l'art. 10 de la Convention, et l'objectif en vue duquel la limitation de son exercice est prévue dans le second paragraphe, a dit que: »Elle ne se trouve pas devant un choix entre deux principes antinomiques, mais devant un principe – la liberté d'expression – assorti d'exceptions, qui appellent une interprétation étroite« (§ 65).

Enfin, la Cour, saisissant l'occasion que lui offraient les affaires *Winterwerp*[44] et *Guzzardi*[45], causes qui concernaient l'application des restrictions au droit à la liberté, prévues au paragraphe premier de l'art. 5, a renouvelé l'expression de sa volonté d'interprétation restrictive des dispositions qui limitent l'exercice des droits et libertés proclamés dans la Convention.

Il y a là, on le voit, une jurisprudence constante dont le principe même trouve sa justification dans l'objet et le but de la Convention.

---

[41] C. E. D. H., 21 février 1975, Série A, n° 18, § 44.
[42] Souligné par nous.
[43] C. E. D. H., 26 avril 1979 (note 18).
[44] C. E. D. H., 24 octobre 1979.
[45] C. E. D. H., 6 novembre 1980.

Die Schiffahrtsfreiheit von Kriegsschiffen nach UNCLOS III

WILHELM KARL GECK *

I. Probleme und Prämissen

1981 erregten zwei Zwischenfälle auf See international Aufsehen. Bei dem ersten ging es um das Eindringen eines getaucht fahrenden Unterseebootes in fremde Gewässer, bei dem zweiten um Übungen eines Flottenverbandes und seiner Flugzeuge innerhalb einer Zone, die der Küstenstaat vor seinem Festland in Anspruch nahm. Die wesentlichen Tatsachen sollen hier insoweit skizziert werden, als sie für die Schiffahrtsfreiheit von Kriegsschiffen von Interesse sind.

Am 27. Oktober 1981 lief das sowjetische Unterseeboot 137 in der Ostsee auf Grund. Das Schiff gehört zu einer Klasse, die im NATO-Code als Whiskey-Klasse bezeichnet wird (»Whiskey on the Rocks«). Es hat eine

---

* Der Verfasser dankt Herrn Assessor *Karl-Heinz Oehler* für die Hilfe bei dem Zusammenstellen der Literatur und für die kritische Durchsicht des Manuskriptes.

Wasserverdrängung von 1350 t (getaucht), konventionellen Antrieb, bis zu 18 Torpedos und eine Besatzung von etwa 54 bis 56 Mann. Die Schäreninsel Sturkö, in deren Nähe das Schiff derart aufsetzte, daß es sich nicht mehr mit eigener Kraft freimachen konnte, liegt etwa 30 sm vor Karlskrona. U 137 geriet innerhalb der schwedischen Eigengewässer *(internal waters)* auf Grund. Diese sind vom schwedischen Recht festgelegt und bilden hier zugleich die Basis- oder Grundlinie der Küstengewässer *(territorial sea)*, die Schweden in einer Breite von 12 sm in Anspruch nimmt. Die äußere Grenze der Binnengewässer (Grundlinie der Küstengewässer) richtet sich hier nach den Verbindungslinien zwischen einer Anzahl von vorgelagerten Inseln [1]. Das ist nach geltendem Völkergewohnheitsrecht und Vertragsrecht zulässig [2].

Der Fall erregte aus mehreren Gründen Aufsehen. Einmal war das sowjetische Kriegsschiff ohne die nach schwedischem Recht schon für die – aufgetauchte – Durchfahrt durch Küstengewässer vorgeschriebene Anmeldung in schwedische Eigengewässer eingedrungen und sogar in ein – völkerrechtlich zulässiges – militärisches Schutz- und Kontrollgebiet um den wichtigen Marinestützpunkt Karlskrona [3]. Das Schutz- und Kontrollgebiet sollte nach schwedischen Vorstellungen sorgfältig gesichert und scharf überwacht sein. Das Unterseeboot drang dort anscheinend gerade zu einem Zeitpunkt ein, als die schwedische Marine einen neuen, streng geheimgehaltenen Torpedotyp erproben wollte. Ferner erklärte der schwedische Ministerpräsident es für sehr wahrscheinlich, daß U 137 mit einem oder mehreren nuklearen Sprengköpfen ausgerüstet war. Schließlich erinnerte der Vorfall die Öffentlichkeit an die vergebliche 14tägige Jagd im Jahre 1980 auf ein nicht näher identifiziertes östliches U-Boot bei dem schwedischen Flottenstützpunkt Muskör und an drei andere Vorfälle im Oktober 1966 [4].

Nach sowjetischen Verlautbarungen war U 137 bei einer routinemäßigen Übungsfahrt in der Ostsee wegen des Ausfalls von Navigationsinstrumenten unbeabsichtigt vom Kurs abgekommen. Diese Verlautbarungen fanden in Schweden kein Verständnis. Nach Aussagen schwedischer Marineoffiziere

---

[1] Siehe Lag om Sveriges Sjöterritörium (vom 3. 6. 1966), Svensk Författningssamling (SFS) 1966 Nr. 374. Dazu die königliche Bekanntmachung vom 3. 6. 1966, SFS 1966 Nr. 375 und die Verordnung vom 18. 12. 1978 zur Änderung dieser Bekanntmachung, SFS 1978 Nr. 961.

[2] Siehe insbes. I.C.J. Reports 1951, S. 128 ff., und Art. 4 der Genfer Konvention von 1958 über das Küstenmeer und die Anschlußzone. Schweden ist nicht Vertragsstaat, wohl aber die Sowjetunion. Diese machte zu Art. 23 der Konvention einen Vorbehalt dahin, daß der Küstenstaat das Recht habe, die Durchfahrt von fremden Kriegsschiffen durch seine *Küsten*gewässer zu regeln.

[3] Einzelheiten bei *Berg*, Das sowjetische U-Boot 137 in schwedischen Hoheitsgewässern, Fragen der Immunität fremder Kriegsschiffe, Zeitschrift für ausländisches öffentliches Recht und Völkerrecht (ZaöRV) 42 (1982), S. 295 ff.

[4] Siehe zu den Vorfällen von 1966 *O'Connell*, International Law and Contemporary Naval Operations, British Year Book of International Law (BYIL) 44 (1970), S. 19 ff., 58.

war das Eindringen in die engen Schärenbuchten von stellenweise sehr geringer Wassertiefe nur mit funktionierenden Instrumenten und mit einer hervorragenden Navigationstechnik möglich. Zur Frage der atomaren Bewaffnung nahm die sowjetische Regierung anscheinend nicht Stellung. Die sowjetische Nachrichtenagentur TASS bezeichnete Angaben über Atomtorpedos an Bord von U 137 als absurd, scheint aber dabei Formulierungen gebraucht zu haben, die das Vorhandensein atomarer Sprengköpfe doch nicht eindeutig ausschlossen. Vorerst endete der Zwischenfall mit einem Protest Schwedens, den die Sowjetunion zurückwies, mit der Freigabe des Bootes nach »neun Tagen Tauziehen« und seinem Abschleppen durch schwedische Schiffe in die Hohe See. Offen ist eine Rechnung für die Bergung von U 137. Das schwedische Außenministerium forderte (umgerechnet) rund 650 000,– DM[5].

Der zweite der aufsehenerregenden Zwischenfälle forderte Menschenleben. Am 18./19. August 1981 führte die 6. US-Flotte im Mittelmeer eine Übung durch, die einige Tage vorher »through notices to airmen and to mariners« bekanntgemacht worden war. Die Übung erstreckte sich in den Golf von Sidra (Große Syrte). Am 19. August schossen amerikanische Flugzeuge, die von dem Flugzeugträger Nimitz aufgestiegen waren, zwei libysche Kampfflugzeuge ab. Die beteiligten Staaten stellten die Entstehung und den

---

[5] Diese Skizze stellt nur auf Fragen der Schiffahrtsfreiheit ab. Sie beruht auf Berichten in den folgenden Zeitungen: Frankfurter Allgemeine Zeitung (FAZ), International Herald Tribune, Le Monde, Neue Zürcher Zeitung, Süddeutsche Zeitung und The Times (London) sowie auf den oben in Anm. 1 angeführten schwedischen Rechtsnormen. Die Pressemeldungen decken sich nicht in allen Punkten. Zum Teil hängt das damit zusammen, daß verschiedene amtliche Stellen unterschiedliche Darstellungen abgaben und sowjetische Äußerungen möglicherweise bewußt unklar formuliert hatten. Die hier berücksichtigten Presseberichte sind durchweg in einem Punkte unzutreffend oder jedenfalls ungenau. Sie sind unzutreffend, soweit sie das Auflaufen der U 137 in schwedischen Küstengewässern – statt Eigengewässern – feststellen. Die Presseberichte sind ungenau, soweit sie den Ausdruck Hoheitsgewässer benutzen, der nicht klar erkennen läßt, ob Eigen- oder Küstengewässer gemeint sind. Hauptsächlich unter immunitätsrechtlichen Gesichtspunkten ist der Fall eingehend untersucht von *Berg* (Anm. 3) mit Übersetzung der Note, in der die schwedische Regierung den Zwischenfall schilderte und wertete. Auch bei der kürzeren Darstellung von *Mössner* stehen diese Gesichtspunkte im Vordergrund: Spionage und Immunität von Kriegsschiffen, Neue Juristische Wochenschrift 35 (1982), S. 1196 ff. *Mössner* erwähnt Schadensersatzansprüche in Höhe von 650 000 $. Diese Angabe ist unzutreffend. – Ein anderer U-Boot-Fall erregte weniger Aufsehen. Der italienische Verteidigungsminister erklärte vor dem Abgeordnetenhaus, am 25. 2. 1982 sei ein atomgetriebenes sowjetisches Angriffs-Unterseeboot der Victor-Klasse in italienische Hoheitsgewässer im Golf von Tarent eingedrungen, um einen Aufklärungsauftrag in der Nähe des Hafens durchzuführen, in dem ein großer Teil der italienischen Kriegsmarine stationiert ist. Das von italienischen Schiffen verfolgte Unterseebot sei in internationale Gewässer entkommen. Die sowjetische Botschaft in Rom bestritt auf einen italienischen Protest hin die Anwesenheit eines sowjetischen Unterseebootes in italienischen Hoheitsgewässern (alles nach Berichten in der FAZ, Saarbrücker Zeitung, Süddeutschen Zeitung vom 2.–6. 3. 1982).

Ort des Zwischenfalls verschieden dar. Nach amerikanischer Darstellung hatten die libyschen Flugzeuge die amerikanischen angegriffen, als diese sich auf einem Übungsflug in dem Golf von Sidra etwa 60 sm vor der libyschen Küste befanden. Nach libyscher Darstellung hatten die amerikanischen Flugzeuge angegriffen, als die libyschen auf einem Routineflug etwa 30 sm vor der libyschen Küste waren [6]. Die unterschiedlichen Angaben sind ein Beispiel für Schwierigkeiten bei der Standortklärung auf und über der See.

Die Frage, wer wen angegriffen hat, muß hier dahinstehen. Es geht nur um den völkerrechtlichen Status des Luftraums. Er hängt von dem Status des unter ihm liegenden Seegebietes ab. Der Luftzwischenfall ist daher zugleich ein Seezwischenfall. Beide Staaten vertraten im Hinblick auf das für den Ort des Zwischenfalls geltende Seerecht entgegengesetzte Rechtsansichten. Über den libyschen Standpunkt berichteten verschiedene Zeitungen unterschiedlich. Nach einer Darstellung nahm Libyen seit 1973 den Golf von Sidra und damit auch den Ort des Zwischenfalls als Eigengewässer mit der Begründung in Anspruch, es handele sich hier um eine historische Bucht. Nach einer zweiten Darstellung beanspruchte Libyen seit 1973 gerade den Golf von Sidra als Küstenmeer. Nach der dritten Darstellung nahm Libyen ganz allgemein – also nicht nur beim Golf von Sidra – Küstengewässer von 200 sm Breite in Anspruch. Aus früheren juristischen Veröffentlichungen ergibt sich, daß Libyen den Golf von Sidra seit 1973 beansprucht und die Durchfahrt ausländischer Schiffe von einer Erlaubnis abhängig macht. Es ist jedoch nicht klar, welchen völkerrechtlichen Status es für den Golf in Anspruch nimmt. Sicher ist allerdings, daß die Vereinigten Staaten und zum mindesten auch Italien Vorbehalte geltend gemacht hatten [7]. Für die Antwort darauf, ob die amerikanischen Flugzeuge 60 (oder auch 30) sm vor der libyschen Küste fliegen durften, ist unerheblich, wie die Begründung für den libyschen Rechtsstandpunkt wirklich lautete. Die Souveränität des Küstenstaates erfaßt nämlich den Luftraum nicht nur über den Eigengewässern mit »nationalen« Buchten, sondern nach allgemeinem Völkergewohnheitsrecht (und Art. 2 der Konvention von 1958 über das Küstenmeer und die Anschlußzone) auch über den Küstengewässern. Die Vereinigten Staaten betrachteten den Golf von Sidra jenseits der traditionellen 3 sm-Grenze des Küstenmeeres als einen Teil der Hohen See. Hier seien Flottenmanöver grundsätzlich zulässig; hier gelte auch die traditionelle Freiheit des Überfliegens. Die Vereinigten Staaten protestierten daher wegen eines unprovozierten Angriffs auf ihre Flugzeuge im internationalen Luftraum.

---

[6] Eine erste Zeitungsnachricht, der Vorfall habe sich nach libyschen Aussagen etwa 6 sm vor der libyschen Küste abgespielt, dürfte entweder eine libysche Äußerung mißverstanden haben oder eine erste libysche Stellungnahme wiedergeben, die bald aufgegeben wurde.

[7] Siehe Revue générale de droit international public (RGDIP) 78 (1974), S. 1177 f.; American Journal of International Law (AJIL) 68 (1974), S. 510 f.; Italian Yearbook of International Law 2 (1976), S. 422 ff.

Pressestimmen äußerten die Vermutung, die Vereinigten Staaten hätten mit der Flottenübung die völkerrechtliche Schiffahrtsfreiheit auf der Hohen See und die ihr entsprechende Überflugsfreiheit augenfällig wahren wollen, da diese Rechte nur durch wirksamen Gebrauch gesichert werden könnten, notfalls in demonstrativer Form. Für diese Vermutung wurde u. a. auf das erhebliche Interesse der Vereinigten Staaten daran hingewiesen, daß die Bewegungsfreiheit ihrer Mittelmeerflotte nicht durch einseitige Akte von Küstenstaaten eingeschränkt wird. Auch seien die amerikanisch-libyschen Beziehungen gespannt gewesen und ein amerikanischer Wunsch, libyschen Ansprüchen Einhalt zu gebieten, unter diesem Gesichtspunkt jedenfalls nicht auszuschließen [8].

Die eben erwähnten Seezwischenfälle stehen nicht isoliert da. *Cable* zählt ohne Anspruch auf Vollständigkeit unter dem nicht ganz deckungsgleichen Begriff *»application of limited naval force«* für die Zeit von 1945–1969 etwa 80 Fälle auf. An ihnen waren Kriegsschiffe von 20–25 Staaten beteiligt. Nach *O'Connell* steigt bei der Anwendung von *coercive sea power* die Kurve an. Er kommt für die Zeit von 1945–1974 auf etwa 100 Fälle, an denen Marineeinheiten von fast 50 Staaten beteiligt waren [9]. Zu den bekannteren Beispielen gehören die Versenkung des israelischen Zerstörers Eilat durch ägyptische Raketen 1967, die »Beschlagnahme« des amerikanischen Aufklärungsschiffs Pueblo 1968 nahe der nordkoreanischen Küste durch nordkoreanische Kriegsschiffe, die Land-, See- und Luftkämpfe zwischen der Volksrepublik China und Südvietnam 1974 um die Paracelinseln in der südchinesischen See und die Fischereistreitigkeiten zwischen Großbritannien und Island 1973 sowie 1975/76 [10].

Es ist wahrscheinlich, daß die Zahl der Seezwischenfälle weiterhin zunehmen wird. Dazu können verschiedene Ursachen beitragen. Die folgenden erscheinen von besonderer Bedeutung.

---

[8] *Oxman* spricht von »successful excercise by the U.S.Navy of the freedoms of the sea in the Gulf of Sidra«, allerdings ohne Darstellung des Zwischenfalls: The Third United Nations Conference on the Law of the Sea: The Tenth Session (1981), AJIL 76 (1982), S. 1 ff. (20). Die Skizze der Vorgänge beruht auf Berichten in den Zeitungen FAZ, The Times, International Herald Tribune vom 20. 8. 1981 sowie Neue Zürcher Zeitung und Le Monde vom 21. 8. 1981.

[9] *Cable,* Gunboat Diplomacy, Political Applications of Limited Naval Force (1971), S. 177 ff.; *O'Connell,* The Influence of Law on Sea Power (1975), S. 2.

[10] Es ist weder möglich noch nötig, die gesamte Literatur aufzuführen, die sich mit den angeführten Seezwischenfällen befaßt – zum Teil sehr eingehend. Einige Belege genügen, aus denen sich Art und Umstände des jeweiligen Geschehens erkennen lassen. Siehe etwa zum *Eilat*-Fall Europa-Archiv 22 (1967), S. Z 251 f.; *O'Connell,* a.a.O., S. 126 ff.; zum *Pueblo*-Fall AJIL 62 (1968), S. 756 f.; RGDIP 72 (1968), S. 168 ff.; 73 (1969), S. 822 ff.; *Cable,* a.a.O., S. 32 ff.; zu den Kämpfen um die Paracelinseln RGDIP 79 (1975), S. 176 ff.; zu dem britisch-isländischen Kabeljaukrieg RGDIP 77 (1973), S. 281 ff., 867 ff., 1225 ff.; 80 (1976), S. 298 ff.; *Janis,* Sea Power and the Law of the Sea (1976), S. 77 ff.

1. Immer mehr Staaten machen ein intensives Interesse an bestimmten Seegebieten oder an bestimmten Arten der Meeresnutzung geltend. Mit konkurrierenden Rechten und Ansprüchen nimmt die Zahl der Konfliktsmöglichkeiten zu. 2. Zur Durchsetzung oder Verteidigung ihrer Rechte oder Ansprüche brauchen die Staaten Seestreitkräfte, wenn auch – je nach der Sachlage – möglicherweise nur sehr »bescheidene«. Seemacht in einem begrenzten Sinne ist jedoch nicht mehr eine Domäne weniger Staaten. Aufrüstung findet heute nicht nur bei Supermächten und mittleren Mächten statt[11]. Die Zahl der militärisch handlungsfähigen Staaten kann auch die Zahl der Zwischenfälle vermehren. 3. Die Verwendung von Kriegsschiffen zu bewaffnetem Einsatz – der nicht zugleich Einsatz von Waffen sein muß – ist häufig weniger riskant als der Einsatz von Landstreitkräften. Die auch im wörtlichen Sinne fließenden Seegrenzen werden nicht für so »heilig« gehalten und nicht so erbittert verteidigt wie Landgrenzen. Ein Konflikt auf See erregt meist weniger Gemüter und diese nicht so tief und dauerhaft. Der maritime Konflikt erscheint leichter isolierbar und auch insofern weniger gefährlich. Den Streitparteien steht eher eine Option zwischen Druck auf den Gegner einerseits und Rückzug andererseits offen als bei vergleichbaren Konflikten auf dem Lande. Daher können die Staaten es auf Zwischenfälle mit Machtproben zur See im allgemeinen eher ankommen lassen als auf Zwischenfälle zu Lande. Das gilt auch für die Hohe See, wo die Anlässe für Konflikte allerdings seltener sind als in Seezonen mit Sonderrechten von Küstenstaaten. 4. Selbst der militärisch klar überlegene Staat ist nicht von vornherein davor sicher, in einen Konflikt verwickelt zu werden. Der zur See Stärkere ist nicht immer willens oder in der Lage, seine überlegene Machtposition auszunützen. Aus den eben erwähnten Gründen fühlt er sich oft weniger als bei Landzwischenfällen »genötigt«, seine Überlegenheit auszuspielen. Es dauerte 11 Monate, bis die Vereinigten Staaten auch nur die Besatzung der »Pueblo« von Nordkorea zurückerhielten. Island konnte den Kabeljaukrieg gegen Großbritannien gewinnen, weil dieses seine Übermacht nicht nutzte – letztlich, um die NATO-Luftbasis in Keflavik nicht zu gefährden[12]. Das gewaltsame Vorgehen Perus im Jahre 1954 gegen Walfangschiffe des Reeders Onassis, die unter der Flagge Panamas weit vor der Küste Perus fuhren, führte zu Protesten des Flaggenstaates Panama sowie Großbritanniens und der Vereinigten Staaten, wo die Schiffe versichert waren. Diese Proteste verhinderten aber nicht, daß Peru bereits im nächsten Jahre mehrere US-amerikanische Schiffe aufbrachte und die Kapitäne zu Geldstrafen verurteilte. Ebenfalls 1955 griff Ecuador zwei US-amerikanische Schiffe auf Hoher See mit Maschinengewehrfeuer an, brachte sie auf und verhängte trotz starker Proteste der Vereinigten

[11] Siehe die Bemerkungen unten im Text zu den Anm. 14–22 a.

[12] Siehe RGDIP 73 (1969), S. 822 ff., zum *Pueblo*-Fall; *Janis* (Anm. 10), S. 77 ff., zum Kabeljaukrieg 1973.

Staaten Geldstrafen von fast 50 000 $ [13]. Derartige Erfahrungen können den Schwächeren von zwei potentiellen Kontrahenten selbst dann ermutigen, wenn er offensichtlich gegen Völkerrecht verstößt. 5. Die Tatsachenaufklärung ist bei See- und Luftzwischenfällen oft besonders schwierig, wie der oben erwähnte Zwischenfall zwischen den Vereinigten Staaten und Libyen sowie der *Pueblo*-Fall beispielhaft zeigen. Wo hat sich ein Zwischenfall abgespielt? Wer hat einen Angriff begonnen? 6. Um so wichtiger ist bei diesen Umständen Klarheit der maßgeblichen Rechtsnormen. Unklarheiten des Seerechts können wesentlich zu einer Zunahme von Seezwischenfällen beitragen. Lücken und Grauzonen im Recht geben dem Staat, der wichtige Interessen zu haben glaubt und meint, er könne sie auch unter politischen, wirtschaftlichen und schließlich militärischen Risiken durchsetzen oder verteidigen, die Möglichkeit, in gutem Glauben oder jedenfalls ohne offensichtliche Rechtsverletzung handeln. Die »Seenahme« durch Küstenstaaten in den letzten Jahrzehnten bietet Beispiele auch für die Nutzung unklarer Rechtslagen.

Nur bei einem kleinen Teil der bekanntgewordenen Seezwischenfälle spielten Fragen der Schiffahrtsfreiheit die entscheidende Rolle. Die Bedeutung der Normen über die Schiffahrtsfreiheit nimmt jedoch ständig zu. Ein wesentlicher Grund ist die räumliche Verengung der Hohen See durch die Einführung von neuen Meereszonen mit Beschränkungen der Meeresfreiheit, die den Küstenstaaten zugute kommen. Ein weiterer wesentlicher Grund ist der Ausbau der Seestreitkräfte in einer Vielzahl von Staaten. Unter Einbeziehung auch der Staaten mit einigen wenigen Küstenschutzbooten nennt *Janis* 1976 über 120 Staaten mit Seestreitkräften. Nach *O'Connell* hatten etwa zu dem gleichen Zeitpunkt 35 Staaten Anti-Schiff-Raketen, unter ihnen Ägypten, Algerien, Indien, Indonesien, Irak, Libyen, Malaysia, Singapur, Syrien, neun kommunistische und fünf lateinamerikanische Staaten [14]. *Weyers* Flottentaschenbuch Warships of the World (1979/81) [15] weist auf die Vielzahl von Seezwischenfällen in den letzten beiden Jahren und den weltweit wachsenden Bedarf an kleineren Kriegs- und Polizeischiffen hin, die zur Kontrolle der neuerworbenen Seegebiete dienen, und führt 127 Staaten an, die über Kriegsschiffe verfügen. Das umfassendste Flottenhandbuch, *Jane*'s Fighting Ships, nennt in der Ausgabe 1978/79 138 Staaten mit Kriegsschiffen irgendeiner Art [16]. Das von *Karkoszka* bearbeitete 6. Kapitel »The expansion of naval Forces« des SIPRI Yearbook 1979 konstatiert mit wachsender Besorgnis das Ansteigen der Seerüstung in praktisch allen Gebieten der Erde und die Teilnahme von zahlreichen Entwicklungsländern an der globalen Expansion der Seestreitkräfte. Die Möglichkeit zum Erwerb von leichten und verhältnis-

---

[13] Siehe *Oda*, New Trends in the Regime of the Seas (I), ZaöRV 18 (1957/58), S. 61 ff., 71 ff.; *Phleger*, Recent Developments Affecting the Regime of the High Seas, Department of State Bulletin 32 (1955), S. 934 ff., 937.

[14] *Janis* (Anm. 10), S. XIII; *O'Connell* (Anm. 9), S. 3.

[15] 55. Jg. hrsg. von *Albrecht* (1979) (insbes. S. 5).

[16] Hrsg. von *Moore* (1978).

mäßig billigen Kriegsschiffen mit hochentwickelter Technik und einer im Vergleich zu früher gewaltigen Feuerkraft gestatte es kleinen oder kaum existierenden Seestreitkräften, in kurzer Zeit zu verhältnismäßig modernen und feuerkräftigen Flotten heranzuwachsen[17]. *Platzöder* spricht von einem »internationalen Rüstungstransfer bisher nicht gekannten Ausmaßes« mit negativen Auswirkungen durch die Einführung der Ausschließlichen Wirtschaftszonen[18].

Die Seestreitkräfte unterscheiden sich schon von der Größe her wie der Hai vom Hering. Außer den (1.) Supermächten der See, den Vereinigten Staaten und der Sowjetunion, haben (2.) Großbritannien und Frankreich *blue water navies*, d.h. Seestreitkräfte, die zu größeren und längeren Operationen fern der eigenen Küste fähig und zudem ebenfalls mit strategischen Atomwaffen ausgerüstet sind[19]. Eine 3. Gruppe von Staaten verfügt über Seestreitkräfte von 80–250 Schiffen mit größeren Einheiten, aber ohne Atomwaffen. Diese Staaten sind Argentinien, Australien, Brasilien, Chile, China, die Bundesrepublik Deutschland, Griechenland, Indien, Italien, Japan, Kanada, Südkorea, Mexiko, die Niederlande, Peru, Portugal, Schweden, Spanien, Taiwan, die Türkei und Venezuela[20]. Diese Seestreitkräfte sind – isoliert betrachtet – nur von regionaler Bedeutung. Als 4. Gruppe nennt *Janis* 29 Staaten mit mindestens einem, höchstens zehn größeren Überwasserkampfschiffen und als 5. Gruppe 72 Staaten mit lediglich kleineren Schiffen in z.T. sehr geringer Zahl[21]. Immerhin können auch kleine Einheiten mit modernen Raketenwaffen selbst gegenüber großen Schiffen ein erhebliches Vernichtungspotential haben. Beispielsweise erreichen moderne Patrouillenboote, die heute mit Torpedos und Raketen ausgerüstet werden können, eine Feuerkraft wie früher nur Fregatten oder Zerstörer[22]. Die Seemacht mittlerer und kleiner Staaten

---

[17] SIPRI Yearbook of World Armaments and Disarmament, hrsg. vom Stockholm International Peace Research Institute (SIPRI Yearbook) 1979, S. 329, 379. Siehe auch SIPRI Yearbook 1981 Teile I und II sowie S. XV.

[18] *Platzöder,* Völkerrechtliche und politische Auswirkungen der Dritten Seerechtskonferenz der Vereinten Nationen auf Nord- und Ostsee (1978), S. 91. Andererseits führt das Ansteigen der Kosten für größere Schiffseinheiten, insbes. Flugzeugträger und atomare Unterseeschiffe, gerade in manchen demokratischen Staaten auch zum Abbau von Seestreitkräften. Ein einprägsames Beispiel bietet Großbritannien.

[19] Die verbreitete Bezeichnung *blue water navy* soll hier mit Hochseestreitkräften umschrieben werden, obwohl das nicht ganz dem traditionellen Verständnis des Begriffs Hochseestreitkräfte entspricht. Die Einteilung der Staaten nach der Bedeutung ihrer Seestreitkräfte in fünf Gruppen folgt *Janis* (Anm. 10). Die Einteilung von *Booth*, Navies and Foreign Policy (1977), S. 113 ff., 120 ff., richtet sich in erster Linie nach der Reichweite der Seestreitkräfte: a) *global navies* der USA und der Sowjetunion, b) *ocean- going navies* Großbritanniens und Frankreichs, c) *contiguous sea navies* mit etwa 60% aller Seestreitkräfte und d) *coastal navies* mit etwa 35% aller Seestreitkräfte.

[20] So *Janis* (Anm. 10), S. 63.

[21] *Ibid.,* S. 64. Siehe zu Einzelheiten die nach Staaten geordneten Übersichten in *Weyers* Flottentaschenbuch und *Jane*'s Fighting Ships (Anm. 15, 16).

[22] So SIPRI Yearbook 1979, S. 330.

darf nicht isoliert gesehen werden. Luftstreitkräfte mit modernen Raketenwaffen können von Landbasen aus in Küstennähe ansonsten weit überlegene Seestreitkräfte fremder Staaten gefährden. Nach *Luttwak* bestand die bis Anfang der sechziger Jahre unbestrittene Luftüberlegenheit der amerikanischen Flugzeugträger im Mittelmeer gegenüber kleineren Anliegerstaaten Anfang der siebziger Jahre nicht mehr, da die kleineren Staaten an Zahl und Qualität der Flugzeuge große Fortschritte gemacht hatten. Das Bild dürfte sich Anfang der achtziger Jahre noch weiter zugunsten der Anliegerstaaten verändert haben[22a].

Die Staaten definieren die Aufgaben ihrer Seestreitkräfte naturgemäß unterschiedlich. Jedoch gibt es zum mindesten in den einzelnen Gruppen von Staaten unübersehbare Gemeinsamkeiten. Die offizielle Zielsetzung der amerikanischen Flotte läßt sich in vier Arten von Aufgaben gliedern: 1. strategisch – atomare – Abschreckung, 2. Seebeherrschung, insbesondere zum Zwecke der Sicherung von überseeischen Verbindungen, d.h. das Offenhalten von Seewegen für sich selbst und Sperrung für den Feind, 3. Machtausübung an fremden Küsten und 4. das Zeigen der Flagge in einer fremden Region als Zeichen von Interesse oder Besorgnis[23]. Der Oberbefehlshaber der sowjetischen Flotte, Admiral Gorschkov, hat die Aufgaben der sowjetischen Seestreitkräfte folgendermaßen umrissen: 1. Verteidigung der Sowjetunion durch strategische Abschreckung und konventionelle Streitkräfte, 2. Sperrung der See für die Flotte der Vereinigten Staaten als einem Instrument ihrer imperialistischen Politik, 3. Demonstration der wirtschaftlichen und militärischen Macht in Friedenszeiten jenseits der Grenzen[24]. Die Aufgaben der britischen Seestreitkräfte wurden 1975 offiziell so definiert: 1. nukleare Abschreckung, 2. Seeverteidigung für Großbritannien und die NATO im östlichen Atlantik und dem Kanalgebiet, 3. – und mit geringerer Betonung – Schutz britischer und NATO-Interessen außerhalb der Gewässer um Großbritannien[25]. Das offizielle französische Selbstverständnis von der Aufgabe der eigenen Seestreitkräfte entspricht im wesentlichen dem britischen. Jedoch betonen die Franzosen wie die Amerikaner und die Sowjets anders als Groß-

---

[22a] *Luttwak,* The Political Uses of Sea Power (1974), S. 49 ff. – Der britisch-argentinische Kampf um die Falklandinseln 1982 hat eindrucksvolle Belege für die im Text geäußerte Erwartung gegeben. Siehe auch unten Anm. 137 a.

[23] *Janis* (Anm. 10), S. 1 ff. Die Sicherung der Küste ist in den Vereinigten Staaten Aufgabe der »Coast Guard«. Siehe allgemein zu den Aufgaben von Seestreitkräften die Klassifizierung bei *Booth* (Anm. 19), S. 17 ff.: »Policing role«, »diplomatic role«, »military role« mit weiteren Differenzierungen.

[24] *Janis* (Anm. 10), S. 23 f. Siehe im einzelnen *Gorškov,* Die Rolle der Flotten in Krieg und Frieden. Zeitschriftenbeiträge (Moskau 1972/73), hier in der Übersetzung von *Arenz* (1975), insbes. S. 150 ff.: Die sowjetische Kriegsflotte als Instrument von Frieden und Freundschaft; *derselbe* (Schreibweise: Gorschkow), Seemacht Sowjetunion (Moskau 1976), hier nach der Übersetzung aus dem Bundessprachenamt (1978), insbes. S. 258 ff., 396 ff.

[25] *Janis* (Anm. 10), S. 39 ff.

britannien die Bedeutung von Flottenpräsenz in einem politisch-militärischen Sinne, auch wenn dies zurückhaltend umschrieben ist[26]. Bei den Staaten der oben angeführten 3. Gruppe ist die seestrategische Konzeption im allgemeinen auf regionale Aufgaben der Küstenverteidigung und der Sicherheit der Schiffahrt in Küstennähe begrenzt[27]. Im Rahmen einer Allianz können sich allerdings unter Umständen überregionale Aufgaben ergeben. Die Aufgaben von Seestreitkräften der 4. und 5. Gruppe sind naturgemäß noch enger. Bei der 5. Gruppe dürfte es sich häufig mehr um Polizeiaufgaben als um eigentliche militärische Aufgaben handeln.

Es liegt auf der Hand, daß unterschiedliche militärische Machtmittel und Gegebenheiten zu unterschiedlichen Ansichten über die Lösung wichtiger seerechtlicher Fragen beitragen, insbesondere über die Rechtsstellung der Küstengewässer und der Meerengen sowie über die Passagerechte, aber auch über neue Fragen wie Archipelgewässer und Ausschließliche Wirtschaftszonen. Die oben skizzierten Erwägungen haben eine erhebliche Bedeutung auf der »Third United Nations Conference on the Law of the Sea« (UNCLOS III) gehabt. Sie sind Hintergrund der unter II.–VI. folgenden Ausführungen, auch wenn dies nicht jeweils im einzelnen dargelegt werden kann. Im Rahmen der Normen über die einzelnen Seegebiete spielt gerade die Schiffahrtsfreiheit von Kriegsschiffen eine wesentliche Rolle[28].

Angesichts der Wichtigkeit der Schiffahrtsfreiheit von Kriegsschiffen im Seevölkerrecht und angesichts des Fehlens einer alle Seezonen einbeziehenden Untersuchung dieser Frage liegt es nahe, die Aussagen von UNCLOS III zu dieser Frage zu untersuchen. Es ist ungewiß, ob die am 30. April 1982 in New York von der Seerechtskonferenz angenommene »Convention on the Law of the Sea« bald in Kraft treten und für wieviele Staaten sie dann bindend sein wird[28a]. Die Annahme des Textes mit 130 gegen nur 4 Stimmen

---

[26] *Janis* (Anm. 10), S. 53 f.

[27] *Janis* (Anm. 10), S. 63 ff. Strategie und Flottenaufbau einzelner regionaler Mächte sind skizziert in Seemacht und Außenpolitik, hrsg. von *Mahncke* und *Schwarz* (1974). S. 260 ff. (Bearbeiter des Kapitels Fischer)

[28] Vgl. etwa *Butler*, The Soviet Union and the Law of the Sea (1971); *Critchley*, Canada's Maritime Forces and the Law of the Sea, International Perspectives (Ottawa) 1978, S. 3 ff., aber auch verstreute Ausführungen im Rahmen einer umfassenderen Betrachtung wie z. B. von *Rembe*, Africa and the International Law of the Sea: A Study of the Contribution of the African States to the Third United Nations Conference on the Law of the Sea (1980).

[28a] Hier wird der letzte von den Vereinten Nationen veröffentlichte Text zugrunde gelegt (A/CONF.62/L.78 vom 28. 8. 1981). Die wenigen Textänderungen von der Frühjahrssitzung 1982 betreffen die im folgenden erwähnten Artikel nicht. Auch die vier zusätzlichen Draft Resolutions zu der Konvention bewirken keine Textänderung. Die Textänderungen sind ersichtlich aus dem vom United Nations Information Service herausgegebenen Press Release SEA/154 vom 30. 4. 1982. Dieses enthält auch Angaben über letzte vergebliche Versuche, die friedliche Durchfahrt von Kriegsschiffen durch Küstengewässer von dem Einverständnis des Küstenstaates oder jedenfalls von einer vorherigen Anmeldung abhängig zu machen oder die Position des

bei 17 Enthaltungen hat den Erwartungen auf die zum Inkrafttreten erforderlichen 61 Ratifizierungen jedenfalls großen Auftrieb gegeben, obwohl die Vereinigten Staaten mit Nein gestimmt und der gesamte Ostblock sowie Großbritannien und die Bundesrepublik sich der Stimme enthalten haben. Sicher ist, daß der Entwurf bereits ein erhebliches Gewicht gewonnen hat und unabhängig von seinem rechtlichen Schicksal die weitere Entwicklung des Seerechts stark beeinflussen wird.

Obwohl die Schiffahrtsfreiheit von Kriegsschiffen nur eines und keineswegs das größte von den vielen Problemen der Konferenz war, hängt ihre Betrachtung mit so vielen Problemkreisen und Regeln zusammen, daß eine Abgrenzung der folgenden Ausführungen notwendig ist. Der Weg soll hier durch die verschiedenen Seegebiete im Sinne von Wassersäulen führen. Das Recht des Festlandsockels, von Meeresboden und Meeresuntergrund im übrigen und damit auch des »Gebiets« im Sinne von Teil XI. der Draft Convention bleibt unberücksichtigt. Die Schiffahrtsfreiheit von Kriegsschiffen läßt sich hier nur generell behandeln, d.h. unter Ausschluß von besonderen Friedens-/atomwaffenfreien-/entmilitarisierten Zonen, und nur unter dem Gesichtspunkt des Friedensvölkerrechts. Damit bleibt etwa der Meeresbodenvertrag von 1971 ebenso außerhalb der folgenden Überlegungen wie das gesamte Kriegsvölkerrecht und Neutralitätsrecht. Ausgeklammert werden muß auch das Atomrecht im übrigen, insbesondere die Frage von Atomversuchen auf Hoher See. Schließlich ergibt sich eine Einschränkung der Untersuchung daraus, daß dem Verfasser die Kompetenz des Seestrategen und des Waffenexperten abgeht, die man im Grunde für dieses Thema brauchte. Aber *O'Connell* hat den Juristen (hoffentlich zu Recht) geraten, sich durch die Lückenhaftigkeit ihrer Kenntnisse und die unvermeidliche Vorläufigkeit ihrer Ergebnisse von der Behandlung seerechtlicher Themen nicht völlig abschrecken zu lassen[29].

Da das Thema trotz dieser Eingrenzungen den Rahmen einer Festschrift zu sprengen droht, ist es im allgemeinen nicht möglich, zeitlich hinter UN-

Küstenstaates auf andere Weise zu verbessern: S. 7, 37 f. sowie unten Anm. 50. Nach dem Press Release unterliegt der am 30. 4. angenommene Text noch einer Überarbeitung durch das Drafting Comittee der Konferenz. Sie soll von Mitte Juli–Mitte August 1982 stattfinden. Man hofft, daß die Unterzeichnung des endgültigen Textes im Dezember 1982 in Caracas stattfinden kann. Aus der Tatsache, daß – nach dem erwähnten Press Release (S. 49) – eine informelle Plenarsitzung der Konferenz bis zum 16. 4. 1982 schon etwa achthundert (!) Empfehlungen des Drafting Comittee gebilligt hatte, lassen sich Schlüsse auf die Aufgaben des Drafting Comittee und auf die bisherige »Kontrolle« durch die Konferenz ziehen. Angesichts der Tatsache, daß ein endgültiger Text nicht in allen Einzelheiten vorliegt und die Unterzeichnung noch aussteht, wird die Konvention trotz der Annahme im folgenden als Draft Convention oder DC abgekürzt. Hier kann von den nach Art. 320 sechs authentischen Texten nur der englische herangezogen werden. Die Materialien sind erschienen unter dem Titel Third United Nations Conference on the Law of the Sea – Official Records, Vol. I (1975)–Vol. XIII (1981); wird fortgesetzt.

[29] *O'Connell* (Anm. 4), S. 20.

CLOS III zurückzugehen. Nur dort, wo es unbedingt erforderlich ist, wird der Rechtszustand vor UNCLOS III einbezogen. »Ältere«, z. T. sogar grundlegende, Literatur und die Vorgeschichte der Konferenz müssen unberücksichtigt bleiben.

Die Darstellung der Schiffahrtsfreiheit von Kriegsschiffen nach der Draft Convention wird viele Auslegungsfragen bei Einzelbestimmungen aufwerfen. Es werden aber auch grundsätzlichere Fragen auftreten, auf die schon hier hingewiesen werden soll: Der Verfasser hat angesichts des Fehlens von allgemein anerkannten Maßstäben für die Beurteilung von Kodifikationen schon früher die Ansicht vertreten, ein Kodifikationszweck sei dann erreicht, wenn ein Kodifikationsvertrag die Rechtsklarheit fördere, inhaltlich sachgerecht sei und einen hinreichend weiten Geltungsbereich habe[30]. Zu den beiden ersten Punkten werden sich mit aller gebotenen Vorsicht in den folgenden Teilen einige Aussagen machen lassen – wenn auch mit unterschiedlicher Präzision. Ebenso sollten einige Aussagen zu der wichtigen Frage möglich sein, der *Hermann Mosler* 1946 seine Bonner Antrittsvorlesung gewidmet hat, »Die Großmachtstellung im Völkerrecht« (veröffentlicht 1949).

UNCLOS III ist im Hinblick auf die erfaßte Materie das bei weitem größte Kodifikationsprojekt, das eine Internationale Organisation jemals unternommen hat. Über das Seerecht hinaus wird die Konvention Auswirkungen auf die zukünftige Gestaltung der Weltwirtschaftsordnung haben. Aber auch in anderer Hinsicht ist UNCLOS III unter den Kodifikationskonferenzen einmalig. Das gilt im Hinblick auf den erfaßten Raum (etwa 70% der Erdoberfläche), den Umfang der Kodifikation (320 Artikel und neun Anhänge), die Zahl der beteiligten Staaten (157 Staaten bei der Frühjahrssitzung 1982), die Größe der Delegationen, die Zahl der Interessengruppen und ihre Überschneidungen, aber auch im Hinblick auf die Vorstadien und die Dauer der Konferenz (seit 1973), vor allem jedoch für die Art des Verfahrens. Alle diese Fragen können hier nicht einmal skizziert werden; zu allen gibt es viele Äußerungen in der juristischen Literatur. Nur ein Punkt soll erwähnt werden. Über die besonderen Interessen an der Schiffahrtsfreiheit gerade von Kriegsschiffen wurde in den offiziellen Gremien der Seerechtskonferenz anscheinend selten so eingehend und deutlich gesprochen, daß man ein genaues Bild nachzeichnen kann. Die Seemächte scheinen diese Interessen nicht mit näheren Begründungen artikuliert, sondern lediglich ihre Existenz deutlich gemacht zu haben. Der außenstehende Beobachter könnte geneigt sein, zwei Redensarten – eine in etwas veränderter Form – auf die Konferenz anzuwenden: Über eine Flotte spricht man nicht, eine Flotte hat man und – im Hinblick auf die Aufgaben der Seestreitkräfte – nie davon sprechen, immer daran denken.

---

[30] *Geck*, Völkerrechtliche Verträge und Kodifikation, ZaöRV 36 (1976), S. 96 ff., 129.

II. Die Schiffahrtsfreiheit von Kriegsschiffen auf Hoher See und in fremden Eigengewässern

1. Der Begriff Kriegsschiffe

Der Begriff der Kriegsschiffe hat sich im Völkergewohnheitsrecht herausgebildet. Er wurde vertraglich zum ersten Mal im Rahmen der Haager Friedenskonferenzen in der Konvention von 1907 betreffend die Umwandlung von Handelsschiffen in Kriegsschiffe definiert[31]. Von dieser Definition gingen die Vorschläge der International Law Commission aus, die zu Art. 8 Abs. 2 der Genfer Konvention über die Hohe See von 1958 führten. Nach Art. 8 Abs. 2 ist ein Kriegsschiff im Sinne der Konvention »a ship belonging to the naval forces of a state and bearing the external marks distinguishing warships of its nationality, under the command of an officer duly commissioned by the government and whose name appears in the Navy List, and manned by a crew who are under regular naval discipline«. Die Genfer Konvention über die Hohe See band am 31. Dezember 1981 57 Staaten. Zu ihnen gehörten die USA, die Sowjetunion, Großbritannien und die Bundesrepublik Deutschland. Trotz der Gültigkeit des Kriegsschiffsbegriffs nur für diese Konvention und trotz der begrenzten Zahl ihrer Vertragsstaaten kann man davon ausgehen, daß die Definition der Kriegsschiffe in Art. 8 Abs. 2 allgemein gelten sollte und auch allgemein so verstanden wurde. Nach der Präambel der Konvention über die Hohe See hat die Seerechtskonferenz der Vereinten Nationen 1958 »the following provisions as generally declaratory of established principles of international law« angenommen. Zu den Vertragsbestimmungen, die nach dieser Absicht nur deklaratorischen Charakter haben, kann man Art. 8 Abs. 2 mit der Definition der Kriegsschiffe zählen.

Auch UNCLOS III befaßte sich mit dem Begriff der Kriegsschiffe. Art. 29 DC schließt sich der Definition der Kriegsschiffe in Art. 8 Abs. 2 der Konvention über die Hohe See im wesentlichen an. Es gibt lediglich drei Abweichungen: An die Stelle der *naval forces* sind die *armed forces* getreten; die »Navy List« ist ersetzt durch die »appropriate service list or its equivalent«; die Besatzung muß nicht mehr »under regular naval discipline« stehen, sondern »under regular armed forces discipline«. Man darf annehmen, daß diese geringfügigen und sachgerechten Änderungen des Kriegsschiffbegriffes sich

---

[31] Verträge der Bundesrepublik Deutschland Serie A: Multilaterale Verträge, Bd. 28 (1969), S. 172. Vgl. zu den Kriegsschiffen im Friedensvölkerrecht zusammenfassend Akademie der Wissenschaften der UdSSR – Institut für Staat und Recht, Modernes Seevölkerrecht: Wissenschaftliche Forschung – Schutz der Meeresumwelt – Handels- und Kriegsschiffahrt (Original 1977, hier in Übersetzung von *Rauch*, 1981), S. 247 ff., im folgenden Wissenschaftliche Forschung; *Geck*, Warships, in: Bernhardt (ed.), Encyclopedia of Public International Law, Instalment 4 (1982), S. 346 ff. Für den Stand vor UNCLOS III siehe *Geck*, Kriegsschiffe, völkerrechtliche Stellung, in: Strupp-Schlochauer, Wörterbuch des Völkerrechts, Bd. 2 (1961), S. 368 ff.

auch unabhängig von dem Schicksal der Draft Convention durchsetzen, d.h. früher oder später zu allgemeinem Völkergewohnheitsrecht werden.

2. Die Schiffahrtsfreiheit auf Hoher See

Die Schiffahrtsfreiheit auf Hoher See galt nach allgemeinem Völkergewohnheitsrecht schon vor 1958 auch und gerade für Kriegsschiffe. Art. 2 Abs. 1 Ziff. 1 der Genfer Konvention von 1958 über die Hohe See ist im Hinblick auf »freedom of navigation« nur deklaratorisch. UNCLOS III hält an dem Grundsatz der Schiffahrtsfreiheit für Kriegsschiffe auf der Hohen See fest. Die in Art. 87 Abs. 1 *lit.* a DC unter anderen Freiheiten angeführte Schiffahrtsfreiheit gilt unbestritten für Kriegsschiffe. Die in Art. 87 Abs. 1 *lit.* b DC vorgesehene Freiheit zum Überfliegen der Hohen See entspricht Art. 2 Abs. 1 Ziff. 4 der Genfer Konvention über die Hohe See und geltendem Völkergewohnheitsrecht. Die Schiffahrtsfreiheit von Kriegsschiffen stößt allerdings auf verschiedene Grenzen.

Art. 88 DC reserviert die Hohe See für friedliche Zwecke. Er ist im Zusammenhang mit dem erst 1980 in den Text aufgenommenen Art. 301 DC über die friedliche Nutzung des Meeres zu sehen. Danach sollen alle Vertragsstaaten sich bei der Ausübung ihrer Rechte und Pflichten gemäß der Konvention jeden Gebrauchs von Gewalt und jeder Gewaltandrohung enthalten, die gegen die territoriale Unversehrtheit oder politische Unabhängigkeit eines Staates gerichtet oder in irgendeiner anderen Weise mit den in der Satzung der Vereinten Nationen verkörperten Grundsätzen des Völkerrechts unvereinbar sind. Unterschiedliche Bewertungen des Art. 301 DC sprechen von »Schlüsselbestimmung für die militärische und sicherheitspolitische Beurteilung des Konventionsentwurfs« oder sagen lediglich »... (it) repeats the Charter requirements without creating new rights or obligations« [32].

Es ist unübersehbar, daß Art. 301 DC dem Art. 2 Abs. 4 Satzung der Vereinten Nationen (SVN) nahezu wörtlich folgt. Art. 2 Abs. 4 SVN ist seit 1945 in Kraft, ohne daß man in ihm jemals ernsthaft ein Hindernis für die Schiffahrt zu nicht-zivilen Zwecken gesehen hat. Die Behauptung, Seestreitkräfte sollten von der Hohen See ausgeschlossen werden, wäre auch im Hinblick auf die Konferenzgeschichte und Art. 95, 100 DC absurd. Zur Schiffahrtsfreiheit von Kriegsschiffen gehören traditionell Nutzungsarten wie Manöver, Übungen, Waffenversuche, Starten und Anbordnehmen von Flugzeu-

---

[32] So einerseits *Rauch* VI, S. 33 f. (*Rauch,* Politische Konsequenzen und Möglichkeiten der Seerechtsentwicklung aus der Sicht der UdSSR, Berichte des Bundesinstituts für ostwissenschaftliche und internationale Studien [1977] [*Rauch* I–VI, 1977–1980]); andererseits mit Hinweis auf die Entstehungsgeschichte *Oxman*, The Third United Nations Conference on the Law of the Sea: The Ninth Session (1980), AJIL 75 (1981), S. 211 ff., 237 f. Die Friedlichkeit der Nutzung erscheint noch in anderen Zusammenhängen. Siehe Art. 141, 143 Abs. 1, 147 Abs. 2 *lit.* d, 155 Abs. 2, 240 *lit.* a, 242 Abs. 1, 246 Abs. 3 DC sowie unten V. 2.

gen. Aus Art. 301 DC läßt sich nicht ableiten, daß friedliche Zwecke im Sinne der Draft Convention nunmehr nur nicht-militärische Zwecke sein sollen. Richtig ist vielmehr die Auslegung, daß Art. 88 in Verbindung mit Art. 301 DC die Schiffahrtsfreiheit von Kriegsschiffen im Frieden nicht enger beschränkt als das bisher geltende Recht, die Satzung der Vereinten Nationen eingeschlossen. Die der Schiffahrtsfreiheit von Kriegsschiffen immanenten Aktivitäten bleiben weiterhin zulässig. Nur so ist Art. 19 Abs. 2 DC verständlich, der bestimmt, daß verschiedene militärische Aktivitäten in den Küstengewässern die Friedlichkeit der Durchfahrt ausschließen[33]. Zur Schiffahrtsfreiheit für Kriegsschiffe gehören bei Erfüllung bestimmter Voraussetzungen auch Waffenübungen mit (bloßen) Warnzonen, und zwar selbst dann, wenn die Übungen der Erprobung von schiffsgefeuerten Interkontinentalraketen (ohne atomare Sprengkörper) dienen[34]. Unzulässig dürfte

---

[33] Herrschende Meinung. *Treves* führt als Grund für die Zulässigkeit militärischer Aktivitäten u.a. Art. 298 Abs. 1 *lit.* b und Art. 19 DC an; er sieht in Art. 301 eine *formulation ouverte*: La notion d'utilisation des espaces marins à des fins pacifiques dans le nouveau droit de la mer, Annuaire français de droit international 26 (1980), S. 687 ff., 693. Mit weiter ausholender Begründung kommt *Rauch* zu dem gleichen Ergebnis: *Rauch* I (Anm. 32), S. 15 ff., 30; *derselbe,* Militärische Aspekte der Seerechtsentwicklung, in: Graf Vitzthum (Hrsg.), Aspekte der Seerechtsentwicklung (1980), S. 75, 77 ff. Das gleiche Ergebnis bei *Burke*, Submerged Passage Through Straits: Interpretations of the Proposed Law of the Sea Treaty Text, Washington Law Review 52 (1977), S. 201; *Wolfrum,* Restricting the Use of the Sea to Peaceful Purposes: Remilitarization in Being?, German Yearbook of International Law 24 (1981), S. 200 ff., 217. Die Sowjetunion hat ihre Haltung vom Ausschluß militärischer Nutzung zum Ausschluß wohl lediglich der militärischen Tätigkeiten geändert, die gegen die Satzung der Vereinten Nationen verstoßen; Nachweise bei *Treves*, a.a.O., S. 691 f. Siehe auch: Akademie der Wissenschaften der UdSSR – Institut für Staat und Recht – Modernes Seevölkerrecht: Die Rechtsordnung der Gewässer und des Bodens des Weltmeeres (Original von 1974, hier in der Übersetzung von *Rauch* 1978), S. 269 f., im folgenden Rechtsordnung. Dort geht der Bearbeiter *Carev* mit Selbstverständlichkeit von der Schiffahrtsfreiheit aller Klassen von Kriegsschiffen aus. Diese müsse natürlich im Rahmen des Völkerrechts, insbesondere des Gewaltverbotes ausgeübt werden. Siehe zu militärischen Aktivitäten in besonderen Seezonen die folgenden Abschnitte, insbesondere V.

[34] So selbst das SIPRI Yearbook 1975, S. 593 ff., im Abschnitt »Security aspects in the Law of the Sea debate« von *Goldblat.* Auf S. 595 wird vorgeschlagen, in Zukunft größere Manöver und Übungen mit Interkontinentalraketen in Seegebieten mit großem internationalem Schiffsverkehr auszuschließen und bis 200 sm Küstennähe von der Genehmigung des Küstenstaates abhängig zu machen – wohlgemerkt *de lege ferenda*. Ein Verzicht auf diese Übungen sei nicht realistisch. Siehe ferner *Jenisch* Das Recht zur Vornahme militärischer Übungen und Versuche auf Hoher See in Friedenszeiten (1970), S. 70 ff., 106, 126 ff. Beispiele für Schießübungen verschiedener Staaten mit Warnzonen, z.T. zur Erprobung von Interkontinentalraketen ohne Sprengköpfe bei *Whiteman*, Digest of International Law, Bd. 4 (1965), S. 619 ff., 552 f.; *Scheuner*, Entwicklungen im Seekriegsrecht seit dem zweiten Weltkrieg, in: Recht über See, Festschrift Stödter, hrsg. von H. P. Ipsen/K. H. Necker [o.J. 1979], S. 287 ff., 306 f.; *Jenisch*, a.a.O., S. 119 ff. Beispiele speziell für sowjetische Versuche mit Warnzonen vor UNCLOS III bei *Whiteman*, a.a.O., S. 623 f., 626; für das Jahr 1976 bei *Rauch* II (Anm. 32), S. 25.

dagegen die Einrichtung von Sperrzonen mit Verboten für die sonstige Schiffahrt sein. Sie bewirken eine Art von temporärer »Seenahme«. Damit kommen sie zumindest in die Nähe des Verbotes von Art. 89 DC, Teile der Hohen See staatlicher Souveränität zu unterwerfen. Die Abgrenzungsschwierigkeiten, die entstehen können, lassen sich hier nicht erörtern. Art. 87 Abs. 2 DC sucht mit einer Generalklausel alle Interessenkonflikte zu erfassen, die bei dem Aufeinandertreffen von verschiedenen Erscheinungsformen der Freiheit der Hohen See entstehen können: Alle Arten der Meeresfreiheit sollen mit gebührender Rücksichtnahme (*due regard*) auf die Interessen anderer Staaten an ihrer Nutzung von Meeresfreiheit ausgeübt werden. Das ist – von einer Nuance abgesehen (*reasonable regard*) – die Regelung des Art. 2 Abs. 2 der Genfer Konvention über die Hohe See von 1958[35].

Bei diesem herkömmlichen Verständnis der Schiffahrtsfreiheit von Kriegsschiffen ist es nicht erforderlich, die Zulässigkeit ihrer eben erwähnten militärischen Aktivitäten erst damit zu begründen, daß die Freiheit der Hohen See mit der Aufzählung in Art. 87 Abs. 1 DC nicht abschließend definiert ist, sondern daß – ebenso wie in der Konvention über die Hohe See von 1958 – durch die Worte *inter alia* weitere Ausprägungen dieser Freiheit anerkannt sind, welche die erwähnten militärischen Aktivitäten erfassen würden. Für den Fall, daß die obige Auslegung der Schiffahrtsfreiheit im Hinblick auf langandauernde Manöver großer Schiffsverbände mit erheblichen Waffenübungen oder auf Warnzonen für die Erprobung von schiffsgefeuerten Interkontinentalraketen auf Widerspruch stoßen sollte, haben die sonstigen Arten der Meeresfreiheit durch die Worte *inter alia* in beiden einschlägigen Konventionen eine Auffang-Funktion. Für die Einrichtung von Warngebieten zur Erprobung landgefeuerter Interkontinentalraketen rücken die weiteren Erscheinungsformen der Meeresfreiheit auf Hoher See jedoch in die »erste Linie« der Begründung.

Neu ist die Pflicht in Art. 87 Abs. 2 DC zur gebührenden Rücksichtnahme (*due regard*) bei der Ausübung der Schiffahrtsfreiheit auf Rechte zu Tätigkeiten hinsichtlich Meeresboden und Meeresuntergrund außerhalb staatlicher Zuständigkeit. In dem für dieses »Gebiet« (*the Area*) maßgeblichen Teil XI der Draft Convention bestimmt allerdings Art. 135, daß der Rechtsstatus der Gewässer über dem »Gebiet« und des Luftraumes über diesen Gewässern unberührt bleibt. So kann man einstweilen davon ausgehen, daß die Ausbeutung des »Gebietes« jedenfalls nicht zu gezielten Einschränkungen der Schiff-

---

[35] Siehe zu dem Fragenkreis, einschließlich der hier ausgeklammerten Atomversuche, zusammenfassend *Jenisch*, a.a.O., insbesondere S. 112 ff. – Der Zusammenstoß eines amerikanischen Atom-Unterseeschiffes mit einem japanischen Frachtschiff im April 1981 wirft diese Rechtsfragen nicht auf: Schiffahrtsfreiheit im engsten Sinne als Bewegungsfreiheit traf hier auf die gleiche Art der Meeresfreiheit. Der Fall erregte vor allem deshalb internationales Aufsehen, weil das amerikanische dem untergehenden japanischen Schiff nicht die erforderliche Hilfe leistete. Siehe *Rousseau*, RGDIP 85 (1981), S. 875 ff.

fahrtsfreiheit führen wird. Natürlich wird die Ausbeutung des »Gebietes« die freie Schiffahrt *in praxi* einschränken, insbesondere durch Installationen. Art. 147 Abs. 2 DC soll diesen Einschränkungen Grenzen ziehen. Nach *lit.* b dürfen die ansonsten zulässigen Installationen nicht errichtet werden, wo sie die Fahrt (*passage*) durch Seewege behindern können, die für die internationale Schiffahrt lebenswichtig sind. Dennoch können die Einschränkungen der Schiffahrtsfreiheit durch Rechte im »Gebiet« *in praxi* unter Umständen erheblich sein. Die Frage läßt sich hier nicht untersuchen. Die nach Art. 77 Abs. 1 DC durch die Zweckbestimmung inhaltlich begrenzten, aber »souveränen« Rechte des Küstenstaates am Festlandsockel beeinflussen nach Art. 78 Abs. 1 DC den Rechtsstatus von Wassersäule und Luftraum über dem Festlandsockel ebenfalls nicht[36]. Auch hier gilt nach Art. 86, 87 Abs. 1 DC die Schiffahrtsfreiheit. Übt der Küstenstaat seine Rechte am Festlandsockel gemäß Art. 80, 81, 85 DC aus, so darf dies nach Art. 78 Abs. 2 DC jedenfalls nicht zu einer unberechtigten Behinderung (*unjustifiable interference*) der Schiffahrtsfreiheit führen. Man muß hoffen, daß sich die mit Art. 78 Abs. 2 DC verbundenen Erwartungen für die Wahrung der Schiffahrtsfreiheit trotz der Unbestimmtheit der Norm bestätigen werden. Zwar war Art. 5 der Genfer Konvention von 1958 über den Festlandsockel auch nicht genauer als die einschlägigen Vorschriften der Draft Convention. Der Stand der Technik gestattet dem Küstenstaat aber schon heute mehr und größere Bohrungen und Installationen als 1958. Massierungen von Bohrungen und Installationen können zu zunehmenden Behinderungen der Schiffahrt führen.

Zusammenfassend läßt sich etwa feststellen: Sieht man das bisherige Recht im Zusammenhang mit dem völkerrechtlichen Gewaltverbot in der Satzung der Vereinten Nationen, so ändert sich die Schiffahrtsfreiheit von Kriegsschiffen auf Hoher See durch die Draft Convention inhaltlich nicht wesentlich. Die Neueinführung des »Gebietes«, auf das ebenso Rücksicht zu nehmen ist wie auf einen erweiterten und intensiver ausgebeuteten Festlandsockel, dürfte allerdings die Konfliktsmöglichkeiten vergrößern. Die Normen für den Ausgleich aufeinanderstoßender Rechte und Interessen ließen sich unter dem Gesichtspunkt der Rechtsklarheit gewiß verbessern. Man darf aber nicht übersehen, daß nicht jede theoretisch mögliche stets auch die praktikable und für die Praxis bessere Formulierung ist und mag daher die oben erwähnten Generalklauseln dem Versuch zu einer speziellen Regelung vorziehen. Größere Nachteile für die Schiffahrtsfreiheit erwachsen daraus, daß die Draft Convention den Bereich der Hohen See einschneidend vermindert. Ohne ausdrückliche Aussage darüber, ob sie aus dem Rechtsbegriff Hohe See ausgeklammert sind, schließt Art. 86 DC zusätzlich zu den Eigen- und den (nunmehr bis zu 12 sm breiten) Küstengewässern, die seit je nicht zur Hohen See gehörten, auch die Archipelgewässer und die Ausschließli-

---

[36] Allerdings wird die Wassersäule über dem Festlandsockel weitgehend zur Anschlußzone/Ausschließlichen Wirtschaftszone gehören. Dazu unten V.

chen Wirtschaftszonen von der automatischen Anwendung aller Normen über die Hohe See aus. Die Tatsache, daß in Teil V der Draft Convention wesentliche Bestimmungen aus dem Recht der Hohen See auf die Ausschließlichen Wirtschaftszonen für anwendbar erklärt werden, gleicht den Verlust an Schiffahrtsfreiheit – isoliert betrachtet – nicht aus (siehe unten V. 2). Die – an anderen Stellen der Konvention noch stärker ausgeprägten – Hin- und Herverweisungen anstelle grundsätzlicher Zuordnungen der jeweiligen Seegebiete sind als politische Kompromisse erklärbar, im Interesse der Rechtsklarheit aber nicht zu begrüßen.

3. Die Schiffahrtsfreiheit in fremden Eigengewässern

Die Souveränität der Staaten über ihre Eigengewässer ist unbestritten. Das Völkergewohnheitsrecht verpflichtet die Staaten nicht, fremden Kriegsschiffen den Zugang zu ihren Eigengewässern zu gewähren. Sofern nicht vertragliche Verpflichtungen oder die gewohnheitsrechtlichen Ausnahmefälle der Seenot oder sonstiger Arten von höherer Gewalt vorliegen [37], dürfen die Küstenstaaten den Zugang fremder Schiffe, insbesondere fremder Kriegsschiffe, zu ihren Eigengewässern nach Belieben von einer Erlaubnis oder vorherigen Anmeldung abhängig machen. Die Befugnis, fremden Kriegsschiffen den Zugang zu den Eigengewässern völlig zu verweigern, schließt auch die Befugnis ein, die Zulassung von Bedingungen abhängig zu machen. Beispiele sind die Gegenseitigkeit, der Verzicht auf militärische Übungen oder auf die Erteilung von Landurlaub für die Besatzung. Die Draft Convention enthält darüber keine Bestimmungen. Sie läßt den gewohnheitsrechtlichen Rechtszustand unberührt, weil sie keine einschlägigen Regelungen trifft. Dieser Grundsatz kommt in der Draft Convention etwa im letzten Absatz der Präambel zum Ausdruck [38]. Die Staatenpraxis bei der Zulassung ist unterschiedlich [39]. Die Zunahme von Schiffen, die durch Kernkraft getrieben oder mit Kernwaffen bestückt sind, sowie Unfälle von Supertankern können zu einer weiteren Verbreitung restriktiver Praktiken beitragen.

Da die Draft Convention für die Schiffahrtsfreiheit von Kriegsschiffen auf der – reduzierten – Hohen See keine grundsätzlichen Änderungen vorsieht und die volle staatliche Souveränität über die Eigengewässer unberührt läßt, richten sich die wesentlichen Fragen auf die Auswirkungen von UNCLOS III in anderen, z. T. rechtlich erst neu geschaffenen Meeresbereichen [40].

---

[37] *Speranskaja*, in: Modernes Seevölkerrecht (Rechtsordnung, Anm. 33), S. 86 Anm. 31 erklärt die Ansicht für beachtenswert, daß die Einfahrt von Atomschiffen in Häfen nach einem (wohl atomaren) Unfall unter Umständen als Ausnahme von dem Gebot der Hilfeleistung verboten sei. Vgl. dazu auch den Fall unten in Anm. 64.

[38] Siehe auch die Hinweise in Art. 25 Abs. 2, Art. 38 Abs. 2 Satz 2 DC.

[39] Siehe dazu *Colombos*, The International Law of the Sea (6. Aufl. 1972), S. 262–264; Modernes Seevölkerrecht (Rechtsordnung, Anm. 33), S. 83 f.

[40] Hier liegt auch der Schwerpunkt bei *Riphagen*, La navigation dans le nouveau droit de la mer, RGDIP 84 (1980), S. 144 ff.

III. Die Schiffahrtsfreiheit von Kriegsschiffen in fremden Küstengewässern und in Meerengen

1. Küstengewässer

Nach allgemeinem Völkergewohnheitsrecht erstreckt sich die Souveränität des Küstenstaates über das Landgebiet und die Eigengewässer hinaus auf die Küstengewässer, ihren Meeresboden und -untergrund sowie auf den Luftraum über ihnen. Art. 1, 2 der Genfer Konvention von 1958 über das Küstenmeer und die Anschlußzone (Küstenmeerkonvention) sind insofern nur deklaratorisch. Inhaltlich entspricht Art. 2 DC diesen Normen. Fest steht auch, daß die Souveränität des Küstenstaates schon gewohnheitsrechtlich gewissen Einschränkungen unterliegt. Zu ihnen gehört das Recht auf friedliche Durchfahrt (*innocent passage*), zumindest für andere Schiffe als Kriegsschiffe[41].

Grundlage der Beratungen über das Durchfahrtsrecht für Kriegsschiffe auf der Seerechtskonferenz von 1958 waren die Vorarbeiten der International Law Commission. Der ILC-Berichterstatter François sah die friedliche Durchfahrt für Kriegsschiffe in Art. 22 seines ersten Entwurfs vor, für sonstige Schiffe in Art. 15. Nach Stellungnahmen einiger Staaten, mehreren Debatten und Änderungen verabschiedete die International Law Commission schließlich einen Konventionsentwurf, der für Kriegsschiffe eine Ausnahme von dem allgemeinen Recht zur friedlichen Durchfahrt vorsah: Gemäß Art. 24 durfte der Küstenstaat die Durchfahrt von Kriegsschiffen durch seine Küstengewässer von seiner vorherigen Einwilligung oder von einer Anmeldung abhängig machen. Er hatte die friedliche Durchfahrt jedoch im Normalfall (*normally*) zu gestatten, sofern seine Regelungsbefugnisse und die Pflichten beachtet wurden, welche die Konvention allen durchfahrenden Schiffen auferlegte[42]. Der zuständige 1. Ausschuß der Genfer Seerechtskonferenz nahm diesen ILC-Vorschlag mit großer Mehrheit an. Dagegen lehnte das Plenum in einer gesonderten Abstimmung mit 45 gegen 27 Stimmen bei 6 Enthaltungen eine Befugnis des Küstenstaates ab, die Durchfahrt von Kriegsschiffen von seiner Einwilligung abhängig zu machen. Bei der Plenarabstimmung über Art. 24 mit der noch verbliebenen Einschränkung des Durchfahrtsrechtes, der Anmeldung, stimmten 43 Staaten für diese Regelung und 24 – unter ihnen die Sowjetunion – dagegen. 12 Staaten enthielten sich der Stimme. Zu den 24 ablehnenden Staaten gehörten mehrere, die bei der voraufgegangenen Abstimmung über die Genehmigungsbedürftigkeit der

---

[41] Die Übersetzung von »*innocent*« *passage* oder *passage* »*inoffensif*« als »friedliche« Durchfahrt ist zu eng. Sie ist auch im Hinblick auf die Artikel der Draft Convention nicht glücklich, die von »friedlicher« Nutzung sprechen. Siehe oben Anm. 32. Die Übersetzung ist aber so eingebürgert, daß Änderungsvorschläge – etwa unschädliche Durchfahrt – sich nicht durchgesetzt haben und weiterhin aussichtslos erscheinen.

[42] Die Pflichten ergaben sich aus Art. 17, 18 des damaligen Entwurfes: YILC 1956 II, S. 258 f.

friedlichen Durchfahrt unterlegen waren und mit der Ablehnung des Anmeldungserfordernisses die Absicht verbanden, das Recht auf friedliche Durchfahrt für Kriegsschiffe ganz zu Fall zu bringen. Da die Anmeldepflicht nicht die notwendige Zweidrittelmehrheit erlangt hatte, wurde auch sie nicht in die Genfer Küstenmeerkonvention aufgenommen[43].

Art. 14 der Küstenmeerkonvention gewährt allen Schiffen das Recht auf friedliche Durchfahrt. Unterseeschiffe müssen nach Art. 14 Abs. 6 aufgetaucht fahren und ihre Flagge zeigen[44]. Abgesehen von der Garantie für die Durchfahrt durch Meerengen (siehe unten III. 2.) ist der Küstenstaat nach Art. 16 Abs. 3 der Küstenmeerkonvention berechtigt, in bestimmten Zonen seines Küstenmeeres die friedliche Durchfahrt fremder Schiffe vorübergehend zu untersagen (*suspend*), wenn dies für den Schutz seiner Sicherheit wesentlich ist. Dabei sind Diskriminierungen untersagt und ist eine vorherige Bekanntmachung geboten. Art. 16 Abs. 3 gilt auch für Kriegsschiffe. Art. 23 der Konvention betrifft nur Kriegsschiffe: Der Küstenstaat darf ein Kriegsschiff auffordern, sein Küstenmeer zu verlassen, wenn das Kriegsschiff die Vorschriften des Küstenstaates über die Durchfahrt durch das Küstenmeer nicht befolgt und eine Aufforderung nicht beachtet, sich nach diesen Vorschriften zu richten.

An der Genfer Seerechtskonferenz hatten sich 86 Staaten beteiligt. Bei der Schlußabstimmung über die Küstenmeerkonvention stimmten 23 Staaten nicht ab. 61 Delegationen sprachen sich für die Konvention aus; zwei enthielten sich; Gegenstimmen wurden nicht abgegeben. Bis zum Jahresende 1981 waren nur 45 Staaten Vertragsstaaten geworden. Immerhin gehören zu ihnen die Vereinigten Staaten, die Sowjetunion und Großbritannien. Jedoch hatten die Sowjetunion, Weißrußland und die Ukraine, Bulgarien, Rumänien, die Tschechoslowakei und Ungarn sowie Kolumbien von vornherein Vorbehalte gegen das Recht von Kriegsschiffen auf erlaubnisfreie friedliche Durchfahrt durch Küstengewässer gemacht[45]. Andere Staaten hatten den Vorbehalten widersprochen. Dennoch erscheinen unter diesen Umständen und angesichts der Vorgeschichte in der International Law Commission und auf der Seerechtskonferenz nicht alle Zweifel daran ausgeräumt, daß zu Beginn von UNCLOS III das nach allgemeinem Völkergewohnheitsrecht bestehende Recht auf friedliche Durchfahrt Kriegsschiffe auch dann einbezog, wenn kein Einverständnis des Küstenstaates oder jedenfalls eine Mitteilung

---

[43] Siehe die Konferenzberichte der Vereinten Nationen A/CONF.13/39, S. 146; A/CONF.13/38, S. 77 f. (Text französisch).

[44] Die übliche Übersetzung von »Submarine« mit »Unterseeboot« erscheint angesichts der großen Tonnagen heute nicht mehr passend. Für die Ohio-Klasse der US-Strategic Missile Submarines sind in »*Jane*'s« (Anm. 16), S. 626, getaucht 18 700 t Wasserverdrängung angegeben.

[45] Siehe zur Sowjetunion insbes. *Butler* (Anm. 28), S. 59 ff., 64. *Butler* hielt noch 1971 nicht für voraussehbar, ob die Sowjetunion als neue See-Großmacht ihre Haltung ändern würde: S. 70. Siehe dazu unten bei Anm. 52–55.

an ihn vorlag. Entscheidend ist die Frage, ob das Recht auf genehmigungs-/anmeldungsfreie friedliche Durchfahrt von Kriegsschiffen schon *vor* der Genfer Seerechtskonferenz von 1958 bestand. Dann war es in der Küstenmeerkonvention bestätigt – trotz aller auf der Konferenz geäußerten Bedenken. Bestand das Durchfahrtsrecht dagegen nicht, so waren die Vorgänge bei der Entstehung der Küstenmeerkonvention und diese selbst nicht geeignet, ein Durchfahrtsrecht über den Kreis der Vertragsstaaten hinaus zu begründen. Es ist nicht möglich, der Frage hier weiter nachzugehen [46].

Auf der Genfer Seerechtskonferenz von 1958 hatte man sich über die Breite der Küstengewässer nicht einigen können. Zahlreiche Staaten hatten an der 3 sm-Grenze festgehalten, sich aber so starken Forderungen nach einer Ausdehnung gegenüber gesehen, daß eine Regelung im Vertrag nicht zustande kam. In UNCLOS III setzte sich Art. 3 DC mit der Höchstbreite von 12 sm durch, bei der es wohl unabhängig vom Schicksal der Draft Convention (jedenfalls erst einmal) bleiben wird. Die friedliche Durchfahrt von Handelsschiffen durch die erweiterten Küstengewässer war nicht umstritten. Wegen der Ausdehnung der Küstengewässer gewannen die Fragen nach dem Recht von Kriegsschiffen auf friedliche Durchfahrt und nach den Bedingungen dieses Rechtes noch an Bedeutung. Beide Fragen waren in UNCLOS III umstritten. Dabei ging es einer Anzahl von Küstenstaaten in erster Linie darum, sich vor politischem Druck durch fremde Kriegsschiffe zu schützen, d.h. um Sicherheitsinteressen. Für die andere Seite waren entgegengesetzte Interessen wesentlich, nämlich die Bewegungsfreiheit ihrer See- und Luftstreitkräfte in Seegebieten von nunmehr beträchtlicher Ausdehnung. In einem weiteren Sinne kann man auch hier von Sicherheitsinteressen sprechen. So wurde die Kontroverse um die Breite der Küstengewässer unter anderen Vorzeichen, aber mit teilweise gleichem Interessenhintergrund weitergeführt [47].

Nicht nur Schweden (oben I.), sondern eine größere Zahl von Küstenstaaten hatten landesrechtliche Vorschriften erlassen, welche die friedliche Durchfahrt fremder Kriegsschiffe durch die Küstengewässer von einer vorherigen Anmeldung oder Erlaubnis abhängig machten [48]. Andere Staaten er-

---

[46] Vgl. etwa *de Vries Reilingh*, Warships in Territorial Waters, their Right of Innocent Passage, Netherlands Yearbook of International Law 2 (1971), S. 29, 36 ff., 48 ff.; *Oppenheim/Lauterpacht*, International Law, Bd. 1: Peace (8. Aufl. 1955), S. 494. Siehe auch unten Anm. 56.

[47] Siehe zur Interessenlage *Bowett*, The Law of the Sea (1967), S. 7 ff. Zusammenfassung von verschiedenen Vorschlägen etwa bei *Rauschning*, Die Durchfahrt durch nationale Gewässer vor den Küsten, in: Bernhardt/Rudolf (Hrsg.), Die Schiffahrtsfreiheit im gegenwärtigen Völkerrecht (1975), S. 39 ff., 46 ff.

[48] Beispiele in Modernes Seevölkerrecht (Rechtsordnung, Anm. 33), S. 105 ff., sowie in der von wechselnden Herausgebern veröffentlichten Sammlung New Directions in the Law of the Sea. Eine Genehmigung verlangen danach etwa die Sowjetunion (Bd. 1, S. 31); Grenada (Bd. 7, S. 33); Somalia (Bd. 7, S. 59); Burma (Bd. 7, S. 357); Pakistan (Bd. 7, S. 478) und die Malediven (Bd. 9, S. 295); eine Ankündigung der Durchfahrt verlangen etwa Guyana (Bd. 7, S. 115); Mauritius (Bd. 7, S. 416) und die Seychellen (Bd. 8, S. 12).

kannten derartige Bestimmungen nicht oder jedenfalls nicht ausdrücklich an[49]. Noch auf der 9. Sitzung von UNCLOS III 1980 setzten sich zuerst etwa 40, später anscheinend sogar 45 Staaten unter Führung Chinas und der Philippinen für einen Zusatz zu dem heutigen Art. 21 DC ein. Er sollte den Küstenstaaten die Befugnis geben, die Durchfahrt durch die Küstengewässer zu regeln im Hinblick auf »(t)he navigation of warships, including the right to require prior authorization or notification for passage through the territorial sea«. Diese Bestrebungen setzten sich jedoch – wie schon 1958 – nicht durch[50].

Es waren die Vereinigten Staaten und die Sowjetunion, die erkennen ließen, daß sie auf dem Recht von Kriegsschiffen zur friedlichen Durchfahrt durch Küstengewässer bestehen würden, und die mit Hilfe insbesondere ihrer Verbündeten erreichten, daß Art. 17 DC das Recht auf friedliche Durchfahrt vorsieht, ohne zwischen Kriegsschiffen und anderen Schiffen zu unterscheiden. Das Ringen um die friedliche Durchfahrt durch »gewöhnliche« Küstengewässer erhielt zusätzliches Gewicht dadurch, daß dabei gleichzeitig wichtige Vorentscheidungen für Küstengewässer mit Meerengencharakter fielen (unten III. 2.). Das entschiedene Eintreten der Vereinigten Staaten für das Recht auf friedliche Durchfahrt von Kriegsschiffen ist schon deshalb nicht überraschend, weil die USA sich zur Bewahrung der Schiffahrtsfreiheit so lange gegen die Ausdehnung der Küstengewässer über 3 sm hinaus gewehrt hatten. Es war nicht nur unter Berücksichtigung ihrer seestrategischen Interessen konsequent, daß die USA die Ausdehnung nur bei Sicherungen für die Schiffahrtsfreiheit auch von Kriegsschiffen akzeptieren wollten[51].

Die Sowjetunion hatte die 12 sm-Breite schon von dem Zaristischen Rußland übernommen und war stets für ihre völkerrechtliche Zulässigkeit eingetreten. Sie hatte gegen das Recht auf friedliche Durchfahrt von Kriegs-

---

[49] Siehe BYIL 49 (1978), S. 395; AJIL 74 (1980), S. 9.

[50] Siehe die Aufzählung der Staaten und der Konferenzdokumente bei *Rauch* VI (Anm. 32), S. 20 f., 53 f. Etwas andere Zahlen bei *Wolf*, Die Dritte Seerechtskonferenz der Vereinten Nationen – Beiträge zur Reform der internationalen Ordnung und Entwicklungstendenzen im Nord-Süd-Verhältnis (1981), S. 215 f., 369 f. Ein letzter Versuch mit gleicher Zielsetzung scheiterte im April 1982, siehe oben Anm. 28a und die Dokumente A/CONF.62/L.97, A/CONF.62/L.117. – Die Draft Convention sieht auch keine militärischen Sicherheitszonen vor, obwohl – nach *Darman* – 1977 32 Staaten irgendwelche Warn-, Überwachungs- oder sonstige Zonen beanspruchten. *Darman*, The Law of the Sea: Rethinking U.S. Interests, Foreign Affairs 56 (1977/78), S. 373 ff., 378. Nach dem Modernen Seevölkerrecht (Wissenschaftliche Forschung, Anm. 31), S. 270, gibt es in der Praxis etwa 20 verschiedene Bezeichnungen für unterschiedliche Sicherheitszonen.

[51] Die Vereinigten Staaten hatten ihre Interessen im Hinblick auf die friedliche Durchfahrt von Kriegsschiffen durch Küstengewässer allerdings zeitweilig anders verstanden. Auf der Haager Kodifikationskonferenz 1930 führten die USA die Gegner des Durchfahrtsrechtes an, während die Sowjetunion sich zurückhielt. Nach 1945 waren die Vereinigten Staaten – inzwischen die stärkste Seemacht der Welt – der Hauptbefürworter und die Sowjetunion der Hauptgegner. So *O'Connell* (Anm. 9), S. 139.

schiffen durch Küstengewässer in der Genfer Küstenmeerkonvention von 1958 die oben erwähnten Vorbehalte gemacht und das Bestehen eines Durchfahrtsrechts nach allgemeinem Völkergewohnheitsrecht entschieden verneint. In dem einschlägigen Kapitel des 1967 vom Institut für Staat und Recht der Sowjetischen Akademie der Wissenschaften veröffentlichten Völkerrechtshandbuches wurden die Versuche bürgerlicher Juristen zur Rechtfertigung eines Durchfahrtsrechts in Parallele zu der aggressiven Politik der imperialistischen Mächte gesehen, die mit ihren Streitkräften den freien Zugang in Küstengewässer anderer Staaten sichern wollten [52]. Noch 1974 erklärte das maßgebliche Werk der Akademie der Wissenschaften ausdrücklich, es gebe im *Völkerrecht* keine Norm, die die ungehinderte Durchfahrt von Kriegsschiffen erlaube. Es berief sich auf die Gesetzgebung vieler Staaten, Arbeiten der International Law Commission, die Beratungen auf der Seerechtskonferenz von 1958 und die Vorbehalte zu der Küstenmeerkonvention. Eine andere Rechtslage würde die Souveränität der Küstenstaaten verletzen und grundlegenden Prinzipien des Völkerrechts widersprechen [53]. Der Aufbau einer globalen sowjetischen Seemacht, die den amerikanischen Seestreitkräften gleichwertig sein sollte, führte in der Frage des Durchfahrtsrechtes zu einer vollständigen Kehrtwendung der sowjetischen Politik [54], ohne die es nicht zu der Regelung von UNCLOS III gekommen wäre. Dabei ging es wahrscheinlich gerade der Sowjetunion noch mehr um die Durchfahrt durch Meerengen innerhalb von Küstengewässern als um die »gewöhnlichen« Küstengewässer [55].

Das Recht auf friedliche Durchfahrt nach Art. 17 DC steht Schiffen aller Staaten zu ohne Unterschied zwischen Handels- und Kriegsschiffen [56].

---

[52] *Kolodkin* und *Molodcov*, Seefriedensrecht, in: Das völkerrechtliche Regime der Territorialgewässer, der Anschlußzone und des Hohen Meeres (Original 1967, hier Übersetzung von *Hermann Weber*, 1973), S. 49 ff.; siehe auch *Rauch* I, S. 78 ff.

[53] Modernes Seevölkerrecht (Rechtsordnung, Anm. 33), S. 107 [Verfasser *Speranskaja*]. Siehe auch die Stellungnahme aus der DDR von *Haalck/Reintanz* 1972, daß die sozialistischen Staaten die Interessen aller um ihre Sicherheit besorgten Länder vertreten, die Wert darauf legten, sich vor fremder Einflußnahme innerhalb ihres Landes zu sichern. *Haalck/Reintanz*, Internationales Seerecht – Leitfaden für Seeoffiziere [o.J. 1972], S. 91.

[54] Siehe zu den Flottenstärken der beiden Supermächte die Übersichten in *Jane*'s Fighting Ships und *Weyers* Flottentaschenbuch Warships of the World sowie die Einleitungsbemerkungen in diesen Bänden (Anm. 15, 16). Das SIPRI Yearbook 1979 hält ein annäherndes Gleichgewicht zwischen den Seestreitkräften von NATO und Warschauer Pakt für gegeben: *Karkoszka*, 6. Kapitel: The expansion of naval forces, S. 329 ff., 354 ff., 370. Bemerkenswert, aber nicht überraschend ist die Zurückhaltung bei der Darstellung der sowjetischen Seeziele und Möglichkeiten in dem Buch von Flottenadmiral *Gorschkow* (Anm. 24), siehe etwa S. 3 ff.

[55] Siehe unten III. 2.

[56] Es ist unverständlich, daß *Maduro*, Passage through international straits: The prospects emerging from the Third United Nations Conference on the Law of the Sea, Journal of Maritime Law and Commerce 12 (1980), S. 85 f., das Recht auf friedliche Durchfahrt nach einem älteren Text der Draft Convention ohne Genehmigung/Anmeldung noch anzweifelt. Ähnliche Zweifel für die Konvention von 1958 trotz des

Art. 18 DC definiert den Begriff Durchfahrt. Art. 20 DC schreibt ebenso wie schon die Küstenmeerkonvention bei »submarines« aufgetauchtes Fahren und das Zeigen der Flagge vor; das gleiche gilt für die – neu aufgenommenen – »other underwater vehicles«. Der Begriff der Friedlichkeit ist in Art. 19 DC eingehender umschrieben als in Art. 14 Abs. 4 der Küstenmeerkonvention. Nach beiden Normen soll sich die friedliche Durchfahrt in Übereinstimmung mit dem jeweiligen Vertrag und anderen Regeln des Völkerrechts vollziehen. In beiden Normen wird die Durchfahrt als friedlich definiert, »so long as it is not prejudicial to the peace, good order or security of the coastal state«. Während die Konvention von 1958 es – von Regelungen für Fischereifahrzeuge abgesehen – bei diesen Generalklauseln bewenden läßt, werden sie in Art. 19 Abs. 2 DC in 12 Unterabsätzen *lit.* a–l und noch mehr Handlungsformen aufgeschlüsselt. Art. 19 Abs. 2 *lit.* g und *lit.* i dürften für Kriegsschiffe kaum Bedeutung haben. Die vorsätzliche und beträchtliche (*serious; lit.* h) Umweltverschmutzung im Widerspruch zur Draft Convention könnte für bestimmte Kriegsschiffe und bestimmte andere Schiffe als Grund für die Disqualifizierung der Durchfahrt eine zu enge Umschreibung sein. Große Gefahren für die Umwelt können durch atomgetriebene oder -bewaffnete Schiffe wie durch Supertanker auch ohne Vorsatz entstehen. Speziell für Kriegsschiffe sind von den Aktivitäten, die eine Durchfahrt durch das Küstenmeer nicht als friedlich erscheinen lassen, die folgenden wichtig: Jede Gewalt oder Gewaltandrohung gegen die Souveränität, territoriale Unversehrtheit oder politische Unabhängigkeit des Küstenstaates; jede Gewalt oder Gewaltandrohung, die in irgendeiner anderen Weise gegen die Grundsätze des Völkerrechts verstößt, die in der Satzung der Vereinten Nationen verkörpert sind; jede Übung mit Waffen irgendeiner Art; jede Tätigkeit zur Sammlung von Informationen zum Nachteil für die Verteidigung oder Sicherheit des Küstenstaates; jede Propaganda mit dem Ziel, die Verteidigung oder Sicherheit des Küstenstaates zu beeinträchtigen; jedes Absetzen, Landen oder Aufnehmen von Luftfahrzeugen oder irgendwelchen militärischen Mitteln (*device*), jede Handlung zur Störung von Kommunikationssystemen oder irgendwelchen anderen Einrichtungen des Küstenstaates; Forschungs- oder Vermessungstätigkeiten sowie *last not least* jede andere Tätigkeit, die nicht unmittelbar mit der Durchfahrt zusammenhängt. Vor allem die letzte Generalklausel reicht sehr weit. Jedenfalls die meisten dieser Tätigkeiten schließen schon nach der Küstenmeerkonvention von 1958 den friedlichen Charakter der Durchfahrt aus, auch wenn sie dort nicht ausdrücklich angeführt sind. Diese

---

ebenfalls klaren Textes bei *Butler* (Anm. 28), S. 65, und *Rojahn*, Die Durchfahrt durch internationale Meerengen, in: Bernhardt/Rudolf (Anm. 47), S. 79, 103, sowie neuestens – unter Berufung auf die Entstehungsgeschichte – bei *Wolfrum* (Anm. 33), S. 230. Realistisch erscheint dagegen die Annahme *Wolfrums,* daß manche Staaten jedes Argument nutzen werden, das ihnen die Draft Convention bietet, um landesrechtliche Bestimmungen über die Genehmigungsbedürftigkeit der Durchfahrt von Kriegsschiffen oder über eine Anmeldepflicht aufrechtzuerhalten (ebenda, S. 233 f.).

Beschränkungen erscheinen unter dem Gesichtspunkt der Schiffahrtsfreiheit im allgemeinen wie speziell für Kriegsschiffe *grundsätzlich* sachgerecht.

Sehr wichtig ist, daß die Aufzählung in Art. 19 Abs. 2 DC objektive, wenn auch teilweise generalklauselartige Maßstäbe setzt, die nur auf das Verhalten des Schiffes abstellen, und daß die Aufzählung abschließend ist. Es sollen derzeit in dieser Frage keine anderen Völkerrechtsregeln eingreifen, obwohl dies nach Art. 2 Abs. 3 DC in Verbindung mit dem letzten Absatz der Präambel denkbar wäre. Für den abschließenden Charakter sprechen schon in Art. 19 Abs. 2 die Worte »in any of the following activities« und das Fehlen des in anderen Konventionsbestimmungen enthaltenen »inter alia« bei der Aufzählung. Wesentlich ist auch die Tatsache, daß angesichts der zahlreichen, überwiegend generalklauselartigen Formulierungen bei vernünftiger Betrachtung ein ungeschriebenes Mehr an Ausschließungsgründen für die Friedlichkeit der Durchfahrt nicht vorstellbar ist, ohne die friedliche Durchfahrt dem Belieben der Küstenstaaten anheimzugeben[57].

Art. 25 Abs. 1 DC ermächtigt den Küstenstaat zu den notwendigen Maßnahmen (*necessary steps*), um eine nicht friedliche Durchfahrt zu verhindern. Der Küstenstaat behält damit die schon in Art. 16 Abs. 1 der Genfer Küstenmeerkonvention enthaltene, generalklauselartig umrissene Befugnis zum ersten Zugriff[58]. Jedoch sind die Voraussetzungen genauer umschrieben, so daß sie jedenfalls nicht im Belieben des Küstenstaates stehen. Im Streitfall muß der Inhalt der unbestimmten Rechtsbegriffe objektiv durch Auslegung ermittelt werden.

Der Küstenstaat darf die friedliche Durchfahrt nach Art. 21 DC für bestimmte Zwecke regeln. Auch hier ist der abschließende Charakter bei der Aufzählung der Regelungsbefugnisse wichtig[59]. Im Hinblick auf Kriegsschiffe können insbesondere die Regelungsbefugnisse zugunsten der Schiffssicherheit und des Seeverkehrs (sehr weit) sowie zum Schutze nicht näher bezeichneter küstenstaatlicher Einrichtungen, der lebenden Hilfsquellen des Meeres und der Umwelt des Küstenstaates erheblich werden. Zu einer (räumlichen) Einengung des Bereichs für die friedliche Durchfahrt kann Art. 22 DC führen. Er ermächtigt den Küstenstaat, Schiffen auf friedlicher Durchfahrt bestimmte Seewege und Systeme zur Trennung des Verkehrs vorzuschreiben, wenn dies im Hinblick auf die Sicherheit der Schiffahrt erforderlich ist. Bestimmte Seewege dürfen insbesondere für Tanker, kernkraftgetriebene Schiffe sowie Schiffe mit spaltbaren, besonders gefährlichen oder schäd-

---

[57] Für den abschließenden Charakter der Aufzählung schon *Stöcker*, Die »Freiheit der Meere« und die Dritte Seerechtskonferenz der Vereinten Nationen – Anmerkungen aus der Sicht der deutschen Handelsschiffahrt, in: Festschrift Stödter (Anm. 34), S. 315 ff., 329. Etwas vorsichtiger *Jaenicke*, Die Dritte Seerechtskonferenz der Vereinten Nationen, Grundprobleme im Überblick, ZaöRV 38 (1978), S. 438, 467. Für eine lediglich beispielhafte Aufzählung *Rauch*, Aspekte (Anm. 33), S. 95.

[58] Siehe aber Art. 30 DC – unten S. 306, 310.

[59] Siehe außer dem Einleitungssatz von Art. 21 auch Art. 24 Abs. 1 DC.

lichen Stoffen vorgeschrieben werden. Zu den Schiffen »carrying nuclear ... substances or materials« dürften auch atomar bewaffnete Schiffe gehören. Damit ermöglicht Art. 22 DC es dem Küstenstaat, die Bewegungsfreiheit zahlreicher und besonders kampfstarker Kriegsschiffe wesentlich einzuschränken. Der Küstenstaat hat bei Ausübung dieser Befugnisse nach Art. 22 Abs. 3 DC allerdings verschiedene Grenzen zu berücksichtigen, die einen Mißbrauch verhindern sollen; er hat auch für gebührende Publizität zu sorgen. Die Formulierungen des Art. 22 Abs. 3 DC lassen dem Küstenstaat einen nicht unerheblichen Spielraum. Zudem bedeutet »Berücksichtigung« natürlich weniger an Bindung als etwa »Beachtung«. Art. 23 DC will die Kontrolle des Küstenstaates über die kraft Antrieb, Bewaffnung oder Ladung besonders gefährlichen Schiffe erleichtern. Schließlich darf der Küstenstaat auf Grund von Art. 25 Abs. 3 DC in von ihm zu spezifizierenden Gebieten seines Küstenmeeres nach angemessener Veröffentlichung die friedliche Durchfahrt zeitweilig aussetzen (*suspend*). Voraussetzung ist, daß dies für den Schutz seiner Sicherheit, einschließlich von Waffenübungen, wesentlich ist. Dabei ist jede Diskriminierung von fremden Schiffen unzulässig.

Angesichts aller dieser einschränkenden Voraussetzungen für die friedliche Durchfahrt und der weiten Regelungsbefugnisse des Küstenstaates ist Art. 24 DC wichtig. Er statuiert zusammenfassend noch einmal, daß alle Einschränkungen durch den Küstenstaat nur in Übereinstimmung mit diesem Vertrag erfolgen dürfen und daß bei allen nach dem Vertrag zulässigen Einschränkungen das Recht auf friedliche Durchfahrt auch *in praxi* gewährleistet bleiben muß. Art. 236 DC über die »sovereign immunity« bewirkt u.a., daß auch die Befugnisse von Küstenstaaten zum Schutz der Meeresumwelt die Schiffahrtsfreiheit von Kriegsschiffen nicht behindern. Wie schon nach Art. 23 der Genfer Küstenmeerkonvention darf der Uferstaat nach Art. 30 DC fremde Kriegsschiffe auffordern, seine Küstengewässer zu verlassen, nunmehr »sofort«. Voraussetzung ist, daß sie die (zulässigen) Normen des Küstenstaates über die Durchfahrt durch die Küstengewässer nicht befolgt und eine entsprechende Aufforderung nicht berücksichtigt haben.

Die Regelung der Schiffahrtsfreiheit für Kriegsschiffe in der Draft Convention läßt sich hier und jetzt nicht in allen Aspekten würdigen. Jedoch lassen sich im Hinblick auf die Küstengewässer einige Gesichtspunkte hervorheben. Ausgangspunkt ist die hier nicht näher belegbare Annahme, daß Meeresfreiheit in Form der Schiffahrtsfreiheit einen Rechtswert darstellt und daß dieser in der Welt souveräner Staaten nicht nur im Hinblick auf Handelsschiffe im weitesten Sinne besteht, sondern – wenn auch z.T. mit anderer Begründung – im Hinblick auf Kriegsschiffe. Bei dieser Betrachtungsweise bedürfen Einschränkungen der Schiffahrtsfreiheit einer Begründung. Sie kann sich aus überwiegenden Interessen von Gruppen von Staaten (Küstenstaaten) oder vorrangigen Gemeininteressen ergeben.

Die Küstenstaaten haben schon wegen der veränderten technischen Gegebenheiten ein legitimes Interesse an einer Erweiterung des Küstenmeeres über die 3 sm-Zone hinaus. Sie sind heute mehr gefährdet als noch in der Mitte des 20. Jahrhunderts. Die ungeheuren Schäden durch die Unfälle von Supertankern sind nicht vergessen. Normierung der Durchfahrt und Aufsicht durch die Küstenstaaten sind zum mindesten eine Ergänzung internationaler Abkommen sowie flaggenstaatlicher Normsetzung und Aufsicht. Über die Frage, ob nicht schon eine Küstenmeerbreite von 6, 8 oder 10 sm genügen würde, mag man streiten. Eine gewisse Ausdehnung der Küstengewässer ist jedenfalls sachgerecht. Insoweit mußte ein Stück Meeresfreiheit geopfert werden. Im übrigen ist für die Beurteilung der Draft Convention im Hinblick auf die Schiffahrtsfreiheit von Kriegsschiffen zu beachten, daß die Breite der Küstengewässer umstritten war, daß eine Anzahl von Staaten Küstengewässer bis zu 200 sm Breite durchzusetzen suchte[60] und daß gegen das Recht von Kriegsschiffen zur friedlichen Durchfahrt immer wieder Bedenken geltend gemacht wurden. Geht man von diesen Prämissen aus, so kann man trotz der Ausdehnung der traditionellen Küstengewässerbreite von 3 sm auf bis zu 12 sm und trotz der weitgehenden Regelungsbefugnisse des Küstenstaates nach der Draft Convention in dieser insofern einen Gewinn für die Schiffahrtsfreiheit von Kriegsschiffen sehen, als sie feste und jedenfalls nicht unerträglich weitergehende räumliche Grenzen vorsieht. Für die Seegebiete, die schon nach altem Recht Küstengewässer waren, bringt die neue Detailregelung der Durchfahrt mehr Rechtssicherheit. Möglicherweise kommt die detailliertere Definition der Friedlichkeit der Durchfahrt eher Drittstaaten zugute als dem Küstenstaat, der nach altem Recht mehr Beurteilungsspielraum oder gar Ermessen hatte. Ähnlich könnte es sogar im Hinblick auf die Regelungsbefugnisse des Küstenstaates sein[61]. Dagegen tritt ein Verlust an Schiffahrtsfreiheit in den Küstengewässern ein, die bisher (unbestritten) zur Hohen See gehörten. Hier wird ein Teil der verhältnismäßig regelungsfreien und konfliktsfreien Hohen See in intensiv geregeltes, eher konfliktsträchtiges Küstenmeer umgewandelt. Das Seegebiet, in dem Unterseeschiffe aufgetaucht fahren müssen, ist erheblich ausgedehnt worden. Der Luftraum zum freien Überflug ist entsprechend beschnitten. Legt man diese Änderungen ohne sonstiges Abwägen des Für und Wider bei der Neuregelung der Küstengewässer zugrunde, so erscheint der Ausgleich inhaltlich vertretbar, den die Draft Convention zwischen den territorialen Souveränitätsansprüchen und den verschiedenen Bedürfnissen von Küstenstaaten einerseits und dem Grundsatz der Schiffahrtsfreiheit mit den entsprechenden Interessen von großen und kleinen Seemächten andererseits vorsieht. Die Draft Convention gibt den Kriegsschiffen grundsätzlich nicht mehr an Schiffahrtsfreiheit als sie jedenfalls nach der Genfer Küstenmeerkonvention von 1958 hatten. Die Definition der Friedlichkeit einer Durchfahrt und die Regelungsbefugnisse der

[60] Nach *Rauch* V (Anm. 32), S. 13, nahmen im August 1979 22 Staaten 3 sm in Anspruch, 76 Staaten 12 sm, 14 Staaten 200 sm und die übrigen zwischen 3 und 200 sm.

[61] Vgl. *Jaenicke* (Anm. 57), S. 467 f.

Küstenstaaten dürften diesen in Friedenszeiten hinreichenden Schutz vor Gefahren und Schäden geben, die der Flaggenstaat nicht verursachen will. Die Konventionsbestimmungen dürften – sofern sie nicht bewußt verletzt werden – die Küstenstaaten auch in dem Punkte hinreichend schützen, der auf der Konferenz immer bedacht wurde, auch wenn man ihn nicht erwähnte, nämlich bei der Sicherung gegen politisch-militärischen Druck durch küstennahe Seestreitkräfte.

Eine Ausnahme von der grundsätzlich positiven Wertung wäre vielleicht bei höchst gefährlichen Schiffen angebracht. Diese in der Draft Convention nicht enthaltene Kategorie müßte enger sein als die des Art. 22 Abs. 2 DC; sie müßte allerdings die Schiffe mit atomarem Antrieb oder atomaren Waffen erfassen. Die furchtbaren Folgen einer Kernwaffenexplosion über bewohntem Gebiet sind verhältnismäßig bekannt[62]. Die Folgen von nuklearen Explosionen in Küstengewässern dürften wahrscheinlich geringer sein. Aber auch sie erscheinen so furchtbar, daß man fragen kann, ob die Öffnung der Küstengewässer für die friedliche Durchfahrt von Atomschiffen jeder Art dem Küstenstaat kein unangemessenes Risiko auferlegt[63]. Die Hoheitsakte, die Art. 21 DC gestattet, könnten bei einer atomaren Katastrophe zu spät kommen. Man kann fragen, ob die oben erwähnten Befugnisse des Küstenstaates, Seewege und Systeme der Verkehrstrennung vorzuschreiben, insoweit hinreichend sind, zumal – anders als bei Handelsschiffen – die besonderen Vorsichtsmaßnahmen auf Grund von völkerrechtlichen Verträgen gemäß Art. 23 DC speziell für höchst gefährliche Kriegsschiffe (noch) gar nicht existieren, vielleicht nie existieren werden[64]. Eine Antwort mit möglicher

---

[62] Siehe etwa den Bericht des VN Generalsekretärs General and Complete Disarmament: Comprehensive Study on Nuclear Weapons (1980), hier in der deutschen Übersetzung die UNO-Studie: Kernwaffen (1982), insbes. S. 69 ff.

[63] Siehe etwa *Reich*, Folgen einer Kernexplosion in der Nordsee, in: C. F. von Weizsäcker (Hrsg.), Kriegsfolgen und Kriegsverhütung (1971), S. 677 ff.

[64] Im Falle eines durch Brand manövrierunfähigen sowjetischen atomar getriebenen Unterseeschiffes war Japan nach dem Recht der Genfer Küstenmeerkonvention von 1958 nur deshalb berechtigt, vor dem Abschleppen durch seine Küstengewässer Auskunft über eine atomare Bewaffnung zu verlangen, weil der sowjetische Vorbehalt zu Art. 23 der Konvention umgekehrt nun zugunsten Japans wirkte. Siehe *Grammig*, The Yoron Jima Submarine Incident of August 1980: A Soviet Violation of the Law of the Sea, Harvard International Law Journal 22 (1981), S. 331 ff. Kurzdarstellung von *Rousseau*, RGDIP 85 (1981), S. 184 f. Nach Art. 22 DC könnte der Küstenstaat im Hinblick auf solche Fälle Seewege und Systeme zur Verkehrstrennung vorschreiben. Als Rechtsgrundlagen für ein Vorgehen nach Art. 21 kommen Abs. 1 *lit.* a, d und f in Betracht. Im Hinblick auf den Seetransport von Produkten der deutschen Chemieindustrie hat *Stöcker* Bedenken gegen Art. 22, 23 DC wegen zu weiter Eingriffsbefugnisse (Anm. 57, S. 333 f.). Diese Fälle sind oben im Text mit dem Begriff der *höchst* gefährlichen Schiffe nur z. T. gemeint. Man müßte trotz der großen Schwierigkeiten in diesem Punkte doch wohl noch weiter differenzieren. Siehe zu Teilaspekten etwa *Pelzer*, Aktuelle internationalrechtliche Probleme der friedlichen Reaktorschifffahrt, in: Bernhardt/Rudolf (Anm. 47), S. 321 ff.; *Rauschning*, Die Reaktorschiffahrt im Lichte der Entwürfe der Dritten Seerechtskonferenz zur Schiffahrtsfreiheit, Jahrbuch für internationales Recht 18 (1975), S. 283 ff.

Tendenz zu einer engeren Beschränkung der Durchfahrt in solchen Fällen brauchte nicht unbedingt eine Vorentscheidung für die gleiche Fragestellung im Recht der Meerengen zu bedeuten. Zwar können die Folgen einer Atomkatastrophe unter Umständen dort noch größer sein als in den »gewöhnlichen« Küstengewässern. In Meerengen dürften die Durchfahrtzeiten aber regelmäßig kürzer und die Gefahren insofern geringer sein. Zudem hat für Drittstaaten die Durchfahrt durch Meerengen im allgemeinen eine viel größere Bedeutung als die durch »gewöhnliche« Küstengewässer.

Vom Blickpunkt dritter Staaten her gesehen kann die Draft Convention bei sachgerechter Anwendung den Staaten, denen lediglich an einer wirklich friedlichen Durchfahrt – im Unterschied etwa zu Flottendemonstrationen – gelegen ist, das für sie Wesentliche geben (im Falle der höchst gefährlichen Schiffe vielleicht sogar mehr, als einem billigen Interessenausgleich entspricht). Die in der Literatur verschiedentlich kritisierte Einführung der Generalklausel des Art. 19 Abs. 2 *lit.* l DC ist nicht nur als politischer Kompromiß verständlich. Die Generalklausel dürfte für die Flaggenstaaten fremder Kriegsschiffe auch erträglich sein, solange sie vom Küstenstaat nicht mißbräuchlich ausgenutzt wird. Ferner ist nicht einzusehen, daß Unterseeschiffen auch in »gewöhnlichen« Küstengewässern das Recht zum untergetauchten Fahren zustehen muß, das ihnen in Meerengen eingeräumt ist. Die »gewöhnlichen« Küstengewässer haben für die Durchfahrt von Kriegsschiffen im allgemeinen nicht annähernd die gleiche Bedeutung wie eine erhebliche Zahl von Meerengen. Diese Aussagen, die natürlich nicht frei von subjektiven Elementen sind, werden im Hinblick auf Kriegsschiffe dadurch erleichtert, daß die Einschränkung ihrer Schiffahrtsfreiheit in den verbreiterten Küstengewässern wegen Art. 30 DC nicht zu unmittelbaren Eingriffen gegen sie führen darf, wie sie bei sonstigen Schiffen zulässig sind.

Da der Teufel bekanntlich im Detail steckt, bleiben Fragen, von denen hier nur einige angesprochen werden können. Ist – zum Beispiel – das unter dem Gesichtspunkt der Interessenabwägung grundsätzlich vertretbare Verbot in Art. 19 Abs. 2 *lit.* l DC von »any other activity not having a direct bearing on passage« im Einzelfall auch praktikabel? Oder: Die Durchfahrt ist nach herkömmlichem Verständnis grundsätzlich nicht friedlich, wenn das Schiff seine Geschütze schußbereit auf die Küsten gerichtet hat. Wie lassen sich die Prinzipien aus dem *Korfu Kanal*-Fall auf Abschußvorrichtungen für Raketen anwenden? Welche Maßstäbe sollen hier gelten[65]? Sind die Befugnisse

[65] Verschiedene Bestimmungen des Art. 19 DC lassen sich kaum auf ihre Einhaltung überwachen – auch beim Überwasserfahren. Grundsätzlich darf das Statuieren einer Rechtspflicht nicht an Schwierigkeiten bei der Überprüfung ihrer Einhaltung scheitern. Andererseits ist es problematisch, Rechtspflichten zu statuieren, wenn die Unmöglichkeit der Überwachung feststeht. Ein Dilemma! – Die Frage nach den rechtlich erlaubten Mitteln für den Küstenstaat, um ein unzulässigerweise untergetaucht fahrendes Unterwasserschiff zum Auftauchen oder Verlassen der Küstengewässer zu zwingen, überschreitet den Raum für diese Arbeit.

des Küstenstaates für »the safety of navigation and the regulation of maritime traffic« (Art. 21 Abs. 1 *lit.* a DC) auch für die Praxis einigermaßen mißbrauchsfrei umrissen? Wie weit kann der Küstenstaat nach Art. 22 Abs. 1 und 2 DC bei der Festlegung von Seewegen und Systemen der Verkehrstrennung innerhalb der ihm von Art. 22 Abs. 3, 24 Abs. 1 DC gezogenen Grenzen gehen? Daß solche Unsicherheiten bleiben, läßt sich allerdings bei einem völkerrechtlichen Vertrag mit weitgehend unumgänglichen Generalklauseln kaum vermeiden. Dem möglichen Mißbrauch von Generalklauseln läßt sich bei heftig umstrittenen völkerrechtlichen Regelungen auch nicht mit der Forderung nach immer mehr Detailregelungen begegnen. Detailregelungen brauchen nicht besser zu sein. Vor allem sind sie bei manchen Materien und Interessenkonstellationen noch weniger konsensfähig als die auslegungsbedürftigen Generalklauseln. In Zweifelsfällen könnte eine Vermutung für den Vorrang entweder der küstenstaatlichen Souveränität oder aber der Schiffahrtsfreiheit in Form der friedlichen Durchfahrt nützlich sein. Die Draft Convention stellt eine derartige Vermutung jedoch nicht auf. Die Küstengewässer gehören traditionell zum Staatsgebiet und unterstehen der staatlichen Souveränität. Das Interesse der Küstenstaaten an ihnen ist grundsätzlich größer als das von Drittstaaten. Die Draft Convention geht von diesen Tatsachen aus. Am Anfang von Teil II – in Art. 2 Abs. 1 – steht die Souveränität. Eine Vermutung zugunsten der friedlichen Durchfahrt ist ausgeschlossen[66]. Umgekehrt ist eine Vermutung zugunsten der staatlichen Souveränität zumindest problematisch. Auch das Recht auf friedliche Durchfahrt hat seine lange Tradition. Art. 2 Abs. 3 DC bestimmt ausdrücklich, daß die Souveränität gemäß der Konvention ausgeübt wird. In dieser ist die friedliche Durchfahrt Ausdruck einer Grundentscheidung und Gegenstand eines eingehenden Unterabschnittes. Art. 24 DC unterstreicht das große Gewicht dieser Grundentscheidung im System.

Eine weitere grundsätzliche Frage wirft Art. 30 DC auf. Schließt er gegenüber Kriegsschiffen die Befugnis des Küstenstaates nach Art. 25 Abs. 1 aus, in seinen Küstengewässern die notwendigen Schritte zu unternehmen, um von vornherein eine Durchfahrt zu verhüten, die nicht friedlich ist? Die Fragen lassen sich vermehren. Sie münden in die Frage nach einer wichtigen Wertung: Sind der Gewinn an Rechtssicherheit[67] durch die abschließenden – wenn auch generalklauselartigen – Spezifizierungen insbesondere in Art. 19–23 DC und damit die – partielle – Minderung der Konfliktsmöglichkeiten größer als die Nachteile aus der Vermehrung von Konfliktsmöglichkeiten, die aus der Umwandlung von Hoher See in Küstengewässer entstehen? Eine Antwort wäre auch dann, wenn die Konvention allgemein angenommen würde, erst nach langjährigen Erfahrungen möglich.

---

[66] Eine Vermutung zugunsten küstenstaatlicher Souveränität nimmt *Reisman*, The Regime of Straits and National Security: An Appraisal of International Lawmaking, AJIL 74 (1980), S. 65, an.

[67] Ihn betont etwa *Jaenicke* (Anm. 57), S. 467 f.

2. Meerengen

In Ermangelung von vertraglichen Sonderregelungen wie z. B. für Bosporus und Dardanellen bestand und besteht nach allgemeinem Völkergewohnheitsrecht für Kriegsschiffe ein Recht auf friedliche Durchfahrt durch fremde Meerengen, die der internationalen Schiffahrt zwischen Teilen der Hohen See dienen[68]. Art. 16 Abs. 4 der Genfer Küstenmeerkonvention von 1958 bestätigt dieses Recht für Meerengen, die der internationalen Schiffahrt zwischen einem Teil der Hohen See und einem anderen Teil der Hohen See oder dem Küstenmeer eines fremden Staates dienen. Die Konvention erklärt jede Suspendierung dieses Rechtes ausdrücklich für unzulässig. Sie behandelt die friedliche Durchfahrt durch Meerengen im übrigen aber wie die friedliche Durchfahrt durch sonstige Küstengewässer. Insbesondere besteht kein Überflugrecht; Unterseeschiffe müssen aufgetaucht fahren.

Mit der Ausweitung der Küstengewässer durch die Draft Convention bis zu maximal 12 sm hat das Durchfahrtsrecht durch Meerengen erheblich an Bedeutung gewonnen. Über die Zahl der internationalen Meerengen innerhalb von Küstengewässern mit 12 sm Breite besteht keine Einigkeit. Überwiegend wird sie auf etwa 120 geschätzt[69]. Ein großer Teil der Meerengen ist für den internationalen Seeverkehr nicht wesentlich. Anderes gilt für die Straßen von Dover (jährlich etwa 130 000 Schiffe ohne Fährschiffe), Gibraltar (jährlich über 150 000 Schiffe), Hormuz, Bab el Mandeb, Malakka und Singapur, Sunda, Lombok, aber auch für eine Anzahl von weniger bekannten Meerengen, durch die (insgesamt) ein Großteil des Welt-Schiffsverkehrs geht.

Ein besonderes Durchfahrtsrecht durch internationale Meerengen in Küstengewässern, das sich von der normalen friedlichen Durchfahrt durch Küstengewässer unterscheidet, war auf der Dritten Seerechtskonferenz speziell für Kriegsschiffe sehr umstritten. In dieser neuen »Meerengenfrage« wurden die Vereinigten Staaten und die Sowjetunion ungeachtet ursprünglicher Mei-

---

[68] Ganz überwiegende Meinung. Siehe statt aller I.C.J. Reports 1949, 1 (28) [*Korfu Kanal*-Fall]; Sondervoten von Alvarez (insbes. S. 46 f.); Azevedo (insbes. S. 101, 103 ff.) und Krylov (insbes. S. 74 f.). Vgl. zum folgenden aus der neuesten Literatur insbes. *Burke* (Anm. 33), S. 193 ff.; *Maduro* (Anm. 56), S. 65 ff.; *Moore*, The Regime of Straits and the Third United Nations Conference on the Law of the Sea, AJIL 74 (1980), S. 77 ff.; *Platzöder*, Meerengen, ZaöRV 38 (1978), S. 710 ff.; *Reisman* (Anm. 66), S. 48 ff.; *Robertson*, Passage Through International Straits: A Right Preserved in the Third United Nations Conference on the Law of the Sea, Virginia Journal of International Law 20 (1980), S. 801 ff.; *Rojahn* (Anm. 56), S. 79 ff., 101 ff. Auch die von *Mangone* herausgegebene Serie International Straits of the World (1978 ff.) weist auf die Bedeutung des Problems hin. Siehe für die strategisch wahrscheinlich wichtigste Meerenge der Serie Bd. 4: *Truver*, The Strait of Gibraltar and the Mediterranean (1980). Die sowjetische Auffassung wird deutlich in Modernes Seevölkerrecht (Rechtsordnung, Anm. 33) Kapitel VIII.

[69] Die Unterschiede in den Schätzungen ergeben sich hauptsächlich aus den Fragen, ob eine Meerenge dem internationalen Seeverkehr dient und wie mehrere, geographisch hintereinander liegende Meerengen zu zählen sind.

nungsverschiedenheiten zu Verbündeten. Das besonders ausgeprägte Interesse der Sowjetunion an der Durchfahrt durch Meerengen hängt natürlich mit der neu erworbenen Stellung als See-Großmacht und den neuerdings weltweiten See-Interessen zusammen und ergibt sich vor allem aus zwei geographischen Gegebenheiten. Für die Zugänge der global einzusetzenden Sowjetflotte zu offenen Meeren sind die Meerengen des Schwarzen Meeres (allerdings unter einem Sonderregime), der Ostsee, bei den Kurilen und dem Japanischen Meer höchst wichtig. Abgesehen von der Barents-See ist das Klima an der langen Eismeerküste ein wesentliches Hindernis für die sowjetischen Überwasser-Seeverbindungen zu den Weltmeeren [70].

Das Interesse der Vereinigten Staaten an den Meerengen ist dank ihrer ungleich günstigeren seestrategischen Lage nicht so groß wie das der Sowjetunion. Daher wurde bei einer Überprüfung der amerikanischen Optionen für UNCLOS III zeitweise auch der Gedanke erwogen, Bestrebungen zu einer Nationalisierung großer Meeresteile, einschließlich von Meerengen, Archipelen und küstennahen Seezonen zu unterstützen, um auf diese Weise die Sowjetunion gewissermaßen einzumauern [71]. Die Politik der Vereinigten Staaten in UNCLOS III richtete sich jedoch aus verschiedenen Gründen darauf, das Recht auf Durchfahrt durch Meerengen (und Überflug) wirksam und für Unterseeschiffe in besonderer Weise zu sichern. Dabei waren wichtig die Aufrechterhaltung der nuklearen Abschreckung, die strategische und wirtschaftliche Bedeutung der Seeverbindungen zum Mittelmeer, Persischen Golf, Roten Meer und Indischen Ozean sowie die Rücksichtnahme auf die NATO-Verbündeten. Eine Rolle spielte aber auch der Wunsch, etwaige Abhängigkeiten zu vermeiden, die sich gezeigt hatten, als Alliierte bei der Mittelostkrise 1973 den amerikanischen Militärflugzeugen den Durchflug durch ihren Luftraum versagt hatten. Hinzu kam die Erwägung, daß für eine Demokratie wie die Vereinigten Staaten Völkerrechtsnormen *in praxi* eine stärkere Bindungswirkung entfalten als für die Staaten des Warschauer Paktes und daß man aus diesem Grunde den Prozeß der Normentstehung in wesentlichen Sachfragen stets besonders ernst nehmen müsse. Hier läßt sich die Frage, ob es bei dem speziellen Schutz für Durchfahrts- (und Überflugs-)rechte durch internationale Meerengen um vitale Interessen der Vereinigten Staaten geht, nicht beantworten. Die Antwort hängt u. a. von strategischen Gesamtkonzeptionen, den Vorstellungen über die Aufgaben der Seestreitkräfte im übrigen und von dem Kräfteverhältnis zwischen den beiden Supermächten und ihren Verbündeten ab. Die tatsächlichen Gegebenheiten und die Ansichten wechseln naturgemäß im Laufe der Zeit – nicht zuletzt wegen der techni-

---

[70] Siehe *Janis* (Anm. 10), S. 25 ff., 30 ff., mit dem Hinweis, daß die obigen Bemerkungen für Unterseeschiffe mit Atomraketen nur bedingt gelten. Vgl. auch *Rauch* VI (Anm. 32), S. 5. In »Seemacht und Außenpolitik« (Anm. 27) sind die Kapitel 10–12 einzelnen Seegebieten gewidmet und ist die geographische Lage der Sowjetunion jeweils berücksichtigt.

[71] Siehe *Rauch* VI (Anm. 32), S. 4 ff.

schen Entwicklung. Das gilt in der »Meerengenfrage« gerade im Hinblick auf die zunehmende Reichweite der atomaren Waffen und auf die Möglichkeiten, unter Wasser fahrende Schiffe zu orten. Die Reichweite der Interkontinentalraketen ist so groß und wird weiterhin derart gesteigert, daß es den USA speziell für die atomare Abschreckung auf die Durchfahrt etwa durch wichtige internationale Meerengen schon jetzt weniger anzukommen braucht als der Sowjetunion [72]. Jedenfalls steuerten die beiden Supermächte ungeachtet gelegentlicher Unterschiede in der Konferenztaktik letztlich in die gleiche Richtung [73]. Ohne ihre hier gleichlaufenden militärischen Interessen wäre das für Kriegsschiffe (und Flugzeuge) günstige Sonderregime über die Meerengen nicht in die Draft Convention aufgenommen worden.

Die Vereinigten Staaten und die Sowjetunion hatten eine Präferenz für ein Recht auf freie Durchfahrt in Anlehnung möglichst an die Schiffahrtsfreiheit auf der Hohen See, d.h. für ein Korridorregime, allerdings unter dem – unvermeidbaren – Verzicht auf bestimmte militärische Aktivitäten während der Durchfahrt. Die Sowjetunion, die selbst ein Küstenmeer von 12 sm Breite beanspruchte und gegen das Recht auf friedliche Durchfahrt von Kriegsschiffen in der Genfer Küstenmeerkonvention von 1958 einen Vorbehalt gemacht hatte, und die Vereinigten Staaten fanden aber nur begrenzte Unterstützung, insbesondere bei Staaten mit größeren Handels- und Kriegsflotten. Manche Küstenstaaten an Meerengen wollten die Durchfahrt speziell von Kriegsschiffen genehmigungsbedürftig machen und die Ausübung in Anlehnung an die Durchfahrt durch »gewöhnliche« Küstengewässer regeln. Die meisten Staaten der sogenannten »Gruppe der 77«, die seit 1964 auf 125 Staaten angewachsen ist (1981), vertraten ebenfalls eher restriktive Vorstellungen. Der Wille, mit den Durchfahrtsrechten für Kriegsschiffe nicht über Art. 16 Abs. 4 der Küstenmeerkonvention von 1958 hinauszugehen, war ver-

---

[72] Siehe die skeptischen Äußerungen zur Bedeutung der Meerengenfrage für die USA von *Osgood*, U.S.Security Interests in Ocean Law, Ocean Development and International Law 2 (1974), S. 1 ff., insbes. S. 13, 15, 17; SIPRI Yearbook 1975, S. 600 ff. *Darman* (Anm. 50), S. 376, hält für »arguable«, daß die Durchfahrt durch Meerengen jedenfalls für die strategische Abschreckung nicht (mehr) nötig ist. Siehe zu der Frage der Ortung von Unterwasserschiffen unten bei Anm. 81. Vgl. als Beispiele für unterschiedliche Meinungsäußerungen im übrigen etwa *Moore* (Anm. 68), S. 78 ff., unter Hinweis auch auf das Allgemeininteresse der Staatenwelt an der Meerengenpassage unter anderen als seestrategischen Gesichtspunkten; *Janis* (Anm. 10), S. 4 ff., 63 ff. (speziell für die Interessen der Küstenstaaten mit bloßen Küstenverteidigungs-Seestreitkräften); *Richardson*, Power, Mobility and the Law of the Sea, Foreign Affairs 58 (1979/80), S. 902 ff., 912 ff.; *Burke* (Anm. 33), S. 197. Viele, vielleicht die meisten der insgesamt zahlreichen Diskussionsbeiträge stammen von Autoren, die in der einen oder anderen Funktion – vom Berater bis zum Delegationsleiter – an UNCLOS III oder den Vorbereitungen beteiligt waren und zu ihrer »eigenen« Politik oder Gegenpositionen Stellung nehmen. Auch bei klaren Aussagen über die großen Linien fehlt – begreiflicherweise – das militärische Detail.

[73] Einleuchtend die Formulierung von *O'Connell*, die größeren Seemächte »prefer a rule of law that does not close the options«: (Anm. 9), S. 99.

breitet. Andererseits war klar, daß jedenfalls die beiden See-Großmächte auf mehr Schiffahrtsfreiheit gerade für Kriegsschiffe bestehen würden. Verschiedene Kompromißvorschläge sollten die Konfrontation auflösen. Die Vereinigten Staaten hatten schon früh das Einverständnis zu 12 sm Küstengewässer als Konzession für die freie Durchfahrt von Kriegsschiffen durch alle internationalen Meerengen signalisiert. Als das nicht genügte, ließen sie und die Sowjetunion eine grundsätzliche Bereitschaft erkennen, auch die Ausschließlichen Wirtschaftszonen zu akzeptieren. Trotz der erheblichen Vorteile für beide Staaten lag die Einrichtung der Ausschließlichen Wirtschaftszonen letztlich doch nicht in ihrem Interesse. Auch bei der Behandlung der Archipele wurde im Hinblick auf die Meerengenfrage erhebliches Entgegenkommen gezeigt. Dieser Preis für das Sonderregime der Meerengen wurde hauptsächlich der Kriegsschiffe wegen gezahlt. Für Handelsschiffe hätte man es wahrscheinlich – wenn auch mit Vorbehalten – bei der friedlichen Durchfahrt bewenden lassen [74].

Das im folgenden skizzierte System der Transitpassage ist ein Kompromiß zwischen einem Korridorregime ähnlich der Schiffahrtsfreiheit auf Hoher See und der friedlichen Durchfahrt durch »gewöhnliche« Küstengewässer (oben II. 2., III. 1.). Das Durchfahrtsrecht durch Meerengen nach der Draft Convention besteht von vornherein nicht in den Fällen der Art. 35, 36. Damit sind grundsätzlich nur Meerengen in Küstengewässern erfaßt. Unberührt bleiben die für einzelne Meerengen ganz oder teilweise maßgeblichen völkerrechtlichen Verträge, z.B. für den Bosporus und die Dardanellen. Die Abgrenzung der Meerengen mit einem langjährigen, vollständigen oder teilweisen Spezialregime für die Durchfahrt führt vereinzelt zu Schwierigkeiten [75]. Das Meerengenregime gilt auch nicht für internationale Meerengen, innerhalb derer noch Wege durch die Hohe See oder durch eine Ausschließliche Wirtschaftszone bestehen, sofern diese Wege im Hinblick auf navigatori-

---

[74] Siehe *Rauch*, Aspekte (Anm. 33), S. 96; *Janis* (Anm. 10), S. 83 f. Vgl. zur Entstehungsgeschichte der Transitpassage *Robertson* (Anm. 68), S. 806 ff. mit Zusammenfassung auch der Kompromißvorschläge. Nach *Jaenicke* (Anm. 57), S. 484, hatten die beiden Großmächte der freien Durchfahrt (Durchflug) zur Sicherung der globalen Bewegungsfreiheit ihrer See- und Luftstreitkräfte unter ihren Konferenzzielen offensichtlich höchste Priorität eingeräumt. *Hollick*, U.S. Foreign Policy and the Law of the Sea (1981), S. 271 ff., erwähnt die Auseinandersetzungen verschiedener amerikanischer Ressorts, die bis zum Weißen Haus führten und bei denen sich das Verteidigungsministerium im Grundsatz durchsetzte, wenn auch nicht mit dem Wunsch nach einem absoluten Transitrecht. Das Buch läßt die Vielzahl der beteiligten Stellen innerhalb (S. 399) und außerhalb des »Government«, interne Kämpfe und auch Führungsschwächen erkennen. Nach einzelnen Beobachtern bestand auch unter dem Gesichtspunkt der Handelsschiffahrt ein wesentliches Interesse an einer besonderen Meerengenpassage. So etwa *Stöcker* (Anm. 57), S. 337 f.

[75] Siehe zu den unterschiedlichen Ansichten über die Ostseezugänge nach dem Kopenhagener Vertrag von 1857 *Rauch* I, S. 84, *Rauch* III, S. 46, *Rauch* VI (Anm. 32), S. 22, und *Jaenicke* (Anm. 57), S. 474. Die Frage ist von großer praktischer Bedeutung.

sche und hydrographische Verhältnisse eine ähnliche Eignung wie die Wege durch die Küstengewässer aufweisen. Diese Ausklammerung zielt darauf, daß Küstenstaaten von der Ausdehnung ihrer Küstengewässer auf 12 sm absehen, so daß in geographischen Meerengen »Hochseekanäle« entstehen[76]. Das besondere Durchfahrtsrecht der Transitpassage besteht dagegen nach Art. 37 DC für Meerengen, die für die internationale Schiffahrt zwischen einem Teil der Hohen See oder einer Ausschließlichen Wirtschaftszone und einem anderen Teil der Hohen See oder einer Ausschließlichen Wirtschaftszone genutzt werden. Auch insoweit gibt es Zweifelsfragen. Durchfahrt (und Überflug) dürfen nach Art. 38 Abs. 2 DC auch von oder zu einem Küstenstaat der Meerenge führen, sofern dessen einschlägige Vorschriften beachtet sind[77].

In den von Art. 37 DC erfaßten Meerengen besteht ein Recht auf Transitpassage nach Art. 38 Abs. 1 (*transit passage*) für alle Schiffe (und Luftfahrzeuge), sofern nicht die Ausnahme des Art. 38 Abs. 1 2. Halbsatz vorliegt. Die Transitpassage ist Ausübung des Rechtes auf Schiffahrtsfreiheit (*freedom of navigation*) und Überflug in Übereinstimmung mit Teil III der Draft Convention; sie besteht lediglich für die kontinuierliche und zügige Durchfahrt (oder Überflug). Aktivitäten, die keine Ausübung des Rechts auf Transitpassage sind, bleiben einschlägigen anderen Bestimmungen der Draft Convention unterworfen.

Die Transitpassage als Ausübung von Schiffahrtsfreiheit ist an bestimmte Voraussetzungen geknüpft und mehr Regelungen unterworfen als die Schiffahrtsfreiheit auf der Hohen See. Das durchfahrende Schiff muß die Pflichten aus Art. 39 DC beachten. Er enthält u.a. ein Gewaltverbot wie Art. 19 Abs. 2 *lit.* a DC und untersagt wie Art. 19 Abs. 2 *lit.* l alle Tätigkeiten, die nicht zur normalen, kontinuierlichen und zügigen Art der Durchfahrt gehören. Forschungs- oder Vermessungstätigkeit bedarf nach Art. 40 DC der Erlaubnis des Küstenstaates. Dieser hat nach Art. 42 DC begrenzte Regelungsbefugnisse, die sämtlich nicht speziell auf Kriegsschiffe zugeschnitten sind. Der Küstenstaat kann gegebenenfalls Ersatz für Verluste und Schäden von dem Flaggenstaat eines Schiffes mit Immunität fordern. Nach Art. 41 DC darf der Küstenstaat Seewege und Systeme zur Verkehrstrennung vorschreiben, die von den durchfahrenden Schiffen zu beachten sind. Bei den meisten dieser Punkte hatten die See-Großmächte in Einzelheiten Konzessionen gemacht. Insgesamt gesehen sind diese Vorschriften erst sinnvoll, wenn man sie als abschließende Regelung der Materie ansieht[78].

---

[76] Siehe dazu *Rauch*, Aspekte (Anm. 33), S. 98 ff. mit Beispielen aus der Praxis und der Befürchtung, eine derartige anfängliche Zurückhaltung werde später zu einer »schleichenden Hoheitsanmaßung« führen.

[77] Das Überflugsrecht für Militärflugzeuge über Meerengen ist neu. Auch an ihm hatten die Vereinigten Staaten und die Sowjetunion besonderes Interesse.

[78] So schon *Jaenicke* (Anm. 57), S. 473; ferner *Moore* (Anm. 68), S. 106 ff., gegen *Reisman* (Anm. 66), S. 69 f. Vgl. auch oben III. 1. bei Anm. 57.

Für die internationalen Meerengen, die nach Art. 38 Abs. 1 DC von der Transitpassage ausgenommen sind oder die einen Teil der Hohen See oder eine Ausschließliche Wirtschaftszone mit dem Küstenmeer eines fremden Staates verbinden[79], soll nach Art. 45 DC das für die Durchfahrt ungünstigere Recht der friedlichen Durchfahrt gelten (oben III. 1.). Die Suspendierung dieses Rechtes ist jedoch ausdrücklich ausgeschlossen (unaufhebbare friedliche Durchfahrt). Der Status von Meerengen, die in Küstengewässern liegen, aber nicht der internationalen Schiffahrt dienen, ist in der Draft Convention nicht besonders geregelt. Daher fallen sie unter das allgemeine Recht der Küstengewässer. Es besteht das Recht auf friedliche Durchfahrt mit der Suspendierungsmöglichkeit gemäß Art. 25 Abs. 3.

Die Transitpassage durch Meerengen ist unter dem Gesichtspunkt der Schiffahrtsfreiheit günstiger als die friedliche Durchfahrt durch Küstengewässer. Zwar schließen die weitgefaßten Generalklauseln in Art. 39 Abs. 1 DC mit dem Verbot jeder Gewaltandrohung und jeder Tätigkeit, die nicht zur üblichen, kontinuierlichen und zügigen Art der Durchfahrt gehört, etwa Schieß- oder sonstige Waffenübungen während der Transitpassage aus, obwohl diese Aktivitäten nicht ausdrücklich erwähnt sind wie in Art. 19 Abs. 2 DC. Überhaupt dürften die meisten Aktivitäten, die nach Art. 19 Abs. 2 die Friedlichkeit der Durchfahrt ausschließen, auch nach Art. 39 Abs. 1 unzulässig sein. Die Umschreibung der Transitpassage und die Pflichten der Schiffe bei der Transitpassage nach Art. 38–40 DC beschränken die Schiffahrtsfreiheit aber jedenfalls nicht mehr als die Aufzählung in Art. 19. Die Regelungsbefugnisse der Küstenstaaten für die Transitpassage nach Art. 42 DC sind enger als die Regelungsbefugnisse nach Art. 21, 25 für die friedliche Durchfahrt durch Küstengewässer. Bei der Transitpassage gibt es keine dem Art. 30 DC entsprechende Norm mit der Befugnis des Küstenstaates, »unbotmäßige« Kriegsschiffe zum sofortigen Verlassen der Küstengewässer aufzufordern. Art. 44 untersagt dem Küstenstaat nicht nur – wie Art. 24 für die friedliche Durchfahrt – jede Hinderung der Transitpassage, sondern verbietet ihm ausdrücklich jede Suspendierung, während die Suspendierung der friedlichen Durchfahrt unter den Voraussetzungen des Art. 25 Abs. 3 zulässig ist.

Bei der Transitpassage gibt es auch keine Parallele zu dem Gebot des Art. 20 DC, daß Unterwasserschiffe aufgetaucht und unter Zeigen ihrer Flagge fahren müssen. Dieser Unterschied ist nicht mit einem Hinweis auf die in Art. 39 Abs. 1 *lit.* c DC erwähnten »normal modes of ... transit« auszuräumen. Die »normal modes« beziehen sich auf die Schiffe, nicht auf eine abstrakte Art der Durchfahrt oder auf die geographischen Gegebenheiten der jeweiligen Meerenge. Für Unterseeschiffe, vor allem für die modernen mit Atomantrieb, ist die Unterwasserfahrt die normale Fahrt. Das Fehlen eines Gebotes zum aufgetauchten Fahren ist auch nicht durch Zufall oder Vergeßlichkeit zu erklären. Diese Frage war bei den Überlegungen und Diskussio-

[79] Siehe *Jaenicke* (Anm. 57), S. 474 f., insbes. die Beispiele in Anm. 72.

nen über die Transitpassage einer der wichtigsten Punkte. Unterseeschiffe sind aufgetaucht ungleich leichter zu orten als bei untergetauchtem Fahren. In Meerengen sind sie besonders leicht zu entdecken. Den atomaren Seemächten, die in ihren Unterwasserschiffen mit Interkontinentalraketen eine unentbehrliche, vielleicht die am besten geschützte Abschreckungswaffe sahen, erschien ihre Bewegungsfreiheit und damit der Schutz vor Ortung von größter Bedeutung. Gerade die Vereinigten Staaten und die Sowjetunion hatten nicht nur auf größere Sicherheit der Transitpassage vor Eingriffsbefugnissen der Küstenstaaten und auf das Überflugsrecht Wert gelegt, sondern auch auf diesen Punkt. Die Diskussionen während der Konferenz und die anschließenden Formulierungen erfolgten in allgemeiner Kenntnis dieser Tatsachen. Sie führten zu der bewußten Auslassung einer Parallelnorm zu Art. 20 DC. Das ist zwar keine ausdrückliche Erklärung über die Zulässigkeit untergetauchten Fahrens, steht aber immerhin in Parallele zu der Schiffahrtsfreiheit in den Normen über die Hohe See. Dort ist das Recht von Unterseeschiffen zum untergetauchten Fahren auch nicht erwähnt, aber dennoch unbestrittener Bestandteil der Schiffahrtsfreiheit in Art. 87 Abs. 1 *lit.* a DC. Damit haben sich in der Sache die größeren Seemächte und ihre Anhänger durchgesetzt, wenn sie auch nicht die günstigere ausdrückliche Formulierung erreichten [80]. Die während der entscheidenden Verhandlungs- und Formulierungsphasen bestehende Interessenlage muß allerdings nicht unbedingt von Dauer sein. Ob die untergetauchte Durchfahrt durch Meerengen die Bedeutung, die ihr auf der Konferenz beigemessen wurde, noch in der Mitte oder am Ende der achtziger Jahre haben wird, kann hier und heute nicht beurteilt werden. Änderungen kann nicht nur eine größere Standortunabhängigkeit der Unterseeschiffe durch noch größere Reichweite der Interkontinentalraketen bewirken. Eine veränderte Situation ergäbe sich auch, wenn die Entwicklung der Ortungstechnik es ermöglicht, getaucht fahrende Unterseeschiffe sicher auszumachen [81].

---

[80] Das obige Ergebnis stimmt mit den wohl überwiegenden Äußerungen in der Literatur überein. Siehe statt aller *Burke* (Anm. 33), S. 200 ff.; *Robertson*, S. 843 ff.; *Moore*, S. 95 ff. (beide Anm. 68) mit weiteren Nachweisen. Zweifel daran, daß die Zulässigkeit untergetauchten Fahrens hinreichend gesichert ist, äußern *Reisman* (Anm. 66). S. 71 ff., und die bei *Robertson* (S. 844 f.) angeführten Autoren. Auch hier ist die Beteiligung/Nicht-Beteiligung der Autoren an UNCLOS III von Interesse. Für den sowjetischen Standpunkt ermöglicht Aufschlüsse ein Artikel von *Molodcov*, der 1977 im offiziellen Organ des Instituts für Staat und Recht der Sowjetischen Akademie der Wissenschaften erschien und vor Beschränkungen von Durchfahrt und Überflug warnt; Einschränkungen dieser Rechte berührten lebenswichtige Interessen der Sowjetunion und seien für diese im Ergebnis unannehmbar. (Wiedergegeben nach *Rauch* II, S. 7 f.; oben Anm. 32). Siehe dazu auch oben bei Anm. 72.

[81] Siehe etwa *Osgood* (Anm. 72), S. 15 ff.; *Herbig*, Die militarisierten Meere, in: Graf Vitzthum (Hrsg.), Die Plünderung der Meere – Ein gemeinsames Erbe wird zerstückelt (1981), S. 79 ff., 84 ff. Im SIPRI Yearbook 1979, S. 428 ff., befaßt sich *Wilkes* eingehend mit den »Developments in detection technology«. Aber schon im SIPRI Yearbook 1975, S. 600 f., meinte *Goldblat*, die hier skizzierte »Meerengenfrage« werde

Günstiger als die bloße friedliche Durchfahrt ist die Transitpassage für die Schiffahrtsfreiheit schließlich im Hinblick auf die Befugnis des Küstenstaates, Seewege und Systeme zur Trennung des Verkehrs vorzuschreiben. Diese müssen allgemein anerkannten internationalen Regeln entsprechen. Vor allem bedürfen sie nach Art. 41 Abs. 4 DC der Annahme durch die zuständige Internationale Organisation (International Maritime Organization IMO). Der Spielraum des Küstenstaates ist damit erheblich kleiner als nach Art. 22 Abs. 3. Schließlich sei auch bei diesem kurzen Vergleich an das Überflugrecht erinnert, das bei der friedlichen Durchfahrt fehlt [82].

Das Meerengenregime der Draft Convention läßt sich hier ebensowenig würdigen wie das Recht der »gewöhnlichen« Küstengewässer. Immerhin lassen sich im Hinblick auf Kriegsschiffe einige Gesichtspunkte hervorheben. Die Möglichkeit zur Ausweitung der Küstengewässer wirkt sich auf die Schiffahrtsfreiheit überall dort nachteilig aus, wo anstelle des Rechtes der Hohen See nunmehr Meerengenrecht gelten soll. Da Art. 16 Abs. 4 der Küstenmeerkonvention von 1958 mit dem Recht von Kriegsschiffen auf friedliche Durchfahrt durch internationale Meerengen Ausdruck auch von allgemeinem, vertragsunabhängigem Völkergewohnheitsrecht ist, bewirkt das Recht auf Transitpassage für die bisherigen Meerengen jedenfalls im Grundsätzlichen kein Mehr an Schiffahrtsfreiheit. Dagegen wird die Schiffahrtsfreiheit durch das Recht zum untergetauchten Fahren erweitert, die Meeresfreiheit auch durch das Recht zum Überfliegen: Beide Rechte bestanden in Meerengen innerhalb fremder Küstengewässer früher nicht. Die Frage, ob die eingehendere Detailregelung für die Transitpassage inhaltlich mehr Rechtssicherheit bietet, ist wahrscheinlich zu bejahen, wenn man mit dem bisherigen Recht über die Durchfahrt durch Meerengen vergleicht. Die Transitpassage führt aber zu mehr Auslegungsproblemen und damit potentiellen Konfliktsmöglichkeiten als die Schiffahrtsfreiheit auf der Hohen See, die ohne die Erweiterung der Küstengewässer in vielen der von der Draft Convention erfaßten Meerengen oder in Teilen von ihnen gelten würde. Andererseits ist im Vergleich zu dem Recht der »gewöhnlichen« Küstengewässer nach der Draft Convention die Begrenzung der küstenstaatlichen Eingriffsmöglichkeiten durch die Transitpassage ein Gewinn für die Schiffahrtsfreiheit und jedenfalls teilweise für die Rechtssicherheit. Ein erheblicher Teil der oben beispielhaft erwähnten Auslegungsprobleme kann aber auch bei der Transitpassage auftreten. Dazu kommen zusätzliche Auslegungsschwierigkeiten [83].

---

mehr von der legalistischen als von der praktischen Sicht her diskutiert. Wahrscheinlich werde ein neuer Seerechtsvertrag erst in Kraft treten, wenn die Unterseeschiffe mit Interkontinentalraketen alle wesentlichen Ziele eines eventuellen Gegners ohne Durchfahrt durch Meerengen erreichen könnten.

[82] Nach *Riphagen* (Anm. 40), S. 167, stehen das Meerengen- wie das Archipelregime insgesamt der Schiffahrtsfreiheit auf der Hohen See offensichtlich näher als dem Regime der Küstengewässer.

Die Auslegungsschwierigkeiten bei Einzelfragen und damit auch die Konfliktsmöglichkeiten ließen sich vermindern, wenn die Draft Convention für Zweifelsfälle eine Vermutung entweder zugunsten der Schiffahrtsfreiheit oder zugunsten der küstenstaatlichen Souveränität aufstellen würde. Der Wortlaut von Teil III läßt eine generelle Vermutung zugunsten der Schiffahrtsfreiheit nicht zu. Am Anfang – in Art. 34 DC – stehen neben dem Hinweis auf die Transitpassage Souveränität und Zuständigkeiten der Küstenstaaten. Zwar waren die großen Seemächte und ihre Verbündeten gegenüber den Meerengenstaaten in der Sache überwiegend erfolgreich. Eine Ausweitung der Einzelaussagen über das Recht zur Transitpassage etwa durch Verweisungen auf das Recht der Hohen See war jedoch nicht zu erreichen. Die Ausführungen oben speziell zum untergetauchten Fahren sind nicht zu verallgemeinern [84]. Andererseits läßt sich angesichts gerade der Art. 34 Abs. 2, 38 und 44 DC mit den ausdrücklichen und sehr gezielten Begrenzungen der küstenstaatlichen Souveränität auch keine Vermutung für diese aufstellen.

Die Frage nach der Sachgerechtigkeit des Meerengenregimes hinsichtlich der Kriegsschiffe stößt sich sofort an dem unterschiedlichen Verständnis von Sachgerechtigkeit. Die Problematik ist hier schwieriger als bei den »gewöhnlichen« Küstengewässern, obwohl im Völkergewohnheitsrecht die (aufgetauchte) Durchfahrt von Kriegsschiffen durch Meerengen fester verankert ist als die durch »gewöhnliche« Küstengewässer. Die größere Problematik der Entwicklung liegt nicht zuletzt in der Waffentechnik und in den sonstigen militärischen Gegebenheiten, die in Zusammenhang mit dem neuen Recht auf untergetauchtes Fahren zu sehen sind. Man müßte die Regeln des Meerengenregimes möglicherweise aufschlüsseln in die über die Transitpassage von Kriegsschiffen im allgemeinen und über das untergetauchte Fahren von Atomträgern sowie das Überflugrecht im besonderen. Für die an der Transitpassage hauptsächlich interessierten Staaten geht es jedoch um die gleichen Ziele. Ein allgemein anerkannter Maßstab für die Sachgerechtigkeit dürfte kaum zu finden, die oben getroffene Feststellung über den Wert der Schiffahrtsfreiheit nicht entscheidend sein. Man kann die Transitpassage lediglich als legalisierte Durchsetzung militärischer Interessen der stärksten Seemächte und ihrer Verbündeten sehen. Man kann sie aber auch sehen als wesentliches Instrument zur Erhaltung des militärischen Gleichgewichts zwischen den Supermächten und ihren Verbündeten und damit zur Vermeidung eines dritten Weltkrieges. Bei der zweiten Alternative könnte man von einer »Sache der ganzen Welt« sprechen, die dem Interesse einer Anzahl von Meerengenstaaten vorgehen müßte. Aber wer kann mit Sicherheit sagen, was die relevanten Freiheiten der Transitpassage insgesamt oder gar was einzelne dieser Freihei-

---

[83] Als Beispiel für die erste Gruppe: Art. 39 Abs. 1 *lit.* b und c ist nicht leichter auszulegen als der fast wortgleiche Art. 19 Abs. 2 *lit.* a und l. Siehe für zusätzliche Auslegungsprobleme *Platzöder* (Anm. 18), S. 65 ff., und – kürzer – (Anm. 68), S. 729 ff.

[84] Vgl. auch *Riphagen* (Anm. 40), S. 167.

ten für dieses militärische Gleichgewicht heute oder in zehn Jahren bedeuten? Wer kann beweisen, daß der Weltfriede tatsächlich von dem militärischen Gleichgewicht zwischen den Supermächten und ihren Verbündeten abhängt? Das Bemühen um militärisches, insbesondere atomares Gleichgewicht trägt zu dem ständigen Wettrüsten jedenfalls bei. Nicht wenige Beobachter sind der Meinung, das ständige Wettrüsten erhöhe die vorhandene Unsicherheit und wirke friedensgefährdend[85]. Von diesem Ausgangspunkt her müßte man vielleicht die Schiffahrtsfreiheit von Kriegsschiffen überhaupt beschränken, jedenfalls aber das Recht der Transitpassage von Schiffen mit Atomwaffen und vor allem ihr untergetauchtes Fahren ablehnen.

Fehlen allgemein anerkannte, sichere Maßstäbe für die Sachgerechtigkeit der Transitpassage von Kriegsschiffen, so läuft die Frage auf eine Interessenabwägung zwischen Staatengruppen hinaus. Bei der Bildung von Gruppen ergeben sich Schwierigkeiten. Meerenge ist nicht gleich Meerenge, Meerengenstaat nicht gleich Meerengenstaat. Es gibt Meerengen, in denen das Recht der Transitpassage von Kriegsschiffen *in praxi* nicht in Anspruch genommen werden wird. Für die Anliegerstaaten ist die Regelung der Transitpassage unwichtig. Auch im übrigen stimmen nicht alle Meerengenstaaten überein. Nicht jeder definiert seine Interessen von seiner geographischen Lage als Meerengenanlieger her. Großbritannien etwa sieht seine Interessen in erster Linie nicht als Küstenstaat der Meerengen von Dover/Calais und Gibraltar, sondern als atomare Seemacht mit einer *blue water navy* und als Mitglied der NATO. Bei anderen Meerengenstaaten kann einerseits die Zugehörigkeit zu einem Militärbündnis, andererseits zur Gruppe der blockfreien Staaten zu ganz verschiedenen Einstellungen führen. Immerhin beantworten die *meisten* Meerengenstaaten die Frage nach der Sachgerechtigkeit der Transitpassage jedenfalls bei isolierter Betrachtung anders als die atomaren Supermächte der Meere und die anderen beiden Staaten mit atomaren Hochseestreitkräften.

Als Ausgangspunkt für einige verallgemeinernde Bemerkungen über die Gruppeninteressen lassen sich die besonderen geographischen Gegebenheiten von internationalen Meerengen und – wiederum verallgemeinert – ihre Bedeutung für den Seeverkehr nutzen. Diese Gegebenheiten legen es nahe, drittstaatliche Interessen an der Durchfahrt stärker zu berücksichtigen als bei den »gewöhnlichen« Küstengewässern. Es erscheint – ohne daß es hier näher ausgeführt werden kann – sachgerecht, jedenfalls Handelsschiffen im weiteren Sinne im Vergleich zu der bloßen friedlichen Durchfahrt durch Küstengewässer verbesserte Durchfahrtsrechte durch internationale Meerengen zu gewähren. Die Zahl der Staaten, die daran interessiert sind, dürfte erheblich größer sein als die Zahl der Meerengenstaaten, die primär entgegengesetzte Interessen haben. Die Regelung der Transitpassage für Handelsschiffe in der

[85] Siehe etwa die Einführung zu der UNO-Studie Kernwaffen (Anm. 62), S. 15 ff.; die SIPRI Jahrbücher oder gar die Ansicht von *Herbig* (Anm. 81), S. 94, »die Richtung des Wettrüstens zielt auf Kriegführung, nicht auf Abschreckung«.

Draft Convention dürfte berechtigten Bedürfnissen der Küstenstaaten *grundsätzlich* genügen. Sie gibt aber wohl auch den Drittstaaten das ihre. Im Hinblick auf das Regime der Transitpassage speziell für Kriegsschiffe ist die Interessenkonstellation anders. Vereinfacht formuliert verläuft die eigentliche Front zwischen den atomaren Supermächten und ihren Verbündeten auf der einen Seite und jenen Meerengenstaaten auf der anderen Seite, die von ihren Interessen her nicht primär zur ersten Gruppe gehören und Meerengen »haben«, bei denen die Durchfahrt von Kriegsschiffen für andere Staaten wesentlich ist. Keine der beiden Gruppen kommt einer Mehrheit in der heutigen Staatenwelt auch nur nahe. Die Frage, ob die küstenstaatlichen Interessen oder die – vereinfacht formuliert – Seemachtinteressen Vorrang haben sollen, läßt sich nicht zuverlässig beantworten. Man kann weder die unmittelbar beteiligten Staaten einfach zählen noch ihre Interessen gewichten. Die Lösung ergibt sich auch nicht aus den Vorstellungen und der Haltung der nicht zu diesen beiden Gruppen gehörenden großen Mehrheit der Staaten, zu der die etwa 30 Binnenstaaten gehören. Die von der Transitpassage nicht unmittelbar betroffenen Staaten sahen die Regelung im Rahmen des seerechtlichen Gesamtkonzeptes, z. T. von ihren Interessen an anderen Regelungen. Viele Staaten waren zuerst unentschieden. Wenn eine deutliche Mehrheit anfänglich die besondere Transitpassage für Kriegsschiffe ablehnte, so waren nicht zuletzt Gesichtspunkte der Ideologie (gegen die großen Seemächte und/oder früheren Kolonialmächte), der Gruppenzugehörigkeit und der Taktik maßgeblich. Als die atomaren Supermächte und ihre Partner zu erkennen gaben, welche Bedeutung sie der Transitpassage beimaßen, als sie mit verschiedenen Mitteln der Diplomatie und mit erheblichen Konzessionen hinsichtlich des Archipelregimes und der Ausschließlichen Wirtschaftszonen Konzessionsbereitschaft in der Frage der Transitpassage zu erreichen suchten, war die Mehrheit der Staaten ohne ein ausgeprägtes unmittelbares Interesse an der Transitpassage für Kriegsschiffe zu einem Entgegenkommen bereit. Welches Gewicht soll man Interessen einräumen, die von den meisten Staaten dem Interesse an den Ausschließlichen Wirtschaftszonen und – bei einzelnen – an den Archipelgewässern nachgeordnet wurden? Wie soll man aus dieser »Gewichtung« der Interessen durch die Staatenmehrheit Rückschlüsse für oder gegen die Sachgerechtigkeit der Regelung ziehen? Jedenfalls von diesem Ausgangspunkt her brauchten sich die Staaten, die sich für die Transitpassage einsetzten und besonderen Anteil an ihrer Aufnahme in die Draft Convention hatten, wohl nicht so sehr unter Begründungszwang zu fühlen, wie es der Fall gewesen zu sein scheint. Von der eben skizzierten Lage aus läßt sich das Regime der Transitpassage für Kriegsschiffe unter den Gesichtspunkten sachgerecht oder sachwidrig heute nicht zuverlässig beurteilen. Es erscheint als Bestandteil eines umfassenden politischen Kräftespiels, nach den Maßstäben der Gegenwart als vertretbarer Kompromiß.

Ist das zutreffend, so bedarf es nicht so sehr der meist eher interessanten als überzeugenden Versuche, die einschlägigen Vertragsbestimmungen durch

Hinweise auf ihren gemeinnützigen Charakter auf eine »höhere Ebene« zu heben. Dafür einige Beispiele: Die Sowjetunion trug auf der Konferenz vor, bei einer bloßen friedlichen Durchfahrt durch Meerengen sei der Welthandel bedroht, an dem alle Staaten ein gemeinsames Interesse hätten, insbesondere die Binnenstaaten [86]. Das »Moderne Seevölkerrecht« der Sowjetischen Akademie der Wissenschaften weist auf die Versuche der Gegner der Schiffahrtsfreiheit durch Meerengen hin, Entwicklungsländer durch Hinweise auf die angeblich existierende Gefahr von seiten der Kriegsflotten der Großmächte zu verwirren. Dabei sei die Freiheit der Schiffahrt durch alle Meerengen, die der internationalen Schiffahrt zur Verbindung von offenen Meeren dienten, doch notwendig für die Befriedigung der Wirtschaftsinteressen; die Schließung von wichtigen internationalen Meerengen könne der Mehrheit der Staaten unvorstellbaren Schaden zufügen [87]. Ein führender sowjetischer Völkerrechtler, *Molodcov*, erklärte unter Berufung auf eine kubanische Argumentation, die Transitpassage sei für Völker in revolutionären und nationalen Befreiungskämpfen wichtig, deren Ausgang nicht selten von der Unterstützung durch befreundete sozialistische Länder abhänge [88]. Differenzierter und aufrichtiger wirken die Ausführungen von *Moore*, der zeitweise Delegationsleiter der USA für UNCLOS III war. Er nennt zwar auch die Bedeutung einer gesicherten Durchfahrt für Handelsschiffe; in bezug auf Öltanker handele es sich nicht nur um eine Kosten-, sondern sogar um eine Existenzfrage. Ferner resultierten aus größeren küstenstaatlichen Kontrollbefugnissen als denen bei der Transitpassage zusätzliche Konfliktsmöglichkeiten. Weiter weist er auf die Gefahr hin, daß küstenstaatliche Kontrollen in Meerengen auf andere Seegebiete wie die Ausschließlichen Wirtschaftszonen übergreifen können (*spill-over threat*). An die Spitze seiner Ausführungen stellt *Moore* aber die Erhaltung des strategischen Gleichgewichts zwischen den Supermächten, die von der Unverwundbarkeit der strategischen Unterseeschiffe als den Trägern der Zweitschlag-Kapazität abhänge. Das strategische Gleichgewicht sei für den Weltfrieden von großer Bedeutung, der Weltfriede liege im Interesse aller Staaten. Nur weil man auf der Konferenz diese Zusammenhänge – *alle* von ihnen – verstanden habe, sei das Transitregime möglich geworden; allein ein Insistieren der Seemächte hätte nie genügt [89]. Dieses Bemühen, in den *offen* ausgesprochenen Interessen der für ein atomares Gleichgewicht auf See eintretenden Staaten *zugleich* das Interesse einer großen Staatenmehrheit aufscheinen zu lassen, wirkt jedenfalls abgewogener als die vorher erwähnten einseitigen Aussagen zur Begründung der Transitpassage.

---

[86] Siehe *Wolf* (Anm. 50), S. 218. Dort auch die ironische Erwiderung Chinas, das von der Sowjetunion erwähnte Anwachsen des Welthandels könne doch kaum von der freien Durchfahrt von Kriegsschiffen oder dem untergetauchten Fahren von Atom-Unterseeschiffen durch Meerengen herrühren.

[87] Rechtsordnung (Anm. 33), S. 243, 224.

[88] Nach *Rauch* II (Anm. 32), S. 7.

[89] *Moore* (Anm. 68), S. 79 ff.

Auch wenn man die Transitpassage für Kriegsschiffe durch Meerengen in der Form der Draft Convention grundsätzlich für einen vertretbaren Kompromiß hält, bleibt die Frage nach dem Schutz vor höchst gefährlichen Schiffen. Oben wurde schon gesagt, daß eine dem Küstenstaat insoweit günstigere Regelung im Bereich der »gewöhnlichen« Küstengewässer nicht unbedingt eine Vorentscheidung für die Meerengen bedeuten muß. In diesen haben Küstenstaaten und Drittstaaten häufig gesteigerte, aber völlig entgegengesetzte Interessen. Diese können hier nicht weiter aufgeschlüsselt werden; sie mögen sich auch in absehbarer Zukunft teilweise ändern [90].

IV. Die Schiffahrtsfreiheit von Kriegsschiffen in Archipelgewässern

Nachdem einige Staaten jahrzehntelang vergeblich einen besonderen Rechtsstatus für Archipelgewässer gefordert hatten, setzte sich dieser in UNCLOS III durch [91]. Ein rechtlicher Sonderstatus von Archipelgewässern wird sich wohl auch dann nicht mehr aufhalten lassen, wenn die Draft Convention nicht allgemein akzeptiert werden sollte. Archipelstaaten im Sinne der Draft Convention können nur Staaten sein, die aus Inseln oder Inselgruppen in einer bestimmten Zusammensetzung bestehen. Ein Staat mit Gebiet auf kontinentalem Festland kommt als Archipelstaat nicht in Betracht. Ausgangspunkt der neuen Regeln ist die Erwägung, daß das Meer zwischen Inseln und Inselteilen eines Archipelstaates nicht die übliche Abgrenzungsfunktion zu fremden Hoheitsgewässern oder zu den Weltmeeren hat, sondern eher der Verbindung zu anderen Teilen desselben Staatsgebietes dient. Als Charakteristikum der Archipelstaaten sieht man einen engen Zusammenhang zwischen Inselgruppen und Seegebieten an. Sie müssen als – annähernd – geographische, wirtschaftliche und politische Einheit erscheinen oder in der Geschichte als derartige Einheiten verstanden worden sein. Wegen dieses – tatsächlichen oder unterstellten – Zusammenhangs dürfen nach Art. 47, 48 DC von den äußersten seewärtigen Punkten der Inselgruppen Grundlinien gezogen werden, die Inseln und innere Seegebiete zu Einheiten zusammenfassen und Aus-

---

[90] Es gibt zu denken, daß ein so sachkundiger Beurteiler wie *Burke*, der die Zulässigkeit des untergetauchten Fahrens von Atom-Unterseeschiffen durch Meerengen in der Draft Convention so zweifelsfrei gewährleistet sieht, gegen dieses Recht unter rechtspolitischen Gesichtspunkten Bedenken hat: »right does not accord with desirable community policy«, *Burke* (Anm. 33), S. 220.

[91] Zur Vorgeschichte statt aller *O'Connell,* Mid-Ocean Archipelagos in International Law, BYIL 45 (1971), S. 1 ff., mit der zutreffenden Voraussage, das Archipelprinzip werde sich wahrscheinlich durchsetzen, wenn das Problem der Durchfahrtsrechte gelöst würde (S. 77). Auf dem neuesten Stand *Sturies,* Archipelgewässer – Zur Entwicklung eines neuen Rechtsbegriffs im Seerecht (1981), S. 102 ff. Siehe zur Interessenlage, S. 65 ff., zur Entstehung der nachkolonialen Archipelansprüche, S. 82 ff. Vgl. auch Modernes Seevölkerrecht (Rechtsordnung, Anm. 33), S. 111 ff.

gangslinie für Küstengewässer, Anschlußzone, Ausschließliche Wirtschaftszone und Festlandsockel sind. Mit anderen Worten: Es ist die Funktion der Grundlinien, die intrainsularen Gewässer zu »vereinnahmen« und die neue Einheit nach außen abzugrenzen. Die Definition und die Abgrenzung der Archipelstaaten und Archipelgewässer in der Draft Convention ergibt sich vor allem aus drei Regeln: Einmal können nach Art. 46 nur Inselstaaten Archipelstaaten sein. Zum anderen dürfen 97% der Grundlinien um die Inseln des Archipels von Punkt zu Punkt nicht länger als 100 sm sein und ist die Länge der verbleibenden 3% auf 125 sm begrenzt. Schließlich muß in einem so abgegrenzten Archipelgebiet das Verhältnis zwischen Wasser- und Landfläche zwischen 1:1 und 9:1 liegen (Art. 47 Abs. 1 und 2 DC). Von den diesen Maßstäben entsprechenden Archipelstaaten sind die wichtigsten die Philippinen, Indonesien, die Bahamas- und die Fidschi-Inseln. Der größte Archipelstaat Indonesien hat nach *Brock* eine Nord-Süd-Ausdehnung von etwa 1300 sm und eine Ost-West-Ausdehnung von etwa 3000 sm, insgesamt etwa 2 300 000 Quadratmeilen Seegebiet[92]. Durch die Archipelgewässer aller dieser Staaten verlaufen internationale Seestraßen von unterschiedlicher, zum Teil großer Wichtigkeit (z. B. Malakka- und Sundastraße)[93].

Art. 49 DC erstreckt die Souveränität des Archipelstaates auf alle Seegebiete innerhalb der nach Art. 47 zulässigen Grundlinien, auf den Luftraum über diesen Seegebieten sowie auf Meeresboden und -untergrund darunter. Nach Art. 86 DC fällt dieses Gebiet nicht unter den VII. Abschnitt der Draft Convention mit dem Recht der Hohen See. Die Besitzzuweisung an Seegebieten, die bisher ganz überwiegend zur Hohen See gehörten, wird dadurch gemildert, daß Teil IV der Draft Convention die Befugnisse des Archipelstaates erheblich begrenzt. Für die Schiffahrtsfreiheit auch von Kriegsschiffen sind vor allem Art. 52–54 DC wichtig. Für den Verkehr zwischen einem Teil der Hohen See oder einer Ausschließlichen Wirtschaftszone einerseits und einem anderen Teil der Hohen See oder einer Ausschließlichen Wirtschaftszone andererseits besteht auf den internationalen Schiffahrtswegen innerhalb der Archipelgewässer und des anschließenden Küstenmeeres das Recht auf Archipelpassage (*archipelagic sea lanes passage*). Die Archipelpassage steht allen Schiffen zu, also auch Kriegsschiffen. Sie umfaßt nach Art. 53 Abs. 3 DC ausdrücklich die kontinuierliche, zügige und unbehinderte Schiffahrt (*navigation*, aber auch den Überflug) in der üblichen Art (*normal mode*). Da die übli-

---

[92] *Brock*, Threats to freedom of navigation, (U.S.)Judge Advocate General's Journal, December 1969–January 1970, S. 75 ff., 78. Die gleichen Zahlen zur Länge und Breite bei *O'Connell*, a.a.O., S. 38. Andere Zahlen bei *Sturies*, a.a.O., S. 87, 90.

[93] Überblick über die Archipelstaaten bei *Sturies* (Anm. 91), S. 84 ff. Art. 46, 47 DC haben weitergehende Forderungen aufhalten können, nach denen 27 Staaten zu Archipelstaaten und etwa 15% der gesamten Meeresoberfläche zu Archipelgewässern geworden wären. Siehe *Platzöder/ Graf Vitzthum*, Zur Neuordnung des Meeresvölkerrechts auf der Dritten Seerechtskonferenz der Vereinten Nationen. Vorgeschichte – Verfahren – Problemfelder – Vorentscheidungen (1974), S. 67.

che Fahrweise von Unterseeschiffen die Unterwasserfahrt ist, unterstreicht die Formulierung *normal mode* die schon für die Transitpassage durch Meerengen festgestellte Zulässigkeit der Unterwasserfahrt [94]. Gegen die Zulässigkeit der Unterwasserfahrt hatten sich insbesondere die Philippinen ausgesprochen. Das Interesse der großen Seemächte an untergetauchter Passage war wegen der Größe und der geographischen Lage einiger Archipelstaaten erheblich. Ein Kompromiß zwischen ihnen und den Archipelstaaten, die wegen ihrer kleinen Zahl und der Vielzahl ihrer Opponenten dringend auf Verbündete angewiesen waren, lag um so näher, als bei der Archipelpassage wegen der größeren Seeräume die Bedenken der Archipelstaaten regelmäßig geringer waren als die, welche sich bei Meerengen gegen das getauchte Fahren vor allem atomgetriebener oder -bewaffneter Unterseeboote vortragen ließen. Da Art. 54 DC die Art. 39, 40, 42 und 44 *mutatis mutandis* für anwendbar erklärt, entspricht die Archipelpassage auch im übrigen weitgehend der Transitpassage durch Meerengen (siehe oben III. 2.). Vor allem sind die Verbote jeder Gewaltandrohung und jeder Tätigkeit wichtig, die nicht zu der normalen Art der kontinuierlichen und zügigen Durchfahrt gehören. Militärische Übungen sind also nach den Art. 53 Abs. 3, 54, 39 Abs. 1 DC ebenso unzulässig wie in Meerengen.

Einschränkungen der Bewegungsfreiheit in Archipelgewässern, die gerade bei Kriegsschiffen wichtig sind, können sich aus der Befugnis der Archipelstaaten ergeben, bestimmte Archipelseewege (*archipelagic sea lanes* und *air routes*) vorzuschreiben, d.h. eine Reihe fortlaufender Linien von den Eingangs- und Ausgangspunkten für die Durchfahrt durch die Archipelgewässer und das anschließende Küstenmeer verbindlich festzulegen. Die durchfahrenden Schiffe dürfen nach jeder Seite nicht mehr als 25 sm von den vorgeschriebenen Fahrtlinien abweichen. Zum Zwecke sicherer Durchfahrt durch enge Kanäle in diesen Seewegen darf der Archipelstaat Regeln über eine Verkehrstrennung erlassen. Zwar sollen grundsätzlich alle für die internationale Schiffahrt üblichen Durchfahrtsstraßen auch Archipelseewege werden. Jedoch ist eine Doppelung von Seewegen ähnlicher Eignung zwischen denselben Eingangs- und Ausgangspunkten nicht erforderlich. Diese Befugnisse der Archipelstaaten könnten die Archipelpassage fremder Schiffe einschneidend beschränken, wenn sie nach Belieben ausgeübt werden dürften. Die Voraussetzungen und die Ausübung sind aber ihrerseits begrenzt, insbesondere durch Art. 53 Abs. 1, 3–6, 8 DC. Nach Art. 53 Abs. 8 müssen die Archipelseewege und die Systeme zur Verkehrstrennung allgemein anerkannten internationalen Regeln entsprechen. Das Bemühen, Willkür des Archipelstaates auszuschließen, zeigt sich auch darin, daß solche Bestimmungen der Annahme (*adoption*) durch die zuständige Internationale Organisation bedürfen (IMO). Das

[94] Für die Zulässigkeit schon *Burke* (Anm. 33), S. 215 ff.; *Sturies* (Anm. 91), S. 128. Die »normal modes ... of transit« in Art. 39 Abs. 1 *lit.* c DC bei Transitpassagen stehen in einem etwas anderen Zusammenhang. Vgl. oben S. 316 f.

ist die Parallele zu Art. 41 Abs. 4 über die Transitpassage durch Meerengen. Diese Regelung ist für die Schiffahrt wesentlich günstiger als Art. 22 Abs. 3 für die friedliche Durchfahrt durch Küstengewässer. Kommt es nicht zu solchen abgestimmten Regelungen, so geht das nicht zu Lasten der Schiffahrtsfreiheit: So lange es – etwa mangels Einverständnisses zwischen Archipelstaat und IMO – an entsprechenden wirksamen Regelungen fehlt, besteht die Archipelpassage auf den Wegen, die üblicherweise für die internationale Schifffahrt benutzt werden[95]. Nach Art. 54 DC gilt für die Archipelpassage auch Art. 44 *mutatis mutandis*. Die Regelungsbefugnis des Küstenstaates darf also nicht zur Behinderung der Archipelpassage führen. Vor allem darf diese ebensowenig suspendiert werden wie die Transitpassage. Bei Fahrten durch Archipelgewässer außerhalb der internationalen Schiffahrtswege, die nicht unter die Archipelpassage fallen, besteht nach Art. 52 DC für alle Schiffe das Recht auf friedliche Durchfahrt entsprechend den oben unter III. 1. skizzierten Normen über die Küstengewässer. Nur die Befugnis des Archipelstaates zur Suspendierung erscheint hier enger: Seine Waffenübungen sind nicht als Suspendierungsgrund genannt.

Auch bei den Archipelen gibt es im Hinblick auf die geographischen, militärischen, wirtschaftlichen und politischen Gegebenheiten und auf die Bedeutung für die Schiffahrt große Unterschiede. Schon sie begrenzen die Möglichkeit für eine zusammenfassende Würdigung der Schiffahrtsfreiheit von Kriegsschiffen in der Draft Convention. Einige allgemeine Erwägungen erscheinen jedoch vertretbar. Oben wurde schon erwähnt, daß die Einführung eines Archipelregimes (mit der Archipelpassage) Ergebnis von Kompromissen im Zusammenhang mit anderen wichtigen Fragen ist, insbesondere mit der Transitpassage durch Meerengen. Die Ausgestaltung des Archipelregimes in der Draft Convention ist aber auch Ergebnis zahlreicher Kompromisse in Einzelfragen. Manche der hier nur erwähnten Normen werfen die gleichen Auslegungsprobleme auf wie die parallelen Regelungen bei der Transitpassage durch Meerengen. Sie können zu den gleichen Konflikten führen. Da die Archipelgewässer nach geltendem Recht ganz überwiegend Hohe See mit dem Grundsatz der Schiffahrtsfreiheit sind, wirkt die Neuregelung mit den oben skizzierten rechtlichen Unklarheiten konfliktvermehrend. Obwohl die Unklarheiten geringer sind als bei der friedlichen Durchfahrt durch Küstengewässer, erscheinen sie gravierender. Die Unklarheiten im Recht der Küstengewässer blieben – z. T. unvermeidlich – bei der Neuregelung für eine bereits bestehende Sonderzone des Meeres. Die Unklarheiten im Recht der Archipelgewässer entstanden erst durch die Einrichtung einer neuen Sonderzone. Auch wegen der Minderung der Rechtssicherheit bedarf das Archipelregime daher einer überzeugenden Begründung.

---

[95] Art. 53 Abs. 12 ist im Zusammenhang so zu verstehen, daß diese Rückverweisung auf die traditionelle Schiffahrtsfreiheit immer dann gilt, wenn die Regelungen nicht entsprechend der Konvention zustande kommen, nicht etwa nur dann, wenn der Küstenstaat sie nicht erlassen will.

Die Draft Convention stellt für die Archipelgewässer ebensowenig wie für die Küstengewässer und Meerengen eine Vermutung auf, die bei Auslegungsschwierigkeiten eingreifen könnte. Die Souveränität der Archipelstaaten kann schon wegen der geographischen Gegebenheiten nicht weiter gehen als die der Meerengenstaaten. Die Entstehungsgeschichte des Archipelregimes spricht ebenfalls dagegen, in den Teilen von Archipelgewässern, die bisher Hohe See sind, unter Berufung auf die neue, aber zugleich begrenzte Souveränität der Archipelstaaten die Archipelpassage zugunsten der Souveränität des Archipelstaates stärker einzuschränken, als das in den Einzelbestimmungen für die jeweilige Einzelfrage vorgesehen ist. Archipelregime und Archipelpassage entstanden gleichzeitig. Ohne Archipelpassage war ein Archipelregime von vornherein nicht akzeptabel. An dem Ausschluß einer Vermutung zugunsten der Souveränität der Archipelstaaten ändert auch Art. 49 Abs. 4 DC nichts. Andererseits schließen Art. 49, 86 mit der – wenn auch durch die Transitpassage eingeschränkten – Souveränität der Archipelstaaten es aus, für Zweifelsfälle eine Vermutung zugunsten einer Angleichung der Archipelgewässer an das Recht der Hohen See aufzustellen. Immerhin stellen die Archipelgewässer für die verschiedenen Formen der Meeresnutzung durch die beteiligten Staaten regelmäßig erheblich mehr Raum zur Verfügung als Meerengen. Das mag helfen, die Zahl der ungewollten Konflikte zu verringern.

Vor allem für die Staaten mit weitgespanntem Seehandel oder größeren Kriegsflotten (insbesondere Unterseeschiffen) und Luftstreitkräften weist die Archipelpassage im Vergleich zu der bloßen friedlichen Durchfahrt durch Küstengewässer in etwa die gleichen Vorteile auf wie die Transitpassage durch Meerengen. Vergleicht man das Recht von Kriegsschiffen auf Archipelpassage dagegen mit ihrer früheren Schiffahrtsfreiheit auf der Hohen See, so ist der Verlust an Schiffahrtsfreiheit beträchtlich. Er wiegt um so schwerer, als es sich bei einigen Archipelen um riesige Seegebiete mit Durchfahrten von großer Bedeutung handelt. Die Einengung der Schiffahrtsfreiheit gilt auch für Handelsschiffe und für eine Vielzahl von Staaten. Die weit ausgreifende Seenahme nützt nur einer kleinen Zahl von Archipelstaaten. Sie schadet der ganz überwiegenden Mehrheit der Staaten mit Schiffahrtsinteressen. Die Seenahme läßt sich nicht damit rechtfertigen, daß der Archipelstaat die wirtschaftlichen Nutzungsmöglichkeiten im Interesse der Allgemeinheit ausüben wird. Im Vordergrund der Begründungen steht eindeutig das Interesse der Archipelstaaten. Die Seenahme wird wesentlich mit wirtschaftspolitischen Interessen der Archipelstaaten begründet, die alle Entwicklungsländer sind[96].

---

[96] Siehe insbes. *Sturies* (Anm. 91), S. 70 ff. mit weiteren Hinweisen. Seine Feststellung, die überwiegende Meinung in der Literatur erkenne das Bedürfnis an Sonderrechten zur Kompensation für geographische Nachteile an, ist mit Vorsicht aufzunehmen. Die meisten angeführten Äußerungen stammen aus Archipelstaaten und/ oder aus der Zeit vor der Durchsetzung von Ausschließlichen Wirtschaftszonen auf der Konferenz. *McDougal/Burke*, The Public Order of the Oceans, A Contemporary International Law of the Sea (1962), S. 411 ff., und *O'Connell* (Anm. 91), S. 75, äußern sich differenzierter.

Aber: Werden diese Interessen nicht weitgehend durch die Erweiterung von Küstengewässern und Anschlußzonen, vor allem aber durch die Einrichtung von Ausschließlichen Wirtschaftszonen befriedigt? Ließen sich – sofern man diese Fragen verneint – die besonderen Interessen der Archipelstaaten nicht durch ergänzende Regeln bei den Ausschließlichen Wirtschaftszonen berücksichtigen, ohne daß es der ausgedehnten Seenahme bedarf? Aber selbst ohne die Einrichtung der Ausschließlichen Wirtschaftszonen kann man mit guten Gründen fragen, ob auf die Archipele zugeschnittene bloße Funktionszonen hier nicht sachgerechter wären als die durch das Archipelregime vorgesehenen Hoheitszonen? Ein weiteres Argument zugunsten der Archipelstaaten sind ihre Sicherheitsbedürfnisse und ihr Interesse an der Erhaltung der maritimen Umwelt [97]. Diese Gründe erscheinen eher geeignet, die weitgehende Verminderung der Hohen See und ihrer Freiheiten jedenfalls zulasten von Kriegsschiffen zu rechtfertigen. Fremde Seestreitkräfte inmitten des eigenen Staatsgebietes, das aus einer größeren Zahl räumlich getrennter Inseln besteht, sind bedrohlicher als fremde Seestreitkräfte vor einer zusammenhängenden Küste, vielleicht selbst in – vergleichsweise kurzen – Meerengen. Die Rivalitäten von atomar getriebenen und bewaffneten Unterseeschiffen und der Systeme zu ihrer Entdeckung können in Archipelgewässern schon im Frieden für die Archipelstaaten zu größeren Gefahren führen als für die Anlieger der Hohen See. Jedenfalls im Hinblick auf den letzten Punkt kann man jedoch fragen, ob nicht schon die Erweiterung der Küstengewässer, ergänzt durch die oben erwähnten Sonderregelungen für höchst gefährliche Schiffe, berechtigten Sicherheitsbedürfnissen der Archipelstaaten genügen würde. Es ist nicht recht einzusehen, daß die Archipelstaaten durch die vorgesehene Seenahme besser gestellt sein müssen als Meerengenstaaten. Für den Schutz der maritimen Umwelt sollte die Einführung der Ausschließlichen Wirtschaftszonen ausreichen. Schließlich wird zugunsten der Archipelstaaten angeführt, daß sie mit besonderen Schwierigkeiten bei der Durchsetzung ihrer Rechtsordnung kämpfen müssen und größere Probleme durch separatistische Bestrebungen zu lösen haben als andere Staaten [98]. Es ist jedoch nicht Aufgabe der Staatenwelt, die Durchsetzung der innerstaatlichen Rechtsordnung von Archipelstaaten durch derartige Opfer an Meeresfreiheit zu erleichtern oder mit solchen Opfern zur Lösung von Integrationsproblemen dieser Staaten beizutragen. Bei dieser Betrachtungsweise bleibt zugunsten des Archipelregimes noch das Argument, daß es für den Raum der Archipelgewässer eine verhältnismäßig übersichtliche und einheitliche Regelung im Vergleich zu dem Rechtszustand bewirkt, der in diesem Seegebiet bestehen würde, wenn es in Eigen-, Küstengewässer, Meerengen, Anschluß-, Ausschließliche Wirtschaftszonen sowie Hohe See aufgeteilt und zudem noch Sonderregelungen unterworfen wäre. Die in der Draft Convention vorgesehene Vereinheitli-

---

[97] *Sturies* (Anm. 91), S. 69 f.
[98] Siehe *Sturies* (Anm. 91), S. 68.

chung bedeutet je nach der geographischen Lage möglicherweise ein Weniger an Regelung für einzelne Seegebiete und – wiederum möglicherweise – trotz der hier angesprochenen Probleme insoweit ein Mehr an Rechtsklarheit und Rechtssicherheit. Das gilt aber nicht im Hinblick auf die erheblich größeren Seegebiete, die nach noch geltendem Recht Hohe See sind und erst durch die Draft Convention einem Sonderzustand unterworfen werden.

Steht man dem Archipelregime nach der Draft Convention so kritisch gegenüber, so wird man es unbedingt für sachgerecht halten, die Befugnisse der Archipelstaaten durch die Gewährleistung der Archipelpassage einzuschränken. Dann läßt sich auch kaum kritisieren, daß Kriegsschiffe nicht schlechter gestellt werden als andere Schiffe, obwohl keineswegs allen an der Archipelpassage für Handelsschiffe (im weiten Sinne) interessierten Staaten an einem besonderen Durchfahrtsrecht auch für Kriegsschiffe gelegen ist. Die Regelung der Archipelpassage speziell für Kriegsschiffe legt den durch die Draft Convention in schwer zu verteidigender Weise räumlich »bereicherten« Archipelstaaten keine unzumutbaren Belastungen auf. Die Aufrechterhaltung eines Stückes der früher umfassenden Meeresfreiheit auch für die Kriegsschiffe kommt unmittelbar zwar nur einer begrenzten Zahl von Staaten zugute. Sie kann letztlich aber der Mehrzahl nützen, die ein Interesse daran hat, daß die Passagerechte von den Archipelstaaten nicht beschnitten werden. Dazu kann die Ausübung dieser Rechte gerade durch Kriegsschiffe beitragen. Auch bei der Archipelpassage und ihren Ausprägungen speziell für Kriegsschiffe wäre jedoch das Problem der höchst gefährlichen Schiffe zweckmäßigerweise noch einmal zu überdenken, obwohl es hier nicht so gravierend ist wie bei Meerengen. Wenn das Urteil von *Sturies* zutrifft, der Weg des Konventionsentwurfes berücksichtige »die wesentlichen Interessen der betroffenen Parteien«[99], dann nur deshalb, weil die Draft Convention die Archipelpassage für alle Schiffe einschließlich der Kriegsschiffe enthält. Vor dem Hintergrund des neuen Archipelregimes ist die Sachgerechtigkeit der Archipelpassage speziell für Kriegsschiffe leichter zu bejahen als die Transitpassage bei den Meerengen. Aber auch hier handelt es sich um einen politischen Kompromiß, der in größeren Zusammenhängen zu beurteilen ist, nicht zuletzt in Verbindung mit der Meerengenfrage. Hält man ihre Regelung vom Allgemeininteresse her für vertretbar, vom Interesse ihrer Befürworter her vielleicht sogar für notwendig, so bleibt zu fragen, ob das Stück Meeresfreiheit, das bei der Einrichtung des Archipelregimes geopfert wurde, ein zu hoher Preis war.

---

[99] Anm. 91, S. 132.

V. Die Schiffahrtsfreiheit von Kriegsschiffen in Anschlußzonen und Ausschließlichen Wirtschaftszonen

Die Küstengewässer einschließlich ihrer Meerengen und die Archipelgewässer unterstehen grundsätzlich der Souveränität der Küsten-/Archipelstaaten, deren Hoheitsbefugnisse durch Durchfahrtsrechte allerdings beträchtlich eingeschränkt sind. Im Unterschied zu diesen Territorialzonen sind die Anschlußzonen und die Ausschließlichen Wirtschaftszonen bloße Funktionszonen, in denen wesentliche Bestimmungen über die Schiffahrtsfreiheit auf der Hohen See selbstverständlich oder kraft ausdrücklicher Verweisungsvorschriften gelten [100].

1. Anschlußzonen

Art. 24 der Genfer Konvention von 1958 über das Küstenmeer und die Anschlußzone ermöglicht dem Küstenstaat die Inanspruchnahme einer seewärts an die Küstengewässer anschließenden Zone [101]. Die äußerste Grenze dieser Zone verläuft seewärts 12 sm parallel zu der Grundlinie, von der aus die Breite der Küstengewässer gemessen wird. Diese Zone ist ausdrücklich als Teil der Hohen See bezeichnet. Hier gilt nach der Küstenmeerkonvention vorbehaltlich der in Art. 24 abschließend aufgeführten küstenstaatlichen Befugnisse das Recht der Hohen See – auch für die Kriegsschiffe und ihre Aktivitäten [102].

Art. 33 DC sieht ebenfalls eine Anschlußzone vor. Ihre zulässige Breite wird – von der gleichen Grundlinie aus – auf 24 sm verdoppelt. Diese Erweiterung der Anschlußzone ist angesichts der Ausdehnung schon der Küstengewässer und der Einführung der Ausschließlichen Wirtschaftszonen nicht unproblematisch. Die Schiffahrtsfreiheit der Handelsschiffe wird durch die küstenstaatlichen Befugnisse in einem weiteren Seegebiet eingeengt, ohne daß ein zwingendes Bedürfnis ersichtlich ist. Für die Schiffahrtsfreiheit speziell von Kriegsschiffen ist wesentlich, daß Art. 33 DC im Vergleich mit

---

[100] Siehe zur Begründung dafür, daß die Ausschließlichen Wirtschaftszonen trotz der dem Küstenstaat zugesprochenen »Souveränität« als Funktionszonen gewertet werden, unten V. 2.

[101] Siehe zur Vorgeschichte und Entwicklung etwa Modernes Seevölkerrecht (Rechtsordnung, Bearbeiter *Serkov*, Anm. 33), S. 131 ff., sowie *Kolodkin* und *Molodcov* (Anm. 52), S. 75 ff. Eingehend zu verschiedenen Arten von Anschlußzonen, die in der Draft Convention zumeist nicht anerkannt und insoweit auch hier nicht berücksichtigt sind, *McDougal/Burke* (Anm. 96), S. 565 ff. Die Verfasser bezeichnen den »brief and enigmatic« Art. 24 der Küstenmeerkonvention von 1958 als mißlungen. Eine für alle Gegebenheiten gleiche, nur 12 sm schmale Anschlußzone entspreche den Bedürfnissen der internationalen Gemeinschaft nicht, sie sei auch wirklichkeitsfern: S. 606.

[102] Siehe *Jenisch* (Anm. 34), S. 88 ff., 93 f. mit weiteren Nachweisen.

Art. 24 der Küstenmeerkonvention von 1958 die Befugnisse der Küstenstaaten jedenfalls nicht inhaltlich erweitert. Diese sind weiterhin lediglich zu den Kontrollen befugt, die erforderlich sind, um Verstöße gegen ihre Zoll-, fiskalischen, Einwanderungs- oder Gesundheitsvorschriften im eigenen Staatsgebiet oder in den Küstengewässern zu verhüten oder zu bestrafen. Diese inhaltlich eng begrenzten Befugnisse geben dem Küstenstaat regelmäßig keine Handhabe, um die Schiffahrtsfreiheit von Kriegsschiffen einzuschränken. Zwar ist die Anschlußzone in Art. 33 DC im Unterschied zu Art. 24 der Genfer Küstenmeerkonvention nicht ausdrücklich als Teil der Hohen See bezeichnet. Nach Art. 86 DC gelten die Bestimmungen über die Hohe See aber für alle Seeteile, die nicht zur Ausschließlichen Wirtschaftszone, den Küstengewässern oder den Eigengewässern eines Staates oder zu den Archipelgewässern eines Archipelstaates gehören. Art. 2 Abs. 1 DC besagt ebenfalls, daß dem Küstenstaat in der Anschlußzone nicht die Souveränität zusteht, die ihm diese Konvention für andere Seegebiete zuerkennt. Somit genießen Kriegsschiffe nach Art. 95 DC auch die vollständige Immunität von jeder Hoheitsgewalt außer der des Flaggenstaates. Die Feststellung, daß Art. 33 DC in der Anschlußzone als solcher die Schiffahrtsfreiheit von Kriegsschiffen nicht einschränkt, bedeutet allerdings keine abschließende Entscheidung über ihre Schiffahrtsfreiheit in diesem Seegebiet. Die Anschlußzonen sind nämlich zugleich Teil der Ausschließlichen Wirtschaftszone, sofern der Küstenstaat eine Ausschließliche Wirtschaftszone in Anspruch nimmt. Das dürfte schon jetzt meistens, in Zukunft vielleicht überall der Fall sein [103].

2. Ausschließliche Wirtschaftszonen [104]

Ebenso wie das Archipelregime sind die Ausschließlichen Wirtschaftszonen trotz ihrer jahrzehntelangen Vorgeschichte als völkerrechtliches Institut Ergebnis erst von UNCLOS III. Während sich die Ausweitung der Anschlußzonen verhältnismäßig problemlos vollzog, war die Einführung der Aus-

---

[103] *Caflisch* sieht das Paradox als gegeben an, daß der Küstenstaat in der Anschlußzone nicht die sonstigen Befugnisse hat, die ihm in der übrigen Ausschließlichen Wirtschaftszone zustehen: Les zones maritimes sous juridiction nationale, leur limites et leur délimitation, RGDIP 84 (1980), S. 68, 85. Diese Auslegung überzeugt nicht. Die Anschlußzone ist vielmehr ein Teil der Ausschließlichen Wirtschaftszone, in dem der Küstenstaat zusätzliche Befugnisse hat.

[104] Siehe zum folgenden insbes. *Dupuy*, L'océan partagé – Analyse d'une négociation (1979), S. 66 ff., 266 ff.; *Extavour*, The Exclusive Economic Zone, A Study of the Evolution and Progressive Development of the International Law of the Sea (1979); *Gündling*, Die exklusive Wirtschaftszone, ZaöRV 38 (1978), S. 616; *Jaenicke* (Anm. 57), S. 478 ff.; *Kovalyov,* The Economic Zone and its Legal Status, International Affairs (Moskau) 1979, S. 58 ff.; *Oxman,* The Third United Nation's Conference on the Law of the Sea: The 1977 New York Session, AJIL 72 (1978), S. 57, 67 ff.; *Rauch,* Aspekte (Anm. 33), S. 75, 77 ff.; *Vargas,* The Legal Nature of the Patrimonial Sea: A First Step towards the Definition of the Exclusive Economic Zone, German Yearbook of International Law 22 (1979), S. 142 ff.; *Wolf* (Anm. 50), S. 221 ff.

schließlichen Wirtschaftszonen so umstritten, daß die Konferenz an dieser Frage zu scheitern drohte. Der Drang zu dieser Seenahme war bei Langküstenstaaten sehr stark, vor allem bei denen, die schon früher eine Ausschließliche Fischerei-, Wirtschaftszone oder gar ein Küstenmeer von 200 sm beansprucht hatten. Diese Bestrebungen stießen zuerst auf grundsätzlichen Widerspruch. Er kam von Staaten mit großen Fernfischereiinteressen (Großbritannien, Japan, Polen, UdSSR) oder größeren See- und Luftstreitkräften (z. B. USA), aber auch von zahlreichen anderen Staaten, meist in Binnen- und Randlagen. 1973/74 zeigten sich die Sowjetunion mit den Ostblockstaaten und die Vereinigten Staaten bereit, den Grundgedanken der Ausschließlichen Wirtschaftszone zu akzeptieren. Die Ausschließliche Wirtschaftszone brachte – für sich betrachtet – den Vereinigten Staaten wegen ihrer günstigen geographischen Lage erheblich mehr Vorteile als der Sowjetunion. Dennoch war ihre Einführung bei einer Gesamtwürdigung auch für die Vereinigten Staaten kein Gewinn. Beide See-Großmächte gingen von der Voraussetzung aus, daß der Charakter der neu zu schaffenden Zone als Hohe See gewahrt werden und ein Ausgleich in anderen strittigen Fragen erfolgen würde, insbesondere bei der Durchfahrt durch Meerengen. Nachdem sich den beiden Großmächten andere wichtige Schiffahrtsstaaten angeschlossen hatten, ließ sich trotz verbleibender erheblicher Widerstände die Aufnahme der Ausschließlichen Wirtschaftszone in den Konventionstext nicht mehr verhindern. Schon 1975 begann die Flut von einseitigen staatlichen Proklamationen über Ausschließliche Wirtschafts- oder jedenfalls Fischereizonen [105]. Es ist zu befürchten, daß die Ausschließlichen Wirtschaftszonen nunmehr im Grundsatz bereits vom Schicksal des Vertrages unabhängig geworden sind, obgleich etwa von sowjetischer wie von US-amerikanischer Seite wiederholt betont worden ist, ihr völkerrechtliches Schicksal hänge von dem der Draft Convention ab.

Art. 55 DC bezeichnet als Ausschließliche Wirtschaftszone ein seewärtiges Gebiet im Anschluß an die Küstengewässer, das der besonderen Rechtsordnung in Teil V der Konvention untersteht. Nach Art. 57 DC darf die Breite der Ausschließlichen Wirtschaftszone gemessen von der Grundlinie für die Küstengewässer 200 sm nicht überschreiten. Die Ausschließlichen Wirtschaftszonen insgesamt erfassen ein riesiges Seegebiet. Unterschiedliche Berechnungsansätze führen zu etwa 30% der gesamten Meeresoberfläche oder 36% der Hohen See. In der Ausschließlichen Wirtschaftszone gibt Art. 56

---

[105] Siehe *Gündling*, a.a.O., S. 623; *Kovalyov*, a.a.O., S. 58. 1976 dehnte der Kongress der Vereinigten Staaten entgegen den Widerständen der Delegationsmitglieder aus dem State Department und dem Department of Defense mit Wirkung von 1977 die Ausschließliche Fischereizone von 12 auf 200 sm aus. Siehe *Hollick* (Anm. 74), S. 353, 357 f. 1976 erging durch das Präsidium des Obersten Sowjets der UdSSR ein Erlaß über vorläufige Maßnahmen zur Erhaltung der lebenden Naturschätze und zur Regulierung der Fischerei in den an die Küste der UdSSR anschließenden Meeresgebieten (bis zu einer Breite von 200 sm). Deutsche Übersetzung bei *Rauch* I (Anm. 32), S. 130 ff.

Abs. 1 *lit.* a DC dem Küstenstaat souveräne Rechte (*sovereign rights*). Sie sind aber durch die Zweckbestimmung begrenzt. Sie bestehen lediglich 1. zur Erforschung, Ausbeutung, Erhaltung und Verwaltung aller lebenden oder nicht lebenden natürlichen Hilfsquellen von Meeresboden, Untergrund und Wassersäule, 2. im Hinblick auf andere Tätigkeiten zur wirtschaftlichen Ausbeutung und Erforschung dieses Gebietes, z.B. durch Energiegewinnung aus Wasser, Strömungen und Wind. Zusätzlich zu dieser inhaltlich begrenzten »Souveränität« gibt Art. 56 Abs. 1 *lit.* b und c DC dem Küstenstaat begrenzte Zuständigkeiten (*jurisdiction*) zur Einrichtung und Nutzung von künstlichen Inseln und anderen Installationen, zur wissenschaftlichen Meeresforschung, zum Schutz und zur Erhaltung der Meeresumwelt sowie andere, in der Draft Convention vorgesehene Rechte (und Pflichten). Die Zuständigkeiten nach Art. 56 Abs. 1 *lit.* b bestehen nur, soweit sie in sonstigen Bestimmungen der Draft Convention spezifiziert sind. Art. 56 Abs. 1 *lit.* b ist also eine ergänzungsbedürftige Kompetenzzuweisung; Art. 56 Abs. 1 *lit.* c ist lediglich eine abstrakte Erwähnung von anderenorts angeführten Zuständigkeiten. Bei der Ausübung aller seiner Befugnisse hat der Küstenstaat die Rechte (und Pflichten) anderer Staaten angemessen zu berücksichtigen (*due regard*). Vor allem gelten für die Ausschließlichen Wirtschaftszonen gemäß Art. 58 Abs. 2 DC die Art. 88–115 DC, d.h. Grundbestimmungen über den Status der Hohen See, und andere einschlägige Völkerrechtsregeln insoweit, als diese Regeln nicht mit den speziellen Bestimmungen in Teil V der Konvention unvereinbar sind. Zu den maßgeblichen Regeln gehört auch Art. 89 DC, nach dem kein Staat rechtsgültig einen Teil der Hohen See seiner Souveränität unterwerfen darf. Erscheinungsformen der Meeresfreiheit auf der Hohen See in Art. 87 DC gelten auch in der Ausschließlichen Wirtschaftszone. Sie sind dort allerdings durch Art. 58 Abs. 1 DC reduziert. Anderen Staaten als dem Küstenstaat bleiben nur die Freiheiten zur Schiffahrt, zum Überflug, Kabel- und Rohrlegen sowie sonstige international rechtmäßige Nutzungsarten der See, die zu diesen Freiheiten in Beziehung stehen und mit der Draft Convention im übrigen vereinbar sind. Beispiele für solche Nutzungsarten sind die mit der Führung von Schiffen, Luftfahrzeugen sowie dem Kabel- und Rohrlegen in Zusammenhang stehenden Tätigkeiten. Damit fehlen von den ausdrücklich in Art. 87 DC angeführten Freiheiten auf der Hohen See außer dem – nach dem Zweck der Ausschließlichen Wirtschaftszone selbstverständlich nicht erwähnten – Fischfang auch der Bau von künstlichen Inseln und Installationen sowie die Meeresforschung. Ferner fehlt in Art. 58 Abs. 1 die Ergänzung zu diesen in Art. 87 DC ausdrücklich genannten Arten der Meeresfreiheit durch die dortigen Worte *inter alia*.

Natürlich liegen die wichtigsten Auswirkungen der Neuschöpfung im Wirtschaftlichen. Jedoch ist es auch im Hinblick auf Kriegsschiffe von großer Bedeutung, ob und wie die Ausschließlichen Wirtschaftszonen die Schiffahrtsfreiheit unmittelbar berühren und wie die Küstenstaaten diese in ihren Zonen einschränken dürfen. Eine beträchtliche Zahl von Staaten verliert durch die Ausschließlichen Wirtschaftszonen den unmittelbaren Zugang zu der von Sonderbefugnissen freien See[106]. Ferner hat unter den Voraussetzungen des Art. 60 DC jeder Küstenstaat die Befugnis zur Errichtung oder Genehmigung von künstlichen Inseln und Installationen mit eigenen – wenn auch kleinen – Sicherheitszonen. Diese sind von allen Schiffen zu respektieren. Künstliche Inseln und Installationen dürfen nach Art. 60 Abs. 7 DC dort nicht eingerichtet werden, wo sie die Nutzung anerkannter Schifffahrtswege beeinträchtigen können, die für die internationale Schiffahrt wesentlich sind. Keine der erwähnten künstlichen Einrichtungen hat den Rechtsstatus von natürlichen Inseln, insbesondere besteht kein eigenes Küstenmeer[107].

Weiterreichende Folgen speziell für Kriegsschiffe ergeben sich bei einem Verständnis des Art. 58 DC, das Manöver, Übungen, Waffenversuche, Starten und Anbordnehmen von Flugzeugen allgemein aus der Schiffahrtsfreiheit ausschließt. Vor allem wegen der Weite des durch die Ausschließlichen Wirtschaftszonen erfaßten Seegebietes ist die Auslegung von Art. 58 (in Verbindung mit Art. 56 DC) wichtig. Die einschlägigen Formulierungen sind ein dilatorischer Formelkompromiß[108]. Der Verweis auf Art. 88 DC in Art. 58 Abs. 2 führt nicht zu einem engen Verständnis der Schiffahrtsfreiheit. Die Beschränkung der Nutzung an der Hohen See auf friedliche Zwecke schließt die eben erwähnten militärischen Tätigkeiten nicht von vornherein aus[109]. Art. 58 Abs. 1 DC gewährleistet unter Bezugnahme auf Art. 87 die Freiheit der Schiffahrt (*freedom of navigation*). Das ist ein wesentlicher Unterschied zu den bloßen Durchfahrtsrechten durch fremde Küstengewässer, Meerengen oder Archipelgewässer (*innocent, transit* oder *archipelagic passage*), in denen die Aktivitäten durchfahrender Schiffe ausdrücklich bestimmten – oben skizzierten – rechtlichen Einschränkungen unterworfen sind. Zwar hat Art. 58 Abs. 1 nicht alle enumerierten Freiheiten des Art. 87 DC und auch nicht das dort zusätzlich zu den enumerierten Freiheiten angeführte *inter alia* übernommen. Jedoch gehören nach der hier vertretenen Meinung Manöver, Übungen, Waffenversuche, Starten und Anbordnehmen von Flugzeugen traditionell zur Schiffahrtsfreiheit von Kriegsschiffen, sofern diese Tätigkeiten nicht in Seegebieten stattfinden, in denen sie gewohnheitsrechtlich oder in Verträgen untersagt sind. Zudem statuiert Art. 58 Abs. 1 DC zusätzlich zur Schiffahrtsfreiheit das Recht auf »other internationally lawful uses of the sea related to these freedoms, such as those associated with the operation of ships, aircraft and submarine cables and pipelines, and compatible with the

---

[106] Nach *Platzöder/Graf Vitzthum* sind es insgesamt 61, nämlich 27 im Nahen Osten und Afrika, 17 in Europa, 11 in Asien und 6 in Lateinamerika: Wirtschaftszonen und Archipelstaaten, Verfassung und Recht in Übersee 7 (1974), S. 289 ff., 292.

[107] Siehe zu diesem Art. 121 DC und *Symmons*, The Maritime Zones of Islands in International Law (1979).

[108] Vgl. *Extavour* (Anm. 104), S. 239 f.; *Gündling* (Anm. 104), S. 624 ff.

[109] Siehe oben II. 2.

other provisions of this Convention«. Diese anderen international rechtmäßigen Nutzungsarten in Beziehung zur Schiffahrtsfreiheit können sich bei Kriegsschiffen sinnvoll nicht allein auf den technisch-nautischen Betrieb, den Vorgang der Fortbewegung und Orientierung beschränken. Die Technik und Nautik zur Fortbewegung sind selbstverständlicher Bestandteil schon der Schiffahrtsfreiheit im engsten Sinne und bedürfen keiner besonderen Erwähnung. Wenn die eben erwähnten Aktivitäten von Kriegsschiffen nicht bereits unmittelbar von der Schiffahrtsfreiheit erfaßt sein sollten, dann gehören sie jedenfalls zu den sonstigen international rechtmäßigen Nutzungsarten [110]. Hätte man die erwähnten traditionellen Aktivitäten von Kriegsschiffen auf der Hohen See bei der Schaffung von derart riesigen Ausschließlichen Wirtschaftszonen untersagen wollen, so hätte man selbst auf einer Konferenz mit den höchst ungewöhnlichen Verfahrensweisen und den Kompromißlösungen von UNCLOS III zu anderen Formulierungen als denen der Draft Convention kommen müssen. Eine Verweisung auf Bestimmungen über die Transitpassage wie die in Art. 54 DC bei der Archipelpassage hätte genügt. Tatsächlich hatte jedoch ein Vorschlag Perus keinen Erfolg, Manöver und Waffenübungen von durchfahrenden Schiffen in der Ausschließlichen Wirtschaftszone von der Einwilligung des Küstenstaates abhängig zu machen [111]. Ebenso waren Versuche gescheitert, Kriegsschiffen in der Anschlußzone alle Tätigkeiten ohne unmittelbare Beziehung zur Fortbewegung zu untersagen [112]. Es ist nicht einzusehen, warum eine derartige Beschränkung nunmehr in der viel größeren Ausschließlichen Wirtschaftszone gelten sollte. Ferner ist zu berücksichtigen, daß die Befugnisse der Küstenstaaten trotz des hier problematischen Ausdrucks *sovereign rights* in Art. 56 Abs. 1 *lit.* a DC erheblich enger sind als seine Souveränität im Küstenmeer sowie den Meerengen und selbst als die des Archipelstaates in den Archipelgewässern. In der Ausschließlichen Wirtschaftszone deckt die »Souveränität« nur Tätigkeiten im Zusammen-

---

[110] Genau umgekehrt allerdings die Auslegung von *Platzöder*: Die eindeutige Sicherung von militärischen Aktivitäten in fremden Wirtschaftszonen würde die Streichung der eben erwähnten Formulierung »related to these freedoms such as . . . this Convention« erfordern, *Platzöder* (Anm. 18), S. 89. Zu dem gleichen *Ergebnis* wie der obige Text kommen dagegen etwa *Jaenicke* (Anm. 57), S. 490; *Dupuy* (Anm. 104), S. 269 f.; *Oxman* (Anm. 104), S. 73; *Richardson* (Anm. 72), S. 916; *Rauch* II, S. 23 f. *Rauch* III, S. 22, 25 (Anm. 32), weist auf die erfolglosen Versuche Brasiliens und Perus hin, die Zulässigkeit von Manövern und Waffenversuchen in einer 200 sm-Zone ausdrücklich vom Einverständnis des Küstenstaates abhängig zu machen. In anderen Veröffentlichungen hält derselbe Verfasser es für zweifelhaft, ob nicht lediglich ein Recht zur »nackten« Militärschiffahrt besteht, d.h. zur bloßen Fahrt mit entsprechend nautisch-navigatorischen Tätigkeiten: *Rauch* I, S. 97, und Aspekte (Anm. 33), S. 88 ff. Auch *Jenisch* scheint hier eine Unklarheit zu sehen, die dem Küstenstaat das Recht zur Entscheidung gibt: Außenpolitik 30 (1979), S. 37 ff., 42 f. Im SIPRI Yearbook 1975, S. 595 (Anm. 17), wird *de lege ferenda* vorgeschlagen, größere Manöver und Waffenübungen in einer 200 sm-Zone genehmigungspflichtig zu machen.

[111] Siehe *Platzöder* (Anm. 18), S. 90.

[112] *Rauch*, Aspekte (Anm. 33), S. 93, 118 Anm. 110.

hang mit den natürlichen Hilfsquellen, d. h. für wirtschaftliche Zwecke, und sind die Befugnisse nach Art. 56 Abs. 1 *lit.* b DC noch enger als die in Abs. 1 *lit.* a. Teil V Draft Convention gestattet den Küstenstaaten eben keine – grundsätzlich – vollständige Seenahme, die nur durch – unterschiedlich ausgestaltete – Durchfahrtsrechte begrenzt ist. Es liegt nahe, daß der »schwächeren« Stellung des Küstenstaates ein größerer Freiheitsraum von Drittstaaten entspricht, da man ja nicht ein ungeschriebenes Archipelregime einführen wollte. An diesem Verständnis der Schiffahrtsfreiheit von Kriegsschiffen ändert auch die Tatsache nichts, daß die Draft Convention die Ausschließliche Wirtschaftszone entgegen den Wünschen zahlreicher Industriestaaten sowie der Gruppe der Binnenländer und geographisch benachteiligten Staaten nicht ausdrücklich als Teil der Hohen See bezeichnet, und daß in Art. 58 Abs. 2 wesentliche Regeln über das Recht der Hohen See nur insoweit auf die Ausschließliche Wirtschaftszone für anwendbar erklärt werden, als diese Regeln mit Teil V Draft Convention vereinbar sind.

Wenn nach dieser Auslegung des Art. 58 DC im Frieden Manöver, Übungen, Waffenversuche, Starten und Anbordnehmen von Flugzeugen nicht von vornherein und allgemein mit den Rechten des Küstenstaates unvereinbar sind, so besteht doch eine Grenze für diese Ausprägungen der Schiffahrtsfreiheit. Nach Art. 58 Abs. 3 müssen Drittstaaten bei der Ausübung ihrer Rechte die Rechte der Küstenstaaten respektieren, soweit sie gemäß dieser Konvention und fortgeltendem sonstigen Völkerrecht bestehen. Mit der Anwendbarkeit der Schiffahrtsfreiheit aus Art. 87 Abs. 1 *lit.* a DC ist zugleich Art. 87 Abs. 2 DC mit der Pflicht zur Rücksichtnahme auf die Interessen aller Staaten an der Ausübung ihrer Freiheiten auf der Hohen See anwendbar. Zwar kann sich ein Küstenstaat auf seine Befugnisse zum Schutze von Installationen oder Fischbeständen nicht berufen, um allgemein übliche Manöver und Übungen oder gar das Starten und Landen von Flugzeugen in jedem beliebigen Seegebiet der Ausschließlichen Wirtschaftszone als Verletzung seiner Rechte zu rügen. Jedoch gibt es – von Atomversuchen ganz abgesehen – Grenzen für Aktivitäten von Kriegsschiffen. So dürfen sich Manöver über zulässige Sicherheitszonen von Installationen überhaupt nicht hinwegsetzen. Es dürfte aber auch ein Mißbrauch der Schiffahrtsfreiheit sein, große Manöver mit Schießübungen in wichtigen Laichgebieten gerade in Laichzeiten durchzuführen [113]. Die Rücksichtnahme in der Ausschließlichen Wirtschaftszone gegenüber dem Küstenstaat und gegenüber dritten Staaten darf nicht geringer sein als die gegenüber allen Staaten auf der Hohen See nach Art. 87 Abs. 2 DC. Die Grenzen für die Einrichtung von Warnzonen durch Drittstaaten bei ihren Manövern sind hier enger. Das schon auf der Hohen See geltende Verbot für die Einrichtung von Sperrzonen gewinnt ein noch größeres Gewicht.

---

[113] *Rauch* III (Anm. 32), S. 29.

Auch in den Ausschließlichen Wirtschaftszonen entstehen für die Schiffahrtsfreiheit Probleme. Es wäre erstaunlich, wenn es bei einem neuen, so hart umstrittenen und schließlich durch Kompromisse verschiedener Art geschaffenen Rechtsinstitut von so großer praktischer Bedeutung anders wäre. Die auslegungsbedürftigen Generalklauseln, die Anlaß zu Meinungsverschiedenheiten geben können, sind hier seltener als im Recht der Archipele, Meerengen und vor allem Küstengewässer. Immerhin stoßen auch in der Ausschließlichen Wirtschaftszone küstenstaatliche Befugnisse, die teilweise nur vage umrissen sind, auf Freiheiten anderer Staaten. Die oben erwähnten Normen geben genügend Beispiele: Wie etwa verhält sich im konkreten Fall das souveräne Recht des Küstenstaates zum Schutze seiner natürlichen Hilfsquellen zu der Schiffahrtsfreiheit von Kriegsschiffen unter Einschluß von großen Manövern oder zu der Errichtung von Warnzonen für den Abschuß oder das Niedergehen von Interkontinentalraketen? Hier fehlt es an der wünschenswerten rechtlichen Klarheit. Art. 59 DC sieht Konfliktslösungen auf der Grundlage der Billigkeit im Lichte aller erheblichen Umstände unter Berücksichtigung der Wichtigkeit der betroffenen Interessen für die Streitparteien wie auch für die internationale Gemeinschaft insgesamt vor. Das ist eine Leerformel. Im Streit um die Einführung einer Ausschließlichen Wirtschaftszone und ihr Regime war keine ausdrückliche Aussage darüber zu erreichen, ob sie noch ein Teil der Hohen See sein soll, der nunmehr unter einem Sonderregime steht, oder aber ganz und gar ein Seegebiet *sui generis*. Im ersten Falle könnte in Konflikten eine Vermutung für die Schiffahrtsfreiheit sprechen, im zweiten Falle vielleicht eine Vermutung für die Befugnisse des Küstenstaates. Die Frage braucht angesichts des sogleich vertretenen Ergebnisses hier nicht vertieft zu werden. Einschlägige Formulierungen des Teiles V mit den verschiedenen Hin- und Herverweisungen dienten nicht zuletzt dem Zweck, die auf der Konferenz so umstrittene Frage in einer Weise zu lösen, die beiden Seiten die Behauptung ermöglichte, sich durchgesetzt zu haben[114]. Nach Ansicht des Verfassers ist die Ausschließliche Wirtschaftszone ein Seegebiet *sui generis*[115]. Diese rechtliche Qualifizierung gestattet es

[114] Siehe *Hollick* (Anm. 74), S. 294 f., 306, 317 (im II. Ausschuß auf der New Yorker Sitzung gab es ein Zahlenverhältnis von etwa 50:50), S. 326 f. *Hollick*, S. 357, führt innenpolitische Gründe für ein Zurückweichen der amerikanischen Delegation auf der Konferenz an.

[115] Der Vorsitzende des II. Ausschusses erklärte auf der New Yorker Sitzung 1976, daran bestehe kein Zweifel (*Hollick*, Anm. 74, S. 317). Die Bemerkung ist aber im Zusammenhang mit dem Wunsch zu sehen, unbedingt eine Einigung durch Kompromiß zu ermöglichen. Ausdrücklich für die juristische Qualifikation als Seegebiet *sui generis* *Gündling* (Anm. 104), S. 653 f. Zu der gleichen Bewertung neigen etwa *Dupuy* (Anm. 104), S. 83; *Stöcker* (Anm. 57), S. 344 f. Die Frage wird eher umschritten als ausdrücklich gestellt bei *Oxman* (Anm. 104), S. 67 ff. (System von Spiegeln), und *Kovalyov* (Anm. 104). Siehe für beide aber unten Anm. 116. *Jaenicke* erklärt die Frage durch den Text vom Juli 1977 für ausgeklammert, ihre Beantwortung für nicht präjudiziert (Anm. 57), S. 489; siehe dort auch S. 485 ff.

aber nicht, in Zweifelsfällen von einer Vermutung für die Zuständigkeit des Küstenstaates auszugehen. Entscheidend ist nach Art. 55 DC der Teil V der Konvention. Ihre übrigen Teile sind nur insoweit maßgeblich, als Teil V sie als erheblich *(relevant)* erkennen läßt. So wie die Zulässigkeit bestimmter militärischer Aktivitäten in der Ausschließlichen Wirtschaftszone oben nicht mit dem Status dieser Zone als besonderem Teil der Hohen See begründet wurde, so lassen sich umgekehrt Befugnisse, die dem Küstenstaat nicht eindeutig zugewiesen sind, nicht mit einer Vermutung zu seinen Gunsten begründen. Die im Wege des Kompromisses eingeführte »Souveränität« ist nicht eine grundsätzlich umfassende Entscheidungsbefugnis auf der Grundlage voller »Gebietshoheit«, also nicht die Souveränität im gebietlichen Kontext nach herkömmlichem Verständnis. *Oxman* trifft dies mit der Feststellung »the sovereign rights are economic rights« [116]. Die Ausschließliche Wirtschaftszone ist ungeachtet der unzutreffenden Formulierung von der »Souveränität« des Küstenstaates bloße Funktionszone. Auch in diesem Zusammenhang ist von Interesse, daß nach Art. 58 Abs. 2 DC vorbehaltlich spezieller Normen in Teil V der Art. 89 DC gilt und daß nach Art. 89 kein Staat rechtsgültig einen Teil der Hohen See seiner Souveränität unterwerfen darf. Wenn die Ausschließliche Wirtschaftszone auch in Art. 86 Satz 1 DC nicht als Teil der Hohen See aufgeführt ist, so sagt doch sogleich Satz 2, daß dies nicht zu einer Verkürzung der Freiheiten führt, die alle Staaten dort gemäß Art. 58 genießen. Jedenfalls ist eine generelle Vermutung für die Lösung von Fällen ausgeschlossen, in denen die Existenz und Reichweite von wirtschaftlichen Befugnissen des Küstenstaates und von nicht-wirtschaftlichen Rechten von Drittstaaten – wie der Schiffahrtsfreiheit von Kriegsschiffen – unklar sind und einander widersprechen. Daran ändert es nichts, daß nach Art. 58 Abs. 2 und 3 DC der Teil V der Draft Convention den Vorrang gegenüber anderen Konventionsbestimmungen hat. Abs. 1 des Art. 58 und Art. 56 stehen ebenfalls in Teil V [117]. Auch wegen des Fehlens einer Aussage über den Charakter der Ausschließlichen Wirtschaftszone insbesondere in Abschnitt V und in Art. 86 DC sowie wegen der – oben skizzierten – Hin- und Herverweisungen in teilweise generalklauselartigen Artikeln (System von Spiegeln) läßt Teil V unter dem Gesichtspunkt der Rechtsklarheit und

---

[116] Anm. 104, S. 68. Ebenso *Dupuy*, S. 83, »zone de souveraineté économic«. Eher noch einschränkender *Kovalyov* »zone of exclusive economic jurisdiction«, S. 64. Siehe auch die Kritik von *Gündling*, S. 625, an der Verwendung des Souveränitätsbegriffes mit dem Ergebnis, es handele sich um eine Funktionalzone (alle Anm. 104).

[117] Wenn *Gündling* (Anm. 104), S. 648, meint, im Konfliktsfall habe das Wirtschaftsregime des Küstenstaates Vorrang, so ist das eine etwas andere Betrachtungsweise. Bei den Ausführungen oben im Text geht es nicht um den Vorrang zwischen klaren Ausprägungen von küstenstaatlicher Wirtschaftshoheit im Verhältnis zu Rechten von Drittstaaten, sondern um die Frage, ob unklare Formulierungen küstenstaatlicher Wirtschaftsfreiheit in der Draft Convention extensiv ausgelegt werden dürfen. *Stöcker* fürchtet, daß sich im Konfliktsfall die Wirtschaftszone gegenüber der Meeresfreiheit durchsetzen wird (Anm. 57), S. 347 f.

Rechtssicherheit zu wünschen übrig. Zwei hervorragende Sachkenner äußern sich dennoch optimistisch oder verhältnismäßig optimistisch. Nach *Kovalyov* haben die Rechte aller Nationen an der Hohen See mit den Rechten des Küstenstaates nichts zu tun. Unvereinbarkeitsfragen könnten nur bei Versuchen rechtswidrigen Handelns auftreten [118]. Diese Annahme ist nicht haltbar. *Oxman* hält die Logik hinter dem »System von Spiegeln« für klar und den Entwurf (in der Fassung von 1977) für eine Verbesserung, die auch Schwierigkeiten mit der staatlichen Gesetzgebung über Sonderzonen ausräumen könne. *Oxman* verschweigt allerdings nicht, daß der Entwurf nicht seinen persönlichen Präferenzen entspreche und weiter verbesserungsfähig sei [119]. Auch bei diesen beiden Beurteilungen mag eine Rolle spielen, daß die Verfasser den Delegationen ihrer Staaten angehört haben. Realistisch erscheint die Befürchtung, daß in weiten Teilen der See Konfliktsmöglichkeiten geschaffen sind, ohne daß ein adäquates Normensystem zu ihrer Lösung zur Verfügung gestellt worden ist (dilatorischer Formelkompromiß) [120].

Die Antwort auf die Frage, ob die Regelung über die Schiffahrtsfreiheit von Kriegsschiffen in den Ausschließlichen Wirtschaftszonen sachgerecht ist, hängt entscheidend von der Vorfrage nach der Sachgerechtigkeit der Ausschließlichen Wirtschaftszonen überhaupt ab. Diese Frage läßt sich hier nicht näher untersuchen. Es möge dem Verfasser gestattet sein, seine Ansicht ohne nähere Begründung vorzutragen. Es ist kein Zufall, daß der in diesem Zusammenhang vielleicht entschiedenste Verteidiger der Meeresfreiheit, die Sowjetunion, auf der 7. Sitzung der Konferenz die Unterstützung einer teilweise ungewöhnlichen Koalition genoß. Ihr gehörten (neben den Ostblockstaaten) zahlreiche Binnenstaaten, zahlreiche Staaten mit ungünstigem Küstenverlauf, die arabischen Staaten und Israel an. Zutreffend erklärte der Heilige Stuhl die ungeheure Seenahme durch die Ausschließlichen Wirtschaftszonen mit dem Zugriff auch auf einen immensen Teil der Meeresschätze für unvereinbar mit allen Vorstellungen über das Meer als gemeinsames Erbe der Menschheit [121]. Die geographischen Gegebenheiten, die Verbindungen von Land und See sprechen bei den Ausschließlichen Wirtschaftszonen nicht annähernd so für die Küstenstaaten wie es bei Küstengewässern einschließlich der Meerengen und selbst bei den Archipelen der Fall ist. Gesichtspunkte einer gerechten Weltwirtschaftsordnung – wie immer man sie sich vorstellen mag – lassen sich ebenfalls nicht als Grund für die Errichtung der Ausschließlichen Wirtschaftszonen heranziehen. Zu den Staaten, welche die größten Meeresräume als Ausschließliche Wirtschaftszonen vereinnahmen werden, gehören die Vereinigten Staaten, Australien, Neuseeland, Kanada,

---

118 *Kovalyov* (Anm. 104), S. 62.

119 Anm. 104, S. 74 f.

120 Siehe von den kritischen Stimmen etwa *Rauch*, Aspekte (Anm. 33), S. 88 f.; *Dupuy* (Anm. 104), S. 84, »Le compromis . . . demeure fragile«.

121 Siehe *Jenisch* (Anm. 110), S. 43, allerdings unter Anführung des Vatikanstaates statt des Hl. Stuhles.

die Sowjetunion, Japan und Großbritannien[122]. Etwa die Hälfte der Fläche aller Ausschließlichen Wirtschaftszonen wird 13 reichen oder wohlhabenden Industriestaaten zugutekommen. Den größten räumlichen Gewinn werden die Vereinigten Staaten haben[123]. Die Einführung des Wirtschaftszonenkonzepts stellt also auch keinen Beitrag zur Entwicklungshilfe dar, sondern wird den Abstand zwischen armen und reichen Staaten noch vergrößern. Zwar gehören zu den begünstigten Langküstenstaaten auch Entwicklungsländer, wenn auch nur wenige der ärmsten. Aber die meisten Binnenstaaten und zahlreiche Staaten mit ungünstigen Küsten sind ebenfalls Entwicklungsländer[124]. Der Gesichtspunkt besserer Erhaltung und schonender Nutzung der natürlichen Hilfsquellen des Meeres könnte angesichts des unbefriedigenden derzeitigen Zustandes die Einrichtung der Ausschließlichen Wirtschaftszonen rechtfertigen. Jedoch berechtigt die Neuregelung, die im wesentlichen in der Nationalisierung der Ressourcen besteht, nicht zu großen Hoffnungen[125]. Eine funktionale, rechtliche oder moralische Legitimation der Tendenz zur Nationalisierung vermag der Verfasser nicht zu erkennen[126].

Geht man von diesen Voraussetzungen aus, so läßt sich bei der Einrichtung der Ausschließlichen Wirtschaftszonen ein Interesse der Küstenstaaten, das den Vorrang vor der Schiffahrtsfreiheit von Handelsschiffen verdient, von vornherein nicht bejahen. Lehnt man die Sachgerechtigkeit der Seenahme mittels Ausschließlicher Wirtschaftszonen in der Form der Draft Convention grundsätzlich ab und sieht man in Übereinstimmung mit jahrhundertealtem Völkerrecht in der Meeresfreiheit die Schiffahrtsfreiheit für Handels- *und* Kriegsschiffe aus plausiblen Gründen einbegriffen, so wird man den Einschränkungen der Schiffahrtsfreiheit von Kriegsschiffen ungeachtet gewisser Unterschiede in der Interessenlage hinsichtlich beider Schiffsarten mit großer Zurückhaltung begegnen. Zudem können die Kriegsschiffe hier in gleichem oder noch stärkerem Maße als in anderen besonderen Meereszonen einen Beitrag zur Wahrung der jeweils verbleibenden Schiffahrtsfreiheit leisten: Es ist zu befürchten, daß die von verschiedenen Staaten schon begonnenen Angriffe auf die nach der Draft Convention verbleibenden Arten der Meeresfreiheit in der Ausschließlichen Wirtschaftszone an Stärke und Häufigkeit zunehmen werden[127]. Die Befürchtungen, daß es trotz Art. 89 DC zu einer schleichenden Ausweitung küstenstaatlicher Zuständigkeiten, einer *creeping jurisdiction*

---

[122] Siehe *Wolf* (Anm. 50), S. 250.

[123] Vgl. dazu die Übersicht bei *Lucchini/Voelckel*, Les états et la mer. Le nationalisme maritime (1977), S. 61 ff.

[124] Vgl. dazu *Apollis*, L'emprise maritime de l'état côtier (1981), S. 15 ff.

[125] Kritisch auch *Jenisch* (Anm. 110), S. 41, und *Wolfrum*, Renationalisierung des Fischereiregimes, in: Graf Vitzthum (Plünderung, Anm. 81), S. 231, 241 f., 245.

[126] So schon *Jenisch* (Anm. 110), S. 41.

[127] So gestehen beispielsweise Burma und die Malediven fremden Schiffen in ihren Ausschließlichen Wirtschaftszonen nur das Recht der friedlichen Durchfahrt zu. Siehe New Directions (Anm. 48) Bd. 7, S. 359, Bd. 9, S. 295.

kommen wird[128], lassen eine Verteidigung der Schiffahrtsfreiheit notwendig erscheinen. Ihre Wahrung auch für Handelsschiffe wird jedenfalls unterstützt, wenn Kriegsschiffen grundsätzlich die gleichen Rechte wie diesen zustehen und sie für die gemeinsamen Rechte in Erscheinung treten. Eine solche Rechtswahrung durch Präsenz wird allerdings nicht dadurch erleichtert, daß einzelne Staaten, die traditionell zu den wichtigsten Verteidigern der Schiffahrtsfreiheit gehören, selbst Langküstenstaaten mit lohnenden Nutzungsmöglichkeiten in Ausschließlichen Wirtschaftszonen sind. Auch die oben[129] erwähnte Aufrüstung von Schwellen- und Entwicklungsländern mit langen Küsten sowie die großen Wirkungsmöglichkeiten für kleine Kriegsschiffe sind kein gutes Omen für die Schiffahrtsfreiheit.

VI. Ausblick

Aufgrund der obigen Übersicht läßt sich die Frage beantworten, wie die unter I. skizzierten Fälle nach der Draft Convention zu entscheiden wären. Ein Kriegsschiff bedürfte zum Einfahren in fremde Eigengewässer weiterhin des Einverständnisses des Küstenstaates, wenn die Einfahrt von diesem nicht generell für alle Staaten oder vertraglich für den Flaggenstaat freigegeben oder schließlich im Einzelfall wegen höherer Gewalt zulässig ist. Damit wäre die schwedische Rechtsansicht im Falle »Whiskey on the Rocks« auch nach der Draft Convention zutreffend.

Kriegsschiffen stünde die friedliche Durchfahrt durch die – nunmehr bis zu einer Höchstbreite von 12 sm ausdehnbaren – fremden *Küsten*gewässer ebenso zu wie jedenfalls nach der Genfer Küstenmeerkonvention von 1958. Genehmigungen oder Anmeldungen dürften nicht gefordert werden. Gewisse Modifizierungen der friedlichen Durchfahrt würden sich auf einen Fall wie den schwedisch-sowjetischen von 1981 nicht auswirken, wenn er sich in Küstengewässern abgespielt hätte. Untergetauchtes Fahren entgegen dem Willen des Küstenstaates wäre in »gewöhnlichen« Küstengewässern weiterhin unzulässig, in internationalen Meerengen dagegen – anders als nach geltendem Recht – bei Beachtung der Voraussetzungen und Regeln für die Transitpassage gestattet.

Im Hinblick auf die völkerrechtliche Beurteilung des amerikanisch-libyschen Zwischenfalls muß hier ungeprüft bleiben, ob es sich bei dem Golf von Sidra um eine historische Bucht handelt. Die Frage dürfte zu verneinen sein. Die Ausdehnung der Küstengewässer durch die Draft Convention bis zu einer Höchstbreite von 12 sm erweitert auch den küstenstaatlichen Luftraum, wirkt sich aber auf den Zwischenfall nicht aus. Hätte sich dieser dage-

---

[128] Siehe z. B. *Dupuy* (Anm. 104), S. 205 f.; *Platzöder/Graf Vitzthum* (Anm. 106), S. 292; *Rauch*, Aspekte (Anm. 33), S. 83.

[129] Siehe unter I. 2. oben bei Anm. 17.

gen über den Küstengewässern abgespielt, so wäre schon das Überfliegen durch fremde Militärflugzeuge rechtswidrig gewesen, da die Draft Convention nicht parallel zu der friedlichen Durchfahrt von Kriegsschiffen ein Recht von Kampfflugzeugen auch zum friedlichen Überflug von Küstengewässern vorsieht. Bei Zwischenfällen zwischen Kampfflugzeugen des Küstenstaates und denen eines dritten Staates, die von einer Kriegsflotte aufgestiegen sind, würde bei einer Entfernung von 30 sm zur Küste nach der Draft Convention das Recht der Ausschließlichen Wirtschaftszone gelten. Folgt man der oben vertretenen Ansicht, so sind in diesem Bereich Flottenmanöver und damit verbundene Luftübungen zulässig. Das Flugrecht erstreckte sich auch auf Luftmanöver, die nicht von Flugzeugträgern ausgehen. Somit dürften Kampfflugzeuge wie die amerikanischen ebensowenig vom Küstenstaat angegriffen werden, wie diese Flugzeuge des Küstenstaates angreifen dürften.

Würde sich im Hinblick auf die Ausgangsfälle nichts ändern, so ist doch festzuhalten, daß die Draft Convention die Schiffahrtsfreiheit von Kriegsschiffen insgesamt gesehen empfindlich beschneidet. Die Befugnis zur Unterwasserpassage (und zum Überflug) stellt einen Gewinn nur in den »alten« Meerengen dar, die auch nach dem noch geltenden Recht nicht Teile der Hohen See sind. Diesem Vorteil stehen erhebliche Verschlechterungen der Schiffahrtsfreiheit zugunsten der Küstenstaaten gegenüber. Sie ergeben sich aus der Tatsache, daß der Raum der Hohen See mit der Meeresfreiheit zugunsten von besonderen Seezonen und damit zugunsten der Küstenstaaten beschnitten wird. Die Gewährleistungen von Transit- und Archipelpassage oder von Schiffahrtsfreiheit im Recht der Ausschließlichen Wirtschaftszone mindern den Grad der Verschlechterung, heben diese aber nicht auf.

Geht man davon aus, daß die Draft Convention grundsätzlich eine umfassende Kodifikation des Meeresrechtes sein (Art. 311 DC und Präambel) und daß Vorbehalte nur insoweit zulässig sein sollen, als sie in der Konvention ausdrücklich gestattet sind (Art. 309 DC), so besteht ein Normensystem, das geschlossen ist, aber eine Vielzahl von Konfliktsmöglichkeiten zuläßt. Die Frage nach der friedlichen Streitregelung hat daher besondere Bedeutung. Sie kann hier nicht näher erörtert werden[130]. Genügen muß der Hinweis, daß die Draft Convention in Art. 286 ff. ein kompliziertes System obligatorischer friedlicher Streitregelung vorsieht. Sie bietet den Vertragsstaaten gewisse Auswahlmöglichkeiten bei den zu verbindlichen Entscheidungen befugten Instanzen. Für Streitfälle, die im Zusammenhang mit der Schiffahrtsfreiheit entstehen können, ist von Art. 286 DC auszugehen: Vorbehaltlich des Abschnittes 3 in Teil XV kann nach einem Mißerfolg bei der Konfliktslösung mit anderen friedlichen Mitteln jede Streitigkeit über Auslegung

[130] Siehe aus der Literatur die neueste größere Arbeit von *Leboux*, La Troisième Conférence sur le droit de la mer et le règlement obligatoire des différends, Canadian Yearbook of International Law 18 (1980), S. 31 ff.; unter Bezugnahme noch auf einen älteren Text, aber mit reichlichen Literaturangaben *Bernhardt*, Die Streitbeilegung im Rahmen der Neuordnung des Seerechts, ZaöRV 38 (1978), S. 959 ff.

und Anwendung der Konvention auf Antrag jeder Streitpartei vor eine Instanz gebracht werden, die nach der Draft Convention zu verbindlichen Entscheidungen zuständig ist. Streitigkeiten, bei denen es um die Ausübung der Souveränität oder der Zuständigkeiten eines Küstenstaates nach der Konvention geht, fallen unter Art. 297 Abs. 1 in Abschnitt 3 des XV. Teiles. Die obligatorische Streitregelung ist – soweit hier erheblich – in zwei Arten von Fällen gegeben. Die erste Art setzt die Behauptung voraus, daß ein Küstenstaat einen Verstoß gegen diejenigen Bestimmungen der Konvention begangen hat, die sich auf die Freiheiten und Rechte zur Schiffahrt und zum Überflug oder zum Legen von Unterseekabeln und -rohrleitungen oder auf andere international zulässige Meeresnutzungen beziehen, wie sie in Art. 58 DC spezifiziert sind[131]. Die zweite Fallgruppe obligatorischer Streitentscheidung liegt bei der Behauptung vor, ein Staat habe bei dem Gebrauch dieser Freiheiten Rechte eines Küstenstaates verletzt, die von der Konvention gegeben oder anerkannt sind. Hier kann nicht geprüft werden, ob diese Bestimmungen alle Streitigkeiten erfassen, die im Zusammenhang mit der Schiffahrtsfreiheit von Kriegsschiffen entstehen können. Soweit derartige Streitigkeiten von der obligatorischen Streitentscheidung erfaßt werden, haben die Vertragsstaaten der Konvention jedenfalls Ausweichmöglichkeiten. Art. 298 Abs. 1 *lit.* b DC gestattet ihnen, jederzeit einseitig Streitigkeiten über militärische Aktivitäten auszuschließen. Ausdrücklich sind hier militärische Aktivitäten von Staatsschiffen und Staatsluftfahrzeugen genannt, die nicht Handelszwecken dienen. Schließlich können durch einseitige Erklärung Streitigkeiten ausgeschlossen werden, bei denen der Sicherheitsrat der Vereinten Nationen die ihm von der Satzung zugewiesenen Aufgaben ausübt, sofern der Sicherheitsrat nicht entscheidet, den Fall von seiner Agenda abzusetzen oder die Streitparteien zur friedlichen Streitregelung gemäß der Draft Convention aufzufordern. Für die Zukunft des verbindlichen Streitentscheidungsverfahrens liegen zwei Prognosen nahe: Es ist anzunehmen, daß die angerufene Entscheidungsinstanz einen erheblichen Teil ihrer Zeit und Arbeitskraft auf Zuständigkeitsfragen verwenden muß. Wichtiger ist die zweite Annahme, daß aus der obligatorischen Streitentscheidung speziell über die Schiffahrtsfreiheit von Kriegsschiffen *in praxi* nicht viel werden wird.

Unter diesen Umständen fragt sich, welche Interessen auf die Dauer den stärkeren Durchsetzungswillen und die größere Durchsetzungskraft haben werden, die an möglichst weitreichender (geographisch) und ungestörter (inhaltlich) Schiffahrtsfreiheit oder die an möglichst weitreichenden (geographisch) und tiefgreifenden (inhaltlich) küstenstaatlichen Befugnissen. Bei ei-

---

[131] Die Formulierung läßt nicht mit letzter Sicherheit erkennen, ob damit auch die Rechte und Freiheiten der Durchfahrt durch Küstengewässer, Meerengen und Archipele Rechtsschutz genießen, wie *Jaenicke* (Anm. 57), S. 507 f., meint, oder ob wegen der Bezugnahme auf Art. 58 DC nur die Schiffahrtsfreiheit in den Ausschließlichen Wirtschaftszonen erfaßt wird, was *Platzöder* (Anm. 68), S. 730 f., für wahrscheinlicher hält.

ner Prognose ist schon deshalb Vorsicht geboten, weil eine Zuordnung der Staaten lediglich zu dieser Zweiheit von Interessen nicht der Wirklichkeit entspricht, diese vielmehr differenziert ist. Zählt man zu den Langküstenstaaten einschließlich der Archipelstaaten wegen der teilweisen Interessengleichheit auch die Meerengenstaaten, so gab es auf UNCLOS III etwa 60 Staaten mit besonderem Interesse an einer möglichst weitgehenden Kontrolle über den Meeresraum vor der Küste. Zu ihnen gehören sowohl Entwicklungsländer als auch Industrieländer. Weitgehend mit diesen 60 Staaten identisch sind die Küstenstaaten, die vor der eigenen Küste fischreiche Meereszonen haben und zur Fernfischerei in größerem Umfang nicht in der Lage sind. Dagegen waren an möglichst weitgehender Schiffahrtsfreiheit die bedeutenden Seemächte und die anderen größeren Schiffahrtsnationen interessiert. Zu den Staaten dieser Gruppe gehören überwiegend Industrie-, aber auch einige Schwellenländer. Teilweise identisch mit diesen 15–20 Staaten sind etwa 10 Staaten mit umfangreicher Fernfischerei [132]. Für die Ausübung ihrer Freiheit zum Fischfang ist die Schiffahrtsfreiheit erste Voraussetzung. An weltweiten Flottenbewegungen waren unmittelbar in erster Linie die beiden atomaren Supermächte interessiert, darüber hinaus am ehesten die beiden anderen atomaren Seemächte mit Hochseestreitkräften, nämlich Großbritannien und Frankreich [133]. Auch in diesem Zusammenhang gilt, daß selbst Langküstenstaaten ihre Interessen nicht nur von der geographisch-wirtschaftlichen Lage her bestimmen. Die Vereinigten Staaten und die Sowjetunion sind Langküstenstaaten *par excellence*, aber zugleich atomare Supermächte mit weltweiten Interessen gerade an der Schiffahrtsfreiheit der Kriegsschiffe. Beide Staaten gehören zudem zu den großen Handelsschiffahrtsstaaten. Für einzelne Staaten kann die Zugehörigkeit zu einem Bündnis eher maßgeblich sein als die Eigenschaft als Küsten- oder Schiffahrtsstaat. Wieder andere Staaten mögen größtes Interesse an der Schiffahrtsfreiheit für Handelsschiffe haben, an der Schiffahrtsfreiheit für Kriegsschiffe aber wenig oder gar nicht interessiert sein, etwa Staaten der billigen Flaggen.

Trotz der hier nur beispielhaft erwähnten Komplikationen bei der Aufstellung von Interessengegensätzen ist festzustellen, daß die Zahl der Staaten, bei denen die Schiffahrtsfreiheit das dominierende Interesse ist, erheblich kleiner ist als die Zahl der Küstenstaaten mit dem Hauptinteresse Wahrung und Ausdehnung von küstenstaatlicher Hoheit. Das Zahlenverhältnis ist für die Vertreter spezifisch küstenstaatlicher Interessen noch günstiger, wenn man auf das Interesse an der Schiffahrtsfreiheit lediglich von Kriegsschiffen

---

[132] Vgl. die Übersicht bei *Wolf* (Anm. 50), S. 104 f.

[133] Gliederungen der Staaten nach ihren verschiedenen, sich teilweise überschneidenden Interessen, auf die hier nicht eingegangen werden kann, finden sich außer bei *Wolf* (Anm. 50), S. 104 ff., z. B. bei *Hollick* (Anm. 74), S. 250 ff.; *Marffy*, La genèse du nouveau droit de la mer – Le Comité des fonds marins (1980), S. 127 ff., 135 ff.; *Stöcker* (Anm. 57), S. 320 f. Eine außerordentlich eingehende Aufschlüsselung staatlicher Interessen mit detaillierten Statistiken bieten *Lucchini/Voelckel* (Anm. 123).

abstellt. Von den etwa 30 Binnenstaaten und den meisten der etwa 25 geographisch benachteiligten Staaten kann man bei diesem Interessenkonflikt aus verschiedensten Gründen kaum eine entschiedene Haltung erwarten, jedenfalls kein besonderes Engagement in Konfliktsfällen [134]. Diese Zahlenverhältnisse legen den Schluß nahe, daß die Konvention selbst als weltweit verbindlicher Vertrag schon vor der Durchführung einer Revisionskonferenz nach Art. 312 DC allerlei Versuchen ausgesetzt sein wird, die küstenstaatlichen Zuständigkeiten auf Kosten der Schiffahrtsfreiheit zu erweitern. Das kann durch offenkundige Vertragsverletzungen geschehen, gegen die kein Vertrag schützt, etwa durch die Ausdehnung von Sonderzonen auf Kosten der Hohen See. Die Erweiterung küstenstaatlicher Befugnisse auf Kosten der Schiffahrtsfreiheit kann aber auch dadurch geschehen, daß man küstenstaatliche Befugnisse extensiv, Bestimmungen zum Schutz der Schiffahrtsfreiheit restriktiv auslegt und Meerengen, Archipelgewässer und die Ausschließlichen Wirtschaftszonen allmählich auf den Status von Küstengewässern herabdrückt [135]. Der beste Schutz dagegen wäre die Erkenntnis, daß die einen wie die anderen Aktionen sich nicht lohnen, vor allem deshalb, weil die größten wirtschaftlichen Nutzungsmöglichkeiten von Meeressäule und Meeresboden/-untergrund in der bis 200 sm breiten Wirtschaftszone und dem Festlandsockel in seiner neuen Ausdehnung liegen [136] und weil die Regelung der Draft Convention den Interessen der Küstenstaaten bereits hinreichend Rechnung trägt. Nach den bisherigen Erfahrungen, die hier nicht im einzelnen aufgeführt werden können, erscheint ein großer Optimismus in diesem Punkte jedoch nicht sonderlich berechtigt, u. a. weil auch vor der 200 sm-Zone noch einiges zu »holen« ist und weil es nicht nur um wirtschaftliche Interessen geht. Rechnet man mit weiteren Expansionsversuchen von Küstenstaaten in Form von konventionswidriger räumlicher Ausdehnung und Reglementierung der bestehenden Sonderzonen, so kommt es wesentlich auf den Widerstand an, der den Küstenstaaten von anderen Staaten entgegengesetzt wird. Die Hoffnungen auf einen entschiedenen Widerstand werden nicht nur durch die eben erwähnten Zahlenverhältnisse gemindert. Wenn diese Anlaß zur Skepsis geben, mag man noch entgegnen, daß zu den Exponenten der Schiffahrtsfreiheit die wichtigsten Seefahrtsstaaten und die größten Seemächte gehören. Skeptisch gegenüber der Hoffnung auf eine wirksame Wahrung der Schiffahrtsfreiheit stimmt jedoch schon ein Blick auf die Bedeutung, welche Staaten mit entgegengesetzten Interessen diesen Interessen beimessen können. Wiederum verallgemeinernd: Zum mindesten manche Küstenstaaten, insbesondere Langküstenstaaten, die nicht zugleich Staaten mit immerhin wesentlichen – wenn auch nicht vorrangigen – Schiffahrts- oder Fernfi-

---

[134] Die Zahlenangaben folgen der Übersicht bei *Wolf* (Anm. 50), S. 109 f.

[135] Vgl. als Beispiele die oben in Anm. 127 angeführten Gesetze.

[136] *Wolfrum* (Anm. 125), S. 239, spricht von 90% des Fischaufkommens, *Rauch*, Aspekte (Anm. 33), S. 86, von nahezu 95%. Siehe sogleich Anm. 137.

schereiinteressen sind oder in Bündnisinteressen ein erhebliches Argument für die Wahrung der Schiffahrtsfreiheit sehen, werden ihren Interessen an der Ausweitung küstenstaatlicher Befugnisse auf Kosten der Schiffahrtsfreiheit einen hohen Rang einräumen. Das ist jedenfalls dann zu erwarten, wenn es ihnen um wesentliche wirtschaftliche, politische oder militärische Interessen, unter Umständen auch um politisches Prestige zu gehen scheint. Unter wirtschaftlichen Gesichtspunkten ist nicht nur der Fischfang in dem Raum von 200 sm Küstennähe wichtig. Zur Erweiterung küstenstaatlicher Befugnisse reizen auch die natürlichen Hilfsquellen des Festlandsockels[137], der bei dieser Untersuchung ausgeklammert ist. Auf der »Gegenseite« dürfte die Verteidigung der Schiffahrtsfreiheit als abstraktes Prinzip heute nur für eine kleine Zahl von Staaten von großer Bedeutung sein. Da Eingriffe in die Schiffahrtsfreiheit unmittelbar immer nur einen oder wenige Staaten treffen, geht es für die anderen um ein bloßes Prinzip. Das Prinzip hat für die eigenen politischen Entscheidungen im allgemeinen weniger Gewicht als für den Küstenstaat das konkrete Gegeninteresse. Einer Großzahl von Staaten – wenn nicht der Mehrzahl – wird eine nicht allzu flagrante und gewalttätige Verletzung der Schiffahrtsfreiheit durch einen Küstenstaat, möglicherweise auch eine Häufung solcher Vorkommnisse, verhältnismäßig gleichgültig sein. Es ist eine Erfahrungstatsache, daß selbst offensichtliche Verletzungen der Völkerrechtsordnung bei nicht unmittelbar betroffenen Staaten selten zu dem Empfinden *tua res agitur* und zu Handlungen führen, die diesem Empfinden auch dann Ausdruck geben, wenn sie unbequem sind oder gar etwas »kosten«. Auch unter den Gesichtspunkten des Ortes der Kontroverse und der Handlungsweise der Beteiligten sind die Verteidiger der Schiffahrtsfreiheit in der ungünstigeren Position. Der Küstenstaat hat die Initiative in einem Seegebiet gewissermaßen vor der eigenen Haustür. Er wählt Ort, Zeitpunkt und Art seines Vorgehens sowie das Objekt dieses Vorgehens. Als unmittelbarer Verteidiger der Schiffahrtsfreiheit kommt grundsätzlich nur der Flaggenstaat in Betracht. Er mag die üblichen diplomatischen Mittel nutzen, unter Umständen mit politischem Druck oder gar mit Repressalien vorgehen, wenn er bereit ist, die natürlich von Fall zu Fall verschiedenen Risiken einzugehen. Von der obligatorischen Gerichtsbarkeit hat er nicht viel zu erhoffen. Gegenwehr an Ort und Stelle dürfte Handelsschiffen fast immer unmöglich sein, Gegenwehr mit Hilfe von Kriegsschiffen dürfte nur wenigen Flaggenstaaten offenstehen: Vor der eigenen Küste kann auch der militärisch schwache Staat dank der modernen Technik mit verhältnismäßig geringen Mitteln seine Macht wirksam entfalten; vor einer Tausende von Meilen entfernten fremden Küste ist schon die mittlere Seemacht in der Handlungsfähigkeit sehr begrenzt und möglicherweise weit schwächeren Kontrahenten unterlegen. Der

---

[137] Beispielsweise befinden sich über 90% der vor den Küsten vermuteten Öl- und Gasvorkommen in den 200 sm-Zonen. So *Rauch*, Aspekte (Anm. 33), S. 86. Nach *Lucchini/Voelckel* (Anm. 123), S. 132, sind es 87%.

oben erwähnte Gesichtspunkt, daß der maritime Konflikt leichter isolierbar, weniger prestigeträchtig und insgesamt nicht so gefährlich erscheint wie ein Konflikt zu Lande, wirkt ebenfalls eher zugunsten der Küstenstaaten (siehe zu allem oben I.)[137a].

Zu diesen für die Wahrung der Schiffahrtsfreiheit wenig Hoffnung erweckenden Erwägungen kommt die Erfahrung, daß auch Staaten, die an der Verteidigung der Schiffahrtsfreiheit jedenfalls bei eigener Betroffenheit grundsätzlich interessiert und zu einer wirksamen Verteidigung in der Lage sind, im konkreten Falle aus Rücksichten verschiedener Art zurückschrecken. Oben wurden schon einige Beispiele genannt[138]. Hier sei ergänzend die Beobachtung eines Sachkenners wie *Darman* wiedergegeben, der große Erfahrungen aus seiner Tätigkeit für die US-Regierung hat. Einzelfälle und Schema territorialer Expansion auf See sind danach ebenso klar wie die Reaktion der Vereinigten Staaten: Die USA hätten dem Grundsatz der Meeresfreiheit Lippendienste gezollt, *in praxi* jedoch die Aushöhlung dieses Grundsatzes immer wieder zugelassen; in der überwältigenden Mehrzahl der Fälle hätten die Vereinigten Staaten die Beziehungen zu dem Kontrahenten für wichtiger gehalten als einen Verlust an Meeresfreiheit[139]. Schiffahrtsfreiheit wird wahrscheinlich nicht so sehr durch Proteste gegen mögliche oder tatsächliche Verletzungen gewahrt wie durch die Ausübung des Rechtes. Angesichts der Wehrlosigkeit von Handelsschiffen kann die Schiffahrtsfreiheit in allen durch den neuen Seerechtsvertrag erfaßten Meereszonen mit einem Sonderstatus am besten durch Kriegsschiffe gewahrt werden. Ihre bloße Durchfahrt durch Küstengewässer, Meerengen, Archipelgewässer und ihre unterschiedlichen

---

[137a] Der unerklärte Krieg um die Falklandinseln im ersten Halbjahr 1982 geht über Seezwischenfälle und das Thema dieser Arbeit natürlich weit hinaus. Er bestätigt jedoch Thesen der Absätze I und VI. Die Schwierigkeiten bei der Rückeroberung der Inseln durch Großbritannien zeigen deutlich die Vorteile des – verhältnismäßig – küstennahen Staates. Die Inseln sind von der argentinischen Küste (Hafen) etwa 380 sm entfernt, von Großbritannien etwa 7000 sm. Die britische Flotte verlor vier Kriegsschiffe, zwei Landungsschiffe und ein Versorgungsschiff; mehrere Schiffe wurden beschädigt. Hier erwies sich die Vernichtungskraft der mit modernen Raketen bewaffneten Flugzeuge, die von Landbasen aus operieren konnten, auch gegenüber einer weit überlegenen Flotte. Nur die Vereinigten Staaten, die Sowjetunion und Frankreich hätten unter ähnlichen Umständen einen Kampf führen können.

[138] Bei Anm. 12.

[139] Anm. 50, S. 378. Von ähnlichen Erfahrungen berichtet *Richardson* (Anm. 72), S. 909, ehemals Under Secretary of State and Secretary of Defense: Auf der einen Seite stehen politische, wirtschaftliche und militärische Rücksichten; die in einer Demokratie wichtige Berücksichtigung der öffentlichen Meinung in der Welt kann dabei ebenso eine Rolle spielen wie die Befürchtung, wichtige Stützpunkte zu verlieren. Ein kleiner Staat, der den Einsatz seiner Streitkräfte vor der Küste nicht scheut, kann zwar überwunden werden, jedoch der Preis dafür ist oft politisch zu hoch. Gegenüber diesen Faktoren hat die Erkenntnis einen schweren Stand, daß mit der Schiffahrtsfreiheit nicht nur ein abstraktes Prinzip zu verteidigen ist, sondern eigene wesentliche Interessen. Bemerkenswert ist die von *Richardson* erwähnte Empfehlung eines Botschafters, Routineflüge in einer 200 sm-Zone vor der Küste einzustellen (Anm. 72), S. 909.

Aktivitäten in Ausschließlichen Wirtschaftszonen können manchmal Grund für eine Beunruhigung der Küstenstaaten sein. Aufenthalt und Aktivitäten der Kriegsschiffe auf allen Meeren und in allen Seegebieten unter Beachtung des gewissenhaft ausgelegten Meeresvertrages bieten aber vielleicht noch am ehesten eine Chance dafür, daß der Abschluß von UNCLOS III nicht zugleich der Auftakt zu einer weiteren Seenahme sein wird[140].

Eine Betrachtung der Schiffahrtsfreiheit von Kriegsschiffen nach der Dritten Seerechtskonferenz der Vereinten Nationen gibt Anlaß zu einer – nur punktuellen – Bemerkung über die Rolle der Großmächte im heutigen Völkerrecht. *Hermann Mosler* hat in seiner Antrittsvorlesung über »Die Großmachtstellung im heutigen Völkerrecht« an Hand von Zeugnissen internationaler Konferenzen und des diplomatischen Verkehrs gezeigt, wie die politische Großmachtstellung in die Völkerrechtsordnung eingedrungen ist[141]. Er hat dargestellt, daß es die Großmächte sind, welche die politischen Geschikke der Welt bestimmen, und daß die Kräfte der zweitrangigen Staaten nur insofern von Belang sind, als sie zusammen mit denjenigen einer Großmacht oder einer Großmächtegruppe in die Waagschale der Weltpolitik geworfen werden[142]. Das war 1946, als es nur 55 Mitglieder der Vereinten Nationen gab. Als die Draft Convention on the Law of the Sea am 30. April 1982 durch Abstimmung angenommen wurde, hatten die Vereinten Nationen 157 Mitglieder. Schon diese Zunahme läßt darauf schließen, daß ein Wandel in der Stellung der Großmächte eingetreten ist, wie immer man Großmächte von anderen Staaten abgrenzen mag. Großmächte sind zwar im Hinblick auf Zahl und Wirkungskraft ihrer Waffen ungleich stärker als 1946. Sie sind insbesondere atomar bewaffnet. Bei ihnen liegt weiterhin die Entscheidung darüber, ob es zu einem dritten Weltkrieg kommen wird. Die Großmächte haben immer noch eine erheblich größere Handlungsfreiheit als alle anderen

---

[140] Das wurde bei den dem Verfasser bekannten Kommentierungen über den amerikanisch-libyschen Zwischenfall im Golf von Sidra durchweg übersehen. Siehe aber *Oxman* (Anm. 8). Andererseits: Überflüge in einer 12 sm-Zone gehören schon heute nicht mehr zu den legitimen Versuchen, Meeresfreiheit zu wahren. Siehe *Rauch* V, S. 13 f. (Anm. 32).

[141] Veröffentlicht 1949; siehe insbes. S. 45.

[142] *Ibid.* – Dem Verfasser hat nicht vorgelegen *George P. Smith II,* Restricting the Concept of Free-Seas: Modern Maritime Law Re-Evaluated (1980). Die nach Drucklegung dieser Abhandlung erschienene Veröffentlichung von *Eitel,* Seerechtsreform und Internationale Politik, befaßt sich mit unseren Fragen nur am Rande: Archiv des öffentlichen Rechts 107 (1982), S. 100 ff. Auf S. 112 spricht der Verfasser von dem bestimmenden Einfluß der beiden Großmächte auf den Gang der Seerechtsreform da, wo sie sich zusammentun, als »Beleg für das *Gewicht politischer* – oder einfach schlichter – *Macht* auch in einer demokratisierten Staatengemeinschaft« (Hervorhebung im Original). Entgegen dieser Äußerung ist an den Ausführungen oben im Text festzuhalten. – Zu spät für irgendeine Berücksichtigung oben hat der Verfasser erhalten: *O'Connell,* Innocent Passage of Warships, in: The Law of the Sea (Thesaurus Acroasium of the Institute of Public International Law and International Relations of Thessaloniki, Bd. 7) (1977), S. 405 ff.

Staaten. Die obige Skizze dürfte – mehr am Rande – aber auch gezeigt haben, daß Großmächte in manchen Konfliktsituationen mit schwächeren Kontrahenten Rücksichten nehmen müssen oder jedenfalls nehmen. Vor allem hat UNCLOS III eine Erfahrung bestätigt, die sich aus der Tätigkeit etwa der »Gruppe der 77« in vielen internationalen Gremien und Konferenzen ergeben hat. Eine Vielzahl von kleinen und mittleren Staaten kann heute bei politischen und völkerrechtlichen Neuregelungen initiierend und sogar mitmaßgeblich sein. Die Archipelgewässer und die Ausschließlichen Wirtschaftszonen sind nicht auf Initiativen der Großmächte zurückzuführen, sondern wurden ihnen eher abgehandelt. Die Großmächte haben ihre Meerengen-Interessen mit einer besonderen Regelung für die Transitpassage, mit Unterwasserfahrt und Überflug zwar durchsetzen können, aber nur unter großen Anstrengungen und mit erheblichen Konzessionen. Der Ausgleich der Interessen im internationalen Verkehr, der ein wesentlicher Zweck der Völkerrechtsordnung ist, vollzog sich auf UNCLOS III auch in der hier erörterten Frage nicht allein zwischen den Großmächten. Allerdings haben die Großmächte noch am ehesten die Möglichkeit, ein so gefährdetes Rechtsinstitut wie die Freiheit der Schiffahrt allgemein und speziell in Gestalt der Schiffahrtsfreiheit für Kriegsschiffe zu verteidigen. Das wäre eine eigenartige »Heilige Allianz«. Ob die Großmächte – und einzelne andere Staaten – dazu willens und in der Lage sein werden, kann nur die Zukunft lehren. 1946 konnte man über die dominierende Stellung der Großmächte nicht mit besonderem Vertrauen schreiben. 1982 mit dem Ende von UNCLOS III und mit der weitgehenden Nationalisierung des Meeres, des gemeinsamen Erbes der Menschheit, ist nicht das Jahr, in dem man mit besonderem Vertrauen über den Aufstieg der anderen Staaten zu schreiben wagt. Alles fließt, gerade auf See. Aber wohin?

La recherche du consensus dans les décisions de la Cour internationale de Justice

André Gros

Peut-on parler de consensus dans le domaine de la juridiction internationale? Dans les présentes observations le mot sera pris dans sa portée réelle contemporaine; un procédé de négociation qui remplace le vote des participants par la systématisation d'une méthode aussi vieille que l'humanité, le *do ut des* magnifié, la concession sur un point pour obtenir un résultat favorable sur un autre, l'ensemble devenant à la fois étendu et fragile puisque toute atteinte à un élément menace l'ensemble. La troisième conférence sur le droit de la mer en est le parfait exemple (projet de convention, A/Conf.62/L.78, 7 septembre 1981; cf. l'analyse du processus dans l'opinion dissidente du juge Oda, *Plateau continental (Tunisie/Jamahiriya arabe libyenne),* arrêt du 24 février 1982, para. 8 à 26).

Rien de neuf dans le procédé, toute négociation diplomatique en particulier est nécessairement recherche d'équilibre entre des demandes de satisfaction d'intérêts étatiques différents et opposés, par l'équilibre d'un compromis; mais le diplomate sait que ce qu'il a négocié subira une épreuve décisive, l'acceptation par son Etat selon les procédures constitutionnelles, l'important pour chacun étant de négocier un texte ratifiable. Or la méthode actuelle de consensus tend à éviter le contrôle de la ratification, du fait que l'ensemble des dispositions arrêtées par consensus sont à prendre ou à laisser; ce n'est pas seulement le vote libre des Etats à chaque étape de la négociation d'un traité qui disparaît mais, en tout cas pour les démocraties, la liberté d'appréciation par la représentation élue au moment de l'approbation du traité. Le consensus, *i.e.*, le compromis, atteint l'appréciation des intérêts étatiques par les organes constitutionnellement responsables de la ratification des traités.

Rien ne paraît, *a priori*, permettre de transposer une telle méthode de négociation dans le processus juridictionnel . . . et cependant la contagion des moeurs internationales, comme toute contagion, ne s'est pas arrêtée à la frontière du judiciaire.

* * *

L'avis consultatif du 20 décembre 1980 (*Interprétation de l'accord du 25 mars 1951 entre l'OMS et l'Egypte,* C.I.J. Recueil 1980, p. 73) a été accueilli par de sérieuses critiques. *Wilhelm Wengler* a relevé que les opinions individuelles émises par huit sur treize des membres de la Cour qui siégeaient montrent une division très large sur l'opportunité d'une réponse positive ou négative à la question posée par l'Organisation mondiale de la Santé (OMS), et en plus sur les motivations que l'organe de l'OMS compétent pour transférer le siège d'un bureau régional pouvait ou devait prendre en considération, le résultat étant que la Cour »pour arriver à une décision presque unanime . . . a transformé, en l'interprétant, la question posée en termes très précis par l'Organisation en la question générale . . . de savoir quels principes et règles sont applicables à la question de savoir à quelles conditions et selon quelles modalités un transfert du Bureau régional de l'Egypte pourrait être effectué« (Neue Juristische Wochenschrift 1981, Fasc. 10, »Un nouvel avis consultatif de la Cour internationale de Justice«, p. 501. Voir aussi un article plus détaillé de *Jean Combacau* »La question du transfert du Bureau régional de l'OMS devant la Cour internationale de Justice« dans l'Annuaire français de droit international, 1980, p. 225–247, qui conclut par le même interrogation sur les limites de la fonction juridictionnelle).

La thèse des critiques de l'avis est que, pour obtenir un »consensus« (12 voix contre 1 sur la question I et 11 voix contre 2 sur la question II), la Cour compromet l'exercice de sa fonction judiciaire en préférant une quasi-unanimité d'apparence, dont le voile est immédiatement levé par 9 opinions, à une faible majorité qui révélerait la division réelle au sein de la Cour.

Il ne s'agit pas d'un problème récent, comme le rappelle *Wengler* dès le début de son commentaire; il est compliqué en apparence par un problème parallèle, l'abondance de l'*obiter dictum* dans des opinions aussi nombreuses que les juges, aussi lorsqu'il n'est pas question de consensus. Ces deux aspects distincts de l'exercice de la fonction judiciaire peuvent se rejoindre dans les regrettables conséquences qu'ils emportent, mais il ne faut pas les confondre.

Il peut être utile aujourd'hui de faire le point sur la question du consensus car il est vrai de dire que la fonction judiciaire, telle qu'elle a été exercée par la Cour, a marqué une nette évolution de ce fait depuis une vingtaine d'années.

Dans son opinion individuelle sur *Certaines dépenses des Nations Unies* (article 17, paragraphe 2, de la Charte), avis consultatif du 20 juillet 1962, p. 198) Sir Gerald Fitzmaurice écrivait:

> »De plus (et ceci est la raison principale qui m'a poussé à écrire une opinion individuelle), il semble que l'avis de la Cour, *tout en traitant minutieusement certains points, s'abstienne intentionnellement d'en aborder d'autres, d'un caractère plus général et comportant des difficultés qui ont embarrassé un certain nombre de ceux qui ont eu à connaître du sujet.* En résumé, l'avis passe sous silence certains points qui m'apparaissent des plus pertinents;

en effet, bien que l'›avis juridique‹ mentionné dans le préambule de la requête porte sur la question du ›financement des opérations des Nations Unies au Congo et au Moyen-Orient‹, j'estime que, même dans ce seul contexte, l'avis n'atteint pas sa pleine utilité s'il ne traite pas de certaines questions plus générales ainsi que d'un ou deux autres points que la Cour a laissés de côté.« (C'est moi qui souligne).

La question est bien posée; que se passe-t-il dans une affaire où les membres de la juridiction ne sont pas d'accord soit sur l'importance, soit sur la pertinence même de certains points de droit ou de fait dans une affaire? Les »difficultés« peuvent-elles être l'objet d'une sorte de consensus de ménagement mutuel des opinions et, en conséquence, voilées ou omises dans le texte d'une décision?

Pour simplifier, il ne peut être question d'examiner le fond de chaque affaire, avis consultatif ou arrêt, où pourraient s'être manifestées ouvertement de telles formes de consensus, chacune présentant des traits particuliers tenant au sujet et à ces difficultés rencontrées quant à la pertinence, aux conséquences des divers points de droit soulevés dans une affaire et à l'opportunité de les traiter, aussi quelques références seulement seront données. Le but est de rechercher quelques cas où l'accord des votes confronté à une division dans les opinions permet de conclure à un consensus sur une base juridique modifiée ou réduite, déclinant de dire tout le droit.

Pour montrer que toute division des opinions ne suppose pas une fuite devant les difficultés d'une affaire et que la multiplication des opinions est, en soi, un problème différent du consensus judiciaire, j'écarterai une affaire à opinions multiples mais sans consensus, l'arrêt de la *Barcelona Traction*. L'arrêt du 5 février 1970 dans l'affaire de la *Barcelona Traction* (C.I.J. Recueil 1970, p. 3 à 357) comporte 49 pages pour l'arrêt et 306 de déclarations, opinions individuelles et dissidente émanant de douze juges sur seize dont onze membres de la Cour sur quatorze. La décision mettait fin à la seconde demande présentée le 19 juin 1962 par la Belgique contre l'Espagne, après qu'elle eut renoncé à poursuivre une première instance introduite en 1958 et rayée du rôle le 10 avril 1961. L'affaire elle-même a donc duré douze ans avec une documentation recueillie dans 11 volumes de la Cour (Mémoires, plaidoiries et documents) totalisant 7215 pages. Il n'est pas question ici de traiter autre chose que le problème de la comparaison entre le contenu des opinions par rapport au vote exprimé sur le dispositif de l'arrêt et d'établir que, s'il y a une certaine recherche de consentement sur le point de droit qui justifie l'arrêt, il n'y a pas en cela consensus. Il faut rappeler que la requête belge tendait à obtenir du Gouvernement espagnol »la réparation du préjudice qui aurait été causé à la société Barcelona Traction du fait d'actes prétendument contraires au droit international commis par des organes de l'Etat espagnol« (C.I.J. Recueil 1970, para. 1) et que cette requête fut rejetée par la Cour pour défaut de qualité pour agir du Gouvernement belge.

Le dispositif de l'arrêt comporte le texte suivant:

»LA COUR
rejette la demande du Gouvernement belge par quinze voix contre une, douze des voix de la majorité se fondant sur les motifs énoncés dans le présent arrêt.« (C.I.J. Recueil 1970, p. 51).

Il semblerait, à lire ceci, que seuls trois juges avaient d'autres motifs de rejeter la requête que ceux exposés dans l'arrêt et ont tenu à le faire savoir. En réalité le compte est plus complexe.

D'après leurs opinions, onze membres de la Cour indiquent comment ils auraient traité l'affaire sur des points non traités ou traités autrement dans l'arrêt. Sans pouvoir les analyser toutes je ferai brièvement le résumé de certaines.

La déclaration commune de Petrén et de Onyeama (C.I.J. Recueil 1970, p. 52) les sépare de l'arrêt sur un point important, le droit de protection diplomatique envers une société et la question du lien effectif. Le Président Bustamante y Rivero aurait voulu que la Cour étudie le problème de l'épuisement des recours internes, ce qu'elle se refusa à faire (*eod. loc.*, p. 56–63) et il indique malgré cela son opinion sur la réponse qu'il convenait, selon lui, de donner.

Sir Gerald Fitzmaurice, dans son opinion (p. 64–113), donne comme en 1962 la raison de tels dépassements de l'objet strict des arrêts de la Cour d'une manière fort claire:

»Il y a en outre plusieurs questions particulières, qui ne sont pas traitées dans l'arrêt de la Cour ou qui n'y sont qu'effleurées, sur lesquelles je voudrais présenter certaines observations. Bien que ces observations ne puissent avoir que le caractère d'*obiter dicta* sans posséder l'autorité d'un jugement, les opinions qui émanent d'une manière ou d'une autre d'une instance judiciaire constituent, tant qu'il ne sera pas possible d'édicter, dans le domaine juridique international, des textes législatifs exprès ayant un effet obligatoire direct, le principal moyen de préciser et de développer le droit de façon quelque peu concrète. Je suis d'accord avec l'ancien juge Sir Hersch Lauterpacht pour estimer que les tribunaux internationaux doivent avoir cette considération présente à l'esprit, ce qui les met dans une situation différente de celle des tribunaux nationaux pour ce qui est d'examiner, ou du moins de commenter, des aspects dépassant la stricte *ratio decidendi* d'une affaire«.

Tanaka va encore plus loin dans son opinion (p. 114–160) puisqu'il y traite le fond et déclare »devoir suivre le cheminement logique qui [lui] est propre et qu'en conscience [il] croit juste« (C.I.J. Recueil 1970, p. 114), en renvoyant à une controverse précédente (C.I.J. Recueil 1966, déclaration du Président Sir Percy Spender, p. 51, et opinion dissidente de Tanaka, p. 262–263).

Philip Jessup (C.I.J. Recueil 1970, p. 161–221) expose des vues favorables au caractère exhaustif des prises de position judiciaires des tribunaux internationaux, reprenant les idées de Sir Hersch Lauterpacht sur ce point et il fait porter son examen sur le problème du droit de protection diplomatique des sociétés qu'il traite en détail dans le cadre de l'affaire devant la Cour.

Morelli (p. 222–242) procède à une analyse approfondie du problème de la qualité substantielle et non procédurale pour obtenir justice et à l'étude des droits reconnus dans le cadre de la protection diplomatique des sociétés.

L'affaire de la *Barcelona Traction* ne pose donc pas de problème de consensus. Comme toute juridiction la Cour a examiné en première ligne le problème de l'intérêt pour agir et, une large majorité s'étant formée, n'a pas eu à aller plus loin. L'importance des questions juridiques soulevées au cours de la procédure a paru telle aux membres de la Cour, même à ceux qui acceptaient la motivation du rejet de la requête, que le volume des opinions a correspondu à leur intérêt pour ces questions. Il y avait consentement sur le rejet de la requête pour défaut d'intérêt, ce qui ne voilait pas les difficultés mais les excluait de l'arrêt; en parler dans les opinions pouvait poser, dans certains cas, le problème traité par le Président Sir Percy Spender dans sa déclaration de 1966 mais aucun problème de consensus destiné à produire une forte majorité ou une quasi-unanimité n'était en cause.

Dans les affaires *Compétence en matière de pêcheries* (C.I.J. Recueil 1974), le problème du consensus apparaît nettement par contre. L'arrêt a décidé que l'Islande n'avait pas le droit d'étendre ses limites de pêche de 12 à 50 milles pour des motifs de conservation; ces motifs n'étaient pas ceux de l'Islande et la Cour a évité de décider que, dans l'état du droit positif, l'extension à 50 milles n'était pas opposable aux Etats qui pêchent. Puis gardant ce thème unique de la conservation, la Cour a construit à l'intention de deux Etats parties à un différend sur la limite des pêcheries une consultation sur les problèmes de conservation comme si leur solution pouvait remplacer la seule décision envisagée dans l'affaire, celle portant sur la licéité d'une extension nouvelle de limites au-delà de 12 milles. Répondre à un différend sur une prétention de juridiction exclusive par des directives pour un accord de conservation n'était pas accomplir la mission de la Cour. L'échec de la solution proposée par la Cour est connu.

C'est le même problème de consensus sur des points sans rapport nécessaire avec l'objet des questions posées à la Cour qui se retrouve dans l'avis consultatif de 1975 sur le *Sahara occidental*. Dans son opinion individuelle (C.I.J. Recueil 1975), le juge Pétren écrivait:

> »Dans ces conditions, on peut se demander si l'interprétation que la Cour a décidé de donner aux questions posées correspond bien aux intentions de l'Assemblée générale ou si elle ne représente pas plutôt une nouvelle formulation de ces questions . . . Dans sa réponse à la première des questions posées par l'Assemblée générale, l'avis consultatif passe à côté de

cet objet de la requête [*i.e.* les prétentions de souveraineté du Maroc et de l'»ensemble mauritanien« sur le territoire du Sahara occidental au moment de sa colonisation] . . . C'est en éludant la question de la souveraineté que . . . l'avis énonce . . .« (p. 111 et 113).

(Voir aussi la déclaration d'Ignacio-Pinto, p. 78 et la mienne, p. 77, para. 12 et 13 renvoyant à un passage de ma déclaration en 1974 dans les affaires *Compétence en matière de pêcheries* »sur cette nouvelle tendance de la Cour à répondre à des problèmes qu'elle se pose plutôt qu'à celui qui lui est posé« (C.I.J. Recueil 1974, p. 148–149)).

On sait que la réponse fournie dans le paragraphe final de l'avis sur le *Sahara occidental* était ainsi rédigé:

> »Les éléments et renseignements portés à la connaissance de la Cour montrent l'existence, au moment de la colonisation espagnole, de liens juridiques d'allégeance entre le sultan du Maroc et certaines des tribus vivant sur le territoire du Sahara occidental. Ils montrent également l'existence de droits, y compris certains droits relatifs à la terre, qui constituaient des liens juridiques entre l'ensemble mauritanien, au sens où la Cour l'entend, et le territoire du Sahara occidental. En revanche, la Cour conclut que les éléments et renseignements portés à sa connaissance n'établissent l'existence d'aucun lien de souveraineté territoriale entre le territoire du Sahara occidental d'une part, le Royaume du Maroc ou l'ensemble mauritanien d'autre part. La Cour n'a donc pas constaté l'existence de liens juridiques de nature à modifier l'application de la résolution 1514 (XV) quant à la décolonisation du Sahara occidental et en particulier l'application du principe d'autodétermination grâce à l'expression libre et authentique de la volonté des populations du territoire (voir paragraphes 54 à 59 ci-dessus).« (C.I.J. Recueil 1975, p. 68, para. 162).

C'est un bon exemple de consensus judiciaire rassemblant deux points de vue opposés dans une apparente unité. Les mots »en revanche« (*on the other hand*, en anglais) sont un voile fragile de la contradiction entre les deux prononcés, l'un sur des liens d'allégeance avec existence de certains droits relatifs à la terre, l'autre sur l'absence de lien de souveraineté territoriale entre le territoire et les deux Etats. Les six années qui nous séparent de l'avis ont montré la vanité de cet essai de concilier l'eau et le feu.

Malgré l'insuccès de cette politique judiciaire dans les affaires précitées, le plus récent arrêt de la Cour, en date du 24 février 1982, est la dernière manifestation de ce penchant à voiler les difficultés majeures d'un différend en essayant de le régler par une conciliation des intérêts en cause sur la base du plus petit dénominateur commun. En adoptant pour partie de sa ligne de délimitation des zones de plateau continental relevant de chacun des deux Etats une ligne établie entre leurs concessions résultant d'actes unilatéraux de l'un et l'autre, et en considérant avec divers arguments que cette ligne était la limite équitable, la Cour n'a pas tranché une question fondamentale: comment l'unilatéral devient-il équitable alors qu'il est l'objet même du différend? (Cf. la première opinion dissidente, para. 18–22 et 23 et l'opinion dissidente de M. Evensen, para. 26).

En 1974 j'écrivais dans une opinion dissidente que »la mission propre de la Cour est toujours de ›régler conformément au droit international les différends qui lui sont soumis‹ (art. 38 du Statut). En introduisant dans les relations internationales une conception selon laquelle la Cour rendrait des décisions selon ce qu'à chaque occasion la majorité penserait être à la fois juste et opportun, une transformation profonde serait opérée . . . Que la nouvelle conception soit à rejeter comme contraire au rôle du juge international me paraît résulter de la simple constatation que celui-ci n'est pas un juge fédéral; les Etats, peu nombreux aujourd'hui, qui viennent à la Cour ne le font pas pour recevoir des conseils, mais pour obtenir la confirmation judiciaire des engagements conventionnels qu'ils ont pris, selon le droit international établi, et en fonction d'une situation bien connue d'eux«. (C.I.J. Recueil 1974, p. 148–149).

En poursuivant la réflexion il faut, semble-t-il, examiner d'abord quels résultats utiles le consensus apparant dans les décisions ou avis de la Cour a apportés pour le règlement des différends ou difficultés soumis par les Etats ou les organisations internationales. Non pas que la fin justifie d'aucune manière le moyen, mais en plus de son caractère déformant du rôle du juge, dans la réalité des choses, le consensus judiciaire n'a rien réglé. L'arrêt sur la *Compétence en matière de pêcheries* n'a pas eu d'effet; l'avis de consensus du Sahara occidental, si minutieusement rédigé pour tout équilibrer dans le paragraphe 162 transposé aussitôt en une sorte de dispositif, a été contredit dès le lendemain et le différend s'est lourdement aggravé depuis lors. La recherche de l'opportunité et du compromis n'ont rien apporté à la justice internationale. Le mal est ancien[1] et il est dénoncé par des juges et des auteurs. Peut-on croire qu'il disparaisse?

Alors que la pratique du consensus n'a en rien amélioré l'efficacité du rôle judiciaire de la Cour, il faut en voir tous les effets dommageables pour

---

[1] *Ch. De Visscher* a écrit à propos de l'affaire du *Droit de passage* qu'une certaine disposition d'esprit »a même conduit la Cour dans l'affaire du Droit de passage sur territoire indien à ne pas faire droit à une conclusion pourtant clairement énoncée par une Partie. La revendication du Portugal, qui pouvait se réclamer d'un principe général de droit aussi bien que de l'équité avait été formulée comme celle d'un droit actuel; l'arrêt l'a située dans le passé. Ce point de vue eut pour conséquence, que la Cour en cette occasion ne s'est pas acquittée pleinement de sa mission; le différend, tel qu'il lui avait été soumis, n'a pas été entièrement vidé. C'est ce que releva dans sa déclaration très ferme le Président de la Cour, Helge Klaestad: ›En se limitant à la situation telle qu'elle existait en juillet 1954, l'arrêt n'a pas résolu l'ensemble du différend tel qu'il a été soumis à la Cour‹«. (De l'équité dans le Règlement arbitral ou judiciaire des litiges du droit international public, p. 16).

l'avenir du contentieux international. Lorsqu'une affaire est complexe et qu'elle soulève de nombreux points de droit, s'il est vrai qu'il n'est pas toujours nécessaire de les trancher tous, certains permettant de donner la solution de l'affaire, il n'est pas bon de prendre comme méthode de travail judiciaire l'effacement systématique des difficultés, la transformation des questions posées, l'accord sur le plus facile et le silence sur le reste. Lorsqu'il s'agit du seul organe juridictionnel permanent dans les relations internationales contemporaines il convient de demander à un tribunal de traiter les problèmes qu'on lui pose, de les régler par une majorité, quelle qu'en soit l'ampleur, la minorité exprimant ses vues. Ce sont deux philosophies qui s'opposent, non seulement de la justice, mais de la vie internationale et le débat n'est pas nouveau. La recherche du consensus tend à faire de la Cour une sorte de conciliateur amiable dont les actes juridictionnels ne seraient plus que des recueils de conseils; ce rôle peut être exercé par d'autres mais le juge dit le droit et rien que le droit.

Die Einschränkung von Grundrechten in einer demokratischen Gesellschaft

Zu den Schrankenvorbehalten der Europäischen Menschenrechtskonvention

Kay Hailbronner

I. Einführung

Der praktische Wert internationaler Menschenrechte bemißt sich nicht zuletzt an der Berechenbarkeit und Überprüfbarkeit ihrer Schranken. Großzügige Gewährungen von Freiheitsrechten, wie z.B. das Recht zum Verlassen des eigenen Landes (vgl. Art. 12 Abs. 2 des Internationalen Paktes über bürgerliche und politische Rechte), stehen solange auf dem Papier, als beinahe uferlose Schrankenvorbehalte das Ausweichen vor den internationalen Verpflichtungen möglich machen. Der internationale Menschenrechtsschutz befindet sich hier in einem Dilemma. Einerseits wäre es wünschenswert, die staatlichen Möglichkeiten zur Einschränkung von Grundrechten möglichst genau zu umschreiben und einer internationalen gerichtlichen Kontrolle zu unterwerfen. Andererseits ist der Verweis auf unbestimmte Schrankenvorbehalte, die auf schützenswerte Rechtsgüter der nationalen Rechtsordnungen verweisen, in vielen Bereichen unverzichtbar, um der Fülle von Konfliktsmöglichkeiten unter Berücksichtigung der jeweiligen nationalen Besonderheiten Rechnung zu tragen. Menschenrechte stehen in einem untrennbaren Zusammenhang mit der gesamten einzelstaatlichen Rechts- und Gesellschaftsordnung.

In der Europäischen Menschenrechtskonvention findet dieses Dilemma im Verweis auf die demokratische Gesellschaft seinen Niederschlag. Die Konvention versucht, durch abgestufte Schrankenvorbehalte ein hohes Maß an Berechenbarkeit der Freiheitsbeschränkungen zu erreichen. Andererseits ist bei einigen von der Konvention gewährleisteten Rechten der Rückgriff auf relativ unbestimmte Vorbehalte der öffentlichen Ruhe und Ordnung, der nationalen Sicherheit, der Rechte und Freiheiten anderer, des wirtschaftlichen Wohles des Landes usw. unentbehrlich, um allen möglichen Konfliktsituationen Rechnung zu tragen. In diesen Fällen sollen aber Beschränkungen nur zulässig sein, soweit sie »in einem demokratischen Staat« (Art. 6 Abs. 1 Satz 2 EMRK) bzw. »in einer demokratischen Gesellschaft« (Art. 8 Abs. 2, Art. 9 Abs. 2, Art. 10 Abs. 2, Art. 11 Abs. 2 EMRK; Art. 2 Abs. 3 und 4 des 4. Zusatzprotokolls zur EMRK) notwendig sind. Der Verweis auf die demokratische Gesellschaft (bzw. den demokratischen Staat) dient daher als »Schrankenschranke«, als Instrument zur Schaffung eines einheitlichen europäischen Mindeststandards menschenrechtlicher Beschränkungen. Mit der Verwendung des Wortes »Schaffung« ist bereits angedeutet, daß es sich nicht lediglich um ein mit den herkömmlichen Mitteln juristischer Auslegung aus der Konvention zu erschließendes europäisches Schrankenkonzept handelt, das von den Organen der Konvention auf die konkreten Fälle anzuwenden wäre. Welche Beschränkungen in einer demokratischen Gesellschaft nach Art der Europaratsstaaten für notwendig gehalten werden, mag nicht nur nach Ort und Zeit unterschiedlich beurteilt werden [1]. Ebenso unterschiedliche Beurteilungen können sich auch daraus ergeben, daß verschiedenartige Verständnisse über die Notwendigkeiten einer demokratischen Gesellschaft zugrunde gelegt werden. Bereits die Diskussion innerhalb der Bundesrepublik Deutschland über die richtige Grundrechtsinterpretation zeigt, zu welch unterschiedlichen Ergebnissen man gelangen kann, je nach dem, ob z.B. bei der Neuordnung des Medienwesens einer liberalstaatlichen oder einer sozialstaatlichen Konzeption der Freiheit des Rundfunks der Vorzug gegeben wird. Es zeigt sich auch hier die aus dem innerstaatlichen Verfassungsrecht wohlbekannte Überschneidung von Inhalt und Grenzen der Grundrechte. Beide sind untrennbar miteinander verknüpft. Wird z.B. aus der Pressefreiheit eine institutionelle Gewährleistung mit der Konsequenz einer staatlichen Schutz- oder Subventionierungspflicht für notleidende Tageszeitungen abgeleitet, so bedeutet dies zugleich notwendig Beschränkungen der Pressefreiheit anderer Zeitungsunternehmen. Dies gilt erst recht, wenn Konventionsrechten »Drittwirkung« gegenüber Privaten beigemessen wird. Kommission und Gerichtshof befinden sich insofern in keiner anderen Situation als ein nationales Verfassungsgericht, das über die Auslegung der Grundrechte zu entscheiden hat. Anders als das nationale Gericht können aber die europäischen Rechtsprechungsorgane nicht auf eine einheitliche Verfassungsordnung mit einem System von verfassungsrechtlich geschützten Rechtsgütern und »Verfassungswerten« zurückgreifen, sondern müssen eine Vielzahl verschiedener Rechtsordnungen und Verfassungskonzeptionen berücksichtigen. Kommission und Gerichtshof haben die schwierige Aufgabe, aus dieser Fülle von Verschiedenheiten und nationalen Eigenheiten die Gemeinsamkeiten einer »demokratischen Gesellschaft« bezüglich des Inhalts und der Schranken

[1] Zu den verschiedenen Grundrechtstheorien vgl. *H. H. Klein*, Die Grundrechte im demokratischen Staat (1972); *E. W. Böckenförde*, NJW 1974, S. 1529 ff.; *A. Bleckmann*, Allgemeine Grundrechtslehren (1979), S. 155 ff.

von Menschenrechten herauszudestillieren. Dies beinhaltet notwendig zugleich eine rechtsschöpferische Tätigkeit, will man die Konvention nicht auf die Funktion einer letztinstanzlichen Pannenhilfe für solche Fälle beschränken, die versehentlich auf der nationalen Ebene nicht bereinigt wurden. Dies kann aber keine Generalvollmacht für die europäischen Menschenrechtsorgane bedeuten, schrittweise eine vereinheitlichte europäische Menschenrechtsordnung dort zu schaffen, wo eine politische Einigung nicht oder noch nicht erreicht werden kann. *Hermann Mosler* hat wiederholt darauf hingewiesen, daß die Straßburger Organe auf die Dauer nur dann effektiv entscheiden können, wenn sie sich mit dem Grundkonsens der Mitgliedstaaten in Einklang befinden[2].

Damit ist das Kernproblem bezeichnet. Es sind die Mitgliedstaaten und nicht die europäischen Menschenrechtsorgane, die über das Ausmaß einer integrierten europäischen Menschenrechtsordnung entscheiden. Wird dies bei einer allzu dynamischen, europafreundlichen Auslegung nicht beachtet, so besteht die Gefahr, daß einzelne Mitgliedstaaten ihre Zustimmung zur Erweiterung der Konvention durch Zusatzprotokolle verweigern oder ihre Unterwerfung unter die Kontrolle der europäischen Organe zurückziehen. Die Konvention kann als völkerrechtlicher Vertrag die nationalstaatliche Souveränität nicht stärker einschränken, als die Europäische Menschenrechtskonvention und der in ihr enthaltene gemeinsame Grundrechtsstandard dies gebieten. Die Konvention kann daher weder jetzt noch in der Zukunft nationalstaatliche Grundrechtsordnungen ersetzen. Andererseits kann die Ermittlung des gemeinsamen Mindeststandards sich nicht darauf beschränken, eine Übereinstimmung der beteiligten Rechtsordnungen bis ins Detail zu verlangen. Dies liefe darauf hinaus, die Konvention weitgehend ihrer Bedeutung zu berauben. Es gilt also, einen Mittelweg zu finden, der die Effektivität der Konvention sichert, aber auch den nationalen Ordnungsvorstellungen einen angemessenen Raum zugesteht. *Hermann Mosler* hat mit seinem wissenschaftlichen und richterlichen Wirken die Richtung dieses Weges gewiesen[3]:

> »L'homogénéité permet l'établissement d'un standard jusqu'à un degré assez élevé, mais ne suffit pas pour assurer l'égalité totale. Les organes de Strasbourg doivent tolérer, par conséquent, la variété des ordres nationaux en tant qu'elle résulte du développement incomplet de l'intégration européenne. L'interprétation ne doit pas nécessairement rester stable. Elle peut être flexible et devenir plus stricte dans la mesure où l'harmonisation des ordres juridiques nationaux fait des progrès«.

---

[2] Vgl. *H. Mosler,* Der Europäische Gerichtshof für Menschenrechte nach zwanzig Jahren, in: Recht als Prozeß und Gefüge, Festschrift für Hans Huber (1981), S. 600; *derselbe,* Bericht über die Ergebnisse des V. Internationalen Kolloquiums über die Europäische Menschenrechtskonvention, Frankfurt vom 9.–12. 4. 1980.

[3] *H. Mosler,* L'influence du droit national sur la Convention européenne des Droits de l'Homme, in: Miscellanea Ganshof van der Meersch, Bd. 1 (1972), S. 521, 533.

Die Zuweisung der Aufgabe einer fortschreitenden Grundrechtsharmonisierung an die Organe der Europäischen Menschenrechtskonvention wirft die Frage auf, nach welchen Methoden die Kriterien des einheitlichen Grundrechtsstandards einer demokratischen Gesellschaft zu ermitteln sind. Der Gerichtshof hat hierzu vor allem in den Fällen *Handyside, Sunday Times* und *Dudgeon* Stellung genommen.

Gestützt auf diese Entscheidungen sollen im folgenden einige Grundfragen zur Auslegung der Schrankenklauseln der Konvention erörtert werden.

II. Die Demokratieklausel in der Entscheidungspraxis der Organe der Europäischen Menschenrechtskonvention

Die Kommission hat der Demokratieklausel erst relativ spät eine eigene Bedeutung zugemessen. Die Notwendigkeit der zur Überprüfung stehenden Menschenrechtsbeschränkung wird in der älteren Praxis häufig nur mit einem formelhaften Hinweis auf die demokratische Gesellschaft untersucht, ohne daß eine selbständige Funktion der Demokratieklausel erkennbar wäre [4]. Die Kommission beschränkt sich in der Regel darauf festzustellen, ob die Einschränkung zum Schutz eines der in der Konvention erwähnten Rechtsgüter (z.B. öffentliche Sicherheit, Ordnung) notwendig war. Als Maßstab wird hierfür der Grundsatz der Verhältnismäßigkeit verwendet. Erstmals im *de Becker*-Fall [5] wird die Demokratieklausel herangezogen, um zu begründen, daß eine Einschränkung des Rechts auf Meinungsfreiheit unverhältnismäßig war. Es ging dabei um die belgische Kollaborateursgesetzgebung, die es erlaubte, einen wegen Gefährdung der staatlichen Sicherheit Verurteilten für immer von jeglicher kulturellen und publizistischen Tätigkeit auszuschließen. Die Kommission kommt zu dem Schluß, daß eine fortdauernde undifferenzierte Beschränkung des Rechts auf freie Meinungsäußerung ohne Rücksicht auf deren politischen Inhalt, die nach ihrer Auffassung im Interesse der öffentlichen Ordnung in der unmittelbaren Nachkriegszeit noch notwendig

---

[4] Zur Praxis von Kommission und Gerichtshof vgl. *Ph. Th. Vegleris,* Valeur et signification de la clause »dans une société démocratique« dans la Convention européenne des Droits de l'Homme, Revue des Droits de l'Homme 1 (1968), S. 219 ff.; *U. Hoffmann-Remy,* Die Möglichkeiten der Grundrechtseinschränkung nach den Art. 8–11 Abs. 2 der Europäischen Menschenrechtskonvention (Schriften zum Völkerrecht, Bd. 49) (1976); *F. Elens,* La notion de démocratie dans la Convention européenne des Droits de l'Homme (unveröffentl. Manuskript, Institut des Hautes Etudes Européennes, Straßburg, 1980–1981).

[5] Beschwerde 214/56, Zulässigkeitsentscheidung vom 9. 6. 1958, abgedr. in Yearbook of the European Convention on Human Rights, Bd. 2 (1958/59), S. 215 ff.; vgl. insbes. den Kommissionsbericht vom 8. 1. 1960, abgedr. in EGMR, Serie B, Nr. 2, 1962, S. 11 ff.

sein konnte, sich nach Stabilisierung der Verhältnisse nicht mehr rechtfertigen läßt[6]. Dieser Schluß erscheint aus heutiger Sicht beinahe selbstverständlich. Die spezifischen Gründe, die den belgischen Gesetzgeber in der Nachkriegszeit zu einer Fernhaltung der Kollaborateure vom gesamten öffentlichen Leben veranlaßt hatten, waren längst weggefallen. Eine Gefahr für eines der in der Konvention enthaltenen Schutzgüter konnte jedenfalls für nicht politische Äußerungen des Beschwerdeführers ernsthaft nicht behauptet werden. Bereits die Anwendung des Notwendigkeitserfordernisses mußte daher zu dem von der Kommission gezogenen Schluß führen. Beachtenswert ist, daß die Kommission es hierbei nicht bewenden läßt, sondern zusätzlich darauf hinweist, daß diejenigen Länder, die eine vergleichbare Gesetzgebung erlassen hätten, keine derart weitgehenden und unflexiblen Sanktionen wie die belgische Gesetzgebung vorgesehen hätten. Aus der Praxis der Europaratstaaten wird der Schluß gezogen, daß solche Einschränkungen, wie sie die belgische Gesetzgebung vorsah, ohne Rücksicht auf den Zeitablauf und die besonderen Umstände, die zu dem strafbaren Verhalten des Beschwerdeführers geführt haben, in einer »demokratischen Gesellschaft« nicht als zum Schutz der öffentlichen Ordnung notwendig angesehen werden.

Im Fall *Handyside*[7] hatte der Gerichtshof zum ersten Mal Gelegenheit, sich intensiv mit dem Begriff der demokratischen Gesellschaft auseinanderzusetzen. Dem Verfahren lag die Bestrafung eines Verlegers wegen Verbreitung eines Nachschlagebuchs für Schüler, das auch sexuelle Fragen ansprach, zugrunde. Unstreitig verfolgte hier die Beschränkung der Meinungsäußerungsfreiheit den gemäß Art. 10 Abs. 2 EMRK rechtmäßigen Zweck des Schutzes der Moral. Über die weitere Frage, ob die konkreten Beschränkungen und Strafdrohungen auch »in einer demokratischen Gesellschaft im Interesse ... der Moral ... unentbehrlich« waren, entspann sich eine aufschlußreiche Methodendebatte. Kommission und Gerichtshof gingen davon aus, daß den nationalen Behörden ein Spielraum bei der Beurteilung der Eingriffsvoraussetzungen des Art. 10 Abs. 2 EMRK zukomme. Der Gerichtshof begründet dies damit, daß das Rechtsschutzsystem nur subsidiären Charakter habe[8]. Grundsätzlich seien die staatlichen Behörden in einer besseren Lage als der nationale Richter, um sich zum genauen Inhalt wechselnder Moralanforderungen sowie zur Notwendigkeit einer Einschränkung oder Strafdrohung zur Durchsetzung dieser Anforderungen zu äußern. Die europäischen Menschenrechtsorgane hätten daher nicht die Aufgabe, sich an die Stelle der zuständigen innerstaatlichen Gerichte zu setzen, sondern seien darauf beschränkt, die Einhaltung der Grenzen dieses Beurteilungsspielraums im Hinblick auf die Konventionserfordernisse zu kontrollieren. Diese Kontrollbefugnis beinhaltet nach Auffassung des Gerichtshofs sowohl die Prüfung der

---

[6] Vgl. den Kommissionsbericht (Anm. 5), S. 127 ff.
[7] Fall *Handyside,* Urteil vom 7. 12. 1976, Serie A, Nr. 24, 1976.
[8] A.a.O. (Anm. 7), S. 22.

Zweckgerichtetheit der strittigen Maßnahme einschließlich der zugrunde liegenden Gesetzgebung als auch ihre Notwendigkeit. Der Gerichtshof bestimmt demnach auf der Grundlage der verfügbaren Unterlagen, ob die von den Behörden zur Rechtfertigung der von ihnen ergriffenen Eingriffsmaßnahmen aufgeführten Gründe im Hinblick auf die Schrankenvorbehalte der Konvention durchgreifen und ausreichen. Die Maßstäbe, die der Gerichtshof für die Abgrenzung des nationalen Beurteilungsspielraums vom europäischen Standard in Bezug auf die Notwendigkeit der Maßnahme als auch hinsichtlich des zu schützenden Rechtsgutes zugrunde legt, werden in der *Handyside*-Entscheidung nicht recht deutlich. Fest steht nur, daß der Gerichtshof davon ausgeht, daß dasjenige, was in einer demokratischen Gesellschaft als zum Schutze der in der Konvention aufgeführten Rechtsgüter als notwendige Beschränkung anzusehen ist, eigenständig und damit europäisch einheitlich zu bestimmen ist; freilich nur als Rahmenordnung, die abweichende Praktiken innerhalb der Mitgliedstaaten prinzipiell zuläßt. Die Mehrheit der Kommission und die britische Regierung hatten im Fall *Handyside* die Auffassung vertreten, der Gerichtshof habe lediglich nachzuprüfen, ob die englischen Gerichtsbehörden nach Treu und Glauben, in sachgerechter Weise und in den Grenzen des den Vertragsstaaten in Art. 10 Abs. 2 EMRK eingeräumten Beurteilungsspielraums gehandelt hätten[9]. Der in der Praxis der Kommission verschiedentlich angewandte Maßstab von *reasonableness and good faith* scheint mit einer Art Vertretbarkeitskontrolle auf einen ausgedehnteren Beurteilungsspielraum der nationalen Behörden hinauszulaufen, als ihn die Mehrheit des Gerichtshofs zubilligen wollte. Auch bei diesem Maßstab stellt sich aber die Frage, welche Kriterien für das, was noch als *reasonable* für konventionsgemäß erachtet wird, gelten. Auch bei diesem Test liegt daher die Vorstellung eines europäischen Standards zugrunde, bei dessen Verletzung die Klassifizierung als *unreasonable* begründet ist.

Weitere Aufschlüsse über die Methode zur Bestimmung des einheitlichen europäischen Standards liefert die Entscheidung des Gerichtshofs im Fall *Sunday Times*[10]. Gestützt auf das *common law*-Konzept des *contempt of court* war der Sunday Times die Veröffentlichung eines Artikels über die thalidomidgeschädigten Kinder, der sich speziell mit der Frage des Verschuldens des englischen Herstellers und seiner Schadensersatzpflicht auseinandersetzte, verboten worden. Die Verbotsverfügung wurde vom House of Lords im wesentlichen mit der Begründung bestätigt, daß die geplante Veröffentlichung ein *prejudgement* eines anhängigen Verfahrens, eine Art *trial by newspaper* bedeute. Ein faires und ordnungsgemäßes Verfahren ist nach Meinung der Lordrichter nur dann gewährleistet, wenn Berichterstattung, welche die vom Gericht zu entscheidende Fragen präjudiziere, untersagt werden kann. Mit einer knappen Mehrheit entschied sich der Gerichtshof für eine Verletzung von

---

[9] A.a.O. (Anm. 7), S. 21 f.

[10] Fall *Sunday Times,* Urteil vom 26. 4. 1979, Serie A, Nr. 30, 1979.

Art. 10 EMRK. Ausgangspunkt der Prüfung, ob die konkrete Einschränkung im Sinne der Konvention notwendig war, ist wiederum die im *Handyside*-Fall entwickelte Lehre vom unüberprüfbaren Beurteilungsspielraum. Diese soll aber nicht bei allen zur Schrankenziehung zugelassenen Rechtsgütern gleich zu bemessen sein. Vielmehr wird unterschieden zwischen mehr »objektiv« und damit »europäisch« zu bestimmenden Rechtsgütern, wie z.B. dem »Ansehen« und der Unparteilichkeit der Rechtsprechung, und stärker subjektiv, d.h. nationalstaatlich geprägten Rechtsgütern, wie z.B. der Moral. Im folgenden distanziert sich der Gerichtshof zunächst von dem durch die Kommissionsmehrheit im *Handyside*-Fall angewandten Test, ob ein Staat *reasonably, carefully and in good faith* gehandelt habe. Für die Prüfung der Frage, ob hinreichende Gründe vorliegen, welche die Einschränkungen der Pressefreiheit zum Schutze des Ansehens der Rechtsprechung in einer demokratischen Gesellschaft notwendig gemacht haben, wird als Maßstab ein europäischer Standard angewandt, den der Gerichtshof aus einer eigenen Wertung der in der Konvention verwandten Begriffe und Abwägung der miteinander kollidierenden Rechtsgüter ableitet. Auch hier wird aber letztlich nicht ganz deutlich, welches die Kriterien für die Gewinnung des europäischen Standards sind. Der Gerichtshof hat im Grundsatz nicht in Abrede gestellt, daß die vom House of Lords unterstellte *interference with the course of justice* bei einer präjudiziellen Berichterstattung gegeben sein könne. Daraus wird aber im Gegensatz zum House of Lords nicht der Schluß gezogen, daß hiermit die Voraussetzungen von Art. 10 EMRK erfüllt seien. Vielmehr prüft der Gerichtshof, ob die Abwägung zwischen dem mit der konkreten Eingriffsmaßnahme zu schützenden Rechtsgut »Ansehen der Gerichtsbarkeit« und dem durch die Beschränkung beeinträchtigten Rechtsgut »Pressefreiheit« richtig vorgenommen wurde, wobei der Gerichtshof davon ausgeht, daß es sich nicht eigentlich um die Herstellung einer Balance zwischen zwei kollidierenden, gleichwertigen Rechtsgütern handle, sondern um den Schutz der Meinungsfreiheit, die ausnahmsweise zum Schutze anderer Rechtsgüter eingeschränkt werden könne. Der Gerichtshof hält sich deshalb für befugt, abweichend von der Wertung des House of Lords, der Meinungsfreiheit bei Publikationen von öffentlichem Interesse ohne Rücksicht auf mögliche präjudizierende Inhalte den Vorrang vor dem Schutz des Ansehens der Rechtsprechung einzuräumen. Grundlegend ist hierfür eine ganz bestimmte Vorstellung von der Rolle der Meinungsfreiheit in einer demokratischen Gesellschaft. Der Gerichtshof betont, daß auch die Gerichte nicht in einem Vakuum tätig werden könnten und sich die vorherige Diskussion von Streitfällen gefallen lassen müßten. Den Medien obliege in einer demokratischen Gesellschaft das Informieren über Angelegenheiten des öffentlichen Interesses, und die Öffentlichkeit habe auch ein Recht auf Information. Die Einschränkung der Pressefreiheit könne deshalb nur dann gerechtfertigt sein, wenn es »absolut sicher« sei, daß eine Veröffentlichung eine Bedrohung des Ansehens der Rechtsprechung bedeute. Man könnte vermuten, daß der Gerichtshof den Maßstab für diesen Standard aus der Praxis der anderen Europaratstaaten entnommen hat, denen derart weitgehende *contempt of court*-Konzepte unbekannt sind. Maßgeblich soll aber nach Auffassung der Gerichte nicht der Blick auf die anderen europäischen Rechtsordnungen sein. Die Zugrundelegung internationaler Standards bedeute nicht, daß absolute Uniformität gefordert sei. Dennoch legt der Gerichtshof offensichtlich großes Gewicht darauf, daß in den anderen Europaratstaaten der Pressefreiheit bei der Berichterstattung über anhängige Gerichtsverfahren große Bedeutung zugemessen werde und daß die *prejudgement*-Doktrin in der Auslegung durch das House of Lords auch in Großbritannien selbst nach Auffassung der englischen Regierung zu überdenken sei.

Angesichts der knappen Gerichtsmehrheit sind für die künftige Praxis des Gerichtshofs die Argumente der dissentierenden Richter[11] von besonderem Interesse. Auch die Minderheit des Gerichts geht davon aus, daß es für die Feststellung der »Notwendigkeit in einer demokratischen Gesellschaft« der Abwägung der möglichen nachteiligen Folgen für die Pressefreiheit und des Schadens für das ordnungsgemäße Funktionieren der Gerichtsbarkeit, wenn die Beschränkung nicht vorgenommen würde, bedarf. Anders als die Mehrheit gehen aber offenbar die dissentierenden Richter von einer echten Güterabwägung, nicht von einem Regel-Ausnahme-Verhältnis aus. Darüber hinaus gestehen sie den nationalen Behörden einen größeren Spielraum bei der Beurteilung der Frage zu, welchen Schaden das Ansehen der Gerichtsbarkeit erleiden kann, wenn die Einschränkung nicht vorgenommen würde, und welche Maßnahmen notwendig sind, um diesen Gefahren zu begegnen. Der Begriff des »Ansehens der Gerichtsbarkeit« sei wie derjenige der »Moral« mit den nationalen Verhältnissen eng verknüpft und könne nicht völlig einheitlich bestimmt werden. Ebenso könne der notwendige Umfang von Einschränkungen zum Schutze dieses Ansehens nach der Rechtsordnung und den Rechtstraditionen des jeweiligen Landes variieren. Innerhalb der Grenzen einer »demokratischen Gesellschaft« sei jedem Staat freigestellt, dasjenige Verfahren zu bestimmen, das ihm am besten zum Schutze des Ansehens der Gerichtsbarkeit geeignet erscheine. Die entscheidende Frage ist damit nach Meinung der Minderheit, ob das House of Lords »gute Gründe zu der Annahme« (*entitled to think*) gehabt habe, die Veröffentlichung des Artikels sei für das Ansehen der Gerichtsbarkeit schädlich. Maßstab dieser Prüfung sei zunächst, ob die nationalen Behörden *in good faith, with due care and in a reasonable manner* bei der Bewertung der Gefahren für die in Art. 10 Abs. 2 EMRK aufgeführten Rechtsgüter gehandelt hätten; darüber hinaus sei aber vorrangiges Ziel der europäischen Kontrolle sicherzustellen, daß in einer demokratischen Gesellschaft die Einschränkungen im richtigen Verhältnis zu dem damit verfolgten legitimen Ziel stünden. Auch die dissentierenden Richter überprüfen deshalb an Hand eigener Vorstellungen über die richtige Zweck-Mittel-Relation die vom House of Lords vorgenommene Abwägung,

[11] A.a.O. (Anm. 10), S. 47 ff.

allerdings unter Respektierung eines breiteren nationalen Spielraums als die Mehrheit des Gerichts.

Im Fall *Dudgeon*[12] hatte sich der Gerichtshof erneut mit der Frage auseinanderzusetzen, inwieweit die nationalen Moralvorstellungen eine auch in einer demokratischen Gesellschaft notwendige Einschränkung der Privatsphäre begründen können. Zur Überprüfung stand die für Nordirland geltende englische Homosexualitätsgesetzgebung, die mit Rücksicht auf die dort herrschenden, besonders strengen Moralregeln die einfache Homosexualität unter Erwachsenen noch als strafbares, wenngleich in der Praxis nicht verfolgtes, Delikt behandelte, während sie in den übrigen Teilen des Vereinigten Königreichs im Sinne eines Jugendschutzes abgeändert worden war. Der Gerichtshof folgt derselben Methode wie in den oben geschilderten Fällen. Die fragliche Einschränkung der Privatsphäre sei nur durch ein dringendes soziales Bedürfnis zu rechtfertigen, das unter Berücksichtigung der Erfordernisse einer demokratischen Gesellschaft festzustellen sei. Als Grundelement der Freiheitsbeschränkung einer demokratischen Gesellschaft wird wiederum ein angemessenes Verhältnis zwischen Einschränkung und damit verfolgtem Ziel verlangt, wobei die nationalen Behörden im Prinzip eher in der Lage seien, eine exakte Einschätzung der Erfordernisse zum Schutze der Moral in dem betreffenden Land zu geben. Der Gerichtshof setzt sich im folgenden eingehend mit den Gründen auseinander, welche die englische Regierung zur Beibehaltung der Strafbestimmung bewogen hatten. Der englischen Regierung wird attestiert, daß sie mit guten Gründen wegen der in Nordirland vorherrschenden besonders strengen gesellschaftlichen Anschauungen eine Freigabe der einfachen Homosexualität abgelehnt habe. Dennoch sei die Einschränkung unverhältnismäßig, da sich insofern kein besonders dringendes soziales Bedürfnis nachweisen lasse. Gegenüber den schweren, nachteiligen Wirkungen für das Leben der Betroffenen ließe sich nämlich keine hinreichende Gefahr eines Schadens für die Gesellschaft feststellen, wenn die Strafbarkeit der Homosexualität aufgehoben würde. Straflosigkeit bedeute nicht Billigung. Maßgeblich für diese Einschätzung ist für den Gerichtshof offensichtlich eine Veränderung der gesellschaftlichen und rechtlichen Normen in der großen Mehrheit der Europaratstaaten in Bezug auf die Strafwürdigkeit der einfachen Homosexualität. Der Gerichtshof betont zwar, daß allein die Tatsache, daß derartige Einschränkungen in anderen Europaratstaaten nicht für notwendig gehalten würden, nicht ausschlaggebend sei, da es verschiedene moralische und gesellschaftliche Normen innerhalb verschiedener kultureller Gemeinschaften selbst innerhalb eines Staates geben könne. Dennoch geht er offenbar davon aus, daß »europaeinheitliche Standards« für die Strafbarkeit der einfachen Homosexualität aus dem Wesen der demokratischen Gesellschaft abgeleitet werden können.

---

12 Fall *Dudgeon,* Urteil vom 23. 10. 1981, Vorabveröffentlichung des Europarats, 1981.

An diesem Punkt setzt die Kritik der dissentierenden Richter an [13]. Sie stellen in Frage, ob die Mehrheitsentscheidung nicht den den nationalen Instanzen überlassenen Beurteilungsspielraum mißachtet. Über die schädlichen Folgen der Aufhebung einer Strafbarkeit für die Moral der nordirischen Gesellschaft läßt sich streiten, wie die dissentierenden Richter eingehend darlegen. Der Richter Walsh weist nach, daß auch die Gefahr einer sittlichen Korrumpierung erwachsener Personen ein schützenswertes Anliegen einer Strafbestimmung sein könne [14]. Darüber hinaus stellt sich die Frage, ob nicht die Freigabe der Homosexualität in einer sittenstrengen Gesellschaftsordnung von der Mehrheit der Bevölkerung als schädlich empfundene gesellschaftspolitische Folgen haben könne. Rechtliche Regeln über die Sexualmoral haben, wie Walsh ausführt, für die Form und Entwicklung der Gesellschaft Konsequenzen und weisen deshalb eine soziale Dimension auf, die den Gesetzgeber zum Handeln berechtigen kann. Die Methode, aus den in der großen Mehrzahl der Europastaaten herrschenden gesellschaftlichen und juristischen Anschauungen verbindliche »*Euronormen*« für alle Mitgliedstaaten in Bezug auf die Strafbarkeit bestimmter Verhaltensweisen abzuleiten, bleibt daher problematisch.

III. Die Demokratieklausel und der einheitliche Grundrechtsstandard

Die Methode des Gerichtshofs zur Bestimmung des Standards der »demokratischen Gesellschaft« hat überwiegend Zustimmung, aber auch Kritik erfahren [15]. *Ermacora* wirft dem Gerichtshof vor, im *Handyside*-Fall seiner Aufgabe, europäische Maßstäbe zu entwickeln, nicht gerecht geworden zu sein [16]. Der Gerichtshof relativiere durch den Rückzug auf innerstaatliche Standards den Konventionsinhalt. Demgegenüber haben *Kelly* und andere vor einer zu strengen Kontrolle über den Gebrauch der verschiedenartigen Beschränkungsmöglichkeiten der Menschenrechte durch die Mitgliedstaaten gewarnt [17]. Insbesondere dürfe über die Europäische Menschenrechtskonvention

---

[13] Vgl. insbes. die *dissenting opinions* der Richter Zekia (Anm. 12), S. 22 ff., Matscher (Anm. 12), S. 26 ff., und Walsh (Anm. 12), S. 31 ff.

[14] A.a.O. (Anm. 12), S. 33.

[15] Vgl. die oben Anm. 4 erwähnten Autoren.

[16] *F. Ermacora,* Richtungsweisendes Handyside-Urteil?, Anmerkung zum Urteil des EGMR vom 7. 12. 1976, EuGRZ 1977, S. 363 f.

[17] Vgl. *J. Kelly,* Die internationale Kontrolle der Beschränkungen und Grenzen der Garantien, Bericht, V. Internationales Kolloquium über die Europäische Menschenrechtskonvention, Frankfurt vom 9.–12. 4. 1980, Europaratsdokument H/Coll. (80) 3; ferner *H. Seibert,* Bericht über das V. Internationale Kolloquium über die Europäische Menschenrechtskonvention, EuGRZ 1980, S. 334, 337.

nicht einer Minderheit die Rechtsvorstellung einer Mehrheit aufgezwungen werden. Die Entwicklung einheitlicher Menschenrechtsstandards sei mit der Aufgabe der Rechtsintegration nicht identisch[18]. Die verschiedenen Ansatzpunkte der Kritik zeigen, daß sich hinter der Auseinandersetzung über die Demokratieklausel divergierende Grundverständnisse über die Konvention und die Rolle des Gerichtshofs bei der Schaffung einer einheitlichen europäischen Menschenrechtsordnung verbergen. Letztlich wird daher die Toleranzschwelle nicht ohne subjektive Wertungen ermittelt werden können. Das kann aber nicht bedeuten, daß der Versuch, nach objektivierbaren Elementen zur Auslegung der Demokratieklausel zu suchen, von vornherein als erfolglos betrachtet werden müßte. Die Rechtsprechung des Gerichtshofs legt vom Sinn dieser Bemühungen Zeugnis ab, wenngleich ein größeres Maß an Berechenbarkeit erwünscht wäre. Die Feststellung *Moslers,* daß die Methode unverzichtbar, aber der Verbesserung fähig sei[19], ist daher Anlaß, die möglichen Verfahrensweisen zur Ermittlung europäischer Standards einer Überprüfung zu unterziehen.

1. Die Bedeutung der Konvention für die Ermittlung des Begriffs »demokratische Gesellschaft«

Die Konvention selbst enthält bekanntlich keine Definition der demokratischen Gesellschaft. Aus der Entstehungsgeschichte, dem Wortlaut und dem Sinn der Konvention können aber Grundsätze abgeleitet werden, die für das Verständnis der in der Konvention enthaltenen Menschenrechte, für die Auslegung der Rechtsgüter, die eine Beschränkung rechtfertigen können und für das Verhältnis zwischen Freiheitsrecht und Beschränkung von Bedeutung sind[20].

Das Menschenrechtsverständnis der Konvention ist geprägt von einer liberalen Grundrechtsauffassung. Gesichert werden soll zunächst die Freiheitssphäre des Einzelnen gegenüber staatlichen Eingriffen. Dies zeigt sowohl die Gesamtheit der Konventionsrechte als auch die Formulierung einzelner Bestimmungen, die zum Teil ausdrücklich Staaten oder öffentliche Behörden als Adressaten ansprechen. Die Konvention beruht also auf einem allen Europaratstaaten eigentümlichen Menschenbild, das Rechte um der Entfaltung des Individuums als Einzelnem oder in der Gemeinschaft mit anderen willen gewährt. Die Freiheit des Einzelnen ist danach nicht von vornherein in die Ziele einer »Volksgemeinschaft« oder »sozialistischen Gemeinschaft« eingebunden und nur zur Verwirklichung einer derartigen Gemeinschaft als wertvoll und

[18] Vgl. *G. Cohen Jonathan,* Cour européenne des Droits de l'Homme (1979), Cahiers de Droit Européen 16 (1980), S. 463, 478.

[19] *Mosler,* Bericht über die Ergebnisse des V. Internationalen Kolloquiums über die Europäische Menschenrechtskonvention (Anm. 2).

[20] Vgl. Nachweise Anm. 4.

schutzwürdig anerkannt[21]. Darüber hinaus basiert die Konvention auf einem Grundkonsens über die politischen Strukturen des Gemeinwesens, die für die beschriebene Entfaltung des Einzelnen unabdingbar sind[22]. Für die Schrankenproblematik folgt daraus, daß sich ein Grundrechtsverständnis verbietet, das dem oben beschriebenen Menschenbild entgegengesetzt wäre. Der Streit um Teilhaberechte, Institutionsgarantien oder Drittwirkung von Menschenrechten wird hierdurch nicht entschieden. Die Konvention schließt keineswegs aus, daß Staaten angesichts der technischen und sozialen Umweltveränderungen möglichen Grundrechtsbedrohungen seitens der Gesellschaft durch Veränderungen der Funktionsweise von Grundrechten Rechnung tragen[23]. Daraus ergeben sich möglicherweise auch Änderungen der Wirkungsweise der in der Konvention enthaltenen Menschenrechte, wenngleich in die Konvention bewußt keine sozialen Grundrechte aufgenommen worden sind; dies aus der Überlegung heraus, daß derartige Rechte nicht hinreichend konkretisierbar sind, um als Grundlage international durchsetzbarer Ansprüche dienen zu können[24]. Ob ein von der Konvention gewährleistetes Menschenrecht im Sinne einer Schutzpflicht, objektiven Gewährleistung, institutionellen Garantie oder mittelbaren Drittwirkung auszulegen ist, wird man daher nur auf Grund einer eingehenden Analyse des betreffenden Rechts unter Heranziehung des Verständnisses dieses Rechts in den Mitgliedstaaten der Konvention entscheiden können[25]. Dabei ist allerdings angesichts dessen, daß auch in den innerstaatlichen Verfassungen die Ableitung positiver Leistungsansprüche gegenüber Gesetzgebung und Verwaltung auf Skepsis stößt, äußerste Vorsicht walten zu lassen[26]. Auch insoweit gilt die Mahnung *Moslers,* daß die europäischen Organe die Fortentwicklung der europäischen Menschen-

---

[21] Vgl. *Bleckmann* (Anm. 1), S. 20; *K. Korinek,* Bedeutung der Grundrechte, Generalbericht (2. Teil), IV. Konferenz der europäischen Verfassungsgerichte in Wien vom 16.–18. 10. 1978, EuGRZ 1978, S. 489 ff.; *O. Luchterhandt,* Politische und soziale Menschenrechte, demokratischer Verfassungsstaat und Völkerrecht im Ost-West-Gegensatz, in: J. Schwartländer (Hrsg.), Menschenrechte und Demokratie (1981), S. 63 ff.

[22] *J. Schwartländer*, Demokratie-Verwirklichung oder Gefährdung der Menschenrechte?, in: Menschenrechte und Demokratie (Anm. 21), S. 189, 209 f.

[23] *M. Sørensen,* Do the Rights Set forth in the European Convention on Human Rights in 1950 have the Same Significance in 1975?, Bericht, IV. Internationales Kolloquium über die Europäische Menschenrechtskonvention, Rom vom 5.–8. 11. 1975, Europaratsdokument H/Coll. (75) 2; vgl. auch *P. Juvigny,* Modern Scientific and Technical Developments and their Consequences on the Protection of the Right to Respect for a Person's Private and Family Life, his Home and Communications, in: A. H. Robertson (Hrsg.), Privacy and Human Rights (1973), S. 129 ff.

[24] *Bleckmann* (Anm. 1), S. 21 ff.

[25] *L. Wildhaber,* To what Extent has the Right to Education Undergone an Evolution?, Bericht, IV. Internationales Kolloquium über die Europäische Menschenrechtskonvention, Rom vom 5.–8. 11. 1975, Europaratsdokument H/Coll. (75) 4, S. 7 ff.; vgl. ferner *Korinek* (Anm. 21), S. 490.

[26] *Bleckmann* (Anm. 1), S. 21 ff.

rechtsordnung nicht in einem Parforceritt entgegen dem europäischen Grundkonsens durchsetzen können [27]. Entsprechendes muß freilich auch für ein bestimmtes Verständnis der Menschenrechte gelten, das möglicherweise nicht von allen Europaratstaaten als Grundlage einer demokratischen Gesellschaft akzeptiert wird. Bedenken bestehen daher z.B. auch gegen eine einseitige Betonung der politisch demokratischen Funktion von Freiheitsrechten mit der Folge einer prinzipiellen Höherrangigkeit einer Inanspruchnahme von Freiheitsrechten für politische Zielsetzungen. Unter diesem Blickwinkel erscheint auch die Feststellung der Mehrheit des Gerichtshofs im Fall *Sunday Times* nicht ganz unbedenklich, daß sich Gerichtsverfahren über öffentliche Angelegenheiten stets auch öffentlicher Diskussion und Berichterstattung stellen müssen [28]. Die Vermutung, daß hier ein demokratietheoretisches Verständnis der gerichtlichen Entscheidungsfindung eines »öffentliche(n) Prozesse(s) gesellschaftlich vermittelter und als solcher zu kontrollierender Konstruktion von Wirklichkeit und richtigem Recht« Pate gestanden hat, ist nicht von der Hand zu weisen [29].

Festzuhalten ist für die Frage nach dem Maß zulässiger Beschränkungen, daß die in den einzelnen Mitgliedstaaten in unterschiedlichem Umfang vorzufindenden sozialen oder demokratietheoretischen Ergänzungen der Freiheitsrechte keine Umwandlung des individualschützenden Charakters der Grundrechte, im Einzelfall oder generell, bewirken dürfen. Soziale und wirtschaftliche Sicherungen dürfen zum bürgerlichen Freiheitsrecht hinzutreten, es aber nicht ersetzen. Die Erkenntnis, daß die Konvention den Staaten innerhalb des aufgezeigten Rahmens die Wahl unter verschiedenen möglichen Grundrechtsverständnissen zur Ergänzung individueller Freiheitssicherung gegenüber dem Staat überläßt, kann allerdings nicht darüber hinwegtäuschen, daß staatliche Schutzmaßnahmen sich auch gegen Dritte richten können und daher Beschränkungen individueller Freiheitsrechte anderer nach sich ziehen können. Wird z.B. die Meinungsfreiheit im Sinne einer konsequenten »inneren Pressefreiheit« verstanden, so folgt daraus eine Beschränkung der Freiheit eines Verlegers, die Zeitung nach eigenen Vorstellungen herstellen und vertreiben zu können [30]. Noch weiter reichende Folgerungen ergeben sich, wenn Vorstellungen unmittelbarer Demokratie auf den gesellschaftlichen Bereich, auf Arbeitswelt und öffentliche Verwaltung erstreckt werden [31]. Die Grenz-

---

[27] Bericht über die Ergebnisse des V. Internationalen Kolloquiums über die Europäische Menschenrechtskonvention (Anm. 2), S. 9.

[28] A.a.O. (Anm. 10), S. 40.

[29] *R. Stürner,* »Fair trial« und öffentliche Meinung, JZ 1980, S. 1, 4 unter Hinweis auf *J. Scherer,* JuS 1979, S. 470, 475.

[30] Zur Bedeutung der Meinungsäußerungsfreiheit für das Medienwesen vgl. *W. Pahr,* The Evolution of the Freedom of Expression, Bericht, IV. Internationales Kolloquium über die Europäische Menschenrechtskonvention, Rom vom 5.–8. 11. 1975, Europaratsdokument H/Coll. (75) 3.

[31] Kritisch zum »Demokratismus« *J. Schwartländer* (Anm. 22), S. 217 ff.

ziehung zwischen zulässiger Ergänzung und unzulässiger Ersetzung individueller Freiheit durch ein sozialstaatliches, demokratisches oder institutionelles Grundrechtsverständnis kann wiederum nur auf der Grundlage der gemeinsamen Überzeugung der Mitgliedstaaten erfolgen. Der Rechtsvergleichung kommt dabei die Aufgabe zu, ungeachtet divergierender Grundrechtsverständnisse in den einzelnen Mitgliedstaaten den individuell freiheitssichernden Kern der Menschenrechte generell und im Einzelfall herauszuarbeiten und gegenüber Ingerenzen durch andere »progressivere« Grundrechtsverständnisse abzusichern. Festzuhalten bleibt andererseits, daß die Notwendigkeit von Grundrechtsbeschränkungen in einer demokratischen Gesellschaft kein einheitliches Schrankensystem, gestützt auf ein bestimmtes Grundrechtsverständnis, beinhaltet. Innerhalb des aufgezeigten Rahmens bleibt es daher jedem Staat überlassen, aus seiner Sicht des richtigen Grundrechtsverständnisses für eine umfassende Grundrechtsverwirklichung Sorge zu tragen und eventuell hieraus Beschränkungen individueller Freiheit abzuleiten.

Die Konvention regelt nur Teilaspekte der Menschenrechtssicherung. Diese Feststellung ist nicht nur für das »richtige« Grundrechtsverständnis von Bedeutung, sondern auch für die Bestimmung des Verhältnisses von Menschenrecht und beschränkendem Rechtsgut. Anders als die innerstaatlichen Verfassungen enthält die Konvention keine umfassende Aufzählung der geschützten Rechtsgüter und öffentlichen Interessen. Auf die öffentlichen Interessen und privaten Rechte anderer wird vielmehr nur generalklauselartig in den Schrankenregelungen Bezug genommen. Daraus ergeben sich Probleme der Abwägung. Aus der Konvention selbst sind Maßstäbe für den Rang des Rechtsgutes, das die Einschränkung rechtfertigen soll, nur ausnahmsweise abzuleiten. Für die Herstellung des richtigen Verhältnisses zwischen Menschenrecht und beschränkendem Rechtsgut ist daher abzustellen auf die innerstaatlichen Verfassungen, soweit nicht konstitutive Prinzipien der »demokratischen Gesellschaft« eine bestimmte Bewertung zwingend vorschreiben. Der Mehrheit des Gerichtshofs im *Sunday Times*-Fall kann deshalb methodisch nicht gefolgt werden, wenn sie darlegt, es ginge nicht eigentlich um die Herstellung einer Balance zwischen zwei miteinander kollidierenden, prinzipiell gleichwertigen Rechtsgütern, sondern nur um die Prüfung, ob ausnahmsweise ein in der Konvention geschütztes Recht durch ein anderes Rechtsgut eingeschränkt werden könne[32]. Der Stellenwert eines beschränkenden Rechtsgutes, sei es in der Schrankenklausel ausdrücklich erwähnt oder nur unter die Generalklausel der öffentlichen Ordnung oder der privaten Rechte anderer subsumierbar, ist daher auch an Hand der vergleichenden Bewertung dieses Rechtsgutes in den innerstaatlichen Verfassungen festzustellen. Danach kann es sehr wohl geboten sein, eine Balance zwischen zwei kollidierenden, gleichwertigen Rechtsgütern im Sinne »praktischer Konkordanz«[33] herzustellen.

[32] A.a.O. (Anm. 10), S. 41.

[33] Vgl. dazu *K. Hesse*, Verfassungsrecht (13. Aufl. 1982), Rdn. 317 ff.

Der *Sunday Times*-Fall zeigt die praktischen Konsequenzen der vom Gerichtshof praktizierten Regel-Ausnahmemethode. Bei Annahme einer prinzipiellen Gleichwertigkeit beider Rechtsgüter hätte der Gerichtshof nicht zur Feststellung gelangen können, die Einschränkung der Pressefreiheit könne nur dann gerechtfertigt werden, wenn »absolut sicher« sei, daß hierdurch das Ansehen der Justiz geschädigt werde. Das Erfordernis der absoluten Sicherheit bedeutet, daß die mögliche Schädigung des Ansehens der Justiz generell hingenommen werden muß gegenüber einer Beeinträchtigung der Freiheit der Meinungsäußerung. Im Einzelfall kann dieses Ergebnis vertretbar sein. Methodisch bedenklich ist es aber, dieses Ergebnis schlechthin aus dem Regel-Ausnahme-Charakter von Menschenrechten und beschränkendem Rechtsgut abzuleiten. Bei einer gravierenden potentiellen Schädigung des Ansehens der Justiz kann eine geringfügige Beeinträchtigung der Freiheit der Meinungsäußerung auch in einer demokratischen Gesellschaft als gerechtfertigt erscheinen, angesichts dessen, daß der Unabhängigkeit und dem Schutz des Ansehens der Justiz auch in den Rechtsordnungen anderer Mitgliedstaaten ein hoher Stellenwert eingeräumt wird. Andererseits kann es geboten sein, bei kollidierenden Rechtsgütern und öffentlichen oder privaten Interessen von einem Regel-Ausnahme-Prinzip auszugehen, mit der Folge, daß nur bei einer nachgewiesenen schweren Beeinträchtigung derartiger Rechtsgüter Beschränkungen der Konventionsrechte zulässig sind. So wird z.B. das wirtschaftliche Wohl des Landes oder die Verteidigung der Ordnung (vgl. Art. 8 Abs. 2 EMRK) in einer demokratischen Gesellschaft nur ausnahmsweise und unter strengsten Voraussetzungen Beschränkungen grundlegender Menschenrechte rechtfertigen können. Methodisch folgt daraus die Notwendigkeit einer Einzelfallabwägung der kollidierenden Rechtsgüter unter Berücksichtigung des Stellenwertes, den das beschränkende Rechtsgut in einer »demokratischen Gesellschaft« genießt. Der Begriff der »Notwendigkeit« wird hierdurch entscheidend beeinflußt. Der vom Gerichtshof zugrunde gelegte Test der *pressing social need* sollte daher nicht als strikt anzuwendende Formel verstanden werden, sondern als flexibler Abwägungsmaßstab.

Ausnahmsweise enthält die Konvention selbst einen Maßstab für die Notwendigkeit bzw. Zulässigkeit von Beschränkungen in demokratischen Gesellschaften. Art. 17 EMRK ist zu entnehmen, daß Tätigkeiten oder Handlungen, die auf die Abschaffung oder unzulässige Beschränkung der Konventionsrechte abzielen, von der Konvention nicht geschützt werden. Die Konvention hat sich insoweit ausdrücklich für das Prinzip der »streitbaren Demokratie« entschieden. Anhänger totalitärer Zielsetzungen sollen demnach die Rechte der Konvention nicht für die Durchsetzung menschenrechtsfeindlicher Zielsetzungen in Anspruch nehmen können[34]. Im Bereich

---

[34] Vgl. *J. E. S. Fawcett,* The Application of the European Convention on Human Rights (1969), S. 251 ff.; *F. G. Jacobs,* The European Convention on Human Rights (1975), S. 210 ff.; *G. L. Weil,* The European Convention on Human Rights (1963), S. 76 ff.

von Art. 17 bedarf es daher keiner (besonderen) Abwägung der miteinander kollidierenden Rechtsgüter.

2. Probleme der Rechtsvergleichung bei der Ermittlung des europäischen Menschenrechtsstandards

Bei der Ermittlung der gemeinsamen demokratischen Überzeugungen kommt der Rechtsvergleichung eine Schlüsselrolle zu. Die Untersuchung der Tätigkeit der europäischen Menschenrechtsorgane läßt allerdings keine eindeutigen Schlüsse auf die zugrunde gelegte rechtsvergleichende Methode zu. Einerseits wird betont, es komme nicht darauf an, ob sich in der Mehrheit der Europaratstaaten vergleichbare Beschränkungen finden ließen. Andererseits wird regelmäßig auf die Rechtslage in den anderen Europaratstaaten hingewiesen und daraus ein Maßstab für die in einer demokratischen Gesellschaft notwendigen Schranken abgeleitet. Dies geschieht meist in der Form eines negativen Tests derart, daß geprüft wird, ob sich vergleichbare Beschränkungen in anderen Europaratstaaten finden lassen. Die Abwesenheit derartiger Beschränkungen indiziert nach dieser Praxis, daß in einer demokratischen Gesellschaft die fragliche Beschränkung nicht für notwendig gehalten wird. Oder es wird auf die Regelpraxis der anderen Staaten verwiesen [35]; der Nachweis, daß auch in den meisten anderen Staaten die fragliche Beschränkung üblich ist, solle den Schluß auf ihre Notwendigkeit in einer demokratischen Gesellschaft erlauben. Auch die Literatur zeigt kein klares Bild über die zugrunde zu legende rechtsvergleichende Methode. Überwiegend wird unter Berufung auf die Rechtsprechung der europäischen Organe die Notwendigkeit rechtsvergleichender Überlegungen hervorgehoben, ohne daß allerdings Leitlinien für deren Bewertungen ersichtlich wären [36].

Dezidiert gegen die Heranziehung der Rechtsvergleichung hat sich *Bleckmann* ausgesprochen, weil die Rechtsvergleichung keinen sicheren Maßstab liefere [37]. Insbesondere sei unklar, welcher Staat als demokratisch anzusehen sei und wieviel Staaten das beschränkende Gesetz kennen müßten. Vielmehr wolle die Menschenrechtskonvention mit der demokratischen Gesellschaft auf absolute Maßstäbe für grundrechtsbeschränkende Gesetze verweisen, die in demokratischen Gesellschaften verwendet würden. Daraus ergebe sich die Anwendung des Verhältnismäßigkeitsprinzips. Der Hinweis auf die demokratische Gesellschaft bedeutet also nach *Bleckmann,* daß bei der erforderlichen Abwägung den Individualrechten die ihnen gebührende Rolle zugesprochen werden muß und daß die Individualrechte nicht stets wie in sozialistischen Gesellschaften hinter das öffentliche Interesse zurückzutreten haben. Den Maßstab für die Abwägung will freilich auch *Bleckmann* aus der Vergleichung des Rechts in rechtsstaatlichen Gesellschaften gewinnen [38].

---

[35] Vgl. z.B. Fall *Dudgeon* (Anm. 12), Fall *Handyside* (Anm. 7).
[36] Vgl. *Jacobs* (Anm. 34), S. 197; *Hoffmann-Remy* (Anm. 4), S. 62 ff.
[37] *Bleckmann* (Anm. 1), S. 19.
[38] A.a.O. (Anm. 37), S. 20.

Diese Auffassung berücksichtigt nicht ausreichend, daß die Konvention mehr erreichen will als nur einen Schutz vor der Anwendung totalitärer Praktiken. Sie ist vielmehr angelegt auf die Verwirklichung einer darüber hinausgehenden einheitlichen europäischen Menschenrechtsordnung, die gerade auch in den demokratischen Ordnungen der Europaratstaaten Maßstäbe und Schranken setzt. Die Präambel der Konvention selbst spricht daher von der Herbeiführung einer größeren Einheit unter den Mitgliedstaaten und der Wahrung *und Entwicklung* der Menschenrechte. Hierzu bedarf es der Rechtsvergleichung nicht nur für die allgemeine Methode der Ermittlung von Inhalt und Schranken in einer demokratischen Gesellschaft. Vielmehr muß die Rechtsvergleichung den Maßstab dafür liefern, ob das Ergebnis der Abwägung im Einzelfall dem von der Konvention geforderten europäischen Standard entspricht oder nicht. Es liegt nahe, bei der Ermittlung des europäischen Standards auf die Rechtsvergleichungsdiskussion innerhalb der Europäischen Gemeinschaften zurückzugreifen [39].

Eine Minimallösung, die einen europäischen Standard nur bei einer mathematischen Übereinstimmung aller Europaratstaaten annehmen würde, scheidet von vornherein aus, da sie die Konvention jeder Bedeutung berauben würde. Aber auch gegen eine Maximallösung, die den europäischen Standard durch die Suche nach der optimalen Grundrechtsverwirklichung ermitteln möchte [40], sprechen erhebliche Bedenken. Anders als bei den Europäischen Gemeinschaften geht es nicht um die Ergänzung einer supranationalen Ordnung durch europäische Standards, sondern um die Gewährleistung einheitlicher Mindestprinzipien in den einzelnen Mitgliedstaaten. Da die Konvention die grundrechtliche Ordnung der einzelnen Mitgliedstaaten weder ersetzen kann noch ersetzen will, verbietet sich jedes System, das der Verschiedenheit der Grundrechtskonzeptionen in den einzelnen Mitgliedstaaten nicht ausreichend Rechnung trägt.

Aus demselben Grund kann nicht auf die numerisch ermittelte Mehrheit der Europaratstaaten abgestellt werden. Die Ermittlung europäischer Standards an Hand einer derart festgestellten Majorität würde die Oktroyierung einer bestimmten Vorstellung von einer demokratischen Gesellschaft für eine Minderheit von Vertragsstaaten bedeuten. Im wesentlichen würden damit die durch bestimmte wirtschaftliche und politische Gemeinsamkeiten in den Europäischen Gemeinschaften verbundenen Staaten den Grundrechtsstandard der Europaratstaaten bestimmen. Für bestimmte Arten von Grundrechtsschranken, wie z.B. für Beschränkungen des Rechts auf Freizügigkeit, ist aber

[39] Vgl. hierzu insbes. *H. Mosler/R. Bernhardt/M. Hilf* (Hrsg.), Grundrechtsschutz in Europa (Beiträge zum ausländischen öffentlichen Recht und Völkerrecht, Bd. 72) (1977); *A. Bleckmann,* Zur Entwicklung europäischer Grundrechte, DVBl. 1978, S. 457 ff.

[40] Vgl. dazu *K. Hailbronner,* Ziele und Methoden völkerrechtlich relevanter Rechtsvergleichung, ZaöRV 36 (1976), S. 190, 219 ff.; zu Bedenken gegen die Maximalisierung vgl. auch *St. Trechsel,* in: Grundrechtsschutz in Europa (Anm. 39), S. 111.

offenkundig, daß die spezifisch EG-einheitlichen Vorstellungen nicht mit denen einer demokratischen Gesellschaft im Sinne der Menschenrechtskonvention gleichgesetzt werden können.

Übrig bleibt die Heranziehung der Rechtsvergleichung im Sinne einer »wertenden« Ermittlung europäischer Standards für die Beschränkung der Konventionsrechte. Da es auf mathematische Übereinstimmung nicht ankommen kann, muß die Rechtsvergleichung lediglich eine »tendenzielle, grundsätzliche« Übereinstimmung ergeben. Der Schluß auf eine Konventionsverletzung ist nur gerechtfertigt, wenn die fragliche Beschränkung »tendenziell übereinstimmend« als nicht notwendig zum Schutze eines in der Konvention genannten Rechtsguts betrachtet wird. Die rechtsvergleichende Untersuchung des Begriffs der demokratischen Gesellschaft ist dabei nicht nur für die Auslegung der beschränkenden Rechtsgüter von Bedeutung (was versteht man unter öffentlicher Ordnung, Moral usw.?), sondern zugleich für die Methode der Schrankenermittlung und ihre Anwendung auf den Einzelfall. Unproblematisch ist zunächst die Heranziehung grundsätzlicher Prinzipien hinsichtlich der Zuordnung von Freiheitsrechten und anderen öffentlichen und privaten Rechtsgütern. Übereinstimmend wird deshalb der Grundsatz der Verhältnismäßigkeit für die Schrankenziehung zugrunde gelegt[41]. Entsprechendes gilt für das Verhältnis von Grundrechtsbeschränkung und -beseitigung. Wesensgehaltsgarantien oder vergleichbare Konzepte lassen sich aus allen Rechtsordnungen der Europaratstaaten ableiten[42]. Problematisch ist die Forderung nach tendenzieller Übereinstimmung dagegen bei der Anwendung des Verhältnismäßigkeitsgrundsatzes im Einzelfall. Soweit hier nicht nur innerstaatliche »Pannen« im Einzelfall auf der Konventionsebene korrigiert werden sollen, geht es häufig um eine Divergenz in der Bewertung des Verhältnisses von Menschenrecht und menschenrechtsbegrenzendem Rechtsgut.

Welche Voraussetzungen müssen vorliegen, damit in diesen Fällen von einem europäischen Standard gesprochen werden kann? Zunächst ist nicht ausreichend, daß eine entsprechende Beschränkung in den übrigen Europaratstaaten nicht vorzufinden ist. Das Fehlen entsprechender Beschränkungen läßt sich unter Umständen aus abweichenden tatsächlichen Verhältnissen, aus vorübergehenden Zweckmäßigkeitserwägungen oder daraus ableiten, daß das Rechtsgut bereits anderweitig ausreichend geschützt wird. Die Rechtsvergleichung hat demgegenüber den Nachweis zu erbringen, daß die fragliche Beschränkung tendenziell übereinstimmend als nicht (oder nicht mehr) notwendig zum Schutze des fraglichen Rechtsguts angesehen wird, sei es, daß die Beschränkung als ungeeignet, als über das Ziel hinausschießend oder als im Verhältnis zum angestrebten Zweck unverhältnismäßig erscheint. Die-

[41] Vgl. zum Grundsatz der Verhältnismäßigkeit *K. Hailbronner,* Generalbericht zum 10. Internationalen Rechtsvergleichungskongreß 1980, Budapest (1982), S. 831 ff.

[42] *Hoffmann-Remy* (Anm. 4), S. 78 ff.

ser Nachweis setzt voraus, daß der behauptete Standard zumindest grundsätzlich allgemeine Anerkennung gefunden hat. Die Abweichung eines einzelnen Mitgliedstaates kann um so eher vernachlässigt werden, als sie auf Umständen beruht, die die Geltung des Grundsatzes nicht in Frage stellen. Erforderlich ist deshalb, daß aus den zugrunde gelegten Rechtsordnungen ein verallgemeinerungsfähiges Prinzip abgeleitet werden kann, das es erlaubt, die zur Debatte stehende Abweichung als bei der Bildung der Menschenrechtsordnung unerheblich außer Betracht zu lassen.

Im übrigen sind bei der Rechtsvergleichung die allgemein zu beachtenden Regeln auch im Rahmen der Ermittlung des Begriffs der demokratischen Gesellschaft zugrunde zu legen [43]. Verglichen werden nicht nur die Rechtsnormen, entscheidend ist vielmehr die Anwendung des Rechts in der Wirklichkeit. Dabei ist besonders darauf zu achten, ob der fragliche Beschränkungseffekt auf Grund anderer Beschränkungsmechanismen, die möglicherweise nicht das konkrete Menschenrecht betreffen, in vergleichbarer Weise erreicht wird. Die Frage, welche die Rechtsvergleichung zu beantworten hat, ist also diejenige nach den Mitteln zur Bewältigung bestimmter Probleme oder Gefährdungen von Rechtsgütern oder öffentlicher Interessen. So wird man z.B. Einschränkungen der politischen Betätigung von Beamten auf Grund von Treuepflichten nicht, gestützt auf europäische Standards, für konventionswidrig erklären können, wenn in anderen Staaten Loyalitätspflichten bereits auf Grund anderer Mechanismen, wie z.B einer nicht überprüfbaren Auswahlpraxis, sichergestellt werden [44]. Der Vergleich hat deshalb nicht nur die fragliche Menschenrechtsbeschränkung einzuschließen, sondern andere Rechtsinstrumente und Verfahrensweisen, durch die ein entsprechender Erfolg erzielt werden kann, einzubeziehen.

Vergleichbarkeitsprobleme stellen sich ferner dann, wenn die fragliche Beschränkung auf bestimmten tatsächlichen Umständen beruht, die in dieser Weise in keinem anderen Land gegeben sind. Hier darf die rechtsvergleichende Betrachtung sich nicht darin erschöpfen, daß entsprechende Beschränkungen in den anderen Europaratstaaten nicht für notwendig gehalten werden. Vielmehr ist zu fragen, ob bei Annahme der besonderen Umstände möglicherweise auch in den anderen Staaten eine Abweichung vom zugrunde gelegten Standard in Betracht kommt. Erforderlich ist also ein hypothetisches Urteil über den anzuwendenden europäischen Standard. Ist eine derartige hypothetische Betrachtung nicht möglich, weil die speziellen Verhältnisse in einem Mitgliedstaat mit der Situation anderer Mitgliedstaaten schlechthin unvergleichbar sind, so wird man in der Regel nicht von der Annahme eines europäischen Standards ausgehen können, sondern von einer Freiheit der Staaten, mit den von der Konvention zugelassenen Mitteln einer besonderen Gefahrensituation zu begegnen.

---

[43] Vgl. dazu *Bleckmann* (Anm. 39).

[44] Vgl. *K. Doehring, A. Bleckmann* u.a., Verfassungstreue im öffentlichen Dienst europäischer Staaten (Schriften zum öffentlichen Recht, Bd. 379) (1980).

Entsprechend ist auch die Frage der Bewältigung neuer Situationen oder Gefährdungen zu lösen. Unter Berufung darauf, daß entsprechende Beschränkungen sich in den Europaratstaaten nicht finden lassen, kann der Weg für neue Gesetzesmodelle nicht versperrt werden[45]. Bei neuen Entwicklungen muß es grundsätzlich Sache jedes Mitgliedstaates bleiben, diesen Entwicklungen, die eventuelle neue Gefährdungen für schutzwürdige Rechtsgüter mit sich bringen, durch neue Schrankenregelungen zu begegnen. Das Urteil auf eine Konventionsverletzung kann hier nur darauf gestützt werden, daß angesichts bestehender Grundprinzipien über die Beseitigung von Gefahren der begangene Weg mit Sicherheit in den anderen Mitgliedstaaten als unverhältnismäßig oder unangemessen angesehen würde. Ist auf Grund einer derartigen Bewertung der in der EMRK aufgeführten öffentlichen und privaten Interessen jedoch eine Erweiterung bestehender Schrankenregelungen möglich oder sogar geboten, so scheidet eine Konventionsverletzung aus. Im Zweifel wird man bei neuartigen Gefährdungssituationen ebenfalls von einer Freiheit der Staaten auszugehen haben, diesen Gefährdungen durch neuartige Gesetzesmodelle Rechnung zu tragen.

Besondere Probleme stellen sich, wenn die inkriminierte Beschränkung nicht auf ein Versagen der zuständigen innerstaatlichen Organe, auf Pannen im Gesetzgebungsverfahren, auf überholte Normen aus vordemokratischen Zeiten oder auf ein Abgleiten in totalitäre Praktiken infolge einer Änderung der politischen Grundstrukturen zurückzuführen ist, sondern auf divergierende Vorstellungen über das Verhältnis von Menschenrecht und beschränkendem Rechtsgut. Die Feststellung einer tendenziellen oder grundsätzlichen Übereinstimmung in diesen Fällen läuft auf die Fiktion eines Grundkonsenses hinaus, da wir es in Wirklichkeit mit abweichenden Vorstellungen zu tun haben. Wenn dennoch ein europäischer Standard zugrunde gelegt werden soll, bedarf es besonderer Rechtfertigung, damit die Minderheitskonzeption eines oder mehrerer Staaten, deren Konventionswidrigkeit behauptet wird, als unbeachtlich außer Betracht bleiben kann. Dieser Rechtfertigungsgrund kann sich einmal daraus ergeben, daß im Rahmen des Europarats bestimmte einheitliche Vorstellungen über Grundrechtsschranken außerhalb der Konvention offiziell niedergelegt worden sind. So wird man z.B. die Resolutionen des Ministerrats der Europaratstaaten über die Pflichten der Polizei in einer demokratischen Gesellschaft[46] oder die Empfehlung des Ministerkomitees über den Zugang der Öffentlichkeit zu Verwaltungsunterlagen in öffentlichen Angelegenheiten[47] als Konkretisierung der Strukturen einer demokratischen Gesellschaft auch für die Schrankenziehung mitberücksichtigen kön-

[45] Vgl. *Bleckmann* (Anm. 1), S. 19.

[46] Parlamentarische Versammlung des Europarates, Straßburg, Entschließung 690 (1979) vom 8. 5. 1979 betr. Pflichten und Status der Polizei in einer demokratischen Gesellschaft, abgedr. in EuGRZ 1979, S. 299 f.

[47] Activities of the Council of Europe in the Field of Human Rights in 1981, Europaratsdokument H (82) 2, S. 39 ff.

nen, auch wenn ein Staat sich innerstaatlich hierzu in Widerspruch befindet. Der Einwand, daß mit derartigen Resolutionen und Empfehlungen gerade keine verbindliche Vereinheitlichung der Rechtsordnung beabsichtigt ist, greift m.E. nicht durch. Es geht nicht darum, diesen Resolutionen eine nicht existierende Verpflichtungskraft zu unterstellen, sondern darum, sie für die Auslegung bestehender Verbindlichkeiten der Menschenrechtskonvention fruchtbar zu machen.

Sind derartige Anhaltspunkte aber nicht vorhanden, so bedarf es für die Annahme eines europäischen Standards angesichts divergierender Konzeptionen einer besonders intensiven Analyse der Gründe und des Ausmaßes der Divergenz. Abgesehen von bereits geschilderten Pannenfällen oder überholten bzw. systemwidrigen Regelungen, die mit einem gewandelten Demokratieverständnis auch des betreffenden Staates selbst nicht mehr in Einklang zu bringen sind, kann die Divergenz auf Besonderheiten der gesellschaftlichen Struktur (z.B. besonders strikte Moralvorstellungen) oder des politisch-juristischen Systems (besondere Stellung der Gerichtsbarkeit in Großbritannien; verfassungsrechtlich garantiertes Berufsbeamtentum in der Bundesrepublik Deutschland) beruhen. Hier ist besondere Vorsicht geboten, wenn die Vorstellungen der Mehrheit für eine Minderheit im Gewande des Begriffs der demokratischen Gesellschaft verbindlich gemacht werden sollen. Im Fall *Dudgeon*[48] hat der Gerichtshof auf den Wandel der Rechtsüberzeugungen in den Mitgliedstaaten hingewiesen, ohne sich damit auseinanderzusetzen, daß neben Nordirland offenbar auch weitere Mitgliedstaaten, wie z.B. Zypern, die Pönalisierung der einfachen Homosexualität für geboten halten. Sicherlich wird man nicht schlechterdings davon ausgehen können, daß die Rücksichtnahme auf die traditionellen Anschauungen der Bevölkerung oder eines Teils der Bevölkerung menschenrechtliche Beschränkungen rechtfertigt. Auch Tradition und ehrwürdige rechtliche Institutionen müssen sich unter der Herrschaft der Konvention daraufhin überprüfen lassen, ob sie auf rational überprüfbaren Gründen, die mit der Konvention in Einklang stehen, beruhen, d.h., ob sie zum Schutz gegen die Gefährdung eines von der Konvention anerkannten Rechtsguts geeignet und erforderlich sind. Zurückhaltung ist dort geboten, wo auf Grund der gesellschaftlichen oder politisch-juristischen Besonderheiten die Prüfung der Verhältnismäßigkeit im engeren Sinne anders ausfällt als bei den anderen Staaten, die dem beschränkenden Rechtsgut einen anderen Rang einräumen, der im Einzelfall zu einem abweichenden Ergebnis bei der Lösung der Kollisionsfrage führen könnte. Je stärker diese Divergenz allgemeine Prinzipien betrifft, desto eher ist die Annahme geboten, daß in Wirklichkeit ein europäischer Standard, der für den Begriff der demokratischen Gesellschaft zugrunde gelegt werden kann, sich noch nicht entwickelt hat. Beispiele für Divergenzen bei der Bewertung der miteinander kollidierenden Rechtsgüter bietet sowohl der Fall *Dudgeon* als auch der *Sun-*

---

[48] A.a.O. (Anm. 12).

day Times-Fall. In beiden Fällen hat der Gerichtshof nicht in Abrede gestellt, daß objektiv nachvollziehbare, an sich konventionsgemäße Gründe der Moral bzw. des Ansehens der Gerichtsbarkeit für eine Beschränkung der jeweiligen Konventionsrechte vorgebracht werden konnten. In beiden Fällen wurde aber für entscheidend gehalten, daß auf Grund einer höheren Bewertung der Konventionsrechte eine mögliche Gefährdung des Schutzguts hingenommen werden mußte.

Verschiedene Anzeichen deuten darauf hin, daß die Mehrheit des Gerichtshofs von einer Befugnis ausgeht, bei der Ermittlung der europäischen Standards – quasi in rechtspolitischem Vorgriff auf die erstrebenswerte Entwicklung – abweichende Regelungen und Konzepte als rückständig, als der »wahren Entwicklung der demokratischen Gesellschaft nicht entsprechend« auszuschalten. Die Aufgabe des Gerichtshofs bestünde nach dieser Konzeption darin, eine angemessene Lösung unter den Mitgliedstaaten unter Berücksichtigung der Entwicklungstendenzen zu finden. Darauf läßt auch die Einbeziehung rechtspolitischer Reformvorstellungen im *Sunday Times*-Fall schließen. Der Schluß von Reformbestrebungen und Entwicklungstendenzen auf einheitliche Strukturen einer demokratischen Gesellschaft ist aber nicht unproblematisch. Man hat zu Recht darauf aufmerksam gemacht, daß dem Straßburger Gerichtshof im Gegensatz zum Luxemburger Gerichtshof nicht die allgemeine Aufgabe einer rechtspolitischen Weiterentwicklung einer integrierten, in politischer Fortentwicklung befindlichen Gemeinschaft zukommt[49]. Die dynamische Auslegung der Menschenrechtskonvention kann daher, anders als bei den Europäischen Gemeinschaften, nicht als ständige Fortentwicklung der Integration, sondern nur als Anpassung der Konvention an gewandelte Umstände, Sitten und Überzeugungen im Rahmen einer vorhandenen Menschenrechtsordnung verstanden werden. Reformbestrebungen oder die Nichtanwendung bestimmter Normen oder auch einfach nur die Überzeugungen in der weit überwiegenden Zahl der Europaratstaaten können ein Anzeichen dafür sein, daß sich derartige einheitliche Grundsätze gebildet haben, hinter denen die tatsächliche Rechtslage in einem oder mehreren Mitgliedstaaten zurückgeblieben ist. Ob dies der Fall ist, muß aber unter Berücksichtigung der kulturellen, gesellschaftlichen und politischen Eigenheiten in allen demokratischen Gesellschaften des Europarats festgestellt werden. Insbesondere dort, wo vergleichbare Regelungen wie die angeblich konventionswidrige Regelung oder Maßnahme auch in anderen Mitgliedstaaten praktiziert werden, muß der Gerichtshof die faktisch präjudizielle Konsequenz seiner Entscheidung auch für diese Staaten in Betracht ziehen. Der Maßstab der demokratischen Gesellschaft ist z.B. nicht der der kontinental-europäischen

[49] *W. J. Ganshof van der Meersch,* Die Bezugnahme auf das innerstaatliche Recht der Vertragsstaaten in der Rechtsprechung des Europäischen Gerichtshofs für Menschenrechte, EuGRZ 1981, S. 481, 483 ff.; Referate von *H. Golsong, Ch. Sasse, P. Pescatore, E. Klein,* in: Grundrechtsschutz in Europa (Anm. 39), S. 7 ff., 51 ff., 64 ff., 133 ff.

Staaten oder der einer gesellschaftspolitisch besonders fortschrittlichen Mehrheit von Europaratstaaten. Die demokratische Gesellschaft der Europaratstaaten muß vielmehr Raum für zahlreiche verschiedene kulturelle und gesellschaftliche Wertvorstellungen lassen. Über abweichende Bewertungen wird man sich deshalb nur hinwegsetzen können, wenn die divergierende Bewertung auf Motiven und Kriterien beruht, die mit den übereinstimmenden Strukturen einer demokratischen Gesellschaft und den Prinzipien der Menschenrechtskonvention nicht mehr in Einklang gebracht werden können. Im Fall *Dudgeon* hätte daher eingehender untersucht werden müssen, ob die zugrunde gelegte Konzeption des Strafrechts als Rechtsgüterschutz unter Ausklammerung privater sexueller Verhaltensweisen, die Dritte nicht schädigen, wirklich als gesichertes Prinzip der Strafrechtsordnung der beteiligten Staaten angesehen werden kann [50]. Entsprechend hätte im *Sunday Times*-Fall näher dargelegt werden müssen, warum die mit der besonderen Stellung der Gerichtsbarkeit in Großbritannien zusammenhängende *contempt of court*-Konzeption mit den Wertvorstellungen und Maßstäben der Konvention nicht in Einklang zu bringen ist. Daß diese Entscheidung letztlich wiederum ein hohes Maß subjektiver Wertung enthält, ergibt sich aus ihrem Wesen. Das letzte Wort über den richtigen Maßstab des Gerichtshofs sprechen die Mitgliedstaaten. Das Maß der Akzeptanz einer Entscheidung ist zugleich ein Indiz dafür, daß es dem Gerichtshof gelungen ist, die Strukturen der demokratischen Gesellschaft in angemessener Weise zu konkretisieren.

3. Die Prüfungsdichte – Beurteilungs- und Ermessensspielraum

Der Begriff des Beurteilungsspielraums *(margin of appreciation)* ist in der Entscheidungspraxis der Kommission und des Gerichtshofs – wie die geschilderten Entscheidungen zeigen – untrennbar mit der Auslegung der unbestimmten Schrankenklauseln der Art. 8–11 und der Notstandsbestimmung des Art. 15 verbunden [51]. Der Gerichtshof leitet den nationalen »Beurteilungsspielraum« vorwiegend aus einem »subsidiären Charakter« der europäischen Kontrolle ab. Im *Handyside*-Fall wird diese Subsidiarität mit der größeren Sachnähe und der besseren Vertrautheit der nationalen Behörden »mit den in

---

[50] Für eine derartige These sprechen verschiedene Entscheidungen der Kommission und nationaler Verfassungsgerichte, vgl. *H.-E. Folz,* Die Europäische Menschenrechtskonvention in der neueren Rechtsprechung des Verfassungsgerichtshofs, in: Völkerrecht und Rechtsphilosophie, Internationale Festschrift für Stephan Verosta (1980), S. 217; vgl. ferner die Entscheidung des österr. Verfassungsgerichtshofs vom 9. 3. 1978, EuGRZ 1978, S. 245 ff.

[51] Zum Beurteilungsspielraum vgl. *A. Bleckmann,* Der Beurteilungsspielraum im Europa- und im Völkerrecht, EuGRZ 1979, S. 485 ff.; *C. C. Morrisson, Jr.,* Margin of Appreciation in European Human Rights Law, Revue des Droits de l'Homme 6 (1973), S. 263 ff.; *Cohen Jonathan* (Anm. 18), S. 481 ff.; *R. Higgins,* Derogations under Human Rights Treaties, BYIL 48 (1976/77), S. 281, 313 ff.; *C. Feingold,* The Little Red Schoolbook, Human Rights Review 3 (1978), S. 21 ff.

den jeweiligen Ländern wirksamen Kräften« begründet. Der Spielraum erstreckt sich in erster Linie auf die Notwendigkeit bzw. Angemessenheit von Einschränkungen. Die innerstaatlichen Behörden sollen frei bleiben in der Wahl der Maßnahmen, die sie in den von der Konvention erfaßten Bereichen zu ergreifen für geeignet halten. Im Fall *Klass*[52] heißt es, es sei nicht Sache des Gerichts, die Bewertung der staatlichen Behörden in der Frage der Ausgestaltung des Überwachungssystems durch irgendeine andere Bewertung dessen, was die beste Politik auf diesem Gebiet sein könnte, zu ersetzen. Neben dem Spielraum bei der Wahl der angemessenen Methoden und Maßnahmen wird aber auch ein Spielraum im Hinblick auf die Subsumtion eines Sachverhalts unter die Schrankenklauseln der Konvention anerkannt. Ob ein festgestellter Tatbestand eine Gefährdung der Moral im Sinne der Konvention bedeutet, ist nach den Feststellungen des Gerichtshofs im *Handyside*-Fall nicht einheitlich zu beurteilen. In der Regel seien die staatlichen Behörden besser in der Lage, die nach Zeit und Ort möglicherweise unterschiedlichen Moralanforderungen festzulegen. Entsprechendes gilt für die anderen in der Konvention aufgeführten unbestimmten Rechtsbegriffe der öffentlichen Ordnung, der internationalen Sicherheit usw. Im *Sunday Times*-Fall hat der Gerichtshof allerdings insoweit differenziert zwischen den eher »objektiv« zu bestimmenden Rechtsgütern der Konvention, wie z.B. Ansehen der Rechtsprechung, und den »subjektiven«, d.h. weit mehr national bestimmten, Begriffen wie z.B. »Moral«[53]. Das Ausmaß der europäischen Kontrolle differiert also danach, ob und inwieweit in größerem zwischenstaatlichem Rahmen objektivierbare Kriterien zur Auslegung eines Konventionsbegriffs vorhanden sind.

In der Literatur ist wiederholt auf die Unbestimmtheit dieses Konzepts hingewiesen worden, wobei allerdings zugleich eingestanden wird, daß der Begriff des Beurteilungsspielraums unverzichtbar sei[54]. Der Gerichtshof hat immer wieder als seine Aufgabe betont, die Einhaltung der Grenzen des Beurteilungsspielraums sicherzustellen. Die Konzeption des Beurteilungsspielraums liefert aber keine Kriterien dafür, wo diese Grenzen verlaufen. *Fawcett* ist der Auffassung, der Beurteilungsspielraum beruhe auf einem Mittelweg zwischen einer rein subjektiv national geformten Begriffsbestimmung durch die Staaten und einer rein objektiven, europaeinheitlichen durch die europäischen Organe[55]. Am Maßstab der demokratischen Gesellschaft sei zu entscheiden, ob die betreffende Beschränkung gerechtfertigt sei. Auch mit dieser Abgrenzung wird freilich nur das Problem anders definiert, nicht aber gelöst. Ähnliche Einwände gelten gegenüber der Theorie von der Vertretbarkeitskontrolle, die von der Kommission und einer Gerichtsminderheit favorisiert

---

[52] Beschwerde 5029/71, vgl. EuGRZ 1977, S. 419 ff.
[53] Kritisch dazu *Kelly* (Anm. 17), S. 8 f.
[54] Vgl. z.B. *Morrisson, Jr.* (Anm. 51), S. 284 ff.; *Feingold* (Anm. 51), S. 31 ff.
[55] A.a.O. (Anm. 34), S. 248.

worden ist[56]. Entscheidend wird der Beurteilungsspielraum hier durch einen *bona fide*-Maßstab bestimmt. Der Gerichtshof hätte demnach nicht zu fragen, ob die fragliche Beschränkung tatsächlich notwendig war, sondern ob die nationalen Behörden venünftige Gründe hatten, dies anzunehmen. Ob solche vernünftigen Gründe angenommen werden können, hängt wiederum von dem Prüfungsmaßstab ab. Stellt man nicht auf rein subjektive Verschuldenskriterien ab – wofür die Konvention keinerlei Anhaltspunkte liefert –, muß auf objektive Maßstäbe zurückgegriffen werden, die letztlich wiederum nur aus den gemeinsamen Grundsätzen einer »demokratischen Gesellschaft« abgeleitet werden können.

Bleibt unter dieser Umständen nur der Rückgriff auf das Letztentscheidungsrecht des Gerichtshofs nach dem Motto »Der nationale Beurteilungsspielraum gilt soweit, wie der Gerichtshof es für richtig befindet«[57]? Man wird diese Frage weder mit einem Ja noch mit einem Nein beantworten können. Zur Bestimmung des nationalen Handlungsspielraums ist zunächst auf die oben dargestellten Methoden der Auslegung der Schrankenklauseln zurückzugreifen. Nationale Handlungsspielräume ergeben sich daraus, daß die Konvention keine umfassende Vereinheitlichung der Rechtsordnungen bewirken will, sondern lediglich auf der Grundlage gemeinsamer Überzeugungen von einer demokratischen Gesellschaft eine einheitliche Menschenrechtsminimalordnung, beschränkt auf den jeweiligen Integrationsstand, begründet. Im Rahmen der Strukturen einer demokratischen Gesellschaft verbleibt Raum für nationale Verschiedenheiten. Dies unterscheidet die »Zweistufigkeit« der Menschenrechtskonvention wesentlich von der einheitlichen EG-Rechtsordnung[58]. Dieser Freiraum läßt sich nicht allgemein umschreiben. Er differiert je nach beschränktem Recht und umfaßt grundsätzlich sowohl die Auslegung der in der Konvention zugelassenen Beschränkungsgründe als auch die Abwägung zwischen den kollidierenden Rechtsgütern. Man wird also zu unterscheiden haben zwischen einem autonom zu bestimmenden »Begriffskern« der öffentlichen Ordnung, Moral, nationalen Sicherheit usw. und einem national geprägten »Begriffshof«. Entsprechendes gilt für das Urteil über die Notwendigkeit bestimmter Maßnahmen zum Schutze eines gefährdeten Rechtsguts. Der Handlungs- oder Ermessensspielraum der nationalen Instanzen ergibt sich also in derartigen Fällen daraus, daß die Konvention das fragliche Handeln zuläßt, weil entgegenstehende europäische Standards nicht festgestellt werden können. Es bedarf keines Hinweises auf die größere Sachnähe der nationalen Behörden oder eine angebliche Subsidiarität der europäischen Kontrolle. Vergleichbar den Grundsätzen des nationalen Rechts über den Ermessensspielraum der Verwaltung trägt die Konvention lediglich der

---

[56] Vgl. z.B. Fall *Sunday Times* (Anm. 10), S. 50 ff., ähnlich *Bleckmann,* (Anm. 51).

[57] *Morrisson, Jr.* (Anm. 51), S. 283 ff.

[58] Vgl. *Mosler* in: Grundrechtsschutz in Europa (Anm. 39), S. 126.

Tatsache Rechnung, daß die Verantwortung für die Gesamtpolitik bei den nationalen Instanzen liegt. *Bleckmann* leitet daraus eine Beschränkung der Gerichtskontrolle ab, wenn es sich um vitale Interessen des Staates oder um Materien handle, die der Natur der Sache nach, wie z.B. die innerstaatliche Organisations- und Kompetenzverteilung in den *domaine réservé* der Staaten fielen[59]. Als eigenständiger Maßstab verwandt ist dieser Gesichtspunkt freilich nicht unproblematisch, da er die Reichweite der Konvention *a priori* durch das unbestimmte Konzept der vitalen Staatsinteressen begrenzt. Die Konvention will aber auch und gerade dort, wo möglicherweise vitale Staatsinteressen berührt sein können, eine einheitliche Menschenrechtsordnung begründen (vgl. Art. 15 EMRK). Sinnvoller ist es daher, auf die oben gekennzeichneten Methoden der Ermittlung europäischer Standards abzustellen.

Vom Handlungsspielraum, der durch das materielle Recht der Konvention vorgegeben ist, kann der eigentliche Beurteilungsspielraum unterschieden werden. Die Entscheidung über beschränkende Maßnahmen nach der Konvention erfordert nicht selten subjektive Wertungsentscheidungen, die nur begrenzt objektivierbar und international überprüfbar sind. Vergleichbar dem Gedanken der situationsgebundenen Prüfungsentscheidung des nationalen Rechts können Entscheidungen nationaler Instanzen eben wegen ihrer Abhängigkeit von einer spezifisch nationalen Situation nur begrenzt überprüft werden. So verhält es sich bei der Ergreifung von Notstandsmaßnahmen »in dem Umfang, den die Lage unbedingt erfordert«[60]. Entsprechendes gilt aber auch allgemein für die Wertung eines Sachverhalts im Hinblick auf einen Schrankenvorbehalt der Konvention. Ob eine beantragte Demonstration Gewaltanwendung provozieren kann, ist auf der Grundlage einer umfassenden Bewertung der Gesamtlage zu beurteilen, zu der die nationalen Instanzen auf Grund ihrer Sachnähe und Kenntnisse der nationalen Verhältnisse besser geeignet sind. Die europäischen Organe haben lediglich festzustellen, ob die nationalen Instanzen die in einer demokratischen Gesellschaft gültigen Kriterien für das Vorliegen einer Störung der öffentlichen Ordnung, Sicherheit, Gesundheit usw. richtig angewendet haben. Der Grund für die Zurückhaltung der europäischen Organe ist hier nicht die von der Konvention zugelassene Bandbreite möglicher staatlicher Reaktionen, sondern die »Nichtvertretbarkeit« der Beurteilung einer bestimmten Situation oder eines Sachverhalts. Diese Nichtvertretbarkeit kann auf der Unwiederholbarkeit der Entscheidungssituation (Notlage), auf mangelnder Sachkunde (technische Feststellung), auf der Komplexität von Entscheidungen (gesetzgeberische

---

[59] *Bleckmann* (Anm. 51), S. 493 ff.

[60] Vgl. dazu *T. Stein,* Die Außerkraftsetzung von Garantien menschenrechtlicher Verträge, Bericht, V. Internationales Kolloquium über die Europäische Menschenrechtskonvention, Frankfurt vom 9.–12. 4. 1980; *J. Velu,* Le contrôle des organes prévus par la Convention des Droits de l'Homme sur le but, le motif et l'objet des mesures d'exception dérogeant à cette convention, in: Mélanges offerts à Henri Rolin (1964), S. 462 ff.

Prognosen) oder auf ihrer Situationsgebundenheit (Anforderungen zum Schutze der Moral) beruhen[61]. In der Praxis lassen sich allerdings Handlungs- und Beurteilungsspielraum der Staaten im hier verstandenen Sinne nur schwer trennen. Dennoch scheint mir die Unterscheidung für eine Herausarbeitung verschiedener Kategorien von Handlungs- und Beurteilungsspielräumen wichtig zu sein. Die im nationalen Recht erarbeiteten Fallgruppen können dazu wesentlich beitragen. Auf die Konturen einer Dogmatik des Beurteilungsspielraums im europäischen Recht kann hier nicht näher eingegangen werden. Festzuhalten bleibt, daß der Begriff der demokratischen Gesellschaft den Staaten Handlungsspielräume gibt, die der europäischen Kontrolle Grenzen setzen.

[61] Vgl. *Bleckmann* (Anm. 51), S. 493 ff.

Europäische Gemeinschaften und internationale Streitbeilegung

MEINHARD HILF

I. Einleitung *

Hermann Mosler hat sich mit der Einordnung der Europäischen Gemeinschaften in den Kreis der traditionellen Völkerrechtssubjekte und der Fortentwicklung und Wirksamkeit der gerichtlichen Streitbeilegung internationaler Konflikte wiederholt auseinandergesetzt [1]. Für ihn handelt es sich dabei nicht um statische Zustände, sondern um zu beobachtende evolutionäre Vorgänge.

Die Gemeinschaften wachsen ungeachtet ihres inneren Zustands derzeit verstärkt in die sie umgebende Völkerrechtsordnung hinein. Drittstaaten und internationale Organisationen beginnen, den Gemeinschaften einen ihren Kompetenzen angemessenen Platz in der Völkerrechtsgemeinschaft zuzuerkennen. Diese Entwicklung führt zwangsläufig zu einer *Verknüpfung* der angesprochenen Fragenkreise: Die Gemeinschaften treten zunehmend in ihren völkerrechtlichen Beziehungen an die Stelle ihrer Mitgliedstaaten bzw. neben diese. Der stetige Ausbau der internen Kompetenzen führt zu parallelen Außenkompetenzen. Bei dem Versuch, in der Falkland-Krise durch gemein-

---

* Wird im Text in Abgrenzung zu den drei Europäischen Gemeinschaften verkürzt von der »Gemeinschaft« gesprochen, so steht diese für die Europäische Wirtschaftsgemeinschaft.

[1] Vgl. *H. Mosler*, Die Entstehung des Modells supranationaler und gewaltenteilender Staatenverbindungen in den Verhandlungen über den Schuman-Plan, in: Festschrift Hallstein (1966), S. 355 ff., sowie *derselbe*, Die Erweiterung des Kreises der Völkerrechtssubjekte, Zeitschrift für ausländisches öffentliches Recht und Völkerrecht (ZaöRV) 22 (1962), S. 1 ff.; zur internationalen Streitbeilegung siehe Judicial Settlement of International Disputes, International Court of Justice, Other Courts and Tribunals, Arbitration and Conciliation (Beiträge zum ausländischen öffentlichen Recht und Völkerrecht, Bd. 62) (1974) sowie zuletzt *derselbe*, Judgments of International Courts and Tribunals, in: Bernhardt (Hrsg.), Encyclopedia of Public International Law, Instalment 1 (1981), S. 111 ff.

sames Handeln zur Streitbeilegung beizutragen [2], bzw. bei der Verteidigung ihrer Handelsinteressen im Rahmen des GATT u.a. mit der Einleitung eines umfassenden Verfahrens gegen Japan [3] haben die Gemeinschaften zuletzt gezeigt, daß sie im Außenbereich handlungsfähiger oder -williger erscheinen als im Innenbereich. Die *Völkerrechtspersönlichkeit* der Gemeinschaften – *Hermann Mosler* würde die umfassenden Begriffe der Völkerrechtsfähigkeit oder -subjektivität wohl noch vorziehen [4] – findet zunehmend Anerkennung und ihren Ausdruck in einem dichten Netz bilateraler und multilateraler Verträge [5] sowie in verschiedenen Formen der Mitwirkung in anderen internationalen Organisationen [6].

Die Aufnahme der genannten völkerrechtlichen Rechtsbeziehungen bringt die Gemeinschaften in die Lage, auf unausweichliche Streitigkeiten über ihre Rechte und Pflichten reagieren zu müssen.

Sind sie bereit, sich dabei einer übergeordneten Gerichtsbarkeit zur Regelung der völkerrechtlichen Streitigkeiten zu unterwerfen, oder streben sie eher nach einer nichtgerichtlichen Streitbeilegung? Fürchten sie Juridiktionskonflikte mit der innergemeinschaftlichen Gerichtsbarkeit? Oder sollten sie eher Vorreiter gerichtlicher Streitbeilegung auch völkerrechtlicher Streitigkeiten sein? Gibt es unter ihren Mitgliedstaaten oder im Recht internationaler Organisationen Verhaltensmuster, an denen sie sich ausrichten können? Oder enthalten bereits die Verträge zur Gründung der Europäischen Gemeinschaften Anhaltspunkte für die Beilegung von völkerrechtlichen Streitigkeiten?

1. Aussagen der Gemeinschaftsverträge

Die Gemeinschaftsverträge enthalten keine ausdrückliche Aussage über Art und Weise internationaler Streitbeilegung. In der Präambel des stark von politischen Impulsen getragenen EGKS-Vertrages rufen sich die Mitgliedstaaten noch zu »angemessenen Anstrengungen« für die Erhaltung des Weltfrie-

---

[2] Vgl. Verordnung (EWG) Nr. 877/82 des Rates vom 16. 4. 1982 zur Aussetzung der Importe aller Erzeugnisse mit Ursprung in Argentinien, ABl. Nr. L 102/1 vom 16. 4. 1982.

[3] Vgl. zum Beschluß des Rates vom 23. 3. 1982 den Bericht »Handelskonflikte werden für Gatt zur Zerreißprobe«, Süddeutsche Zeitung Nr. 85 vom 14. 4. 1982, S. 21.

[4] Vgl. u. a. unter dem Stichwort Völkerrechtsfähigkeit, in: Strupp-Schlochauer, Wörterbuch des Völkerrechts Bd. 3 (1962), S. 665 f.

[5] *H. Krück*, Völkerrechtliche Verträge im Recht der Europäischen Gemeinschaften (Beiträge zum ausländischen öffentlichen Recht und Völkerrecht, Bd. 70) (1977), S. 182 ff., listet bis Ende 1976 389 bi- und multilaterale Verträge auf; das gemeinschaftseigene CELEX-Dokumentationssystem weist von 1977 bis März 1982 weitere 311 Verträge aus.

[6] Vgl. das von der Kommission der Europäischen Gemeinschaften periodisch herausgegebene Heft »La Communauté européenne, les organisations internationales et les accords multilatéraux« (Stand: 1. 1. 1980).

dens auf. Im EWG-Vertrag (EWGV) wird an gleicher Stelle das Ziel der Handelspolitik nur noch mit der »Beseitigung der Beschränkungen im zwischenstaatlichen Wirtschaftsverkehr« angegeben; immerhin enthält der siebente Erwägungsgrund einen Hinweis auf die »Grundsätze der Satzung der Vereinten Nationen«, allerdings nur im Hinblick auf die Förderung des Wohlstandes der überseeischen Länder. Aus einer Gesamtschau der den Gemeinschaften zugewiesenen materiellen Einzelkompetenzen im auswärtigen Bereich zum Abschluß von Abkommen und zur Pflege der Zusammenarbeit mit Drittstaaten und internationalen Organisationen (u.a. Art. 113, 229 ff. EWGV) läßt sich eine Offenheit und Grundentscheidung der Gründer für eine intensive internationale Zusammenarbeit ableiten[7]. Dies schließt sicherlich, wie unten näher zu zeigen ist, die Verpflichtung der Gemeinschaften zu einer friedlichen Beilegung etwaiger völkerrechtlicher Streitigkeiten mit ein. Eine genauere Festlegung auf bestimmte Mittel der Streitbeilegung fehlt indes. Die Rechtswahrungsfunktion des Gerichtshofs der Europäischen Gemeinschaften bezieht sich im wesentlichen nur auf die Auslegung und Anwendung der Gemeinschaftsverträge (Art. 164 EWGV), er wird nur in dem durch die Gemeinschaftsrechtsordnung gesteckten Rahmen tätig[8]. Das die Gemeinschaften bindende Rechtsstaatsprinzip, das grundsätzlich eine gerichtliche Entscheidung von Streitigkeiten mit einschließt, ist in den Verträgen in bezug auf den Innenrechtsraum der Gemeinschaften verwirklicht. Hier kommt es zunehmend zur Auslegung völkerrechtlicher Verträge durch den Gerichtshof vor allem im Rahmen des Vorlageverfahrens[9]. Auch kann der Gerichtshof mit der Erstattung von Gutachten über die Vereinbarkeit beabsichtigter Abkommen mit dem Vertrag befaßt werden (Art. 228 Ziff. 1 Abs. 2 EWGV). Diese Verfahren betreffen aber insgesamt nur Streitfälle, die die Anwendung völkerrechtlicher Verträge im Rechtsraum der Gemeinschaft betreffen. Über eine entsprechende Ausstrahlung des Rechtsstaatsprinzips auch in den Bereich der völkerrechtlichen Außenbeziehungen hinein sagen die Verträge indessen nichts aus. Zwar hält Art. 181 EWGV einen möglichen Zugang zum Gerichtshof auch für völkerrechtliche Streitigkeiten bereit, falls es der Gemeinschaft gelingen sollte, in einem »öffentlich-rechtlichen«, d.h. unter Umständen auch völkerrechtlichen, Vertrag eine Schiedsklausel zugunsten der Zuständigkeit des Gerichtshofs zu vereinbaren. Nur ist dies bisher der Gemeinschaft – von einem Sonderfall abgesehen[10] – nicht gelungen.

---

[7] Vgl. zur »Internationalisierung der Verfassung«, die die Mitgliedstaaten ebenso wie die Gemeinschaften betrifft, die Hinweise bei *U. Everling,* Sind die Mitgliedstaaten der Europäischen Gemeinschaften noch Herren der Verträge?, oben S. 173 ff.

[8] EuGH 30. 4. 1974 – *Haegemann,* Rs. 181/73 – Slg. 1974, S. 449, 460 (Rdnr. 2/6).

[9] Vgl. dazu unten die Übersicht bei *P. Pescatore,* Die Rechtsprechung des Europäischen Gerichtshofs zur innergemeinschaftlichen Wirkung völkerrechtlicher Abkommen, S. 661 ff.

[10] Vgl. zum Assoziationsabkommen mit Griechenland unten S. 403.

Eine entsprechende Unterwerfung ist von vornherein weder von einem völkerrechtlichen Partner zu erwarten wie auch nicht von der Gemeinschaft im Hinblick auf die Gerichtsbarkeit eines ihrer Vertragspartner [11].

Eine gewisse Erweiterung über den Rechtsraum der Gemeinschaften hinaus deutet Art. 182 EWGV an. Hier wird der Gerichtshof auch für Streitigkeiten der Mitgliedstaaten untereinander für zuständig erklärt. Die entsprechende schiedsvertragliche Zuständigkeitsvereinbarung darf sich indes nicht auf beliebige Streitigkeiten zwischen den Mitgliedstaaten beziehen, sondern nur auf solche, die »mit dem Gegenstand dieses Vertrages in Zusammenhang« stehen. Art. 182 EWGV erlaubt einen Zugriff der Mitgliedstaaten auf die Zuständigkeit des Gerichtshofs. Dieser kann im Wege der Organleihe von den Mitgliedstaaten zur Streitbeilegung herangezogen werden. Die Mitgliedstaaten können aber auch den Gerichtshof damit betrauen, im Falle einer Streitigkeit im Rahmen eines zwischen den Mitgliedstaaten geschlossenen Abkommens das zur Regelung des Falles zuständige Schiedsgericht zu berufen [12].

Die Mitgliedstaaten haben von der Möglichkeit der Organleihe vor allem in bezug auf Abkommen im Rahmen des Art. 220 EWGV mehrfach Gebrauch gemacht [13] und so zu erkennen gegeben, daß sie in diesem weit abzusteckenden Rahmen eine gerichtliche Streitbeilegung in bezug auf die ent-

---

[11] Entsprechend schließt das Bundesverfassungsgericht (BVerfG) aus dem Sinn und Zweck der Ermächtigung nach Art. 24 Abs. 1 GG, die auf eine Übertragung von Hoheitsrechten auf funktionsfähige zwischenstaatliche Einrichtungen gerichtet sei, daß hinsichtlich der organisatorischen Gestaltung des Rechtsschutzes dem Gesetzgeber ein weites Ermessen zustehe: ein verfassungsrechtliche Pflicht, stets einen Rechtsschutz durch innerstaatliche (deutsche) Gerichte zu gewährleisten, bestehe nicht: BVerfG 23. 6. 1981, »Eurocontrol«, Deutsches Verwaltungsblatt (DVBl.) 97 (1982), S. 186.

[12] So in Art. 29 Abs. 2 des Übereinkommens vom 19. 4. 1972 über die Gründung eines Europäischen Hochschulinstituts, Amtsblatt der Europäischen Gemeinschaften (ABl.) Nr. C 29/1 vom 9. 2. 1976.

[13] Vgl. die Zusatzprotokolle vom 3. 6. 1971 zu den Übereinkommen über die gerichtliche Zuständigkeit und Vollstreckung gerichtlicher Entscheidungen sowie über die gegenseitige Anerkennung von Gesellschaften und juristischen Personen, Bundesgesetzblatt (BGBl.) 1972 II, S. 845 und 857; in dem Internen Abkommen vom 20. 11. 1979 zur Durchführung des Lomé II-Abkommens (ABl. Nr. L 347/206 vom 22. 12. 1980) haben die Mitgliedstaaten die Zuständigkeit des Gerichtshofs für alle Streitigkeiten vereinbart, die sich zwischen ihnen ergeben; das Übereinkommen über das Europäische Patent für den Gemeinsamen Markt vom 15. 12. 1975 überträgt ebenfalls dem Gerichtshof die einheitliche Auslegung (ABl. Nr. L 17/1 vom 26. 1. 1976); dem Übereinkommen über das auf vertragliche Schuldverhältnisse anzuwendende Recht vom 19. 6. 1980, dem noch nicht alle Mitgliedstaaten beigetreten sind, ist bisher kein Protokoll über die Zuständigkeit des Gerichtshofs beigefügt worden, obgleich auch für diese Konvention die einheitliche Auslegung die Grundlage für ihre Wirksamkeit ist: vgl. die gemeinsame Erklärung zur weiteren Prüfung dieser Frage in ABl. Nr. L 266/17 vom 9. 10. 1980. – Vgl. im einzelnen *I. Schwartz*, Übereinkommen zwischen den EG-Staaten: Völkerrecht oder Gemeinschaftsrecht, in: Festschrift Grewe (1981), S. 551 ff.

sprechenden völkerrechtlichen Rechtsbeziehungen wünschen. Eine Rechtspflicht der Mitgliedstaaten zum Abschluß der genannten Schiedsvereinbarungen läßt sich aus Art. 182 EWGV nicht ableiten, auch nicht aus dem Grundsatz der Gemeinschaftstreue, der entsprechende Verhaltenspflichten zur Wahrung der Rechtseinheit zwischen dem Gemeinschaftsrecht und dem es umgebenden Völkervertragsrecht der Mitgliedstaaten untereinander begründen könnte. Der Annahme einer solchen Rechtspflicht steht insbesondere Art. 219 EWGV entgegen, der ausdrücklich die Mitgliedstaaten nur verpflichtet, »Streitigkeiten über die Auslegung oder Anwendung dieses Vertrages nicht anders als hierin vorgesehen« zu regeln. Die von den Mitgliedstaaten bisher abgeschlossenen Schiedsvereinbarungen auf der Grundlage des Art. 182 EWGV bilden dennoch für die Gemeinschaft selbst einen Teil des Verhaltensmusters, daß die Mitgliedstaaten insgesamt im Hinblick auf die Wahl der Streitbeilegungsmittel in bezug auf ihre Völkerrechtsbeziehungen bieten.

2. Verhaltensmuster der Mitgliedstaaten

Sind die Gemeinschaften vor die Frage gestellt, wie sie ihre völkerrechtlichen Streitigkeiten oder ob sie diese gerichtlich beilegen sollen, so wäre ein eindeutiges entsprechendes Verhaltensmuster der Mitgliedstaaten nicht ohne Bedeutung. Die Gemeinschaften haben wesentliche verfassungsrechtliche Grundüberzeugungen der Mitgliedstaaten zu beachten, soweit sich dies mit Ziel und Struktur der Gründungsverträge in Einklang bringen läßt[14]. Sollte sich in den Rechtsordnungen der Mitgliedstaaten eine durchgängige verfassungsrechtliche Pflicht nachweisen lassen, völkerrechtliche Streitigkeiten vorwiegend gerichtlich entscheiden zu lassen, so wäre davon auszugehen, daß die Mitgliedstaaten auch von den Gemeinschaften eine gerichtliche Streitbeilegung erwarten.

a) Bereits das *Grundgesetz* läßt eine entsprechende verfassungsrechtliche Pflicht nicht erkennen. Art. 24 Abs. 3 GG verpflichtet den Bund, »zur Regelung zwischenstaatlicher Streitigkeiten ... Vereinbarungen über eine allgemeine, umfassende, obligatorische, internationale Schiedsgerichtsbarkeit« beizutreten. Die Kumulierung der Vorbedingungen würde bei einer strengen Auslegung diesen Absatz vollständig seines realen Sinngehaltes entleeren. Es wird im Völkerrecht weder eine alle Staaten bzw. alle Streitigkeiten umfassende Gerichtsbarkeit geben. Eine weite Auslegung könnte allenfalls eine Grundlage für den erfolgten Beitritt der Bundesrepublik Deutschland zum Statut des Internationalen Gerichtshofs abgeben bzw. die Abgabe einer Unterwerfungserklärung nach Art. 36 Abs. 2 des Statuts fordern. Die Stärkung der internationalen Gerichtsbarkeit liegt in der Zielrichtung dieser Verfas-

[14] EuGH 17. 12. 1970 – *Internationale Handelsgesellschaft,* Rs. 11/70 – Slg. 1970, S. 1125, 1135.

sungsvorschrift, indes fordert sie kaum eine Vorreiterrolle, die über die übliche Praxis der anderen Staaten hinausgeht. Bei lediglich 47 Staaten, die sich unter teils sehr weitreichenden Vorbehalten der obligatorischen Gerichtsbarkeit des Internationalen Gerichtshofs unterworfen haben, kann von einer »allgemeinen« bzw. »umfassenden« Gerichtsbarkeit nicht gesprochen werden[15].

Ein weitergehender Ansatzpunkt läßt sich in bezug auf die von Art. 24 Abs. 1 und 2 GG zugelassene Übertragung von Hoheitsrechten gewinnen. Unterliegt innerstaatlich die Ausübung jeglicher den Einzelnen betreffenden Hoheitsgewalt der richterlichen Kontrolle (Art. 19 Abs. 4 GG), so steht eine Übertragung der Hoheitsgewalt auf zwischenstaatliche Einrichtungen von vornherein unter dem Vorbehalt einer wirksamen gerichtlichen Kontrolle[16]. Eine Weiterübertragung von Hoheitsrechten durch die Europäischen Gemeinschaften auf völkerrechtliche Einrichtungen setzt aus dieser Sicht ebenfalls eine richterliche Kontrolle durch Organe der jeweiligen Einrichtung voraus, sofern es den Gemeinschaften nicht gelingt, die Zuständigkeit des Gerichtshofs der Europäischen Gemeinschaften zu vereinbaren. In dem Entwurf eines Übereinkommens der Gemeinschaft mit der Schweiz über die Errichtung eines Europäischen Stillegunsfonds für die Binnenschiffahrt ist zur Kontrolle der dem Fonds zu übertragenden Befugnisse ein »Gericht des Fonds« mit umfassenden Zuständigkeiten vorgesehen[17]. Diese Wahrungspflicht hinsichtlich des rechtsstaatlichen Standards folgt aus dem Grundgedanken der Art. 164 ff. EWGV ebenso wie aus dem die Gemeinschaften bindenden Rechtsstaatsprinzip, das sich auf die gemeinsamen verfassungsrechtlichen Überlieferungen der Mitgliedstaaten gründet.

b) Läßt nach allem das Grundgesetz den zuständigen Verfassungsorganen grundsätzlich freie Hand, wie sie die Streitregelung in den von ihnen abzuschließenden völkerrechtlichen Verträgen ausgestalten wollen – mit Ausnahme der Fälle der Übertragung von Hoheitsgewalt –, so verwundert es kaum, wenn sich auch in den Verfassungen der *übrigen Mitgliedstaaten* keine strengeren Bindungen hinsichtlich der Wahl der Mittel zur friedlichen Streitbeilegung nachweisen lassen. Neben allgemeinen Bekenntnissen zum friedlichen Zusammenleben der Völker bzw. zur »Entwicklung einer internationa-

---

[15] Vgl. im Ergebnis *Th. Maunz*, in: Maunz/Dürig/Herzog/Scholz, Grundgesetz, Kommentar, Bd. 2 (Lieferung 1964), Art. 24, Rdnr. 30 f., sowie *O. Rojahn*, in: v. Münch, Grundgesetz-Kommentar, Bd. 2 (1976), Art. 24 Rdnr. 53 ff. Anderer Ansicht dagegen insbes. *W. Niedermeier*, Der Beitritt der Bundesrepublik Deutschland zum Statut des Internationalen Gerichtshof (Augsburger Schriften zum Staats- und Völkerrecht, Bd. 2) (1975), S. 78. – Zu den Unterwerfungserklärungen siehe International Court of Justice, Yearbook 1980/1981, S. 45.

[16] So zutreffend *C. Tomuschat*, in: Bonner Kommentar (Zweitbearbeitung 1981), Art. 24 Rdnr. 58.

[17] Vgl. Art. 41 ff. des Übereinkommens; der paraphierte Text ist im ABl. Nr. C 208/2 vom 3. 9. 1976 abgedruckt; zu dem Übereinkommen, das auf Grund des Gutachtens des EuGH vom 26. 4. 1977, Nr. 1/76, Slg. 1977, S. 741 ff., nicht in Kraft gesetzt werden konnte, vgl. unten S. 415.

len Rechtsordnung«[18] enthalten die Verfassungen keine nähere Bestimmung der Streitbeilegungswege. Lediglich in Art. 29 Abs. 2 der irischen Verfassung wird das »Festhalten am Grundsatz der friedlichen Erledigung internationaler Streitfragen durch schiedsrichterliche oder richterliche Entscheidung auf internationaler Ebene« bekräftigt. Diese Aussage geht weit über die Festlegung in Art. 24 Abs. 3 GG hinaus. Sie ist indes in der Regelungsintensität begrenzter, da sie keine Pflicht begründet, sondern nur einen Grundsatz bekräftigt, der nicht in allen Fällen beachtet werden muß.

Für die Europäischen Gemeinschaften ist insgesamt keine gemeinsame verfassungsrechtliche Überlieferung der Mitgliedstaaten erkennbar, die sie auf bestimmte Arten der internationalen Streitbeilegung festlegen würde. Die Völkerrechtspraxis der Mitgliedstaaten könnte dagegen ein aussagekräftiges Verhaltensmuster ausweisen, dem sich die Gemeinschaften kaum entziehen könnten.

c) Aus der gemeinsamen Mitgliedschaft der Mitgliedstaaten in den Vereinten Nationen lassen sich Hinweise für völkerrechtliche Pflichten entnehmen, die auch für die Gemeinschaften Bedeutung haben. Die in Art. 2 Ziff. 3 der Charta verankerte Pflicht zur friedlichen Streitbeilegung sowie die in Art. 33 Abs. 1 der Charta aufgefächerte Pflicht zum Bemühen um eine Streitbeilegung durch verschiedene Mittel, die von dem Verhandlungsweg bis hin zur gerichtlichen Entscheidung reichen, dürften als allgemeiner Grundsatz des Völkerrechts anerkannt sein und insoweit auch unmittelbar die Gemeinschaften binden. Darüber hinaus sind alle Mitgliedstaaten dem Statut des Internationalen Gerichtshofs beigetreten. Sie haben jedoch nicht alle die fakultative Unterwerfungserklärung nach Art. 36 Abs. 2 des Statuts abgegeben[19].

Überblickt man die Beteiligungs*praxis* der Mitgliedstaaten hinsichtlich der multilateralen Schiedsvertragswerke, so wird eine generelle Zurückhaltung deutlich: dies gilt für das 1. Haager Abkommen zur friedlichen Erledigung internationaler Streitfälle vom 29. Juli 1899 (zwei Mitgliedstaaten[20]), für das 2. Haager Abkommen vom 18. Oktober 1907 (sieben weitere Mitgliedstaaten[21]), für die Genfer Generalakte vom 28. April 1949 (vier Mitgliedstaaten[22]) sowie insbesondere für das Europäische Übereinkommen vom

---

[18] Art. 58 Abs. 2 der niederländischen Verfassung.

[19] Nach dem Stand vom 31. 7. 1981 haben lediglich Belgien, Dänemark, Luxemburg, die Niederlande und das Vereinigte Königreich eine Unterwerfungserklärung abgegeben: vgl. Anm. 15.

[20] Griechenland und Italien.

[21] Belgien, Dänemark, Bundesrepublik Deutschland, Frankreich, Luxemburg, Niederlande und Vereinigtes Königreich; Irland ist trotz seiner oben angeführten Verfassungsbestimmung nicht beigetreten.

[22] Belgien, Dänemark, Luxemburg und Niederlande; zur Beteiligung Frankreichs und des Vereinigten Königreichs vgl. *F. A. Freiherr von der Heydte,* General Act for the Pacific Settlement of International Disputes, in: Bernhardt (Hrsg.), Encyclopedia of Public International Law, Instalment 1, (1981), S. 62 ff.

29. April 1957 zur friedlichen Beilegung von Streitigkeiten (sieben Mitgliedstaaten[23]).

Der Umfang der Beteiligung der Mitgliedstaaten an den angeführten Vertragswerken läßt für die Europäischen Gemeinschaften kaum Rückschlüsse zu. Ein Beitritt steht ohnehin jeweils nur Staaten offen. Selbst wenn den Gemeinschaften in ihrer Sonderstellung ein Beitrittsrecht eingeräumt würde, wären sie auf Grund der unterschiedlichen Haltung ihrer Mitgliedstaaten kaum in der Lage, die entsprechenden Beitritte zu vollziehen. Die Gemeinschaften würden die Zurückhaltung der Mitgliedstaaten gegenüber allgemeinen Schiedsverträgen teilen.

Im Gegensatz zur Praxis der Mitgliedstaaten, allgemeine Schiedsverträge zurückhaltend zu beurteilen, ist auf zahlreiche Verträge der Mitgliedstaaten hinzuweisen, die besondere Klauseln zur Streitbeilegung einschließen. Dabei dürfte eine Streitbeilegung im Wege der Verhandlung oder anderer nichtgerichtlicher Mittel im Vordergrund stehen. Gerichtliche Streitbeilegungsklauseln sind am ehesten in Verträgen geringerer Bedeutung zu beobachten[24]. Sollte sich in bezug auf einen Vertragsgegenstand, der auch die Gemeinschaften berührt, feststellen lassen, daß alle Mitgliedstaaten sich für eine (schieds-)gerichtliche Streitbeilegung in den entsprechenden bi- oder multilateralen Verträgen entschieden haben, so läge damit ein Verhalten vor, das auch auf die Gemeinschaften zurückwirken würde. Sind alle Mitgliedstaaten der Europäischen Konvention zum Schutze der Menschenrechte und Grundfreiheiten (EMRK) beigetreten und haben sie sich der obligatorischen Gerichtsbarkeit des Europäischen Gerichtshofs für Menschenrechte unterworfen oder haben sie alle – bis auf Griechenland – die Individualbeschwerde zur Europäischen Kommission für Menschenrechte zugelassen, so kann dieses übereinstimmende Verhalten die Gemeinschaften zumindest veranlassen, die von ihnen ausgeübte Hoheitsgewalt gleichermaßen den Rechtsschutzorganen der Konvention zu unterwerfen. Der Hinweis des Gerichtshofs der Europäischen Gemeinschaften im *Nold*-Fall (1974)[25] auf verschiedene internationale Verträge,

---

[23] Es fehlen Frankreich, Griechenland und Irland; die übrigen Mitgliedstaaten haben zum Teil in gesonderten Erklärungen die obligatorische Schiedsgerichtsbarkeit (Kapital III) ausgeschlossen: vgl. *K. Ginther,* European Convention for the Peaceful Settlement of Disputes, in: Bernhardt (Hrsg.), Encyclopedia of Public International Law, Instalment 1 (1981), S. 57.

[24] Vgl. – auch zum folgenden – *H. v. Mangoldt,* Die Schiedsgerichtsbarkeit als Mittel internationaler Streitschlichtung (Beiträge zum ausländischen öffentlichen Recht und Völkerrecht, Bd. 63) (1974), S. 54.

[25] EuGH 14. 5. 1974 – *Nold,* Rs. 4/73 – Slg. 1974, S. 491: Nach dem Wortlaut der Entscheidung (Rdnr. 13) scheint der Gerichtshof die Mitgliedschaft *aller* Mitgliedstaaten nicht für erforderlich zu halten, solange diese wenigstens »am Abschluß beteiligt« waren. Hierbei könnte es sich auch schon um eine Beteiligung an der Ausarbeitung und um die Unterzeichnung handeln. »Abschluß« dürfte indes im Gegensatz zum späteren Beitritt die rechtswirksame Teilnahme an den betreffenden internationalen Verträgen von Anfang an meinen.

namentlich die EMRK, mag sich im konkreten Zusammenhang ausschließlich auf deren *materiellen* Gehalt und deren Ausdruck als »gemeinsame Verfassungsüberlieferungen der Mitgliedstaaten« bezogen haben. Er dürfte aber ebenso in verfahrensrechtlicher Hinsicht die Gemeinschaften veranlassen, sich vergleichbaren Streitbeilegungs- bzw. Rechtsschutzverfahren zu unterwerfen, sofern dies im jeweiligen Vertragszusammenhang möglich ist.

In der sonstigen Vertragspraxis der Mitgliedstaaten dürfte es schwer sein, übereinstimmende Verhaltensmuster festzustellen. Sollte sich ein solcher Nachweis für oder gegen die Vereinbarung einer schiedsgerichtlichen Streitbeilegung in bestimmten Vertragskategorien (etwa Handels-, Luftverkehrs-, Fischerei- oder Niederlassungsverträge) nachweisen lassen, so dürfte diese Ausgangslage auch die Gemeinschaften veranlassen, in der von ihr zu gestaltenden Vertragspraxis auf die Vereinbarung übereinstimmender Regelungen hinzuwirken. Praxisvergleichende Untersuchungen dürften am ehesten bei Verträgen auf Übereinstimmungen stoßen, die in einem institutionellen Rahmen geschlossen werden (z.B. GATT, ICAO, IAEO). Eine grundsätzlich positive Einstellung zu einer gerichtlichen Streitbeilegung in bezug auf alle völkerrechtlichen Vertragsbeziehungen – wie sie etwa von der Schweiz vertreten wird[26] – läßt sich für keinen der Mitgliedstaaten der Gemeinschaften feststellen. Den Gemeinschaften steht aus dieser Sicht der gleiche Ermessensspielraum offen wie den Mitgliedstaaten, die je nach Vertragsgegenstand oder -partner eines der in Art. 33 Abs. 1 der Charta vorgesehenen Streitbeilegungsverfahren wählen.

3. Die Praxis internationaler Organisationen

Die Praxis internationaler Organisationen ist für die Gemeinschaften weit weniger aussagekräftig. Wenn schon die Mitgliedstaaten sich weitgehend freie Hand lassen zur Wahl der Streitbeilegungsmittel – sieht man einmal von den Fällen der Übertragung von Hoheitsbefugnissen und des Schutzes der Menschenrechte ab –, so ist damit die Praxis der von ihnen gegründeten internationalen Organisationen vorgezeichnet. Dennoch könnte eine ständige Praxis im Einzelfall (Sitzstaatsabkommen, Abkommen über finanzielle Hilfe[27]) zumindest Auswirkungen auf die Vertragspolitik der Gemeinschaften haben.

Die International Law Commission hat sich in ihren Vorarbeiten über Fragen der Verträge zwischen Staaten und internationalen Organisationen bzw. zwischen zwei oder mehreren internationalen Organisationen hinsichtlich der Schiedsgerichtsbarkeit sehr zurückhaltend gezeigt: In Art. 66 des

---

[26] Vgl. v. *Mangoldt* (Anm. 24), S. 49 f., sowie *F. Münch,* Zur schweizerischen Initiative für die friedliche Beilegung von Streitigkeiten in Helsinki und Montreux, 1973 und 1978, in: Festschrift Bindschedler (1980), S. 385, 387 ff.

[27] Vgl. näher bei *N. Wühler,* Arbitration Clause in Treaties, in: Bernhardt (Hrsg.), Encyclopedia of International Public Law, Instalment 1 (1981), S. 34 ff.

nach der ersten Lesung erstellten Entwurfs ist diese nur für bestimmte Streitigkeiten zwischen Staaten vorgesehen. Dagegen hält Art. 66 Ziff. 2 und 3 für Fälle, an denen internationale Organisationen beteiligt sind, lediglich ein formalisiertes Vermittlungsverfahren bereit. Die Europäische Wirtschaftsgemeinschaft hat zum Ausdruck gebracht, daß sie auch für die zuletzt genannten Fälle ein verbindliches Schiedsgerichtsverfahren für angemessen halte[28].

Zusammengefaßt bestehen für die Gemeinschaften kaum deduktive Vorgaben zur Gestaltung der Streitbeilegung in den von ihr zu schließenden Verträgen auf Grund der Gemeinschaftsverträge selbst, der gemeinsamen Rechtsüberzeugungen bzw. Vertragspraxis der Mitgliedstaaten oder auf Grund des bisher auf die internationalen Organisationen anwendbaren Rechts. Der folgende Nachweis der Vertragspraxis der Gemeinschaften wird allgemeine Aussagen über die Rolle der Gemeinschaften bei der Vereinbarung internationaler Streitbeilegungsverfahren und über rechtliche Folgerungen hieraus für den jeweiligen Vertrag bzw. seine Auslegung im Gemeinschaftsrecht ermöglichen.

II. Die Vertragspraxis der Europäischen Gemeinschaften

Die Untersuchung der völkerrechtlichen *Vertragspraxis* der Europäischen Gemeinschaften schränkt bereits die möglichen Verfahren der internationalen Streitbeilegung in mehrfacher Hinsicht ein. Nicht erfaßt werden Maßnahmen zur Regelung von Streitigkeiten, die in der Ausübung wirtschaftlichen Drucks bestehen. Die Gemeinschaften haben in der Afghanistan- und Polenkrise wirtschaftliche Sanktionen ergriffen sowie in der Falklandkrise mit der Unterbindung aller Importe aus Argentinien versucht, zur Beilegung entstandener Streitigkeiten beizutragen[29].

Vorliegend sollen die Verfahrenswege von Interesse sein, die im Rahmen bi- und multilateraler Verträge den Gemeinschaften offenstehen, um Streitigkeiten über die Vertragsauslegung und -anwendung beizulegen. Ausgeklammert bleiben dabei die überwiegend privatrechtlichen Verträge, über die in der Regel – wie ausdrücklich für die Verträge der Europäischen Investionsbank in Art. 29 Abs. 1 ihrer Satzung bestimmt – die zuständigen einzelstaatlichen Gerichte zu entscheiden haben[30].

---

[28] Vgl. United Nations, General Assembly, Doc. A/CN. 4/350 Add. 2 vom 23. 3. 1982, S. 10, sowie zum Entwurf Yearbook of the International Law Commission 1980, Bd. 2, Doc. A/CN. 4/SER. A/1980/Add. 1 (part 2), S. 64, 66 ff.

[29] Vgl. hierzu die Darstellungen bei *E. U. Petersmann,* Internationale Wirtschaftssanktionen als Problem des Völkerrechts und des Europarechts, Zeitschrift für vergleichende Rechtswissenschaft 80 (1981), S. 1 ff., und *M. Schröder,* Wirtschaftssanktionen der Europäischen Gemeinschaften gegegenüber Drittstaaten dargestellt am Beispiel des Iran-Embargos, German Yearbook of International Law 23 (1980), S. 111 ff.

[30] Art. 29 Abs. 2 der Satzung erlaubt es der Investitionsbank, in ihren Verträgen Schiedsverfahren vorzusehen, für die u. a. auch der Gerichtshof der Europäischen Ge-

Im Vordergrund stehen der wirtschaftlichen Ausrichtung der Gemeinschaften gemäß die nicht-gerichtlichen Verfahren der Streitbeilegung. Vereinbarungen über eine gerichtliche Streitbeilegung, sind bisher die Ausnahme. Sie sind eher in Verträgen aus jüngerer Zeit anzutreffen und bieten vor allem der rechtlichen Betrachtung Ansatzpunkte. Die Beteiligung der Gemeinschaften an multilateralen Abkommen *neben* den Mitgliedstaaten (gemischte Abkommen) wirft insbesondere in den Streitbeilegungsverfahren Sonderfragen der geteilten Verantwortlichkeit bzw. Verfahrensherrschaft auf, die zunächst weitgehend ausgeklammert werden sollen.

4. Nicht-gerichtliche Verfahren der Streitbeilegung

a) Zweiseitige Verträge

Die weit überwiegende Anzahl aller von den Gemeinschaften geschlossenen *zweiseitigen* Verträge sieht ausschließlich Konsultationen oder Verhandlungen zwischen den Vertragsparteien vor, um Streitigkeiten über die Auslegung oder Anwendung des Vertrages beizulegen. In den meisten Handels- oder Zusammenarbeitsabkommen der Europäischen Wirtschaftsgemeinschaft sind die Verhandlungen in den institutionellen Rahmen eines »gemischten Ausschusses« eingebunden. Dient dieser Ausschuß in einer Reihe von Abkommen nur als Forum für Verhandlungen, ohne daß der Ausschuß sich zu dem Streit oder einem etwaigen Verhandlungsergebnis zu äußern hätte, so ist dem Ausschuß in anderen Fällen die Befugnis übertragen, »Empfehlungen« an die Vertragsparteien zu richten, die sich auf die Verwaltung des Abkommens beziehen. Die »Empfehlungen« geben dabei dem Streitbeilegungsverfahren keine erkennbar erhöhte Wirksamkeit, da sich der Ausschuß aus Vertretern beider Vertragsparteien zusammensetzt, die Empfehlungen nur im gegenseitigen Einvernehmen abgeben (sog. »Vertragsorgane«). Gegen den Willen einer der Vertragsparteien kann keine bindende Auslegung oder Anwendung der Abkommen festgelegt werden. Der Gerichtshof der Europäischen Gemeinschaften hat im *Polydor*-Fall auf diese institutionelle »Schwäche« des Freihandelsabkommens der EWG mit Portugal[31] hingewiesen und daraus Folgerungen für die Auslegung des Abkommens abgeleitet: Während die Gemeinschaft über Verfahren oder Einrichtungen (*instruments*) verfüge, um die einheitliche Anwendung des Gemeinschaftsrechts sicherzustellen, sei in dem Abkommen mit Portugal kein vergleichbares Verfahren vereinbart. Bereits aus diesem Grund hätte die in Frage gestellte unmittelbare Anwendbarkeit einer Abkom-

---

meinschaften für zuständig erklärt werden kann. Vergleichbar sieht Art. 20 des Beschlußes des Rates vom 29. 6. 1976 über die Assoziation der überseeischen Länder und Gebiete mit der EWG (ABl. Nr. L 176/8 vom 1. 7. 1976) vor, daß Streitigkeiten der Länder und Gebiete mit Unternehmen einem Schiedsgerichtsverfahren unterworfen werden, das vom Rat erlassen bzw. bis dahin der Vergleichs- und Schiedsordnung der Internationalen Handelskammer unterstellt wird.

[31] EuGH 9. 2. 1982 – *Polydor,* Rs. 270/80 – hekt. S. 36 (Rdnr. 20).

mensbestimmung, die sich gleichlautend im EWG-Vertrag finde, verneint werden können [32].

Die aufgezeigte institutionelle Schwäche kann umfassenden Abkommen über gegenseitige Wirtschaftsbeziehungen angemessen sein. Ziel der Abkommen ist es, zum gegenseitigen Vorteil Beziehungen zu entwickeln und aufrechtzuerhalten. Dies setzt voraus, daß die Vertragspartner in der Lage bleiben, auf auftretende wirtschaftliche Schwierigkeiten etwa durch die Anwendung von Schutzklauseln angemessen zu reagieren. Nicht eine starre juristische Konfliktlösung, sondern der jeweils im gegenseitigen Einvernehmen ausgehandelte Kompromiß kann am ehesten Garant für das weitere reibungslose Funktionieren der Abkommen sein, in dem beide Parteien ihren erstrebten Vorteil finden.

b) Mehrseitige Verträge

Diese Ausgangslage trifft auch für die *mehrseitigen* Abkommen mit wirtschaftlicher Zielsetzung zu, an denen die Gemeinschaft beteiligt ist. Sowohl im GATT einschließlich der Sonderabkommen als auch in den internationalen Rohstoffabkommen geht es im Streitfalle um die Beurteilung wirtschaftlicher Sachverhalte mit Austauschcharakter, die sich kaum für eine gerichtliche bindende Entscheidung durch ein unabhängiges Organ eignen [33]. Dennoch können die zwischen zwei Vertragsparteien bilateral ausgehandelten Auslegungen Auswirkungen auf das Gesamtgeflecht der in dem multilateralen Abkommen verwobenen Interessen einer großen Anzahl von Staaten haben. Die Gemeinschaft sieht sich daher in der Regel in ihren multilateralen Abkommen in Streitbeilegungsverfahren eingebunden, die den Vertragsorganen mitunter eine einflußreiche Rolle zuweisen, ohne ihnen zugleich eine richterliche Funktion zuzuerkennen.

Im Vordergrund stehen derzeit die Verfahren im Rahmen des GATT. Die Gemeinschaft, die im Rahmen ihrer Kompetenzen die früher von den Mitgliedstaaten im Anwendungsbereich des GATT ausgeübten Befugnisse übernommen und einen Teil der ergänzenden GATT-Abkommen im eigenen

---

[32] Gleichlautend sehen Art. 14 Abs. 2 des Freihandelsabkommens vom 22. 7. 1972 (ABl. Nr. L 301/164 ff. vom 31. 12. 1972) und Art. 30 EWGV die Abschaffung bzw. das Verbot »mengenmäßiger Einfuhrbeschränkungen sowie aller Maßnahmen gleicher Wirkung« vor. Zur Streitfrage vgl. *J. V. Louis,* in: Le droit de la Communauté économique européenne, Bd. 12: *Louis/Brückner,* Relations extérieures (1980), S. 189, für den im Ergebnis die unmittelbare Anwendbarkeit zu bejahen gewesen wäre. Der Gerichtshof hat die Frage der unmittelbaren *Anwendbarkeit* nicht abschließend entscheiden müssen.

[33] Zu den besonderen Bedingungen der Wirksamkeit der nicht-integrierten internationalen Wirtschaftsorganisationen, die ein gerichtliches Streitbeilegungsverfahren weitgehend ausschließen, vgl. *G. Malinverni,* Le règlement des différends dans les organisations internationales économiques (Institut Universitaire de Hautes Etudes Internationales, Collection de Droit International, 3) (1974), S. 89 ff., 197 ff.

Namen abgeschlossen hat[34], hat eine eher zurückhaltende Haltung zu den im GATT-Rahmen bestehenden Streitbeilegungsverfahren erkennen lassen.

Art. XXII und XXIII GATT sehen zur Regelung von Meinungsverschiedenheiten in erster Linie Verhandlungen zwischen den betroffenen Parteien und in einem weiteren Stadium die Einschaltung der Vertragsparteien zur Prüfung der Streitfragen vor. Sie sind durch die Praxis der GATT-Vertragsparteien im Laufe der Zeit zu einem mehrstufigen Verfahren insbesondere mit der Einsetzung von Sondergruppen *(panels)* fortentwickelt worden[35]. Diese Verfahren dienen in erster Linie der Vermittlung und der Suche einvernehmlicher Lösungen, sie weisen aber dennoch durch die Einschaltung unabhängiger Sachverständiger und durch die Möglichkeiten des GATT-Rats, Empfehlungen, Weisungen oder Ermächtigungen auszusprechen, justizförmige Züge auf.

Die Entwicklung des GATT-Verfahrens ist im Hinblick auf seine Verrechtlichung durch den Streit der Verfechter entgegengesetzter Schulen geprägt. Die Vereinigten Staaten scheinen einen *legalistic approach* zu verfolgen[36] und haben entsprechend die überwiegende Zahl der Verfahren anhängig gemacht, nicht zuletzt auf Grund ihrer nationalen Handelsgesetzgebung, die Privaten einen gerichtlich durchsetzbaren Einfluß auf die von der Regierung zu betreibenden GATT-Verfahren einräumt[37]. Die Gemeinschaft scheint dagegen eher einem *non-legalistic approach* zuzuneigen. Dies kommt nicht nur anläßlich der Tokyo-Runde in der Initiative zum Ausdruck, an Stelle einer grundlegenden Reform zunächst die zuvor eingefahrenen Verfahrens-

---

[34] Grundlegend EuGH 12. 12. 1972 – *International Fruit Company,* Rs. 21–24/72 – Slg. 1972, S. 1219, 1228 f., sowie den Beschluß des Rates vom 10. 10. 1979 über den Abschluß der multilateralen Übereinkommen, die im Rahmen der Tokio-Runde ausgehandelt worden sind, ABl. Nr. L 71/1 vom 17. 3. 1980; aus der neueren Literatur vgl. *J. H. J. Bourgeois,* Le GATT et le Traité CEE, Diritto Comunitario e degli Scambi Internazionali 19 (1980), S. 31 ff.; *derselbe,* The Tokyo Round Agreements on Technical Barriers and on Government Procurement in International and EEC Perspective, Common Market Law Review 19 (1982), S. 5 ff.

[35] Vgl. die Übersichten bei *R. E. Hudec,* GATT Dispute Settlement after the Tokyo Round: An Unfinished Business, Cornell International Law Journal 13 (1980), S. 145 ff.; *D. E. de Kieffer,* GATT Dispute Settlements: A New Beginning in International and U.S. Trade Law, Northwestern Journal of International Law and Business 2 (1980), S. 317 ff.; *E. U. Petersmann,* International Governmental Trade Organizations – GATT and UNCTAD, in: International Encyclopedia of Comparative Law (Offprint 1981), S. 3, 24 ff.; *W. Meng,* Streitbeilegung im GATT, ZaöRV 41 (1981), S. 69 ff. mit Nachweisen aus der älteren Literatur.

[36] *Hudec,* a.a.O., S. 151.

[37] Gemäß Sect. 302 (a) des Trade Act 1974, in der geänderten Fassung des Trade Agreements Act 1979, kann jede interessierte Partei über den United States Trade Representative den Präsidenten ersuchen, im Hinblick auf die vorgetragenen Beschwerden Maßnahmen zu ergreifen: zum Verfahren, das Fristen vorsieht und gegebenenfalls zur Einleitung der GATT-Verfahren führen kann, vgl. im einzelnen *de Kieffer* (Anm. 35), S. 327 ff.

praktiken informell festzuschreiben[38], sondern auch in der Zurückhaltung (die im Einzelfall aber auch auf die Uneinigkeit der Mitgliedstaaten zurückgeführt werden kann!), das GATT-Verfahren aktiv zur Durchsetzung ihrer Interessen einzusetzen. In den bisher über 100 eingeleiteten Streitverfahren steht die Gemeinschaft lediglich in vier Verfahren, davon erstmals in dem *U.S. DISC tax legislation*-Fall und zuletzt in dem eingeleiteten Verfahren gegen Japan, auf der »Kläger«seite, dagegen aber in über 20 Verfahren auf der »Beklagten«seite[39].

Die bisher ersichtliche Haltung der Gemeinschaft bei der Handhabung der anhängigen Verfahren, die sich offenbar gegen eine Verrechtlichung der Verfahren wendet und in die bezeichenderweise die juristischen Dienste der Kommission und des Rates nicht formell eingeschaltet werden, dürfte zwar von einer Vielzahl der GATT-Partner geteilt werden. Sie entspricht auch dem Grundanliegen des GATT, einen Interessenausgleich herbeizuführen, der durch notwendigerweise »einseitige« und zugleich bindende Rechtsentscheidungen kaum gefunden werden könnte. Dennoch führt die seit dem Abschluß der Tokyo-Runde ständig wachsende Zahl von Verfahren notwendigerweise zu der Herausbildung eines Fall-»Rechts« durch die ständig wiederkehrende Deutung und Anwendung der in den Abkommen verwendeten Begriffe. Die einzelnen *panels* werden in ihrer Entscheidungsfreiheit durch das zuwachsende Fallmaterial eingeengt. Zugleich dürften sich Verfahrensgrundsätze verfestigen und sowohl im eigentlichen GATT-Verfahren als auch in den vergleichbar ausgestalteten Verfahren der Sonderübereinkommen allgemeine Anerkennung finden. Und ebenso dürfte es auf längere Zeit für die Vertragsparteien schwieriger werden, sich den Verfahren bzw. den getroffenen Feststellungen der *panels* bzw. der Vertragsparteien zu entziehen.

Der Gerichtshof der Europäischen Gemeinschaften hat das Streitverfahren in den Art. XXII und XXIII GATT in dem Fall *International Fruit Company* (1972)[40] ungewöhnlich eingehend beschrieben und auf Grund der erkennbaren Flexibilität und Geschmeidigkeit der Vorschriften gefolgert, daß

---

[38] In dem »Memorandum d'accord concernant les notifications, les consultations, le règlement des différends et la surveillance« vom 28. 11. 1979 (L/4907) wird unterstrichen, daß die Verfahren nach Art. XXIII Ziff. 2 »ne devraient pas être conçus ni considérés comme des actes contentieux, et que, si les différends surviennent, toutes les parties contractantes engageront ces procédures en toute bonne foi dans un effort visant à régler ces différends«: Instruments de base et documents divers, Suppl. No. 26 (Genf 1980), S. 231, 233.

[39] Die Zahlenangaben ergeben sich auf Grund der Übersichten bei *Hudec* (Anm. 35), S. 200 f., sowie *derselbe*, The GATT Legal System and World Trade Diplomacy (1975), S. 275 ff., und aus den seither veröffentlichten Angaben in GATT Activities in 1980 (Genf 1981), S. 43 ff., und GATT Activities in 1981 (Genf 1982), S. 42 ff. In dem Verfahren der EWG gegen Japan sind bisher nur Konsultationen gemäß Art. XXIII Ziff. 1 eingeleitet worden; Japan bereitet bereits ein gegenläufiges Verfahren vor: siehe VWD-Europa vom 18. 5. 1982, S. II/1.

[40] Vgl. oben Anm. 34.

sich der Einzelne vor Gericht grundsätzlich nicht auf das materielle GATT-Recht berufen könne, dieses also in der Gemeinschaftsrechtsordnung nicht als unmittelbar anwendbar angesehen werden könne. Nachfolgende Entscheidungen des Gerichtshofs lassen es indes als offen erscheinen, ob diese Folgerung unter allen denkbaren Umständen zutreffen wird. Zumindest dürfte der Gerichtshof dann Bestimmungen die unmittelbare Anwendbarkeit im Innenrechtsraum der Gemeinschaft zuerkennen, wenn zuvor in dem für alle Vertragsparteien verbindlichen Streitverfahren der Inhalt der GATT-Vorschriften eindeutig festgelegt worden ist. In dem im Jahre 1981 entschiedenen Fall *Dürbeck*[41] prüft der Gerichtshof eingehend die Vereinbarkeit von Schutzmaßnahmen der Gemeinschaft gegenüber Apfelimporten aus Chile an Hand der Feststellungen der GATT-Sondergruppe (*panel*): Diese habe »lediglich« festgestellt, daß die Gemeinschaft bei der Berechnung der zum Import noch zugelassenen Mengen ein falsches Referenzjahr herangezogen habe. Dieser Verstoß – so der Gerichtshof – sei nicht geeignet, die Gültigkeit der Schutzmaßnahmen der Gemeinschaft in Frage zu stellen. In gewichtigeren Fällen könnte der Gerichtshof daher einer verletzten Bestimmung des GATT eine unmittelbare Anwendbarkeit dargestalt zuerkennen, daß diese dem widersprechenden sekundären Gemeinschaftsrecht vorgehen würde. Eine *zunehmende Verrechtlichung* der GATT-Verfahren sowie eine *Verdichtung der Spruchpraxis* der Sondergruppen zum materiellen GATT-Recht könnte den Gerichtshof veranlassen, seine Rechtsprechung über die Qualifikation des GATT-Rechts im Rahmen der Gemeinschaftsrechtsordnung zu überprüfen. Voraussetzung wäre jedoch, daß eine *gleichmäßige Befolgung des so verfestigten GATT-Rechts seitens der GATT-Partner nachgewiesen werden könnte, die es rechtfertigen würde, von einer dem Gegenseitigkeitsprinzip entsprechenden Praxis zu sprechen.*

Eher im Hintergrund der genannten GATT-Verfahren steht das – zumindest aus der Sicht der Gemeinschaft beurteilt – wirksamere Überwachungs- und Klageverfahren im Rahmen des Welttextilabkommens (Vereinbarung über den internationalen Handel mit Textilien)[42]. Die Gemeinschaft nimmt an den Arbeiten und Entscheidungen des »Textile Surveillance Body« in unterschiedlichen Rollen teil, das zum einen eine ständige Überwachungsaufgabe über alle Maßnahmen der Vertragsparteien ausübt (*review procedure*), aber auch zuständig ist, Streitfälle zwischen den Vertragsparteien im Wege der Vermittlung, des Vergleichs oder letztlich der Entscheidung zu regeln, deren Wirksamkeit auf der Autorität des Gremiums beruht (*complaint procedure*)[43]. Beide Verfahren vollziehen sich in einem wesentlich engeren Rahmen als die

---

[41] EuGH 5. 5. 1981 – *Dürbeck,* Rs. 112/80 – hekt., Rdnr. 45 f.; vgl. hierzu auch *Bourgeois,* Common Market Law Review 19 (1982), S. 26 Anm. 63.

[42] Art. 11 der Vereinbarung über den internationalen Handel mit Textilien vom 20. 12. 1973, ABl. Nr. L 118 vom 30. 4. 1974; die im Dezember 1981 abgeschlossenen Verhandlungen über ein neues Textilabkommen haben die Vorschriften über die Streitbeilegung nicht geändert.

[43] Vgl. Art. 11 Abs. 13 bzw. 5 der Vereinbarung.

allgemeinen GATT-Verfahren, allgemeine Handelsinteressen der Parteien werden selten berührt. Als arbeitsintensives und aufwendiges ständiges Verfahren eignet es sich kaum zur Übernahme in das allgemeine GATT-Vertragswerk[44]. Die Gemeinschaft soll an den wenigen Streitverfahren, die bisher anhängig gemacht worden waren, nicht beteiligt gewesen sein[45].

Die nicht-gerichtlichen Streitbeilegungsverfahren der internationalen Rohstoffabkommen, denen die Gemeinschaft in jüngerer Zeit beigetreten ist, spielen bisher im Vergleich zu den GATT-Verfahren in der Gemeinschaftspraxis keine hervorzuhebende Rolle[46]. Der in allen einschlägigen Abkommen vorgesehene Weg über Konsultationen, Vergleichsverfahren unter Vermittlung der Organisation und schließlich Entscheidung über eine behauptete Verletzung des Abkommens durch den Rat der Organisation nach Einholung eines Gutachtens einer *ad hoc*-Beratergruppe, hat bisher zu keiner bedeutenden Spruchpraxis geführt. Auch in diesen Abkommen wird der Versuch der Verständigung über die gerichtliche Streitbeilegung gestellt.

5. Streitbeilegung durch schiedsgerichtliche Verfahren

Insgesamt stehen in der Vertragspraxis der Gemeinschaften wie auch der Mitgliedstaaten die nicht-gerichtlichen Streitbeilegungsverfahren der Zahl nach eindeutig im Vordergrund. In relativ seltenen Fällen sind jedoch Schiedsklauseln vereinbart worden: Die vertraglich vorgesehene Möglichkeit, zu einer bindenden und endgültigen Regelung von Streitfällen zu gelangen, kann auf die Auslegung und Anwendung eines Vertrages zurückwirken.

Eine erste Durchsicht der von den Gemeinschaften abgeschlossenen und zugänglichen Verträge hat 31 Verträge ergeben, die ein gerichtliches Schiedsverfahren zur Regelung von Streitigkeiten aufweisen. Es lassen sich nur Vermutungen darüber anstellen, warum die Gemeinschaften gerade in diesen Fällen bereit waren oder veranlaßt worden sind, in die Aufnahme eines Schiedsverfahrens einzuwilligen. Es werden im wesentlichen die gleichen Gründe sein, die auch die Mitgliedstaaten in ihren bi- und multilateralen Vertragsbe-

---

[44] So die Einschätzung bei *Hudec* (Anm. 35), S. 145, 168 ff., sowie *G. H. Perlow*, The Multilateral Supervision of International Trade: Has the Textiles Experiment Worked?, American Journal of International Law (AJIL) 75 (1980), S. 93, 103 ff.

[45] Vgl. indes die kritischen Anmerkungen zur Haltung der Gemeinschaft (»EEC's challenge to its legal authority«) und zum Verlust des für die Wirksamkeit des Streitbeilegungsverfahrens notwendigen Grundübereinstimmung zwischen den Partnerstaaten (»breakdown of substantive consensus«) *Perlow*, a.a.O., S. 130 f.

[46] Vgl. etwa Art. 57 ff. des Internationalen Kakao-Übereinkommens von 1980, Beschluß des Rates (EWG) vom 30. 6. 1981 über die vorläufige Anwendung, ABl. Nr. L 313/1 ff. vom 31. 10. 1981; Art. 40 des Internationalen Olivenöl-Übereinkommens 1979, Beschluß des Rates (EWG) vom 12. 11. 1979, ABl. Nr. L 327/1 ff. vom 24. 12. 1979; Art. 56 des Internationalen Naturkautschuk-Übereinkommens 1979, Beschluß des Rates (EWG) vom 22. 4. 1980, ABl. Nr. L 213/1 ff. vom 16. 8. 1980 sowie die entsprechenden Abkommen über Zinn (vgl. ABl. Nr. L 222/1 ff. vom 14. 8. 1976) und Kaffee (ABl. Nr. L 309/1 ff. vom 10. 11. 1976).

ziehungen bestimmen[47]. Auffällig ist gleichwohl, daß die Mehrzahl der entsprechenden Verträge und Abkommen erst aus jüngerer Zeit datieren.

Dabei werfen zweiseitige Verträge im Hinblick auf die Vereinbarung eines Schiedsverfahrens weniger Gestaltungsfragen auf als multilaterale Verträge und solche, an denen außer den Gemeinschaften die Mitgliedstaaten einerseits und eine Mehrzahl von Vertragspartnern andererseits beteiligt sind. Die Verträge der drei Gemeinschaften stehen dabei in unterschiedlichen rechtlichen Zusammenhängen.

a) Schiedsverfahren in Verträgen der EWG

aa) Zweiseitige Verträge

Wenige zweiseitige Verträge der Wirtschaftsgemeinschaft sehen ein Schiedsverfahren zur Streitbeilegung vor. Es handelt sich vor allem um Assoziations- und Kooperationsabkommen, die zum Teil auf eine engere institutionelle Bindung an die Gemeinschaft gerichtet sind, sowie um einen Teil der Fischereiabkommen, die überwiegend präzise Rechtsregime festschreiben.

Das *Assoziationsabkommen* mit Griechenland vom 9. Juli 1961[48] hält den Vertragsparteien insgesamt vier verschiedene Wege der Streitbeilegung offen. Art. 67 bestimmt im einzelnen:

»(1) Jede in Artikel 65 bezeichnete Vertragspartei kann den Assoziationsrat mit jeder Streitigkeit in bezug auf Anwendung oder Auslegung dieses Abkommens befassen, soweit sie die Gemeinschaft, einen Mitgliedstaat der Gemeinschaft oder Griechenland betrifft.

(2) Der Assoziationsrat kann die Streitigkeit durch Beschluß beilegen; er kann ferner beschließen, die Beilegung der Streitigkeit dem Gerichtshof der Europäischen Gemeinschaften oder irgendeinem anderen bestehenden Gericht zu unterbreiten.

(3) Kann der Assoziationsrat die Streitigkeit nicht gemäß Absatz (2) beilegen oder hat er nicht gemäß Absatz (2) das zur Beilegung dieser Streitigkeit berufene Gericht benannt oder hat das danach benannte Gericht die Streitigkeit nicht beigelegt, so kann eine Partei der anderen Partei die Bestellung eines Schiedsrichters notifizieren; die andere Partei ist verpflichtet, binnen zwei Monaten einen zweiten Schiedsrichter zu bestellen. Für die Anwendung dieses Verfahrens gelten die Gemeinschaft und die Mitgliedstaaten zusammen als eine Streitpartei.

Ein dritter Schiedsrichter, der das Amt des Vorsitzenden ausübt, wird nach Maßgabe des Absatzes (4) bestellt.

Die Schiedssprüche ergehen mit Stimmenmehrheit.

. . .«.

---

[47] Vgl. die Übersicht über die politisch-psychologischen Vorbedingungen für den Abschluß von Schiedsverträgen bei *v. Mangoldt* (Anm. 24), S. 173 ff.

[48] ABl. Nr. L 26 vom 18. 2. 1963.

In den folgenden Absätzen wird zunächst der Präsident des Gerichtshofs der Europäischen Gemeinschaften als dritter Schiedsrichter benannt. Nach Ablauf von fünf Jahren soll im Streitfall der Präsident des Internationalen Gerichtshofs den dritten Richter aus dem Kreise von Persönlichkeit bestellen, die in den OECD-Staaten hohe richterliche Funktionen ausüben.

Eine nahezu übereinstimmende Regelung weist das Assoziationsabkommen mit der Türkei vom 12. September 1963 auf (Art. 25) [49]. In keinem anderen Abkommen ist es der Gemeinschaft gelungen, die Zuständigkeit des Gerichtshofs der Europäischen Gemeinschaften als eine der vertraglich vorgesehenen Streitbeilegungsinstanzen zu vereinbaren. Die Zielrichtung der Abkommen auf einen späteren Beitritt Griechenlands und der Türkei wird die Vertragsparteien bestimmt haben, Fragen, die in besonderer Weise die Rechtsordnung der Gemeinschaft berühren, durch übereinstimmenden Beschluß dem Gerichtshof der Gemeinschaften unterbreiten zu können. Dieses ausgefeilte Streitregelungsverfahren dürfte den Gerichtshof der Gemeinschaften bestärkt haben, davon auszugehen, daß die Parteien auf eine vollständige und gegenseitige Erfüllung aller Pflichten des Abkommens zielten und damit eine in ihren Rechtsordnungen unmittelbare Anwendbarkeit einer Reihe von Vertragsbestimmungen herbeiführen wollten [50].

Unter den 13 *Kooperationsabkommen* der Gemeinschaft enthalten lediglich die Abkommen mit Tunesien vom 25. April 1976, mit Algerien vom 26. April 1976 und mit Marokko vom 27. April 1976 ein *ad hoc*-Schiedsverfahren. Es soll eingreifen, falls es dem Kooperationsrat nicht gelingt, einen Streitfall über die Auslegung der Abkommen beizulegen. Der Kooperationsrat bleibt einvernehmlich für die Bestellung des dritten Schiedsrichters zuständig, so daß die Vertragsparteien die Bildung des Schiedsgerichts in der Hand behalten. Diese Kooperationsabkommen haben zuvor bestehende Assoziationsabkommen (sog. Entwicklungsassoziationen [51]) abgelöst und führen den angestrebten festen Rahmen der gegenseitigen Beziehungen unter einer weniger belasteten Bezeichnung weiter. Die übrigen Kooperationsabkommen sind ihrem Inhalt nach eher Handelsabkommen und verweisen daher auf den Verhandlungsweg, um im Streitfall eine beide Seiten zufriedenstellende Lösung zu erreichen.

Von den neun *Fischereiabkommen* der Gemeinschaft, die nach der Ausdehnung der Fischereizonen auf 200 sm mit verschiedenen Drittstaaten abgeschlossen worden sind, enthalten allein die Abkommen mit Schweden, Finnland (nicht in Kraft getreten) und der Republik Senegal besondere Schieds-

---

[49] ABl. Nr. L 217 vom 29. 12. 1964.

[50] Vgl. EuGH 29. 4. 1982 – *Pabst & Richarz,* Rs. 17/81 – hekt., S. 35 ff.

[51] Zu den unterschiedlichen Inhalten der Assoziationsabkommen der Gemeinschaft vgl. *E. U. Petersmann,* Struktur und aktuelle Rechtsfragen des Assoziationsrechts, ZaöRV 33 (1973), S. 266 ff.

bzw. Schlichtungsverfahren[52]. Es lassen sich wiederum nur Vermutungen anstellen, warum die Gemeinschaft teils Konsultations- und teils Schiedsverfahren zugestimmt hat oder zustimmen mußte. Sowohl gegenüber der Republik Senegal als etwa auch gegenüber Norwegen war und ist die Gemeinschaft der eher fordernde Vertragspartner. Die Erwägungsgründe enthalten hierzu keine Anhaltspunkte[53]. Die Ergebnisse der Dritten Seerechtskonferenz, die das Fischereiregime innerhalb der Wirtschaftszonen nur eingeschränkt den verschiedenen Wegen der gerichtlichen Streitbeilegung unterstellt, dürften zumindest künftig die Gemeinschaften veranlassen, nur noch in entsprechender Eingrenzung der Aufnahme von Schiedsgerichtsklauseln in ihre Verträge zuzustimmen.

bb) Mehrseitige Verträge

Die Verträge der Europäischen Wirtschaftsgemeinschaft mit mehreren Vertragspartnern lassen zwei Gruppen erkennen. Die auf der Grundlage des Art. 238 EWGV (Assoziierungen) geschlossenen Verträge sind, auch wenn an ihnen eine Vielzahl von Drittstaaten beteiligt ist, zweiseitig strukturiert: die eingesetzten Vertragsorgane fassen ihre Beschlüsse »im gegenseitigen Einvernehmen der Gemeinschaft einerseits und der AKP-Staaten andererseits«[54]. Diese Struktur prägt seit dem Jaunde I-Abkommen alle aufeinanderfolgenden Verträge, die auch bei der Gestaltung der vereinbarten Streitbeilegungsverfahren eine Fortentwicklung erkennen lassen. Die übrigen mehrseitigen Abkommen der Europäischen Wirtschaftsgemeinschaft sind echte multi-

[52] Vgl. jeweils Art. 7 und den Anhang der Fischereiabkommen mit Schweden vom 21. 3. 1977 (ABl. Nr. L 226/2 vom 29. 8. 1980), mit Finnland (Vorschlag der Kommission vom 23. 2. 1979, ABl. Nr. C 69/6 vom 13. 3. 1979) und der Republik Senegal vom 15. 6. 1979 (ABl. Nr. L 226/16 vom 29. 8. 1980). Die vereinbarten Schiedsgerichtsverfahren stimmen weitgehend überein, insbesondere legen sie fest, daß Hoheitsrechte der Vertragsparteien in den jeweiligen Zonen nicht Gegenstand eines Schiedsgerichtsverfahrens sein können. Während die Abkommen mit Schweden und Finnland den Präsidenten des Internationalen Gerichtshofs mit der gegebenenfalls erforderlichen Ernennung des dritten Schiedsrichters betrauen, enthält das Abkommen der Republik Senegal keine entsprechende Klausel, so daß das Verfahren von einer Partei – trotz vereinbarter Fristsetzung – blockiert werden kann.

[53] Vgl. die übrigen Abkommen mit Dänemark/Faröer vom 15. 3. 1977, mit Guinea-Bissau vom 27. 2. 1980, mit Norwegen – ohne Datum – (jeweils ABl. Nr. L 226/ 11, 33 und 47 vom 29. 8. 1980) sowie mit Kanada – ohne Datum – (ABl. Nr. L 379/ 53 vom 31. 12. 1981), mit Spanien vom 15. 4. 1980 (ABl. Nr. L 322/3 vom 28. 11. 1980) und mit den Vereinigten Staaten vom 15. 2. 1977 (ABl. Nr. L 141/1 vom 9. 6. 1977): Dieser Ausschuß der gerichtlichen Streitbeilegung könnte bereits auf Art. 297 Abs. 3 (a) der Draft Convention on the Law of the Sea, UN-Doc. A/ CONF. 62/L.78 vom 28. 8. 1981 zurückzuführen sein, da hier die gerichtlichen Wege der Streitbeilegung in bezug auf das Fischereiregime weitgehend ausgeschlossen werden.

[54] Vgl. etwa Art. 167 des Lomé II-Abkommens vom 31. 10. 1980, ABl. Nr. L 347/55 vom 22. 12. 1980.

laterale Verträge, denen die Gemeinschaft in der Regel erst später beigetreten ist und an denen oft zugleich die Mitgliedstaaten beteiligt sind. Diese Vielzahl der Parteien führt zu schwierigen Aufgaben der Vertragsgestaltung. Die in dem Entwurf für eine Seerechtskonvention gefundenen Lösungen fallen in ihrer Vielgestalt aus dem bisher bekannten Rahmen.

Das erste mit den assozierten Staaten Afrikas und Madagaskar geschlossene Abkommen von *Jaunde I* vom 20. Juli 1963 [55] betont die Funktion des Streitbeilegungsverfahrens im Rahmen der Vertragserfüllung: Es weist in Art. 51 ein *ständiges* »Schiedsgericht der Assoziation« aus, das aus fünf Mitgliedern bestehen und tätig werden soll, wenn weder der Assoziationsrat noch die Parteien einen Streitfall auf andere Weise beilegen können. Die Schiedsrichter sind ernannt, und das – zunächst nicht veröffentlichte – Statut des Schiedsgerichts ist auf seinen Vorschlag hin erlassen worden. Erstaunlicherweise hat der Assoziationsrat im Jahre 1963 den Präsidenten des Gerichtshofs der Europäischen Gemeinschaften als fünften Schiedsrichter und zum Präsidenten des Schiedsgerichts bestellt [56]. Diese Wahl unterstreicht die enge Bindung des Abkommenswerkes an die Gemeinschaft und läßt einen besonderen Rechtscharakter erkennen [57]. Sie würde indes nicht mehr dem Selbstverständnis der AKP-Staaten unter den Lomé-Abkommen entsprechen. Streitfälle sind nicht anhängig gemacht worden.

Das Folgeabkommen von *Jaunde II* vom 29. Juli 1969 [58] übernimmt gleichlautend die Vorstellung eines ständigen Schiedsgerichts (Art. 53), dessen bisher bestehende Satzung nunmehr von den Vertragsparteien selbst im Protokoll Nr. 8 zu dem Abkommen festgeschrieben worden ist [59]. Art. 9 des Protokolls legt den Sitz des Schiedsgerichts an den Ort des Gerichtshofs der Gemeinschaften, dessen Kanzlei zugleich als Kanzlei des Schiedsgerichts tätig werden soll. In Anhang VIII des Abkommens vereinbarten die Vertragsparteien ausdrücklich, vor Anrufung des Assoziationsrats bzw. des Schiedsgerichts ein Vermittlungsverfahren einzuleiten, in dem es allen »interessierten Parteien« erlaubt sein sollte, ihre Rechte geltend zu machen [60].

Die Vertragsparteien sind im nachfolgenden Abkommen von *Lomé I* vom 28. Februar 1975 [61] von der Konzeption eines ständigen Schiedsgerichts abgerückt, da offensichtlich der institutionelle Aufwand den auftretenden Bedürfnissen nicht entsprach. Zur Vollzugssicherung erschien den Parteien ein *ad*

---

[55] ABl. Nr. 1431/64 vom 11. 6. 1964.

[56] Vgl. *D. Vignes,* L'association des Etats africains et malgaches à la C.E.E. (1970), S. 32.

[57] Vgl. EuGH 5. 2. 1976 – *Bresciani,* Rs. 87/75 – Slg. 1976, S. 129, 140 f., zum besonderen Rechtscharakter des Abkommens, das bewußt ungleiche Verpflichtungen der Parteien festlege und auf dessen Bestimmungen sich dennoch unter Umständen die Gemeinschaftsangehörigen unmittelbar vor Gericht berufen könnten.

[58] ABl. Nr. L 282 vom 28. 12. 1970.

[59] ABl. Nr. L 282/25 ff. vom 28. 12. 1970.

[60] A.a.O., S. 41.

[61] ABl. Nr. L 25 vom 30. 1. 1976.

hoc-Schiedsverfahren ausreichend, in dem die Gemeinschaft und die Mitgliedstaaten als eine Partei angesehen werden (Art. 81). Fakultativ wurde den Parteien der Zugang zu einem vorgängigen Vermittlungsverfahren eingeräumt, in dem die einzigen beiden Streitfälle, die während der Laufzeit des Abkommens zu regeln waren, beigelegt werden konnten [62].

Wiederum leicht verändert ist das Streitverfahren im derzeit geltenden Lomé II-Abkommen vom 31. Oktober 1979 [63]: In einem gestuften Verfahrensablauf ist zunächst nach erfolglosen Verhandlungen der Ministerrat zur Behandlung des Streitfalles zuständig, er *kann* auf Antrag einer der Parteien das Vermittlungsverfahren einleiten oder letzten Endes einen Schiedsrichter bestellen, dem die Parteien innerhalb von zwei Monaten je einen Schiedsrichter zur Seite stellen. Der Schiedsspruch hat innerhalb von 18 Monaten zu ergehen. Auch dieses Verfahren hat bisher – bis auf die erste Stufe – nicht eingreifen müssen [64]. Es behält weiterhin die Schwäche, daß die Bestellung aller Schiedsrichter vom Willen der Parteien abhängig bleibt. Grundlegende Änderungen sind in den weiteren Abkommen nicht zu erwarten, zumal das Verfahren neben seiner Präventivwirkung keine praktische Bedeutung erlangt hat. Das am 24. September 1969 geschlossene Assoziationsabkommen mit drei ostafrikanischen Staaten enthält übereinstimmend eine *ad hoc*-Schiedsklausel mit der Befugnis des Assoziationsrats, den dritten Schiedsrichter zu bestimmen [65].

Neben den »Assoziationsabkommen« weist eine Reihe *echter* mehrseitiger Abkommen, an denen sich die Europäische Wirtschaftsgemeinschaft beteiligt hat oder deren Beitritt bzw. Abschluß sie anstrebt, Schiedsverfahren in vielfältigen Varianten auf. Die Art der Klauselgestaltung, die die Gemeinschaft in der Regel nicht oder nicht bestimmend beeinflussen kann, hängt dabei vor allem von der Praxis derjenigen internationalen Organisation ab, in deren Rahmen das jeweilige Abkommen ausgehandelt worden ist.

Eine Reihe der im Rahmen des *Europarats* vereinbarten Übereinkommen betrifft Bereiche, die in die Zuständigkeit der Gemeinschaft fallen. Die Übereinkommen sehen entsprechend in ihrem Text oder in einem Zusatzprotokoll den Beitritt der Gemeinschaft vor. Das dreigestufte Grundmodell der Streit-

---

[62] Zu dem Ergebnis eines Vermittlungsverfahrens vgl. den Bericht der Kommission über das Stabex-System der Jahre 1975–1979 in SEK(81) 1104 vom 13. 7. 1981.

[63] ABl. Nr. L 347/57 vom 22. 12. 1980; siehe zur Regelung der Verwaltungskosten des Schiedsgerichts Art. 3 des Protokolls Nr. 2 zum Abkommen, a.a.O., S. 137; Art. 11 des Protokolls Nr. 3, a.a.O., S. 138 f., eröffnet dem Rat der AKP-Minister sowie der Europäischen Investitionsbank die Möglichkeit, hinsichtlich der ihnen eingeräumten Vorrechte und Immunitäten in einem Schiedsverfahren nach Art. 176 des Lomé II-Abkommens als Partei aufzutreten.

[64] Die Drohung bereits, das Vermittlungsverfahren einzuleiten, hat die Gemeinschaft Anfang 1982 veranlaßt, einer lange Zeit umstrittenen und seit dem 1. 5. 1981 überfälligen Neufestsetzung der Zuckerpreise für AKP-Zucker zuzustimmen: Vgl. die Verordnung (EWG) Nr. 1000/82 des Rates in ABl. Nr. L 118/2 vom 1. 5. 1982.

[65] Vgl. Art. 28 des Arusha-Abkommens, ABl. Nr. L 282 vom 28. 12. 1970.

beilegungsregelung, das in einige der Übereinkommen aufgenommen worden ist, geht von dem Versuch des jeweiligen ständigen Organs der Übereinkommen aus, Streitfälle gütlich beizulegen. Haben anschließende Verhandlungen der Streitparteien keinen Erfolg oder können sie sich nicht auf ein abweichendes Verfahren einigen, so wird auf Antrag einer Partei der Streit einem Schiedsverfahren unterworfen: Dieses weist Fristen zur Bestellung der Schiedsrichter aus und erklärt ergänzend den Präsidenten des Europäischen Gerichtshofs für Menschenrechte für zuständig, im Karenzfall die Schiedsrichter zu ernennen.

Dieses flexible und zugleich effektive Verfahren ist u. a. in dem Übereinkommen zur Erhaltung der europäischen freilebenden Tiere und wildwachsenden Pflanzen und ihrer natürlichen Lebensräume vom 19. September 1979 vorgesehen [66]. In einem Zusatz wird angestrebt, in einem Streitfall Klarheit darüber herbeizuführen, ob die Gemeinschaft selbst, einer ihrer Mitgliedstaaten oder beide zusammen zur Aufnahme des jeweiligen Rechtsstreits zuständig sind. Erfolgt von seiten der Gemeinschaft keine entsprechende Notifizierung, so gelten die Gemeinschaft und der betreffende Mitgliedstaat für die Zwecke des Schiedsverfahrens als eine Streitpartei [67].

Im Rahmen der *Umweltschutzpolitik* ist die Gemeinschaft drei internationalen Übereinkommen zum Schutze der Meere sowie dem Übereinkommen zum Schutz des Rheins gegen chemische Verunreinigung beigetreten, die jeweils im Anhang detaillierte Schiedsverfahren enthalten. Diese sind weitgehend gleichlautend formuliert, unterscheiden sich aber bei der Frage, ob die Verfahren für die Parteien fakultativ oder obligatorisch sein sollen. Art. 22 des Übereinkommens vom 16. Februar 1976 zum Schutz des Mittelmeeres vor Verschmutzung (Barcelona-Übereinkommen) [68] setzt für die Durchführung des Schiedsverfahrens das Einverständnis der Streitparteien voraus, sieht aber zugleich vor, daß die Parteien im voraus eine sie bindende Unterwerfungserklärung abgeben können. Ein obligatorisches Schiedsverfahren sichert dagegen die Durchführung des Übereinkommens vom 21. Februar 1974 zur Verhütung der Meeresverschmutzung vom Lande aus (Art. 21 des Übereinkommens von Paris) sowie des Übereinkommens vom 29. April 1963 zum Schutze des Rheins (Art. 15 des Übereinkommens von Bern) [69].

---

[66] Vgl. Beschluß des Rates vom 3. 12. 1981, ABl. Nr. L 38/1 vom 10. 2. 1982. Eine vereinfachte Variante des Verfahrens ist in Art. 45 des Europäischen Übereinkommens über den Schutz der Tiere auf dem internationalen Transport vom 13. 12. 1968 aufgenommen, dem die Gemeinschaft bisher in Anwendung des beigefügten Zusatzprotokolls vom 10. 5. 1979 nicht beigetreten ist: vgl. BGBl. 1973 II, S. 722.

[67] Siehe Art. 18 Abs. 3 sowie unten Anm. 113.

[68] Vgl. Beschluß des Rates vom 25. 7. 1977 über den Abschluß des Übereinkommens in ABl. Nr. L 240/1 vom 19. 9. 1977 sowie den Anhang A zum Übereinkommen auf S. 10.

[69] Vgl. Beschluß des Rates vom 3. 3. 1975 (ABl. Nr. L 194/5 vom 25. 7. 1975) zum Übereinkommen von Paris, in dem Art. 8 bestimmt: »La Communauté économi-

Schließlich findet sich eine äußerst flexible Streitregelung in dem Übereinkommen über die Erhaltung der lebenden Meeresschätze der Antarktis vom 20. Mai 1980, dem die Gemeinschaft nach seinem Inkrafttreten beigetreten ist[70]. Die im Streitfall vorgeschriebenen Konsultationen (Art. XXV) sollen dazu führen, die Streitigkeit »durch Verhandlung, Untersuchung, Vermittlung, Vergleich, Schiedsverfahren, gerichtliche Beilegung oder sonstige friedliche Mittel ihrer Wahl« beizulegen. Gelingt dies nicht, wird mit Zustimmung der Parteien der Fall dem Internationalen Gerichtshof oder einem Schiedsverfahren unterworfen, für das laut Anlage zu dem Übereinkommen der Generalsekretär des Ständigen Schiedsgerichtshofs gegebenenfalls mit der Ernennung des zweiten oder auch dritten Schiedsrichters beauftragt wird. Ein besonderer Hinweis auf Streitigkeiten, in die die Gemeinschaft einbezogen sein könnte, fehlt, zumal die Gemeinschaft auf die Möglichkeit, den Streitfall dem Internationalen Gerichtshof zu unterbreiten, nicht eingehen könnte.

In dieser Vielfalt der Verfahrenswege kommt der Antarktisvertrag dem im April 1982 festgelegten Text für eine Seerechtskonvention nahe, wird aber in dieser Hinsicht von diesem übertroffen. Da die Seerechtskonvention schiedsgerichtliche und neu zu schaffende gerichtliche Wege der Streitbeilegung kennt, soll sie im Zusammenhang mit der Erörterung der Verfahren echter Gerichtsbarkeit einbezogen werden.

b) Schiedsverfahren in Verträgen der EGKS

Die fehlende Außenhandelskompetenz der Europäischen Gemeinschaft für Kohle und Stahl wirft bei der Aushandlung der Verträge der Wirtschaftsgemeinschaft regelmäßig die Frage auf, ob der Vertrag auf der Grundlage des EWG-Vertrages die Produkte für Kohle und Stahl mit erfaßt oder ob die Mitgliedstaaten für die Kohle- und Stahlprodukte ein gesondertes völkerrechtliches Abkommen schließen sollen. Auch kennt die Praxis nach dem EGKS-Vertrag eine Reihe gemischter Abkommen. Die bisher weitgehend im Schatten des juristischen Interesses liegende Praxis ist nicht eindeutig. Die Entscheidung für den jeweiligen Weg dürfte oft auf hier nicht zu erörternden

---

que européenne, comme toute partie contractante à la convention, est habilitée à agir comme partie requérante ou appelée devant le tribunal arbitral«. Kompetenzkonflikte der Gemeinschaft mit ihren Mitgliedstaaten sind im Text nicht angesprochen, in der Schlußakte (im Amtsblatt nicht abgedruckt) wird es der Gemeinschaft aufgegeben, sich über die interne Verteilung der Kompetenzen zu äußern: vgl. auch unten Anm. 97. Zum Beitrittsbeschluß des Rates zum Übereinkommen von Bern vgl. ABl. Nr. L 240/51 vom 19. 9. 1977 sowie unten Anm. 113. – Im Transportbereich hat der Rat mit Verordnung (EWG) Nr. 2112/78 (ABl. Nr. L 252/1 vom 14. 9. 1978) die Teilnahme der Gemeinschaft am Genfer Zollübereinkommen über den internationalen Warentransport mit Carnets TIR vom 14. 11. 1975 beschlossen, das in Art. 57 eine obligatorische Schiedsklausel aufweist.

[70] Beschluß des Rates vom 4. 9. 1981, ABl. Nr. L 252/26 vom 5. 9. 1981.

rechtspolitischen Beweggründen beruhen [71]. Es leuchtet indes nicht ein, wenn die Streitbeilegungsverfahren je nach dem EWG- und EGKS-Bereich unterschiedlich gestaltet werden. In den Abkommen zwischen den Mitgliedstaaten der EGKS und Algerien, Marokko und Tunesien [72] werden die Bestimmungen über die Streitbeilegung in den Kooperationsabkommen der EWG sinngemäß für anwendbar erklärt. Dagegen weicht das Abkommen der Mitgliedstaaten mit den Lomé-Staaten über die Waren, die in die Zuständigkeit der EGKS fallen, von dem (gemischten) Lomé II-Abkommen ab [73] und beschränkt sich im Streitfall auf Konsultationen, anstatt auch für diesen Bereich das Schiedsverfahren des Art. 176 des Lomé II-Abkommens zu übernehmen. Die jeweiligen Erwägungsgründe ergeben über die unterschiedlichen Verfahrensweisen keinen Aufschluß.

c) Schiedsverfahren in Verträgen der Euratom-Gemeinschaft

Die völkerrechtliche Vertragspraxis der Euratom-Gemeinschaft umfaßt wenige Abkommen. Sie betreffen vor allem Fragen der Zusammenarbeit und der nuklearen Sicherheit. Die ersten zwei Abkommen über Zusammenarbeit mit den Vereinigten Staaten vom 8. November 1958 und mit Kanada vom 6. Oktober 1959 [74] beschränkten das Streitbeilegungsverfahren auf die Durchführung von Konsultationen. Seit Beginn der siebziger Jahre enthalten dagegen *alle* veröffentlichten Abkommen Schiedsklauseln unterschiedlicher Gestaltung, die durchgehend Vorsorge für den Fall treffen, daß eine Partei mit der Ernennung eines Schiedsrichters säumig ist. So wird der Präsident des Internationalen Gerichtshofs, der Generalsekretär der Vereinten Nationen oder etwa der Generalsekretär der OECD ersucht, die fälligen Ernennungen vorzunehmen [75]. Diese Änderung der Vertragspraxis dürfte der gesteigerten inter-

---

[71] Die Vertragspraxis folgt weitgehend dem Weg über den Abschluß gesonderter Abkommen: Vgl. z. B. die Parallelabkommen über die EGKS-Produkte zu den Freihandelsabkommen mit den Rest-EFTA-Staaten sowie das Lomé-Abkommen. Dagegen beziehen sich die im Rahmen der Tokio-Runde von der EWG abgeschlossenen nicht tarifären Sonderübereinkommen auch auf EGKS-Produkte: vgl. ABl. Nr. L 71/1 ff. vom 17. 3. 1980 sowie hierzu *Bourgeois* (Anm. 34), S. 22 f.; in der Literatur wird weitgehend, im Anschluß an *K. Carstens,* Die Errichtung des Gemeinsamen Marktes in der Europäischen Wirtschaftsgemeinschaft, Atomgemeinschaft und Gemeinschaft für Kohle und Stahl, ZaöRV 18 (1957/58), S. 459, 521, vertreten, daß die handelspolitischen Kompetenzen auf dem Sektor Kohle und Stahl auf die Organe der EWG übergegangen sind. – Ein gemischtes Abkommen besteht beispielshalber mit Österreich über die Einführung direkter internationaler Eisenbahntarife (allerdings ohne Schiedsklausel): Abkommen vom 26. 7. 1957, ABl. EGKS Nr. 6 vom 20. 2. 1958.

[72] Vgl. die Abkommen in ABl. Nr. L 263/121 vom 27. 9. 1978, Nr. L 264/121 sowie Nr. L 265/121, jeweils vom 27. 9. 1978.

[73] Vgl. ABl. Nr. L 347/183 vom 22. 12. 1980.

[74] Vgl. ABl. Nr. 17 vom 19. 3. 1959 sowie ABl. Nr. 60 vom 24. 11. 1959.

[75] Vgl. z. B. im folgenden zum Übereinkommen über den Objektschutz von Kernmaterial (Art. 17) bzw. die Änderung des vorstehenden Abkommens mit Kanada vom 16. 1. 1978, das nunmehr eine Schiedsklausel mit einem ergänzenden Ernen-

nationalen Aufmerksamkeit für alle Vorgänge entsprechen, die den Brennstoffkreislauf betreffen. Die verschiedenen Aspekte der Versorgungs- und Planungssicherheit, der Sorge um die Nichtverbreitung von Atomwaffen sowie der nuklearen Sicherheit dürften dieser Absicherung der Einhaltung der Vertragspflichten zugrunde liegen.

Der Gerichtshof der Europäischen Gemeinschaften hat sich in dem Beschluß vom 14. November 1978 mit dem Entwurf eines Übereinkommens über den Objektschutz von Kernmaterial, kerntechnischen Anlagen und Nukleartransporten befaßt und unter anderem festgestellt, daß die von der IAEO entworfene Streitregelung einen dreistufigen Weg über Verhandlungen, ein Schiedsverfahren und im Falle der Erfolgslosigkeit bis hin zu einer Klageerhebung vor dem Internationalen Gerichtshof vorsieht[76]. Auf Grund der Klärung des Gerichtshof der Gemeinschaften, daß das Übereinkommen zum Teil in die Zuständigkeit der Gemeinschaft fällt, hat die Kommission am 10. Dezember 1979 dem Rat eine Beschlußvorlage über einen Übereinkommenstext übersandt, der hinsichtlich der möglichen Beteiligung der Gemeinschaft angepaßt worden war: Die Anrufung des Internationalen Gerichtshofs steht nach der neuen Fassung *alternativ* neben dem Schiedsverfahren, so daß es die Gemeinschaft vermeiden kann, eine Sonderrolle in Anspruch zu nehmen. Die Kommission hat folgerichtig bei der am 26. Oktober 1979 in Wien erfolgten Unterzeichnung erklärt, daß die Gemeinschaft nur an das alternativ vorgesehene Schiedsverfahren gebunden sei, und zugleich diejenigen Artikel des Übereinkommens benannt, die nicht in den Kompetenzbereich der Gemeinschaft fallen[77].

Das in den Abkommen mit Schweden vom 10. Mai 1976[78] und der Schweiz[79] über die Beteiligung an dem Fusionsforschungsprogramm der Gemeinschaft gewählte Schiedsverfahren weist wiederum dem Präsidenten des Internationalen Gerichtshofs für den Karenzfall das Ernennungsrecht des bzw. der Schiedsrichter zu. Die Abkommen führen zur Beteiligung an der Versuchsanlage »JET«, die in der Rechtsform eines gemeinsamen Unternehmens nach Art. 46 ff. Euratom-Vertrag gegründet worden ist. Im Hinblick auf Streitigkeiten zur Auslegung oder Anwendung der Satzung haben sich beide Drittstaaten bereit gefunden, sich einem Schiedsgerichtsverfahren zu unterwerfen, das bei fortbestehenden Meinungsverschiedenheiten letztlich auf

---

nungsrecht durch den Generalsekretär der OECD enthält: ABl. Nr. L 65/19 vom 8. 3. 1978.

[76] Beschluß Nr. 1/78 nach Art. 103 EAG-Vertrag, Slg. 1978, S. 2151, 2170; die Gemeinschaft ist durch Beschluß des Rates vom 9. 6. 1980 dem Übereinkommen beigetreten: ABl. Nr. L 149/41 vom 17. 6. 1980.

[77] Vgl. KOM(79)727 endg. vom 10. 12. 1979, S. 3 und Anhang 3; zur Offenlegung der Kompetenzabgrenzung zwischen der Gemeinschaft und ihren Mitgliedstaaten vgl. unten S. 423.

[78] ABl. Nr. L 162/28 vom 23. 6. 1976.

[79] Ohne Abschlußdatum in ABl. Nr. L 242/2 vom 4. 9. 1978.

ein Ernennungsrecht durch den Präsidenten des Gerichtshofs der Europäischen Gemeinschaften hinausläuft[80].

Diese Bereitschaft der Unterwerfung erklärt sich aus der gleichberechtigten Eingliederung in die Struktur des JET-Unternehmens, das enge institutionelle Bindungen zur Organisationsstruktur der Euratom-Gemeinschaft aufweist[81]. Insgesamt hebt sich der Euratom-Bereich mit seinem begrenzten Regelungsgegenstand auch in der Gestaltung der Streitbeilegungsverfahren deutlich von der Vertragspraxis der EWG ab, die die unterschiedlichsten Interessenlagen im auswärtigen Verkehr zu berücksichtigen hat.

6. Streitbeilegung durch echte Gerichte

In den völkerrechtlichen Beziehungen der Gemeinschaften können bestehende Gerichte zur Beilegung von Streitigkeiten eingeschaltet werden. Die Gemeinschaften können sich indes auch mit ihren Vertragspartnern dahingehend verständigen, daß sie ein neues Gericht einsetzen, um Streitfälle im Zusammenhang mit dem jeweiligen Vertragswerk entscheiden zu lassen. Bereits bestehende Gerichte sind in diesem Zusammenhang in erster Linie der Gerichtshof der Europäischen Gemeinschaften, der Europäische Gerichtshof für Menschenrechte sowie der Internationale Gerichtshof. Beispiele für neu zu errichtende Gerichtsinstanzen sind das Gericht des Europäischen Stillegungsfonds sowie das »International Tribunal for the Law of the Sea«.

a) Der *Gerichtshof der Europäischen Gemeinschaften* kommt grundsätzlich zur Entscheidung über Streitigkeiten mit Drittstaaten nicht in Betracht. Dies wurde eingangs näher dargelegt. Seiner Funktion als gemeinschaftsinternes Gericht, dem die Wahrung des Gemeinschaftsrechts obliegt, wird die völkervertragsrechtliche Streitschlichtung nicht gerecht. Bis auf Griechenland und die Türkei – im Rahmen der auf einen Beitritt abzielenden Assoziationsabkommen[82] – hat sich bisher kein Drittstaat bereit erklärt, sich zumindest alternativ der Gerichtsbarkeit des Gerichtshofs zu unterwerfen. Realistischer ist dagegen die Bildung gemischter Kammern oder eines Schiedsgerichts, an denen Richter des Gerichtshofs der Europäischen Gemeinschaften neben den vom Vertragspartner benannten Richtern beteiligt sind und die so für eine harmonisierende Auslegung des Gemeinschaftsrechts und der völkerrechtlichen Rechtsbeziehungen der Gemeinschaften sorgen können[83].

---

[80] Vgl. Art. 25 JET-Satzung vom 30. 5. 1978, ABl. Nr. L 151/10 vom 7. 6. 1978.

[81] Vgl. *M. Hilf,* Die Organisationstruktur der Europäischen Gemeinschaften (Beiträge zum ausländischen öffentlichen Recht und Völkerrecht, Bd. 79) (1982), S. 102 ff.

[82] Siehe oben Anm. 48 f.

[83] Dieser bereits von *P. Pescatore,* Les relations extérieures des Communautés européennes, Recueil des Cours (RdC) 103 (1961 II), S. 235, gewiesene Weg ist von der Kommission in ihrem Vorschlag einer Verordnung zum Abschluß des Übereinkommens über den Stilllegungsfonds (vgl. oben Anm. 17) aufgegriffen worden: gemäß Art. 6 sollen insgesamt sechs für das Gericht des Fonds zu berufende Richter vom

b) Ein mit Art. 181 EWGV vergleichbarer kompromissarischer Weg zur Vereinbarung der Zuständigkeit des *Europäischen Gerichtshofs für Menschenrechte* besteht nicht. Seine Zuständigkeit ist auf die Auslegung und Anwendung der EMRK begrenzt. Nur im Falle eines Beitritts der Gemeinschaften zur EMRK würde sich in ihren Verhältnissen zu Drittstaaten, die zugleich Unterzeichnerstaaten der EMRK sind, die Zuständigkeit des Europäischen Gerichtshofs für Menschenrechte ergeben können. Die Kommission der Europäischen Gemeinschaften strebt einen solchen Beitritt an [84], sie wird insoweit vom Wirtschafts- und Sozialausschuß [85] sowie auch vom Europäischen Parlament [86] unterstützt. Je nach den noch offenen Modalitäten eines Beitritts würden die Gemeinschaften sowohl in der Rolle des Beschwerdeführers als auch in der Rolle des beschwerdeführenden vertragsschließenden Teils (Art. 48 EMRK) Drittstaaten vor dem Europäischen Gerichtshof gegenüberstehen können. Die Kommission verspricht sich von einem Beitritt der Gemeinschaften unter anderem eine Stärkung ihrer internationalen Rechtspersönlichkeit, die in der Unterwerfung durch eine gemeinschaftsexterne Gerichtsinstanz zum Ausdruck kommen soll [87].

c) Auch das Statut des *Internationalen Gerichthofs* schließt für die Gemeinschaften den Zugang sowohl für das streitige Verfahren als auch für das Gutachtenverfahren aus [88]. Die vorstehend angeführten multilateralen Verträge, an denen die Gemeinschaften beteiligt sind und die die Zuständigkeit des

---

Gerichtshof aus seiner Mitte bestellt werden: ABl. Nr. C 208/3 vom 3. 9. 1976. Der Gerichtshof selbst hat in seinem Gutachten 1/76 (oben Anm. 17) Bedenken gegen diese Besetzung geäußert, da es nach dem Inhalt des Übereinkommens durchaus zu Vorlageverfahren mit übereinstimmenden Fragestellungen kommen könne, die die jeweils dann beteiligten Richter in eine unzulässige Doppelbeteiligung bringen würden: Slg. 1977, S. 761 f.; vgl. auch unten Anm. 94.

[84] Vgl. Memorandum der Kommission vom 4. 4. 1979 betreffend den Beitritt zur EMRK, Bull. EG, Beilage 2/79.

[85] Europäische Grundrechte-Zeitschrift (EuGRZ) 8 (1981), S. 63, sowie ABl. Nr. C 353/1 ff. vom 31. 12. 1980.

[86] Vgl. zur Zeit den Entwurf eines Berichts des Rechtsausschusses vom 20. 4. 1982 (Berichterstatter: G. Gonella), PE 74 231/rev., sowie die Entschließung vom 27. 4. 1979, ABl. Nr. C 127/69 f. vom 21. 5. 1979.

[87] Vgl. Anm. 84, S. 12. Die Regierungen der Mitgliedstaaten haben sich bisher nicht eindeutig festgelegt; aus der Literatur, in der die zustimmenden Äußerungen überwiegen, vgl. *C. D. Ehlermann/E. Noël,* Der Beitritt der EG zur EMRK, Schwierigkeiten – Rechtfertigung, in: Das Europa der Zweiten Generation, Gedächtnisschrift Sasse, Bd. 2 (1981), S. 685 ff.; eher ablehnend *L. N. Brown/J. McBride,* Observations of the Proposed Accession by the European Community to the European Convention of Human Rights, The American Journal of Comparative Law 29 (1981), S. 691 ff.

[88] Vgl. Art. 34 Ziff. 1 und Art. 65 des Statuts. Zu den Möglichkeiten, im Rahmen eines verschleierten Prozesses zwischen Staaten Rechtsfragen internationaler Organisationen zum Gegenstand des Verfahrens zu machen, vgl. im folgenden sowie bei *T. Furukawa,* Le double rôle de la Cour internationale de justice à l'égard des organisations internationales: protection et contrôle, in: Mélanges Reuter (1981), S. 293, 296 ff.

Internationalen Gerichtshofs vorsehen, berücksichtigen den begrenzten Zugang, indem sie diese Zuständigkeit lediglich alternativ neben anderen Wegen der Streitbeilegung bestimmt haben.

Dennoch nimmt das Statut von der Existenz auch anderer internationaler Organisationen außerhalb der Familie der Vereinten Nationen Kenntnis und bezieht diese ergänzend in anhängige Verfahren ein, soweit es der Sach- und Rechtsaufklärung dienen kann: Der Internationale Gerichtshof kann die Gemeinschaften als »öffentlich-rechtliche internationale Organisationen« um Auskunft über die vor ihn gebrachten Angelegenheiten ersuchen; der Gerichtshof »soll« solche Auskünfte auch entgegennehmen, wenn sie aus eigener Initiative vorgelegt werden (Art. 34 Ziff. 2). Denkbar ist auch, daß die Gemeinschaftsverträge oder die von ihnen geschlossenen völkerrechtlichen Verträge in Verfahren vor dem Gerichtshof eine Rolle spielen, etwa bei der Abgrenzung der Kompetenzbereiche zwischen den Gemeinschaften und ihren Mitgliedstaaten. In diesen Fällen würden die Gemeinschaften gemäß Art. 34 Ziff. 3 von den gesamten Unterlagen des schriftlichen Verfahrens in Kenntnis gesetzt und könnten eine schriftliche Stellungnahme abgeben, die auch in der jeweiligen mündlichen Verhandlung mit einem Vertreter der Gemeinschaften erörtert werden würde[89]. Auch im Gutachtenverfahren sieht Art. 66 des Status die Möglichkeit vor, daß der Gerichtshof an die Gemeinschaften als »internationale Organisation« herantritt, sofern diese über die Gutachtenfrage schriftlich oder auch mündlich Aufschluß geben können.

Ebenso wie die Parteifähigkeit ist für die Gemeinschaften der Weg der Intervention gemäß Art. 62 und 63 des Statuts verschlossen[89a]. Wollten sich die Gemeinschaften gegen völkerrechtswidriges Verhalten von Drittstaaten zur Wehr setzen oder wollten Drittstaaten gegen die Gemeinschaften vorgehen, so bliebe nur ein gemeinsames Vorgehen der Mitgliedstaaten als Streitgenossen[90] bzw. ein Verfahren gegen diese. Auch die Betrauung nur eines Mitgliedstaates mit der Führung eines Verfahrens im Interesse der übrigen Mitgliedstaaten und damit auch im Interesse der Gemeinschaften wird in weiter Auslegung des Statuts für zulässig gehalten[91], wenngleich auf diese Weise der auf Staaten begrenzte Zugang zu den streitigen Verfahren vor dem Gerichtshof aufgehoben werden würde.

Eine Klage gegen die Gesamtheit der Mitgliedstaaten dürfte nicht zu ihrem gewünschten Ziel führen. Zwar bleiben die Mitgliedstaaten auf Grund

---

[89] Vgl. Art. 69 der Verfahrensordnung vom 14. 4. 1978.

[89a] Rechtsfragen, die sich für die Intervention nach Art. 62 des Statuts stellen, sind erstmals vom Internationalen Gerichtshof im Fall *Plateau continental* (Tunisie/Jamahiriya arabe libyenne), requête à fin d'intervention, Urteil vom 14. 4. 1981, Recueil des arrêts 1981, S. 3 ff.; siehe hierzu *K. Oellers-Frahm,* Die Intervention nach Art. 62 des Statuts des Internationalen Gerichtshofs, ZaöRV 41 (1981), S. 579 ff.

[90] Vorgesehen im Statut im Zusammenhang mit der Benennung der *ad hoc*-Richter in Art. 31 Ziff. 5.

[91] So *Pescatore* (Anm. 83), S. 228, im Anschluß an *E. Hambro,* The Jurisdiction of the International Court of Justice, RdC 76 (1950 I), S. 160 f.

ihres Gründungsaktes gegenüber Drittstaaten verantwortlich; dies gilt indes nur solange, als die Drittstaaten die Gemeinschaften noch nicht als selbstverantwortliches Völkerrechtssubjekt anerkannt haben. Selbst wenn eine solche Anerkennung fehlen sollte, so könnte ein Verfahren gegen alle Mitgliedstaaten nur anhängig gemacht werden, wenn diese sich insgesamt unter den gleichen Bedingungen der Gerichtsbarkeit des Internationalen Gerichtshofs unterworfen haben. Dies ist – wie oben gezeigt wurde – bisher nicht der Fall gewesen und ist auch nicht gem. Art. 38 Abs. 5 der Verfahrensordnung im Wege des *forum prorogatum* ohne weiteres zu erwarten. Hieran würde auch eine gemeinsame Klage der Mitgliedstaaten gegen einen Drittstaat scheitern, der Rechte der Gemeinschaften verletzt hat. Eine gemeinsame Unterwerfungserklärung aller Mitgliedstaaten im Hinblick auf die völkerrechtlichen Streitbeilegungen der Gemeinschaften unter die Gerichtsbarkeit des Internationalen Gerichtshofs ist zwar denkbar [92], hätte aber angesichts der unterschiedlichen Haltung der Mitgliedstaaten zur obligatorischen Gerichtsbarkeit keine Aussicht auf Verwirklichung und würde sich weder gegenüber Staaten auswirken können, die die Gemeinschaften anerkannt haben, noch gegenüber Staaten, die sich nicht der Gerichtsbarkeit des Gerichtshofs unterworfen haben. Die Rechtskraft des Urteils würde sich überdies nicht auf die Gemeinschaften erstrecken (Art. 59 Statut). Diesen angedeuteten Schwierigkeiten ließe sich erst mit einer wohl wünschbaren, aber nicht absehbaren Öffnung des Zugangs zu den streitigen Verfahren vor dem Internationalen Gerichtshof auch für internationale Organisationen begegnen, die in weitem Umfang völkerrechtliche Vertragschließungsbefugnisse an Stelle ihrer Mitgliedstaaten wahrnehmen [93].

d) Im Ergebnis bietet sich für die Gemeinschaften nur der Weg der *Neugründung eines ständigen Gerichts* an, um die Einhaltung der von ihnen abgeschlossenen völkerrechtlichen Verträge einer echten Gerichtsbarkeit zu unterstellen. Der zusammen mit der Schweiz unternommene Versuch, ein ständiges Gericht im Rahmen des Übereinkommens über die Errichtung eines Europäischen Stillegungsfonds für die Binnenschiffahrt einzusetzen, ist vorerst gescheitert. Der Gerichtshof der Europäischen Gemeinschaften hatte in einem Gutachten festgestellt, daß gegen diesen neuartigen Weg der Streitschlichtung nichts einzuwenden sei, daß also der Wirtschaftsgemeinschaft die erforderliche Organisationskompetenz zur mitwirkenden Gründung zustehe [94]: ein angepaßtes Rechtsschutzverfahren liege in der Beteiligung eines Drittlandes begründet. Der Gerichtshof hat im Ergebnis aus anderen Grün-

---

[92] Angesprochen bei *Pescatore,* a.a.O., S. 230.

[93] Reformansätze in dieser Richtung werden immer wieder erörtert, vgl. bei *L. Gross,* Review of the Rôle of the International Court of Justice, AJIL 66 (1972), S. 479, 487.

[94] EuGH 26. 4. 1977 – Gutachten 1/76 – Slg. 1977, S. 741, 761 f.; als Rechtsgrundlage diente allein Art. 75 EWGV, der die Kompetenzen der Gemeinschaft im Bereich der Verkehrspolitik festlegt.

den den Übereinkommensentwurf mit dem Gemeinschaftsrecht für unvereinbar erklärt. Das vorläufige Scheitern dieses Unterfangens ist indes nicht auf den Ausgang des Gutachtenverfahrens zurückzuführen, sondern auf die veränderten wirtschaftlichen Rahmenbedingungen, die die beteiligten Kreise nicht mehr auf den Abschluß des Abkommens drängen lassen.

Ein weiteres Mitwirken der Europäischen Wirtschaftsgemeinschaft an der Gründung neuer, dem jeweiligen Vertragszusammenhang angepaßter Gerichtsinstanzen liegt im Rahmen der Dritten *Seerechtskonferenz* vor[95]. Der im April 1982 mit Mehrheit beschlossene Text der neuen Seerechtskonvention kennt vier alternative Streitbeilegungsverfahren für den Fall, daß es den Streitparteien nicht gelingt, durch Verhandlung, Vermittlung oder anderweitige einverständliche Regelung den Streit beizulegen. Es handelt sich nach Art. 287 des Konventionstextes um die Gerichtsbarkeit des neu zu schaffenden »International Tribunal for the Law of the Sea«, des Internationalen Gerichtshofs, eines allgemeinen Schiedsgerichts und mehrerer Sonderschiedsgerichte, die für die einzelnen von der Konvention abgedeckten Sachgebiete zuständig sein können. Die Konferenz hat in ihrer Schlußphase über einen Beitritt der internationalen Organisationen entschieden[96]. Im Anhang IX zur Konvention werden die Vorschriften der Konvention *mutatis mutandis* für die internationalen Organisationen für anwendbar erklärt. Der Zugang zum Internationalen Gerichtshof wird für sie gemäß Art. 7 Ziff. 1 ausgeschlossen. Die internationalen Organisationen und damit auch die Gemeinschaft sind aufgefordert, bei der Unterzeichnung eine Liste der Sachgebiete aufzustellen, für die die Organisationen an der Stelle ihrer Mitgliedstaaten zuständig sind. Die Gemeinschaft hatte sich gegen die Abgabe einer solchen Erklärung – wie auch schon in früheren Fällen[97] – vergeblich gewandt, denn beim verflochtenen Stand der Integration ist jede summarische Erklärung unscharf und trifft kaum die Trennungslinie der Kompetenzbereiche, die sich ständig verändert. Entsprechend der Aufteilung der Kompetenzen kann es zu Streitfällen kommen, in denen die Gemeinschaft zusammen mit ihren Mitgliedstaaten beteiligt ist oder auch zu Fällen, die ausschließlich die Gemeinschaft betreffen. Art. 7 Ziff. 3 des Anhangs IX versucht den durch die unterschiedliche Wahl der Streitbeilegungsinstanzen drohenden möglichen Verwicklungen durch die subsidiäre Verpflichtung auf die Schiedsgerichtbarkeit zu lösen[98].

---

[95] Vgl. zum folgenden *R. Bernhardt,* Law of the Sea, Settlement of Disputes, in: Bernhardt (Hrsg.), Encyclopedia of Public International Law, Instalment 1 (1981), S. 133 ff.; *derselbe,* Die Streitbeilegung im Rahmen der Neuordnung des Seerechts, ZaöRV 38 (1978), S. 959 ff., sowie *L. Caflisch,* Le règlement judiciaire et arbitral des différends dans le nouveau droit international de la mer, in: Festschrift Bindschedler (1980), S. 351 ff.

[96] Vgl. Art. 1*bis* in Verbindung mit Art. 305 Ziff. 1 (e) unter Hinweis auf Anhang IX.

[97] Vgl. oben Anm. 66 und 77.

[98] »Where an international organization and one or more of its States members are joint parties to a dispute, or parties in the same interest, the organization shall be

Die insgesamt den Unterzeichnerstaaten belassene Gestaltungsfreiheit, die zu einer Unübersichtlichkeit der im konkreten Fall gegebenen Streitbeilegungsverfahren führt, sollte es der Gemeinschaft ermöglichen, bei einem Beitritt die ihrem Kompetenzbereich angemessenen Streitbeilegungswege zusammen mit den Mitgliedstaaten oder an ihrer Stelle zu wählen.

III. Rechtliche Folgerungen aus der Ausgestaltung der Streitbeilegungsverfahren

7. Erkenntnisse aus der Vertragspraxis

a) Die nachgewiesene Vertragspraxis vermittelt kein einheitliches Bild, das eindeutige rechtliche Folgerungen aufdrängen könnte. Zwar sind nahezu alle *Verfahren der Streitschlichtung* bis auf den den Gemeinschaften versperrten Weg zum Internationalen Gerichtshof vertreten. Ein systematisches Vertragsmuster oder eine einheitliche Vertragspolitik hinsichtlich der Wahl der Streitbeilegungsverfahren ist indes auf den ersten Blick nicht zu erkennen.

Immerhin gibt es einzelne Vertragsgruppen, die sich abheben. So fand sich in keinem typischen Handelsabkommen ein gerichtliches Streitbeilegungsverfahren. Ein gemeinsames Erwägen und Verhandeln der wirtschaftlichen Balance dieser Abkommen soll über Meinungsverschiedenheiten hinweghelfen. Dagegen enthalten alle »echten« Assoziationsabkommen, die die Vertragspartner in eine institutionelle Verbindung und Nähe zur Gemeinschaft bringen sollen, zumindest eine *ad hoc*-Schiedsklausel, wenn nicht sogar ein ständiges Schiedsgericht. Das gleiche trifft für die neueren Abkommen nach dem Euratom-Vertrag zu, die auf der Verläßlichkeit der Vertragserfüllung aufbauen.

Eine eher uneinheitliche Vertragsgestaltung läßt sich in jenem weiten Bereich der Verträge feststellen, die weder den reinen Warenaustausch noch die institutionelle Annäherung an die Gemeinschaft zum Vertragsgegenstand haben. Während für die Freihandelsabkommen mit den Rest-EFTA-Staaten nach längeren Überlegungen generell eine nicht-gerichtliche Streitbeilegung gewählt wurde, ist der Außenstehende angesichts der uneinheitlichen Vertragspraxis etwa bei den Fischereiabkommen eher ratlos. Die Frage der Streitbeilegung steht in der Regel nicht im Mittelpunkt von Verhandlungen. Als typische Randklausel ist sie von einem Vertragspartner, der auf eine be-

---

deemed to have accepted the same procedures for the settlement of disputes as the States members; provided that where a State member has only chosen the International Court of Justice under article 287, the organization and the State member concerned shall be deemed to have accepted arbitration in accordance with annex VII, unless the parties to the dispute otherwise agree«. Zu den hierbei auftretenden rechtstechnischen Schwierigkeiten bei der Besetzung der Richterbank vgl. im folgenden S. 418 f., 425.

stimmte Ausgestaltung besonders drängt, in ihrem Inhalt unter Umständen leichter zu beeinflussen als echte Schlüsselfragen des Vertrages. So haben Schweden und der Senegal auf die Aufnahme von Schiedsklauseln in ihre Fischereiabkommen gedrängt, während Norwegen offensichtlich auf Grund der der Gemeinschaft überlegenen Verhandlungsposition eine solche Lösung nicht anstrebte. Insgesamt ist auf seiten der Gemeinschaften bisher hinsichtlich der Gestaltung der Streitbeilegungsklauseln keine eindeutige Vertragspolitik zu erkennen. Die Sorge möglicher Jurisdiktionskonflikte, die auf dem Nebeneinander der nationalen, gemeinschaftsrechtlichen und einer zusätzlichen internationalen Gerichtsbarkeit beruhen kann, soll etwa bei der Entscheidung für die Verhandlungsklausel in den EFTA-Abkommen mit ausschlaggebend gewesen sein.

b) Überblickt man die in die vorliegende Untersuchung einbezogenen 34 Verträge der Gemeinschaften, die ein schiedsgerichtliches Streitbeilegungsverfahren enthalten, so fällt die *Vielfalt der Klauseln* auf. Nahezu alle Schiedsklauseln sehen ein *ad hoc* einzuleitendes Verfahren vor. Das ständige Schiedsgericht nach den Jaunde-Abkommen ist eine Ausnahme geblieben. In der Regel begrenzen die Vertragsparteien ihre Verpflichtungen auf den Versuch, einen Streit durch Verhandlungen oder durch Vergleich beizulegen. Die gemischten Vertragsorgane bilden hierzu den Rahmen.

Die meisten Schiedsklauseln folgen dem typischen Muster, das auch den Verträgen der Mitgliedstaaten zugrunde liegt: Drei Schiedsrichter sind zu ernennen, die in Anwendung des jeweiligen Vertrages bzw. ergänzend des Völkerrechts zu entscheiden haben. Eine Entscheidung *ex aequo et bono* ist in keiner Klausel vorgesehen. Sofern überhaupt etwas zu dem Verfahren gesagt ist, so wird dem Schiedsgericht die Verfahrensherrschaft zuerkannt. Die Entscheidungen sind bindend, die Kosten werden in der Regel geteilt, nur im Jaunde II-Abkommen erklärte sich die Gemeinschaft bereit, einer Kostenregelung *ex aequo et bono* zuzustimmen.

Für die Vertragspartner der Gemeinschaften werfen die gemischten Verträge die schwierigsten Fragen auf. In den zweiseitig angelegten Handels-, Kooperations- und Assoziationsabkommen ist bei einer gleichzeitigen Beteiligung der Mitgliedstaaten festgelegt, daß die Gemeinschaft und die Mitgliedstaaten im Falle der Durchführung des Schiedsverfahrens als eine Partei gelten. Im Rahmen der neuen Seerechtskonvention ist im Anhang IX in Art. 4 Ziff. 4 ausdrücklich festgehalten:

> »The participation by such international organizations shall in no case entail an increase of the representation that their States members, which are States parties, would otherwise be entitled to, including rights in decision-making«.

Internationale Organisationen, also auch die Gemeinschaft, sollen nicht schlechter stehen, als wenn ihre Mitgliedstaaten allein in einen Streitfall verwickelt worden wären; sie sollen aber auch keinen zusätzlichen Repräsentationsgewinn erzielen. Nur wenn Fragen in den Kompetenzbereichen der Gemeinschaft und ihrer Mitgliedstaaten im Streit sind, ist eine zusätzliche Vertretung gerechtfertigt: Die Zahl der von allen beteiligten Streitparteien gemeinsam zu ernennenden Schiedsrichter hat in diesem Fall stets größer zu sein (n + 1) als die Zahl der von jeder Partei autonom zu ernennenden Schiedsrichter (n) [99]. Hier zeigen sich die besonderen Schwierigkeiten bei der sich in der Tendenz – trotz der gleichzeitigen Verdichtung der Gemeinschaftskompetenzen – ausweitenden Zahl der gemischten Abkommen.

Die Effektivität einer Schiedsklausel hängt entscheidend davon ab, ob die Parteien die Möglichkeit haben, die Ernennung der Schiedsrichter zu blockieren, oder ob Vorsorge getroffen ist, im Falle der Obstruktion durch eine Partei das Schiedsgericht dennoch durch eine »neutrale« Instanz einsetzen zu lassen. Einigen sich die Partner in den vorgeschalteten Verhandlungen nicht, so werden sie auch kaum eine Übereinstimmung über die Zusammensetzung des Schiedsgerichts erzielen.

Nur ein geringer Teil der Klauseln legt Fristen zur Ernennung der Schiedsrichter fest. Etwa die Hälfte der von den Gemeinschaften vereinbarten Klauseln laufen auf ein Bestimmungsrecht durch eine neutrale Instanz hinaus: in 10 Abkommen ist der Präsident des Internationalen Gerichtshofs berufen, im Karenzfall einer Partei den von ihr zu benennenden Schiedsrichter zu bestellen oder im Falle der Nichteinigung den dritten Schiedsrichter zu berufen. Eine entsprechende Zuständigkeit der jeweiligen gemischten Vertragsorgane hat nicht die gleiche Wirkung, weil diese Organe stets im Einvernehmen der beteiligten Partner entscheiden, eine Obstruktion einer Partei also nicht überwunden werden kann.

Für die Gemeinschaften mag die bisherige Vertragspraxis Anlaß sein, bei der Aushandlung künftiger Verträge die Ausgestaltung der Schiedsklauseln näher zu bedenken. Allein die Effektivität des Schiedsverfahrens muß indes nicht immer dem Vertragsinteresse der Parteien entsprechen: Beiden Vertragspartnern kann daran gelegen sein, das Schiedsverfahren nur im Einvernehmen zur Anwendung zu bringen.

8. Folgerungen aus der Vereinbarung von Schiedsgerichtsverfahren

Das Gestaltungsermessen der in den Gemeinschaften für den Vertragsschluß zuständigen Organe ist bei der Wahl zwischen dem Verhandlungsweg oder dem Weg zur Streitbeilegung über ein Schiedsverfahren weitgehend frei. Die Wahl wird in Anbetracht des institutionellen Aufwandes nur dann auf die Bildung oder den Beitritt zu einer Instanz echter obligatorischer Gerichtsbarkeit fallen, wenn die Gemeinschaften Hoheitsbefugnisse übertragen (Stillegungsfonds) oder sich einer internationalen Kontrolle (EMRK) unterstellen

---

[99] Vgl. Art. 3 (g) des Anhangs VII zur Seerechtskonvention; hierzu *B. H. Oxman,* The Third United Nations Conference on the Law of the Sea: The Tenth Session (1981), AJIL 76 (1982), S. 1 ff., 16.

wollen. Streben die Organe der Gemeinschaften dagegen den Einbau eines Schiedsverfahrens in einen ihrer völkerrechtlichen Verträge an, dann werden sie mit in Erwägung zu ziehen haben, ob sie damit einen Jurisdiktionskonflikt mit der gemeinschaftseigenen Gerichtsbarkeit heraufbeschwören (a)), ob dieser Weg besser als der Weg gegenseitiger Verständigung der Vollzugssicherung dienen kann (b)), welche Auswirkungen die Wahl auf die Auslegung des Vertrages haben wird (c)), welche Sonderlagen im Falle der gleichzeitigen Beteiligung der Mitgliedstaaten zu beachten sind (gemischte Verträge) (d)) und wie die Verfahrensherrschaft unter den Gemeinschaftsorganen aufzuteilen sein wird (e)).

a) Jurisdiktionskonflikte

Die von den Gemeinschaften abgeschlossenen Verträge unterliegen im Rahmen ihrer Rechtsordnung der Auslegung des Gerichtshofs im Verfahren der Vorabentscheidung u. a. nach Art. 177 EWGV. Unterschiedlichen Auslegungen durch nationale Gerichte kann auf diesem Wege wirksam begegnet werden. Ein zur Auflösung von Fragen der Auslegung und Anwendung eines Vertrages eingesetztes Schiedsgericht wäre als völkerrechtliche Instanz an die Auslegungen des Gerichtshofs der Gemeinschaften nicht gebunden. Eine befürchtete Verwirrung bei der Vertragshandhabung durch die Schaffung einer den Gemeinschaften überlagerten und unter Umständen zu widersprechenden Auffassungen gelangenden Schiedsinstanz könnte die Organe der Gemeinschaften veranlassen, die völkerrechtliche Vertragsauslegung eher in die Hände der Vertragspartner mit dem Ziel einvernehmlicher Vertragsauslegung zu legen.

Die Vorgeschichte der von der Wirtschaftsgemeinschaft ausgehandelten Freihandelsabkommen mit den EFTA-Staaten zeigt, daß zwischen den Vertragsparteien keine Einigung über die Einrichtung einer gemeinsamen Schiedsinstanz und damit über das Ziel einer übereinstimmenden Rechtsentwicklung in bezug zur Gemeinschaftsrechtsordnung zustande gekommen ist[100]. Selbst wenn sich die Wirtschaftsgemeinschaft mit den Rest-EFTA-Staaten auf eine *ad hoc*-Schiedsklausel geeinigt hätte, so hätte sie jedenfalls keine Jurisdiktionskonflikte befürchten müssen. Ein Spruch des Schiedsgerichts würde den geschlossenen völkerrechtlichen Vertrag mit bindender Wirkung für beide Partner einheitlich auslegen. Diese Auslegungen, denen die Vertragspartner beim Abschluß des Vertrages antizipierend zugestimmt haben, würden auch für den Gerichtshof der Europäischen Gemeinschaften als Organ der Gemeinschaft im Sinne des Art. 228 Ziff. 2 EWGV verbindlich sein. Dieser hätte ebenso wie die innerstaatlichen Gerichte der Vertragspartner die Verpflichtung, seine Vertragsauslegung der auf der völkerrechtlichen Ebene erfolgten Auslegung anzupassen. Einem Jurisdiktionskonflikt

---

[100] Vgl. *E. J. Mestmäcker,* Gerechte Wettbewerbsbedingungen in den Freihandelsabkommen der EG, in: Festschrift Zweigert (1981), S. 681, 691.

dürfte damit vorgebeugt sein. *Ohne* die Vereinbarung einer Schiedsinstanz besteht zumindest theoretisch viel eher die Gefahr, daß mangels einer gerichtlichen Auslegung auf der völkerrechtlichen Ebene der Gerichtshof der Gemeinschaften den Vertrag anders deuten könnte als die Gerichte der Vertragspartner[101]. Sollten jedoch die Vertragsparteien im bloßen Verhandlungswege zu einer vereinbarten Auslegung kommen, so wäre auch diese für die Gerichte beider Seiten verbindlich. Trotz der möglicherweise verwirrenden Verfahrensvielfalt würde im Ergebnis die Vereinbarung eines Schiedsverfahrens eher Jurisdiktionskonflikte ausräumen denn begründen.

b) Vollzugssicherung

Die Aufnahme einer verbindlichen Schiedsregelung kann sowohl auf der völkerrechtlichen Ebene als auch im Rahmen der Organe der Gemeinschaften zur Vollzugssicherung des Vertrages beitragen. Sind lediglich Konsultationen oder Verhandlungen zur Konfliktlösung vereinbart, so behält jeder Partner die Möglichkeit, die Erreichung der im Vertrag vereinbarten Ziele zu blockieren oder zu verzögern. Eine Eskalation einseitiger Maßnahmen wie das Ergreifen der in den Freihandelsabkommen mit den Rest-EFTA-Staaten vorgesehenen Schutzklauseln bis hin zur Kündigung des Vertrages wegen nicht auflösbarer Auslegungsdivergenzen wären bei dem Ausbleiben einer Verhandlungslösung denkbar.

Die Präventivwirkung eines drohenden Schiedsverfahrens hat sich u. a. während der zuvor erwähnten Verhandlungen über die Neufestsetzung der Preise für AKP-Zucker, die im April 1982 erfolgte, gezeigt. Der wiederholte Hinweis auf das im Lome II-Abkommen festgelegte Schiedsverfahren hat zu der Einigung im Rat der Gemeinschaft sowie auf dem Verhandlungsweg mit den AKP-Staaten beigetragen[102]. Die vollzugssichernde Funktion eines vereinbarten Schiedsverfahrens kann damit *doppelt,* d. h. sowohl im völkerrechtlichen als auch im innergemeinschaftlichen Bereich wirken. Diese Folgerung könnte auch für wirtschaftlich komplexe Streitfälle gezogen werden, für die es schwerfällt, eindeutige rechtliche Maßstäbe zu finden. Dennoch ist je nach der Art der Vertragsbeziehungen zu differenzieren: So dürfte im Rahmen des GATT weitgehend Übereinstimmung bestehen, daß sich nicht alle Streitfälle gerichtlich auflösen lassen, ohne dem Vertragswerk insgesamt Schaden zuzufügen. Als *wrong cases,* die sich nicht für eine gerichtliche Streitbeilegung eignen, werden im GATT-Rahmen die Fälle der verbreiteten Regelverletzungen

---

[101] Diese Gefahr dürfte in der Praxis gering sein: Im Hinblick auf die Deutung des auch in den EFTA-Abkommen enthaltenen Verbotes der »Maßnahmen gleicher Wirkung« hat der Gerichtshof der Gemeinschaften u. a. im *Polydor*-Fall (vgl. oben Anm. 31), hekt., S. 29 ff., das gleiche Auslegungsergebnis erzielt wie die obersten Gerichte in der Schweiz und in Österreich (Nachweise in EuGH, a.a.O., S. 17, sowie bei *Mestmäcker,* a.a.O., S. 686).

[102] Vgl. oben Anm. 64.

in wirtschaftlich schwierigen Zeiten genannt, die stillschweigende Hinnahme der Verletzung von nicht mehr funktionsfähigen Regeln (Landwirtschaftsbereich, Selbstbeschränkungsabkommen) sowie Fälle, in denen Grundsatzfragen der jeweiligen Rechtssysteme aufgeworfen werden [103].

c) Rückwirkung auf die Vertragsauslegung

Die Vereinbarung eines Schiedsverfahrens kann die Vertragsauslegung beeinflussen. Steht hinter den Vertragspartnern eine Instanz, die gegebenenfalls die Vertragshandhabung für beide Seiten verbindlich festlegt, so ist zugleich sichergestellt, daß die Vertragspartner im gegenseitigen Verhältnis zu einer übereinstimmenden Vertragsanwendung gelangen. Der Gerichtshof der Europäischen Gemeinschaften hat sich bei der Frage, ob eine Bestimmung eines völkerrechtlichen Vertrages der Gemeinschaften als unmittelbar anwendbar verstanden werden muß, unter anderem von der institutionellen Struktur der jeweiligen Verträge leiten lassen. Hat ein Vertrag Vertragsorgane eingesetzt, die jederzeit eine einheitliche Anwendung des Vertrages sicherstellen können, so stehen der Annahme der unmittelbaren Anwendung weniger Bedenken entgegen als in Fällen, in denen der Gerichtshof nicht mit Sicherheit davon ausgehen darf, daß der Vertragspartner der Gemeinschaften den jeweiligen Vertrag ebenfalls als unmittelbar anwendbar versteht. Die Qualifizierung einer Vertragsnorm als unmittelbar anwendbar hängt sicher in erster Linie von der Natur und Reichweite der jeweils vereinbarten Rechte und Pflichten ab. Die Annahme einer solchen Qualifizierung hat weitreichende Folgen: Der Gemeinschaftsbürger kann sich auf die völkerrechtliche Vertragsbestimmung zumindest auch gegenüber entgegenstehendem sekundärem Recht der Gemeinschaften berufen, und der Gerichtshof kann der Rechtsstellung des Einzelnen zur Durchsetzung verhelfen, ohne daß stets sichergestellt wäre, daß der völkerrechtliche Vertragspartner in seiner Rechtsordnung den Vertrag mit der gleichen Intensität anwenden würde. Jedenfalls bietet die Vereinbarung eines Schiedsverfahrens dem Gerichtshof einen Anhaltspunkt dafür, daß die Vertragspartner ein stabiles Vertragsverhältnis schaffen wollten, in dem eine übereinstimmende Vertragsanwendung weitgehend sichergestellt werden soll. Der Gerichtshof hat denn in seiner bisherigen Rechtsprechung völkerrechtlichen Vertragsbestimmungen nur in solchen Verträgen eine unmittelbare Anwendbarkeit bestätigt, die durch die Absicherung durch ein Schiedsverfahren ein stabiles Vertragsverhältnis erkennen lassen: Dies trifft für das Jaunde I-Abkommen [104] ebenso zu wie für das Assoziationsabkommen mit Griechenland [105], nicht indes für das Freihandelsabkommen mit Portugal [106]

[103] Zu den Fällen der *ordinary noncompliance,* der *inoperative rules* sowie der Fälle des *over-taxing the procedure* vgl. *Hudec,* (Anm. 35), S. 159 ff.

[104] EuGH 5. 2. 1976 – *Bresciani,* Rs. 87/75 – Slg. 1976, S. 129.

[105] Vgl. zuletzt EuGH 29. 4. 1982 – *Pabst & Richarz,* Rs. 17/81 – hekt., S. 35 ff.

[106] Im Verfahren *Polydor* (in dem der Gerichtshof die Frage der unmittelbaren Anwendbarkeit letztlich nicht entscheiden mußte: vgl. oben Anm. 31 f.) hatten neben

bzw. für das GATT[107]. Die Vereinbarung eines Schiedsverfahrens dürfte im Ergebnis zumindest *ergänzend* ein wichtiges Auslegungselement für den Gerichtshof sein, um einer Vertragsbestimmung im Rahmen der Gemeinschaftsrechtsordnung eine unmittelbare Anwendung zu bestätigen[108].

Die Verrechtlichung des Verfahrens der völkerrechtlichen Streitbeilegung führt damit indirekt im Gemeinschaftsbereich zu einer stärker hervortretenden Rolle der rechtsprechenden Gewalt bei der Vertragshandhabung.

d) Schiedsverfahren in gemischten Verträgen

Die Durchführung von Schiedsgerichtsverfahren wirft in gemischten Verträgen besondere rechtliche Gestaltungsprobleme auf. In der bisherigen Vertragspraxis finden sich nur in Ausnahmefällen vorgeschriebene Lösungswege.

Geht man von der Vorstellung vermischter Kompetenzbereiche aus[109], so könnte das Schiedsgerichtsverfahren entweder die Gemeinschaft oder ihre Mitgliedstaaten oder beide zusammen erfassen. Indes stößt diese Trennung auf praktische und rechtliche Schwierigkeiten. Sowohl die Gemeinschaften als auch die Mitgliedstaaten zögern, den Vertragspartnern eine Aufstellung über die jeweiligen Kompetenzbereiche zu geben[110]: Der Graubereich an der Trennungslinie dürfte kaum zuverlässig abzugrenzen sein, die evolutionäre Verschiebung der Kompetenzbereiche steht jeder momentanen Festschreibung entgegen.

Im Falle der aktiven Betreibung eines Schiedsverfahrens stellt sich die Abgrenzungsfrage in gleicher Weise wie im Falle der Klage eines Drittstaates gegen die Gemeinschaften und/oder ihre Mitgliedstaaten. Art. 4 des Internen Abkommens zur Durchführung des Lomé II-Abkommens ist ein Versuch, wenigstens gemeinschaftsintern einen Weg zu finden, um jedem Mitglied-

---

der Kommission die fünf sich am Verfahren beteiligenden Mitgliedstaaten auf die unterschiedlichen Ziele und Wirkungen des EWG-Vertragswerks gegenüber dem Freihandelsabkommen mit Portugal hingewiesen, die u. a. auch in der institutionellen Ausgestaltung deutlich werden.

[107] Vgl. oben Anm. 34 zum Fall *International Fruit Company* (1972) sowie die zur Zeit anhängigen Vorlagen des italienischen Kassationsgerichtshofs in den Rs. 266–269/81.

[108] Daß die Existenz eines Schiedsverfahrens nicht stets die Frage der unmittelbaren Anwendbarkeit bestimmter Vertragsnormen präjudiziert, wurde bereits gezeigt: In dem Verfahren EuGH 24. 11. 1977 – *Razanatsimba,* Rs. 65/77 – Slg. 1977, S. 2229, hat die Frage der unmittelbaren Anwendbarkeit des Art. 62 des Lomé I-Abkommens eine Rolle gespielt, das ein – allerdings nicht erzwingbares – Schiedsverfahren enthält (vgl. oben Anm. 61): Der Gerichtshof konnte die Frage offenlassen, von Generalanwalt Reischl wurde die Frage verneint.

[109] Zutreffend gegen die Annahme einer rechtlichen *Verschmelzung* der Kompetenzbereiche in gemischten Abkommen: *K. M. Meessen,* Das Abkommen von Lomé als gemischter Vertrag, EuR 15 (1980), S. 36, 49.

[110] Siehe oben S. 416.

staat die Verteidigung der Interessen seines Kompetenzbereichs zu ermöglichen:

> »Hält ein Mitgliedstaat in Bereichen, die unter die Zuständigkeit der Mitgliedstaaten fallen, die Anwendung des Artikels 176 des Abkommens für erforderlich, so konsultiert er vorher die anderen Mitgliedstaaten.
> Hat der AKP-EWG-Ministerrat zum Vorgehen des in Absatz 1 genannten Mitgliedstaats Stellung zu nehmen, so entspricht die Haltung der Gemeinschaft der des betreffenden Mitgliedstaats, es sei denn, daß die im Rat vereinigten Vertreter der Regierungen der Mitgliedstaaten etwas anderes beschließen« [111].

Die Konsultation nach Abs. 1 sollte richtigerweise auch die Gemeinschaft (Kommission) einschließen, da gerade die Abgrenzung der Kompetenzbereiche in Frage stehen kann.

Im Verfahren selbst sollte der beklagte Drittstaat nicht einwenden können, daß die Klage des Mitgliedstaates sich auf einen Sachbereich beziehe, der in die ausschließliche Zuständigkeit der Gemeinschaft falle. Nähme das Schiedsgericht hierzu Stellung, so würde es nach dem EWG-Vertrag die Kompetenzbereiche abgrenzen und könnte dabei zu Ergebnissen gelangen, die dem Gemeinschaftsrecht widersprächen. Diese für die Gemeinschaften untragbare Folge kann nur vermieden werden, wenn man entweder bei der Festschreibung des Verfahrens dem beklagten Drittstaat im Wege der »Streitverkündung« das Recht einräumt, vor der Einlassung zur Sache von der Gemeinschaft und dem betreffenden Mitgliedstaat gemeinsam eine für das Verfahren bindende Erklärung über die Aufteilung der Kompetenzbereiche zu verlangen, so daß das Schiedsgericht über die Zuständigkeitsfrage nicht mehr aus eigener Kompetenz zu entscheiden hätte. Oder der Streitfall müßte ohne Prüfung der gemeinschaftsinternen Kompetenzaufteilung zur Sache entschieden werden in der Annahme, daß die Gemeinschaften im Falle der Kompetenzanmaßung des Mitgliedstaats im internen Vertragsverletzungsverfahren nach Art. 169 EWGV – oder im umgekehrten Verhältnis durch Klage des Mitgliedstaats gemäß Art. 173 Abs. 1 EWGV – eine *vor* der Endentscheidung des Schiedsgerichts erfolgende Klärung herbeigeführt haben würden.

Wird die Gemeinschaft oder einer ihrer Mitgliedstaaten verklagt, so könnten diese grundsätzlich wiederum nicht einwenden, die Auswahl des Klagegegners entspreche nicht der internen Kompetenzverteilung zwischen der Gemeinschaft und ihren Mitgliedstaaten [112]. Diese sollten sich allein oder gemeinsam der Klage des Drittstaates stellen müssen, solange sie ihm gegen-

---

[111] ABl. Nr. L 347/207 vom 22. 12. 1980.

[112] Vgl. Art. 27 Satz 1 Wiener Vertragsrechtskonvention: »A party may not invoke the provisions of its internal law as justification for its failure to perform a treaty«. Auszunehmen dürften die Fälle sein, in denen die interne Kompetenzverteilung eindeutig ist und diese dem Drittstaat bekannt ist bzw. bekannt sein müßte.

über nicht gemeinsam eine verbindliche Erklärung über die Abgrenzung der Kompetenzbereiche abgegeben haben [113].

Eine generelle, im Vertrag selbst oder anläßlich des Beitritts der Gemeinschaft zu einem Abkommen abgegebene Erklärung hilft im konkreten Streitfall nicht viel weiter. Wie gezeigt, verlangt der im April 1982 angenommene Text des Anhangs IX der Seerechtskonvention in Art. 5 Ziff. 1 von vornherein von den sich beteiligenden internationalen Organisationen die Hinterlegung einer generellen Erklärung, die die Kompetenzbereiche zwischen der Organisation und ihren Mitgliedstaaten abgrenzt. Liegt eine solche Erklärung vor, so muß sich zwar ein klagender Drittstaat in einem späteren Gerichtsverfahren die hierin summarisch festgeschriebene Kompetenzabgrenzung entgegenhalten lassen. Indes dürfte die Kompetenzfrage im konkreten Streitfall trotz des Vorliegens der generellen Erklärung Zweifelsfragen aufwerfen. Eine völlig zufriedenstellende Lösung ist bisher in der Vertragspraxis nicht gefunden worden. In jedem Fall sollte die Gemeinschaft bereit sein, im Zweifelsfall gegenüber dem Drittstaat eine subsidiäre Verantwortlichkeit für die Erfüllung des gemischten Abkommens zu übernehmen.

Für die Besetzung der Schiedsinstanz wird der allgemeine Grundsatz zur Anwendung kommen, daß die gemeinsame Beteiligung der Gemeinschaft und ihrer Mitgliedstaaten nicht zu einer übermäßigen Vertretung auf der Richterbank führen darf. In der Seerechtskonvention kommt dieser Gedanke im Anhang IX in Art. 4 Ziff. 3 zum Ausdruck.

e) Gemeinschaftsinterne Verfahrensherrschaft

Die vorstehenden Hinweise machen auf den unsicheren Boden aufmerksam, auf dem sich die Gemeinschaften besonders im Falle gemischter Abkommen bei der Wahrnehmung ihrer völkerrechtlichen Außenkompetenzen bewegen. Die Vereinbarung von Schiedsverfahren hat – wie zu zeigen war – in erster Linie eine Präventivfunktion, die Vertragsverletzungen verhindern soll. Es wird auch auf absehbare Zeit nicht zur Durchführung eines Schiedsverfahrens kommen, das wiederum eine Fülle von Gestaltungsfragen im gemein-

---

[113] In einer exemplarischen Klausel sieht Anhang B des Übereinkommens zum Schutz des Rheins gegen chemische Verunreinigung vom 3. 12. 1976 (ABl. Nr. L 240/51, 60 vom 19. 9. 1977 unter Ziffer 8) eine Klärung der Kompetenzfrage für ein etwaiges Schiedsverfahren vor: »Im Fall von Streitigkeiten zwischen zwei Vertragsparteien, von denen nur eine ein Mitgliedstaat der Europäischen Wirtschaftsgemeinschaft ist – die ihrerseits selbst Vertragspartei ist –, richtet die andere Partei den entsprechenden Antrag gleichzeitig an diesen Mitgliedstaat und an die Gemeinschaft, die dieser Partei gemeinsam innerhalb einer Frist von zwei Monaten nach Eingang des Antrags mitteilen, ob der Mitgliedstaat, die Gemeinschaft oder der Mitgliedstaat und die Gemeinschaft gemeinsam als Streitpartei auftreten. Ergeht eine solche Mitteilung nicht innerhalb dieser Frist, so gelten der Mitgliedstaat und die Gemeinschaft für die Anwendung dieses Anhangs als ein und dieselbe Streitpartei. Das gleiche gilt, wenn der Mitgliedstaat und die Gemeinschaft gemeinsam als Streitpartei auftreten«.

schaftsinternen Bereich aufwerfen würde. Der vorerwähnte Art. 4 des Internen Abkommens zur Durchführung des Lomé II-Abkommens versucht, die Koordinierung im Rahmen eines gemischten Abkommens zu regeln. Hinsichtlich der Beteiligung der Gemeinschaft an einem Schiedsverfahren ist das zuständige Organ zur Bestellung des Schiedsrichters bzw. zur Bestimmung des Schiedsverfahrens zu bestimmen. Es spricht viel dafür, dem Rat diese Zuständigkeit im Zusammenwirken mit der Kommission zuzuerkennen. Der Rat schließt die Verträge für die Gemeinschaft (EWG) ab, die Kommission ist für die Aushandlung und Durchführung der Verträge zuständig. Die Durchführung eines Schiedsverfahrens, das auf eine verbindliche Auslegung des Vertrages hinausläuft, steht dabei dem Vertragsschluß näher. Überdies haben sich die Mitgliedstaaten in den Verträgen die Ernennung der innerhalb der Gemeinschaften tätigen Richter vorbehalten [114]. Aus institutioneller Sicht wäre es aber auch sachgerecht, wenn der Rat in der jeweiligen Verordnung zum Abschluß der Verträge festlegen würde, daß gegebenenfalls der Gerichtshof der Europäischen Gemeinschaften den zu bestellenden Schiedsrichter ernennen soll [115]. Für diesen Weg können auch die wenigen Abkommen herangezogen werden, in denen der Gerichtshof der Europäischen Gemeinschaften in das Streitbeilegungsverfahren eingeschaltet ist [116]. Je nach der zu regelnden Sachmaterie dürfte indes der Rat auf der mit der Bestellung des Schiedsrichters verbundenen Beeinflussungsmöglichkeit des Verfahrens bestehen.

Für die Durchführung von Konsultationen oder Verhandlungen dürfte dagegen der Kommission die sachnähere Kompetenz zustehen. So nimmt sie im GATT-Rahmen die im Rahmen der Streitbeilegungsverfahren zu erfüllenden Funktionen wahr, wobei sie allerdings ständig im Kontakt mit einem Ausschuß der Vertreter der Mitgliedstaaten steht.

IV. Schlußfolgerungen

Die Gemeinschaften haben sich in ihren bisher geschlossenen Verträgen gegenüber einer gerichtlichen Streitbeilegung eher zurückhaltend gezeigt, wenn auch in jüngerer Zeit eine Tendenz zugunsten der Vereinbarung zumindest

---

[114] Vgl. Art. 167 EWGV.

[115] Für diese Lösung gibt es in der Gemeinschaftspraxis Beispiele: Die Kommission hat vorgeschlagen, daß die Richter des Verwaltungsgerichts erster Instanz in Personalsachen nicht von den Regierungen der Mitgliedstaaten, sondern vom Gerichtshof der Europäischen Gemeinschaften ernannt werden sollen (vgl. ABl. Nr. C 225/6,9 vom 22. 9. 1978); ebenso soll der Gerichtshof gemäß Art. 6 des Vorschlags der Kommission für eine Verordnung des Rates (EWG) zum Abschluß des Übereinkommens über den Stillegungsfonds (oben Anm. 17) aus seiner Mitte die sechs von den beteiligten sechs Mitgliedstaaten zu bestellenden Richter ernennen.

[116] Vgl. oben S. 412.

von *ad hoc*-Schiedsgerichtsklauseln zu erkennen ist. Die gegenüber der International Law Commission gegebene Anregung, auch für Streitfälle, an denen internationale Organisationen beteiligt sind, ein obligatorisches schiedsgerichtliches Verfahren vorzusehen, bringt diese Tendenz erkennbar zum Ausdruck.

Die bisherige Zurückhaltung dürfte, von der Zahl der Abkommen her gesehen, nicht zuletzt daran liegen, daß ein Großteil der Abkommen im Rahmen der gemeinsamen Handelspolitik abgeschlossen worden ist. Von der Natur ihres Vertragsgegenstandes her kann hier eine Verhandlungslösung eher zu einem befriedigenden Ausgleich der beteiligten Handelsinteressen führen denn eine eher nur den Einzelfall beurteilende Schiedsgerichtsentscheidung. Der mittelbare Einfluß der Mitgliedstaaten bleibt indes bei der Verhandlungslösung beträchtlich: So bietet gerade der Weg der Streitschlichtung über ein *ad hoc* durchzuführendes Schiedsgerichtsverfahren den Gemeinschaften die Möglichkeit, ihre Rechtspersönlichkeit im völkerrechtlichen Vertragsbereich zu stärken, als diese Verfahren die Gemeinschaften zu einer einheitlichen Teilnahme mit einer einheitlichen Rechtsposition veranlassen.

Eine gerichtliche Streitentscheidung entspricht der rechtsstaatlichen Tradition der Gemeinschaften und der Mitgliedstaaten. Letztere lassen sich indes über die innerstaatliche Bedeutung des Rechtsstaatsprinzips hinaus nicht zwangsläufig und generell auf ein gerichtliches Streitbeilegungsverfahren in ihren Außenbeziehungen festlegen. Für die Gemeinschaften sollten die vorstehenden Übersichten und Überlegungen zumindest eine Offenheit im Grundsatz *für* die Vereinbarung von Schiedsgerichtsverfahren ergeben: Dies gilt gerade, weil die Gemeinschaften in einem Großteil ihrer Vertragsbeziehungen als die wirtschaftlich und politisch stärkere Seite auftreten, die sich dem Wunsch eines Vertragspartners nach der Vereinbarung eines Schiedsverfahrens nicht entziehen sollte.

Indes sollte im Einzelfall eine Reihe weiterer Faktoren in die Überlegungen einfließen wie die Funktion des jeweiligen Vertrages im Rahmen der gegenseitigen Beziehungen, die angestrebte Eindeutigkeit und Schärfe der festgelegten vertraglichen Verpflichtungen, das Risiko und die voraussichtlichen Folgen etwaiger Vertragsverletzungen sowie die Qualität der Beziehungen zu dem jeweiligen Vertragspartner. Die Gemeinschaften werden nur dann ihren außenpolitischen Spielraum für die Bestimmung des jeweiligen Streitbeilegungsverfahrens bewahren, wenn sie keine generelle Linie für oder gegen ein gerichtliches Verfahren festlegen, die ihnen entgegengehalten werden könnte.

Das schiedsgerichtliche *ad hoc*-Streitbeilegungsverfahren dürfte insgesamt durch seine Flexibilität und die Vermeidung eines institutionellen Aufwands den Bedürfnissen der Vertragspraxis der Gemeinschaften am ehesten gerecht werden. Eine Öffnung des Zugangs zum Internationalen Gerichtshof auch für die Gemeinschaften, die in weiten Bereichen der Außenzuständigkeiten an die Stelle ihrer Mitgliedstaaten getreten sind, wird sich bei einem weiteren positiven Verlauf der Integration aufdrängen, sofern der Internationale Gerichtshof seine Rolle im Bereich der internationalen Streitbeilegung verstärken kann. In unausweichlichen Einzelfällen werden sich die Gemeinschaften durch ein gemeinsames Vorgehen ihrer Mitgliedstaaten behelfen können. Den Gemeinschaften stehen mit den verschiedenen Arten der dargestellten Verfahren Wege offen, um zur effektiven Durchsetzung des Rechts in ihren internationalen Beziehungen beizutragen.

The Legal Status of the International Seabed

The Controversy about the Legality of National Legislation on Deep Sea Mining

GÜNTHER JAENICKE

After years of marine and technological research and development, the recovery of minerals (so-called polymetallic or manganese nodules) from the deep seabed has reached the stage of technical and commercial feasibility [1]. Although at present world market conditions for the metals contained in the nodules (manganese, nickel, copper and cobalt) do not appear to be very attractive for embarking on large-scale mining ventures on the deep seabed, it has nevertheless become more and more urgent to provide for an appropriate legal regime which should govern these mining activities: First, if further investments in this field are to be encouraged in the interest of industrial development, the operator of a mine site on the international seabed will need secure and generally recognized rights of tenure for the financing of his activites; Second, if the industry goes ahead with mining activities on the deep seabed, it will be indispensable to exercise control over such activities in order to provide for the rational mangement of the respective resources, to protect the legitimate uses by others of the high seas, and to prevent damage to the marine environment. The legal regime which has to be developed to this end, should preferably be a universally applied international regime; however, as long as such an international regime is not forthcoming it falls upon the States to exercise the necessary control over deep sea mining activities of their nationals.

The efforts to establish an international legal regime which would govern the exploration and exploitation of the resources of the international

[1] *M. A. Dubs* (Director Ocean Resources Department, Kennecott Copper Corp.), Statement on behalf of the American Mining Congress before the United States House of Representatives Committee on Merchant Marine and Fisheries (Subcommittee on Oceanography) on 22 May 1979; United Nations Department of International Economic and Social Affairs, Seabed Mineral Resource Development: Recent Activities of the International Consortia, U.N. Doc. ST/ESA/107 (1980) and Addendum (1982).

seabed commenced in 1970 with the »Declaration of Principles governing the Sea-Bed and the Ocean Floor, and the Subsoil thereof, beyond the Limits of National Jurisdiction« by the General Assembly of the United Nations (hereafter referred to as the »Declaration of Principles«)[2]. By this Declaration the General Assembly declared the international seabed and its resources to be the »common heritage of mankind« and formulated principles and guidelines for an international legal regime which should be established by an international treaty of a universal character with respect to the exploration and exploitation of these resources[3]. At the same time the General Assembly decided to convene in 1973 the Third United Nations Conference on the Law of the Sea which should primarily deal with the establishment of an international regime – including appropriate international machinery – for the international seabed area, and also with a broad range of related law of the sea issues[4]. The General Assembly instructed its Committee on the Peaceful Uses of the Sea-Bed and the Ocean Floor beyond the Limits of National Jurisdiction (hereafter referred to as »Seabed Committee«) which had been constituted in 1967 for the study of matters relating to the international seabed and had been transformed into a standing committee in 1968, to prepare draft treaty articles for an international seabed regime[5]. When the Conference convened in 1973[6], the Seabed Committee had not succeeded in preparing an agreed set of draft articles: While all could accept that the international seabed area (the limits of which were still undefined) and its resources should be considered the »common heritage of mankind«[7], widely

[2] Res. No. 2749 (XXV) of 17 December 1970. The text of this Resolution is reproduced in 10 International Legal Materials (I.L.M.) (1971) 220. The Resolution was adopted by a vote of 108 in favour, none against, with 14 abstentions.

[3] The legal effect of the Declaration will be discussed later at p. □.

[4] Res. No. 2750 C (XXV) of 17 December 1970. The text of this Resolution is reproduced in 10 I.L.M. (1971) 226.

[5] Para. 6 of Res. No. 2750 C (XXV).

[6] Meanwhile the mandate of the Conference had been considerably enlarged and fundamentally changed so as to comprise all matters relating to the law of the sea without focusing primarily on the legal regime of the international seabed as originally envisaged by Res. No. 2750 C (XXV) of 17 December 1970. By Res. No. 3067 (XXVIII) of 16 November 1973 (reproduced in Third United Nations Conference on the Law of the Sea, Official Records, Vol. I, p. vii) the General Assembly redefined the mandate for the Conference as follows: »... the mandate of the Conference shall be to adopt a convention dealing with all matters relating to the law of the sea, taking into account the subject-matters listed in paragraph 2 of General Assembly Resolution 2750 C (XXV) and the list of subjects and issues relating to the law of the sea formally approved on 18 August 1972 by the Committee on the Peaceful Uses of the Sea-Bed and the Ocean Floor beyond the Limits of National Jurisdiction and bearing in mind that the problems of ocean space are closely interrelated and need to be considered as a whole...«.

[7] This was in line with the policy statement of the President of the United States of 23 May 1970 where he, *inter alia,* proposed that »...all nations adopt as soon as possible a treaty under which they would renounce all national claims over the natural

divergent views were held how this principle should be reflected in the structure, functions and powers of the international organization (named »International Seabed Authority«) which was to be established. The developing States, organized in the Group of 77, held the view that the International Seabed Authority should have comprehensive powers and discretion in regulating deep sea mining and eventually, after a transitory stage of allowing national activities on behalf of the International Seabed Authority, assume the sole competence to conduct exploitation activities in the international seabed area. The industrialized States, including the eastern european socialist States, and among them in particular those whose nationals were actively planning to engage in deep sea mining activities, maintained that the International Seabed Authority should merely be endowed with well-defined regulatory powers, and that States and their nationals should have direct, nondiscriminatory assured access to the resources of the international seabed under a licensing or concession system. In view of these diametrically opposed views the Seabed Committee was only able to prepare sets of alternative articles which reflected the different views of the members of the Committee[8]. At the Third United Nations Conference on the Law of the Sea, it took nine sessions of protracted and laborious negotiations (1974–1980) to arrive at a text on an international seabed regime which was incorporated as Part XI in the draft of the comprehensive Convention on the Law of the Sea which had been prepared by the Conference[9]. The provisions in Part XI of the Draft Convention and in the Annexes thereto tried to bridge the cleavage between the opposed views by devising an elaborate »parallel system« under which States and their nationals as well as the International Seabed Authority through an international »Enterprise« would be accorded access to the resources of the international seabed. It was anticipated, after the end of the 9th Session of the Conference, that this text would have good prospects to be adopted by consensus once some still outstanding issues had been resolved. One important outstanding issue was the question how the interests of those national companies or consortia could be protected which had already made sizeable investments in the preparation of deep sea mining activities, including prospecting for suitable mine sites and carrying out pilot projects.

---

resources beyond the point where the high seas reach a depth of 200 meters (218.8 yards), and would agree to regard these resources as the common heritage of mankind. . .« (9 I.L.M. 807 (1970). This was coupled with a proposal that exploration and exploitation of the resources between the 200 m line and the edge of the continental margin would be regulated by coastal States as trustees for the international community, and beyond the continental margins by an international authority.

[8] See Report of the Committee on the Peaceful Uses of the Sea-Bed and the Ocean Floor beyond the Limits of National Jurisdiction (Sub-Committee I), GAOR 28th Sess., Suppl. 21 (Doc. A/9021), Vol. I, pp. 19–37; Vol. II, pp. 39–162.

[9] Draft Convention on the Law of the Sea – Doc. A/Conf. 62/L. 80, 27 August 1980.

The hopes for a consensus in this matter were, however, shattered at the beginning of the 10th Session of the Conference in March 1981 when the United States announced that quite a number of key provisions on the international seabed regime in the present Draft Convention were unacceptable to them and that they would suspend their participation in the negotiations of the Conference until they had made a thorough review of the text in order to find out whether they could continue to participate in the Conference and what changes they would seek to make the text acceptable to them. The United States were not the only ones who had serious difficulties with the international legal regime as presently devised. Already at the end of the 9th Session (1980) other industrialised States which were interested in deep sea mining (Belgium, France, the Federal Republic of Germany, Italy, Japan, the Netherlands and the United Kingdom) had criticised the text of the Draft Convention relating to the international seabed regime in several respects[10]. In particular, the Federal Republic of Germany had, in a written Statement submitted to the Conference before the beginning of the 10th Session[11], listed its concerns with some key provisions of the text. At the 11th Session (from 8 March to 30 April 1982) a last effort was made to achieve a consensus on the international seabed regime. Although some progress was made, primarily in devising a scheme for the protection of »pioneer investors«[12], the Conference ultimately failed to reach the much desired con-

[10] See the statement made by the Federal Republic of Germany (A/Conf. (A/Conf. 62/SR. 135, p. 18), France *(ibid.,* p. 26), United Kingdom (A/Conf. 62/SR. 137, p. 24), Italy *(ibid.,* p. 30), Netherlands (A/Conf. 62/SR. 138, p. 23), Japan (A/Conf. 62/SR. 139, p. 2), and Belgium *(ibid.,* p. 36) in the Plenary Meetings from 25 to 27 August 1980.

[11] Doc. A/Conf. 62/WS/16. This Statement had been submitted to the Secretariat of the Conference on 25 February 1981, even before the announcement of the United States to review the text, but was released as a conference document as late as on 10 March 1981.

[12] The scheme which was finally adopted by the Conference in form of a Resolution (Resolution II governing Preparatory Investment in Pioneer Activities – Doc. A/Conf. 62/L. 132/ Add. 1, as amended by A/Conf. 62/L. 132/Add. 1/Corr. 1 and L. 141/Add. 1) and linked with the conventional regime by Art. 308 para. (5) in the Final Clauses of the Convention. It originated from proposals made by the western industrialised countries which wanted to ensure that the consortia which had made sizeable investments in the preparation of seabed mining will be accorded assured access to the international seabed when mining will become commercially feasible. Under the scheme as it emerged from the negotiations of the 11th Session 4 consortia which are at present active in deep seabed mining and which are made up of private companies having the nationality of the following States: Belgium, Canada, Federal Republic of Germany, Italy, Japan, Netherlands, United Kingdom and United States, as well as 4 States: France, India, Japan and the Soviet Union which also claim to have invested in deep sea mining activities through their respective State enterprises, may register with the Preparatory Commission (which will function until the entry into force of the Convention) as »pioneer investors« as soon as the respective States have signed the Convention and certify that their enterprises have spent at least 30 Million

sensus. On 30 April 1982 the Conference adopted the Convention on the Law of the Sea only by a majority vote without the concurrence of the major industrialised States: the United States voted against; Belgium, the Federal Republic of Germany, Italy, Luxembourg, the Netherlands, the United Kingdom, the Soviet Union and the other eastern bloc States (except Romania) abstained[13]. Thus, the prospects for the early establishment of a universally applicable international seabed regime are not very promising; it remains uncertain whether and when the United States and the other States which have abstained from voting for the Convention, will nevertheless eventually sign and ratify the Convention, and it is equally uncertain to what extent their attitude will influence the attitude of others in the ratification process which requires 60 ratifications for the entry into force of the Convention. We have to face the fact that an international seabed regime which according to the Declaration of Principles should be based on a convention of a universal character, is still missing and will probably be missing for some time to come.

The advanced stage of the technological development of deep sea mining and the already heavy investment of industry in this field as well as the dim

---

U.S. $ in pioneer activities prior to 1 January 1983; in addition, developing States or their companies may also qualify as »pioneer investors« until 1 January 1985 under the same conditions. The status as »pioneer investor« allows exploration activities until the entry into force of the Convention, but no commercial exploitation; however, after the entry into force of the Convention and the establishment of the International Seabed Authority »pioneer investors« will have the right to an exploitation contract and a production authorization with priority over other applicants within the limits of the production ceiling prescribed by Art. 151 of the Convention provided that in the meantime the respective certifying States have ratified the Convention. This priority will not apply in relation to the operating organ of the International Seabed Authority (the »Enterprise«) until the »Enterprise« has also got production authorizations for two mine sites. It is estimated that 10 mine sites will eventually be available to satisfy the needs of all »pioneer investors«. No »pioneer investor« may be accorded more than one mine site; however, the rights of a »pioneer investor« may devolve upon its successor in interest. For further details the reader must be referred to the above-mentioned document.

[13] The Convention was adopted by 130 votes in favour to 4 votes against, with 17 abstentions. The other 3 negative votes came from Israel, Turkey and Venezuela for reasons not connected with the international seabed regime; similarly, Albania and Ecuador did not take part in the voting for other reasons. Of the 17 abstaining States, 15 of them abstained because of their discontent with certain aspects of the international seabed regime; only two States (Spain and Thailand) abstained for other reasons. France and Japan voted for the Convention; however, Japan declared that it would study the situation in the light of the uncertainty surrounding the effectiveness of the seabed regime before deciding whether to sign the Convention, and France while admitting that its vote had been influenced by the positive aspects of the scheme for the protection of »pioneer investors«, declared that its positive vote did not prejudge its decision whether to sign the Convention. China which had also voted in favour, expressed the opinion that it was not very optimistic that the Convention could achieve its objectives.

prospects of an early entry into force of an international legal regime for deep seabed mining were the main factors which motivated some highly industrialized States whose industries are active in this field, to enact national legislation for the regulation of deep sea mining activities of their nationals. On 28 June 1980 the United States enacted the Deep Seabed Hard Minerals Resources Act[14]. On 16 August 1980 the Federal Republic of Germany[15], on 28 July and 23 December 1981 the United Kingdom[16] and the French Republic[17], and finally on 18 April 1982 the Soviet Union[18] followed with similar legislation. Other industrialized States with deep sea mining interests (Belgium, Italy, Japan) are also considering action in this respect. This legislation is, according to the legislative intent clearly expressed in the text of the respective laws, essentially of an interim nature to cover the time until a convention on the international seabed regime will enter into force for the respective States; however, it cannot be excluded that the national legislation will remain longer in operation than originally anticipated after the Conference has failed to arrive at a consensus on a universally acceptable convention.

The common features of the above-mentioned national laws are the following:

(1) Natural and juridical persons under national jurisdiction are prohibited to explore or exploit mineral resources (manganese nodules) on the international seabed except under a specific licence to be issued for this purpose by the competent national authorities; no exploitation licence may be issued before 1988 in order to leave time for the entry into force of an international seabed regime.

(2) The afore-mentioned licences will confer on the licencee an exclusive right under national law to undertake exploration or (after 1 January 1988) exploitation activities within a defined mine site; reciprocal recognition of such exclusive rights is envisaged vis-à-vis States which have enacted equivalent parallel legislation, and reciprocity agreements may be concluded for the purpose of resolving conflicts of priority or overlap of applications for mine sites.

---

[14] Public Law 96–283, 94 Stat. 553. The text is reproduced in 19 I.L.M. 1003 (1980). The Regulation for the execution of this Act (U.S. Federal Register, Vol. 46, No. 178 of 15 September 1981, pp. 45890–45920) appear at 20 I.L.M. 1228 (1981).

[15] Gesetz zur vorläufigen Regelung des Tiefseebergbaus of 18 August 1980 (BGBl. 1980 I 1457), amended on 12 February 1982 (BGBl. 1982 I 136).

[16] Deep Sea Mining (Temporary Provisions) Act 1981 of 28 July 1981 (Statutes 1981 Chapter 53). The text is reproduced in 20 I.L.M. 1217 (1981).

[17] Loi No. 81–1135 du 23 décembre 1981 sur l'exploration et l'exploitation des ressources minérales des grands fonds marins (Journal Officiel de la République Française, 1981, p. 3499).

[18] Decree of the Presidium of the USSR Supreme Soviet of 18 April 1982 on »Provisional Measures to Govern Activities of Soviet Entities in the Exploration and Exploitation of Seabed Mineral Resources beyond the Limits of the Continental Shelf«.

(3) By granting exclusive exploration or exploitation rights which are valid under the national laws of the licencing and reciprocating States, the licencing State exercises merely its jurisdiction over its nationals or persons otherwise subject to its jurisdiction, but does not purport to exercise thereby sovereignty or sovereign rights over areas or resources of the international seabed[19].

(4) Strict control is to be exercised by the national authorities over exploration or exploitation activities under licences for the purpose of protecting the marine environment and the legitimate uses by others of the high seas which might be affected thereby; this is of particular importance as long as no international regime is effective in this respect.

(5) A tax-like contribution of 0,75% of the fair market value of the commercially recoverable metals contained in the respective resource has to be paid by the licencee into a fund which is set aside and might eventually be used for purposes in accordance with the international regime to be established.

(6) This legislation will be superseded by the provisions of a convention which provides for an international seabed regime, as soon as such a convention enters into force for the licencing State.

To assess the legal effects of the national licencing schemes correctly, it should be emphasized that the national legislation as described above neither purports to confer exclusive mining rights on the licencee in the sense that they would be enforceable against third States, nor does it have the international legal competence to do so. The exclusivity of the exploration or exploitation rights within the geographical coordinates of the specific mine site as defined in the licence can only be asserted as a subjective legal right within the realm of jurisdiction of the licencing and reciprocating States, or any other State which will be prepared, by agreement or otherwise, to recognize such an exclusive right. This does not mean, however, that exploration or exploitation activities on the international seabed, if authorized by a State, are without legal protection under general international law: Once it is assumed, as the States which have such national legislation on deep seabed mining have repeatedly emphasized[20], that exploration or exploitation activities with

[19] The United States legislation contains a special disclaimer in this respect in Sect. 3 para. (a) of the Deep Sea Hard Minerals Resources Act. The legislation of the Federal Republic of Germany contains a similar reference in the declaration of the purposes of this legislation in § 1 subpara. (1) of the Deep Sea Mining Law. The same disclaimer is contained in Art. 1 para. (2) of the French Law of 23 December 1981 and in para. 2 of the Soviet Decree of 18 April 1982.

[20] The United States Act contains in Sect. 2 para. (12) the following statement: »It is the legal opinion of the United States that exploration for and recovery of hard mineral resources of the deep seabed are freedoms of the high seas subject to a duty of reasonable regard to the interests of other States in their exercise of those and other freedoms recognized by general principles of international law«. Under the terms of the Act »hard minerals« are meant to be manganese nodules.

respect to manganese nodules are permissible uses under the regime of the high seas, such activities if authorized by a State und undertaken with due regard to the lawful uses by others of the high seas, enjoy the same protection under general international law as any other lawful activity under the regime of the high seas and may not be interfered with[21]. Thus, under this assumption the legal protection of nationally licenced activities on the international seabed would not seem to be substantially inferior in comparison with the legal situation under a universally applicable international regime. The problem, however, is precisely that there are divergent views as to whether the recovery of manganese nodules from the international seabed is *per se* a permissible use under the present regime of the high seas or whether, as many developing States strongly assert, exploration and exploitation of the resources of the seabed and its subsoil are prohibited or at least not permissible under present international law until authorized under an international legal regime to be established. It follows therefrom that the legal protection of exploration and exploitation activities on the international seabed which the industry needs before embarking an such a costly venture, is to some extent precarious under the system of parallel national legislations; and this is not due to any inherent inadequacy of such legislation but rather to the uncertainty of its underlying assumption that States and their nationals may lawfully exploit resources of the international seabed under present international law without having to wait for a specific authorization of the international community in this respect[22].

This demonstrates the relevancy of the question which will be discussed in this article: What is the present legal status of the international seabed and its subsoil beyond the limits of the national continental shelves? To what extent is this area under the high seas open to all nations for exploiting its resources under present international law? While the States which have enacted national legislation on the recovery of manganese nodules from the international seabed, maintain that exploration and exploitation activities in this respect are lawful under the principle of the freedom of the high seas, the developing States have, at the Law of the Sea Conference, severely criticized this legislation and its underlying legal assumptions as being contrary to fun-

---

[21] The problem how to safeguard site-specific mining activities on the seabed against intruders, has been specifically adressed by *Goldie* in several articles (see below note 29).

[22] The Resolution attached to the Convention on the Law of the Sea which would allow so-called »pioneer investors« (national companies or international consortia which have already made sizeable investments in site-specific prospection and development of recovery techniques) to go on with the exploration of their chosen »sites« until the entry into force of the Convention under the conditions laid down in this Resolution (see above note 12), will eliminate for those States which avail themselves of the option under this Resolution, the unfavourable impact of the controversy about the legality of national deep sea mining activities under national legislation.

damental principles of international law and therefore illegal. The developing States have rested their case mainly on the following arguments[23]:

(1) The first argument was that the principle of the freedom of the high seas did not apply to the resources of the international seabed and its subsoil; and that, therefore, no right existed for States or their nationals to appropriate these resources without authorization by the international community.

(2) The second argument was that the resources of the international seabed and its subsoil had been declared the common heritage of mankind and that this principle had been generally accepted by the international community and thereby acquired the character of general international law, or even of international *jus cogens*; and that it followed therefrom that these resources were the property of the international community and could only be exploited under the control and management of an international authority but not by States individually.

(3) The third argument was that by taking part in the third Conference on the Law of the Sea which had the mandate to establish an international regime for the exploration and the exploitation of the resources of the international seabed, all participating States had assumed an obligation to negotiate in good faith such a treaty; and that, therefore, national legislation which provides for unilateral mining activities on the international seabed and may thereby create *faits accomplis,* were not in conformity with this obligation.

In the following pages I shall confine myself to the consideration of the first and second line of argument. The third argument which relates to questions of treaty-making procedure and is not germane to the substantive issues that are at stake here, will not be considered in the present context; it will have to be judged in the light of the situation which had developed after so many years of fruitless negotiation at the Conference and, in any event, will have lost its relevancy after the conclusion of the Conference.

The controversy whether and to what extent resources of the international seabed may lawfully be exploited by States and their nationals under the existing legal regime of the high seas or whether such exploitation must await authorization by a convention of universal application, dates back to the time of the discussions in the Seabed Committee and in the General Assembly of the United Naitons in 1969 which culminated in the debate on the so-called Moratorium Resolution of the United Nations General Assembly of 15 December 1969[24]. This Resolution, which was adopted by a majority

---

[23] The legal arguments of the Group of 77 against the legality of the national legislation were first expounded in a letter dated 24 April 1979 from the Chairman of the Group of 77 addressed to the President of the Conference (Doc. A/Conf. 62/77). In subsequent letters of 24 August 1979 (Doc. A/Conf. 62/89) and 29 August 1980 (Doc. A/Conf. 62/106) the arguments were further elaborated.

[24] Res. 2574 D (XXIV). As to the discussion which preceded the vote on this Resolution see U.N. Doc. A/PV. 1833 (1969) and A/C.1/ PV. 1683, 1708–1709 (1969).

vote against the votes of nearly all industrialized States (including the Soviet Union and its allies) [25], declared that pending the establishment of an international seabed regime

> »... (a) States and persons, physical or juridical, are bound to refrain from all activities of exploitation of the resources of the area of the seabed and ocean-floor, and the subsoil thereof, beyond the limits of national jurisdiction;
>
> (b) No claim to any part of the area or its resources shall be recognized«.

This Resolution, however, did not make law. Apart from the fact that such resolutions of the General Assembly have by themselves no legally binding effect on the member States of the United Nations, the negative votes of the industrialized States which were specifically addressed by the terms of this Resolution, preclude any argument to the effect that its content might have affirmed or contributed to the formation of international law. As the International Court of Justice had emphasized in its Judgment of 20 February 1969 in the *North Sea Continental Shelf* Cases:

> »In order to form a new rule of customary international law, state practice must be both extensive and virtually uniform, including the practice of those States whose interests are specifically affected«.

In 1970, when the Seabed Committee was engaged in formulating the guiding principles for an international seabed regime, the same group of developing States tried to incorporate the content of the Moratorium Resolution in the Declaration of Principles on the international seabed, but met again with the firm resistance of those industrialized States which were interested in deep seabed mining and relied on their right to exploit the resources of the international seabed under existing international law as long as no agreement on an international regime had been reached. In order to break the impasse the Chairman of the Seabed Committee proposed a compromise formula which allowed both sides to interpret it according to their own view. This formula found its way into the Declaration of Principles of 17 December 1970 and read as follows:

---

[25] The Resolution was adopted by 62 votes in favour, 28 votes against and 28 abstentions. The negative votes came from Australia, Austria, Belgium, Bulgaria, Byelorussia, Canada, Czechoslovakia, Denmark, France, Ghana, Hungary, Iceland, Ireland, Italy, Japan, Luxembourg, Malta, Mongolia, Netherlands, New Zealand, Norway, Poland, Portugal, South Africa, Soviet Union, Ukraine, United Kingdom and United States. The abstaining States were Burma, China, Cuba, El Salvador, Greece, Indonesia, Iran, Israel, Ivory Coast, Laos, Lebanon, Liberia, Libya, Madagascar, Malawi, Nigeria, Philippines, Romania, Saudi Arabia, Sierra Leone, Spain, Sudan, Swaziland, Syria, Togo, Turkey, United Arab Republic (Egypt), and Upper Volta.

» . . .

3. No State or person, natural or juridical, shall claim, exercise, or acquire rights with respect to the area or its resources incompatible with the international regime to be established and the principles of this Declaration.

4. All activities regarding the exploration and exploitation of the resources of the area and other related activities shall be governed by the international regime to be established«[26].

It is obvious that these principles do not refer to the existing legal regime under international law, but to a regime which is "to be established". On the other hand, the somehow ambiguous wording of these principles allows both sides to interpret them in conformity with their respective views: If one starts from the premise that existing international law does not prohibit the exploitation of the resources of the international seabed, the wording of the provisions cited above may logically be interpreted in the sense that deep sea mining activities shall be brought under an exclusive international legal regime once such a regime will have been established but that until then no restrictions apply. If, however, one starts from the opposite assumption that there is no right to exploit the deep seabed resources under existing international law, the wording of the provisions cited above may also logically be interpreted in the sense that deep sea mining activities may only be undertaken in conformity with and under the governance of the international regime to be established, that is after its establishment. It is beyond doubt that the above-cited principles were typical compromise formulae and were meant as such; they were not designed to decide the moratorium issue, but rather to leave that issue open. Thus, nothing can be gained for our purpose by going deeper into the controversial interpretation of these paragraphs of the Declaration of Principles, and we cannot hope to find there an answer to our question whether deep seabed mining activities are permissible under existing international law.

In 1972, the issue of a moratorium for deep seabed mining was revived again when in the course of the discussions in the Seabed Committee some developing States proposed a »decision« of the Seabed Committee to reaffirm that such a moratorium was already in effect[27]. The proposed »decision« re-

---

[26] As already mentioned above in note 2 the Declaration was finally adopted without opposition by 108 votes to none, with 14 abstentions. The vote was not recorded, but it is known that the abstentions came from the Soviet Union and its allies who were at that time not very favorable to the setting up of an international organisation for the control of deep sea mining activities.

[27] U.N.Doc.A/AC. 138/L. 11/Rev. 1 (reproduced in the Report of the Seabed Committee to the General Assembly on its 1972 Sessions – U.N.Doc.A/8721, p. 69). The proposal had been submitted by Algeria, Brazil, Chile, China, Iraq, Kenya, Kuwait, Mexico, Peru, Venezuela, Yemen and Yugoslavia. It had been opposed by the United States, the United Kingdom, Japan, the Soviet Union and others.

lied in its preamble on both the Moratorium Resolution as well as on the Declaration of Principles, and read in its operative part as follows:

> »The Committee on the Peaceful Uses of the Sea-Bed and the Ocean Floor beyond the Limits of National Jurisdiction...
> *Calls* upon all States engaged in activities in the sea-bed area beyond the limits of national jurisdiction, in conformity with the provisions of the two resolutions cited above to cease and desist from all activities aiming at commercial exploitation of the sea-bed area and to refrain from engaging directly or through their nationals in any operations aimed at the exploitation of the area before the establishment of the international regime;
> *Reaffirms* that prior to the establishment of the international regime no claims on any part of the area or its resources based on past, present or future activities will be recognized«.

At the discussion of this proposal, the developing States and the industrialized States again expressed opposite views of whether existing international law permitted deep seabed mining; in particular, the industrialized States opposed the view that the Declaration of Principles already implied such a moratorium[28]. The proposed »decision« was (among other proposals) included in the Report of the Seabed Committee to the United Nations General Assembly, but in view of the opposition was neither adopted by the Seabed Committee nor later taken up by the General Assembly. At the Law of the Sea Conference, the idea of a moratorium resolution by the Conference has sometimes loomed in the background, but has not been actively pursued any more in view of the consensus procedure in the Conference and the foreseeable opposition of the deep sea mining States.

Thus, the proceedings in the United Nations and the aforementioned Resolutions of the General Assembly do not allow the conclusion that thereby an express or tacit agreement had been achieved to the effect that no national mining activities on the international seabed should be allowed until the entering into force of an international regime. In this context, it is fitting to recall that the States which have enacted national legislation on seabed mining, have already reasonably responded to the demand for a moratorium by prohibiting exploitation of international seabed resources until 1 January 1988.

Having disposed of the moratorium argument we have now to address ourselves to the central issue: What is the present legal status of the international seabed and its resources? In the legal literature this subject has found increasing attention in recent years[29]. But, opinions among writers are

[28] U.N.Doc.A/AC. 138/SR. 71–89 (1972). The discussion is also summarized in the Report of the Seabed Committee (above note 27), p. 8–10.

[29] *Anand,* The Legality of the Interim Seabed Mining Regimes, Foreign Affairs Reports (Feb. 1980); *Arrow,* The Proposed Regime for Unilateral Exploitation of

as sharply divided as between States on the legality of national mining activities on the international seabed. We will have to review the arguments of both sides and try to assess their respective merits.

Most commentators approach the issue of the legal status of the international seabed in two phases: At first they ask what the legal situation had been before the recovery of manganese nodules from the international seabed had become an international issue; next they ask whether and to what extent the widespread, if not universal acceptance of the concept qualifying these resources as the »common heritage of mankind« has changed the legal situation. It may well be doubted whether this differentiation perceives the matter in the true perspective. There are good reasons to assume that the essence of the common heritage concept has always been inherent in the international character of areas and resources beyond national jurisdiction in the sense that these areas and resources may not be brought under national sovereignty, but remain open to be used or exploited by all States and their nationals[30]; this holds true for all areas and resources beyond national jurisdiction, be it in the

---

Deep Seabed Mineral Resources by the United States, Harvard International Law Journal 1980, p. 337; *Biggs,* Deep Seabed Mining and Unilateral Legislation, Journal of Maritime Affairs 1980, p. 223; *Burton,* Freedom of the Seas: International Law Applicable to Deep Seabed Mining Claims, Stanford Law Review 29 (1979), p. 1135; *Goldie,* Mining Rights and the General International Law Regime of the Deep Ocean Floor, Brooklyn Journal of International Law 2 (1975), p. 21; Customary International Law and Deep Seabed Mining: A Range of Perspectives, Syracuse Journal of International Law and Commerce 6 (1979), p. 173; A Selection of Books Reflecting Perspectives in the Seabed Mining Debate, The International Lawyer 15 (1981), p. 293 and p. 445; *Graf Vitzthum,* Die Bemühungen um ein Regime des Tiefseebodens, Zeitschrift für ausländisches öffentliches Recht und Völkerrecht (ZaöRV) 38 (1978), p. 745; *Jennings,* The United States Draft Treaty on the International Seabed Area: Basic Principles, International and Comparative Law Quarterly 20 (1970), p. 433; *Kronmiller,* The Lawfulness of Deep Seabed Mining (New York 1980); *Larsen,* The United States Position on the Deep Seabed, Suffolk Transnational Law Journal 1979, p. 1; *Luard,* The Control of the Seabed: A New International Issue (London 1974); *Orego Vicuna,* Les législations nationales pour l'exploitation des fonds des mers et leur incompatibilité avec le droit international, Annuaire français de droit international 1978, p. 810; *Petersmann,* Rechtsprobleme der deutschen Interimsgesetzgebung für den Tiefseebergbau, ZaöRV 41 (1981), p. 267; *Saffo,* Has the General Assembly Created a Law to Govern Sea-bed Mining?, Tulane Law Review 1979, p. 492; *R. Young,* The Legal Regime of the Deep Sea Floor, American Journal of International Law 62 (1968), p. 641.

[30] In this context, the question whether the seabed and its subsoil are *res communis* or *res nullius* has been debated. I do not think that these notions are particularly helpful for defining the legal status of the international seabed and its subsoil. As there is general agreement that no part of it may be validly appropriated by a State or subjected to its sovereignty, the international seabed and its subsoil cannot be considered *res nullius* and open to annexation. On the other hand, the assertion that the international seabed is *res communis* does not explain or define convincingly whether and under what conditions property rights may be aquired in respect of the resources taken therefrom. As to the facets of this problem see *Goldie* (above note 29), pp. 304 ff.

oceans, in the Antarctica or in outer space. At this juncture it may be recalled that the common heritage concept is neither by its denomination, nor by its substance a new concept; the similar wording has, in fact, already been used in the discussion of fishery matters at the 1930 Codification Conference where it had been used as an argument to demonstrate the need for international conservation measures for the protection of fish stocks in international waters. At its closure the Conference unanimously adopted a Recommendation concerning the Protection of Fisheries which affirmed the importance and necessity for the safeguarding of »riches constituting the common patrimony« [30a].

The common heritage concept became a useful argument again when it was realized that under the provisions of the 1958 Continental Shelf Convention which had made the seaward limit of the national continental shelf dependent on the »exploitability« of the resources, States could virtually claim continental shelf rights to the whole of the seabed under the high seas. The unattractive prospect that under the exploitability criterion all seabed areas under the oceans might eventually be divided up among the coastal States, led Ambassador Pardo of Malta to take his famous initiative which called for a definite outer limit of national continental shelf claims and the preservation of the seabed and its resources beyond this limit as a »common heritage of mankind« to be used and exploited for the benefit of the international community as a whole, taking into consideration the needs of the poor countries [31]. Thus, the concept of the common heritage of mankind was put forward in the first place for the defence of the international seabed against the claims of coastal States for exclusive national jurisdiction over extensive seabed areas under the oceans.

The common heritage concept is certainly not a mere empty shell devoid of any substance. It carried with it the expectation that any legal regime

---

[30a] Actes de la Conférence pour la Codification du Droit international – S.d.N. Doc. C. 351. M. 145. 1930. V. – Vol. I, p. 169.

[31] Pardo's Statement is *in extenso* reproduced in the Official Records of the United Nations General Assembly, 22th Session, First Committee, 1515th and 1516th Meeting (1 November 1967 – Doc.A/C. 1/PV. 1515 and 1516). The main points of his initiative were the following: »*First,* the seabed and the ocean floor are a common heritage of mankind and should be used and exploited for peaceful purposes and for the exclusive benefit of mankind as a whole. The needs of poor countries. . . . should receive preferential consideration in the event of financial benefits being derived from the exploitation of the seabed and ocean floor for commercial purposes.
Second, claims to sovereignty over the seabed and ocean floor beyond present national jurisdiction, as presently claimed, should be frozen until a clear definition of the continental shelf is formulated.
Third. . . to draft a comprehensive treaty to safeguard the international character of the seabed and ocean floor beyond present national jurisdiction; and. . . to provide for the establishment of an international agency which ill ensure that national activities undertaken in the deep seas and on the ocean floor will conform to the principles and provisions incorporated in the proposed treaty«.

which was to govern the exploitation of the resources of the international seabed, should be rational, equitable and considerate of the interests of all States, including the needs of developing countries. But the common heritage concept did not entail any specific legal regime, neither in respect of its substantive rules, nor in respect of the institutional machinery for their implementation. That is left, as the Declaration of Principles clearly envisages, to later agreement among States. When the concept of the common heritage of mankind was accepted by the industrialized countries as well as by the developing countries in para. (1) of the Declaration of Principles as the basic concept for the international seabed regime, it was the object and purpose of the Declaration to give practical effect to this concept and to agree on the political guidelines how such an international regime should be established and what should be the main requirements for that regime. Therefore, it would go beyond the limits of proper interpretation if one would purport to draw from the acceptance of the common heritage concept in the first paragraph of the Declaration legal conclusions which do not follow from the following paragraphs of the Declaration, and if such an interpretation would moreover be at variance with the understanding under which the Declaration had been adopted. In particular, nothing in those paragraphs of the Declaration which specify the main contures of the future regime[32], supports the

---

[32] The relevant paras. 5 to 15 of the Declaration of Principles read as follows:

»5. The area shall be open to use exclusively for peaceful purposes by all States whether coastal or land-locked, without discrimination, in accordance with the international régime to be established.

6. States shall act in the area in accordance with the applicable principles and rules of international law including the Charter of the United Nations and the Declaration on Principles of International Law concerning Friendly Relations and Co-operation among States in accordance with the Charter of the United Nations, adopted by the General Assembly on 24 October 1970, in the interests of maintaining international peace and security and promoting international co-operation and mutual understanding.

7. The exploration of the area and the exploitation of its resources shall be carried out for the benefit of mankind as a whole, irrespective of the geographical location of States, whether land-locked or coastal, and taking into particular consideration the interests and needs of the developing countries.

8. The area shall be reserved exclusively for peaceful purposes, without prejudice to any measures which have been or may be agreed upon in the context of international negotiations undertaken in the field of disarmament and which may be applicable to a broader area. One or more international agreements shall be concluded as soon as possible in order to implement effectively this principle and to constitute a step towards the exclusion of the sea-bed, the ocean floor and the subsoil thereof from the arms race.

9. On the basis of the principles of this Declaration, an international régime applying to the area and its resources and including appropriate international machinery to give effect to its provisions shall be established by an international treaty of a universal character, generally agreed upon. The régime shall, *inter alia,* provide for the orderly and safe development and rational management of the area and its resources and for expanding opportunities in the use thereof and ensure the equitable sharing

interpretation that the common heritage concept necessarily implied some sort of common ownership or property right of the international community in the resources of the international seabed and that consequently exploitation of these resources had to await specific legal authorization by the international community, let alone the difficult question whether it would be admissible to employ such private law concepts in public international law.

---

by States in the benefits derived therefrom, taking into particular consideration the interests and needs of the developing countries, whether land-locked or coastal.

10. States shall promote international co-operation in scientific research exclusively for peaceful purposes:

(a) By participation in international programmes and by encouraging co-operation in scientific research by personnel of different countries;

(b) Through effective publication of research programmes and dissemination of the results of research through international channels;

(c) By co-operation in measures to strengthen research capabilities of developing countries, including the participation of their nationals in research programmes.
No such activity shall form the legal basis for any claims with respect to any part of the area or its resources.

11. With respect to activities in the area and acting in conformity with the international régime to be established, States shall take appropriate measures for and shall co-operate in the adoption and implementation of international rules, standards and procedures for, *inter alia:*

(a) Prevention of pollution and contamination, and other hazards to the marine environment, including the coastline, and of interference with the ecological balance of the marine environment;

(b) Protection and conservation of the natural resources of the area and prevention of damage to the flora and fauna of the marine environment.

12. In their activities in the area, including those relating to its resources, States shall pay due regard to the rights and legitimate interests of coastal States in the region of such activities, as well as of all other States which may be affected by such activities. Consultation shall be maintained with the coastal States concerned with respect to activities relating to the exploration of the area and the exploitation of its resources with a view to avoiding infringement of such rights and interests.

13. Nothing herein shall affect:

(a) The legal status of the waters superjacent to the area or that of the air space above those waters;

(b) The rights of coastal States with respect to measures to prevent, mitigate or eliminate grave and imminent danger to their coastline or related interests from pollution or threat thereof resulting from, or from other hazardous occurrences caused by, any activities in the area, subject to the international régime to be established.

14. Every State shall have the responsibility to ensure that activities in the area, including those relating to its resources, whether undertaken by governmental agencies, or non-governmental entities or persons under its jurisdiction, or acting on its behalf, shall be carried out in conformity with the international régime to be established. The same responsibility applies to international organizations and their members for activities undertaken by such organizations or on their behalf. Damage caused by such activities shall entail liability.

15. The parties to any dispute relating to activities in the area and its resources shall resolve such dispute by the measures mentioned in Article 33 of the Charter of the United Nations and such procedures for settling disputes as may be agreed upon in the international régime to be established«.

In view of the fact that the Declaration, as shown above, has purposely avoided settling the issue of a moratorium for mining activities on the international seabed, and in view of the fact that the Declaration has been adopted on the basis of that understanding, it would seem inconsistent therewith to give the common heritage concept a legal meaning which would result in such a moratorium.

Thus, whether or not the common heritage concept already applies to the international seabed it does not lead us much further in defining its legal status under present international law than that the international seabed and its resources are »international« in the sense that no State, as the Declaration in its para. (2) correctly states, »shall claim or exercise sovereignty or sovereign rights over any part thereof«.

It has sometimes been disputed whether the international seabed (and/or its subsoil) are part of the high seas. If this question would be answered in the negative, the legal regime of the high seas would *a priori* not apply to the respective resources. However, the Declaration of Principles has taken a definite stand on this issue: In the preamble of the Declaration it had been expressly recognized that

> »... the existing legal regime of the high seas does not provide substantive rules for regulating the exploration of the aforesaid area and the exploitation of its resources«.

This can only be interpreted to the effect that, in principle, the legal regime of the high seas embraces also the exploitation of the resources of the »aforesaid area«, that is »the sea-bed and ocean floor, and the subsoil thereof«. The recognition that there are no substantive rules for the regulation of the exploitation of the respective resources could, however, mean two different things: Either the principles and rules governing the high seas are too general for a satisfactory regulation of this new activity; or, there is, in fact, a »legal vacuum« or a »lacuna« in the juridical sense which has to be filled out be creating new law through treaty or state practice. Leaving aside for the moment the „legal vacuum" alternative, we shall first consider what legal principles and rules of the high seas regime could be relevant in the present context.

The divergent views as to the applicability of the principles and rules of the high seas regime can be summarized as follows:

(a) One side argues that activities aimed at the exploitation of the resources of the international seabed and its subsoil belong to the freedoms of the high seas because there is sufficient analogy to other established high seas freedoms, and that there is at least no rule of international law which specifically prohibits the exploitation of the seabed.

(b) The other side argues that activities aimed at the exploitation of the resources of the international seabed are not comparable to any of the recognized high seas freedoms, and it must at least be proved that such activities have been recognized by the international community as freedoms of the high seas.

The arguments which have been advanced by both sides in support of their line of thought, are not fully convincing. The following critical comments may be appended:

1) It is difficult to find sufficient similarity between the exploitation of the resources of the international seabed or its subsoil and such other freedoms of the high seas as fishing, laying of cables and pipelines, scientific research, or deployment of installations in order to draw a convincing analogy to them. Exploitation of the mineral resources of the international seabed and its subsoil affects, in general, non-renewable resources and is more intensive and exclusive in its character than the aforementioned freedoms. Only in the case of the recovery of manganese nodules, an analogy to fishing activities may suggest itself: Manganese nodules are extremely abundant on the seabed in all the deep oceans and their recovery requires merely special dredging techniques and no fixed mining installations. Scientists even maintain that manganese nodules are constantly in formation and may thus not be considered as falling into the category of non-renewable resources.

2) Even if the analogy to other recognized high seas freedoms may be difficult to accept, it does not follow therefrom that the exploitation of the resources of the international seabed and its subsoil is *a priori* disqualified from being a permissible activity under the regime of the high seas merely because it is new and without precedent. On the other hand, the somewhat outdated maxim that all is permitted what is not specifically prohibited by a rule of international law, is also not a good argument under modern perceptions of the international legal order. The situation may rather be described as follows: New activities or uses on or under high seas do not necessarily require prior recognition by the international community before they may qualify as a permissible activity or use under the regime of the high seas; otherwise new activities or uses could never lawfully be developed. It has not been the practice of States to seek prior recognition before the development of new uses of the sea.

3) This conclusion is in harmony with Art. 2 [33] of the High Seas Convention of 1958 and the corresponding Art. 87 [34] of the Convention adopted

---

[33] Art. 2 reads as follows: »The high seas being open to all nations, no State may validly purport to subject any part of them to its sovereignty. Freedom of the high seas is exercised under the conditions laid down by these articles and by other rules of international law. It comprises, *inter alia,* both for coastal and non-coastal States:

(1) Freedom of navigation;
(2) Freedom of fishing;
(3) Freedom to lay submarine cables and pipelines;
(4) Freedom to fly over the high seas.

by the Law of the Sea Conference on 30 April 1982. The content of these Articles has never been in dispute and is considered as expressing general international law. If one reads these Articles carefully, it should be noticed that a clear distinction has to be made between the fundamental principle of the »freedom of the high seas« and the specific »freedoms« which are either those expressly mentioned in these Articles or other »freedoms« which have been developed under the concept of the »freedom of the high seas«. It follows clearly from the wording of these Articles and has never been in dispute that the list of the recognized high seas freedoms contained therein is not exhaustive, and that other activities or uses of the sea may qualify as »freedoms« (which is merely a synonymous expression for permissible activities and uses) if they are compatible with the fundamental principle of the freedom of the high seas. Thus, it cannot very well be maintained, that the exploitation of the resources of the international seabed and its subsoil is *a priori* not permissible under the regime of the high seas simply because it cannot rely on prior recognition. But it must nevertheless first be ascertained whether or not this new activity or use is compatible with the principle of the freedom of the high seas and is exercisable in conformity therewith.

4) We have now to examine whether there are criteria by which the compatibility with the concept inherent in the principle of the freedom of the high seas can be determined. It seems rather difficult to find reliable criteria for this purpose: Neither Art. 2 of the High Seas Convention of 1958 nor the corresponding Art. 87 of the new Convention on the Law of the Sea provide us with any criteria in this respect. Therefore, the listing of established high seas freedoms in the above-mentioned Articles and the general recognition by the international community of other permissible uses as such freedoms is certainly the best way to overcome this difficulty, but that does not solve the question of new activities in respect of which is apparently no

---

These freedoms, and others which are recognized by the general principles of international law, shall be exercised by all States with reasonable regard to the interests of other States in their exercise of the freedom of the high seas«.

[34] Art. 87 reads as follows:

»(1) The high seas are open to all States, whether coastal or landlocked. Freedom of the high seas is exercised under the conditions laid down by this Convention and by other rules of international law. It comprises, *inter alia,* both for coastal and land-locked States:

(a) freedom of navigation;

(b) freedom of overflight;

(c) freedom to lay submarine cables and pipelines, subject to Part VI;

(d) freedom to construct artificial islands and other installations, permitted under international law, subject to Part VI;

(e) freedom of fishing, subject to the conditions laid down in section 2;

(f) freedom of scientific research, subject to Parts VI and XIII.

(2) These freedoms shall be exercised by all States with due regard for the interests of other States in their exercise of the freedom of the high seas, and also with due regard for the rights under this Convention with respect to activities in the Area«.

international consensus as to their permissibility under the regime of the high seas[35]. There seem to be, however, some considerations which may be relevant to the question of compatibility with the concept of the freedom of the high seas:

a) It is no longer disputed that States may validly exploit the resources of the seabed and subsoil within the limits of their continental shelf jurisdiction. It follows therefrom that such activities on the seabed are by their nature not prohibited *per se.*

b) It has been recognized that exploitation of the resources of the seabed may take place under the continental shelf regime although this area and the waters above have been and are still to a certain extent, functionally as well as geographically, considered part of the high seas. It follows therefrom that exploitation of the resources of the seabed as such is not incompatible with the regime of the high seas and the exercise by others of the freedoms therein.

c) It is inherent in the concept of the high seas which shall be »open to all nations«, that activities undertaken there should by their very nature be exercizable by all nations on an equal footing. While this will depend *in concreto* on the nature of the resource and the technique of its recovery, manganese nodules on the seabed, by their abundance in terms of exploitable »sites«, do allow participation by all nations in their recovery. The international community may consider it desirable, as has been expressed in the Declaration of Principles, to regulate the access to this resource in view of the unequal distribution of financial and technical capabilities among the nations and the needs of developing countries; but this can be no precondi-

---

[35] It might be argued that under the terms of Art. 2 of the High Seas Convention of 1958 only those activities and uses of the sea may qualify as high seas »freedoms« which have already found general recognition, and that other activities and uses have an inferior legal status or no legal status at all under the legal regime of the high seas. This might be inferred from the – rather unclear – wording of Art. 2 where it refers to the »other freedoms recognized under the general principles of law«. This proviso which has now been ommitted in Art. 87 of the new Convention on the Law of the Sea was not inserted for the purpose of restricting the freedom of the high seas to those activities and uses which were generally accepted as »recognized freedoms« by the international community. Unfortunately, the official records of the 1958 Conference on the Law of the Sea do not shed much light on the interpretation of this part of Art. 2, which resulted from an amendment by the United Kingdom delegation (see Official Records, Vol. IV (Second Committee) pp. 37–56; A/Conf. 13/ C.2./L. 68); this amendment was proposed as a compromise in view of certain proposals which were designed to restrict or prohibit naval manoeuvres in the vicinity of foreign coasts and nuclear weapons tests on the high seas. The correct interpretation would rather be that the words »recognized under the general principles of law« were meant to emphasize that new activities and uses must not only be compatible with the concept of the freedom of the high sea, but must also not violate other rules of international law. This interpretation would be in harmony with the present wording of Art. 87 of the new Convention on the Law of the Sea according to which the freedom of the high seas is to be exercised »under the conditions laid down by this Convention and by other rules of international law«.

tion for the existence of the originary right to undertake such exploitation just as the regulation of navigation, fishing, dumping of waste or other uses of the sea was no precondition for the enjoyment of these freedoms and rights.

d) The recovery of manganese nodules does not require for its practical performance the exercise of sovereignty or sovereign rights by the home State of the operator over the »site« from which the nodules will be recovered. The national legislation of the seabed mining States expressly disclaims any intention in this respect. It is true that economic recovery of nodules in an international area requires that the operations can be undertaken undisturbed within an area or »site« of certain dimensions, and that it would be desirable to regulate by generally accepted international rules the reasonable size of such »sites«, criteria for the priority in exploiting them, and the mutual recognition of an exclusive right to such a »site«. This would be achieved in the most optimal way by a convention of universal application, but could also be achieved to a large extent through the application of general principles of law.

To conclude the discussion of this point, there seems to be no cogent reason to consider the recovery of manganese nodules from the international seabed as being incompatible with the concept of the freedom of the high seas. We must certainly take note that there is general agreement to establish an international regime of universal application for the exploitation of the resources of the international seabed, but it cannot convincingly be argued that pending the effective establishment of such a regime national activities aimed at the recovery of manganese nodules must be considered as unlawful under the legal regime of the high seas.

We have finally to consider the – probably more helpful – alternative that the exploitation of the resources of the international seabed, and in particular the recovery of manganese nodules is not regulated by the existing principles and rules of the regime of the high seas, but that there is rather a »legal vacuum« or a »lacuna« in international law with respect to such activities.

Activities of men which have become feasible by the advancement of science and the development of technology, and in particular activities for the exploitation of new resources which are needed by the world economy or are useful to mankind, will be developed and pursued irrespective of whether there are legal rules for their regulation. There is neither a legal rule that prohibits new activities nor the opposite rule that any new activity remains lawful as long as it is not prohibited or regulated by a specific rule. It falls upon the States which constitute the legal order in the world, to assume the responsibility, either acting internationally by international agreement or acting nationally within the sphere of their national jurisdiction, to regulate new activities as far as it is necessary or desirable for the common good. Thus, in those areas of men's activities where there is as yet no rule of international law in existence, it is the practice of States, either acting nationally or acting in concert by international agreement, which forms and develops the respective rules of international law for the regulation of new activities. This practice may at first be diffuse, disparate or even divergent, and it will take some time before rules have been developed which command general recognition throughout the international community. Until this stage is reached, the question whether a new activity is lawful or not, is besides the point and cannot convincingly be answered either way. It is not unreasonable to assume that this is exactly the present situation with respect to the exploitation of the resources of the international seabed. The historical development of this issue within the United Nations which has been described above in some detail, is sufficient evidence of the fact that there exist no generally accepted rules of international law which would decide the lawfulness of the exploitation of the international seabed and its subsoil. Under these circumstances the Declaration of Principles must be regarded as an agreed effort to create the necessary legal framework for the regulation of the exploitation of the international seabed and its subsoil; but, until the establishment of such a special international regime, activities aimed at the exploitation of international seabed resources cannot be branded as unlawful if undertaken with due regard to the rights of other States under the regime of the high seas.

Finally it cannot be argued that national activities aimed at the recovery of manganese nodules from the international seabed are *per se* unlawful because they tap a resource which is »international«, and not part of the national domain of the individual State. Such an argument is untenable because it would neglect the development of international law with respect to the exploitation of the resources of the seabed and its subsoil under the continental shelf doctrine. This body of international law has been developed on the premise that the exploitation of the resources of the seabed and its subsoil is a permissible activity by States – or under the continental shelf doctrine by the respective coastal State – and may be extended to the depths of the ocean as far as the »exploitability« of these resources would admit. Only the question up to what limit such exploitation should be the privilege of the coastal State or should remain open to all States, was left in abeyance at the 1958 Conference on the Law of the Sea. If coastal States were entitled to claim exploitation rights under the continental shelf doctrine over the adjacent areas of the deep seabed, including the right to recover manganese nodules which happen to be within these areas, no cogent reason can be found why States should not equally be entitled to undertake activities in this respect in the areas where coastal States do not or cannot claim exclusive rights of exploitation.

To sum up: It would certainly be in the interest of the international community as well as in the interest of the deep sea mining industry to have a universally applicable international legal regime which would govern the exploitation of the resources of the international seabed. But as long as the regime which has been provided for in the new Convention on the Law of the Sea, cannot command universal recognition, present international law provides no convincing legal argument for considering national activities aimed at the recovery of manganese nodules from the international seabed as unlawful. The interim legislation by which States in the meantime regulate the recovery of manganese nodules by persons under their jurisdiction, cannot, therefore, be accused of violating international law.

Intervention under Article 62 of the Statute of the International Court of Justice

EDUARDO JIMÉNEZ DE ARÉCHAGA

1. The judgment of the International Court of Justice of 14 April 1981 rejecting the application of Malta for permission to intervene in the proceedings between Tunisia and Libya concerning the delimitation of their continental shelf, was based on two grounds. The first of these grounds was that »the object of Malta's request, as declared and defined in its application, falls altogether outside the scope of the form of intervention for which Article 62 provides«[1]. The second ground invoked in the judgment as a basis for rejecting the application was that Malta, in the Court's view, did not succeed »in showing the existence of an interest of a legal nature which may be affected by the decision in the case, in the meaning of Art. 62 of the Statute«[2].

In view of its negative conclusions as to these two grounds, the Court did not find it necessary to examine the issue of whether the existence of a jurisdictional link constitutes another of the requirements for allowing an intervention[3].

2. While the Court examined the first two grounds jointly, considering them to be closely connected, it would seem that from a logical point of view the question whether the Court was confronted in that case with a request for what it has described as »a genuine intervention«[4], has priority with respect to the two other issues, which may only arise subsequently, namely, the existence of a legal interest and of a jurisdictional link. The logical independence between the first two questions is confirmed by the fact that the absence of a »genuine intervention« was the ground which ensured the unanimity of the decision, while the inexistence of a legal interest was not accepted by Judges Oda and Schwebel, who stated in their separate opinions that they disagreed from the Court's conclusion on this point[5].

---

[1] I.C.J. Reports, 1981, p. 12.
[2] *Ibid.*
[3] I.C.J. Reports, 1981, p. 20, para. 36.
[4] *Haya de la Torre* case, I.C.J. Reports, 1951, p. 77.
[5] I.C.J. Reports, 1981, p. 31–32 and 37.

I. The Concept of a »Genuine Intervention«

3. As indicated in the Court's judgment, Malta »did not seek to be admitted as a veritable party to the proceedings, having a status on a footing of complete equality with the Parties to the case, but was seeking the procedural position of a ›participant‹ by way of intervention« [6].

4. The immediate answer to this request is that Art. 62 of the Statute does not permit this form of intervention on a »non-party« basis. To accept the role in the proceedings sought by Malta would introduce into the Statute a distinction which neither the text nor the *travaux préparatoires* of Art. 62 authorize.

5. The distinction made in the Statute is that between intervention under Art. 62 and intervention under Art. 63. The Permanent Court of International Justice stated in the first sentence of its first Judgment:

> »The Statute of the Permanent Court of International Justice provides for two sets of circumstances and two different forms in which intervention is possible on the part of States which are not from the outset concerned in a suit brought before it« [7].

Art. 62 does not make any distinction between the rights and obligations which a State may have under this provision. Consequently, there are no statutory grounds for distinguishing intervention as a party and on a »non-party« basis. According to Art. 62 of the Statute and to the Rules of Court implementing it, a State allowed to intervene is entitled to submit written and oral observations including submissions. Art. 49 (4) and 60 (2) of the Rules of Court not only assume, but clearly say that a State authorized to present submissions which the Court must decide in its judgment constitutes a *party* to the proceedings. This seems obvious, for how can a State take part in judicial proceedings and make submissions to a Court without becoming a party to these proceedings?

Art. 62, therefore, requires the intervener to become a party to the case, in the full sense of the term, equally bound by the judgment under Arts. 59 of the Statute and 94 of the United Nations Charter. It must be remarked that both these provisions only apply to a State which has been a »party« to the judicial proceedings. If, under Art. 63 of the Statute, the intervening party becomes bound by the Court's interpretation of the treaty, the intervening State, under Art. 62, should *a fortiori* be bound by the Court's judgment to the extent of its intervention. Art. 63, para. 2, cannot be interpreted *a contrario* since its purpose is not to extend, but to restrict the effects of *res judicata* only to the »interpretation of the Convention«.

6. From a substantive point of view, the basic difficulty resulting from the procedural situation sought by Malta is the privileged position it would

---

[6] I.C.J. Reports, 1981, p. 8, para. 12.

[7] *S.S. »Wimbledon«*, P.C.I.J., Series A, No. 1, p. 12.

have achieved with respect to the two original parties. The intervener on a non-party basis would submit its views on the legal issues arising in the case and would thus influence the outcome of the litigation by means of its pleadings and submissions, but Arts. 59 of the Statute and 94 (1) of the Charter would not apply to it. A first unacceptable consequence of allowing this form of intervention would be that the eventual decision of the Court, including its *ratio decidendi*, if it were favourable to Malta's submissions and contentions – for instance by applying in that area equidistance, or a certain criterion of proportionality – might later be invoked as *res judicata* by the intervener against the original parties, either in litigation or on negotiations. But, on the other hand, if the judgment was unfavourable to Malta's submissions, none of the original parties could invoke it conclusively against Malta, since the intervener had disclaimed the *status* of a party, and it is only to a party that Arts. 59 of the Statute and 94 (1) of the Charter apply. This would constitute a flagrant violation of the basic principles of fair dealing and equality in the administration of justice.

7. In this case the Court was confronted with the unprecedented situation of a prospective litigant who attempted to infiltrate the judicial proceedings between two parties – who happen to be its future adversaries – in order to influence the outcome of their litigation by making submissions and contentions identical or very similar to those it would advance later in future litigations or negotiations with them. Not only it would thus have the advantage of advocating these views twice, but it would be doing so, in the first instance, without running the risk, inherent to all judicial proceedings, of losing its case.

8. A second unacceptable consequence of this limited form of intervention is that the intervener would not expose itself to counter claims or opposing submissions that could be made by the original parties against it. Thus, for instance, Tunisia and Libya would have to defend the closing lines of their Gulfs of Gabes and Sirte against Malta's contentions, but they would be denied the possibility of opposing any base-lines which, on similar or different criterion, might have been established by Malta. In other words, while Tunisia and Libya maritime claims and boundaries would become vulnerable to legal attack and to adverse submissions from Malta, the reverse would not be true. Thus, the intervener on a non-party basis would have the best of two worlds: the advantages of judicial litigation without any of its risks. This inadmissible situation brings to mind Paulo's sentence: »the power to condemn necessarily entails the power to absolve«.

II. The Legal Interest Affected by the Decision

9. Two questions related *inter se* have been the subject of debate since the establishment of the Permanent Court:

a) whether a general interest in the development of the principles and rules of International Law may suffice for permission to intervene under Art. 62, or a direct interest of a material character in the actual outcome of the dispute is required and
b) whether a valid link of jurisdiction between the would be intervener and the original parties is an essential condition for the granting of permission to intervene.

These two questions will be taken up separately in the light of the *travaux préparatoires* of the Statute and the Rules and of the jurisprudence of both Courts.

a) Travaux préparatoires of the Statute

10. When the draft Statute of the Permanent Court prepared by the Advisory Committee of Jurists was brought to the consideration of the Council of the League of Nations, the United Kingdom delegate, Mr. Balfour (as he was then) submitted a note with certain observations.

One of his observations was the following one:

> »There is another point on which I speak with much diffidence. It seems to me that the decision of the Permanent Court cannot but have the effect of gradually moulding and modifying international law. This may be good or bad; but I do not think this was contemplated by the Covenant; and in any case there ought to be some provision by which a State can enter a protest, *not* against any particular decision arrived at by the Court, but against any ulterior conclusions to which that decision may seem to point.
> If this danger has any reality, it becomes doubly formidable from the fact that at present the three most populous Western States – the United States, Germany and Russia – are not Members of the League, and cannot be expected to take their views on international law from the Court's decision« [8].

It seems clear that for Mr. Balfour, the present Art. 62 of the draft Statute did not afford sufficient protection for the question he raised, and consequently that in his view there was no right to intervene based merely on a general interest in the development or modification of International Law.

11. This observation was examined in the report submitted to the League's Council by M. Bourgeois, a well-known document which led to the final approval of the Statute.

[8] Note on the Permanent Court of International Justice, submitted by Mr. Balfour to the Council of the League of Nations, Brussels, October 1920, published in: Documents concerning the action taken by the Council of the League under Article 14 of the Covenant, Document No. 28, p. 38 (emphasis in the original).

In his report, M. Bourgeois dealt, *inter alia,* with the question whether

»the right of intervention in its various aspects and in particular the question whether the fact that the principle implied in a judgment may affect the development of International Law in a way which appears undesirable to any particular State may constitute for it a sufficient basis for any kind of intervention in order to impose the contrary views held by it with regard to this principle« [9].

12. Under the title »The Right of Intervention«, M. Bourgeois' report begins with a transparent allusion to Mr. Balfour's observation:

»The observations in the draft project of The Hague by one of our colleagues drew attention to the following case: it might happen that a case appearing unimportant in itself might be submitted to the jurisdiction of the Court, and that the Court might take a decision on this case, laying down certain principles of international law which, if they were applied to other countries, would completely modify the principles of the traditional law of this country, and which might therefore have serious consequences. The question has been raised whether, in view of such an alternative, the States not involved in the dispute should not be given the right of intervening in the case in the interest of the harmonious development of the law« [10].

13. If the draft Statute had contained a recognition of the right of intervention of a third State merely on the basis of a concern of principle over the points of law involved in a case, and without the need of a concrete interest and a jurisdictional link, then the obvious and categorical answer to Mr. Balfour's observation would have been to point out the existence of Art. 62 of the Statute as constituting the solution to the problem. However, M. Bourgeois stated something different. He said that:

»The Hague Jurists have not moreover disregarded the necessity of bearing in mind considerations which, *if not exactly identical, are* at least in *the same order of ideas.* They have, indeed, given to *some* litigant States the right to intervene in a case where any interest of a judicial nature which may concern them is involved« [11].

14. Furthermore, M. Bourgeois, not only abstained from saying that the answer to Mr. Balfour's preoccupation is to be found in Art. 62 of the Statute, but he proposed instead the addition to the draft Statute of two new articles. These two new articles were clearly designed to solve Mr. Balfour's

[9] Documents concerning the action taken by the Council of the League under Article 14 of the Covenant, Extract from the Procès-verbal of the Tenth Session of the Council, Document No. 30, p. 46.

[10] *Ibid.,* p. 50.

[11] *Ibid.,* p. 50 (emphasis added).

observation, which greatly preoccupied at that time the great Powers, who feared that the existence of the Court would diminish the predominant influence they then exercised in the development of international law.

These two new articles, added by the Council of the League to the draft of the Advisory Committee of Jurists, are *mutatis mutandis* the present Arts. 57 and 59 which read:

> »57. If the judgment does not represent in whole or in part the unanimous opinion of the judges, any judge shall be entitled to deliver a separate opinion«.
>
> »59. The decision of the Court has not binding force except between the parties and in respect of that particular case«.

The need felt by M. Bourgeois and the League's Council to insert these two important new articles in the Statute drafted by the Advisory Committee of Jurists demonstrates that Art. 62 was not then understood as giving an answer to the preoccupations of the Great Powers, which had been voiced by Mr. Balfour.

15. Thus, an examination of M. Bourgeois' report, and of the League's Council action upon it, which constitute one of the most important *travaux préparatoires* of the Statute, leads to the view that Art. 62 does not authorize intervention merely on the basis of a general interest in the development of International Law. On the contrary, as Judge Anzilotti has put it, »the Article in question only contemplated cases in which the States desiring to intervene had an actual legal interest in the dispute« [12]. And the views expressed by Judge Anzilotti are of significance in respect of M. Bourgeois' report because, as recognized by him, Judge Anzilotti, before coming to the Court, in his capacity as legal advisor of the Council, had greatly assisted M. Bourgeois in the preparation of the report [13].

b) The S.S. »Wimbledon« case

16. In the *S.S. Wimbledon* case, the question of the jurisdictional link did not arise, since the applicant States, the respondent and the intervener had accepted the jurisdiction of the Court under Art. 386 of the Treaty of Versailles and Art. 37 of the Statute of the Court.

However, this case is illustrative of the factors characterizing the existence of »an interest of a legal nature which may be affected by the decision in the case«.

17. In the application the original parties made two submissions, asking the Court to decide that

[12] Acts and Documents concerning the organization of the Court, Series D, No. 2, pp. 90 and 87.

[13] Documents concerning the action taken by the Council of the League under Article 14 of the Covenant, Document No. 30, p. 42.

»the German authorities wrongfully refused, on 21 March 1921, free access to the Kiel Canal of the steamship »*Wimbledon*«, and that the German Government shall make reparation for the loss incurred by the afore-mentioned vessel in consequence of this action, a loss which is estimated at francs 174.082^{86}, together with interest at 6% per annum as from 20 March 1921« [14].

18. The request by the Polish Government to intervene on the side of the Applicants only invoked as the basis for intervening the fact that »Poland was one of the parties to the Treaty of Peace of Versailles« and

»the refusal in question constitutes a violation of the rights and material advantages guaranteed to Poland by Article 380 of the Treaty of Versailles« [15].

No mention was made in that document of any damage suffered by Poland and no submission was made asking the Court to condemn Germany to indemnify Poland.

19. It was in view of this application that the United Kingdom Government made its observations, considering

»that as the question at issue in the case is that of the proper construction of a treaty to which Poland is a party, Poland is entitled as of right to intervene in the proceedings in accordance with Article 63 of the Statute of the Court. The British Government notes, however, that the Polish application is based on Article 62 of the Statute, not on Article 63, and also that it is not an application for permission to intervene as a third party, as provided in Article 62, but is an application to be allowed to intervene on the side of the British, French, Italian, and Japanese Governments« [16].

20. In the light of these observations, the United Kingdom Government continued to say:

»Article 62 relates only to States alleging an interest of a legal nature (un intérêt d'ordre juridique) which may be affected by the decision in the case. Apart from the general Polish interest in the interpretation of the Treaty of Versailles, it is not clear that the Polish government possesses any interest of a legal nature in the matters at issue in the ›*Wimbledon*‹ case. Indeed, the Polish application itself states the grounds of the application in a way which shows that it is based merely on the injury which the Polish Government would suffer from any restriction of its rights under Article 380 of the Treaty of Versailles. It is, therefore, submitted

---

[14] P.C.I.J., Series A, No. 1, p. 8.

[15] *Ibid.*, p. 9.

[16] P.C.I.J., Series C, No. 3, vol. 1, p. 107.

that an application of this nature would more properly be made under Article 63«[17].

The British observations ended by saying that:

»as the position which will be enjoyed by a State allowed to intervene under Article 62 appears to be identical in all respects with that of a State which intervenes under Article 63 so far as concerns the binding nature of the final decision given by the Court, it is submitted that it would be preferable that the intervention of the Polish Government in the ›*Wimbledon*‹ case should take place under Article 63 and not under Article 62«[18].

21. What was the attitude of the Polish Government with respect to these observations? In a communication made to the Court by its agent, this Government stated that

»being the consignee of the cargo, considers that it has incontestably a legal interest in the ›*Wimbledon*‹ case. A certain quantity of rolling-stock belonging to the Polish State railways was waiting idle for two weeks for the unloading of the vessel, which was prevented by the German Government from arriving in time at the port of destination. The Polish Government, however, does not intend to ask the German Government for special damages for this loss«[19].

The agent continued by saying that

»the Polish Government, however, having in view of the fact that it is a signatory of the Treaty of Versailles, a right to intervene under Article 63, intends to avail itself of this right and does not insist that the reasons for intervention which it has given in regard to Article 62 should be taken into consideration«[19].

22. The Court took account of this statement and also that the Polish Government

»states that it does not intend to ask the German Government for any special damages for the prejudice caused to it in the case of the ›*Wimbledon*‹«[20].

It accepted Poland's intervention under Article 63 and concluded:

»The attitude thus adopted renders it unnecessary for the Court to consider and satisfy itself whether Poland's intervention in the suit before it

---

[17] *Ibid.*
[18] *Ibid.*, p. 108.
[19] *Ibid.*, p. 118.
[20] P.C.I.J. Series A, No. 1, p. 13.

> is justified by an interest of a legal nature, within the meaning of Article 62 of the Statute«[21].

23. The inference which may be drawn from this case is that the key element in determining whether a State

> »has an interest of a legal nature which may be affected by the decision in the case«,

is whether or not that State is presenting to the Court submissions *in its own and direct interest,* which are different from those of the original applicants and go beyond them. The Court considered it unnecessary to pronounce on the request under Art. 62 only after Poland had declared that it did not intend to ask the defendant for any special damages for the prejudice caused to it. Since there was no actual Polish request for damages before the Court, the importance thus attributed to the attitude of not intending to ask for »any special damages for the prejudice caused to it« reveals the importance assigned to this element, for determining the existence of an interest of a legal nature liable to be affected by the decision. In other terms, for the Permanent Court Art. 62 requires the assertion by means of an independent submission of a specific claim of the would-be intervener which may be affected by the operative part of the Court's Judgment. And this did not occur with Malta's interest which only concerned the potential implications of the Court's reasoning in a judgment to be delivered in a dispute between two other States.

c) The application of Malta

24. In its judgment of 14 April 1981 the International Court of Justice took a similar position: it refused to find the existence of a legal interest which might be affected by the decision of the case on the ground that it was not the object of the Maltese application »to obtain any form of ruling or decision from the Court concerning its continental shelf boundaries with either or both of those countries«.

The Court stated that »the interest of a legal nature invoked by Malta does not relate to any legal interest of its own directly in issue as between Tunisia and Libya in the present proceedings or as between itself and either one of those countries. It concerns rather the potential implications of reasons which the Court may give in its decision in the present case on matters in issue as between Tunisia and Libya with respect to the delimitation of their continental shelves for a subsequent delimitation of Malta's own continental shelf«[22].

---

[21] *Ibid.*

[22] I.C.J. Reports, 1981, p. 12, para. 19.

III. The Requirement of a Jurisdictional Link

25. The question of the need of a jurisdictional link in order to allow intervention was discussed by the members of the Permanent Court in their Preliminary Session and was thoroughly and ably examined by Counsel in the oral proceedings concerning the Maltese application. Yet the Court did not find it necessary to go into this issue, although the three separate opinions deal with it.

a) The discussion in the Permanent Court

26. When the judges of the Permanent Court met in 1922 for preparing the Rules of Court a marked divergence of views appeared as to the need of a jurisdictional link in order to allow intervention.

Several Judges, such as Lord Finlay, Judges Weiss, Oda, Loder and Moore were of the view that no such requirement was needed. The latter referred »to the fact that no great Power had up to the present accepted the optional clause«, adding:

> »that for this reason it was all the more desirable to give these Powers the right to come before the Court in suits instituted by the smaller Powers, for the purpose of obtaining the decision of the Court on main principles of international law«.

Mr. Moore »could not see how the condition of reciprocity in regard to compulsory jurisdiction could nullify the provisions of an Article of the Statute, i.e., Article 62« [23].

27. On the other hand Judges Anzilotti, Huber, Altamira, Negulesco and Yovanovitch had an opposite view [24]. Judge Huber said »that a State which was not bound by the reciprocity clause could not be allowed to intervene in a case between States which had accepted it« [25]. Judge Anzilotti stated that »the right of intervention could only exist either in virtue of an agreement between the two original parties or when the parties to the case, as well as those who desired to intervene, had accepted the optional clause with regard to the compulsory jurisdiction of the Court«. And Judge Altamira pointed out »that the scheme of the Jurists of 1920 was based on the principle of the compulsory jurisdiction of the Court. When this principle was modified by the Assembly, the text of certain articles had unfortunately not been brought into line with the new principle which had been introduced« [26].

---

[23] Acts and Documents concerning the Organization of the Court, Series D, No. 2, p. 91.

[24] *Ibid.,* pp. 87, 89, 88, 93 and 381.

[25] *Ibid.,* p. 87.

[26] *Ibid.,* p. 89.

28. The Scandinavian Judges Nyholm and Beichmann took the position that it was not for the Court to legislate in its Rules of Procedure, in regard to the right of intervention. »The question whether or not this right existed in particular cases must be left to the decision of the Court, as each concrete case arose« [27]. When adopting the Rules of Procedure »it is only necessary to determine the general mode of procedure to be followed by States requesting permission to intervene« [28].

In view of the divergence of opinions this suggestion was followed, on the basis of a presidential ruling [29].

b) The Fiji case

29. The Fiji request for intervention in the *Nuclear tests* cases (*Australia* v. *France*, and *New Zealand* v. *France*) raised, for the first time, the issue of the jurisdictional link.

Fiji was not a party to the 1928 General Act, as were Australia, New Zealand and France, and had not accepted the optional clause.

Yet it intended to intervene in the case in support of the applicants, thus joining in their submissions that France's conduct was inconsistent with international law, should be discontinued, and violated the rights of Fiji. If Fiji, as an intervener, obtained such a declaration, it could afterwards claim damages against France on the basis of *res judicata* of the declaratory Judgment.

30. This possibility of Fiji becoming an intervener entitled to invoke rights against France and placing this country in the position of a respondent State, appeared as clearly inadmissible, in view of the lack of any jurisdictional link between France and Fiji.

Such a possibility would violate three basic principles of the Statute:

a) the principle of consent to jurisdiction, namely that »the Court can only exercise jurisdiction over a State with its consent« [30], and this applies both to jurisdiction *ratione personae* and *ratione materiae*.
   France had never consented to the exercise of the Court's jurisdiction with respect of France at the behest of Fiji;
b) the principle of equality of the position of the parties before the Court. Fiji would be making submissions and claims and suing France before the Court, while France could never make any claim or sue Fiji before the Court;
c) the principle of strict reciprocity of rights and obligations among States parties to the Statute, which have accepted the Court's compulsory juris-

---

[27] Judge Nyholm, *ibid.*, pp. 88 and 352.
[28] *Ibid.*, p. 352.
[29] *Ibid.*, p. 29. This was the ruling made by President Loder and not the individual view mentioned by Judge Oda in his separate opinion at p. 26.
[30] I.C.J. Reports, 1954, p. 32.

diction. This is the reason why para. 2 of Art. 36 provides that the acceptance of compulsory jurisdiction must always be *»in relation to any other State accepting the same obligation«*. A corollary of this principle of reciprocity is that a State cannot be obliged to submit to the jurisdiction of the Court in a larger measure than the other party is itself subject. This is the reason why the two declarations must coincide and any of the parties may invoke the reservations of the other. A State which cannot be brought to Court as a defendant by another State can neither become an applicant vis-à-vis that State nor an intervener against that same State, entitled to make independent submissions in support of an interest of its own.

c) Malta's application to intervene

31. In the hearings concerning Malta's application, Counsel for Malta when examining the *Fiji* case, distinguished this case from the Maltese one and observed that Malta did not try to intervene for the purpose of asserting a right against either party to the case[31]. Thus, he implicitly admitted that if the intervening State has »the purpose of asserting a right against the respondent«, then the requirement of an independent jurisdictional link between intervener and respondent applies.

32. It is true that Malta did not attempt to intervene for the purpose of asserting directly a right of its own, which may be affected by the decision of the case; it did not invoke a right which is directly »in issue« for Malta (»pour *lui* en cause«) as the French text of Art. 62 requires. Yet, what Malta attempted to do with its intervention was to assert its own rights, but in an indirect and roundabout way, by means of a decision applying to other parties and binding them alone.

33. The oral proceedings demonstrated that if Malta's intervention was allowed, it would make submissions to the Court advocating the line of equidistance and opposing certain baselines established by Libya and Tunisia.

These submissions were of a concrete and precise character and their effect would have resulted in concrete and precise boundary lines. These Maltese submissions would not, in this respect, be less precise than the submissions filed by the original parties of the case. Mr. Lauterpacht, for instance, clearly suggested that Malta, in its submissions, would favour a line of equidistance as the appropriate one for continental shelf delimitation in that region. In this respect it is interesting to recall that neither Tunisia nor Libya had advocated the line of equidistance.

Equally, Counsel for Malta advanced the contention that it opposed the closing lines for gulfs established by Tunisia in Gabes and by Libya in Sirte. This latter baseline was not in issue in the case between Libya and Tunisia.

---

[31] CR 81/3, p. 51.

These submissions signify that Malta, if allowed to intervene, would have made what in German law is described as an »agressive intervention«, a type of intervention by a third party who elects to make independent contentions and arguments without supporting the case of either party to the principal proceedings[32].

34. If allowed to intervene, and if successful in opposing the baselines of Tunisia and Libya, Malta would have achieved a result which really constituted an assertion of its own claims or a recognition of its own rights, supported by a judgment that could later be invoked by Malta against the original parties.

35. And this would, in effect, place Tunisia and Libya in the position of respondents against Malta's claims, obliged to defend their own baselines against Maltese legal attacks. With respect to the jurisdictional link the question then is: how could Malta place both Libya and Tunisia in the position of defendants against its claims and submissions, when neither Tunisia nor Libya had consented to the Court's exercise of jurisdiction with respect to Maltese submissions against any of them?

It seems indisputable that the requirement of consent for the exercise of jurisdiction both *ratione materiae* and *ratione personae* is a fundamental principle of the Statute which cannot be ignored by a literal and isolated reading of Art. 62. How can a State, merely by invoking Art. 62, bring another to court and make submissions against it, when it cannot do so as a principal party?

---

[32] *V. S. Mani,* International Adjudication; Procedural Aspects (The Hague 1980), p. 258.

Paralleles Tätigwerden von Sicherheitsrat und Internationalem Gerichtshof bei friedensbedrohenden Streitigkeiten

Zu Fragen der Zuständigkeit und Organtreue

ECKART KLEIN

I. Problemstellung

Erhaltung und Sicherung des Friedens sind heute das zentrale Thema des Völkerrechts. Die einschlägigen Vorschriften sind rechtsexistenziell: Wo aufgrund der bestehenden Bedingungen ein Krieg zur Vernichtung aller Staaten führen kann, müssen Kriegsverhütung und Friedenssicherung einen notwendigen Bestandteil des Völkerrechts bilden, will es Rechtsordnung sein; denn zur Rechtsordnung gehört der Schutz ihrer Träger, gehört die Abwehr von sie zerstörenden Einwirkungen [1].

Das Gewaltverbot und die Pflicht der Staaten, ihre Streitigkeiten friedlich beizulegen, entsprechen dieser Logik. Ihr untertan ist auch die Satzung der Vereinten Nationen (SVN), indem sie diese Staatenverpflichtungen in den Mittelpunkt rückt (Art. 2 Ziff. 3, 33) und ihren vornehmsten Organisationszweck gerade in der Effektuierung dieser Pflichten sieht (Art. 1 Ziff. 1).

Die Indienststellung der Organe der Vereinten Nationen (Art. 7 SVN) zugunsten dieses Zwecks ergibt sich hieraus von selbst. Besondere Bedeutung kommt dabei dem Sicherheitsrat, der Generalversammlung und dem Internationalen Gerichtshof zu. Die diesbezüglichen Zuständigkeiten ergeben sich aus der Satzung als der Verfassung der Organisation. Insoweit zeigt sich jedoch, daß keine klare Kompetenzverteilung in dem Sinn erfolgt ist, daß immer nur ein Organ ausschließliche Befassungs- und Entschließungskompe-

---

[1] Vgl. dazu *H. Mosler,* The International Society as a Legal Community (1980), S. 18; *V. Bruns,* Das Völkerrecht als Rechtsordnung, Teil I, ZaöRV 1 (1929), S. 1 ff. (26).

tenz im Hinblick auf eine bestimmte Streitigkeit hätte. Paralleles[2] Tätigwerden der genannten Organe ist danach denkbar. Eine ausdrückliche Regelung besteht nur im Verhältnis von Sicherheitsrat zur Generalversammlung, wonach diese gehindert ist, Empfehlungen zu Fragen zu verabschieden, die noch Verhandlungsgegenstand des Rates sind[3]. Diese Zuständigkeitsbeschränkung der Versammlung ist an der dem Sicherheitsrat von den Mitgliedstaaten übertragenen »primären Verantwortung« für die Aufrechterhaltung des Weltfriedens und der internationalen Sicherheit (Art. 24 Abs. 1 SVN) orientiert; sie schließt abweichende und damit die Autorität der Beschlüsse des Rates untergrabende Entschließungen der Generalversammlung aus. Gegenstand der folgenden Ausführungen ist, ob sich diese primäre Verantwortung des Sicherheitsrats für den Weltfrieden auch dem Internationalen Gerichtshof gegenüber als Jurisdiktionsschranke durchzusetzen vermag oder ob umgekehrt ein Tätigwerden des Gerichtshofs als des »Hauptrechtsprechungsorgans« der Vereinten Nationen (Art. 92 SVN) zu einer Zuständigkeitsbeschränkung des Sicherheitsrats führen kann[4]. Dabei wird von der Erörterung sonstiger Jurisdiktionsfragen abgesehen, vielmehr vorausgesetzt, daß die Organe bezogen auf den Streitfall grundsätzlich zuständig sind; problematisiert wird nur, ob darin durch die vorangehende oder gleichzeitige Befassung des anderen Organs eine Änderung eintreten kann. Da das Ergebnis dieser Untersuchung im wesentlichen negativ sein wird, also von der Zulässigkeit parallelen Tätigwerdens von Sicherheitsrat und Internationalem Gerichtshof prinzipiell auszugehen ist, hat sich die Erörterung der weiteren und interessanteren Frage anzuschließen, ob und inwieweit eine Pflicht dieser Organe besteht, divergierende Entscheidungen zu vermeiden oder in sonstiger Weise das Verfahren vor dem jeweils anderen Organ zu berücksichtigen[5]. Dies führt in die wichtige und weitgehend noch unbearbeitete Problematik der Organtreue und Organloyalität in internationalen Organisationen hinein.

Die aufgeworfenen Fragen sind nicht nur akademisch, sie haben sich vielmehr konkret in der Praxis gestellt. Als Richter am Internationalen Gerichtshof war *Hermann Mosler* unmittelbar mit ihnen befaßt.

II. Ausgangsfälle

Zur Illustration der behandelten Probleme besonders geeignet sind der *Korfu-Kanal*-Fall, der *Ägäis-Festlandsockel*-Fall und der Fall der in *Teheran festgehaltenen amerikanischen Geiseln.*

---

[2] Dieser Begriff ist dem der »konkurrierenden« Tätigkeit vorzuziehen, da sich mit letzterem – jedenfalls nach deutschem Verständnis – doch ein Exklusivitätsanspruch verbindet; vgl. den Begriff der »konkurrierenden Gesetzgebung«, Art. 72 GG.

[3] Art. 12 SVN; siehe dazu auch unten in und bei Anm. 59. Auf A/RES/377(V) – Uniting for Peace-Resolution – ist freilich hinzuweisen.

[4] Siehe unten III.

[5] Siehe unten IV.

1. Korfu-Kanal-Fall

Nachdem am 22. Oktober 1946 zwei Zerstörer der britischen Marine durch albanische Minen im Korfu-Kanal schwer beschädigt und viele Personen dabei getötet und verletzt worden waren und nachdem Albanien sich nicht geneigt gezeigt hatte, dem Wiedergutmachungsanspruch des Vereinigten Königreichs zu entsprechen, wurde am 10. Januar 1947 der Sicherheitsrat der Vereinten Nationen auf Antrag Großbritanniens mit dieser Streitigkeit befaßt[6]. Albanien, zu dieser Zeit noch nicht Mitglied der Organisation, akzeptierte die Einladung nach Art. 32 SVN und übernahm insoweit die einem Mitglied obliegenden Verpflichtungen[7]. Ein britischer Entschließungsentwurf[8], der die Feststellung enthielt, Albanien habe im Korfu-Kanal ein nicht notifiziertes Minenfeld gelegt, scheiterte am Veto der Sowjetunion[9]. Hingegen kam es am 9. April 1947 zur Verabschiedung der Entschließung 22 (1947), mit der – einem weiteren britischen Vorschlag gemäß – der Sicherheitsrat den Parteien empfahl, ihren Streit unmittelbar dem Internationalen Gerichtshof zu unterbreiten[10]. Im Hinblick auf die daraufhin vom Vereinigten Königreich einseitig beim Gerichtshof anhängig gemachte Klage bejahte dieser zunächst seine Jurisdiktion[11]; in der Hauptsache wurde später auf der Grundlage eines zwischen beiden Staaten vereinbarten Kompromisses entschieden[12].

Die Empfehlung des Sicherheitsrats, die streitigen Fragen dem Internationalen Gerichtshof vorzulegen, entsprach Art. 36 Abs. 3 SVN[13]. Die Anwendung dieser Bestimmung erlaubte es dem Rat, sich jeder weiteren Meinungsäußerung sachlicher oder prozeduraler Art zu enthalten und eine parallel geschaltete Befassung beider Organe mit dem Streitfall zu vermeiden. Die Frage einer Berücksichtigung der Entschließung des Sicherheitsrats durch den Gerichtshof konnte sich daher nur insoweit stellen als klärungsbedürftig erschien, ob der Verweis auf den Internationalen Gerichtshof *per se* zuständigkeitsbegründend wirken konnte[14].

---

[6] UN Doc. S/247.

[7] Decision of 20 January, 1947.

[8] SCOR 1947, 120th meeting, S. 567.

[9] *Ibid.*, S. 589.

[10] »... that the United Kingdom and Albanian Governments should immediately refer the dispute to the International Court of Justice in accordance with the provisions of the Statute of the Court«.

[11] *Affaire du détroit de Corfou*, Arrêt sur l'exception préliminaire, ICJ Reports 1948, S. 15.

[12] ICJ Reports 1949, S. 4 – *Affaire du détroit de Corfou*, Arrêt (Fond), und S. 244 – Arrêt (Fixation du montant des réparations).

[13] Vgl. etwa *S. D. Bailey*, The Procedure of the Security Council (1975), S. 246.

[14] Siehe dazu die *op. ind. collective*, ICJ Reports 1947/48, S. 31 f.

2. Ägäis-Festlandsockel-Fall

Wegen der sich zuspitzenden Lage in der Ägäis[15] rief Griechenland, gestützt auf Art. 35 SVN, am 10. August 1976 den Sicherheitsrat an mit dem Ersuchen, die Türkei zur Beendigung der behaupteten Verletzungen der souveränen Rechte Griechenlands am Kontinentalschelf der Ägäis aufzufordern[16]. Am selben Tag erhob Griechenland zudem Klage beim Internationalen Gerichtshof, mit der es um Feststellung der beiderseitigen Kontinentalschelfgrenzen ersuchte. Zugleich beantragte Griechenland beim Gerichtshof den Erlaß vorläufiger Maßnahmen; unter anderem sollten beide Regierungen aufgefordert werden, von allen weiteren militärischen Maßnahmen abzusehen, welche die friedlichen Beziehungen gefährden könnten[17].

In seiner Entschließung 395 vom 25. August 1976 rief der Sicherheitsrat die Parteien zu äußerster Zurückhaltung auf (Ziff. 1). Zu den aufgeworfenen Rechtsfragen nahm er selbst nicht Stellung, drängte jedoch beide Seiten dazu, die direkten Verhandlungen mit dem Ziel einer gegenseitig akzeptablen Lösung wiederaufzunehmen (Ziff. 3) und dabei der Beiträge eingedenk zu sein, die richterliche Erörterung, insbesondere durch den Internationalen Gerichtshof, für die Streitbeilegung bedeuteten (Ziff. 4). Anders als im *Korfu-Kanal*-Fall erfolgte also keine unmittelbare Verweisung der Parteien auf den Internationalen Gerichtshof nach Art. 36 Abs. 3 SVN[18]; die genannten Ziffern 3 und 4 der Resolution stellen in der Zusammenschau vielmehr Kompromißformeln dar, die auf die bereits erhobene Klage Griechenlands ebenso Bedacht nehmen wollen wie auf die während der Sicherheitsratsdebatte deutlich gewordene Weigerung der Türkei, die Zuständigkeit des Gerichtshofs in der vorliegenden Streitsache anzuerkennen[19].

Am 11. September 1976 lehnte der Internationale Gerichtshof die beantragten vorläufigen Maßnahmen ab[20]. In dem hier interessierenden Punkt berief er sich dabei auf die vorangegangene Entschließung des Sicherheitsrats 395 (1976): Da beide Regierungen die Nützlichkeit dieser Entschließung anerkannt hätten und da nicht zu vermuten sei, »that either State will fail to heed its obligations under the Charter of the United Nations or to fail to heed the recommendations of the Security Council addressed to them with respect to their present dispute«, bestehe keine Notwendigkeit, vorläufige

---

[15] Dazu *E. Klein*, Tätigkeit der Vereinten Nationen in völkerrechtlichen Fragen (Berichtszeitraum: 1. 1. 1976–31. 12. 1976), ArchVR 17 (1978), S. 371 ff. (397).

[16] S/12167.

[17] Vgl. zu den Anträgen ICJ Reports 1976, S. 3 (4 f.).

[18] Grundlage der Ratsentschließung sind vielmehr Art. 33 Abs. 2 und 35 SVN; vgl. auch *L. Gross*, The Dispute between Greece and Turkey Concerning the Continental Shelf in the Aegean, AJIL 71 (1977), S. 31 ff. (36).

[19] Vgl. UN Chronicle 1976 No. 8, S. 22 ff. – Hieraus erklärt sich auch die Annahme der Resolution im Konsensusverfahren.

[20] *Aegean Sea Continental Shelf*, Interim Protection, ICJ Reports 1976, S. 3.

Maßnahmen zu beschließen[21]. Die Entscheidung macht deutlich, daß der Gerichtshof sich nicht etwa wegen der eine entsprechende Aufforderung enthaltenden Sicherheitsratsresolution für unzuständig gehalten hat, an die Parteien zu appellieren, friedensgefährdende Aktionen zu unterlassen; doch schloß er aus der Existenz der Entschließung und der Vermutung der Rechtstreue der Parteien auf die fehlende Dringlichkeit der beantragten vorläufigen Maßnahme und damit auf das Fehlen einer nach Art. 41 IGH-Statut zur Begründung eines solchen Antrags notwendigen Voraussetzung[22]. Der Mäßigungsappell des Sicherheitsrats ist somit in die Argumentation des Gerichtshofs nur als bloßer »Umstand« (*circumstances*) im Sinne von Art. 41 des Statuts eingegangen; zur Sache hat sich der Gerichtshof hingegen nicht geäußert[23]. Diese Haltung ist in einzelnen Separatvoten kritisiert worden[24]; man hat dort den argumentativen Rückzug auf die Sicherheitsratsentschließung und die damit sichtbar gewordene »Verweigerungshaltung« bedauert und einen eigenständigen Beitrag des Gerichtshofs zur Beruhigung der gespannten Lage für richtiger gehalten.

3. US-Geiseln in Teheran-Fall

Zu einer der eben geschilderten ähnlichen verfahrensrechtlichen Situation kam es im Fall der amerikanischen Geiseln in Teheran. Am 4. November 1979 hatten, ohne von iranischen Sicherheitskräften behindert worden zu sein, iranische Staatsangehörige das amerikanische Botschaftsgebäude in Teheran besetzt und ca. 50 Mitglieder des diplomatischen und konsularischen Personals gefangen genommen. Am 9. November 1979 beantragten die Vereinigten Staaten gem. Art. 35 SVN eine dringliche Sitzung des Sicherheitsrats; ihr Ergebnis war eine vom Rat autorisierte Erklärung seines Präsidenten, in der der Iran zur sofortigen Freilassung der Geiseln aufgefordert wurde[25]. Die Nichtbeachtung dieses Appells und die sich daraus ergebende Zuspitzung der Lage ließen den Generalsekretär am 25. November 1979 von seiner Kompetenz nach Art. 99 SVN Gebrauch machen, eine erneute Sitzung des Sicherheitsrats einzuberufen[26]. Am 27. November 1979 wurde der Appell des

[21] *Ibid.*, S. 13. – Am 19. 12. 1978 verneinte der IGH mit zwölf gegen zwei Stimmen seine Jurisdiktionskompetenz in der Sache, ICJ Reports 1978, S. 3.

[22] So auch *K. Oellers-Frahm*, Der Beschluß des Internationalen Gerichtshofs im griechisch-türkischen Streit um den Festlandsockel in der Ägäis, ZaöRV 37 (1977), S. 620 ff. (636 f.).

[23] Der IGH ließ die Frage offen, ob Art. 41 IGH-Statut überhaupt eine entsprechende eigenständige Kompetenz gibt; dagegen Richter Mosler (*sep. op.*), ICJ Reports 1976, S. 24 ff., der aber zugleich gegen eine künstliche Trennung der beantragten vorläufigen Maßnahmen Stellung nimmt.

[24] Richter Lachs (*sep. op.*), ICJ Reports 1976, S. 20 f.; Elias (*sep. op.*), *ibid.*, S. 27 ff. (S. 30: »language of refusal«); Tarazi (*op. ind.*), *ibid.*, S. 32 ff.; Stassinopoulos (*op. diss.*), *ibid.*, S. 35 ff. (38).

[25] S/13615, 13616.

[26] S/13646.

Präsidenten vom 9. November wiederholt[27]. Noch bevor der Sicherheitsrat eine formelle Entschließung faßte, erhoben die Vereinigten Staaten am 29. November beim Internationalen Gerichtshof Klage gegen den Iran auf Freilassung der Geiseln und auf Wiedergutmachung; zugleich stellten sie den Antrag, im Wege der vorsorglichen Maßnahmen gemäß Art. 41 IGH-Statut die sofortige Freilassung der Geiseln zu verfügen[28].

Am 4. Dezember erging Entschließung 457 (1979), in der unter Hinweis auf die schon vorangegangenen Aktivitäten der Vereinten Nationen in dieser Sache und auf die den Weltfrieden gefährdende Spannung zwischen beiden Staaten der Sicherheitsrat den Iran zur sofortigen Freilassung der Geiseln aufforderte, beide Parteien aufrief, ihre verbleibenden Streitfragen mit friedlichen Mitteln zu lösen, sowie den Generalsekretär um seine guten Dienste ersuchte. In der Debatte wurde auf die bereits anhängigen Verfahren vor dem Internationalen Gerichtshof zunächst kein Bezug genommen; doch gab nach der einstimmig erfolgten Annahme der amerikanische Delegierte McHenry zu Protokoll, daß die Entschließung 457 (1979) friedliche Bemühungen im Rahmen anderer UN-Organe nicht ersetzen wolle. Weder die USA noch ein anderes Mitglied seien der Auffassung, daß die Entschließung eine irgendwie geartete präjudizielle Auswirkung auf das an den Internationalen Gerichtshof gerichtete Ersuchen habe, einstweilige Maßnahmen zu beschließen[29]. Diese zweifellos durch die oben dargestellten Vorgänge im *Ägäis*-Fall provozierte Stellungnahme blieb unwidersprochen[30].

Am 15. Dezember 1979 entsprach der Internationale Gerichtshof dem amerikanischen Antrag auf Erlaß vorsorglicher Maßnahmen[31]. In der Entscheidung wurde auf die gleichzeitige Befassung des Sicherheitsrats und Resolution 457 (1979) nur beiläufig hingewiesen, und zwar zum Beleg dafür, daß die Geiselnahme entgegen der iranischen Auffassung nicht als sekundäre Angelegenheit zu behandeln, sondern in sich geeignet sei, den Weltfrieden und die internationale Sicherheit ernsthaft zu bedrohen[32]. Allerdings war in der mündlichen Verhandlung (am 10. Dezember) der amerikanische Bevollmächtigte vom Präsidenten des Gerichtshofs nach den Auswirkungen der Sicherheitsratsresolution befragt worden. In der Antwort wurde versucht, den vorliegenden Fall vom *Agäis*-Fall zu differenzieren; zusätzlich wurde auf die oben erwähnte Stellungnahme des amerikanischen Delegierten im Zusammenhang mit Entschließung 457 (1979) verwiesen. Vor allem wurde auf die

---

[27] S/13652.

[28] Die Anträge finden sich in ILM 18 (1979), S. 1464 ff. und 1482 ff.; in beiden Fällen machten die USA darauf aufmerksam, daß der Sicherheitsrat vom Generalsekretär auf der Grundlage von Art. 99 SVN einberufen worden war.

[29] Vgl. den Wortlaut der Erklärung in ICJ Public Sitting, Dec. 10, 1979, Verbatim Record (Doc. CR 79/1, 1979), S. 39.

[30] UN Chronicle 1980/1, S. 5 ff. (12 f.).

[31] *United States Diplomatic and Consular Staff in Tehran,* Provisional Measures, ICJ Reports 1979, S. 7 ff.

[32] *Ibid.*, S. 15.

Funktion des Internationalen Gerichtshofs abgestellt, die ihm unterbreiteten Streitigkeiten dem Völkerrecht gemäß zu entscheiden[33].

In der Entschließung 461 vom 31. Dezember 1979 erneuerte der Sicherheitsrat seinen dringenden Aufruf an den Iran und drohte die Verhängung von Zwangsmaßnahmen an. Die einstweilige Anordnung des Gerichtshofs fand Erwähnung in der Präambel und in Ziffer 2 des operativen Teils der Resolution, wo beklagt wurde, daß der Iran die Geiseln entgegen Entschließung 457 (1979) und der Order des Internationalen Gerichtshofs nach wie vor festhalte. Auch in der Ratsdebatte wurde immer wieder auf die Entscheidung des Gerichtshofs vom 15. Dezember Bezug genommen, ohne daß aber jedoch auch dabei die Gleichzeitigkeit der Tätigkeit der UN-Organe problematisiert worden wäre[34].

Zu einer weiteren Sicherheitsratsresolution ist es nicht gekommen, nachdem die Verhängung von Zwangsmaßnahmen am Veto der Sowjetunion gescheitert war (Januar 1980)[35]. Am 24. Mai 1980 erging jedoch das Urteil des Gerichtshofs in der Hauptsache[36]. Hier finden sich im Kontext der Zuständigkeitserörterung für unsere Problematik einschlägige Hinweise. Der Gerichtshof verwies zunächst auf seine Order vom 15. Dezember 1979, in der er die Einbettung der aufgeworfenen Rechtsfragen in einen weiteren politischen Zusammenhang als nicht ausreichend angesehen hatte, seine Zuständigkeit zu verneinen[37]. Noch wichtiger sind die sich anschließenden Ausführungen darüber, ob die vom Generalsekretär in Durchführung seiner guten Dienste eingesetzte *fact finding*-Kommission die Jurisdiktionskompetenz des Internationalen Gerichtshofs beeinträchtigt haben konnte[38]. An dieser Stelle macht der Gerichtshof darauf aufmerksam, daß weder er selbst am 15. Dezember 1979 noch der Sicherheitsrat am 31. Dezember 1979 sich durch die gleichzeitige Ausübung ihrer jeweiligen Funktionen irritiert gezeigt hätten. Dazu bestehe auch kein Anlaß, da es im Verhältnis von Sicherheitsrat zu Internationalem Gerichtshof keine Art. 12 SVN entsprechende Norm gebe: »The reasons are clear. It is for the Court, the principal judicial organ of the United Nations, to resolve any legal questions that may be in issue between parties to a dispute; and the resolution of such legal questions by the Court may be an important, and sometimes decisive, factor in promoting the peaceful settlement of the dispute«[39]. Es folgt der Hinweis auf Art.36 Abs. 3 SVN. Von

[33] Verbatim Record (Anm. 29), S. 37 ff. Vgl. dazu auch *L. Gross*, The Case Concerning United States Diplomatic and Consular Staff in Tehran: Phase of Provisional Measures, AJIL 74 (1980), S. 395 ff. (402 ff.).

[34] UN Chronicle 1980/2, S. 18 ff.

[35] *E. Klein*, Tätigkeit der Vereinten Nationen in völkerrechtlichen Fragen (Berichtszeitraum: 1. 1. 1979–31. 12. 1979), ArchVR 19 (1981), S. 287 ff. (317).

[36] *United States Diplomatic and Consular Staff in Tehran*, Judgment, ICJ Reports 1980, S. 3 ff.

[37] *Ibid.*, S. 19.

[38] *Ibid., S. 20 ff.*

[39] *Ibid.*, S. 22.

hier aus wird im Ergebnis gleichfalls verneint, daß die Errichtung der Kommission mit der gleichzeitigen Durchführung des gerichtlichen Verfahrens unvereinbar wäre. Für einen Willen der Parteien, wonach die Kommission Ersatzfunktion für den Internationalen Gerichtshof hätte übernehmen sollen, ergaben sich keinerlei Anhaltspunkte [40].

Der *Geiselnahme*-Fall ist wegen der mehrfachen Befassung von Sicherheitsrat und Internationalem Gerichtshof in besonders guter Weise geeignet, die hier untersuchte Problematik zu verdeutlichen. Kein Organ hat sich in seinem *procedere* von dem Vorgehen des anderen Organs behindert gesehen. Wegen der Ablehnung im *Ägäis*-Fall, vorläufige Maßnahmen zu beschließen, ist dies vor allem für den Internationalen Gerichtshof sowohl bezüglich der Entscheidung im einstweiligen Anordnungsverfahren wie in der Hauptsache hervorzuheben; dabei ist allerdings zu berücksichtigen, daß der Gerichtshof im Jahr 1980 anders als 1976 – und mit Grund – davon ausgehen konnte, daß die vorangegangene Sicherheitsratsentschließung zu keiner Beruhigung des Streits geführt hatte, so daß nach wie vor die »Umstände« den Erlaß vorläufiger Maßnahmen als erforderlich erscheinen ließen. Auf der anderen Seite haben sowohl Sicherheitsrat als auch Internationaler Gerichtshof die Erkenntnisse des jeweils anderen Organs ohne weiteres in die eigenen Überlegungen einbezogen.

III. Zuständigkeitsproblematik

Die erste zu erörternde Frage geht dahin, ob sich die Befassung mit einer zwischenstaatlichen Streitigkeit oder die inhaltlich-sachliche Stellungnahme hierzu durch Sicherheitsrat oder Internationalen Gerichtshof wegen der vorangegangenen oder gleichzeitigen Befassung des jeweils anderen Organs den Einwänden der Litispendenz oder der *res iudicata* ausgesetzt sehen. Die eben vorgestellten Fälle deuten nicht darauf hin; dieses Ergebnis bedarf jedoch noch der dogmatischen Absicherung [41].

1. Problem der Litispendenz

a) Das im nationalen Recht wohlbekannte Institut der Rechtshängigkeit ist mit der zunehmenden Zahl internationaler Organe, die sich mit Streitfällen befassen, auch für den internationalen Bereich problematisiert worden. Für

---

[40] *Ibid.*, S. 23. – Zum Problem auch *E. Zoller*, L'affaire du personnel diplomatique et consulaire des Etats-Unis à Téhéran, RGDIP 84 (1980 II), S. 973 ff. (989 ff.).

[41] Vgl. hierzu vor allem die wichtigen Studien von *D. Ciobanu*, Litispendence between the International Court of Justice and the Political Organs of the United Nations, in: L. Gross (ed.), The Future of the International Court of Justice (1976), S. 209 ff., und *S. Rosenne*, The Law and Practice of the International Court, Bd. 1 (1965), S. 83 ff.; ferner *G. Ténékidès*, L'exception de litispendance devant les organismes internationaux, RGDIP 36 (1929), S. 502 ff.

die Übertragung der im nationalen Recht entwickelten Gedanken eignen sich vor allem Konstellationen, in denen zwei oder gar noch mehr richterliche Gremien parallel oder nacheinander mit einer Streitigkeit befaßt werden. Zum Teil treffen die einschlägigen Rechtsgrundlagen selbst Vorkehrungen gegen derartige parallele Befassungen [42]. Wo dies nicht der Fall ist, muß auf allgemeine Überlegungen zurückgegriffen werden. In diesem Sinn hat der Ständige Internationale Gerichtshof gegenüber dem Einwand der Streithängigkeit vor dem deutsch-polnischen Schiedsgericht auf dem Vorliegen von drei Voraussetzungen bestanden, die zum Erfolg dieses Einwands notwendig sind: Es müsse sich um die identische Klage (*identical action*) und um dieselben Parteien (*same parties*) vor einem artgleichen Gericht (*court of the same character*) handeln; im konkreten Fall hat der Ständige Internationale Gerichtshof diese Voraussetzungen für nicht gegeben erachtet [43].

b) Untersucht man das Litispendenzproblem [44] nun im Verhältnis von Sicherheitsrat und Internationalem Gerichtshof und geht man auch hier von den Voraussetzungen der Identität des Streitgegenstandes, der Parteien und der Gleichartigkeit der Organe aus, so stellen sich bald unüberwindbare Zweifel daran ein, daß die gleichzeitige Befassung zur Errichtung von Jurisdiktionsschranken geeignet ist.

Zugunsten einer Identität des Streitgegenstandes könnte freilich zunächst der Mißerfolg aller Versuche sprechen, politische und rechtliche Streitigkeiten sinnvoll zu unterscheiden und zum Zuordnungskriterium für Sicherheitsrat bzw. Internationalen Gerichtshof zu machen. Ein sachlicher Gegensatz zwischen rechtlichen und politischen Streitigkeiten existiert nicht [45]. Prinzipiell ist jede zwischenstaatliche Streitigkeit sowohl politisch als auch auf rein rechtlicher Basis lösbar. Seine Hauptbedeutung hat das Begriffspaar »rechtliche und politische Streitigkeiten« daher auch gar nicht im hier diskutierten Verhältnis von Sicherheitsrat zu Internationalem Gerichtshof, sondern bei der Prüfung, ob die Jurisdiktion des Gerichtshofs als Fall der sogenannten obligatorischen Gerichtsbarkeit im Sinne von Art. 30 Abs. 2 Statut gegeben ist, soweit dies von einer Streitseite bestritten wird; wo beide Seiten übereinstimmend den Streitfall dem Internationalen Gerichtshof unterbreiten (Art. 36 Abs. 1 Statut), bedarf es der entsprechenden Qualifikation als *legal*

---

42 Siehe etwa Art. 5 Abs. 2 *lit.* a) Optional Protocol zum UN Pakt über bürgerliche und politische Rechte und die sich hieran anschließenden Interpretationsschwierigkeiten.

43 CPJI Sér. A No. 6 (1925), S. 20 – *Intérêts allemands en Haute-Silésie* (Compétence).

44 *Ph. Bretton*, L'affaire des »otages« américains devant la Cour Internationale de Justice, Journal du Droit International (Clunet) 107 (1980), S. 787 ff. (800), spricht von der *exception de recours parallèle.* Dieser Ausdruck macht deutlicher, um was es hierbei geht, als der Begriff der Litispendenz, den man eher geneigt ist, nur auf die richterliche Streithängigkeit zu beziehen.

45 Dazu *G. Dahm*, Völkerrecht, Bd. 2, S. 493 f.; *R. Higgins*, Policy Considerations and the International Judicial Process, ICLQ 17 (1968), S. 58 ff. (74).

dispute von vornherein nicht[46]. Immerhin hat über Art. 36 Abs. 3 SVN die angesprochene Differenzierung auch für die Beziehung Sicherheitsrat – Internationaler Gerichtshof Bedeutung erhalten. Kontur hat der Begriff der »rechtlichen Streitigkeit« aber dadurch nicht gewonnen. Zwar läßt sich auch insoweit an eine Orientierung an Art. 36 Abs. 2 Statut denken; ausschlaggebend kann sie jedoch nicht sein, weil je nach der Einstellung der Parteien Streitigkeiten sowohl außerhalb des Katalogs von Art. 36 Abs. 2 Statut richterlicher Streiterledigung zugänglich als auch Materien innerhalb dieses Katalogs dieser Erledigungsmethode nicht offen sein können. Zudem obliegt die Einschätzung hierüber dem Sicherheitsrat[47]; gerade diese Einschätzungsfreiheit zeigt, daß weder eine vorgegebene noch eine durch Befassung eines Organs erworbene Natur des Streits (als rechtlich oder politisch) in der einen oder anderen Richtung exklusive Zuständigkeiten zu begründen vermag.

Dennoch kann letztlich von der Identität einer vor den Sicherheitsrat gebrachten Streitigkeit mit einem dem Internationalen Gerichtshof unterbreiteten Streitfall nicht ausgegangen werden. Grund hierfür ist die Unterschiedlichkeit des Begehrens (Streitgegenstand): Vor dem Internationalen Gerichtshof geht es um die Klärung der umstrittenen Fragen auf der Grundlage des Rechts und im Fall des Art. 41 Statut um den vorläufigen Schutz des bedrohten Rechts vor Verschlechterung[48]. Vor dem Sicherheitsrat geht es stets um die im Interesse der Friedenswahrung notwendige Streitbeilegung als solche[49]. Diese sicherlich zugespitzte Unterscheidung besagt weder, daß Erwägungen der Friedenssicherung oder Friedenserhaltung vor dem Internationalen Gerichtshof, noch daß rechtliche Überlegungen vor dem Sicherheitsrat keine Rolle spielten; doch unbestreitbar sind die Akzente deutlich anders gesetzt[50].

Das zweite Merkmal der Identität der Streitparteien ist weniger problematisch. Immerhin ist zu beachten, daß auch am Streit nicht unmittelbar beteiligte Mitgliedstaaten der Vereinten Nationen den Sicherheitsrat anrufen können (Art. 35 SVN); diesem Recht von Drittstaaten läßt sich schwerlich mit dem Litispendenz-Einwand begegnen. Wenn aber Dritte insoweit nicht gehindert sind, ist Abweichendes für die Streitparteien selbst wohl kaum zu vertreten.

Am eindeutigsten führt das Erfordernis der Jurisdiktionshomogenität[51] zu einer negativen Antwort auf die Frage nach der rechtlichen Relevanz der

---

[46] *Rosenne* (Anm. 41), S. 94; nicht unbestritten.

[47] *N. Bentwich/A. Martin*, A Commentary on the Charter of the United Nations (1950), S. 83.

[48] Der Fall der *ex aequo et bono*-Entscheidung (Art. 38 Abs. 2 Statut) bleibt hier außer Betracht.

[49] Welche Entscheidungsbefugnisse der Sicherheitsrat im Verhältnis zum IGH hat, ist unten (nach Anm. 52) näher darzustellen.

[50] *R. Higgins*, The Place of International Law in the Settlement of Disputes by the Security Council, AJIL 64 (1970), S. 1 ff. (15 ff.).

Litispendenz im hier erörterten Zusammenhang. Dafür sind zwei Gründe maßgebend. Der eine ergibt sich aus der verschiedenartigen Charakterisierung der beiden Organe: Der Sicherheitsrat wird als politisches, der Internationale Gerichtshof als richterliches Organ qualifiziert. Hinter diesen Bezeichnungen verbergen sich materielle Unterschiede, die hier nur stichwortartig zu nennen sind: Zusammensetzung, (Un-) Abhängigkeit, (nicht) justizförmiges Verfahren, unterschiedlich intensive Verwiesenheit auf das Völkerrecht[52].

Der zweite gegen die Jurisdiktionshomogenität von Sicherheitsrat und Internationalem Gerichtshof sprechende Grund ist die Andersartigkeit ihrer Entscheidungsbefugnis, die in der unterschiedlichen Zuordnung von verbindlicher Entscheidung und Streitinhalt zum Ausdruck kommt. Nur der Gerichtshof kann verbindlich zur Streitsubstanz selbst entscheiden; so kann er etwa – *inter partes* – mit bindender Kraft (Art. 94 Abs. 1 SVN; 59 Statut) die Zugehörigkeit eines umstrittenen Territoriums zum Gebiet der einen oder anderen Streitpartei oder die rechtliche Verpflichtung zu die Streitsubstanz betreffenden Handlungen oder Unterlassungen aussprechen. Die Entscheidungsbefugnis des Sicherheitsrats ist anders geregelt: Soweit der Rat Beschlüsse streitsubstantieller Natur faßt, kann er nur in der Rechtsform der Empfehlung handeln[53]. Für den Bereich des Kapitels VI SVN ist dies unbestritten[54]. Aber auch im Bereich des Kapitels VII SVN bleibt für streitinhaltsbezogene Beschlüsse nur die in Art. 39 und 40 SVN auch ausdrücklich oder implizit zur Verfügung gestellte rechtlich unverbindliche Empfehlung. Verbindlichkeit kommt Ratsbeschlüssen im Rahmen des Kapitels VII SVN nur insoweit zu, als diese den Konflikt der friedlichen Beilegung offen halten oder wieder zugänglich machen wollen[55]. Eine den Streit inhaltlich erledigende Instanz ist der Sicherheitsrat folglich nicht; er kann z.B. weder unmittelbar den Status eines umstrittenen Territoriums definieren noch seine streitsubstanziellen Empfehlungen zwangsweise durchsetzen[56]. Von der Satzung ist die verbindliche politische Streiterledigung nicht vorgesehen; ihr ist das Souveränitätsprinzip im Weg gestanden[57].

---

[51] Vgl. *Ténékidès* (Anm. 41), S. 523: »En effet la notion de litispendance est intimement liée à une question d'homogénéité juridictionnelle«.

[52] Vgl. dazu *Mosler* (Anm. 1), S. 286; *Rosenne* (Anm. 41), S. 74 ff.; *Ciobanu* (Anm. 41), S. 211 ff.; *A. Verdross/B. Simma,* Universelles Völkerrecht (2. Aufl. 1981), S. 137.

[53] So zu Recht *M. Krökel,* Die Bindungswirkung von Resolutionen des Sicherheitsrates der Vereinten Nationen gegenüber Mitgliedstaaten (1977), S. 87 f.

[54] Als Rechtsgrundlagen kommen in Betracht Art. 35, 37, 38 SVN.

[55] *Krökel* (Anm. 53), S. 79 ff., 97.

[56] *M. Virally,* L'organisation mondiale (1972), S. 418.

[57] Vgl. hierzu mit weiteren Nachweisen *E. Klein,* Statusverträge im Völkerrecht (Beiträge zum ausländischen öffentlichen Recht und Völkerrecht, Bd. 76) (1980), S. 106 f.

Aus alledem folgt, daß im Bereich der verbindlichen streitsubstanzbezogenen Regelung divergierende Entscheidungen von Sicherheitsrat und Internationalem Gerichtshof gar nicht auftreten können, eine unlösbare Pflichtenkollisionsituation für die Beteiligten daher nicht entstehen kann. Nicht ausgeschlossen sind freilich den Streitinhalt betreffende Divergenzen schlechthin[58]. Das damit aufgeworfene Problem ist aber nicht mehr ein solches der Litispendenz; denn gerade die dargestellten Unterschiede in der Entscheidungsbefugnis verbieten es, Sicherheitsrat und Internationalen Gerichtshof als rechtlich gleichartige Organe zu qualifizieren.

Dieses Resultat wird auch abgesichert durch einen Umkehrschluß aus Art. 11 und 12 SVN, die eben nur im Verhältnis der politischen Organe Generalversammlung und Sicherheitsrat Anwendung finden[59]. Ferner bietet auch Art. 36 Abs. 2 SVN für die Gegenargumentation keine Handhabe; denn weder der Wortlaut noch die Entstehungsgeschichte noch die nachfolgende Praxis weisen darauf hin, daß der Sicherheitsrat gegenüber einem Streit *sub iudice* gehindert wäre, sich einzuschalten[60]. Schließlich ist noch darauf hinzuweisen, daß es dem Sinn des Gebots der friedlichen Streitbeilegung nicht entsprechen kann, die Prozeduren und Methoden solcher Streiterledigung einzuschränken, bevor der Streit auch wirklich faktisch beigelegt ist[61]. Im Ergebnis stellt sich daher das Litispendenzproblem im Verhältnis Sicherheitsrat und Internationalem Gerichtshof nicht[62].

---

[58] So könnte der Sicherheitsrat als streitbeilegende Maßnahme die Abtretung eines umstrittenen Gebiets von Staat A an Staat B empfehlen (vgl. *Dahm* (Anm. 45), S. 368), während der IGH *ex iure* zur Feststellung gelangt, das Gebiet gehöre zu A.

[59] Siehe oben bei Anm. 39. Ferner: Richter Stassinopoulos (*op. diss.*), ICJ Reports 1976, S. 35 ff. (38); *L. Gross,* The International Court of Justice and the United Nations, RdC 120 (1967 I), S. 313 ff. (328); *Rosenne* (Anm. 41), S. 73. Übrigens ist bemerkenswert, daß Art. 12 SVN der Generalversammlung nicht die deliberierende Befassung, nur die Beschlußfassung verbietet. – Wesentlich stärker ist die Anbindung der Hilfsorgane geregelt; vgl. etwa zum Verhältnis Generalversammlung und Sicherheitsrat zur Interimskommission *D. Chr. Dicke/H.-W. Rengeling,* Die Sicherung des Weltfriedens durch die Vereinten Nationen (1975), S. 138 f. – Ein interessantes Beispiel hierarchischer Ordnung in einer internationalen Organisation findet sich jetzt in Art. 6 Abs. 5 des Vertrages vom 18. 6. 1981 über die Errichtung der Organisation der ostkaribischen Staaten (ILM 20 (1981), S. 1166 ff.).

[60] Ausführlich dazu *Ciobanu*(Anm. 41), S. 220 ff.

[61] Dieses Argument bezieht sich auch auf kritische Anmerkungen zu den vom StIGH (Anm. 43) aufgestellten Voraussetzungen eines relevanten Litispendenzeinwandes, mit denen die starke nationalrechtliche Orientierung beklagt wird, vgl. etwa *J. A. W. Verzijl,* The Jurisprudence of the World Court, Bd. 1 (1965), S. 24 f.

[62] So auch die ganz herrschende Meinung: Neben den oben Anm. 41 Genannten vgl. noch *Gross* (Anm. 59), S. 328 ff.; *derselbe* (Anm. 18), S. 37 f.; *H. Fakher,* The Relationships among the Principal Organs of the United Nations (1951), S. 64 f.

2. Res iudicata-Einwand

Aufgrund der vorstehenden Überlegungen ist auch der Einwand der *res iudicata* im hier erörterten Zusammenhang ohne Bedeutung[63]. Dem entspricht die Rechtsprechung des Weltgerichtshofs: Der Ständige Internationale Gerichtshof hat einen solchen sich auf eine Völkerbundratsentschließung beziehenden Einwand wegen des unterschiedlichen Jurisdiktionscharakters zurückgewiesen[64]. Entgegen der Ansicht von *Ciobanu*[65] weist auch das *Namibia*-Gutachten des Internationalen Gerichtshofs in keine andere Richtung: Wenn dort nämlich der Gerichtshof die Einschätzung des Sicherheitsrats übernimmt, bei der Diskussion der Resolution S/276 (1970) sei eine *situation* erörtert worden und kein *dispute* im Sinne von Art. 32 SVN, so ist dies nicht die Folge einer vom Internationalen Gerichtshof für bindend erachteten Entscheidung des Sicherheitsrats; ausdrücklich wird nämlich darauf hingewiesen, daß diese Frage deshalb nicht mehr geprüft zu werden brauche, weil Südafrika dieser Einschätzung durch den Sicherheitsrat nicht widersprochen habe[66].

Vom Sicherheitsrat her gesehen wäre es bei seiner »primary responsibility for the maintenance of international peace and security« abwegig, Befassung und Beschlußfassung deshalb nicht zulassen zu wollen, weil der Gerichtshof zwar bereits entschieden, diese Entscheidung aber nicht tatsächlich zur Streitbeilegung geführt hat. Eine ganz andere Frage ist, inwieweit der Sicherheitsrat die vorangegangene Entscheidung des Internationalen Gerichtshofs bei seiner eigenen Entschließung zu berücksichtigen hat – und *vice versa*. Darauf ist alsbald zurückzukommen[67].

3. Kann der Sicherheitsrat die Anrufung des Internationalen Gerichtshofs verbieten?

Ist nach allem die parallele, aber auch hintereinander geschaltete Befassung von Sicherheitsrat und Internationalem Gerichtshof im Hinblick auf dieselbe Streitsache grundsätzlich zulässig, bleibt doch noch zu bedenken, ob nicht unter besonderen Bedingungen die an sich – zumindest im Rahmen von Art. 36 Abs. 6 Statut – gegebene Jurisdiktionskompetenz des Internationalen Gerichtshofs durch eine Entscheidung des Sicherheitsrats beschränkt werden kann. Richter Alvarez hat eine solche Konstellation bejaht: »Si une affaire

---

[63] So etwa auch *Dahm* (Anm. 45), S. 515; zur Relevanz des Konzepts innerhalb der Judikatur des IGH vgl. *S. Rosenne*, The Law and Practice of the International Court, Bd. 2 (1965), S. 623 ff.

[64] CPJI Sér. A No. 15, S. 29 – *Droits des minorités en Haute-Silésie (Ecoles minoritaires).*

[65] *Ciobanu* (Anm. 41), S. 232.

[66] ICJ Reports 1971, S. 16 ff. (22 f.) – *Legal Consequences for States of a Continued Presence of South Africa in Namibia.*

[67] Siehe unten IV. 2. und 3.

soumise à la Cour vient menacer la paix du monde, le Conseil de Sécurité peut s'en saisir et mettre fin à la compétence de la Cour. La compétence du Conseil résulte de la nature de l'Organisation internationale créée par la Charte et des attributions du Conseil«[68]. Dieser Ansicht hat *Rosenne* widersprochen: »Such an automatic consequence can only occur where the title under which the Court is exercising jurisdiction contains an express reservation to the jurisdiction of the Court in the event of the dispute being submitted to the Security Council«[69].

Die Auffassung von Alvarez geht zunächst insofern zu weit, als sie bei jedem friedensgefährdenden Streit dem Sicherheitsrat die übergeordnete Befassungskompetenz einräumt. Eine solche Hierarchie im Verhältnis zum Internationalen Gerichtshof besteht aber – wie oben ausgeführt – gerade nicht[70]. Friedensgefährdende zwischenstaatliche Streitigkeiten können dem Internationalen Gerichtshof also gewiß nicht *ratione materiae* entzogen werden. Allerdings kann nicht übersehen werden, daß möglicherweise gerade die Anrufung des Gerichtshofs zu einer gefährlichen Eskalation der Auseinandersetzung führt; dies ist natürlich nur denkbar in Fällen, in denen sich die eine Streitpartei dem Gang zum Gerichtshof, aus welchem Grund auch immer, entziehen möchte. Daß dies nicht nur eine hypothetische Konstruktion ist, belegen Streitigkeiten wie der *Ägäis-Festlandsockel*-[71] und der *Beagle-Kanal*-Fall[72]. Um den Ausbruch einer militärischen Konfrontation zu verhindern, ist es vorstellbar, daß der Sicherheitsrat gemäß Art. 39 oder 40 SVN den Parteien verbindlich aufgibt, in dieser Situation auf die Anrufung des Internationalen Gerichtshofs zu verzichten[73]. Weil der Sicherheitsrat selbst eine substanzielle Streitlösung nicht mit die Parteien bindender Wirkung vorlegen kann[74], kann dieses Verbot allerdings nur vorübergehender Natur sein.

---

[68] Richter Alvarez (*op. diss.*), ICJ Reports 1952, S. 124 (134).

[69] *Rosenne* (Anm. 41), S. 73. In diesem Sinn wohl auch *O. Audéoud,* La Cour internationale de Justice et le règlement des différends au sein des organisations internationales, RGDIP 81 (1977), S. 945 ff. (967 f.); unklar *Ciobanu* (Anm. 41), S. 228 f.

[70] Dazu auch *H. Steinberger,* The International Court of Justice, in: H. Mosler/R. Bernhardt (eds.), Judicial Settlement of International Disputes (Beiträge zum ausländischen öffentlichen Recht und Völkerrecht, Bd. 62) (1974), S. 193 ff. (203).

[71] Vgl. die Äußerungen der Türkei in der S/RES/395 (1976) vorausgehenden Ratsdebatte (oben bei und in Anm. 19).

[72] Die argentinische Regierung gab am 21. 1. 1982 folgende Erklärung ab: »Die Regierung der Republik Argentinien ist zu der Auffassung gelangt, daß das Generalabkommen für eine juristische Lösung der Kontroversen, das am 5. April 1972 mit der Republik Chile unterzeichnet wurde, in seiner gegenwärtigen Form weder den Interessen der argentinischen Nation entspricht, noch das geeignete Instrument für eine Lösung der zwischen beiden Ländern entstandenen Zwischenfälle ist. Ein Beweis hierfür liegt in der Tatsache, daß es absolut ungeeignet war, die schwerste Krise zu überwinden, und daß es während der Vorfälle sich als notwendig erwies, auf die Vermittlung des Papstes zurückzugreifen. . .«.

[73] Entsprechendes gälte für die Unterbrechung eines laufenden Verfahrens.

[74] Dazu oben III. 1. b) zur Entscheidungsbefugnis.

Der Internationale Gerichtshof wäre an eine solche Sicherheitsratsentschließung insoweit gebunden, als sie seine Zuständigkeit *ratione personae* berührt: Wie er nämlich *inter partes* vereinbarte Jurisdiktionsvorbehalte zur Kenntnis zu nehmen hat, muß er auch von außen an die Parteien herangetragene und diese bindende Handlungsbeschränkungen beachten.

Zugegebenermaßen erreicht die Untersuchung hier einen kritischen Punkt – nicht weil sich nun doch eine Art Unterordnung des Internationalen Gerichtshofs unter den Sicherheitsrat herausgestellt hätte (was nicht der Fall ist), sondern weil der Rekurs auf das Völkerrecht, der in der Anrufung des Gerichtshofs zu sehen ist, als Friedensgefährdung charakterisiert zu werden scheint. Sicher wäre eine solche Argumentation so nicht haltbar[75]. Die objektive rechtliche Würdigung schließt jedoch nicht aus, daß aus der Anrufung des Internationalen Gerichtshofs eine tatsächliche, den Weltfrieden gefährdende Eskalation entstanden sein kann – und sei es allein wegen der unangemessenen Reaktion der anderen gerichtsunwilligen Partei. Die Satzung der Vereinten Nationen stellt die Friedenserhaltung in den Mittelpunkt ihrer Bemühungen. Dabei kommt dem Völkerrecht eine wichtige, aber keine ausschließliche Funktion zu. Dieser mangelnde Exklusivitätsanspruch des Völkerrechts, der den wichtigsten Unterschied zur Rechtsfunktion in einer freiheitlich organisierten Staatsordnung markiert, führt dazu, daß – wo notwendig – der Sicherheitsrat zur Beseitigung der militärischen Konfliktgefahr über das Bestehen von Rechtspositionen (zeitweilig) hinweggehen kann[76].

IV. Loyalitätsgebot und Berücksichtigungspflicht

1. Allgemeines

Mit der Parallelität des Tätigwerdens[77] von Sicherheitsrat und Internationalem Gerichtshof tritt der Ordnungsgesichtspunkt der Koordination[78] ins Blickfeld. Da ein gemeinsames übergeordnetes Koordinationsorgan fehlt, muß sie von Sicherheitsrat und Gerichtshof selbst geleistet werden. Sie vollzieht sich im Weg der funktionalen Kooperation[79].

Dogmatisch läßt sich die Kooperationspflicht aus dem Loyalitätsgebot herleiten, dem die Organe der Vereinten Nationen im Verhältnis zueinander unterliegen. Die gemeinsame Organstellung (Art. 7 SVN) und die Verpflichtung aller Organe auf die Zwecke und Grundsätze der Vereinten Nationen

---

[75] Vgl. etwa Prinzip II Abs. 5 Satz 2 der Friendly Relations-Deklaration, A/RES/2625 (XXV), Annex.

[76] Vgl. zu dieser Frage auch noch unten IV. 3. a) und bei Anm. 134.

[77] *Rosenne* (Anm. 41), S. 87, spricht von »functional parallelism«.

[78] Richter Elias (*sep. op.*), ICJ Reports 1976, S. 27, bezeichnet Sicherheitsrat und IGH als »co-ordinate principal organs of the United Nations«.

[79] *Rosenne* (Anm. 41), S. 70; *Steinberger* (Anm. 70).

(Art. 1 und 2 SVN)[80] bilden die positiv-rechtlichen Ansätze zur Entwicklung einer organisationsspezifischen Organtreue. Wie im innerstaatlichen Bereich die »Verfassungsorgantreue« letztlich mit der Herstellung staatlicher Einheit zu begründen ist[81], die ihrerseits Voraussetzung ist für die Verwirklichung der Staatszwecke, gründet sich die Organloyalität internationaler Organisationen auf die von den Mitgliedstaaten intendierte Verwirklichung des Organisationszwecks, die eine Zusammenfassung der Organisationskapazitäten fordert, zumindest aber die gegenseitige Kompetenzverhinderung und Lahmlegung verbietet. Gerade im Bereich internationaler Organisationen ist der Auslegungstopos der Zweckverwirklichung keine unbekannte Größe; von hier aus erklären sich z.B. die *effet utile-* und *implied powers*-Lehre[82]. Derselbe Gedanke wirkt sich im Binnenverhältnis der Organe aus. Dabei darf es freilich nicht – ebensowenig wie bei Anwendung der genannten Lehren – zu einer das »institutionelle Gleichgewicht« erschütternden Kompetenzaufweichung oder -verschiebung kommen. Organtreue kann nur eine am Organisationszweck und am Organisationsgefüge orientierte gegenseitige Rücksichtnahme bei der Ausübung der Organkompetenzen bedeuten.

Der rechtliche Zusammenhang, in dem die Organtreue steht, verbietet es, sie als bloß unverbindliche Maxime, als eine Angelegenheit des politischen Stils abzutun[83]. Sie ist vielmehr Rechtsgebot. Aus ihr folgt die Pflicht, das Verfahren und die verfahrensabschließenden Entscheidungen des jeweils anderen Organs im Rahmen des eigenen Verfahrens und der eigenen Entscheidungsfindung zu berücksichtigen. Dieses Berücksichtigungsgebot beinhaltet jedenfalls die Anerkennung der tatsächlichen Befassung und Entscheidung des anderen Organs[84], die sich zunächst einmal als Zurkenntnisnahme auswirkt. Darüber hinaus sind die Organe verpflichtet, Befassung und Befassungsergebnis der anderen Organe in die eigenen Erwägungen einzustellen mit dem Ziel, divergierende Ergebnisse zu vermeiden. Für diesen Einstellungsvorgang erweist sich der Organisationszweck sowohl als Grundlage wie als Grenze. Eine inhaltliche Bindung im Sinne einer die eigene Entscheidung prägenden obligatorischen Zugrundelegung kommt daher nur in Be-

---

[80] Einprägsam die Formulierung des Richters Azevedo (*op. ind.*), ICJ Reports 1950, S. 79 ff. (80) – *Interpretation of Peace Treaties*: »La Cour, désignée comme organ principal et ainsi plus dans l'engrenage de l'O.N.U., doit faire de son mieux pour collaborer avec les autres organes en vue d'atteindre les buts et principes visés«. Vgl. auch Richter Tarazi (*op. ind.*), ICJ Reports 1976, S. 32 ff. (33).

[81] Dazu *W.-R. Schenke,* Die Verfassungsorgantreue (1977), S. 26 ff.

[82] Im Überblick hierzu *I. Seidl-Hohenveldern,* Das Recht der Internationalen Organisationen einschließlich der Supranationalen Gemeinschaften (3. Aufl. 1979), Rdr. 1602 ff.

[83] Vgl. *Gross* (Anm. 59): »good sense, restraint and tact of political organs«.

[84] *Rosenne* (Anm. 41), S. 71: »(I)t follows that when one principal organ has to deal with a decision taken by another principal organ, it has to take that decision as it is, having no power to change the *meaning* of that decision as determined by the organ making it«. – Man könnte insoweit von »Tatbestandswirkung« sprechen.

tracht, wenn dadurch die im Interesse der Friedenswahrung eingeräumte Eigenständigkeit der organspezifischen Kompetenz unangetastet bleibt; dies ist gleich – für Gerichtshof und Sicherheitsrat getrennt – näher darzustellen. Demgegenüber ist ein Nichtbefassungsgebot [85] aus dem Rechtsgrundsatz der Organtreue nicht herzuleiten. Der Verwirklichung des Organisationszwecks wird vielmehr durch sachlich übereinstimmende bzw. sich ergänzende Entscheidungen besser gedient sein als durch möglicherweise als Distanzierung zu verstehende Nichtbefassung [86].

2. Berücksichtigungspflicht des Internationalen Gerichtshofs

a) Der Internationale Gerichtshof hat, judiziert er nicht *ex aequo et bono,* seiner Rechtsprechung das die Parteien in ihrem Verhältnis zueinander bindende Recht zugrunde zu legen; sofern sie Mitglieder der Vereinten Nationen sind, also vor allem die Satzung. Von hier aus leitet sich auch die Bindung der Parteien an bestimmte Entschließungen des Rates ab. Bei der Entscheidung am Maßstab des Rechts darf der Internationale Gerichtshof diese auf die Rechtsposition der Parteien unmittelbar einwirkenden Resolutionen nicht übergehen [87].

Wie gezeigt wurde, beziehen sich freilich die bindenden Ratsentschließungen gerade nicht auf die Streitsubstanz [88]. Insoweit besteht daher keine Pflicht des Internationalen Gerichtshofs, die Resolutionen des Rats zu beachten, jedenfalls nicht über das Instrument der Parteibindung. Pflichtenkollisionen können in diesem Bereich für die Streitparteien nicht auftreten. Auch die Verweisung der Parteien auf den Internationalen Gerichtshof nach Art. 36 Abs. 3 (unmittelbar oder in Verbindung mit Art. 37 Abs. 2 SVN) ist für den Gerichtshof nicht *ipso iure* zuständigkeitsbegründend [89], ebensowenig wie seine Zuständigkeit durch die Empfehlung des Sicherheitsrats (gem. Art. 33 Abs. 2 oder 36 Abs. 1 SVN) [90], eine andere Form der Streitlösung zu suchen, beeinträchtigt werden kann.

b) Von alledem unberührt bleibt das aus der Organtreuepflicht herzuleitende Berücksichtigungsgebot, dem der Internationale Gerichtshof gegenüber den Akten der anderen Organe, hier des Sicherheitsrats, unterliegt. Daraus erwächst jedoch kein Hindernis, eine unabhängige, eigenverantwortliche

---

[85] Etwa verbrämt als *judicial* oder *political self-restraint.*

[86] In diese Richtung ging der Vorwurf des Richters Elias (Anm. 24) bei der Ablehnung der einstweiligen Anordnung im *Ägäis*-Fall (»language of refusal«) – freilich zu Unrecht, da der Gesichtspunkt der Organloyalität nicht klare Zuständigkeitsgrenzen (Art. 41 Statut) übergehen darf.

[87] Siehe dazu schon oben III. 3.

[88] Dazu oben bei und in Anm. 53; vgl. auch *Higgins* (Anm. 50), S. 11 ff.

[89] Darauf weist die *op. ind.* im *Korfu-Kanal*-Fall richtig hin (Anm. 14). – Übrigens ist eine bindende Verweisung der Parteien auf den IGH auch nicht nach Kap. VII SVN möglich.

[90] Zum Unterschied dieser Rechtsgrundlagen vgl. *Bentwich/Martin* (Anm. 47).

Rechtsprüfung vorzunehmen, ja nicht einmal die Ermächtigung zu einem solchen Verzicht. Das Prinzip der Organtreue kann nämlich die Kompetenzverteilung nicht ändern. Damit reduziert sich jedoch die Berücksichtigungspflicht nicht auf Null; denn bei seiner Rechtsfindung wird der Internationale Gerichtshof eine als Rechtsüberzeugung des Sicherheitsrats feststehende Auffassung im Rahmen vertretbarer Erwägungen in Betracht ziehen müssen. Allerdings ginge es zu weit, für Ratsresolutionen eine Vermutung zugunsten ihrer Völkerrechtsmäßigkeit eingreifen zu lassen, da anders als im innerstaatlichen Recht, wo die staatlichen Organe an das Recht gebunden sind, eine derartige Rechtsfestlegung für die politischen Organe der Vereinten Nationen, soweit sie streitbeilegend tätig werden, nicht besteht. Letztlich sind daher sachliche Divergenzen auch unter dem Aspekt des Berücksichtigungsgebots nicht ausgeschlossen und hinzunehmen.

c) Das Problem einer Berücksichtigungspflicht des Internationalen Gerichtshofs ist vor allem im Zusammenhang mit dem Erlaß oder Nichterlaß vorläufiger Maßnahmen aktuell geworden; zu verweisen ist auf den *Ägäis-Festlandsockel-* und den *Iran*-Fall [91]. Ausgangspunkt einer Diskussion hat auch hier die einschlägige Kompetenzvorschrift zu sein (Art. 41 Statut) [92]. Die entscheidende Frage, die dabei aufgetaucht ist, lautet, welchen Einfluß eine vorangegangene fallbezogene Sicherheitsratsentschließung auf den Anordnungsgrund (Dringlichkeitsproblem, »if... circumstances so require«) hat.

Der bloße Hinweis auf die Existenz einer derartigen Resolution ist weder ausreichend, um die Dringlichkeit der beantragten vorläufigen Maßnahme zu bejahen, noch um sie zu verneinen; denn die Resolution kann einerseits zwischenzeitlich Früchte getragen haben, so daß eine Anspruchsverschlechterung *lite pendente* nicht zu besorgen ist, sie kann andererseits aber auch ihr Ziel, die Spannung zu vermindern, verfehlt haben, so daß die Rechtsgefährdung der antragstellenden Streitpartei unverändert fortbesteht. Auch hier ist daher auf der Grundlage der Kompetenzvorschrift eine eigenständige Prüfung des Internationalen Gerichtshofs erforderlich [93], die weder in der einen noch in der anderen Richtung durch das Berücksichtigungsgebot obsolet wird. Der Gerichtshof hat freilich im Sinne des oben Gesagten die Sicherheitsratsentschließung in seine Rechtserwägungen einzubeziehen [94], und er hat, wenn dies der Rat festgestellt hat, von der im Zeitpunkt des Ergehens der Resolution bestehenden friedensgefährdenden Situation auszugehen; denn diese festzustellen, obliegt wieder dem Sicherheitsrat bei der Ausübung seiner eigenen Kompetenz. Der Internationale Gerichtshof ist kein – einem

---

[91] Siehe oben II. 2. und 3.

[92] Dabei ist der Auffassung recht zu geben, die in Art. 41 Statut keine eigenständige, sondern nur eine »akzessorische« Kompetenz sieht; vgl. dazu Richter Mosler, ICJ Reports 1976, S. 24.

[93] Vgl. auch *Oellers-Frahm* (Anm. 22), S. 636.

[94] So auch *M. Bettati*, L'affaire du plateau continental de la mer Egée devant la Cour internationale de Justice, AFDI 22 (1976), S. 99 ff. (114).

staatlichen Verfassungsgericht vergleichbarer – Letztinterpret der »Verfassung« der Organisation [95].

Diese vom Gerichtshof selbständig und eigenverantwortlich vorzunehmende Prüfung fiel – mit guten Gründen – im *Ägäis*-Fall anders aus als im iranischen *Geisel*-Fall [96]. Der Unterschied liegt nicht (jedenfalls soweit in der Entscheidung von außen erkennbar) darin, daß im ersten Fall [97] ein erheblich stärkeres Abstellen auf die Sicherheitsratsentschließung erfolgt wäre als im zweiten, sondern in der anderen Einschätzung der Dringlichkeitssituation, wie sie sich dem Internationalen Gerichtshof auf Grund der Reaktion der Streitparteien auf die beiden Ratsresolutionen darstellte. Insoweit ist einerseits auf die Äußerungen von Regierungsmitgliedern beider Seiten zur Entschließung 395 (1976) zu verweisen [98], andererseits auf die eindeutige negative Haltung des Iran gegenüber dem Appell des Sicherheitsrats [99]. Demgegenüber ist der Hinweis des amerikanischen Delegierten, Resolution 457 (1979) wolle kein Ersatz für Maßnahmen anderer UN-Organe sein [100], ohne rechtliche Bedeutung. So wie die Kompetenz des Gerichtshofs von anderen Organen grundsätzlich nicht beschränkt werden kann, kann sie auch nicht erweitert werden. Die Dringlichkeit der Schutzmaßnahme ist zwingende Voraussetzung ihres Erlasses. Ein inhaltlicher Bruch in der Rechtsprechung des Gerichtshofs ist daher nicht erkennbar [101].

Freilich kann man fragen, ob es dem Internationalen Gerichtshof auch bei negativer Prüfung des Dringlichkeitserfordernisses möglich oder sogar geboten ist, sich dem vorangegangenen Appell des Sicherheitsrats anzuschließen und somit seine Autorität mit der des Rats zu vereinen [102]. Wie ich meine, ist auch diese Frage von der Kompetenz des Gerichtshofs her zu beantworten: Der Internationale Gerichtshof ist zuständig zum Erlaß vorläufiger Maßnahmen, sofern die Voraussetzungen vorliegen. Er ist nicht zuständig zur Abgabe allgemeiner, seien es politisch, seien es rechtlich begründeter Erklärungen. Auch hier hilft die Organtreue nicht über die Kompetenzschranke hinweg [103].

Aus diesen Erwägungen folgt schließlich auch, daß der Internationale Gerichtshof im Rahmen seiner Zuständigkeit nach Art. 41 Statut inhaltlich

---

95 *Steinberger* (Anm. 70), S. 203.

96 Die tatsächliche Entwicklung hat dies *ex post* bestätigt.

97 Diesen Vorwurf erhebt der Sache nach *U. Villani,* Le misure cautelari nell'affare della piattaforma continentale del mare Egeo, Riv. Dir. Int. 60 (1977), S. 128 ff. (131).

98 Siehe oben bei Anm. 21.

99 *Klein* (Anm. 35), S. 300 f., 316 f.

100 Siehe oben bei Anm. 29 und 30.

101 Anders *Gross* (Anm. 33), S. 403 f.

102 In diese Richtung weisen die *sep. op.* der Richter Lachs, Elias und Tarazi, ICJ Reports 1976, S. 19 f., 27 ff., 31 ff.

103 Unverständlich daher auch *Gross* (Anm. 18), S. 44 f., der sich angesichts Ziff. 4 der Resolution 395 (1976) über die Zurückhaltung des IGH wundert.

über die Anordnungen oder Empfehlungen des Sicherheitsrats hinausgehen oder auch dahinter zurückbleiben darf. Hätte etwa der Sicherheitsrat im *Geisel*-Fall die beteiligten Staaten nur zur Besonnenheit und zum Verzicht auf militärische Maßnahmen aufgefordert, wäre es dem Gerichtshof nicht verwehrt gewesen, seine vorläufigen Maßnahmen so zu formulieren, wie er es tatsächlich getan hat. Umgekehrt wäre es dem Gerichtshof rechtlich unbenommen geblieben, seinerseits und anders als der Sicherheitsrat nicht auf der Freilassung der Geiseln zu bestehen, wenn er darin eine unzulässige Vorwegnahme der Hauptsacheentscheidung gesehen hätte [104]; diese Rechtsauffassung wäre zwar im konkreten Fall nicht richtig, aber als prinzipielle Überlegung – trotz der anders lautenden Sicherheitsratsentschließung – nicht zu beanstanden gewesen.

3. Berücksichtigungspflicht des Sicherheitsrats

Die eben erörterten Fragen stellen sich natürlich auch umgekehrt für den Sicherheitsrat im Hinblick auf die Entscheidungen des Internationalen Gerichtshofs.

a) Die Antwort hierauf verlangt zunächst nach einer Klärung des Stellenwerts rechtlicher Erwägungen im Entscheidungsfindungsprozeß des Rates. Man hat eine derartige Orientierung für nachgerade schädlich gehalten; im *Ägäis*-Fall wurde darauf hingewiesen, daß der Sicherheitsrat kaum der Platz sei »to analyze such complex issues of international law« [105]; als die Suez-Intervention 1956 im Sicherheitsrat diskutiert wurde, erklärte der britische Premierminister, es sei »not particularly helpful«, im Sicherheitsrat über die rechtlichen Aspekte des Falles nachzudenken [106]. Man wird wohl dem Dilemma, in dem sich der Sicherheitsrat insoweit offenbar befindet, am ehesten gerecht mit der Unterscheidung von Maßnahmen, die unmittelbar die kriegerische Konfrontation ausschalten wollen (*peace-keeping*), und solchen, die eine inhaltliche Streitlösung erstreben (*peace-making*) [107].

Bei Maßnahmen des *peace-keeping* steht die Verhinderung des Ausbruchs eines militärischen Konflikts oder die Wiederherstellung der Waffenruhe ganz im Vordergrund. Es geht dabei zunächst nicht um Recht oder Unrecht, sondern um die Verhinderung einer sonst vielleicht nicht mehr aufzuhaltenden Entwicklung. Denkbare verbindliche Entscheidungen, z.B. die Aufforderung zur sofortigen Kampfeinstellung, müssen sich notgedrungen an beide Seiten richten; nach der Philosophie der Satzung sind Zwangsmaßnahmen im

---

[104] Zum Problem vgl. ICJ Reports 1979, S. 16; *Gross* (Anm. 33), S. 404 f.
[105] UN Doc. S/PV. 1953 (Aug. 25, 1976), S. 14 f.
[106] Hierzu mit weiteren Beispielen *Higgins* (Anm. 50), S. 9 f.
[107] Die Unterscheidung findet ihre unmittelbare Rechtfertigung im Wortlaut des Art. 1 Ziff. 1 SVN, wo auf das Völkerrecht erst im Kontext der Streitlösung, nicht bei der unmittelbaren Beseitigung der Friedensgefährdung bzw. der Aggressionsakte hingewiesen wird; siehe auch unten bei Anm. 110.

Sinne von Kapitel VII SVN auch gegen den Angegriffenen, keineswegs nur gegen den Aggressor selbst möglich[108]. Dabei ist freilich das »Auswahlermessen« des Sicherheitsrats bezüglich des Adressaten von Zwangsmaßnahmen nicht völlig frei, sondern hat sich an der Bereitschaft, die Anordnungen des Sicherheitsrats zu befolgen, auszurichten. Letztlich steht ja sogar – nimmt man die Satzung ernst – das Selbstverteidigungsrecht nach Art. 51 SVN unter dem Vorbehalt eines Tätigwerdens des Sicherheitsrats[109].

Ein doch etwas anderes Bild ergibt sich dort, wo der Sicherheitsrat um substanzielle Streitlösung (*peace-making*) bemüht ist. Die hier angestrebte dauerhafte Lösung läßt das Völkerrecht zu einem sehr viel bedeutsameren Faktor werden. Aber auch hier ist es nicht Aufgabe des Sicherheitsrats, aufgrund der Rechtslage über Ansprüche der Streitparteien zu entscheiden, sondern den Streit dadurch zu lösen, daß er die Parteien zu einer selbständig vereinbarten Übereinkunft bewegen kann oder selbst substantielle Vorschläge macht. Allerdings verweist Art. 1 Ziff. 1 SVN auch den Sicherheitsrat (als Organ der Vereinten Nationen) auf die Beachtung des Völkerrechts bei der Erledigung internationaler Streitigkeiten[110]. Diese Beachtung erfolgt jedoch, eben anders als beim Internationalen Gerichtshof, nicht im Sinne einer Rechtsfindung – Aufspüren und Interpretation der einschlägigen Rechtsnormen und Fallsubsumtion –; das Völkerrecht bietet vielmehr nur die Grenze, nicht die Basis der auf den Streitinhalt bezogenen Ratsempfehlungen[111]. Das bedeutet, daß die Vorschläge des Sicherheitsrats nicht das Völkerrecht wiedergeben müssen, ihm aber auch nicht widersprechen dürfen; der Lösungsvorschlag muß völkerrechtskompatibel[112] sein, d.h. er darf nichts beinhalten, was die Parteien nicht selbst vereinbaren können. Die substantielle Empfehlungskompetenz des Rates (z.B. nach Art. 37 Abs. 2, 38 SVN) findet daher ihre Grenze am *ius cogens* und völkerrechtlichen *ordre public*[113]. Die empfohlene Regelung muß im Dispositionsbereich der Parteien liegen. Auch der Si-

---

[108] Dies macht auch der Wortlaut der Art. 39 ff. SVN deutlich. – In eine andere Richtung weist allerdings der sog. Entwurf der Manila-Deklaration über die friedliche Regelung von Streitigkeiten (GAOR, 35. Sess., Suppl. No. 33 [A/35/33]), in dem die Forderung aufgestellt ist, daß »cease-fire injunctions should in all cases be based on full respect for the territorial integrity of States which required withdrawal to international boundaries or to recognised cease-fire lines«.

[109] Auf die Sonderrolle des Sicherheitsrats weisen etwa auch hin Art. 73 Wiener Vertragsrechtskonvention, Art. 2–4 der Aggressionsdefinition (A/RES/3314 (XXIX), Annex) und Prinzip I Abs. 10 der Friendly Relations- Deklaration (Anm. 75).

[110] *Higgins* (Anm. 50), S. 1.

[111] Ebenso schon *Dahm* (Anm. 45), S. 368.

[112] So *Higgins* (Anm. 50), S. 16.

[113] Zu Unterscheidung und Inhalt beider Kategorien *H. Mosler,* Ius cogens im Völkerrecht, Schweiz. Jahrbuch für Internationales Recht 25 (1968), S. 9 ff.; *derselbe* (Anm. 1), S. 17 ff.

cherheitsrat besitzt keine Verfügungsbefugnis über die zwingenden Normen des Völkerrechts [114].

Untersuchungen zeigen, daß die Praxis des Sicherheitsrats im wesentlichen diesen Überlegungen entspricht. Der Rat hat in der Regel die relevanten rechtlichen Gesichtspunkte in seine Erwägungen einbezogen, ohne sich freilich immer von ihnen leiten zu lassen [115]; meist vermeidet er auf rechtliche Gesichtspunkte gestützte Verurteilungen, doch gibt es hiervon die immer wiederkehrenden Ausnahmen [116]. Im *Namibia*-Fall hat der Sicherheitsrat sogar beim Internationalen Gerichtshof implizit um die rechtliche Bestätigung seiner Aktion nachgesucht; explizit ging es um die sich aus der Beendigung des Mandats ergebenden Rechtsfolgen [117].

b) Fragt man nun konkret nach der Bedeutung der Entscheidungen des Internationalen Gerichtshofs für die Entschließungen des Sicherheitsrats, so kann sich aus der Verbindlichkeit der Gerichtsentscheidungen für die Parteien [118] – anders als im umgekehrten Fall [119] – kein Argument für die Berücksichtigung durch den Sicherheitsrat ergeben [120]. Denn der Sicherheitsrat ist eben gerade nicht auf die Ermittlung der Rechtslage fixiert. Eine Berücksichtigungspflicht folgt jedoch aus dem durch die Kompetenzzuweisung konkretisierten Prinzip der Organtreue. Danach hat der Sicherheitsrat die Entscheidung des Gerichtshofs als die für die Streitparteien verbindliche Rechtsfeststellung zur Kenntnis zu nehmen; dem Sicherheitsrat ist es nicht möglich, diese Entscheidung aus seiner eigenen Diskussion mit der Begründung auszuklammern, sie sei nicht Rechtens. Dies besagt andererseits nicht, daß der Rat seine Entschließungen an der Entscheidung des Internationalen Gerichtshofs auch inhaltlich ausrichten müßte. Hierzu bedarf es näherer Erläuterung.

Hat der Gerichtshof vorläufige Maßnahmen zur Sicherung von streitbefangenen Ansprüchen erlassen (Art. 41 Statut), so muß sie der Sicherheitsrat in seine Lageeinschätzung insoweit einbeziehen, als er auch von da aus das

---

[114] Hier ist vor allem auf die durch das Selbstbestimmungsrecht gezogene Schranke hinzuweisen; vgl. auch *H. Mosler*, Völkerrecht als Rechtsordnung, ZaöRV 36 (1976), S. 6 ff. (37).

[115] Vgl. *T. J. Kahng*, Law, Politics, and the Security Council (2. Aufl. 1969), S. 227 ff.; *O. Schachter*, The Relation of Law, Politics and Action in the United Nations, RdC 109 (1963 II), S. 165 ff.; *M. Schaefer*, Die Funktionsfähigkeit des Sicherheitsmechanismus der Vereinten Nationen (Beiträge zum ausländischen öffentlichen Recht und Völkerrecht, Bd. 77) (1981), S. 381 f.

[116] Vor allem gegenüber Südafrika und Israel; vgl. *O. Schachter*, The Quasi-Judicial Rôle of the Security Council and the General Assembly, AJIL 58 (1964), S. 960 ff. (961); *Higgins* (Anm. 50), S. 18.

[117] ICJ Reports 1971, S. 16 ff.

[118] Art. 94 Abs. 1 SVN, 59 Statut. – Die Verbindlichkeit erstreckt sich auf einstweilige Anordnungen gem. Art. 41 Statut; vgl. dazu *K. Oellers-Frahm*, Die einstweilige Anordnung in der internationalen Gerichtsbarkeit (Beiträge zum ausländischen öffentlichen Recht und Völkerrecht, Bd. 66) (1975), S. 107 ff.

[119] Siehe oben IV. 2. a).

[120] *Steinberger* (Anm. 70), S. 203.

Verhalten der Streitparteien näher beurteilen kann; denn kommen die Parteien der einstweiligen Anordnung nach, so wird in aller Regel eine friedensgefährdende Situation nicht mehr bestehen. Im *Geisel*-Fall hat der Sicherheitsrat bei Verabschiedung der Resolution 461 (1979) von eben dieser Grundlage aus gehandelt[121]. Der Sicherheitsrat ist aber frei, bei der tatsächlichen Einschätzung der Friedensgefährdung vom Ergebnis des Internationalen Gerichtshofs abzuweichen und deshalb keine oder inhaltlich abweichende Entschließungen zu fassen[122], die freilich nicht – soweit sie ihrerseits verbindlich sind – die vom Sicherheitsrat ja nicht selbst mit bindender Wirkung herbeizuführende materielle Streitlösung präjudizieren dürfen.

Auch die die Streitsubstanz betreffenden Urteile des Gerichtshofs bewirken keine sachliche Einengung der gleichfalls auf inhaltliche Streiterledigung gerichteten Ratsempfehlungen; eine solche Wirkung stünde mit der Verantwortung des Sicherheitsrats für die Aufrechterhaltung und Sicherung des Weltfriedens und der internationalen Sicherheit nicht in Einklang. Denn es ist unbestritten und unbestreitbar, daß die (richtig oder falsch) gefundene rechtliche Lösung nicht notwendig die wirkliche Pazifizierung nach sich zieht, sondern möglicherweise eine Verschärfung des Konflikts heraufbeschwört[123]. In dieser Situation ist es Aufgabe des Sicherheitsrats, den Rekurs auf Waffengewalt zu verhindern und den Streit friedlicher Beilegung offenzuhalten[124]. Hierzu können sich dann gerade Initiativen empfehlen, die von dem vom Internationalen Gerichtshof gefundenen Rechtsstandpunkt abweichen.

Art. 94 Abs. 2 SVN vermittelt für diesen Zusammenhang wichtige Einsichten. Diese Vorschrift verweist die vor dem Internationalen Gerichtshof obsiegende Partei für die Durchsetzung des Urteils auf den Sicherheitsrat, der dadurch zum eigentlichen Vollstreckungsorgan wird[125]. Aus dieser Funktion ergibt sich, daß der Rat, wenn er auf dieser Rechtsgrundlage handelt, nur (bindende) Entscheidungen oder Empfehlungen erlassen kann, die inhaltlich dem Gerichtsurteil entsprechen[126]. Allerdings steht es im Ermessen des

---

[121] Siehe oben bei Anm. 34.

[122] Vgl. auch oben nach Anm. 103 für den umgekehrten Fall.

[123] Vgl. *Mosler* (Anm. 1), S. 285: »If a vital interest is involved, the dispute will hardly be settled by judicial procedure although international law may provide the legal basis for a decision«; ferner: *M. A. Kaplan/B. Katzenbach,* The Political Foundations of International Law (1961), S. 278; *H. v. Mangoldt,* Tradition in der zwischenstaatlichen Streitbeilegung?, in: Tradition und Fortschritt im Recht, Festschrift zum 500-jährigen Bestehen der Tübinger Juristenfakultät (1977), S. 435 ff. (484 f.); *Steinberger* (Anm. 70), S. 206 f.

[124] *Virally* (Anm. 56), S. 432.

[125] Dazu vgl. *K. Oellers-Frahm,* Zur Vollstreckung der Entscheidungen internationaler Gerichte im Völkerrecht, ZaöRV 36 (1976), S. 654 ff. (665 ff.); *H. Mosler,* Judgments of International Courts and Tribunals, in: Bernhardt (ed.), Encyclopedia of Public International Law, Instalment 1 (1981), S. 111 ff. (117).

[126] *Rosenne* (Anm. 41), S. 152; *Krökel* (Anm. 53), S. 94. Anders: *H. Kelsen,* The Law of the United Nations (1951), S. 539, und *Dahm* (Anm. 45), S. 564.

Rates, ob er überhaupt von dieser Kompetenz Gebrauch machen und wie er auf den die Ausführung des Urteils verweigernden Staat einwirken will. Schon hierin zeigt sich, daß es dem Rat überlassen bleiben soll, wie er die tatsächliche Streitlösungskapazität des Urteils einschätzt [127]. Daneben aber – und dies ist wesentlich – bleiben dem Rat seine sonstigen Kompetenzen gem. Kapitel VI und VII SVN in vollem Umfang erhalten [128]. So beruht etwa die Ratsentschließung 461 (1979) im Teheraner *Geisel*-Fall nicht auf Art. 94 Abs. 2 SVN, ist also nicht als Vollstreckungsmaßnahme begreifbar [129]. Aus der Tatsache, daß in diesem Fall der Resolutionsinhalt der Sache nach der einstweiligen Anordnung des Internationalen Gerichtsnof entsprach [130], kann auch nicht darauf geschlossen werden, daß inhaltliche Übereinstimmung über ein Tätigwerden nach Art. 94 Abs. 2 SVN hinaus geboten ist [131]. Der Sicherheitsrat nämlich behält die Freiheit, insoweit – also im Bereich seiner Zuständigkeit jenseits von Art. 94 SVN – über den Inhalt seiner (substantiellen oder prozeduralen) Empfehlungen oder seiner (nicht-substantiellen) bindenden Entscheidungen in eigener Verantwortung zu bestimmen.

Die sich hierbei ausdrückende politische Lagebeurteilung [132] ist jedoch nicht völlig frei von rechtlichen Erwägungen. Die Existenz des Urteils und die hierin enthaltenen juristischen Feststellungen sind vom Rat als solche hinzunehmen, dürfen also nicht als völkerrechtswidrig qualifiziert werden und sind hineinzunehmen in die dem Rat obliegende Würdigung, wie der Streitfall nun friedlich zu lösen ist, nachdem die unterlegene Partei dem Urteil nachzukommen ablehnt. Die Gerichtsentscheidung ist also ein wichtiges Element bei dieser vorzunehmenden Einschätzung [133].

An dieser Stelle ist nochmals auf die Funktion des Völkerrechts bei der Streiterledigung zurückzukommen [134]. Es wurde schon erwähnt, daß der Si-

---

[127] Zum bisher einzigen und gescheiterten Fall einer Heranziehung von Art. 94 Abs. 2 SVN, dem *Anglo-Iranian Oil Co.* Case – ICJ Reports 1951, S. 89 –, vgl. *Bailey* (Anm. 13), S. 247 ff.; *Rosenne* (Anm. 41), S. 156 ff.

[128] *L. M. Goodrich/E. Hambro/A. P. Simons,* Charter of the United Nations (3. Aufl. 1969), S. 556 f.; *Rosenne* (Anm. 41), S. 153.

[129] Dies ergibt sich schon aus dem fehlenden Antrag der USA; auf welcher Grundlage die Resolution beruht, ist unklar; in Betracht kommt Art. 35, auch Art. 39 und 40 (im Sinne von Empfehlungen) sind zu erwägen. – Vgl. auch *L. C. Green,* The Tehran Embassy Incident – Legal Aspects, ArchVR 19 (1980/81), S. 1 ff. (12). Zum Problem, wieweit Art. 94 Abs. 2 SVN für die Vollstreckung einstweiliger Anordnungen taugt, vgl. *J. P. A. Bernhardt,* The Provisional Measures Procedure of the International Court of Justice Through U.S. Staff in Tehran: Fiat Iustitia, Pereat Curia?, Virginia Journal of International Law 20 (1980), S. 557 ff. (607 ff. mit weiteren Nachweisen).

[130] Siehe oben bei Anm. 34.

[131] Anders *Fakher* (Anm. 62), S. 67 f.

[132] Zur Politikdominanz in der post-adjudikativen Phase vgl. *Rosenne* (Anm. 41), S. 115 ff., 152.

[133] *Rosenne* (Anm. 41), S. 153.

[134] Siehe auch schon oben IV. 3. a).

cherheitsrat als Organ der Vereinten Nationen dem Völkerrecht nicht indifferent gegenüberstehen kann (vgl. Art. 1 Ziff. 1 SVN)[134]. Darüber hinaus ist zu beachten, daß die Lösung von zwischenstaatlichen Streitigkeiten auf der Grundlage des Völkerrechts jedenfalls dem langfristigen Interesse der Staaten eher entspricht als die Berücksichtigung noch so drängender kurzfristiger Interessen[135]. Die auf der Grundlage der Gegenseitigkeit entstandenen Normen des allgemeinen Völkerrechts wie der Völkerrechtsverträge stellen ja meist die Normativierungen von auch den Streitparteien akzeptablen Vorstellungen dar, an deren Beachtung und Erhaltung jeder Staat ein prinzipielles Interesse hat[136], da dies zur Vorhersehbarkeit des politischen Handelns und zur Rechtssicherheit beiträgt, systemimmanentes Verhalten fördert[137] und damit zur Stabilisierung der internationalen Beziehungen führt. Dies kann nicht ausschließen, daß Friedenswahrung davon abweichende Schritte verlangt. Der Sicherheitsrat hat aber der Vermutung, daß die Einhaltung des Völkerrechts der Friedenswahrung in besonderer Weise dienlich ist, dadurch Rechnung zu tragen, daß er über diese Überlegungen nicht in freier Willkür hinweggeht, sondern ernsthaft bei seiner eigenen Entscheidung berücksichtigt.

V. Abschließende Bemerkungen

Die dem Prinzip der Organtreue zu entnehmende Berücksichtigungspflicht bedeutet nicht, daß ein Organ bloßes Vollzugs- oder Akklamationsorgan des anderen Organs ist; es zielt nicht auf eine Verunklarung der jeweiligen Organidentität und -verantwortung. Eine Veränderung der innerorganisatorischen Kompetenzstrukturen kann durch das Berücksichtigungsgebot nicht bewirkt, das institutionelle Gleichgewicht[138] darf nicht erschüttert werden. Dennoch ist das Berücksichtigungsgebot mehr als eine bloße Stilfrage; es beinhaltet veritable Rechtspflichten, die von der Unzulässigkeit der Druckausübung[139] auf das andere Organ bis zur unmittelbaren Einbeziehung von dessen Entscheidungen in den eigenen Entscheidungsfindungsprozeß reichen. Die weithin fehlende Möglichkeit, solche Pflichten zu effektuieren, sollte an ihrer Charakterisierung als Rechtspflichten nicht hindern. Je höhere Anforderungen an die Ordnungsfunktion des Völkerrechts gestellt werden müssen, desto mehr bedarf dieses der dogmatischen Durchdringung und Konkretisierung.

---

135 Dazu *Schachter* (Anm. 115), *passim; Gross* (Anm. 59), S. 435.

136 *Schachter* (Anm. 116), S. 965; *Klein* (Anm. 57), S. 54 f.

137 *Higgins* (Anm. 50), S. 14.

138 Auf die Bedeutung des Gleichgewichts im Verhältnis Sicherheitsrat zur Generalversammlung macht aufmerksam *Ch. Chaumont,* L'équilibre des organes politiques des Nations Unies et la crise de l'organisation, AFDI 11 (1965), S. 428 ff.

139 Ein Beispiel stellt die Reaktion der politischen Gremien der Vereinten Nationen gegenüber dem *Südwestafrika*-Urteil von 1966 (ICJ Reports 1966, S. 6 ff.) dar; dazu *Gross* (Anm. 59), S. 325 f.

The Threshold in Law-Making

MANFRED LACHS

It gives me great pleasure to join others in honouring *Hermann Mosler*, not only a distinguished teacher and judge who has enriched the international law of our generation, but also a generous colleague and friend.

The richness and complexity of modern life, the ever-increasing contacts between States, the development and activity of international organizations, the needs of political and economic relations, these are some of the key factors contributing to the continuous creation of new rules of international law. They have also been decisive factors in the development of and changes in the law-making process.

Need I re-emphasize that the essence of the problem remains the consent of parties whether they be two, three or fifty in number, to be bound by obligations accepted by them and to respect rights conferred upon their partners. If we accept this point of departure, a specifically defined form is no longer of prime importance. This has been recognized for a long time. The novelty is in the wealth of manifestations created by international life, the new possibilities and formulae arising out of new methods of co-operation [1].

Notwithstanding the many changes, it is a truism to state that treaties remain the principal instrument shaping inter-State relations; their role has even expanded to a hitherto unknown degree, and international organizations, in particular the United Nations, have made a large contribution to that development; not only in playing the passive role of a depositary, but also by taking the initiative of preparing texts. Thus their contribution to the elaboration of treaties of worldwide importance cannot be overestimated.

Yet, even in a domain so well established as treaty law, more particularly on the multilateral level, new techniques have been resorted to with regard to the procedures and methods. Though the classic procedure of the diplomatic conference and its concomitant formalities remains the method most frequently employed – in principle it is used by the International Law Com-

---

[1] Cf. my article, Some Reflections on Substance and Form in International Law in a Changing Society, in: Essays in Honour of Philip Jessup (New York 1972), pp. 99 ff.

mission – traditional procedures are on the way out. They are frequently replaced by new methods; for example, a succession of discussions, comments and drafts within the United Nations, so that one might say that *it is a series of conferences which takes place,* until governments arrive at a text which will be acceptable to them.

Naturally there is a clear difference between the preliminary consultations and the final stage. Sometimes it is the Legal Committee of the General Assembly which plays the role of a diplomatic conference (this was the case in respect of the Convention on Special Missions, prepared by the Legal Committee and approved by the General Assembly at its 24th Session in 1969). Usually the text of a treaty is discussed and approved by a Committee and opened for signature after its adoption by the plenary of the General Assembly. It is hardly necessary to recall that one of the first, important examples of this method was the Convention for the Prevention and Punishment of the Crime of Genocide drafted by the Legal Committee in 1948 and approved by the General Assembly that same year. Many others have followed the same pattern.

A different method was followed in drawing up a new Convention on Narcotics. The drafts were prepared by two distinct organs: the Narcotics Commission and the Social Committee of the Economic and Social Council. A special conference was next convened in New York in 1961, and at that meeting the text of this Convention was adopted.

Still another method for the finalization of treaties is to set up *ad hoc* bodies to deal exclusively with problems arising in the specific field in question. Disarmament provides an example; first it was the Disarmament Committee, later transformed into the Conference of the Disarmament Committee, which dealt, for instance, with the issue of nuclear tests, preparing the Nuclear Text Ban Treaty (Moscow, 5 August 1963). It was this same Committee which drew up the draft treaty on the non-proliferation of nuclear weapons, discussed at several sessions of the General Assembly, and finally approved on 12 June 1968.

A similar procedure was adopted in regard to all questions in the field of outer space. The first Treaty on the Principles concerning the Activities of States in the Exploitation and Utilisation of Space, the Moon and Other Celestial Bodies was worked out in the Legal Sub-Committee of the United Nations Committee for the Peaceful Uses of Outer Space, submitted to the Committee of the whole, next to the General Assembly, adopted by the latter on 18 December 1966 and opened for signature on 27 January 1967. (The same procedure was followed in regard to four other conventions and treaties in this domain). But it also happens that committees are set up with a view to the drafting of a single instrument. This was the case in the Convention on the Recovery Abroad of Maintenance, for the purposes of which a special committee was appointed by the Secretary-General of the United Nations. (The draft was adopted by a special conference, convened in accordance with a resolution of the Economic and Social Council and opened for signature on 20 June 1956.) Likewise with the Recognition and Enforcement of Foreign Arbitral Awards: a special committee was set up and prepared a draft Convention which was considered by a conference convened by the United Nations in 1958. Thus new treaty-making procedures have evolved within the framework of international organizations[2].

These changes in negotiation techniques together reflect the new needs of our time, the adaptation of the processes to the subject matter of a treaty and the circumstances and interests of the potential parties to it. The novelty of the procedure arises out of the existence of permanent frameworks, which create permanent organs or may establish others for the elaboration of one or of a whole series of instruments and within which States co-operate. Their common feature is the continuous character of the negotiations, even if sittings are held periodically, until the treaty is finally born, unless of course, the negotiations break down, which they frequently do for extra-legal reasons. In this type of law-making the question of the threshold indicated in the title of these reflections does not arise: for with the entry into force of the treaty the rights and obligations provided by it become law – the threshold is crossed. This is also the case with bilateral treaties.

However, ever more frequently other methods are resorted to: in some the first step constitutes only a stage, remaining outside the sphere of law, in others the threshold of law is crossed, yet without a *lex perfecta* being born: this is to follow later. Still others constitute a stage in shaping customary law, thus being a different source of law. These methods differ also in their origins: some have been known for a long time, others have only recently come to the fore.

The most frequent contemporary illustration is to be seen in resolutions adopted by organs of international organizations. Some are unquestionably binding, their effects arising out of the constitution or other basic documents of the organization in question. More frequently they are recommendations only. However, even so, could it have been intended that resolutions of the General Assembly or of similar bodies should be devoid of any legal value? Even if the emphasis is laid on other aspects of these recommendations, can it be presumed that they are devoid of all effect[3]? I do not propose to enter the vast arena of these resolutions, which has been the subject of many com-

---

[2] Cf. my article, La contribución de las Naciones Unidas al derecho internacional, 1945–1970, in: La ONU-Dilema a los 25 años, Foro Internacional (Mexico 1970), pp. 339 ff. The International Labour Organisation provides a much older and very typical illustration of an organization within which conventions (and recommendations) are elaborated.

[3] For whatever »the content of the recommendation and the nature and circumstances of the majority . . . it is nevertheless a legal act of the principal organ of the United Nations which Members of the United Nations are under a duty to treat with a degree of respect appropriate to a resolution of the General Assembly«, Lauterpacht, ind. op., I. C. J. Reports 1955, p. 120.

mentaries and controversies. It may suffice to conclude that in some, but only some, circumstances they may become binding, on grounds of their subject matter, the special circumstances in which they were adopted, the will of the parties to be bound by them, the character of the agreement reached or the events that followed their adoption. Thus, apart from the thousands of treaties now in force and registered with the United Nations, there are scores of resolutions which may be binding, creative of or determining law[4]. But even where they are not binding, they stand on the threshold of law and may lead to the creation of binding rules.

I need scarcely recall some of the well-known instances, such as the Convention for the Prevention and Punishment of the Crime of Genocide, adopted on 9 December 1948, which was preceded by the resolution adopted two years earlier by the General Assembly on 11 December 1946; the Universal Declaration of Human Rights of 10 December 1948, which preceded the two International Covenants on Human Rights (International Covenant on Economic, Social and Cultural Rights, and the International Covenant on Civil and Political Rights (1966); the Declaration on the Elimination of all Forms of Racial Discrimination (of 1964) preceded the Convention on the same subject concluded in 1966; finally the Treaty on Outer Space (opened for signature on 27 January 1967), was preceded by the Declaration on the Legal Principles concerning the Activity of States in the Exploitation and Use of Outer Space, adopted by the General Assembly in 1963.

The same may be said of some similar resolutions and declarations of other organs of the United Nations or other international organizations. But even if it is evident that such texts can, and often do, constitute stages in the law-making process, the question remains: what is their legal value from the moment of their adoption and up to the time when (if ever) they are replaced by treaties? Greater precision is required in the definition of their status: does it lie within or outside the sphere of law? Here we face a motley picture, for many are the varieties of such texts. Some have confirmed or reiterated principles recognized by States. With regard to them the question of the threshold does not arise: the law already in force has found in them only its confirmation[5]. Close to these may be declarations of what *Hermann Mosler* describes as »general principles«, which are not born in the usual law-creating process but are part of the law; in the absence of their recognition the international community would deny the law's very existence; these may find expression in General Assembly resolutions[6]. At times such resolutions have

---

[4] Cf. *Schwarzenberger*, A Manual of International Law (London 1967), p. 26.

[5] Cf. again the GA resolution on genocide: it defined it as »a crime under international law« (96/I of 11 Dec. 1946). In dealing with the Convention which followed the adoption of the resolution, the ICJ held that »the principles underlying the Convention are principles which are recognized by civilized nations as binding on States even without any conventional obligation« (I. C. J. Reports 1951, p. 23). Thus the same applies to the resolution.

[6] Cf. *H. Mosler*, Völkerrecht als Rechtsordnung, 36 ZaöRV, 1976, p. 42.

served to interpret, sharpen or bring up to date the provisions of the Charter; at other times they have combined two or three functions: the confirmation of a pre-existing rule but with a new interpretation, and the crystallization of a new rule, setting in motion further developments. Some have been specifically defined as »a stage in the progressive development and codification[7]«.

The role of others in the development of principles of international law may be the result of further action following their adoption and initiated by them. Thus it has been held that:

> »The subsequent development of international law in regard to non-self-governing territories, as enshrined in the Charter of the United Nations, made the principle of self-determination applicable to all of them... A further important stage in this development was the Declaration on the Granting of Independence to Colonial Countries and Peoples (General Assembly resolution 1514 (XV) of 14 December 1960), which embraces all peoples and territories which ›have not yet attained independence‹« [8].

But in most cases, as I pointed out, they constitute the first step in the process of law-creation, a view I suggested long ago[9]. Their status has been compared to non-ratified treaties, such as the 1909 London Declaration or the Code on Air Warfare drawn up in 1923. It has rightly been argued:

> »... qu'ayant participé ainsi à l'énoncé de ces règles, un Etat membre ne pourrait prétendre que d'autres règles contraires résultent de la coutume internationale et qu'il entend les appliquer... Au fond, on peut estimer que cette ›étape‹ consisterait à établir une législation ›matérielle‹, n'acquérant force obligatoire que par une action supplémentaire des Etats intéressés« [10].

While it is true that, in principle, States have accepted the idea that the fact of delegates raising their hands or pronouncing the word »yes«, and thus voting for a certain resolution, does not create a legal obligation, it is nonetheless also true that States may be bound by such a resolution and that, in any case, they cannot be prevented from regarding such a resolution as

---

[7] Cf. Res. 2181 (XXI) on »The Declaration on Principles of International Law concerning Friendly Relations and Co-operation among States«. The Preamble and para. 8 refer specifically to it as a stage. (Final Text of Res. 2625 of 24 Oct. 1970).

[8] *Legal Consequences for States of the Continued Presence of South Africa in Namibia (South West Africa) notwithstanding Security Council Resolution 276 (1970),* Advisory Opinion, I. C. J. Reports 1971, p. 31, para. 52.

[9] The International Law of Outer Space, 113 RdC 1964 III, p. 97.

[10] *Suzanne Bastid,* Observations sur une ›étape‹ dans le développement progressif et la codification des principes du droit international, in: En Hommage à Paul Guggenheim, (Geneva 1968), p. 145.

binding them. Thus the text in question may immediately enter the sphere of law[11].

Next, there has appeared on the horizon another type of instrument called a »code of conduct« or »guideline«. It started with a code of conduct concerning shipping, elaborated by Unctad. It was followed by a similar code of conduct, concerning multinational enterprises, worked out by the OECD. »The observance of the guidelines«, it was said, »is voluntary and not legally enforced. However, they should help to ensure that the operations of these enterprises are in harmony with national policies of the country where they operate and to strengthen the basis of mutual confidence between enterprise and State«[12]. Similar documents are now being elaborated by the United Nations.

Thus we have instruments that are not binding but are intended to encourage States to comply with them. This leads me to confront the key problem of these reflections: when can it be said that an action of a State has crossed the threshold into the realm of law?

In the past the problem presented no difficulties: life moved more slowly, law could be formed at greater leisure. Nowadays, on the other hand, one or more treaties are concluded in one of the world's capitals every day; every day diplomatic correspondence produces new agreements and also new differences of view; changes continue, there is a regular succession of varied events. Need I recall that the threshold is clearly crossed once the treaty has entered into force? But to some degree it is crossed by the mere signature, prior to ratification even when one is required[13]. Moreover, some instruments, even those which remain unratified for a long time by the States in question, are held binding if they initiated or confirmed State practice[14].

The problem becomes more difficult once we are bound to operate by way of presumptions, or rely on implied consent. States sometimes deliberately seek a less conspicuous method for creating legal links and do so for various reasons. Then it is up to those interpreting their action to ascertain whether this constituted only a stage which continues to remain outside the sphere of law, or whether the threshold of law has been crossed.

---

[11] For a general analysis of the subject cf. *B. Simma,* Zur völkerrechtlichen Bedeutung von Resolutionen der UN-Generalversammlung in Völkerrecht und Außenpolitik, in: Fünftes deutsch-polnisches Juristen-Kolloquium, vol. 2 (Baden-Baden 1981), pp. 45–76.

[12] Para. G, Declaration on International Investment in Multinational Enterprises, 21 June 1976.

[13] Cf. Art. 18, Vienna Convention on the Law of Treaties. Cf. also earlier I. C. J. Reports 1951 (the Court held that without ratification »signature« does not make the signatory State a party to the Convention. Nevertheless it establishes a provisional status in favour of that State), p. 28.

[14] Cf. Convention relating to the Conflict of Nationality Laws (The Hague 1930); I. C. J. Reports 1955, p. 23.

While the international community has so far codified only the Law of Treaties »concluded between States in written form and governed by international law«[15], it was explicitly laid down that this »shall not affect« the legal force of »international agreements not in written form«[16]. Thus it is well known that sometimes contractual links arise out of oral statements made by Heads of States, Prime Ministers of Foreign Ministers. Need I recall a well-known declaration of a Foreign Minister which found its consequences confirmed by the Permanent Court of International Justice[17]? Several more illustrations are known from the past[18]. But it may be of interest to recall a recent case in which an oral promise to prohibit poppy-plantation made in a conversation by the Prime Minister of Turkey to the President of the United States (March 1972) was invoked by the United States Government when, two years later, Turkish legislation in several provinces permitted such plantation[19].

Similar problems arise with regard to unilateral declarations. Their binding force should problably be submitted to a more severe test, since they are acts of one State only, attracting no obligations on the part of any other. Yet some of them, on matters of no small importance, have had legal effects. The threshold can be presumed to have been crossed when the intention of the State concerned leaves no room for doubt«. It is from the actual substance of these statements«, as it was held, »and from the circumstances attending their making that the legal implications of the unilateral act must be deduced«[20].

Here, as elsewhere, matters of form have receded into the background; that a statement be made in writing is no longer a condition of its validity: »nor is it today required that it be made within the context of international negotiations«[21]. Thus, for instance, with regard to »declarations accepting its compulsory jurisdiction«, the Court held that »the sole relevant question is whether the language employed in any given declaration does reveal a clear intention«[22].

Again, it is not a question of imposing upon a State an obligation against its will. It is only one of ascertaining what that will was at the time the de-

---

[15] Vienna Convention on the Law of Treaties 1969, Art. 21 (a).

[16] Art. 3 (a) of the Vienna Convention; nor, it was added, shall this affect »the application to them of any of the rules set forth in the present Convention to which they would be subject under international law independently of the Convention« (Art. 3 (b)).

[17] Cf. P. C. I. J. Series A/B, 53, 1933, pp. 22 ff.

[18] For example, the oral proposal made by the Chargé d'affaires of the Austro-Hungarian monarchy to the Foreign Minister of France for the extension of a trade agreement between the two States and the latter's oral acceptance of that offer (11 May 1883).

[19] Cf. Neue Zürcher Zeitung, 8 July 1974. Cf. also 79 RGDIP 1975, pp. 542 ff.

[20] I. C. J. Reports 1975, p. 269, para. 51.

[21] I. C. J. Reports 1974, p. 267, para. 43.

[22] I. C. J. Reports 1961, p. 32.

claration was made, thus of preventing the possibility of giving an act a *post facto* interpretation which would be contrary to the original one. This risk is present on both sides: for the State involved, and for those dealing with it.

In all these cases, multi-, bi- or unilateral, the identification of the will and intention of the State or States concerned at the critical moment must today take into account the circumstances of international intercourse in the contemporary world. The daily contacts of States, the growing network of their relations, determine the degree of rigidity or strictness of the tests, which must take account of the necessity for a balanced relationship – rights corresponding to obligations are involved.

To continue the theme, it is worth stressing (*ut supra*) that the threshold into the sphere of law may be crossed not only by a written or oral agreement, but also by »a very definite, very consistent course of conduct on the part of a State«[23]. This raises again the question of the rigidity of the test. Perhaps the qualification »very consistent« is not so easy to verify; it might be too severe. Hence while, in cases of testing a rule, »criteria of frequency, continuity and uniformity are involved«[24], »too much importance need not be attached to . . . a few uncertainties or contradictions, real and apparent«[25]. As to the obligation of an individual State incurred by its consistent behaviour, frequency is certainly an important test: its intensity does, of course, depend on the number of opportunities a State may have to act or refrain from acting, in accordance with what may be claimed to be a »rule of law«. Continuity is not always applicable. Thus, tests vary from case to case. Sometimes it should be rather strict; on other occasions less severe evidence may suffice: the whole complex of circumstances should be taken into account to produce the evidence which is to be deemed adequate and unequivocal.

These reflections may lead into a much wider enquiry. However, the present observations may suffice, as I write to honour a jurist who values brevity.

The conclusion I arrive at is that the recent developments have undoubtedly contributed to widening the possibilities of passing the threshold (and hence have increased the frequency with which it is passed). While some claim that this results in weakening the law by softening what should be hard and strictly defined rules, the sometimes easy passage of the threshold helps the expansion of law and its strengthening by other methods. These rules are born – as I indicated earlier – in circumstances in which States for one reason or another are unwilling or unable to sign a solemn treaty or to subscribe formally to a rule. Yet they may be anxious to establish a rule which may become binding otherwise – immediately or in the future.

Thus States make promises: to comply with provisions that they themselves do not openly characterize as rules of law, or to follow patterns of con-

[23] I. C. J. Reports 1969, p. 25, para. 28, Cf. also *Giuliano*, Diritto Internazionale (Milan 1974) part 6, pp. 335 ff.

[24] Cf. my opinion, I. C. J. Reports 1969, pp. 22 ff.

[25] I. C. J. Reports 1951, p. 138.

duct which are perhaps not enforceable but gradually grow into the body of law.

»A quoi servent toutes les chartes, conventions, déclarations, proclamations, protocoles que se jurent de siècle en siècle les gouvernements, s'il faut périodiquement négocier pour acquérir la certitude provisoire que cette fois-ci est la bonne?« wrote *Robert Escarpit* in *Le Monde* some time ago. The irony of this cannot escape the reader.

This is the inevitable result of the changing conditions of life itself. It is a consequence of historical evolution that a certain number of principles and rules have become intergral parts of our heritage and bind every State; others are in the process of evolution and still others have yet to be born. Thus the *corpus juris gentium* grows. Obviously each act or action »must be considered within the general framework of the security of international intercourse, and the confidence and trust which are so essential in relations among States«[26].

As the rapid succession of events makes it imperative to adapt law to the needs of life – because it is in their very close harmony with life that the rules of law can best play their part, so must the tests evolve as to whether a given principle, rule of conduct, permission or prohibition has crossed the threshold of law. The conclusion seems inescapable. In view of the expansion and diversification of the law-making processes we witness a corresponding growth of variety and flexibility of tests concerning the crossing of the threshold.

As the law's task is to facilitate life and relations among States, so is it the jurist's duty to help in clarifying, and even simplifying, the journey of rules of conduct in crossing the threshold from a stage outside into the realm of law[27].

---

[26] I. C. J. Reports 1974, p. 269.

[27] As was wisely said by a colleague and friend whose premature demise we continue to mourn: »The lawyer's task is not only to interpret the resulting consensus but also to make understandings between States as flexible an instrument as possible in order to encourage agreement. He does no service to the establishment of order if he adopts an either-or posture.« *R. R. Baxter,* International Law in ›Her Infinite Variety‹, 29 ICLQ, part 4, October 1980, p. 565.

Versäumnisverfahren in der internationalen (Schieds-)Gerichtsbarkeit und souveräne Gleichheit

HANS V. MANGOLDT

Fragen der internationalen Gerichtsbarkeit bilden nicht nur einen Schwerpunkt im bisherigen wissenschaftlichen Oeuvre *Hermann Moslers* – vor dem Hintergrund seiner Überzeugung, daß richterliche und schiedsrichterliche Einrichtungen der Sache des Friedens in der internationalen Gemeinschaft dienen [1]. Er war und ist auch berufen, dieser Sache des Friedens in besonders ausgezeichneter Weise als internationaler Richter zu dienen. Und er hat es vermocht, in der jüngeren Generation den Gedanken der friedlichen Streitbeilegung durch internationale richterliche Instanzen zu stärken. So möchte es angemessen erscheinen, wenn ich in der ihm gewidmeten Festschrift ein Thema aus der jüngeren Entwicklung der internationalen Gerichtsbarkeit aufgreife, das gerade erst auch seine besorgte Aufmerksamkeit gefunden hat [2].

I

In der modernen Neuentwicklung der internationalen Schiedsgerichtsbarkeit seit dem Jay-Vertrage [3] ist recht bald der Gedanke aufgetreten, daß eine Partei versuchen könnte, sich einer einmal vereinbarten Schiedsgerichtsbarkeit in einer konkreten Streitigkeit dadurch zu entziehen, daß sie sich weigert, die vorgesehenen oder üblichen Verfahrenshandlungen vorzunehmen, insbesondere am schriftlichen bzw. mündlichen Verfahren teilzunehmen und

[1] Judicial Settlement of International Disputes (Beiträge zum ausländischen öffentlichen Recht und Völkerrecht, Bd. 62) (1974), S. 3 ff., 8; *Mosler,* RdC 140 (1974 IV), S. 300; *derselbe,* The International Society as a Legal Community (1980), S. 294.

[2] Vgl. *Mosler,* Nichtteilnahme einer Partei am Verfahren vor dem Internationalen Gerichtshof, in: Festschrift Schlochauer (1981), S. 439 ff.

[3] Vgl. *Ralston,* International Arbitration from Athens to Locarno (1929), S. 190 ff.

Anträge zu stellen, oder daß sie einfach nicht erscheint. Jede dieser Verhaltensweisen – ich möchte sie zusammenfassend als Versäumung bezeichnen – erscheint schon auf den ersten Blick als geeignet, das Verfahren zu gefährden und dem hinter der Einigung über das Verfahren stehenden Gedanken die Grundlage zu entziehen, daß es die Streitigkeit entpolitisieren kann, daß es zur Lösung des Streites führen kann auf Grund der Anwendung objektiver Normen durch eine nicht am Ausgang des Verfahrens interessierte, also unparteiische Instanz[4]. So mag es als nur zu natürlich gelten, wenn für den »Ernstfall« versucht wird, mit besonderen Verfahrensbestimmungen für die Säumigkeit einer Partei »zu retten, was zu retten ist«.

1. Jedenfalls wird bereits 1955 im Kompromiß zwischen Großbritannien und Portugal im *Croft*-Fall vorgesehen, daß der Schiedsrichter *ex parte* – d.h. auf Betreiben bloß der tätigen Streitpartei – entscheiden könne, wenn die andere mit ihrer Antwort auf das Vorbringen der tätigen säumig sein sollte[5]. Ebenso ist in Kompromissen zwischen Chile und den USA im *Macedonien*-Fall[6], zwischen Großbritannien und Portugal im *Yuille, Shortridge et Cie*-Fall[7] und zwischen der Republik Südafrika (Transvaal) und Griquasland wegen einer Grenzstreitigkeit[8] ein solches *ex parte*-Verfahren vorgesehen worden[9].

Allerdings ist es in keinem der genannten Verfahren wirklich zur Abwesenheit einer Partei gekommen. Mit Recht betont *Witenberg*[10], daß es sich durchweg um Fälle der isolierten Schiedsgerichtsbarkeit gehandelt hat, um Verfahren, die auf Grund jeweils eines Kompromisses durchgeführt worden waren, also auf der Grundlage einer auf den konkreten Streitfall bezogenen konkreten Einigung der streitenden Parteien über die Durchführung des Schiedsverfahrens. Mit Recht macht er weiter darauf aufmerksam, daß selbst in solchen Fällen die Versäumung einer Partei denkbar wäre, wenngleich im Grunde paradox und außergewöhnlich, weil es nur schwer verstanden werden könnte, wenn Parteien die Durchführung eines Verfahrens beeinträchtigen, das sie gerade erst in voller Kenntnis der besonderen Umstände eines konkreten Streitfalles vereinbart haben. Isolierte Schiedsgerichte, einmal von den

---

[4] Vgl.. *Mosler* (Anm. 1).

[5] Vgl. *Stuyt,* Survey of International Arbitrations (1972), Nr. 50; *Witenberg,* L'organisation judiciaire – la procédure et la sentence internationales (1937), S. 382 f.; Hinweise auch in: Commentary on the Draft Convention on Arbitral Procedure, UN Publ. 1955, V 1, zu Art. 20, S. 78. – Es ist danach unrichtig, wenn im Bericht *J. B. Scott,* The Project of a Permanent Court of International Justice and Resolutions of the Advisory Committee of Jurists (1920), S. 124 (zu Art. 52 – später Art. 53 IGH-Statut), die Auffassung vertreten wird, die Anwesenheit beider Parteien im Verfahren sei für die Schiedsgerichtsbarkeit »essential«.

[6] *Stuyt* (Anm. 5), Nr. 59.

[7] *Stuyt* (Anm. 5), Nr. 67.

[8] *Stuyt* (Anm. 5), Nr. 92.

[9] Hinweise auch bei *Witenberg* und im Commentary on the Draft Convention on Arbitral Procedure (Anm. 5).

[10] (Anm. 5), S. 381, 384 f.; siehe in diesem Zusammenhang auch *Fitzmaurice,* BYIL 51 (1980), S. 99 f.

streitenden Parteien gebildet und mit der Erledigung der konkreten Streitigkeit betraut, dürfen eben zumeist auf das fortdauernde Einverständnis der Parteien mit der Durchführung des Verfahrens bis zu einer Entscheidung rechnen.

2. Ganz anders liegt es bei der institutionellen Schiedsgerichts- und bei der internationalen Gerichtsbarkeit in den Fällen, in denen die eine Seite allein und ohne das vorherige Einverständnis der anderen eine Streitigkeit auf Grund einer früheren, allgemeineren Gerichtsbindung der Parteien anhängig machen kann. Hier kann später im konkreten Streitfalle, der bei Eingehung der allgemeineren Streitbeilegungsbindung in seinen Konturen nicht absehbar gewesen sein mag, der anderen Seite die Bereitschaft fehlen, den vereinbarten unparteiischen Dritten einzuschalten.

a) Als Anzeichen kommen verschiedene Verhaltensweisen in Betracht: Die unwillige Partei mag ihre Mitwirkung bei der Bildung eines früher vereinbarten institutionellen Schiedsgerichts zurückhalten und so zur Anwendung gegebenenfalls vereinbarter Vorschriften nötigen, die eine Einsetzung des Schiedsgerichts durch Dritte im Wege der Ersatzbestellung zulassen; oder sie mag sich einer an sich notwendigen einvernehmlichen Anrufung der richterlichen Instanz entziehen und dadurch zur Anwendung etwa vereinbarter Vorschriften über die Ersetzung der »für den Normalfall« erforderlichen Schiedsordnung [11] nötigen; oder sie mag sich gegen eine einseitige Anrufung der richterlichen Instanz mit zahlreichen, sehr intrikate Fragen aufwerfenden und den Weg zur Entscheidung der Hauptsache stark behindernden, letztlich aber nicht tragfähigen vorgängigen Einreden zur Wehr setzen – insbesondere im Verfahren vor dem IGH bisweilen zu beobachten; oder sie mag, ohne Rücksicht auf das anhängige Verfahren, mit ihren dem Antragsteller nachteiligen Einwirkungen auf den Streitgegenstand fortfahren und eine bereits ergangene richterliche einstweilige Regelung mißachten. Oder die unwillige Partei verweigert eben ganz oder zum Teil nach der Rechtshängigkeit vorgesehene oder übliche Prozeßhandlungen – wird also säumig. Dankbar sind auch mehrere dieser Verhaltensweisen, nacheinander bzw. gleichzeitig – wie jüngst in verschiedenen Verfahren vor dem IGH.

Säumigkeit einer Partei ist also vor allem eine Möglichkeit, mit der man bei der Durchführung institutioneller (Schieds-)Gerichtsbindungen rechnen muß, das Versäumnisverfahren ein Bereich des internationalen Prozeßrechts, dessen Komplementärverhältnis zur obligatorischen richterlichen Streitbeilegung nicht übersehen werden kann [12]. Dies belegt auch die Praxis konkreter zwischenstaatlicher gerichtlicher Verfahren.

---

[11] Muster: Art. 53 des Abkommens zur friedlichen Erledigung internationaler Streitfälle von 1907; *Martens* NRG 3e sér. tome 3, S. 360; RGBl. 1910, 5.

[12] Es ist darum erklärbar, daß das Übereinkommen von 1907 (vgl. Anm. 11), das eine obligatorische Schiedsgerichtsbarkeit trotz Schaffung der Schiedsrichterliste des Ständigen Schiedshofes gerade nicht begründet, die Möglichkeit der Untätigkeit einer Partei nur in einem Fall und ganz am Rande berücksichtigt (Art. 69).

b) Der erste Fall, der ohne Mitwirkung einer Seite durchgeführt werden mußte, ist der von Costa Rica durch einseitigen Antrag beim Zentralamerikanischen Gerichtshof am 27. März 1916 anhängig gemachte Streit mit Nicaragua wegen dessen sog. Bryan-Chamorro-Vertrages mit den USA [13]. Die für beide Seiten verbindliche Gerichtsbindung ergab sich aus Art. 1 der Konvention über die Errichtung eines Zentralamerikanischen Gerichtshofes [14] und war außerordentlich weit. Nach Art. 14 der Konvention hatte die einseitige Antragsschrift zu umfassen »all the points of fact and law relative to the matter, and all pertinent evidence«. Nach den vom Gerichtshof auf der Grundlage des Art. 26 der Konvention beschlossenen Verfahrensregeln [15] bedurfte der Antrag der Annahme durch den Gerichtshof, ehe die Antragsschrift zur Beantwortung im schriftlichen Verfahren dem Antragsgegner zuzustellen war.

Der Gerichtshof hat diese Annahmeentscheidung am 1. Mai 1916 getroffen, die Zustellung der Antragsschrift an Nicaragua angeordnet, die übliche Beantwortungsfrist festgesetzt und eine einstweilige Anordnung dahingehend erlassen, daß der *status quo* aufrechterhalten bleiben solle. An diese Anordnung hat sich Nicaragua jedoch nicht gehalten.

In der Annahmeentscheidung hat der Gerichtshof bereits seine Gerichtsbarkeit und die Zulässigkeit des Verfahrens geprüft und bejaht, wenngleich nur vorläufig, schon wegen des Nicaragua bis dahin noch nicht formell gewährten rechtlichen Gehörs. Er verfügt aber bereits über die Texte von Noten Nicaraguas an Costa Rica, die Nicaragua dem Gerichtshof informell mitgeteilt hatte und in denen Nicaragua Costa Rica gegenüber die Gerichtsbarkeit bestritten und die *impropriety* des anhängig gemachten Verfahrens geltend gemacht hat; die Annahmeentscheidung stellt dazu fest, »that of the merits of these allegations it is not possible to judge without a hearing« [16].

Aus dem weiteren Verfahrensgang ist von besonderem Interesse, daß Nicaragua die vom Gericht gesetzte Frist zur Antragsbeantwortung hat verstreichen lassen. Der Gerichtshof hat daraufhin Art. 15 der Konvention angewandt, der bestimmt:

> »If the term designated shall have expired without answer having been made to the complaint, the Court shall require the complainant or complainants [Übersetzungsfehler!, gemeint sind: defendant or defendants, der Verf.!] to do so within a further term not to exceed twenty days, after the expiration of which and in view of the evidence presented and of

---

[13] Vgl. *Stuyt* (Anm. 5), Nr. G 1.

[14] Verbindlicher spanischer Text in AJIL 2 (1908), Suppl., S. 231; englisch auch in Papers Relating to the Foreign Relations of the United States, 1907, S. 697.

[15] Ordinance of procedure, Art. 6, 73; englischer Text in AJIL 8 (1914), Suppl., S. 194.

[16] Englische Übersetzung der Annahmeentscheidung in: US Foreign Relations Papers (Anm. 14), 1916, S. 841; dazu auch die Entscheidung zur Hauptsache, AJIL 11 (1917), S. 181, 210 f.; vgl. ferner *Hudson,* AJIL 26 (1932), S. 774.

such evidence as it may ex officio have seen fit to obtain, the Tribunal shall render its decision in the case, which decision shall be final«.

Innerhalb der dieser Vorschrift gemäß gewährten Nachfrist hat der Gerichtshof schließlich eine offizielle Stellungnahme Nicaraguas erhalten. Der nicaraguanische Außenminister nimmt darin zunächst in einigen Punkten zu Sachargumenten Costa Ricas zur Hauptsache Stellung – »wishing to be understood by the foregoing as implying nothing more than a demonstration of courtesy towards this Court« [17]. Im übrigen konzentriert er sich ganz auf die nach der Auffassung Nicaraguas fehlende Gerichtsbarkeit des Gerichtshofes in der Sache und stellt abschließend fest, »that his Government believes, and ... so declares, that it is under no obligation to reply to the complaint ... because it cannot admit, even conditionally, the competency of the Court to take cognizance of, and to decide, that complaint«. Und weiter: »That not for a moment could Nicaragua consider – except for the purpose of rejecting them – the baseless statements, nor the proceeding attacking her sovereignty and dignity«; »that if the Court insists upon taking cognizance of the cause in violation of the provisions of the convention that brought it into existence, the Government of Nicaragua protests in most solemn manner that Nicaragua has proceeded within her rights, that the Central American Court is without jurisdiction, and, in the event of a decision adverse to her, Nicaragua declares that she will be unable to abide by it« [18].

Trotz dieser Hinweise, die als höchst zweifelhaft erscheinen lassen, ob Nicaragua die Stellungnahme als formelle Antragsbeantwortung verstanden wissen wollte, hat der Gerichtshof sie als solche gedeutet und die Beantwortungsfrist für abgelaufen erklärt [19]. Im mündlichen Verfahren ist Nicaragua nicht vertreten gewesen, und der Vertreter Costa Ricas hat sich in seiner Argumentation und Beweisführung auf die bereits in der Antragsschrift unterbreiteten Punkte beschränkt [20]. Das Gericht hat in seiner Endentscheidung zunächst die gerade berichtete Stellungnahme des nicaraguanischen Außenministers als peremptorische Einrede gegen die Gerichtsbarkeit gedeutet, sich mit ihr – und auch den in verschiedenen inoffiziellen Stellungnahmen Nicaraguas dazu geäußerten Argumenten [21] – auseinandergesetzt, die Einrede zurückgewiesen und sodann weitgehend zugunsten Costa Ricas in der Hauptsache entschieden. Auch bei der Entscheidung in der Sache hat es sich mehrfach mit Argumenten Nicaraguas auseinandergesetzt, die Nicaragua nur durch informelle Mitteilungen dem Gerichtshof zur Kenntnis gebracht hatte,

---

[17] Text nach der offiziellen englischen Übersetzung durch die costaricanische Gesandtschaft in Washington, AJIL 11 (1917), S. 181, 204; im Text teilweise divergierend die amerikanische Übersetzung in US Foreign Relations Papers (Anm. 14), 1916, S. 874.

[18] AJIL 11 (1917), S. 206 f.

[19] *Ibid.*, S. 207.

[20] *Ibid.*, S. 207 f.

[21] Vgl. *ibid.*, S. 210 f.; dazu auch *Hudson*, AJIL 26 (1932), S. 774.

teils durch die Bekanntgabe der schon erwähnten Noten, welche Nicaragua an Costa Rica wegen des Streitfalles gerichtet hatte[22]. Wie in der Stellungnahme des nicaraguanischen Außenministers schon angekündigt, hat der Antragsgegner das Urteil nicht respektiert, genau so wenig wie vorher die einstweilige Anordnung des Gerichtshofes[23].

Letztlich hat dieses Verhalten Nicaraguas den Anlaß für das Scheitern des Gerichtshofes gegeben[24]. Um ganz sicher zu gehen, daß er seine Tätigkeit nicht fortsetzen könne, hat Nicaragua den Gründungsvertrag 1917 gekündigt und dazu gegenüber den USA in einer Note ausgeführt, das Gericht sei degeneriert »into a center of lively intrigues«[25].

c) Moderne Parallelen zu diesem Vorgang braucht man angesichts der jüngsten Entwicklung beim IGH nicht lange zu suchen:

Zum ersten ist seit der Anrufung des IGH gegen Island durch das Vereinigte Königreich und durch die Bundesrepublik Deutschland 1972 wegen der einseitigen Ausdehnung der Fischereigewässer kein beklagter Staat mehr regulär, d.h. durch Prozeßhandlungen im Rahmen der Bestimmungen des Statuts und der Verfahrensordnung vertreten gewesen, wenn es sich um eine einseitige Klage handelte, die auf eine allgemeinere Gerichtsbindung des Beklagten gestützt war[26]. Neben diesen beiden Fischerei-Fällen sind weiter zu nennen: die beiden Streitigkeiten um die französichen Atomwaffenversuche im Pazifik[27], der pakistanisch-indische Kriegsgefangenenstreit[28], der griechisch-türkische Streit um die Abgrenzung des Festlandsockels in der Ägäis[29] und der US-iranische Streit um die Behandlung des diplomatischen und konsularischen Personals in Teheran[30].

---

[22] Vgl. *ibid.,* S. 182–184, 217, 225.

[23] Vgl. den Bericht von *Hudson,* AJIL 26 (1932), S. 773 ff., 775; *derselbe,* La Cour permanente de Justice internationale (1936), S. 61 ff.

[24] Vgl. auch *Tomuschat,* in: Judicial Settlement (Anm. 1), S. 320 f.

[25] *Hudson,* AJIL 26 (1932), S. 785, 781; *derselbe,* La Cour permanente (Anm. 23), S. 78, 71 f.

[26] Fishery's Jurisdiction case *(UK* v. *Iceland)*: Order ICJ Rep. 1972, S. 184; Jurisdiction of the Court, Judgement, ICJ Rep. 1973; S. 3; Merits, Judgement, ICJ Rep. 1974, S. 3. Fishery's Jurisdiction *(F. R. Germany* v. *Iceland)*: Order, ICJ Rep. 1972, S. 191; Jurisdiction of the Court, Judgement, ICJ Rep. 1973, S. 49; Merits, Judgement, ICJ Rep. 1974, S. 175.

[27] Nuclear Tests case *(Australia* v. *France)*: Order, ICJ Rep. 1973, S. 99; Order, ICJ Rep. 1973, S. 338; Judgement, ICJ Rep. 1974, S. 253. *New Zealand* v. *France*: Order, ICJ Rep. 1973, S. 135, 341; Judgement, ICJ Rep. 1974, S. 457.

[28] Case concerning Trial of Pakistani Prisoners of War *(Pakistan* v. *India)*: Order, ICJ Rep. 1973, S. 328, 344, 347.

[29] Aegean Sea Continental Shelf case *(Greece* v. *Turkey)*: Order, ICJ Rep. 1976, S. 3, 42; 1977, 3; Judgement, ICJ Rep. 1978, S. 3.

[30] United States Diplomatic and Consular Staff in Tehran: Order, ICJ Rep. 1979, S. 23; Judgement, ICJ Rep. 1980, S. 3. – Zu allen Verfahren *Sinclair,* ICLQ 1981, S. 340 ff.; *Fitzmaurice,* BYIL 51 (1980), S. 89 ff.

Zum zweiten sind, soweit der IGH in den genannten Streitfällen einstweilige Regelungen getroffen hat, diese vom Beklagten nicht beachtet worden[31].

Zum dritten gibt es in allen genannten Fällen Stellungnahmen der beklagten Staaten, teilweise direkt dem Gerichtshof gegenüber, teilweise in der Öffentlichkeit, aber wohl auch für diesen bestimmt, in denen jeweils insbesondere die Gerichtsbarkeit, z.T. näher begründet, bestritten[32] und deshalb die Teilnahme am Verfahren abgelehnt worden ist. In allen Fällen hat der Gerichtshof diese Stellungnahmen bei seinen jeweiligen Entscheidungen berücksichtigt.

Zum vierten haben Entscheidungen, soweit sie gegen den jeweiligen Beklagten ergangen sind oder für diesen als nachteilig gedeutet werden konnten, auf sein unmittelbares Verhalten keinen Einfluß gehabt[33].

Schließlich ist fünftens in den *Atomtest*-Fällen, den einzigen, in denen im Kreis der hier erörterten eine allgemeine Unterwerfung des Beklagten unter die Gerichtsbarkeit des IGH nach der Fakultativklausel vorlag[34], die Behandlung der Angelegenheit für Frankreich Grund gewesen, seine Unterwerfung zu beenden. Der französische Außenminister begründet dazu am 26. Januar 1974 in der Kammer auf Anfrage:

> [On]»ne pouvait ... qu'être surpris de la façon dont la Cour s'est abstenue d'admettre qu'elle était manifestement incompétente pour connaître de l'instance intentée contre la France par l'Australie et la Nouvelle-Zélande à propos des essais nucléaires en Polynésie française. Le Gouvernement a noté que la Cour, ayant pris cette position, a indiqué des mesures conservatoires devant demeurer en vigueur jusqu'à ce que le jugement sur la compétence soit, tout à loisir, intervenu. Même en tenant compte de ce que ces mesures conservatoires n'ont aucun caractère contraignant, le Gouvernement a dû conclure que la conjonction de ces deux procédés juridiques (refus de reconnaître une incompétence manifeste, indication de mesures conservatoires) tendait à anéantir la valeur des réserves les plus précises assortissant l'acceptation de la juridiction de la

---

[31] *UK* v. *Iceland,* ICJ Rep. 1972, S. 12; *F. R. Germany* v. *Iceland,* ICJ Rep. 1972, S. 30; *Australia* v. *France,* ICJ Rep. 1973, S. 99; *New Zealand* v. *France*, ICJ Rep. 1973, S. 135; *USA* v. *Iran,* ICJ Rep. 1979, S. 7.

[32] Irregulär vorgebracht möglicherweise, um dem Konzept des *forum prorogatum* zu entgehen, um nicht durch Einlassung eine fehlende Gerichtsbarkeit zu begründen; dazu *Gross,* AJIL 71 (1977), S. 56 ff.

[33] Fischereifälle, Geiselnahme in Teheran, jüngste Wiederaufnahme der französischen Kernwaffenversuche im Pazifik und dagegen ICJ Rep. 1974, S. 272, § 60, wonach Frankreich offenbar an die Auffassung des Gerichtshofes gebunden sein sollte, daß es künftig völkerrechtlich verpflichtet sei, die streitigen Atomwaffenversuche zu unterlassen.

[34] Darüber hinaus war als Jurisdiktionsgrundlage auch die Genfer Generalakte herangezogen worden; auch diese hat Frankreich gekündigt.

Cour de la part des Etats, et donc à faire bon marché de la volonté de ceux-ci, qui est pourtant la condition de base de l'exercice de la juridiction internationale. Le Gouvernement s'est donc convaincu que la sauvegarde qu'il pensait trouver dans les réserves qualifiant la déclaration française ne pouvait l'être, en réalité, que dans un retrait pur et simple de cette déclaration, . . .«.

Auch nach den Endentscheidungen des IGH in den *Atomtest*-Fällen, mit denen jeweils die Sache für erledigt erklärt worden ist, hat Frankreich nach Auskunft des französischen Außenministers in der Kammer keinen Anlaß gefunden, seine neue, negative Position dem Gerichtshof gegenüber zu überprüfen [35].

In diesen beiden letztgenannten Fällen ist die Parallele zur Entwicklung beim Zentralamerikanischen Gerichtshof nach seiner Behandlung des Streits zwischen Costa Rica und Nicaragua besonders deutlich: Dort wie hier sieht der Beklagte seine rechtlich geschützten Interessen als ganz ungewöhnlich gravierend beeinträchtigt, da sich nach seiner Auffassung das Gericht über die Grenzen seiner Jurisdiktion hinweggesetzt hat, Grenzen, die zwar informell, aber mit um so größerem Nachdruck, dem Gerichtshof gegenüber postuliert worden waren. In beiden Fällen darum für den Beklagten die Entscheidung, die in seiner Sicht schwerwiegende Verletzung seiner Souveränität mit der Beendigung seiner Unterwerfung unter die betroffene Gerichtsbarkeit zu beantworten – und mit der Nichtachtung des Richterspruchs.

d) Man sieht bereits an dieser Begründung, daß hinter der Nichtteilnahme des Beklagten bei der Einleitung eines Schieds- oder Gerichtsverfahrens in den verschiedenen eingangs berichteten Techniken und schließlich beim schriftlichen bzw. mündlichen Verfahren selbst noch keine prinzipielle Völkerrechtsfeindlichkeit im Verhältnis zu antragstellendem Staat und Gericht oder die Absicht zu stecken braucht, einmal eingegangene und eigentlich als bestehend erkannte vertragliche Schieds- oder Gerichtsbindungen nicht zu honorieren. Vielmehr kann ein beklagter Staat, der sich so vehement der Durchführung eines Verfahrens widersetzt, gerade darauf Wert legen wollen, daß ihm gegenüber das Recht geachtet werde, genauer: die Grenzen seiner Unterwerfung, außerhalb deren er nicht verpflichtet ist, im Verhältnis zum antragstellenden Staat seine Streitigkeit vor dritten Instanzen auszubreiten und in der Sache entscheiden zu lassen.

Andererseits hat der jetzt zur richterlichen Streitbeilegung unwillige Staat allerdings früher bestimmte Verpflichtungen im Verhältnis zu der antragstellenden Seite und zu dem angerufenen (Schieds-)Gericht übernommen und diesem in allgemeinerer Form entweder stillschweigend oder durch ausdrückliche Regelung gewisse Verfahrens- und Entscheidungsrechte zugestan-

---

[35] Vgl. J. O., Assemblée, Détats, 9. 3. 1974, S. 1088 f., Nrn. 7930, 7980; 1. 2. 1975, S. 381 f., Nr. 15949. Dazu auch *H. Thierry,* AFDI 1974, S. 286 ff. Ferner *Sur,* RGDIP 79 (1975), S. 976 f., mit weiteren Hinweisen.

den für den Fall, daß eine Sache in einer näher bestimmten Weise vor das Gericht gebracht werden sollte. Aus dem kategorischen Gegenüber von Beharren des Beklagten auf seinem Recht zur Selbstbeurteilung der Grenzen der eigenen Unterwerfung einerseits und richterlichem Anspruch, die ihm früher durch fortbestehende vertragliche Regelung auch gegenüber dem Beklagten zugestandenen Rechte und Befugnisse zu nutzen, können natürlich eine Reihe von Schwierigkeiten entstehen; insbesondere daraus, daß das Gericht unter Umständen aus den für seine Tätigkeit bestehenden Vorschriften ableiten kann, auch gegenüber einem säumigen Staat richterlich tätig werden zu dürfen.

Besonders akzentuiert erscheinen die Schwierigkeiten aus dem Gegeneinander von souveräner Selbstbeurteilung der Grenzen der eigenen Unterwerfung und diesbezüglicher richterlicher Entscheidungskompetenz, wenn schließlich der unterlegene Beklagte wegen des Fehlens besonderer Vorschriften für die Feststellung der Nichtigkeit internationalrichterlicher Entscheidungen die Nichtigkeitsfrage selbst beurteilt und entscheidet und sich danach von einer Beachtung der richterlichen Entscheidung befreit glaubt.

e) Bestimmungen für den Fall der Versäumung finden sich in einer Reihe prozeßrechtlicher Instrumente. Auf Art. 15 der Gründungskonvention des Zentralamerikanischen Gerichtshofes habe ich oben (b) bereits hingewiesen. Sodann ist Art. 53 IGH-Statut zu nennen, die einzige Bestimmung im Prozeßrecht des IGH, welche die Folgen der Säumigkeit einer Partei regelt. Auch das Projekt eines Prisengerichtshofes aus dem Jahre 1907 sah die Möglichkeit des Verfahrens ohne die säumige Partei vor und bestimmte[36]: »La Cour décide d'après les éléments d'appréciation qu'elle a à sa disposition«.

So wichtige Regelungen wie der deutsch-schweizerische Schiedsvertrag vom 3. Dezember 1921, die Genfer Generalakte vom 26. September 1928/28. April 1949, der Bogotá Pakt vom 30. April 1948 und das Europäische Übereinkommen zur friedlichen Beilegung von Streitigkeiten vom 29. April 1957 enthalten dagegen Vorschriften über das Versäumnisverfahren nicht, anders als die Model Ways on Arbitral Procedure, die in Art. 25 – stark am IGH-Statut und dem zentralamerikanischen Muster orientiert – bestimmen[37]:

»1. Whenever one of the parties has not appeared before the tribunal, or has failed to present its case, the other party may call upon the tribunal to decide in favour of its case.
2. The arbitral tribunal may grant the defaulting party a period of grace before rendering the award.
3. On the expiry of this period of grace, the tribunal shall render an award after it has satisfied itself that it has jurisdiction. It may only de-

[36] Art. 40, Hinweis bei *Stauffenberg,* Statut et règlement de la Cour permanente de justice internationale (1934), S. 288.

[37] AJIL 53 (1959), S. 239; zu Unterschieden gegenüber dem Statut, *Sinclair,* ICLQ 1981, S. 339.

cide in favour of the submissions of the party appearing, if satisfied that they are well founded in fact and in law«.

Weitere Regelungen über das Versäumnisverfahren sind vor allem für richterliche Instanzen getroffen worden, die zwischen verschiedenen Individuen oder zwischen Individuum und Staat judizieren sollten, besonders hinzuweisen auf die Tribunaux arbitraux mixtes nach dem ersten Weltkrieg [38]. Diese sollen hier aber wegen der prinzipiellen Unterschiede beim Verfahren zwischen Völkerrechtssubjekten im Rahmen der souveränen Gleichheit der Parteien auf der einen Seite und Verfahren zwischen Staat und Individuum oder zwischen verschiedenen Individuen vor völkerrechtlichen Instanzen auf der anderen Seite nicht weiter verfolgt werden, auch wenn sich (Schieds-)Gerichte in dem letztgenannten Bereich immer wieder am zwischenstaatlichen Verfahren orientiert haben [39].

II

Insbesondere die Auslegung und Anwendung des Art. 53 IGH-Statut hat in jüngerer Zeit zu Kontroversen geführt, sowohl im Gerichtshof selbst, sichtbar durch Sondervoten, als auch in der Literatur. Um auf sie eingehen zu können, bedarf es der Erinnerung an einige Grundgegebenheiten der internationalen Gerichtsbarkeit, die auch bei der Entstehung des Art. 53 IGH-Statut eine wesentliche Rolle gespielt haben.

1. Zu den Fundamentalprinzipien des Völkerrechts gehört es [40], daß die Unterwerfung unter ein von den betroffenen Staaten für bestehende oder künftige Streitigkeiten frei vereinbartes Streitbeilegungsverfahren, und auch die Einleitung eines derart vereinbarten Verfahrens im konkreten Falle, nicht als unvereinbar mit der souveränen Gleichheit der Staaten angesehen werden können [41]. Jede mit der souveränen Gleichheit der Staaten vereinbarte interna-

---

[38] Weitere Hinweise bei *Guyomar,* Le défaut des parties à un différend devant les jurisdictions internationales (1960), S. 88 ff.; *Hallier,* Völkerrechtliche Schiedsinstanzen für Einzelpersonen und ihr Verhältnis zur innerstaatlichen Gerichtsbarkeit (Beiträge zum ausländischen öffentlichen Recht und Völkerrecht, Bd. 35) (1962), S. 112 ff.; Commentary on the Draft Convention (Anm. 5), zu Art. 20, S. 79.

[39] So z.B. der Einzelschiedsrichter Dupuy im *Topco and Calasiatic* v. *Libyen*-Fall, vgl. *Lalive,* JDI (Clunet) 104 (1977), S. 325, 329, 330; siehe auch den Schiedsspruch in der Hauptsache, ILR S. 53, 389, 431 ff., 436; vgl. ferner *Sinclair,* ICLQ 1981, S. 339 f., 353 ff.; *Stuyt,* RGDIP 82 (1978), S. 425.

[40] Vgl. Declaration on Principles of International Law concerning Friendly Relations and Co-operation among States in accordance with the Charter of the United Nations, A/RES/2625 (XXV), General Part § 3.

[41] So die Riphagen-Formel in der Friendly Relations Declaration, Principle that states shall settle their international disputes by peaceful means etc., Abs. 5.

tionale Gerichts- oder Schiedsgerichtsbarkeit beruht also auf dem übereinstimmenden Willen der betroffenen Staaten – und nur auf ihm. Zugleich darf freilich die Vereinbarung eines verbindlichen *third party settlement* für einen bestimmten konkreten oder für eine größere Klasse von künftigen möglichen Streitfällen als besonders aufgeklärte Ausübung der Souveränität der beteiligten Staaten angesehen werden[42].

Der Sache nach sind solche Vereinbarungen völkerrechtliche Verträge, fallen also unter das völkerrechtliche Grundprinzip: *pacta sunt servanda.* Beide Parteien sind damit prinzipiell verpflichtet, auf Verlangen der jeweils anderen Seite solche (Schieds-)Gerichtsbindungen zu honorieren – es sei denn, sie hätten die Verpflichtung einer nur ihrer eigenen Beurteilung zugänglichen Beschränkung unterworfen, wie im Falle der berühmten Ehren- und Interessenklauseln zu Beginn dieses Jahrhunderts oder der *self-judging reservation* nach dem Muster des Connally-Amendment, das bei der Unterwerfung nach der Fakultativklausel des IGH-Statuts so sehr Schule gemacht hat[43].

a) Honoriert nun eine Seite – ohne über die genannte Rückzugsmöglichkeit zu verfügen – ihre Verpflichtung nicht, z.B. indem sie gebotene Mitwirkungshandlungen bei der Bildung des Gerichts oder bei der Unterbreitung des Streitfalles unterläßt, so kann es – abstrakt gesehen – keinem Zweifel unterliegen, daß sie damit ihre völkerrechtlichen Verpflichtungen gegenüber der anderen Partei des Streitfalles verletzt[44].

Ebenso abstrakt gesehen ist anderes nur anzunehmen, wenn die Parteien bei Eingehung der Streitbeilegungsverpflichtung vorausgesehen haben, daß eine Partei zur Einsetzung des Gerichts und Anhängigmachung des Verfahrens erforderliche Mitwirkungshandlungen unterlassen könnte, und mit Rücksicht darauf Regeln für eine Ersatzbestellung oder Ersetzung der Schiedsordnung oder eine einseitige Anrufungsmöglichkeit geschaffen haben. Gibt es solche zusätzlichen Regeln, sind im Prinzip zwei Wege vereinbart, die zur Rechtshängigkeit eines Streitfalles führen können, kann es folglich *nicht als rechtswidrig* angesehen werden, wenn die unwillige Partei die andere nötigt, die Rechtshängigkeit im Wege des vielleicht komplizierteren und zeitraubenderen Alternativverfahrens der Ersatzbestellung etc. herbeizuführen.

Freilich können schon damit für die unwillige Partei *Nachteile* im Verfahren verbunden sein: Daß die Ersatzbestellung der Richterbank zu einer Besetzung des Schiedsgerichts führt, die mit der Denkweise und den Positionen des Unwilligen weniger vertraut ist und ihnen darum vielleicht weniger Affi-

---

[42] So der Berichterstatter der Spezialkommission für die friedliche Streitbeilegung auf der Haager Friedenskonferenz 1899, vgl. bei *v. Mangoldt,* in: Tradition und Fortschritt im Recht, Festschrift zum 500jährigen Bestehen der Tübinger Juristenfakultät (1977), S. 448, 449 f., 461, 464 ff.

[43] Vgl. *v. Mangoldt* (Anm. 42), S. 485; siehe auch *Fitzmaurice,* BYIL 51 (1980), S. 100 f.

[44] Vgl. dazu das IGH-Gutachten zur Interpretation der Friedensverträge von 1947, ICJ Rep. 1950, S. 65, 75 ff., 77; 221, 226 ff., 228 f.

nität entgegenbringen kann; oder daß die Formulierung des Streitgegenstandes in der ersetzten Schiedsordnung weniger den Interessen des Unwilligen entspricht, als das bei seiner Mitwirkung hätte der Fall sein können. Doch auch diese möglicherweise im konkreten Verfahren eintretenden Nachteile für den Unwilligen sind von seiner freiwilligen Unterwerfung mitumfaßt, da er die genannten Techniken zur praktischen Durchsetzung einer Streitbeilegungsverpflichtung ja selbst mitkontrahiert hatte. Daß sie ihn im Einzelfalle treffen können, ist nichts anderes als der Preis dafür, daß ihm selbst im umgekehrten Falle mit Hilfe dieser Techniken die Durchführung eines Verfahrens ermöglicht würde, das er unbedingt wünscht – also Ergebnis der Reziprozität.

b) Doch nur abstrakt gesehen ist dies alles so einfach. Der Unwillige könnte seine mangelnde Mitwirkungsbereitschaft gerade darauf stützen, daß kein Fall gegeben sei, der unter die vereinbarte Streitbeilegungsverpflichtung fällt. Hat er Recht, kann er durch seine Untätigkeit weder eine Verpflichtung zu einem verbindlichen *third party settlement* verletzen, noch im Prinzip die Voraussetzungen für die Anwendung des alternativen Mechanismus' einer Begründung der Rechtshängigkeit ohne seine Mitwirkung schaffen. Die Frage kann darum bei einer derartigen Position des Schiedsunwilligen nur lauten, wie Klarheit darüber geschaffen werden soll, ob der konkrete Streitfall unter die Streitbeilegungsverpflichtung fällt oder nicht. Hier einfach einer der beiden Positionen den Vorrang einzuräumen, scheidet wegen der souveränen Gleichheit beider Parteien und der damit verbundenen prinzipiellen rechtlichen Gleichwertigkeit beider Auffassungen im fundamentalen Relativismus der anarchischen Völkerrechtsordnung von vornherein aus.

Ein Lösungsansatz ergibt sich jedoch aus einer anderen Grundregel internationaler (Schieds-)Gerichtsbarkeit.

Hätten die Parteien den konkreten Streitfall gemeinsam der fraglichen richterlichen Instanz unterbreitet und hätte der unwillige Staat bei dieser Gelegenheit geltend gemacht, daß die Streitigkeit nicht unter die von der anderen Seite herangezogene Streitbeilegungsverpflichtung falle, so wäre es zunächst Aufgabe des Richters geworden, darüber zu entscheiden, ob ein Klauselfall gegeben ist. Denn die *Feststellung der eigenen* Gerichtsbarkeit, die *jurisdiction on jurisdiction*, gehört zu den völkerrechtlich wohlbegründeten und immer wieder auch vertraglich geregelten (insbes. Art. 36 [6] IGH-Statut) Befugnissen des zwischenstaatlichen Richters[45]. Schließlich kann es keinem Zweifel unterliegen, daß Parteien, die eine Schiedsverpflichtung ohne *self-judging reservation* eingegangen sind, den fundamentalen Relativismus im Anwendungsbereich der Klausel gerade vermeiden, den Dritten auch über die

[45] Vgl. z.B. das französisch-chilenische Schiedsgericht, RIAA Bd. 15, S. 99 f.; ferner den IGH im *Nottebohm*-Fall, Preliminary Objection, ICJ Rep. 1953, S. 111, 119; weitere Hinweise bei *v. Mangoldt,* Die Schiedsgerichtsbarkeit als Mittel internationaler Streitschlichtung (Beiträge zum ausländischen öffentlichen Recht und Völkerrecht, Bd. 63) (1974), S. 100, 102.

Auslegung und Anwendung der Schiedsverpflichtung entscheiden lassen wollten, darüber, ob eine konkrete Streitigkeit darunter fällt oder nicht.

Beide Staaten, auch der schiedsunwillige, würden in einem solchen Falle Partei eines streitigen Verfahrens; in diesem wäre zunächst allerdings nur festzustellen, ob der konkrete Streit unter die Klausel fällt, und es kann, wenn der Richter seine Gerichtsbarkeit verneint, herauskommen, daß sich der schiedsunwillige wegen der konkreten streitigen Hauptsache nicht als Partei vor eine internationale Instanz ziehen zu lassen braucht. An seiner Parteistellung im Verfahren um das Bestehen der Gerichtsbarkeit änderte ein solches Ergebnis jedoch überhaupt nichts.

Nicht grundsätzlich anders wird die Lage, wenn der Unwillige bereits gegenüber dem Schiedsbegehren der anderen Seite das Fehlen einer Verpflichtung zur richterlichen Streitbeilegung geltend macht *und* wenn die Parteien bei der in ihrer Auslegung und Anwendung streitigen Gerichtsbindung die Möglichkeit vorgesehen haben, die im allgemeinen bis zur Rechtshängigkeit des Verfahrens erforderlichen Mitwirkungshandlungen im Falle ihrer Unterlassung zu ersetzen. Der staatliche Verzicht auf den unbedingten Vorrang der souveränen Selbstbeurteilung, ob ein Klauselfall vorliegt oder nicht, ist hier nicht geringer als in den Fällen, in denen nach der vereinbarten Streitbeilegungsregelung beide Parteien gemeinsam das Schiedsgericht bilden und den Streitfall anhängig machen müssen, sondern größer: Haben die Parteien mit Bestimmungen über etwa notwendige Ersatzbestellungen und über die etwa notwendige Ersetzung einer fehlenden Schiedsordnung prinzipiell *schon einer Partei allein* die Möglichkeit zugestanden, einen Streitfall vor einen internationalen Richter zu bringen, so steht dahinter der Gedanke, daß die Jurisdiktionseinrede, über die der berufende Schiedsrichter erst im Verfahren selbst entscheiden könnte, nicht soll das von einer Seite gewünschte Verfahren torpedieren können, wenn die Einrede gleichsam vorzeitig erhoben wird.

Ein Dritter, der wegen eines konkreten Streitfalles unter Berufung auf ein möglicherweise einschlägiges jurisdiktionsbegründendes Instrument gebeten wird, dort vorgesehene Handlungen vorzunehmen, um die fehlende Mitwirkung der schiedsunwilligen Partei zu ersetzen, wird darum im Auge behalten müssen, daß die streitenden Parteien grundsätzlich der in diesem Instrument vereinbarten Gerichtsinstanz die Entscheidung überlassen wollten, wie es für den konkreten Fall auszulegen ist, ob er darunter fällt oder nicht. Denn an der allgemeinen Zuständigkeit des früher einmal für künftige Streitfälle vereinbarten Richters, die Grenzen seiner Gerichtsbarkeit selbst zu bestimmen, ändern die zur Sicherung des Verfahrens beigegebenen Ersatzbestellungsregelungen etc. nichts.

Allein der um Ersatzregelung angegangene Dritte leitet seine eigene Kompetenz ebenfalls aus dem vom unwilligen Staat als nicht anwendbar bezeichneten Instrument ab. Auch er kann darum eigentlich die übertragene Kompetenz nur ausüben, wenn er selbst geklärt hat, ob ein Anwendungsfall des Streitbeilegungsinstrumentes vorliegt oder nicht; prinzipiell müßte er also dieselben Tatsachen- und Rechtsfragen geprüft haben, die später das in der möglicherweise einschlägigen Streitbeilegungsverpflichtung vorgesehene Gericht im Rahmen seiner *jurisdiction on jurisdiction* zu prüfen hätte. Damit würde jedoch praktisch die ganze *jurisdiction on juristiction,* die auszuüben dem endlich berufenen Richter zugewiesen ist, in den möglicherweise vorausgehenden Abschnitt der Ersatzregelungen verschoben[46]. Dies kann aber nicht der Sinn einer Ersatzregelungskompetenz sein, deren Funktion es ist, gegen den Willen des Schiedsunwilligen ein Verfahren zu ermöglichen, in dem dann freilich später zunächst einmal zu prüfen sein wird, ob ein Fall vorgängiger Schiedsbindung nach der in seinen Anwendungsgrenzen bestrittenen Streitbeilegungsbindung vorliegt oder nicht. Man wird deshalb die Zuständigkeit des um Ersatzregelungen angegangenen Dritten darauf beschränken müssen zu prüfen, ob es ausgeschlossen ist, daß ein Klauselfall vorliegt. Ist dies nicht offensichtlich ausgeschlossen, wird die erbetene Ersetzungshandlung vorzunehmen und dadurch der in dem streitigen Instrument vorgesehenen Gerichtsinstanz die Entscheidung zu ermöglichen sein, ob der Unwillige zur richterlichen Streitbeilegung verpflichtet ist. In der Praxis ist auch so verfahren worden[47].

Ist auf diese Weise endlich der Streitfall trotz der Untätigkeit des schiedsunwilligen Staates vor die richterliche Instanz gebracht, deren Gerichtsbarkeit er verneint, entscheidet sie nicht anders über die Grenzen ihrer eigenen Gerichtsbarkeit als im Falle einvernehmlicher Anrufung.

Auch hier wird darum der Schiedsunwillige Partei des Verfahrens wenigstens bis zur Entscheidung darüber, ob der konkrete Streitfall unter die herangezogene Streitbeilegungsverpflichtung fällt oder nicht. Denn noch mehr

---

[46] Zu den damit verbundenen Problemen für den Teilsaspekt der Ersatzbestellung von Schiedsrichtern vgl. *Wengler,* Revista española de Dorecho internacional 25 (1972), S. 425 ff.

[47] So hatte in der portugiesisch-jugoslawischen Poststreitigkeit (RIAA Bd. 12, S. 339 ff.) Jugoslawien die schiedsrichterliche Streitbeilegung mit der Begründung verweigert, es handele sich nicht um eine Poststreitigkeit, also nicht um einen Klauselfall nach Art. 31 der Weltpostkonvention vom 11. 7. 1952, und mit dieser Begründung die Bestellung seines Schiedsrichters verweigert (*ibid.,* S. 345). Die in der kompromissarischen Klausel vorgesehene Stelle hat gleichwohl die Ersatzbestellung vorgenommen (*ibid.*, S. 346 f.), und das auf diese Weise gebildete Schiedsgericht hat sodann seine Gerichtsbarkeit prüfen können und in dem wesentlichen Punkt, der jugoslawischen Auffassung entsprechend, verneint (*ibid.,* S. 347 f.). Umgekehrt hat sich ganz entsprechend der IGH im Fall des *Schiedsspruchs des Königs von Spanien zwischen Honduras und Nicaragua* (ICJ Rep. 1960, S. 192) ganz ungewöhnlich zurückgehalten, auf die Behauptung einzutreten, der frühere Richter habe seinerzeit seine Kompetenzen überschritten; der IGH hat es so im wesentlichen vermieden, die *jurisdiction on jurisdiction* dieses Schiedsrichters nachträglich an sich zu ziehen (nicht ganz so zurückhaltend dagegen im Gutachten über die Friedensverträge, ICJ Rep. 1950, S. 65, 75 ff., aber es handelte sich eben auch »nur« um ein Gutachtenverfahren, in dem bindende Entscheidungen für die streitenden Parteien und Eingriffe in die Zuständigkeit eines möglichen Schiedsgerichts, über die Grenzen seiner Gerichtsbarkeit zu entscheiden, nicht denkbar waren).

als bei richterlichen Verfahren, die nach den zugrundeliegenden Instrumenten nur einvernehmlich eingeleitet werden können, hat die Vereinbarung einer mit Sicherungen für ihre Funktionsfähigkeit versehenen Streitbeilegungsklausel die Aufgabe, es wenigstens zu einem geregelten Verfahren zwischen den streitenden Staaten darüber kommen zu lassen, ob die Streitbeilegungsklausel Anwendung zu finden hat.

Das aber setzt die Parteistellung beider streitenden Staaten voraus; ihnen in Bezug auf das Verfahren eine unterschiedliche Rechtsstellung zuzuweisen, wäre mit ihrer souveränen Gleichheit unvereinbar[48] – es sei denn, es lägen deutlich erkennbare, zwischen den Beteiligten frei vereinbarte abweichende Regelungen vor. Im übrigen muß in diesem Zusammenhang an die üblichen, auch in Art. 59 IGH-Statut zum Ausdruck gebrachten subjektiven Grenzen der Rechtskraft erinnert werden, die es grundsätzlich ausschließen, daß ein internationaler Richter Feststellungen über die eigene Gerichtsbarkeit mit Bindungswirkung gegenüber nicht als Partei am Verfahren beteiligten Staaten trifft[49].

c) An dieser Parteistellung des Schiedsunwilligen im Verfahren, in dem der Richter zunächst feststellen muß, ob ihm überhaupt die Kompetenz zugewiesen ist, über die konkrete völkerrechtliche Angelegenheit der streitenden Staaten verbindlich zu entscheiden, oder ob in einer solchen Entscheidung eine Kompetenzanmaßung, eine Verletzung der Souveränität des Schiedsunwilligen, liegen würde, kann sich selbst dadurch nichts ändern, daß der Unwillige keine Verfahrenshandlungen vornimmt, möglicherweise nicht einmal indirekt seine Auffassungen zu erkennen gibt. Jedenfalls eine Streitbeilegungsklausel, die es ermöglicht, daß ein Staat eine Sache selbst gegen den Willen der anderen Seite vor die in der Klausel prinzipiell vorgesehene Instanz bringen kann, läßt nur die Auslegung zu, daß die Vertragschließenden unbedingt auf ihre ursprüngliche Zuständigkeit zur Beurteilung der Grenzen ihrer Unterwerfung verzichten und die *jurisdiction on jurisdiction* der früher für bestimmte künftige Streitfälle allgemein vereinbarten Instanz zuweisen wollten. Jede andere Auslegung würde dem Vertragszweck nicht gerecht, der hinter den unter Umständen recht aufwendigen Regelungen zur Sicherung der Streitbeilegungsabrede sichtbar wird: den prinzipiell vereinbarten Richter allein über die Anwendungsgrenzen der Jurisdiktionsklausel entscheiden zu lassen, wenn er nach ihren Bestimmungen schon auf das alleinige Betreiben eines der im Streite befindlichen Staaten gültig angerufen werden kann.

Weitere Sicherungen der Durchführbarkeit der *jurisdiction on jurisdiction*, insbesondere solche gegen Unterlassungen der unwilligen Partei nach gültiger Anrufung der prinzipiell vorgesehenen richterlichen Instanz[50], sind nur in

---

[48] Siehe auch unten bei 4.
[49] Dazu auch nachfolgend bei c) mit Anm. 53.
[50] Siehe oben I. 2. e).

seltenen Fällen ausdrücklich vorgesehen worden; umgekehrt ist mir kein Fall bekannt, in dem eine internationale richterliche Instanz aus dem bloßen Fehlen von Vorschriften über das Verfahren bei Säumigkeit einer Seite gefolgert hätte, es dürfe nicht wirksam entscheiden. Es überrascht darum auch nicht, wenn im beratenden Juristenausschuß bei der Vorbereitung des Art. 53 des (St)IGH-Statuts die Auffassung vertreten worden ist, die ganze Vorschrift sei überflüssig, wenn man dem Gericht keine weiteren Befugnisse geben wolle als sonst im »normalen« Verfahren [51].

Nicht zu billigen ist es danach, wenn Richter Gros in seinen Sondervoten zu den *Atomtest*-Fällen [52] annimmt, daß es im Verfahren nur noch eine Partei gebe, sobald die Nichtteilnahme des unwilligen Staates am Verfahren offen feststehe. Selbst wenn man diese Äußerung auf den Fall begrenzt, daß der Untätige die Gerichtsbarkeit verneint, bleibt wenigstens der Streit um die Anwendbarkeit der möglicherweise jurisdiktionsbegründenden Klausel zwischen den betroffenen Staaten zu erledigen. Wie schon oben (b) am Ende) bemerkt, verträgt diese Erledigung keine unterschiedliche Stellung der Betroffenen im Prozeß.

Mit Recht hat *Mosler* [53] außerdem darauf aufmerksam gemacht, daß auch Art. 53 IGH-Statut in einer Prozeßlage, in der über die Gerichtsbarkeit noch nicht entschieden zu sein braucht, von beiden betroffenen Staaten als »Parteien« spricht.

Eine Ausnahme kommt höchstens in den – hier aber wohl nicht einschlägigen – Fällen in Betracht, in denen nach der Praxis des IGH [54] oder nach der neuen Vorschrift des Art. 38 [5] Verfahrensordnung 1978 [55] die Eintragung in das allgemeine Register nicht erfolgt oder eine erfolgte Eintragung zu streichen ist – eine Möglichkeit analog derjenigen in dem oben I. 2. b) berichteten Verfahren vor dem Zentralamerikanischen Gerichtshof, die Annahme einer Klage durch Entscheidung abzulehnen.

Es ist darum auch nur sehr eingeschränkt richtig, wenn vor allem im französischen Schrifttum formuliert wird: »L'on ne peut parler de carence que dans le cas où il existe à la charge de l'Etat défaillant une véritable obligation de soumettre le différend considéré à un juge, c'est-à-dire lorsqu'il a ac-

---

[51] Und wenn noch im Bericht des Ausschusses ganz unproblematisch davon ausgegangen wird, das Gericht könne auch bei Säumnis einer Partei gültig seine Gerichtsbarkeit feststellen; zum Ganzen *Stauffenberg* (Anm. 36), S. 389; siehe auch die abweichende Meinung Barwick in den *Atomtest*-Fällen, ICJ Rep. 1974, S. 398. – *Fitzmaurice* hat dies bei seinen radikalen Vorschlägen in BYIL 51 (1980), S. 97, 93, 107, 113, 116, 120 f., nicht berücksichtigt.

[52] ICJ Rep. 1973, S. 118 f.; 1974, 290 ff.

[53] Festschrift Schlochauer, S. 447 f.; siehe auch *Oellers-Frahm,* ArchVR 1979, S. 390 f.; im Ergebnis wohl ebenso *Fitzmaurice,* BYIL 51 (1980), S. 89/90, 98 Anm. 1, 121.

[54] Siehe auch *Fitzmaurice*, BYIL 51 (1980), S. 90.

[55] Dazu *Oellers-Frahm,* ArchVR 1979, S. 313.

cepté à un moment quelconque de recourir à une juridiction«[56]. Denn der gerichtsunwillige Staat hat wenigstens seine in der fraglichen (Schieds-)Gerichtsklausel übernommene Verpflichtung zu achten, dem Richter seine Zuständigkeit zur Entscheidung über die eigene Gerichtsbarkeit zu belassen; und in dem Verfahren, das zur Prüfung dieser Frage notwendig ist, kann er sehr wohl säumig sein.

d) Allerdings wird man selbst hierin keine Pflichtverletzung, keine Verletzung der Streitbeilegungsbindung sehen können, so wenig wie bei einem Verhalten, das die tätige Partei zur Anrufung Dritter nötigt, damit das Verfahren durchgeführt werden kann (vgl. oben a)). Auch hier bleibt ja die Möglichkeit, das Verfahren durchzuführen, auch hier ergibt sich letztlich aus dem für die Jurisdiktion maßgeblichen Instrument die Möglichkeit einer Alternative zu dem üblichen Verfahren unter aktiver Teilnahme beider Parteien[57].

Dies festzuhalten ist deshalb wichtig, weil es sich damit verbietet, den Säumigen unter dem Gesichtspunkt einer Verletzung völkerrechtlicher Verpflichtungen mit besonderen prozessualen Nachteilen im Versäumnisverfahren zu belasten[58].

2. Ist Säumigkeit einer Partei prinzipiell auch möglich in einem Verfahren, oder Verfahrensabschnitt, in dem allein über das Bestehen der Gerichtsbarkeit entschieden wird, ist nicht zu erkennen, warum für das in diesem Falle erforderliche *procedere* anderes gelten soll, als wenn die Versäumung in der Hauptsache eintritt. Es ist darum auch nicht zu erkennen, warum Art. 53 IGH-Statut, dessen Formulierungen auf die Behandlung der Hauptsache zugeschnitten sind, in der Substanz seiner Regelung nicht auch für ein abgetrenntes Verfahren zur Feststellung der Gerichtsbarkeit – oder besonderer Zulässigkeitsfragen – gelten soll[59]. Wird das durch einseitigen Antrag anhängig gemachte Verfahren nicht in zwei Teile gespalten, Gerichtsbarkeit/Zulässigkeitsfragen und Hauptsache, ist überhaupt nicht zu erkennen, warum die

---

[56] *Favoreu,* AFDI 1974, S. 253, 258, im Anschluß an *Guyomar* (Anm. 38), S. 20 Anm. 26; *Goy,* JDI (Clunet) 101 (1974), S. 279, 291 f.; *Eisemann,* AFDI 1973, S. 356 f.; dazu *Stuyt,* RGDIP 82 (1978), S. 423, 431 f.

[57] Ebenso *Mosler,* in: Festschrift Schlochauer, S. 443; Gros, Sondervotum zu den *Atomtest*-Fällen (einstweilige Regelung), ICJ Rep. 1973, S. 116 f./151 f., 118/152 f.; bei Lachs, Sondervotum in *Teheran*-Fall, ICJ Rep. 1980, S. 48, bleibt unklar, ob er eine Rechtspflicht meint, wenn er davon spricht, daß die Staaten das Erscheinen vor Gericht »schuldeten«.

[58] Zum Problem der Nachteile des Versäumnisverfahrens für *beide* Parteien und zur souveränen Gleichheit noch näher unten bei 4.

[59] Vgl. IGH, *Fischerei*-Fälle (Gerichtsbarkeit), ICJ Rep. 1973, S. 7–8/54; ferner *Mosler,* in: Festschrift Schlochauer, S. 441, 451; *Oellers-Frahm,* ArchVR 1979, S. 389; *Sur,* RGDIP 79 (1975), S. 983. Abweichend anscheinend *Fitzmaurice,* BYIL 51 (1980), S. 97, 93, 107, 113, 116, 120 ff.; zu dessen stark an nationalen Vorbildern orientierter Betrachtungsweise auch noch unten 4. mit Anm. 77. – Ebenfalls unten bei 4. mit Anm. 92 auch zu dem besonderen Problem, ob (insbes. Verfahrens-)Einreden nur berücksichtigt werden können, wenn sie vom Beklagten formell erhoben worden sind.

Behandlung der Jurisdiktionsfrage anderen Kriterien folgen sollte als die des materiellen Anspruches. Es gibt aber keinen rechtlich zwingenden Grund, aus dem die Behandlung der vorgängigen Fragen und der Fragen der Hauptsache unbedingt getrennt werden müßte.

Zwar ist es abstrakt gesehen richtig, daß sich eine Partei bei bestrittener Jurisdiktion nicht auf eine Verhandlung und Entscheidung der Hauptsache einzulassen braucht, wenn die Gerichtsbarkeit nicht besteht. Wird sie aber schließlich innerhalb der Entscheidung zur Hauptsache rechtlich fehlerfrei bejaht, so bestand sie eben auch schon zu dem Zeitpunkt, zu dem das Gericht – ohne Vorabentscheidung über die Gerichtsbarkeit – von den Parteien die Verhandlung der Hauptsache erwartete, wurde also schon zu diesem Zeitpunkt dem Untätigen nichts zugemutet, was mit seiner Souveränität unvereinbar wäre.

Freilich wird ein Gericht, damit konfrontiert, daß ein Staat in einem konkreten Streitfall das Bestehen der Gerichtsbarkeit – in formeller Prozeßhandlung oder informell – bestreitet, gut daran tun, zunächst die Jurisdiktionsfrage zu entscheiden[60]. Dies ist wohl auch der Gedanke hinter dem 1922 schließlich nicht durchgeführten Vorhaben im StIGH, bei Einreden gegen die Gerichtsbarkeit und bei Säumigkeit einer Partei unbedingt über das Bestehen der Gerichtsbarkeit vorab zu entscheiden[61]. Er steht ebenso hinter Art. 79 (insbes. 7) Verfahrensordnung 1978[62]. Dort ist aber auch hinreichend berücksichtigt, daß eine vorgängige Entscheidung über die Gerichtsbarkeitsfrage nicht immer möglich sein wird, z.B. bei den sog. qualifizierten Prozeßvoraussetzungen, wie etwa, wenn sich der geltend gemachte materielle Anspruch und die Gerichtsbarkeit aus demselben Vertrag ergeben, dessen Wirksamkeit der Beklagte bestreitet.

3. Allerdings kann der untätige Beklagte nicht beides haben: Beschränkung des Verfahrens zunächst auf die Frage der bestrittenen Gerichtsbarkeit und zugleich die Verpflichtung des Klägers, im ersten Schriftsatz nach Einreichung der Klage den gesamten Fall auszubreiten, also auch die Anträge und Argumentationen zur Hauptsache, damit der Beklagte frei entscheiden könne, ob er säumig werden oder bleiben wolle – wie dies das gemeinsame Sondervotum der Richter Bengzon und Jiménez de Aréchaga zu den verfahrensleitenden Anordnungen des IGH in den *Fischerei*-Fällen gegen Island[63] unter Berufung auf »general principles of law recognized in national legislations concerning default proceedings« fordert.

---

[60] Dies lehren nicht nur die oben I. 2. b), c) berichteten Empfindlichkeiten Nicaraguas und Frankreichs. – Vgl. im übrigen auch *Sur,* RGDIP 79 (1975), S. 1010.

[61] Vgl. die Hinweise im Sondervotum Gros zu den *Atomtest*-Fällen, ICJ Rep. 1974, S. 299; ferner *Guyomar,* Commentaire du Règlement de la Cour internationale de Justice (1973), S. 357; *Guggenheim,* ZaöRV 19 (1958), S. 134.

[62] Art. 67 (7) Verfahrensordnung 1972; dazu *Eisemann,* AFDI 1973, S. 372 f.

[63] ICJ Rep. 1972, S. 185/186, 192/193. Ähnlich die Regelung des Art. 14 Konvention zur Gründung des Zentralamerikanischen Gerichtshofes, vgl. oben I. 2. b).

Damit ist nicht nur die Frage nach den Details der Durchführung eines Versäumnisverfahrens gestellt, sondern auch die, wie weit es zulässig ist, im internationalen Prozeß auf Erfahrungen aus dem nationalen Bereich mit der Säumnis einer Partei zurückzugreifen.

a) Der beratende Juristenausschuß hat sich bei der Erarbeitung des Art. 53 (St)IGH-Statut, wie mit Recht häufig in der Literatur hervorgehoben wird, geweigert, dies zu tun. Der Bericht erwägt sowohl für den Fall, daß eine Partei nur in der Hauptsache nicht vertreten ist, als auch für den Fall, daß eine Partei überhaupt nicht vertreten ist, der Gerichtshof jedoch seine Gerichtsbarkeit bejaht, folgendes:

> »On peut se demander si, alors, il n'y a pas lieu d'adjuger au demandeur ses conclusions sans exiger de preuves. Certains systèmes de procédure interne l'admettent; mais il ne faut pas oublier qu'ici les plaideurs sont des Etats, et qu'il est particulièrement grave de prononcer une condamnation contre eux, lorsqu'ils dénient à la Cour le droit de les juger. Pour rendre acceptable aux susceptibilités de la souveraineté le jugement par la Cour, même au refus de comparaître, il était nécessaire d'entourer la condamnation qui frapperait l'Etat défaillant, de toutes les garanties désirables pour lui donner l'autorité morale et, par suite, en assurer le respect en attendant d'en faciliter l'exécution«.

Der Ausschuß hat deshalb bei der Formulierung des Versäumnisverfahrens vor dem IGH das englische und amerikanische Verfahrensrecht zum Vorbild genommen, denn dort waren die schärfsten Anforderungen an die Begründetheit der Anträge der tätigen Partei festzustellen, ehe ein Urteil gegen die untätige ergehen kann [64]. Allgemeine, aus den verschiedenen nationalen Verfah-

---

[64] Vgl. die Hinweise bei *Scott* (Anm. 5), S. 125 f.; für das gegenwärtige US-Recht 28 U.S.C.A. § 1251 Note 109; *Moore's* Federal Practice § 55.02 [1]; 28 U.S.C.A. Rules of the Supreme Court, Rule 9 § 8; 28 U.S.C.A. Federal Rules of Civil Procedure, Rule 55, Notes 5 und 8; zum englischen Recht *Bunge, J.,* Das englische Zivilprozeßrecht (1974), S. 96 ff.; siehe aber auch *Linke,* Die Versäumnisentscheidungen im deutschen, österreichischen, belgischen und englischen Recht (1973), S. 79 f. – Nicht unerheblich großzügiger ist demgegenüber das französische Prozeßrecht in den jetzt maßgeblichen Art. 467–479 des Nouveau Code de procédure civile, J. O. 9. 12. 1975 annexe: Bei Nichterscheinen des Beklagten wird nach Art. 472 in der Sache entschieden, der Richter aber »ne fait droit à la demande que dans la mesure où il l'estime régulière, recevable et bien fondée«. Dazu näher *Vincent,* Procédure civile (19. Aufl. 1979), Nrn. 503 ff., 511 ff., insbes. 515. Damit ist wenigstens eine summarische Prüfung des Falles des Beklagten gemeint, es können auch Beweise erhoben werden. Und in Fällen, in denen öffentliches Interesse besteht, ist die Pflicht des Richters, sich von der wirklichen Sach- und Rechtslage zu überzeugen, noch sehr viel strenger gefaßt. Zur älteren französischen Rechtslage vgl. das Dekret vom 28. 8. 1972, J. O. 1972, S. 9300, Art. 41 ff.; *Dalloz,* Répertoire de procédure civile, mise à jour, 1974, »jugement par défaut«. Weiter *Guyomar* (Anm. 38), S. 3 ff.; siehe auch *E. Michellet,* Juris-Classeur de procédure civile, Bd. 2, Fasc. 538, jugements par défaut §§ 97 ff. – Wieder anders das deutsche Recht mit seiner Geständnisfiktion des § 331 Zivilprozeßordnung zu Lasten

rensrechten stammende Rechtsgrundsätze werden damit gerade nicht aufgenommen; schon der in der wiedergegebenen Begründung des Ausschusses herausgestellte Gegensatz zwischen verschiedenen nationalen Verfahrensrechten zeigt, daß der Ausschuß solche Grundsätze nicht zu finden vermag[65]. Die dem widersprechende Bezugnahme der Richter Bengzon und Jiménez de Aréchaga erweist sich nach alledem in der gewählten Allgemeinheit als nicht tragfähig.

Es kann darum in diesem Zusammenhang nicht eindringlich genug davor gewarnt werden, mit dem Begriff Versäumnis/*default*/*défaut* nationale »Hintergedanken« einströmen zu lassen, etwa die Vorstellung, der Säumige könne wegen einer Pflichtverletzung[66] »bestraft« werden[67].

b) Unter dem konkreten verfahrensrechtlichen Aspekt, unter dem sich Bengzon und Jiménez de Aréchaga auf allgemeine Rechtsgrundsätze des Versäumnisverfahrens beziehen, ist diese Bezugnahme ebenfalls nicht tragfähig. Für ein System wie das amerikanische, welches im Falle der Versäumnis zunächst die gerichtliche Feststellung dieses Tatbestandes vorsieht, die säumige Partei damit von jedem weiteren Vorbringen ausschließt und sodann den Antrag auf Versäumnisurteil freistellt[68], mag es selbstverständlich sein, daß ein Versäumnisurteil nicht auf Grund eines Vorbringens erlassen werden darf, welches nach der Feststellung der Versäumung liegt[69]. Nationale Rechte hingegen, die diese Zweiteilung des Verfahrens nicht kennen, wie z.B. das deutsche, können auch im Versäumnisfalle weiteres Parteivorbringen zulassen. Nur daß dieses dem Untätigen angemessen mitgeteilt und er nochmals geladen werden muß, ein Versäumnisurteil zugunsten der tätigen Partei also nicht im Termin der ersten Versäumung erlassen werden kann[70]. Was diesen

---

des säumigen Beklagten; aber auch hier besteht nicht für alle Fälle der Säumigkeit des Beklagten eine einheitliche Rechtslage, insbesondere entfällt die Geständnisfiktion, wenn es um von Amts wegen zu berücksichtigende Fragen geht; aus diesem Grunde – vgl. § 86 Verwaltungsgerichtsordnung (VwGO) – gibt es im Prozeß vor den allgemeinen Verwaltungsgerichten ein Versäumnisverfahren im technischen Sinne des deutschen Prozeßrechts überhaupt nicht, wird in allen Fällen der Säumigkeit streitig und nach Prüfung der relevanten Fragen *ex officio* – auch der Tatsachenfragen – entschieden, vgl. § 102 Abs. 2 VwGO.

[65] Zur Verschiedenheit der nationalen Prozeßrechte im Versäumnisverfahren siehe auch den Einzelschiedsrichter Lagergren im *BP* v. *Libyen*-Fall, ILR 53, S. 297, 311 f.; vgl. ferner *Lalive*, JDI (Clunet) 104 (1977), S. 329 f.

[66] Für entsprechende US-amerikanische Überlegungen vgl. *Bouvier*'s Law Dictionary (3rd revision, 1914), »default«; *Black*, Law Dictionary (4th ed.) »default«.

[67] Dazu schon oben II. 1. d); vgl. ferner *v. Mangoldt*, ZaöRV 40 (1980), S. 562 ff.

[68] Dazu die Hinweise oben in Anm. 65.

[69] Analoge Überlegungen scheinen auch für das oben berichtete Verfahren vor dem Zentralamerikanischen Gerichtshof maßgebend gewesen zu sein.

[70] Vgl. *Baumbach*/*Lauterbach*/*Albers*/*Hartmann*, ZPO-Kommentar (40. Aufl.) § 331 ZPO Anm. 4 C a; zur Rechtslage nach französischem Recht vgl. insoweit *Dalloz*, Répertoire de procédure civile, jugement par défaut, § 46.

Regelungen zugrunde liegt, ist nichts anderes als der Gedanke der Gleichheit der Parteien vor dem Richter in seiner Ausprägung des *audiatur et altera pars*[71]. Auch die untätige Partei kann nicht hinsichtlich allen künftigen Vorbringens der erscheinenden als geständig angesehen werden. Prozessual ist die Gleichheit nur gewahrt, wenn vor der richterlichen Entscheidung der säumigen Partei wenigstens Gelegenheit zur Stellungnahme zu demjenigen Vorbringen gegeben war, das der Richter zu erwägen haben würde. Dies freilich setzt die Unterrichtung des Säumigen in den prozessual vorgeschriebenen Formen voraus.

4. Gleichheit ist das maßgebende Prinzip auch für das zwischenstaatliche Versäumnisverfahren[72] – aber nicht als allgemeiner Rechtsgrundsatz, auch in den Ausprägungen nicht, wie z.B. bei dem *audiatur et altera pars*[73]. Vielmehr geht es um das völkerrechtliche Fundamentalprinzip, »that international disputes shall be settled on the basis of the sovereign equality of states«[74]. Dieser Völkerrechtsgrundsatz gilt auch für das Verfahren[75]. Eine ungleiche prozessuale Behandlung der Parteien duldet er nicht[76]. Und er verträgt auch keine irgendwie »aufgelockerte« Handhabung, ganz besonders nicht bei Säumnis einer Partei[77].

Den von einer Versäumung betroffenen internationalen Richter kann das allerdings in erhebliche Verlegenheit bringen.

Einerseits hat er, wenn eine genügend »wasserdichte« Formulierung des jurisdiktionsbegründenden Instruments zu der Annahme nötigt, daß auch bei Untätigkeit einer Partei entschieden werden soll, über die Klage der tätigen zu verhandeln und bei Bejahung der Gerichtsbarkeit auch in der Sache zu entscheiden – statt eine Entscheidung abzulehnen. Eine irgendwie geartete Bestrafung des Tätigen wegen des Verhaltens des Säumigen darf somit nicht eintreten.

Andererseits hat der Richter zu beachten, daß die Parteien mit dem jurisdiktionsbegründenden Instrument ihr fundamentales Recht der Selbstbeurteilung aufgegeben, dem prinzipiellen Vorrang jeweils der eigenen Betrachtung der streitigen Angelegenheit abgeschworen und die Erledigung dem Unpar-

---

[71] Dazu statt vieler *Maunz/Dürig/Herzog/Scholz,* Art. 3 GG, Rdnr. 384 a *lit.* f aa, Rdnr. 392 ff.

[72] Gros betont es zu Recht in seinem Sondervotum zu den *Atomtest*-Fällen, ICJ Rep. 1974, S. 290–292.

[73] In der Literatur wird das Prinzip freilich als allgemeiner Rechtsgrundsatz aus den nationalen Rechten auch für das Versäumnisverfahren nutzbar gemacht, vgl. *Bin Cheng,* General Principles of Law (1953), S. 290 ff., 296 f.

[74] Friendly Relations Declaration, vgl. oben Anm. 40/41.

[75] Vgl. *v. Mangoldt* (Anm. 42), S. 477, mit weiteren Hinweisen.

[76] Siehe auch den 1. Bericht *Scelle* zum schiedsgerichtlichen Verfahren für die International Law Commission, YILC 1950 II, S. 133 § 64.

[77] Ebenso das Sondervotum Gros in den *Atomtest*-Fällen (einstweilige Regelung), ICJ Rep. 1973, S. 118/152 f.; darum sind auch die Überlegungen von *Fitzmaurice,* BYIL 51 (1980), S. 97, 93, 107, 113, 116, 120 f., zu verwerfen.

teiischen übertragen haben. Dies schließt es aus, daß der Richter einfach die Position einer Partei übernehmen kann, nötigt ihn, ungeachtet der Untätigkeit des Säumigen die wirkliche Sach- und Rechtslage zu erforschen und darauf seinen Spruch zu gründen [78]. Auch eine Bestrafung des Säumigen wegen seiner Untätigkeit ist also prinzipiell ausgeschlossen. Sie wäre im übrigen wenig tunlich, wie der beratende Juristenausschuß mit Recht festgestellt hat, wenn eine Partei, deren Untätigkeit auf vehementer Verneinung der Gerichtsbarkeit und auf der Befürchtung beruht, mit eigenen wesentlichen Rechten »zu kurz« zu kommen, zu derjenigen aufgeklärten Ausübung der eigenen Souveränität zurückgeführt werden soll, die in der ursprünglichen Unterwerfung lag.

Nun ist weiter der internationale Richter bei Säumigkeit einer Seite für die Würdigung der Sach- und Rechtslage in dem streitigen Falle zunächst ganz auf das Vorbringen der tätigen angewiesen. Zudem ist diese jedenfalls prinzipiell im normalen Verfahren berechtigt [79], ihren Vortrag und ihre Beweise zu ergänzen, wenn sie feststellt, daß »ihr Fall« in der bisherigen Konstruktion beim Gerichtshof keine Aussicht auf Erfolg hat. Gerade in diese Lage kann sie aber auch kommen, wenn der Richter im Versäumnisverfahren *proprio motu* Fragen aufwerfen darf und unter Umständen muß [80], die der bisherigen Konstruktion des Falles des Tätigen die Grundlage entziehen. Soll also, und in welchen Grenzen, das Gericht für den untätigen Beklagten, »without espousing the cause of the defendant, ... nevertheless, act as its counsel«? Der Bericht von *J. B. Scott* über das Projekt eines ständigen internationalen Gerichtshofes [81] nimmt dies an [82]. Soll weiter auf entsprechendes »Vorbringen« des Richters hin die tätige Partei noch im mündlichen Verfahren ihr Vorbringen dahingehend ergänzen dürfen, daß ihre Klage zulässig bzw. begründet wird? Man wird hier mehr als nur erhebliche Zweifel haben müssen.

Einmal bringt der Richter mit einem solchen Vorgehen seine Unparteilichkeit in Gefahr – zumindest aus der Perspektive der tätigen Partei. Dies ist einer weiteren Bereitschaft zur Unterwerfung der Staaten unter den internationalen Richter nicht eben förderlich. Zum anderen stellt sich die Frage der souveränen Gleichheit insofern, als dem untätigen Staat wenigstens Gelegenheit zur Stellungnahme gegenüber neuem Vorbringen zur Sach- und Rechts-

---

[78] Dazu *Guyomar* (Anm. 38), 22–24.

[79] Beim IGH im Rahmen der Art. 52 ff. Verfahrensordnung 1972 (Art. 56 ff. Verfahrensordnung 1978).

[80] Dazu insbes. *Witenberg* (Anm. 5), S. 386, unter Hinweis auf die im zwischenstaatlichen Prozeß fehlende Möglichkeit eines Einspruchs gegen ein Versäumnisurteil. Ausnahmen davon sind bei Verfahren zwischen Staat und Privaten festzustellen, vgl. *Hallier* (Anm. 38), S. 114.

[81] (Anm. 5), S. 125.

[82] Ansätze dazu wohl auch beim IGH, *Fischerei*-Fälle, ICJ Rep. 1973, S. 7–8/54; *Ägäis*-Fall, ICJ Rep. 1978, S. 18 f. (§§ 42, 43), siehe aber auch die sehr vorsichtigen Formulierungen dort S. 19 § 45 und S. 20 § 47; dazu *Bettati*, AFDI 1978, S. 318.

lage bleiben muß, das ihm bisher nicht regulär bekannt gegeben worden war [83]. Dies ganz besonders auch deshalb, weil an sich »gut gemeintes« »Vorbringen« des Richters gegen Konstruktionen der tätigen Partei wegen unbeabsichtigter und unbedachter Zusammenhänge Verschlechterungen der Position des Säumigen zur Folge haben könnte. Wird aber nach dem Sach- und Rechtsgespräch zwischen tätiger Partei und Richter in der mündlichen Verhandlung dem Säumigen erneut Gelegenheit zur Stellungnahme gegeben, kann dies nicht nur die Entscheidung der Sache innerhalb vertretbarer Zeit unmöglich machen; vielmehr kann darin auch eine Art Bestrafung des Tätigen liegen, der bei Tätigkeit auch der anderen Seite darauf rechnen dürfte, daß unter bestimmten Umständen verspätetes Vorbringen zurückgewiesen und unbeachtet bleiben müßte [84].

Was bietet sich in dieser Lage als Ausweg an, wenn ein internationaler Richter eine Klage bei Untätigkeit des Beklagten *ex officio* zu prüfen hat? Vor allem [85], daß er sich auf bekannt gewordene, auch der tätigen Partei bekannte [86], Positionen und Argumente des Säumigen stützt [87]. Weder stellen sie für den Säumigen etwas Neues in dem eben behandelten Sinne dar, noch läßt sich sagen, daß die tätige Partei nicht gerade mit diesen Positionen und Argumenten zu rechnen brauchte und daß sie sich nicht auf sie hätte einstellen können [88].

Zwar ist ein solches richterliches *procedere* bei den bisherigen Verfahren, denen eine Partei unter Berufung auf die angeblich fehlende Gerichtsbarkeit fernblieb, bei denen sie aber gleichwohl informell ihre Rechtspositionen und teilweise auch Begründungen wissen ließ, heftig kritisiert worden – von Ver-

---

[83] Siehe auch *Scelle*, YILC 1950 II, S. 133 § 64; *Bin Cheng* (Anm. 73), S. 296; anderes könnte höchstens gelten, wenn nachträglich nur Notorisches in den Prozeßstoff eingeführt worden war.

[84] Zum Zusammenhang verspätetes Vorbringen/»gleiche« Möglichkeit zur Stellungnahme auch die Sondervoten Read und Azevedo im *Asyl*-Fall, ICJ Rep. 1950, S. 351, 352. – Dazu auch Art. 52, 67 Verfahrensordnung 1972 (Art. 56, 79 Verfahrensordnung 1978). In Formfragen (auch Fristfragen) weigert sich der IGH allerdings mit Recht, dieselbe Strenge walten zu lassen wie manche nationalen Rechte (dazu unten Anm. 92), um die Feststellung der wirklichen Wahrheit nicht ungebührlich zu behindern. Dazu auch *Jiménez de Aréchaga*, AJIL 67 (1973), S. 19.

[85] Aber nicht nur! – doch sollen die möglichen weiteren Aspekte eines »espousing of the cause of the defendant« (vgl. oben bei Anm. 81), insbesondere die Grenzen selbständiger Einführung von Prozeßstoff in das Versäumnisverfahren durch das Gericht, hier nicht näher untersucht werden.

[86] Durch formelle Mitteilung, möglicherweise auch durch Allgemeinkundigkeit.

[87] Siehe auch das Sondervotum de Castro in den *Atomtest*-Fällen, ICJ Rep. 1974, S. 372.

[88] Damit argumentiert auch der IGH in den *Atomtest*-Fällen, ICJ Rep. 1974, S. 263, 264 bzw. 467 f. – Anderes gilt möglicherweise bei verspätetem »Vorbringen« des Säumigen, vgl. IGH, *Korfu-Kanal*-Fall (Entschädigung), ICJ Rep. 1949, S. 237 ff., 244 ff., 248; *Nottebohm*-Fall (zweite Phase), ICJ Rep. 1955, S. 6; *Ägäis*-Fall, ICJ Rep. 1978, S. 7; dazu *Mosler*, in: Festschrift Schlochauer, S. 453; *Sinclair*, ICLQ 1981, S. 349 f.; siehe auch *Oellers-Frahm*, ArchVR 1979, S. 387 f., 390.

tretern der tätigen Partei ebenso wie aus der Perspektive der untätigen, und vor allem in der Literatur [89]. Die souveräne Gleichheit beider Parteien *und* die neutrale Stellung des Richters wird so in der mißlichen Lage, in welche der Richter durch die Säumigkeit gebracht ist, aber noch am ehesten gewahrt [90].

Hatte sich die *tätige Partei* in der Präsentation ihres Falles nicht genügend auf die informell vorgebrachte Position der untätigen eingerichtet und verliert sie deshalb ihren Fall, ohne vorher vom Richter darauf hingewiesen worden zu sein, daß ihm das informelle Vorbringen der säumigen Seite als relevant erscheint, so mag sie sich überrascht und in gewisser Weise durch das Verhalten des Untätigen auch bestraft fühlen. Aber sie hat ja mit dem jurisdiktionsbegründenden Instrument nicht nur die Möglichkeit geschaffen, daß im *ex parte*-Verfahren der Richter die Tragfähigkeit des Falles *ex officio* untersucht, sie hat sich auch noch im konkreten Fall dieser Prüfungsbefugnis *ex officio* unterworfen, indem sie trotz Säumigkeit der anderen Seite das Verfahren weiter betrieb. Und sie hätte wohl auch in einem kontradiktorisch durchgeführten Verfahren keinen Anspruch auf einen Hinweis des Richters, wenn dieser auf Grund der Einlassungen des Beklagten oder schon auf Grund von Mitteilungen des Klägers über das Verhalten des Beklagten zu der Auffassung gelangt, der klägerische Antrag sei aus prozessualen Gründen oder in der Sache abzuweisen [91]. Noch weniger kann schließlich der im Versäumnisverfahren unterlegene Kläger einen Anspruch darauf gehabt haben, in der Sache zu obsiegen, nur weil der Beklagte ihm zustehende peremptorische oder dilatorische Einreden unterlassen haben und dadurch z.B. die Gerichtsbarkeit prorogiert oder das bisherige Fehlen der Zulässigkeitsvoraussetzung erfolgloser Verhandlungen vor Anrufung des Richters unschädlich gemacht haben könnte [92].

---

[89] Vgl. besonders bei *Fitzmaurice,* BYIL 51 (1980), S. 94 ff., 115 ff.; *Cot,* AFDI 1973, S. 256; *Bettati,* AFDI 1978, S. 316; *Sinclair,* ICLQ 1981, S. 345 ff. Aus Sondervoten: Fitzmaurice, *Fischerei*-Fälle (jurisdiction), ICJ Rep. 1973, S. 35; Gros, *Atomtest*-Fälle (einstweilige Regelung), ICJ Rep. 1973, S. 116 f./151.

[90] Vgl. auch *Oellers-Frahm,* ArchVR 1979, S. 389.

[91] Art. 61, 62 Verfahrensordnung 1978 (Art. 57, 59, Verfahrensordnung 1972, dazu *Guyomar* – Anm. 61 –, S. 290 ff.) enthalten zwar ein Hinweis- und Fragerecht für den IGH bzw. einzelne Richter, begründen aber keine entsprechende Pflicht.

[92] Dies hat *Oellers-Frahm,* ArchVR 1979, S. 392, nicht berücksichtigt. – Auch der IGH prüft *proprio motu* Zulässigkeitsfragen im Versäumnisverfahren neben der eigenen Gerichtsbarkeit, vgl. die *Atomtest*-Fälle, ICJ Rep. 1974, S. 259 ff./463 ff.; siehe auch Petrén, Sondervoten zur einstweiligen Regelung in den *Atomtest*-Fällen, ICJ Rep. 1973, 124/159; *Teheran*-Fall, ICJ Rep. 1980, S. 18 § 33; sehr viel vorsichtiger im *Ägäis*-Fall, ICJ Rep. 1978, S. 20 § 47, wohl angesichts der besonderen Umstände des Falles, vgl. dazu das Sondervotum Stassinopoulos, *ibid.*, S. 74 f.; vgl. ferner *Zoller,* RGDIP 84 (1980), S. 981. Der Hauptgrund für dieses »liberale« Verhalten des IGH dürfte in seiner ständigen Rechtsprechung zu finden sein, nach der er nicht gehalten ist, »d'attacher à des considérations de forme la même importance qu'elles pouvaient avoir dans le droit interne«, *Mavrommatis*-Fall, CPJI série A 2, S. 34; *Nord-Kamerun*-Fall, ICJ Rep. 1963, S. 28; *Korfu-Kanal*-Fall (preliminary objection), ICJ Rep. 1948,

In der Durchführung eines Versäumnisverfahrens können gewiß auch *Nachteile für die tätige Partei* umschlossen sein; sie zu tragen, hat die tätige Partei im Interesse eines gesicherten Obligatoriums jedoch übernommen [93], soweit sie mit der Durchführung eines *ex parte*-Verfahrens notwendig verbunden sind, welches der souveränen Gleichheit der Parteien verpflichtet bleiben muß.

Der Eindruck, daß die Technik insbesondere des IGH, informelles Vorbringen des Säumigen zu berücksichtigen, nur der tätigen Partei nachteilig sei, ist im übrigen zwar verbreitet, aber wohl unrichtig. Die von französischer Seite zu hörende Rüge, daß das außerprozessuale Verhalten Frankreichs in den *Atomtest*-Fällen nicht hätte berücksichtigt werden dürfen [94], mag das veranschaulichen. Gerade die Behandlung dieser Fälle durch den IGH macht *Risiken* sichtbar, *denen der Säumige* in Bezug auf die Beurteilung seines außerprozessualen Verhaltens *ausgesetzt* ist, weil er es nicht durch nuancierte Argumentation im Rahmen des Verfahrens verständlich macht. Die bloß informelle »Einbringung« von Positionen und Argumenten des Säumigen in ein gerichtliches Verfahren bleibt wohl doch eher fragmentarisch bis kursorisch und verkürzt darum allzuleicht eine an sich vielleicht gar nicht schlecht zu verteidigende Sache des Untätigen. Man wird nicht erwarten können, daß der internationale Richter durchweg in solchen Fällen auch nur in der Lage ist, die notwendigen Verfeinerungen bei seiner Prüfung *ex officio* vorzunehmen [95]. Zwischenstaatliche Streitverfahren sind nicht selten vor allem durch die Besonderheit der konkreten Situation in einer Weise beherrscht, daß maßgebliche Gesichtspunkte sachgerecht nur von der interessierten Seite in das richterliche Verfahren eingeführt und verständlich gemacht werden können. Erst dann aber können sie den Ausgang des Verfahrens angemessen beeinflussen. Daß dies so ist, daß bei der Streitbeilegung vor allem das Konkrete der Situation die Bereinigung der Streitigkeit beherrscht, zeigt sich nicht zuletzt noch darin, daß es nach wie vor an einer befriedigenden Theorie zur Präzedenzwirkung internationalrichterlicher Entscheidungen fehlt – selbst für die Recht-

---

S. 27; *Ägäis*-Fall, ICJ Rep. 1978, S. 18 § 42; Sondervotum de Castro, ICJ Rep. 1974, S. 372; vgl. auch *Bettati,* AFDI 1978, S. 318; ferner *v. Mangoldt,* ZaöRV 40 (1980), S. 559 ff. – Diese Flexibilität um des Zieles der Feststellung der wirklichen Wahrheit Willen wird bei der auf einem ungewöhnlich engen und formalen Verständnis der Verfahrensordnung aufbauenden Kritik von *O'Connell* und *Fitzmaurice,* BYIL 51 (1980), S. 96 f., nicht berücksichtigt.

[93] Dazu näher auch schon oben bei II. 1. a), d), 3. a) am Ende.

[94] So aber der IGH, ICJ Rep. 1974, S. 263/467 f. Dagegen *Sur,* RGDIP 79 (1975), S. 986 f., 978 f.; siehe auch das gemeinsame Sondervotum Onyeama, Dillard, Jiménez de Aréchaga, Waldock, ICJ Rep. 1974, S. 322.

[95] Siehe auch das Sondervotum Gros zu den einstweiligen Regelungen in den *Atomtest*-Fällen, ICJ Rep. 1973, S. 118 f./153; ebenfalls Gros zu den *Fischerei*-Fällen (Merits), ICJ Rep. 1974, S. 126/234, 237; Sondervotum Lachs im *Teheran*-Fall, ICJ Rep. 1980, S. 48.

sprechung des IGH[96]. Entschließt sich darum eine Partei, dem gegen sie anhängig gemachten Verfahren vor einem internationalen Richter fernzubleiben und ihren Fall nur inoffiziell zu »vertreten«, so kann dies ihre Position nicht unerheblich verschlechtern.

Daß der Richter trotzdem das Verfahren durchführt und am Ende entscheidet, verletzt aber auch gegenüber dem Säumigen nicht die souveräne Gleichheit, denn dieser war ja früher selbst aufgeklärt genug gewesen, eine Streitbeilegungsregelung zu vereinbaren, die auch gegen ihn, nicht nur zu seinen Gunsten, die Möglichkeit der Durchführung des richterlichen Verfahrens bei Untätigkeit einer Seite vorsah, um auf diese Weise zu einem wirklichen Obligatorium zu gelangen[97].

III

Die bisherige Verfahrensweise der internationalen Gerichte, welche sich mit der Säumigkeit einer Seite bei gleichzeitigem informellem »Vortrag« auseinandersetzen mußten, dürfte inzwischen eine gefestigte Rechtsprechung gebildet haben. Es gehört zu den Leistungen der internationalen Gerichtsbarkeit, daß sie mit der gewählten Art der Einbeziehung informellen Vorbringens – aufs Ganze gesehen – nicht nur die richterliche Unparteilichkeit und damit letztendlich das richterliche Prestige zu wahren verstanden und die Balance zwischen der tätigen und der säumigen Partei zu halten vermocht hat, sondern daß es ihr auch gelungen sein dürfte, die Einbeziehung der informellen Argumentation des Säumigen in einer Weise vorzunehmen, welche nicht als Einladung zur Säumigkeit mißverstanden werden kann. Vielleicht vermag dies die Untätigkeit des Beklagten auf die Dauer zu einer bloßen Episode bei der Verwirklichung des alten staatlichen Bestrebens zu machen, nicht alle früher eingegangenen Bindungen an eine richterliche Streitbeilegung honorieren zu müssen. Was dagegen stets bleiben wird, ist am Ende die Frage, ob eine richterliche Entscheidung gegen eine Partei, die das Verfahren unter allen Umständen nicht wollte, auch durchgesetzt werden kann. Auch diese Frage hat etwas mit der Integrität des richterlichen Verfahrens zu tun, welches zu der Verurteilung geführt hatte; noch mehr aber hängt die Antwort von der Fähigkeit des Richters ab, sein Ergebnis dem Unterlegenen auch politisch schmackhaft zu machen. Und diese Fähigkeit ist wohl nicht zuletzt eine Funktion der Bereitschaft, auf beide Parteien gleichgewichtig einzugehen.

---

[96] Dazu *Rosenne,* The Law and Practice of the International Court, Bd. 2 (1965), S. 611 ff.; *Tunkin,* Theory of International Law (1974), S. 179 ff.; *v. Mangoldt* (Anm. 42), S. 469, 480.

[97] Dazu näher auch schon oben bei II. 1. a), d), 3. a) am Ende.

Staatliche Aufklärungsansprüche und Völkerrecht

F. A. Mann

I

In zahlreichen Zusammenhängen verleiht das Recht wohl aller Länder, insbesondere auch der Bundesrepublik Deutschland, dem Staat und seinen Organen Anspruch auf Aufklärung von Tatsachen durch Vorlegung von Urkunden oder mündliche Aussage. Im Steuerrecht gelten die §§ 90, 93 und 97 Abgabenordnung (AO), durch die den »Beteiligten« Mitwirkungspflichten auferlegt werden; zu den Beteiligten zählen nach § 78 AO auch alle diejenigen, an die die Finanzbehörde den Verwaltungsakt richten will oder gerichtet hat. Nach § 90 Abs. 2 AO obliegt den »Beteiligten« die Pflicht zur Aufklärung des Sachverhalts und zur Beschaffung der Beweismittel, wenn der Sachverhalt sich auf Vorgänge außerhalb der Bundesrepublik bezieht; in einem solchen Fall haben die Beteiligten »alle für sie bestehenden rechtlichen und tatsächlichen Möglichkeiten auszuschöpfen« und können sich nicht auf eine von ihnen selbst geschaffene Unmöglichkeit berufen. Oder man denke an § 46 Gesetz gegen Wettbewerbsbeschränkungen (GWB), nach dem die Kartellbehörde von »Unternehmen oder Vereinigungen von Unternehmen« und von den »nach Gesetz oder Satzung zur Vertretung berufenen Personen« Auskunft verlangen kann; auch hier umfaßt diese Vorschrift mündliche Auskunft sowie die Vorlegung von Urkunden. Die Verletzung solcher Pflichten wird in der Regel als Ordnungswidrigkeit bestraft.

Nach der Vorstellung des Gesetzgebers richten sich Bestimmungen dieser Art – wie alles Recht – vornehmlich an diejenigen, die er in erster Linie, vielleicht sogar ausschließlich im Auge hat, nämlich an Personen, die seiner Territorialhoheit unterworfen sind, zuweilen auch an seine eigenen Staatsangehörigen, wo immer sie sich aufhalten mögen. Aber der Wortlaut derartiger Bestimmungen ist so allgemein gefaßt, daß er sich auf alle Personen zu beziehen scheint, ohne Rücksicht auf ihren Aufenthalt oder ihre Staatsangehörigkeit. Die »Unternehmen« in § 46 GWB scheinen alle Unternehmen in der ganzen Welt zu sein. Diese Lesart geht jedoch gewiß zu weit. Der deutsche Steuergesetzgeber beabsichtigt auch sicherlich nicht, ein ausländisches Unternehmen wie etwa die Ford Corporation in Detroit mit ihrem gesamten Vermögen oder Gewinn in Deutschland zu besteuern. Man darf sogar davon ausgehen, daß er noch nicht einmal beabsichtigt, auf den Teil des Gewinns der Ford Corporation eine Steuer zu erheben, der aus Exporten nach der Bundesrepublik erwächst. Der Wortlaut bedarf also einer einschränkenden Auslegung. Sie kann nur dadurch gefunden werden, daß man die Reichweite von Bestimmungen der erwähnten Art am Völkerrecht mißt, sie also völkerrechtskonform auslegt. Im Steuerrecht hat man davon gesprochen, daß Konflikte zwischen inländischem und ausländischem Recht im Einzelfall »nach den Ermessensgrundsätzen der Verhältnismäßigkeit und der Zumutbarkeit« zu lösen sind[1]. Aber das ist im Grunde eine Redensart, die der Diskretion weiten Raum läßt und die Unsicherheit keineswegs beseitigt. Und wenn man § 46 GWB gemäß § 98 Abs. 2 GWB dahin auslegt, daß er überall, aber auch nur da Anwendung findet, wo sich ein Vorgang auf die Bundesrepublik auswirkt, so ersetzt man das ursprüngliche Problem mit einem anderen; denn es bleibt die Frage, welche der vielen in Betracht kommenden (direkten oder indirekten) Wirkungen rechtlich relevant sind. Auch darauf kann wiederum nur das Völkerrecht Antwort geben; denn das (auf bedauerlichen amerikanischen Einfluß zurückgehende) Wirkungsprinzip als solches ist völkerrechtlich von zweifelhafter Rechtmäßigkeit[2] und kann nur dadurch rechtmäßig werden, daß man ihm einen völkerrechtskonformen Inhalt gibt.

Die Grenzen staatlicher Aufklärungsansprüche werden durch die völkerrechtliche Jurisdiktionslehre gezogen[3]. Diese hat bekanntlich zwei Aspekte. Danach fehlt dem Staat das Recht zur Handlung (*enforcement jurisdiction*) im Ausland, in jedem Fall insoweit, als es sich nicht um seine eigenen Staatsangehörigen handelt[4]. Genau so wie ohne Zustimmung des territorialen Souveräns der Staat durch seine Beamten im Ausland keine Verhaftungen oder Durchsuchungen vornehmen oder sein Notar keine Urkunden aufnehmen kann, so ist es ihm versagt, eine Anordnung zur Vorlegung von Urkunden etwa in der Weise zuzustellen, daß er einen eingeschriebenen Brief an den Ausländer im Ausland richtet; das hat ein amerikanisches Gericht gerade vor kurzem in einer ungewöhnlich sorgfältig begründeten, auf Völkerrecht ge-

---

[1] Dazu *Tipke-Kruse*, Abgabenordnung (10. Aufl. 1980), § 90 Anm. 6. Daß bei Auslandsbeziehungen erhöhte Mitwirkungspflichten bestehen, entspricht alter und ständiger Rechtsprechung. Vgl. zuletzt Entscheidungen des Bundesfinanzhofs 118, 553, 557 (21. 1. 1976).

[2] Zu diesem Problem vgl. insbes. *Meessen*, Völkerrechtliche Grundsätze des Internationalen Kartellrechts (1975), und neuerdings *Meng*, ZaöRV 41 (1981), S. 469 mit zahlreichen Hinweisen auf neuere Literatur.

[3] Dazu grundsätzlich *Mann*, The Doctrine of Jurisdiction in International Law, RdC 111 (1964 I), S. 1, oder *id.*, Studies in International Law (1973), S. 1.

[4] Darüber *Mann*, RdC 111 (1964 I), S. 110 ff.

stützten, wegweisenden Entscheidung ausgesprochen, die sich auf die durch eingeschriebenen Brief erfolgte Zustellung eines Vorlegungsbefehls durch die amerikanische Kartellbehörde an eine französische Gesellschaft bezog[5]. Wesentlich schwieriger dagegen liegen die Fälle, in denen der Staat nicht im Ausland handelt, sondern in denen allein die Regelungsbefugnis (*legislative jurisdiction*) gegenüber ausländischen Tatbeständen in Frage steht. Danach ist der Ausgangspunkt, daß jene Aufklärungsansprüche auch im Inland nur in dem Umfang geltend gemacht werden können, ja bestehen, in dem der zu beurteilende Tatbestand in den Rahmen der legislativen Jurisdiktion des Staates fällt, d.h. von der Gesetzgebung des Staates so erfaßt werden kann und soll, daß in die inneren Angelegenheiten fremder Staaten nicht eingegriffen, ihre Hoheitsgewalt nicht verletzt wird. Diese Hoheitsgewalt ist – wie diejenige aller Staaten – grundsätzlich territorial begrenzt, wenn auch Ausnahmen, wie etwa das strafrechtliche Weltrechtsprinzip, vorkommen und zulässig sind. Man hat davon gesprochen, daß es die Binnenbeziehung ist, die es dem Staat gestattet, seine Regelungsbefugnis über seine territoriale Souveränität hinaus zu erstrecken. Was damit gemeint ist, ist ein sachgemäßes rechtliches Interesse an der Regelung des Tatbestands, ein Anknüpfungspunkt, der *reasonable* erscheint[6], der genügend eng ist, um es dem Staat zu erlauben, ohne Verletzung fremder Hoheitsrechte, den Tatbestand zu regeln, wobei ein bloßes

---

[5] *Federal Trade Commission* v. *Compagnie de Saint Gobain - Pont à Mousson*, 636 F. 2d 1300 (1980). Die Entscheidung, die mit den Erwägungen von *Mann*, a.a.O., S. 118, im Einklang steht, ist schlechthin überzeugend. Sie legt mit Recht auf die Tatsache Gewicht, daß die französische Regierung gegen die Zustellung durch eingeschriebenen Brief protestiert hatte. Der Europäische Gerichtshof entschied im entgegengesetzten Sinn (*Geigy* gegen *EWG Kommission*, Amtliche Sammlung 18 [1972] 825, 826, Abschnitt c). Wie der Entscheidung erster Instanz in dem amerikanischen Fall zu entnehmen ist (493 F. Supp. 286 (1980), Anm. 7), hat Pierre Mayer die Entscheidung des Europäischen Gerichtshofs als »unhaltbar« bezeichnet. Sie beruht in der Tat auf einem Fehlschluß, auf einem *non sequitur* einzigartigen Ausmaßes: »Den Akten ist zu entnehmen, daß die Behörden des vorliegend beteiligten Drittstaates [Schweiz] gegenwärtig keine praktikable Möglichkeit der Zustellung auf dem Hoheitsgebiet dieses Staates anerkennen, die sie nach innerstaatlichem Recht als wirksam ansehen. Der Gemeinschaft kann daher [*sic*] nicht unter Berufung auf das Völkerrecht die Befugnis abgesprochen werden, das Erforderliche zu tun, um die Wirksamkeit ihrer Maßnahmen ... auch dann zu gewährleisten, wenn der Urheber dieser Handlungen seinen Sitz in einem Drittstaat hat«. Zu anderen Aspekten dieses für den Juristen geradezu beschämenden Urteils siehe *Mann*, International and Comparative L.Q. 1973, S. 35.

[6] Dieser Ausdruck, hier mit »sachgemäß« übersetzt und der Sache nach mit den 1964 gemachten Vorschlägen übereinstimmend, wird nunmehr von dem Tentative Draft eines revidierten, vom American Law Institute herausgegebenen Foreign Relations Law of the United States (1981) verwendet: Zwar hält § 402 leider daran fest, daß Regelungsbefugnis besteht, wenn »conduct outside its territory has or is intended to have substantial effect within its territory«. Jedoch hat ein Staat keine Regelungsbefugnis in Bezug auf »conduct, relations, status or interests of persons or things having connections with another state or states when the exercise of such jurisdiction is unreasonable« (§ 403).

wirtschaftliches Interesse gewiß nicht ausreicht[7]. In keinem Fall aber kann seine Durchsetzungsbefugnis weiter reichen als seine Regelungsbefugnis: Er kann, um zu dem obigen Beispiel zurückzukehren, keine Anordnung an die amerikanische Ford Corporation richten, wenn er nicht den zugrunde liegenden Tatbestand zu regeln befugt ist, – er kann keine Anordnung auf Vorlegung von Urkunden erlassen, wenn er nicht den Vorgang, auf den sich die Urkunden beziehen, besteuern kann[8]. Im Ergebnis kann also die Aufklärungsanordnung aus doppeltem Grund unzulässig sein: einmal weil sie unter Überschreitung der *Handlungs*befugnis des inländischen Staates und unter Verletzung der ausländischen Gebietshoheit einem Ausländer im Ausland zugestellt wird; sodann weil sie zwar im Inland erlassen und zugestellt wird, aber auf einer Bestimmung beruht, die über die inländische *Regelungs*befugnis hinausgeht und einen allein im Ausland verankerten Tatbestand erfaßt. Um Mißverständnisse zu vermeiden, sei nur die Selbstverständlichkeit hinzugefügt, daß gegenüber dem inländischen Unternehmen, mag es Haupt- oder Zweigniederlassung oder Tochtergesellschaft sein, der inländische Gesetzgeber völlig frei ist[9], soweit nicht Art. 25 Grundgesetz (GG) eingreift, und daß das ausländische Unternehmen, das im Inland in relevantem Zusammenhang tätig wird, insoweit der inländischen Hoheitsgewalt unterworfen ist[10].

Man mag einwenden, daß eine Jurisdiktionslehre des soeben skizzierten Inhalts völkergewohnheitsrechtlich auf zu schwachen Füßen steht, weil sie nicht von einer umfassenden und eindeutigen Staatenpraxis getragen werde. Darauf kann an dieser Stelle nur zweierlei erwidert werden. Einmal gibt es, wenn man die gesamte Staatenpraxis überblickt, wenn man also alle in Betracht kommenden Gebiete wie Steuer-, Kartell-, Straf-, Wertpapierhandels-, Schiffahrts-, zivilprozessuales Zuständigkeits-, aber auch internationales Privatrecht (um nur die wichtigsten Gebiete zu nennen) überblickt, ein äußerst

---

[7] Das ist der Hauptgrund, aus dem im Kartellrecht das Wirkungsprinzip im amerikanischen Sinn, das der in der vorigen Anm. erwähnte Vorentwurf aufrechterhält, kein zulässiger Anknüpfungspunkt ist. Vgl. dazu unten.

[8] Zu dem im Text gebildeten Fall unzulässiger Besteuerung vgl. *Mann*, RdC 111 (1964 III), S. 103, und zu der Abgrenzung von Regelungs- und Durchsetzungsbefugnis *ibid.*, S. 131.

[9] Deshalb ist die amerikanische Entscheidung *Securities and Commission* v. *Minas de Artemisa S.A.*, 50 F.2d 215 (1945), in keiner Weise bemerkenswert; denn die beklagte mexikanische Gesellschaft war angesichts der tatsächlichen Feststellungen in Wahrheit in den Vereinigten Staaten ansässig.

[10] Wenn deshalb eine schweizerische Bank an der New Yorker Börse Aktien kauft, so gilt für sie nicht nur das Recht der New Yorker Börse, sondern im Rahmen der Tätigkeit in New York auch das Recht der Vereinigten Staaten. Das ist der einfache Gesichtspunkt, der es der Bank versagt, sich auf ihr Bankgeheimnis nach schweizerischem Recht zu berufen. In dem Fall *Securities & Exchange Commission* v. *Banca della Svizzera Italiana*, F. Supp. (1981) kam der Richter zwar zu diesem Ergebnis, aber stützte es darauf, daß er »carefully balanced the interests at stake« und »our vital national interest« die Offenlegung verlangte. Nirgends hat der *ordre public* so extreme Formen angenommen wie in Amerika. Vgl. dazu unten unter IV.

umfangreiches Beweismaterial. Sodann handelt es sich um nichts anderes als die Anwendung traditioneller und alteingesessener Grundsätze auf eine spezielle Fallgruppe: Die Lehre von der Gebietshoheit des Staates, die sowohl für den handelnden wie für den duldenden Staat gilt, erfordert allseitige Rücksichtnahme und Nichteinmischung in einem Sinn und einem Umfang, dessen Ermittlung im gegebenen Zusammenhang Sache der Wissenschaft und der Rechtsprechung ist. An dem Ausgangspunkt: Gebietshoheit wird niemand zweifeln; seine Bedeutung und Anwendung in einem speziellen Fragenkomplex ist Sache der Rechtsfindung. Dabei kann es kaum einen Zweifel über den einfachsten, wenn auch unrealistischen Fall geben: Wenn man sich ein deutsches Gesetz vorstellt, das eine Steuer von allen ausländischen Gesellschaften zu erheben beabsichtigt, die in die Bundesrepublik exportieren, so ist es ganz offensichtlich, daß es völkerrechtswidrig und selbst in Deutschland unbeachtlich wäre (Art. 25 GG). Keinem Staat steht es zu, das Verhalten von Ausländern im Ausland zu regeln. Man wende nicht ein, ein solches Gesetz sei *brutum fulmen*, undurchführbar und sinnlos; denn die besteuerte ausländische Gesellschaft mag Vermögen in der Bundesrepublik haben, so daß die Steuerbehörde ihre Forderung durchsetzen könnte. Die Gültigkeit oder Ungültigkeit des Gesetzes ist also höchst relevant. Ist es ungültig, so ist auch ein dem staatlichen Aufklärungsanspruch entsprechender Vorlegungsbefehl unbeachtlich. Wäre das Gesetz gültig, so würde dasselbe von einer Aufklärungsanordnung gelten.

Nun liegen jedoch die in Betracht kommenden Sachverhalte nicht immer so klar und eindeutig. Viele Abwandlungen sind denkbar und haben in der Praxis zu Schwierigkeiten Anlaß gegeben. Ihnen wird man sich nunmehr zuzuwenden haben.

II

Der einfachste Tatbestand liegt dann vor, wenn die inländische Behörde einer ausländischen Gesellschaft, die im Inland weder eine Zweigniederlassung noch eine Tochtergesellschaft unterhält, Befehle über ihr Verhalten im Ausland erteilt: Die deutsche Steuerbehörde ordnet an, daß eine ausländische Gesellschaft ihre Bücher vorzulegen hat. Eine solche Maßnahme erscheint unzulässig, »weil der deutsche Gesetzgeber keine Kompetenz hat, das Tun und Lassen von Ausländern im Ausland zu regeln« [11].

Die Frage der Vereinbarkeit der inländischen Anordnung mit der ausländischen Gebietshoheit wäre nur dann anders zu beurteilen, wenn es möglich wäre, dem kartellrechtlichen Wirkungsprinzip im Sinn der amerikanischen

---

[11] So neuestens Finanzgericht Düsseldorf, Recht der Internationalen Wirtschaft (RIW/AWD) 1981, S. 717.

Rechtsprechung allgemeine Bedeutung beizulegen, ja es auch nur für das Kartellrecht anzuerkennen, – ob die deutsche Rechtsprechung zu § 98 Abs. 2 GWB die gleiche Richtung nehmen wird, ist vorläufig ungewiß und hoffentlich unwahrscheinlich. Nach amerikanischer Auffassung erfaßt die amerikanische Kartellgesetzgebung jede Absprache zwischen Ausländern im Ausland, die sich in Amerika rein wirtschaftlich auswirkt, es sei denn die Ausländer weisen nach, daß diese Auswirkung unbeabsichtigt ist [12]. Seit etwa 30 Jahren haben amerikanische Gerichte und Verwaltungsbehörden sich aufgrund dieses Satzes für berechtigt gehalten, Ausländern im Ausland die Vorlegung von Urkunden aufzugeben, die sich im Ausland befanden und sich auf Vorgänge im Ausland bezogen. Zahlreiche Staaten haben gegen diese Maßnahmen auf diplomatischem Weg protestiert. Zahlreiche Staaten, darunter auch die Bundesrepublik Deutschland, haben gesetzgeberische Maßnahmen ergriffen, um es ihren Staatsangehörigen zu verbieten, die amerikanischen Befehle zu befolgen [13]. Viele Belege tun dar, daß die Staatengemeinschaft die amerikanische Haltung für völkerrechtswidrig hält. Ein Vorfall aus jüngster Zeit ist so aufschlußreich, daß er dargestellt zu werden verdient.

Im Jahr 1977 war Großbritanniens höchstes Gericht, das House of Lords, mit einem Antrag der Westinghouse Electric Corporation befaßt, aufgrund dessen die englische Gesellschaft Rio Tinto Zinc Corporation Ltd. und ihre Konzerngesellschaften Urkunden gemäß einem Gesetz vorlegen sollten, das in Ausführung des Haager Übereinkommens vom 18. März 1970 über die Beweisaufnahme im Ausland in Zivil- und Handelssachen erlassen worden war. Ohne auf die Komplikationen des Falles und die zahlreichen Entscheidungen und Zwischenfälle einzugehen, zu denen er sowohl in England wie in Amerika geführt hat, ist es hier lediglich relevant, darauf hinzuweisen, daß vor dem House of Lords der englische Staat vertreten durch den Attorney-General (Generalanwalt) intervenierte, um dem Antrag der amerikanischen Gesellschaft entgegenzutreten; der Attorney-General – und dies muß ganz

---

[12] Das ist der wesentliche Inhalt des Prinzips, das durch den Fall *United States* v. *Aluminium Company of America*, 148 F.2d 416 (1945), in die amerikanische Rechtsprechung eingeführt wurde.

[13] Vgl. dazu unten Abschnitt IV. Es ist genugtuend, in dem in Anm. 6 erwähnten amerikanischen Vorentwurf Hinweise auf einen Teil der durch die amerikanische Praxis hervorgerufenen internationalen Proteste zu finden. Sie schließen mit der vielleicht allzu optimistischen Feststellung ab: »In response to these and similar reactions, the United States has begun to modify its assertions of jurisdiction in some areas« (S. 91, 92). Es ist ferner erfreulich, bei *Atwood & Brewster*, Antitrust and American Business Abroad (2. Aufl. 1981), § 6.21, ein gewisses Verständnis für die Unangemessenheit, wenn auch nicht die Völkerrechtswidrigkeit, der amerikanischen Praxis zu finden. Im § 18.04 erkennen sie überdies an: »the arguments for adjustment seem to us compelling«. Es ist ein erster Schritt in diese Richtung, wenn in § 18.19 Rücksichtnahme auf die Interessen anderer Staaten verlangt wird: »Many would argue that international norms require more, but few would claim that they require any less«. Die erstere Gruppe umfaßt fast die ganze Welt außerhalb der Vereinigten Staaten.

besonders hervorgehoben werden – konnte erklären, daß Kanada, Australien und Frankreich seiner Intervention beitraten. Im Laufe der umfang- und aufschlußreichen Erklärung, die zur Ablehnung der amerikanischen Anträge führte, wurde zur amerikanischen Kartellgesetzgebung formell wie folgt Stellung genommen [14]:

> (1) The anti-trust laws of the United States of America (»U.S.«) should not provide jurisdiction for U.S. courts to investigate non-U.S. companies and non-U.S. individuals in respect of their actions outside the U.S., although the U.S. claims to have such jurisdiction. (2) For the purposes of United Kingdom sovereignty the U.K. does not recognise any such investigation as having any validity or as being proper. (3) The matters set out above are rendered a fortiori by virtue of the penal character of the anti-trust laws. (4) Any use of the U.S. anti-trust laws for the above purposes, except with the authority of the U.K., is an invasion of and prejudicial to U.K. sovereignty. (5) In the instant case no such authority exists.

Diese Haltung, die England seit vielen Jahren gegenüber den Vereinigten Staaten wie auch – vor 1973 – gegenüber der EWG vertreten hatte [15], bedeutet die Zurückweisung der Wirkungslehre amerikanischer Prägung. Ihre Völkerrechtswidrigkeit wird klar zum Ausdruck gebracht, richtig begründet und erhält durch den Beitritt von drei einflußreichen Staaten wesentliche Unterstützung, zumal bekannt ist, daß auch viele andere Staaten die amerikanische Praxis ablehnen.

Es erscheint jedoch unwahrscheinlich, daß die amerikanische Praxis sich ändern wird. Auf der einen Seite können die Vereinigten Staaten sich auf gewisse diplomatische Erfolge berufen, wie insbesondere ein weitgehend unbekanntes und unbeachtetes Abkommen, das im Jahre 1976 die Vereinigten Staaten und die Bundesrepublik – leider – geschlossen haben und das auf Austausch von Information hinausläuft [16]. Auf der anderen Seite darf man für den Augenblick der Tatsache kein großes Gewicht beilegen, daß in zwei neueren Entscheidungen die Gerichte Gesichtspunkte zusammengefaßt haben, die für die Frage der Ausübung der Gerichtsbarkeit zu dem Test der »beabsichtigten Wirkung« hinzukommen sollen [17]. Aber dabei handelt es sich um kaum mehr als Worte, die bisher ohne Einfluß geblieben sind. Das wurde klar, als in einem der Westinghouse Electric Corporation Fälle die Re-

---

[14] (1978) A.C. 547, 591, 592.

[15] Vgl. die Note, die bei *E. Lauterpacht*, British Practice in International Law 1967, 58–67; Report of the 54th Conference of the International Law Association (The Hague 1970), S. 185–188, und *Brownlie*, Principles of International Law (3. Aufl. 1979), S. 310–313 abgedruckt ist.

[16] BGBl. 1976 II, 1711.

[17] *Timberlane Lumber Co.* v. *Bank Bank of America*, 549 F.2d 597 (1976); *Mannington Mills Inc.* v. *Congoleum Corporation*, 595 F.2d 1287 (1979).

gierungen von Australien, Kanada, Südafrika und Großbritannien intervenierten, um die amerikanische Gerichtsbarkeit zu bestreiten, und ein drittes Gericht die unveränderte Reichweite des Wirkungsprinzips im hergebrachten Sinn hervorhob[18]. Die Geschäftswelt wird sich also auf die fortdauernde Anwendung dieses von fast allen Staaten abgelehnten und allein deshalb völkerrechtswidrigen Prinzips und damit auf die Gefährdung ihres amerikanischen Vermögens einzustellen haben, – eine Gefährdung, die selten so folgenreich war wie in dem *Westinghouse*-Fall, in dem das amerikanische Vermögen der betroffenen ausländischen Gesellschaften jahrelang gesperrt war, in dem ungeheure Prozeßkosten aufgewendet wurden und der schließlich mit einem Vergleich endete, der den nichtamerikanischen Gesellschaften rund $ 40 Millionen kostete. Im Lauf seines Urteils hat Lord Wilberforce[19] mit Recht festgestellt: »It is axiomatic that in anti-trust matters the policy of one State may be to defend what it is the policy of another State to attack«. Die völkerrechtswidrige Praxis der Vereinigten Staaten wird aller Wahrscheinlichkeit nur dadurch beseitigt werden können, daß die nichtamerikanische Welt sich nicht auf die – aussichtslose – Verteidigung beschränkt, sondern deutliche Gegenmaßnahmen ergreift und überdies die Verurteilung der Vereinigten Staaten durch ein internationales Gericht größter Autorität zu erreichen versucht. Wäre die Zuständigkeit gegeben, so könnte eine Verurteilung schwerlich ausbleiben; denn die bloße wirtschaftliche Auswirkung im Inland eines Verhaltens von Ausländern im Ausland kann unter keinem Gesichtspunkt ein Anknüpfungspunkt für die inländische Regelungsbefugnis sein. Man braucht sich nur die Folgen einer entgegengesetzten Auffassung für den internationalen Wirtschaftsverkehr vorzustellen, und man wird sie ablehnen müssen: man nehme den heute durchaus denkbaren Fall an, in dem etwa in den Vereinigten Staaten ein Zinssatz von mehr als 10% im Jahr verboten ist. In der Bundesrepublik werden 20% bezahlt. Deshalb werden große Geldmengen (vor allem arabischen Ursprungs) von Amerika abgezogen und in die Bundesrepublik verlegt. Sollten amerikanische Banken berechtigt sein, von deutschen Banken Schadensersatz wegen Verletzung des amerikanischen Gesetzes zu verlangen?

III

So kommt man zu der allgemeinen Frage, ob es völkerrechtlich und damit für die Anwendung und Auslegung staatlicher Anordnungen des Inlands über das Verhalten einer ausländischen Gesellschaft im Ausland ausreicht, daß sie im Inland eine Zweigniederlassung oder Tochtergesellschaft hat oder

---

18 *Westinghouse Electric Corporation* v. *Rio Algom Ltd.*, 617 F.2d 1248 (1980).

19 *In re Westinghouse Uranium Contract,* (1978) A.C. 547, 617.

daß das Mitglied eines Organs, etwa des Vorstands oder Aufsichtsrats, oder auch nur ein Angestellter sich im Inland aufhält und die Zustellung an die inländische Organisation erfolgt ist.

1. Was die beiden ersteren Fälle angeht, so wäre eine Maßnahme gegen das inländische Unternehmen (Zweigniederlassung oder Tochtergesellschaft) nur dann zulässig, wenn der Anspruch gegen das ausländische Unternehmen und das darauf gestützte Aufklärungsverlangen völkerrechtlich gerechtfertigt ist. Was nicht genügt, ist die bloße Tatsache, daß im Inland Vermögen vorhanden ist, das für die Schulden des ausländischen Unternehmens haftet. Wenn es sich um Ansprüche handelt, die gegen das inländische Unternehmen gerichtet und zu deren Verfolgung im Ausland liegende Urkunden notwendig oder erwünscht sind, so ist es Sache des inländischen Schuldners, sie zu beschaffen; völkerrechtliche Grenzen bestehen hier nicht. Deshalb ist z.B. dem Bundesfinanzhof in vollem Umfang darin beizutreten, wenn er in den Preisen, die eine deutsche GmbH ihrer schweizerischen Muttergesellschaft berechnet, eine versteckte Gewinnausschüttung sieht und es der deutschen GmbH zur Last legt, daß Urkunden nicht vorgelegt werden, die sich angeblich im Besitz der Muttergesellschaft befinden und deren Herausgabe diese verweigert[20]. Hätte dagegen die deutsche Steuerbehörde die schweizerische Muttergesellschaft wegen empfangener Gewinne besteuert und wäre darin eine Überschreitung der deutschen Steuerhoheit zu sehen (was wahrscheinlich nicht behauptet werden könnte), so wären Zwangsmaßnahmen gegen die deutsche Tochter unzulässig.

Diese Meinung steht jedoch im Widerspruch mit einer Bemerkung des Kammergerichts in einer Kartellsache sowie mit der amerikanischen Praxis. Das Kammergericht[21] war mit einem Verfahren gegen die deutsche Tochtergesellschaft der englischen British Petroleum Ltd. befaßt. In seinem Beschluß führte es aus:

> »Da die BP einen derartigen, auf den deutschen Inlandsmarkt einwirkenden Mißbrauch gegebenenfalls mitverursacht hätte, kann die Kartellbehörde gemäß § 46 i.V.m. § 98 II GWB die erforderlichen Auskünfte auch von dem ausländischen Konzernunternehmen – eventuell durch die Beschwerdeführerin – verlangen«.

Das ist die reine Theorie der wirtschaftlichen Auswirkung. Die deutsche Kartellbehörde kann von der englischen Gesellschaft nichts verlangen. Sie kann von der deutschen Tochter Auskünfte verlangen und es ihr anlasten, wenn sie nicht erteilt werden. Aber sie hat keine Hoheitsgewalt in England. Ausgangspunkt der amerikanischen Lehre ist die von den meisten Staaten abgelehnte Lehre, nach der eine ausländische Gesellschaft, die in Amerika eine

---

[20] Bundessteuerblatt 1981 II, 482 (16. 4. 1980).

[21] Entscheidung vom 14. 5. 1974, Wirtschaft und Wettbewerb, Entscheidungssammlung zum Kartellrecht, 1974 I, S. 1467.

Zweigniederlassung oder Tochtergesellschaft hat, selbst in Amerika präsent ist und der Zuständigkeit der amerikanischen Gerichte und Verwaltungsbehörden unterliegt[22]. Ist das der Fall, so zieht man den weiteren, noch bedenklicheren Schluß, daß lediglich wegen der Existenz der »*in personam* jurisdiction« die Vorlegung von im Ausland befindlichen Urkunden angeordnet werden kann[23]. Von dieser – angeblichen – Befugnis wird in aller Regel Gebrauch gemacht, es sei denn, daß schwerwiegende, vor allem strafrechtliche Sanktionen im Ausland der Erfüllung der Aufklärungspflicht entgegenstehen und nicht aus der Welt geschafft werden können oder ein international anerkanntes Mittel anderer Art, wie etwa ein Rechtshilfeersuchen, zur Verfügung steht. Der Gesichtspunkt, der entscheidend gegen die amerikanische Praxis spricht, ist die völlige Verkennung der Grenzen staatlicher Regelungsbefugnis. Sie ist gewiß nicht allein mit dem persönlichen Gerichtsstand gegeben, vor allem wenn dieser seinerseits auf einem denkbar schwachen Fundament beruht. Man kann nicht ohne Verletzung ausländischer Souveränität einem Ausländer ein Verhalten im Ausland vorschreiben, wenn nicht zwischen dem vorgeschriebenen Verhalten und dem Inland eine enge rechtliche Beziehung besteht[24].

2. Daß der bloße Aufenthalt eines Organs der ausländischen Gesellschaft im Inland dem inländischen Gesetzgeber nicht die Befugnis zur Regelung der Angelegenheiten einer ausländischen Gesellschaft gibt, sollte kaum eines Wortes bedürfen. Aber auch hier gibt die amerikanische Rechtsprechung zu Bedenken Anlaß.

Auf der einen Seite gibt es eine erfreuliche Entscheidung des Berufungsgerichts in Washington, D.C.: Wie erwähnt, erklärte sie die Zustellung einer

---

[22] In Wahrheit ist nach amerikanischer Auffassung für die Zuständigkeit sehr viel weniger erforderlich: Es genügt, wenn die ausländische Gesellschaft *is doing business* – ein Begriff, der außerordentlich weit ausgelegt wird, so weit, daß allein darin oft eine Völkerrechtswidrigkeit zu sehen ist, weil die Staatenpraxis die amerikanische Auslegung verwirft. Ein gutes Beispiel aus neuerer Zeit ist der Fall *Zenith Radio Corporation* v. *Matsuahita Elec Indus Co.*, 402 F. Supp. 262, 327, wo die Anknüpfungspunkte aufgeführt werden. In der Praxis ist auch eine Muttergesellschaft in den Vereinigten Staaten tätig, wenn sie dort eine Tochtergesellschaft hat, mit deren Hilfe ihre Waren »into the stream of commerce« geraten. Vgl. *World-Wide Volkswagen Corporation* v. *Woodson*, 444 U.S. 286, 297 (1980). *Mann*, RdC 111 (1964 III), S. 60 ff., hat es leider unterlassen, diesem Problem nachzugehen.

[23] »It is no longer open to doubt that a federal court has the power to require the production of documents located in foreign countries if the court has in *personam* jurisdiction of the person in possession or control of the material«. *United States* v. *First National City Bank*, 396 F.2d 897 (1968) mit weiteren Nachweisen.

[24] Man erwähnt in diesem Zusammenhang häufig die Entscheidung *Ings* v. *Ferguson*, 282 F.2d 149 (1960), aber sie ist nicht einschlägig. In einem Prozeß zwischen A und B wurde der New Yorker Filiale einer kanadischen Bank die Aufforderung (subpoena) zugestellt, in Kanada befindliche Urkunden der Hauptniederlassung zu beschaffen und vorzulegen. Das Gericht entschied, daß dies im Wege des Rechtshilfeverfahrens zu erfolgen habe.

Anordnung der Federal Trade Commission durch nach Paris gerichteten eingeschriebenen Brief für unzulässig. Darüber hinaus hatte jedoch die Kartellbehörde die Anordnung nicht nur dem Anwalt der französischen Gesellschaft in Washington, sondern auch ihrem »general delegate in the United States« sowie dessen Tochter in New York zugestellt. Auch diese Zustellungen wurden für ungültig erklärt[25]. Auf der anderen Seite gibt es jedoch den für manche geradezu grotesken Fall *United States* v. *Field*[26]. Field, ein kanadischer Staatsangehöriger, war Leiter einer Bank auf den britischen Cayman Inseln in Westindien. Er befand sich im Warteraum des Flughafens in Miami, um nach Cayman, seinem Wohnsitz, zurückzufliegen, als ihm die Anordnung zugestellt wurde, im Rahmen einer strafrechtlichen Untersuchung gegen Dritte wegen angeblicher Steuerdelikte über Konten und Geschäfte seiner Bank auszusagen. Nach dem Recht von Cayman stellte die bloße Tatsache der Aussage über Bankgeheimnisse für Field die Gefahr einer Strafverfolgung dar. Das Berufungsgericht für den 5. Circuit wies seine Beschwerde zurück. Field, dem die Anordnung gewiß ordnungsgemäß in Miami zugestellt worden war, sollte als Organ seiner Bank aussagen. Die – vom Gericht nicht gestellte – entscheidende Frage war deshalb die, ob den Vereinigten Staaten ein Recht auf Aufklärung gegen die Cayman Bank zustand. Eine verneinende Antwort ist unabweisbar. Wenn man sich ein amerikanisches Gesetz vorstellt, in dem ausländische Banken im Ausland verpflichtet werden, amerikanischen Strafuntersuchungsbehörden Auskunft zu geben, so wird die Völkerrechtswidrigkeit des Vorgehens und der Entscheidung offensichtlich. Die Entscheidung sollte für viele ein warnendes Beispiel sein: sie zeigt, was jedem, der sich in Amerika auch nur vorübergehend aufhält, passieren kann.

IV

Es bleibt schließlich der umgekehrte Fall, in dem die Hauptniederlassung (oder Muttergesellschaft) sich im Inland, eine Filiale oder Tochtergesellschaft im Ausland befindet: Der Fall eines sich im Inland aufhaltenden Organs des ausländischen Unternehmens liegt genau so wie der soeben behandelte Tatbestand und bedarf keiner gesonderten Erörterung.

1. Was den Fall der ausländischen Zweigniederlassung angeht, so kann grundsätzlich kein Zweifel daran bestehen, daß dem inländischen Staat sowohl die Regelungs- wie die Durchsetzungsbefugnis zusteht, ja völkerrechtliche Beschränkungen schwerlich bestehen; denn Haupt- und Zweigniederlassung bilden eine Einheit. Gewiß gilt dies nicht durchweg. So hat die Filiale einer Bank eine gewisse Selbständigkeit, indem wahrscheinlich die Hauptnie-

[25] Oben Anm. 5.
[26] 532 F.2d 404 (1976).

derlassung nicht für die Depositen der Filiale haftet[27] und indem ein gegen die Hauptniederlassung erlassener Arrest nicht die ausländische Filiale bindet[28]. Dennoch wird für den Zweck des staatlichen Aufklärungsanspruchs wohl von der Einheit des Unternehmens auszugehen sein. In einem solchen Fall hat ein amerikanisches Gericht deshalb mit Recht erklärt: »Thus the task before us, as Citibank concedes, is not one of defining power but of developing rules governing the proper exercise of power«[29]. Der Fall betraf das Verlangen, im Rahmen einer kartellrechtlichen Untersuchung Urkunden vorzulegen, die sich im Besitz der Frankfurter Filiale der First National City Bank befanden und die Geschäftsgebahrung der C. F. Böhringer & Söhne GmbH betrafen, und ähnliche Fälle sind oft entschieden worden. Die Gerichte haben dem Aufklärungsverlangen fast immer stattgegeben. Auf die Gesichtspunkte, die die Ausübung ihres Ermessens bestimmt haben, wird zurückzukommen sein.

2. Die Dinge liegen wesentlich schwieriger, wenn es sich um ein Aufklärungsverlangen handelt, das zwar an die Muttergesellschaft gerichtet ist, sich aber auf Urkunden oder mündliche Auskünfte bezieht, die allein die ausländische Tochtergesellschaft erteilen kann und soll. Wäre das Aufklärungsverlangen ausschließlich an die inländische Muttergesellschaft gerichtet, so kann seine Gültigkeit nicht am Völkerrecht gemessen werden. Wäre es ausschließlich an die ausländische Tochtergesellschaft gerichtet, so wäre es völkerrechtlich unzulässig. Daß – nach deutscher und richtiger – Auffassung die Muttergesellschaft eine Mitwirkungspflicht intensivsten Umfangs hat, kann nicht bezweifelt werden, sie darf jedoch nicht überspannt werden.

In den Vereinigten Staaten ergibt sich eine besondere Schwierigkeit aus der Tatsache, daß nach amerikanischer Auffassung eine 100%ige Tochterge-

---

[27] Die amerikanischen Entscheidungen sind in der aus anderen Gründen aufgehobenen Entscheidung *United States* v. *First National City Bank*, 321 F.2d 14; 325 F.2d 1020 (1964), zusammengestellt. Anders jedoch *Vishipco Line* v. *Chase Monhattan* Bank, 660 F.2d 854 (1981). Vgl. unten Anm. 28. Der im Text aufgestellte Satz gilt in den meisten Ländern, insbesondere in England und Amerika nicht für Versicherungsgesellschaften.

[28] Vgl. für England Lord Denning in *Power Curber International Ltd.* v. *National Bank of Kuwait* (1981), 1 W.L.R. 1233, 1241. Es ist also unmöglich, der Entscheidung des Supreme Court der Vereinigten Staaten in dem Fall *Omar* zuzustimmen: *United States* v. *First National City Bank*, 379 U.S. 378 (1965). Omar S.A., eine Aktiengesellschaft in Uruguay, hatte ein Bankkonto bei der Filiale der Beklagten in Montevideo. Der amerikanische Fiskus glaubte Steueransprüche gegen Omar zu haben und erwirkte einen Arrest gegen die Hauptniederlassung der Bank in New York. Diese verteidigte sich mit der Begründung, Omar habe kein Vermögen in New York. »Respondent has actual practical control over its branches . . . The branch bank's affairs are, therefore, as much within the reach of the in personam order entered by the District Court as are those of the Home Office. Once personal jurisdiction of a party is obtained, the District Court has authority to order it to ›freeze‹ property under its control, whether the property is within or without the United States«.

[29] So die in Anm. 23 erwähnte Entscheidung.

sellschaft fast immer als Zweigniederlassung, als bloßer Teil der Muttergesellschaft gilt, der Schleier der Rechtspersönlichkeit also beiseite geschoben wird. Die dadurch hervorgerufene Problematik wird durch eine Entscheidung des Berufungsgerichts (9. Circuit) beleuchtet[30].

Im Jahr 1970 erwarb die Vetco Inc. sämtliche Aktien einer schweizerischen Aktiengesellschaft, Vetco AG. Im Jahr 1976 führten die Wirtschaftsprüfer Deloitte Haskins & Sells eine Betriebsprüfung des Konzerns durch. Sie kamen zu dem Ergebnis, daß Vetco Inc. zusätzliche Steuer schuldete und machten der amerikanischen Steuerbehörde freiwillige Anzeige. Diese übergab die Sache jedoch ihrer Strafabteilung. Sie erließ eine Anordnung, aufgrund deren die in dem Urteil als »Swiss records« bezeichneten, jedoch im einzelnen nicht aufgeführten Urkunden vorzulegen waren. Demgegenüber verteidigte sich Vetco Inc. mit der Berufung auf den der internationalen Rechtsprechung wohlbekannten § 273 des schweizerischen Strafgesetzbuchs. Danach wird derjenige mit Gefängnis bestraft, der Geschäftsgeheimnisse einer fremden amtlichen Stelle oder ihrem Agenten zugänglich macht. Das Gericht wies diese Verteidigung mit eingehenden, wenig durchsichtigen Erwägungen zurück, die das Gericht selbst wie folgt zusammenfaßte[31]:

> »We conclude that the United States has a powerful interest in obtaining the summoned documents and that Switzerland has a small interest in insisting that they not be produced. Appellants have not shown that production pursuant to a court order enforcing an IRS summons would impose hardship on them«.

Der Fall gibt zu vielen Fragen Anlaß. Ein Teil der Unsicherheit beruht auf der fehlenden Identifizierung der »summoned documents«. Es scheint, daß die Urkunden schweizerischen Tochtergesellschaften amerikanischer Unternehmen gehörten, die vermutlich schweizerische Vorstandsmitglieder hatten. Deshalb entstand die vom Gericht völlig übergangene Frage, wie die amerikanische Vetco Inc. die Urkunden rechtmäßig in ihren Besitz schaffen konnte. Sollte eine wahrscheinlich zum Scheitern verurteilte Klage gegen die schweizerischen Töchter erhoben werden? Sollten die schweizerischen Vorstandsmitglieder abberufen, Amerikaner ernannt und von diesen die Urkunden gestohlen werden? Sollte Vetco Inc. die schweizerischen Gesellschaften liquidieren? Indem das Gericht die Rechtspersönlichkeit der schweizerischen Gesellschaft ignorierte (und damit übrigens gegen die von ihm nicht erwähnte, jedoch völlig eindeutige Standardklausel von Art. II (1) (c) des amerikanisch-schweizerischen Doppelbesteuerungsabkommens verstieß)[32], schien es

---

[30] *United States* v. *Vetco Inc.*, 644 F.2d 1324 (1981).

[31] S. 1333.

[32] Danach gelten, wie in wahrscheinlich allen Doppelbesteuerungsabkommen, Tochtergesellschaften nicht als ständige Niederlassungen der Muttergesellschaft, sondern werden als selbständige Rechtssubjekte behandelt.

anzunehmen, daß die Organe von Vetco Inc. die Urkunden einfach an sich nehmen und nach Amerika transportieren konnten. Ähnliches schien von den Wirtschaftsprüfern erwartet zu werden, denn sowohl diese wie Vetco Inc. hatten $ 500 pro Tag zu zahlen, an dem die Urkunden nicht bereitgestellt waren.

Die Entscheidung beruht auf einer dem Völkerrecht widersprechenden Überspannung des Gedankens, daß die Rechtspersönlichkeit einer Tochtergesellschaft für Zwecke des amerikanischen Steuerrechts nicht anerkannt wird. Die Vereinigten Staaten können zwar Vetco Inc. so besteuern, als existiere die schweizerische Tochter nicht, aber der Beschluß des amerikanischen Gerichts läuft darauf hinaus, daß die Rechtspersönlichkeit der Tochter *in der Schweiz* ignoriert wird und irgend jemand in der Schweiz eine strafbare Handlung begehen muß. In der Anordnung einer solchen Maßnahme, die allen Regeln des internationalen Privatrechts widerspricht, liegt eine Mißachtung schweizerischer Hoheitsgewalt. Es mag sein, daß in der Entscheidung auch eine Verletzung amerikanischen Verfassungsrechts gesehen werden könnte, indem die Gebote des internationalen Privatrechts verkannt worden sind, aber das ist eine Frage, die hier weder relevant ist noch erörtert werden kann.

Zum Glück besteht Grund zu der Annahme, daß die Gerichte der Bundesrepublik den Geboten des Völkerrechts wesentlich verständnisvoller gegenüber stehen werden. So hat das Kammergericht der ausländischen Tochtergesellschaft eines deutschen Unternehmens den Erwerb einer ausländischen Betriebsstätte einer ausländischen Gesellschaft gestattet, obwohl dadurch eine »relativ unbedeutende weitere Wettbewerbsbeschränkung« im Inland (§ 98 Abs. 2 GWB) eintrat. Der Grund war der, daß eine Untersagung des Erwerbs eine Verletzung des Völkerrechts bedeutet, »dem Grundsatz sinnvoller Anknüpfung im internationalen Verwaltungsrecht und dem völkerrechtlichen Nichteinmischungsprinzip« widersprochen hätte. Das Prinzip der rein wirtschaftlichen Wirkung ist in diesem Konflikt unterlegen[33].

V

Die Entscheidungen, die zu dem im Vorstehenden behandelten Fallgestaltungen ergangen sind, insbesondere die amerikanischen Entscheidungen gehen im allgemeinen nicht den Weg, der hier empfohlen wurde. Vor allem in Amerika glaubt man, mit dem von *Grossfeld*[34] – einem der wenigen deut-

---

[33] Entscheidung vom 26. 11. 1980 (*Bayer*), RIW/AWD 1981, S. 406 (407) mit kritischer, aber nicht überzeugender Anmerkung von *Markert*.

[34] Inländische Auskunftspflichten und ausländische Auskunftsverbote, in: Festschrift für Karl Michaelis (1972), S. 118, 127 ff. Dort auch weitere Nachweise über die recht beschränkte deutsche Literatur. *Grossfeld* spricht wie auch manch anderer von den Geboten »internationaler Courtoisie« (*comity*). Das ist keine rechtlich faßbare Institution. Es gibt Gebote des Völkerrechts. Es gibt ferner Gebote des nationalen *ordre public*. Es gibt nichts Drittes. Vgl. auch *Meng* (Anm. 2), S. 508.

schen Bearbeiter des Problems – übernommenen Gedanken einer Interessenabwägung, eines *balancing of interests* weiterzukommen. Eine solche Argumentation führt natürlich stets zur Anwendung der *lex fori*, denn das eigene Interesse wird stets für überwiegend gehalten. So sehen die amerikanischen Gerichte – selbstverständlich – das Interesse des amerikanischen Fiskus als so vordringlich an, daß die Begehung einer strafbaren Handlung in der Schweiz keine Rolle spielt[35]: »The interest of the United States in the enforcement of the summons outweighs the contrary Swiss interest«. So sieht ein anderes amerikanisches Gericht das amerikanische Interesse an der Aufklärung von angeblichen Verstößen gegen das amerikanische Kartellgesetz als so überragend an, daß das deutsche Recht und Interesse wenigstens dann unberücksichtigt bleiben kann, wenn nicht die Bundesrepublik oder die Bundesbank interveniert[36]: Während Antitrustgesetze »have long been considered cornerstones of this nation's economic policies«, unterliegt das Bankgeheimnis in Deutschland »severe limitations«, so daß die Schlußfolgerung gerechtfertigt ist, »that the legislature did not value the public interest in bank secrecy as highly as it did the duty of secrecy of doctors and attorneys«. Solche unjuristischen, rein politischen Erwägungen mögen am Platz sein, wenn es sich um die Ausübung richterlichen Ermessens oder auch eines Rechts handelt. Sie sind wertlos, wenn es um die Existenz des Rechts geht. Diese kann nicht von Interessenabwägung abhängen.

Die unausweichlichen Folgen der in Amerika geübten Willkür sind Abwehrmaßnahmen der betroffenen Staaten, Maßnahmen, die keinen Zweifel an dem Interessenvorrang lassen und die auf eine höchst ungesunde und unerfreuliche Art von Kriegsführung mit juristischen Waffen hinauslaufen. Seit dem Ende der 50er Jahre gibt es kaum ein Land in der westlichen Welt, das nicht Gesetze mit dem Ziel erlassen hat, die Erteilung von Auskünften an ausländische (sprich: amerikanische) Gerichte oder Behörden zu unterbinden. Dazu gehören nicht nur die Bundesrepublik[37], sondern auch Holland, Schweden, Norwegen, Finnland, Belgien, Indien[38], ferner Kanada, Australien und Südafrika[39], vor allem aber auch Frankreich[40] und Großbritannien[41] mit völlig neuen und verfeinerten Gesetzen.

Dabei ist allerdings keineswegs sicher, daß diese Gesetze, welche die für die Entstehung einer Völkerrechtsregel so wichtige Staatenpraxis zum Aus-

---

[35] So die in Anm. 30 erwähnte Entscheidung auf S. 1323.

[36] So die Anm. 23 erwähnte Entscheidung.

[37] Verordnung vom 14. 12. 1966, BGBl. 1966 II, 1542.

[38] Nachweis bei *Mann*, Conflict of Laws and Public, RdC 132 (1971 I), S. 164.

[39] Nachweise bei *Meng* (Anm. 2), S. 492.

[40] Gesetz vom 16. 7. 1980, dazu *Herzog*, AJIL 75 (1981), S. 382, und *Névot*, Revue critique de droit international privé 70 (1981), S. 421.

[41] Protection of Trading Interests Act 1980, der ein Gesetz aus dem Jahr 1964 ablöst. Dazu *Lowe*, AJIL 75 (1981), S. 257. Die Bedenken, die *Meng* (Anm. 2), S. 495, 510 f., gegen Art. 6 des Gesetzes von 1980 erhebt, sind unbegründet. Sie übersehen, daß es sich um eine Retorsionsmaßnahme gegen Völkerrechtsverletzungen handelt.

druck bringen, in den Vereinigten Staaten Beachtung finden werden. Sie können häufig erst zur Anwendung kommen, wenn der Betroffene sich an seine Regierung gewandt hat. Aber gerade dadurch kann er in Amerika dem Vorwurf ausgesetzt sein, daß er es an einem *good faith effort*, an dem echten Bestreben fehlen ließ, die notwendige Erlaubnis zu erhalten. Hier mag für die Betroffenen ein Dilemma entstehen, für das amerikanische Gerichte wahrscheinlich wenig Verständnis aufbringen werden. Wie die Entscheidung im Fall *Vetco Inc.* zeigt, ist selbst die Strafbarkeit der Aufklärungsmaßnahme im Ausland in Amerika nicht mit Sicherheit als Entschuldigungsgrund anerkannt, obwohl – paradoxerweise – im Zivilprozeß, auf den der Betroffene sich freiwillig eingelassen, den er vielleicht sogar angestrengt und mit dessen Verfahrensregeln er sich abzufinden hat, unter solchen Bedingungen die ausländische Partei weitgehend entschuldigt sein kann [42].

Eine Würdigung der gesamten Situation läßt nur einen Schluß zu. Man sollte den Weg zum Recht, in diesem Fall zum Völkerrecht und zur völkerrechtskonformen Gesetzesauslegung wieder finden und die Verfolgung von Interessen, also von Politik, den Gerichten ersparen.

---

[42] Soweit ersichtlich, ist die Verweigerung einer nach ausländischem Recht erforderlichen Erlaubnis nur in einem Fall als Entschuldigungsgrund behandelt worden: *In re Westinghouse Electric Corporation*, 563 F.2d 992 (1977). Die Entscheidung ist deshalb schwer verständlich, weil sie eine Anordnung (subpoena) für ungültig erklärte, die in einem in Amerika schwebenden Verfahren der Rio Algom Corporation, einer im Staat Delaware gegründeten Gesellschaft mit Sitz in Kanada, in Utah zugestellt worden war, wo sie eine Betriebsstätte unterhielt. Die Urkunden befanden sich in Kanada. Hier wurde also eine amerikanische Gesellschaft von der Vorlegungspflicht befreit. Die amerikanische Rechtsprechung greift immer wieder auf den Fall *Société Internationale* v. *Rogers*, 357 U.S. 197 (1958), zurück, in dem es sich jedoch um etwas ganz anderes handelte. Die Klägerin, eine schweizerische Gesellschaft, verlangte vom amerikanischen Staat Herausgabe der während des Krieges beschlagnahmten, angeblich ihr gehörigen Aktien der General Aniline & Film Corporation, die der Verwalter feindlichen Vermögens beschlagnahmt hatte. Die schweizerische Bundesanwaltschaft beschlagnahmte die vorzulegenden Urkunden. Obwohl die Klägerin sich ernstlich um Freigabe der Urkunden bemüht hatte, aber »nur« 190 000 Urkunden vorlegen konnte, wurde von den unteren Gerichten die Klage wegen Verletzung der Vorlegungspflicht ohne Sachverhandlung abgewiesen. Der Supreme Court hielt aus verfassungsrechtlichen Gründen dieses Verfahren für unzulässig und verlangte eine Verhandlung zur Sache aufgrund des vorliegenden Materials, wobei im Wege der Beweiswürdigung das Fehlen rechtserheblichen Materials zu berücksichtigen war. Der Fall lag also völlig anders als die hier behandelten Fälle. Ein vergleichbarer Fall dagegen war *Arthur Andersen & Co.* v. *Finesilver*, 546 F.2d 336, aber dort wurde die Berücksichtigung schweizerischen Rechts abgelehnt.

Vertragsauslegung durch Vertragsrechtsvergleichung in der Judikatur internationaler Gerichte, vornehmlich vor den Organen der EMRK *

FRANZ MATSCHER

I. Einleitung; zur Rolle der Rechtsvergleichung im Völkerrecht

Daß die Rechtsvergleichung auch im Völkerrecht eine bedeutende Rolle spielt, ja daß gewisse, dem Interpreten völkerrechtlicher Normen gestellte Aufgaben ohne Rückgriff auf die rechtsvergleichende Methode überhaupt nicht lösbar sind, ist heute unbestritten [1], mag auch in der Lehre gelegentlich darüber geklagt werden, daß in den Entscheidungen und in der sonstigen Praxis internationaler Gerichte und Organisationen nur selten zum Ausdruck kommt, daß ein bestimmtes Problem auf rechtsvergleichender Grundlage gelöst wurde; jedenfalls hat es den Anschein, daß Methodenfragen der Rechtsvergleichung dabei kaum je ausdrücklich zur Sprache kommen würden [2]. Indes ist auch in der Lehre die Erforschung geeigneter Kriterien eines methodischen Vorgehens, die der jeweiligen Nutzanwendung der Rechtsvergleichung im Völkerrecht angepaßt sein müßten, über Ansätze nicht hinausgekommen.

Das Hauptanwendungsgebiet der Rechtsvergleichung im Völkerrecht wird vornehmlich in der Feststellung des Vorhandenseins eines allgemeinen Rechtsgrundsatzes (nach Art. 38 Abs. 1 *lit.* c IGH-Statut) in einer bestimmten Materie gesehen. Dazu ist eine wenigstens überblicksartige vergleichende Betrachtung der wesentlichsten nationalen Rechtsordnungen unerläßlich.

* Ohne nähere Angabe der Instanz zit. Entscheidungen sind solche des EGMR.

[1] *Rabel*, Rechtsvergleichung und internationale Rechtsprechung, RabelsZ 1 (1927), S. 5; *Bothe*, Die Bedeutung der Rechtsvergleichung in der Praxis der internationalen Gerichte, ZaöRV 36 (1976), S. 280.

[2] *Hailbronner*, Ziele und Methoden völkerrechtlich relevanter Rechtsvergleichung, ZaöRV 36 (1976), S. 190.

Aber auch wer den allgemeinen Rechtsgrundsätzen nicht den Charakter einer eigenständigen Völkerrechtsquelle zugestehen will, diese vielmehr nur als Aspekte allgemein anerkannter Grundsätze des Völkerrechts betrachtet, die sich aus der *opinio iuris sive necessitatis* geübten Praxis der Staaten (nach Art. 38 Abs. 1 *lit.* b IGH-Statut) oder aus der Judikatur internationaler Gerichte (nach Art. 38 Abs. 1 *lit.* d IGH-Statut) herausschälen lassen oder die überhaupt nur Ausdruck allgemeiner Vernunftregeln sind[3], wird nicht umhin können, die Staatenpraxis und die Judikatur nationaler und internationaler Gerichte rechtsvergleichend zu untersuchen. Diesen Gedanken weiterspinnend möchte ich überhaupt festhalten, daß auch die Konstatierung eines allgemein anerkannten Völkerrechtssatzes (gem. Art. 38 Abs. 1 *lit.* b IGH-Statut) nach einer vergleichenden Betrachtung der Praxis der Staaten, in ihrem gegenseitigen Verkehr, unter Einbeziehung der nationalen Rechtsordnungen und der von ihnen geschlossenen Verträge – deren Regelungen unter Umständen Ausdruck des Bestehens oder des Fehlens eines solchen Völkerrechtssatzes sind[4] – verlangt. Auch zur Lösung des Problems der Vertragskollisionen (oder Vertragskonflikte)[5] – Art. 30 Wiener Vertragsrechtskonvention (WVK) – ist eine Gegenüberstellung und Vergleichung der angeblich miteinander konkurrierenden Vertragsbestimmungen erforderlich.

Neuerer Erkenntnis entspricht die Hervorhebung der Rolle der Rechtsvergleichung für das Recht der internationalen Organisationen[6]. Daß schließlich rechtsvergleichende Studien zur Vorbereitung von Kodifikationsvorhaben unerläßlich sind, liegt auf der Hand[7]. Beispielhaft sei auf die Arbei-

---

[3] Vgl. darüber *Seidl-Hohenveldern*, Die Rolle der Rechtsvergleichung im Völkerrecht, in: Festschrift Verdross (1960), S. 255 ff.; *Hailbronner* (Anm. 2), S. 205; *Sereni*, Principi generali di diritto e processo internazionale (1955), S. 7 ff.

[4] *Scheuner*, Internationale Verträge als Elemente der Bildung von völkerrechtlichem Gewohnheitsrecht, in: Festschrift Mann (1977), S. 409; *Doehring*, Gewohnheitsrecht aus Verträgen, ZaöRV 36 (1976), S. 77; *Geck*, Völkerrechtliche Verträge und Kodifikation, ZaöRV 36 (1976), S. 139; *Strebel*, Quellen des Völkerrechts als Rechtsordnung, ZaöRV 36 (1976), S. 312, spricht von der »indirekten Wirkung« von Verträgen.

[5] Zahlreiche Vertragskonflikte ergeben sich im Privat- und Prozeßrecht im Hinblick auf die überaus rege, nicht immer ausreichend koordinierte bi- und multilaterale Kodifikationstätigkeit in diesen Materien. Vgl. dazu *Meessen*, Staatsverträge zum IPR, Berichte der Deutschen Gesellschaft für Völkerrecht 16 (1975), S. 55 ff.; *Majoros*, Les conventions internationales en droit international privé (1976); *Volker*, Konventionskonflikte im IPR (1977); *Matscher*, Étude des règles de compétence judiciaire dans certaines conventions internationales, RdC 161 (1978 III), S. 177.

[6] *Zemanek*, Was kann die Vergleichung staatlichen öffentlichen Rechts für das Recht der internationalen Organisationen leisten?, ZaöRV 24 (1964), S. 452; *Ress*, Die Bedeutung der Rechtsvergleichung für das Recht internationaler Organisationen, ZaöRV 36 (1976), S. 227.

[7] Vgl. dazu etwa *Ago*, La codification du droit international et les problèmes de sa réalisation, in: Festschrift Guggenheim (1968), S. 93; *Zemanek*, Die Bedeutung der Kodifizierung des Völkerrechts für seine Anwendung, in: Festschrift Verdross (1971), S. 565; *Geck* (Anm. 4), S. 96.

ten der Haager Privatrechtskonferenz, des UNIDROIT, des Europarats und der International Law Commission hingewiesen [8].

Die Aktualität des Phänomens zeigt sich u.a. auch darin, daß der X. Internationale Kongreß für Rechtsvergleichung (Budapest, 1978) dem Problem eine eigene Sektion (IV. A. 1) gewidmet hat und daß ein Teil der Budapester Referate, zusammen mit anderen einschlägigen Arbeiten, in einer eigenen Publikation (*Butler* [Hrsg.], International Law in Comparative Perspective [1980] [9]), gesammelt und systematisch geordnet, veröffentlicht wurde und ferner, daß auch auf dem Kolloquium, welches das Heidelberger Max-Planck-Institut für ausländisches öffentliches Recht und Völkerrecht im Jahr 1975, aus Anlaß seines 50jährigen Bestehens, veranstaltet hatte (»Völkerrecht als Rechtsordnung – Grundlagen und Quellen« [10]), eine Reihe von grundlegenden Referaten der Rolle der Rechtsvergleichung im Völkerrecht gewidmet war.

Es ist das Verdienst des Jubilars, die Bedeutung der Rechtsvergleichung im Zusammenhang mit Aufgaben hervorgehoben zu haben, mit denen sich neuere Erscheinungsformen von internationaler Gerichtsbarkeit in zunehmendem Maße konfrontiert sehen [11].

Indes ist es nicht mein Anliegen, die Rolle der Rechtsvergleichung bei der Feststellung, sondern bei der Auslegung von in ihrem Bestand gesicherten Völkerrechtsnormen [12], letztlich also bei der Auslegung von Normen des

---

[8] *Dutoit*, Comparative Law and Public International Law, in: Butler (zu Anm. 9), S. 75 ff.

[9] Dazu *Matscher*, Österreichische Zeitschrift für öffentliches Recht 31 (1980), S. 362.

[10] ZaöRV 36 (1976); siehe bereits den Bericht von *Steinberger*, Rechtsvergleich und Gerichtsschutz des Einzelnen gegenüber der vollziehenden Gewalt: Ein Weg zur Gewinnung von Völkerrecht?, in: Gerichtsschutz gegen die Exekutive, Bd. 3: Rechtsvergleichung – Völkerrecht (Beiträge zum ausländischen öffentlichen Recht und Völkerrecht, Bd. 52) (1971), S. 273.

[11] *Mosler*, Rechtsvergleichung vor völkerrechtlichen Gerichten, in: Festschrift Verdross (1971), S. 381.

[12] Wenn auch theoretisch zwischen Rechts*findung* und Rechts*auslegung* – als getrennte Anliegen – unterschieden werden kann (in diesem Sinne *Sereni*, Diritto internazionale, Bd. 1 [1956], S. 178; *Haraszti*, Some Fundamental Problems of the Law of Treaties [1973], S. 25 f.; *Yasseen*, L'interprétation des traités d'après la Convention de Vienne sur le droit des traités, RdC 151 [1976 III], S. 9 f. Auch die grundlegende Arbeit von *Bernhardt*, Die Auslegung völkerrechtlicher Verträge, insbesondere in der neueren Rechtsprechung internationaler Gerichte [Beiträge zum ausländischen öffentlichen Recht und Völkerrecht, Bd. 40] [1963], geht von der Prämisse aus, daß die Anwendbarkeit der auszulegenden Vertragsnormen an sich außer Streit steht [S. 31, zu Anm. 176]); *Lachs*, La preuve et la Cour Internationale de Justice, in: Perelman/Foriers (Hrsg.), La preuve en droit (1981), S. 116 f., weist auf die beiden Tätigkeiten innewohnenden gemeinsamen Züge hin und spricht von »interprétation au sens large« (welche die Rechtsfindung einschließt) und »interprétation au sens étroit«; ähnlich *Kutscher*, Thesen zu den Methoden der Auslegung des Gemeinschaftsrechts, aus der Sicht eines Richters (in: EuGH, Begegnung von Justiz und Hochschule, 27.–28. 9. 1976), I-7.

Völkervertragsrechts (nach Art. 38 Abs. 1 *lit.* a IGH-Statut) zu untersuchen. Infolge der zunehmenden »Vertragsdichte« verwundert es nicht, daß auch zu diesem Zweck in erhöhtem Maße rechtsvergleichend vorgegangen wird.

II. Die Rolle der Rechtsvergleichung bei der Auslegung völkerrechtlicher Verträge

Wer heute von der Auslegung völkerrechtlicher Verträge spricht oder darüber schreibt, muß von den in den Art. 31–33 WVK getroffenen Regelungen ausgehen. Mögen diese Bestimmungen im wesentlichen auch nur eine Kodifizierung der bisherigen Praxis darstellen, so sind sie schon deshalb von ausschlaggebender Bedeutung, weil sie klarstellen, daß es sich bei ihnen um Rechtsnormen handelt [13]. Im übrigen hat sich der Europäische Gerichtshof für Menschenrechte (Gerichtshof) (auf dessen Judikatur sich die vorliegende Arbeit konzentrieren wird) bereits vor dem Inkrafttreten der WVK grundsätzlich – »unbeschadet aller einschlägigen Vorschriften der Organisation« (Art. 5 WVK) – auf deren Auslegungsregeln gestützt [14].

Halten wir schließlich fest, daß die WVK zu den mehr rechtspolitischen als rechtsdogmatischen Fragen einer statischen bzw. historischen oder dynamischen bzw. evolutiven Auslegung, einer restriktiven oder extensiven Auslegung, keine klar in eine bestimmte Richtung weisende Aussagen macht. Eindeutig ist nur der Vorrang, den sie dem Vertragstext als solchem, als Ausgangspunkt und Grundlage einer jeglichen Auslegung, einräumt [15]. Zwischen einer Auslegung rechtsgeschäftlicher und normativer Verträge unterscheidet sie nicht. Die von ihr aufgestellten Regeln erheben vielmehr Anspruch auf Allgemeingültigkeit [16].

Auch zur Auslegung von Verträgen, d.h. zur Feststellung des Inhalts vertraglich festgelegter Rechte und Pflichten, kann die Rechtsvergleichung in mannigfacher Hinsicht ein probates Hilfsmittel darstellen. Arbeitshypothese

---

[13] *Verdross/Simma*, Universelles Völkerrecht (1976), S. 390. Über den rechtlichen Standort der völkerrechtlichen Auslegungsregeln im allgemeinen siehe *Bernhardt* (Anm. 12), S. 25 ff.

[14] *Golder*, 21. 2. 1975, Série A Nr. 18, §§ 29, 30=EuGRZ 1975, 93.

[15] *Bernhardt*, Interpretation and Implied (Tacit) Modification of Treaties, ZaöRV 27 (1967), S. 496 f.; *Köck*, Vertragsinterpretation und Vertragsrechtskonvention (1976), S. 91; *Yasseen* (Anm. 12), S. 16 f. Im Fall *Wemhoff*, 27. 6. 1968, Série A Nr. 7, S. 23 (also noch vor dem Zustandekommen der WVK), hatte der Gerichtshof betont, daß es sich bei der EMRK um einen normativen Vertrag handle, der nicht restriktiv, sondern nach objektiven und teleologischen Kriterien anzulegen sei; dazu *Mosler*, L'influence du droit national sur la Convention Européenne des Droits de l'Homme, in: Festschrift Ganshof van der Meersch, Bd. 1 (1972), S. 527. Eine Auslegung nach WVK-Grundsätzen würde allerdings zum gleichen Ergebnis geführt haben.

[16] *Bernhardt* (Anm. 15), S. 494 f.

– wie bei jeder Auslegung von Rechtsnormen – ist dabei die Annahme einer Lücke oder Unklarheit im anwendbaren Vertragstext[17].

Hier muß noch eine weitere Abgrenzung des Themas aufgezeigt werden: Die vorliegende Untersuchung befaßt sich nicht mit dem schon angedeuteten (Anm. 5) Problem der Konkurrenz von Verträgen. Mag auch die Prüfung der Anwendbarkeit eines bestimmten Vertrags – im Verhältnis zu anderen Verträgen – einen Aspekt der Auslegung der konkurrierenden Verträge darstellen[18], und mag es auch zutreffen, daß es zu diesem Zweck erforderlich ist, diese Verträge zueinander in Beziehung zu setzen, wird dabei keine rechtsvergleichende Operation im eigentlichen Sinn vorgenommen[19]. Zur Lösung des hier angesprochenen Problems ist vielmehr generell nach Art. 30 WVK[20] vorzugehen. Teilweise enthalten die Verträge auch spezielle Konfliktsregeln; vgl. etwa, was den hier interessierenden Bereich anbelangt, Art. 25 des Europäischen Niederlassungsabkommens vom 13. Dezember 1955[21], Art. 6 der Europäischen Vereinbarung vom 6. Mai 1969 über die an Verfahren vor der Europäischen Kommission für Menschenrechte (Kommission) und dem Gerichtshof teilnehmenden Personen[22], insbesondere aber Art. 60 Europäische Menschenrechtskonvention (EMRK).

Was nun die hier allein behandelte Auslegung von Verträgen im engeren Sinne anbelangt, können einmal die zur Zeit des Vertragsabschlusses in den Rechtsordnungen der Mitgliedstaaten aufscheinenden Begriffe Ausdruck der »üblichen Bedeutung« (nach der allgemeinen Auslegungsregel des Art. 31 Abs. 1 WVK) sein, in der die gleichen, im Vertrag verwendeten Begriffe zu verstehen sind. Nicht nur wenn ein Vertrag, mehr oder weniger ausdrücklich, auf das Recht einer bestimmten Vertragspartei[23] oder der Vertragsparteien

---

[17] *Hailbronner* (Anm. 2), S. 199.

[18] Es gilt auch in diesem Zusammenhang, was in Anm. 12 ausgeführt wurde.

[19] Vgl. *Capotorti*, Widerspruch in der innerstaatlichen Rechtsordnung zwischen der EMRK und anderen völkerrechtlichen Abkommen, in: Menschenrechte im Staatsrecht und im Völkerrecht, Zweites Internationales Kolloquium über die Europäische Konvention zum Schutze der Menschenrechte und Grundfreiheiten, Wien 1965 (1967), S. 69; *Frowein*, Probleme des allgemeinen Völkerrechts vor der EKMR, in: Festschrift Schlochauer (1981), S. 289.

[20] *Verdross/Simma* (Anm. 13), S. 334 f., 397 ff. Die zit. Bestimmung spricht von der »Anwendung« (*application*) von Verträgen. Der Begriff der Anwendung – im Gegensatz zur Auslegung – von Verträgen wird von *Bernhardt* (Anm. 12), S. 32 ff. (und desgleichen in der vorliegenden Arbeit) allerdings in einem anderen Sinn verwendet (»Auslegung« = Bestimmung der Tragweite, des Gehalts, einer Norm; »Anwendung« = Subsumtion eines zu beurteilenden Sachverhalts unter die festgestellte Norm).

[21] European Treaty Series Nr. 19.

[22] dBGBl. 1977 II, 1445.

[23] »Rechtmäßig« und auf »gesetzlich vorgeschriebener Weise« durch ein »zuständiges Gericht« (nach Art. 5 EMRK): *Winterwerp*, 24. 11. 1979, Série A Nr. 33, §§ 36 ff., §§ 44 ff. = EuGRZ 1979, 653 f., 654 f.; *Guzzardi*, 6. 11. 1980, Série A Nr. 39, § 102.

überhaupt[24] (oder gar auf einem weiteren Kreis von Staaten typische Institutionen) verweist, ist nämlich im Zweifel davon auszugehen, daß diese Begriffe in dem Sinn zu verstehen sind, der ihnen in der Rechtssprache der Partnerstaaten gemeiniglich zukommt[25].

Freilich darf nicht übersehen werden, daß insbesondere multilaterale Verträge vielfach autonome Begriffe kreieren. So ist es heute ständige Rechtsprechung der Organe der EMRK, daß etwa die Begriffe »zivilrechtliche Ansprüche und Verpflichtungen«[26], »Anklage«[27], »strafrechtlich«[28], »angemessene Frist«[29] (Art. 6 Abs. 1 EMRK); »Schuld«[30] (Art. 1 Abs. 2 EMRK); Achtung des »Privat- und Familienlebens«[31], des »Privatlebens, der Wohnung und des Briefverkehrs«[32] (Art. 8 EMRK) nicht allein unter Bezug auf das innerstaatliche Recht des belangten Staats interpretiert werden dürfen, son-

---

[24] In einer »demokratischen Gesellschaft ... notwendig« (etwa nach Abs. 2 der Art. 8–11 EMRK oder in Abs. 3 des Art. 2, 4. Zusatzprotokoll): *Handyside*, 7. 12. 1976, Série A Nr. 24, § 53 ff.=EuGRZ 1979, 388; dazu Sondervotum *Mosler* §§ 2, 3=EuGRZ 1979, 51 ff.; *Sunday Times*, 27. 10. 1978, Série A Nr. 30, §§ 58 ff.=EuGRZ 1979, 388 ff.; dazu gemeinsames abweichendes Sondervotum, §§ 7 ff.=EuGRZ 1979, 392 f.; *Young u.a.* 13. 8. 1981, Série A Nr. 44 II, §§ 62 ff.=EuGRZ 1981, 562; *Dudgeon*, 22. 10. 1981, Série A Nr. 45, §§ 48 ff.; dazu abweichendes Sondervotum *Matscher*. Wenn der Vertrag, wie hier, auf bestimmte, in den Rechtsordnungen der Vertragspartner typenmäßig ausgebildete Institutionen verweist (*Hailbronner* [Anm. 2], S. 222 f.), liegt ein Phänomen vor, das von dem der Bildung vertragsautonomer Kategorien (darüber unten Anm. 28–35) kaum zu trennen ist. Dafür spricht allein schon die Erwägung, daß das Element der »Notwendigkeit« – mögen auch die Voraussetzungen ihres Zutreffens nach den jeweiligen Verhältnissen in den einzelnen Ländern zu beurteilen sein – ein vertragsautonomer Begriff ist, während die Bezugnahme auf die »demokratische Gesellschaft« einen Verweis auf bestimmte, in den Rechtsordnungen der Vertragspartner (und darüber hinaus) ausgebildete Institutionen darstellt. Aber auch sonst mag es mitunter zweifelhaft sein, ob ein Vertrag, etwa die EMRK, auf das Recht eines oder mehrerer Vertragsstaaten verweist oder ob er zur Bildung eines autonomen, dann aber für Fälle aus allen Vertragsstaaten gleich anwendbaren Begriffs tendiert (dazu *Mosler* [Anm. 11], S. 397 f.).

[25] *Schiesser*, 4. 12. 1979, Série A Nr. 34, § 28=EuGRZ 1980, 205.

[26] *Ringeissen*, 16. 7. 1971, Série A Nr.13, S. 39, § 94; *König*, 28. 6. 1978, Série A Nr. 27, § 88=EuGRZ 1978, 415; *Le Compte u.a.*, 23. 6. 1981, Série A Nr. 43, §§ 44 ff.=EuGRZ 1981, 552.

[27] *Neumeister*, 27. 6. 1968, Série A Nr. 8, S. 41, § 18 i.V.m. S. 35, § 28.

[28] *Engel*, 8. 6. 1976, Série A Nr. 22, §§ 80 ff.=EuGRZ 1976, 231 ff.; *Deweer*, 27. 2. 1980, Série A Nr. 35, § 42=EuGRZ 1980, 671.

[29] *Wemhoff*, 27. 6. 1968, Série A Nr. 7, S. 22 ff., §§ 6 ff., *Neumeister* (Anm. 27), S. 41 ff., §§ 17 ff.; *König* (Anm. 26), §§ 97–99=EuGRZ 1978, 417; *Buchholz*, 6. 5. 1981, Série A Nr. 42, §§ 47–49=EuGRZ 1981, 494; siehe auch *Peukert*, Die überlange Verfahrensdauer in der Rspr. der Straßburger Instanzen, EuGRZ 1979, 261.

[30] *Adolf*, 26. 3. 1982, Série A Nr. 49, §§ 35 ff.

[31] *Marckx*, 13. 6. 1979, Série A Nr. 31, § 31=EuGRZ 1979, 455; *Dudgeon* (Anm. 24), § 41.

[32] *Klass u.a.*, 6. 9. 1978, Série A Nr. 28, § 41=EuGRZ 1979, 284; dazu (und zu Anm. 31) *Stolzlechner*, Der Schutz des Privat- und Familienlebens (Art. 8 MRK) im Licht der Rechtsprechung des VfGH und der Straßburger Instanzen, Österreichische Juristen-Zeitung 1980, S. 85.

dern im Sinne der Konvention zu verstehen sind. Dabei sind, wie sich der Europäische Gerichtshof einmal treffend ausgedrückt hat[33], »die Zielsetzungen und die Systematik des Übereinkommens sowie die allgemeinen Rechtsgrundsätze, die sich aus der Gesamtheit der innerstaatlichen Rechtsordnungen ergeben«, zu berücksichtigen. Demnach ist selbst hier das Recht der betreffenden Staaten nicht ohne Belang[34], und das Ergebnis der »eigenständigen« oder »autonomen« Auslegung darf nicht ein Begriff sein, der von der Begriffsbildung in den Rechtsordnungen der Mitgliedstaaten des Übereinkommens völlig abstrahiert[35, 36]. Ein gewisses Maß an rechtsvergleichender Argumentation ist dazu also erforderlich[37]. Lediglich wenn der Vertrag einen eigenständigen, übernationalen Begriffsapparat kreiert, erübrigt sich ein Rückgriff auf die Rechtsordnungen der Partnerstaaten[38].

Ferner können die nach Abschluß des Vertrags in den Rechtsordnungen der Vertragsstaaten eingeführten Regelungen, besonders soweit es sich um solche handelt, die zur Ausführung des Vertrags erlassen wurden, Emanation einer »nachfolgenden Praxis« (im Sinne von Art. 31 Abs. 3 *lit.* b WVK) sein, die gleichfalls das Verständnis der Partner betreffend die im Vertrag getroffenen Regelungen zum Ausdruck bringt.

Sowohl in einem wie im anderen Beispiel gelangt der Interpret zum Sinn einer Vertragsnorm mittels einer rechtsvergleichenden Analyse der entsprechenden Institutionen der Vertragsstaaten. Aber auch auf alle damit in Zusammenhang stehenden Probleme soll hier nicht näher eingegangen werden. Ich möchte mich vielmehr darauf beschränken, der Frage nachzugehen, inwieweit eine Vergleichung mit Verträgen, die mit denselben oder auch mit anderen Partnern geschlossen wurden, ein geeignetes Mittel zur Vertragsauslegung sein kann.

III. Probleme der Auslegung der EMRK

Hauptaufgabe der Organe der EMRK – vornehmlich der Kommission und des Gerichtshofs – ist die Auslegung und Anwendung (vgl. Anm. 20) der

---

[33] Rechtssache *LTU Lufttransportunternehmen* gegen *Eurocontrol*, Urteil vom 14. 10. 1976, Slg. 1976, 1550.

[34] *König* (Anm. 26), § 89 = EuGRZ 1978, 415.

[35] Sondervotum *Matscher* zu *König* (Anm. 26), 45 f. = EuGRZ 1978, 422.

[36] Zur »autonomen Auslegung« vgl. auch *Evrigenis*, Reflections on the National Dimension of the ECHR, in: Kolloquium Athen, 1978, S. 72 f.

[37] *Ganshof van der Meersch*, Die Bezugnahme auf das innerstaatliche Recht der Vertragsstaaten in der Rechtsprechung des EGMR, EuGRZ 1981, 481 ff.

[38] Vgl. etwa die Definition des Begriffs der Konnexität in Art. 22 Abs. 3, Übereinkommen über die gerichtliche Zuständigkeit und die Vollstreckung gerichtlicher Entscheidungen in Zivil- und Handelssachen; dazu *Martiny*, Autonome und einheitliche Auslegung im Europäischen und im Internationalen Zivilprozeßrecht, RabelsZ 1981, S. 427.

Konvention. Da sich diese in weitem Maße unbestimmter Begriffe bedient (Anspruch auf »faires Verfahren« [Art. 6 Abs. 1], auf Achtung des »Privat- und Familienlebens« [Art. 8], auf »wirksame Beschwerde« [§ 13], »in Übereinstimmung mit den Allgemeininteressen des Staates erforderlich« [Art. 1 Abs. 2, 1. Zusatzprotokoll] usw.), sieht sich der Interpret oft vor schwierige Auslegungsfragen gestellt.

Andererseits verhindern solche flexiblen Formulierungen der Konvention und ihrer Zusatzprotokolle eine Versteinerung der Begriffe; sie laden vielmehr zu einer evolutiven oder dynamischen, den sich wandelnden gesellschaftlichen Vorstellungen und Bedürfnissen angepaßten Auslegung und Anwendung der Konvention ein [39]. Aber auch wenn man die besondere Funktion der Organe der EMRK berücksichtigt, die in der Nähe derjenigen eines Verfassungsgerichts einzuordnen ist, bleibt deren Aufgabe die Rechts*anwendung* und nicht die Rechts*findung*. Leitlinie sollte dabei sein, wie *Sereni* dazu sinngemäß ausgeführt hat [40], daß es nicht die Aufgabe des Interpreten ist, den Inhalt einer Norm von sich aus weiterzuentwickeln, sondern vielmehr die Änderungen nachzuvollziehen, die der Inhalt einer Norm durchgemacht hat.

Von der Möglichkeit einer evolutiven Interpretation machen die Organe der EMRK einen ausgiebigen Gebrauch, ja sie betrachten diese als die der EMRK am meisten angepaßte Auslegungsmethode [41]; dabei gehen sie mitunter bis hart an den Rand dessen heran, was man noch als Vertragsauslegung im Rechtssinn bezeichnen kann. Um so angebrachter erschiene es mir – auch im Interesse der Autorität der zu treffenden Entscheidung, vor allem aber um aufzuzeigen, daß es sich um eine rechtlich begründete Überlegung und nicht um einen Willkürakt handelt [42] – die befolgte Interpretationsmethode offen-

---

[39] *Sørensen*, Les droits inscrits en 1950 dans la Convention européenne des Droits de l'Homme ont-ils la même signification en 1975?, in: Actes du 4e Colloque international, Rom 1975, S. 83; er verweist dabei (S. 88) insbes. auf das *Namibia*-Gutachten vom 21. 6. 1971, ICJ Rep. 1971, S. 31; *Mosler*, Problems of Interpretation in the Case Law of the ECHR, in: Gedächtnisschrift Panhuys (1980), S. 149, insbes. S. 158f.; *Ganshof van der Meersch*, De la mise en oeuvre d'une sauvegarde collective des Droits de l'Homme en droit international – La Convention européenne, in: Festschrift Dehousse, Bd. 1 (1979), S. 197.

[40] *Sereni* (Anm. 12), S. 182.

[41] *Tyrer*, 25. 4. 1978, Série A Nr. 26, § 31 = EuGRZ 1979, 164; *Marckx* (Anm. 31), § 41 = EuGRZ 1979, 457; *Dudgeon* (Anm. 24), § 60.

[42] Vgl. zu der nicht nur im Mittelalter, sondern bis weit in die Neuzeit herauf häufig unterbliebenen Begründung der von moralischen (Papst) oder politischen (monarchische oder republikanische Staatsoberhäupter) Autoritäten gefällten Schiedssprüche, vornehmlich, um diese nicht einer Kritik zu unterziehen, *Matscher*, Die Begründung der Entscheidungen internationaler Gerichte, in: Sprung/König (Hrsg.), Entscheidungsbegründung in europäischen Verfahrensrechten und im Verfahren vor internationalen Gerichten (1974), S. 448 ff.

zulegen und zu rechtfertigen[43]. Und wenn der Interpret dazu nicht in der Lage ist, würde er sich vielleicht dessen bewußt werden, daß der Weg, den da oder dort zu begehen er sich anschickt, bedenklich ist. Andererseits könnte in vielen Fällen die Überzeugungskraft einer Entscheidung durch die Aufnahme rechtsvergleichender Argumente in die Begründung nur gewinnen[44]. In diesem Zusammenhang möchte ich an die weise Bestimmung des § 272 österr. Zivilprozeßordnung erinnern, die in Abs. 1 den Richter zur freien Beweiswürdigung ermächtigt, ihn in Abs. 3 aber verpflichtet, »die Umstände und Erwägungen, welche für ... (seine) ... Überzeugung maßgeblich waren ... in der Begründung anzugeben«.

Demgegenüber tritt die andere, gerichtlichen oder gerichtähnlich strukturierten Organen gleichfalls eigentümliche Aufgabe, nämlich die Feststellung strittiger Tatsachen, für die Kommission weitgehend, für den Gerichtshof fast zur Gänze, in den Hintergrund[45].

IV. Die Auslegung der EMRK mit Hilfe der Vertragsrechtsvergleichung in der Praxis von Kommission und Gerichtshof

Eine Durchsicht der Entscheidungen der Kommission und des Gerichtshofs zeigt, daß beide Organe bei der Herausarbeitung des Sinnes und bei der Auslotung der Tragweite der im Grundrechtskatalog der Konvention und der Zusatzprotokolle statuierten Rechte des Individuums – und der korrespondierenden Verpflichtungen der Konventionsstaaten – nicht selten die Normen dieses Katalogs mit den Bestimmungen anderer, zwischen den Vertragsstaaten geschlossener Verträge vergleichen bzw. sie an diesen messen. Bei den letzteren kann es sich um solche von gleichfalls grundrechtlichem Inhalt oder auch um Vertragsbestimmungen handeln, die denselben Gegenstand, wie die Konventionsnorm, betreffen, diesen aber unter einem anderen Gesichtspunkt regeln oder die, in einem bestimmten Zusammenhang, zumindest mit den gleichen Begriffen operieren wie die fraglichen Konventionsnormen.

In der Folge soll – ohne jeden Anspruch an Vollständigkeit – eine Auswahl von Beispielen aus diesem Bereich aufgezeigt werden. Ergänzend möchte ich auf Hypothesen hinweisen, wo eine solche Vertragsrechtsvergleichung

---

[43] Zur Bedeutung der Entscheidungsbegründung für das Vertrauen in die Rechtsprechung, insbes. in der internationalen Gerichtsbarkeit, vgl. *Matscher*, a.a.O., S. 442, 446.

[44] Zur Überzeugungskraft der Rechtsvergleichung siehe *Ress* (Anm. 6), S. 234, insbes. Anm. 13.

[45] Vgl. die Referate von *Melchior* und *Matscher* bei dem vom Institut International d'Etudes des Droits de l'Homme, Triest (29.–30. 11. 1980), organisierten Round Table-Gespräch über »La CEDH: les pouvoirs d'enquête de la Commission et de la Cour« (im Druck).

– zur Stützung oder zur Widerlegung der Meinung der Konventionsorgane oder von in der Lehre zu Problemen der Konventionsauslegung geäußerten Ansichten – meines Erachtens brauchbare Argumente geliefert hätte oder liefern würde. Dabei soll nicht chronologisch, sondern nach der Systematik der Konvention und der Zusatzprotokolle vorgegangen werden.

1. Zur Auslegung des Begriffs der »Zwangs- oder Pflichtarbeit« nach Art. 4 Abs. 2 und 3 EMRK wurde im Verfahren vor der Kommission von den Beschwerdeführern und von den belangten Regierungen in mehreren Fällen die ILO-Konvention 1930 betreffend die Zwangs- oder Pflichtarbeit (Nr. 29) herangezogen, da die auszulegenden EMRK-Bestimmungen auf die entsprechenden Regelungen der ILO-Konvention zurückgehen[46]. Ferner hat sich die Kommission zur Auslegung der Begriffe »Sklaverei« und »Leibeigenschaft« (nach Art. 4 Abs. 1 EMRK) und zu deren Abgrenzung von den verwandten Begriffen der »Zwangs- oder Pflichtarbeit« in einem Fall auf Art. 1 der Zusatzkonvention vom 30. April 1956 über die Beseitigung der Sklaverei berufen[46a].

2. Im Fall *Guzzardi*[47] ging es u. a. um die Frage, ob die Zuweisung eines Zwangsaufenthalts (*soggiorno obbligatorio*) an ein als *mafioso* qualifiziertes Individuum auf einem Teil einer Insel (Bewegungsspielraum auf einer Fläche von ca. 2,5 km^2), mit Unterwerfung unter verschiedene Überwachungsmaßnahmen, einen »Freiheitsentzug« nach Art. 5 Abs. 1 EMRK oder lediglich eine »Bewegungs- und Aufenthaltsbeschränkung« nach Art. 2 Abs. 1, 4. Zusatzprotokoll, das vom belangten Staat (Italien) aber nicht ratifiziert wurde, darstellt. Wenn auch allseits anerkannt wurde, daß zwischen »Freiheitsentzug« und »Bewegungs- oder Aufenthaltsbeschränkung« nur ein gradueller Unterschied besteht, hat sich die Mehrheit des Gerichtshofs für das erstere entschieden. In zwei Sondervoten wurde aber darauf hingewiesen, daß der Begriff des Freiheitsentzugs gem. Art. 5 Abs. 1 EMRK nach einer engen Auslegung verlangt, weil sonst die Notwendigkeit der zusätzlichen Regelung des Art. 2 Abs. 1, 4. Zusatzprotokoll nicht einsichtig wäre[48]. Würde man ferner gewisse Formen von Bewegungs- und Aufenthaltsbeschränkung als »Freiheitsentzug« qualifizieren, liefe das darauf hinaus, auf Umwegen auch diejenigen Staaten den Verpflichtungen des Art. 2 Abs. 1, 4. Zusatzprotokoll zu unterwerfen, welche das letztgenannte Instrument nicht ratifiziert haben.

---

[46] *Iversen* gegen *Norwegen*, Zulässigkeitsentscheidung vom 17. 12. 1963, Annuaire de la Convention européenne des Droits de l'homme 6 (1963), S. 329 f.; dazu *Fawcett*, The Application of the European Convention on Human Rights (1969), S. 43 ff. Ähnlich im Fall *Gussenbauer* gegen *Republik Österreich*, Zulässigkeitsentscheidung vom 22. 3. 1972, Annuaire de la Convention européenne des Droits de l'homme 15 (1972), S. 455, 459.

[46a] *Van Droogenbroek*, Beschwerde Nr. 7906/77, Zulässigkeitsentscheidung vom 5. 7. 1979, Décisions et Rapports 17, 66, und Bericht vom 9. 7. 1980, § 79.

[47] 6. 11. 1980, Série A Nr. 39, insbes. § 92.

[48] Kurz angedeutet im Sondervotum *Matscher*, S. 62; ausführlich erörtert im Sondervotum *Fitzmaurice*, S. 51 f.

3. *W. Pahr* hat in einem Aufsatz[49] die Meinung vertreten, eine Staatenimmunität in Zivilrechtssachen verstoße gegen das von Art. 6 Abs. 1 EMRK garantierte Recht auf Zugang zu den Gerichten[50]; die Notwendigkeit, diesen Staat vor seinen eigenen Gerichten zu belangen, sei wiederum mit dem gleichfalls von Art. 6 Abs. 1 garantierten Recht auf *fair trail* nicht vereinbar.

Es soll nun nicht verkannt werden, daß in Zivilrechtssachen die Immunität der Staaten (und ebenso die der Diplomaten) eine rechtspolitisch durch nichts zu rechtfertigende Erscheinung darstellt; zutreffenderweise wurde sie deshalb in den letzten Jahrzehnten beträchtlich eingeschränkt. Indes braucht auf dieses Problem hier nicht näher eingegangen zu werden. Ich möchte lediglich darauf hinweisen, daß die Immunität des Staates, selbst in Angelegenheiten, die von den Organen der Konvention als Zivilrecht im Sinne des Art. 6 Abs. 1 EMRK qualifiziert werden[51], in weitem Ausmaß auch nach dem Verständnis der europäischen Staaten als Völkergewohnheitsrechtssatz anerkannt und darüber hinaus in dem auf Initiative der III. Europäischen Justizministerkonferenz (1964) vorbereiteten und am 16. Mai 1972 in Basel zur Unterzeichnung durch die Mitgliedstaaten des Europarats aufgelegten Europäischen Übereinkommen über die Staatenimmunität[52] festgeschrieben worden ist. Kann aber etwas, was nach einer europäischen Konvention erlaubt ist, nach einer anderen untersagt sein[53]?

4. Im Fall *Marckx* war die Vereinbarkeit von Regelungen des belgischen Zivilrechts über die Stellung des außerehelichen Kindes mit dem in Art. 8 EMRK statuierten Recht auf Achtung des Familienlebens zu prüfen. Der Gerichtshof anerkannte, daß zur Zeit der Ausarbeitung der Konvention mehrere Rechtsordnungen von Konventionsstaaten ähnliche Regelungen enthielten wie die heute inkriminierten Bestimmungen des belgischen Rechts. Im Sinne des Grundsatzes der evolutiven Interpretation sei diese aber im Lichte der heute herrschenden Konzeptionen auszulegen. Tatsächlich hätten die meisten europäischen Rechtsordnungen eine Entwicklung in diesem Sinn durchgemacht[54]. Auf derselben Linie lägen auch das von der Commission in-

---

[49] Die Staatenimmunität und Art. 6 Abs. 1 der EMRK, in: Festschrift Modinos (1968), S. 223, 231 f.

[50] *Golder* (Anm. 14), § 36=EuGRZ 1975, 97 f.; *Deweer* (Anm. 28), § 49=EuGRZ 1980, 672 f. mit Anm. *Espenhain*, EuGRZ 1981, 16; vgl. ferner *Eissen*, Le droit à un tribunal dans la jurisprudence de la Commission, in: Festschrift Ganshof van der Meersch, Bd. 1 (1972), S. 455.

[51] Vgl. die Urteile *König* (Anm. 26), *Le Compte* (Anm. 26).

[52] European Treaty Series Nr. 74.

[53] Dazu *Matscher*, Die Verfahrensgarantien der EMRK in Zivilrechtssachen, Österreichische Zeitschrift für öffentliches Recht 31 (1980), S. 20 Anm. 75.

[54] Von einer noch höheren Warte aus gesehen *Schwind*, Wandlung von Ehe und Familie im Spiegel internationaler Abkommen, in: Festschrift Beitzke (1979), insbes. S. 727 ff. Seine Ausführungen hätten auch zur Stützung der Argumentation des EGMR herangezogen werden können, als dieser im Fall *Airey* aus dem von Art. 8 EMRK garantierten Recht auf Achtung des Privat- und Familienlebens auch den An-

ternationale de l'état civil ausgearbeitete Brüsseler Übereinkommen vom 12. September 1962 über die Feststellung der mütterlichen Abstammung unehelicher Kinder[55] (am 23. April 1964 in Kraft getreten; von Belgien zwar unterzeichnet, bisher aber nicht ratifiziert) sowie das Europäische Übereinkommen vom 15. Oktober 1975 über die Rechtsstellung der außerhalb der Ehe geborenen Kinder[56] (in Kraft getreten am 11. August 1978; von Belgien bisher nicht unterzeichnet).

Während aber die belangte belgische Regierung in der Tatsache, daß das erstgenannte Übereinkommen bislang nur 8 Unterschriften und 4 Ratifikationen und das zweitgenannte 10 Unterschriften und gleichfalls 4 Ratifikationen erhalten hat, ein Indiz dafür erblickte, daß diese Übereinkommen nur sehr bedingt Ausdruck einer europäischen *communis opinio* wäre, betrachtete der Gerichtshof allein das Faktum der Ausarbeitung dieser Übereinkommen zwischen (überwiegend) europäischen Staaten als ein Argument für die Weiterentwicklung der in Europa über die Stellung des außerehelichen Kindes vorherrschenden Konzeptionen, die auch auf die Auslegung des Art. 8 EMRK nicht ohne Einfluß bleiben könnte[57].

Indessen könnte man mit Hilfe des Europarat-Übereinkommens 1975 aber auch anders argumentieren: Art. 14 Abs. 1 gestattet jedem Staat die Anmeldung von drei Vorbehalten. Einer davon könnte daher auch die Modalitäten der (in Art. 2 geregelten) Feststellung der außerehelichen Mutterschaft betreffen. Kann nun aber ein in einem zwischen Europarat-Staaten geschlossenen Übereinkommen zulässiger Vorbehalt mit den Verpflichtungen aus der gleichfalls zwischen den Europarat-Staaten geschlossenen EMRK in Widerspruch stehen? Mit anderen Worten, ist es denkbar, daß in einem europäischen Übereinkommen ein Vorbehalt als zulässig angesehen wird, der mit einer europäischen Grundrechtsnorm nicht vereinbar wäre[58]?

5. Im Fall *Klass u. a.* stand die Vereinbarkeit des im Sinne des (auf Grund des Art. 10 Abs. 2 GG erlassenen) Gesetzes vom 13. August 1968 zur Beschränkung des Brief-, Post- und Fernmeldegeheimnisses (»G 10«), dBGBl. I, 949, eingerichteten Überwachungssystems mit dem von Art. 8 EMRK garantierten Recht auf Achtung des Privatlebens, der Wohnung und des Briefverkehrs zur Diskussion.

Im Verfahren wurde zwar von der belangten Regierung vorgebracht, daß das inkriminierte Gesetz u.a. erlassen worden war, um die entsprechenden

---

spruch auf eine effektive Ermöglichung einer gerichtlichen Eheauflösung herausgelesen hat (Urteil vom 9. 10. 1979, Série A Nr. 32, §§ 31–33 = EuGRZ 1979, 629; siehe aber auch die abweichenden Meinungen der Sondervoten *O'Donoghue*, S. 23, und *Vilhjalmsson*, S. 27 (beide EuGRZ 1979, 631).

[55] dBGBl. 1965 II, 23.

[56] European Treaty Series Nr. 85.

[57] 13. 6. 1979, Série A Nr. 31, insbes. § 41 = EuGRZ 1979, 457.

[58] In ähnlicher Weise hat auch der EGMR in den Fällen *Nationale Belgische Polizeigewerkschaft* und *Schwedischer Lokomotivführerverband* (Text zu Anm. 60) argumentiert.

Kontrollrechte, die ursprünglich von den Besatzungsmächten wahrgenommen worden waren und die sie sich auch noch in Art. 2 des Deutschland-Vertrags (zwischen den Vereinigten Staaten von Amerika, dem Vereinigten Königreich, der Französischen Republik und der Bundesrepublik Deutschland) vom 26. Mai 1952, in der durch das Pariser Protokoll vom 23. Oktober 1954 geänderten Fassung, dBGBl. 1955 II, 385, vorbehalten hatten, auf deutsche Stellen zu übertragen. Am 27. Mai 1968 hatten die drei Mächte sinngemäß erklärt, daß das »G 10« ihrem Sicherheitsbedürfnis entspreche und beschlossen, nach dessen Inkrafttreten ihre Kontrollrechte erlöschen zu lassen[59]. Weitere Konsequenzen wurden aus diesem Sachverhalt aber weder von der belangten Regierung noch in den Entscheidungsgründen des Gerichtshofs gezogen. Im Ergebnis schloß der Gerichtshof, daß das inkriminierte Überwachungssystem unter den gegebenen Umständen als im Sinne des Art. 8 Abs. 2 »notwendig« betrachtet werden kann und daß es auch den anderen, von dieser Bestimmung aufgestellten Erfordernissen Rechnung trägt. Unter diesen Umständen erübrigte es sich, diese »Notwendigkeit« auch im Lichte des in den oben erwähnten Abkommen und Erklärungen der Alliierten zum Ausdruck kommenden Rechtsstandpunkts zu prüfen.

6. In mehreren Gewerkschaftsfällen wurde von der Kommission und vom Gerichtshof die Tragweite des Rechts auf Koalitionsfreiheit, »einschließlich des Rechts, zum Schutze (der eigenen) Interessen Gewerkschaften zu bilden und diesen beizutreten«, im Lichte der entsprechenden Bestimmungen der Europäischen Sozialcharta 1961, der UN-Menschenrechtsdeklaration 1948, der UN-Weltpakte 1966 und der ILO-Konventionen 1948 (Nr. 87) und 1949 (Nr. 98) ausgelegt.

In den Fällen *Nationale Belgische Polizeigewerkschaft* und *Schwedischer Lokomotivführerverband* ging es um die Frage, ob das von Art. 11 Abs. 1 EMRK garantierte Recht auf gewerkschaftliche Freiheit auch das Recht auf Konsultierung bei Tarifverhandlungen einschließt. In diesem Zusammenhang wurde auf die Bestimmung des Art. 6 Abs. 1 Europäische Sozialcharta hingewiesen, welche die Konventionsstaaten (lediglich) verpflichtet, »gemeinsame Beratungen zwischen Arbeitnehmern und Arbeitgebern zu fördern«. Aus der vorsichtigen Formulierung der Sozialcharta schloß der Gerichtshof, daß auch diese ein echtes Recht auf Konsultierung nicht anerkennt. Im übrigen könne jeder Staat durch Vorbehalt gem. Art. 20 die aus Art. 6 Z. 1 Sozialcharta erfließende Verpflichtung nicht annehmen. Daraus lasse sich des weiteren schließen, daß ein Recht auf Konsultierung auch nicht aus Art. 11 Abs. 1 EMRK abgeleitet werden könne, da nicht anzunehmen sei, daß die Sozialcharta 1961 einen Rückschritt gegenüber der Konvention 1950 darstellt[60].

---

[59] 6. 9. 1978, Série A Nr. 28; siehe die zusammenfassende Sachverhaltsabschilderung in §§ 14 und 15 = EuGRZ 1979, 279, und den Kommissionsbericht vom 9. 3. 1977 zur Beschwerde Nr. 5029/71, § 44.

[60] 27. 4. 1975, Série A Nr. 19, § 38 = EuGRZ 1975, 464, bzw. 6. 2. 1976, Série A Nr. 20, § 39 = EuGRZ 1976, 64 f.

Vermerkt sei, daß die Kommission die Bestimmung des Art. 11 EMRK auch mit den entsprechenden Regelungen des Art. 8 Abs. 1 Weltpakt für wirtschaftliche, soziale und kulturelle Rechte 1966 bzw. des Art. 22 Abs. 1 Weltpakt für bürgerliche und politische Rechte 1966 sowie mit den ILO-Konventionen 1948 über die Vereinigungsfreiheit und den Schutz des Vereinigungsrechts (Nr. 87) und 1949 betreffend die Anwendung der Prinzipien des Rechts auf Organisation und auf Kollektivverhandlungen (Nr. 98) in Beziehung gesetzt hat[61].

Im Fall *Schmidt* und *Dahlström* kam der Gerichtshof zum Ergebnis, daß das Recht auf Koalitionsfreiheit nach Art. 11 Abs. 1 EMRK kein Recht auf bestimmte Behandlung seitens des Staates, insbesondere kein Recht auf den Genuß rückwirkender Geltung von Vorteilen, z. B. von Lohnerhöhungen, die einem neuen Tarifvertrag entfließen, einschließt. Ein solches Recht sei ebensowenig in der Europäischen Sozialcharta erwähnt[62]. Auch in diesem Fall hat die Kommission ihre rechtsvergleichenden Untersuchungen weiter ausgedehnt und in ihre Erwägungen zusätzlich die ILO-Konvention 1949 (Nr. 98) einbezogen[63].

Im jüngst entschiedenen *Closed shop*-Fall (Young, James und Webster) wurde, zur Klärung des von den Beschwerdeführern geltend gemachten Rechts auf negative Gewerkschaftsfreiheit, die Bestimmung des Art. 11 EMRK vom Gerichtshof auch im Lichte des Art. 20 Abs. 2 der UN-Menschenrechtsdeklaration 1948 beurteilt. Seitens der belangten britischen Regierung und des als *amicus curiae* zugelassenen Vertreters des Trades Union Congress wurden im Verfahren auch die Bestimmungen der Art. 8 Abs. 1 Weltpakt für wirtschaftliche, soziale und kulturelle Rechte 1966, Art. 22 Abs. 1 Weltpakt für bürgerliche und politische Rechte 1966, Art. 5 Europäische Sozialcharta sowie die ILO-Konventionen 1948 (Nr. 87) und 1949 (Nr. 98) zu Vergleichszwecken herangezogen[64]. Der Gerichtshof ist aber nicht näher darauf eingegangen.

Festgehalten sei schließlich, daß die Kommission bereits im Fall *X* gegen *Republik Irland* die Tragweite der von Art. 11 EMRK garantierten Gewerkschaftsfreiheit (*in concreto* ging es darum, ob diese ungebührlich eingeschränkt worden war) an der ILO-Konvention 1948 (Nr. 87) sowie an der Bestimmung des auf die genannte Konvention verweisenden Art. 22 Abs. 3 Weltpakt für bürgerliche und politische Rechte 1966 gemessen hat[65].

---

[61] Kommissionsbericht vom 27. 5. 1974, §§ 63 ff., § 75, Série B, Bd. 17 (1976), 49 ff., 54 bzw. §§ 65 ff., § 78, Série B, Bd. 18 (1977), 42 ff., 48.

[62] 6. 2. 1926 Série A Nr. 21, § 34 = EuGRZ 1976, 70.

[63] Kommissionsbericht vom 17. 4. 1974, §§ 61 ff., Série B, Bd. 19 (1977), S. 38 ff.

[64] 13. 8. 1981, Série A Nr. 44, § 52; siehe auch Kommissionsbericht vom 14. 12. 1979 zu den Beschwerden Nr. 7601/76 und 7806/77, §§ 114, 115.

[65] Zulässigkeitsentscheidung vom 1. 2. 1971, Annuaire de la Convention européenne des Droits de l'homme 14 (1971), S. 221 f.; dort findet sich auch der interessante Hinweis, daß – mit Ausnahme der Türkei – sämtliche EMRK-Mitglieder die ILO-Konvention 1948 (Nr. 87) ratifiziert hätten.

7. Zur Auslegung des Art. 2, 1. Zusatzprotokoll (Beachtung der Elternrechte bei der Ausübung des Rechts auf Erziehung) wurde in den belgischen Sprachenfällen im Verfahren vor der Kommission und vor dem Gerichtshof[66] von den Beschwerdeführern u.a. auch die UNESCO-Konvention 1960 gegen die Diskriminierung in Erziehungsangelegenheiten herangezogen. Die belangte belgische Regierung hat sich dagegen ausgesprochen, da dieses internationale Instrument mit der Rechtslage nach der EMRK, welche eine internationale gerichtliche Kontrolle vorsieht, nicht verglichen werden könnte und die Mitgliedstaaten der EMRK nur die dort klar umrissenen Verpflichtungen übernommen hätten[67]. Kommission und Gerichtshof sind auf die Frage der Vergleichbarkeit der beiden Instrumente nicht weiter eingegangen.

8. Auch in anderem Zusammenhang ist es denkbar, zur Auslegung von Konventionsbestimmungen auf andere, zwischen Mitgliedstaaten des Europarats geschlossene, dieselben oder ähnliche Materien zum Gegenstand habende Übereinkommen zurückzugreifen.

So besteht etwa ein auf der Hand liegendes Naheverhältnis zwischen der den Schutz des Privatlebens garantierenden Bestimmung des Art. 8 EMRK und dem Europarat-Übereinkommen vom 28. Januar 1981 über den Datenschutz (siehe insbesondere dessen Art. 1)[68] oder zwischen den von Art. 1 des Entwurfs eines 6. Zusatzprotokolls normierten Verfahrensgarantien für den Fall der Ausweisung[69] und den entsprechenden Bestimmungen des Art. 3 des Europäischen Niederlassungsabkommens (Anm. 21) sowie des Art. 13 Weltpakt für bürgerliche und politische Rechte 1966, oder ferner zwischen dem von Art. 8 garantierten Recht auf Achtung des Privatlebens (im Lichte des behaupteten Rechts auf staatliche Anerkennung von »Geschlechtsumwandlungen«) und den Bestimmungen der entsprechenden, in Ausarbeitung befindlichen CIEC-Konvention[70], oder schließlich zwischen dem Recht auf Eigentum nach Art. 1, 1. Zusatzprotokoll 1952 und den entsprechenden Schutzbestimmungen des Art. 23 Europäisches Niederlassungsabkommen (Anm. 21), der Flüchtlingskonvention 1951 und der Staatenlosenkonvention 1954.

---

[66] 23. 7. 1968, Série A Nr. 6, S. 26, § 4.

[67] Kommissionsbericht vom 24. 6. 1965 Série B/1 (1965–1967), S. 96 f., 163, 212; dazu *Opsahl*, La Convention et le respect de la famille et la protection des droits des parents et des tuteurs familiaux dans l'éducation des enfants, in: Vie privée et Droits de l'Homme, Zweites Kolloquium Brüssel, 1970 (1973), S. 294 f.

[68] European Treaty Series Nr. 108.

[69] CEDH (80) 28.

[70] Die Frage hätte im belgischen Fall *van Oosterwijek* von Interesse sein können. Im Verfahren vor dem EGMR war sie allerdings nicht mehr relevant, weil dieser die Zulässigkeit der Beschwerde – wegen Nichterschöpfung des innerstaatlichen Instanzenzugs nach Art. 26 EMRK – verneint hat (6. 11. 1980, Série A Nr. 40=EuGRZ 1981, 275).

V. Zur dogmatischen Eingliederung der Vertragsrechtsvergleichung in das System der Auslegungsregeln der WVK

In den unter IV. erwähnten Beispielen wurden die Bestimmungen der EMRK an denen anderer, zur Gänze oder zum Teil zwischen den EMRK-Mitgliedern geschlossenen Verträgen gemessen. Allen diesen Fällen von Vertragsrechtsvergleichung lag, ganz allgemein gesprochen, der Gedanke zugrunde, daß die in einem Vertrag verwendeten Begriffe im Zweifel in dem Sinn zu verstehen sind, in welchem sie auch in anderen, gleiche oder ähnliche Materien betreffende Verträgen verwendet werden.

Ist eine solche Vorgangsweise erlaubt, ich meine, entspricht sie anerkannten Auslegungsgrundsätzen? Trotz der Häufigkeit des Phänomens hat sich die Lehre bislang nur wenig mit der Frage befaßt. In aller Regel beschränkt sie sich auf die kurze Feststellung, daß die Vergleichung mit anderen Verträgen ein taugliches Auslegungsmittel sein kann [71]. Dabei wird dieser Vorgang vielfach auch mit der Analogie gleichgesetzt [72].

In der Folge möchte ich nun versuchen, die Vertragsrechtsvergleichung als Mittel der Vertragsauslegung in das System der Auslegungsregeln der WVK – welche, wie wir bereits gesehen haben, auch für die Auslegung der EMRK gelten (vgl. zu Anm. 14) – einzugliedern. Dabei soll auch auf die wenigen differenzierteren Aussagen der Lehre zu diesem Problem spezifisch eingegangen werden.

1. Nach der allgemeinen Auslegungsregel des Art. 31 Abs. 1 WVK sind die von einem Vertrag verwendeten Begriffe in ihrer »üblichen Bedeutung« und »in ihrem Zusammenhang« zu verstehen. Man kann nun wohl davon ausgehen, daß die übliche Bedeutung eines Ausdrucks nicht nur sein umgangssprachlicher Sinn, sondern vor allem der Sinn ist, in dem er in den zwischenstaatlichen Beziehungen, insbesondere in Verträgen zwischen Staaten, verwendet wird. Das wird vor allem dann zutreffen, wenn die verschiedenen Verträge den gleichen Gegenstand haben [73], besonders aber, wenn sie nach ei-

---

[71] Beispielhaft bereits *Rivier*, Lehrbuch des Völkerrechts (1889), S. 333; *Ehrlich*, L'interprétation des Traités, RdC 24 (1928 IV), S. 115. Für die neuere Literatur vgl. *Oppenheim/Lauterpacht*, International Law Bd. 1 (8. Aufl. 1967), S. 954; *Rousseau*, Droit international public, Bd. 1 (1970), S. 275 ff.; *Soubeyrol*, L'interprétation internationale des traités et la considération de l'intention des parties, Journal du droit international 85 (1958), S. 692; *Dutoit* (Anm. 8), S. 71.

[72] Vgl. *Zajtay*, Sur le raisonnement par analogie comme méthode d'interprétation du droit, in: Festschrift Aubin (1979), S. 305. Vorbehalte gegen die Argumentation *per analogiam* im Völkervertragsrecht meldet *Bernhardt* (Anm. 12), S. 181 f., an.

[73] Vgl. die Begriffe der Zwangs- oder Pflichtarbeit nach der ILO-Konvention 1930 (Anm. 29) und nach Art. 4 Abs. 2 und 3 EMRK (oben zu Anm. 6), der Koalitionsfreiheit nach der UN-Menschenrechtsdeklaration 1948, der Europäischen Sozialcharta 1961, den ILO-Konventionen 1948 (Nr. 87) bzw. 1949 (Nr. 98), den beiden Weltpakten 1966 und nach Art. 11 Abs. 1 EMRK (oben zu Anm. 60–65).

nem einheitlichen Grundkonzept ausgerichtet sind [74]. Insofern stellt der Vergleich mit einem anderen Vertrag, in dem ein bestimmter Begriff vielleicht klargestellt ist, ein brauchbares Argument dafür dar, daß dieser Begriff im auszulegenden Vertrag auch im gleichen Sinn zu verstehen ist [75]. Ganz eindeutig erscheint mir die Situation, wo ein Vertrag ausdrücklich auf einen anderen verweist [76].

Das vorstehende gilt meines Erachtens aber nicht nur für den Fall, daß die beiden Verträge in einem gewissen sachlichen Zusammenhang – der bis zur gegenseitigen Interdependenz gehen kann [77] – stehen [78]. Dort kann sogar noch das zusätzliche – stärkere – Argument herangezogen werden, daß Vertragsnormen im Zweifel so auszulegen sind, daß Konflikte nach Möglichkeit vermieden werden, Verpflichtungen aus mehreren Verträgen sich vielmehr harmonisch ergänzen sollen. Damit wird gleichzeitig die Auslegungsregel des Art. 31 Abs. 3 *lit.* c WVK angesprochen, derzufolge »neben dem Zusammenhang« (also gleichwertig mit diesem) »alle einschlägigen Regeln des Völkerrechts, die in den Beziehungen zwischen den Parteien anwendbar sind« berücksichtigt werden sollen. Es ist unbestritten, daß dazu nicht nur die Regeln des Völkergewohnheitsrechts und die allgemeinen Rechtsgrundsätze, sondern auch die Regeln anderer, zwischen denselben Partnern in Geltung stehender Verträge zu verstehen sind [79]. Hier zeigt sich auch eine Nahtstelle zwischen dem Problem der Vertragsauslegung im engeren Sinne und dem der Vertragsanwendung in dem Sinne, in dem der Ausdruck in Art. 30 WVK verwendet wird (vgl. dazu oben, zu Anm. 20).

Freilich steht es den Vertragsparteien frei, ein und dieselbe Frage in anderen, verschiedene Ziele anstrebenden Verträgen auch verschieden zu regeln [80], so daß ein und derselbe Begriff in den kontroversen Verträgen nicht immer dasselbe bedeuten muß. Soweit aber eine solche Differenzierungsabsicht

---

[74] Die Literatur der sozialistischen Länder verweist in diesem Zusammenhang auf die im wesentlichen gleichartigen Verträge zwischen sozialistischen Staaten über Freundschaft, gegenseitigen Beistand und Zusammenarbeit, so z. B. *Haraszti* (Anm. 12), S. 148; siehe auch *Schweisfurth*, Der internationale Vertrag in der modernen sowjetischen Völkerrechtstheorie (1968), S. 282 f.; Ähnliches gilt aber gleichfalls etwa für die Rechtshilfeverträge in Zivil- und Strafsachen, für die Auslieferungsverträge usw., auch zwischen ideologisch verschiedenen Partnern.

[75] *Hailbronner* (Anm. 2), S. 222.

[76] *Bernhardt* (Anm. 12), S. 88.

[77] *Ch. De Visscher*, Problèmes d'interprétation judiciaire en droit international public (1963), S. 102 ff.

[78] Ein klassisches Beispiel ist die vom Österr.-Deutschen Schiedsgericht vorgenommene Auslegung des Begriffs »zu caritativen Zwecken« nach Art. 20 Vermögensvertrag, dBGBl. 1958 II, 129=öBGBl. 1958/119, in Übereinstimmung mit Art. 22 Z. 13 des Österr. Staatsvertrags 1955, öBGBl. 1955/152; dazu *Seidl-Hohenveldern*, The Austrian-German Arbitral Tribunal (1972), S. 87 f.

[79] *Bernhardt* (Anm. 15), S. 500; *Haraszti* (Anm. 12), S. 146 Anm. 181.

[80] Darauf weist *Bernhardt* (Anm. 12), S. 87 f., hin; siehe ferner *Haraszti* (Anm. 12), S. 147 f.

nicht aus dem Vertrag hervorgeht, erscheint die hier erörterte Interpretationsmethode durchaus legitim, mag es auch richtig sein, daß nur ein vorsichtiger Gebrauch davon gemacht werden darf.

2. Wenn ferner Art. 31 Abs. 1 WVK von der Auslegung der in einem Vertrag verwendeten Begriffe »im Zusammenhang« spricht, meint er in erster Linie, wie auch aus der Aufzählung in Art. 31 Abs. 2 hervorgeht, den Zusammenhang bzw. die systematische Interpretation im engeren Sinne. In der Tat ist im Zweifel nicht davon auszugehen, daß – bei sorgfältiger Redaktion – innerhalb des geschlossenen Systems eines Vertrags mit ein und demselben Begriff an verschiedenen Stellen Verschiedenes ausgedrückt werden soll, und insbesondere nicht, daß eine widersprüchliche Regelung gewollt ist. Die einzelnen Normen und Teile eines Systems sind vielmehr als sinnhaft aufeinander bezogen aufzufassen. Ein Vertrag mit seinen Anhängen, Zusatzprotokollen, Erklärungen usw. stellt nun – für seinen Bereich – ebenso ein in sich geschlossenes »System« dar wie eine staatliche Rechtsordnung.

Die Völkerrechtsordnung als Ganzes bildet aber kein geschlossenes System, auch wenn die internationale Gemeinschaft als eine Rechtsgemeinschaft angesprochen werden kann [81]. Sie ist weniger durch hierarchische Unterordnung als durch die Gleichordnung ihrer Glieder gekennzeichnet. Durch Verträge und Kodifikationen – eine rationalisierte Form von Rechtssetzung [82] – wird aber eine Koordinierung, zumindest in Teilbereichen, angestrebt. Ist ein gewisser Grad an »Vertragsdichte« erreicht, insbesondere wenn es auch zur Bildung gemeinsamer Organe gekommen ist, geht die Koordination in eine Integration über. In den erfaßten Teilbereichen kann man meines Erachtens von einem koordinierten bzw. integrierten System sprechen [83]. Das trifft – vornehmlich für den Bereich der Grundrechte – auch für die Europarat-Staaten zu: Die Präambel zur Satzung beruft sich auf die »Verbundenheit mit den geistigen und sittlichen Werten, die das gemeinsame Erbe ihrer Völker sind« (Abs. 3), und spricht anschließend (Abs. 4) von den europäischen Ländern, die »von demselben Geist beseelt sind« und zwischen denen »eine engere Verbindung« hergestellt werden soll. Denselben Gedanken greift auch die EMRK auf, die in Abs. 5 der Präambel erneut von den europäischen Staaten spricht, »die vom gleichen Geiste beseelt sind und ein gemeinsames Erbe an geistigen Gütern, politischen Überlieferungen, Achtung der Freiheit und Vorherrschaft des Gesetzes« besitzen [84].

---

[81] *Mosler*, The International Society as a Legal Community, RdC 140 (1974 IV).

[82] *Scheuner* (Anm. 4), S. 409.

[83] Insofern stellen die Europarat-Staaten, um mit *Mosler* (Anm. 15), S. 526, und mit *Hailbronner* (Anm. 2), S. 195, zu sprechen, gleichzeitig eine regionale Ordnung und einen ideologisch geprägten Völkerrechtskreis dar. Daß Homogenität aber auch außerhalb regionaler Verbände bestehen kann, hat *Bleckmann*, Völkergewohnheitsrecht trotz widersprüchlicher Praxis?, ZaöRV 36 (1976), S. 402 ff., nachgewiesen.

[84] In den Fällen *Nationale Belgische Polizeigewerkschaft* (Anm. 60), § 38, *Schwedischer Lokomotivführerverband* (Anm. 60), § 39; *Marckx* (Anm. 31), § 41=EuGRZ

Von einem solchen koordinierten oder integrierten System kann aber angenommen werden, daß es in sich widerspruchsfrei ist [85], so daß eine systematische Interpretation im weiteren Sinne bzw. eine methodologische Interpretation gerechtfertigt erscheint [86]. Freilich gilt auch in diesem Zusammenhang, daß es den Gliedern des Systems freisteht, einzelne Fragen untereinander auch unterschiedlich zu regeln.

Mit der systematischen oder methodologischen Interpretation hat sich vor allem die sozialistische Völkerrechtslehre befaßt [87]. Sie wird dort als Unterfall der logischen Interpretation betrachtet oder auch auf die Auslegungsregel des Art. 31 Abs. 3 *lit.* c WVK gestützt. Dabei wird teilweise in Zweifel gezogen, ob hier überhaupt noch von Rechtsvergleichung gesprochen werden kann [88], weil eine solche die Vergleichung von Begriffen und Institutionen aus verschiedenen Rechtsordnungen zum Gegenstand hat, die methodologische Interpretation aber gerade von der Prämisse der »Systemeinheit« ausgeht. Auch hinsichtlich der Auslegung, durch Gegenüberstellung, von Verträgen der Europäischen Gemeinschaften wird behauptet, daß hier keine Rechtsvergleichung vorliege, weil das Gemeinschaftsrecht eine Einheit darstellt [89]. Aber das sind meines Erachtens letztlich nur terminologische Probleme [90].

Ist eine systematische Vergleichung nur mit Verträgen zwischen denselben Partnern gestattet [91]? Gewiß kann man von einem »System« nur bei Identität der Parteien sprechen. Aber auch wo eine solche Identität nicht gegeben

---

1979, 457, wurde ausdrücklich darauf hingewiesen, daß es bei den zu Vergleichszwekken herangezogenen Verträgen um solche des Europarats bzw. um Verträge handelt, denen auch alle europäischen Staaten angehören. Im Fall *X* gegen *Republik Irland* (Anm. 65) wurde hervorgehoben, daß die zu Vergleichszwecken herangezogene ILO-Konvention 1948 (Nr. 87) von allen MRK-Mitgliedern – mit Ausnahme der Türkei – ratifiziert worden war.

[85] Vgl. zu Anm. 53 und 58.

[86] Siehe auch *Bülck*, Vertragsauslegung, in: Strupp-Schlochauer, Wörterbuch des Völkerrechts, Bd. 3 (1962), S. 551.

[87] Siehe die Zusammenfassung bei *Schweisfurth* (Anm. 74), S. 282 f.

[88] *Haraszti*, The Comparative Method in International Law, in: Butler (zu Anm. 9), S. 121 f.

[89] *Kutscher* (Anm. 12), I-24.

[90] Im Sinne von *Zweigert*, Rechtsvergleichung, in: Strupp-Schlochauer, Wörterbuch des Völkerrechts, Bd. 2 (1962), S. 80, würde es sich hier um interne (im Gegensatz zur externen) Rechtsvergleichung handeln.

[91] Die Problematik hat eine gewisse Ähnlichkeit mit derjenigen, die sich bei der Feststellung von Völkergewohnheitsrecht oder von allgemeinen Rechtsgrundsätzen stellt, wo gefragt wird, ob ein Rechtsgrundsatz in allen Rechtsordnungen anerkannt sein muß, um als ein »allgemeiner« zu gelten. Mit Recht wird in diesem Zusammenhang betont, daß eine Anerkennung in den wesentlichsten Rechtsordnungen ausreicht, ein Stillschweigen einzelner nicht stört und daß nur bei widersprüchlichen Auffassungen darüber einem Rechtsgrundsatz der Charakter der Allgemeinheit nicht mehr zukommt; vgl. *Bernhardt* (Anm. 12), S. 156 f.; *Hailbronner* (Anm. 2), S. 221; *Zemanek* (Anm. 7), S. 459, 470 f.; *Dutoit* (Anm. 8), S. 68, 73.

ist, wird man immerhin das (schwächere) Argument der »üblichen Bedeutung« im obigen Sinne heranziehen können. Ähnliches gilt für Verträge, die zwar zwischen denselben Partnern ausgearbeitet wurden, die aber noch nicht für alle in Kraft getreten sind, und, im geringeren Ausmaß, für Verträge, die im Stadium des Entwurfs steckengeblieben sind. Gerade bei den letzteren ist aber das Argument nicht von der Hand zu weisen, daß die Staaten die vorbereitete Regelung schließlich doch nicht gewollt haben, so daß solche nicht verbindlich gewordene Regelungen kaum ein geeignetes Mittel zur Erkundung des anderen Verträgen innewohnenden Sinns sein können [92].

3. Unbestritten ist es, daß ein früherer Vertrag, der einem späteren (nachweislich) zum Vorbild gedient hat, zur Auslegung des letzteren herangezogen werden kann. Das geschieht sogar sehr häufig vor den und seitens der Organe der EMRK [93]. Diese Art der Auslegung liegt auf einer ähnlichen Ebene wie die Berufung auf die *travaux préparatoires;* sie stellt eine legitime Auslegungsmethode nach Art. 32 WVK dar.

4. Spätere Verträge, insbesondere soweit sie zwischen denselben Partnern geschlossen wurden, können Ausdruck einer »nachfolgenden Praxis« sein; eine Berufung auf sie ist daher gleichfalls ein geeignetes Interpretationsmittel nach Art. 31 Abs. 3 *lit.* b WVK [94].

5. Frühere – aber doch in einem gewissen zeitlichen Zusammenhang – besonders aber gleichzeitig geschlossene Verträge zwischen denselben Partnern können als Ausdruck der »allgemeinen Lebensordnung« betrachtet werden, in welche ein Vertrag eingebettet und in deren Licht er im Zweifel zu verstehen ist [95]. Dieses Argument wird besonders zu einer statischen bzw. historischen [96] Auslegung herangezogen werden können. Umgekehrt sind neuere Verträge Ausdruck der jetzt herrschenden Verhältnisse und Konzeptionen (in deren Rahmen die EMRK – nach der Spruchpraxis ihrer Organe – auszulegen ist [97]); sie können daher für eine evolutive Interpretation brauchbare Argumente liefern.

---

[92] Solche Verträge entfalten auch nicht die von *Strebel* (Anm. 4), S. 312, angesprochene »indirekte Wirkung«, bzw. sie besitzen, um mit *Geck* (Anm. 4), S. 139, zu sprechen, in aller Regel keine »Ausstrahlungskraft«. Im Fall *Marckx* (Anm. 31) hat es der EGMR allerdings abgelehnt, aus der Tatsache, daß die zu Vergleichszwecken herangezogenen zwei Übereinkommen nur eine geringe Anzahl von Ratifikationen aufweisen konnten, Konsequenzen in die Richtung zu ziehen, daß sie nicht Ausdruck der heute im Bereich der europäischen Staaten herrschenden Konzeptionen wären.

[93] Vgl. etwa zu Anm. 46; besonders ausführlich das Sondervotum *Kellberg*, dem *Nørgaard* beigetreten ist, zum Kommissionsbericht vom 14. 12. 1979 betr. die Beschwerden 7601/76, 7806/77 (*Closed shop*-Fall; siehe oben zu Anm. 64), S. 44. Siehe ferner *Spencer*, L'interprétation des Traités par les Travaux préparatoires (1935), S. 72 ff.

[94] *Karl*, Vertrag und spätere Praxis im Völkerrecht (Habilitationsschrift, Salzburg, 1981 [im Druck]).

[95] Dazu *Bernhardt* (Anm. 12), S. 134 ff.

[96] In diesem Sinne *Bülck* (Anm. 86), S. 500.

[97] Vgl. bei Anm. 41.

Ist es möglich, zwischen den von der Vertragsrechtsvergleichung, in ihren verschiedenen Anwendungsformen, gebotenen Auslegungshilfen eine »Rang-« oder »Wertordnung« aufzustellen? Generell sei gesagt, daß die Vertragsauslegung – so wie jede Art von Normenauslegung – ein komplexer Vorgang ist, in dem sich normative mit psychologischen, dogmatische mit weltanschaulichen und politischen Elementen mischen[98] und der sich nicht durch mechanische Regeln steuern läßt. Die Heranziehbarkeit bzw. der »Wert« der verschiedenen Auslegungsmittel wird sich letztlich daher nur im Einzelfall beurteilen lassen. Dazu kommt, daß ein und dieselbe Erscheinung Argumente verschiedenster Art – und nicht nur solche gegensätzlicher Natur[99] – liefern kann: Ein früherer Vertrag kann dem späteren, auszulegenden, als Vorbild gedient haben; auch wo das nicht der Fall ist, können die von ihm verwendeten Begriffe Ausdruck der »üblichen Bedeutung« sein, in der die des späteren gleichfalls zu verstehen sind, bzw. der frühere Vertrag kann Ausdruck der gleichen »allgemeinen Lebensordnung« sein, in die beide Verträge eingebettet sind. Oder: die systematische bzw. methodologische Interpretation kann sowohl auf den »Zusammenhang« als auch auf das Gebot der Berücksichtigung der in den Beziehungen zwischen den Vertragsparteien anzuwendenden »einschlägigen Völkerrechtsregeln« gestützt werden usw.

Aus dem vorhergehenden möchte ich den Schluß ziehen, daß die Aufstellung einer starren, allgemeingültigen Rangordnung zwischen den einzelnen, von der Vertragsrechtsvergleichung gebotenen Auslegungshilfen nicht möglich ist. Unberührt bleibt aber das – letztlich auch dem System der WVK zugrunde liegende (siehe auch Anm. 15) – Prinzip[100], daß sich jede Auslegung von geschriebenen Rechtsnormen in erster Linie an den (Vertrags)Text zu halten hat; dazu gehören auch der engere und der weitere Zusammenhang, in dem die auszulegende Vertragsbestimmung steht; ferner, daß die Umstände, die außerhalb des Vertragstextes liegen, zu ihm selbst aber doch eine gewisse nähere Beziehung aufweisen, wie die Begleitumstände des Vertragsabschlusses, Vorarbeiten, herangezogene Vorbilder, frühere und spätere Praxis, erst in zweiter Linie zu berücksichtigen sind. Noch loser ist die Bezie-

---

[98] *Ch. De Visscher*, Théories et réalités en Droit international public (3. Aufl. 1960), S. 317 ff.

[99] Ein Vertrag, besonders aber eine Mehrheit von Verträgen über dieselbe Materie, kann Ausdruck des Vorhandenseins eines Völkergewohnheitsrechtssatzes sein, bzw. es läßt sich daraus schließen, daß eine ähnliche, im auszulegenden Vertrag nicht ausdrücklich geregelte Frage im gleichen Sinn, wie in den anderen Verträgen, zu beurteilen ist; aus der Tatsache, daß nur eine bestimmte Materie geregelt oder daß nur zwischen bestimmten Partnern eine vertragliche Regelung darüber getroffen wurde, kann aber ebenso geschlossen werden, daß außerhalb der ausdrücklichen vertraglichen Regelung das Bestehen einer entsprechenden völkergewohnheitsrechtlichen Norm nicht anerkannt wird bzw. daß die getroffene Regelung nur für die von ihr ausdrücklich erfaßte Materie oder nur zwischen den Partnern gilt, welche diese Regelung ausdrücklich getroffen haben (vgl. auch zu Anm. 4, ferner zu Anm. 57).

[100] *Bernhardt* (Anm. 12), S. 186 ff.

hung zur »allgemeinen Lebensordnung« in die der Vertrag eingebettet ist; sie kann daher nur eine ganz subsidiäre Auslegungshilfe bieten. Nach diesem Prinzip ist auch der »Stellenwert« der einzelnen von der Vertragsrechtsvergleichung, in ihren verschiedenen Anwendungsformen, gebotenen Auslegungshilfen zu beurteilen.

The Legislative Rôle of the World Court in an Era of Transition

EDWARD MCWHINNEY

Judge *Mosler*'s concern with the judicial process considerably antedates his election to the World Court and his prior election to the European Court of Human Rights. It is a long-standing interest, from the academic world, reflected in two major symposium volumes – one on constitutional review in national legal systems[1], and the other on the judicial settlement of international disputes[2], in which Judge *Mosler* was a principal in the intellectual planning and organisation. The juxtaposition of those two legal volumes, the one comparative and municipal, the other international, confirms the basic truth that international adjudication is, historically, deeply rooted in particular national legal experiences and national legal thought-ways, both as to its procedural forms and practice and also as to its overriding philosophy and political *raison d'être* or justification. It is both a problem for international adjudication today when so much of international law procedure, institutions, and substantive principles is under attack today on account of its alleged ethnocentricity or cultural relativity, but also an opportunity in so far as the particular national legal systems from which international adjudication borrowed so heavily in its earliest, formative years do not present one single monolithic bloc of experience but rather a plurality of different patterns and traditions offering alternative choices for application and adjustment to rapidly changing political conditions in the World Community.

An age of transition (or revolution) such as we live in today, in the World Community, in which the passage from an older, established public

---

[1] Verfassungsgerichtsbarkeit in der Gegenwart, Länderberichte und Rechtsvergleichung – Constitutional Review in the World Today, National reports and comparative studies – La Juridiction Constitutionnelle à l'époque contemporaine, Exposé de la situation dans différents pays et étude comparée (Beiträge zum ausländischen öffentlichen Recht und Völkerrecht, Vol. 36) (1962).

[2] Judicial Settlement of International Disputes. International Court of Justice, Other Courts and Tribunals, Arbitration and Conciliation (Beiträge zum ausländischen öffentlichen Recht und Völkerrecht, Vol. 62) (1974).

order system is clear, but where what is not so clear is the predominant ideological orientation and basic social and economic organisation of the new system emerging in its place and also its degree of tolerance of intellectual cross-currents and divergencies[3], has difficulty in maintaining consensus as to its own ordering principles. These ordering principles – what *Pound*, borrowing from *Kohler*, identified as jural postulates of a civilisation period[4], and what contemporary jurists tend to call the goal values, are the keys to the legal resolution of interests-conflicts in concrete problem-situations where legal doctrine and past jurisprudence are ambiguous or provide no clear-cut answer. The difficulty is that the goal values advanced today to order and control present disputes are likely to be those of yesterday or the day before yesterday, and hardly likely to be acceptable to or compatible with the newer social and economic forces challenging the older international order now in process of disappearance. An international adjudicatory institution runs the risk, in an era of transition, of ruling in vain, and issuing its judicial *fiat* with the political likelihood that, in the absence of international enforcement machinery whose own efficacy is, of course, predicated upon an existing substantial societal consensus, its ruling will be ignored or flouted by the party adversely affected. It is better, in these circumstances, to approach the *rôle* of international adjudication today less in terms of a conception of timeless absolutes of legal theory and legal organisation, good for all historical seasons and all places, than of what *Radbruch* characterised as legal antinomies[5] – different, sometimes directly competing principles and practice, that offer alternative models for international legal development for the future, and the ultimate choice among which will be determined by the ordinary historical-dialectical process of empirical, trial and error testing, and political give-and-take and bargaining, and pragmatic compromise. In considering antimonies in international adjudication and in Court-based settlement or resolution of international conflicts generally, in the current era of transition, some key questions arise.

First, is judicial settlement of disputes *ethocentric* in character, and more specifically Western, in the special sense of being rooted in distinctive Western legal-cultural experience; or is it more nearly universal in character and potentiality of application today? If it is the former, of course, it will politically rise and fall with those major Western legal systems whose political leaders institutionalised it, as part of the post-World War I »victors' settle-

---

[3] See the new UNESCO series, »Nouveaux défis au droit international«, and its lead-off volume, – *Mohammed Bedjaoui*, Pour un nouvel ordre économique international (1979). And see also the author's Conflict and Compromise. International Law and World Order in a Revolutionary Age (1981).

[4] *J. Kohler*, Lehrbuch der Rechtsphilosophie (1908); *Roscoe Pound*, Introduction to American Law (1919); *Pound*, Social Control through Law (1942); *J. Stone*, The Province and Function of Law. Law as Logic, Justice, and Social Control (1946).

[5] See generally *G. Radbruch*, Rechtsphilosophie (4th ed. 1950).

ment« at Versailles, in the old Permanent Court, and then continued it, after 1945, with the successor International Court of Justice. Within the special legal community of the United Nations, there has been a tendency to denigrate the International Court, and international adjudication in general, as Western, or more precisely »Western Imperialist« in character. One remembers, here, the angry Third World characterisation of the one-vote majority judgment of the Court in *South West Africa, Second Phase* in 1966[6] as a »White Man's«, »colonialist«, verdict after the equally close judgment, the other way, in *South West Africa, Preliminary Objections* in 1962[7] had seemed to point the way to a more creative, historically deterministic, judicial interpretation that would seek to re-write and up-date the old League of Nations-era institution of the Mandate so as to accord with the contemporary international law imperative principle of self-determination of peoples and, above all, of colonial peoples. These quick political reactions and criticisms in the U.N. General Assembly to *South West Africa, Second Phase* have given way, by now, to legally more nuanced evaluations that take account, also, of »technical«, procedural factors in the Court's 1966 ruling, and of the degree of relevance of existence of a proper, adjectival law base before a court can or should rule on substantive legal issues. There is also a political recognition that while the Court, *qua* institution, is permanent, including in its institutional *rôle* its capacity to alter or change International Law for the future, its actual membership composition is not; and that the accidents of particular judicial personalities on the Court at any time and thus the Court's ideological orientation, can be corrected, albeit somewhat slowly, through the staggered system of replacement of one third of the Court's membership, every three years, in the periodic elections of judges by the Security Council and General Assembly. A less overtly political and more intrinsically legal (strictly, constitutional-legal) objection to the Court's post-1945, tendencies to an expansionist, legislative *rôle* has been the Soviet and Soviet-bloc insistence that it must function as a court *stricto sensu*[8]. On this argument, the Court must respect its own jurisdictional requirements and limit itself to narrow technical constructions, avoiding large, »policy«-style interpretations or the venturing unnecessarily upon broad, »political« questions. The antinomy, here, is between the Soviet, internal (Municipal, constitutional) law rejection of judicial policy-making as élitist and basically undemocratic in character, with the concomitant insistence on the duty of judicial deference to the

---

[6] *South West Africa,* Second Phase, Judgment, I.C.J. Reports 1966, p. 6.

[7] *South West Africa,* Preliminary Objections, Judgment, I.C.J. Reports 1962, p. 319.

[8] The special Soviet and other national conceptions of the World Court and of international adjudication are discussed in the author's The World Court and the Contemporary International Law-Making Process (1979), at p. 13 *et seq.* As to comparative attitudes to the *rôle* of adjudication in Municipal, national law, see generally the author's Judicial Review (4th ed. 1969).

other, more »popular« (executive-legislative) institutions of government, and the more overtly »American«, free-wheeling, legislative approach to the judicial function. The antinomy is not in fact a pure East-West (Soviet-Western) one, for the Soviet doctrinal position in regard to the non-policy *rôle* of the World Court is paralleled in the internal (Municipal, constitutional) law thinking of Anglo-Saxon, Common Law countries other than the United States and has its protagonists, even within the United States, in the school of judicial self-restraint associated in particular with the late Mr. Justice Felix Frankfurter of the United States Supreme Court[9]. Continental European Civil Law-based thinking on international adjudication and the limits of the World Court's law-making *rôle* seems also to be more nuanced than the »American« approach and, in fact, quite close to the Soviet approach in some more recent Court jurisprudence.

This leads on to a second question having to do with the main alternative, technical conceptions of the judicial process, national and international. The basic antinomy is between strict construction – »grammatical«, »logical«, »historical« interpretation, on the one hand, and broad, purposive »teleological« interpretation, on the other; and it is a distinction known, under slightly different rubrics, to Common Law and Civil Law based legal systems alike. The Legal Realists of the Common Law World, in insisting on the fallacies of the logical form inherent in legal reasoning and the impossibility, therefore, of divorcing a »logical« from a »policy« interpretation of statute or past judicial precedent, find their analogues in the French Civil Law and Gény's *libre recherche scientifique* and the German Civil Law's *Freirechtsschule*. The distinction between the logical, orthodox, positivist, »black-letter law«, strict interpretation on the one hand, and a creative, legislative, policy interpretation on the other, is not an absolute one. Rather than being thought of as a dichotomy, the different approaches are better conceived as points upon a continuum; though the difference of degree can be a considerable one. The two basic approaches are well represented on the International Court of recent years, though the main individual protagonists seem, very often, to be less the consequence of objective factors of national legal culture and national legal thought-ways, than the product of casual, subjective factors of pre-election training and professional exposure and hence of the accidents of judicial nomination and judicial elections. There will be frequent intellectual cross-currents, in other words. Thus, within the Anglo-Saxon Common Law special legal community on the Court at the one time period, Sir Percy Spender and Sir Gerald Fitzmaurice[10] find themselves on the opposite side of the Court to Philip Jessup[11], in the decision on *South West Africa, Second Phase,* and with diametrically opposing philosophical con-

[9] *Ibid.*

[10] I.C.J. Reports 1966, p. 6, at p. 8.

[11] *Ibid.* at p. 323.

ceptions of the judicial *rôle*. Is the crucial difference, here, one between a »British« and an »American« Common Law approach, or one between the Equity Bar approach (which Spender and Fitzmaurice both reflected in their training and experience), and the more generalist, policy approach of a more generalist, non-legal, State Department adviser? How does one explain the close intellectual parallels between the strict, procedural, adjectival law approach of Judges Spender and Fitzmaurice in *South West Africa, Second Phase,* and the substantial technical emphasis, in the same case, of the two Soviet bloc members of the Court, Judge Winiarski (who joined with the Court majority, in the official Opinion of Court signed by Sir Percy Spender, as President) [12] and Judge Koretsky who, though dissenting, did so in a Dissenting Opinion written on a deliberatley restrained, technical, procedural, adjectival law argumentation [13], – unless it be on the basis that they had a common conception of the Court *qua* institution and of the obligations and limits of any judicial legislative function, that transcended elements of national legal origins or of casual national political advantage in the instant case? One comes very close, in such circumstances, to classical notions, common to most legal systems, of the independence of the judiciary and of the collegial responsibility of individual judges to the larger institution of which they are part.

When one examines the actual record since *South West Africa, Second Phase,* one is struck by the fact that, in spite of the angry recriminations in the General Assembly following on the Court decision, there has been very little direct follow-up in those ways directly envisaged, under the U.N. Charter, for changing Court membership and thus any particularist, ideological orientation claimed to be discerned in its decisions and objected to by the General Assembly majority of the day. The triennial elections to the Court, since the Court decision was handed down in the Summer of 1966, have followed the same general patterns of voting that have operated in the Security Council and General Assembly in the judicial selections since the creation of the United Nations after World War II: the Gentlemen's Agreement involving a »regional« distribution and allocation of membership that is balanced in geographical and general political-ideological terms, and even in strictly technical, linguistic and legal-systemic terms, and that is the stronger, perhaps, for resting on pure custom or convention and not being formally codified in positive law terms, as, for example, with the complicated »regional group« representation formula for the expanded International Law Commission, stipulated in the General Assembly Resolution of 20 November, 1981 amending the Commission Statute [14]. The informal, Gentlemen's

[12] *Ibid.* at p. 8.
[13] *Ibid.* at p. 237.
[14] U.N. General Assembly A/RES/36/39, 20 November 1981: Enlargement of the International Law Commission: amendments to articles 2 and 9 of the Statute of the Commission.

Agreement as to the Court gives what may be considered, with a tribunal of only 15 members, an overly preponderant weight to the five Permanent Members of the Security Council that is no longer warranted, in the case of several at least of them, by the political facts-of-life of effective power in the World Community of today, however it may have been at San Francisco in 1945. Thus, except for the one case of China, and that only from the Court elections of 1966 onwards when the historical anomaly of nationalist China's continuing to occupy the Chinese seat in the United Nations to the exclusion of the People's Republic, had become acute, the »Big Five« countries of 1945 have always had their judicial nominees elected or reelected to the Court; though the noticeable slippage of votes for the French, and more particularly for the British, nominees for the most recent, 1981 elections[15], suggests that any polite custom of automatic election in these two national cases may be in doubt in the future. Except, however, in the one negative instance, in the election of 1966 that immediately followed on the *South West Africa, Second Phase* decision, when the exceptionally well-qualified (in U.N. and general terms) Australian candidate, Sir Kenneth Bailey, was beaten because, (it is generally thought), of resentment against his fellow-countryman, Sir Percy Spender's, deciding part in that case with his second, tie-breaking vote, as President, political-ideological considerations do not appear to have been preponderant in weighting the final voting for judges. The ultimate choices seem to have turned, rather, on somewhat marginal geographical-regional factors, such as the exact number of seats to be allocated to Latin America or to Continental Western Europe. The clear Third World majority in the United Nations General Assembly that united so effectively in the mid-1970s to enact the abstract juridical framework of a New International Economic Order, has been unable or unwilling, or simply not sufficiently interested, to make common cause in marshalling its voting numbers so as to try to place a Third World-leaning, »activist« core of judges, committed to the progressive development of international law through judicial legislation, – on the Court. The innovations, in terms of judicial personnel, have occurred incrementally, through the nuanced choice of one individual national candidate rather than another, within a particular regional bloc grouping, rather than through any attempt at direct political break-through by displacement of the existing political balance between such regional groups.

More overt reactions to a conceived new political *rôle* for the Court seem, in fact, to have come from the traditional, long-term supporters of the Court

---

[15] United Nations, Department of Public Information, Press Release GA/6478, 5 November 1981. While the French and British nominees, in that order, headed the polls in both the General Assembly and the Security Council, the actual votes obtained, even with two outstanding candidates, fell short of the proportions of previous elections: for the French nominee, 127 out of 155 votes cast in the General Assembly, and 11 out of 15 votes cast in the Security Council; for the British nominee, 107 out of 155 cast in the General Assembly, and 10 out of 15 cast in the Security Council.

and of the cause of judicial settlement in general. While the negative reactions, in *Certain Expenses of the United Nations*, in 1962[16], of the French judge, and also of the two Soviet bloc judges[17], might reasonably have been expected to the rather precipitate and somewhat heavy-handed U.S. initiative, in terms of Article 19 of the Charter, to deprive France and the Soviet Union of their votes in the General Assembly because of their non-payment of the special financial assessments for the U.N. military operations in the Middle East and the Congo, rather more surprising perhaps was the thinly veiled criticism in some Western circles of the *rôle* of the non-Western European judges in their special opinions in what was essentially an intra-Western European dispute of the nature that used to constitute the agenda of the old, League of Nations-era, Permanent Court of International Justice, namely the *North Sea Continental Shelf* judgment of 1969[18]. For both in its limited (Western European) range of parties, and also its limited technical (essentially non-philosophic or non-policy) issue presented for judicial interpretation, *North Sea Continental Shelf* came closest of all to that more traditional, »legal« problem, – ripe for judicial determination rather than political decision-making – that used to characterise the work of the »old« Court in its original, European »family compact« era of its history; and there was therefore no call for the interpolation of non-Western European opinions or parties extraneous to the original legal dispute between the original legal parties. The Court itself, in response to and in recognition of this »special legal community« – whether political-ideological or geographical-regional – aspect of its historical work-load, has moved in recent years to sponsor recourse to special regional chambers constituted within its general jurisdiction, as a means of encouraging recourse to international adjudication. But traditional friends of the Court who share the same, older, Western European, legal *Weltanschauung*, like Great Britain and France, have anticipated the Court's new policy by bypassing its jurisdiction altogether in favour of the creation of special *ad hoc* Commissions for the arbitrament of their own »special legal community« disputes, without the distracting presence of judges stranger to their intra-Western European family. Such anticipatory, defensive moves, while no doubt politically understandable, are regrettable, since placing too little faith in the plenary jurisdiction of the Court and in the Court's own capacity for creative adjustment and accommodation of its own special problem-solving techniques and processes to a wide range of problems, – the older, technical-legal ones as well as the newer policy ones.

In fact, since *South West Africa, Second Phase,* in 1966, the Court seems to have moved with considerable flexibility and pragmatism in a sea of new

---

[16] *Certain Expenses of the United Nations,* Advisory Opinion of 20 July 1962, I.C.J. Reports 1962, p. 151.

[17] Winiarsky J., (dissenting), *ibid.,* at p. 230; Koretsky J., (dissenting), *ibid.*, at p. 268.

[18] *North Sea Continental Shelf,* Judgment, I.C.J. Reports 1969, p. 3.

problems, some of these of a high political character. The legal counter to the earlier political charges of an inherent Western judicial bias in *South West Africa, Second Phase*, was supplied, decisively, in the *Namibia (South West Africa)* Advisory Opinion of 1971 [19], when the Court, by overwhelming judicial majorities, recognised the »termination of the Mandate« originally granted by the old League of Nations to the present Republic of South Africa over South West Africa (Namibia), and also the »illegality of South Africa's presence in Namibia«, but also, at the same time, made imaginative, innovatory extensions into the »new« International Law in both its substantive and its procedural aspects. In this regard, the Court reached out for the concept of Inter-Temporal Law and the time dimension in legal interpretation, to conclude that the Mandate granted over the former Germany colony (Namibia) in 1919, whatever its original character, must be interpreted today in an evolutionary rather than a static way, particularly in respect to that »sacred trust of civilisation« that mandates connoted and whose ultimate objective was the self-determination and independence of the peoples concerned [20].

The intellectual boldness with which the Court majority approached both substantive legal issues and also new procedural issues, including herein judicial fact-finding and the possibilities of taking judicial notice of matters of general public knowledge and already within the public domain [21] – including those issues of »accumulation of expressions of condemnation of *Apartheid* ... proof of the pertinent contemporary international standard« [22], on which Judge Jessup had stood alone in his dissenting opinion in *South West Africa, Second Phase,* five years earlier – should not lead us to conclude that the Court will now venture into a form of »policy-making in the air« without proper regard for observance of preliminary, procedural or adjectival law requirements. Thus in *Nuclear Tests (Australia* v. *France)* in 1974 [23] which, on first sight, might have seemed to furnish the Court with a ready-made and golden opportunity to place itself on the side of the Angels by ruling that high-altitude atmospheric nuclear tests violated »new« international law norms as to international environmental protection and the control of pollution and nuclear fall-out, the Court correctly refused to enter on discussion of any such substantive legal issues once the original dispute had become moot [24] with France's (in the Court's view, legally binding) unilateral

---

[19] *Legal Consequences for States of the Continued Presence of South Africa in Namibia (South West Africa)* notwithstanding Security Council Resolution 276 (1970), Advisory Opinion, I.C.J. Reports 1971, p. 16.

[20] *Ibid.,* at p. 27 *et seq.*

[21] *Ibid.,* at p. 50.

[22] Jessup J., (dissenting), I.C.J. Reports 1966, p. 6, at p. 441.

[23] *Nuclear Tests (Australia* v. *France),* Judgment of 20 December 1974, I.C.J. Reports 1974, p. 253.

[24] *Ibid.,* at p. 271–272.

declaration of intention to cease such high altitude nuclear tests altogether[25]. The Court's approach, here, fully accords with all the canons of judicial self-restraint and avoidance of substantive rulings where the preliminary adjectival law conditions are not met or no longer exist, found in municipal, national legal experience, including that of the United States.

On the other hand, in the *Western Sahara* Advisory Opinion of 1975[26], the preliminary, procedural issues being properly satisfied, the Court, again by lop-sided judicial majorities, was not afraid to re-examine »old« international categories as to territorial sovereignty and the acquisition by European colonial powers of title to lands previously occupied by non-European, indigenous or aboriginal peoples (*terra nullius*). Some of the specially concurring opinions, in particular, go beyond the necessities of the case to dispose of the concept of *terra nullius* itself, thereby rejecting the purely Western, ethnocentric conception of the legal modes of establishing sovereignty over territory. The application of the doctrine of Inter-Temporal Law to the Western Sahara situation, on this particular judicial approach, would seem to involve an inter-cultural aspect, – the spatial as well as the temporal element in the international legal space-time dimension[27].

The culmination of the Court's intellectual evolution, since *South West Africa, Second Phase,* in 1966, and its increasing confidence in making use of the judicial process (including the techniques of judicial self-restraint) in problem-solving, having a high political character, of the sort that is likely increasingly to come to it in an era of general transition in the World Community, should have been demonstrated in *United States Diplomatic and Consular Staff in Tehran (United States* v. *Iran)* in 1980[28]. Unfortunately for the Court and for the more general cause of the progressive development of International Law adumbrated in Article 13 (1) (a) of the Charter, the substantive legal conflicts involved were not clearly presented and there was no direct joinder of legal issue due to the failure of one of the parties either to be represented in the hearings before the Court or even, failing an appearance, to file a written memorandum as France had done in *Nuclear Tests (Australia* v. *France)* before it. That case, brought by the United States against Iran, and in respect to which the Iranian Government chose not to be represented before the Court, had arisen from the taking over of the United States Embassy in Tehran, and the forcible detention of the U.S. diplomatic and consular personnel on the premises, by Iranian demonstrators in November, 1979, im-

---

[25] *Ibid.,* at p. 267.

[26] *Western Sahara,* Advisory Opinion, I.C.J. Reports 1975, p. 12.

[27] Institut de Droit International, »Le problème dit du droit intertemporel dans l'ordre international«, (Report of 11th Commission), Annuaire de l'Institut de Droit International 55 (1973), p. 1; and see, generally, the author's The World Court and the Contemporary International Law-Making Process (1979), at p. 93 *et seq.*

[28] *United States Diplomatic and Consular Staff in Tehran (United States of America* v. *Iran),* Judgment of 24 May 1980, I.C.J. Reports 1980, p. 42.

mediately after the Carter Administration of the United States' decision to admit the ex-Shah of Iran from Mexico to the United States for medical treatment. In its interim order of December, 1979[29], the World Court took note of a letter received from the Iranian Foreign Minister contending that the American hostages issue represented only –

»a marginal and secondary aspect of an overall problem, one such that it cannot be studied separately, and which involves, *inter alia*, more than 25 years of continual interference by the United States in the internal affairs of Iran«[30].

The Iranian Government's contention, at this preliminary stage, was that the taking of the American diplomatic and consular personnel in Tehran, as hostages, could not be –

»divorced from its proper context, namely the whole political dossier of the relations between Iran and the United States over the last 25 years . . ., in particular the *coup d'état* of 1953 stirred up and carried out by the [United States secret service agency] C.I.A., the overthrow of the lawful national government of Dr. Mossadegh, the restoration of the Shah and of his *régime* which was under the control of American interests, and all the social, economic, cultural, and political consequences of the direct interventions in our internal affairs . . .«[31].

Were these issues of fact in respect to which the Court could, and should, have taken formal judicial notice, in the spirit of its innovatory approach to judicial fact-finding, already adverted to, in *Namibia (South West Africa)* in 1971? Perhaps the Court was too cautious or too traditional in its approach, in *United States Diplomatic and Consular Staff in Tehran*, in comparison to *Namibia (South West Africa)*. But *Namibia (South West Africa)* had arisen under the Advisory Opinion jurisdiction of the Court where, by definition, there were no parties of the record, on whom the burden of carrying the proof of facts in issue could be imposed by the Court; and even if one stripped away the veil and recognised that, in reality, it was a contest between the Mandatory Power, white South Africa, and the indigenous black inhabitants of the Mandated territory, the latter, not yet being recognised as a State, were a legal »non-person« and hardly in a position to discharge the burden of proof of disputed facts. In such a situation, in *Namibia (South West Africa)*, the Court took its own initiative and moved boldly to fill the gap as to fact-finding. By comparison, in *United States Diplomatic and Consular Staff in Tehran*, the party claiming to be adversely affected by alleged past illegal

---

[29] *United States Diplomatic and Consular Staff in Tehran,* Provisional Measures, Order of 15 December 1979, I.C.J. Reports 1979, p. 7.

[30] *Ibid.,* at p. 11.

[31] *Ibid.*

interventions in its internal affairs by the U.S. Government and its agencies, Iran, had always been in a position to appear before the Court and to make oral or written submissions, or to file a written memorandum together with supporting evidence or averments of fact, in attempted legal justification of the seizing of U.S. diplomatic and consular personnel in Tehran. In the absence of any such initiative by the directly interested party, would there be any reason or justification for the Court's departing from the normal *contra proferentem* rule as to facts in issue? Whether the presence before the Court of sufficiently evidenced proof of past systematic U.S. interventions in Iranian internal affairs could have, in itself, provided a legal warrant or excuse for the forcible taking over of diplomatic and consular premises and the seizing of their personnel, remains open to question, after the Court's ruling in *United States Diplomatic and Consular Staff in Tehran.* The principles assuring the protection of foreign ministers were among the most ancient in customary International Law, and they had been confirmed and codified, only a few years before the Court ruling, in the Vienna Convention on Diplomatic Relations of 1961 and the Vienna Convention on Consular Relations of 1963, which had provided their own in-built remedies against abuses of diplomatic privileges and immunities, by way of espionage or interference in the internal affairs of the receiving State, through the declaring as *persona non grata* and thus compelling the recall of offending foreign diplomatic personnel. It might be suggested that the effect of the two Vienna Conventions, referred to, is to raise to the level of imperative principles of International Law – *jus cogens* – respect for and protection of foreign diplomatic premises and their personnel; and if the Court had decided in these terms it would have given a welcome finality and freedom from legal doubt to its ruling. The Court does not, however, go as far as that, and so the larger issue remains open; though it must be said that it would require a very considerable extension of the old International Law category of Reprisals to reach the legal conclusion seemingly contended for by the Iranian Government in justification of its actions against the United States and its diplomatic premises and personnel in the instant case[32]. In the three Separate Opinions filed in the case, and particularly the Dissenting Opinion, the contested issues of fact as to alleged prior U.S. illegal interventions in Iranian internal affairs seem to be considered as at least relevant to Iran's obligation to make general reparation, apart from the specific return of the diplomatic personnel and premises and property. Judge Lachs, in his Separate Opinion, felt that the Court majority should have applied »sound judicial economy« and not gone into the issue of reparation apart from return of personnel, premises, and property. Judge Tarazi, in his Dissenting Opinion, took affirmative judicial notice of the U.S.

---

[32] Compare the analysis by *E. Zoller*, L'affaire du personnel diplomatique et consulaire des Etats-Unis à Téhéran, Revue générale de droit international public 80 (1980), p. 973, at 1012 *et seq.*

Government's actions in Iran in behalf of the Shah, in 1953; and he also considered as being in the public domain the U.S. Government's granting of permission to the ex-Shah to enter the United States in October, 1979, for the U.S. Government, in his view, was perfectly aware that this – »might have untoward consequences«. Any Iranian Government responsibility thus had to be envisaged, in Judge Tarazi's opinion – »in the context of the revolution which took place in that country and brought about, as it were, a break with a past condemned as oppressive«. Judge Tarazi was also very critical, as was Judge Morozov in his own separate Dissenting Opinion, of various U.S. Government actions before and after the Court was seized with jurisdiction in the case and which, in Judge Tarazi's view, – »constituted an encroachment on the functions of the Court«, – namely, the freezing of Iranian economic assets in the U.S. and the abortive U.S. military operation of April, 1980. In short, the three Separate Opinions properly introduce additional legal nuances not fully present in the official Opinion of the Court and important to the progressive development of International Law doctrine for the future, and this without derogating from the principle – unanimously reaffirmed by the members of the Court – of the inviolability of diplomatic personnel and premises. This is just what Special Opinions are supposed to do, for, with appropriate balance and restraint in their formulation, they maintain the collegial quality of the Court at the same time as permitting experimental ventures into judicial law-making for the future.

* * *

In sum, if the actual fact-situation involved in *United States Diplomatic and Consular Staff in Tehran* is likely to prove unique, there is no reason to believe that it will not be followed by other cases, involving a similar high political content; and which will end up before the Court, in default of political action in the other, more overtly and avowedly political international arenas (the Security Council, the General Assembly), as the parties concerned increasingly turn to the judicial process for resolution of their conflicts.

In surveying the work of the Court since the political watershed, 8-to-7, decision rendered in 1966 in *South West Africa, Second Phase,* it seems clear that the Court has displayed a considerable degree of initiative and ability to innovate, as to both substantive legal principles and also as to adjectival, procedural law elements. The Court's commitment to the cause of the progressive development of International Law is sufficiently evidenced in its leading cases since that time, as is also its recognition of the legislative, policy-making character inherent in adjudication by a final tribunal. But such judicial policy-making as it has developed in its detailed jurisprudence has not been any simple, or simplistic, »legislation in the air«, but has proceeded upon a proper respect for preliminary jurisdictional requirements, and for procedural forms and processes. The Court has been aware, at all times, of the sensible limits of judicial fact-finding, and also of its obligations not to decide in vain by rendering judgment on an issue that is premature or that may already have become moot. In short, it has always understood that it is, in the end, a court; and it has tried, therefore, to function as a »responsible magistrature« and not as a professionally untrammelled »academy of jurists«[33]. After some early tendencies to judicial Byzantinism, as reflected in a plethora of Separate Opinions – Specially Concurring and Dissenting – in a number of leading cases, producing »no-clear-majority« decisions whose actual majority *ratio decidendi* was often obscure, the Court seems to have settled down into a proper respect for its own collegial character and responsibilities, and of the limits, therefore, of permissible tone and language and also the actual numbers of dissonance within its own ranks. This latter is part of the larger concept of judicial self-restraint which includes obligations of mutual respect as to one's own colleagues, as well as the more obvious duty of comity and deference in relation to the other, political institutions of the World Community. All this seems to augur well as to the exercise of the legislative capacity of the Court in the present, difficult era of transition in the World Community. The political composition of the Court is bound to change, through the operation of the triennial elections and the increased political attention given to them, in ways that make the Court more representative, for better or for worse, of the turbulent cross-currents of opinion within the current General Assembly majorities. But the Court itself has done its best to assist the general cause of change, through creative adaptation of its own internal decision-making processes and procedures, and through not being afraid to test new doctrine.

[33] The dichotomy is that of ex-President Manley Hudson of the old, League of Nations-era, Permanent Court of International Justice: *Hudson,* The Permanent Court of International Justice, 1920–1942, at p. 511; cited by Fitzmaurice J., (dissenting), in *Namibia (South West Africa),* I.C.J. Reports 1971, p. 16, at p. 303.

Internationaler Schutz der Menschenrechte und Völkerrechtsunmittelbarkeit

SHIGEKI MIYAZAKI

I. Veränderungen in Staat und Völkergemeinschaft

Nach Meinung der Sozialpsychologen denkt nicht der Einzelmensch, sondern die Gesellschaft; das Sozialbewußtsein prägt das Denken des Einzelnen. Dieses traditionelle, vorgefaßte Bewußtsein bestimmt den Gedanken, daß die Völkergemeinschaft aus Staaten bestehe und die internationalen Beziehungen lediglich zwischenstaatliche Verhältnisse seien. Lange nahm man an, die rechtliche Mediation des Einzelmenschen durch den Staat sei im Völkerrecht vollständig, der Einzelne könne folglich weder Subjekt im Völkerrecht noch Träger völkerrechtlicher Rechte und Pflichten sein, d. h. eine Völkerrechtsunmittelbarkeit stehe ihm daher nicht zu [1].

Das klassische Völkerrecht kannte wohl einige Regeln zum Schutz der Menschenrechte, sie waren aber nicht mehr als ein bloßer Reflex der Rechte und Pflichten des Staates, der allein als Subjekt des Völkerrechts galt. Das Völkerrecht erfährt neuerdings eine grundsätzliche Umwandlung als Folge einer Änderung der Staatenstruktur, als Folge wachsender gegenseitiger Abhängigkeit (*interdependance*) zwischen Staaten, als Folge vermehrter transnationaler Beziehungen und als Folge einer entsprechenden Vorstellung im Bewußtsein der Menschen.

1. Früher konnte man den Staat mit dem »König« gleichsetzen, denn dieser hatte die echte Souveränität in seinem Staatsbereich, wie es Ludwig XIV. mit dem Schlagwort *l'Etat, c'est moi* ausdrückte. Der Staat des Mittelalters war nichts anderes als ein König mit den ihn umgebenden Gefolgsleuten und dem Söldnerheer in seinem Dienst.

Der Einzelne war dabei nicht Mitglied des Staates, sondern nur Objekt der Regierungsgewalt. Er wurde als das Nebenstück des Lehens und als Quelle des Tributs sowie der Blutsteuer betrachtet. Erst seit der französi-

---

[1] *F. A. von der Heydte*, Völkerrecht, Bd. 1 (1958), S. 32.

schen Revolution begann man damit, den Einzelnen in das System einer Staatsgliederung einzufügen.

Das Staatsverhalten orientierte sich immer am nationalen Interesse; früher bestand dies wohl nur im Vorteil des Königs, gegenwärtig spricht man vom Vorteil des betreffenden Volkes, weil zur Zeit die meisten Staaten von einem demokratischen Selbstverständnis getragen werden, die Interessen der Staatsangehörigen also im Vordergrund stehen – sei es nun eine Republik oder eine Monarchie. Diese Veränderung wirkt sich auch aus auf die Konstruktion der internationalen Beziehungen und die völkerrechtliche Rechtssetzung.

Im 19. Jahrhundert wurden alle internationalen Verträge im Namen des Königs geschlossen. Im Gegensatz dazu steht der Vertrag über die Ächtung des Krieges (Briand-Kellogg-Pakt) vom 27. August 1928; ihn erklärten »die Hohen Vertragschließenden Parteien *im Namen ihrer Völker*«; ähnlich beginnt die Charta der Vereinten Nationen: »*Wir, die Völker* der Vereinten Nationen, entschlossen . . .«.

2. In der gegenwärtigen Struktur der Koexistenz kann kein Staat isoliert existieren, denn die internationale gegenseitige Abhängigkeit ist immer stärker geworden: auf staatlicher Regierungsebene, auf innerstaatlicher Verwaltungsebene und auf der nichtstaatlichen privaten Ebene. Durch internationale Zusammenarbeit auf den Gebieten von Wirtschaft, Sozialpolitik, Kultur, Erziehung und Gesundheitspflege usw. wird das Bewußtsein der gegenseitigen Abhängigkeit der Völker der Erde mehr und mehr bestärkt.

3. Der Kontakt der Einzelnen untereinander über die nationalen Grenzen hinweg, der zuvor durch die eigene Regierung vermittelt werden mußte, ist inzwischen zu einer transnationalen unmittelbaren Beziehung erwachsen. Wir leben sowohl als Angehörige irgendeines Staates, gleichzeitig aber auch als Mitglieder der Menschheit insgesamt. Nichtstaatliche Organisationen, ICRC, multinationale Unternehmen usw. werden jetzt zu Rechtsträgern der internationalen Gesellschaft und erlangen dadurch internationalrechtliche Existenz (*entity*) als Rechtsträger der Völkergemeinschaft[2].

Massenmedien sichern die dafür notwendige Kommunikation; sie liefern nicht nur Informationen über Tatsachen in aller Welt, sondern vermitteln auch Wertschätzungen und das Wertgefühl dazu. Besonders intensiv wirkt hier die Rundfunk- und Fernsehsendung, die ohne Grenzkontrolle jeden unmittelbar anspricht.

4. Auf diese Weise bildet sich ein gemeinsames Menschheitsgefühl neben dem Nationalgefühl, und es verstärkt sich nach und nach.

Schwere Verletzungen der Menschenwürde und der Menschenrechte können jetzt den internationalen Frieden und die internationale Sicherheit gefährden, da sie gemeinschaftliche Interessen der gegenwärtigen Menschheit be-

---

[2] *H. Mosler,* Die Erweiterung des Kreises der Völkerrechtssubjekte, Berichte der Deutschen Gesellschaft für Völkerrecht, Heft 4 (1961), S. 44.

treffen. Zudem haben sich die Staaten verpflichtet, die Menschenrechte und Grundfreiheiten allgemein zu achten, zu verwirklichen und durchzusetzen.

Repräsentative Auffassungen der Völkerrechtswissenschaft beginnen damit, diese zwangsläufigen Zusammenhänge zu erkennen und die Bedeutung der Menschenrechte für die Sicherung des Weltfriedens herauszustellen [3].

II. Die Menschenrechte in der Gegenwart

Der Schutz der Menschenrechte erfordert eine Garantie, damit menschenwürdiges Leben und menschliche Behandlung praktisch so verwirklicht werden, wie sie rechtlich gewährleistet sein sollen.

1. Das Menschenleben an sich ist das erste und wichtigste Schutzgut. Jeder Mensch hat das Recht auf Leben, Freiheit und Sicherheit der Person. Angriffe auf das Leben und die Person, namentlich Tötung jeder Art, Verstümmelung, grausame Behandlung, Folter und die Festnahme von Geiseln sind jederzeit und überall verboten. Auch im Krieg, der besonders als Atomkrieg die Lebensbedingungen zerstört, bleiben die oben erwähnten Verhaltensweisen in Bezug auf Personen, die nicht unmittelbar an den Feindseligkeiten teilnehmen, verboten.

Neuerdings werden diese Probleme besonders im internationalen humanitären Recht behandelt. Das Recht auf eine menschenwürdige Lebenshaltung, auf Gesundheit und Wohlbefinden einschließlich Nahrung, Kleidung, Wohnung, ärztliche Versorgung und notwendige Leistungen der sozialen Fürsorge soll gewährleistet sein. Damit der Einzelne menschlich leben kann, ist mehr erforderlich als die Erhaltung seiner biologischen Existenz: der Mensch muß als Mensch wirklich »menschlich« leben können.

Schon in seiner Botschaft an den Kongreß vom 6. Januar 1941 verkündete Präsident Roosevelt vier Freiheiten als Grundlagen für die Zukunft und für die Sicherheit der Welt: die Rede- und Glaubensfreiheit sowie die Freiheit vor Furcht und Not.

Art. 1 der Allgemeinen Erklärung der Menschenrechte vom 10. Dezember 1948 bestimmt: »Alle Menschen sind frei und gleich an Würde und Rechten geboren. Sie sind mit Vernunft und Gewissen begabt und sollen einander im Geist der Brüderlichkeit begegnen«. Dieser Erklärung nach hat jeder Mensch Anspruch auf die oben erwähnten Rechte und Freiheiten ohne irgendeine Unterscheidung wie etwa nach Rasse, Farbe, Geschlecht, Sprache, Religion, politische und sonstige Überzeugung, nationale oder soziale Herkunft, nach Eigentum, Geburt oder sonstigen Umständen. Die in dieser Erklärung verkündeten Rechte und Freiheiten, die hauptsächlich als die traditionellen bürgerlichen Menschenrechte bekannt sind, werden bereits auf der

---

[3] *E. Friesenhahn,* Menschenrechte, in: Strupp-Schlochauer, Wörterbuch des Völkerrechts, Bd. 2 (1961), S. 504.

staatlichen Ebene, insbesondere im Verfassungsrecht von fast allen Staaten geschützt.

Diese Rechte und Freiheiten sind gewöhnlich von der Staatsgewalt zu achten, sie sollen aber auch von anderen Einzelnen oder von sozialen Gruppen geachtet werden. Schutz vor Sklaverei, Zwangsarbeit und Folter, vor sonstigem Freiheitsentzug, die Glaubens-, Gewissens-, Meinungs- und Informationsfreiheit, die Versammlungs- und Vereinigungsfreiheit, der Anspruch auf eine gerichtliche Nachprüfung von Freiheitsentzug innerhalb angemessener Frist und das Recht auf politische Mitwirkung im Heimatstaat unmittelbar oder durch frei gewählte Vertreter sind für ein »menschliches« Leben notwendig. Das reicht aber noch nicht aus; weiter müssen folgende Rechte gewährleistet sein: das Recht auf Bildung, und zwar auf den allgemeinen, fachlichen und beruflichen Unterricht; das Recht auf Arbeit, auf freie Berufswahl, auf angemessene und befriedigende Arbeitsbedingungen; das Recht auf angemessene Entlohnung eigener Arbeit und Tätigkeit; das Recht auf Schutz des privaten Lebens (*privacy*); das Recht auf Schutz vor willkürlicher Berichterstattung über eigene Handlungen, der Anspruch auf Richtigstellung solcher Berichte (*right of correction*); das Recht, am kulturellen Leben der Gemeinschaft frei teilzunehmen, sich der Künste zu erfreuen und den wissenschaftlichen Fortschritt und seine Wohltaten zu genießen; ferner das Recht auf Familienleben.

Wirklich wirkungsvoll kann man diese Rechte und Freiheiten nur mit Hilfe von Maßnahmen des eigenen Staates nicht gewährleisten; das ist erst auf der Ebene internationaler Garantien möglich, weil Wissenschaft, Musik, Literatur, Religion, private Liebhabereien, Liebe, Freundschaft, Freude, Trauer, Krankheit usw. vor Landesgrenzen nicht mehr haltmachen. Der Lebensbereich vieler Menschen verbreitet sich so weit über die Grenzen hinaus, daß sie als ein transnationales Netz menschlicher Beziehungen erscheinen, wie wir es schon beschrieben haben.

Ubi societas ibi ius. Die Ausbildung eines transnationalen Rechts in einer transnationalen Völkergemeinschaft befindet sich vorerst noch in den Anfängen. Zunehmend aber breitet sich die Erkenntnis aus, daß der Einzelne eine begrenzte Völkerrechtssubjektivität als ein Träger völkerrechtlicher Rechte und Pflichten gewonnen hat.

III. Internationale Verträge für den Schutz der Menschenrechte

1. Schon vor dem zweiten Weltkrieg gab es einige Verträge, die die Verpflichtung der Staaten zur Achtung bestimmter Menschenrechte enthielten. Dabei wurde die Sicherung der im Vertrag umschriebenen Menschenrechte

des Einzelnen als Anspruch des Vertragsstaats auf Vertragserfüllung definiert, und ihre Verletzung wurde als Vertragsverletzung gewertet [4].

Zu erwähnen sind hier:
- das Abkommen betreffend die Sklaverei vom 25. September 1926,
- das Übereinkommen über Zwangs- oder Pflichtarbeit vom 28. Juni 1930,
- die Internationale Übereinkunft zur Unterdrückung des Frauen- und Kinderhandels vom 30. September 1921.

2. Nachdem der zweite Weltkrieg unsagbares Leid über die Menschheit gebracht hatte, richten sich die Bemühungen um den Schutz der Menschenrechte gleichermaßen auf die staatliche wie auf die internationale Ebene.

Die Erfahrungen mit totalitären Regimen, die nicht dem Menschen dienen, sondern ihn unterjochen, lehren, daß nur rechtsstaatliche Regierungen, die den Einzelnen als Persönlichkeit achten und ihm Menschenrechte zuerkennen, wahre Autorität haben und damit die friedliche Ordnung in internationalen Beziehungen verbürgen können [5].

In der UN-Charta bekräftigen die Völker der Vereinten Nationen ihren Glauben an die Grundrechte des Menschen, an Würde und Wert der menschlichen Persönlichkeit, an die Gleichberechtigung von Mann und Frau. Nach Art. 55 der Charta haben die Vereinten Nationen die allgemeine Achtung und Verwirklichung der Menschenrechte und der Grundfreiheiten für alle Menschen ohne Unterschied der Rasse, des Geschlechts, der Sprache oder der Religion zu fördern. Alle Mitgliedstaaten verpflichten sich, gemeinsam und für sich selbst mit der Organisation zusammenzuarbeiten, um dieses Ziel zu erreichen [6].

Die Allgemeine Erklärung der Menschenrechte war wohl nicht als unmittelbar verpflichtendes Vertragsinstrument konzipiert, wurde aber als das von allen Völkern und Nationen zu erreichende gemeinsame Ideal verkündet. Sie hat deshalb als allgemeine Richtschnur für zahlreiche spätere Verträge über den Schutz der Menschenrechte und für nationale Verfassungen weltweite Wirkungen entfaltet [7]. Zu erwähnen sind hier [8]:
- die Konvention über die Verhütung und Bestrafung des Völkermordes vom 9. Dezember 1948,
- das Abkommen über die Rechtsstellung der Flüchtlinge vom 28. Juli 1951,
- das Übereinkommen über die politischen Rechte der Frau vom 31. März 1953,
- das Übereinkommen über die Rechtsstellung der Staatenlosen vom 28. September 1954,

---

[4] *von der Heydte* (Anm. 1), S. 268. *M. Grassi,* Die Rechtsstellung des Individuums im Völkerrecht (1955), S. 127 ff.

[5] *Friesenhahn* (Anm. 3), S. 503.

[6] *H. Guradze,* Der Stand der Menschenrechte im Völkerrecht (1956), S. 101 ff.

[7] *E. Menzel/K. Ipsen,* Völkerrecht (2. Aufl. 1979), S. 123.

[8] *J. Delbrück,* Menschenrechte und Grundfreiheiten im Völkerrecht (1972), S. 68.

- das Übereinkommen über die Abschaffung der Zwangsarbeit vom 25. Juni 1957,
- das Übereinkommen über die Diskriminierung in Beschäftigung und Beruf vom 25. Juni 1958,
- das Übereinkommen gegen Diskriminierung im Unterrichtswesen vom 14. Dezember 1960,
- das internationale Übereinkommen zur Beseitigung jeder Form von Rassendiskriminierung vom 7. März 1966,
- das Protokoll über die Rechtsstellung der Flüchtlinge vom 31. Januar 1967,
- das Übereinkommen über die Unterdrückung und die Bestrafung des Verbrechens der Apartheid vom 30. November 1973,
- das Übereinkommen zur Beseitigung jeder Form von Frauendiskriminierung vom 18. Dezember 1979.

Als wichtigste sind aber vor allem zu nennen: die Konventionen der Vereinten Nationen zum Schutz der Menschenrechte von 1966, nämlich 1. der Internationale Pakt über wirtschaftliche, soziale und kulturelle Rechte, 2. der Internationale Pakt über bürgerliche und politische Rechte und 3. das Fakultativprotokoll zum Internationalen Pakt über bürgerliche und politische Rechte vom 16. Dezember 1966, als universale Vertragswerke, 4. die Europäische Konvention zum Schutz der Menschenrechte und Grundfreiheiten vom 4. November 1950 (Rom-Konvention) sowie 5. die Europäische Sozialcharta vom 18. Oktober 1961 und schließlich 6. die Amerikanische Menschenrechtskonvention vom 22. November 1969, als regionale Vertragswerke.

Diese sechs Konventionen sind die Nachfolger der Allgemeinen Menschenrechtserklärung der Vereinten Nationen von 1948. Die ersten zwei Internationalen Menschenrechtspakte haben erneut festgestellt, daß die Mitgliedstaaten sich verpflichteten, in Zusammenarbeit mit den Vereinten Nationen die allgemeine Achtung und Verwirklichung der Menschenrechte und Grundfreiheiten durchzusetzen.

IV. Formeln zur Bestimmung des Schutzes der Menschenrechte

Die Konventionen sehen verschiedene Bestimmungen zum Schutz der Menschenrechte vor; sie lassen sich folgendermaßen klassifizieren in Bestimmungen, die

1) lediglich Staaten die Achtung der Menschenrechte vorschreiben,
2) auch Einzelnen die Befugnis verleihen, sich gegenüber entgegenstehendem staatlichen Recht unmittelbar auf sie zur Durchsetzung der Menschenrechte zu berufen,
3) den Staaten oder Einzelnen vorschreiben, Angriffe auf die Menschenrechte zu verhüten und zu bestrafen,

4) den internationalen Organen die Möglichkeit eröffnen, Verletzungen der Menschenrechte durch Vertragsstaaten zu prüfen,
5) Einzelnen die Möglichkeit eröffnen, den Staat wegen Verletzungen der Menschenrechte vor internationalen Organen zur Verantwortung zu ziehen [9].

1. Einer größeren Zahl von ehemaligen Verträgen zufolge, die die Lebenssachverhalte des Einzelnen zum Gegenstand hatten, ist die Verpflichtung der Staaten zur Achtung bestimmter Menschenrechte anerkannt.

Ein Staat kann *einseitig* in einem Vertrag einem anderen Staat gegenüber rechtlich bindend die Achtung der Menschenrechte aller Gewaltunterworfenen versprechen. Gewöhnlich versprechen die Staaten aber *einander* lediglich gegenseitig, den Schutz der Menschenrechte zu gewährleisten.

Der bisherigen Ansicht nach erreichten dabei die Menschenrechte noch nicht die völkerrechtliche Geltung im vollen Sinne, weil diese Rechte nur durch Vermittlung des Staates, und zwar durch Ratifikation oder Transformation zum innerstaatlichen Recht erhoben werden konnten.

a) Diplomatischer Schutz. Das Recht des Staates, seine Staatsbürger im Ausland oder gegen die Einwirkung fremder Staatsgewalten zu schützen, nennt man das Recht über den diplomatischen Schutz oder diplomatische Protektion. Der Staatsbürger kann von seinem eigenen Heimatstaat einen solchen Schutz beanspruchen; dem Staat obliegt eine Pflicht zum Schutze auf staatlicher Ebene. Auf der internationalen Ebene hat der in seinem Staatsangehörigen verletzte Staat das Recht auf diplomatischen Schutz. Der verletzte Einzelne hat indessen keinen Anspruch gegen den Verletzungsstaat auf völkerrechtlicher Ebene und bleibt damit nur Objekt eines rechtlichen Schutzes. Der Staat des Verletzten kann aus politischen Gründen sogar darauf verzichten, sein Recht über den diplomatischen Schutz zu verwirklichen [10].

b) Humanitäre Intervention. Alle Mitgliedstaaten der Vereinten Nationen legen ihre internationalen Streitigkeiten durch friedliche Mittel so bei, daß der Weltfriede, die internationale Sicherheit und die Gerechtigkeit nicht gefährdet werden. Schwere Verletzungen der Menschenrechte gefährden aber die internationale Sicherheit und die Gerechtigkeit [11].

Die Staaten können auch Interesse an der Behandlung haben, die andere Staaten ihren eigenen Staatsangehörigen angedeihen lassen. Der Internationale Gerichtshof hat in seinem Gutachten vom 28. Mai 1951 über das Völkermordabkommen bestätigt, daß »the principles underlying the Convention (elementary considerations of humanity) are principles which are recognized

---

[9] *Friesenhahn* (Anm. 3), S. 507.

[10] *K. Doehring,* Die Pflicht des Staates zur Gewährung diplomatischen Schutzes (Beiträge zum ausländischen öffentlichen Recht und Völkerrecht, Bd. 33) (1959), S. 7 ff.

[11] *A. Verdross/B. Simma,* Universelles Völkerrecht, Theorie und Praxis (1976), S. 584.

by civilized nations as binding on states, even without any conventional obligation«[12]. Die Geltendmachung der humanitären Intervention hängt auch vom Willen des betreffenden Staates ab.

2. Inzwischen wurden sog. *self-executing*-Verträge geschlossen, die unmittelbar, d.h. ohne die Transformation zum staatlichen Recht, von den innerstaatlichen Instanzen (Gerichte und Verwaltungsbehörden) anzuwenden sind. Es gibt außerdem noch Verträge, die neben der *self-executing*-Bestimmung auch noch *non self-executing*-Bestimmungen enthalten. Es ist ferner möglich, daß einige Bestimmungen eines Vertrags für einen Vertragsstaat unmittelbar anwendbar sind, während sie es für andere nicht sind, weil nicht nur der Text des Vertrags, sondern auch die innerstaatliche Rechtsordnung des betreffenden Vertragsstaates unmittelbar die Anwendung durch innerstaatliche Instanzen vorsehen, um die Bestimmung als *self-executing* qualifizieren zu können[13].

Wäre der Einzelne dabei das Subjekt der Tat oder der Unterlassung in einer solchen Bestimmung, dann wäre er auch Subjekt im Sinne des Vertragswerks, d.h. auf völkerrechtlicher Ebene der Rechtsträger.

Die Vorschrift des Art. 7 des Internationalen Paktes über bürgerliche und politische Rechte bestimmt: »Niemand darf der Folter oder grausamer unmenschlicher oder erniedrigender Behandlung oder Strafe unterworfen werden«. Die Vorschrift des Art. 3 der Europäischen Menschenrechtskonvention enthält den gleichen Ausdruck. Die Vorschrift des Art. 9 (1) des besagten Paktes sieht vor: »Jedermann hat ein Recht auf persönliche Freiheit und Sicherheit«.

Die Vorschrift des Art. 12 EWG-Vertrag bestimmt zudem: »Die Mitgliedstaaten werden untereinander weder neue Einfuhr- oder Ausfuhrzölle oder Abgaben gleicher Wirkung einführen noch die in ihren gegenseitigen Handelsbeziehungen angewandten erhöhen«. Der Europäische Gerichtshof hat festgestellt, daß diesem Artikel *self-executing* Charakter zukommt[14].

3. Es gibt Verträge, in denen der Einzelne unmittelbar auf Grund des Völkerrechts verpflichtet und dann auch zur Verantwortung gezogen werden kann; als wichtigste sind zu nennen: die Verträge über den Völkermord, über die Luftpiraterie, über den verbrecherischen Angriff gegen Diplomaten und andere Vertreter fremder Staaten sowie über die Kriegsverbrecher. Vielleicht könnte man den Vertrag über die Apartheidsverbrecher hinzufügen.

a) Der Völkermord ist das schwerste Verbrechen nach dem internationalen Vertrag – abgesehen von den Kriegsverbrechen. Die Vorschriften der Art. 1 und 2 der Konvention über die Verhütung und Bestrafung des Völker-

---

[12] *Meyer-Lindenberg*, Die Menschenrechte im Völkerrecht, Berichte der Deutschen Gesellschaft für Völkerrecht, Heft 4 (1961), S. 88. CIJ Recueil 1951, S. 23.

[13] Menschenrechte im Staatsrecht und im Völkerrecht, hrsg. von der Rechts- und staatswissenschaftlichen Fakultät an der Universität Wien (1967), S. 26.

[14] A.a.O., S. 27.

mordes vom 9. Dezember 1948 sehen vor, daß Völkermord, ob im Frieden oder im Krieg begangen, ein Verbrechen gemäß internationalem Recht ist und daß jeder Vertragsstaat zu dessen Verhütung und Bestrafung verpflichtet ist. Der Vertrag sieht ein internationales Strafgericht vor, das bis jetzt noch nicht errichtet wurde.

b) Zur Bekämpfung der Luftpiraterie finden sich zur Zeit drei multilaterale Verträge: das Tokyoter Abkommen über strafbare und bestimmte andere an Bord von Luftfahrzeugen begangene Handlungen vom 14. September 1963, das Haager Übereinkommen zur Bekämpfung widerrechtlicher Inbesitznahme von Luftfahrzeugen vom 16. Dezember 1970 und die Montrealer Konvention zur Bekämpfung rechtswidriger Handlungen gegen die Sicherheit der Zivilluftfahrt vom 23. September 1971 [15].

c) Die Konvention über die Verhütung und Bestrafung von Angriffen auf international geschützte Personen, einschließlich der Diplomaten vom 14. Dezember 1973 sieht vor: Verbrecherische Angriffe auf ausländische Staatsoberhäupter, Regierungschefs, Außenminister, Diplomaten, andere Staatenvertreter und Funktionäre internationaler Organisationen, die nach Völkerrecht Anspruch auf erhöhten Schutz haben, sowie auf Angehörige dieser Personen sind zu verhindern und zu bestrafen. Täter der erwähnten Verbrechen können wohl nicht unmittelbar auf Grund dieser Verträge verfolgt werden, sondern erst auf Grund staatlicher Strafnormen, die in Erfüllung völkerrechtlicher Pflichten der Vertragsstaaten erlassen wurden [16]. Diese Verbrechen gehören aber zu jenen völkerrechtlichen Straftaten, die im Bewußtsein der Völker als strafbares Unrecht gelten. Daneben kann der Einzelne freilich unmittelbar auf Grund des Völkerrechts wegen Verletzung völkerrechtlichen Kriegsrechts bestraft werden.

4. Eine echte völkerrechtliche Berechtigung des Einzelnen liegt nur vor, wenn die Menschenrechte unmittelbar durch eine Völkerrechtsnorm geschützt, ihre Verletzung vor internationalen Instanzen geprüft und die Einhaltung der Menschenrechte durch diese Instanzen kontrolliert werden kann.

V. Verfahren zur Sicherung von Menschenrechten

Drei Arten von Verfahren zur Sicherung von Menschenrechten durch internationale Instanzen lassen sich unterscheiden: 1. die Berichtsverfahren, 2. die Staatenbeschwerdeverfahren und 3. die Verfahren der Individualbeschwerde [17].

---

[15] *Verdross/Simma* (Anm. 11), S. 224.
[16] *Verdross/Simma,* a.a.O., S. 225.
[17] *Delbrück* (Anm. 8), S. 51. *Menzel/Ipsen* (Anm. 7), S. 128.

1. Die Berichtsverfahren

a) Nach Art. 40 des Internationalen Paktes über bürgerliche und politische Rechte von 1966 sind die Vertragsstaaten verpflichtet, Berichte über Maßnahmen zur Verwirklichung der Konventionsrechte zu erstatten, die von einem unabhängigen Menschenrechtskomitee aus 18 Staatsangehörigen der Vertragsstaaten geprüft werden [18].

Ähnliche Berichtsverfahren zur Kontrolle der Einhaltung von Verpflichtungen der Staaten auf dem Gebiet der Menschenrechte findet man in einer Reihe von internationalen Verträgen. Zu nennen sind:

b) Die Berichte der Mandatsmächte an den Völkerbund. Sie werden von einem ständigen Mandatsausschuß (9 Mitglieder) geprüft. Dieses war das erste der Berichtsverfahren (Art. 22 ix der Satzung des Völkerbundes vom 28. Juni 1919).

c) Die Berichte der Mitgliedstaaten der Internationalen Arbeitsorganisation (ILO), die von der Konferenz der ILO geprüft werden (Art. 22 und 23 ILO-Verfassung vom 28. Juni 1919 und 9. Oktober 1946).

d) Die Berichte der Treuhandverwaltungsmächte, die von der Generalversammlung und dem ihrer Autorität unterstehenden Treuhandrat geprüft werden (Art. 87 (a) UN-Charta und Rule 74 der Geschäftsordnung des Treuhandrates).

e) Die Berichte (Erklärungen) der Vertragsstaaten der Europäischen Menschenrechtskonvention (EMRK), die vom Europarat geprüft werden (Art. 57 EMRK von 1950) [19].

f) Die Berichte der Vertragsstaaten des Übereinkommens gegen Diskriminierung im Unterrichtswesen, die von der Generalkonferenz der UNESCO geprüft werden (Art. 7 dieses Übereinkommens vom 14. Dezember 1960).

g) Die Berichte der Vertragsparteien der Europäischen Sozialcharta, die von einem Sachverständigenausschuß des Europarats geprüft werden (Art. 21, 24 dieser Charta vom 18. Oktober 1961).

h) Die Berichte der Vertragsstaaten des Internationalen Übereinkommens zur Beseitigung jeder Form von Rassendiskriminierung, die von einem Ausschuß für die Beseitigung der Rassendiskriminierung geprüft werden (Art. 9 dieses Übereinkommens vom 7. März 1966).

i) Die Berichte der Vertragsstaaten des Internationalen Paktes über wirtschaftliche, soziale und kulturelle Rechte, die vom UN-Wirtschafts- und Sozialrat geprüft werden (Art. 16 dieses Paktes vom 16. Dezember 1966).

j) Die Berichte der Vertragsstaaten des Übereinkommens über die Unterdrückung und die Bestrafung des Verbrechens der Apartheid, die von einer

---

[18] *W. K. Geck,* Der internationale Stand des Schutzes der Freiheitsrechte: Anspruch und Wirklichkeit, ZaöRV 38 (1978), S. 207. *Delbrück* (Anm. 8), S. 52.

[19] *Z. M. Nedjati,* Human Rights under the European Convention (1978), S. 4 und 6.

Gruppe aus 3 Mitgliedern der UN-Menschenrechtskommission geprüft werden (Art. 7 dieses Übereinkommens vom 30. November 1973).

k) Die Berichte der Vertragsstaaten des Übereinkommens zur Beseitigung jeder Form von Frauendiskriminierung, die von einem Ausschuß für die Beseitigung der Frauendiskriminierung geprüft werden (Art. 18 dieses Übereinkommens vom 18. Dezember 1979).

In allen diesen Fällen handelt es sich aber nicht um echte völkerrechtliche Berechtigungen des Einzelnen selbst, sondern nur um staatliche Pflichten zur Sicherung der Menschenrechte auf Grund dieser Verträge, obwohl die internationalen Kontrollgremien auf eben diese Sicherung hin angelegt sind.

2. Die Staatenbeschwerde

a) Nach Art. 41 des Internationalen Paktes über bürgerliche und politische Rechte von 1966 kann ein Vertragsstaat, der einen anderen einer Verletzung der in diesem Pakt niedergelegten Menschenrechte für schuldig hält, ein Menschenrechtskomitee benachrichtigen (*communication*), damit dieses die Umstände prüft – sofern der betroffene Vertragsstaat sich ausdrücklich diesem Verfahren unterworfen hat.

Ähnliche Verfahren über die Staatenbeschwerde sind in einer Reihe internationaler Menschenrechtsverträge vorgesehen.

b) Mitgliedstaaten der ILO können beim Internationalen Arbeitsamt Klage gegen andere Mitglieder erheben, sofern diese nach Ansicht des klagenden Staates die Durchführung eines von beiden Teilen ratifizierten Übereinkommens nicht in befriedigender Weise sicherstellen (Art. 26 I ILO-Verfassung). Die Klage kann unter Umständen von einem Untersuchungsausschuß geprüft werden. Der Bericht des Ausschusses wird dann dem Verwaltungsrat der ILO und jeder an dem Streitfall interessierten Regierung mitgeteilt; seine Veröffentlichung wird veranlaßt. In diesem Falle bedarf es keiner besonderen Erklärung des verklagten Mitgliedstaates, um die Zuständigkeit des Ausschusses als obligatorisch anzuerkennen.

c) Vertragsstaaten der EMRK können eine Beschwerde an die Europäische Kommission für Menschenrechte wegen jeder angeblichen Verletzung der Bestimmungen der EMRK durch einen anderen Vertragsstaat richten. Die Kommission soll mit guten Diensten auf eine gütliche Einigung hinwirken; mangels einer solchen kann der Vertragsstaat, der die Kommission mit dem Fall befaßt hat oder gegen den sich die Beschwerde richtet, sowie die Kommission selbst den Fall dem Europäischen Gerichtshof für Menschenrechte vorlegen.

Es bedarf keiner Zustimmung des verklagten Staates, um die Zuständigkeit der Kommission zu begründen, seine Zustimmung ist jedoch für die Zuständigkeit des Gerichtshofs erforderlich. Wird die Frage nicht innerhalb eines Zeitraums von drei Monaten dem Gerichtshof vorgelegt, so entscheidet der Ministerausschuß des Europarats, ob die Konvention verletzt worden ist.

Der Vertragsstaat übernimmt die Verpflichtung, die Entscheidung des Gerichtshofs zu befolgen, ihre Durchführung wird vom Ministerausschuß überwacht (Art. 24, 53 und 54 EMRK).

d) Jeder Vertragsstaat, der das Protokoll über die Errichtung einer Schlichtungs- und Vermittlungskommission zum Übereinkommen gegen Diskriminierung im Unterrichtswesen unterzeichnet hat, kann gegen jeden anderen Vertragsstaat wegen einer Verletzung des Übereinkommens eine Schlichtungs- und Vermittlungskommission anrufen; er kann den Fall dieser Kommission vorlegen, wenn er durch gegenseitige Verhandlung nicht zur Befriedigung der beteiligten Staaten bereinigt werden konnte (Art. 12 dieses Protokolls vom 10. Dezember 1962).

e) Jeder Vertragsstaat des Internationalen Übereinkommens zur Beseitigung jeder Form von Rassendiskriminierung kann gegen jeden anderen Vertragsstaat wegen einer Verletzung des Übereinkommens einen Ausschuß für die Beseitigung der Rassendiskriminierung anrufen. Zudem kann eine *ad hoc*-Kommission, die aus fünf Mitgliedern besteht, mit einmütiger Zustimmung der Streitparteien ernannt werden; sie bietet den beteiligten Staaten ihre guten Dienste an, um auf der Grundlage der Achtung dieses Übereinkommens eine gütliche Beilegung herbeizuführen (Art. 12 dieses Übereinkommens).

f) Jeder Vertragsstaat der Amerikanischen Menschenrechtskonvention, der einen anderen Vertragsstaat der Verletzung der in dieser Konvention niedergelegten Menschenrechte für schuldig hält, kann diesen vor die Amerikanische Kommission für Menschenrechte zitieren; die Kommission wird ihre guten Dienste zu einer gütlichen Einigung leisten; mangels einer solchen können die Streitparteien oder die Kommission selbst den Fall dem Amerikanischen Gerichtshof für Menschenrechte vorlegen. Dieses Verfahren ist dem der EMRK sehr ähnlich. Es bedarf aber einer Zustimmung des verklagten Staates sowohl für die Zuständigkeit der Kommission wie auch für die Zuständigkeit des Gerichtshofs (Art. 41, 45, 49, 61 und 62 dieser Konvention).

3. Die Individualbeschwerde

Das Verfahren, mit dem ein Einzelner seine Menschenrechte vor internationalen Instanzen gegen einen diese Rechte verletzenden Staat, einschließlich seines Heimatstaats, verfolgen kann, ist sicher das fortschrittlichste.

a) Das Fakultativprotokoll zum Internationalen Pakt über bürgerliche und politische Rechte sieht vor, daß der Einzelne Mitteilungen (*communications*) über Verletzungen der Menschenrechte an einen Ausschuß für Menschenrechte richten kann. Der Ausschuß soll die erhaltene Mitteilung prüfen und das Ergebnis diesem Staat und dem betreffenden Einzelnen mitteilen (Art. 5 dieses Protokolls).

Ähnliche Verfahren über die Individualbeschwerde befinden sich in einigen internationalen Verträgen.

b) Der zentralamerikanische Gerichtshof, der durch den Vertrag der zentralamerikanischen Republiken (Costa Rica, Guatemala, Honduras, Nicaragua und El Salvador) vom 20. Dezember 1907 errichtet wurde, war für Streitfälle zwischen einer Regierung und einem Einzelnen zuständig, wenn sie die angebliche Verletzung eines völkerrechtlichen Vertrages betrafen. Die Klage eines Einzelnen bedurfte keiner Unterstützung durch die Regierung seines Staates. Das Gericht war bis 1918 tätig und hat insgesamt fünf Individualklagen verhandelt (Art. 2 dieses Vertrags).

c) Gemischte Schiedsgerichte, die auf Grund des Versailler Friedensvertrages errichtet wurden, entschieden über die Entschädigung für die den Staatsangehörigen der alliierten und assoziierten Mächte – während des ersten Weltkrieges – durch außerordentliche Maßnahmen zugefügten Schäden. Außerdem erkannte das Schiedsgericht für Oberschlesien auf Grund des deutsch-polnischen Abkommens vom 15. Mai 1922 die Klagemöglichkeit des Einzelnen gegen seinen eigenen Staat an. Nach dem zweiten Weltkrieg wurden einige völkerrechtliche Schiedsinstanzen für den rechtlichen Schutz einzelner Personen eingerichtet.

d) Auf Grund einiger nach dem ersten Weltkrieg geschlossener Minderheitenschutzverträge hatte der individuelle Minderheitenangehörige das Recht, sich bei Verletzung von Menschenrechten an den Völkerbund zu wenden; Generalsekretär, Dreierausschuß und Völkerbundsrat hatten in abgestufter Zuständigkeit die Sache zu prüfen. Dem verletzenden Staat konnten konkrete Hilfsmaßnahmen auferlegt werden, wenn die Beschwerde begründet war.

e) Berufsverbände von Arbeitnehmern oder Arbeitgebern können eine Beschwerde an das Internationale Arbeitsamt richten, sofern irgendein Mitgliedstaat der ILO die Durchführung eines Übereinkommens, dem er beigetreten ist, nicht in befriedigender Weise sichergestellt hat. Die Beschwerde wird von dem Verwaltungsrat der ILO geprüft; er hat sie der betreffenden Regierung zu übermitteln und diese zu einer Erklärung aufzufordern. Wenn die Antwort nicht zufriedenstellend ausfällt, hat der Verwaltungsrat das Recht, die Beschwerde und die Antwort zu veröffentlichen (Art. 24 und 25 der ILO-Verfassung, besonders ihre Verbesserung vom 25. Juni 1953).

f) Einwohner von Mandatsgebieten konnten Beschwerden (*petition*) einem ständigen Mandatsausschuß vorlegen, sofern sie mit der Verwaltung der Mandatare unzufrieden waren. Die Beschwerde wurde zuerst dem Mandatar zugeleitet, der sie mit seiner Meinungsäußerung diesem Ausschuß vorlegte.

g) Einwohner von Treuhandgebieten können ihre Beschwerden (*petition*) an den Treuhandrat richten, der sie entgegenzunehmen und in Konsultation mit der Verwaltungsmacht zu prüfen hat (Art. 87b der UN-Charta und Art. 76–93 der Geschäftsordnung des Treuhandrats).

h) Das System der Individualbeschwerde an die Europäische Kommission für Menschenrechte ist das fortschrittlichste Verfahren für den internationalen Schutz von Menschenrechten bis auf den heutigen Tag [20].

Natürliche Personen, nichtstaatliche Organisationen oder Personenvereinigungen, die sich durch eine Verletzung der in der Europäischen Menschenrechtskonvention anerkannten Rechte durch einen der Vertragsstaaten geschädigt fühlen, können ihre Beschwerde (Gesuch) über den Generalsekretär des Europarats an die Europäische Kommission für Menschenrechte richten; diese prüft die Angelegenheit und versucht, einen freundschaftlichen Ausgleich auf der Grundlage der Achtung der Menschenrechte zu erreichen.

Das alles setzt aber voraus, daß der Vertragsstaat, gegen den sich die Beschwerde richtet, eine Erklärung abgegeben hat, wonach er die Zuständigkeit der Kommission in dieser Angelegenheit anerkannt hat (Art. 25 EMRK).

i) Das System der Individualbeschwerde vor der Amerikanischen Kommission für Menschenrechte ist dem vorgenannten der Europäischen Kommission ähnlich. Allerdings sieht die AMRK den Einzelnen, Gruppen von Einzelnen und nichtstaatliche Organisation (NGO) als die Beschwerdeführer an.

Es bedarf hier auch keiner Zustimmung des verklagten Staates für die Zuständigkeit der Kommission (Art. 44 AMRK) [21].

j) Einwohner von Kolonialgebieten können ihre Beschwerden einem besonderen Ausschuß auf Grund der Resolution der UN-Generalversammlung 1514 (XV) vom 14. Dezember 1960 vorlegen.

k) Die UN-Kommission für Menschenrechte behandelte bis zum Jahre 1970 nur allgemeine Fragen und Angelegenheiten der Förderung der Menschenrechte.

Seit der Resolution des Wirtschafts- und Sozialrats 1503 (XLVII) vom 27. Mai 1970 darf nun die Kommission auch Individualbeschwerden offiziell in Empfang nehmen und behandeln. Sie behandelt insbesondere Beschwerden der Einwohner von Kolonialgebieten und Gewaltherrschaftsgebieten [22].

l) Einwohner im Gebiet der Vertragsstaaten des Internationalen Übereinkommens zur Beseitigung jeder Form von Rassendiskriminierung können ihre Beschwerden gegen einen die Rechte dieses Vertrags verletzenden Staat an den Ausschuß für die Beseitigung der Rassendiskriminierung richten. Die Zustimmung des verklagten Staates ist für die Zuständigkeit der Kommission (Art. 14 dieses Übereinkommens) erforderlich.

---

[20] *H. Wiebringhaus,* Die Rom-Konvention für Menschenrechte in der Praxis der Straßburger Menschenrechtskommission (1950), S. 106 ff. *A. H. Robertson,* Human Rights in Europe (2. Aufl. 1977), S. 26 ff. *Derselbe,* Die Menschenrechte in der Praxis des Europarates (1972), S. 137 ff.

[21] *A. P. Schreiber,* The Inter-American Commission on Human Rights (1970), S. 13 ff.

[22] *Menzel/Ipsen* (Anm. 7), S. 122.

Die erwähnten Individualbeschwerdeverfahren haben verschiedene Formen. Die fünf verschiedenen Stufen des Verfahrens durch die internationalen Instanzen sind so zu klassifizieren:

1) Die Beschwerde, die nur als Information von einer internationalen Instanz empfangen wird, ohne daß ein weiteres Verfahren folgt;

2) die Beschwerde, die von einer internationalen Instanz zwar geprüft wird; das Prüfungsergebnis wird aber weder den Beteiligten mitgeteilt noch sonst veröffentlicht;

3) die Beschwerde, die von einer internationalen Instanz geprüft wird; das Prüfungsergebnis wird in einem allgemeinen Bericht veröffentlicht;

4) die Beschwerde, die in einem gerichtsförmigen oder gerichtsähnlichen Verfahren auf eine Menschenrechtsverletzung hin geprüft wird; das Ergebnis wird den Beteiligten mitgeteilt und eventuell in einem konkreten Bericht veröffentlicht;

5) die Beschwerde, die von einer internationalen Instanz in einem gerichtsförmigen Verfahren geprüft wird; die Entscheidung wird von den Beteiligten als obligatorisch anerkannt und ihre Durchführung von einem internationalen Organ überwacht.

Neuerdings befassen sich einige Forscher mit den »nichtförmlichen« Sicherungsverfahren der Menschenrechte[23]. Diese Verfahren sind eine Art politischer Verfahren, in denen der Schutz der Menschenrechte von internationalen Gremien, z. B. der Generalversammlung und dem Sicherheitsrat der UN, gewährleistet werden soll.

VI. Gegenwärtige Probleme

Mehrere Schriftsteller der Völkerrechtswissenschaft haben gelegentlich gemeint, die Einzelmenschen könnten die ihnen zustehenden Menschenrechte nur so lange genießen, als die Staaten durch Verträge den Schutz der Menschenrechte gewährleisten. Die Rechte des Einzelnen könnten nur durch Vereinbarung der Vertragsstaaten durchgesetzt und durch diese weiter aufgehoben werden. Die oben erwähnten Sicherungsverfahren für Menschenrechte seien keine Ausnahme davon[24].

Man darf dabei allerdings nicht übersehen, daß es einige zwingende Normen (*ius cogens*) im Völkerrecht gibt.

Was im einzelnen Inhalt zwingender Normen im Völkerrecht ist, konnte bisher noch nicht eindeutig geklärt werden. In der Beratung über das Vertragsrecht in der Internationalen Law Commission der Vereinten Nationen haben einige Mitglieder erwähnt, daß Menschenrechte, einschließlich des

---

[23] *Delbrück* (Anm. 8), S. 51. *Menzel/Ipsen* (Anm. 7), S. 130.

[24] *I. Seidl-Hohenveldern,* Völkerrecht (2. Aufl. 1969), S. 147, para. 655.

Verbotes des Sklavenhandels und Selbstbestimmung der Völker, zum zwingenden Völkerrecht (*ius cogens*) gehörten.

Daß Mitgliedstaaten der Vereinten Nationen sich verpflichtet hatten, die allgemeine Achtung und Verwirklichung der Menschenrechte und Grundfreiheiten durchzusetzen, haben mehrere internationale Gremien anerkennend festgestellt.

Der Staat kann auch durch Vereinbarungen die Menschenrechte, die zum zwingenden Normbestand gehören, nicht wieder aufheben, weil von ihnen keine Abweichung erlaubt ist.

Bestimmungen über die Menschenrechte sind nicht erst auf Grund der Zustimmung von Vertragsstaaten zur Wirksamkeit gelangt, sondern besitzen Geltung bereits wegen ihrer Übereinstimmung mit dem grundlegenden zwingenden Normenbestand. Man hat auch die Auffassung vertreten, nur der Staat könne als aktives Subjekt internationale Verträge schließen, der Einzelne bleibe dagegen das passive Subjekt im Völkerrecht, das selbst nicht handeln könne.

Niemand hat indessen gemeint, nur das Gesetzgebungsorgan könne als aktives Subjekt im Staatsrecht tätig werden, denn der Einzelne kann zwar gewöhnlich nur mittelbar durch das Parlament, tatsächlich aber auch durch die Interessengruppen an der Gesetzgebung mitwirken. Die internationalen Verträge werden in der Regel von Bevollmächtigten der Staaten, die an den diplomatischen Konferenzen teilnehmen, auch zuweilen von Arbeitgebervertretern, Arbeitnehmervertretern und Regierungsvertretern, die an der Allgemeinen Konferenz der ILO teilnehmen, oder Vertretern in der UN-Generalversammlung geschlossen.

Die Zustimmung eines Staates zur Bindung an einen Vertrag wird im allgemeinen durch Ratifikation zustande gebracht: Die Regierung ratifiziert mit der Zustimmung ihres eigenen Gesetzgebungsorgans den Vertrag. Der Einzelne kann dabei ebenfalls nur mittelbar, nämlich durch das Organ (Regierungsvertreter, eventuell Arbeitgeber- oder Arbeitnehmervertreter oder Abgeordnete) an der Vertragsschließung teilnehmen – tatsächlich besteht noch eine gewisse Mitwirkungsmöglichkeit durch Interessengruppen. Heutzutage kann eine Regierung die öffentliche Meinung, d. h. den Willen des Einzelmenschen, sowohl an der Vertragsschließung als an der Gesetzgebung mitzuwirken, nicht mehr völlig ignorieren.

Wer meint, daß ohne Verträge und ohne eine Organisation von Instanzen Menschenrechte geschützt werden könnten, begeht einen Irrtum. Die früheren Lehren beschäftigen sich zu sehr mit rechtlichen Mechanismen in der Völkergemeinschaft; man soll jetzt stärker die Funktionen betrachten. Von einem funktionalen Standpunkt aus erweist sich, daß die Menschenrechte in der Völkergemeinschaft verhältnismäßigen Schutz genießen. Aber nicht die Vorschriften der Verträge oder die internationalen Instanzen gewähren diesen Schutz, sondern dieser begründet sich im nichtstaatlichen Bereich in transnationalen Lebensverhältnissen von Einzelmenschen.

In diesem Fall kann man der Ansicht sein, daß es sich um die Menschenrechte auf innerstaatlicher Ebene handelt. Die Rechte stammen aber nicht nur aus dem Staatsrecht, sondern auch aus dem Völkerrecht, sei es aus *self-executing* Verträgen, sei es aus dem *ius cogens*.

Zur Zeit werden Menschenrechte eigentlich durch die rechtlichen Überzeugungen geschützt, nämlich das Bewußtsein der Völker, weil Diplomaten, Bevollmächtigte, Regierungsvertreter, Richter des Internationalen Gerichtshofs, Mitglieder der NGO und Einzelmenschen sich ihrem Bewußtsein entsprechend verhalten, das mit dem allgemeinen Bewußtsein der Völkergemeinschaft, d. h. der Weltöffentlichkeit, verknüpft ist.

Die großen Hindernisse gegen die Verwirklichung der Menschenrechte, d.h. Krieg, Kolonialherrschaft, Despotismus, Armut usw., können erst mit der internationalen und transnationalen Solidarität der Völker wirklich beseitigt werden.

Zum internationalen Schutz der Menschenrechte kommt es jetzt auf die folgenden Maßnahmen an:

1. Unterricht über die Menschenwürde in allen Schulen und in der Gesellschaft
2. Aufklärung über Menschenrechtssicherung durch die Massenmedien
3. Erweckung öffentlichen Bewußtseins über die Bedeutung der Menschenwürde
4. Beförderung der Tätigkeit der NGO zur Pflege der Menschenrechte, z. B. ICRC, Amnesty International und Mitwirkung an ihrer Tätigkeit
5. Einrichtung von internationalen oder nichtstaatlichen Untersuchungs- und Beobachtungs-Instanzen zum Kampf gegen Angriffe auf die Menschenrechte
6. Erweiterung der Beobachtungs- und Prüfungszuständigkeit der internationalen Instanzen für Menschenrechte
7. Förderung des Beitritts zum Fakultativprotokoll zum Internationalen Pakt über bürgerliche und politische Rechte vom 16. Dezember 1966
8. Errichtung eines universalen Gerichtshofs für Menschenrechte (ähnlich den Gerichtshöfen für Menschenrechte in EMRK oder AMRK).

Alle diese Maßnahmen umfaßt die Präambel der Satzung der UNESCO:

> »Da die Verletzung der Menschenrechte im Geist des Menschen entsteht, muß die Verteidigung der Menschenrechte im Geist des Menschen ihren Anfang nehmen«.

Observations sur la hiérarchie des sources du droit international

RICCARDO MONACO

1. L'ordre juridique international, du fait de sa formation et de sa nature de caractère social, n'est pas une entité statique; au contraire il évolue sans cesse au cours des années. Dernièrement cet ordre a évolué d'une façon tellement rapide que les mutations qu'il a subi sont d'une ampleur exceptionnelle. D'où une conséquence bien visible et constatable: l'ordre international actuel présente des caractères sensiblement différents de ceux qu'il possédait autrefois.

On peut d'ailleurs remarquer que les mutations qui se sont produites au sein de l'ordre international n'ont pas seulement eu pour objet les secteurs périphériques du droit international, c'est-à-dire les secteurs nouveaux spécialisés qui appartiennent désormais au domaine juridique international; il s'agit là surtout de certaines organisations internationales nouvelles, dont l'activité dépasse les frontières habituelles de l'ordre international.

Il faut reconnaître en outre que l'ensemble de ces développements nouveaux a également touché la structure de cet ordre et aussi, par conséquent, le régime des sources.

En ce qui concerne les sources du droit international, ces changements de structure ont donné lieu à certaines manifestations juridiques qui dépassent ou du moins tendent à dépasser le cadre des sources de caractère traditionnel. A cet égard on entend parler souvent de sources atypiques, c'est-à-dire de sources qui s'ajouteraient aux sources traditionnelles.

Ce fait nécessiterait, à lui seul, un examen approfondi: mais, à cet égard, qu'il me soit permis de renvoyer à une analyse récente que j'ai faite et qui constitue, à mon avis, les prémisses de ce que je voudrais exposer ici [1].

---

[1] Je me réfère à mon étude Fonti e pseudo fonti del diritto internazionale, Rivista di diritto internazionale 1978, p. 740 ss., qui, étant destinée aux Mélanges offerts à Adolfo Miaja de la Muela, figure également dans Estudios de derecho internacional en l'honneur dudit Professeur, vol. 1 (Madrid 1979), pp. 411 ss.

2. Devant un mouvement doctrinal qui est devenu très important dernièrement, il me semble qu'il existe un point qui apparaît particulièrement apte à éclaircir les idées dans ce domaine. Il s'agit de la tâche tendant à établir une hiérarchie aussi exacte que possible entre les différentes sources.

En effet, il ne suffit pas d'avoir une connaissance exacte de chacune des sources; il est nécessaire en outre de mettre chacune d'elles à la place qui lui appartient dans une échelle hiérarchique.

A partir du moment où le nombre des sources tend à augmenter, il apparaît nécessaire de les graduer selon un ordre de priorité: ainsi il devient possible d'apprécier la portée et l'efficacité de chacune d'elles par rapport aux autres.

Tout cela oblige d'autre part à considérer les sources traditionnelles d'une façon distincte par rapport aux sources nouvelles, qui se sont manifestées à la suite de l'évolution récente de l'ordre juridique international.

Une analyse tendant à fixer chacune des sources ainsi que sa portée exacte n'est pas nécessaire au stade actuel de la doctrine et de la pratique, qui ont déjà clarifié les éléments d'appréciation auxquels est liée toute recherche de ce genre.

Ce qui au contraire me semble indispensable c'est de fixer les critères d'après lesquels on peut arriver à la graduation des sources. A cet égard on rencontre les expressions les plus variées, telles que hiérarchie des sources, priorité des sources, prédominance d'une source à l'égard d'une autre, utilisation graduelle des sources par le juge ou par l'interprète: de toutes façons, l'idée qui se trouve à la base de notre sujet ne doit pas être entendue dans un sens strict.

Nous nous en tenons donc à l'expression hièrarchie des sources[2]: et alors le concept qui se rapproche particulièrement de celui de hiérarchie est l'idée de la coordination de plusieurs activités exercées dans le même domaine. Cela signifie que l'on peut retenir la notion de hiérarchie sans pourtant tenir compte de la supériorité ou au contraire de la subordination d'une personne, d'une fonction, d'une structure par rapport à d'autres.

C'est pour cela que le problème de la hiérarchie entre les différentes sources normatives peut être ramené à celui des relations entre les sources elles-

---

[2] Dans presque tous les travaux en matière de sources on trouve des allusions plus ou moins précises au problème de la hiérarchie, mais les contributions destinées spécialement à la solution de ce problème ne sont pas tellement nombreuses: voir surtout *Bos,* The Recognized Manifestations of International Law: a New Theory of Sources, German Yearbook of International Law 20 (1977), pp. 9 ss.; *idem,* The Hierarchy among the Recognized Manifestations (sources) of International Law, in: Estudios de derecho internacional en l'honneur de Miaja de la Muela, *op. cit.*, pp. 363 ss.; *Akehurst,* The Hierarchy of the Sources of International Law, British Year Book of International Law 47 (1974–1975), pp. 273 ss.

mêmes[3]. Et ces relations peuvent être de nature différente, étant donné qu'il faut tenir compte de la valeur que chacune des sources possède par rapport aux autres.

3. Il y a donc certaines prémisses à notre recherche: elles dominent ces relations, au-delà même du problème qui est celui des sources du droit international.

Tout d'abord une source peut être supérieure et donc prédominante à l'égard d'autres sources; de sorte que, dans ce cas on peut à juste titre parler de hiérarchie, pas tellement dans un sens absolu, mais au sens réciproque: dans une telle hypothèse, la source supérieure peut remplacer ou bien modifier l'autre, tandis qu'elle ne peut être ni remplacée ni modifiée par cette dernière.

Les effets découlant d'une telle situation se produisent différemment compte tenu du fait que les deux sources concernent la même matière ou une matière distincte. Dans le premier cas la source supérieure est toujours une source primaire, tandis que la source inférieure apparaît comme une règle secondaire. Dans le deuxième cas chacune des deux sources peut régler la matière qui est de sa propre compétence par des normes primaires et indépendantes: ce qui fait que tout chevauchement entre les normes respectives est exclu.

Il se peut au contraire que plusieurs sources soient équivalentes: et alors, même dans cette hypothèse, il faut distinguer le cas de la concurrence par rapport à celui de la séparation entre les sources. Si les deux sources peuvent régler la même matière, étant donné leur équivalence, il n'est pas question de normes primaires et secondaires, mais seulement de normes antérieures et postérieures. Par conséquent, les normes plus récentes, bien que provenant d'une source différente, remplacent et modifient les normes plus anciennes, et peuvent, à leur tour, être modifiées par d'autres normes qui, dans l'avenir, émanent de la même source qui avait produit les normes plus récentes.

Au cas où les matières faisant l'objet de deux sources équivalentes sont séparées, une telle séparation s'impose également à l'une et à l'autre, de sorte qu'est exclu tout empiètement de l'une de sources dans le domaine qui est réservé à l'autre.

---

[3] Une distinction qui assez souvent figure dans la doctrine est celle qui sépare les sources primaires des sources secondaires: d'ailleurs elle se reflète sur les normes qui sont produites par l'une et par l'autre catégorie de sources, de sorte qu'on parle aussi de normes primaires et de normes secondaires. On pourrait douter de ce parallélisme: en tout état de cause il est reconnu dans le domaine du droit interne. En ce qui concerne l'ordre juridique italien voir surtout *Giannini,* Lezioni di diritto amministrativo, vol. 1 (Milan 1950), p. 59; *Amato,* Rapporti tra norme primarie e secondarie (Milan 1962); *Lavagna,* Istituzioni di diritto pubblico (2e éd., Turin 1973), p. 175; pour une synthèse concernant les différentes significations qu'on peut attribuer à cette distinction voir *Pizzorusso,* Delle fonti del diritto, in: Commentario del Codice civile dirigé par Scialoja et Branca (Bologne, Rome 1977), p. 41.

Voilà donc un certain nombre de règles de caractère logique, qui peuvent être utilisées même à l'égard des normes juridiques internationales[4].

4. Avant de continuer sur le chemin qui nous amènera à fixer la hiérarchie des sources du droit international il nous semble opportun de frationner notre recherche, dans le sens qu'un examen particulier s'impose en ce qui concerne trois catégories de sources, qui figurent parmi les plus discutables, à cause aussi du fait qu'elles dépassent le cadre des sources traditionnelles.

Il s'agit tout d'abord des principes généraux de droit, à l'égard desquels il faut préciser qu'ils ne sont pas les principes qu'on trouve dans tout ordre juridique interne, mais les principes propres à l'ordre juridique international. Cette distinction est nécessaire, parce que, nonobstant les conceptions divergentes qui résultent parfois de la doctrine et de la jurisprudence, ces principes ont un caractère autonome, étant donné qu'ils trouvent leur origine dans l'ordre international lui-même, de sorte qu'ils opèrent exclusivement dans cet ordre[5].

---

[4] En exposant les critères qui figurent ci-dessus, je me suis inspiré surtout de la ligne logique qui a été suivie à cet effet par *Zanobini,* Gerarchia e parità fra le fonti, in: Scritti giuridici in onore di Santi Romano, vol. 1 (Padou 1940), p. 595. Ces critères, qui ont été énoncés afin de mettre un ordre logique de priorité entre les sources du droit étatique, sont certainement valables aussi dans le domaine du droit international.

[5] La distinction dont il s'agit a été précisée d'une façon très nette et convaincante par *Rousseau,* Droit international public, vol. 1 (Paris 1970), pp. 371 ss., qui sépare, sur la base d'une analyse exhaustive qui trouve son fondement dans la jurisprudence internationale, les principes généraux du droit en tant que catégorie qui est inhérente à n'importe quel ordre juridique international ou interne, des principes qui, étant communs aux différents ordres juridiques internes, sont empruntés à l'ordre juridique international (art. 38 par. 3 du Statut de la Cour internationale de Justice) et aussi des principes qui sont spéciaux à l'ordre juridique international et qui, par conséquent, peuvent être appelés, à juste titre, principes généraux du droit international. Cette tripartion apparaît, à mon avis, indispensable lorsqu'on essaie de suivre un ordre logique dans le classement des sources du droit international: ce qui n'arrive pas toujours, à cause aussi du fait que ces trois catégories sont très proches l'une de l'autre et que, dans cette matière, l'approche pratique prévaut assez souvent sur la vision systématique. C'est pour cela que même les auteurs qui ont consacré le plus d'attention aux principes généraux dans leur ensemble, comme *Verdross,* dans ses nombreuses et précieuses contributions, et dernièrement dans l'ouvrage Die Quellen des Universellen Völkerrechts (Fribourg en Br. 1973), pp. 120 ss., et *Guggenheim,* Traité de droit international public, vol. 1 (Genève 1953), pp. 152 et 156, ne tiennent pas compte de ladite tripartion. Au surplus *Guggenheim* déclare que les principes généraux du droit reposent soit sur le droit conventionnel, soit sur le droit coutumier, ce qui déprive ces principes de leur autonomie en tant que source. – Par contre il faut signaler l'analyse approfondie de *Barberis,* Fuentes del derecho internacional (La Plata 1973), p. 280, qui tend à séparer le processus de création des principes généraux de celui à travers lequel se manifestent les normes coutumières. – Un exemple trés important des recherches qui seraient nécessaires pour clarifier l'ensemble des problèmes qui doivent être résolus dans ce domaine est donné par *Schlesinger,* Research on the General Principles of Law Recognized by Civilized Nations, American Journal of International Law 51

Il n'est pas toujours facile de fixer des principes généraux; à cet égard l'aide fondamentale provient de la jurisprudence internationale. De cette base on peut déduire une série de principes généraux: certains se réfèrent aux relations entre les différents ordres juridiques; tel est le cas, par exemple, du principe en vertu duquel le traité a une priorité sur la loi interne; d'autres concernent l'exercice des compétences internationales de l'Etat et, parmi eux, figurent le principe de la continuité de l'Etat et celui de son indépendance; d'autres encore dominent la matière de la responsabilité internationale de l'Etat et celle de la conduite de la guerre en mer [6].

En principe, il y a accord presque unanime pour reconnaître l'existence de la catégorie des principes généraux; mais le point qui nous intéresse est celui de classer ces principes dans la hiérarchie des sources. Pour l'instant il faut toutefois attendre d'être arrivés à la fin de notre examen des sources controversées.

En tout état de cause il faut d'ores et déjà remarquer que d'un côté la jurisprudence se borne à attribuer aux principes généraux la fonction de source subsidiaire, alors que, d'un autre côté, la doctrine la plus avancée n'hésite pas à affirmer qu'aux principes généraux appartient une position de premier plan en tant que source du droit international.

Mais ici encore il faut procéder avec beaucoup de prudence car les bases de départ de la jurisprudence et de la doctrine diffèrent sensiblement. En effet, tandis qu'on a tendance à s'en tenir de plus en plus aux principes généraux du droit reconnus par les nations civilisées, citées dans l'art. 38 du Statut de la Cour Internationale de Justice en renforçant leur rôle, d'un autre côté, on rapproche de plus en plus les principes généraux de l'ordre international des principes constitutionnels de l'ordre lui-même [7].

(1957) p. 734. De cette étude il ressort très clairement que le champ d'application de ces principes dépasse largement l'idée initiale posée par le Statut de la Cour. – D'ailleurs il ne faut pas oublier les connexions possibles entre les principes généraux de l'ordre international et les principes généraux de certains ordres particuliers: voir, à cet égard, *Meessen,* Zur Theorie allgemeiner Rechtsgrundsätze des internationalen Rechts, Jahrbuch für internationales Recht 17 (1974), pp. 283 ss.

[6] *Rousseau* (note 5), pp. 393–394, sur la base concrète de la jurisprudence internationale essaie de déterminer ainsi le contenu de ces principes généraux.

[7] La doctrine n'accepte pas toujours la catégorie des principes constitutionnels de l'ordre international en tant que source destinée à trouver sa place à côté des autres sources traditionnelles. En tout état de cause j'estime que la complexité croissante de l'ordre international a fait apparaître l'existence de certaines règles de structure qui sont préexistantes à la fonction normative et qui, en tant que telles, sont inhérentes à l'ordre international. Ce concept a surtout été développé par la doctrine italienne: voir *Romano S.,* Corso di diritto internazionale (Padoue 1939), p. 31; *Fedozzi,* Introduzione al diritto internazionale e parte generale (Padoue 1939), p. 50; *Ottolenghi,* Corso di diritto internazionale (Turin 1951), p. 42; *Bentivoglio,* Interpretazione del diritto e diritto internazionale (Pubblicazioni dell'Università di Pavia, Studi nelle scienze giuridiche e sociali, n. 119), p. 254; *Monaco,* Cours général de droit international public, Recueil des Cours 125 (1968 III), p. 131. En ce qui concerne la doctrine étrangère, je me bor-

D'après cette dernière tendance, l'origine des principes généraux se ramènerait seulement à l'ordre international, de sorte qu'il n'existernait plus aucun lien entre ces principes et les autres, dont l'origine et la justification se trouvent au sein des ordres juridiques internes des différents Etats.

Dans ce domaine les recherches de la doctrine ont beaucoup progressé, et c'est pour cela qu' à l'heure actuelle on pense à un ensemble de principes, qui ne seraient pas seulement ceux indiqués dans le Statut de la Cour, mais qui couvriraient, sous le titre de principes généraux, une réalité plus vaste et plus directement rattachée à l'ordre international.

En tout état de cause, à partir du moment où l'on doit établir une hiérarchie entre les sources du droit international, il devient impossible de laisser de côté les principes généraux de l'ordre international en tant que tels, parce que désormais ils expriment les exigences de structure et de fonctionnement de l'ordre international dans sa réalité actuelle.

5. Il faut passer maintenant à une autre catégorie de sources, dont l'importance est telle, que l'on peut déduire aussi des conséquences sur l'ordre hiérarchique selon lequel les sources peuvent être placées: il s'agit des règles juridiques créées par des actes émanant d'organes d'institutions internationales[8].

Pour être au clair de ce problème complexe, il faut tout d'abord remarquer que ces actes ne peuvent pas être ramenés à une catégorie de portée générale. En effet, les organisations internationales, dont les organes ont des pouvoirs normatifs directs, sont de nature très variée, de sorte qu'ici peuvent seulement être considérées celles qui, à travers de tels actes, peuvent avoir une influence sur la production juridique internationale, telle qu'elle se manifeste dans le cadre universel.

Par conséquent ne peuvent faire l'objet de notre recherche les actes qui créent des normes destinées à devenir efficaces seulement dans l'ordre juridique de l'organisation ou qui donnent vie à des règles de droit particulier visant les ordres juridiques extérieurs à l'organisation.

En d'autres termes, le problème se pose seulement pour les actes qui créent des normes juridiques relevant du droit international général: en pratique l'examen est limité aux résolutions des organes des Nations Unies. Cependant, ces actes n'ont pas tous la même efficacité juridique. Cela implique

---

nerai à citer *Sørensen,* Les sources du droit international (Copenhague 1946), p. 116, et *Schwarzenberger,* The Fundamental Principles of International Law, Recueil des Cours 87 (1955 I), p. 204. – Sur l'ensemble de cette conception, voir l'analyse approfondie de *Giuliano,* Diritto internazionale, vol. 1 (Milan 1974), pp. 217–225.

[8] Sur l'ensemble des problèmes qui se posent à l'égard des actes normatifs qui sont créés par des organisations internationales, ainsi que sur l'efficacité des règles qui en découlent, je me permets de renvoyer à mon exposé, tel qu'il figure dans *Monaco,* Les principes régissant la structure et le fonctionnement des organisations internationales, Recueil des Cours 156 (1977 III) pp. 141 ss.

que, parmi les très nombreuses résolutions, il faut retenir seulement celles qui donnent lieu à des effets juridiques obligatoires.

Compte tenu de cela, il n'y a pas lieu de considérer les résolutions qui ont un caractère interorganique: celles-ci obligent un organe inférieur à observer les instructions d'un organe supérieur, de sorte que ces règles peuvent devenir efficaces seulement au sein de l'ordre interne des Nations Unies et non pas à l'égard des Etats membres.

Au contraire il existe un nombre limité de résolutions qui, émanant de l'Assemblée générale ou du Conseil de sécurité dans l'exercice de certaines compétences statutaires et ayant force obligatoire à l'égard des Etats membres, ont un caractère juridique international.

6. Ce qui s'impose davantage à notre attention c'est le fait que le moyen de la résolution a été utilisé à des fins diverses de celles en vue desquelles elle avait été originairement conçue: en effet, tandis que la résolution ordinaire édicte elle-même les normes émanant de la volonté des organes des Nations Unies, dans d'autres cas elle est devenue le moyen par lequel sont adoptés des textes organiques destinés à être introduits dans le contexte de l'ordre international général.

Il s'agit des déclarations de principes, qui assez souvent ont marqué un moment important de l'oeuvre confiée à l'Assemblée générale des Nations Unies[9]. A cet égard il faut distinguer certaines déclarations qui, tout en ayant une grande portée politique, ont toutefois un objet limité et, de ce fait, n'ont pas d'influence sur l'ordre juridique international, et d'autres déclarations qui touchent aux bases mêmes de la communauté internationale, en tant qu'ordre qui se développe sans cesse, et qui par conséquent tendent à donner une nouvelle structure à cet ordre.

---

[9] Ce point est, à mon avis, fondamental en ce qui concerne l'admissibilité des déclarations de principes en tant que sources: à cet égard, étant donné les limites qui sont imposées à la présente étude, je m'abstiens de faire de longues citations. Je voudrais seulement me référer à mon article Fonti e pseudo fonti del diritto internazionale (note 1), pp. 752 ss., où il y a une tentative de présenter une synthèse des problèmes très délicats qui se posent à cet égard. En tout état de cause, il faut être très prudent dans ce domaine: ce qui explique que même les auteurs qui semblent être les plus inclins à élargir la catégorie des sources du droit international comme *Bos,* The Recognized Manifestations (note 2), pp. 24–25, acceptent comme sources seulement certaines décisions des organisations internationales. Il faut avouer qu'avec cette affirmation on reste beaucoup trop dans le vague. – Etant donné que le problème le plus important dans ce domaine est celui de l'efficacité des déclarations de principes, je trouve que l'on peut conclure avec l'affirmation qu'on lit juste à la fin du précieux petit volume de *Verdross* (note 5), p. 140: »die in einer Deklaration der Generalversammlung enthaltenen Normen eine völkerrechtliche Verbindlichkeit nur dadurch erlangen können, daß sie entweder ins völkerrechtliche Gewohnheitsrecht oder ins Vertragsrecht eingehen oder als allgemeine Rechtsgrundsätze anerkannt werden. Die Behauptung, das außerhalb dieser Rechtsquellen noch andere formelle Rechtsquellen ersten Grades in der Völkerrechtsordnung bestehen, ist daher unhaltbar«.

Il suffit de se référer à certains exemples, tels que la Déclaration qui énonce les principes de droit international relatifs aux relations amicales et à la coopération entre les Etats, approuvée par la résolution 2625 (XV) du 24 octobre 1970; la Charte des droits et devoirs économiques des Etats adoptée par la résolution 3281 (XXIX) du 14 décembre 1974.

La nature juridique de telles déclarations est très controversée, ce qui a conduit à des résultats contradictoires. Cela à cause du caractère très varié des matières faisant l'objet de ces actes, de sorte qu'il en découle des conséquences très différentes à l'égard du droit international en vigueur.

En fin de compte, les déclarations présupposent qu'à la suite de leur émanation on arrive à poser des règles juridiques correspondant à leur contenu et aux motifs qui les ont inspirées; cela prouve qu'elles ne sont pas capables, en tant que telles, de donner vie à des normes juridiques.

Cela s'applique aussi aux déclarations de principes, qui ne sont pas adoptées par des résolutions de l'Assemblée générale des Nations Unies, mais sont approuvées en tant que conclusions de conférences internationales tenues sous l'égide des Nations Unies, ou bien résultent de négociations diplomatiques qui se sont réalisées en dehors des Nations Unies.

Ces déclarations, quant à leur forme, diffèrent des actes en vertu desquels une organisation internationale manifeste sa volonté: en effet, ces derniers actes sont formés d'après les procédures concernant la manifestation de la volonté de l'organisation. Ce qui, au contraire, n'arrive pas au sein des conférences internationales qui échappent au système des Nations Unies. A part cela, il n'y a pas de différence entre les deux catégories de déclarations. En effet, les résolutions des organisations internationales, y compris celles en question, doivent être approuvées par les organes compétents de l'institution, afin de devenir efficaces à l'égard des Etats membres; de la même manière les documents formels emanant d'une conférence internationale n'ont pas, par eux-mêmes, force obligatoire, parce qu'ils ont besoin d'être approuvés, selon les procédures étatiques requises, par les Etats participants.

7. Ainsi, nous sommes parvenus à avoir une idée assez précise des principes généraux et des actes émis par des organisations internationales, en tant que catégories possibles de sources du droit international.

Il s'agit en outre de définir une troisième catégorie de normes pouvant entrer en ligne de compte afin de fixer la hiérarchie des sources: ce sont les normes faisant partie du *jus cogens*[10]. Il faut tout d'abord remarquer que cette

---

[10] Sur l'ensemble des problèmes qui se posent à l'égard du *jus cogens* qu'il me soit permis de renvoyer à l'exposé fait dans mon Cours général de droit international public, Recueil des Cours 125 (1968 III), pp. 202–212, ainsi qu'à mon article, déjà cité (note 1), Fonti e pseudo fonti del diritto internazionale, pp. 744–745. Nonobstant que la catégorie des normes de *jus cogens* ait reçu une consacration officielle dans l'art. 53 de la Convention de Vienne sur les traités, plusieurs difficultés relatives à la définition de ce concept demeurent irrésolues. Ce qui d'ailleurs amènera la doctrine et la pratique à faire des efforts ultérieurs en vue de parvenir à une définition de ce concept qui ne

catégorie de règles a des limites incertaines et cela pour deux raisons essentielles: premièrement parce qu'il est presque impossible de définir le contenu de la catégorie, c'est-à-dire d'établir quelles sont les règles ayant force impérative et celles qui ne l'ont pas; deuxièmement parce qu'il est presque impossible d'affirmer *a priori* à quel niveau normatif se situent les règles de *jus cogens*. En effet, il se peut qu'une règle d'origine conventionnelle se transforme, à un moment donné, en une règle de *jus cogens*; d'autre part on ne peut pas nier qu'une règle juridique coutumière devienne une norme impérative ayant la même efficacité que les autres règles de *jus cogens*.

Il existe encore d'autres difficultés et incertitudes qui proviennent du fait que les règles de *jus cogens* peuvent avoir une influence plus ou moins marquée sur les autres normes: coutumières, conventionnelles, principes généraux. A cause de cela placer exactement une règle de *jus cogens* dans l'échelle hiérarchique des sources devient toujours difficile; mais, pour l'instant, nous ne sommes pas encore arrivés au moment qui viendra lorsqu'on fera une comparaison entre les différentes sources.

8. Le problème le plus délicat qui se pose au départ est celui de fixer les relations entre les normes de *jus cogens* et les normes constitutionnelles de l'ordre international. Entre autres, il faut savoir si une telle distinction est vraiment possible et si elle doit être maintenue. Il faut rappeler qu'une règle constitutionelle est certainement de caractère impératif et, en tant que telle, elle ne comporte pas de dérogation. Toutefois, il peut exister une règle de *jus cogens* qui ne figure pas parmi les règles constitutionnelles et à laquelle on ne peut néanmoins déroger. Dans ces conditions on peut bien imaginer deux catégories différentes de règles impératives: celles constitutionnelles et celles im-

---

soit pas seulement fondée sur une règle conventionnelle. – Certaines difficultés étaient d'ailleurs déjà présentes à l'esprit de ceux qui ont poursuivi le travail qui a abouti à la Convention de Vienne: voir, parmi les différentes analyses à cet égard, *Paul,* The Legal Consequences of Conflict between a Treaty and an Imperative Norm of General International Law, Österr. Zeitschrift für öffentliches Recht 21 (1971), p. 18, *Fahmi,* Peremptory Norms as General Rules of International Law, Österr. Zeitschrift für öffentliches Recht 22 (1971), pp. 388–393. Dans les traités les plus récents on peut quand même remarquer que l'élaboration doctrinale avance: voir, par exemple, *Verdross/Simma,* Universelles Völkerrecht (Vienne 1976), p. 266, où on trouve une précision importante dans le sens que le *jus cogens* couvre un domaine plus restreint que celui du droit international universel; de sorte que celui-ci, étant de nature dispositive, devient applicable seulement si les parties ne sont pas tombées d'accord pour y déroger. Par conséquent, toute tentative d'identifier le *jus cogens* avec le droit international général n'est pas admissible. – Sur un plan différent, on doit tenir compte d'une autre remarque: voir *Quoc Dinh/ Daillier/Pellet,* Droit international public (Paris 1980), p. 196, qui à juste titre observent que l'introduction dans l'ordre international du concept du *jus cogens* présuppose que la structure actuelle du droit international soit changée: »c'est dans la perspective du développement progressif du droit international qu'il faut se placer pour apprécier cet événement juridique, qu'est la reconnaissance de l'existence du *jus cogens:* dans l'édification des bases contitutionnelles écrites de la Communauté internationale, il faut bien un commencement et il réside dans la solution de principe adoptée par la Convention de Vienne«.

pératives au sens strict du mot. Mais on pourrait se demander si on doit les considérer comme deux catégories de normes, ou bien s'il ne serait pas plus conforme à la logique de les réunir dans le même secteur normatif. Voilà donc les incertitudes que l'existence des règles de *jus cogens* fait naître et auxquelles il faudra bien faire face au cours de notre recherche.

9. Après l'examen, nécessairement rapide, que nous avons consacré à trois catégories de sources, c'est-à-dire, les principes généraux de l'ordre international, les actes émanant d'organisations internationales et les régles de *jus cogens,* on peut laisser de côté, pour l'instant, les problèmes particuliers qui concernent ces sources.

Ainsi, l'examen exposé ci-après, qui vise les sources de caractère classique, devient plus simple. Bien entendu, tout cela comporte une réflexion ultérieure à l'égard du point fondamental que nous devons traiter, c'est-à-dire, la fixation de la hiérarchie des sources. Les sources de caractère classique, nous le savons bien, sont l'accord et la coutume. Elles sont marquées par leur nature, qui se rattache à la volonté ou à la pratique des Etats; en tout état de cause, sans revenir à des discussions doctrinales bien connues tendant à séparer le phénomène accord du phénomène coutume, il faut tenir compte du fait que les deux sources, contrairement aux autres, se rattachent directement au comportement des Etats.

En ce qui concerne le problème de l'efficacité réciproque de l'accord et de la coutume, on admet qu'ils se placent sur le même plan: il s'ensuit que les problèmes relatifs à la priorité entre les normes respectives trouvent leur solution par application des principes bien connus de la priorité dans le temps des normes et des relations entre normes générales et normes particulières.

En ce qui concerne les relations entre normes générales créées par voie d'accord et normes particulières de formation coutumière, il faut appliquer le principe précité, en vertu duquel, en présence de normes ayant la même efficacité, c'est la priorité dans le temps qui prévaut. Ainsi, un accord multilatéral peut abroger ou modifier des règles coutumières qui, après son entrée en vigueur, ont continué à être appliquées entre un certain nombre d'Etats participants. Au contraire, si après l'entrée en vigueur d'un accord multilatéral, entre un certain nombre d'Etats participants surgissent des règles coutumières, qui sont en contradiction avec des normes de l'accord, les règles coutumières ne pourront déroger aux régles conventionnelles que si ces dernières sont de nature dispositive.

Si au contraire les normes de l'accord multilatéral ont un caractère impératif, les règles créées par des coutumes particulières ne pourront pas déroger aux normes conventionnelles: cela sur la base du même principe, en vertu duquel les normes posées par des accords particuliers ne peuvent pas déroger à des règles conventionnelles de caractère impératif.

Il faut maintenant voir quelles sont les solutions possibles, lorsque la comparaison se fait entre normes générales posées par des coutumes et normes particulières émanant de coutumes ou d'accords.

Lorsque les normes particulières (peu importe qu'elles proviennent d'une coutume ou d'un accord), sont antérieures à une coutume générale, cette dernière peut y déroger totalement ou partiellement, dans la mesure où elles sont incompatibles avec la coutume générale. Ceci découle du fait que la formation d'une coutume entre les mêmes sujets qui, antérieurement, ont donné vie à des accords particuliers, produit une mutation essentielle au sein de la production normative préexistante. Par conséquent, les règles les plus récentes prévalent sur celles qui les ont précédées, non pas parce que la coutume est une source supérieure à l'accord, mais tout simplement parce que dans ce cas on doit appliquer le critère chronologique. Enfin, en matière de relations entre une coutume générale et des normes particulières qui se sont formées plus récemment, on se trouve en présence de la même situation, déjà indiquée, qui oppose le droit conventionnel général au droit particulier [11].

10. Voilà donc que le moment est venu de voir comment on peut établir un ordre hiérarchique entre les sources du droit international. A cet égard on pourrait tout d'abord se demander si, dans l'ordre international, des règles existent qui visent d'une façon particulière l'échelle prioritaire des sources. Il semble que la réponse doit être négative: en effet, tout au plus, certains critères, et non des règles juridiques, pourraient être déduits de la doctrine et de la jurisprudence, mais, ces critères, en tant que tels, ne pourraient pas dépasser une simple fonction d'orientation.

Cela dit, il faut remarquer qu'assez souvent on se réfère à une règle qui paraît être la seule utilisable dans cette matière: il s'agit de l'art. 38 du Statut de la Cour internationale de Justice, dont le libellé est bien connu, mais qui, toutefois, mérite d'être reproduit ci-dessous: la Cour applique:

a) les conventions internationales, soit générales, soit spéciales, établissant des règles expressément reconnues par les Etats en litige;
b) la coutume internationale, comme preuve d'une pratique générale, acceptée comme étant le droit;
c) les principes généraux de droit reconnus par les nations civilisées.

---

[11] Voir, sur cette matière, mon exposé dans Manuale di diritto internazionale pubblico (Turin 1971) pp. 163 ss., ainsi que le résumé des critères applicables dans mon Cours général de droit international public (note 7), pp. 213 ss. – Evidemment les critères qui peuvent être adoptés en vue de la coordination entre normes ayant une efficacité différente sont utilisables aussi pour établir une hiérarchie entre les sources respectives. Et cet ordre de priorité peut amener à la coordination que l'on veut obtenir. Voilà donc pourquoi *Verdross/Simma* (note 10), p. 334, ramènent ladite coordination au rang qui appartient à chacune des normes à l'intérieur de la source dont elle découle. Ce qu'il faut en outre remarquer c'est que entre l'accord et la coutume sont valables, d'après les auteurs dernièrement cités, les mêmes critères qui sont applicables à l'intérieur d'une même source. Selon *Bos* (note 2), p. 366, en appliquant les règles bien connues de la coordination entre accord et coutume on arriverait au résultat que: »The rank or treaty and custom may vary from one case to another«. Pour mieux dire, au stade de l'application de la norme il ne s'agit plus d'établir le rang des deux sources, étant donné qu'elles ont un rang équivalent, mais d'appliquer la norme choisie sur la base de certains critères.

Cet article a donné lieu à des interprétations très variées et parfois opposées, sur lesquelles il n'y a pas lieu à revenir ici. En tout état de cause, à l'exception de certaines opinions qui sont restées isolées, la doctrine s'est orientée dans le sens d'exclure que l'art. 38 établisse un ordre hiérarchique entre les sources qu'il indique [12].

L'interprétation la plus fondée est, au contraire, celle qui constate dans l'art. 38 l'existence d'un ordre des sources, qui indique au juge le rang auquel le juge lui-même place les sources applicables au cas d'espèce. Ce qui est très différent d'une hiérarchie au sens propre du mot: celle-ci en réalité imposerait des résultats obligatoires en présence de plusieurs sources concurrentes, entre lesquelles il faudrait choisir.

11. Avant d'essayer d'établir une hiérarchie, qui, à défaut de règles spécifiques, serait fondée uniquement sur la pratique et sur l'appréciation logique de tous les éléments entrant en ligne de compte, il faut encore vérifier si les sources faisant l'objet de notre examen sont seulement celles prises en considération jusqu'ici, c'est-à-dire les principes généraux, les accords, la coutume et les principes généraux de droit communs aux Etats, ou bien si d'autres sources méritent d'être retenues.

Il y a des auteurs qui placent le droit naturel à côté des autres sources; toutefois ils ne le qualifient pas comme une source primaire, mais comme une source complémentaire à l'égard des autres [13].

---

[12] Dans ce sens voir *Parry,* The Sources and Evidences of International Law (Manchester 1965), p. 28; *Fitzmaurice,* The Future of Public International Law and of the International Legal System in the Circumstances of Today, Institut de Droit international, Livre du Centenaire (Bâle 1973), p. 325; *Bos* (note 2), p. 368. – Cette conclusion est aussi tirée des travaux préparatoires de l'art. 38. En effet, comme le rappelle *Akehurst* (note 2), p. 274, le projet du Statut de la Cour permanente de Justice internationale, tel qu'il fut préparé par le Comité des juristes de la S.d.N., précisait que les sources indiquées par le premier paragraphe de l'art. 38 devraient s'appliquer en *ordre successif,* alors que cette mention fut supprimée dans la rédaction définitive. Voir en outre, pour une interprétation qui ne tient pas compte du rang des normes *Heinz,* Zur Systematik der Völkerrechtsquellen und der Einteilung des Völkerrechts, Österr. Zeitschrift für öffentliches Recht 21 (1971), pp. 197 ss. D'après *Vallindas,* General Principles of Law and the Hierarchy of the Sources of International Law, in: Problèmes fondamentaux du droit international, Festschrift für Jean Spiropoulos (Bonn 1957), p. 426, l'ordre des sources indiqué par l'art. 38 est applicable à ce qui *normalement,* mais pas toujours, se produit dans cette matière. Par conséquent l'ordre devrait être corrigé, en plaçant au premier rang la coutume, au deuxième les principes généraux de droit, et au troisième les accords internationaux. Mais ici, encore une fois, comme d'ailleurs d'autres auteurs, *Vallindas* semble répéter la confusion, que nous avons tâchée d'éviter, entre les principes généraux du droit international et les principes évoqués dans l'art. 38.

[13] *Bos,* Hierarchy among the Recognized Manifestations (note 2), n'hésite aucunement à placer le droit naturel *complémentaire* parmi les sources du droit international, plus précisément parmi »The Recognized Manifestations of International Law«. – Le droit naturel ainsi évoqué occuperait la cinquième place de l'échelle hiérarchique, après certaines décisions d'organisations internationales, les traités, la coutume et les principes généraux du droit reconnus par les nations civilisées. – Pour comprendre

Compte tenu des origines de la science du droit international, qui trouve son fondement dans les conceptions de droit naturel, ainsi que de la formation de ce droit, qui s'est longuement inspirée des principes supérieurs dominant l'évolution de la pratique, tout cela s'explique aisément. Toutefois, à l'heure actuelle, on peut certainement douter qu'il existe une telle source: en effet, la codification sur le plan positif des règles du droit international ainsi que la fixation par des normes écrites des règles du droit coutumier général font obstacle à toute subsistance du droit naturel.

Ce n'est qu'en donnant une prédominance à l'élément rationnel dans la formation et l'application du droit que l'on pourrait parvenir à un pareil résultat: ce qui semble précisément démenti par la tendance à laquelle nous venons de faire allusion.

En effet le droit naturel est encore considéré comme une source au sein des systèmes juridiques dans lesquels le droit créé par les juristes l'emporte sur le droit d'origine législative. Ce qui explique pourquoi le droit naturel joue toujours un rôle important dans les Pays de *common law*; de sorte que même les auteurs de droit international, dont la formation est de *common law*, ont tendance à considérer le droit naturel comme une source de l'ordre juridique international.

Au cas où le droit naturel serait une source, son emprise sur l'ordre international devrait être d'une très grande ampleur et même illimitée: ce qui signifierait qu'il ne se poserait pas seulement comme une source complémentaire par rapport aux principes généraux de droit normalement reconnus: en effet, ces derniers ont une origine et une nature positive, de sorte qu'ils ne pourraient pas être intégrés par des principes découlant du droit naturel.

De toute façon, si vraiment on voulait rattacher les principes du droit naturel à l'art. 38 du Statut de la Cour internationale de Justice et, par conséquent, aux principes visés par cet article, une telle assimilation ne serait pas admissible.

En effet, l'opinion communément acceptée affirme que l'art. 38 a en vue les principes qui, étant reconnus par les Etats, font déjà partie de l'ordre international positif. Ce qui prouve tout d'abord que ces principes sont distincts du droit naturel et rend en outre évident qu'ils ne peuvent pas être intégrés par d'autres principes, qui, ayant une origine rationnelle, sont de nature tout à fait différente des règles du droit international positif.

---

l'idée dont s'est inspiré l'auteur, il faut revenir à son travail de base, que nous avons cité plusieurs fois, c'est-à- dire The Recognized Manifestations (note 2), p. 44. Dans le but d'éviter tout malentendu sur ce point très délicat, mais, en même temps, très important, je m'exprimerai avec les mots employés par l'auteur: »complementary natural law as the present writer sees it has much to do with a ›natural order‹ in a context of world geography. In order to avoid confusion between this type of natural order and the *preexisting* natural order, the former will be referred to, here, as ›natural geographical order‹«. L'idée de ce droit naturel particulier serait fondée en bonne substance sur la pratique et la jurisprudence internationale actuelle en matière d'espace aérien, maritime, terrestre.

12. Un autre problème analogue se pose à l'égard de la jurisprudence internationale: en effet on pourrait se demander si une telle jurisprudence peut apparaître comme source de droit international. Sur ce point la dispute doctrinale a une longue histoire, parce qu'elle remonte à l'époque où fut rédigé l'art. 38 du Statut de la Cour. Dans cet article on lit que la Cour applique les décisions judiciaires. . . comme moyen auxiliaire de détermination des règles de droit, sous réserve toutefois de l'art. 59 du même Statut, qui dispose que la décision de la Cour n'est obligatoire que pour les parties en litige et dans le cas qui a été décidé. L'interprétation de ces dispositions, telle qu'elle résulte d'une longue expérience doctrinale et pratique, est que, comme d'ailleurs le dit l'art. 38 lui-même, les décisions judiciaires valent seulement comme moyen apte à constater l'existence de certaines normes [14].

Il en découle donc que la jurisprudence ne peut pas être considérée comme moyen créateur de normes juridiques internationales, et, par conséquent, en tant que source, parce que son efficacité se trouve être limitée par la fonction, seulement indirecte, que le Statut vise.

Cela dit, on peut comprendre pourquoi l'art. 38, à cause de son caractère général et du fait qu'il figure dans un texte dont la portée concrète résulte de l'évolution de la jurisprudence internationale, peut être interprété de façon évolutive. Voilà pourquoi une partie de la doctrine persiste à considérer la jurisprudence comme une source, tout au moins auxiliaire.

Sur ce point on pourrait encore évoquer la doctrine qui, en traitant des sources du droit international, opère une distinction importante, en séparant les sources qui créent du droit, des sources dont la fonction est simplement la connaissance ou la constatation du droit: ainsi il serait possible de placer la jurisprudence dans cette deuxième catégorie.

D'ailleurs nous estimons qu'on ne peut pas s'attacher à un autre argument, qui n'est décisif qu'en apparence, consistant à reconnaître que la

---

[14] *Barberis* (note 5), pp. 263–308, consacre un chapitre tout entier à cette matière. Il s'agit d'une analyse approfondie et conduite avec une rigueur logique parfaite. Mais nous estimons que c'est précisément le point de départ qui prête à critique: l'auteur en effet, d'après le titre lui-même, qu'il a pris comme prémisse de sa recherche: *La jurisprudencia internacional como fuente de derecho,* tend, tout au long de son exposé, à démontrer que la jurisprudence peut être utilisée en tant qu'élément créateur de normes juridiques. Ce qui reste encore à démontrer. Le point qui en tout état de cause mérite beaucoup d'attention est celui, à l'égard duquel on peut constater l'accord unanime de la doctrine, que la jurisprudence joue un rôle fondamental dans la formation des règles coutumières: mais c'est toujours la coutume, une fois formée, qui acquiert la valeur de source et non pas la jurisprudence.

Cela dit, et tout en reconnaissant qu'aussi le domaine des sources évolue dans le temps, en ce qui concerne la jurisprudence il est préférable de s'en tenir à l'opinion classique. Celle-ci est exprimée avec toute clarté par *Verdross* (note 5), p. 135, qui nie à la jurisprudence la valeur de source formelle du droit international; or, à part les discussions qu'on peut faire en faveur ou contre les sources matérielles, c'est bien aux sources formelles que nous croyons devoir nous en tenir lorsqu'on parle de leur hiérarchie.

jurisprudence est un élément de très grande importance dans le processus de formation des règles coutumières. A cet égard il faut remarquer que les coutumes elles-mêmes sont des sources au sens propre, et non pas les éléments qui concourent à leur formation. Autrement on pourrait faire le même raisonnement en ce qui concerne la pratique internationale et les autres faits qui participent à la création des coutumes.

Une partie de la doctrine, qui retient la jurisprudence comme source, ne se réfère pas en général aux décisions des organes judiciaires internationaux, mais seulement à certaines décisions. Ce qui toutefois ne change en rien la solution du problème: en effet il est évident que certaines décisions seulement sont importantes en tant que preuve de l'activité de juge tendant à interpréter le droit. D'ailleurs dans les ordres juridiques internes on a aussi la même situation, étant donné que seuls les arrêts des juridictions suprêmes font jurisprudence: mais celle-ci n'acquiert pas la qualité de source.

13. Nous en arrivons ainsi à nous confronter, dans une situation plus claire, au problème de la hiérarchie des sources; en effet la sélection faite nous indique que la graduation doit être établie entre celles qui sont les véritables sources du droit international; cela signifie que les sources atypiques ne doivent pas entrer en ligne de compte.

Ainsi l'ordre de priorité doit être fixé entre les règles constitutionnelles, les normes de *jus cogens,* les normes créées en vertu d'accords, les normes découlant de coutumes, les principes généraux du droit.

Il nous semble au départ pouvoir faire un choix important: c'est-à-dire que les règles constitutionnelles peuvent être assimilées aux règles de *jus cogens,* parce que les deux catégories sont formées par des règles impératives et, par conséquent, auxquelles on ne peut déroger.

A cet égard il faut encore éclaircir la situation, parce qu'on parle aussi de règles fondamentales, qui seraient en tout cas semblables aux règles constitutionnelles: or, il s'agit de préciser les relations possibles entre ces règles fondamentales et celles appartenant au *jus cogens.* Sur le plan de la théorie les solutions seraient plutôt rationnelles, étant basées sur la simple logique. Mieux vaudrait s'adresser à la pratique et il serait mieux encore de pouvoir utiliser une décision judiciaire internationale.

La Cour internationale de Justice a eu récemment l'occasion de traiter ce problème lorsqu'elle a été saisie de l'affaire relative au personnel diplomatique et consulaire des Etats-Unis à Téhéran. En effet dans l'ordonnance émise par la Cour le 15 décembre 1979 on lit (par. 41) qu'à partir du moment où un Etat entretient des relations diplomatiques ou consulaires avec un autre Etat, il »ne saurait manquer de reconnaître les obligations impératives qu'elles comportent et qui sont maintenant codifiées dans les conventions de Vienne de 1961 et 1963, auxquelles l'Iran et les Etats-Unis sont parties«. Dans l'arrêt qui a suivi, du 24 mai 1980, la Cour a réaffirmé non seulement les principes posés dans l'ordonnance, mais a aussi rappelé (par. 88) le caractère impératif des obligations juridiques incombant au gouvernement de l'Iran. Il y a eu un auteur qui, sur la base des expressions utilisées par la Cour, s'est demandé si la Cour »n'a pas fait subrepticement référence à une norme de *jus cogens,* sans aller jusqu' à employer un mot ou un concept ayant suscité et suscitant encore tant de controverses et de passions«[15]. En outre ce même auteur se demande pourquoi la Cour, ayant employé les termes »fondamental« et »impératif«, n'a pas saisi dans l'affaire de Téhéran une occasion idéale pour jeter les bases d'une approche concrète de l'insaisissable *jus cogens,* étant donné que les deux conventions de Vienne comprennent des règles auxquelles on ne peut déroger et qui, en tant que telles, ne peuvent être modifiées que par des règles ayant le même caractère.

En ce qui me concerne, je partage l'opinion de ceux qui interprètent les allusions faites par la Cour à certaines règles fondamentales et, par conséquent, impératives, comme étant seulement des références qui ne voulaient pas donner une base formelle à la définition de certaines normes comme faisant partie du *jus cogens*[16].

L'importance réelle des arguments employés par la Cour consiste dans sa convinction que, désormais, la logique doit amener non seulement le juge mais aussi l'interprète à rapprocher les règles fondamentales de l'ordre international – qu'on peut appeler d'ailleurs aussi constitutionnelles – des règles impératives: on peut en effet affirmer que les unes et les autres ont comme caractère prédominant le fait qu'on ne peut y déroger.

Ce qui nous amène à dire qu'en décidant quelles sont les normes qui doivent être mises à la première place dans l'ordre hiérarchique, il n'y a pas lieu d'indiquer deux catégories de normes: en effet, il est impossible d'établir si les normes fondamentales priment celles de *jus cogens,* ou vice versa. Ce qui nous amène à placer au premier rang les unes et les autres, sans tenir compte de leur qualification respective.

---

[15] Voir *Bretton,* L'affaire des otages américains devant la Cour internationale de justice, Journal du droit international 1980, pp. 819 ss. Au contraire l'évaluation faite par *Zoller,* L'affaire du personnel diplomatique et consulaire des Etats-Unis à Téhéran, Revue générale de droit international public 84 (1980), p. 1024, est plus équilibrée: en effet cet auteur constate tout simplement que la Cour s'est abstenue de résoudre le problème de l'éventuelle appartenance au *jus cogens* des règles en question.

[16] *Morelli,* Norme c.d. fondamentali e norme cogenti, Rivista di diritto internazionale 64 (1981), p. 510. En réalité, comme le souligne *Morelli,* les différentes qualifications données aux règles sur les immunités diplomatiques par l'arrêt de la Cour, ne concernent pas ces règles dans leur valeur formelle. Elles ont pour objet seulement les exigences et les intérêts que ces mêmes règles ont en vue et, par conséquent, se réfèrent non à la règle en tant que telle, mais visent uniquement son contenu. En d'autres termes, si la substance des règles tend à sauvegarder des intérêts fondamentaux ou d'importance capitale, il n'en découle pas la nature impérative des règles elles-mêmes: celles- ci restent donc en dehors du domaine du *jus cogens,* dont le concept est axé sur l'idée absolue de l'impossibilité de toute dérogation de la part des sujets destinataires.

14. En deuxième position peuvent figurer les normes créées par accord international; à celles-ci il faut reconnaître une priorité, comme nous indique la tradition, sur les normes découlant de la coutume. Comme nous l'avons déjà expliqué, dans les relations réciproques entre le deuxième rang – celui des accords – et le troisième rang – celui des normes coutumières – on applique les critères déjà mis en lumière. Il est évident que, dans chacun des cas concrets, on parviendra à la solution appropriée en vertu de l'interprétation des éléments particuliers propres à l'espèce.

Restent encore à résoudre certains problèmes, qui seulement à première vue semblent simples: il s'agit du cas où dans un accord international des règles de *jus cogens* sont comprises, et de celui de l'efficacité des coutumes générales à l'égard de règles de *jus cogens.* Nous estimons que dans ces cas aussi les solutions à adopter doivent être recherchées à travers l'interprétation des normes dans leur ensemble et de l'efficacité respective des normes elles-mêmes.

Au quatrième rang se placent les principes généraux de l'ordre juridique international, auxquels peuvent être assimilés, en ce qui concerne leur rôle en tant que source, les principes généraux de droit reconnus par les nations civilisées.

A cet égard on pourrait observer que l'assimilation n'est pas possible, parce qu'il s'agit de deux catégories différentes de principes généraux. En effet, tandis que les premiers ont leur origine et leur fondement dans l'ordre international, les deuxièmes sont élaborés et admis dans les ordres internes des Etats et acquièrent leur efficacité à condition d'être communs à l'ordre juridique de plusieurs Etats.

Toutefois, à l'heure actuelle il faut évaluer ces phénomènes en tenant compte de l'évolution récente de l'ordre international. En effet les principes cités à l'art. 38 du Statut de la Cour sont indentifiés et appliqués au fur et à mesure que la jurisprudence internationale avance, tandis que les principes généraux de l'ordre international résultent de la structure récente de cet ordre, à partir du moment où elle s'est consolidée: cela explique pourquoi la doctrine les admet et la pratique les reconnaît.

En regardant vers l'application dudit art. 38 et vers la formule qu'il contient, on peut affirmer que si cet article avait dû être conçu à l'heure actuelle, et non il y a plusieurs années, on aurait maintenant dans le Statut une formule différente, dans laquelle seraient compris les principes généraux de l'ordre juridique international. En ce qui concerne la place qui revient aux principes généraux, on doit remarquer qu'ils ne peuvent pas avoir une position hiérarchique supérieure à la quatrième. En tout état de cause il n'est pas admissible qu'un principe général prime une norme écrite contenue dans une convention internationale, ni qu'il prime une règle coutumière.

Il faut reconnaître en outre que les principes généraux ont une fonction complémentaire par rapport aux autres sources, ce qui fait qu'ils ne peuvent pas occuper, dans l'échelle hiérarchique des sources, une position différente de celle qui leur appartient.

Bemerkungen zum ius cogens

FRITZ MÜNCH

1. Unser Jubilar hat bei einem Rückblick auf seine Leitung des Max-Planck-Instituts für ausländisches öffentliches Recht und Völkerrecht bestritten, eine »Schule« begründet zu haben. Über die Grundlagen des Völkerrechts habe jeder sein eigenes Credo, und dieses wirke selten und wenig auf die praktische Arbeit [1]. So hat denn im Institut ein undogmatisches Klima geherrscht, gewiß zum Nutzen und zur Freude der Mitarbeiter.

Das hindert aber nicht, daß *Hermann Mosler* sich seine Gedanken über Grundfragen des Völkerrechts gemacht hat. Eine davon ist das *ius cogens,* das er in mehreren Veröffentlichungen behandelt [2].

Ganz kurz zusammengefaßt, bejaht er das *ius cogens* so:

Auch die unorganisierte Staatengemeinschaft [3] hat eine Verfassung. Mindestens muß es eine Norm über das Entstehen des Rechts geben. Außerdem sichern sachliche Normen ein Mindestmaß von Einheitlichkeit im Verhalten. Logischerweise haben diese Normengruppen höheren Rang und sind *ius cogens. Mosler* scheut sich nicht, auf die Möglichkeit einer Begründung aus Naturrecht hinzuweisen, scheint aber eher dem neuen Realismus der Italiener zuzuneigen, die die Grundlage des Völkerrechts auf die Beobachtung des Wesens der Völkerrechtsgemeinschaft stützen.

---

[1] Auf einer Festveranstaltung des Instituts am 27. 2. 1981 laut der aus diesem Anlaß vervielfältigten Broschüre, S. 27 ff., insbes. S. 28, 30, 33.

[2] *H. Mosler,* Der »Gemeinschaftliche Ordre Public« in europäischen Staatengruppen, Revista de Derecho Internacional 21 (1968), S. 523–535 = Estudios de Derecho Internacional, Homenaje a D. Antonio de Luna (1968), S. 372–384; *derselbe,* Ius cogens im Völkerrecht, Schweizerisches Jahrbuch für internationales Recht 25 (1968), S. 9–40 (der Vortrag ist im April 1969 gehalten und nach der Abfassung der Vertragsrechtskonvention vom 23 5. 1969 ergänzt worden); *derselbe,* Völkerrecht als Rechtsordnung, ZaöRV 36 (1976), S. 6–47; *derselbe,* The International Society as a Legal Community, RdC 140 (1974 IV), S. 1–320; selbständig veröffentlicht 1980 mit Ergänzungen.

[3] In *Moslers* Beitrag zur Festschrift de Luna steht S. 372 (=Revista S. 523), daß ein *ordre public* nur in einer organisierten Rechtsgemeinschaft bestehen kann. Jedoch erwähnt er S. 374 (525) den »Internationalen Ordre public« der Völkerrechtsgemeinschaft und erinnert S. 383 f. (534 f.) an das *droit public européen.*

Obwohl die Vertragsrechtskonvention nicht ohne Widerspruch gegen das *ius cogens* zustande gekommen und noch nicht allgemein formell angenommen ist, hat sich der Gedanke weit verbreitet. Kein Lehrbuch und keine Monographie kommt daran vorbei, daß die Staatengemeinschaft immerhin bei der Schlußabstimmung am 22. Mai 1969 in Wien einen Konsens von Gewicht ausgedrückt hat.

Dieser Beitrag soll die Erörterung für und wider das *ius cogens* nicht nachzeichnen; die Schrift von *Jerzy Sztucki* [4] gibt bei aller grundsätzlichen Kritik einen guten Überblick. Vielmehr soll er auf Punkte eingehen, die bisher vernachlässigt erscheinen.

Hierher gehört zunächst das Verhältnis des *ius cogens* im Völkerrecht zu den gleichnamigen und ähnlichen Erscheinungen im innerstaatlichen Recht. Seine Aufklärung kann immer noch zum Verständnis der Probleme beitragen. Sodann ist nicht genug auf das zwingende Recht in den Rotkreuzabkommen von 1949 geachtet worden. Endlich sollte am Katalog der zwingenden Normen gearbeitet werden, damit ihr Sinn deutlicher erscheine.

2. Ein Tadel am *ius cogens* der Wiener Vertragsrechtskonvention (WVRK) von 1969 geht dahin, daß seine Definition in Art. 53 von der Wirkung her aufgestellt ist[5], aber nicht sagt, wie es sich in Erscheinungsform und Inhalt auszeichnen soll.

Dies ist ein Mangel, der dem älteren [6] und dem gegenwärtigen deutschen Recht auch anhaftet. Man findet überhaupt keine Definition im Gesetz, sondern nur eine Sammlung zwingender Vorschriften[7], die aus der Praxis ergänzt wird, ohne daß eine deskriptive Definition, nicht einmal eine einheitliche Motivation zu konstruieren wäre. Man zählt zum zwingenden Recht das Sachenrecht wegen der beschränkten Zahl von Institutionen, das Familienrecht, die Formvorschriften, den Grundsatz von Treu und Glauben[8], Verfahrens-, Grundbuch- und Wertpapierrecht[9]. Nicht nur das BGB überläßt es in den meisten Fällen der Auslegung, den zwingenden oder nichtzwingenden Charakter der Vorschrift zu erkennen [10], sondern auch die anderen Gesetze halten es so.

---

[4] *J. Sztucki,* Jus Cogens and the Vienna Convention on the Law of Treaties. A Critical Appraisal (1974). Diese Kritik stößt ihrerseits auf Kritik: *A. Bleckmann* in der Besprechung ZaöRV 35 (1975), S. 684 ff.

[5] Das andere Merkmal, daß zwingendes Recht nur durch neues zwingendes Recht geändert werden kann, hilft nicht weiter und wird hier nicht erörtert.

[6] *F. Förster,* Preußisches Privatrecht (6. Aufl. 1892), S. 41.

[7] *Enneccerus/Nipperdey,* Allgemeiner Teil des bürgerlichen Rechts, Bd. 1 (1952) 1, S. 189.

[8] *Staudinger,* Kommentar zum BGB (11. Aufl.) Einleitung vor § 854 Anm. 11–13; (10./11. Aufl.) Einleitung zum Familienrecht Anm. 21; § 242 Anm. A 141 ff.

[9] *M. Rehbinder,* Strict Law and Considerations of Equity in the Legal System of the Federal Republic of Germany, in: R.A. Newman (Hrsg.), The Unity of Strict Law (1978), S. 374, 382; ferner weithin das Arbeitsrecht.

[10] So *B. Windscheid,* Lehrbuch des Pandektenrechts (T. Kipp), Bd. 1 (9. Aufl. 1906), S. 127.

Sogar wo ein Schritt zur Vervollständigung der ganz formalen Bestimmung des zwingenden Rechts getan ist, gilt er als unzulänglich. Der Code civil, Art. 6, nennt als zwingend die Vorschriften des *ordre public* und der guten Sitten. Man hält das für zu vage, außerdem gehen Formvorschriften und der Schutz der Schwächeren nicht notwendig auf den *ordre public* zurück[11]. Dem Art. 14 der Allgemeinen Vorschriften zum Niederländischen Recht, der dem Art. 6 Code civil nachgebildet ist, wird derselbe Mangel an Präzision vorgeworfen[12].

Nun hat das alles nicht gehindert, daß Rechtsprechung und Lehre neben die Vorschriften, die positiv für zwingend erklärt sind, noch zahlreiche andere stellen konnten[13]. Man gewinnt sogar im jeweiligen nationalen Recht ein Bild davon, was den Gesetzgeber ausdrücklich oder implizit bewogen hat, Vorschriften zu zwingenden zu erheben. Aber diese Motivationen sind sehr mannigfach [14], und je mehr man sie begreift, um so weniger leuchtet die Möglichkeit ein, die Idee ins Völkerrecht zu übernehmen, weil dort die Sachlagen nicht vorkommen.

Zudem beginnt beim Blick in die Vergangenheit und in andere Rechtssysteme das *ius cogens,* wie es hier verstanden worden ist (Abweichungen durch Privatwillkür ausschließend), zu verschwimmen. Denn zwingend hat man auch genannt diejenigen Normen des Gemeinen Rechts, die entgegen der sonst geltenden Rangordnung dem Partikularrecht vorgehen[15]. Das ist antiquiert, aber der weitverbreitete Begriff des *strict law* konkurriert mit dem zwingenden Recht und meint meist den Gegensatz zur *equity* oder, außerhalb des angelsächsischen Bereichs, zum *ius aequum*[16]. Er meint auch, zumal in Frankreich, eng zu interpretierendes Recht[17]. Das alles möchte *Rehbinder*[18]

---

[11] *G. Marty/P. Raynaud,* Droit Civil, Bd. 1 (1961), S. 163 § 99.

[12] *P. Scholten,* in: Mr. C. Asser's Handleiding tot de Beoefening van het Nederlandsch Burgerlijk Recht, Algemeene Deel (2. Aufl. 1934), S. 31.

[13] Im soeben erwähnten niederländischen Recht gelingt es *J.L.M. Elders,* Strict Law and Dutch Law, in: Newman (Anm. 9), S. 169 ff., aus der Praxis eine Anzahl Beispiele zwingenden Rechts beizubringen. Auch fast jeder andere Beitrag dort kann das Gesagte belegen.

[14] *Dernburg,* Pandekten, Bd. 1 (3. Aufl. 1892), S. 69, spricht von den notwendigen Folgerungen aus der Eigenart eines Rechtsinstituts, von Rücksichten des allgemeinen Wohls, von rechtspolizeilichen Gründen (wir würden heute sagen »rechtspolitischen«) und von guten Sitten. *Rehbinder* (Anm. 9), S. 389, gibt mehrere Rechtfertigungen für zwingendes Verfahrensrecht. *R.A. Newman,* The General Principles of Strict Law, in: Newman (Anm. 9), S. 397 ff., faßt die Ergebnisse zusammen, die die Untersuchung von 14 Materien (S. 420) erbracht hat.

[15] *R. Sohm,* Institutionen des Römischen Rechts, in: Sohm/Mitteis/Wenger (noch in der 17. Aufl. 1949), S. 24 f.

[16] *Sohm,* a.a.O.; Newman, The Unity of Strict Law (Anm. 9), S. 11, 35, 161, 356; *Windscheid* (Anm. 10), S. 119, scheidet das Paar strenges/billiges Recht vom zwingend/nachgiebigen.

[17] Newman, a.a.O., S. 91 Anm. 2, 109, 139, 215; *F. Somló,* Juristische Grundlehre (1917), S. 387 f. – er polemisiert freilich gegen den Gedanken, siehe auch S. 380.

noch in den zwingenden Formvorschriften des gegenwärtigen deutschen Rechts sehen. Wieder muß man zweifeln, ob diese Dinge im zwischenstaatlichen Bereich, im gegenwärtigen Völkerrecht, von Belang sind.

Weniger Bedenken hingegen möchte man aus der Meinung herleiten, daß *ius cogens* dem angelsächsischen Juristen unfaßbar sei. Hemmungen dieser Art mögen einigen Stimmen gegen das *ius cogens* zugrunde liegen. Aber es ist doch *Sir Humphrey Waldock* gewesen, der in der International Law Commission das *ius cogens* in die Kodifikationsarbeiten eingebracht hat [19], und *D. B. Walters* widerspricht entschieden einer solchen Inkompatibilität [20].

3. Die vielfach zerstreuten zwingenden Sätze im Privatrecht bilden kein organisches Ganzes und lassen sich allenfalls auf einige Tendenzen zurückführen, die unabhängig voneinander sind. Sie bilden keine Ordnung. Man sollte darum die Vorstellung eines höheren Rechts innerhalb des Völkerrechts aus einem anderen Denkbereich herleiten [21].

Einigermaßen zusammenhängende Ordnungen höheren Rangs findet man in den starren Verfassungen einzelner Staaten, insbesondere in dem Amendments-Katalog der USA, in den Grundrechten des Grundgesetzes oder der italienischen Verfassung. Man nennt diese Bestimmungen zwar nicht *ius cogens,* und sie richten sich nicht eigentlich gegen Abmachungen Einzelner; aber diese Analogie des völkerrechtlichen *ius cogens* zum Verfassungsrecht sollte zum Verständnis dessen genügen, was die Befürworter des *ius cogens* im Völkerrecht wollen. Die Redewendung vom *ordre public international* ist denn auch, so gesehen, ganz angemessen, und sogar der Anklang an den *ordre public* des internationalen Privatrechts ist nützlich, sofern man ihn nicht als Gleichsetzung mißversteht. So hat denn *M. Katz* als Leiter des Kolloquiums von Lagonissi in seinem Schlußwort verschiedene Andeutungen aufgegriffen, nach denen der gegen *ius cogens* verstoßende Vertrag dem verfassungswidrigen Gesetz vergleichbar sei [22].

In einer gleichzeitigen Erörterung des internationalprivatrechtlichen *ordre public* ist auch ein Schritt zum völkerrechtlichen *ius cogens* getan worden. *G. Jaenicke* hat in seinem Bericht an die Deutsche Gesellschaft für Völkerrecht 1965 [23] gefordert, man müsse Völkerrechtswidrigkeiten in einem fremden Ge-

---

[18] (Anm. 9), S. 371 ff., 376–379, 386–388 – er scheint stellenweise zu weit zu gehen. Aber der Zusammenstoß des Grundsatzes von Treu und Glauben mit dem zwingenden Recht ist in der Tat ein Problem, siehe *Staudinger* (Anm. 8), § 242 Anm. A 17, 18, 142, 144.

[19] Er selbst macht darauf aufmerksam, daß das englische Recht keinen angemessenen Ausdruck für das *ius cogens* habe, YILC 1963 I, S. 62 § 25. *Pal,* ebenda, S. 69 § 31, hat noch mehr Schwierigkeiten.

[20] In: Newman (Anm. 9), S. 140 ff.

[21] Davon geht auch *E. Suy* aus, The Concept of Jus Cogens in International Law, Bericht an das Kolloquium 1966 in Lagonissi (1967), S. 22.

[22] Ebenda, S. 113.

[23] Zur Frage des internationalen ordre public, Berichte der Deutschen Gesellschaft für Völkerrecht, Heft 7 (1967), S. 80 ff., 88 ff.

setz mit dem Mittel des *ordre public* abwenden[24], und gelangt in diesem Zusammenhang auf die Lehre vom völkerrechtlichen *ius cogens.*

Offenbar führen in dieser Materie die Worte nicht durch die sichere Pforte zum Tempel der Gewißheit. Die weitgehende Übereinstimmung in der Sache ist 1963 in der International Law Commission von vielen Vorbehalten gegen den Ausdruck »*ius cogens*« begleitet gewesen, und bezeichnenderweise erscheint er schließlich nur als Klammerdefinition in den Überschriften zu Art. 53 und 64 WVRK. Auch *international public order* hat in den Beratungen manche Gegner gehabt, so daß man heute noch um ein passendes Wort verlegen ist. »Verfassungssätze der Völkerrechtsgemeinschaft« wäre brauchbar, aber wahrscheinlich dort unverständlich, wo die Verfassung nur als Staatsorganisationsstatut aufgefaßt wird. Indessen werden Verfassungen mit Grundrechtskatalogen und Sozialsystemen immer häufiger, und das müßte zusammen mit den beiden Menschenrechtspakten der Vereinten Nationen von 1966 das Verständnis für völkerrechtliches *ius cogens* wecken.

Sicherlich kann man heute die alte Konstruktion verlassen, nach der nur ein Vertrag oder ein vertragsähnlicher Konsens völkerrechtliche Bindungen erzeugt. Der Normenkomplex über völkerrechtliche Verträge bewies eigentlich schon aus seinem Dasein seine Unabhängigkeit von einer Quelle im Vertragsrecht[25]. Nimmt man ihn als Gewohnheitsrecht, so führt die Frage nach dessen Verbindlichkeit zu einem Zirkel, wenn man nicht Sätze über die Existenz und das Entstehen von Völkerrecht als ein Kapitel der Verfassung der Völkergemeinschaft ansieht. Neben dieses treten materiellrechtliche Verfassungselemente, aus der Struktur der Gemeinschaft sich ergebende Sätze, Gestaltungsgrundsätze der Völkerrechtsordnung und rechtslogische Grundsätze.

Das alles hat *Hermann Mosler* beschrieben[26]; wenn man es systematisieren würde, entstünde ein sehr umfangreicher Text. Es würde sich eher um eine neue Begründung des Völkerrechts und die Reform der Quellenlehre handeln, wie sie viele Italiener vorgenommen haben. *Mosler* ist anscheinend auch von da beeinflußt [27].

Es ist aber nicht gesagt, daß diese höherrangigen und logischen Sätze auch zwingende sein müßten. Es bleibt immer übrig, aus der Völkerrechtsordnung die zwingenden Bestandteile hervorzuheben. Dies kann nun nicht wie beim Entstehen einer starren Verfassung geschehen, wo die in die Verfassung übernommenen Sätze und Institute positiv ihren Rang erhalten. Es ist

---

[24] Für den Fall der völkerrechtswidrigen Enteignung im Ausland ist das schon vorher dargelegt worden: *J.H.W. Verzijl,* The Relevance of Public and of Private International Law Respectively for the Solution of Problems Arising from Nationalization of Enterprises, ZaöRV 19 (1958), S. 531, 537 ff., und *F. Münch,* Les effets d'une nationalisation à l'étranger, RdC 98 (1959 III), S. 411, 458 ff.

[25] Dies Argument wird im Bewußtsein verblassen, aber doch richtig bleiben, wenn die Vertragsrechtskonvention universales Vertragsrecht werden sollte.

[26] Völkerrecht als Rechtsordnung, ZaöRV 36 (1976), S. 31 ff., 44 ff.

[27] Wenn er auch nicht dem *diritto spontaneo* anhängen möchte, a.a.O., S. 44.

unzulänglich, das völkerrechtliche *ius cogens* auf die Satzung der Vereinten Nationen zu erweitern oder zu beschränken, schon deshalb, weil die Satzung nicht im Gedanken an ein *ius cogens* geschrieben worden ist. Was an wichtigen Grundsätzen dort genannt ist, ist politisches Programm für die Vereinigung, die universal werden wollte, es aber noch nicht war. Es hat also die Satzung nicht die Dignität, das *ius cogens* des allgemeinen Völkerrechts darzustellen.

Eine Überlegung könnte helfen: Verfassungen, besonders starre, haben oft die Funktion eines *contrat social*. Ein neuer Staat kann in ihr aufklärerisch Prinzipien des idealen Staatswesens niederlegen; eine siegreiche Revolution kann ihr Programm aufstellen; nach einer mißglückten Revolution kann ein Volk einen Kompromiß finden, nach dem es in Ruhe weiterleben will. Dabei sind häufig Sätze materiellen Rechts wichtiger als Elemente des Staatsaufbaus.

Einen derartigen Maßstab an die Verfassung der Völkerrechtsgemeinschaft legen bringt gewiß die Gefahr subjektiver Werturteile. Wenn man aber die bisherige Erörterung über mögliche Sätze des *ius cogens* betrachtet, muß man das für ein notwendiges Übel halten [28] und eben darauf bauen, daß im Verlauf einer solchen Diskussion sich mit der Zeit eine Mehrheitsmeinung herausbildet und daß man die Sätze und Institutionen eines notwendigen Kerns der Ordnung für die Staatengemeinschaft erkennt.

4. Das *ius cogens* des humanitären Kriegsrechts ist nicht eben Strukturelement, aber doch wichtig und sollte geeignet sein, einige Aspekte des Problems zu beleuchten. Es wird wenig herangezogen; *Sztucki* [29] hält es deshalb nicht für ein geeignetes Beispiel, weil es in den Rotkreuzabkommen von 1949 selbst umschrieben wird, also zur Materie der mehrfachen Verträge über dieselbe Materie gehört (Art. 30 und 41 WVRK), und nach der Meinung der International Law Commission solche Fälle nicht eigentlich das *ius cogens* der höheren Ordnung darstellen [30].

Nun gilt aber doch das humanitäre Kriegsrecht in seinen überlieferten Sätzen trotz der Kodifikation als allgemeines Völkerrecht, und daher kann man Beschränkungen der Vertragsfreiheit auf diesem Gebiet heranziehen.

Das System der vier Rotkreuzabkommen vom 12. August 1949 hat mehrere Besonderheiten. Die eine besteht darin, daß angegeben ist, welche Bestimmungen vor Abänderungen geschützt sein sollen, und zwar geschieht das jeweils an zwei Stellen.

Art. 6 des I.–III. Abkommens (Art. 7 des IV. Abkommens) ermöglicht Sonderabkommen unter den Parteien, erhebt also die Abkommen nicht eigentlich schlechthin zu unabänderlichem Recht. Abs. 1 Satz 2 schließt aber

---

[28] Alle Begründungen für ein breites oder für ein enges *ius cogens*, die man gehört hat, sind rein subjektiv, auch wo sie sich auf die objektiven Bedürfnisse der Völkergemeinschaft richten wollen.

[29] (Anm. 4), S. 38 f., siehe auch S. 16.

[30] Kommentar der ILC zu Art. 50 ihres letzten Entwurfs zur WVRK, YILC 1966 II, S. 248 Ziff. 2.

aus, daß solche Sonderabkommen die Lage und die Rechte der geschützten Personen beeinträchtigen. Das Recht der Abkommen ist also ein zwingendes Minimum. Gemäß dem folgenden Art. 7 (8) können die geschützten Personen nicht auf die Rechte aus den Abkommen und den möglichen Sonderabkommen verzichten. Das ist eine Verstärkung des *ius cogens*-Effekts, der normalerweise nur im Ausschluß eines abweichenden Vertrages zwischen Staaten besteht – sie hat in der Materie einen guten Sinn, und an diese Konsequenz aus dem *cogens*-Gedanken müßte man auch anderwärts denken.

In den Teilen über die Durchführung der Abkommen sind jeweils [31] die schweren Verletzungen aufgezählt, die mit Strafe bedroht werden müssen, die aber auch insofern *ius cogens* sind, als nach dem jeweils folgenden Artikel keine Vertragspartei sich oder andere von den Verantwortlichkeiten befreien kann, die aus solchen Verletzungen hervorgehen. Es ist also nicht möglich, beim Friedensschluß aufzurechnen oder auf Ersatz oder Verfolgung der Untaten zu verzichten [32]. Auch das ist eine notwendige Folgerung des *ius cogens,* die zu verallgemeinern wäre; denn als *ius cogens* im üblichen Sinne sind die Abkommen schon in den Art. 6 bzw. 7 gedacht.

Eine Wirkung von *ius cogens* findet sich noch in Art. 47 des IV. Abkommens zum Schutze von Zivilpersonen; sie beschränkt mögliche Vereinbarungen zwischen der Besatzungsmacht und den Behörden des besetzten Gebiets. Sinngemäß müßte man in den letzten Worten des Artikels auch Beschränkungen eventueller Verwaltungs- oder Abtretungsverträge zwischen Besatzungsmacht und besetztem (oder teilweise besetztem) Staat sehen.

Endlich untersagen die Rotkreuzabkommen allgemein Vergeltungsmaßnahmen gegen geschützte Personen [33]; auch dies kann eine Facette des *ius cogens* sein [34].

So ist schon 1949 auf einem Teilgebiet des Völkerrechts ein Modell von *ius cogens* mit ziemlich ausführlichen Bestimmungen errichtet worden. Es ist aber damals nicht so genannt worden; der Ausdruck erscheint nicht in den Vorarbeiten und Verhandlungen [35] und auch nicht im quasi-amtlichen Kommentar (siehe Anm. 32).

---

[31] I. Abkommen Art. 50; II. Abkommen Art. 51; III. Abkommen Art. 130; IV. Abkommen Art. 147.

[32] *J. S. Pictet,* Les Conventions de Genève du 12 août 1949, Commentaire, Bd. 1 (1952), S. 420, und an den entsprechenden Stellen der Bände 2–4 (1959, 1958, 1956); Actes de la Conférence diplomatique de Genève de 1949, Bd. 2 B (1951), S. 128.

[33] I. Abkommen Art. 46; II. Abkommen Art. 47; III. Abkommen Art. 13 Abs. 3; IV. Abkommen Art. 33 Abs. 3.

[34] *C. Dominicé,* Représailles et droit diplomatique, in: Festschrift für Hans Huber (1981), S. 552, deutet das vorsichtig an und erwähnt das humanitäre Kriegsrecht. Man kann das sicherlich verallgemeinern.

[35] Actes de la Conférence diplomatique de Genève de 1949. In den Erörterungen, Überschriften, Randtiteln und im Register ist vielmehr die Rede von *clause de sauvegarde,* Art. 6 (bzw. 7) Abs. 1 Satz 2; *droit acquis* und *inaliénabilité,* Art. 7 (bzw. 8); *responsabilité des Parties contractantes,* Art. 51, 52, 131, 148; und *intangibilité* bzw. *inviolabilité des droits,* Art. 47 des IV. Abkommens.

Das sollte aber nicht hindern, vom heutigen Sprachgebrauch aus dies Etikett anzubringen. Man hat damals recht spontan Völkerrechtsnormen in den Stand von *ius cogens* erhoben aus der Notwendigkeit, solche verstärkten Sicherungen ins humanitäre Kriegsrecht einzubauen, nachdem schlechte Erfahrungen in der Praxis gemacht worden waren.

Während der Beratungen gab es manche sachlichen Einwendungen [36], die aber heute nach der universalen Annahme der Konventionen [37] als erledigt zu gelten haben. Dogmatische Schwierigkeiten, wie sie mancherorts gegen das *ius cogens* der WVRK noch empfunden werden, sind gar nicht laut geworden, und das ist ein Grund, warum das zwingende humanitäre Kriegsrecht besondere Beachtung verdient.

5. Im Verfahren über die einstweilige Anordnung gegen den Iran wegen der amerikanischen Geiseln hat der Internationale Gerichtshof das Diplomatenrecht als Quelle von *obligations impératives* bezeichnet, die auch nicht aus Gründen der Repressalie außer acht gelassen werden können [38]. *C. Dominicé* hat das unterstrichen und mit dem humanitären Kriegsrecht in Parallele gesetzt [39].

6. Die Wiener Vertragsrechtskonvention von 1969 hat nicht das günstige Schicksal gehabt wie die Rotkreuzabkommen von 1949. Sie ist erst am 27. Januar 1980 zwischen 35 Staaten in Kraft getreten und gilt Ende 1981 unter 40 Staaten, unter denen sich nur ein ständiges Mitglied des Sicherheitsrates befindet. Sie wird dennoch seit ihrem Abschluß vielfach in Auseinandersetzungen verwendet. Der Internationale Gerichtshof und einzelne Richter in ihren Voten zitieren sie [40].

In den letzten Abstimmungen fanden die einschlägigen jetzigen Art. 53, 64 und 71 Mehrheiten von 87 bzw. 84 bzw. 87 gegen 8 bzw. 8 bzw. 5 Stimmen bei 12 bzw. 16 bzw. 12 Enthaltungen. Die Gesamtabstimmung ergab 79 Annahmen der WVRK, 1 Ablehnung und 19 Enthaltungen [41].

Wenn man die Liste der gebundenen Staaten durchsieht – sie ist nicht sehr eindrucksvoll –, so vermerkt man eine gewisse Divergenz zwischen der formalen Verbindlichkeit der WVRK und ihrem praktischen Gewicht. Sie

---

[36] Von den Briten besonders. Ein französischer Delegierter wollte es nicht verdammt sehen, daß im ersten Weltkrieg elsässische Soldaten des deutschen Heeres, die in französische Gefangenschaft gerieten, ins französische Heer traten.

[37] Nach dem Fundstellennachweis B zum BGBl. per 31. 12. 1981 sind 151 Staaten gebunden.

[38] CIJ Rec. 1979, S. 20 par. 41.

[39] Siehe Anm. 34, dort S. 547, 550.

[40] Nachweise im Register der Fontes Iuris Gentium Series A Sectio I Tomus 6 (1978), der bis 1975 geht. Hinzu kommen Stellen im *Ägäis*-Fall, CIJ Rec. 1978, S. 56, 68, 69, und im Gutachten über die WHO-Stelle Alexandria, CIJ Rec. 1980, S. 91 f., 94 f., 96.

[41] United Nations Conference on the Law of Treaties, Second Session, Official Records (1969), Meetings, S. 106 Ziff. 65, S. 125 Ziff. 80, S. 127 Ziff. 14, S. 207 Ziff. 51.

ist weithin als gelungene Formulierung des bestehenden Gewohnheitsrechts anerkannt; die Frage wäre die, ob das auch von ihrem *ius cogens* gilt.

Ius cogens hat im Schrifttum[42] und in der internationalen Gerichtsbarkeit[43] eine ältere Geschichte als die Kodifikation des Vertragsrechts. Sie geht zurück auf die älteren Vorstellungen über die Ungültigkeit sittenwidriger völkerrechtlicher Verträge und Schückings Voten in den Fällen *Wimbledon* 1923 und *Oscar Chinn* 1934. Vor allem bestätigen die oben Ziff. 4 erörterten Rotkreuzabkommen von 1949, daß die Völkerrechtsgemeinschaft klare Vorstellungen von *ius cogens* schon damals hatte und ein solches System einrichtete.

7. Der Umfang des zwingenden Völkerrechts ist bisher sehr undeutlich dargestellt. Es gibt viele Formeln und Kategorien und viele Versuche zu Katalogen[44]. Man kommt nicht weit, wenn man vom *ius cogens* des Zivilrechts ausgeht. Der Gedanke, daß man gewissermaßen den heutigen stillschweigenden *contrat social* der Völkergemeinschaft erkunden müsse, scheint fruchtbarer, bedarf aber noch der Ausarbeitung.

Das Gewaltverbot hat sicherlich viele Stimmen für seine Einreihung ins *ius cogens*. Bei ihm wird klar, daß man die Bedeutung von *ius cogens* abwandeln muß. Es wäre bloßer Schulfall, an Verträge zu denken, die das Gewaltverbot aufheben wollten. Seine zwingende Macht müßte sich darin äußern, daß der mit Gewalt herbeigeführte Zustand rechtsunwirksam ist und auch nicht durch Verträge legalisiert werden kann. Es müßte auch die Pflicht der Staaten folgen, mit Druckmitteln die Wiederherstellung des vorigen Zustandes zu erwirken. Die Vereinten Nationen hätten auf dem Papier diese Möglichkeit; die Erfahrungen lehren aber, daß die Organisation nicht funktioniert. Darum war der Versuch interessant – er betraf freilich einen anderen Unrechts-Sachverhalt –, in den Gutachten über Südwestafrika eine Rechtspflicht der Staaten zu konstruieren, durch eine Reihe von Sanktionen auf die Beseitigung eines rechtswidrigen Zustandes hinzuwirken.

In den Vorschlägen, das *ius cogens* zu umschreiben, taucht auch das Selbstbestimmungsrecht auf. Es ist deswegen interessant, weil die Forderung der Selbstbestimmung praktisch gegen einen bestehenden, äußerlich sehr oft legal erscheinenden Zustand gerichtet wird. Rechnet man es zum *ius cogens*, so behauptet man, daß es bestehende Souveränitäts- und Vertragsverhältnisse breche.

So müßte das zwingende Selbstbestimmungsrecht, das einem Staate zukommt, hegemoniale Positionen, die aus Regionalpakten und ideologischen

---

[42] *Suy* (Anm. 21), S. 26–49; *Jaenicke* (Anm. 23), S. 82 Anm. 7; *Sztucki* (Anm. 4), S. 54–59.

[43] *Suy*, ebenda, S. 60–64; *Sztucki*, ebenda, S. 12–22; siehe ferner Fontes Iuris Gentium Series A Sectio I Tomus 6, S. 17, 32 f., 269, und den oben bei Anm. 38 erwähnten Fall der Geiseln im Iran.

[44] *Sztucki* (Anm. 4), S. 76–79, 79–81, 81–84. Er versucht sogar eine regionale Analyse der Ansichten S. 89–93.

Bündnissen abgeleitet werden, wie etwa die Breschnew-Doktrin, aufheben können. Das Selbstbestimmungsrecht eines Volkes würde jetzt, wirksamer als das Nationalitätenprinzip des vorigen und jetzigen Jahrhunderts, die Gebietsverteilungen auflösen, die ihm widersprechen.

Die Dekolonisierung, seit 1945 im Gange und nun fast abgeschlossen, erscheint im Zusammenhang nicht als gewaltsame Erhebung und nicht als politischer Vorgang allein, sondern als Verwirklichung eines zwingenden Satzes der Selbstbestimmung gegenüber den Rechtstiteln, auf denen der Kolonialstatus beruht hatte[45]. Die Verkündung des Rechts in der Satzung der Vereinten Nationen und in Art. 1 der Menschenrechtspakte der Vereinten Nationen vom 19. Dezember 1966 ist von einer Praxis begleitet gewesen, die über einer Milliarde Menschen in mehr als 60 neuen Staaten eine neue politische Existenz verschafft hat.

Der Internationale Gerichtshof hat im Gutachten vom 21. Juni 1971 über Südwestafrika doch wohl die Selbstbestimmung eines Kolonialvolks als Rechtssatz aufgefaßt[46]; zwar nicht so, daß das Mandat hinfällig würde, aber daß in den Mandatsvertrag eine neue Verpflichtung zur Verwirklichung der Selbstbestimmung einfließt, deren Verletzung zur Aufhebung des Mandats berechtigt. Im Gutachten vom 16. Oktober 1975 über die West-Sahara ist die Selbstbestimmung zunächst auch nur Prinzip genannt, aber dann mehrmals Recht[47]. Mehrere Richter haben sich noch schärfer in diesem Sinne ausgesprochen.

Zugegeben, daß in der Verwirklichung dieses Selbstbestimmungsrechts Mängel vorgekommen sind und dem Betrachter leicht Spott auf die Lippen kommt, so möchte er doch nicht deswegen an seinem Rechtscharakter zweifeln. Es halten zwar erfahrungsgemäß die Lehrbücher an den überlieferten Vorstellungen fest und gehen nur langsam und widerwillig zu neuen Auffassungen über. Die Spezialliteratur ist immer eher bereit, neue Entwicklungen zu verfolgen und an einem bestimmten Punkt die Geltung eines neuen Rechts festzustellen.

---

[45] Daß die Kolonisierung sehr oft auf Verträgen zwischen der Metropole und den Fürsten oder Häuptlingen der Kolonie beruht, hat *C. H. Alexandrowicz* mehrmals dargetan: Treaty and Diplomatic Relations between European and South Asian Powers in the Seventeenth and Eighteenth Centuries, RdC 100 (1960), S. 207; *derselbe,* An Introduction to the History of the Law of Nations in the East Indies (16th, 17th and 18th centuries) (1967); *derselbe,* The Afro-Asian World and the Law of Nations, RdC 123 (1968), S. 177; *derselbe,* The European-African Confrontation, A Study in Treaty Making (1973); *E. Hertslet,* The Map of Africa by Treaties (1909). Für die deutschen Schutzgebiete besonders: *F. Schack,* Das deutsche Kolonialrecht in seiner Entwicklung bis zum Weltkriege (1923), S. 4–12; *Martens* NRG2, Bd. 4, S. 473 ff.; Bd. 11, S. 461 ff., 479 ff.

[46] CIJ Rec. 1971, S. 31. Wenn auch zunächst in Ziff. 52 die Selbstbestimmung nur ein Prinzip genannt ist, so soll es doch anwendbar sein, und in Ziff. 53 wird sie in die Rechtsentwicklung einbegriffen.

[47] CIJ Rec. 1975, S. 33 Ziff. 59, S. 36 Ziff. 70 und 71, S. 67 Ziff. 161.

Beim Selbstbestimmungsrecht in der Dekolonisierung ist auch der Rang als *ius cogens* einleuchtend[48], weil, wie oben gesagt, dieses Recht die früheren Titel der Kolonialmacht beseitigt hat. Man kann sogar die weithin verbreitete Meinung, das Selbstbestimmungsrecht berechtige zum bewaffneten Kampf – nicht nur der ringenden Bevölkerung, sondern auch anderer Staaten, die ihr zu Hilfe kommen –, zum Beweis eines Charakters als *ius cogens* nehmen; und es ist sogar die These aufgestellt worden, eine solche Bevölkerung und ihre Helfer dürfen sich des Terrors bedienen.

Nun erweist sich der Charakter des Selbstbestimmungsrechts als zwingend am Beispiel der Dekolonisierung. Aber zu behaupten, daß es nur Kolonialvölkern zustehe, ist ein Kurzschluß, der sich nur auf einige Entschließungen der Vereinten Nationen stützt und die grundsätzliche Bedeutung dieses Rechts mißkennt. Offenbar hat man Geister gerufen, die lästig werden, und versucht nun Einschränkungen anzubringen, wie die Integrität der neuen Staaten – ihre Grenzen sind aber doch von der Kolonialmacht geschaffen – oder die Ergebnisse des Krieges. Dabei hat die Entschließung des Sicherheitsrates 242 mit großer Einmütigkeit den Satz aufgestellt, daß durch Krieg kein Gebiet erworben werden kann, und daran haben die Vereinten Nationen seitdem immer festgehalten, wohlgemerkt im Fall der Palästinenser, die kein Kolonialvolk sind. Darum steht das zwingende Selbstbestimmungsrecht auch den Bevölkerungen zu, deren rechtmäßige Gebiete (wenn sie vertrieben worden sind) in andere Hand kommen sollen und gekommen sind. Diese Variante der Selbstbestimmung besagt, daß kein Gebiet ohne Zustimmung der rechtmäßig hingehörenden Bevölkerung seinen Herrn wechseln kann – in der Formulierung der französischen Verfassung von 1958 Art. 53 Abs. 3.

8. Weil *ius cogens* verschiedentlich bereits lebendig gewesen ist, kann man ihm nicht den Nichtrückwirkungsartikel 4 WVRK entgegensetzen. Insbesondere erweist sich die Selbstbestimmung seit 1945 als wirksam, unabhängig vom Text der WVRK. Es steht also nichts im Wege, ältere Verträge und Gebietsregelungen am Maßstab des Selbstbestimmungsrechts der Bevölkerung zu messen.

Es ist dies z.B. geschehen hinsichtlich der Regelungen in Mittel- und Osteuropa[49]. Es gibt auch Beanstandungen als *nul et non avenu* gegen das Abkommen von Madrid vom 14. November 1975 über die West-Sahara[50] und

---

[48] *A. Rigo Sureda,* The Evolution of the Right of Self-Determination (1973), S. 353: »... self-determination has become a peremptory norm«.

[49] *F. Münch,* Ostverträge und neues ius cogens, in: Ostverträge, Berlin-Status, Münchener Abkommen, Beziehungen zwischen der BRD und der DDR, Symposium März 1971 (Veröffentlichungen des Instituts für Internationales Recht an der Universität Kiel, Nr. 66) (1971); *G. Meier,* Der Begriff des bewaffneten Angriffs, ArchVR 16 (1973–75), S. 375, 385, über den Freiheitskampf gegen die Sowjetunion.

[50] Bericht für die Association Internationale des Juristes Démocrates, Revue de droit contemporain 1977 Nr. 1, S. 170 (g).

gegen die Abmachungen von Camp David über Palästina[51]. Das sog. Tribunal Permanent des Peuples hat im Jahre 1981 den sowjetischen Angriff auf Afghanistan[52] und die Besetzung von Ost-Timor durch Indonesien[53] als Verletzungen des Selbstbestimmungsrechts verdammt.

Man wird solche Stimmen, die nicht isoliert sind, sondern weiter verbreitete Tendenzen widerspiegeln, nicht einfach als apokryph abtun können; sie zeigen jedenfalls, wie die Selbstbestimmung einen Rang in der öffentlichen Meinung erlangt hat. Die Regionalisten und Föderalisten Europas[54] fördern seit langem den Gedanken, auch wenn sie ihm nicht in konkreten Fällen so zugespitzten Ausdruck geben.

9. Am Selbstbestimmungsrecht läßt sich eine Bedeutung des *ius cogens* besonders klar dartun: Es kann und soll ein Instrument des *peaceful change* sein. Alle tiefere Einsicht in die Voraussetzungen eines Friedens führt zur Forderung nach einer guten Ordnung der Staatengemeinschaft; denn die bloße Abwesenheit von Krieg ist ein prekärer Zustand. Die gegenwärtige Weltpolitik beweist es zur Genüge, da zwar ein Friedenszustand vorgetäuscht wird, in Wahrheit fortgesetzt irgendwo Krieg tobt, weil die Abwicklung des letzten Krieges und die Vorgänge seit 1945 keine Ordnung hinterlassen haben.

Solche Gedanken haben schon die Zeit nach dem ersten Weltkrieg beherrscht. Solange *peaceful change* ein rein politisches Konzept war, bestand die Gefahr, mit ihm Unruhe erst zu schüren. Nimmt man aber *ius cogens* in dem weiten Sinn einer auch materiellen Grundordnung der Staatengemeinschaft, so ergibt sich die Verbindlichkeit der Änderung als bestehendes Recht mit einem doch größeren Nachdruck als je vorher. Es müßte leichter sein, bei den weniger unmittelbar Beteiligten Stimmen und Einfluß zu gewinnen, wenn die Änderung nicht als subjektives Petitum, sondern als Teil der objektiv notwendigen Ordnung erscheint.

---

[51] Kolloquium in Paris der Association Internationale des Juristes Démocrates, 29./30. 9. 1979, Revue de droit contemporain 1979 Nr. 2, S. 144 ff.

[52] Session Stockholm, 1.–3. 5. 1981, Sentence (vervielfältigt), S. 37.

[53] Session Lisbonne, 19.–21. 6. 1981, Sentence (vervielfältigt), S. 40 f.

[54] *G. Héraud,* Modèle pour une application générale du droit d'autodétermination, L'Europe en formation, Nr. 236 (1980), S. 96–118.

Intervention in the International Court of Justice

Articles 62 and 63 of the Statute

Shigeru Oda

Introduction

It was more than thirty years ago that the author first visited the Max-Planck-Institut für ausländisches öffentliches Recht und Völkerrecht in Heidelberg, which was then located at the foot of the Philosophenweg. Since then he has been closely associated with the Institute and has pursued research work there on several occasions, owing a great deal of his achievements to the help and kindness of Professor *Hermann Mosler*, then Director of the Institute. It is much to his regret that the schedule of the International Court of Justice has not permitted him to write a new article for the occasion of Judge Mosler's seventieth birthday.

Part I is a result of the research which the author, as a Judge of the International Court of Justice, undertook while preparing his note in the case concerning the *Continental Shelf (Tunisia/Libyan Arab Jamahiriya) Application of Malta for Permission to Intervene,* and Part II is the view once expressed by him in his separate opinion appended to the Judgment which the Court delivered on 14 April 1981 in the above-mentioned case.

Part I

Historical Backgrounds of Articles 62 and 63 of the Statute of the International Court of Justice

1. Convention for the Peaceful Settlement of International Disputes

When the Statute of the Permanent Court of International Justice was drafted in 1920 the following provision of the 1907 Convention for the Peaceful

Settlement of International Disputes[1], which was inherited, with some minor modifications, from Art. 56 of the 1899 Convention for the Peaceful Settlement of International Disputes[2], was at hand. Art. 84 of the 1907 Convention provides as follows:

> »La sentence arbitrale n'est obligatoire que pour les Parties en litige. Lorsqu'il s'agit de l'interprétation d'une Convention à laquelle ont participé d'autres Puissances que les Parties en litige, celles-ci avertissent en temps utile toutes les Puissances signataires. Chacune de ces Puissances a le droit d'intervenir au procès. Si une ou plusieurs d'entre Elles ont profité de cette faculté, l'interprétation contenue dans la sentence est également obligatoire à leur égard«.

In the 1907 Convention as well as the 1899 Convention, States other than those in dispute were entitled to intervene only in the case of interpretation of a multilateral treaty to which they were parties.

2. Advisory Committee of Jurists of The Hague in 1920

At its second meeting in London in February 1920, the Council of the League of Nations decided to appoint an Advisory Committee of Jurists for the purpose of preparing plans for the establishment of the Permanent Court of International Justice, provided for in Art. 14 of the Covenant of the League of Nations, and of presenting a report to the Council.

The Advisory Committee of Jurists consisted of ten members, of which Baron Descamps of Belgium served as President, and met from 16 June to 24 July 1920 in the Peace Palace at The Hague[3].

The Secretariat of the Committee was entrusted with the preparation of a tabular synopsis of the different existing schemes[4]. The synopsis, consisting of three parts, included a section on the right of intervention, as Section VI of Part C – Procedure[5]. Various drafts were referred to, among which some introduced the new idea that not only in a case of the interpretation of a multilateral convention, but also when its interests were affected, a third State might intervene in a litigation before the Court. A Plan of the Five

---

[1] Deuxième Conférence Internationale de la Paix, La Haye, 15 juin–18 octobre 1907. Actes et documents, tome premier, Séances plénières de la conférence, Ministère des affaires étrangères (The Hague 1907).

[2] Conférence Internationale de la Paix, La Haye, 18 mai–29 juillet 1899, Ministère des affaires étrangères (new ed., The Hague 1907).

[3] Permanent Court of International Justice, Advisory Committee of Jurists, Procès-Verbaux of the Proceedings of the Committee, June 16–July 24 1920 with Annexes (The Hague 1920).

[4] *Ibid.*, pp. 51–99.

[5] *Ibid.*, pp. 93–95.

Neutral Powers, which had been drafted by jurists of Norway, Denmark, Sweden, Holland and Switzerland in February 1920 [6], provided as follows:

»Article 48 – 1. Lorsqu'un différend soumis à la Cour touche les intérêts d'un Etat tiers, celui-ci a le droit d'intervenir au procès«.

The draft of a convention respecting a Permanent International Court of Justice drawn up by a Swedish Governmental Commission in 1919 [7] had a similar provision, namely:

»Article 21 – Lorsqu'un différend soumis à la Cour est relatif à une convention internationale générale ou concerne à d'autres égards les intérêts d'un Etat tiers qui n'est pas Partie dans le litige, ce dernier aura le droit d'intervenir dans l'affaire.

La Partie qui a saisi la Cour d'un litige est tenue d'en donner avis à l'Etat qui, aux termes du premier alinéa du présent article, a le droit d'intervenir dans l'affaire«.

Furthermore, the draft scheme of a convention concerning an international judicial organization drawn up by the three committees nominated respectively by the Governments of Denmark, Norway and Sweden [8] had the following provision:

»Article 31 – Lorsqu'une affaire soumise à la Cour porte sur l'interprétation d'une convention internationale générale ou universelle, ou si elle concerne d'une autre manière les intérêts d'un Etat tiers, ce dernier aura le droit d'intervenir dans l'affaire.

Les Etats tiers doivent être avertis par la Partie qui a intenté l'affaire«.

A text of a draft scheme for the establishment of a Permanent Court of International Justice had been prepared by the drafting committee [9]. When Art. 2 of Chapter 3 – Procedure of the draft scheme concerning application to the Court (later Art. 40 of the Statute) was under discussion at the twenty-eighth meeting on 20 July [10], Lord Phillimore called attention to Art. 23 (later Art. 63 of the Statute) and raised the question of the right of intervention. He thought that para. 4 of Art. 2 concerning the notification of an application to the Court to the Members of the League of Nations through the Secretary-General would be useless if this right were not conceded to a wider extent than had been done in Art. 23, which, as quoted below, only mentioned general treaties. Mr. Fernandes proposed the following wording:

---

[6] Permanent Court of International Justice, Advisory Committee of Jurists, Documents presented to the Committee relating to existing plans for the establishment of a Permanent Court of International Justice, pp. 300–323.

[7] *Ibid.*, pp. 236–251.

[8] *Ibid.*, pp. 168–201.

[9] Permanent Court of International Justice, Advisory Committee of Jurists, Procès-Verbaux of the Proceedings of the Committee, pp. 561–571.

[10] *Ibid.*, pp. 587–594.

»Le droit d'intervention est reconnu aux Etats ayant un intérêt légitime, soit d'assister une des parties en cause, en raison d'un droit conjoint, soit d'exclure le demandeur ou le défendeur. Dans le premier cas, l'intervenant agira dans les mêmes délais accordés à la partie assistée. Dans le second cas, l'opposition sera formée dans les délais et selon les formes déterminées par la Cour«.

Mr. Hagerup, for his part, proposed the wording of Art. 21 of the plan of the three Scandinavian countries, according to which, as mentioned above, the right to receive notification and the right of intervention were not limited to Members of the League of Nations. Art. 2 as originally drafted was adopted.

The draft scheme prepared by the drafting committee provided, with regard to the right of intervention, as follows[10a]:

»Article 23 – Lorsqu'il s'agit de l'interprétation d'une convention à laquelle ont participé d'autres Etats que les parties en litige, le Greffe, au vu de la requête introductive d'instance, avertit sans délai toute puissance signataire.
Chacun des Etats a la faculté d'intervenir au procès, et, s'il l'exerce, devient partie en litige au point de vue de la chose jugée«.

This provision was adopted after some discussion, which was summarized in the official documents[11] as follows:

»*Lord Phillimore* explained the right of intervention as existing in English Law. He emphasised, in particular, the fact that in England, an intervening party could only associate itself with the defendant.
M. Loder gave an explanation of the same institution as it existed in Dutch Law. This admits of intervention both on the side of the plaintiff and of the defendant.
The President explained that the right of intervention in a case concerning a general Convention amounted to an extension of the principle of *res judicata*. A distinction must be drawn between intervention based upon an existing interest, and intervention for a purely potential reason.
M. Hagerup said he thought the wording of Article 23 was satisfactory as far as it went, but that it was incomplete. An addition was necessary, which might be worded as follows:
›Chacun des Etats signataires a la faculté d'intervenir au procès et, s'il exerce cette faculté, il devient partie au litige au point de vue de la chose jugée. Il en est de même de tout Etat dont les intérêts sont touchés par le procès‹.
M. Loder suggested Article 48 of the Five Powers' plan in the same connection.

10a *Ibid.*, p. 571.
11 *Ibid.*, pp. 592–594.

Lord Phillimore suggested the following wording:
›Lorsqu'un Etat tiers pense qu'un différend soumis à la Cour touche ses intérêts, cet Etat peut former une requête aux fins d'admission à l'intervention; et la Cour, si bon lui semble, y fera droit‹.
M. Fernandes agreed with Lord Phillimore on principle, but wished to make the right of intervention dependent upon certain conditions; for instance, it should be stated that the interests affected must be legitimate interests.
The President thought that the solution of the question of intervention should be drawn from common law. He proposed a wording based on this idea:
›Lorsqu'un Etat estime que, dans un différend, il peut être porté atteinte à ses droits, cet Etat peut adresser à la Cour une requête aux fins d'intervention, et la Cour peut y donner satisfaction. . .‹.
M. Adatci suggested to amend the wording proposed by M. Loder, by replacing the word ›right‹ by the word ›interest‹.
M. de Lapradelle wished to know whether the Court, with its normal composition, would be competent to hear the suits of third parties, supposing that the latter were not represented on the Court by a judge of their nationality.
It was agreed to leave this question open.
The President proposed the following new wording:
›Lorsqu'un Etat estime que dans un différend un intérêt d'ordre juridique le concernant est en cause, il peut adresser à la Cour une requête, à fin d'intervention. La Cour décide‹.

(English text – Should a State consider that it has an interest of a legal nature, which may be affected by the decision in the case, it may submit a request to the Court to be permitted to intervene. It will be for the Court to decide upon this request.)

The President's formula was adopted, subject to amendments in wording[12]. It would be a separate article to be inserted before the original Article 23. The latter Article was adopted in the following terms, borrowed from M. Asser and based on Article 84 of the Convention of 1907:

›Lorsqu'il s'agit de l'interprétation d'une Convention à laquelle ont participé d'autres Etats que les parties en litige, le greffe avertit sans délai tous les signataires.
Chacun d'eux a le droit d'intervenir au procès et s'il exerce cette faculté, l'interprétation contenue dans la sentence est également obligatoire à son égard‹.

---

[12] The text which the Drafting Committee prepared afterwards was slightly different from this text because of the suggestion of Mr. de Lapradelle, *ibid.*, p. 644.

(English text – Whenever the construction of a Convention to which States other than those concerned in the case are parties is in question, the Registrar shall notify all such States forthwith.
Every State so notified has the right to intervene in the proceedings; but if it uses this right, the construction given by the judgment shall be as binding upon it as upon the original parties to the dispute).

M. Hagerup objected to the expression ›will take force of *res judicata*‹ (au point de vue de la chose jugée) in the original wording of Art. 23«.

The text was thus revised by the drafting committee [13] and Art. 23 read as follows:

»Adopté (avec une addition à insérer avant l'article)...
›Lorsqu'il s'agit de l'interprétation d'une convention à laquelle ont participé d'autres Etats que les parties en litige, le Greffe, en vue de la requête introductive d'instance, avertit sans délai toute puissance signataire.
Chacun des Etats a la faculté d'intervenir au procès, et s'il exerce cette faculté il devient partie en litige au point de vue de la chose jugée‹«.

Thus the text concerning the right of intervention which was adopted after the first reading was as follows [14]:

»Article 60 (new) – Lorsqu'un Etat estime que dans un différend un intérêt d'ordre juridique est pour lui en cause, il peut adresser à la Cour une requête, afin d'intervention.
La Cour décide.
(English text – Should a State consider that it has an interest of a legal nature in a certain case, it may submit a request to the Court to be permitted to appear as a third party.
The Court shall decide).
Article 61 (formerly Article 23) – Lorsqu'il s'agit de l'interprétation d'une convention à laquelle ont participé d'autres Etats que les parties en litige, le Greffe avertit sans délai tous les signataires.
Chacun d'eux a le droit d'intervenir au procès et s'il exerce cette faculté, l'interprétation contenue dans la sentence est également obligatoire à son égard«.

As indicated in the text revised by the drafting committee as mentioned above, Art. 60 was quite new.

At its thirty-first meeting on 22 July 1920 [15], the committee devoted itself to the second reading of the text provisionally adopted at the previous

---

[13] *Ibid.*, p. 637.
[14] *Ibid.*, pp. 659–670.
[15] *Ibid.*, pp. 645–652.

meetings and distributed on the previous day, and Arts. 60 and 61 were adopted without alteration.

At the thirty-second meeting on 23 July [16], all the members of the committee signed the copy of the text adopted in the second reading on the previous day. At this meeting Mr. de Lapradelle, as the chairman of the drafting committee, began to present its report. The reading of the report was completed at the thirty-fourth meeting on 24 July [17]. The report by Mr. de Lapradelle reads, in the part dealing with Arts. 60 and 61, as follows [18]:

> »Lastly, the question of the right of intervention, a question which was not even mentioned in the project of the Court of Arbitral Justice of 1907, which was necessarily very short for reasons already given, and which question was only incidentally dealt with by the Convention for the Pacific Settlement of International Disputes of 1899 in connection with the force of the *res judicata* and the compulsory character of sentences by arbitration (Art. 56; Art. 84 of the 1907 Convention), is dealt with explicitly in this plan.
>
> There are three possibilities: a party may wish to take sides with the plaintiff or the defendant; a party may claim certain exclusive rights; or a party may request that one of the two contesting States should withdraw on the ground that it is not the real *dominus* of the right which it claims. In this latter case intervention tends to become exclusion, but as a rule a State is content to take joint action with one of the parties: should this be allowed? The Committee replies in the affirmative, but on condition that an interest of a legal nature is involved. The Court is to decide whether the interest is legitimate and consequently whether the intervention is admissible. To refuse all right of intervention, might have unfortunate results. The essential point is to limit it to cases in which an interest of a legal nature can be shown, so that political intervention will be excluded, and to give the Court the right of decision.
>
> Further there is one case in which the Court cannot refuse a request to be allowed to intervene; that is in questions concerning the interpretation of a Convention in which States, other than the contesting parties, have taken part; each of these is to have the right to intervene in the case. If such a State uses this right, the interpretation contained in the sentence becomes binding between it and the other parties to the case.
>
> Where collective treaties are concerned, general interpretations can thus be obtained very quickly, which harmonise with the character of the Convention«.

---

16 *Ibid.*, p. 671.
17 *Ibid.*, pp. 689–691.
18 *Ibid.*, pp. 693–749.

3. The Council of the League of Nations

At its eighth session held in San Sebastian from 30 July to 5 August 1920[19], Mr. Bourgeois, Representative of France, read a report by the Advisory Committee of Jurists. The Council passed a resolution dated 5 August 1920[20] by which the report was approved and the draft prepared by the Committee, and the report presented to the Committee by its rapporteur was sent to the governments of State Members of the League of Nations[21].

The tenth session of the Council was held from 20 to 28 October 1920 in Brussels[22]. Mr. Bourgeois presented a report on the Permanent Court of International Justice which was adopted by the Council on 27 October[23]. The part relevant to the right of intervention reads as follows:

> »The observations in the draft project of The Hague by one of our Colleagues draw attention to the following case: it might happen that a case appearing unimportant in itself might be submitted to the jurisdiction of the Court, and that the Court might take a decision on this case, laying down certain principles of international law which, if they were applied to other countries, would completely modify the principles of the traditional law of this country, and which might therefore have serious consequences. The question has been raised whether, in view of such an alternative, the States not involved in the dispute should not be given the right of intervening in the case in the interest of the harmonious development of the law, and otherwise after the closure of the case, to exercise, in the same interest, influence on the future development of law. Such action on the part of a non-litigant State would moreover have the advantage of drawing attention to the difficulty of making certain States accept such and such a new development of jurisprudence.
>
> These considerations undoubtedly contain elements of great value. The Hague Jurists have not moreover disregarded the necessity of bearing in mind considerations which, if not exactly identical, are at least in the same order of ideas. They have, indeed, given to non-litigant States the right to intervene in a case where any interest of a judicial nature which may concern them is involved.
>
> Moreover, Article 61 of the Draft lays down that: ›Whenever the construction of a convention in which States other than those concerned in the case, are parties, is in question, the Registrar shall notify all such

---

[19] League of Nations, Permanent Court of International Justice, Documents concerning the action taken by the Council of the League of Nations under Article 14 of the Covenant and the adoption by the Assembly of the Statute of the Permanent Court [see also League of Nations, Official Journal No. 8 (1920), pp. 12 ff.].

[20] *Ibid.*, p. 20.

[21] *Ibid.*, p. 26.

[22] *Ibid.*, p. 42.

[23] *Ibid.*, pp. 45–50.

States forthwith. Every State so notified has the right to intervene in the proceedings: but if it uses this right, the construction given by the judgment will be as binding upon it as upon the original Parties to the dispute.‹

This last stipulation establishes, in the contrary case, that if a State has not intervened in the case the interpretation cannot be enforced against it.

No possible disadvantage could ensue from stating directly what Article 61 indirectly admits. The addition of an Article drawn up as follows can thus be proposed to the Assembly:

›*The decision of the Court has no binding force except between the Parties and in respect of that particular case*‹.

(French text – La décision de la Cour n'est obligatoire que pour les parties en litige et dans le cas qui a été décidé).

The observations presented had, however, a much wider bearing: The question of giving to the various legal systems represented by the various States the possibility of collaborating with the Court in the development of international law. It must first of all be noted that the Court will contain representatives of the different judicial systems into which the world is divided and that the judgments of the Court will therefore be the result of the co-operation of entirely different thought and systems[24]«.

The draft scheme for the institution of the Permanent Court of International Justice, mentioned in Art. 14 of the Covenant of the League of Nations, presented to the Council by the Advisory Committee of Jurists as amended by virtue of the decisions of the Council, read as follows[25]:

»Article 57*bis*. La décision de la Cour n'est obligatoire que pour les parties en litige et dans le cas qui a été décidé.
Article 60. (identical to the previous text)
Article 61. (identical to the previous text)«.

4. The Third Committee of the First Assembly of the League of Nations

The Third Committee of the First Assembly of the League of Nations met from 17 November to 11 December 1920 under the chairmanship of Mr. Bourgeois of France[26]. A sub-committee was set up under the Third Committee and held ten meetings from 24 November to 7 December 1920 under the chairmanship of Mr. Hagerup of Norway[27]. The draft scheme presented

[24] *Ibid.*, p. 50.
[25] *Ibid.*, pp. 54–60.
[26] *Ibid.*, pp. 84–110.
[27] *Ibid.*, pp. 111–158.

by the sub-committee contained Arts. 59, 62 and 63, which were identical to Arts. 57*bis*, 60 and 61 of the Brussels text[28]. The report by Mr. Hagerup on behalf of the sub-committee which accompanied the draft scheme did not carry any substantive explanations on Arts. 59, 62 and 63[29].

The Third Committee, at its sixth meeting on 9 December[30], discussed the plan for the Permanent Court of International Justice submitted by the sub-committee and these articles were adopted without discussion.

5. First Assembly of the League of Nations

The First Assembly of the League of Nations took up the report of the Third Committee on the Permanent Court of International Justice at its twentieth and twenty-first plenary meetings, both held on 13 December[31]. Without much detailed discussion on the provisions of the draft scheme, and particularly on Chapter III dealing with procedure, there was no proposed amendment and no one asked to make any remarks. A resolution concerning the establishment of a Permanent Court of International Justice was unanimously adopted at the twenty-first plenary meeting[32].

6. Statute of the Permanent Court of International Justice

The Statute adopted by the First Assembly of the League of Nations read as follows:

> »Article 59. La décision de la Cour n'est obligatoire que pour les parties en litige et dans le cas qui a été décidé.
> Article 62. Lorsqu'un Etat estime que dans un différend un intérêt d'ordre juridique est pour lui en cause, il peut adresser à la Cour une requête, à fin d'intervention.
> La Cour décide.
> (English text – Should a State consider that it has an interest of a legal nature which may be affected by the decision in the case, it may submit a request to the Court to be permitted to intervene as a third Party.
> It will be for the Court to decide upon this request).
> Article 63. Lorsqu'il s'agit de l'interprétation d'une convention à laquelle ont participé d'autres Etats que les parties en litige, le Greffe les avertit sans délai.
> Chacun d'eux a le droit d'intervenir au procès, et s'il exerce cette faculté, l'interprétation contenue dans la sentence est également obligatoire à son égard«.

---

[28] *Ibid.*, pp. 214–221.
[29] *Ibid.*, pp. 206–213.
[30] *Ibid.*, pp. 100–104.
[31] *Ibid.*, pp. 225–256.
[32] *Ibid.*, p. 256.

The Statute was amended by a Protocol concerning the Revision of the Statute of the Permanent Court of International Justice which formed the subject of the Resolution of the Assembly of the League of Nations of 14 September 1929. The Protocol came into force on 1 February 1936. None of the three articles as mentioned above were subject to revision. Art. 40 (formerly Art. 2 of Chapter 3 of the draft scheme) was amended to include »and also any States entitled to appear before the Court«.

7. Statute of the International Court of Justice

A Committee of Jurists designated by the United Nations met in Washington from 9 to 19 April 1945 for the purpose of preparing and submitting to the San Francisco Conference a draft statute of an International Court of Justice[33]. In fact, in the drafting committee there was almost no discussion on Arts. 59, 62 and 63. As of 15 April 1945, the text of the Statute of the Permanent Court of International Justice – revisions prepared by the Drafting Committee – was prepared[34], and, according to an explanation given by Mr. Jessup of the United States[35], this text was based on the American draft[36]. The only change in the relevant provisions was that in Art. 62 the expression »as a third Party« was deleted.

At the tenth meeting on 16 April[37], Mr. Read of Canada, as the chairman of the drafting committee, gave some explanation of the draft, as follows:

> »The phrase ›as a third Party‹ was eliminated from Article 62 (1) as misleading«.

This was the only explanation concerning the relevant provisions.

The report on the draft statute of an International Court of Justice was submitted by the Committee to the San Francisco Conference on 20 April 1945[38]. The only relevant explanation was as follows[39]:

> »Articles 58 to 64 contain no change in the French text; the formal emendations made in the English text of ... Article 62, paragraph 1 (elimination of the words: ›As a third Party‹) do not change the sense thereof«.

---

[33] Documents of the United Nations Conference on International Organization (San Francisco 1945) vol. 14 (United Nations Committee of Jurists).

[34] *Ibid.*, pp. 485–500.

[35] *Ibid.*, p. 210.

[36] *Ibid.*, pp. 323–347.

[37] *Ibid.*, pp. 204–215.

[38] *Ibid.*, pp. 648–680.

[39] *Ibid.*, p. 676.

Part II

Interpretations of Articles 62 and 63 of the Statute of the International Court of Justice

1. Article 62 of the Statute

(i) Intervention »as a party«?

The records of the proceedings of the Advisory Committee of Jurists of 1920 which prepared the Statute of the Permanent Court of International Justice shed little light on what kind of functions a third State permitted to intervene under Art. 62 of the Statute (which was identical to Art. 62 of the Statute of the International Court of Justice as far as the French text is concerned) can exercise, and on what kind of effects may flow from its intervention. Although the Rules of Court adopted in 1922 at the preliminary session of the Permanent Court of International Justice contained provisions governing the application for permission to intervene, they did not deal with the scope of intervention, or the way in which the intervention of a third party, once granted, should be conducted. The Permanent Court of International Justice and its successor left such questions of intervention to be decided in the light of the particular circumstances of each case. In sixty years, there has hardly been a case before the Court in which Art. 62 could be said to have been a key issue.

The English text of Art. 62 of the Statute of the Permanent Court of International Justice spoke of intervention »as a third party«, and these words were omitted when the Statute of the International Court of Justice was drafted in 1945 by the United Nations Committee of Jurists. From the outset, the French text of the Statute of the Permanent Court of International Justice did not contain any phrase corresponding to »as a third party«. Art. 62 of the Statute, when redrafted for the present Court in 1945, did not undergo any change as far as the French text was concerned, and the relevant part of the report of the Committee is quoted above.

It is true that both the English and the French texts of the Statute of the Permanent Court of International Justice are authentic, as expressly mentioned in the Protocol of Signature of that Statute. On the other hand, the Preface to the *Procès-verbaux* of the Proceedings of the Advisory Committee of Jurists clearly indicated that[40]:

> »As all the members of the Committee, with the exception of Mr. Elihu Root, spoke in the French language, the English text of the Procès-Verbaux is to be looked upon as a translation, except in so far as concerns the speeches and remarks of Mr. Root«.

---

[40] *Op. cit.* (note 9), p. IV.

The reason why, in 1920, the phrase »as a third party« was introduced into the English text, as a translation from the French text, is not known. At all events, this introduction would not seem to have been explicable on the basis of the change in the French text from »un intérêt d'ordre juridique le concernant est en cause« to »un intérêt d'ordre juridique est *pour lui* en cause«. There is in the records of the discussions no suggestion that in 1920 the drafters had specifically in mind the idea of intervention »as a party«. Given this want of information, it does not seem justified to draw conclusions about the meaning of intervention »as a third party« based essentially on the English text of the Statute.

It is far from clear that participation *qua* party is a *conditio sine qua non* of the institution of intervention. Moreover, the question of whether or not the institution of intervention under Art. 62 of the Statute requires the participation of a third State solely »as a party« is closely interrelated with two further questions: first, whether or not a jurisdictional link which connects the intervening State with the original litigant States in the principal case should be required; and, second, whether or not the judgment of the Court in the principal case should also be binding upon the intervening State. It is difficult to discuss the institution of intervention without taking into account these two further questions, which are so closely interrelated with the nature of the institution under Art. 62.

(ii) Jurisdictional link and binding force of the Judgment

It is arguable that a jurisdictional link between the intervening State and the original parties to the case would be required if the intervening State were to participate as a full party, and that, in such a case, the judgment of the Court would undoubtedly be binding upon the intervening State. Such a right of intervention is basically similar to that provided for in the municipal law of many States. As a result of the participation of the third party as a full party in the principal case, the case will become a litigation among three parties. In the case of municipal law, of course, the link of jurisdiction between the third party seeking intervention and the original litigants is not at issue. This municipal institution has existed for many years to protect the right of a third party which might otherwise be affected by the litigation between two other parties and to promote economy of litigation. In such circumstances two or three causes of action concerning the same set of rights or obligations are dealt with as a single case.

Similarly, before the International Court of Justice, there may be cases in which the third State seeking intervention to secure its alleged right, which is involved in the very subject-matter of the original litigation, is linked with the original litigant States by its acceptance of the compulsory jurisdiction of the Court under the optional clause of the Statute or through a specific treaty or convention in force, or by special agreement with these two States. In such cases the third State may participate as a plaintiff or a defendant or as an independent claimant. Probably, in fact, this third State would in such circumstances also be entitled to bring a separate case on the same subject before the Court. On the other hand, participation in the proceedings by a third State as a full party without having any jurisdictional link with the original parties, while remaining immune from the binding force of the judgment, would certainly be tantamount to introducing through the back door a case which could not otherwise have been brought before the Court because of lack of jurisdiction. This seems inadmissible prima facie, because the jurisdiction of the International Court of Justice is based on the consent of sovereign States and is not otherwise compulsory.

Nevertheless, it is by no means clear that the only hypothesis contemplated when the draft of Art. 62 was under discussion was the hypothesis of the intervening State being connected by a jurisdictional link with the original litigants in the principal case. When the Permanent Court of International Justice met in 1922 for its preliminary session to discuss, among other things, the Rules of Court, the Committee on Procedure prepared questionnaires in which the Court was asked, in connection with intervention: »Have third parties interested in a case the right of intervention only when the original parties to a dispute have accepted the compulsory jurisdiction of the Court?« [41]. The Court in 1922 was divided in its answer and did not come to any definite conclusion. Yet it must be noted that the President, Judge Loder, ruled at the seventeenth meeting on 24 February 1922 that he

> »could not take a vote upon a proposal the effect of which would be to limit the right of intervention (as prescribed in Article 62) to such States as had accepted compulsory jurisdiction. If a proposal in this sense were adopted, it would be contrary to the Statute« [42].

The possibility in respect of Art. 62 of a somewhat broader scope of overall interpretation is traceable in the proceedings of the preliminary session of the Permanent Court of International Justice. In this respect, it may be pertinent to quote from the *Summary of Previous Discussions on the Question of the Right of Intervention,* submitted by Judge Beichmann, also at the seventeenth meeting on 24 February 1922. In the circumstances of Art. 62, he said:

> »no State has a right to intervene, but may only ask the Court for permission to do so; permission shall only be given if the Court considers that the State in question has an interest of a legal nature in the case. This condition, however is not necessarily the only one, and its fulfilment does not necessarily involve the right of intervention. Even though

[41] P.C.I.J., Series D, No. 2, p. 291.
[42] *Ibid.,* p. 96.

the Court is of opinion that this condition is fulfilled, it may refuse the request.
Article 62 of the Statute lays down that the question shall be decided in each particular case as it arises; there is therefore no need to adopt any decision at the moment either with regard to the interpretation of the words ›interest of a legal nature which may be affected by the decision‹, or with regard to the question whether the right of intervention is subject to other conditions of a legal nature, for example, the acceptance of the compulsory jurisdiction of the Court by the original parties and the party desiring to intervene, or the consent of the original parties. The question whether, when the right to intervene has been admitted and exercised, the intervening State is to be bound by the judgment, as well as the original parties, must also remain open.
Nevertheless, the discussion has shown that intervention may be based on other grounds: the intervening State may have a subjective right, which is incompatible with the claims of the original parties or of one of them, or again it may be to the interest of the intervening State that opinions contrary to its own should not prevail as regards the rules to be applied. The last named reason for intervention might be regarded as sufficient, at all events in the circumstances contemplated in Article 63. The question whether this reason would also suffice in other circumstances remains open« [43].

The situation where a right *erga omnes* is at issue between two States, but a third State has also laid a claim to that right, is a hypothesis which here merits consideration. For instance, in the case of the sovereignty over an island, or the delimitation of a territorial boundary dividing two States, with a third party also being in a position to claim sovereignty over that island or the territory which may be delimited by this boundary, or in a case in which a claim to property is in dispute, an unreasonable result could be expected if a jurisdictional link were required for the intervention of the third State. If this link is deemed at all times indispensable for intervention, the concept of intervention in the International Court of Justice will inevitably atrophy. Accordingly if the third State does not have a proper jurisdictional link with the original litigant States, it can nevertheless participate, but not as a party within the meaning of the term in municipal law. The role to be played by the intervening State in such circumstances must be limited. It may assert a concrete claim against the original litigant States, but that claim must be confined to the scope of the original Application or Special Agreement in the principal case. The intervening State cannot seek a judgment of the Court which directly upholds its own claim. The scope of the Court's judgment will also be limited: it will be bound to give judgment only within the scope

[43] *Ibid.*, p. 349.

of the original Application or Special Agreement. The intervening State cannot, of course, escape the binding force of the judgment, which naturally applies to it to the extent that its intervention has been allowed. The intervening State will have been able to protect its own right merely in so far as the judgment declines to recognize as countervailing the rights of either of the original two litigant States. On the other hand, to the extent that the Court gives a judgment positively recognizing rights of either of the litigant States, the intervening State will certainly lose all present or future claim in conflict with those rights. In this light, it does not seem tenable to argue that unless the intervener participates as a party on an equal footing with the original litigant States, it would unreasonably benefit without putting itself in any disadvantageous position.

2. Article 63 of the Statute

(i) No jurisdictional link and intervention not »as a party«

Intervention in the International Court of Justice is not necessarily limited to the situation concerning some well-defined right which is in dispute between litigant States. Relevant in this respect is Art. 63 of the Statute. The subject-matter of the dispute between the original parties in the case of Art. 63 will certainly be concrete rights claimed by both sides. But if any third State were to intervene, it would be because that third State was concerned with the interpretation of the convention falling to be construed in the judgment of the Court, but not with the subject-matter itself. This kind of intervention is unique in international law and, unlike Art. 62, was borrowed from the provisions of Art. 84 of the 1907 Convention for the Pacific Settlement of International Disputes, which was inherited, with some minor modifications, from Art. 56 of the 1899 Convention for the Pacific Settlement of International Disputes. This was confirmed by the President of the Advisory Committee of Jurists in 1920[44], although in fact no extensive discussions on this point have been reported from that time.

In the application of Art. 63, no jurisdictional link is apparently required between the intervening State and the original litigant States. The third State may participate in the case, but not »as a party« on an equal footing with the original litigant States because the object of the intervention is not necessarily connected with the claims of the original parties. The third party participates, but not as a plaintiff or defendant or even as an independent claimant. This seems to be clear from some precedents of the Court. In the *Haya de la Torre* case, the delivery of Haya de la Torre, who was enjoying asylum at the Colombian Embassy in Peru, was the subject-matter of the case, in which Cuba was not directly concerned. There is no reason to maintain that Cuba's intervention was assumed to be a participation »as a party« in the sense the

---

[44] *Op. cit.* (note 9), p. 594.

author has described above (although in the list of participants in the case Cuba was mentioned as the »intervening party«). In fact, Cuba's participation consisted simply in presentation of its interpretation of the Havana Convention. Similarly, in the *S. S. »Wimbledon«* case, the subject-matter was not the cargo in which Poland was interested but the right of access of the vessel in question to the Kiel Canal. In neither case was the intervention thought to be conditional on the presentation of any concrete claim against both or either of the original litigant States.

(ii) Binding force of the Judgment

The judgment of the Court will certainly be binding upon the litigant States, but all that will be binding upon the intervening State is, as para. 2 of Art. 63 provides, »the construction [of a convention] given by the judgment«. In other words, the intervening State will be bound by the Court's interpretation of the convention if it becomes involved in a case involving the application of that instrument.

In this respect it seems pertinent to examine the meaning of Art. 59 of the Statute, which provides for the binding force of the judgments of the Court, particularly since the meaning of that Article is sometimes discussed in connection with Art. 63. Art. 59 was not contained in the draft prepared by the Advisory Committee of Jurists in June/July 1920. It stemmed from comments of the British delegate at the Council of the League of Nations in October 1920. Mr. Balfour submitted a note on the Permanent Court of International Justice, a passage of which read[45]:

> »There is another point on which I speak with much diffidence. It seems to me that the decision of the Permanent Court cannot but have the effect of gradually moulding and modifying international law. This may be good or bad; but I do not think this was contemplated by the Covenant; and in any case there ought to be some provision by which a State can enter a protest, *not* against any particular decision arrived at by the Court, but against any ulterior conclusions to which that decision may seem to point«.

The report of Mr. Léon Bourgeois of France, who had also once submitted a report on the draft scheme of the Advisory Committee of Jurists at the Council meetings at San Sebastian in August, was presented at the Council on 27 October 1920. It starts with these words: »The following are the points which I propose that you should consider: . . .«, and continues:

---

[45] Permanent Court of International Justice, Documents concerning the action taken by the Council of the League of Nations under Article 14 of the Covenant, p. 38.

»8. The right of intervention in its various aspects, and in particular the question whether the fact that the principle implied in a judgment may affect the development of international law in a way which appears undesirable to any particular State may constitute for it a sufficient basis for any kind of intervention in order to impose the contrary views held by it with regard to this principle«[46].

Apparently taking into account the observation which had been made by Mr. Balfour, the report continued in connection with the institution of intervention in the case of the construction of a convention, as follows:

»This last stipulation establishes, in the contrary case, that if a State has not intervened in the case the interpretation cannot be enforced against it. No possible disadvantage could ensue from stating directly what Article 61 [now Article 63] indirectly admits. The addition of an Article drawn up as follows can thus be proposed to the Assembly: ›*The decision of the Court has no binding force except between the Parties and in respect of that particular case*‹ [now Article 59]«[47].

It may accordingly be concluded that the drafters of the Statute apprehended that the interpretation which the Court would place on international law would be shaped by prior judgments of the Court, and that, by adding this provision, they intended to inhibit the extension of a modified interpretation of international law to those States which had not participated in the case.

If Art. 59 is interpreted against this background, it does not add much to what was contemplated under Art. 63, and thus has no direct bearing on it. It may be asked, however, what significance it may have to state, as implied by Art. 63, that the construction of a convention will not be binding on States not party to a case before the Court. For regardless of such a postulate there is little doubt that, in a case where the construction of a particular convention is in dispute, the construction placed upon it by the Court in a previous case will tend to prevail. It is submitted that in this sense there will not be much difference between those States which have intervened in a case and those States which have not intervened, so far as the practical effect of the Court's construction of an international convention is concerned. It is questionable whether the intention of the founders – *i.e.,* not to make the interpretation of a convention by the Court binding upon the States which have not participated in the case – was really given effect by the formulation of Art. 59.

3. Case of the interpretation of the principles and rules of international law

If an interpretation of a convention given by the Court is necessarily of concern to a State which is a party to that instrument, though not a party to the case, there seems to be no convincing reason why the Court's interpretation of the principles and rules of international law should be of less concern to a State. If, therefore, the interpretation of an international convention can attract the intervention of third States under Art. 63 of the Statute, it may be asked why the interpretation of the principles and rules of international law should exclude a third State from intervening in a case. Lack of jurisdiction is not a sufficient reason for preventing a State from intervening as a non-party in a principal case in which the application of the principles and rules of international law is at issue, for the interpretation given by the Court of those principles and rules will certainly be binding on the intervening State. What is more, as in the case of Art. 63, the provisions of Art. 59 do not in fact guarantee a State which has *not* intervened in the principal case any immunity from the subsequent application of the Court's interpretation of the principles and rules of international law.

The author is not of course suggesting that such an intervention would fall within the meaning of Art. 63 of the Statute. He is simply saying that such a type of intervention – i.e., non-party intervention in the case in which a jurisdictional link is absent, but the interpretation given by the Court is binding – was introduced under Art. 63. And if such a type of intervention is therefore possible, Art. 62, if looked at in the light of Art. 63, can also be viewed as comprehending this form of intervention as well, providing that the interest of a legal nature is present. That is to say, intervention under Art. 62 encompasses the hypothesis where a given interpretation of principles and rules of international law is sought to be protected by a non-party intervention. In this hypothesis, the mode of intervention may be the same as under Art. 63, so that the third State neither appears as a plaintiff or defendant nor submits any specific claim to rights or titles against the original litigant States.

It may be objected that the States which may be affected by the interpretation of such principles and rules by the Court will be without number, and that, if an interpretation of the principles and rules of international law can open the door of the Court to all States as interveners, this will invite many future instances of intervention. This problem should be considered from the viewpoint of future judicial policy, and more particularly from the viewpoint of the economy of international justice. Yet this cannot be the reason why a request for intervention which is actually pending should be refused when the requesting State claims that its legal interest may be affected by the Court's rulings on the principles and rules of international law. The possibility of an increasing number of cases invoking Art. 63 may likewise not be avoided. The fact that in the past Art. 63 has been rarely invoked does not guarantee that the situation will remain unchanged in the future. Thus the problem is related not only to Art. 62, but also to Art. 63.

However, unlike Art. 63 dealing with the case of the interpretation of an international convention, Art. 62 comprises certain restrictions. Para. 2 of Art. 62 provides that: »It shall be for the Court to decide upon this request«.

This means that the Court has certain discretionary powers to allow or not to allow any requesting State to intervene in the litigation. Still more important is the restriction of para. 1 of Art. 62. This paragraph requires the State requesting intervention to show that »it has an interest of a legal nature which may be affected by the decision in the case«. Thus any danger of expansive application of Art. 62 will certainly be restricted by the Court's exercising its discretionary power, more particularly to determine whether the requesting State has such an interest.

Menschenrechte und Minderheitenschutz

Zu den Arbeiten des Institut de Droit international in der Zwischenkriegszeit

KARL JOSEF PARTSCH

I

Es ist erstaunlich, daß es im Schrifttum gelegentlich immer noch so dargestellt wird, als reichten nicht nur die Wurzeln eines positivrechtlichen Menschenrechtsschutzes, sondern auch der Menschenrechtsbewegung nicht weiter zurück als bis in den zweiten Weltkrieg. Erst die Kriegszielerklärungen der alliierten Mächte hätten dazu geführt, daß ein völlig neues Element in die Entwicklung des modernen Völkerrechts eingeführt worden sei, das in der Charta der Vereinten Nationen einen Niederschlag fand [1].

Das kann nicht nur daran liegen, daß die Erschütterungen des Rechtsbewußtseins vor und während des zweiten Weltkrieges die Erinnerung an frühere Ansätze zu dieser Entwicklung überdeckten, sondern muß auch noch andere Gründe gehabt haben, denen hier nachgegangen sei.

II

1. Die wichtigsten systematischen Vorarbeiten für einen internationalen Menschenrechtsschutz wurden in der Zeit zwischen den beiden Weltkriegen

---

[1] So etwa *D. W. Greig*, International Law (London 1970), S. 614, aber auch schon *P. N. Drost*, Human Rights as Legal Rights (Leiden 1951), S. 20; unlängst *V. Pechota*, The Development of the Covenant on Civil and Political Rights, in: L. Henkin (ed.), The International Bill of Rights (New York 1981), S. 32 ff. Dagegen die sorgfältigen historischen Untersuchungen von *F. Ermacora*, Menschenrechte in der sich wandelnden Welt, Bd. 1 (Wien 1974), S. 335–407 (mit Dokumenten).

(1921–1929) im Institut de Droit international geleistet. Sie führten zur Annahme einer »Déclaration des droits internationaux de l'homme« auf der Sitzung des Institut in New York am 12. Oktober 1929 [2].

Hermann Mosler, dem dieser Beitrag gewidmet ist, hat sie in den Zusammenhang mit den Arbeiten des Institut auf diesem Gebiet gestellt [3] und dabei die New Yorker Erklärung als »das herausragende Ereignis auf diesem Gebiet (S. 99), als »grundlegende Stellungnahme, die von keiner weiteren ... an Bedeutung erreicht wurde«, bezeichnet (S. 86).

Sie fand im zeitgenössischen Schrifttum einen starken Widerhall [4]. Daß sie auch einen Völkerrechtler, welcher die Anerkennung von Menschenrechten ausschließlich in den Bereich des Naturrechts – einem angeblichen Recht der Menschheit, fern der Völkerrechtsordnung – verwies, zum Nachdenken anregte, zeigt eine Notiz, die *Dionisio Anzilotti* als Merkposten für eine Neuauflage seines Lehrbuches zu seinen Ausführungen über die Stellung des Individuums im Völkerrecht zu Papier brachte: »Tener conto delle risoluzioni dell'Istituto di diritto internazionale, nella sessione di ... 1929« [5].

Auch nach 1945 haben noch einzelne Teilnehmer an den Diskussionen im Institut an die Erklärung von 1929 erinnert [6], obwohl sonst der Widerhall eher schwach war.

2. Auf der anderen Seite hat aber das Institut, als es 1947 in Lausanne seine Arbeiten wiederaufnahm und als erstes Thema die internationalen Menschenrechte »als Grundlage einer Wiederherstellung des Völkerrechts« diskutierte [7], nicht wieder an die früheren Arbeiten angeknüpft. Der Bericht von *Charles De Visscher* zitiert nur ganz beiläufig den ersten Satz der Präambel der Erklärung von New York: »Considérant que la conscience juridique du monde civilisé exige la reconnaissance à l'individu de droits soustraits à toute atteinte de la part de l'Etat« [8]. Als in den Beratungen zwei Mitglieder, die an den Beratungen in New York teilgenommen hatten, auf die Erklärung zurückkamen und anregten, sie unter Berücksichtigung der inzwischen eingetretenen Entwicklung zu revidieren [9], fanden sie keine Zustimmung. Ziel der

[2] Annuaire de l'Institut de Droit international (Annuaire) 35 (1929 II), S. 298.

[3] Das Institut de Droit international und die völkerrechtliche Stellung der menschlichen Person, in: Justitia et Pace, Festschrift zum 100jährigen Bestehen des Institut de Droit International, hrsg. von W. Wengler (Berlin 1974), S. 77–100.

[4] Nachweise bei *H. Mosler*, a.a.O., S. 86 f. (Anm. 22–24), und *L. Oppenheim/H. Lauterpacht*, International Law, Bd. 1 (8. Aufl., London 1955), S. 738 Anm. 1.

[5] Corso di diritto internazionale, hrsg. von der Società Italiana per l'Organizzazione Internazionale, Bd. 1 (4. Aufl., Padua 1955), S. 120 Anm. 7.

[6] So vor allem *A. Verdross*, in: UNESCO, Um die Erklärung der Menschenrechte (Zürich, Wien, Konstanz 1951), S. 338; *H. Wehberg*, Stichwort »Institut de Droit International«, in: Strupp-Schlochauer, Wörterbuch, Bd. 2 (1961), S. 24.

[7] Annuaire 41 (1947), Session de Lausanne, Bericht *Ch. De Visscher*, S. 1–13; Beratungen, S. 152–190.

[8] A.a.O., S. 6.

[9] *Alvarez*, a.a.O., S. 163, und *Mandelstam,* S. 165.

gegenwärtigen Arbeiten sei nicht die Ausarbeitung eines Kataloges zu schützender Rechte – deren gäbe es schon mehrere –, sondern »de favoriser le déclenchement d'un courant d'opinions en faveur de la protection des droits fondamentaux«[10]. So hatte in der Tat der Berichterstatter die Zielsetzung der Arbeiten in Lausanne umschrieben, indem er sich auf die *mission primordiale* des Institut nach seinem Statut (Art. 1 Abs. 1) berief, den Fortschritt des Völkerrechts zu fördern,

> »1° En travaillant à formuler les principes généraux de la science, de manière à répondre à la conscience juridique du monde civilisé«[11].

Dieser Thematik ist die Entschließung in ihrer endgültigen Fassung[12] auch treugeblieben. »Es handelt sich nicht um einen Vorentwurf für eine internationale Konvention, sondern um eine Grundsatzerklärung, die die geistigen Werte, die gleichzeitig die wesentlichsten und die bedrohtesten sind, herausstellen soll«; die »Erklärung vermeidet konsequenterweise die Anknüpfung an die New Yorker Erklärung von 1929«[13].

3. Es gibt eine ganze Reihe von Gründen dafür, daß die Arbeiten der zwanziger Jahre auf diesem Gebiet im Jahre 1947 nicht auf derselben Linie fortgeführt wurden. Damals trat das Institut zum ersten Mal nach dem Ende des Krieges zusammen. Ganz allgemein bestand in dieser Zeit das Gefühl der Notwendigkeit eines Neubeginns, verbunden mit einer Kritik an den Errungenschaften der Vorkriegszeit. Das zeigte sich auf internationaler Ebene bei den Beratungen über die Ausgestaltung der neuen Weltorganisationen wie auch innerstaatlich bei verschiedenen Verfassungsberatungen.

Fraglich erscheint hingegen, ob eine neue Startposition schon damals auf einen grundlegenden Wandel der allgemeinen dogmatischen Grundlagen des Völkerrechts zurückgeführt werden konnte. Denn für eine Neudefinition der Stellung des Individuums im Völkerrecht lag in der Charta vorwiegend ein Auftrag vor. Andererseits war man bei den Arbeiten, die zur Erklärung von New York führten, in allen dogmatischen Fragen unendlich vorsichtig und behutsam vorgegangen, um denkbaren Kontroversen aus dem Wege zu gehen[14].

Allerdings ist zu fragen, ob die Erklärung von 1929 nicht schon deshalb als Arbeitsgrundlage ausschied, da sie eng mit politischen und rechtlichen Vorstellungen verknüpft war, die nach dem zweiten Weltkrieg fragwürdig

---

[10] *Ch. De Visscher*, a.a.O., S. 165.

[11] A.a.O., S. 1.

[12] A.a.O., S. 258.

[13] *Mosler* (Anm. 3), S. 92; dennoch ist es erstaunlich, daß *Ch. De Visscher* in seinem Sonderbericht zur 100-Jahr-Feier »La contribution de l'Institut de Droit international au développement du droit international«, Livre du Centenaire 1873–1973 (Basel 1973), S. 131, kein Wort der Wertung für die in der Zwischenkriegszeit geleistete Arbeit fand.

[14] *Mosler* (Anm. 3), S. 92.

geworden waren. Zu denken ist dabei an das System des Schutzes nationaler Minderheiten, das mit der Gründung der Vereinten Nationen – will man es sehr milde ausdrücken – stark in den Hintergrund trat, wenn es nicht sogar verworfen wurde [15].

Daß die Arbeiten im Institut in den zwanziger Jahren vom Minderheitenschutz ausgingen und auf ihm aufbauten, erscheint eindeutig. Es ist jedoch die Frage, ob sich das Ergebnis dieser Arbeiten, die New Yorker Erklärung, aus der Verbindung mit dem Minderheitenschutzsystem zu lösen vermochte, indem sie als Menschenrechtsinstrument eigene Konturen gewann, so daß sie nicht als Bestandteil des Minderheitenschutzsystems dessen Schicksal zu teilen hatte.

4. Der ursprüngliche Untersuchungsauftrag lautete: »La protection internationale des droits de l'homme, du citoyen et des minorités«. Menschenrechte und Minderheitenschutz waren also miteinander kombiniert.

Als Berichterstatter wurde ein ausgesprochener Kenner und Spezialist des Minderheitenrechts gewonnen: *André Mandelstam* (1869–1949), Schüler von *F. von Martens*, seit 1898 Dragoman der russischen Botschaft bei der Pforte. Er hatte 1912/13 die Verhandlungen mit der türkischen Regierung über die Gewährung einer gewissen Autonomie an die Armenier geführt und sich als Autor des »Mandelstam-Plans« – unter dieser Bezeichnung erscheint er in der deutschen diplomatischen Korrespondenz – einen Ruf als Befürworter der Rechte ethnischer Minderheiten erworben. Sein uneigennütziges Eintreten für die Rechte des armenischen Volkes, Opfer des ersten großen Völkermordes im 20. Jahrhundert, zeigte, daß er sich persönlich von humanitären Beweggründen leiten ließ und nicht – wie ihm nicht selten unterstellt – von der machtpolitischen Erwägung, das Testament Peters des Großen zu verwirklichen, Rußland den Zugang zu den Meerengen zu eröffnen [16].

---

[15] *H. Raschhofer*, Artikel »Minderheiten«, in: R. Wolfrum (Hrsg.), Handbuch der Vereinten Nationen (München 1977), S. 286–292; *R. Hauser*, International Protection of Minorities, in: Israel Yearbook of Human Rights, Bd. 1 (1971), S. 92–102 mit einem Diskussionsbeitrag von *H. Mosler*, a.a.O., S. 405–407, der dazu auffordert, dem Minderheitenproblem wissenschaftlich wieder stärkere Beachtung zu schenken. Dies geschah z. B. seitens *F. Capotorti*, Study on the Right of Persons Belonging to Ethnic, Religious and Linguistic Minorities, UN Doc. E/CN.4/Sub.2/384/Rev.1 (1979). Auf eine neue Behandlung des Problems durch *L. B. Sohn* ist noch zurückzukommen.

[16] Sein umfangreiches Schrifttum über das Problem begann mit einer 1917 in Lausanne und Paris veröffentlichten Flugschrift: »Le sort de l'Empire ottoman«. In einer weiteren Publikation: La Société des Nations et les Puissances devant le problème Arménien (Paris 1926), übte er harte Kritik an den früheren Verbündeten Rußlands. Neue Aktenpublikationen veranlaßten ihn, in der Schrift: Das armenische Problem im Lichte der Völker- und Menschenrechte (mit einem Anhang: Die deutsch-armenischen Beziehungen während der armenischen Reformaktion, S. 91–149) (Kieler Vorträge und Einzelschriften, Heft 15) (1931) frühere Schilderungen der deutschen Haltung zum Problem zu revidieren und zu differenzieren. Vgl. den schönen Nachruf von *A. Makarov*, Annuaire 43 (1950), S. 482–484, und *K. J. Partsch*, Die Ar-

Bald nach Aufnahme der Vorarbeiten zur Ausführung des ihm 1921 in Rom erteilten Auftrages erkannte er den grundlegenden Unterschied zwischen dem Minderheitenschutz des Völkerbundsystems und einem allgemeinen Schutz von Menschenrechten, der auch dem Mehrheitsvolk zugute kommt. Während der Minderheitenschutz nur an eine Diskriminierung auf Grund ganz bestimmter Kriterien – wie Rasse, Religion und Sprache – anknüpft, ist der Menschenrechtsschutz ein allgemeines politisches Problem. Auf Grund dieser Erkenntnis entschloß er sich, vorab den Minderheitenschutz zu behandeln und das breitere Problem des Menschenrechtsschutzes für eine spätere Behandlung zurückzustellen[17]. Damals (d.h. im Juli 1925) sah es fast so aus, als sei die Behandlung des allgemeinen Menschenrechtsschutzes im Institut auf unbestimmte Zeit vertagt. Jedenfalls wurde die mit diesen Arbeiten betraute Kommission so umbenannt, daß sie nur noch die Arbeiten zum Minderheitenschutz deckte[18].

Schon im September 1925 führte jedoch eine Diskussion in der Völkerbundsversammlung zu einer Wende. Es ging dabei um die Minderheiten, nicht um die Menschenrechte. Litauen hatte beantragt, einen besonderen Ausschuß damit zu betrauen, eine Konvention über die Erstreckung des Minderheitenschutzes auf alle Völkerbundsmitglieder auszuarbeiten, und fand die Unterstützung mehrerer auch durch Minderheitenverträge gebundener Mitglieder, die sich gegen eine ungleiche Behandlung sträubten. Die Großmächte, insbesondere Frankreich und Großbritannien, aber auch China, widersprachen einer Generalisierung des Minderheitenschutzes aber so heftig, daß Litauen seinen Antrag zurückziehen mußte[19].

4. Der Berichterstatter im Institut zog aus diesem Vorgang die Konsequenz, das Menschenrechtsproblem aus dem System des Minderheitenschutzes herauszulösen, um auf diese Weise die menschenrechtlichen Errungenschaften dieses Systems – er nennt es das »juristische Minimum« – auf alle Mitglieder der Staatengemeinschaft zu erstrecken, ohne daß dabei die Frage auftaucht, in welchen Staaten es nationale Minderheiten gibt, wie sie zu definieren sind und ob sie nicht etwa durch ein Schutzsystem erst künstlich geschaffen und konstituiert worden sind. Daß er dabei den Kreis der lediglich zu proklamierenden Menschenrechte auf die schon in den Minderheitenschutzverträgen und -erklärungen enthaltenen Staatenverpflichtungen beschränkte, geschah sicher aus taktischen Gründen. Er wollte dem Einwand ausweichen, durch eine derartige Erklärung in den Vorbehaltsbereich nationa-

---

menierfrage und das Völkerrecht in der Zeit des ersten Weltkrieges, insbesondere das Wirken von *A. Mandelstam*, in: E. A. von Renesse (Hrsg.), Der erste Völkermord des 20. Jahrhunderts (1982).

[17] Einleitung zum 1. Bericht, Annuaire 32 (1925), S. 246.

[18] Annuaire 32 (1925), S. 538; *H. Wehberg* äußerte deshalb ernsthafte Zweifel, ob die Kommission den Auftrag wieder auf das allgemeine Menschenrechtsproblem erweitern dürfe: Annuaire 34 (1928), S. 312.

[19] Annuaire 34 (1928), S. 276–280.

ler Zuständigkeiten für innere Angelegenheiten einzudringen. Denn wenn gewisse Materien, welche das Verhältnis von Staaten zu ihren eigenen Angehörigen berühren, in Verträgen geregelt seien, könnten sich die Vertragspartner auf beiden Seiten nicht mehr darauf berufen, sie gehörten noch zum *domaine réservé*[20].

Diesen Erwägungen folgt der Vor-Entwurf für eine Menschenrechtserklärung, den *André Mandelstam* auf der Sitzung in Stockholm vorlegte[20] und der außer einer Präambel nur zwei Artikel umfaßte. Die Präambel bezog sich nachdrücklich auf die 1919 und 1920 abgeschlossenen Minderheitenschutzverträge, die ein Minimum von Individualrechten enthielten, welche das gegenwärtige Völkerrecht anerkenne und die nun auf alle Staaten zu erstrecken seien.

Die darauf folgenden zwei Artikel des operativen Teils[21] sind wortwörtlich aus dem Vertrag zwischen den alliierten und assoziierten Mächten mit Polen vom 28. Juni 1919[22] entnommen. An die Stelle der polnischen Regierung beziehungsweise der Einwohner Polens sind lediglich »die vertragschließenden Teile« bzw. »die Angehörigen der vertragschließenden Teile« getreten.

Die enge Verbindung zwischen Minderheitenschutz und Menschenrechtserklärung ist also in aller Offenheit dargetan, um die Autorität des positivierten Minderheitenschutzes zu nützen.

5. Bei diesem Vor-Entwurf ist es freilich nicht geblieben. Eine tiefgreifende Umgestaltung erfuhr die Präambel. Von den Minderheitenschutzverträgen ist nicht mehr ausdrücklich die Rede. Nationale Verfassungsentwicklungen in den Vereinigten Staaten und Frankreich werden in den Vordergrund gerückt. Aus damaliger Sicht war das taktisch nicht ungeschickt. Die Staaten, welche so massive Einwände gegen eine Generalisierung des Minderheitenschutzes erhoben hatten, sollten nicht darauf gestoßen werden, daß hier in verdeckter Form nichts anderes geschehe. Erst im vorletzten Absatz dieser Präambel heißt es dann: »Que, d'autre part, un certain nombre de traités stipule la reconnaissance des droits de l'homme«. Das war nur eine verdeckte Bezugnahme.

Auch der operative Teil wurde weiter entwickelt. Daß die drei ersten Artikel in der endgültigen Fassung durchgehend mit dem Passus beginnen: »Il est du devoir de tout Etat de reconnaître à tout individu le droit égal . . .«, folgt der ursprünglichen Konzeption, daß durch diese Proklamation das Individuum nicht zum Völkerrechtssubjekt erhoben werden soll[23], während der

[20] *Ibid.*, *S. 291.*

[21] Sie betreffen den Schutz von Leben und Freiheit (Art. 1 Abs. 1), die Religionsfreiheit (Art. 1 Abs. 2) jedes Einwohners, die Rechtsgleichheit bei Genuß privater und öffentlicher Rechte durch Staatsangehörige (Art. 2).

[22] *Martens*, NRG 3, Bd. 13, S. 505.

[23] 2. Bericht *Mandelstam*, Annuaire 34 (1928), S. 287: »Le moment n'est, peut-être, pas encore venu de conférer à l'individu la qualité de *sujet* de ces droits . . .«.

Religionsartikel in seiner ursprünglichen Fassung[24] früher mit den Worten begann: »Les habitants de chacun des Etats contractants auront le droit . . .«, was auch abweichend von dieser Konzeption gedeutet werden konnte.

Daß als Schutzobjekt nun das Individuum (*individu*) und nicht mehr der Einwohner (*habitant*) erscheint, soll die Probleme der juristischen Personen ausklammern[25], stellt aber keine Hinwendung zu einer menschenrechtlichen Terminologie dar.

Ein ähnliches praktisches Ziel der Eingrenzung hat in Art. 4 der endgültigen Fassung[26] die Wahl des Wortes *nationaux* statt *ressortissants*, um nämlich die Eingeborenenbevölkerung von Kolonialgebieten vom Genuß des dort erwähnten Rechts auszuschließen[27].

Auch die Ergänzungen führen nicht aus der Gedankenwelt des Minderheitenschutzes heraus. So lehnt sich der eingefügte Sprachenartikel (endgültige Fassung: Art. 3) an Art. 7 Abs. 3 des Polenvertrages an, ist freilich schwächer. Auch das Verbot der Entziehung der Staatsangehörigkeit (endgültige Fassung: Art. 6) hat enge Beziehungen zum Minderheitenschutz. Er soll davor schützen, daß die Angehörigen der Minderheit diejenigen Rechte einbüßen, die ihnen nur zustehen, so lange sie dem Staatsverband des Aufenthaltsstaates angehören[28].

Allenfalls könnte man sagen, daß die Aufnahme des Geschlechts unter die verbannten Kriterien aus dem Bereich des Minderheitenschutzes herausfällt. Damals war das ohnehin weitgehend ein Postulat, das weit in die Zukunft wies.

6. So ergibt also die Analyse nicht nur des Vorentwurfes des Berichterstatters, sondern auch seiner Weiterentwicklung während der Debatten im Institut, daß die ganze Erklärung in der Gedankenwelt des Minderheitenschutzes verbleibt und keine Rede davon sein kann, daß sie sich aus dieser Gedankenwelt gelöst habe. So ist es nicht erstaunlich, daß sie als ein Instrument des Minderheitenschutzes aus dem System des Völkerbundes angesehen wurde und dasselbe Schicksal erlitt wie andere Instrumente dieses Rechtsgebiets.

III

Will man Bedeutung und Tragweite eines Instrumentes des Rechtslebens würdigen, darf man nicht nur seiner Wirkung als unmittelbarer Arbeitsgrundlage für spätere Entwicklungen nachgehen. Es kann auch mittelbar wei-

---

[24] Vor-Entwurf: Art. 1 Abs. 2, endgültig Abs. 2.
[25] Annuaire 35 (1929 I), S. 120, und dazu *Mosler* (Anm. 3), S. 91.
[26] Vor-Entwurf: Art. 2 Abs. 2.
[27] Annuaire 35 (1929 I), S. 124–127.
[28] Annuaire 36 (1931 I), S. 545: Bericht *André Mandelstam*.

tergewirkt haben. Ein wichtiges Beispiel für eine derartige mittelbare Wirkung ist etwa der Einfluß des Gedankengutes der Paulskirche nicht nur auf ihr zeitlich nachfolgende Verfassungswerke in Deutschland, sondern auch auf die schlichte Gesetzgebung, etwa die Reichsjustizgesetze.

In dem hier gegebenen Rahmen sind nur einige mehr aphoristische Bemerkungen darüber möglich, wie sich die Erklärung von 1929 im Lichte des gegenwärtigen Rechts des internationalen Menschenrechtsschutzes darstellt.

1. Die außerordentliche Zurückhaltung, welche im Institut hinsichtlich einer *Anerkennung des Individuums als Rechtssubjekt* des Völkerrechts geübt wurde, erscheint nicht nur im Lichte der Situation in den zwanziger Jahren gerechtfertigt. Trotz aller Bemühungen ist die Entwicklung auf diesem Gebiet nicht sehr viel weitergekommen, auch wenn man die Feststellung, selbst eine beschränkte Völkerrechtsunmittelbarkeit des Individuums stecke »heute noch in den Kinderschuhen« [29], als leicht unterkühlt ansehen mag. Schon vor zwanzig Jahren konnten durchaus überzeugende Ansätze dafür nachgewiesen werden, daß das Individuum zum Träger völkerrechtlicher Rechte und Pflichten geworden sei [30], und diesen Nachweisen wären heute zahlreiche neue hinzuzufügen.

2. Die New Yorker Erklärung enthält eine *Generalklausel für Leben, Freiheit und Eigentum* (Art. 1) und nur wenige Einzelrechte (Art. 2–4 und 6). Ein Übergang von einer solchen Generalklausel zum Katalogsystem schon in der Universellen Erklärung der Menschenrechte von 1948, dann aber verfeinert und differenziert in den Menschenrechtspakten von 1966 wie auch in den regionalen Menschenrechtsinstrumenten wird überwiegend als wesentliche Errungenschaft angesehen. Das soll keineswegs grundsätzlich geleugnet werden. Freilich hat das Katalogsystem den Nachteil, daß ständig das Bestreben auftritt, einmal etablierte Kataloge von Menschenrechten weiter aufzuspalten und mit neuen Rechten anzureichern. Andererseits zeigt die Judikatur auf nationaler wie internationaler Ebene die steigende Tendenz, in den Katalogen versteckte Generalklauseln zu benützen. Auch im Vertragsrecht gibt es Rückfälle in das System von Generalklauseln (wie Art. 5 des Internationalen Übereinkommens zur Beseitigung der Rassendiskriminierung von 1965, wo der Katalog nur zur Umschreibung des Anwendungsgebietes einer Generalklausel dient, ohne einzelne Rechte zu definieren). Das gibt doch zu denken und sollte davor warnen, schon einen Übergang zum Katalogsystem in jedem Falle als Vorteil oder Fortschritt anzusehen. Damit soll nicht geleugnet werden, daß die inzwischen geschaffenen Kataloge einen Zuwachs an Rechtsklarheit und Rechtssicherheit gebracht haben. Das sollte jedoch nicht überschätzt werden.

---

[29] *Verdross/Simma,* Universelles Völkerrecht (1976), S. 222.

[30] Stichwort »Individuum im Völkerrecht«, in: Strupp-Schlochauer, Wörterbuch, Bd. 2 (1961), S. 15.

3. Die *Trias der Rechtsgüter Leben – Freiheit – Eigentum* ist inzwischen durch die Trias »Leben – Freiheit – Sicherheit« abgelöst worden. Zwar enthält die Universelle Erklärung der Menschenrechte noch eine Eigentumsgarantie mit einem Konfiskationsverbot (Art. 17), doch schon bei der Ausarbeitung der Europäischen Menschenrechtskonvention gab es Schwierigkeiten, an ihr festzuhalten, während sie in den Menschenrechtspakten völlig fortfiel. Auf der anderen Seite wuchs das Bedürfnis, soziale Menschenrechte zu statuieren und zu sichern, selbst wenn dadurch klassische Freiheitsrechte geschmälert wurden. Diese Tendenz setzt sich immer stärker durch. Von dieser Entwicklung war freilich die Erklärung von 1929 noch nicht berührt.

4. Das *Diskriminierungsverbot* hat die Erklärung stark in den Vordergrund gerückt (Art. 1 und 5). Es soll den Gleichheitssatz absichern, indem es gewisse Unterscheidungskriterien absolut für unzulässig erklärt. Diese Tradition hat sich weitgehend durchgesetzt. Es konnte freilich nicht ausbleiben, daß sich die gebannten Kriterien mit dem Wandel sozialer Auffassungen veränderten. So wurden Geburtsvorteile (*naissance*) schon während der Arbeiten im Institut fallen gelassen, da sie für überwunden galten. Es ist fraglich, ob das schon damals realistisch war. So blieben nur Staatsangehörigkeit, Rasse, Sprache und Religion als untersagte Kriterien. Dazu kam das Geschlecht. Die Staatsangehörigkeit gehört eigentlich nicht in diese Liste[31], zumal wenn das Eigentum unter den geschützten Gütern erscheint. Deswegen haben sowohl die Universelle Erklärung (Art. 2 Abs. 1) wie die Pakte (Art. 2) die Staatsangehörigkeit durch *national or social origin*, ein faktisches, kein rechtliches Unterscheidungsmerkmal, ersetzt.

5. Die Erklärung verzichtet im allgemeinen auf eine *Definition* der gewährten Rechte und auch auf eine *Umschreibung der zulässigen Einschränkungen*[32]. Nur bei der Religionsfreiheit gibt es eine Schrankenklausel. Ihre Ausübung soll nur zulässig sein, falls sie nicht unvereinbar ist mit dem *ordre public* und den guten Sitten. Diese Schranke wurde sehr weit ausgelegt. Auf ihrer Grundlage könne in einem Lande, das eine Staatsreligion kenne, die öffentliche Ausübung einer anderen unterbunden werden.

Die Einschränkung von Menschenrechten durch Rücksichten auf den *ordre public* ist ein altes und dennoch außerordentlich schwieriges Thema. Es geht zurück bis auf die französische Erklärung der Menschenrechte von 1789 und ist von dem amerikanischen Präsidenten W. Wilson durch seinen ersten Entwurf eines Völkerbundspaktes (Art. 21) wieder in die moderne Diskussion eingeführt worden. Meine Sympathie gehört den Engländern, die sich seit jeher gegen diese Formel wandten und nur eine *prevention of disorder* zulassen wollten, was sehr viel enger ist und keine uferlose Einschränkung von

[31] So schon *Valloton*, Annuaire 34 (1928), S. 369.

[32] *Mandelstam* meinte damals, dafür sei die Zeit noch nicht reif, das sollte einer künftigen Konvention überlassen bleiben: Annuaire 35 (1929 II), S. 119.

Menschenrechten auf eine beliebige politische Ideologie – wie den Sozialismus – erlaubt[33].

6. Zum Schluß nur einige wenige Bemerkungen zu den in der Erklärung erwähnten Einzelrechten.

a) Die Garantie der *Religionsfreiheit* (Art. 2) war recht schwach ausgestaltet. Kein Unterschied wurde gemacht zwischen dem Recht, eine Religion zu haben, sich zu ihr zu bekennen, und dem Recht, sie auszuüben. Erst später hat sich durchgesetzt, daß diese Unterscheidung nötig ist, um die Schranken dieser beiden Rechte sinnvoll zu bestimmen. So wie die Einschränkung ausgedrückt ist, wirkt sie auch für das Recht, eine Religion zu haben, und ist daher bedenklich.

b) Das gleiche Recht jedes Individuums, *die Sprache eigener Wahl* zu benützen und in ihr unterrichtet zu werden, soll geachtet werden, ohne daß ausgesprochen wäre, in welchen Bereichen und Beziehungen diese Rechte gelten sollen, wer für die Erziehung sorgt. Es ist aber aus den parallelen Bestimmungen der Minderheitenschutzerklärung zu entnehmen, daß der Staat weder verpflichtet sein soll, besondere Unterrichtsanstalten dafür zur Verfügung zu stellen noch private Schulen zu subventionieren. Da der Gebrauch der Sprache eigener Wahl nicht auf einen bestimmten Personenkreis beschränkt ist, kann wohl davon ausgegangen werden, daß das Recht hierzu allgemein gilt, also auch gegenüber den Behörden und Gerichten des Aufenthaltsstaates.

Auf diesem Gebiet ist eher ein Rückschritt zu verzeichnen. Ein allgemeines Recht, die Sprache eigener Wahl zu benützen, ist in den Menschenrechtspakten nicht gewährt. Es gilt nur für die Angehörigen bereits bestehender ethnischer oder sprachlicher Minderheiten und nur »gemeinsam mit anderen Angehörigen ihrer Gruppe«, nicht auch im Verkehr mit Angehörigen des Mehrheitsvolkes oder mit Behörden oder Gerichten des Staates[34]. Von der Erziehung in der eigenen Sprache ist weder in dem Pakt für bürgerliche und politische Rechte noch in dem Schulartikel des Sozialpaktes (Art. 13) die Rede.

Die in dem Sprachenartikel der Erklärung von 1929 geweckten Hoffnungen sind also nicht erfüllt, teilweise sogar enttäuscht worden. Freilich ist dabei zu berücksichtigen, daß dieser Sprachenartikel auf einer spezifischen Errungenschaft des Minderheitenschutzsystems des Völkerbundes beruht. Andererseits ist es aber auch bezeichnend, daß gegenwärtig den linguistischen Minderheiten noch nicht einmal die Rechte gewährt werden sollen, die 1929 allen Menschen zugedacht waren.

---

[33] Zu dem Problem allgemein *A. Kiss*, Permissible Limitations of Rights, in: L. Henkin (ed.), The International Bill of Rights (New York 1981), S. 290 ff. (299–302).

[34] Art. 27 des Paktes über bürgerliche und politische Rechte und dazu jetzt *L. B. Sohn*, The Right of Minorities, in: L. Henkin (ed.), a.a.O., S. 270 ff.

c) Das besondere *Diskriminierungsverbot* hinsichtlich privater und öffentlicher Rechte der Staatsbürger (Art. 4) wurde seinerzeit sehr engherzig ausgelegt. Zu den »öffentlichen Rechten« sollten staatsbürgerliche Rechte nicht gehören. Zugang zu Ämtern, Funktionen und Ehrenstellungen wurden aus den erwähnten Beispielen gestrichen, um insbesondere eine Garantie des Frauenstimmrechts auszuschließen. Diese stark zeitbedingte Schwelle wurde inzwischen übersprungen, und Art. 25 des Paktes über bürgerliche und politische Rechte gewährt die staatsbürgerlichen Rechte allen Bürgern, Männern und Frauen ohne Diskriminierung und läßt nur Beschränkungen zu, die »vernünftig« (*reasonable*) sind[35].

d) Ein Verbot willkürlicher *Ausbürgerungen* besteht zwar nach der amerikanischen Menschenrechtskonvention (Art. 20 Abs. 3), nicht aber nach den universell geltenden Pakten. Insofern ist die New Yorker Erklärung den bisher erreichten Resultaten vorausgeeilt.

IV

Würdigt man den der Erklärung von 1929 zugrundegelegten Standard mit dem inzwischen auf der Weltebene erreichten, so kann zwar nicht übersehen werden, daß diese Erklärung zwar in vieler Hinsicht weit hinter dem gegenwärtig erreichten Standard zurückbleibt, da sie für so wichtige Rechtsgüter wie Leben und Freiheit nur eine Generalklausel mit einem Diskriminierungsverbot kennt, ohne zu definieren, welche Einzelgarantien unter diese Generalklausel fallen. Auch wenn Abs. 4 der Präambel auf die »equal protection of the law-clause« der amerikanischen Verfassung verweist, ist daraus doch nicht ohne alle Zweifel zu entnehmen, daß die Generalklauseln für Leben und Freiheit und für den Genuß aller bürgerlichen und politischen Rechte auch eine Garantie für ein Gerichtsverfahren enthält, das rechtsstaatlichen Erfordernissen genügt.

Andererseits reichen die in der Erklärung ausdrücklich garantierten Einzelrechte teilweise an das Ergebnis heran, welches inzwischen in jahrzehntelangen Bemühungen erreicht wurde, einzelne von ihnen gehen sogar darüber hinaus und visieren ein Ziel an, das im positiven Völkerrecht noch nicht erreicht ist.

Die hohe Einschätzung, welche *Hermann Mosler* der Erklärung zuteil werden ließ, ist schon deswegen berechtigt.

---

[35] Dazu *K. J. Partsch*, Freedom of Conscience and Expression and Political Freedoms, in: L. Henkin (ed.), a.a.O., S. 238–244.

Die Rechtsprechung des Europäischen Gerichtshofs zur innergemeinschaftlichen Wirkung völkerrechtlicher Abkommen

PIERRE PESCATORE

So wie große Künstler ihre Perioden gehabt haben, gibt es auch bei *Hermann Mosler* eine »Europäische Periode«; er war unter den Juristen, die der ersten Europäischen Gemeinschaft, der EGKS, ihre rechtliche Gestalt gegeben haben, und er ist zu dieser Zeit mit Veröffentlichungen hervorgetreten, die zum grundlegenden Bestand des Gemeinschaftsrechts gehören[1]. Danach hat er sich wieder dem Gebiet seiner eigentlichen Berufung, dem Völkerrecht, zugewandt, dennoch aber in seinen persönlichen Arbeiten wie auch in der Leitung des Heidelberger Instituts den Konnex mit dem Integrationsrecht nicht verloren. Es mag daher angemessen erscheinen, mit dieser zwischen Gemeinschaftsrecht und Völkerrecht gelegenen Untersuchung einen Beitrag zu seiner Festschrift zu leisten.

Drei einleitende Bemerkungen sollen Rahmen und Tragweite dieser Studie verdeutlichen.

Es geht hier nicht darum, die Beziehungen zwischen Gemeinschaftsrecht und Völkerrecht in ihrer ganzen Tragweite darzulegen. Das hat vor kurzem *Olivier Jacot-Guillarmod* in seiner Dissertation über »Droit communautaire et droit international public« getan[2]. Ich habe mich darauf beschränkt, einen

---

[1] Siehe die »Publikationen zum Europarecht« im Verzeichnis der Schriften von *Hermann Mosler.*

[2] Etudes suisses de droit internal (Genève 1979). Siehe noch zum selben Thema: *Ch.-E. Held,* Les accords internationaux conclus par la Communauté économique européenne (1977), S. 182 ff.; *A. Bleckmann,* Die Position des Völkerrechts im inneren Rechtsraum der Europäischen Gemeinschaften, Jahrbuch für Internationales Recht 18 (1975), S. 300–319; *K. M. Meessen,* The Application of Rules of Public International Law within Community Law, Common Market Law Review 13 (1976), S. 485–501; *J.-P. Pietri,* La valeur juridique des accords liant la Communauté économique européenne, Revue trimestrielle de droit européen 12 (1976), S. 51–75 und 194–214; *J.-V. Louis,* Mise en oeuvre des obligations internationales de la Communauté dans les ordres juridiques de la Communauté et de ses Etats membres, Revue belge de droit inter-

Aspekt dieses vielschichtigen Komplexes, der mir besonders gut bekannt ist, intensiv zu untersuchen, nämlich die Rechtsprechung des Europäischen Gerichtshofs zur Anwendung völkerrechtlicher Verträge im Bereich des Gemeinschaftsrechts. Es gibt derzeit genau 42 Fälle, in denen der Europäische Gerichtshof sich auf völkerrechtliche Abkommen bezogen hat, dazu noch 37 Urteile sehr spezieller Art, die sich auf die Anwendung des sogenannten »Brüsseler Zolltarifschemas« beziehen[3]. Wenn diese Fälle auch nur wenige Prozente der gesamten Rechtsprechung des Gerichtshofs darstellen, so ist es immerhin ein nicht unrepräsentatives Anschauungsmaterial, das es auszuwerten gilt.

Um ein Mißverständnis von vornherein auszuschalten, möchte ich darauf hinweisen, daß ich in dieser Studie als »völkerrechtliche Abkommen« nur solche Abkommen ins Auge gefaßt habe, die in den Bereich der *auswärtigen* Beziehungen der Gemeinschaft fallen oder zumindest einem die Gemeinschaft überragenden völkerrechtlichen Kontext angehören. Außer Betracht bleiben daher die Verträge zur Gründung der Gemeinschaft selbst, die Beitrittsverträge sowie Abkommen zwischen den Mitgliedstaaten, wie etwa das Brüsseler Zuständigkeits- und Vollstreckungsabkommen, die allesamt zum internen Recht der Gemeinschaft gehören. Außer Betracht bleiben ebenfalls zweiseitige oder mehrseitige Abkommen zwischen Mitgliedstaaten sowie Abkommen zwischen einzelnen Mitgliedstaaten und Drittstaaten, die gelegentlich in der Rechtsprechung, besonders zur Frage der sozialen Sicherheit, eine Rolle gespielt haben. Hier geht es also ausschließlich um Abkommen, welche die Gemeinschaft in einen weiteren völkerrechtlichen Zusammenhang einbinden.

Zum Thema selbst noch eine dritte Vorbemerkung. Wenn man die relevanten Urteile durchsieht, so drängt sich auf den ersten Blick die Feststellung auf, daß der Gerichtshof, der sich in anderen Materien durch mutige und kategorische Stellungnahmen hervorgetan hat und grundsätzlichen Aussagen nicht ausgewichen ist, an das Problem der Wirkung völkerrechtlicher Abkommen im Gemeinschaftsbereich nur sehr vorsichtig herangegangen ist. Man sucht hier vergebens nach Formeln, wie man sie in der Rechtsprechung

---

national 13 (1977), S. 122–143; *L. Citarella,* La rilevanza del diritto internazionale nell'ordinamento comunitario, Studi Riccardo Monaco (1977), S. 123–140; *E. C. Pinto,* I trattati internazionali nel diritto communitario, Rivista di diritto internazionale 63 (1980), S. 313–329. Ich erlaube mir, auch auf zwei eigene Publikationen hinzuweisen: External Relations in the Case-Law of the Court of Justice of the European Communities, Common Market Law Review 16 (1979), S. 615–645, ergänzt durch eine »Contribution to the Discussion« in: Division of Powers between the European Communities and their Member States in the Field of External Relations, Europa Institute, University of Amsterdam (1981), S. 69–75.

[3] Siehe die dieser Studie beigefügten Listen: A. Urteile zur gemeinschaftsinternen Wirkung völkerrechtlicher Abkommen; B. Urteile mit Bezugnahme auf die im Rahmen des Brüsseler Abkommens zum Zolltarifschema ergangenen Tarifavise. Die jeweils in Frage stehenden völkerrechtlichen Abkommen sind in den betreffenden Urteilen näher bezeichnet.

etwa zur unmittelbaren Anwendbarkeit und zum Vorrang des Gemeinschaftsrechts oder zur Vertragsfähigkeit und internationalen Rechtspersönlichkeit der Gemeinschaft findet. Auf dem hier ins Auge gefaßten Gebiet gibt es kein *Van Gend & Loos*-, kein *Costa ./.ENEL*-, kein *AETR-Urteil*[4]. Vielmehr hat sich der Gerichtshof in allen hier besprochenen Fällen in seinen Aussagen auf das zur Fallösung jeweils Notwendige beschränkt; statt die Grundsatzfragen aufzugreifen, hat er es in den meisten Fällen vorgezogen, diese zu überspringen und damit bestenfalls implizite zu lösen. Diese Verfahrensweise zeigt, daß er die typische Haltung eines internen, nicht eines internationalen Gerichts einnimmt. Es ist offensichtlich, daß er sich – obwohl er seine Tätigkeit in einem mehrstaatlichen Kontext entfaltet – in erster Linie als eine systemimmanente Gerichtsbarkeit empfindet. Wenn dieser Pragmatismus auch die nach Grundsätzen verlangende juristische Doktrin nicht befriedigen mag, so hat er doch seine guten Gründe[5].

Immerhin hat der Gerichtshof, wenn in gewissen Fällen völkerrechtliche Abkommen ihm ein zweckdienliches Argument an die Hand gaben, zur Lösung der gestellten Probleme auf diese Abkommen zurückgegriffen, ohne deswegen – von Ausnahmen abgesehen – grundsätzliche Fragen aufzuwerfen. Wir werden sehen, daß diese Bezugnahmen in der Rechtsprechung in sehr verschiedenartigen Zusammenhängen und zu sehr unterschiedlichen Zwekken auftauchen. Dabei wird sich zeigen, daß die prinzipiellen Fragen, ob völkerrechtliche Abkommen innerhalb der Gemeinschaft »anwendbar« sind und ob sie etwa eine »unmittelbare Wirksamkeit« entfalten, ungenügend nuancierte Fragestellungen sind. Das Recht im allgemeinen und das Gemeinschaftsrecht im besonderen sind viel zu komplex, als daß man mit solchen apriorischen Fragestellungen auskommen könnte[6].

---

[4] *Van Gend & Loos,* 26/62 vom 5. 2. 1963, Slg. S. 1, ist richtungweisend für die unmittelbare Anwendbarkeit des Gemeinschaftsrechts, *Costa ./. ENEL,* 6/64 vom 5. 7. 1964, Slg. S. 1141, für den Vorrang des Gemeinschaftsrechts; das Urteil *Kommission ./. Ministerrat* im sog. Europäischen Straßenverkehrsabkommensfall, besser bekannt unter der französischen Bezeichnung »AETR«, 22/70 vom 31. 3. 1971, ist grundlegend für die internationale Vertragsfähigkeit der Gemeinschaft.

[5] Diese Zurückhaltung hat mancherlei Gründe, von denen der eindeutigste wohl in der Sorge besteht, das Gemeinschaftsrecht durch Einführung völkerrechtlicher Wertmaßstäbe nicht desintegrieren zu lassen. Daß diese Besorgnis nicht illusorisch ist, mag man aus der Studie von *R. Bernhardt,* Das Recht der europäischen Gemeinschaften zwischen Völkerrecht und staatlichem Recht, in: Festschrift Bindschedler (1980), S. 229–239, ersehen. Daraus geht nämlich hervor, daß Dinge wie: formlose Änderung und Aufhebung der Verträge, Außerkraftsetzung des Gemeinschaftsrechts durch widersprechende staatliche Gesetze und, in »gravierenden Konfliktsituationen«, der Vorrang der staatlichen Macht vor dem Recht im Völkerrecht immerhin erwägenswerte Fragen sind. Wenn solche Denkweisen in der Tat für völkerrechtliche Argumentation repräsentativ sind, muß man verstehen, daß der Gerichtshof es vermeidet, ein mit solchen Ideen befrachtetes trojanisches Pferd in das Gemeinschaftsrecht einzuführen.

[6] Hier ist auf die Analyse von *A. Koller* in seinem Buch: Die unmittelbare Anwendbarkeit völkerrechtlicher Verträge und des EWG-Vertrages im innerstaatlichen

I. Welche Verträge hat der Gerichtshof anzuwenden?

Bevor wir die Frage der Wirkung völkerrechtlicher Abkommen im Rahmen des Gemeinschaftsrechts untersuchen können, müssen wir vorab feststellen, unter welchen Voraussetzungen solche Abkommen in der Gemeinschaft anwendbar sind. Wir werden sehen, daß im Rahmen des Gemeinschaftsrechts nicht nur solche Abkommen in Kraft stehen, die von der Gemeinschaft als solcher abgeschlossen worden sind. Auch andere völkerrechtliche Verträge können im Rahmen des Gemeinschaftsrechts Relevanz haben, sei es als Folge einer völkerrechtlichen Sukzession oder Substitution der Gemeinschaft in bestehende Verträge der Mitgliedstaaten oder sei es auf Grund der gebotenen Rücksichtnahme auf andere völkerrechtliche Bindungen dieser Staaten.

1. Eigene Abkommen der Gemeinschaft

In Betracht zu ziehen und anzuwenden hat der Gerichtshof in erster Linie die von der Gemeinschaft selbst im Rahmen ihrer internationalen Vertragsfähigkeit abgeschlossenen Abkommen. Während langer Jahre waren solche Abkommen Ausnahmeerscheinungen, heute gehören sie zur laufenden Praxis der Gemeinschaft. Es handelt sich dabei im wesentlichen um drei oder vier große Gruppen von Verträgen: Handelsabkommen, Abkommen über wirtschaftliche Hilfe und Zusammenarbeit, Assoziierungsabkommen (wobei, vom Standpunkt der Gemeinschaft aus, auch die Jaunde- und Lomé-Abkommen einzubeziehen sind) und schließlich Freihandelsabkommen, die bereits eine erste Stufe der internationalen Wirtschaftsintegration darstellen [7].

Bei solchen Abkommen ist in erster Linie zu untersuchen, ob sie für die Gemeinschaft in Kraft sind, d. h. ob die Gemeinschaft auf Grund eines regel-

---

Bereich (Bern 1971) (S. 58 ff.) und auf den Aufsatz von *M. Zuleeg*, Die innerstaatliche Anwendbarkeit völkerrechtlicher Verträge am Beispiel des GATT und der Europäischen Sozialcharta, ZaöRV 35 (1975), S. 341–363 (344 ff.), zu verweisen, wo der Unterschied zwischen »Geltung« und »Anwendbarkeit« von völkerrechtlichen Verträgen hervorgehoben wird. Wie wir sehen werden, bezeichnet diese Unterscheidung genau den Punkt, wo sich die monistischen und die dualistischen Doktrinen trennen. Stellungnahmen wie die oben zitierten Artikel von *Bleckmann* und *Pinto* (Anm. 2), in denen die Anwendung des Völkerrechts im inneren Rechtsraum der Gemeinschaft sogleich in ein dualistisches Denkschema hineingezwängt wird, tragen weder der Komplexität des Problems, noch der Rechtsprechung des Gerichtshofs Rechnung.

[7] Diese Praxis hat sich als Folge des *AETR*-Urteils vom 31. 3. 1971 entwickeln können, da dieses Urteil feststellt, daß die Gemeinschaft eine völkerrechtliche Vertragsfähigkeit nicht nur im Rahmen der ihr ausdrücklich durch die Art. 113 (Zoll- und Handelsabkommen) und 238 (Assoziierungsabkommen) des EWG-Vertrags zugewiesenen Vertragsschließungskompetenz, sondern im gesamten Bereich der von diesem Vertrag der Gemeinschaft zugewiesenen materiellen Zuständigkeiten besitzt. Siehe zu dieser Frage *C. Flaesch-Mougin*, Les accords externes de la CEE, essai d'une typologie (1979) und die regelmäßigen Chroniken von *J. Raux*, Revue trimestrielle de droit européen.

rechten Abschlußverfahrens völkerrechtlich gebunden ist. In dieser Beziehung haben sich bisher keinerlei Streitfälle ergeben. Immerhin ist zweierlei festzuhalten.

Zunächst einmal, daß der Gerichtshof keinen Unterschied in der Wirkungsweise dieser Abkommen gemacht hat, je nachdem ob sie von der Gemeinschaft in ausschließlicher Zuständigkeit oder parallel mit den eigenen Mitgliedstaaten abgeschlossen worden waren. Abkommen, die nach dem sogenannten »gemischten Verfahren« von der Gemeinschaft und den Mitgliedstaaten gleichlaufend abgeschlossen worden sind, binden die Mitgliedstaaten, vom Standpunkt des Gemeinschaftsrechts aus gesehen, auf Grund des Abschlusses durch die Gemeinschaft. Relevant aus der Sicht des Gemeinschaftsrechts ist mit anderen Worten nur der Abschluß durch die Gemeinschaft; der Abschluß durch die Mitgliedstaaten mag in bezug auf solche Verpflichtungen, die außerhalb der Zuständigkeit der Gemeinschaft liegen, seine Bedeutung haben, für den gemeinschaftseigenen Bereich gilt nur der Abschlußakt der Gemeinschaft. So wird im *Bresciani*-Urteil vom 5. Februar 1976 bezüglich des Jaunde-Abkommens gesagt: »Das Abkommen wurde sowohl von den Mitgliedstaaten als auch von der Gemeinschaft geschlossen und war daher nach Artikel 228 (des EWGV) für diese Parteien verbindlich«[8].

Eine weitere Frage ergibt sich aus der Form des Abschlußaktes, den der Ministerrat wählt: Aus Gründen, die nicht ohne weiteres ersichtlich sind, bediente er sich in früherer Zeit der Form der »Entscheidung« und wählte er neuerdings meist die Form der »Verordnung«. Der Gerichtshof hat diese Frage zur Bestimmung der Wirkung völkerrechtlicher Abkommen offensichtlich als unerheblich erachtet und in allen Fällen ausschließlich auf eine Analyse des Abkommensinhaltes abgestellt, ohne die Frage nach der Natur des Abschlußaktes überhaupt aufzuwerfen. Eine in dieser Beziehung durch den London Court of Appeal in dem jüngsten *Polydor*-Fall gestellte Frage ist im Urteil vom 9. Februar 1982, weil unerheblich, ohne Antwort geblieben. In der Tat ist es nicht denkbar, daß die Wirkungsweise eines völkerrechtlichen Abkommens durch die Wahl der Form des Abschlußaktes durch eine der Vertragsparteien einseitig modifiziert werden kann. Aus der Wahl der Verordnungsform kann man allenfalls auf den Willen schließen, den Abkommensbestimmungen gemeinschaftsintern eine optimale Wirksamkeit zu sichern; welches aber diese Wirkung ist, bleibt offen.

2. Substituierung der Gemeinschaft in völkerrechtliche Abkommen der Mitgliedstaaten

Höchst bemerkenswert vom völkerrechtlichen Standpunkt aus ist eine zweite Kategorie von Verträgen, genauer: von multilateralen Abkommen, die von

[8] Dieser Punkt wurde noch einmal bekräftigt im Gutachten 1/76 vom 26. 4. 1977 (Slg. S. 741) betreffend das geplante Abkommen mit der Schweiz über die Schaffung eines Stillegungsfonds für die Rheinschiffahrt (Nr. 7 der Begründung).

den Mitgliedstaaten vor Schaffung der EWG auf Gebieten abgeschlossen worden waren, die später in den Zuständigkeitsbereich der Gemeinschaft überführt wurden. Die Praxis zeigt, daß es dabei besonders um zwei Verträge geht, einerseits das GATT-Abkommen von 1947, zu dessen »Contracting Parties« sämtliche Mitgliedstaaten gehören, andererseits das Brüsseler Abkommen vom 15. Dezember 1950 über das Zolltarifschema; bekanntlich ist der gemeinsame Zolltarif auf dieser sogenannten »Brüsseler Nomenklatur« aufgebaut. Es muß darauf hingewiesen werden, daß es sich in beiden Fällen um »lebendige«, d.h. mit handlungsfähigen Organen versehene Vertragssysteme handelt, die somit evolutionsfähige Verpflichtungen enthalten. Das GATT ist heute nicht mehr, was es im Jahre 1947 war; es wurden in der Zwischenzeit in diesem Rahmen ungezählte Verhandlungen über den Zolltarif und über andere handelspolitische Fragen geführt, und es kam dabei zu Abschlüssen, die entweder die Form von Protokollen, von Verhaltenscodices oder von Entscheidungen der »Contracting Parties« annahmen; es müssen also im Rahmen des GATT nicht nur die ursprünglichen Vertragsregeln ins Auge gefaßt werden, sondern auch die Arbeiten der GATT-Organe und ihr Ergebnis in Form von Protokollen oder Entscheidungen. Ähnlich, wenn auch auf viel engerer Basis verhält es sich im Rahmen der Brüsseler Konvention über die Zollnomenklatur. In diesem Rahmen funktioniert unter der Autorität des Rats für Zusammenarbeit auf dem Gebiet des Zollwesens ein Ausschuß zur Auslegung des Tarifschemas, der laufend sogenannte »Tarifavise« zur Klärung des ursprünglichen Tarifschemas abgibt[9].

Auf dem Gebiet der Fischerei ist man ähnlichen Problemen begegnet. Es gibt hier Abkommen zur Regelung der Fischerei in den verschiedenen Meereszonen, wobei naturgemäß die Abkommen für den Nord-Ost- und den Nord-Westatlantik aus geographischen Gründen für die Gemeinschaft von vorrangigem Interesse sind. Im Rahmen dieser Abkommen wurden internationale Fischereikomissionen eingerichtet, die ein Entscheidungsrecht in Form von »Empfehlungen« ausüben. Von dem Augenblick an, in dem die Gemeinschaft Anfang der 70er Jahre ihre Befugnisse auf dem Gebiet der Fi-

[9] Es handelt sich hier um zwei Abkommen, die am 15. Dezember 1950 in Brüssel unterzeichnet wurden, nämlich die Convention establishing a Customs Cooperation Council und die Convention on Nomenclature for the Classification of Goods in Customs Tariffs. Durch das erstgenannte Abkommen wird ein Customs Cooperation Council geschaffen; Art. III des zweiten Abkommens bestimmt folgendes: »(a) The Council shall supervise the operation of the present Convention with a view of securing uniformity in its interpretation and application. (b) To this end, the Council shall establish a Nomenclature Committee . . .«. Art. IV überträgt dem Nomenclature Committee die Funktion, u.a. »(c) to prepare explanatory notes as a guide to the interpretation and application of the Nomenclature«. Augenblicklich ist das Abkommen über die Zollnomenklatur für 50 Staaten in Kraft; nach Informationen des Customs Cooperation Council wird es durch insgesamt 148 Staaten *de iure* oder *de facto* angewandt, wobei die Gemeinschaft jeweils für 10 Staaten zählt.

scherei wahrgenommen hat, stellte sich auch hier die Frage nach der Wirkung solcher Abkommen innerhalb des Gemeinschaftsbereichs [10].

Zu erwähnen ist in diesem Zusammenhang zunächst ein frühes Urteil des Gerichtshofs, vom 27. Februar 1962, *Kommission ./. Italien,* der sogenannte »*Radioröhren*-Fall«. Hier stellt der Gerichtshof fest, daß in den innergemeinschaftlichen Verhältnissen die GATT-Regeln durch die Regeln des EWG-Vertrags ausgeschaltet werden; ein Mitgliedstaat kann sich daher nicht auf die GATT-Bestimmungen berufen, um seinen gemeinschaftsrechtlichen Verpflichtungen zu entgehen.

Was die auswärtigen Handelsbeziehungen anbelangt, so finden sich die entscheidenden Aussagen im Urteil vom 12. Dezember 1972, *International Fruit Cy.* Dieses Urteil, bekräftigt durch das *Schlüter*-Urteil vom 24. Oktober 1973, ist in seinen Ausdrücken noch etwas zögernd; geklärt und gefestigt wurde die Rechtsprechung etwas später im Urteil vom 19. November 1975, *Nederlandse Spoorwegen.* In diesen Urteilen hat der Gerichtshof festgestellt, daß die Gemeinschaft, weil ihr die Kompetenzen auf dem zoll- und handelspolitischen Gebiet übertragen worden sind, in der Ausübung der Rechte und Pflichten des GATT den Mitgliedstaaten »substituiert« wurde. Der Gerichtshof hat sich hier, wenn er auch den Ausdruck selbst vermeidet, von der völkerrechtlichen Doktrin der Staatensukzession leiten lassen: Die von den Mitgliedstaaten an die Gemeinschaft abgegebenen zoll- und handelspolitischen Kompetenzen wurden gewissermaßen *cum emolumento et onere* übertragen; spätestens ab Inkraftsetzung des gemeinsamen Zolltarifs steht es der Gemeinschaft zu, die sich aus dem GATT ergebenden Befugnisse und die entsprechenden völkerrechtlichen Pflichten einheitlich für die gesamte Gemeinschaft wahrzunehmen [11].

Die Tragweite dieser Stellungsnahme kann nur aus den Fragen verstanden werden, die in diesen Vorlageverfahren von einem niederländischen und einem deutschen Gericht an den Gerichtshof herangetragen worden waren. Man darf nämlich nicht vergessen, daß auch nach der Übernahme der entsprechenden Befugnisse durch die Gemeinschaft die Mitgliedstaaten ihre Position als »Contracting Parties« des GATT nicht aufgegeben haben und daß sich damit die Frage stellte, ob nicht die nationalen Gerichte in ihrer Eigenschaft als Gerichte der ursprünglichen Vertragsparteien befugt waren, die GATT-Bestimmungen in eigener Zuständigkeit auszulegen und gegebenenfalls Maßnahmen der Gemeinschaft (wie etwa handelspolitische Schutzmaßnahmen im *International Fruit Cy*-Fall oder die Erhebung von sogenannten

---

[10] Näheres dazu im *Kramer*-Urteil vom 14. 7. 1976.

[11] Diese Frage ist in der juristischen Literatur noch kaum behandelt worden. Siehe dazu *H. Krück,* Völkerrechtliche Verträge im Recht der Europäischen Gemeinschaften (Beiträge zum ausländischen öffentlichen Recht und Völkerrecht, Bd. 70) (1977), S. 137 ff., der allerdings hier eine Sukzessionswirkung nicht annehmen möchte, sowie meine oben (Anm. 2) erwähnte Studie in Common Market Law Review 16 (1979), S. 637 f.

Ausgleichsbeträgen auf Schweizerkäse im *Schlüter*-Fall) auf ihre GATT-Konformität zu prüfen. Besonders deutlich ist in dieser Hinsicht die Antwort im *Nederlandse Spoorwegen*-Urteil, wo der Gerichtshof unterstreicht, daß die Wirkung der GATT-Bestimmungen nur noch aus der Sicht der Rechtsordnung der Gemeinschaft bestimmt werden kann und nicht mehr aus der der Rechtsauffassungen der verschiedenen Mitgliedstaaten. In dem zuletzt genannten Urteil wird dieselbe Konzeption auf das Brüsseler Abkommen über das Zolltarifschema übertragen. Hier wird darauf hingewiesen, daß die Zollavise des im Rahmen dieses Abkommens funktionierenden Ausschusses in der Gemeinschaft zu beachten sind, damit die Einheitlichkeit der Anwendung der Zollnomenklatur in sämtlichen Vertragsstaaten gewährleistet wird [12].

3. Anwendung sonstiger Abkommen

Zu den eigenen Verträgen der Gemeinschaft gehören demnach nicht nur solche, die von der Gemeinschaft als eigener Rechtsperson abgeschlossen wurden, sondern auch diejenigen, bezüglich derer sich die Gemeinschaft auf Grund der erfolgten Kompetenzübertragungen den Mitgliedstaaten substituiert hat. Die Rechtsprechung zeigt indes, daß damit der Kreis der innerhalb der Gemeinschaft anwendbaren Verträge nicht erschöpft ist. Mehrmals hatte der Gerichtshof mit Verträgen zu tun, an denen die Mitgliedstaaten beteiligt sind, für die aber in Ermangelung klarer Kompetenzregeln die oben beschriebene Substitutionswirkung nicht in Betracht kommen konnte. Den Zufälligkeiten der Rechtsfälle entsprechend kamen hier die folgenden völkerrechtlichen Verträge ins Spiel: im Zusammenhang mit dem freien Waren- und Dienstleistungsverkehr so verschiedenartige Abkommen wie das New Yorker Abkommen über den vorübergehenden Import von Kraftfahrzeugen, das Florenzer Abkommen über den Import von Gegenständen erzieherischer, wissenschaftlicher oder kultureller Art, das Pflanzenschutzabkommen der FAO, das Straßburger Abkommen über die Pflichtversicherung von Kraftfahrzeugen, ja sogar das Weltpostabkommen und ein Abkommen gegen die Verbreitung unsittlicher Veröffentlichungen; in anderen Zusammenhängen handelt es sich um die Europäische Menschenrechtskonvention und verschiedene Abkommen der Internationalen Arbeitsorganisation.

Wie hat sich nun der Gerichtshof derartigen Abkommen gegenüber verhalten? Es ist bemerkenswert, daß er, wann immer eine Berufung auf solche Abkommen erfolgte, nicht das getan hat, was man als erste Reaktion eines

---

[12] Die Notwendigkeit einer einheitlichen Zollnomenklatur wird hier, wohlverstanden, im Hinblick auf sämtliche Vertragsstaaten des Brüsseler Abkommens erwähnt. Mit anderen Worten erfüllen diese Tarifavise in der Gemeinschaft eine doppelte Funktion, da sie einerseits die Einheitlichkeit der Anwendung der Brüsseler Zollnomenklatur, andererseits die Einheitlichkeit der Anwendung des gemeinsamen Zolltarifs gewährleisten helfen. Es sei vermerkt, daß auch innerhalb der Gemeinschaft ein Ausschuß geschaffen wurde, der Stellungnahmen zur Auslegung des gemeinsamen Zolltarifs ausarbeitet.

Gerichts erwartet hätte: Er hat nicht die Frage gestellt, ob die betreffenden Abkommen auch für sämtliche Mitgliedstaaten in Kraft sind. Das war jedenfalls für eines der erwähnten Abkommen, nämlich die Europäische Menschenrechtskonvention, problematisch, solange die französische Republik diesem Abkommen noch nicht beigetreten war: Bekanntlich ist dieser Beitritt erst im Laufe des Jahres 1974 erfolgt.

Welchen Status die einzelnen Mitgliedstaaten im Rahmen dieser Abkommen auch haben, der Gerichtshof hat sich offenbar von der Auffassung leiten lassen, daß diese Verträge an sich, d. h. als Elemente des Völkerrechts, auch für das Gemeinschaftsrecht Relevanz haben können. Die so in Kauf genommene formalrechtliche Unsicherheit hat er dadurch abgefangen, daß er, und zwar zum ersten Mal in seinem *Nold*-Urteil vom 14. Mai 1974, zur Frage des internationalen Status der Menschenrechte ausgesagt hat, internationale Rechtsakte, an denen die Mitgliedstaaten »mitgewirkt« haben oder denen sie »beigetreten sind«, könnten »Hinweise geben, die im Rahmen des Gemeinschaftsrechts zu berücksichtigen« seien. Diese Methode der freien Verwertung von Elementen der Europäischen Menschenrechtskonvention hat er danach in einer ganzen Reihe von Urteilen praktiziert, um Lösungen, die im Rahmen des Gemeinschaftsrechts erarbeitet worden waren, auch völkerrechtlich zu untermauern. Nicht anders ist er bei der Verwertung von Abkommen der Internationalen Arbeitsorganisation verfahren, so im Fall des Abkommens Nr. 100 über Lohngleichheit weiblicher und männlicher Arbeitnehmer, des Abkommens Nr. 69 über Arbeitsvermittlung durch Private und schließlich des Abkommens Nr. 20 über die Nachtarbeit in den Bäckereien.

Es liegt auf der Hand, daß diese unformalistische Betrachtungsweise eine günstige Ausgangsbasis für die Einbeziehung völkerrechtlicher Abkommen in die Begründungen des Gerichtshofs geschaffen hat.

II. Welche innergemeinschaftliche Wirkung haben völkerrechtliche Abkommen?

Nachdem wir gesehen haben, welche völkerrechtlichen Abkommen der Gerichtshof als Bestandteile seiner juristischen Begründungen in Betracht zieht, bleibt zu untersuchen, welche innergemeinschaftlichen Wirkungen die auf diese Weise ins Gemeinschaftsrecht eingeführten Rechtselemente entfalten. Wir wollen hier die Urteile vorwegnehmen, in denen mehr grundsätzliche Stellungnahmen zum Ausdruck kommen, um dann im einzelnen zu sehen, in welchen Zusammenhängen und mit welcher Wirkung völkervertragsrechtliche Bestimmungen vom Gerichtshof in Betracht gezogen wurden. Wir werden dabei feststellen, daß diese Elemente ihren Effekt in sehr verschiedenartigen Perspektiven entfalten und daß man hier nicht von einer einheitlichen »Anwendbarkeit« sprechen kann.

1. Völkerrechtliche Abkommen als »integrierende Bestandteile« des Gemeinschaftsrechts

Die erste grundsätzliche Stellungnahme finden wir im *Haegeman*-Urteil vom 30. April 1974[13]. Hier stellte sich zum erstenmal die Frage, ob ein Abkommen der Gemeinschaft, in diesem Falle das Athener Abkommen zur Assoziierung Griechenlands, der Auslegungskompetenz des Gerichtshofs nach Art. 177 EWGV unterliegt. Der Gerichtshof sagt dazu folgendes:

> »Wie sich dem Beschluß vom 26. 9. 1961 entnehmen läßt, wurde das Athener Abkommen vom Rat gemäß Artikel 228 des Vertrags geschlossen. Dieses Abkommen stellt somit für die Gemeinschaft die Handlung eines Gemeinschaftsorgans im Sinne des Artikels 177 Absatz 1 b) dar. Die Bestimmungen des Abkommens bilden seit dessen Inkrafttreten einen integrierenden Bestandteil der Gemeinschaftsrechtsordnung. In dem durch diese Rechtsordnung gesteckten Rahmen ist der Gerichtshof zur Vorabentscheidung über die Auslegung dieses Abkommens befugt«.

In diesen paar Zeilen sind außerordentlich viele Aussagen zusammengepackt. Es sei daran erinnert, daß das Athener Abkommen zu den sogenannten »gemischten Abkommen« gehört, weil daran sowohl die Gemeinschaft als auch die Mitgliedstaaten beteiligt waren. Da das Abkommen durch die Gemeinschaft abgeschlossen wurde, ist es von seiten der Gemeinschaft als Organakt einer Institution anzusehen, genauer: als Akt des Ministerrats, der ja die Vertragsschließungskompetenz besitzt. Als solches unterliegt das Abkommen der Auslegungskompetenz des Gerichtshofs nach Art. 177. Auf Grund von Art. 228 des Vertrags ist das Abkommen zugleich für die Mitgliedstaaten bindend, und die Auslegung wird somit auch für den internen Bereich der Mitgliedstaaten maßgebend. Allerdings ist diese Auslegung nur in dem durch die Rechtsordnung der Gemeinschaft »gesteckten Rahmen«, d. h. gemeinschaftsintern wirksam; eine so erlangte Auslegung kann also dem beteiligten Drittstaat, in diesem Fall Griechenland, nicht entgegengehalten werden. Die für die Zukunft wichtigste Aussage aber findet sich in dem Satz: »Die Bestimmungen des Abkommens bilden seit dessen Inkrafttreten einen integrierenden Bestandteil der Gemeinschaftsrechtsordnung«. Es ist hier zunächst festzuhalten, daß der Gerichtshof nicht etwa auf die Förmlichkeiten des Abschlusses, sondern auf die Tatsache des Inkrafttretens des Abkommens abstellt. Seit diesem Zeitpunkt ist das Abkommen Teil der Gemeinschaftsrechtsordnung; seine Wirkungen bestimmen sich also nach Maßgabe dieser Rechtsordnung und nicht nach nationalen Maßstäben, wie das noch einmal

---

[13] Siehe dazu die Besprechungen von *Tizzano,* Foro Italiano 99 (1974), S. 311 f.; *R. Adam,* Rivista di diritto internazionale 58 (1975), S. 584–589; *R. Kovar,* Journal du droit international 103 (1976), S. 195–197; *K. Iliopoulos,* Revue hellénique de droit international 29 (1976), S. 193–211.

in dem wenig später ergangenen *Nederlandse Spoorwegen*-Urteil vom 9. November 1975 besonders unterstrichen wird. Diese Aussage zeigt, daß der Gerichtshof dem Gedanken einer unmittelbaren Anwendbarkeit völkerrechtlicher Abkommen zuneigt, sonst hätte er nicht den Ausdruck »integrierender Bestandteil« verwendet.

Das zweite Urteil mehr grundsätzlicher Art ist am 5. Februar 1976 in der Rechtssache *Bresciani* ergangen [14]. In dieser Rechtssache war von einem italienischen Gericht die Frage gestellt worden, ob eine Bestimmung des Jaunde-Abkommens über die Abschaffung von Abgaben zollgleicher Wirkung in dem Sinne unmittelbar gelte, daß es den Gemeinschaftsangehörigen das subjektive Recht verleihe, keine Abgaben zollgleicher Wirkung an einen Mitgliedstaat zu entrichten. Einmal mehr handelt es sich hier um ein sogenanntes »gemischtes Abkommen«, das sowohl von den Mitgliedstaaten als auch von der Gemeinschaft abgeschlossen worden war. Auf Grund von Art. 228 EWGV ist dieses Abkommen für die Gemeinschaft wie auch für die Mitgliedstaaten verbindlich. Trotz mangelnder Gegenseitigkeit im Rahmen des Jaunde-Abkommens ist der Gerichtshof der Ansicht, daß die in Frage stehende Bestimmung unmittelbar anwendbar ist; dieses Ungleichgewicht entspricht der besonderen Eigenart des Verhältnisses der Gemeinschaft zu den ihr assoziierten Staaten. Die in Frage stehende Bestimmung »ist inhaltlich bestimmt, und sie unterliegt weder einem stillschweigenden noch einem ausdrücklichen Vorbehalt seitens der Gemeinschaft«; sie begründet daher ein Recht der Gemeinschaftsangehörigen, sich vor Gericht darauf zu berufen. Zur Erhärtung dieser Schlußfolgerung weist der Gerichtshof darauf hin, daß das Jaunde-Abkommen, was die Abschaffung von Abgaben zollgleicher Wirkung anbelangt, ausdrücklich auf Art. 13 EWGV verweist. Aus der Begründung geht aber hervor, daß nicht diese Verweisung, sondern die inhaltliche Analyse der einschlägigen Bestimmung des Jaunde-Abkommens die wahre *ratio decidendi* darstellt.

2. Völkerrecht als Maßstab zur Prüfung der Gültigkeit von Organakten

Der Gerichtshof verwendet völkerrechtliche Abkommen als Maßstab der Prüfung der Gültigkeit von Organakten im Sinne von Art. 177 EWGV. Zu diesem Punkt gibt ebenfalls das bereits erwähnte *United Fruit Cy*-Urteil vom 12. Februar 1972 Aufschluß [15]. In diesem Fall war von einem niederländischen

---

[14] Siehe dazu die Besprechungen von X., The Journal of Business Law 1976, S. 299 f.; *G. White*, European Law Review 1 (1976), S. 402–406; *L. Citarella*, Rivista di diritto europeo 16 (1976), S. 248–251.

[15] Dieses Urteil hat zu einer Flut von Besprechungen Anlaß gegeben: *E. Millarg*, Europarecht 8 (1973), S. 145–155; *G. Meier*, AWD des Betriebsberaters 19 (1973), S. 376 f.; *S. A. Riesenfeld*, AJIL 67 (1973), S. 504–508; *P. J. G. Kapteyn*, Sociaal-Economische Wetgeving 21 (1973), S. 491–498, und The International Lawyer 8 (1974),

Gericht die Frage gestellt worden, ob die Prüfung der Gültigkeit von Handlungen der Organe der Gemeinschaft im Sinne von Art. 177 EWGV auch die Prüfung ihrer Gültigkeit nach Völkerrecht umfasse. Darauf antwortete der Gerichtshof wie folgt:

»Diese die Zuständigkeit des Gerichtshofs regelnde Vorschrift enthält keine Einschränkung hinsichtlich der Gründe, mit denen die Gültigkeit der genannten Handlungen in Frage gestellt werden könnte. Da diese Zuständigkeit sich somit auf sämtliche Ungültigkeitsgründe erstreckt, die diesen Handlungen anhaften können, muß der Gerichtshof auch prüfen, ob ihre Gültigkeit dadurch beeinträchtigt werden kann, daß sie einer Regel des Völkerrechts widersprechen«.

In dem genannten Rechtsfall ging es um die Gültigkeit handelspolitischer Schutzmaßnahmen, die an Hand von Art. XI des GATT-Abkommens beurteilt werden sollten. Die Antwort des Gerichtshofs ist aber der gestellten Frage entsprechend ganz allgemein: Die Gültigkeit von Organakten kann am Maßstab irgendeiner Regel des *Völkerrechts* geprüft werden; neben Völkervertragsrecht können hier also auch Entscheidungen internationaler Organisationen (wie sie ja gerade im GATT eine Rolle spielen) sowie Regeln des ungeschriebenen Völkerrechts in Betracht kommen. Diese Feststellung ist für die Systematik des Gemeinschaftsrechts sehr bedeutsam. Da die Gültigkeitsprüfung nach Art. 177 im systematischen Zusammenhang mit der Rechtmäßigkeitskontrolle nach Art. 173 und ganz allgemein im Zusammenhang mit der »Wahrung des Rechts« nach Art. 164 des Vetrags gesehen werden muß, macht diese Entscheidung des Gerichtshofs klar, daß auch das Völkerrecht in allen seinen Formen als integrierender Bestandteil der Rechtsordnung der Gemeinschaft und der gemeinschaftsrechtlichen »Gesetzmäßigkeit« angesehen werden muß, jedenfalls im Rahmen der auswärtigen Beziehungen.

Die Anwendung des im Fall *International Fruit Cy* entwickelten Grundsatzes führt allerdings in diesem Urteil selbst und später in dem gleichgelagerten *Schlüter*-Urteil vom 24. Oktober 1973 nicht zu dem Ergebnis, das man auf den ersten Blick aus den erwähnten Prämissen ableiten möchte (Bindung der Gemeinschaft an die GATT-Regeln und Prüfung der Organakte am Maßstab des Völkerrechts). Der Gerichtshof fügt nämlich eine weitere Bedingung hinzu: »Falls die Ungültigkeit vor einem staatlichen Gericht geltend gemacht wird, ist weiterhin erforderlich, daß diese Bestimmung ein Recht

---

S. 74–82; *A. Giardina,* Rivista di diritto internazionale e processuale 9 (1973), S. 582–600; *J. Rideau,* Cahiers de droit européen 9 (1973), S. 465–483; *H. G. Schermers,* Ars Aequi 23 (1974), S. 58–64, und Common Market Law Review 12 (1975), S. 77–90; *A. Del Vecchio,* Il diritto negli scambi internazionali 12 (1973), S. 274–278; *A. M.,* Rassegna dell'Avvocatura dello Stato 26 (1974), S. 414–418; *M. Waelbroeck,* Journal of World Trade Law 8 (1974), S. 614–623; *G. Scarangella Arpino,* Jus Gentium 9 (1975), S. 239–253; *R. Kovar,* Revue du Marché Commun 1974, S. 345–360. Siehe dazu auch den in Anm. 6 zit. Aufsatz von *Zuleeg.*

der Gemeinschaftsangehörigen begründen kann, sich vor Gericht auf sie zu berufen«, mit anderen Worten, es muß nachgewiesen sein, daß die in Frage stehende Abkommensbestimmung unmittelbar wirksam ist. Dies wird aber vom Gerichtshof im Fall des GATT verneint, jedenfalls was Art. XI und im Fall *Schlüter* Art. II anbelangt. Diese Schlußfolgerung zieht der Gerichtshof nicht aus einer Analyse der betreffenden Artikel, sondern ganz allgemein aus »Sinn, Aufbau und Wortlaut« des Abkommens, wobei er alle jene Bestimmungen hervorhebt, die das GATT als ein Forum handelspolitischer Verhandlungen auf zwischenstaatlicher Ebene mit mannigfaltigen Entpflichtungs- und Rücknahmemöglichkeiten charakterisieren.

Diese Stellungnahme des Gerichtshofs zur innergemeinschaftlichen Anwendbarkeit oder genauer: Nichtanwendbarkeit des GATT hat zu zahlreichen Kommentaren Anlaß gegeben, wobei auch kritische Stimmen nicht fehlen. Das ist nicht verwunderlich, wenn man bedenkt, daß bereits Generalanwalt Mayras in seinen Schlußanträgen darauf hingewiesen hatte, daß zu dieser Frage widersprüchliche Stellungnahmen der Gerichte mindestens zweier Mitgliedstaaten bestehen: In Italien werden die GATT-Bestimmungen auf Grund eines gesetzlich erfolgten Anwendungsbefehls als unmittelbar anwendbar angesehen, während die deutschen Gerichte einen gegenteiligen Standpunkt eingenommen haben[16]. Eine Wahl war hier unausweichlich. *Kapteyn, Schermers* und *Waelbroeck* haben gerade an diesem Aspekt der *International Fruit Cy-* und *Schlüter*-Urteile lebhafte Kritik geübt. Sie werfen dem Gerichtshof die Anwendung eines »Doppelstandards« vor, je nachdem, ob es sich um das interne Gemeinschaftsrecht oder um völkerrechtliche Regeln handelt, und bedauern, daß durch diese Stellungsnahme die Regeln des GATT in ihrer Wirksamkeit unnötigerweise geschwächt worden seien. *Waelbroeck* bedauert, daß der Gerichtshof nicht den nationalen Gerichten gefolgt sei, die eine großzügigere Haltung in bezug auf die unmittelbare Wirkung der GATT-Bestimmungen eingenommen haben, und spricht die Hoffnung aus, der Gerichtshof möge auf diese Urteile zurückkommen.

Besonders hervorzuheben ist die Kritik des italienischen Autors *Scarangella Arpino,* der dem Gerichtshof vorhält, er habe das Problem der unmittelbaren Anwendbarkeit der GATT-Bestimmungen von sich aus aufgeworfen, ohne daß diese Frage von den vorlegenden Gerichten überhaupt gestellt war; überdies habe er sich dazu auf Grund von globalen und sachfremden Erwägungen ausgesprochen, die mit der gestellten Frage – nämlich die Tragweite der Art. II und XI des GATT-Abkommens – nichts zu tun hätten. Mit dieser kritischen Bemerkung ist das zentrale Methodenproblem angesprochen, das sich aus der gesamten hier besprochenen Rechtsprechung ergibt. Wie wir in der Folge noch besser sehen werden, ist der Gerichtshof im übrigen tatsächlich stets auf analytische Weise verfahren, d.h. er hat die Frage der innerge-

[16] Zur Rechtsprechung in den Vereinigten Staaten siehe die soeben zit. Studien von *Waelbroeck* und *Zuleeg.*

meinschaftlichen Wirkung völkerrechtlicher Abkommen nicht aus apriorischen Erwägungen, sondern aus einer Substanzanalyse der einschlägigen Einzelbestimmungen beantwortet. Bereits in dem wenig später ergangenen *Nederlandse Spoorwegen*-Urteil vom 19. November 1975, das sich durch seine feste Federführung auszeichnet, hebt er mit verschiedenen Redewendungen den verbindlichen Charakter der GATT-Bestimmungen hervor, ohne noch einmal auf die Frage der unmittelbaren Anwendbarkeit einzugehen. In dem jüngsten *Dürbeck*-Urteil vom 5. Mai 1981, das sich genau wie *International Fruit Cy* auf handelspolitische Abwehrmaßnahmen bezieht, wird die Argumentation des früheren Urteils nicht mehr erwähnt; hier begnügt sich der Gerichtshof mit der schlichten Feststellung, die betreffenden Abwehrmaßnahmen hätten im Rahmen der GATT-Gremien zu einem Gedankenaustausch Anlaß gegeben und seien dort in der Sache nicht beanstandet worden.

Der Vollständigkeit halber seien hier noch die Urteile *Schroeder* und *Wünsche* vom 6. Februar bzw. 12. Juli 1973 erwähnt, in denen es um die Vereinbarkeit einer von der Gemeinschaft verfügten Importbeschränkung für landwirtschaftliche Produkte aus Griechenland mit den Bestimmungen des Athener Assoziierungsvertrags geht; das *Nold*-Urteil vom 14. Mai 1975 zur Vereinbarkeit einer neuen Handelsregelung für Kohle mit einer Bestimmung der Europäischen Menschenrechtskonvention über das Eigentumsrecht; das *Panasonic*-Urteil vom 26. Juni 1980 zur Vereinbarkeit von Kontrollmaßnahmen im Rahmen des europäischen Wettbewerbsrechts mit dem Recht auf Schutz der Privatsphäre und dem Recht auf Schutz des Briefgeheimnisses, so wie diese Rechte durch die Europäische Menschenrechtskonvention gewährleistet sind. Im *Arbelaiz-Emazabel*-Urteil vom 8. Dezember 1981, das sich auf die Rechte spanischer Fischer in den seit der Ausdehnung der Fischereigrenzen auf 200 Seemeilen der Kontrolle der Gemeinschaft unterliegenden Gewässern bezieht, geht es um die Vereinbarkeit der von der Gemeinschaft eingeführten Fangbeschränkungen mit gewissen Fischerei-Abkommen; in ihm wird zur Begründung ganz allgemein auf die »völkerrechtliche Entwicklung« auf dem Gebiet der Hochseefischerei Bezug genommen. In diesen Fällen bestätigt der Gerichtshof die Gültigkeit der angefochtenen gemeinschaftsrechtlichen Maßnahmen.

3. Völkerrechtliche Abkommen als Maßstab zur Prüfung der Vereinbarkeit nationaler Gesetze mit dem Gemeinschaftsrecht

Gelegentlich werden die Maßstäbe des Völkerrechts vom Gerichtshof angewandt, um die Kompatibilität nationaler Gesetzgebungsakte mit Gemeinschaftsrecht zu prüfen. Die profilierteste Entscheidung in dieser Hinsicht ist zweifellos das Urteil *Kommission ./. Niederlande* vom 12. Juli 1977. Hier wies der Gerichtshof eine Vertragsverletzungsklage wegen der Erhebung einer Abgabe bei der Durchführung sanitärer Kontrollen in den Niederlanden bei der Ausfuhr von pflanzlichen Erzeugnissen unter Berufung auf ein völkerrechtliches Abkommen ab. Es stellte sich nämlich heraus, daß diese Kontrollen im Rahmen des FAO-Abkommens über den Pflanzenschutz durchgeführt wurden; der Gerichtshof stellte fest, daß sämtliche Mitgliedstaaten dem System dieses Abkommens angehören und daß dieses daher im Rahmen der Gemeinschaft eine gemeinsame Regel zur Förderung des internationalen Handels mit Pflanzen darstelle.

In einer ganzen Anzahl anderer Urteile, die im Rahmen von Vorlageverfahren ergangen sind, dient die Berufung auf völkerrechtliche Abkommen dazu, die Konformität nationaler Gesetzgebungsakte oder Praktiken mit dem Gemeinschaftsrecht oder mit den von der Gemeinschaft eingegangenen Verpflichtungen zu beurteilen. Diese Urteile werden in chronologischer Reihenfolge erwähnt; einige davon sind sehr deutlich, bei anderen taucht das Argument nur am Rande oder als eine Art *obiter dictum* auf.

In dem bereits erwähnten *Haegeman*-Urteil vom 30. April 1974 wird die Erhebung einer Importabgabe auf griechischen Wein durch Belgien als mit dem Assoziierungsabkommen vereinbar bezeichnet. Im *Charmasson*-Urteil vom 10. Dezember 1974 hingegen wird eine französische Importregelung für Bananen als unvereinbar mit einem internationalen Einverständnis bezeichnet, das sich aus parallelen Entscheidungen des im Rahmen des Jaunde-Abkommens gebildeten Assoziierungsrates und des Rates der Gemeinschaft ergeben hat. Im *Rutili*-Urteil vom 28. Oktober 1975 wird eine fremdenpolizeiliche Maßnahme der französischen Verwaltung als mit einem Maßstab unvereinbar erklärt, der in der Europäischen Menschenrechtskonvention mehrfach verwendet wird: nämlich dem Maßstab, was mit den Erfordernissen einer »demokratischen Gesellschaft« vereinbar ist. Im *Kramer*-Urteil vom 14. Juli 1976 wird eine niederländische Schutzmaßnahme auf dem Gebiet der Hochseefischerei als mit dem Gemeinschaftsrecht vereinbar erklärt, weil sie auf Grund einer im Rahmen der Fischereikonvention für den Nord-Atlantik getroffenen Entscheidung in Kraft gesetzt worden war. Im *Van Ameyde*-Urteil vom 9. Juni 1977 wird ausgeführt, daß das von der Straßburger Konvention über die Pflichtversicherung von Kraftfahrzeugen eingeführte System der »grünen Karte« vom Gemeinschaftsrecht anerkannt wird und daß daher die in einem Mitgliedstaat in diesem Rahmen getroffenen Bestimmungen im Prinzip nicht mit dem EWG-Vertrag in Widerspruch stehen. In den Urteilen *Cayrol* (vom 30. November 1977) und *Rivoira* (vom 28. März 1979) werden französische für den Import spanischer Trauben geltende Beschränkungen unter dem Gesichtspunkt des mit Spanien abgeschlossenen Handelsabkommens untersucht und als mit diesem konform erklärt. In der Rechtssache *Van Wesemael* hatte sich die belgische Regierung, um die Wirkung einer in Frankreich erteilten Lizenz zur Stellenvermittlung im Theater- und Musikgewerbe auszuschließen, auf das Abkommen Nr. 96 der Internationalen Arbeitsorganisation über Stellenvermittlung berufen; auf Grund einer Analyse der einschlägigen Bestimmung dieser Konvention weist der Gerichtshof in seinem Urteil vom 8. Januar 1979 diesen Einwand als nicht stichhaltig zurück. Im *Henn & Darby*-

Urteil vom 14. Dezember 1979 wird im Zusammenhang mit einem Importverbot für unzüchtige Filme und Magazine in England ausgeführt, daß die Anwendung des Genfer Abkommens zur Bekämpfung der Verbreitung unzüchtiger Veröffentlichungen und der einschlägigen Bestimmungen des Weltpostvereins den Bestimmungen des Gemeinschaftsrechts über den freien Warenverkehr nicht widerspricht. Im *Pecastaing*-Urteil vom 5. März 1980 wird die Frage gestreift, ob die Bestimmungen der belgischen Gesetzgebungen über das Klagerecht von Ausländern im Fall von Ausweisungen der Europäischen Menschenrechtskonvention entsprechen. Im *Chatain*-Urteil vom 24. April 1980 wird festgehalten, daß die Anwendung besonders strenger Strafbestimmungen des Zollrechts eines Mitgliedstaats, in diesem Falle Frankreichs, eine gegen das Freihandelsabkommen mit der Schweiz verstoßende »Maßnahme gleicher Wirkung« wie ein Importverbot darstellen könnte. Im *Carciati*-Urteil vom 9. Oktober 1980 wird eine italienische sich auf vorübergehenden Import von Kraftfahrzeugen beziehende Maßnahme, die im Rahmen des New Yorker Abkommens getroffen worden war, als mit dem Gemeinschaftsrecht vereinbar erklärt. Im *Oebel*-Urteil vom 4. Juli 1981 wird das in der Bundesrepublik angewandte Nachtbackverbot als mit dem Gemeinschaftsrecht vereinbar bezeichnet; dabei wird auf die Übereinstimmung dieser Gesetzgebung mit dem Abkommen Nr. 20 der IAO hingewiesen.

4. Völkerrechtliche Abkommen als Auslegungshilfe

In anderen Fällen werden Bestimmungen völkerrechtlicher Verträge vom Gerichtshof als Auslegungshilfe zur Verdeutlichung oder Fortentwicklung des Gemeinschaftsrechts verwandt. Hier ist vor allem auf die reichhaltige Rechtsprechung zur innergemeinschaftlichen Geltung der Tarifavise des im Rahmen des Brüsseler Abkommens über die Zollnomenklatur eingerichteten Ausschusses zu verweisen. Anfangs war der Gerichtshof bezüglich der Rezeption dieser Avise etwas zögernd. So sagt er in dem ersten zu dieser Frage am 8. Dezember 1970 auf Vorlage des Bundesfinanzhofs ergangenen *Bakels*-Urteil, daß der gemeinsame Zolltarif auf dem »Brüsseler Zolltarifschema« begründet ist und daß deshalb die unter Aufsicht des Rates für die Zusammenarbeit auf dem Gebiet des Zollwesens ergangenen Erläuterungen und Avise, solange noch keine gemeinschaftsrechtlichen Erläuterungen ergangen sind, als ein »maßgebliches Erkenntnismittel« für die Auslegung des gemeinsamen Zolltarifs anzusehen sind. Später hat der Gerichtshof, wie bereits erwähnt, im *Nederlandse Spoorwegen*-Urteil vom 19. November 1975 darauf hingewiesen, daß die Gemeinschaft die Verpflichtungen aus dem Brüsseler Abkommen übernommen hat und an diese Verpflichtungen gebunden ist. Daher haben die im Rahmen dieses Abkommens ergangenen Tarifavise auch für die Gemeinschaft eine Relevanz, die mit folgenden Worten beschrieben wird:

> »Derartige Tarifavise sind zwar für die vertragsschließenden Teile gewiß nicht verbindlich, stellen jedoch Auslegungshilfen dar, denen umso grö-

ßeres Gewicht zukommt, als sie von einer Stelle herrühren, die von den vertragsschließenden Teilen damit beauftragt ist, die einheitliche Auslegung und Anwendung des Tarifschemas sicherzustellen«[17].

In den zahlreichen auf den gemeinsamen Zolltarif und seine Positionen sich beziehenden Auslegungsfragen, die dem Gerichtshof laufend unterbreitet wurden, werden immer wieder die Brüsseler Erläuterungen und Tarifavise ins Feld geführt, und sie werden auch vom Gerichtshof beachtet. Damit sind diese zu einer Art *de facto*-Rechtsquelle für die Auslegung des gemeinsamen Zolltarifs geworden. Die einzige Entscheidung, in der hiervon Abstand genommen wird, ist ein Urteil der ersten Kammer des Gerichtshofs (*Chem-Tec* vom 11. Juli 1980); in ihm wird ein Vorbehalt aktualisiert, der sich bereits in den Urteilen *Bakels* und *Nederlandse Spoorwegen* findet, insofern nämlich als die Tarifavise von Natur aus nicht verbindlich sind und deshalb vom Gerichtshof bei offensichtlichem Widerspruch mit dem Sinn der Nomenklatur abgelehnt werden können.

Auch in anderen Sparten der Rechtsprechung spielen internationale Abkommen eine Rolle als Auslegungshilfe. So werden im Urteil *Markus & Walsh* vom 15. Oktober 1969 zur Auslegung einer Agrarverordnung, die eine Verweisung auf das GATT enthielt, der Begriff der »Konsolidierung« in der Praxis des GATT und eine Fußnote zur bekannten Konzessionsliste XL herangezogen. In einem ganz anderen Zusammenhang wird durch das *Defrenne*-Urteil vom 15. Juli 1978 der Ausdruck »gleiche Arbeit« in Art. 119 EWGV im Sinne des Abkommens Nr. 100 der IAO – erweiternd – ausgelegt.

5. Völkerrechtliche Abkommen als Grundlage von subjektiven Rechten

Erscheinen in den meisten bisher besprochenen Fällen die Bestimmungen völkerrechtlicher Abkommen als mittelbare Stadien einer juristischen Argumentation, als Auslegungshilfen, wenn nicht gar als Lückenbüßer, so sind aus der Rechtsprechung des Gerichtshofs solche Urteile besonders hervorzuheben, in

---

[17] Die Frage stellt sich, ob mit dieser Formel ein Vorbehalt erledigt ist, der sich im *Bakels*-Urteil findet und demzufolge die Tarifavise nur »bei Fehlen einschlägiger von der Gemeinschaft erlassener Bestimmungen« gelten. In der Tat trägt jede gemeinschaftseigene Auslegung die Gefahr einer uneinheitlichen Anwendung der weltweit rezipierten Nomenklatur in sich. Man muß dabei bedenken, daß diese Nomenklatur nicht nur für die Anwendung des Zolls, sondern auch für die Außenhandelsstatistik relevant ist. Aus diesen Erwägungen müßte sich im Konfliktsfall ein Vorrang der »Brüsseler Erläuterungen« vor den gemeinschaftsinternen Erläuterungen ergeben. Dieses Problem ist im *Cleton*-Urteil vom 4. 10. 1979 angesprochen, aber nicht gelöst worden. Immerhin wird hier darauf hingewiesen, daß in den Vorbemerkungen zu den Erläuterungen des Zolltarifs der Gemeinschaft ausdrücklich festgestellt wird, »daß diese die Erläuterungen des Rates für die Zusammenarbeit auf dem Gebiet des Zollwesens nicht ersetzen, sondern lediglich vervollständigen sollen« (Nr. 13); damit ist aber praktisch der Vorrang der »Brüsseler Erläuterungen« anerkannt worden.

denen von Rechten und Pflichten der Einzelnen gesprochen wird, sei es im Rahmen des Gemeinschaftsrechts selbst oder sei es, und das ist noch bemerkenswerter, in Form von Ansprüchen oder Pflichten den Mitgliedstaaten gegenüber. In dieser Perspektive zeigt sich nämlich das, was man gewöhnlich als »unmittelbare Wirksamkeit« völkerrechtlicher Abkommen bezeichnet, sei es im Rahmen des Gemeinschaftsrechts selbst oder sei es als Ausfluß des Gemeinschaftsrechts im Rahmen des nationalen Rechts der Mitgliedstaaten.

Hier ist indes eine Vorbemerkung notwendig. Man spricht gewöhnlich von »unmittelbarer Wirkung« nur in solchen Fällen, in denen die Bezugnahme auf ein völkerrechtliches Abkommen zur Anerkennung subjektiver Rechte führt. Meiner Ansicht nach ist das eine zu enge Auffassung. Auch die Verneinung solcher Rechte impliziert die unmittelbare Wirksamkeit, wenn sie auf Grund einer Auseinandersetzung *mit dem materiellen Inhalt* eines völkerrechtlichen Vertrages erfolgt. Denn das Eingehen auf die Substanz einer solchen Bestimmung zeigt immerhin die Bereitschaft des Richters, Rechte anzuerkennen, falls diese sich aus der sachlichen Analyse ergeben. Dies um so mehr, als im Prozeß dem Recht jeder der Parteien eine irgendwie geartete Pflicht auf Leistung oder Unterlassung der anderen Partei gegenübersteht; anders gesagt: Der positiven Rechtsstellung des einen entspricht eine negative Rechtsstellung des anderen und umgekehrt.

Man muß also unter diesem Blickwinkel alle Entscheidungen ins Auge fassen, in denen der Gerichtshof auf völkerrechtliche Abkommen eingegangen ist, um die Rechtsposition Einzelner zu bestimmen, auch wenn die Entscheidung für den Rechtssuchenden im Ergebnis negativ ausgefallen ist. Die im folgenden aufgeführten Entscheidungen sind wiederum chronologisch geordnet: So kann man am besten die praktische Entwicklung des Gedankens einer »unmittelbaren Wirkung« völkerrechtlicher Abkommen erkennen.

Hier sind zunächst wieder die Urteile zur Auslegung des gemeinsamen Zolltarifs zu erwähnen, da es hier jeweils um sehr konkret bestimmte Positionen einzelner Importeure geht, deren Rechte und Pflichten gegenüber der Zollverwaltung bestimmt werden. Zum selben Ergebnis führt das *Interfood*-Urteil vom 26. April 1972, in dem auf ein im Rahmen des GATT abgeschlossenes Zollabkommen Bezug genommen wird, um die Einordnung einer Ware in den gemeinsamen Zolltarif zu bestimmen. In den Urteilen vom 7. Februar 1973 (*Schroeder*) und 12. Juli 1973 (*Wünsche*) geht es um die Kompatibilität von Einfuhrbeschränkungen der Gemeinschaft gültig für Tomatenkonzentrat und Erzeugnisse aus der Verarbeitung von Obst und Gemüse mit einer Bestimmung des Assoziierungsabkommens mit Griechenland; da diese Beschränkungen als durch das Abkommen gerechtfertigt erscheinen, ergeben sich keine Rechte der Importeure. Auf derselben Linie liegt das *Haegeman*-Urteil vom 30. April 1974, in dem die Erhebung einer Abgabe auf den Import von griechischen Weinen in Belgien gerechtfertigt wird. Die Berufung der *Firma Nold* auf die Europäische Menschenrechtskonvention, unternommen um eine neue Handelsregelung auf dem Bereich des Kohlesektors zu

Fall zu bringen, wird im Urteil vom 14. Mai 1974 als nicht stichhaltig bezeichnet. Hingegen verhilft die Berufung auf eine Entscheidung des Assoziierungsrates im Rahmen des Jaunde-Abkommens im bereits erwähnten *Charmasson*-Urteil vom 10. Dezember 1974 dem Kläger dazu, den freien Import von Bananen in Frankreich durchzusetzen. Das ebenfalls schon erwähnte *Bresciani*-Urteil vom 5. Februar 1975 erkennt ausdrücklich das Recht Einzelner an, sich auf die Bestimmung des Jaunde-Abkommens zur Eliminierung von Abgaben gleicher Wirkung wie Zollrechte zu berufen, in diesem Fall, um einer von Italien erhobenen Importabgabe zu entgehen. Im Falle *Rutili* wird durch Urteil vom 28. Oktober 1975 das Recht des freien Aufenthalts des Klägers in Frankreich unter anderem auf eine Bestimmung der Europäischen Menschenrechtskonvention gestützt. Im Falle *Kramer* wird die Anwendung nationaler Maßnahmen zum Schutz der Fischereibestände durch die Niederlande und damit die Rechtmäßigkeit eines Strafverfahrens gegen Rechtsbrecher auf Grund des Abkommens über die Fischerei im Nord-Ost-Atlantik bestätigt. Im *Prais*-Urteil vom 27. Oktober 1976 wird im Rahmen der Rechtsprechung über Beamtenfragen die Achtung der religiösen Rechte der europäischen Beamten auf Grund der Europäischen Menschenrechtskonvention als eine Selbstverständlichkeit bejaht. In den Urteilen *Cayrol* und *Rivoira* vom 30. November 1977 bzw. 28. März 1979 wird die Rechtmäßigkeit von französischen Importbeschränkungen auf spanische Trauben zu Lasten der Importeure gerechtfertigt. Im *Defrenne*-Urteil vom 5. Juli 1978 wird der Begriff der »gleichen Arbeit« zur Definierung der Rechte weiblicher Arbeitnehmer aus einem Abkommen der Internationalen Arbeitsorganisation in das Gemeinschaftsrecht übernommen. Im *Hauer*-Urteil vom 13. Dezember 1979 wird das erste Protokoll zur Europäischen Menschenrechtskonvention über den Schutz des Privateigentums zur Aufzeigung der Grenzen des Eigentumsrechts, also zu Lasten der Klägerin, angeführt. Durch das *Chatain*-Urteil vom 24. April 1980 wird es dem im Hauptverfahren Angeklagten möglich gemacht, sich auf eine Bestimmung des Freihandelsabkommens mit der Schweiz zu berufen, um besonders schwerwiegenden Strafbestimmungen der französischen Zollgesetzgebung zu entgehen. Im *Panasonic*-Urteil vom 26. Juni 1980 verweist der Gerichtshof auf die Grenzen, die sich im Rahmen der Europäischen Menschenrechtskonvention in bezug auf den Schutz der Privatsphäre und des Briefgeheimnisses aus Gesichtspunkten des öffentlichen Interesses ergeben, um die Kontrollbefugnisse der Kommission einem Unternehmen gegenüber auf dem Gebiet des Wettbewerbsrechts sicherzustellen. Im Urteil vom 8. Dezember 1981 (*Arbelaiz-Emazabel u. a.*) werden aus dem Londoner Fischereiabkommen Vorbehalte angeführt, die es möglich machen, spanischen Fischern die gemeinschaftlichen Fischereibeschränkungen aufzuerlegen. Im *Polydor*-Urteil vom 9. Februar 1982 legt der Gerichtshof die Bestimmungen des Freihandelsabkommens mit Portugal über die Eliminierung der Importbeschränkungen im Zusammenhang mit der Wahrung gewerblicher Schutzrechte dahin aus, daß diese Bestimmungen dem Inhaber des Schutzrechts im Vereinigten Königreich und nicht dem Importeur von in Portugal vermarkteten und dort aufgekauften Tonträgern zugute kommen.

Wie ersichtlich ist all diesen Fällen gemeinsam, daß die Bestimmungen völkerrechtlicher Abkommen in dem einen oder anderen Sinne angewendet werden, um individuelle Rechtspositionen zugunsten oder zu Lasten Einzelner zu definieren.

III. Welches ist die rechtstheoretische Stellung des Gerichtshofs?

Durch seine Rechtsprechung zur Anwendung des Gemeinschaftsrechts hat sich der Luxemburger Gerichtshof einen Namen als konsequenter Vertreter des rechtlichen »Monismus« gemacht[18]. Diese Grundauffassung drückt sich in den zahlreichen Aussagen zur bindenden Wirkung der Verträge, zur Irreversibilität der von den Mitgliedstaaten eingegangenen Verpflichtungen, zur sogenannten »Drittwirkung« des Gemeinschaftsrechts gegenüber Privaten, zur unmittelbaren Anwendbarkeit und zum Vorrang des Gemeinschaftsrechts aus[19]. Die Frage ist, ob solche Auffassungen sich auch auf die auswärtigen Beziehungen übertragen lassen. Im Gegensatz zu den Stellungnahmen zur Wirkungsweise des internen Gemeinschaftsrechts, wo die wesentlichen Akzente seit Beginn der 60er Jahre eindeutig gesetzt wurden, ist der Gerichtshof in der Tat, wie bereits eingangs bemerkt, an das Problem der inner-

---

[18] Die Anwendung dieses Prädikats mag verwundern: Ist nicht im Gemeinschaftsrecht die Alternative Monismus/Dualismus überwunden? Der Wahl dieser Terminologie liegt eine rechtstheoretische Auffassung zugrunde. Meiner Ansicht nach stellt sich diese Alternative wo immer autonome Rechtssysteme in Berührung kommen: ob Völkerrecht und Landesrecht; ob nationale Rechtssysteme unter sich; ob Staatsrecht und Kirchenrecht, ob Bundesrecht und Staatsrecht; ob Gemeinschaftsrecht und nationales Recht ... Überall stellt sich die Frage, ob solche Rechtsordnungen als zusammenhängende und gegenseitig geöffnete oder als unzusammenhängende und geschlossene Systeme angesehen werden; das erste ist »Monismus«, das zweite »Dualismus«. Diese Zusammenschau zeigt, daß es nicht möglich ist, hier Monist und dort Dualist zu sein. Dieses Problem stellt sich also auch im Rahmen des Gemeinschaftsrechts; hier ist der Standpunkt »getrennter« Rechtsordnungen längst überwunden zugunsten eines »kooperativen« Verhältnisses mit Vorrang des Gemeinschaftsrechts. Siehe dazu meinen Schlußvortrag auf dem V. Internationalen Kongress für Europarecht, Berlin 1970, Das Zusammenwirken der Gemeinschaftsrechtsordnung mit den nationalen Rechtsordnungen (Kölner Schriften zum Europarecht, Bd. 13) *in fine.*

[19] Ich habe die diesbezügliche Rechtsprechung eingehend in meiner Studie: Aspectos judiciales del »acervo communitario«, Revista de instituciones europeas 8 (1981), S. 331–366, französische Fassung unter dem Titel: Aspects judiciaires de l'»acquis communautaire«, Revue trimestrielle de droit européen 16 (1981), S. 617–651, dargestellt.

gemeinschaftlichen Wirkung völkerrechtlicher Bindungen nur sehr behutsam herangegangen.

Ich möchte zunächst versuchen, diese Rechtsprechung in methodischer Hinsicht auf ihre Ergebnisse zu untersuchen. Man möge mir erlauben, daran einige persönliche Erwägungen anzuschließen, die außerhalb des Bereichs richterlicher Motivationen liegen.

1. Pragmatischer Monismus als Grundhaltung

Wie sich aus der Betrachtung der angeführten Rechtsprechung ergibt, hat der Gerichtshof, von wenigen Ausnahmen abgesehen, die Frage der Wirkung von völkerrechtlichen Abkommen innerhalb des Gemeinschaftsbereichs auf pragmatische Weise gelöst. Ohne eine dogmatische Vorfrage, etwa in bezug auf die »unmittelbare Wirksamkeit« der in Betracht kommenden völkerrechtlichen Verträge aufzuwerfen, ist er jeweils – mit Ausnahme der oben erwähnten Urteile *International Fruit Cy* und *Schlüter* – unvermittelt auf die Frage eingegangen, ob die vor ihm behandelten Abkommen für die Gemeinschaft oder in der Gemeinschaft oder für ihre Mitgliedstaaten in Kraft sind und ob sie für die Lösung des gestellten Problems relevant sind. Wie wir gesehen haben, wird zu der Frage, ob derartige Abkommen in Kraft stehen, ein sehr flexibler Maßstab angewendet: insofern als es für die eigenen Abkommen nicht auf die Rechtsnatur des Abschlußaktes ankommt, insofern auch als die Übernahme völkerrechtlicher Rechte und Pflichten durch Sukzession auf Grund der Kompetenzübertragung auf die Gemeinschaft formlos angenommen wird, insofern schließlich als die bindende Wirkung völkerrechtlicher Verpflichtungen sogar in einem erweiterten Bereich angenommen wird, wo eine solche Substitutionswirkung nicht vorliegen kann.

Eng damit verflochten ist die Frage nach der Relevanz völkerrechtlicher Verpflichtungen in materiellrechtlicher Hinsicht. Daß der Gerichtshof die völkerrechtlichen Verträge nicht unbesehen hinnimmt, erhellt daraus, daß in einer ganzen Anzahl von Fällen die von den Parteien ins Feld geführten Abkommen, gegebenenfalls nach vertiefter Analyse ihres Inhalts, als zur Lösung der gestellten Frage nicht relevant ausgeklammert wurden. Man kann dafür eine ganze Reihe von Urteilen anführen. So das *Lensing*-Urteil vom 11. Februar 1973, in dem der Gerichtshof schlicht feststellt, daß Waren aus Guinea nicht in den Genuß der Vorteile des Jaunde-Abkommens kommen, und zwar einfach deswegen, weil Guinea diesem Abkommen niemals beigetreten ist. So in den Urteilen *EMI* vom 15. Juni 1976 und *Bouhelier* vom 1. Oktober 1979, aus denen hervorgeht, daß die angerufenen Handels- oder Freihandelsabkommen mit Spanien, Österreich, Schweden und der Schweiz zur Sache nichts zu bestellen hatten. Ähnlich in den Urteilen *Pecastaing* vom 5. März 1980 und *Heinz van Landewyck* vom 29. Oktober 1980, wo erklärt wird, warum die Europäische Menschenrechtskonvention für die gestellte Problematik unergiebig ist. Besonders interessant ist in dieser Hinsicht das *Razanatsimba*-Urteil vom 24. November 1977, wo ausgeführt wird, daß das Lomé-Abkommen einem Staatsangehörigen aus Madagaskar kein Anrecht in bezug auf die Niederlassung als Rechtsanwalt in Frankreich verleiht; diese negative Auslegung folgt aus einer eingehenden Auseinandersetzung mit der auf das Niederlassungsrecht sich beziehenden Klausel des Abkommens. Eine alternative Lösung hätte ja darin bestehen können, das Abkommen in Bausch und Bogen als »nicht unmittelbar anwendbar« zu erklären, womit sich der Gerichtshof weitere Bemühungen zur Ausdeutung der betreffenden Abkommensvorschrift erspart hätte.

Zu dem eben Gesagten ist anzumerken, daß der Gerichtshof nur in den wenigsten Fällen die Frage nach der Anwendbarkeit und der Relevanz von Abkommensbestimmungen ausdrücklich gestellt und erwogen hat. In den meisten Fällen wird dies als selbstverständlich vorausgesetzt. Ein beredtes Beispiel für diese Verfahrensweise ist das *Polydor*-Urteil vom 9. Februar 1982. Hier hatte der London Court of Appeal ausdrücklich die Frage nach der unmittelbaren Anwendbarkeit des Freihandelsabkommens mit Portugal gestellt; die Parteien hatten zu dieser Frage längere Ausführungen gemacht, und die am Verfahren beteiligten Regierungen hatten gerade auf diesen Aspekt das Schwergewicht gelegt. Auch die Kommission hatte nach anfänglichem Zögern dargelegt, die Frage der unmittelbaren Anwendbarkeit des Abkommens habe eine »grundlegende Bedeutung für die Stellung der Gemeinschaft in ihren internationalen Beziehungen, für ihre Autonomie, ihre Identität und die Möglichkeit, ihre Interessen und die Interessen der Mitgliedstaaten, für die sie Verantwortung übernommen habe, zu schützen«. Zudem werfe die Frage der unmittelbaren Wirkung »das Problem der richtigen Verteilung der Befugnisse und Zuständigkeiten zwischen dem Gesetzgeber und der Exekutive auf der einen und der Judikative auf der anderen Seite im Rahmen der Rechtsordnung der Gemeinschaft auf«. Aus diesem Grunde dürfe »der Begriff der unmittelbaren Wirkung, wie er im Gemeinschaftsrecht entwickelt worden sei, nicht als solcher auf die internationalen Beziehungen der Gemeinschaft übertragen werden«[20]. Ohne diese Auffassungen in seiner Begründung auch nur zu erwähnen, ist der Gerichtshof in die Analyse der Substanz des Freihandelsabkommens eingetreten, und er hat aus dieser Sicht das gestellte Problem – nämlich die Frage der Erschöpfung von gewerblichen Schutzrechten durch Vermarktung von geschützten Produkten in einem der Freihandelsstaaten – gelöst. Am Ende seines Urteils erklärt der Gerichtshof, daß damit die Frage nach der unmittelbaren Wirkung des Abkommens gegenstandslos geworden sei.

Auf dieselbe Weise war er schon in dem *Bresciani*-Urteil vom 5. Februar 1976 verfahren; hier war vom vorlegenden italienischen Gericht die Frage nach der unmittelbaren Anwendbarkeit einer Bestimmung über die Eliminierung zollgleicher Abgaben ausdrücklich gestellt worden. Auch hier ist der

---

[20] Diese Zitate sind dem Tatbestand des Urteils entnommen.

Gerichtshof unmittelbar in eine Analyse der angerufenen Bestimmung im Rahmen des Systems dieses Abkommens eingetreten; das entscheidende Argument wird aus der Substanz der in Frage stehenden Abkommensbestimmung hergeleitet, indem gesagt wird, diese Bestimmung unterliege »weder einem stillschweigenden noch einem ausdrücklichen Vorbehalt seitens der Gemeinschaft«, und sie begründe daher ein Recht der Einzelnen, sich vor Gericht darauf zu berufen.

Man sieht also, daß der Gerichtshof seine Stellungnahme über die Wirkung völkerrechtlicher Abkommen im Gemeinschaftsbereich – immer vorausgesetzt, daß diese Abkommen für die Gemeinschaft in Kraft und zur Sache relevant sind – meist aus einer materiellrechtlichen Analyse einschlägiger Bestimmungen gewonnen hat. Das pragmatische Vorgehen führte demnach zu Lösungen, die im Ergebnis der Haltung des Gerichtshofs zur Wirkungsweise des internen Gemeinschaftsrechts entsprechen. Nur in den oben besprochenen Urteilen *United Fruit Cy* und *Schlüter* hat der Gerichtshof, wie bereits gesagt, das gestellte Problem auf Grund der Vorfrage gelöst, ob die angerufenen Bestimmungen des GATT überhaupt unmittelbar anwendbar sind. Betrachtet man die Begründung näher, so stellt man fest, daß die verwendeten Argumente auf systematischen Erwägungen beruhren, die mit dem vorliegenden Sachproblem (Gültigkeit einer handelspolitischen Schutzmaßnahme bzw. Erhebung des Währungsausgleichs) direkt nichts zu tun hatten. Ob nicht die sachliche Analyse der angerufenen Bestimmungen zu genau demselben praktischen Ergebnis geführt hätte, bleibt damit eine offene Frage.

Bei einem Vergleich dieser beiden methodischen Linien der Rechtsprechung zeigt sich also, daß hier das pragmatische Vorgehen zu Lösungen führt, die man in der Tat als »monistisch« bezeichnen kann, während die apriorische Methode in eher »dualistische« Vorstellungen mündet.

2. Monismus, Dualismus, rechtliches Gehör und »bona fides«

Die klassische Unterscheidung zwischen Monismus und Dualismus wird heutzutage meist als überholt bezeichnet. Niemand will noch Monist oder Dualist sein, wir haben es nur noch mit »gemäßigten Monisten« oder »gemäßigten Dualisten« oder zwischen beiden Extremen vermittelnden Haltungen zu tun, wenn nicht gar dieser Theorienstreit kurzerhand als nutzlos abgetan wird[21]. Der Schreiber dieser Zeilen ist indes der Ansicht, daß diese Diskussion keineswegs überholt ist, mehr noch, daß die Entscheidung für die eine oder die andere dieser Auffassungen sogar in »gemäßigter« Form eine keineswegs wertneutrale Stellungnahme ist. Die Durchsetzungskraft des Völkerrechts hängt nämlich ganz entscheidend von der Auffassung all derjenigen zu diesem grundlegenden Dilemma ab, die mit der Anwendung dieses Rechts zu tun haben.

---

[21] Das ist z. B. die Schlußfolgerung von *Ch. Rousseau,* Droit international public, Bd. 1 (1970), S. 48.

Da indes über diese Begriffe einige Verwirrung herrscht, sei es mir erlaubt zu sagen, war hier mit Monismus und Dualismus gemeint ist. Für mich besteht die »monistische« Auffassung darin, daß ein völkerrechtlicher Vertrag, der gültig abgeschlossen wurde und somit völkerrechtlich in Kraft ist[22], auch im internen Bereich – der Gemeinschaft wie des Staates – anwendbar ist, und daß er damit auch vom Richter zu seiner Urteilsfindung herangezogen werden kann[23]. Dabei ist jeweils eine Vorfrage und eine Sachfrage zu prüfen. Die Vorfrage besteht darin, ob ein ins Feld geführter Vertrag tatsächlich in Kraft ist, d.h. ob er in gehöriger Form abgeschlossen und veröffentlicht wurde. Das letztere Erfordernis ergibt sich in der Gemeinschaft entweder aus einer analogen Anwendung von Art. 191 EWGV über die Veröffentlichung von Rechtsakten oder aus einem allgemeinen rechtsstaatlichen Gedanken derart, daß in unseren demokratischen Rechtssystemen keine unveröffentlichten oder unzugänglichen Vorschriften rechtens sein können. Die Prüfung der Sachfrage geht dahin, ob die angerufene Vertragsbestimmung inhaltlich gesehen näher bestimmte Rechtsfolgen im Hinblick auf eine gegebene Sachlage entfalten kann. Ob solche Rechtsfolgen bestehen und worin sie genau bestehen, kann nicht aus apriorischen Erwägungen, sondern nur aus einer vertieften Analyse jeder einzelnen Abkommensbestimmung nach den Regeln juristischer Auslegungskunst geschlossen werden.

Wenn ich recht verstehe, was demgegenüber mit der »dualistischen« Auffassung gemeint ist, so besteht sie darin, zwischen den Abschluß eines völkerrechtlichen Abkommens und seine Anwendung das Erfordernis eines irgendwie gearteten internen Anwendungsbefehls oder Umsetzungsaktes einzuschieben. Nach dieser Auffassung bedarf es also noch eines formellen oder materiellrechtlichen Vollzugsbefehls des Gesetzgebers. Da aber im System der Gemeinschaftsverfassung für den Erlaß eines solchen Befehls keine Vorsorge getroffen ist und ein solcher Befehl daher nicht ergehen kann und in der Tat nie ergeht, würden völkerrechtliche Abkommen als solche niemals anwendbar sein können. Man müßte allenfalls warten, bis ihre Konsequenzen auf gewisse Verordnungen übertragen würden; das geschieht in der Tat laufend beim Zolltarif und ist auch z.B. bei der Feststellung des Zollwerts und bei der Übernahme des Antidumpingcodex des GATT geschehen[24]. Aber es

---

[22] Also nach *A. Koller* (Anm. 6), in »Geltung« ist.

[23] Ich verstehe also Art. 228 des EWGV dahin, daß ein Abkommen, das vertragskonform abgeschlossen wurde und die Gemeinschaft »bindet«, damit alle Voraussetzungen erfüllt, um auch gemeinschaftsintern angewendet zu werden. Welches genau die Wirkungen eines solchen Abkommens sind, kann nur aus der Auslegung der einzelnen Bestimmungen, selbstverständlich mit Rücksichtnahme auf System und Zweck des gesamten Abkommens, erschlossen werden. Genau so ist der Gerichtshof in dem mehrmals zitierten *Polydor*-Urteil verfahren.

[24] Siehe dazu die Präambeln der Zollwert-VO 1224/80 vom 28. 5. 1980 (Amtsblatt der EG 1980, L 134, S. 1) und der Dumping-VO 3017/79 vom 20. 12. 1979, (*ibid.*, 1979, L 339, S. 1). Dasselbe Verfahren wurde in der VO 1798/75 (mehrmals

gibt auch andere Materien, wo Abkommensbestimmungen ihrer Natur nach darauf angelegt sind, als solche ihre Wirkung zu entfalten; das ist so bei sämtlichen Gleichbehandlungsgeboten oder Diskriminierungsverboten und ganz generell allen völkerrechtlichen Verbotsnormen. Solche Normen sind wegen ihres allgemeinen und vorsorglichen Charakters nicht geeignet, zu konkreten Durchführungsbestimmungen Anlaß zu geben; sie lassen sich erst im Fall ihrer Verletzung konkretisieren; derartige Normen werden daher durch das Erfordernis eines gesonderten Anwendungsbefehls – der in einer allgemeinen Form sinnlos ist – praktisch stillgelegt, soweit es um ihre interne Auswirkung geht. Hier kann nur die Anwendung der Abkommensbestimmungen selbst zu einem konkreten Resultat führen.

Betrachtet man nun diesen Theorienstreit und seine praktische Auswirkung vom Standpunkt der richterlichen Funktion, so kann man nicht verkennen, daß die dualistische Haltung in die Gefahrenzone der Rechtsverweigerung führt. Denn was heißt es, wenn ein Rechtssuchender sich vor Gericht auf einen in Kraft getretenen, d.h. völkerrechtlich gültig abgeschlossenen, also bindenden Vertrag beruft, die Berücksichtigung dieses Vertrages ihm aber wegen mangelnder »unmittelbarer Anwendbarkeit« verweigert wird? Was heißt es, wenn der Richter es unterläßt, die Bestimmungen eines solchen Vertrages daraufhin zu untersuchen, ob sich daraus gegebenenfalls Elemente zur Lösung des ihm vorliegenden Rechtsstreits ergeben? Das kommt nahe an Rechtsverweigerung heran, denn hier lehnt es der Richter ab, eine Rechtsregel anzuwenden, die sonder Zweifel in Kraft ist, die bindet, die relevant ist und die bei näherem Zusehen gegebenenfalls Wirkungen für ein strittiges Rechtsverhältnis erzeugen könnte[25].

Diese Erwägung zeigt, warum ein Gericht wie der Europäische Gerichtshof, das mit seiner Aufgabe, das Recht zu wahren und die Rechtmäßigkeit zu schützen, ernst macht, fast zwangsläufig zu monistischen Auffassungen und Haltungen kommen mußte: Wer nach dem Grundsatz des rechtlichen

---

geändert) über die Zollbefreiung von erzieherischen, wissenschaftlichen und kulturellen Gegenständen, in Durchführung des Florenzer UNESCO-Abkommens angewandt (siehe Amtsblatt 1975, L 184, S. 1).

[25] Besonders gravierend ist diese Haltung im Fall eines auf die staatsinterne Anwendung zugeschnittenen Abkommens wie der Europäischen Menschenrechtskonvention, deren Anwendung in immerhin sieben Vertragsstaaten (Dänemark, Irland, Island, Malta, Norwegen, Schweden, Vereinigtes Königreich, Zypern) wegen der herrschenden »dualistischen« Vorstellungen lahmgelegt ist, ohne daß die betreffenden Staaten etwas tun würden, um diesem Zustand ein Ende zu bereiten. Siehe dazu die übereinstimmenden Informationen und Wertungen von *P. Seidel,* Der Rang der europäischen Menschenrechtskonvention in den Mitgliedstaaten, Deutsches Verwaltungsblatt 1975, S. 747–757; *A. Drzemczewski,* The Domestic Status of the European Convention on Human Rights, Legal Issues of European Integration 1977, S. 1–85; *H. Golsong,* L'effet direct, ainsi que le rang en droit interne de la Convention européene des droits de l'homme, in: Les recours des individus devant les instances nationales en cas de violation du droit européen, Institut d'Etudes Européennes (1978), S. 59–83.

Gehörs lebt, wer sich nicht mit der Ausrede des *non liquet* seiner Richterpflicht entziehen will, der kann nicht die Augen vor gegebenen Lösungselementen nur deswegen verschließen, weil diese Elemente aus einer völkerrechtlichen Quelle fließen. Denselben Gedanken kann man vom Richter auf das Gemeinwesen selbst übertragen, d. h. auf den Staat oder, in unserem Fall, auf die Gemeinschaft. Von diesem mehr globalen Standpunkt aus gesehen, führen nämlich die dualistischen Auffassungen zu einer Verletzung der Grundregel aller völkerrechtlichen Beziehungen, nämlich der *bona fides*. Denn worin besteht letzten Endes der Effekt dualistischer Auffassungen? Darin, daß der Staat (oder hier die Gemeinschaft) zwar eine völkerrechtliche Bindung eingeht, sich zugleich aber die Entscheidung vorbehält, ob und wie er dieser Bindung in der internen Rechtsordnung eine angemessene Wirksamkeit gibt. Man bindet sich völkerrechtlich und verhält sich gegebenenfalls intern dennoch so, als sei man nicht gebunden – das ist mit dem Grundsatz von Treu und Glauben im völkerrechtlichen Verkehr nicht zu vereinbaren. Diese Erwägungen zeigen, daß die geistigen Väter des Dualismus mit ihren abstrakten Anschauungen über die Verschiedenheit der Normadressaten, über die Trennung der inneren und äußeren Rechtssphäre des Staates, über die Undurchdringlichkeit der staatlichen Rechtsordnungen[26] der Wirksamkeit des Völkerrechts einen schlechten Dienst erwiesen haben[27]. Die dualistische Theorie fördert geradewegs die Zwiespältigkeit des staatlichen Handelns. Als solche ist sie mit dem Grundsatz völkerrechtlicher *bona fides* unvereinbar.

In diesem Widerstreit rechtstheoretischer und im letzten rechtsphilosophischer Auffassungen, hat der Europäische Gerichtshof, soweit das Verhältnis zwischen Gemeinschaftsrecht und nationalem Recht in Frage steht, eine eindeutige Stellung zugunsten monistischer Denkweisen bezogen. Hier sind in der Tat die Postulate der *bona fides communitaria,* der Respektierung rechtsstaatlicher Prinzipien, des rechtlichen Gehörs und der Rechtsgewäh-

[26] *Bleckmann,* in seinem oben (Anm. 6) zit. Artikel, verwendet in dieser Beziehung das befremdende Bild eines »Souveränitätspanzers«.

[27] Es ist hier nicht der Platz, auf dieses klassische Thema einzugehen. Ich erlaube mir, stellvertretend, auf die vertiefte Behandlung dieser Problematik durch *R. Quadri,* in seinem Cours général de droit international public, RdC 113 (1964 III), S. 280 ff., zu verweisen. Ich möchte dazu nur bemerken, daß der Dualismus nicht nur theoretische, sondern auch staatsrechtliche Wurzeln hat. Hinzuweisen ist dabei besonders auf den Dualismus britischer Prägung, der eine Konsequenz der für das Vereinigte Königreich charakteristischen Spaltung zwischen Krone und Parlament ist: Die Krone schließt die Verträge ab, hat aber keine Gesetzgebungsgewalt; das Parlament erläßt Gesetze, hat aber keinen Anteil am *treaty making power*; überdies gibt es kein eigentliches Amtsblatt, in dem die Verträge gesetzlich bekanntgemacht werden könnten. Somit kann ein Vertrag »als solcher« niemals im Vereinigten Königreich in Kraft treten. Im Fall der Europäischen Gemeinschaften wurde dieses Hindernis erfolgreich durch den *European Communities Act* überwunden.

rung realisiert[28]. Das Problem der innergemeinschaftlichen Wirkung völkerrechtlicher Abkommen ist, wie wir gezeigt haben, in demselben Geist gelöst worden, soweit der Gerichtshof bisher *pragmatisch* an dieses Problem herangegangen ist; insoweit ist von der Anwendung eines Doppelstandards dualistischer Prägung nichts zu merken. Aber im Grundsätzlichen scheint die Frage noch nicht endgültig entschieden, hier scheint der Vorbehalt dem Völkerrecht gegenüber noch nicht überwunden. Der Gerichtshof wird also noch sagen müssen, welches im Grunde genommen sein »Selbstverständnis« ist: Ob er seine vornehmliche Aufgabe darin sieht, die innere Ordnung der Gemeinschaft zu pflegen, oder ob er sich es darüber hinaus angelegen sein läßt, die Gemeinschaft auch in judizieller Hinsicht in die weiteren völkerrechtlichen Zusammenhänge hineinzustellen, soweit diese »justitiabel« sind. So wie es die Kommission im *Polydor*-Fall treffend dargelegt hat, geht es hier in der Tat um die Abgrenzung von Politik und Recht; die Frage ist nur die, wo genau diese Grenze gezogen wird.

A. Urteile zur gemeinschaftsinternen Wirkung völkerrechtlicher Abkommen

			Slg. S.
27. 2. 1962	10/61	Kommission ./. Italien (GATT)	1
15. 10. 1969	14/69	Markus & Walsh (GATT)	349
9. 7. 1970	26/69	Kommission ./. Französische Republik (Assoziierungsabkommen Tunesien)	565
26. 4. 1972	92/71	Interfood (GATT)	231
12. 12. 1972	21–24/73	International Fruit Cy (GATT)	1219
7. 2. 1973	40/72	Schröder (Assoziierungsabkommen Griechenland)	125
12. 7. 1973	59/72	Wünsche (*idem*)	791
24. 10. 1973	9/73	Schlüter (GATT)	1135
11. 12. 1973	147/73	Lensing (Jaunde-Abkommen)	1543
30. 4. 1974	181/73	Haegeman (Assoziierungsabkommen Griechenland)	449
14. 5. 1974	4/73	Nold (Europäische Menschenrechtskonvention)	491
10. 12. 1974	48/74	Charmasson (Jaunde-Abkommen)	1383
28. 10. 1975	36/75	Rutili (Europäische Menschenrechtskonvention)	1219
19. 11. 1975	38/75	Nederlandse Spoorwegen (GATT, Brüsseler Abkommen Zolltarifschema)	1439

[28] Die deutlichsten Ausprägungen dieser Haltung in neuerer Zeit findet man in zwei Urteilen: *Simmenthal,* 106/77, vom 9. 3. 1978 (Slg. S. 629), zur Unmöglichkeit der Inkraftsetzung von gemeinschaftsrechtswidrigen Gesetzen in den Mitgliedstaaten, und *Ursula Becker,* 8/81, vom 19. 1. 1982, zur innerstaatlichen Reflexwirkung der Bindung des Staats durch Richtlinien.

5. 2. 1976	87/75	Bresciani (Jaunde-Abkommen)	129
15. 6. 1976	51/75	EMI (Freihandelsabkommen u. a.)	811
14. 7. 1976	3, 4 u. 6/76	Kramer u. a. (Fischereiabkommen)	1279
27. 10. 1976	130/75	Prais (Europäische Menschenrechtskonvention)	1589
9. 6. 1977	90/76	Van Ameyde (Straßburger Abkommen für Pflichtversicherung)	1091
12. 7. 1977	89/76	Kommission ./. Niederlande Pflanzenschutzabkommen	1355
24. 11. 1977	65/77	Razanatsimba (Lomé-Abkommen)	2229
30. 11. 1977	52/77	Cayrol (Handelsabkommen Spanien)	2261
2. 2. 1978	72/77	Universiteitskliniek Utrecht (Florenzer UNESCO-Abkommen)	189
15. 7. 1978	149/77	Defrenne (Abkommen Nr. 100 der IAO)	1365
18. 1. 1979	110 u. 111/78	Van Wesemael (Abkommen Nr. 96 der IAO)	35
28. 3. 1979	179/78	Rivoira (Handelsabkommen Spanien)	1147
4. 10. 1979	141/78	Französische Republik ./. Vereinigtes Königreich (Fischereiabkommen)	2923
11. 10. 1979	225/78	Bouhelier (Freihandelsabkommen u. a.)	3151
13. 12. 1979	44/79	Hauer (Europäische Menschenrechtskonvention)	3727
14. 12. 1979	34/79	Henn & Darby (Weltpostabkommen u. a.)	3795
5. 3. 1980	98/79	Pecastaing (Europäische Menschenrechtskonvention)	691
24. 4. 1980	65/79	Chatain (Freihandelsabkommen Schweiz)	1345
26. 6. 1980	136/79	Panasonic (Europäische Menschenrechtskonvention)	2033
9. 10. 1980	823/79	Carciati (New Yorker Abkommen über Import von Motorfahrzeugen)	2773
29. 10. 1980	209, 215, 218/78	Heintz van Landewyck u. a. (Europäische Menschenrechtskonvention)	3125
5. 5. 1981	112/80	Dürbeck (GATT)	1095
14. 7. 1981	155/80	Oebel (Abkommen Nr. 20 der IAO)	1993
8. 12. 1981	180 u. 266/80	Crujeiras und Yurrita (Fischereiabkommen)	
8. 12. 1981	181/80	Arbelaiz-Emazabel (*idem*)	
9. 2. 1982	270/80	Polydor (Freihandelsabkommen Portugal)	
29. 4. 1982	17/81	Pabst & Richarz (Assoziierungsabkommen Griechenland)	

B. Urteile mit Bezugnahme auf die im Rahmen des Brüsseler Abkommens zum Zolltarifschema ergangenen Tarifavise

*Bemerkung: nur die mit * bezeichneten Urteile enthalten grundsätzliche Stellungnahmen; alle anderen Urteile enthalten eine bloße Bezugnahme auf die Tarifavise oder eine Wiederholung der »Bakels-Formel«.*

			Slg. S.
18. 6. 1970	72/69	Bremer Handelsgesellschaft	427
18. 6. 1970	74/69	Krohn	427
8. 12. 1970 *	14/70	Bakels	1001
8. 12. 1970	28/70	Otto Witt	1021
14. 7. 1971	12/71	Henck	743
14. 7. 1971	13/71	Henck	767
14. 7. 1971	14/71	Henck	779
24. 11. 1971	30/71	Siemens	919
15. 12. 1971	21/71	Brodersen	1069
15. 12. 1971	77/71	Gervais-Danone	1127
23. 3. 1972	36/71	Henck	187
14. 12. 1972	38/72	Arend van de Poll	1329
9. 10. 1973	12/73	Muras	963
7. 11. 1973	49/73	Fleischer	1199
22. 11. 1973	128/73	Past	1277
8. 5. 1974	183/73	Osram	477
29. 5. 1974	185/73	König	607
25. 9. 1975	28/75	Baupla	989
23. 2. 1975	35/75	Matisa	1205
19. 11. 1975 *	38/75	Nederlandse Spoorwegen	1439
18. 2. 1976	98–99/75	Carstenskeramik	241
8. 4. 1976	106/75	Merkur-Außenhandel	531
22. 9. 1976	22/76	Import Gadgets	1371
23. 3. 1977	86/76	Gervais-Danone	619
26. 5. 1977	108/76	Kloeckner-Ferromatik	1047
27. 10. 1977	23/77	Westfälischer Kunstverein	1985
15. 12. 1977	60/77	Fuss	2453
15. 12. 1977	63/77	Poppe	2473
9. 3. 1978	111/77	Bleiindustrie	659
16. 3. 1978	104/77	Oelschläger	791
28. 3. 1978	158/78	Biegi	1103
2. 5. 1979	137/78	Henningsen Food Inc.	1707
4. 10. 1979 *	11/79	Cleton	3069
13. 3. 1980	111/79	Caterpillar	773
11. 7. 1980 *	798/79	Chem-Tec	2639
16. 10. 1980	816/79	Mecke	3029
16. 7. 1981	159/80	Wünsche	2161

Limites constitucionales y jurídico internacionales en la competencia penal internacional del Estado español

Glosa breve al Art. 13 de la Constitución de 1978

José Puente Egido

I. Consideraciones generales

1. En 1917 *J. Kohler* concluía su prólogo a la clásica obra »Internationales Strafrecht« con la siguiente frase: »Die neue Zeit verlangt einen neuen Geist, aber der neue Geist kann nur wirken, wenn er sich erfüllt mit dem historisch Gegebenen«. Sirva aquí también esta idea de lema y orientación de esta glosa que en modesto homenaje me atrevo a ofrecer al venerado Maestro. Armoniza además, a lo que se me alcanza, con lo que entiendo ha sido y es constante en el talante científico de *H. Mosler*. Lejos de él tomar a la Historia como una permanente y nunca acabada autorrealización del Espíritu del Mundo, según lo exigiría la ortodoxia hegeliana. No; en el *Mosler* jurista el recurso a la Historia es sobre todo método, criterio epistemológico, porque la historia en la proyección vertical, al igual que la comparación de sistemas e instituciones jurídicas en la proyección horizontal, permiten superar las contingencias de cualquier forma jurídica positiva para apuntar hacia los ideales del Derecho. Con un cierto matiz suareciano – que él no tendría inconveniente en aceptar – me atrevo a decir que se trata de penetrar en la esencia de las cosas a través de las manifestaciones concretas y recortadas de su existencia [1].

No encaja mal tampoco esta idea introductoria para ilustrar la orientación que se quiere dar a este breve comentario al art. 13 de nuestra Constitución. De una Constitución, ciertamente, elaborada de modo mediocre y con un resultado en conjunto también de mediocridad. Pero de la que en el tema con-

---

[1] Vid., entre otros, este ejemplo: *H. Mosler,* The International Society as a Legal Community (Sijthoff and Noordhoff 1980), pp. 6 ss.

creto que nos ocupa, y no obstante las imperfecciones que más abajo se indicarán, se puede decir, por fortuna, que el resultado es bueno. Dicho *cum granu salis*, hasta llegaría a sugerir que la obra ha superado aquí al artista o la profecía a los profetas. Cualquiera que sea la suerte que el futuro desarrollo constitucional reserve a este texto, hay en él una positiva aportación en la línea del respeto a la regla de Derecho.

2. Toda constitución moderna ofrece y está respondiendo a un concreto modelo de relación entre el Orden jurídico internacional y el respectivo orden jurídico interno. Conocidamente estas normas constitucionales se agrupan en dos grandes categorias: las de *procedimiento,* que no solo fijan los cauces de penetración de las normas internacionales en el Derecho interno sino también su valor normativo en él; y las reglas *materiales o de fondo,* que suponen otras tantas limitaciones a la libertad originaria de cada Estado. Que algunas de estas normas sean expresión y duplicación de los límites trazados por el orden internacional a esa actividad del Estado, mientras que otras reglas constitucionales van más allá de este marco y limitan la competencia de los órganos estatales en ámbitos en los que todavía el Derecho internacional no se ha interferido, es asunto que por ahora dejamos al margen de nuestras consideraciones.

Pues bien, la vigente Constitución española se caracteriza indiscutiblemente por una amplia apertura hacia el orden internacional en esa doble vertiente de reglas de procedimiento y reglas de fondo, y el art. 13 se encuadra en esta última categoría.

Una breve referencia al proceso de elaboración de este precepto en las Cortes constituyentes permite captar mejor su sentido y alcance, así como encuadrarlo sistemáticamente en el conjunto. Las Cortes partieron de una buena base, por cuanto el primitivo texto del art. 12 del Anteproyecto, que pasó a ser el 13 del texto definitivo, respondía bastante bien a los datos que la técnica jurídica ofrece hoy en materia de constitucionalización del régimen de extranjeros, el de la extradición y el asilo territorial, que son las cuestiones en él contempladas. Sistemáticamente se encuadró el precepto, ya desde el primer momento, en el Titulo II (De los derechos y deberes fundamentales) y dentro de ese título en el Capítulo primero, que trata de los españoles y extranjeros. Pero como vicio de origen, fuente de errores en los ulteriores debates, del que no ha sido purgado el texto definitivo, está la heterogeneidad de sus contenidos. Así como la confusión de la regulación de la extradición como institución jurídica, en contraposición, con la vieja concepción de acto de Estado o de gobierno, generadora de derechos y deberes individuales (y de ahi su adecuada colocación sistemática) y como mandato del constituyente al legislador ordinario, y en general a los demás órganos del Estado, que en cuanto limitación de esas competencias podía haber sido encuadrada en otra parte. Pero no nos detenemos ahora en el tema de la competencia penal internacional del que trataremos después.

Perturbadora ha sido, en efecto, la amalgama de dos cuestiones que como la extradición y la extranjería temática y practicamente tienen relación, pero son conceptualmente tan diferentes. La constitución española se orienta, en principio también, por la clásica distinción de derechos y libertades fundamentales (*Menschenrechte*) y derechos y deberes ciudadanos (*Bürgerrechte*) y precisamente por ello pretendió agrupar a los primeros en la Sección primera del Cap. 2°, Tit. I, y a los segundos en la Sección 2ª del mismo capítulo y título; pero lo hizo sin congruencia ni sistema porque en la primera sección está el art. 23 que habla de »los ciudadanos. . .« o el art. 19 y el 29 que hablan de »los españoles. . .«, mientras que en la sección segunda se garantizan derechos que como la libertad para contraer matrimonio (art. 32) o el de propiedad privada (art. 33) parecen – según el desafortunado texto literal – estar garantizados únicamente a los ciudadanos españoles. . . Vestigia terrent. . .!

Si por los efectos hay que juzgar las causas, dicho queda el juicio que merece a una crítica imparcial el tono del debate parlamentario. La ponencia había presentado un texto que, ciertamente, introducía un criterio excesivamente restrictivo en cuanto a la admisión del extranjero residente en España al goce y ejercicio de derechos políticos. Pero en el extremo opuesto se situaron las enmiendas de los partidos de izquierda, socialistas y comunistas, que llevaban a una práctica equiparación del extranjero al nacional en el goce de los derechos políticos sin restricción alguna[2]. Y a esta confusión, que estuvo presente en todo el debate parlamentario, se debe también la falta de claridad con la que se trató el tema de la extradición e, indirectamente, el de la competencia internacional del Estado español en materia penal. Veamos cómo.

Hubo coincidencia en todos los grupos políticos en exigir constitucionalmente que la extradición solo sea otorgada en base a las disposiciones de un tratado o de la ley. Introducción, pues, en esta materia del principio de legalidad, en un campo precisamente donde por tradición y práctica internacional e interna no se ha exigido como consecuencia del respeto a la regla de Derecho, el control jurisdiccional de los actos de la administración o del gobierno. La hubo también, a no dudarlo, aunque no aluden a ello los diarios de Cortes,

---

[2] Fueron numerosas las intervenciones en este sentido, las más de las veces utilizando peregrinos argumentos sin consistencia científica alguna. Ambos grupos políticos tendían a la eliminación del inciso »solamente los españoles son titulares de derechos políticos«. En esa equiparación del extranjero al nacional se trataba unas veces de poner al Derecho constitucional español con lo que se decía era linea inequívoca de evolución en el Derecho europeo (Constitución Española. Trabajos Parlamentarios, Madrid 1980 [en adelante Const. Esp.], I, pp. 960 ss.; otras se trataba de conseguir un argumento que sirviera de baza en una futura negociación internacional »para poder exigir posteriormente una igualdad de trato (en favor de) los ciudadanos españoles que residen en el extranjero, a los cuales se les niega normalmente este tipo de derechos. . .« (!!!!), *ibid.*, p. 956. Se llegó incluso a »profetizar« que en el año 1981 todos los trabajadores emigrantes en Europa participarían en las elecciones municipales. ¡Nada menos! *ibid.*, II, p. 1998. – En contra, las intervenciones jurídicamente mucho más fundadas de O. Alzaga, *ibid.*, p. 2002, y Fraga Iribarne, pp. 2005 y 2006.

sobre que la extradición como instituto jurídico quedaba bajo el régimen general de las garantias individuales. Pero donde los espíritus se separaron fué en el tema de los límites constitucionales a la facultad de entregar de los órganos estatales españoles y, sobre todo, en el concepto de »delito político« y del »delito de terrorismo« como excepción a la excepción. En los partidos comunista y socialista hubo una clara tendencia a debilitar la obligación internacional de entrega mediante la eliminación de la referencia expresa a los delitos de terrorismo (versión moderna y actualizada de la claúsula del atentado)[3], según deseos de los socialistas, o mediante la »gasificación« del concepto, ya en si mismo impreciso, de delito político por la sustitución de »delitos de intencionalidad política«. Esto haría mucho más dificil el control jurisdiccional si, como es de esperar, por la via de los recursos judiciales la calificación en última instancia de qué sea y qué no sea un delito político a efectos de extradición será resuelta por instancias judiciales[4]. En algún caso bien pudo ocurrir que la »singularidad« de la propuesta se explique no tanto por un exceso doctrinario del programa político del grupo proponente, sino por desconocimiento técnico de los datos del problema[5].

Grave ha sido, finalmente, que en la discusión parlamentaria solo muy tardíamente y en forma casual apareciera la relación íntima en que el tema de

---

[3] Si en la claúsula del atentado (y al margen de todo planteamiento egoista de autoprotección entre quien la negocia y resulta asi protegido) había fundamento para una especial protección de lo que todavía en el siglo pasado era base esencial del poder político, hoy la base de la protección deberá ser extendida respecto de conductas terroristas criminosas, porque también el poder político se ha extendido y difuminado en las comunidades políticas modernas.

[4] Los comunistas proponían la sustitución del término »delito político« por el de »delitos de intencionalidad política« (Const. Esp. I, p. 974) y los comunistas y socialistas pidieron la eliminación del texto constitucional de la mención expresa del terrorismo. Los socialistas solicitaban, además, la supresión de la extradición de todos aquellos delitos que en la ley penal extranjera estuvieren penados con la pena capital y, en general, la no entrega de delincuentes a paises cuyos sistemas jurídicos no reconozcan esos »principios penales reconocidos en los textos internacionales«. Razón: »sería absurdo que a una persona que se encuentra en territorio español y que, como creo que quedó claro ayer, está protegida por las garantías de la Convención europea... etc.« (Const. Esp. I, p. 969). Somos incorregibles en nuestro doctrinarismo. Si la propuesta hubiera triunfado el texto constitucional habría declarado a España refugio seguro de asesinos, violadores... etc. procedentes de paises (que son los más) que no reconocen esas garantías.

[5] Por el representante de la Minoria catalana se propuso que la extradición fuera denegada »cuando la pena solicitada por la acusación fuere superior a la correspondiente por el mismo delito de acuerdo con la legislación española« (Const. Esp. I, p. 972) lo que equivalía a exigir que las leyes penales extranjeras, cualesquiera que fueren las circunstancias de la comisión del delito por el que se reclamaba la entrega, se plegaran a las exigencias de la ley penal española que podría tener como único título para intervenir el de ser la ley del pais de refugio. Se abandonaba con ello uno de los principios más firmes del Derecho penal internacional cual es el de la *lex loci.* Se hubiera introducido así una jurisdicción exorbitante, contraria a elementales normas del orden jurídico internacional y fuente permanente de conflictos internacionales.

la extradición está con el de los límites competenciales del Estado español en materia penal. No se le mencionó para nada en la discusión parlamentaria en el Congreso de los Diputados, y, cuando apareció de modo marginal en las discusiones de la Cámara Alta, el grado de desconocimiento e incompetencia que sobre él muestran tener los »Padres de la Patria« llega al límite del sonrojo nacional[6].

Verdad es, pues, que la obra ha resultado ser superior a sus artífices, y su fórmula final expresa en parte lo que la práctica interna e internacional española había incorporado ya en el tema de la extradición y de la fijación de la competencia internacional[7], y en parte la retoca mejorándola. El mencionado precepto constitucional, con sus zonas de luz y de sombras, ofrece ciertamente una base sólida para un futuro desarrollo constitucional puesto que recoje bastante bien valores que pueden y deben ser plasmados en una constitución democrática moderna. Acierto ha sido la constitucionalización de la extradi-

---

[6] Vid. Const. Esp. pp. 3192–3193. La sugerencia la formuló el Sr. Villar Arregui quien con agudeza, aunque sin precisión conceptual y terminológica, y siguiendo una propuesta de Azcárate, planteó bien la conexión entre la extradición y las reglas de la LOPJ (Ley Organica del Poder Judicial, de 1870) sobre la competencia del juez penal español en la persecución y castigo de un delito cometido en el extranjero. ¿Quid del francés -se preguntaba- que comete un delito de robo con asesinato en Francia, penado allí con la pena capital y se refugia en territorio español? El juez español no lo puede juzgar porque carece de competencia, el gobierno español no lo puede entregar porque el delito está castigado en el pais requirente con la pena de muerte.

La contestación que recibió de sus compañeros del Senado no pudo ser más singular. No nos resistimos a dejar de transcribir una, porque apenas se pueden acumular tantos despropósitos jurídicos en menos palabras: »Por otra parte, me parece que también que no es correcto, que no es totalmente cierto que ese francés no pueda ser juzgado en España, porque existe una institución, seguramente llena de deficiencias y que ocasiona grandes retrasos, que es el auxilio judicial, y a través del auxilio judicial, primero por la via diplomática y luego de la via diplomática a la Magistratura, el francés podrá ser juzgado en España. Por lo menos podrán hacerse todos los actos procesales – por ejemplo la declaración de testigos, la prueba del delito – para que el Tribunal francés pase su correspondiente sentencia. Este es el procedimiento normal de auxilio judicial. . .«.

[7] En el debate parlamentario hubo un olvido, motivado quizá por razones políticas. Nadie en él aludió al antecedente inmediato español de este precepto que estaba en el art. 30 de la Const. republicana de 1931 y por el que se prohibia que el Estado español concluyera tratados de extradición donde se permitiese la extradición de delincuentes político-sociales. Formulado como si se tratara de un derecho individual, constituía en realidad un mandato al Gobierno y Cortes ordinarias, defecto en el que se ha incurrido de nuevo. Tampoco nadie pareció acordarse de que la prohibición de claúsulas en tratados de extradición que prevean entrega de delincuentes políticos es obra de una vieja ley de 4 de Diciembre de 1855.

El mencionado art. 30 no gozó de buena opinión entre los constitucionalistas de entonces, ni suscitó entusiasmos en el momento de su aprobación. Si hemos de dar crédito al antiguo Presidente de la República (*N. Alcalá-Zamora,* Los Defectos de la Constitución [Madrid 1936], pp. 99 ss.) fué obra de la improvisación y hasta de un malentendido. Vid. también *Pérez-Serrano,* La Constitución española (Madrid 1932), p. 150.

ción, primero como garantía constitucional de una voluntad de cooperación de España en la tarea solidaria de la represión internacional del delito, después (y esto es quizá lo más importante) entendiéndola como un acto sustancialmente jurisdiccional, sometido a las garantías constitucionales de un Estado de Derecho y no como un acto de soberanía o de política exterior en el que la persona humana esté sujeta o pueda estarlo a las contingencias políticas del momento. Pero una voluntad de cooperación que debe encontrar sus límites allí donde el Estado deja de ser tutelador y garante de valores sociales comunes y corre el riesgo de transformarse en juez y parte en el enjuiciamiento y castigo de los delitos políticos [8].

Positiva es la inclusión del asilo territorial (art. 13,4) como un mandato del legislador constitucional sobre el ordinario para su ulterior reglamentación [9], aunque de él solo nos ocuparemos aquí tangencialmente. Idea esta que armoniza con las realidades del pluralismo de la actual comunidad internacional. Desarmar el brazo de un poder estatal prepotente – de cualquier poder estatal y no solo el de los Estados no democráticos – [10] no significa necesariamente alinearse en el bando de la subversión, sino – quizá – un rasgo de humanidad que intenta moderar los posibles excesos de un poder siempre en riesgo de incurrir en violencia estructural o legal. El asilo territorial, como institución del Derecho internacional general, si es así entendido, equilibra en

---

[8] Para la adecuada comprensión de esta afirmación tengamos en cuenta los siguientes elementales principios. Según la Teoria General del Estado y del Derecho internacional el *ius puniendi* del Estado se asienta en el monopolio de la fuerza y el Derecho y en ser garante (por lo tanto), en términos de obligación, del orden social. Desde la exigencia del principio de la división de poderes y de la *rule of law* el derecho del Estado moderno a castigar no tiene su base en el delito, esto es, en la acción u omisión incriminada, sino en la sentencia judicial condenatoria. En otras palabras, en el moderno Estado de Derecho las garantías individuales en el proceso penal, como superación del procedimiento penal del Estado policial, son elemento esencial de ese Estado. Tal evolución se inició en el Derecho interno español, material y procesal, a partir de 1837, con las grandes reformas administrativas. Ahora bien, la penetración de estas normas elementales en la relación exterior, o por mejor decir, en aquellos usos y prácticas (más bien que normas) en materias como la extradición, auxilio judical, competencia internacional etc. ha sido mucho más lenta y tardía.

[9] También aquí los grupos de izquierda, con sus enmiendas al texto del anteproyecto, eran partidarios de fórmula más perentoria y precisa. El texto final sobre el asilo, aprobado por la conjunción de los partidos de la derecha, remite en buena parte la materia a la regulación del legislador ordinario, pero no es exacto que se procediera a una »desconstitucionalización« del derecho de asilo, sin eficacia alguna sobre la legislación en materia de extranjeros, según se afirmó por algún parlamentario (Const. Esp. I, p. 984). Las sentencias más recientes de nuestro T. S., recogen en efecto, ese cambio en el derecho de extranjeria, del que es base y fundamento nuestro art. 13, ya que la esperada nueva Ley de Extranjería no se ha promulgado, ni ha pasado de la primera fase de elaboración.

[10] Los autores de la propuesta que en las Constituyentes del 31 pasó a ser el art. 30 de la Constitución, la formularon con una confesada intención béligerante. Se trataba de impedir que el Estado degenerase en poder dictatorial y entregara delincuentes político-sociales a otros Estados.

algo las insatisfacciones y carencias de una sociedad en la que todavía está muy presente el Estado como formidable máquina de poder [11].

II. Régimen de la extradición en España: Perspectivas históricas, naturaleza y problemas principales

1. Hasta la Ley de Extradición de 1958 el Derecho español careció de una ordenación sistemática de esta institución; sus normas internas eran fraccionarias y preferentemente procesales. La ley se elaboró pensando sobre todo en la extradición pasiva y con la intención manifestada de que sirviera de pauta en el ajuste de futuros tratados [12], lo que no fué obstáculo para que la misma ley reconozca el carácter prevalente del tratado sobre las normas internas, incluida la propia ley de extradición.

Independientemente de lo que después diremos, el juicio de conjunto que merece este intento de sistematización, es positivo. Se regoje en ella el principio de especialidad, el de *non bis in idem,* se excluye al delito político como susceptible de entrega, o los puramente militares, se prohibe la entrega del reo de un delito castigado con pena de muerte según la ley penal del Estado reclamante mientras éste Estado no dé seguridades de que tal pena le será conmutada. Pero la ley es un tanto renuente al no garantizar de modo estricto el principio de legalidad y, sobre todo, al otorgar un amplísmo margen de intervención al gobierno quien sobre el delito político »apreciará libremente en cada caso el carácter de la infracción« (art. 6). Es igualmente excesivo el principio de competencia real que reconoce en el Estado extranjero requirente permitiendo al Gobierno atender la petición de entrega de un delincuente, que no es nacional suyo cuando »se trate de delitos que la ley española sometería a la competencia de Tribunales españoles, aún siendo cometido en el extranjero por extranjeros«.

La Ley de Enjuiciamiento Criminal (L. E. Crim.), así como alguna otra ley especial de menor interés para nosotros, habían regulado ya mucho antes de 1958 los aspectos procesales de la extradición activa que, salvo casos excepcionales, y una vez agotada la via judicial con decisión favorable, debe ser

---

[11] En el Derecho intermedio parece haber sido doctrina común que la competencia del juez penal estaba regida exclusivamente por tres fueros, el de la comisión del delito, el del origen o el del domicilio. Vid. *Carlevalii,* De iudiciis. . . etc., quaestio 840, quien sienta esta regla: »domiciliarius vel originarius unius Regni in illo delinquens et fugiens ad alliud, non potest puniri in Regno ad quod confugit«, y da la razón: »forum non sortitur, ac per consequens puniri non potest: nemo enim potest puniri nisi a iudice competente«.

Ya en plena etapa absolutista, el tratado hispano-francés de extradición de 1765 contenía una curiosa claúsula que combinaba el asilo eclesiástico que todavía subsistía en España, pero no en Francia, con la extradición entre ambos paises. Vid. art. 5.

[12] En expresión del texto legal, art. 2,2: »Los tratados de extradición que se concierten en el futuro se ajustarán, en lo posible, a las prescripciones de la propia ley«.

obtenida por via diplomática (arts. 823 ss.). Entre esas reglas destaca por su interés el art. 826 que contiene, dentro de la sencillez de sus tres normas, una armónica combinación de las reglas comunes de la extradición con las de la competencia penal internacional en el Derecho español.

Pero la fuente principal normativa de la institución de la extradición ha estado tradicionalmente en el sistema español, y aun sigue estando, no en la ley interna sino en los tratados, especialmente los bilaterales[13]. A los que por una posición muy arraigada en nuestro sistema jurídico no se les considera simplemente como una »ley del Reino«, sino con jerarquía normativa superior a la ley.

Es, pues, sobre este conjunto de reglas sobre el que deberá proyectarse ahora la nueva normativa constitucional. Con la brevedad que este comentario exije diremos ahora simplemente que en razón del carácter normativo inmediato del texto constitucional, el efecto derogatorio sobre las reglas jurídicas inferiores que sean contrarias es directo[14]. Y, puesto que el mandato constitucional afecta a todos los órganos del Estado, la constitución exije una adaptación de la ley del 1958 al precepto constitucional por via legislativa ordinaria, una interpretación jurisprudencial por los tribunales ordinarios y, enventualmente, por la via extraordinaria, del Tribunal Constitucional, en armonía con el texto constitucional. En materia de acuerdos internacionales se impone una adaptación de los ahora en vigor a esas reglas superiores, asi como el respeto de estas mismas reglas en la conclusión de futuros acuerdos.

2. El antecedente convencional más antiguo del que se tiene noticia cierta en el Derecho español es una Pragmática de los Reyes Católicos por la que se aprobaba y hacía extensiva a todo el Reino de Castilla una concordia concertada entre los jueces de Galicia y Portugal que estableció una intensa colaboración en la persecución de delincuentes a uno y otro lado de la Raya. Notable es en ella, medido al menos con las categorías del moderno Derecho de la extradición, la regla siguiente: »el natural de estos reinos que pase a Portugal y mate en él a alguna persona. . . etc. sea entregado a las justicias de Portugal«. Dos posibles vias de explicación se nos ocurren; o bien se trata de una regla muy especial que explicaría la vecindad y frecuente relación entre los habitantes de ambos reinos, o bien no es más que la concrección práctica de la norma cuasi universal y exclusiva de la *lex loci commissi,* sin duda fuertemente arraigada en los Derechos hispánicos de entoces, pero que más que una singularidad ibérica era probablemente común a todo el occidente cristiano[15].

---

[13] Vid. Ministerio de Justicia, Convenios de Extradición (Madrid 1980).

[14] Vid. respecto de Italia *A. Cassese,* en: Commentario de la Costituzione. Principi Fondamentali (1975), pp. 556 ss. y decisiones jurisprudenciales italianas en contrario citadas.

[15] Como información general de la doctrina común en el siglo XVII, de nuevo *Carlevalii, op. cit.* (nota 11) y en especial Tit I, Disp. II, Sect. II, De remissionibus reorum.

Si esto fuera así, y a partir de un momento que el estado actual de nuestros conocimientos histórico-jurídicos no permiten precisar, debió de producirse una completa inversión en la competencia penal de las monarquías absolutas, precisamente por obra y gracia de este poder. Indicio de esta evolución sería la regla contenida en el tratado de paz de 1500 entre los mismos monarcas y Luis XII de Francia que Dumont III, 2 recoge: Se prometen en ella estos reyes la recíproca entrega de los súbditos que habiendo cometido en el reino un delito de lesa majestad o cualquier otro que no haya sido previamente castigado, se refugien en el otro reino[16]. Bien es verdad que todavía ahi el lugar de la comisión y el vinculo de fidelidad parecen ir en paralelo, pero en la evolución futura, al menos en el Derecho español, debió prevalecer este segundo sobre el primero. Esto es lo que parece deducirse de un texto bastante claro, aunque de modo indirecto:

»Habiendo llegado a la Real noticia que en diferentes paises extranjeros, quando algunos de los vasallos de S. M., asi soldados como paysanos, transeuntes o domiciliados en ellos delinquen contra sus leyes y bandos públicos, se les forma proceso por las Justicias ordinarias. . . sin remitir los delinquentes a los Tribunales Españoles, se manda que las justicias de estos Reynos procedan siguiendo la regla de la reciprocidad contra los extranjeros transeuntes o domicialidos de qualquiera nación. . . conforme a las leyes del Reyno. . . sin permitir se forme sobre ello competencia alguna«[17].

Sea de ello lo que fuere, lo que evidentemente este texto que acabamos de citar muestra es lo que ya sabemos. Con el cambio revolucionario en ideas e instituciones que introduce en Europa la Revolución Francesa se inicia un nuevo proceso de territorialización de la ley, particularmente las de Derecho público. A la soberanía del monarca, que en sus formas jurídicas tiene en lo personal todavía determinados rasgos feudales, se sustituye la nacional, que en lo territorial se configura sin excepciones ni fisuras en la medida en que el Estado moderno se identifica con su territorio.

Con el constitucionalismo se abre una nueva fase en el Derecho convencional de la extradición, aunque no de modo inmediato. El primer tratado

---

Vid. la Pragmática de 20 de Mayo de 1499 que fué ampliada por Felipe II y Carlos III y posó después a la Nueva Rec. y más tarde a la Novis., Leyes 3,4, y 5 Tit. 36, lib. 12. Vid. *Ortiz de Zúñiga,* Elementos de Práctica Forense, vol. 2 (1843), p. 208–10. Los testimonios que hemos podido recoger sobre este punto están contestes en afirmar el constante uso que a lo largo de tres siglos debió hacerse de este convenio, las más de las veces, mediante comunicación directa de tribunal a tribunal sin utilizar la via diplomática, v. gr. Pacheco en los debates parlamentarios durante la discusión del primer código penal (vid. más adelante nota No. 37) sobre los datos que él aporta de su experiencia como juez de comisión.

[16] Sobre la referencia al tratado entre Fernando e Isabel y Luis XII de Francia vid. *Kohler, op. cit.,* p. 150.

[17] Es una Real Cédula de Carlos IV de 1782, vid. *Pérez y López,* Teatro de la Legislación Universal de España e Indias, T. 13, pp. 416–417.

concertado por España en materia de extradición en el pasado siglo fué uno con Portugal en 1823, pero se sitúa de un modo bastante atípico porque no es un normal tratado de extradición sino instrumento de colaboración entre ambos paises frente a las agitaciones de los movimientos absolutistas[18]; que no debió, por lo demás, ser muy eficaz y que la perdió sin duda por completo del lado español en la declaración general de nulidad que en la restauración absolutista hizo Fernando VII de los actos »del gobierno llamado constitucional«. El comienzo de la nueva fase del Derecho de extradición hay que verlo exactamente en la mitad del siglo pasado, coincidiendo, además, con la nueva reglamentación del régimen de extranjería. El primer tratado de extradición que España ajustó sobre estos nuevos planteamientos fué el hispano-francés de 1850. Durante la segunda mitad del pasado siglo y el primer decenio del actual se fueron concluyendo tratados bilaterales de extradición con los principales paises europeos y americanos, que en parte continúan hoy todavía en vigor[19].

Supera el estrecho marco de este trabajo intentar siquiera hacer un análisis valorativo de la masa normativa que esos acuerdos contienen, y aunque desde un punto de vista doctrinal sería arriesgado deducir reglas generales por las básicas coincidencias que muchos presentan, séanos permitido formular algunos rasgos característicos. Durante todo este periodo el tratado de extradición sigue siendo un acto esencialmente reservado a la competencia del ejecutivo y los actos concretos de ejecución, cuando se trata de extradición pasiva, están abandonados por completo a la Administración y Gobierno. Pero la misma materia penal aproxima necesariamente la extradición a las garantías básicas del Derecho penal moderno. No parece que las Cámaras parlamentarias hayan ejercido un control directo e inmediato sobre el gobierno en esta materia, pero por la via consultiva, especialmente del Ministerio de Estado (hoy de Asuntos Exteriores) o del de Justicia o del propio Gobierno, el Consejo de Estado fué formulando doctrina que ha dado más tarde paso al control jurisdiccional. Hoy este control es una firme adquisición en nuestro sistema jurídico. El paso previsible ahora es la constitucionalización de esta materia a través del recurso de amparo.

3. Desde el momento en que se supera la idea de que el tratado de extradición es un acto internacional estrictamente interestatal, aparece de inmediato su naturaleza compleja. Evidentemente que el tratado es y seguirá siendo una manifestación de la voluntad de cooperación de los Estados en la represión internacional del delito, pero la conexión necesaria entre las figuras delic-

[18] Este tratado presenta la singularidad de ofrecer uno de los pocos ejemplos que la práctica internacional tiene del derecho de *hot pursuit* terrestre. Por dicho acuerdo ambos gobiernos se otorgaban recíprocamente el derecho de persecución de *partidas de facciosos* »sin que la entrada por tal motivo se considerase violación de territorio«.

[19] Durante este periodo España concertó algo más de un medio centenar de tratados de extradición. Vid. Guia Diplomática y Consular de España, Año 1908, pp. 447–449.

tivas que contempla como susceptibles de extradición y las respectivas leyes penales de los Estados contratantes (al igual que ocurre con los convenios de Derecho internacional privado y los respectivos Derechos privados internos) lleva al acuerdo internacional las garantías de la ley penal interna y de modo primordial el principio de la legalidad. Pero el tratado de extradición en cierto sentido no es más que un aspecto de un problema del que la competencia penal internacional es la otra cara. ¿Es que acaso sea en la posible »fisura«, en el engarce entre uno y otro aspecto; donde se instala la institución del asilo territorial? En verdad un Estado no dará refugio a un delincuente extranjero justiciable ante sus tribunales como reo de un delito sobre el que sus tribunales poseen competencia. No lo deberá dar tampoco cuando en virtud de un compromiso internacional esté obligado a la entrega.

Atención especial nos merece el tratado de extradición como tutelador del principio de legalidad. Si nos atenemos a la redacción del art. 827 L.E. Crim. procederá la petición de extradición activa no solo cuando exista tratado o lo prevea el Derecho escrito, sino también cuando lo permita el Derecho consuetudinario del pais requerido o cuando sea procedente en virtud de la reciprocidad. ¿Cómo se compadece esta norma con la legalidad en delitos y penas?

Para captar mejor el tema merece la pena traer a colación nuestra experiencia histórica del pasado siglo. Todavía durante la vigencia del viejo tratado de 1765 con Francia, en 1838 el Ministro de Estado español y el Embajador francés en Madrid convinieron mediante canje de notas adicionar la quiebra fraudulenta a la lista de delitos susceptibles de extradición que, ciertamente, aquel tratado no había contemplado. El acuerdo fué tomado, no en abstracto, sino a la vista de una petición concreta de extradición por quiebra que el gobierno francés había formulado al español. Este hecho fué causa de una fuerte protesta del Tribunal Supremo español y el acuerdo logrado por el canje de notas fué anulado, pero este incidente no fué obstáculo a que en 1851, vigente ya el nuevo tratado de 1850 por el mismo procedimiento se incluyera »la tentativa de asesinato, manifestada por un principio de ejecución« en el régimen del tratado. El Gobierno español, que no debía tener seguridad sobre la regularidad del procedimiento seguido, acudió en consulta al Consejo de Estado, quien tachó de irregular este proceder a causa de la falta de solemnidades y publicidad del acuerdo y por ser lesivo del derecho de asilo [20].

---

[20] Sin embargo, años más tarde el propio Consejo de Estado rectificó un tanto esta doctrina. El gobierno francés solicitó del español la entrega de un súbdito suyo, eclesiástico, presunto reo de un delito de atentado contra el pudor en una joven menor de 13 años. El gobierno galo reconocía que el supuesto no estaba comprendido en el art. 2 del convenio de extradición, pero ofrecía reciprocidad para casos futuros y semejantes. Pasado el expediente a dictamen, este Organo fué de parecer que podía, no obstante, accederse a la entrega por razones de moralidad y conveniencia, cuidando el gobierno español de precisar exactamente el alcance y obligación de la reciprocidad ofrecida.

Pero la doctrina del propio Consejo de Estado fué, a pesar de todo, vacilante durante un largo periodo de tiempo. Ciertamente, si este Alto Organismo en un primer momento parece seguir el criterio gubernamental para considerar a los tratados de extradición como actos meramente intergubernamentales, posteriormente, y por aplicación de reglas básicas en el Derecho administrativo interno, sienta el criterio de que en el procedimiento administrativo interno español, a efectos de extradición, la persona reclamada es parte interesada y ha de dársele audiencia en él, permitiéndole hacer oposición a la petición de entrega formulada por el gobierno extranjero contra el presunto delicuente. Pero aún todavía en los años veinte de este siglo, este Organo seguía entendiendo que las peticiones de extradición no incluidas en los supuestos previstos en un tratado internacional debían ser contempladas a la vista de la legislación nacional y no con relación al pacto, y que quedaban, por lo tanto, reservadas a la libre apreciación del gobierno requerido. Los criterios a tener en cuenta para esa decisión serían sobre todo el de apoyo y mutua correspondencia entre los paises, el de la mayor sumisión posible del presunto delincuente a la jurisdicción de los tribunales del lugar de la comisión del delito y el del asilo teritorial[21]. No obstante, ya en esas fechas, en opiniones minoritarias pero fuertes, tal y como lo demuestran los votos separados de algunos consejeros, se fué sentando la doctrina de que »los actos diplomáticos no pueden ser objeto de interpetación extensiva« y que la »enumeración de los delitos en los tratados de extradición es vinculante para los respectivos gobiernos, y por lo tanto limitativa«.

El paso decisivo lo ha dado más tarde la jurisprudencia del Tribunal Supremo (T. S.), reforzado en la posición que ya había tomado por la ley de 1958, y que se concreta en estas tres notas: a) Prioridad del tratado sobre la ley interna española y por lo tanto, sobre la ley de extradición y el propio Código penal. b) En tanto que reglas incorporadas al Derecho interno españolas, las estipuladas en los acuerdos de extradición crean derechos y deberes de carácter individual. c) Siendo el principio de la especialidad una regla básica – porque de otra forma se desconocería el principio del *nullum crimen* – hay lugar a la casación y por consiguiente a la declaración de nulidad de lo actuado cuando en la sentencia del tribunal de instancia esta regla no haya sido respetada[22].

---

[21] Tal fué el caso de un interesante asunto al que más adelante hacemos referencia, nota No. 26. Y en la misma linea se inserta también una petición del gobierno argentino al español para que éste le entregara un súbdito suyo, reo de un delito de estafa cometido en Buenos Aires y que se había refugiado en España. El gobierno argentino reconocía que tal delito no figuraba en la lista de los previstos en el tratado de extradición concluido por ambos paises. Este es sin duda el origen del cambio de notas que incluyó el delito de estafa en el régimen del tratado de extradición hispano-argentino de 1881, vid. Gaceta, 27 de Marzo 1935.

[22] Vid. S.T.S., 5 de Nov. 1968, Aranzadi, No. 4529 y Auto, 5 de Nov. 1976, Aranzadi No. 4632. Por el primer fallo se estimó el recurso de casación de quien había sido condenado por la Audiencia como autor de un delito de hurto de uso de vehículo

4. En el art. 13 de la Constitución no se ha establecido como causa de exclusión de la extradición pasiva la ciudadanía española del reclamado. El silencio puede ser interpretado de dos modos: ha sido práctica tan arraigada en el Derecho español la no entrega de nacionales que no se hizo necesaria la prohibición expresa; o también, el silencio es intencionado y trata de no obstaculizar una posible evolución futura del Derecho de la extradición que tiende a dulcificar el rigor y automatismo, al menos en favor de aquellos casos en los que la competencia del Estado extranjero aparezca sólidamente fundada[23].

Se aparta, pues, el texto constitucional de una antigua y bien arraigada práctica y regla en el Derecho interno español. El nivel de información histórica que hoy poseemos permite afirmar que por lo menos a partir de la segunda mitad del pasado siglo tanto la administración activa como la consultiva en España entendieron que era regla del Derecho de gentes la no entrega de los propios nacionales que les vinculaba a ellos en cualquier caso, con tanta mayor razón si la no entrega por esta causa estaba prevista en el tratado de extradición[24]. De ella no se ha separado España (ni siquiera en el tratado con la Gran Bretaña, hoy denunciado) hasta época reciente[24]. Es decir, como una consecuencia de las nuevas ideas políticas del s. XIX, se introduce un fuerte principio de competencia internacional *ratione personae* que distorsiona el de la

---

de motor. Formulada la petición de entrega a las autoridades alemanas, sobre la base del tratado de extradición hispano-alemán de 1878, éstas alegaron por sendas notas verbales su creencia de que no se trataba de un delito de hurto, sino de uso indebido de vehículo. Transferido el individuo a España a causa de otros delitos, la Audiencia le condenó también por el hurto. El Tribunal Supremo anuló esta sentencia.

En el segundo caso se declara igualmente la nulidad de la sentencia de la Audiencia Nacional por la que se condenaba a un súbdito francés entregado por las autoridades belgas en virtud del tratado de extradición hispano-belga de 1870 y que habia sido juzgado por »delito anterior y no contenido en el Convenio«. El Tribunal Supremo razona así su fallo: »La naturaleza de estas disposiciones son de orden público, tanto sustantivo como procesal, en su observancia se compromete la seriedad y prestigio del Estado español y la seguridad del tráfico juridico internacional. Y como en tales supuestos de inobservancia no existe otro medio que la nulidad »ex officio; . . la doctrina de esta Sala ha proclamado que aunque el recurso de nulidad no existe en la L. E. Crim. es remedio excepcional aplicable a los supuestos indicados«.

[23] Así p. ej.: tratado hispano-norteamericano de 1970, art. 4; hispano-mexicano de 1978 art. 7,1; hispano-italiano, art. 25,1. En todos ellos se preve la denegación de entrega de los nacionales como una facultad que se reserva el gobierno, lo que implica que el supuesto general, del que la regla es excepción, es el contrario.

[24] Poco despues de la celebración del tratado hispano-francés de 1850 el gobierno galo reclamó la entrega de un español originario que durante unos años fué soldado en la Legión extranjera francesa y estaba acusado de la comisión de un delito de asesinato en territorio francés. La consulta dirigida por el gobierno español al Consejo de Estado fue sobre si el servicio de armas en un ejército extranjero había hecho perder a este español su ciudadanía. En su contestación el Consejo negó la perdida de ésta y afirmó la obligación del gobierno español de no acceder a la entrega por entender que era un principio de Derecho internacional, recogido además en el tratado, pero también por ser conforme a la constante práctica en Francia.

lex loci commissi, particularmente en un sistema tan fuertemente arraigado a él como lo era el español[25].

Quizá por eso en esa práctica internacional se advierte una cierta laxitud en la apreciación de la ciudadanía española cuando el requerido por la demanda de extradición no era español originario, o no era claramente español, aunque pudiera ser considerado súbdito[26]. Pero, a su vez, esta flexibilidad ha sido conseguida con un muy insatisfactorio grado de seguridad jurídica incompatible con elementales principios jurídicos. Por fortuna esta carencia está hoy superada, en primer lugar, gracias a la jurisprudencia del Tribunal Supremo[27], y hoy también a una mejor y más cuidada técnica en la redacción de los tratados de extradición[28].

---

[25] Tal fué el caso de la solicitud de entrega formulada por el embajador francés a principios de este siglo respecto de un argelino. Kaid Slimane, antiguo soldado indígena en Argelia quien después de robar y asesinar a un soldado mulsulmán se refugió en el Marruecos español. En el momento de la reclamación prestaba servicios como policía indígena en la guarnición de Melilla. El dictamen del Consejo de Estado fué favorable a la entrega, interpretando un tanto restrictivamente el art. 1 del tratado de extradición hispano-galo de 1877.

Pero en otras ocasiones su criterio ha estado más matizado. Asi fué en otra reclamación de Francia sobre la base del mencionado tratado. Solicitaba el embajador galo que le fuera entregado un tal Montana, natural de los Valles de Andorra como presunto reo de un delito de asesinato y del que el mencionado diplomáticio presumía que »las autoridades españolas no tendrían dificultad alguna en su entrega por ser Andorra un pais independiente de España«. Es decir, un criterio exactamente contrario al de Francia que no entrega a ningún andorrano por considerar que son súbditos franceses. El parecer entonces del Consejo Real (antiguo Consejo de Estado) fué que el gobierno debía dar largas al asunto para no prejuzgar indirectamente la naturaleza de las relaciones de España con Andorra.

[26] Un caso verdaderamente límite fué el asunto resuelto en 1911. En virtud de Real Decreto se otorgó a un súbdito belga la nacionalidad española con fecha de 19 de Julio de 1910, la inscripción de esa concesión se hizo en el Registro Civil con fecha de 12 de Agosto de ese mismo año. El encargado de negocios de la Embajada francesa, en virtud de nota de 26 de Julio de 1910 solicitó la detención provisional de ese individuo por encontrarse sumariado por el juzgado de instrucción del Tribunal del Sena a causa de un delito de estafa cometido en 1909. Por ulterior nota de Abril del año siguiente la Embajada formalizó la petición de entrega en virtud de condena por sentencia firme. El gobierno español notificó el asunto al belga, pero éste se desentendió.

Consultado el Consejo de Estado, éste fue de parecer que debia accederse a la entrega por estimar que, tanto si se tomaba como criterio el de la fecha de la comisión del delito (Resolución de Oxford del IDI) como la de la petición de extradición (discutible) el interesado no era en ninguna de ellas ciudadano español.

Lo menos satisfactorio de todo este asunto fué la negativa del Consejo a dar audiencia al interesado por entender que el tratado no lo permitía.

[27] Vid. S. del T. S. 3 de Febrero 1969, Aranzadi, No 501.

[28] »La condición de nacional será apreciada en el momento de la decisión sobre la extradición«, precisan hoy los tratados con México e Italia, arts. 7,1, y 25, 1 respectivamente.

5. Sobre el tema de la exclusión del delito político de la extradición, la Constitución, por el contrario, se pone en linea con la legislación y práctica anterior. Pero esto no quiere sugerir que la mención constitucional sea superfetatoria; muy por el contrario. La exigencia constitucional exigirá que se matice y perfile el alcance de la prohibición, tarea que se inició durante la vigencia de la Constitución de 1931, pero quedó en los meros balbuceos. Seáme permitido extraer de los problemas prácticos surgidos en esa experiencia algunos puntos de mayor interés.

Según quedó ya dicho, el mandato constitucional es absoluto, no condicionado ni aplazado; afecta pues a todas las normas contrarias cualesquiera que sea su naturaleza y origen o la actividad del órgano estatal que regulen. Pero sobre esa claridad inicial se acumulan después toda una serie de interrogantes: ¿Están incluidos en la prohibición los delitos conexos? Puesto que el texto constitucional para nada alude a la claúsula del atentado está incluida ésta en los delitos de terrorismo. ¿Afecta el texto constitucional con sus permisiones y prohibiciones al régimen convencional del auxilio judicial, que no lleva necesariamente a la extradición? La fijación del concepto mismo de »delito político« es un problema en si. Nosotros hemos sugerido ya, en atención al sistema general de garantías de la constitución, que el órgano más calificado en última instancia es el Tribunal de Garantías Constitucionales, cuyo control será tanto más necesario cuanto que el extranjero, sujeto a un procedimiento de extradición, puede hacer valer ante este Tribunal derechos y libertades constitucionalmente garantizados, pero sobre todo si se hace objeto de extradición a ciudadanos españoles. Pero no está en nuestro proposito negar lo que en la extradición hay de »acto de relación exterior« e, incluso, de contingencia y oportunidad política en la acción pública del Ministerio Fiscal en la persecución del delito.

6. Dentro todavía del tema de la extradición, haremos unas referencias a ciertos problemas jurídicos que nacen de la extrecha conexión que se establece entre los acuerdos de extradición y el Derecho interno (sobre todo el Derecho penal de los Estados contratantes) y en el que la práctica española ofrece algunos casos de interés.

El problema más conocido es el de la adaptación necesaria a que los tratados de extradición vigentes deben ser sometidos, bien mediante una revisión formal, o meramente interpretativa, como consecuencia de la modificación de los respectivos códigos o leyes penales en los Estados contratantes. La dificultad de esta adaptación está en relación directa con las reformas en las leyes penales internas[29]. Conocido es también, aunque menos frecuente, la asime-

---

[29] Aqui reside una de las principales ventajas del método de eliminación, en contraste con el de enumeración. A la hora de configurar los tratados de extradición, la adaptación del tratado a la reforma de la ley penal interna por el primero se opera de modo automático. Vid. *H. Grützner,* Auslieferung, Strupp-Schochauer, Wörterbuch des Völkerrechts, vol. 1 (1960), pp. 115 ss.

tría en determinados tratados entre las obligaciones asumidas por una y otra parte[30].

Pero donde la cuestión presenta aspectos verdaderamente interesantes es cuando esta intensa relación entre los distintos sistemas jurídicos adquiere rasgos conflictuales. Examinaremos dos casos:

a) Aplicación de ley extranjera: Prescindamos ahora de si cuando los tribunales de un determinado pais han de tener en cuenta normas de Derecho público extranjero simplemente realizan una »toma en consideración« o proceden a una verdadera aplicación; es una *quaestio disputata* que no hace a nuestro caso. Pero indiscutiblemente, reglas como la de la doble incriminación o la de la *lex mitior,* o la de la prescripción posible del delito por el que se solicita la extradición exigen que el tribunal o, en general, los órganos de decisión conozcan y apliquen normas penales extranjeras.

Pues bien, resolviendo un recurso de casación el T. S. Sala 2ª (Penal) formuló un pronunciamiento en materia de aplicación y prueba de ley penal extranjera que se aparta de los criterios matenidos por la Sala Civil del mismo Tribunal:

> »Aunque carezcan de la estimación de derecho interno, las normas penales u ordenamiento jurídico sancionador de otro pais, no pueden estimarse será ›quaestio facti‹, sometida a la prueba de las partes, como si se tratara de un derecho de su única disposición, sino que por afectar a derechos públicos de interés preponderante, la precisión de su existencia, vigencia y aplicabilidad corresponde de oficio a los Tribunales de lo penal«[31].

En fecha muy anterior a este pronunciamiento, el Consejo de Estado formuló doctrina parecida sobre la aplicación de normas de Derecho público francés por los órganos de la Adminstración española, en el siguiente y curioso caso:

De nuevo Francia reclamaba la entrega de un presunto delincuente, ciudadano francés asentado en Madrid como comerciante al que se le acusaba de un delito de encubrimiento de robo. En la audiencia en el expediente de extradición – que esta vez si se le concedió al interesado – éste alegó que no se trata-

[30] Durante la negociación del tratado de extradición anglo-español de 1878, el negociador español ya advirtió la dificultad que para el delito político nacía de que en Inglaterra no podía ser fijado por criterio gubernamental, sino que en éste como en cualquier otro caso de extradición la fijación en el Derecho inglés de la naturaleza del delito es de la exclusiva competencia del *jury.* Una observación equivalente hizo también respecto del *habeas corpus.* Por esto, sin duda, el referido tratado regulaba minuciosamente, y de modo inusual para los tratados de esa época, el procedimiento a seguir para la extradición en cada uno de los dos paises. En estas razones está quizas la causa de las dificultades a las que Fraga Iribarne aludía en el debate parlamentario de las Cortes que elaboraron el art. 13, haciendo valer su experiencia como Embajador de España en Londres.

[31] S. de. T. S. 10 de Abril de 1970, Aranzadi No. 1669.

ba de un delito de robo, sino de hurto y que el hurto no estaba comprendido en la lista del tratado como susceptible de extradición. Alegaba, además, que el hecho caia bajo un decreto de amnistía que el Gobierno francés había dado en favor de todos los que, como él, habían sido soldados en la Guerra Europea. En el tratado de extradición hispano-francés, art. II No. 25, se habia equiparado efectivamente »vol« a »robo« y por eso la Sección de lo Contencioso del Ministerio de Estado era de parecer que había que atenerse a »los términos claros y precisos del tratado, aunque resulte desigualdad en su aplicación para las partes contratantes«. El Consejo de Estado, por el contrario, estimó que aunque era cierto que la ambivalencia del término francés »vol« podría inducir a incluir indistintamente los dos conceptos del Derecho español, la falta de precisión en dicho idioma, no obstante, »no debe inducir a las autoridades españolas a colocarse en situación de inferioridad en la defensa de los acogidos en su territorio«. En relación con la cuestión de si la ley francesa de amnistía de 29 de abril de 1921 alcanzaba o no al delito incriminado, declaró que »respetando la jurisdicción y potestad del Estado francés para hacer aplicación de los preceptos de dicha ley, y limitando el examen de la cuestión al punto concreto del derecho que a España corresponde para aplicar el art. II del citado convenio. . . es indiscutible el derecho de España para negar la pretensión del gobierno francés«.

b) Aspectos conflictuales en la reserva de derechos civiles que efectuan los tratados de extradición: Como claúsula de estilo que se encuentra en la mayor parte de los tratados suscritos por España se ha solido pactar una obligación independiente de entrega de efectos y dinero hallados sobre el delincuente y, como decía el tratado hispano-francés de 1765 de modo expresivo »con particularidad si el delincuente fuese ladrón, todo el dinero y efectos que hubiese robado«. La independencia de esta obligación de entrega la expresan bien aquellos tratados que especifican que cuando fuere física o jurídicamente imposible la entrega de la persona, subsiste, sin embargo, la de los efectos [32].

Pues bien, elemento esencial en esa obligación es la claúsula de reserva que la misma norma convencional establece en favor de los derechos que los terceros hubieran podido haber adquirido sobre los mencionados efectos [33]. En otros términos, la obligación de entrega no es absoluta, está condicionada al respeto de los derechos legítimos de terceros y con ello se efectua imperceptiblemente una remisión conflictual: la tercería sobre esos bienes deberá ser juzgada de acuerdo con el Derecho privado (incluido el Internacional pri-

[32] En el tratado hispano-búlgaro de 17 de Julio de 1930, notable por su perfección jurídico-técnica, se distingue bien el derecho de propiedad privada sobre esos efectos (interés particular) y el interés público que sobre ellos pueda recaer como piezas de convicción, y se reserva el legítimo derecho de propiedad, pero sin liberar al Estado de la obligación de la entrega provisional, con garantía de devolución.

[33] El art. 42 del tratado hispano-italiano de 1973 hace ya expresa referencia a la obligación de la parte requerida de ocupación y entrega de dichos efectos *en la medida prevista por su legislación.*

vado) del pais que realiza la entrega. Ahora bien, si se acepta la moderna tendencia de la autonomía absoluta de las normas penales (en el supuesto dudoso de que este campo pudiera también ser incluido en esa autonomía), la solución sería la contraria. La misma naturaleza de la reserva parece inclinar a caracterizar el derecho así configurado como un derecho real sobre cosas (*ius in rem)* y no obligacional (*ius ad rem).*

Los tratados establecen pura y simplemente la reserva sin fijar criterio alguno respecto del organo llamado a enjuiciar el mejor derecho sobre esos bienes, mas si tenemos en cuenta que, en última instancia se trata de decidir sobre tituladad en derecho de propiedad, cualquiera que sea la competencia de los organos administrativos en el procedimiento de extradición, el asunto deberá ser remitido a los tribunales de justicia, que es donde según los principios básicos y comunes en los distintos sistemas jurídicos occidentales se resuelven los litigios sobre este derecho [34].

III. La competencia penal internacional en el derecho español

1. La competencia internacional de los tribunales españoles en materia civil, penal etc. es uno de los temas marginales, menos cuidados por la doctrina. A

[34] Todos estos interesantes problemas confluyeron en un curioso caso. A lo largo de los años treinta surgió una diferencia entre el Gobierno español y francés que hasta se pensó podría ser objeto de un recurso de interpretación ante el TPJI por los discrepantes criterios en la inteligencia del art. 8 del tratado de extradición de 1877.

En efecto, en Octubre de 1928 la policía francesa detuvo a los autores de un delito de estafa cometido contra un banco español en España. En aplicación del referido tratado, el gobierno francés entregó a los delincuentes, pero no las cantidades monetarias que se les ocuparon y que para entonces constituían una respetable suma. Las autoridades francesas se fundaban para proceder así en el art. 8 del referido tratado. Los delincuentes eran estafadores que habían cometido estafas tambien en bancos de Londres y Ginebra. La Audiencia de Alicante, que era la competente en el enjuiciamiento del delito de estafa cometido en España, no solo había declarado en su sentencia la culpabilidad de los acusados, sino que la propiedad de las cantidades dinerarias que se les habían ocupado correspondian al banco español estafado. El gobierno francés para fundar su derecho a retenerlas alegaba no solamente el art. 8 sino también la regla del Derecho civil francés (válida igualmente en el Derecho español, añadimos nosotros) de que tratándose de cosas fungibles para que se reconzca la propiedad sobre ellas es preciso que quien las reclama las especifique. A ello oponía al banco español la conexión en que estaba el delito con sus consecuencias civiles (sistema español de acumulación de acciones civiles y penales en pleitos criminales) y de que tratándose de distintas estafas había en ellas un mero concurso real y no un delito continuado, por lo tanto, no podía aplicarse el dinero de esta estafa a la indemnización de las víctimas de las otras (tesis, al parecer, de Jiménez de Asúa cuya opinión había sido solicitada en dictamen). Se aducía también por parte española que »la reserva de un derecho no declarado *a priori* no puede implicar dilacción en el cumplimiento de un trámite obligado«, lo que sin duda es cierto dentro del Derecho administrativo o procesal español, pero inidóneo en el plano de la relación jurídica internacional.

caballo entre distintas disciplinas, dispares entre si, ha sido y es mancomún de muchos, y por lo tanto no considerado como propio por nadie. Probablemente por esta misma razón las normas que la regulan presentan esas mismas deficiencias. Formuladas de manera fragmentaria y asistemática, se encuentra dispersas en distintas leyes especiales y sin que el legislador haya establecido la necesaria relación entre el ámbito de aplicación espacial de las reglas materiales y las correspondientes normas de procedimiento. Tanto la competencia internacional civil como la procesal estuvieron por igual bajo el perturbador influjo del fuero de extranjería[35, 35 bis].

Así pues, en el tema general de la competencia internacional (tanto de la civil como de la penal) nos movemos todavía entre zonas de luz y oscuridad. Conocemos todavía bastante mal las líneas básicas del tema en la legislación y práctica del s. XVIII, como antecedente inmediato a las reformas legislativas del siglo pasado[36]. Pero puesto que existió fuero especial de extranjería, evidente es que al extranjero (por eso tal competencia especial meramente interna) se le encausaba y juzgaba en España según leyes penales españolas. Ahora bien, ¿Como se armoniza tal proceder con las normas internas españolas (vid. pp. 712 ss.) o las obligaciones pactadas en acuerdos internacionales sobre la entrega de extranjeros, delicuentes en España, a su respectivas soberanías para el adecuado castigo?

En otra linea de pensamiento lo que si parece evidente es que la justicia administrada en España »en nombre del Rey«, lo fué en el sentido estricto de la delegación y que los casos de extranjería, como asuntos que afectaban a la relación internacional estaban especialmente reservados a su acción directa. Fué sin duda el principio de separación de poderes, y quizá aún más el de le-

---

[35] En otra parte nos hemos ocupado de estas carencias en la fijación de la competencia internacional en los tribunales civiles españoles (*J. Puente Egido,* Derecho Internacional Privado Español [Barcelona 1981], pp. 925 ss.). No es este el lugar para volver sobre este aspecto del problema, baste con decir como complemento de lo que diremos respecto de la competencia penal, que los tribunales civiles en el pasado siglo por una defectuoso comprensión de la entonces nueva normativa (ciertamente fraccionaria) sobre competencia internacional, y sobre todo olvido del Derecho anterior afirmaron, en ocasiones, una competencia exorbitante que pudo hacer incurrir al Estado español en responsabilidad internacional. Y, sin embargo, su radical incompetencia v. gr. en acciones de embargo intentadas por extranjeros respecto de buques extranjeros surtos en puertos españoles, estaba ya bien señalada en ese Derecho. Vid. p. ej. *Tapia,* Librería de Jueces. . . etc., vol. 5 (1829), p. 73: »Las naves extranjeras que traen a estos reinos mercaderías y bastimentos, tampoco deben ser ejecutadas. . . etc.« y lo seguía recogiendo bien el primer código de comercio de 1829 art. 605, y aunque con menos claridad también lo contempla el actualmente vigente.

[35 bis] Un curioso caso de arresto de un marino inglés en el puerto de Málaga a mediados del pasado siglo como presunto autor de un delito, y que dió lugar a un conflicto competencia entre el juez de extranjeros y el ordinario, muestra que una de las razones del establecimiento del fuero especial de extranjeros era la celeridad en la administración de justicia.

[36] Vid. *A. Cornejo,* Diccionario histórico y forense del Derecho Real de España (Madrid 1779), pp. 362 ss.

galidad, los que impulsaron al legislador español a formular las normas de competencia internacional penal, bien que en fecha relativamente tardía, que siguen hoy todavía en vigor.

Así se deduce ciertamente de los datos históricos siguientes: Fué inmediatamentes después de la instauración del periodo constitucional cuando con ocasión del primer código penal, el legislador español formuló las primeras reglas de competencia internacional. Pero este código sufrió la suerte general de toda la legislación constitucional con la reimplantación del poder absoluto, y las sucesivas reformas que sufrió la legislación penal en el pasado siglo dejaron fuera del código las referentes a la validez de la ley penal en el espacio [37]. Y en esta situación de anonimia se estuvo hasta que fueron formuladas las aludidas normas en 1870 con ocasión de la promulgación de la LOPJ. Parece bastante seguro que el origen de tales normas está en un incidente diplomático con Rusia, el asunto Bella, que por su interés merece ser contado aquí en detalle.

Este español avencindado en Odesa fué perseguido por las autoridades rusas como presunto reo de quiebra fraudulenta en la que una de las principales víctimas había sido otro español, también avecindado en aquella ciudad. El español huyó y se refugió en Constantinopla donde el representante diplomático ruso en Turquía siguió solicitando la colaboración de las autoridades turcas y del encargado de negocios español para su captura y conducción a Odesa. El Ministro de Estado español, que cursó instrucciones a nuestro representante en Constantinopla en este sentido, consiguió alegando el art. 6 del Tradado de 1782 con la Sublime Puerta que, capturado Bella por las autoridades turcas, se le remitiera detenido a Barcelona donde quedó a disposición del Regente de la Audiencia de Barcelona. Conocido este hecho por las autoridades rusas, la legación de este pais solicitó la entrega, pero le fué denegada por

[37] El Código penal de 1822 (Vid. Colección de los Decretos y Ordenes Generales expedidos por las Cortes, T. 9 [Madrid 1922], p. 212) fijó por primera vez como principio general la aplicación de la ley española a españoles y extranjeros por todos los delitos que cometieren en territorio español »salvas las exceciones estipuladas en los tratados existentes con otras Potencias«. La cláusula de reserva de tratado apuntaba, según lo muestran los debates parlamentarios, a lo estipulado entonces con Portugal y Marruecos sobre entrega a esos Estados de súbditos suyos autores de delitos cometidos en España. Según las explicaciones que la Comisión redactora dió a las Cortes se trataba: a) De establecer la generalidad de la ley penal española »sin distinción alguna«, y como exigencia del principio de igualidad. b) De imponer la aplicación de la ley española a los delitos cometidos por los españoles en el extranjero cuando fueren detenidos en territorio español o fueren entregados a España. Vid. Diario de Cortes, Ses. 1 de Dic. de 1821, p. 1062.

Las normas sobre validez espacial de la ley penal española fueron suprimidas en las sucesivas reformas del Código y no vuelven a aparecer hasta la de 1928, pero sin transcendencia, dada la corta vida que tuvo esta reforma. La que se fectuó en 1932, las volvió a suprimir para someterlas a una mayor elaboración en una reforma futura más profunda, que todavía no ha tenido lugar. El actualmente vigente, contiene en la parte especial y referida a una categoría concreta de delitos alguna regla de validez espacial.

el Ministerio de Estado español en razón a la ciudadanía española indiscutible del reclamado. Entendía el Ministerio español que siendo las Audiencias de Barcelona y Mallorca tribunales de alzada respecto de las decisiones judiciales que los cónsules de España en Levante dieran en ejercicio de sus funciones judiciales, el Cónsul español de Odesa debía remitir el fallo de los tribunales rusos de Odesa para deducir el tanto de culpa y encausar a Bella.

Pero planteado judicialmente el asunto en España, el juez de primera instancia decidió inhibirse y la inhibición fué confirmada por la Audiencia. En su decisión entendía la Audiencia de Barcelona que la detención de Bella estaba motivada por una medida gubernativa del Ministerio de Estado, pero que no no habiendose dictado auto de prisión por juez ni tribunal alguno que dependiera de su jurisdicción, no podía retener al detenido porque de otro modo violaría los arts. 3 y 4 de la Constitución (1869). Por decisión de la propia Audiencia el detenido quedaba a disposición del gobernador de Barcelona, quien entendiendo que la detención gubernativa lesionaba una garantía constitucional, le concedió libertad bajo fianza.

La situación resultaba cada vez más embarazosa para el gobierno español. Los indicios racionales de culpabilidad eran manifiestos, y presionaban en favor de la entrega no solo la legación rusa en Madrid, sino también el cónsul español de Odesa y los representantes diplomáticos de España en Rusia y Turquía, ya que en la posible impunidad de Bella veían ellos comprometido en buen nombre de España. El Gobierno español entendía que no podía acceder a la extradición sin infringir elementales normas internas e internacionales, que, por lo demás, tenía que respetar la independencia de los tribunales españoles; y estos a su vez entendían que no podían dictar auto de prisión si no había juez manifiestamente competente, y no lo había. Se estaba, pues, ante una verdadera laguna legal a la que puso remedio la ley que en aquellos momentos se estaba preparando y mayor relación de proximidad tenía con la materia [38]. Este parece ser el origen de las normas de competencia penal internacional en la LOPJ.

Y fué en esta situación, inmediatamente anterior y posterior a la promulgación de la LOPJ, como el caso paso a consulta del Consejo de Estado, quien en un bien equilibrado dictamen, puso de manifiesto las irregularidades y carencias en que habían incurrido los diversos órganos de la administración, incluidos los de justicia, y sugirió una vía de solución, a la que nosotros, brevemente, desearíamos formular unas observaciones.

Promulgada la nueva ley – estimaba este Alto Cuerpo Consultivo – no había ya incompetencia de los tribunales españoles sobre hechos delictivos cometidos por ciudadanos españoles en el extranjero contra otros españoles o contra extranjeros, siempre que medie querella de la parte ofendida. No hay, pues, impulsión de oficio, ni bastaba la reclamación de la representación di-

---

[38] Los datos que figuran en este expediente, indican muy bien el modo en que se enjuició el tema antes y depués de la reforma legislativa aludida.

plomática rusa en Madrid. Con ello reflejaba el mismo espíritu liberal de la ley, que hoy – al menos en parte – ha sido superado por la jurisprudencia de nuestro T. S., como después se dirá. La circunstancia de que la LOPJ tuviera un efecto retroactivo no le planteaba dificultad alguna puesto que estimaba que la nueva ley ni creaba penalidad nueva, ni definía como delitos hechos que con anterioridad no estuvieran ya comprendidos en nuestras leyes penales. Lo que establecía simplemente era la competencia de los tribunales españoles y siendo estas normas reglas procesales, a ellas no le alcanza la regla estricta de la irretroactividad de la ley penal. Refleja aquí el Consejo un error de apreciación, muy extendido en nuestro sistema, en el que sobre todo los jueces son proclives a la calificación »procesal« de normas que, por afectar a principios básicos en la recta administración de justicia o de garantías fundamentales del proceso o de formación de la voluntad en órganos colegiados, son evidentemente reglas sustanciales.

¿Cuál fué el origen de esta laguna en el sistema jurídico español? Las cosas debieron ocurrir más o menos de esta forma. La competencia de la *lex loci commissi* como criterio de fijación de la competencia penal está firmemente arraigada en la coinciencia jurídica hispana. La tenían nuestros fueros medievales, la recogen Las Partidas, las dos Recopilaciones, y debió estar todavía muy firme en el siglo XVIII [39]. Pero el proceso de territorialización que el Estado moderno opera, sobre todo cuando se transforma en constitucional, modifica esta conexión en un doble sentido: a) La conexión a la *lex loci* no es conexión directa a la ley del lugar del delito, sino a la ley del Estado en cuyo territorio nacional el acto delictivo fué cometido [40]. b) El paso de la noción de súbdito a ciudadano, con derechos y deberes constitucionalmente garantizados, hace de la pertenencia del individuo a esa comunidad – de la que el derecho constitucional a permanecer en su teritorio no es más que un corolario – un derecho fundamental, del que dispone libremente el titular, pero que ha de ser respetado por toda autoridad o poder público. En otros términos, el constitucionalis-

[39] Vid. *Carlevalii* (nota 11), Saldaña en sus anotaciones a la obra de *F. von Liszt, Tratado de Derecho Penal*, vol. 2 (Madrid 1916), p. 184. En la información pública que las Cortes de 1820 abrieron en los trabajos de preparación del primer código penal diversas instituciones se proclamaron abiertamente en favor de la *lex loci commissi*, Audiencia de Valladolid y de Extremadura, Universidad de Salamanca, Colegio de Abogados de Madrid etc. Vid. Diario de Cortes *loc. cit.* En esta misma linea de pensamiento está la opinión disidente del juez español Altamira en el asunto del *Lotus*, Recueil des Arrêts Ser. A, No. 10 etc. etc.

[40] La obligatoriedad de la ley penal sin excepción a todos los habitantes del territorio nacional *(nemo legem ignorare censetur)* se impuso en el debate parlamentario del primer código penal como una exigencia del orden interno constitucional, pero no sin cierta resistencia por parte de algunos parlamentarios en los que todavía pesaba la tradicional distinción en el pensamiento jurídico-moral hispano de las acciones *mala in se* y *mala quia prohibita* a efectos del presumible efecto en el conocimiento o desconocimiento por parte del extranjero en el cumplimiento de la ley penal española y graduación, por consiguiente, de la responsabilidad penal en que estuviere incurso el infractor. Vid. Diario de Cortes (nota No. 37).

mo no solo impone por la regla de la igualdad ante la ley la sumisión estricta de todos a la ley penal nacional, sino que establece un principio de competencia personal igualmente absoluto respecto de los actos delictivos de los propios ciudadanos, presuntos delincuentes por actos cometidos en el extranjero y acogidos a su territorio nacional.

2. Principios de competencia penal internacional. Dentro del reducido marco de esta reflexión haremos una breve referencia a los principales:

a) Principio territorial: Doctrina y jurisprudencia estan contestes en afirmar que el criterio primordial en la fijación de la competencia penal internacional es el principio territorial [41]. Al margen de las reglas contenidas en leyes especiales, que confirman y modulan este criterio [42], las normas básicas que lo establecen son el art. 8 del Código civil, el 29 del R. D. de 17 de Nov. 1852 y de modo indirecto también los arts. 333 y 335 LOPJ. La jurisprudencia confirma, y en cierto sentido amplía esa orientación territorialista de la ley.

Ahora bien, aunque sin demasiadas consecuencias prácticas todavía, esta fijación de la competencia no ha sido suficientemente esclarecida por referencias a exigencias precisas del orden jurídico internacional. Muy por el contrario, en algún fallo nuestro Alto Tribunal ha entendido que la limitación que de esa regla se sigue es una *auto* y no una *heterolimitación:* »El principio de la territorialidad de las leyes penales, consagrado en al art. 8 del C. c. no tiene más limitaciones que las que el mismo Estado, en el ejercicio de su soberanía ha querido imponerse y que en el ordenamiento jurídico actual están contenidas en el art. 334 de la LOPJ, limitaciones que han de tener interpretación restrictiva. . . etc.« [43].

---

[41] De esencial lo califica la S. del T.S. de 20 de Dic. de 1980 y considera a los demás principios complementarios de éste con los que se trata de »colmar las lagunas de impunidad que comportaría una aplicación exclusiva del principio territorial«. Otra S. de 10 de Abril de 1970 decia: »Como emanación de la soberanía de los Estados nacionales, por la coordinación indispensable de su imperio y de los regímenes existentes en la comunidad internacional, y por la lesión y agravio más directo que el delito causa en el orden público interno del pais donde se actúa, se impone con indudable prevalencia en los distinto derechos positivos penales el principio básico de territorialidad de la ley«.

[42] Vid. referencias actualizadas en *Rodriguez Devesa,* Derecho Penal Español, vol. 1 (Madrid 1981), pp. 210 ss.

[43] Evidentemente, la expresión es inexacta por cuanto no es ni puede ser el art. 334 (Inmunidades diplomáticas) la única limitación a la jurisdicción territorial del Estado. En el caso concreto se trataba de negar el privilegio de inmunidad jurisdiccional a la esposa del agregado naval de la Embajada norteamericana en Madrid, encausada por delito de imprudencia. El concepto restrictivo que aparece en este auto fué modificado más tarde en decisión de 9 de mayo de 1974 en una causa de adulterio seguida contra un miembro del MAAG-Spain, personal de la ayuda técnica que por el Convenio de Ayuda y Defensa Mutua de 26 de Septiembre 1953 entre España y los Estados Unidos de América del Norte se les asimiló al personal diplomático. A mayor abundamiento eran aplicables ahora también los arts. 31, c y 37,2 del la Convención de Viena de 1961, ratificada por España el 21 de Nov. 1967 y por los E. U. de América del Norte el 13 de Nov. 1972.

De aquí quizá una cierta tendencia expansiva, no obstante la confesada territorialidad, hacia ámbitos penales en los que la tipicidad de la regla penal y del hecho delictivo depende ampliamente, bien del Derecho privado o bien del Derecho extranjero (privado o del lugar de la comisión del hecho o de la omisión de la acción en la que consiste el delito). Es el caso de delitos que como el adulterio la bigamia o el abandono de familia están en directa conexión con el estado y condición de las personas y que hacen planear algunas dudas sobre la estricta territorialidad de las leyes penales. La generalidad de la ley penal debe subsistir, evidentemente, como garantía del orden territorial que corresponde a todo poder estatal; se trataría más bien, dentro de esa generalidad de modularla, personalizando más la norma penal para atender al pluralismo interno de las sociedades modernas[44].

En las pocas ocasiones en las que el T. S. ha sido invitado en los recursos de casación a atenuar el rigor estricto de la ley penal española como ley territorial (recursos, ciertamente, endebles desde el punto de vista de la argumentación doctrinal), el T. S. ha apelado a principios de moral universal o de derecho natural queriendo dejar patente que la obligación que impone la norma penal española y por la que se castiga al reo no depende de la particular concepción del legislador que no hace sino concretarla[45]. Por otra via en otra ocasión acudió al caracter de delito permanente en el abandono de familia para justificar el castigo de españoles residentes en Suiza, cuyo domicilio estaba también en aquel pais[46].

Todas estas carencias en la aplicación del principio territorial pueden y deben ser enfrentadas desde las limitaciones que las reglas generales del Derecho internacional impongan a la competencia teritorial de los Estados; es decir, en un sentido equivalente al que tiene la más reciente doctrina alemana con ocasión de algunas decisiones jurisprudenciales[47].

---

[44] *Donnedieu de Vabres,* Les principes modernes du droit pénal international (París 1928), p. 78, admitió ya que normas penales como las que fijan el límite de la responsabilidad penal y otras »sont des lois de *garantie individuel.* Aussi estimons- nous que la doctrine et la jurisprudence aujourd'hui constantes qui les appliquent sans discernement aux nationaux et aux étrangers, qui bornent leur application aux limites du territoire, sont entachées d'erreur«.

[45] S. del T.S. 18 de Febrero 1964. Aranzadi No. 831.

[46] S. 22 de Octobre 1975, Aranzadi No. 3974, no aparece claro del apuntamiento dónde estaba domiciliado el matrimonio en el momento de la imposición de la pena. El delito de abandono de familia, según tiene declarado la jurisprudencia, se comete allí donde se deja de prestar el auxilio a que se está obligado y en lo que consiste el delito ¿Vale igualmente este punto de conexión cuando el lugar de la prestación es el extranjero? ¿Sigue siendo igualmente delito cuando la *lex loci* (que es la del lugar de la prestación) siendo extranjera no considera delito el abandono de familia? Acaso por esta razón el T. S. hechó mano de la noción de delito permanente.

[47] El proyecto originario de constitución existía una regla equivalente al art. 25 de la Ley Fundamental de Bonn, que fué después suprimida.

b) Principio de la competencia personal: Los arts. 339 y 340 de la LOPJ expresan bien que esta competencia es excepción y complemento de la territorial, y así lo ha entendido de modo constante la jurisprudencia [48].

Si la competencia teritorial exige una fijación precisa y exacta del concepto de territorio, la personal postula igualmente una precisión del vínculo que liga a la persona individual con el Estado que pretende tener jurisdicción sobre ella y ya dijimos algo de este problema cuando hablamos de la extradición. Añadamos aquí que, por lo menos en el Derecho español, la nacionalidad del sujeto activo del delito no basta por si misma a fundar competencia, sino que entran en juego otras circunstancias como la nacionalidad del sujeto pasivo, el bien jurídico lesionado, la ley del luegar de la comisión y la querella de parte. Evidentemente esta es una competencia muy condicionada y en eso se diferencia radicalmente de la anterior, que es absoluta.

En contra de lo que vimos fué doctrina del Consejo de Estado, el T. S. tiene resuelto que el momento decisivo para apreciar la nacionalidad del delincuente a efectos de competencia no es el de la comisión del hecho sino el de su represión [49].

La aplicación extraterritorial de la ley se opera por dos vias: 1) Delitos cometidos por español contra español. 2) Delitos cometidos por español contra extranjero. Estos son los supuestos contemplados por los arts. 339 y 340 LOPJ respectivamente, bajo condiciones que la doctrina procesal española – quizá por infuencia de la italiana – ha considerado meramente como condiciones de procedibilidad. La diferencia sustancial entre uno y otro radica en que mientras que en el segundo supuesto es preciso que la ley del lugar de la comisión considere a la acción u omisión como *delito grave* (grave según la valoración que haga la ley española), exigencia, pues, de la doble incriminación, no ocurre lo mismo con el primer caso porque ahi basta la mera incriminación por la ley española. Se está, pues, ante un caso, *químicamento puro,* de competencia personal ya que la proyección extraterritorial de la ley española se efectúa por la doble via de la nacionalidad del sujeto activo y del sujeto pasivo del delito. ¿Es esta regla compatible con la ordenación de las competencias penales de los Estados que realiza el Derecho internacional? ¿Reponde al modelo de relaciones intersistemáticas y de limitación de competencias estatales ínsito en la vigente constitución?

---

[48] S. 10 de Abril 1970, Aranzadi No. 1669.

[49] S. 3 de Febrero 1969 No. 501. Según *Jiménez de Asúa,* Tratado de Derecho Penal, vol. 2, 4ª edic., p. 850, el tema había sido ya resuelto por una Resolución de la Fiscalía de 10 de Febrero de 1911. Se trataba de una italiana que cometió delito en Buenos Aires contra ciudadano español y adquirió después la nacionalidad española por matrimonio con español. La idea rectora en la argumentación del T.S. es evitar la impunidad, ante la imposibilidad de la extradición por la nacionalidad española del encausado. El argumento prueba demasiado, puesto que no debería jugar allí donde la extradición no se solicite. En todo caso consideramos que esta tesis parece incompatible con el principio de legalidad, hoy además garantizado constitucionalmente.

Toda extensión de competencia penal de las leyes de un Estado incriminando conductas de individuos que las ponen y realizan por entero en el territorio de otro (aunque sean ciudadanos del primero) debe mantener un cierto grado de conexión con el Estado que incrimina para que dicha competencia no resulte exorbitante. Pero aún aquí hay que reconocer grados: la acción puede afectar a los vínculos de fidelidad del ciudadano con su propio pais y ahí la conexión es manifiesta. Puede ser una acción u omisión simplemente permitida por las leyes estatales del lugar de la realización, y entonces habrá lugar a muchas matizaciones; y, finalmente, puede tratarse de conductas impuestas por las leyes del Estado local, y aqui la excusa debe ser absolutoria. En todo caso entra en juego una garantía individual de seguridad jurídica a la que no es ajeno el sistema protector de la Constitución y el orden jurídico internacional[50].

Tambien una reciente sentencia del T. S. muestra las deficiencias que en esta via tiene nuestro sistema. Dos españoles se trasladaron al extranjero y, concretamente a un pais abortista, donde se le practicó a la mujer el aborto. De retorno a España ambos fueron encausados y se les impusieron penas aplicando el art. 411,2 del Código penal. El recurso de casación fué formulado por infracción de ley. El art. 339 de la LOPJ había sido infringido, porque, exigiendo ésta como condicion de procedibilidad, que el delito cometido por español en el extranjero sea contra otro español, difícilmente podía sostenerse que el feto, que no había nacido, sin adquirir la condición de sujeto, hubiera podido poseer nacionalidad española. La mujer sobre la que se practicó el aborto no podía ser a la vez sujeto activo y pasivo del delito. El T. S. salió del apuro con una sentencia escasamente convincente. Apeló a dos argumentos: Utilizó sobre todo la noción de fraude a la ley. El traslado a pais extranjero cuyas leyes permisivas toleran prácticas abortivas habría constituido en realidad un fraude a la ley española [51]. El argumento es inconsistente; una de dos, o el ámbito de la ley penal española es extraterritorial y entonces no hay fraude, sino pura y simple infracción, o no tiene tal eficacia extraterritorial y entonces no hay fraude pura y simplemente porque no habría delito. ¡Excelente ejemplo de la buena causa defendida con malos argumentos! En una segunda via el T. S. se fué por el principio de la protección real, el de los bienes jurídicos protegidos »el de la vida del feto y el del Estado que tiene un interés ético, familiar o simplemente demográfico en la conservación del nasciturus« [52].

---

[50] Altamira en su voto particular *loc. cit.* hacía ya observar agudamente la relación en que está la buena aplicación del principio territorial con el respeto a los derechos de la persona humana (p. 104).

[51] Por esto *Kohler, op. cit.*, p. 90, y *Oehler,* Internationales Strafrecht, pp. 143 y 144, rechazan el principio de la personalidad activa y el segundo funda la competencia en estos y caso equivalentes en el de la solidaridad entre Estados.

[52] Un riesgo de deslizamiento hacia la competencia real, que ya había advertido *Kohler, op. cit.*, pp. 93 ss., quien cita un caso de la jurisprudencia belga muy semejante al que ahora presentamos. Vid. S. T.S. 20 de Dic. 1980.

Los referidos preceptos de la LOPJ exigen que el presunto delincuente sea habido en territorio nacional. Es decir, está excluido el juicio en rebeldía. Pero la ley nada especifica sobre el modo de su presencia, no exige que esta sea voluntaria o en virtud de un procedimiento regular de extradición. Aunque no parece haber antecedentes, en caso de que la presencia en territorio español fuera debida a una detención ilegal, habría probablemente fundamento para negar la competencia[53].

Tambien estas normas hablan de que medie querella del ofendido o »cualquier persona que pueda hacerlo con arreglo a las leyes«. En el pensamiento liberal de la época estaba la idea de la necesidad de que la acción pública en esta clase de delitos cometidos en el extranjero hubiera impulsión de parte, y a esa regla se atuvo la primera interpretación[54]. Aprovechando la ambigüedad de los términos entrecomillados, el T. S. ha ido a una interpetación extensiva y hoy acepta que el querellante sea el propio Ministerio Fiscal, salvo en los delitos en que por ley española se necesita instancia de parte[55].

---

[53] Como se sabe los paises anglosajones e Israel, caso Eichmann, tienen criterio distinto.

[54] Así lo indican las primeras Memorias de la Fiscalía del Tribunal Supremo Vid. Memoria de 1890, p. 45. Delito de estafa cometido por español en Portugal: »No habiéndose querellado el ofendido, sino simplemente habiendo presentado denuncia, debe cesar el procedimiento de oficio«.

[55] S. del T.S. de 1 de Dic. 1979, Aranzadi, No. 4518.

Die »Einzelfallbezogenheit« in der Rechtsprechung des Europäischen Gerichtshofs für Menschenrechte

Georg Ress

1. Die dreifache Problemstellung

Der Einzelfall, an Hand dessen vom Europäischen Gerichtshof für Menschenrechte (EGMR) der Verstoß gegen die Europäische Menschenrechtskonvention (EMRK) festgestellt wird, kann in dreifacher Hinsicht zum rechtlichen Problem im Verfahren vor den Straßburger Organen werden: als Entscheidungsmaxime (»Einzelfallgerechtigkeit«), als objektive und subjektive Grenze der Rechtskraft (Streitgegenstand des Verfahrens und Problem der *inter partes*- bzw. *erga omnes*-Wirkung) sowie als Richtschnur für das Verfahren nach Art. 54 EMRK, in welchem das Ministerkomitee die Durchführung der Urteile überwacht.

Gerechtigkeit als Maxime der Rechtsprechung erhält besonderes Gewicht, wenn es um die behauptete Verletzung von Menschenrechten geht. Wer wollte ernsthaft in Zweifel ziehen, daß dem Richter aufgegeben ist, Gerechtigkeit gerade im konkreten Einzelfall zu finden – man könnte fragen: Kann es eine andere Gerechtigkeit geben als die auf den konkreten Einzelfall bezogene? Doch was bedeutet es, wenn im Zusammenhang mit der richterlichen Tätigkeit auch internationaler Spruchkörper von »*Einzelfallgerechtigkeit*« als Entscheidungsmaxime gesprochen wird? Die Erfassung einer derartigen Entscheidungsmaxime gehört, wie *Hermann Mosler* erst jüngst festgestellt hat[1], zu den »schwierigsten Fragen« richterlicher Rechtsfindung. Ihre Bedeutung zeigt sich vor allem bei der Abgrenzung des staatlichen Beurteilungsspielraums (*marge d'appréciation* bzw. *margin of appreciation*) von Bindungen auf Grund autonomer Begriffe oder Begriffselemente der EMRK. Der Gerichtshof muß die Frage, ob der staatliche Bewertungs- und Handlungsspiel-

[1] *H. Mosler,* Der Europäische Gerichtshof für Menschenrechte nach zwanzig Jahren, in: Recht als Prozeß und Gefüge, Festschrift für Hans Huber (1981), S. 601.

raum überschritten worden ist, an Hand der konkreten Beschwerde prüfen[2]. Dadurch entsteht, wie *H. Mosler*[3] schreibt, »das aus der nationalen Verfassungsgerichtsbarkeit bekannte Problem der Einzelfallgerechtigkeit, die auf der Anwendung unbestimmter Rechtsbegriffe beruht«. Das Ergebnis sei so lange schwer voraussehbar, als die Begriffe noch nicht in einer Vielzahl von Fällen konkretisiert worden sind. Diese Methode sei – trotz der Kritik, die sie gefunden habe[4] – unverzichtbar. Der Gerichtshof bemühe sich, sie mit Sorgfalt anzuwenden, wie die Begründungen, die im *Sunday Times*-Fall[5] von der schwachen Mehrheit im Urteil und der großen Minderheit in der gemeinsamen abweichenden Meinung (11:9) für ihre Auffassungen in diesem Punkt abgegeben worden sind, zeigen.

Mit der »Einzelfallgerechtigkeit« hängt ein weiteres Problem, nämlich das der unmittelbaren Wirkung der Urteile des EGMR zusammen. Die Konzentration auf den Einzelfall führt nicht nur zu der Frage, inwieweit Aspekte der Einzelfallgerechtigkeit selbst zum Beurteilungsmaßstab gehören. Vielmehr führt diese Konzentration auf den Einzelfall auch zu der Frage, worauf sich die Feststellung einer Konventionsverletzung *in concreto* bezieht, was also 1) Urteilsgegenstand ist und 2) worauf sich die Verpflichtung der Staaten, sich in allen Fällen, an denen sie beteiligt sind, nach der Entscheidung des Gerichtshofs zu richten (Art. 53 EMRK), bezieht. Welche Verpflichtung erwächst demnach den Vertragsstaaten aus Art. 53 EMRK? Ist diese Verpflichtung nur auf die Wiedergutmachung im konkreten Fall gerichtet oder schließt sie die Verpflichtung ein, ähnlichen Fällen in Zukunft vorzubeugen?

---

[2] Vgl. z.B. EGMR vom 13. 8. 1981 im Fall *Young, James und Webster* (Europäische Grundrechte-Zeitschrift [EuGRZ] 1981, 559 ff.), in dem es um die Verletzung von Art. 11 EMRK durch nachträgliche *closed shop*-Regelungen geht. Der EGMR hebt »zum wiederholten Male hervor, daß er sich in Verfahren, die auf Individualbeschwerden zurückgehen, soweit als möglich auf die Problemstellungen zu beschränken hat, die sich aus dem konkret zu entscheidenden Sachverhalt ergeben, wobei allerdings der allgemeine Zusammenhang solcher Problemstellungen nicht aus den Augen verloren werden darf (vgl. u.a. Fall *Guzzardi,* Urteil vom 6. 11. 1980, Série A, Nr. 39, S. 31–32, § 88=Human Rights Law Journal 1980, 273). Im Einklang mit diesem Grundsatz ist der Gerichtshof nicht berufen, in dem vorliegenden Fall das *closed shop*-System als solches auf seine Vereinbarkeit mit der Konvention zu überprüfen oder zu jeder seiner einzelnen Auswirkungen oder Zwangsmaßnahmen Ausführungen zu machen; der Gerichtshof beschränkt seine Überlegungen vielmehr auf diejenigen Auswirkungen, die dieses System im Hinblick auf die Beschwerdeführer gehabt hat«.

[3] (Anm. 1), S. 601 f.

[4] Die Diskussion über dieses Problem hatte schon auf dem Fünften Internationalen Kolloquium über die Europäischen Menschenrechtskonvention (vom 9.–12. 4. 1980 in Frankfurt a.M.) eine Rolle gespielt. Vgl. dazu die Bemerkungen in meinem Bericht über »Die Wirkungen der Urteile des Europäischen Gerichtshofs für Menschenrechte im innerstaatlichen Recht und vor innerstaatlichen Gerichten«, in: Europäischer Menschenrechtsschutz – Schranken und Wirkungen, Verhandlungen des 5. Int. Kolloquiums über die EMRK in Frankfurt/M. (1982), mit Schlußbericht von *H. Mosler,* S. 355 ff.

[5] EuGRZ 1979, 386.

Wirkt sich die »Individualisierung« des Verfahrens demnach nicht nur auf die Entscheidungsmaxime, sondern auch auf die Folgeverpflichtungen der Vertragsstaaten aus? Beides sind methodisch nicht zusammengehörende Fragestellungen: gleichwohl wirkt sich zur Beantwortung beider der Sinn und Zweck des gesamten, durch die EMRK eingerichteten Schutzsystems aus.

Daran schließt sich ein weiteres Auslegungsproblem an: Korrespondiert die vom Ministerkomitee nach Art. 54 EMRK zu überwachende Durchführung des Urteils der Verpflichtung der Parteien nach Art. 53 EMRK? Umfaßt die »Durchführung« auch Maßnahmen, die der Herstellung eines konventionsgemäßen Zustandes in der Zukunft dienen – also z. B. Maßnahmen des Gesetzgebers, auch wenn nur die Konventionswidrigkeit eines Urteils oder eines Verwaltungsaktes festgestellt wurde?

2. Die Einzelfallgerechtigkeit

a) Einzelfallgerechtigkeit als Entscheidungsmaßstab?

Nach der »Entscheidungsmethode« [6] im *Sunday Times*-Fall genügt es nicht, daß der beanstandete Eingriff unter die in Art. 10 Abs. 2 EMRK aufgeführten Ausnahmen fällt; der Gerichtshof muß vielmehr überzeugt sein, daß der Eingriff unter Berücksichtigung aller erheblichen Tatsachen und Umstände des ihm unterbreiteten Falles notwendig war. Unter Berücksichtigung dieser Umstände kommt der Gerichtshof dann zu dem Schluß, daß der beanstandete Eingriff (Urteil des House of Lords, welches eine einstweilige Anordnung gegen die Sunday Times wegen *contempt of court* bestätigte) [7] »nicht einem sozialen Bedürfnis entsprach, das hinreichend zwingend war, um das öffentliche Interesse an der Freiheit der Meinungsäußerung im Sinne der Konvention zu überwiegen«. Diese *Abwägung mittels* und nur *für* den konkreten Einzelfall [8] – im Fall *Airey* [9] in den Bereich der »further realisation of social and economic

---

[6] So der EGMR in EuGRZ 1979, 391 (Nr. 67).

[7] Streitig war, ob sich die Beschwerde nur gegen diese Entscheidung des House of Lords oder auch gegen die »weiten und unbestimmten Vorschriften (des common law) über contempt of court« richtete (vgl. Nr. 42, 43 des Urteils). Der EGMR ging davon aus, daß die Sachprüfung der Kommission sich *ratione materiae* auf die Frage beschränkte, ob die Regeln des *contempt of court*, so wie sie vom House of Lords der Bestätigung der einstweiligen Anordnung zugrunde gelegt worden seien, die Beschränkung nach Art. 10 Abs. 2 rechtfertigen. Daher habe er nur zu prüfen, ob die Konvention durch das Urteil des House of Lords verletzt worden sei.

[8] Vgl. auch die Entscheidung im *Marckx*-Fall, EuGRZ 1979, 454, 460 (Nr. 27, 58).

[9] *Airey*-Fall, EuGRZ 1979, 629 (Nr. 26). Interessant und bezeichnend ist vor allem folgende Feststellung des Gerichts: »Es wäre falsch, das Ergebnis – daß die Möglichkeit, persönlich vor dem High Court aufzutreten, Frau Airey kein wirksames Recht auf Zugang gibt – zu verallgemeinern. Dieses Ergebnis gilt nicht für alle Fälle, die ›zivilrechtliche Ansprüche und Verpflichtungen‹ betreffen, noch für jeden der an ihnen Beteiligten. Unter bestimmten Umständen wird die Möglichkeit, ohne Beistand

rights« fortgesetzt – könnte den Übergang zur »Einzelfallgerechtigkeit«, zum *case by case approach* ohne generalisierende und – was noch problematischer wäre – generalisierungsfähige Aussage bedeuten. Problematisch erscheint eine derartige Tendenz deshalb, weil die Kontinuität der Rechtsprechung des EGMR, auch die Verallgemeinerungsfähigkeit ihrer Aussagen wesentliche Elemente für die Bereitschaft der Gerichte der Vertragsstaaten sind, ihr zu folgen. Je mehr der EGMR auf den Einzelfall, die besonderen Umstände und die spezifische soziale Entwicklung abstellt, um so eher besteht die Gefahr, daß auch die staatlichen Gerichte geneigt sind, in jedem Fall das Besondere zu sehen und sich nicht von der Rechtsprechung des EGMR leiten zu lassen [10].

In seinem Schlußbericht auf dem Frankfurter Menschenrechtskolloquium hatte *Hermann Mosler* seinerzeit bemerkt, hinter der Einzelfallgerechtigkeit, »diesem Wortungetüm, das es leider in unserer Sprache gibt«, verberge sich folgende Überlegung. Im Einzelfall, der bei Gericht anhängig ist, seien unbestimmte Begriffe der gesetzlichen Tatbestände auszulegen und *in concreto* anzuwenden. Solche Bestimmungen kenne die Konvention ebenso wie das nationale Recht in vielen Fällen, beispielsweise die Begriffe demokratische Gesellschaft, nationale Sicherheit, öffentliche Ruhe und Ordnung, wirtschaftliches Wohl des Landes, notwendige Maßnahmen zur Verteidigung der Ordnung und zur Verhinderung von strafbaren Handlungen, Schutz der Gesundheit und der Moral. Im Einzelfall müsse entschieden werden, ob eine dieser Klauseln im konkreten Fall die Maßnahme, über die sich der Beschwerdeführer in Straßburg beklagt, rechtfertigt, mit anderen Worten, ob die Konvention verletzt sei oder nicht. Diese Bewertung der unbestimmten Begriffe und die Abgrenzung der Grundrisse der Konvention legten dem Gerichtshof eine besondere Verantwortung auf, die durch die Stichworte: Beurteilungsspielraum der Staaten bei der durch die Konvention erlaubten Einschränkung von geschützten Rechten einerseits und Bestimmung des Inhalts allgemeiner unbestimmter Begriffe an Hand von Einzelfällen andererseits umschrieben werde [11].

Das Problem der Einzelfallgerechtigkeit als Entscheidungsmaßstab ist in der Tat aus der Rechtsprechung des Bundesverfassungsgerichts (BVerfG) [12],

---

eines Rechtsanwalts persönlich vor Gericht auftreten zu können, die Anforderungen von Art. 6 Abs. 1 erfüllen; es kann Fälle geben, in denen eine solche Möglichkeit selbst zum High Court einen angemessenen Zugang darstellt. Tatsächlich wird viel von den jeweiligen Umständen abhängen«.

[10] Vgl. meinen Bericht auf dem Fünften Internationalen Menschenrechtskolloquium (Anm. 4), S. 40.

[11] A.a.O. (Anm. 4), S. 357 f.

[12] Die »Umstände des (Einzel)Falles« werden vom BVerfG bei der Grundrechtsinterpretation im Rahmen der sog. »Güterabwägung« maßgeblich herangezogen (vgl. BVerfGE 7, 198 [210]; 35, 5 [11]. In der Literatur ist diese Methode als Etikettenschwindel bezeichnet worden, da eine als Ausgleich abstrakter Rechtsgüter angekün-

aber auch des Bundesgerichtshofs (BGH) [13] bekannt und als allgemeines juristisches Erkenntnisproblem diskutiert worden [14]. Die Spannung zwischen Abstraktheit der Rechtsanwendungsregeln und Konkretheit der Rechtsanwendungsaufgabe ist zwar, wie *Göldner* [15] treffend feststellt, in jeder Methodenlehre notwendig angelegt. Sie prägt sich aber »um so schärfer aus, je weiter die Rechtsanwendungsdogmatik vom Maßstab der Logik zu demjenigen der Teleologik übergeht und damit auf den spezifizierenden Ableitungszusammenhang logischer Deduktion verzichtet« [16]. In der Rechtsprechung kommt dieser Zusammenhang deutlich in dem Satz des BVerfG zum Ausdruck, daß für den Richter das Gebot sachgerechter Entscheidung im Rahmen der Gesetze unter dem Blickpunkt materialer, wertorientierter Gerechtigkeit bestehe [17]. Zu diesem Satz hat *Ernst Wolf* [18] kritisch angemerkt, daß nach der idealistischen Philosophie gerecht ist, was im einzelnen Fall mit der absoluten Rechtsidee übereinstimmt. Die »materiale Gerechtigkeit« sei absolute Einzelfallgerechtigkeit. Eine solche »Gerechtigkeit« gäbe es unter Menschen nicht. Der Versuch, sie dennoch zu praktizieren, führte zur autoritären und totalitären Willkürherrschaft. Diese Kritik an der idealistischen Philosophie führt *Ernst Wolf* zu der Schlußfolgerung, daß die darin entstandene »Idee« einer absoluten Einzelfallgerechtigkeit zur Auflösung von Gesetz und Recht und damit zur Auflösung der gerichtlichen Rechtserkenntnis führe [19]. Die dadurch bedingte Prozeßschwemme könne nicht durch unvermeidliche fehlerhafte prozeßrechtliche Vereinfachungsgesetze, sondern nur durch Rückkehr zu den Grundlagen der gerichtlichen Urteile überwunden werden.

Was auf der einen Seite als richterliche Willkürherrschaft, erscheint auf der anderen Seite als unverzichtbares Element des Rechtsanwendungsvorganges.

---

digte »Güterabwägung« in Wahrheit zu einer Interessenabwägung der Beteiligten im Einzelfall umfunktioniert wurde (vgl. *Harald Schneider,* Die Güterabwägung des Bundesverfassungsgerichts bei Grundrechtskonflikten [1979], S. 23 mit weiteren Nachweisen). In der Tat geht damit, wie *W. Knies* (Schranken der Kunstfreiheit als verfassungsrechtliches Problem [1967], S. 41) dargelegt hat, ein jedenfalls bei abstrakter Abwägung vorhandener »letzter Rest regulativer Substanz« verloren.

[13] Vgl. beispielsweise Entscheidungen des Bundesgerichtshofs in Zivilsachen (BGHZ) 3, 308 ff.; 17, 266 f.; 21, 319 ff.; 26, 349 ff.; 30, 203 ff.; siehe *U. Diederichsen,* Zur Begriffstechnik richterlicher Rechtsfortbildung im Zivilrecht, in: Festschrift für Franz Wieacker (1978), S. 325 ff.

[14] Vgl. insbes. *Esser,* Vorverständnis und Methodenwahl in der Rechtsfindung (1972), S. 65 ff.; *M. Kriele,* Theorie der Rechtsgewinnung (2. Aufl. 1967), S. 177 ff.; *K. Larenz,* Die Bindung des Richters an das Gesetz als hermeneutisches Problem, in: Festschrift für Ernst Rudolf Huber (1973), S. 291 ff.; *R. Fischer,* Die Weiterbildung des Rechts durch die Rechtsprechung (1971).

[15] *D. C. Göldner,* Verfassungsprinzip und Privatrechtsnorm bei der verfassungskonformen Auslegung und Rechtsfortbildung (1969), S. 82.

[16] *Göldner,* a.a.O., S. 24 ff., 312 These 11.

[17] BVerfGE 42, 78.

[18] Das Wesen des gerichtlichen Urteils, in: Gedächtnisschrift für Rudolf Bruns (1980), S. 236.

[19] A.a.O., S. 237.

Es dürfte daher von Interesse sein zu klären, was damit gemeint ist, wenn von »Einzelfallgerechtigkeit« als Entscheidungsmaxime gesprochen wird, ob es überhaupt sinnvoll ist, diesen Begriff zu verwenden, oder ob er nicht eher zu Mißdeutungen Anlaß gibt, und schließlich, ob bei einer internationalen gerichtlichen Entscheidungsinstanz gegenüber der sogenannten »Einzelfallgerechtigkeit« nicht stärkere Zurückhaltung – verglichen mit nationalen Entscheidungsinstanzen – am Platz ist.

b) Einzelfallgerechtigkeit als Teilbereichsgeneralisierung und Korrekturentscheidung

Um den Begriff der Einzelfallgerechtigkeit sinnvoll zu verwenden, muß er im Rahmen des Straßburger Kontrollsystems an die spezifischen Eigenheiten einer internationalen Gerichtsbarkeit, welche sich tendenziell zu einer »gemeineuropäischen Verfassungsgerichtsbarkeit« entwickelt, angepaßt werden[20]. Selbstverständlich besteht bei der Auslegung unbestimmter Rechtsbegriffe der EMRK und ihrer Anwendung im Einzelfall – ebenso wie im nationalen Recht – jene in der Theorie häufig beschriebene Wechselbeziehung[21] zwischen einzelnen Auslegungs- und Rechtsanwendungsschritten. Dabei darf jedoch der konkrete Einzelfall nicht der einzige rationale Steuerungsfaktor für den Auslegungsvorgang sein, sondern es müssen zumindest für einen Teilbereich die gedanklichen Alternativen anderer Fallkonstellationen erwogen und die Generalisierbarkeit der konkreten Aussage in diesem Rahmen überprüft werden.

Eine derartige Tendenz zur Generalisierung der Auslegung unbestimmter Rechtsbegriffe der EMRK ist der Rechtsprechung des EGMR auch nicht fremd[22]. Sie liegt z. B. der Argumentation im Fall *Engel*[23] zugrunde, wo der EGMR festgestellt hat, daß Disziplinarverfahren grundsätzlich nicht in den Anwendungsbereich des Art. 6 Abs. 1 EMRK fallen, daß es aber Fälle geben könne, in denen sich hinter einem vom innerstaatlichen Recht als disziplinarisch bezeichneten Vorwurf in Wirklichkeit eine strafrechtliche Anklage verbirgt. Insoweit würde sich die Überprüfungsbefugnis des Gerichtshofes, wie wohl im Einzelfall die Art des Verstoßes dem Staat die Anwendung des Disziplinar- statt des Strafrechts gegen den Betroffen gestattet, »im Regelfall als

[20] Vgl. *J. A. Frowein*, Die Europäische Menschenrechtskonvention in der neueren Praxis der Europäischen Kommission und des Europäischen Gerichtshofs für Menschenrechte, EuGRZ 1980, 231, 237; siehe auch *G. Ress*, Wechselwirkungen zwischen Völkerrecht und Verfassung bei der Auslegung völkerrechtlicher Verträge (Berichte der Deutschen Gesellschaft für Völkerrecht, Heft 23) (1982), S. 9, 29 ff.

[21] Vgl. *Kriele* (Anm. 14), S. 64 ff.

[22] Vgl. *U. Scheuner*, Die Fortbildung der Grundrechte in internationalen Konventionen durch die Rechtsprechung – Zur Rechtsprechung des Europäischen Gerichtshofs für Menschenrechte, in: Staatsrecht – Völkerrecht – Europarecht, Festschrift für H.-J. Schlochauer (1982), S. 916 ff.

[23] Série A Nr. 22, S. 35 Ziff. 82 = EuGRZ 1976, 221 (232).

illusorisch erweisen, wenn man nicht auch den Schweregrad der Sanktion berücksichtigen würde, die dem Betroffenen droht«. Ebenso bemüht sich der EGMR bei der Auslegung des Begriffs »zivilrechtlich« (*de caractère civil/civil rights*) um eine Generalisierung: So hat er im Fall *Le Compte* nicht auf den Einzelfall abgestellt, sondern mit vielleicht unscharfen und zweideutigen Begriffen (wie Richter Matscher kritisiert)[24] die Abgrenzung zu umschreiben versucht. Begriffe wie: »für den Ausgang des Verfahrens unmittelbar bestimmend« (Nr. 46 und 47 des Urteils) bzw. »unmittelbarer und nachhaltiger Eingriff in die Ausübung des fraglichen Rechts« (Nr. 49 des Urteils) bieten zweifellos mehr an regulativer Substanz, als es die bloße Behauptung wäre, Verfahren, die mit einem zeitweiligen oder endgültigen Verbot der Berufsausübung enden können, seien stets zivilrechtlich.

Aus der theoretischen Diskussion wird man zwei Aspekte der »Einzelfallgerechtigkeit« aufnehmen können: Einmal den Einzelfall als Korrektur eines allgemeinen, in den Einzelheiten noch variablen Begriffsverständnisses[25]; zum anderen als Teilbereich einer noch zu findenden abstrakten Gesamtlösung. *W. Henke* hat, was den ersten Gesichtspunkt angeht, mit Recht darauf hingewiesen, daß »die verhältnismäßig groben Einteilungen der Rechtsordnungen« – und dies ist maßgeblich auch für die vagen Generalbegriffe wie Menschenwürde[26], öffentliche Ordnung, demokratische Gesellschaft, Gesundheit, Eigentum, usw. – immer wieder »der Korrektur unter dem Gesichtspunkt der Einzelfallgerechtigkeit (bedürfen), die nur von Fall zu Fall und darum nur von der Rechtsprechung und bloß nachzeichnend und vage von der Rechtswissenschaft vorgenommen werden kann«[27]. Eine derartige Einzelfallgerechtigkeit als »Korrekturlösung« setzt freilich voraus, daß vorab ansatzweise eine abstrakte Begriffsbestimmung entwickelt wird, welche geeignet ist, die *rationalen* Anforderungen eines internationalen Wertsetzungs- und Steuerungsmechanismus für die Vertragsstaaten der EMRK zu erfüllen.

---

[24] In seiner abweichenden Meinung, EuGRZ 1981, 537, schreibt Richter Matscher, daß die vom EGMR gewählten Formulierungen »weder dazu geeignet (sind), die Vertragsstaaten bei der Anpassung ihrer Rechtsordnungen an die von den Organen der Konvention aufgestellten Anforderungen zu leiten, noch die Rechtssuchenden über die Erfolgsaussichten einer Beschwerde aufzuklären. Kurz, sie fördern nicht die Rechtssicherheit, die in den Urteilen des Gerichtshofs zu Recht wiederholt beschworen worden ist«.

[25] Diese Abgrenzung kann auch negativ sein und besagen, daß eine staatliche Maßnahme *nicht* ein (möglicherweise) geschütztes Recht verletzt (vgl. *Paton*-Fall, Entscheidung der Menschenrechtskommission vom 13. 5. 1980, EuGRZ 1981, 20 ff.).

[26] Siehe dazu *A. Verdross,* La dignité de la personne humaine base des Droits de l'homme, Österreichische Zeitschrift für öffentliches Recht 1980, S. 271.

[27] *W. Henke,* Das subjektive Recht im System des öffentlichen Rechts, Die Öffentliche Verwaltung 1980, S. 622, der sich auf *B. Heusinger,* Rechtsfindung und Rechtsbildung im Spiegel richterlicher Erfahrung (1975) bezieht mit dem Bemerken, das Werk lasse, ohne gegen sie zu polemisieren, die Fruchtlosigkeit der rechtslogischen und der meisten rechtsmethodischen Bemühungen im Kontrast sehr deutlich hervortreten.

Der zweite Aspekt der Einzelfallgerechtigkeit – Teilbereichslösung einer noch zu entwickelnden abstrakten Gesamtlösung – bildet im Grunde die Umkehrung des ersten Aspekts. Der Einzelfall wird induktiv zum Aufbau der abstrakten Begriffsbestimmung genutzt[28]. Bedenklich stimmt – wie auch in der Judikatur des Bundesverfassungsgerichts[29] – nur eine Vorgangsweise, bei welcher das Gericht seine Aussagen in der Nachzeichnung der besonderen Umstände des Einzelfalles erschöpft, ohne für einen Teilbereich die für den hermeneutischen Subsumtionsvorgang unerläßliche allgemeine Aussage in Form des Obersatzes bzw. des Maßstabes für eine »Abwägung« zwischen zwei Werten zu entwickeln[30]. Einzelfallgerechtigkeit ist mit einer derartigen »Maßstabslosigkeit« nicht zu verwechseln.

Die Frage, ob für eine internationale Gerichtsbarkeit die Verwendung des Begriffs der Einzelfallgerechtigkeit am Platz ist, wird man daher im Sinne der von *H. Mosler* getroffenen Feststellung beantworten können, daß Einzelfallgerechtigkeit ein unerläßliches Element auch des internationalen Rechtsfindungsprozesses an Hand der EMRK ist, wenn die beiden Aspekte – Korrekturentscheidung bzw. Teilbereichsgeneralisierung – beachtet werden.

3. Erga omnes-Wirkung der Feststellungsurteile des EGMR?

a) Allgemeine Erwägungen

Gegenstand des Verfahrens vor dem EGMR ist die Frage der Konventionsverletzung an Hand eines konkreten Einzelfalles und gegebenenfalls der Anspruch auf eine gerechte Entschädigung (Art. 50 EMRK). Gleichwohl rückt das Individuum, das zu Beginn des Verfahrens vor der Kommission im Mittelpunkt steht, im Laufe des Verfahrens in den Hintergrund, da es vor dem EGMR nicht mehr Partei ist. Damit hängt zusammen, daß aus einer Quasi-Verfassungsbeschwerde (im Individualbeschwerdeverfahren) unversehens, wie *Wildhaber* ausführt, ein Kontrollsystem wird, das Bezüge zur abstrakten Normenkontrolle aufweist. »Denn im Grunde ringt man vor dem EGMR an Hand konkreter Einzelfälle um die Konventionskonformität und Revisionsbedürftigkeit staatlicher Gesetze ganz allgemein«[31].

---

[28] Vgl. dazu Beispiele aus der Judikatur des BVerfG bei *Schneider* (Anm. 12), S. 187 ff. Ein Beispiel aus der Judikatur des EGMR ist die »Verletzteneigenschaft« auf der Grundlage der »Besonderheiten« des *Klass*-Falles (EuGRZ 1979, 278).

[29] Nachweise bei *Schneider* (Anm. 12), S. 196 ff.

[30] Problematisch in dieser Hinsicht insbes. die *Sunday Times*-Entscheidung des EGMR (EuGR 1979, 386; vgl. dazu kritisch *U. Scheuner* [Anm. 22], S. 925, sowie *F. A. Mann,* Contempt of Court in the House of Lords and the European Court of Human Rights, Quarterly Review 1979, S. 348 ff.).

[31] *L. Wildhaber,* Erfahrungen mit der Europäischen Menschenrechtskonvention, Schweizerischer Juristenverein, 1979, S. 316. Zutreffend ist seine Feststellung, daß die

Diese Eigentümlichkeit des Kontrollsystems bildet einen der Ansätze für die These, daß den Urteilen, mit denen der EGMR eine Konventionsverletzung feststellt[32], Wirkung über den konkreten Einzelfall und die Parteien des Verfahrens hinaus zukomme[33]. Der EGMR stelle nur fest, was über den konkreten Fall hinaus nach Art. 1 EMRK ohnedies Verpflichtung der Vertragsparteien sei. Eine ähnliche Argumentation wird auch im Hinblick auf Urteile des EuGH vorgetragen[34].

Eine derartige Erstreckung der Bindungswirkung ist trotz des »objektiven« oder *ordre public*-Charakters des Konventionsschutzes nicht überzeugend. Art. 53 EMRK stellt die Bindung ausdrücklich auf die an dem Verfahren beteiligten *Parteien* ab (subjektive Rechtskraft *inter partes*)[35]. *Objektiv* wird die Rechtskraft durch die konkrete Frage nach der Konventionsverletzung einer oder mehrerer *bestimmter* staatlicher Maßnahmen begrenzt. Es wird also nicht über die Auslegung der EMRK *in abstracto* (»ein für alle mal«) entschieden[36].

Als entscheidender Angelpunkt der folgenden Argumentation erweist sich dabei die Frage, welche staatliche(n) Maßnahme(n) wirklich Gegenstand der gerichtlichen Feststellung der Konventionswidrigkeit sind. Die damit verbundene objektive Begrenzung der Bindungswirkung führt zu Problemen, weil ein Vertragsstaat bei *künftigen* gleichartigen Maßnahmen theoretisch keiner Verpflichtung aus Art. 53 EMRK unterliegt. So hat die belgische Cour de Cassation dem EGMR die Gefolgschaft bei der Auslegung von Art. 6 Abs. 1 EMRK in den Fällen *Simonis*[37] und *Guchez*[38] am 21. Januar 1982 in einer Weise versagt, die eine grundsätzliche Stellungnahme erfordert.

---

klassischen völkerrechtlichen, zwischenstaatlichen Elemente des Kontrollsystems desto stärker hervortreten, je näher eine Beschwerde an die mögliche Festellung der Konventionsverletzung heranrückt.

[32] Gleiches müßte konsequenterweise auch für die negativen Feststellungsurteile (keine Konventionsverletzung) gelten.

[33] Vgl. die Diskussion auf dem Frankfurter Kolloquium (Anm. 4).

[34] *A. Bleckmann* (Europarecht [3. Aufl. 1980], S. 156) hält es in den Verfahren nach Art. 169 und 170 EWG-Vertrag für angemessen, alle Organe der EWG und alle Mitgliedstaaten an die ergangenen Urteile zu binden, da sie auf einem völlig objektiven Verfahren beruhen und Kommission und Mitgliedstaaten im Interesse der europäischen Rechtsordnung klagen (»Wächter«-Funktion).

[35] *H. Mosler,* Supra-National Judicial Decisions and National Courts, Hastings International and Comparative Law Review 4 (1981), S. 425 ff. (456 f.).

[36] *Bleckmann* (Anm. 34), S. 157, vertritt für das Vorlageverfahren nach Art. 177 EWG-Vertrag demgegenüber die – mich nicht überzeugende – Ansicht, die »Auslegung« müsse gegenüber jedermann wirken, weil nur so die Einheitlichkeit der europäischen Rechtsordnung gesichert werde. Die herrschende Lehre geht mit Recht nur von der Bindung des vorlegenden Gerichts aus und hält Vorlagen *derselben* Rechtsfragen für zulässig.

[37] Entscheidung Nr. 6458 vom 21. 1. 1982 in der Sache *Louis Simonis ./. Ordre des Médecins,* Journal des Tribunaux 1982, S. 446.

[38] Entscheidung Nr. 6445 vom 21. 1. 1982 in der Sache *Henry Guchez./.Ordre des Architects,* Journal des Tribunaux 1982, S. 428 mit *observations* von *J. Salmon* (S. 446).

b) Bindungen an die Auslegung im Fall Le Compte?

Im Fall *Le Compte*[39] hatte der EGMR entschieden, daß Art. 6 Abs. 1 EMRK unter bestimmten Voraussetzungen (nachhaltiger Eingriff in die Berufsausübung) auch auf Disziplinarverfahren Anwendung findet. Die belgische Cour de Cassation stand demnach – nach den Ausführungen des Procureur général F. Dumon – vor der Frage, ob sie ihre Rechtsprechung ändern solle, entweder

1. weil sie von der Auslegung des Art. 6 Abs. 1 EMRK durch den EGMR überzeugt worden sei oder
2. weil sie an dessen Auslegung gebunden sei oder
3. weil sie der Ansicht sei, »que, en raison de la mission d'application et aussi d'interprétation que ladite Convention internationale, ratifiée par la Belgique et dont les dispositions font partie du droit positif de notre pays, confie à la haute juridiction internationale, il importe eu égard à l'indispensable sécurité juridique, de même qu'à la souhaitable uniformité d'interprétation dans tous les états liés par ladite Convention, de se rallier à l'interprétation donnée par l'arrêt du 23 juin 1982«.

Der Procureur général führt zur Beantwortung der Frage, ob die nationalen Gerichte frei sind, der Rechtsprechung des EGMR zu folgen, die Unterscheidung zwischen relativer und absoluter Bindung ein. Auf der einen Seite müsse man die *conséquences particulières* betrachten, die sich für einen Vertragsstaat der Konvention ergeben, dem gegenüber der EGMR eine Konventionsverletzung festgestellt habe, und welche eventuell seine Verurteilung zu einer angemessenen Entschädigung nach sich ziehe. Diese Wirkung bezeichnet er als »autorité relative ou *in specie* des arrêts de la cour«. Ihr stehe als denkbare Alternative die *autorité absolue, erga omnes* gegenüber, welche eine Bindungswirkung aller nationalen Gerichte der Mitgliedstaaten an die vom EGMR gegebene Auslegung auslösen würde; sie würde sich darauf erstrecken, »de respecter et d'appliquer l'interprétation se dégageant de ses arrêts, comme c'est le cas en vertu du principe du ›stare decisis‹ en Angleterre et dans d'autres pays dites de ›Common Law‹«[40].

Der Procureur général stellt auch die Frage, ob den belgischen Staat eine völkerrechtliche Verantwortlichkeit träfe, wenn seine Gerichte in einem *konkreten Einzelfall* die Bindung, die sich aus einem Spruch des EGMR ergibt, mißachte (»de nier l'autorité in specie, qui se dégage d'un arrêt de la Cour de Strasbourg«) oder sich von dessen Rechtsprechung, die er in anderen Fällen entwickelt hat, nicht leiten ließe. Dies sind in der Tat zwei völlig verschiedene Probleme. Im entschiedenen Einzelfall (d.h. soweit eine staatliche Maß-

---

[39] Urteil im Fall *Le Compte, Van Leuven, De Meyere* vom 23. 6. 1981, Journal des Tribunaux 1981, S. 622 ff. mit *observations* von *Pierre Lambert*. Vgl. auch EuGRZ 1981, 551 ff., und European Human Rights Reports 1982, S. 1 ff.

[40] Journal des Tribunaux 1982, S. 438 ff.

nahme unmittelbar für konventionswidrig erklärt wird) ist nicht nur der Staat, sondern sind m. E. auch dessen Gerichte bei Gegenstandsidentität gebunden [41]. Die nationalen Gerichte, welche im Zusammenhang mit der festgestellten Konventionsverletzung zur Durchsetzung eines nationalen Schadensersatzanspruchs angerufen werden, können die Verletzung der Konvention im konkreten Fall nicht mehr bezweifeln [42]. Sie können freilich Tatbestandsmerkmale, welche im Verfahren in Straßburg nicht zur Entscheidung standen, prüfen und sind an die Tatsachenfeststellungen des EGMR nicht gebunden. Von dieser Rechtsfortentwicklung ist die Präzedenz- oder Orientierungswirkung der Auslegung für künftige Fälle deutlich zu trennen [43].

Die Zweifel, die vom Procureur général an einer *autorité erga omnes* geäußert werden, sind überzeugend. Auch Art. 177 EWG-Vertrag führt, trotz der zaghaften Ansätze in der neueren Judikatur des Gerichtshofs der Europäischen Gemeinschaften (EuGH) [44], nicht zu einer *erga omnes*-Wirkung. Dabei ist freilich zwischen den Interpretationsentscheidungen und den Entscheidungen über die Gültigkeit von Akten zu unterscheiden – eine Unterscheidung, die der Procureur général nicht trifft. Den Interpretationsentscheidungen kommt Rechtskraftwirkung nur für den vorgelegten Fall, der Gültigkeitsentscheidung jedoch »Rechtskraftwirkung« – besser: Gestaltungswirkung – für alle künftigen Entscheidungen zu, da der für nichtig erklärte Akt damit aus dem Rechtsbestand des europäischen Gemeinschaftsrechts ausgeschieden ist.

Eine Bindung an Interpretationsentscheidungen internationaler Gerichte im Sinne der *stare decisis*-Regel gehört nicht zu den allgemeinen Prinzipien des Völkerrechts [45]. Art. 63 des Statuts des Internationalen Gerichtshofs (IGH) [46] ist insoweit sicher ein gutes Beispiel für die relative Rechtskraftwirkung.

---

[41] Vgl. dazu im einzelnen meinen Bericht (Anm. 4), S. 13 ff. *Mosler* (Anm. 35), S. 456, ist demgegenüber der Auffassung, die EMRK gebiete nicht, »that the judgments of the Strasbourg Court be treated as res iudicata before the national courts of the State against which the judgment is given«. Anders *D. Schindler,* Die innerstaatlichen Wirkungen der Entscheidungen der europäischen Menschenrechtsorgane, in: Festschrift Max Guldner (1973), S. 285: Stellt der EGMR z. B. die Konventionswidrigkeit eines Verwaltungsaktes oder tatsächlichen, dem Staat zurechenbaren Handelns fest, »so ist die Widerrechtlichkeit der Schadenszufügung (vom nationalen Richter) zweifellos zu bejahen«. Meines Erachtens wäre es wenig sinnvoll, wenn der nationale Richter *dieselbe* Frage (der Konventionswidrigkeit) nochmals – gegebenenfalls in anderem Sinne – entscheiden dürfte.

[42] In meinem Bericht (Anm. 4, S. 16) habe ich vorgeschlagen, die Bindungswirkung auf die für den Beschwerdeführer *günstige* Feststellung der Konventionswidrigkeit zu beschränken.

[43] Vgl. dazu *Mosler* in seinem Schlußbericht (Anm. 4), S. 366.

[44] Siehe Neue Juristische Wochenschrift (NJW) 1980, S. 508.

[45] Vgl. *Mosler,* Judgements of International Courts and Tribunals, in: Encyclopedia of Public International Law, Instalment 1 (1981), S. 116.

[46] Art. 63 Abs. 2 sieht vor, daß die in einem Urteil des IGH enthaltene Auslegung eines Vertrages eine Vertragspartei, die nicht Streitpartei ist, nur bindet, wenn sie dem Verfahren beigetreten ist. Gleichwohl verbirgt sich dahinter ein »extended

c) Umfang der Rechtskraftwirkung inter partes und Umfang der (indirekten) Präzedenz- oder Orientierungswirkung

Man hat den Eindruck, daß die vom Procureur général zitierten Äußerungen von *H. Rolin*[47] und von *Sir Vincent Evans*[48] die Unterscheidung zwischen der (relativen) Rechtskraftwirkung im konkreten Fall und der Orientierungswirkung nicht völlig klar durchführen. *Rolin* legte Gewicht darauf, daß die Entscheidungen »de la Cour exercent en fait une autorité indirecte sur les organes législatifs et judiciaires nationaux« – von einer Bindung im Sinne einer »règle du précédent« oder im Sinne einer »force obligatoire pour les gouvernements et les juridictions d'ordre interne des Etats parties à la Convention« könne jedoch keine Rede sein. Im gleichen Sinne zieht *Sir Vincent Evans* die Schlußfolgerung, daß »the jurisprudence of the European Court is not legally binding on the Courts and other authorities in the countries concerned, but will have persuasive authority«. Es gibt also keine Wirkung *erga omnes,* auch keine *stare decisis*-Bindung, sondern nur eine, wie auch *Mosler* sie genannt hat, *Orientierungswirkung* für die Gerichte der Mitgliedstaaten[49].

Diese Orientierungswirkung ist von der Rechtskraft *inter partes* klar zu unterscheiden. Ob die Behauptung von *J. Velu*[50], daß »l'autorité de la chose jugée par la Cour ne peut, en cas de contestation devant le juge belge, être envoquée que par et contre l'une des parties au litige devant la Cour« wirklich zutreffend ist, erscheint zweifelhaft, weil vor dem EGMR der Beschwerdeführer keine Parteistellung hat, gleichwohl aber das Urteil zu seinen Gun-

concept of the scope of the res iudicata«: *S. Rosenne,* The Law and Practice of the International Courts, Bd. 2 (1965), S. 628 f.; vgl. auch IGH Urteil vom 13. 6. 1951, *Haya de la Torre*-Fall, ICJ Reports 1951, S. 76 f. Nicht völlig geklärt scheint selbst beim IGH die Funktion von Feststellungsurteilen im Lichte von Art. 59 des Statuts. Im *Northern Cameroon*-Fall (ICJ Reports 1963, S. 34) führte der IGH aus: »If in a declaratory judgment it (sc. the Court) expounds a rule of customary law or interprets a treaty which remains in force its judgment has a continuing applicability«. *Rosenne,* a.a.O., S. 621, bemerkt dazu – etwas sibyllinisch, wie mir scheint –: »Here the Court stressed what it termed the forward reach of the judgment as establishing once and for all and with binding force as between the parties a legal position which cannot again be called into question in so far as the legal effects ensuing therefrom are concerned«. Im Gegensatz zu den Feststellungsurteilen des EGMR haben diese IGH-Urteile die Vertragsauslegung *selbst* zum Feststellungsgegenstand. Die Ausführungen sind daher nicht auf Feststellungsurteile des EGMR übertragbar.

[47] Rapport sur l'autorité des arrêts et des décisions des organes de la Convention européenne des droits de l'homme, Revue des droits de l'homme, Human Rights Journal 6 (1973), S. 746.

[48] Bericht in den »Actes du Colloque sur la Convention européenne des droits de l'homme par rapport à d'autres instruments internationaux pour la protection des droits de l'homme« – Conseil de l'Europe – Direction des droits de l'homme – Strasbourg 1979, S. 193.

[49] Vgl. Schlußbericht (Anm. 4), S. 366.

[50] *J. Velu,* Les effets directs des instruments internationaux en matière des Droits de l'homme – Prolegomena – collection sous la direction de A. Vandeplas et J. P. Sprentels (1981), S. 143, no. 32.

sten Feststellungen trifft. Man kann deshalb von einer Rechtskraftwirkung nach internem Recht auch zu seinen Gunsten sprechen[51]. Sicher haben die Urteile eine sogenannte Tatbestandswirkung (*effet direct de fait*)[52].

Wenn bezweifelt wird, ob *Rigaux*[53] und *Bossuyt*[54] die Probleme der Rechtskraftwirkung *erga omnes* oder *inter partes* zutreffend erörtert haben – so der Vorwurf von *F. Dumon*[55] –, dann ist nicht minder zweifelhaft, ob die Ausführungen des Procureur général selbst wirklich zum Kern der Sache vorgedrungen sind. Die Entscheidungen des EGMR haben – grundsätzlich nicht anders als die Entscheidungen eines obersten nationalen Gerichts – über den Einzelfall hinausgehend bezüglich der gegebenen Auslegung nur Überzeugungs- bzw. Orientierungswirkung. Die Rechtskraft im eigentlichen Sinne bleibt auf den entschiedenen Fall *inter partes* beschränkt.

Eine Abweichung von der vom EGMR gegebenen Auslegung in einem *neuen* Fall ist daher den Gerichten der Mitgliedstaaten nicht verwehrt, auch wenn eine solche Abweichung erst nach Ausschöpfung aller Argumente des EGMR und Darlegung der »besseren« Rechtsmeinung Platz greifen darf. Die Gerichte der Vertragsstaaten haben die vom EGMR vertretene Auslegung vorrangig zu beachten (Vorrang jedenfalls gegenüber bisherigen Urteilen nationaler, auch höchster nationaler Gerichte zur gleichen Rechtsfrage)[56]. Es ist von Bedeutung, wie der Procureur général und die Cour de Cassation diesen Grundsatz verstehen und anwenden.

Der EGMR hat die Aufgabe der Auslegung und der Kontrolle der Anwendung der EMRK, und daher wirkt sich seine Rechtsprechung in gleicher Weise wie die eines obersten Bundesgerichtshofs auf die nachgeordneten Gerichte aus. Wenn seine Rechtsprechung nicht auf diese Weise einwirken würde, »son institution n'aurait pas l'effet qui a été sa raison d'être et non seulement la sécurité juridique mais aussi l'application coordonnée de la Convention dans tous les Etats liés par celle-ci – coordination ou harmonisation conditionnant son efficacité et l'égalité de traitement de tous les citoyens – seraient gravements compromises«. Mit Recht weist *F. Dumon* darauf hin, daß sich diese Gesichtspunkte um so mehr aufdrängen, weil es sich um eine internationale Gerichtsbarkeit handelt, welche die gleiche Rechtsanwendung in allen Vertragsstaaten sicherstellen muß. Wenn daher ein nationales Ge-

---

[51] Siehe dazu *Mosler* (Anm. 35), S. 457, sowie meinen Bericht (Anm. 4), S. 14 f.

[52] Vgl. *Ress, ibid.*, S. 17; vgl. auch *J. Velu*, Les voies de droit ouvertes aux individus devant les instances nationales en cas de violation de la Convention européenne des droits de l'homme, in: Les recours des individus devant les instances nationales, Institut d'Etudes européennes (1978), S. 241, nach Anm. 171.

[53] *F. Rigaux*, La loi condamnée, Journal des Tribunaux, 22. 9. 1979, no. 70, 72, 73, 74, 76.

[54] *M. J. Bossuyt*, The direct applicability of international instruments on human rights, Revue belge de droit international 1980, S. 317 ff., 323–325.

[55] (Anm. 40), S. 444 Anm. 29.

[56] Zu diesem allgemeinen Grundsatz über die »Pflicht zur vorrangigen Beachtung der Urteile des EGMR« vgl. meinen Bericht (Anm. 4), S. 22.

richt von der Auslegung, die der EGMR zur EMRK entwickelt hat, abweicht, dann muß es dabei nicht nur die eigenen nationalen Rechtskonzeptionen des betreffenden Rechtsgebiets im Auge haben, »mais aussi à celles des juridictions de ces autres Etats et spécialement à celles des Cours suprêmes ou les plus hautes de ceux-ci« [57]. Man fragt sich freilich, wie dies geschehen soll. Denn der Mitgliedstaat, der in einem Verfahren verurteilt worden ist, wird höchstwahrscheinlich in seinem internen Recht inzwischen Vorsorge dafür getroffen haben, daß die vom EGMR vertretene Auslegung sich in Zukunft in dem nationalen Recht niederschlägt – sei es durch Änderung der Gesetzgebung oder der Verwaltungspraxis oder zumindest durch eine entsprechende Mitteilung der Entscheidung des EGMR an die nationalen Gerichte [58]. Für diesen Staat ist die Abweichung der Rechtsprechung der nationalen Gerichte eines anderen Staates von der Auslegung durch den EGMR zumindest problematisch – im Hinblick darauf, daß der EGMR möglicherweise die von ihm vorgenommene Auslegung in einem künftigen Fall ändern könnte. Diese Möglichkeit ist aber nie auszuschließen. Eine Änderung der Auslegungspraxis durch den EGMR wirft jedoch, anders als bei einem obersten Gerichtshof eines Staates, Anpassungsprobleme in verschiedenen nationalen Rechtsordnungen auf, die um ein Erhebliches die Anpassungsprobleme innerhalb eines einzelnen Staates, auch eines Bundesstaates, übersteigen. Aus diesem Grunde wird ein internationales Gericht seine Rechtsprechung nur in den allerseltensten Fällen ändern. Seine Autorität beruht wesentlich auf der Kontinuität seiner Rechtsprechung. Und wie ein nationales Gericht, welches von der Auslegung durch den EGMR abweicht, die rechtlichen Bedingungen in den anderen Mitgliedstaaten dabei mitberücksichtigen soll, bleibt unbeantwortet. Soll es berücksichtigen, inwieweit in anderen Staaten vergleichbare Probleme der Anpassung an die Auslegung durch den EGMR bestehen [59], wieweit die

---

[57] Proc.gén. *F. Dumon* (Anm. 40), S. 444.

[58] Vgl. die im Anschluß an die Entscheidung im Fall *Tyrer* (EuGRZ 1979, 162 ff.) vom Chief Justice gegebene Information an alle für die Verhängung und Ausführung von Prügelstrafen maßgeblichen Stellen. Im Fall *H. O'Callaghan* (Urteil vom 6. 10. 1981) versuchte der Berufungsrichter einer neuerlichen Verurteilung zur Prügelstrafe auszuweichen. Er ging davon aus, daß »a decision of the European Court on Human Rights is binding on us in respect of the question as to whether or not a particular act is a breach of the Convention or not. Consequently, in our view, further sentences of birching if carried out, would render the United Kingdom and the Isle of Man in breach of an International Treaty obligation«. Nun »bindet« das im Fall *Tyrer* ergangene Urteil den Richter nicht im Fall *O'Callaghan,* da es nicht um *dieselbe* staatliche Maßnahme geht. Richtig ist aber, daß eine erneute Verurteilung zur Prügelstrafe wiederum eine Konventionsverletzung darstellt – aber nicht wegen der Urteilsbindung, sondern wegen der Bindung an die EMRK selbst.

[59] Insofern erscheint der Hinweis auf die Entscheidung des französischen Conseil d'Etat vom 27. 10. 1978, *Debout* Leb. 395–397 mit Schlußanträgen *Daniel Labetoulle* bedeutsam. Siehe dazu *Chr. Autexier,* Frankreich und die Europäische Konvention zum Schutze der Menschenrechte und Grundfreiheiten nach der Unterwerfungserklärung (Art. 25) vom 2. 10. 1981, ZaöRV 42 (1982), S. 327 ff., 345 ff.

Staaten sich schon in ihrer internen Praxis angepaßt haben, wieweit Änderungen auf der Verfassungsebene in anderen Staaten gegebenenfalls notwendig werden?

Zugunsten einer Abweichung von der Interpretation, die der EGMR gegeben hat, spricht meiner Ansicht nach nicht die Tatsache, daß die Auslegung nicht einstimmig, sondern unter *Beifügung abweichender Meinungen* gegeben wurde [60]. Natürlich zeigt die abweichende Meinung andere Interpretationsmöglichkeiten auf. Man wird auch zugeben müssen, daß sie den Überzeugungscharakter des Spruchs mindert. Es erscheint mir aber fragwürdig, ob man aus der Existenz abweichender Meinungen den Schluß ziehen kann, daß es den nationalen Gerichten dadurch erleichtert wird, »de penser qu'un revirement de jurisprudence n'est pas exclu«. Die auch im innerstaatlichen Recht umstrittene Zulassung von abweichenden Meinungen bei höchsten Gerichten [61] mag zwar in einem *stare decisis*-System als unproblematisch erscheinen. Außerhalb dieses Systems rechtlicher Bindung an Präjudizien führt sie zu einer gewissen Minderung der Überzeugungskraft der Urteile [62]. Gleichwohl bleibt die Frage, ob aus den »opinions contraires, opposées ou divergentes de certains juges« [63] wirklich die Schlußfolgerung zu ziehen ist, daß den nationalen Gerichten dadurch die Abweichung von den Auslegungen der EMRK durch den EGMR erleichert oder in solchen Fällen sogar weniger problematisch werde. Denn stets bleibt nicht nur das Gewicht der abweichenden Meinungen, sondern auch ihre numerische Erheblichkeit zu beachten [64].

Schließlich soll die *Rechtssicherheit* ein Gesichtspunkt sein, welcher die nationalen Gerichte daran hindern könnte, von der von ihnen bisher eingeschlagenen Rechtsprechung (Auslegung) abzuweichen. Der Procureur général ist der Ansicht, daß die Gerichte dann ihre eigene bisherige Rechtsprechung fortsetzen sollten, wenn sie überzeugt sind oder wenn es »vorhersehbar« ist, daß eine Änderung der Interpretation, die der EGMR gegeben hat, bei einem nächsten Fall zu erwarten ist [65]. Die in dieser Argumentation zutage tretende Umkehrung der Pflicht zur vorrangigen Beachtung (Vorrang der bisherigen nationalen Judikatur) [66] ist mit dem Sinn des internationalen Kontrollsy-

---

[60] So aber *Dumon* (Anm. 40), S. 30.

[61] Zur Zweckmäßigkeit des *dissenting vote* beim Bundesverfassungsgericht siehe *K. Doehring,* Staatsrecht der Bundesrepublik Deutschland (2. Aufl. 1980), S. 235.

[62] Wie weise es war, beim EuGH ein *dissenting vote* nicht zuzulassen, zeigt sich an der Integrationskraft, die von den Urteilen des EuGH ausgeht.

[63] *Dumon* (Anm. 40), S. 30.

[64] Wenn im Fall *Le Compte* – einer Plenarentscheidung von 20 Richtern – vier abweichende und eine zum Teil abweichende Meinung angefügt sind (Anwendbarkeit von Art. 6 Abs. 1 EMRK auf den Fall: 15 : 5), dann steht diese Gewichtung im anderen Verhältnis wie etwa im Fall *Sunday Times* (11 : 9 zur Frage der Verletzung von Art. 10).

[65] (Anm. 40), S. 444 f.

[66] Die »Regel« wird von *Dumon, ibid.,*wie folgt gefaßt: »Par ailleurs, celle-ci (sc. les Cours suprêmes nationales) responsables de la sécurité juridique ont, à mon avis,

stems unvereinbar. Übersehen wird auch, daß die nationale Rechtssicherheit eingebettet ist in eine übergreifende Konzeption der Rechtssicherheit, welche Bestandteil der EMRK ist [67].

Erst als vierten Gesichtspunkt führt der Procureur général den *Umfang der konkreten Rechtskraftwirkung inter partes* ein. Bemerkenswerterweise war im Fall *Le Compte* gerade gegenüber Belgien eine Verletzung von Art. 6 Abs. 1 EMRK vom EGMR festgestellt worden (fehlende Öffentlichkeit des Disziplinarverfahrens gegen die Beschwerdeführer vor dem Berufsrat der belgischen Ärztekammer) [68]. Damit war eine unmittelbare Bindung Belgiens und der belgischen Gerichte aber nur soweit eingetreten, wie die konkrete Urteilsfeststellung reichte. Diese Feststellung beschränkte sich darauf, daß »die Sache der Beschwerdeführer nicht von einem mit umfassender Entscheidungsbefugnis ausgestatteten Gericht öffentlich gehört worden ist«. Die Konventionswidrigkeit der königlichen Anordnung vom 6. Februar 1970 [69], welche die Öffentlichkeit in solchen Verfahren vollständig und ausnahmlos ausschließt, wird dagegen vom EGMR nicht ausdrücklich festgestellt [70].

Art. 53 EMRK bedeutet auch nach der Auffassung von *Dumon,* daß der betroffene Staat sich einerseits bemühen muß, die Wirkungen der von ihm getroffenen Entscheidung – einer legislativen, exekutiven oder gerichtlichen – auszulöschen, wenn der EGMR festgestellt hat, daß *diese* Entscheidung im Widerspruch zu einer oder mehreren Bestimmungen der Konvention steht. Der Staat muß deshalb »éventuellement modifier sa législation ou sa réglementation nationale constitutant l'infraction à ladite convention ou ayant été éventuellement la cause de la violation des règles de celle-ci par une juridic-

---

l'impérieux devoir, avant de décider qu'elles abandonneront leur propre jurisprudence que, par définition, elles estimaient fondée, dès lors qu'elles l'ont adoptée, de se demander s'il n'y a pas possibilité voire risque prévisibles d'une modification prochaine de l'interprétation donnée par la Cour de Strasbourg«.

[67] Vgl. EGMR im *Marckx*-Fall, EuGRZ 1979, 460.

[68] Die Öffentlichkeit des Revisionsverfahrens vor der belgischen Cour de Cassation vermochte nach Ansicht des EGMR den festgestellten Mangel nicht zu heilen, weil die Cour die Rechtssachen »nicht in tatsächlicher Hinsicht« überprüft und entscheidet (Nr. 60 – EuGRZ 1981, 554). Deshalb schlägt Richter Pettiti in seiner *concurring opinion* als Lösung, »die am besten den Respekt vor der Tradition der Berufsstände und der Gerichtsbarkeit der Kammern mit der Beachtung der Regeln des gerechten Verfahrens nach der Europäischen Konvention verbindet«, eine Verfahrensgestaltung vor, »die ein abschließendes, zu umfassender Sach- und Rechtsprüfung führendes Rechtsmittel vorsieht und die Öffentlichkeit auf diesen Verfahrensabschnitt beschränkt« (EuGRZ 1981, 558). Dieser Vorschlag zeigt im übrigen, daß mehr als eine Möglichkeit besteht, das Disziplinarverfahrensrecht in solchen Fällen konventionskonform zu gestalten.

[69] Die Anordnung war nicht Verfahrensgegenstand und nur mittelbar Angriffsobjekt der Beschwerdeführer.

[70] Dies wäre an sich möglich gewesen. Art. 95 der Verfassung und Art. 23 der königlichen Anordnung Nr. 79, wonach der Kassationshof die Rechtssachen nicht in tatsächlicher Hinsicht überprüft, hätten nicht aus den gleichen Gründen beanstandet werden können.

tion nationale«[71]. Eine derartige formale Pflicht zur Änderung der Gesetzgebung kann sich als überflüssig darstellen, wenn sich diese Wirkung unmittelbar schon aus der nationalen Rechtsordnung – wie in Belgien – ergibt[72]. Geht man, wie in Belgien, vom Vorrang des Völkerrechts (einschließlich der völkerrechtlichen Verträge) vor dem nationalen Recht aus, so würden durch eine entsprechende Auslegung des Art. 6 Abs. 1 EMRK die entgegenstehenden nationalen Normen verdrängt, d.h. im konkreten Fall nicht angewendet. Diese Stellungnahme unterstreicht die Verpflichtung der Mitgliedstaaten, die festgestellte Konventionswidrigkeit zu beseitigen. Die Verpflichtung des Staates beschränkt sich nicht darauf, eine angemessene Entschädigung, zu der der Staat nach Art. 50 EMRK verurteilt werden kann, zu bezahlen[73].

d) Eigenständige Völkerrechtsverletzung durch Nichtbefolgung der Auslegung des EGMR im Falle Le Compte?

Es wäre zu erwägen, ob der Völkerrechtsverletzung, die ein Mitgliedstaat durch Nichteinhaltung der EMRK begeht, dadurch eine *zusätzliche,* eigenständige Völkerrechtsverletzung hinzugefügt wird, daß er einer in einem Urteil des EGMR vertretenen Auslegung der EMRK nicht folgt[74]. Da aber keine über die EMRK hinausgehende strikte verfahrensrechtliche Bindung erkennbar ist, scheidet eine derartige *formale* zusätzliche Rechtsverletzung wohl aus.

Allein die Tatsache, daß die Cour de Cassation der Auslegung der EMRK durch den EGMR nicht folgt, hat daher die völkerrechtliche Verantwortlichkeit Belgiens nicht *notwendigerweise* zur Folge[75].

Jedoch: gibt es für derartige Abweichungen nationaler Gerichte nicht eine *evidente* Grenze? Muß die Abweichung – wie *Salmon*[76] bemerkt – nicht auf völlig außergewöhnliche Umstände beschränkt bleiben? Hätte der EGMR nach dem Fall *Le Compte* in weiteren Entscheidungen seine Rechtsprechung schon bekräftigt, so hätte auch der Procureur général der belgischen Cour de Cassation davon abgeraten, sich dieser Auslegung nicht anzuschließen; denn in diesem Fall wäre die Verantwortlichkeit des Staates evident. Der Procureur général hat freilich nicht beachtet, daß der Fall *Le Compte* wegen der Grund-

---

[71] (Anm. 40), S. 445, vgl. dazu meinen Bericht (Anm. 4), S. 4 ff.

[72] Vgl. Cour de Cassation 27. 5. 1971, Pas. 1971, I, 886; 14. 1. 1977, Pas. 1976, I, 539; 26. 9. 1978, Pas. 1979, I, 126.

[73] Vgl. *Mosler, separate opinion* im Fall *De Wilde,* Ooms und Versyp, Urteil zu Art. 50 vom 10. 3. 1972, Série A, Bd. 14, S. 17.

[74] Vgl. dazu meine Schlußbemerkung (Anm. 4). Die völkerrechtliche Verantwortlichkeit Belgiens, von der *Salmon* (Anm. 38), S. 450, ausgeht, bezieht sich auch nur darauf, daß die Beschlüsse der Cour de Cassation vom 21. 1. 1982 gegen Art. 6 Abs. 1 EMRK verstoßen »et qu'il est prévisible que si la Cour européenne est saisie de ces espèces, elle confirmera sa jurisprudence et constatera ces infractions qui . . . engagent la responsabilité de la Belgique«.

[75] Vgl. *Dumon* (Anm. 40), S. 445.

[76] (Anm. 39), S. 448.

satzfrage schon vom Plenum des EGMR entschieden wurde. Es war »une démarche mûrement réfléchie«[77]. Aus dem Sinn des EMRK-Kontrollsystems scheint mir zu folgen, daß die Rechtsprechungsfreiheit der nationalen Gerichtsbarkeit in einer solchen Situation auf eine *limite évidente* stößt. Die »*Einzelfallbezogenheit*« findet in dieser tendenziellen Generalisierung der Auslegungswirkung ihre praktische[78] – und, soweit darin sich eine *Pflicht* zur vorrangigen Beachtung der durch den EGMR gegebenen Auslegung manifestiert, auch rechtliche – Begrenzung[79].

4. Die subjektiven und objektiven Grenzen der Rechtskraft der Urteile des EGMR und das Verhältnis von Art. 53 zu Art. 54 EMRK

a) Die Überwachungsfunktion nach Art. 54 EMRK

Die Befugnis des Ministerkomitees, die Durchführung (*exécution*) der Urteile gemäß Art. 54 EMRK zu überwachen, hat im Anschluß an die Urteile in den Fällen *Marckx* und *Luedecke* die Frage aufgeworfen, ob diese rechtliche Überwachungsfunktion sich nur auf die Behebung der *subjektiven* Beschwer im *Einzelfall* bezieht oder auch *objektiv* die Beseitigung des konventionswidrigen Zustandes in einer Weise erfaßt, daß dadurch *künftigen* Verstößen vorgebeugt wird. Im Mittelpunkt steht demnach wiederum das Problem der *Einzelfallbezogenheit*: Wieweit darf das Ministerkomitee die Herstellung eines konventionsgemäßen Zustandes nicht nur *für*, sondern eher *anläßlich* des Einzelfalles auch im Hinblick auf Erlaß, Aufhebung oder Änderung staatlicher Akte, vor allem: gesetzgeberischer Akte überwachen?

---

[77] *Salmon* (Anm. 38), S. 450.

[78] *Dumon* (Anm. 40), S. 445, beklagt, daß mit der Anpassung an die vom EGMR gegebene Auslegung »nous attribuerions, *en réalité*, à la jurisprudence de la Cour des droits de l'homme une autorité erga omnes, contraignante, qu'elle n'a pas«.

[79] Auffällig ist, daß die Entscheidungen der Cour de Cassation vom 21. 1. 1982 in den Fällen *Guchez* und *Simonis* auf die Probleme, die sich aus der Abweichung von dem *Le Compte*-Urteil des EGMR ergeben, mit keinem Wort eingehen, sich freilich mit den Argumenten, wenn auch in knapper Form, auseinandersetzen. Die Cour de Cassation schließt sowohl aus dem Sinn als auch aus dem Umfang des Begriffs »contestation sur des droits et obligations de caractère civil« als auch aus der Natur und dem Gegenstand von Disziplinarverfahren, »que les auteurs de la Convention de sauvegarde des droits de l'homme n'ont pu avoir eu l'intention de soumettre celle-ci à la publicité que prescrit l'art. 6 § 1er de cette Convention«. Der Schluß auf den historischen Parteiwillen bedeutet nicht, daß die Cour de Cassation sich von einer lediglich historischen und subjektiven Auslegung leiten läßt. Vielmehr stehen neben dem Text und dem sinnhaften Zusammenhang der Regelung in Art. 6 EMRK auch und vor allem der Zweck der dort geregelten Verfahren und Verfahrensgarantien im Vordergrund der Betrachtung, die folglich objektiven Charakter hat.

Da es sich bei den Urteilen des EGMR um Feststellungsurteile handelt, stellt sich die Frage, welche Wiedergutmachungs- bzw. Beseitigungspflichten sich daraus für die Vertragsstaaten ergeben. Für diese Betrachtung ist zu unterscheiden a) die Verpflichtung aus der Konvention selbst (Art. 1 in Verbindung mit Art. 57 EMRK) und b) die Verpflichtung, die sich aus der Rechtskraft des Urteils (Art. 52 EMRK in Verbindung mit Art. 53 EMRK) ergibt. Da es sich um Feststellungsurteile handelt, besagt die Rechtskraft der Entscheidung zwischen den Parteien an sich nur, daß mit bindender Wirkung feststeht, daß eine Konventionsverletzung stattgefunden hat. Indem Art. 53 bestimmt, daß die Parteien des Verfahrens sich nach der Entscheidung des Gerichtshofs »zu richten« haben, wird eine Verhaltenspflicht statuiert, welche der ohnedies nach allgemeinem Völkervertragsrecht bestehenden Pflicht entspricht, den Vertragsbruch (seinen auslösenden Akt und die Folgen) zu beseitigen. Den Staaten steht dabei ein gewisser Ermessensspielraum zu [80].

Auf diese Verpflichtung nach Art. 53 EMRK bezieht sich die Befugnis des Ministerausschusses, die »Durchführung« der Urteile zu überwachen. Diese Überwachung ist in Bezug auf Verfahren und die Anordnung konkreter Überwachungsmaßnahmen nicht näher spezifiziert [81]. Es spricht jedoch viel für die These [82], daß die »Durchführung« auch die zur Vorbeugung

[80] Siehe im einzelnen meinen Bericht (Anm. 4), S. 4 ff.

[81] Die vom Ministerkomitee 1976 erlassenen Regeln zur Anwendung von Art. 54 EMRK (abgedruckt im Anhang zur »Collection of Resolutions adopted by the Committee of Ministers in application of article 54« – H (81)5) schreiben nur vor, daß das Ministerkomitee »invite l'Etat intéressé à l'informer des mesures prises à la suite de cet arrêt, eu égard à l'obligation qu'il a de s'y conformer selon l'article 53 de la Convention«. Diese Wendung bringt wenig Aufschluß. Man wird m. E. – schon wegen der vergleichbaren Ausgangslage – auf die Regelungen in Art. 32 Abs. 2 bis 4 EMRK in analoger Anwendung zurückgreifen dürfen.

Nach Nr. 5 der im Juni 1979 erlassenen Regeln des Ministerkomitees für die Anwendung des Art. 32 EMRK (vgl. Anhang des Dok. H. (81) 4) sollen Ratschläge, Anregungen oder Empfehlungen sein »closely related to the violation« (bzw. »directement trait à la violation« – die englische Fassung ist wohl weiter). Die Regel 5 legt außerdem fest, daß diese Ratschläge nicht bindend sind, da sie keine Entscheidungen im Sinne von Abs. 4 des Art. 32 darstellen. Diese durch den Wortlaut des Art. 32 nicht unbedingt getragene Auffassung beruht offensichtlich auf der Erwägung, daß die Vertragsstaaten bezüglich der von ihnen zu ergreifenden »befriedigenden Maßnahmen« (vgl. Art. 32 Abs. 3 EMRK) einen Ermessensspielraum haben, der durch bindende Beschlüsse des Ministerkomitees nicht eingeschränkt werden soll. Von Interesse ist die Wendung, daß sich diese Ratschläge nur »genau bzw. unmittelbar auf die Verletzung« beziehen dürfen. Die Frage, welcher Akt des Vertragsstaates eigentlich die Verletzung ausmacht, wird damit zu einem entscheidenden Anknüpfungspunkt.

[82] Dafür spricht auch die Praxis der Staaten, welche im Anschluß an einen ihnen gegenüber festgestellten Konventionsverstoß ihre Gesetzgebung änderten oder andere generelle Anweisungen erließen, um künftige Verstöße auszuschließen (vgl. Collection H (81) 5) im *Golder*-Fall: »instructions for an immediate introduction of new procedures in all prison department establishments« – sowie Einbringung eines Gesetzes zur Änderung der Prison rules 1974. Im Fall *Engel* hatte die Änderung der Gesetzgebung schon zuvor stattgefunden, so daß »violations of Art. 5 and 6 of the Conven-

künftiger Konventionsverstöße zu ergreifenden generellen Maßnahmen erfaßt, wobei freilich das Ministerkomitee in den Ermessensspielraum der betreffenden Staaten nicht eingreifen darf. Zur Begründung sei darauf hingewiesen, daß die Individualbeschwerde häufig[83] den Charakter einer »verdeckten Rechtssatz-Beschwerde« haben wird, d.h. die Beschwerde richtet sich nicht gegen die unrichtige Anwendung eines an und für sich konventionsgemäßen innerstaatlichen Gesetzes im Einzelfall, sondern mittelbar gegen die Konventionswidrigkeit des Gesetzes, welches den konventionswidrigen Einzelakt (Urteil oder Verwaltungsakt) determiniert.

b) Vergleich mit der Verfassungsbeschwerde

In diesem Zusammenhang ist ein Vergleich mit dem Verfassungsbeschwerdeverfahren vor dem BVerfG angebracht. Nach § 95 Abs. 1 Satz 1 Bundesverfassungsgerichtsgesetz (BVerfGG) stellt das BVerfG zunächst fest, welche Vorschrift des Grundgesetzes und durch welche Handlung oder Unterlassung sie verletzt wurde. Wird der Verfassungsbeschwerde gegen Verwaltungsakte oder Gerichtsurteile stattgegeben, die auf einer grundrechtswidrigen Rechtsnorm beruhen, so »ist die Feststellung nach § 95 Abs. 1 Satz 1 auch auf das Gesetz zu erstrecken«[84]. Begründet wird diese Auffassung mit dem »untrennbaren Zusammenhang« zwischen Verfassungswidrigkeit des Gesetzes und des Einzelaktes sowie mit der »objektiven Funktion« der Verfassungsbeschwerdeentscheidung, die insoweit über die konkrete Beschwerde des Antragstellers hinaus zu einer Klärung der Verfassungsrechtslage beitragen soll[85]. Positiv-rechtliche Anhaltspunkte für diese Aus-

---

tion found by the Court cannot occur under the legislation currently in force«. Im Fall *Tyrer* informierte die Britische Regierung, »in so far as the decision of the Court raises a general issue because of the risk of further applications«, daß sie alle für die Verhängung von Prügelstrafen zuständigen Amtspersonen davon unterrichtet habe, »that the effect of the judgment is that judicial corporal punishment must now be held to be in breach of the European Convention«. Hier hat sich das Ministerkomitee (offenbar angesichts der Rechtsstellung der Isle of Man) mit dieser Erklärung ohne Änderung der Gesetzgebung zufrieden gegeben; im Fall *Sunday Times* hat die britische Regierung einen Gesetzesvorschlag eingebracht, »to amend the English law regards Contempt of Court (and thus) to prevent further conflict in this respect with the provisions of the European Convention of Human Rights as interpreted by the Court in this case«. Im *Airey*-Fall wurde dem irischen Parlament ein neues »Scheme of Civil Legal Aid and Advice« vorgelegt. Im Fall *Le Compte* schließlich ist inzwischen eine »proposition de loi relative à la publicité des procédures disciplinaires contre les médecins, pharmaciens, architects et médecins vétérinaires« eingebracht worden (Vorschlag R. Lallemand, Senat, 12. 3. 1982).

[83] *E. Schumann,* Verfassungs- und Menschenrechtsbeschwerde gegen richterliche Entscheidungen (1963), S. 254, spricht sogar von einer »generellen Geneigtheit« zur verdeckten Rechtssatzbeschwerde.

[84] *Schmidt-Bleibtreu,* in: Maunz/Schmidt-Bleibtreu/Klein/Ulsamer, BVerfGG, § 95 Rn. 17.

[85] Zum objektiven Zweck der Verfassungsbeschwerde siehe *Schumann* (Anm. 83), S. 91 ff., 117 f.

weitung der Feststellungsformel enthalten die im Verfassungsbeschwerdeverfahren entsprechend anwendbaren §§ 78 Satz 2, 82 Satz 1 BVerfGG[86]. Nach § 95 Abs. 1 Satz 2 BVerfGG kann das Gericht aussprechen, daß auch jede Wiederholung der beanstandeten Maßnahme das Grundgesetz verletzt. Da *die* beanstandete Maßnahme nie wiederholt, sondern nur die gleiche nochmals gesetzt werden kann, handelt es sich um eine die objektiven Grenzen der Rechtskraft überschreitende Feststellung[87]. Diese Befugnis unterstreicht die über den Individualrechtsschutz hinausgehende objektive Funktion der Verfassungsbeschwerdeentscheidung, »die nicht allein darauf gerichtet ist, einen in der Vergangenheit ausgebrochenen konkreten Streit unter den Beteiligten zu entscheiden, sondern für die Zukunft allgemein mit Bindung für alle eine streitige Rechtsfrage zu entscheiden und dadurch Wiederholung gleichartiger Streitigkeiten auszuschließen«[88]. Es ist deshalb nicht überraschend, daß *Lechner*[89] die Problematik des § 95 Abs. 1 Satz 2 darin sieht, daß die »beanstandete Maßnahme« nicht immer rechtstypisch eindeutig erfaßbar sei und den jeweiligen Umständen möglicherweise entscheidende Bedeutung zukommen könne. Das Problem besteht in der Tat darin zu bestimmen, ob eine Wiederholung *der* beanstandeten Maßnahme vorliegt[90].

Nach Ansicht von *Schmidt-Bleibtreu* und anderen[91] bedarf es nach der Rechtsprechung des BVerfG zu § 31 Abs. 1 BVerfGG keines Ausspruchs über das Wiederholungsverbot (§ 95 Abs. 1 Satz 2), um eine über den Einzelfall hinausgehende Bindungswirkung der Verfassungsbeschwerdeentscheidung herbeizuführen, da auch ohne einen solchen Ausspruch »die sich aus dem Tenor und den tragenden Gründen der Entscheidung ergebenden Grundsätze für die Auslegung der Verfassung von den Gerichten und Behörden in allen künftigen Fällen beachtet werden müssen«[92].

Diese Grundsätze können nicht ohne weiteres auf die Interpretation von Art. 53 EMRK übertragen werden, nicht nur, weil den Urteilen des EGMR keine kassatorische Wirkung zukommt, sondern weil eine dem § 95 Abs. 1 Satz 2 BVerfGG vergleichbare Regelung fehlt.

Gleichwohl wird man folgendes festhalten dürfen: Hat der EGMR festgestellt, daß eine »Entscheidung oder Maßnahme einer gerichtlichen oder

---

[86] Vgl. *Schmidt-Bleibtreu* (Anm. 84); BVerfGE 18, 288 [300]; 24, 75 [103].

[87] *Schumann* (Anm. 83), S. 91.

[88] *Geiger,* BVerfGG (1952), § 95 Anm. 4.

[89] *Lechner,* BVerfGG 1973, § 95 zu Abs. 1 b; so auch *Scherer,* Die Verfassungsbeschwerde (Diss. 1959), S. 234 f.

[90] Die gleichen Probleme bestehen bei der Verwaltungsgerichtsbarkeit: siehe *G. Ress,* Die Entscheidungsbefugnis in der Verwaltungsgerichtsbarkeit (1968), S. 106 f., 121 f.

[91] (Anm. 84), Rn. 20 unter Hinweis auf *Redelberger,* NJW 1953, S. 363; *Spanner,* Das Bundesverfassungsgericht (1972), S. 152.

[92] BVerfGE 19, 377 [392]; 20, 56 [87].

sonstigen Behörde« – und dies kann auch ein Gesetz sein[93] – gegen die EMRK verstößt, so ist der Staat verpflichtet, diese Maßnahmen und die von ihr ausgehenden Wirkungen zu beseitigen und gegebenenfalls Genugtuung zu leisten. Mit dieser sich aus dem objektiven Umfang der Rechtskraft ergebenden Verpflichtung wäre es unvereinbar, daß der Staat entweder die konventionswidrige Maßnahme aufrechterhält (und nur Genugtuung leistet)[94] oder die in der Vergangenheit schon aufgehobene Maßnahme mit der Begründung erneut erläßt, die Auslegung der EMRK durch den EGMR überzeuge ihn nicht.

Bei Einzelakten (Urteilen und Verwaltungsakten) erstreckt sich die Feststellungswirkung mittelbar aber auch auf alle jene generellen Staatsakte (Gesetze, Verordnungen, Verwaltungsanweisungen), welche den Verstoß gegen die EMRK im rechtlich-normativen Zusammenhang untrennbar determinieren. Beruht der konkrete Akt auf einem konventionswidrigen Gesetz und wird diese Konventionswidrigkeit in den Gründen der Entscheidung (wenn auch nur verklausuliert) erkennbar, dann erstreckt sich mittelbar die Feststellung der Konventionswidrigkeit auch auf die Akte (Gesetzesbestimmungen), welche diese Konventionswidrigkeit determinieren. Der EGMR – am Einzelfallvollzug orientiert – hat diesen Zusammenhang nicht immer klar zugeben wollen. Im *Marckx*-Urteil[95] steht, es sei nicht Aufgabe des Gerichtshofs, eine abstrakte Prüfung der angegriffenen Gesetzestexte durchzuführen; er prüfe, ob deren Anwendung auf die Beschwerdeführerinnen mit der Konvention in Einklang steht. Dabei wird zunächst verdeckt, daß die Konventionswidrigkeit gerade nicht im Anwendungsvorgang einer an sich konventionsgemäßen innerstaatlichen Norm, sondern in dieser selbst begründet ist! Des-

---

[93] Vgl. *Marckx*-Fall, EuGRZ 1979, 460 (Nr. 58); *Mosler* (Anm. 45), S. 116; mein Bericht (Anm. 4), S. 5 ff. Noch viel deutlicher ist der *Klass*-Fall ein – von einem denkbaren Vollzug ausgehendes – reines Gesetzesprüfungsverfahren gewesen (EuGRZ 1979, 278 ff.), bei dem der EGMR die vom BVerfG entwickelten Voraussetzungen für die Zulässigkeit einer unmittelbar gegen ein Gesetz gerichteten Vefassungsbeschwerde *mutatis mutandis* übernommen hat (Nr. 36).

[94] Vgl. dazu die Auseinandersetzung mit *W. Schmid,* Die Wirkungen der Entscheidungen der europäischen Menschenrechtsorgane (1974), S. 86, 154 ff., in meinem Bericht (Anm. 4), S. 5 f. (Anm. 26 und 36). Leistet der Staat dem Beschwerdeführer Genugtuung (*satisfaction équitable*) im Sinne von Art. 50 EMRK, so erlischt damit nicht seine Verpflichtung, sich nach dem Urteile »zu richten«, d. h. den konventionsgemäßen Zustand herzustellen. Die Überwachungsfunktion des Ministerkomitees findet nicht *ipso iure* ihr Ende, wenn dem Beschwerdeführer gemäß Art. 50 EMRK Genüge getan wurde, oder nach Ansicht des EGMR im konkreten Fall kein Anlaß zur Anwendung von Art. 50 EMRK besteht. So besteht z. B. im Fall *Marckx* die Verpflichtung Belgiens, durch Mittel seiner Wahl, d. h. aber doch wohl gesetzgeberische Maßnahmen, »die ihm aus Art. 53 obliegende Verpflichtung zu erfüllen« (Nr. 58 des Urteils, EuGRZ 1979, 460), fort, auch wenn sich die gerechte Entschädigung für die Beschwerdeführerinnen in »der Feststellung mehrerer Verletzungen ihrer Rechte« (Nr. 68 *loc.cit.*) erschöpfte.

[95] EuGRZ 1979, 460 (Nr. 58).

halb läßt der EGMR auch sofort eine etwas gewundene, den ersten Satz in Frage stellende Passage folgen: »Allerdings läßt sich nicht vermeiden[96], daß seine Entscheidung – woran kein Zweifel besteht – über den spezifischen Fall hinaus Wirkungen haben wird, um so mehr, als die hervorgetretenen Verletzungen ihre *unmittelbare Wurzel* in den genannten Gesetzestexten haben und nicht in einzelnen Vollzugsmaßnahmen«. Bedeutsam ist der folgende Satz: Die Entscheidung des EGMR »kann jedoch nicht durch sich selbst die strittigen Vorschriften aufheben oder ändern! Sie hat im wesentlichen Feststellungscharakter . . .«. Darin kommt deutlich zum Ausdruck, daß sich die Feststellung mittelbar auch auf die »strittigen Vorschriften« bezieht – nur ist deren Änderung bzw. Aufhebung Sache des betroffenen Staates.

In § 95 Abs. 3 Satz 2 BVerfGG ist für das BVerfG bei dieser Konstellation die Konsequenz gezogen worden, daß das Gericht ein Gesetz auch dann für nichtig erklären kann, wenn die Verfassungsbeschwerde sich gegen eine konkrete Entscheidung richtet und ihr stattgegeben wird, weil die aufgehobene Entscheidung auf einem verfassungswidrigen Gesetz beruht[97]. Der Umfang der Kassationsbefugnis des BVerfG in § 95 Abs. 3 hängt von einem ursächlichen Zusammenhang zwischen der Verfassungswidrigkeit des Gesetzes »und dem Erfolg der Verfassungsbeschwerde« ab, d. h. der konkrete Einzelakt ist deshalb verfassungswidrig, weil das Gesetz verfassungswidrig ist[98]. Diese Vorschrift dient »nicht primär dem Rechtsschutzinteresse des Beschwerdeführers, dessen Beschwer durch die Aufhebung der ihm nachteiligen grundrechtswidrigen Entscheidung in den meisten Fällen beseitigt sein dürfte. Sie soll vielmehr über den subjektiven Rechtsschutz hinaus eine Klärung der objektiven Rechtslage herbeiführen. Die von § 95 Abs. 3 Satz 2 geforderte Kausalbeziehung zwischen der Grundrechtswidrigkeit des Einzelaktes und der des Gesetzes trägt dabei dem Gedanken Rechnung, daß das BVerfG nicht von sich aus, sondern nur auf einen Anstoß von außen hin zur Normenkontrolle befugt sein soll«[99].

Zwischen der Verfassungsbeschwerde und der Menschenrechtsbeschwerde bestehen, wie die Regelungen zeigen, sicher eine Reihe von Unterschieden. Beiden Beschwerdeverfahren ist jedoch eigen, daß sie neben der Funktion, die subjektive Beschwer zu beseitigen, auch eine *objektive* Funktion haben. Beide dienen der grundsätzlichen Klärung von Rechtsfragen, was die Kommission für die Menschenrechtsbeschwerde (sowohl Staaten- als auch

---

[96] Man fragt sich: Warum sollte dies auch vermieden werden? Hätte ein positives Feststellungsurteil im Fall *Klass* sich nicht zur Konventionswidrigkeit des G 10 aussprechen sollen?

[97] Vgl. dazu *Schmidt-Bleibtreu* (Anm. 84), § 95, Rd. 33; vgl. auch *Schumann,* Verfassungs- und Menschenrechtsbeschwerde (1963), S. 82 ff.

[98] Nichtigerklärungen nach § 95 Abs. 3 Satz 2 BVerfGG enthalten u. a. BVerfGE 7, 377 [444]; 9, 83 [89]; 10, 201 [221] usw. (siehe *Schmidt-Bleibtreu* [Anm. 84], S. 25 Anm. 2).

[99] *Schmidt-Bleibtreu,* a.a.O., S. 25.

Individualbeschwerde) seit dem *Pfunders*-Fall[100] stets betont hat und was sich auch aus dem Art. 28 *lit.* b (freundschaftlicher Ausgleich der Angelegenheit nur auf der Grundlage der Achtung der Menschenrechte, wie sie in dieser Konvention niedergelegt sind) ergibt. Sofern andere Verfahren, wie z. B. die staatsrechtliche Beschwerde nach Schweizer Recht, die Prüfungskompetenz des Bundesgerichts stärker auf den angegriffenen Hoheitsakt und die behauptete Rechtsverletzung beziehen[101], kann dieses nicht ohne weiteres auch für das Verfahren nach der EMRK[102] behauptet werden. Während dem schweizerischen staatsrechtlichen Beschwerdeverfahren der Grundsatz *iura novit curia* fremd ist, ist er Bestandteil des EMRK-Verfahrens. Dem schweizerischen Recht ist die in § 95 Abs. 3 Satz 2 BVerfGG vorgesehene Möglichkeit der Nichtigerklärung einer nur mittelbar angegriffenen Rechtsnorm fremd; damit ist nichts für die Parallelität zum EMRK-Verfahren ausgesagt, in welchem es zwar keine Kassationsbefugnis des EGMR, wohl aber entsprechende Verhaltenspflichten der Parteien gibt. Diese (neben der subjektiven) *objektive* Funktion[103] spricht dafür, daß das Ministerkomitee auch die daraus sich ergebenden Folgen überwachen darf, d. h., ob der betroffene Staat die »unmittelbare Wurzel« der Konventionsverletzung beseitigt hat. Der Überwachungsfunktion des Ministerkomitees nach Art. 54 EMRK kommt demnach eine Aufgabe zu, welche im Verfassungsbeschwerdeverfahren durch gerichtliche Aussprüche (nach § 95 Abs. 3 Satz 2 BVerfGG) und wesentlich »unflexibler« erfüllt wird.

c) Unterschiedliche Konzeptionen des EMRK-Kontrollsystems?

Genau gesehen müßte einer derartigen Überwachung der objektiven Funktion der EGMR-Urteile auch eine Verpflichtung der an einem Verfahren beteiligten Parteien korrespondieren, sich nach diesem objektiven Gehalt der Urteile des EGMR »zu richten«, also eine Verpflichtung, alle die Konven-

---

[100] Entscheidung vom 11. 1. 1961, Yearbook of the European Convention on Human Rights 1961, S. 138 ff. Siehe dazu *Waldock,* EuGRZ 1979, 599; *Eissen,* AFDI 1963, S. 727 ff.; kritisch dazu *H. Walter,* Die europäische Menschenrechtsordnung (Beiträge zum ausländischen öffentlichen Recht und Völkerrecht, Bd. 53) (1970), S. 48 ff.

[101] *Schmidt-Bleibtreu* (Anm. 84), § 95 Rn. 6.

[102] Vgl. dazu den Streit um die Geltung des Satzes *iura novit curia*, der nach *Fleiner/Giacometti* (Schweizerisches Bundesstaatsrecht, 1949 [1965], S. 898 Anm. 77) in dem schweizerischen Verfahren nicht gilt. Zur EMRK siehe dagegen *H. Walter,* Der Grundsatz iura novit curia im Europäischen Menschenrechtsverfahren. Zur Teilabweisung von Individualbeschwerden wegen offensichtlicher Unbegründetheit, ZaöRV 28 (1968), S. 561 ff.; der EGMR hat die Geltung des Grundsatzes für die EMRK im *Neumeister*-Fall klargestellt (bei *Walter,* S. 576).

[103] *R.-J. Dupuy,* AFDI 1957, S. 449 ff., 470, geht zu weit, wenn er behauptet, das Individuum erhebe die Individualbeschwerde gemäß Art. 25 EMRK weniger zur Verteidigung eines subjektiven Rechts als zur Wahrung der objektiven Legalität. Gegen diese Überzeichnung *Walter,* (Anm. 100), S. 64 ff., und derselbe, Übermäßige Objektivierung des Europäischen Menschenrechtsschutzes?, ZaöRV 35 (1975), S. 120.

tionswidrigkeit im konkreten Fall *determinierenden* Staatsakte zumindest *ex nunc* zu beseitigen oder außer Anwendung zu bringen, mithin die Beschwer nicht nur *in concreto,* sondern auch *in abstracto* im Hinblick auf die Wiederholung der gleichen beanstandeten Maßnahme zu beseitigen. Geht man von einer derartigen, die Rechtskraftwirkung im engeren Sinne erweiternden Verpflichtung und Überwachung aus, so bedeutet dies nicht, daß diese Verpflichtung gegenüber dem Beschwerdeführer besteht. Es handelt sich um eine objektive Verpflichtung [104], deren Einhaltung der Ministerausschuß zu überwachen hat, weil erst dadurch die »befriedigenden Maßnahmen« (analog Art. 32 Abs. 3 EMRK) herbeigeführt werden [105].

Mosler hat in seiner persönlichen Meinung im Fall *Koç* [106] die Auffassung vertreten, Art. 25 EMRK könne nicht dahin verstanden werden, daß er dem Beschwerdeführer ein rechtliches Interessse zubillige, »daß in seinem *Einzelfall* die Unvereinbarkeit der Gesetzgebung eines Staates oder des auf dieser Gesetzgebung beruhenden Urteils eines nationalen Gerichts festgestellt wird, es sei denn, daß diese Feststellung für die Entscheidung über die Verletzung der Konvention in dem konkreten Fall des Beschwerdeführers notwendig ist«. *Mosler* deutet die Argumentation des Gerichtshofs im *Luedicke*-Fall dahingehend, daß sie davon auszugehen scheine, der Einzelne, der die Kommission anrufe, habe nicht nur einen *Anspruch* darauf, daß die in Bezug auf ihn begangene Verletzung beseitigt und ihm gegebenenfalls eine gerechte Entschädigung zuerkannt werde, sondern darüber hinaus einen Anspruch darauf, daß der Gerichtshof über die Frage entscheidet, ob die Gesetzgebung und die Gerichtsentscheidungen, die die Verletzung veranlaßt haben, mit der Konvention vereinbar sind.

Ausgehend vom Umfang des subjektiven Anspruchs des Beschwerdeführers, wird dieser Argumentation zu folgen sein. Gleichwohl bleibt die Frage, ob angesichts des objektiven Charakters des Verfahrens die Bedenken berechtigt sind.

Im Fall *Koç* trat die Konventionsverletzung in einer Kostenfestsetzung gegen den Beschwerdeführer zutage, welche auf einem Schwurgerichts-Urteil beruhte, das in Anwendung des Gerichtskostengesetzes erging. Insoweit lag eine verdeckte Rechtssatzbeschwerde vor. Sicher wurde die Beschwer für den Beschwerdeführer mit der Erklärung der Bundesregierung, die Beitreibung der Kosten aus dem Schwurgerichts-Urteil werde künftig unterbleiben und außergerichtliche Kosten des Beschwerdeführers würden gegebenenfalls ge-

---

[104] Gegen die objektive Verpflichtungstheorie hat sich mit beachtlichen Gründen *Walter* (Anm. 100), S. 48 ff., gewandt.

[105] Bei mittelbar konventionswidrigen Gesetzen stellt sich freilich die Frage, ob der Staat nicht abwarten kann, daß von den Gerichten und Verwaltungsbehörden in Zukunft eine konventionskonforme Auslegung des Gesetzes praktiziert werden wird (vgl. oben Anm. 82 zum Fall *Tyrer*). Problematisch daran ist, daß das Ministerkomitee unter Umständen für diesen Nachweis Fristen setzen müßte oder seine Aufgabe in eine Art »Dauerüberwachung« überginge.

[106] EuGRZ 1979, 42.

tragen, nicht beseitigt, denn wegen anderer Folgen konnte dem Beschwerdeführer eine angemessene Entschädigung nach Art. 50 EMRK zustehen. Diese Erklärung der Bundesregierung ist sicher kein prozessuales Anerkenntnis, wie dies aus anderen Verfahren geläufig ist. Denn die Regierung hat nicht den Beseitigungs- und Wiedergutmachungsanspruch des Beschwerdeführers anerkannt, sondern nur ein Versprechen gegeben. Sie hat auch nicht den konkreten Akt – Kostenfestsetzung und Gerichtsurteil – aufgehoben.

Mir scheint, daß sich hinter dieser Streitfrage zwei grundverschiedene Konzeptionen des EMRK-Kontrollsystems verbergen: Wenn man davon ausgeht, daß Streitgegenstand des Beschwerdeverfahrens einschließlich des Verfahrens vor dem Gerichtshof der Anspruch des Beschwerdeführers auf Beseitigung der Rechtsverletzung und auf eine eventuelle angemessene Entschädigung ist, dann liegt es auf Grund dieses subjektiv-rechtlichen Bezugspunktes nahe, die Erledigung des Verfahrens (Streichung aus der Liste) stets dann anzunehmen (und danach auch den Umfang der Verpflichtung der Parteien nach Art. 53 EMRK zu bestimmen), sobald und soweit die individuelle Beschwer beseitigt ist. Diese Beseitigung mag im Einzelfall differenziert sein (Verzicht auf Kosten, Begnadigung, Aufhebung des Strafurteils), erstreckt sich jedoch nicht auf weitere, den konkreten Akt determinierende Rechtsakte (Gesetze). Der Staat mag sich dazu entschließen, daß er – da er die Auslegung der EMRK durch den EGMR nicht teilt – das Gesetz aufrechterhält und auf eine Änderung der Rechtsprechung in einem nächsten Fall hoffen. Bei dieser Art der Einzelfallbezogenheit bliebe für eine erweiterte Überwachungsfunktion des Ministerkomitees nach Art. 54 EMRK kein Raum.

Stellt dagegen die Feststellung der Konventionswidrigkeit eines bestimmten staatlichen Verhaltens den eigentlichen Verfahrensgegenstand dar (wie von Art. 50 erster Halbsatz EMRK nahegelegt), so steht bei der Auslegung von Art. 53 EMRK die Beseitigung der festgestellten Konventionswidrigkeit im Mittelpunkt der staatlichen Verpflichtung und erst in zweiter Linie die Beseitigung der individuellen Beschwerde. In diesem Fall wäre Streitgegenstand des Verfahrens die Frage, ob der Staat konventionswidrig gehandelt hat. Diese Frage wäre natürlich auch zu prüfen, wenn der individuelle Beseitigungs- und Wiedergutmachungsanspruch Streitgegenstand wäre, sie wäre jedoch nur ein Element (Vorfrage) im Rahmen einer andersartigen Hauptfrage.

Es liegt nahe, das Verfahren vor der Kommission und dem Gerichtshof weder ausschließlich in der einen noch in der anderen Richtung zu deuten, sondern ausgehend von der Bezogenheit auf die konkrete Beschwer in der Person des Beschwerdeführers *zusätzliche* objektive, auf Rechtsklärung angelegte Elemente anzunehmen, wie z. B. die Geltung des Grundsatzes *iura novit curia,* und die objektive Verpflichtung der Staaten, die als konventionswidrig festgestellten Staatsakte, einschließlich der Gesetze, zu ändern, außer Anwendung zu setzen oder sonstwie »befriedigend« zu beheben. An diese zusätzlichen objektiven Elemente darf auch die Überwachung durch das Ministerkomitee anknüpfen.

L'ordre juridique international et les traités des organisations internationales

PAUL REUTER

Le dédicataire de ces »Mélanges« a consacré comme tous les internationalistes confirmés une part non négligeable de ses méditations et de ses oeuvres à l'ordre juridique international et aux organisations internationales. Les réflexions qui suivent ont pour objet de rendre hommage à sa pensée en esquissant quelques-uns des rapports qui naissent entre cet ordre et une nouvelle catégorie d'actes juridiques, les traités auxquels les organisations internationales sont parties. On présentera d'abord le cadre dans lequel le problème se pose et l'on formulera ensuite trois propositions qui, sans épuiser cette matière, en expriment des aspects essentiels.

* * *

L'existence d'organisations internationales intergouvernementales au nombre de plus de deux cents a posé de nombreux problèmes qui touchent les caractères de l'ordre juridique international; ainsi en est-il des actes unilatéraux et de la responsabilité internationale. On se limitera néanmoins à la question des traités de ces organisations parce que c'est la plus riche et celle qui, après les travaux de la Commission du droit international, est aujourd'hui la mieux connue [1].

Il y a bien des façons de définir un ordre juridique [2], mais l'on ne voudrait pas au seuil même de cette étude ouvrir des discussions doctrinales qui re-

---

[1] On se reportera aux Rapports de cette Commission publiés dans son Annuaire depuis 1972.

[2] On se dispensera d'aborder ici la question de la définition des organisations internationales parce qu'il est clair que l'on n'envisage ici que les organisations internationales qui peuvent conclure, si limitée que soit cette capacité, des traités. Cette remarque suffit aux fins de cet essai comme elle a suffi à la Commission du droit international dans la préparation du projet d'articles sur ces traités. Ceci n'ôte nullement son intérêt à l'étude de la notion d'organisation internationale en général.

tarderaient l'examen du sujet. C'est pourquoi on se contentera de rappeler qu'un ordre juridique est d'abord un ordre, c'est-à-dire un ensemble d'éléments »mis en ordre« les uns par rapport aux autres de manière à constituer une certaine unité. Quand on parle de l'ordre juridique international on a présent à l'esprit l'ordre juridique international *général*, c'est-à-dire l'ensemble des normes internationales générales ordonnées entre elles par les règles fondamentales portant notamment sur la création des normes, leur interprétation, leur application et sanction ainsi que sur leur respective hiérarchie. Comme on le sait il y a aussi, à côté de l'ordre juridique international, les ordres juridiques des Etats, plus ou moins distincts et séparés de l'ordre international selon des conceptions monistes ou dualistes, qui sont elles-mêmes à un moment donné de l'histoire et pour un Etat donné l'expression de l'intensité et de la nature des relations internationales.

L'introduction des organisations internationales dans ce schéma classique conduit à des problèmes nouveaux. En effet le développement de ces dernières les amène à exercer des fonctions normatives d'abord à l'égard de leurs agents administratifs, ensuite à l'égard des Etats qui en sont membres et enfin parfois à l'égard de particuliers qui sont sous la juridiction de ces Etats. Ainsi apparaît progressivement un ordre juridique propre à chaque organisation. Cet ordre est selon le cas plus ou moins riche en règles générales, – plus ou moins homogène (comment mettre dans la même catégorie les normes concernant le statut des agents administratifs de l'organisation et celles obligeant directement les Etats membres?), – plus ou moins rigoureusement défini et hiérarchisé. Comment désigner cet ordre juridique propre à une organisation?

On a parfois employé à ce sujet l'expression »ordre juridique interne de l'organisation«. On comprend les raisons qui justifient l'emploi d'un terme qui accentue ou au moins souligne la personnalité de l'organisation et suggère par antithèse que celle-ci a des relations *externes* ce qui est comme on le verra bientôt un point très important. Cependant un tel vocabulaire est chargé de références analogiques. Ce sont essentiellement les Etats qui possèdent un droit »interne« et ce droit national présente des caractères et une structure bien déterminés et sensiblement identiques pour tous les Etats contemporains (sources variées avec prédominance des sources écrites, – rareté des lacunes, – hiérarchie précise et rigoureuse, – sanction par un contrôle judiciaire généralisé etc.). Le vocabulaire en cause n'est donc pas neutre. Qualifier le droit propre à une organisation internationale de droit »interne«, c'est suggérer que ce droit n'est pas sans analogie avec un ordre juridique étatique et s'éloigne plus ou moins des caractères de l'ordre juridique international. Au contraire le qualifier de »droit international particulier« c'est indiquer que, sous réserve de quelques traits spécifiques il participe aux caractères de cet ordre juridique international. Il s'agit là d'une question de principe et de vocabulaire, mais la jurisprudence offre déjà des cas où le choix d'une analyse ou d'une autre entraine des conséquences pratiques importantes[3].

Aussi la pratique conventionnelle, quand il s'agit de désigner ce droit propre à une organisation internationale dans l'article d'une convention ou d'un projet de convention a évité cette expression pour recourir à une terminologie plus anodine et descriptive, ainsi la Convention sur la représentation des Etats dans leurs relations avec les organisations universelles de 1975 ou le Projet d'articles de la Commission du droit international sur les traités auxquels des organisations internationales sont parties[4]. Ces deux textes se réfèrent aux »règles de l'Organisation« et cette expression »s'entend notamment des actes constitutifs de l'Organisation, de ses décisions et résolutions pertinentes et de la pratique bien établie de l'Organisation«.

Ces définitions montrent toutes les incertitudes et toutes les ambiguïtés qui peuvent peser d'un point de vue théorique et pratique sur la notion de »droit propre à une organisation«. En effet, non seulement ce droit propre ainsi défini reste incertain dans sa nature, mais même son étendue est ambigüe. L'acte constitutif d'une organisation internationale fait en effet partie selon ces définitions des »règles de l'organisation«; or l'acte constitutif de l'organisation est un traité entre Etats auxquel l'Organisation qu'il créée n'est pas partie, alors que les actes mentionnés dans ces définitions sont des actes unilatéraux de l'Organisation. Autrement dit et pour reprendre un vocabulaire inventé par les Communautés européennes, le droit propre d'une Organisations internationale comprend à la fois un »droit originaire« (les actes constitutifs) et un »droit dérivé« (toutes les normes issues des organes de l'Organisation).

Telles sont les observations qui constituent le cadre général des ordres juridiques par rapport auxquels il faut situer maintenant les traités auxquels des organisations internationales sont parties. On est ainsi à même de faire quelques observations de principe que l'on énoncera sous forme de propositions que l'on va exposer successivement en les commentant. Elles peuvent être énoncées de la manière suivante:

1) L'ordre juridique international contient et même contient nécessairement, des règles concernant les traités des organisations internationales.

---

[3] Dans les avis consultatifs que la CIJ a rendus sur le caractère définitif de certains jugements du Tribunal administratif des Nations Unies et du Tribunal administratif de l'Organisation internationale du travail, mais surtout dans le premier, la Cour rencontrait la question de savoir si la théorie de l'excès de pouvoir de l'arbitre, théorie propre à l'ordre juridique international s'étendait aussi aux jugements d'un tribunal compétent pour résoudre les litiges entre l'Organisation et ses agents. La Cour a répondu négativement: l'ordre juridique propre à une organisation, au moins en ce qui concerne les rapports de l'Organisation et de ses agents ne subit pas l'attraction de l'ordre juridique international parce qu'il présente une structure différente (cf. l'avis du 13 juillet 1954, Recueil, p. 56, et l'avis du 23 octobre 1956, p. 97).

[4] La définition qui figure à l'article 1 n° 34 de la Convention de 1975 est reprise dans le Projet d'article 2 j) de la Commission du droit international. La Convention de Vienne sur le droit des traités se référait dans son art. 5 à »toute règle pertinente de l'Organisation«, mais sans définir cette expression.

2) Tous les actes conventionnels des organisations internationales ne relèvent pas de l'ordre juridique international.

3) Les rapports entre les traités des organisations internationales et l'ordre propre à chacune d'entre elles doivent être aménagés.

* * *

La première proposition ci-dessus énoncée ne sera guère contestée et semble bénéficier de la force de l'évidence. Cependant elle mérite d'être précisée et approfondie.

Il y a plusieurs voies pour l'établir. La plus simple est de se référer à la pratique internationale. Les traités auxquels des organisations internationales sont parties se comptent par milliers et sont soumis dans la pratique à un régime qui d'une manière générale ressemble beaucoup à celui des traités entre Etats. Historiquement, quand la Commission du droit international a commencé ses premiers travaux sur le droit des traités, les rapporteurs spéciaux chargés d'étudier la matière ont tous accepté le principe d'un certain parallèlisme entre les traités entre Etats et les traités auxquels des organisations internationales sont parties. Quand la préparation du Projet d'articles sur le droit des traités est entré, à partir de 1962, dans sa phase active, la Commission a commencé par tenter de rédiger son projet en englobant les traités des organisations internationales et c'est, non point tant pour des raisons de principe, que pour des raisons pratiques et notamment par souci de garder une rédaction directe et relativement simple qu'elle a décidé de se limiter aux traités entre Etats. La Conférence de Vienne sur le droit des traités a retrouvé le problème dès son ouverture en 1968, et c'est pour ne pas retarder la conférence, mais non pour des raisons de principe, qu'elle a écarté une suggestion des Etats-Unis tendant à inclure les traités des organisations dans le projet sous examen; cette conférence a renvoyé la question, étant donné son importance, à la Commission du droit international[5]. Mais pendant ces longues années il n'a jamais été mis en doute par aucun Etat que la règle *pacta sunt servanda* s'appliquait aussi aux traités des organisations internationales et par conséquent que ces traités relevaient de l'ordre juridique international[6].

En dehors de cette approche historique et fortement positiviste, on peut établir l'existence de règles générales régissant les traités des organisations in-

---

[5] En mentionnant le fait que »les organisations internationales ont des pratiques diverses à cet égard« la Conférence se bornaite à se référer à la capacité et aux formes de la conclusion, mais non au régime général de ces traités.

[6] Sur cet aspect historique cf. Premier Rapport sur la question des traités conclus entre Etats et organisations internationales ou entre deux ou plusieurs organisations internationales, Annuaire de la Commission du droit international 1972 vol. II, p. 187.

ternationales sur la base d'une autre considération. Les traités des organisations internationales sont conclus par toutes les parties à de tels accords avec l'intention et la conviction que ces traités obligent. Or la force d'un engagement conventionnel découle nécessairement de l'ordre juridique dans lequel il est conclu. Il est manifeste que ce n'est ni dans l'ordre juridique d'un Etat ni dans »les règles d'une organisation« que l'on peut trouver, sauf ce qui sera dit au sujet de la deuxième proposition énoncée ci-dessus, la source de la force obligatoire d'un tel traité. Les Parties à un tel traité n'ont voulu se soumettre à aucun système juridique propre à l'un d'eux, ou propre à un autre Etat non partie au traité. Seul un ordre universel auquel les intéressés sont directement soumis pouvait satisfaire leur volonté.

Ce dernier raisonnement implique que l'ordre juridique international contient les règles générales nécessaires au régime applicable à de tels traités; l'existence de telles règles est en ce sens *nécessaire*. Sans doute peut-on faire observer que pour ces traités, à la différence de ce qui regarde les traités entre Etats, il est difficile de citer des précédents coutumiers. Mais même en se plaçant sur ce terrain l'objection n'est pas convaincante, parce qu'il existe des principes généraux du droit international public, applicables aux traités entre Etats et que ces principes généraux fournissent la base des règles que l'ordre juridique international doit appliquer aux traités des organisations internationales[7]. En effet pour sa plus large part le droit des traités n'est que l'expression des mécanismes classiques du consensualisme sous réserve du *jus cogens* que la Convention de Vienne a expressément mentionné.

On est donc conduit à reconnaître que le »droit des organisations internationales« entendu comme un droit valable pour toutes les organisations internationales porte sur le droit des traités, de même que sur quelques autres secteurs qui concernent les relations des organisations internationales avec d'autres sujets de droit international comme la responsabilité. En contrepartie seules les règles de chaque organisation régissent sa structure, ses organes, ses fonctions et ses pouvoirs, en un mot tout ce qui correspond à ce que l'on pourrait en quelque sorte appeler son »droit constitutionnel«. Ceci est d'autant plus vrai qu'il n'existe pas de principe d'égalité des organisations internationales analogue à celui de l'égalité des Etats et que les organisations,

---

[7] Si l'on considère par ailleurs les traités qui ont comme parties plusieurs Etats et une ou plusieurs organisations internationales, la Convention de Vienne sur le droit des traités apporte un singulier renfort. Son art. 3 n'exclut pas l'application de ses dispositions à des accords conclus entre des Etats et d'autres sujets de droit international ou entre ces autres sujets de droit international des règles qu'elle énonce; de plus il prévoit l'application de ces mêmes règles aux relations entre Etats nées d'un traité conclu entre des Etats et d'autres sujets de droit international. La dissociation des relations entre les différentes parties à un traité pour les soumettre à des règles différentes (non prévues à l'accord) n'est possible que dans d'étroites limites; aussi bien une telle disposition implique-t-elle d'une certaine manière l'identité du régime des traités entre Etats et des traités des organisations internationales.

entièrement suspendues à leurs fonctions et au traité qui les créée sont fondamentalement différentes les unes des autres.

Deux précisions doivent encore être apportées en ce qui concerne cette saisie par l'ordre juridique international des traités conclus par les organisations. Cette saisie appelle une restriction et une extension.

La restriction tient au fait bien connu et commun aux Etats et aux organisations que le droit constitutionnel propre à chacun d'entre eux et à chacune d'entre elles intervient également dans le domaine de la conclusion des traités en déterminant quelles personnes peuvent représenter l'Etat et les organisations et les engager définitivement. Il y là un problème classique et délicat d'ajustement entre l'ordre international et les ordres constitutionnels. Il a été réglé pour les Etats par les art. 7 et 46 de la Convention de Vienne sur le droit des traités; pour les organisations internationales la Commission du droit international a adapté ces solutions en renforçant quelque peu la protection assurée au respect des règles constitutionnelles propres à chaque organisation [8].

Une extension de l'emprise de l'ordre juridique international apparaît dans la mesure où le droit international général se saisit de données qui sont généralement abandonnées à des normes particulières propres à chaque organisation. Ainsi le régime des privilèges et immunités est en principe déterminé cas par cas pour chaque organisation; il y a donc à ce sujet des règles particulières propres à chaque organisation, et l'on ne peut pas dire que l'ordre juridique international règle directement cette question d'une manière générale, bien que l'on puisse établir l'existence de certains principes généraux en cette matière [9]. On peut imaginer toutefois que par un mécanisme de codification ou par un lent développement de la pratique apparaisse dans ce domaine un droit uniforme qui devienne valable pour toutes les organisations.

On avait déjà signalé cette possibilité dans une étude antérieure [10]; la jurisprudence récente de la CIJ en a suggéré un exemple qui mérite d'être signalé [11].

En effet il s'agissait de savoir dans quelles conditions le siège du bureau régional de l'Organisation mondiale de la santé (OMS) peut être transféré hors du pays où il est fixé et alors qu'un certain nombre d'accords relatifs à ce

---

[8] Annuaire de la CDI 1975 vol. II, p. 30, et 1979 vol. II 2è partie, p. 170, ainsi que notre étude Quelques réflexions sur la notion de »pratique internationale«, spécialement en matière d'organisations internationales, à paraître dans les »Mélanges en l'honneur de G. Sperduti«.

[9] *J. Duffar,* Les privilèges et immunités des organisations internationales intergouvernementales; à paraître en 1982.

[10] *P. Reuter,* Sur quelques limites du droit des organisations internationales, Festschrift für Rudolf Bindschedler (1980), p. 491.

[11] *Interprétation de l'accord du 25 mars 1951 entre l'OMS et l'Egypte, avis consultatif,* CIJ Recueil 1980, p. 73 et surtout p. 94 §§ 45 ss. Pour une analyse critique de l'avis cf. l'étude de *J. Combacau,* La question du transfert du bureau régional de l'OMS davant la CIJ, Annuaire français de droit international 1980, p. 225.

bureau ont été conclus avec le pays hôte. Le problème pouvait être placé sur le plan du droit particulier conventionnel, c'est-à-dire sur le plan des accords pertinents. A défaut d'indication en ce sens dans ces derniers on pouvait remonter à une règle générale du droit des traités traitant de la dénonciation des traités ne contenant pas de dispositions relatives à la dénonciation (art. 56 de la Convention de Vienne de 1969; projet d'article identique pour les traités des organisations internationales). Mais cette règle ne contient qu'une directive très souple. Sans exclure toute indication tirée de ces deux données la Cour a accordé une importance essentielle à une autre méthode en tirant de l'examen d'un grand nombre d'accords de siège des précédents qui permettaient de poser une règle générale décrivant les droits et obligations des deux parties dans une situation de ce genre:

»Telles sont, selon la Cour, les implications des principes et règles juridiques généraux applicables en cas de transfert du siège d'un bureau régional hors du territoire d'un Etat hôte«.

Il est certain que les règles communes dégagées par la Cour ne sont pas d'une excessive précision (§ 43):

». . . les relations juridiques entre l'Egypte et l'Organisation sont devenues et demeurent celles d'un Etat hôte et d'une organisation internationale, c'est-à-dire des relations dont l'essence même consiste en un ensemble d'obligations réciproques de coopération et de bonne foi«.

Et l'on peut penser que les »précisions« apportées par la Cour à l'obligation de consultation, de négociation et de délai dans la réalisation du transfert sont dépourvues de toute spécificité. Mais elles n'en témoignent pas moins d'une tendance à développer un droit commun de toutes les organisations internationales, formé cette fois non plus de règles *nécessaires* inhérentes à l'existence d'un ordre juridique international, mais simplement de règles *possibles et utiles.*

* * *

La deuxième proposition »Tous les actes conventionnels des organisations internationales ne relèvent pas de l'ordre juridique international« ne sera contestée par personne. Elle ne constitue qu'une application d'un célèbre *dictum* de la Cour internationale de Justice dans son avis de 1949 sur *la réparation des dommages subis au service des Nations Unies*[12]:

». . . la Cour arrive à la conclusion que l'Organisation est une personne internationale . . . Cela n'implique même pas que tous les droits et de-

[12] Recueil 1949, p. 179.

voirs de l'Organisation doivent se trouver sur le plan international, pas plus que tous les droits et devoirs d'un Etat ne doivent s'y trouver placés«.

Il y a donc des actes conventionnels d'une organisation qui ne relèvent pas de l'ordre juridique international. Les problèmes que soulève une telle éventualité sont nombreux et difficiles; on en donnera quelques exemples avant de s'attacher au plus intéressant d'entre eux.

Il est bien connu qu'une organisation peut placer un acte conventionnel dans l'ordre juridique d'un Etat quelconque. Elle le fera normalement pour des marchés courants de fournitures et de services conclus avec des entreprises privées. S'il s'agit d'un acte conventionnel avec un sujet de droit international, autre organisation ou Etat, on pourra se demander parfois si les Statuts des parties à cet acte conventionnel ouvrent un choix à celles-ci et, en cas de réponse affirmative, comment se détermine dans les cas douteux l'identification de l'ordre juridique choisi. Ainsi une organisation internationale peut emprunter de l'argent à une autre organisation internationale qui exerce des fonctions bancaires ou à un Etat et un tel acte conventionnel pourrait constituer un traité ou un contrat placé sous une loi nationale déterminée. Mais ces problèmes existent aussi pour les actes conventionnels entre Etats[13]. De même on pourrait discuter la question qui se pose pour des Etats ou des personnes privées de savoir si un contrat peut échapper à toute loi étatique et être soumis par exemple aux principes généraux du droit, ou aux principes communs à un ensemble de systèmes juridiques déterminés. Ces questions sont aujourd'hui bien connues et l'on n'insistera pas davantage sur elles.

En revanche il n'en est pas de même de la question de savoir si un acte conventionnel peut être placé dans l'ordre juridique propre d'une organisation.

Au premier abord la question peut sembler sinon inintelligible, au moins obscure[14]. Il est nécessaire pour commencer de considérer à quelles situations concrètes elle pourrait correspondre.

Aucune organisation, pour le moment tout au moins, n'a reçu compétence pour élaborer un code complet régissant les engagements conventionnels et offrant ainsi aux intéressés une *proper law* pour des contrats à conclure. Une telle organisation serait très proche d'une union fédérale. Quoique moins caractérisée, une organisation qui offrirait à des particuliers, ressortissants de ses Etats membres un contrat spécifique, par exemple un type

---

[13] Par exemple l'affaire des *Cargaisons déroutées*, Recueil des sentences arbitrales, vol. 12, p. 65.

[14] Le rapporteur spécial pour la question des traités des organisations internationales à la Commission du droit international avait par l'intermédiaire du Secrétaire général des Nations Unies adressé à un ensemble d'organisations un questionnaire qui se référait notamment à ce problème. Ces organisations ont dans l'ensemble indiqué que cette question ne se posait pas pour elles.

de société commerciale déterminé apparaîtrait sans doute également engagée dans la voie de l'»intégration«. En revanche on voit plus fréquemment d'autres situations dans lesquelles ce ne sont pas des relations commerciales de droit privé qui sont en cause, mais bien des relations qui semblent relever du droit international public, mais sont en étroite relation avec les règles propres de l'organisation.

Un premier exemple est constitué par ce que l'on appelle parfois des »accords interservices«; il s'agit d'accords qui ont pour parties différents organes d'une organisation ou divers services administratifs d'une même organisation bénéficiant d'un minimum d'indépendance les uns à l'égard des autres[15]. On peut douter du caractère »conventionnel« de tels actes juridiques émanant d'organes et de services qui ne possèdent pas une personnalité indépendante. Mais même en admettant qu'ils possèdent une telle personnalité, il apparaît bien qu'il ne s'agit pas de »conventions« régies par des règles *externes* à l'organisation. En réalité il s'agit d'actes juridiques dont la licéité et les effets dépendent uniquement du droit propre à chaque organisation.

D'une manière plus générale un acte conventionnel conclu entre un ou plusieurs Etats et une organisation peut être en rapports étroits avec les règles pertinentes de l'organisation lorsqu'il est conclu par un ou plusieurs Etats membres *en leur qualité de membre* de cette organisation, c'est-à-dire lorsque cet acte a un étroit rapport avec les obligations et les droits des Etats membres. Dans ce cas cet accord ne manquera pas de subir dans son régime, certains effets de cet étroit rapport.

Quels peuvent être ces effets?

Ils dépendent évidemment de l'étroitesse de ce rapport. Les règles de l'organisation serviront, le cas échéant, à interpréter cet accord en vertu d'un principe général consacré par la Convention de Vienne sur le droit des traités (art. 31 § 3c). Mais il faut aller plus loin; cet accord peut avoir des liens si étroits avec les règles pertinentes de l'organisation qu'il apparaît n'être conclu que pour l'*exécution* d'une telle règle ou d'un acte de l'organsation. Dans ce cas cet accord est conditionné et subordonné à cette règle ou à cet acte. Cet accord rentre dans l'hiérarchie des actes juridiques propres à l'organisation avec toutes les conséquences qui en découlent et c'est pourquoi on pourrait considérer qu'il fait partie de l'ordre juridique propre à l'organisation. Ceci ne veut nullement dire que les règles générales du droit des traités, règles nécessaires relevant directement de l'ordre juridique international ne s'appliquent pas à lui, mais qu'en plus il est soumis aux obligations résultant de l'ordre juridique propre d'une organisation déterminée.

On peut en donner un exemple qui n'est pas imaginaire. A la suite d'un conflit armé le Conseil de sécurité des Nations Unies prescrit un cessez-feu et

---

[15] Le droit interne des Etats connaît également à côté des actes conventionnels conclus entre des collectivités ou services personnalisés, des actes conventionnels entre services ou organes non personnalisés.

pose des règles concernant les obligations des forces armées situées dans une zone déterminée; il donne également mandat au secrétaire général des Nations Unies de conclure au nom des Nations Unies un accord avec un des Etats en présence, notamment en ce qui concerne le contrôle du cessez le feu par des éléments d'une force de maintien de la paix. L'accord ainsi conclu est entièrement subordonné à la décision qu'a prise le Conseil de sécurité, il ne peut modifier la décision, ni interdire au Conseil de revoir sa décision ou d'y mettre fin si la situation l'exige. On ne peut considérer que cet accord est entièrement extérieur à l'organisation, il fait partie du système juridique de l'organisation.

La Commission du droit international a rencontré ce problème quand elle a voulu adapter au cas des organisations internationales une règle sans malice de la Convention de Vienne sur le droit des traités qui dispose:

> Art. 27: »Une partie ne peut invoquer les dispositions de son droit interne comme justifiant la non-exécution d'un traité. Cette règle est sans préjudice de l'article 46«.

Certains ont pu craindre que si l'on disposait que »Une organisation internationale ne peut invoquer les dispositions des règles de l'organisation comme justifiant la non-exécution d'un traité«, les accords conclus par l'organisation pour mettre en oeuvre une décision de l'organisation n'immobilisent l'exercice par celle-ci des compétences qui lui sont attribuées. La Commission du droit international s'est efforcée d'adopter pour un projet d'art. 27 une disposition qui dissipe cette crainte[16].

Ces considérations et ces exemples ramènent à la question générale soulevée au début, de l'existence et des caractères d'un ordre juridique propre aux organisations internationales et de l'insertion des traités conclus par l'organisation dans cet ordre juridique.

Dans leurs observations sur le Projet d'articles élaboré par la Commission du droit international certains Etats se sont demandé s'il ne convenait pas de faire un sort particulier à ces accords qu'une organisation conclut avec ses propres membres. C'était reprendre le problème que l'on vient d'examiner en lui donnant une forme très large. En vérité si certains accords conclus par une organisation apparaissent en rapport étroit avec le droit propre de l'Organisation, il ne faut pas généraliser cette situation. S'il n'en était pas ainsi on serait conduit à dire en pratique que les organisations universelles (avec les Nations Unies en tête) ne concluent que des accords internes, sauf avec quelques Etats non membres. Ce serait d'ailleurs donner à ces relations internes de l'organisation une importance et un caractère qu'elles n'ont pas et conférer

[16] Le texte adopté en première lecture dispose »2. Une organisation internationale partie à un traité ne peut invoquer des règles de l'Organisation comme justifiant la non-exécution du traité à moins que l'exécution du traité dans l'intention des parties ne soit subordonnée à l'accomplissement des fonctions et pouvoirs de l'Organisation«.

par exemple un caractère différent à un accord de siège conclu par les Nations Unies avec les Etats-Unis d'Amérique et le même accord conclu avec la Suisse.

En réalité il faut dans l'analyse théorique faire ici comme ailleurs preuve d'esprit de mesure. Les organisations internationales engendrent un ordre juridique propre, c'est certain. Pour la grande majorité des organisations en mettant à part des unions d'Etats aussi particulières que les Communautés européennes et en excluant certains secteurs de cet ordre juridique comme ceux qui concernent les rapports avec leurs agents, ces ordres juridiques propres restent des ordres juridiques rattachés au droit international général. Les traités conclus par les organisations internationales avec des Etats restent des traités relevant de l'ordre juridique international, même s'ils sont conclus avec des Etats membres. Mais ils trouvent également leur place dans l'ordre juridique de l'organisation et ils peuvent dans leur interprétation, dans leur insertion parmi les autres actes juridiques de l'organisation être fortement marqués par les règles de l'organisation. Mais n'est-ce pas conforme à une indication déjà donnée par la Convention de Vienne sur le droit des traités, non pas à propos des traités des organisations internationales, mais à propos de traités entre Etats en rapport avec l'activité des organisations internationales, dans son art. 5:

> »La présente Convention s'applique à tout traité qui est l'acte constitutif d'une organisation internationale et à tout traité adopté au sein d'une organisation internationale, sous réserve de toute règle pertinente de l'organisation«.

Autrement dit encore, l'ordre juridique propre d'une organisation est une réalité, mais qui n'a encore aujourd'hui et généralement qu'un contenu d'une assez faible densité. En mettant en oeuvre cette notion on doit tenir compte de ce caractère; tout en restant soumis aux règles de l'ordre juridique international, les traités des organisations internationales sont, avec des effets divers, partie intégrante des règles de chaque organisation.

* * *

On vient de voir que d'une manière générale le contenu des ordres juridiques propres à chaque organisation restait suffisamment rudimentaire pour que les rapports entre les traités des organisations internationales et cet ordre juridique ne soulèvent pas en règle générale de difficultés pratiques considérables. Cependant on a indiqué au début de cette étude, dans une troisième proposition que »ces rapports doivent être aménagés«. En effet dès qu'une organisation se voit attribuer des pouvoirs réels, dès qu'elle est quelque chose de plus qu'un lieu de rencontre où les Etats discutent des questions d'intérêt commun, l'ordre juridique propre des organisations gagne en richesse et en poids et la place dans cet ordre juridique des accords conclus par l'organisation doit être précisée.

On vient d'en donner un exemple concernant les Nations Unies, mais le problème devrait être l'objet de solutions plus systématiques dans des organisations exerçant sur les Etats membres, et même parfois sur leurs ressortissants une emprise plus forte. C'est là une question très difficile qui appellerait de longs développements et que l'on ne peut évoquer ici que par de brèves allusions.

1) Tout d'abord certains problèmes d'application des traités d'une organisation internationale peuvent se poser dans les mêmes termes que pour les Etats. Prenons un exemple simple et pas trop imaginaire. Deux organisations internationales concluent un traité concernant les droits pécuniaires de leurs fonctionnaires lorsque ceux-ci passent en cours de carrière d'une organisation à une autre. Par ailleurs les fonctionnaires de ces organisations peuvent porter leurs différends devant un tribunal administratif. Un tel traité crée-t-il au bénéfice de ces agents des droits que ceux-ci peuvent invoquer directement, même en l'absence d'une mesure »interne« de l'organisation le prévoyant? Le Tribunal administratif compétent peut-il considérer que ce traité est applicable à une réclamation sans une mesure de promulgation ou de publication particulière? On peut imaginer des exemples plus compliqués pour des organisations gérant certains intérêts économiques communs à des Etats et ayant le droit de conclure des traités dans ces matières.

2) Ensuite l'insertion des traités conclus par l'organisation dans un ordre juridique propre de l'organisation se heurte à une difficulté sérieuse en ce qui concerne les effets de cette insertion à l'égard des Etats membres. En effet il a été soutenu devant la Commission du droit international et en doctrine que par rapport aux traités conclus par une organisation, les Etats membres étaient des tiers et ces traités ne pouvaient produire à leur égard des effets que sous les mêmes conditions que celles prévues en termes assez rigoureux en ce qui concerne les obligations par la Convention de Vienne sur le droit des traités (art. 34 ss.). Si les traités conclus par l'organisation n'obligent que celle-ci, quel est l'intérêt pratique de l'insertion de ces traités dans l'ordre juridique de cette organisation?

Il est bien certain que les Etats membres d'une organisation, tout en acceptant que celle-ci soit liée par un traité conclu par elle, n'acceptent pas de l'être eux-mêmes. Mais il y a des cas où l'objet des accords conclus par une organisation est tel qu'il n'est pas possible que les Etats membres prétendent les ignorer. Le problème est alors de déterminer les hypothèses et les conditions d'un effet des traités d'une organisation internationale sur ces Etats membres. Il n'est pas concevable par exemple qu'une organisation qui gère une union douanière et qui, à ce titre, peut conclure des traités de commerce dont l'exécution matérielle sera confiée aux agents de ses Etats membres puisse se voir opposer par un Etat membre la qualité de tiers. Il y a donc un aménagement à réaliser. La Convention de Vienne sur le droit des traités a posé la règle générale que des obligations à l'égard des Etats tiers ne pouvaient naître que d'un consentement exprès donné par écrit. Bien qu'une autre opinion ait été défendue par quelques membres de la Commission du droit international, on peut estimer qu'une règle aussi stricte, assimilant complètement les membres d'une organisation à des tiers en ce qui concerne tous les traités d'une organisation devrait être assouplie. Mais la question est restée jusqu'à présent pendante devant la Commission.

C'est donc le concept de »personnalité juridique« qui n'est qu'un concept abstrait et qui a été tant invoqué pour favoriser le développement des organisations internationales, qui, par un retour des choses qui n'a rien de surprenant se dresse ici comme un obstacle au développement des relations extérieures des organisations internationales. Ceci montre une fois de plus la mesure avec laquelle il faut user de concepts trop généraux.

The Meaning of »Authentic Text« in Modern Treaty Law

SHABTAI ROSENNE *

»[C]omme les autres branches du droit, le droit international emprunte son vocabulaire à un langage étranger au droit, mais il opère ensuite une transformation plus ou moins profonde du sens des mots. En second lieu, eu facteur propre au droit international, l'absence de langue proprement internationale, vient d'accélérer la formation d'un vocabulaire particulier. En troisième lieu les contradictions profondes qui dominent les structures de la société internationale actuelle conduisent, surtout dans le vocabulaire le plus récent, à une ambiguïté et à une incertitude dont les raisons ne sont pas accidentelles« [1].

Most modern treaties, multilateral and bilateral, are drawn up in more than one language, the treaty stating that the texts in the named languages are all »authentic« (or »authoritative«). Our purpose here is in a preliminary way to examine the question whether in this context the adjective possesses some specific *juridical* content or is it merely descriptive of a fact. We are not concerned with language versions howsoever prepared which are not so designated, or with the case where the parties have agreed that in the event of an irreconcilable lack of concordance between the authentic texts, a named text (whether or not »authentic«) should prevail for purposes of interpretation. Coincidence in time of two events has prompted this brief survey of what may be a more complex problem than hitherto commonly assumed. Those were the unexpected difficulties – both substantive and quantitative – encountered in this sphere by the Drafting Committee of the Third United Nations Conference on the Law of the Sea (hereafter UNCLOS III) [2], and

---

* This article, completed in November 1981, expresses the personal views of the author alone.

[1] *P. Reuter,* Quelques Réflexions sur le vocabulaire du droit international, Mélanges offerts à Monsieur le Doyen Trotabas (Paris 1970), 423, 424.

[2] Rule 56 of the Rules of Procedure provides that Arabic, Chinese, English, French, Russian and Spanish shall be the languages of the Conference, A/CONF. 62/30/Rev. 3. Art. 320 of the Draft Convention on the Law of the Sea, not yet examined by the Drafting Committee, provides that texts of the Convention in those six languages »are equally authentic«, A/CONF. 62/L. 78.

the somewhat unexpected approach of the International Court of Justice to the interpretation of Art. 62 of its quinquilingual Statute in the judgment of 14 May 1981 in the *Application to Intervene* by Malta in the *Tunisia/Libya Continental Shelf* case[3].

Regardless of one's doctrinal views, the point of departure for this examination must be Art. 33 of the Vienna Convention on the Law of Treaties of 23 May 1969, in force from 27 January 1980. That provision reads:

> *Interpretation of treaties authenticated in two or more languages*
> 1. When a treaty has been authenticated in two or more languages, the text is equally authoritative in each language, unless the treaty provides or the parties agree that, in case of divergence, a particular text shall prevail.
> 2. A version of a treaty in a language other than one of those in which the treaty was authenticated shall be considered an authentic text only if the treaty so provides or the parties so agree.
> 3. The terms of the treaty are presumed to have the same meaning in each authentic text.
> 4. Except where a particular text prevails in accordance with paragraph 1, when a comparison of the authentic texts discloses a difference of meaning which the application of articles 31 and 32 [General rule of interpretation, and Supplementary means of interpretation, respectively] does not remove, the meaning which best reconciles the texts, having regard to the object and purpose of the treaty, shall be adopted[4].

Noteworthy in that text is the constant repetition of the concept of »authentication«. As a stage in the conclusion of a treaty, authentication has only recently been identified. Not without hesitation it was formally introduced in Art. 10 of the Vienna Convention which lays down that the text of a treaty is established as *authentic and definitive* (italics supplied) either by such procedure as may be provided for in the text of the treaty or agreed upon by the States participating in its drawing up; or failing such procedure, by signature, signature *ad referendum* or initialling by the representatives of those States of the text of the treaty or of the Final Act of a conference *incorporating* (italics supplied) the text[5]. It is, moreover, closely related to the

[3] I.C.J. Reports 1981, 3.

[4] Originally Art. 29 of the draft articles on the law of treaties of the International Law Commission, Yearbook of the International Law Commission (hereafter YBILC), 1966, vol. II, 173 at 224. Reproduced in United Nations Conference on the Law of Treaties, Official Records, Documents of the Conference, 3 at 43. For the history of this article in the Conference, see A/CONF. 39/14, paras. 276–282, *ibid.*, at 151. For its final adoption at the 14th plenary meeting, see *ibid.*, Second Session, 58.

[5] In the case of UNCLOS III, Rule 39 of the Rules of Procedure contains a specific provision regarding »the adoption of the text of the Convention on the Law of the Sea« – a rare instance of this. In normal United Nations practice, the full text of a multilateral treaty in all its authentic texts is incorporated in the Final Act, which is

traditional »adoption« of the text of a treaty, with which Art. 9 of the Vienna Convention deals. The text of Art. 33 was carefully prepared. The International Law Commission required specific information on how multilingual treaties are drawn up in United Naitons practice, and it placed that information alongside the standard literature concerning the interpretation of multilingual treaties[6]. Art. 33 heavily emphasizes the role of the process of authentication and the incorporation of the authenticated texts in the Final Act as the starting point for multilingual interpretation.

In this article, a brief examination will first be made of some recent jurisprudence of four international tribunals selected at random: the International Court of Justice, the Arbitral Tribunal and Mixed Commission for the Agreement on German External Debts[7], the Court of Justice of the European Communities (which may be in a special position in this regard) and the European Court of Human Rights – the order is deliberate. That examination, cursory though it may be, will attempt to establish the extent to which it is possible to pinpoint the relationship between the preparation of the multilingual instrument and its judicial interpretation, and whether there are other factors to be considered. This will be followed by a brief survey of the treatment of concordance problems in UNCLOS III. It is hoped that this double examination, of randomly chosen jurisprudence and of practice in a major international conference faced with a draft nearly 500 articles long in six languages and drawn from many sources, will permit some conclusions to be drawn on the main issue posed at the outset, whether the word »authentic« possesses any juridical content or whether it is merely descriptive to serve a political purpose.

* * *

Although the Protocol of Signature and the Statute of the Permanent Court of International Justice were authentic in two languages, English and French, and by Art. 39 of the Statute those two languages were the official languages of that Court, it seems that no serious question arose in that

---

customarily signed by the President and Executive Secretary of the Conference and in many cases also by delegations present. But in itself the Final Act has no legal significance, as once was explained by the Secretariat, YBILC 1970, vol. I, 99 (1061st meeting).

[6] See discussion at the Commission's 767th meeting, YBILC 1964, vol. I, 298 (Articles then numbered 72 and 73). For the Secretariat's memorandum, see *ibid.,* 1966, vol. II, 104 (A/CN. 4/187). For the initial reaction of the Special Rapporteur, Sir Humphrey Waldock, see *ibid.,* 51, 101 (A/CN. 4/186). For subsequent discussion at the Commission's 874th meeting, see *ibid.,* 1966, vol. I, Part 2, 208.

[7] The London Agreement of 27 February 1953, 333 United Nations Treaty Series 3.

Court's judicial (as opposed to administrative) work in which interpretation of the Statute as a bilingual operation was germane. In fact the main work of drafting that Statute had been done in French, and a note prefaced to the published *Procès-Verbaux* of the 1920 Advisory Committee of Jurists states that since all the members of the Advisory Committee except one spoke in French, »the English text of the Procès-Verbaux is to be looked upon as a translation, except for the words of that member«[8]. Be that as it may for that preliminary stage, all subsequent work in the Council and Assembly of the League of Nations was technically bilingual, and in due course some attempt was made to establish a degree of concordance between the two languages[9]. Despite this (in some respects a reflection of the bitter struggle then being waged for formal recognition of English as a »diplomatic language«), this concordance was well known to be defective, and with the first steps to reorganize the Court after the Second World War, at the Washington Committee of Jurists (April 1945) efforts were made to improve it[10]. However, at the subsequent San Francisco Conference itself a further major change was introduced with the decision that the Statute would be an integral part of the Charter of the United Nations and that both instruments would be adopted in five texts, Chinese, English, French, Russian and Spanish, each »equally authentic«[11]. These events prompted the leading commentator on the Permanent Court to make the following observation:

---

[8] Permanent Court of International Justice, Advisory Committee of Jurists, Procès-Verbaux of the Proceedings of the Committee (The Hague, 1920) iv. It may be noticed that both in the Protocol of Signature and in Art. 39 of the Statute the two languages are not listed in alphabetical order, as would now be customary – an indication, *inter alia,* that the French text of the Statute may legitimately be examined first. In Art. 51 of the 1978 revised Rules of the present International Court of Justice the unusual order of listing first used in 1920 is maintained. Nevertheless, in the Advisory Committee of Jurists, the English text of the Covenant of the League of Nations was taken as the basis of the Committee's work, *M. Hudson,* The Permanent Court of International Justice, 1920–1942 (New York 1943), 190, note 19.

[9] League of Nations, Permanent Court of International Justice, Documents concerning the action taken by the Council of the League of Nations under Article 14 of the Covenant and the Adoption by the Assembly of the Statute of the Permanent Court of International Justice (1921), 206 (Report and Draft Scheme presented to the Assembly by the Third Committee).

[10] In this respect, note the observation by Judge Read on the lack of concordance in Art. 48 of the Statute in his dissenting opinion in the *Norwegian Loans* case, I.C.J. Reports 1957, 9, 79 at 81.

[11] 1 UNICIO 631, 15 *ibid.,* 335. Although special arrangements were made for the preparation and examination of the Chinese, Russian and Spanish texts of the Charter, in all probability they (and to some extent the French) were translations hurriedly prepared from the English negotiated text of the Charter: as for the Statute, it is understood that the Chinese, Russian and Spanish texts were also prepared from the English text as it had been revised at the Washington Committee of Jurists. In 1981, the Drafting Committee of UNCLOS III, in its work on Part XV and Annexes V, VI, VII and VIII on the Draft Convention on the Law of the Sea (Informal Text),

»Though only the English and French versions constituted the text of the old Statute, the text of the new Statute consists of versions in Chinese, English, French, Russian and Spanish. As the five versions are »equally authentic« under Article 111 of the Charter, each of them is to serve in the process of interpretation. Yet the historical fact that English and French were employed as the working languages at San Francisco, and the provision (in Article 39 of the Statute) that »the official languages of the Court shall be French and English,« may be thought to justify a process of interpretation which proceeds, *in the beginning at any rate* [italics supplied], on the basis of the versions in these two languages«[12].

Apart from the addition of three authentic texts of the Statute to the existing two, and leaving aside the quinquilingual Charter itself serving as the constitutional setting for the Statute, the principal innovations at San Francisco consisted of a few new provisions introduced then into the Statute and, as mentioned a more deliberate effort to improve the concordance between the different texts, especially the English and French texts. The practice of the majority opinions of the Court since 1946 when a question of interpretation of the Charter or Statute has arisen shows a clear tendency only to rely on the English and French texts. On the other hand, individual opinions are freer in examining also all the other texts.

The issue arose early, in the first *Admission* case, concerning the interpretation of Art. 4 of the Charter. The Court in its opinion could not »find any conflict between« the English and French texts. But the dissenting judges examined all five texts, and seem to have found some differences of nuance at least between some of them[13]. More closely in relation to the Statute itself, one of the new provisions inserted in 1945 is Art. 36, para. 5, the meaning of which was in issue in the *Aerial Incident of 27 July 1955* case. The majority

Doc. A/CONF. 62/WP. 10/Rev. 3, encountered difficulties due to unhappy renderings of the Statute in languages other than French, both the 1920 version as far as concerns English, and the 1945 version as far as concerns Chinese, Russian and Spanish (there is no authentic United Nations version of the Charter and Statute in Arabic, that in common use and circulated by the Office of Public Information of the United Nations [O.P.I. 511] being an earlier translation prepared by the Government of Egypt. Having regard to the San Francisco decisions, it seems that both the United Nations and the Court should publish a quinquilingual edition of the Charter and Statute, to which the customarily used Arabic text should be annexed for convenience and for legal accuracy.

[12] *M. O. Hudson,* The Twenty-fourth Year of the World Court, 40 American Journal of International Law 1, 14 (1946) (footnotes in original omitted).

[13] *Admission of Members into the United Nations,* advisory opinion, I. C. J. Reports 1947–1948, 57, 62. And see the joint dissenting opinion of Judges Basdevant, Winiarski, Sir Arnold McNair and Read, *ibid.,* 82, 86, and the dissenting opinion of Judge Krylov, 107 at 110, 112 (where the issue seems to have been treated with a touch of wry humour). On that opinion, see further *E. McWhinney,* Conflict and Compromise: International Law and World Order in a Revolutionary Age (New York 1981), 39.

opinion again only had regard for the English and French texts notwithstanding that at least one of the judges forming part of the majority did find some slight drafting differences between these two texts[14]. More important is the observation found in the joint dissenting opinion of Judges Sir Hersch Lauterpacht, Koo and Sir Percy Spender, who examined all five texts:

»The fact that the Chinese, Russian and Spanish texts ... approximate to the English text does not invalidate or weaken the obvious meaning of the French text. Those three texts were translated from the English version, whereas the French text was that of one of the two working languages adopted at the San Francisco Conference. However... [t]here is no question ... of giving preference to the French text. The French text is no more than an accurate translation of the English text as generally understood. Or, rather, in so far as it appears that the *final* version [italics supplied] was first formulated in the French language, the English text is no more than an accurate translation from the French« [15].

One of the provisions which, after having been amended during the original drafting process in 1920 in both languages, was further amended in English only at the Washington Committee of Jurists was Art. 62, para. 1, on intervention. The English text of the 1920 version of the Statute ended with the words »intervene as a third party«. The Washington Committee of Jurists in 1945 found that the words »as a third party« were misleading, and that to drop the phrase »would not change« the sense of the article[16]. These two explanations may not be entirely consistent with each other; and neither indicates what the Committee thought the »sense« of the provision was, or in what respect the English text was »misleading«, something which was not clarified in the 1920 *travaux préparatoires* either, nor is it a matter of comment in the standard literature on the Permanent Court. The question of the interpretation of the text as revised in English only in 1945 was one of the main issues discussed in the judgment of 14 April 1981 in connection with

---

[14] I.C.J. Reports 1959, 127, 144. See also the opinions of Judges Badawi and Zoričić, *ibid.,* 148–150, 152–154. It may be noted that no question of multilingual interpretation arose in the *Barcelona Traction* (preliminary objections) case in so far as it concerned another provision, Art. 37, newly inserted into the Statute in 1945, *ibid.,* 1964, 6.

[15] *Ibid.,* 1959, 161, 162. See also Judge *ad hoc* Goitein, *ibid.,* 195, 197.

[16] 14 UNCIO 211, 497, 501, 513, 648, 676, 821, 849. For a question by the Secretariat as to the necessity for harmonization of the new versions of that provision, see 18 *ibid.,* 667, 668. For approval at the San Francisco Conference by Committee IV/1, see 13 *ibid.,* 821, 849, 854 (English) and 881 (French). But it might be noticed that the two most authoritative commentaries on the Statute of the Permanent Court, one in French and one in English, do not contain any detailed discussion of these aspects. Statut et Règlement de la Cour permanente de Justice internationale, Eléments d'interprétation, ed. by B. Schenk Graf von Stauffenberg (Berlin 1934) 435–442; *M. O. Hudson,* The Permanent Court of International Justice 1920–1942 (New York 1943) 208, 419–424, although that author does point out that the French version of the Statute includes no equivalent of the phrase »as a third party« (p. 420, note 78).

Malta's application to intervene in the *Tunisia-Libya Continental Shelf* case[17]. The issue concerned the words »interest of a legal nature which may be affected«, »un intérêt d'ordre juridique qui est pour lui en cause«. That phrase had originally been drafted by the Advisory Committee in 1920 and subsequently became incorporated unchanged as Art. 62 of the Statute of the Permanent Court[18], and some saw a connection between that phrase and the words dropped from the English text in 1945, perhaps matching, even if not exactly corresponding to the French words *pour lui*, although this is not mentioned in the published records, and any such »concordance« would appear unlikely. As a result of the work of the Washington Committee of Jurists and the San Francisco Conference, the French text of Art. 62 of the present Statute corresponds to the French text of Art. 62 of the Statute of the Permanent Court, and the English text differs from it, that difference however supposedly not changing its sense, or at least not being intended to change the sense whatever that might have been[19].

In the *Application to Intervene* judgment the present Court, interpreting Art. 62, briefly described the drafting of the provision in 1920 and the revision of the English text in 1945[20]. In all these instances, the references were made to the original drafting of Art. 62: subsequent stages, including their adoption as authentic texts in 1920 and again in 1945, were not mentioned. This omission in a judgment delivered in 1981, after the Vienna Convention on the Law of Treaties had entered into force, is particularly striking; as regards the present Statute, it makes no mention of the addition of three new »authentic texts« alongside the other two, even if only translations, and in applying modern techniques of multilingual interpretation it passes in silence over the stages of the adoption of the text, authentication, and the role authentication is now intended to play in the interpretation of a multilingual treaty under the Vienna Convention[21].

* * *

---

[17] *Supra,* note 3.

[18] *Op. cit. supra* (note 8), pp. 571, 587, 592–594, 621, 643, 657, 669, 745. But no real statement of interpretation appears in the records.

[19] For details, see note 16 *supra.* No comments on Art. 62 were made before the Washington meeting. 14 UNCIO 444. On this change *Hudson* commented: »The deletion of the concluding phrase in the English version of paragraph 1 of the Article brings it into correspondence with the French version«. *Supra,* note 12 at 42.

[20] Judgment, para. 22. A slightly fuller account was given by Judge Oda in his separate opinion (para. 3). On the other hand, in his separate opinion Judge Schwebel referred only to the English text (p. 26).

[21] In this connection, it is interesting to notice that the President of the Court, Sir Humphrey Waldock, had himself been the special rapporteur of the International Law Commission and later expert consultant for the United Nations Conference on the Law of Treaties, and seven of the Members of the Court together with one Judge *ad hoc* had been members of the International Law Commission or of their delegations at the Vienna Conference, or had been otherwise closely associated with the preparation of the Vienna Convention.

The authentic texts of the German External Debts Agreement of 1952 are English, French and German, and the organs established by that treaty have had several times to employ multilingual interpretation techniques. Here a change is noticeable in the jurisprudence, after the Vienna Conference had adopted Art. 33 of the Convention on the Law of Treaties.

In *Swiss Confederation* v. *Federal Republic of Germany* (No. 1) decided on 3 July 1958[22] the question arose of the meaning of the expressions »place of payment«, »Zahlungsort« and »que le paiement serait fait à l'étranger« appearing in the relevant provision of the Agreement. Here the Arbitral Tribunal refused to apply conflicts of law rules but turned to the practice of the International Court, according to which »words and phrases are to be given their normal, natural, and unstrained meaning in the context in which they occur«[23]. It then proceeded to examine the meanings of those expressions in the different legal systems involved, and reached the conclusion that it preferred the natural meaning of the term to a technical »and artificial« meaning advanced by the respondent. The Tribunal went on to say that it was confirmed in its opinion when it examined the origin of the provision in question and the legal position of the creditors as it was at the time when the Agreement was negotiated. This brought the Tribunal to a close analysis of preceding legislation of the Military Government and then of the proceedings at the London Conference itself. In turn this led to the following dictum:

> »The application of the principle of interpreting treaties according to the natural sense of the words is therefore particularly appropriate to this case where the natural meaning seems to coincide with the intention of the parties as deduced from the circumstances of the case«[24].

Another question of multilingual interpretation then arose, but it concerned the private law contract and not the Agreement itself. Nevertheless the Tribunal did say that the language of the provision of the Agreement only became unclear or obscure »when there is imported into the meaning of the word *Zahlungsort* in the German text the unique and restricted definition given under German law to the word *Leistungsort*«. That the Tribunal, after close examination of the circumstances, refused to do[25].

---

[22] Arbitral Tribunal and Mixed Commission for the Agreement on German External Debts, Reports of Decisions and Advisory Opinions 1958, 7 (in German, English and French, with no indication of which text is »authentic« or »authoritative«). English text reproduced in 35 International Law Reports 33. The Arbitral Tribunal was composed of Daehli, Michelson, Barabdon, Wolff, Richard, Phenix, von Caemmerer, Makarov and Hinderling (the last two being arbitrators *ad hoc*).

[23] English version, p. 38; International Law Reports at 58. The Tribunal also here relied upon the Granada resolution of the Institute of International Law on the interpretation of treaties, 46 Institut de Droit international, Annuaire (1956) 364.

[24] *Op. cit.* at pp. 48 and 66. A joint dissenting opinion was appended by Barabdon, Wolff, von Caemmerer and Makarov.

[25] See also *Rommenhöller, Rotterdam* v. *Schuh, Bad Nederbreisig,* decided on 7 March 1961. *Ibid.,* 1961/1, 1. The Mixed Commission was composed of Daehli, Michelson and Euler.

More important, and probably more controversial, was the fuller reasoning adopted by the same Tribunal in 1980 in the so-called *Young Loan* case[26], which largely turned on the proper interpretation of an allegedly obscurely drafted clause in the London Agreement. Here the Tribunal thoroughly examined the legislative history of that provision, and also the method of preparation of the authentic texts of the London Agreement. The compromissory clause under which the Tribunal operated required the Tribunal to apply the generally accepted rules of international law. Regarding the international law of treaties, the Tribunal observed that the provisions of the Vienna Convention did not apply in the present proceedings on account of Art. 4 thereof[27], »but nevertheless it is widely held in jurisprudence and legal literature that the Convention properly reflects both the present and the past state of international treaty law since, as regards interpretation at least, it is restricted to the codification of customary law in force« (para. 16). Referring to Art. 31–33 of the Vienna Convention, and especially to Art. 33, para. 1, the Tribunal »directly inferred« that the English text of the London Agreement carried

> »no special interpretative weight merely because the Agreement was largely – and undisputedly, as far as the disputed clause is concerned – drafted in that language and discussed in English by the Committees concerned with it on the basis of English drafts. The Tribunal takes the view that the habit occasionally found in earlier international practice of referring to the basic or original text as an aid to interpretation is now, as a general rule, incompatible with the principle, incorporated in Article 33 (1) of the VCT [Vienna Convention on the Law of Treaties], of the equal status of all authentic texts in plurilingual treaties[28]. The interpretational maxim of the special importance or precedence – whatever form it may take – of the original text would relegate the other authentic texts again to the status of subordinated translations (para. 17)«[29].

---

[26] *Belgium, France, Switzerland, the United Kingdom and the United States of America* v. *Federal Republic of Germany,* award of 16 May 1980. *Ibid.,* 1980, 5. The English version is reproduced in 59 International Law Reports 495. The Tribunal was composed of Castrén, Arndt, Robinson, Maier, Bathurst, Monguilan and Kewenig, and it reached its decision by four to three (Robinson, Bathurst and Monguilan, who appended a joint dissenting opinion).

[27] But see our article »The Temporal Application of the Vienna Convention on the Law of Treaties«, 4 Cornell International Law Journal (1970), 1.

[28] The Tribunal here referred to *M. Hilf,* Die Auslegung mehrsprachiger Verträge, Eine Untersuchung zum Völkerrecht und zum Staatsrecht der Bundesrepublik Deutschland (Beiträge zum ausländischen öffentlichen Recht und Völkerrecht, Bd. 61) (Berlin 1973).

[29] In fact, the suggestion to include a rule to fall back on the language in which the treaty was drafted if it was otherwise impossible to reconcile divergences was discussed in the International Law Commission, especially at its 884th meeting, on the initiative of Verdross. That discussion, at which the President of the Tribunal was

The Tribunal then examined whether the conditions of the Vienna Convention might still be met if an attempt to determine the »objectified« (*objektivierten*, *volonté ›objective‹*) will of the parties, as expressed in the text of the treaty, were based on the normal significance of the terms used at the time the treaty was concluded – a particular manifestation of the intertemporal law. But this did not help matters, and the Tribunal reached the conclusion that a possible discrepancy between the German version on the one hand, and the English and French versions on the other, could not be reconciled. For that reason the system of contextual interpretation within the meaning of Art. 31, para. 1, of the Vienna Convention would have to be applied (paras. 19, 20). Close examination of the London Agreement, both linguistically and in its total setting, led the majority to a certain conclusion which it then proceeded to test against the »object and purpose« of the London Agreement, also in accordance with Art. 33, para. 1, of the Vienna Convention. In accordance with para. 3 of that same article, this was followed by examination of the subsequent agreements between the contracting parties on interpreting the London Agreement and the subsequent practice in its application »from which a consensus between the parties regarding the interpretation of specific parts of the treaty might be deduced« (para. 31). Having done all that, the Tribunal had recourse to the *travaux préparatoires* on the basis laid down in Art. 32 of the Vienna Convention[30], in the process also availing itself of the testimony of some of the principal negotiators[31]. All of this led the majority to prefer the German text. The Tribunal then went on to clinch its argument by relying on Art. 33, para. 4, of the Vienna Convention, with its further reference to the »object and purpose« of the London Agreement, leading it once again to support the German contentions (paras. 39–41).

Needless to say, this reasoning was contested on every point by the joint dissenting opinion, since this was one of the rare instances in which the deci-

---

present, did not go so far as the Tribunal's pronouncement would suggest, and by not codifying the proposal the Commission merely did not think that this was a »rule« which ought to be included in the codified law of treaties, YBILC 1966, vol. I, Part 2, 271. It is curious that the joint dissenting opinion which is copious (and, with respect, convincing) in its citations to rebut the majority view, did not refer to that discussion.

[30] Para. 34 of the Award contains an important attempt to define the concept of *travaux préparatoires,* something which the International Law Commission was careful not to do. Analysis of that definition does not come within the scope of this article: but we find it necessary to express a formal reservation of our position thereon.

[31] With respect, we consider that testimony of negotiators (unless contemporary with the negotiation) may not be a reliable factor in the interpretation of a treaty. Negotiators may be able to recollect their own intentions, but that is a matter of secondary importance since they would rarely have full and direct knowledge of all the undisclosed reasons that had prompted their Government's instructions. See our observations, based on personal experience, at the 678th meeting of the International Law Commission. YBILC 1963, vol. I, 32.

sion itself depended almost exclusively on the answer given to the issue of plurilingual interpretation (more frequently the issues of plurilingual interpretation have an incidental or marginal relationship to the main issues of a case). We believe that this case, viewed from the angle of the development of the jurisprudence of international multilingual interpretation in the light of contemporary multilingual drafting practices, may open more questions than it solves, and that it cannot be accepted as the last word on the subject. One of the questions which it raises, and which we have seen raised in other contexts, is the real, juridical, meaning of the formal statement in a treaty that certain language texts are »authentic«. This question is important both intrinsically and in the light of Art. 10 and 33 of the Vienna Convention, to which we will return in due course.

* * *

The jurisprudence of the Court of Justice of the European Communities that has come to our notice[32] at first sight does not fully correspond to this pattern or with the underlying premisses of the Vienna Convention. However, all this jurisprudence concerns not the interpretation of a multilingual treaty, not even the Treaty of Rome itself, but apparent linguistic divergences between the multilingual texts of internal Community legislation of one

[32] The following cases have been examined: Case 825/79 *Folci* v. *Amministrazione delle Finanze dello Stato* [1980] ECR 3053; Case 9/79 *Koschniske* v. *Raad van Arbeid* [1979] *ibid.*, 2717; Case 16/78 *Choquet* [1978] *ibid.*, 2293; Case 76/77 *Dufour* [1977] *ibid.*, 2485; Case 55/77 *Maris, ibid.*, 2327; Case 30/77 *Regina* v. *Bouchereau, ibid.*, 1999; Case 108/76 *Klöckner-Ferromatic* v. *Oberfinanzdirektion Munich, ibid.*, 1047; Case 80/76 *North Kerry Milk Products Ltd.* v. *Minister for Agriculture and Fisheries, ibid.*, 425; Case 82/76 *Hoechst* v. *Hauptzollamt Frankfurt-am-Main, ibid.*, 335; Case 21/76 *Bier* v. *Potasse d'Alsace* [1976] *ibid.*, 1735; Case 14/76 *Sarl* v. *Bouyer, ibid.*, 1497; Cases 98/75, 99/75 (joined) *Carstens Keramik GmbH* v. *Oberfinanzdirektion Frankfurt-am-Main, ibid.*, 241; Cases 40–48, 50, 54–56, 111, 113, 114/73 (joined) *Vereniging Suiker et al.* v. *Commission* [1975] *ibid.*, 1663; Case 35/75 *Matisa* v. *Hauptzollamt Berlin, ibid.*, 1205; Case 66/74 *Farrauto* v. *Bau Berufsgenossenschaft, ibid.*, 157; Case 6/74 *Moulijn* v. *Commission* [1974] *ibid.*, 1287; Case 183/73 *Osram* v. *Oberfinanzdirektion Frankfurt-am-Main. ibid.*, 477; Case 6/72 *Euoropemballage et al.* v. *Commission* [1973] *ibid.*, 215; Case 45/72 *Meroal* v. *Fonds national de Retrait* [1972] C.J. C.E. *Recueil* 1225; Case 41/69 *Chemiefarma* v. *Commission* [1970] *ibid.*, 661; Case 29/69 *Stauder* v. *Ulm* [1969] *ibid.*, 419; Case 19/67 *Sociale Versekeringsbank* v. *van der Vecht* [1967] ECR 345; Case 6/67 *Guerra* v. *Institut National d'Assurance, ibid.*, 219; Case 1/60 *Acciaieria Ferriera di Roma* v. *High Authority* [European Steel and Coal Community] [1960] *ibid.*, 165, reproduced in 27 International Law Reports 27. We are indebted to Mme A–M Blois, Head of the Information Service of the Court of Justice of the European Communities for bringing this jurisprudence to our attention, and to the computerized data retrieval system which greatly facilitated our examination of it.

form or another. According to our information, there is no jurisprudence concerning the multilingual interpretation of the Treaty of Rome itself, the language history of which, in this respect has its own peculiarities[33]. There are, however, complex political and psychological reasons why, it appears, Community law shows on the surface a marked tendency to distance itself from general international law, and from international conventional law in particular (such as the Vienna Convention on the Law of Treaties), at least in its formal exposition[34]. Thus we find in that jurisprudence little interest in the *travaux préparatoires* of the instruments in question, and no mention at all of the combined processes of adoption and authentication, neither in the judgments of the Court nor in the submissions of the Advocates-General[35]. The reason was expressed several years ago in the following instructive terms:

> »Les Communautés européennes constituent un nouvel ordre juridique de droit international, au profit duquel les Etats ont limité dans des domaines restreints leur droits souverains, et dont les sujets sont non seulement les Etats membres mais également leurs ressortissants. Une des caractéristiques de l'intégration partielle réalisés par les traités européens est la superposition des normes communautaires aux systèmes internes des membres«[36].

This notwithstanding, and keeping in mind the automatic consequences that must flow from the legislative effect in internal law of the Member

---

[33] 294 United Nations Treaty Series 3. English translation prepared by the International Committee for the Common Market and Euratom in vol. 298, 3. The original Treaty of Rome was drawn up in four authentic texts – Dutch, French, German and Italian, and it is widely accepted that the other texts are a translation from the French. By Art. 160 of the Act concerning the Conditions of Accession and Adjustments to the Treaties of 22 January 1972, as amended by the Decision of the Council of the European Communities of 1 January 1973 adjusting the documents concerning the accession of new Member States to the European Communites (Official Journal of the European Communities, No. L. 2, 1 January 1973, 1), texts of the Treaties drawn up in the Danish, English and Irish languages »shall be authentic under the same conditions as the original texts of the Treaties. . .«. On that occasion a new authentic English text was prepared by the United Kingdom Government and is now in general use. The 1972 Act also included Norwegian as an authentic text, but this was dropped by the adjustment decision of 1973. For the current texts (in force before the entry of Greece into the Communities), see Treaties establishing the European Communities (Luxembourg 1978). Separate editions exist in Danish, Dutch, English, French, German, Irish and Italian.

[34] See on this in particular, *O. Jacot-Guillarmod,* Droit communautaire et droit international public (Geneva 1979).

[35] On these submissions, a reflection of the practice of some Continental jurisdictions, see Art. 166 of the Treaty of Rome.

[36] *F. Bonn*, Les problèmes juridico-linguistiques dans les Communautés européennes, 68 Revue général du Droit international public (1964), 708, 716.

States of much if not all of the Community legislation which imposes, to a far greater extent than is necessary in much of general international law, the need for identicalness in legal effect rather than harmonization or concordance of texts in a juridical sense, close inspection of the many dicta of the European Court indicates that in the long run, multilingual interpretative problems are approached in ways not very different from those found in international tribunals of the more usual type. Two selections may be given to illustrate this. Thus, to take one of the early cases, the Court stated:

> »Attendu que, lorsqu'une décision unique est addressée à tous les Etats membres, la nécessité d'une application et dès lors d'une interprétation uniformes exclut que ce texte soit considéré isolément dans une de ses versions, mais exige qu'il soit interprété en fonction, tant de la volonté réelle de son auteur que du but poursuivi par ce dernier, à la lumière notamment des versions établies dans toutes ces langues« [37].

This is not very different from the fuller construction given in one of the latest cases, where that Court said:

> »In fact, the wording of the passage in question, considered solely in the Dutch version, is capable of giving the impression that the term used refers exclusively to a person of the female sex.
> However, the need for a uniform interpretation of Community regulations makes it impossible for that passage to be considered in isolation and requires that it should be interpreted and applied in the light of the versions existing in the other official languages.
> A comparison with the other versions of the provision in question reveals that, in all the other versions, a word has been used which includes equally male and female workers.
> This interpretation is borne out on the one hand by the purpose of the provision ... and on the other by the principle of the equal treatment of male and female workers in the field of social security« [38].

Thus we find in this jurisprudence heavy emphasis placed on the process of comparison of the different language versions as what might be called the mechanical aspect of multilingual interpretation, coupled with equally heavy emphasis on the object and purpose of the author of the instrument in question, supplying a dynamic element to this kind of interpretative process, all taking place under the general umbrella of community integration. This substantiates the view that clearly in this respect the Court of Justice has not

---

[37] *Stauder* v. *Ulm, supra,* note 32, at 425. Note the interesting inversion of »application« and »interpretation« in that *attendu.*

[38] *Koschniske* v. *Raad van Arbeid, supra,* note 32, at 2724. Indeed so constant is this jurisprudence that it would be otiose here to analyse all the cases listed in that note.

gone beyond the margin of appreciation possessed by every court in the choice of criteria for interpretation, consolidated in the Vienna Convention[39].

* * *

At first glance, the European Court of Human Rights supplies a large quantity of jurisprudence on the questions under discussion here, but without making any unexpected innovations[40]. The usual rule is set forth in what might be called classical terms in the *Sunday Times* case as follows:

> »confronted with versions of a law-making treaty which are equally authentic but not exactly the same, the Court must interpret them in a way that reconciles them as far as possible and is most appropriate in order to realize the aim and achieve the object of the treaty«[41].

In some of the earlier cases, notably in *Ringeisen* and *Golden*, the Court had closer regard for the detailed rules of the Vienna Convention but this does not appear to be consistent practice.

Cursory perusal of these judgments might lead to a facile conclusion that the concordance of the English and French versions of the European Convention on Human Rights[42] was not given high priority during the negotiating and drafting stages. However, this may not be an entirely correct way of looking at things, and closer inspection may reveal that what the Court has treated as problems of multilingual *interpretation* are in reality something much deeper, and relate to the reaching of decisive conclusions, for the purpose of the case at hand, about conceptual differences between the different systems of law involved, both in their substantive and in their adjectival

[39] *Jacot-Guillarmod, op. cit. supra* (note 34), at 117. Further at p. 216, where the Vienna Convention is nevertheless termed »une base de référence privilégiée«. See also *T. C. Hartley,* The Foundations of European Community Law (Oxford 1981), 47.

[40] The principal jurisprudence of this Court relating to multilingual interpretation is found in the following cases: *Sunday Times* (1979), Ser. A, No. 30, para. 48, 58 International Law Reports 491; *Golden* (1975), Ser. A, No. 18, paras. 30–36, 57 International Law Reports 200; *Ringeisen* (1971), Ser. A, No. 13, para. 110, 51 International Law Reports 442; *Vagrancy* (1971), Ser. A, No. 12, 56 International Law Reports 351; *Delcourt* (1970), Ser. A, No. 11, para. 25, 47 International Law Reports 169; *Belgian Languages* (1968), Ser. A, No. 9, 45 International Law Reports 114; *Wemhoff* (1968), Ser. A, No. 7, para. 8, 41 International Law Reports 281.

[41] Examination of the French text of this judgment indicates that the Court was here picking up the language of Art. 33 of the Vienna Convention, without directly citing it. This is lost in the rather heavy English formulation.

[42] 213 United Nations Treaty Series 221. See also Council of Europe, European Convention on Human Rights, Collected Texts (Strasbourg 1978). On discrepancies between the two versions, see, *inter alia, M. Tabory,* Multilingualism in International Law and Institutions (Alphen aan den Rijn 1980), 166, note 318.

aspects, and not mere linguistic or terminological absence of formal concordance. This opens up new perspectives, which traditional concepts of the interpretation of multilingual treaties will not be able to solve. The *Golden* case in particular can be read as shifting the weight of the interpretational exercise from the effort to extract some common meaning from apparent divergences between the authentic texts of the Convention to a search for the underlying concept to which the authors of the Convention desired to give expression, namely, in that case the concept of the »rule of law« as it is taken to exist generally and generically in Western Europe. In that respect, an unexpressed element, which may be denominated »Europeanism«, may be found to exist here and to supply elements not directly contemplated by the Vienna Convention, much in the same way that, as seen, this is an extra-judicial factor influencing the comparable jurisprudence of the Court of Justice of the European Communities.

* * *

Before looking at the drafting problems, and more particularly those of concordance of UNCLOS III, it is to be noted that an uneasy feeling that received techniques of multilingual treaty drafting (mostly inherited from the bilingual practice of the League of Nations) might not be adequate began to surface in the course of the work of the International Law Commission on the law of treaties[43]. In 1964, responding to complaints in the Sixth Committee and elsewhere about lack of concordance between the English and French versions of the draft articles and the Spanish version, the Commission requested its Drafting Committee to assume responsibility also for the Spanish text[44]. At the Vienna Conference itself, while formally, as in previous cases, concordance was within the especial province of the competent services of the Secretariat operating under the general control of the Conference Drafting Committee nevertheless a slight change was introduced. It was

---

[43] Cf. Doc. A/CN. 4/187, supra, note 6. For some earlier instances of difficulties experienced in the United Nations, see the memorandum of the Secretariat entitled Resolutions of the General Assembly concerning the Law of Treaties (A/CN. 4/154), YBILC 1963, vol. II, 1; same, Summary of the Practice of the Secretary-General as Depositary of Multilateral Agreements (ST/LEG/7, 1959, mimeographed), paras. 2–9; *Sh. Rosenne,* United Nations Treaty Practice, 86 Recueil des Cours (1954 II) 281, 381.

[44] 772nd and 777th meetings, YBILC 1964, vol. I, 2, 77; 1964, vol. II (A/5809, para. 5) 174. For our appraisal of the consequences of that decision, see The Law of Treaties: A Guide to the Legislative History of the Vienna Convention (Leyden & Dobbs Ferry, 1970), 36: »... frequently brought to light questions of substance, sometimes quite unsuspected, requiring further elucidation«. See also our »Terminaison des traités collectifs: Rapport provisoire«, 52 Institut de Droit international, Annuaire (1967), vol. 1, 24, 39.

explained by the Committee of the Whole in one of its reports to the Conference in the following terms:

> »At the 59th meeting of the Committee of the Whole, on 8 May 1968, the Chairman of the Drafting Committee explained the procedure followed by the Drafting Committee in preparing the texts of the articles in the different official languages [Chinese, English, French, Russian and Spanish[45]]. He stated that those members of the Drafting Committee whose mother tongue was Chinese, Russian or Spanish carefully studied the text of the International Law Commission's draft prepared in their language and submitted from time to time to the Drafting Committee corrections to the syntax or terminology. The Drafting Committee then referred such corrections to the Conference's language services so that the latter could ensure that they did not affect the versions in other languages. When this had been ascertained, the corrections were incorporated in the text of the language to which they related. The text, so corrected, was to be found in the reports submitted by the Drafting Committee to the Committee of the Whole«[46].

Later the Commission developed special techniques for producing concordant Russian texts of its draft articles[47], and it is to be assumed that similar arrangements exist for the preparation of the Arabic and Chinese texts of the draft articles (but not necessarily the commentaries) produced by the Commission. However, it is with UNCLOS III that the problems of ensuring concordance between six authentic texts have exploded as a heavy substantive matter, demanding, time-consuming, expensive and tedious, making heavy calls on all those concerned with it, and requiring a high degree of inventiveness, especially on the part of the Secretariat, in order to ensure what the Chairman of the Drafting Committee has called »linguistic concordance to the extent possible, and juridical concordance in all cases«[48].

The functions of the Drafting Committee of UNCLOS III – consisting of 23 elected members[49] – are set forth in Rule 53 of the Rules of Procedure as follows:

---

[45] The United Nations subsequently prepared an Arabic translation of the Vienna Convention for the United Nations Conference on Succession of States in Respect of Treaties (A/CONF. 80/WP. 2, 1977, mimeographed).

[46] A/CONF. 39/14, para. 16 (d), Official Records, *supra* (note 4), 108.

[47] Not always appreciated by the Secretariat. See the protest of Professor Ushakov, and the Commission's decision, at its 1330th meeting. YBILC 1975, vol. I, 131. On this incident, see *V. Evintov,* Osob'ennost'i sostavl'enniya raznoyazychnykh tekstov sovremennykh m'ezhdunarodnykh dogovorov, 1979 Soviet Year-Book of International Law 161, 166.

[48] A/CONF. 62/L. 67/Rev. 1, para. 8 (mimeographed).

[49] For those elections, see UNCLOS III, Official Records, vol. I, 18, 19 (7th plenary meeting).

»[W]ithout reopening substantive discussion on any matter, [to] formulate drafts and give advice on drafting as requested by the Conference or by a Main Committee, co-ordinate and refine the drafting of all texts referred to it, without altering their substance, and report to the Conference or to the Main Committee as appropriate. It shall have no power of or responsibility for initiating texts«.

This partly reflects the traditional role of drafting committees, but at the same time it hesitantly points the way to new directions, with its reference to the co-ordination and refinement of the texts referred to it. But when those Rules of Procedure were drawn up in 1974, no one had any idea of how the Conference would proceed to develop texts and »refer« them to the Drafting Committee. It is in that respect that the novelty of the work of this Drafting Committee originates. For unlike previous conferences, articles were not referred to the Drafting Committee individually after debate and decision (in principle at least) in one of the Main Committees. Entirely new techniques were evolved to deal with the substance, and these, in their turn, affected the techniques used by the Drafting Committee. For in the end, not individual articles, but the Convention as a whole in various stages of its evolution, was referred as a whole to the Drafting Committee.

As is known, the Conference worked in a fairly systematic manner, through three Main Committees and through the Informal Plenary institutionalized as a Main Committee (for the topics of the settlement of disputes, the final clauses, and some miscellaneous matters not allocated to or coming within the scope of the work assigned to one or other of the Main Committees). In this way, the various parts and sections of the text were put together step by step. Initially this Committee-stage work found expression in a series of individual working papers known as the Informal Single Negotiating Text and the Revised Single Negotiating Text, in four separate parts[50]. In July 1977 these papers were consolidated for the first time in a document entitled Informal Composite Negotiating Text. This itself went through three revisions before the issuance of the Draft Convention on the Law of the Sea in August 1981[51]. It was only after the Informal Composite Negotiating Text first appeared that it became possible to make a start on studying the text of the proposed Convention as a whole and that the Drafting Committee could commence to assess the work that it would be called upon to do. It was rapidly discovered that this work would fall into two broad categories, partly overlapping – »harmonization« in the sense of ensuring internal consistency of terminology and presentation in each authentic

[50] UNCLOS III, Official Records, vol. IV, 137 (A/CONF. 62/WP. 8, Parts I, II and III); vol. V, pp. 111, 185 (A/CONF. 62/WP. 9 & Rev. 1, Part IV).

[51] A/CONF. 62/WP. 10, Official Records, vol. VIII, 1; Rev. 1 (April, 1979), Rev. 2 (April, 1980), Rev. 3 (August 1980) mimeographed. A/CONF. 62/L. 78 + Corr. 1–3 (mimeographed).

text taken individually, and »concordance« (in the traditional sense) as between each and all of those authentic texts; and for concordance the Committee accepted as its yardstick the principle of the integrity of the text of the single treaty in all its language versions, on the basis of Art. 33 of the Vienna Convention. The Drafting Committee soon found that it would need to establish subsidiary organs – which it is entitled to do under Rule 50 of the Rules of Procedure of UNCLOS III. Accordingly, in 1978 it set up six »Language Groups«, membership in which is open to all participating delegations, whether or not members of the Drafting Committee. The Chairmen of the Language Groups, designated »Co-Ordinators«, sitting together received the reports of the Language Groups. The Co-Ordinators' conclusions were submitted to the Drafting Committee and, subject to the decisions of the Drafting Committee, came before the Conference sitting as the Informal Plenary, where the final decisions were taken[52]. The period

[52] For the establishment of the Language Groups, see the reports of the Chairman of the Drafting Committee in Official Records, vol. IX, 101 and vol. X, 199 (A/CONF. 62/RCNG. 2). The Language Groups organized their work according to their convenience, being serviced by the UNCLOS Secretariat. Their initial reports appear in an informal series of documents which, after various methods had been tried, became standardized in 1980 under the symbols ALGDC, CLGDC, ELGDC, FLGDC, RLGDC and SLGDC, all of which were translated into the other five languages. In 1982, for reasons of economy and speed, these translations were abolished. The consequences were not encouraging. Those reports, with translations, were subsequently consolidated in a series of working papers, one for each article for which revisions were proposed by any one of the Language Groups, under the symbol DC/Article x, and in that form the reports of the Language Groups were laid before the Co-Ordinators. The conclusions of the Co-Ordinators were consolidated in another series of working papers under the symbol CG/WP.–, usually relating to a larger unit of the text – a section or a Part. The conclusions reached on these documents constituted the report of the Drafting Committee to an informal meeting of the Conference. In 1981 the Drafting Committee completed most of the work on Arts. 2–149, 186–299 and Annexes I, II and V, A/CONF. 62/L. 67/Rev. 1 + Add. 1, 3–16 and A/CONF. 62/L. 75 + Add. 1–13. The Draft Convention (L. 78) incorporates all the recommendations to date of the Drafting Commitee accepted by the Conference. Most of the harmonization exercise was conducted on the basis of the ICNT and Revs. 1 and 2, and that of concordance on Rev. 3, *supra* (note 51). It has been estimated that the Drafting Committee and its subsidiary organs up to 1981 between them produced something like a quarter of a million (sic) pages of documentation. *E. Rauch,* Politische Konsequenzen und Möglichkeiten der Seerechtsentwicklung nach New York 1981 (VII), Berichte des Bundesinstituts für ostwissenschaftliche und internationale Studien 23–1981 (Cologne 1981), 47. In 1981 alone it has been reported that the Language Groups together held something like 700 meetings, the Co-Ordinators 75, the Drafting Committee 30, and the Conference, meeting in informal meetings of the plenary, some 15 meetings for consideration of the reports of the Drafting Committee, which submitted a total of something like 1500 recommendations on the articles with which it dealt. An important innovation, which greatly facilitated the work as a whole, was the computerization of the text in some of the languages. Much time was saved, especially in the English Language Group, through the use of computer runs of recurring words and phrases. For the decision of the Con-

1978–1980 was spent mainly on the harmonization aspect (including grammar, syntax and spelling), the application of the conclusions being deferred to the article-by-article reading as part of the concordance aspect. The Chairman of the Drafting Committee later explained the slow pace of the work, especially on the settlement of disputes provisions, by their »volume and complexity« which compelled the Drafting Committee to deal with two main problems raising juridical, technical and linguistic issues, which he described as the internal co-ordination of the various provisions of this Part [Part XV, Arts. 279–299 and Annexes V, VI, VII and VIII]; and the co-ordination between these provisions and their counterparts in Part XI, section 5 [Arts. 186–191] of the draft[53].

In this complicated process, every article, in each language, was given three or four readings, and differences which revealed themselves in the Co-Ordinators' meetings were resolved by a process of »consultations« and other less formal methods, in which the Chairman of the Drafting Committee and members of his delegation played a leading role.

This unaccustomed, heavy and expensive system of work is the consequence of several extraordinary circumstances, the repetition of which in identical terms is unlikely. These unusual features included the excessively politicized composition of the Drafting Committee itself, membership in which was not based on the professional or linguistic competence of the representatives of the States elected to the Committee (China, France and the United Kingdom were not members of the Drafting Committee); the extreme difficulty of the actual negotiating process much of which, undertaken by representatives not working in their mother tongue, was conducted outside the formal framework of the Conference and of which no official records have been kept in the Conference itself; the absence of a formal single »basic text« in all or most of the official languages of the Conference and the addition of one language during the Conference itself[54]. As a result, the Drafting Committee found itself obliged to undertake tasks which, from some points of view, may be regarded as going beyond what is normally expected

ference requesting the Drafting Committee to commence this work, see Official Records, vol. IX, 21.

[53] A/CONF. 62/L. 75, para. 4. Later he also stated in all frankness that »the draft provisions contained many technical defects which the Drafting Committee had the duty to remedy«, A/CONF. 62/SR. 152, p. 6 (provisional version).

[54] Arabic was added as a language of the Conference in 1975. Official Records, vol. IV, 3. In itself, at that stage of the work, this was not a complicating factor. The complications arose later, when it was realized that no authentic Arabic texts existed of most of the international documents, including the Charter of the United Nations and the Statute of the International Court, the 1958 Conventions on the Law of the Sea and various relevant documents of the Intergovernmental Maritime Consultative Organization, out of which the new Convention has emerged. A/CONF. 62/L. 39, Official Records, vol. XI, 102. This later induced a *cri de coeur* from the Arabic Language Co-Ordinator, reproduced in A/CONF. 62/L. 67/ Rev. 1, Annex 3.

of a Drafting Committee of a Diplomatic Conference, and more akin to those performed regularly by the Drafting Committee of the International Law Commission which frequently acts as an informal negotiating body and as the real generator of the texts of treaty articles[55]. Also, the Committee was not pressed for time in the early stages, in fact not until 1982.

While, therefore, the kind of work which this Drafting Committee has performed may be regarded as exceptional, the very magnitude of its tasks throws a searchlight on the nature of the problems which have to be faced in the drafting of a multilingual treaty of any size and importance. For short treaties, or for treaties prepared with the thoroughness with which the International Law Commission treats some of the concordance problems involved in the draft articles which it prepares, little of this behind-the-scences work will be publicized, it being rare for the reports of the Drafting Committee to consist of more than the texts actually adopted, together with such additional comments as the Committee has authorized to accompany those texts. The complex process adopted in UNCLOS III is a highly sophisticated return to the process of preparing the multilingual Charter at the San Francisco Conference, and it represents the most complete illustration of what is involved in giving effect to the principle of the integrity of a multilingual treaty in all its authentic texts.

The real test of the efficacy of this process is yet to come, however. The Drafting Committee has been able to produce a single text out of a heterogeneous series of origins – a text most of which is reasonably (though probably not entirely) harmonious and concordant. In the absence of *travaux préparatoires* of the normal kind, enabling the legislative history of any given text to be traced through the record, the reports of the Drafting Committee on its article-by-article examination of this Convention ought to assume a leading place in any process of interpretation in the future. It may well happen, for instance, that the true intention of the Conference can be established by noting a change made in the Chinese text only, and the relevant report of the Chinese Language Group will show whether the change was purely a linguistic one, or whether its effect was to introduce greater clarity into the text of the Convention as a whole. But this, in turn, will impose quite exceptional burdens on the interpretation process, and it remains to be seen whether and how it will accept this challenge.

One of the consequences of all this effort should be noted at this stage. When the Conference finally adopts the text of this Convention, within the

---

[55] This *modus operandi* of the Drafting Committee, especially during the second part of the tenth session of the Conference, was the object of some criticism in the Conference, principally on the part of the East European Group and the Group of 77. See discussions at the 64th meeting of the General Committee and the 152nd plenary meeting, A/CONF. 62/BUR/SR. 64, A/CONF. 62/SR. 152. However, that criticism cannot be isolated from the political circumstances in which the whole of the tenth session of the Conference met.

meaning of Art. 9 of the Vienna Convention, the Convention will have been drawn up in six authentic texts, all of which were prepared and examined by delegations to the Conference in an »open-ended« process, and are not translations prepared by the Secretariat and approved by the Conference after perusal by members of the Drafting Committee personally familiar with one or more of those languages. There is therefore an enormous difference between the preparation of this Convention and the preparation of the Charter of the United Nations or even the Vienna Convention on the Law of Treaties. Against that background, the methodological approach adopted by the International Court in the *Application to Intervene* case towards the interpretation of the quinquilingual Statute of the Court would probably not be appropriate. The unknown factor today is how this fundamental difference in the drafting process will find expression in the interpretative process, whether by States, by international organizations or by international tribunals, as well as in the technical literature which the new Convention on the Law of the Sea is bound to generate.

Before leaving this topic, there is interest in noting the first impression views of Governments as expressed in the Sixth Committee of the General Assembly in the on-going »Review of the Multilateral Treaty-making Process«. In 1980 the Secretary-General submitted a questionnaire which raised two issues: (*a*) should treaties continue to be formulated simultaneously in all languages in which their text is to be authentic, or should they originally be formulated in only one or two languages, with additional versions established by a special procedure later; (*b*) if negotiation in multiple languages is to continue, should the example of UNCLOS III be followed, of establishing a subgroup for each language, whose co-ordinators should meet from time to time to resolve any interlingual or general questions about the text[56]. Parenthetically it may be observed even now that question (*b*) in that form gives a mistaken view of the shape of the issue, for two reasons at least: (1) it confuses the language of *negotiation* (which may be whatever is common between the persons conducting the negotiation, unless they are accompanied by interpreters, which for the most part they were not) with the language in which the result of that negotiation is written down, in the case of UNCLOS III usually English; (2) at a time when in UNCLOS III was still under way, it opened the door to subjective and impressionable responses from which the emotion of personal frustrations and disappointments could not be entirely removed. The immediate reactions to those questions cannot, therefore, be taken as necessarily the last word on the matter.

---

[56] Doc. A/35/312 (1980), para. 63. On this general review, see General Assembly resolutions 32/48, 8 December 1977, 35/162, 15 December 1980, and Documents A/35/312 and A/36/553 (1981). It is to be observed that the figures given *supra* (note 52) illustrate the rapid loss of importance of the Drafting Committee itself in relation to the two »subsidiary organs« which it created.

The replies covered a wide range. Orally, in the Sixth Committee itself in 1980 the need was stressed to continue the current practice of formulating treaties simultanously in all the languages in which the treaty was to be authentic. At the same time warnings were uttered that the precise procedure to be followed would have to depend on the merits of each individual treaty, UNCLOS III only being a model in cases of a similar nature and not serving as a precedent[57]. In written comments, one country favoured the practice of adopting multilateral treaties in the six working languages of the General Assembly in the interest of »avoiding cultural hegemonies«. Another warning of a more fundamental character was directed to »the establishment of additional versions which might result in official translations which altered the meaning in some respect«. The polarities of opinion thus expressed will probably make general political agreement on this issue unattainable in the foreseeable future; but the warnings against over-hasty generalization of the very special procedure of UNCLOS III are well taken.

In assessing the special procedure adopted in UNCLOS III, it is important not to lose sight of what many regard as the flawed composition of the Drafting Committee, thanks to the preference given in its election to political factors rather than the professional skills of the representatives chosen to sit on it. The real issue remains, how is interlingual concordance best to be effected – indeed one might even go further and ask what *is* interlingual concordance? – and that is hardly given to a single answer generally valid for all cases.

* * *

We can now return to the primary purpose of this article.

It will perhaps come as a surprise that, so far as we have been able to ascertain, notwithstanding the repeated preoccupation of the international community, especially since the League of Nations, with practical aspects of the law of treaties, there seems to be no formal definition of »authentic text«. There is some discussion on what is not an authentic text despite a formal provision in the treaty stating that a given language text is »equally authentic« (the word »equally« in this context is probably pleonastic). Mostly this has concerned the Chinese texts of conventions adopted since 1948. These discussions were always highly colored by the political aspects of anything to do with China before the regularization of the status of the People's Republic in the United Nations in 1971 (this question has not arisen since). The best known is the issue of the revision or correction (the precise designation was a matter of controversy) of the Chinese text of the Convention on the Prevention and Punishment of the Crime of Genocide, brought to the

[57] A/36/553, Section II *passim,* and para. 23.

General Assembly by the Government of (Nationalist) China in 1951 and leading, after two unusual debates in the Sixth Committee, to resolutions 605 (VI), 1 February 1952, and 691 (VII), 21 December 1952[58]. A less known incident occurred in a conference held in 1972, when a representative enquired whether any steps were being taken to ensure that the authentic Chinese text »was in complete correspondence with the other texts«. The legal adviser to that Conference gave the following answer:

> »[S]ince the People's Republic of China had not been represented at the Conference, arrangements had been made to produce the Chinese text at a later stage, *after the original of the Protocol had been transmitted to United Nations Headquarters for deposit* [italics supplied]. That procedure had been adopted for reasons of economy, in order to avoid the expense of bringing a staff of Chinese translators and calligraphers to Geneva to prepare the Chinese text«[59].

This is the only instance known to the writer in which the *travaux préparatoires* clearly state that a text designated as authentic was not in existence when the treaty itself was adopted.

The most authoritative attempt to arrive at an international understanding over the meaning of »authentic« seems to have taken place during the tenth session of the International Labour Conference in 1927. There the issue arose of the preparation of German texts of draft conventions and recommendations, German not being an official language of the International Labour Conference at the time. In the course of a long and complicated discussion in the Standing Orders Committee and the Drafting Committee, the following series of propositions emerged:

> »(1) An official translation cannot be confused with an authentic text, since a text becomes authentic only by reason of its adoption by the Conference. Translations drawn up by the Director of the International Labour Office, on his own responsibility, would obviously not be authentic.
>
> (2) In the event of a dispute before the Governing Body or the Permanent Court of International Justice, only the French and English texts would be taken into consideration. The amendment itself provided that the official translations could be regarded as valid for the application of the Conventions and Recommendations only by ›the Governments concerned‹ and only ›in their respective countries‹. Moreover, the translations would contain the full text of the final Article in the two au-

[58] See on this *Yuen-li Liang,* The Question of Revision of a Multilingual Treaty Text, 47 American Journal of International Law (1953) 263. And see the Sectetariat's memorandum A/CN. 4/154, paras. 144–154, *supra* (note 43).

[59] United Nations Conference to consider amendments to the Single Convention on Narcotic Drugs, 1961 (1972), Official Records, vol. II, 64 (14th plenary meeting).

thentic texts of every Convention, which states that only the French and English texts shall be authentic.

(3) It was realised that it was impossible to avoid all mistakes in translation, but it was pointed out that such mistakes would probably be no more frequent in the translations drawn up by the Director of the International Labour Office than in those at present prepared by the Governments concerned upon their own responsibility.

(4) It was also pointed out that the proposed amendment would make it possible for States sharing the same language to agree upon a common text.

(5) Finally, it was emphasised that in no case would the Director impose a text upon a Government. As the text of the amendment definitely states, the translations would be made only ›at the request of interested Governments‹. Moreover, the last sentence of the amendment very definitely ensures the freedom of each Government. Even if a State Member had requested a translation from the Director of the Office, it would still remain free to attach no legal value to it. It would be open to the Governments themselves to consider such translations ›as authoritative in their respective countries for the application of the Conventions and Recommendations‹ «[60].

That seems also to be the philosophy underlying the conclusions of the International Law Commission and the Vienna Conference on the Law of Treaties in the matter of the interpretation of multilingual treaties, and the legal function of the authentic texts in that process. The consequence would be that for a text of a treaty to be »authentic« in the legal sense, the first condition would be that it should have been adopted by the treaty-adopting body itself. If that was not the case, then for all its political and psychological importance (exaggerated in an age of spreading nationalism and fears of cultural hegemony), the designation of a text as »authentic« carries in itself little if any weight. This does not exclude a later Conference from changing the authentic texts of an existing treaty, as from an agreed date, such as has occurred with the 1944 Convention on Civil Aviation or with the Treaty of Rome. Where a text designated »authentic« formed part of the instrument as adopted, then the drafting history of that text, as distinct from the legislative history of the provision or of the treaty as a whole, should be examined to establish what steps were taken to ensure interlingual concordance. In this sense, the somewhat categoric statement in the *Young Loan* case that the

[60] League of Nations, International Labour Conference, Tenth Session, Geneva 1927, vol. I – First, Second and Third Parts at 551. Note in this connection the Buenos Aires Protocol of 24 September 1968 on the trilingual text of the ICAO Convention (Chicago 1944). See *G. F. FitzGerald,* The Development of the Authentic Trilingual Text of the Convention on Civil Aviation, 64 American Journal of International Law (1970) 364.

Vienna Convention has abolished »the status of subordinated translations« may not be a complete reflection of the real intention of Art. 33. In the case of languages not widely known or not expected to be widely known, or those used by a very small number of States or even by one State only, their concordance with the texts of what are sometimes called »universal« languages may not be a matter of overriding importance in an interpretative exercise not involving those »rare« languages. The real difficulties arise when a delicate political compromise is reflected in a slight absence of concordance between the texts of those »universal« languages, such as where moods and tenses of verbs are skillfully used, when similar moods and tenses do not commonly exist in some other languages. Experience in UNCLOS III, especially with the authentic Arabic text, suggests that attempts to align a text on one only of the »universal« texts may upset delicately balanced compromises, or lead to unexpected political difficulty. The present writer encountered a similar experience when he prepared a Hebrew translation of the Vienna Convention itself[61].

The international jurisprudence here surveyed does serve to indicate that with the exception of the *Young Loan* case and some (but not all) of the post-1969 decisions of the European Court of Human Rights, international tribunals have not shown any clear tendency to read too much into Art. 33, to regard it as being innovative, or to examine closely the interlingual history of a multilingual treaty. This parallels the relative disinterest shown by participants in international conferences in the drafting problems of multilingual treaties. UNCLOS III is believed to be the first major conference in which the problem of language concordance has been treated with such unusual care. But the context and circumstances of that Conference, including the virtual absence of time pressures, were unusual and are unlikely to be repeated, at all events not on so large and so long a scale. The transfer of UNCLOS III practice to a conference lasting but a few weeks, working on a »basic text« in which concordance problems have already had some treatment, would be almost impossible.

However, it does seem that when faced with a real or alleged lack of concordance between authentic texts of a multilingual treaty, the general principle of the integrity of the single treaty in all its authentic texts would require more careful examination of all the language versions, even if they were prepared by the Secretariat and receive their status as »authentic« texts from the rules of procedure. It is in that sense that the multilingual aspects of the judgment in the *Application to Intervene* case sit uneasily with the more adventurous precedent of the 1948 *Admission* case.

At first sight, it might be thought that Art. 33 of the Vienna Convention contains a larger element of »progressive development« than has hither-

---

[61] *Sh. Rosenne/M. Zohory*, The Vienna Convention of 23 May 1969 on the Law of Treaties (the »Treaty on Treaties«) (Jerusalem 1978).

to been assumed, and that may be found to impose on the average practitioner demands that are too high, and could lead to results different from what the authors of a multilingual treaty ever intended. The fact that a text is designated »authentic« implies no more than that it *may* – perhaps even should – be *one* of the elements mobilized in the interpretative process. But neither the word »authentic« nor Art. 33 of the Vienna Convention takes us very far in answering the question of *how* it should be used. The authors of the Vienna Convention probably assumed that they were doing no more than codifying the law in a widely drawn statement of principle, and any attempt to read more into it than that is likely to lead to a deformation not only of the art of multilingual interpretation, but also of standard and empirically developing multilingual conference practice.

»Linguistic concordance to the extent possible, and juridical concordance in all cases« – reported the Chairman of the Drafting Committee to UNCLOS III. Perhaps the ultimate aim is to reach something even more elusive – conceptual concordance in all cases, if necessary at the expense of formal linguistic concordance.

Die Europäische Stiftung und die deutschen Bundesländer

Walter Rudolf

I

Der Jubilar gehört zu den ersten Wissenschaftlern, die nach Gründung der Bundesrepublik Deutschland das Thema »Auswärtige Gewalt und Bundesländer« intensiv behandelt haben [1]. Das Thema blieb in den 33 Jahren seit dem Entstehen der Bundesrepublik aktuell. Gerade im Bereich der auswärtigen Gewalt ist die *Triepel*'sche Erkenntnis, daß es im Bundesstaat keine Statik gibt, sondern nur einen Waffenstillstand ohne dauernden Frieden, besonders augenfällig, worauf der Jubilar in seinem Diskussionsbeitrag auf der Bonner Staatsrechtslehrertagung 1953 hingewiesen hat [2]. Solche »Waffenstillstände« sind durch die Lindauer Absprache vom 14. November 1957 [3] für den völkerrechtlichen Bereich und die EG-Erklärung des Bundeskanzlers vom 19. September 1979 [4], zu welcher der Vorsitzende der Ministerpräsidentenkonferenz mit Schreiben vom 26. September 1979 im Namen der Regierungschefs der Länder das Einverständnis erklärte, für den europäischen Sektor erreicht worden. Versuche, zu verfassungsrechtlich normierten Fixierungen – nach *Triepels* Bild zu einem dauernden Frieden – zu gelangen, wie sie die Enquete-Kommission Verfassungsreform vorgeschlagen [5] und die Enquete-Kommis-

---

[1] *H. Mosler*, in: Festschrift für Richard Thoma (1950), S. 129 ff., 150; sowie Diskussionsbeitrag in Veröffentlichungen der Vereinigung der Deutschen Staatsrechtslehrer (VVDStRL) 12 (1954), S. 240 ff.; Zeitschrift für ausländisches öffentliches Recht und Völkerrecht (ZaöRV) 16 (1955/56), S. 1 ff.

[2] VVDStRL 12 (1954), S. 240.

[3] Text: ZaöRV 20 (1959/60), S. 116 Anm. 102, und *R. Morawitz*, Die Zusammenarbeit von Bund und Ländern bei Vorhaben der Europäischen Gemeinschaft (1981), S. 85 f.

[4] Text: Gemeinsames Ministerialblatt Nr. 28 vom 15. 10. 1980, und *Morawitz* (Anm. 3), S. 102 ff.

[5] Schlußbericht der Enquete-Kommission Verfassungsreform, BT-Drucks. 7/5924, Kapitel 4 (Internationale Beziehungen), S. 228 ff.

sion Auswärtige Kulturpolitik erörtert, aber alsbald aufgegeben hat[6], sind Episoden geblieben. Bei Wahrung ihrer divergierenden Rechtsauffassungen haben sich Bund und Länder miteinander arrangiert, was nicht ausschließt, daß aus Gründen der Praktikabilität oder durch Veränderungen in der Entwicklung des Verhältnisses von Bund und Ländern Korrekturen notwendig werden können.

Im EG-Bereich liegen die Schwierigkeiten im Bund-Länder-Verhältnis darin, daß der Bund durch den EGKS-Vertrag und die Römischen Verträge auch »Hoheitsrechte« der Länder auf die Europäischen Gemeinschaften »übertragen« hat, wozu er nach der herrschenden Meinung[7] die durch Art. 24 I GG gedeckte Kompetenz besessen hat. In der Mehrzahl der Fälle sind die Europäischen Gemeinschaften freilich auf Gebieten rechtsetzend tätig, die, gäbe es die Gemeinschaft nicht, in der Bundesrepublik Deutschland in die Kompetenz des Bundes fallen würden. Insoweit sind die deutschen Länder nur betroffen, als sie durch den Bundesrat an der Rechtsetzung mitwirken oder zum Vollzug europäischer Rechtsakte verpflichtet sind[8]. Art. 2 des Zustimmunsgesetzes zu den Römischen Verträgen sieht für diese Fälle Unterrichtungspflichten der Bundesregierung vor[9], auf deren Einhaltung auch gegenüber dem Parlament nach wie vor von den zu informierenden Verfassungsorganen geachtet wird, wie noch in jüngster Zeit eine parlamentarische Anfrage der die Bundesregierung tragenden Fraktionen zeigt[10].

Es gibt allerdings auch eine Reihe von Materien, für welche die Europäischen Gemeinschaften Kompetenzen beanspruchen, die zur ausschließlichen Regelungszuständigkeit der Länder gehören würden[11]. Diese Fälle werden nach dem durch die EG-Erklärung des Bundeskanzlers vom 19. September 1979 und § 85a Gemeinsame Geschäftsordnung (GGO) II vorgesehenen Verfahren behandelt, soweit es um Maßnahmen im integrierten Bereich der

---

[6] Schlußbericht der Enquete-Kommission Auswärtige Kulturpolitik, BT-Drucks. 7/4121, S. 39 f., Nr. 202 ff. Vgl. dazu *W. Rudolf*, Zeitschrift für Kulturaustausch 26 (1976), S. 11 ff.

[7] Vgl. die Nachweise bei *O. Rojahn*, in: I. von Münch, Grundgesetz-Kommentar, Bd. 2 (1976), Rdn. 39 f.; *A. Bleckmann*, Recht der Internationalen Wirtschaft/Außenwirtschaftsdienst des Betriebsberaters (RIW/AWD) 1978, S. 144 ff.; *R. Riegel*, Deutsches Verwaltungsblatt (DVBl.) 1979, S. 246.

[8] Vgl. dazu *W. Rudolf*, in: Festschrift für Hans-Jürgen Schlochauer (1981), S. 123 ff.

[9] BGBl. 1957 II, S. 753.

[10] Kleine Anfrage der Fraktionen der SPD und FDP zur Struktur der Regierungstätigkeit im Bereich der Europäischen Politik vom 5. 3. 1982, BT-Drucks. 9/1428, dazu die Antwort des Parlamentarischen Staatssekretärs des Bundesministers für Wirtschaft vom 29. 3. 1982, BT-Drucks. 9/1520. Vgl. auch die Große Anfrage der Fraktion der CDU des Landtags Rheinland-Pfalz vom 17. 10. 1980, LT-Drucks. 9/968, und die Antwort des Ministers der Justiz vom 22. 12. 1980, LT-Drucks. 9/1122.

[11] *H. E. Birke*, Die deutschen Bundesländer in den Europäischen Gemeinschaften (1973), S. 17 ff.

EG geht. Demgegenüber sind europäische Aktivitäten außerhalb des integrierten Bereichs im Rahmen der europäischen politischen Zusammenarbeit der EG-Staaten dann, wenn Kompetenzen und Interessen der Länder involviert sind, gemäß der Lindauer Absprache zu behandeln. Dabei können Fälle in einer Gemengelage zwischen europäischer Integration und europäischer Koordination auftreten, die sowohl gemäß dem Verfahren der Lindauer Absprache wie nach der EG-Erklärung des Bundeskanzlers zu behandeln sind. Aktuelle Beispiele hierfür sind die Initiative der Bundesregierung für eine Europäische Akte[12] und die am 29. März 1982 von den EG-Staaten vereinbarte Errichtung einer Europäischen Stiftung.

---

[12] Mit der von der italienischen Regierung unterstützten Initiative der Bundesregierung für eine Europäische Akte wird das Ziel verfolgt, die Idee der politischen Einigung Europas neu zu beleben. Vgl. *H. D. Genscher*, Europäische Zeitung 32 Nr. 3, März 1982, S. 1. Die Europäische Union solle schrittweise verwirklicht werden, indem »eine enge kulturelle Zusammenarbeit zwischen den Mitgliedstaaten« beabsichtigt sei, »um das Bewußtsein der kulturellen Gemeinsamkeit als Teil der europäischen Identität zu fördern, dabei zugleich die Vielfalt der eigenständigen Überlieferungen auszuschöpfen, um den Austausch der gegenseitigen Erfahrungen, insbesondere auch unter der Jugend zu intensivieren«. Es sollen ein Rat der für kulturelle Zusammenarbeit zuständigen Minister und ein Rat der Justizminister gebildet werden. Der Rat der für kulturelle Zusammenarbeit zuständigen Minister solle regelmäßig einen Meinungsaustausch über eine enge Zusammenarbeit im kulturellen Bereich führen, um damit eine weitgehende Harmonisierung der Standpunkte in kulturellen Fragen zu ermöglichen. Bei diesen Beratungen könnten sich die Mitgliedstaaten nach Maßgabe ihrer verfassungsrechtlichen Bestimmungen durch die jeweils zuständigen Minister vertreten lassen. Im Bereich der kulturellen Zusammenarbeit wird ferner die Unterstützung durch ein ausbaufähiges Sekretariat der Europäischen Politischen Zusammenarbeit vorgeschlagen.

Die Europäische Akte wird von der Bundesregierung als Willensbekundung von »hoher Intensität« gekennzeichnet. Es handelt sich dabei nicht um einen Vertrag, sondern mehr um eine Absichtserklärung. Die Möglichkeit, daß durch eine politische Aktion ohne rechtliche Wirkungen die innerstaatliche Kompetenzverteilung zwischen Bund und Ländern umgangen werden könnte, wurde von den Bundesländern gesehen, die entsprechende Bedenken äußerten. Als der für die Konferenz für kulturelle Zusammenarbeit zuständige Vertreter könne nur der Präsident der Kultusministerkonferenz in Betracht kommen. Bayern lehnte diesen Vorschlag insoweit zunächst gänzlich ab, da es keine Notwendigkeit für dieses neu zu schaffende Organ sah, weil auf diesem Gebiet bereits die EG und der Conseil de Coopération Culturelle beim Europarat wirkten und die Europäische Stiftung demnächst auch zuständig sei.

Die Regierungschefs der Bundesländer haben das Ziel der Initiative der Bundesregierung in einer Stellungnahme vom 4. 3. 1982 gleichwohl begrüßt. Bei der Besprechung des Bundeskanzlers mit den Regierungschefs der Länder am 5. 3. 1982 herrschte Übereinstimmung darüber, daß Bund und Länder in Kultur- und Bildungsfragen, soweit sie innerstaatlich in die ausschließliche Zuständigkeit der Länder fallen, sowie bei allen sonstigen Fragen, die wesentliche Interessen der Länder berühren, zu einer engen und vertrauensvollen Zusammenarbeit verpflichtet sind. Als zuständige Stellen für die innerstaatliche Abstimmung wurden von Bundesseite das Auswärtige Amt und von Länderseite die Bayerische Staatskanzlei benannt.

II

Der Plan zur Errichtung einer Europäischen Stiftung geht zurück auf den Tindemans-Bericht vom 7. Januar 1976, der im Auftrag der Regierungschefs der EG-Staaten erstattet wurde[13]. In seinem Bericht schlug Tindemans in einem Kapitel über das »Europa der Bürger« die Errichtung einer Europäischen Stiftung mit dem Ziel vor, das Verständnis zwischen den Völkern der Gemeinschaft zu verbessern[14]. Es reiche nicht aus, daß die europäische Schicksalsgemeinschaft eine Realität sei; sie müsse auch als solche anerkannt werden. Den Europäern von morgen müsse die europäische Realität als eine persönliche und konkrete Erfahrung vor Augen geführt und ihnen eine gründliche Kenntnis unserer Sprachen und unserer Kultur vermittelt werden, weil hieraus das gemeinsame Erbe erwachse, das die Europäische Union schützen müsse.

Der Vorschlag Tindemans wurde allgemein begrüßt[15]. Allerdings wurden auch Bedenken laut: Man befürchtete, daß nationales Proporzdenken, unzureichende Besetzung der Spitzenpositionen, zu wenig Finanzmittel, Bürokratisierungstendenzen, zu viele Mitspracherechte zu vieler Stellen und das Abgleiten der Programme »auf das Niveau einer mehr oder weniger veredelten oder kulturell verbrämten Öffentlichkeitsarbeit« die hochgesteckten Ziele, die mit der Europäischen Stiftung verfolgt werden, verfehlen könnten, bevor noch das Projekt dieser Stiftung genauer konturiert war[16].

Vom Erscheinen des Tindemans-Berichts dauerte es fast zwei Jahre, bis der Europäische Rat am 5. und 6. Dezember 1977 einen Grundsatzbeschluß zur Errichtung einer Europäischen Stiftung faßte. Das Europäische Parlament begrüßte diesen Grundsatzbeschluß und forderte die Kommission auf, »in kürzester Frist diesbezügliche Vorschläge auf der Basis von Art. 235 EWG-Vertrag vorzulegen«[17], nachdem die Kommission dem Rat der EG bereits Vorschläge zum Aufgabenbereich, zu den Zielen, zur Struktur und zur Finanzierung dieser Stiftung gemacht hatte[18].

---

[13] Bull. EG, Beilage 1/1976.

[14] Vgl. dazu *H. Schneider/W. Wessels* (Hrsg.), Auf dem Weg zur Europäischen Union?, Diskussionsbeiträge zum Tindemans-Bericht (1977); *R. Picht* (Hrsg.), Kulturpolitik für Europa (1977).

[15] Vgl. etwa *Th. Jansen*, in: R. Picht (Anm. 14), S. 113 ff.; *D. Klose,* in: R. Picht, a.a.O., S. 174 ff.; *H. Arnold,* EG-Magazin 9/79, S. 9; *P. Karpenstein,* Europarecht (EuR) 1979, S. 300, 305.

[16] *Arnold* (Anm. 15).

[17] Entschließung des Europäischen Parlaments zur Gründung einer Europäischen Stiftung zur Mobilisierung der öffentlichen Meinung zugunsten der Ziele und Politiken Europas, Abl. EG Nr. C 85 vom 10. 4. 1978, S. 23. – Auf seiner Tagung im Juni 1982 hat das Europäische Parlament bedauert, daß die Tätigkeit der Stiftung den Charakter einer zwischenstaatlichen Zusammenarbeit erhalten hat, und energisch dagegen protestiert, daß es nicht, wie gefordert, vorher zu Aufgabenbereich, Zielen, Struktur, Finanzierung und parlamentarischer Kontrolle der Stiftung konsultiert wurde. Vgl. BT-Drucks. 9/1817 vom 30. 6. 1982.

[18] KOM (78) 51 endg.

Bei seinem Treffen am 7. und 8. April 1978 legte der Europäische Rat [19] Ziele, Aufgaben und Strukturen der Europäischen Stiftung in den Grundzügen fest. Er entschied auch, daß die Stiftung durch eine Vereinbarung zwischen den im Rat der EG vereinigten Mitgliedstaaten errichtet werden solle. Von einer Beteiligung der Gemeinschaft an der Vereinbarung ging er stillschweigend aus. In die Ziele der Stiftung sollten Angelegenheiten eingeschlossen sein, welche die Gemeinschaft angehen und in deren Zuständigkeit fallen. Die Gemeinschaftsorgane sollten sich an der Benennung der Mitglieder des Stiftungsrats beteiligen. Der Haushalt der Gemeinschaft sollte die Stiftung finanzieren.

Die Kommission der EG legte am 8. Mai 1978 als Arbeitsdokument der Dienststellen der Kommission den Entwurf eines Übereinkommens über die Gründung einer Europäischen Stiftung vor [20]. In diesem Arbeitsdokument wurde festgestellt, daß die Errichtung einer Stiftung mit eigener Rechtspersönlichkeit nach nationalem und internationalem Recht eine internationale Konvention verlange. Damit war ausgesprochen, daß die Stiftung nicht als bloßes Hilfsorgan auf Grund primären Gemeinschaftsrechts geschaffen werden kann, weil die Ermächtigungsgrundlage des Art. 235 EWGV dafür nicht ausreicht [21].

Die Frage, ob die Europäische Stiftung auf Art. 235 EWGV gestützt werden kann oder durch völkerrechtlichen Vertrag unter Beteiligung der EG geschaffen werden muß, richtet sich danach, ob der Stiftung Aufgaben im Rahmen der Ziele des Gemeinsamen Marktes übertragen werden sollen oder ob ihre Aufgabenstellung darüber hinausreicht. Diese Frage ist auch für die innerstaatliche verfahrensmäßige Beteiligung der Bundesländer von Bedeutung [22].

Nach dem Vertragsentwurf vom 8. Mai 1978 sollte Aufgabe der Stiftung sein, zu einer besseren Verständigung zwischen den Völkern der Gemeinschaft beizutragen und eine bessere Kenntnis des kulturellen Erbes Europas in seiner Mannigfaltigkeit und Einheit sowie ein größeres Verständnis der europäischen Integration zu fördern. Als Zielsetzungen wurden genannt die Förderung der Kontakte und des Gedankenaustauschs zwischen Personen und Gruppen verschiedener Länder der Gemeinschaft, die Verbesserung des Sprachunterrichts, die Vertiefung der Kenntnisse der Staatsangehörigen der Gemeinschaft von der geschichtlichen, kulturellen und sozialen Entwicklung der Vertragsstaaten sowie von ihrem gemeinsamen Erbe und der europäi-

---

[19] Zur rechtlichen Charakterisierung des Europäischen Rats vgl. *L. Tindemans*, in: Mélanges Fernand Dehousse, Bd. 2 (1979), S. 167 ff.; *W. Wessels*, Der Europäische Rat, Stabilisierung statt Integration? (Europäische Studien des Instituts für Europäische Politik, Bd. 13) (1980); *H. P. Ipsen*, EuR 1981, S. 309 ff.

[20] SEK (78) 1930 endg.

[21] Zu den Hilfsorganen vgl. *M. Hilf*, ZaöRV 36 (1976), S. 551 ff.; *M. Schweitzer/W. Hummer*, Europarecht (1980), S. 75 ff.; *H. Weis*, EuR 1980, S. 273 ff.

[22] Vgl. unten V.

schen Integration und die Förderung einschlägiger Forschungsarbeiten [23]. Die Stiftung solle direkte Tätigkeiten durchführen oder Tätigkeiten anderer Einrichtungen richtungweisend beeinflussen oder unterstützen. Sie solle dafür sorgen, daß ihre Tätigkeiten diejenigen anderer Organe oder Einrichtungen auf nationaler, bilateraler oder internationaler Ebene ergänzen [24], ohne daß sich ihre Tätigkeiten mit den in den bestehenden oder künftigen Programmen der Gemeinschaft vorgesehenen Tätigkeiten überschneiden dürften [25]. Im Rahmen der ihr zugewiesenen Aufgaben habe die Stiftung mit den Organen und Einrichtungen zusammenzuarbeiten, die im gleichen Bereich oder in ähnlichen Bereichen tätig sind und den Wunsch äußern, sie zu unterstützen [26]. Die vorrangigen Tätigkeiten solle die Stiftung in einem Programm festlegen, das sich über mehrere Jahre erstrecken könne. Bei der Ausarbeitung des Programms solle das besondere Augenmerk den Interessen und Bedürfnissen der Jugend gelten [27].

Der Entwurf sah vor, daß die Stiftung und ihr Personal die zur Erfüllung ihrer Aufgaben erforderlichen Vorrechte und Immunitäten nach Maßgabe eines Protokolls über die Vorrechte und Immunitäten der Europäischen Stiftung genießen, das Bestandteil des Übereinkommens werden solle und in 18 Artikeln ebenfalls im Entwurf ausgearbeitet wurde [28]. Da Paris als Sitz der Stiftung vorgesehen war [29], solle diese mit der französischen Regierung ein Sitzabkommen schließen [30]. In jedem Vertragsstaat solle die Stiftung die umfassendste Rechtsfähigkeit haben, die juristischen Personen nach dem innerstaatlichen Recht zuerkannt wird. Insbesondere solle sie unbewegliche und bewegliche Sachen erwerben oder veräußern, Verträge abschließen und vor Gericht auftreten können [31].

Als Organe der Europäischen Stiftung sah der Entwurf einen 36köpfigen Stiftungsrat und einen aus elf Personen bestehenden Exekutivausschuß vor [32]. Der Stiftungsrat solle zur Hälfte aus Mitgliedern bestehen, die von den Mitgliedstaaten, zu einem Viertel aus Mitgliedern, die vom Rate der EG mit qualifizierter Mehrheit auf Vorschlag der Kommission nach Anhörung des Europäischen Parlaments ernannt werden. Das restliche Viertel solle von den so ernannten Mitgliedern kooptiert werden [33]. Die Amtszeit solle sechs Jahre

---

[23] Art. 2 Abs. 2 des von der EG-Kommission erarbeiteten Entwurfs vom 8. 5. 1978.
[24] Art. 3 I des Entwurfs.
[25] Art. 3 II des Entwurfs.
[26] Art. 4 I des Entwurfs.
[27] Art. 5 I und II des Entwurfs.
[28] Art. 6 I des Entwurfs.
[29] Art. 20 I des Entwurfs.
[30] Art. 6 II des Entwurfs.
[31] Art. 7 des Entwurfs.
[32] Art. 8 ff. des Entwurfs.
[33] Art. 10 I des Entwurfs.

betragen[34]. Dem Stiftungsrat solle die Leitung der Stiftung obliegen[35]. Der auf drei Jahre bestellte Exekutivausschuß, der aus dem Präsidenten und den drei Vizepräsidenten des Stiftungsrats und sieben vom Stiftungsrat unter seinen Mitgliedern benannten Personen bestehen solle, war als Leitungsorgan der Stiftung vorgesehen[36]. Ein vom Stiftungsrat ernannter Generalsekretär solle den Exekutivausschuß bei seinen Organisations- und Verwaltungsaufgaben unterstützen und die laufenden Geschäfte leiten[37].

Recht knapp waren die Finanzbestimmungen gehalten[38], über die in der Folgezeit noch intensiv und kontrovers verhandelt wurde. Im Prinzip wurde vorgeschlagen, die Finanzmittel der Stiftung aus einem Zuschuß der EG für die Erstausstattung, etwaigen zusätzlichen Zuschüssen der EG und freiwilligen Zuwendungen aus öffentlicher und privater Hand zu besorgen. Bei Streitigkeiten der Vertragsparteien untereinander oder mit der Stiftung über Anwendung und Auslegung des Gründungsabkommens war die Entscheidung des Gerichtshofs der EG vorgesehen[39].

Dem Entwurf des Übereinkommens wurde schließlich noch ein zusätzlicher Entwurf einer Vereinbarung zwischen den Vertragspartnern beigefügt, dessen Zweck sein sollte, dazu beizutragen, daß die Stiftung ihre Tätigkeit möglichst schnell aufnehmen kann. Er sah vor, einen Ausschuß zu bilden, dessen Mitglieder von den Vertragsparteien und von der EG-Kommission benannt werden und der »durch Untersuchungen und Überlegungen die Errichtung und Arbeit der Europäischen Stiftung vorbereiten« solle.

Das Arbeitsdokument der Dienststellen der EG-Kommission wurde in der Gruppe »Allgemeine Fragen« des Rats der EG im Mai und Juni 1978 detailliert überprüft. Als Ergebnis der Überprüfung wurden ein überarbeiteter erweiterter Entwurf für ein Abkommen über die Gründung einer Europäischen Stiftung, ein Entwurf für ein Abkommen über die vorläufige Anwendung des genannten Abkommens und die Hauptpunkte eines Entwurfs für einen Beschluß des Rats der EG ausgearbeitet[40]. Bei den Beratungen sprachen sich die dänische und die französische Delegation gegen eine Beteiligung der EG an dem Projekt mit der Begründung aus, daß der EG eine Zuständigkeit zuerkannt würde, die sie nach den Römischen Verträgen nicht habe. Die Gruppe stellte fest, daß ein autonomer Beschluß des Rates notwendig sei, gleichgültig, ob die EG Vertragspartei des Übereinkommens werde oder nicht. Der Beschluß müsse wegen der Ernennung eines Teils der Mitglieder des Stiftungsrates, des Ernennungsverfahrens und der Bereitstellung von EG-Haushaltsmitteln auf Art. 235 EWGV gestützt werden. Keine Einigung konnte

[34] Art. 10 II des Entwurfs.
[35] Art. 11 I des Entwurfs.
[36] Art. 12 u. 13 des Entwurfs.
[37] Art. 14 des Entwurfs.
[38] Art. 15–19 des Entwurfs.
[39] Art. 24 des Entwurfs.
[40] Dok.R/1432/78 (FE 22) vom 5. 6. 1978.

die Gruppe bezüglich der Anzahl der Mitglieder des Stiftungsrats erzielen. Während die Kommission an der Zahl von 36 Mitgliedern festhielt, erschien einigen Delegationen diese Zahl zu hoch. Ein Kompromißvorschlag, der 17 Mitglieder vorsah, wurde von der deutschen und der niederländischen Delegation abgelehnt. Über die Finanzierung der Stiftung konnten noch keine übereinstimmenden Vorschläge vorgelegt werden. Auch über das Verzeichnis der Tätigkeiten der Stiftung bestanden noch Meinungsverschiedenheiten. Die Prüfung des Protokolls über die Vorrechte und Immunitäten wurde zurückgestellt.

Seit Sommer 1978 blieb die Gründung der Stiftung in der Schwebe, bis im Herbst 1981 das Projekt energisch weiterverfolgt wurde. Im November 1981 konnte in der Gruppe »Allgemeine Fragen« des EG-Rats in drei bisher strittigen Fragen weitgehende Einigung erzielt werden, nachdem der Ausschuß der Ständigen Vertreter des EG-Rats zuvor Leitlinien aufgestellt hatte. Erstens war man sich grundsätzlich einig, daß unbeschadet der Gemeinschaftskompetenzen die Ziele der Stiftung dem vom Europäischen Rat im April 1978 genehmigten Text entsprechen müßten. Eine als Hinweis dienende Liste der Tätigkeiten solle in das Übereinkommen aufgenommen werden. Allerdings machten vier Delegationen einen einstweiligen Vorbehalt gegen eine von der Gruppe aufgestellte Liste. Zweitens stimmten alle Delegationen der Gründung der Stiftung durch ein Übereinkommen zwischen den im Rat vereinigten Vertretern der Mitgliedstaaten zu. Die EG könne sich dadurch an der Stiftung beteiligen, daß sie einen finanziellen Beitrag leiste und in einer noch festzulegenden Weise im Rat der Stiftung vertreten werde. Drittens bestand grundsätzlich Übereinstimmung über die Finanzierung der Stiftung, deren Erstausstattung in einem über fünf Jahre hinweg geleisteten jährlichen Beitrag der Gemeinschaft bestehen solle, über dessen Höhe allerdings keine Einigung erzielt wurde. Verfahrenstechnisch wurden drei Lösungsvorschläge erörtert. Favorisiert wurde der Abschluß eines Abkommens zwischen Stiftung und EG zur Festlegung der Einzelheiten der Zusammenarbeit, wobei Rechtsgrundlage eines solchen Abkommens ein auf Art. 235 EWGV beruhender Rechtsakt wäre. Außerdem könne die Stiftung ermächtigt werden, zusätzliche finanzielle Beiträge, insbesondere von den Regierungen der Mitgliedsländer, entgegenzunehmen. Der Vorschlag, die französische Regierung solle der Stiftung kostenlos ein Grundstück in Paris sowie die für den Betrieb der Stiftung erforderlichen Gebäude zur Verfügung stellen und deren Unterhaltung übernehmen, stieß auf den Vorbehalt der französischen Delegation. Nicht gelöst blieben das Problem der Zahl der Mitglieder des Verwaltungsrats und des Exekutivausschusses sowie die Frage der Vorrechte und Immunitäten. Die griechische Delegation meldete einen einstweiligen allgemeinen Vorbehalt an.

In mehreren Sitzungen des Ausschusses der Ständigen Vertreter und der Arbeitsgruppe im Dezember 1981 und im Januar, Februar und März 1982 konnten weitere Fortschritte erzielt werden. Am 11. Dezember 1981 erstellte die Arbeitsgruppe einen überarbeiteten Entwurf eines Gründungsabkommens, der auf dem Entwurf von 1978 basierte und hinsichtlich der Zahl der Mitglieder des Stiftungsrates die große Lösung von 40 Mitgliedern vorsah[41]. Um möglichst bald ein Ergebnis zu erzielen, stellte die Arbeitsgruppe gemäß dem Mandat, das ihr der Ausschuß der Ständigen Vertreter tags zuvor erteilt hatte, die noch offenen Fragen weiterzuprüfen, am 19. Januar 1982 ein Vier-Punkte-Arbeitsprogramm auf, das die Fertigstellung des Gründungsabkommens, des Beschlusses des Rates zur Ermächtigung der Kommission, mit der Stiftung zu verhandeln, der Vorlage eines von der Kommission dem Rat zu unterbreitenden Entwurfs eines Briefwechsels zwischen der Gemeinschaft und der Stiftung und des Beschlusses des Rates zur Genehmigung des vorgenannten Briefwechsels vorsah. Von einer Einigung war man noch weit entfernt, zumal in der Sitzung des Ausschusses der Ständigen Vertreter am 28. Januar 1982 die deutsche und die griechische Delegation einen einstweiligen allgemeinen Vorbehalt eingelegt hatten.

Im Februar 1982 hat die Gruppe die Artikel des Entwurfs des Gründungsabkommens erneut geprüft und überarbeitet, wobei Bestimmungen über den Rat und den Exekutivausschuß einer doppelten Überarbeitung unterzogen wurden. Als Arbeitsergebnis der Gruppe vom 4., 12. und 24. Februar 1982 wurden zwei Textentwürfe vorgelegt, nämlich ein Entwurf eines Abkommens über die Errichtung der Europäischen Stiftung und ein Entwurf für eine Vereinbarung über die Einsetzung eines Vorbereitenden Ausschusses. Eine zweite Lesung einiger Artikel des Entwurfs für ein Abkommen über die Errichtung der Europäischen Stiftung fand Anfang März 1982 statt. Um auch die Frage der Vorrechte und Immunitäten weiter voranzutreiben, legte der Vorsitz des EG-Rates am 3. März 1982 einen Vermerk mit dem Text eines Protokolls der Gruppe »Allgemeine Fragen« zur Prüfung vor.

In seiner Sitzung vom 18. März 1982 stellte der Ausschuß der Ständigen Vertreter fest, daß die meisten der noch offenen Fragen geregelt werden konnten. Offen blieben nur noch die Fragen der Unvereinbarkeit der Tätigkeit der Mitglieder des Rates der Stiftung, der Zusammensetzung des Vorbereitenden Ausschusses, der Einstellung eines Programmdirektors und einer Erklärung über die Regelung für die Stiftung, ihre Mitglieder und ihr Personal. Auch hinsichtlich des Briefwechsels zwischen der EWG und der Stiftung bestanden noch Bedenken der italienischen Delegation. Zu dem letzten Punkte übersandte der Präsident der Kommission Thorn dem Präsidenten des Rats der EG am gleichen Tage eine tags zuvor erarbeitete Empfehlung für einen Beschluß zur Ermächtigung der Kommission zu Verhandlungen über einen Briefwechsel zwischen der EWG und der Europa-Stiftung.

Am 22. März 1982 erzielte der Rat überraschend Einigkeit über den Abkommenstext und die Schlußakte zum Abkommen. Auf Drängen des Rats-

---

[41] Durch den Beitritt Griechenlands zur EG mußte die Zahl der Stiftungsratsmitglieder von 36 auf 40 erhöht werden.

vorsitzenden Tindemans wurden alle Vorbehalte zu einzelnen Bestimmungen der Entwürfe zurückgezogen, so daß die Außenminister der EG-Mitgliedstaaten am 29. März 1982 im Rahmen der Festsitzung aus Anlaß des 25. Jahrestages der Römischen Verträge das Gründungsabkommen, eine Vereinbarung über die Einsetzung eines Vorbereitenden Ausschusses und eine Schlußakte unterzeichnen konnten.

III

1. Das eine Präambel und 29 Artikel umfassende Abkommen über die Errichtung einer Europäischen Stiftung regelt im ersten Kapitel Grundsätze und Ziele der Stiftung. Danach besitzt die Stiftung Rechtspersönlichkeit, verfolgt keinen Erwerbszweck und hat ihren Sitz in Paris[42]. Sie hat die Aufgabe, zu einer besseren Verständigung zwischen den Völkern der EWG beizutragen und eine bessere Kenntnis des kulturellen Erbes Europas in seiner großen Vielfalt und seiner Einheit zu fördern sowie ein größeres Verständnis für die europäische Integration zu entwickeln[43]. Diese Aufgabe hat sie nach den in Art. 5 festgelegten Leitlinien zu erfüllen[44]. Die Tätigkeiten der Stiftung er-

---

[42] Art. 1 und 7. Die Stiftung schließt gemäß Art. 6 mit der französischen Regierung ein Sitzabkommen ab. Frankreich stellt der Stiftung gemäß Art. 21 unentgeltlich ein Gelände in Paris sowie die für die Arbeit der Stiftung erforderlichen Gebäude zur Verfügung, deren Instandhaltung es übernimmt.

[43] Art. 2.

[44] Art. 5 lautet: »Die Stiftung erstellt das Programm mit den vorrangigen Tätigkeiten und den Einzelheiten ihrer Beteiligung.
Die Tätigkeiten, welche die Stiftung im Rahmen ihrer in Artikel 2 definierten Aufgabe durchführen kann, können unter anderem folgende Ziele haben:
- Zu fördern sind – in erster Linie bei den Völkern der Gemeinschaft – das Verständnis für den Europa-Gedanken sowie die Verbreitung von Informationen über den Aufbau Europas einschließlich der Informationen über die Länder der Gemeinschaft und ihre Geschichte;
- zu untersuchen ist, mit welchen Mitteln die Länder der Gemeinschaft unter Berücksichtigung der heutigen Entwicklung der Gesellschaft und der Technik ihr gemeinsames kulturelles Erbe bewahren und weiterentwickeln können;
- zu fördern sind das Erlernen der Sprachen der Länder der Gemeinschaft und die praktische Nutzung der so erworbenen Kenntnisse;
- zu unterstützen ist der Austausch von Personen innerhalb der Gemeinschaft einschließlich des berufsbezogenen Austausches sowie des Austausches im Zusammenhang mit Tätigkeiten, durch die das Verständnis für die Gemeinschaft gefördert werden soll;
- es sind insbesondere Programme aufzustellen und zu fördern, die den Interessen und Bedürfnissen der Jugend Rechnung tragen;
- zu fördern ist sowohl innerhalb als auch außerhalb des Gebiets der Gemeinschaft die kulturelle Ausstrahlung der Gemeinschaft, und zwar insbesondere durch Unterstützung kultureller und sonstiger Vorhaben, um das Wesen der Gemeinschaft und die Zusammenarbeit zwischen ihren Völkern in volkstümlicher und ansprechender Weise vor Augen zu führen«.

gänzen diejenigen anderer Organe und Einrichtungen, die auf nationaler, bilateraler oder multilateraler Ebene in den unter die Zuständigkeit der Stiftung fallenden Bereichen tätig sind, ohne sich dabei mit den in den Programmen der Gemeinschaft vorgesehenen Tätigkeiten zu überschneiden[45]. Auf Wunsch anderer Organe und Einrichtungen kann die Stiftung mit diesen im gleichen Bereich oder in ähnlichen Bereichen zusammenarbeiten[46]. Mit Vorrang sind indirekte Tätigkeiten durchzuführen[47], doch kann die Stiftung auch unmittelbar tätig werden, wenn andere Organe oder Einrichtungen bei solchen Vorhaben zur Durchführung nicht in der Lage sind[48]. Die Stiftung handelt in völliger Unabhängigkeit und trägt für eine ausgewogene Durchführung ihrer Tätigkeiten Sorge[49].

Im zweiten Kapitel wird die Struktur der Stiftung geregelt. Organe sind gemäß Art. 8 der Rat und der Exekutivausschuß, die von einem Generalsekretär unterstützt werden. Aus dem Wortlaut folgt, daß der Generalsekretär kein selbständiges Organ der Stiftung ist, sondern nur amtswalterische Funktionen wahrzunehmen hat, die in Art. 15 näher umschrieben sind[50]. Der Rat besteht aus 40 Mitgliedern, nämlich je zwei von den Vertragsstaaten im gegenseitigen Einvernehmen ernannten, zehn von der EG bestimmten und zehn von den so bestellten 30 Mitgliedern kooptierten[51]. Mindestens die Hälfte der kooptierten Mitglieder wird unter Persönlichkeiten von Organen oder Einrichtungen ausgewählt, die in den gleichen Bereichen tätig sind wie die Stiftung[52]. Die Amtszeit beträgt vier Jahre mit der Möglichkeit einmaliger Verlängerung[53]. Der Präsident, der ein von den Mitgliedstaaten ernanntes Mitglied sein muß, und die beiden Vizepräsidenten werden für die Dauer von zwei Jahren bei möglicher einmaliger Wiederwahl gewählt[54]. Die Mitglieder des Rats, in den nur hochgestellte Persönlichkeiten, die jede Gewähr für Unabhängigkeit bieten, berufen werden, üben ihr Mandat weisungsfrei in voller Unabhängigkeit aus[55]. Es besteht Inkompatibilität mit dem Amt eines Regierungsmitglieds und eines Mitglieds der EG-Kommission[56]. Nach dem

---

[45] Art. 3 I.

[46] Art. 4: »Die Stiftung arbeitet mit den Organen und Einrichtungen zusammen, die in dem gleichen Bereich oder in ähnlichen Bereichen tätig sind und den Wunsch haben, sie zu unterstützen«.

[47] Art. 3 II.

[48] Art. 3 III.

[49] Art. 3 V.

[50] Erstellung der Vorentwürfe des Tätigkeitsprogramms und des jährlichen Haushaltsplans für den Exekutivausschuß, Verwaltung und Durchführung der Tätigkeiten der Stiftung nach den Weisungen des Rates und Exekutivausschusses, Vorschläge an den Exekutivausschuß für die Einstellung und Entlassung des Personals.

[51] Art. 10 § 1 I.

[52] Art. 10 § 1 II.

[53] Art. 10 § 2.

[54] Art. 10 § 3 I und II.

[55] Art. 9 I und II.

[56] Art. 9 III.

Wortlaut von Art. 9 III sind nur Mitglieder der Bundesregierung als Ratsmitglied ausgeschlossen; Mitglieder von Landesregierungen können danach zum Ratsmitglied berufen werden. Die Funktionen des Rats regelt Art. 11 [57]. Beschlüsse des Rats, der alle sechs Monate oder auf Antrag von wenigstens einem Drittel seiner Mitglieder vom Präsidenten einberufen werden muß, werden mit der absoluten Mehrheit der Mitglieder gefaßt [58]. Art. 12 sieht vor, daß die Vertragsstaaten zu gegebener Zeit im gegenseitigen Einvernehmen die Bestimmungen über die Tagegelder der Mitglieder des Rats sowie die Regeln für ein Statut für das Personal der Stiftung erlassen. In diesen Regeln muß das Verfahren für die Beilegung von Streitigkeiten zwischen der Stiftung und ihren Bediensteten festgelegt werden.

Die allgemeine Verwaltung der Stiftung obliegt gemäß Art. 14 dem Exekutivausschuß [59]. Dieser besteht aus zehn Mitgliedern, je einem aus jedem Vertragsstaat. Der Präsident und die beiden Vizepräsidenten des Rats sind automatisch Mitglieder, wobei der Präsident des Rats gleichzeitig den Vorsitz im Exekutivausschuß führt. Die übrigen sieben Mitglieder werden vom Rat aus seinen Reihen gewählt. Dabei ist im Rahmen des Möglichen dafür Sorge zu tragen, daß die drei Kategorien von Mitgliedern im Exekutivausschuß im gleichen Verhältnis vertreten sind [60]. Auch der Exekutivausschuß beschließt mit der absoluten Mehrheit der Mitglieder [61].

---

[57] Art. 11 lautet: »Der Rat hat die oberste Leitung der Stiftung und legt die allgemeinen Leitlinien für sie fest.
Dem Rat obliegt es zu diesem Zweck insbesondere,
- das Programm zur Bestimmung der Rangfolge der Tätigkeiten der Stiftung aufzustellen,
- den jährlichen Haushaltsplan aufzustellen und die Rechnungen abzuschließen,
- die internen Vorschriften für die Arbeitsweise der Stiftung zu erlassen,
- über die Annahme von Schenkungen, Subventionen und sonstigen Zuwendungen zu entscheiden,
- den Generalsekretär der Stiftung zu ernennen und die Dauer seiner Amtszeit festzulegen«.

[58] Art. 10 § 5.

[59] Art. 14 lautet:
»(1) Der Exekutivausschuß ist für die allgemeine Verwaltung der Stiftung zuständig.
(2) Er stellt das Tätigkeitsprogramm der Stiftung auf und unterbreitet es dem Rat.
(3) Er stellt den Entwurf des jährlichen Haushaltsplans und gegebenenfalls die Entwürfe mehrjähriger finanzieller Vorausschätzungen auf und unterbreitet sie dem Rat.
(4) Er bereitet die Arbeiten des Rates vor.
(5) Er sorgt für die Ausarbeitung und Durchführung des Tätigkeitsprogramms.
(6) Er nimmt auf Vorschlag des Generalsekretärs die Einstellung und Entlassung des Personals der Stiftung vor«.
Gemäß Art. 23 erstellt der Exekutivausschuß spätestens zum 31. März den jährlichen Gesamtbericht über die Tätigkeiten der Stiftung und übermittelt ihn dem Rat zur Genehmigung.

[60] Art. 13 § 1 III.

[61] Art. 13 § 3 Satz 2.

Das dritte Kapitel trifft Regelungen über die Finanzen der Stiftung. Die Finanzmittel stammen einmal aus einem Beitrag der Gemeinschaft vorbehaltlich eines Gemeinschaftsbeschlusses und zum anderen aus freiwilligen Zuwendungen aus öffentlicher und privater Hand. Schenkungen und Zuwendungen unter Auflagen, die mit den Aufgaben der Stiftung unvereinbar sind, darf die Stiftung nicht annehmen[62]. Der Rat erläßt die Finanzvorschriften und stellt nach deren Maßgabe alljährlich den Haushaltsplan auf[63], der vom Exekutivausschuß auszuführen ist[64]. Die Finanzkontrolle obliegt dem Rechnungshof der EG[65].

Art. 24 regelt, daß Streitigkeiten der Vertragsstaaten untereinander oder eines oder mehrerer Vertragsstaaten mit der Stiftung über Anwendung oder Auslegung des Abkommens, die nicht innerhalb von sechs Monaten im Wege von Verhandlungen beigelegt werden können, einem Schiedsverfahren unterworfen werden. In diesem Fall bestimmt der Präsident des Europäischen Gerichtshofs auf gemeinsamen Antrag der streitenden Parteien oder, wenn ein solcher nicht vorliegt, auf Antrag einer der streitenden Parteien entsprechend den Modalitäten einer von den Vertragsstaaten nach Anhörung des Gerichtshofs erstellten Verfahrensordnung die Schiedsinstanz, die den betreffenden Streit beilegen soll. Die Vollstreckung von Entscheidungen der Schiedsinstanz obliegt den Vertragsstaaten und der Stiftung.

Das Abkommen gilt gemäß Art. 25 für das europäische Hoheitsgebiet der Vertragsstaaten, die französischen überseeischen Departements und die französischen überseeischen Gebiete. Gemäß Anhang 3 gilt das Abkommen nach einer Erklärung der Bundesregierung auch für das Land Berlin, sofern nicht die Bundesregierung gegenüber den übrigen Vertragsparteien binnen drei Monaten nach Inkrafttreten des Abkommens eine gegenteilige Erklärung abgibt. Auf die Kanalinseln und die Insel Man findet das Abkommen dann Anwendung, wenn die britische Regierung eine entsprechende Erklärung abgibt. Keine Anwendung findet das Abkommen auf die Färöer und auf Grönland, doch kann die dänische Regierung dem Depositar notifizieren, daß das Abkommen auf die genannten Gebiete angewendet werden soll.

Der Beitritt eines neuen Mitgliedstaates der Gemeinschaft zu diesem Abkommen erfolgt durch Hinterlegung einer Beitrittsurkunde beim Depositar[66].

Das Abkommen, das der Ratifizierung bedarf[67], wird auf unbestimmte Dauer geschlossen und tritt einen Monat nach der Hinterlegung der Ratifika-

---

[62] Art. 16.

[63] Art. 17 und 18.

[64] Art. 19.

[65] Art. 20.

[66] Art. 27.

[67] Art. 28. – In Anhang 1 verpflichten sich die Unterzeichnerstaaten, sofort die internen Verfahren zur parlamentarischen Genehmigung und Ratifizierung einzuleiten, um die Durchführung des Abkommens und die effektive Einrichtung der Stiftung binnen kürzester Frist zu ermöglichen.

tionsurkunden des zuletzt ratifizierenden Staates in Kraft[68]. Depositar ist die französische Regierung. Vertragssprachen sind Dänisch, Deutsch, Englisch, Französisch, Griechisch, Irisch, Italienisch und Niederländisch[69]. Diese Sprachen sind als Amtssprachen der Gemeinschaft auch Sprachen der Stiftung[70].

2. Die Vereinbarung über die Einsetzung eines Vorbereitenden Ausschusses wurde in der Erwägung abgeschlossen, daß schon von der Unterzeichnung des Abkommens an unverzüglich mit den Vorarbeiten insbesondere zur Erleichterung des materiellen und administrativen Aufbaus der Stiftung und zur Vorbereitung ihrer Tätigkeit begonnen werden muß. Der Vorbereitende Ausschuß setzt sich aus je einer von den einzelnen Unterzeichnerstaaten des Abkommens und von der EG-Kommission ernannten Persönlichkeiten zusammen. Aus dem Kreise dieser Persönlichkeiten bestellen die Unterzeichnerstaaten im gegenseitigen Einvernehmen und in enger Zusammenarbeit mit der EG-Kommission eine Person, der die Koordinierung und Verfolgung der Tätigkeiten des Ausschusses obliegt. Der Ausschuß tritt binnen drei Monaten nach der Unterzeichnung des Abkommens auf Einladung der französischen Regierung zusammen[71]. Die EG-Kommission wird ersucht, die Sekretariatsgeschäfte des Vorbereitenden Ausschusses zu übernehmen. Er übt seine Tätigkeit in Paris aus und findet sein Ende zum Zeitpunkt der Ernennung der von den Regierungen benannten Mitglieder des Stiftungsrates.

3. In der Schlußakte wird festgestellt, daß die Vertreter der Unterzeichnerstaaten die Texte des Abkommens über die Errichtung einer Europäischen Stiftung und der Vereinbarung über die Einsetzung eines Vorbereitenden Ausschusses sowie die in den Anlagen 1 und 2 enthaltenen Erklärungen angenommen und die Erklärung der Bundesrepublik Deutschland über die Geltung des Abkommens in Berlin zur Kenntnis genommen haben. Anhang 1 enthält Erklärungen zum Abkommen über die Errichtung einer Europäischen Stiftung[72]. In Anhang 2 wird eine Erklärung zu der Regelung für die Stiftung, ihre Mitglieder und ihr Personal abgegeben. Darin wird niedergelegt, daß die Unterzeichnerstaaten binnen vier Monaten nach Unterzeichnung des

---

[68] Art. 26 §§ 1 und 2. Die Stiftung wird auf der ersten Tagung des Rats errichtet und nimmt auf dieser Tagung ihre Tätigkeit auf (Art. 28 § 3).

[69] Art. 29.

[70] Art. 22.

[71] Der Ausschuß muß insbesondere die öffentlichen und privaten Einrichtungen bestimmen, mit denen die Stiftung zusammenarbeiten könnte, die Möglichkeiten der Zusammenarbeit der Stiftung mit diesen Einrichtungen vorbereiten und mit der Suche nach auswärtigen Finanzierungsquellen beginnen. Er muß ferner der Stiftung durch Erarbeitung von Vorschlägen für einen Programmentwurf die Annahme ihres ersten Tätigkeitsprogramms erleichtern und Schritte unternehmen, die im Hinblick auf den Abschluß des Sitzabkommens zwischen der Stiftung und der Regierung des Gastlandes erforderlich sind.

[72] Anhang 1 enthält Erklärungen zu Art. 4 (Zusammenarbeit mit anderen Einrichtungen), Art. 16 (Finanzierung der Stiftung) und Art. 26 § 2 (schnellstmögliche Ratifikation).

Abkommens eine Regelung für die Stiftung, deren Mitglieder und deren Personal aushandeln werden, wobei sie den durch die Tätigkeit der Stiftung bedingten Bedürfnissen und Interessen sowie der Unabhängigkeit der Stiftung Rechnung tragen. Die Erklärung stellt mithin ein *pactum de contrahendo* dar. Sie war notwendig, da eine Einigung über die Frage der Vorrechte und Immunitäten für die Stiftung und ihr Personal nicht rechtzeitig erzielt werden konnte.

Die Erklärung in Anhang 2 zählt beispielhaft die zu prüfenden Fragen auf: die Immunität der Stiftung bei der Ausübung ihrer Tätigkeit, die Unverletzlichkeit der Räumlichkeiten, Gebäude und Archive, die Befreiung von verwaltungsmäßigen und gerichtlichen Zwangsmaßnahmen sowie von direkten und indirekten Steuern und Ein- und Ausfuhrbeschränkungen, die Veröffentlichungs- und Informationsfreiheit und besondere Devisen- und Wechselkursregelungen. Für die Mitglieder des Rates sollen die Immunität gegen gerichtliche Verfolgung wegen Handlungen bei der Ausübung ihrer Tätigkeit und Verwaltungserleichterungen bei Reise, Aufenthalt und Devisenverkehr geprüft werden. Bezüglich der an der Arbeit der Stiftung beteiligten Personen erstreckt sich die Prüfung nur auf Verwaltungserleichterungen bei Reise, Aufenthalt und Devisenverkehr. Für den Generalsekretär und das Personal sollen die Immunität gegen gerichtliche Verfolgung wegen Handlungen bei der Ausübung ihrer Amtstätigkeit, Vorzugsbehandlung bei Einwanderung und Anmeldung, Erleichterungen im Bereich der Währung und des Devisenverkehrs und das Recht zur Ein- und Ausfuhr von Mobiliar, Kraftwagen und persönlicher Habe geprüft werden. Schließlich sollen auch das System der sozialen Sicherheit und die Steuerregelung für den Generalsekretär und das Personal der Stiftung festgelegt werden.

IV

Die Europäische Gemeinschaft ist durch das Abkommen über die Errichtung der Europäischen Stiftung insoweit betroffen, als sie gemäß Art. 10 § 1 I ein Viertel der Mitglieder des Stiftungsrats bestellen und auf Grund von Art. 16 I Nr. 1 Finanzmittel für die Stiftung aufbringen wird [73]. Eine unmittelbare Verpflichtung der EG durch den Vertrag ist freilich nicht möglich, weil die EG nicht Vertragspartner ist. Die einschlägigen Bestimmungen der Art. 10 § 1 I und 16 I Nr. 1 stehen deshalb unter dem Vorbehalt von Beschlüssen der Gemeinschaft.

[73] Der finanzielle Beitrag der EG beläuft sich auf 4 Mio. ECU in den ersten drei Jahren mit der Maßgabe, daß im ersten Jahr 1 Mio., in den beiden folgenden Jahren je 1,5 Mio. ECU zu zahlen sind. Danach soll eine Überprüfung dieses Beitrags erfolgen. Die Arbeiten des Vorbereitungsausschusses werden durch getrennte, von der EG-Kommission zur Verfügung gestellte Sondermittel finanziert werden.

Die Formulierungen in Art. 10 § 1 I »vorbehaltlich eines etwaigen Beschlusses der Gemeinschaft« und Art. 16 I Nr. 1 »vorbehaltlich eines Beschlusses dieser Gemeinschaft« deuten darauf hin, daß ein bloßer Ratsbeschluß, zehn Mitglieder in den Stiftungsrat zu entsenden und Finanzmittel für die Stiftung zur Verfügung zu stellen, ausreicht, um die genannten Bestimmungen realisieren zu können. Aus dem bloßen Text des Gründungsvertrages folgt jedenfalls nicht, daß ein Abkommen zwischen der Gemeinschaft und der Stiftung notwendig ist, um die genannten Bestimmungen des Gründungsvertrages wirksam werden zu lassen. Ein solcher Vertrag könnte ohnehin kaum unmittelbar auf Art. 228 EWGV gestützt werden, da der Gemeinschaft zumindest teilweise die Kompetenzen auf dem Gebiet der Stiftungsaufgaben fehlen. Auf der anderen Seite dürfte auch ein Beschluß gemäß § 235 EWGV allein nicht ausreichen; denn Inhalt eines Beschlusses könnte nur die einseitige Verpflichtung der Gemeinschaft zur Realisierung der genannten Bestimmungen des Gründungsvertrags sein. Eine einseitige Erklärung würde aber nicht ausreichen, zumal sie einseitig widerrufen werden könnte. Da die EG selbst nicht Partner des Gründungsvertrages der Stiftung ist, was trotz teilweise fehlender Kompetenzen möglich wäre [74], sie durch diesen Vertrag aber vorbehaltlich ihrer Zustimmung berechtigt und verpflichtet wird, kann sie Rechte und Verpflichtungen gegenüber der Stiftung nur durch Übereinkunft mit der Stiftung begründen. Rechtliche Bedenken aus dem Gemeinschaftsrecht gegen eine solche Übereinkunft bestehen nicht, weil es sich nur um eine rechtstechnische Frage handelt, ob die Gemeinschaft durch ein »gemischtes Abkommen« mit den Gründungsstaaten [75] oder durch eine bilaterale Übereinkunft mit der Stiftung verpflichtet oder berechtigt wird. Die gemeinschaftsrechtliche Kompetenz folgt aus Art. 235 EWGV.

Ein Abkommen der Gemeinschaft mit der Europäischen Stiftung ist erst möglich, wenn die Stiftung existent geworden ist, also gemäß Art. 26 § 3 erst mit der ersten Tagung des Stiftungsrats. Verhandlungen zwischen der Gemeinschaft und der Stiftung können also erst nach diesem Zeitpunkt beginnen. Trotzdem können Verhandlungsrichtlinien schon jetzt aufgestellt werden. Der EG-Rat hat demgemäß durch Beschluß, der im – am 25. März 1982 eingeleiteten und am 31. März abgeschlossenen – schriftlichen Verfahren genehmigt wurde, die erforderlichen Voraussetzungen für die Realisierung der Art. 10 § 1 I und 16 I Nr. 1 des Gründungsabkommens geschaffen, indem er an die EG-Kommission das Mandat erteilte, nach Gründung der Stiftung mit dieser einen Schriftwechsel zu führen.

Inhalt dieses Ratsbeschlusses vom 25./31. März ist die Ermächtigung der EG-Kommission zu Verhandlungen über einen Briefwechsel zwischen der

[74] Sog. »gemischtes Abkommen«. Vgl. *Schweitzer/Hummer* (Anm. 21), S. 137.

[75] *Schweitzer/Hummer* (Anm. 21), S. 137, begründen die Zulässigkeit solcher »gemischter Abkommen« »aus praktischen, rechtspolitischen und rechtlichen Bedürfnissen«.

Europäischen Wirtschaftsgemeinschaft und der Europäischen Stiftung entsprechend dem Vorschlag des Vorsitzenden der EG-Kommission vom 17. März 1982. Im einzigen Artikel des Beschlusses wird diese Ermächtigung in der Form ausgesprochen, daß zugleich der Entwurf eines Schreibens an die Europäische Stiftung als Anlage beigefügt wird, aus dem sich ergibt, daß die Gemeinschaft von der Möglichkeit des Gründungsabkommens für die Stiftung Gebrauch zu machen gedenkt, zehn Mitglieder für den Stiftungsrat zu bestellen und einen Finanzbeitrag von 4 Mio. ECU zu gewähren. Der Entwurf enthält auch die Einverständniserklärung der Stiftung, daß die Verwendung der Gemeinschaftsmittel entsprechend den für den Gesamthaushalt der Gemeinschaften geltenden Bestimmungen von der Kommission und dem Rechnungshof geprüft werden kann. Gemäß dem Entwurf wird der Präsident des Stiftungsrats um die Bestätigung seines Einverständnisses mit dem Inhalt des Schreibens gebeten.

Nicht ausdrücklich geregelt bleibt die Frage, wie der Präsident des Stiftungsrates, mit dem der Briefwechsel geführt werden soll, gewählt werden kann, wenn zwei Viertel der Ratsmitglieder noch nicht bestellt sind, weil deren Bestellung überhaupt erst nach der Vereinbarung mit der EG möglich ist. Die von der EG zu benennenden Mitglieder können nämlich vor dem Briefwechsel noch nicht Mitglieder des Stiftungsrates sein, und auch die kooptierten Mitglieder können erst in den Rat einziehen, wenn neben den 20 von den Staaten bestellten die von der EG benannten Ratsmitglieder im Amt sind.

Der erste Präsident des Stiftungsrates könnte zunächst nur mit der Mehrheit der Stimmen der von den Staaten bestellten Ratsmitglieder gewählt werden. Anderenfalls bliebe nur die Möglichkeit, daß die EG Mitglieder entsendet, die sich an der Kooptation des letzten Viertels der Mitglieder beteiligen, bevor die Vereinbarung zwischen der EG und der Stiftung über die Entsendung der von der EG benannten Mitglieder getroffen wurde. Um die von der EG entsandten Mitglieder rechtzeitig vor der Wahl des Stiftungsratspräsidenten zu bestellen, reicht der Beschluß vom 23. März 1982 jedenfalls nicht aus. Im übrigen bedarf der endgültige Abschluß der Übereinkunft zwischen der Gemeinschaft und der Stiftung nach erfolgreichem Abschluß der Verhandlungen zwischen dem Präsidenten der Kommission und dem Präsidenten des Stiftungsrates bei entsprechender Anwendung von Art. 228 EWGV eines weiteren Ratsbeschlusses[76]. Ein solcher Ratsbeschluß dürfte auch dann nicht entbehrlich sein, wenn das Abkommen zwischen EG und Stiftung dem in der Anlage zum Ratsbeschluß vom 25./31. März beigefügten Entwurf eines Schriftwechsels exakt entsprechen wird.

Außerdem beschloß der EG-Rat Protokollerklärungen, nach denen – einem Wunsch der deutschen Delegation entsprechend – ein Haushalts- und

[76] »Gedoppelter Ratsbeschluß«. Vgl. *Schweitzer/Hummer* (Anm. 21), S. 136.

Finanzausschuß eingesetzt werden kann und die Absicht der EG-Kommission begrüßt wird, den Vorbereitungsausschuß zu finanzieren.

V

Da bereits das Arbeitsdokument der Dienststellen der EG-Kommission vom 8. Mai 1978 in den Artikeln über die Ziele der Stiftung und die Zusammensetzung des Stiftungsrats Bestimmungen auf Gebieten enthielt, die in die Kompetenz der Länder fallen oder deren wesentliche Interessen berühren, hat das Auswärtige Amt der Ständigen Vertragskommission der Länder gemäß der Lindauer Absprache diesen Konventionsentwurf zur Prüfung übersandt. Zu einer Abstimmung der deutschen Haltung nahm an einer vom Auswärtigen Amt geladenen erweiterten Ressortbesprechung am 23. Mai 1978 auch ein Vertreter Nordrhein-Westfalens als Berichterstatter für die Ständige Vertragskommission der Länder teil. Durch ihren Vertreter äußerten die Länder starke Bedenken gegen Formulierungen der Stiftungsaufgaben. Wegen der Mitwirkung der Länder müßten im Stiftungsrat zwei Stimmen für die Bundesrepublik Deutschland zugestanden werden, von denen eine von den Ländern zu benennen sei. Auf einer Sitzung der Ständigen Vertragskommission am 29. Juni 1978 äußerte der Berichterstatter aus Nordrhein-Westfalen die Auffassung, daß das Übereinkommen über die Gründung der Europäischen Stiftung der Zustimmung der Länder nach Nr. 3 der Lindauer Absprache bedürfe, da Kompetenzen der Länder berührt werden. Die Länder würden dem Abkommen nur zustimmen, wenn sie am Stiftungsrat personell beteiligt werden. Im übrigen gehe die Ständige Vertragskommission davon aus, daß den Ländern aus der Stiftung keine finanziellen Belastungen erwachsen.

Auch die Ständige Konferenz der Kultusminister der Länder in der Bundesrepublik Deutschland wurde sofort nach der Vorlage des Arbeitsdokuments der Dienststellen der Kommission vom 8. Mai 1978 über die Angelegenheit informiert. Ein Vertreter des Generalsekretariats der Kultusministerkonferenz nahm an der Ressortbesprechung im Auswärtigen Amt am 23. Mai 1978 teil. Auf ihrer 188. Plenartagung am 8./9. Juni 1978 beschloß die Kultusministerkonferenz, daß die Länder auf jeden Fall in den Stiftungsgremien vertreten sein müßten. Die Ständige Vertragskommission der Länder und die Kultusministerkonferenz waren somit bereits in einem frühen Stadium mit der Angelegenheit befaßt.

In den dafür vorgesehenen Koordinierungsgremien der Länder sind die materiellen und organisatorischen Bestimmungen des Entwurfs eingehend beraten und die Bedingungen entwickelt worden, die Voraussetzung für die Zustimmung der Länder zu einem Gründungsvertrag für eine Europäische Stiftung sein müßten. Im Jahre 1981 ist auch der Beobachter der Länder bei den Europäischen Gemeinschaften entsprechend dem neuen Länderbeteiligungsverfahren bei Maßnahmen der EG auf Gebieten, die zur Gesetzgebungskompetenz der Länder gehören oder deren wesentliche Interessen berühren, informiert worden. Die Einschaltung des Beobachters der Länder bei der EG war deshalb notwendig, weil die EG, ohne Partner des Gründungsvertrages der Stiftung zu sein, sich an der Benennung der Mitglieder des Stiftungsrats beteiligen und zunächst die Hauptlast der Finanzierung der Stiftung tragen wird.

Nachdem seit November 1981 die Verhandlungen über die Errichtung der Stiftung Fortschritte machten, wurden die Ständige Vertragskommission der Länder, die Kultusministerkonferenz und der Beobachter der Länder bei den EG über alle Verhandlungsergebnisse des Ausschusses der Ständigen Vertreter und der Gruppe »Allgemeine Fragen« informiert. Angesichts der überraschend schnellen Einigung der Vertragspartner auf Grund der Initiative des Ratsvorsitzenden Tindemans konnten die Länder aber weder im Rahmen der Lindauer Absprache und der Kultusministerkonferenz ihr Einverständnis zu den endgültigen Texten rechtzeitig vor der Vertragsunterzeichnung erklären noch sich entsprechend dem Verfahren nach der EG-Erklärung des Bundeskanzlers vor der Beschlußfassung des EG-Rats abschließend abstimmen.

Wie in einer Stellungnahme der Ständigen Vertragskommission vom 14. Januar 1982 festgestellt wurde, sind die Länder vor allem an vier Punkten essentiell interessiert, nämlich an einer personellen Vertretung im Stiftungsrat, an einer genauen Begrenzung der Aufgaben der Stiftung in Art. 5 des Vertrages (sog. Indikative Liste), an der Gleichberechtigung der deutschen Sprache und am Ausschluß jeder finanziellen Belastung. Da diese Bedingungen zum Zeitpunkt der überraschenden Einigung mit den anderen Vertragspartnern erfüllt waren, hat die Bundesregierung ohne das erklärte Einverständnis der Länder die Vertragsdokumente am 29. März 1982 unterzeichnet. Denn bereits Ende Februar 1982 hatte die Ständige Vertragskommission der Länder dem Auswärtigen Amt mitgeteilt, daß die Wiederherstellung der sog. großen Lösung bei der Zusammensetzung des Stiftungsrats begrüßt werde und gegen die Indikative Liste keine grundsätzlichen Bedenken bestünden. Bei der Beschlußfassung auf der Tagung des Rats am 22. März 1982 hat der deutsche Vertreter auf die innerstaatliche Kompetenzzuweisung an die Länder hingewiesen, indem er die Erklärung abgab, daß die Bundesregierung alles versuchen werde, um die innerstaatlichen Voraussetzungen für die Annahme des Abkommens zu schaffen und insbesondere die Zustimmung der Länder zu erhalten. Damit waren die anderen Vertragspartner über die noch fehlende Zustimmung der Länder informiert.

Weil ihr kein endgültiger Vertragstext vorlag, hielt die Ständige Vertragskommission der Bundesländer noch am 18. März 1982 auf ihrer 171. Sitzung die Angelegenheit wegen des unfertigen Zustands des Vertragswerks nicht für entscheidungsreif. Als der endgültige Text mit Schreiben vom 23. März 1982 vom Auswärtigen Amt übermittelt wurde, blieb keine Zeit mehr, die Zustimmung der Länder bis zur Unterzeichnung am 29. März einzuholen.

Die 172. Tagung der Ständigen Vertragskommission der Länder fand erst am 22. April 1982 mit der Europäischen Stiftung als Tagesordnungspunkt[77] statt.

In eine zeitlich noch größere Bedrängnis gerieten Bund und Länder, um gemäß der EG-Erklärung des Bundeskanzlers zu verfahren. Der Beobachter der Länder bei den Europäischen Gemeinschaften war zwar über alle Vorgänge bis zur Beschlußfassung des EG-Rats Ende März 1982 gemäß § 85 a I GGO II durch den Bundesminister für Wirtschaft bzw. unmittelbar informiert, doch konnte eine eingehende und umfassende Darstellung des Länderstandpunkts nicht vor dem Hintergrund eines endgültigen Textes stattfinden. Der Ratsbeschluß vom 25./31. März 1982 basierte auf dem vom Präsidenten der Kommission vorgeschlagenen Text vom 17. März. Schon dieser Text konnte auf keinen Fall rechtzeitig nach dem dafür vorgesehenen Verfahren über den Beobachter der Länder den zuständigen Länderressorts zur Kenntnis gebracht und von den Ländern erörtert werden, um untereinander zu einer einheitlichen Haltung zu gelangen und den einvernehmlichen Standpunkt dem Bund mitzuteilen. Über den möglichen Inhalt eines EG-Ratsbeschlusses nach Art. 235 EWGV waren die Länder durch ihren Beobachter erst mit Schreiben vom 9. März informiert worden, als dieser ein »Non-Paper« der EG-Kommission vom 19. Februar 1982 über die am 10. Februar definierte Haltung der Kommission zum Abschluß eines Übereinkommens zwischen der Kommission und der Stiftung übermittelte. Hinsichtlich der Finanzierung der Stiftung ging dieses »Non-Paper«, das übrigens nur in französischer Sprache vorlag, noch von anderen Vorstellungen aus als der Ratsbeschluß vom 25./31. März. Eine Koordination der Länder über einen endgültigen Text eines Ratsbeschlusses nach dem dafür vorgesehenen Verfahren[78] konnte also angesichts der überraschenden Einigung der Vertragspartner über das Gründungsabkommen und des sofort gefaßten Ratsbeschlusses unter gar keinen Umständen stattfinden.

VI

Am Beispiel der Europäischen Stiftung lassen sich Licht- und Schattenseiten der Bund-Länder-Beziehungen sowohl im völkerrechtlichen als auch im europarechtlichen Bereich deutlich aufzeigen. Zunächst ist festzustellen, daß der Bund während des ganzen Verfahrens bemüht war, die Länder absprachegemäß unverzüglich zu informieren[79] und deren Wünsche gegenüber

---

[77] Neben dem Tagesordnungspunkt »Verschiedenes« war die Europäische Stiftung ursprünglich als einziger Tagesordnungspunkt vorgesehen.

[78] Beschluß der Ministerpräsidenten der Länder vom 5. 7. 1979, Text: *Morawitz* (Anm. 3), S. 107. Vgl. *Rudolf* (Anm. 8), S. 133 f.

[79] Neben die Information der Ständigen Vertragskommission der Länder und des Sekretariats der Kultusministerkonferenz tritt auch die Direktbenachrichtigung der Bonner Vertretungen der Länder durch das Auswärtige Amt. Zum Verhältnis Landesvertretungen und Ständiger Vertragskommission vgl. *H. Laufer/J. Wirth,* Die Landesvertretungen in der Bundesrepublik Deutschland (1974), S. 170 f.

den auswärtigen Verhandlungspartnern und im EG-Rat zu berücksichtigen. Die Zusammenarbeit zwischen Bund und Ländern verlief reibungslos sowohl hinsichtlich des völkerrechtlichen Sektors im Verhältnis des Auswärtigen Amts zur Ständigen Vertragskommission der Länder und zur Kultusministerkonferenz als auch bezüglich des europäischen Bereichs im Verhältnis des Bundesministers für Wirtschaft und anderer Stellen des Bundes zum Beobachter der Länder bei den Europäischen Gemeinschaften. Ebenso funktionierte die Zusammenarbeit der Länder auf den verschiedenen Ebenen und in den dafür vorgesehenen Koordinierungsgremien. Die Länder konnten sich vor allem auf vier essentielle Forderungen einigen, und der Bund konnte diese Forderungen gegenüber den neun Vertragspartnern durchsetzen.

Die vorgesehenen Verfahren funktionierten auch dann noch, als die Verhandlungen Anfang dieses Jahres in ihr entscheidendes Stadium traten. Jedenfalls konnten die Interessen der Länder gewahrt werden, weil die Länder im Bereich ihrer Kompetenzen ihren Willen zur Geltung bringen konnten. Erst die völlig überraschende Einigung der Vertragspartner am 22. März 1982 überforderte die vorgesehenen Verfahren, als sich der Bund über die fehlende Zustimmung der Länder zur Unterzeichnung des Gründungsabkommens im europäischen Interesse hinwegsetzte; denn noch am 18. März sah sich die Ständige Vertragskommission der Länder wegen des fehlenden verbindlichen Textes zu einer Stellungnahme nicht in der Lage.

1. Muß deshalb hinsichtlich des *völkerrechtlichen Sektors* das Urteil revidiert werden, daß sich das Verfahren nach der Lindauer Absprache eingespielt zu haben scheint[80], daß sich die Lindauer Absprache mehr oder weniger bewährt habe[81]? In ihrem Schlußbericht hatte die Enquete-Kommission Auswärtige Kulturpolitik zwar festgestellt, daß die Praxis nach der Lindauer Absprache im allgemeinen funktioniere[82], daß sie aber zeitraubend sei, was bei den bilateralen Beziehungen noch hingenommen werden könne, während der unverhältnismäßig hohe Zeitaufwand – erst recht auf längere Sicht – kaum den Voraussetzungen genüge, die eine sich verstärkende und vertiefende multilaterale Zusammenarbeit im kulturellen Bereich erfordere[83]. Das Beispiel der Europäischen Stiftung zeigt, daß die zeitaufwendige Koordination und Artikulation der Länder in der Tat dann nicht möglich ist, wenn der Bund im Interesse einer raschen Einigung mit den Vertragspartnern zu einer sofortigen Entscheidung gedrängt ist. Der Fall der Europäischen Stiftung zeigt aber auch, daß der Bund jedenfalls auf der völkerrechtlichen Ebene voll handlungsfähig blieb – allerdings auf Kosten der Länder, denen, wie die Ständige Vertragskommission noch vier Tage vor der plötzlichen Einigung der Ver-

---

[80] *W. Rudolf,* Archiv des Völkerrechts 13 (1966), S. 62.

[81] *W. Rudolf,* in: Festschrift für Hubert Armbruster (1976), S. 70.

[82] Schlußbericht der Enquete-Kommission Auswärtige Kulturpolitik, BT-Drucks. 7/4121, S. 39, Nr. 202.

[83] Schlußbericht der Enquete-Kommission Auswärtige Kulturpolitik, BT-Drucks. 7/4121, S. 39 f., Nr. 204. Vgl. *Rudolf* (Anm. 6), S. 13.

tragspartner ausdrücklich feststellte, wegen eines fehlenden Vertragstextentwurfs die erforderliche Koordination und Artikulation nicht möglich war. Da es bei völkerrechtlichen Abkommen auf die genaue Formulierung jeder einzelnen Bestimmung ankommt, ist die Auffassung der Ständigen Vertragskommission der Länder sicher nicht unbegründet gewesen.

Das Problem wird freilich dadurch entschärft, daß hinsichtlich des Gründungsabkommens kein die Kompetenzen der Länder berührendes *fait accompli* geschaffen wurde, weil das Abkommen der Ratifikation bedarf. Aus der Erklärung des deutschen Delegationsleiters anläßlich der Einigung über den Vertragstext am 22. März 1982, daß der Bund bemüht sein werde, insbesondere die Zustimmung der Länder zu dem Abkommen zu erhalten – womit auf völkerrechtlicher Ebene auf die innerstaatlichen Kompetenzabscheidungen hingewiesen wurde –, folgt, daß jedenfalls ohne Zustimmung der Länder der Abschluß des Abkommens nicht möglich sein wird. Rechtlich gesehen bleibt es den Ländern unbenommen, wegen der ihre Kompetenzen betreffenden Bestimmungen des Abkommens dieses anzunehmen oder abzulehnen. Dabei sollte nicht verkannt werden, daß eine Ablehnung zu Unstimmigkeiten des Bundes mit den anderen Vertragspartnern führen würde, die der Bund vor dem Hintergrund einer stagnierenden Europapolitik politisch kaum riskieren könnte. Die Zustimmung der Länder dürfte allerdings keine Schwierigkeiten bereiten, da deren Interessen im wesentlichen gewahrt sind, zumal auch der deutsche Vertreter im Vorbereitenden Ausschuß im Benehmen mit den Ländern tätig werden wird.

Was den völkerrechtlichen Sektor betrifft, ist in der gegenwärtigen Situation kein anderes Verfahren als das nach der Lindauer Absprache möglich. Es berücksichtigt die Kompetenzverteilung zwischen Bund und Ländern, ohne die verfassungsrechtliche Frage nach der Zuständigkeit zum Vertragsschluß bei Verträgen über Gegenstände der Landesgesetzgebung zu präjudizieren. Wegen der Zeitaufwendigkeit des Verfahrens muß der Bund bei raschen Entscheidungen auf der völkerrechtlichen Ebene abwägen, ob er die Einigung mit den Vertragspartnern auch ohne Zustimmung der Länder trifft und damit einen möglichen Konflikt mit den Ländern riskiert oder ob er ohne Zustimmung der Länder die Unterzeichnung verweigert und damit den Konflikt mit den ausländischen Vertragspartnern in Kauf nimmt. In Anbetracht der nicht möglichen Einigung auf einen gemeinsamen verfassungsrechtlichen Standpunkt muß der Bund von Fall zu Fall entscheiden, ob er sich zugunsten der Einigkeit mit den auswärtigen Vertragspartnern oder zugunsten der Länder erklären will. Bei der Europäischen Stiftung fiel die Entscheidung zugunsten der Einigung mit den Vertragspartnern leicht, weil die Länder prinzipiell mit dem Inhalt des Gründungsabkommens einverstanden waren, ohne daß sie ihr Einverständnis zu einem endgültigen Entwurf eines Abkommenstextes erklärt hatten.

2. Was den *europarechtlichen Sektor* betrifft, zeigt das Beispiel der Europäischen Stiftung, daß sich die Einrichtung des Ständigen Beobachters der Län-

der bei den Europäischen Gemeinschaften bewährt hat [84]. Der Beobachter der Länder war in jedem Stadium der Beratungen im integrierten und im koordinierten Bereich über den Sachstand informiert und hat seinerseits die vorgesehenen Anlaufstellen, vor allem auch die im konkreten Falle zuständige Kultusministerkonferenz unterrichtet. Jedenfalls bedarf es insoweit derzeit keiner weiteren Institutionalisierungen, auch keiner Ständigen Vertretungen der Länder bei den Europäischen Gemeinschaften, die ohnehin *de lege lata* nicht zulässig wären [85]. Das Verfahren, wie es nach der EG-Erklärung des Bundeskanzlers und dem Beschluß der Ministerpräsidentenkonferenz vom 5. Juli 1979 vorgesehen ist, funktionierte bis zu dem Augenblick der überraschenden Einigung der Vertragspartner am 22. März 1982. Vor dem Ratsbeschluß vom 25./31. März 1982 war eine vorherige Koordination und Artikulation der Länder über den genauen Inhalt dieses Beschlusses aus Zeitgründen unmöglich.

Geht man davon aus, daß es wegen des »gedoppelten« Beschlußverfahrens eines weiteren Beschlusses des EG-Rats gemäß Art. 235 EWGV bedarf, um den Schriftwechsel zwischen der Gemeinschaft und der Stiftung zu sanktionieren, dann ist die fehlende Koordination und Artikulation der Länder zu einem Ratsbeschluß vom 25./31. März 1982 praktisch ohne erhebliche Bedeutung, weil der EG-Rat durch seinen Beschluß nur der EG-Kommission das Mandat zu Verhandlungen mit der Stiftung entsprechend dem in der Anlage beigefügten Entwurf eines Schriftwechsels erteilt hat. Vor einem endgültigen Beschluß des EG-Rats über die Beteiligung an der Ernennung von Mitgliedern des Stiftungsrats und über die Finanzierung können die Länder noch ihre Haltung koordinieren und artikulieren. Gerade im integrierten Bereich liegt der endgültige Text noch nicht vor, auch wenn es unrealistisch ist anzunehmen, daß der dem Ratsbeschluß als Anlage beigefügte Entwurf eines Schriftwechsels zwischen der EG und der Stiftung noch verändert werden wird.

Die bisherige Praxis des Verfahrens gemäß der EG-Erklärung des Bundeskanzlers vom 19. September 1979 und dem Beschluß der Ministerpräsidentenkonferenz vom 5. Juli 1979 läßt nur den Schluß zu, daß dieses Verfahren bei überraschenden Entscheidungen des EG-Rates nicht ausreicht. Im übrigen liegen zu wenige Erfahrungen vor, um das Verfahren bereits abschließend beurteilen zu können [86]. Verfahrensmängel sind jedenfalls nicht offenkundig geworden, was möglicherweise darauf zurückzuführen ist, daß das Verfahren erst kurze Zeit erprobt wurde und die Länder nur in verhältnismä-

---

[84] Zum Beobachter der Länder vgl. *Birke* (Anm. 11), S. 53 f.; *K. Oberthür,* Integration 1978, S. 62; *Rudolf* (Anm. 8), S. 134; *Morawitz* (Anm. 3), S. 31 ff.

[85] *W. Strohmeyer,* Bayerische Verwaltungsblätter 1982, S. 207.

[86] Zum Verfahren bei EG-Vorlagen vgl. *D. Blumenwitz,* in: Gedächtnisschrift für Christoph Sasse, Bd. 1 (1981), S. 223 ff., 226 f.; *Rudolf* (Anm. 8), S. 131 ff.; *Morawitz* (Anm. 3), S. 69 ff.

ßig wenigen Fällen im Rahmen dieses Verfahrens beteiligt werden, weil Landeskompetenzen und -interessen nicht allzu häufig berührt sind.

Jedoch läßt sich schon jetzt feststellen, daß die Information des Beobachters der Länder durch den Bund und die Benachrichtigung der Anlaufstellen durch den Beobachter der Länder schnell und reibungslos funktionierten. Die Abstimmung der Länder untereinander bedarf aber eines längeren Zeitraums. Soweit ersichtlich, werden Entscheidungen des EG-Rats normalerweise so vorbereitet, daß für die Koordinierung und Artikulierung der Länder genügend Zeit bleibt. Bei äußerst kurzfristigen und überraschenden Entscheidungen des Rats – wie im Falle der Europa-Stiftung – ist die Koordinierung der Länder zeitlich nicht möglich. Das bedeutet, daß sich das deutsche Ratsmitglied in einem solchen Falle entscheiden muß, ob es ohne die Stellungnahme der Länder dem Ratsbeschluß zustimmen oder ob es im Interesse der Länder dem Ratsbeschluß seine Zustimmung verweigern will.

3. Eines Hinweises bedarf die *Information des Bundesrats,* in welchem die Länder an der Gesetzgebung und der Verwaltung des Bundes mitwirken, gemäß Art. 2 des Zustimmungsgesetzes zu den Römischen Verträgen[87]. Das Unterrichtungsverfahren wurde durch eine Ressortabsprache vom 12. Juni 1981 so gestaltet, daß sämtliche vom Sekretariat des Rats der EG eingehenden Vorschläge zu Verordnungen und Richtlinien des EG-Rats sowie Vorschläge zu offensichtlich in die innerdeutsche Gesetzgebung eingreifenden Entscheidungen des Rats vom Bundesministerium für Wirtschaft an den Chef des Bundeskanzleramtes zur Weiterleitung an den Bundestag und Bundesrat zu schicken sind. Das Zuleitungsschreiben muß dem in der EG-Vorlage vorgesehenen Titel der Maßnahme, das Datum der Übersendung an den Ratspräsidenten, einen Hinweis, ob die EG-Vorlage vor ihrer Verabschiedung dem Europäischen Parlament vorgelegt wird, und einen Hinweis auf den voraussichtlichen Zeitpunkt der Verabschiedung im Rat enthalten. Dabei ist zu vermerken, ob dieser Zeitpunkt noch nicht abzusehen, ob mit einer alsbaldigen Verabschiedung oder mit der unmittelbar bevorstehenden Beschlußfassung durch den Rat zu rechnen ist. Im Falle der Europäischen Stiftung reichte die Zeit zwischen der Einigung der Parteien am 22. März und der Einleitung des Beschlußverfahrens im Rat am 25. März nicht zur Unterrichtung über die unmittelbar bevorstehende Beschlußfassung aus.

VII

Die Errichtung der Europäischen Stiftung wirft schließlich die Frage auf, ob die Landtage am Verfahren, die innerstaatlichen Voraussetzungen für ei-

[87] *U. Oetting,* Bundestag und Bundesrat im Willensbildungsprozeß der Europäischen Gemeinschaften (1973), S. 41 ff., 106 ff.; *Ch. Sasse,* in: Der Bundesrat als Verfassungsorgan und politische Kraft (1974), S. 351 ff.; *Oberthür* (Anm. 84), S. 60; *Riegel* (Anm. 7), S. 246; *Blumenwitz* (Anm. 86), S. 217 f.; *Rudolf* (Anm. 8), S. 124 ff.; *Morawitz* (Anm. 3), S. 27 ff.

nen Abschluß des Gründungsabkommens zu schaffen, zu beteiligen sind. Eine solche Beteiligung ist immer dann erforderlich, wenn Gegenstände der *Gesetzgebung* des Landes betroffen sind[88]. Das dürfte bei der Europäischen Stiftung bisher nicht der Fall sein, weil das Gründungsabkommen, die Vereinbarung über die Einsetzung eines Vorbereitenden Ausschusses und die Schlußakte keine Regelungen betreffen, die nicht allein durch die Exekutive innerstaatlich vollzogen werden können. Dies gilt auch für die Erklärung in Anlage 2 der Schlußakte, binnen vier Monaten nach Vertragsunterzeichnung bestimmte Fragen der Privilegien und Immunitäten zu prüfen. Da es lediglich um eine »Prüfung« geht, also noch keine konkreten Normen über Exemtionen von landesrechtlichen Bestimmungen und über Befreiungen von Steuern, die der Landesgesetzgebung unterliegen, geschaffen wurden, bedarf es der Zustimmung der Landtage in den dafür verfassungsrechtlich vorgesehenen Formen nicht.

Im Bundesrat sind die den Landesregierungen angehörenden Mitglieder nicht an Weisungen der Landtage gebunden[89], so daß von den Landtagen kein Einfluß auf den Abschluß des Gründungsvertrages der Europäischen Stiftung ausgeübt werden kann, wenn man von den allgemeinen Kontrollmöglichkeiten des Parlaments gegenüber der Regierung absieht[90].

Eine andere Frage ist, ob Unterrichtungspflichten der Landesregierungen gegenüber den Landtagen oder deren Ausschüssen bestehen. Die Antwort hängt von der Regelung in den einzelnen Ländern ab. Am weitesten geht die Informationsbereitschaft der Landesregierung von Rheinland-Pfalz. Auf Grund eines Briefwechsels zwischen dem Ministerpräsidenten und dem Landtagspräsidenten vom 6./11. Juli 1977 wird der jeweils zuständige Ausschuß des Landtags über Bundesratssachen in den Fällen von Verfassungsänderungen, mit denen Kompetenzen der Länder auf den Bund oder vom Bund auf die Länder verlagert werden sollen, und sonstigen Verfassungsänderungen, Gesetzen oder Gesetzesänderungen von herausragender landespolitischer Bedeutung, die wesentliche Interessen des Landes unmittelbar berühren, informiert[91]. Durch einen Schriftwechsel zwischen Ministerpräsident und Landtagspräsident vom 20. Juni/24. September 1980 wurde diese Informationsbereitschaft auf Beschlüsse der Fachministerkonferenzen ausgedehnt, soweit sie sich auf Gegenstände beziehen, die der verfassungsrechtlichen Entschei-

---

[88] Vgl. *Rudolf* (Anm. 81), S. 68 f.

[89] *J. Linck,* DVBl. 1974, S. 861, 863, mit weiteren Literaturangaben. Vgl. auch den Schlußbericht der Enquete-Kommission Verfassungsreform, BT- Drucks. 7/5924, S. 101 f.

[90] *Linck* (Anm. 89), S. 863 ff., mit weiteren Literaturangaben. Vgl. auch *P. G. Schneider*, Beteiligung der Landesparlamente beim Zustandekommen von Staatsverträgen und Verwaltungsabkommen der Bundesländer (Diss. Mainz 1978), S. 153 ff.; *J. Linck,* Die Öffentliche Verwaltung 1979, S. 116 ff.; *M. Hofmann*, Die Bundesrepublik Deutschland – ein gouvernementaler Bundesstaat? (Diss. Mainz 1980), S. 77 ff.

[91] LT-Drucks. 8/2307.

dungskompetenz des Landtags unterliegen[92]. Diese Unterrichtung erstreckt sich auf Beschlüsse, die dem Landtag empfehlen, bestimmte Gesetze einzubringen, bestimmte Staatsverträge abzuschließen und bestimmte sonstige Vorhaben durchzuführen, deren Verwirklichung im Landeshaushalt zu Mehr- oder Minderausgaben von jährlich über 1 Mio. DM führen würde. In der Antwort des Ministeriums der Justiz vom 22. Dezember 1980 auf eine Große Anfrage der Fraktion der CDU über Europapolitik wird festgestellt, daß der Landtag von Rheinland-Pfalz derzeit durch die Landesregierung über wichtige EG-Vorhaben unterrichtet wird, und zwar während der Kommissionsphase, soweit ein ausdrücklicher Beschluß einer Fachministerkonferenz zu dem EG-Vorhaben gefaßt wird, und nach Beginn der Ratsphase, sobald ein entsprechender Entwurf im Bundesrat ansteht[93]. Damit wird bei EG-Vorhaben eine über die Regelung in anderen Bundesländern hinausgehende Unterrichtung des Landtags von Rheinland-Pfalz eingeräumt. Darüber hinaus erklärte sich die Landesregierung zu einer weiteren Prüfung bereit, ob und wieweit, insbesondere bei wichtigen EG-Vorhaben, die ganz oder in einzelnen Bestimmungen in die ausschließliche Gesetzgebungskompetenz der Länder fallen würden, eine zusätzliche Unterrichtung des Landtags vorgenommen werden kann. Die Möglichkeit einer vor der Beratung im Bundesrat liegenden Information im Einzelfall werde davon abhängen, ob sich das probeweise eingeführte Verfahren nach der EG-Erklärung des Bundeskanzlers und dem Beschluß der Ministerpräsidentenkonferenz vom 5. Juli 1979 als praktikabel erweise.

Eine Informationspflicht gegenüber dem Landtag über das Gründungsabkommen über die Europäische Stiftung und den Ratsbeschluß vom 25./31. März 1982 besteht auch in Rheinland-Pfalz nicht, da bisher keine Gesetzgebungskompetenzen des Landtags berührende Normen geschaffen wurden. Ob die erfolgte Prüfung der Vorrechte und Befreiungen eine Unterrichtungspflicht nahelegen könnte, hängt bei Zugrundelegung der genannten Briefwechsel zwischen Ministerpräsident und Landtagspräsident davon ab, ob es sich hier um eine bevorstehende Gesetzesänderung von herausragender landespolitischer Bedeutung handelt, die wesentliche Interessen des Landes unmittelbar berühren. Dies dürfte für die Befreiung der Stiftung, ihrer Organe und ihres Personals von den der Landesgesetzgebung unterliegenden Steuern zu verneinen sein. Die Landtage werden also erst im Zusammenhang mit der Erteilung des innerstaatlichen Rechtsanwendungsbefehls[94] für landesgesetzliche Kompetenzen betreffende Vorschriften eines noch auszuhandelnden

[92] LT-Drucks. 9/946.

[93] LT-Drucks. 9/1122, S. 6 f.

[94] An der Auffassung, daß der innerstaatliche Anwendungsbefehl auch einen Transformationseffekt hat, wird festgehalten. *W. Rudolf,* Völkerrecht und deutsches Recht (1967), S. 167 ff., 205 ff.

Abkommens über Privilegien und Immunitäten gemäß den landesverfassungsrechtlichen Bestimmungen beteiligt werden [95].

Die Tätigkeiten der Europäischen Stiftung werden den sog. »gesetzesfreien« Raum betreffen. Bei der Stiftungsarbeit wird es in erster Linie um die Entwicklung neuer Impulse und Projekte für schon vorhandene nationale, bilaterale und europäische Institutionen gehen [96]. Der Stiftung wird zunächst eine Anregungs- und Clearingfunktion zukommen [97]. Ob die Stiftung den hohen Anforderungen, die an sie bei geringen eigenen Mitteln gestellt werden, gerecht werden kann, wird nicht zuletzt von der personellen Besetzung des Stiftungsrats abhängen. Daß die Länder der Bundesrepublik Deutschland ein Mitglied im Stiftungsrat stellen werden, ist Ausdruck eines gewiß komplizierten, aber funktionierenden kooperativen Föderalismus. Europäische kulturelle Zusammenarbeit ist ohne die Länder nicht möglich – nicht nur wegen der innerstaatlichen Kompetenzverteilung, sondern auch wegen der Verteilung von fachlicher Kompetenz in einer in Vergangenheit und Gegenwart dezentralisierten reichen und lebendigen Kulturlandschaft.

---

[95] Vgl. *F. Regehr,* Die völkerrechtliche Vertragspraxis in der Bundesrepublik Deutschland (1974), S. 202 ff.; *Rudolf* (Anm. 81), S. 68 f.; *Schneider* (Anm. 90), S. 32 ff.

[96] Art. 3 II des Gründungsvertrages.

[97] *Picht* (Anm. 14), S. 113 ff.

Creativity and Objectivity in International Tribunals

OSCAR SCHACHTER

In view of Professor *Mosler*'s distinguished role as an international judge and his many contributions to better understanding of the judicial process in international affairs, it seems appropriate to offer some observations on the perplexing problems of creativity and objectivity in international tribunals.

»The act of the Court is a creative act in spite of our conspiracy to represent it as something else«[1]. This comment of *Brierly* reminds us how reluctant international lawyers have been to recognize that judges perform a creative role through their decisions and pronouncements. This is not surprising in view of the dominance of positivist doctrine and the fragility of institutions for judicial settlement.

In actuality, international lawyers generally welcome the decisions of judicial tribunals, especially those of the International Court of Justice, as the touchstone of law. Many have quoted with approval the oracular-sounding dictum of Justice Cardozo in a U.S. Supreme Court case of 1933: »International law or the law that governs between states, has at times, like the common law within states, a twilight existence during which it is hardly distinguishable from morality or justice, till at length the *imprimatur* of a court attests to its jural quality«[2]. The *imprimatur* of a court has a nearly irresistible attraction for lawyers, especially those raised in the tradition of the common law. The fact that Art. 38 of the Statute of the Court refers to judicial decisions only as a »subsidiary means« of determining rules of law has had little effect on the deference shown to the judgments and opinions of the International Court and of international arbitral bodies. The judicial decisions of national courts though more cautiously treated, have also been widely used as highly persuasive declarations of international law. This has been as true among »civil law« jurists as among common lawyers. The great value

---

[1] *J. L. Brierly,* The Judicial Settlement of International Disputes, in: The Basis of Obligation in International Law 98 (1958).

[2] *New Jersey* v. *Delaware*, 291 U.S. 383 (1933).

attached to the collections of case-law by *Politis, de La Pradelle* and *Viktor Bruns*, as well as to the Annual Digest of *Lauterpacht* and more recently to the U.N. Reports of International Arbitral Awards is indicative of the lawyers attitude to judicial decisions. Such decisions are treated as authoritative evidence of law to a greater degree than are the primary sources of custom and treaty.

Why is this so? For one thing, judicial decisions often produce a measure of certainty where previously confusion and obscurity existed. They usually invoke principles that have long been familiar to the profession. Their style in many cases is magisterial, pronouncing their judgments with assurance and certainty. The very fact that there have been so few decisions of the International Court of Justice – about two a year on the average since 1948 – has enhanced their interest. Each one is closely examined, dissected, quoted and pondered for its implications. The individual opinions of judges, increasingly used, have added to this process by their more thorough and detailed reasoning. Sharp divisions in the Court (for example, the 1966 judgment on *South West Africa*) reduce the authority of such decisions while those that are unanimous or nearly so are rarely questioned. It is not irrelevant that the main »audience« for the decisions – those who pass upon their legal significance – are members of a closely-linked professional group, »the invisible college« of international lawyers which includes most of the judges. Many of the judges are highly esteemed for their learning and judicial wisdom. Their views are almost always highly valued.

Exceptions do occur. The 1966 judgment in the case brought by Ethiopia and Liberia against South Africa is a notorious example. In that case an evenly divided Court (the President having 2 votes) decided on a technical issue of *locus standi* to avoid a judgment on the merits of South Africa's conformity to its mandate[3]. That judgment was widely criticized, and the Court later in an advisory opinion ruled that South Africa did in fact violate its legal obligations[4].

The Court cases on *South West Africa* raised more sharply than most cases the issue of objectivity – to what extent is the Court governed by »political« or more generally by »subjective« rather than »objective« factors? These issues can be found in a great many cases. Does the inevitable process of selectivity and abstraction mean that the reasoning of the Court involves judgments of international relations, social needs, or political philosophies? These are questions that have been discussed at length and often with subtlety by international lawyers and notably by many of the great judges themselves. Many of them recognize the creative element in the judicial function; they do not think of themselves as phonographs playing the recording of custom and treaty. When we read opinions closely we become aware of the dis-

---

[3] *South West Africa, Second Phase,* I.C.J. Reports 1966, p. 6.

[4] *Advisory Opinion on Namibia,* I.C.J. Reports 1971, p. 16.

cretionary elements in the art of judging: the selection of »relevant« facts, the need to give specific meaning to broad, undefined concepts, the subtle process of generalizing from past cases, the uses of analogy and metaphor, the inevitable choices between competing rules and principles. As we know these elements can be found as well in domestic courts, most notably in constitutional litigation. But they are more conspicuous and troublesome in international law. One reason is the large area of uncertainty in international law, especially customary law, and not infrequently on the meaning of basic treaty law. *Sir Hersch Lauterpacht*, in 1955 after some years on the International Law Commission, and for many years the editor of *Oppenheim,* the leading English-language treatise, was moved to say:

> »Once we approach at close quarters practically any branch of international law we are driven, amidst some feeling of incredulity, to the conclusion that although there is as a rule a consensus of opinion on one broad principle – even this may be an overestimate in some cases – there is no semblance of agreement in relation to specific rules and problems«[5].

The sweeping character of this statement may be qualified today especially in the light of the several codification treaties concluded since 1955. But it still remains pertinent to many issues of international law.

A second reason for underlining the differences between judicial discretion in international and domestic courts relates to the absence of a legislative organ. This difference interestingly has given rise to two opposing viewpoints. One holds that the domestic court can leave it to its legislature to remedy gaps or obscurities in the law whereas the international community

> »is peculiarly dependent on its international tribunals for the development and clarification of the law, and for lending to it an authority more substantial and less precarious than can be drawn from often divergent or uncertain practices of states, or even from the opinions of individual publicists whatever their repute«[6].

An opposing view is that the absence of an international legislative organ and the uncertainties of the treaty process require more restraint by international tribunals since they are not subject to the review and corrective processes of legislative bodies. A court which moves too far and too fast, which appears to create new law and obligations, will in the absence of a legislature be more powerful than states would wish it to be. Its fragile and precarious jurisdiction dependent on the continued consent of states would then be

---

[5] *H. Lauterpacht,* Codification and Development of International Law, 49 A.J.I.L. 16, 17 (1955).

[6] *G. Fitzmaurice,* Hersch Lauterpacht – the Scholar as Judge, 37 British Year Book of International Law (B.Y.I.L.) 14–15 (1961).

placed at great risk. One might add to these cautionary admonitions doubts whether judges are qualified to adapt law to new needs, particularly when no legislature exists to correct judicial errors in the light of later knowledge. *Brierly* said judges are the last people who ought to have the duty of changing the law thrust upon them, for their whole training disqualifies them from deciding disputes by a balancing of interests[7].

Discussions of judicial creativity – a topic endlessly fascinating to judges and scholars – are often concluded with Polonius-like homilies: the judge must recognize the need for development but also the importance of stability and certainty; he »must hold a just middle way between excess of valour and excess of caution«; he has a duty to decide on the basis of general principles applicable to all but he should also defer to the particular configuration of facts in the case at hand. He is advised to »balance« the competing conceptions of law and liberty (sovereignty): thus legal norms should not be extended by analogy or abstraction to restrict the freedom of state action; on the other hand, the scope of sovereignty depends on law, hence legal obligation cannot be determined by emphasis on sovereign rights. Admonitions of this kind do not seem especially helpful for deciding particular controversies, since they do not – cannot – tell the judge how much weight to give to one or the other of the »polarities« in the given circumstances. Their value lies rather in emphasizing the dialectical element in the judicial function and the danger of allowing one-sided abstractions to determine specific cases. They point to the necessity of considering opposing principles together in order to reach a just and rational solution. A felicitous expression of this idea can be found in the following comment by an American scholar discussing the opposing principles in U.S. constitutional law:

> »These abstractions, arrayed in intransigent hostility like robot sentinels facing each other across a border, can become useful guardians on either hand in the climb to truth if they can be made to march together. Somehow the life blood of the concrete problem tempers the mechanical arrogance of abstractions« [8].

International judges, even more than constitutional arbiters, have good reason to »temper the mechanical arrogance of abstractions with the life blood of the concrete problem«. By concentrating its decision on the particular circumstances of a case, and »not yielding to the charm of generalizing abstraction«, the international tribunal is less likely to be perceived as excessively creative. It does not then seem to be stretching the »consent« of the parties (whether that consent was *ad hoc* or given for a broad category of

---

[7] *Brierly, op. cit.*(note 1) at p. 105 note 34 . For similar views sceptical of the ability of courts to deal with changing conditions and conflicts of interest, see *J. Stone*, Non-Liquet and the Function of Law in the International Community, 35 B.Y.I.L. 149 (1959).

[8] *P. Freund*, 69 Harv.L.Rev. 803 (1965).

cases). Nor does it threaten the interests of other states with »judicial legislation« that limits their sovereign rights. Understandably, therefore, international judicial decisions are often closely fitted to the individual configurations of facts. At the same time, the felt necessity for authoritative and persuasive reasons demands the formulation of principles and often the pairing and weighing of competing norms. Two of the notable judgments of the International Court – those in the *Corfu Channel* Case[9] and the *Anglo-Norwegian* Case[10] – exhibit the way in which principles are linked to the detailed and special facts of each dispute. The judgments are premised on those facts. Yet both judgments are also »principled« in that they rely on a variety of general norms and concepts to reach their individualized judgments. Both judgments thus have become important contributions to clarification and development of the law.

We return to our earlier query. If the judges must select and generalize, if they need to weigh and balance conflicting rules, to decide what facts are relevant, what principles are to prevail, whether to apply deductive logic rather than inductively rely on precedent, is it not obvious that their personal attitudes, political outlook or ideological affinities will play an important, if not decisive role? The answer is of course affirmative[11], but in giving that answer it is important to draw distinctions both of kind and of degree.

It is difficult to deny, in view of the record, that individual judges are, by and large, influenced by the views of their own governments. This has been most evident – and is indeed taken for granted – when a judge is a national of a state party to the case. Rarely has a judge differed from a stand taken by his government in litigation[12]. The provision for *ad hoc* judges for parties who do not have a judge of their own nationality is, at least in part, recognition of that expected partisanship. It is also generally assumed that most judges will share the views of their governments on issues in which that government has taken a position or has an obvious interest even though it is not a party to a case. Predictions in these cases may be hazardous for some judges but in broad terms, the distribution of votes in many cases can often be related to political or ideological groupings. Some caution is called for when the technical or jurisprudential question before the Court does not fall easily into the pattern drawn by different governmental policies. In some of these cases, judicial opinions cannot readily be linked to state interests.

---

9 I.C.J. Reports 1949, p. 4.

10 I.C.J. Reports, 1951, p. 8.

11 Professor *Mosler* has observed »Where there is no common agreement as to the substantive rules and especially where there is no complete agreement as to the sources of general international law, the subjective views of the individual judges assume increased importance«, *Mosler*, The International Society as a Legal Community, 140 Recueil des Cours (1974), p. 293.

12 *I. R. Suh,* Voting Behavior of National Judges in International Courts, 63 A.J.I.L. 224 (1969).

The fact that judges often reflect particular state interests is of course at variance with the ideal of objectivity of the judicial function. Yet it is not unreasonable to regard the reflection of national or group interests as appropriate and advantageous for an international court in a divided and heterogeneous world. The Statute of the Court expressly calls for heterogeneity among the judges and it is of course acknowledged in practice that the different political as well as regional groups should be given due representation. There is inevitably an irreducible political element in the Court, an element that means that different special interests – national, regional, ideological, philosophical – will have their champions on the bench. Their inarticulate major premises may be more important for their decisions than either deductive reasoning from general principles or inductive reasoning from precedent and practice.

Although international lawyers, by and large, deplore the influence of »non-legal« political factors (some attribute the reluctance of states to use the Court to that factor), the governments show little disposition to make the Court more »objective«. Indeed, the selection of judges has probably become more politicized in recent years than it was some decades ago[13]. The distribution of seats in the International Court of Justice (mainly arranged by gentlemen's agreements among the different »political-regional groups«) reflects the political orientation of states more than the abstract representational criteria of Art. 9 of the Statute (i.e. »the main forms of civilization and the principal legal systems of the world«). The political element is even more evident in the background of the individuals elected to the Court. More and more come from the ranks of governmental officials. Those who do not have generally served as representatives of their governments in United Nations meetings or other international conferences[14]. This holds true for nearly all the judges; it runs across political and regional lines. One consequence is that many of the judges find themselves sitting in cases in which they previously had partisan roles or were otherwise active when the disputes were in U.N. bodies.

The Court, faced with this phenomenon, has not considered it necessary that a judge refrain from sitting on a case because he took part in earlier U.N. proceedings concerned with that case – not even if he played a leading partisan role at that time. Art. 17 of the Statute, which bars a judge from participating in the decision of any case in which he had previously served as counsel or advocate for one of the parties »or in any other capacity«, has been construed in practice as not applicable to partisan activity as delegates in the U.N. or similar bodies.

---

[13] *Mosler, op. cit.* (note 11) at p. 293.

[14] *S. Rosenne,* The Composition of the Court, in: The Future of the International Court of Justice (L. Gross ed., 1976), p. 377, 388 ff.

These facts about the composition of the Court can only reinforce the impression drawn from the decisions and the votes of the judges that politics and governmental interests have a significant impact on the judicial decisions. This has been decried as a lack of the requisite judicial objectivity and as a good reason why states shun the Court. But one may also see the political element in more positive terms. First, the governmental experience of the judges enables them to look behind the legal abstractions to the underlying interests and attitudes that may be affected by the potential decisions. If one acknowledges that law is not autonomous, there is good reason for judges to take account of consequences in the real world. That they do so more readily in terms of their own particular governmental interests is a fact of life[15]. There is a second, less parochial, consequence of political experience. That experience, especially when obtained in international conferences, often facilitates the task of harmonizing competing views so as to achieve general agreement or at least a majority consensus. For one thing, the judges realize that unlike academic scholars, it is their task to make collective decisions; their own opinion, no matter how well reasoned, is not in itself sufficient. Prior governmental experience contributes a sophistication about the ways of achieving consensus. That task may sometimes call for recognition of a common interest; the preservation of diplomatic inviolability, as in the *Hostages* Case[16], is a good example. In other cases, the »common interest« may be difficult to perceive amidst conflicting demands. Then the experienced judge will seek not optimal solutions, but those that are seen to do least harm to contending principles and interests. I do not mean to suggest that all judicial decisions should reflect an outcome that completely harmonizes or accommodates clashing positions. There are nearly always winners and losers; judicial settlement is significantly different in its function from conciliation. It remains, nonetheless, true that the art of judicial decision in an international court for a heterogeneous community generally involves a weighing and balancing of competing principles. The court should seek to give as much effect to each of these principles (assuming their authoritative character as law) as is consonant with the primary task of deciding the particular case.

But we must also bear in mind that the judicial task of »harmonization« is different from political negotiation and legislation. The judges must find the justification for their decisions in authoritative legal norms, the *lex lata*, no matter how cognizant they are of consequences and conflicting interests. The fact that they face the kinds of choices I have underlined above does not mean that they can make such choices as naked choices of policy. They must give reasons that are grounded in law – that is, in principles and processes

---

[15] *Ibid.* at p. 389.

[16] Case *Concerning the United States Diplomatic and Consular Staff in Tehran*, I.C.J. Reports 1980, p. 3.

that have been accepted as law generally by the international community or by the relevant group of states. They will therefore often present their conclusions in a way that obscures the act of choice and stresses the »objective« character of their decision. *Brierly*'s comment – quoted earlier – about »our conspiracy« to represent a decision of the Court »as something other than the creative act it truly is« recognizes how much we lawyers like to have the court appear as if it did nothing but declare positive law based on verifiable facts of practice and agreement. But it is futile to expect that this desire and the »conspiracy« based on it will effectively hide the reality of choice and »creativity« from the governments.

Thus when some judges stress that they wish only to state the law as it is – and »not to arrogate a creative power which does not belong« to the court[17] – and when they reach this conclusion in order to limit the jurisdiction of the court, one cannot help but observe that their conclusion generally accords with the attitude of their own government (or a closely allied government), an attitude that strongly favors sovereignty and the right of each state to interpret the scope of its obligations. The coincidence of the judge's view of »objective« law and his government's policy does not mean they are necessarily wrong; it does, however, lead to some scepticism about their emphatic claim of objectivity. Such scepticism is increased when many other judges find substantial legal basis for a contrary opinion.

Despite the desire of some scholars that the Court revert to positive law in the »classic sense«, it seems clear when we look at the decisions that strict positivism is a chimerical goal. The necessities of choice, the normative complexities of the law, the inadequacies of precedent and the great diversity of facts make it impossible to assume that the »objective« facts of prior state practice or the words of treaties will be enough for decisions when states differ substantially on the law. However, this conclusion does not in itself imply that the Court should always be expansive or »developmental« in its decisions. On occasion they may do so successfully. The *Corfu Channel* Case[18] and the *Advisory Opinion on Reparation for Injuries Suffered in the Service of the U.N.*[19] are examples of such development where the Court looked not to state practice in the strict sense but searched for underlying principles that reflected the purposes of states in adopting the U.N. Charter and the new post-1945 international law. However, these decisions, it must be noted, met the test of general acceptability by the community of states. If that test were not met, the decision would be seen as a partisan or »political« act and its effect on future law would be dubious.

I would suggest that it is not judicial wisdom to exhaust fully the processes of judicial reasoning in every case or to express principles »in their full

---

[17] I.C.J. Reports, 1974, at p. 37 (opinion of Judge Ignacio-Pinto).
[18] I.C.J. Reports, 1949, p. 4.
[19] I.C.J. Reports, 1949, p. 174.

generality« (*Lauterpacht*). »Individualization« rather than generalization will often be the appropriate way for an international court to cope with the inevitable choices it must make. As long as the judicial »web is woven in small knots« (as *Cardozo* commented)[20], international judges can meet changing conditions without seeming to arrogate to themselves an authority they have not been given. But those »small knots« can strengthen the fabric of law, probably more so than attempting to stretch that fabric through undue generalization. The decisions once made and widely accepted become elements in the evolution of law. They may be seen at first as limited in scope, as »molecular«, but they are nonetheless treated as authoritative and they stand ready to be invoked and relied on in future situations where one party or another considers their reasoning as helpful to a new problem. They mediate the underlying positivist data presented to them in the claims of the parties and produce the reasoned and principled decisions that now become accepted as the expression of authoritative law.

[20] *Benjamin N. Cardozo,* The Nature of the Judicial Process 69 (1921).

International Organizations as Members of Other International Organizations

HENRY G. SCHERMERS

A. The EEC as a Model

Gradually, an increasing number of public functions are being taken over by international organizations. There are many reasons for this. Some are financial: the launching of a space vehicle being too expensive for a small country, several countries together may do it in an international organization. Another reason for delegating public tasks to an international organization may be that no State can singly perform that task, such as making the rules for air navigation above the high seas, and possibly the exploitation of the ocean floor. Then there are tasks which States could perform but which need so much international coordination that international organizations can perform them better. Development aid is probably a good example of this.

The more international organizations participate in international administration, the stronger the need to bind them to rules of international law and to have them participate in international relations.

The European Economic Community (EEC) is regarded by some as a first step towards a European federal State. It, then, is an intermediary phase for a stronger union, important for Europe but not of general interest to the international lawyer. One may strongly doubt whether this view is correct. First, it is not at all sure that the EEC will become a European federal State in the forseeable future and, secondly, even if it may do so, then this does not necessarily mean that this is a unique phenomenon. On the contrary, it seems likely that other groups of States will pass through similar phases when they find it necessary to form larger units. Therefore, we submit that the EEC is a forerunner and that other organizations of States will in future meet many of the same problems which the EEC meets now. That makes it worthwhile to consider the problems which have confronted the EEC in its functioning under international law. Once we have established that a body like the EEC can participate in international legal relations, we will be confronted with the question whether and under what conditions it can also become a member of an international organization.

B. International Organizations as Subjects of International Law

Traditionally, international law has often been seen as a legal system between States. Entities other than States could not be subjects of it. This view has never been universally accepted, but even if it may have had widespread support, it no longer applies to modern international law. In the international system of our time, those international organizations which have a legal identity independent of the States which have established them participate in much the same way as States[1]. Many international organizations have concluded treaties; some of them, and in particular the EEC, have done more and have even participated in other international organizations.

According to the EEC Treaty, the Community has a common outer tariff and exercises a common trade policy *vis-à-vis* third States. It concludes trade agreements. In the beginning some States, particularly in Eastern Europe, objected to the Community's participation in multilateral treaties, but the need for Community participation in the commodity agreements was so strong that the legal objections were set aside.

Nowadays, the EEC's competence to act under international law is generally accepted. Most States have made treaties with the Community; over a hundred of them have accredited diplomats to the Community. The power of the Community to operate under international law can no longer be seriously disputed. This means that other similar entities would also be able to participate in international relations, in the same way.

C. Division of Competence between Organization and Members

International treaty-making has two aspects. In the first place, it performs an international political role. By making a treaty with another State, a relationship is built which may alleviate international tensions and promote mutual good will. The making of a treaty with another State can be seen as a token of friendship. This aspect of treaty-making is part of international relations and comes under the competence of the Ministry of Foreign Affairs. As most

[1] See *Hermann Mosler,* The International Society as a Legal Community (1980), p. 50.

treaty-making is finally executed by civil servants of this Ministry, it may well be that the importance of this aspect is often overstressed.

In the second place, treaty-making plays a part in the legislative machinery. Often, rules of law cannot effectively operate when they are limited to the national territory of a State. They need to be extended to other countries and this can be done only through treaties. European road transport offers a good example of this. All States in the region considered it necessary to set a maximum limit on the time which a lorry driver may sit at the wheel of his vehicle. Continuous driving for too long a time creates a danger for road safety. Without some international agreement, however, a lorry driver could drive up to the maximum time in one country, cross the border and continue driving up to the maximum time set by the next country. In such a case, an international arrangement is needed in order to create one maximum driving time for a number of States. That one combined maximum driving time is what the European Road Transport Agreement intended to achieve.

Submitting that this second aspect of treaty-making, the legislative, is the most important one, the Commission of the European Communities has always taken the position that the legislator competent to make internal rules should also be the one who must make a treaty with third States. As the Community had made internal legislation on European road transport, it should also have been able to make the European Road Transport Agreement with other European States. The Council of the Communities, on the other hand, saw the European Road Transport Agreement as an instrument of foreign policy and for that reason submitted that only the Member States were competent to enter into this agreement. According to the Council's submission, which was most strongly defended by France, only trade agreements could be entered into by the Community. From the fact that the EEC Treaty contained a provision for these agreements only (Art. 113 of the EEC Treaty), it was derived that other international agreements were outside the competence of the Community.

The Court of Justice of the European Communities, however, supported the position of the Commission, at first with some restriction[2], but later in its entirety. From the legal personality of the Community (Art. 210) it derived the Community's capacity to enter into international commitments. It considered the Community competent to enter into international agreements in all fields where it has the competence to legislate internally[3].

After some years, the Council has in fact accepted the Community's power to enter into treaties other than trade agreements, but it defines these fields restrictively.

---

[2] *ERTA* case (22/70), 31 March 1971, [1971] ECR 263.

[3] *Laying-up Fund Opinion* (1/76), 26 April 1977, [1977] ECR 741.

D. Mixed Agreements

1. Definition

Not all treaties are confirmed to one specific subject. Often, several matters are regulated in one treaty. It may then happen that some of these matters fall within the competence of the Communtiy while others are part of the exclusive domain of the Member States. Who should then be the party to the treaty?

For this situation a special construction was developed: the mixed agreement. A mixed agreement is a treaty to which both the Community and its individual Member States are parties. Ideally, the provisions of such a treaty should be split up, and the Community should become a party to that part of the treaty which covers subjects in respect of which the Community is competent, whilst the Member States should be parties to the parts of the treaty regulating matters under the competence of the Member States. In most cases, however, such a division of treaty obligations proves to be impossible, mainly for two reasons: a) many treaties do not permit adherence to only one part of the treaty provisions, and b) the Community does not want to delimit clearly what provisions of a treaty come under Community law, as it does not want to freeze its competences. Often a subject not covered by Community law today may be covered by Community law tomorrow. When a division of treaty obligations is impossible, both the Community and its Member States adhere to the entire treaty.

In practice, the Community, its Member States, and the other parties to the treaties have not taken the legal problems involved very seriously. Both the Community and its Member States have adhered to the treaties without any clear restriction as to the extent of their treaty obligations. So far, this has not led to significant problems. The Community and the Member States each perform the obligations which come under their respective competences, and no problems have arisen on borderline cases.

2. Legal effect for individuals

There is no uniform opinion on the effect of treaty obligations in the national legal order. In France, the Netherlands and Greece, a treaty is part of the national legal order and, when its provisions have direct effect, they can be invoked by individuals. In the Federal Republic of Germany, Italy and Belgium, treaties are approved by law and that same law makes them part of the national legal order. This also leads to individuals being able to invoke the provisions of the treaty, even though officially they do not invoke the treaty but rather the national law approving it. In Britain and Ireland, separate legislation is needed in order to make a treaty effective within the national legal order. For the Community itself, Art. 228 (2) of the EEC Treaty provides that agreements between the Community and Third States are binding on the Member States. The Court of Justice has ruled that treaties entered into by the Community form part of Community law. They therefore can be invoked by individuals before their national courts [4].

In most Member States, the individuals have obligations and rights under mixed agreements in two ways. They can invoke these agreements as part of their national legal order and, as well, as part of the Community legal order. Although this may give rise to some procedural questions, it will not normally affect the substantive rights individuals may derive from the agreements. This double coverage does not lead to more rights than would single coverage. In some cases, however, problems will arise.

3. Problems as to mixed agreements

a) Incomplete mixed agreements

Some mixed agreements have been concluded to which the Community and only some of its Member States are parties, whilst the other Member States do not participate. Examples are the Convention of Barcelona of 16 February 1976 on the Protection of the Mediterranean Sea against Pollution [5], the Convention of Bonn of 3 December 1976 for the Protection of the Rhine against Chemical Pollution [6], the Memorandum of Understanding on Nuclear Science and Technological Information of 9 September 1974, concluded between the United States on the one hand and Euratom plus six of its then nine Members on the other [7], and the agreement of 3 April 1973 between the International Atomic Energy Agency on the one hand and Euratom and seven of its Member States on the other [8].

A similar situation is foreseen in the Convention on International Liability for Damage caused by Space Objects of 29 March 1972 which provides that international organizations may adhere on the condition that at least a *majority* of their Members are parties to the Convention [9].

Such an agreement is not necessarily incomplete. When all rights and obligations of the agreement relate to either the Community or to some specific Member States, than an agreement covering both the Community and those Member States forms a complete unit. The agreement, for example, of Switzerland on the one hand, and the Community plus France, the Federal

---

[4] See, e.g., *Second Haegeman* case (181/73), 30 April 1974, [1974] ECR 449; *Bresciani* case (87/75), 5 Feb. 1976, [1976] ECR 129.

[5] O.J. 1977, L 240/3.

[6] O.J. 1977, L 240/39.

[7] O.J. 1974, L 307/19; Trb. 1974, 238; Trb. 1975, 93.

[8] Trb. 1973, 97; Trb. 1974, 30; UN Juridical Yearbook 1976, p. 123; Bull. EC 2-1977, p. 61, point 2. 2. 26.

[9] GA Res. 2777 (XXVI), Art. 22, para. 3.

Republic of Germany, Luxembourg and the Netherlands on the other, on chemical pollution of the Rhine covers all parties which might pollute the Rhine as well as those which use the water of the river. The other Member States of the Community did not have to participate, as the treaty could not affect them.

In most cases, however, the non-participation of one or more Member States will cause a gap in the treaty. The Community is unable to fulfill all rights and obligations, and the remaining ones are covered only in respect to the territories of some Member States, but not in respect to the territories of the others. We call these mixed agreements, therefore, incomplete mixed agreements.

The question arises whether such incomplete mixed agreements are permissible under international law. We must distinguish several possibilities.

(i) A bilateral mixed agreement (that is, a mixed agreement with only one Third State) may be incomplete if the other party accepts such incompleteness. When the US negotiated with Euratom on the exchange of nuclear information, it knew that France and the United Kingdom would not participate. Nonetheless, it was willing to accept the agreement with Euratom and its other Member States. In such a case, there can be no objection against the incomplete mixed agreement.

(ii) When it is possible to adhere to a part of a treaty without accepting other parts, then the Community may adhere to those parts which come under its competence without adhering to the rest of the treaty. For purposes of allowing the Community to become a party, its inability to accept all parts of the treaty is irrelevant. Likewise, the Member States may or may not adhere to the other parts of the treaty, or to some of them. The freedom to adhere only to parts makes all forms of mixed agreements possible whithout harming the treaty as a whole or the other parties.

(iii) If a treaty permits reservations to such an extent that parties may exclude parts of the treaty, then again there will be no problem. The Community adhering to only a part of the obligations and some Member States not adhering to the other parts can be seen as equivalent to the reservation that these other parts of the treaty will not be binding in the whole territory of the Community.

(iv) A real problem arises when a treaty is concluded as a unit to which no reservations or restrictions are permitted. This will usually be the case when different obligations are bound together in a package deal. In a convention on the law of the sea, for example, most developing nations want to reserve wide fishing zones to the coastal States, and most industrialized nations want some guarantee for free shipping through all waters. The compromise might be a convention accepting wide fishing zones but also the obligation to allow innocent passage of foreign ships through their waters. If such a compromise is reached, States will not be entitled to adhere only to the articles on fishing, nor will they be permitted to make reservations which in fact exclude either the fishing or the shipping articles. If to such a convention the Community, or any other international organization competent only for fishing, could adhere without all Members at the same time adhering, then a split would occur which the treaty-makers did not want to create: the inhabitants of the Community could profit from the fishing rights granted by the Convention because of the Community being a party, while foreign ships would not have the treaty rights of innocent passage, as that would be governed by the competence of Member States which may not be a party to the Convention. International organizations should be entitled to become parties to this kind of package-deal agreement only if all their Members also become parties. And vice versa: Members could be party to the treaty only if the organization also adhered, because the Members are equally unable to undertake all obligations on their own. This means also that adherence to the agreement should be a condition for the admission of new Members to the organization, unless the organization and its existing Members are prepared to withdraw from the agreement when they admit a new Member State which is unwilling to adhere to the mixed agreement.

b) Liability

For the other parties to a mixed agreement it may be difficult to know whether the Community or a Member State is liable for a particular violation of the treaty. When under a new convention on the law of the sea a fishing vessel returning from the fishing grounds would fish in the territorial waters of a foreign State, the Community would be liable, being competent on matters of fishing. But would it also be liable if a merchant vessel fished in the waters of a foreign State? And what would be the case if the merchant vessel were State owned, or if it were a warship? Obviously, there will always be marginal situations in which it is not clear whether the Community or the Member State should accept responsibility.

During the law of the sea negotiations, the Community was asked to indicate clearly which parts of the new treaty would be covered by its competence and which parts would be covered by the responsibility of the Member States. The Community felt unable to make a clear and final statement to this effect, as its competence is not static. Through a further development of the Common Market, subjects which previously were covered by the competences of the Member States may become the competence of the Community. The Community, therefore, could not freeze its position by making an official statement on the limits of its competences with respect to the law of the sea.

The example given above shows that a declaration of competence by the Community would not solve all problems. Such a declaration could indicate what parts of a law of the sea convention would be covered by the competences of the Community and what parts would be the domain of the Member States; it could not solve the factual problems of whether an actual violation was a violation of the one part or of the other.

The solution of the question of the liability, therefore, should be sought in a procedure, rather than in a partition of the treaty obligations. There are several possibilities.

(i) The most appropriate procedure would be to allow third parties to bring all claims to the Community, leaving it to the latter to pass those claims on to Member States which are outside the Community's competence. For third parties this would be the easiest procedure, as it would relieve them of having to choose between the Community and Member States as well as of having to decide which Member State is liable in a case where a violation is committed by nationals of several Member States acting together. This latter situation may present itself more frequently the further that integration has progressed.

(ii) For the Member States, however, it may be difficult to accept that all questions of competence are decided by the Community. For them, therefore, it may be more acceptable if all claims are addressed to the States, leaving it to them to decide who is competent. This solution seems even less acceptable. Not only does it leave the decision on the competence to one of the parties involved, but it also creates the risk that different Member States may solve the same question of competence in a different way.

(iii) One could leave the choice to the Third State involved, allowing it to apply to either the Communtiy or the Member State, leaving it to them to pass the claim on if that would be appropriate. This would, however, leave the decision to a large extent to a third party who is neither competent nor able to take such a decision. None of the disadvantages of either of the two previously mentioned possibilities would be eliminated.

(iv) One could request the other treaty partners to address their claims to the Member States, and at the same time oblige these Member States to request a preliminary ruling of the Court of Justice on the question of whether the Community or the Member State is liable. This would be an appropriate solution if all claims would go to the courts of the Member States. For them it is part of their normal task to refer questions of Community law to the Community Court. The preliminary ruling obtained would at the same time guarantee uniform interpretation throughout the Community and an objective legal decision as to the competence. In practice, however, international claims are usually not presented to the courts but to the Ministry of Foreign Affairs of the State concerned. That Ministry cannot request a preliminary ruling, and permitting it to do so would upset the judicial system of the Community.

(v) The most acceptable solution seems to be to have all claims addressed to the Council of the Community. There, all Member States together could solve the question of the competence. If the Council comes to an unanimous decision, then all interests seem sufficiently guaranteed. For the Third

States, however, there would be the risk that the Council would not be able to agree within a reasonable period of time and that no decision on the competence would be taken. The claim would then get bogged down in the procedure. This risk can be overcome by the adoption of a rule of Community law obliging the Council to decide this sort of question within a specific period of time. If the Council would then be unable to do so, the question could be passed on to the Court of Justice under Art. 175 of the EEC Treaty, which contains a remedy against the inaction of Community institutions.

Thus a fair balance of all interests would be reached. The Third State would be enabled to bring its claim in a reasonably easy way, the Member States would get the opportunity to solve the question of competence by mutual consultation in the Council of the Communities, and a deadlock would be prevented by the Court of Justice as the most appropriate organ in a case where the Member States cannot agree but where some decision must be taken within a reasonable period of time.

c) Parliamentary approval

When national parliaments have to approve a mixed agreement, they normally want to know to what obligations their State will be committed. In principle, the whole mixed agreement will become binding on the national population, either through the ratification of the Communtiy or through that of the Member State. In some respects it may, therefore, not be of great importance for parliamentary approval which provisions will bind the Member States and which the Community.

In most Member States, parliamentary approval of a treaty is a condition for binding the State to the treaty. Logically, parliaments then want to know to what obligations their State will be bound, apart from the obligations of the Community. To meet this wish of the national parliaments, some indication should be given as to the division of competences when a mixed agreement is entered into. This indication does not need to be as specific as a declaration of competences to third parties for the sake of the Community's liability. It may, therefore, be given without necessarily freezing the relationship between the Community and its Member States. In an indication on the division of competences for the purpose of parliamentary approval, room can be left for future developments.

E. Independent Partnership in Treaties

In a mixed agreement, an international organization and its Member States are both parties, each for that part of the agreement which concerns its competence. The obligations of the international agreement are split up, divided between the organization and its Members.

In exceptional cases, it is possible that an international organization as well as its Member States can be full parties to all provisions of an international agreement. The agreement is then not mixed, but the international organization becomes an extra, full party. This situation will arise when the Community's competence has not affected the full competence of the Member States, but is newly created in addition to that competence. The following examples may illustrate this situation.

(i) The United Nations may establish international military forces. For these forces it could adhere to the Geneva Red Cross Conventions. Such adherence would in no way affect the obligations of the Member States under these conventions. The UN, in addition to its Members, would be bound as an independent partner to the Conventions.

(ii) The European Communities, or any other international organization exerting power, may bind itself to conventions on human rights. It will then be obliged to respect those human rights in the exercise of its powers. That will not affect the rights and duties of the Member States. They will continue to be bound by all provisions of the same conventions. The Community would become an additional independent partner to such a convention.

(iii) When the Community sends its own diplomats to foreign countries, it could – with some restrictions – become a party to the Vienna Convention on Diplomatic Relations. That would not affect the position of the Member States under that convention. Their diplomatic relations are neither increased, nor diminished by the adherence of the Community. Again, this would be an example of the Community becoming an independent partner to a convention.

As the treaties to which international organizations are independent parties are not mixed agreements, the problems of mixed agreements do not arise. An international organization can be an independent partner to a treaty, without its Members being parties or when only some of its Members are parties. That affects neither the position of the organization nor that of the Member States.

F. Membership of International Organizations

1. Sorts of membership

To some extent, the problems of international organizations becoming Members of other international organizations run parallel to the problems of international organizations becoming parties to treaties. Since virtually all public international organizations are established by treaties, this close relationship is quite natural.

In the case of international organizations, however, an extra dimension is added to the treaty relationship. Members of international organizations not only obtain rights and undertake obligations; they also form part of the organization itself and are elements of it. The most important organs of all international organizations are the Members themselves. This leads us to the question of whether international organizations are suitable to form part of other international organizations.

As far as knowledge, competence to participate or input in general are concerned, there can be no doubt. All large international organizations have more competence in the field in which they operate and more able people than many governments. Membership of other international organizations can greatly benefit the content of new proposals and the policy of an international organization.

With respect to the execution of the tasks of international organizations, the situation is more complex. Many recommendations of international organizations can be implemented only by States. In the case of the European Communities (or other similar bodies) participating in international organizations, we can distinguish the same categories as in the case of multilateral treaties. In some fields the Community has taken over the powers of its Member States so that only the Community can be Member of the organizations in those fields. In other instances, both the Community and its Member States carry part of the responsibilities of the organization so that a »mixed membership« would be needed. And, finally, there may be situations where membership of the Community and of its Member States may exist independent of each other.

2. The Community replacing its Member States

The European Economic Community has a common customs tariff and a common commercial policy. Consequently, all international agreements on customs duties are to be made by the Community with the exclusion of the individual Member States. In international organizations concerning tariffs and trade, the Community should replace its Members. In fact, the Community functions as a Member of the General Agreement on Tariffs and Trade (GATT) and of several Commodity Councils. The other parties to the GATT have recognized the Community as a Member [10].

This replacement does not offer great problems. So far, the Community has not formally adhered to GATT. It operates through the Member State which holds the presidency of the Council. Formally, therefore, there are still ten Community Members to GATT, who follow one policy, expressed by the delegation holding the presidency of the Council. That delegation contains the representatives of the Commission. The other nine delegations do not actively participate, but they retain their right to vote.

---

[10] *Third International Fruit Company* case (21–24/72), Dec. 1972, [1972] ECR 1219, at p. 1227.

The fact that one economic unit in GATT holds ten votes could be objected to by the other parties to GATT. For the Community itself it causes no problems. The other parties have not objected, partly because the political and legal problems involved are too complicated, so that one rather leaves things as they are, and partly because they do not seem to care enough to protest. In fact, ten votes in GATT do not give the Community an influence out of proportion to its economic importance. In comparison to most other Members of GATT, ten votes for the Community still means an underrepresentation. Only with respect to the USA and Japan, which are also underrepresented in GATT, are the ten votes out of proportion.

In international organizations, one partner holding more than one membership is not a novelty. When in the nineteenth century the Universal Postal Union and the International Telegraph Union were established, separate membership was granted to the colonies of the larger States precisely with the purpose of giving these States more voting power. The same happened after the Second World War when the Byelorussian and Ukrainian Soviet Republics obtained separate membership in the United Nations and in several specialized agencies.

In practice, there are two different forms in which the combined votes of the EEC Member States can be expressed [11]. The International Tin Agreement provides that the Community has no voting rights, but that the voting rights of the Member States can be exerted collectively whenever the Community is competent [12]. The International Cocoa Agreement, on the other hand, provides that the Community itself can vote, by expressing the collective votes of the Member States [13]. The latter provision better reflects the position of the Community as the actual Member of the organization and should, therefore, be preferred in this kind of agreement.

An eleventh membership for the Community itself would be a *novum* for which little ground can be found in an organization such as GATT where, in fact, only one commercial policy is represented.

3. Independent membership

The best example of an international organization in which the European Community could be an independent Member in addition to its Member States is the organization created by the European Convention on Human Rights. Though this organization is incorporated in the Council of Europe, it can be seen as a separate unit composed of a Court and a Commission and serviced by the Secretariat of the Council of Europe in much the same way as

---

[11] See *Manfred A. Dauses,* Die Beteiligung der Europäischen Gemeinschaften an multilateralen Völkerrechtsübereinkommen, 14 Europarecht (1979), pp. 138–170, at pp. 159, 160.

[12] International Tin Agreement, Art. 50, O.J. 1972, L 90 1–32 at p. 23.

[13] International Cocoa Agreement 1975, Art. 4, para. 2., Trb. 1976, 123.

several international organizations are serviced by the Secretariats of other organizations (e.g., the International Development Association by the Secretariat of the International Bank for Reconstruction and Development).

The parties to the European Convention on Human Rights have undertaken to secure a number of rights and freedoms to everyone within their jurisdiction [14]. But complaints can be brought only for violations committed by a party to the Convention [15]. When an individual alleged that his rights were violated by the European Community, his claim was held inadmissible as the Community was not a party to the Convention [16]. Apparently, the responsibility of the States parties to the Convention is understood as not including acts committed by the Community. Therefore, there would be reason for allowing, even for encouraging, the Community to adhere to this Convention in order to have its activities brought under the control of the Convention.

The Community would then be an independent party to the Convention, with its obligations concerning matters other than the obligations of the Member States, namely Community acts, and with the Member States responsible for domestic acts. As far as the treaty obligations are concerned, this would cause no problems; but what about the organizational aspects of the European Convention?

The Convention provides that the European Commission of Human Rights consists of a number of members equal to that of the parties to the Convention and that no two members may be nationals of the same State. In practice, this means that every participating State has one of its own nationals in the Commission. By excluding more than one national of the same State, the treaty-makers wanted to avoid any pretext that one State could be treated more favourably than others. In a sensitive field such as the protection of human rights this seems a wise precaution. Nonetheless, it also has the disadvantage that a Party which has no national available cannot make use of a national of another Party. When Liechtenstein had no national for the European Court of Human Rights, which contains a similar restriction, it had to appoint someone from outside the Council of Europe. Should the Community adhere to the Convention, it might also have to appoint a Human Rights Commissioner from outside the Council of Europe. As the EEC Commission has some non-European nationals on its staff, this would be possible without necessarily having to look for an outsider. It could also be submitted that the lawyers working for the Community have become Community lawyers to such an extent that they may be considered as Community nationals for the purpose of the European Convention on Hu-

---

[14] European Convention on Human Rights (ECHR), Art. 1.

[15] ECHR, Art. 25.

[16] Application No. 8030/77 (Confédération Française Démocratique du Travail), 10 July 1978, 13 Decisions and Reports, 231.

man Rights. The appointment of a Commissioner for the Community would, therefore, not lead to insurmountable problems.

Participation in the other organs of the European Convention on Human Rights would not be required, as these organs do not depend on the parties to the European Convention. The European Court of Human Rights is composed of one judge for each Member State of the Council of Europe; the Committee of Ministers for the Convention is the Committee of Ministers of the Council of Europe. According to the Convention, the Community would be entitled to appoint a judge *ad hoc* to the Chamber of the Court discussing a case in which it is involved[17]. In the Committee of Ministers, the Member States can present the views of the Community. Whenever this would be useful, the Committee of Ministers could also hear an observer from the Community.

In other international organizations to which the Community could adhere as an independent Member, the situation may be more complicated. Usually, all Members of an international organization are represented in its principal organ. There, their views can be expressed and votes will be taken. In theory, there can be no objection to attributing an extra vote to the Community in organizations where it is an independent Member, because it has an independent responsibility. In that case, the governmental powers within the Community have been distributed between eleven governments. The power of ten of them are geographically limited, those of the eleventh are functionally limited. Legally, the situation is the same as if each Member State of the Community had sacrificed a little part of its territory to form an eleventh State, which would then become an independent Member of the international organization.

The Community should be treated as any other Member, once it has become an additional Member of an international organization. This means that on the one hand, it could be elected in all organs of the organization whilst on the other hand, it should pay its full share of the financial contribution.

4. Mixed membership

As in the case of agreements, most difficulties arise in the mixed situation where a part of the responsibilities is performed by the Community and another part by the Member State. None of them can alone exert the full powers of membership.

It is not impossible for an international organization to provide for membership which is restricted to particular subject matters. The World Meteorological Organization admits Members which are not fully sovereign. Dependent territories which have their own meteorological service can become Members of the organization, but they are entitled to fully participate in me-

[17] ECHR, Art. 43.

teorological matters only. They may not vote on amendment of the constitution, membership, relations with other international organizations, and election of the president and vice-presidents of the organization and of the members of the Executive Committee[18]. Similarly, the constitution of an international organization could provide that customs unions can be Members but that they will have voting rights only on matters concerning tariffs and trade, whilst Member States of such customs unions can also be Members but will have voting rights only on matters other than tariffs and trade. It is clear, however, that this kind of arrangement can easily lead to problems in cases which only partially affect tariffs and trade and in cases where the effect is disputed.

Mixed membership should, therefore, be avoided. The European Communities (and similar international organizations) should become Members of other international organizations only if they have full competence to perform the duties of membership. They should abstain from joining international organizations whose task extends in a significant way beyond the task of the Community (or similar organizations).

Mixed membership is not needed in an international organization which operates in principle in the same field as the Community but which may occasionally perform functions which involve political aspects outside the competence of the Community, such as the acceptance of new Members. In such cases the Community can become a Member with the exclusion of its Member States and the Community can make an internal rule providing that voting on that kind of political items will occur only after binding consultation with the Member States.

In all other cases of mixed responsibility, the Member States of the Community should be the Members of the other organization. They should admit representatives of the Commission in the delegation of the Member State which holds the presidency of the Council, and the Member States can make an agreement allowing only that delegation to speak on matters falling under the competence of the Community, in much the same way as at present in the GATT. The representation of the Community in the GATT works sufficiently well and should be preferred to any form of mixed membership.

In a case of mixed membership, the financial burden of the membership should be divided between the Community and its Member States in the same proportion as their competences in the organization are divided.

---

[18] Constitution of the World Meteorological Organization, Art. 11.

Rechtsfragen grenzüberschreitender Wasserstraßen am Beispiel des Rhein-Main-Donau-Schiffahrtsweges

HANS-JÜRGEN SCHLOCHAUER

I

Die Frage nach einer Rechtsordnung für transnationale vom Meere aus schiffbare Ströme ist seit Regelungen in den Friedensverträgen nach dem ersten Weltkrieg in der Völkerrechtsliteratur [1] erörtert und – in Anknüpfung an die Erklärung der französischen Revolutionsregierung von 1792 und das Reglement in den Schlußakten des Wiener Kongresses 1815 [2] – seit der Mitte des vorigen Jahrhunderts für internationalisierte Ströme [3] durch Konventionen

---

[1] *A. Lederle,* Das Recht der internationalen Gewässer (1920); *H. Triepel,* Internationale Wasserläufe (1931); *B. Winiarski,* Principes généraux du droit fluvial international, Recueil des Cours 45 (1933 III), S. 79 ff.; *A. Lederle,* Die Zukunft der deutschen internationalen Ströme, Zeitschrift für Völkerrecht 20 (1936), S. 65 ff.; *W. Wegener,* Probleme der völkerrechtlichen Ordnung der internationalen Binnenschiffahrt in geschichtlicher und systematischer Bedeutung (1939); *O. Gönnenwein,* Die Freiheit der Flußschiffahrt (1940); *J. Bärmann,* Die Freiheit der europäischen Binnenschiffahrt (1950); *H. Thalmann,* Grundprinzipien des modernen zwischenstaatlichen Nachbarrechts (1951); *J. J. Lador-Lederer,* Vom Wasserweg zur internationalen Gemeinschaft, Die Friedens-Warte 53 (1955), S. 243 ff.; *B. Vitányi,* The International Regime of River Navigation (1979).

[2] *R. Rie,* Der Wiener Kongreß und das Völkerrecht (1957); *A. Nussbaum,* Geschichte des Völkerrechts (1960), S. 206 ff.; *J. H. W. Verzijl,* International Law in Historical Perspective, Bd. 3 (1970), S. 103 ff.

[3] Insbesondere Rhein und Donau. – Rhein-Regelung in Wiener Schlußakte 1815, Mannheimer Rheinschiffahrtsakte 1831 und Revidierte Mannheimer Rheinschiffahrtsakte 1868 mit Revision 1963 sowie Zusatzregelung 1972. Donau-Regelung in Pariser Frieden 1856, Donauschiffahrtsakte 1857 und 1865, Pontus-Vertrag 1871, Berliner Vertrag 1878, Londoner Vertrag 1883, Pariser Donauakte 1921 und Konvention über die Regelung der Schiffahrt auf der Donau der Ostblock-Anliegerstaaten 1948 mit Beitritt Österreichs 1960.

geregelt worden. In neuerer Zeit sind für einzelne internationale Flüsse bilaterale Abkommen getroffen worden, die vergleichbare Grundsätze aufstellen [4]. Für nicht durch Vertrag »unmittelbar internationalisierte«, sogenannte konventionelle Ströme ist eine generelle Regelung durch die Barcelona-Konvention von 1921 und das integrierende »Statute on the Regime of navigable Waterways of International Concern« gefunden worden [5]. Diese Rechtsordnung für grenzüberschreitende »mittelbar internationalisierte« Wasserläufe ist von einer nur geringen Zahl der Konferenzstaaten ratifiziert worden. Der Versuch ist praktisch gescheitert, aber theoretisch von Bedeutung geblieben. Auch für interozeanische Kanäle [6] wurde eine einheitliche Rechtsordnung angeregt. Der Rechtsstatus der Kanäle ist aber unterschiedlich durch Verträge, also »konventionell« geregelt.

II

Das Projekt eines Rhein-Main-Donau-Schiffahrtsweges geht zurück auf den im Jahre 793 von Karl dem Großen unternommenen Versuch (»Fossa Carolina«), eine Verbindung zwischen Rhein und Donau zu schaffen [7]. Der erste Teil eines schiffbaren Wasserlaufes zwischen Main und Donau entstand nach zehnjähriger Bauzeit im Jahre 1846. Dieser von Ludwig I. angeregte »Ludwigs Kanal« führte von Bamberg über Nürnberg nach Dietfurt an der Altmühl und sollte das nordbayerische Gebiet dem südlichen »Altbayern« annähern. Dem Unternehmen lagen politische Motive und ein für jene Epoche typischer nationalromantischer Gedanke zugrunde. Der wirtschaftliche

[4] Sankt-Lorenz-Strom als Grenzfluß zwischen den Vereinigten Staaten von Amerika – Kanada und Verbindung Atlantischer Ozean – Große Seen; 1959 als Seeweg eröffnet. – Verträge Großbritannien – Vereinigte Staaten von Amerika in Anknüpfung an im Jay-Vertrag 1794 geregelte Grenzstreitigkeiten seit 1854, Boundary-Waters-Treaty 1909 und Agreement Vereinigte Staaten von Amerika–Kanada 1941; dazu *L. Chevrier,* The St. Lawrence Seaway (1959); *T. H. Hills,* The St. Lawrence Seaway (1959); *R. R. Baxter,* The Law of International Waters (1964), S. 91 ff.; *R. R. Baxter,* Documents of the St. Lawrence Seaway (1960). – Ferner Columbia-Flußvertrag Vereinigte Staaten von Amerika – Kanada 1961, Text: Archiv des Völkerrechts (AVR) 13 (1966/67), S. 332 ff. – Chamizal-Vertrag Vereinigte Staaten von Amerika – Mexiko 1963, Text: AVR, a.a.O., S. 359 ff. – Abmachungen Frankreich – Spanien über Garonne 1963 und Bidassoa 1964, Text: AVR 12 (1964/65), S. 433 ff.

[5] *F. Corthésy,* Etude de la Convention de Barcelone sur le régime des voies navigables d'intérêt international (1927); *H. Krüger,* Verkehrskonferenz von Barcelona, in: Strupp-Schlochauer, Wörterbuch des Völkerrechts, Bd. 3 (1962), S. 153 ff.

[6] Insbesondere Kieler Kanal, Panama Kanal und Suez Kanal, die das Gebiet nur eines Staates durchlaufen. – Neuestens Konflikt Argentinien – Chile über den Beagle Kanal; dazu *K. Hernekamp,* Der argentinisch-chilenische Grenzstreit am Beagle Kanal (1980); *Brigitte Bollecker-Stern,* L'arbitrage dans l'affaire du canal de Beagle entre l'Argentine et le Chile, Revue Générale de Droit International Public 83 (1979), S. 7 ff.

[7] *H. H. Hofmann,* Kaiser Karls Kanalbau (1969).

Nutzen des Fahrwassers war gering, da der Rhein und die auf deutschem Gebiet verlaufende Donau noch nicht hinreichend kanalisiert waren; belastend wirkten zudem die bis 1866 geltenden hohen Rheinzölle.

Die Schiffbarmachung des Mains und der oberen Donau schritten in wechselvoller Entwicklung fort. Der Staatsvertrag zwischen dem Deutschen Reich und den Ländern vom März 1921 überführte die größeren Wasserstraßen der Länder auf das Reich, das deren Ausbauarbeiten fortzuführen hatte, und machte den Main von der Mündung bis Bamberg und die Donau von Kelheim bis zur österreichischen Grenze bei Passau zu Reichswasserstraßen. Im Juni 1921 schlossen das Deutsche Reich und Bayern einen Vertrag über den Ausbau der Rhein-Donau-Wasserstraße. Hierfür wurde eine Rhein-Main-Donau-Aktiengesellschaft gegründet, die nach einem Konzessionsvertrag vom Dezember 1921 die Verpflichtung übernahm, den Schiffahrtsweg von Aschaffenburg bis zur Grenze bei Passau entsprechend dem Juni-Vertrag auszubauen. Die Finanzierung ist nach einer Zwischenregelung vom Februar im Vertrag zwischen der Bundesrepublik Deutschland und dem Freistaat Bayern vom September 1966 dahin geregelt worden, daß der Rhein-Main-Donau-Aktiengesellschaft zum Bau der Strecke Nürnberg–Regensburg für die über Eigenmittel und am Kapitalmarkt aufgenommene Fremdmittel hinausgehenden Kosten unverzinsliche Darlehen (Verhältnis Bundesrepublik–Bayern 2:1) zu gewähren waren; für die von der Gesellschaft im Auftrag des Bundes auszubauende Donaustrecke Regensburg–Vilshofen hatte Bayern einen Beitrag von 50% der vom Bund aufgewendeten Haushaltsmittel zu leisten.

Die Großschiffahrtsstraße besteht, abgesehen vom Rhein, zum Teil aus kanalisierten Teilen des Mains und der Donau, zum Teil aus künstlichen Wasserstraßen. Sie gliedert sich nach drei in Entwicklung, Aufbauplan und Rechtsstellung unterschiedlichen Teilstücken: 1. den Main von Mainz bis Bamberg; 2. das sogenannte Kanalstück von Bamberg über Nürnberg und Dietfurt (Ludwigs-Kanal, Altmühltal) bis Kelheim, das zum Teil ausgebaut und in der zweiten Hälfte bis 1986 fertigzustellen ist; 3. die Donau von Kelheim über Regensburg, Straubing und Vilshofen bis zur österreichischen Landesgrenze bei Passau.

III

Der wirtschaftliche Nutzen und die politische Zweckmäßigkeit einer Rhein-Main-Donau-Großwasserstraße sind seit jeher umstritten. Das Unternehmen hat zwar als »gesamteuropäisches Problem« [8] sowie als erste Schaf-

[8] *W. Böttger,* Die Rhein-Main-Donau-Wasserstraße als gesamteuropäisches Problem (1975); *E. Seidel,* Die Großschiffahrtsstraße Rhein-Main-Donau, eine wirtschaftliche Idee und ihre Wirklichkeit (1960).

fung einer Kanalverbindung zwischen zwei internationalisierten Strömen Beachtung gefunden und ist für die Energieförderung sowie die wasserwirtschaftliche Entwicklung bedeutsam.

Die Fertigstellung des Verbindungsweges ist in neuester Zeit jedoch zweifelhaft geworden. Die Finanzmisere, der die Bundesregierung seit 10 Jahren in steigendem Maße, mit Kulmination in der Gegenwart, erlegen ist, hat die Frage entstehen lassen, ob die Kosten für den erst zur Hälfte fertiggestellten Abschnitt des sogenannten Kanalstückes von der Bundesregierung aufgebracht werden können. Sie fühlt sich zwar an die Verträge mit Bayern von 1921 und 1966 gebunden, hat aber Bundesfinanzminister Matthöfer beauftragt, mit der bayerischen Regierung über die Beendigung der Bauarbeiten zu verhandeln [9]. Minister Matthöfer hat in Erwägung gezogen, jene Beträge im Haushalt zu streichen und zum Ausgleich finanzielle Unterstützung beim Ausbau des bayerischen Straßen- und Schienennetzes anzubieten [10]. Bundesverkehrsminister Hauff hat zugestimmt und dabei den Schiffahrtsweg das »dümmste Bauwerk seit dem Turmbau von Babel« genannt [11]. Eine Einstellung des Kanalbaues ist andererseits als »größter Schildbürgerstreich aller Zeiten« bezeichnet worden [12] und in Bayern aus regionalwirtschaftlichen Gründen auf Widerstand gestoßen [13]. Diese Vorgänge haben alte Kontroversen über die »fixe Kanalidee« wieder aufleben lassen und zu neuen Erwägungen unter wirtschaftlichen, politischen und kulturellen Gesichtspunkten geführt.

Die Unwirtschaftlichkeit des Schiffahrtsweges wird kaum bestritten. Die Ausbaukosten haben bisher vier Milliarden betragen und sind für die Fertigstellung auf etwa zwei Milliarden errechnet. Der Nutzen-Kostenfaktor würde auch bei Verwendung von Erträgen der großen Zahl von Kraftwerken für den Kanalbau höchstens 0,5 betragen; zu berücksichtigen sind auch die Unterhaltungskosten der Kanalstrecke von jährlich 20 Millionen. Die wirt-

---

[9] Frankfurter Allgemeine Zeitung vom 24. 7. 1981 und 28. 1. 1982.

[10] A.a.O., 24. 7. 1981.

[11] A.a.O., 2. 12. 1981.

[12] So Präsident der Kammer der gewerblichen Wirtschaft für Oberösterreich Trauner, a.a.O., 30. 12. 1980.

[13] Ministerpräsident Strauß und der Bayerische Wirtschaftsminister Jaumann erklärten, daß Bayern den Bund nicht aus seinen Vertragspflichten entlassen werde, a.a.O., 1. sowie 30. 12. 1981 und 28. 1. 1982. – Während der Drucklegung wird bekannt das dem Bund für Umwelt und Naturschutz erstattete Gutachten von *B. Simma* „Möglichkeiten zur Aufhebung und Änderung der Rhein-Main-Donau-Verträge" (1982, 65 S. Maschinenschrift). *Simma* hält die Verträge wegen grundlegender Veränderung der bei Vertragsabschluß bestehenden Verhältnisse für unzumutbar und gibt dem Bund einen Anpassungsanspruch. Im übrigen sei die Nichtigkeit der Verträge von 1921 und 1966 anzunehmen, da sie in Widerspruch zum Prinzip der Gesetzmäßigkeit der Verwaltung und zu gesetzlichen Vorschriften (Finanzierungsregelung, Bürgschaftsverpflichtung, Betrauung der Rhein-Main-Donau-AG als Bauherrn) ständen.

schaftspolitischen Aussichten des Unternehmens, für das ein vordringliches Interesse kaum besteht, haben sich seit den Vertragschlüssen mit Bayern von 1921 und 1966 erheblich verschlechtert.

Eine durchgängig schiffbare Wasserstraße brächte der Deutschen Bundesbahn und der bundesdeutschen Binnenschiffahrt Nachteile. Die ohnedies hochverschuldete Bundesbahn würde eine Minderung ihrer Frachteinnahmen treffen, da die Tarife der Binnenschiffahrt um fünfzig Prozent niedriger sind. Die westdeutsche Binnenschiffahrt würde andererseits durch die Dumpingmethode der östlichen Staatshandelsschiffahrt, wie sie die »rote Flotte« in der Seeschiffahrt ausübt, Einbußen erleiden. Maßnahmen, die eine solche Dumpingkonkurrenz ausschließen, erscheinen wenig aussichtsreich. Die Öffnung des Schiffahrtsweges würde überdies den Staatsreedereien des Ostblocks ermöglichen, in das westeuropäische Wasserstraßennetz einzudringen. Dadurch würde die Forderung der Sowjetunion nach Internationalisierung des Verbindungsweges zwischen der »roten« Donau – und dem Main-Rhein-Flußsystem eine Grundlage erhalten.

Unter kulturellem Aspekt wird geltend gemacht, daß der weitere Ausbau der Kanalstrecke zwei der reizvollsten deutschen Landschaften, das bisher nur zum Teil betroffene Altmühltal und das hier verlaufende Donautal, vernichten würde[14]. Andererseits würde die Stillegung der Arbeiten eine gigantische Bauruine im Kanalstück entstehen lassen.

IV

Die völkerrechtliche Problematik, die eine durch Fertigstellung der Kanalstrecke geschaffene Rhein-Main-Donau-Verbindung als eine für den durchgehenden Schiffsverkehr geeignete Wasserstraße entstehen läßt, ist in der Fachliteratur übereinstimmend erkannt worden. Die Fragen sind von *Jaenicke, Zemanek und Kippels*[15] unter Einbeziehung politischer und wirtschaftlicher Aspekte untersucht worden. Hier mag auf diese alle Faktoren beachtenden Arbeiten verwiesen und – unter Ausklammerung von Fragen der Kollision mit Europäischem Gemeinschaftsrecht – versucht werden, das Wesentli-

---

[14] *G. Schlösser*, Das Altmühltal (1969).

[15] *G. Jaenicke*, Die neue Großschiffahrtsstraße Rhein-Main-Donau. Eine völkerrechtliche Untersuchung (Völkerrecht und Außenpolitik, Bd. 21) (1973); *K. Zemanek*, Die Schiffahrtsfreiheit auf der Donau und das künftige Regime der Rhein-Main-Donau-Schiffahrtsstraße. Eine völkerrechtliche Untersuchung, Österreichische Zeitschrift für Öffentliches Recht, Suppl. 4, 1976; *K. W. Kippels*, Der völkerrechtliche Status des zukünftigen Europakanals und seine Auswirkungen auf das Rhein-Donauregime (Schriften zum Völkerrecht, Bd. 62) (1978). – Früher *P. Krzizanowski*, Die Rhein-Main-Donau-Verbindung nach gegenwärtigem Völkerrecht (1965); *P. Krzizanowski*, Die Rechtslage des Rhein-Main-Donau-Verbindungsweges, AVR 14 (1969/70), S. 343 ff.

che zu markieren. Es geht um zwei Fragen: den Status der bisher erst zum Teil geschaffenen Main und Donau verbindenden Kanalstrecke und die unter Berücksichtigung der Rhein-Donau-Regime denkbare Regelung für die Großschiffahrtsstraße.

Die Kanalstrecke Bamberg–Kelheim verläuft im ganzen auf dem Gebiet der Bundesrepublik Deutschland, hat also keinen grenzüberschreitenden Charakter. Sie ist keine internationale Wasserstraße, deren Kriterium darin liegt, daß sie mehrere Staaten durchläuft oder zwei Staaten voneinander trennt. Da das Kanalstück aber zwei unter internationalen Regimen stehende Ströme verbindet, kann es fraglich sein, ob die Bundesrepublik Deutschland kraft ihrer Gebietshoheit über diesen Wasserlauf als nationales Gewässer verfügen kann. Obsolet geworden ist die Bestimmung des Artikels 353 des Friedensvertrages von Versailles von 1919 [16], nach der im Falle des Baues eines Großschiffahrtsweges Rhein-Donau das Deutsche Reich verpflichtet wurde, »auf diesen Schiffahrtsweg die in den Artikeln 332 bis 338 niedergelegte Ordnung zur Anwendung zu bringen«, wonach die Bundesrepublik Deutschland die Kanalstrecke als internationalisiert hätte behandeln müssen. Sie unterliegt nicht den Regelungen der Mannheimer Rheinschiffahrtsakte oder der – hinfällig gewordenen – Donauakte von 1921, die sich im übrigen nur auf die von Natur aus schiffbaren Teile der Donau bezog. Da das auf der Barcelona-Konferenz von 1921 verabschiedete Statut für Ströme von internationalem Interesse nicht in Kraft getreten ist, läßt sich aus jener angestrebten Regelung keine »mittelbare« Internationalisierung des Main-Donau-Kanals herleiten. Auch die »Helsinki-Rules« der International Law Association vom Jahre 1966 [17], nach denen für Wasserläufe internationalen Charakters ohne weiteres eine Internationalisierung angenommen oder vereinbart werden müsse, haben sich trotz weitgehender theoretischer Anerkennung nicht zu Völkergewohnheitsrecht entwickelt. Es enthält keinen derartigen Grundsatz. Die Bundesrepublik Deutschland ist also nicht veranlaßt, die innerhalb ihres Staatsgebiets gelegene Kanalstrecke für ein internationales Schiffahrtsregime freizugeben und auf ihr Schiffen anderer Staaten Verkehrsrechte einzuräumen [18]. Die Bundesregierung könnte aber mit Rhein- und Donau-Anliegerstaaten Vereinbarungen über die Nutzung dieses Wasserweges treffen. Nach einer allgemein geltenden Auffassung kann der Anlieger einer Wasserstraße im Anschluß an

---

[16] *H.-J. Schlochauer,* Die Einwirkungen des Krieges auf den Bestand völkerrechtlicher Verträge, Deutsche Rechts-Zeitschrift 1. Jg. (1946), S. 161 ff.

[17] Die International Law Association hat auf ihrer Konferenz in Helsinki 1966 die in 37 Artikel gefaßten »Helsinki-Rules« für Uses of Waters in International Rivers als wünschenswerte Regelung des internationalen Flußrechtes angenommen und den Vereinten Nationen zur Ausarbeitung eines Konventionsentwurfes für die Mitgliedstaaten vorgeschlagen, International Law Association, Report of the Fifty-second Conference Helsinki 1966, S. 447 ff.

[18] So auch *Jaenicke* (Anm. 15), S. 108; *Vitányi* (Anm. 1), S. 191; *Zemanek* (Anm. 15), S. 60.

grenzüberschreitende Ströme den Uferstaaten angrenzender Flußläufe Abreden über im einzelnen festzulegende Schiffahrtsfreiheit nicht versagen.

Eine internationale Regelung für die Rhein-Main-Donau-Großschiffahrtsstraße stößt auf Schwierigkeiten. Das Kanalstück Bamberg–Kelheim liegt zwischen internationalisierten Flußsystemen, die nach den vertraglichen Regelungen Unterschiede aufweisen.

Auf dem Rhein gelten die Mannheimer Rheinschiffahrtsakte von 1868 [19], nach denen die Schiffahrtsfreiheit den Fahrzeugen aller Nationen unter Gleichbehandlung der Staatsangehörigen, Schiffe sowie Güter mit den Rheinuferstaaten zu gewähren und eine Abgabe nur zulässig ist, soweit sie für die Stromunterhaltung verwendet wird [20].

Auf dem Main von Mainz bis Bamberg gibt es keine Schiffahrtsfreiheit, da das internationale Regime des Rheins sich nicht auf dessen Nebenflüsse erstreckt; auf sie wirken indirekt nur die Grundsätze der Inländergleichbehandlung und der Abgabenfreiheit.

Der Rechtsstatus der Donau ist seit der Suspendierung ihrer Internationalisierungsbestimmung infolge des zweiten Weltkrieges diffus [21]. Es bestanden unterschiedliche Regelungen: Für die Seedonau mit der Europäischen Donaukommission galten die auf Artikel 15 des Pariser Friedens von 1856 als »Teil des europäischen öffentlichen Rechts« [22] zurückgehenden Donauschiffahrtsakte von 1857 und 1865, ferner der Pontusvertrag von 1871, der Berliner Vertrag von 1878 sowie der Londoner Vertrag von 1883. Der Status der Binnendonau mit der Internationalen Donaukommission war in Artikel 347 des Versailler Vertrages und der Donauakte von 1921 geregelt. Nach Kontroversen zwischen den Siegermächten des zweiten Weltkrieges kam es zu der Belgrader Konferenz von 1948 über die Regelung der Schiffahrt auf der Donau. Beteiligt waren Frankreich, Großbritannien, die Vereinigten Staaten von Amerika, die Sowjetunion, das Gastland Jugoslawien sowie die fünf östlichen Donauuferstaaten. Der von der Sowjetunion eingebrachte Entwurf einer

---

[19] *M. Fleischmann,* Völkerrechtsquellen (1905), S. 81 ff. – Der Vertrag gilt in der revidierten Fassung von 1963 (in Kraft 1967) – Text: AVR 14 (1969/70), S. 439 ff. – mit Zusatzprotokoll 1972.

[20] Dazu *U. Scheuner,* Das internationale Recht der Rheinschiffahrt und der nationale Binnenverkehr (1954); *H. Kraus/U. Scheuner,* Rechtsfragen der Rheinschiffahrt (1956); *U. Scheuner,* Rhein, in: Strupp-Schlochauer, Wörterbuch des Völkerrechts, Bd. 3 (1962), S. 117 mit Literaturnachweisen. – Zur Rheinschiffahrtszentralkommission und den Donaukommissionen *H.-J. Schlochauer,* Der Rechtsschutz gegenüber der Tätigkeit internationaler und übernationaler Behörden (1952), S. 12 ff.

[21] Dazu *A. Lederle,* Die Donau und das internationale Schiffahrtsrecht (1928); *G. Martius,* Die Entwicklung des zwischenstaatlichen Donauschiffahrtsrechtes, AVR 1 (1948/49), S. 233 ff.; *I. Seidl-Hohenveldern,* Donau, in: Strupp-Schlochauer, Wörterbuch des Völkerrechts, Bd. 1 (1960), S. 393 ff.; *St. Gorove,* Law and Politics of the Danube (1964).

[22] Text: *Martens,* Nouveau Recueil Général de Traités, Bd. 15, S. 770. – Vgl. auch Anm. 3.

Konvention über die »Schiffahrt auf der Donau von Ulm bis zum Schwarzen Meer«[23] ist mit sieben Stimmen der Ostblockstaaten gegen die Stimme der Vereinigten Staaten angenommen worden; Frankreich und Großbritannien blieben der Abstimmung fern. Die anderen Signatarstaaten der Donaukonvention von 1921 wurden nicht hinzugezogen. Die »Konvention« 1948 ist völkerrechtswidrig, da sie eine nur suspendierte Regelung über die Donauschiffahrt ohne Zustimmung aller Signatare beseitigte und Bestimmungen über eine Donaustrecke innerhalb der Bundesrepublik Deutschland enthielt, die nicht am Vertragsabschluß beteiligt war. Die westeuropäischen Signatare der Donaukonvention von 1921 haben die Frage ihrer Fortgeltung – nach anfänglichem Zögern wohl aus politischen Gründen – weder dem hierfür zuständigen Ständigen Schiedshof noch dem Internationalen Gerichtshof unterbreitet. Die Konvention wird *de facto* von den Ostblock-Anliegerstaaten und von der Bundesrepublik Österreich, die ihr im Jahre 1960 beigetreten ist, angewendet, abgesehen von Einwirkungen auf die im Hoheitsgebiet der Bundesrepublik Deutschland verlaufende Donaustrecke.

Für die Donau von Kelheim bis zur österreichischen Grenze bei Passau gilt weder die Donaukonvention von 1921 noch der Belgrader Vertrag von 1948. Die Bundesrepublik Deutschland kann aber anderen Donauanliegerstaaten Schiffahrtsrechte nach dem Prinzip der Gegenseitigkeit gewähren.

Die verschiedenen Rechtsverhältnisse auf den Stromstrecken bedingen Konfliktmöglichkeiten hinsichtlich der durchgehenden Benutzung der Großschiffahrtsstraße.

Würde die Bundesrepublik Deutschland bei einer weiter steigenden Tendenz zur Ostpolitik der Donaukonvention von 1948 beitreten, gälten deren Bestimmungen für die »deutsche« Donau von der österreichischen Grenze bis Kelheim. Die Bundesregierung müßte alsdann überlegen, unter welchen Bedingungen das anschließende Kanalstück Kelheim–Bamberg der internationalen Schiffahrt eröffnet werden sollte. Da im Verlauf einer künftigen Großschiffahrtsstraße die Donau- und Rhein-Regime nebeneinander bestehen, würden ihre Unterschiede sich infolge der verschiedenen Wirtschaftssysteme der Anliegerländer zuungunsten der Rheinuferstaaten auswirken[24].

Zemanek hat drei Lösungen zur Gestaltung der Rechtslage auf der Großschiffahrtsstraße umrissen[25]. Die »Maximallösung« als einheitliches neues Regime für den Stromverlauf von der Rhein- bis zur Donaumündung ist unrealistisch, da eine Änderung der Rheinschiffahrtsakte und der Belgrader Donaukonvention nicht zu erwarten ist. Dies gilt auch für die »optimale Lösung«, in einem multilateralen Vertrag unter Aufrechterhaltung der Regime von Rhein und Donau sowie unter Einbeziehung der Kanalstrecke ein inte-

---

[23] Text der Konvention vom 18. 8. 1948: *Schweitzer/Rudolf*, Friedensvölkerrecht (2. Aufl. 1979), S. 171 ff.

[24] So *Jaenicke* (Anm. 15), S. 103; *Zemanek* (Anm. 15), S. 60.

[25] *Zemanek* (Anm. 15), S. 62 ff.

grales Kommunikationssystem zu vereinbaren. Am aussichtsreichsten erscheint die »Minimallösung« mit Widmung des Kanalstückes zum internationalen Verkehr ohne Harmonisierungsvereinbarungen. Sie würde den Rhein- und Donaustaaten den Zugang zum anderen Strom bei Fortbestehen der konventionellen Regelungen ermöglichen. Zusätzlich müßten bilaterale oder multilaterale Abkommen zwischen den Rhein- und Donauuferstaaten über die Bedingungen getroffen werden, unter denen »die Schiffe des einen Staates im Gebiet des anderen Staates Transporte durchführen dürfen und welche Rechte und Vorteile ihnen dabei gewährt werden«[26], um zu einem Gleichgewicht der Nutzungsmöglichkeiten zu gelangen.

---

[26] *Jaenicke* (Anm. 15), S. 104. – Zu eventuell neu entstehenden, hier nicht erörterten Kabotage-Fragen vgl. *Jaenicke* (Anm. 15), S. 34, und *Kippels* (Anm. 15), S. 180, 241.

Zur authentischen Interpretation von Gesetzen

HANS SCHNEIDER

Um den Wortlaut und Sinn des Gesetzes gegen Verfälschung, Entstellung oder Mißverständnisse zu sichern, haben sich frühere Gesetzgeber ausdrücklich jede Änderung, Streichung oder Vermehrung selbst vorbehalten. Damit sollte in erster Linie einem anderen Gesetzgeber die spätere Änderung verwehrt und daneben einer richterlichen Rechtsfortbildung vorgebeugt werden. Der Vorbehalt einer ausschließlichen Änderung durch den Gesetzgeber ist in der deutschen Territorialgesetzgebung des 16.–18. Jahrhunderts allenthalben zu finden, ja er stellt in der europäischen Gesetzgebung dieser Zeit eine verbreitete Erscheinung dar [1]. Das liegt im Zuge der Entwicklung des absoluten Staates.

I

Mit diesem Änderungsvorbehalt steht die Wiederaufnahme des Gedankens in Zusammenhang, daß die maßgebliche Auslegung des Gesetzes allein dem Gesetzgeber zustehe. Wie Kaiser Justinian die Abfassung von Kommentaren zu seinem Corpus juris bei Strafe verboten und sich selbst die Beantwortung etwaiger Zweifelsfragen vorbehalten hatte (533 n. Chr.), so enthalten französische und deutsche Gesetze des absolutistischen Zeitalters einen Interpretations-Vorbehalt zugunsten des Gesetzgebers [2]. In dieser Richtung ging das in Preußen von Cocceji ausgearbeitete Projekt eines Corpus Fridericianum (1749) ganz justinianisch vor: Nach dem Entwurf der Vorrede dazu sollte das Gesetzbuch in deutscher Sprache verfaßt werden, »damit ein Jeder, der einen Prozeß habe, selbst nachsehen und, ob er Recht oder Unrecht habe, daraus erlernen könne . . . Damit dann ferner Private, insonderheit Professoren, keine Gelegenheit haben, dieses Landrecht durch eigenmächtige In-

[1] *G. Immel*, in: Handbuch der Quellen u. Literatur der neueren europäischen Privatrechtsgeschichte (hrsg. von Coing), Bd. 2, Teilbd. 2 (1976), S. 86 f.

[2] *Immel*, a.a.O., S. 90 f.

terpretation zu corrumpiren«[3]. Deswegen sollte es bei Strafe verboten werden, einen Kommentar zu schreiben[4]. Dazu ist es freilich nicht gekommen.

Das Allgemeine Landrecht für die Preußischen Staaten (ALR) von 1794[5] – auf dessen einleitende §§ 74 und 75 sich bekanntlich der Aufopferungsanspruch bis zur Gegenwart gründet – enthielt einen Interpretationsvorbehalt zugunsten der königlichen Gesetzeskommission: »Findet der Richter den eigentlichen Sinn des Gesetzes zweifelhaft, so muß er, ohne die prozeßführenden Parteien zu benennen, seine Zweifel der Gesetzescommission anzeigen, und auf deren Beurteilung antragen«[6]. Diese Regelung hat sich freilich schon nach kurzer Zeit als nicht praktikabel erwiesen[7]. Die bereits seit 1781 unter Vorsitz des Großkanzlers bestehende Gesetzkommission sollte in erster Linie neue Gesetze vorbereiten und war deswegen kaum im Stande, die von den Gerichten nur zu gern vorgelegten Zweifelsfragen authentisch zu beantworten. Schon vier Jahre nach Verkündung des ALR wurde daher bestimmt, daß Anfragen im Laufe eines Prozesses nicht mehr stattfinden sollten, die Richter nur verpflichtet seien, die vermeintliche Dunkelheit eines Gesetzes dem Chef der Justiz zum Zweck der Klärung bei einer künftigen Gesetzgebung zu melden. Das Recht der authentischen Interpretation ging 1806 auf den Justizminister über und die Gesetzkommission scheint damals »bei der allgemeinen Stockung der Staatstätigkeit außer Aktivität gekommen und nach der Rückkehr der Ordnung nicht wieder zusammenberufen zu sein«[8]. Die Rolle der Kommission als höchstem Gremium (vor der Entscheidung des Königs) zur Prüfung von Gesetzentwürfen hat später der 1818 eingerichtete Staatsrat übernommen[9].

Zum Allgemeinen Landrecht für die Preußischen Staaten sind noch bis 1847 mehrere generelle »Deklarationen« ergangen, in denen der König »zur Beseitigung von Zweifeln über die Auslegung« einzelner Vorschriften eine authentische Interpretation bekanntgab (in der seit 1810 bestehenden Geset-

---

[3] Zitat nach *A. Stölzel*, Brandenburg-Preußens Rechtsverwaltung und Rechtsverfassung, Bd. 2 (1888), S. 209.

[4] Ob dieser Vorschlag die eigene Meinung des Projektverfassers wiedergab, darf bezweifelt werden. Samuel Freiherr von Cocceji (1679–1755), ein gebürtiger Heidelberger, Sohn eines Professors der Rechte und selbst Ordinarius in der Fakultät seines Vaters (Frankfurt an der Oder), kam erst 1714 in den Justizdienst nach Berlin, wurde 1723 Präsident des Kammergerichts, dann Staatsminister und 1730 Curator aller preußischen Universitäten. Ein Feind der Wissenschaft war er gewiß nicht, aber er hatte genug Gründe ein *ius certum* zu erstreben.

[5] Dem heutigen Interessenten ist das ALR in der von *H. Hattenhauer* besorgten Textausgabe (1970) leicht zugänglich.

[6] § 47 Einl. zu ALR.

[7] Dazu *C. Bornhak*, Preußische Staats- und Rechtsgeschichte (1903), S. 261.

[8] *L. v. Rönne* (langjähriger Vizepräsident des Appellationsgerichts), Das Staatsrecht der preußischen Monarchie, Bd. 1 (4. Aufl. 1881), S. 74/75.

[9] Dazu *H. Schneider*, Der preußische Staatsrat (1952), insbes. S. 148 f. über die legislative Tätigkeit bis 1847.

zessammlung veröffentlicht). Diese allgemeinverbindlichen Deklarationen ergingen stets auf Antrag des Staatsministeriums und nach Begutachtung durch den Staatsrat, in dem auch Savigny und andere bedeutende Juristen saßen.

Nachdem Preußen eine Verfassungsurkunde erhalten hatte und nunmehr die Gesetzgebung der Mitwirkung von beiden Kammern des Parlaments bedurfte, sind in der zweiten Hälfte des 19. Jahrhunderts einige Gesetze erlassen worden, in denen erklärt wurde, wie ein früheres Gesetz – richtig verstanden – auszulegen sei. Solchen Interpretationsgesetzen wurde von der damaligen Rechtslehre rückwirkende Kraft (auf den Zeitpunkt des Inkrafttretens des ursprünglichen Gesetzes) beigemessen [10]. Ein spätes Beispiel für eine solche authentische Interpretation bietet das im Jahre 1906 erlassene preußische »Gesetz zur Deklarierung des Kommunalabgabengesetzes« von 1893; es sollte die Rechtsprechung des Oberverwaltungsgerichts korrigieren, hat aber zu neuen Streitfragen geführt [11]. Da Zweifel für die Zukunft nie auszuschließen sind und stets darüber gestritten werden kann, ob das Interpretationsgesetz das ursprüngliche Gesetz (konstitutiv) abgeändert oder nur (deklaratorisch) seine Anwendung in der schon vorher gebotenen Lesart bestätigt hat (was für die Frage seiner Rückwirkung Bedeutung besitzt), sind solche authentischen Interpretationen von seiten des Gesetzgebers aus der Mode gekommen.

---

[10] *R. v. Mohl,* Staatsrecht des Königreichs Württemberg, Bd. 2 (2. Aufl. 1840), S. 99; *H. v. Schulze-Gaevernitz,* Preußisches Staatsrecht, Bd. 2 (2. Aufl. 1890), S. 50; *Meyer/Anschütz,* Lehrbuch des deutschen Staatsrechts (6. Aufl. 1905), S. 555.

[11] Die §§ 7, 20 und 27 preuß. Kommunalabgabengesetz (KAG) besagten, daß Gebühren, direkte Gemeindesteuern und Steuern vom Grundbesitz »nach festen Normen und Sätzen« bzw. »nach festen und gleichmäßigen Grundsätzen« zu erheben sind. In Bezug auf die Gebühren war hinzugefügt: »Eine Berücksichtigung Unbemittelter ist nicht ausgeschlossen«. Das preußische Oberverwaltungsgericht (OVG) hatte daraus geschlossen, daß für Benutzungsgebühren eine gleichmäßige Verteilung nach dem Maß der Nutzung erfolgen müsse und *nur* bei Unbemittelten eine Ermäßigung zulässig sei (OVG 36, 92). Dagegen hatten sich die Kommentatoren des KAG (*Nöll/Freund, Adickes* und *Strutz*) gewandt mit dem Argument, die Unbemittelten-Klausel sei exemplifikativ zu verstehen; ihre Kritik wurde vom Herrenhaus aufgenommen und führte zu dem Deklarationsgesetz vom 24. 6. 1906 (Gesetzessammlung, S. 376), dessen einziger Paragraph bestimmte: Die §§ 7, 20, 27 KAG »stehen einer Abstufung der Gebühren und Steuersätze nicht entgegen. Insbesondere ist es zulässig, die Gebührensätze nach Maßgabe der Leistungsfähigkeit bis zur gänzlichen Freilassung abzustufen und einzelne Grundstücksarten oder Besitzgruppen mit verschiedenen Sätzen zu den Steuern vom Grundbesitze heranzuziehen«. Das OVG erkannte zwar an, daß dieser authentischen Interpretation rückwirkende Kraft beizulegen sei (50, 107), blieb aber bei einer restriktiven Auslegung (50, 60), indem es eine Degression von Benutzungsgebühren nur bei minderer Leistungsfähigkeit des Schuldners, nicht bei regelmäßigem Massenbezug zuließ. Das Deklarationsgesetz von 1906 hat in den ehemals preußischen Gebieten in der Bundesrepublik Deutschland bis zum Inkrafttreten neuer Kommunalabgabengesetze der Länder fortgegolten.

II

In der Gegenwart werden unerwünschte Ergebnisse einer streng am Gesetz orientierten Rechtsprechung durch offene Umgestaltung der gesetzlichen Bestimmungen korrigiert. Der Gesetzgeber verändert sein Gesetz. Dabei sind ihm freilich – was die Geltung für die vergangene Zeit angeht – Grenzen gezogen. Weder dürfen rechtskräftige Urteile nachträglich beseitigt werden, noch darf eine verschlechternde Rückwirkung jemanden nachträglich belasten, der damit nicht zu rechnen brauchte. Die erste einschlägige Entscheidung des Bundesverfassungsgerichts [12] hat diese Grenze (noch dazu in einem dubiosen Fall) überspitzt zu dem Leitsatz verdichtet: »Der Gesetzgeber darf ein von der höchstrichterlichen Rechtsprechung zutreffend angewandtes Gesetz nicht rückwirkend ändern, um die Rechtsprechung für die Vergangenheit ins Unrecht zu setzen und zu korrigieren«. Die kargen Entscheidungsgründe rechtfertigen eine so weitgehende These nicht. Es geht um Fragen des Vertrauensschutzes, nicht um ein Ins-Unrecht-Setzen der Rechtsprechung (soweit rechtskräftig entschiedene Fälle unberührt gelassen werden). Der Gesetzgeber hat sich deswegen hernach nicht gescheut, generelle Korrekturen vorzunehmen, die sich auch auf Vergangenes erstrecken. So sind die Rechtsfolgen von Verstößen gegen gewisse Verfahrens- und Formvorschriften des Bundesbaugesetzes (Verstöße, die nach der bisherigen Judikatur zur Nichtigkeit von zahlreichen Bebauungsplänen geführt hätten) nachträglich gemildert worden [13]. In ähnlicher Weise hat der Landesgesetzgeber von Baden-Württemberg Fehler, die wegen einer Verletzung von Befangenheitsregeln die Beschlüsse eines Gemeinderates nichtig gemacht hätten, für »unbeachtlich« erklärt, wenn die Fehler nicht binnen eines Jahres gerügt werden [14], was sich auch auf alte Fehler im laufenden Jahr bezog. Die höchstrichterliche Rechtsprechung zum Beurkundungsgesetz hat zu einer raschen Korrektur durch ein Ergänzungsgesetz vom 20. Februar 1980 (BGBl. I, S. 157) geführt. Danach ist ein notariell beurkundetes Rechtsgeschäft nicht deshalb nichtig, weil – wie der Bundesgerichtshof zuvor angenommen hatte [15] – in der Niederschrift auf eine öffentliche Urkunde nur Bezug genommen, sie aber nicht beigefügt oder nicht verlesen worden ist, wie das Beurkundungsgesetz es befahl. Diese vom Gesetzgeber vorgenommene Korrektur seiner

---

[12] BVerfGE 18, 429 (Erstattung einer zu Unrecht gezahlten Verschollenheitsrente).

[13] § 155 a Bundesbaugesetz 1976 (*Sartorius* I, 300) in Verbindung mit Art. 3 § 12 des Änderungsgesetzes vom 18. 8. 1976 (BGBl. I, S. 2221), wonach eine Gemeinde die Wirkung des § 155 a auch nachträglich herbeiführen konnte.

[14] § 18 Abs. 6 Gemeindeordnung Bad.-Württ. 1977 (*G. Dürig,* Gesetze des Landes Baden-Württemberg, 56).

[15] BGH Urteil vom 23. 9. 1977, 23. 2. und 27. 4. 1979, NJW 1978, S. 102, und 1979, S. 1495 f.

früheren Regelung kommt einem Interpretationsgesetz nahe. Die Novelle ist zwar nicht rückwirkend in Kraft gesetzt worden, sie bezieht sich aber auch und gerade auf früher vorgenommene, nach der BGH-Rechtsprechung zum ursprünglichen Gesetz also ungenügende Beurkundungsfälle. Das rechtsstaatlich geschärfte Gewissen unserer Gerichte hat prompt zu verfassungsrechtlichen Bedenken gegen diese nachträgliche Heilung von Beurkundungsmängeln geführt[16], wobei groteskerweise geltend gemacht wurde, das »Vertrauen« einer Vertragspartei in den Eintritt der Nichtigkeitsfolge des falsch beurkundeten Rechtsgeschäfts hätte nicht enttäuscht werden dürfen. Schlitzohrigkeit sollte also belohnt werden?

In der Ebene des Verfassungsrechts hat sich die authentische Interpretation durch Gesetz noch einen bescheidenen Platz erhalten. Während der Geltung der Reichsverfassung von 1919 sind mehrfach Reichsgesetze, deren vollständige Vereinbarkeit mit der Reichsverfassung umstritten war, mit den für Verfassungsänderungen vorgeschriebenen Mehrheiten beschlossen worden. In der Verkündungsformel hieß es dann »... nachdem zur Vermeidung von Zweifeln festgestellt ist, daß die Erfordernisse verfassungsändernder Gesetzgebung für die §§ ... erfüllt sind«. Das Gesetz war damit auf jeden Fall gültig; aber besaß es auch den Rang eines Verfassungssatzes? Bedurfte es also zur späteren Änderung oder Aufhebung jener Paragraphen wiederum der gesteigerten Mehrheiten? Um solchen verdeckten Verfassungsänderungen oder Durchbrechungen zu begegnen, verlangt das Grundgesetz für die Bundesrepublik Deutschland, daß jede Verfassungsänderung im Wortlaut des Grundgesetzes zum Ausdruck kommt (Art. 79 Abs. 1 Satz 1 GG).

Dieses Text-Gebot ist bei verfassungsrelevanten Vertragsgesetzen zu völkerrechtlichen Abkommen kaum zu erfüllen. Deswegen soll es »zur Klarstellung«, daß die Bestimmungen des Grundgesetzes dem Abschluß und dem Inkraftsetzen von gewissen völkerrechtlichen Verträgen (Friedensregelung, Abbau von Besatzungsrecht, Verteidigung) nicht entgegenstehen, genügen, daß das Grundgesetz durch eine solche »Klarstellung« ergänzt wird. Diese verschrobene Konstruktion (Art. 79 Abs. 1 Satz 2) ist 1954 erfunden worden, um das Zustandekommen einer damals geplanten Europäischen Verteidigungsgemeinschaft nicht an selbstfabrizierten, verfassungsgesetzlichen Hin-

---

[16] Vorlagebeschluß des Landgerichts Düsseldorf vom 28. 5. 1980, NJW 1980, S. 1800 mit Anm. von *Dempewolf*. Zutreffende Bemerkungen dazu bei *Hartmut Bauer*, NJW 1980, S. 2552. Der BGH hat eine Verfassungswidrigkeit in dem Beurkundungsänderungsgesetz nicht erblickt (NJW 1981, S. 228). Den im oben zit. Leitsatz von BVerfGE 18, 429 geäußerten Gedanken (es sei unzulässig, die Rechtsprechung nachträglich ins Unrecht zu setzen) haben die Richter am BGH nicht angeführt. Sie lassen sich wohl lieber vom Gesetzgeber als vom Bundesverfassungsgericht belehren. Wenn das BVerfG die obersten Fachgerichte korrigiert, schließt dies die Feststellung ein, daß deren Richter die Verfassung verkannt oder deren Gewichtung nicht genügend berücksichtigt haben. Wenn der Gesetzgeber durch Änderung einschlägiger Vorschriften korrigiert, so liegt darin die Anerkennung, daß die Gerichte zuvor gesetzestreu entschieden haben.

dernissen auf seiten der Bundesrepublik Deutschland scheitern zu lassen. Damals wurde zugleich eine entsprechende Deklaration in das Grundgesetz aufgenommen (Art. 142a)[17]. Die Einleitungsformel zu diesem verfassungsändernden Gesetz ist übrigens mit einer motivierenden Präambel nach Weimarer Vorbild versehen worden: »Zur Klarstellung von Zweifeln über die Auslegung des Grundgesetzes hat der Bundestag mit Zustimmung des Bundesrates unter Einhaltung der Vorschrift des Artikels 79 Abs. 2 GG das folgende Gesetz beschlossen«[18]. Der Vertrag über die Europäische Verteidigungsgemeinschaft kam aber hernach wegen des politischen Widerstandes der französischen Nationalversammlung nicht zustande. Damit blieb ein wesentlicher Teil der »Klarstellung« in Art. 142a gegenstandslos. Der ganze Artikel ist 1968 aus dem Grundgesetz entfernt worden[19]. Die Möglichkeit, mit Hilfe des Art. 79 Abs. 1 Satz 2 GG gewisse völkerrechtliche Verträge als verfassungskonform zu interpretieren (wozu selbstverständlich die in Abs. 2 verlangten ⅔ Mehrheiten notwendig sind), ist dem Bundesgesetzgeber erhalten geblieben. Bisher hat er davon keinen weiteren Gebrauch gemacht.

III

Die allgemein verbindliche und daher maßgebliche Interpretation des Grundgesetzes steht im übrigen dem Bundesverfassungsgericht zu. Aber es ist nicht nur der Verfassungstext, den das Gericht dabei interpretiert: Wenn nämlich das Bundesverfassungsgericht ein einfaches Gesetz auf seine Vereinbarkeit mit dem Grundgesetz überprüft und dabei das einfache Gesetz in einem bestimmten Sinn so auslegt, daß es nur in diesem Verständnis dem Inhalt der Verfassung nicht widerspricht, so bindet auch diese verfassungskonforme Auslegung eines einfachen Gesetzes alle Regierungen und Parlamente, Behörden und Gerichte in der Bundesrepublik (§ 31 Abs. 1 BVerfGG). Die verfassungskonforme Auslegung ist daher maßgeblich und darf als authentische Interpretation eines einfachen Gesetzes durch das Bundesverfassungsgericht bezeichnet werden[20].

---

[17] Gesetz zur Änderung des GG vom 26. 3. 1954 (BGBl. I, S. 45): »Die Bestimmungen dieses Grundgesetzes stehen dem Abschluß und dem Inkrafttreten der am 26. und 27. Mai 1952 in Bonn und Paris unterzeichneten Verträge ... mit ihren Zusatz- und Nebenabkommen, insbes. dem Protokoll vom 26. Juli 1952, nicht entgegen«.

[18] *G. Wild*, Ausfertigung von Gesetzen und RechtsVO (Heidelberg 1969), S. 48, hebt zutreffend hervor, daß die gesetzgebenden Körperschaften in diesem Falle den Einleitungssatz in ihre Beschlußfassung miteinbeziehen mußten. Sonst ist es nicht Sache des Parlaments, die Eingangsworte festzulegen.

[19] 17. Gesetz zur Ergänzung des GG vom 24. 6. 1968 (BGBl. I, S. 709).

[20] Ein Musterbeispiel dafür bietet der Beschluß des Plenums des BVerfG vom 11. 6. 1980 (E 54, 277), dessen Entscheidungsformel lautet: »§ 554b Abs. 1 der Zivil-

Einer solchen maßgeblichen Auslegung sind auch Mantelgesetze zugänglich, mit denen der Bundestag unter Mitwirkung des Bundesrates die Zustimmung zu Staatsverträgen gemäß Art. 59 Abs. 2 GG erteilt. Zwar kann nicht der Vertrag als solcher den eigentlichen Gegenstand einer verfassungsgerichtlichen Prüfung bilden, sondern nur das dazu ergangene Zustimmungsgesetz[21], aber: »da der materiell-rechtliche Gehalt eines solchen Gesetzes sich erst aus dem Vertrag ergibt, dem zugestimmt wird, hat das Bundesverfassungsgericht hier zu prüfen, ob die gesetzgebenden Körperschaften einem Vertrag dieses Inhalts zustimmen durften«[22]. Im Hinblick auf diese Bindung des Zustimmungsgesetzes im Sinne des Art. 59 Abs. 2 GG an den Vertrag hat sich das Bundesverfassungsgericht im Urteil vom 31. Juli 1973 über den Grundlagen-Vertrag nicht gescheut zu schreiben, Gegenstand des (von der Bayerischen Staatsregierung beantragten) Normenkontrollverfahrens sei »das Vertragsgesetz vom 6. Juni 1973 (BGBl. II S. 421) und der in ihm in Bezug genommene Vertrag samt Zusatzprotokoll«[23]. Folgerichtig hat das Gericht auch den am 21. Dezember 1972 unterzeichneten Vertrag über die Grundlagen der Beziehungen zwischen der Bundesrepublik Deutschland und der Deutschen Demokratischen Republik am Maßstab des Grundgesetzes überprüft, um feststellen zu können, ob der Bundestag dem Vertragswerk in der Form eines Gesetzes zustimmen durfte. Das Ergebnis dieser Prüfung hat das Gericht in folgender Entscheidungsformel zusammengefaßt[24]:

> »Das Gesetz zu dem Vertrag vom 21. Dezember 1972 zwischen der Bundesrepublik Deutschland und der Deutschen Demokratischen Republik über die Grundlagen der Beziehungen zwischen der Bundesrepublik Deutschland und der Deutschen Demokratischen Republik vom 6. Juni 1973 (BGBl. II S. 421) ist in der sich aus den Gründen ergebenden Auslegung mit dem Grundgesetz vereinbar«.

In den Gründen des Urteils heißt es unter Hinweis auf Art. 16 und 116 GG, der Vertrag müsse, »um verfassungskonform zu sein«, so ausgelegt werden, daß die Deutsche Demokratische Republik für die Bundesrepublik nicht zum Ausland geworden ist, daß jeder Bürger der Deutschen Demokratischen Republik, der in den Schutzbereich der Bundesrepublik gelangt, als Deutscher wie jeder Bürger der Bundesrepublik behandelt wird und daß jede Verkürzung des verfassungsrechtlichen Schutzes durch den Vertrag oder eine Vereinbarung zur Ausführung des Vertrages grundgesetzwidrig wäre.

---

prozeßordnung i.d.F. vom 8. Juli 1975 (BGBl. I, S. 1683) ist nicht dahin auszulegen, daß die Annahme von Revisionen, die im Endergebnis Aussicht auf Erfolg besitzen, abgelehnt werden darf«.

[21] BVerfGE 12, 288 (Überleitungsvertrag); 29, 358 (deutsch-niederländischer Finanzvertrag).

[22] BVerfGE 12, 288 (Überleitungsvertrag) unter Bezugnahme auf 1, 410 (Bonner Verträge 1952) und 4, 162 (Saar-Statut).

[23] BVerfGE 36, 13 (Grundlagen-Vertrag).

[24] BVerfGE 36, 2.

Die hier vorgenommene verfassungskonforme Auslegung eines Staatsvertrages als einer Voraussetzung für die verfassungsrechtliche Unbedenklichkeit des Zustimmungsgesetzes im Sinne des Art. 59 Abs. 2 GG ist innerstaatlich für die Bundesrepublik Deutschland bindend. Als eine authentische Interpretation, die auch den Vertragspartner zu binden vermöchte[25], kann sie freilich nicht angesehen werden.

[25] Dazu *R. Bernhardt,* Die Auslegung völkerrechtlicher Verträge (Beiträge zum ausländischen öffentlichen Recht und Völkerrecht, Bd. 40) (1963), S. 44/45; *I. Seidl-Hohenveldern*, Völkerrecht (4. Aufl. 1980), S. 87/88.

Europabekenntnis und Wiedervereinigungsgebot des Grundgesetzes

THEODOR SCHWEISFURTH

Herausragendes Kennzeichen der »Europa«-Politik der deutschen Bundesregierung im Jahre 1981 war eine neue Initiative zur Errichtung einer »Europäischen Union« der EG-Staaten. Im Januar 1981 nutzte der Bundesminister des Auswärtigen, Genscher, die Eröffnung der Ausstellung »Die Nazarener in Rom« zur Verkündung der Forderung nach einem »neuen Anlauf in der Einigungspolitik«: »Die politische Form, in der die Wert- und Schicksalsgemeinschaft der Völker Europas den klarsten Ausdruck finden kann, ist die Europäische Union«[1]. Der Bundesaußenminister legte dem Bundeskabinett eine »europapolitische Initiative« vor, auf deren Grundlage das Kabinett in seiner Sitzung am 18. September 1981 »die Grundlinien der deutschen Europapolitik« erörterte und im Anschluß daran den Bundesaußenminister ermächtigte, »mit den EG-Partnern über eine politische Grundsatzerklärung zur Europäischen Union zu verhandeln«[2]. Dem Bundesaußenminister gelang es, die Unterstützung Italiens für den Plan einer Europäischen Union zu gewinnen; dies führte zu einer deutsch-italienischen Initiative, der Vorlage des »Entwurfs einer Europäischen Akte«[3]. Diesem Entwurf zufolge soll ein »vereintes Europa« in Gestalt einer »Europäischen Union« geschaffen werden, wobei die »Europäische Akte« selbst diese »Europäische Union« nicht begründen, sondern nur ein »Beitrag zu (ihrer) Errichtung« sein soll. Der Entwurf sieht ein »gemeinsames Vorgehen auf dem Gebiet der Sicherheitspolitik« und eine »gemeinsame Außenpolitik« der zehn EG-Staaten vor. Als wichtigste institutionelle Neuerung ist vorgesehen, die Entscheidungsstrukturen der EG und der Europäischen Politischen Zusammenarbeit (EPZ) unter der Verantwortung des Europäischen Rates zusammenzuführen; der Europäische Rat soll das »politische Lenkungsorgan« der EG und der EPZ werden. Der Entwurf

---

[1] Bulletin des Presse- und Informationsamtes der Bundesregierung (Bulletin) vom 23. 1. 1981 (Nr. 5), S. 42.

[2] Bulletin vom 24. 9. 1981 (Nr. 82), S. 723.

[3] Bull. EG 11-1981, S. 99 ff.

sieht vor, daß die »Europäische Akte« fünf Jahre nach ihrer Unterzeichnung überprüft wird, und zwar mit dem Ziel, die bis dann erreichten Fortschritte des europäischen Einigungswerkes in einem »Vertrag über die Europäische Union« zusammenzufassen. Der Bundesaußenminister hat dann noch zusätzlich das Europäische Parlament aufgefordert, bis zu den nächsten europäischen Wahlen 1984 den Entwurf für eine »europäische Verfassung« vorzulegen [4].

Die neue »europapolitische Initiative« der Bundesregierung wird, wie auch immer ihre Erfolgsaussichten beurteilt werden mögen, sicher von vielen Befürwortern der Integrationspolitik begrüßt, sie macht andererseits, wenn man vor sie das Prisma des Grundgesetzes hält, auch nachdenklich. Zielt der »Entwurf einer Europäischen Akte« auf ein »vereintes Europa« ab, so nimmt er ganz offenkundig die entsprechende Formel aus der Präambel des Grundgesetzes auf; die Bundesregierung will, so der Eindruck, mit ihrer neuen Initiative dem »Europabekenntnis« des Grundgesetzes entsprechen, der diesbezüglichen Zielsetzung der Präambel des Grundgesetzes gerecht werden. Nun enthält aber die Präambel des Grundgesetzes bekanntlich eine doppelte Zielsetzung: neben dem Europabekenntnis steht das Wiedervereinigungsgebot. Beide Zielsetzungen der Präambel erfahren, so ein weiterer Eindruck, in der derzeitigen Verfassungspraxis eine unterschiedliche Behandlung: Der neuen »europapolitischen Initiative« kontrastiert eine völlige Inaktivität der Bundesregierung auf dem Felde der Wiedervereinigungspolitik. Die Gründe für diese Inaktivität sind bekannt; auf der Basis der außenpolitischen Diagnose, der internationale Bezugsrahmen, in den »die deutsche Frage« eingebettet ist, sei petrifiziert, erscheint der Bundesregierung eine Aktivität auf diesem Felde sinnlos, weil von vornherein zum Mißerfolg verdammt. Diese Lagebeurteilung hat zur Konsequenz, daß die Bundesregierung den weiten Ermessensspielraum, den das Wiedervereinigungsgebot den zu politischem Handeln berufenen Organen der Bundesrepublik Deutschland einräumt [5], nicht für ein Handeln zur Herbeiführung der Einheit Deutschlands nutzt, sondern auf ein Unterlassen reduziert. Das bloße »Offenhalten der deutschen Frage« kann wohl kaum als aktive Wiedervereinigungspolitik qualifiziert werden.

Angesichts dieses Befundes einer unterschiedlichen Behandlung zweier Verfassungsgebote in der Praxis eines obersten Bundesorgans erhebt sich die Frage, ob eine solche Ungleichbehandlung verfassungsrechtlich legitimiert ist. Darf die Bundesregierung sich an der Weiterführung der »europäischen Integration« bis hin zur Bildung einer »Europäischen Union« beteiligen, ja sogar aktiv führend sich dafür einsetzen, während sie das Wiedervereinigungsgebot des Grundgesetzes weiterhin in einem »Dornröschenschlaf« [6] be-

---

[4] Frankfurter Allgemeine Zeitung vom 15. 12. 1981.

[5] Entscheidungen des Bundesverfassungsgerichts (BVerfGE) 5, 127 f.

[6] Vgl. *G. Ress*, Das Wiedervereinigungsgebot des Grundgesetzes, in: Fünf Jahre Grundvertragsurteil des Bundesverfassungsgerichts, hrsg. von G. Zieger (Schriften zur Rechtslage Deutschlands, Bd. 1) (1979), S. 265 ff.

läßt? Schickt der Bund sich nicht an, die ihm vom Grundgesetz übertragene »Integrationsgewalt« (Art. 24 GG) zu überschreiten, wenn er mit seiner neuen »europapolitischen Initiative« eine »Europäische Union« anstrebt trotz Fortbestandes der Spaltung Deutschlands?

In der wissenschaftlichen Literatur bleibt die Frage des Verhältnisses von Wiedervereinigungsgebot und Europabekenntnis (oder »europäischer Option« des Grundgesetzes) in der Regel unerörtert[7]. Wo es angesprochen wird, bleibt es im allgemeinen bei knappen Bemerkungen; so wird es beispielsweise als ein »Wechselverhältnis« des Inhalts bezeichnet, daß das Wiedervereinigungsstreben das Europa-Bekenntnis nicht ignorieren dürfe wie umgekehrt die Europa-Politik die Einheitswahrung zu beachten habe[8]; oder es wird ein Konflikt zwischen Europapolitik und Wiedervereinigungsgebot für möglich gehalten, bei dessen Eintritt der Bund sich »um eine Harmonisierung zwischen beiden Geboten bemühen« müsse[9]. Wie diese Harmonisierung zu geschehen habe, bleibt offen.

Nach Kenntnis des Verfassers hat sich lediglich *Karl Doehring* vor einiger Zeit ausführlich mit der doppelten Zielsetzung der Präambel des Grundgesetzes befaßt. Er schreibt von einer »Gegensätzlichkeit beider Ziele« der Präambel, die bei Erlaß des Grundgesetzes im Jahre 1949 »sicherlich nicht in gleichem Maße evident« gewesen sei wie heute, »oder man wollte sie nicht sehen«[10]. Eine von *Doehring* seinerzeit formulierte Frage ist durch die neue »europapolitische Initiative« der Bundesregierung in ein neues Stadium der Aktualität getreten, die Frage nämlich, »wieweit die funktionelle Integration der Europäischen Gemeinschaften ausgeweitet und vervollständigt werden kann, ohne daß eine weitere Souveränitätsaufgabe die Autonomie der Bundesrepublik zur Herstellung der Wiedervereinigung Deutschlands entscheidend aufhebt«[11]. *Doehrings* Überlegungen über die von ihm »angedeutete *contradictio*« der doppelten Zielsetzung der Präambel kommen zu dem Ergebnis, daß »angesichts der rechtlichen Gebundenheit der Bundesrepublik an das Wiedervereinigungsgebot deren Beteiligung an einer ihre völkerrechtliche Souveränität entscheidend beeinträchtigenden Vereinigung Europas nur dann zulässig (wäre), wenn eine das Wiedervereinigungsgebot als verfassungsrechtliches Gebot aufhebende Verfassungsänderung vorgenommen würde«[12].

[7] Typisch *Maunz/Dürig/Herzog/Scholz*, Grundgesetz, Kommentar (1980), wo es unter Präambel Rdnr. 24 heißt, daß das Streben nach einem »vereinten Europa« überwiegend als »Pflicht der Organe der Bundesrepublik« angesehen wird; und unter Präambel Rdnr. 26 wird die Förderung der Wiedervereinigung als ein Verfassungsgebot bezeichnet; wie sich beide Verfassungsgebote zueinander verhalten, wird nicht diskutiert.

[8] Grundgesetz, Kommentar, Bd. 1, hrsg. von I. v. Münch (1974), S. 22.

[9] *A. Bleckmann,* Grundgesetz und Völkerrecht (1975), S. 238.

[10] *K. Doehring,* Die Wiedervereinigung Deutschlands und die europäische Integration als Inhalte der Präambel des Grundgesetzes, Deutsches Verwaltungsblatt 1979, S. 633 ff., hier S. 634

[11] *Ibid.,* S. 637

[12] *Ibid.,* S. 639 (634).

Wenn in diesen Zitaten von Widerspruch (*contradictio*) die Rede ist und Wege zu seiner Auflösung aufgezeigt werden, so stellt sich die Frage, ob es sich dabei tatsächlich um einen Widerspruch im Grundgesetz oder nicht vielmehr um einen Widerspruch zwischen Verfassungsrecht und Verfassungspraxis handelt.

Das Verfassungsrecht

Der deutsch-italienische »Entwurf einer Europäischen Akte« versteht unter der »Vereinigung Europas« bzw. unter »vereintes Europa« den immer weiter fortzuführenden Zusammenschluß der EG-Staaten; die Verträge über die Europäischen Gemeinschaften werden als juristische Basis der »Vereinigung Europas« angesehen. Das wird besonders deutlich im dritten Teil des »Entwurfes einer Europäischen Akte«, der die »Perspektive« behandelt und dabei auch »andere europäische Staaten« ins Blickfeld nimmt, die sich an der Vereinigung Europas beteiligen könnten, und zwar durch einen Beitritt zu den Europäischen Gemeinschaften [13]. Entspricht diese Identifizierung des »vereinten Europa« mit den Europäischen Gemeinschaften dem Europabegriff des Grundgesetzes?

Das Grundgesetz spricht an zwei Stellen von Europa. In der Präambel ist vom Willen des deutschen Volkes die Rede, als gleichberechtigtes Glied »in einem vereinten Europa« dem Frieden der Welt zu dienen; nach Art. 24 Abs. 2 GG kann sich der Bund einem System gegenseitiger kollektiver Sicherheit einordnen, das eine friedliche und dauerhafte Ordnung »in Europa« und zwischen den Völkern der Welt herbeiführen und sichern soll. Was meint das Grundgesetz hier mit »Europa«?

Sind auch bei der Auslegung der Verfassung die in der Rechtsdogmatik allgemein geltenden Auslegungsmittel einzusetzen und ist dabei mit einer schlichten Verbalinterpretation zu beginnen [14], die den »natürlichen Wortsinn« zu ermitteln hat, so ist der natürliche Sinn des Wortes »Europa« jener, der durch die Geographie und jahrhundertealte Geschichte und Politik geprägt worden ist. Geographisch bezeichnet das Wort »Europa« den Raum, der durch die Fixpunkte Spitzbergen im Norden, Kreta im Süden, Island im Westen und Ural im Osten eingegrenzt wird; geschichtlich-politisch meint das Wort die in diesem Raum existente Staatengesellschaft. Würde das Grundgesetz bei der Verwendung des Wortes »Europa« von diesem »natürlichen Wortsinn« abweichen wollen, würde es einen engeren Raum, eine engere Staatengruppe bezeichnen wollen, so müßten sich in seinem Text eingrenzende Zusätze finden; das ist nicht der Fall, in der Präambel und in Art. 24 GG ist schlicht von »Europa« die Rede.

[13] Bull. EG 11-1981, S. 103.

[14] *K. Stern,* Das Staatsrecht der Bundesrepublik Deutschland, Bd. 1 (1977), S. 102 und 34.

Auch eine Interpretation des Wortes »Europa« im Kontext des Art. 24 Abs. 2 GG zeigt (vor dem Hintergrund seiner Entstehungszeit), daß der Europa-Begriff des Grundgesetzes den historisch geprägten Inhalt des Wortes meint. Kann der Bund sich nach Art. 24 Abs. 2 Satz 1 in ein System der »kollektiven Sicherheit« einordnen, so macht Art. 24 Abs. 2 Satz 2 deutlich, daß »hierbei« in erster Linie an eine durch ein kollektives Sicherheitssystem herbeizuführende friedliche und dauerhafte Ordnung »in Europa« gedacht ist. Es gehört nun zum Wesen eines »kollektiven Sicherheitssystems« daß seine Mitgliedschaft nicht nur aus »Verbündeten« (wie bei einem System kollektiver Selbstverteidigung) besteht, sondern daß es potentielle Gegner einschließt[15]. Zur Entstehungszeit des Grundgesetzes waren diese potentiellen Gegner schon hinreichend identifiziert: die Sowjetunion und die mit ihr durch bilaterale Pakte verbündeten Staaten Osteuropas. Auf sie *auch* zielte das avisierte kollektive Sicherheitssystem ab; auch daraus folgt, daß »Europa« in Art. 24 Abs. 2 Satz 2 mehr umfaßt als nur die westlich der schon zur Entstehungszeit des Grundgesetzes durch Mitteleuropa und Deutschland verlaufenden Trennungslinie belegenen Staaten.

Auch eine historische Auslegung ergibt, daß der Europa-Begriff des Grundgesetzes nicht mit dem EG-Europa identifiziert werden kann, und zwar aus dem simplen Grund, weil es dieses EG-Europa zur Zeit der Entstehung des Grundgesetzes noch gar nicht gab; welche (juristische) Gestalt das »vereinte Europa« annehmen würde, war 1948/49 noch völlig offen. Diese simple historische Erinnerung muß natürlich mit dem Einwand rechnen, daß, obwohl es das EG-Europa noch nicht gab, sich 1948/49 aber doch bereits die Konturen der Spaltung Europas in einen West- und einen Ostblock so stark herausgebildet hätten, daß die Verfassungsväter mit der europäischen Option des Grundgesetzes realistischerweise nur eine Vereinigung der Bundesrepublik Deutschland mit den Staaten jenes Teils Europas im Auge gehabt haben konnten, in dem der zu reorganisierende deutsche Staat selbst belegen war, also mit den Staaten des westlichen Europa; die europäische Option würde sich damit auf die EG *pro futuro* bezogen haben, so daß die Feststellung, das EG-Europa habe es zur Entstehungszeit des Grundgesetzes noch nicht gegeben, doch eine allzu simple historische Auslegung und damit nicht stichhaltig sei. Diesen Einwand entkräftet aber vollständig eine historisch-genetische, also eine auf die Materialien zum Grundgesetz zurückgreifende Interpretation.

Zitiert sei hier jeweils ein Vertreter der stärksten Fraktionen im Parlamentarischen Rat. In seiner Sitzung am 8. September 1948 sagte das Mitglied des Parlamentarischen Rates Dr. Süsterhenn (CDU): »Die Konfliktsituation

[15] Vgl. *U. Scheuner*, Kollektive Sicherheit, in: Strupp-Schlochauer, Wörterbuch des Völkerrechts, Bd. 2 (1961), S. 242: »Ein wirksames System kollektiver Sicherheit besteht nur dort, wo es in der Staatengemeinschaft oder in einer Region alle Staaten umschließt und zu gemeinsamer Abwehr auch gegen die Mitglieder der Verbindung verpflichtet«.

zwischen den Alliierten, der verhängnisvolle Riß zwischen Ost und West, der quer durch Deutschland und Europa geht, hindert uns an der Durchführung unseres Vorhabens« (nämlich der Beteiligung auch der Ostzone an der Grundgesetzschöpfung). »Eine Fortdauer des gegenwärtigen politischen und wirtschaftlichen Chaos würde nicht nur eine Gefährdung der Existenz unseres deutschen Volkes bedeuten, sondern ganz Europa, das sich bereits am Rande des Abgrundes befindet, mit in unser Verderben hineinziehen. Nur die sofortige und effektive Konsolidierung Westeuropas und die Schaffung einer politischen Ordnungs- und einer wirtschaftlichen Wohlstandssphäre in Westdeutschland werden in der Lage sein, auf die auch dem abendländischen Kulturbereich angehörenden Völker Ostmittel- und Südeuropas soviel Anziehungskraft auszuüben, daß ihre demnächstige Wiedereingliederung in die europäische Konföderation ermöglicht wird« [16].

In der Sitzung des Parlamentarischen Rates am 8. Mai 1949 sagte sein Mitglied Dr. Schmid (SPD): »Dann wollen wir diese Einheit des deutschen Volkes aus Gründen europäischer Solidarität; denn Europa kann nur geschaffen werden, wenn das ganze Deutschland dabei ist. Der Riß, der heute durch Deutschland geht, geht in Wirklichkeit durch Europa, und Europa unter Anerkennung dieses Risses schaffen würde bedeuten, daß man Europa zu Kleineuropa degradiert ... Indem wir unsere Einheit fordern, sind wir uns bewußt, daß wir ein Anliegen Europas erfüllen. Unsere Forderung ist eine europäische Forderung« [17].

Aus diesen beiden Zitaten ist eindeutig zu entnehmen, daß bei den Vätern des Grundgesetzes der historisch überkommene Inbegriff Europa lebendig war, d.h. »ganz Europa«, das neben Westeuropa zumindest noch »die auch dem abendländischen Kulturbereich angehörenden« Staaten »Ostmittel- und Südeuropas« einschloß, also Polen, die Tschechoslowakei und Ungarn sowie Rumänien, Jugoslawien, Bulgarien und Griechenland, sowie selbstverständlich jenen Teil Deutschlands, der zur Zeit der Schaffung des Grundgesetzes »die Ostzone« war. Daraus folgt, daß das EG-Europa nicht identisch ist mit dem (räumlichen) Europabegriff des Grundgesetzes, dieser ist umfassender als jenes.

Der Europabegriff des Grundgesetzes hat über den räumlichen hinaus auch noch einen institutionellen Inhalt. Der Präambel des Grundgesetzes zufolge will das deutsche Volk als gleichberechtigtes Glied nicht schlicht »in Europa« dem Frieden der Welt dienen, sondern in einem »vereinten« Europa. Der institutionelle Inhalt des Europabegriffs ist allerdings recht konturlos oder man könnte sagen »offen«, denn »vereint« besagt nur, daß irgendwelche die Staaten Europas vereinigende Institutionen geschaffen werden sollen, sagt über die Art dieser Institutionen aber nichts aus. Richtig schreibt dazu *Doehring* in seiner oben erwähnten Schrift: »Verschiedene Deutungen dieses so

---

16 Parlamentarischer Rat, Stenographischer Bericht, 2. Sitzung, S. 19.
17 *Ibid.*, 10. Sitzung, S. 199.

unbestimmten Begriffs sind denkbar. Europa könnte als »vereint« gelten, wenn es sich um die Herstellung einer internationalen Organisation handelt. Das gleiche könnte gelten, wenn es um die Bildung eines Staatenbundes geht oder um diejenige eines Bundesstaates«[18]. Daß das Grundgesetz hinsichtlich der Art der »Vereinigung Europas« keinerlei Festlegungen enthält, folgt aus der Natur des Regelungsgegenstandes: Eine »Vereinigung Europas« ist nur unter Mitwirkung der anderen Staaten möglich – eine einseitige gewaltsame »Neuordnung Europas« ist der Bundesrepublik Deutschland sowohl faktisch unmöglich wie rechtlich untersagt, Art. 25 und 26 GG –, erst aus der Mitwirkung der anderen Staaten können sich die Institutionen eines »vereinten« Europas herauskristallisieren, das »Europabekenntnis« des Grundgesetzes kann daher nur ein institutionell flexibles Angebot sein.

Einen institutionell einigermaßen konkretisierten Vorschlag zur »Vereinigung Europas« enthält das Grundgesetz allerdings doch. Auf dem Sektor der sicherheitspolitischen »Vereinigung Europas« bietet das Grundgesetz zur Herbeiführung einer »dauerhaften Ordnung in Europa« ein »System gegenseitiger kollektiver Sicherheit« an, Art. 24 Abs. 2 GG. Ein solches System wird institutionell durch die Schaffung einer internationalen Organisation der europäischen Staaten verwirklicht.

Über Art. 24 Abs. 2 GG hinaus eröffnet das Grundgesetz noch einen spezifischen Weg, auf dem die »Vereinigung Europas« institutionell herbeigeführt werden kann: die Übertragung von Hoheitsrechten auf »zwischenstaatliche Einrichtungen« nach Art. 24 Abs. 1 GG. Auch dabei kann es sich wegen der Mitwirkungsbedürftigkeit der anderen europäischen Staaten um nicht mehr als ein Angebot zur Schaffung mit Hoheitsrechten ausgestatteter »zwischenstaatlicher Einrichtungen« handeln, und daher ist Art. 24 Abs. 1 GG auch nur als Kann-Vorschrift formuliert. Die verfassungsrechtliche Eröffnung des Weges zur Übertragung von Hoheitsrechten auf »zwischenstaatliche Einrichtungen« gehört aber nun nicht mehr zum institutionellen Inhalt des Europabegriffs des Grundgesetzes; dies folgt aus dem Wortlaut des Art. 24 Abs. 1 GG, dem ein spezifischer Bezug zur Region Europa nicht entnommen werden kann. Die Ermächtigung aus Art. 24 Abs. 1 GG bezieht sich nicht speziell auf zwischenstaatliche Einrichtungen europäischer Provenienz, oder sagen wir besser nicht nur auf sie, sondern sie erstreckt sich über diese hinaus auf überregionale und sogar universelle »zwischenstaatliche Einrichtungen«[19]. Art.24 Abs. 1 GG stellt daher nur eine Methode (die »Übertragung von Hoheitsrechten«) für eine generelle »Öffnung der deutschen

[18] *Doehring* (Anm. 10), S. 636.

[19] Vgl. Stern (Anm. 14), S. 389; *Maunz/Dürig/Herzog/Scholz* (Anm. 7), Art. 24 Rdnr. 21. *Tomuschat*, in: Bonner Kommentar (Lieferung April 1981) Art. 24 Rdnr. 4: Art. 24 Abs. 1 ist nicht lediglich »eine (untergeordnete) Ausführungsnorm zum ›Europaauftrag‹ der Präambel«. Siehe auch Rdnr. 43.

Staatlichkeit«[20] zur Verfügung, formuliert aber weder eine spezifische »europäische Option« – diese ist in der Präambel und in Art. 24 Abs. 2 GG enthalten –, noch verweist er den das »vereinte Europa« anstrebenden Bund zwingend auf den von ihm der deutschen auswärtigen Gewalt eröffneten Weg.

Wie verhält sich nun das Europabekenntnis des Grundgesetzes, seine »europäische Option«, zum Wiedervereinigungsgebot? Besteht eine »Gegensätzlichkeit beider Ziele« der Präambel des Grundgesetzes, ein Zielkonflikt zwischen zwei Verfassungsgeboten?

Da der Europa-Begriff des Grundgesetzes sich auf den ganzen Kontinent bezieht bzw. da er doch zumindest auch die »dem abendländischen Kulturbereich angehörenden Staaten« Ostmittel- und Südeuropas sowie die seinerzeitige Ostzone Deutschlands umfaßt, ist ein Zielkonflikt zwischen Europa-Bekenntnis und Wiedervereinigungsgebot im Grundgesetz nicht vorhanden. Denn in dem so umschriebenen Kreis europäischer Staaten ist das ganze Deutschland einbezogen; eine »Vereinigung« dieses Europas schlösse die Wiedervereinigung Deutschlands nicht nur nicht aus, sondern würde sie geradezu notwendig im Gefolge haben oder doch zumindest eine gute Voraussetzung für die Erreichung auch des Verfassungszieles der Wiedervereinigung Deutschlands darstellen. Auf diese Weise bilden Wiedervereinigungsgebot und Europa-Bekenntnis eine Einheit, oder, wenn man es so will, ein »Wechselverhältnis«, oder, um einen der oben zitierten Verfassungsväter noch einmal zu bemühen: »Europa kann nur geschaffen werden, wenn das ganze Deutschland dabei ist«. Geht man von dem Bezug des Europa-Bekenntnisses des Grundgesetzes auf ganz Europa aus, kann von einer »Gegensätzlichkeit beider Ziele« der Präambel nicht die Rede sein. Was die doppelte Zielsetzung der Präambel betrifft, so waren »die Überlegungen bei Erlaß des Grundgesetzes« keineswegs »unklar«, wie *Doehring* vermutet[21]; die Väter des Grundgesetzes (vgl. die obigen Zitate) strebten eindeutig kein auf Westeuropa beschränktes »Kleineuropa« an, in das später nur der andere Teil Deutschlands, die Ostzone, einbezogen werden sollte[22]; das Ziel war vielmehr eine »europäische Konföderation«, die – wie auch immer sie institutionell konkret gestaltet werden würde – sich jedenfalls über den ganzen Kontinent erstrecken sollte und damit ganz Deutschland in ganz Europa einbeziehen sollte.

Unklar waren auch nicht die Wege, die die Väter des Grundgesetzes zu diesen Zielen einschlagen wollten, allerdings bestanden darüber sehr unter-

---

[20] *H. P. Ipsen*, Europäisches Gemeinschaftsrecht (1972), S. 52; *K. Vogel*, Die Verfassungsentscheidung des Grundgesetzes für eine internationale Zusammenarbeit (1964).

[21] *Doehring* (Anm. 10), S. 634.

[22] *Doehring, ibid.*, charakterisiert die Zielsetzung der »Politik der Stärke« doch als zu bescheiden, wenn er schreibt, sie wurde »als Möglichkeit aufgefaßt, die DDR aus dem Ostblock zu lösen und danach Gesamtdeutschland in Europa« (sprich Westeuropa) »zu integrieren«. Die »Politik der Stärke« zielte auf das *roll-back* der Sowjetunion aus ihrem gesamten ostmitteleuropäischen Glacis.

schiedliche Vorstellungen: Die einen meinten, die Ziele seien am ehesten über die »sofortige Konsolidierung Westeuropas« mit »demnächstiger Wiedereingliederung« der Staaten Osteuropas, bewirkt durch die »Anziehungskraft« dieses Westeuropa, erreichbar; die anderen hielten diese kleineuropäische Zwischenlösung für eine Sackgasse auf dem Weg zu einem »vereinten Europa«, sie wollten dieses Ziel auf dem Weg über die Wiederherstellung der Einheit Deutschlands erreichen. So war nicht eine »Gegensätzlichkeit beider Ziele« der Präambel bei Erlaß des Grundgesetzes im Jahre 1949 »nicht in gleichem Maße evident« wie heute (*Doehring*); nicht in gleichem Maße evident wie heute war nur, daß die kleineuropäische Zwischenlösung kein Weg war, der zum Ziel des »vereinten Europa« und der Wiedervereinigung Deutschlands führte.

Zu der Annahme eines Widerspruches zwischen beiden Zielsetzungen der Präambel des Grundgesetzes kann man nur gelangen auf der Grundlage eines verengten, auf Westeuropa oder gar nur auf das EG-Europa beschränkten Europabegriffs, der nicht der Europabegriff des Grundgesetzes ist. Dieser verengte Europabegriff steht hinter den Überlegungen, die *Doehring* über eine angebliche *contradictio* in der doppelten Zielsetzung der Präambel angestellt hat; *Doehring* stellte sich zwar die Frage, was »der vage Ausdruck ›vereintes‹ Europa« in der Präambel bedeuten könnte, beantwortete sie aber dann nur im Hinblick auf das Adjektiv »vereintes«[23]; »Europa« ist für ihn das Europa der EG-Staaten[24], und die die Souveränität der Bundesrepublik möglicherweise entscheidend beeinträchtigende weitere »Vereinigung Europas« bedeutet für ihn der engere Zusammenschluß der EG-Staaten in »eine politische Union Europa in Gestalt eines Bundesstaates«[25].

Die Verfassungspraxis

Die Verfassungspraxis in bezug auf das Europabekenntnis und das Wiedervereinigungsgebot des Grundgesetzes folgte den Vorstellungen jener Verfassungsväter, die die Ansicht vertreten hatten, beiden Präambelzielen würde man am ehesten näherkommen durch »die sofortige und effektive Konsolidierung Westeuropas« (Süsterhenn). Für die institutionelle Form des »vereinten Europa« wurde dabei jene vom Grundgesetz zur Verfügung gestellte Methode gewählt, die die intensivste »Vereinigung« ermöglicht: die Methode der Übertragung von Hoheitsrechten auf »zwischenstaatliche Einrichtungen« nach Art. 24 Abs. 1 GG. Auf dem Sektor der Sicherheitspolitik erwies

[23] *Doehring, ibid.*, S. 636 f.

[24] *Ibid.*, S. 634, wird das besonders deutlich in dem Satz: »Ein vereinigtes Europa unter Ausschluß der DDR scheint ebenfalls realisiert werden zu können«. Denkt man an die Vereinigung ganz Europas, kann diese natürlich nicht »unter Ausschluß der DDR«, eines im Zentrum Europas liegenden Staates, realisiert werden.

[25] *Ibid.*, S. 639.

sich diese Methode infolge der Ablehnung der Europäischen Verteidigungsgemeinschaft durch die französische Nationalversammlung allerdings als nicht verwendbar. Der ersatzweise erfolgte Beitritt der Bundesrepublik Deutschland zur NATO war ein doppelter Verzicht, dem Europabekenntnis des Grundgesetzes auf sicherheitspolitischem Gebiet zu entsprechen: Anstelle des von Art. 24 Abs. 2 GG avisierten Aufbaus einer »dauerhaften Ordnung in Europa« durch Errichtung eines kollektiven Sicherheitssystems trat der Beitritt der Bundesrepublik Deutschland zu einem kollektiven Selbstverteidigungssystem, dessen Mitgliedschaft nicht nur außereuropäische Staaten einschließt, sondern das sogar gegen andere europäische Staaten gerichtet ist. Die sicherheitspolitische »europäische Option« des Grundgesetzes in Art. 24 Abs. 2 gehört seitdem zu den »unerfüllten« Teilen des Grundgesetzes.

Die »Vereinigung Europas« wurde dann auf wirtschaftlichem Sektor gesucht, und hier war sie, sogar über die Methode des Art. 24 Abs. 1 GG erfolgreich, partiell zumindest. Aus der Gründungszeit der Europäischen Wirtschaftsgemeinschaft ist in unserem Zusammenhang folgendes bemerkenswert: Was den Europabegriff in der Verfassungspraxis betrifft, so sind in dieser Gründungszeit bereits deutlich erste Anzeichen eines Abrückens vom umfassenden Europabegriff des Grundgesetzes erkennbar. Als kritisiert wurde, daß die zu schaffende Europäische Wirtschaftsgemeinschaft der sechs EWG-Gründungsstaaten nur ein »Kleineuropa« schaffen werde, wurde seitens der die Bundesregierung tragenden Fraktion erwidert, »daß sich die Einigung in diesem Raum auch auf das übrige Europa fördernd auswirken« werde[26]; unter »übriges Europa« wurden dabei aber nur die OEEC-Staaten verstanden, und diese Gruppe wurde als »das gesamte Europa« bezeichnet[26]; die osteuropäischen Staaten hatten in diesem Europabegriff schon keinen Platz mehr[27].

Was das Wiedervereinigungsgebot des Grundgesetzes betrifft, so wurde die durch das Beschreiten des »kleineuropäischen« Weges für dieses Gebot entstehende Problematik von allen Beteiligten gesehen. Die Bundesregierung

---

[26] Vgl. den Debattenbeitrag des Abgeordneten Dr. Furler (CDU/CSU), 2. BT – 200. Sitzung, Sten. Ber., Bd. 35, S. 11 349.

[27] Ganz deutlich wird dies z. B. in folgender Passage eines Debattenbeitrages des Abgeordneten Dr. Hellwig (CDU/CSU): »Ich glaube es ist nicht ganz richtig, wenn bei diesen Sechs nur von einem »Klein-Europa« gesprochen wird. Immerhin handelt es sich dabei auch in seiner wirtschaftlichen Bedeutung um den Kern des kontinentalen Europas. Im wesentlichen stehen folgende Gruppen draußen: Einmal Großbritannien zufolge seiner Empire-Orientierung, dann die skandinavischen Länder, die einen ganz bestimmten abgeschlossenen Lebens- und Wirtschaftsraum für sich darstellen, weiter Südeuropa mit den Ländern Griechenland und Türkei und schließlich Portugal und Spanien. Das Sonderproblem Österreich ist in dem Fall sicher zu beachten; aber gerade Österreich hat sich in den letzten Monaten auch wiederholt sehr positiv für die Zollunion für den Gemeinsamen Markt ausgesprochen, und es wird, wenn nicht alle Zeichen trügen, doch zu einem Zusammenschluß mit dem Gemeinsamen Markt kommen«, 2. BT – 200. Sitzung, Sten. Ber., Bd. 35, S. 11 362.

erklärte, es sei »eine tief beklagenswerte Tatsache, daß Deutschland in die neue europäische Gemeinschaft eingefügt wird, belastet mit der schweren politischen Hypothek unserer erzwungenen Teilung«; die Bundesregierung habe »daher bei Abfassung des Vertragswerkes ihr ganzes Bestreben darauf gerichtet, alle Möglichkeiten einer Wiedervereinigung Deutschlands offenzuhalten und die künstliche Spaltung nicht zu vertiefen« [28]; zum Zwecke dieses »Offenhaltens« hatte die Bundesregierung während der EWG-Verhandlungen erklärt, sie gehe »von der Möglichkeit aus, daß im Fall der Wiedervereinigung Deutschlands eine Überprüfung der Verträge über den Gemeinsamen Markt und Euratom stattfindet« [29]. Mit dieser Erklärung hatte die Bundesregierung dafür Sorge getragen, daß im Falle der Wiedervereinigung Deutschlands seine Mitgliedschaft neu verhandelt werden kann [30]; hinsichtlich der Frage, ob eine solche Situation nach Gründung des Gemeinsamen Marktes überhaupt noch je eintreten werde, vertraute die Bundesregierung, wie schon ein Teil der Gründungsväter des Grundgesetzes, auf die »Anziehungskraft« [31] der zu schaffenden Gemeinschaft.

Daß die Beteiligung der Bundesrepublik Deutschland an der EWG gerade auch die Lösung der deutschen Frage fördern werde, war eines der wesentlichen Argumente der Befürworter dieser Beteiligung; so wurde die Überzeugung zum Ausdruck gebracht, »daß das Vertragswerk ein Meilenstein sein wird auf dem Wege sowohl zu einem einigen Europa als auch zu einem einigen Deutschland« [32]. Deutlich war damit gesagt, daß die kleineuropäische Politik beide Ziele der Präambel des Grundgesetzes nicht nur nicht in einen Gegensatz bringen werde, sondern daß beide durch sie gefördert würden. Andere waren da skeptischer, sie sahen die Gefahr heraufziehen, daß die kleineuropäische Einigungspolitik in einen Konflikt mit dem Wiedervereinigungsgebot des Grundgesetzes geraten könnte. So wurde gesagt, man werde »das ungute Gefühl nicht los, daß sich durch den Beitritt des einen Teils Deutschlands zu einer westeuropäischen Gemeinschaft die beiden Teile Deutschlands auseinanderentwickeln werden« [33]; so wurde gefragt, »ob die europäische Wirtschaftsunion die Verhältnisse hinsichtlich der Wiedervereinigung versteinert. Ich wage die Frage nicht zu entscheiden. Es ist möglich, daß die Bundesregierung recht hat und die Tatsache, daß nun ein starkes Klein-Europa entsteht, Rußland zu einer anderen Haltung gegenüber der Wiedervereinigung veranlaßt. Das ist möglich; ich weiß es aber nicht. Es ist

[28] So Staatssekretär Dr. Hallstein, 2. BT – 200. Sitzung, Sten. Ber., Bd. 35, S. 11 332.

[29] *Ibid.*

[30] Vgl. dazu *P. Dagtoglou*, Recht auf Rücktritt von den Römischen Verträgen?, in: Festschrift für Ernst Forsthoff (1972), S. 77 ff.

[31] So Staatssekretär Dr. Hallstein (Anm. 28), S. 11 334.

[32] So der Abgeordnete Dr. Elbrächter (DP (FVP)), 2. BT – 200. Sitzung Sten. Ber., Bd. 35, S. 11 360.

[33] So der Abgeordnete Margulies (FDP), *ibid.*, S. 11 352.

auch das Gegenteil möglich, daß durch diesen Vertrag eine solche Versteinerung eintritt, daß an eine Wiedervereinigung überhaupt nicht mehr gedacht werden kann«[34]. Wieder andere hielten die von der Bundesregierung verfolgte, man könnte sagen mittelbare Wiedervereinigungspolitik *qua* kleineuropäischen Zusammenschluß für wenig erfolgversprechend, sie vertrauten nicht schlicht auf die »Anziehungskraft« der zu schaffenden Gemeinschaft und verlangten deshalb zusätzlich eine »aktive freiheitliche Wiedervereinigungspolitik«, der die EWG nicht entgegenstehen dürfe[35]; auf diese Weise sollte durch eine doppelseitige Politik der doppelten Zielsetzung der Präambel des Grundgesetzes entsprochen werden.

Allen hier zitierten Aussagen ist, so läßt sich wohl feststellen, gemeinsam, daß dem engen Zusammenhang zwischen Wiedervereinigungsgebot und »Europäischer Option« des Grundgesetzes auch in der Verfassungspraxis Rechnung getragen werden müsse, so unterschiedlich auch die Auffassungen waren, in welcher Weise das zu geschehen habe. In den seit dem Gründungsjahr der EWG inzwischen abgelaufenen fünfundzwanzig Jahren ist der Zusammenhang zwischen beiden Verfassungsgeboten in der Verfassungspraxis verloren gegangen.

Folgende Feststellungen lassen sich wohl kaum bestreiten: Die in die »Anziehungskraft« der EWG gesetzten Erwartungen haben sich nur zum Teil erfüllt; das »Klein-Europa« der Sechs hat sich zwar zur Zehnergemeinschaft erweitert, weitere Beitrittskandidaten befinden sich im »Sog« der EG; dabei handelt es sich aber ausschließlich um westeuropäische Staaten. Keiner hegt heute mehr die Erwartung, daß die »Anziehungskraft« der EG auch zu einer Vereinigung mit den »auch dem abendländischen Kulturbereich angehörenden Völkern Ostmitteleuropas« führen werde; dem wirken Kräfte entgegen, die die »Anziehungskraft« der EG neutralisieren. Dieser Abschied von der noch im Gründungsjahr der EWG gepflegten Illusion der Machbarkeit eines durch »supranationale« Institutionen »vereinten Europas«, dem auch der sowjetisch besetzte Teil Deutschlands und andere Staaten Ostmitteleuropas sich anschließen würden, hat im Laufe der Jahre im Hinblick auf das Europabekenntnis und das Wiedervereinigungsgebot des Grundgesetzes gravierende Wirkungen in der Verfassungspraxis gezeitigt.

Der Europabegriff hat sich in der Praxis der Bundesregierung in doppelter Weise verengt. An die Stelle des räumlich-geographisch umfassenden Europabegriffs des Grundgesetzes wurde ein Europabegriff gesetzt, der sich nur noch auf die EG-Staaten, auf andere europäische Staaten allenfalls noch erstreckt, soweit es sich um EFTA- oder neutrale Staaten handelt; die Beziehungen zu den europäischen Staaten des Rates für Gegenseitige Wirtschaftshilfe (RGW) rubriziert die Bundesregierung nicht mit »Europapolitik«, sondern mit »Ostpolitik«. Zu notorisch ist diese Verengung des Europabegriffs

---

[34] So der Abgeordnete Stegner (GB/BHE), *ibid.*, S. 11 367 f.

[35] So der Abgeordnete Dr. Arndt (SPD), *ibid.*, S. 11 372.

in der Praxis der Bundesregierung, als daß es dafür hier eines ausführlichen Nachweises bedürfte; nur exemplarisch sei verwiesen auf die eingangs erwähnte Kabinettsitzung am 18. September 1981, bei der die Erörterung der »Grundlinien der deutschen Europapolitik« sich nur auf »Erhaltung und Ausbau der Europäischen Gemeinschaft« erstreckt[36]; und exemplarisch verwiesen sei auch auf den »Entwurf einer Europäischen Akte« (Dritter Teil: Perspektive); wenn diese Akte »andere europäische Staaten« zur Beteiligung an der »Vereinigung Europas«, d.h. zum Beitritt zu den EG, aufruft, so sind damit doch nur jene gemeint, »welche die in dieser Akte enthaltenen Wertvorstellungen und Ziele teilen«[37] – damit bleiben die europäischen RGW-Staaten bei der von der Bundesregierung angestrebten »Vereinigung Europas« draußen vor der Tür.

Auch hinsichtlich seines institutionellen Inhalts ist der Europabegriff in der Praxis der Bundesregierung verengt im Vergleich zum Europabegriff des Grundgesetzes, oder vielleicht besser gesagt erstarrt. Ist der Europabegriff des Grundgesetzes hinsichtlich der institutionellen Mittel der »Vereinigung« flexibel[38], so hat die bei der Gründung der EWG getroffene Wahl der von Art. 24 Abs. 1 GG zur Verfügung gestellten Methode in der nachfolgenden Praxis dazu geführt, daß unter »Vereinigung« Europas nur noch die Form der intensiven Integration verstanden wird, wie sie von den EG-Staaten praktiziert wird und fortentwickelt werden soll. Alle anderen möglichen, lockeren Formen, in denen ein »vereintes Europa« hergestellt werden könnte, sind der Praxis aus dem Gesichtsfeld geraten.

Der verengte Europabegriff in der Praxis der Bundesregierung wirkt dann wiederum auf wissenschaftliche Kommentierungen des Grundgesetzes zurück, so wenn gesagt wird, die europäische Option des Grundgesetzes habe »sich inzwischen durch die Gründung der Europäischen Gemeinschaften ... realisiert«[39]; eine solche Aussage erfolgt vor dem Hintergrund eines auf die EG-Staaten reduzierten Europabegriffs; vor dem Hintergrund des Europabegriffs des Grundgesetzes hätte es zu heißen »teilrealisiert«.

Hinsichtlich des Wiedervereinigungsgebotes führte die alsbald fällige Einsicht in die Wirkungslosigkeit der »Anziehungskraft« der EWG auf die RGW-Staaten und insbesondere auf den sowjetisch besetzten Teil Deutschlands in der Verfassungspraxis zu einer Abkoppelung des Wiedervereinigungsgebotes von der europäischen Option des Grundgesetzes. Hatte im EWG-Gründungsjahr die Bundesregierung, wie oben zitiert, »die Tatsache, daß Deutschland in die neue europäische Gemeinschaft eingefügt wird, belastet mit der schweren politischen Hypothek unserer erzwungenen Teilung«, noch als »tief beklagenswerte Tatsache« bezeichnet[40], so findet die noch im-

[36] Bulletin vom 24. 9. 1981 (Nr. 82), S. 723.
[37] Bull. EG 11-1981, S. 103.
[38] Vgl. oben S. 863.
[39] *Stern* (Anm. 14), S. 351.
[40] Vgl. oben S. 867.

mer vorhandene »Tatsache unserer erzwungenen Spaltung« in dem neuerlichen Bemühen der Bundesregierung, die Integration der EG in Richtung auf eine »Europäische Union« zu vertiefen, keinerlei Erwähnung mehr; daß der doch immerhin von der deutschen Bundesregierung ausgehende »Entwurf einer Europäischen Akte« sich durch absolutes Stillschweigen zu dieser Tatsache auszeichnet, macht anschaulich, wieweit die Praxis die europäische Option des Grundgesetzes vom Wiedervereinigungsgebot losgelöst hat.

Die aus der Einschätzung der internationalen Lage sich ergebende Einschätzung der Realisierungschancen beider Präambelgebote durch die verschiedenen Bundesregierungen führte zu unterschiedlichen Realisierungsbemühungen. Während das Präambelziel eines »vereinten Europa« – wenn auch nur bezogen auf Westeuropa – mit großer Intensität verfolgt wurde, zu den uns allen bekannten Ergebnissen geführt hat, und, wie die Vorlage des »Entwurfs einer Europäischen Akte« zeigt, auch weiterhin verfolgt wird, wurde eine »aktive freiheitliche Wiedervereinigungspolitik« von keiner Bundesregierung betrieben.

Alle Bundesregierungen beschränkten sich auf eine Politik des »Offenhaltens« der Wiedervereinigung Deutschlands, denn auch die aktivere Deutschlandpolitik der von der sozialliberalen Koalition getragenen Bundesregierungen war und ist nur eine Politik des Erträglichmachens, nicht aber der Überwindung des *status quo* der Teilung des Landes.

Die Abkopplung des Wiedervereinigungsgebotes von der europäischen Option des Grundgesetzes in der Verfassungspraxis, die intensive Verfolgung der westeuropäischen Integration im Rahmen der EG und die gleichzeitige Inaktivität bei der Verfolgung des Wiedervereinigungsgebotes haben nun mittlerweile zu einer Situation geführt, von der sich zutreffend sagen läßt, daß, »je stärker die Bundesrepublik in die westeuropäische Integration eingebunden ist, um so schwerer sich das Ziel der Wiedervereinigung erreichen« läßt[41]. Unzutreffend ist diese Situation jedoch beschrieben, wenn zu ihr gesagt wird, es sehe »derzeit so aus, als seien beide Ziele der Präambel des Grundgesetzes noch niemals in so starken Widerspruch zueinander getreten, wie das heute der Fall ist«[42]; Verfassungsziele können zueinander nachträglich, d. h. zeit unveränderter Geltung einer ausformulierten Verfassung, gar nicht zueinander in Widerspruch treten, entweder es bestand von Anfang an ein Widerspruch zwischen ihnen oder es bestand keiner. Nachträglich zueinander in Widerspruch treten können die Verfassungsgebote mit der Verfassungspraxis, oder besser gesagt umgekehrt – und genau dies ist bei dem uns hier interessierenden Problem der Fall. Zwischen europäischer Option und Wiedervereinigungsgebot des Grundgesetzes bestand und besteht kein Widerspruch. Wenn sich das Ziel der Wiedervereinigung um so schwerer erreichen läßt, je stärker die Bundesrepublik Deutschland in die westeuropäische

[41] So *Tomuschat* (Anm. 19), Rdnr. 6.
[42] So *Doehring* (Anm. 10), S. 634.

Integration eingebunden ist bzw. sich immer stärker einbindet, dann besteht ein Widerspruch zwischen der Verfassungspraxis der westeuropäischen Integration und dem Wiedervereinigungsgebot des Grundgesetzes.

Verfassungsrechtliche Grenzen und Möglichkeiten der Europapolitik

Daß sich die die *derzeitige* Wiedervereinigungspolitik und Europapolitik betreffende Verfassungspraxis der Bundesregierung in einem rechtlich schon nicht mehr tolerablen Widerspruch zum Grundgesetz befände, ist meines Wissens noch von niemandem, insbesondere auch noch nicht von dem authentischen Interpreten des Grundgesetzes, dem Bundesverfassungsgericht, behauptet worden. Im Hinblick auf das Wiedervereinigungsgebot sieht das Bundesverfassungsgericht sogar die derzeitige Nichtverfolgung einer »aktiven freiheitlichen Wiedervereinigungspolitik« offenbar noch innerhalb der weiten Ermessensgrenzen, die das Wiedervereinigungsgebot auszeichnen; zwar hat es im Grundlagenvertragsurteil das Wiedervereinigungsgebot u.a. dahin präzisiert, daß »alle Verfassungsorgane verpflichtet sind, in ihrer Politik auf die Erreichung dieses Ziels hinzuwirken«, es hat sich aber enthalten, die wiedervereinigungspolitische Inaktivität, also das Nichthinwirken der Bundesregierung zu rügen, weil die Abschätzung der Chancen dieser Politik und damit die Entscheidung über die Entwicklung von Aktivität Sache der Bundesregierung und der sie tragenden parlamentarischen Mehrheit sei[43].

Was die westeuropäische Integrationspolitik betrifft, so wird sie über jeden Zweifel erhaben als grundgesetzkonform angesehen. So zweifelsfrei ist dies indessen nicht, und zwar aus folgendem Grunde: Wie die ursprüngliche These von der »Anziehungskraft« des zu schaffenden EWG-Europas zeigt, war die auf »Klein-Europa« gerichtete Einigungspolitik durchaus als Zwischenstufe gedacht zur Herstellung eines größeren »vereinten Europa«, von der selbst die RGW-Staaten nicht ausgenommen waren; damit entsprach die EWG-Gründungspolitik durchaus dem (geographischen) Inhalt des Europabegriffs des Grundgesetzes. Mit der durch gesteigerten Realitätssinn erfolgten Einschränkung der These von der »Anziehungskraft« des geschaffenen EG-Europas auf westeuropäische Staaten, einer Einschränkung, die ihre jüngste Verdeutlichung im »Entwurf einer Europäischen Akte« gefunden hat, sind die europäischen RGW-Staaten von der »europäischen« Einigungspolitik nunmehr ausgenommen; die so in der Praxis erfolgte Einschränkung des Europabegriffs stimmt mit dem Europabegriff des Grundgesetzes nunmehr nicht überein, so daß westeuropäische Integrationspolitik heute wohl kaum noch als Europapolitik im Sinne der »europäischen Option« des Grundgesetzes verstanden werden kann.

---

[43] BVerfGE 36, 18.

Als Begründung für diese Einschränkung des Europabegriffs durch die Praxis kann nun darauf hingewiesen werden, daß eine Integration der europäischen RGW-Staaten in die EG infolge der unterschiedlichen politischen und wirtschaftlichen Strukturen nicht möglich sei. Dieser Hinweis ist sicher zutreffend. Die hieraus gezogene Schlußfolgerung, daß sich deshalb europäische Einigungspolitik auf westeuropäische Staaten beschränken müsse, ist aber nicht zwingend. Denn dieser Hinweis koppelt die Politik zur Herbeiführung eines »vereinten Europa« mit einer bestimmten, nämlich der bei der Vereinigung der EG-Staaten verfolgten, für die deutsche Beteiligung durch Art. 24 Abs. 1 GG zur Verfügung gestellten Vereinigungsmethode; und er geht von der – unausgesprochenen – Vorstellung aus, daß nur mittels dieser Vereinigungsmethode ein »vereintes Europa« geschaffen werden könne. Indem der Ermächtigungscharakter des Art. 24 Abs. 1 GG in eine für deutsche Politik zur Herbeiführung eines »vereinten Europa« quasi zwingende Vorschrift ummodelliert wird, wird das institutionell flexible Angebot im »Europabekenntnis« des Grundgesetzes auf *ein* Angebot reduziert und damit zur Erstarrung gebracht. Wenn sich aber nun – wie das heute der Fall ist – herausstellt, daß die Benutzung der Vereinigungsmethode der »Integration« über Art. 24 Abs. 1 GG nicht zu einem im Sinne des Grundgesetzes den ganzen Kontinent erfassenden »vereinten Europa« führen kann, dann drängt sich doch die Frage auf, ob die Benutzung dieser Vereinigungsmethode im Hinblick auf den umfassenden Europabegriff des Grundgesetzes noch weiter als grundgesetzkonform betrachtet werden kann. Die deutschen »Vereiniger Europas« verabsolutieren die Vereinigungsmethode der »Integration«, sie erreichen damit zwar eine enge Vereinigung europäischer Staaten, nicht aber das grundgesetzliche Ziel eines »vereinten Europa«. Ist es im Sinne der europäischen Option des Grundgesetzes, zugunsten der Vereinigung einer Gruppe europäischer Staaten einer bestimmten Vereinigungsmethode den Vorrang zu geben gegenüber dem grundgesetzlichen Ziel eines »vereinten Europa«? Entspräche es nicht eher der europäischen Option des Grundgesetzes, wenn sich die Verfassungspraxis flexibler verhalten und sich weniger wirksamen Vereinigungsmethoden zuwendete, bei denen dann zwar eine weniger intensive Vereinigung als die im Rahmen der EG mögliche in Kauf genommen werden müßte, die aber dann doch ein »vereintes Europa« und nicht bloß die Vereinigung eines Teils europäischer Staaten ermöglichen würde?

Diese Fragen sollen hier nur einmal aufgeworfen werden. Zur Frage, an welche flexibleren Methoden der Herbeiführung eines »vereinten Europa« vielleicht gedacht werden könnte, sollen am Schluß dieser Überlegung noch einige Bemerkungen gemacht werden.

Wenn wir uns jetzt der Frage der verfassungsrechtlichen Grenzen und Möglichkeiten der *künftigen* die Europapolitik betreffenden Verfassungspraxis zuwenden, so kann es sich dabei auch nicht um den Versuch einer Beantwortung, sondern nur gleichsam um eine Auffächerung dieser Frage handeln. Auch nach Vorlage des »Entwurfs einer Europäischen Akte« sind die Konturen der künftigen »europapolitischen« Verfassungspraxis noch zu ungenau, als daß sie im einzelnen am Grundgesetz gemessen werden könnten. Es soll lediglich die wohl nicht zu weit greifende Annahme unterlegt werden, daß der angestrebte »Vertrag über die Europäische Union« auf die Etablierung eines von den EG-Staaten zu bildenden Bundesstaates abzielt. Schon vor der neuerlichen »europapolitischen Initiative« der Bundesregierung meinte auch *Doehring*, »daß der bisher eingeschlagene Weg über die partielle wirtschaftliche Integration unter partieller Souveränitätsaufgabe bei folgerichtiger Beurteilung zu einem Bundesstaat zu führen scheint, wenn überhaupt ein weiteres Integrationsziel angestrebt wird oder erreichbar ist« [44].

Zweifel an der verfassungsrechtlichen Zulässigkeit einer deutschen Beteiligung an der Errichtung eines (west-)europäischen Bundesstaates könnten sich schon aus der »europäischen Option« des Grundgesetzes selbst ergeben. Hier schließt der Wortlaut des Grundgesetzes, das instititutionell flexible Angebot eines »vereinten« Europa, auch die Form einer bundesstaatlichen Vereinigung ein. Es ist aber fraglich, ob das Grundgesetz eine so weitreichende Vereinigung Europas intendiert, wenn man wieder auf die Materialien seiner Entstehung zurückgreift. Wenn einer der intensivsten Befürworter einer »westeuropäischen Zwischenlösung« nur von einer angestrebten »europäischen Konföderation« sprach [45], so ist dies ein Indiz dafür, daß die Verfassungsväter doch wohl nur an eine Vereinigung Europas auf völkerrechtlicher Stufe gedacht, eine bundesstaatliche Vereinigung aber nicht im Auge gehabt haben. Ein weiteres Indiz dafür ist, daß das Grundgesetz dort, wo es einen institutionell einigermaßen konkretisierten Vorschlag zur Vereinigung Europas enthält, nämlich auf dem Sektor der Sicherheitspolitik in Art. 24 Abs. 2, auch nur eine Vereinigung auf völkerrechtlicher Stufe (ein kollektives Sicherheitssystem) ins Auge faßt. Läßt sich hierauf nicht die These stützen, das Grundgesetz lasse eine Vereinigung Europas nur auf völkerrechtlicher Stufe zu, woraus folgte, daß die Eingliederung der Bundesrepublik Deutschland in einen europäischen Bundesstaat die Grenzen der »europäischen Option« des Grundgesetzes überschritte?

Zweifel an der verfassungsrechtlichen Zulässigkeit einer deutschen Beteiligung an der Errichtung eines (west-)europäischen Bundesstaates ergeben sich aber vor allem aus dem Wiedervereinigungsgebot. In dieser Hinsicht ist das Grundlagenvertragsurteil des Bundesverfassungsgerichts noch kaum gewürdigt worden. Nach der authentischen Interpretation des Wiedervereinigungsgebotes durch das Bundesverfassungsgericht ist der breite Raum politischen Ermessens dann eindeutig überschritten, wenn eine Maßnahme des Ge-

---

[44] *Doehring* (Anm. 10), S. 637 Vgl. im übrigen zur Entstehung und noch vorhandenen Unbestimmtheit des Begriffs »Europäische Union« *H. v. d. Groeben/H. Möller* (Hrsg.), Die Europäische Union als Prozeß, Bd. 1, (1980), S. 149 ff. und S. 213 ff.

[45] Vgl. oben Anm. 16.

setzgebers der Wiedervereinigung »rechtlich oder tatsächlich« offensichtlich entgegensteht[46].

Wieweit die Eingliederung der Bundesrepublik Deutschland in einen (west-)europäischen Bundesstaat der Wiedervereinigung tatsächlich entgegenstehen würde, ist eine schwer zu beantwortende Frage. Immerhin sei daran erinnert, daß schon im EWG-Gründungsjahr im Deutschen Bundestag zu bedenken gegeben wurde, daß man über die Frage der rechtlichen Reversibilität des EWG-Vertrags gar nicht zu reden brauche, »weil die sukzessiv vorgenommene Durchführung des Vertrages eine Verschmelzung der nationalen Volkswirtschaften zu einer europäischen bedeutet, die sich nicht zurückdrehen läßt, also irreversibel ist«[47]. In bezug auf die durch den EWG-Vertrag eingeschränkte Autonomie der Bundesrepublik Deutschland in wesentlichen wirtschaftlichen Materien der Staatsgestaltung hat *Doehring* gesagt, daß es müßig sei, heute darüber nachzudenken, ob dieser so hergestellte Rechtszustand schon gegen das Wiedervereinigungsgebot verstoße, denn das Bundesverfassungsgericht habe darüber geschwiegen, obwohl es genügend Gelegenheit gehabt hätte, die Vereinbarkeit festzustellen[48]; gleiches läßt sich über den faktischen Zustand der inzwischen eingetretenen weitgehenden Verschmelzung der nationalen Volkswirtschaften sagen. Die Frage stellt sich aber für die Zukunft neu, wenn in einem (west-)europäischen Bundesstaat weitere wesentliche Materien der Staatsgestaltung (Sicherheitspolitik, Außenpolitik) der Autonomie der Bundesrepublik Deutschland entglitten. Könnte die Bundesrepublik Deutschland der Verpflichtung, auf die Erreichung des Zieles der Wiedervereinigung Deutschlands hinzuwirken[49], als Gliedstaat eines europäischen Bundesstaates überhaupt noch entsprechen? Könnte, welche rechtlichen Kautelen in einem Unionsgründungsvertrag auch immer vereinbart werden würden, und trotz auch dann wieder zu erwartender Versicherungen der Bundesregierung, »auf einen Zustand des Friedens in Europa hinzuwirken, in dem das deutsche Volk in freier Selbstbestimmung seine Einheit wiedererlangt«, der Beitritt zu einem (west-)europäischen Bundesstaat als eine verhüllte Aufgabe der »Wiederherstellung der staatlichen Einheit (Deutschlands) als politisches Ziel«[50] gedeutet werden, weil eine Konzentration der Bundesorgane auf Errichtung und Ausbau eines (west-)europäischen Bundesstaates diese rein faktisch von dem Bemühen um Wiederherstellung der Einheit Deutschlands auf Jahre oder Jahrzehnte hinaus ablenken würde, mit der Folge, daß diese Einheit *ad calendas graecas* verschoben werden würde? Wie könnten »alle Verfassungsorgane« dann auch nur noch der aus dem Wiedervereinigungsgebot fließenden Forderung, »den Wiedervereinigungs-

---

[46] BVerfGE 36, 17.
[47] 2. BT – 200. Sitzung, Sten. Ber., Bd. 35, S. 11 367 (Abgeordneter Stegner (GB/BHE)).
[48] *Doehring* (Anm. 10), S. 637.
[49] BVerfGE 36, 18.
[50] BVerfGE 36, 17.

anspruch im Innern wachzuhalten und nach außen beharrlich zu vertreten«[51], mit einigermaßen ernstzunehmender Glaubwürdigkeit entsprechen?

Zu der Frage, inwieweit eine Eingliederung der Bundesrepublik Deutschland in einen (west-)europäischen Bundesstaat der Wiedervereinigung rechtlich entgegenstehen würde, lassen sich aus dem Grundlagenvertragsurteil deutlichere Hinweise entnehmen. Die Grenze des Ermessens würde dann überschritten, wenn die Bundesrepublik »einen mit dem Grundgesetz unvereinbaren Rechtstitel schafft oder sich an der Begründung eines solchen Rechtstitels beteiligt, der ihr bei ihrem Streben nach diesem Ziel (der Wiedervereinigung) entgegengehalten werden kann«[52]. Bei der Interpretation der »in einem inneren Zusammenhang mit dem Wiedervereinigungsgebot« stehenden Bestimmung des Art. 23 GG hat dann das Bundesverfassungsgericht die Ermessensgrenze des Wiedervereinigungsgebotes noch deutlicher konturiert; Art. 23 beinhaltet ein »rechtlich Offensein« der Bundesrepublik Deutschland gegenüber dem erstrebten Zuwachs um andere Teile Deutschlands, und dieses liege »spezifisch darin, daß sie, die Bundesrepublik, rechtlich *allein* Herr der Entschließung über die Aufnahme der anderen Teile ist, sobald diese sich dafür entschieden haben beizutreten. Diese Vorschrift verbietet also, daß sich die Bundesregierung *vertraglich in eine Abhängigkeit begibt*, nach der sie rechtlich nicht mehr *allein*, sondern nur noch im Einvernehmen mit dem Vertragspartner die Aufnahme verwirklichen kann«[53].

Diesen Maßstab angelegt, müßte ein westeuropäischer Unionsgründungsvertrag, sollte sich die Bundesrepublik Deutschland daran beteiligen können, Klauseln enthalten, die das »rechtlich Offensein« der Bundesrepublik Deutschland perpetuierten und garantierten. Zwei Klauseln dieser Art wären denkbar: ein Sezessionsrecht der Bundesrepublik Deutschland für den Fall der Wiedervereinigung oder eine Beitrittsklausel des Inhalts, daß über einen Beitritt der DDR zur »Europäischen Union« allein die Bundesrepublik Deutschland zu entscheiden hätte, nicht aber auch die übrigen Unionsmitglieder. Zum Sezessionsrecht hat schon *Doehring* ausgeführt, daß »diese Methode nicht ungefährlich für das Ziel der Wiedervereinigung Deutschlands wäre«, weil die Regierung eines europäischen Bundesstaates es unterlaufen könnte[54]. Was eine spezifische »DDR-Beitrittsklausel« betrifft, so ist mir keine bundesstaatliche Verfassung bekannt, die die Entscheidung über den Beitritt eines Drittstaates ausschließlich einem ihrer Gliedstaaten überließe. Ein Vertrag über die Europäische Union würde insoweit ein völliges Novum schaffen; das wäre zwar theoretisch denkbar, aber es erscheint mir im höchsten Grade unwahrscheinlich, daß eine spezifische »DDR-Beitrittsklausel« politisch durchsetzbar wäre. Ein deutscher Verzicht auf eine solche Klausel

---

[51] BVerfGE 36, 18.
[52] *Ibid.*
[53] BVerfGE 36, 28 (Hervorhebung im Original).
[54] *Doehring* (Anm. 10), S. 639.

aber wäre meines Erachtens ein »schwerwiegender Einwand gegen die verfassungsrechtliche Zulässigkeit einer Europäischen Union« [55].

Die sich aus dem Wiedervereinigungsgebot ergebenden Zweifel an der verfassungsrechtlichen Zulässigkeit einer deutschen Beteiligung an der Errichtung eines (west-)europäischen Bundesstaates lassen sich auch nicht mittels Art. 24 Abs. 1 GG aus dem Weg räumen. Der Beitritt der Bundesrepublik Deutschland zu einem (west-)europäischen Bundesstaat »durch Gesetz«, das gemäß Art. 24 Abs. 1 GG erginge, wäre allein schon deswegen nicht möglich, weil ein Bundesstaat nicht als »zwischenstaatliche Einrichtung« im Sinne dieses Artikels verstanden werden kann. »Zwischenstaatliche Einrichtungen« sind nur solche Staatenverbindungen, deren Begründung oder Fortentwicklung nicht zur Aufhebung der Völkerrechtsunmittelbarkeit ihrer Mitglieder führen, der Beitritt der Bundesrepublik Deutschland zu einem Bundesstaat hätte aber diesen Effekt.

Zusätzlich sind der Integrationsgewalt gerade durch die Bindung der Verfassungsorgane an das Wiedervereinigungsgebot Grenzen gesetzt. Die Ansicht, daß Art. 24 Abs. 1 GG eine »Bresche«, eine »offene Flanke« des Grundgesetzes bilde, »daß er mit anderen Worten eine vorweggenommene Änderung der Verfassung darstelle und infolgedessen keinesfalls an deren sonstige Vorschriften gebunden sei« [56], sollte spätestens mit dem »Solange«-Beschluß des Bundesverfassungsgerichts vom 29. Mai 1974 [57] überwunden sein, unabhängig von der Kritik, die dieser Beschluß im konkreten Fall erfahren hat [58, 59]. Art. 24 Abs. 1 GG, so hat das Bundesverfassungsgericht ausgeführt, »muß wie jede Verfassungsbestimmung ähnlich grundsätzlicher Art im Kontext der Gesamtverfassung verstanden und ausgelegt werden. Das heißt, er eröffnet nicht den Weg, die Grundstruktur der Verfassung, auf der ihre Identität beruht, ohne Verfassungsänderung« zu ändern [60]. Ist nun nicht wie der Grundrechtsteil auch das Wiedervereinigungsgebot »ein unaufgebbares, zur Verfassungsstruktur des Grundgesetzes gehörendes Essentiale der geltenden Verfassung der Bundesrepublik Deutschland«, das zu relativieren Art. 24 GG »nicht vorbehaltlos« gestattet [61]? Ist doch das »Grundgesetz rechtlich auf die Ziele der Wiedervereinigung Deutschlands angelegt« [62], und

---

[55] So *Ress* (Anm. 6), S. 286 Anm. 64, zustimmend zu früheren einschlägigen Ausführungen des Verfassers, vgl. *Th. Schweisfurth*, KSZE und Deutsche Frage, Deutschlandarchiv 1977, S. 941 f.

[56] *Maunz/Dürig/Herzog/Scholz* (Anm. 7), Art. 24 Rdnr. 17.

[57] BVerfGE 37, 271 ff.

[58] Vgl. hierzu z. B. die Stellungnahme zum »Solange«-Beschluß vom 29. 5. 1974 von *M. Hilf*, *E. Klein* und *A. Bleckmann*, Sekundäres Gemeinschaftsrecht und deutsche Grundrechte, ZaöRV 35 (1975), S. 51 ff.

[59] Vgl. auch *Doehring* (Anm. 10), S. 638, es ergebe sich aus dem »Solange«-Beschluß jedenfalls mittelbar, daß der Anwendung des Art. 24 GG verfassungsrechtliche Grenzen gesetzt seien.

[60] BVerfGE 37, 279.

[61] BVerfGE 37, 280.

[62] So *Doehring* (Anm. 10), S. 635.

hat doch das Bundesverfassungsgericht im Grundlagenvertragsurteil den »in einem inneren Zusammenhang mit dem Wiedervereinigungsgebot« stehenden Art. 23 GG »zu den zentralen Vorschriften« gezählt, »die dem Grundgesetz sein besonderes Gepräge geben«[63]. Stellt Art. 24 Abs. 1 GG auch eine Möglichkeit für eine intensive Verflechtung der Bundesrepublik Deutschland im Rahmen einer internationalen Kooperation zur Verfügung, so weist er doch keinen Weg, der um das Wiedervereinigungsgebot herumführt. Wenn dieses »nicht vorbehaltlos« relativiert werden darf, wie sich dies meines Erachtens aus den angeführten Judikaten ergibt, dann ist die auswärtige Gewalt von Verfassungs wegen gehalten, auch bei der Fortführung der EG-Integration (über Art. 24 Abs. 1) entsprechende »Vorbehalte«, wie sie oben angesprochen worden sind (Sezessionsrecht, »DDR-Beitrittsklausel«), zu machen – womit wir wieder bei deren Problematik angelangt wären[64].

Gegen alle bislang hier vorgetragenen Zweifelsfragen der verfassungsrechtlichen Zulässigkeit der deutschen Beteiligung an einem (west-)europäischen Bundesstaat könnte nun grundsätzlich eingewendet werden, sie basierten auf einer zu einseitigen Betonung der »subjektiven« Auslegung des Grundgesetzes, d.h. einer Auslegung, die sich zu stark an die Intention der Verfassungsväter halte; sie berücksichtigen überhaupt nicht die Idee des »Verfassungswandels«[65]. Würde eine Auslegung des Grundgesetzes vorgenommen, die sich mehr »am Verhalten der Verfassungsorgane im Laufe der Verfassungsentwicklung«[66] und an den seit Inkrafttreten des Grundgesetzes eingetretenen Veränderungen der internationalen Lage orientierte, dann könnte man zu dem Ergebnis kommen, daß das Wiedervereinigungsgebot inhaltlich auf die Pflicht zur Pflege »normaler« Beziehungen zur DDR auf der Basis des Grundlagenvertrages und der Europabegriff des Grundgesetzes geographisch auf Westeuropa reduziert sei.

---

[63] BVerfGE 36, 28.

[64] Pressemeldungen zufolge (Der Tagesspiegel vom 22. 9. 1981) soll die Bundesregierung beabsichtigen vorzuschlagen, daß die »Europäische Akte« nicht die Form eines völkerrechtlichen Vertrages bekommen, sondern als »gemeinschaftliche verbindliche Willensbekundung der Regierungen der Mitgliedstaaten« der EG angenommen werden soll; die KSZE-Schlußakte schwebt hier offensichtlich als Vorbild vor. Sollte dies tatsächlich geschehen, würde die Entscheidung über einen weiteren wichtigen Schritt zu einem (west-)europäischen Bundesstaat dem parlamentarischen Mitverantwortungsbereich entzogen werden. Es ergäbe sich dann die Frage, ob es verfassungsrechtlich zulässig ist (eventuelle Umgehung von Art. 24 Abs. 1 bzw. 59 Abs. 2 GG), daß die Exekutive in einer geradezu schicksalhaften außenpolitischen Grundorientierung politisch vollendete Tatsachen schaffen kann, ohne daß das Für und Wider in einer parlamentarischen Debatte ausdiskutiert werden kann; vgl. zu dieser Problematik »einer apokryphen internationalen Rechtsetzung« grundsätzlich *C. Tomuschat*, Der Verfassungsstaat im Geflecht der internationalen Beziehungen, Veröffentlichungen der Vereinigung der Deutschen Staatsrechtslehrer (VVDStRL) Heft 36 (1978), S. 7 ff., hier S. 32 ff.

[65] Vgl. dazu *R. Bernhardt*, Deutschland nach 30 Jahren Grundgesetz, VVDStRL Heft 38 (1980), S. 7 ff., hier S. 11 f. und die dort angegebene Literatur.

[66] *Doehring* (Anm. 10), S. 636.

Die Idee des Verfassungswandels hat in unserem Zusammenhang meines Erachtens wenig Überzeugungskraft. *Doehring* hat auf diesen Einwand zutreffend erwidert, daß trotz aller Wandlungen im Faktischen und im Rechtlichen (Abschluß des Grundlagenvertrages) das Wiedervereinigungsgebot des Grundgesetzes »nach wie vor« besteht[67]. Kann und muß das Verfassungsverständnis sich auch ändern, weil die Verfassungsauslegung »auch auf neue Entwicklungen und Notwendigkeiten reagieren« muß, so doch nur »innerhalb gewisser Grenzen und bei gleichbleibendem Wortlaut«[68]. Bemerkenswert ist, daß das Bundesverfassungsgericht nur von einem »Bedeutungswandel« gesprochen hat, den eine Verfassungsbestimmung erfahren kann, wenn in ihrem Bereich neue, nicht vorausgesehene Tatbestände auftauchen[69]; »Bedeutungswandel« ist weitaus weniger als ein Obsoletwerden. Ein Bedeutungswandel des Wiedervereinigungsgebotes scheint inzwischen eingetreten zu sein, soweit es sich um das zur Wiedervereinigung führende *procedere* handelt; denn wenn das Bundesverfassungsgericht darunter einen »längeren Prozeß« versteht, »der zunächst in einem der dem Völkerrecht bekannten verschiedenen Varianten einer Konföderation« enden kann[70], so ist hier, unter Berücksichtigung neuer inzwischen aufgetauchter Tatbestände – nämlich der sozialen, politischen und ökonomischen Auseinanderentwicklung beider Teile Deutschlands – ein Weg »in Richtung auf die Verwirklichung der Wiedervereinigung des deutschen Volkes in einem Staat«[71] ins Blickfeld gekommen, an den die Verfassungsväter sicher noch nicht gedacht hatten. Ein solcher Bedeutungswandel des Wiedervereinigungsgebotes läßt aber seinen Kern, das Ziel, Deutschland staatlich zu reorganisieren, unbeührt.

Grundsätzlich wird vor der Verwendung der Idee des Verfassungswandels als Auslegungsmaxime mit Recht gewarnt, weil sie »die Gefahr des Abdankens des Verfassungsrechts vor dem Einbruch des Faktischen« bedeutet[72]. Und gerade in bezug auf das Wiedervereinigungsgebot und die europäische Option des Grundgesetzes trifft zu, was *Ress* zu bedenken gegeben hat: »Mit dem rechtsstaatlichen Verfassungsverständnis ist eine Interpretation, welche auf das jeweils »Zeitgemäße« und damit politisch »Wünschbare« abstellt, nicht vereinbar, weil sie die normative Kategorie der Verfassung und damit ihren Rechtscharakter zugunsten einer politischen Realanalyse vernachlässigt, wenn nicht gar aufgibt«[73].

Was die verfassungsrechtlichen Möglichkeiten der Fortführung der deutschen Europapolitik betrifft, so sollen die vorstehenden Überlegungen mit folgenden Bemerkungen ihren Abschluß finden. Der derzeit bestehende

---

[67] *Ibid.*
[68] So *Bernhardt* (Anm. 65), S. 12.
[69] BVerfGE 2, 401.
[70] BVerfGE 36, 25 f.
[71] *Ibid.*
[72] So *Stern* (Anm. 14), S. 115.
[73] *Ress* (Anm. 6), S. 266.

Widerspruch zwischen der Verfassungspraxis der westeuropäischen Integration und dem Wiedervereinigungsgebot (sowie im Grunde auch dem Europabekenntnis) des Grundgesetzes läßt sich meines Erachtens nur auf zweierlei Weise auflösen. Zum einen kann der bisher eingeschlagene Weg der westeuropäischen Integration fortgesetzt werden, nunmehr entlang der im »Entwurf einer Europäischen Akte« enthaltenen Leitlinien und ausmündend in die Gründung eines westeuropäischen Bundesstaates. Wird dieser Weg weiter verfolgt, dann ist, so das zutreffende Ergebnis der Überlegungen *Doehrings*[74], dies der Bundesrepublik Deutschland angesichts der Bindung an das Wiedervereinigungsgebot nur möglich, wenn eine dieses Gebot aufhebende Verfassungsänderung vorgenommen wird.

Die Verfassungspraxis könnte mit der Verfassung aber auch dadurch wieder zur Kongruenz gebracht werden, daß die Verfassungspraxis sich wieder an der umfassenden europäischen Option des Grundgesetzes orientiert und diese wieder an das Wiedervereinigungsgebot ankoppelt. Hinsichtlich des Wiedervereinigungsgebotes würde das bedeuten, daß die Verfassungspraxis – wenn schon eine »aktive freiheitliche Wiedervereinigungspolitik« derzeit noch immer nicht möglich erscheint – das Wiedervereinigungsgebot »nach der negativen Seite hin«[75] streng zu beachten hätte, d. h. jeden weiteren Schritt zur westeuropäischen Integration sorgfältig daraufhin prüft, ob er nicht eine Maßnahme darstellt, die die Wiedervereinigung rechtlich hindert oder faktisch unmöglich macht.

Hinsichtlich der europäischen Option des Grundgesetzes würde dies bedeuten, daß die Verfassungspraxis wieder den auf ganz Europa bezogenen Europabegriff des Grundgesetzes ins Blickfeld nimmt und sich von der Beschränkung auf die Vereinigungsmethode der »Integration« über Art. 24 Abs. 1 GG befreit. Wäre aber nun nicht eine solche auf ganz Europa bezogene Europapolitik zum Scheitern verurteilt, weil »die Vision eines zwischen kommunistischen und kapitalistischen Staaten gebildeten vereinigten Europas ohnehin utopisch (erscheint), es sei denn, Grundprinzipien des bestehenden Verfassungsrechts der Bundesrepublik werden aufgegeben«[76]?

Utopisch ist eine solche Vision in der Tat – allerdings nur, wenn man sich die Vereinigung Europas ausschließlich in den Formen vorstellt, wie sie in der EG Gestalt angenommen haben. Besinnt man sich aber darauf, daß Art. 24 Abs. 1 GG kein »aus sich selbst heraus legitimiertes Ziel darstellt«[77], sondern nur *eine* der nach dem Grundgesetz möglichen Vereinigungsmethoden, dann ist ein »vereintes Europa«, das sowohl kommunistische wie kapitalistische Staaten einschließt, durchaus keine Utopie. Die Gründung gesamteuropäischer internationaler Organisationen ist doch denkbar, ohne daß ei-

---

[74] *Doehring* (Anm. 10), S. 639.
[75] BVerfGE 5, 128.
[76] So *Doehring* (Anm. 10), S. 634.
[77] Vgl. *Tomuschat* (Anm. 19), Rdnr. 5.

ner der beteiligten Staaten Grundprinzipien seines Verfassungsrechts aufgeben müßte. Sachgebiete europäischer Kooperation sind beispielsweise in der KSZE-Schlußakte zur Genüge aufgelistet. Würden die derzeit stagnierenden Verhandlungen zwischen den EG und dem RGW fortgeführt, so könnten auch sie in eine institutionelle Zusammenarbeit zwischen diesen beiden europäischen Wirtschaftsorganisationen bzw. zwischen deren Mitgliedstaaten ausmünden. Für eine ferner liegende Zukunft wäre sogar ein »vereintes Europa« auf sicherheitspolitischem Gebiet denkbar, womit die auf diesem Sektor formulierte »europäische Option« des Grundgesetzes – Art. 24 Abs. 2 – endlich doch noch realisiert werden würde. Beide Wege sind gangbar. Die Bundesrepublik Deutschland muß sich nur, wird die westeuropäische Integrationspolitik fortbetrieben bis zur Gründung eines westeuropäischen Bundesstaates, für den einen oder anderen entscheiden. Der heute schon bestehende und sich durch eine konsequente Politik in Richtung auf die Gründung einer »Europäischen Union« noch verstärkende Widerspruch zwischen der Verfassungspraxis der westeuropäischen Integration und den Geboten der Präambel des Grundgesetzes ist jedenfalls in einem Rechtsstaat auf die Dauer verfassungsrechtlich nicht tolerabel.

Der Rückgriff auf die Mitgliedstaaten in Internationalen Organisationen

IGNAZ SEIDL-HOHENVELDERN

Zwischenstaatliche Organisationen verdanken ihre Existenz als Völkerrechtssubjekte dem dahin zielenden ausdrücklich[1] oder stillschweigend[2] geäußerten Willen ihrer Mitgliedstaaten. Dieser ist in aller Regel in einem völkerrechtlichen Vertrag niedergelegt, der gleichzeitig die Verfassung der Organisation darstellt.

1. Gelten nun auch für diesen Vertrag uneingeschränkt Art. 54 und 59 des Wiener Übereinkommens über das Recht der Verträge[3] als bloße Kodifizierung eines Völkergewohnheitsrechtes? Bleiben also die Mitgliedstaaten auch hier Herren des Vertrages? Können sie ihn also durch einen von allen Mitgliedstaaten geschlossenen späteren Vertrag beenden oder abändern, ohne die Vorschriften einzuhalten, die der als Verfassung der Organisation beschlossene frühere Vertrag für seine Beendigung oder Abänderung vorsieht? Können hier die Mitglieder, unter völliger oder teilweiser Nichtachtung der Tatsache, daß sie durch den Gründungsvertrag ein Gebilde geschaffen haben, das eine eigene, von den ihrigen verschiedene Völkerrechtspersönlichkeit besitzt, ihren nunmehrigen neuen gemeinsamen Willen unter Rückgriff auf die Tatsache durchsetzen, daß die Organisation ihr Bestehen ja nur dem dahin gehenden früher geäußerten Willen der Mitgliedstaaten verdankt?

2. Kann ein solcher Änderungs- oder Endigungswille auch von Staatenvertretern geäußert werden, die gleichzeitig Mitglieder eines Organs der betreffenden Organisation sind?

3. Sollen aber nur die Mitgliedstaaten selbst durch den Schleier der eigenen Rechtspersönlichkeit der Organisation hindurch greifen können oder sollte nicht ein solcher Rückgriff auch Dritten möglich sein, um eine Haf-

---

[1] Vgl. Art. 6 Abs. 2 EGKS-Vertrag.

[2] *Seidl-Hohenveldern,* Das Recht der Internationalen Organisationen einschließlich der Supranationalen Gemeinschaften (3. Aufl. 1979), S. 45.

[3] Österr. Bundesgesetzblatt (ÖBGBl.) Nr. 40/1980, vgl. unten Text nach Anm. 5.

tung der Mitgliedstaaten für ein nach Völkerrecht oder innerstaatlichem Recht rechtswidriges Verhalten der Organisation geltend zu machen, in Ermangelung oder in Ergänzung einschlägiger Vorschriften im Recht der Organisation?

I

Was den Rückgriff auf den gemeinsamen Willen der Mitgliedstaaten zur Änderung oder Beendigung des Gründungsvertrages betrifft, halten wir mit *Mosler*[4] zunächst einmal fest, daß in den Verträgen zur Errichtung internationaler Verbände sich vertragliche und satzungsrechtliche Elemente unterscheiden lassen. Im ersten Bereich haben die Staaten gegenseitige Rechte und Pflichten aus dem Vertrag, im zweiten Rechte und Pflichten im Verhältnis zur Organisation als deren Mitglieder. Welches dieser Elemente überwiegt nun im Konfliktsfall? Laut Art. 5 des Wiener Übereinkommens über das Recht der Verträge findet dieses Übereinkommen auf jeden Vertrag Anwendung, durch den eine internationale Organisation errichtet wird, sowie auf jeden Vertrag, der innerhalb einer Organisation geschlossen wird, unbeschadet der hierfür von dieser Organisation erlassenen Regeln. Der letzte Halbsatz kann und will aber gar nicht für diesen besonderen Fall die alte Streitfrage, wer eher war, die Henne (die Staaten) oder das Ei (die Organisation), zugunsten der Regeln der Organisation lösen. Dafür spricht zunächst einmal schon die Überlegung, daß, um dem Vergleich treu zu bleiben, zwar die Staaten die Organisationen schaffen, Staaten aber ihre Existenz im allgemeinen nicht von Organisationen ableiten. Die Staaten stellen also gegenüber den Internationalen Organisationen ein *prius* dar[5]. Art. 5 des Übereinkommens will lediglich klarstellen, daß *innerhalb* einer Organisation deren Sonderregelungen z.B. über das Verhältnis zu anderen Verträgen (Art. 103 UN-Satzung, Art. 234 EWG) Vorrang vor den allgemeinen Regeln des Art. 59 des Übereinkommens haben.

Nur in diesem Sinn ist auch das Rechtsgutachten des Ständigen Internationalen Gerichtshofes vom 28. August 1928[6] über die Auslegung des Griechisch-Türkischen Abkommens vom 1. Dezember 1926 zu verstehen. Das Griechisch-Türkische Abkommen vom 30. Januar 1923[7] hatte eine Gemisch-

---

[4] National- und Gemeinschaftsinteressen im Verfahren des EWG-Ministerrates. Die Beschlüsse vom 29. 1. 1966, ZaöRV 26 (1966), S. 5.

[5] So *H. Mosler,* Die Erweiterung des Kreises der Völkerrechtssubjekte, Berichte der Deutschen Gesellschaft für Völkerrecht, H. 4 (1961), S. 56.

[6] Series B, Nr. 16, hierzu *E. Lauterpacht*, The Development of the Law of International Organization by the Decisions of International Tribunals, RdC 152 (1976 IV), S. 404–405.

[7] LNTS 32, S. 77, Art. 11.

te Kommission als »mainly an administrative body« errichtet. Sie bestand aus elf Mitgliedern, je vier davon wurden von Griechenland und der Türkei und drei vom Völkerbundsrat ernannt. Die Kommission sollte den in diesem Abkommen vorgesehenen Bevölkerungsaustausch überwachen. Das Griechisch-Türkische Abkommen vom 1. Dezember 1926[8] bestimmte sodann, daß Fragen hinsichtlich der Zuständigkeit dieser Kommission einer Schiedsentscheidung des Präsidenten des Griechisch-Türkischen Schiedsgerichtes unterbreitet werden sollten, erwähnte aber nicht, *wer* zur Einholung dieser Schiedsentscheidung zuständig sein sollte. Im konkreten Fall hatte Griechenland versucht, dies im Alleingang zu tun.

Dem Ständigen Internationalen Gerichtshof lag allerdings die Frage vor, ob ein Mitgliedstaat oder beide Mitgliedstaaten gemeinsam an Stelle der Kommission das Schiedsverfahren in Gang setzen konnten. Das verneinte der Ständige Internationale Gerichtshof mit folgenden Worten: »To accord to individual members of an organization constituted as a corporate body any right to take action of any kind outside the sphere of proceedings within that organization, would be clearly contrary to an accepted principle of law. In the absence of an express provision, therefore, it is impossible to accept such an abnormal state of things«.

Auch hier ist ein solches Vorgehen nur *im Rahmen der Organisation* »abnormal« zu nennen. Die Frage war in diesem Sinne zu verstehen und konnte im gegenständlichen Fall auch gar nicht anders beantwortet werden. Die Kommission bestand ja nicht nur aus Vertretern der Türkei und Griechenlands, sondern auch aus drei Vertretern des Völkerbundsrates. Schon aus diesem Grunde hätte auch ein Einvernehmen der beiden Staaten nicht dem Willen der Kommission gleichgesetzt werden können.

Wenn aber wirklich alle an einer Organisation Beteiligten ein Abweichen von deren Gründungsvertrag beschließen, dann können sie dies als Herren des Vertrages tun. Es erscheint mir unangebracht, diesen Ausfluß des souveränen Willens der Mitgliedstaaten mit der Begründung zu leugnen, der Abschluß des Gründungsvertrages stelle für die Mitgliedstaaten eine unwiderrufliche Selbstbindung ihrer Souveränität dar[9]. Der Gedanke der Selbstbindung hat zur Linderung zumindest der ärgsten Auswüchse der Theorie der absoluten Souveränität gewiß gute Dienste geleistet[10]. Dieser Satz besteht aber nicht um seiner selbst willen. Ein Staat bindet sich durch einen Vertrag jeweils gegenüber einem oder mehreren anderen Völkerrechtssubjekten als Partner dieses Vertrages. Nun sind aber in dem hier zu behandelnden Fall ja alle Partner des Gründungsvertrages einverstanden, ihn durch den späteren Vertrag zu ersetzen. Sie entlassen sich also gegenseitig aus ihrer früheren

---

[8] LNTS 68, S. 21.

[9] So aber *A. Bleckmann,* Internationale Beamtenstreitigkeiten vor nationalen Gerichten, Rechtsgutachten für die Union Syndicale, Section Eurocontrol (1981), S. 100.

[10] *A. Verdross/B. Simma,* Universelles Völkerrecht (1976), S. 66.

Selbstbindung. Im vertragsrechtlichen Bereich geht also der spätere Vertrag dem Gründungsvertrag mit der gleichen mathematischen Sicherheit vor, demzufolge eins minus eins gleich null ist.

Im satzungsrechtlichen Bereich sieht es allerdings, wie oben dargestellt [11], anders aus. Kann dadurch aber rechtlich und tatsächlich diese mathematische Zwangsläufigkeit entkräftet werden? Gewiß, Internationale Organisationen gehören heute unbestrittenermaßen zum Kreis der Völkerrechtssubjekte [12]. Die mit dem Gründungsvertrag von den Mitgliedstaaten ins Leben gerufene Organisation ist aber nicht Partner dieses Vertrages, sondern nur ein drittes, durch diesen Vertrag gegründetes Völkerrechtssubjekt. Das Völkerrecht kennt zwar Verträge zu Gunsten Dritter, die diesem Dritten eine eigene Rechtsposition einräumen [13]. Ein solches Recht könnte ihm auch durch einen *actus contrarius* der Partner des betreffenden Vertrages nicht entzogen werden. So verdankt z. B. die Republik Zypern ihre Existenz dem Londoner Vertrag vom 19. Februar 1959 zwischen Griechenland, der Türkei und Großbritannien [14]. Dennoch könnten diese drei Staaten durch einen zwischen ihnen abgeschlossenen künftigen Vertrag den durch den Vertrag vom 19. Februar 1959 geschaffenen Staat nicht wieder beseitigen, also etwa die Insel zwischen Griechenland und der Türkei nicht aufteilen. In diesem Beispiel zeigt sich aber ein grundlegender Unterschied in der Qualität der Völkerrechtssubjektivität von Staaten und Internationalen Organisationen. Die Völkerrechtspersönlichkeit der Staaten besteht, einmal entstanden, absolut. Der so geschaffene Staat genießt alle Rechte, die das Völkerrecht den Staaten einräumt, einschließlich seines Rechtes auf Existenz und Nichteinmischung, auch gegenüber seinen Gründungsstaaten. Eine Internationale Organisation ist dagegen dergestalt aus dem Willen ihrer Mitgliedstaaten abgeleitet, daß sie nur die Rechte genießt, die ihr die Mitgliedstaaten ausdrücklich oder stillschweigend einräumen, und auch diese nur so lange, als die Mitgliedstaaten nicht gemeinsam ihren Willen ändern [15]. Von einer hierarchischen Unterwerfung der Mitgliedstaaten unter die Satzung der Organisation kann also nur im Bereich des Rechts der Organisation selbst die Rede sein. Vom allgemeinen Völkerrecht her steht aber der gemeinsame Wille der Mitgliedstaaten über der Satzung. Sie bleiben hier also Herren über ihr Geschöpf.

Dies beweist nachstehender kurzer Blick auf die Praxis, deren Ergebnis im übrigen schon auf Grund folgender Überlegung nahezu selbstverständlich ist. Bei allem Respekt vor der eigenständigen Völkerrechtssubjektivität der Organisation und vor den innerhalb des Rechts der Organisation bestehen-

---

[11] Vgl. oben Text bei Anm. 5–8.

[12] *Mosler* (Anm. 5), S. 39 ff.

[13] Vgl. insbes. die Entscheidung des StIGH im *Genfer Freizonen*-Fall vom 7. 6. 1932, PCIJ Series A/B Nr. 46.

[14] Jahrbuch des öffentlichen Rechts 9 (1960), S. 286 ff., *L. Dischler,* Die Zypernfrage (1960), S. 151.

[15] *Seidl-Hohenveldern* (Anm. 2), S. 39–40, ebenso *Mosler* (Anm. 5), S. 59–60.

den satzungsrechtlichen Bindungen bleibt doch unbestritten, daß den aus Vertretern der Mitgliedstaaten bestehenden Organen eine sehr entscheidende Rolle bei der Willensbildung jeder Organisation zukommt. Selbst wenn man der hier vertretenen Vorstellung von den Mitgliedstaaten als Herren des Vertrages nicht folgen wollte, hätten es die Mitgliedstaaten in der Hand, durch einvernehmliches, wenn auch satzungswidriges Verhalten dieser ihren Weisungen unterworfenen Organe die Rechtswirklichkeit in der Organisation derart zu beeinflussen, daß das nunmehr von ihnen gewollte Verhalten als Gewohnheitsrecht der Organisation deren ursprünglich gesatztes Recht abändern könnte[16]. Ich verweise als Beispiel auf die gewohnheitsrechtlich eingeführte Regel, daß trotz des Wortlauts des Art. 27 Abs. 3 UN-Satzung eine Stimmenthaltung einer vetoberechtigten Großmacht im Sicherheitsrat nicht als Ausübung ihres Vetorechts angesehen wird. Im Extremfall hätten es die Mitgliedstaaten in der Hand, durch gemeinsame völlige Untätigkeit der Staatenvertretungsorgane in der Organisation deren Tätigkeit gänzlich zum Erliegen zu bringen. Die Untätigkeit einer Organisation kann aber genau so gut wie ein *actus contrarius*[17] das Ende einer Organisation bewirken. Mit diesem Argument hat *Kelsen*[18] das Ende des Völkerbundes bejaht, obwohl seiner Ansicht nach kein *actus contrarius* vorlag, da ja nicht alle Mitgliedstaaten des Völkerbundes dem Auflösungsbeschluß zugestimmt hatten. So sind auch die Internationale Oderkommission ebenso wie die Internationale Elbekommission und die Organisation Internationale Africaine et Malgache ohne förmlichen Auflösungsbeschluß, lediglich durch ein späteres Verhalten ihrer Mitgliedstaaten, nämlich durch deren Untätigkeit, untergegangen[19].

II

Man kann die Durchsetzung des Willens der Mitgliedstaaten darauf zurückführen, daß sie die Herren über den Gründungsvertrag geblieben sind oder darauf, daß sie durch ein einvernehmliches Verhalten ihrer Vertreter in

---

[16] *Lauterpacht* (Anm. 6), S. 461 und 465.

[17] Laut einer Stellungnahme der Völkerrechtsdirektion des Eidgenössischen Politischen Departements kann eine Internationale Organisation auch anders als durch *actus contrarius*, so auch durch eine stillschweigende Billigung ihres Endens zu bestehen aufhören (Schweiz. Jahrbuch für intern. Recht 36 [1980], S. 168). Die Stellungnahme beruft sich hierbei darauf, daß die UN-Völkerrechtskommission in ihrem Entwurf zu dem späteren Art. 54 des Wiener Übereinkommens über das Recht der Verträge vom Erfordernis des *actus contrarius* abgerückt ist. (Vgl. *R. G. Wetzel/D. Rauschning*, Die Wiener Vertragsrechts-Konvention, Materialien zur Entstehung der einzelnen Vorschriften (1978), S. 386.

[18] *H. Kelsen*, The Law of the United Nations (1951), S. 595.

[19] *H. G. Schermers*, International Institutional Law (2. Aufl. 1980), S. 812.

der Organisation innerhalb des Rechtes der Organisation satzungsänderndes Gewohnheitsrecht geschaffen haben. In der Praxis wird die Wahl zwischen diesen beiden Deutungen oft nicht leicht sein. Dies ist darauf zurückzuführen, daß die als Staatenvertreter in die Organe der Organisation entsandten Personen zwei Rollen spielen können. Als Mitglieder eines Organs der Organisation sind sie an deren Satzung grundsätzlich gebunden. Verhalten sich aber alle Mitglieder eines solchen Organs ständig in einer mit der Satzung unvereinbaren Weise, schaffen sie dadurch satzungsintern Gewohnheitsrecht, das insoweit an Stelle der Satzung tritt. Einfacher ist es aber anzunehmen, daß die Staatenvertreter hier in ihrer anderen Eigenschaft, nämlich nicht als Mitglieder des Organs der Organisation, sondern eben als Vertreter ihrer Staaten auftreten. Ihre Willenseinigung stellt dann eben einen im vereinfachten Verfahren geschlossenen Vertrag dar. Vom Völkerrecht her gesehen, ist ein solcher Vertrag gleichwertig mit jedem im klassischen Vertragsabschlußverfahren abgeschlossenen Vertrag[20], also auch mit dem Gründungsvertrag der Organisation.

Nur innerstaatlich könnten sich Bedenken ergeben, wenn der Gründungsvertrag mit Zustimmung der gesetzgebenden Körperschaften zustande käme, wohingegen eine solche Zustimmung wohl kaum vorliegen dürfte, wenn ein Beschluß z. B. des Ministerrats einer Organisation in ein im vereinfachten Verfahren zwischen den Außenministern der Mitgliedstaaten abgeschlossenes Regierungsübereinkommen umgedeutet wird. Infolge der funktionellen Verdoppelung der beschlußfassenden bzw. vertragschließenden Personen wird es oft gar nicht leicht sein festzustellen, in welcher ihrer beiden Funktionen sie tätig werden wollten[21]. Die Luxemburger Beschlüsse vom 28. und 29. Januar 1966[22] sollen nach Auskunft des Rates der EG nicht auf einer Regierungskonferenz, sondern auf einer außerordentlichen Tagung des Rates zustandegekommen sein. Diese Auskunft ließ dennoch die Rechtsfrage offen, ob es sich dabei um einen Beschluß des Rates handelte[23]. *Mosler* sieht die Beschlüsse zu Recht als eine Vereinbarung der Regierungen über das Verfahren im Rat an[24]. Klarer kommt dies bei dem Beschluß der Außenminister der EG-Staaten vom 30. Mai 1980 über die Verringerung der von Großbritannien auf Grund des EWG-Vertrages an den EWG-Haushalt abzuführenden Summen zum Ausdruck, der der Bestätigung durch die Regierungen der Mit-

---

[20] So konnte der Notenwechsel zwischen der Deutschen Botschaft in Paris und dem Französischen Außenminister vom 8. 3. 1939 die wörtliche Auslegung des § 2 Ziff. 2 der Anlage hinter Art. 79 des Versailler Vertrages, der die französischen Gerichte bis dahin gefolgt waren, für nicht mehr anwendbar erklären. Vgl. *Lichter*, Die Staatsangehörigkeit nach deutschem und ausländischem Recht (2. Aufl. 1955), S. 390.

[21] Vgl. etwa *Fouchard/Mourgeon*, Le fonds d'entraide et de garantie des emprunts du Conseil de l'Entente, Clunet 1969, S. 31.

[22] *Mosler,* (Anm. 4), S. 1–4.

[23] *Mosler,* a.a.O., S. 5.

[24] *Mosler,* a.a.O., S. 18 und 27.

gliedstaaten bedurfte[25]. *Mosler* verweist auf weitere Beispiele aus dem Recht der Europäischen Gemeinschaften[26] und der Vereinten Nationen[27]. Der Internationale Gerichtshof hat außersatzungsmäßige Änderungen der UN-Satzung in seinen Rechtsgutachten über *Südwest-Afrika* gutgeheißen, soweit diese Änderungen nicht mit den Hauptzielen und dem Grundaufbau der Vereinten Nationen unvereinbar wären[28]. Eine Entscheidung der Mitgliedstaaten vom 28. November 1979 haben Art. XVIII A und C das GATT *de facto* abgeändert[29]. Auch aus der Praxis der EUROCONTROL lassen sich einschlägige Beispiele anführen[30]. Die Praxis bestätigt also die Möglichkeit der Mitgliedstaaten jeder Organisation, trotz der eigenen Völkerrechtspersönlichkeit der Organisation auch ohne förmliche Änderung der Satzung ihren gemeinsamen späteren Willen selbst dann durchzusetzen, wenn er mit der Satzung unvereinbar wäre.

III

Können nun aber auch Dritte trotz der eigenen Völkerrechtspersönlichkeit der Organisation auf deren Mitgliedstaaten zur Geltendmachung von Haftungsansprüchen zurückgreifen? Wenn den Mitgliedstaaten, wie wir gesehen haben, die Möglichkeit bleibt, durch gemeinsames Vorgehen ihren späteren Willen selbst entgegen der Satzung durchzusetzen, erscheint deren Haftung für das Handeln der Organisation nur recht und billig. Der Schleier der Rechtspersönlichkeit einer juristischen Person kann nach allgemeinen Rechtsgrundsätzen sowohl zum Schutz der Rechte ihrer Mitglieder als auch Dritter durchstoßen werden. Was der IGH im *Barcelona Traction*-Fall[31] diesbezüglich für nach innerstaatlichem Recht errichtete Gesellschaften ausführte, sollte auch für Internationale Organisationen gelten, da die Beziehungen zwischen Organisation und Mitgliedstaaten denen zwischen einer juristischen Person und ihren Mitgliedern entsprechen.

So entspricht es völlig der Rechtsansicht des IGH im *Barcelona Traction*-Fall[32], daß im Falle der Beendigung der Existenz einer Organisation[33] deren Aktivvermögen auf die Mitglieder übergeht, die aber auch für deren Passiven

[25] Keesing's Archiv der Gegenwart 1980, S. 23579.
[26] *Mosler* (Anm. 4), S. 19–20.
[27] *Mosler*, a.a.O., S. 14–17.
[28] ICJ Reports 1950, S. 137, und ICJ Reports 1955, S. 74.
[29] Annuaire français de droit international 1979, S. 595.
[30] *I. Seidl-Hohenveldern,* Die Immunität Internationaler Organisationen in Dienstrechtsstreitfällen, Rechtsgutachten für Eurocontrol (1981), S. 69–70.
[31] Entscheidung vom 5. 2. 1970, ICJ Reports 1970, S. 39, paras. 56 und 57=ILR 46, S. 213.
[32] A.a.O., S. 41, para. 66=ILR 46, S. 215.
[33] Vgl. oben bei Anm. 17–19.

zumindest mit dem übernommenen Vermögen haften. So ist das Vermögen des Völkerbundes anteilsmäßig an die Mitglieder ausgezahlt worden [34], die nicht den Vereinten Nationen beitraten. So hat Österreich ein Zusatzprotokoll der Belgrader Donaukonvention vom 18. August 1948 [35] für völkerrechtswidrig gehalten und daher nicht ratifiziert, demzufolge lediglich die Aktiva nicht aber auch die Passiva der Europäischen Donaukommission und der Internationalen Donaukommission auf die neue Donaukommission übergehen sollten [36].

Ein Rückgriff auf die Haftung der Mitgliedstaaten besteht ferner jedenfalls dann zu Gunsten von Drittstaaten, wenn sie die Völkerrechtspersönlichkeit der Organisation nicht oder nicht mehr anerkennen. Abgesehen vom Fall einer solchen Anerkennung kann die Gründung einer Organisation und die darin liegende Schaffung einer Völkerrechtspersönlichkeit nur für deren Mitgliedstaaten rechtsverbindlich sein. Durch den Abschluß des Gründungsvertrages können die Mitgliedstaaten die Rechtslage nicht zu Lasten dritter Staaten verändern und an Stelle der Staaten ein Haftungsobjekt treten lassen, gegen das schon infolge seines, verglichen mit den Mitgliedstaaten wesentlich geringeren Vermögens Haftungsansprüche dritter Staaten schwerer durchsetzbar wären. So haben Italien und Hitler-Deutschland bei der Schweiz gegen die von ihnen als feindselig empfundenen Sendungen des auf Schweizer Gebiet errichteten Völkerbundsenders Radio-Nations protestiert [37]. Von einer solchen Haftung kann sich ein Sitzstaat auch nicht etwa durch Bestimmungen wie Abschnitt 46 des Sitzabkommens vom 11. Dezember 1957 zwischen Österreich und der Internationalen Atomenergie-Organisation (IAEO) [38] freizeichnen, demzufolge: »Außer derjenigen völkerrechtlichen Verantwortlichkeit, die sich aus der Mitgliedschaft der Republik Österreich bei der IAEO ergeben könnte, erwächst der Republik Österreich aus der Tatsache, daß sich der Amtssitzbereich der IAEO innerhalb ihres Gebietes befindet, keine völkerrechtliche Verantwortlichkeit für Handlungen oder Unterlassungen der IAEO oder für Handlungen oder Unterlassungen der Angestellten der IAEO im Rahmen ihrer Funktionen«. Eine solche Bestimmung kann nur als Rechtsgrundlage künftiger, sich aus der Geltendmachung solcher Haftungsansprüche ergebender Regreßansprüche des Sitzstaates gegen die übrigen Mitgliedstaaten dienen.

---

[34] *H. Aufricht*, A Guide to League of Nations Practice (1951), S. 638–639, gibt diesen Liquidationsbericht wieder.

[35] Österr. BGBl. Nr. 118/1956.

[36] Nr. 37 der Beilagen zu den Stenographischen Protokollen des Österreichischen Nationalrats, IX. Gesetzgebungsperiode, S. 28 laut *Pichler*, Die Donaukommission und die Donaustaaten. Kooperation und Integration (Wien 1973), S. 58.

[37] *E. F. Ranshofen-Wertheimer*, The International Secretariat (1945), S. 197.

[38] ÖBGBl. Nr. 82/1958; UNTS 339, S. 110.

[39] Hierzu *I. Seidl-Hohenveldern*, Gemeinsame zwischenstaatliche Unternehmen, in: Liber amicorum Bernhard C. Aubin (1977), S. 193–216.

Was die Haftung der Mitgliedstaaten einer Organisation oder gemeinsamer zwischenstaatlicher Unternehmen[39] im innerstaatlichen Recht betrifft, die nicht finanzieller Natur sein wird, so entscheidet über das Bestehen einer solchen Haftung im Verhältnis zwischen den Mitgliedstaaten das Satzungsrecht der Organisation. Für einen Drittstaat sollte auch hier ein Rückgriff auf die Mitgliedstaaten in gewissen Grenzen möglich sein, die meines Erachtens die Tatsache genügend berücksichtigen, daß jede Organisation im innerstaatlichen Recht nicht nur der Mitgliedstaaten Rechtspersönlichkeit genießt. Diese Rechtspersönlichkeit ergibt sich anders als die Völkerrechtspersönlichkeit nicht direkt, wenn auch meist nur konkludent[40], aus dem Gründungsvertrag, der ja für Drittstaaten *res inter alios acta* ist, sondern daraus, daß deren internationales Privatrecht auf den Gründungsvertrag als das »nationale Recht« der Organisation verweist[41].

Adam[42] neigt der Ansicht zu, daß die Mitgliedstaaten eines Unternehmens für dieses voll haften. Die Unternehmen seien kraft Völkerrecht errichtet, das anders als das innerstaatliche Recht keine Haftungsbeschränkungen bei Mitgliedern juristischer Personen kenne. *Shihata*[43] ist hier vorsichtiger. Er möchte darauf abstellen, was die Mitgliedstaaten insoweit berücksichtigten und dritten Staaten als ihre Absicht zu erkennen gaben. Eine unbeschränkte Haftung der Mitgliedstaaten könnte aber dann bestehen, wenn das Unternehmen, unbeschadet seiner eigenen Völkerrechtspersönlichkeit, als Organ[44] der Staaten auftrete. *Eli Lauterpacht* hält die Frage nach dem Bestehen einer subsidiären Haftung der Mitgliedstaaten im Fall der Zahlungsverweigerung der Organisation für offen[45]. Mit *Ginther*[46] möchte ich aber eine solche Haftung an Hand der von ihm untersuchten Entstehungsgeschichte von Abkommen über die Haftung für die Schäden aus dem Betrieb von Atomschiffen und Weltraumfahrzeugen sowie über Grundsätze betreffend die Erforschung und Nutzung des Weltraums bejahen, die nach *Eli Lauterpacht*[47] aber noch nicht deklaratorisch bestehendes Völkerrecht wiedergeben. Eine solche Haftung erscheint mir besonders in solchen Fällen nur recht und bil-

---

[40] *H. Mosler,* The International Society as a Legal Community, RdC 140 (1974 IV), S. 66–67.

[41] *Seidl-Hohenveldern* (Anm. 2), S. 88–89, sowie in: Liber amicorum Bernhard C. Aubin, S. 209–210.

[42] *H. Adam,* Les organismes internationaux spécialisés, Bd. 1 (1965), S. 129–130.

[43] *I. Shihata,* Role of Law in Economic Development, The Legal Problems of International Public Ventures, Revue égyptienne de droit international 25 (1969), S. 125.

[44] Vgl. *K. H. Böckstiegel,* Der Durchgriff auf den Staat (1972), S. 19, und *F. A. M. Riad,* L'enterprise publique ou semi-publique en droit international privé, RdC 108 (1963 I), S. 648–649.

[45] *Lauterpacht* (Anm. 6), S. 412–413.

[46] *K. Ginther,* Die völkerrechtliche Verantwortlichkeit Internationaler Organisationen gegenüber Drittstaaten (1969), S. 172–173.

[47] (Anm. 6), S. 412–413.

lig[48], in denen, wie z. B. im Fall des gemeinsamen zwischenstaatlichen Unternehmens Eurochemic, die Mitgliedstaaten aus politischen Gründen auf die Fortführung des Unternehmens bestanden hatten, obwohl dessen wirtschaftlich hoffnungslose Lage jeden ordentlichen Kaufmann zur sofortigen Liquidation verpflichtet hätte[49]. Die Mitgliedstaaten der Eurochemic haben dann schließlich auch die Haftung für deren Schulden weit über ihre sich aus dem Gründungsvertrag ergebenden Verpflichtungen übernommen[50]. Sie haben dabei aber gegenüber den Gläubigern der Eurochemic keine direkte Haftung übernommen, sondern durch Zahlungen an die Eurochemic diese in die Lage versetzt, diesen Verpflichtungen nachzukommen[51]. Mit einer solchen Einschränkung scheint mir einerseits der Tatsache des Bestehens einer gesonderten Rechtspersönlichkeit der Organisation bzw. des Unternehmens und andererseits der realen Wirtschafts- und Machtlage, die sich hinter dem Schleier dieser Rechtspersönlichkeit verbirgt, hinreichend Rechnung getragen.

---

[48] Zustimmend *E. Libbrecht*, Entreprises à caractère juridiquement international (1973), S. 287.

[49] *von Busekist,* Der Werdegang der Eurochemic – eine Bilanz, Atomwirtschaft 1980, S. 264.

[50] *Libbrecht* (Anm. 48), S. 287.

[51] Allgemein für eine solche Regelung *Ginther* (Anm. 46), S. 125. Im gleichen Sinn *Reuter*, Sixth report on the question of treaties concluded between States and International Organizations or between two or more International Organizations, Yearbook of the International Law Commission 1977 Bd. II, Teil 1, S. 131, para. 10, und *I. Brownlie*, Principles of Public International Law (3. Aufl. 1979), S. 685.

Remarks on the Interpretation of the United Nations Charter

Krzysztof Skubiszewski

This article is not a full and systematic exposition of problems relating to the interpretation of the Charter of the United Nations. The treatment of the subject is selective and the contribution limits itself to four questions: (1) the effect of the constitutional nature of the Charter on its interpretation, (2) the role of preparatory work, (3) some comments on the practice that can be resorted to in aid of interpretation, and (4) binding interpretation.

1. Constitutional Nature of the Charter

The Charter of the United Nations is not a model of flawless drafting. Several of its provisions are unclear and more than one meaning can be attributed to them[1]. This difficulty can be solved, and the choice of the right meaning can be facilitated by interpreting the Charter in the light of its object and purpose. Does one get a better perception of what they actually are if one looks at the Charter as a constitution?

According to *Lord McNair,* the Charter is »something very much more than a treaty«: it is a constitution[2]. This feature of treaties establishing international organizations has been commented upon in legal writings[3], includ-

[1] *Lazare Kopelmanas*, L'Organisation des Nations Unies, vol. 1 (Paris 1947), fasc. 1, p. 253; *Hans Kelsen,* The Law of the United Nations: A Critical Analysis of Its Fundamental Problems (London 1950), p. XIII. See also the pioneering commentary by *Leland M. Goodrich/Edvard Hambro,* Charter of the United Nations: Commentary and Documents (Boston 1946), in particular pp. 300–302.

[2] Conference on international law held under the auspices of the David Davies Memorial Institute, quoted by *F. A. Vallat,* The Competence of the United Nations General Assembly, RdC 97 (1959 II), p. 203, at p. 250.

[3] *Goodrich/Hambro* (note 1), p. 48; *Shabtai Rosenne,* Is the Constitution of an International Organization an International Treaty?, Commnicazioni e Studi 12 (1966), p. 21; *Riccardo Monaco,* Le caractère constitutionnel des actes institutifs d'Organisations internationales, in: Mélanges offerts à Charles Rousseau (Paris 1974), p. 153.

ing the writings on the Charter; the problem has often been discussed in connection with its interpretation [4].

Judge Charles De Visscher observed [5]:

> »It is an acknowledged rule of interpretation that treaty clauses must not only be considered as a whole, but must also be interpreted so as to avoid as much as possible depriving one of them of practical effect for the benefit of others. This rule is particularly applicable to the interpretation of a text of a treaty of a constitutional character like the United Nations Charter, above all when, as in this case, its provisions create a well-defined international régime, and for that reason may be considered as complementary to one another«.

The learned judge referred to the trusteeship system. It is one of the many constitutional elements of the Charter. Constitutional regulation pervades the whole instrument. The Charter establishes the concept of the Organization, lays down the basic principles to which it shall conform, and defines its functions and powers, and, in particular, distributes them among its different organs. These characteristics bear a certain analogy to the national constitution as the organic and fundamental law of the country. Some of the definitions of the constitution could easily encompass the Charter. It is an instrument in which »the structure and mechanism« as well as »the fundamental rules to regulate the motions« of the Organization have been prescribed [6], and it is also the »supreme law« [7] for the United Nations.

However, these similarities do not suffice to make it possible to approach the interpretation of the Charter along the lines and according to the patterns of national constitutions. The Charter remains a treaty concluded by States, and what pertains to its interpretation is governed by the law of trea-

---

[4] E.g., *Kopelmanas* (note 1), p. 295 and *pasim; Alf Ross,* Constitution of the United Nations: Analysis of Structure and Function (Copenhagen 1950), pp. 30–40; *Charles De Visscher,* L'interprétation judiciaire des traités d'organisation internationale, Rivista di Diritto Internazionale 41 (1958), p. 177; *J. F. McMahon,* The Court of the European Communities: Judicial Interpretation and International Organization, BYIL 37 (1961), p. 320, reprinted in: *J. E. S Fawcett/Rosalyn Higgins* (eds.), International Organization: Law in Movement (London, New York, Toronto 1974), p. 1, pp. 23–24; *Sir Humphrey Waldock,* General Course on Public International Law, RdC 106 (1962 II), p. 1, at p. 26 ff.; *Philippe Cahier,* Le droit interne des organisations internationales, RGDIP 67 (1963), p. 563, at p. 576; *Ruth B. Russell,* Changing Patterns of Constitutional Development, International Organization 19 (1965), p. 410, at pp. 410–413; *Finn Seyersted,* United Nations Forces in the Law of Peace and War (Leyden 1966), pp. 350–351.

[5] Diss. op., *International Status of South-West Africa,* ICJ Reports 1950, p. 186, at p. 187.

[6] Cf. the definition of constitution in *Eakin* v. *Raub,* Pa., 12 Serg. and R. 330, 347.

[7] Cf. *Rison* v. *Farr,* 24 Ark. 161, 166, 87 Am. Dec. 52. For Members the Charter constitutes the supreme law with regard to their treaty obligations, Art. 103.

ties, unless the Charter itself lays down rules that deviate from that law[8]. Analogies derived from the place, role and development of constitutional law in the State, if at all relevant, call for great caution. The United Nations is composed of States which represent various views on the function of the constitution in their national spheres, in particular the adaptation, through interpretation, of the constitutional framework to the changing circumstances and exigencies of life. The United Nations would probably confront an insoluble problem, if it had to choose, by searching for analogies among the manifold and differing attitudes of national systems. The meaning of the word »constitution« changes when transposed from the domestic to the international scene; it does not automatically carry with it the introduction of domestic patterns into the interpretation of the law of international organization.

For all that, it cannot be denied that the constitutional nature of the treaty has an influence on its interpretation. Judge Alejandro Alvarez has made the following comment in connection with the interpretation of the Charter: »an institution, once established, acquires a life of its own, independent of the elements which have given birth to it, and it must develop, not in accordance with the views of those who created it, but in accordance with the requirements of international life«[9]. »A treaty or a text that has once been established acquires a life of its own«[10]. The concept of the United Nations »as a living organism«[11] and a »dynamic instrument«[12] was behind some of the interpretations of the Charter by the Secretary-General, Dag Hammarskjöld. The constitutional perspective, no doubt, facilitated this approach. For constitutions »have certain underlying and implicit premises, which are literally extra-constitutional, but which provide a »higher-law« rationale to justify choices between competing principles«[13].

Obviously, the perception of the Charter as a constitution does not entail the power to extend, alter or disregard its provisions under the guise of interpretation. Nonetheless the emphasis on the purposes of the Organization favours the teleological method and the application of the doctrine of implied powers. The Organization is supposed to possess a competence, or competences, which are not explicitly stated in the Charter but which follow logically and necessarily from the U.N.'s purposes and aims. The teleological approach belongs to the »creative or dynamic methods of interpretation« and

---

[8] Cf. Art. 5 of the Vienna Convention on the Law of Treaties.

[9] Ind. op., *Conditions of Admission of a State to Membership in the United Nations (Article 4 of the Charter),* ICJ Reports 1948, p. 67, at p. 68.

[10] Diss. op., *Competence of the General Assembly for the Admission of a State to the United Nations, ibid.,* 1950, p. 12, at p. 18.

[11] GAOR, 14th Sess., Supp. No. 1A, A/4132/Add. 1.

[12] GAOR, 16th Sess., Supp. No. 1A, p. 1.

[13] *Oscar Schachter,* The Relation of Law, Politics and Action in the United Nations, RdC 109 (1963 I), p. 165, at p. 195.

its use has been admitted in the interpretation of the constitutions of international organizations [14], including the Charter [15].

The teleological method minimizes the importance of the intention of the parties and, consequently, of the preparatory work. This method, however, produces its own difficulties, for it shifts the old problems of interpretation onto another plane. The factor of intention cannot always be divorced from the object and purposes of the Charter [16].

Some view the Charter »as a constitution designed to give effect to certain purposes and principles« [17]. The Members agree on the purposes and principles as long as they are expressed in general and abstract language. They begin to differ when the purposes and principles are invoked to indicate a specific course of conduct. In that case the purposes and principles may become an additional source of interpretative controversy. And often it is the other way round: the interpreter must first look into the detailed provisions to elucidate the meaning of purposes and principles [18]. The latter are not always a reliable to guide interpretation. Judge Sir Percy Spender notes that »[t] he stated purposes of the Charter should be the prime consideration in interpreting its text«. Yet »the Court is to look [. . .] at the terms of the Charter itself. What does it provide to carry out its purposes« [19]? Judge Gros also saw a limit to the role of the purposes and principles: »The degree of solidarity accepted in an international organization is fixed by its constitution. It cannot be subsequently modified through an interpretation based on purposes and principles which are always very broadly defined [. . .]« [20].

---

[14] *G. G. Fitzmaurice,* The Law and Procedure of the International Court of Justice: Treaty Interpretation and Certain Other Points, BYIL 28 (1951), p. 1, at p. 8, note 1. *Ervin P. Hexner,* Teleological Interpretation of Basic Instruments of Public International Organizations, in: Law, State and International Legal Order (Knoxville 1964), p. 119.

[15] An example is the interpretation in *Reparation for Injuries Suffered in the Service of the United Nations,* ICJ Reports 1949, p. 174.

[16] *G. G Fitzmaurice,* The Law and Procedure of the International Court of Justice 1951-4: Treaty Interpretation and Other Treaty Points, BYIL 33 (1957), p. 201, at p. 209.

[17] *Vallat* (note 2), p. 269.

[18] *Kopelmanas* (note 1), p. 285: »certains articles de la Charte ne contiennent que l'énoncé de tendances générales, dont les précisions et les règles pratiques d'application doivent être cherchées dans d'autres parties du texte. Cela est surtout vrai du Préambule et du premier Chapitre consacré aux buts et aux principes de l'Organisation [. . .]«.

[19] Sep. op., *Certain Expenses of the United Nations (Article 17, paragraph 2, of the Charter),* ICJ Reports 1962, p. 182, at p. 185.

[20] Diss. op., *Legal Consequences for States of the Continued Presence of South Africa in Namibia (South West Africa) notwithstanding Security Council Resolution 276 (1970), ibid.,* 1971, p. 323, at p. 341.

2. Preparatory Work

If the constitutional nature of the Charter carries weight in the process of interpretation (para. 1), the role of preparatory work will tend to diminish. But irrespective of this point, the proper value of preparatory work in our problem is debatable.

Preparatory documents relating to the Charter shed light on the intentions of the original Members. In today's Organization they are a minority of 51 States[21]. The intention of the present majority cannot be explained on the basis of what was said at San Francisco. A great number of Members were not present there, and most of them did not exist as States when the Conference convened. Hence the study of the preparatory work is irrelevant if the purpose is to ascertain the intention of the whole membership in its present composition. On the other hand, such a study is of historical value; it permits the appreciation of the trends prevailing at the birth of the United Nations and the identification of national attitudes towards the Organization at the moment of its foundation.

The Members which have been admitted to the Organization by virtue of Art. 4 have accepted the final result of the San Francisco deliberations, *i.e.* the Charter itself. Do they thereby accept, as has been suggested, »any interpretation [. . .] that may be given on the basis of the records of the preparatory work when such records have already been made public« [22]? In view of the frequent inconclusiveness of the preparatory documents the opinion quoted is, perhaps, of limited practical importance. However, it is open to doubt on its own merits. As one Government observed, »the States Members of the United Nations were parties to the Charter and not to the twenty-two volumes of preparatory work of the San Francisco Conference, in which the majority of the present Members had not even participated« [23]. Judge Alvarez argued that the value of the *travaux préparatoires,* in particular of those relating to international organizations, was reduced for several reasons: the preparatory work contained contradictions; the decision of States to sign a treaty was not influenced by the preparatory work »with which, in many cases, they were unacquainted«; finally, »the texts should continue to be in harmony with the new conditions of social life«. »It is therefore necessary [. . .] to look ahead [. . .] and not to look back, or have recourse to *travaux préparatoires*«.

---

[21] One original Member, Poland, was not represented at the San Francisco Conference.

[22] *Pollux,* The Interpretation of the Charter, BYIL 23 (1946), p. 54, at p. 72.

[23] A/AC. 125/SR.66 (Romania). A less radical view was taken by *Kopelmanas* (note 1), p. 308: »En adhérant à l'Organisation, [les nouveaux Members] ont pu avoir connaissance de tous les documents de la Conférence de San Francisco et apprécier les possibilités d'interprétation qu'ils contiennent«. He did not deny that the new Members might always try to impose an interpretation which did not conform to the *travaux préparatoires.* Cf. also *Goodrich/Hambro* (note 1), p. 49.

In interpreting a treaty »we must have regard to the exigencies of contemporary life, rather than to the intentions of those who framed it« [24]. Judge Sir Percy Spender said: »It is hardly the intention of those States which originally framed the Charter which is important except as that intention reveals itself in the text. What is important is what the Charter itself provides«. The learned judge did not list the *travaux préparatoires* among the materials the analysis of which would be the first task of the International Court of Justice while interpreting the Charter [25].

3. Practice

Practice by Member States, including practice of organs composed by them, must be taken into account as an aid in interpreting of the Charter. However, the value and importance of practice will depend on its nature.

A »practice which is of a peaceful, uniform and undisputed character accepted in fact by all current Members« may be said to resolve the interpretative problem at issue [26]. It can be said that universal acceptance gives the practice the rank of authoritative interpretation (Paragraph 4 below) [27]. An example is practice relating to voting in the Security Council: an abstention by a permanent member is not regarded as a veto, though a different conclusion would follow from the ordinary meaning of the term »concurring votes« used by Art. 27, para. 3. But practice by only some of the Member States, even if expressed through a U.N. organ in a valid resolution, is nothing more than practice of those States and of the organ concerned. It is not equivalent to subsequent conduct of the parties to the Charter establishing their agreement on its interpretation. Such a practice throws light on the attitude of the majority and remains simply their interpretation. As to the organ, the point was well made by Judge Sir Percy Spender in his Separate Opinion in the *Expenses* case. He argued that it was »not possible to equate ›subsequent conduct‹ with the practice of an organ of the United Nations« because, first, an organ was not a party to the Charter and, second, the organ was but a mechanism through which the Members expressed their views and

---

[24] *Op. cit.* (note 10), p. 23. See also Judge Azevedo's diss. op., *ibid.*, p. 22, at pp. 23 and 30.

[25] *Loc. cit.* (note 19).

[26] Judge Sir Percy Spender (note 19), p. 195.

[27] According to the United States, Res. 2625 (XXV) »may be an authoritative interpretation of international law, adopted as it was unanimously and stated as it was by many Members to be such – at any rate, if it is supported by state practice«, GAOR, 32nd sess., 6th Com., 11 November 1977, reprinted in AJIL 72 (1978), p. 377. *Salo Engel*, Procedures for the *De Facto* Revision of the Charter, ASIL Proceedings, vol. 59, 1965, p. 108, at p. 116, writes of practice which »clearly establishes the understanding of all Members« regarding Charter interpretation.

through which they acted. The practice established within an organ had the same probative value in the process of interpretation as the conduct of the Members that comprised the organ's majority; that practice had no »particular probative value of its own«[28]. The learned judge made the latter statement in regard to the interpretation of their functions by the General Assembly and the Security Council. In this writer's view, it applies, with even stronger reason, to the interpretation of those Charter provisions which give expression to the rules of general international law.

The General Assembly has often interpreted these provisions. Some of its resolutions are evidence of the practice followed within the Organization. But the Assembly is also competent to give an abstract interpretation of the Charter[29], in particular it can elicit the meaning of the principles and other provisions of the Charter irrespective of any specific problem or problems submitted to it. Occasionally, Governments and writers have differed as to whether a declaration by the General Assembly has constituted an interpretative instrument or was something else. Cases in point are the Universal Declaration of Human Rights[30] and the Declaration on the Granting of Independence to Colonial Countries and Peoples[31]. Resolution 1803 (XVII) on permanent sovereignty over natural resources belonged, according to one view, to the »interpretations of the fundamental provisions of the Charter«[32]. Several Governments regarded the Declaration on Principles of International Law as an exercise in Charter interpretation[33].

Usually the resolutions which contain interpretation of the Charter are recommendations. However, there is room for a distinction between, on the one hand, the resolution itself and, on the other, the interpretative assertions

---

[28] *Loc. cit.* (note 19), p. 192.

[29] For the competence of the I.C.J. to deal with an interpretative question couched in abstract terms, see the *Admission case* (note 9), pp. 61 and 74 (in favour of abstract interpretation) and p. 108 (against).

[30] Res. 217 A (III). In favour of the interpretation thesis: A/C.3/SR.92 and E/SR.215, p. 18 (France); GAOR, 18th sess., 6th Com., 811th Mtg, 14 November 1963, p. 162, para. 8 (Poland); *ibid.*, p. 164, para. 22 (Egypt); Judge Tanaka, diss. op., *South West Africa* cases, Second Phase, ICJ Reports 1966, p. 250, at p. 293. Against the interpretation thesis: GAOR, 18th sess., 6th Com., 817th Mtg, 21 November 1963, p. 204, para. 13 (Australia).

[31] Res. 1514 (XV). In favour of the interpretation thesis: GAOR, 17th sess., 6th Com., 760th Mtg, 15 November 1962, p. 133, para. 18 (Ceylon); *ibid.*, para. 22 (Poland). GAOR, 18th sess., 6th Com., 811th Mtg, 14 November 1963, p. 162, paras. 8 and 9 (Poland), *ibid.*,p. 164, para. 22 (Egypt). Against: cf. doubts expressed by France, the United Kingdom and the United States during the debate on the resolution, GAOR, 15th sess., Pl.; see also the Australian position, *loc. cit.* (note 30).

[32] Egypt, *loc cit.* (note 30).

[33] Res. 2625 (XXV) A/AC.119/SR.10, p. 4 (Australia). GAOR, 25th sess., 6th Com., p. 9, paras. 35 and 36 (Australia), p. 13, para. 22 (Denmark), p. 14, para. 33 (Hungary), p. 22, para. 18 (Israel), p. 24, para. 31 (Greece), p. 25, para. 38 (Mali), p. 27, para. 5 (Peru) and p. 28, para. 10 (Mongolia).

which accompany recommendations submitted to Governments. These assertions, as Professor Oscar Schachter maintains, are not recommendatory [34].

4. Binding Interpretation

Binding interpretation of the Charter can either take the form of authoritative (»authentic«) interpretation by the contracting parties or be incorporated in an organ decision that is obligatory for its addressees.

The authoritative (»authentic«) interpretation is the responsibility of the contracting parties, *i.e.* all the U.N. Member States. Once they unanimously, and without exception, agree on an interpretation, it is binding. The report of Committee IV/2 of the San Francisco Conference states that »if an interpretation made by any organ of the Organization or by a committee of jurists is not generally acceptable it will be without binding force« [35]. The term »generally« must be understood to refer to all the Members, and not to their majority. The majority would be empowered to give a binding interpretation only by virtue of a specific provision of the Charter, but such a provision is lacking (the possibility inherent in Arts. 108 and 109 is discussed below). On the other hand, the acceptance need not be express [36], but if it is implicit, there ought to be no doubt that it has been given.

All the Members being represented in the General Assembly, the question arises whether unanimous resolutions formulate or reveal the authoritative interpretation of the Charter. A unanimous resolution can constitute evidence of the acceptance of the interpretation by the Member States. But the mere fact of unanimity is not identical with acceptance in the sence of the

---

[34] *Oscar Schachter,* The Relation of Law, Politics and Action in the United Nations, RdC 109 (1963 I), p. 165, at p. 185, *idem.,* The Generation Gap in International Law, ASIL Proceedings, vol. 63, 1969, p. 230, at pp. 231 and 233. In his Written Statement in the *Namibia* case the U.N. Secretary-General referred to the »contestation« of the legal force of the G.A. resolutions. But the interpretation contained therein could not be called in question: »no such contestation could justify any State in refusing to apply to the international Territory of Namibia a contemporary interpretation of the United Nations Charter formally adopted and upheld by an overwhelming majority of member States«, ICJ Pleadings, vol. 1, p. 75, at p. 87, para. 50. In his oral statement before the Court the representative of the Secretary-General said: »Included among the international rules which are binding on the administration of the international territory of Namibia are declarations and resolutions formally adopted by principal organs of the United Nations which represent generally accepted interpretations and applications of the provisions of the United Nations Charter, and which either are of general application, or are stated to have specific reference to the situation of Namibia«, *ibid,* vol. 2, p. 30, at p. 36.

[35] UNCIO, vol. 13, p. 710. *Engel* (note 27), p. 109, assumes that here »acceptable« means probably »accepted«, for »mere acceptability cannot give binding force to an interpretation [. . .]«.

[36] *Kopelmanas* (note 1), pp. 272–273.

San Francisco report. To find it one must look beyond the resolution [37] and ask the question: did the Members aim at arriving at an authoritative interpretation? Likewise, the resolution may embody the agreement between the Member States regarding the interpretation of the Charter (Art. 31, para. 3 (a) of the Vienna Convention on the Law of Treaties) [38]. Again, one must prove that the Members have expressed their agreement in the resolution.

Some of the declarations on human rights [39], colonial countries [40] and principles of international law [41] (referred to in Paragraph 3 above) have occasionally been regarded as having the status of authoritative interpretations. So was Res. 3314 (XXIX) on the definition of aggression [42]. The real answer depends on the meaning and significance of the individual votes: did they reflect the intention to give an authoritative interpretation [43]?

At any rate, unanimity of the contracting parties is one of the essentials of authoritative (»authentic«) interpretation. Yet the report of Committee IV/2 seems to envisage the possibility of a majority decision. It states that

---

[37] *Alfred Verdross,* Kann die Generalversammlung der Vereinten Nationen das Völkerrecht weiterbilden?, ZaöRV 26 (1966), p. 690, at p. 695, denies the possibility of authoritative interpretation by the Assembly. *Engel* (note 27), p. 116, speaks of the necessity to make a determination »in each particular case«. Some writers mention the recognition of the interpretation (as embodied in a resolution) by States. For other writers, however, the resolution itself is sufficiently conclusive.

[38] *Suzanne Bastid,* Observations sur une »étape« dans le développement progressif et la codification des principes du droit international, in: Recueil d'études de droit international en hommage à Paul Guggenheim (Genève 1968), p. 132, at pp. 136 and 144; *Gaetano Arangio-Ruiz,* The Normative Role of the General Assembly of the United Nations and the Declaration of Principles of Friendly Relations with an Appendix on The Concept of International Law and the Theory of International Organisation, RdC 137 (1972 III), p. 419, at pp. 515–517.

[39] Cf. the French position (note 30). In the *South West Africa* cases, Second Phase, Ethiopia and Liberia maintained that »the formal acts of the constituent organs of the United Nations have produced an authoritative construction of Articles 55 (c) and 56 of the Charter, inter alia, such that the practice of apartheid is legally impermissible«, ICJ Pleadings, vol. 9, p. 347. *Ian Brownlie,* Principles of Public International Law (2nd ed., Oxford 1973), p. 674, and *Louis B. Sohn,* The Development of the Charter of the United Nations: The Present State, in: Maarten Bos (ed.), The Present State of International Law and Other Essays written in honour of the Centenary Celebration of the International Law Association 1873–1973 (Deventer 1973), p. 39, at p. 54, classify the Declaration under the head of authoritative interpretation.

[40] E.g., *Brownlie* (note 39).

[41] GAOR, 25th sess., 6th Com., p. 18, para. 8 (Iraq), p. 26, para. 45 (Cyprus) and p. 35, para 7 (India). Cf. also note 27 above.

[42] In favour of this view: GAOR, 29th sess., 6th Com. 1478th Mtg, 16 October 1974, p. 75, para. 14 (Iraq). Against: *ibid.,* 1477th Mtg, 15 October 1974, p. 70, para. 18 (United Kingdom) and 1480th Mtg, 18 October 1974, p. 93, para. 55 (Israel).

[43] See remarks and distinctions by *Milton Katz,* The Relevance of International Adjudication (Harvard University Press 1968), pp. 116 and 119–120. Cf. also *Heribert Golsong,* Das Problem der Rechtsetzung durch internationale Organisationen (insbesondere im Rahmen der Vereinten Nationen), Berichte der Deutschen Gesellschaft für Völkerrecht, vol. 10 (1971), p. 1, at p. 23.

»in cases where it is desired to establish an authoritative interpretation as a precedent for the future, it may be necessary to embody the interpretation in an amendment to the Charter. This may always be accomplished by recourse to the procedure provided for amendments«. Amendments and alterations enter into force for all the Members when ratified by two-thirds of them, including the five permanent members of the Security Council (Arts. 108 and 109). Thus any interpretation that is incorporated in, and takes the form of, an amendment to, or alteration of, the Charter would be binding, though it might not have been consented to by all[44]. But this procedure is in fact to little avail. The history of Charter revision shows that amendments and alterations are agreed upon rarely, slowly and with difficulty. To date interpretation has never been their subject. Besides, the procedure itself is one for modifying, not interpreting, the Charter. It may, therefore, be concluded that the Charter is subject to the requirement of unanimity inherent in the rule of authoritative interpretation[45].

When turning to examine the possibility of a binding interpretation being made by the Organization itself (as distinguished from all its Members), it must be said at the outset that the Charter conferred no such function of interpretation on any U.N. organ.

If an organ is competent to adopt a binding decision, and this decision contains an interpretative statement, the latter shares in the mandatory nature of the act. But it applies to the specific problem under consideration and is not binding for other organs[46]. If the act is non-obligatory, so is its interpretation. However, in Professor Schachter's opinion (referred to in Paragraph 3 above), interpretative statements made in recommendatory resolutions are not necessarily non-binding.

In discussing binding interpretation by U.N. organs, one thinks primarily of the International Court of Justice. Now, though the Court was made the principal judicial organ of the United Nations, it did not receive any general

---

[44] The non-concurring Member is free to withdraw from the Organization if it finds itself unable to maintain its membership under the revised Charter. See report of Committee I/2 of the San Francisco Conference, UNCIO, vol. 1, pp. 619–620, vol. 6, p. 246 and vol. 7, pp. 327–329.

[45] Ejus est interpretari legem cujus est condere. *Ioan Voïcu,* De l'interprétation authentique des traités internationaux (Paris 1968). According to the Hague Court, »it is an established principle that the right of giving an authoritative interpretation of a legal rule belongs solely to the person or body who has power to modify or suppress it«, Advisory Opinion regarding the Delimitation of the Polish-Czechoslovakian Frontier (*Question of Jaworzina*), PCIJ, Series B, No. 8, p. 37. This principle does not permit the inference that the majority of two-thirds (including all the permanent members of the Security Council) can authoritatively interpret the Charter, unless they have recourse to Arts. 108 and 109.

[46] If the United Nations could be a party in a case before the International Court of Justice, the Court's interpretation, contained in the judgment, would then be binding for all organs of the Organization, although the obligatory effect of the interpretation so given would remain limited to that particular case.

competence to interpret the Charter. In this respect its position differs from some other courts, especially from the Court of the European Communities, which have been endowed with the power to interpret constitutional instruments. Whenever the I.C.J. explains the meaning of the Charter in contentious proceedings, its interpretation is, of course, binding, but only within the limits defined by Art. 59 of the Statute. An interpretation given in an advisory opinion could become binding through incorporation into a mandatory decision of a U.N. organ or through its acceptance as being obligatory by all the Members.

As to the latter point, there is nothing in the Charter or in general international law to substantiate the view[47] that the interpretation of the International Court of Justice will on principle be acceptable to all the Members and will always constitute an obligatory solution of conflicts among individual interpretations. Contrary to some expectations or wishes which go back to the beginnings of the United Nations, the Court has not become the organ that would normally and regularly clarify the provisions of the Charter. In 1947, by Res. 171 (II), the Assembly recommended that »the difficult and important points of law« within the jurisdiction of the Court which had arisen in the course of the activities of the U.N. organs and which involved »questions of principle« that it was desirable to have settled, »including points of law relating to the interpretation of the Charter«, should be referred to the Court for an advisory opinion. At the same time, the Assembly put to the Court a question relating to the interpretation of Art. 4, para. 1. The question was couched in abstract terms, though it had its origin in the divergencies of views among the particular Governments on the admission of certain States to the Organization. The fate of the resulting advisory opinion[48] did not augur well for the effectiveness of the Court's interpretative activity. The opinion of the Court has not been followed by the Security Council, *i.e.* by the organ for the guidance of which it was meant. The non-compliance with the opinion marked a departure from the practice of the League of Nations, which never failed to apply an opinion of the Permanent Court of International Justice. The League, however, was more prudent, as it did not implicate the Hague Court in disputes of interpretation which reflected the basic political divisions among the Members. Of course, the competence of the Court in such disputes remains beyond doubt, because interpretation of treaties is always a legal question. But the advisability of taking recourse to the Court is another matter: »it may be dangerous to the authority and to the very existence of a court to subject it to the test of having to pass upon a question which is regarded by a strong group of states as unsuitable for settlement in this way«[49]. In fact, not even a strong group is necessary to

---

[47] Expressed by *Kopelmanas* (note 1), p. 274.
[48] ICJ Reports 1948, p. 57.
[49] *Pollux* (note 22), p. 66

frustrate the Court's interpretation: the opposition of one permanent member of the Security Council will suffice. The Soviet attitude in the *Admission* case is an illustration. In that instance divergencies of interpretation reflected a political conflict, and its solution could not have been left to the Court[50]. And the final disposal of the membership problem was arranged »in diametrical opposition to the Court's opinion«[51]. The decision in *Certain Expenses of the United Nations (Article 17, paragraph 2, of the Charter)*[52] is another well-known instance of the Court's interpretation not prevailing. Where the issue touches upon an acute political controversy, the Court's view, though accepted by the organ which has asked for it, will not automatically be complied with by another organ, nor will it change the attitude of those who oppose the interpretation adopted by the Court. Wherever, on the other hand, the load of political friction was none or minimal, the interpretation of the Court would be acceptable to the Members and would, therefore, settle the legal point at issue. An example is the advisory opinion on the *Reparation for Injuries Suffered in the Service of the United Nations*[53]. However, more often than not the interpretation of the Charter concerns vital interests which the States are not willing to submit to any judicial control[54].

---

[50] Cf. *Shabtai Rosenne,* On the Non-Use of the Advisory Competence of the International Court of Justice, BYIL 39 (1963), p. 1, at p. 31, *Michel Virally,* L'Organisation mondiale (Paris 1972), pp. 152–154 and 214; *Serge Sur,* L'Interprétation en droit international public (Paris 1974), pp. 126 ff. See also the observations by *Paul Reuter,* in: Institut de Droit International, Livre du Centenaire 1873–1973, Évolution et perspective du droit international (Basle 1973), pp. 395–396, and by Judge Gros, sep. op., *Nuclear Tests* case, ICJ Reports 1974, p. 276, at p. 297.

[51] *Rosenne* (note 50), p. 41.

[52] ICJ Reports 1962, p. 151.

[53] *Ibid.* 1949, p. 174. Another instance is the interpretation of Art. 4, para. 2, *ibid.* 1950, p. 4.

[54] *Goodrich/Hambro* (note 1), pp. 20 and 301. *McMahon* (note 4) speaks of »the acute need for an informal, flexible, and an expeditious method of interpretation that will militate against the concept of judicial control«. See, generally, *Ludwik Ehrlich,* Interpretacja traktatów (Warsaw 1957); *Rudolf Bernhardt,* Die Auslegung völkerrechtlicher Verträge (Beiträge zum ausländischen öffentlichen Recht und Völkerrecht, Bd. 40) (Köln, Berlin 1963); *J. Rideau,* Juridictions internationales et contrôle du respect des traités constitutifs des organisations internationales (Paris 1969); *Sur* (note 50), pp. 315 ff.

Comment concevoir, dans sa spécificité, l'adhésion des Communautés européennes à la Convention de Rome sur la sauvegarde des Droits de l'Homme et des libertés fondamentales

GIUSEPPE SPERDUTI

1. Le Mémorandum de la Commission des Communautés européennes en date du 4 avril 1979

Le problème de l'adhésion des Communautés européennes à la Convention de Rome sur la sauvegarde des Droits de l'Homme et des libertés fondamentales a été soulevé dans une étude approfondie par la Commission des Communautés, étude qui a pris forme d'un Mémorandum, portant la date du 4 avril 1979. L'importance de cette initiative est témoignée par ses larges répercussions dans l'opinion publique et dans les milieux des savants. Il suffit de rappeler ici la première rencontre internationale qui a eu lieu à Venise, Ile S. Giorgio, les 9–11 novembre 1979, convoquée par le »Comité consultatif italien pour les droits de l'homme« et avec la participation de nombreuses personnalités de différents pays européens [1].

Il importe de remarquer ceci: si l'on veut acheminer ledit problème de l'adhésion vers une forme pratique et convenable de solution, il ne faut pas strictement s'en tenir au schéma classique de l'acte d'adhésion, mais il faut concevoir cet acte avec les spécifications et les adaptations que la situation exige.

---

[1] Les Actes de la Rencontre ont paru en 1981: S.I.O.I., Europa e diritti umani, p. 283.

2. La spécificité des liens à établir entre le système de la Convention de Rome et le système des Traités communautaires

On essayera de préciser dans le présent contexte la spécificité des liens dont paraît souhaitable la création entre le système de la Convention de Rome et le système des Traités communautaires, en vue d'accomplir une oeuvre à la fois rentable et soustraite aux risques des innovations radicales.

Il s'agit, en effet, de:

a) renforcer la protection des droits de l'homme en Europe et en favoriser le développement ultérieur;

b) viser à la concordance de la jurisprudence, qui risquerait d'être compromise si les deux juridictions européennes, la Cour de Strasbourg et la Cour de Luxembourg, étaient appelées à se prononcer en pleine indépendance l'une de l'autre par rapport aux mêmes règles matérielles sur les droits de l'homme;

c) tâcher d'atteindre ces objectifs en évitant de créer des procédures dont la complexité et la longueur pourraient affecter le droit fondamental de toute personne à ce que sa cause soit jugée »dans un délai raisonnable« (Convention de Rome, art. 6, § 1).

3. Un exemple des conséquences fâcheuses que pourrait entraîner le recours strict au schéma classique de l'adhésion aux traités internationaux

Le dernier point qu'on vient d'évoquer demande une attention toute particulière. Il faut prendre soin d'écarter la forme classique de l'adhésion, à savoir un acte d'adhésion des Communautés qui mettrait celles-ci sur le même plan des Etats par rapport à la Convention et porterait atteinte notamment à la nature supranationale et internationale de la Cour de Justice des Communautés par suite d'une sorte de déclassement de la Cour au rôle d'un tribunal interne.

Sur le même point il convient d'ajouter quelques remarques en se référant à l'arrêt le plus récent concernant la protection des droits de l'homme, qui a été prononcé par la Cour de Justice des Communautés.

C'est l'arrêt Hauer n° 44/79, *Interdiction de nouvelles plantations de vignes*, pris en date du 13 décembre 1979. L'arrêt a fait suite à une demande qu'avait adressée à la Cour le Verwaltungsgericht de Neustadt an der Weinstraße et par laquelle cette juridiction tendait à obtenir une décision à titre préjudiciel sur un règlement du Conseil, le règlement n° 1162/75 du 17 mai 1976, portant des mesures visant à adapter le potentiel viticole des pays membres aux besoins du marché. Il s'agissait, entre autres, d'examiner »la compatibilité des dispositions du règlement n° 1162/76 avec les règles relatives à la protection des droits fondamentaux« et ce du fait que certains doutes manifestés par le Verwaltungsgericht devaient »être compris comme mettant en cause la validité du règlement«. S'agissant notamment d'examiner lesdites dispositions sous l'angle de leur compatibilité avec le respect dû au droit de propriété, la Cour s'est référée au premier Protocole joint à la Convention de Rome, dont l'art. 1 porte justement sur la protection du droit de propriété, ainsi qu'aux »indications résultant des règles et pratiques constitutionnelles des neuf Etats membres«. Elle a conclu à la justification de la restriction à l'usage de la propriété apportée par le règlement susdit: à cet égard elle a fait état des »objectifs d'intérêt général poursuivis par la Communauté«. Bref, à son avis, ledit règlement »ne porte pas atteinte à la substance du droit de propriété tel qu'il est reconnu et garanti dans l'ordre juridique communautaire«.

Dans le présent contexte, il importe notamment d'observer ceci: à supposer que cet arrêt de la Cour de Justice des Communautés soit intervenu après adhésion de celle-ci à la Convention de Rome, on aurait par ce même arrêt atteint le stade de l'épuisement des voies de recours à l'intérieur de la CEE, de sorte que si le requérant devant la juridiction allemande continuait de se croire victime d'une violation de son droit de propriété à cause à la fois de l'interdiction par le règlement communautaire de nouvelles plantations de vignes et du refus par les autorités allemandes de l'autoriser à procéder à la plantation d'une vigne dans un terrain de sa propriété, une procédure ultérieure lui resterait ouverte pour demander le respect de son droit, procédure à engager devant la Commission européenne des Droits de l'Homme et susceptible à son tour d'aboutir aussi à une procédure devant la Cour européenne. Il n'est pas besoin de souligner l'aspect peu pratique d'une telle chaîne de procédures.

Au demeurant, une chaîne tout autant lourde peut être envisagée par rapport à la situation telle qu'elle existe aujourd'hui, ce qui amène encore une fois à se demander si le droit – ci-dessus rappelé – de toute personne à ce que sa cause soit entendue »dans un délai raisonnable« n'en résulterait pas gravement affecté. Les Etats membres des Communautés européennes demeurent, en principe, responsables des mesures prises par eux, en contraste avec la Convention de Rome, à titre de mesures auxquelles les astreint leur qualité de membres communautaires. Il s'ensuit que la personne affectée par l'une de ces mesures et qui se considérerait victime de la violation de l'un des droits que lui reconnaît la Convention, devrait être considérée comme habilitée à introduire une requête devant la Commission européenne des Droits de l'Homme et déclencher ainsi une procédure susceptible d'aboutir à son tour, dans les conditions visées dans l'art. 48 de la Convention européenne, à une procédure ultérieure devant la Cour établie par cette même Convention.

4. Nécessité de garder aux institutions communautaires et aux institutions créées par la Convention de Rome les caractères qui leur sont propres

Il faut admettre qu'une notion même vague d'adhésion à la Convention de Rome n'est pas dépourvue d'un certain éclat politique. Mais il faut également dire que cet éclat peut être d'obstacle à s'apercevoir des aspects pratiques du procédé, notamment des complications auxquelles aboutit, par le prolongement des procédures, une forme inappropriée de raccord entre le Traité et la Convention de Rome.

L'idée à retenir est qu'il faut s'efforcer de garder aux institutions communautaires, d'une part, et aux institutions créées par la Convention de Rome, d'autre part, les caractères qui leur sont propres.

L'adhésion de type classique des Communautés européennes à la Convention de Rome affecterait d'une façon marquée la Cour de Justice des Communautés dans son statut d'instance judiciaire supranationale, car ses décisions, à l'instar de celles des tribunaux nationaux, pourraient désormais donner lieu à des requêtes devant des instances internationales.

Or, il est certain que l'interprétation finale et conclusive de la Convention de Rome, quelles que soient les circonstances qui la provoquent, doit être réservée à la Cour européenne. En effet, l'interprétation des règles par lesquelles la Convention énonce les droits de l'homme est étroitement liée à la définition des obligations incombant aux Parties Contractantes et finit par se confondre avec celle-ci. Il est donc évident que la définition par la voie judiciaire de ces obligations relève de la compétence de la Cour européenne. Par contre, il n'y a pas véritablement des raisons pour ne pas s'en tenir au Traité instituant les Communautés européennes en ce qui concerne la mission de la Cour de Justice des Communautés. En d'autres termes, il faut prendre garde de ne pas porter atteinte au statut fondamental de la Cour. La notion à retenir peut s'exprimer ainsi: la Cour de Justice assure par ses propres décisions le respect du droit au niveau communautaire, quitte à se coordonner avec la Cour européenne des Droits de l'Homme lorsque, pour statuer sur la validité d'un acte communautaire, il importe de résoudre auparavant un problème d'interprétation de la Convention de Rome.

5. Un dessein de Protocole d'adhésion

Voici en quels termes – signalés, bien sûr, à titre d'indication de principe – on peut envisager la rédaction d'un Protocole d'adhésion, conçu de manière à satisfaire les différentes exigences qui se présentent.

– Article 1. – »Par acte d'adhésion, les Communautés Européennes déclarent leur acceptation pleine et entière des principes sur la reconnaissance des droits de l'homme et des libertés fondamentales, qu'énoncent la Convention de

Rome du 4 novembre 1950 et les Protocoles additionnels à cette Convention«.
– Article 2. – »Lorsque la question se pose pour la Cour de Justice des Communautés de savoir si une mesure prise par l'une des institutions des Communautés est compatible avec un droit reconnu dans ladite Convention et que pour résoudre cette question il importe d'interpréter la Convention, la Cour de Justice s'adresse, de son propre chef ou sur demande de l'une des parties au différend, à la Cour européenne des Droits de l'Homme afin que celle-ci, assistée par la Commission européenne des Droits de l'Homme, interprète les dispositions pertinentes de la Convention. L'interprétation donnée par la Cour européenne lie la Cour de Justice«.
– Article 3. – »La Cour de Justice annule tout acte d'une institution communautaire soumis à son jugement si elle conclut à l'incompatibilité de l'acte avec la Convention européenne«.

Freedom of the Press and of Broadcasting and Prior Restraints

Aspects of the case law of the Constitutional Court of the Federal Republic of Germany

HELMUT STEINBERGER

In its decision in the *Handyside* case the European Court of Human Rights stated that Art. 10 sec. 2 of the European Convention on Human Rights[1] leaves to the Contracting Parties a limited »margin of appreciation« with regard to restrictions on the exercise of the freedoms guaranteed by Art. 10 sec. 1 of the Convention[2]. It accorded this power of appreciation both to the domestic legislator and to other domestic bodies called upon to interpret and apply the laws, in particular, to domestic courts. This principal approach was confirmed by the first decision on *Sunday Times*[3]; the strong dissenting minority in this case did not question it but would rather have extended its scope. The Court clearly stressed that this domestic power of appreciation was not unlimited but was under the control of the institutions of the Convention. It had, in particular, to control the purpose of a restrictive law or measure as well as its necessity in a democratic society, including the pro-

---

[1] Art. 10: »1. Everyone has the right to freedom of expression. This right shall include freedom to hold opinions and to receive and impart information and ideas without interference by public authority and regardless of frontiers. This Article shall not prevent States from requiring the licensing of broadcasting, television or cinema enterprises. 2. The exercise of these freedoms, since it carries with it duties and responsibilities, may be subject to such formalities, conditions, restrictions or penalties as are prescribed by law and are necessary in a democratic society, in the interests of national security, territorial integrity or public safety, for the prevention of disorder or crime, for the protection of health or morals, for the protection of the reputation or rights of others, for preventing the disclosure of information received in confidence, or for maintaining the authority and impartiality of the judiciary«.

[2] Judgment of 7 Dec. 1976, Publications of the European Court of Human Rights, Series A, no. 24; Europäische Grundrechte-Zeitschrift (EuGRZ) 1977, pp. 38 ff.; see, in particular, secs. 48, 49, 52, 54, 57, 62 of the Judgment.

[3] Judgment of 26 April 1979, Series A, no. 30; EuGRZ 1979, p. 386; see, in particular, sec. 59 of the Judgment.

portionality between the restriction and its lawful purpose, according to the autonomous international standards as laid down by the Convention. To expound the various limits, aspects and implications of this domestic power of appreciation, which goes hand in hand with the power of control of the institutions of the Convention, will be the task of the case law of the Strasbourg institutions. From the two judgments mentioned a few limited conclusions can already be derived.

At first it would appear that the assessment of whether there exist pressing social needs which might require imposing by law restrictions of the kinds mentioned in Art. 10 sec. 2 of the Convention falls within the margin of appreciation of a Contracting Party. A second conclusion would be that the extent of the power of appreciation is not the same with regard to the various legitimate aims of restrictions under Art. 10 sec. 2; it may vary with the respective aims[4]. Accordingly, a more extensive European control may correspond to a less discretionary domestic power of appreciation[5]. As the most important conclusion, at least for judicial practice, would probably appear that the majority opinion in *Sunday Times* rejected the idea to limit the European standard of control with regard to the domestic power of appreciation to a *standard of arbitrariness, i.e.* to control merely whether the domestic bodies in imposing the restriction acted reasonably, carefully and in good faith[6]. This will be of particular practical importance with regard to the European control of the finding and evaluation by the domestic courts of the facts and circumstances of a measure[7]. The dissenting opinions of Judge

---

[4] While in *Handyside* the Court found that it was not possible to find in the laws of the Contracting Parties a uniform European conception of »the requirements of morals« in the sense of Art. 10 sec. 2, it considered in *Sunday Times* the aim of maintaining »the authority« of the judiciary to be a far more objective notion, domestic law and practice, as well as provisions, like Art. 6, in the Convention itself revealing a fairly substantial measure of common ground in this area.

[5] *Sunday Times* case, sec. 59, at pp. 36 f.

[6] *Ibid.*

[7] This approach led the majority in *Sunday Times* not to limit itself to a consideration whether the evaluation of the procedural circumstances of the claims against Distillers by the House of Lords had been unreasonable but to replace it by its own evaluation which was the decisive step to be able to find the injunction against Sunday Times to be a disproportionate and, accordingly, not »necessary« restriction in the sense of Art. 10 sec. 2 of the Convention – a rather keen and risky approach indeed. One might ask if it was not more adequate if the Court restrained its control of the findings and evaluation of the facts and circumstances to the test whether such findings and evaluations by the domestic courts were *based* on considerations *apparently* (evidently) unreasonable in the light of the standards of the Convention or on considerations which fundamentally misconceived (and thereby violated) these standards. Such a test would enable the Court to meet any manipulation of the statement of facts by the domestic courts. Like the Constitutional Court of the Federal Republic of Germany the European Court of Human Rights is not a »court of revision«; one important consequence of this is that both Courts are not bound to the statements of facts by other courts. It is a question of judicial wisdom and *self*-restraint which credit

Mosler in the *Handyside* case as well as of nine judges in the *Sunday Times* case demonstrate the serious difficulties the Court had and will have to master in this context. The reasons for these difficulties are quite complex. One results from the fact that in order to control the margin of appreciation left to a Contracting Party the Court will have to consider and assess the interactions of the relevant domestic legal institutions, of substantive and procedural law, of legislative, administrative and judicial practice (actions as well as omissions) surrounding the measure attacked before it [8]. A »condition« or »formality«, rather innocent on the face of it, in combination with other legal provisions or practices may turn out to be a hard core intrusion into a fundamental right or liberty. Other measures, restrictive at first glance, in the context of actual practice may turn out to be no more than reasonable regulations, rather than restrictions, of a fundamental right. Much more serious will be the fact that restraints on a conventional right or freedom almost always already have been the result of a balancing by the national authorities of the conventionally protected right or freedom against constitutionally or otherwise legally protected interests; to consider and weigh them in the terms of Art. 10 sec. 2 of the Convention will require to posite them on the scale of the relevant interests within the whole cultural context of the respective democratic society while at the same time maintaining that the Convention embodies common fundamental legal values and standards.

In *Handyside* and *Sunday Times* the Court related the margin of appreciation to the question of the »necessity« of restrictive measures. This will include the assessment – though under European control – whether such measures were prompted by a pressing social need, whether the reasons for a measure were relevant and sufficient, and whether the measure was required for, appropriate and proportional to the aims mentioned in Art. 10 sec. 2 of the Convention. »Necessity«, nevertheless, is a relational notion. In the context of Art. 10 sec. 1 it is related to the freedoms guaranteed under sec. 1, on the one side, and to the aims recognized to legitimate and justify restrictions »in a democratic society« under sec. 2, on the other side. The question, therefore, arises whether all the notions related by the term »necessary« carry with them certain margins of (domestic) appreciation which do (and must) not encroach upon the essential substance of the international standards laid down by the Convention and which may, if they do not so encroach, justify restrictions by a Contracting Party. In other words: do even notions like

---

they may accord to other courts in this context. With regard to the practice of the Federal Constitutional Court in this context see *Hans Joachim Faller*, Zu den Eingriffsmöglichkeiten des Bundesverfassungsgerichts bei Rügen der Verletzung von Meinungs- und Pressefreiheit (im Verfassungsbeschwerdeverfahren gegen Gerichtsentscheidungen), in: Presserecht und Pressefreiheit, Festschrift für Martin Löffler (München 1980), pp. 43 ff.

[8] It seems that such considerations brought Judge Mosler to dissent from part of the *Handyside* decision.

»democratic society« or »prescribed by law«, for instance the problem of the certainty of the »law« in the sense of Art. 10 sec. 2[9], leave margins of domestic appreciation? Even the concepts of the freedoms themselves guaranteed by Art. 10 sec. 1? Would, for instance, a concept of democracy as it is expressed in Arts. 21 II[10], 9 II[11] and 18 of the Basic Law of the Federal Republic of Germany[12] and which deviates from the more Kelsonian, relativistic types of concepts of democracy prevailing in other Contracting Parties fall within or outside such a margin? Would the Court venture to evaluate historical experience of a certain society differently from the framers of its constitution? Consequences on the domestic as well as on the international plane might be very far-reaching.

The European Court of Human Rights may have to face these questions as pertinent cases might be brought before it. The following paper will try to outline a few limited aspects of the constitutional law and protection of freedoms covered by Art. 10 of the Convention as they result from the case law of the Constitutional Court of the Federal Republic of Germany being one of the domestic judicial bodies to which a power of appreciation might lie[13].

Art. 5 of the Basic Law of the Federal Republic of Germany provides[14]:

---

[9] The concurring opinions by judges Zekia, O'Donoghue and Evrigenis in *Sunday Times* appear to leave a very small margin of appreciation, if any at all, to domestic courts in qualifying national laws as being sufficiently certain.

The certainty of laws is a constitutional requirement under the Basic Law of the Federal Republic of Germany. It has been developed by the case law of the Federal Constitutional Court and is frequently relied upon in cases before this Court. It may raise very complex questions, as for instance in the *Fast Breeder (Kalkar)* case where the Court with specific regard to the requirement of the certainty of the law in question (§ 7 sec. 3 of the Statute on Atomic Energy) developed the principle of »dynamic protection of fundamental rights«, see Entscheidungen des Bundesverfassungsgerichts (Decisions of the Federal Constitutional Court (Tübingen 1952 ff.) (BVerfGE) vol. 49, pp. 89, 133 ff. (1978).

Notions like »public order and security« as used in German statutes dealing with police power at first sight may appear to be overly broad and vague; when considered however, against the background of the case law of the competent courts they will stand even the most rigorous (domestic) tests of »certainty«.

[10] Under 21 II political parties which by reason of their aims or the behaviour of their adherents, seek to impair or abolish the free democratic fundamental order or to endanger the existence of the Federal Republic of Germany, shall be unconstitutional. The Federal Constitutional Court shall decide on the question of unconstitutionality.

[11] Under Art. 9 II associations, the purposes or activities of which conflict with criminal laws or which are directed against the constitutional order or the concept of international understanding, are prohibited.

[12] See the concept of democracy as developed by the Federal Constitutional Court in BVerfGE 2, 1 (12 f.) (1952); 5, 85 (140) (1956); 44, 125 (145) (1977).

[13] For a broader treatment see *W. Schmitt Glaeser*, Die Meinungsfreiheit in der Rechtsprechung des Bundesverfassungsgerichts, Archiv des öffentlichen Rechts 97 (1972), pp. 60 ff., 276 ff.

[14] »(I) Everyone shall have the right freely to express and disseminate his opinion by speech, writing and pictures and freely to inform himself from generally acces-

»(I) Jeder hat das Recht, seine Meinung in Wort, Schrift und Bild frei zu äußern und zu verbreiten und sich aus allgemein zugänglichen Quellen ungehindert zu unterrichten. Die Pressefreiheit und die Freiheit der Berichterstattung durch Rundfunk und Film werden gewährleistet. Eine Zensur findet nicht statt.
(II) Diese Rechte finden ihre Schranken in den Vorschriften der allgemeinen Gesetze, den gesetzlichen Bestimmungen zum Schutze der Jugend und in dem Recht der persönlichen Ehre.
(III) Kunst und Wissenschaft, Forschung und Lehre sind frei. Die Freiheit der Lehre entbindet nicht von der Treue zur Verfassung«.

I

Art. 5 is one cornerstone of a constitutional concept drafted to secure the dignity and liberty of man and a democratic, open society in which intellectual freedom is considered not only as an end but also as a means to attain and preserve the end. Art. 5 operates within an inter-related set of other fundamental rights and liberties, constitutional principles, rules and standards, institutional and procedural devices. With the concept of the spiritual nature and personal dignity of man as an individual at its core it includes the rights and liberties of the *forum internum* of man – freedom of thought and of belief, the right to form and to hold opinions on any subject, the right to privacy and to be let alone – as well as the rights and liberties covering the social nature of man – the right to communicate through any medium, the right to inquire and to inform oneself from generally accessible sources; the right to combine with others in assembly or in association to form and to articulate opinions, receive communications, provide information, pursue common goals; the right to free movement and to freely choose one's occupation and profession, one's place of work and of training.

These fundamental rights and liberties must not only equally be protected, they also have to be reconciled with the rights and liberties of other persons and groups as well as with other individual and social interests recognized by the constitution[15]. The task of delimitation and reconciliation of

---

sible sources. Freedom of the press and freedom of reporting by means of broadcasts and films are guaranteed. There shall be no censorship.

(II) These rights are limited by the provisions of the general laws, the provisions of law for the protection of youth, and by the right to inviolability of personal honour.

(III) Art and science, research and teaching, shall be free. Freedom of teaching shall not absolve from loyalty to the constitution«.

If not indicated otherwise, Articles cited in the following are those of the Basic Law of the Federal Republic of Germany of May 23, 1949 (Bundesgesetzblatt [BGBl.] 1949, p. 1), as amended.

[15] BVerfGE 7, 198 (210) (1958); 20, 162 (176) (1966); 28, 191 (201 f.) (1970).

these rights, liberties and interests in practice raises the most difficult questions, e.g. as to the competence and legal forms of action (legislative acts of parliament; regulatory powers of the executive; administrative actions) or as to the standards of proportionality and reasonableness, of overbreadth and vagueness of the respective actions.

Rich historical experience, moreover, demonstrates that, as a rule, freedom of expression, and in particular freedom of the press, cannot effectively be secured unless it is guarded by impartial and independent courts armed with adequate competences. This requires a set of procedural devices such as the right of access to the courts, a full and fair hearing and adequate powers of the courts to adjudicate the facts as well as the law. Judicial protection on the other side will only be effective if the other branches of government respect and execute the decisions of the courts which neither command the sword nor the purse of society [16]. The best safeguards will consist in a courageous public constantly supporting and defending the idea of freedom of expression. The press can take a leading role in this effort thereby supporting and defending the ideas upon which its own freedom rests.

It would go beyond the limits of this paper to expound the indicated constitutional framework in which freedom of the press as guaranteed by Art. 5 I 2 of the Basic Law is embedded and operates. Suffice it to indicate the following: Arts. 1 to 17 of the Basic Law recognize a traditional set of fundamental rights and liberties which by express provision of Art. 1 III bind the three branches of governmental power as directly enforceable law. If according to the Basic Law a fundamental right or liberty may be restricted by or pursuant to a law, such law must apply generally and not solely to an individual case; in no case may the essential content of the restricted fundamental right or liberty be encroached upon, Art. 19 I, II. Should any person's right (constitutional or other) be violated by public authority, recourse to the courts shall be open to him, Art. 19 IV. Whoever abuses the freedom of expression, in particular the freedom of the press, in order to combat the free democratic basic order, shall forfeit these basic rights; such forfeiture and the extent thereof shall only by pronounced by the Federal Constitutional Court, Art. 18 [17]. Judicial power is vested in the judges only and exclusively, Art. 92. Independence of the judges is guaranteed by the Constitution, Arts. 97, 98. Any person who claims that one of his fundamental rights or liberties or of

---

[16] See *Alexander Hamilton,* The Federalist Papers No. 78.

[17] Since 1953, when the Federal Constitutional Court began its work, only two cases had to be decided by the Court under Art. 18; they were dismissed, see BVerfGE 11, 282 (1960) and 38, 23 (1974). In the latter case the Federal Government had brought an action before the Court under Art. 18 against the editor of a newspaper considered as rightist. The Court dismissed the action because the Government had not answered the extensive factual and legal statements of the opponent nor had it put forward arguments relating to the question of the present danger of the opponent.

his rights under certain other provisions of the Basic Law has been violated by public authority (legislative, administrative or judicial), after exhaustion of other judicial remedies may enter a complaint of unconstitutionality with the Federal Constitutional Court, Art. 93 I 4a. Under the Basic Law German courts have the power of judicial review of legislative acts; if, however, a court considers unconstitutional a law the validity of which is relevant to its decision, it may not declare such law void but must stay the proceedings and obtain a decision of the competent constitutional court (State or federal, as the case may be) on the constitutional issue of the validity of the law in question, Art. 100 I.

Legislative competences in the field of broadcasting (except the technical side) rest with the States; in the fields of press and film they are divided between the Federation and the States: the Federation is competent to enact skeleton provisions on the general status of the press and the film industry, Art. 75; until now the Federation has not made use of this »framework« competence; the other competences rest with the States, Art. 30; they have enacted press laws. The Basic Law, nevertheless, is the supreme law of the land; the States, therefore, are bound by Art. 5.

II

1. In its decisions under Art. 5 the Federal Constitutional Court from the very beginning has underlined two principal dimensions of the freedom of expression[18]: One being the individual dimension; it protects one of the most immediate issues of the spiritual nature of man, flowing from personal individuality, which seeks dialogue. Without this freedom it is denied to man to live in accordance with his nature. Therefore, it is one of the most precious human rights[19]. The other dimension is social: Freedom of expression is at the very basis of ordered liberty and democracy because it makes possible the permanent public discussion and struggle of opinions on subjects of general and political importance essential to democracy[20]. This fundamental liberty, therefore, is of a comprehensive nature[21].

Freedom of the press, which is a special freedom within the scope of the freedoms of expression, secures a *subjective* basic right directed against coer-

---

[18] BVerfGE 7, 198 (208) (1958).

[19] BVerfGE 7, 198 (208) (1958).

[20] BVerfGE 5, 85 (205) (1956); 7, 198 (212, 219) (1958); 12, 113 (125) (1961); 20, 56 (98); 162 (174 ff.) (1966); 25, 256 (265) (1969); 36, 193 (204) (1973); 50, 234 (239) (1979).

[21] BVerfGE 33, 1 (15) (1972). The fact that the decisions of the Court under Art. 5 in nearly every case dealt with the sociopolitical dimension of the freedoms of expression and of information does not imply that the Court would deny their fundamental importance in non-political spheres. It rather depended on the kinds of cases actually brought before it, and demonstrates the actual points of collision.

cive State action in press related contexts. Yet Art. 5 I 2 goes beyond mere recognition of a subjective right; it embodies an *objective* constitutional guarantee of the institution of a »free press«[22]. This institutional element may enable and even oblige the government to enact protective measures to preserve the institution of »free press« should developments occur which seriously endanger it, e.g. by excessive concentration destroying the free market place of ideas[23].

2. Exercise of the freedom of the press does not dispense from the obligations resulting from the provisions of »the general laws, the provisions of law for the protection of youth, and by the right to inviolability of personal honour«, Art. 5 II. Aside from limitations which are covered by provisions of the Basic Law itself, like Arts. 9 II, 21 II and 18 this means that restraints on the exercise of the freedom of the press are constitutional *only* if they are covered by the provisions of Art. 5 II.

In daily practice most legal problems will arise with regard to what qualifies as a »general law« capable of limiting the exercise of the freedom of the press. Obviously the notion of »general laws« cannot mean »any law« otherwise freedom of the press would be at the mercy and discretion of the legislature and would be granted only »according to law«. In this context three points are to be stressed:

a) It is well established by the decisions of the Constitutional Court that »general laws« are those laws which do not outlaw the content of a certain opinion as such, are not directed against such opinion but are neutral as to content and serve, without regard to any particular opinion, to protect a legal good equally recognized by the Constitution[24].

b) Even if a law meets this requirement and is, therefore, constitutionally capable of limiting the exercise of the freedom of the press, it must not encroach (or be applied so as to encroach) upon the essential content of the freedom of the press, Art. 19 II.

---

[22] BVerfGE 10, 118 (121) (1959); 20, 162 (176) (1966). The relation and interaction between these two components may raise difficult problems which will not be pursued here; for an excellent critical treatment see *W. Schmitt Glaeser* (note 13), pp. 82 ff. Principally the institutional component must only serve to strengthen the subjective fundamental right (see BVerfGE 7, 198 (205) (1958) – for other fundamental rights, see BVerfGE 24, 367 (389) (1968); 50, 290 (337) (1979) but not to accord privileges to the press nor to dissolve the subjective right by way of imposing additional restraints upon it derived from pretended institutional necessities or »public« obligations resulting from them.

[23] In the opinion of the Administrative Court and of the Administrative Court of Appeals of Berlin financial subsidies to newspapers would require a law which must contain so precise provisions that any discretionary power of the executive in administering and distributing such subsidies would be banned; see the decisions in Neue Juristische Wochenschrift (NJW) 1974, p. 330; 1975, p. 1940.

[24] BVerfGE 7, 198 (207 ff.) (1958); 21, 271 (280) (1967); 26, 186 (205) (1969); 28, 175 (185 f.) (1970); 28, 282 (292) (1970); 33, 52 (66) (1972).

c) Such laws themselves must be interpreted and applied restrictively with a view to leaving the exercise of the freedom of the press (as well as other basic rights under Art. 5 I) unrestrained as far as possible and compatible with the constitutionally protected aim of such laws. The Constitution is not only the supreme law of the land and thereby a measuring-scale for the validity of statutory provisions; the provisions must be interpreted in conformity with the Constitution and may receive, within the frame provided by their wording, an additional content from it[25]. This requires a kind of dialectical method of interpretation and a general as well as a concrete balancing of the constitutionally protected values involved[26]. »Absolutist« doctrines within this context have found no support within the German judiciary[27]. If, however, a restriction is neither covered by Art. 5 II nor by other provisions of the Basic Law, there is no room for any balancing whatsoever: it is unconstitutional.

Art. 5 II applies likewise to the other freedoms guaranteed by Art. 5 I, *i.e.* to the freedoms of the expression of opinions, of information, of film and of broadcasting. A large body of decisions has been developed dealing with the problems of »general laws«. Generally speaking (and excepting works of art and science, teaching and research[28]) the author, publisher, editor, reporter, journalist, distributor and shopkeeper who sells books, magazines, newspapers, tapes or records, the producer of a motion picture, the movie theater, the reader of a book or paper, the viewer of a film, in exercising their respective fundamental rights are subject, e.g. to the law of contracts and of torts[29], to property law, the law against unfair competition, copyright law and other laws protecting intellectual property, to labour law, social insurance law and tax law[30], to criminal[31] and to administrative[32] law, etc. Most laws qualify as »general laws« in the sense of Art. 5 II as they are not directed against a certain opinion, do not discriminate against or unreasonably restrict any specific press-, film- or information-related behaviour, or institution or entity. A vast

---

[25] BVerfGE 51, 304 (323) (1979).

[26] BVerfGE 7, 198 (210 f.) (1958); 21, 239 (243) (1967); 28, 191 (202) (1970); 33, 52 (66) (1972); 43, 130 (139) (1976).

[27] Every judge knows of the extreme difficulties of balancing competing constitutional interests; he might well long for »absolute« standards; in practice, however, these hardly ever prove to be workable unless one simply ignores other interests protected as well by the Constitution.

[28] Which are covered by the special provision of Art. 5 III and are not subject to the limitations of Art. 5 II.

[29] BVerfGE 7, 198 (211, 214) (1958); 7, 230 (234) (1958).

[30] Federal Financial Court (BFH), NJW 1974, pp. 1157, 1158.

[31] BVerfGE 20, 162 (177) (1966); 21, 239 (1967); 28, 175 (1970); 28, 191 (200) (1970); 42, 143 (149) (1976); 47, 130 (143) (1978); 198 (232) (1978); – 2 BvR 215/81 – dec. of 26 May 1981, 57, 250 (270) (1981).

[32] BVerfGE 28, 36 (47) (1970); 28, 282 (289) (1970); Administrative Court of Appeals (OVG) Koblenz, Deutsches Verwaltungsblatt (DVBl.) 1966, p. 578.

body of decisions has emerged from adjudicating questions of »general laws«. Only a few cases can be mentioned here:

The appeal by the chairman of the Hamburg Press Club to the German public, owners of movie theaters and film distributors to be prepared to boycott a motion picture produced by a film production company and directed by Veit Harlan, who had directed an anti-semitic film during the period of nationalsocialism, under the circumstances of the case was covered by the freedom of expression and could not be restrained by claims of omission under § 826 of the Civil Code[33]. §§ 99, 100 of the Criminal Code which penalized publishing State secrets were considered to be »general laws«, limiting the exercise of the freedom of the press, as they were not directed against specific opinions, although such provisions in the light of Art. 5 I had to be interpreted restrictively[34].

In 1961, a weekly magazine published a fabricated »exclusive interview« with Soraya, the former empress of Iran; the interview dealing with her private affairs had never taken place. She sued for damages under §§ 823, 847 of the Civil Code. Against the judgment of the civil courts in her favour the respondent entered a constitutional complaint with the Federal Constitutional Court contending, among other arguments, that the judgment violated his right to freedom of the press. The Court, while stating that the notion »press« required broad interpretation covering not only »serious« publications but also the »rainbow« (*i.e.* yellow) press, confirmed that § 823 protecting the right of personal integrity against tort was a »general law« in the sense of Art. 5 II limiting the exercise of the freedom of the press[35].

In 1972, the country-wide public television corporation »Second German Television« intended to broadcast a documentary film on a criminal affair that had actually taken place in 1969; four soldiers sleeping in their barracks had been murdered by three men. One of the participants, the plaintiff, in 1970 was sentenced to six years imprisonment for assisting in the murder. The documentary intended to show the relations, in part homosexual, between the three men, the planning and execution of the crime, the investigation of the deed and a blackmail affair connected with it. At the beginning of the film, the plaintiff's picture was to be shown, after that his role was to be represented by an actor; his real name was to be used constantly during the film. Arguing that his rights to personal integrity, to his name and his portrait were in danger of being violated by the documentary, if broadcast, he petitioned the civil courts for an injunction which was denied. Upon his constitutional complaint against these decisions the Federal Constitutional Court granted the injunction and ordered the television corporation not to

---

[33] BVerfGE 7, 198 (1958); see also BVerfGE 12, 113 (1961); 24, 278 (1968); 54, 129 (137) (1980).

[34] BVerfGE 20, 162 (170) (1966) – *Spiegel* case; 21, 239 (1967); 28, 175 (1970) – *Porst* case; 57, 250 (270) (1981) – *Cremer* case.

[35] BVerfGE 34, 269 (283 f.) (1973).

broadcast the documentary to the extent that it dealt with the plaintiff. This decision is most typical for the Court's approach:

The Court undertook to balance the fundamental rights of the plaintiff to personal integrity and privacy, in particular, his constitutionally protected interest in resocialization from a criminal act and in returning to a normal life in society after he had served his sentence, with the freedom of broadcasting enjoyed by the television corporation. While underlining the constitutive function of the freedom of broadcasting for a democratic society which in general protects all kinds of broadcasts, the Court held that as soon as other constitutionally protected rights may be violated by it, the conflicting interests generally and *in concreto* must be weighed against each other. In general, neither the right to personal integrity, including the right to one's own portrait, protected by Arts. 1 and 2, nor the freedom of broadcasting prevail over each other without more. If *in concreto* they conflict, their common relation to the protection of the dignity of man must be taken into consideration. With a view to all the relevant circumstances of a case the intensity of the intrusion into the personal sphere must be weighed against the public interest in information on the subject. For *contemporaneous* reports on serious crimes this public interest in general prevails. However, even then regard must be had to the innermost personal privacy and to the proportionality of the intervention; this may not always permit publicly naming, portraying or otherwise identifying the person concerned. Once there is no longer a *contemporaneous* report the weight of the protection of personal integrity increases. It prohibits broadcasts which, irrespective of the point on time, deal with the perpetrator of a crime and his privacy, at least if they cause substantially new or additional injury, in particular, if they endanger the rehabilitation of the person concerned. In the opinion of the Court, this was the case here, several years after the plaintiff had been tried and was presently serving his sentence[36].

In 1976, the author of a political pamphlet which contained harsh attacks concerning the personal political past of a member of the Federal Diet was sentenced because of defamation under §§ 186, 187 a of the Criminal Code. His constitutional complaint was successful. The Criminal Court had argued that these provisions, in order to protect personal honour, required an extensive interpretation. The Constitutional Court considered this to be incompatible with Art. 5 I because it did not take due account of the fact that the pamphlet's contents contributed to the struggle of opinions in a question of essential concern to the public[37].

At the occasion of the funeral of the president of the Court of Appeals of Berlin who had been murdered by terrorists, a well-known commentator over television charged the German author *Heinrich Böll* with having helped to

---

[36] BVerfGE 35, 202 (1973).

[37] BVerfGE 43, 130 (1976).

prepare the soil in which the seeds of terrorism could germinate. He quoted Mr. Böll as having publicly declared that the State fought by the terrorists was a dung-heap, that he discerned only remnants of rotten power defended by rat-like rage, that the State persecuted the terrorists in a chase without quarter. Mr. Böll sued for damages in the civil courts alleging of violation of his honour. He argued that the quotations were either false or taken out of the context where they had a different meaning than that imputed by the commentary. The Federal Supreme Court for Civil Cases denied the claim; although the citations were not correct, the commentary was nevertheless widely justified by a reasonable interpretation of Böll's publications and therefore remained within the limits of the right to free expression of opinion. Following a constitutional complaint, the Federal Constitutional Court reversed the judgment. Against wrong, falsified or distorted reproduction of his statements a person could rely on his fundamental right to personal integrity recognized by Arts. 1 I and 2 I of the Basic Law. False quotations were not covered by Art. 5 I. The function of the freedom of expression in a democratic society did not require such protection. Value judgments in the interest of contending public opinions enjoyed protection by Art. 5 I regardless of their contents; the same was not true, however, of false factual assertions[38].

The Law on the distribution of printed publications endangering youth, which forbids certain methods of publicly advertising these publications was considered constitutional with regard to the express provision for the protection of youth contained in Art. 5 II[39].

In 1966, the Constitutional Court had indicated that persons in press-related occupations might claim, directly based on their basic right to freedom of the press, a right to refuse to testify in criminal proceedings going beyond the statutory right to do so provided for in § 53 sec. 1 no. 5 of the Code of Criminal Procedure[40]. The statute of 25 July 1975[41] took this consideration into account and extended the statutory right to refuse testimony to the author, transmitter, or informant of contributions and records as well as to those providing information for press activities to the extent that these contributions, records, and informations were destined for the editorial parts of periodicals or broadcasts. A three-judge panel of the Constitutional Court held that this privilege did not extend to pictures made by the reporters of a daily newspaper themselves[42].

---

[38] BVerfGE 54, 208 (219) (1980).

[39] BVerfGE 11, 234 (238) (1960); insofar as the Law stipulated an irrefutable presumption that pictures of nudity were always endangering the youth it was considered an unreasonable (disproportionate) restriction and – to that extent – declared void.

[40] BVerfGE 20, 162 (187 ff., 191 f., 195 f., 216 ff.) (1966) – *Spiegel* case.

[41] BGBl. 1975 I, p. 1973.

[42] Decision of 4 March 1981 – 2 BvR 195/81 – EuGRZ 1981, p. 314. The reporters themselves had photographed persons actually participating in violent tur-

As may be discerned from this body of decisions, freedom of the press as well as freedom of film and broadcasting under the Basic Law do not just mean freedom from *prior* restraints. They also protect against restrictions *after* publication which do not qualify as »general laws« in the sense indicated above.

3. Under the legal order of the Federal Republic of Germany no print media licencing requirements for authors, publishers, editors, reporters, journalists, correspondents or distributors exist. No authorities exist for the issuance, renewal or revocation of licences, nor are there any standards or procedures for such purposes. No such requirements exist either for newspapers, magazines, books, motion pictures, pictures, or plays. The Constitutional Court, while recognizing a »public« function of the press[43] underlined that this function could not be accomplished by the public authorities. The unrestrained foundation and establishment of press enterprises as well as free access to press-related trades, occupations, or professions are essential to a free press operating on principles of private economy and organized under private law. Public power in general must not intervene in the intellectual and economic competition between the elements concerned[44].

The Press Laws of the States, in complying with this principal requirement of federal constitutional law, uniformly provide that press-related activity, including the establishment of publishing enterprises or any other press-related trade must not be subjected to licensing[45]. Besides requiring an

---

moil. On application by the prosecutor, the District Court of Hannover ordered the pictures seized. Constitutional complaint was denied.

[43] BVerfGE 20, 162 (175) (1966). From the whole body and lines of decisions of the Court it would appear that this »public function« (*öffentliche Aufgabe*) does not mean that the (political) press exerted any »public power« or »public mandate«; the term rather intends to give a sociological description of the effects of the (political) press on public opinion in a democratic Society (comp., e.g., BVerfGE 44, 125 (140 f.) (1977). This is not to deny that the Court developed some limited constitutional consequences out of this public function, like an obligation on public authorities to inform the press (BVerfGE 20, 162 (172) (1966), the constitutional protection of editorial secrets, and the collection and dissemination of information and opinions (BVerfGE 10, 118 (121) (1959)) as well as specific responsibilities, like correct reporting (BVerfGE 12, 113 (130) (1961)) or responsibilities in the field of national security (BVerfGE 20, 162 (212) (1966)). But in substance the Court has withstood the danger of deriving from this public function of the press the admissibility of restraints on the exercise of the subjective basic right of the freedom of the press going beyond restraints covered by Art. 5 II or by other provisions of the Basic Law. The decision BVerfGE 42, 53 (62) (1976), for instance, rejected the argument that the private press had an obligation to stay neutral during an election campaign; under present statutory law which, the Court stressed, meets no constitutional objection, a newspaper was free to refuse to print political advertisements or letters to the editor, even if it was the only newspaper in the region concerned.

[44] BVerfGE 20, 162 (175 f.) (1966) – *Spiegel* case.

[45] See, for instance, § 2 of the Press Law of the State of Bavaria.

imprint for printed products[46] they provide that the editor-in-charge (*verantwortlicher Redakteur*) must have his permanent residence in the Federal Republic of Germany, must not under a judicial decision have lost the capacity to hold public offices or to gain functions through public elections[47], be at least 21 years of age, and possess full legal capacity[48]. These provisions, although especially related to the press, are not to be considered incompatible with the freedom of the press because they are neutral in content and not directed against the formation or holding of a certain opinion; they do not unreasonably restrain press-related activities, and serve to secure other constitutionally protected interests, in particular, they serve to clarify the legal responsibility where cases giving rise to claims concerning a publication are concerned. The provisions do not operate as a prior restraint as no prior approval or licence by a public authority is required for an editor-in-charge to exercise his occupation or to be hired.

4. Even if a limitation of the freedoms of expression meets the requirements of Art. 5 II, one specific kind of restraint is constitutionally prohibited: censorship. Its prohibition in Art. 5 I 3 applies to print media, motion pictures and broadcasts (radio, television).

The notion censorship in Art. 5 I 3 means prior (preventive) censorship. It comprises restrictive measures by public authorities prior to the production, publication, or dissemination of an intellectual product, in particular, to make its production, or dissemination dependent on the prior examination or approval of its intellectual contents. The mere existence of censorship would paralyse intellectual life; therefore, even »general laws« in the sense of Art. 5 II must not provide for censorship[49].

A few decisions of the Constitutional Court may be mentioned in this context.

The Statute on the dissemination of publications endangering youth restricts or prohibits certain methods of distributing and advertising publications which have immoral contents or praise crime, war, or racial hatred. Such publications are registered; the register is published. Registration and publication authority is rested in a federal administrative authority whose members are appointed by the Minister of the Interior and who are not bound by instructions. The authority decides on application by the Minister

---

[46] *I.e.* the indication of the name, the firm and the address of the printer and editor. Periodical press products must indicate the name and the address of the responsible editor (*Redakteur*); see, for instance, § 8 of the Press Law of the State of Baden-Württemberg.

[47] See § 45 of the German Criminal Code.

[48] See, for instance, § 9 of the Press Law of the State of Baden-Württemberg. The State Laws differ to some extent in the additional requirement that the editor-in-charge must not be protected against criminal prosecution by sovereign or parliamentary immunity or indemnity.

[49] BVerfGE 33, 52 (72) (1972).

for Family Affairs and by the administrative authorities for the protection of youth. Its decisions may be brought before the administrative courts. The law, however, does not make publication, advertising, or dissemination dependent upon prior control or approval. Its provisions, therefore, do not violate the prohibition of censorship in Art. 5 I 3; the provisions restricting advertising are covered as constitutional by Art. 5 II [50].

In connection with this Statute the Constitutional Court, in another decision, stressed that the registration system provided for in the law must be interpreted and applied so as to ensure that the judicial control by the courts must not be curtailed [51].

A law of 1965 empowers the postal service and the customs authorities to secure those goods which for reasons of State security may not be imported into the country. The plaintiff had asked friends in Eastern Germany to send him daily newspapers published there. The newspapers were confiscated; the district court approved the action. On constitutional complaint the decision was reversed. The Constitutional Court based its decision not on the prohibition of censorship in Art. 5 I 3 but on the plaintiff's freedom of information protected by Art. 5 I 2 [52]. If interpreted and applied in conformity with the freedom of information the law did not violate Art. 5 I 1 [53].

In most State Press Laws it is provided [54] that official documents on pending criminal court proceedings must not be published before they have been dealt with in the public hearing of the trial or before the proceedings have been terminated, except with the permission of the competent authority. Operation of this provision clearly demonstrates what is not meant by preventive censorship. The provision does not oblige, a newspaper reporting on the trial for instance, to submit the proofs prior to publication to the competent authority – that would constitute censorship. If publication of the official document takes place without permission the publisher/editor will be held liable for it – this does not constitute censorship in the sense of Art. 5 I 3. § 184 of the German Criminal Code penalizing, *inter alia*, the publication and dissemination of pornographic publications operates in a similar way; it does not constitute censorship but provides sanctions *after* the incriminating behaviour has occurred. Nor do injunctions or other decisions by courts ordering the attachment of a publication or orders enjoining publication or any further distribution, based on »general laws« in the sense of Art. 5 II, constitute censorship in the sense of Art. 5 I 3.

---

[50] BVerfGE 11, 234 (238) (1961); 30, 336 (347 ff.) (1971).

[51] BVerfGE 51, 304 (1974).

[52] BVerfGE 27, 71 (1969); prohibition of censorship protects the author, publisher and, possibly, the distributor (but not the reader as such) of a publication, BVerfGE 27, 88 (102) (1969).

[53] BVerfGE 27, 88 (1969).

[54] See, for instance, § 5 I of the Law of Baden-Württemberg.

Other cases demonstrate that not only censorship in the sense of Art. 5 I 3 but also other kinds of prior restraint can be unconstitutional. When a reporter of a newspaper was excluded from the public hearing by the judge because the paper – not even the reporter himself – had previously published an article critical of the judge, the Constitutional Court considered the exclusion to have violated both the paper's and the reporter's right to freedom of the press which protected the field of preparatory press work in general, to the gathering of information in particular[55].

Other decisions have stressed the importance protecting the sources of information of the press for the fulfilment of its function[56]. § 102 of the Law on Organization of Enterprises of 1972 provides that the workers' council, an institution elected by the employees, must be consulted by the management before an employee may be dismissed. The Law in § 118 at the same time provides that it is not applicable to enterprises which directly and preponderantly serve the purposes of reporting or expressing opinions falling under Art. 5 I 2 of the Basic Law, to the extent that the nature of the enterprise does not permit such an application. This exception applies, in particular, to newspapers. The publisher of a Cologne daily newspaper dismissed an editor (*Redakteur*) without having consulted the workers' council of the enterprise. The plaintiff sued against his dismissal. He succeeded before the Federal Supreme Court of Labour Law. The Constitutional Court confirmed the decision. It pointed out, however, that the freedom of the press not only prohibits any direct influence by State action on the editorial tendency of a press product; the State also must not enact regulations subjecting or opening up the press to outside non-State influences incompatible with the freedom of the press. § 118 of the Law on Organization of Enterprises recognized this constitutional objection. In conformity with Art. 5 I 2 it was to be interpreted so as to permit the workers' council to put forward social arguments only against the dismissal of an editor/reporter but excluding any arguments relating to the editorial tendency or policy[57].

In contrast to the Weimar Constitution the Basic Law prohibits censorship in the indicated sense also with regard to films. For reasons of space the related constitutional problems will not be dealt with here[58]. Only one case is to be mentioned.

---

[55] BVerfGE 50, 234 (240) (1979).

[56] BVerfGE 20, 162 (176, 187) (1966) – *Spiegel* case; 36, 193 (204) (1973).

[57] BVerfGE 52, 283 (1979). The Federal Supreme Court for Labour Law in a decision of 30 June 1981 – 1 ABR 26/79 – pronounced that the workers' council of a publishing enterprise under the Law on the Organisation of Enterprises had the right – without having to show a special case or interest – to inspect the salary list also with regard to the salaries of editors. The Supreme Court refuted the argument that this would violate the protection of editional tendency accorded by the Law, as the enterprise was not restrained to granting salaries in line with the tendency.

[58] The German motion picture industry has founded a private association. Within this association a Board on Voluntary Selfcontrol of the Film Industry (FSK)

The Law of 24 May 1961 [59] prohibits the importation of motion pictures into the country whose contents are capable of acting as means of propaganda against the free democratic basic order or against the idea of international understanding. Under § 5 sec. 2 anybody importing motion pictures into the country, has to submit a copy to a federal administrative authority within one week. By executive regulation of the Federal Government it may

---

was established. Its commissions consist of members part of whom are nominated by public authorities and churches, the other part by the film industry, the voting power being distributed in approximately the same proportions. The board examines all motion pictures to be publicly shown in the Federal Republic of Germany, including documentary as well as advertising films and excepting scientific films and films for instruction in public or publicly recognised educational and scientific institutions. It decides whether a film may be shown in public or not, or whether any cuts or changes will have to be made before being shown, whether it is suitable for youth, and whether it must not be shown on certain holidays. The substantive standards for the examination of films have been established by the board's »principles of voluntary self-control«, as amended. They define the task of the FSK as preventing films from exerting negative influences in the field of morals, religion, or politics. Accordingly, no film should be produced, distributed, or publicly shown capable

a) of violating moral or religious feelings, in particular of having a demoralizing or brutalizing effect;
b) of advancing national-socialistic, militaristic, imperialistic, chauvinistic, or racial hatred tendencies;
c) of endangering the relations of Germany with other States or of degrading its standing in foreign countries;
d) of endangering or abusing the constitutional foundations and the principles of the rule of law of the German people as a whole and within the Federal States;
e) of distorting historical facts by explicitly propagandistic or tendentious perspectives.

No film must be refused a showing for personal reasons of taste and aesthetics in art. The decisive criteria is the effect of a film on the viewer, that is, the average normal viewer. Effect means the passive receptive as well as the active effects.

There is a review procedure against the decisions of the Working Commission of the Board, exercised in the first instance by its Main Commission and, finally, by its Legal Commission.

Enforcement of the decisions of the Board is secured by an agreement of the members of the associations of the producers, distributors, and owners of movie theaters to observe the decisions. Mutual interest and reciprocal economic dependence up to the present time appear to have ensured that the system has worked; outsiders are supposed to have no chance economically of evading the system.

Except with regard to the system's operation aimed at protecting youth, the question of the compatibility of the system of voluntary selfcontrol with Art. 5 I has been much debated in German legal discussion. It has been defended with the argument that the persons and enterprises concerned with regard to the participation of representatives of public authority in the system have renounced, by voluntary agreement, their right not to be subjected to censorship. It has been attacked on the ground that Art. 5 I protects the »institution of the free process of communication« not permitting the renunciation of freedom from censorship, in particular, if voluntary renunciation can only be evaded at the cost of financial ruin. The constitutionality of the FSK system up to now has not been tested in the Constitutional Court.

[59] BGBl. 1961 I, p. 607.

be determined that motion pictures from certain countries of origin are not subject to the obligation to submit. By regulation of 12 October 1961 obligation to submit motion pictures from most countries of origin has been waived, except for the countries of Eastern Europe. Obligation to submit films to the administrative authority does not legally exclude the motion picture being shown publicly. Therefore, § 5 sec. 2 of the Law as such does not constitute censorship in the sense of Art. 5 I 3. The provision must, however, be considered in context with other provisions which subject transactions to administrative approval on the acquisition from foreign persons [60] of the rights to show films publicly. The combinated effect of these provisions gave rise to the question whether this constituted factual censorship [61]. Administrative approval of the acquisition, however, is not dependent on the intellectual content of the motion picture, which must not be examined by the authority under § 17 AWG/§ 48 AWV. It has been argued, therefore, that factual censorship does not take place [62]. By interlocutory procedure [63] the question was brought before the Constitutional Court whether § 5 of the Law violated, *inter alia*, Art. 5 of the Basic Law. The Court considered § 5 to be a »general law« in the sense of Art. 5 II of the Basic Law. Nor did § 5 violate the prohibition of censorship in Art. 5 I 3 as this constitutional provision meant only preventive censorship. With regard to motion pictures censorship meant a general prohibition to make uncontrolled films accessible to the public, together with an obligation to submit them to a public authority prior to public showing which examined their content according to censorial standards and, as a result of such examination, either did or did not allow their public showing [64]. The provisions attacked did not provide for preventive censorship; the fact that the economic changes for a motion picture might be diminished due to sanctions imposed upon public showing did not constitute censorship. The situation was similar to criminal sanctions imposed after publication. It was up to the individual to run that risk, while prior censorship prevented even the possibility of being prepared to take that risk. The obligation in § 5 sec. 2 of the Law to submit a copy of the motion picture did not legally exclude its public showing. Nor was the required approval of transactions to acquire the right to show a motion picture publicly dependent upon the examination of its content; because of that no

---

[60] § 17 Außenwirtschaftsgesetz (AWG) and § 48 Außenwirtschaftsverordnung (AWV).

[61] In this sense *M. Rehbinder*, Film und Recht (1967), pp. 47, 50.

[62] *Bernd Rieder,* Die Zensurbegriffe des Art. 118 Abs. 2 der Weimarer Reichsverfassung und des Art. 5 Abs. 1 Satz 3 des Bonner Grundgesetzes (Berlin 1970), pp. 180 ff., 183.

[63] The plaintiff in the administrative court had brought a motion picture from the Leipzig Fair in East Germany with him and had shown it publicly several times in the Federal Republic.

[64] BVerfGE 33, 52 (72) (1972).

factual censorship was implied[65]. The Court, nevertheless, thought it feasible that under different circumstances the combined effect of provisions could lead to unconstitutional *de facto* censorship.

Tax laws of the States provide for tax reduction if a motion picture has been rated »valuable« or »especially valuable« by the Board of Film Evaluation. The Board exists since 1957. It has been established by a treaty concluded between the States of the Federal Republic of Germany[66]. According to the treaty its purpose is to obtain uniform records for the taxation and for the promotion of valuable motion pictures. The Board does not decide whether a motion picture may be shown in public or not; its decision, therefore, cannot be qualified as censorship in the sense of Art. 5 I 3 of the Basic Law. A plaintiff whose motion picture had not been rated so as to fulfil the requirements for a tax reduction had argued before the administrative courts that by its economic impact the rating system corresponded to factual censorship. The Federal Supreme Court for Administrative Law followed this argument to some extent in saying that it would be a circumvention of the prohibition of censorship if the tax law operated in a way so as to prevent economically the public showing of motion pictures which had not obtained a favourable rating by the Board[67]. The issue has not been brought before the Constitutional Court. It would appear that if the economic impact of tax law provisions would practically exclude the showing of a motion picture Art. 5 I 2 and 5 II might be violated, but it would not constitute censorship in the sense of Art. 5 I 3.

III

The freedoms guaranteed by Art. 5 I of the Basic Law include the freedom of broadcasting, that is to transmit as well as to receive by radio and television broadcasts destined for the general public. The fact, however, that the frequencies technically available for such purposes have been limited[68] has meant that broadcasting under the laws of the Federal Republic of Germany has been organized in the form of public law establishments.

1. Under Art. 73 no. 7 of the Basic Law, the Federation has the exclusive power to legislate on the telecommunication services. By decision of the Federal Constitutional Court this was interpreted to cover only the *technical* side

---

[65] *Ibid.*, pp. 74 ff.

[66] For the text, see Hessischer Staatsanzeiger 1957, p. 1071.

[67] BVerwGE 23, 194 (199) (1966).

[68] In part due to international treaties, such as the treaty of Malaga – Torremolinos of 25 October 1973 (BGBl. 1976 II, p. 1090) and the European Convention of 22 January 1965 (BGBl. 1969 I, p. 1940).

of telecommunication[69]; it does not include the power to legislate on the organisation of broadcasts or of broadcasters[70] which rests with the States, Art. 30.

With the exception of two federal establishments[71] all radio and television media operating under German law have been established by laws of the States or by inter-state treaties[72]. Two television systems are broadcasting nationwide; the (first) »German Television« system (ARD) based on a cooperation agreement of 9/10 June 1950 between the broadcasting establishment of the States, and the »Second German Television« system (ZDF) based on the Treaty of 5 June 1961 between the States of the Federation[73]. They are financed by fees having a public law nature[74] and, to a smaller part, by commercials[75].

2. In their performance of their legally defined tasks the broadcasting systems are subject to *legal* surveillance by the respective States, that is surveillance as to their operating within the legal order, in particular, in their financial and budgetory affairs; but they are not subject to a censorship by the State with regard to their programming policy or to the contents of their broadcasts[76]. The organizational structures of these establishments – which are decisive to the freedom of broadcasting under a public law system – vary to some extent but essentially follow the same pattern[77].

---

[69] The Federal Law on Telecommunication (BGBl. 1977 I, p. 460) reserves the right to install and maintain (technical) licences. Administration of this law, in particular, the granting of broadcasting licences and the allotting of frequencies rests with the Federal Postal Administration.

[70] BVerfGE 12, 205 (1961).

[71] »Deutschlandfunk« and »Deutsche Welle« the main purpose of which is to transmit broadcasts destined to listeners in foreign countries and Germany as a whole; established by Law of 29 November 1960 (BGBl. 1961 I, p. 862).

[72] See the State Laws of Bavaria, Baden-Württemberg, Berlin, Bremen, Hesse, Northrhein-Westphalia and the Saarland; the Treaties on the North-German Broadcasting System between the States of Hamburg, Lower Saxony and Schleswig-Holstein of 26 February 1955, the Treaty on the South West Broadcasting System between the States of Baden-Württemberg and Rheinland-Palatinia of 27 August 1951, and the Treaty on the »Second German Television« System of 6 June 1961 concluded between all the States of the Federation; for texts see *G. Herrmann*, Rundfunkgesetze (2d ed., Cologne et al. 1977).

[73] In addition each broadcasting establishment of the States runs a »Third Programme« of a regional nature.

[74] Presently based on inter-State treaties of 5 December and 5 July 1973 and uniform State regulations.

[75] The television systems may broadcast commercials only for a limited time-space per day; the two federal stations do not broadcast commercials. Commercials are provided by private companies founded by the establishments; their profits go to the establishments.

[76] Factual influence by the State School Administrations may occur with the radio broadcasts destined for educational purposes in public schools.

[77] Usually there are three organs: the general council (Rundfunkrat), the administrative council (Verwaltungsrat), and the superintendent (Intendant); with some establishments, in addition, there exists a program council (Programmbeirat).

3. In 1960, when only one countrywide television system (ARD) existed, the Federal Government under Chancellor Konrad Adenauer tried to establish a second system by way of a private company the shares of which were to be held by the Federation and the States. When the State Governments, however, declined to join the company the Federal Government decided to go it alone by itself taking over all the shares. This was attacked by the States of Hamburg and of Hesse in the Federal Constitutional Court, mainly on the grounds that it invaded the powers of the States unconstitutionally. The Court decided that the Federal Government, by founding the television company, had trespassed upon the powers of the States, its constitutional obligation of friendly co-operation with the States, and the freedom of broadcasting as guaranteed by Art. 5 I of the Basic Law[78].

In its judgment of 16 June 1981[79] the Constitutional Court declared unconstitutional and void the provisions of a State law which provided for the admission of private broadcasting – the only State law until now which had provided so[80]. Both decisions went to develop constitutional principles in the field of broadcasting:

While the present system of public law establishments under the given technical conditions of limited facilities (frequencies) was compatible with Art. 5 I of the Basic Law, the Constitution did not exclude private broadcasting establishments. Any system, nevertheless, would have to meet certain constitutional requirements.

In the opinion of the Court the subjective as well as the objective components of the freedom of broadcasting serve a system of freedom of expression. This requires, on the one side, that radio and television must not be subjected to dominance and influence by the State. But this negative, defen-

---

The general council is to represent the public. Membership varies between 11 and 66 members. They are chosen either from social groups (economic, like trade unions, social, religious, or cultural groups, representing also women, youth, sports etc.) or by the State – mostly by the State parliaments –; the State Governments, in general, are represented by one member only. Representation has been criticized as having been influenced too strongly by party politics. The general council is competent to create all the other organs of the establishment. It elects the members of the administrative council, and elects and dismisses or confirms the election by the administrative council of the superintendent. It may advise the superintendent on questions of program policy and on other questions.

The administrative council controls the current economic and technical affairs administered by the superintendent who needs the approval of the council for certain kinds of acts.

The superintendent, as a rule, is elected by the general council, in some cases by the general and the administrative council. After election he is engaged on the basis of a private law contract. The superintendent represents the establishment to the outside world and directs it. He has the power to determine programs and enjoys nearly complete independence in this respect.

[78] BVerfGE 12, 205 (1961).

[79] BVerfGE 57, 295 (1981).

[80] Law on Broadcasting of 7 June 1967 of the Saarland.

sive dimension alone would not ensure a free and comprehensive system of freedom of expression. This required, moreover, a positive legal construction which secures that the plurality of existing opinions may have their voices represented in the media as broadly and comprehensively as possible. In view of possible collisions of various fundamental rights – such as the rights to comprehensive and correct information arising out of the freedom of information together with the freedom of expression of those who produce a program or have a hearing in it – it was the task of the legislator to harmonize and delimit the legal positions concerned. Art. 5 I required that the *legislature itself by statute* determine all the essentials necessary to ensure the freedom of broadcasting; this determination must not be left or delegated to the executive branch nor to standing rules adopted by the broadcasting establishment or provided for in contracts. In particular, the legislature had to ensure that the media will not be delivered up to one or more single groups within society, but that all relevant groups should have a voice in the programs broadcast over the media as a whole, so as to ensure the existence of a »marketplace of opinions« in which the plurality of views will not be abridged. In spite of the technological developments one had to take into account the possibilities of the concentration of power determining opinions as well as the dangers of abuse (*i.e.* the unilateral exercise of influence over public opinion). In view of this situation it would not meet the constitutional requirement to exclude governmental interventions only and to leave the media to the *laissez-faire laissez-passer* of market forces. It was the responsibility of the legislature to ensure the existence of a system of broadcasting in which – taken as a whole – the plurality of opinions will be represented.

The Court did not definitively enumerate the essential requirements which a statute on private broadcasting establishments would have to meet. It pointed out, however, a few minimum requirements. One was that the legislature must decide on the basic order and organizational model of broadcast media whether to secure the plurality of opinions by the *internal* structure of an establishment (as it is the intention of the presently existing model of public law establishments) or by an (*external*) plurality of establishments which on the whole represent the plurality of opinions. The legislature, moreover, would have to provide binding guidelines that ensure a minimum of balanced contents of the programs, of fairness, and of mutual respect. In a system of »external plurality« the individual establishment would not be obliged to present a balanced program but to pertinent, comprehensive and correct information. Supervision by the State was permissible *only* to secure the observance of the legal rules enacted to guarantee the freedom of broadcasting; to that extent, however, supervision was necessary in order to protect that freedom.

If the legislature chose a system comprising private broadcasting establishments it would have to enact criteria for the admission and for the denial of access to the system and for a due procedure to decide upon these questions. Such a licensing procedure must serve only the purpose to guarantee the freedom of broadcasting. The law itself would have to determine the criteria for admission or denial. If there exist only limited facilities for broadcasts which do not allow all applications to be taken up rules will have to be enacted for allotting the available facilities meeting the requirements of the equal protection clause of Art. 3 I of the Basic Law; it must not be left to accidents, or *laissez faire – laissez passer*, nor to the unbound discretion of the executive. If a sharing of facilities, for instance the allocation of times for broadcasting is not feasible, or if the legislator has chosen a system in which only licences for »full programs« will be granted, the law must determine criteria for admission which guarantee equal chances for applicants; the degree of realization of chances must be determined by objective, pertinent criteria which can reasonably be met by an individual application.

4. Under the present legal order no licensing system has been enacted and no licensing procedures exist. Under the existing laws of the States responsibility for the selection and the contents of programs, as a rule, rests with the superintendent. In determining the principal programming policy he has to consult the general council. Programming is not subject to any instructions from outside the establishment, in particular, it is not subject to prior examination or approval by executive authorities.

Freedom of broadcasting constitutionally, is not a professional privilege conferred on persons employed or engaged by the public broadcasting establishments; no speaker, reporter, commentator, or editor as such may claim from Art. 5 I 2 of the Basic Law to have his views broadcast; he may express them freely in the same manner and under the same circumstances as anybody else[81]. Partly on the basis of statutory or treaty law, partly on a voluntary basis, the presently existing broadcasting establishments grant limited access to State governments, political parties, churches and religious communities, trade unions and associations of employers to present their views during appropriate broadcasting time[82]. In particular, political parties during election campaigns are granted limited free time to present electoral spots.

During the 1975 election campaign in the State of Northrhein-Westphalia, the »Communist Party of Germany/Marxist-Leninist« was granted broadcasting time by the West German broadcasting system for an electoral broadcast. The party submitted a manuscript for radio broadcast. The establishment refused to broadcast it because the manuscript contained several phrases which, in its opinion, transgressed the limits of freedom of expression and possibly constituted criminal acts; the manuscript contained such

---

[81] It is not a constitutional question whether such a claim might result from a contract with the establishment, or from treaty or statutory law.

[82] See, e.g., § 6 of the Treaty on the South West Broadcasting System; sec. 3 provides that political parties and organizations which seek to impair or abolish the free democratic basic order or to endanger the existence of the Federal Republic of Germany shall have no right to use the establishment.

phrases as: »The bourgeoise parliament is a corrupt gossipping stall which, as Lenin says, does not deserve anything else but be chased apart by the revolutionary masses of the people[83]«.

In 1976, the »Communist Federation of West Germany« (KBW) was granted free broadcasting time for electoral broadcasts in the election campaign for the Federal Diet. On September 8, 1976 it submitted a manuscript to be broadcast by the South West Broadcasting System. The establishment refused to broadcast the following two phrases: »That is so because the capitalist class is ruling; it is in its power to send the police and the army against striking workers«; and: »Parliamentary elections are a cheating manoeuvre to accomplish this. They serve to suppress the working class and the people«. The establishment considered these phrases to violate §§ 90a and 90b of the Criminal Code.

The »Communist Party of Germany« (KPD) during the same election campaign submitted an electoral broadcast script to the Second German Television System the text of which contained words like »terror by the State«, »establishment of proletarian dictatorship«, and »decaying bourgeois parliamentarism«. The establishment refused to broadcast the item.

Applications for injunctive relief by the respective parties in all three cases were denied by the administrative courts. Their constitutional complaints before the Federal Constitutional Court were successful[84]. The Court considered the refusal to broadcast the material to be a violation of the fundamental right of political parties to have equal chances in elections. The fact that the manuscript contained phrases hostile to the Constitution did not justify their refusal. Art. 21 II of the Basic Law, indeed, provided that political parties which, by reason of their aims or the behaviour of their adherents, seek to impair or abolish the free democratic basic order, shall be unconstitutional. The power to declare a party unconstitutional, however, is confined to a decision by the Federal Constitutional Court solely and exclusively, as expressly stated in Art. 21 II 2 of the Basic Law. The complainant parties had not been so declared. Accordingly, the refusal could not be based on their being unconstitutional; provisions to the contrary in various State laws and treaties on broadcasting establishments could not prevail over the Constitution. The broadcasting establishment, on the other hand, was not constitutionally debarred from examining whether a submitted manuscript was in fact intended for electoral purposes, and whether it violated criminal statutes. Examination and refusal on these grounds did not constitute censorship in the sense of Art. 5 I 3. Such examination was not intended to suppress, according to censorial standards, public expression of certain opinions; its purpose, rather, was to control whether the texts fulfilled the exclusive conditions under which limited access to the media was granted, that is to say

[83] For the complete texts, see BVerfGE 47, 198 (201 ff.) (1978).
[84] BVerfGE 47, 198 (1978).

that they were electoral broadcasts, and, moreover, to secure that the text did not evidently support the commission of criminal acts. The establishments could be denied such power only if the political parties had an unlimited right of access to the establishment's broadcasts. Such a right, however, did not exist either under statutory or under constitutional law.

5. Present laws and inter-State treaties provide binding principles and guidelines for the program and for the programming policy[85]. But they do not provide for prior restraints in order to enforce these principles or guidelines. The superintendent may be discharged if he violates his contract, or on any other lawful reason[86]. This might be used as a powerful means to enforce the principles and guidelines which are also binding upon the superintendent; but it does not constitute a prior restraint. If the establishment should violate »general laws« in the sense of Art. 5 II of the Basic Law, recourse to the courts is open for the persons or institutions concerned.

6. The public law system of radio and television as it presently exists in the Federal Republic of Germany has been passionately attacked and defended in public debate. It has been criticized for presenting an unbalanced scale of opinions which was due to the fact that certain establishments were safely in the hands of leftist editors; it therefore needed competition by private establishments; technological and technical developments (cable television etc.) no longer justified a »public law monopoly« but required the recognition in practice of the subjective basic right of free broadcasting; granting this freedom only »according to law« served to disqualify it as a fundamental right. It is mainly defended by the argument that with private establishments there is an inherent danger of delivering the media in the hands of a few powerful groups and diminishing the quality of programs. After the decision of the Federal Constitutional Court of 16 June 1981, several State governments have announced plans to prepare laws which would provide for the possibility of licensing private broadcasting establishments. If they are enacted, one may expect new kinds of prior restraints problems to arise; they will have to be resolved in accordance with Art. 5 of the Basic Law.

---

[85] The texts vary but are more or less the same in substance. § 5 of the Treaty on the South West Broadcasting System, e.g., provides that the programs of the system must carry by the spirit of democratic freedom and understanding among peoples, that the system must not unilaterally serve the purposes of a government, a political party, a church, a religious community, an ideology, a profession or a pressure group; news and reports must be correct in their contents and fair in their reproduction; doubts as to the reliability of news must be mentioned; news is to be separated from comment and opinion. – If a news or a fact has been broadcast contrary to truth the system may be requested to broadcast an opposing view if a legitimate interest can be shown by persons or institutions affected; § 7 of the Treaty.

[86] § 16 sec. 5 of the Law on the Hessian Broadcasting System, for instance, provides that the establishment may at any time dispense with the services of the superintendent with the consequence that he is discharged from his position (while retaining his salary); this requires a two thirds majority by the general council and the administrative council of the establishment.

Droit des traités et droits de l'homme

Erik Suy

L'interprétation et l'application des conventions internationales relatives aux droits de l'homme posent des problèmes de tout ordre aux instances internationales appelées à en vérifier la mise en oeuvre, mais il est très rare que ces instances se réfèrent au droit des traités tel qu'il est codifié dans la Convention de Vienne de 1969. Sans doute le droit des traités régit-il de manière générale les conventions concernant les droits de l'homme mais son application dans ce contexte soulève quelques questions spécifiques qui méritent de retenir notre attention. Sans prétendre les examiner de manière exhaustive, nous voudrions faire quelques observations sur les questions suivantes: les normes sur la protection des droits de l'homme relèvent-elles du *jus cogens*? La violation d'une convention sur les droits de l'homme par un Etat partie laisse-t-elle une possibilité de dénonciation pour les autres parties? Quel est l'effet des réserves apportées à ces conventions?

I. Droits de l'homme et jus cogens

La Convention de Vienne sur le droit des traités contient deux articles relatifs aux normes impératives du droit international général (*jus cogens*). L'art. 53 stipule que tout traité qui, au moment de sa conclusion, est en conflit avec le *jus cogens* est nul. L'article se poursuit par une définition assez générale de ce qu'il faut entendre par norme impérative du droit international général. L'art. 64 de son côté dispose que »Si une nouvelle norme impérative du droit international général survient, tout traité existant qui est en conflit avec cette norme devient nul et prend fin«.

La notion de *jus cogens* a fait couler beaucoup d'encre ces derniers temps et le Professeur *Reuter* dit justement qu'elle »est apte, par son mystère et ses contradictions, à servir de véhicule à toutes sortes d'aspirations à la fois vives et confuses (décolonisation, développement, désarmement) et à mobiliser des considérations idéologiques; à ce titre elle intéresse plus la sociologie politique que le droit« [1].

---

[1] *P. Reuter,* Introduction au droit des traités (1972), p. 1437.

Le *jus cogens* est trop complexe pour qu'on puisse l'invoquer à la légère. Il a été dit et écrit que le droit des peuples à disposer d'eux-mêmes en fait partie et nous sommes enclins à partager ce point de vue. Mais doit-on en conclure, comme certains l'ont fait, que les règles internationales protégeant les droits de l'homme relèvent également du *jus cogens*, puisqu'elles sont de nature assez semblable à celles qui garantissent le droit des peuples à disposer d'eux-mêmes?

Dans l'arrêt qu'elle a rendu le 5 février 1970 dans l'affaire de la *Barcelona Traction*, la Cour internationale de Justice parle notamment de la distinction essentielle à établir entre les obligations des Etats envers la communauté internationale dans son ensemble et celles qu'ils ont vis-à-vis d'un autre Etat. Et la Cour poursuit: »Par leur nature même les premières concernent tous les Etats. Vu l'importance des droits en cause, tous les Etats peuvent être considérés comme ayant un intérêt juridique à ce que ces droits soient protégés; les obligations dont il s'agit sont des obligations *erga omnes*.

Ces obligations découlent par exemple, dans le droit international contemporain, de la mise hors-la-loi des actes d'agression et du génocide mais aussi des principes et des règles concernant les droits fondamentaux de la personne humaine, y compris la protection contre la pratique de l'esclavage et de la discrimination raciale« [2].

Plusieurs auteurs affirment que les droits de l'homme appartiennent à la catégorie des normes impératives auxquelles aucune dérogation n'est possible. Le Professeur *McDougal* écrit que la Déclaration universelle »is now hailed ... as established customary law, having the attributes of *jus cogens* and constituting the heart of a global bill of rights« [3]. Et il poursuit: »It should be no cause for surprise that the great bulk of contemporary human rights prescriptions ... should be widely regarded today ... as among the principles clearly identifiable as *jus cogens*« [4].

Quant au Professeur *Ago*, il écrit que font partie du *jus cogens* »les règles fondamentales concernant la sauvegarde de la paix et notamment celles qui interdisent le recours à la force ou à la menace de la force: les règles fondamentales de caractère humanitaire (interdiction du génocide, de l'esclavage, de la discrimination raciale, protection des droits essentiels de la personne humaine en temps de paix et en temps de guerre) etc. . . .« [5].

En examinant la doctrine, on constate que la plupart des auteurs considèrent les règles relatives à la protection des droits de l'homme comme relevant du *jus cogens* mais tous restent plutôt vagues, utilisant des formules telles que »droits essentiels«, »règles fondamentales«, etc. ... Il est évident

---

[2] C.I.J. Recueil 1970, p. 32.

[3] *M. McDougal,* Human Rights and World Public Order (1980), p. 274.

[4] *Idem,* p. 345.

[5] *R. Ago,* Droit des traités à la lumière de la Convention de Vienne, RdC 134 (1971 III), p. 324, note 37.

que d'un point de vue pratique, cette situation est loin d'être satisfaisante. Peut-être les brèves observations qui suivent contribueront-elles à jeter quelque lumière sur cette question.

Il convient de rappeler tout d'abord que la notion de *jus cogens* en droit international apparaît dans le contexte du droit des traités: le traité contraire au *jus cogens* est nul, comme l'est le contrat contraire à l'ordre public. D'un point de vue théorique donc, on peut affirmer qu'un traité international par lequel deux ou plusieurs Etats se seraient mis d'accord pour violer une prescription fondamentale concernant les droits de l'homme serait nul au sens de l'art. 53 de la Convention de Vienne sur le droit des traités. Mais n'est-ce pas là une pure hypothèse d'école? Peut-on concevoir que deux ou plusieurs Etats s'entendent pour violer le droit à la vie ou d'autres droits »essentiels«? Le génocide peut certes être le fait d'un Etat – mais est-il imaginable que deux ou plusieurs Etats concluent un accord pour le pratiquer? Si l'on s'en tient à la vraisemblance et si on limite le concept du *jus cogens* à sa portée originale, c'est-à-dire à son rôle assez restreint d'élément annulateur dans le droit des traités, on arrivera à la conclusion que, dans la pratique, la notion n'a guère d'application dans le contexte des droits de l'homme, à moins d'élargir le concept de »droits de l'homme« au point de l'assimiler au droit à l'autodétermination des peuples ou encore au »droit au développement«. Ces deux concepts dépassent toutefois, à notre avis, celui des droits de l'homme au sens où on l'entend en doctrine. Par droits de l'homme nous entendons les droits et libertés *individuels* qu'ils aient un caractère civil, politique, social, économique ou culturel.

Sans doute, un Etat peut-il se rendre coupable, sur le plan législatif et administratif par exemple, de violations flagrantes des droits de l'homme, mais le concept de nullité de son comportement, qui nous paraît être l'essence même de la notion de *jus cogens*, est dépourvu de sens en l'espèce. Le comportement en question peut être considéré comme illicite, voire constituer un crime international, mais le concept de *jus cogens* en tant que critère d'annulation d'un engagement international n'a pas d'application ici.

Si l'on veut donc parler de *jus cogens* dans l'hypothèse envisagée, ce ne peut être qu'en redéfinissant la notion en lui attribuant un sens tout à fait différent de celui qu'elle a dans le droit des traités.

Nous partageons entièrement la distinction faite par la Juge *H. Mosler* entre le *jus cogens* et l'ordre public de la communauté internationale. »The term *ius cogens* refers to restrictions on freedom of contract which are imposed on all members of the international community in their mutual relations . . . *Ius cogens* is the counterpart of *ius dispositivum.* The terms apply only to agreements« [6]. En revanche, la notion plus large d'ordre public de la communauté internationale »consists of principles and rules the enforcement of which is of such vital importance to the international community as a whole that any

---

[6] *H. Mosler,* The International Society as a Legal Community (1980), p. 19.

unilateral action or any agreement which contravenes these principles can have no legal force . . .«[7].

Une deuxième remarque s'impose. Si certains auteurs ont pu considérer les droits de l'homme comme relevant du *jus cogens* c'est probablement parce que plusieurs instruments internationaux, tout en donnant aux Etats la possibilité de déroger dans certaines circonstances et sous certaines conditions à leurs dispositions, interdisent cependant toute dérogation à certains articles – donc à certains droits considérés comme »essentiels«. L'art. 4 du Pacte international relatif aux droits civils et politiques en est un exemple typique de même que l'art. 15 de la Convention européenne des droits de l'homme et des libertés fondamentales. Relèveraient donc de l'ordre public de la communauté internationale les principes auxquels aucune dérogation n'est autorisée.

L'interdiction formelle de toute dérogation nous paraît en effet être un critère objectif permettant d'identifier une règle relevant de l'ordre public de la communauté internationale. En appliquant ce critère, non seulement on sera amené à exclure du champ de cet ordre public tous les droits de l'homme auxquels il est expressément permis de déroger, mais on y fera aussi rentrer toutes les normes auxquelles il n'est pas permis de déroger qu'elles aient trait ou non aux droits de l'homme. Bien que l'effet de traités contraires au *jus cogens* ait été établi par la Convention de Vienne de 1969 sur le droit des traités, la doctrine et la pratique ne se sont pas encore suffisamment penchés sur l'effet d'une mesure législative ou administrative contraire à l'ordre public de la communauté internationale. Il ne suffit pas de constater que ces mesures entrainent la responsabilité internationale de l'Etat car telle pourrait être la conséquence de la violation de n'importe quelle obligation internationale. La conséquence d'une mesure dérogeant à l'ordre public de la communauté internationale devrait être plus lourde et impliquer, outre sa nullité absolue, l'obligation de tous les Etats de ne pas reconnaître ces mesures . . . tout en laissant ouverte la question de savoir comment une norme impérative donnée pourrait être remplacée par une autre norme ayant le même caractère.

Un autre critère objectif semble résider dans l'attribution du caractère de crime international à la violation de certaines normes. Cette idée a été développée par le Professeur *Ago* dans son projet sur la Responsabilité internationale des Etats, et elle a été reprise par la Commission du droit international. Dans son commentaire, la Commission lance cependant un avertissement en disant qu'il »serait erroné de vouloir en conclure simplement que toute violation découlant d'une norme impérative . . . serait un crime international«. On peut admettre que les obligations dont la violation constitue un crime seront »normalement« des obligations découlant d'une règle impérative. On ne peut cependant pas nier que la catégorie des obligations internationales pour lesquelles aucune dérogation n'est admise est plus vaste que celles des obligations dont la violation est nécessairement un crime in-

[7] *Idem*, p. 18.

ternational« [8]. Par exemple, le principe de la liberté de la haute mer est considéré comme une norme du *jus cogens,* mais sa violation n'est pas un crime, et la violation des droits de l'homme n'est, en général, pas un crime sauf »une violation grave et à une large échelle d'une obligation internationale d'importance essentielle pour la sauvegarde de l'être humain, comme celles interdisant l'esclavage, le génocide, l'*apartheid*« [9].

En résumé, le *jus cogens* en tant que notion de droit international et, plus particulièrement, du droit des traités, a trait à la nullité d'accords internationaux. Or, un accord international par lequel des Etats s'obligeraient à violer certains droits de l'homme »essentiels« est peu vraisemblable. En second lieu, on ne peut parler de norme impérative que pour la catégorie restreinte des droits civils et politiques qui ne sont susceptibles d'aucune dérogation. Des dérogations à de tels droits ne sont guère concevables dans la pratique que par voie de mesures nationales, ce qui situerait le problème dans le contexte de la responsabilité internationale de l'Etat et non pas dans celui du droit des traités. Enfin, cette responsabilité se situerait sur un plan plus élevé si la dérogation se manifestait par une violation d'un droit de l'homme qualifiée de crime international. Somme toute, certains droits de l'homme appartiennent au droit impératif, c'est-à-dire absolu, sans avoir pour autant le caractère de *jus cogens* tel que l'envisage le droit des traités.

II. Extinction et suspension des traités de caractère humanitaire

L'art. 60 de la Convention de Vienne sur le droit des traités se rapporte à l'extinction d'un traité ou à la suspension de son application comme conséquence de sa violation. Le par. 2 dispose que la violation substantielle d'un traité multilatéral par l'une des parties permet de prendre plusieurs actions, soit suspension, soit abrogation. Et la sec. 5 précise que les possibilités de suspension ou d'abrogation »ne s'appliquent pas aux *traités de caractère humanitaire*, notamment aux dispositions excluant toute forme de représailles à l'égard des personnes protégées par lesdits traités«.

Ici se posent un certain nombre de problèmes, le premier étant de savoir si la notion de »traités de caractère humanitaire« recouvre également la catégorie des conventions sur les droits de l'homme. La réponse à cette question doit être trouvée dans les travaux préparatoires. Lors de la 61ème séance de la Commission plénière de la première session de la Conférence, le représentant de la Suisse, M. Bindschedler, proposa l'amendement oral suivant à ce qui était alors l'art. 57 du projet:

---

[8] Annuaire de la C.D.I., 1976, vol. II (Deuxième partie), p. 111, par. 62.
[9] Art. 19 du projet d'articles sur la responsabilité des Etats, *idem,* p. 88.

»Les règles précédentes ne s'appliquent pas à des conventions humanitaires conclues entre ou avec des Etats non liés par des conventions multilatérales pour la protection de la personne humaine, qui excluent les représailles contre les personnes. Des accords de cette nature doivent être observés en tout état de cause« [10].

En présentant cet amendement, M. Bindschedler s'est référé aux traités du type des Conventions de Genève pour la protection des victimes de guerre qui devraient, selon lui, être à l'abri d'une extinction ou d'une suspension de nature à mettre en danger la vie même des individus. Il s'est en outre référé à d'autres conventions également importantes, concernant le statut des réfugiés, la répression de l'esclavage, l'interdiction du génocide, et la protection des droits de l'homme en général, affirmant que même leur violation par une partie ne devait pas avoir pour conséquence de frapper des individus innocents [11].

L'expert-conseil à la Conférence, Sir Humphrey Waldock, a fait observer qu'un grand nombre de conventions humanitaires contenaient des clauses autorisant à les dénoncer par une simple notification et a ajouté qu'il paraîtrait dès lors étrange d'exclure toute possibilité de riposte à une violation substantielle. Les délégations britannique et libanaise se sont en outre posé la question de savoir à quel traité la formule »humanitaire« devait s'appliquer.

Lors de la deuxième session de la Conférence (en 1969), la délégation suisse proposa un amendement formel [12]. En la présentant, M. Ruegger a declaré que parmi les conventions humanitaires figuraient, outre les conventions de Genève de 1949, d'autres conventions, notamment celles concernant la protection des droits de l'homme en général [13]. L'amendement suisse fut adopté sans opposition.

Aux termes du par. 5 de l'art. 60 donc, il n'est pas permis à une ou plusieurs parties à un traité multilatéral de caractère humanitaire, de suspendre ou de terminer »les dispositions relatives à la protection de la personne humaine«. Si l'on admet que les conventions sur les droits de l'homme sont comprises parmi les »conventions de caractère humanitaire«, on est amené à la conclusion que toute violation substantielle d'une convention sur les droits de l'homme autorise les autres parties à suspendre l'application du traité ou à mettre fin à celui-ci, sauf pour ce qui est des dispositions concernant la protection physique de la personne. Est-ce là un résultat satisfaisant?

Nous pensons pour notre part qu'il ne fallait pas introduire cet élément à l'art. 60, car il est maintenant clair qu'on a adopté le principe de la suspension ou de l'extinction des traités de caractère humanitaire, ce que l'on n'aurait pas

---

[10] Conférence des Nations Unies sur le droit des traités, Première session, 1968. Documents officiels, A/CONF.39/11, p. 385.

[11] *Idem,* p. 385.

[12] A/CONF.39/L.31.

[13] A/CONF.39/11/Add.1, p. 119.

dû faire. En outre, l'art. 60, même compte tenu des restrictions du par. 5, risque d'entrer en conflit avec l'art. 4 du Pacte international relatif aux droits civils et politiques qui ne permet aucune dérogation à une série d'articles et de droits, alors que selon le par. 5, seules les dispositions relatives à la protection physique de la personne c'est-à-dire les art. 6, 7 et 8 ne seraient pas susceptibles de dérogation.

Qui plus est, si l'application de l'art. 60 de la Convention de Vienne produit un resultat logique dans le cas de traités multilatéraux du type classique qui imposent des obligations aux Etats dans leurs relations mutuelles, il n'en va pas de même s'agissant des traités de caractère humanitaire qui imposent des obligations aux Etats à l'égard des individus de telle sorte que l'extinction ou la suspension des accords humanitaires n'affecterait l'Etat qui les viole que d'une façon tout à fait marginale et sanctionnerait en realité les individus.

Rappelons à cet égard, que, dans son avis consultatif du 21 juin 1971, sur les conséquences juridiques pour les Etats de la présence continue de *l'Afrique du Sud en Namibie*, la Cour internationale de Justice a dit que les Etats avaient l'obligation de s'absentir d'invoquer ou d'appliquer les traités ou dispositions des traités conclus par l'Afrique du Sud au nom de la Namibie ou en ce qui la concerne. Cette obligation, a dit la Cour, »ne peut s'appliquer à certaines conventions, comme les conventions de caractère humanitaire dont l'inexécution pourrait porter préjudice au peuple namibien« [14].

Un troisième argument milite en faveur du maintien en vigueur des conventions à caractère humanitaire – y compris les conventions relatives aux droits de l'homme – même après des violations substantielles: la plupart des conventions humanitaires contiennent des stipulations concernant le contrôle de leur application. Dans les conventions relatives aux droits de l'homme, il s'agit de comités ou de commissions indépendantes auxquels se superposent parfois des instances judiciaires. N'est-il pas dans l'intérêt de la partie qui prétend être la victime d'une violation d'obtenir réparation en ayant recours à ces mécanismes plutôt que de dénoncer la convention?

Cela dit, on peut soutenir que dans la mesure où la Convention contient une clause de dénonciation – et tel est le cas pour les Conventions de Genève de 1949 et les Protocoles additionnels de 1977 – la dénonciation pour cause de violation substantielle doit être possible. La Convention sur la prévention et la répression du crime de génocide de 1948 contient également des dispositions sur la dénonciation (art. XIV et XV). Il en est de même pour la Convention internationale sur l'élimination de toutes les formes de discrimination raciale (art. 21), et la Convention internationale sur l'élimination et la répression du crime d'*apartheid* (art. XVI).

Si l'on admet la possibilité d'une dénonciation ou d'une suspension des conventions à caractère humanitaire, il faudra sans doute le faire de la façon la

---

[14] C.I.J. Recueil 1971, p. 55, Sec. 122.

plus restrictive en exigeant tout d'abord que soient respectées les clauses de dénonciation qui, dans les Conventions de Genève de 1949 et les protocoles de 1977, sont très précises. Il devra en outre être entendu que la dénonciation ou la suspension ne saurait s'appliquer aux dispositions relatives à la protection *physique* de la personne humaine et ne saurait par surcroît avoir d'effet que sur l'Etat ayant violé la Convention – ce qui signifie qu'au bout du compte, ce sont les ressortissants de cet Etat qui seraient affectés!

Cette dernière remarque fait apparaître l'absurdité de la procédure de dénonciation pour violation substantielle d'une convention de caractère humanitaire. Quoi qu'il en soit, l'interprétation du par. 5 de l'art. 60 est loin d'être simple, et les quelques remarques qui précèdent ne visent qu'à mettre en lumière les difficultés que pourrait engendrer son application.

III. Les réserves

Rappelons tout d'abord le système des réserves dans la Convention de Vienne de 1969. Dans l'art. 19, la Convention énonce le principe général selon lequel un Etat peut - au moment de signer, de ratifier, d'accepter, d'approuver un traité, ou d'y adhérer – formuler une réserve, à moins que la réserve ne soit interdite par le traité ou ne soit incompatible avec l'objet et le but du traité; l'art. 20 contient des dispositions très détaillées sur l'acceptation des réserves et les objections aux réserves; l'art. 21 traite des effets juridiques des réserves et des objections aux réserves.

On se limitera ici à quelques problèmes particuliers touchant les réserves au Pacte international relatif aux droits civils et politiques.

Tout d'abord, les clauses finales du Pacte ne contiennent aucune disposition relative aux réserves. La question avait cependant été discutée au cours de l'élaboration des Pactes, et plusieurs propositions à cet effet ont été introduites. Le rapport de la Commission des droits de l'homme indique toutefois:

> »Plusieurs représentants ont fait observer que le droit de tout Etat contractant à formuler des réserves à un traité multilatéral était désormais un principe du droit international accepté à condition que ces réserves ne soient pas incompatibles avec l'objet et le but du traité. Il était également bien établi que toute objection élevée contre une réserve par un autre Etat contractant empêchait l'entrée en vigueur du traité entre l'Etat qui faisait la réserve et l'Etat qui formulait l'objection, à moins que celui-ci n'ait exprimé une intention contraire«.

Cette conclusion est, en fait, celle de la Convention de Vienne de 1969.

Les réserves étant admises, on analysera maintenant quelques-unes des réserves faites au Pacte. Il y a tout d'abord des réserves *ratione loci* visant à écarter l'application de la totalité ou d'une partie du Pacte dans certains territoires, et qu'on peut négliger. Viennent ensuite des réserves *ratione temporis*, qui reprennent en substance les dérogations autorisées dans l'art. 4 du Pacte.

On sait que les conditions dans lesquelles les Etats peuvent se prévaloir des dérogations visées à l'art. 4 sont strictement définies:

- Les dérogations doivent être proportionnées aux exigences de la situation;
- Elles doivent être compatibles avec les normes de droit international;
- Elles ne doivent pas être fondées sur une discrimination sociale, raciale ou religieuse;
- Elles doivent être limitées dans le temps;
- Elles ne doivent pas déroger aux art. 6 (droit à la vie), 7 (interdiction de la torture), 8 (esclavage), 11 (prison pour dettes contractuelles), 15 (non-rétroactivité des lois pénales), 16 (personnalité juridique) et 18 (liberté de pensée).

Le Chili, la Colombie, le Nicaragua, le Royaume-Uni et l'Uruguay se sont prévalus de la possibilité offerte par l'art. 4.

On notera cependant qu'avant l'entrée en vigueur du Pacte le Gouvernement de la République fédérale d'Allemagne avait indiqué les conditions dans lesquelles il appliquerait le principe de la non-rétroactivité des lois pénales (art. 15, par. 1), et qu'après l'entrée en vigueur du Pacte:

- Le Royaume-Uni s'est réservé le droit de ne pas appliquer l'art. 11 (emprisonnement pour dettes contractuelles) à Jersey;
- L'Italie a aussi indiqué les conditions dans lesquelles elle appliquerait les dispositions de l'art. 15, par. 1;
- La Trinité-et-Tobago a également donné son interprétation de l'art. 15, par. 1, et a formulé une réserve au par. 2 de l'art. 4;
- La Norvège a fait une réserve à l'art. 6, par. 4 (droit de grève).

Ceci pose les problèmes suivants: un Etat peut-il formuler des réserves à des articles auxquels nulle dérogation ne peut être apportée? De telles réserves ne sont-elles pas interdites et ne sont-elles pas contraires au but et à l'objet du Pacte? Quelle doit être l'attitude à adopter par le dépositaire?

Dans la mesure où l'art. 4 peut être assimilé à une clause de réserve, on peut considérer que les parties ont clairement exprimé leur volonté d'interdire certaines réserves, et qu'aller à l'encontre de cette intention serait incompatible avec le droit des traités.

Toutefois, dans la mesure où l'art. 4 n'est pas une clause de réserve, mais seulement une modalité d'application du Pacte, le problème est celui de l'interprétation de la volonté des parties.

A cet égard, lors des discussions au sujet des réserves à la Commission des droits de l'homme[15], on a fait valoir que certaines dispositions contenues dans la Trosième Partie du Pacte étaient essentielles et qu'on ne devrait pas

---

[15] Commission des droits de l'homme, Dixième session, 23 février–16 avril 1954; Conseil économique et social, Documents officiels, Dix-huitième session, suppl. 7 (E/2573, E/CN.4/705), p. 32, par. 279.

autoriser de réserves aux dispositions auxquelles l'art. 4 ne permettait pas de déroger, même au cas où un danger exceptionnel menacerait l'existence de la nation. Cependant, les discussions relatives à l'art. 4, tant à la Commission des droits de l'homme qu'à la Troisième Commission de l'Assemblée générale, n'apportent guère d'éclaircissements sur ce point. Il semblerait plutôt qu'en dehors des dispositions de cet article on ait envisagé de dresser une liste des articles auxquels aucune réserve ne pourrait être faite, mais l'examen de cette question fut renvoyé au moment des discussions sur les réserves. Toutefois, il se dégage des opinions émises à ce moment là, que toute réserve portant sur les articles essentiels énumérés à l'art. 4 devait être présumée comme incompatible avec l'objet et le but du Pacte.

Ne s'agit-il pas en fait d'un faux problème résultant d'une confusion entre conditions d'application (possibilité pour un Etat partie de modifier l'application du Pacte) et conditions de participation (obligation pour un Etat partie de respecter un minimum de normes, même en cas de danger public). L'objet de l'art. 4 n'est pas d'interdire les réserves à certains articles, mais de prévoir l'impossibilité d'exécution temporaire au sens de l'art. 61 de la Convention de Vienne qui stipule: »... Si l'impossibilité est temporaire, elle peut être invoquée seulement comme motif pour suspendre l'application du traité«. Prévoir, conformément aux conditions énoncées dans l'art. 4, la possibilité de mesures dérogeant aux obligations du Pacte sauf à celles prévues dans certains articles, n'équivaut pas exactement à autoriser des réserves sauf à certains articles. Sinon, on dénaturerait la volonté des parties qui, en l'absence de toute disposition sur les réserves, a été de laisser les Etats libres de subordonner leur participation au Pacte à toutes les réserves, sauf objection des parties. Une réserve à un article, même essentiel, peut ne pas être considérée en soi comme incompatible avec le but et l'objet du traité (par exemple la réserve de portée limitée du Royaume-Uni en ce qui concerne la non-application de l'art. 15, par. 1, à Jersey). Cette question importante n'est pas seulement théorique: par le biais de réserves à des dispositions qui doivent être présumées plus importantes que les autres (sinon essentielles) et auxquelles il ne peut être dérogé même au motif de circonstances exceptionnelles, les Etats ont la possibilité de vider le Pacte d'une grande partie de sa substance. Il appartient, certes, à l'organe de contrôle de l'application du Pacte, c'est-à-dire au Comité des droits de l'homme institué en application de l'art. 28 du Pacte, de faire »toutes observations générales qu'il jugerait appropriées« (art. 40, par. 4) et d'en faire rapport à l'Assemblée générale des Nations Unies (art. 45). Mais il ne s'agit là que d'un contrôle *a posteriori*.

En premier lieu, le problème s'est posé et se pose pour le dépositaire dans les termes suivants:

– Pour l'entrée en vigueur du Pacte en vertu de son art. 49, un instrument assorti d'une réserve à l'un des articles auxquels il n'est pas permis de déroger devait-il être compté au nombre des instruments requis à cet effet?

– Un instrument de ratification, d'approbation, d'acceptation ou d'adhésion comportant une réserve semblable peut-il être reçu en dépot?

Avant l'entrée en vigueur du Pacte, la Norvège, en ratifiant (le 13 septembre 1972), avait formulé une réserve relative à l'art. 6, par. 4 (droit de grève) auquel il ne peut être dérogé[16]. L'instrument de ratification de la République fédérale d'Allemagne, transmis pour dépôt le 17 décembre 1973, était accompagné d'un certain nombre de réserves dont celle-ci: »... Le paragraphe 1 de l'article 15 du Pacte sera appliqué comme suit: dans le cas d'un adoucissement des dispositions pénales en vigueur, dans certains cas exceptionnels précis, le droit en vigueur antérieurement reste applicable à des actes commis avant la modification de la loi«.

C'était là une déclaration qui, bien que limitée à des cas exceptionnels et précis, constituait une réserve au principe fondamental de la non-rétroactivité des lois pénales reconnu à l'art. 15 du Pacte et auquel aucune dérogation ne pouvait être apportée, même en cas de circonstances exceptionnelles. Le Secrétaire général a reçu l'instrument en dépôt à la date de sa transmission et a communiqué aux parties le texte de ladite réserve sans attirer leur attention sur le problème de droit que la réserve soulevait. De même, sans consulter les parties, le Secrétaire général a compté l'instrument en question au nombre des 35 instruments requis pour l'entrée en vigueur du Pacte. L'attitude du dépositaire se justifie, entre autres, par la considération que la République fédérale était le 25ème Etat à ratifier ou adhérer et qu'il était raisonnable de prévoir qu'il s'écoulerait encore un peu de temps avant l'entrée en vigueur (dépôt de 35 instruments) du Pacte[17]. Ainsi les parties contractantes ou signataires avaient le temps nécessaire pour faire objection à la réserve dans le cas où elles l'auraient estimée interdite et contraire au but et à l'objet du Pacte. Il n'y eut en fait aucune objection.

Après l'entrée en vigueur du Pacte, le 23 mars 1976, d'abord le Royaume-Uni, en ratifiant le 20 mai 1976, se réservait entre autres le droit de ne pas appliquer l'art. 11 (emprisonnement pour dettes contractuelles) à Jersey. Cet article est l'un de ceux auxquels il ne peut être dérogé en dehors de circonstances exceptionnelles.

Les 15 septembre et 8 décembre 1978, respectivement, l'Italie, en ratifiant, et la Trinité-et-Tobago, en adhérant, ont formulé une déclaration interprétant la dernière phrase de l'art. 15 (»... si, postérieurement à cette infraction la loi prévoit une peine plus légère, le délinquant doit en bénéficier«) comme s'appliquant aux procédures en cours – aux affaires pendantes. Bien que formulée en termes similaires, la déclaration de la Trinité-et-Tobago dans l'instrument figurait sous la rubrique »réserve«. A la demande du Secrétariat, le Gouvernement de la Trinité-et-Tobago a alors indiqué qu'il s'agissait seulement d'une déclaration interprétative.

---

[16] Le Gouvernement norvégien a retiré cette réserve le 21 novembre 1979.

[17] Le dépôt de l'instrument de ratification de la République fédérale a été effectué le 17 décembre 1973; le Pacte est entré en vigueur le 23 mars 1976.

Plus certaine, néanmoins, reste la réserve de ce Gouvernement précisément au par. 2 de l'art. 4, qui énonce les articles auxquels il n'est pas permis de déroger: »... se réserve de ne pas appliquer intégralement les dispositions du paragraphe 2 de l'article 4 du Pacte, car, aux termes de l'article 7, par. 3, de la Constitution, le Parlement peut valablement adopter des lois en contradiction avec les articles de la Constitution«. Conformément à sa pratique dépositaire, le Secrétaire général a annoncé l'entrée en vigueur du Pacte pour la Trinité-et-Tobago, sauf les conséquences juridiques que les Etats estimeraient devoir attacher à la réserve.

Les réserves *ratione materiae* relatives au Pacte relatif aux droits civils et politiques sont nombreuses et très diverses. Sur 69 Etats parties à ce Pacte[18], 21 ont formulé des réserves: République fédérale d'Allemagne, Australie, Autriche, Barbade, Danemark, Finlande, France, Gambie, Guyane, Inde, Islande, Italie, Japon, Mexique, Norvège, Nouvelle-Zélande, Pays-Bas, Royaume-Uni, Suède, Trinité-et-Tobago et Venezuela. La formulation de ces réserves rend parfois leur qualification par les Etats (réserve ou simple déclaration interprétative) et la détermination de leur portée assez délicates. Par exemple les réserves suivantes:

– »Les articles ... seront appliqués de la manière suivante: ...« (République fédérale d'Allemagne);
– »Les articles ... seront appliqués pour autant qu'ils ne portent pas atteinte à ...« (Autriche).
– »... Déclare que la teneur de (telle ou telle disposition du Pacte) ne correspond pas à sa législation actuelle ...« (Finlande).

Il en va de même pour certaines déclarations interprétatives:

– »... L'article ... est interprété comme n'excluant pas la distinction entre ...«;
– »... Les dispositions de ... sont considérées comme étant compatibles avec ...«;
– »... L'article ... ne saurait faire obstacle ...«.

La déclaration des Pays-Bas à ce sujet illustre cette incertitude:

»... Bien que les réserves énoncées soient en partie de caractère interprétatif, il a décidé de formuler dans tous les cas des réserves plutôt que des déclarations interprétatives étant donnés que si cette dernière formule était utilisée il pourrait être mis en doute que les textes du Pacte permettent les interprétations proposées«.

A titre d'information on relèvera que ces réserves et ces déclarations portent en général sur des dispositions précises d'articles se rapportant soit à la procédure pénale (recours devant les tribunaux, régime pénitentiaire, détention, droit de grève), soit à la liberté de circulation, à la discrimination sous toutes ses formes, à la liberté de réunion, d'opinion, ou encore à l'accès aux fonctions publiques.

---

[18] Au 1er janvier 1982.

Cependant parmi les réserves de nature à affecter la substance même des Pactes, il faut mentionner les réserves dites constitutionnelles, c'est-à-dire celles qui subordonnent l'application des dispositions des Pactes à leur conformité avec la constitution de l'Etat auteur de la réserve. Ainsi, les réserves de l'Inde précisaient que les dispositions des art. 9, 12, 19 (alinéa 3), 21 et 22 seraient appliqués de manière à se conformer aux dispositions des art. 22 (alinéa 3 à 7) et 19 de la constitution de l'Inde[19].

Il est très évident qu'un gouvernement ne peut engager l'Etat sur le plan international au-delà des compétences qui lui sont dévolues par la constitution. Il est non moins évident qu'en droit international classique les engagements de l'Etat sur le plan international transcendent les dispositions de sa législation interne, voire de sa constitution[20]. Une semblable réserve, en tout état de cause, apparaît superflue. Elle s'apparente davantage à une déclaration de portée politique à moins de l'analyser comme le déplacement du fardeau de la preuve pour déterminer la portée éventuelle de la réserve.

Ces quelques réflexions à propos des réserves ne sauraient se terminer sans qu'on souligne la position parfois difficile du dépositaire lorsqu'il se trouve confronté à des réserves de toute évidence inadmissibles en l'absence de toute clause relative aux réserves dans un accord international. A l'Organisation des Nations Unies, l'Assemblée générale a prié le Secrétaire général, en tant que dépositaire, de laisser à chaque Etat, en l'absence de clauses, le soin de tirer les conséquences juridiques des réserves et des objections aux réserves[21]. Cela signifie que toutes les réserves reçues sont communiquées aux Etats pour information et éventuelle réaction, et il appartient donc aux Etats de réagir et, éventuellement, d'objecter aux réserves. On peut toutefois se demander si cette procédure ne contient pas un risque très grave pour ce qui est des réserves interdites *prima facie.* En effet, la pratique démontre que les Etats qui objectent à des réserves sont très peu nombreux et les réserves considérées risquent alors d'avoir plein effet, chose normale pour les réserves admissibles mais qui pourrait avoir des conséquences dommageables dans les autres cas. Faute d'objections, les réserves inadmissibles pourraient sortir leurs effets contre la volonté originale des auteurs de l'accord, ce qui équivaudrait à amender indirectement celui- ci. Cette situation mériterait qu'à l'avenir on se penche sur l'effet de ces réserves et qu'on précise le rôle du dépositaire en la matière.

---

[19] A la demande du Secrétaire général, le Gouvernement de l'Inde lui a communiqué le texte des dispositions constitutionnelles sur lesquelles portaient les réserves.

[20] Art. 46 de la Convention de Vienne: »1. Le fait que le consentement d'un Etat à être lié par un traité a été exprimé en violation d'une disposition de son droit interne concernant la compétence pour conclure des traités ne peut être invoqué par cet Etat comme viciant son consentement, à moins que cette violation n'ait été manifeste et ne concerne une règle de son droit interne d'une importance fondamentale«.

[21] Résolutions de l'Assemblée générale 598 (VI) du 12 janvier 1952 et 1452 B (XIV) du 7 décembre 1959.

Protection of Minorities under Article 27 of the International Covenant on Civil and Political Rights

Christian Tomuschat

International protection of human rights, today an elaborate network of legal norms and mechanisms of implementation [1], has one of its most noticeable roots in the system for the protection of minorities as it had evolved in Central Europe after the First World War [2]. Thus, the needs of minorities have contributed to opening up that new dimension of international law outside the traditional field of inter-State relationships which *Hermann Mosler* has so cogently analyzed, namely, the extension of the circle of subjects of international law[3]. Due to the belief that human rights for everyone could replace specific devices for the sole benefit of minorities, for many decades after the Second World War, interest in minority protection was weak. However, new momentum was gained when in 1976 the International Covenant on Civil and Political Rights (CCPR) came into force. According to Art. 27 of that instrument[4], »persons belonging to ... ethnic, religious or linguistic minorities« shall not be denied certain freedoms. Being designed to find world-wide acceptance, Art. 27 had necessarily to be drafted in a summary manner. Consequently, any attempt exactly to state its scope and meaning encounters considerable difficulties. Not surprisingly, therefore, commentators

[1] For the United Nations see the impressive compilation United Nations Action in the Field of Human Rights (1980), for the Council of Europe see Bilan de la Convention européenne des droits de l'homme, Council of Europe Doc. DH (79) 1.

[2] Compare, for instance, *G. Dahm,* Völkerrecht, vol. 1 (1958), p. 421; *H. Mosler,* Das Institut de Droit International und die völkerrechtliche Stellung der menschlichen Person, in: W. Wengler (ed.), Iustitia et Pace (1974), pp. 77–100, at 87, 90.

[3] Die Erweiterung des Kreises der Völkerrechtssubjekte, Berichte der Deutschen Gesellschaft für Völkerrecht no. 4 (1961), pp. 39–75, at 62–63; The International Society as a Legal Community, RdC 140 (1974 IV), pp. 1–320, at 70–72.

[4] Henceforth all unspecified references to articles are made to articles of the CCPR.

have criticized Art. 27 for the »scantiness« of its substance[5] which indeed contrasts sharply with the wealth of the substantive as well as procedural safeguards laid down in the minority treaties which had been concluded after the First World War[6].

I. Introduction

1. Although one may easily agree that Art. 27 is a rather succinct provision which raises more problems than it really resolves, its existence alone is noteworthy enough. Indeed, while it is commonly said that at the regional level more effective and elaborate protection can be and is generally afforded than on a world-wide scale, the contrary is true here. None of the regional treaties or draft instruments deals specifically with the rights of (members of) minorities. The European Convention for the Protection of Human Rights and Fundamental Freedoms (ECHR) confines itself to listing »association with a national minority« as a forbidden ground of discrimination (Art. 14), while the American Convention on Human Rights (ACHR)[7] abstains from even mentioning the word »minority«[8]. The same negative result emerges from a reading of the African Charter of Human and Peoples'

---

[5] *F. Münch,* Der Minderheitenartikel im Menschenrechtstext der Vereinten Nationen, in: Th. Veiter (ed.), System eines internationalen Volksgruppenrechts (1972) part 2, pp. 64–72, at 65; see also *J. W. Brügel,* Internationaler Minderheitenschutz nach dem Zweiten Weltkrieg, Europa-Archiv 421–430, at 424 (1972): »offenkundige Unzulänglichkeit dieser Bestimmung«; *Y. Dinstein,* Collective Human Rights of Peoples and Minorities, 25 ICLQ 102–120, at 118 (1976): »a minimum – rather than a maximum – of rights«; *F. Ermacora,* Möglichkeiten der Ausgestaltung des Minderheitenschutzes im Rahmen der Vereinten Nationen, in: System eines internationalen Volksgruppenrechts, *supra,* pp. 54–63, at 56; *J. D. Humphrey,* The United Nations Sub-Commission on the Prevention of Discrimination and the Protection of Minorities, 62 AJIL 869–888, at 870 (1968): »relatively weak article 27«; *T. Modeen,* The International Protection of National Minorities in Europe (1969), p. 108: »no real advance on the U.N. Universal Declaration of Human Rights«; *J. Robinson,* International Protection of Minorities. A Global View, 1 Israel Yearbook on Human Rights 61–91, at 89 (1971): »This is a classic example of restrictive toleration of minorities«.

[6] For a comprehensive review of those treaties see, e.g., *F. Capotorti,* Study on the Rights of Persons Belonging to Ethnic, Religious and Linguistic Minorities, UN Doc. E/CN.4/Sub.2/384/Rev. 1, pp. 17–26; *M Ganji,* International Protection of Human Rights (1962), pp. 45–77; the implementation at the international level has been analyzed by *C. Gütermann,* Das Minderheitenschutzverfahren des Völkerbundes (1979).

[7] Of 22 November 1969, in force since 18 July 1978, International Legal Materials, vol. 9 (1970), pp. 99–121.

[8] It would appear, however, that implicitly at least Art. 1 (1) ACHR prohibits discrimination on the ground of a person's ties to a minority by outlawing distinctions based, *inter alia,* on race, language, religion, national or social origin, birth or any other social condition.

Rights[9]. And in spite of its Art. X labelled »Rights of Minorities« not even the recent Islamic Universal Declaration of Human Rights[10] changes the general pattern because the substantive scope of that provision is confined to religious minorities under muslim rule which, as it appears, are thereby enjoined to refrain from applying coercion against their members[11].

Recently, however, new efforts have been undertaken with a view to improving the status of minorities. A first, though rather modest step was accomplished in 1975 when the Final Act of the Conference on Security and Co-Operation in Europe[12] in its »Declaration on Principles Guiding Relations between Participating States« took account of the hardship suffered by national minorities in certain countries by stating what under the régime of the rule of law would be self-evident, namely, that persons belonging to such minorities shall enjoy »equality before the law« and shall be secured »the full opportunity for the actual enjoyment of human rights and fundamental freedoms«. Essentially, these propositions come down to a ban against overt or covert discrimination of individuals who do not form part of the majority of the population, and their real significance is additionally affected by the obvious weakness of the mechanism of implementation. More courageous initiatives have been ultimately started in Western European political fora. The Parliamentary Assembly of the Council of Europe, focusing specifically on the language problem, has formulated rather far-reaching proposals in its recommendation 928 (1981) of 7 October 1981[13]. In concrete terms, however, this recommendation constitutes but a suggestion which the Committee of Ministers may take up or may, if not flatly decline, defer *sine die* by transmitting it to a study group of experts. An analogous fate could easily happen to the Community Charter for Regional Languages and Cultures and Charter of Rights of Ethnic Minorities[14] adopted by the European Parliament on 16 October 1981, shortly after the Parliamentary Assembly of the Council of Europe had concluded its deliberations. The Charter mainly

---

[9] International Legal Materials, vol. 21 (1982), pp. 59–68.

[10] Of 22 September 1981. Its author, the Islamic Council, is a non-governmental organization. The instrument, therefore, possesses only a rather limited representative value.

[11] Art. X Rights of Minorities: a) The Qur'anic principle »There is no compulsion in religion« shall govern the religious rights of non-Muslim minorities. b) In a Muslim Country religious minorities shall have the choice to be governed in respect of their civil and personal matters by Islamic Law, or by their own laws.

[12] For the English text see International Legal Materials, vol. 14 (1975), pp. 1292–1325, at 1295: »The participating States on whose territory national minorities exist will respect the right of persons belonging to such minorities to equality before the law, will afford them the full opportunity for the actual enjoyment of human rights and fundamental freedoms and will, in this manner, protect their legitimate interests in this sphere«.

[13] Printed as a separate document of the Council of Europe.

[14] C 287 Official Journal of the European Communities 106 (9 November 1981).

contains an appeal to national governments to promote the use of minority languages in all spheres of life (operative paragraph 1), but it lacks any binding force and does not purport to establish machinery for securing a permanent control over the implementation of its provisions. Thus, it looks more like a momentaneous and perhaps short-lived interest in minority rights which does not really deserve its pretentious name – which also would have required a more careful drafting than has actually gone into that instrument.

2. Seen against this general background, the merits of Art. 27 appear in a much clearer light. Although undeniably somewhat vague and too succinct[15], Art. 27 goes beyond a purely hortatory declaration in that it constitutes, from the formal point of view, hard, binding treaty law. A closer look at its contents will reveal, furthermore, that its substance is not as poor as is often suggested. In fact, it seems to have frightened the French Government to such an extent that France felt prompted to enter a rather enigmatic reservation[16] when it ratified the CCPR in 1980. It was the first country to do so.

Minorities do not only enjoy the protection of Art. 27. In a country bound by the CCPR or by an analogous regional instrument and respecting the obligations undertaken thereunder, only few problems remain unsettled. In particular, freedom of thought, conscience and religion (Art. 18), freedom of opinion (Art. 19) and freedom of association (Art. 22) define areas of autonomy safeguarded against Government interference which generally enable minorities to preserve and maintain their own identities. The prohibition of discrimination prevents measures directly aimed at placing minorities in an unfavourable position as compared with the remainder of the population[17]. Within the modern nation-State, however, minorities are structurally and permanently put in jeopardy by forces of cultural assimilation. It is this threat which Art. 27 seeks to combat.

3. As to the means and methods of interpretation to be resorted to in attempting to elicit answers to open questions, it goes without saying that the principles laid down in the Vienna Convention on the Law of Treaties

---

[15] *F. Capotorti,* Die Rechte der Angehörigen von Minderheiten, Vereinte Nationen 113–118, at 113 (1980): »völlig unbestimmt«.

[16] »Le Gouvernement français déclare, compte tenu de l'article 2 de la Constitution de la République française, que l'article 27 n'a pas lieu de s'appliquer en ce qui concerne la République«, see UN Doc. CCPR/C/2/Add. 5, p. 2 (1982). The introductory words of Art. 2 of the French constitution of 1958 to which apparently reference has been made read: »La France est une République indivisible. . .«.

[17] Compare, for instance, *F. Capotorti,* The Protection of Minorities under Multilateral Agreements on Human Rights, 2 The Italian Yearbook of International Law 3–32, at 4–5 (1976); *P. Pernthaler,* Der Schutz der ethnischen Gemeinschaften durch individuelle Rechte (1964) *passim,* esp. at p. 28.

(VCLT)[18] must be applied. In the first place, it is the terms of the treaty on which attention has to be focused (Art. 31 (1) VCLT). More often than not, however, words are ambiguous, even if read in their context. In such circumstances, the general question arises as to what extent the *travaux préparatoires* can be used. *B. G. Ramcharan*[19] has recently criticized the present writer for emphasizing, he thinks, to an excessive degree this source in connection with an article on equality and non-discrimination according to Art. 2 (1) and 26[20].

It can certainly not be denied that the preparatory work has been relegated by Art. 32 VCLT to a secondary place. Art. 32 VCLT provides that it constitutes but a supplementary means of interpretation. It can be foreseen, furthermore, that in the future the practice of the Human Rights Committee (HRCee) established under Art. 28 will acquire an ever greater weight. At the present time, however, all these observations cannot detract from the basic fact that, failing the literal method of interpretation, the only additional clues as to the intention of the States parties are provided by the official UN documents covering the different drafting stages[21]. To leave aside that material would be tantamount to refusing even to take note of available knowledge. Such neglect would appear to be all the less appropriate since during the deliberations within the relevant UN bodies occasionally a clear consensus emerged, while the practice of States since the entry into force of the CCPR has produced no such clear-cut results. In contradistinction to traditional treaties regarding typical inter-State relationships, the practice of States in implementing the CCPR escapes any precise and accurate description. Millions of cases to be handled every day according to the precepts of that instrument can neither be identified nor summarized or compared between themselves. There are few, if any occasions on which the States parties to the CCPR could be viewed as acting together in such a way as to manifest their common understanding of a specific provision of the CCPR. At any rate, the regular meetings convened for election purposes in accordance with Art. 30 have never been used as procedural devices for solving difficult issues of construction of the CCPR.

---

[18] International Legal Materials, vol. 8 (1969), pp. 679–713; bilingual (English-French) edition in 29 Zeitschrift für ausländisches öffentliches Recht und Völkerrecht 711–760 (1969).

[19] *B. G. Ramcharan,* Equality and Nondiscrimination, in: L. Henkin (ed.), The International Bill of Rights (1981), pp. 246–269, 463–470, at 466 note 27.

[20] *C. Tomuschat,* Equality and Non-Discrimination under the International Covenant on Civil and Political Rights, in: Staatsrecht – Völkerrecht – Europarecht, Festschrift für Hans-Jürgen Schlochauer, ed. I. von Münch (1981), pp. 691–716.

[21] For an account of the drafting history see, in particular, the *Capotorti* report (supra note 6), pp. 31–34; *F. Ermacora,* Der Minderheitenschutz in der Arbeit der Vereinten Nationen (1964), pp. 38–65; *L. Sohn,* The Rights of Minorities, in: L. Henkin (ed.), The International Bill of Rights (1981), pp. 270–289, at 273–274.

II. Scope of Application Ratione Personae

1. a) Concerning the holders of the rights under Art. 27, no doubts can exist. Protection is not afforded to minority groups as such, but rather to »persons« belonging to minorities. This formulation cannot be viewed just as an accident of drafting. To conceive of minority protection in individualistic terms fits well into the general pattern of the CCPR which, apart from the right of self-determination, addresses itself directly to the individual concerned (»everyone«). In addition, during the drafting process delegates engaged in the work on the CCPR were well aware of the divergence between the two conflicting conceptions. The individualistic approach (»persons belonging to ... minorities«) had already been adopted in the very first text, the draft outline of an International Bill of Human Rights prepared by the Division of Human Rights of the UN Secretariat (Art. 46) [22]. Later, the USSR attempted to alter this general conceptual line by suggesting that the obligation of States should be framed in such a way as to exist vis-à-vis minority communities as such [23], but its motions were rejected by clear majorities. During the final discussion in the Third Committee of the General Assembly it was expressly noted by the representative of India that the draft article under examination did not apply to »minorities considered as groups« [24]. Finally, in legal writings authors have unanimously acknowledged the specific wording by which the beneficiaries of Art. 27 are defined [25]. In this connection, it has rightly been stressed that using the general label of protection of minorities amounts to an incorrect method of expression which is justifiable only in a broader sense by focusing on the practical effects of the provision [26].

b) Since it is the relationship tying a person to a minority group which bestows it with the rights under Art. 27, it cannot be avoided asking the question what criteria are available to classify a group of persons as a mi-

---

[22] UN. Doc. E/CN. 4/21 Annex A, p. 8.

[23] UN Doc. E/CN.4/L.222, reproduced in Report of the Ninth Session of the Commission on Human Rights, ESCOR 16th Sess. Suppl. 8, Annex III, p. 55 para. 13: »The State shall ensure to national minorities the right to use their native tongue and to possess their national schools, libraries, museums and other cultural and educational institutions«.

[24] UN Doc. A/C.3/SR.1103, p. 215 para. 39.

[25] Compare, for instance, the *Capotorti* report *(supra* note 6), p. 35 paras. 206–210, and *Capotorti (supra* note 17), p. 18; *E.-I. A. Daes,* Protection of Minorities under the International Bill of Human Rights and the Genocide Convention, in: Festschrift für Pan. J. Zepos, vol. 2 (1973), pp. 35–86, at 62; *Ermacora (supra* note 21), p. 61; *Modeen* (*supra* note 5), p. 108; *Münch* (*supra* note 5), p. 70; *Y. M. Rechetov,* Minority Rights in Contemporary International Law, 31 Revue Hellénique de droit international 154–164, at 159–161; *A. H. Robertson,* Human Rights in the World (1972), p. 96; *Sohn* (*supra* note 21), p. 274; *C. Zanghi,* in: Presidenza del Consiglio dei Ministri (ed.), La protezione internazionale dei diritti dell'uomo (1979), p. 72.

[26] See the *Capotorti* report (*supra* note 6), p. 35 para. 206.

nority. As usual, the Covenant itself does not give any more detailed explanation apart from setting forth that it contemplates only »ethnic, religious or linguistic« minorities. Those qualifications, nonetheless, tend to demonstrate that certainly not each and every group of people may claim to be a minority. The common characteristic feature of ethnic, religious and linguistic minorities is that they form stable units with a notable degree of historical continuity. To belong to an ethnic or linguistic community is a distinction mark which pervades the entire existence of an individual and cannot easily be changed at will. In the case of religious minorities, the degree of stability is less accentuated, all the less so since according to Art. 18 everyone is free to leave the community in which he was born and to join another group of his preference. Nonetheless, it may be assumed that the drafters proceeded from a general conception of religious communities according to which such communities are historical entities which their members normally adhere to for their entire life-span.

The text itself providing no further clues as to the scope and meaning of the concept of minorities, it is legitimate to turn to the drafting history in order to elucidate what other criteria should guide the construction of Art. 27. In that respect, one has to acknowledge above all that the Third Committee of the General Assembly, where Art. 27 got its final approval through a debate on the relevant substantive issues, based its deliberations on UN Doc. A/2929, the annotations on the draft International Covenants on Human Rights[27], which aptly summarize the work done at the previous drafting stages, in particular by the Human Rights Commission (HRCion)[28]. Concerning the present Art. 27 some of the observations contained therein are extremely clear-cut, leaving no room for doubt. »It was agreed«, says the passage which is relevant here, »that the article should cover only separate or distinct groups, well-defined and long-established on the territory of a State«[29]. This statement, which the annotations relate to the opening clause (»In those States . . exist«), faithfully reflects the proceedings of the HRCion. Indeed, the text on which the HRCion reached agreement, which had been prepared by the Sub-Commission on Prevention of Discrimination and Protection of Minorities[30], had started with the words: »Persons belonging . . .«[31]. It was at the insistence of the Chilean delegation that the opening

[27] GAOR 10th Sess. Annexes, Agenda item 28, Part II.

[28] See UN Doc. A/C.3/SR. 1103, p. 213 para. 7.

[29] UN Doc. A/2929 (*supra* note 27), p. 63 para. 184.

[30] Adopted at its 57th meeting on 23 January 1950, see UN Doc. E/CN.4/Sub. 2/SR. 57, p. 3.

[31] Apart from the opening clause, the text was identical with the present wording of Art. 27, see Report of the Third Session of the Sub-Commission on the Prevention of Discrimination and the Protection of Minorities to the Commission on Human Rights, UN Doc. E/CN.4/358, p. 22, also reproduced in UN Doc. E/2447 (*supra* note 23), p. 55 para. 15.

clause was added. Although the earlier version of its amendment had conveyed the underlying idea much more clearly[32], its later streamlining was in no way intended to detract from its original meaning[33].

Consonant with these views, during the debates held in the Third Committee several speakers declared their full support for the basic approach to the concept of minorities as it had been chosen by the HRCion[34]; no representative advanced a different understanding of the provision. In addition, delegates from a number of countries took care to emphasize that the text under review should not be capable of being invoked by »new« minorities[35]. It appears, therefore, that the conclusions implicit in the restriction of Art. 27 to ethnic, religious and linguistic minorities are fully borne out by the drafting history. Along the same lines, it has been emphasized by the Special Rapporteur of the Sub-Commission on Prevention of Discrimination and Protection of Minorities, *Francesco Capotorti,* that a minority necessarily presupposes stable characteristics which differ sharply from those of the rest of the population[36].

The legal requirement of a definite historical identity does not exclude diffused minorities from the protection of Art 27. The text of the CCPR does not indicate that additional conditions must be fulfilled. Although frequently minorities will be concentrated in certain geographical areas, they may also live largely dispersed among the majority of the population. In the case of religious communities this could even be the rule. It should not be overlooked, furthermore, that lack of territorial cohesion between members of a minority may be the direct result of measures of persecution which took place in the past. To exempt the responsible States from observing Art. 27 would, in such circumstances, amount to rewarding injustices inflicted upon the aggrieved communities. Likewise, a narrow interpretation would encour-

---

32 »In those States in which stable and well defined ethnic, religious or linguistic minorities have long been established«, see UN Doc. E/CN.4/L.261, original version; for its later modification see UN Doc. E/CN.4/SR.371, p. 4.

33 Compare the relevant statements of the Chilean representative, UN Doc. E/CN.4/SR.370, pp. 5, 11, referring to »well-established historical minorities« in the sense that they »had been formed by historical processes«. Along the same lines: Uruguay, *ibid.,* SR.369, p. 7, Philippines, *ibid.,* SR.370, p. 10.

34 Brazil, UN Doc. A/C.3/SR.1103, p. 213 paras. 11, 12; Chile, *ibid.,* p. 214 para. 18; *ibid.,* SR.1104, p. 221 para. 23; Pakistan, *ibid.,* p. 220 para. 16.

35 Chile, UN Doc. A/C.3/SR.1103, p. 214 para. 19; Venezuela, *ibid.,* p. 215 para. 228; India, *ibid.,* p. 215 para. 36; Pakistan, *ibid.,* SR.1104, p. 220 para. 16; Nicaragua, *ibid.,* p. 222 para. 36.

36 *Loc. cit. (supra* note 6), p. 96 para. 566, and *supra* note 17, p. 16. On the other hand, *Capotorti* lacks full consistency in pointing out that, where a minority exists, Art. 27 is applicable to it, irrespective of the date of its formation. The problem, however, is to define the characteristics of a minority. Although minorities evolve in time and history, they cannot be formed in the same way as an association under Art. 22. The relevant point is also missed by *P. Thornberry,* Is there a Phoenix in the Ashes? – International Law and Minority Rights, 15 Texas International Law Journal 421–458, at 448 (1981), who rejects the distinction between old and new minorities.

age governmental policies designed to displace persons belonging to minorities from their original homes.

c) As far as size is concerned, first of all the upper limit has to be determined. Without inquiring into every detail of this complex issue, it may be stated quite unequivocally that, contrary to an isolated voice in legal doctrine[37], a majority can never constitute a minority in the sense contemplated by Art. 27. Indeed, if, as in South Africa, a majority lives in a state of oppression, its members do not require specific protection »in order to enjoy their own culture«, the only need they really have is not to be withheld their basic democratic rights as enunciated in Art. 25. As soon as the principle »one man, one vote« is secured, the majority may then successfully assert and effectuate its political and cultural aspirations[38].

In some countries which are made up of different linguistic communities organized in autonomous sub-units, be they qualified as States, provinces, regions or otherwise, the main group of the population may be in a minority position in such sub-units. This is true, for instance, of Canada where the province of Québec is largely inhabited by French-speaking Canadians, of Switzerland and the USSR, where the Russians constitute only a small, though increasing part of the population in some of the Union Republics. Conflicts are likely to arise in such circumstances if at the lower level, by virtue of its powers in matters of education, the sub-unit concerned attempts to enforce school policies designed to impose the »minority«-language as the only admissible language in public educational institutions, eventually even prohibiting the establishment of private schools where instruction would be given in the language dominant at the national level. In the *Belgian Linguistic* cases, the European Court of Human Rights had to deal with a situation of that type. The major bone of contention was the strict separation of Belgium into different linguistic regions to the effect that public or Government-subsidized education was made available only in the local language. Although the ECHR, as has already been pointed out, lacks a specific guarantee for the benefit of minorities, it is worthwhile noting that the European Court of Human Rights denied a right to be taught in one's own language. In the Court's view, such a right could not be derived from the first sentence of Art. 2 of the [First Additional] Protocol to the ECHR which sets forth that »no person shall be denied the right to education«[39].

---

[37] *Dinstein* (*supra* note 5), p. 112; *contra* the *Capotorti* report (*supra* note 6), p. 96 para. 566, and *Capotorti (supra* note 17), p. 1516; *E. H. Pircher,* Der vertragliche Schutz ethnischer, sprachlicher und religiöser Minderheiten im Völkerrecht (1979), p. 27; *Sohn (supra* note 21), p. 279.

[38] A viable solution presupposes, however, entrenched constitutional safeguards for the protection of the rights of the white minority.

[39] Case *Relating to Certain Aspects of the Laws on the Use of Languages in Education in Belgium* (merits), judgment of 23 July 1968, Publications of the European Court of Human Rights, Ser. A, vol. 8, pp. 31 and 35.

If one looks at this issue from the standpoint of the aggrieved individual, protection seems to be warranted. Language being a constituent part of everyone's identity, he runs the risk of being structurally alienated from his family if he receives education in a language which is not his by his personal origins. Parents may find that their children, through the medium of another language whose influence they cannot escape, evolve characteristics which open up a gap of mutual misunderstanding. On the other hand, starting from the premise that Art. 27 should at least concurrently protect the group as such, one has to acknowledge that the general configuration is quite a different one if the »regional minority« forms part of a national majority. If Art. 27 has been conceived of in order for minorities to survive and to preserve their identity, such reasons play no rôle here. On the contrary, educational language requirements as described above normally serve as a means of protecting the true minority against pressures of assimilation originating from the global majority. Furthermore, individuals have the possibility to return to those parts of the national territory where their own language is being practised. To be sure, practical obstacles may often prevent returning to the land of one's mother-tongue. In addition, the general tension between the principles of national unity and freedom of movement, on the one hand, and territorial linguistic separation, on the other, is all too obvious. But one cannot fail to observe that during the drafting process the difficult problems inherent in the existence of bilingual or plurilingual States with different national communities were never discussed. All references to factual situations to be dealt with concerned true minorities, that is to say, minorities defined by comparison with global population numbers. The present author therefore concludes that Art. 27 does not address itself to the issue of »regional minorities«.

Further reflection shows that, indeed, protective needs differ fundamentally from each other. Within the framework of the nation-State which has crystallized as the creation of a leading cultural and linguistic group, ethnic and linguistic minorities have to struggle to maintain a continuous, daily fight against being totally drawn into the melting pot[40]. By contrast, as far as pluri-national (pluri-ethnic, pluri-linguistic) States like Canada, Switzerland or the USSR with one preponderant component are concerned, the crucial problem is to strike a balance between the requirements of national unity, on one hand, and ethnic and cultural diversity, on the other. The delicate equilibrium existing in such States needs institutional safeguards in order to resist the prevailing majority culture which inevitably has a dominant impact on every kind of organized State activity. In such circumstances, territorial unity

[40] In legal writings, authors are mostly well aware of this general background, compare, for instance, *O. Kimminich,* The Function of the Law of Ethnic Groups in the International System, 23 Law and State 37–51, at 49 (1981); *Pernthaler* (*supra* note 17), pp. 20–21.

reflected in the school language may be an indispensable means of safeguarding the minority against erosion by the outside pressures to which it is permanently exposed[41]. It would, at any rate, be a somewhat amazing result if minority protection could be undermined in the name of minority rights.

The conclusion which has emerged in the course of the above considerations, namely that the concept of minority cannot be defined in the abstract, without any regard to the substantive rights granted by Art. 27, finds a further confirmation when one examines the question as to whether a lower limit exists[42]. Does the minority need to be composed of a certain minimum member of persons in order to be respected as such? The question would be extremely pertinent if Art. 27 would require States to take positive steps designed to ensure the existence and the well-being of minorities. If, on the other hand, Art. 27 contains mainly a negative obligation in the sense that governments are obligated to refrain from interfering with the culture, religion and language of minority groups, a definition becomes almost superfluous. States would then not be charged with burdens to be measured also in budgetary terms, but would mainly be required to let minority communities develop their own characteristics. Essentially, sheer passivity would then be sufficient to comply with Art. 27. A closer examination of the problem shall, therefore, be deferred.

2. Though obvious from the text of the provision, it should be repeated that Art. 27 confines itself to specific categories of minorities, viz. ethnic, religious or linguistic communities. Normally no major difficulties in identifying religious or linguistic groups will arise. Religious minorities are united in a common creed, linguistic communities have in common a language spoken by all of its members. To define an ethnic minority, however, may prove to become a considerably more complex issue since the ethnic origin of a person, except in cases of obvious racial differences related to colour, does not always reveal itself unmistakably even to an outside observer if it is not supported by an additional linguistic element. Art. 27 relies heavily on the cultural ties existing between a group of people in that it guarantees the enjoyment by ethnic communities of their own culture.

National minorities have not been brought within the scope of Art. 27. Proposals put forward to that effect by the USSR[43] were rejected in the HRCion. No clear justification had been provided by the USSR for its suggestion. In a contradictory manner, it was asserted, on the one hand, that »the term ›ethnic or linguistic groups‹ was not nearly as far-reaching as ›national minorities‹«, while, on the other, it was pointed out that the adoption of the Sub-Commission's text »would have the effect of protecting

---

[41] See also *M. J. Bossuyt,* The Direct Applicability of International Instruments on Human Rights, 15 Revue belge de droit international 317–343, at 329 note 54.

[42] See the *Capotorti* report (*supra* note 6), pp. 12 para. 56, 96 para. 566.

[43] See *supra* note 23.

groups which might never become national minorities«[44]. Only the second statement is in conformity with the true state of affairs[45]. Indeed, though the concept of national minority may lack precision, it is clear nonetheless that it adds an additional subjective criterion to the relevant objective characteristics. Only few ethnic minorities would qualify as national minorities. National minorities are groups which have become conscious of their own identity to such an extent that they seek to become masters of their own fate at least partially. Any reference to a national minority thus introduces a definite political element. For that reason, too, the boundary line between a people in the sense contemplated by Art. 1 and a minority would tend to become blurred. It is probably the awareness of this danger which prompted the majority of States to omit any reference to national minorities.

3. a) During the preparation of Art. 27 it was sometimes asserted that the present Art. 27 did not apply to aliens[46]. The same view has been expressed in statements of legal doctrine[47], in particular by the *Capotorti* report[48]. At first glance this position seems to possess rather solid foundations. Immigrant aliens constitute, as a kind of model, a type of community which lacks the characteristic of stability which, as has been shown earlier, is one of the key-criteria on which Art. 27 hinges. Nonetheless, closer examination gives rise to serious doubts. In the first place, one cannot fail to observe that the word employed is »persons«, not nationals. This choice accords with the general spirit of the CCPR which aims to do away with the traditional distinction between nationals and aliens[49]. Although international protection of human rights is mainly designed to extend the benefits of international guarantees to nationals who, under the traditional rules, were deprived of any kind of protection, the reverse tendency is also reflected in the CCPR, for instance in Art. 12 (1) which grants freedom of movement and freedom to take residence within the national territory also to aliens, and in Art. 19, 21 and

---

[44] UN Doc. E/CN.4/SR.369, p. 13.

[45] France rightly criticized the Soviet draft for its narrow character, see UN Doc. E/CN.4/SR. 370, p. 7.

[46] Discussion in the Third Committee of the General Assembly in 1961: India, UN Doc. A/C.3/SR.1103, p. 215 para. 38; Iraq, *ibid.*, SR.1104, p. 219 para. 7.

[47] See *Modeen* (*supra* note 5), p. 108; other authors state that, in general, minorities comprise only nationals, see, e.g., *Pircher (supra* note 37), p. 25; *contra: Münch (supra* note 5), p. 72; *M. Tabory,* Language Rights as Human Rights, 10 Israel Yearbook on Human Rights 167–223, at 182 (1980); with some slight doubts also *E. Schwelb,* Some Aspects of the International Covenants on Human Rights of December 1966, in: A. Eide/A. Schou (eds.), International Protection of Human Rights (1969), pp. 103–129, at 122. For a critical assessment of the exclusion of foreign workers see *D. Gerdes,* Minderheitenschutz – eine internationale Rechtsnorm auf der Suche nach ihrem Gegenstand, Vereinte Nationen 126–131, at 130 (1980).

[48] *Supra* note 6, p. 12 para. 57, p. 96 para. 566; see also *Capotorti (supra* note 17), p. 16. The report is, however, incorrect in stating that it is »generally accepted« that members of a minority must be nationals of the State concerned.

[49] This is also underlined by *Schwelb (supra* note 47), p. 122.

22, which extend certain politically sensitive freedoms to »everyone«. In fact, what are important are the historical roots of the group concerned. Its existence may not be of recent origin, but must go back into the past. This can also be true of persons who are not nationals of the State concerned, for instance where a State, on whatever grounds, has acquired territory which had originally belonged to another State, but does not automatically grant its nationality to all persons living in this new part of the country [50]. Another situation not to be overlooked is the arbitrary discrimination against a specific ethnic group, whose members, although their ancestors have come to the country concerned decades or even centuries ago, are denied nationality with the intention of keeping them under the tight discipline of aliens legislation, reserving the right to expel them eventually.

b) Similar considerations apply to the question as to whether immigrants are capable of claiming the benefits of Art. 27. During the deliberations not only by the HRCion [51], but also by the General Assembly [52] it was emphasized time and again by representatives of Latin American countries in particular that immigrants would not come within the purview of Art. 27. Speakers pointed to the past where, during the existence of the German Third Reich, German immigrants had forcefully asserted their »rights« to privileged treatment [53]. In general, it was felt that the recognition of immigrants as minorities would pose a threat to national unity.

Again, the basic premise must be the general agreement, reflected in the opening words of Art. 27, that only such groups qualify as minorities that are »well-defined and long-established« on the territory of a State. Normally, immigrants will mix rather quickly with the remainder of the population. Their main aim is not to stay an isolated group within their new mother-country but rather to become an integral part of the national community which by their free will they have chosen. Consequently, they generally lack the characteristic of stability, being destined instead to gradually disappear as a distinct group. But things may develop along different lines. In multi-ethnic societies immigrants may indeed preserve and keep their specific identity over generations. It must be acknowledged, therefore, that immigrants are

---

[50] See also ECOSOC Res. 502 F (XVI) of 3 August 1953, ESCOR 16th Sess. Suppl. 1, p. 12, and Brazil, UN Doc. A/C.3/SR.1103, p. 213 para. 12.

[51] Chile, UN Doc. E/CN.4/SR.368, pp. 10, 14; Uruguay, *ibid.*, SR.369, p. 7, SR.370, p. 4, joined by Australia, *ibid.*, SR.369, p. 11; *contra*: France, *ibid.*, SR.368, p. 16: »It was always possible that sooner or later certain groups of immigrants might claim minority status«; Philippines, *ibid.*, SR.370, p. 9 (drawing attention to the purely negative dimension of the rights to be granted).

[52] Brazil, UN Doc. A/C.3/SR.1103, p. 214 paras. 12–13, supported by Chile, *ibid.*, p. 214 para. 18; Australia, *ibid.*, p. 214 para. 25; Spain, *ibid.*, p. 215 para. 35; Ecuador, *ibid.*, p. 215 para. 44; Chile, *ibid.*, SR.1104, p. 221 para. 24.

[53] Brazil, UN Doc. A/C.3/SR.1103, p. 214 para. 13; Chile, *ibid.*, p. 214 para. 22; Ecuador, *ibid.*, p. 215 para. 44.

not as such debarred from becoming historical entities in the sense contemplated by Art. 27[54].

c) Finally, it has been maintained that indigenous communities do not meet the requirements of a minority. In the HRCion, it was said by the Chilean representative that »it would be sheer romanticism« to treat those communities as minorities[55]. Even more harshly, the Australian delegate pointed out that the aborigines »had no separate competing culture of their own, for as a group they had only reached the level of food-gatherers«, and that the only possible solution for the problem of those people »lay in assimilation«[56].

Similar observations were later repeated in the General Assembly[57] where also Brazil stated that Indians could not correctly be called minorities[58]. Essentially, these statements reveal a naive feeling of cultural superiority. According to this view, indigenous ways of life, characterized by backwardness and primitivity, are simply doomed to disappear and should be swept away as vestiges of an archaic past which have no right to existence in a modern society. Besides being politically unwise, those interpretations have no legal foundation in Art. 27. Indeed, while immigrants mostly lack historical roots in the social network of the national community concerned, indigenous communities most perfectly embody the type of »well-defined and long-established« groups that the drafters generally had in mind. In addition, most of them do not only have in common one of the three specific features taken into account by the relevant provision, but overwhelmingly the ethnic specificity is supplemented by a common language and a common religion. The sole fact that during the deliberations some delegations expressed a different understanding does in no way alter the objective meaning of the text of the treaty.

The view that indigenous communities cannot be excluded from the protection afforded by Art. 27, has found an additional confirmation in the practice of the HRCee. When taking up the reports of Ecuador[59], Chile[60], Co-

---

[54] The same view is held by *Capotorti (supra* note 17), p. 16; *Tabory (supra* note 47), p. 182, and – but without a specific reference to Art. 27 – *Pircher (supra* note 37), p. 26. The *Capotorti* report also rejects the standpoint of the countries of immigration, see *supra* note 6, p. 12 para. 58.

[55] UN Doc. E/CN.4/SR.368, p. 10.

[56] *Ibid.,* SR.369, p. 11.

[57] Chile, UN Doc. A/C.3/SR.1103, p. 214 para. 21; *ibid.,* SR. 1104, p. 221 para. 23; Australia, *ibid.,* SR.1103, p. 214 para. 26.

[58] UN Doc. A/C.3/SR.1104, p. 220 para. 19.

[59] No information had been furnished in the first report under Art. 40, UN Doc. CCPR/C/1/Add. 8. Questions were put by members Evans, UN Doc. CCPR/C/SR.31, p. 9 para. 40; Tarnopolsky, *ibid.,* p. 9 para. 42. For the answer of the Government see UN Doc. CCPR/C/SR.32, p. 5–6 paras. 19, 22 and 25. The second report confined itself to stating that »problems of racism or racial discrimination have never existed in Ecuadorian society«, UN Doc. CCPR/C/1/Add. 29, p. 11 para. 37; for the ensuing questions and answers see UN Doc. CCPR/C/SR.118, p. 8 paras. 42–45.

lombia[61] and Venezuela[62], members always asked questions about the legal status and factual situation of Indians, and during the examination of the Canadian report this issue received particular emphasis[63]. Not a single member of the HRCee felt that Art. 27 should be interpreted restrictively *ratione personae.* In the *Lovelace* case concerning a native Indian woman having lost her legal status as an Indian under the Canadian Indian Act by virtue of her marriage to an ordinary Canadian, the HRCee has furthermore ruled quite explicitly that Indian tribes unreservedly meet the requirements set forth in Art. 27[64].

Finally, one may also point to the *raison d'être* of Art. 27. However weak its safeguards may be, they are due to the insight that in the modern nation-State with a leading national group, minority cultures are structurally threatened with extinction. It is precisely this situation which obtains in the Western hemisphere where the modern States in their present configuration were created almost exclusively by settlers of European origin who have forcefully

---

[60] In its report under Art. 40, UN Doc. CCPR/C/1/Add. 25, p. 53, the Chilean Government had simply asserted that »in Chile there are no ›minorities‹ within the meaning of article 27 of the Covenant«. This judgement was rejected by members Tomuschat, UN Doc. CCPR/C/SR.128, p. 10 para. 37; Tarnopolsky, *ibid.,* p. 13 para. 49. For the answer of the Chilean delegation see *ibid.,* SR.130, pp. 12–13 para. 47. Though maintaining its stance in principle, the representative stated that »the Government was adopting the necessary steps to guarantee Chileans of Mapuche origin the full enjoyment of their rights while respecting not only the individual characteristics of Mapuche life but also protecting the right of the Mapuches to follow their own customs and use their own language«.

[61] The Government of Colombia, with explicit reference to Art. 27, gave detailed information on the status of minorities in its report under Art. 40, UN Doc. CCPR/C/1/Add. 50, pp. 64–67, although it was stated that indigenous groups »cannot be regarded as an ethnic minority«. Questions were but by members Bouziri, UN Doc. CCPR/C/SR.222, p. 5 para. 23; Tomuschat, *ibid.,* p. 7 para. 34; Tarnopolsky, *ibid.,* p. 8 para. 45; Koulishev, *ibid.,* p. 10 para. 55; Hanga, *ibid.,* p. 11 para. 70; Graefrath, *ibid.,* SR.223, p. 3 para. 12. For the answer see *ibid.,* SR.226, pp. 7–8 para. 35: »His Government was currently making every effort to establish institutions which would preserve the cultural integrity of the indigenous population while encouraging their integration«.

[62] The Venezuela Government had acknowledged the existence of indigenous communities as minorities in the sense envisaged by Art. 27 in its report under Art. 40, UN Doc. CCPR/C.6/Add. 3, p. 24. Questions were put by members Bouziri, UN Doc. CCPR/C/SR.248, p. 9 para. 51; Hanga, *ibid.,* SR.249, p. 3 para. 11; Tarnopolsky, *ibid.,* p. 6 para. 26; Koulishev, *ibid.,* p. 9 para. 50; Janča, *ibid.,* p. 10 para. 60; Graefrath, *ibid.,* p. 12 para. 69. For the answer see *ibid.,* SR.252, p. 6 para. 35.

[63] For the information provided as to the federal law see UN Doc. CCPR/C/1/Add. 43 (vol. I), pp. 107–108. Questions were put by members Hanga, UN Doc. CCPR/C/SR.206, p. 2 para. 3; Bouziri, *ibid.,* p. 4 para. 12; Evans, *ibid.,* p. 12 para. 51; Movchan, SR. 207, p. 2 para. 4; Koulishev, *ibid.,* p. 7 para. 28; Diéye, *ibid.,* p. 9 para. 37. For the extensive Canadian answer see SR.211, pp. 4–6 paras. 11–23.

[64] Published in the Fifth Report of the HRCee, GAOR 36th Sess. Supp. 40, p. 166; German translation: 8 Europäische Grundrechte-Zeitschrift 522 (1981).

permeated the national texture with the specific values inherited from their ancestors, not infrequently to the detriment of autochthonous populations [65]. More than many other communities, which have successfully met the challenges of the industrialized society of today, Indian tribes are badly in need of protection [66].

In sum, no valid arguments can be found to deny the benefit of Art. 27 to indigenous communities [67].

4. a) Art. 27 does not specify when a person can be considered to »belong« to a minority. On this issue, a subjective and an objective doctrine compete with one another [68]. Even superficial reflection makes it clear, however, that no uniform solution would be appropriate. To adhere to a religious community is, in the spirit of the CCPR as reflected in Art. 18, a determination which the individual should be free to make without any outside coercion. Consequently, the fact that a person has adhered to, or has withdrawn from, a religious community simply has to be acknowledged by the State party concerned, as long as indeed a true religious minority is in issue [69]. On the other hand, ethnic groups are social entities characterized by common descent and kinship, in other words, by objective elements, however difficult an accurate definition of ethnicity may prove. No person can, therefore, become a member of an ethnic minority just by declaring his sympathy or personal liking for that group. It should not be overlooked, however, that a minority

---

[65] The study »Of the Problem of Discrimination Against Indigenous Populations«, UN Doc. E/CN.4/Sub.2/476 with Add. 1–6, submitted to the Sub-Commission on Prevention of Discrimination and Protection of Minorities by the Special Rapporteur José R. Martínez Cobo, provides in that respect a most illustrative picture.

[66] For a summary assessment see, for instance, *P. Gerber/M. Münzel,* Nicht Wiederbelebung, sondern Wandel des Lebendigen – Indianische Minderheitenbewegungen heute, Vereinte Nationen 122–126 (1980).

[67] The same view has been expressed by *Capotorti (supra* note 17), p. 17. In this connection, reference should also be made to the ILO Convention 107 concerning the Protection and Integration of Indigenous and other Tribal and Semi-Tribal Populations in Independent Countries, of 26 June 1957, UNTS vol. 328, p. 248. Its acceptance by the community of States is hitherto quite unsatisfactory; it has only been ratified by 27 States (December 1981).

[68] In its judgment on *Rights of Minorities in Upper Silesia,* of 28 April 1928 [1928] PCIJ Ser. A, no. 15, p. 32, the Permanent Court of International Justice relied mainly on objective criteria (»a question of fact and not solely one of intention«). The *Capotorti* report (*supra* note 6), p. 15 paras. 78–81, provides no clear answer, but refers only to the views which have been expressed on the issue. Recently, the subjective approach has been suggested by *Modeen (supra* note 5), p. 24, while objective criteria are preferred by *Pircher (supra* note 37), pp. 28–29. For further discussion see *Y. de Montigny,* L'O. N. U. et la protection internationale des minorités depuis 1945, 13 Revue Juridique Thémis 389–447, at 405–407 (1978); *F. H. E. W. du Buy,* Einige Gedanken über Probleme bei der Interpretation des Minderheitenbegriffs, 20 Literatur-Spiegel 25–50, at 34–47 (1977).

[69] The religious quality of artificially created sects may doubtless be questioned in many instances.

in the same way as a people goes through a continuous process of transformation. Birth, marriage and death affect its composition and existence. When there are no rules of membership established by the minority itself, the relevant facts of life must be taken as the decisive criteria. Children born from mixed marriages should be respected in their wish either to remain associated with the minority or to integrate into the mainstream of national life. In the case of a more distant connection, no other guideline can be perceived than the actual behaviour of the individual himself. It emerges from the innate logic of the matter itself that subjective elements play an important role in supplementing the pattern derived from objective criteria.

b) In this connection, the question arises as to whether national legislative bodies are entitled to make binding determinations on the requirements for the recognition of a person's »belonging« to a minority. In principle, such power should not be called into question. Although the concepts employed in the CCPR, as the outcome of an international legislative process, mostly constitute autonomous products and not just simple references to notions adopted in domestic law, it may be quite legitimate for the competent national authorities to concretize the scope and meaning of formulae enshrined in the CCPR with a view to facilitating their implementation. Provided that such determinations are made *bona fide* and confine themselves to drawing a boundary line in an area which is objectively open to doubt, they may be binding also upon the HRCee as the body called upon to monitor at the international level the observance of the CCPR [70].

The *Lovelace* case referred to above [71], however, has already shown the difficulty of relying on national legislation. The Canadian Indian Act of 1869 as presently in force, the relevant domestic statute in that dispute, does not purport to circumscribe the right to a specific linguistic or cultural status, but confines itself to specifying who is entitled to preferential treatment with regard to land ownership, hunting, fishing, Government grants in aid etc. It would appear to be obvious that concerning such supportive measures tight control is required. No one can acquire the right to receive a share of public funds just because he claims them. Under Art. 27 no such privileges are provided for. As will still be pointed out in more detail, no more is granted *grosso modo* than not to be subjected to assimilation policies. For that reason, in the *Lovelace* case the HRCee did not feel debarred from ruling that the applicant, although legally a non-Indian, could invoke Art. 27 since ethnically she

[70] It is certainly not suggested that States should be free either to grant or to withhold recognition of a minority or of a person's belonging to a minority. Legal writers are almost unanimous in rejecting such power, see, for instance, the *Capotorti* report *(supra* note 6), pp. 96–97 para. 570; *Capotorti (supra* note 17), p. 18; *Pircher (supra* note 37), p. 218; *Thornberry (supra* note 36), p. 448, because otherwise Art. 27 could easily be deprived of any substance; the only voice to the contrary is *Rechetov (supra* note 25), p. 164.

[71] See *supra* note 64.

was a pure Indian. The conclusion is, therefore, that there may exist different legal categories of members of a minority, a classification according to domestic standards not necessarily prejudging the personal sphere of application of Art. 27.

5. Coming back to the initial question as to whether Art. 27 constitutes an individual or a collective right, a final word of clarification seems to be warranted. In spite of its general profile as an individual guarantee, the beneficial effects which the provision produces extend also to the groups to which the persons concerned are associated. It is said explicitly that the rights which members of a minority enjoy are designed to be exercised »in community with the other members of their group«. By logical implication it can be inferred, therefore, that the existence of the three types of minorities mentioned in Art. 27 is also protected as such[72]. Otherwise, the rights accorded to persons belonging to a minority could easily become devoid of any substance. Endeavours to do away with an existing minority would infringe the rights of everyone pertaining to that minority. Viewed in the perspective of practical impact, therefore, the difference between the actual individual right and a truly collective right is a rather slight one. It nevertheless remains that the group as such is not the holder of the rights under Art. 27[73] so that, for instance, it is legally prevented from initiating proceedings under the Optional Protocol.

III. Substantive Contents of Rights

Concerning the substantive rights granted to individuals forming part of a minority, the text of Art. 27 is worded rather cautiously. In a negative formulation, it is set forth that no one shall be denied the right to enjoy his culture, to profess and practise his own religion or to use his own language.

1. A basic requirement for a clear understanding of Art. 27 is to know exactly for what general purpose the provision has been drafted. In fact, some commentators believe that Art. 27 is just redundant, confining itself to outlawing discrimination without providing any kind of special protection to minorities[74]. In this connection, reference is often made to the advisory opinion of the Permanent Court of International Justice of 6 April 1935 on

[72] See in the same sense *Pernthaler (supra* note 17), pp. 32 ff.; *Pircher (supra* note 37), p. 224; *Sohn (supra* note 21), p. 275; a cautious reflection of the same idea also by *Capotorti (supra* note 17), p. 19.

[73] Erroneously, *R. Arnold,* Probleme der Individual- und Kollektivrechte in Völkerrecht, in: Studien zum Staats- und Völkerrecht, Festschrift für Hermann Raschhofer (1977), pp. 27–42, at 40, speaks of »kumulative(n) Rechtssubjektivität des Einzelnen und der Gruppe«.

[74] In particular *Ermacora (supra* note 21), p. 62; *idem,* Volksgruppenrecht und Minderheitenschutz im regionalen Rahmen, in: F. Wittmann/S. Graf Bethlen (eds.), Volksgruppenrecht. Ein Beitrag zur Friedenssicherung (1980), pp. 115–125, at 124.

Minority Schools in Albania which has highlighted that distinction between non-discrimination (equality in law) and special protection (equality in fact):

> »The idea underlying the treaties for the protection of minorities is to secure for certain elements incorporated in a State, the population of which differs from them in race, language or religion, the possibility of living peaceably alongside that population and co-operating amicably with it, while at the same time preserving the characteristics which distinguish them from the majority, and satisfying the ensuing special needs.
>
> In order to attain this object, two things were regarded as particularly necessary, and have formed the subject of provisions in these treaties.
>
> The first is to ensure that nationals belonging to racial, religious or linguistic minorities shall be placed in every respect on a footing of perfect equality with the other nationals of the State.
>
> The second is to ensure for the minority elements suitable means for the preservation of their racial peculiarities, their traditions and their national characteristics.
>
> These two requirements are indeed closely interlocked, for there would be no true equality between a majority and a minority if the latter were deprived of its own institutions, and were consequently compelled to renounce that which constitutes the very essence of its being as a minority«[75].

There can be no doubt that the same reasoning also applies to the present-day situation. The needs of minorities have not changed over the past decades since the Permanent Court gave its opinion.

2. Contrary to the views referred to above, the starting-point of any legal analysis should be the fact that Art. 27 does not speak of non-discrimination at all. In the HRCion and in the Third Committee of the General Assembly, no such classification was made[76]. These findings may not be lightly discarded. Nor can a justification for placing a kind of corrective interpretation on

---

[75] [1935] PCIJ Ser. A/B, No. 64, p. 17.

[76] In the HRCion the following statements were delivered: Great Britain (Hoare), UN Doc. E/CN.4/SR.369, p. 5: »He agreed also that the provisions in the draft covenants relating to non-discrimination applied to minorities. Clearly the Sub-Commission had been aware of the purport of those provisions and had therefore confined itself to the other problem of what protection, on the assumption that minorities enjoyed equality in respect of the rights enunciated in the covenants, was needed to enable them to preserve their identity«; *ibid.*, p. 14: »The Sub-Commission's text was not concerned with non-discrimination, which was already fully secured by article 2. It was an additional provision to ensure that minorities, besides enjoying the other rights laid down in the Covenant, should also be allowed to use their own language, profess and practice their own religion and enjoy their own culture«; France, *ibid.*, SR.370, p. 7: »The Sub-Commission's draft took account of the fact that minorities possessed special rights additional to other human rights«. See also the *Capotorti* report *(supra* note 6), p. 41 para. 242, and *Capotorti (supra* note 17), p. 4.

the text be seen. Arts. 2 (1) and 26 sufficiently take care of the need of minorities not to be discriminated against at the juridical level of legal equality. One of the forbidden grounds of distinction is status, which certainly covers the fact of belonging to a minority; in addition, race, colour, language and religion are mentioned specifically. Accordingly, the main thrust of Art. 27 lies elsewhere. The text itself suggests quite unequivocally that specific collective goods, viz. minority cultures and languages, have been brought within the ambit of the CCPR which otherwise would enjoy no protection within the framework of this instrument. Focusing on the general aim of the article, it can be stated that a definite, though modest step has been taken with a view to ensuring equality in fact as defined in the opinion of the Permanent Court of 1935.

While cultural rights enjoy protection under Art. 15 of the International Covenant on Economic, Social and Cultural Rights (CESCR), their inclusion in the CCPR greatly strengthens their impact. The obligation of States parties under the CCPR are generally commitments to be strictly complied with whereas the obligations deriving from the CESCR have the much weaker nature of promotional obligations. In addition, the individual complaints procedure under the Optional Protocol is only open to claims that rights under the CCPR have been violated. In the *Lovelace* case already referred to, the HRCee has explicitly recognized that Art. 27 is suitable to be invoked by individuals by means of an individual communication.

3. Concerning the substantive contents of Art. 27, the most important question is whether the provision simply contains a prohibition of interference or whether States are obligated to take positive action in favour of minorities. There is no need to dwell on the enormous difference between the two alternative interpretations.

It should be noted to begin with that the negative formulation was chosen deliberately. During the ultimate drafting stage before the HRCion in 1953, a text proposed by the USSR was couched in the following terms: »The State shall insure to national minorities the right to use their native tongue and to preserve their national schools, libraries, museums and other cultural and educational institutions«[77]. In the debate on the merits of that proposal as compared with the draft submitted by the Sub-Commission, the pros and cons of either solution were extensively discussed. While the USSR and its allies most eloquently defended their proposal, delegations from other parts of the world cautioned against imposing too heavy a burden on States. Thus, the British representative stated that besides non-discrimination the only further duty which seemed necessary was to emphasize the importance of toleration; in its negative form, the obligations provided for in the text submitted by the Sub-Commission could easily be complied with and were not onerous[78]. The Chilean delegate pointed out that a provision like that

[77] *Supra* note 23.
[78] UN Doc. E/CN.4/SR.369, p. 6.

suggested by the USSR would have its convenient place in the CESCR[79], and France underscored the importance of setting forth minimum guarantees acceptable to all countries[80]. Interestingly enough, when in 1961 the final discussions took place in the Third Committee of the General Assembly, the USSR did not attempt again to claim that the rights to be enjoyed by minorities should be strengthened. No formal amendment to that effect was placed before the Third Committee. Although the Mexican delegation stressed that a positive formulation should have been preferred[81], no concrete steps were taken to alter the text. Indeed, Mr. Baroody, the representative of Saudi-Arabia, explained his full satisfaction with the draft as it stood[82].

Under these circumstances, it is difficult to see a convincing justification for giving a broader construction to Art. 27 by requiring States to give positive assistance to minorities[83]. One may easily agree with the Special Rapporteur of the Sub-Commission, Francesco Capotorti, that the negative formulation of Art. 27 »does not meet the requirements of the situation«[84]. But this statement is just a maxim of legal policy which does not rest on fully reflected foundations. Above all, one has to bear in mind that, as one of the components of the CCPR, Art. 27 is designed to find world-wide application. Thus, account has to be taken of the specific problems of Third World countries. As far as Africa is concerned, it has been reported, for instance, that in Nigeria not less than 250 native languages exist[85]. It is simply unrealistic to assume that the competent public authorities could take affirmative action for the benefit of all of those linguistic communities. On the other hand, there is no obstacle of any kind which would prevent authorities from tolerating the use of those languages and their manifold dialects. The same is true of the cultural life of the different communities. Stretching the scope of Art. 27 to encompass positive obligations could lead in the last analysis to an outright breakdown of its guiding value and hence to a total loss of credi-

---

79 *Ibid.*, p. 15.

80 *Ibid.*, SR.370, pp. 7–8.

81 UN Doc. A/C.3/SR.1104, p. 219 para. 3.

82 *Ibid.*, p. 221 para. 32.

83 Most authors indeed advocate a cautious interpretation, see *G. Bennett*, Aboriginal Rights in International Law (1978), p. 52; *Dinstein (supra* note 5), p. 118; *L. C. Green*, Human Rights and Canada's Indians, 1 Israel Yearbook on Human Rights 156–190, at 188 (1971); *Modeen (supra* note 5), p. 108; *Montigny (supra* note 68), p. 435; *Pircher (supra* note 37), p. 220; *Rechetov (supra* note 25), p. 159; *Robinson (supra* note 5), p. 91.

84 *Loc. cit. (supra* note 6), pp. 98–99 para. 588; see also *supra* note 15, p. 114, and *supra* note 17, pp. 22, 25. His views are shared by *Sohn (supra* note 21), p. 285; *B. Vukas*, General International Law and the Protection of Minorities, 8 Human Rights Journal 41–70, at 45 (1975); *R. C. A. White*, Self-Determination: Time for a Re-Assessment?, 28 Netherlands International Law Review 147–170, at 168 (1980) (but without any real examination of the issue).

85 *H. Dröge*, Murtala Muhammeds Vermächtnis. Ein Beitrag zur jüngsten Geschichte Nigerias, in: J. Kroneck/Th. Oppermann (eds.), Im Dienste Deutschlands und des Rechtes. Festschrift für Wilhelm G. Grewe (1981), pp. 39–58, at 45.

bility. Art. 27 will be more effective if it is restricted to a hard core of obligations easily to be complied with.

Only in an indirect way can an obligation to take positive steps arise. If and when a State provides financial support to members of majority groups in respect of activities coming within the scope of Art. 27, it is then required by virtue of the principle of non-discrimination to extend analogous treatment to persons belonging to an ethnic, linguistic or religious minority since non-discrimination applies to each and every right enshrined in the CCPR[86]. In fact, to exclude such persons from benefits granted to all other citizens would amount to a denial of those rights which apparently are effective only if their exercise is being subsidized out of public funds. In case the country concerned has at the same time ratified the CESCR, additional confirmation for this conclusion can be derived from Art. 15 in connection with Art. 2 (2) of that Covenant.

To sum up, it can hardly be doubted that Art. 27 was not intended to provide direct rights to positive State action; on the other hand, it should not be overlooked that it is well suited to give rise to derivative rights to positive State action. This conclusion implies at the same time that generally there is no need to trace a lower limit for minorities.

4. a) The needs of being protected against State interference through a specific provision are not the same for all kinds of minorities. First of all, one may openly daresay that religious communities are sufficiently safeguarded by Art. 18. Indeed, one fails to see what supplementary guarantee Art. 27 adds to the substance of Art. 18. All available evidence tends to show that the drafters resorted to a duplication simply because they felt that within the context of a provision on minorities religious communities should not be left out[87].

b) A different picture emerges with regard to ethnic minorities whose specific culture Art. 27 is designed to support, their existence as such being placed under the legal shield of the Genocide Convention and of the general rules contained in Art. 6. While the modern nation-State which is based on a strict separation between itself and the churches does not breathe a specific philosophy or *Weltanschauung* which permeates all of its activities, in cultural terms normally it is far from being neutral. Almost inevitably the leading national group(s) convey its (their) peculiar cultural values through the Constitution and through the work of the entire machinery of State agencies to all strata of the population. The mostly Government controlled mass media, in particular radio and TV, continuously exert additional pressures of adaptation and assimilation.

The cultural gap between the dominant society and certain minorities may even reach such dimensions that the mere fact of coming into contact

---

86 For analogous considerations see *Thornberry (supra* note 36), p. 450.

87 Compare also the *Capotorti* report *(supra* note 6), p. 38 para. 227.

with modern forms of civilization is likely to threaten if not the existence of those minorities, at least their cultural survival as a distinct group. Examples are readily at hand. Suffice it to point to the aborigines in Australia and to certain Indian tribes in the vastness of the Amazonas region. The most radical solution, in such instances, would obviously consist of declaring it unlawful to bring to an end the isolation of such minority groups. Nonetheless, it has to be admitted quite frankly that here the limits of what Art. 27 contains in substance are attained and even exceeded. While Art. 27 constitutes in itself a rejection of the naive belief that the numerical majority should be entitled to impose its value system by force, it is still the product of societies which are profoundly convinced that everyone should have his chance to live in a modernized environment instead of leading an existence in backwardness and ignorance. Indeed, Governments would be harshly criticized if they refrained from taking any action, leaving instead the endangered tribes undisturbed. In this perspective, in particular, Art. 13 CESCR recognizes the right of everyone to education [88]. Consequently, Art. 27 cannot be interpreted as an absolute obstacle to establishing any contacts with an ethnic minority.

c) Nonetheless analogous considerations apply to the protection of language groups. The protective needs of the minorities concerned may even be more accentuated here. Language is an omnipresent element in almost every form of State action. Even if such action uses a specific national language merely as its form and instrument, it has an substantive impact outside its main target, namely the promotion of the language actually employed and the gradual elimination of the language(s) (almost) never resorted to by State agencies. The most deeply going influence is obviously exercized by educational institutions. To a lesser extent, but almost as intensely, the media and the press contribute to strenghtening the leading national language.

5. In trying to clarify some of the concrete legal consequences flowing from the prohibitions contained in Art. 27, the basic premise should again be spelled out that minority protection under the CCPR is intended to secure the cultural survival of minority groups. Art. 27 shall provide them with the legal foundations for preserving their specific identity. It is in this general light that the following considerations will be developed.

a) First and foremost, persons belonging to ethnic and linguistic minorities have the right to employ their language between themselves, not only in private conversations, but also in public [89]. It would run counter to the CCPR to ban the use of a minority language under the pretext that it consti-

[88] Compare also the UNESCO Convention against Discrimination in Education, of 15 December 1960, UNTS vol. 429, p. 93.

[89] *J.-W. Bruegel,* A Neglected Field: The Protection of Minorities, 4 Human Rights Journal 413–442, at 433 (1971); *Daes (supra* note 25), p. 62; *Modeen (supra* note 5), p. 108; *Münch (supra* note 5), p. 70; *Pircher (supra* note 37), pp. 222–223; *Tabory (supra* note 47), p. 183.

tuted seditious activity. Nor could it be reconciled with Art. 27 to enact binding guide-lines on the use of minority languages in the press. Whenever a minority group wishes to publish its own newspapers or other periodicals, it is free to do so subject to the general conditions laid down in the relevant press laws. To prohibit any written expression of a minority language would amount to a hardly concealed attempt to do away with that language. At the present stage of evolution of modern societies, a language of purely oral expression would be definitely doomed to extinction. Only writings are capable of stimulating sustained interest in a language, promoting its intrinsic creative forces and adapting it to the ever changing cultural milieu. In the light of these considerations, it would appear rather short-sighted to maintain that the language guarantee in Art. 27 expresses just a banality.

Likewise, minorities may not be prevented from setting up their own institutions without which cultural life would lack any solid foundations. In that respect, museums and libraries are albeit secondary, though indispensable components of a comprehensive strategy for the right to enjoy one's own culture and language. Although such activities may be viewed as coming within the purview of Art. 19, the explicit reference to culture in Art. 27 assumes at least an important function of clarification [90].

b) The most complex questions are connected with the issue as to whether cultural and linguistic minorities are endowed with the right to establish their own schools. Here, the internal sphere of the communal life of minorities is exceeded. School systems imprint an indelible hallmark on human life. Children brought up in minority schools will for their entire life bear the traces of that education, but the same is true of State owned schools which do not take account of a child's ethnic and linguistic descent. Whereas one type of school is well suited to maintain the cultural and linguistic heritage of the minority concerned, running the risk, however, of alienating the child from the dominant groups in society, the other one will be more efficient in securing equal opportunities for minority children within the cultural and political system shaped by the majority although it may easily end up in an outright strategy of assimilation by destruction. It would appear, therefore, that given changing circumstances a delicate balance has to be struck. The present writer feels that no clear-cut and uniform answers can be given. Only one thing is certain, namely that effective opportunities must be given to children belonging to a minority to acquaint themselves fully with the language and culture of their ancestors. It is of secondary importance how this result is arrived at. If States do not permit, at the elementary school level, the creation of establishments by the minority concerned itself [91], they

---

[90] The *Capotorti* report (*supra* note 6), pp. 37–38 paras. 218–224, pp. 99–100 paras. 591–599, relies too much on Government support without developing sufficiently the autonomous cultural rights of ethnic minorities.

[91] Under Art. 13 (3) CESCR, no such prohibition would be permissible.

at least are duty-bound to allow for sufficient programs to be taught within the framework of the general school system. To impede any organized and systematic language education outside the parental home would be tantamount to pursuing eradication policies and would thus constitute a denial in the sense envisaged by Art. 27. For a minority language to remain a living social network, it needs support from other sources than parents, relatives and friends. In particular, the capabilities of expression in writing would in no way be developed. Generally, such discriminated languages are necessarily condemned to go through a progressive process of degradation. A somewhat more differentiated answer than normally provided is therefore called for in response to the introductory question of this section [92].

In order to dispel any misunderstandings, it may be repeated that claims to subsidization may not be directly based on Art. 27. Only if other groups of the population receive funds for their cultural purposes must minorities be treated alike.

c) Nor does Art. 27 require States to admit minority languages as an official language before courts and administrative agencies [93]. Again, this is not a new issue. Already the very first draft of an International Bill of Human Rights had provided (Art. 46) that persons belonging to a minority should have the right »to use their own language before the courts and other authorities and organs of the State« [94]. Later, a USSR proposal for insertion of a new article on minorities in the Universal Declaration on Human Rights suggested that everyone should have the right to use his tongue »in the courts and in other official premises« [95]. If the present text does not mention the relationship between the individual and the State, this omission, therefore, is not due to forgetfulness or ignorance of the problem.

For the sake of completeness it should be observed, however, that in extreme cases the guarantees enshrined in Art. 14 should apply. An accused person has a right to be assisted by an interpreter »if he cannot understand or speak the language used in court« (para. 3 (f)). This guarantee does not depend on the nationality of the accused. Its beneficiaries are also nationals, irrespective of whether they are supposed to be able to understand and to speak

---

[92] During the drafting process, a right of minorities to set up their own schools was affirmed in relation to Art. 27 by the Philippines, UN Doc. E/CN.4/SR.370, p. 9. In the same sense: *Dinstein,* Cultural Rights, 9 Israel Yearbook on Human Rights 58–81, at 68 (1979); *Sohn (supra* note 21), p. 284. Doubts have been expressed by *Daes (supra* note 25), p. 62; *Modeen (supra* note 5), p. 108. A negative conclusion is suggested by *M. Lebel,* Le choix de la langue d'enseignement et le droit international, 9 Revue juridique Thémis 221–248, at 236–237; *Pircher (supra* note 37), pp. 222–223.

[93] Compare also *Daes (supra* note 25), p. 63; *Modeen (supra* note 5), p. 108; *Pircher (supra* note 37), p. 222; *Tabory (supra* note 47), p. 183; the only voice to the contrary is *Sohn (supra* note 21), p. 284.

[94] *Supra* note 22.

[95] UN Doc. A/800, contained in UN Doc. A/C.3/308/Rev. 2, GAOR 3rd Sess. 1st part, 3rd Committee, Annexes, p. 45.

the main national language. Concerning private law suits, Art. 14 does not explicitly mention an analogous right. However, there could obviously be no question of a »fair hearing« in the sense determined by Art. 14 (1) (1) if proceedings were held in a language inaccessible to one of the parties without translation or interpretation. Just to mention an example: Actions to title for land initiated against Indian tribes in an American country could not be concluded validly if the significance of such court proceedings would not be grasped by the defendants due to linguistic difficulties.

6. In principle, the rights under the Covenant are directed against State interference. Public authorities should refrain from intruding into the personal sphere of individual citizens. This is essentially also true of Art. 27. Indeed, the main threats to minorities can be traced to action undertaken by State authorities. However, the existence and cultural autonomy of minorities may also be jeopardized by societal forces. It is a matter of common knowledge, for instance, that Indian communities almost everywhere in the Western hemisphere have suffered great harm from the inimical, if not outright genocidal activities of land-seeking settlers. The CCPR is well aware of such instances where human rights are not threatened by public power, but by uncontrolled groups in the society. The most illustrative example in point is Art. 6 (1) with its firm ruling that the right to life »shall be protected by law«, a proposition supplemented by the general precept that States are duty-bound not only to respect, but also to »ensure« the rights under the CCPR (Art. 2 (1)). In its latest report to the General Assembly the HRCee has underscored the importance of this particular aspect of the commitments under the CCPR with particular regard to the principle of equality of sexes [96], stating, however, that the obligation to ensure the enjoyment of rights »relates to all rights set forth in the Covenant« [97]. Although one may nurture legitimate doubts as to the actual scope of this new dimension of obligations undertaken by virtue of the CCPR, it appears to be obvious that Art. 27 denotes something of extreme fragility which, in the same vein as Art. 6 or Arts. 7 and 8, deserves protection against any kind of outside interference. States are thus required to protect minorities also against threats originating from private individuals.

7. Finally, it should be pointed out that minority rights under Art. 27 have to be distinguished from the right of self-determination which, pursuing to Art. 1, is accorded to »peoples«. In the view of the drafters there was a clear dividing-line between minorities, which had to respect the existing framework of the States in which they were living, and peoples, which could assert claims to establish their own State [98]. This fundamental divergency has

[96] General Comment 4/13, para. 2, [Fifth] Report of the Human Rights Committee, GAOR 36th Sess. Supp. 40, p. 109.

[97] General Comment 3/13, para. 1 (*supra* note 96).

[98] Thus, in the Third Committee of the General Assembly, the duty of loyalty incumbent on minorities was stressed by several delegations: Iraq, UN Doc. A/C.3/

furthermore found an explicit and clear expression in the text of the CCPR itself. Its logical implication is that Art. 27 cannot be invoked with a view to advancing claims for national independence[99].

There exists, however, a point of intersection between the two provisions. Minorities have been defined here, in consonance with the *travaux préparatoires,* as »separate or distinct groups, well-defined and long-established on the territory of a State«. This definition greatly reduces the number of groups that qualify as beneficaries of the rights provided for in Art. 27. At the same time, it brings them close to the characteristics possessed by a people under Art. 1[100]. Since in any event it is highly controversial what distinctive criteria make a community a people in the sense of Art. 1[101], the status of a minority may thus easily become a kind of prelude to asserting rights of self-determination, all the more so since a people would doubtless be entitled to content itself with claiming the rights under Art. 27. It is this fluidity between Arts. 1 and 27 which makes many States reluctant to grant *bona fide* the rights set forth in Art. 27. In fact, it is a historical experience that minorities, once they have attained a status of cultural autonomy, frequently attempt to proceed to realizing political independence[102].

Again, such aspirations are mostly quite understandable in view of the general attitude of intolerance which is still the dominant feature of the nation-State based on a dominant State-nation[103]. Majorities can only expect that degree of loyalty which they themselves are prepared to manifest vis-à-vis minorities living among them. Loyalty is a reciprocal bond of trust and

---

SR.1104, p. 219 para. 6; Pakistan, *ibid.,* p. 220 para. 17; Saudi Arabia, *ibid.,* p. 221 para. 32; United Arab Republic, *ibid.,* p. 222 para. 51. Beforehand, in UN Doc. A/2929 *(supra* note 27), p. 63 para. 186, which reflects the earlier discussions, it had been stressed that Art. 27 (then Art. 25) should not encourage disruptive tendencies »dangerous for the unity of the State«.

[99] See the *Capotorti* report (*supra* note 6), p. 98 para. 581, and *Capotorti* (*supra* note 17), pp. 21, 25; *A. Cassese,* The Self-Determination of Peoples, in: L. Henkin (ed.), The International Bill of Rights (1981), pp. 92–113, at 96; *Münch (supra* note 5), p. 67; *Pircher (supra* note 37), p. 219; *Sohn (supra* note 21), p. 276.

[100] According to views expressed by *Th. Veiter,* Nationalitätenkonflikt und Volksgruppenrecht im 20. Jahrhundert (1977), pp. 160–188, almost every ethnic and linguistic minority would constitute a people in the sense of Art. 1.

[101] It has been asserted, for instance, that Indian tribes in North America are holders of the right of self-determination: *J. H. Clinebell/J. Thomson,* Sovereignty and Self-Determination: The Rights of Native Americans under International Law, 27 Buffalo Law Review 669–714 (1978).

[102] Noted also by *Thornberry (supra* note 36), p. 456.

[103] It is certainly not unreasonable to conclude that a minority, which is systematically being denied its basic rights of existence, acquires a right of self-determination as a defense of last resort, see *L. Condorelli,* Droit des minorités et garantie des droits des peuples: les risques de la Déclaration, in: A. Cassese/E. Jouve (eds.), Pour un droit des peuples (1978), pp. 129–135, at 131; *M. Sornarajah,* International Colonialism and Humanitarian Intervention, 11 Georgia Journal of International and Comparative Law 45–77 (1981); *White* (*supra* note 84), p. 168.

solidarity[104] and should not be confounded with the duty to obey a unilateral *diktat*. In no way, however, can this general political background detract from the legal substance of Art. 27.

IV. Limitations

The text of Art. 27 does not mention any limitation clause. Some authors have drawn the conclusion from this omission that the rights provided to members of minorities may not be restricted by national legislative acts[105]. But the results of a purely textual interpretation become less convincing when scrutinized in the light of a more systematic approach.

Concerning the rights of religious minorities, it has been rightly observed during the different drafting stages that Art. 27 runs parallel to Art. 18 and that, consequently, freedom to manifest one's religion should be liable to the conditions laid down in Art. 18 (3), irrespective of whether the individuals concerned invoke one or the other rule as the basis of their freedom[106]. It could hardly make sense to assert that Art. 27 constitutes »harder« law than Art. 18 although in substance both provisions refer to one and the same type of religious activity.

Inherent limitations of the right to enjoy one's own culture may mainly flow from Art. 5. Traditional cultures are not accepted as such, also with their negative and perhaps even abhorrent sides, but only to the extent that they fit into the value-system of the CCPR. Thus, for instance, sacrificing human lives in accordance with archaic rites would not be condoned by the CCPR. It is up to the national authorities to enact legislation designed to prevent such customs which negate the rights of other individuals.

As far as linguistic communities are concerned, it is less less evident that there may exist instances in which restrictions of the right to use one's own language vis-à-vis the other members of the group would prove legitimate. Generally, rights under the CCPR may be limited only in order to combat clear and present dangers to the common good or to the rights of others[107].

---

[104] Rightly underscored by *Veiter (supra* note 100), p. 106. Compare also the forceful criticism of a duty of loyalty by *Brügel (supra* note 5), pp. 428–429.

[105] *Pircher (supra* note 37), p. 221; *A. Fenet,* Un regrettable pas en arrière: Le concept de peuple minoritaire dans la Déclaration universelle des droits des peuples, in: Peuples et Etats du Tiers Monde (1978), pp. 109–130, at 123; *contra: Sohn (supra* note 21), p. 285.

[106] Netherlands, UN Doc. A/C.3/SR.1103, p. 214 para. 17; China, *ibid.,* p. 214 para. 24; India, *ibid.,* p. 215 para. 40; Belgium, *ibid.,* p. 216 para. 47; Turkey, *ibid.,* SR.1104, p. 220 para. 20.

[107] For a clear expression of this basic principle see the *Landinelli Silva* case decided by the HRCee in 1981 *(supra* note 96), pp. 130–133, at 133 para. 8.4; see also the conclusions reached by *A. Ch. Kiss,* Permissible Limitations on Rights, in: L. Henkin (ed.), The International Bill of Rights (1981), pp. 290–310, at 308.

One fails to see, however, that the use of a minority language solely could ever amount to such a threat. Frequently, linguistic minorities are seen as undermining national unity. But if indeed loyalty is not maintained towards the State and if the existence of the national community is being questioned, such activities have nothing to do with the language employed by dissidents. It is obvious that language may in such circumstances become a symbol. Nonetheless, to prohibit the use of a minority tongue would even in these circumstances exceed the limits of what the CCPR allows in the light of its basic principle of proportionality.

V. International Implementation

For reasons of completeness, a short glance may be thrown on recent efforts to implement Art. 27 by a Declaration to be finally adopted by the General Assembly. The *Capotorti* report had recommended the preparation of such a declaration with a view to defining in more elaborate detail the consequences flowing from Art. 27, since this provision required careful interpretation in order to be fully understood [108]. In fact, accepting this proposal, the Sub-Commission recommended the drafting of such an instrument at its thirteenth session in 1977 [109]. After the HRCion had endorsed this proposal [110], work started on the basis of a Yugoslav draft [111]. Proceedings are still continuing within the HRCion which has not even reached agreement on the preamble where many key words as yet are contained between square brackets [112].

If the original intention had been to concretize the substantive contents of Art. 27, it is manifest that this line has been abandoned. Divergencies are already present over the title itself. While Art. 27 only refers to ethnic, religious and linguistic minorities, a new current of opinion suggests the additional inclusion of national minorities [113]. Furthermore, many of the draft articles do not relate the relevant rights to »persons belonging to . . . minorities«, but to minorities themselves [114]. But the most significant change has been introduced regarding the nature and the substance of the rights at stake.

---

[108] *Loc. cit. (supra* note 6), p. 102 para. 617. The same view has been expressed by *Dinstein (supra* note 5), p. 118; *Thornberry (supra* note 36), p. 450.

[109] Res. 5 (XXX) of 31 August 1977, UN Doc. E/CN. 4/1261, p. 45.

[110] Res. 14 A (XXXIV) of 6 March 1978, ESCOR 1978 Suppl. 4, p. 119.

[111] UN Docs. E/CN.4/1261, p. 45; E/CN.4/L.1367/Rev. 1; revised version: E/CN.4/Sub. 2/L.734.

[112] The latest document available is a report by a working-group of the HRCion, UN Doc. E/CN.4/L.1579.

[113] UN Doc. E/CN.4/L.1579, Annex; pertinent criticism by *Capotorti (supra* note 15), p. 116.

[114] The dangers of such an orientation have been eloquently underlined by *Capotorti (supra* note 15), pp. 114–116.

While Art. 27 grants to members of minorities freedom to preserve and develop their own cultural identity, it is now intended to impose on States the burden of promoting the characteristics of the groups concerned (Art. 1 of the draft)[115]. No prophetical gifts are needed to forecast that such a considerable extension of the obligations incumbent upon States will either be rejected – as exceeding by far the actual capabilities of Governments – or will simply be accepted with the mental reservation not to take those new commitments too seriously. For that reason alone, it must be feared that instead of strengthening Art. 27, the adoption of the draft declaration would on the contrary adversely affect its binding substance.

In other parts the draft declaration ostensibly transgresses the boundary line between binding law and pure policy statements. Its Art. 3 (2) sets forth that »it is essential« (sic) to create favourable conditions for the development of minorities. In addition, Art. 4 of the draft contains propositions which to some extent are factual statements, while the remainder part is couched in hortatory language. On the other hand, it can hardly be seen that the contours of Art. 27 would become any more precise through the adoption of the draft declaration. In particular, no attempt has been made to specify what the consequences of the right to use one's own language are in practice[116]. It appears doubtful, therefore, whether any meaningful progress would be brought about[117]. To change this situation, a tremendous and deliberate effort would be required in order to do away with cloudy terminology and come down instead to the real and difficult issues. In its present version, the draft declaration seems more an exercise in legal actionism than a contribution to enhancing real enjoyment of human rights.

VI. Outlook

Many authors have complained that Art. 27 does not provide sufficient protection to minorities. In particular, it has been criticized that no account is taken of minorities as such, Art. 27 granting only individual and not collective rights[118].

---

[115] All references are to the latest version contained in UN Doc. E/CN.4/Sub.2/L.734.

[116] Exemplary in this respect is Art. 6 of the recent Declaration on the Elimination of All Forms of Intolerance and of Discrimination Based on Religion or Belief, adopted by the General Assembly on 25 November 1981, UN Doc. A/RES/36/35.

[117] The same negative evaluation is made by *Capotorti (supra* note 15), pp. 116–117. More positive is a Yugoslav voice, *B. Vukas,* La proposition yugoslave d'une Déclaration sur les droits des minorités nationales, ethniques et religieuses, 25 AFDI, 281–294 (1979); similarly, *Sohn (supra* note 21), p. 289, and *Thornberry (supra* note 36), p. 455, strike a hopeful note.

[118] Compare, in particular, *Kimminich (supra* note 40), p. 47; *Pernthaler (supra* note 17), pp. 30–31; *Veiter (supra* note 100), p. 139.

The present author does not share those criticisms. Collective rights mostly are affected by an inherent deficiency in that they place the group over and above the individual[119]. Responsible citizens, however, need no guardian who takes care of their rights. Fundamental rights and freedoms are a matter of personal concern where, in principle, representation has no legitimacy[120]. On the other hand, no serious defects can be perceived. Since each and every person belonging to a minority is endowed with the rights under Art. 27, their joint action is perfectly lawful – subject to the condition only to respect, within the minority, everyone's personal autonomy. It certainly cannot be denied that minorities need common institutions to defend their rights and interests. But to provide for the establishment of such mechanisms designed to shape the collective will of minorities cannot be attempted in an instrument of potential world-wide application.

In addition, the fact that Art. 27 is related to the individual components of a minority brings it within the scope of the Optional Protocol, whereas collective rights like the right to self-determination are not capable of being asserted through the channel of proceedings for the examination of individual communications. The availability of the Optional Protocol should in no way be underrated. The decision of the HRCee in the *Lovelace* case affords an illustrative example of how far the legal consequences of Art. 27 can reach. In the perspective of procedural efficiency the present attempts to shift emphasis more on the collective aspects of minority protection could lead to an improvement of the real situation of minorities if it were envisaged to set up a specific monitoring body modelled upon the example of the Committee of 24 whose task it was to bring to an end colonialism in all its forms. No such intention has emerged, however, in the course of the deliberations in the Sub-Commission and the HRCion.

Protection of minorities is not a technical matter which can be successfully accomplished simply by drawing up carefully worded texts. More often than not, relationships between a leading national group and existing minorities are marred by tensions which go far back into the past. Due to a vicious circle of negative mutual experiences, majorities distrust the loyalty of minorties, while minorities resent the intolerant way majorities have oppressed themselves or their ancestors. Almost everywhere in the world, a considerable amount of confidence-building is required before Art. 27 can take its full effect.

---

[119] This is also acknowledged by *Pernthaler (supra* note 17), pp. 29–30.

[120] The present writer agrees with *Capotorti (supra* note 15), pp. 114–115, and *supra* note 17, p. 19.

La Monarquía hispánica de la Casa de Austria como forma de Estado*

Antonio TRUYOL SERRA

El Estado moderno, que como es sabido va apareciendo en la baja Edad Media y se consolida a partir de la época del Renacimiento y de la Reforma protestante, se presenta como un fenómeno histórico diferenciado en la evolución política de la Europa occidental, con rasgos típicos comunes entre los cuales cabe recordar la centralización del poder en una instancia territorial suprema (»soberana«) que monopliza el uso legítimo de la fuerza frente al pluralismo de poderes de la sociedad feudal y no reconoce superior, la constitución de un cuerpo de funcionarios técnicamente cualificados y de un ejército permanente al servicio de dicha instancia central, la asunción por ésta de un poder legislativo propio con un campo de acción cada vez mayor.

Sin embargo, en el marco de estos rasgos comunes, la evolución en cuestión presenta peculiaridades de un país a otro o de una región a otra del continente. Si el Estado, a lo largo del Antiguo Régimen, es esencialmente de carácter monárquico, no faltan formas republicanas, aristocráticas u oligárquicas (Venecia, Génova, cantones suizos). Pero las mismas monarquías se diversifican según su mayor o menor grado de centralización, su carácter meramente dinástico-territorial (Alemania, Italia) o ya dinástico-nacional (Francia, Inglaterra, Escocia, Portugal, Polonia) o, como veremos, dinástico-supranacional (España, Austria).

El objeto que hoy nos ocupa va a consistir en fijarnos ante todo en las peculiaridades de la monarquía española, tanto en el orden institucional como en el doctrinal, especialmente en sus diferencias con respecto a la monarquía francesa; diferencias que, si por un lado resultan de su respectivo pasado, han gravitado a su vez decisivamente sobre su respectivo futuro, hasta hoy.

---

* Este artículo se basa esencialmente en una conferencia dada bajo el título »Dos conceptos de la monarquía en el siglo XVII: España y Francia«, el 22 de febrero de 1980, en la Biblioteca Española de París, dentro del ciclo »Les relations entre l'Espagne et la France au cours de l'histoire: le XVIIe siècle«, organizado por el entonces agregado cultural de la Emabajada don Gonzalo Puente Ojea.

En Francia, como en el Península Ibérica (y empleamos aquí esta última expresión con toda intención, en lugar de »España«), se da en las respectivas realezas lo que caracteriza el Estado moderno. En la monarquía de Isabel de Castilla y la de Fernando de Aragón como en la de Carlos VIII y Luis XII, en la de Francisco I y de Enrique II como en la de Carlos de Gante y los Felipes en los siglos XVI y XVII, encontramos la misma lucha (llena de vicisitudes) contra las fuerzas feudales, especialmente nobiliarias, la misma tendencia a fortalecer el poder real, apoyado en un cuerpo de legistas o letrados, una burocracia y unas fuerzas armadas propias.

Pero esta lucha y esta tendencia se desarrollaron desde un principio en contextos muy diferentes, sin la consideración de los cuales no puede comprenderse la configuración de la monarquía en uno y otro Estado en los siglos XVI y XVII.

La diferencia fundamental, en la que hemos de hacer hincapié, por ser el quicio sobre el cual gira el concepto mismo de monarquía en los dos países, es el siguiente. Acabamos de hablar de Francia y de la Península Ibérica, y lo hemos hecho, porque en Francia hay un reino y un rey de Francia, mientras que en la Península Ibérica, la *Hispania* medieval, hay varios reinos y reyes; con la peculiaridad de que los reyes pueden serlo de uno o más reinos a la vez, que en este caso no eran siempre los mismos y ni siquiera estaban siempre situados todos en la Península.

Se trata de un hecho harto conocido, pero que quisiéramos examinar especialmente desde el punto de vista jurídico-constitucional – el único que nos explica su trascendencia, no siempre debidamente valorada, y que por lo demás sigue actuando sobre aspectos significativos de la actualidad española de los días que vivimos.

Francia como reino nació de una decisión política concreta, casi podríamos decir de una vez, con la división del Imperio franco entre los hijos de Luis el Piadoso (Tratado de Verdún, 843), como »Francia occidental«, lo mismo que Alemania (»Francia oriental«), junto a una zona intermedia cuyo núcleo fuera la Lotaringia e Italia.

Desde un principio no hubo sino un reino de Francia y un rey de Francia. El poder de este rey pudo ser modesto y encontrar resistencias en los grandes señores feudales; su persona pudo, en algún momento de crisis, resultar cuestionable. Pero la realeza como tal fue única.

Y no vamos a insistir en la tantas veces subrayada continuidad de la acción de los Capetos en la dirección de un progresivo acrecentamiento y fortalecimiento del *domaine royal* y de las prerrogativas de la realeza, que logró aglutinar en torno a ella a la nación, cuya unidad expresara (Felipe II Augusto; Luis IX (San Luis); Felipe IV el Hermoso; Luis XI).

Especialmente característico de esta evolución es el hecho de que los reyes de Francia rodeasen su dinastía, su corona y su persona de un halo religioso del que nunca participarían los reyes españoles. Si Luis XI se hizo reconocer por la Sante Sede el título de *Roi très chrétien*, ello significó algo más que un título honorífico: elevaba al rey a un rango superior, que encontraba en el *sacre* y en los poderes taumatúrgicos de que se le suponía provisto su manifestación más singular y asombrosamente duradera, hasta entrados los tiempos modernos.

En la Península Ibérica, en cambio, la invasión musulmana destruyó la unidad (frágil de suyo) de la monarquía visigoda. Sólo sobrevivieron del naufragio, en las cordilleras del norte, núcleos cristianos separados y autónomos, que pronto iniciarían trayectorias paralelas en cuanto entes políticos independientes: el reino de Asturias (718), que después de la repoblación de León (856) se transformaría en el reino astur-leonés o simplemente leonés; los condados pirenáicos del este, que se integraban en la *Marca Hispanica* del Imperio carolingio y alcanzaron una independencia de hecho en los siglos IX–X, y de los que surgiría, en torno al condado de Barcelona, el principado de Cataluña; entre ambos complejos, Navarra, convertida en reino en 905. Del reino de León se desgajó el condado de Castilla, erigido en reino en 1035, mientras nacía, el mismo año, entre Navarra y los condados de Cataluña, el reino de Aragón. De una ulterior secesión emergió en el extremo oeste el reino de Portugal (1137–43).

A diferencia de lo que ocurriera en Francia, tenemos, pues, en España, desde el principio, una pluralidad de reinos y principados distintos e independientes, debiendo mencionarse al lado de los indicados los señoríos de Guipúzcoa, Alava y Vizcaya.

Entre unas entidades y otras se produjeron uniones dinásticas, unas variables y esporádicas, y otras duraderas. Entre éstas hay que señalar la de los señoríos de Vizcaya (1179) y de Guipúzcoa y Alava (a partir de 1200) con Castilla, sobre la base de la conservación de un régimen jurídico peculiar que duraría hasta el siglo XIX y en parte hasta el XX; pero debemos mencionar sobre todo dos uniones que dejarían una huella mayor en todo el futuro peninsular: la unión dinástica del reino de Aragón y de Cataluña (1137) y la de los reinos de Castilla y León (1229), constituyéndose así las Coronas de Castilla (Castilla y León) y de Aragón.

Con ello llegamos a un concepto peculiar del sistema dinástico-estatal español: el de Corona, claramente diferenciado del de Reino. Como ha escrito un destacado historiador del derecho español, »en virtud de unos u otros títulos los reyes españoles van reuniendo bajo su poder desde el siglo XVII al XVIII un conjunto cada vez más numeroso de Reinos y Principados, no sólo en la Península, sino también fuera de ella. Tales conjuntos constituyen formas políticas más amplias que las de los antiguos Reinos, que se designan en la baja Edad Media como Coronas«, o, más tarde, también Monarquía, sobre la base de que todos los Reinos y Señoríos que integran una Corona forman una unidad indivisible, que como tal se transmite hereditariamente[1]. Los ter-

---

[1] *A. García Gallo,* Manual de historia del derecho español, tomo 1 (7ª ed. revisada, Madrid 1977), §§ 1203 y 1205.

ritorios integrados en una Corona o Monarquía conservan siempre su nombre y rango, pero no siempre se hallan en la misma situación.

Esta presenta dos modalidades, propias de la Corona de Castilla y de la Corona de Aragón respectivamente.

La primera modalidad es la de »Reinos unidos«, que se dió en la Corona de Castilla desde mediados del siglo XIV, fecha a partir de la cual los Reinos se consideran unidos no sólo en cuanto reconocen a un mismo Rey, sino también en cuanto unos y otros se funden en un todo principal, y por ello todos los súbditos tienen la misma »naturaleza«, las instituciones políticas son comunes a todos ellos y todos se rigen por un mismo derecho (excepto en lo que respecta al de cada ciudad). Ese régimen común se extiende a los territorios musulmanes que se conquistan (Córdoba, Murcia, Jaén, Sevilla, Niebla, etc.) y a las islas Canarias. Unicamente Guipúzcoa y Alava, y luego Vizcaya al unirse definitivamente a la Corona en 1379, conservan un régimen propio. Este será también el caso cuando Fernando el Católico conquiste Navarra y la incoropore a la Corona de Castilla (1512–15).

En cambio, la segunda modalidad de unión, la de los »Reinos separados«, se dió en la Corona de Aragón y en Navarra desde un principio. Aquí lo que hay es una unión personal de los Reinos, que conservan plenamente su propio régimen, la propia »naturaleza« de sus habitantes, por lo cual éstos son »extranjeros« en los demás Reinos de la Corona. Lo interesante es que incluso los nuevos territorios conquistados a los musulmanes (Mallorca y Valencia) se constituyeron en Reinos distintos e independientes, con organización propia. Lo mismo ocurrió con Navarra y los territorios franceses dependientes del Rey de ésta. Y también como separados, se integraron a la Monarquía sucesivamente Sicilia, Cerdeña, Nápoles (vinculadas a la Corona de Aragón), Flandes, el Franco Condado, Milán y, más tarde, bajo Felipe II, Portugal (1580–1640)[2].

En contra de lo que a veces se piensa, el matrimonio de Isabel de Castilla y Fernando de Aragón no significó la unidad formal, constitucional, de España: fué una unión personal de las dos Coronas, una de las cuales constaba de tres Reinos y un Principado separados en la Península (Aragón, Cataluña, Mallorca, Valencia) y otros en Italia. Navarra quedó al margen hasta 1512. El segundo matrimonio de Fernando con Germana de Foix pudo perpetuar la separación provocada por la muerte de Isabel, de haber sobrevivido al monarca

---

[2] Para una consideración específica de los reinos y territorios europeos no peninsulares de la Monarquía española en el conjunto de la historia de su derecho, véase *J. M. Pérez-Prendes*, Curso de historia del derecho español, tomo 1 (Introducción y parte general) (Madrid 1978), cap. 23 A) Italia; B) Portugal; C) Países Bajos, y *R. Gibert,* Historia general del derecho español (Granada 1968); reprod. fotot. (Madrid 1978), los capítulos (no numerados) »Dominación española en los Países Bajos«, »Reino hispánico de Cerdeña«, »Dominación española en Italia«, »Reino hispánico de Sicilia«, »Reino hispánico de Nápoles«, »El Milanesado, feudo del Imperio«.

el hijo que le diera Germana. Al morir sin heredero de ella, en 1516, la unión personal (pero sólo unión personal) se hizo definitiva, al convertirse Carlos I en rey en las dos Coronas.

En este año nace así la Monarquía hispánica, que va a constituir durante dos siglos una forma política peculiar: una unión de Estados distintos y separados (Reinos y Principados) agrupados en dos Coronas (Castilla y Aragón), e incluso tres (con Portugal) durante sesenta años, estando parte de dichos Reinos, Principados y Estados (ya se les va llamando así) no ya fuera de España, sino también de Europa (las Indias, incorporadas, con peculiaridades resultantes de su lejanía y situación, a Castilla).

Unida dicha Monarquía al Imperio bajo Carlos V, se distingue de él, aunque su diferenciación sea menos aparente por la común estructura supranacional de ambos. Separados a raíz de la abdicación de Carlos, y vinculado el Imperio a la rama menor de los Austrias a partir de Fernando I, se perfila nítidamente la Monarquía hispánica bajo Felipe II y sus sucesores, alcanzando entonces no sólo su configuración singular, sino también su ápice. Iniciado su ocaso en la segunda mitad del siglo XVII, perdurará en lo esencial hasta el cambio dinástico que daría lugar a la llamada Guerra de Sucesión de España. En una palabra, la Monarquía hispánica, en el sentido estricto que aquí le damos, es la forma política clásica propia de la Casa de Austria.

En esta fase plenaria de su constitución, el Rey se convierte en Monarca, en rector de la Monarquía formada por todos sus reinos y principados, por todos sus Estados. Pero se da la situación paradójica de que, si bien el Príncipe procede como Monarca o rector de la Monarquía, sólo tiene poder en cuanto rey o príncipe de cada reino o principado, y este poder varía de uno a otro según su respectiva constitución política.

Aquí tenemos la diferencia esencial entre la noción de la Monarquía francesa y la de la Monarquía española. La primera es una monarquía nacional (sin duda el prototipo de la monarquía nacional), que se ha constituido a base de *rattachements* o *réunions* de territorios que se van fundiendo políticamente en el núcleo expansivo del reino (incluso cuando lo son de poblaciones periféricas, nacionalmente más o menos heterogéneas, como en el caso de Bretaña, Alsacia, el Rosellón o Córcega). La segunda es una monarquía supranacional (por decirlo con un término actual que nos parece adecuado), que en el límite podría ser universal y lo era potencialmente – y de ahí que también se la califique de tal: de Monarquía universal.

Esta diferencia se refleja claramente en el plano jurídico-constitucional. La realeza, en Francia como en España, se había rodeado, para las tareas de gobierno cada vez más complejas, de Consejos. Pero en la Monarquía española cada reino o corona, grupo de reinos o principados, tenía su consejo propio: a los de Castilla, de Aragón y de Navarra se sumaron los de Indias (1520–24), de Italia (1555) – cuyos asuntos quedaban así desgajados de la jurisdicción del Consejo de Aragón –, de Portugal (1582), de Flandes (1582). Ello dificultaba la acción unitaria del monarca, puesto que la resolución de un mismo asunto en los diferentes reinos tenía que tramitarse a través de los respectivos Consejos. Y para soslayar tal obstáculo, los monarcas fueron fortaleciendo viejos órganos o creando otros nuevos para una acción conjunta: así, se reforzó la posición de los Secretarios del Rey; se creó el Consejo de Estado o de la Monarquía (1520) sin intervención en el gobierno de los reinos; se establecieron los Consejos de Inquisición (1478), Cruzada (1534) y Hacienda (1593). Y la necesidad de disponer de un ejército sustraído a la legislación de cada reino, o sea, mercenario, y del dinero que ello requería, llevó a los monarcas a préstamos personales sobre la base de la explotación de los recursos americanos.

En resumidas cuentas, vemos que la Monarquía universal española se entiende en el sentido de la Monarquía de Dante, de »gobierno de uno« extensivo (al menos potencialmente) a toda la Cristiandad o gran parte de ella, para distinguirla del Sacro Imperio (alemán), de finalidades en principio análogas. Esta analogía hace que a veces los autores hablen también del Imperio español, aunque oficialmente no se use el término. Y como la Monarquía española asumió la defensa de la catolicidad frente a la Reforma y su difusión entre los infieles, se designó asimismo como »Monarquía católica«, lo cual era otra manera de decir Monarquía universal.

Una diferencia característica de la monarquía de Francia y de España, es lo relativo a la lengua.

Oigamos a un expositor autorizado sobre la actitud de la realeza francesa ante la lengua: »Le premier signe de cette réalité consciente de l'Etat monarchique est l'unité de langue. Le français, c'est essentiellement la langue du roi, celle que l'on parle dans les contrées où il emmène sa cour, Isle de France, Touraine et Poitou. Vers 1450 il est encore maltraité dans le Nord et ignoré dans le Midi. Vers 1500 il descend le Rhône et diffuse dans le Massif Central et le Languedoc. Vers 1550 il a conquis la grande majorité du territoire. Sa qualité de langue du roi est définie par le fait que les parlements de province l'emploient dans leurs actes officiels, à Toulouse dès 1444, à Bordeaux dès 1465, à Aix dès 1560. L'Ordonnance de Villiers-Cotterêts (août 1539) vient donc à son heure, comme tous les actes de nos rois, pour officialiser une situation déjà mûre, en rendant obligatoire l'usage exclusif du ›langage maternel français‹ dans tous les actes juridiques privés ou publics« [3].

La Monarquía hispánica, en cambio, era por naturaleza plurilingüe. Sus distintas partes conservaban sus lenguas, por lo general comunes a varias de ellas. Y los monarcas no sólo respetaron tal diversidad, sino que la consideraban normal. Otro autorizado conocedor de la época, Luis Díez del Corral, lo ha subrayado, recurriendo al testimonio del propio rey Felipe IV. Felipe IV, rey esteta, tradujo al castellano nada menos que la Historia de Italia de *Guic-*

[3] *P. Mesnard,* capítulo 1 del artículo »L'Etat«, Encyclopédie française, tomo X (París 1966), p. 71.

ciardini[4]. Tan significativo como el hecho en sí es la justificación que le da el monarca; no sólo aduce lo mucho que el gobernante puede aprender de los libros de los grandes historiadores, sino – y ello es lo más interesante aquí – la perfección que supone traducir »lengua tan copiosa como la italiana«, la cual es »necesario que la sepa quien posee tantos Estados en aquellas provincias«. Y añade el profesor Díez del Corral: »No sólo el italiano aprendió el Monarca, sino también las lenguas de España: la aragonesa, catalana y portuguesa; y también la francesa, para poder comunicarse con vasallos tan estimados como los de Flandes, pensando hacerlo con los germánicos a través del italiano, lengua calificada por el monarca en su prólogo como la más usada y casi vulgar en Alemania y en todos los Estados hereditarios de ella, que por tantos títulos y tantas razones de sangre y públicas me tocan«[5].

No podemos dejar de suscribir su comentario cuando resalta que el esfuerzo lingüístico de Felipe IV es muy significativo para apreciar lo que en su estimación y en su esencia era la Monarquía de España, y también cuando añade que »todo esto debe ser recordado para llegar a comprender lo que era la pluralista y tolerante [yo precisaría: en este aspecto; pues en el religioso no lo era, como tampoco lo era propiamente la generalidad de los Estados de entonces] Monarquía de los Austrias españoles, y lo que son, al presente, los regionalismos lingüísticos de nuestra península«[6]. Tal política, unida a la relativa a las franquicias tradicionales, explica, entre otras cosas, la fidelidad del Franco Condado a la Monarquía de España, y el encarnizamiento de su resistencia a las tropas de Luis XIV, que sin embargo hablaban su misma lengua.

A decir verdad, la tendencia de las monarquías de la época al absolutismo no fué ajena a la Monarquía de España. Y ello explica que con el tiempo creciese en ella la importancia de Castilla. Porque ya hemos visto que en la Corona de Castilla, el reino resultante de la unión con León y de las ulteriores conquistas (salvo en el caso de Navarra) se había realizado en cierto modo según una línea que recuerda la del reino de Francia. Y es un hecho que en Castilla la Monarquía encontró (tras conflictos iniciales) más facilidades para su acción, sobre todo la exterior. Finalmente, Castilla hizo suya la causa pancatólica de la Monarquía hispánica. De ahí el deseo de los monarcas de tomarla como modelo. Las dificultades de los tiempos de la guerra de los Treinta años movieron al conde-duque de Olivares a proponer a Felipe IV incrementar la centralización del poder en el célebre memorial del que son estas frases: »Tenga V. M. por el negocio más importante de su Monarquía el hacerse rey de España: quiero decir, Señor, que no se contente V. M. con ser rey de Portu-

[4] Historia de Italia, por *Francesco Guicciardini*, traducida del italiano con la vida del autor, por D. Felipe IV, Rey de España, en la »Biblioteca clásica«, 6 tomos (Madrid 1889–1890).

[5] La Monarquía hispánica en el pensamiento político europeo. De Maquiavelo a Humboldt (Madrid 1975), pp. 14–15.

[6] *Ibid.*, p. 15.

gal, de Aragón, de Valencia, conde de Barcelona, sino que trabaje y piense, con consejo mudado y secreto, por reducir estos reynos de que se compone España, al estilo y leyes de Castilla sin ninguna diferencia, que si V. M. lo alcanza, será el Príncipe más poderoso del Mundo«.

Como dice un autor, »el consejo de Olivares es bien claro: lo que él intenta es suprimir la personalidad jurídica y política de los reinos, conventir esa plural y compleja realidad de los reinos y naciones (en el sentido de entonces) de España, en una realidad jurídica y política homogénea y unificada sobre la base del derecho de Castilla (. . .)« [7].

Es bien sabido que los esfuerzos de Olivares en esta dirección llevaron al fracaso. El año de 1640 fue el del levantamiento de Cataluña y de Portugal, que conduciría a la secesión definitiva de este último. Hará falta el advenimiento de los Borbones en la persona de Felipe V, y la guerra de Sucesión de España, para que el sueño del Conde-Duque alcance realidad. Pero esta realidad tuvo la forma de medidas en el fondo vindicativas. Alegando que los reinos de la Corona de Aragón habían roto el vínculo de fidelidad que las ligaba al Monarca (pues habían reconocido al nuevo rey, tomando luego partido por el archiduque Carlos), Felipe V llevó a cabo, en cuatro decretos escalonados entre 1707 y 1716, una revolución constitucional por la que quedaba abolido en su totalidad el derecho particular, público y privado, del reino de Valencia, y el derecho público de Aragón, Mallorca y Cataluña, pero quedando detenido el derecho privado subsistente en su estado de entonces. Se impuso el castellano en la administración. Y lo más importante: lo que subsistía del régimen anterior, se entendía de »nuevo establecido« (Decretos »de nueva planta«).

Que se trataba de medidas discriminatorias se pone de manifiesto por el hecho de que en relación con Navarra, Alava, Guipúzcoa y Vizcaya no se impuso ninguna disposición legal unificadora, quedando en vigor sus respectivos regímenes.

Por otra parte, tras la pérdida del Franco Condado (paz de Nimega, 1678), de los reinos y principados italianos y de los Países Bajos (paces de Utrecht y Rastatt), la Monarquía española quedaba reducida a lo que es hoy España (menos temporalmente Menorca), además de las Indias.

Con ello llegaba a su fin la Monarquía de España como forma peculiar de sociedad política. El cambio operado ha sido muy certeramente caracterizado por el profesor *García Gallo* en esos términos: »(. . .) Ahora, al equipararse a Castilla los Reinos de Aragón, Valencia y Mallorca y el Principado de Cataluña – desgajados ya los territorios europeos (en las paces de Utrecht y Rastatt) –, toda la *Monarquía* queda fundida y el sistema político de Castilla, al extenderse a los otros territorios, se identifica con el de la *Monarquía.* Se sigue dando el nombre de *Reinos* a las regiones – como ya antes se hacía con las de la Corona de Castilla –, pero estos Reinos no tienen significación política. Polí-

---

[7] *F. Tomás y Valiente,* Manual de historia del derecho español (Madrid 1979), p. 238.

tica y jurídicamente, los Reinos desaparecen y sólo queda, triunfante, la *Monarquía*; aquéllos se convierten en meras *provincias*. Todos los súbditos de ella tienen la misma *naturaleza* o nacionalidad y los mismos derechos y obligaciones. El poder real es el mismo en todas partes y las Cortes no se reúnen por Reinos, sino de toda la *Monarquía*. Los *secretarios del despacho* o ministros tampoco se distribuyen su competencia por Reinos, sino por asuntos (exteriores, eclesiásticos, guerra, etc.). El *Consejo Real de Castilla* actúa sobre todos los Reinos y desaparecen los anteriores de éstos. Sólo la lengua castellana es la oficial. Se intensifica la aplicación y el estudio del Derecho nacional« [8].

Esta descripción nos muestra a la Monarquía española alineada con la francesa, por cierto bajo un rey de la misma familia que gobernaba a Francia. Una peculiaridad política hispana había dejado de existir, pero sin por ello dejar de gravitar sobre el futuro en la memoria colectiva [9].

Sólo nos queda dedicar unas palabras finales al impacto de la Monarquía de Francia y la Monarquía de España sobre el pensamiento de la época.

Aquí también observamos una diferencia entre ambas. En Francia la elaboración doctrinal del concepto de la monarquía prepara la evolución de la misma o corre pareja con ella.

El punto de referencia más significativo es *Jean Bodin*, cuyos Six livres de la République (1576) habían perfilado con tal relieve la *monarchie royale*. La *monarchie royale* de *Bodino* no es todavía la monarquía absoluta. Sobre el supuesto de la noción de soberanía, se caracteriza por el hecho de que la autoridad se concentra cada vez más en manos del Rey, pero queda limitada por el derecho, los estamentos y cuerpos constituidos, y los consejos.

El sesgo absolutista se produce después, por ejemplo con *Guy Coquille* (Institution du droit des Français, 1603), quien desliga al monarca de su Consejo y proclama en cierto modo la soledad de su poder: »*Le roi est monarque et n'a pas de compagnon en sa majesté royale*«; más todavía con *Cardin Le Bret* (De la souveraineté du Roi, 1632), en quien la soberanía, más que la forma de la república, tiende a convertirse en la propiedad del monarca, lo que permitiría identificar el soberano al Estado. Por otra parte, *Le Bret* afirma la procedencia directa del poder de los Reyes de Dios, que les hace sus representantes en la tierra. También *Bossuet* (Politique tirée de l'Ecriture Sainte) está en la línea del absolutismo monárquico, si bien busca al mismo un contrapeso moral en

---

[8] Obra citada, tomo 1, § 1232 (las cursivas son del autor).

[9] Merece señalarse que con ocasión de la expedición a Cerdeña promovida por Alberoni en 1717, Felipe V promulgó también para la isla un »decreto de nueva planta« con fecha 24 de noviembre del mismo año, desarrollado por una real cédula de 16 de febrero de 1719: disposiciones que instauraban una organización semejante en lo fundamental a la catalana (también se establecía la obligatoriedad del castellano en las causas y pleitos de la Audiencia); pero ya no hubo lugar a su aplicación y cumplido desarrollo. Véase sobre el particular *J. L. Bermejo Cabrero,* Un Decreto más de nueva planta, Revista del Departamento de Derecho político, Universidad Nacional de Educación a Distancia, Madrid, n°5 (1979–1980), pp. 129–144.

la conciencia del rey. El galicamismo era otro factor coadyuvante en esta dirección.

La relación entre la Monarquía hispánica y la doctrina política es muy distinta, por la estructura peculiar de ésta.

La doctrina política española se ocupa preferentemente, bajo la influencia de la tradición medieval, de los deberes y la función del rey. El pensamiento escolástico, que en España revive con tanto vigor en el siglo XVI y a comienzos del XVII, se refiere a la república en general. En ella el titular del poder es la comunidad en su conjunto, que lo entregó al rey, el cual sólo puede ser desposeído del mismo si se convierte en tirano. De ahí la importancia del tema (por su incidencia en las monarquías no católicas) en estos autores, principalmente en *Juan de Mariana.* La admisión de un poder indirecto del Papa en lo temporal contribuía asimismo a una línea antiabsolutista, mitigada en cuanto a su efecto interno por el carácter católico de los reyes españoles.

Estas discusiones doctrinales no guardan relación con la Monarquía española *stricto sensu.* Tampoco es del todo específica al respecto la reflexión de los tratadistas políticos, que se ocupan ante todo de la *»razón de Estado«,* o sea, de la relación entre los medios y los fines en el gobierno de los Estados. Hablan de »monarquía« como gobierno del mundo, sobre la base de la distinción entre rey y monarca, recibida de la baja Edad Media. Hablan también de Imperio indistintamente; de una Monarquía y un Imperio al servicio de los ideales de la Contrarreforma.

Sin embargo, la Monarquía Hispánica no podía dejar de suscitar la reflexión de los teóricos del Estado. Pero fué esencialmente fuera de España. *Díez del Corral* ha subrayado el hecho, estudiando tal reflexión, precursora en *Maquiavelo* (quien señaló con claridad la distinta génesis de la monarquía en Francia y en España) y en *Guicciardini,* actual en *Campanella,* e histórica en *Montesquieu* y en *Guillermo de Humboldt.*

Entre estos pensadores, sin duda el más significativo, por contemporáneo y por la verdadera obsesión con que una y otra vez vuelve sobre el tema, es *Tomás Campanella,* al que hace años dediqué algunos estudios, a los que para más detalles me remito [10].

No deja de ser paradójico que la teoría específica de la Monarquía hispánica en el momento mismo de su máximo despliegue se deba a un hombre que, acusado de participación en una conspiración contra dicha monarquía en Calabria, permaneció preso en sus cárceles napolitanas veintisiete años, entre 1599 y 1626, y que luego, tras unos años de estancia en Roma bajo la vigilancia de la Inquisición, buscaría refugio en la Francia de Richelieu, en cuya capital murió.

De hecho, la teoría campanelliana de la monarquía hispánica aparece asociada a su teoría de la monarquía universal pontificia, que intentaba dar nueva

---

[10] Trabajos refundidos en mi libro Dante y Campanella. Dos visiones de una sociedad mundal (Madrid 1968).

vigencia a la pretensión de una *plenitudo potestatis* de algunos Papas medievales y sus defensores doctrinales, los llamados curialistas. De ahí que fuera expuesta principalmente en dos obras que se complementan: El De Monarchia Hispanica (escrito hacia 1598, editado en una versión alemana en 1620 y en latín en 1640) y el De Monarchia Messiae (escrito en la cárcel en 1605, editado en 1633).

Lo esencial del pensamiento de *Campanella* en este punto consiste en su idea de una monarquía universal pontificia, de la que la monarquía hispánica sería el brazo secular. La monarquía hispánica se le presentaba en efecto como un conjunto de pueblos que se complementaban unos a otros. Esta concepción desembocaba en una interpretación providencialista de la historia, según la cual, si bien el tesón y la paciencia de los españoles les había hecho acreedores al imperio, el crecimiento rápido e inesperado de éste hasta su dimensión mundial era fruto de un superior designio, que se valió de la política matrimonial de sus monarcas y del descubrimiento del Nuevo Mundo. Y de ahí que en sus Discorsi ai principi d'Italia (escritos en 1593, pero cuya licencia de publicación denegó paradójicamente el Virrey en 1624), incitase *Campanella* a sus compatriotas a que lo apoyasen.

Es sabido que en sus últimos años, vería *Campanella* en Francia a la nación más idónea para asumir el caudillaje político al servicio del Pontificado, en su obra conocida, por las primeras palabras de su texto, como »Le monarchie delle nazioni« (escrita en París en 1635, permaneció inédita hasta el siglo XIX). Pero lo que podemos llamar el »episodio francés« del postrer *Campanella* no era capaz de romper ya el indisoluble vínculo anteriormente establecido entre *Campanella* y la Monarquía hispánica. La monarquía nacional francesa no estaba en condiciones por su misma naturaleza de asumir una empresa universal, que por lo demás, en el nuevo mundo político diversificado de los Estados soberanos, ya fueran de base nacional o de base dinástica, se revelaría finalmente inviable.

También es paradójico otro hecho. La obra de *Campanella* sobre la Monarquía española fué muy leída en Europa, incluso y acaso sobre todo por quienes temían el poderío español, buscando en el libro los arcanos de su política [11]. En cambio, su eco en España fué nulo. Sólo cabe asociar a su concepción

---

[11] Ya hemos indicado que la Monarchia di Spagna salió primero, por obra de *Christoph Besold,* en traducción alemana de 1620, reeditada en 1623. Ya el título señalaba que se trataba de poner al descubierto los manejos del Rey de España para conseguir la dominación del mundo: Von der spanischen Monarchie, erst und ander Theyl, oder aussführliches Bedencken welcher Massen von dem König in Hispanien der gantzen Weltbeherrschung, sowol insgemein als auff jedes Königreich unnd Land besonders, allerhand Anstalt zu machen (...). La edición de 1630 es todavía más explícita, refiriéndose a Alemania, Francia y especialmente los Países Bajos como objetivos de la red de la dominación española y dirigida »bienintencionadamente a todos los altoalemanes y bajoalemanes evangélicos para aviso fiel«: Spanisch Angelhacken, das ist Discours Fr. Campanellae, darinnen er dem König von Hispanien Mittel und Wege

a un autor, Fray *Juan de Salazar,* cuya »Política española« (Logroño 1619) desarrolla, con un espíritu ingenuamente triunfalista, algunos de los temas de *Campanella* en orden a la monarquía española[12]. *Salazar* estuvo en Roma y debió conocer algunos de los manuscritos de *Campanella* que circulaban, cuando escribió allí su libro. Cita una vez a *Campanella,* sin dar la referencia. Pero mientras el providencialismo campanelliano era dinámico y favorecía la acción, el de *Salazar* desemboca en un quietismo político que prefigura el que poco después invadirá la gran máquina de la monarquía misma.

A modo de conclusión, podríamos decir que no es anómalo el que la teoría de la Monarquía de España o de las Españas (como también se decía) fuese desarrollada del modo más coherente por quien era »natural« de uno de los reinos no españoles de la Monarquía. Ello corresponde en cierto modo a lo que podríamos llamar la lógica de una Monarquía que, más allá de la Península y sus archipiélagos, »disponía de súbditos en las regiones más adelantadas de Europa, dependientes de la suprema instancia regia con diversos y aborígenes títulos políticos que no eran formalmente distintos del que tenían los valencianos o los gallegos«, por servirnos de las palabras de *Díez del Corral*[13]. Si la habían servido fielmente políticos y militares como el piamontés Mercurino de Gattinara (1465–1530), Nicolás Perrenot de Granvelle (h. 1468–1550) y su hijo Antonio, el cardenal (1517–1586), del Franco Condado, el romano Alejandro Farnesio (1545–1592), el genovés Ambrosio Spínola (1569–1630) y el portugués Francisco de Melo (1597–1651), nada de extraño tenía que interpretase cabalmente su estructura y su función (por así decirlo), con el adecuado impulso barroco, el calabrés Tomás Campanella (1568–1639).

---

zeiget wie er Deutschland, Franckreich insonderheit aber Niederland angeln und in seine Gewalt bringen könne. Allen evangelischen Hoch- und Niederdeutschen zu treuer Warnung wolmeinend in deutsche Sprache versetzt. El título de una versión inglesa posterior no es menos significativo: Thomas Campanella an Italian Friar and Second Machiavel. His Advice to the King of Spain for attaining the universal Monarchy of the World (...). Translated into English by Ed. Chilmead, and published for awakening the English to prevent the approaching ruine of their Nation, With an admonitorie Preface by William Prynne of Lincolnes-Inne Esquire, Londres, sin fecha. El prefacio lleva la de 1659. Si tenemos en cuenta que es el año de la paz de los Pirineos, no resultaba muy perspicaz el traductor al valorar el peligro de la Monarquía española para su país. Desde entonces hasta la paz de Utrecht, el peligro vendría de la Francia de Luis XIV. En cuanto a la versión latina, se multiplican las ediciones a partir de 1640.

[12] Véase le edición de *M. Herrero García,* con estudio preliminar (Madrid 1945). Cf. asimismo *A. Montoro-Ballesteros,* Fray Juan de Salazar, moralista político (1619) (Madrid 1972); y en relación directa con nuestro tema: *H. Mechoulan,* Juan de Salazar lecteur de *La Monarchie Espagnole* de T. Campanella dans *La Politique Espagnole,* en: Ethno-Psychologie. Revue de Psychologie des Peuples, 1/1973, pp. 103–120.

[13] Obra citada, p. 11.

Tal fué la Monarquía hispánica como forma política peculiar, que según el autor al que acabamos de citar »no puede calificarse como medieval ni como plenamente moderna, sino como característica del Renacimiento y del primer Barroco: es decir, de la Alta Edad Moderna, que se extiende desde principios del siglo XVI a mediados del XVII«. Nosotros añadiríamos que, desde el punto de vista constitucional y el jurídico-internacional (siempre estrechamente unidos aquí, por la índole de la Monarquía hispánica), se extiende en realidad hasta comienzos del siglo XVIII. Entonces, con los Decretos de Nueva Planta y los Tratados de Utrecht y Rastatt, tuvo fin ese singular cuerpo político que fué la Monarquía hispánica o Monarquía católica, regida por la Casa de Austria y que, mientras en la Península, al extinguirse la dinastía, se transformaría, bajo la Casa de Borbón, según el esquema francés, vería rasgos esenciales suyos pasar duraderamente a la Monarquia austríaca regida por la otra rama de la Casa de Habsburgo, baluarte del catolicismo frente al Islam en los Balcanes y multinacional hasta finalmente en la época del auge de las nacionalidades. Porque, como se ha dicho desde una perspectiva de altos vuelos, »esta Casa de Austria ha dado su nombre a un imperio supranacional dirigido primero desde Toledo y a partir de Felipe II desde Madrid y que se extendió a las cuatro partes del mundo conocidas, y a la otra comunidad de pueblos del centro europeo, dirigido por Viena« [14].

Hemos tratado de caracterizar brevemente el edificio político levantado por la Casa de Austria a partir de la realidad española legada por los Reyes Católicos. Cae fuera de este empeño la actractiva tarea de analizar comparativamente el que la Casa edificara desde la realidad de sus tierras patrimoniales originarias, en esa Alemania de la que fué cabeza hasta la ascensión de Prusia y que finalmente se constituiría en Estado nacional sin ella, vinculada como estaba a una idea supranacional; una Austria que, transformada en Austria-Hungría como consecuencia de ello, vería, como la Monarquía hispánica, su Imperio amenazado tanto por la evolución del contexto internacional como la de sus tensiones internas, cuya pacífica superación constituía su específica tarea [15].

---

[14] *H. Benedikt,* Die Casa D'Austria, das Reich und Europa. Entwicklung des österreichischen Universalismus bis 1866, en la obra colectiva Spectrum Austriae, dirigida por O. Schulmeister con la colaboración de J. C. Allmayer-Beck y A. Wandruszka (Viena 1957), p. 108.

[15] Cabe remitir en parte al trabajo citado en la nota anterior y a otros dos de la misma obra: *H. Hantsch,* Der Völkerstaat an der Donau, pp. 151 ss., y *R. Kann,* Nationalitätenproblem und Nationalitätenrecht, pp. 195 ss. Sin perder de vista naturalmente las diferencias del entorno cultural y político en las respectivas épocas.

Apéndice

Cuadro cronológico de la Monarquía hispánica de los Reyes Católicos y de los Austrias desde el punto de vista histórico-constitucional

1469: matrimonio del infante Fernando de Aragón con Isabel de Castilla	
1474: Isabel I, reina de Castilla	
1476: victoria de Isabel y Fernando en Toro contra Portugal y los partidarios de la infanta Juana en Castilla	
1478: La Inquisición	
1479: Fernando V, rey de Aragón. Casa de Foix en Navarra	1479: Tratado de las Alcaçovas entre Castilla y Portugal: reconocimiento por Portugal de los derechos de Castilla sobre la islas Canarias; renuncia de Castilla a pretensiones sobre la costa de Africa y otros archipiélagos
1492: rendición de Granada. Expulsión de los judíos	1492: descubrimiento de América por Cristóbal Colón al servicio de Castilla
	1493: Tratado de Barcelona entre Fernando y Carlos VIII de Francia: devolución a Aragón del Rosellón y la Cerdaña. Bulas de Alejandro VI de delimitación de las zonas de evangelización de Castilla y Portugal en Ultramar
	1494: Tratado de Tordesillas entre Castilla y Portugal: corrección de la »línea de demarcación« pontificia en favor de Portugal
1496: matrimonio de la princesa Juana y Felipe el Hermoso	
1500: Juana y Felipe, proclamados príncipes herederos	1500: Pedro Alvares Cabral descubre el Brasil, que pasa a depender de Portugal en virtud del tratado de Tordesillas
1504: muerte de Isabel; Juana I, »la Loca«, reina de Castilla; Fernando, gobernador general; fin de la unión personal de Castilla y Aragon	1504: incorporación a la Corona de Aragón del reino de Nápoles
1505: concordia de Salamanca: gobierno conjunto de doña Juana, su consorte don Felipe y don Fernando. Felipe une el ducado de Borgoña. Fernando se casa con Germana de Foix	

1506: concordia de Villafáfila: don Felipe, rey; incapacitación de doña Juana; don Fernando se retira a Aragón. Muerte de Felipe; regencia provisional de Ximénez de Cisneros; regencia de don Fernando	
1512: Fernando ocupa la Alta Navarra, incorporándola a la Corona de Castilla en 1515	
1515: mayoría de edad del príncipe Carlos; regencia de Cisneros (1515/1517) en su ausencia	
1516: muerte de Fernando. Carlos I, rey de Aragón y de Castilla. Unión definitiva de las dos coronas	
1519: Carlos, emperador: Carlos V	
	1519/21: conquista de Méjico por Hernán Cortés
1520/21: sublevación de las Comunidades (Castilla)	
1521/23: sublevación de las Germanías (Valencia, Mallorca)	
1526: matrimonio de Carlos con Isabel de Portugal († en 1539)	1526: Tratado de Madrid: adquisición del Milanesado y del ducado de Borgoña (ésta, no consumada)
	1531/35: conquista del Perú por Pizarro
	1546: unión del ducado de Milán a la Monarquía hispánica
	1549: Carlos separa a los Países Bajos del Imperio y los une a la Monarquía hispánica
1556: abdicación de Carlos. Empieza a reinar Felipe II en los territorios de la Monarquía hispánica, que se separa del Imperio. Unión del Franco Condado a aquélla	1556: abdicación de Carlos. Le sucede en Austria y Alemania su hermano menor Fernando, rey de Hungría y de Bohemia, elegido emperador en 1558 (Fernando II). El Franco Condado se une a la Monarquía hispánica
1558: muerte de Carlos V	
	1565/66: sublevación de los Países Bajos
1580: unión de la Corona de Portugal a las de Castilla y Aragón en la persona de Felipe II (I en Portugal) en régimen de distinción	
	1581: los Estados Generales de los Países Bajos deponen en La Haya a Felipe II
1591: rebelión foral de Aragón	
1598: Felipe III (1598 – 1621). Privanza del duque de Lerma	
1609: expulsión de los moriscos	1609: tregua de doce años con las Provincias Unidas

1621: Felipe IV (1621 – 65). Privanza del conde-duque de Olivares; política centralizadora

1621: reanudación de la guerra con las Provincias Unidas

1640: sublevación de Cataluña (que durará hasta 1652) y Portugal (que terminará con su independencia)

1643: caída de Olivares

1647: sublevación de Sicilia

1648: tratado de Muenster con las Provincias Unidas; reconocimiento de su independencia. Sublevación de Nápoles

1652: fin de la guerra de Cataluña; vuelta a la situación institucional anterior

1659: Paz de los Pirineos: cesión del Rosellón y la Cerdaña

1665: Carlos II (1665/1700); regencia de Mariana de Austria (h. 1675)

1668: Tratado de Lisboa: reconocimiento de la independencia de Portugal

1674: sublevación de Messina

1679: Paz de Nimega: cesión del Franco Condado

1700: testamento de Carlos II nombrando heredero al duque de Anjou. El duque de Anjou, rey: Felipe V de Borbón

1701: comienzo de la guerra de Sucesión

1705: el archiduque Carlos de Austria se hace proclamar rey de España en la Corona de Aragón

1707: Decreto relativo a la abolición del régimen particular de los reinos de Aragón y de Valencia

1711: Decreto por el que se restablece el derecho privado de Aragón

1711: el archiduque Carlos de Austria, Emperador (Carlos VI)

1713/14: Paces de Utrecht y Rastatt: reconocimiento de Felipe V como rey de España y de las Indias españolas; cesión a Austria de los reinos y territorios de Italia (salvo Sicilia, que pasó a Saboya) y de los Países Bajos; cesión de Gibraltar y Menorca a Inglaterra

1715: Decreto de abolición del régimen particular de Mallorca, salvo en lo concerniente al derecho privado

1716: Decreto de abolición del régimen particular de Cataluña, salvo en lo concerniente al derecho privado

Ein Recht auf innerparteiliche Opposition?

Ein Beitrag zur Auslegung der Art. 20 Abs. 2 und 21 GG

DIMITRIS TH. TSATSOS*

---

* Für wertvolle, anregende Diskussion und kritische Beurteilung der Gedankengänge dieses Beitrages danke ich meinen Mitarbeitern am Lehrgebiet für Deutsches und Ausländisches Staatsrecht und Staatslehre an der Fernuniversität Hagen *Martin Morlok* und *Wolfgang Czepluch*.

§ 1 Einleitung

I. Der Anspruch, Bürger zu sein

1. Der Mensch, seine Würde und seine Freiheitsrechte stehen im Zentrum nicht nur des wissenschaftlichen Interesses, sondern ganz besonders auch der internationalen gerichtlichen Tätigkeit *Hermann Moslers,* der als Richter am Europäischen Gerichtshof für Menschenrechte in Straßburg und am Internationalen Gerichtshof in Den Haag der Gerechtigkeit in hervorragender Weise gedient hat.

2. Die hier angestellten Überlegungen fallen nicht außerhalb dieser großen Thematik. Ganz im Gegenteil: Der Mensch wird erst durch seine politische Dimension zum Bürger, wenn er den Willen und die *institutionelle* Möglichkeit hat, am politischen Geschehen nicht nur als Zuschauer dabei zu sein, sondern als Akteur *mitzuwirken.* Setzen wir jenen Willen voraus – und klammern ihn somit aus diesen Überlegungen aus – hängt die *Möglichkeit zur Mitwirkung* von der konkreten Gestalt ab, die das positive Recht und die normrelevante Wirklichkeit der politischen Ordnung jeweils geben.

3. Im parlamentarischen parteienstaatlichen Regierungssystem, so wie es sich durch die deutsche Geschichte in der Grundgesetzordnung konkretisiert hat, vollzieht sich die politische Willensbildung nicht ausschließlich, aber doch *grundsätzlich* in und durch die politischen Parteien. Dieser Gegebenheit konnte sich der Verfassungsgeber von 1949 nicht verschließen, und deshalb nahm er eine diese Gegebenheit sanktionierende Bestimmung in Art. 21 Abs. 1 Grundgesetz (GG) auf: »Die Parteien wirken bei der politischen Willensbildung des Volkes mit«.

4. Geht man von dieser faktischen Priorität der politischen Partei aus und berücksichtigt man die Oligopolstellung der wenigen – in der Bundesrepublik Deutschland vier – Parteien, die sich etablieren konnten, wird die Enge des Raumes sichtbar, der für die ideologischen und politischen Optionen der mitwirkungswilligen Bürger zur Verfügung steht. Wie noch zu zeigen sein wird, darf deshalb der politische und ideologische Pluralismus der Gesellschaft nicht nur im Verhältnis zwischen den Parteien seinen Ausdruck finden, sondern muß sich auch in den Parteien realisieren, will man mit dem demokratischen Prinzip des Art. 20 Abs. 2 GG Ernst machen.

5. Die Möglichkeit des Bürgers, nach Art. 20 Abs. 2 GG seine politische Dimension zu verwirklichen, hängt damit entscheidend von den rechtlichen und faktischen Möglichkeiten ab, die er als Mitglied in einer Partei hat. Nach diesen Möglichkeiten bemißt sich der Wirklichkeitsgehalt des Volkssouveränitätsprinzips des Art. 20 Abs. 2 GG.

6. Die Überlegungen dieses Beitrags treffen freilich nicht die Gesamtproblematik: »Rechtsstellung des Parteimitgliedes«, wie sie im Grundgesetz (Art. 21 Abs. 1 Satz 3) und im Gesetz über die politischen Parteien vor allem

in den §§ 6, 9, 10 und 15 verwurzelt ist [1]. Dieser Beitrag will lediglich einige verfassungsrechtliche und verfassungspolitische Aspekte ansprechen, die für eine rechtliche Beurteilung innerparteilicher Oppositionsbildung relevant sein können.

II. Bestandsaufnahme

1. Rechtsfragen fallen nicht vom Himmel. Sie entstehen im Entwicklungsprozeß des Gemeinlebens. Gerade die Aktualität der hier angesprochenen Thematik zeigt, wie stark die Praxis Rechtsprobleme aufdeckt und prägt.

2. War es über Jahre hinaus sehr selten, daß andere als die vier sogenannten etablierten Parteien in den gesetzgebenden Körperschaften vertreten sein könnten, so scheint der Einzug der »Grünen Partei« in mehrere Landtage bzw. Abgeordnetenhäuser – trotz der Restriktionen durch die 5%-Klausel und die neue politische Bewegungen benachteiligende Wahlkampfkostenerstattung – die Wandlung des bundesrepublikanischen Vier- in ein Fünf-Parteiensystem möglich gemacht zu haben.

3. Auch wenn man diese Erscheinungen noch nicht abschließend verfassungsrechtlich und politisch bewerten kann, läßt sich gleichwohl bereits jetzt feststellen, daß die aufgezeigte Entwicklung auch das innerparteiliche Gefüge der vier großen Parteien in der Bundesrepublik Deutschland nachhaltig beeinflußt, und zwar in zwei Richtungen:

a) Das Auftreten der »ökologischen« Bewegungen hat die Diskussion darüber ausgelöst, wie das Entstehen einer solchen, ernst zu nehmenden parlamentarischen Konkurrenz zu erklären ist, wo doch alle vier großen Parteien immer für sich in Anspruch nahmen, Volksparteien zu sein, mit der erklärten Absicht, alle im Volke vorhandenen politischen Richtungen abzudecken und deshalb für jedermann wählbar zu sein [2].

Letztlich geht es um die Frage nach der Integration oder Abgrenzung im Verhältnis zu politischen Gruppierungen, deren Programmatik sich nicht in allen Punkten mit der offiziellen Parteilinie deckt. Mit dem Mittel der Integration könnte man zwar einer drohenden Abwanderung weiterer bisheriger Stammwähler begegnen, andererseits würde gleichzeitig das innerparteiliche Konfliktpotential in den Sachdebatten erhöht, und die Partei insgesamt erschiene nach außen noch mehr als im Inneren uneinig und zerrissen [3].

---

[1] Gesetz über die politischen Parteien vom 24. 7. 1967, zuletzt geändert durch Gesetz vom 21. 12. 1979 (im folgenden PartG), BGBl. I, S. 2358.

[2] Zum Begriff der Volkspartei vgl. *O. Kirchheimer*, Politische Vierteljahresschrift 1965, S. 20 ff.

[3] Gerade die Sozialdemokratische Partei Deutschlands, welche durch die sich augenblicklich abzeichnende Entwicklung große Stimmenverluste hinzunehmen hat, sieht sich einem erhöhten Handlungsdruck ausgesetzt. Sinnfälliger Ausdruck hierfür ist die parteiinterne Diskussion um die sog. »Löwenthal-Thesen«; siehe *R. Löwenthal*, Die Neue Gesellschaft (1981), S. 1085 ff.

b) Die Mutation des Parteiensystems hat jedoch gleichzeitig auch die Position derjenigen verändert, die bereits bisher in der Darstellung ihrer eigenen politischen Position die offizielle Parteipolitik in Frage stellten. Barg diese Kritik an der offiziellen Parteilinie bisher auch stets die Gefahr in sich, von der Partei isoliert – möglicherweise sogar ausgeschlossen – zu werden, und bedeutete dies für den hiervon Betroffenen das »politische Todesurteil« [4], so ist die augenblickliche Entwicklung gekennzeichnet durch eine Stärkung der Position des einzelnen Parteimitgliedes ohne Änderung seiner parteigesetzlich und -satzungsmäßig bestimmten Rechtsstellung. Obwohl innerparteiliche Auseinandersetzungen – wegen des möglicherweise in der Öffentlichkeit entstehenden Bildes fehlender Geschlossenheit – gescheut werden, müssen sich die Parteien in stärkerem Maße mit Meinungen beschäftigen, die den offiziellen Kurs der Partei auch außerhalb der Parteitagsdiskussion in Frage stellen.

Beredtes Zeugnis für solche Entwicklungen ist beispielsweise die Rubrik *pro* und *contra* in der SPD-Mitglieder-Zeitung »Sozialdemokrat – Magazin«, in welcher jeweils zwei Parteimitglieder zu einer kontroversen Sachfrage aus ihrer Sicht in unterschiedlicher Weise Stellung beziehen.

4. Allenthalben zeichnet sich hierdurch eine parteiinterne Entwicklung ab, die die Unantastbarkeit des gerade von den Parteien immer wieder beschworenen Geschlossenheitsprinzips [5] in Frage stellt. Parteimitglieder, ob einfache, Funktions- oder Mandatsträger, begnügen sich nicht mehr damit, auf Parteitagen und Versammlungen die offizielle Parteipolitik in Frage zu stellen, sondern versuchen, ihren abweichenden Meinungen insbesondere dadurch Ausdruck und mithin auch Gewicht zu verleihen, daß sie gleichgesinnte Mitglieder um sich versammeln, sich organisieren und sich ebenso wie die »offizielle Parteimeinung« über die Medien artikulieren.

Der Blick in das politische Tagesgeschehen macht deutlich, daß letztlich alle Parteien von diesen internen Entwicklungen berührt werden.

Die Auseinandersetzung um den sogenannten »Nato-Doppelbeschluß« oder um die »Kernenergiefrage« in der SPD lassen sich hier ebenso nennen wie die Entwicklung in der FDP, welche zu dem sogenannten »Kölner Kongreß« Anfang des Jahres 1982 führte. Ähnliche Erscheinungen zeigen sich beispielsweise in der CDU im Verhältnis der Partei zu ihren Sozialausschüssen. Verfolgt man die politische Bewertung dieser parteiinternen Vorgänge durch die offiziellen Parteigremien, aber auch in den entsprechenden Kommentaren der Tagespresse, so fällt durchgehend die überaus negative Bewer-

---

[4] So *H. P. Bull*, Zeitschrift für Rechtspolitik (ZRP) 1971, S. 196.

[5] *E. Stein*, Staatsrecht (6. Aufl. 1978), S. 139 ff.; *B. Zeuner*, Innerparteiliche Demokratie (1969), S. 14; *R. Wolfrum*, Die innerparteiliche demokratische Ordnung nach dem Parteiengesetz (1974), S. 146; *W. Henke*, in: Bonner Kommentar, Rz. 55 zu Art. 21; *K.-H. Seifert*, Die politischen Parteien im Recht der Bundesrepublik Deutschland (1975), S. 214 ff.; *D. Hahn*, Innerparteiliche Demokratie (Diss. Köln 1973), S. 49.

tung dieser Erscheinungen auf. Man spricht nicht von parteiinternem Pluralismus oder innerparteilicher Meinungsfreiheit, sondern von fehlender glaubwürdiger Orientierung, vom Bild der Uneinigkeit und Zerrissenheit [6].

Das Infragestellen der offiziellen Parteipolitik wird als ein die Partei gefährdender Vorgang gewertet, da mit dem Prinzip der Geschlossenheit und Einheit der Partei Effektivität und Glaubwürdigkeit untrennbar verbunden seien.

5. Innerparteiliche Oppositionsbildungen stellen nicht nur eine Eigenart des bundesrepublikanischen Parteiensystems dar [7]. Zahlreiche politikwissenschaftliche Untersuchungen unterstreichen die Aktualität und Internationalität der Erscheinung [8]. Trotz vorliegender politikwissenschaftlicher Erkenntnisse wurden innerparteiliche Oppositionsbildungen parteiverfassungs- und parteienrechtlich bisher kaum beachtet [9].

§ 2 Zum Begriff der innerparteilichen Opposition

I. Herrschende Sprachverwirrung

1. Die rechtliche Auseinandersetzung mit innerparteilichen Oppositionsbildungen litt von Anbeginn an unter der offensichtlichen Unklarheit über den Untersuchungsgegenstand selbst [10].

---

[6] So beispielsweise Bundeskanzler Helmut Schmidt nach »Süddeutsche Zeitung« vom 27./28. 3. 1982.

[7] Organisierte innerparteiliche Gruppenbildungen finden sich insbesondere in den britischen und italienischen Parteien. Die britische Labour Party war seit jeher durch die Existenz eines gut durchorganisierten sog. linken »Flügels« gekennzeichnet. Kern dieser oppositionellen linkssozialistischen Gruppierung innerhalb der Partei war die sog. »Tribune-Gruppe«, welche im Gegensatz zu den lediglich punktuell aufbrechenden Protesten gegen die Parteiführung eine Gruppierung mit eigenen Zielen und Mitgliedern darstellte. Die innerparteilichen Auseinandersetzungen entzündeten sich insbesondere an Fragen der Nationalisierung von Industriebranchen, der Atombewaffnung, der Zugehörigkeit zum Atlantikpakt sowie dem Beitritt Großbritanniens zur Europäischen Gemeinschaft. Guter Überblick bei *J. Hartmann*, Parteienforschung (1979), S. 40 ff.

[8] *R. Rose*, Parties, Factions and Tendencies in Britain, Political Studies 1964, S. 33 ff.; *A. J. Stern/S. Tarrow/M. F. Williams*, Factions and Opinion Groups in European Mass Parties, Comparative Politics 1970/1, S. 529 ff.; *G. Sartori*, Party and Party Systems: A Framework for Analysis 1976, S. 71 ff.; *R. Zariski*, Party Factions and Comparative Politics, Midwest Journal of Political Science 1960, S. 27 ff.

[9] Politikwissenschaftliche Untersuchungen im deutschsprachigen Raum: *A. Ashkenasi*, Reformpartei und Außenpolitik (1968); *B. Zeuner* (Anm. 5), S. 113 ff.; *J. Raschke*, Innerparteiliche Opposition (1974); *derselbe*, Organisierter Konflikt in westeuropäischen Parteien (1977); *derselbe*, Demokratisierung durch innerparteilichen Gruppenwettbewerb, Aus Politik und Zeitgeschichte, 1975 Heft 14, S. 11 ff.; *J. Seifert*, Frankfurter Hefte 1960, S. 765 ff.; *D. Preuße*, Gruppenbildungen und innerparteiliche Demokratie (1981).

[10] Solche Defizite wären zumeist dann zu vermeiden gewesen, wenn man sich bei der wissenschaftlichen Beschäftigung insbesondere der politikwissenschaftlichen Er-

Im folgenden soll daher zunächst – stets im Hinblick auf die verfassungsrechtliche Relevanz – unter Zuhilfenahme parteiensoziologischer Erkenntnisse untersucht werden, welche strukturellen Merkmale innerparteiliche Gruppierungen aufweisen müssen, damit sie »innerparteiliche Opposition« im hier verstandenen Sinne darstellen.

2. Der Begriff der Opposition in der modernen Verfassungslehre ist ein Produkt der Parlamentarismusterminologie. Er kennzeichnet einen notwendigen Bestandteil parlamentarischer Demokratie. Die Möglichkeit und das Recht zur Bildung einer Opposition ist Wesensmerkmal der freiheitlich-demokratischen Grundordnung des Grundgesetzes, wie sie in der Rechtsprechung des Bundesverfassungsgerichts ihren Ausdruck gefunden hat [11]. In diesem Sinne kennzeichnet Opposition jedoch nur das Vorhandensein einer oder mehrerer parlamentarischer Gruppen, die Regierung und Mehrheitsfraktionen politisch bekämpfen und abzulösen versuchen [12].

3. Die Begriffsbildung *»innerparteiliche Opposition«* ist demgegenüber weit weniger gebräuchlich. Gruppenbildungen innerhalb der Parteien werden vielmehr als »Gruppierung«, »Richtung«, »Fraktion«, »Faktion« oder auch »Flügel« bezeichnet. Es erscheint daher zunächst gerechtfertigt, den Begriff »innerparteiliche Opposition« zu präzisieren, um damit den Untersuchungsgegenstand deutlich herauszustellen und gleichzeitig von vergleichbaren verfassungsrechtlichen Fragestellungen abzugrenzen. Unter innerparteilichen Oppositionsbildungen im Rahmen der hier aufgezeigten Problematik sollen im folgenden solche innerparteilichen Formierungen verstanden werden, welche sich aus Parteimitgliedern unterschiedlicher Parteiorgane oder Untergliederungen (Gebietsverbände, Jugendorganisationen, Vereinigungen, Arbeitsgemeinschaften, Fraktionen, Parteiausschüsse) zusammensetzen, die eine von den jeweiligen Parteimehrheiten und/oder Beschlüssen abweichende politische Zielsetzung gegenüber der Partei als auch der Öffentlichkeit vertreten und hierdurch versuchen, die Mehrheiten in der Partei entweder personell abzulösen bzw. die Politik der Mehrheiten zumindest zu ändern [13].

II. Wesentliche Merkmale innerparteilicher Opposition

1. Die rechtliche Anerkennung von Konflikten bedeutet *per se* Akzeptanz der Opposition. Jeder Mehrheitsentscheidung in den verschiedenen, im Parteiengesetz vorgesehenen Organen (Mitgliederversammlung, Vertreterversamm-

---

kenntnisse zur innerparteilichen Opposition bedient hätte; daß dies offensichtlich nicht geschehen ist, bezeugt einmal mehr den Hang vieler Staatsrechtler zum Rückzug auf den sog. »überpolitischen Standpunkt«, *G. Stuby*, Der Staat 1969, S. 303 ff. (305).

[11] Entscheidungen des Bundesverfassungsgerichts (BVerfGE) 2, S. 1 ff. (12 f.).

[12] Beispielhaft für einen formalen Oppositionsbegriff: *K. Stern*, Das Staatsrecht der Bundesrepublik Deutschland, Bd. 1 (1977), S. 821.

[13] Die Definition lehnt sich an diejenige von *Raschke*, Organisierter Konflikt (Anm. 9), S. 22 f., an.

lung, Arbeitsgemeinschaften, Parteitag) oder auch in den Fraktionen, die sich in den Vertretungskörperschaften im Bund, den Ländern oder den Kommunen bilden, steht eine oppositionelle Minderheit gegenüber. Es versteht sich von selbst, daß solche Oppositionen parteirechtsimmanent sind, denn die durch Abstimmungsniederlagen formierte Minderheit legitimiert geradezu die getroffene Mehrheitsentscheidung, da sie dem demokratischen Grundkonsens über die Beachtung der Verbindlichkeit von Mehrheitsentscheidungen Ausdruck verleiht. Man könnte sogar sagen, daß die Funktion satzungsmäßig entstehender Minderheiten bereits im Zeitpunkt ihrer Entstehung auch wieder endet, da durch die Beschlußfassung selbst der innerparteiliche Willensbildungs- und Diskussionsprozeß beendet worden ist; demgegenüber erstrebt innerparteiliche Opposition eine über diesen Zeitpunkt hinausgehende Wirkung.

Sie findet sich mit der für sie ungünstigen Beschlußfassung zwar ab, versucht jedoch, die demokratisch ermittelten Mehrheiten zu ihren Gunsten zu ändern. Trotz grundsätzlicher Zustimmung zur Legitimität der Mehrheitsentscheidung nimmt sie auch weiterhin einen von der offiziellen Parteilinie abweichenden Standpunkt ein. Besonders deutlich wird der Unterschied zwischen innerparteilicher Opposition und den durch Abstimmung entstehenden Minderheiten jedoch, wenn man berücksichtigt, daß besonderes Kennzeichen innerparteilicher Oppositionsbildungen eine parakonstitutionelle Zusammensetzung ist. Bei entsprechenden personellen Konsequenzen scheint es jedoch nicht ausgeschlossen, daß eine innerparteiliche oppositionelle Gruppierung auch aus einer solchen formierten Minderheit heraus entstehen kann.

2. Während die Meinungsbildung innerhalb der Parteien sich auf unterschiedlichen Entscheidungsebenen durch Beschlußfassung nach innerparteilicher Diskussion der Sach- und Personalfragen entwickelt, bildet sich die innerparteiliche Opposition in ihrer personellen Zusammensetzung neben den satzungs- und parteirechtlich vorgesehenen Untergliederungen und ist in diesem Sinne parakonstitutionell. Wegen dieses besonderen Merkmales bedarf es zur effektiven politischen Zielverwirklichung eines gewissen Grades an innerer Organisation, welche ihrerseits nur bei Zusammenschlüssen von einer gewissen zeitlichen Dauer möglich ist. Sogenannte »*ad hoc*-Gruppierungen« erfüllen daher ebenfalls nicht die Merkmale einer innerparteilichen Opposition; sie bilden sich lediglich zu bestimmten begrenzten Sachfragen und lösen sich nach der Entscheidung der jeweiligen Frage auch schnell wieder auf.

3. Es erscheint jedoch zweifelhaft, ob man hieraus die positive Schlußfolgerung ziehen darf, daß innerparteiliche Opposition über ein einheitliches politisches Gesamtkonzept verfügen müsse, daß innerparteiliche Opposition sozusagen »die Partei in der Partei« darstellt. In der Literatur ist insoweit teilweise von »innerparteilichem Parteiensystem« oder »innerparteilicher Partei« die Rede [14].

---

[14] *Zeuner* (Anm. 5), S. 113 ff.

Wegen der deutlichen Eingrenzung des Parteibegriffs durch das Parteiengesetz selbst erscheint es jedoch unzweckmäßig, innerparteiliche Opposition als »Partei in der Partei« zu umschreiben; dies könnte den Eindruck hervorrufen, es handele sich bei innerparteilicher Oppositionsbildung doch um eine Art Vorstufe der Parteispaltung, wenn auch minderen Ranges. Daß dies gerade nicht die Zielsetzung innerparteilicher Opposition ist, wird deutlich an der Tatsache, daß innerparteiliche Opposition grundsätzlich auch für sich die Verbindlichkeit des jeweiligen Grundsatzprogrammes anerkennen muß, will sie nicht die Grenze zum parteischädigenden Verhalten verwischen. Innerparteiliche Opposition will sich dementsprechend aus ihrer Sicht nicht als Alternativgruppierung zur Gesamtpartei darstellen, sondern lediglich als Teil *innerhalb* der Partei, welche durch das Aufzeigen von alternativen Sach- und Personallösungen die Mehrheiten in der Partei abzulösen und/oder zu ändern versucht.

4. Innerparteiliche Oppositionsgruppierungen unterscheiden sich aber auch von »Richtungen« und »Strömungen« in einer Partei. Zwar verfolgen auch letztere Ziele, die mit denen der Gesamtpartei bzw. den dort maßgebenden Mehrheiten nicht übereinstimmen. Im Gegensatz zur innerparteilichen Opposition wird jedoch hier eine abweichende Meinung lediglich artikuliert und nicht auch noch organisiert [15]. Eine personelle Konsistenz innerparteilicher Richtungen und Strömungen ist nur in einem beschränkten Umfange vorhanden, weshalb beispielsweise gleiches Abstimmungsverhalten innerhalb solcher Gruppierungen in größerem Maße als rein gefühlsmäßige Entscheidung zu werten sein wird denn als bewußtes und gewolltes Zusammenwirken gegen die herrschende Parteimeinung. Von den Parteien selbst werden Richtungen und Strömungen auch weitestgehend toleriert, solange sie nicht den Versuch einer internen Organisation machen.

Teilweise wird zur Kennzeichnung der hier dargestellten innerparteilichen Richtungsgruppierungen auch der Begriff »Fraktion« verwandt [16]. Wegen seiner besonderen Verwendung im Parlamentsrecht sollte dieser Begriff, um Verwechslungen zu vermeiden, zur Kennzeichnung innerparteilicher Gruppierungen keine Verwendung finden.

5. Zwischen innerparteilichen Oppositionsgruppierungen und der Gesamtpartei müssen sachlich-politische Zielunterschiede bestehen. Dies schließt selbstverständlich im Einzelfall nicht aus, daß innerparteiliche Opposition mit ihrem Tätigwerden auch personalpolitische Ziele verfolgt. Durch dieses besondere Strukturmerkmal soll jedoch verhindert werden, daß beispielsweise der lediglich um Parteiämter geführte organisierte Gruppenkampf oder der Versuch, spezielle regionale oder lokale Forderungen durchzusetzen, bereits als innerparteiliche Opposition gewertet werden kann [17].

---

[15] *Raschke,* Organisierter Konflikt (Anm. 9), S. 22.
[16] *Sartori* (Anm. 8).
[17] *Raschke,* Organisierter Konflikt (Anm. 9), S. 23.

Zusammenfassend ist festzuhalten:

6. Im Sprachgebrauch hat sich eine Reihe von Begriffen eingebürgert, mit welchen Gruppierungen bezeichnet werden, die eine von der offiziellen Parteimeinung abweichende Auffassung sowohl im Parteiinnern als auch nach außen vertreten. Die hiermit umschriebenen innerparteilichen Erscheinungen stellen jedoch keine innerparteiliche Opposition im Sinne der hier gegebenen Definition dar.

7. Wenn gleichwohl im folgenden der bereits oben erwähnte Begriff der Faktion und der Begriff innerparteilicher Opposition synonym verwandt werden, so soll hiermit der Tatsache Rechnung getragen werden, daß auch die politikwissenschaftliche Forschung beide Begriffe zur Kennzeichnung der gleichen innerparteilichen Erscheinungen gebraucht [18]: Faktion und innerparteiliche Opposition sind keine legalisierten parteilichen Suborganisationen, sondern Gebilde, die neben den parteigesetzlichen und parteisatzungsrechtlich vorgesehenen Organisationen entstehen. Sie stellen mithin auch keine ständige *Institution*, sondern eine organisierte, auf Zeit angesetzte *Aktion* dar.

III. Innerparteiliche Opposition und Interessengruppen in der Partei

1. Die Betrachtung und verfassungsrechtliche Einordnung innerparteilicher Opposition stoßen auf die Erscheinung innerparteilicher Interessengruppen. Diese Gruppierungen zur bevorzugten Durchsetzung ganz bestimmter Interessen sind zur Konturierung des Phänomens innerparteilicher Opposition im folgenden näher zu betrachten. Beispiele solcher parteiinterner Gruppierungen zur Verfolgung partikularer Interessen sind die Sozialausschüsse der christlichen-demokratischen Arbeitnehmerschaft oder die Wirtschaftsvereinigung in der CDU, die Fachausschüsse in der FDP oder die verschiedenen Arbeitsgemeinschaften innerhalb der SPD, hier insbesondere die Arbeitsgemeinschaft der Selbständigen, die der Arbeitnehmer, aber auch die der sozial-demokratischen Juristen.

2. Diese Interessengruppierungen innerhalb einer Partei weisen eine Reihe von *Gemeinsamkeiten* mit den Erscheinungen innerparteilicher Opposition auf. Wie diese verfolgen die Interessenvereinigungen und ihre Mitglieder innerhalb ihrer Partei Ziele, die nicht völlig mit denen der Gesamtpartei bzw. der dort maßgebenden Mehrheit übereinstimmen. Sie verfolgen Ziele, die erklärtermaßen zwar im Rahmen der politischen Orientierung ihrer Partei stehen und verwirklicht werden sollen, die aber durchaus als eigenständig erkannt und akzeptiert sind. Die Institutionalisierung solcher Interessengruppen bringt zum Ausdruck, daß ihre Mitglieder ihre Partikularinteressen nicht völlig wegen der politischen Zielsetzungen der Partei aufgeben. Mit der satzungsmäßigen Anerkennung von solchen Interessenvereinigungen [19] wird im

---

[18] *Raschke*, a.a.O., S. 22.

[19] Siehe beispielsweise §§ 38 f. Statut der CDU, zuletzt geändert am 5. 11. 1981.

Gegenteil offiziell gebilligt, daß die Partei in sich heterogene Strömungen zusammenfaßt, die sogar in einem mehr oder weniger akuten ständigen Konflikt zueinander stehen. In der satzungsmäßigen Billigung solcher Partikularinteressenvertretung innerhalb der Parteien kommt deren Volksparteicharakter zum Ausdruck. Alle Parteien spiegeln in relativ breiter Weise die gesellschaftlichen Interessengegensätze in sich selbst wider. Die vorhandenen gesellschaftlichen Interessen werden nicht direkt auf verschiedene Parteien abgebildet, sondern über wenige Parteien – nach Möglichkeit über alle zugleich – in den politischen Entscheidungsprozeß eingebracht.

3. Innerparteiliche Opposition wie Interessengruppen innerhalb einer Partei haben weiter gemeinsam, daß beide ihr Tätigkeitsfeld nicht nur innerhalb der Parteigremien haben, sondern auch bewußt und gezielt nach außen werben. Das Handlungsfeld beider Gruppierungen beschränkt sich also nicht auf den innerparteilichen Raum, sondern erstreckt sich in die fachspezifische oder allgemeine Öffentlichkeit hinein, wo es zu einem Konkurrenzverhältnis mit der werbenden Darstellung der Partei als ganzer kommen kann. Die für innerparteiliche Opposition kennzeichnende organisatorische Festigung ist bei den parteiinternen Interessengruppen in besonderem Maße vorhanden. In aller Regel sind sie sogar satzungsmäßig anerkannt und auch mit materiellen Resourcen durch die Partei ausgestattet. Dieser organisatorische Fundus führt dazu, daß innerparteiliche Opposition sich nicht selten an existierende Interessenvereinigungen innerhalb der Partei anlehnt und versucht, diese Substrukturen der Partei für die eigenen Ziele einzusetzen. Im Erscheinungsbild der Parteien können sich so innerparteiliche Oppositionsgruppierungen und die Organisationen spezifischer parteiinterner Interessenvertretung verwischen.

4. Die Möglichkeit der Anlehnung einer innerparteilichen Opposition an solche Interessengruppen läßt eine der *Funktionen* solcher parteiinterner Interessengruppen sichtbar werden: Sie eröffnen einen zweiten Weg der Willensbildung innerhalb der Partei[20].

Für die derart vertretenen Interessen bedeutet diese innerparteiliche Repräsentanz eine Verstärkung ihrer Einwirkungsmöglichkeiten auf die politischen Entscheidungen. Die Vertretung in mehreren, gar sämtlichen politischen Parteien erhöht insgesamt die Wirksamkeit einer solchen Einflußnahme – jedenfalls für die bei diesem Spiel beteiligten Interessen.

Für das politische System sind die interessenbegründeten Einflußnahmen oft wegen der darin liegenden Sachkunde auf dem jeweils angesprochenen Gebiet wertvoll[21]. Weiter begründet die Tatsache, daß die starken Interessen der Gesellschaft sich in allen Parteien artikulieren und daß sämtliche Parteien mit diesen in gewisser Weise verbunden sind, für die einzelne Partei wie für

---

[20] So *Zeuner* (Anm. 5), S. 94 f.

[21] Vgl. zu diesem Aspekt z. B. *H. H. von Arnim*, Gemeinwohl und Gruppeninteressen (1977), S. 138 ff.

das politische System insgesamt ein größeres Maß an Unabhängigkeit gegenüber den einzelnen Interessengruppen. Interessen wirtschaftlicher Art und politische Programmatik können dieserart in stärkerem Maße differenziert werden. Die eigene Rationalität partikularistischer Interessenverfolgung einerseits, der Verfolgung und Durchsetzung politischer Ziele andererseits kann sich so besser entfalten.

5. Die Darstellung wichtiger Interessen innerhalb der Parteien wirkt weiterhin integrativ. Die Interessenverbände werden durch ihre innerparteiliche Vertretung dazu motiviert, ihre eigenen Ziele nicht ohne Rücksicht auf die im politischen Raum sonst noch zu beachtenden Ziele zu verfolgen. Zugleich wird die Schärfe der gesellschaftlichen Konflikte dadurch gemildert, daß Interessengegensätze und Unterschiede der politischen Präferenz nicht immer parallel liegen, sondern sich kreuzweise überschneiden können. Der politische Gegner teilt möglicherweise die gleichen beruflichen Interessen, oder der Interessenantagonist gehört der gleichen politischen Partei an. Damit wird verhindert, daß Interessenkonflikte und politische Auseinandersetzungen eine disruptive Schärfe gewinnen und möglicherweise nicht mehr mit den friedlichen, verfahrensmäßigen Mitteln einer Konkurrenzdemokratie lösbar sind[22]. Wegen dieser relativen Befriedungsfunktion einer Nichtidentität von Interessengegensätzen und politischen Auffassungsunterschieden können Konflikte ohne Gefährdung des inneren Friedens verstärkt ausgetragen werden. Das mindestens teilweise Auseinanderfallen von eher ökonomisch orientierten Interessenkonflikten einerseits und »eigentlich« politischen Konflikten erhöht das Maß an Konfliktfähigkeit, das eine Gesellschaft hat[23].

6. Die Tatsache, daß im Gegensatz zu den Interessenvereinigungen innerhalb einer Partei interne Oppositionsgruppen als solche keinen satzungsmäßigen Ausdruck gefunden haben, weist auf wichtige *Unterschiede* dieser beiden in ihrer Erscheinung oft ähnlichen Formen politischer Assoziation hin. Interessengruppierungen innerhalb einer Partei sind mit satzungsmäßiger Legitimität ausgestattet; sie haben regelmäßig eigene Organe. Nach dem Selbstverständnis der Partei haben solche Untergruppierungen einen legitimen Platz innerhalb des größeren Ganzen der Partei. Demgegenüber ist innerparteiliche Opposition in einem organisatorischen Sinne nicht anerkannt.

7. Ein weiterer äußerlich greifbarer Unterschied zwischen Formierungen innerparteilicher Opposition und innerparteilichen Interessengruppen liegt darin, daß Interessengruppen häufig ihr hauptsächliches organisatorisches

---

[22] Das Mit- und Gegeneinander der Verfolgung von ökonomischen Interessen und politischen Zielen bezieht sich damit auf die zentrale Verfassungsfunktion der politischen Einheitsbildung, hierzu *K. Hesse*, Grundzüge des Verfassungsrechts der Bundesrepublik Deutschland (13. Aufl. 1982), § 1 II.

[23] Zu Konflikten innerhalb von Gesellschaften siehe grundlegend *L. Coser*, The Function of Social Conflicts (1956); in Deutschland hat vor allem *R. Dahrendorf* auf die heilsame Wirkung von Konflikten aufmerksam gemacht, siehe insbes. Die Funktionen sozialer Konflikte, in: Pfade aus Utopia (1967), S. 263 ff.

Fundament außerhalb der Partei haben: in einem spezifischen Interessenverband. Die Beziehung zwischen diesem und der parteiinternen Gruppierung ist deswegen besonders eng. Die parteiinterne Interessengruppe kann, selbstverständlich mit Modifizierungen, als Verlängerung dieses Interessenverbandes in die politische Partei hinein verstanden werden. Bei innerparteilichen Oppositionsbewegungen ist die Situation eher umgekehrt gelagert. Ihr primäres Entstehungsgebiet und ihr vorzügliches Handlungsfeld ist der politische Raum innerhalb der Partei. Auch wenn sie in die allgemeine Öffentlichkeit hineinzuwirken versucht und Interessenverbände zu ihren Verbündeten machen möchte, so bleibt die Wurzel innerparteilicher Opposition wegen ihres genuin politischen Charakters innerhalb der Partei.

8. Diese Unterschiede im Erscheinungsbild lassen sich auf eine grundsätzliche Unterscheidung von Interessengruppen einerseits und innerparteiliche Opposition andererseits zurückführen. Diese Unterscheidung ist analytischer Natur. Sie drückt theoretisch erfaßte Erkenntnisse aus, wird wegen der Mehrzahl der an jeder realen Gruppierung mitwirkenden Faktoren aber nur selten in theoretischer Reinheit an den tatsächlichen politischen Organisationsformen abzulesen sein.

Die Orientierung politisch tätiger Gruppen läßt sich verstehen als auf einem Kontinuum angesiedelt, das sich zwischen rein materiellen Interessen am einen Ende und »ideologischen« Zielvorstellungen am anderen Ende erstreckt. »Interessen« in diesem Sinn bedeuten ein Begehren nach materiellen Vorteilen und deren Sicherung für die eigene Gruppe. »Ideologische Vorstellungen« beziehen sich demgegenüber nicht unmittelbar auf materielle Bestände. »Ideologisch« soll dabei keine pejorative Bedeutung haben; damit soll nur gekennzeichnet werden, daß es um Zielvorstellungen geht, die grundsätzlich ihre Rechtfertigung in sich selbst sehen, die selbstzweckhaft befolgt werden. Es geht mithin hierbei um moralische – in einem weiteren Sinne – Vorstellungen darüber, wie das Gemeinwesen eingerichtet werden soll. Dabei geht es typischerweise nicht um ein Mehr oder Weniger, sondern um ein grundsätzliches So oder Anders. Diese »ideologischen Zielvorstellungen« bezeichnen also Übereinkünfte in normativen Fragen.

9. Diese Entgegensetzung knüpft an die theoretische Unterscheidung von Arbeit und Interaktion, wie sie *Jürgen Habermas* entwickelt hat, an. Die Reproduktion des menschlichen Lebens setzt die Befriedigung grundsätzlich anthropologisch angelegter Bedürfnisse voraus, wozu Handlungen in technischer Einstellung notwendig sind. Zugleich kann die Reproduktion des menschlichen Lebens nur in Gemeinschaft mit anderen erfolgen und setzt immer schon bestimmte normativ eingerichtete soziale Ordnungsmuster voraus und zudem kulturelle Vorstellungen über die menschlichen Bedürfnisse. Handlungstheoretisch kann – bezogen auf soziales Handeln – strategisches Handeln kommunikativem Handeln entgegengesetzt werden[24].

Bemerkenswerterweise hat die empirisch orientierte Parteien- und Organisationsforschung es für nützlich erachtet, mit einer entsprechenden Unterscheidung zu arbeiten und durch gemeinsame Interessen konstituierte Gruppen von genuin politischen Vereinigungen abzugrenzen[25].

10. Auf dieser theoretischen Basis lassen sich die Interessengruppen dem Pol materieller Interessen der Bedürfnisbefriedigung, die Erscheinung innerparteilicher Opposition dem Pol ideologischer Zielvorstellungen zuordnen. Interessen sind in einem grundsätzlichen Sinn kompatibel, was auch bedeutet, daß sie kompromißfähig sind. Interessengegensätze können häufig durch eine Steigerung der Menge der zur Verteilung kommenden Güter geschlichtet werden. Unter Bedingungen wirtschaftlichen Wachstums sind Interessenkonflikte deswegen grundsätzlich lösbar.

Demgegenüber sind Auseinandersetzungen über die richtigen politischen Zielvorstellungen häufig sehr viel grundsätzlicher und gehen um miteinander unvereinbare Vorstellungen. Kompromisse sind dann nur aus Gründen – selbst wieder politischer – Einsicht in die Notwendigkeit der Aufrechterhaltung einer politischen Ordnung nötig, die auch zwischen den Anhängern verschiedener Überzeugungen aufrecht erhalten werden soll. Kompromisse sind in diesem Bereich also nur sekundärer Natur[26].

11. Für den vorliegenden Zusammenhang ist es wichtig festzuhalten, daß partikulare Interessen konstituiert werden durch ihre gesellschaftliche Verortung, etwa aus der Existenz als Landwirt, als abhängiger Arbeiter oder als Angehöriger des selbständigen Mittelstandes. Interessen in diesem Sinne entstehen innerhalb gesellschaftlicher Strukturen. Mit der zunehmenden Verflechtung von staatlicher Tätigkeit und wirtschaftlichem Geschehen werden diese Interessenkonflikte nicht nur im Wege des ökonomischen Konkurrenzkampfes ausgetragen, sondern auch zum Gegenstand politischer Entscheidung. Damit wird die politische Sphäre eine wichtige Arena der ökonomischen Interessenauseinandersetzungen. Ihren eigentlichen Ursprung haben diese Interessen aber nicht in diesem politischen Bereiche im engeren Sinne.

Anders die ideologischen Zielvorstellungen. Sie werden im politischen Diskurs erst entwickelt und artikuliert. Während die innerparteilichen Interessengruppen ihre anderswo formierten Interessen voraussetzen und nur ver-

---

[24] Diesen Grundgedanken hat *J. Habermas* zunächst in »Arbeit und Interaktion« (zuerst 1967) vorgestellt, abgedruckt in: Technik und Wissenschaft als »Ideologie« (1968), S. 9 ff.; aus der gleichen Zeit ist dann natürlich das große Werk »Erkenntnis und Interesse« (1968) zu nennen. Spätere Ausarbeitungen finden sich etwa in »Legitimationsprobleme im Spätkapitalismus« (1973), z.B. S. 14 ff. Siehe zuletzt auch: Theorie des kommunikativen Handelns, 2 Bde. (1981), z.B. Bd. 1, S. 384 ff.

[25] Siehe z.B. *S. M. Lipset/M. Trow/J. Coleman*, Union Democracy (1956) zit. nach Taschenbuchausgabe (1962), S. 77 ff., 306 ff.; *Zeuner* (Anm. 5), S. 93 ff.; *G. Teubner*, Organisationsdemokratie und Verbandsverfassung (1978), S. 198 ff., 303 ff.

[26] Deswegen besteht auch eine Tendenz, daß Auseinandersetzungen um Grundfragen der richtigen Ziele schärfer werden als Interessenkonflikte im materiellen Bereich. Paradigmatisch dafür sind in der europäischen Geschichte die Religionskriege – wenngleich diese freilich sich oft genug mit Interessenauseinandersetzungen überlagerten.

suchen, diese über die Parteien in den politischen Entscheidungsprozeß einzubringen, haben die alternativen politischen Zielvorstellungen der innerparteilichen Opposition ihren Ursprungsort innerhalb der innerparteilichen Diskussion. Sie sind genuin politisch und Kernelement des Geschehens in einer Partei.

§ 3 Verfassungsrechtliche Ansatzpunkte eines Rechts auf innerparteiliche Opposition

I. Einleitung

1. Die rechtliche Einschätzung einer Bildung und organisatorischen Verfestigung innerparteilicher Opposition hat ihren obersten Bezugspunkt in Art. 21 GG. Die in Art. 21 Abs. 1 Satz 1 GG liegende Anerkennung der Rolle der politischen Parteien bei der politischen Willensbildung des Volkes ist eine wichtige Errungenschaft des Grundgesetzes. Ein hervorragendes Merkmal dieser Bestimmung ist der in ihr zum Ausdruck kommende Realismus, der die verfassungsrechtliche Konsequenz daraus zieht, daß unter den Verhältnissen moderner großer politischer Gemeinwesen Demokratie nur über die Aktivität politischer Parteien verwirklichbar ist. Diesem Realismus muß insofern Rechnung getragen werden, als auch die Funktionsvoraussetzungen für die politischen Parteien und ihre Tätigkeit unter den Schutz des Grundgesetzes gestellt werden.

2. Einmal kann die Tätigkeit innerparteilicher Oppositionsgruppen die rechtlich vorausgesetzte Handlungsfähigkeit der Partei wie auch die Orientierungsmöglichkeiten der Bürger und ihre daran geknüpften politischen Handlungschancen in Mitleidenschaft ziehen. Damit könnten die Parteien in eine Entwicklungsrichtung gedrängt werden, die sie vom Bild relativ stabiler, handlungsfähiger Verbände entfernte, das § 2 Abs. 1 PartG für den Rechtsbegriff der politischen Partei zeichnet.

Zugleich muß aber auch berücksichtigt werden, daß die Effektivität der Parteien und ihrer Arbeit eine Leistungsfähigkeit im Interesse freiheitlicher politischer Willensbildung ist. Die Bezeichnung der Parteien als »Verfassungsorgane« darf nicht darüber hinwegtäuschen, daß es Aufgabe der Partei ist, die Vielfalt der politischen Ansichten aufzunehmen und darzustellen. Pluralismus und Dynamik gehören zu den wesentlichen Grundeigenschaften des politischen Prozesses, die durch die verfassungskräftige Anerkennung der politischen Parteien geschützt werden. Die Effektivität der politischen Parteien muß also an diesen Zielen gemessen werden.

Schlagkräftigkeit und Geschlossenheit einer Partei dürfen nicht über die Vielfalt und Entwicklungsoffenheit der politischen Meinungsbildung gestellt werden. So wie die Konkurrenzdemokratie zwischen den Parteien von Konflikten und Spannungen lebt, so muß dies grundsätzlich aber auch für den Binnenbereich einer Partei gelten.

3. Die Verpflichtung der Parteien auf demokratische Grundsätze durch Art. 21 Abs. 1 Satz 3 GG macht dies unmißverständlich klar. Das Gebot innerparteilicher Demokratie ist deswegen eine weitere Grundlage für die rechtliche Beurteilung innerparteilicher Oppositionsbildung. Dabei wäre es freilich ein Fehler, die Effektivität der Parteien gegen das Gebot innerparteilicher Demokratie auszuspielen. Tatsächlich sind beide Imperative aufeinander bezogen: Im Hinblick auf das Ziel eines freiheitlichen demokratischen politischen Willensbildungsprozesses.

Für Einzelfragen sagt die Verpflichtung auf die »demokratischen Grundsätze« als solche noch nicht sehr viel aus. Die Bestimmung verlangt erhebliche Konkretisierung. Für vorliegenden Zusammenhang stellt sich dabei insbesondere die Frage, inwiefern die Bestimmungen des Parteiengesetzes und die Vorschriften der einzelnen Satzungen, die in bestimmten Teilen eindeutig als Ausfluß des Demokratiegebotes zu begreifen sind, noch Raum lassen für einen Rückgriff auf das Gebot innerparteilicher Demokratie selbst. Mit anderen Worten, es geht darum, in welchem Maße die innerparteiliche Demokratie in den Kompetenz- und Verfahrensbestimmungen aufgegangen und von diesen absorbiert worden ist, respektive inwieweit daneben und gegebenenfalls dagegen ein unmittelbarer Rekurs auf die »demokratischen Grundsätze« möglich ist.

4. Als ein weiterer möglicher Maßstab zur rechtlichen Bewertung innerparteilicher Opposition kommen die Grundrechte in Frage. Hier stellt sich einmal die Problematik der Grundrechtsgeltung innerhalb der Parteien; sodann geht es darum, die Position und die Handlungsmöglichkeiten der Mehrheit der Partei zu achten.

5. Der rechtliche Befund hinsichtlich der Bildung und Betätigung innerparteilicher Opposition erschließt sich demnach nicht dem ersten Blick. Die normativen Vorgaben sind wegen des in ihnen selbst angelegten Bezugs zur Wirklichkeit nicht selbstgenügsam, aus sich heraus zu interpretieren. Der speziell bei politisch geprägten Materien gesteigerte Wirklichkeitsbezug des Rechts und die besondere Qualität der politischen Parteien als dynamische Elemente der Demokratie verlangen es, die tatsächlichen Bestimmungsgründe und Funktionen innerparteilicher Opposition mehr in den Blick zu nehmen, um so informiert den Gehalt der verfassungsrechtlichen Vorgaben präzisieren zu können. Das soll im Anschluß an diese juristischen Überlegungen (II, III und IV dieses Paragraphen) unter § 4 versucht werden.

II. Die Institutionalisierung der politischen Parteien

1. Die das Grundgesetz für die Bundesrepublik Deutschland durchziehende demokratische Idee findet ihren besonderen Ausdruck darin, daß nach Art. 20 GG alle Staatsgewalt vom Volke ausgeht.

2. Wichtigstes Instrument zur Einlösung dieses Anspruches sind im demokratischen Verfassungsstaat die politischen Parteien. Das Volk als Inbegriff einer Vielzahl ungeordnet nebeneinanderstehender Individuen würde sich ohne die Institutionalisierung der politischen Parteien als weniger handlungsfähig erweisen. In dieser Erkenntnis hat bereits der Verfassungsgesetzgeber die politischen Parteien mit einem eigenen Verfassungsartikel bedacht, dessen besondere Bedeutung auch in seiner Stellung unmittelbar im Anschluß an Art. 20 zum Ausdruck kommt. Die genannte Verfassungsbestimmung zählt die Existenz politischer Parteien anders als die früheren deutschen Verfassungen zu den unverzichtbaren, integrierenden Bestandteilen des Verfassungssystems der Bundesrepublik Deutschland. Sie werden damit aus der Reihe der »bloßen« Vereinigungen nach Art. 9 GG besonders hervorgehoben. Die Hinwendung der Verfassung zur parteienpluralistisch organisierten Demokratie hat teilweise sogar zur verfassungstheoretischen Bejahung einer Identifizierung der Parteien mit dem politisch aktiven Volk geführt[27]. Das verfassungsrechtliche Postulat der Mitwirkung der Parteien bei der politischen Willensbildung des Volkes nach Art. 21 Abs. 1 Satz 1 GG soll nicht nur die Entscheidung des Grundgesetzes für ein offenes Mehrparteiensystem durch die bloße Existenz mehrerer vorhandener politischer Parteien sicherstellen, sondern gleichzeitig darüber hinaus den Parteien besondere Aufgaben im politischen Gesamtsystem zur eigenverantwortlichen Erfüllung zuweisen. Während Art. 21 selbst lediglich von der »Mitwirkung« spricht, lassen sich die einzelnen Elemente dieses verfassungsrechtlichen Kataloges insbesondere dem § 1 Abs. 2 PartG entnehmen[28].

Im Rahmen der hier untersuchten Fragestellung nach der verfassungsrechtlichen Einordnung und Bewertung innerparteilicher Oppositionsbildungen kann auf die einzelnen Parteifunktionen nicht im Detail eingegangen werden.

3. Die besonderen Funktionszuweisungen an die politischen Parteien haben jedoch zur Folge, daß jede *binnenstrukturelle Erscheinung*, wie beispielsweise innerparteiliche Oppositionsbildungen, – aber auch in diesem Zusammenhang die Rechtsstellung des einzelnen Parteimitgliedes – normativ nur bewertet werden können unter Berücksichtigung der verfassungsrechtlich erwünschten Optimierung parteilicher Funktionserfüllung.

Anders gefaßt könnte man sagen, daß derjenige, der Erscheinungen wie innerparteiliche Opposition zu rechtfertigen sucht, die Darlegungslast dafür trägt, daß solche Erscheinungen die Erfüllung der den politischen Parteien *qua* Verfassung zugewiesenen Aufgaben nicht beeinträchtigt.

---

[27] *G. Leibholz,* Strukturprobleme der modernen Demokratie (3. Aufl. 1967), S. 146; zusammenfassend hiergegen *P. Haungs*, Zeitschrift für Parlamentsfragen 1973, S. 502 ff.; auch *M. Kriele,* Einführung in die Staatslehre (1. Aufl. 1975), S. 321 ff.

[28] Zu den Parteifunktionen *A. Mintzel/H. Schmitt,* Politische Bildung 1981, S. 3 ff.

4. Die Mitwirkung der Parteien an der politischen Willensbildung läßt sich grundsätzlich nach drei Funktionen unterscheiden:

a) Wichtigste Funktion der politischen Parteien ist ihr Beitrag zur Durchführung von Wahlen. Sie bilden Kandidaten heran, stellen diese zur Wahl auf und präsentieren sie dem Wahlvolk. Diese – ein Höchstmaß an Organisation erfordernden – Aufgaben können von anderen Gruppierungen nicht geleistet werden. Im Wege der öffentlichen Wahlkampfführung wirken sie auf die Wahlbereitschaft der Bevölkerung ein und erhöhen damit – wenn auch nur mittelbar – die demokratische Legitimation staatlicher Herrschaft.

b) Neben den sich im gesellschaftlichen Raum bildenden Verbänden obliegt den Parteien die Strukturierung gesellschaftlicher Meinungs- und Willensbildungstendenzen. Sie tragen zur Bildung einer öffentlichen Meinung bei und kanalisieren gleichzeitig die ständige Einflußnahme des Volkes auf die politischen Entscheidungsorgane. Integrierender Bestandteil der demokratischen Grundordnung sind die politischen Parteien auch insofern, als sie die gruppenmäßig von den Interessenverbänden an sie herangetragenen Sonderinteressen zunächst filtrieren und in ihrer Bedeutung zu sonstigen Interessengruppen relativieren, um sie sodann bei der institutionalisierten staatlichen Willensbildung in den Parlamenten und Regierungen zur Geltung zu bringen.

c) Die unmittelbare Beteiligung der Parteien am staatlichen Willensbildungsprozeß zeigt sich insbesondere auch darin, daß die Entscheidungsorgane selbst, Regierung und Parlamente, in ihrer personellen Zusammensetzung aus Personen bestehen, deren politische Existenz der jeweiligen Partei, welcher sie angehören, auf das engste verbunden ist.

5. Der Verfassungsgesetzgeber hat sich jedoch nicht nur mit der Inkorporation politischer Parteien in das verfassungsrechtliche Gesamtgefüge sowie der Zuweisung verfassungsrechtlicher Aufgaben begnügt, sondern darüber hinaus gleichzeitig durch Art. 21 Abs. 3 GG den Gesetzgeber ermächtigt, die funktionsgerechten Folgen jener Institutionalisierung zu normieren. Dem Interesse an der Effektivität der Parteien im Hinblick auf ihr Tätigwerden im Rahmen der demokratischen Ordnung hat das Parteiengesetz insofern Rechnung getragen, als zwei besondere Begriffsmerkmale bei der Definition der politischen Partei besonders hervorgehoben worden sind: Einmal die *Einflußmöglichkeit* politischer Parteien auf die Willensbildung und zum anderen die *Festigkeit* der Organisation. Beide Merkmale münden in eine Forderung nach Geschlossenheit politischer Parteien, die dem verfassungsrechtlichen Effektivitätspostulat als funktionssichernd und -erhaltend zuzuordnen ist[29]. In diesem Zusammenhang ist auch die Möglichkeit der Parteien zu sehen, Mitglieder, welche der Partei durch ihr Verhalten Schaden zufügen, auszuschließen (§ 10 Abs. 4 PartG).

[29] Zum Geschlossenheitsprinzip siehe Anm. 5.

Der Gesetzgeber ist offensichtlich davon ausgegangen, daß die Erfüllung der Parteifunktionen nur unter der Voraussetzung innerer Geschlossenheit effektiv erfolgen kann. Obwohl der Begriff Geschlossenheit im Parteiengesetz selbst keine Erwähnung findet, hat er für den Bereich der innerparteilichen Willensbildung längst den Charakter einer normativen und nicht nur deskriptiven Kategorie angenommen. Hieraus wird gefolgert, daß die Partei als politischer Aktionsverband nur dann im Wettbewerb mit den anderen Parteien bestehen kann und imstande ist, den Wähler vor klare Wahlalternativen zu stellen, wenn der Konsens im Innern gewahrt ist. Jede andere Entscheidung würde gleichzeitig eine Entwertung des demokratischen Wahlaktes darstellen, denn eine Wahlentscheidung zugunsten der nicht als geschlossene Einheit auftretenden Partei würde das politisch Gewollte nicht mehr klar zutage treten lassen[30]. Nähme man deshalb eine Oppositionsbildung, welche über die Ebene und außerhalb der institutionalisierten Parteiorgane und Parteiorganisationen hinaus die einmal beschlossene Parteipolitik in Frage stellt, hin, so würde man das Prinzip der Geschlossenheit der politischen Parteien preisgeben und damit einen Vorgang legitimieren, welcher sich hemmend auf die einzelnen Parteifunktionen auswirken würde.

6. Gegen die Zulässigkeit innerparteilicher Oppositionsbildungen könnte auch die Tatsache sprechen, daß die innere Ordnung der Parteien Gegenstand der abschließenden Regelungen in den §§ 6–16 des PartG ist. Der Schutz innerparteilicher Minderheiten klingt hier lediglich bei § 15 Abs. 3 PartG an, so daß die einzig realistische Einflußmöglichkeit von Minderheiten nur in dem Versuch liegen kann, in den einzelnen Gebietsverbänden zur Mehrheit zu werden, um auf die innerparteiliche Willensbildung sodann einwirken zu können. Zwar können nach § 8 Abs. 2 PartG die jeweiligen Parteisatzungen weitere der Willensbildung dienende Einrichtungen bzw. Organe vorsehen; tatsächlich ist dies jedoch in keiner Parteisatzung bisher erfolgt.

III. Das Gebot innerparteilicher Demokratie

1. Der ideologische und politische Konflikt ist dem Wesen der politischen Partei immanent. Das kommt unmißverständlich in der Konzeption des Grundgesetzes über die politische Partei zum Ausdruck, denn das Demokratiegebot für die innere Ordnung der politischen Parteien nach Art. 21 Abs. 1 Satz 3 GG präzisiert das Konfliktlösungsverfahren für den innerparteilichen politischen Prozeß und setzt somit den Konflikt voraus. Durch die Herstellung und Gewährleistung innerer Parteifreiheit soll gleichzeitig die Freiheit des politischen Prozesses an seiner Quelle gesichert werden[31].

---

[30] A.a.O., S. 139; *Wolfrum* (Anm. 5), S. 141 ff., insbes. S. 145 f.; *Zeuner* (Anm. 5), S. 16.

[31] *Hesse* (Anm. 22), S. 72. Zum ganzen Problem der demokratischen Willensbildung vgl. *Ute Müller,* Die demokratische Willensbildung in den politischen Parteien

Die Vorschrift richtet sich zunächst gegen alle Tendenzen, die versuchen, die freie politische Willensbildung innerhalb der Parteien zu beeinträchtigen. Fraglich ist allerdings, ob hieraus bereits der positive Schluß gezogen werden kann, daß Faktionsbildungen verfassungsrechtlich keinen Bedenken begegnen.

2. Die Übertragung eines Rechts auf Bildung einer Opposition aus dem staatlichen Organisationsbereich auf den innerparteilichen Raum würde jedoch voraussetzen, daß der Demokratiebegriff in Art. 21 Abs. 1 Satz 3 GG mit dem aus Art. 20 GG identisch ist. Eine solche prinzipielle Gleichschaltung zwischen dem Bereich staatlicher Organisation und innerparteilichem Bereich begegnet jedoch erheblichen Bedenken. Die Tatsache, daß Art. 21 Abs. 1 Satz 3 GG nur von den *demokratischen Grundsätzen* spricht, macht deutlich, daß eine schematische Übertragung demokratischer Institutionen des Grundgesetzes auf das Innenleben der politischen Parteien nicht beabsichtigt war, sondern nur unter Berücksichtigung der Parteifunktionen und den besonderen Eigenarten parteiinterner Willensbildung vorgenommen werden kann [32]. Allerdings könnte man sich mit gutem Recht auch auf den Standpunkt stellen, daß das Recht zur Ausübung einer Opposition sozusagen als *demokratisches Minimum* aus dem Bereich staatlicher Organisation auch auf den innerparteilichen Bereich übertragbar ist [33].

3. Bereits die Parteirechtskommission hat gefordert, daß die Willensbildung innerhalb der Parteien sich »von unten nach oben« entwickeln soll [34]. Wegen des zunehmenden Umfangs sowie der wachsenden Komplexität politisch-gesellschaftlicher Problemstellungen kann eine wirksame Willensbildung von unten nach oben in den Parteien jedoch nur in den Formen der Kooperation und Kommunikation zwischen den Mitgliedern stattfinden. Demokratische Willensbildung heißt deshalb auch Akzeptanz der Gruppenbildung. Entscheidungsalternativen in der parteiinternen Willensbildung setzen

---

(1967) und *E. Friesenhahn,* Die Stellung der politischen Parteien in der Verfassung, Hefte der Vereinigung für den Gedankenaustausch zwischen deutschen und italienischen Juristen, Nr. 4, S. 14 ff.

[32] *I. v. Münch,* Grundgesetz-Kommentar, Bd. 2 (1976), Rz. 45 zu Art. 21; *Henke,* a.a.O., Rz. 45; *D. Th. Tsatsos/M. Morlok,* Parteienrecht (1982), S. 36 ff.; *H. Trautmann,* Innerparteiliche Demokratie im Parteienstaat (1975), S. 170 ff. (174).

[33] *K.-H. Seifert* (Anm. 5), insbes. S. 15, 217 ff., 267 ff.; *J. Seifert* (Anm. 9); *H.-P. Schneider,* Die parlamentarische Opposition im Verfassungsrecht der Bundesrepublik Deutschland, Bd. 1 (1974), S. 214, 314; *H. See,* Volksparteien im Klassenstaat (1969), S. 43; *Trautmann,* a.a.O., S. 130 ff.; *F. J. Bröder,* Blätter für deutsche und internationale Politik 1970, S. 1165 ff., insbes. S. 1168. Die Akzeptanz innerparteilicher Oppositionsbildungen liegt offensichtlich auch im Verständnis des Bundesverfassungsgerichts von innerparteilicher Demokratie; siehe BVerfGE 2 (1 ff.), 42 f.: »Dieser Kollektivausschluß ist nicht nur eine Verletzung der Mitgliedschaftsrechte, er ist zugleich ein unzulässiges Mittel, um jede Opposition gegen die Parteiführung mundtot zu machen«.

[34] Rechtliche Ordnung des Parteienwesens; Bericht der vom Bundesminister des Inneren eingesetzten Parteienrechtskommission (1957), S. 157.

deshalb auch die »Publizität des internen Parteigeschehens« voraus[35]. Die Notwendigkeit innerparteilichen Gruppenwettbewerbs spricht aus verfassungsrechtlicher Sicht daher für die Zulässigkeit innerparteilicher Oppositionsbildungen[36].

Lösungen, welche die Akzeptanz davon abhängig machen würden, ob die sachlich-politischen Zielvorstellungen denen der Gesamtpartei bzw. den dort herrschenden Mehrheiten entsprechen, würden einerseits die normative Funktion des Geschlossenheitsprinzips überschätzen und gleichzeitig den erstrebten Willensbildungsprozeß umkehren, da dann innerparteiliche Machteliten über die Existenzberechtigung solcher Gruppierungen entscheiden würden.

4. Unter Berücksichtigung des Erfordernisses innerparteilicher demokratischer Ordnung sind auch die Möglichkeiten der Parteien zur Durchführung von Ausschlußverfahren zu betrachten[37]. Hierzu sind die Ausschlußgründe des § 10 Abs. 4 PartG demokratiekonform zu interpretieren und das Verfahren selbst ist unter Beachtung demokratischer Grundsätze, sprich Öffentlichkeit, durchzuführen[38].

IV. Die Grundrechte

1. Auch die Frage der Grundrechtsgeltung könnte für die hier angesprochene Problematik relevant sein. Über die Frage der Grundrechtsgeltung im innerparteilichen Bereich ist in der verfassungsrechtlichen Literatur oft und mit unterschiedlichen Ergebnissen gestritten worden[39].

Der in der Faktionsbildung liegende bewußte und gewollte Zusammenschluß von Parteimitgliedern zur Durchsetzung von der jeweiligen Parteienmehrheit abweichender sachlich-politischer Zielvorstellungen könnte verfassungsrechtlich dann unbedenklich sein, wenn sich die einzelnen abweichenden Parteimitglieder zur Rechtfertigung ihres Handelns auf die Grundrechte aus den Art. 5, 8 bzw. 9 GG berufen könnten. Die Grundrechtsgel-

---

[35] *J. Habermas,* Strukturwandel der Öffentlichkeit (9. Aufl. 1978), S. 248.

[36] *Zeuner* (Anm. 5), S. 113 ff.; *Raschke,* Innerparteiliche Opposition (Anm. 9), S. 273; *derselbe,* Demokratisierung (Anm. 9), S. 17 ff.

[37] *P. Häberle,* Struktur und Funktion der Öffentlichkeit im demokratischen Staat, in: derselbe, Die Verfassung des Pluralismus (1979), S. 145. Ausführlich zur Ausschlußproblematik *E. Lengers,* Rechtsprobleme bei Parteiausschlüssen (Diss. Bochum 1973) und die in der folgenden Anm. angegebene Literatur.

[38] Zum Parteiausschlußverfahren siehe *G. P. Strunk,* Parteiausschlußverfahren wegen innerparteilicher Opposition (Diss. Köln 1974); sowie jüngstens *K. H. Hasenritter,* Parteiordnungsverfahren (1981).

[39] Die Meinungen führen von einer gänzlichen Ablehnung der Grundrechtsgeltung zwischen Partei und ihren Mitgliedern, so beispielsweise bei *Henke* (Anm. 5), Rz. 52, über eine bedingte Geltung, wie z. B. bei *K. Hesse,* Veröffentlichungen der Vereinigung der Deutschen Staatsrechtslehrer, Bd. 17, S. 11 ff. (32), *W. Luthmann,* Deutsches Verwaltungsblatt (DVBl.) 1962, S. 166 ff., bis zu einer unbeschränkten Geltung, wie z. B. bei *H. Lenz/Ch. Sasse,* Juristenzeitung 1962, S. 233 ff.; *G. Dux,* DVBl. 1966, S. 553 ff.

tung im innerparteilichen Bereich begegnet jedoch ebensolchen Bedenken wie die unbesehene Übernahme demokratischer Elemente aus dem staatlichen Organisationsbereich auf den Bereich der innerparteilichen Ordnung. Die Problematik soll hier auch nicht weiter verfolgt werden, da die Grundstrukturen der Problemstellung noch dem herkömmlichen Trennungsdenken verhaftet sind. Tatsächlich geht es weniger darum, die Partei und ihre Mitglieder in einen Gegensatz zu bringen, der durch die Geltung der Grundrechte im Sinne eines *status negativus* zugunsten der Mitglieder schützend zu strukturieren wäre, als vielmehr darum, die Parteienfreiheit auch nach innen hin für die Mitglieder und Vereinigungen von Mitgliedern zu effektivieren. Die Grundrechte sind daher weniger in ihrem abwehrenden Charakter zu sehen, als in ihrer Bedeutung für die Praktizierung von Demokratie[40]. Die negatorische Seite darf jedenfalls nicht in den Vordergrund gestellt werden.

2. Die Tatsache, daß die innerparteiliche Ordnung demokratischen Grundsätzen entsprechen muß, stellt jedoch sicher, daß bestimmte Grundrechtsprinzipien in den parteiinternen Raum transformiert werden. Hierbei kann es sich jedoch nur um solche Prinzipien handeln, die für den freien innerparteilichen Willensbildungsprozeß konstituierend sind. In diesem Sinne sind die Meinungs- und Vereinigungsfreiheit elementare, wesentliche Voraussetzungen innerparteilicher Demokratie. Nicht nur das Recht, seine Meinung frei zu äußern, sondern auch die Möglichkeit zur freien, sozialen Gruppenbildung[41] müssen im innerparteilichen Raum ihre Entsprechung finden.

Gerade weil die Parteien Träger und Mittler des freien Willensbildungsprozesses sind, haben sie zuvörderst in ihrem internen Bereich darauf bedacht zu sein, daß jegliche Willensbildung sich frei von Repressionen und demokratisch vollziehen kann.

3. Gleichwohl lassen sich aus der Anerkennung innerparteilicher Meinungsfreiheit zur Problematik der Zulässigkeit innerparteilicher Oppositionsbildungen keine weiteren Konsequenzen ziehen. Innerparteiliche Oppositionsbildungen setzen zwar insoweit die Möglichkeit jedes Parteimitgliedes voraus, sich innerhalb der Partei – mithin gegenüber der Parteiöffentlichkeit – zu artikulieren. Faktionen jedoch beabsichtigen nicht nur, diese »parteiintern« in Anspruch zu nehmen, sondern versuchen, die parteiinterne Auseinandersetzung vor der gesellschaftlichen Öffentlichkeit auszutragen. Noch in einem weiteren Punkte geht die Frage nach der Zulässigkeit innerparteilicher Oppositionsbildungen über das Problem innerparteilicher Meinungsfreiheit hinaus: Meinungsfreiheit im angedeuteten Sinne der Gewährleistung freier politischer Auseinandersetzung kann nur eine Funktion zur Vorbereitung abschließender Beschlußfassung in der jeweiligen Parteiorganisation haben. Sol-

---

[40] Zum Zusammenhang von Grundrechten und Demokratie vergleiche *P. Häberle*, Die Wesensgehaltgarantie des Art. 19 Abs. 2 Grundgesetz (2. Aufl. 1972), S. 17 ff.; die aktivbürgerliche Seite der Grundrechte hat *Trautmann* (Anm. 32), S. 181 ff., für die Grundrechtsproblematik im parteiinternen Raum herausgearbeitet.

[41] Hierzu *Hesse* (Anm. 22), S. 168.

che Beschlüsse werden in den nach demokratischen Formprinzipien aufgebauten Parteien durch Mehrheitsentscheid gefunden. Der Mehrheitsentscheid beendet zunächst die freie innerparteiliche Meinungsäußerung insoweit, als nunmehr der Mehrheitswille sowohl den Bürgern gegenüber verdeutlicht werden als auch in das staatliche Entscheidungshandeln eingehen soll. Die besondere Bedeutung von Faktionsbildungen liegt jedoch gerade darin, daß diese auch nach abschließender Entscheidung ihren von der Parteimehrheit abweichenden Willen nach innen und außen zu vertreten versuchen.

§ 4 Ursachen innerparteilicher Oppositionsbildung

I. Einleitung

1. Die Frage nach den Ursachen einer Bildung von innerparteilichen Oppositionsgruppen zielt auf die Faktoren, die verantwortlich dafür sind, daß politische Meinungsunterschiede und Interessengegensätze innerhalb von bestehenden Parteien ausgetragen werden und nicht, wie eigentlich zu erwarten wäre, in der Form einer Konkurrenz zwischen verschiedenen Parteien.

2. Die Suche nach den Ursachen hat Einfluß auf die normative Bewertung innerparteilicher Oppositionsbildungen, denn die parteirechtliche Beurteilung innerparteilicher Erscheinungsformen hat ständig zu beachten, daß Parteien politische Gebilde sind und daher stets unter Berücksichtigung der gesellschaftlichen und verfassungspolitischen Dynamik zu betrachten sind.

Dementsprechend läßt sich eine komplexe Erscheinung wie innerparteiliche Opposition auch nur unter Einbeziehung parteiensoziologischer und politikwissenschaftlicher Erkenntnisse verfassungsrechtlich bewerten. Im Rahmen eines kurzen Beitrags können Ursachen von Faktionsbildung jedoch nur sehr kursorisch dargestellt werden.

II. Parteiexterne Faktoren

1. Die Hauptursache für diese Zentrierung des politischen Handelns auf den Innenbereich einer politischen Partei – und zwar auch dann, wenn es um rivalisierende Auffassungen geht – liegt im faktisch bestehenden *weitgehenden Monopol* der wenigen existierenden politischen Parteien auf relevantes politisches Handeln. Wer politisch etwas bewirken will, wird dies in aller Regel über die bestehenden großen Parteien versuchen. Die rechtlichen und die faktischen Hindernisse für eine Parteikonkurrenz und eine gleiche Chance auf Parteineugründung legen es nahe, den Versuch der Durchsetzung eigener Vorstellungen auf dem Weg über die Einflußnahme in einer bestehenden Partei zu unternehmen. Demgegenüber erscheint es dornenreich und schwieriger, sich mit einer eigenständigen Parteineugründung Gehör zu verschaffen.

Die großen Parteien haben Medienzugang, die Vermittler der öffentlichen Meinung sind auf diese Parteien und ihre Stellungnahmen ausgerichtet.

Vor allem haben die bestehenden großen Parteien Mandatsträger in den institutionalisierten Entscheidungsgremien. Über politische Ziele wird dort verbindlich entschieden. Einfluß auf diese Mandatsträger läßt sich aber vorwiegend über deren Partei gewinnen. Die Parteibasis von Mandatsträgern zu sein, verspricht mehr Aussicht auf politische Durchsetzung der eigenen Ziele als eine Basis ohne parlamentarische Vertretung zu sein.

2. Unter den Ursachen dieser faktischen Dominanz der bestehenden großen Parteien ist zunächst handgreiflich die 5%-Klausel des Wahlrechts. Sie legt die »Repräsentationsschwelle« [42] höher und begründet die Gefahr, daß die für eine kleinere Partei abgegebenen Stimmen im Endergebnis nicht zählen. Viele Wähler, jedenfalls bei Wahlen, die sie für wichtig halten, werden ein solches Risiko nicht eingehen wollen. Ehe eine Partei mit einiger Sicherheit die 5%- Hürde nimmt, hat sie mit einem erheblichen faktischen Nachteil zu rechnen.

Ein weiterer rechtlich abgesicherter Vorteil der bestehenden größeren Parteien liegt im Schlüssel für die Vergabe von Sendezeiten in Rundfunk und Fernsehen vor Wahlen. Diese Zeiten werden gemäß § 5 Abs. 1 PartG nach dem bisherigen Erfolg der Parteien vergeben und benachteiligen damit Neuankömmlinge auf dem Markt der Parteien. Die bisher Erfolgreichen gehen mit einem Wettbewerbsvorteil ins neue Rennen.

3. Ein weiterer Faktor für die Stärkung innerparteilicher Gruppierungen liegt im Parteiensystem. Während im Zweiparteiensystem die Chancen von Faktionsbildungen wegen der scharfen Konkurrenzsituation zwischen den Parteien relativ gering sind, begünstigt ein Parteiensystem wie das der Bundesrepublik Deutschland wegen der Notwendigkeit von Regierungs- und Wahlbündnissen die Herausbildung innerparteilicher Gruppierungen. Innerhalb einer Partei können sich mehrere Richtungen organisatorisch verfestigen, welche unterschiedliche Bündnisse ihrer Partei anstreben und sich daher in ihren programmatischen Aussagen den jeweils avisierten Bündnisparteien annähern.

4. Eher im Bereich der politischen Kultur liegt eine Abneigung weiter Kreise gegen eine Parteienzersplitterung. Die Einheit der Parteien wird als hohes Gut betrachtet, um dessentwillen auch erhebliche Kompromisse in der Sache verlangt werden. Diese Einstellung ist ungeachtet ihrer historischen Verwurzelung und ungeachtet der Frage ihrer aktuellen Berechtigung jedenfalls zunächst einmal als Faktum vorhanden und hat ihren Einfluß.

5. Ein weiterer Grund für die Hinwendung zu den bestehenden großen Parteien liegt in der Logik politischen Handelns begründet. Politisches Handeln will etwas bewirken, und zwar in aller Regel möglichst schnell. Diese besondere Ausrichtung auf Effektivität legt es nahe, sich der bestehenden Einflußkanäle und Durchsetzungsstrategie zu bedienen, d.h. aber im wesentlichen, den Weg über die bestehenden großen Parteien zu gehen.

---

[42] Ausdruck bei *Raschke*, Organisierter Konflikt (Anm. 9), S. 175.

Freilich ist diese Anlehnung an die bestehenden politischen (*input-*) Strukturen damit verbunden, daß die darin liegenden Einseitigkeiten und Verzerrungen mit übernommen werden. Neue politische Gruppierungen könnten demgegenüber unbelastet von solchen Vorgaben sich in größerer Freiheit bewegen und politisch stärker innovatorisch wirken. Die Tendenz, politische Ziele innerhalb und über die bestehenden großen Parteien zu erreichen, hat auch darin eine Ursache, daß die zu aktivem politischem Handeln motivierten Bürger sich bereits verstärkt in den politischen Parteien befinden. Wenn diese Bürger jetzt ein aktuelles politisches Anliegen verfechten wollen, so ist es selbstverständlich, dies über ihre Partei zu tun. Für die relativ starke Zentrierung politischer Aktivitäten auf die bestehenden Parteien sind zu einem nicht unerheblichen Ausmaß auch selbstselektive Effekte in der politischen Aktivbürgerschaft verantwortlich. Nicht zufällig leben Parteineugründungen zu einem erheblichen Teil von der nachwachsenden jungen Generation.

6. Bringt man die verschiedenen Bestimmungsgründe für die Formen politischen Handelns auf einen gemeinsamen Nenner, so muß festgestellt werden, daß die bestehende politische Infrastruktur die Realisierung politischer Anliegen bricht. Jedes politische Handeln hat unter den historisch gewordenen Bedingungen der gegebenen Struktur sich zu orientieren und zu bewähren. Es gibt keinen Neubeginn auf einer politischen *tabula rasa.* Auch neue Kräfte mit neuen Zielen haben aus den bestehenden Strukturen mit ihren Verwerfungen heraus zu handeln. Die politische Infrastruktur in der Bundesrepublik wird nun wesentlich von den bestehenden großen Parteien geprägt. Politisches Handeln kommt nicht umhin, sich an diesen Parteien zu orientieren. Auch gegenüber den von diesen Parteien bisher vorwiegend verfolgten Zielen abweichende Auffassungen werden zu einem erheblichen Teil versuchen, über diese Parteien zum Erfolg zu gelangen.

III. Parteiinterne Faktoren

1. Die Entstehungsbedingungen für Faktionen sind auch abhängig von den jeweiligen innerparteilichen Strukturbedingungen. Parteien mit großer Konflikttoleranz werden die Bildung innerparteilicher Oppositionsgruppierungen eher hinzunehmen bereit sein.

2. Auch die Größe der Partei kann Auswirkungen haben: Kleine Parteien, wie beispielsweise die Freien Demokraten, werden geneigt sein, innerparteiliche Gruppierungen eher zu dulden, da sie sich ansonsten eines Teils ihrer personellen Rekrutierungsbasis begeben würden.

3. Demgegenüber wird sich eine große und mächtige Parteibürokratie im Hinblick auf Faktionsbildungen eher negativ auswirken. Jede Art von Organisation bedeutet eine Verfestigung, bedeutet die Ausbildung von Schwerpunkten der Tätigkeit und umgekehrt dazu die Ausblendung anderer Aspekte. Durch Organisation bekommen verschiedene Interessen und Ziele unter-

schiedliche Durchsetzungschancen innerhalb der Partei. Unter dem Stichwort des »ehernen Gesetzes der Oligarchie«[43] sind diese Erscheinungen seit Robert Michels diskutiert worden. Neben sonstigen strukturellen Vorteilen für die bestehende Ausrichtung hat die Spitze einer Partei die verschiedensten Mittel in der Hand, ihre Auffassungen zu unterstützen. Dazu zählen die offiziellen Kommunikationsmittel der Partei, also Rundschreiben, Zeitungen und ähnliches, der organisatorische Apparat mit hauptberuflichen Funktionären, aber auch persönliches Ansehen und Prestige sowie der Hinweis auf die im parlamentarischen Raum sich ergebenden zwingenden Notwendigkeiten. Insbesondere neue politische Anliegen stoßen auf unverhältnismäßig große Hindernisse »im normalen Geschäftsgang« der Parteien. Verfechter solcher Anliegen werden deswegen danach trachten, sich durch organisatorische Zusammenarbeit bessere Durchsetzungschancen zu verschaffen. Eine Zusammenarbeit dieser Anhänger quer zu den satzungsmäßigen Organen liegt hier besonders nahe. Die Anhänger neuer Anliegen verfügen noch nicht über satzungsmäßige Gremien oder Arbeitskreise, sachliche und personelle Stränge müssen erst geknüpft werden. Die Hebung der kommunikativen Effektivität für Minderheitsgruppen mit neuen Anliegen läßt es deswegen ratsam erscheinen, daß diese sich um geeignete Organisationsformen innerhalb ihrer Partei bemühen.

4. Auch das Hinzutreten neuer politischer Bewegungen, wie beispielsweise augenblicklich das der sogenannten »Grünen«, begünstigt das Entstehen innerparteilicher Oppositionsbildungen in den bereits etablierten Parteien. Die Gefahr, daß die im innerparteilichen Meinungsstreit unterlegenen Mitglieder und damit möglicherweise auch Wähler zu der neuen Bewegung »abwandern«, stärkt die Bereitschaft zur Akzeptanz organisierter abweichender Meinungsbildung im Parteiinnern. Dies trifft selbst auf die Parteien zu, die ihrer historischen Entwicklung nach eher auf Einheit und Geschlossenheit bedacht sind.

§ 5 Auswirkungen innerparteilicher Opposition in verfassungsrechtlicher Sicht

I. Einleitung

1. Die bisherigen Überlegungen haben deutlich gemacht, daß innerparteiliche Oppositionsbildungen, welche sich gegen die offizielle Parteipolitik wenden, die Konsequenz der faktischen Rolle und Position politischer Parteien im politischen Gesamtsystem der Bundesrepublik Deutschland darstellen. Gleich-

[43] Stichwortgebend *R. Michels*, Zur Soziologie des Parteiwesens in der modernen Demokratie (1925); aus der neueren Diskussion in erster Linie *F. Naschold,* Organisation und Demokratie (1969), *Teubner* (Anm. 25).

zeitig konnte gezeigt werden, daß die verfassungsrechtlichen Rahmenbedingungen des Parteiensystems Faktionsbildungen außerordentlich begünstigen.

Die verfassungsrechtliche Bewertung von Faktionen bewegt sich ständig im Spannungsfeld zwischen innerparteilichem Pluralismus und parteilichem Geschlossenheitsprinzip.

2. Wie an Hand der einschlägigen Verfassungsgrundsätze nachgewiesen werden konnte, würde eine einseitige Auflösung dieses Spannungsverhältnisses zugunsten des Geschlossenheitsprinzips der Vielgestaltigkeit parteilicher Funktionen und den hieraus folgenden normativen Bewertungsgrundsätzen innerparteilicher Strukturen nicht gerecht werden. Bedauerlicherweise ist eine solch undifferenzierte Betrachtungsweise gerade von seiten der Parteien selbst ständig zu beobachten.

3. Wenngleich der Begriff »innerparteiliche Opposition« oder »Faktionsbildung« – soweit ersichtlich – kaum verwandt wird, sondern stattdessen von Uneinigkeit und Zerrissenheit gesprochen wird, besteht über die Parteigrenzen hinweg Einigkeit darüber, daß solche organisierten Erscheinungen in mehrfacher Hinsicht für die Partei selbst nachteilig sein könnten: Daß innerparteiliche Auseinandersetzungen Wählerverluste und Vertrauensschwund in der Bevölkerung herbeiführen, gehört zum ständigen ideologischen Repertoire der Parteiführungen[44].

Diese einseitigen und auch empirisch nur schwer nachvollziehbaren Feststellungen zu den tatsächlichen Auswirkungen innerparteilicher Opposition lassen es neben der rein rechtlichen Bewertung und Einordnung geboten erscheinen, abschließend noch zu der Frage Stellung zu nehmen, wem Faktionsbildungen letztlich schaden bzw. nutzen und in welchen Bereichen die verfassungsrechtliche Zulässigkeit innerparteilicher Oppositionsbildungen Änderungen zu bewirken vermag.

II. Auswirkungen auf die Parteimitglieder

1. Die Etablierung innerparteilicher Oppositionsgruppen bedeutet eine Steigerung der Chancen auf aktive Beteiligung am Parteigeschehen. Besonders gilt dies für die Anhänger von Minderheitsauffassungen. Diese sehen sich nämlich sonst vor dem Problem, daß sie in Abstimmung und Wahlen der Mehrheit unterliegen und am höherstufigen Parteigeschehen nicht oder nicht mehr angemessen repräsentiert teilnehmen können. Auch kann ein in der jeweiligen Basiseinheit voraussehbares Unterliegen leicht dazu führen, Tendenzen der Resignation aufkommen zu lassen und zur politischen Apathie der Minderheitskreise zu führen[45]. Die Parteiarbeit, die auf die Stabilisierung

---

[44] So gelten Parteitage insbesondere dann als erfolgreich, wenn möglichst wenig Widerspruch gegen die Parteiführungen bestand, wobei dann weniger der sachliche Gegenstand als die Tatsache eines Streites interessierte.

[45] Dies kann seinerseits dazu führen, daß auch die Vertreter der Mehrheitsauffassung mangels Notwendigkeit und Reiz einer politischen Beteiligung sich aus der aktiven Parteiarbeit zurückziehen.

und den Ausbau der Oppositionsgruppe zielt, bietet ein sinnvolles Ziel weiterer Arbeit. Die systematische Zusammenarbeit mit Gleichgesinnten bestärkt die eigene Auffassung und fördert die Entwicklung der politischen Programmatik. Die Zusammenarbeit von oppositionellen Kräften erhöht deren Gewicht, steigert die Aufmerksamkeit für ihre Anliegen und erhöht ihre Erfolgschancen, was alles wiederum einen motivierenden Effekt auf die einzelnen Mitglieder hat.

2. Politische Beteiligung des Bürgers setzt voraus, daß er sich für Vorhaben engagieren kann, die ihm sinnvoll erscheinen, daß er in überschaubarer Zeit Erfolge oder Mißerfolge persönlich erleben kann, daß er auch persönliche Erfolgserlebnisse dabei hat und nicht zuletzt die Unterstützung durch eine Gruppe, die kommunikativ seine Wertausrichtung bestärkt und ihm in persönlicher Hinsicht Anerkennung verschafft.

Die tatsächlichen Voraussetzungen dafür, daß Parteimitglieder sich auch aktiv am Parteigeschehen beteiligen und darüber hinaus Einfluß gewinnen, sind von der Forschung eingehend untersucht worden[46]; deren Ergebnisse können hier nicht dargestellt werden. Entscheidend ist im vorliegenden Zusammenhang jedenfalls der Punkt, daß die einfachen Mitglieder, die einer Minderheitsauffassung anhängen, nur auf der untersten Stufe der Hierarchie sich betätigen und Einfluß nehmen können und nur Gratifikationen auf dieser Ebene bekommen können – aber eben kontrolliert durch die Anhänger der Mehrheit.

Die Einbettung einer politischen Betätigung für Minderheitspositionen in Gruppen von Gleichgesinnten ist eine notwendige Voraussetzung für die Stabilisierung politischer Beteiligung überhaupt. Politische Aktivität der Bürger ist immer bedroht von der – verglichen mit anderen Tätigkeiten – relativ geringen Erfolgswirksamkeit der Beiträge des Einzelnen, jedenfalls vor der Langfristigkeit möglichen Erfolgs. Konkurrierende Handlungsfelder, etwa der Beruf oder die Familie, können vergleichsweise rascher und auch deutlicher greifbare Erfolge vermitteln. Eine andauernde politische Beteiligung der Bürger muß deswegen den Aktiven (gemessen an den letztlich verfolgten politischen Zielen) sekundäre Erfolge bieten. Dazu zählen in erster Linie die Unterstützung durch Gruppen von Gleichgesinnten, aber auch Erfolgserlebnisse, als Gruppe wie als Einzelner; dazu zählt auch die Betrauung mit bestimmten Aufgaben für die eigene Gruppe. Die Ausbildung einer parteiinternen Opposition erhöht die Zahl der Ämter bzw. »Quasi-Ämter«, die eine Partei vergeben kann, und kann damit Prestigebedürfnisse befriedigen. Persönlicher Prestigegewinn als Antriebsmotiv für die politische Betätigung in

[46] Aus der klassischen Forschung zu den Voraussetzungen politischer Beteiligung vgl. *H. Eulau/S. J. Eldersveld/M. Janowitz*, Political Behavior (1956); *R. E. Lane*, Political Life (1959); *L. Milbrath*, Political Participation (1956); *J. Steiner*, Bürger und Politik (1968); *H. Nie/G. B. Powell/K. Prewitt*, American Political Science Review 63 (1969), S. 361 ff., 808 ff.

einer Partei mag kein besonders ehrenwerter Grund sein, in tatsächlicher Hinsicht ist er aber ein besonders wichtiger und wirksamer[47].

Auch steigert die politische Betätigung selbst, gegebenenfalls auch in Oppositionsgruppen, ihrerseits soziale Fertigkeiten und Persönlichkeitseigenschaften, die wiederum einen positiven Einfluß auf die politische Betätigung haben[48].

3. Eine durch Oppositionsgruppen bewirkte und verstärkte Betätigung in den politischen Parteien ermöglicht sachlich eine Vergrößerung der Chance für die Mitglieder, ihre Auffassungen und Interessen in den politischen Entscheidungsprozeß einzubringen. Zugleich ist dies auch von expressiver Bedeutung für die Mitglieder. Diese emotional-expressive Seite des politischen Engagements darf nicht unterschätzt werden. Gerade der mündige Bürger mit seinem Anspruch, selbst politisch mitzubestimmen, darf nicht nur auf geschlossene Türen stoßen. Die eigene politische Aktivität mit erfahrbaren Ergebnissen des eigenen Tuns wird letztlich von den normativen Grundlagen einer der Würde des Einzelnen verpflichteten Demokratie gefordert. Er soll nicht nur passives Objekt der Apparate und der Parteigrößen sein, sondern soll selbst die Chance zu verantwortlichem Gestalten haben, was in modernen Großgesellschaften nur im Rahmen von Untergruppierungen von Parteien für eine größere Zahl von Bürgern möglich ist. Gerade weil die Demokratie im staatlichen Bereich stark repräsentativ ausgestaltet ist, sind kompensatorisch dazu Möglichkeiten direkten Engagements in den Parteien notwendig – auch und gerade für die Anhänger von Minderheitspositionen[49].

4. Die *normativen Vorgaben des Grundgesetzes* sprechen für eine Dynamik des politischen Systems, die zu einer Steigerung der direkten politischen Mitwirkungsmöglichkeiten der Einzelnen führt. Der Grundsatz der Menschenwürde realisiert in politischer Hinsicht für den Einzelnen das Prinzip der Volkssouveränität und begründet letztlich die politischen Betätigungsrechte der Bürger, mit Hinblick auf die Parteien, die innerparteiliche Demokratie und die innerparteilichen Handlungschancen. Die politikwissenschaftliche wie die juristische Beschäftigung mit den Parteien darf freilich auch keiner Partizipations- und Oppositionseuphorie für den innerparteilichen Bereich verfallen. Die Funktionsfähigkeit der Parteien darf nicht aus dem Blick verloren werden, dies gebietet schon Art. 21 Abs. 1 Satz 1 GG; im Hinblick auf die Mitglieder darf auch nicht außer acht gelassen werden, daß die Anhänger

---

[47] Zur Rolle von Prestige-Bedürfnissen und der Prestige-Verteilung als Instrument sozialer Steuerung vgl. umfassend *G. J. Goode*, The Celebration of Heroes (1979).

[48] Vgl. z. B. *Nie/Powell/Prewitt* (Anm. 46), S. 808 ff.

[49] Vgl. dazu *Tsatsos/Morlok* (Anm. 32), § 3 II: »Je stärker sich die repräsentativen Elemente in Verfassungsrecht und Verfassungswirklichkeit entwickeln, desto systemnotwendiger wird es, daß sich die Demokratieverwirklichung im innerparteilichen Raum im Sinne der Stärkung der Mitglieder- und Basismöglichkeiten (*Elemente der unmittelbaren Demokratie*) gestaltet«.

der Mehrheitsauffassung ihrerseits nicht demotiviert werden. Sie haben als Mehrheit offizielle Bestimmungschancen, und sie müssen diese auch wahrnehmen können, falls nicht ihre politische Betätigung sinnlos erscheinen soll. Auch ist zu beachten, daß manche Parteimitglieder eher von Gemeinschaftsgefühlen angesprochen werden als von innerparteilichen Auseinandersetzungen und Machtkämpfen. Ob solche Einstellungen allerdings realistisch sind und ob sie bewahrenswert sind, ist eine andere Frage.

Insgesamt ist für die Perspektive der Mitglieder festzuhalten, daß die innerparteiliche Oppositionsformierung die tatsächlichen Voraussetzungen zu erhöhter politischer Betätigung verbessert und daß das Verfassungsrecht einer solchen Beteiligung grundsätzlich positiv gegenübersteht. Vor allem ist zu betonen, daß die Perspektive der Mitglieder von Verfassungs wegen mit großer Bedeutung ausgestattet ist. Wegen dieses Gewichts der Mitgliederperspektive dürfen die in ihr gewonnenen Argumente für eine innerparteiliche Opposition nicht den Funktionserfordernissen der Partei als Organisation geopfert werden. Verfassungsrechtlich wird dies dadurch garantiert, daß die Parteienfreiheit und das Recht auf chancengleiche politische Betätigung auch ein *Individualrecht* der einzelnen Parteimitglieder ist, nicht nur ein Recht der Partei als solcher.

III. Auswirkungen auf die Parteien

1. Das bewußt weitgefaßte Tätigkeitsmerkmal »Mitwirkung bei der politischen Willensbildung« nach Art. 21 Abs. 1 GG gewinnt erst dann stärkere Konturen, wenn man an die Vielzahl von politischen Parteien denkt, die wegen des Wahlsystems keine parlamentarische Mitwirkungsmöglichkeit haben.

Die politische Entscheidung selbst ist staatlicher Herrschaftsakt der von der Verfassung hierzu bestimmten Staatsorgane. Demgegenüber bewegt sich die Tätigkeit der Parteien zunächst in einer der politischen Entscheidung vorgelagerten Phase. In diesem Stadium stellen die Parteien ihre in einem inneren Verfahren mehrheitlich getroffenen Aussagen der gesellschaftlichen Öffentlichkeit als Sachentscheidungsalternative zur Verfügung.

2. Im Hinblick auf die Artikulation von politischen Ansichten und Interessen ist die Aktivierung von Minderheitsmeinungen in Form organisierter innerparteilicher Opposition positiv zu sehen. Es wäre unnatürlich und ein Zeichen von Behinderung des politischen Kommunikationsprozesses, wenn Fragen, zu denen man mit guten Gründen verschiedener Auffassung sein kann, nicht auch in einer freiheitlichen Partei entsprechend kontrovers diskutiert werden und sich hierzu auch Faktionen bilden könnten. Dies gilt inbesondere für die Verhältnisse in den meisten modernen westlichen Demokratien, in welchen der tatsächliche Interessen- und Meinungspluralismus nicht in einer Vielzahl von Parteien abgebildet wird, sondern über relativ wenige

Parteien repräsentiert werden soll. Parteiinterne Opposition ist damit die notwendige Konsequenz der Entwicklung zu den sogenannten Volksparteien [50].

3. Die den politischen Parteien im gesamtgesellschaftlichen Prozeß zukommende Integrationsfunktion kann von der rechtlichen Akzeptanz parteilicher Oppositionsbildungen nur profitieren, Gesellschaftliche Gruppierungen, die zunächst außerhalb der Parteien und diesen kritisch oder sogar feindlich gegenüberstehen, können nur dann effektiv in den Meinungsbildungsprozeß mit einbezogen werden, wenn sie Hoffnung auf eine Veränderung der Strukturen innerhalb der Parteien haben können. Die Anziehungskraft politischer Parteien wird daher nach außen hin dadurch größer, daß die Anhänger recht verschiedener Auffassungen in ihnen eine Heimstatt finden können. Dies gilt gerade auch für Strömungen, die ihrer Natur nach zunächst als oppositionelle in Erscheinung treten: Historisch relativ neue Probleme, die in der bestehenden politischen Struktur noch keinen fixierten organisatorischen Ausdruck gefunden haben, werden zwangsläufig zunächst als Minderheitspositionen und als Opposition gegen die bisherige Praxis, die die neuen Probleme noch nicht beachtet hat, auf der politischen Bühne erscheinen.

4. Demgegenüber sind die Auswirkungen von Faktionsbildungen in bezug auf die Einflußnahme politischer Parteien auf die Entwicklung in den Parlamenten und Regierungen eher vorsichtig zu betrachten. Die Probleme ergeben sich insbesondere aus dem satzungsorganübergreifenden Charakter innerparteilicher Oppositionsgruppierungen. Das Charakteristikum von Faktionen besteht gerade darin – wie wir bereits oben erklärt haben –, daß es sich hierbei nicht um Suborganisationen der jeweiligen Partei handelt, sondern um Gruppierungen, die sich *neben* den partei- und satzungsrechtlich institutionalisierten Gremien herausbilden. Es besteht daher die Möglichkeit, daß Faktionen personell und sachlich bis in die Parlamentsfraktionen bzw. die Regierungen hineinreichen.

Gerade im Bereich der Fraktionen aber kommt dem Geschlossenheitsprinzip eine besondere Qualität zu. Seine Funktionalität besteht in diesem Bereich nicht nur im Hinblick auf die Selbstdarstellung der Partei, sondern unmittelbar auf die Stabilität des parlamentarischen Regierungssystems. Die Gefahr innerparteilicher Oppositionsbildungen in Regierungsfraktionen besteht insbesondere in einer nachlassenden Unterstützung der von der Fraktion zu tragenden Regierung, möglicherweise sogar in der Ablösung des Regierungschefs selbst.

5. Jede Öffnung der Parteien zur Gesellschaft hin wird jedoch darauf gerichtet sein, das erfolgreiche Bestehen im innerparteilichen Wettbewerb durch größtmögliche Erhöhung des Wählerpotentials zu sichern. Gerade in diesem Tätigkeitsbereich wird dem geschlossenen Auftreten der Partei besondere Bedeutung beigemessen. Der Versuch, einen Zusammenhang zwischen Wahlverhalten und Faktionsbildungen herzustellen, scheint jedoch insbeson-

---

[50] Dazu klassisch *Kirchheimer* (Anm. 2).

dere im Hinblick auf empirische Wahlanalysen mehr auf Vermutungen angewiesen zu sein denn auf wissenschaftlich nachvollziehbare Erkenntnisse[51]. Auf der anderen Seite darf außerdem nicht übersehen werden, daß die Akzeptanz innerparteilicher Oppositionsbildungen von seiten der Parteien eine Parteispaltung – und damit einen noch höheren Wählerschwund – zu verhindern vermag.

6. Die Entscheidung des Grundgesetzes für einen demokratischen Pluralismus in Art. 20 Abs. 1 und 2 soll insbesondere auch durch die Existenz eines pluralistischen Parteiensystems gesichert werden. Nur die Konkurrenz der Parteien untereinander kann die politische Willensbildung des Volkes offenhalten. Gefahren für die Offenheit des politischen Prozesses ergeben sich insbesondere aus der ständigen programmatischen Angleichung der Parteien untereinander. In dem Maße jedoch, in dem sich die Konkurrenz der Parteien untereinander zurückentwickelt, erfordert die Pluralismussicherung eine verstärkte Hinwendung zu den innerparteilichen Strukturen.

7. Der zentrale Maßstab für die rechtliche Beurteilung der Wirkungen innerparteilicher Oppositionsgruppen ist daher im Verfassungsgebot zur innerparteilichen Demokratie, Art. 21 Abs. 1 Satz 3 GG, zu suchen.

Zu den wesentlichen Elementen innerparteilicher Demokratie gehört sicher, daß der Willensbildungsprozeß von unten nach oben verläuft und damit letztlich von den Mitgliedern und den unteren Gruppierungen der Partei bestimmt wird und nicht von der Parteispitze. Die Mitglieder und Gruppierungen von Mitgliedern auf verschiedenen Ebenen müssen die Möglichkeit haben, Ansichten in effektiver Weise zu artikulieren und Vorschläge zur Diskussion und Abstimmung stellen zu können. Gesichert werden muß daher zunächst die Rechtsposition des einzelnen Parteimitglieds. Daneben darf jedoch nicht verkannt werden, daß für den tatsächlichen Geschehensablauf in einer Partei Gruppen von Mitgliedern die ausschlaggebende Einheit darstellen. Eine wirklichkeitsorientierte Betrachtung der Problematik innerparteilicher Demokratie wird sich deswegen weit stärker als bisher um die Ausdrucks- und Handlungsmöglichkeiten von parteiinternen Gruppierungen zu kümmern haben.

Die sozialwissenschaftliche Organisationsforschung hat mehrfach belegt, daß die innerorganisatorische partizipative Demokratie nicht zureichend gesichert wird durch die herkömmlichen institutionellen Vorkehrungen. Die Möglichkeit wirksamer Beteiligung der Mitglieder muß dadurch gesichert werden, daß zwischen der Ebene der einzelnen Mitglieder und der Parteispitze die Zwischenebene der innerparteilichen Gruppierungen berücksichtigt wird[52]. Im Hinblick auf das innerparteiliche Demokratiegebot aus Art. 21

---

[51] Beispiele für das besonders erfolgreiche Abschneiden als uneinig und zerstritten geltender Parteien bei *Hasenritter*, Aus Politik und Zeitgeschichte 1982, Heft 14, S. 23 f.

[52] Klassisch hierfür ist die Studie von *S. M. Lipset/M. Trow/J. Coleman* (Anm. 25); ein wesentliches Ergebnis dieser Studie über die amerikanische Druckge-

Abs. 1 Satz 3 GG bedeutet dies, daß die parteiinternen organisatorischen und verfahrensmäßigen Ausgestaltungen derart sein müssen, daß innerparteiliche Gruppierungen als solche ausdrucks- und handlungsfähig sind und ein innerparteilicher Wettbewerb von verschiedenen Gruppierungen möglich wird[53].

An dieser Stelle ist zunächst nur die Feststellung wichtig, daß die Bildung innerpaterteilicher Oppositionsgruppen als legitim und im Interesse innerparteilicher Demokratie als notwendig anerkannt wird.

IV. Auswirkungen auf das politische System

1. Die Bedeutung innerparteilicher Opposition für *das politische System* ergibt sich aus der wesentlichen Rolle innerparteilicher Opposition für das Funktionieren innerparteilicher Demokratie.

Wie in § 4 gesehen, steigt die Wichtigkeit innerparteilicher Opposition mit dem Grad der Beschränkungen des Wettbewerbs zwischen den bestehenden Parteien, den tatsächlichen Hindernissen einer Parteineugründung und den realistisch gesehenen Möglichkeiten einer Fortsetzung effektiver politischer Arbeit nach einem Parteiaustritt bzw. Parteiausschluß.

2. Im Interesse der Demokratie im Gemeinwesen überhaupt muß das Parteiensystem selbst flexibel bleiben. Das bedeutet unter den die bestehen-

---

werkschaft ist, daß die innerverbandliche Demokratie durch den Wettbewerb von internen Gruppen erst wirksam wird. Siehe z. B. S. 15: ». . . democracy is most likely to become institutionalized in organizations whose members form organized or structured subgroups which while maintaining a basic loyality to the larger organization constitute relatively independent and autonomous centers of power within the organization. Or to put it in another way, democracy is strengthened when members are not only related to the larger organization but are also affiliated with or loyal to subgroups within the organization.« – In Deutschland hat insbes. *F. Naschold*, Organisation und Demokratie (1969), die einschlägige Forschung dargestellt und rezipiert, siehe besonders S. 33 ff., 90 ff.; z. B. S. 91: »Denn wer eine Demokratisierung der Gesamtorganisation will, muß ›Fraktionsbildung‹ wollen, wenn das Bekenntnis zu demokratischen Strukturen sich nicht zur reinen Außendarstellung verkehren soll . . . es für die innerparteiliche Demokratisierung entscheidend wichtig ist, die Beziehungen zwischen breiter Mitgliedschaft und Parteiführung durch den Einbau einer strategisch richtig plazierten Zwischenebene zu institutionalisieren«. Insbesondere hebt er hervor, daß institutionelle Regelungen nur richtig wirksam werden, wenn sie in sozialer und psychischer Hinsicht abgesichert sind durch den Aufbau von informellen Gruppen. Speziell für die rechtswissenschaftliche und rechtspolitische Diskussion hat *Gunther Teubner*, Organisationsdemokratie und Verbandsverfassung (1978), diese Thematik aufgegriffen, weitergeführt und für rechtswissenschaftliche Fragestellungen spezifiziert. Auch er bestätigt die ganz entscheidende Rolle von verbandsinternen Koalitionen mit einer relativen Autonomie bei der Verwirklichung innerverbandlicher Demokratie. Rechtliche Normierungen, die bloß beim einzelnen Mitglied ansetzen, bleiben deswegen relativ folgenlos. Gerade das Verbandsrecht hat seine Aufmerksamkeit den innerverbandlichen Gruppierungen zuzuwenden, . . . Hier besonders S. 197 f.

[53] Welche einzelnen Regelungen dazu tauglich und nötig sind, kann im Rahmen dieses Beitrages nicht erörtert werden, bedarf vor allem auch fundierter parteiensoziologischer und -psychologischer Untersuchungen.

den Parteien favorisierenden Bedingungen, daß die bestehenden Parteien sensibel bleiben für die Wünsche und Befürchtungen, Vorstellungen und Interessen der Bevölkerung, die sich in einer ständig sich wandelnden Welt gleichfalls ändern. Die institutionellen Festschreibungen innerhalb der Parteien müssen immer wieder in Frage gestellt werden können, sollen die Parteien wirksame Instrumente des Ausdrucks und auch der Formulierung der politischen Anliegen des Volkes sein. Wenn aus den skizzierten Gründen Parteineugründungen geringere Chancen haben, so muß auf der Grundlage von Art. 21 Abs. 1 Satz 1 und 3 GG die Chance bestehen, daß durch die alten Parteien neue Politik gemacht wird.

3. Die politische Situation ist gekennzeichnet von einem starken Pluralismus der Interessen und politischen Auffassungen. Die Legitimation der politischen Ordnung hängt nun davon ab, daß die Spannweite des faktischen politischen Pluralismus wenigstens einigermaßen abgebildet wird auf die wesentlichen politischen Handlungseinheiten: die Parteien. Wenn die Spannweite der von den Parteien insgesamt vertretenen Interessen und Ziele erheblich geringer ist als die der im Volk tatsächlich vorhandenen, so führt dies zu einer fundamentalen Legitimationskrise. In den bestehenden Volksparteien können die heterogenen politischen Vorstellungen aber nur dann effektiven Ausdruck finden, wenn es dort Raum gibt für die Betätigung von politischen Gruppen, die bei Beibehaltung einer übergeordneten Loyalität zu ihrer Partei verschiedene politische Ziele verfolgen können. Die Funktionsfähigkeit und die Legitimität der politischen Ordnung verlangen, daß der bestehende *Außenpluralismus* in einem adäquaten *Innenpluralismus* in den wenigen großen Parteien repräsentiert wird.

4. Verstärkt wird die Notwendigkeit zur innerparteilichen Vielfalt und Offenheit durch die Dynamik der Verhältnisse. Der Wandel ist mit der geschichtlichen Natur des Menschen schlechthin gegeben. Eine der Wurzeln des Demokratiegebotes dürfte gerade darin zu sehen sein, daß demokratische Strukturen die Anpassung an veränderte Umstände erleichtern.

Gerade in den letzten Jahren ist unter dem Schlagwort »Wertewandel« eine Verschiebung in der Wertvorstellung und Motivationsstruktur nicht unerheblicher Teile der jüngeren Generation festzustellen[54]. Auch für die Zukunft ist davon auszugehen, daß die jeweils junge Generation aus ihren je subjektiv erlebten Erfahrungen nicht vorhersehbare politische Einstellungen entwickelt, die sie politisch dargestellt und verwirklicht sehen möchte. Die Wichtigkeit und Dringlichkeit bestimmter Anliegen kann sich daraus verändern. Für solche Prozesse müssen die politischen Parteien offen sein.

5. Das bedeutet in organisatorischer Hinsicht die Ermöglichung der Tätigkeit von verschiedenen Gruppen innerhalb einer Partei. Im Hinblick auf die – immer notwendige und zulässige! – Grenzziehung des innerhalb einer

---

[54] Wichtig *R. Inglehart*, The Silent Revolution (1977); *H. Klages/P. Kmieciak* (Hrsg.), Wertwandel und gesellschaftlicher Wandel (1979).

spezifischen Partei Tolerablen heißt dies aber auch, daß die Kriterien der Identität einer Partei, die als Kriterien der Abgrenzung dieser Partei dienen, sich wandeln können und müssen! Konkret bedeutet dies, daß der Inhalt der »Grundsätze der Partei«, der nach § 10 Abs. 4 PartG Maßstab für schwere Verfehlungen und damit ein Parteiausschlußgrund sein kann, sich ändern kann und muß. Gleiches gilt für die weitere Ausschlußvoraussetzung des schweren Schadens, der einer Partei zugefügt wird. Gerade im Hinblick auf die vielfältige politische Entwicklung kann nicht immer ohne weiteres gesagt werden, daß eine Auflockerung der Geschlossenheit der Partei für diese einen Schaden darstellt.

Parteiideologien und Schadenskriterien wandeln sich und müssen sich wandeln können. Dies ist die Folge der Tätigkeit von parteiinternen Oppositionsgruppen, wie umgekehrt die Anerkennung einer solchen Veränderbarkeit eine wichtige Voraussetzung für die Akzeptierung von parteiinterner Opposition ist.

6. Die Transparenz des parteiinternen Geschehens könnte zunächst das Argument stützen, eine in der politischen Öffentlichkeit betriebene oder ausgetragene innerparteiliche Opposition gegen die institutionell legitimierte Parteiführung könnte das notwendige Bild der Geschlossenheit der Partei stören. Ein solches Argument aber steht und fällt mit einem überholten Parteiverständnis, das von der Unterstellung ausgeht, die Partei sei ein ideologisch und politisch homogenes Gebilde, in dem Konflikt und Meinungskampf ein pathologischer, mit dem Begriff der Partei unvereinbarer Zustand sei. Ein so eng verstandenes »Geschlossenheitsprinzip« kann der Partei den ihr von der Verfassung gegebenen Charakter als Forum souveräner politischer Willensbildung entscheidend schwächen. Anders ausgedrückt: Art. 20 Abs. 2 GG läßt nicht zu, daß eine bestimmte Ideologie oder eine bestimmte Politik ein für alle Mal inhaltlich determiniert wird. Vielmehr unterliegen solche Inhalte dem Nomos des Wandels. Nicht die *Bezeichnung* einer politischen Partei z. B. als »liberal« oder »sozialdemokratisch« ist das Verbindliche, sondern der *Inhalt* jener Ideologien. Der Streit um solche Inhalte aber bleibt ewig aktuell und gibt somit der Partei, die ihn sucht, die Möglichkeit, eine Institution der Volkssouveränität nach Art. 20 Abs. 2 GG zu bleiben.

7. Wird die Notwendigkeit und Berechtigung des politischen Pluralismus und Wandels auch innerhalb einer Partei anerkannt, so kann auch aus dem Blickwinkel des politischen Systems insgesamt einer innerparteilichen Opposition grundsätzlich nicht widersprochen werden. Nur gruppenmäßig nach außen abgegrenzte und nach innen verstärkte Kommunikation verleiht solchen oppositionellen Bestrebungen Effektivität und Durchsetzungschancen.

8. Freilich muß auch ein Interesse des gesamtpolitischen Systems anerkannt werden, eine Parteienzersplitterung zu verhindern. Die Diskussion darüber braucht hier nicht aufgerollt zu werden. *De lege lata* ist dieses Interesse durch die 5%-Klausel in den Wahlgesetzen anerkannt. Eine Anmerkung zu

diesem Komplex scheint aber nötig. In Deutschland wird häufig vor einer Parteienzersplitterung unter Hinweis auf den Untergang der Weimarer Republik gewarnt, wofür die Aufspaltung der demokratischen Kräfte eine wesentliche Mitursache gewesen sei. Dem ist entgegenzuhalten, daß die Vielzahl der Parteien nicht *per se* die Schwierigkeiten der Weimarer Republik begründete, sondern vielmehr deren Unfähigkeit und Unwillen, miteinander tragfähige Bündnisse einzugehen. Parteienzersplitterung an sich kann damit nicht als so großes Übel angesehen werden, vielmehr die Verfestigung der politischen Gegensätze, die Tiefe der Gräben, welche die verschiedenen politischen Lager trennt. Gerade unter dem Dach einer Partei, welches doch eine gewisse Einheitlichkeit stiften dürfte, scheinen im Hinblick auf dieses Problem der politischen Einheitsbildung die Chancen einer immer neu notwendigen Verständigung und Kompromißbildung größer zu sein. Gerade im Hinblick auf die Weimarer Erfahrungen dürfte deswegen eine starke interne Auffächerung der Parteien als der gegenüber einer Vielzahl von Parteien empfehlenswertere Weg erscheinen.

9. Schließlich sind auch die Gegebenheiten der politischen Kultur eines Landes veränderbar. In dieser Hinsicht ist ein pädagogisches Moment der Anerkennung innerparteilicher Opposition hervorzuheben: Wesentlich für das Funktionieren einer Demokratie ist die Fähigkeit und Bereitschaft weiter Teile der politisch interessierten Bevölkerung, politisch Andersdenkende grundsätzlich zu akzeptieren, mit Konflikten zu leben und nicht jeden Gegner zu verteufeln (etwa als »Verfassungsfeind« oder »Parteischädling«). Meinungsverschiedenheiten zu ertragen auf der Basis von Gemeinsamkeiten – das ist ein Essential der Demokratie, im großen wie im kleinen. Die Praktizierung und Tolerierung innerparteilicher Opposition kann somit für die politische Kultur wie für die einzelnen Bürger und Parteimitglieder wichtige Lernerfahrungen darstellen.

§ 6 Schlußbemerkung

1. Es besteht ein Recht auf Bildung, Teilnahme, Organisierung und öffentliche Artikulierung einer innerparteilichen Opposition.

2. Gemeint sind Oppositionen, die nicht durch das parteigesetzliche oder parteisatzungsrechtliche Organisationsrecht legitimiert sind. Solche Oppositionen gehen über den Rahmen und die Strukturen institutionalisierter Parteiorgane hinaus. Sie entstehen im Rahmen der großen gesellschaftlichen Konflikte, ideologischer wie politischer Art, die parteiübergreifend und somit *nicht nur unter* den Parteien, sondern auch und immer *stärker in den Parteien* ausgetragen werden.

3. Dieses Recht resultiert aus der Funktionsbestimmung der politischen Parteien durch das Grundgesetz.

a) Das Prinzip der Volkssouveränität (Art. 20 Abs. 2 GG) verwirklicht sich zu einem wesentlichen Teil durch den Pluralismus nicht nur zwischen, sondern auch *in* den Parteien; damit ist den Parteimitgliedern die rechtliche Möglichkeit eröffnet, an der Austragung der großen Konflikte unserer Zeit effektiv teilzuhaben.

b) Das Parteimitglied hat aber nur dann in seinem organisierten Verband wirklich Einfluß, wenn es sich mit Gleichdenkenden organisatorisch zusammenschließen darf.

c) Das positive Parteiorganisationsrecht darf nicht dahin interpretiert werden, daß jede Tätigkeit unzulässig sei, die über die Aktivität in den satzungsmäßigen Gremien hinausgeht.

4. Das Recht auf innerparteiliche Opposition verdrängt nicht den normativen Bestand und die Notwendigkeit des Schutzes der Partei im internen Bereich. Die Prinzipien der Disziplin, der Geschlossenheit und der Homogenität stehen sicher in einer gewissen Spannung zur Anerkennung der innerparteilichen Opposition; sie sind aber mit ihr nicht unvereinbar.

5. Die verfassungsrechtliche Anerkennung innerparteilicher Oppositionsbildungen schließt jedenfalls aus, daß solche Aktivitäten *per se* als Ausschlußgrund gewertet werden.

6. In dem Umfang und unter den Voraussetzungen, in denen nach diesen Überlegungen Teilnahme und Mitwirkung an innerparteilicher Opposition zulässig sind, kann diese Aktivität weder als »Gefährdung der Partei« noch als »Verletzung der Satzung« interpretiert werden.

7. Das Geschlossenheits- und Homogenitätsideal im modernen Parteiwesen, auch in der Bundesrepublik Deutschland, wird – so fürchte ich – dann zu einem untauglichen Instrument von Machtpolitik einer verlegenen Parteispitze, wenn es in der Legitimationsbasis der Partei kriselt.

Der Angriffskrieg im System des Deutschen Bundes (1815–1866)

STEPHAN VEROSTA

Im Ersten Pariser Frieden vom 30. Mai 1814 erklärten die Vertragsstaaten – Großbritannien, Österreich, Preußen, Rußland und Frankreich – ihr Verlangen, »den langwierigen Erschütterungen von Europa und dem Unglücke der Völker durch einen festen, aber eine richtige Verteilung der Kräfte unter die Mächte gegründeten, und in seinen Bestimmungen die Gewährleistung für seine Dauer enthaltenden Frieden ein Ende zu machen«; Art. 6 dieses Vertrages bestimmte: »Die Staaten Deutschlands werden unabhängig und durch ein föderatives Band vereinigt sein«. Schon im Allianzvertrag von Chaumont vom 1. März 1814 hatten Großbritannien, Österreich, Preußen und Rußland es als Ziel dieses Defensivbündnisses bezeichnet, »das Gleichgewicht in Europa aufrechtzuerhalten, die Ruhe und Unabhängigkeit der Mächte zu sichern und die Überfälle (*envahissements*) zu verhindern, die seit so vielen Jahren die Welt verwüstet haben«. Die Präambel zum Zweiten Pariser Frieden vom 20. November 1815 stellte fest, daß »die verbündeten Mächte durch ihre vereinten Anstrengungen und den Erfolg ihrer Waffen Frankreich und Europa vor den Zerrüttungen, womit sie durch das letzte frevelhafte Unternehmen Napoleon Bonapartes und die zur Unterstützung desselben in Gang gebrachten revolutionären Maßregeln bedroht waren, gerettet haben«; sie wiederholt die Verurteilung von Angriffs- und Eroberungskriegen in der Absicht, »die ehemaligen durch den verderblichen Einfluß der Revolution und des Eroberungssystems lange Zeit gestörten Verhältnisse wieder anzuknüpfen«. Der Friedensvertrag sollte den verbündeten Mächten gerechte Schadloshaltung für das Vergangene und befriedigende Gewährleistung für die Zukunft sichern. Die verbündeten Mächte qualifizierten damit neuerlich und abschließend die Kriege der französischen Revolutionsregierungen und des napoleonischen Regimes samt den folgenden Annexionen als völkerrechtswidrige Angriffs- und Eroberungskriege.

Für die Aufrechterhaltung der im Zweiten Pariser Frieden und in der Wiener Kongreßakte vom 9. Juni 1815 bis ins Detail formulierten neuen

Friedensordnung Europas sorgte das Europäische Konzert der vier siegreichen Mächte (Quadrupelallianz), dem bald das Frankreich unter den restaurierten Bourbonen zugezogen wurde (Quintupelallianz). In dieses europäische System war noch die vom Zaren Alexander I. initiierte sogenannte Heilige Allianz vom 26. September 1815 eingefügt. Dieses Europäische Konzert war kein organisierter Staatenbund, aber in Krisenfällen traten die fünf Mächte zu Staatenkonferenzen zusammen, die den Streitfall beilegten oder ihre Beschlüsse auch mit militärischen Mitteln durchsetzten. Das Europäische Konzert wurde daher wiederholt als ein Vorläufer des Völkerbundes angesehen. Angriffskriege mit dem Ziel der Eroberung fremden Gebietes und mit militärischer Gewalt herbeigeführte Änderungen des *status quo* in Europa waren völkerrechtswidrig. Gebietsveränderungen in Europa bedurften der Zustimmung des Europäischen Konzerts, das den Frieden und den *status quo* in Europa durch 45 Jahre weitgehend aufrechterhalten hat.

Der Deutsche Bund als regionale internationale Organisation

Im Rahmen des Systems des Europäischen Konzerts spielte der Deutsche Bund, dem die beiden Kongreßmächte Österreich und Preußen angehörten, eine bedeutende Rolle. In der Präambel seiner Gründungsurkunde, der Bundesakte vom 8. Juni 1815, heißt es:

> »Die souveränen Fürsten und freien Städte Deutschlands, den gemeinsamen Wunsch hegend, den VIten Artikel des Pariser Friedens vom 30. Mai 1814 in Erfüllung zu setzen, und von den Vorteilen überzeugt, welche aus ihrer festen und dauerhaften Verbindung für die Sicherheit und Unabhängigkeit Deutschlands und die Ruhe und das Gleichgewicht Europas hervorgehen würden, sind übereingekommen, sich zu einem beständigen Bunde zu vereinigen«.

Dieser multilaterale völkerrechtliche Vertrag war ein Annex der Wiener Kongreßakte und samt seinen nachfolgenden Ausführungsbestimmungen ein Teil des europäischen Völkerrechts. Er begründete im Deutschen Bund (Conféderation Germanique) einen Bund von 38 Staaten, nach der heutigen Terminologie eine regionale internationale Organisation mit politischen Zwecken, innerhalb des Europäischen Konzerts, dem überdies die Vormächte der Confoederatio Germanica, Österreich und Preußen angehörten. Als Zweck des Staatenbundes definiert Art. I der Bundesakte die Erhaltung der äußeren und inneren Sicherheit Deutschlands und der Unabhängigkeit und Unverletzlichkeit der einzelnen deutschen Staaten.

Hauptorgan dieser internationalen Organisation war die Bundesversammlung, später Bundestag genannt, der entweder als Plenum oder als engere Versammlung (engerer Rat) tagte. Im Plenum hatte jeder Mitgliedstaat mindestens eine Stimme; wegen ihrer Bevölkerungszahl hatten die Staaten Österreich, Preußen, Sachsen, Bayern, Hannover und Württemberg je vier Stimmen, Kurhessen, das Großherzogtum Hessen, Holstein (vertreten durch den König von Dänemark) und Luxemburg (vertreten durch den König der Niederlande) je drei Stimmen; Braunschweig, Mecklenburg-Schwerin und Nassau je zwei Stimmen, insgesamt also 69 Stimmen. Für ihre Beschlüsse im Plenum war Zweidrittelmehrheit erforderlich. Im engeren Rat hatten die 14 Mitgliedstaaten, die im Plenum mehr als eine Stimme besaßen, je eine Stimme, die 24 anderen Mitgliedstaaten traten zu sechs Kurien zusammen, die je eine Stimme führten. Der engere Rat zählte also siebzehn Stimmen. Für Beschlüsse war einfache Stimmenmehrheit vorgesehen; bei Stimmengleichheit stand dem Vorsitzenden die Entscheidung zu. Hinsichtlich der Annahme oder Abänderung der Grundgesetze, von organischen Bundesreinrichtungen, von *jura singulorum* oder Religionsangelegenheiten konnte weder im engeren Rat, noch im Plenum ein Beschluß durch Stimmenmehrheit gefaßt werden (Art. VII der Bundesakte; im folgenden BA). In diesen Angelegenheiten konnte also auch der kleinste Mitgliedstaat eine Beschlußfassung verhindern (Vetorecht). Die Bundesakte selbst konnte als multilateraler Vertrag nur durch einen Beschluß der Vertragsstaaten der Wiener Kongreßakte geändert werden; da ein solcher Zusatzvertrag nie abgeschlossen wurde, blieb die BA ungeändert für die Mitglieder und die auswärtigen Vertragsstaaten in Geltung.

Auf der Bundesversammlung, mit Sitz in Frankfurt am Main, waren die Mitgliedstaaten durch weisungsgebundene »Bundestaggesandte« vertreten, die eine ständige Gesandtenkonferenz darstellten. Die Bundesversammlung konnte sich auf eine bestimmte Zeit, jedoch nicht länger als vier Monate, vertagen (Art. VII BA). Sie beschloß die notwendigen Durchführungsbestimmungen, die Grundgesetze oder organische Gesetze genannt wurden (Art. VII und VIII BA). Ein solches Grunddokument ist die »Wiener Schlußakte« vom 8. Juni 1820. Sie wurde in Anwesenheit der beim Deutschen Bund beglaubigten diplomatischen Vertreter der anderen Kongreßmächte beschlossen, ohne daß seitens irgendeines der Kongreßstaaten ein Einspruch erfolgt wäre. Die Wiener Schlußakte entsprach also sowohl dem Rahmen der Wiener Kongreßakte, insbesondere auch der deutschen Bundesakte als Annex der Wiener Kongreßakte, wie dem geltenden positiven Völkerrecht.

Für die Erarbeitung von »Grundsätzen in bezug auf des deutschen Bundes völkerrechtlichen Verhältnisse des Kriegs und Friedens« hatte die Bundesversammlung einen Ausschuß eingesetzt, der aus Vertretern Bayerns, Hannovers, Mecklenburgs, Österreichs und Preußens bestand; dieser legte der Bundesversammlung im September 1819 ein Dokument mit dem obigen Titel vor. Auf dessen Grundlage wurden die auf »des deutschen Bundes völkerrechtlichen Verhältnisse des Kriegs und Friedens« bezüglichen Bestimmungen der Wiener Schlußakte vom 8. Juni 1820 ausgearbeitet.

Der Deutsche Bund war ein eigenes Völkerrechtssubjekt. Schon Art. X BA führt ausdrücklich die auswärtigen militärischen Verhältnisse des Bundes an. Art. XI Abs. 1 spricht von einem Bundeskrieg. Art. XXXV der Wiener Schlußakte (WSA) gibt dem Bund »als Gesamtmacht das Recht, Krieg, Frieden, Bündnisse und andere Verträge zu beschließen; unter Bezugnahme auf die in Art. II der BA ausgesprochenen Zwecke übt der Bund aber diese Rechte nur zu seiner Selbstverteidigung, zur Erhaltung der Selbständigkeit und äußeren Sicherheit Deutschlands und der Unabhängigkeit und Unverletzbarkeit der einzelnen Bundesstaaten aus«. Die Bundesversammlung hatte »als Organ der Gesamtheit des Bundes für die Aufrechterhaltung friedlicher und freundschaftlicher Verhältnisse mit den auswärtigen Staaten Sorge zu tragen« (Art. 50 Abs. 1 WSA) und daher »die von fremden Mächten bei dem Bunde beglaubigten Gesandten anzunehmen und, wenn es nötig befunden werden sollte, im Namen des Bundes Gesandte an fremde Mächte abzusenden« (Art. 50 Abs. 2), ferner »in eintretenden Fällen Unterhandlungen für die Gesamtheit des Bundes zu führen und Verträge für denselben abzuschließen« (Art. 50 Abs. 3). Schließlich oblag es der Bundesversammlung, auf Verlangen von Regierungen einzelner Mitgliedstaaten für dieselben die Verwendung des Bundes bei fremden Regierungen und in gleicher Art auf Verlangen fremder Staaten die Dazwischenkunft des Bundes bei einzelnen Mitgliedstaaten eintreten zu lassen (Art. 50 Abs. 4 WSA).

Der Deutsche Bund war also eine regionale internationale Organisation mit völkerrechtlicher Rechts- und Handlungsfähigkeit, mit aktivem und passivem Gesandtschaftsrecht, mit dem Recht, mit auswärtigen Staaten Verträge zu verhandeln und abzuschließen, auf Ersuchen zwischen Bundesmitgliedern und auswärtigen Staaten zu vermitteln und zu schlichten, Verteidigungskriege zu führen und Frieden zu schließen. Die Völkerrechtswissenschaft hat dem System des Deutschen Bundes als einer regionalen internationalen Organisation im Rahmen des Europäischen Konzerts wenig Aufmerksamkeit geschenkt. Der Akzent lag stets auf der »Verfassung Deutschlands«, was noch heute in der Rechtsgeschichte maßgebend ist. Im folgenden soll auf die Verdienste des Deutschen Bundes hingewiesen werden, die alle Fragen um Krieg und Frieden zu regeln suchten. Die dabei von der Bundesversammlung ausgearbeiteten, dem positiven Völkerrecht entsprechenden Regeln nehmen vielfach Formulierungen vorweg, wie sie im Völkerbundpakt, in der Charta der Vereinten Nationen und im Nordatlantikpakt vorhanden sind.

Die Hauptzwecke des Deutschen Bundes waren die Erhaltung der äußeren und inneren Sicherheit Deutschlands und die Unabhängigkeit und Unverletzlichkeit der deutschen Staaten. Da bei einem Staatenbund das gute Verhältnis der Mitgliedstaaten die Grundlage für die äußere Sicherheit und Selbstbehauptung der Gesamtheit der regionalen internationalen Organisation und ihrer souveränen Mitglieder ist, sei zunächst auf die Rechte und Pflichten der Mitgliedstaaten des Deutschen Bundes untereinander eingegangen.

Die innere Sicherheit des Deutschen Bundes – der Gewaltverzicht der Mitgliedstaaten untereinander

Die Präambel der Bundesakte beginnt mit den Worten »die *souveränen* Fürsten und freien Städte Deutschlands«. Die 38 Mitgliedstaaten des Deutschen Bundes nahmen jedenfalls die Qualifikation »souverän« für sich in Anspruch. Jeder Vertrag und Bündnisvertrag beschränkt zwar nicht die Rechts- und Handlungsfähigkeit der Vertragspartner, wohl aber ihre Handlungsfreiheit. Der Vorgänger des Deutschen Bundes, das Römische Reich deutscher Nation, war auch ein Staatenbund, aber die Mitgliedstaaten hatten weitgehende Handlungsfreiheit; sie konnten auch Bündnisse mit auswärtigen Staaten eingehen und gegeneinander Krieg führen. Die Bundesakte schränkt in Art. XI Abs. 2 und 3 die Handlungsfreiheit aber ganz erheblich ein:

> »Die Bundesglieder behalten zwar das Recht der Bündnisse aller Art, verpflichten sich jedoch, in keine Verbindungen einzugehen, welche gegen die Sicherheit des Bundes oder einzelner Bundesstaaten gerichtet wären.
> Die Bundesglieder machen sich ebenfalls verbindlich, einander unter keinerlei Vorwand zu bekriegen, noch ihre Streitigkeiten mit Gewalt zu verfolgen, sondern sie bei der Bundesversammlung anzubringen. Dieser liegt alsdann ob, die Vermittlung durch einen Ausschuß zu versuchen, und falls dieser Versuch fehlschlagen sollte, und demnach eine richterliche Entscheidung notwendig würde, solche durch eine wohlgeordnete Austrägalinstanz zu bewirken, deren Aussprache die streitenden Teile sich sofort zu unterwerfen haben«.

Die Mitgliedstaaten des Deutschen Bundes hatten untereinander auf jede Kriegführung und Gewaltanwendung verzichtet und sich verpflichtet, ihre Streitfälle von der Bundesversammlung schlichten zu lassen oder sich dem richterlichen Entscheid eines *ad hoc*-Schiedsgerichtes zu unterwerfen. Gegen einen Mitgliedstaat, der den Vermittlungsbeschluß der Bundesversammlung oder das Urteil der »Austrägalinstanz« nicht durchführte, war als kollektive Zwangsmaßnahme die Bundesexekution mit militärischer Gewaltanwendung vorgesehen. Derartige überaus weitreichende Beschlüsse waren unter den wichtigen Kompetenzen des Art. VI der Bundesakte nicht angeführt und konnten daher mit einfacher Mehrheit des engeren Rates oder des Plenums gefaßt werden.

Diese radikale Lösung für die innere Sicherheit des Deutschen Bundes auf Kosten der Handlungsfreiheit seiner Mitgliedstaaten zeigte die Möglichkeit an, daß aus dem Deutschen Bund durch Errichtung »supranationaler« Organe allmählich ein Bundesstaat entstehen konnte. Der Schweizer Staatenbund hat diese Entwicklung genommen. Nach vorübergehender Beherrschung durch Frankreich hat die »Tagsatzung« – die Generalversammlung der Gliedstaaten (Kantone) des Schweizer Staatenbundes – aus Gründen der

Selbsterhaltung gegenüber den die Schweiz umgebenden Großmächten 1848 die Umwandlung in einen Bundesstaat vollzogen; hierbei blieben Reservatrechte der sich weiter souverän nennenden Gliedstaaten des Schweizer Bundesstaates erhalten. Die Wiener Kongreßmächte haben diesen Übergang des Schweizer Staatenbundes von der völkerrechtlichen Ebene auf die staatsrechtliche des Bundesstaates schließlich, nach Beilegung des »Neuenburger Handels«, hingenommen und die dauernde Neutralität der regionalen internationalen Organisation auf den Bundesstaat Schweiz übertragen.

Die äußere Sicherheit des Deutschen Bundes

Bundesakte und Wiener Schlußakte setzen in ihren Bestimmungen über die auswärtigen Beziehungen des Deutschen Bundes und seiner Mitgliedstaaten das positive Völkerrecht und dessen Beobachtung voraus. Die Sicherheit des Deutschen Bundes konnte aber nicht nur durch den Angriff eines fremden Staates verletzt oder durch Kriegsvorbereitungen und andere Maßnahmen von außen her bedroht werden, sondern auch durch völkerrechtswidriges Verhalten einzelner Mitgliedstaaten; vom Bund als Gesamtheit wurde ein derartiges Verhalten nicht angenommen, seine Regelung blieb dem allgemeinen Völkerrecht und den Beschlüssen des Europäischen Konzerts vorbehalten. Dagegen enthält die Wiener Schlußakte Bestimmungen, welche die Mitgliedstaaten zum Völkerrecht gemäßen Verhalten verpflichten und bei Völkerrechtsverletzungen seitens eines Mitgliedstaates dem Ausland gegenüber ein bestimmtes Verfahren anordnen.

Die Mitgliedstaaten sind nicht nur verpflichtet, das Völkerrecht zu beobachten und insbesondere keinen Angriffskrieg gegen einen fremden Staat zu führen sowie sich aller (provokatorischen) Handlungen zu enthalten, die einen fremden Staat zur Anwendung von Gewalt veranlassen könnten; Art. 36 WSA bestimmt:

> »Dagegen sind die einzelnen Bundesstaaten verpflichtet, von ihrer Seite weder Anlaß zu dergleichen Verletzungen zu geben, noch auswärtigen Staaten solche zuzufügen«.

Bei einem Streitfall kann sowohl der fremde Staat wie der Mitgliedstaat die Sache bei der Bundesversammlung anhängig machen. Der erste Fall ist in Art. 36 Abs. 2 WSA geregelt:

> »Sollte von seiten eines fremden Staates über eine von einem Mitglied des Bundes ihm widerfahrende Verletzung bei der Bundesversammlung Beschwerde geführt und diese begründet befunden werden, so liegt der Bundesversammlung ob, das Bundesglied, welches die Beschwerde veranlaßt hat, zur schleunigen und genügenden Abhilfe aufzufordern, und mit dieser Aufforderung, nach Befinden der Umstände, Maßregeln, wodurch

> weiteren friedestörenden Folgen zur rechten Zeit vorgebeugt werde, zu verbinden«.

Wenn ein Mitgliedstaat in einem Streitfall mit einem fremden Staat den Bund um Vermittlung ersucht, so hat nach Art. 37 WSA die Bundesversammlung

> »den Ursprung solcher Irrung und das wahre Sachverhältnis sorgfältig zu prüfen. – Ergibt sich aus dieser Prüfung, *daß dem Bundesstaate das Recht nicht zur Seite steht*, so hat die Bundesversammlung denselben von der Fortsetzung des Streites ernstlich abzumahnen, und die begehrte Dazwischenkunft zu verweigern, auch erforderlichen Falls zur Erhaltung des Friedensstandes geeignete Mittel anzuwenden«.

Die Gründung des Deutschen Bundes durch die Mächte des Wiener Kongresses, einschließlich Österreichs und Preußens, bezweckte nicht nur die Sicherheit des Deutschen Bundes und jedes einzelnen seiner Mitgliedstaaten, sondern auch die Sicherung des europäischen Gleichgewichts und des Friedens in Europa. Daher erklärt sich die der Bundesversammlung in den angeführten Artikeln der Wiener Schlußakte eingeräumte Kompetenz, bei Anrufung in Streitfällen im Sinne der Beobachtung des Völkerrechtes und der Erhaltung des Friedens, auf die einzelnen Mitgliedstaaten einzuwirken. Der Deutsche Bund und seine Mitgliedstaaten sollten stark genug sein, jeden Angriff von außen abzuwehren, zugleich aber auch zu verhindern, daß ein Mitgliedstaat eine riskante und zum Krieg führende Politik betreibe. Diese doppelte Funktion des Schutzes des Deutschen Bundes und seiner Mitgliedstaaten und der Erhaltung des Friedens in Mitteleuropa hat der Deutsche Bund bis 1864 erfüllt.

Wenn trotz dieser vorbeugenden Bestimmungen es doch zur Androhung oder Anwendung von Gewalt durch einen fremden Staat gegen ein Bundesmitglied kam, sollten die der Schutzfunktion des Deutschen Bundes entsprechenden Bestimmungen zur Anwendung kommen. Der Angriffskrieg spielt auch hier eine zentrale Rolle.

Aggression von außen gegen einen Mitgliedstaat

Jeder Angriff eines fremden Staates gegen einen Mitgliedstaat ist Angelegenheit des ganzen Bundes; Art. 36 Abs. 1 lautet:

> »Da in dem elften Artikel der Bundesakte alle Mitglieder des Bundes sich verbindlich gemacht haben, sowohl ganz Deutschland, als jeden einzelnen Bundesstaat gegen jeden Angriff in Schutz zu nehmen, und sich gegenseitig ihre sämtlichen unter dem Bunde begriffenen Besitzungen zu garantieren, so kann kein einzelner Bundesstaat von Auswärtigen verletzt werden, ohne daß die Verletzung zugleich und in demselben Maße die Gesamtheit des Bundes treffe«.

Wenn das zum Bund gehörige Gebiet eines Mitgliedstaates von einer feindlichen Macht feindlich überfallen wird, tritt sofort der Kriegszustand ein, und es muß unverzüglich zu den erforderlichen Verteidigungsmaßnahmen geschritten werden (Art. 38 WSA); diese werden vom engeren Rat mit einfacher Mehrheit beschlossen. Dagegen bestimmt Art. 40: Sieht sich der Bund zu einer förmlichen Kriegserklärung genötigt, so kann solche nur in der vollen Versammlung nach der für dieselbe vorgeschriebene Stimmenmehrheit von zwei Dritteilen beschlossen werden. Die Mitgliedstaaten waren zur Stellung des in der »Kriegs-Verfassung« des Bundes festgesetzten Bundes-Kontingentes an Truppen und Material verpflichtet; doch konnte jedes Mitglied eine größere Macht stellen, als sein Bundeskontingent betrug (Art. 44 WSA).

Da einzelne Mitgliedstaaten außerhalb des Bundesgebietes Gebiet (»Besitzungen«) besaßen – so Österreich in Galizien, der Bukowina, Dalmatien und Lombardei-Venetien, Preußen in Hinterpommern, Ost- und Westpreußen, Posen sowie Neuchâtel – mußte der Fall einer fremden Aggression gegen solche Gebiete geregelt werden; Art. 47 lautet:

> »In den Fällen, wo ein Bundesstaat in seinen außer dem Bunde gelegenen Besitzungen bedroht oder angegriffen wird, tritt für den Bund die Verpflichtung zu gemeinschaftlichen Verteidigungsmaßregeln, oder zur Teilnahme und Hilfeleistung nur in sofern ein, als derselbe, nach vorhergehender Beratung durch Stimmenmehrheit in der engeren Versammlung, *Gefahr für das Bundesgebiet* erkennt«.

Wenn ein Bundesstaat mit Besitz außerhalb des Bundesgebietes »in seiner Eigenschaft als Europäische Macht« einen Krieg beginnt, wird der Bund nicht berührt, er bleibt neutral (Art. 46 WSA).

Gefahr einer Aggression gegen einen Mitgliedstaat

Das Verfahren des Deutschen Bundes bei Bedrohung eines Mitgliedstaates oder der Gesamtheit des Bundes durch eine feindliche Aggression ist in der Wiener Schlußakte eingehend geregelt; Art. 38 bestimmt zunächst:

> »Wenn aus der Anzeige eines Bundesstaats, oder aus anderen zuverlässigen Angaben, Grund zu der Besorgnis geschöpft wird, daß ein einzelner Bundesstaat, oder die Gesamtheit des Bundes, von einem feindlichen Angriffe bedroht sei, so muß die Bundesversammlung sofort die Frage, ob die Gefahr eines solchen Angriffes wirklich vorhanden ist, in Beratung nehmen, und darüber in der kürzest-möglichen Zeit einen Ausspruch tun«.

Weder der Fall eines Angriffs gegen den Deutschen Bund als solchen noch seiner Bedrohung durch einen Angriff ist jemals eingetreten. Die vereinigte Macht Österreichs, Preußens und der anderen Bundesstaaten sowie das Interesse der anderen Mächte des Europäischen Konzerts an der Aufrechterhaltung des *status quo* in Mitteleuropa haben solche Gedanken gar nicht aufkommen lassen.

Für den Fall der Feststellung der Bedrohung eines Mitgliedstaates bestimmt Art. 38 WSA weiter:

> »Wird die Gefahr anerkannt, so muß gleichzeitig mit diesem Ausspruche wegen der in solchem Fall unverzüglich in Wirksamkeit zu setzenden Verteidigungsmaßregeln, ein Beschluß gefaßt werden. Beides, jener Ausspruch und dieser Beschluß, ergeht von der engeren Versammlung, die dabei nach der in ihr geltenden absoluten Stimmenmehrheit verfährt«.

Der vom engeren Rat gefaßte Beschluß »über die Wirklichkeit der Gefahr eines feindlichen Angriffes« verpflichtet nach Art. 41 WSA sämtliche Mitgliedstaaten zur Teilnahme an den vom Bundestag notwendig erachteten Verteidigungsmaßregeln. Auch die vom Plenum allenfalls ausgesprochene Kriegserklärung verpflichtet alle Mitgliedstaaten zur unmittelbaren Teilnahme an dem konkreten Bundeskrieg. Ein solcher Fall trat 1848 ein, als die dänische Krone den Mitgliedstaat Holstein mit Dänemark vereinigen wollte. Die Bundesversammlung stellte am 4. April 1848 die Bedrohung Holsteins fest, und Preußen wurde mit den vom Bund für nötig erachteten Gegenmaßnahmen beauftragt. Diese weiteten sich in der Folge zu einem Bundeskrieg aus, der durch den Friedensvertrag des Deutschen Bundes mit Dänemark vom 8. Juli 1850 in Berlin beendet wurde.

Für den Fall, daß die Bedrohung nicht oder nicht ausreichend befunden wird, bestimmt Art. 42 WSA:

> »Wenn die Vorfrage, ob Gefahr vorhanden ist, durch die Stimmenmehrheit verneinend entschieden wird, so bleibt nichts desto weniger denjenigen Bundesstaaten, welche von der Wirklichkeit der Gefahr überzeugt sind, unbenommen, gemeinschaftliche Verteidigungsmaßregeln untereinander zu verabreden.«

In den Fällen, in denen ein Mitgliedstaat in seinen außer dem Bunde gelegenen Besitzungen bedroht wird, greift der Bund nach Art. 47 WSA nur ein, wenn nach vorhergehender Beratung durch Stimmenmehrheit in der engeren Versammlung *Gefahr für das Bundesgebiet* festgestellt wird.

Das Bundesgebiet kann auch durch einen Krieg zwischen fremden Staaten bedroht werden; hierzu sagt Art. 45 WSA: »Wenn in einem Kriege zwischen auswärtigen Mächten oder in anderen Fällen Verhältnisse eintreten, welche die Besorgnis einer Verletzung der Neutralität des Bundesgebietes veranlassen, so hat die Bundesversammlung ohne Verzug im engeren Rate die zur *Behauptung der Neutralität* erforderlichen Maßregeln zu beschließen. Der Deutsche Bund sollte in seiner Friedensfunktion durch bewaffnete Neutralität die nun einmal kriegführenden auswärtigen Mächte vor jedem Mißbrauch des Bundesgebietes abschrecken oder abhalten.

Die Art. 48 und 49 handeln schließlich vom Friedensschluß in einem Bundeskrieg. Art. 48 wiederholt die Bestimmung des Art. XI Abs. 2 der Bundesakte, daß kein Bundesmitglied bei einmal erklärtem Bundeskrieg einseitige Unterhandlungen mit dem Feinde eingehen noch einseitig Waffenstillstand oder Frieden schließen darf, und hebt hervor, daß diese Bestimmung für sämtliche Bundesstaaten, sie mögen außerhalb des Bundes Besitzungen haben oder nicht, gleich verbindlich ist.

Wenn in einem Bundeskrieg der Bund als Völkerrechtssubjekt in Unterhandlungen über den Abschluß eines Friedens oder eines Waffenstillstandes eintritt, hat die Bundesversammlung zu spezieller Leitung derselben einen Ausschuß zu bestellen, zu den Unterhandlungen selbst aber eigene Bevollmächtigte zu ernennen und mit den gehörigen Instruktionen zu versehen. Die Annahme und Bestätigung eines Friedensvertrages kann nur im Plenum erfolgen (Art. 49).

Zusammenfassend kann gesagt werden, daß die Wiener Schlußakte in völkerrechtlicher Hinsicht ein ausgezeichnetes Dokument darstellt, das ganz modern wirkt, wenn man von einzelnen sprachlichen Ausdrücken und Wendungen der Entstehungszeit absieht. Das ist nicht erstaunlich, weil die besten Köpfe der 38 Mitgliedstaaten die »Grundsätze« und die Wiener Schlußakte ausgearbeitet, mitredigiert und jedenfalls vor Annahme genau geprüft haben; so war z. B. Metternich nicht nur Staatsmann und Diplomat, sondern ein sehr guter Jurist und hat die Wiener Schlußakte bis 1848 gehandhabt und angewendet. In der Wiener Schlußakte kommt dem Ausdruck »Angriff« zentrale Bedeutung zu; er ist aber weder in der Wiener Kongreßakte noch in den beiden Pariser Friedensverträgen, noch in der Bundesakte und der Wiener Schlußakte definiert, ebensowenig wie er in den Bündnisverträgen zwischen 1858 und 1914, dem Völkerbundpakt und in der Charta der Vereinten Nationen umschrieben wird. Es ist zur Auslegung daher das allgemeine Völkerrecht heranzuziehen, wobei auf die Entstehungszeit der genannten Dokumente – 1813–1820 – Bedacht zu nehmen ist.

Die mehr oder minder absoluten Monarchien hatten sich im 18. Jahrhundert stets bemüht, ihre »Kabinettskriege« als in und zur Durchsetzung völkerrechtlicher Ansprüche geführt erscheinen zu lassen und auf diese Weise zu »rechtfertigen«; damit anerkannten sie für den europäischen Bereich weiterhin das Rechtsverbot von Eroberungen mit bloßer Gewalt. Allerdings waren dabei manche »Rechtsgründe« weit herbeigeholt worden, nur dürftig die eigene Sucht nach Vergrößerung des eigenen Staatsgebietes, etwa im schlesischen Krieg, verbergend. Gegen diese »heuchlerische« Praxis des Ancien Régime hatte die französische Nationalversammlung in wiederholten Resolutionen seit 1790 erklärt, daß »ein Volk nicht das Recht hat, das Eigentum eines anderen Volkes zu überfallen, noch das Recht, es seiner Freiheit und seiner natürlichen Vorteile zu berauben«. Diese Rechtsgrundsätze galten in Europa seit der Herausbildung des Staatensystems; nun aber sollte jeder Angriffs- und Eroberungskrieg verboten und nur die Abwehr eines völkerrechtswidrigen Angriffs als Verteidigungskrieg erlaubt sein. Diese am Naturrecht orientierten Sätze wurden mit dem anderen Gedankengut der Französischen Revolution in Europa überaus populär. Kant als Bewunderer der Französischen Revolution hat sie weitgehend in seine Rechtslehre (1797) übernommen. Die Kongreßmächte verurteilten in den Präambeln der beiden Pariser Friedensverträge die Eroberungskriege und gewaltsamen Annexionen Frankreichs unter dem Directoire, Konsulat und Empire und wollten die neue Friedensordnung gegen jede gewaltsame Veränderung sichern, vor allem was den territorialen Besitzstand anlangt. Angriffs- und Eroberungskriege sollten in Europa grundsätzlich verboten sein. In diesen Regeln trafen sich, aus verschiedenen geistigen Ursprüngen stammend, Gedanken der frühen Französischen Revolution und der konservativen Restauration. Unter »Angriff« wurde daher immer der erste feindliche Einfall in fremdes Gebiet verstanden, in der Absicht, den bisherigen territorialen Besitzstand gewaltsam zu verändern. Dessen Aufrechterhaltung war einer der Hauptzwecke des Europäischen Konzerts und in dessen Rahmen des regionalen Deutschen Bundes. Die Abwehr eines Angriffes in legitimer Selbsthilfe bis zum Verteidigungskrieg war völkerrechtskonform. Aber auch bei dieser Lösung blieb noch ein erheblicher Raum für subjektive Bewertungen, etwa im Fall der Bedrohung mit Angriff und der Abwehrhandlung des Bedrohten. Im Zweifelsfall hatte eine Instanz mit moralischer Autorität – und Macht – zu entscheiden, etwa die Mehrheit des Europäischen Konzerts (z. B. im Krimkrieg) oder im Rahmen des grundsätzlich defensiv konzipierten Deutschen Bundes der engere Rat (mit Stimmenmehrheit). Über diese Lösung sind weder die späteren Bündnisverträge noch der Völkerbundpakt, noch die Charta der Vereinten Nationen hinausgekommen. In der Zeit des Völkerbundes sind einige Tatbestände der Aggression umschrieben worden, und die Generalversammlung der Vereinten Nationen hat 1976 mit einer Resolution eine Definition der Aggression angenommen. Diese Definitionen sind als Vorlage für konkrete Bündnisverträge und als Richtlinien (*guide-lines*) für den Sicherheitsrat und – bei dessen Schweigen – für die einzelnen Staaten gewiß sehr nützlich. Eine restlose Lösung ist wohl nicht in allen Fällen möglich; nicht immer liegen die Sachverhalte so eindeutig wie beim präventiven Angriffskrieg 1914 und bei den Aggressionen der Achsenmächte zwischen 1931 und 1943.

Der Deutsche Bund wurde nicht von auswärtigen Mächten überwältigt, wie dies der preußisch-deutschen Großmacht 1918 und 1945 widerfahren ist. Er zerbrach, weil Preußen sich dem Mehrheitsbeschluß des engeren Rates nicht fügte und in dem anschließenden Bundes-Exekutionskrieg gegen die Mehrheit der deutschen Staaten und der Mehrheit des deutschen Volkes siegreich blieb. Nach außen hat der Deutsche Bund von 1815 bis 1866 den Besitzstand aller Mitgliedstaaten gewahrt und zum Frieden in Europa wesentlich beigetragen.

Autoren der Festschrift

BALLREICH, HANS, Dr. iur., Wissenschaftlicher Mitarbeiter am Max-Planck-Institut für ausländisches und internationales Patent-, Urheber- und Wettbewerbsrecht, München

BARBERIS, JULIO A., Profesor de Derecho Internacional Público de la Universidad de Buenos Aires; Membre de la Cour Permanente d'Arbitrage

BASTID, SUZANNE, Professeur émérite à l'Université de droit, d'économie et de sciences sociales de Paris; Membre de l'Institut de Droit international

BAYER, HERMANN-WILFRIED, Dr. iur., Professor an der Ruhr-Universität Bochum; Honorar-Professor an der Universität Tübingen

BERNHARDT, RUDOLF, Direktor am Max-Planck-Institut für ausländisches öffentliches Recht und Völkerrecht; Professor an der Universität Heidelberg; Richter am Europäischen Gerichtshof für Menschenrechte

BLECKMANN, ALBERT, Dr. iur., Docteur en droit, Professor an der Universität Münster

BOTHE, MICHAEL, Dr. iur., Professor an der Universität Hannover

DOEHRING, KARL, Professor an der Universität Heidelberg; Direktor am Max-Planck-Institut für ausländisches öffentliches Recht und Völkerrecht; Membre de l'Institut de Droit international

ELIAS, TASLIM O., President of the International Court of Justice; Associate Member of the Institute of International Law

EVERLING, ULRICH, Professor Dr., Richter am Gerichtshof der Europäischen Gemeinschaften, Luxemburg

EVRIGENIS, DIMITRIOS, Professeur à l'Université de Thessaloniki; Directeur du Centre de droit économique international et européen; Juge à la Cour européenne des Droits de l'Homme; Associé de l'Institut de Droit international

FITZMAURICE, SIR GERALD [† 7. 9. 1982], Member of the English Bar; Queen's Counsel (Q.C.); Knight Grand Cross of the Order of St. Michael & St. George (G.C.M.G.); Master of the Bench of Gray's Inn; Doctor of Law *honoris causa* in the Universities of Cambridge, Edinburgh and Utrecht; formerly Member and past President of the United Nations International Law Commission; formerly Chief Legal Adviser (Jurisconsult) in the Foreign Office, London; Judge of the International Court of Justice 1960–1973; Judge of the European Court of Human Rights 1974–1980; Member of the Permanent Court of Arbitration 1964–1980; Member of the Institute of International Law

FROMONT, MICHEL, Professeur de droit public; Directeur de l'Institut de droit comparé de l'Université de Dijon

FROWEIN, JOCHEN ABR., Dr. iur., M.C.L., Direktor am Max-Planck-Institut für ausländisches öffentliches Recht und Völkerrecht; Professor an der Universität Heidelberg; Vizepräsident der Europäischen Kommission für Menschenrechte

GANSHOF VAN DER MEERSCH, WALTER J., Juge à la Cour européenne des droits de l'homme; Procureur général ém. à la Cour de cassation; Membre de l'Académie royale de Belgique; Professeur hre. à l'Université de Bruxelles

GECK, WILHELM KARL, Dr. iur., M.A., Professor an der Universität des Saarlandes, Saarbrücken; Richter am Saarländischen Verfassungsgerichtshof

GROS, ANDRÉ, Juge à la Cour internationale de Justice; Membre de l'Institut de Droit international; Professeur Honoraire des Facultés de Droit; Jurisconsulte Honoraire du Ministère des Affaires Etrangères

HAILBRONNER, KAY, Dr. iur., LL.M. (McGill, Montreal), Professor an der Universität Konstanz

HILF, MEINHARD, Dr. iur. habil., Professor an der Universität Bielefeld

JAENICKE, GÜNTHER, Dr. iur., em. Professor an der Universität Frankfurt am Main

JIMÉNEZ DE ARÉCHAGA, EDUARDO, Former President of the International Court of Justice; Professor of International Law, Montevideo Law School; Member of the Institute of International Law

KLEIN, ECKART, Dr. iur. habil., Professor an der Universität Mainz

LACHS, MANFRED, Judge and former President of the International Court of Justice; Member of the Institute of International Law

v. MANGOLDT, HANS, Dr. iur. habil., Professor an der Universität Tübingen

MANN, F. A., Professor Dr. Dr. h.c., Mitglied der britischen Akademie und des Institut de Droit international

MATSCHER, FRANZ, Universitätsprofessor DDr., Salzburg; Richter am Europäischen Gerichtshof für Menschenrechte

McWhinney, Edward, Queen's Counsel; Professor of International Law and Relations, Simon Fraser University (Vancouver, Canada); Membre de l'Institut de Droit international

Miyazaki, Shigeki, Dr. iur., Professor an der Meiji Universität (Tokio)

Monaco, Riccardo, Professeur à l'Université de Rome; Doyen de la Faculté de sciences politiques; Membre de l'Institut de Droit international

Münch, Fritz, Dr. iur., em. apl. Professor an der Universität Bonn; Membre de l'Institut de Droit international

Oda, Shigeru, Professor Dr., Judge of the International Court of Justice

Partsch, Karl Josef, em. Professor an der Universität Bonn; Mitglied des Ausschusses zur Beseitigung der Rassendiskriminierung bei den Vereinten Nationen

Pescatore, Pierre, Richter am Gerichtshof der Europäischen Gemeinschaften; Professor an der Universität Lüttich und am Universitätszentrum Luxemburg; Membre de l'Institut de Droit international

Puente Egido, José, Catedrático de Derecho Internacional U.N.E.D. Madrid

Ress, Georg, Dr. iur., Dr. rer. pol., Professor, Direktor des Europa-Instituts der Universität des Saarlandes; stellvertretendes Mitglied des Verfassungsgerichtshofs des Saarlandes

Reuter, Paul, Professeur émérite à l'Université de droit, d'économie et de sciences sociales de Paris, membre de la Commission du droit international; membre associé de l'Académie Royale de Belgique; Membre de l'Institut de Droit international

Rosenne, Shabtai, Fellow, Jewish Academy of Arts and Sciences (New York); Member, Institute of International Law; Honorary member, American Society of International Law; Ambassador-at-Large and Chairman, Israel Delegation to the Third United Nations Conference on the Law of the Sea

Rudolf, Walter, Dr. iur., Professor an der Universität Mainz; Staatssekretär im Ministerium der Justiz, Mainz; Associé de l'Institut de Droit international

Schachter, Oscar, Hamilton Fish Professor of International Law and Diplomacy, Columbia University (N.Y.); Co-editor in chief, American Journal of International Law; Member, Institut de Droit international

Schermers, Henry G., Professor of Law at the University of Leiden

Schlochauer, Hans-Jürgen, Dr. iur., Professor an der Universität Frankfurt am Main; Membre de la Cour Permanente d'Arbitrage

SCHNEIDER, HANS, em. Professor an der Universität Heidelberg; Honorarprofessor an der Universität Stuttgart

SCHWEISFURTH, THEODOR, Dr. iur., Privatdozent, wissenschaftlicher Referent am Max-Planck-Institut für ausländisches öffentliches Recht und Völkerrecht; Lehrbeauftragter an der Rechts- und Wirtschaftswissenschaftlichen Fakultät der Universität Bayreuth

SEIDL-HOHENVELDERN, IGNAZ, Dr. Dr. h.c., Professor an der Universität Wien; Membre de l'Institut de Droit international

SKUBISZEWSKI, KRZYSZTOF, Professor of Law, Polish Academy, Warsaw; Member, Institute of International Law

SPERDUTI, GIUSEPPE, Professeur de droit international à l'Université de Rome; Membre de l'Institut de Droit international; Président du Comité consultatif italien des droits de l'homme

STEINBERGER, HELMUT, Dr. iur. utr., Richter des Bundesverfassungsgerichts; Auswärtiges Wissenschaftliches Mitglied des Max-Planck-Instituts für ausländisches öffentliches Recht und Völkerrecht; Professor an der Universität Mannheim

SUY, ERIK, Docteur en droit, Docteur ès sciences politiques, Professeur à l'Université de Louvain (KUL); Secrétaire général adjoint et Conseiller juridique de l'ONU; Associé de l'Institut de Droit international

TOMUSCHAT, CHRISTIAN, Dr. iur., Professor, Direktor des Instituts für Völkerrecht der Universität Bonn; Mitglied des Menschenrechtsausschusses der Vereinten Nationen

TRUYOL SERRA, ANTONIO, Catedrático de la Universidad Complutense, Madrid; Magistrado del Tribunal Constitucional; Associé de l'Institut de Droit international

TSATSOS, DIMITRIS TH., Professor an der Pantios Hochschule für Politische Wissenschaft Athen und an der Fernuniversität Hagen; Rechtsanwalt am Staatsgerichtshof Athen

VEROSTA, STEPHAN, Professor an der Universität Wien; vorm. Botschafter; Mitglied der Österreichischen Akademie der Wissenschaften und des Institut de Droit international

Publikationen von Hermann Mosler

(ohne Buchbesprechungen und Diskussionsbeiträge)

I. Selbständige Publikationen und Abhandlungen in Zeitschriften und Sammelwerken

1. Die Intervention im Völkerrecht (Neue Deutsche Forschungen Abt. Staats-, Verwaltungs-, Kirchen-, Völkerrecht und Staatstheorie, Bd. 12) (1937)
2. Zur Auslegung von Art. 7 § 3 des deutsch-polnischen Wiener Abkommens über Staatsangehörigkeits- und Optionsfragen vom 30. August 1924, ZaöRV 7 (1937), S. 832–840
3. Die Änderungen im Statut der Deutschen Reichsbahn und Reichsbank, ZaöRV 7 (1937), S. 457–461
4. Die Konferenz zur internationalen Bekämpfung des Terrorismus (November 1937), ZaöRV 8 (1938), S. 99–107
5. Die Revision des Haager Abkommens über die Anwendung des Genfer Abkommens auf den Seekrieg, ZaöRV 8 (1938), S. 282–291
6. Die Verwaltung im Elsaß und in Lothringen zur Zeit des Waffenstillstandes, Reichsverwaltungsblatt 61 (1940), S. 501–506
7. Das französische Prisenverfahren im gegenwärtigen Kriege, ZaöRV 10 (1940), S. 480–494
8. Die Anwendung der belgischen Weltkriegsgesetzgebung seit der Mobilmachung, Reichsverwaltungsblatt 62 (1941), S. 7–8
9. Der Konflikt über die gerichtliche Nachprüfung der Verordnung der Generalsekretäre in den belgischen Ministerien, ZaöRV 11 (1943), S. 610–620
10. Der Einfluß der Rechtsstellung Deutschlands auf die Kriegsverbrecherprozesse, Süddeutsche Juristenzeitung 2 (1947), S. 362–370

Abkürzungen: Ann. IDI = Annuaire de l'Institut de Droit International; ArchVR = Archiv des Völkerrechts; BT-Drs. = Bundestagsdrucksache; DÖV = Die Öffentliche Verwaltung; I.C.J. Reports = Reports of the International Court of Justice; JBl. = Juristische Blätter (Wien); JZ = Juristenzeitung; NJW = Neue Juristische Wochenschrift; RdC = Académie de Droit International, Recueil des Cours; Staatslexikon = Staatslexikon. Recht, Wirtschaft, Gesellschaft, hrsg. von der Görres-Gesellschaft (6. Aufl. 1957 ff.); ZaöRV = Zeitschrift für ausländisches öffentliches Recht und Völkerrecht

11. Rechtsgutachten des Instituts für internationales Recht und Politik an der Universität Bonn über die Frage, inwieweit die Besatzungsmächte nach geltendem Völkerrecht berechtigt sind, im Privateigentum stehende industrielle Produkte und Industrieanlagen wegzunehmen oder zu zerstören (zusammen mit R. Thoma und E. Friesenhahn). Erstattet im Auftrag der Stadt Essen (1947), 18 S.
12. Die religiös-sittlichen Grundlagen des Völkerrechts, Der katholische Gedanke (1947), S. 96–106
13. Wirtschaftskonzessionen bei Änderung der Staatshoheit. Eine völkerrechtliche Studie zum Hoheitswechsel und zur Hoheitsausübung auf fremdem Staatsgebiet (1948), 256 S.
14. Die Kriegshandlung im rechtswidrigen Kriege, Jahrbuch für internationales und ausländisches öffentliches Recht 2/3 (1948), S. 335–357
15. Die Großmachtstellung im Völkerrecht (Schriften der Süddeutschen Juristenzeitung, Heft 8) (1949), 52 S.
16. Die völkerrechtliche Wirkung bundesstaatlicher Verfassungen. Eine Untersuchung zum Völkerrecht und zum vergleichenden Verfassungsrecht, in: Festschrift für Richard Thoma (1949), S. 129–172
17. Der Vertrag über die Europäische Gemeinschaft für Kohle und Stahl. Entstehung und Qualifizierung, ZaöRV 14 (1951), S. 1–45
18. Die Wendung zum supranationalen Gedanken im Schumanplan, Recht, Staat, Wirtschaft 3 (1951), S. 245–259
19. Die Sicherung der Menschenrechte durch die internationale Rechtsordnung, in: Die Menschenrechte in christlicher Sicht, 2. Beiheft zur Herder-Korrespondenz (1953), S. 34–49
20. Menschenrechte und Verfassungskompromiß, in: Die Menschenrechte in christlicher Sicht, 2. Beiheft zur Herder-Korrespondenz (1953), S. 60–73
21. Die auswärtige Gewalt im Verfassungssystem der Bundesrepublik Deutschland, in: Völkerrechtliche und staatsrechtliche Abhandlungen, Carl Bilfinger zum 75. Geburtstag am 21. Januar 1954 gewidmet (Beiträge zum ausländischen öffentlichen Recht und Völkerrecht, Bd. 29) (1954), S. 243–299
22. Kulturabkommen des Bundesstaats. Zur Frage der Beschränkung der Bundesgewalt in auswärtigen Angelegenheiten, ZaöRV 16 (1955), S. 1–34
23. Das Völkerrecht in der juristischen Ausbildung, NJW 23 (1955), S. 860–861
24. Der Konkordatsprozeß (Veröffentlichungen des Instituts für Staatslehre und Politik e.V. Mainz, Bd. 7) (1956), Ausführungen über das Ermächtigungsgesetz (S. 1321–1325, 1341–1343), über den Grundsatz der Bundestreue (S. 1583–1601), über die Transformation des Reichskonkordats von 1933 in Deutsches Recht (S. 1349–1354) und über seine Fortgeltung für die Bundesrepublik Deutschland (S. 1382–1389)

25. Proposals for an International Code Regulating the Handling of Property of Enemy Nationals, Heidelberg 1956, maschinenschriftliche Vervielfältigung, 41 gez. Bl. 4°, und in: International Bar Association, 6th Conference Report, Oslo, July 1956 (The Hague 1957), S. 351–364
26. Überstaatliche Organisationen. Hessische Hochschulwochen für staatswissenschaftliche Fortbildung 1955, Bad Sooden-Allendorf (1956), S. 167–188
27. Aufgaben und Grenzen der organisierten Forschung des Völkerrechts, in: Aus der deutschen Forschung der letzten Dezennien, Festschrift für Dr. Ernst Telschow (1956), S. 258–266
28. Internationale Organisation und Staatsverfassung, in: Rechtsfragen der internationalen Organisation, Festschrift für Hans Wehberg zu seinem 70. Geburtstag (1956), S. 273–300
29. Zur Anwendung der Grundsatzartikel des Vertrages über die Europäische Gemeinschaft für Kohle und Stahl, ZaöRV 17 (1957), S. 407–427
30. Das Völkerrecht in der Praxis der deutschen Gerichte (Juristische Studiengesellschaft Karlsruhe, Schriftenreihe Heft 32/33) (1957), 48 S.
31. La Organización internacional y la distribución de competencias. Cursos y Conferencias de la Escuela de Funcionarios Internacionales, Bd. 1 (1957), S. 533–603
32. L'application du droit international public par les tribunaux nationaux, RdC 91 (1957), S. 623–709
33. Auswärtige Gewalt, in: Staatslexikon, Bd. 1 (1957), Sp. 753–760
34. Vermögensschutz im geltenden Völkerrecht, Wirtschaftliche Nachrichten der Industrie- und Handelskammer für die Stadtkreise Essen, Mülheim (Ruhr) und Oberhausen zu Essen 11 (1957), S. 81–83 (Zusammenfassung eines Vortrags und wörtliche Auszüge)
35. Die Aufnahme in internationale Organisationen, ZaöRV 19 (1958), S. 275–317
36. Europäische Gemeinschaft für Kohle und Stahl, Organisation und völkerrechtliche Stellung, in: Staatslexikon, Bd. 3 (1959), Sp. 128–130
37. Grundgesetz der Bundesrepublik Deutschland, in: Staatslexikon, Bd. 3 (1959), Sp. 1113–1122; Ergänzungsband 2 (1970), Sp. 224–229
38. Organisation und Verfahren des Europäischen Gerichtshofs für Menschenrechte, ZaöRV 20 (1960), S. 415–449
39. Geschichte des Max-Planck-Instituts für ausländisches öffentliches Recht und Völkerrecht, Jahrbuch der Max-Planck-Gesellschaft 1961 Teil II, S. 687–703
40. Kriegsbeginn, in: Wörterbuch des Völkerrechts, hrsg. von H.-J. Schlochauer in Zusammenarbeit mit H. Krüger, H. Mosler, U. Scheuner, Bd. 2 (2. Aufl. 1961), S. 326–329
Kriegsende, *ibid.,* S. 333–337
41. Potsdamer Abkommen, in: Staatslexikon, Bd. 6 (1961), Sp. 431–437

42. Die Erweiterung des Kreises der Völkerrechtssubjekte, Berichte der Deutschen Gesellschaft für Völkerrecht, Heft 4 (1961), S. 39–83; erweiterte Fassung in ZaöRV 22 (1962), S. 1–48
43. Thoma, in: Staatslexikon, Bd. 7 (1962), Sp. 972–973
44. Die Gewährleistung des Völkerrechts durch die nationale Verfassung, dargestellt am Grundgesetz der Bundesrepublik Deutschland, Problèmes Contemporains de Droit Comparé (Institut Japonais de Droit Comparé [Tokyo], Bd. 1) (1962), S. 157–178
45. Staatliche Zuständigkeit, in: Wörterbuch des Völkerrechts, hrsg. von H.-J. Schlochauer in Zusammenarbeit mit H. Krüger, H. Mosler, U. Scheuner, Bd. 3 (2. Aufl. 1962), S. 317–323 (zusammen mit H.-O. Bräutigam) Völkerrechtsfähigkeit, *ibid.*, S. 665–677
46. Das Heidelberger Kolloquium über Verfassungsgerichtsbarkeit, Ziel – Methode – Ergebnis, in: Verfassungsgerichtsbarkeit in der Gegenwart, Länderberichte und Rechtsvergleichung – Constitutional Review in the World Today, National reports and comparative studies – La Juridiction Constitutionnelle à l'époque contemporaine, Exposés de la situation dans différents pays et étude comparée (Beiträge zum ausländischen öffentlichen Recht und Völkerrecht, Bd. 36) (1962), S. IX–XLIII
47. Die Beendigung des Kriegszustands mit Deutschland nach dem zweiten Weltkrieg, bearbeitet mit einer Studiengruppe des Max-Planck-Instituts für ausländisches öffentliches Recht und Völkerrecht von H. Mosler und K. Doehring (Beiträge zum ausländischen öffentlichen Recht und Völkerrecht, Bd. 37) (1963), X, 486 S. (besonders S. 454–486: Die Beendigung des Kriegszustands mit Deutschland und die Entwicklung des völkerrechtlichen Kriegsbegriffs)
48. Weimarer Verfassung, in: Staatslexikon, Bd. 8 (1963), Sp. 518–526
49. Réflexions sur la personnalité juridique en droit international public, in: Problèmes de droit des gens, Mélanges Henri Rolin (1964), S. 228–251
50. The Protection of Human Rights by International Legal Procedure, Georgetown Law Journal 52 (1964), S. 800–823
51. Kritische Bemerkungen zum Rechtsschutzsystem der Europäischen Menschenrechtskonvention, in: Festschrift für Hermann Jahrreiss (1964), S. 289–317
52. Die Verfassung des Deutschen Reiches vom 11. 8. 1919, Reclams Universal-Bibliothek Nr. 6051 (1964; Neuaufl. 1977)
53. National- und Gemeinschaftsinteressen im Verfahren des EWG-Ministerrates. Die Beschlüsse der außerordentlichen Tagung des EWG-Rates in Luxemburg vom 28. Januar 1966, ZaöRV 26 (1966), S. 1–32
54. European Law – Does it Exist?, Current Legal Problems 19 (1966), S. 168–191
55. Die Entstehung des Modells supranationaler und gewaltenteilender Staatenverbindungen in den Verhandlungen über den Schuman-Plan, in: Pro-

bleme des europäischen Rechts, Festschrift für Walter Hallstein (1966), S. 355–386

56. Wer ist aus dem Reichskonkordat verpflichtet?, in: Gedächtnisschrift für Hans Peters (1967), S. 350–374
57. Das Heidelberger Kolloquium über die Haftung des Staates für rechtswidriges Verhalten seiner Organe, in: Haftung des Staates für rechtswidriges Verhalten seiner Organe, Länderberichte und Rechtsvergleichung – Liability of the State for Illegal Conduct of its Organs, National reports and comparative studies – La responsabilité de l'Etat pour le comportement illégal de ses organes, Exposé de la situation dans différents pays et étude comparée (Beiträge zum ausländischen öffentlichen Recht und Völkerrecht, Bd. 44) (1967), S. IX–XXIII, 805–807
58. Das Notstandsrecht in Frankreich, Belgien und den Niederlanden sowie Ausführungen über die Anwendung rechtsvergleichender Methoden auf die Gestaltung des deutschen Notstandsrechts, BT-Drs., 5. Wahlperiode, Rechtsausschuß (12. Ausschuß), Innenausschuß (6. Ausschuß), Protokoll Nr. 60/76, S. 1–17, 24
59. Der „Gemeinschaftliche Ordre Public" in europäischen Staatengruppen, in: Estudios de Derecho Internacional, Homenaje a D. Antonio de Luna (1968), S. 372–384 (=Revista Española de Derecho Internacional 21 (1968), S. 523–535)
60. Repertorien der nationalen Praxis in Völkerrechtsfragen – eine Quelle zur Erschließung des Allgemeinen Völkerrechts?, in: Recueil d'études de droit international en hommage à Paul Guggenheim (1968), S. 460–489
61. Europarecht – ein neues Fach der Rechtswissenschaft?, Jahrbuch der Aristotelischen Universität Thessaloniki 1968, S. 419–438
62. Begriff und Gegenstand des Europarechts, ZaöRV 28 (1968), S. 481–502
63. La procédure de la Cour internationale de Justice et de la Cour Européenne des Droits de l'Homme, in: Liber Amicorum Discipulorumque René Cassin, Bd. 1 (1969), S. 196–212
64. Europäisches Recht, in: Staatslexikon, Ergänzungsband 1 (1969), Sp. 849–856
65. Das Heidelberger Kolloquium über Gerichtsschutz gegen die Exekutive, Gegenstand und Methode, in: Gerichtsschutz gegen die Exekutive – Judicial Protection against the Executive – La protection juridictionnelle contre l'exécutif (Bde. 1–3) (Beiträge zum ausländischen öffentlichen Recht und Völkerrecht, Bd. 52) (1969), Bd. 1, S. XIII-LX
66. Ius Cogens im Völkerrecht, Schweizerisches Jahrbuch für internationales Recht 25 (1968), S. 9–40
67. Münchner Abkommen, in: Staatslexikon, Ergänzungsband 2 (1970), Sp. 686–692 (zusammen mit J. Jurina)
68. Rechtsvergleichung vor völkerrechtlichen Gerichten, in: Internationale Festschrift für Alfred Verdross zum 80. Geburtstag (1971), S. 381–411

69. Das Österreichisch-Deutsche Schiedsgerichtsverfahren nach dem Finanz- und Ausgleichsvertrag vom 27. 11. 1961, ZaöRV 32 (1972), S. 36–70
70. L'influence du droit national sur la Convention européenne des Droits de l'Homme, in: Miscellanea Ganshof van der Meersch (1972), S. 521–543
71. Nationale Gerichte als Garanten völkerrechtlicher Verpflichtungen, Zum Begriff des Gerichts in der neuesten Rechtsprechung des Europäischen Gerichtshofs für Menschenrechte, in: Liber amicorum J. J. M. van der Ven (1972), S. 381–390
72. Judicial Settlement of International Disputes, International Symposium, Heidelberg, July 10–12, 1972, ZaöRV 32 (1972), S. 543–564 (zusammen mit R. Bernhardt, nicht gezeichnet)
Deutsche Fassung in: Mitteilungen aus der Max-Planck-Gesellschaft 1972, S. 431–459
73. Gleichheit der Eltern beim Erwerb der Staatsangehörigkeit der Kinder, in: Festschrift für U. Scheuner zum 70. Geburtstag (1973), S. 473–491
74. Das Institut de Droit International und die völkerrechtliche Stellung der menschlichen Person, in: Iustitia et Pace, Festschrift zum 100jährigen Bestehen des Institut de Droit International (1974), S. 77–100
75. Problems and Tasks of International Judicial and Arbitral Settlement of Disputes Fifty Years after the Founding of the World Court, in: Judicial Settlement of International Disputes, International Court of Justice – Other Courts and Tribunals – Arbitration and Conciliation (Beiträge zum ausländischen öffentlichen Recht und Völkerrecht, Bd. 62) (1974), S. 3–15
76. Das Max-Planck-Institut für ausländisches öffentliches Recht und Völkerrecht, zusammen mit R. Bernhardt, Max-Planck-Gesellschaft, Berichte und Mitteilungen, Heft 2 (1975), 45 S.
77. Control of Administrative Activities by Administrative Courts and Non-Judicial Means (particularly by ombudsmen), Paper delivered at the South African Law Conference Cape Town 1975, S. 38–65
78. Völkerrecht als Rechtsordnung, ZaöRV 36 (1976), S. 6–49
79. Das Max-Planck-Institut für ausländisches öffentliches Recht und Völkerrecht, Heidelberger Jahrbücher, Bd. 20 (1976), S. 53–78
80. The International Society as a Legal Community, RdC 140 (1974 IV), S. 1–320
81. Antrittsrede vor der Heidelberger Akademie der Wissenschaften, Jahrbuch der Heidelberger Akademie der Wissenschaften 1975 (1976), S. 117–120
82. Ergebnisse und Perspektiven des Kolloquiums über Grundrechtsschutz in Europa, zusammen mit R. Bernhardt, in: Grundrechtsschutz in Europa, Europäische Menschenrechts-Konvention und Europäische Gemeinschaften (Beiträge zum ausländischen öffentlichen Recht und Völkerrecht, Bd. 72) (1977), S. 205–223

83. Souveräne Gleichheit und universelle Staatengemeinschaft im Völkerrecht der Gegenwart, Jahrbuch der Heidelberger Akademie der Wissenschaften 1976 (1977), S. 77–90
84. Remarques sur la superposition du droit des organisations internationales au droit des gens interétatique, in: Estudios de Derecho Internacional, Homenaje al Profesor Miaja de la Muela, Bd. 1 (1979), S. 429–439
85. The International Court of Justice at its Present Stage of Development, Dalhousie Law Journal 5 (1979), S. 545–567
86. Aktuelle Aspekte des Verfahrensrechts des Internationalen Gerichtshofs, in: Völkerrecht und Rechtsphilosophie, Internationale Festschrift für Stephan Verosta zum 70. Geburtstag (1980), S. 249–259
87. Problems of Interpretation in the Case Law of the European Court of Human Rights, in: Essays on the Development of the International Legal Order, in memory of H. F. van Panhuys (1980), S. 149–167
88. The International Society as a Legal Community. Revised edition of No. 80 *supra* (Alphen aan den Rijn 1980), XIX, 327 S.
89. Der Europäische Gerichtshof für Menschenrechte nach zwanzig Jahren, in: Recht als Prozeß und Gefüge, Festschrift für Hans Huber (1981), S. 595–607
90. Nichtteilnahme einer Partei am Verfahren vor dem Internationalen Gerichtshof, in: Staatsrecht – Völkerrecht – Europarecht, Festschrift für Hans-Jürgen Schlochauer (1981), S. 439–456
91. Supra-National Judicial Decisions and National Courts, Hastings International and Comparative Law Review 4 (1981), S. 424–472
92. Judgments of International Courts and Tribunals, in: Bernhardt (ed.), Encyclopedia of Public International Law, Instalment 1 (1981), S. 111–118
93. Austro-German Arbitration Award under the Treaty of Finance and Compensation of 1961, in: Bernhardt (ed.), Encyclopedia of Public International Law, Instalment 2 (1981), S. 27–28
94. Verhandlungen des Fünften Internationalen Kolloquiums über die Europäische Menschenrechtskonvention in Frankfurt (Main), Schlußbericht, in: Europäischer Menschenrechtsschutz, Schranken und Wirkungen, hrsg. von Irene Maier (1982), S. 355–369

II. Würdigungen und Nachrufe

1. Zum 70. Geburtstag Hans Wehbergs (15. Dezember 1955), JZ 1955, S. 764–765
2. Richard Thoma zum Gedächtnis, DÖV 1957, S. 826–828
3. Richard Thoma, ArchVR 6 (1957), S. 385–386

4. Zum wissenschaftlichen Werk von Hans Peters, Jahresbericht der Görres-Gesellschaft für 1956 (1957), S. 46–55
5. Georg Schreiber, ZaöRV 22 (1962), S. IX
6. Ernst Reibstein, ZaöRV 26 (1966), S. 493–494
7. Adolf Schüle, ZaöRV 27 (1967), S. 657–659
8. Alexander N. Makarov zum 80. Geburtstag am 4. 8. 1968, ZaöRV 28 (1968), S. 475–476
9. Friedensnobelpreis für René Samuel Cassin, Ruperto Carola 46 (1969), S. 6–7
10. Übergabe der Internationalen Festschrift zum 80. Geburtstag von Alfred Verdross; Rede, gehalten am 20. 3. 1971 in der Salzburger Residenz (Auszug), JBl. 93 (1971), S. 411–414
11. Erich Kaufmann zum Gedächtnis, ZaöRV 32 (1972), S. 235–238
12. Erich Kaufmann, Mitteilungen aus der Max-Planck-Gesellschaft 1972, S. 413–418
13. Alexander N. Makarov 1888–1973, ZaöRV 33 (1973), S. 443–446
14. Alexandre N. Makarov, Ann. IDI 55 (1973), S. 903–905
15. Georg von Gretschaninow 1892–1973, ZaöRV 33 (1973), S. 836–837
16. Rudolf Smend 1882–1975, ZaöRV 35 (1975), S. 721–722
17. Im Spannungsfeld von Recht und Politik. Zur 100. Wiederkehr des Geburtstages von Carl Bilfinger, Ruperto Carola 31 (1979), S. 119
18. Im Dienste der Wahrheit und des Menschen. Laudatio auf Egon Schwelb, in: Dag-Hammarskjöld-Ehrenmedaille, Deutsche Gesellschaft für die Vereinten Nationen, 1979, Sadruddin Aga Khan, Egon Schwelb (1980), S. 28–33 (englische Übersetzung S. 34–38)

III. Sondervoten zu Entscheidungen internationaler Gerichte

1. *Europäischer Gerichtshof für Menschenrechte*

a) Affaires De Wilde, Ooms et Versyp, Arrêt du 10 mars 1972 (article 50), Série A, Vol. 14, pp. 17–20
b) Affaire Handyside, Arrêt du 7 décembre 1976, Série A, Vol. 24, pp. 32–35

2. *Internationaler Gerichtshof*

a) Aegean Sea Continental Shelf, Interim Protection, Order of 11 September 1976, I.C.J. Reports 1976, pp. 25–27
b) Interpretation of Agreement of 25 March 1951 between the WHO and Egypt, Advisory Opinion, I.C.J. Reports 1980, pp. 125–130
c) Application for Review of Judgment No. 273 of the United Nations Administrative Tribunal, Advisory Opinion, I.C.J. Reports 1982, pp. 379–388

IV. Herausgebertätigkeit

1. *Publikationen des Max-Planck-Instituts für ausländisches öffentliches Recht und Völkerrecht (seit 1954)*

Zeitschrift für ausländisches öffentliches Recht und Völkerrecht
Beiträge zum ausländischen öffentlichen Recht und Völkerrecht
Fontes Iuris Gentium (Series A Sectio I und Sectio II)
Materialien für Forschung und Praxis des ausländischen öffentlichen Rechts und Völkerrechts
Public International Law. A Current Bibliography of Articles
Nouveau Recueil Général de Traités. Continuation du Grand Recueil de G. Fr. de Martens. Vol. 41, Préface à la Troisième (dernière) Livraison, préparée par O. Steiner et A. Maier

2. *Andere Publikationen*

Erich Kaufmann, Gesammelte Schriften, Bde. 1–3 (1960), zusammen mit A. H. von Scherpenberg, W. Strauss, E. Kordt, F. A. Frhr. von der Heydte, K.-J. Partsch
Wörterbuch des Völkerrechts, hrsg. von H.-J. Schlochauer in Zusammenarbeit mit H. Krüger, H. Mosler, U. Scheuner, Bde. 1–3 (2. Aufl. 1960–1962)
Gedächtnisschrift für Hans Peters. Zusammen mit H. Conrad, H. Jahrreiss, P. Mikat, H. C. Nipperdey, J. Salzwedel (1967)
Internationale Festschrift für Alfred Verdross zum 80. Geburtstag, zusammen mit R. Marcic, E. Suy, K. Zemanek (1971)
Themistokles Tsatsos, Perí Politeías. Staatstheoretische Studien. Mit einer Einführung von H.-G. Gadamer. Zusammen mit H. Schneider, H. Strebel (1972)

Max-Planck-Institut für ausländisches öffentliches Recht und Völkerrecht

Beiträge zum ausländischen öffentlichen Recht und Völkerrecht

R. Bernhardt, K. Doehring, J. Abr. Frowein

Bde. 27–59 erschienen im Carl Heymanns Verlag KG Köln, Berlin;
ab Band 60 im Springer-Verlag Berlin, Heidelberg, New York

80 Jutta *Stoll:* **Vereinbarungen zwischen Staat und ausländischem Investor.** Rechtsnatur und Bestandsschutz. 1982. XI, 166 Seiten (6 Seiten English Summary). Geb. 49,– DM

79 Meinhard *Hilf:* **Die Organisationsstruktur der Europäischen Gemeinschaften.** Rechtliche Gestaltungsmöglichkeiten und Grenzen. 1982. XVIII, 442 Seiten (12 Seiten English Summary, 13 Seiten Résumé français). Geb. 98,– DM

78 Ludwig *Weber:* **Die Zivilluftfahrt im Europäischen Gemeinschaftsrecht.** 1981. XXVIII, 428 Seiten (6 Seiten English Summary). Geb. 114,– DM

77 Michael *Schaefer:* **Die Funktionsfähigkeit des Sicherheitsmechanismus der Vereinten Nationen.** 1981. XXI, 471 Seiten (10 Seiten English Summary). Geb. 105,– DM

76 Eckart *Klein:* **Statusverträge im Völkerrecht.** Rechtsfragen territorialer Sonderregime. 1980. XIV, 395 Seiten (10 Seiten English Summary). Geb. 86,– DM

75 **Die Koalitionsfreiheit des Arbeitnehmers/The Freedom of the Worker to Organize/ La liberté syndicale des salariés.** Rechtsvergleichung und Völkerrecht/Comparative Law and International Law/Droit comparé et droit international public. 1980, XLIX, 1536 Seiten. (In 2 Bänden, die nur zusammen abgegeben werden.) Geb. 360,– DM

74 Hans *Reinhard:* **Rechtsgleichheit und Selbstbestimmung der Völker in wirtschaftlicher Hinsicht.** Die Praxis der Vereinten Nationen. 1980. XIII, 373 Seiten. Geb. 86,– DM

73 Theodor *Schweisfurth:* **Sozialistisches Völkerrecht?** Darstellung – Analyse – Wertung der sowjetmarxistischen Theorie vom Völkerrecht „neuen Typs". 1979. XIV, 615 Seiten. Geb. 108,– DM

72 **Grundrechtsschutz in Europa.** Europäische Menschenrechts-Konvention und Europäische Gemeinschaften. Internationales Kolloquium, veranstaltet vom Max-Planck-Institut für ausländisches öffentliches Recht und Völkerrecht, Heidelberg 1976. 1977. VIII, 248 Seiten (3 Seiten englische, 20 Seiten französische Zitate). Geb. 62,– DM

71 Georg *Ress:* **Die Rechtslage Deutschlands nach dem Grundlagenvertrag vom 21. Dezember 1972.** 1978. XIV, 436 Seiten. Geb. 66,– DM

70 Hans *Krück:* **Völkerrechtliche Verträge im Recht der Europäischen Gemeinschaften.** Abschlußkompetenzen – Bindungswirkung – Kollisionen. 1977. XIII, 210 Seiten. Geb. 78,– DM

69 Michael *Bothe:* **Die Kompetenzstruktur des modernen Bundesstaates in rechtsvergleichender Sicht.** 1977. XIV, 350 Seiten. Geb. 130,– DM

68 Ulf-Dieter *Klemm:* **Die seewärtige Grenze des Festlandsockels.** Geschichte, Entwicklung und lex lata eines seevölkerrechtlichen Grundproblems. 1976. XIV, 285 Seiten. Geb. 98,– DM

67 Torsten *Stein:* **Amtshilfe in auswärtigen Angelegenheiten.** 1975. XII, 193 Seiten. Geb. 78,– DM

66 Karin *Oellers-Frahm:* **Die einstweilige Anordnung in der internationalen Gerichtsbarkeit.** 1975. X, 168 Seiten. Geb. 72,– DM

65 Werner *Morvay:* **Souveränitätsübergang und Rechtskontinuität im Britischen Commonwealth.** Ein Beitrag zur Lehre von der Staatensukzession. 1974. VIII, 116 Seiten. Geb. 58,– DM

64 Hartmut *Schiedermair:* **Der völkerrechtliche Status Berlins nach dem Viermächte-Abkommen vom 3. September 1971.** 1975. VII, 223 Seiten. Geb. 67,– DM

63 Hans *von Mangoldt:* **Die Schiedsgerichtsbarkeit als Mittel internationaler Streitschlichtung.** Zur Beilegung von Rechtsstreitigkeiten auf der Grundlage der Achtung vor dem Rechte. 1974. XII, 214 Seiten. Geb. 60,– DM

62 **Judicial Settlement of International Disputes.** International Court of Justice. Other Courts and Tribunals. Arbitration and Conciliation. An International Symposium. 1974. XII, 572 Seiten. Geb. 94,– DM

61 Meinhard *Hilf:* **Die Auslegung mehrsprachiger Verträge.** Eine Untersuchung zum Völkerrecht und zum Staatsrecht der Bundesrepublik Deutschland. 1973. XI, 249 Seiten. Geb. 81,60 DM

60 Helmut *Steinberger:* **Konzeption und Grenzen freiheitlicher Demokratie.** Dargestellt am Beispiel des Verfassungsrechtsdenkens in den Vereinigten Staaten von Amerika und des amerikanischen Antisubversionsrechts. 1974. XV, 646 Seiten. Geb. 178,– DM

59 Michael *Bothe:* **Das völkerrechtliche Verbot des Einsatzes chemischer und bakteriologischer Waffen.** Kritische Würdigung und Dokumentation der Rechtsgrundlagen. 1973. XIV, 397 Seiten. Kartoniert. 96,– DM

58 Kay *Hailbronner:* **Der Schutz der Luftgrenzen im Frieden.** 1972. VIII, 117 Seiten. Kartoniert. 28,– DM

57 Bernhard *Schlüter:* **Die innerstaatliche Rechtsstellung der internationalen Organisationen** unter besonderer Berücksichtigung der Rechtslage in der Bundesrepublik Deutschland. 1972. XIII, 200 Seiten. Kartoniert. 52,– DM

56 Jörg P. *Müller:* **Vertrauensschutz im Völkerrecht.** 1971. XVI, 275 Seiten, Kartoniert. 53,– DM

55 Christian *Tomuschat:* **Die Aufwertung der Deutschen Mark.** Staats- und völkerrechtliche Überlegungen zur Neufestsetzung der Währungsparität im Jahre 1969. 1970. VII, 49 Seiten. Kartoniert. 12,– DM

54 **Demonstrationen und Straßenverkehr.** Landesberichte und Rechtsvergleichung. Belgien, Bundesrepublik Deutschland, Frankreich, Großbritannien, Italien, Niederlande, Österreich, Schweden, Schweiz, USA. 1970. LXII, 210 Seiten. Kartoniert. 30,– DM

53 Hanfried *Walter:* **Die Europäische Menschenrechtsordnung.** Individualrechte, Staatenverpflichtungen und ordre public nach der Europäischen Menschenrechtskonvention. 1970. X, 150 Seiten. Kartoniert. 26,– DM

52 **Gerichtsschutz gegen die Exekutive – Judicial Protection against the Executive – La protection juridictionnelle contre l'exécutif.** Bde. 1 und 2: Länderberichte. 1969/70. LIV, 1258 Seiten. Bd. 3 [Deutsche Ausgabe]: Rechtsvergleichung – Völkerrecht. 1971. XI, 400 Seiten; Bd. 3 [English Edition]: Comparative Law – International Law. 1971. XII, 400 Seiten. (In Gemeinschaftsverlag mit Oceana Publications, Inc., Dobbs Ferry, N.Y. 10522, USA.) Bde. 1–3 Leinen. 220,– DM

51 Gundolf *Fahl:* **Der Grundsatz der Freiheit der Meere in der Staatenpraxis von 1493–1648.** Eine rechtsgeschichtliche Untersuchung. 1969. IX, 143 Seiten. Kartoniert. 28,– DM